Exploration
of Halley's Comet

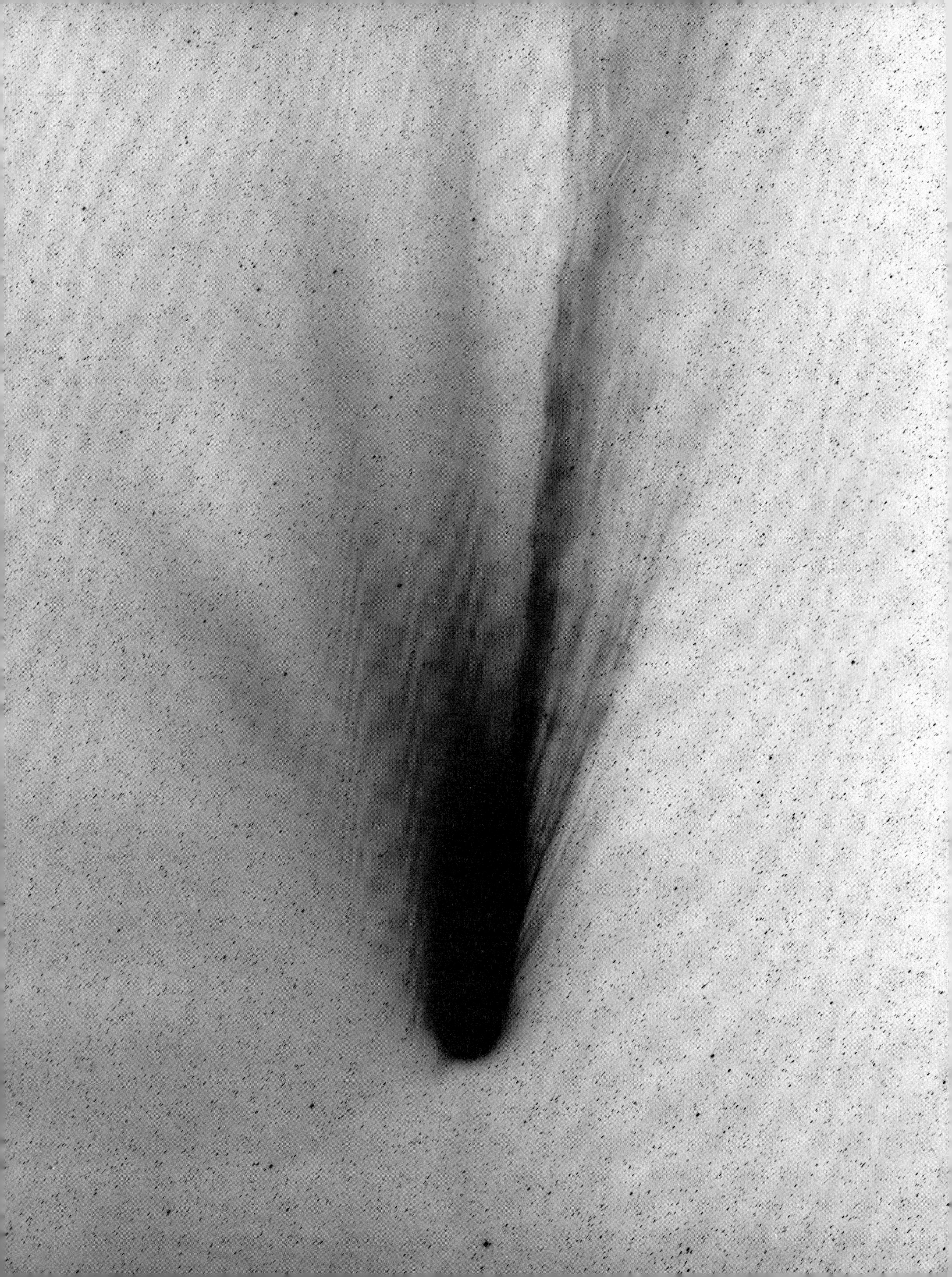

M. Grewing F. Praderie
R. Reinhard (Eds.)

Exploration of Halley's Comet

Springer-Verlag Berlin Heidelberg New York
London Paris Tokyo

Professor Dr. Michael Grewing
Astronomisches Institut der Universität, Keplerstraße 17
D-7400 Tübingen, Fed. Rep. of Germany

Professor Dr. Françoise Praderie
Observatoire de Meudon, F-92195 Meudon Principal Cedex, France

Dr. Rüdeger Reinhard
Space Science Department of ESA/ESTEC, Keplerlaan 1,
NL-2201 AZ Noordwijk, The Netherlands

Cover photograph: Comet Halley. Digitized false-colour representation of a photograph taken at the 80/120 cm Schmidt telescope of the Calar Alto Observatory, Spain on March 16, 1986 (K. Birkle, Max-Planck-Institut für Astronomie).

Frontispiece: Gas tails of Halley's Comet as observed on March 8, 1986. The main gas tail has a wiggly structure and there are many narrow streamers on the south side. At least seven broad, diffuse dust tails are also visible on the opposite side (Photo European Southern Observatory).

Pages 1-936 of this book originally appeared as the journal **Astronomy and Astrophysics, Volume 187, Number 1 and 2** (ISSN 0004-6361) Springer-Verlag Berlin, Heidelberg 1987

ISBN 3-540-17344-7 Springer-Verlag Berlin Heidelberg New York
ISBN 0-387-17344-7 Springer-Verlag New York Berlin Heidelberg

Library of Congress Cataloging-in-Publication Data
ESLAB Symposium (20th: 1986: Heidelberg, Germany). Exploration of Halley's comet: proceedings of the 20th ESLAB Symposium, Heidelberg, October 27–31, 1986.
M. Grewing, F. Praderie, R. Reinhard, (eds). p. cm.
"Originally ... appeared as the journal Astronomy and Astrophysics, volume 187, number 1 and 2" – T.p. verso.
Includes index.
1. Halley's comet – Congresses. 2. Space plasmas – Congresses. 3. Comets – Congresses. I. Grewing, M. (Michael), II. Praderie, F. (Françoise), III. Reinhard, R. IV. Title.
QB723.H2E75 1986 523.6′4–dc19 88-12371

Printed in Germany

Typesetting and printing: Brühlsche Universitätsdruckerei, Giessen
Binding: Konrad Triltsch, Graphischer Betrieb, Würzburg
2156/3150-543210

Preface

The 1985/86 apparition of Halley's Comet turned out to be the most important apparition of a comet ever. It provided a worldwide science community with a wealth of exciting new discoveries, the most remarkable of which was undoubtedly the first image of a cometary nucleus.

Halley's Comet is the brightest periodic comet, and the most famous of the 750 known comets. With its 76-year period, its recent appearance was truly a "once-in-a-lifetime" observational opportunity. The 1985/86 apparition was the thirtieth consecutive recorded apparition.

Five apparitions ago, the English astronomer Edmond Halley discovered the periodicity of "his" comet and correctly predicted its return in 1758, a triumph for science best appreciated in the context of contemporary views, or rather fears, about comets at that time.

The increasingly rapid progress in technological development is very much apparent when one compares the dominant tools for cometary research during Halley's next three apparitions: in 1835 studies were made based on drawings of the comet; in 1910 photographic plates were used; while in March 1986 an armada of six spacecraft from four space agencies approached the comet and carried out in situ measurements, 1 AU from the Earth. In 1910, nobody could have dreamed that this was possible, and today it is equally difficult to anticipate what scientists will be able to achieve in 2061.

The 1985/86 apparition began with the record-breaking recovery of the comet at 11 AU from the Sun on 16 October 1982, 3½ years before perihelion. It was not until late 1985, however, that the comet was close enough to the Sun to become fully active. By then, 1000 professional astronomers from more than 50 countries, coordinated in the International Halley Watch (IWH), were turning their telescopes towards Halley to provide almost continuous coverage over a wide range of wavelengths. With the in situ measurements from the six space probes and other remote observations from space, from rockets and high-flying airplanes, this was the largest observing campaign ever on an astronomical object.

Because cometary science is truly multidisciplinary and because comets contain the most pristine solar-system material, the little "dirty snowballs" are among the most fascinating objects in the solar system. Scientists from such different fields as astronomy, physics (in particular plasma physics), chemistry, meteoritic research and even history have all worked together in their studies of comets. They have applied such diverse techniques as in situ measurements, remote observations, theoretical analysis, laboratory simulation and computer modelling, yielding the wide variety of results presented in this book.

The combined efforts of thousands of scientists around the globe generated a huge amount of unique scientific data which has now undergone a preliminary evaluation. Many of the results obtained were presented at a special symposium on the "Exploration of Halley's Comet" held in Heidelberg in October 1986. This symposium was attended by more than 500 scientists. An important goal of the symposium was to stimulate discussion between the scientists of the various communities, in particular, the experimenters of the Halley fly-by Spacecraft and the astronomers using ground-based telescopes. While the main emphasis of the symposium was of course on Halley, discussions and papers on other comets were also included, bearing in mind the need to put the Halley results into proper perspective. A total of 370 invited and contributed oral and poster papers were presented, most of which were published in the symposium proceedings published in December 1986 in an ESA Special Publication (SP-250).

Following the meeting in Heidelberg, which provided an opportunity for the first full-scale overview of the analysis work being performed by many different groups, the manuscripts were revised, rewritten and subsequently subjected to the refereeing process of Astronomy and Astrophysics. This process could not have succeeded without the enormous help that the editors received from a large number of scientists. The editors owe special thanks to Drs. M. F. A'Hearn, C. Arpigny, I. Axford, H. Balsiger, M. Combes, J. Crovisier, A. Dollfus, S. Drapatz, P. Eberhardt,, Th. Encrenaz, H. Fechtig, P. D. Feldman, M. Festou, E. Gérard, R. H. Giese, E. Grün, D. Hovestadt, W. F. Huebner, W.-H. Ip, K. Jockers, A. D. Johnstone, H. U. Keller, R. Knacke, L. Kohoutek, P. Lämmerzahl, Y. Langevin, J. M. Le Borgne, B. A. Lindblad, A. Mangeney, B. G. Marsden, D. A. Mendis, F. M. Neubauer, M. Niedner, H. Pedersen, J. Rahe, A. Richter, H. U. Schmidt, R. Schwenn, Z. Sekanina, A. Tokunaga, M. Wallis, H. A. Weaver, R. M. West, F. L. Whipple, K. Wilhelm, and S. Wyckoff, who all accepted particularly large shares of the refereeing work and whose expert advice was instrumental in ensuring the high scientific standard of the papers.

This book contains (p. 1–936) all the papers that originally appeared in the Comet Halley dedicated issue of the journal *Astronomy and Astrophysics* (Vol. 187, Number 1 and 2) and, in addition, a summary of the scientific results presented at the Heidelberg conference and five papers providing background information on the various space missions to Halley's Comet and the International Halley Watch.

Michael Grewing *Rüdeger Reinhard*
Françoise Praderie

Contents

New information on comet P/Halley as depicted by Giotto di Bondone and other Western artists
Olson, R.J.M., Pasachoff, J.M. 1

Encounters with comets: discoveries and puzzles in cometary plasma physics
Galeev, A.A. 12

The pick-up of cometary protons by the solar wind
Neugebauer, M., Lazarus, A.J., Altwegg, K., Balsiger, H., Goldstein, B.E., Goldstein, R., Neubauer, F.M., Rosenbauer, H., Schwenn, R., Shelley, E.G., Ungstrup, E. 21

Alfvénic turbulence in the solar wind flow during the approach to comet P/Halley
Johnstone, A.D., Coates, A.J., Heath, J., Thomsen, M.F., Wilken, B., Jockers, K., Formisano, V., Amata, E., Winningham, J.D., Borg, H., Bryant, D.A. 25

General features of comet P/Halley: solar wind interaction from plasma measurements
Rème, H., Sauvaud, J.A., d'Uston, C., Cros, A., Anderson, K.A., Carlson, C.W., Curtis, D.W., Lin, R.P., Korth, A., Richter, A.K., Mendis, D.A. 33

The Sakigake/Suisei encounter with comet P/Halley
Hirao, K., Itoh, T. 39

Waves in the magnetic field and solar wind flow outside the bow shock at comet P/Halley
Johnstone, A., Glassmeier, K., Acuna, M., Borg, H., Bryant, D., Coates, A., Formisano, V., Heath, J., Mariani, F., Musmann, G., Neubauer, F., Thomsen, M., Wilken, B., Winningham, J. 47

Solar wind flow through the comet P/Halley bow shock
Coates, A.J., Johnstone, A.D., Thomsen, M.F., Formisano, V., Amata, E., Wilken, B., Jockers, K., Winningham, J.D., Borg, H., Bryant, D.A. 55

Macroscopic perturbations of the IMF by P/Halley as seen by the Giotto magnetometer
Raeder, J., Neubauer, F.M., Ness, N.F., Burlaga, L.F. 61

Low-frequency magnetic field fluctuations in comet P/Halley's magnetosheath: Giotto observations
Glaßmeier, K.H., Neubauer, F.M., Acuña, M.H., Mariani, F. 65

Fine structure of the magnetic field in comet P/Halley's coma
Yeroshenko, Y.G., Styashkin, V.A., Riedler, W., Schwingenschuh, K., Russel, C.T. 69

Giotto magnetic-field results on the boundaries of the pile-up region and the magnetic cavity
Neubauer, F.M. 73

Identification of boundaries in the cometary environment from ac electric field measurements
Mogilevsky, M., Mikhailov, Y., Molchanov, O., Grard, R., Pedersen, A., Trotignon, J.G., Béghin, C., Formisano, V., Shapiro, V., Shevchenko, V. 80

Dust observations of comet P/Halley by the plasma-wave analyser
Trotignon, J.G., Béghin, C., Grard, R., Pedersen, A., Formisano, V., Mogilevsky, M., Mikhailov, Y. 83

Comparative study of the low-frequency waves near comet P/Halley during the Vega-1 and Vega-2 flybys
Savin, S., Avanesova, G., Balikhin, M., Wozniak, D., Wronowski, P., Klimov, S., Krawczyk, Z., Nozdrachev, M., Orlowski, D., Sokolov, A., Juchniewicz, J. 89

Plasma flow in the cometosheath of P/Halley during the encounter of Suisei
Takahashi, S., Terasawa, T., Mukai, T., Kitayama, M., Miyake, W., Hirao, K. 94

MHD waves detected by ICE at distances $\geqq 28\ 10^6$ km from comet P/Halley: Cometary or solar wind origin?
Tsurutani, B.T., Brinca, A.L., Smith, E.J., Thorne, R.M., Scarf, F.L., Gosling, J.T., Ipavich, F.M. 97

Plasma-tail activity at the time of the Vega encounters
Niedner, M.B., Jr., Schwingenschuh, K. 103

Observations of cometary plasma-wave phenomena
Scarf, F.L., Coroniti, F.V., Kennel, C.F., Gurnett, D.A., Ip, W.-H., Smith, E.J. 109

Hydromagnetic waves associated with cometary water group ions: Sakigake observation
Yumoto, K., Saito, T., Nakagawa, T. 117

Plasma properties from the upstream region to the cometopause of comet P/Halley: Vega observations
Verigin, M.I., Gringauz, K.I., Richter, A.K., Gombosi, T.I., Remizov, A.P., Szegö, K., Apáthy, I., Szemerey, I., Tátrallyay, M., Lezhen, L.A. 121

Observations of heavy energetic ions far upstream from comet P/Halley
Sanderson, T.R., Wenzel, K.-P., Daly, P.W., Cowley, S.W.H., Hynds, R.J., Richardson, I.G., Smith, E.J., Bame, S.J., Zwickl, R.D. 125

Spatial distribution of water-group ions near comet P/Halley observed by Suisei
Mukai, T., Miyake, W., Terasawa, T., Kitayama, M., Hirao, K. 129

An interpretation of the ion pile-up region outside the ionospheric contact surface
Ip, W.-H., Schwenn, R., Rosenbauer, H., Balsiger, H., Neugebauer, M., Shelley, E.G. 132

Description of the main boundaries seen by the Giotto electron experiment inside comet P/Halley-solar wind interaction region
d'Uston, C., Rème, H., Sauvaud, J.A., Cros, A., Anderson, K.A., Carlson, C.W., Curtis, D., Lin, R.P., Korth, A., Richter, A.K., Mendis, A. 137

In-situ observations of a bi-modal ion distribution in the outer coma of comet P/Halley
Thomsen, M.F., Feldman, W.C., Wilken, B., Jockers, K., Stüdemann, W., Johnstone, A.D., Coates, A., Formisano, V., Amata, E., Winningham, J.D., Borg, H., Bryant, D., Wallis, M.K. 141

The composition and radial dependence of cometary ions in the coma of comet P/Halley
Korth, A., Richter, A.K., Mendis, D.A., Anderson, K.A., Carlson, C.W., Curtis, D.W., Lin, R.P., Mitchell, D.L., Rème, H., Sauvaud, J.A., d'Uston, C. 149

Pick-up ions at comet P/Halley's bow shock: observations with the IIS spectrometer on Giotto
Wilken, B., Johnstone, A., Coates, A., Borg, H., Amata, E., Formisano, V., Jockers, K., Rosenbauer, H., Stüdemann, W., Thomson, M.F., Winningham, J.D. 153

Ion temperature and flow profiles in comet P/Halley's close environment
Schwenn, R., Ip, W.-H., Rosenbauer, H., Balsiger, H., Bühler, F., Goldstein, R., Meier, A., Shelley, E.G. 160

The composition and dynamics of cometary ions in the outer coma of comet P/Halley
Balsiger, H., Altwegg, K., Bühler, F., Fuselier, S.A., Geiss, J., Goldstein, B.E., Goldstein, R., Huntress, W.T., Ip, W.-H., Lazarus, A.J., Meier, A., Neugebauer, M., Rettenmund, U., Rosenbauer, H., Schwenn, R., Shelley, E.G., Ungstrup, E., Young, D.T. 163

Expansion velocity and temperatures of gas and ions measured in the coma of comet P/Halley
Lämmerzahl, P., Krankowsky, D., Hodges, R.R., Stubbemann, U., Woweries, J., Herrwerth, I., Berthelier, J.J., Illiano, J.M., Eberhardt, P., Dolder, U., Schulte, W., Hoffman, J.H. 169

Giotto-IMS observations of ion-flow velocities and temperatures outside the magnetic cavity of comet P/Halley
Goldstein, B.E., Neugebauer, M., Balsiger, H., Drake, J., Fuselier, S.A., Goldstein, R., Ip, W.-H., Rettenmund, U., Rosenbauer, H., Schwenn, R., Shelley, E.G. 174

The dependence of mass resolution and sensitivity of the PUMA instrument on the energy spread of ions produced by hypervelocity impacts
Sagdeev, R.Z., Kissel, J., Evlanov, E.N., Fomenkova, M.N., Inogamov, N.A., Khromov, V.N., Managadze, G.G., Prilutski, O.F., Shapiro, V.D., Shutyaev, I.Y., Zubkov, B.V. 179

Spatial distribution of heavy ions in comet P/Halley's coma
Vaisberg, O.L., Zastenker, G., Smirnov, V., Khazanov, B., Omelchenko, A., Fedorov, A., Zakharov, D. 183

Quasi-periodic features and the radial distribution of cometary ions in the cometary plasma region of comet P/Halley
Gringauz, K.I., Verigin, M.I., Richter, A.K., Gombosi, T.I., Szegö, K., Tátrallyay, M., Remizov, A.P., Apáthy, I. 191

Electronic spectroscopy and relaxation of some molecular cations of cometary interest
Leach, S. 195

Possible models on disturbances of the plasma tail of comet Halley during the 1985–1986 apparition
Saito, T., Saito, K., Aoki, T., Yumoto, K. 201

Structure and dynamics of the plasma tail of comet P/Halley. I. Knot event on December 31, 1985
Saito, T., Yumoto, K., Hirao, K., Minami, S., Saito, K., Smith, E. 209

Structure and dynamics of the plasma tail of comet P/Halley. II. Kink event on January 10–11, 1986
Tomita, K., Saito, T., Minami, S. 215

Hot ions observed by the Giotto ion mass spectrometer at the comet P/Halley contact surface
Goldstein, R., Young, D.T., Balsiger, H., Buehler, F., Goldstein, B.E., Neugebauer, M., Rosenbauer, H., Schwenn, R., Shelley, E.G. 220

Photographic observations of tail-formation activities of comet P/Halley in November 1985
Liu, Z.L. 225

The outburst of comet P/Halley on December 12, 1985
Watanabe, J., Kawakami, H., Tomita, K., Kinoshita, H., Nakamura, T., Kozai, Y. 229

Structure and dynamics of plasma-tail condensations of comet P/Halley 1986 and inferences on the structure and activity of the cometary nucleus
Celnik, W.E., Schmidt-Kaler, T. 233

Observations of the coma of comet P/Halley and the outburst of 1986 March 24–25 (UT)
Rettig, T.W., Kern, J.R., Ruchti, R., Baumbaugh, B., Baumbaugh, A.E., Knickerbocker, K.L., Dawe, J. 249

Observations of ions in comet P/Halley with a focal reducer
Jockers, K., Geyer, E.H., Rosenbauer, H., Hänel, A. 256

Two disconnection events in comet P/Halley and possible solar causes
Lundstedt, H., Magnusson, P. 261

Activity of the plasma tail of comet P/Halley in March 1986
Wu, M.C., Qiu, P.Z. 264

The cause of two plasma-tail disconnection events in comet P/Halley during the ICE-Halley radial period
Brosius, J.W., Holman, G.D., Niedner, M.B., Brandt, J.C., Slavin, J.A., Smith, E.J., Zwickl, R.D., Bame, S.J. 267

Energy spectra of energetic ions in the vicinity of comet P/Giacobini-Zinner
Richardson, I.G., Cowley, S.W.H., Moore, V., Staines, K., Hynds, R.J., Sanderson, T.R., Wenzel, K.-P., Daly, P.W. 276

Plasma structures in comets P/Halley and Giacobini-Zinner
Brand, J.C., Niedner, M.B., Jr. 281

Analysis of the electron measurements from the Plasmag-1 experiment on board Vega-2 in the vicinity of comet P/Halley
Gringauz, K.I., Remizov, A.P., Verigin, M.I., Richter, A.K., Tàtrallyay, M., Szegö, K., Klimenko, I.N., Apàthy, I., Gombosi, T.I., Szemerey, T. 287

The upstream region, foreshock and bow shock wave at comet P/Halley from plasma electron measurements
Anderson, K.A., Carlson, C.W., Curtis, D.W., Lin, R.P., Rème, H., Sauvaud, J.A., d'Uston, C., Korth, A., Richter, A.K., Mendis, D.A. 290

Stochastic Fermi acceleration of ions in the pre-shock region of comet P/Halley
Gribov, B.E., Kecskeméty, K., Sagdeev, R.Z., Shapiro, V.D., Shevchenko, V.I., Somogyi, A.J., Szegö, K., Erdös, G., Eroshenko, E.G., Gringauz, K.I., Keppler, E., Marsden, R.G., Remizov, A.P., Richter, A.K., Riedler, W., Schwingenschuh, K., Wenzel, K.-P. 293

Measurements of low energy electrons and spacecraft potentials near comet P/Halley
Pedersen, A., Grard, R., Trotignon, J.G., Beghin, C., Mikhailov, Y., Mogilevsky, M. 297

Charge exchange of solar wind ions in the coma of comet P/Halley
Shelley, E.G., Fuselier, S.A., Balsiger, H., Drake, J.F., Geiss, J., Goldstein, B.E., Goldstein, R., Ip, W.-H., Lazarus, A.J., Neugebauer, M. 304

Fluid simulation of comet P/Halley's ionosphere
Baumgärtel, K., Sauer, K. 307

Unusual characteristics of electromagnetic waves excited by cometary newborn ions with large perpendicular energies
Brinca, A.L., Tsurutani, B.T. 311

Far-ultraviolet objective spectra of comet P/Halley from sounding rockets
Opal, C.B., McCoy, R.P., Carruthers, G.R. 320

IUE observations of comet P/Halley: evolution of the ultraviolet spectrum between September 1985 and July 1986
Feldman, P.D., Festou, M.C., A'Hearn, M.F., Arpigny, C., Butterworth, P.S., Cosmovici, C.B., Danks, A.C., Gilmozzi, R., Jackson, W.M., McFadden, L.A., Patriarchi, P., Schleicher, D.G., Tozzi, G.P., Wallis, M.K., Weaver, H.A., Woods, T.N. 325

Some diatomic molecules from comet P/Halley's UV spectra near spacecraft flybys
Wallis, M.K., Krishna Swamy, K.S. 329

Activity of comet P/Halley on March 23–25, 1986: IUE observations
McFadden, L.A., A'Hearn, M.F., Feldman, P.D., Roettger, E.E., Edsall, D.M., Butterworth, P.S. 333

Cometary MHD and chemistry
Wegmann, R., Schmidt, H.U., Huebner, W.F., Boice, D.C. 339

Atomic hydrogen production rates for comet P/Halley from observations with Dynamics Explorer 1
Craven, J.D., Frank, L.A. 351

The spectrum of P/Halley's coma obtained with an objective-prism
Florsch, A., Marcout, J., Traversa, G. 357

Comet P/Halley neutral gas density profile along the Vega-1 trajectory measured by the Neutral Gas Experiment
Curtis, C.C., Fan, C.Y., Hsieh, K.C., Hunten, D.M., Ip, W.-H., Keppler, E., Richter, A.K., Umlauft, G., Afonin, V.V., Dyachkov, A.V., Erö, J., Jr., Somogyi, A.J. 360

Low-resolution maps of comet P/Halley in principal atomic and molecular species
Kidger, M.R., Acosta, J.A., Garzón, F., Prieto, M., Gómez, R. 363

Pioneer Venus measurements of H, O, and C production in comet P/Halley near perihelion
Stewart, A.I.F. 369

Anisotropy of the neutral gas distribution of comet P/Halley deduced from NGE/Vega-1 measurements
Hsieh, K.C., Curtis, C.C., Fan, C.Y., Hunten, D.M., Ip, W.-H., Keppler, E., Richter, A.K., Umlauft, G., Afonin, V.V., Erö, J., Jr., Somogyi, A.J. 375

The atomic carbon distribution in the coma of comet P/Halley
Woods, T.N., Feldman, P.D., Dymond, K.F. 380

Carbon-isotope ratio in PUMA 1 spectra of P/Halley dust
Šolc, M., Vanýsek, V., Kissel, J. 385

Study of the isotopic features of Swan bands in comets
Krishna Swamy, K.S. 388

Kinematic properties of the neutral gas outflow from comet P/Halley
Larson, H.P., Mumma, M.J., Weaver, H.A. 391

The spectrum of comet P/Halley between 0.9 and 2.5 μm
Maillard, J.P., Crovisier, J., Encrenaz, T., Combes, M. 398

Anisotropic non-stationary gas flow dynamics in the coma of comet P/Halley
Kömle, N.I., Ip, W.-H. 405

Infrared investigation of water in comet P/Halley
Weaver, H.A., Mumma, M.J., Larson, H.P. 411

The ortho-para ratio of water vapor in comet P/Halley
Mumma, M.J., Weaver, H.A., Larson, H.P. 419

The 2.7 μm water band of comet P/Halley: interpretation of observations by an excitation model
Bockelée-Morvan, D., Crovisier, J. 425

Curves of growth of emission lines in cometary spectra. Implications for H_2O and OH bands of comet P/Halley
Krasnopolsky, V.A., Tkachuk, A.Y. 431

The D/H ratio in water from comet P/Halley
Eberhardt, P., Dolder, U., Schulte, W., Krankowsky, D., Lämmerzahl, P., Hoffman, J.H., Hodges, R.R., Berthelier, J.J., Illiano, J.M. 435

Improved gas-kinetic treatment of cometary water sublimation and recondensation: application to comet P/Halley
Crifo, J.F. 438

Detection of OH rotational emission from comet P/Halley in the far-infrared
Stacey, G.J., Lugten, J.B., Genzel, R. 451

18-cm wavelength radio monitoring of the OH radical in comet P/Halley (1982i)
Gérard, E., Bockelée-Morvan, D., Bourgois, G., Colom, P., Crovisier, J. 455

10.7 GHz continuum observations of comet P/Halley
Falchi, A., Gagliardi, L., Palagi, F., Tofani, G., Comoretto, G. 462

Rotational structure of the (2,0) Phillips band of C_2 in comet P/Halley
Appenzeller, I., Münch, G. 465

OH radio observations of comet P/Halley
Schloerb, F.P., Claussen, M.J., Tacconi-Garman, L. 469

Observations of HCN in comet P/Halley
Schloerb, F.P., Kinzel, W.M., Swade, D.A., Irvine, W.M. 475

The CO and N_2 abundance in comet P/Halley
Eberhardt, P., Krankowsky, D., Schulte, W., Dolder, U., Lämmerzahl, P., Berthelier, J.J., Woweries, J., Stubbemann, U., Hodges, R.R., Hoffman, J.H., Illiano, J.M. 481

Resolution of the [O I]+NH_2 blend in comet P/Halley
Arpigny, C., Magain, P., Manfroid, J., Dossin, F., Danks, A.C., Lambert, D.L. 485

A rotational-state population analysis of the high-resolution IUE observation of CS emission in comet P/Halley
Prisant, M.G., Jackson, W.M. 489

Search for methane in comet P/Halley
Drapatz, S., Larson, H.P., Davis, D.S. 497

Evidence for methane and ammonia in the coma of comet P/Halley
Allen, M., Delitsky, M., Huntress, W., Yung, Y., Ip, W.-H., Schwenn, R., Rosenbauer, H., Shelley, E., Balsiger, H., Geiss, J. 502

Detection of parent molecules in comet P/Halley from the IKS-Vega experiment
Moroz, V.I., Combes, M., Bibring, J.P., Coron, N., Crovisier, J., Encrenaz, T., Crifo, J.F., Sanko, N., Grigoryev, A.V., Bockelée-Morvan, D., Gispert, R., Nikolsky, Y.V., Emerich, C., Lamarre, J.M., Rocard, F., Krasnopolsky, V.A., Owen, T. 513

Detection of a new emission band at 2.8 μm in comet P/Halley
Tokunaga, A.T., Nagata, T., Smith, R.G. 519

Photometry of P/Halley (1982i)
Sterken, C., Manfroid, J., Arpigny, C. 523

Polarimetry of comet P/Halley: continuum versus molecular bands
Le Borgne, J.F., Leroy, J.L., Arnaud, J. 526

Photometric observations of comet P/Giacobini-Zinner
Schleicher, D.G., Millis, R.L., Birch, P.V. 531

Circular polarization near the nucleus of comet P/Halley
Metz, K., Haefner, R. 539

Spectrophotometry of comet P/Halley. I. Flux, column density and emission gradients within the coma in the emission bands and the continuum
Sivaraman, K.R., Babu, G.S.D., Shylaja, B.S., Rajamohan, R. 543

Spectrophotometry of comet P/Halley at wavelengths 275–710 nm from Vega-2
Moreels, G., Clairemidi, J., Parisot, J.P., Zucconi, J.M., Bertaux, J.L., Blamont, J.E., Hersé, M., Krasnopolsky, V.A., Moroz, V.I., Krysko, A.A., Tkachuk, A.Y., Gogoshev, M., Gogosheva, T., Werner, R., Spasov, S. 551

The visual brightness behavior of P/Halley during 1981–1987
Green, D.W.E., Morris, C.S. 560

The spectral behavior of P/Halley at large heliocentric distance in light of the Giotto/Vega results
Belton, M.J.S., Spinrad, H., Wehinger, P.A., Wyckoff, S., Yeomans, D.K. 569

Periodicities in the light curve of P/Halley and the rotation of its nucleus
Festou, M.C., Drossart, P., Lecacheux, J., Encrenaz, T., Puel, F., Kohl-Moreira, J.L. 575

Photometry of comet P/Halley at near post-perihelion phases
Neckel, T., Münch, G. 581

Observations of comet P/Halley at minimum phase angle
Meech, K.J., Jewitt, D.C. 585

Chinese observations of comet P/Halley in China and abroad
Gong (Kung), S.M., Wu, G.J., Chen, P.S., Zhang, X.F., Sun, S.S. 594

Thermal infrared imaging of comet P/Halley
Campins, H., Telesco, C.M., Decher, R., Ramsey, B.D. 601

Low resolution mapping of comet P/Halley in the near-infrared
Lázaro, C., Garzón, F., Arévalo, M.J. 605

Infrared monitoring of comet P/Halley
Lorenzetti, D., Moneti, A., Stanga, R., Strafella, F. 609

Airborne and groundbased spectrophotometry of comet P/Halley from 5–13 micrometers
Bregman, J.D., Campins, H., Witteborn, F.C., Wooden, D.H., Rank, D.M., Allamandola, L.J., Cohen, M., Tielens, A.G.G.M. 616

The near-infrared polarization and color of comet P/Halley
Brooke, T.Y., Knacke, R.F., Joyce, R.R. 621

The 3.2–3.6 μm emission features in comet P/Halley: spectral identifications and similarities
Knacke, R.F., Brooke, T.Y., Joyce, R.R. 625

Airborne spectrophotometry of P/Halley from 16 to 30 microns
Herter, T., Campins, H., Gull, G.E. 629

Photometry of comet P/Halley from 40 to 160 microns
Campins, H., Joy, M., Harvey, P.M., Lester, D.F., Ellis, H.B., Jr. 632

Airborne spectrophotometry of P/Halley from 20 to 65 microns
Glaccum, W., Moseley, S.H., Campins, H., Loewenstein, R.F. 635

Comet P/Halley near-nucleus phenomena in 1986
Larson, S., Sekanina, Z., Levy, D., Tapia, S., Senay, M. 639

The sunward spike of Halley's comet
Sekanina, Z., Larson, S.M., Emerson, G., Helin, E.F., Schmidt, R.E. 645

Complex refractive index of grain material deduced from the visible polarimetry of comet P/Halley
Mukai, T., Mukai, S., Kikuchi, S. 650

Infrared emission from P/Halley's dust coma during March 1986
Hanner, M.S., Tokunaga, A.T., Golisch, W.F., Griep, D.M., Kaminski, C.D. 653

The dust tail of comet P/Halley in April 1986
Lamy, P.L., Pedersen, H., Vio, R. 661

Albedo maps of comets P/Halley and P/Giacobini-Zinner
Hammel, H.B., Telesco, C.M., Campins, H., Decher, R., Storrs, A.D., Cruikshank, D.P. 665

Polarimetry of grains in the coma of P/Halley. I. Observations
Dollfus, A., Suchail, J.-L. 669

Polarimetry of comet P/Halley
Kikuchi, S., Mikami, Y., Mukai, T., Mukai, S., Hough, J.H. 689

Calibration of the DIDSY-IPM dust detector and application to other impact ionisation detectors on board the P/Halley probes
Göller, J.R., Grün, E., Maas, D. 693

Dust in comet P/Halley from Vega observations
Mazets, E.P., Sagdeev, R.Z., Aptekar, R.L., Golenetskii, S.V., Guryan, Yu, A., Dyachkov, A.V., Ilyinskii, V.N., Panov, V.N., Petrov, G.G., Savvin, A.V., Sokolov, I.A., Frederiks, D.D., Khavenson, N.G., Shapiro, V.D., Shevchenko, V.I. 699

Properties of dust in comet P/Halley measured by the Vega-2 three-channel spectrometer
Krasnopolsky, V.A., Moroz, V.I., Krysko, A.A., Tkachuk, A.Y., Moreels, G., Clairemidi, J., Parisot, J.P., Gogoshev, M., Gogosheva, T. 707

Dust distribution of comet P/Halley's inner coma determined from the Giotto Radio-Science Experiment
Edenhofer, P., Bird, M.K., Brenkle, J.P., Buschert, H., Kursinski, E.R., Mottinger, N.A., Porsche, H., Stelzried, C.T., Volland, H. 712

The dust distribution within the inner coma of comet P/Halley (1982i): encounter by Giotto's impact detectors
McDonnell, J.A.M., Alexander, W.M., Burton, W.M., Bussoletti, E., Evans, G.C., Evans, S.T., Firth, J.G., Grard, R.J.L., Green, S.F., Grun, E., Hanner, M.S., Hughes, D.W., Igenbergs, E., Kissel, J., Kuczera, H., Lindblad, B.A., Langevin, Y., Mandeville, J.-C., Nappo, S., Pankiewicz, G.S.A., Perry, C.H., Schwehm, G.H., Sekanina, Z., Stevenson, T.J., Turner, R.F., Weishaupt, U., Wallis, M.K., Zarnecki, J.C. 719

The dust coma of comet P/Halley: measurements on the Vega-1 and Vega-2 spacecraft
Simpson, J.A., Rabinowitz, D., Tuzzolino, A.J., Ksanfomality, L.V., Sagdeev, R.Z. 742

Spatial and mass distribution of low-mass dust particles ($m < 10^{-10}$ g) in comet P/Halley's coma
Vaisberg, O.L., Smirnov, V., Omelchenko, A., Gorn, L., Iovlev, M. 753

First statistical analysis of 5000 mass spectra of cometary grains obtained by PUMA 1 (Vega-1) and PIA (Giotto) impact ionization mass spectrometers in the compressed modes
Langevin, Y., Kissel, J., Bertaux, J.-L., Chassefière, E. 761

Comet P/Halley: implications of the mass distribution function for the photopolarimetric properties of the dust coma
Lamy, P.L., Grün, E., Perrin, J.M. 767

An attempt to evaluate the structure of cometary dust particles
Smirnov, V.N., Vaisberg, O.L., Anisimov, S. 774

Systematics of the "CHON" and other light-element particle populations in comet P/Halley
Clark, B.C., Mason, L.W., Kissel, J. 779

Secondary electron emission induced by gas and dust impacts on Giotto, Vega-1 and Vega-2 in the environment of comet P/Halley
Grard, R.J.L., McDonnell, J.A.M., Grün, E., Gringauz, K.I. 785

Dust environment of comet P/Halley: a review
Sekanina, Z. 789

Charging of dust particles in comets and in interplanetary space
Notni, P., Tiersch, H. 796

Evaporating grains in P/Halley's coma
Wallis, M.K., Rabilizirov, R., Wickramasinghe, N.C. 801

Comet P/Halley's nucleus and its activity
Keller, H.U., Delamere, W.A., Huebner, W.F., Reitsema, H.J., Schmidt, H.U., Whipple, F.L., Wilhelm, K., Curdt, W., Kramm, R., Thomas, N., Arpigny, C., Barbieri, C., Bonnet, R.M., Cazes, S., Coradini, M., Cosmovici, C.B., Hughes, D.W., Jamar, C., Walaise, D., Schmidt, K., Schmidt, W.K.H., Seige, P. 807

Electrostatic charging and fragmentation of dust near P/Giacobini-Zinner and P/Halley
Boehnhardt, H., Fechtig, H. 824

Evolution of comet P/Halley in early march 1986 as observed from Vega pictures
Abergel, A., Bertaux, J.L. 829

The spatial distribution of dust jets seen during the Vega 2 flyby
Sagdeev, R.Z., Smith, B., Szegö, K., Larson, S., Tóth, I., Merényi, E., Avanesov, G.A., Krasikov, V.A., Shamis, V.A., Tarnapolski, V.I. 835

Temperature and size of the nucleus of comet P/Halley deduced from IKS infrared Vega-1 measurements
Emerich, C., Lamarre, J.M., Moroz, V.I., Combes, M., Sanko, N.F., Nikolsky, Y.V., Rocard, F., Gispert, R., Coron, N., Bibring, J.P., Encrenaz, T., Crovisier, J. 839

Fine dust structures in the emission of comet P/Halley observed by the Halley Multicolour Camera on board Giotto
Thomas, N., Keller, H.U. 843

Detailed analysis of a surface feature on comet P/Halley
Schwarz, G., Craubner, H., Delamere, A., Göbel, M., Gonano, M., Huebner, W.F., Keller, H.U., Kramm, R., Mikusch, E., Reitsema, H., Whipple, F.L., Wilhelm, K. 847

The cometary nucleus: current concepts
Whipple, F.L. 852

Composition measurements and the history of cometary matter
Geiss, J. 859

Modeling P/Halley before and after the encounters
Divine, N., Newburn, R.L., Jr. 867

Post-perihelion brightening of comet P/Halley: Springtime for Halley
Weissman, P.R. 873

P/Halley dust characteristics: a comparison between Orionid and Eta Aquarid meteor observations and those from the flyby spacecraft
Hughes, D.W. 879

Radiation formation of a non-volatile comet crust
Johnson, R.E., Cooper, J.F., Lanzerotti, L.J., Strazzulla, G. 889

Astrometric positions of comet P/Halley
Barbieri, C., Kranjc, A., Scardia, M., Cremonese, G. 893

A new approach to investigations of the long-term motion of comet P/Halley
Sitarski, G., Ziolkowski, K. 896

High-order librations of Halley-type comets
Carusi, A., Kresák, L., Perozzi, E., Valsecchi, G.B. 899

Dormant phases in the aging of periodic comets
Kresák, L. 906

The dynamical lifetime of comet P/Halley
Olsson-Steel, D.I. 909

Galactic tides affect the Oort cloud: an observational confirmation
Delsemme, A.H. 913

The P/Halley meteor showers in 1985–1986
Hajduková, M., Hajduk, A., Cevolani, G., Formiggini, C. 919

The spectra of meteors from comet P/Halley
Halliday, I. 921

Meteoroids from comet P/Halley. The comet's mass production and age
Hajduk, A. 925

The 1985 return of the Giacobinid meteor stream
Lindblad, B.A. 928

The meteor stream associated with comet P/Grigg-Skjellerup
Lindblad, B.A. 931

Meteor contribution by short-period comets
Štohl, J. 933

Associations between ancient comets and meteor showers
Kresáková, M. 935

Supplement

Exploration of Halley's Comet: symposium summary
Mendis, D. A. 939

The Giotto mission to Halley's Comet
Reinhard, R. 949

The Vega mission to Halley's Comet
Sagdeev, R.Z. 959

The Sakigake/Suisei missions to Halley's Comet
Hirao, K., Itoh, T. 965

The International Cometary Explorer
Brandt, J. C., Farquhar, R. W., Maran, S. P., Niedner, M. B. Jr., von Rosenvinge, T. T. 969

The International Halley Watch
Edberg, S. J., Newburn, R. L. Jr., Rahe, J. 981

Author-Title Index

Abergel, A., Bertaux, J.L.: Evolution of comet P/Halley in early March 1986 as observed from Vega pictures 829
Acosta, J.A., see Kidger, M.R., et al. 363
Acuña, M.H., see Glaßmeier, K.H., et al. 65
Acuña, M., see Johnstone, A.D., et al. 47
Afonin, V.V., see Curtis, C.C., et al. 360
Afonin, V.V., see Hsieh, K.C., et al. 375
A'Hearn, M.F., see Feldman, P.D., et al. 325
A'Hearn, M.F., see McFadden, L.A., et al. 333
Alexander, W.M., see McDonnell, J.A.M., et al. 719
Allamandola, L.J., see Bregman, J.D., et al. 616
Allen, M., Delitsky, M., Huntress, W., Yung, Y., Ip, W.-H., Schwenn, R., Rosenbauer, H., Shelley, E., Balsiger, H., Geiss, J.: Evidence for methane and ammonia in the coma of comet P/Halley 502
Altwegg, K., see Balsiger, H., et al. 163
Altwegg, K., see Neugebauer, M., et al. 21
Amata, E., see Coates, A.J., et al. 55
Amata, E., see Johnstone, A.D., et al. 25
Amata, E., see Thomsen, M.F., et al. 141
Amata, E., see Wilken, B., et al. 153
Anderson, K.A., Carlson, C.W., Curtis, D.W., Lin, R.P., Rème, H., Sauvaud, J.A., d'Uston, C., Korth, A., Richter, A.K., Mendis, D.A.: The upstream region, foreshock and bow shock wave at comet P/Halley from plasma electron measurements 290
Anderson, K.A., see d'Uston, C., et al. 137
Anderson, K.A., see Korth, A., et al. 149
Anderson, K.A., see Rème, H., et al. 33
Anisimov, S., see Smirnov, V.N., et al. 774
Aoki, T., see Saito, T., et al. 201
Apáthy, I., see Gringauz, K.I., et al. 191
Apáthy, I., see Gringauz, K.I., et al. 287
Apáthy, I., see Verigin, M.I., et al. 121
Appenzeller, I., Münch, G.: Rotational structure of the (2,0) Phillips band of C_2 in comet P/Halley 465
Aptekar, R.L., see Mazets, E.P., et al. 699
Arévalo, M.J., see Lázaro, C., et al. 605
Arnaud, J., see Le Borgne, J.F., et al. 526
Arpigny, C., Magain, P., Manfroid, J., Dossin, F., Danks, A.C., Lambert, D.L.: Resolution of the [OI]+NH_2 blend in comet P/Halley 485
Arpigny, C., see Feldman, P.D., et al. 325
Arpigny, C., see Keller, H.U., et al. 807
Arpigny, C., see Sterken, C., et al. 523
Avanesov, G.A., see Sagdeev, R.Z., et al. 835
Avanesova, G., see Savin, S., et al. 89
Babu, G.S.D., see Sivaraman, K.R., et al. 543
Balikhin, M., see Savin, S., et al. 89
Balsiger, H., Altwegg, K., Bühler, F., Fuselier, S.A., Geiss, J., Goldstein, B.E., Goldstein, R., Huntress, W.T., Ip, W.-H., Lazarus, A.J., Meier, A., Neugebauer, M., Rettenmund, U., Rosenbauer, H., Schwenn, R., Shelley, E.G., Ungstrup, E., Young, D.T.: The composition and dynamics of cometary ions in the outer coma of comet P/Halley 163
Balsiger, H., see Allen, M., et al. 502
Balsiger, H., see Goldstein, B.E., et al. 174
Balsiger, H., see Goldstein, R., et al. 220
Balsiger, H., see Ip, W.-H., et al. 132
Balsiger, H., see Neugebauer, M., et al. 21
Balsiger, H., see Schwenn, R., et al. 160
Balsiger, H., see Shelley, E.G., et al. 304
Bame, S.J., see Brosius, J.W., et al. 267
Bame, S.J., see Sanderson, T.R., et al. 125
Barbieri, C., Kranjc, A., Scardia, M., Cremonese, G.: Astrometric positions of comet P/Halley 893
Barbieri, C., see Keller, H.U., et al. 807
Baumbaugh, A.E., see Rettig, T.W., et al. 249
Baumbaugh, B., see Rettig, T.W., et al. 249
Baumgärtel, K., Sauer, K.: Fluid simulation of comet P/Halley's ionosphere 307
Béghin, C., see Mogilevsky, M., et al. 80
Béghin, C., see Pedersen, A., et al. 297
Béghin, C., see Trotignon, J.G., et al. 83
Belton, M.J.S., Spinrad, H., Wehinger, P.A., Wyckoff, S., Yeomans, D.K.: The spectral behavior of P/Halley at large heliocentric distance in light of the Giotto/Vega results 569
Bertaux, J.L., see Abergel, A. 829
Bertaux, J.-L., see Langevin, Y., et al. 761
Bertaux, J.L., see Moreels, G., et al. 551
Berthelier, J.J., see Eberhardt, P., et al. 435
Berthelier, J.J., see Eberhardt, P., et al. 481
Berthelier, J.J., see Lämmerzahl, P., et al. 169
Bibring, J.P., see Emerich, C., et al. 839
Bibring, J.P., see Moroz, V.I., et al. 513
Birch, P.V., see Schleicher, D.G., et al. 531
Bird, M.K., see Edenhofer, P., et al. 712
Blamont, J.E., see Moreels, G., et al. 551
Bockelée-Morvan, D., Crovisier, J.: The 2.7 μm water band of comet P/Halley: interpretation of observations by an excitation model 425
Bockelée-Morvan, D., see Gérard, E., et al. 455
Bockelée-Morvan, D., see Moroz, V.I., et al. 513
Boehnhardt, H., Fechtig, H.: Electrostatic charging and fragmentation of dust near P/Giacobini-Zinner and P/Halley 824
Boice, D.C., see Wegmann, R., et al. 339

Bonnet, R.M., see Keller, H.U., et al. 807
Borg, H., see Coates, A.J., et al. 55
Borg, H., see Johnstone, A.D., et al. 25
Borg, H., see Johnstone, A.D., et al. 47
Borg, H., see Thomsen, M.F., et al. 141
Borg, H., see Wilken, B., et al. 153
Bourgois, G., see Gérard, E., et al. 455
Brandt, J.C., Niedner MB, Jr.: Plasma structures in comets P/Halley and Giacobini-Zinner 281
Brandt, J.C., Farquhar, R.W., Maran, S.P., Niedner MB, Jr., von Rosenvinge, T.T.: The International Cometary Explorer 969
Brandt, J.C., see Brosius, J.W., et al. 267
Bregman, J.D., Campins, H., Witteborn, F.C., Wooden, D.H., Rank, D.M., Allamandola, L.J., Cohen, M., Tielens, A.G.G.M.: Airborne and groundbased spectrophotometry of comet P/Halley from 5–13 μm 616
Brenkle, J.P., see Edenhofer, P., et al. 712
Brinca, A.L., Tsurutani, B.T.: Unusual characteristics of electromagnetic waves excited by cometary newborn ions with large perpendicular energies 311
Brinca, A.L., see Tsurutani, B.T., et al. 97
Brooke, T.Y., Knacke, R.F., Joyce, R.R.: The near-infrared polarization and color of comet P/Halley 621
Brooke, T.Y., see Knacke, R.F., et al. 625
Brosius, J.W., Holman, G.D., Niedner, M.B., Brandt, J.C., Slavin, J.A., Smith, E.J., Zwickl, R.D., Bame, S.J.: The cause of two plasma-tail disconnection events in comet P/Halley during the ICE-Halley radial period 267
Bryant, D.A., see Coates, A.J., et al. 55
Bryant, D.A., see Johnstone, A.D., et al. 25
Bryant, D., see Johnstone, A.D., et al. 47
Bryant, D., see Thomsen, M.F., et al. 141
Buehler, F., see Goldstein, R., et al. 220
Bühler, F., see Balsiger, H., et al. 163
Bühler, F., see Schwenn, R., et al. 160
Burlaga, L.F., see Raeder, J., et al. 61
Burton, W.M., see McDonnell, J.A.M., et al. 719
Buschert, H., see Edenhofer, P., et al. 712
Bussoletti, E., see McDonnell, J.A.M., et al. 719
Butterworth, P.S., see Feldman, P.D., et al. 325
Butterworth, P.S., see McFadden, L.A., et al. 333
Campins, H., Joy, M., Harvey, P.M., Lester, D.F., Ellis HB, Jr.: Photometry of comet P/Halley from 40 to 160 μm 632
Campins, H., Telesco, C.M., Decher, R., Ramsey, B.D.: Thermal infrared imaging of comet P/Halley 601
Campins, H., see Bregman, J.D., et al. 616
Campins, H., see Glaccum, W., et al. 635
Campins, H., see Hammel, H.B., et al. 665
Campins, H., see Herter, T., et al. 629
Carlson, C.W., see Anderson, K.A., et al. 290
Carlson, C.W., see d'Uston, C., et al. 137
Carlson, C.W., see Korth, A., et al. 149
Carlson, C.W., see Rème, H., et al. 33
Carruthers, G.R., see Opal, C.B., et al. 320
Carusi, A., Kresák, L., Perozzi, E., Valsecchi, G.B.: High-order librations of Halley-type comets 899
Cazes, S., see Keller, H.U., et al. 807
Celnik, W.E., Schmidt-Kaler, T.: Structure and dynamics of plasma-tail condensations of comet P/Halley 1986 and inferences on the structure and activity of the cometary nucleus 233
Cevolani, G., see Hajduková, M., et al. 919
Chassefière, E., see Langevin, Y., et al. 761
Chen, P.S., see Gong , et al. 594
Clairemidi, J., see Krasnopolsky, V.A., et al. 707
Clairemidi, J., see Moreels, G., et al. 551
Clark, B.C., Mason, L.W., Kissel, J.: Systematics of the „CHON“ and other light-element particle populations in comet P/Halley 779
Claussen, M.J., see Schloerb, F.P., et al. 469
Coates, A.J., Johnstone, A.D., Thomsen, M.F., Formisano, V., Amata, E., Wilken, B., Jockers, K., Winningham, J.D., Borg, H., Bryant, D.A.: Solar wind flow through the comet P/Halley bow shock 55
Coates, A.J., see Johnstone, A.D., et al. 25
Coates, A., see Johnstone, A.D., et al. 47
Coates, A., see Thomsen, M.F., et al. 141
Coates, A., see Wilken, B., et al. 153
Cohen, M., see Bregman, J.D., et al. 616
Colom, P., see Gérard, E., et al. 455
Combes, M., see Emerich, C., et al. 839
Combes, M., see Maillard, J.P., et al. 398
Combes, M., see Moroz, V.I., et al. 513
Comoretto, G., see Falchi, A., et al. 462
Cooper, J.F., see Johnson, R.E., et al. 889
Coradini, M., see Keller, H.U., et al. 807
Coron, N., see Emerich, C., et al. 839
Coron, N., see Moroz, V.I., et al. 513
Coroniti, F.V., see Scarf, F.L., et al. 109
Cosmovici, C.B., see Feldman, P.D., et al. 325
Cosmovici, C.B., see Keller, H.U., et al. 807
Cowley, S.W.H., see Richardson, I.G., et al. 276
Cowley, S.W.H., see Sanderson, T.R., et al. 125
Craubner, H., see Schwarz, G., et al. 847
Craven, J.D., Frank, L.A.: Atomic hydrogen production rates for comet P/Halley from observations with Dynamics Explorer 1 351
Cremonese, G., see Barbieri, C., et al. 893
Crifo, J.F.: Improved gas-kinetic treatment of cometary water sublimation and recondensation: application to comet P/Halley 438
Crifo, J.F., see Moroz, V.I., et al. 513
Cros, A., see d'Uston, C., et al. 137
Cros, A., see Rème, H., et al. 33
Crovisier, J., see Bockelée-Morvan, D. 425
Crovisier, J., see Emerich, C., et al. 839
Crovisier, J., see Gérard, E., et al. 455
Crovisier, J., see Maillard, J.P., et al. 398
Crovisier, J., see Moroz, V.I., et al. 513
Cruikshank, D.P., see Hammel, H.B., et al. 665
Curdt, W., see Keller, H.U., et al. 807
Curtis, C.C., Fan, C.Y., Hsieh, K.C., Hunten, D.M., Ip, W.-H., Keppler, E., Richter, A.K., Umlauft, G., Afonin, V.V., Dyachkov, A.V., Erö J, Jr., Somogyi, A.J.: Comet P/Halley neutral gas density profile along the Vega-1 trajectory measured by the Neutral Gas Experiment 360
Curtis, C.C., see Hsieh, K.C., et al. 375
Curtis, D.W., see Anderson, K.A., et al. 290
Curtis, D.W., see Korth, A., et al. 149
Curtis, D.W., see Rème, H., et al. 33
Curtis, D., see d'Uston, C., et al. 137
Daly, P.W., see Richardson, I.G., et al. 276
Daly, P.W., see Sanderson, T.R., et al. 125
Danks, A.C., see Arpigny, C., et al. 485
Danks, A.C., see Feldman, P.D., et al. 325

Davis, D.S., see Drapatz, S., et al. 497
Dawe, J., see Rettig, T.W., et al. 249
Decher, R., see Campins, H., et al. 601
Decher, R., see Hammel, H.B., et al. 665
Delamere, A., see Schwarz, G., et al. 847
Delamere, W.A., see Keller, H.U., et al. 807
Delitsky, M., see Allen, M., et al. 502
Delsemme, A.H.: Galactic tides affect the Oort cloud: an observational confirmation 913
Divine, N., Newburn RL, Jr.: Modeling P/Halley before and after the encounters 867
Dolder, U., see Eberhardt, P., et al. 435
Dolder, U., see Eberhardt, P., et al. 481
Dolder, U., see Lämmerzahl, P., et al. 169
Dollfus, A., Suchail, J.-L.: Polarimetry of grains in the coma of P/Halley. I. Observations 669
Dossin, F., see Arpigny, C., et al. 485
Drake, J.F., see Shelley, E.G., et al. 304
Drake, J., see Goldstein, B.E., et al. 174
Drapatz, S., Larson, H.P., Davis, D.S.: Search for methane in comet P/Halley 497
Drossart, P., see Festou, M.C., et al. 575
d'Uston, C., Rème, H., Sauvaud, J.A., Cros, A., Anderson, K.A., Carlson, C.W., Curtis, D., Lin, R.P., Korth, A., Richter, A.K., Mendis, A.: Description of the main boundaries seen by the Giotto electron experiment inside comet P/Halley-solar wind interaction region 137
d'Uston, C., see Anderson, K.A., et al. 290
d'Uston, C., see Korth, A., et al. 149
d'Uston, C., see Rème, H., et al. 33
Dyachkov, A.V., see Curtis, C.C., et al. 360
Dyachkov, A.V., see Mazets, E.P., et al. 699
Dymond, K.F., see Woods, T.N., et al. 380
Eberhardt, P., Dolder, U., Schulte, W., Krankowsky, D., Lämmerzahl, P., Hoffman, J.H., Hodges, R.R., Berthelier, J.J., Illiano, J.M.: The D/H ratio in water from comet P/Halley 435
Eberhardt, P., Krankowsky, D., Schulte, W., Dolder, U., Lämmerzahl, P., Berthelier, J.J., Woweries, J., Stubbemann, U., Hodges, R.R., Hoffman, J.H., Illiano, J.M.: The CO and N_2 abundance in comet P/Halley 481
Eberhardt, P., see Lämmerzahl, P., et al. 169
Edberg, S.J., Newburn RL, Jr., Rahe, J.: The International Halley Watch 981
Edenhofer, P., Bird, M.K., Brenkle, J.P., Buschert, H., Kursinski, E.R., Mottinger, N.A., Porsche, H., Stelzried, C.T., Volland, H.: Dust distribution of comet P/Halley's inner coma determined from the Giotto Radio-Science Experiment 712
Edsall, D.M., see McFadden, L.A., et al. 333
Ellis HB, Jr., see Campins, H., et al. 632
Emerich, C., Lamarre, J.M., Moroz, V.I., Combes, M., Sanko, N.F., Nikolsky, Y.V., Rocard, F., Gispert, R., Coron, N., Bibring, J.P., Encrenaz, T., Crovisier, J.: Temperature and size of the nucleus of comet P/Halley deduced from IKS infrared Vega-1 measurements 839
Emerich, C., see Moroz, V.I., et al. 513
Emerson, G., see Sekanina, Z., et al. 645
Encrenaz, T., see Emerich, C., et al. 839
Encrenaz, T., see Festou, M.C., et al. 575
Encrenaz, T., see Maillard, J.P., et al. 398
Encrenaz, T., see Moroz, V.I., et al. 513
Erdös, G., see Gribov, B.E., et al. 293
Erö J, Jr., see Curtis, C.C., et al. 360
Erö J, Jr., see Hsieh, K.C., et al. 375
Eroshenko, E.G., see Gribov, B.E., et al. 293
Evans, G.C., see McDonnell, J.A.M., et al. 719
Evans, S.T., see McDonnell, J.A.M., et al. 719
Evlanov, E.N., see Sagdeev, R.Z., et al. 179
Falchi, A., Gagliardi, L., Palagi, F., Tofani, G., Comoretto, G.: 10.7 GHz continuum observations of comet P/Halley 462
Fan, C.Y., see Curtis, C.C., et al. 360
Fan, C.Y., see Hsieh, K.C., et al. 375
Farquhar, R.W., see Brandt, J.C., et al. 969
Fechtig, H., see Boehnhardt, H. 824
Fedorov, A., see Vaisberg, O.L., et al. 183
Feldman, P.D., Festou, M.C., A'Hearn, M.F., Arpigny, C., Butterworth, P.S., Cosmovici, C.B., Danks, A.C., Gilmozzi, R., Jackson, W.M., McFadden, L.A., Patriarchi, P., Schleicher, D.G., Tozzi, G.P., Wallis, M.K., Weaver, H.A., Woods, T.N.: IUE observations of comet P/Halley: evolution of the ultraviolet spectrum between September 1985 and July 1986 325
Feldman, P.D., see McFadden, L.A., et al. 333
Feldman, P.D., see Woods, T.N., et al. 380
Feldman, W.C., see Thomsen, M.F., et al. 141
Festou, M.C., Drossart, P., Lecacheux, J., Encrenaz, T., Puel, F., Kohl-Moreira, J.L.: Periodicities in the light curve of P/Halley and the rotation of its nucleus 575
Festou, M.C., see Feldman, P.D., et al. 325
Firth, J.G., see McDonnell, J.A.M., et al. 719
Florsch, A., Marcout, J., Traversa, G.: The spectrum of P/Halley's coma obtained with an objective-prism 357
Fomenkova, M.N., see Sagdeev, R.Z., et al. 179
Formiggini, C., see Hajduková, M., et al. 919
Formisano, V., see Coates, A.J., et al. 55
Formisano, V., see Johnstone, A.D., et al. 25
Formisano, V., see Johnstone, A.D., et al. 47
Formisano, V., see Mogilevsky, M., et al. 80
Formisano, V., see Thomsen, M.F., et al. 141
Formisano, V., see Trotignon, J.G., et al. 83
Formisano, V., see Wilken, B., et al. 153
Frank, L.A., see Craven, J.D. 351
Frederiks, D.D., see Mazets, E.P., et al. 699
Fuselier, S.A., see Balsiger, H., et al. 163
Fuselier, S.A., see Goldstein, B.E., et al. 174
Fuselier, S.A., see Shelley, E.G., et al. 304
Gagliardi, L., see Falchi, A., et al. 462
Galeev, A.A.: Encounters with comets: discoveries and puzzles in cometary plasma physics 12
Garzón, F., see Kidger, M.R., et al. 363
Garzón, F., see Lázaro, C., et al. 605
Geiss, J.: Composition measurements and the history of cometary matter 859
Geiss, J., see Allen, M., et al. 502
Geiss, J., see Balsiger, H., et al. 163
Geiss, J., see Shelley, E.G., et al. 304
Genzel, R., see Stacey, G.J., et al. 451
Gérard, E., Bockelée-Morvan, D., Bourgois, G., Colom, P., Crovisier, J.: 18-cm wavelength radio monitoring of the OH radical in comet P/Halley (1982i) 455
Geyer, E.H., see Jockers, K., et al. 256
Gilmozzi, R., see Feldman, P.D., et al. 325
Gispert, R., see Emerich, C., et al. 839
Gispert, R., see Moroz, V.I., et al. 513
Glaccum, W., Moseley, S.H., Campins, H., Loewenstein, R.F.: Airborne spectrophotometry of P/Halley from 20 to 65 μm 635

Glaßmeier, K.H., Neubauer, F.M., Acuña, M.H., Mariani, F.: Low-frequency magnetic field fluctuations in comet P/Halley's magnetosheath: Giotto observations 65
Glaßmeier, K.H., see Johnstone, A.D., et al. 47
Göbel, M., see Schwarz, G., et al. 847
Gogoshev, M., see Krasnopolsky, V.A., et al. 707
Gogoshev, M., see Moreels, G., et al. 551
Gogosheva, T., see Krasnopolsky, V.A., et al. 707
Gogosheva, T., see Moreels, G., et al. 551
Goldstein, B.E., Neugebauer, M., Balsiger, H., Drake, J., Fuselier, S.A., Goldstein, R., Ip, W.-H., Rettenmund, U., Rosenbauer, H., Schwenn, R., Shelley, E.G.: Giotto-IMS observations of ion-flow velocities and temperatures outside the magnetic cavity of comet P/Halley 174
Goldstein, B.E., see Balsiger, H., et al. 163
Goldstein, B.E., see Goldstein, R., et al. 220
Goldstein, B.E., see Neugebauer, M., et al. 21
Goldstein, B.E., see Shelley, E.G., et al. 304
Goldstein, R., Young, D.T., Balsiger, H., Buehler, F., Goldstein, B.E., Neugebauer, M., Rosenbauer, H., Schwenn, R., Shelley, E.G.: Hot ions observed by the Giotto ion mass spectrometer at the comet P/Halley contact surface 220
Goldstein, R., see Balsiger, H., et al. 163
Goldstein, R., see Goldstein, B.E., et al. 174
Goldstein, R., see Neugebauer, M., et al. 21
Goldstein, R., see Schwenn, R., et al. 160
Goldstein, R., see Shelley, E.G., et al. 304
Golenetskii, S.V., see Mazets, E.P., et al. 699
Golisch, W.F., see Hanner, M.S., et al. 653
Göller, J.R., Grün, E., Maas, D.: Calibration of the DIDSY-IPM dust detector and application to other impact ionisation detectors on board the P/Halley probes 693
Gombosi, T.I., see Gringauz, K.I., et al. 191
Gombosi, T.I., see Gringauz, K.I., et al. 287
Gombosi, T.I., see Verigin, M.I., et al. 121
Gómez, R., see Kidger, M.R., et al. 363
Gonano, M., see Schwarz, G., et al. 847
Gong , Kung), S.M., Wu, G.J., Chen, P.S., Zhang, X.F.: Chinese observations of comet P/Halley in China and abroad 594
Gorn, L., see Vaisberg, O.L., et al. 753
Gosling, J.T., see Tsurutani, B.T., et al. 97
Grard, R.J.L., McDonnell, J.A.M., Grün, E., Gringauz, K.I.: Secondary electron emission induced by gas and dust impacts on Giotto, Vega-1 and Vega-2 in the environment of comet P/Halley 785
Grard, R.J.L., see McDonnell, J.A.M., et al. 719
Grard, R., see Mogilevsky, M., et al. 80
Grard, R., see Pedersen, A., et al. 297
Grard, R., see Trotignon, J.G., et al. 83
Green, D.W.E., Morris, C.S.: The visual brightness behavior of P/Halley during 1981–1987 560
Green, S.F., see McDonnell, J.A.M., et al. 719
Gribov, B.E., Kecskeméty, K., Sagdeev, R.Z., Shapiro, V.D., Shevchenko, V.I., Somogyi, A.J., Szegö, K., Erdös, G., Eroshenko, E.G., Gringauz, K.I., Keppler, E., Marsden, R.G., Remizov, A.P., Richter, A.K., Riedler, W., Schwingenschuh, K., Wenzel, K.-P.: Stochastic Fermi acceleration of ions in the pre-shock region of comet P/Halley 293
Griep, D.M., see Hanner, M.S., et al. 653
Grigoryev, A.V., see Moroz, V.I., et al. 513
Gringauz, K.I., Remizov, A.P., Verigin, M.I., Richter, A.K., Tátrallyay, M., Szegö, K., Klimenko, I.N., Apáthy, I., Gombosi, T.I., Szemerey, T.: Analysis of the electron measurements from the Plasmag-1 experiment on board Vega-2 in the vicinity of comet P/Halley 287
Gringauz, K.I., Verigin, M.I., Richter, A.K., Gombosi, T.I., Szegö, K., Tátrallyay, M., Remizov, A.P., Apáthy, I.: Quasiperiodic features and the radial distribution of cometary ions in the cometary plasma region of comet P/Halley 191
Gringauz, K.I., see Grard, R.J.L., et al. 785
Gringauz, K.I., see Gribov, B.E., et al. 293
Gringauz, K.I., see Verigin, M.I., et al. 121
Grün, E., see Göller, J.R., et al. 693
Grün, E., see Grard, R.J.L., et al. 785
Grün, E., see Lamy, P.L., et al. 767
Grün, E., see McDonnell, J.A.M., et al. 719
Gull, G.E., see Herter, T., et al. 629
Gurnett, D.A., see Scarf, F.L., et al. 109
Guryan Yu, A., see Mazets, E.P., et al. 699
Haefner, R., see Metz, K. 539
Hajduk, A.: Meteoroids from comet P/Halley. The comet's mass production and age 925
Hajduk, A., see Hajduková, M., et al. 919
Hajduková, M., Hajduk, A., Cevolani, G., Formiggini, C.: The P/Halley meteor showers in 1985–1986 919
Halliday, I.: The spectra of meteors from comet P/Halley 921
Hammel, H.B., Telesco, C.M., Campins, H., Decher, R., Storrs, A.D., Cruikshank, D.P.: Albedo maps of comets P/Halley and P/Giacobini-Zinner 665
Hänel, A., see Jockers, K., et al. 256
Hanner, M.S., Tokunaga, A.T., Golisch, W.F., Griep, D.M., Kaminski, C.D.: Infrared emission from P/Halley's dust coma during March 1986 653
Hanner, M.S., see McDonnell, J.A.M., et al. 719
Harvey, P.M., see Campins, H., et al. 632
Heath, J., see Johnstone, A.D., et al. 25
Heath, J., see Johnstone, A.D., et al. 47
Helin, E.F., see Sekanina, Z., et al. 645
Herrwerth, I., see Lämmerzahl, P., et al. 169
Hersé, M., see Moreels, G., et al. 551
Herter, T., Campins, H., Gull, G.E.: Airborne spectrophotometry of P/Halley from 16 to 30 μm 629
Hirao, K., Itoh, T.: The Sakigake/Suisei encounter with comet P/Halley 39
Hirao, K., Itoh, T.: The Sakigake/Suisei missions to comet P/Halley 965
Hirao, K., see Mukai, T., et al. 129
Hirao, K., see Saito, T., et al. 209
Hirao, K., see Takahashi, S., et al. 94
Hodges, R.R., see Eberhardt, P., et al. 435
Hodges, R.R., see Eberhardt, P., et al. 481
Hodges, R.R., see Lämmerzahl, P., et al. 169
Hoffman, J.H., see Eberhardt, P., et al. 435
Hoffman, J.H., see Eberhardt, P., et al. 481
Hoffman, J.H., see Lämmerzahl, P., et al. 169
Holman, G.D., see Brosius, J.W., et al. 267
Hough, J.H., see Kikuchi, S., et al. 689
Hsieh, K.C., Curtis, C.C., Fan, C.Y., Hunten, D.M., Ip, W.-H., Keppler, E., Richter, A.K., Umlauft, G., Afonin, V.V., Erö J, Jr., Somogyi, A.J.: Anisotropy of the neutral gas distribution of comet P/Halley deduced from NGE/Vega-1 measurements 375
Hsieh, K.C., see Curtis, C.C., et al. 360
Huebner, W.F., see Keller, H.U., et al. 807
Huebner, W.F., see Schwarz, G., et al. 847
Huebner, W.F., see Wegmann, R., et al. 339

Hughes, D.W.: Comet P/Halley dust characteristics: a comparison between Orionid and Eta Aquarid meteor observations and those from the flyby spacecraft 879
Hughes, D.W., see Keller, H.U., et al. 807
Hughes, D.W., see McDonnell, J.A.M., et al. 719
Hunten, D.M., see Curtis, C.C., et al. 360
Hunten, D.M., see Hsieh, K.C., et al. 375
Huntress, W.T., see Balsiger, H., et al. 163
Huntress, W., see Allen, M., et al. 502
Hynds, R.J., see Richardson, I.G., et al. 276
Hynds, R.J., see Sanderson, T.R., et al. 125
Igenbergs, E., see McDonnell, J.A.M., et al. 719
Illiano, J.M., see Eberhardt, P., et al. 435
Illiano, J.M., see Eberhardt, P., et al. 481
Illiano, J.M., see Lämmerzahl, P., et al. 169
Ilyinskii, V.N., see Mazets, E.P., et al. 699
Inogamov, N.A., see Sagdeev, R.Z., et al. 179
Iovlev, M., see Vaisberg, O.L., et al. 753
Ip, W.-H., Schwenn, R., Rosenbauer, H., Balsiger, H., Neugebauer, M., Shelley, E.G.: An interpretation of the ion pile-up region outside the ionospheric contact surface 132
Ip, W.-H., see Allen, M., et al. 502
Ip, W.-H., see Balsiger, H., et al. 163
Ip, W.-H., see Curtis, C.C., et al. 360
Ip, W.-H., see Goldstein, B.E., et al. 174
Ip, W.-H., see Hsieh, K.C., et al. 375
Ip, W.-H., see Kömle, N.I. 405
Ip, W.-H., see Scarf, F.L., et al. 109
Ip, W.-H., see Schwenn, R., et al. 160
Ip, W.-H., see Shelley, E.G., et al. 304
Ipavich, F.M., see Tsurutani, B.T., et al. 97
Irvine, W.M., see Schloerb, F.P., et al. 475
Itoh, T., see Hirao, K. 39
Itoh, T., see Hirao, K. 965
Jackson, W.M., see Feldman, P.D., et al. 325
Jackson, W.M., see Prisant, M.G. 489
Jamar, C., see Keller, H.U., et al. 807
Jewitt, D.C., see Meech, K.J. 585
Jockers, K., Geyer, E.H., Rosenbauer, H., Hänel, A.: Observations of ions in comet P/Halley with a focal reducer 256
Jockers, K., see Coates, A.J., et al. 55
Jockers, K., see Johnstone, A.D., et al. 25
Jockers, K., see Thomsen, M.F., et al. 141
Jockers, K., see Wilken, B., et al. 153
Johnson, R.E., Cooper, J.F., Lanzerotti, L.J., Strazzulla, G.: Radiation formation of a non-volatile comet crust 889
Johnstone, A.D., Coates, A.J., Heath, J., Thomsen, M.F., Wilken, B., Jockers, K., Formisano, V., Amata, E., Winningham, J.D., Borg, H., Bryant, D.A.: Alfyénic turbulence in the solar wind flow during the approach to comet P/Halley 25
Johnstone, A.D., Glaßmeier, K.H., Acuña, M., Borg, H., Bryant, D., Coates, A., Formisano, V., Heath, J., Mariani, F., Musmann, G., Neubauer, F., Thomsen, M., Wilken, B., Winningham, J.: Waves in the magnetic field and solar wind flow outside the bow shock at comet P/Halley 47
Johnstone, A.D., see Coates, A.J., et al. 55
Johnstone, A.D., see Thomsen, M.F., et al. 141
Johnstone, A.D., see Wilken, B., et al. 153
Joy, M., see Campins, H., et al. 632
Joyce, R.R., see Brooke, T.Y., et al. 621
Joyce, R.R., see Knacke, R.F., et al. 625
Juchniewicz, J., see Savin, S., et al. 89
Kaminski, C.D., see Hanner, M.S., et al. 653
Kawakami, H., see Watanabe, J., et al. 229
Kecskeméty, K., see Gribov, B.E., et al. 293
Keller, H.U., Delamere, W.A., Huebner, W.F., Reitsema, H.J., Schmidt, H.U., Whipple, F.L., Wilhelm, K., Curdt, W., Kramm, R., Thomas, N., Arpigny, C., Barbieri, C., Bonnet, R.M., Cazes, S., Coradini, M., Cosmovici, C.B., Hughes, D.W., Jamar, C., Malaise, D., Schmidt, K., Schmidt, W.K.H., Seige, P.: Comet P/Halley's nucleus and its activity 807
Keller, H.U., see Schwarz, G., et al. 847
Keller, H.U., see Thomas, N. 843
Kennel, C.F., see Scarf, F.L., et al. 109
Keppler, E., see Curtis, C.C., et al. 360
Keppler, E., see Gribov, B.E., et al. 293
Keppler, E., see Hsieh, K.C., et al. 375
Kern, J.R., see Rettig, T.W., et al. 249
Khavenson, N.G., see Mazets, E.P., et al. 699
Khazanov, B., see Vaisberg, O.L., et al. 183
Khromov, V.N., see Sagdeev, R.Z., et al. 179
Kidger, M.R., Acosta, J.A., Garzón, F., Prieto, M., Gómez, R.: Low-resolution maps of comet P/Halley in principal atomic and molecular species 363
Kikuchi, S., Mikami, Y., Mukai, T., Mukai, S., Hough, J.H.: Polarimetry of comet P/Halley 689
Kikuchi, S., see Mukai, T., et al. 650
Kinoshita, H., see Watanabe, J., et al. 229
Kinzel, W.M., see Schloerb, F.P., et al. 475
Kissel, J., see Clark, B.C., et al. 779
Kissel, J., see Langevin, Y., et al. 761
Kissel, J., see McDonnell, J.A.M., et al. 719
Kissel, J., see Sagdeev, R.Z., et al. 179
Kissel, J., see Šolc, M., et al. 385
Kitayama, M., see Mukai, T., et al. 129
Kitayama, M., see Takahashi, S., et al. 94
Klimenko, I.N., see Gringauz, K.I., et al. 287
Klimov, S., see Savin, S., et al. 89
Knacke, R.F., Brooke, T.Y., Joyce, R.R.: The 3.2–3.6 μm emission features in comet P/Halley: spectral identifications and similarities 625
Knacke, R.F., see Brooke, T.Y., et al. 621
Knickerbocker, K.L., see Rettig, T.W., et al. 249
Kohl-Moreira, J.L., see Festou, M.C., et al. 575
Kömle, N.I., Ip, W.-H.: Anisotropic non-stationary gas flow dynamics in the coma of comet P/Halley 405
Korth, A., Richter, A.K., Mendis, D.A., Anderson, K.A., Carlson, C.W., Curtis, D.W., Lin, R.P., Mitchell, D.L., Rème, H., Sauvaud, J.A., d'Uston, C.: The composition and radial dependence of cometary ions in the coma of comet P/Halley 149
Korth, A., see Anderson, K.A., et al. 290
Korth, A., see d'Uston, C., et al. 137
Korth, A., see Rème, H., et al. 33
Kozai, Y., see Watanabe, J., et al. 229
Kramm, R., see Keller, H.U., et al. 807
Kramm, R., see Schwarz, G., et al. 847
Kranjc, A., see Barbieri, C., et al. 893
Krankowsky, D., see Eberhardt, P., et al. 435
Krankowsky, D., see Eberhardt, P., et al. 481
Krankowsky, D., see Lämmerzahl, P., et al. 169
Krasikov, V.A., see Sagdeev, R.Z., et al. 835
Krasnopolsky, V.A., Tkachuk, A.Y.: Curves of growth of emission lines in cometary spectra. Implications for H_2O and OH bands of comet P/Halley 431

Krasnopolsky, V.A., Moroz, V.I., Krysko, A.A., Tkachuk, A.Y., Moreels, G., Clairemidi, J., Parisot, J.P., Gogoshev, M., Gogosheva, T.: Properties of dust in comet P/Halley measured by the Vega-2 three-channel spectrometer 707
Krasnopolsky, V.A., see Moreels, G., et al. 551
Krasnopolsky, V.A., see Moroz, V.I., et al. 513
Krawczyk, Z., see Savin, S., et al. 89
Kresák, L.: Dormant phases in the aging of periodic comets 906
Kresák, L., see Carusi, A., et al. 899
Kresáková, M.: Associations between ancient comets and meteor showers 935
Krishna Swamy, K.S.: Study of the isotopic features of Swan bands in comets 388
Krishna Swamy, K.S., see Wallis, M.K. 329
Krysko, A.A., see Krasnopolsky, V.A., et al. 707
Krysko, A.A., see Moreels, G., et al. 551
Ksanfomality, L.V., see Simpson, J.A., et al. 742
Kuczera, H., see McDonnell, J.A.M., et al. 719
Kursinski, E.R., see Edenhofer, P., et al. 712
Lamarre, J.M., see Emerich, C., et al. 839
Lamarre, J.M., see Moroz, V.I., et al. 513
Lambert, D.L., see Arpigny, C., et al. 485
Lämmerzahl, P., Krankowsky, D., Hodges, R.R., Stubbemann, U., Woweries, J., Herrwerth, I., Berthelier, J.J., Illiano, J.M., Eberhardt, P., Dolder, U., Schulte, W., Hoffman, J.H.: Expansion velocity and temperatures of gas and ions measured in the coma of comet P/Halley 169
Lämmerzahl, P., see Eberhardt, P., et al. 435
Lämmerzahl, P., see Eberhardt, P., et al. 481
Lamy, P.L., Grün, E., Perrin, J.M.: Comet P/Halley: implications of the mass distribution function for the photopolarimetric properties of the dust coma 767
Lamy, P.L., Pedersen, H., Vio, R.: The dust tail of comet P/Halley in April 1986 661
Langevin, Y., Kissel, J., Bertaux, J.-L., Chassefière, E.: First statistical analysis of 5000 mass spectra of cometary grains obtained by PUMA 1 (Vega-1) and PIA (Giotto) impact ionization mass spectrometers in the compressed modes 761
Langevin, Y., see McDonnell, J.A.M., et al. 719
Lanzerotti, L.J., see Johnson, R.E., et al. 889
Larson, H.P., Mumma, M.J., Weaver, H.A.: Kinematic properties of the neutral gas outflow from comet P/Halley 391
Larson, H.P., see Drapatz, S., et al. 497
Larson, H.P., see Mumma, M.J., et al. 419
Larson, H.P., see Weaver, H.A., et al. 411
Larson, S., Sekanina, Z., Levy, D., Tapia, S., Senay, M.: Comet P/Halley near-nucleus phenomena in 1986 639
Larson, S.M., see Sekanina, Z., et al. 645
Larson, S., see Sagdeev, R.Z., et al. 835
Lázaro, C., Garzón, F., Arévalo, M.J.: Low resolution mapping of comet P/Halley in the near-infrared 605
Lazarus, A.J., see Balsiger, H., et al. 163
Lazarus, A.J., see Neugebauer, M., et al. 21
Lazarus, A.J., see Shelley, E.G., et al. 304
Le Borgne, J.F., Leroy, J.L., Arnaud, J.: Polarimetry of comet P/Halley: continuum versus molecular bands 526
Leach, S.: Electronic spectroscopy and relaxation of some molecular cations of cometary interest 195
Lecacheux, J., see Festou, M.C., et al. 575
Leroy, J.L., see Le Borgne, J.F., et al. 526
Lester, D.F., see Campins, H., et al. 632
Levy, D., see Larson, S., et al. 639
Lezhen, L.A., see Verigin, M.I., et al. 121
Lin, R.P., see Anderson, K.A., et al. 290
Lin, R.P., see d'Uston, C., et al. 137
Lin, R.P., see Korth, A., et al. 149
Lin, R.P., see Rème, H., et al. 33
Lindblad, B.A.: The meteor stream associated with comet P/Grigg-Skjellerup 931
Lindblad, B.A.: The 1985 return of the Giacobinid meteor stream 928
Lindblad, B.A., see McDonnell, J.A.M., et al. 719
Liu, Z.L.: Photographic observations of tail-formation activities of comet P/Halley in November 1985 225
Loewenstein, R.F., see Glaccum, W., et al. 635
Lorenzetti, D., Moneti, A., Stanga, R., Strafella, F.: Infrared monitoring of comet P/Halley 609
Lugten, J.B., see Stacey, G.J., et al. 451
Lundstedt, H., Magnusson, P.: Two disconnection events in comet P/Halley and possible solar causes 261
Maas, D., see Göller, J.R., et al. 693
Magain, P., see Arpigny, C., et al. 485
Magnusson, P., see Lundstedt, H. 261
Maillard, J.P., Crovisier, J., Encrenaz, T., Combes, M.: The spectrum of comet P/Halley between 0.9 and 2.5 μm 398
Malaise, D., see Keller, H.U., et al. 807
Managadze, G.G., see Sagdeev, R.Z., et al. 179
Mandeville, J.-C., see McDonnell, J.A.M., et al. 719
Manfroid, J., see Arpigny, C., et al. 485
Manfroid, J., see Sterken, C., et al. 523
Maran, S.P., see Brandt, J.C., et al. 969
Marcout, J., see Florsch, A., et al. 357
Mariani, F., see Glaßmeier, K.H., et al. 65
Mariani, F., see Johnstone, A.D., et al. 47
Marsden, R.G., see Gribov, B.E., et al. 293
Mason, L.W., see Clark, B.C., et al. 779
Mazets, E.P., Sagdeev, R.Z., Aptekar, R.L., Golenetskii, S.V., Guryan Yu, A., Dyachkov, A.V., Ilyinskii, V.N., Panov, V.N., Petrov, G.G., Savvin, A.V., Sokolov, I.A., Frederiks, D.D., Khavenson, N.G., Shapiro, V.D., Shevchenko, V.I.: Dust in comet P/Halley from Vega observations 699
McCoy, R.P., see Opal, C.B., et al. 320
McDonnell, J.A.M., Alexander, W.M., Burton, W.M., Bussoletti, E., Evans, G.C., Evans, S.T., Firth, J.G., Grard, R.J.L., Green, S.F., Grün, E., Hanner, M.S., Hughes, D.W., Igenbergs, E., Kissel, J., Kuczera, H., Lindblad, B.A., Langevin, Y., Mandeville, J.-C., Nappo, S., Pankiewicz, G.S.A., Perry, C.H., Schwehm, G.H., Sekanina, Z., Stevenson TJ: The dust distribution within the inner coma of comet P/Halley (1982i): encounter by Giotto's impact detectors 719
McDonnell, J.A.M., see Grard, R.J.L., et al. 785
McFadden, L.A., A'Hearn, M.F., Feldman, P.D., Roettger, E.E., Edsall, D.M., Butterworth, P.S.: Activity of comet P/Halley on March 23–25, 1986: IUE observations 333
McFadden, L.A., see Feldman, P.D., et al. 325
Meech, K.J., Jewitt, D.C.: Observations of comet P/Halley at minimum phase angle 585
Meier, A., see Balsiger, H., et al. 163
Meier, A., see Schwenn, R., et al. 160
Mendis, A., see d'Uston, C., et al. 137
Mendis, D.A.: Exploration of Halley's Comet: symposium summary 939
Mendis, D.A., see Anderson, K.A., et al. 290
Mendis, D.A., see Korth, A., et al. 149
Mendis, D.A., see Rème, H., et al. 33

Merényi, E., see Sagdeev, R.Z., et al. 835
Metz, K., Haefner, R.: Circular polarization near the nucleus of comet P/Halley 539
Mikami, Y., see Kikuchi, S., et al. 689
Mikhailov, Y., see Mogilevsky, M., et al. 80
Mikhailov, Y., see Pedersen, A., et al. 297
Mikhailov, Y., see Trotignon, J.G., et al. 83
Mikusch, E., see Schwarz, G., et al. 847
Millis, R.L., see Schleicher, D.G., et al. 531
Minami, S., see Saito, T., et al. 209
Minami, S., see Tomita, K., et al. 215
Mitchell, D.L., see Korth, A., et al. 149
Miyake, W., see Mukai, T., et al. 129
Miyake, W., see Takahashi, S., et al. 94
Mogilevsky, M., Mikhailov, Y., Molchanov, O., Grard, R., Pedersen, A., Trotignon, J.G., Béghin, C., Formisano, V., Shapiro, V., Shevchenko, V.: Identification of boundaries in the cometary environment from ac electric field measurements 80
Mogilevsky, M., see Pedersen, A., et al. 297
Mogilevsky, M., see Trotignon, J.G., et al. 83
Molchanov, O., see Mogilevsky, M., et al. 80
Moneti, A., see Lorenzetti, D., et al. 609
Moore, V., see Richardson, I.G., et al. 276
Moreels, G., Clairemidi, J., Parisot, J.P., Zucconi, J.M., Bertaux, J.L., Blamont, J.E., Hersé, M., Krasnopolsky, V.A., Moroz, V.I., Krysko, A.A., Tkachuk, A.Y., Gogoshev, M., Gogosheva, T., Werner, R., Spasov, S.: Spectrophotometry of comet P/Halley at wavelengths 275–710 nm from Vega-2 551
Moreels, G., see Krasnopolsky, V.A., et al. 707
Moroz, V.I., Combes, M., Bibring, J.P., Coron, N., Crovisier, J., Encrenaz, T., Crifo, J.F., Sanko, N., Grigoryev, A.V., Bockelée-Morvan, D., Gispert, R., Nikolsky, Y.V., Emerich, C., Lamarre, J.M., Rocard, F., Krasnopolsky, V.A., Owen, T.: Detection of parent molecules in comet P/Halley from the IKS-Vega experiment 513
Moroz, V.I., see Emerich, C., et al. 839
Moroz, V.I., see Krasnopolsky, V.A., et al. 707
Moroz, V.I., see Moreels, G., et al. 551
Morris, C.S., see Green, D.W.E. 560
Moseley, S.H., see Glaccum, W., et al. 635
Mottinger, N.A., see Edenhofer, P., et al. 712
Mukai, S., see Kikuchi, S., et al. 689
Mukai, S., see Mukai, T., et al. 650
Mukai, T., Miyake, W., Terasawa, T., Kitayama, M., Hirao, K.: Spatial distribution of water-group ions near comet P/Halley observed by Suisei 129
Mukai, T., Mukai, S., Kikuchi, S.: Complex refractive index of grain material deduced from the visible polarimetry of comet P/Halley 650
Mukai, T., see Kikuchi, S., et al. 689
Mukai, T., see Takahashi, S., et al. 94
Mumma, M.J., Weaver, H.A., Larson, H.P.: The ortho-para ratio of water vapor in comet P/Halley 419
Mumma, M.J., see Larson, H.P., et al. 391
Mumma, M.J., see Weaver, H.A., et al. 411
Münch, G., see Appenzeller, I. 465
Münch, G., see Neckel, T. 581
Musmann, G., see Johnstone, A.D., et al. 47
Nagata, T., see Tokunaga, A.T., et al. 519
Nakagawa, T., see Yumoto, K., et al. 117
Nakamura, T., see Watanabe, J., et al. 229
Nappo, S., see McDonnell, J.A.M., et al. 719
Neckel, T., Münch, G.: Photometry of comet P/Halley at near post-perihelion phases 581
Ness, N.F., see Raeder, J., et al. 61
Neubauer, F.M.: Giotto magnetic-field results on the boundaries of the pile-up region and the magnetic cavity 73
Neubauer, F.M., see Glaßmeier, K.H., et al. 65
Neubauer, F.M., see Neugebauer, M., et al. 21
Neubauer, F.M., see Raeder, J., et al. 61
Neubauer, F., see Johnstone, A.D., et al. 47
Neugebauer, M., Lazarus, A.J., Altwegg, K., Balsiger, H., Goldstein, B.E., Goldstein, R., Neubauer, F.M., Rosenbauer, H., Schwenn, R., Shelley, E.G., Ungstrup, E.: The pick-up of cometary protons by the solar wind 21
Neugebauer, M., see Balsiger, H., et al. 163
Neugebauer, M., see Goldstein, B.E., et al. 174
Neugebauer, M., see Goldstein, R., et al. 220
Neugebauer, M., see Ip, W.-H., et al. 132
Neugebauer, M., see Shelley, E.G., et al. 304
Newburn RL, Jr., see Divine, N. 867
Newburn RL, Jr., see Edberg, S.J., et al. 981
Niedner MB, Jr., Schwingenschuh, K.: Plasma-tail activity at the time of the Vega encounters 103
Niedner MB, Jr., see Brandt, J.C. 281
Niedner MB, Jr., see Brandt, J.C., et al. 969
Niedner, M.B., see Brosius, J.W., et al. 267
Nikolsky, Y.V., see Emerich, C., et al. 839
Nikolsky, Y.V., see Moroz, V.I., et al. 513
Notni, P., Tiersch, H.: Charging of dust particles in comets and in interplanetary space 796
Nozdrachev, M., see Savin, S., et al. 89
Olson, R.J.M., Pasachoff, J.M.: New information on comet P/Halley as depicted by Giotto di Bondone and other Western artists 1
Olsson-Steel, D.I.: The dynamical lifetime of comet P/Halley 909
Omelchenko, A., see Vaisberg, O.L., et al. 183
Omelchenko, A., see Vaisberg, O.L., et al. 753
Opal, C.B., McCoy, R.P., Carruthers, G.R.: Far-ultraviolet objective spectra of comet P/Halley from sounding rockets 320
Orlowski, D., see Savin, S., et al. 89
Owen, T., see Moroz, V.I., et al. 513
Palagi, F., see Falchi, A., et al. 462
Pankiewicz, G.S.A., see McDonnell, J.A.M., et al. 719
Panov, V.N., see Mazets, E.P., et al. 699
Parisot, J.P., see Krasnopolsky, V.A., et al. 707
Parisot, J.P., see Moreels, G., et al. 551
Pasachoff, J.M., see Olson, R.J.M. 1
Patriarchi, P., see Feldman, P.D., et al. 325
Pedersen, A., Grard, R., Trotignon, J.G., Béghin, C., Mikhailov, Y., Mogilevsky, M.: Measurements of low energy electrons and spacecraft potentials near comet P/Halley 297
Pedersen, A., see Mogilevsky, M., et al. 80
Pedersen, A., see Trotignon, J.G., et al. 83
Pedersen, H., see Lamy, P.L., et al. 661
Perozzi, E., see Carusi, A., et al. 899
Perrin, J.M., see Lamy, P.L., et al. 767
Perry, C.H., see McDonnell, J.A.M., et al. 719
Petrov, G.G., see Mazets, E.P., et al. 699
Porsche, H., see Edenhofer, P., et al. 712
Prieto, M., see Kidger, M.R., et al. 363
Prilutski, O.F., see Sagdeev, R.Z., et al. 179
Prisant, M.G., Jackson, W.M.: A rotational-state population analysis of the high-resolution IUE observation of CS emission in comet P/Halley 489

Puel, F., see Festou, M.C., et al. 575
Qiu, P.Z., see Wu, M.C. 264
Rabilizirov, R., see Wallis, M.K., et al. 801
Rabinowitz, D., see Simpson, J.A., et al. 742
Raeder, J., Neubauer, F.M., Ness, N.F., Burlaga, L.F.: Macroscopic perturbations of the IMF by P/Halley as seen by the Giotto magnetometer 61
Rahe, J., see Edberg, S.J., et al. 981
Rajamohan, R., see Sivaraman, K.R., et al. 543
Ramsey, B.D., see Campins, H., et al. 601
Rank, D.M., see Bregman, J.D., et al. 616
Reinhard, R.: The Giotto mission to Halley's Comet 949
Reitsema, H.J., see Keller, H.U., et al. 807
Reitsema, H., see Schwarz, G., et al. 847
Rème, H., Sauvaud, J.A., d'Uston, C., Cros, A., Anderson, K.A., Carlson, C.W., Curtis, D.W., Lin, R.P., Korth, A., Richter, A.K., Mendis, D.A.: General features of comet P/Halley: solar wind interaction from plasma measurements 33
Rème, H., see Anderson, K.A., et al. 290
Rème, H., see d'Uston, C., et al. 137
Rème, H., see Korth, A., et al. 149
Remizov, A.P., see Gribov, B.E., et al. 293
Remizov, A.P., see Gringauz, K.I., et al. 191
Remizov, A.P., see Gringauz, K.I., et al. 287
Remizov, A.P., see Verigin, M.I., et al. 121
Rettenmund, U., see Balsiger, H., et al. 163
Rettenmund, U., see Goldstein, B.E., et al. 174
Rettig, T.W., Kern, J.R., Ruchti, R., Baumbaugh, B., Baumbaugh, A.E., Knickerbocker, K.L., Dawe, J.: Observations of the coma of comet P/Halley and the outburst of 1986 March 24–25 (UT) 249
Richardson, I.G., Cowley, S.W.H., Moore, V., Staines, K., Hynds, R.J., Sanderson, T.R., Wenzel, K.-P., Daly, P.W.: Energy spectra of energetic ions in the vicinity of comet P/Giacobini-Zinner 276
Richardson, I.G., see Sanderson, T.R., et al. 125
Richter, A.K., see Anderson, K.A., et al. 290
Richter, A.K., see Curtis, C.C., et al. 360
Richter, A.K., see d'Uston, C., et al. 137
Richter, A.K., see Gribov, B.E., et al. 293
Richter, A.K., see Gringauz, K.I., et al. 191
Richter, A.K., see Gringauz, K.I., et al. 287
Richter, A.K., see Hsieh, K.C., et al. 375
Richter, A.K., see Korth, A., et al. 149
Richter, A.K., see Rème, H., et al. 33
Richter, A.K., see Verigin, M.I., et al. 121
Riedler, W., see Gribov, B.E., et al. 293
Riedler, W., see Yeroshenko, Y.G., et al. 69
Rocard, F., see Emerich, C., et al. 839
Rocard, F., see Moroz, V.I., et al. 513
Roettger, E.E., see McFadden, L.A., et al. 333
Rosenbauer, H., see Allen, M., et al. 502
Rosenbauer, H., see Balsiger, H., et al. 163
Rosenbauer, H., see Goldstein, B.E., et al. 174
Rosenbauer, H., see Goldstein, R., et al. 220
Rosenbauer, H., see Ip, W.-H., et al. 132
Rosenbauer, H., see Jockers, K., et al. 256
Rosenbauer, H., see Neugebauer, M., et al. 21
Rosenbauer, H., see Schwenn, R., et al. 160
Rosenbauer, H., see Wilken, B., et al. 153
Ruchti, R., see Rettig, T.W., et al. 249
Russel, C.T., see Yeroshenko, Y.G., et al. 69
Sagdeev, R.Z.: The Vega mission to P/Halley comet 959
Sagdeev, R.Z., Kissel, J., Evlanov, E.N., Fomenkova, M.N., Inogamov, N.A., Khromov, V.N., Managadze, G.G., Prilutski, O.F., Shapiro, V.D., Shutyaev, I.Y., Zubkov, B.V.: The dependence of mass resolution and sensitivity of the PUMA instrument on the energy spread of ions produced by hypervelocity impacts 179
Sagdeev, R.Z., Smith, B., Szegö, K., Larson, S., Tóth, I., Merényi, E., Avanesov, G.A., Krasikov, V.A., Shamis, V.A., Tarnapolski, V.I.: The spatial distribution of dust jets seen during the Vega-2 flyby 835
Sagdeev, R.Z., see Gribov, B.E., et al. 293
Sagdeev, R.Z., see Mazets, E.P., et al. 699
Sagdeev, R.Z., see Simpson, J.A., et al. 742
Saito, K., see Saito, T., et al. 201
Saito, K., see Saito, T., et al. 209
Saito, T., Saito, K., Aoki, T., Yumoto, K.: Possible models on disturbances of the plasma tail of comet P/Halley during the 1985–1986 apparition 201
Saito, T., Yumoto, K., Hirao, K., Minami, S., Saito, K., Smith, E.: Structure and dynamics of the plasma tail of comet P/Halley. I. Knot event on December 31, 1985 209
Saito, T., see Tomita, K., et al. 215
Saito, T., see Yumoto, K., et al. 117
Sanderson, T.R., Wenzel, K.-P., Daly, P.W., Cowley, S.W.H., Hynds, R.J., Richardson, I.G., Smith, E.J., Bame, S.J., Zwickl, R.D.: Observations of heavy energetic ions far upstream from comet P/Halley 125
Sanderson, T.R., see Richardson, I.G., et al. 276
Sanko, N.F., see Emerich, C., et al. 839
Sanko, N., see Moroz, V.I., et al. 513
Sauer, K., see Baumgärtel, K. 307
Sauvaud, J.A., see Anderson, K.A., et al. 290
Sauvaud, J.A., see d'Uston, C., et al. 137
Sauvaud, J.A., see Korth, A., et al. 149
Sauvaud, J.A., see Rème, H., et al. 33
Savin, S., Avanesova, G., Balikhin, M., Wozniak, D., Wronowski, P., Klimov, S., Krawczyk, Z., Nozdrachev, M., Orlowski, D., Sokolov, A., Juchniewicz, J.: Comparative study of the low-frequency waves near comet P/Halley during the Vega-1 and Vega-2 flybys 89
Savvin, A.V., see Mazets, E.P., et al. 699
Scardia, M., see Barbieri, C., et al. 893
Scarf, F.L., Coroniti, F.V., Kennel, C.F., Gurnett, D.A., Ip, W.-H., Smith, E.J.: Observations of cometary plasma-wave phenomena 109
Scarf, F.L., see Tsurutani, B.T., et al. 97
Schleicher, D.G., Millis, R.L., Birch, P.V.: Photometric observations of comet P/Giacobini-Zinner 531
Schleicher, D.G., see Feldman, P.D., et al. 325
Schloerb, F.P., Claussen, M.J., Tacconi-Garman, L.: OH radio observations of comet P/Halley 469
Schloerb, F.P., Kinzel, W.M., Swade, D.A., Irvine, W.M.: Observations of HCN in comet P/Halley 475
Schmidt, H.U., see Keller, H.U., et al. 807
Schmidt, H.U., see Wegmann, R., et al. 339
Schmidt, K., see Keller, H.U., et al. 807
Schmidt, R.E., see Sekanina, Z., et al. 645
Schmidt, W.K.H., see Keller, H.U., et al. 807

Schmidt-Kaler, T., see Celnik, W.E. 233
Schulte, W., see Eberhardt, P., et al. 435
Schulte, W., see Eberhardt, P., et al. 481
Schulte, W., see Lämmerzahl, P., et al. 169
Schwarz, G., Craubner, H., Delamere, A., Göbel, M., Gonano, M., Huebner, W.F., Keller, H.U., Kramm, R., Mikusch, E., Reitsema, H., Whipple, F.L., Wilhelm, K.: Detailed analysis of a surface feature on comet P/Halley 847
Schwehm, G.H., see McDonnell, J.A.M., et al. 719
Schwenn, R., Ip, W.-H., Rosenbauer, H., Balsiger, H., Bühler, F., Goldstein, R., Meier, A., Shelley, E.G.: Ion temperature and flow profiles in comet P/Halley's close environment 160
Schwenn, R., see Allen, M., et al. 502
Schwenn, R., see Balsiger, H., et al. 163
Schwenn, R., see Goldstein, B.E., et al. 174
Schwenn, R., see Goldstein, R., et al. 220
Schwenn, R., see Ip, W.-H., et al. 132
Schwenn, R., see Neugebauer, M., et al. 21
Schwingenschuh, K., see Gribov, B.E., et al. 293
Schwingenschuh, K., see Niedner MB, Jr. 103
Schwingenschuh, K., see Yeroshenko, Y.G., et al. 69
Seige, P., see Keller, H.U., et al. 807
Sekanina, Z.: Dust environment of comet P/Halley: a review 789
Sekanina, Z., Larson, S.M., Emerson, G., Helin, E.F., Schmidt, R.E.: The sunward spike of Halley's comet 645
Sekanina, Z., see Larson, S., et al. 639
Sekanina, Z., see McDonnell, J.A.M., et al. 719
Senay, M., see Larson, S., et al. 639
Shamis, V.A., see Sagdeev, R.Z., et al. 835
Shapiro, V.D., see Gribov, B.E., et al. 293
Shapiro, V.D., see Mazets, E.P., et al. 699
Shapiro, V.D., see Sagdeev, R.Z., et al. 179
Shapiro, V., see Mogilevsky, M., et al. 80
Shelley, E.G., Fuselier, S.A., Balsiger, H., Drake, J.F., Geiss, J., Goldstein, B.E., Goldstein, R., Ip, W.-H., Lazarus, A.J., Neugebauer, M.: Charge exchange of solar wind ions in the coma of comet P/Halley 304
Shelley, E.G., see Balsiger, H., et al. 163
Shelley, E.G., see Goldstein, B.E., et al. 174
Shelley, E.G., see Goldstein, R., et al. 220
Shelley, E.G., see Ip, W.-H., et al. 132
Shelley, E.G., see Neugebauer, M., et al. 21
Shelley, E.G., see Schwenn, R., et al. 160
Shelley, E., see Allen, M., et al. 502
Shevchenko, V.I., see Gribov, B.E., et al. 293
Shevchenko, V.I., see Mazets, E.P., et al. 699
Shevchenko, V., see Mogilevsky, M., et al. 80
Shutyaev, I.Y., see Sagdeev, R.Z., et al. 179
Shylaja, B.S., see Sivaraman, K.R., et al. 543
Simpson, J.A., Rabinowitz, D., Tuzzolino, A.J., Ksanfomality, L.V., Sagdeev, R.Z.: The dust coma of comet P/Halley: measurements on the Vega-1 and Vega-2 spacecraft 742
Sitarski, G., Ziolkowski, K.: A new approach to investigations of the long-term motion of comet P/Halley 896
Sivaraman, K.R., Babu, G.S.D., Shylaja, B.S., Rajamohan, R.: Spectrophotometry of comet P/Halley. I. Flux, column density and emission gradients within the coma in the emission bands and the continuum 543
Slavin, J.A., see Brosius, J.W., et al. 267
Smirnov, V.N., Vaisberg, O.L., Anisimov, S.: An attempt to evaluate the structure of cometary dust particles 774
Smirnov, V., see Vaisberg, O.L., et al. 183
Smirnov, V., see Vaisberg, O.L., et al. 753
Smith, B., see Sagdeev, R.Z., et al. 835
Smith, E.J., see Brosius, J.W., et al. 267
Smith, E.J., see Sanderson, T.R., et al. 125
Smith, E.J., see Scarf, F.L., et al. 109
Smith, E.J., see Tsurutani, B.T., et al. 97
Smith, E., see Saito, T., et al. 209
Smith, R.G., see Tokunaga, A.T., et al. 519
Sokolov, A., see Savin, S., et al. 89
Sokolov, I.A., see Mazets, E.P., et al. 699
Šolc, M., Vanýsek, V., Kissel, J.: Carbon-isotope ratio in PUMA 1 spectra of P/Halley dust 385
Somogyi, A.J., see Curtis, C.C., et al. 360
Somogyi, A.J., see Gribov, B.E., et al. 293
Somogyi, A.J., see Hsieh, K.C., et al. 375
Spasov, S., see Moreels, G., et al. 551
Spinrad, H., see Belton, M.J.S., et al. 569
Stacey, G.J., Lugten, J.B., Genzel, R.: Detection of OH rotational emission from comet P/Halley in the far-infrared 451
Staines, K., see Richardson, I.G., et al. 276
Stanga, R., see Lorenzetti, D., et al. 609
Stelzried, C.T., see Edenhofer, P., et al. 712
Sterken, C., Manfroid, J., Arpigny, C.: Photometry of P/Halley (1982i) 523
Stevenson TJ, see McDonnell, J.A.M., et al. 719
Stewart, A.I.F.: Pioneer Venus measurements of H, O, and C production in comet P/Halley near perihelion 369
Štohl, J.: Meteor contribution by short-period comets 933
Storrs, A.D., see Hammel, H.B., et al. 665
Strafella, F., see Lorenzetti, D., et al. 609
Strazzulla, G., see Johnson, R.E., et al. 889
Stubbemann, U., see Eberhardt, P., et al. 481
Stubbemann, U., see Lämmerzahl, P., et al. 169
Stüdemann, W., see Thomsen, M.F., et al. 141
Stüdemann, W., see Wilken, B., et al. 153
Styashkin, V.A., see Yeroshenko, Y.G., et al. 69
Suchail, J.-L., see Dollfus, A. 669
Sun, S.S., see Gong , et al. 594
Swade, D.A., see Schloerb, F.P., et al. 475
Szegö, K., see Gribov, B.E., et al. 293
Szegö, K., see Gringauz, K.I., et al. 191
Szegö, K., see Gringauz, K.I., et al. 287
Szegö, K., see Sagdeev, R.Z., et al. 835
Szegö, K., see Verigin, M.I., et al. 121
Szemerey, I., see Verigin, M.I., et al. 121
Szemerey, T., see Gringauz, K.I., et al. 287
Tacconi-Garman, L., see Schloerb, F.P., et al. 469
Takahashi, S., Terasawa, T., Mukai, T., Kitayama, M., Miyake, W., Hirao, K.: Plasma flow in the cometosheath of P/Halley during the encounter of Suisei 94
Tapia, S., see Larson, S., et al. 639
Tarnapolski, V.I., see Sagdeev, R.Z., et al. 835
Tátrallyay, M., see Gringauz, K.I., et al. 191
Tátrallyay, M., see Gringauz, K.I., et al. 287
Tátrallyay, M., see Verigin, M.I., et al. 121
Telesco, C.M., see Campins, H., et al. 601
Telesco, C.M., see Hammel, H.B., et al. 665
Terasawa, T., see Mukai, T., et al. 129

Terasawa, T., see Takahashi, S., et al. 94
Thomas, N., Keller, H.U.: Fine dust structures in the emission of comet P/Halley observed by the Halley Multicolour Camera on board Giotto 843
Thomas, N., see Keller, H.U., et al. 807
Thomsen, M.F., Feldman, W.C., Wilken, B., Jockers, K., Stüdemann, W., Johnstone, A.D., Coates, A., Formisano, V., Amata, E., Winningham, J.D., Borg, H., Bryant, D., Wallis, M.K.: In-situ observations of a bi-modal ion distribution in the outer coma of comet P/Halley 141
Thomsen, M.F., see Coates, A.J., et al. 55
Thomsen, M.F., see Johnstone, A.D., et al. 25
Thomsen, M., see Johnstone, A.D., et al. 47
Thomson, M.F., see Wilken, B., et al. 153
Thorne, R.M., see Tsurutani, B.T., et al. 97
Tielens, A.G.G.M., see Bregman, J.D., et al. 616
Tiersch, H., see Notni, P. 796
Tkachuk, A.Y., see Krasnopolsky, V.A. 431
Tkachuk, A.Y., see Krasnopolsky, V.A., et al. 707
Tkachuk, A.Y., see Moreels, G., et al. 551
Tofani, G., see Falchi, A., et al. 462
Tokunaga, A.T., Nagata, T., Smith, R.G.: Detection of a new emission band at 2.8 μm in comet P/Halley 519
Tokunaga, A.T., see Hanner, M.S., et al. 653
Tomita, K., Saito, T., Minami, S.: Structure and dynamics of the plasma tail of comet P/Halley. II. Kink event on January 10–11, 1986 215
Tomita, K., see Watanabe, J., et al. 229
Tóth, I., see Sagdeev, R.Z., et al. 835
Tozzi, G.P., see Feldman, P.D., et al. 325
Traversa, G., see Florsch, A., et al. 357
Trotignon, J.G., Béghin, C., Grard, R., Pedersen, A., Formisano, V., Mogilevsky, M., Mikhailov, Y.: Dust observations of comet P/Halley by the plasma-wave analyser 83
Trotignon, J.G., see Mogilevsky, M., et al. 80
Trotignon, J.G., see Pedersen, A., et al. 297
Tsurutani, B.T., Brinca, A.L., Smith, E.J., Thorne, R.M., Scarf, F.L., Gosling, J.T., Ipavich, F.M.: MHD waves detected by ICE at distances $\geqq 28\ 10^6$ km from comet P/Halley: Cometary or solar wind origin? 97
Tsurutani, B.T., see Brinca, A.L. 311
Tuzzolino, A.J., see Simpson, J.A., et al. 742
Umlauft, G., see Curtis, C.C., et al. 360
Umlauft, G., see Hsieh, K.C., et al. 375
Ungstrup, E., see Balsiger, H., et al. 163
Ungstrup, E., see Neugebauer, M., et al. 21
Vaisberg, O.L., Smirnov, V., Omelchenko, A., Gorn, L., Iovlev, M.: Spatial and mass distribution of low-mass dust particles ($m < 10^{-10}$ g) in comet P/Halley's coma 753
Vaisberg, O.L., Zastenker, G., Smirnov, V., Khazanov, B., Omelchenko, A., Fedorov, A., Zakharov, D.: Spatial distribution of heavy ions in comet P/Halley's coma 183
Vaisberg, O.L., see Smirnov, V.N., et al. 774
Valsecchi, G.B., see Carusi, A., et al. 899
Vanýsek, V., see Šolc, M., et al. 385
Verigin, M.I., Gringauz, K.I., Richter, A.K., Gombosi, T.I., Remizov, A.P., Szegö, K., Apáthy, I., Szemerey, I., Tátrallyay, M., Lezhen, L.A.: Plasma properties from the upstream region to the cometopause of comet P/Halley: Vega observations 121
Verigin, M.I., see Gringauz, K.I., et al. 191
Verigin, M.I., see Gringauz, K.I., et al. 287
Vio, R., see Lamy, P.L., et al. 661
Volland, H., see Edenhofer, P., et al. 712
von Rosenvinge, T.T., see Brandt, J.C., et al. 969
Wallis, M.K., Krishna Swamy, K.S.: Some diatomic molecules from comet P/Halley's UV spectra near spacecraft flybys 329
Wallis, M.K., Rabilizirov, R., Wickramasinghe, N.C.: Evaporating grains in P/Halley's coma 801
Wallis, M.K., see Feldman, P.D., et al. 325
Wallis, M.K., see Thomsen, M.F., et al. 141
Watanabe, J., Kawakami, H., Tomita, K., Kinoshita, H., Nakamura, T., Kozai, Y.: The outburst of comet P/Halley on December 12, 1985 229
Weaver, H.A., Mumma, M.J., Larson, H.P.: Infrared investigation of water in comet P/Halley 411
Weaver, H.A., see Feldman, P.D., et al. 325
Weaver, H.A., see Larson, H.P., et al. 391
Weaver, H.A., see Mumma, M.J., et al. 419
Wegmann, R., Schmidt, H.U., Huebner, W.F., Boice, D.C.: Cometary MHD and chemistry 339
Wehinger, P.A., see Belton, M.J.S., et al. 569
Weissman, P.R.: Post-perihelion brightening of comet P/Halley: Springtime for Halley 873
Wenzel, K.-P., see Gribov, B.E., et al. 293
Wenzel, K.-P., see Richardson, I.G., et al. 276
Wenzel, K.-P., see Sanderson, T.R., et al. 125
Werner, R., see Moreels, G., et al. 551
Whipple, F.L.: The cometary nucleus: current concepts 852
Whipple, F.L., see Keller, H.U., et al. 807
Whipple, F.L., see Schwarz, G., et al. 847
Wickramasinghe, N.C., see Wallis, M.K., et al. 801
Wilhelm, K., see Keller, H.U., et al. 807
Wilhelm, K., see Schwarz, G., et al. 847
Wilken, B., Johnstone, A.D., Coates, A., Borg, H., Amata, E., Formisano, V., Jockers, K., Rosenbauer, H., Stüdemann, W., Thomson, M.F., Winningham, J.D.: Pick-up ions at comet P/Halley's bow shock: observations with the IIS spectrometer on Giotto 153
Wilken, B., see Coates, A.J., et al. 55
Wilken, B., see Johnstone, A.D., et al. 25
Wilken, B., see Johnstone, A.D., et al. 47
Wilken, B., see Thomsen, M.F., et al. 141
Winningham, J.D., see Coates, A.J., et al. 55
Winningham, J.D., see Johnstone, A.D., et al. 25
Winningham, J.D., see Thomsen, M.F., et al. 141
Winningham, J.D., see Wilken, B., et al. 153
Winningham, J., see Johnstone, A.D., et al. 47
Witteborn, F.C., see Bregman, J.D., et al. 616
Wooden, D.H., see Bregman, J.D., et al. 616
Woods, T.N., Feldman, P.D., Dymond, K.F.: The atomic carbon distribution in the coma of comet P/Halley 380
Woods, T.N., see Feldman, P.D., et al. 325
Woweries, J., see Eberhardt, P., et al. 481
Woweries, J., see Lämmerzahl, P., et al. 169
Wozniak, D., see Savin, S., et al. 89
Wronowski, P., see Savin, S., et al. 89
Wu, G.J., see Gong , et al. 594
Wu, M.C., Qiu, P.Z.: Activity of the plasma tail of comet P/Halley in March 1986 264
Wyckoff, S., see Belton, M.J.S., et al. 569
Yeomans, D.K., see Belton, M.J.S., et al. 569
Yeroshenko, Y.G., Styashkin, V.A., Riedler, W., Schwingenschuh, K., Russel, C.T.: Fine structure of the magnetic field in comet P/Halley's coma 69

Young, D.T., see Balsiger, H., et al. 163
Young, D.T., see Goldstein, R., et al. 220
Yumoto, K., Saito, T., Nakagawa, T.: Hydromagnetic waves associated with cometary water group ions: Sakigake observation 117
Yumoto, K., see Saito, T., et al. 201
Yumoto, K., see Saito, T., et al. 209
Yung, Y., see Allen, M., et al. 502
Zakharov, D., see Vaisberg, O.L., et al. 183
Zastenker, G., see Vaisberg, O.L., et al. 183
Zhang, X.F., see Gong , et al. 594
Ziolkowski, K., see Sitarski, G. 896
Zubkov, B.V., see Sagdeev, R.Z., et al. 179
Zucconi, J.M., see Moreels, G., et al. 551
Zwickl, R.D., see Brosius, J.W., et al. 267
Zwickl, R.D., see Sanderson, T.R., et al. 125

Astron. Astrophys. 187, 1–11 (1987)

ASTRONOMY AND ASTROPHYSICS

New information on comet P/Halley as depicted by Giotto di Bondone and other Western artists

Roberta J.M. Olson [1] **and Jay M. Pasachoff** [2]

[1] Art Department, Wheaton College, Norton, MA 02766, USA
[2] Williams College-Hopkins Observatory, Williamstown, MA 01267, USA

Received January 26, accepted March 25, 1987

Summary. Artists' depictions of comets provide the only visual evidence of historical comets, most notably of Halley's Comet. In this paper we discuss the visual evidence of comet P/Halley at several passages through that of 1301 and compare it with descriptions and modern images. Since it was first recognized that Giotto di Bondone painted a comet in place of the Star of Bethlehem and suggested that this was a portrait of the 1301 apparition of comet Halley (Olson, 1979), a great deal of new information has come to light. We present a synopsis of the textual, visual, and astronomical evidence to support the theory that when Giotto painted his comet in the Scrovegni Chapel he was reflecting his viewing of Comet Halley in 1301.

Key words: comet P/Halley – art – historical – Giotto

1. Introduction

Artists' representations of comets provide the only visual evidence of historical comets, most notably Halley's Comet, before the advent of astronomical photography in the later nineteenth century. As such, they can furnish unique information about past apparitions that supplements astronomical observations and texts.

Artists' depictions also reveal contemporary attitudes toward celestial phenomena and reflect current scientific thought. There has been particular interest in the theory first advanced by one of us (Olson, 1979) that the object rendered as the Star of Bethlehem in a fresco by the Italian painter Giotto di Bondone was based on the 1301 apparition of P/Halley. Prior to that time none of the literature on Giotto had even mentioned that his Star was a comet. This identification led the European Space Agency to name its spacecraft to P/Halley "Giotto." The evidence for the identification is complex and indirect but strong; we argue the case below. Halley's Comet and other objects in the same series of frescoes were among the first to be painted in a naturalistic Renaissance manner, and the effect of the apparition of a bright comet on Giotto may have helped to push his art toward a more radical naturalism and thus played an important role in determining the future of art.

Send offprint requests to: J.M. Pasachoff

2. Discussion

2.1. *Halley's Comet–684*

While this stylized woodcut of 1493 (Fig. 1) purports to record the 684 A.D. apparition of P/Halley, the prickly form is repeated for subsequent comets in the text, and thus merely provides visual punctuation for the chronicle. This exact form is repeated once more; a related more horizontal form appears once; a form that is nearly but not quite its mirror image appears six times; and a form with a similar tail but with a cloud-like configuration for a head appears three times. This image is not, therefore, the earliest representation of P/Halley itself, and is significant only because it calls our attention to the fact that the apparition was recorded in a European source used by the later compiler. Chinese records also briefly report the apparition.

2.2. *Halley's Comet–1066*

The Chinese recorded this apparition as having a broom-like vapor, while the Korean Koryo-sa describes it as being as large as

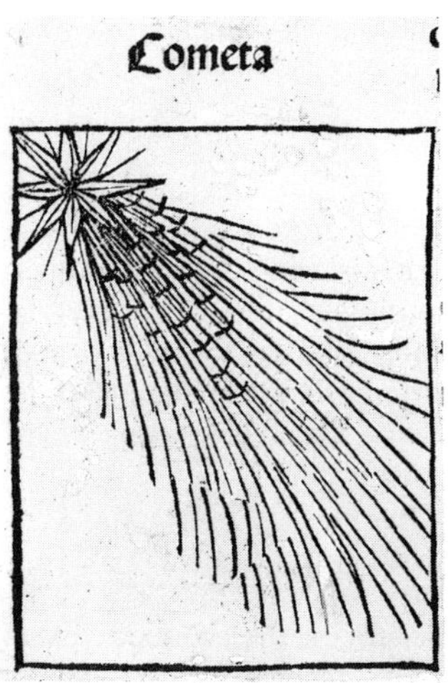

Fig. 1. *Halley's Comet of A.D. 684*, from Hartmann Schedel, the *Nuremberg Chronicle*, 1493, Latin edition. New York, Rare Books and Manuscripts Division, The New York Public Library, Astor, Lenox, and Tilden Foundations

Fig. 2. *Halley's Comet of 1066*, detail of the Bayeux Tapestry, 1073–83. Bayeux Town Hall. With special authorization of the Ville de Bayeux

the moon, which western sources also note (either because of a common source or similar rhetorical judgments, or as absolute judgments of its size). The cathedral archives at Viterbo, Italy, record it as having a tail streaming like smoke up to nearly half the sky. Other westerners describe it rhetorically as "a torch of the sun" or a "flaming beam" – language that parallels the Romanesque stylizations of the Bayeux Tapestry (Fig. 2). These simplified cartoon-like forms endow the image with a symbolic power worthy of the accompanying inscription: *Isti mirant stellă* ("they are in awe of the star"). The comet, resembling a primitive rocket ship, does reveal its anatomical parts, albeit in stylized form: the coma, center of condensation, plasma tail, etc.

2.3. Halley's Comet–1145

Halley's Comet of 1145 is shown in this marginal notation by the monk Eadwine (Fig. 3a). Even though it is a line drawing, there is an attempt to show a head (including a nine-ray center of condensation) and some structure in the tail. In April 1986, Comet Halley exhibited a fan tail (Fig. 3b) similar to Eadwine's line drawing, suggesting that the apparently stylized drawing might have been roughly based on the comet's shape.

2.4. Halley's Comet–1222

A comet (*STELLA COMMETA*) occurs on the right side of an archivolt of Piacenza Cathedral (Fig. 4), usually dated c. 1222 on stylistic grounds. The date should be considered the *terminus post quem*, for 1222 was the year of a spectacular apparition of P/Halley. It would explain why the comet is featured in such an unusually prominent position, balancing the star (*STELLA*), symbolizing all the fixed stars of heaven, found on the opposite side of the moon, the hand of God, and the sun. Datei (1983) posited that a fresco fragment of c. 1250 in the Palazzo della Ragione, Mantua, preserves the 1222 apparition of P/Halley, but this is very unlikely (see Olson and Pasachoff, 1986).

2.5. Halley's Comet–1301

P/Halley was recorded in 1301 in Europe by several individuals, including the eminent historian Giovanni Villani, who describes "great rays of fumes behind [con grandi raggi di fumo dietro]." According to him (Villani, c. 1308–48), the comet remained visible until January 1302 ("durò infino al gennajo"), although it is generally agreed that it remained visible from at least 16 September to about 1 November. Its perihelion was October 25th.

The comet's close approach to Earth and its resulting spectacular nature may partially explain why this is the first apparition of Halley's, or of any other comet for that matter, to have inspired a visually convincing portrait. While the time was artistically and intellectually ripe for such an accomplishment, it took the proclivities of Giotto di Bondone, the Florentine pioneer of naturalism, to observe the naked-eye apparition, translate his sensual experience into this image, and thus shatter the schematic artistic conventions for representing comets.

Giotto painted his comet in place of the customary Star of Bethlehem/Epiphany in the Adoration of the Magi scene in the Scrovegni Chapel in Padua about 1303–5, as part of an ambitious fresco cycle (Fig. 5). This in no way implies a belief that Halley's Comet was the historical Star of Bethlehem/Epiphany. Hughes

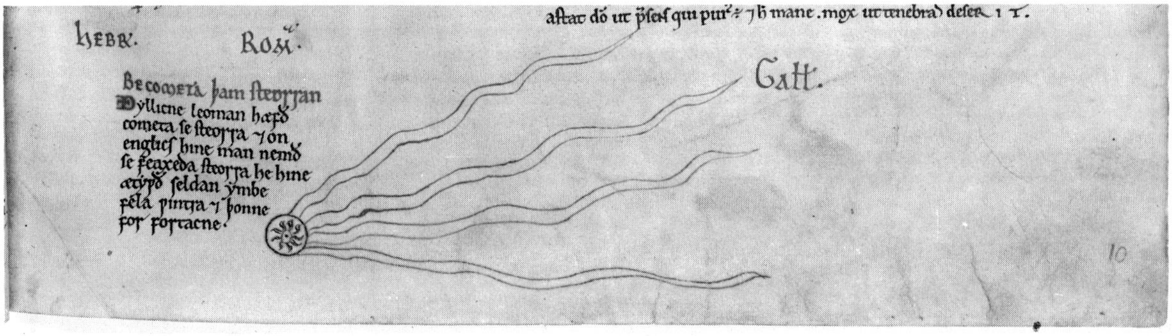

Fig. 3a. Eadwine, *The 1145 Apparition of Halley's Comet*, *The Eadwine* (*Canterbury*) *Psalter*, Ms. R. 17.1, Fol. 10r. Cambridge, Trinity College Library. Courtesy of the Master and Fellows of Trinity College

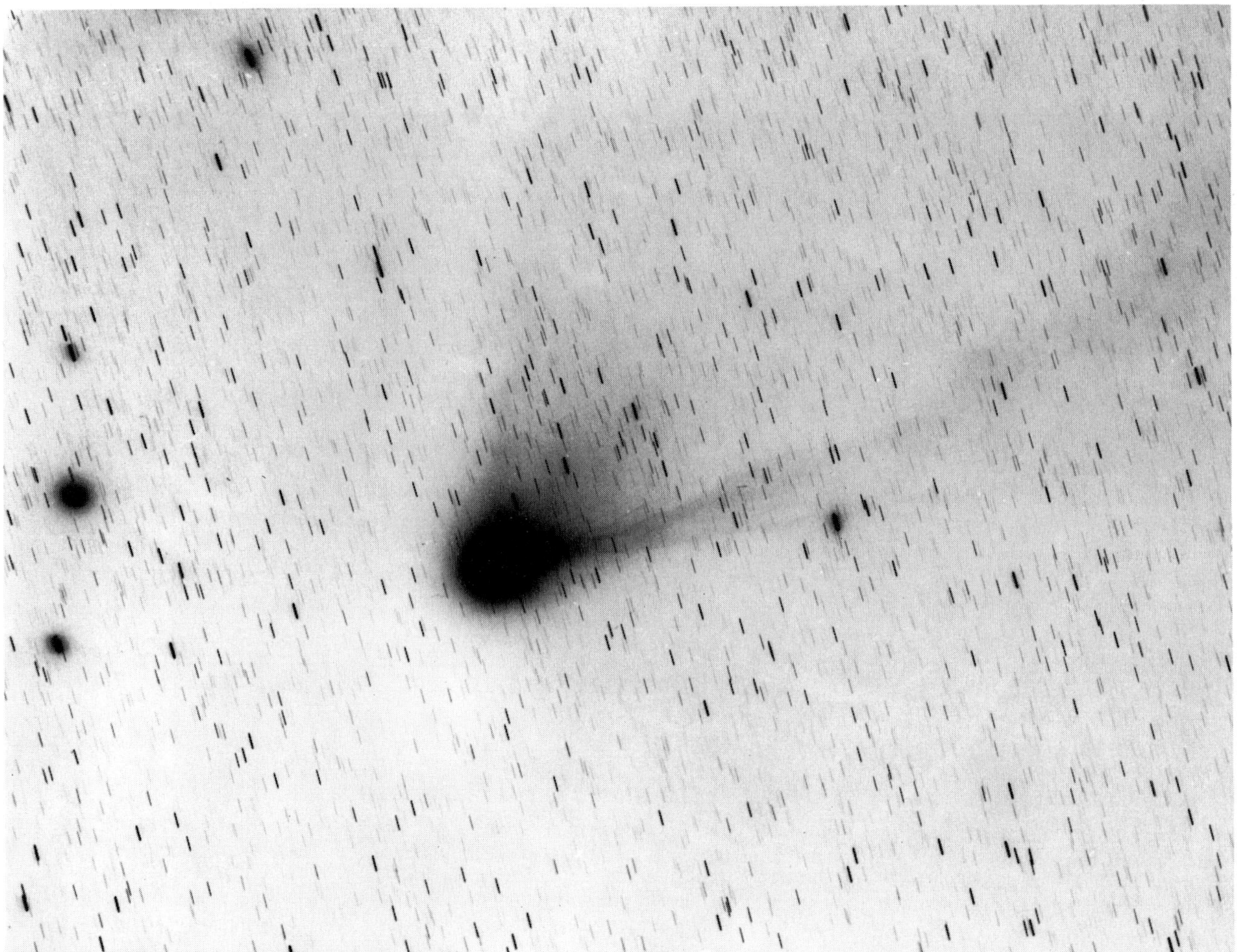

Fig. 3b. *Halley's Comet in April 1986*, photographed at Mauna Kea by Dale P. Cruikshank, Marc Buie, Heidi Hammel, and Alex Storrs

(1979) summarizes the history of the identification of the Star of Bethlehem/Epiphany with astronomical events, concluding that it was an actual physical object explicable by scientific law. (To further complicate the matter two different stars/lights are indicated in the *Apocrypha* – one for the Nativity and a separate one for the Adoration.)

The Star is usually depicted as a tiny stylized object, often with rays descending from it that point to the Christ Child to signify God's blessing of the event (Fig. 6). A rarer visual tradition for representing the Star of Bethlehem/Epiphany as an unusual star existed in Italy before Giotto, as seen in an eleventh-century manuscript page with its bizarre heavenly configuration or in a

Fig. 4. Attributed to Niccolò, detail of the Zodiacal Cycle on the central west portal, Piacenza Cathedral, c. 1222. F. lli Manzotti

Fig. 5a. Giotto, *The Adoration of the Magi*, 1303–5, fresco, 78 3/4 by 72 3/4 in. Padua, The Scrovegni Chapel. Alinari/Art Resource, New York

Fig. 5b. Giotto, *Halley's Comet of 1301*, detail of Fig. 5a

Fig. 5c. *Halley's Comet in 1986*, photographed by David Malin, © 1986 Anglo-Australian Telescope Board

Fig. 6. Duccio, detail from the *Maesta*: *Nativity with the Prophets Isaiah and Ezekiel*, tempera on panel, 1308–11. Washington D.C., The National Gallery of Art, Andrew W. Mellon Collection (1937.1.8)

Fig. 7a. *Initial with the Adoration of the Magi from the Gradual of St. Stefano*, Codex 123, F. 40, 51v, 11th. c. Rome, Biblioteca Angelica. Garrison Collection, Conway Library, Courtauld Institute of Art.

Fig. 7b. *Adoration of the Magi*, Codex 541, 13th c. Padua, Biblioteca del Seminario

Fig. 8. Follower of Giotto, *Nativity*, c. 1316, fresco. Assisi, San Francesco, Lower Church. Alinari/Art Resource, New York

Fig. 9. Contemporary of Giotto, Copy of *The Adoration of the Magi*, tempera on panel. Whereabouts unknown

thirteenth-century manuscript with its more standard star (Figs. 7a and 7b). Significantly, neither depicts a comet.

By contrast, Giotto's fresco contains the first unambiguous comet in an episode from the Infancy of Christ cycle. Moreover, Giotto's revolutionary, large comet is more or less anatomically correct; that is, it shows the head and tail of the comet as it might have appeared to the naked eye. It is thus viscerally convincing.

Giotto created its center of condensation by building up layers of red, yellow, and gold pigment in the comet's head. Beneath the glowing center of condensation and reaching out into the coma are traces of an eight-point gilt star. This configuration may be either a *pentimento*, a first more conservative idea for the Star that the artist converted into a comet, or may be an integral part of his visualization of the coma. (Comets have frequently been described as having a star-like head.) The strong red coloration may be partially due to the flaking tempera pigments and red bole adhesive. But red was also an important ingredient in the artist's original palette. It is significant that historical comets, including Halley's, have been described as having a red cast.

The image's immediacy is heightened by Giotto's method of painting the sweeping arc of the tail. Giotto's technique thus successfully imparts the impression the comet might have given to the naked eye. No contemporary description provides this kind of veracity, strongly suggesting that Giotto's image could not have been rendered without careful firsthand observation. Giotto's magnificent comet dominates the Adoration scene. Because he included a historical comet, which his viewers would have also seen, he enhanced the contemporary impact of his painting and encouraged them to identify with the strong emotion and significance of the scene. For they too had recently witnessed the awesome apparition of P/Halley in 1301.

By including the naturalistic comet in this Adoration scene, Giotto was true to his humanistic method demonstrated most tightly in the Scrovegni Chapel. That is, in each scene of the fresco cycle he painted one or two familiar objects with such authenticity that his audience could immediately associate with their tangibility. Furthermore, each of these objects usually had a profound symbolic dimension. So too with the comet.

Though there has been much debate over what the Star of Bethlehem might have been – a conjunction, a comet, a supernova, etc. – it may well be that there was no actual astronomical event. The idea of the Epiphany Star (epi = above; phanos = appearance) certainly owes its origin to the pagan belief that the birth of a king was signaled by a comet, and follows a complex literary, theological, and astrological tradition. Depictions of the Star usually follow the written traditions (including Matthew 2:2; 2:9–10, the only Evangelist to mention the Magi's Star) rather than making astronomical statements. Giotto's choice of a naturalistic depiction would not have disagreed with any fourteenthcentury beliefs. It seems to follow Origen, the early Church Father, who is the only writer to actually state (*Against Celsus*, I, 59) that the Star was a comet (Olson, 1979, and Olson and Pasachoff, 1986). Thus an important textual source exists as a theological justification for Giotto's image.

Giotto was able to paint such a progressive, daring image of the comet because of the private status of the Scrovegni Chapel and the nature of the city of Padua. Because it was a family chapel, more artistic and theological license was allowed. In planning the

Fig. 10. Follower of Giotto, *Adoration of the Magi*, tempera on panel. New York, The Metropolitan Museum of Art, John Stewart Kennedy Fund, 1911 (11.126.1) (The Museum experts believe the comet is a nineteenth-century repaint; there is a *pentimento* of a star just above and to the left.)

program of the cycle Giotto consulted at least one theological advisor, although his identity is unknown. Furthermore, the Scrovegni Chapel was located in Padua, the university town where Galileo would eventually hold a chair. Already during Giotto's time the university was noted for the study of mathematics and was a center for nascent astronomy. Certainly Giotto and/or his patron and his theological advisor had access to some of the latest theories about the heavens.

It has been suggested (Bellinati, 1981, 1985) but not proven that Pietro d'Abano, the philosopher, medical doctor, and astrologer, was influential on both Giotto's naturalism and on his specific image of the comet. Since Giotto was a Florentine, his naturalistic aesthetic was formed long before his sojourn to paint for Scrovegni and reflects instead the ambience of Florence and Rome. D'Abano, following Aristotle and Ptolemy, had written that comets are dry and hot atmospheric exhalations and that "After a great fire, the material loses its red color and is tinged black." Bellinati believes that this brief description rather than the actual apparition of P/Halley in 1301 accounts for Giotto's powerful comet. He also does not account for the technical problems of the fresco that have resulted in its present appearance. His correlation is problematical, especially since d'Abano is not

Fig. 11. Jacopo Gradenigo, *Scenes of the Infancy of Christ*, Ms. 78 C. 18, Fol. 4r. Berlin, SMPK, Kupferstichkabinett

known to have returned to Padua from Paris until 1306 or later, probably too late for an encounter with Giotto (Gaudenzio, 1962).

More significant for Giotto's image of the comet is the apparent theological controversy of Giotto's time surrounding the nature of the Star. It is, for example, already encountered in a vernacular form in Jacobus de Voragine's *Golden Legend* of 1298. The churchman F. Giordano da Rivalto (born 1260) wrote that the Star of Bethlehem was not like other stars in the sky, for it did not follow the same path as the other stars. Rather, it was a new star. He specifically states that it was *not* a comet, revealing that the nature of the Star of Bethlehem was a hotly debated topic at the time when Giotto painted his image, and that Origen's linkage of that Star with a comet was well-known.

Throughout Italy in the popular tradition – i.e., in Nativity (*presepio*) scenes erected in churches and sacred plays (*sacre rappresentazioni*) thought to have influenced Giotto in the Scrovegni Chapel – the Star is today frequently shown as a comet. Parts of these practices date back to before Giotto's time, but it is difficult to trace the exact features and origins of the ephemeral events.

Giotto's comet was such an advanced artistic/scientific statement that it was followed by less than a handful of tiny, timid imitations by his school, but by enough of them to suggest that Giotto's comet had a very special impact. While these diminutive comets echo the literary/theological traditions, they do not preserve Giotto's naturalistic features, proving that Giotto's inspiration was his own experience of a comet and that his model was Halley's Comet of 1301. For example, followers of Giotto painted two minuscule comets in the Nativity (Fig. 8) and Adoration of the Magi scenes in the Lower Church of San Francesco, Assisi, about 1316. The singular nature of Giotto's image is underlined by looking at another fourteenth-century artist's copy of his Adoration scene (Fig. 9), and at a panel by a follower of Giotto depicting the Adoration (Fig. 10), wherein stylized comets appear instead of Giotto's naturalistic portrait. A grand but diverse stylized comet with an angel occurs in an Infancy of Christ scene in a manuscript by the Paduan Jacopo Gradenigo of around 1399 (Fig. 11); the angel-comet derives from the Infancy Gospel of the *Apocrypha*.

Giotto's interest in convincing celestial phenomena was bequeathed to one of his most talented pupils, Taddeo Gaddi, who is recorded to have been partially blinded by an eclipse. A reflection of Taddeo's physical and emotional trauma is found in his Annunciation to the Shepherds scene in Santa Croce, Florence, where an unusual light, emanating from the angel and synonymous in medieval-Renaissance symbolism with God, permeates the fresco, and one shepherd shades his eyes to gaze up at it in an attempt to penetrate its meaning. The quality of the light and the shepherd's gesture convincingly argue that this preserves Gaddi's autobiographical experience of an eclipse (Smart, 1977).

Based on documentary evidence, the recent art historical literature dates the Scrovegni frescoes, and certainly the section in which the Adoration occurs, before the 25 March 1305 consecration of the chapel (e.g. Bellinati, 1967, and Stubblebine, 1969). The consecration occurred on the feast of the Annunciation, the festival of the Annunciate Virgin, to whom the chapel is dedicated.

Between at least 1300 and 1305 there were no other spectacular comets to rival P/Halley in 1301, although Maffei (1984) raises the possibility of a second comet in late 1301. He bases his theory on old evidence, usually not given much credence, citing: (1) Pingrés later eighteenth-century work listing a second comet that appeared in December 1301 for only 15 days; and (2) Villani's claim that the 1301 comet appeared in September and lasted until January. Maffei believes that both Villani and Giotto may have conflated the two comets Pingré posited. A similar situation occurred in 1910, when Halley's Comet (1910II) was conflated with the earlier Great Comet (1910I). The so-called "second comet" of 1301 has never been confirmed and its apparition is questionable. (In a forthcoming study, we will evaluate Pingré's sources.) Many of the reports of comets during Giotto's time are unreliable and false, pure inventions or misunderstandings of other celestial phenomena. Lancaster-Brown (1985), in an undeveloped essay, also suggested a second comet in 1301. He questionably based his suggestion once again on much later authorities (J. Hevelius, *Cometografia*, 1668; S. Lubieniecki, *Theatrum Cometicum*, 1666–68; N. Stryck, *Algemeene Geographie*, 1740; A.-G. Pingré, *Comètographie*, 1783), who cite each other as evidence as well as other later sources to compound the confusion. While Lancaster-Brown stated that there is no way to determine which comet influenced Giotto, he questionably concluded that a second comet, appearing around Christmastime, would be better suited to Giotto's purpose. It seems to us that the time of the year when Giotto saw a comet would not have had a bearing on his including it. Its duration and brillance would have been far more important.

Marsden (1986), in his catalogue of comets whose orbits are known, as well as comet catalogues from the early part of this century list only P/Halley for the year 1301. (Modern catalogues list only those comets whose orbits have been calculated.) Though it remains possible that a second comet could have appeared and disappeared in 1301 without returning to have its orbit calculated, it seems more likely to us that the appearance of Comet Halley at different times in the morning and evening sky were noted as two comets, as frequently occurred until into the seventeenth century.

3. Discussion

Despite a literary tradition and contemporary trends, Giotto could not have painted his unprecedented portrayal of the Star as the vivid comet he rendered without experiencing firsthand a spectacular apparition. No written word or visual prototype can explain its revolutionary nature. The literary tradition and the fledgling zeal of the time for the observation of natural phenomena, rather, created a receptivity in both Giotto and his audience for such a seminal image.

Naturalism was resuscitated in the visual arts for the first time since Antiquity by Giotto about the time of the 1301 apparition of Halley's Comet (Boccaccio, c. 1350, and Vasari, 1568). The excitement of its apparition led Giotto to paint his comet with great immediacy, stimulating a more naturalistic style and imparting astronomical realism. Analogous but less intense responses are found with other objects that Giotto painted from life in the frescoes. Certainly of all the physical objects that the artist included, the comet was the most unusual and challenging. It could be that Halley's Comet itself inspired Giotto to take greater strides in his new naturalistic manner of painting. We present a fuller discussion of artists' depictions of Halley's Comet in Olson and Pasachoff (1986) and discuss the assessment of the accuracy of artistic depictions by comparing them with photographs in Pasachoff and Olson (1986).

Giotto's comet was, surprisingly, not equalled in painting until the nineteenth century and not surpassed until the advent of astronomical photography in the late nineteenth century. Giotto's galvanizing image received the highest compliment imaginable when the European Space Agency used the name "Giotto" for their mission to encounter the very comet that Giotto saw 685 years earlier.

Acknowledgements. Some of the photographs are courtesy of The National Air and Space Museum, Smithsonian Institution, Washington, D.C., from the book *Fire and Ice: A History of Comets in Art* by Roberta J.M. Olson.

This work has been supported in part by a grant from the National Geographic Society; by the Dudley Award of the Dudley Observatory.

We thank Brian G. Marsden and Donald K. Yeomans for calculating orbits for us. We appreciate the assistance of Elizabeth Stell, Scott McGuire, and Joan Silva with the manuscript. We are further grateful to Robert Volz of the Chapin Library of Williams College for his assistance and to Professor Ulrich Middeldorf for his suggestion.

References

Bellinati, C.: 1967, *La capella di Giotto all'Arena*, Padua, Tipografia del Seminario di Padova, 9–10

Bellinati, C.: 1981, *Jesus*, III [December], 50 ff

Bellinati, C., et al: 1985, *Magico Appuntamento/Halley e Giotto nel Cosmo*, Castel di Godego, *mp*/edizioni, 81–85

Boccaccio, G.: 1976 (c. 1350 original), *Decameron*, Florence, L'Accademia della Crusca, 6th day, 5th story

Datei, E.: 1983, *Astronomia* **23** [June], 30–34

Gaudenzio, L.: 1962, *Pietro d'Abano*, Abano, Azienda Autonoma di Soggiorno e Turismo

Hughes, D.: 1979, *The Star of Bethlehem: An Astronomer's Confirmation*, New York, Walker and Company

Lancaster-Brown, P.: 1985, *New Scientist*, 13 June, 29

Maffei, P.: 1984, *La Cometa di Halley*, Milan, Biblioteca EST/Mondadori, 217–227

Marsden, B.G.: 1986, *Catalogue of Cometary Orbits*, Cambridge, Mass., International Astronomical Union, 8

Olson, R.J.M.: 1979, *Scientific Amer.* **240** [May], 160–170

Olson, R.J.M., Pasachoff, J.M.: 1986, in *Proceedings of the 20th ESLAB Symp.: Exploration of Halley's Comet*, III, Paris ESA SP-250, 201–213

Pasachoff, J.M., Olson, R.J.M.: 1986, *Bull. Am. Astron. Soc.* **18**, 792

Pingré, A.-G., 1783, *Comètographie*, I, Paris, Imprimerie Royale, 423

Smart, A.: 1977, *Taddeo Gaddi, Orcagna and the Eclipses of 1332 and 1339*, in *Studies in Late Medieval and Renaissance Painting in Honor of Millard Meiss*, I, New York, New York University Press, 403–414

Stubblebine, J.: 1969, *Giotto: The Arena Chapel Frescoes*, New York, W.W. Norton and Co., 72–73

Vasari, G.: 1906 (1568 original), *Le Vite*, I, ed. G. Milanesi, Florence, G.C. Sansoni, 369–409

Villani, G.: 1802 (c. 1308–48 original), *Istorie Fiorentine*, IV, Milan, Società Tipografica, 71

Astron. Astrophys. 187, 12–20 (1987)

ASTRONOMY AND ASTROPHYSICS

Encounters with comets: discoveries and puzzles in cometary plasma physics

A. A. Galeev

Space Research Institute, Profsoyuznaya 84/32, SU-117810 Moscow, GSP-1, USSR

Received February 9, accepted May 30, 1987

Summary. The encounters of various cometary probes with comets led to the discovery of a number of phenomena that were not envisaged by the existing theories based mainly on the magnetohydrodynamic description of the solar wind interaction with cometary plasma: the generation of strong magnetohydrodynamic turbulence by the loading of solar wind by cometary ions, the discovery of thin plasma structures in an inner coma that correlate with the bursts of plasma waves and the anomalously thin boundaries in a solar wind flow near comets that indicate the importance of collective plasma processes for the formation of these structures and boundaries. The paper reviews the observations of these phenomena from spacecraft and the kinetic theories of cometary plasma that were developed recently and that are able to provide a natural explanation of observed puzzling phenomena.

Key words: comets – magnetohydrodynamics – shock waves – solar wind – turbulence

1. Introduction

The theoretical models of solar wind interaction with comets were to a great extent based on the results of space studies of solar wind interaction with planets. However, Biermann et al. (1967) pointed out that in contrast to the solar wind interaction with planets the interaction with freely expanding cometary atmospheres starts already at distances of millions of kilometers from the tiny cometary nucleus. This is due to the solar wind loading by photoionized atoms from the cometary atmosphere. The theoretical studies of solar wind flow near comets were carried out mainly in the framework of the hydrodynamical description of a plasma (Wallis, 1973; Schmidt and Wegman, 1982). The main result of these studies was the determination of the positions of the cometary bow shock and the region of solar wind stagnation in the process of loading behind the shock.

The incorporation of collisions of plasma particles with atoms and molecules from the cometary atmosphere and the influence of magnetic stresses on plasma flow in theoretical models has led to the prediction of two other boundaries, the collisionpause and the contact surface that are closer to the nucleus of a comet. The ion-neutral friction heavily breaks the loaded solar wind at the collisionpause (Biermann, 1974). On the other hand, the magnetic field line tension stops the cometary photoplasma collisionally coupled to the expanding cometary gas (Ip and Axford, 1982).

All four boundaries were identified in the data of plasma and field measurements aboard various spacecraft that encountered Halley's comet. But the changes of the chemical properties and physical parameters of the plasma across some of these boundaries were unpredictably dramatic. This is in particular true for the cometopause that is formed in the region of solar wind stagnation due to its mass loading (Gringauz et al., 1986a, b) and separates the solar wind flow dominated by protons from the slowly moving cometary plasma. Another thin boundary separates the heavily loaded solar wind that is strongly braked by ion-neutral friction from the stagnant cometary plasma produced by the photoionization of cometary gas between this boundary and the contact surface (Balsiger et al., 1986; Ip et al., 1986b) (see Fig. 1). The extremely small thickness of these boundaries in comparison with any collisional length indicates the importance of collisionless interaction of different plasmas in cometary comae.

Moreover, it was discovered (Smith et al., 1986) that the loading of the solar wind by a relatively small number of cometary ions makes it turbulent in a huge area around the comet. This also requires the revision of the thory of solar wind loading to include the pitch angle scattering of cometary ions picked up by the solar wind that violate the adiabatic approximation (Wallis and Ong, 1975) usually used to describe this component of solar wind plasma near comets.

The revised theory of solar wind flow near comets is used below (Sect. 2) to define the characteristic plasma boundaries in the outer coma where the solar wind loading by cometary ions plays a dominant role. We show that two intermediate boundaries between the bow shock and the cometopause discovered by the Giotto plasma team (Balsiger et al., 1986; Rème et al., 1986; Johnstone et al., 1986) and later revealed in the Vega plasma data (Savin et al., 1986) are naturally emerging in the theory that takes into account the effect of charge-exchange cooling of picked-up cometary ions on the solar wind flow (Galeev et al., 1985b). The structure of the cometary bow shock discussed in Sect. 3 is of special interest for astrophysics since it represents a new type of collisionless shock that differs from both the planetary bow shock and the cosmic-ray shock. Then in Sect. 4 we briefly discuss possible mechanisms of formation of the thin boundary between the piled up solar wind and the stagnant cometary plasma above the contact surface. Finally, in Sect. 5, we review the observational evidences for the Alfvén's critical ionization velocity effect in operation in the inner comae of the comets Halley and Giacobini-Zinner. The latter could be a natural solution of the long standing

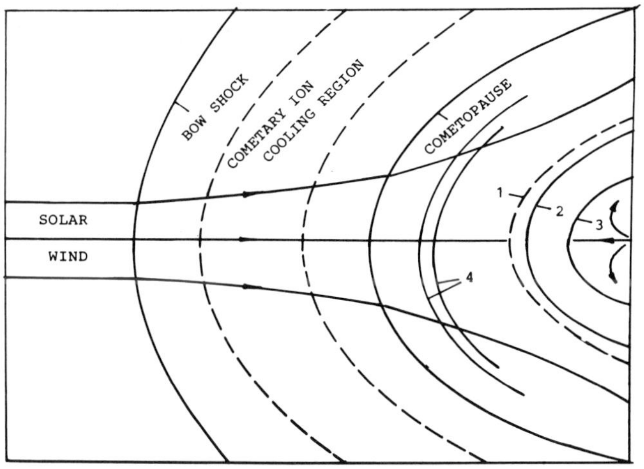

Fig. 1. A schematic view of plasma structures in a cometary coma. The structures in the inner coma are: *1* the collisionpause, *2* the plasma pile up region, *3* the contact surface, *4* the receding plasma envelopes. The directions of solar wind flow and cometary plasma expansion are indicated by arrows on flow lines

problem of the anomalous ionization of cometary atmospheres (Wurm, 1963).

2. Solar wind flow near comets

As we have mentioned above, the solar wind flow near comets is to a large extent controlled by its loading by cometary ions. In the case of solar wind flow across the interplanetary magnetic field the newly born cometary ions start to rotate around the local magnetic field line, frozen into the solar wind plasma. Thus the solar and cometary plasma are tied up by the magnetic field lines threading them and do not move relatively to each other. This is the case when the adiabatic approximation is used to describe the picked-up cometary ions (Wallis et al., 1975). When the solar wind flows obliquely to the magnetic field the newly born cometary ions are drifting across the magnetic field with the plasma, but are sliding along the magnetic field lines relative to the solar wind plasma. The counterstreaming flow of solar wind and cometary plasma along the magnetic field lines as well as their cyclotron motion is an energy source for a variety of plasma instabilities (Winske et al., 1985). However, the well known fact of the possible development of these instabilities was exploited only to justify the assumption of prompt assimilation of cometary ions by the solar wind (Ip and Axford, 1981). Meanwhile the reaction of cometary ions on the development of the above mentioned instabilities is very specific and cannot be reduced only to their assimilation by the solar wind. Namely, the low frequency Alfvén waves excited by the newly born cometary ions scatter these ions mainly in pitch-angles and do not change their energy (Sagdeev et al., 1986).

Therefore, cometary ions remain to be a special (supra-thermal) component of solar wind plasma near comets and should be described only by means of kinetic theory. Such a theory (Wallis and Ong, 1975) was first formulated using the assumption that the magnetic moment of picked up cometary ions is conserved. Moreover, only the case of solar wind flow across the magnetic field was considered in detail. Then the equation for the distribution of cometary ions $f_{0\mathrm{i}}(x,\mu)$ has the form

$$A^{-1}\frac{\partial}{\partial x}(uAf_{0\mathrm{i}})=\frac{Q\exp(-r/v_{\mathrm{g}}\tau)}{4\pi r^2 v_{\mathrm{g}}\tau}\,\delta\left(\mu-\frac{m_{\mathrm{i}}u^2}{2B}\right), \tag{1}$$

where A is the area of the plasma flow tube, μ the magnetic moment of cometary ions, $u(x)$ the hydrodynamic velocity of the plasma along the Sun-comet line, x the coordinate along this line, Q the gas production rate of a comet, v_{g} the velocity of a spherically symmetric expansion of cometary gas, τ the ionization time of cometary gas, B the magnetic field strength, m_{i} the mass of cometary ions, and r the distance to the nucleus of the comet ($r=|x|$ since we consider here the plasma flow only along the Sun-comet line).

To get a close set of equations we add here the hydrodynamic equations for the velocity and density of loaded solar wind

$$A^{-1}\frac{d}{dx}(\varrho uA)=\frac{Qm_{\mathrm{i}}}{4\pi r^2 v_{\mathrm{g}}\tau}\exp(-r/v_{\mathrm{g}}\tau), \tag{2}$$

$$A^{-1}\frac{d}{dx}(\varrho u^2A)=-\frac{d}{dx}P, \tag{3}$$

where ϱ is the mass density of loaded solar wind, m_{i} the ion mass and the main contribution to the plasma pressure, P comes from the cometary ions

$$P=\int_0^\infty \mu B(x)f_{0\mathrm{i}}(x,\mu)\,d\mu. \tag{4}$$

The magnetic field here is considered to be frozen into the plasma in the unshocked solar wind, where $A=\mathrm{const}$:

$$d[u(x)B(x)]/dx=0. \tag{5}$$

However, recognizing that the conservation of the magnetic moment of cometary ions is certainly violated by plasma instabilities, we also consider here the kinetic theory of solar wind loading based on the assumption of strong pitch angle scattering of cometary ions (Sagdeev et al., 1986). Following this assumption we neglect a small anisotropy of the pitch angle distribution of cometary ions that is necessary to maintain the plasma instability and thus continuously generate Alfven waves. Then the equation for the isotropic part of velocity distribution of cometary ions $f_{0\mathrm{i}}(x,v)$ takes the form:

$$A^{-1}\frac{\partial}{\partial x}(uAf_{0\mathrm{i}})=\frac{Q\exp(-r/v_{\mathrm{g}}\tau)}{16\pi^2 r^2 v_{\mathrm{g}}\tau}u^{-2}\delta(v-u)+A^{-1}\left(\frac{duA}{dx}\right)\frac{\partial}{3v^2\,\partial v}(v^3 f_{0\mathrm{i}}). \tag{6}$$

Here the last term on the right hand side describes the adiabatic cooling of cometary ions[1]. The solar wind flow is still described by the hydrodynamic Eqs. (2) and (3) where the pressure is given by the following expression

$$P=(4\pi m_{\mathrm{i}}/3)\int_0^\infty v^4 f_{0\mathrm{i}}(x,v)\,dv. \tag{7}$$

2.1. Flow of the unshocked solar wind

The solar wind flow ahead of the cometary bow shock can be considered as one-dimensional, i.e. $A=\mathrm{const}$ in Eqs. (1)–(7). Then it can be easily shown that these equations are reduced to the magnetohydrodynamic equations for a plasma with $\gamma=2$ in the adiabatic approximation and $\gamma=5/3$ in the limit of strong pitch

[1] This term was not included in the paper by Galeev et al. (1986b). So the bow shock position was calculated there incorrectly

angle scattering. Thus we recover the results of Biermann et al. (1967). As computer simulations in these two limits (Schmidt and Wegman, 1982; Baranov et al., 1986) show, the bow shock with Mach number $M = 2$ is formed in front of the comet. The position R_s of the shock is defined by the relation following from Eq. (2):

$$\varrho_1 u_1/\varrho_\infty u_\infty = 1 + (m_i Q/4\pi v_g^2 \tau^2 \varrho_\infty u_\infty)\,\psi(R_s/v_g\tau), \quad (8)$$

where $\psi(\xi) = \xi^{-1}\exp(-\xi) + \mathrm{Ei}(-\xi)$, $\mathrm{Ei}(-\xi) = \int_{-\infty}^{\xi} e^{-\xi}\xi^{-1}\,d\xi$, $\varrho_1 u_1$ and ϱ_∞, u_∞ are the mass density and velocity of solar wind at the shock and at infinity, respectively. The parameters of the loaded solar wind before and behind the shock are found from the corresponding magnetohydrodynamic equations (Biermann et al., 1967; Wallis, 1973) for the case of $\gamma = 2\,(\gamma = 5/3)$:

$$\varrho_1 u_1/\varrho_\infty u_\infty = 1.185\,(1.323);\ u_1/u_\infty = 0.75\ \ (0.657);$$
$$P_1/\varrho_\infty u_\infty^2 = 0.11\ \ (0.13);\ \ u_2/u_\infty = 0.375\,(0.287)$$
$$P_2/\varrho_\infty u_\infty^2 = 0.555\ (0.62). \quad (9)$$

Figure 2 shows two bow shocks whose stand-off distances of 4.8 10^5 km and 3 10^5 km were calculated in these approximations with the help of expression (8) for the case of the Vega-1 encounter with comet Halley [$n_\infty = 12\ \mathrm{cm}^{-3}$, $u_\infty = 510\ \mathrm{km\,s}^{-1}$, $Q = 1.3\ 10^{30}\ \mathrm{s}^{-1}$, $v_g\tau = 2\ 10^6$ km according to Gringauz et al. (1986a) and Galeev et al. (1986b)]. The shape of the bow shock front is taken from computer simulations for the case of plasma with $\gamma = 5/3$ (Baranov et al. 1986). Here we have marked also all the identified bow shock crossings by spacecraft Vega-1 and Vega-2 (Klimov et al., 1986; Galeev et al., 1986b), Suisei (Mukai et al., 1986), and Giotto (Johnstone et al., 1986; Neubauer et al., 1986; Balsiger et al., 1986; Rème et al., 1986). We see that the adiabatic approximation certainly fails due to the pitch angle scattering of cometary ions by excited Alfvén waves. So below we limit ourselves to the case of strong pitch-angle scattering described by Eqs. (2), (3), (6), and (7).

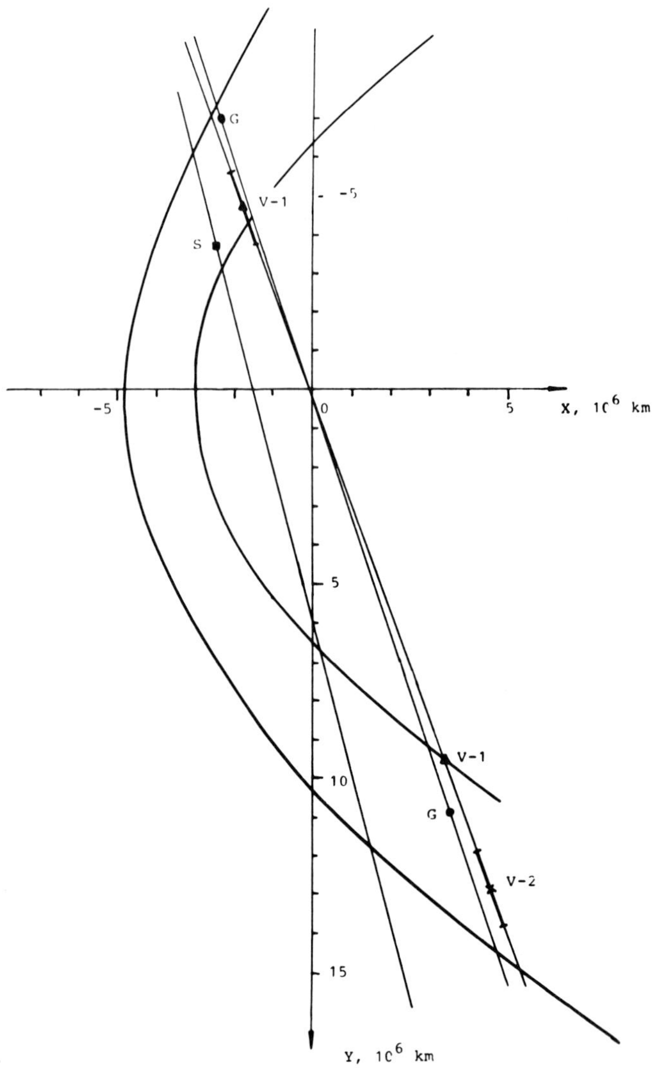

Fig. 2. The trajectories of the Vega-1 (V-1), Vega-2 (V-2), Giotto (G) and Suisei (S) spacecraft with identified bow shock crossing points indicated. The uncertainties of bow shock crossing points are marked by the error bars. The positions of bow shock fronts of paraboloid shapes were calculated in the adiabatic approximation and the approximation of strong pitch angle scattering of cometary ions (the shock is closer to the nucleus)

2.2. Plasma boundaries behind the bow shock front

To solve the set of Eqs. (2), (3) and (6), (7) behind the bow shock we assume that here the plasma is incompressible, i.e. $uA = \mathrm{const}$. This assumption is based on the results of computer simulation (Schmidt and Wegman, 1982) and was used to find the analytic solutions of the hydrodynamic equations for the plasma flow behind the shock (Wallis, 1973). As a result we obtain the equation of plasma motion (Galeev et al., 1985b):

$$u\,d\varrho u/dx = -(u^2/3)\,d\varrho/dx - dP_i/dx. \quad (10)$$

Here the first term of the right hand side represents the pressure gradient of locally born cometary ions. The gradient of the pressure P_i of cometary ions born in the unshocked solar wind and later energized at the shock results from the charge exchange within the cometary atmosphere

$$P_i = P_{i,2}\exp\left[-\sigma_{ct}\bar{u}\tau(\varrho - \varrho_2)/m_i\right], \quad (11)$$

where $P_{i,2}$ is the pressure of cometary ions behind the shock front, σ_{ct} the charge exchange cross-section, $\bar{u}$ the thermal velocity of cometary ions. If we neglect the charge exchange cooling of cometary ions we find from Eqs. (2) and (10) that the solar wind loading by cometary ions results in the fast stagnation of the solar wind flow behind the shock (Wallis, 1973):

$$\varrho = \varrho_2(u_2/u)^{3/4} = \varrho_2\left[1 - \frac{R_L}{r}\left(1 - \frac{r}{R_s}\right)\right]^{-3}, \quad (12)$$

where the position of the stagnation point was calculated assuming $R_L \ll v_g\tau$:

$$R_L \cong m_i Q/12\pi v_g \varrho_2 u_2. \quad (13)$$

However, there are two effects temporarily preventing the stagnation: the tension of magnetic field lines frozen in the solar wind moving fast along the flanks of the interaction region (Schmidt and Wegman, 1982) and the reduction of the cometary ion pressure due to charge exchange processes (Galeev et al., 1985b). The observations (Riedler et al., 1986; Neubauer et al., 1986) show that the second effect dominates and the magnetic field simply follows the plasma compression.

Thus we take into account only charge exchange cooling of cometary ions. At large distances from the nucleus this cooling is slow and we expand the exponential in Eq. (11). Then we obtain from Eq. (10):

$$(u/u_2)^2 - \alpha = (\varrho_2/\varrho)^{8/3}(1 - \alpha), \quad (14)$$

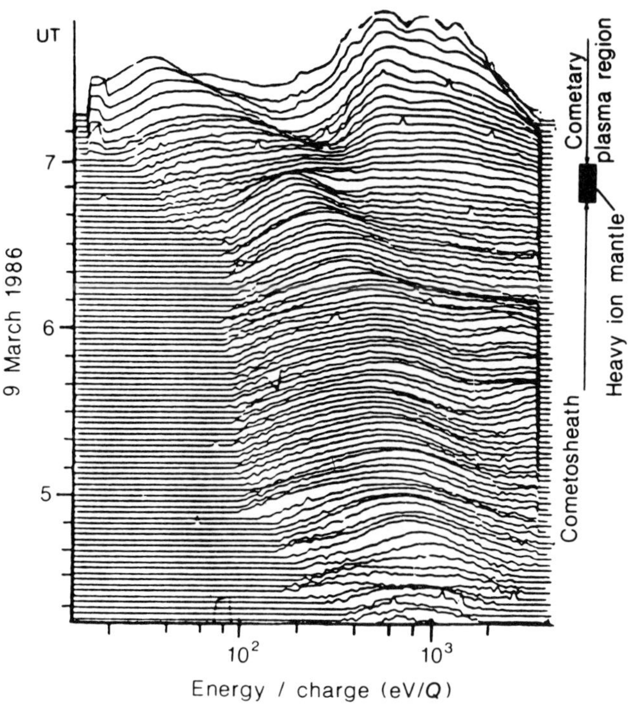

Fig. 3. High-time resolution ion energy spectra (2-min average) measured by the Vega-2 Cometary Ram Analyser (CRA) during encounter with comet Halley (from Gringauz et al., 1986a)

where $\alpha = 3\sigma_{ct}\bar{u}P_{i,2}/4m_i u_2^2$. Combining Eqs. (12) and (14) we find that at the distance

$$R_1 = R_L/(1-\alpha^{1/8}+R_L/R_s) \cong 5R_L \tag{15}$$

the solar wind deceleration is stopped and further on its speed is maintained by the reduction of cometary ion pressure at the level $u=\sqrt{\alpha}u_2$. Here the density of cometary ions can be estimated from Eq. (2):

$$\varrho/\varrho_2 = (3R_L/\sqrt{\alpha r})(1-r/R_1)+\alpha^{-3/8}; \tag{16}$$

closer to the nucleus at the distance

$$R_2 \cong 3\sigma_{ct}\tau\varrho_2\bar{u}R_L/m_i\sqrt{\alpha} \cong 1.7R_L, \tag{17}$$

where according to expression (11) most of energetic cometary ions disappear, the plasma flow starts to be decelerated again until it stagnates at the distance R_L from the cometary nucleus. Making the numerical estimates in Eqs. (15) and (17), we have assumed that the ionization of the cometary atmosphere is produced mainly by charge exchange of solar wind protons with cometary neutrals (Ip, 1986). Thus we put $n_\infty u_\infty \sigma_{ct}\tau = 1$. Besides that we neglected the heating of solar wind protons at the subshock inside the cometary bow shock (see next section) and estimated the values $\bar{u}=1.3u_\infty$ and $P_{i,2}\cong P_2$ from relations (9).

The positions of the plasma boundaries where the plasma flow velocity changes its behavior as described above are easily found from measurements aboard Vega-2 displayed in Fig. 3: $R_1 = 7\ 10^5$, $R_2 = 3.5\ 10^5$, $R_L = 1.7\ 10^5$. Thus the observed positions of these boundaries at flanks agree with the estimates (16) and (17). However, their thickness is unexpectedly small. For example, the last boundary (cometopause) that separates the loaded solar wind from the plasma dominated by cometary ions has the width $\cong 10^4$ km (Gringauz et al., 1986a, b). The solar wind measurements aboard Giotto (Johnston et al., 1986) show that the flux of cometary ions picked up by unshocked solar wind almost dissappears at the second boundary at $5.6\ 10^5$ km from the nucleus with the width $2\ 10^4$ km. Though the positions of the boundaries under discussion are controlled by the elementary processes (photoionization, charge-exchange) their sharpness indicates that their structure is determined by the collisionless interaction of solar and cometary plasma. The enhancement of the plasma wave intensity at all these boundaries (Savin et al., 1986) is the consequence of this interaction.

3. Bow shock structure

Computer simulations by Galeev et al. (1985a) show that the cometary ions contribute significantly to the bulk kinetic energy of the plasma and its internal energy influences strongly the structure of cometary bow shock. The results of simulations of the shock propagating perpendicular to the magnetic field are shown in Fig. 4. Here the profiles of the plasma mass density ϱ, the mass density of cometary ions ϱ_i, and the magnetic field strength are plotted in the upper part of the figure. The distribution of cometary ions and protons in velocity space (the v_x, v_y plane is perpendicular to B) is displayed in the lower part of Fig. 4. This shock was stationary in a loaded proton plasma flow with Mach number $M=2$, expected at the subsolar point of the cometary bow shock. The mass density of cometary ions was about 30% of that of protons. We see that the cometary ions due to their large velocity of cyclotron motion leak easily forward from behind the shock front and are turned back by the Lorentz force. The incoming solar wind in its turn is decelerated by cometary ions. As a result the plasma is compressed and the magnetic field is enhanced at the foot of the shock in a region with roughly the size of the cyclotron radius of the cometary ions ($\cong 10^4$ km). The subshock is formed at the end of the foot and the magnetic field and plasma density overshoot behind it. Let us note finally that cometary ions leaking forward are accelerated in the foot by the selfconsistent electric field of the plasma flow and thus a beam of ions bunched on the phase of their cyclotron motion is clearly seen in the lower part of Fig. 4.

The behavior of the plasma parameters across the perpendicular bow shock observed during the bow shock crossing by Vega-1 and the Giotto spacecraft agrees in general with the results of computer simulations. Figure 5 shows (from top to bottom): the velocity and temperature of the solar wind, the spectral amplitudes of ion flux and electric field oscillations with the frequency $f=1.5$ Hz that is somewhat below the frequency of lower-hybrid resonance and the magnetic field strength measured by the PLASMAG, APV-N and MISCHA instruments near the inbound crossing of the bow shock by Vega-1, March 6, 1986 (Galeev et al., 1986b). The crossing of the bow shock is identified here most easily by the sharp increase in intensity of the plasma waves at 3.46 UT when the spacecraft was at a distance of $(9.9\pm0.1)\,10^5$ km from the nucleus of comet Halley. These lower hybrid plasma waves are excited by the beam of cometary ions leaking forward from behind the subshock front similarly to the excitation of these waves at the quasiperpendicular Earth bow shock by the beam of protons reflected from the shock. Exactly this similarity suggested the idea (Klimov et al., 1986) that Vega-1 had crossed the quasiperpendicular shock that was verified by magnetic field data later (Galeev et al., 1986b). The velocity of the solar wind shown in Fig. 4 is the first moment of the ion energy

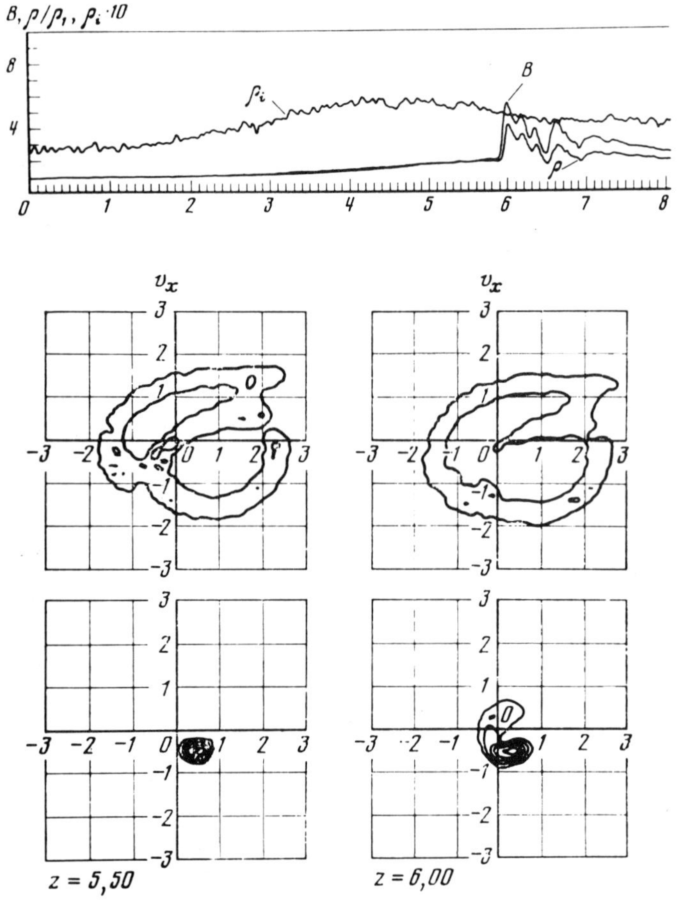

Fig. 4. The profiles of plasma mass density ϱ, cometary ion density ϱ_i, magnetic field strength B (top panel) and the distribution of heavy cometary ions and protons in the shock front obtained by the numerical simulations (from Galeev et al., 1985a)

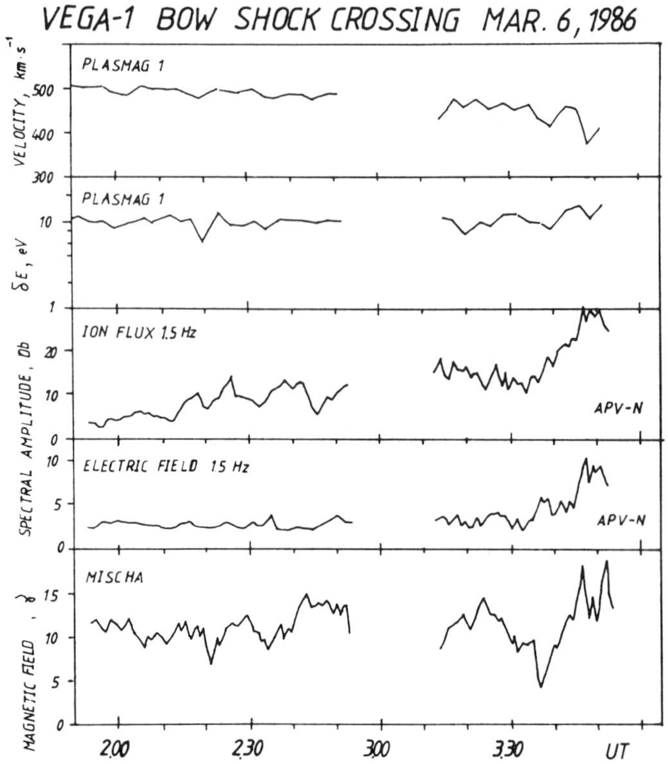

Fig. 5. The behavior of plasma and field parameters across the quasiperpendicular bow shock according to the measurements aboard the Vega-1 spacecraft (from Galeev et al., 1986b)

distribution measured by the wide angle plasma analyzer. It also drops at the moment of the plasma wave intensity enhancement. As to the jump of the magnetic field at the shock front it is masked in Fig. 5 by the strong MHD turbulence of a loaded solar wind. Let us note that the time resolution of plasma and magnetic field measurements aboard Vega-1 was too low (2.5 min to resolve the bow shock front.

The plasma and magnetic field measurements aboard Giotto have much higher time resolution and provide a detailed picture of the plasma behavior in the front of the cometary bow shock. During the inbound crossing of the bow shock by the Giotto spacecraft, the magnetic field (Neubauer et al., 1986) and the density of suprathermal electrons (Anderson et al.) were slowly building up at the foot of the bow shock with a width of 33 000 km (approximately 4 gyroradii of oxygen ions) and the velocity jump of the solar wind protons indicates that the subshock with a width of 500 km (approximately 2 gyroradii of protons) is formed at the end of the foot (Coates et al., 1986). It is interesting to note that the large density enhancement in cometary ions at the beginning of the foot (Wilken et al., 1987) causes the local depletion of the solar wind density and the drop of its velocity and magnetic field strength due to the outflow of protons along magnetic field lines and the mass loading of the solar wind. The signatures of the same phenomena are seen in the profiles of solar wind velocity and magnetic field shown in Fig. 5. The suprathermal electrons in the foot of the cometary shock are accelerated by the lower-hybrid waves generated by the beam of cometary ions similarly to the case of the Earth bow shock (Vaisberg et al., 1983). In turn these electrons generate whistle waves detected near the cometary bow shocks during the encounters of the ICE spacecraft with the Giacobini-Zinner comet (Scarf et al., 1986) and Vega-1,2 with Halley's comet (Grard et al., 1986).

Thus the theory of quasiperpendicular cometary shocks and in-situ measurements are in a good agreement. As to the observation of quasiparallel shocks near comet Halley observed by Vega-2 (Klimov et al., 1986; Galeev et al., 1986b) it needs further study and comparison with computer simulations.

4. Plasma boundaries in the inner comae

The plasma measurements aboard the Giotto spacecraft (Balsiger et al., 1986; Neubauer et al., 1986) have shown that the loaded solar wind does not come into close contact with the cometary plasma expanding with the neutral gas from the nucleus of the comet. The solar wind is stopped at the distance 1.5 10^4 km to the nucleus by the collisions of ions with the atoms of cometary gas evaporated from the nucleus, and the interplanetary magnetic

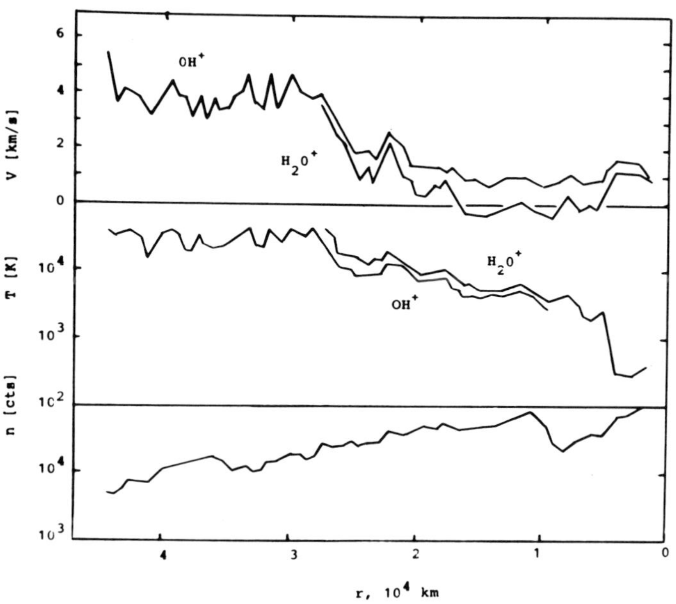

Fig. 6. The velocity and temperature of cometary H_2O^+ and OH^+ ions in the ram directions and the total plasma density measured in the inner coma of comet Halley by the IMS-HIS instrument aboard the Giotto spacecraft (from Schwenn et al., 1987)

field penetrating into the inner coma with the solar wind pushes the expanding cometary plasma closer to the nucleus (4700 km). Thus a wide magnetic barrier filled with the stagnant cometary photoplasma is formed between these two colliding plasmas. The profiles of plasma density, ion temperature and velocity in the ram direction in this region are shown in Fig. 6 (from Schwenn et al., 1986). We see that the plasma density is continuous and the ion temperature and velocity experience large changes across the contact surface separating the stagnant plasma in the magnetic barrier and the unmagnetized cometary plasma (see Fig. 1). The continuity of the plasma density indicates that both plasmas are in photoionization equilibrium

$$\frac{Qm_i}{4\pi v_g r^2 \tau_{ph}} = \alpha \varrho_i^2 / m_i, \tag{18}$$

where $\alpha \cong 1.3\ 10^{-7}\ T_e^{-1/2}\ \mathrm{cm^3\,s^{-1}}$ is the recombination rate, T_e the electron temperature in electron volts (Huebner and Fikani, 1982), $\tau_{ph} = 5.4\ 10^6$ s the photoionization time (Huebner and Carpenter, 1979). The ion temperature in the magnetic barrier is higher because the cometary atoms photoionized in the stagnant plasma gyrate around the magnetic field line with the relative velocity of gas and plasma $(v_g + u)$. Finally the expansion of unmagnetized cometary plasma is stopped at the contact surface by the magnetic field line tension (Ip and Axford, 1982).

On the other side of the magnetic barrier, near the collisionpause the plasma density after reaching its maximum suddenly drops to the lower value. This can be explained by the continuous inflow of cometary ions with the solar wind into the plasma pile up region and the recombination of stagnant plasma. However, Ip et al. (1986) pointed out that the density drop at this boundary is most probably due to the decrease of the electron temperature which makes the plasma recombination faster. The effect of plasma pile up between the collisionpause and the contact surface was verified by the measurements aboard the Vega-1 spacecraft (Vaisberg et al., 1987). The measured value of the maximum plasma density $n_i = 4\ 10^3\ \mathrm{cm^{-3}}$ is higher than that calculated with the help of Eq. (18) for the case of photoionization equilibrium: $n_i \cong 1.4\ 10^3\ \mathrm{cm^{-3}}$. Making this calculation we have assumed that the plasma density spikes on both sides of the tail plasma sheet of the Giacobini-Zinner comet discovered by Meyer-Vernet et al. (1986) are produced by the plasma outflow from the plasma pile up region into the tail. This permitted us to use the electron temperature $T_e \cong 1$ eV measured in these spikes as the best estimate of electron temperature in the pile up region. Thus the cooling of electrons to a temperature of 1 eV could result in the plasma density drop between the plasma pile up region and the magnetic barrier filled with stagnant plasma. However, the mechanism of rapid cooling of electrons in the cometary plasma flow is still unidentified. Let us note finally that the measured positions of the collisionpause and the contact surface agree with those theoretically estimated (Biermann, 1974; Ip and Axford, 1982; Galeev and Lipatov, 1984).

5. The role of collective plasma effects in the ionization of cometary atmospheres and the formation of plasma structures

The problem of anomalous ionization of cometary atmospheres is one of the most intriguing problems of cometary physics and the critical ionization velocity phenomenon by Alfven is the most natural mechanism of anomalous ionization for the magnetized solar wind flow through the cometary atmosphere. However, to arrive at a well reasoned conclusion regarding the role of this particular mechanism it is necessary to measure simultaneously all three plasma components participating in the process: cometary ions, electrons and plasma waves. Presently this analysis (Klimov et al., 1986; Galeev et al., 1986c; Ip, 1986; Haerendel, 1986) is based on the measurements of one or at most two of these parameters. Moreover, as was shown by Galeev et al. (1986c), the plasma density profile in the inner coma is very structured so possibly the anomalous ionization of cometary gas takes place only in thin structures. Thus very detailed data are needed for such kind of analysis.

To illustrate our point let us consider the profiles of the density of heavy cometary ions and the intensity of plasma waves in the lower-hybrid range of frequencies obtained in the PLASMAG and APV-N experiments aboard the Vega-2 spacecraft. These measurements were performed with high time resolution: the energy distribution of ions is measured every second and 64 measurements of one electric field component per second provide a wave form. Figure 7 shows the profiles of the ion density with the 5-second averaging and the spectral energy density of electric field oscillations integrated over the frequency range 1 Hz ÷ 32 Hz (Savin et al., 1986). We see that the build up of ion density is intermittant: the periods of slow density growth due to photo-ionization are interrupted by the periods of faster density increase. It is natural to explain these fast density increases by the impact ionization of gas by suprathermal electrons (note that the temperature of thermal electrons here is certainly lower than the ionization potential of cometary atoms according to the measurements aboard the ICE spacecraft in the inner coma of comet Giacobini-Zinner by Meyer-Vernet et al., 1986). The comparison of the spatial behavior of these parameters displayed in Fig. 7 shows that there is a significant enhancement of the plasma wave intensity at the beginning of each period of an anomalously fast density growth at least at the distances beyond 3 10^4 km from the nucleus of the comet. These density variations could not be explained by the convection of plasma nonuniformities from the

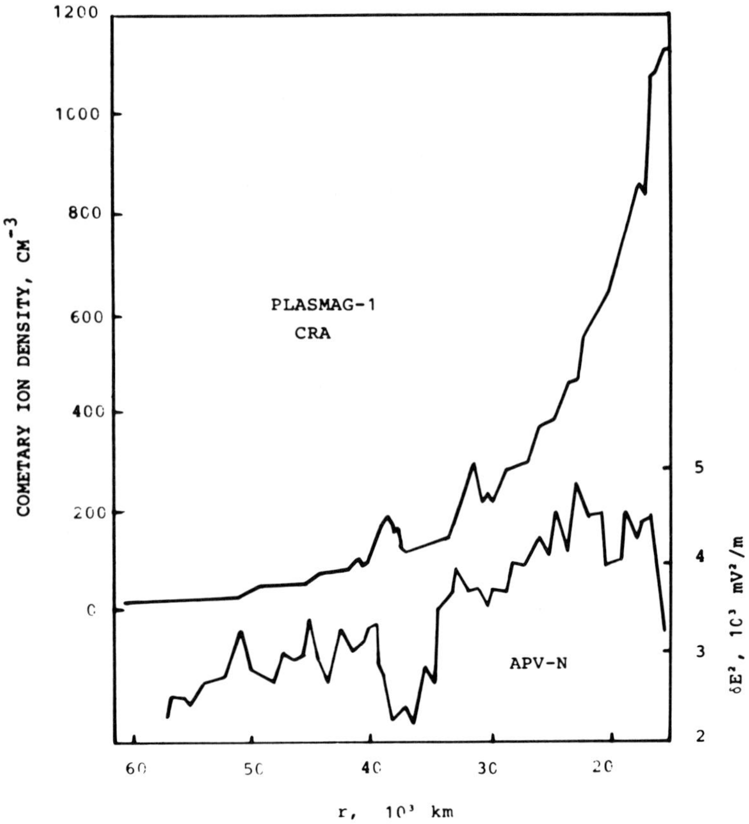

Fig. 7. The profile of heavy ion density (top curve) and the electric field intensity integrated over the range of frequencies 1 ÷ 32 Hz as measured aboard Vega-2 (from Savin et al., 1986)

cometosheath solar wind where they are generated by the ion pick-up processes into the inner comae. The latter follows from the fact that in the convecting nonuniformities there are no energy sources for the excitation of plasma waves. Only the newly born ions provide such an energy source for plasma instability.

Let us also note that the inner cometary coma is filled with the whistler radiation (Scarf et al., 1986; Grard et al., 1986). These waves are most probably excited by the suprathermal electrons accelerated by the lower hybrid waves like it happens in the Io torus (Coroniti et al., 1980; Galeev and Khabibrachmanov, 1985). Thus the intensity of whistler waves could be a good measure of the flux of suprathermal electrons. Unfortunately, the high time resolution data are not available for the detailed comparison with the plasma density and the lower hybrid wave intensity.

There were no direct measurements of the plasma flow velocity aboard the Vega spacecraft. But indirect estimates give the value $u \cong 15 \div 30\,\mathrm{km\,s^{-1}}$ (Galeev et al., 1986c; Vaisberg et al., 1987) that is not much higher than the critical Alfven velocity $v_{cr} = 11\,\mathrm{km\,s^{-1}}$ for water dominated cometary gas. In these conditions the problem of the efficiency of energy transfer from the newly ionized atoms to suprathermal electrons via lower-hybrid plasma waves becomes the key factor to decide whether the development of the critical ionization phenomenon is theoretically possible (see discussion by Galeev and Lipatov, 1984; Galeev et al., 1986c; Haerendel, 1986). It was shown that the high efficiency ($=2/3$) is reached only in the limit of a fast development of the instability when the energy relaxation of newly ionized atoms can be considered as one-dimensional (Formisano et al., 1986). The estimates of the growth rate of the instability produced by an ion beam formed of newly ionized atoms (Galeev et al., 1986c) have led us to the conclusion that it becomes greater than the ion cyclotron frequency (the condition for the one-dimensional energy relaxation of newly born ions) when the following inequality is satisfied:

$$v_{ion}\, n_{Te}/n_i > \omega_{ci}\, u \ln\Lambda/\pi V_{Ae}, \tag{19}$$

where $v_{ion} = \sigma_{ion} v_e Q/4\pi v_g r^2$ is the ionization frequency, v_e and n_{Te} the velocity and density of suprathermal electrons, $\ln\Lambda$ the Coulomb logarithm, u the bulk velocity of plasma, V_{Ae} the electron Alfven velocity. Using the values of the plasma and gas parameters measured aboard the Vega spacecraft, we find that this inequality is satisfied in the inner coma where the two large density enhancements are observed (see Fig. 7). Checking this inequality, we have assumed that the density of suprathermal electrons is approximately equal to their averaged density $n_{Te} \cong 6\,\mathrm{cm^{-3}}$ measured by the RPA1 instrument aboard the Giotto spacecraft at distances larger than 4.5 10^4 km to the comet nucleus (Rème, private communication). Let us note that the density of suprathermal electrons according to these measurements and the plasma bulk velocity drop to low values closer to the nucleus. We can expect that the corresponding sharp decrease of the plasma velocity during the Vega encounter with comet Halley takes place even closer to its nucleus since the dynamic pressure of the solar wind was about 5 times higher at this time.

We cannot yet identify the mechanisms of interruption of the instability under discussion though it is very similar to the interruption of the beam-plasma discharge which is frequently observed in laboratory and space experiments. But the resulting enhancements of the cometary plasma density along the line of sight will be viewed by the outside observer as the receding envelope structures. These structures could also manifest themselves in the fine plasma structures in the tail of comet.

6. Conclusions

The existing theory of the solar wind interaction with a comet is able to explain the global behavior of plasma parameters in the cometary coma. Nevertheless we need further work to find the mechanisms of formation of thin plasma boundaries and spatial structures.

Acknowledgements. The author is grateful to Academician R.Z. Sagdeev for constant support and enlightning discussions.

References

Alfvén, H.: 1957, *Tellus* **9**, 92

Alfvén, H.: 1954, On the Origin of the Solar System, Oxford University Press

Anderson, K.A., Carlson, C.W., Curtis, D.W., Lin, R.P., Rème, H., Sauvaud, J.A., d'Uston, C., Korth, A., Richter, A.K., Mendis, D.A.: 1986, 20th ESLAB, Symp., ESA SP-250, Vol. **I**, p. 259

Balsiger, H., Altwegg, K., Bühler, F., Geiss, J., Ghielmetti, A.G., Goldstein, B.E., Goldstein, R., Huntress, W.T., Ip, W.-H., Lazarus, A.J., Meier, A., Neugebauer, M., Rettenmund, U., Rosenbauer, H., Schwenn, R., Sharp, R.D., Shelley, E.G., Ungstrup, E., Young, D.T.: 1986, *Nature* **321**, 330

Bame, S.J., Anderson, R.C., Asbridge, J.R., Baker, D.N., Feldman, W.C., Fuselier, S.A., Gosling, J.T., McComas, D.J., Thomsen, M.F., Young, D.T., Zwickl, R.D.: 1986, *Sciences* **232**, 356

Baranov, V.B., Zaitsev, N.A., Lebedev, M.G.: 1986, *Astron. Zurn.* **63**, 170

Biermann, L., Brosowski, B., Schmidt, H.U.: 1967, *Solar Phys.* **1**, 254

Biermann, L.: 1974, *Solar Wind III*, ed. C.T. Russell, IGPP UCLA Publication, Los Angeles, USA, p. 396

Coates, A.J., Johnstone, A.D., Thomsen, M.F., Formisano, V., Amata, E., Wilken, B., Jockers, K., Winningham, J.D., Borg, H., Bryant, D.A.: 1986, 20th ESLAB Symp., ESA SP-250, Vol. **I**, p. 263

Coroniti, F.V., Scarf, F.L., Kennel, C.F., Kurth, W.S., Gurnett, D.A.: 1980, *Geophys. Res. Letters* **7**, 45

Formisano, V., Galeev, A.A., Sagdeev, R.Z.: 1982, *Planet. Space Sci.* **30**, 491

Galeev, A.A., Lipatov, A.S.: 1984, *Adv. Space Res.* **4**, No. 9, 229

Galeev, A.A., Lipatov, A.S., Sagdeev, R.Z.: 1985a, *Zhurn. Exper. Teor. Fiz.* **89**, 1495

Galeev, A.A., Cravens, T.E., Gombosi, T.I.: 1985b, *Astrophys. J.* **289**, 807

Galeev, A.A., Sagdeev, R.U., Shapiro, V.D., Shevchenko, V.I., Szego, K.: 1986a, in *Plasma Astrophysics*, ed. T.D. Guyenne, ESA SP-251, p. 307

Galeev, A.A., Gribov, B.E., Gombosi, T., Gringauz, K.I., Klimov, S.I., Oberz, P., Remizov, A.P., Riedler, W., Sagdeev, R.U., Savin, S.P., Sokolov, A.Y., Shapiro, V.D., Shevchenko, V.I., Szegö, K., Verigin, M.I., Yeroshenko, Y.G.: 1986b, *Geophys. Res. Letters* **13**, 841

Galeev, A.A., Gringauz, K.I., Klimov, S.I., Remizov, A.P., Sagdeev, R.Z., Savin, S.P., Sokolov, A.Y., Verigin, M.I.: 1986c, *Geophys. Res. Letters* **13**, 845

Galeev, A.A., Khabibrakhmanov, I.K.: 1985, *Pis'ma Astron. Zhurn.* **11**, 292

Grard, R., Pedersen, A., Trotignon, J.-G., Beghin, C., Mogilevsky, M., Mikhailov, Y., Molchanov, O., Formisano, V.: 1986, *Nature* **321**, 290

Gringauz, K.I., Gombosi, T.I., Remizov, A.P., Apathy, I., Szemerey, I., Verigin, M.I., Denchikova, L.I., Dyachkov, A.V., Keppler, E., Klimenko, I.N., Richter, A.K., Somogyi, A.J., Szegö, K., Szendo, S., Tatrallyay, M., Varga, A., Vladimirova, G.A.: 1986a, *Nature* **321** 282

Gringauz, K.I., Gombosi, T.I., Tatrallyay, M., Verigin, M.I., Remizov, A.P., Richter, A.K., Apathy, I., Szemerey, I., Dyachkov, A.V., Balakina, O.V., Nagy, A.F.: 1986b, *Geophys. Res. Letters* **13**, 613

Huebner, W.F., Carpenter, C.W.: 1979, *Solar Photo Rate Coefficients*, Los-Alamos Scientific Laboratory Report LA-8085-MS

Huebner, W.F., Fikani, M.M.: 1982, *Comet Halley, Parameters Study I*, Los-Alamos Scientific Laboratory Report LA-9346-MS

Ip, W.-H., Axford, W.I.: 1982, in *Comets*, ed. L.L. Wilkening, Univ. of Arizona Press, Tucson, Arizona, p. 588

Ip, W.-H., Axford, W.I.: 1986, *Nature* (in press)

Ip, W.-H.: 1986, *Adv. Space Res.*, **5**, No. 12, 47

Ip, W.-H., Schwenn, R., Rosenbauer, H., Balsiger, H., Neugebauer, M., Shelley, E.G.: 1987, *Astron. Astrophys.* **187**, 132

Johnstone, A., Coates, A., Kellock, S., Wilken, B., Jockers, K., Rosenbauer, H., Studemann, W., Weiss, W., Formisano, V., Amata, E., Cerulli-Irelli, R., Dobrowolny, M., Terenzi, R., Edigi, A., Borg, H., Hultqvist, B., Winningham, J., Gurgiolo, C., Bryant, D., Edwards, T., Feldman, W., Thomsen, M., Wallis, M., Biermann, L., Schmidt, H., Lust, R., Haerendel, G., Paschmann, G.: 1986, *Nature* **321**, 344

Klimov, S.I., Savin, S.P., Aleksevich, Y., Avanesova, G., Balebanov, V.M., Balihin, M., Galeev, A.A., Gribov, B.E., Nozdrachev, M., Smirnov, V., Sokolov, A., Vaisberg, O., Oberc, P., Krawczyk, Z., Grzedrielski, S., Juchnewicz, J., Nowak, K., Orlowskii, D., Parfianovich, B., Wozniak, D., Zbyszynki, Z., Voita, Y., Triska, P.: 1986, *Nature* **321**, 292–293

Krankowsky, D., Lämmerzahl, P., Herrwerth, I., Woweries, J., Eberhardt, P., Dolber, U., Herrmann, U., Schulte, W., Berthelier, J.J., Illiano, J.M., Hodges, R.R., Hoffman, J.H.: 1986, *Science* **232**, 326

Meyer-Vernet, N., Couturier, P., Hoang, S., Perche, C., Steinberg, J.L., Fainberg, J., Meetre, C.: 1986, *Science* **232**, 370

Mukai, T., Miyake, W., Terasawa, T., Kitayama, M., Hirao, K.: *Nature* **321**, 299

Neubauer, F.M., Glassmeier, K.H., Pohl, M., Raeder, J., Acuna, M.H., Barlaga, L.F., Ness, N.F., Musmann, G., Mariani, F., Wallis, M.K., Ungstrup, E., Schmidt, H.U.: 1986, *Nature* **321**, 352

Rème, H., Sauvaud, J.A., d'Uston, C., Cotin, F., Cros, A., Anderson, K.A., Carlson, C.W., Curtis, D.W., Lin, R.P., Mendis, D.A., Korth, A., Richter, A.K.: 1986, *Nature* **321**, 249

Riedler, W., Schwingenschuh, K., Yeroshenko, Y.G., Styashkin, V.A., Russell, C.T.: 1986, *Nature* **321**, 288

Sagdeev, R.Z., Shapiro, V.D., Shevchenko, V.I., Szegö, K.: 1986, *Geophys. Res. Letters* **13**, 85

Savin, S.P., Avanesova, G., Balichin, M., Klimov, S., Sokolov, A., Oberc, P., Orlowskii, D., Krawczyk, Z.: 1986, 20th ESLAB Symp., ESA SP-250, Vol. **3**, p. 433

Scarf, F.L., Coroniti, F.V., Kennel, C.F., Gurnett, D.A., Ip, W.H., Smith, E.J.: 1986, *Science* **232**, 377

Schmidt, H.U., Wegman, R.: 1982, in *Comets*, ed. Wilkening, L.L., Univ. of Arizona Press, Tucson, Arizona, p. 538

Schwenn, R., et al.: 1987, *Astron. Astrophys.* **187**, 160
Smith, E.J., Tsurutani, B.T., Slavin, J.A., Jones, D.E., Siscow, G.L., Mendis, D.A.: 1986, *Science* **232**, 382
Vaisberg, O.L., Zastenker, G.N., Smirnov, V.N., Khazanov, B., Omelchenko, A., Fedorov, A., Zakharov, D.: 1987, *Astron. Astrophys.* **187**, 183
Vaisberg, O.L., Galeev, A.A., Zastenker, G.N., Klimov, S.I., Nozdrachev, M.N., Sagdeev, R.Z., Sokolov, A.Y., Shapiro, V.D.: 1983, *Zhurn. Eksper. Teor. Fiz.* **85**, 1232
Wallis, M.K.: 1973, *Planetary Space Sci.* **21**, 1647
Wallis, M.K., Ong, R.S.: 1975, *Planetary Space Sci.* **23**, 713
Wilken, B., Jockers, K., Studemann, W., Rosenbauer, H., Thomsen, M., Johnstone, A.D., Coates, A.J., Formisano, V., Amata, E.T., Winningham, J.D., Borg, H.: 1987, *Astron. Astrophys.* (this volume)
Winske, D., Wu, Y.Y., et al.: 1975, *J. Geophys. Res.* **90**, 2713
Wurm, K.: 1963, in *The Moon, Meteorites and Comets*, eds. E.B.M. Middlehurst, G.P. Kuiper, Univ. of Chicago Press, Chicago, p. 573

Astron. Astrophys. 187, 21–24 (1987)

ASTRONOMY AND ASTROPHYSICS

The pick-up of cometary protons by the solar wind

M. Neugebauer[1], A.J. Lazarus[2], K. Altwegg[3], H. Balsiger[3], B.E. Goldstein[1], R. Goldstein[1], F.M. Neubauer[4], H. Rosenbauer[5], R. Schwenn[5], E.G. Shelley[6], and E. Ungstrup[7]

[1] Jet Propulsion Laboratory, California Institute of Technology, Pasadena, CA 91109, USA
[2] Massachusetts Institute of Technology, Cambridge, MA 02139, USA
[3] Physikalisches Institut, University of Bern, Bern, Switzerland
[4] Institut für Geophysik und Meteorologie der Universität zu Köln, D-5000 Köln, Federal Republic of Germany
[5] Max-Planck-Institut für Aeronomie, D-3411 Katlenburg-Lindau, Federal Republic of Germany
[6] Lockheed Palo Alto Research Laboratory, Palo Alto, CA 94304, USA
[7] Danish Space Research Institute, Lyngby, Denmark

Received February 18, accepted May 11, 1987

Summary. The HERS detector of the Ion Mass Spectrometer on the Giotto spacecraft measured the 3-dimensional distribution of picked-up cometary protons over a distance of ~ 8 million km upstream of the bow shock of comet P/Halley. The protons were observed to be elastically scattered out of their original cycloidal trajectories such that they were nonuniformly distributed over a spherical shell in velocity space. The shell radius (relative to its expected radius) and thickness increased as the bow shock was approached. Down-stream of the shock, the cometary protons could not be distinguished from the heated solar wind protons.

Key words: comets – plasmas – interplanetary medium

1. Introduction

The first part of a comet sensed by the solar wind is its extensive corona of atomic hydrogen. The comet-solar wind interaction begins when these H atoms are ionized, either by photoionization or by charge-exchange collisions with the solar wind, and the newly ionized protons respond to the electric and magnetic fields in the solar wind. This paper briefy summarizes some of the principal features of this initial interaction and pickup of cometary protons.

2. Instrumentation

The data presented here were obtained by the High Energy Range Spectrometer (HERS) of the Giotto Ion Mass Spectrometer (IMS) experiment. The instrument design and performance have been described by Balsiger et al. (1986a, b). With this instrument, it was possible to study separately the velocity distributions of each ion species (i.e. each narrow range of mass/charge ratio, M/Q). Only proton data are considered here.

During the period of interest, HERS mapped the distribution of protons every fourth spacecraft spin (i.e., 4 s out of every 16 s). A separate channel electron multiplier detector (CEM) counted the flux of protons in each of four elevation angle bins, which viewed the protons incident at angles with respect to the spacecraft spin direction of 15° to 30° (CEM D), 30° to 45° (CEM C), 45° to 60° (CEM B), and 60° to $\sim 70°$ (CEM A). The spacecraft spin axis was closely aligned with the relative velocity vector between the spacecraft and the comet and was pointed toward the comet. During each four-second spin, the proton counts were sorted into 31 azimuth bins (11°.25 each) and 64 quasi-logarithmically spaced energy bins, ranging from 10 to 4024 eV. The total field of view for protons was thus $349° \times \sim 55°$.

Send offprint requests to: M. Neugebauer

3. Preliminary results

The method used to identify protons of cometary origin is illustrated in Fig. 1. Shown as small plots A, B, C, and D are polar plots of contours in velocity space of the proton phase-space density, $f(v)$, observed with the four CEM detectors A, B, C, and D, respectively, during the interval 0803 to 0908 spacecraft event time (SCET), March 13. Each plot represents the distribution on a conical surface; the horizontal axis corresponds to flow in the plane defined by the spacecraft spin vector and the direction to the Sun while positive vertical values correspond to flow with a northward component. Figure 1 A shows data from CEM A, whose field of view was furthest from the spin axis and included the solar wind. The peak of the solar-wind distribution is indicated by an "x" in the center of the roughly circular contours. At this time, the solar wind speed was 397 km s^{-1} relative to the spacecraft. The distortion of the contours at low energy is due to inaccuracies in the preliminary algorithms used for processing the low-energy proton data. Three contour levels represent a one-decade change in phase-space density. With increasing displacement from the "x" at the peak of the solar-wind distribution in Fig. 1 A, the phase-space density decreased monotonically across 16 contour levels (>5 orders of magnitude); the outer four contour intervals have been shaded, with a darker shade representing a higher value of $f(v)$. Two still lower density levels have been left unshaded in Fig. 1 A. Even further outward from the "x", at the bottom of the frame (corresponding to flow with a southward component), the phase-space density climbed again, through four contour levels, as again indicated by shading. This

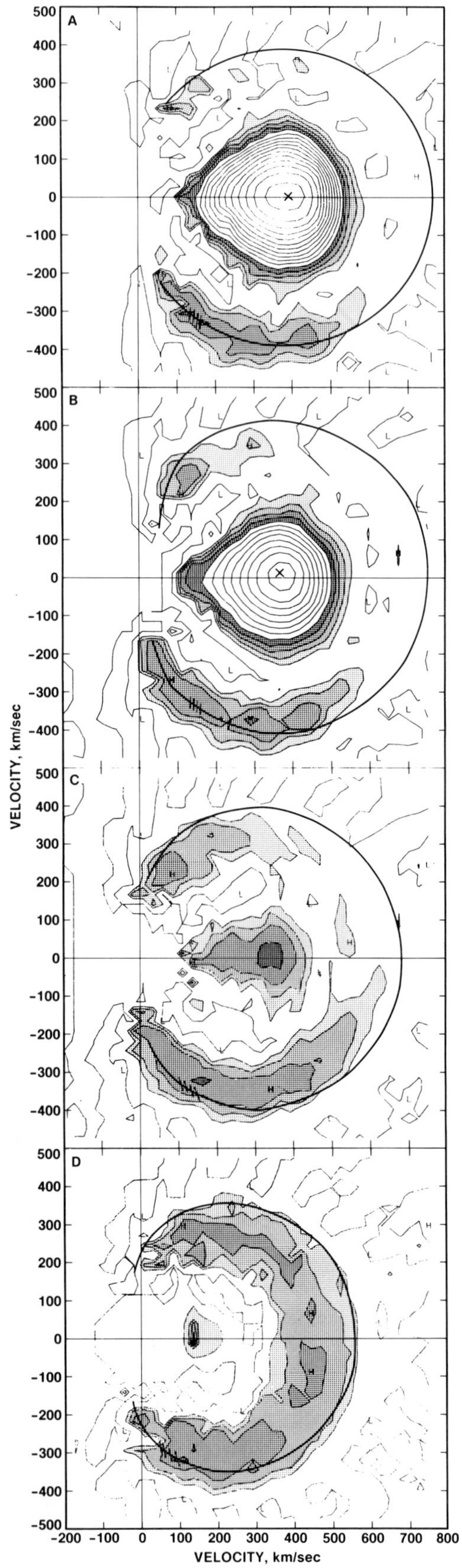

Fig. 1. Phase space density contours of CEMs A–D for the period 0803–0908 UT, March 13, 1986

outer arc is due to the pick-up protons. Because the phase-space density of the pick-up protons was about four orders of magnitude less than that at the peak of the solar-wind proton distribution, it is necessary to sum the data over many spins; 237 spins of proton data were used to generate Fig. 1. During the time that the data in Fig. 1 were obtained, the direction of the interplanetary magnetic field remained relatively steady with solar-ecliptic latitude and longitude angles of +26 and 13°, respectively. The expected location in velocity space of picked-up cometary protons can be calculated as follows: The photo-dissociation of cometary water leads to H atoms with speeds, a, of 8 and 20 km s^{-1} (Keller, 1976). Since both these values are considerably less than the 43.8 km s^{-1} heliocentric speed of the comet, which was in turn considerably less than the solar-wind speed, the velocity of the picked-up protons is relatively insensitive to the precise value of the magnitude of a. For the calculations which follow, it is assumed that $a = 11$ km s^{-1}, in keeping with the results of analysis of Lyman-alpha observations of comet Halley by Suisei (Kaneda et al., 1986).

Since the velocity of the spacecraft was directed towards the comet and the relative speed between the two was 68.4 km s^{-1}, the neutrals appeared to come from the direction of the comet with velocity $v_n = (68.4 + a)$ km s^{-1}. In a reference frame moving with the solar wind, the velocity of a newly created proton would be $\boldsymbol{v}_r^* = \boldsymbol{v}_n - \boldsymbol{v}_{sw}$, where $\boldsymbol{v}_n$ and $\boldsymbol{v}_{sw}$ are the velocities of the neutrals and the solar wind, respectively, in the spacecraft frame. Resolution of the velocity $\boldsymbol{v}_r^*$ into components perpendicular and parallel to the interplanetary magnetic field allows calculation of the protons' helical motion (a ring in velocity space in the solar-wind reference frame). The trace of the ring of freshly ionized hydrogen through the instrument field of view can be calculated. The intersections of this ring with the fields of view of the four CEMs are shown by cross-hatching in Fig. 1.

It is clear from Fig. 1 that the location of the pick-up protons in velocity space includes but extends beyond the (cross-hatched) locations expected for newly created ions spiralling around the local magnetic field. There is evidence that the pick-up protons have been scattered in pitch angle so that they occupy a large part of a spherical shell with radius v_r^*, centered at the solar-wind velocity. The scattering was not sufficiently intense, however, to fill the shell uniformly. The curves showing the intersections of this shell with the conical surfaces mapped out by the four CEMs are also indicated in Fig. 1. The population of picked-up protons is spread out from the expected ring location along these curves.

The band of pick-up protons is broader for CEM-D than for the other detectors because the shell is nearly tangent to the field of view of this sensor.

A similar set of four diagrams is given in Fig. 1 of Neugebauer et al. (1986) for the time interval 1436–1552 SCET, March 13, when the field direction was not quite so steady as in the interval shown in Fig. 1 above. In that case, the shells were nearly completely filled.

At this distance of nearly 4 million km from the comet, the thickness of the shell was roughly consistent with the energy resolution of the instrument. There is no strong evidence for energy diffusion or heating of the protons in the shell; this question will be examined in greater detail in future work. In summary, until the spacecraft was quite close to the shock, the picked-up protons were nonuniformly distributed over a spherical shell in velocity space, with the shell center and radius determined by the velocities of the solar wind and the comet. This result is consistent with the two-dimensional measurements made by Suisei (Terasawa et al., 1986).

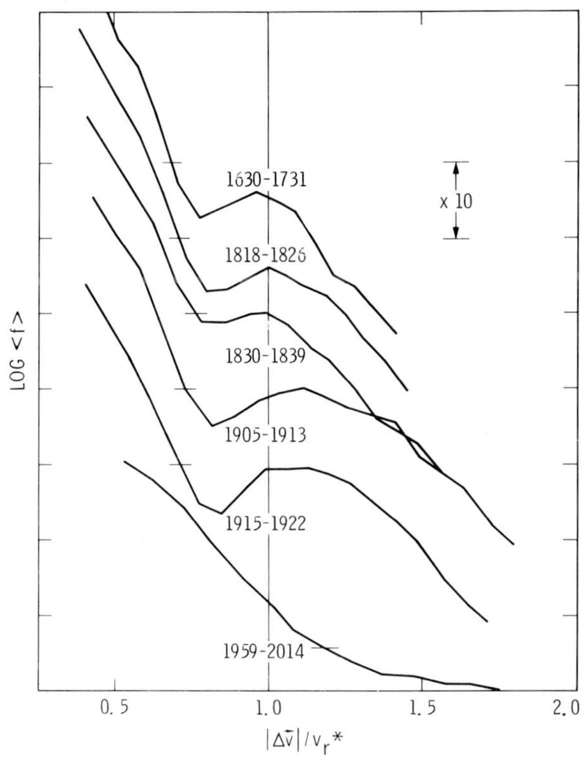

Fig. 2. Average phase space density as a function of displacement in velocity space from the solar-wind bulk speed normalized by $v_r^* =$ the radius of the shell in velocity space expected for protons picked up locally. The numbers on each curve indicate the time intervals (SCET) during which each spectrum was measured. The spacecraft crossed the bo shock during the interval 1922–1931 SCET. The short, horizontal lines on each spectrum denote equal values of $\langle f(v) \rangle$

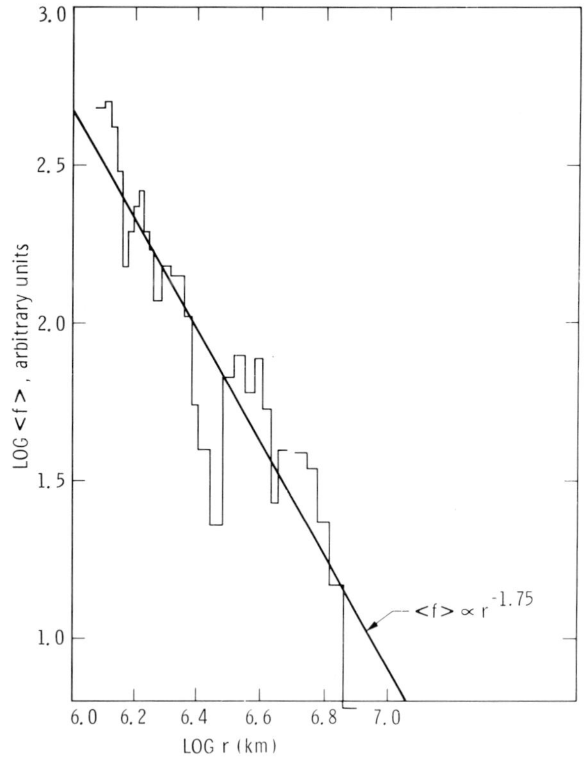

Fig. 3. Average phase space density of the part of the shell of pick-up protons within the field of view of the HERS instrument versus distance from the comet nucleus

Better statistics on the radius (in velocity space) and the thickness of the shell of pick-up protons can be obtained by calculating the distribution of the magnitudes of the vectors $\Delta \boldsymbol{v} = \boldsymbol{v} - \boldsymbol{v}_{sw}$. Figure 2 shows the average $f(v)$ plotted versus $|\Delta \boldsymbol{v}|/v_r^*$ for several intervals when the solar-wind velocity was relatively steady. These curves were obtained by sorting the proton counts obtained in each energy-azimuth-elevation bin into bins of $|\Delta \boldsymbol{v}|$, each bin being 20 km s^{-1} wide, then averaging the contents of each $|\Delta \boldsymbol{v}|$ bin, and finally dividing $|\Delta \boldsymbol{v}|$ by v_r^* before plotting. In the ideal case of a perfectly calibrated instrument with very narrow energy and angular resolution, if all the newly ionized protons had been picked up at the same solar wind speed and had no energy changes thereafter, this plot would show a sharp spike at $|\Delta \boldsymbol{v}|/v_r^* = 1$, superimposed on the flank of the solar-wind distribution function.

Each curve in Fig. 2 is labelled according to the times (in SCET) of the first and last data included. Upstream of the shock crossing at $\sim$1930 SCET, the data show rather broad peaks near $|\Delta \boldsymbol{v}|/v_r^* = 1.0$, with the positions and the widths of the peaks increasing as the comet was approached. Far upstream of the shock, the radius of the shell appeared to be somewhat less than expected. It is possible that this effect is not real, but that it is due to a combination of instrumental factors which are currently under investigation. The widths of the peaks in the first three curves in Fig. 2 are not significantly broader than the widths expected from instrumental smear; future modeling efforts will allow quantitative comparisons of these distributions to those expected for pick-up shells of infinitesimal thickness. There is no question, however, about the reality of the increasing radius (relative to the expected value of v_r^*) and of the increasing thickness of the shell as the shock is approached.

These changes can be ascribed to a combination of effects. First, the instrument always detected cometary protons ionized over a range of distances upstream of the spacecraft. The "older" protons, which were picked up further upstream where the solar-wind speed was presumably higher, should be on a larger-radius shell than should the "young" protons picked up locally. In addition, the protons in the shell would be adiabatically heated as the solar wind was decelerated and compressed (Ip and Axford, 1986). Although there is a mixture of ages of pick-up protons at all distances from the shock, the effect becomes much more pronounced close to the shock where the pickup of heavier ions, such as C^+ and O^+, increases the rates of mass loading and slowing of the solar wind. Separation of the effect of superposition of shells of protons with different ages from the effect of adiabatic acceleration will be very difficult to model because the Giotto spacecraft approached the comet from a direction nearly perpendicular to the solar-wind streamlines instead of along a single streamline. Although the solar-wind velocity did, on the average, decrease as the bow shock was approached, this decrease was not smooth. It can be seen from Fig. 1 of Johnstone et al. (1986) that in this upstream region, the spacecraft crossed many discontinuities separating plasmas of different speeds.

Two additional effects which could broaden the shell symmetrically – both inward and outward from its nominal position – are energy diffusion and an increased level of variation in $\boldsymbol{v}_{sw}$ associated with an increased amplitude of turbulence close to the bow shock. Neither of these effects were apparently very important outside the bow shock.

It is apparent from the bottom curve in Fig. 2 that inside the shock, the shell of pick-up protons could no longer be distinguished from heated solar-wind protons. At this time, protons were detected in all energy channels of the instrument, which included velocities as high as $\sim 850\ \mathrm{km\,s^{-1}}$. The presence of protons with energies ~ 16 times the local bulk flow energy is indirect evidence for the continuing presence of protons picked up at higher speeds and of additional energization processes as well.

Pick-up protons were first detected at ~1700 UT on March 12, the day before encounter when the spacecraft was nearly 8 million km from the comet. The dependence of the number of picked-up protons on distance from the comet can be crudely estimated from Fig. 3, which shows the integral under the pick-up-proton peaks (in curves like those shown in Fig. 2) plotted versus distance from the comet nucleus. Two effects are evident in Fig. 3. First, there is a general dropoff associated with the expected decrease in the number of H atoms with distance from the comet. A least-squares fit to the data produces an average dropoff proportional to $r^{-1.75}$. Second, there are very large fluctuations about the average dropoff rate caused by (a) spatial inhomogeneities in atomic H arising from time variations in cometary activity, (b) perhaps some variation in the rate of charge exchange collisons between H atoms and the temporally and spatially varying solar wind, and (c) variations in the direction of the interplanetary magnetic field, which causes changes in how the cycloidal motion of picked-up protons cuts through the field of view of the instrument. Further work on untangling these various effects and on determining the rate of pitch-angle and energy scattering is in progress.

Acknowledgements. The research described in this paper was carried out, in part, by the Jet Propulsion Laboratory, California Institute of Technology, under a contract with the US National Aeronautics and Space Administration. The work at MIT was partially supported by NASA grant NGL 22-009-015.

References

Balsiger, H., Altwegg, K., Bühler, F., Fischer, J., Geiss, J., Meier, A., Rettenmund, U., Rosenbauer, H., Schwenn, R., Benson, J., Hemmerich, P., Sager, K., Kulzer, G., Neugebauer, M., Goldstein, B.E., Goldstein, R., Shelley, E.G., Sanders, T., Simpson, D., Lazarus, A.J., Young, D.T.: 1986a, ESA SP-1077, p. 129

Balsiger, H., Altwegg, K., Bühler, F., Geiss, J., Ghielmetti, A.G., Goldstein, B.E., Goldstein, R., Huntress, W.T., Ip, W.-H., Lazarus, A.J., Meier, A., Neugebauer, M., Rettenmund, U., Rosenbauer, H., Schwenn, R., Sharp, R.D., Shelley, E.G., Ungstrup, E., Young, D.T.: 1986b, *Nature* **321**, 330

Ip, W.-H., Axford, W.I.: 1986, *Planet. Space Sci.* **34**, 1061

Johnstone, A.D., Coates, A.J., Heath, L., Thomsen, M.F., Wilken, B., Jockers, K., Formisano, V., Amata, E., Winningham, J.D., Borg, H., Bryant, D.: 1986, 20th ESLAB Symp., ESA SP-250, Vol. **I**, p. 25

Kaneda, E., Hirao, K., Shimizu, M., Ashihara, O.: 1986, *Geophys. Res. Letters* **13**, 833

Keller, H.U.: 1976, *Space Sci. Rev.* **18**, 641

Neugebauer, M., Lazarus, A.J., Altwegg, K., Balsiger, H., Goldstein, B.E., Goldstein, R., Neubauer, F.M., Rosenbauer, H., Schwenn, R., Shelley, E.G., Ungstrup, E.: 1986, 20th ESLAB Symp., ESA SP-250, Vol. **I**, p. 19

Terasawa, T., Mukai, T., Miyake, W., Kitayama, M., Hirao, L.: 1986, *Geophys. Res. Letters* **13**, 837

Astron. Astrophys. 187, 25–32 (1987)

ASTRONOMY AND ASTROPHYSICS

Alfvénic turbulence in the solar wind flow during the approach to comet P/Halley

A.D. Johnstone[1], A.J. Coates[1], J. Heath[1], M.F. Thomsen[2], B. Wilken[3], K. Jockers[3], V. Formisano[4], E. Amata[4], J.D. Winningham[5], H. Borg[6], and D.A. Bryant[7]

[1] Mullard Space Science Laboratory, Holmbury St. Mary, Dorking, Sŭrrey RH5 6NT, UK
[2] Los Alamos National Laboratory, Los Alamos, NM 87545, USA
[3] Max Planck Institut für Aeronomie, D-3411 Katlenburg-Lindau, Federal Republic of Germany
[4] Istituto di Fisica dello Spazio Interplanetario, I-00044 Frascati, Italy
[5] Southwest Research Institute, San Antonio, TX 78284, USA
[6] Kiruna Geophysical Institute, Umeå University, S-90187 Umeå, Sweden
[7] Rutherford Appleton Laboratory, Chilton, Didcot, Oxon, OX11 OQX, UK

Received February 23, accepted April 2, 1987

Summary. The Johnstone Plasma Analyser on Giotto operated almost continuously for two days before the encounter with comet Halley. Throughout this period it made observations of the proton and alpha particle distributions in the solar wind with 8 seconds time resolution. As the comet was approached, fluctuations were observed in all the primary bulk parameters, i.e. density, temperature and flow velocity, of both distributions at levels above the usual solar wind turbulence. We present here a survey of the data from a distance of $5\,10^6$ km from the nucleus up to the cometary foreshock at $1.4\,10^6$ km. Early in this period the variance was higher than in the solar wind on a similar day but the activity may have been associated with a crossing of the heliospheric current sheet. From a distance of $2.7\,10^6$ km the power level gradually increased, as the mass loading of the solar wind increased, to levels well above normal solar wind turbulence. We estimate that up to 16% of the free energy associated with the implanted cometary ion distribution went into the wave energy, similar to the proportion which has been estimated by quasilinear theory. Although the solar wind convection time through the mass-loaded region is short compared with the development time for a shell distribution in the cometary ions estimated from numerical simulation, the observations indicate that the cometary ion distribution is still basically ring-like.

Key words: comets – interplanetary medium – plasmas

1. Introduction

As had been anticipated on theoretical grounds (Ip and Axford, 1982) low frequency turbulence is generated in the solar wind flow by the implantation of heavy cometary ions upstream from the cometary bow shock. Turbulence was detected in the magnetic field (Tsurutani and Smith 1986 a,b) and in the electron distributions (Gosling et al., 1986) at comet Giacobini-Zinner by the ICE spacecraft. Similar turbulence has been detected by Giotto (Neubauer et al., 1986) and Vega (Riedler et at., 1986) at comet Halley. We report here the first observations of the turbulence in the solar wind protons and alpha particles where the fluctuations are seen not only in the plasma velocity as expected for simple Alfven waves but also in the density and temperature. By observing the effect of mass-loading on the velocity of the solar wind, we can infer the density of the cometary ions and estimate the free energy associated with the implantation process. In a companion paper (Johnstone et al., 1987) a detailed comparison of the waves in the magnetic field and solar wind is made for a short (one hour) section of the data.

2. Instrumentation

The Johnstone Plasma Analyser (JPA) instrument on Giotto consisted of two sensors, the Fast Ion Sensor (FIS) and the Implanted Ion Sensor (IIS), (Johnstone et al., 1986a). The data presented here comes from FIS only although the information on the cometary ion distributions obtained by IIS will be referenced. The FIS has an electrostatic energy analyser covering the range $10\,\text{eV}\,q^{-1}$ to $20\,\text{keV}\,q^{-1}$ 16 times per 4s spin of the spacecraft. It covers the angular range from 20° to 160° with respect to the spin axis so that within one revolution it obtains essentially the complete distribution of positive ions. On alternate spins it uses another energy sweep, covering one quarter of the normal range four times as often, for the 45° sector centred on the solar direction. In this solar wind mode it is able to distinguish protons and alpha particles in the energy spectrum, if their temperature is low enough that the distributions do not overlap. The sensor therefore obtains a complete energy and angular distribution of both species in 0.25 s every 8 s. The parameters are derived by a moment integration over the energy distribution and by fitting gaussian distributions to the polar angle distribution and the azimuthal distribution separately. From the width of the distribution in these three cuts, three different values of the temperature, related to the temperature anisotropy in the solar wind, are obtained. In this analysis we use the mean value.

Send offprint requests to: A.D. Johnstone

3. Observations

An overview of the data is shown in Fig. 1 (protons) and Fig. 2 (alpha particles). A quick comparison of the two figures shows that they both have the same general features. The x-component of the velocity (from the comet towards the sun in HSE (Halley Solar Ecliptic) coordinates) shows a steady decline in magnitude throughout the plot. The proton density is variable, decreasing in the first half, increasing in the second half. Some of these variations, and the short term changes also visible are likely to be caused by changes in the upstream solar wind flow before it becomes affected by the comet. In the last third of the plot the level of fluctuations increases considerably above solar wind level in a way which must be due to the comet. The simultaneous speed decrease in the bottom panel and the density increase are clearly the effect of mass loading. In order to obtain the solar wind background level so that a reasonable comparison could be made with conditions in the absence of a comet, data from the cruise phase were searched in order to find a day when the density, temperature and velocity were similar. The closest conditions occurred on day 65, 1986, when the spacecraft was 30 million km from the comet. The results are shown in Fig. 3. There are fluctuations in all the parameters but none which correspond in magnitude to the fluctuations in the last third of Fig. 1. In order to make the comparison more quantitative the r.m.s. deviation of each parameter in the 45 values obtained over a 6 min period has been calculated.

The r.m.s. deviations have been normalised to the mean density, mean temperature, or mean total speed for the three velocity components. The results are plotted in Figs. 4,5,6 corresponding to the data samples in Figs. 1,2,3 respectively. This calculation is similar to one carried out by Gosling et al. (1986) on solar wind electron parameters during the ICE encounter with comet Giacobini-Zinner. The effective frequency range of our r.m.s. deviation runs from 2.7 mHz (corresponding to the 6 min period) to 62.5 mHz (the Nyquist frequency at the FIS sample rate). The corresponding figures for Gosling et al. (1986) are 3.3 mHz to 20.8 mHz. These frequency ranges cover gyrofrequencies of ions with mass/charge values ranging from 32 to 2 in the mean field of 6 nT near Halley and 36 to 6 in the mean field of 8 nT near Giacobini-Zinner respectively. The mass range of the main proportion of cometary ions is well-covered in both cases.

In reviewing the observations first note that on day 65 the r.m.s. deviation only briefly exceeds 1.4% in the velocity components and 10% in the density. On day 72 the proton and alpha particle results often exceed these levels but are similar to each other. Since the calculation of the alpha particle parameters is based on much smaller count rates (by a factor of at least 20) a higher level of statistical noise would be expected. The difference in the r.m.s. deviation of the velocity components is not detectable demonstrating that the variability is not produced by statistical noise – a conclusion which is reinforced by the results from day 65. The variability in the alpha density is much higher than that in the proton density and, in this case, is probably caused by statistical effects.

The period from 0500 SCET (Spacecraft Event Time) to 1900 SCET on day 72 can be divided in four sections with different characteristics as shown in Fig. 4. The first section, during which

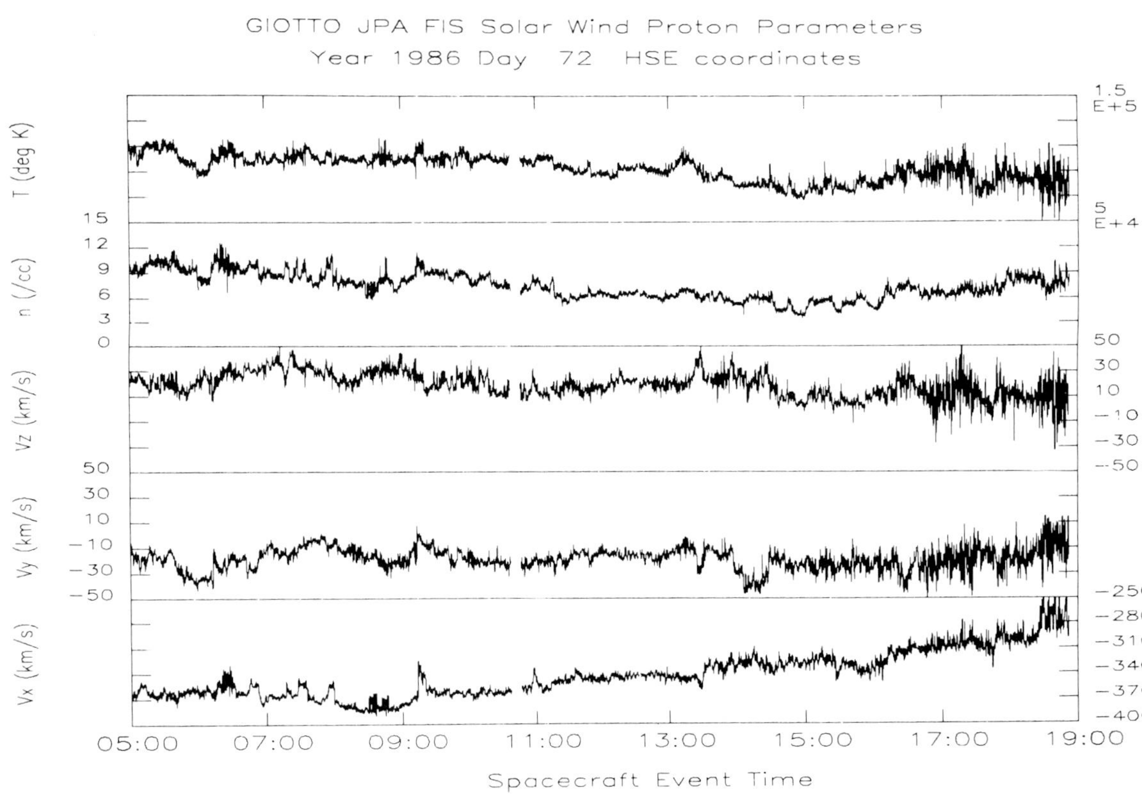

Fig. 1. Solar wind proton parameters from 5 million to 1.25 million km from the nucleus of comet Halley with the full 8 s time resolution. The bottom panel shows the decrease in the solar wind flow speed caused by the mass loading by cometary ions. Simultaneously the fluctuations increase in all components

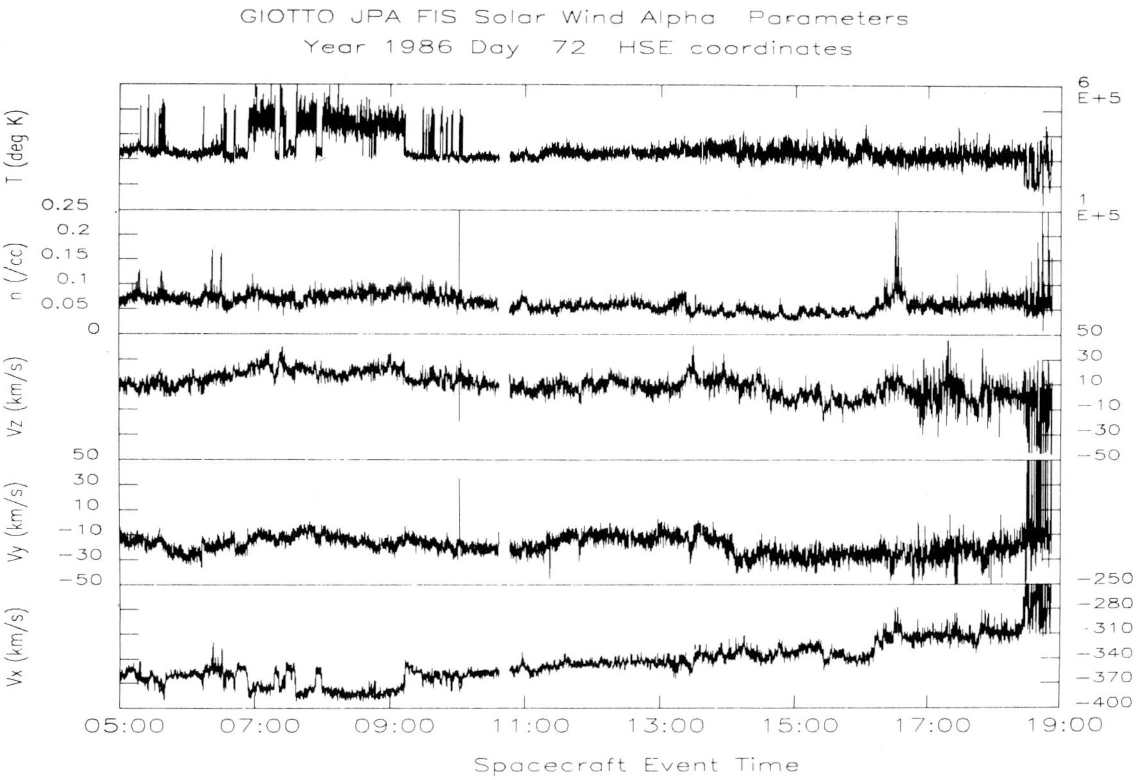

Fig. 2. Solar wind alpha particle parameters over the same time period as Fig. 1

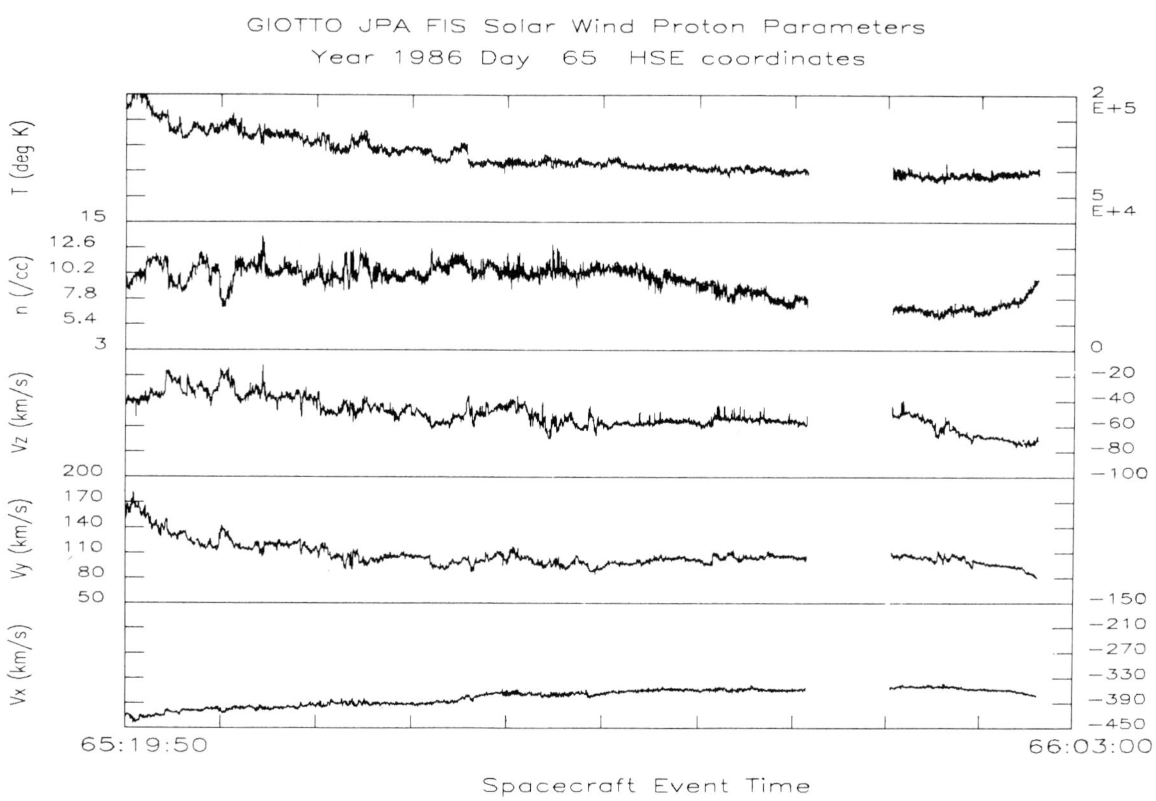

Fig. 3. Solar wind proton parameters at 8 s time resolution obtained on day 65, 30 million km from the comet, when the values of the density, velocity and temperature most closely corresponds to those during the encounter

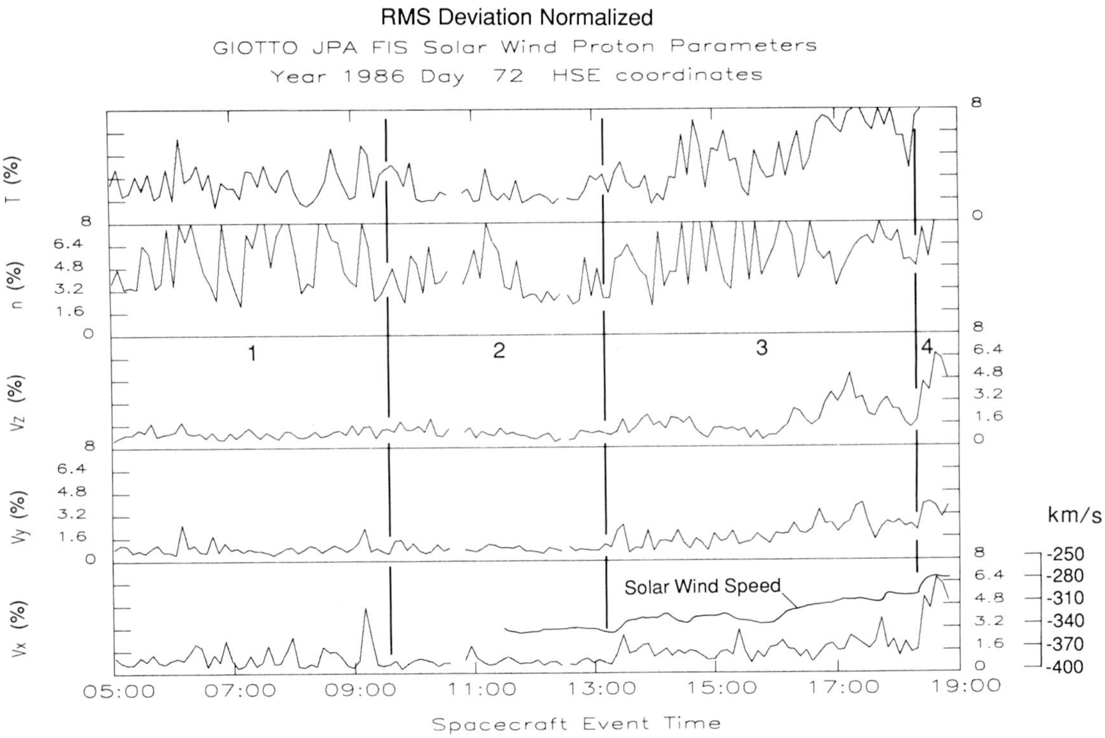

Fig. 4. The normalised r.m.s. deviation of the solar wind proton parameters for the data shown in Fig. 1

the spacecraft moves from 4.7 to 3.6 million km from the nucleus, has higher than normal r.m.s. deviation in all parameters but most noticeably in the V_x component and the density. The plot is spiky in character and the individual peaks seem to be associated with discrete events in the solar wind, seen in particular in the plot of the proton V_x component in Fig. 1. At the present time it is not possible to be sure that this activity is caused by the presence of the comet because during this period the spacecraft was close to and crossed the heliospheric current sheet (H. Rème, F. Neubauer, private communication) and the activity

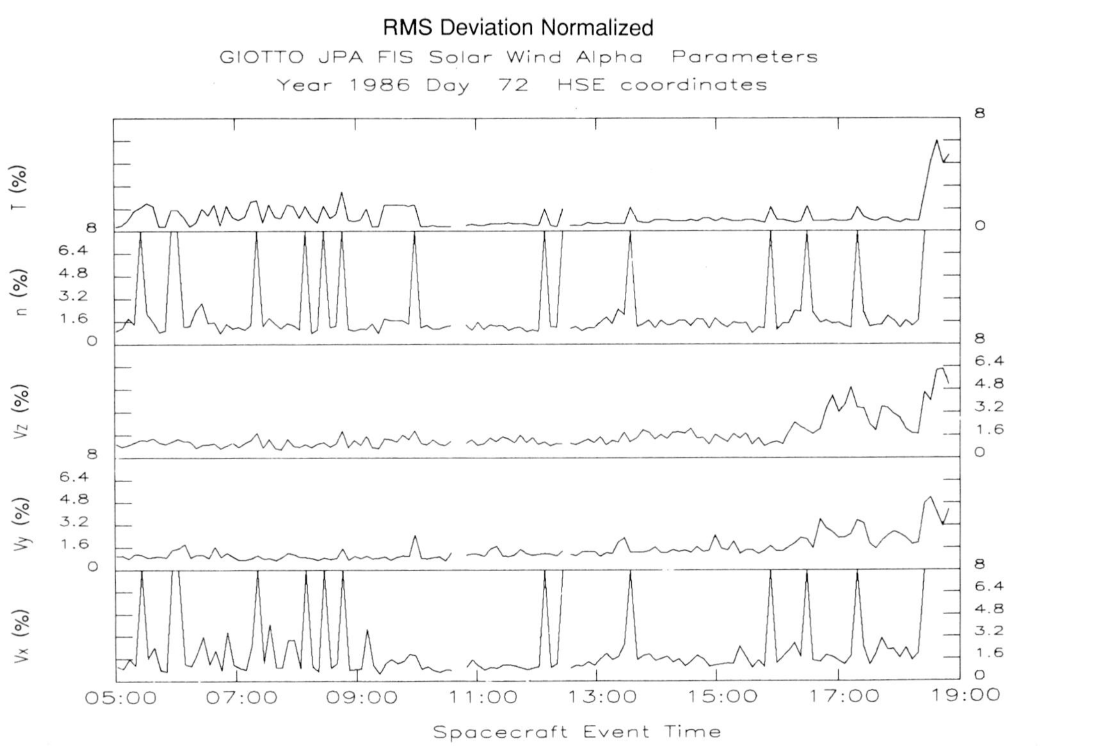

Fig. 5. The normalised r.m.s. deviation of the solar wind alpha particle parameters for the data shown in Fig. 2

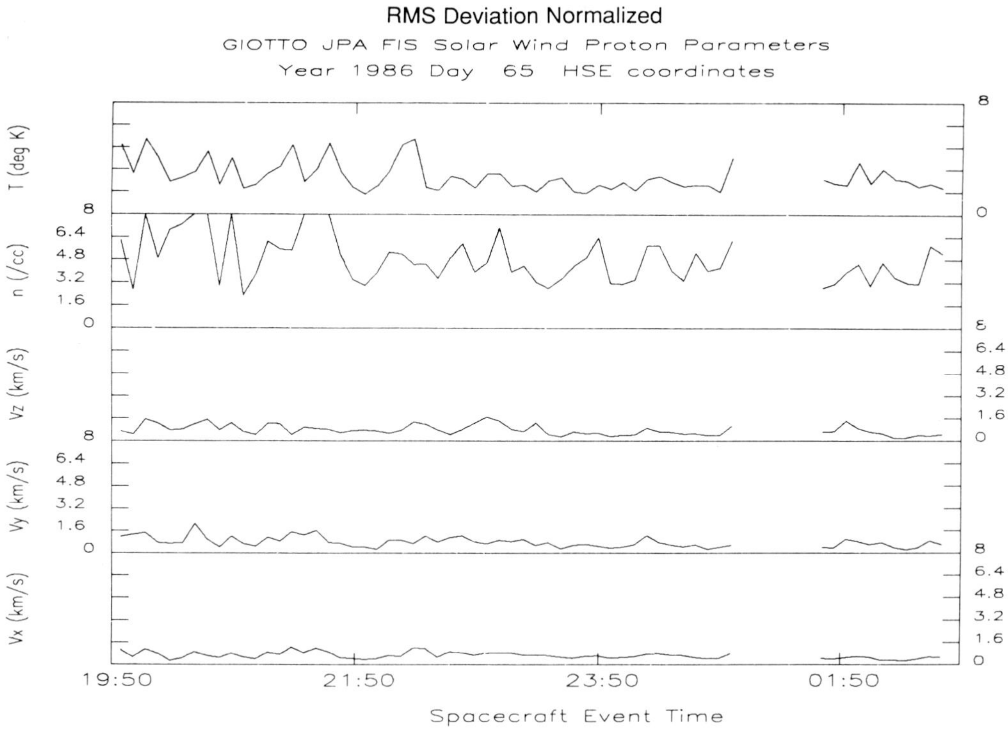

Fig. 6. The normalised r.m.s. deviation of the solar wind proton parameters for the data shown in Fig. 3

may be related to that structure. Further investigation is required to study this event in detail and to relate its occurrence to the observed proximity of the current sheet determined by other instruments.

In the second section of Fig. 4, from 3.6 to 2.7 10^6 km from the nucleus, there is no activity above the solar wind background. The power spectra are indistinguishable from those taken on day 65.

In the third section the activity starts to increase steadily in all parameters from a distance of 2.7 million km. In the first half of the period the highest level of activity is in the density. It is apparently associated with a series of discrete, step-like changes in density seen in Fig. 1. The r.m.s. deviation in the x and y velocity components increases uniformly through the period but the z component reaches a peak at 1700 SCET and then declines. At 1825 SCET the spacecraft reaches the foreshock (Sect. 4 in Fig. 4) and the activity in all parameters increases sharply. At this time the character of the velocity variations changes, as can be seen in a companion paper (Coates et al., 1987). In the bottom panel of Fig. 4, which shows the r.m.s. deviation of the x component of velocity, the mean x velocity has been superimposed for comparison. It is clear that as the mass-loading reduces the speed, the level of activity increases whether expressed as a normalised value as here, or as an absolute level. The activity in the third section is clearly the anticipated turbulence generated by the implanted cometary ions. A power spectrum of the turbulence in the three velocity components is given in Fig. 7. It shows power levels at least an order of magnitude above background levels at frequencies above the heavy ion gyrofrequency. At the lowest frequencies the power at this time is predominantly in the V_z component. If the waves are being generated at the gyrofrequency in the space-craft frame as theory predicts (Tsurutani and Smith, 1986b) then this plot suggests that the waves may be linearly polarised when generated but become isotropic as the power cascades to higher frequencies through non-linear interactions. This supposition is supported by the preliminary conclusion of an analysis of plasma and magnetic field data (Johnstone et al., 1987) at comet Halley and by the magnetic field measurements made by ICE at comet Giacobini-Zinner (Tsurutani and Smith, 1986a).

The power-law spectral slope at $f^{-1.7}$, is close to that found for the magnetic turbulence at Giacobini-Zinner i.e. $f^{-5/3}$ (Tsurutani and Smith, 1986b). Although no measurements of the solar wind proton distribution were made by ICE, the fluctuations in the solar wind electron density have been analysed by Gosling et al. (1986). Just outside the Giacobini-Zinner bow wave the r.m.s. deviation reached levels between 10% and 25% of the mean density. At the Halley foreshock the proton density fluctuations were approximately 8% of the mean density. It is also interesting to compare the wave amplitude in velocity components outside the foreshock with other characteristic velocities. The total r.m.s. deviation of the three components is $\langle v^2 \rangle^{1/2} = 17\,\mathrm{km\,s^{-1}}$. Since the average thermal velocity is approximately $38\,\mathrm{km\,s^{-1}}$ and the Alfvén velocity V_A is $55\,\mathrm{km\,s^{-1}}$, even outside the foreshock the turbulence should be characterised as strong. Any instrument which averages the velocity distribution over periods of the order of two minutes would observe the wave activity as an increase in temperature.

4. Discussion

When new cometary ions are created in the solar wind they are immediately picked up into cycloidal orbits which combine gyration in the magnetic field at a velocity $u_s \sin \phi$ with a drift motion perpendicular to the field, also $u_s \sin \phi$, where ϕ is the angle between the magnetic field and the solar wind flow direction. In

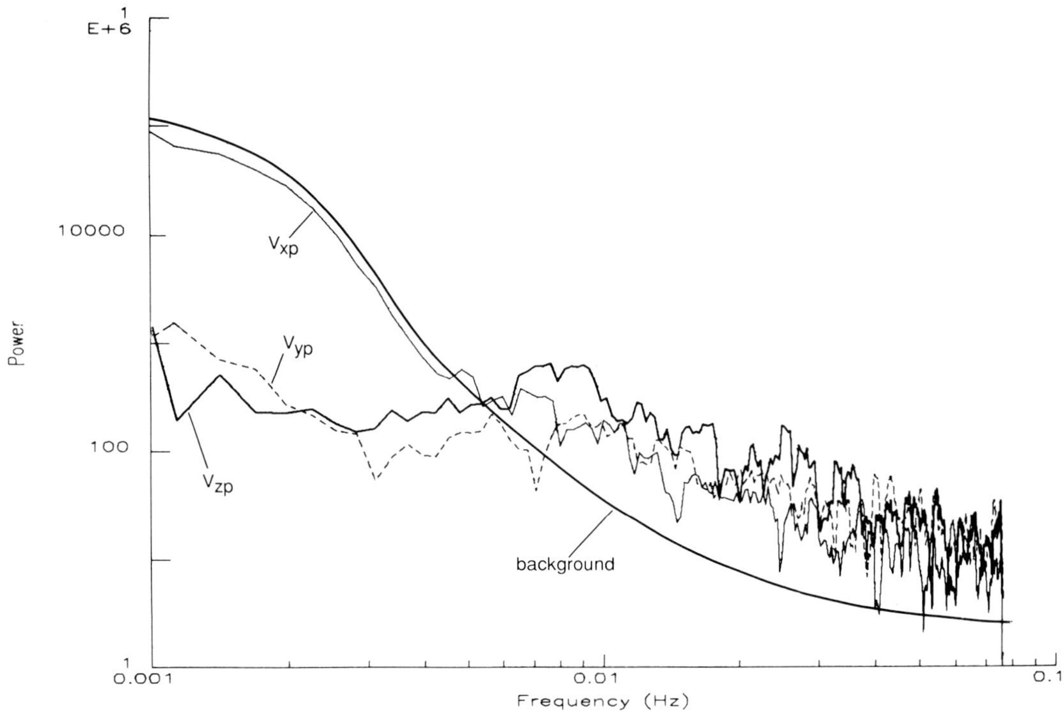

Fig. 7. The power spectrum of the three velocity components between approximately 1648 and 1800. Increased power is observed at all frequencies above the gyrofrequency of the heavy implanted cometary ions

the frame moving with the solar wind, the ions are gyrating and moving sunward along the magnetic field with a velocity $u_s \cos\phi$. This motion parallel to the field is the source of the free energy for the electromagnetic ion beam instability (Wu and Davidson, 1972; Gary et al., 1984) thought to be responsible for the hydromagnetic turbulence. As a result of the turbulence the ions are scattered in pitch angle and thereby pick up momentum parallel to the magnetic field from the solar wind. The free energy per ion is W, where

$$W = \frac{1}{2} m_i u_s^2 \cos^2\phi$$

At any point in the flow the local density of cometary ions is composed of all those ions which have been created upstream and have then drifted along a path, not necessarily identical with the solar-wind flow, to that point. The waves they generate propagate mainly along field lines in the solar wind frame at velocities of the order of the Alfven speed. At the same time they convect downstream with the solar wind at a velocity of the order of 300 km s^{-1}. To a first approximation therefore, we can relate the different types of measurement made at one point by the spacecraft by ignoring the parallel motion of ions and waves in the solar-wind frame. The cometary ions themselves load the solar wind and decrease its velocity; the local wave energy density is derived from the free energy of these cometary ions. Hence we use the decrease in solar-wind velocity to obtain first the cometary ion density, and then the free energy for comparison with the wave energy density. Previous estimates of the cometary ion density have been based on an integration of the upstream ionisation rate (Tsurutani and Smith, 1986a; Yumoto et al., 1986; Galeev et al., 1986) using theoretical parameters.

To obtain the cometary ion density we use the equation (Galeev et al., 1985).

$$\frac{d}{dx}\left(\rho u^2 + p_i + \frac{B^2}{2\mu_0}\right) = 0 \tag{1}$$

The magnetic term can be ignored in this upstream region where there is little change in the mean magnitude. The pressure p_i is dominated by the gyratory motion of the heavy implanted ions i.e.

$$p_i = \frac{1}{2} n_i m_i u_i^2 \tag{2}$$

The ions are implanted over a range of distances upstream where the solar wind velocity varies from u_s, the local measured value, to u_∞, the undisturbed value far upstream. We assume that the mean perpendicular velocity u_i can be taken as

$$u_i = \alpha u_s \sin\phi \tag{3}$$

where α is a number between 1 and u_∞/u_s and is intended ultimately to give an estimate of the accuracy. The momentum flux is

$$\rho u^2 = \rho_s u_s^2 + \rho_i u_i^2 \tag{4}$$

The mass flux in the solar wind must be constant i.e.

$$\rho_\infty u_\infty = \rho_s u_s \tag{5}$$

as has been confirmed experimentally by Coates et al. (1987). Combining equations (1) to (5) we obtain the following expression for the free energy

$$\frac{1}{2} \rho_i u_s^2 \cos^2\phi = \rho_\infty u_\infty (u_\infty - u_s)/3\alpha^2 \tan^2\phi$$

The wave energy density is $\rho_s\langle v_s^2\rangle$ where we have allowed for an equal amount of magnetic wave energy. Setting the wave energy density equal to a fraction K of the free energy density we obtain

$$\frac{\langle v_s^2 \rangle}{u_s^2} = \frac{K}{3\alpha^2 \tan^2 \phi} \frac{u_\infty}{u_s} - 1 \tag{6}$$

The term on the LHS is the sum of the three velocity components of the normalised variances plotted in Fig. 4, which is plotted in Fig. 8 against the velocity term $(u_\infty/u_s)-1$. The upstream conditions are obtained from a measurement in Sect. 2 of Fig. 4 where $u_\infty = 345\,\mathrm{km\,s^{-1}}$ and $n_\infty = 6.4$ protons/cm^3 (Coates et al., 1987). Figure 8 shows that the wave energy density is not simply proportional to the free energy density but that it varies up to an upper limit (represented by the line on the plot) which is proportional to the free energy density. The variability could be the result of changes in the field direction ϕ as shown by Eq. (6). Taking $\phi = 56°$ (Wilken et al., 1986), a value obtained from the energy spectrum of implanted ions in the upstream region, gives

$$K = 0.16 \text{ to } 0.26$$

where the range comes from allowing α to vary between 1 and 1.27. On the basis of quasi-linear theory, which may not be valid in the strong turbulence here, Galeev (1986) has estimated that

$$K = (V_A/u_\infty)(1 + \cos^4 \phi/3)$$

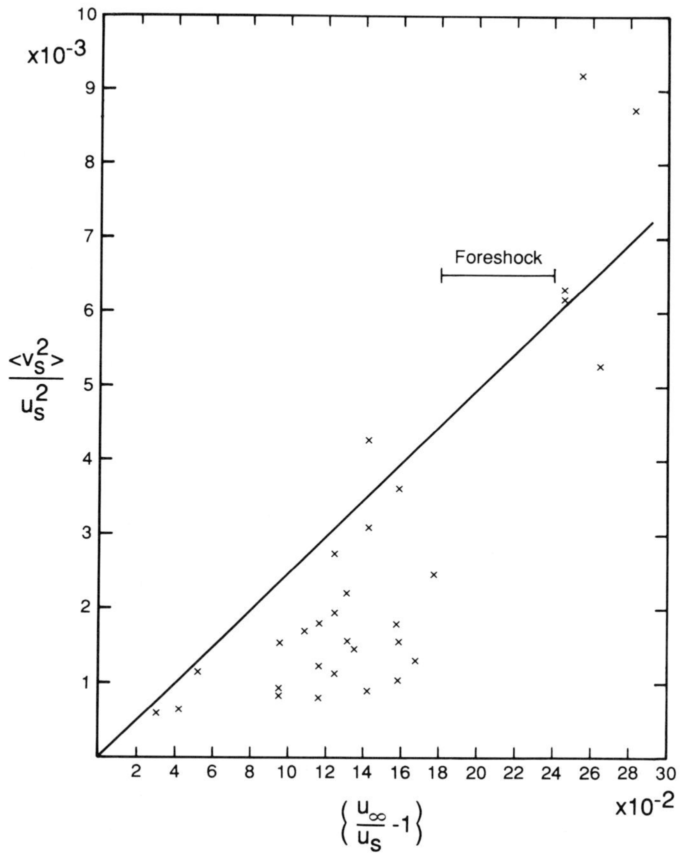

Fig. 8. A plot of wave energy density measured by the normalised velocity variance against the free energy density which is proportional to the normalised velocity decrease. The gap in the distribution of points is due to the decrease in mean velocity on passing through the outer limit of the foreshock at 1825 SCET

Substituting measured values in this expression gives $K = 0.16$ which is within the range of values obtained from the observed waves. Galeev's calculation is based on the assumption that the pitch angle distribution of the cometary ions has formed a shell from the initial ring distribution (Wu et al., 1986) with just sufficient anisotropy provided by newly injected ions to maintain the intensity of the waves against the loss process of induced scattering by solar wind protons (Sagdeev et al., 1986). The angular distribution of the cometary ions is far from shell-like outside the foreshock (Johnstone et al., 1986b; Wilken et al., 1986). The preliminary analysis given by these authors indicates only a partially-filled shell distribution. The time scale τ for the formation of a shell, obtained by Wu et al (1986) from numerical simulation, as a function of wave amplitude is

$$\tau = B/\Omega_i \Delta B = V_A/\Omega_i \langle v^2 \rangle^{1/2}$$

where we have assumed that the waves are Alfven waves to obtain the relationship with the r.m.s. deviation. Conditions immediately upstream from the foreshock give a value of

$$\tau \sim 80\,\mathrm{s}$$

In this time the solar wind has convected a distance of the order of 24,000 km, much less than the scale length for the injection of cometary ions which is of order 10^6 km. The implication of this comparison is that the mechanism is less effective at pitch angle scattering than has been suggested by the numerical simulations.

5. Conclusions

The generation of Alfvenic turbulence in the solar wind has been directly linked here with mass-loading of the solar wind. The wave spectrum reaches strong turbulence levels outside the foreshock comparable with estimates based on quasi-linear theory. The waves appear to be less effective at pitch angle scattering than has been suggested by some numerical simulations.

Acknowledgements. The work at Mullard Space Science Laboratory and the Rutherford Appleton Laboratory was supported by the UK Science and Engineering Research Council. The work at the Max Planck Institut für Aeronomie was supported by the Max-Planck-Gesellschaft zur Förderung der Wissenschaften and by the Bundesminister für Forschung und Technologie under Grant 01-OF-112-4. The work at the Southwest Research Institute was supported by NASA contract Number NASW-3574. The work at the Kiruna Geophysical Institute was supported by the Swedish Board for Space Activities. The work at the Istituto di Fisica dello Spazio Interplanetario was supported by the Piano Spaziale Nazionale of the Consiglio Nazionale delle Ricerche. The work at Los Alamos was performed under the auspices of the US Dept. of Energy and was supported by NASA through order number W-15164.

References

Coates, A.J., A.D. Johnstone, M.F. Thomsen, V. Formisano, E. Amato, B. Wilken, K. Jockers, J.D. Winningham, H. Borg, D.A. Bryant: 1987, *Astron. Astrophys.* (submitted)

Galeev, A.A., T.E. Cravens, T.I. Gombosi: 1985, *Astrophys. J.* **289**, 807

Galeev, A.A.: 1986, 20th ESLAB Symp. ESA SP-250 Vol. I p. 3

Galeev, A.A., et al.: 1986, *Geophys. Res. Letters* **13**, 841

Gary, S.P., C.W. Smith, M.A. Lee, M. Goldstein, D.W. Forslund, (1984). *Phys. Fluids* **27**, 1852

Gosling, J.T., J.R. Asbridge, S.J. Bame, M.F. Thomsen, R.D. Zwickl: 1986, *Geophys. Res. Letters* **13**, 267

Ip., W.H, Axford, W.I.: 1982, in *Comets*, ed. L.L. Wilkening, Univ. Arizona Press, Tucson, Arizona, p. 588

Johnstone, A.D., J.A. Bowles, A.J. Coates, A.J. Coker, S.J. Kellock, J. Raymont, B. Wilken, W. Studemann, W. Weiss, R. Cerulli-Irelli, V. Formisano, E. de Giorgi, P. Perani, M. de Bernardi, H. Borg, S. Olsen, J.D. Winningham, D.A. Bryant: 1986a, ESA SP-1077, p. 15

Johnstone, A., A. Coates, S. Kellock, B. Wilken, K. Jockers, H. Rosenbauer, W. Studemann, W. Weiss, V. Formisano, E. Amata, R. Cerulli-Irelli, M. Dobrowolny, R. Terenzi, A. Egidi, H. Borg, B. Hultquist, J. Winningham, C. Gurgiolo, D. Bryant, T. Edwards, W. Feldman, M. Thomsen, M.K. Wallis, L. Biermann, H. Schmidt, R. Lust, G. Haerendel, G. Paschmann: 1986b, *Nature* **321**, 344

Johnstone, A.D., K.H. Glassmeier, M. Acuna, H. Borg, D. Bryant, A. Coates, V. Formisano, J. Heath, F. Mariani, G. Musmann, F. Neubauer, M. Thomsen, B. Wilken, J. Winningham (1987). *Astron. Astrophys.* (submitted)

Neubauer F.M., K.H. Glassmeier, M. Pohl, J. Raeder, M.H. Acuna, L.F. Burlaga, N.F. Ness, G. Musmann, F. Mariani, M.K. Wallis, E. Ungstrup, H.U. Schmidt (1986). *Nature* **321**, 352

Riedler W, K. Schwingenschuh, Y.G. Yeroshenko, V.A. Styashkin, C.T. Russell (1986). *Nature* **321**, 288

Sagdeev, R.Z., V.D. Shapiro, V.I. Shevchenko, K. Szego. (1986). *Geophys. Res. Letters* **13**, 85

Tsurutani, B.T., E.J. Smith (1986a). *Geophys. Res. Letters* **13**, p. 259

Tsurutani, B.T., E.J. Smith (1986b). *Geophys. Res. Letters* **13**, p. 263

Wilken B, K. Jockers, A.D. Johnstone, A. Coates, J. Heath, V. Formisano, E. Amata, J.D. Winningham, M. Thomsen, D.A. Bryant, H. Borg: 1986, *Adv. Space Res.* **6**, 1, p. 337

Wu, C.S., R.C. Davidson: 1972, *J. Geophys. Res.* **77**, 5399

Wu, C.S., D. Winske, J.D. Gaffey: 1986, *Geophys. Res. Letters* **13**, 865

Yumoto, K., T. Saito, T. Nakagawa: 1986., *Geophys. Res. Letters* **13**, 825

Astron. Astrophys. 187, 33–38 (1987)

ASTRONOMY AND ASTROPHYSICS

General features of comet P/Halley: solar wind interaction from plasma measurements

H. Rème [1], J.A. Sauvaud [1], C. d'Uston [1], A. Cros [1], K.A. Anderson [2], C.W. Carlson [2], D.W. Curtis [2], R.P. Lin [2], A. Korth [3], A.K. Richter [3], and D.A. Mendis [4]

[1] Centre d'Etude Spatiale des Rayonnements, CNRS-UPS, BP 4346, F-31029 Toulouse Cedex, France
[2] Space Sciences Laboratory, UC-Berkeley, Berkeley, CA 94720, USA
[3] Max-Planck Institut für Aeronomie, Postfach 20, D-3411 Katlenburg-Lindau, Federal Republic of Germany
[4] EECS, UC-San Diego, La Jolla, CA 92093, USA

Received February 5, accepted April 15, 1987

Summary. The Giotto RPA-COPERNIC plasma experiment identified several regions in which the solar wind interaction with the cometary plasma displayed characteristic features:

– Beginning ~4.6 10^6 km from the comet there is an upstream region with sporadic connection to the comet,

– An electron foreshock is present up to 2.5 10^5 km away from the bow shock,

– A bow shock is detected at 1.15 10^6 km,

– Between the bow shock and the cometopause the outer regions can be divided into 3 parts,

– A cometopause is found at ~1.35 10^5 km and a density decrease is detected at ~4.5 10^4 km from the comet.

The detailed plasma features associated with these regions are successively described.

Key words: cometary ionospheres – cometary plasma – solar wind

1. Introduction

The RPA-COPERNIC (Rème Plasma Analyzer – Complete Positive ion, Electron and Ram Negative Ion measurements near comet Halley) plasma experiment on the Giotto spacecraft is a collaboration between the Centre d'Etude Spatiale des Rayonnements, Toulouse, France; the Max-Planck Institute für Aeronomie, Lindau, W. Germany; the University of California, Berkeley, USA; and the University of California, San Diego, USA.

This experiment (Rème et al., 1986a) consists of two detectors: the RPA1-EESA (Electron Electrostatic Analyzer), which provides three-dimensional measurements of the distribution of electrons from 10 eV to 30 keV (Rème et al., 1986b), and the RPA2-PICCA (Positive Ion Cluster Composition Analyzer), which provides mass analysis of positively charged cold cometary ions from mass 10 to 213 amu, with $\Delta M = 0.4$ up to 50 amu and $\Delta M = 1$ above (Korth et al., 1986b). The EESA measures the full electron distribution every half spin of the spacecraft (2 seconds). The detection system consists of 17 anodes defining 16 angular sectors, giving an angular resolution of $\sim 22.5° \times 4°$, and the ram sector. By varying the analyzer high voltage, a 39-step energy spectrum is obtained every 250 ms. The data are compressed by computing onboard the axis of symmetry of the electron distribution, which generally coincides with the magnetic field direction, and then sorting the electron data into pitch angle bins about that direction. The magnetic field direction thus obtained, and the binned electron data are telemetered down every 2 seconds, while full 3-D distributions are usually returned twice every 16 s. The PICCA scans the entire mass range up to every 3.2 s (~220 km along the trajectory of Giotto relative to comet Halley).

The first results from the plasma electron analyzer were published very soon after the encounter (Rème et al., 1986b). A more complete and more accurate description of the interaction of the solar wind with the cometary plasma can be given now.

Figure 1 summarizes 3 electron parameters as measured by EESA during the last 43 hours before the Giotto encounter with comet Halley: n_e is the density of 10 eV–30 keV electrons calculated every 2 s from the omni-directional energy spectra, T_e is the electron temperature estimated from the maximum slope of the distribution function above about 50 eV and the heat flux ratio is the ratio of minimum to maximum heat flux computed parallel and antiparallel to the magnetic field direction taken as the axis of symmetry of the electron momentum flux tensor. The data gap on day 72, around 05.00 UT (Earth Received Time), coincides with the last manoeuvre of the spacecraft before the encounter. On day 71 and on day 72 until 04.00 UT (Earth Received Time), there are no significant variations of n_e and T_e. In fact for several days before closest approach, the solar wind was relatively quiet. Its flow speed was $< 400\ \mathrm{km\,s^{-1}}$ and its density was $\sim 5\ \mathrm{cm^{-3}}$. However, increases in the heat flux ratio are observed several times, the first time more than 10^7 km away from the comet nucleus. Are these effects due to the presence of the comet? Comparing the electron data with the magnetometer data (Neubauer et al., 1986a) these increases of the heat flux ratio are found to be clearly correlated with changes in B_x (the sunward or antisunward B direction). During this period Giotto is very near the interplanetary current sheet and the effects of the heat flux ratio are probably of interplanetary and not cometary origin.

After the last manoeuvre, on day 72, after 05.15 UT (Earth Received Time), the data show the influence of the comet. A large turbulent region extends upstream from the bow shock for

Send offprint requests to: H. Rème

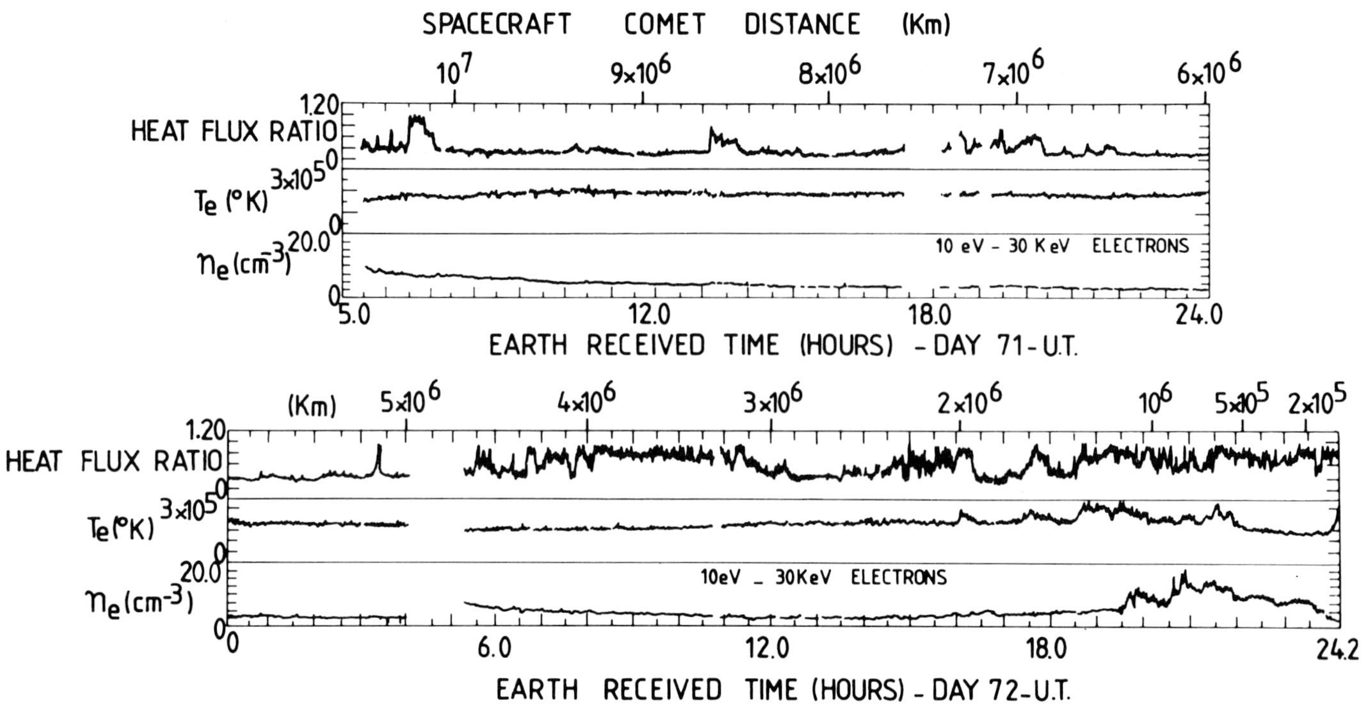

Fig. 1. Electron parameters as measured during the last 43 hours before the Giotto encounter with comet Halley: n_e is the density of 10 eV–30 keV electrons, T_e is the electron temperature estimated from the slope of the energy spectra near 70 eV and the heat flux ratio is the ratio of minimum to maximum heat flux computed parallel and antiparallel to the magnetic field direction, taken as the symmetry axis of the electron momentum flux tensor. Times given are UT Earth Received Times ($\sim$8 minutes later than the spacecraft event times)

$\sim$4.6 10^6 km from the nucleus. Here fluctuations in the electron distribution occur, characterized by abrupt variations in the electron heat fluxes with the heat flux ratio reaching a value of unity when the measurement region seems magnetically connected with the comet. A heat flux ratio of unity means equal forward and reverse heat fluxes, i. e. that electrons may be streaming back along the interplanetary magnetic field, in the direction opposite to the heat flux from the solar corona. This region, with sporadic connections to the comet, is the *upstream region* which extends from 4.6 10^6 to 1.4 10^6 km from the comet.

Figure 2 summarizes some plasma data measured by the RPA-COPERNIC experiment during the last 6.2 hours before the Giotto encounter with comet Halley when the spacecraft went from the upstream region to the closest approach to the comet: n_e is the density of 10 eV–30 keV electrons; n_e (hot) the density of 0.8–3.6 keV electrons; T_e the electron temperature estimated from the slope of the energy spectra near 70 eV; V_e the electron bulk velocity calculated from the electron distribution. At the top panel the RPA2-PICCA count rate in the ram direction is shown for 250–500 eV/q ions.

In its trajectory, the spacecraft successively goes through different regions or boundaries indicated in Fig. 2.

2. The foreshock

An electron foreshock is present from 1.4 10^6 to 1.15 10^6 km from the comet. There is a slight increase in n_e and n_e (hot), an increase of $\geqq 50\%$ in the temperature, a decrease in the bulk velocity which becomes quite variable, and some ions are detected in the ram direction. The heat flux ratio (Fig. 1) rises to nearly unity.

Spikes of $\sim$1 keV electrons appear in front of the shock. These hard electrons may have been accelerated in the same way electrons are accelerated at the Earth's bow shock, to energies of 1 to 10 keV.

3. The bow shock

The shock is detected at 1.15 10^6 km from the comet with a relatively abrupt jump in density. The flow direction determined from the plasma electron distribution function changes abruptly at 19.30 UT (Earth Received Time, i.e. $\sim$19.22 UT Spacecraft Event Time) and the flow speed begins a sharp decline, thus defining the location of the shock quite precisely and lending support to the belief that comet Halley indeed possesses a bow shock. No later than 90 s after this, the electron distribution function becomes isotropic. More details on the bow shock are given in Anderson et al. (1987). Taking into consideration a flaring ratio $\simeq 2$ of the bow shock (Mendis et al., 1986) and the fly-by geometry (the Giotto trajectory is at an angle of $\sim 107°$ to the Sun-comet line) the estimate of the stand-off distance of the bow shock (that is, its distance from the nucleus in the sunward direction) is $\sim$4 10^5 km, which corresponds to a total production rate of neutrals, $\dot{Q}_n \simeq 3.8\ 10^{29}\ s^{-1}$ if the shock Mach number is 2 and $\dot{Q}_n \simeq 5.6\ 10^{29}\ s^{-1}$ if the shock Mach number is 1.5 according to the model of Mendis et al. (1986).

4. The outer regions of the cometary environment

After the crossing of the shock, and before the cometopause, three distinct regions seem evident and are called the "outer regions".

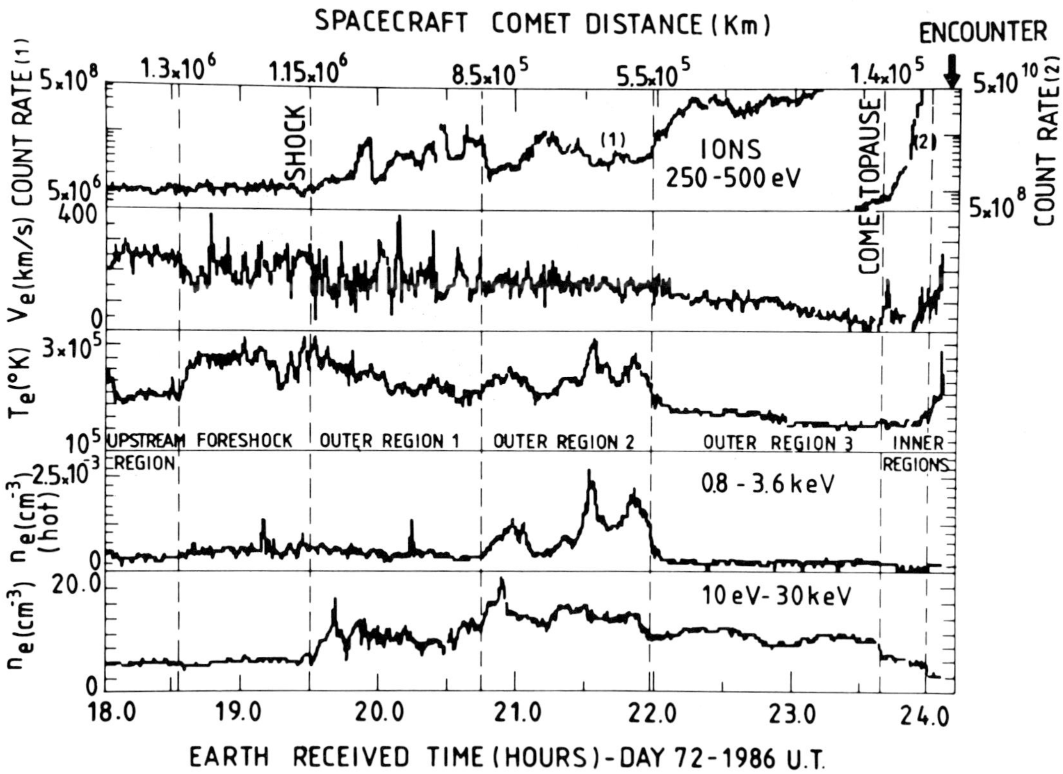

Fig. 2. Plasma data during the last 6.2 hours before the Giotto encounter with comet Halley. From bottom to top panels: n_e, the density of 10 eV–30 keV electrons, n_e (hot), the density of the 0.8–3.6 keV electrons, T_e, the electron temperature estimated from the slope of the energy spectra near 70 eV, V_e, the electron bulk velocity calculated from the electron distribution and the count rate of RPA 2–PICCA in the ram direction for 250–500 eV/q ions (note a change of scale from count rate (1) to count rate (2) at time ~23.5). Times given are UT Earth Received Times (~8 minutes later the spacecraft event times)

Outer region 1 extends from the shock to 8.5 10^5 km from the comet. In this region there is a large increase in the electron density with rapid and strong fluctuations on time scales of the order of one minute. The electron temperature and bulk velocity also show fluctuations. The average temperature decreases toward the comet whereas the average density tends to increase. In this region, the electron distribution function may fluctuate rapidly from quiet solar wind-like to nearly isotropic, which explains the variations of the heat flux ratio between 0.45 and 1. Some keV electrons are detected but their fluxes decrease toward the comet. Inside this region, low energy ions (protons) are detected in the ram direction but their fluxes seem anticorrelated with the electron bulk velocity.

Outer region 2 is crossed at distances between 8.5 10^5 and 5.5 10^5 km and is characterized by larger electron densities, and by larger variations of electron density and temperature that occur on a slower time scale. The average electron bulk velocity is almost constant. The fluxes of the ions detected in the ram direction are anticorrelated with the electron density and temperature. But the main feature of this region is probably the detection of significant fluxes of keV electrons as seen in the plot of n_e (hot) in Fig. 2. Figure 3 shows, for the same time interval as Fig. 2, the total 10 eV–30 keV electron density and electron density data for 6 different energy ranges: 10–40 eV, 40–73 eV, 78–160 eV, 170–760 eV, 0.8–3.6 keV, and 3.9–30 keV. The fluxes of 0.8–3.6 keV electrons detected in this outer region 2 are not present after it, and are much smaller before it. The 3.9–30 keV electron density stays at the background level. Except for the highest energy level, all the densities are maximum in this region.

Figure 4 gives the normalized 0.8–3.6 keV electron pitch angle distribution for the same time interval as Figs. 2 and 3, deduced from the computation of the symmetry axis of the electron momentum flux tensor. It is interesting to note that the 0.8–3.6 keV electron fluxes display a tendency to be magnetic field-aligned. Due to these unexpected high energy aligned electron fluxes this region is called the *mystery region.*

These outer regions 1 and 2 were not distinguished in Rème et al. (1986b), where they were called the transition region.

At about 5.5 10^5 km from the comet, Giotto penetrates into *outer region 3*. The boundary between region 2 and 3 is described in d'Uston et al. (1987). Region 3 extends from 5.5 10^5 km to 1.35 10^5 km from the comet. In this region the average 10 eV–30 keV electron density is smaller than before and relatively constant, and the electron temperature and bulk velocity decrease smoothly. The heat flux ratio is everywhere greater than 0.57 showing that the electron distribution functions are more isotropic than in the other outer regions. In fact the electron distributions become stable and monotonically cool. On the contrary, low energy ions are increasingly detected in the ram direction, as seen in Figure 2. Due to the much smaller fluctuations of the electron properties in this region, it can be called a *quiet region*. Here the high energy electron density is small (smaller than in all the regions encountered before, in Fig. 2).

At 1.35 10^5 km, Giotto goes through a new boundary, the cometopause, described in detail by d'Uston et al. (1987). There is a sudden decrease in electron density and heat flux ratio, associated with minima in the electron temperature and bulk velocity, which tends to zero, sometimes after the beginning of the

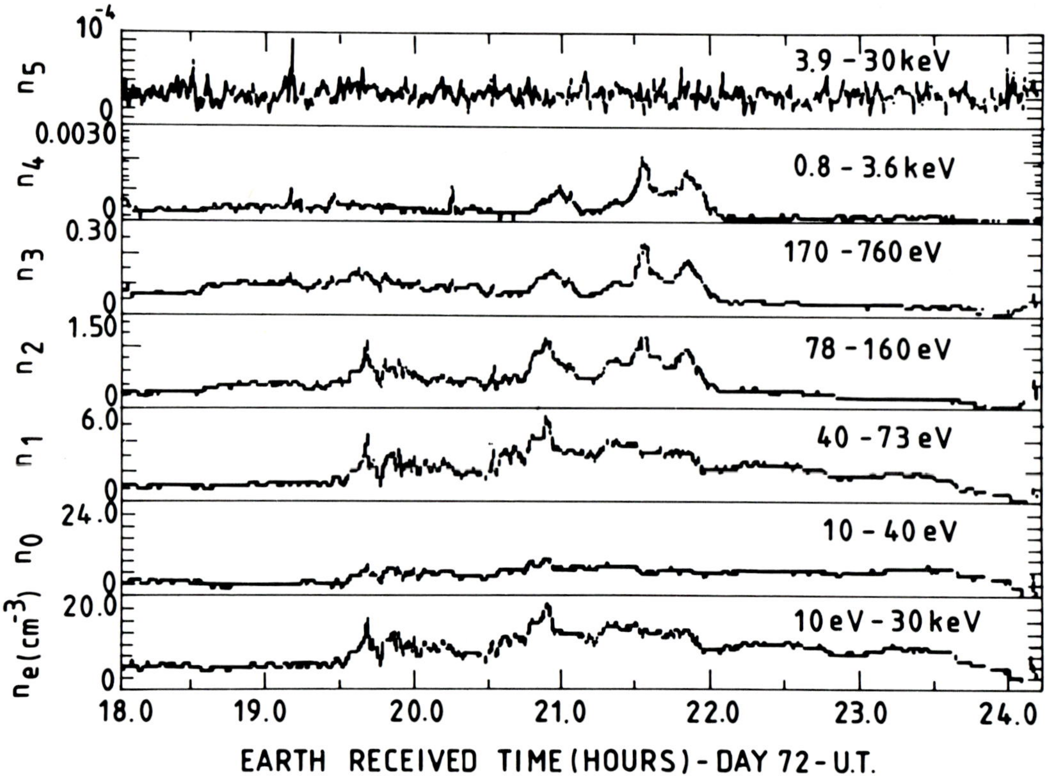

Fig. 3. Electron density data for various energy ranges during the last 6.2 hours before the Giotto encounter with comet Halley. From bottom to top panels: n_e, the total 10 eV–30 keV density; n_0, the 10–40 eV density; n_1, the 40–73 eV density; n_2, the 78–160 eV density; n_3, the 170–760 eV density; n_4, same as n_e (hot) in Fig. 2, the 0.8–3.6 keV density and n_5, the 3.9–30 keV density. Times given are UT Earth Received Times

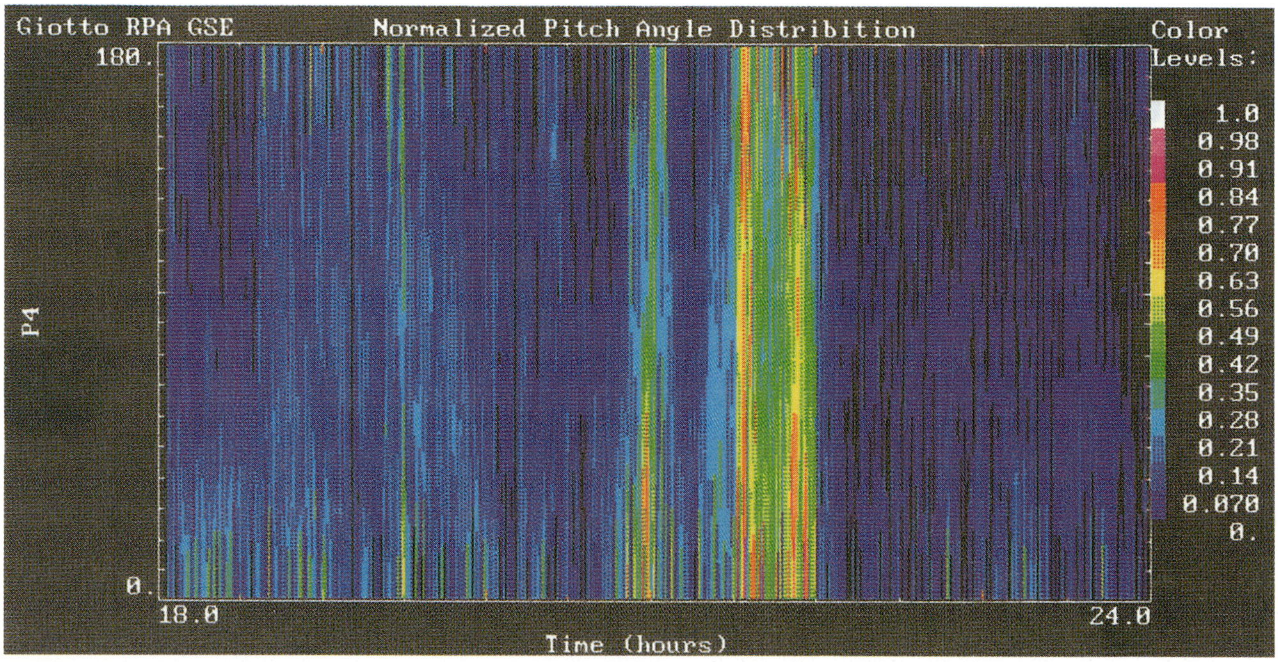

Fig. 4. Normalized 0.8–3.6 keV electron pitch angle distribution during the last 6 hours before the Giotto encounter with comet Halley deduced from the computation of the symmetry axis of the electron momentum flux tensor. Times are UT Earth Received Times

Table 1. Summary of comet Halley's plasma environment

Giotto – comet distance		Main effects
~4.6 10^6–1.4 10^6 km	Upstream region	Heat flux ratio, sporadic connections with the comet
1.4 10^6–1.15 10^6 km	Foreshock region	T_e increase – some ram ions
1.15 10^6 km	Bow shock	V_e decrease – n_e increase
1.15 10^6–8.5 10^5 km	Outer region 1	Large n_e increase, n_e, T_e, V_e fluctuations, anticorrelations: V_e–n_i (ram ions)
8.5 10^5–5.5 10^5 km	Outer region 2	Energetic electrons, anticorrelations of n_e and T_e with n_i (ram ions)
5.5 10^5–1.35 10^5 km	Outer region 3	n_e, T_e, V_e decrease, ram ions
1.35 10^5 km	Cometopause	Cometary ions, n_e decrease $V_e \rightarrow 0$
4.5 10^4 km	2nd sharp density decrease	
~4.7 10^3 km	Ionopause [a]	

[a] Balsiger et al., 1986; Krankowsky et al., 1986

pile-up magnetic region (Neubauer et al., 1986b). After the cometopause, there is a rapid increase, in the ram direction, of the cold ion density while water-group ions are first resolved well above background just before the cometopause (Korth et al., 1986a). By calculation this boundary coincides with the place where the contaminated solar wind is rapidly decelerated by collisions with outflowing cometary neutrals (Mendis et al., 1986) at a sunward distance from the nucleus of ~5 10^4 km. From the cometopause to the time of the encounter, Giotto is in the inner regions; there is a fast increase in the cometary ion fluxes detected in the ram direction. Up to 4.5 10^4 km from the nucleus the electron density decreases. At this distance from the comet a second sharp density decrease is detected. After, the 10 eV–30 keV electron density has a value as low as 2 cm^{-3} showing that cometary electrons have much smaller energies to ensure neutrality with the large number of cometary ions. All along these inner regions keV electron fluxes are very small, close to background.

Measurements in the last few minutes before closest approach are contaminated by intense neutral gas bombardment and frequent dust impacts with the spacecraft and the detectors. Thus the analysis of these data, which are clearly of great interest, is not yet finished and they will not be discussed here.

A summary of the comet Halley plasma environment is given in Table 1.

5. Conclusion

The interaction of the solar wind with comet Halley is very complex. A well defined bow shock appears to be found with features quite different from that of P/Giacobini-Zinner. Calculations indicate that the positions of the main transitions (shock, cometopause) are consistent with the total gas production rate of the comet and the associated mass loading and neutral drag on the inflowing solar wind. There are, however, several other sharp features, whose origins, whether temporal or spatial, are unknown and which have not been identified with any obvious physical process, although the neutral interaction with inflowing contaminated solar wind plasma may play a role in all these transitions. An important feature is the detection of accelerated keV electrons upstream from the comet, in the well-defined electron foreshock and in outer regions 1 and 2 with much higher intensities in outer region 2.

Acknowledgements. We have had useful discussions on the interplanetary magnetic field and particle data with M H Acuña. This work was supported by CNES under Grant No. 1212, By NASA Contract NASW-3575 and by the Max-Planck Gesellschaft für Förderung der Wissenschaften and by the Bundesminister für Forschung und Technologie under Grant No. 01 0F 052.

References

Anderson, K.A., Carlson, C.W., Curtis, D.W., Lin, R.P., Rème, H., Sauvaud, J.A., d'Uston, C., Korth, A., Richter, A.K., Mendis, D.A.: 1987, **187**, 290

Balsiger, H., Altwegg, K., Bühler, F., Geiss, J., Ghielmetti, A.G., Goldstein, B.E., Goldstein, R., Huntress, W.T., Ip, W.-H., Lazarus, A.J., Meier, A., Neugebauer, M., Rettenmund, U., Rosenbauer, H., Schwenn, R., Sharp, R.D., Shelley, E.G., Ungstrup, E., Young, D.T.: 1986, *Nature* **321**, 330

Korth, A., Richter, A.K., Anderson, K.A., Carlson, C.W., Curtis, D.W., Lin, R.P., Rème, H., Sauvaud, J.A., d'Uston, C., Cotin, F., Cros, A., Mendis, D.A.: 1986, *Adv. Space Res.* **5**, 221

Korth, A., Richter, A.K., Loidl, A., Güttler, W., Anderson, K.A., Carlson, C.W., Curtis, D.W., Lin, R.P., Rème, H., Cotin, F., Cros, A., Médale, J.L., Sauvaud, J.A., d'Uston, C., Mendis, D.A.: 1987, *J. Phys. E*: Sci. Instrum. (in press)

Krankowsky, D., Lämmerzahl, P., Herrwerth, I., Woweries, J., Ebarhardt, P., Dolder, U., Herrmann, U., Schulte, W., Berthelier, J.J., Illiano, J.M., Hodges, R.R., Hoffman, J.H.: 1986, *Nature* **321**, 326

Mendis, D.A., Smith, E.J., Tsurutani, B.T., Slavin, J.A., Jones D.E., and Siscoe, G.L.: 1986, *Geophys. Res. Letters* **13**, 239

Neubauer, F.M., Acuña, M.H., Burlaga, L.F., Franke, B., Gramkow, B., Mariani, F., Musmann, G., Ness, N.F., Schmidt, H.U., Terenzi, T., Ungstrup, E., Wallis, M.: 1986a, ESA SP-*1077*, 1

Neubauer, F.M., Glassmeier, K.H., Pohl, M., Roeder, J., Acuña, M.H., Burlaga, L.F., Ness, N.F., Musmann, G., Mariani, F., Wallis, M.F., Ungstrup, E., Schmidt, H.U.: 1986b, *Nature* **321**, 352

Rème, H., Cotin, F., Cros, A., Médale, J.L., Sauvaud, J.A., d'Uston, C., Korth, A., Richter, A.K., Loidl, A., Anderson, K.A., Carlson, C.W., Curtis, D.W., Lin, R.P., Mendis, D.A.: 1986a, ESA SP-**1077**, 33

Rème, H., Sauvaud, J.A., d'Uston, C., Cotin, F., Cros, A., Anderson, K.A., Carlson, C.W., Curtis, D.W., Lin, R.P., Mendis, D.A., Korth, A., Richter, A.K.: 1986b, *Nature* **321**, 349

Rème, H., Cotin, F., Cros, A., Médale, J.L., Sauvaud, J.A., d'Uston, C., Anderson, K.A., Carlson, C.W., Curtis, D.W., Lin, R.P., Korth, A., Richter, A.K., Loidl, A., Mendis, D.A.: 1987, *J. Phys. E: Sci. Instr.* (in press)

d'Uston, C., Rème, H., Sauvaud, J.A., Cros, A., Anderson, K.A., Carlson, C.W., Curtis, D.W., Lin, R.P., Korth, A., Richter, A.K., Mendis, D.A.: 1987, this issue

Astron. Astrophys. 187, 39–46 (1987)

ASTRONOMY AND ASTROPHYSICS

The Sakigake/Suisei encounter with comet P/Halley

K. Hirao [1] **and T. Itoh** [2]

[1] Department of Aeronautics and Astronautics Tokai University 1117, Kitakaname, Hiratsuka, 259–12 Japan
[2] The Institute of Space and Astronautical Science 4-6-1, Komaba, Meguro-ku, Tokyo, 153 Japan

Received March 2, accepted May 19, 1987

Summary. Two spacecraft, Sakigake and Suisei, which were launched by the Institute of Space and Astronautical Science of Japan, flied successfully by the comet P/Halley at minimum distances of 7 million and 150 thousand kilometers respectively. The scientific objectives of these spacecraft were to explore the environment of the comet directly taking the opportunity of the re-revisiting of the first class comet, Halley. Plasma and neutral hydrogen environments were selected for scientific exploration. This article gives the overview of recent scientific results of our missions.

Key words: Halley's comet – encounter – plasma – hydrogen cloud – dust

1. Introduction

After one year discussion among scientists and engineers, including those of other universities, the Institute of Space and Astronautical Science (ISAS) decided to send two spacecraft to study the environment of the comet Halley by fly-by. From the expected capability of newly designed launch vehicle called M-3 SII, the weight of spacecraft was limited to 140 kg. The scientific experiments in the two spacecraft were discussed under this constraint and it was decided to have plasma experiments and one ultraviolet camera. These missions were decided in the ISAS in 1979, six years before the first launch. The first spacecraft called MS-T5, was named Sakigake after launch on 7 January 1985 UTC. The second one was called Planet-A and renamed Suisei after its launch on 18 August 1985 UTC.

Sakigake could not approach the comet Halley closer than several million kilometers due to its launch window and was targetted to fly-by the comet at distance of 7 million kilometers to measure the solar wind interaction with the cometary plasma. The predominant interaction of cometary gas with the solar wind was found even in such a distant place. On the contrary, Suisei should go up to a very close encounter with the comet. It could not have any dust protection due to its weight constraint, and it was launched initially into an orbit to approach the comet within 200 thousand kilometers. The orbit was corrected later to approach the comet at 150 thousand kilometers. Suisei took a lot of UV pictures of the neutral hydrogen corona around the comet and also found quite a complicated situation around the comet.

Send offprint requests to: K. Hirao

Although comprehensive papers are submitted in this symposium some results are described in this overview, which was presented as an invited poster paper at the Heidelberg Symposium.

2. Scientific experiments

In this section, the experiments are introduced briefly.

2.1. Scientific payloads on board Sakigake

The scientific objective of Sakigake was to measure the solar wind conditions blowing to the comet at rather distant places, 7 million kilometers at its closest approach as was mentioned previously. For this purpose, three experiments were mounted, as described below.

2.1.1. Plasma wave probe (PWP)

PWP experiment consists of two measurements of both electric as well as magnetic fields. Electric field is measured in two frequency ranges, one is called as HF range between 5 kHz to 200 kHz by sweeping mode and the other is called as ELF range between 70 Hz to 2.8 kHz in 16 multichannel mode. The sensor used is a dipole antenna of 10 m tip-to-tip length. The magnetic field is measured in the above ELF range by using search coil as its sensor. PWP payload is shown in Fig. 1.

2.1.2. Interplanetary magnetic field measurement (IMF)

Three axis ring-core type fluxgate magnetometer is installed to measure the interplanetary magnetic field in the frequency range between 0 to 3 Hz. The sensor is mounted on the tip of a pantagraph boom of 2 m long from the wall of the spacecraft. The dynamic range is 64 nT with resolution of 0.03215 nT. The instrument is shown in Fig. 2.

2.1.3. Solar wind ion detector (SOW)

The sensor is a modulation Faraday cup with four grids. The effective area is around 70 cm and the dynamic range of ion current is $10^{-11} \sim 10^{-8}$ A. The field of view of the cup is a 120° cone. By means of this detector, flow direction, bulk velocity, density and temperature of the solar wind can be measured.

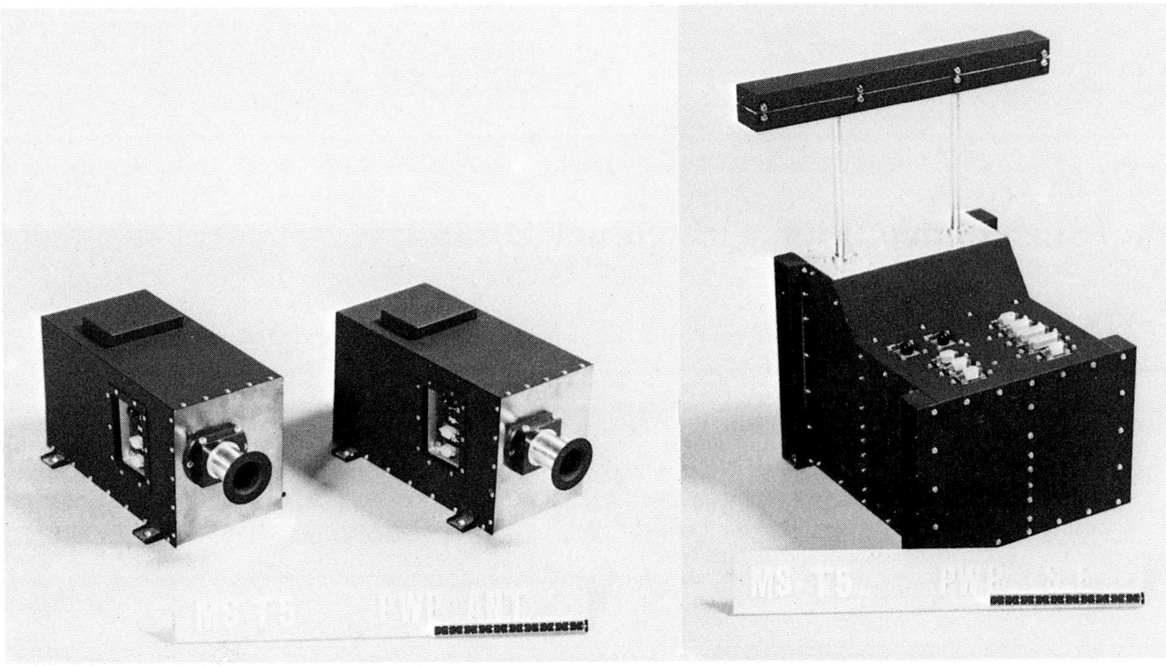

Fig. 1. Plasma wave probe on board Sakigake. On the left are two elements of dipole antenna and on the right electronics is shown with search coil

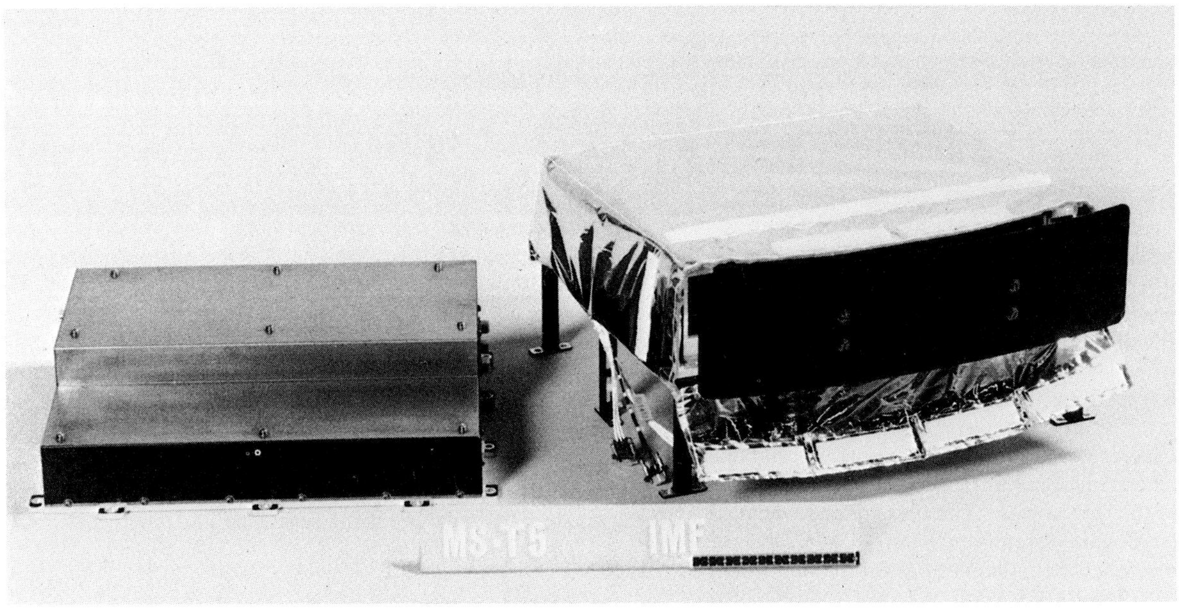

Fig. 2. Magnetometer on board Sakigake. On the left: electronics, on the right: sensor and boom system

The spacecraft is spin stabilized to keep its axis perpendicular to the ecliptic plane and its azimuthal angle is identified by so-called sun-pulse from the sun sensor.

2.2. Scientific payloads onboard Suisei

Scientific objective of Suisei was to study the feature of the hydrogen cloud around the comet and the interaction between solar wind and cometary plasma, an UV camera and an energy analyzer of charged particles were mounted. The distance of closest approach was decided to be 150 thousand kilometers to avoid any serious damage due to the hitting of dust particles.

2.2.1. Ultraviolet imager (UVI)

UVI consists of a vacuum-ultraviolet telescopic mirror lens, ultraviolet image intensifier and CCD detector with 122×153 pixels. Its field of view is $1.^\circ 85 \times 1.^\circ 96$. To avoid the blurr of the image due to the spinning motion of the spacecraft, even as low as 0.2 RPM during imaging, CCD was driven in the mode of spin-synchronized shift. The imager can be operated for imaging as well as photometry. A microprocessor is used for imaging operation and also for data compression. The CCD is cooled by a radiation cooling method. The instrument is shown in Fig. 4.

Fig. 3. Solar wind ion detector on board Sakigake

Fig. 4. Ultraviolet imager on board Suisei. Top: camera system with cylindrical mirror part; bottom: cooling radiator for the CCD

2.2.2. Energy analyser of charged particles (ESP)

The sensor consists of 270° spherical electrostatic analyzer with fan-shaped collimeter of $5° \times 60°$ field of view, and microchannel plate with five discrete anodes. The observable range of energy is from 30 eV/q to 16 keV/q in 96 steps, those are equally spaced in logarithmic scale. The energy resolution is $\Delta E/E = 0.06$. The instrument is shown in Fig. 5. ESP can be operated fully in the spinning rate of 6 RPM. Due to this constraint and telemetry capability, ESP was operated exclusively with UVI.

Fig. 5. Energy spectrum analyzer on board Suisei

3. Results of Sakigake during closest approach to the comet

3.1. PWP observation

Sakigake spacecraft passed through a region of hydrogen coma of comet Halley within the minimum distance of 6.99 10^6 km from the nucleus of the comet at 04:17:51 UT on March 11, 1986. The radio and plasma waves detector on board Sakigake with a 10 m tip-to-tip antenna has detected plasma waves whose origins belong to the comet as indicated in the top figure of Fig. 6 as the dynamic spectrum. In the middle, the total plasma wave intensity integrated over the frequency range from 5 kHz to 195 kHz is indicated with relative scale for each corresponding position indicated by arrows. Structures in the dynamic spectrum suggest that the radio waves were coming from the travelling shock waves formed in the distance range from 3.0 10^5 to 6.0 10^5 km from the nucleus. In Fig. 7, the enhancement of the plasma waves in the multichannel records of ELF plasma wave is shown as well as some typical spectra. From this figure, one can presume that the spacecraft was passing through the plasma turbulences associated with the ion-pick up effects by the solar wind.

3.2. IMF observation

From the analysis of the magnetic field variation during the closest approach of the spacecraft to the comet, long-period HM waves corresponding with the local "water group" ion cyclotron frequency were detected. These waves have an amplitude of about 2 nT and propagate approximately along the ambient field. It is presumed that the hydromagnetically turbulent region extended

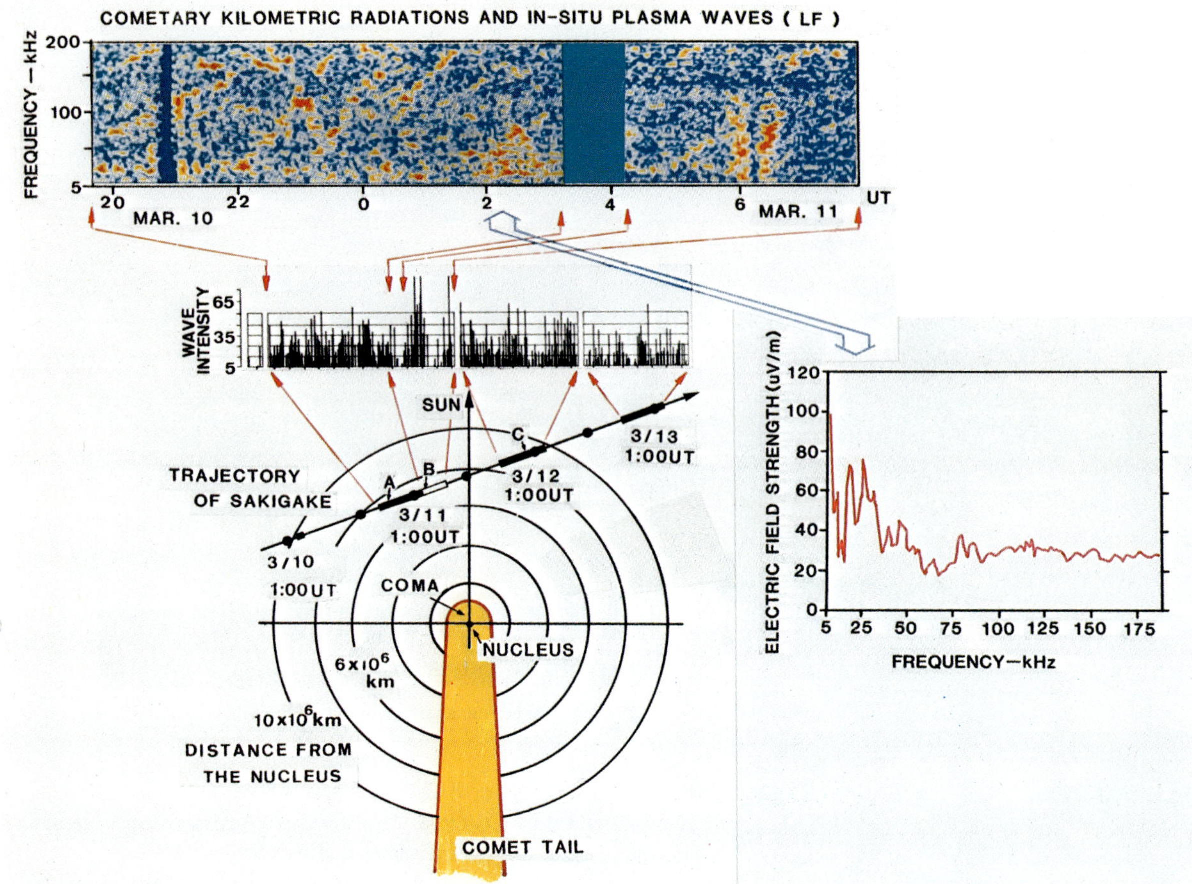

Fig. 6. Kilometric radiation observed by PWP during the closest encounter

up to 9 10^6 km from the nucleus of the comet at that time. The heliospheric neutral sheet was almost parallel to the ecliptic plane on the day of closest approach. It is guessed that this situation might be the reason of no distinct DE event in Halley's ion tail during this period.

3.3. SOW observation

Figure 8 shows velocity and flow direction of solar wind during the closest encounter along the trajectory of Sakigake. On the top of the figure, solar wind velocities are plotted with base line of 350 km s^{-1}. Solar wind parameters on March 10 and 11 are shown in Fig. 9. From top to bottom are: flow direction, wind velocity, ion density and ion temperature.

4. Results from Suisei during its closest approach to the comet

4.1. UVI observation

UVI took a lot of Lyman-α images of the hydrogen coma of the comet Halley, starting middle of November 1985 and ending middle of April 1986. During the closest approach of Suisei, UVI was operated in the photometer mode and obtained the density distribution of hydrogen molecule versus cometocentric distance in the hydrogen coma. From these observations, it was recognized that the comet shows a strong breathing throughout all time with a regular period of 2.2 days which corresponds with the spin period of the nucleus. An example of image is shown in Fig. 10 with pseudo colours. This image were taken at 11:00 UT on February 28, 1986 from the distance of 0.339 AU. The picture dimensions are 1.8 10^6 km in vertical and 1.7 10^6 km in horizontal scales. This image contains a slight stray light effect. The obtained density profile is almost inversely proportional to the square of the cometocentric distance. However some fine structures are superposed to this distribution, suggesting the correspondence with the breathing mentioned previously.

4.2. ESP observation

The Suisei probe passed the nucleus on the sunward side at a minimum distance of 151,000 km at 13:06 UT on 1986 March 8, as shown in Fig. 11. The length and direction of the arrows in the figure indicate the speed and direction of the solar wind plasma. Inside the bow shock the solar wind is decelerated and deflected around the obstacle "comet". Dashed lines around the closest approach represent the magnetic field directions estimated from the symmetry axes of the pickup shells, suggesting the magnetic field draping around the comet. Left and middle diagrams in the top of the figure show examples of the phase space density of ions in the cometosheath, while the right one shows an example outside the bow shock. Horizontal and vertical axes give the direction of $+V_x$ and $+V_y$. Different colours represent different intensities, red being the highest intensity. Ringlike structures are seen in these

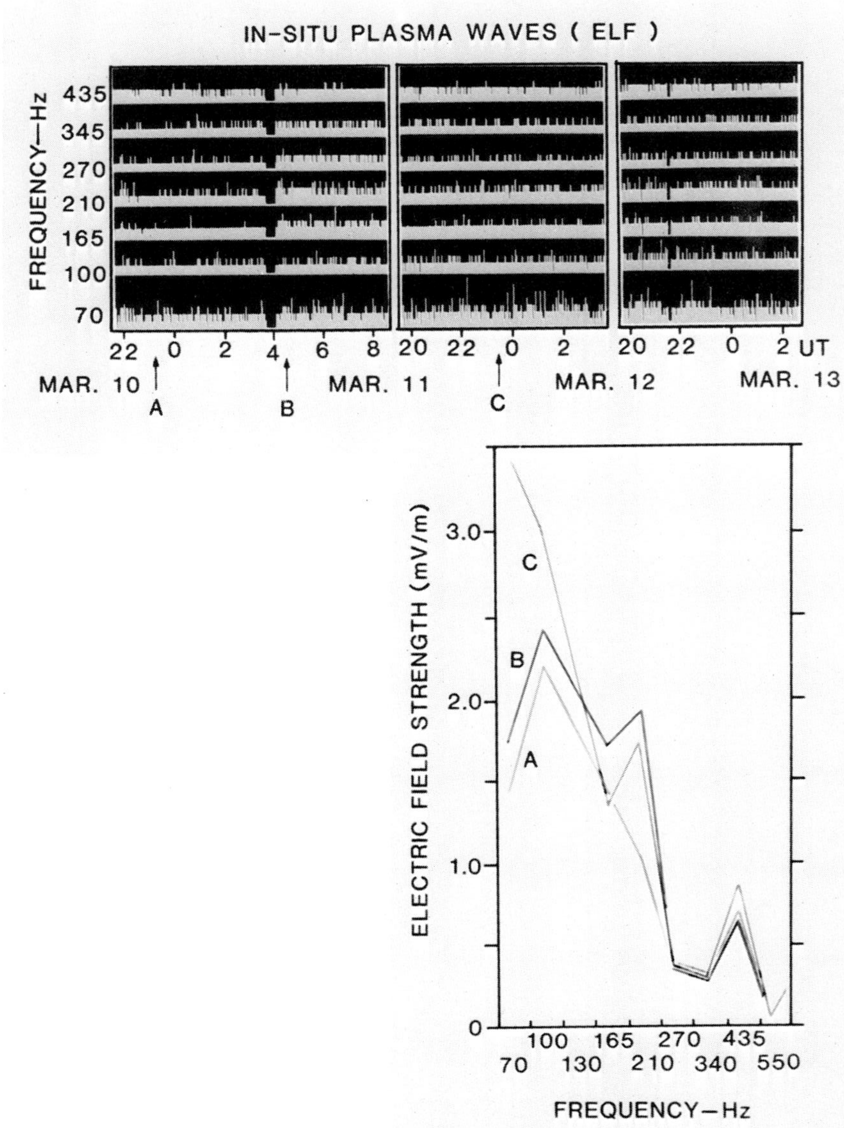

Fig. 7. ELF Plasma waves during the encounter of Sakigake

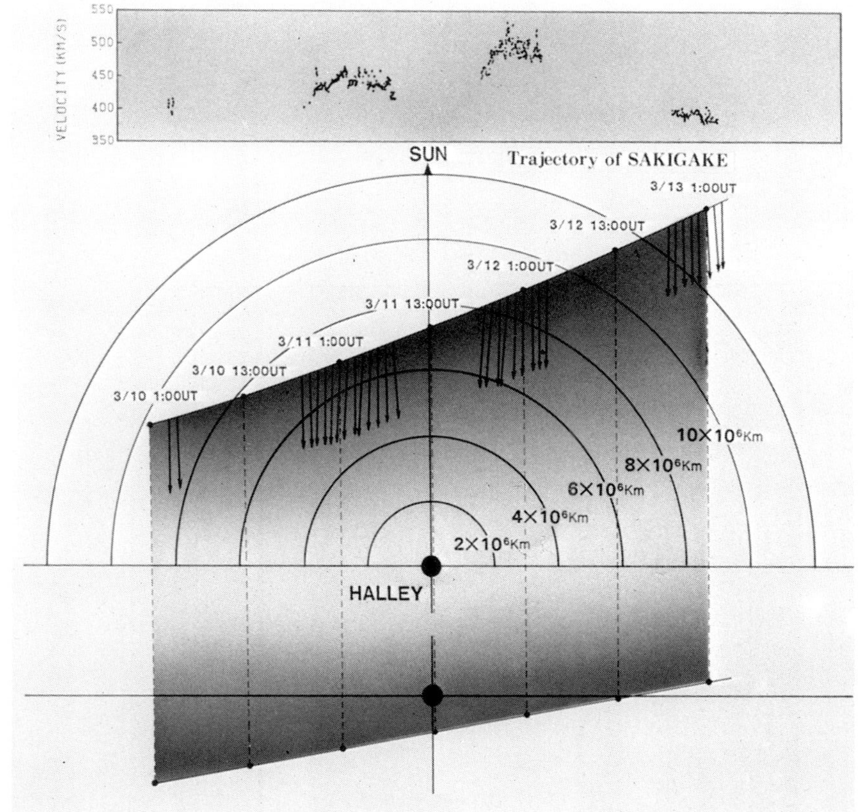

Fig. 8. Solar wind velocity and direction during encounter.

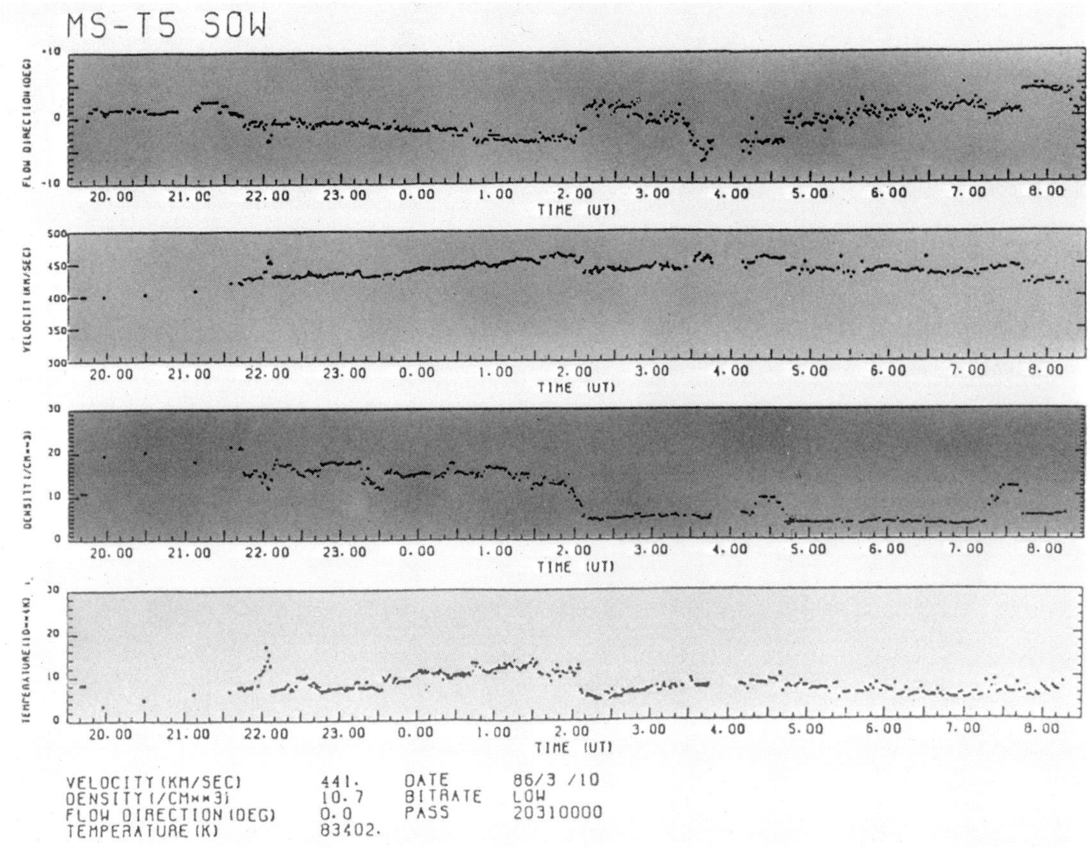

Fig. 9. Solar wind parameters during the encounter

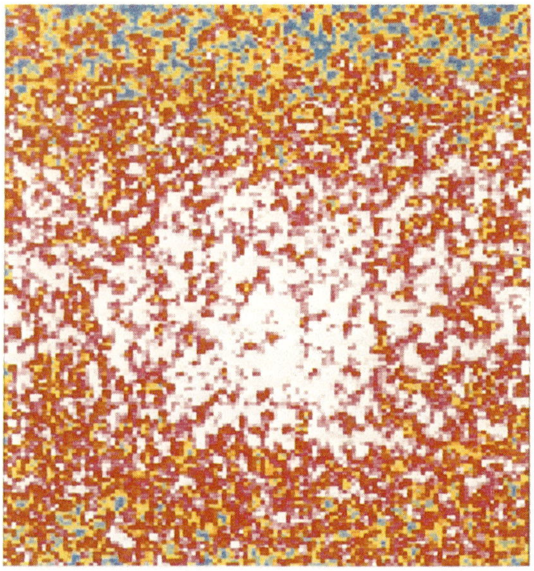

Fig. 10. An example of UV image of the hydrogen coma of Halley's comet in pseudo colour. White is the brightest in Lyman-α line

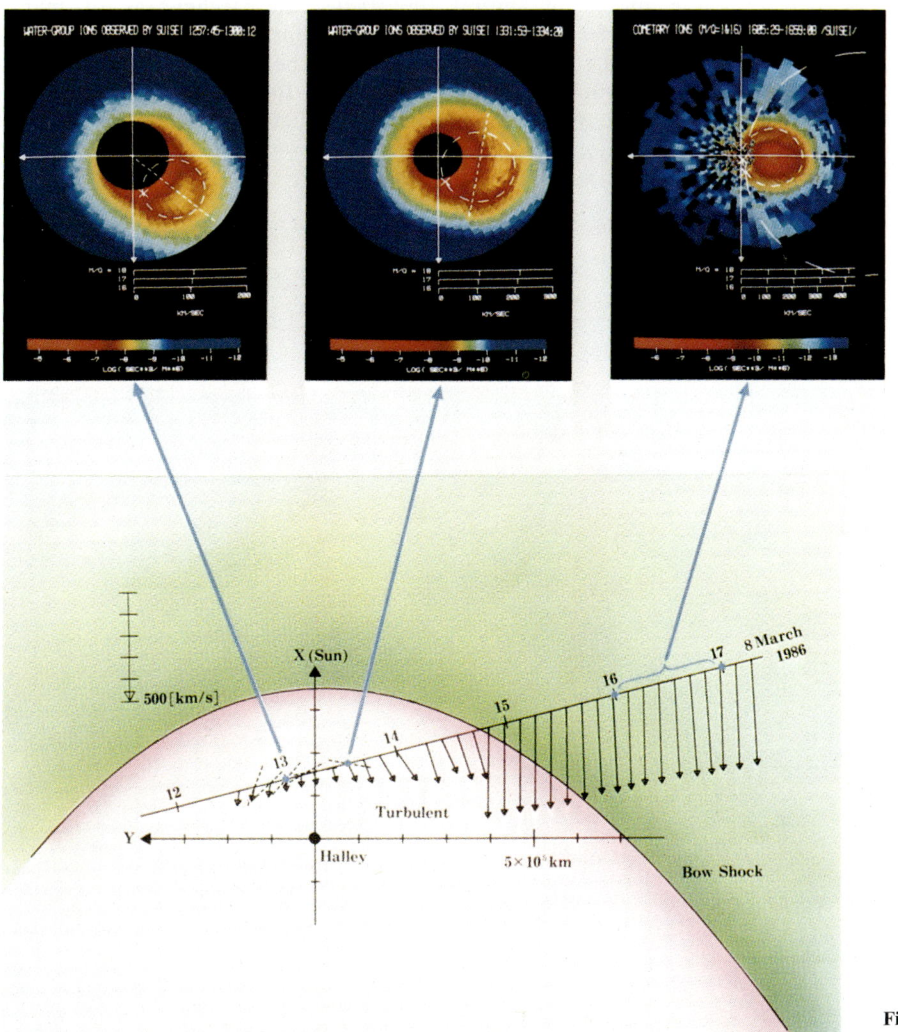

Fig. 11. Plasma environment and cometary ion trace during the closest approach of Suisei

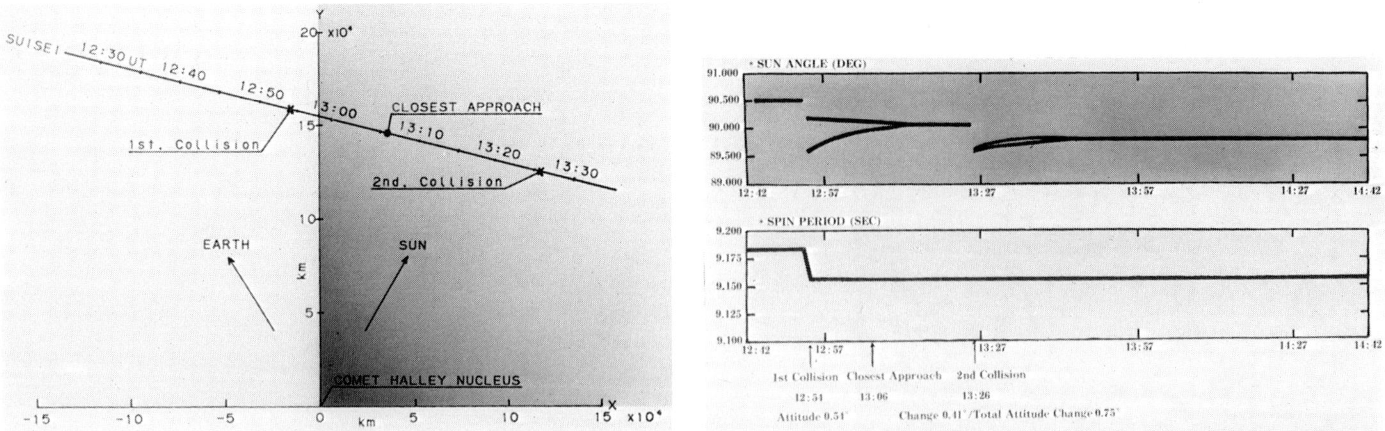

Fig. 12. History and attitude changes of spacecraft at the time of dust particle hitting

examples. They are interpretated as representing the ecliptic cross section of the pickup shells for water-group ions (left and middle) and H^+ (right, small circle) and O^+ (right, large circle). Flow turbulence is observed as short-term fluctuations of the anisotropy direction in the outer cometosheath.

5. Collision with big dust

Suisei was attacked by big dusts before and after its closest approach to the comet as is shown in Fig. 12. The collision was detected by the abrupt change of attitude of the spacecraft. The dusts are estimated to have a weight of several milligrams. Those dusts did not make any damage to the Suisei function.

Acknowledgements. The explorations of Halley's comet was achieved thanks to the splendid cooperation of members of the Institute of Space and Astronautical Science, scientists and engineers of several universities and technicians of many industries in Japan. We express our sincere thanks to their elaborate works for this project. The cooperations with other international agencies included in the IACG was also quite fruitful. We also appreciate the full support of the Ministry of Education, Science and Culture.

Astron. Astrophys. 187, 47–54 (1987)

ASTRONOMY AND ASTROPHYSICS

Waves in the magnetic field and solar wind flow outside the bow shock at comet P/Halley

A. Johnstone[1], K. Glassmeier[2], M. Acuna[3], H. Borg[4], D. Bryant[5], A. Coates[1], V. Formisano[6], J. Heath[1], F. Mariani[7], G. Musmann[8], F. Neubauer[2], M. Thomsen[9], B. Wilken[10], and J. Winningham[11]

[1] Mullard Space Science Laboratory, University College London, Holmbury St. Mary, Dorking, Surrey RH5 6NT, UK
[2] Universtät zu Köln, Institut für Geophysic, D-5000 Köln 41, Federal Republic of Germany
[3] Laboratory for Extraterrestrial Physics, NASA/GSFC, Greenbelt, MD 20771, USA
[4] Kiruna Geophysical Institute, Umeå University, S-90187 Umeå, Sweden
[5] Rutherford Appleton Laboratory, Chilton, Didcot, Oxon, OX11 OQX, UK
[6] Istituto di Fisica dello Spazio Interplanetario, I-00044 Frascati, Italy
[7] Dipartimento di Fisica, University of Rome, I-00185 Roma, Italy
[8] Institut für Geophysik und Meteriologie, TU Braunschweig, D-3300 Braunschweig, Federal Republic of Germany
[9] Los Alamos National Laboratory, Los Alamos, NM 87545, USA
[10] Max Planck Institut für Aeronomie, D-3411 Katlenburg-Lindau, Federal Republic of Germany
[11] Southwest Research Institute, San Antonio, TX 78284, USA

Received February 23, accepted April 2, 1987

Summary. The existence of hydromagnetic waves in the mass-loaded solar wind upstream from the bow shock of a comet is well-established both for comet Giacobini-Zinner and for comet Halley. Whereas previous reports have been concerned either with the magnetic field observations or with plasma observations, here we combine observations of the magnetic field with the solar wind proton and alpha particle distributions. This allows the three possible modes of propagation for these waves to be separated. The magnetic component is predominantly transverse to the magnetic field and linearly polarised. The flow vector also has a substantial amount of power parallel to the magnetic field. A examination of the pressure variations shows that slow magnetosonic waves are more common than the fast mode.

Key words: comets – interplanetary medium – plasmas

1. Introduction

The low frequency hydromagnetic turbulence, expected on theoretical grounds to be found upstream from a comet (Ip and Axford, 1982), has been found at comet Giacobini-Zinner (Tsurutani and Smith, 1986a,b) and at comet Halley (Neubauer et al., 1986b; Riedler et al., 1986). So far, most of the studies have been based on magnetic field measurements although ICE measured oscillations in the electron distribution (Gosling et al., 1986) and turbulence in the solar wind ion distributions at comet Halley has been studied in a companion paper (Johnstone et al., 1987). A full study of the turbulence requires the use of both magnetic field and plasma observations to establish the wave modes, to determine the important plasma parameters such as phase velocities and to study the ion implantation process which provides the free energy to drive the turbulence. In this paper we report a preliminary study of this type based on a detailed comparison of one hour of data selected arbitrarily from the period outside the foreshock of comet Halley when it was known that there was a significant level of wave activity. In this hour the spacecraft travels from 2 million km to 1.75 million km from the nucleus.

Send offprint requests to: A.D. Johnstone

2. Instrumentation

The plasma data was acquired by the Fast Ion Sensor (FIS) of the JPA instrument on Giotto (Johnstone et al., 1986). FIS is a 270° electrostatic energy analyser covering the range 10 eV/q to 20 keV/q. On alternate spins at 8 s intervals, it sweeps the energy over a narrower range for a 45° sector centred on the solar direction in order to make measurements of the solar wind with high angular and energy resolution. Each measurement is made completely within 0.25 s. From these samples of the distribution five parameters, density, three components of the velocity and the temperature, are calculated. The temperature is obtained from the width of the distribution along the three instrument measurement axes, energy, polar and azimuthal angle. Three different values are obtained because of the anisotropy in the distribution. The value used here is the mean of the three values.

The magnetic field was measured by the MAG instrument (Neubauer et al., 1986a). The magnetic field components, obtained at rate of 28 vectors per second, have been averaged over eight seconds centred on the solar wind observation.

3. Observations

The two data sets are displayed in Figs. 1 and 2 and both show variations with periods of the order of two minutes as well as long term (ie. >10 min) trends.

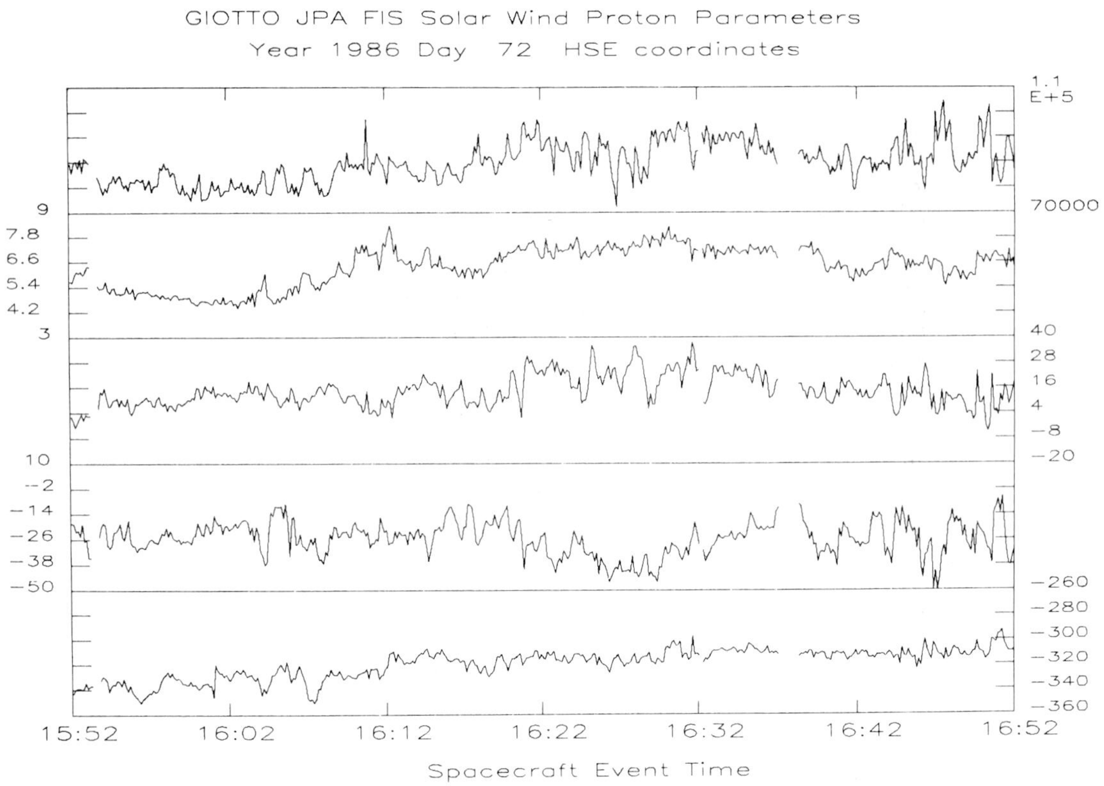

Fig. 1. Solar wind proton parameters measured by the JPA-FIS sensor on Giotto with a time resolution of one sample every 8 seconds. The Halley Solar Eclipse (HSE) co-ordinate system has the x-axis parallel to the comet-sun direction, and the z-axis perpendicular to the ecliptic, northward

Other than this general similarity it is difficult to see any closer relationship between the two sets of data. We ignore the long term trends on the assumption that they are introduced by conditions in the solar wind well upstream from the comet and examine the short period variations in detail. The long term trends are separated from the high frequency fluctuations by passing the data throuch a numerical filter with Gaussian weighting and a cut-off at 3 mHz. From the low-pass filtered values of $\boldsymbol{V}$ and $\boldsymbol{B}$ a new co-ordinate system was defined, which we have called the FLD (Field-Line-Draping) system. It has the y-axis

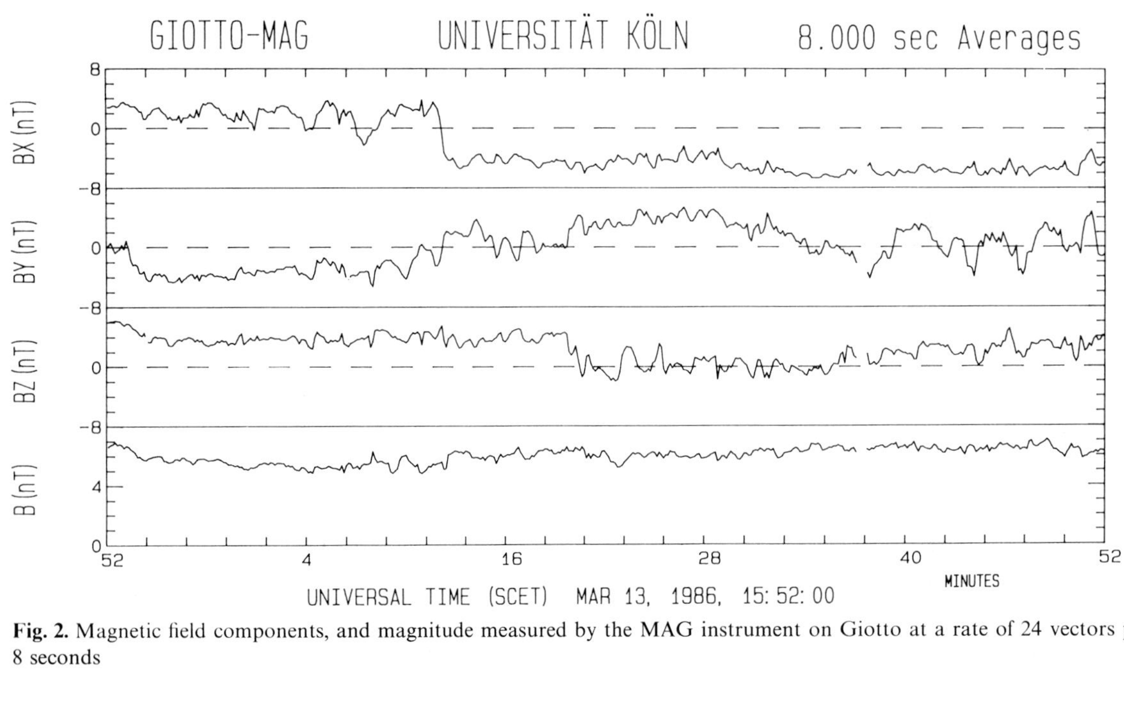

Fig. 2. Magnetic field components, and magnitude measured by the MAG instrument on Giotto at a rate of 24 vectors per second, averaged over 8 seconds

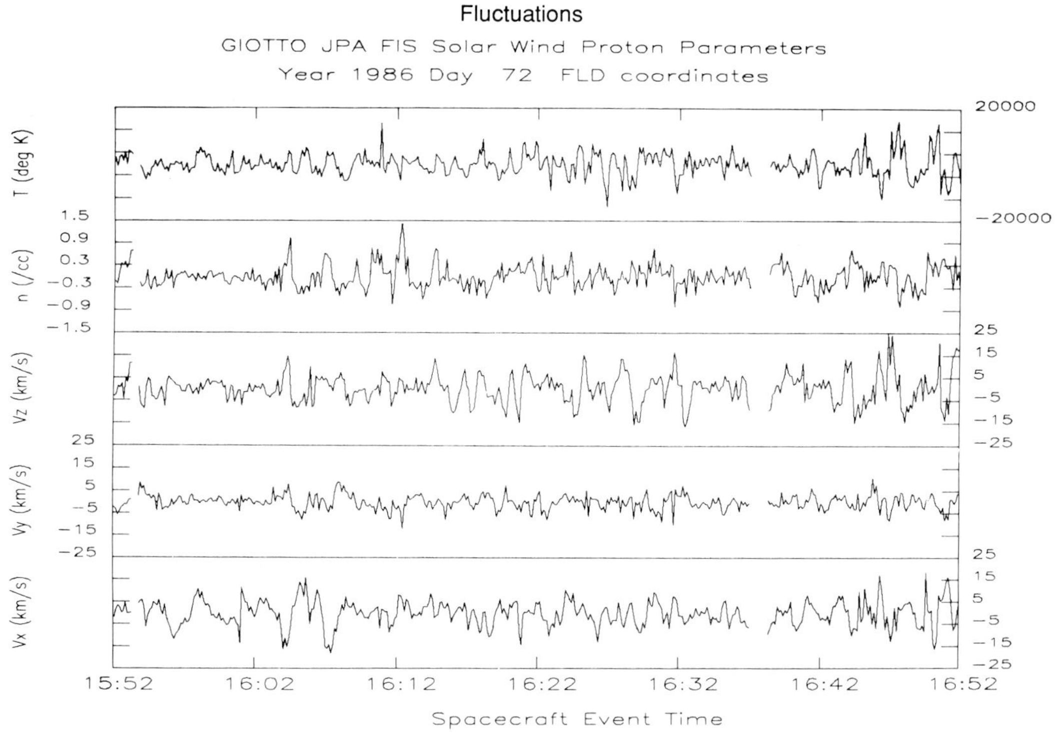

Fig. 3. Fluctuations in the solar wind proton parameters, after being passed through a high pass filter with a cut-off at 3 mHz, plotted in FLD co-ordinates

parallel or anti parallel to the magnetic field, $\boldsymbol{B}$, the z-axis parallel to the electric field, $-\boldsymbol{V} \times \boldsymbol{B}$ and the x-axis forming a right-hand set but oriented to point in the general sunward direction. This system is useful for examining mass loading effects in the draped field around a comet. The disturbance vectors were transformed into this system and are shown plotted in Figs. 3 and 4 for the solar wind protons and the magnetic field respectively. In addition, Fig. 5 shows the solar wind alpha particles.

The first point to note is that the magnetic fluctuations are predominantly transverse to the mean magnetic field, i.e. in the

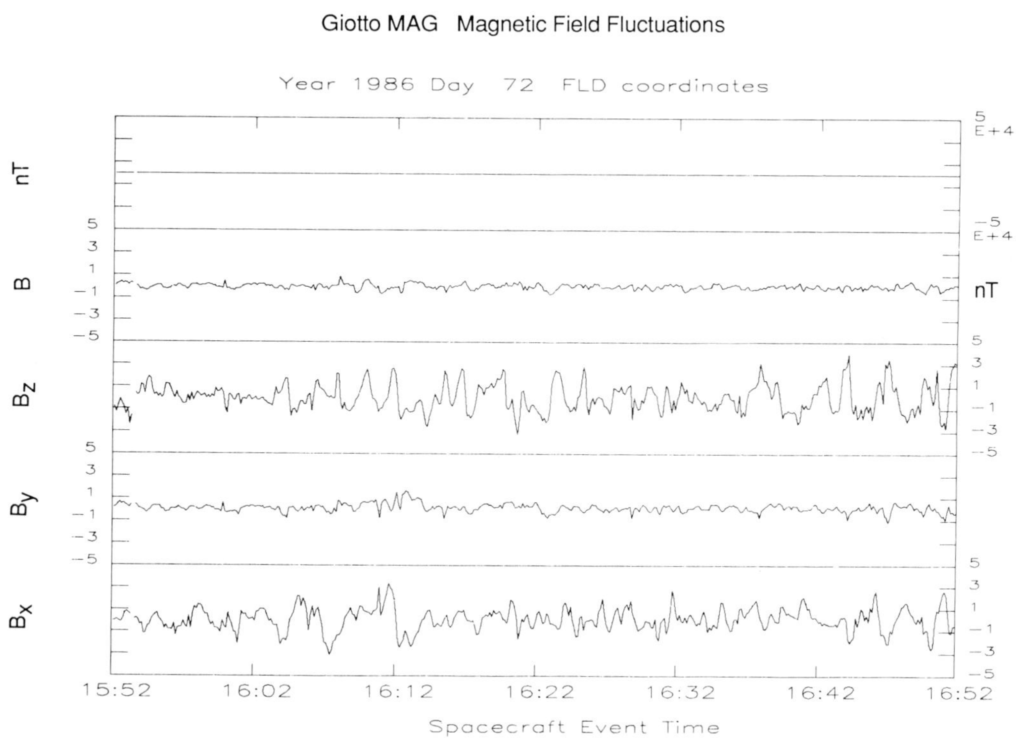

Fig. 4. Fluctuations in the magnetic field components processed in the same way as the solar wind parameters in Fig. 3

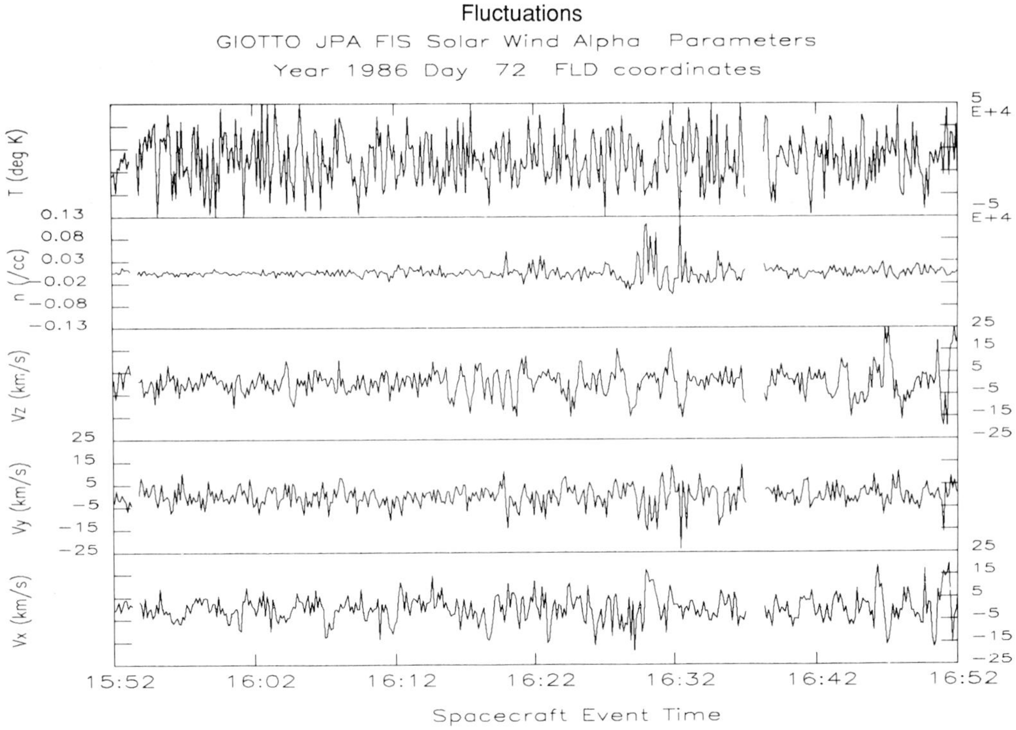

Fig. 5. Fluctuations in the solar wind alpha particle parameters processed in the same way as the proton parameters in Fig. 3

x and z components. The peak amplitude in the transverse components is of the order of $4\,nT$ while in the field-aligned component and the total field strength it is only $1\,nT$. This is borne out by the power spectral analysis of the magnetic field shown in Fig. 6 where the power in the field-aligned component is more than an order of magnitude smaller over the complete frequency range. In the proton velocities the field-aligned component is smaller than the transverse components but not as small as for the magnetic field. This time the power spectrum, Fig. 7, shows that power in the field aligned velocity is an order of magnitude

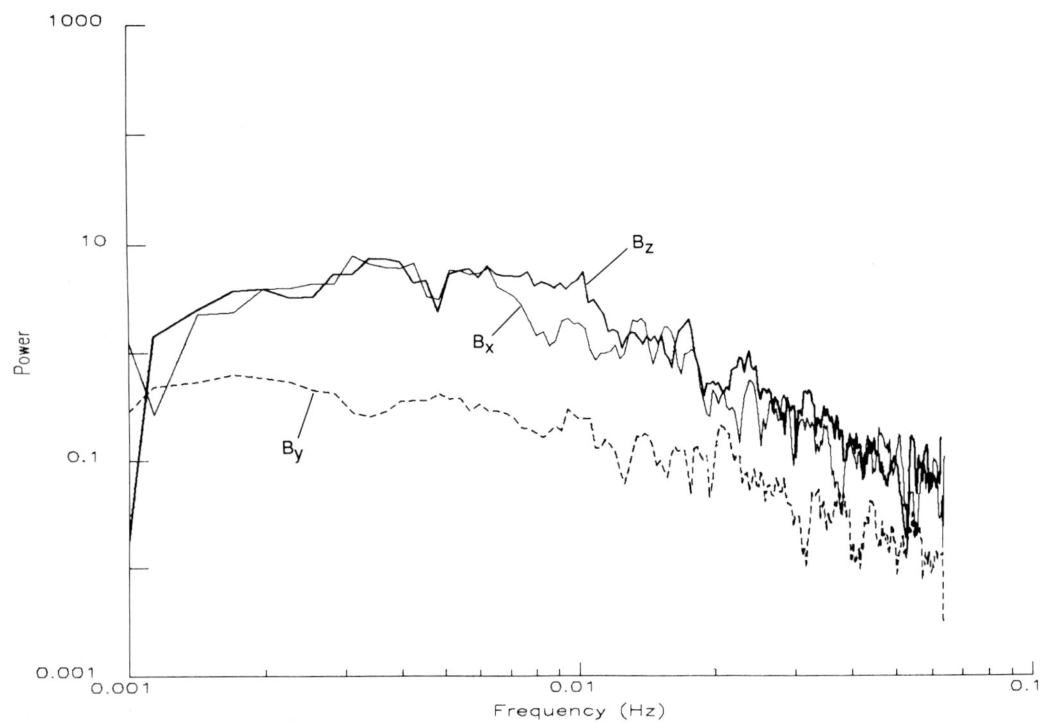

Fig. 6. The power spectrum of the magnetic field components (in FLD co-ordinates) for the data shown in Fig. 4. Note that the power in the field-aligned component is more than an order of magnitude smaller than in the transverse components over the entire frequency range

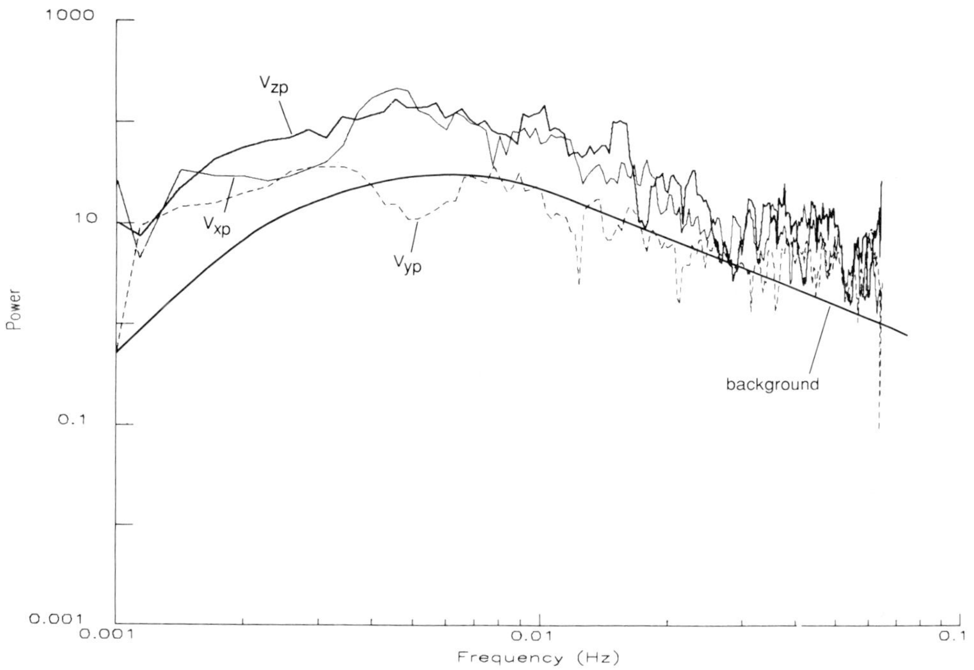

Fig. 7. The power spectrum of the solar wind proton velocity components (in FLD co-ordinates) for the data in Fig. 3
Also shown is the power spectrum measured in the solar wind four hours before when there was no apparent cometary wave activity

smaller over the frequency range covering the gyrofrequencies of the heavy cometary ions. If, as now seems well-established, the waves are generated by an electromagnetic ion beam instability (Wu and Davidson, 1972; Gary et al., 1984), the waves are generated close to resonance in the frame moving with the solar wind ie.

$$\omega - \boldsymbol{k}.\boldsymbol{V}_s = \Omega_i$$

when the waves are Doppler-shifted into the spacecraft frame then $\omega = \Omega_i$. Fig. 7 suggests that the power is generated in the components transverse to the magnetic field and that, as the power cascades to higher frequencies the energy is also coupled into particle motion parallel to the field. Over the upper half of the frequency range the wave power in the velocities is essentially isotropic.

When the different plots are compared in detail some interesting features emerge. The correlation is extremely close between the alpha particles and the protons with both sets of velocity components showing identical variations except for a group of waves between 1603 SCET (Space Craft Event Time) and 1609 SCET. During this period the amplitude of the alpha velocity components is smaller, by a factor of approximately 2/3 than the proton velocity components.

There are two periods, in Figs. 5 and 6, between 1552 SCET and 1609 SCET and between 1642 SCET and 1652 SCET, when there is excellent correlation between the magnetic field components and the velocity components transverse to the mean magnetic field. At other times the correlation is either not as good or even non-existent. If we interpret these variations as simple Alfven waves in a plasma with isotropic pressure then the ratio of the amplitudes b,v should be

$$b = v(\mu_0 \rho)^{1/2}$$

The ratio is measured to be

$$\frac{b}{v} = \frac{7\,nT}{37.6\,\mathrm{km\,s^{-1}}}$$

which leads to a density of 14 proton masses per cm^3. The measured density of solar wind ions at this time is $7\,cm^{-3}$ and the estimated density of cometary ions is 1 proton mass per cm^{-3} (Coates et al., 1987). The difference between the measured and deduced values of the density is probably the result of the considerable pressure anisotropy generated by the cometary ions. The magnetic and velocity components are in phase, therefore, the waves are propagating antiparallel to the magnetic field, i.e. sunward in the frame moving with the solar wind.

The polarisation of the magnetic and velocity waves in these two trains has been examined by generating hodograms (Fig. 8) for the plane perpendicular to the mean magnetic field.

The two periods for which the hodograms have been done are those with good correlation between the plasma and magnetic waves visible in Figs. 3 and 4. There is no dominant polarisation in either sample but the general impression is of a linear component, as has been found by other investigators (Tsurutani and Smith, 1986a) rather than the circular polarisation expected on theoretical grounds (Wu and Davidson, 1972). The magnetic and plasma hodograms are quite similar to each other in each sample but differ between samples. The axis of polarisation lies between 30° and 40° to the z axis (the $\boldsymbol{V} \times \boldsymbol{B}$ direction) in the top two panels and between 50° and 60° in the bottom panels. At the present time there is no accepted explanation for the linear polarisation (Gary and Winske, 1986).

The contribution from slow and fast magnetosonic waves in this assemblage of wave trains can be deduced by comparing the particle P_p and magnetic P_b pressure terms. In the fast mag-

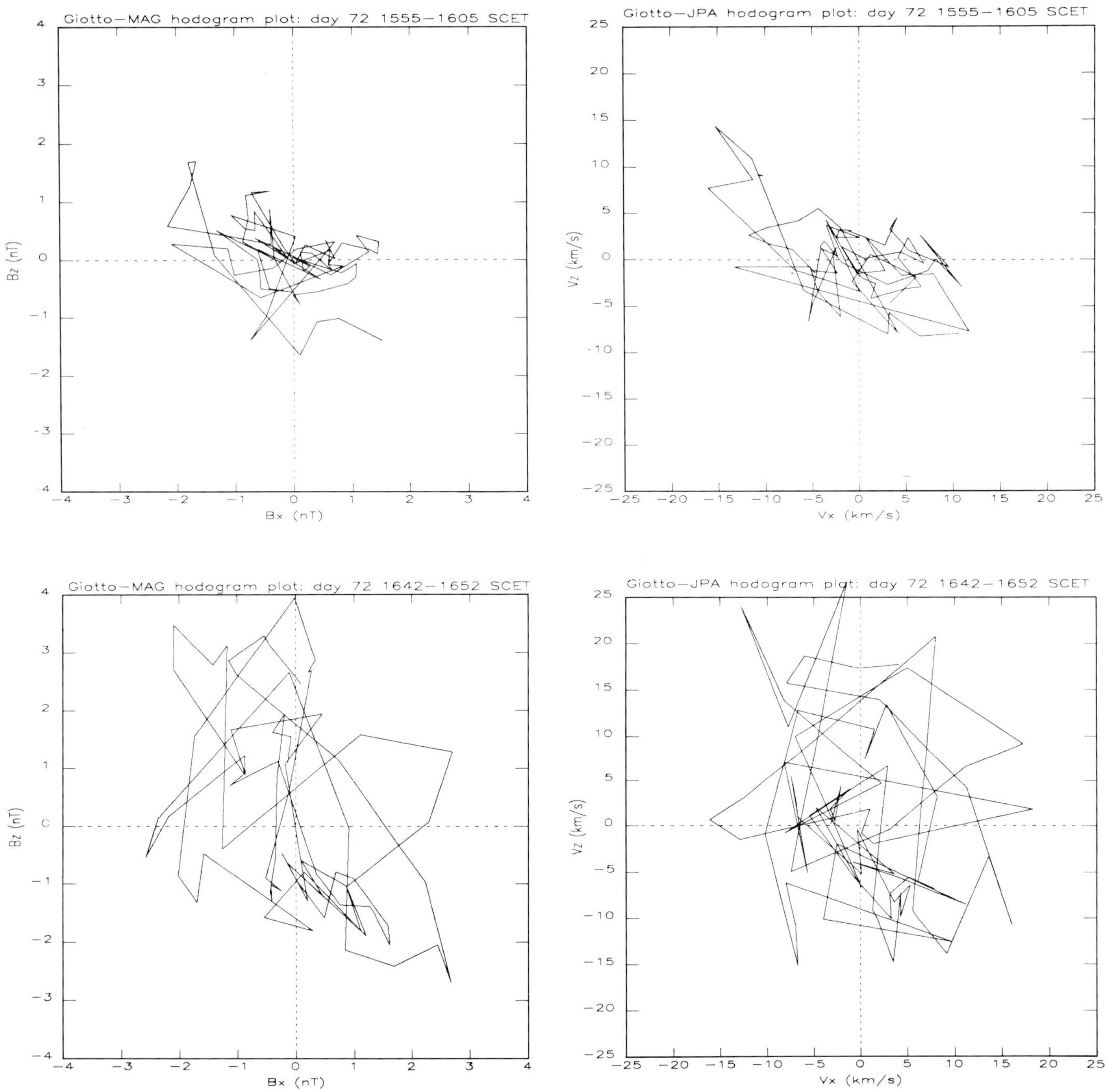

Fig. 8. Hodograms of the magnetic vector and flow vector in the plane perpendicular to the mean magnetic field for the two wave trains with good correlation

netosonic mode the magnetic and particle pressure terms vary in phase; in the slow mode they are out of phase (Southwood and Hughes, 1983; Roberts, 1985). In Fig. 9, the five pressures P_a, P_b, P_p, $P_b + P_p$, $P_b - P_p$ have been plotted to the same linear scale. P_a is the alpha particle pressure. It is possible to find, by comparison of the various plots, examples of anticorrelation, implying slow mode waves, for example in the period from 1607 SCET to 1619 SCET. The amplitude of the magnetic pressure variations is about 0.6 the amplitude of the particle pressure variations here. Most of the time the correlation is not good. No corresponding evidence has yet been found for the existence of fast-mode waves.

The free energy for the electromagnetic ion beam instability is derived from the difference in velocity parallel to the magnetic field between the solar wind protons and the implanted cometary ions. Assuming that all the cometary ions have pitch angles of 90° the difference in velocity v_b (the beam velocity in the theoretical approach (Gary et al., 1984)) can be calculated from the solar wind velocity u_s and the angle between the flow and the field ϕ, i.e.

$$v_b = u_s \cos \phi$$

Figure 10 shows the reduced beam velocity v_b/v_A, the Alfven velocity v_A, and the slow-mode velocity for propagation parallel

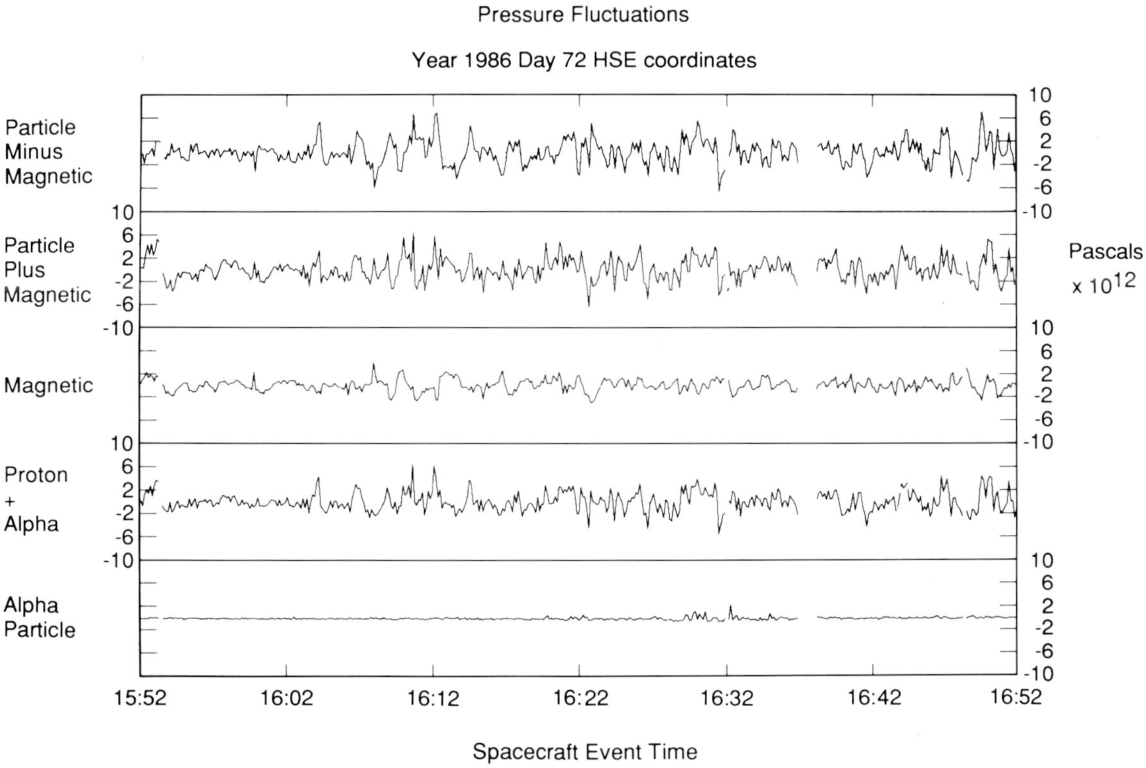

Fig. 9. The variation of the magnetic pressure and the particle pressure

to the field, calculated from the smoothed values of the field and plasma parameters. Over the one hour interval the reduced beam speed varies from -2 through zero to $+5$ as the direction of the magnetic field changes. According to theory the growth rate of the resonant instability has a threshold at a reduced beam velocity of the order of 1 to 3 and is strongly dependent on its value up to 10 (Gary et al., 1984). The non-resonant instability has no threshold but the growth rate is also dependent on the beam velocity. The observed wave intensity might therefore depend on the field direction unless the wave spectrum is saturated. There is no obvious correlation between the two in this short sample of data which is all that has so far been examined.

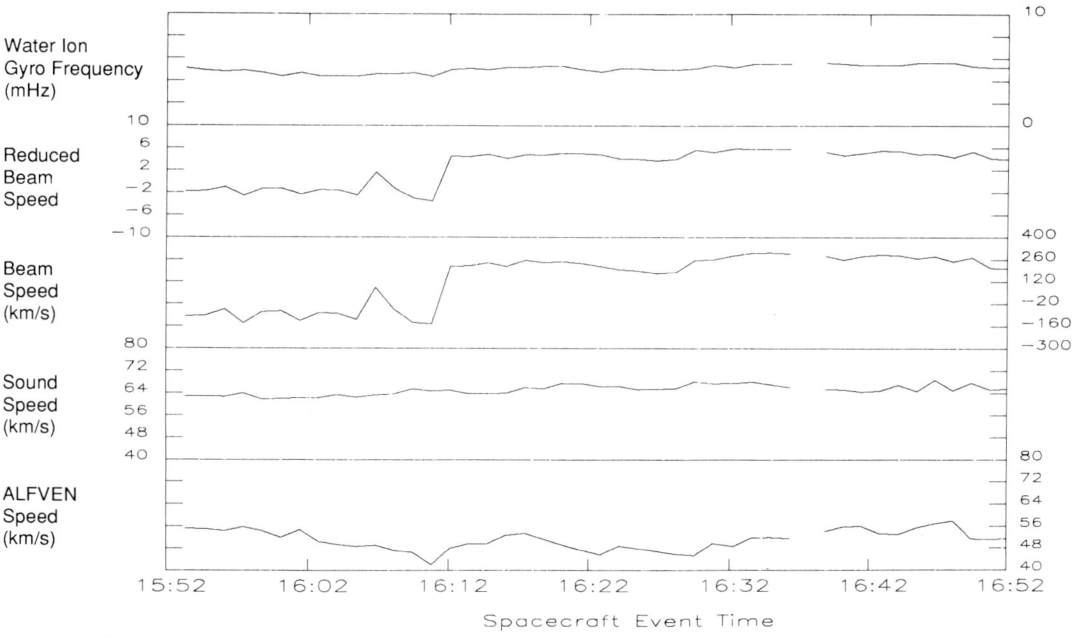

Fig. 10. The reduced beam velocity v_b/v_A, the Alfven speed v_a and the slow mode phase velocity calculated from the smoothed values of magnetic field and plasma parameters

4. Conclusion

The main results of this preliminary analysis of the magnetic and particle turbulence upstream from the bow shock of comet Halley are:

1. The waves in the magnetic field are predominantly transverse to the field but there are large fluctuations in the flow velocity parallel to the field as well as the transverse velocity waves.

2. There are examples of Alfven wave and slow magnetosonic waves in the frequency range 3 mHz to 20 mHz which encompasses the gyrofrequencies of the heavy cometary ions.

3. The Alfven waves are propagating antiparallel to the magnetic field, i.e. upstream in the frame moving with the solar wind.

4. The waves seem to be generated in the heavy ion gyrofrequency band and then cascade through non-linear interactions to higher frequencies. At the higher frequencies the waves in the flow are isotropic while those in the field remain mainly transverse.

5. The waves in the heavy ion gyrofrequency band are predominantly linearly polarised.

Acknowledgements. The work at Mullard Space Science Laboratory and the Rutherford Appleton Laboratory was supported by the UK Science and Engineering Research Council. The work at the Max Planck Institut für Aeronomie was supported by the Max-Planck-Gesellschaft zur Förderung der Wissenschaften and by the Bundesminister für Forschung und Technologie. The work at the Southwest Research Institute was supported by NASA Contract Number W-3574. The work at the Kiruna Geophysical Institute was supported by the Swedish Board for Space Activities. The work at the Istituto di Fisica dello Spazio Interplanetario was supported by the Piano Spaziale Nazionale of the Consiglio Nazionale delle Ricerche. The work at the Universität zu Köln was supported by the Bundesminister für Forschung und Technologie. The work at Los Alamos was performed under the auspices of the US Department of Energy and was supported by NASA through order number W-15164.

References

Coates, A.J., A.D. Johnstone, M.F. Thomsen, V. Formisano, E. Amata, B. Wilken, K. Jockers, J.D. Winningham, H. Borg, D.A. Bryant: 1987, *Astron Astrophys.* (*submitted*)

Gary, S.P., C.W. Smith, M.A. Lee, M. Goldstein, D.W. Forslund: 1984, *Phys. Fluids* **27**, 1852

Gary, S.P., D. Winske: 1986, *J. Geophys. Res.* **91**, 13699

Gosling, J.T., J.R. Asbridge, S.J. Bame, M.F. Thomsen, R.D. Zwickl: 1986, *Geophys. Res. Letters* **13**, 267

Ip., W.H, Axford, W.I.: 1982, in *Comets*, ed. L.L. Wilkening, Univ. Arizona press, Tucson, Arizona, p. 588

Johnstone, A.D., J.A. Bowles, A.J. Coates, A.J. Coker, S.J. Kellock, J. Raymont, B. Wilken, W. Studemann, W. Weiss, R. Cerulli-Irelli, V. Formisano, E. de Giorgi, P. Perani, M. de Bernardi, H. Borg, S. Olsen, J.D. Winningham, D.A. Bryant: 1986, ESA SP-1077, p. 15

Johnstone, A.D., A.J. Coates, J. Heath, M.F. Thomsen, B. Wilken, K. Jockers, V. Formisano, E. Amata, J.D. Winningham, H. Borg, D.A. Bryant: 1987, *Astron. Astrophys.* **187**, 25

Neubauer, F.M., M.H. Acuna, L.F. Burlaga, B. Franke, B. Gramkow, F. Mariani, G. Musmann, N.F. Ness, H.U. Schmidt, T. Terenzi, E. Ungstrup, M. Wallis: 1986a, ESA SP-1077, p. 1

Neubauer, F.M., K.H. Glassmeier, M. Pohl, J. Raeder, M.H. Acuna, L.F. Burlaga, N.F. Ness, G. Musmann, F. Mariani, M.K. Wallis, E. Ungstrup, H.U. Schmidt: 1986b, *Nature* **321**, 352

Riedler W, K. Schwingenschuh, Y.G. Yeroshenko, V.A. Styashkin, C.T. Russell: 1986, *Nature* **321**, 288

Roberts B.: 1985, Magnetohydrodynamic Waves in Solar System Magnetic Fields, ed. E.R. Priest, D. Reidel, Dordrecht, p. 37

Southwood, D.J., W.J. Hughes: 1983, *Space Sci. Rev.* **35**, 301

Tsurutani, B.T., E.J. Smith: 1986a, *Geophys. Res. Letters* **13**, p. 259

Tsurutani, B.T., E.J. Smith: 1986b, *Geophys. Res. Letters* **13**, p. 263

Wu, C.S., R.C. Davidson: 1972, *J. Geophys. Res.* **77**, 5399

Astron. Astrophys. 187, 55–60 (1987)

ASTRONOMY AND ASTROPHYSICS

Solar wind flow through the comet P/Halley bow shock

A. J. Coates [1], A. D. Johnstone [1], M. F. Thomsen [2], V. Formisano [3], E. Amata [3], B. Wilken [4], K. Jockers [4], J. D. Winningham [5], H. Borg [6], and D. A. Bryant [7]

[1] Mullard Space Science Laboratory, UCL, Holmbury St. Mary, Dorking, Surrey, RH5 6NT, UK
[2] Los Alamos National Laboratory, Los Alamos, NM 87545, USA
[3] Istituto di Fisica dello Spazio Interplanetario, I-00044 Frascati, Italy
[4] Max-Planck-Institut für Aeronomie, D-3411 Katlenburg-Lindau, Federal Republic of Germany
[5] Southwest Research Institute, San Antonio, TX 78284, USA
[6] Kiruna Geophysical Institute, Umea University, S-90187 Umea, Sweden
[7] Rutherford Appleton Laboratory, Chilton, Didcot, Oxon, OX11 OQX, UK

Received February 13, accepted May 29, 1987

Summary. The bow shock of a comet is formed by the interaction of three different particle populations; solar wind ions, cometary ions and electrons. We follow the behaviour of the solar wind protons through the Giotto inbound shock crossing at comet P/Halley. A foreshock boundary is seen at $\sim 1.4\ 10^6$ km where the level of solar wind fluctuations increases substantially. The shock itself is seen some $2.5\ 10^5$ km closer. It is a complex structure with high-amplitude waves in the cometary ion foot of the shock preceding a permanent drop in the solar wind speed. The width of the shock structure is $\sim 40{,}000$ km. There is evidence for a further excursion back into the cometary ion foot after the initial shock crossing. The cometary ion density is inferred from the solar wind speed changes and is compared with measurements.

Key words: comets – bow shock – solar wind – Giotto – Johnstone Plasma Analyzer

1. Introduction

The interaction of the solar wind with a comet near to the sun occurs over huge regions of space. Neutral atoms and radicals drift away from the nucleus constrained little by gravitational forces. When ionization of the neutrals occurs, the resultant ion and electron are subject to the electromagnetic forces in the solar wind. They are initially accelerated in the electric field, the ions gaining momentum at the expense of the solar wind. The latter is slowed by this process of ion pick-up in the cometary coma, and it has been shown (Biermann et al., 1967) that if the rate of mass loading causes the mass flux to exceed a critical value, a bow shock must form at the critical point upstream of the comet.

One of the main questions for the plasma instruments on the cometary spacecraft was whether a bow shock did indeed form. There was some doubt both before (Wallis and Dryer, 1985) and immediately after (Bame et al., 1986; Smith et al., 1986a) the ICE flyby of comet Giacobini-Zinner whether a shock forms on the flanks of a small comet. However, a recent coordinated study of data from all the ICE instruments (Smith et al., 1986b) has confirmed the existence of a weak shock. At Halley many investigators from the various probes have reported shock transitions. In the case of Giotto the shock signature on the inbound leg was much more clear than on the outbound leg, probably due to the geometry of the interplanetary magnetic field (Neubauer et al., 1986). In this paper we present measurements of the protons in the solar wind at the Giotto inbound bow shock crossing.

Send offprint requests to: A. J. Coates

2. Instrument

We use data from the Fast Ion Sensor (FIS) of the Johnstone Plasma Analyser (JPA) experiment (Johnstone et al., 1986) on Giotto. This is a 270° electrostatic E/q ion analyser with a full energy range of $10\,\mathrm{eV}/q$ to $20\,\mathrm{keV}/q$. The analyser works simultaneously in two modes, the "solar wind mode" and the "3D mode" (see Fig. 1). Only data from the solar wind mode are used here. This mode has a $45° \times 45°$ field of view, the centre of which was arranged to be the solar direction at encounter. The energy range of this mode is 1/4 of the full energy range; the highest energy in this reduced range is selected on-board by an autoranging procedure which keeps the peak count rate, namely the solar wind proton peak, in the reduced energy range. If the peak is lost, for example if it leaves the field of view of the solar wind mode, then the autoranging initiates a cycle of the energy start value through 8 possible values until the peak is found again. Counts are accumulated and then telemetered in 8 azimuthal angle and 4 polar angle bins, for 30 logarithmically spaced energy levels. Each measurement takes 500 ms and is repeated every 8 s.

Bulk parameters are calculated from the counts arrays on the ground. It is usually possible to compute the moments of the proton and alpha particle distributions separately. Only the proton parameters are used here. During the bow shock crossing the flow at times approached the edge of the angular field of view of the solar wind mode. For this reason in the plots presented here all density and temperature measurements (absolute and relative) are untrustworthy after 1915 spacecraft event time (SCET); all the velocities presented here should however be accurate [1].

[1] Note that since the initial publication of this work in ESA SP-250 (Coates et al., 1986) we have discovered an error in the analysis procedure so that the sign of v_z component is reversed

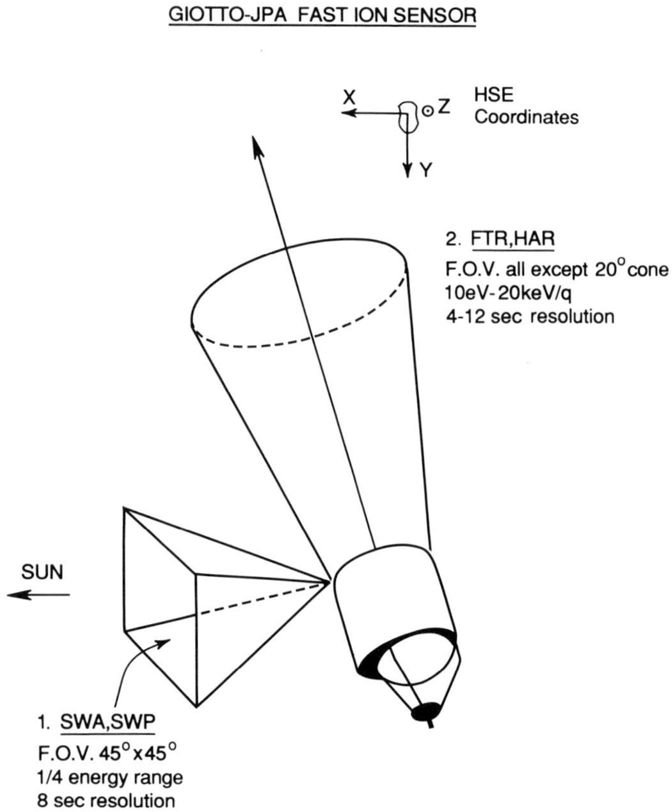

Fig. 1. Fields of view of the Giotto-JPA Fast Ion Sensor (FIS). The solar wind mode is marked 1 and the 3D mode marked 2. The HSE coordinate system is also illustrated

3. Results

We present results for times on March 13 from 1800 SCET (before the foreshock) to 2030 SCET (well into the sheath). Figure 2 shows the solar wind proton bulk parameters for this period. The bottom three panels show the proton velocity in the Halley-centred Solar Ecliptic (HSE) coordinate system defined in Fig. 1.

First we follow the development of v_x, the dominant solar wind velocity component. Until 1825 (marked F), the speed is relatively constant at $\sim 300\,\mathrm{km\,s^{-1}}$, having slowed throughout the afternoon of March 13 due to the mass-loading process (Johnstone et al., 1987). Between 1825 and 1915 SCET the spacecraft is in the foreshock region (Rème et al., 1986) where we observe that the average speed is somewhat slower and significant fluctuations are present. At 1915 the speed recovers to its pre-foreshock value. At 1921 the speed begins a decrease, which becomes more rapid at 1922 (marked S1), reaching a minimum of $<200\,\mathrm{km\,s^{-1}}$ at 1923:30 SCET. After this it recovers almost to its pre-shock level although there is clear evidence for further waves. At 1930:30 however the speed decreases in one 8 s sample, and subsequently its average value remains low. We call this point, which is marked S2, the "permanent" speed decrease and the region between S1 and S2 the cometary ion foot (Coates et al., 1987). After this time, the autoranging cycling mentioned above occurs a few times (visible as gaps in this plot) since the peak of the distribution reached the edge of the solar wind mode field of view. The cycling stopped at 2015 when the start energy level was set to its lowest value by ground command.

The v_y and v_z components in Fig. 2 show that in the foreshock significant directional changes occur in the velocity vector. The speed dip following S1 is accompanied by a marked deflection in flow, at least to the edge of the field of view of this mode. The density trace shows that in the foreshock $\Delta n/n$ is low, $\sim 10\%$, in agreement with the electron measurements (Rème et al., 1986).

Figure 3 shows, on the same time scale, the variation of some parameters derived from the quantities in Fig. 2. From the bottom, the panels are (1) polar angle θ, (2) azimuthal angle ϕ, (3) normalised magnitude of the solar wind velocity vector $\boldsymbol{u}$, (4) normalised mass flux ϱu and (5) normalised momentum flux ϱu^2 of the solar wind protons. The normalisation conditions at infinity are taken as those prevailing at 1200 on March 13, ie $345\,\mathrm{km\,s^{-1}}$ and $6.4\,\mathrm{cm^{-3}}$ for the velocity and density, respectively. The magnitude of velocity behaves similarly to the dominant v_x component discussed above. At the foreshock (marked F) the mass flux changes slightly (although it is essentially conserved up until the bow shock) but the momentum flux shows a pronounced decrease. The width of the shock structure S1–S2 is approximately 40,000 km along the spacecraft track. Subsequent to the shock there is a further recovery of the speed between ~ 1940 and 1945. It may be that this corresponds to an excursion of the spacecraft back into the cometary ion foot of the shock (cf. Formisano et al., 1986). Again the parameters in the top two panels which both involve density are not trustworthy beyond ~ 1915.

Figure 4 shows a projection of the solar wind proton velocity vector into two planes: the x–y plane (upper panel) and the z–y plane (lower panel). Each vector is plotted along the spacecraft trajectory, with gaps at times of the autoranging cycles. Closest approach is towards the left in this figure. The points S1 and S2 are identified on the plot; at S1 the angular change is sharp while at S2 the speed change is sharp.

4. Discussion

The entire region S1–S2 has been identified as the cometary bow shock (Coates et al., 1986, 1987). We have noted that its width is 40,000 km if it is caused by the spacecraft traversing a purely spatial structure. This may correspond to a cometary ion foot of the shock. The major pick-up ions at the bow shock are from the water group (Wilken et al., 1987) and the thickness of this foot is approximately 5 water group ion gyroradii upstream from S2, using the upstream values of velocity and magnetic field. However due to the geometry of the shock crossing this value may be somewhat less in the shock normal direction (Coates et al., 1987). For comparison, the width of the low-speed spike just after S1 is approximately 1.5 water group ion gyroradii. The width of the proton speed change S2, however, is less than 550 km which is less than 2 solar wind proton gyroradii.

From the solar wind proton measurements we can infer the cometary ion density. Assuming that the total momentum flux ϱu^2 is conserved, then

$$\varrho_\infty u_\infty^2 = \varrho_s u_s^2 + \varrho_i u_i^2 ,$$

where ∞ refers to the solar wind well upstream of the comet, s refers to the local solar wind and i to the cometary ions. Now assuming $u_i = u_s$ (in the solar wind frame) and $\varrho_s u_s$, the proton mass flux, is conserved (see Fig. 3), then we have

$$\varrho_i = \varrho_\infty u_\infty (u_\infty - u_s)/u_s^2$$

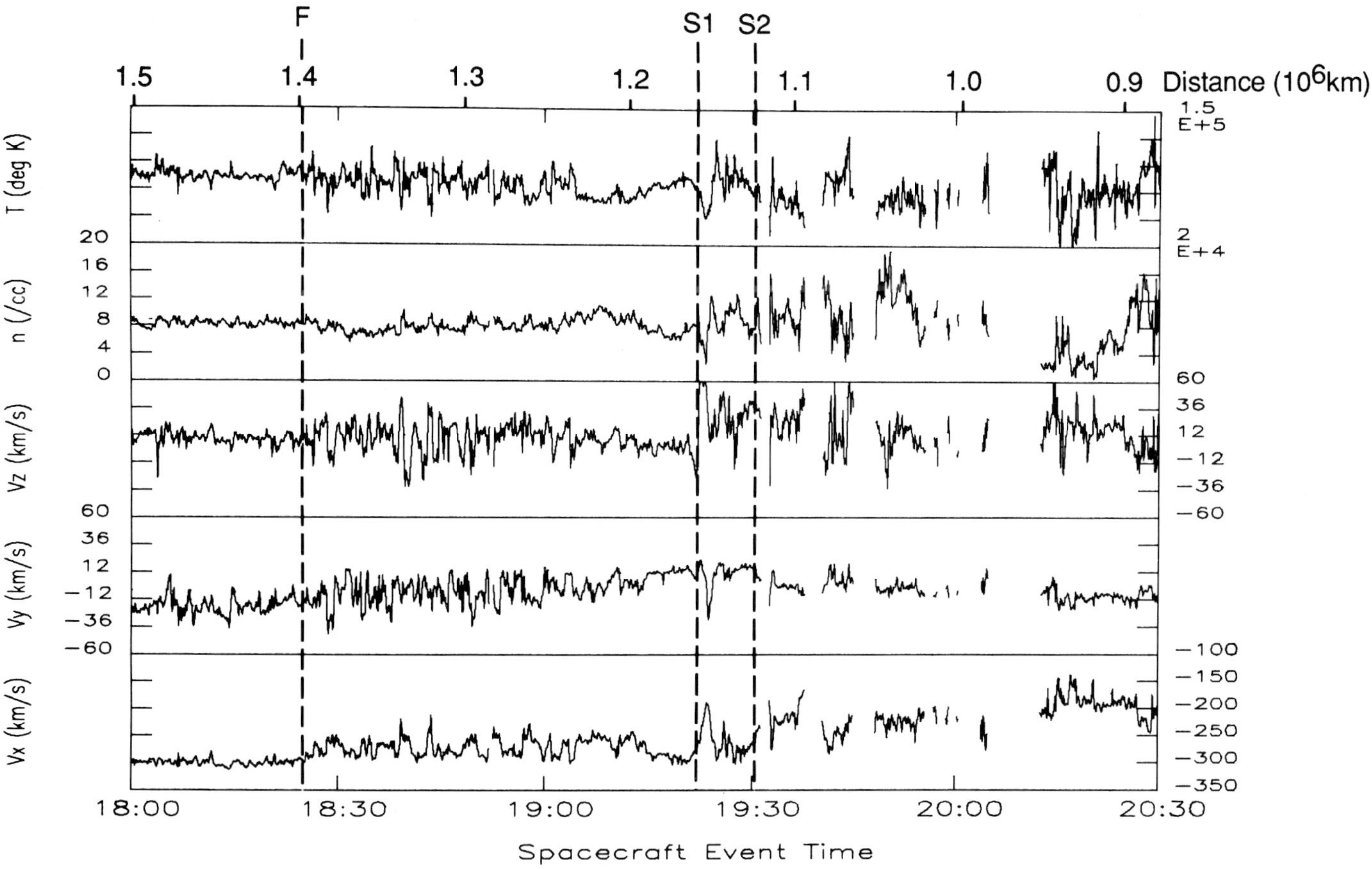

Fig. 2. Solar wind proton bulk parameters in the bow shock region. The foreshock is marked F, the beginning of the low-speed dip is marked S1 and the "permanent" speed change is marked S2. The shock structure extends from S1 to S2

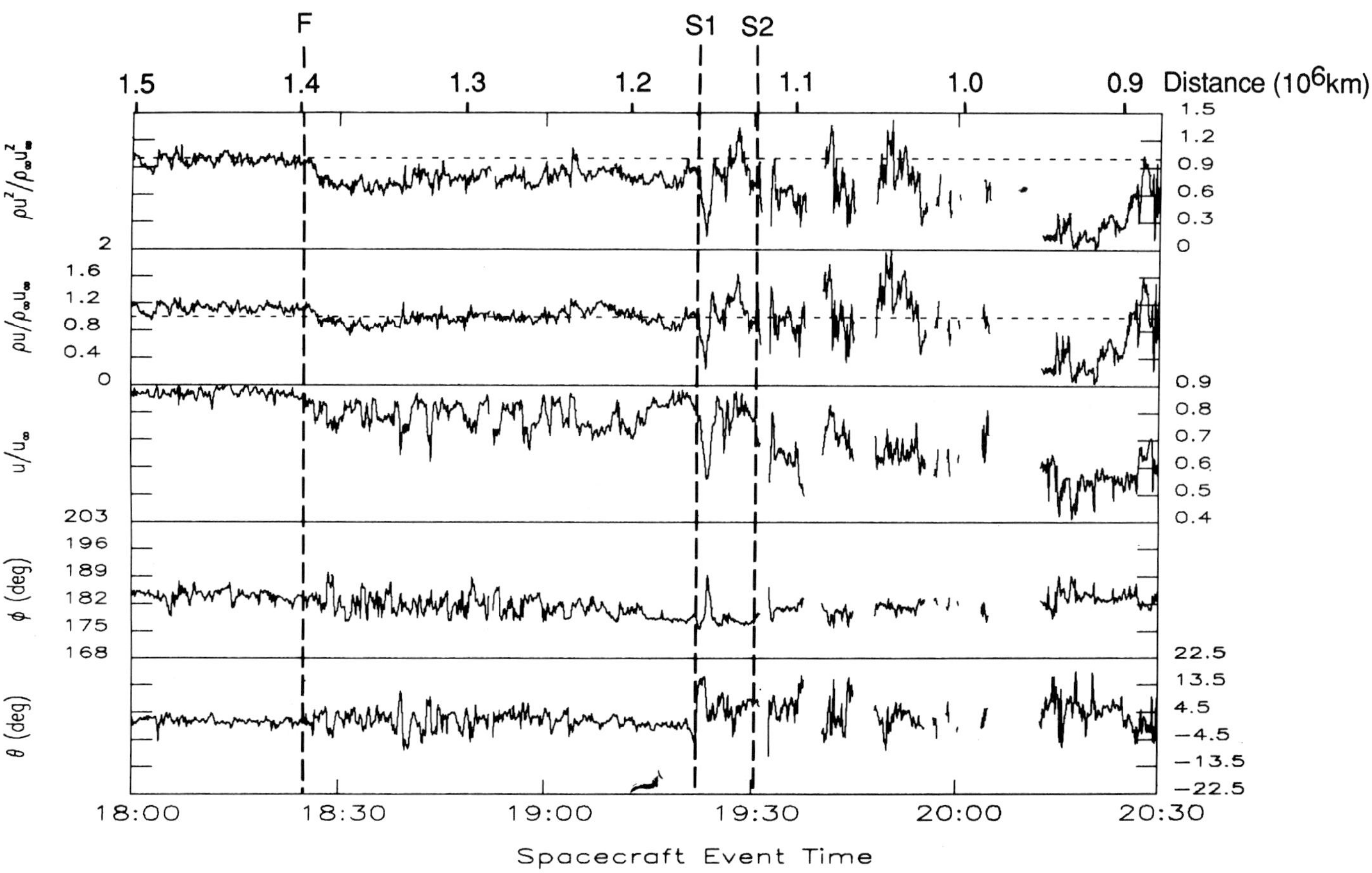

Fig. 3. Parameters derived from Fig. 2 (see text). The values are normalised using the upstream values at 1200 on 13 March, ie $u_\infty = 345\ \mathrm{km\,s^{-1}}$ and $n_\infty = 6.4\ \mathrm{cm^{-3}}$

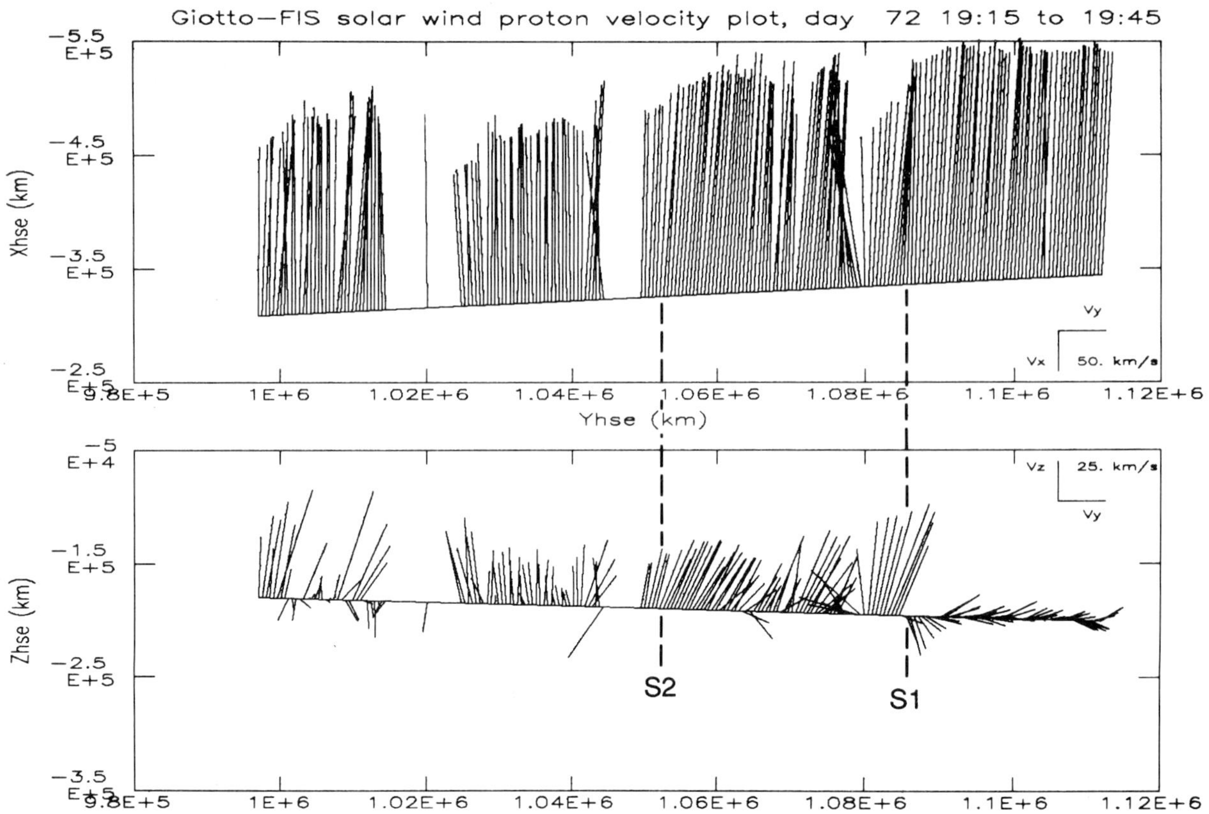

Fig. 4. Projection of the solar wind proton velocity along the spacecraft trajectory. The upper panel shows the x–y plane, ie the view from above the ecliptic plane. The lower panel shows the z–y plane, ie the view from the sun. The vectors are plotted once every 8 s according to the velocity scale in each plot

or

$$n_i = n_\infty u_\infty (u_\infty / u_s - 1)/18\, u_s \,. \tag{1}$$

Ion implantation upstream of the solar wind mass element under consideration, where the local solar wind speed is different, is ignored in Eq. 1. It has been shown in a fuller treatment elsewhere that the error caused by this assumption is low (Johnstone et al., 1987). The assumption $u_i = u_s$ here essentially means that the implanted ion distributions are shell-like in velocity space and thus the bulk velocity is that of the solar wind. According to measurements (Wilken et al., 1987) this is reasonable in the vicinity of the shock but may not be true further upstream where the distributions are more ring-like.

In Fig. 5 we plot the values of n_i inferred from the solar wind speed changes using single-point measurements of the solar wind speed substituted in Eq. (1). The cometary water group ion density from the Implanted Ion Sensor (IIS) (Wilken et al., 1987) is shown for comparison. The inferred densities follow the measurements well in a relative sense while the absolute values differ by a factor ~5. Although the inferred values are not expected to follow the detailed wave structure apparent in the solar wind data, it is significant that at the sharp solar wind speed decrease following S1 the inferred plot has a peak in density which is borne out by the IIS measurements. Indeed detailed inspection shows that this is true for many of the samples where the IIS shows a density peak in this period. The absolute IIS measurements may be out by a factor of 2 or more (Coates et al., 1987; Wilken et al., 1987) since the density calculation omits particles below 1.2 keV. Further work is needed to reconcile the two data sets in absolute terms but the initial impression is encouraging.

Using the measured solar wind mass flux $\varrho_s u_s$ and the inferred cometary ion mass flux $\varrho_i u_i$ we can construct the total mass flux. For a shock to form this quantity should reach a critical value of 4/3 according to the simplified 1-d model of Biermann et al. (1967). Note that more realistic models reduce this value a little (e.g. Schmidt and Wegmann, 1982). The total mass flux is plotted in the top panel of Fig. 6 over a wider time period than the previous figures. The horizontal broken line shows the critical value. Clearly the shock (~1921–1931 SCET) occurs near to the predicted position. The figure also shows the inferred cometary ion density, the solar wind speed and the measured cometary water group ion density over the same period. The extended time period shows more clearly the qualitative agreement between the inferred and the measured cometary ion density.

The foreshock region (~1825–1915 SCET) also stands out in Fig. 6. It is curious that at the foreshock boundary (1825 SCET) the changes in the solar wind protons, the cometary ions and the electrons (Remè et al., 1986) occur simultaneously and suddenly. The proton speed decreases, its fluctuation level increases and the cometary ion density increases. This region will be the subject of further study.

5. Conclusion

We have examined the high resolution solar wind proton measurements in the vicinity of the comet Halley bow shock. The

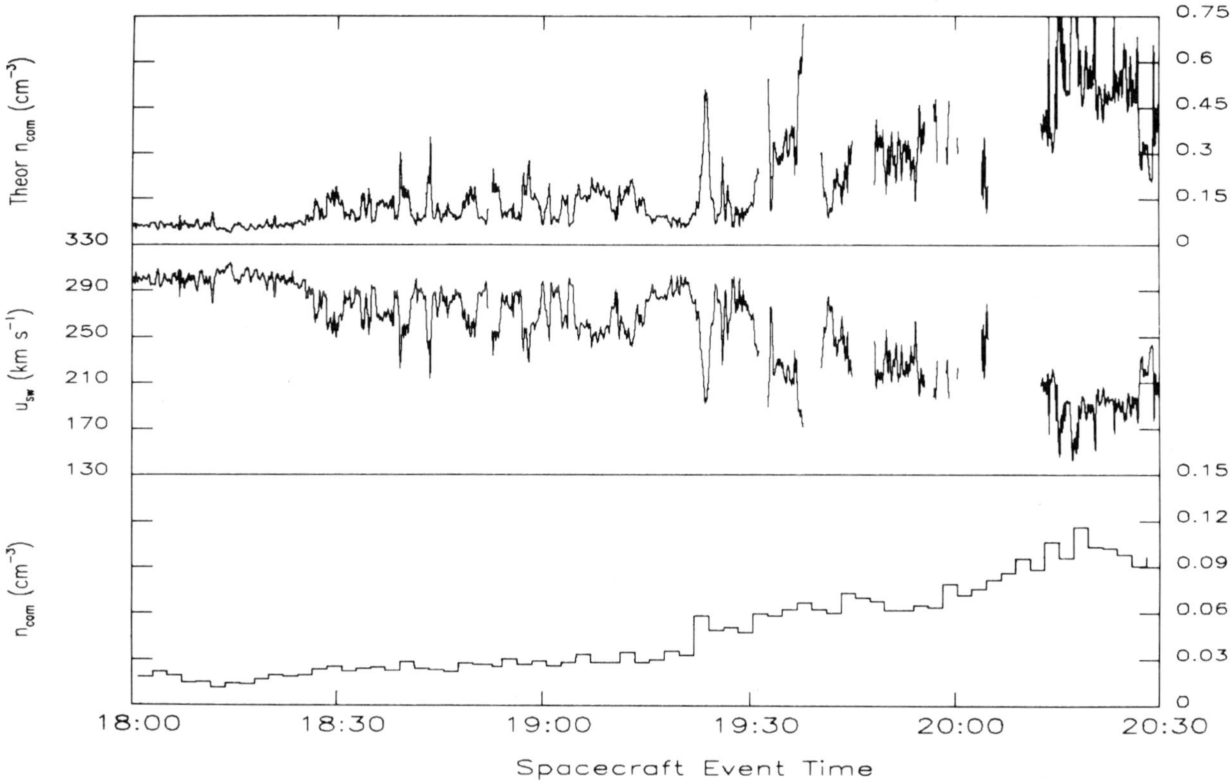

Fig. 5. Top panel, inferred cometary ion density using measurements substituted into Eq. (1). Middle panel, solar wind speed. Bottom panel, IIS measurements of the cometary water group ion density. The plot is on the same time scale as Figs. 1 and 2

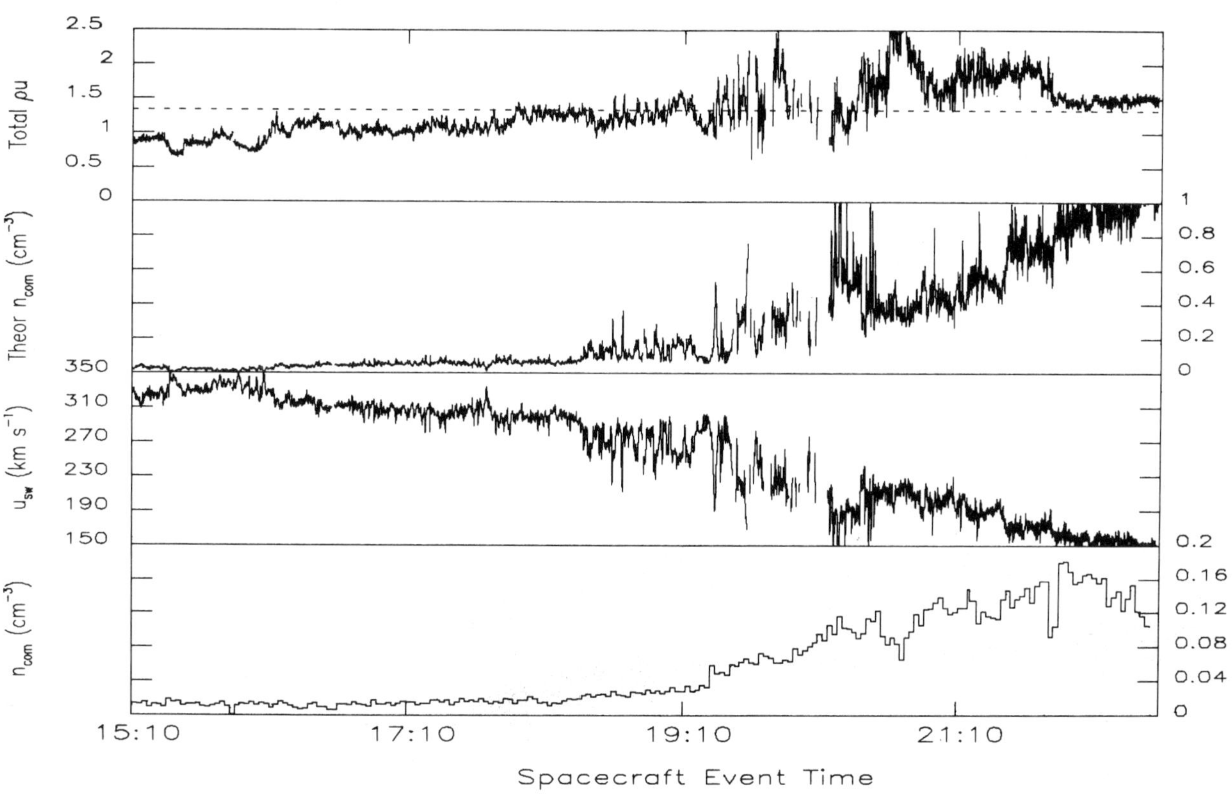

Fig. 6. Top panel. total normalised mass flux, calculated as the addition of the measured solar wind mass flux $\varrho_s u_s$ to the calculated cometary ion mass flux $\varrho_i u_i$. The horizontal broken line shows the critical value 4/3. Second panel, inferred cometary ion density. Third panel, solar wind speed. Fourth panel, IIS measurements of the cometary water group ion density

structure is complicated with a short-lived decrease preceding a more permanent decrease in speed. The solar wind speed decreases have been used to infer the cometary ion density with some success. The shock appears to form at a critical value of ion mass flux as predicted.

Acknowledgements. This work was supported by the Science and Engineering Research Council, the US Department of Energy, NASA order number W-15, 164, the Piano Spaziale Nazionale of the Consiglio Nazionale delle Ricerche, the Bundesministerium für Forschung und Technologie under grant 01 0F 112-4, NASA contract NASW-3574, and the Swedish Board for Space Activities.

References

Bame, S.J., Anderson, R.C., Asbridge, J.R., Baker, D.N., Feldman, W.C., Fuselier, S.A., Gosling, J.T., McComas, D.J., Thomsen, M.F., Young, D.T., Zwickl, R.D.: 1986, *Science* **232**, 356

Biermann, L., Brosowski, B., Schmidt, H.U.: 1967, *Solar Phys.* **1**, 254

Coates, A.J., Johnstone, A.D., Thomsen, M.F., Formisano, V., Amata, E., Wilken, B., Jockers, K., Winningham, J.D., Borg, H., Bryant, D.A.: 1986, 20th ESLAB Symp., ESA SP-250, Vol. **I**, p. 263

Coates, A.J., Lin, R.P., Wilken, B., Amata, E., Anderson, K.A., Borg, H., Bryant, D.A., Carlson, C.W., Curtis, D.W., Formisano, V., Jockers, K., Johnstone, A.D., Korth, A., Mendis, D.A., Rème, H., Richter, A.K., Rosenbauer, H., Sauvaud, J.A., Studemann, W., Thomsen, M.F., d'Uston, C., Winningham, J.D.: 1987, *Nature*, **327**, 489

Formisano, V., Amata, E., Wilken, B., Jockers, K., Johnstone, A., Coates, A., Heath, J., Thomsen, M., Winningham, J.D., Borg, H., Bryant, D.A.: 1986, *Adv. Space Res.* **6**, 229

Johnstone, A.D., Bowles, J.A., Coates, A.J., Coker, A.J., Kellock, S.J., Raymont, J., Wilken, B., Studemann, W., Weiss, W., Cerulli-Irelli, R., Formisano, V., de Giorgi, E., Perani, P., de Bernardi, M., Borg, H., Olsen, S., Winningham, J.D., Bryant, D.A.: 1986, ESA SP-**1077**, p. 15

Johnstone, A.D., Coates, A.J., Heath, J., Thomsen, M.F., Wilken, B., Jockers, K., Formisano, V., Amata, E., Winningham, J.D., Borg, H., Bryant, D.A.: 1987, *Astron. Astrophys.* (this volume)

Neubauer, F.M., Glassmeier, K.H., Pohl, M., Raeder, J., Acuna, M.H., Burlaga, L.F., Ness, N.F., Musmann, G., Mariani, F., Wallis, M.K., Ungstrup, E., Schmidt, H.U.: 1986, *Nature* **321**, 352

Rème, H., Sauvaud, J.A., d'Uston, C., Cotin, F., Cros, A., Anderson, K.A., Carlson, C.W., Curtis, D.W., Lin, R.P., Mendis, D.A., Korth, A., Richter, A.K.: 1986, *Nature* **321**, 349

Schmidt, H.U., Wegmann, R.: 1982, in *Comets*, ed. L.L. Wilkening, Tucson, University of Arizona Press, p. 538

Smith, E.J., Tsurutani, B.T., Slavin, J.A., Jones, D.E., Siscoe, G.L., Mendis, D.A.: 1986a, *Science* **232**, 382

Smith, E.J., Slavin, J.A., Bame, S.J., Thomsen, M.F., Cowley, S.W.H., Richardson, I.G., Hovestadt, D., Ipavich, F.M., Ogilvie, K.W., Coplan, M.A., Sanderson, T.R., Wenzel, K.-P., Scarf, F.L., Vinas, A.F., Scudder, J.D.: 1986b, 20th ESLAB Symp., ESA SP-250, Vol. **3**, p. 461

Wallis, M.K., Dryer, M.: 1985, *Nature* **318**, 646

Wilken, B., Jockers, K., Studemann, W., Rosenbauer, H., Thomsen, M., Johnstone, A.D., Coates, A.J., Formisano, V., Amata, E., Winningham, J.D., Borg, H.: 1987, *Astron. Astrophys.* (this volume)

Astron. Astrophys. 187, 61–64 (1987)

ASTRONOMY
AND
ASTROPHYSICS

Macroscopic perturbations of the IMF by P/Halley as seen by the Giotto magnetometer

J. Raeder [1], **F. M. Neubauer** [1], **N. F. Ness** [2], **and L. F. Burlaga** [2]

[1] Institut für Geophysik und Meteorologie, Universität zu Köln, Albertus-Magnus-Platz, D-5000 Köln 41, Federal Republic of Germany
[2] Goddard Space Flight Center, Code 690, Greenbelt, MD 20771, USA

Received February 13, accepted May 11, 1987

Summary. Giotto magnetic field data have been used to analyze the macroscopic field structure in the vicinity of comet P/Halley. During the Giotto flyby at P/Halley the IMF showed quite a stable "away" polarity. Draping of magnetic field lines is clearly observed along the outbound leg of the trajectory. Inside the magnetic pile-up region the field reverses its polarity several times. A symmetry of oppositely magnetized sheets with respect to the nucleus is found and can be explained in terms of convected IMF features.

Key words: Giotto magnetometer observations – magnetic field draping – P/Halley – solar wind interaction

1. Introduction

Some initial results obtained by the Giotto magnetometer in situ measurements have already been published elsewhere (Neubauer et al., 1986). An overview of the magnetic field structure and the major transitions may be found in the paper by Neubauer (1987). In this report we want to show and discuss the macroscopic structure of the magnetic field in the vicinity of comet P/Halley.

Alfvén (1957) first proposed the model of draped magnetic field lines around a cometary obstacle in the solar wind. This model has subsequently been refined by numerical modelling of the solar wind – comet interaction, for a prominent example see Schmidt and Wegmann (1982). These models usually assume a homogeneous interplanetary magnetic field (IMF) and plasma as well as no turbulence or waves which may be caused by the interaction of pick-up ions with the solar wind. Since these ideal conditions are not met in the real solar wind – comet interaction, the measured field will be characterized by IMF structures as well as by features due to the solar wind – comet interaction.

These considerations make an interpretation difficult in spite of the well known nature of solar wind disturbances which have been studied extensively in the past. In addition the solar wind variations which may be particulary complex at sector boundaries (Behannon et al., 1981) will be strongly modified by the interaction with the comet.

On the other hand the magnetic field variations at P/Halley are quite untypical for the solar wind and can therefore be considered as basically due to the comet – solar wind interaction.

Send offprint requests to: F. M. Neubauer

We shall then investigate this interaction in two ways. First we shall show, that the large scale structure between the inbound and the outbound foreshock region clearly confirms the draping picture. Second, we use directional discontinuities (DD's) of large angular spread which are obviously part of sector boundaries swept into the cometary plasma region as tracers.

2. Large-scale magnetic field draping

Figure 1 shows 4 min averaged magnetic field unit vectors along the Giotto trajectory, decimated by 2 and projected onto the X–Y plane of the Halley centered Solar Ecliptic system (HSE). Most of the wavelike fluctuations are smoothed out by the averaging (see also Glassmeier et al., 1987). The plot covers the time interval from 15 UT on March 13 to 9 UT on March 14 (all times are SpaceCraft Event Time, SCET). We used unit vectors to make a good representation of directional variations possible, thus the length of a vector is proportional to the cosine of the elevation angle between the vector and the ecliptic. Vectors with an arrow tip point southward, whereas the other point to the north of the ecliptic. We also show parabolic fittings of the bowshock and the pile-up boundary as they are derived in Neubauer (1987).

2.1. *Inbound observations*

The first transition boundary is given by the cometary bowshock. Before crossing the bowshock at approximately 19:23 UT on March 13, there is no indication of a major macroscopic disturbance of the IMF by the comet visible. This may be masked by strong fluctuations of cometary or interplanetary origin, however. The field is characterized by large fluctuations, but the dominant direction is outward.

Crossing the bowshock the dominant IMF direction persists, but the fluctuation level increases. Some sheets with a towards polarity are present and similar structures appear in the solar wind before entering the sheath, so they are probably due to convected IMF features. There is no draping visible in the whole sheath. Any deflection of the magnetic field lines is small and is thus masked by the fluctuations. The boundary of the magnetic pile-up region occurs at 23:30 UT. This boundary is associated with a decrease in variability in magnetic field direction.

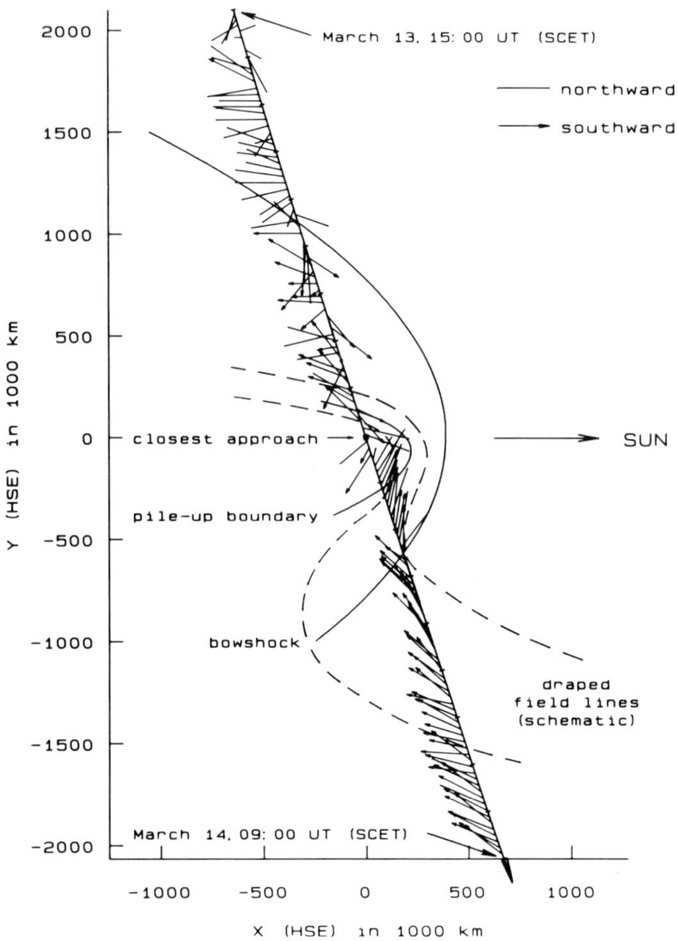

Fig. 1. Four-minute averaged magnetic field unit vectors, decimated by 2 and projected onto the *X–Y* plane of the Halley Solar Ecliptic System (HSE) from 15 UT on March 13 to 9 UT on March 14. The length of the vectors is proportional to the cosine of the elevation angle with respect to the *X–Y* plane. Ticmarcs on the trajectory are 1 h apart. The bowshock and the pile-up boundary are indicated together with two schematically drawn draped field lines

2.2. Outbound observations

Following the trajectory from 9 UT on March 14 backwards, the gross structure of the IMF is different compared with the corresponding parts on the inbound leg. The field is smoother and shows less fluctuations. Outside the bowshock the field points steadily away from the sun at an angle between 135° and 165°. Between 2:40 UT and 2:20 UT the field is smoothly deflected towards the sun to an angle of about 60°. The very broad bowshock transition seems to occur between 3:05 UT and 2:30 UT, i.e. between 750 000 km and 600 000 km from the comet, respectively. The rotation of the IMF in Fig. 1 occurs partly during the broad bowshock structure. Because the deflection persists in the sheath with a gradually decreasing angle, we conclude that this magnetic field deflection clearly reveals draping. Some draped field lines are sketched in the plot. One should observe, that the field along the trajectory is not a snapshot of a spatial structure, but influenced by temporal variations of the IMF during the encounter too. Obviously the IMF conditions before and after the encounter are quite similar, at least with respect to the field polarity. Thus the observed field along the trajectory coincides with the ideal model of a stationary and spatial structure very well.

The outbound pile-up boundary is not very distinct, due to the rather gradual decrease of the field magnitude and some data gaps. Although detailed studies of this boundary still have to be carried out, we consider the discontinuity at 1:07 UT as the best candidate. Well inside this boundary we observe different structures in the magnetic field, which we will discuss next.

3. Observations in the pile-up region

An overview of the magnetic field in the inner region is shown in Fig. 2. Vectors are represented in the same fashion as in Fig. 1. The comet-sun line, also representing the positive *X*-axis in the HSE system gives some orientation. The time interval is from 23:20 UT on March 13 to 00:31 UT on March 14 and covers most of the pile-up region. A more detailed section is shown in Fig. 3.

On both the inbound and the outbound leg the field consists of sheets, in a way, that the magnetic field directions of adjacent sheets are almost antiparallel. In addition, there is a one to one relationship between the sheets inbound and outbound. Every sheet on either side of the nucleus corresponds to a sheet on the other side. While the magnetic field direction of one sheet is sunward, the field direction of the corresponding sheet points away from the sun. Hence two corresponding sheets on either side of the nucleus may be connected, forming a layer of plasma folded around the nucleus, as indicated in Fig. 3 and precisely as expected in the draping picture.

In Figs. 2 and 3 the discontinuities separating these sheets are labelled with small letters on the inbound leg and correspondingly with capital letters outbound. The innermost discontinuities "a" and "A" are the ionopause. Some minor structures are not shown, so the labelling in the figure has gaps. Many of the field reversals have been analyzed using the minimum variance method and have been named directional discontinuities (DD). Due to camera disturbances inbound and data gaps outbound some of them could not be analyzed with confidence. The intersections of the tangential planes with the *X–Y* plane are shown in Figs. 2 and 3. No significant changes in the field magnitude are associated with these discontinuities, except for the ionopause and the discontinuity "k" where the pile-up begins. The minimum variance normal magnetic field components show, that many of these DD's may be tangential discontinuities. The main characteristics, i.e. the minimum variance directions, eigenvalue ratios, normal field components and deflection angles are shown in Table 1. In analyzing a strong directional transition we have systematically subdivided the corresponding time interval into subintervals of varying length. Inclusion in Table 1 reguired fulfilment of two conditions. All of the subintervals must approximately yield the same values of ϕ_n and θ_n. The subinterval with the largest value of $\lambda_2/\lambda_3 > 15$ is then shown. The deflection angles in the table should therefore be considered as lower limits, because of the different and in some cases complex structure of these discontinuities.

The strong symmetry between the inbound and outbound leg shows, that these structures are spatial in the comet rest frame, rather than temporal changes during the interval, in which Giotto crossed the inner region. We suggest, that these plasma layers in the pile-up region, characterized by their magnetic field direction, arise from features of the IMF as they are convected into the pile-up region. The solar wind flow in the pile-up region is rapidly slowed and deflected by the mass loading and the ionosphere,

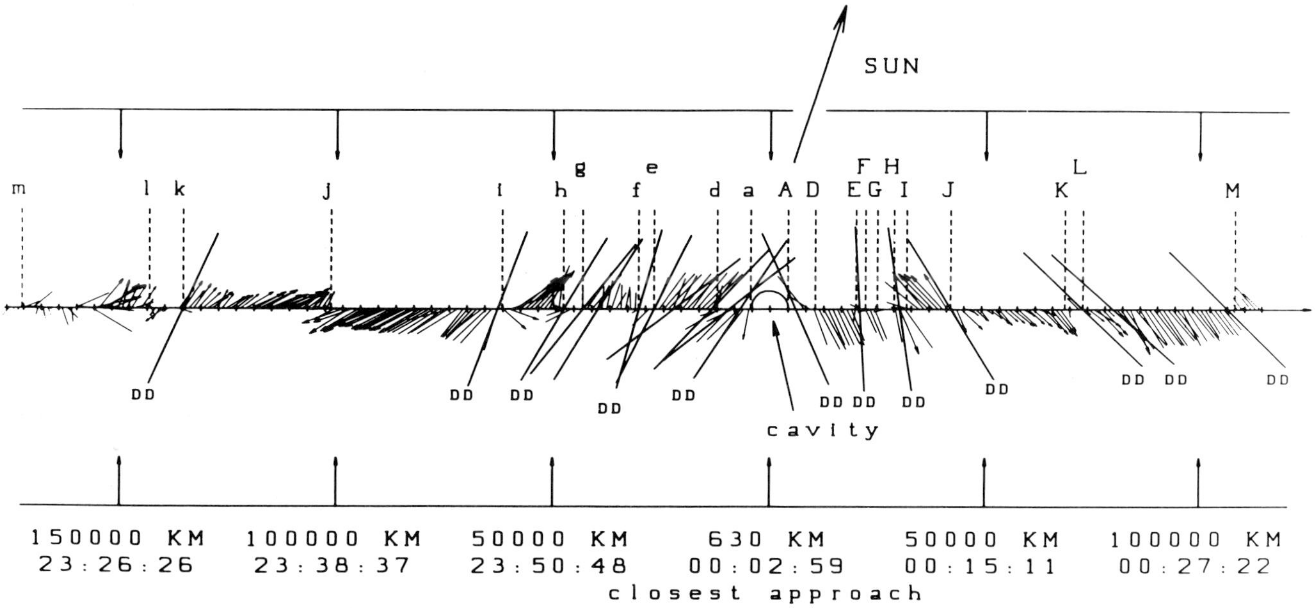

Fig. 2. Projections of 4 s average unit vectors, decimated by 4 onto the X–Y plane of the HSE system from 23:20 UT on March 13 to 00:31 UT on March 14. The length of the vectors indicate their elevation angle with respect to the X–Y pplane as in Fig. 1. The tics along the trajectory are all 60 s. Dashed lines mark the visible directional discontinuities. The intersection of the normal planes of directional discontinuities, which have been analyzed with the minimum variance method are labelled with "DD"

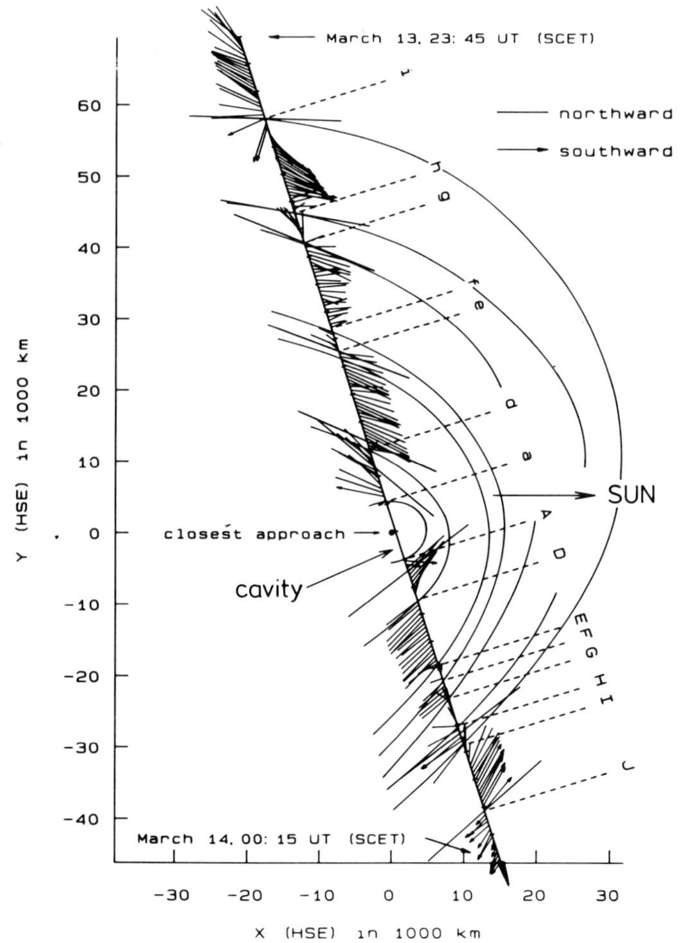

Fig. 3. Section of Fig. 2 from 23:45 UT March 13 to 00:15 UT March 14. The 4 s averages are decimated by 2. The magnetic layers with different polarity are sketched here

acting as an obstacle. Thus the magnetic field, frozen into the plasma will be enhanced by compression of flux tubes and stretching of field lines. Convected structures of the IMF will for the same reason be compressed when entering the pile-up region. These compressed IMF structures, namely the sheets seen by Giotto, will then move slowly through the pile-up region until they slip over the field free cavity and are accelerated into the ion tail. The vicinity to the heliospheric current sheet suggests that we are seeing part of the cascade of strong directional changes which sometimes are associated with the sector boundaries (Behannon et al., 1981).

It should be noted here, that Vega observations show a similar structure around closest approach, but with only one field reversal on either side of the nucleus (Riedler et al, 1986). The authors suggest, besides our interpretation above, that these discontinuities may be related to a diffuse layer related to the contact surface (ionopause), because the normal vectors of the discontinuities roughly point to the nucleus. Because the layers seen by Giotto extend to greater distances and most of their normals do not point to the nucleus, we conclude, that these features are not directly related to the contact surface.

The observation of the rather surprising number of field reversals may have implications for the model of disconnection events (DE). It is assumed, that a DE is driven by the penetration of a sector boundary (field reversal) into the cometary magnetic barrier and enhanced reconnection (Brandt, 1982). The question then arises, why many layers with opposite polarity can coexist deep in the pile-up region. Further detailed and comparative studies will be necessary towards a better understanding.

4. Conclusion

During the Giotto flyby at P/Halley, magnetic field data reveal magnetic field draping very clearly along the outbound leg and particularly close to the comet.

Table 1. Characteristics of the discontinuities in Figs. 2 and 3

Time (SCET)		ϕ_N (°)	θ_N (°)	λ_2/λ_3 (–)	$\|B_Z\|/B$ (–)	ω (°)	Label
86:03:13:	23:25:00	–	–	–	–	150	–
	23:30:00	84	– 3	13	0.051	32	k
	23:38:19	–	–	–	–	98	j
	23:47:42.5	83	–33	42	0.057	155	i
	23:51:25	77	0	78	0.146	144	h
	23:52:28.5	69	–11	139	0.121	96	g
	23:53:20	76	–11	50	0.103	35	–
	23:55:36	–88	4	26	0.003	33	f
	23:56:23	86	– 3	95	0.066		–
	23:56:27	81	– 8	37	0.094	94	e
	23:57:32.5	55	–17	19	0.070	35	–
	23:59:42	64	–16	55	0.031	38	d
86:03:14:	00:00:40	56	–15	40	0.101	53	–
	00:04:04	–54	– 5	271	0.338	107	–
	00:04:10	–56	18	112	0.177	117	–
	00:04:24	–49	0	26	0.048	60	–
	00:04:28	–42	8	107	0.014	35	–
	00:05:30	–	–	–	–	57	D
	00:08:02	–69	6	189	0.014	20	E
	00:09:03	–	–	–	–	44	G
	00:10:01	–	–	–	–	30	H
	00:10:19	–64	0	128	0.198	54	–
	00:10:44	–29	9	34	0.034	126	I
	00:13:13	–41	– 8	63	0.107	95	J
	00:20:43	–26	–10	55	0.218	30	L
	00:22:26	–24	4	28	0.016	52	–
	00:29:00	–27	– 4	358	0.097	46	M

Notes: ϕ_N is the longitude of the normal vector ($\phi_N = 0$ means sunwards), θ_N is the latitude. λ_2/λ_3 is the ratio of the intermediate to the least eigenvalue of the covariance matrix, $|B_Z|/B$ is the ratio of the absolute value of the magnetic field component in the normal direction to the mean magnitude

Inside the pile-up region the field consists of layers of opposite polarity, which arise from convected IMF structures.

Field reversals, convected into the pile-up region are not destroyed by diffusional effects or field line merging but persist even close to the ionopause.

The coexistence of adjacent sheets with opposite polarity close to the comet make a causal relationship between IMF polarity changes and disconnection events questionable.

Acknowledgements. We thank the Bundesminister für Forschung und Technologie (BMFT) and the National Aeronautics and Space Agency (NASA) for funding this investigation.

References

Alfven, H.: 1957, *Tellus* **9**, 92

Behannon, K.W., Neubauer, F.M., Barnstorf, H.: 1981, *J. Geophys. Res.* **86**, 3273

Brandt, J.C.: 1982, in *Comets*, ed. L.L. Wilkening, p. 519

Glassmeier, K.H., Neubauer, F.M., Acuna, M.H., Mariani, F.: 1987 (this issue)

Neubauer, F.M., Glassmeier, K.H., Pohl, M., Raeder, J., Acuna, M.H., Burlaga, L.F., Ness, N.F., Musmann, G., Mariani, F., Wallis, M.K., Ungstrup, E., Schmidt, H.U.: 1986, *Nature* **321**, 352

Neubauer, F.M.: 1987 (this issue)

Riedler, W., Schwingenschuh, K., Yeroshenko, Y.G., Styashkin, V.A., Russell, C.T.: 1986, *Nature* **321**, 288

Schmidt, H.U., Wegmann, R.: 1982, in *Comets*, ed. Wilkening, L.L. p. 538

Astron. Astrophys. 187, 65–68 (1987)

ASTRONOMY AND ASTROPHYSICS

Low-frequency magnetic field fluctuations in comet P/Halley's magnetosheath: Giotto observations

K.H. Glaßmeier[1], F.M. Neubauer[1], M.H. Acuña[2], and F. Mariani[3]

[1] Institut für Geophysik und Meteorologie, Universität zu Köln, Albertus-Magnus-Platz, D-5000 Köln 41, Federal Republic of Germany

[2] Laboratory for extraterrestrial Physics, NASA, Goddard Space Flight Center, Code 690 Greenbelt, MD 20771, USA

[3] II Universita Degli Studi di Roma, Dipartimento di Fisica, 4 Via Orazio Raimondo, J-00173 Roma, Italy

Received February 5, accepted March 25, 1987

Summary. The interaction region between comets and the solar wind is characterized by large amplitude, low frequency magnetic field fluctuations, both within the upstream region as well as in the magnetosheath. Magnetosheath observations of the magnetic field experiment onboard Giotto indicate values of $\delta B/B \simeq O(1)$. Power spectral peaks appear at frequencies of $\simeq 10\,\mathrm{mHz}$ with the spectrum exhibiting a power law dependence with an exponent of the order 2. Radial variation of the fluctuation level does not clearly increase with decreasing distance from the cometary nucleus as observed by the magnetometer onboard Vega-1 and as expected from quasilinear theory. The entrance into the cometary bow shock is furthermore characterized by an order of magnitude increase of the fluctuation level, both on the in- and outbound pass of Giotto.

Key words: comet – comet-solar wind interaction – magnetosheath – magnetic field turbulence – fluctuation level

1. Introduction

The pick-up of freshly ionized atoms and molecules of cometary origin by the electromagnetic fields of the solar wind is one of the basic physical processes of importance for the solar wind-comet interaction. The "free" energy of the freshly ionized particle distributions is the reservoir for excitation of low-frequency magnetic field fluctuations as observed by the recent flybys at comets P/Giacobini-Zinner and P/Halley (Smith et al., 1986; Galeev et al., 1986; Neubauer et al., 1986). Due to recent theoretical work (e.g. Winske et al., 1985; Sagdeev et al., 1986) these fluctuations are invoked primarily by ion cyclotron and fire-hose type instabilities.

It is the purpose of the present brief report to give a first overview on low-frequency fluctuations as observed by the Giotto magnetic field experiment. Special attention is paid to those fluctuations detected within the magnetosheath, i.e. in the region between the cometary bowshock and the pile-up region (e.g. Neubauer et al., 1986). Two accompanying papers (Acuña et al., this volume; Johnstone et al., this volume) discuss in more detail the characteristics of low-frequency waves in the region upstream of the cometary bowshock.

Send offprint requests to: K.H. Glaßmeier

2. Overview

An overview of the magnetic field observations during the encounter of Giotto with comet P/Halley is given in Fig. 1. The upper three panels show 256 s averaged values of the magnetic field magnitude, the azimuth as well as the elevation angle in a *H*alley-*S*olar-*E*cliptic (HSE) coordinate system with the x-axis pointing towards the sun along the comet-sun line, the z-axis pointing approximately towards the north of the ecliptic plane, and the y-axis completing the right hand triad. Basic features of the comet-solar wind interaction region as reflected in the magnetic field data and as discussed in more detail in work by Neubauer et al. (1986), Neubauer (this volume), or Raeder et al. (this volume) can be identified. These are the inbound bowshock crossing at about a distance of $1.1\,10^6$ km from the nucleus, the occurrence of a significant pick-up boundary at 135,000 km, or the basically asymmetric behaviour of the magnitude along the inbound and outbound trajectory, showing a much smoother decay on the outbound pass.

The lower two panels indicate the level of low-frequency fluctuations. The lower one of these two displays the logarithm of the mean of the trace of the power spectral density matrix as computed from 1 s averaged data and data intervals covering 256 s. The upper panel exhibits the Pythagorean sum of the root-mean-squared deviations of the three vector components, determined over 256 s intervals, too. Both measures of the fluctuation level show no clear increase with decreasing radial distance to the nucleus, as suggested by the quasi-linear theory of Sagdeev et al. (1986) and as observed in Vega-1 data by Galeev et al. (1986) and Galeev (1987). This aspect will be discussed in more detail further down.

To further characterize the observed fluctuations the variance, normalized to the squared mean magnetic field, has been determined for both, the transverse as well as compressional field fluctuations. Values between 0.25 and 0.75 have been found with some indication of the compressional variations always being slightly smaller than the transverse ones. These values show that we are dealing with strong low-frequency turbulence.

Power spectral analysis, illustrated by examples in Fig. 2, reveals that throughout the magnetosheath the spectra for both,

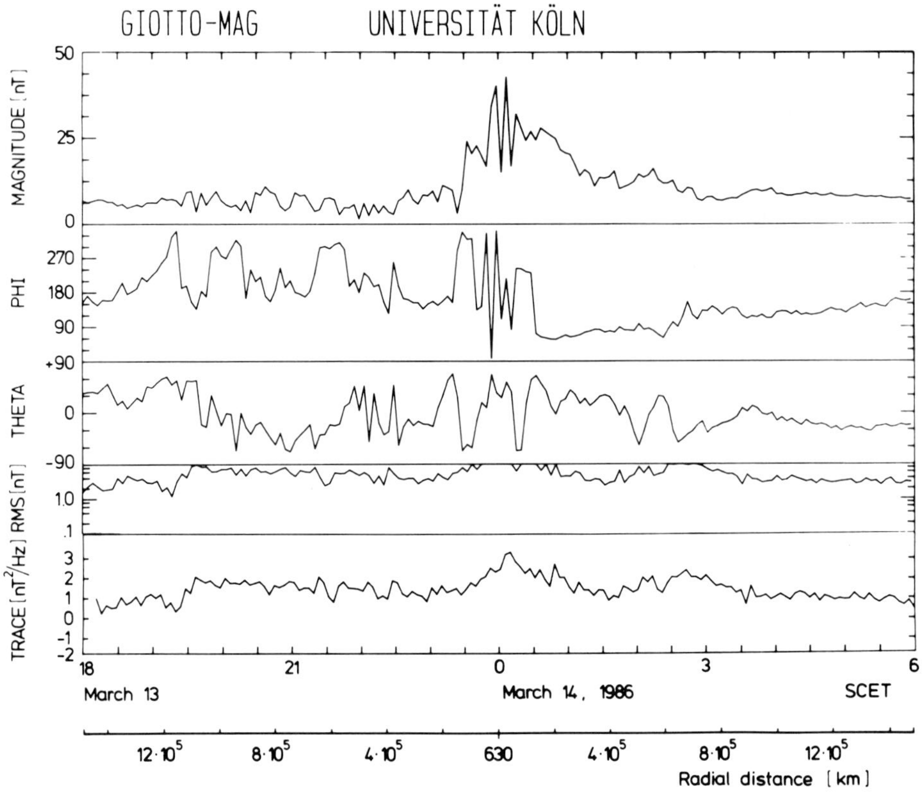

Fig. 1. Overview on magnetic field observations of the Giotto magnetic field experiment near comet P/Halley. The upper three panels display magnetic field magnitude, azimuth, as well as elevation angle in the HSE coordinate system. The bottom two panels indicate the level of fluctuations, using a root-mean-square parameter as well as the mean value of the trace of the spectral density matrix as an indicator

transverse and compressional components, can well be described by a power law $P(f) \sim f^{-\alpha}$ in the frequency range 10–1000 mHz. Typical values are $\alpha = 1.8$–2.1, i.e. usually higher than suggested by a Kolmogoroff spectrum.

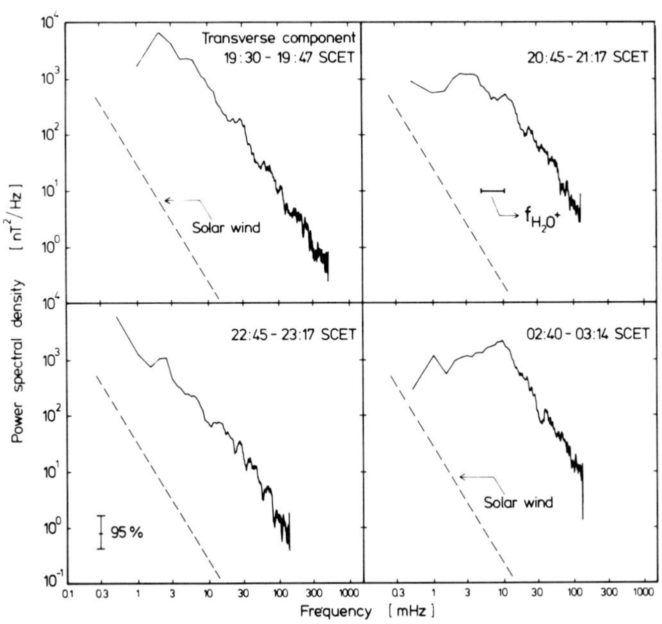

Fig. 2. Power spectra of the transverse magnetic field component for different intervals, covering the in and outbound bowshocks as well as the magnetosheath. The dashed lines indicate typical solar wind spectra. The gyro-frequency of water ions is indicated in the upper right panel. A 95% confidence interval, based on the assumption of a purely statistical process, is given in the bottom left panel

The dashed lines in Fig. 2 represent solar wind spectra at about 1 AU for 'stable' interplanetary conditions, i.e. for a period of minimum solar wind activity, where the stream structure was characterized by stable recurrent high speed streams (Denskat and Neubauer, 1982). The power level is typically about two orders of magnitude higher as observed in the solar wind, in agreement with observations near comet P/Giacobini-Zinner (Tsurutani and Smith, 1986a).

For frequencies less than about 10 mHz the spectra often are flat as shown, for example, in the right two panels of Fig. 2. This suggests that significant wave power is generated at frequencies of about 10 mHz and is distributed to higher frequencies by a turbulence cascade process. It should be noted that 10 mHz corresponds to the gyro-frequency of water group ions in a field of about 10 nT. Thus heavy ion cyclotron unstable waves are a likely candidate to deliver the necessary energy for the turbulence cascade, as noted earlier by Tsurutani and Smith (1986a).

Beside this more general description of the magnetic field observations in Halley's magnetosheath, a more detailed visual inspection of the observations reveals not only turbulent, i.e. chaotic structures, but also exhibits a variety of different quasi-periodic wave forms as noted by Glaßmeier et al. (1987). As an example, Fig. 3 shows such a quasi-periodic wave train, with about 70 s period variations present in all three components as well as the magnitude. Closer inspection of this wave event reveals that it is predominantly a compressional wave with smaller transverse contributions. Clear wave steepening with associated high-frequency waves and partial rotations as observed near P/Giacobini-Zinner is not seen in Fig. 3. Thus this magnetosheath wave event differs significantly from wave events seen both, in the upstream region of comets P/Giacobini-Zinner (Tsurutani and Smith, 1986b) and P/Halley (Acuña et al., Johnstone et al., both this volume).

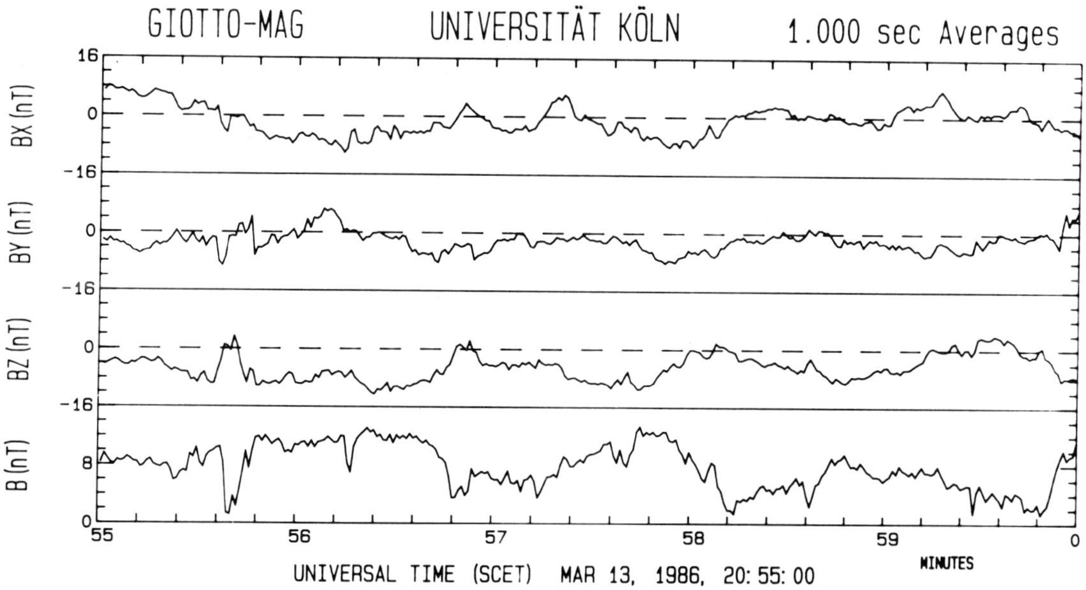

Fig. 3. Typical example of a quasi-coherent wave event as seen in the magnetosheath. Shown are the three components of the magnetic field in the HSE coordinate system as well as the magnitude

3. Radial variation of the fluctuation level

Large amplitude magnetic field variations as observed by the recent cometary flybys will significantly influence the newborn, pick-up ions. Ion acceleration (e.g. Ip and Axford, 1986) and in particular strong pitch-angle scattering will occur. Wave growth, non-linear wave decay, and pitch angle scattering constitute a delicately balanced process which might be described in terms of quasi-linear theory as presented by Sagdeev et al. (1986). Their results indicate a clear dependence of the magnetic field fluctuation level on the density of newborn ions, i.e. the level is expected to increase with decreasing distance from the nucleus. Quasi-linear theory has subsequently been used by Galeev et al. (1986) and Galeev (1987) to analyse magnetic field fluctuations observed by the magnetic field experiment onboard Vega-1 during its encounter with P/Halley. The observed decrease of the fluctuation level with radial distance is, according to Galeev (1987), in "wonderful" agreement with quasi-linear theory by Sagdeev et al. (1986).

However, as already visible in the bottom two panels of Fig. 1 a similar clear increase of the fluctuation level with decreasing distance from the nucleus is not observed in the Giotto magnetic field data within the cometary magnetosheath.

Determination of the level of fluctuation in a highly turbulent data set is a delicate problem as it requires a careful determination of a mean or ambient magnetic field relative to which the turbulence level is determined. This is especially true for the case of the Giotto magnetic field data as there are prominent macroscopic spatial magnetic field variations visible in the data (e.g. Neubauer, this volume; Raeder et al., this volume; see also upper panel of Fig. 1). For example, the overall increase of the magnetic field magnitude when approaching the nucleus gives rise to a significant route-mean-square deviation which might be interpreted as a measure of δB, the fluctuation level.

To demonstrate the problem, Fig. 4 shows radial profiles of the fluctuation level for both, the inbound and the outbound pass of the Giotto trajectory. In the upper panel of Fig. 4 data segments of length 1024 s of 1 s averaged data have been used to determine $\delta B^2 = (\delta B_x^2 + \delta B_y^2 + \delta B_z^2)/3$, after the mean value has been subtracted. For most parts of the interaction region out to about a radial distance of 1.4 10^6 km the level is fairly constant with a typical value of about 100nT^2, for both the inbound and outbound pass. Only very close to the comet nucleus, i.e. for a radial distance less than about 200000 km, a clear increase up to values of 100nT^2 is observed. Comparison with the upper panel

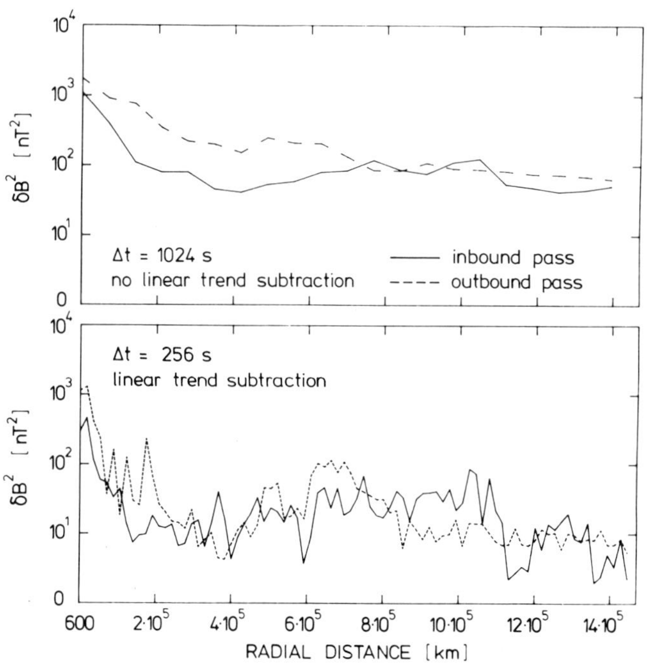

Fig. 4. Radial variation of the magnetic field fluctuation level for both, the inbound and the outbound pass. In the upper panel δB^2 has been determined from data segments 1024 s long, after the mean has been subtracted. In the bottom panel 256 s data segments and subtraction of both, the mean and the linear trend have been used

of Fig. 1 shows a clear correlation of this increase with the overall change of the magnitude of B.

In the lower panel of Fig. 4 a different way of determination of the fluctuation level has been chosen. Instead of using the mean field, determined from 1024 s duration intervals, as a measure of the ambient field, the data interval has been shortened to 256 s. The ambient field is determined by computing the mean value as well as the linear trend over this interval. δB^2 is subsequently calculated from the residual field.

The resulting radial variation of the fluctuation level exhibits more structure than the curves shown in the upper panel of Fig. 4. The overall level in the lower panel is significantly smaller than in the upper one, suggesting that contributions to the variance by the overall increase of the ambient magnetic field have successfully been removed. On the inbound pass the level is about $10 \mathrm{nT}^2$ in the upstream region with a short interval of smaller fluctuations right before the entry into the cometary bowshock at a distance of about $1.2\,10^6$ km on the inbound pass. Entry into the cometary magnetosheath is characterized by a pronounced increase of the level by about an order of magnitude. The level stays constant up to about a radial distance of 400,000 km, when a slight decrease of the level occurs which lasts until about 200,000 km before closest approach. The following increase is probably still due to contributions of the overall increase of the field magnitude due to prominent magnetic field draping near the comet. It should be noted that no non-linear trends have been subtracted from the observations.

Radial variation on the outbound pass exhibits essentially the same behaviour as on the inbound pass. Lower level fluctuations occur in the upstream region, and entrance into the cometary magnetosheath at around 700,000 km is again characterized by a clear increase of the level. A region of slightly decreasing level around 400,000 km appears as on the inbound pass. Thus both radial profiles of fluctuation levels on the inbound and outbound pass exhibit similar spatial variation.

4. Summary and outlook

Magnetic field observations in the magnetosheath of comet P/Halley are characterized by large-amplitude, low-frequency fluctuations in both, magnitude as well as the field components. Values of $\delta B/B \simeq O(1)$ have been observed. Power spectral densities typically exhibit a power law dependence $f^{-\alpha}$ with $\alpha \simeq 1.8$–2.1. Occasionally spectra flatten out at frequencies below 10 mHz. Spectral peaks appear at frequencies corresponding to water group ion gyro-frequencies, and suggest that the observed magnetic field turbulence is due to pick-up cometary ions by the solar wind plasma.

Closer inspection of the radial variation of the magnetic field fluctuations as seen by the Giotto magnetic field experiment reveals a clear discrepancy between Giotto and Vega-1 observations at P/Halley. In the latter case a clear decrease with radial distance is reported (e.g. Galeev, 1987), in accord with quasi-linear theory.

In the Giotto magnetic field data the radial variation of the fluctuation level is characterized by a sudden increase at the inbound bowshock, a nearly constant level down to a distance of 400,000 km, and a slight decrease until about 250,000 km, when the level increases again.

Thus the magnetosheath fluctuations are characterized by a significantly higher level as compared to the upstream region. This suggests that not only the pick-up process contributes to the observed fluctuations, but yet unanalysed processes in the cometary bowshock region and magnetosheath might be as well important.

Finally, it must be noted that magnetosheath fluctuations exhibit clear compressional character, different from the predominantly transverse waves observed in the upstream region of cometary bowshocks (Tsurutani and Smith, 1986; Acuna et al., this volume; Johnstone et al., this volume). Recently Ip and Axford (1986) discussed in some detail the importance of second order Fermi acceleration for ion acceleration processes in comet-solar wind interaction regions taking into account a turbulent ensemble of Alfvén waves. The compressional magnetic field observations within P/Halley's magnetosheath, however, makes transit-time damping (e.g. Fisk, 1976) another attractive acceleration mechanism, the relevance of which should be subject of future research.

Acknowledgement. We thank the Bundesminister für Forschung und Technologie (FRG) and the National Aeronautics and Space Agency (USA) for financial support of this investigation.

References

Denskat, K.U., Neubauer, F.M.: 1982, *J. Geophys. Res.* **87**, 2215

Fisk, L.A.: 1976, *J. Geophys. Res.* **81**, 4633

Galeev, A.A. et al.: 1986, *Geophys. Res. Letters* **13**, 841

Galeev, A.A.: 1987, 20th ESLAB Symp. ESA SP-250, Vol. I, p. 3

Glaßmeier, K.H., Neubauer, F.M., Acuña, M.H., Mariani, F.: 1987, 20th ESLAB Symp. ESA SP-250, Vol. III, p. 167

Ip, W.-H., Axford, W.I.: 1986, *Planet. Space Sci.* **34**, 1061

Neubauer, F.M. et al.: 1986, *Nature*, **321**, 352

Sagdeev, R.Z., Shapiro, V.D., Shevchenko, V.I., Szegö, K.: 1986, *Geophys. Res. Letters* **13**, 85

Smith, E.J., Tsurutani, B.T., Slavin, J.A., Jones, D.E., Siscoe, G.L., Mendis, D.A.: 1986, *Science* **232**, 382

Tsurutani, B.T., Smith, E.J.: 1986a, *Geophys. Res. Letters* **13**, 259

Tsurutani, B.T., Smith, E.J.: 1986b, *Geophys. Res. Letters* **13**, 263

Winske, D., Wu, C.S., Li Y.Y., Mou, Z.Z., Guo, S.Y.: 1985, *J. Geophys. Res.* **90**, 2713

Astron. Astrophys. 187, 69–72 (1987)

ASTRONOMY AND ASTROPHYSICS

Fine structure of the magnetic field in comet P/Halley's coma

Y.G. Yeroshenko [1], **V.A. Styashkin** [1], **W. Riedler** [2], **K. Schwingenschuh** [2], **and C.T. Russel** [3]

[1] IZMIRAN, SU-142092 Troitsk, USSR

[2] Space Research Institute, Austrian Academy of Sciences, Inffeldgasse 12, A-8010 Graz, Austria

[3] Institute of Geophysics and Planetary Physics University of California, Los Angeles, CA 90024, USA

Received February 19, accepted June 19, 1987

Summary. During the encounter with comet P/Halley the magnetic field measurements aboard Vega-1 and Vega-2 have revealed short magnetic field depressions (MFD) at times of a generally rather smooth rise or fall of the magnetic field. A maximum MFD magnitude of 40 nT and a maximum duration of 50 s were observed at distances of $50 \div 70\ 10^3$ km. Most likely these events are due to spatial plasma irregularities inside the coma region and not to temporal field variations. The observed effects may be caused by cometary plasma rays.

Key words: magnetic field depressions – plasma irregularities – cometary rays

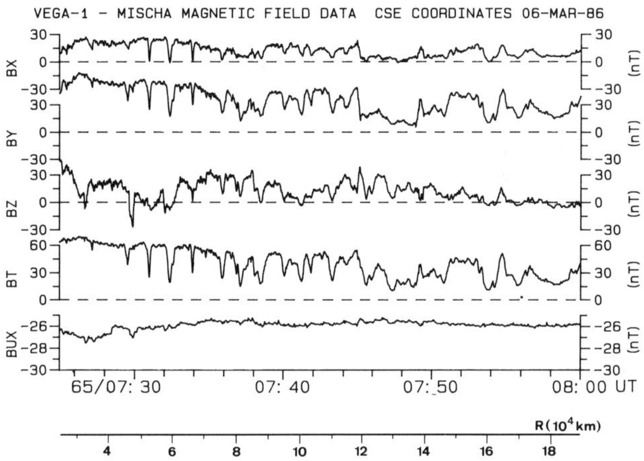

Fig. 1. Magnetic field variations inside comet Halley's coma observed by Vega-1 during the outbound phase of the encounter. The field components (B_x, B_y, B_z) and B-total (B_t) of the outboard sensor are presented in cometocentric solar ecliptic (CSE) coordinates. B_{ux} is the difference in instrument coordinates between the data of the outboard B_x sensor and the inboard B_u (∥ to B_x) single-axis sensor. B_{ux} is plotted with a 15 times larger vertical resolution as the components of the main sensor

1. Introduction

A main feature of the magnetic field along Vega's trajectories inside Comet Halley's coma is a gradual enhancement of B-total with decreasing distance to the nucleus (Riedler et al., 1986a). Preliminary data analysis of the Vega-1/2 magnetic field measurements has also revealed different boundaries:

A bow shock, an outer boundary of the magnetic barrier, behind which a steeper increase of the magnetic field strength is observed and an inner boundary where possibly a "storage" of the previous magnetic field (~ 8 h) before could be seen (Riedler et al., 1986a; Schwingenschuh et al., 1986). Unexpected depressions of the magnetic field superimposed on the background of the increasing field were first assumed to be spacecraft effects, such as, interactions of the solar panels with dust particles (Riedler et al., 1986b). However, a more thorough data analysis showed that plasma irregularities in Halley's coma could be a reason of the magnetic field depressions (MFD). The characteristics of 14 short field depressions are described and an attempt is made to explain the physical nature of these events.

2. Observations

The first MFD's were detected on the outbound leg of the Vega-1 orbit after closest approach (Fig. 1). Similar events were observed on the inbound leg during the Vega-2 encounter (Fig. 2). Due to

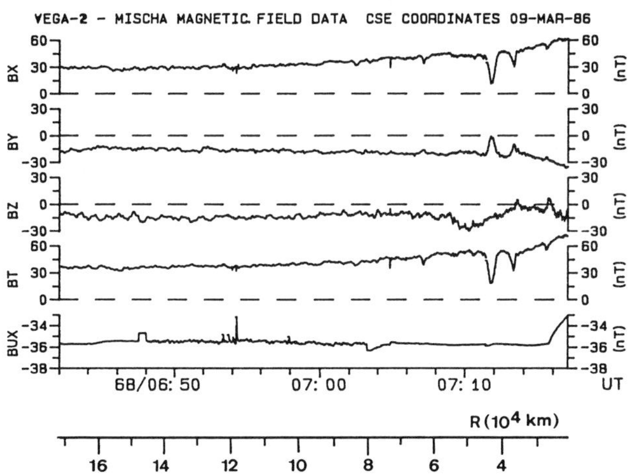

Fig. 2. Magnetic field variations inside comet Halley's coma (Vega-2 inbound data). Labels are the same as in Fig. 1

Send offprint request to: K. Schwingenschuh

the failure in the three-axial sensor of the Vega-2 magnetometer no MFD information on the outbound leg is available.

A decrease of two components (B_x and B_y) is the most obvious feature of about 50 % of MFD events. A simultaneous decrease of all three components was observed from a distance 70 10^3 km and further out on the outbound leg of Vega-1.

Detailed characteristics of the most pronounced MFD's can be seen from Figs. 3 and 4. B_{ux} denotes the difference of B_x and data from the fourth magnetometer sensor, which is situated 1 m closer to the spacecraft (SC) on the same boom parallel to B_x. As can be seen, there is practically no correlation of B_{ux} with the MFD's in the magnetic field components, indicating the absence of a significant gradient during these events. Thus, the MFD's do not seem to be caused by SC currents.

The main parameters of 14 MFD's and their distribution along the Vega orbits are presented in Table 1 and Fig. 5. It can be seen that magnitude and duration of the MFDs depend on the position along the SC trajectories: the biggest and longest depressions are observed at distances of 50 ÷ 70 10^3 km from the comet's nucleus. Minimum variance analyses of eleven Vega-1 and three Vega-2 events (see Table 1) revealed a typical decrease of

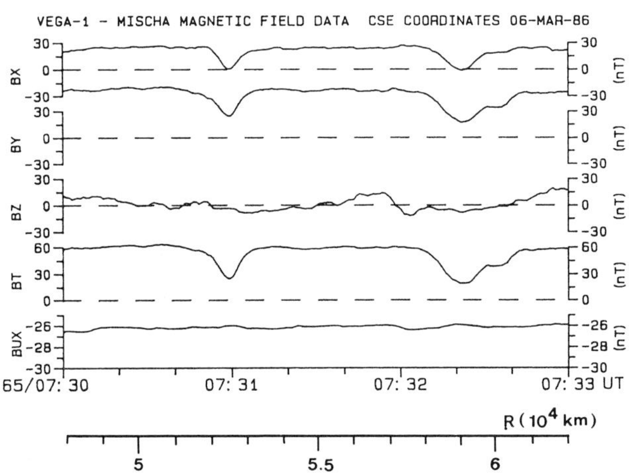

Fig. 3. Detailed structure of the maximum field depressions observed by Vega-1. Labels are the same as in Fig. 1

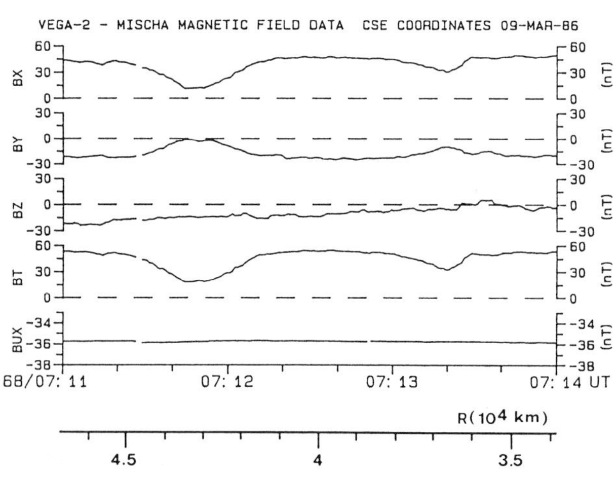

Fig. 4. High resolution plot of Vega-2 MFDs. These events occur at the same time as the plasma density enhancements (Vega-2 PLASMAG data) which have been associated with Critical Ionization Velocity (CIV) events (Galeev et al., 1986)

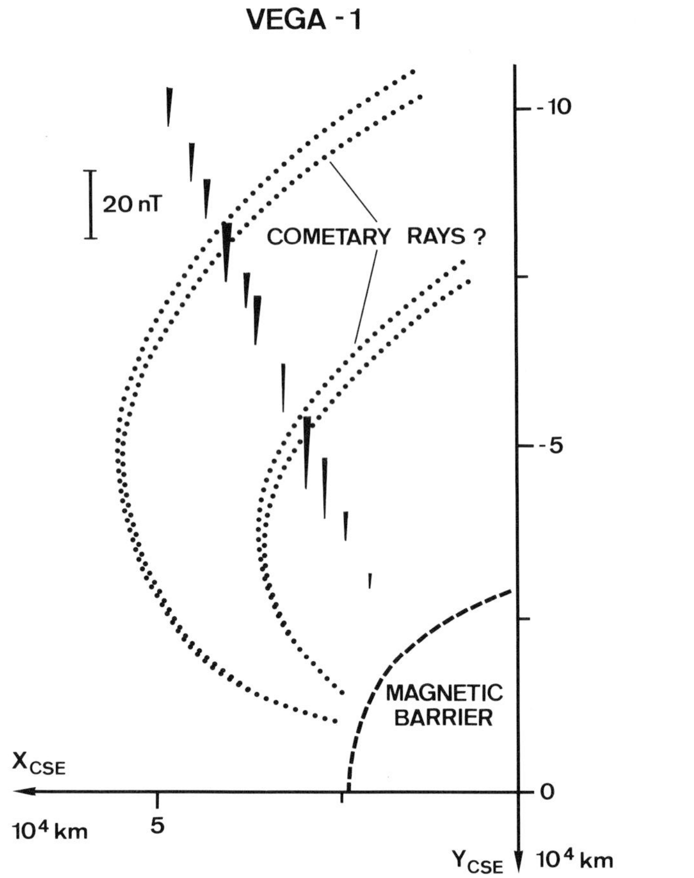

Fig. 5. Spatial MFD distribution along the Vega-1 trajectory. 1/2 ΔB_t is shown. The dashed curves indicate the possible paths of cometary rays

the magnetic field which is not parallel to the undisturbed magnetic field vector. Figure 6 shows an example of the analysis. The perpendicular component of all events is in the order of 10 nT. A pure diamagnetic effect should have no significant components perpendicular to the magnetic field direction.

3. Discussion

Comparison of MFDs with simultaneous plasma measurements aboard Vega-2 showed that the field decreases are rather well correlated with short increases of the plasma density. The maximal magnitude of the MFDs coincides with maximal plasma density increases. The latter were explained by critical ionization velocity effects (Galeev et al., 1986).

Magnetic field depressions were also observed by PVO above the ionosphere of Venus (Russell et al., 1982). The duration of magnetic field "dips" in the case of Venus was ~90 s and the magnitude ~20 nT. Contrary to the MFDs (see the minimum variance analysis in Fig. 5) these cloud events show also a distinct rotation of the magnetic field vector. Spatially distributed, so-called "plasma clouds" have been deduced which were correlated with magnetic field depressions (Brace et al., 1982). There is some trend of flank concentration of "plasma clouds". A possible similarity of these "plasma clouds" detected above the venusian ionopause to rays in cometary tails of Type I was pointed out by Russell et al. (1982).

Table 1.

t(UT)	ΔB(nT)	X_{CSE}	Y_{CSE}	Δt(s) at level $0.5\,\Delta B$	$R = V.\Delta t = 79{,}2.\Delta t$ (km)
		(10^3 km)			
Vega 1, 1986 March 6					
7:27:14	− 7	21	− 31	5	396
7:29:34	−18	25	− 40	9	713
7:31:01	−35	27	− 48	10	792
7:32:24	−40	29.5	− 54	18	1425
7:33:56	−27	33	− 62	8	634
7:35:55	−23	36.5	− 71	14	1109
7:37:08	−20	38	− 75	13	1030
7:38:30	−30	41	− 81.5	25	1980
7:40:08	−20	43.5	− 88	14	1109
7:41:15	−22	46	− 94	20	1584
7:43:17	−22	49	−101.5	20	1584
Vega 2, 1986 March 9					
					$R = 76{,}8.\Delta t$ (km)
7:12:00	−34	−7	34	25	1920
7:13:20	−20	−5	28	12	922
7:15:30	− 7	0	19	9	691

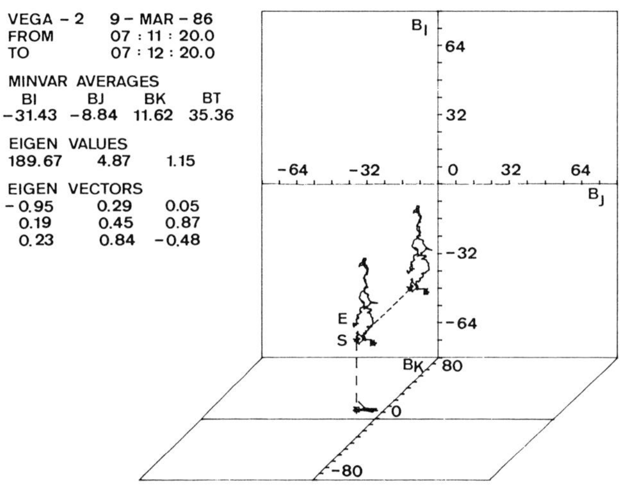

Fig. 6. Minimum variance analysis of the Vega-2 MFD event at 7:11:20 (S)–7:12:20 (E) UT at a distance to the nucleus of 39000 km. B_I is the maximum, B_J the median and B_K the minimum variance direction. The 3D trace of the magnetic field vectors in the minimum variance system (B_I, B_J, B_K) as well as the projections onto the $B_I B_J$ and $B_J B_K$ plane have been plotted. The MFD event is characterized by a change of the magnetic field along B_I with a very small perpendicular rotation. B_I is not parallel to the undisturbed magnetic field ($B_T = 55$ nT). There is an 11 nT perpendicular component and the angle between the undisturbed magnetic field vector and the maximum variance direction is about 20°

Plasma density enhancements were also observed by ICE near comet Giacobini-Zinner which are possibly related to cometary rays (Bame et al., 1986).

There are thus reasons to assume that the MFDs observed by the Vega-1/2 magnetometers could be the effect of plasma concentrations, intersected by the SC along their orbits inside Halley's coma. If the plasma irregularities mentioned above are assumed to be cometary plasma rays, their expansion at the periphery can either be explained as a real expansion of rays or as intersections of the plasma ray by the SC at different angles due to the ray bending at the coma day side [see "fountain effect" in Wurm (1986)].

The fact that the maximal magnitudes of MFDs are observed not at the closest distance to the nucleus may be caused either by ion acceleration in cometary rays (Wurm, 1968) or by the nonresonant ion instability. The latter reason was used to explain the similar plasma fluctuations during the comet Giacobini-Zinner fly-by (Gosling et al., 1986). However, additional analysis incorporating plasma data is necessary for any final conclusion on the dependence of the magnitude of the MFDs on the distance from the cometary nucleus.

Due to the Vega magnetic field experiment all MFDs were observed upstream the outer boundary of the magnetic barrier. Note that MFDs were detected closer to the comet with Vega-2 than with Vega-1. Most probably this is the result of the higher (factor 1.6) solar wind dynamic pressure during the Vega-2 encounter compared with Vega-1 (Gosling et al., 1986). The boundaries in the inner coma could be compressed by the enhanced solar wind dynamic pressure.

The suggested hypothesis of cometary ray effects could explain the absence of MFDs at the inbound leg of the Vega-1 orbit simply by an asymmetrical distribution of cometary rays.

4. Conclusions

Irregular field depressions with amplitudes up to 40 nT and durations up to 50 s were observed by the magnetometers aboard the Vega-1/2 SCs inside comet Halley's coma at distances $\sim 20 \div 100\ 10^3$ km from the nucleus.

The MFDs are distributed along the SC trajectory in such a way that their magnitudes and durations increase with increasing distance to the nucleus and reach a maximum at $\sim 50\ 10^3$ km.

An asymmetrical distribution of magnetic field "dips" relative the subsolar point of the SC orbit was observed by Vega-1.

It is suggested that these events were caused by the filamentary structure of plasma irregularities or by cometary plasma rays.

Acknowledgements. The authors are grateful for discussion and useful comments to A.A. Galeev, K.I. Gringauz, M.I. Verigin, S.I. Klimov, V. Formisano, R. Grard, A. Pedersen, J. Luhmann, M.M. Mogilevsky, Y.M. Michailov, R.P. Rijnbeek, and G. Schelch.

References

Bame, S.J., Anderson, R.C., Asbridge, J.R., Baker, D.N., Feldman, W.C., Fuselier, S.A., Gosling, J.T., McComas, D.J., Thomsen, M.F., Young, D.T., Zwickl, R.D.: 1986 *Science* **232**, 356

Brace, L.H., Theiss, R.F., Hoegy, W.R.: 1982, *Planet. Space Sci.* **30**, 29

Galeev, A.A., Gringauz, K.I., Klimov, S.I., Remizov, A.P., Sagdeev, R.Z., Savin, S.P., Sokolov, A.Y., Verigin, M.I., Szegö, K.: 1986, *Geophys. Res. Letters* **13**, 845

Gosling, J.T., Asbridge, J.R., Bame, S.J., Thomsen, M.F., Zwickl, R.D.: 1986, *Geophys. Res. Letters* **13**, 267

Gringauz, K.I., Gombosi, T.I., Remizov, A.P., Apathy, I., Szemerey, I., Verigin, M.I., Denchikova, L.I., Dyachkov, A.V., Keppler, E., Klimenko, I.N., Richter, A.K., Somogyi, A.J., Szegö, K., Szendrö, S., Tatrallyay, M., Varga, A., Vladimirova, G.A.: 1986, *Nature* **321**, 282

Riedler, W., Schwingenschuh, K., Yeroshenko, Y., Styashkin, V.A., Russell, C.T.: 1986a, *Nature* **321**, 288

Riedler, W., Schwingenschuh, K., Yeroshenko, Y., Styashkin, V.A., Russell, C.T.: 1986b, *Pis'ma Astron. Zh.* **12**, 647 (in Russian)

Russell, C.T., Luhmann, J.G., Elphic, R.C., Scarf, F.L., Brace, L.H.: 1982, *Geophys. Res. Letters* **9**, 45

Schwingenschuh, K., Riedler, W., Schelch, G., Yeroshenko, Y.G., Styashkin, V.A., Luhmann, J.G., Russell, C.T., Fedder, J.A.: 1986, Adv. Space Res. **6**, No. 1, 217

Wurm, K.: 1968, *Icarus* **8**, 287

Astron. Astrophys. 187, 73–79 (1987)

ASTRONOMY AND ASTROPHYSICS

Giotto magnetic-field results on the boundaries of the pile-up region and the magnetic cavity

F.M. Neubauer

Institut für Geophysik und Meteorologie der Universität zu Köln, Albertus-Magnus-Platz, D-5000 Köln 41, Federal Republic of Germany

Received January 30, accepted May 12, 1987

Summary. The application of the minimum-variance-technique to the magnetic field measurements obtained during the Giotto encounter at comet P/Halley has led to the following results on the inner plasma boundaries in the magnetoplasma surrounding the comet:

(a) Modelling the boundary of the magnetic pile-up region firstly by a half-sphere cylinder combination and secondly by a paraboloid with the axes of both surfaces aligned with the aberrated solar wind has led to stand-off distances towards the aberrated solar wind of 143 000 km and 215 000 km, respectively. These can be considered as extreme case without flaring and with strong flaring, respectively.

The axis of symmetry had a transverse offset of about 80 000 km from the nucleus.

(b) The ionopause i.e. the outer boundary of the magnetic cavity region had either a smooth surface of radius 4600 km displaced from the nucleus by 1500 km or, alternatively, it had no average displacement but strong ripples aligned with the aberrated solar wind direction and a wavelength of at least 200 km. Combinations are also possible.

The ionopause must be a tangential discontinuity thereby containing further evidence of the null magnitude of the cavity field.

Key words: Giotto magnetometer observations – plasma boundaries at comet Halley – P/Halley solar wind interaction – ionopause of comet P/Halley

1. Introduction

In this paper we shall continue to use measurements by the Giotto magnetometer experiment to help disentangle the physical nature of the plasma interaction between the solar wind and comet Halley. More specifically, we shall investigate the inner boundaries of the plasma flow regions with regard to their physical nature and shape in three dimensions. The study presented in this paper relies almost exclusively on magnetometer data and will have to be complemented by joint investigations with the various plasma experiments on Giotto in the near future.

The Giotto magnetometer experiment has been discussed in detail elsewhere (Neubauer et al., 1987, 1986a). We repeat only the most important properties. In addition to data obtained during cruise the instrument has provided continuous vector magnetic field data at a rate of 28.24 vectors per second from the late hours of March 11, 1986 through the encounter of comet P/Halley until the early hours of March 15, 1986. In analysing the data two special sources of magnetic interferences have to be kept in mind. Even after the data processing there are residual magnetic field variations at the spin frequency and its harmonics in the rotating spacecraft frame due to the despin motor of the spacecraft. In addition, at certain times the motors of the Halley Multicolor Camera (HMC) cause disturbances the nature of which is well understood, however. The interval from March 13, 20:03 UT to the encounter was more or less affected by HMC disturbances. All times used in this paper are spacecraft event times (SCET). Also we note that a dust impact event just before closest approach initiated a damped nutation of 0°.9 nutation angle implying an initial attitude change of up to 1°.8 (Reinhard, 1986).

Since the attitude of the spacecraft has been determined before the event and after the damping out of the nutation the directional error of the magnetic field vectors in Halley centered solar ecliptic coordinates (HSE-coordinates) is always $\leqq 0°.9$ during this time period.

In order to make the geometrical results of this paper easily comparable with other studies we use the fly-by trajectory solution with a minimum distance of 605 km at 00:03:02 SCET on March 14, 1986. Note that this orbit solution is slightly different from the one used in the complementary report by Neubauer (1986c). In addition a small timing error has been found and corrected in the mean-time.

In the next section we shall then introduce the three major boundaries seen in the magnetometer data with emphasis on the pile-up boundary and the ionopause boundary. In Sect. 3 we shall use the minimum variance technique (Sonnerup and Cahill, 1967) to determine the normals of the pile-up boundary and attempt to model its approximate shape. In Sect. 4 we shall present analogous results on the magnetic cavity boundary i.e. the ionopause.

2. Overview of magnetic field boundaries subdividing the magnetoplasma surrounding comet Halley

Three major magnetic field boundaries have been identified during the Giotto encounter of comet Halley both inbound and outbound (Neubauer et al., 1986b). Figure 1 gives an overview of minute averages of the magnetic field vectors observed from March 13, 1986 18:00 SCET until March 14, 1986 10:00 SCET.

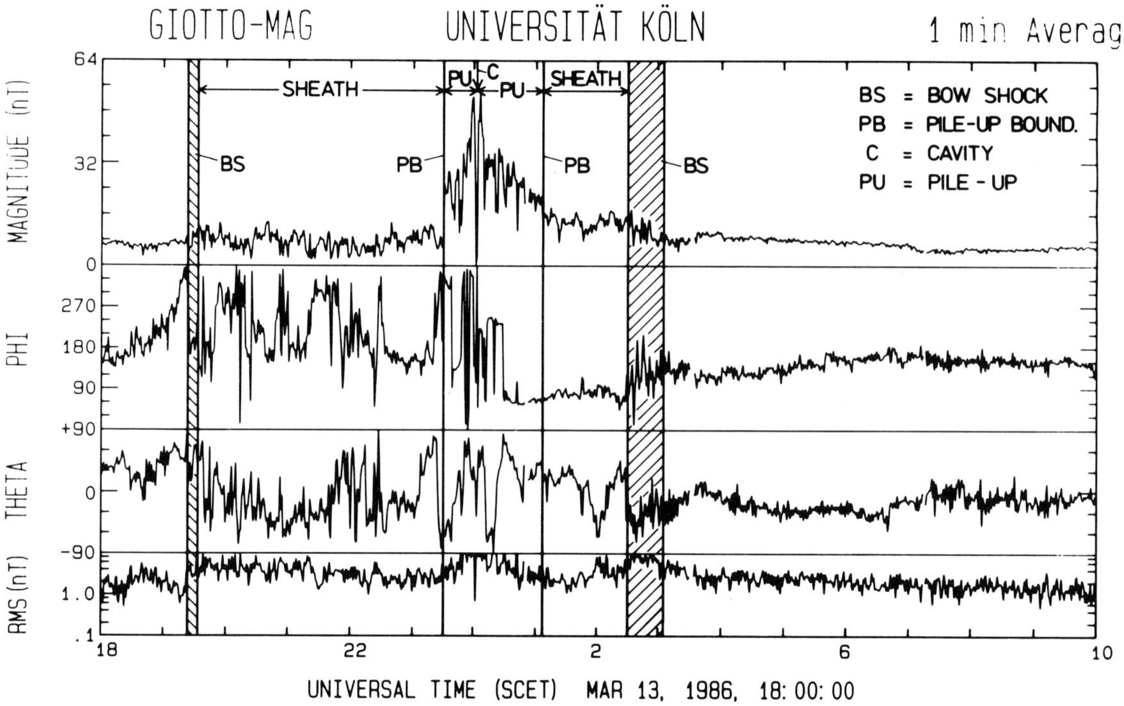

Fig. 1. One minute average vector magnetic field observations in Halley centered solar ecliptic coordinates. The definition of ϕ and θ is standard i.e. θ is the elevation angle with respect to the X, Y-plane and ϕ the azimuth of the projection with $\phi = 0$ towards the sun. RMS is the pythagorean mean of the component rms-values

The magnetic field variations define four spatial regions inbound and outbound. The outermost region is the upstream wave region separated from the "sheath" by the bow shock. The sheath region is separated from the magnetic pile-up region by the pile-up boundary. The innermost boundary, i.e. the ionopause or contact surface, separates the magnetic pile-up region from the magnetic cavity characterized by magnetic field zero. It can only be seen in Fig. 1 as a sharp dip in magnitude and requires better time resolution as in Fig. 3.

A detailed discussion of the bow shock is deferred to a later paper. At this point we mention only that in (Neubauer et al., 1986b) its shape has been approximated by the axisymmetric paraboloid

$$X = 3.9\ 10^5\ \text{km} \left(1 - \frac{Y^2 + Z^2}{(7.8\ 10^5\ \text{km})^2}\right), \tag{1}$$

where the justification for the paraboloid came from theory. This is in approximate agreement with the bow shock shape suggested by Rème and collaborators (1986).

The second major boundary is the outer boundary of the magnetic pile-up region. It appears as a distinct boundary only in the Giotto data but not in the data of the Vega-1 and Vega-2 fly-by's (Riedler et al., 1986; Schwingenschuh et al., 1986). In the Giotto data the very sharp increase in magnetic field magnitude inbound is particularly noteworthy. The outbound crossing is less clear. Hence it may require special conditions in the solar wind or with regard to cometary activity for a sharp boundary to form. We also note that during the ICE-encounter at comet Giacobini-Zinner very pronounced magnetic field maxima were observed at the inbound and outbound tail boundaries (Slavin et al., 1986). We tentatively suggest that their tail boundary is the downstream counterpart of our pile-up boundary.

The third major boundary has only been seen in the Giotto data because of the particularly close approach to the comet. It is the ionopause, which is the outer boundary of the magnetic cavity with a field of essentially zero magnitude. It could also be defined as the inner boundary of the plasma region aroung the comet magnetically connected to the interplanetary medium and finally the sun. If the contrast in magnetic fields between both side of a boundary is considered, the ionopause certainly is the most dramatic boundary.

We shall now discuss the inner two boundaries in detail.

3. The pile-up boundary

In this section we shall investigate the geometry of the pile-up boundary (PB) i.e. the boundary between the sheath and the innermost magnetized plasma region. The boundary is very pronounced on the inbound part of the Giotto trajectory at 23:30 UT on March 13, 1986 and less pronounced outbound where it has been tentatively identified at 1:07 UT on March 14.

Although we consider the discontinuity at 1:07 the best candidate for the outbound pile-up boundary we cannot completely exclude the possibility that the PB is located elsewhere and that this discontinuity represents a solar wind structure convected into the cometary environment. Even for the inbound pile-up boundary it remains to be seen whether the magnetic field pile up as a basic ingredient of the cometary flow system is amplified by solar wind inhomogeneities. We note that the boundary can also be called the isodyne for magnetic field values around 20 nT.

Table 1 gives the locations of the inbound and outbound pile-up boundary and the ionopause in HSE-coordinates. In addition the results of the minimum variance analyses applied to these

Table 1. Geometry of inner boundaries in the plasma environment of comet P/Halley

Time on March 13 or 14	HSE-coordinates in 10^3 km			ϕ_n	θ_n	λ_2/λ_3	$\lvert B_n \rvert/B$
	X	Y	Z	(deg)			
Inbound pile-up boundary 23:30:00	−39.5	127.6	−23.1	84.4	−3.4	12.5	0.05
Inbound ionopause 00:01:54.35	−0.83	4.47	−1.03	86.3	−19.2	18.8	0.05
Outbound ionopause 00:03:58.85	1.69	−3.53	0.40	280.2	−21.9	155.4	0.015
Outbound pile-up boundary 01:07:00	78.5	−246.5	44.0	313.2	2.1	19.4	0.18

boundaries are given. The minimum variance technique has proved to be very powerful for the characterisation and identification of plasma boundaries in the past. The angles ϕ_n and θ_n characterising the minimum variance direction $\boldsymbol{n}$ are also given together with the ratio of the intermediate eigenvalue λ_2 to the minimum eigenvalue λ_3 as a quality measure. Finally, the average component of the magnetic field in the minimum variance direction $\boldsymbol{n}$ divided by the maximum magnetic field is given.

Before we can use the entries of Table 1 for the construction of a boundary surface we have to note that the inbound pile-up boundary normal is not too well determined whereas the outbound λ_2/λ_3 is quite acceptable. Here we have used $\lambda_2/\lambda_3 \geqq 15$ for a minimum variance direction to be acceptable. We do not assign standard errors to these normals since in terms of absolute values these are almost meaningless. A more appropriate but also more difficult question would be to answer the question of the influence of accompanying wave fields, ripples on the surface etc.

Inspection of Fig. 1 shows in addition that the magnetic structure on the inbound and outbound parts of the encounter trajectory is quite different on a scale of tens of minutes (Raeder et al., 1987). Since the upstream magnetic field is neither parallel nor perpendicular to the solar wind speed $\boldsymbol{V}_s$ with aberration taken into account properly an asymmetry is expected because of the asymmetry in the magnetic forces due to the draped magnetic field.

In addition the fly-by trajectory is not perpendicular to the solar wind direction. Taking the undisturbed solar wind speed to be 350 km s^{-1} (Balsiger et al., 1986; Johnstone et al., 1986) and the solar wind vector to be radial in the inertial frame we obtain an aberration angle of $\alpha = 6°$. The angle between the trajectory and the aberrated solar wind speed is then 78°.7.

In addition the different appearance of the inbound and outbound data may be due to temporal variations in the incident solar wind. Moreover ripples may be present on the boundaries leading to a deflection of the normals. Even if the latter two effects should be absent we would expect a boundary with no axisymmetry. In the case of stationary solar wind flow the pile-up boundary should be symmetric only with respect to the plane spanned by $\boldsymbol{V}_s$ and the interplanetary magnetic field $\boldsymbol{B}_s$ i.e. as a function of $\boldsymbol{Z}$. These considerations suggest that the boundary surface may be quite complicated.

On the other hand the minimum variance normals indicating the orientations of tangential planes to the true curved boundary at the two crossing points represent one of the few examples of direct observational information on the magnetoplasma away from the one-dimensional encounter trajectory. Since a three-dimensional boundary model is also of interest for comparison with ground-based observations it is worthwhile to attempt fitting simple model boundary surfaces to these limited data. The simpliest shape is a halfsphere towards the sun with a cylinder in the tailward direction. A best fit is obtained with a vector $\boldsymbol{x}_0 = (X_0, Y_0, Z_0)$ for the center of the sphere given by

Model 1

$$X_0 = -6.7\ 10^4\ \text{km}$$
$$Y_0 = -8.7\ 10^4\ \text{km}$$
$$Z_0 = \ \ 1.3\ 10^4\ \text{km} \tag{2}$$

radius $R = 2.19\ 10^5$ km.

The boundary is shown in the plane $z = Z_0$ in Fig. 2 (Mod. 1). We have also fitted a paraboloid which is axisymmetric with respect to the aberrated solar wind direction to the observations.

This surface is described in a coordinate system with the axes X', Y', Z' with X' paralle to $-\boldsymbol{V}_s$ and Z' parallel to Z. We obtain

$$X' = X'_0 - f\,[(Y' - Y'_0)^2 + (Z' - Z'_0)^2]$$

with

Model 2

$$X'_0 = \ \ 2.15\ 10^5\ \text{km}$$
$$Y'_0 = -8\ \ \ 10^4\ \text{km}$$
$$Z'_0 = \ \ 1.4\ \ 10^4\ \text{km} \tag{3}$$

$f = 5.252\ 10^{-6}\ \text{km}^{-1}$.

The semi-latus rectum for the paraboloid is 202 000 km.

The stand-off distances i.e. the maximum values of X' are 143 000 km and 215 000 km for models 1 and 2, respectively. These models can also be considered as the case of no flaring of the cometary tail and a case with strong flaring. Earth-based observations of the cometary tail at the Giotto encounter may further help in constraining the geometry.

A further question is that of the physical nature of the boundary. The quantities $\lvert B_n \rvert/B_{max}$ in Table 1 indicate that from the minimum variance analysis alone the pile-up boundaries could

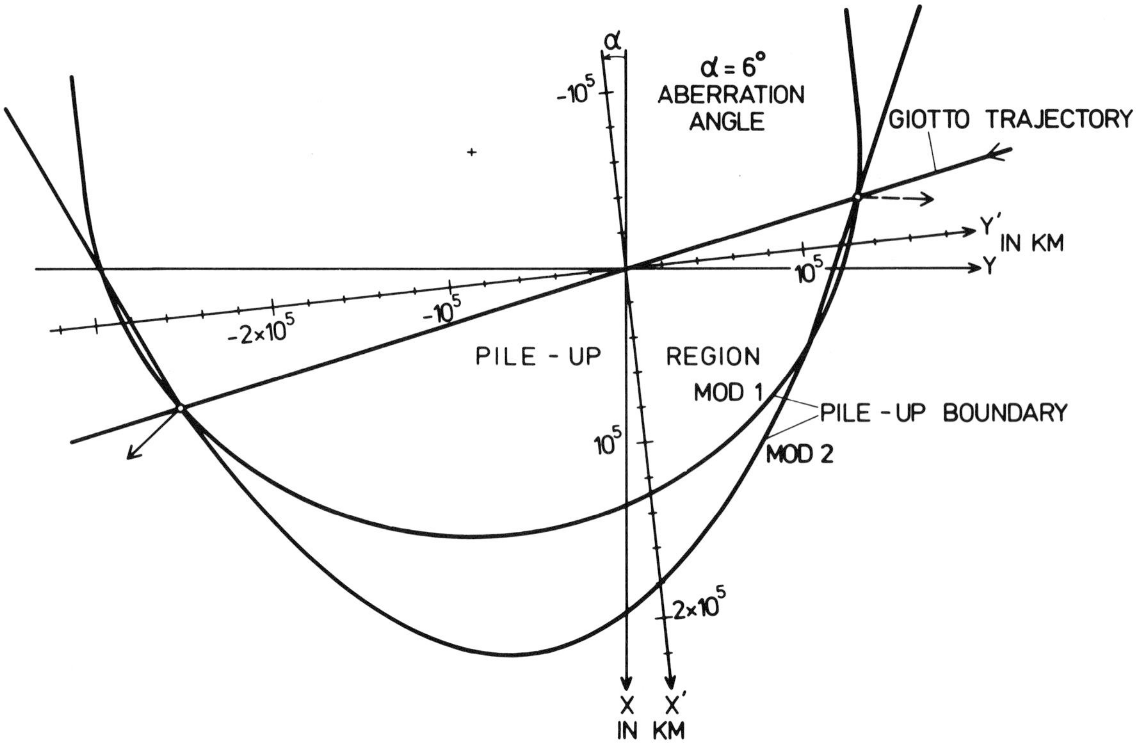

Fig. 2. Giotto trajectory projected on the Halley centered solar ecliptic X, Y-plane with X towards the sun. Also shown is the X', Y'-coordinate system aberrated by 6°. The crossing points of the magnetic pile-up boundary are given together with the normals projected on the X, Y-plane. Two best fit models for the pile-up boundary are also shown in their planes of maximum extent i.e. $Z_0 = 1.3\ 10^4$ km for Model 1 and $Z_0 = 1.4\ 10^4$ km for Model 2 (paraboloid) (see text)

be either tangential discontinuities or slowly propagating rotational discontinuities with strongly differing plasma anisotropies on both sides (Hudson, 1970; Neubauer, 1970) required because of the large change in magnetic field magnitude. Since stationary tangential pressure balance structures like tangential discontinuities do not allow plasma transport through their structure they could occur at most intermittently as parts of nonstationary flow in the comet frame but not as standing boundaries. This is because the magnetic field observed in the pile-up region can only be due to transport of interplanetary magnetic field lines into the pile-up region. A rotational discontinuity allows such transport of plasma and magnetic flux. Hence the pile-up boundary could only be of the strongly anisotropic rotational type, if it is present for an extended time period as a standing boundary.

4. The ionopause

The situation seems more favorable for the ionopause. Table 1 shows that its normals are very well determined. Because of the suspicion that a remaining despin motor disturbance field may affect the minimum variance results we have removed this spacecraft field contribution particularly carefully using data from several spin rotations in the cavity. Only minor changes have resulted from this procedure thereby confirming the accuracy of the results. Both boundaries show a similar structure and similar thicknesses. Figure 3 indicates that the vector magnetic field variations just before the inbound ionopause and just after the outbound ionopause are very similar.

This is except for the observation that the outbound variations appear to be somewhat compressed compared with the inbound variations. Apart from the somewhat different scales on both sides the relations

$$B_X(t - t_E) \approx -B_X(t_E - t)\,, \tag{4a}$$

$$B_Y(t - t_E) \approx \quad B_Y(t_E - t)\,, \tag{4b}$$

$$B_Z(t - t_E) \approx \quad B_Z(t_E - t)\,, \tag{4c}$$

are fulfilled approximately, where t_E is the time of closest approach. In other words B_X, B_Y, and B_Z are approximately odd, even and even functions of time from closest approach, respectively. These observations are therefore fully consistent with theoretical expectations based on draping of field lines over the ionopause where the somewhat different scales are due to the somewhat different angles with respect to the incoming flow direction.

This evidence suggests that on the small scale of four second averages of Fig. 3 the magnetic structure around closest approach is much more symmetric than on larger scales shown in Fig. 1. Therefore the assumption of a simple surface like an ellipsoid or hyperboloid for the ionopause seems to be justified. However, it turns out that the fly-by trajectory with the normals observed at the entry and exit crossing points of the ionopause is not very favorable for the observations. The shape of the boundary in the $\pm X'$-direction is only weakly constrained by the data, as a sensitivity analysis on the errors in $\boldsymbol{n}_1$ and $\boldsymbol{n}_2$ shows. Hence we fit a sphere with unknown center $\boldsymbol{x}_0$ to the observations. We obtain

$$\begin{aligned} X_0 &= -\ 170\,\text{km} & X_0' &= -\ 110\,\text{km} \\ Y_0 &= 570\,\text{km} & Y_0' &= 580\,\text{km} \\ Z_0 &= 1350\,\text{km} & Z_0' &= 1350\,\text{km} \end{aligned} \tag{5}$$

radius $R = 4610$ km.

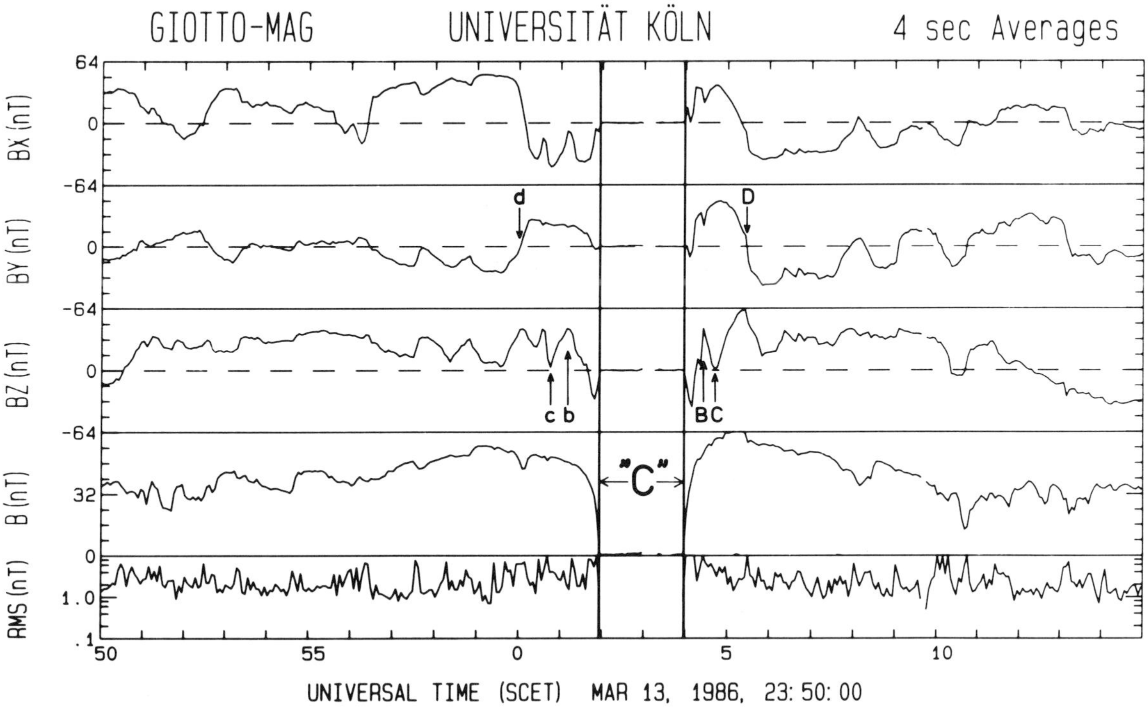

Fig. 3. Spin average vector magnetic field observations in Halley centered solar ecliptic coordinates from March 13, 23:50:00 UT to March 14, 00:15:00 UT. "C" denotes the magnetic field cavity region where the rms-values have been omitted. The letter combinations b, B, c, C and d, D point to the symmetry of the magnetic field

One could imagine that the true shape is described reasonably well by this sphere at positive and not too negative values of X' apart from the physical expectation that because of the anisotropy introduced by the magnetic field the cross-section at constant X' should not be circular.

The cross-section of the sphere in the plane $X'=0$ is shown in Fig. 4 together with the minimum variance normals. Table 2 gives the crossing points and the normals in the primed coordinate system.

The most striking feature of the solution given by Eq. (5) is the large value of the offset towards positive Z'-values. We could imagine several resons for deviations of the geometric center of the cavity from the cometary nucleus in the Y', Z'-plane:

– The neutral gas flow deviates from radial symmetry close to the nucleus because of the day-night-asymmetry, the distribution of sources and possible evaporation of volatiles from the dust particles of jets.

– Plasma sources not associated with solar UV-radiation may cause an asymmetry in the cavity plasma pressure although the large mean free path of hot electrons in the outer cavity combined with their high mobility not constrained by a magnetic field should smooth out local differences.

– Asymmetric pressure from the magnetized plasma region outside the ionopause may be important.

An alternative is an ionopause which is centered on average but distorted by ripples due to instabilities driven by velocity shear across the ionopause, a small density gradient across the ionopause, a small density gradient across the ionopause or the far effect of the density jump at about 10 000 km observed by the ion mass spectrometer experiment (Balsiger et al., 1986). These interesting possibilities have been investigated by Ershkovich and Flammer (1986). In this case the unusual normal directions are a consequence of ripples on the ionopause. In this interpretation the locations $\boldsymbol{x}'_1$ and $\boldsymbol{x}'_2$ in Table 2 are due to the average ionopause plus the displacement by the ripples. Unfortunately both could only be separated if a model of the undisturbed surface or the ripples existed.

If we assume that the unperturbed surface is circular at any X' the projections of $\boldsymbol{n}'_2$ and $\boldsymbol{n}'_2$ on the Y', Z'-plane are most diagnostic for the boundary ripples. These normal projections should be aligned with the vectors $\boldsymbol{x}'_\perp=(0, Y', Z')$ in case of no ripples. The projections of the observed normals on the Y', Z' plane deviate by 6°6 and 28°2 from these idealized directions for the inbound and outbound crossings, respectively. Figure 4 illustrates the geometry. Note that the normals $\boldsymbol{n}'_1$ and $\boldsymbol{n}'_2$ deviate by 9°1 and 3°9 only, respectively, from the Y', Z'-plane.

In this interpretation we conclude that the ripples on the ionopause must be approximately aligned with the aberrated solar wind direction. The good planarity of the magnetic field variations in the ionopause fine structure suggests a wavelength of more than 200 km on the boundary compared with a circumference of 27 000 km.

Finally we note that the very good planarity of the ionopause structure also suggests that the theory of one-dimensional magnetoplasma structures and discontinuities is fully applicable. In this theory a discontinuity with the large contrast in magnetic field magnitude between both sides can only be a tangential discontinuity characterized by $B_n=0$. Since the normals $\boldsymbol{n}_1$ and $\boldsymbol{n}_2$ are approximately aligned with the spin-axis direction, this result indicates a zero magnetic field component in the spin-axis direction. This is important, because this component is measured least accurately in an absolute sense. This is then additional confirmation that the magnetic field must truly be zero in the cavity.

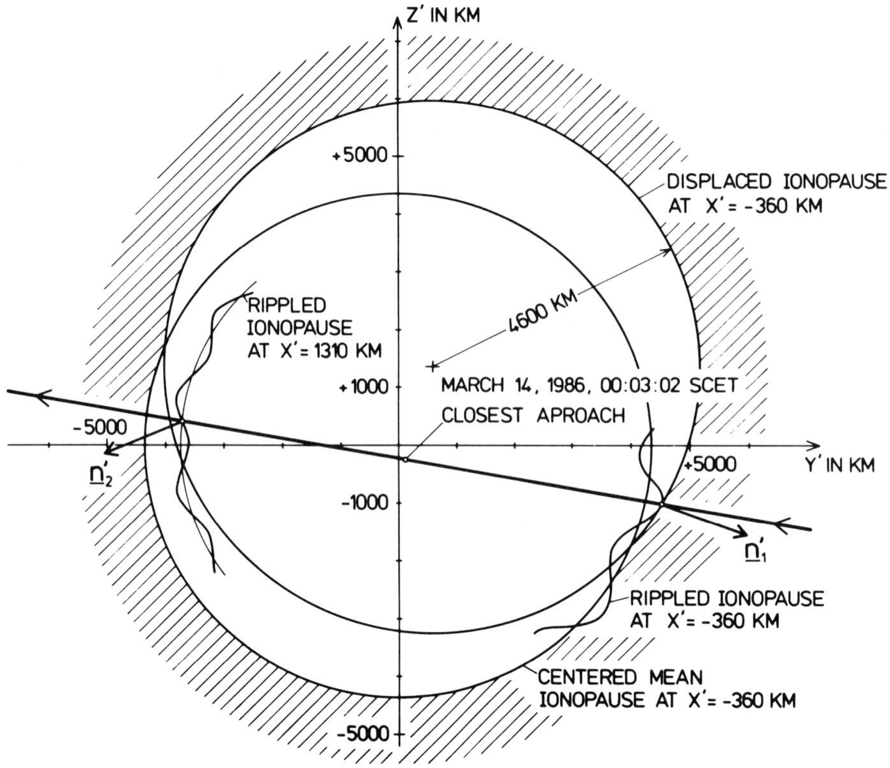

Fig. 4. Giotto encounter trajectory projected on Halley centered solar ecliptic Y', Z'-plane. The ionopause crossing points are shown together with the boundary normals. A best fit sphere displaced from the origin is shown for $X'=0$ under the assumption of ionopause smoothness. Also shown is the centered sphere fitting the observations at $X'=0$ and $X'=1310$ km with ripples on the ionopause indicated

Table 2. Ionopause crossings in aberrated coordinate system

Inbound:	$\boldsymbol{x}_1' = (-0.36,\ 4.53,\ -1.03)\ 10^3$ km
	$\boldsymbol{n}_1' = (0.159,\ 0.931,\ -0.329)$
Outbound:	$\boldsymbol{x}_2' = (1.31,\ -3.69,\ 0.40)\ 10^3$ km
	$\boldsymbol{n}_2' = (0.673,\ -0.926,\ -0.373)$

5. Conclusions

Using the magnetic field data obtained by the Giotto magnetometer experiment during the encounter at comet Halley on March 13/14, 1986, we have investigated the most conspicuous boundaries in the inner plasma region surrounding comet Halley i.e. the pile-up boundary and the ionopause. The normal vectors at the boundary crossings have been determined using the minimum variance technique. In particular, the magnetic pile-up boundary which will probably become the tail boundary far downstream can be described by a paraboloid along the direction of the aberrated solar wind but somewhat offset from the line through the nucleus towards the $-Y_{\mathrm{HSE}}$ direction is strong flaring is allowed. The stand-off distance is 215 000 km. For no flaring i.e. a half-sphere cylinder shape the stand-off distance is 143 000 km. The results on the ionopause surface are inconsistent with a boundary, which has simple symmetry properties. For example, a centered elliptical cross-section is not possible in planes perpendicular to the aberrated solar wind direction. From these observations alone the surface must be characterized by an undeterminate combination of a surface displaced from the origin and superimposed ripples. If no ripples are assumed to be present, it could be characterized by a sphere of radius 4600 km displaced mainly towards the $+Z_{\mathrm{HSE}}$-direction by about $+1500$ km. If only ripples are assumed to explain the observations, they must be aligned with the aberrated solar wind direction and have wavelengths greater than 200 km on the ionopause. The choice between these two extreme cases could come from stability arguments.

In the future the magnetic field structure near the boundaries will be used to draw further conclusions. In addition the interesting fine structure of the boundaries will be investigated.

Acknowledgements. We gratefully acknowledge the contributions by the members of the Giotto magnetometer team and the Giotto project team. This work was supported financially by the Bundesministerium für Forschung und Technologie.

References

Balsiger, H., Altwegg, K., Bühler, F., Geiss, J., Ghielmetti, A.G., Goldstein, B.E., Goldstein, R., Huntress, W.T., Ip, W.H., Lazarus, A.J., Meier, A., Neugebauer, M., Rettenmund, U., Rosenbauer, H., Schwenn, R., Sharp, R.D., Shelley, E.G., Ungstrup, E., Young, D.T.: 1986, *Nature* **321**, 330

Ershkovich, A.I., Flammer, K.R.: 1986, *EOS* **67**, 1144

Hudson, P.S.: 1970, *Planet. Space Sci.* **18**, 1161

Johnstone, A., Coates, A., Kellock, S., Wilken, B., Jockers, K., Rosenbauer, H., Stüdemann, W., Weiss, W., Formisano, V., Amata, E., Cerulli-Irelli, R., Dobrowolny, M., Terenzi, R., Egidi, A., Borg, H., Hultquist, B., Winnigham, J., Gurgiolo, C., Bryant, D., Edwards, T., Feldman, W., Thomsen, M., Wallis, M.K., Biermann, L., Schmidt, H., Lüst, R., Haerendel, G., Paschmann, G.: 1986, *Nature* **321**, 344

Neubauer, F.M.: 1970, *Z. Phys.* **237**, 205

Neubauer, F.M., Acuña, M.H., Burlaga, L.F., Franke, B., Gramkow, B., Mariani, F., Musmann, G., Ness, N.F., Schmidt, H.U., Terenzi, T., Ungstrup, E., Wallis, M.: 1986a, *ESA* SP-1077, p. 1

Neubauer, F.M., Glassmeier, K.H., Pohl, M., Raeder, J., Acuña, M.H., Burlaga, L.F., Ness, N.F., Musmann, G., Mariani, F., Wallis, M.K., Ungstrup, E., Schmidt, H.U.: 1986b, *Nature* **321**, 352

Neubauer, F.M.: 1986c, 20th ESLAB Symp., ESA SP-250, Vol. **I**, p. 35

Neubauer, F.M., Acuña, M.H., Burlaga, L.F., Franke, B., Gramkow, B., Mariani, F., Musmann, G., Ness, N.F., Schmidt, H.U., Terenzi, R., Ungstrup, E., Wallis, M.: 1987, accepted by J. Physics *E*

Raeder, J., Ness, N., Neubauer, F.M.: 1987 (this volume)

Reinhard, R.: 1986, *Nature* **321**, 313

Rème, H., Sauvaud, J.A., d'Uston, C., Cotin, F., Cros, A., Anderson, K.A., Carlson, C.W., Curtis, D.W., Lin, R.P., Mendis, D.A., Korth, A., Richter, A.K.: 1986, *Nature* **321**, 349

Riedler, W., Schwingenschuh, K., Yeroshenko, Y.G., Styashkin, V.A., Russell, C.T.: 1986, *Nature* **321**, 288

Schwingenschuh, K., Riedler, W., Schelch, G., Yeroshenko, Y.G., Styashkin, V.A., Luhmann, J.G., Russell, C.T., Fedder, J.A.: 1986, *Adv. Space Res.* (in press)

Slavin, J.A., Smith, E.J., Tsurutani, B.T., Siscoe, G.L., Jones, D.E., Mendis, D.A.: 1986, *Geophys. Res. Letters* **13**, 3, 282

Sonnerup, B.U., Cahill, L.J.: 1967, *J. Geophys. Res.* **72**, 171

Astron. Astrophys. 187, 80–82 (1987)

ASTRONOMY AND ASTROPHYSICS

Identification of boundaries in the cometary environment from ac electric field measurements

M. Mogilevsky[1], **Y. Mikhailov**[1], **O. Molchanov**[1], **R. Grard**[2], **A. Pedersen**[2], **J.G. Trotignon**[3], **C. Béghin**[3], **V. Formisano**[4], **V. Shapiro**[5], **and V. Shevchenko**[5]

[1] IZMIRAN, 142092 Troitsk, USSR

[2] Space Science Department, ESA/ESTEC, Keplerlaan, 2200 AG, Noordwijk, The Netherlands

[3] Laboratoire de Physique et de Chimie de l'Environnement, CNRS, F-45045 Orléans, France

[4] Istituto Plasma Spazio, CNR, 00044 Frascati, Italy

[5] Space Research Institute, Profsoyuznaya 84/32, 117810 Moscow, GSP-7, USSR

Received February 18, accepted April 30, 1987

Summary. Electric fields are measured with the APV-V experiment in the frequency range 8 Hz–300 kHz. The field amplitude increases significantly, first at a distance of $2\,10^5$ km, then at distances of $1.2–1.5\,10^5$ km, and $5–7\,10^4$ km from the nucleus. These phenomena have been observed both on Vega-1 and Vega-2. The electric field measurements are compared with data obtained from dust and plasma experiments; possible mechanisms responsible for the existence of these boundaries are discussed.

Key words: comet Halley – Vega – electric field – boundaries – cometopause – plasma

1. Introduction

The plasma wave analyser, APV-V carried by the Vega-1 and Vega-2 spacecraft measured ac electric fields in the frequency range 8 Hz–300 kHz during the encounters with comet Halley. As this kind of measurements was performed in a dusty environment of a comet for the first time, the following questions arise:

(1) How are the measurements influenced by dust?

(2) Can we distinguish natural wave phenomena from signals induced by dust impacts?

Several regions have already been tentatively identified in the cometary environment on the basis of plasma wave data (Grard et al., 1986a); the purpose of this paper is to investigate in more detail these instabilities and to understand the mechanism responsible for their spectral characteristics.

2. Experimental results

The measurements of ac electric fields by the APV-V experiment were performed with a parallel spectral analyzer which consisted of 16 filters; the experiment is described elsewhere (Sagdeev et al., 1985; Grard, 1986).

The signal levels observed on Vega-1 and Vega-2 are in general similar (Fig. 1a,b). We shall therefore consider the Vega-2 measurements (Fig. 1b) as typical and will follow their evolution up to the point of closest approach. In Figs. 1a,b we present some filter bank data obtained during the encounter. The original data have time resolution of 1 ps^{-1} (Parameter A) and 2 ps^{-1} (Parameters H-P). This data were averaged over 64 seconds for Vega-1 and 32 seconds for Vega-2. The statistical significance of the averaged data points is better than 10%.

Ignoring small scale fluctuations, we observe a regular increase of amplitude in all filters from a cometocentric distance of $2\,10^5$ km (Grard et al., 1986a, 1986b). The signal grows more rapidly at several boundaries. The first gradient is found at a distance of $\sim 1.6\,10^5$ km from the nucleus for frequencies ~ 100 Hz. The second gradient appears in the band 30 Hz–4 kHz at a distance of $\sim 1.3\,10^5$ km, and it is the most pronounced in the range 0.8–1.8 kHz. The third gradient is visible in the range 0.4–200 kHz at a distance of $7.0\,10^4$ km and is best defined between 8 and 36 kHz. The fourth gradient is observed only in the higher frequency range, around 200 kHz, at a distance of $\sim 2.0\,10^4$ km.

The shape of the spectrum evolves during the approach towards the nucleus. Figure 2 shows examples of electric field spectra at various distances from the nucleus. At distances larger than $1.6\,10^5$ km, the spectral amplitude above 1 kHz drops monotonically with increasing frequency f as $f^{-\alpha}$, with $1 < \alpha < 2$. Two spectral maxima are observed at frequencies of the order of 30 Hz and 200–400 Hz, in the spatial range $7.0\,10^4–1.5\,10^4$ km and $1.6\,10^5–1.0\,10^4$ km, respectively.

Two classes of layers, or boundaries, can therefore be differentiated within the cometary environment:

a) Transitions where strong signal gradients are detected (four such boundaries are observed on Vega-2; the second and third ones are also clearly seen on Vega-1 but closer to the nucleus).

b) Layers where maxima in spectral intensity are observed for frequencies of 30 Hz and 200–400 Hz (these regions are located in the vicinity of the previous boundaries but do not coincide with them).

It has already been reported in the literature that the impacts of dust particles with probes and spacecraft give rise to an intense electric signal in a broad frequency band. This effect has been used, for example, to investigate the dust distribution in the

Send offprint requests to: M. Mogilevsky

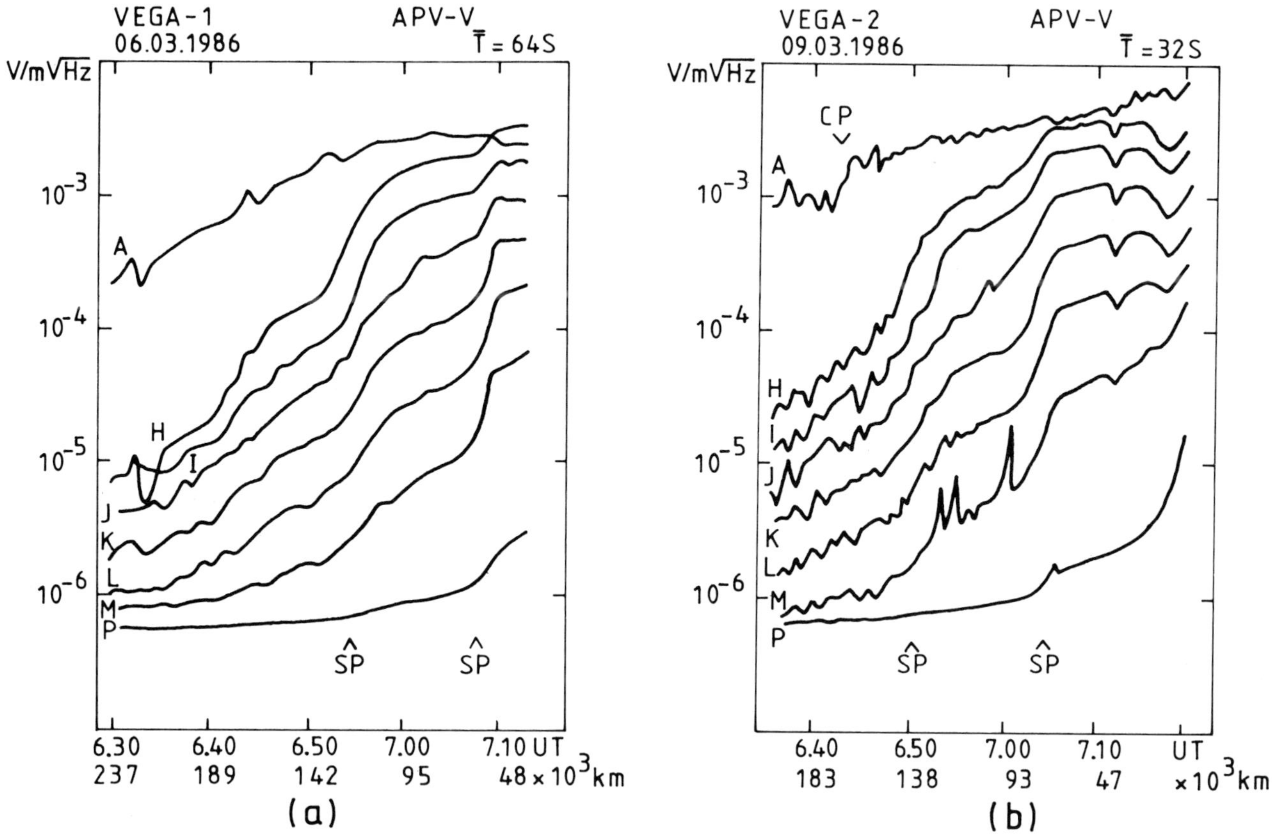

Fig. 1a and b. Electric field intensities measured by the filter banks on Vega-1 (64 s average) and Vega-2 (32 s average). The central frequencies of the filters are: A–8 Hz, H–869 Hz, I–1.8 kHz, J–3.7 kHz, K–7.7 kHz, L–15 kHz, M–36 kHz, P–203 kHz. The labels SP indicate the boundaries detected by the SP-1 experiment; the label CP gives the position of the cometopause

environment of Giacobini-Zinner (Scarf et al., 1986). The signal received by the APV-V antenna has been compared with the dust flux measured by the SP-1 experiment (Vaisberg et al., 1986). The locations of count rate gradients given by this instrument are labelled SP in Fig. 1; they coincide reasonably well with the boundaries where the APV-V electric signal increases. It should be mentioned, however, that (a) the impact rate increases by two orders of magnitude between 1.3 10^5 and 0.7 10^5 km, whereas the frequency of the spectral maxima does not vary by more than a factor of 2.5 and (b) the dust impact rate increases at 1.5 10^4 km while the amplitude of the spectral maxima decreases.

We can therefore tentatively conclude that the increase of the electric signal above 1 kHz is probably governed by the interaction between the spacecraft and the dust environment while the existence of maxima in the spectrum is due to some other reason which is not connected with the dust.

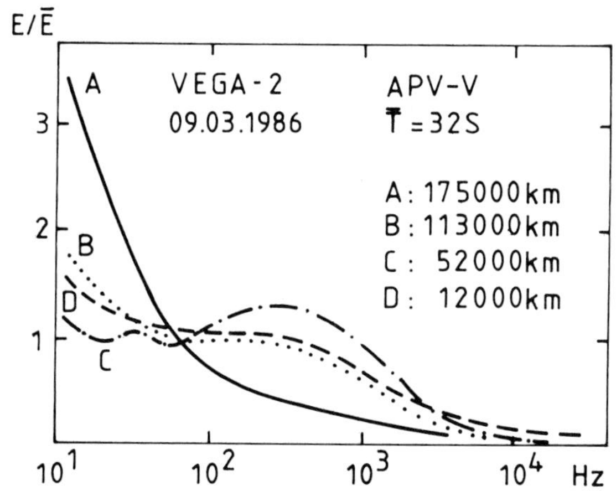

Fig. 2. Electric field spectra (32 s average) measured on Vega-2 at various distances from the nucleus. The amplitude of the electric field component is normalized to the average amplitude $\bar{E} = \sum_{i=1}^{n} E_i$ where n is the number of filters

The second spectral peak shifts from one filter to the next, namely from the 150–300 Hz to the 300–600 Hz channel when the quasistatic magnetic field measured by the MISCHA instrument increases by a factor of 2.5 to 3 (Riedler et al., 1986). One can therefore assume, within the limitations imposed by the available spectral resolution, that the peak frequency is approximately proportional to the ambient magnetic field.

Finally, comparing the APV-V and PLASMAG results, it is found that the discontinuity labelled CP in Fig. 1b is related to the crossing of the cometopause (Gringauz et al., 1986a, 1986b).

3. Discussion

In order to interpret the wave emission seen in the frequency range 200–400 Hz, we suggest a mechanism which has already been proposed to explain phenomena observed at the front of the Earth's bow shock, namely the "fan" instability (Vaisberg et al., 1983).

A fraction of the downstreaming solar wind ion flux is decelerated and reflected by the cometary plasma, which leads to the excitation of electrostatic oscillations at frequencies around the lower hybrid frequency (30 Hz). Part of the instability energy is then absorbed by electrons, which results in the formation of an anisotropic suprathermal tail with an energy range of the order of a few keV. The collisionless relaxation of the anisotropic velocity distribution of the suprathermal electrons finally gives rise to oscillations with frequencies several times lower than the electron cyclotron frequency (200–400 Hz). These findings are corroborated in part by the APV-N experimenters (Klimov et al., 1986) which also report instabilities around the lower hybrid frequency in a region of space where a spectral maximum at 200–400 Hz is seen by APV-V.

Making use of a result given by Vaisberg et al. (1983), the frequency in the second spectral maximum is estimated to be:

$$\omega_{max} = \frac{1}{4}\,\omega_{ce}\,\frac{v_{min} f_0[\varepsilon_e^{1/2}/m_e^{1/2}]}{\int dv_{||} f_0(v_{||})} \cong \frac{v_{min}\omega_{ce}}{4(\varepsilon_e/m_e)^{1/2}},$$

where ω_{ce} is the electron cyclotron angular frequency, $v_{min} = 2\omega_{ce}/\bar{k}$ is the minimum electron resonant velocity, f_0 is the electron distribution function, ε_e is the mean kinetic energy of the suprathermal electrons, m_e is the electron mass, $v_{||}$ is the projection of the velocity vector along the magnetic field and $\bar{k}$ is the wave vector.

With reference to the results obtained with the PLASMAG experiment (Gringauz et al., 1986a, 1986b) we shall assume that electrons with energies of a few keV are observed at distances of $1.5\,10^4$ km from the nucleus. Substituting this value for the suprathermal electron energy and taking $f_{ce} = \omega_{ce}/2\pi = 1400$ Hz, we obtain $f_{max} = \omega_{max}/2\pi \cong 300$ Hz, in fair agreement with the measurements. The electric field amplitude at the frequency of the emission maximum E, is given by

$$\left\langle \frac{E^2}{8\pi} \right\rangle \sim n_{Te}\varepsilon_e(\omega_{ce}/\omega_{pe})^2(\omega_{max}/\omega_{ce}),$$

where n_{Te} is the mean density of suprathermal electrons and ω_{pe} is the electron plasma angular frequency. Using the suprathermal electron density given by Gringauz et al. (1986), we obtain $E \sim 0.3$–$1.0\,\mathrm{mV\,m^{-1}\,Hz^{-1/2}}$, which is also in satisfactory agreement with the APV-V experimental results.

Acknowledgements. The authors thank D. Klinge, A. Butler, K. Hjortnaes, J.-F. Karczewski, J.-M. Moreau, B. Novikov and his team for technical support and A. Dyatchkov for data processing. Thanks are due to A. Galeev, K. Gringauz, E. Eroshenko, O. Vaisberg, S. Klimov and S. Savin for useful discussions.

References

Grard, R.: 1986, *Field, Particle and Wave-Experiments on Cometary Missions*, K. Schwingenschuh, W. Riedler, eds., Technical University of Graz, p. 131

Grard, R., Pedersem, A., Knott, K., Trotignon, J.G., Beghin, C., Formisano, V., Mogilevesky, M., Mikhailov, Y.: 1986, *Adv. Space Res.* **5**, 175

Grard, R. Pedersen, A., Trotignon, J.G. Beghin, C., Mogilevsky, M., Mikhailov, Y., Molchanov, O., Formisano, V.: 1986b, *Nature* **321**, 290

Gringauz, K.I., Gombosi, T.I., Remizov, A.P. et al.: 1986a, *Nature* **321**, 282

Gringauz, K.I., Gombosi, T.I., Tatrallyay, M. et al.: 1986b, *Geophys. Res. Letters* **13**, 613

Klimov, S., Savin, S., Aleksevich, Y. et al.: 1986, *Nature* **321**, 292

Riedler, W., Schwingenschuh, K., Eroshenko, E. et al.: 1986, *Nature* **321**, 288

Sagdeev, R.Z. et al.: 1985, *Venus-Halley Mission*, Louis-Jean, Gap, p. 127

Scarf, F.L., Coroniti, F.V., Kennel, C.F., Gurnett, D.A., Ip, W.-H., Smith, C.T.: 1986, *Science* **232**, 377

Vaisberg, O.L., Galeev, A.A., Zastenker, G.N. et al.: 1983, *Zh. Eksp. Teor. Fiz.* **85**, 1232

Vaisberg, O.L., Smirnov, V.N., Gorn, L.S., Iovlev, M.V., Balikhin, M.A., Klimov, S.I., Savin, S.P., Shapiro, V.D., Shevchenko, V.I.: 1986, *Nature* **321**, 274

Astron. Astrophys. 187, 83–88 (1987)

ASTRONOMY AND ASTROPHYSICS

Dust observations of comet P/Halley by the plasma-wave analyser

J. G. Trotignon [1], C. Béghin [1], R. Grard [2], A. Pedersen [2], V. Formisano [3], M. Mogilevsky [4], and Y. Mikhailov [4]

[1] Laboratoire de Physique et Chimie de l'Environnement, CNRS, 3 A Avenue de la Recherche Scientifique, F-45071 Orléans Cédex 02, France
[2] Space Science Department, ESA, ESTEC, Postbus 299, NL-2220 AG Noordwijk, The Netherlands
[3] Istituto Plasma Spazio, CNR, CP 27, I-00044 Frascati (Roma), Italy
[4] IZMIRAN, SU-142092 Troitsk, Moscow Region, USSR

Received February 12, accepted May 11, 1987

Summary. This paper reports the electric field observations made with the high-frequency plasma-wave analyser, during the Vega spacecraft encounters with comet Halley on 1986, March 6 and 9. Electric fields are measured in the bandwidth 0–300 kHz with an antenna made of two spheres separated by a distance of 11 m. As the nucleus is approached, the average amplitude and bandwidth of the AC electric field increase dramatically; at frequencies above 1 kHz, the power law index of the electric field spectrum is about $-3.4 (E^2 \sim f^{-3.4})$, a value close to those reported elsewhere for random electric pulses induced by dust particles. Moreover, a strong correlation is found between spikes in the quasi-static electric field and small-scale variations in dust count rates for masses less than 10^{-12} g. Finally we identify several large scale gradients in the AC electric field intensity within 140,000 km from the nucleus which are believed to be associated with enhancements of dust jet activity. The locations of these gradients are used to estimate the apparent rotation period of the cometary nucleus and the radial velocity of the small dust particles.

Key words: comet Halley – Vega 1 – Vega 2 – electrostatic noise – dust particle impacts – electric field intensity gradients

1. Introduction

Although the high frequency plasma wave (APV-V) experiment on the Vega spacecraft has been designed to measure wave-electric fields and to study the plasma in the environment of comet Halley, the preliminary results of which are reported in Grard et al. (1985, 1986b), Pedersen et al. (1986) and Mogilevski et al. (1986), it also reacted to impacts of dust particles. It is known that intense wide band noise due to small micron-sized dust particles hitting the antennas and/or the spacecraft has been also detected by the Voyager plasma and radio wave experiments in the ring planes of Saturn and Uranus (Aubier et al., 1983; Gurnett et al., 1983; Meyer-Vernet et al., 1986b) and by ICE (International Cometary Explorer) during the flyby of comet Giacobini-Zinner (Gurnett et al., 1986; Meyer-Vernet et al., 1986a; Scarf et al., 1986). In this paper, we present the electric field data recorded with the high-frequency plasma-wave analyser flown on board Vega 1 and Vega 2 during the comet Halley flybys on 1986, March 6 and 9. We will restrict our study to the data collected within 140,000 km from the nucleus, where the number of dust grains increases substantially. The electric field was measured in the frequency range from DC up to 300 kHz using two spheres, 10 cm in diameter, which were mounted at the tips of two carbon fibre booms attached to the outer solar panels; the two spheres were separated by a distance of 11 m. The electronics box was mounted on the outer part of the spacecraft body (Grard et al., 1986a, b).

Send offprint requests to: J. G. Trotignon

2. Small scale features and correlation with dust flux

2.1. *Difficulties encountered in interpreting the spectral characteristics in terms of plasma waves*

In Fig. 1, we have plotted the Vega 2 equivalent electric field spectra averaged over 2-min intervals against universal time during 40 min preceding closest approach. The intensity and bandwidth of the electric field increase dramatically in the vicinity of the nucleus. The peak electric field spectral intensity is of the order of a few mV m^{-1} Hz$^{-1/2}$ and the high-frequency part of the spectrum from about 1 kHz up to 300 kHz decreases monotonically with a power law index of about $-3.4 (E^2 \sim f^{-3.4})$. The part of the spectrum above the electron cyclotron frequency can hardly be interpreted in terms of plasma waves. As a matter of fact, electromagnetic waves, wherever they are generated, cannot propagate when their frequency lies between the electron cyclotron frequency f_{ce} and the plasma frequency f_p, i.e. here approximately in the range 1–100 kHz (Grard et al., 1985).

We can also rule out the observation of highly Doppler-shifted ion-acoustic turbulence in the same frequency range. Indeed, the highest possible frequency for such electrostatic waves is the lower hybrid frequency, f_{LHR}. The maximum value of the wave vector $\boldsymbol{k}$ is approximately given by:

$$\boldsymbol{k}_{max} \sim 2\pi f_{LHR}/V_{Ti},$$

where V_{Ti} is the ion thermal velocity and:

$$f_{LHR} \sim f_{ce}/M_i^{1/2},$$

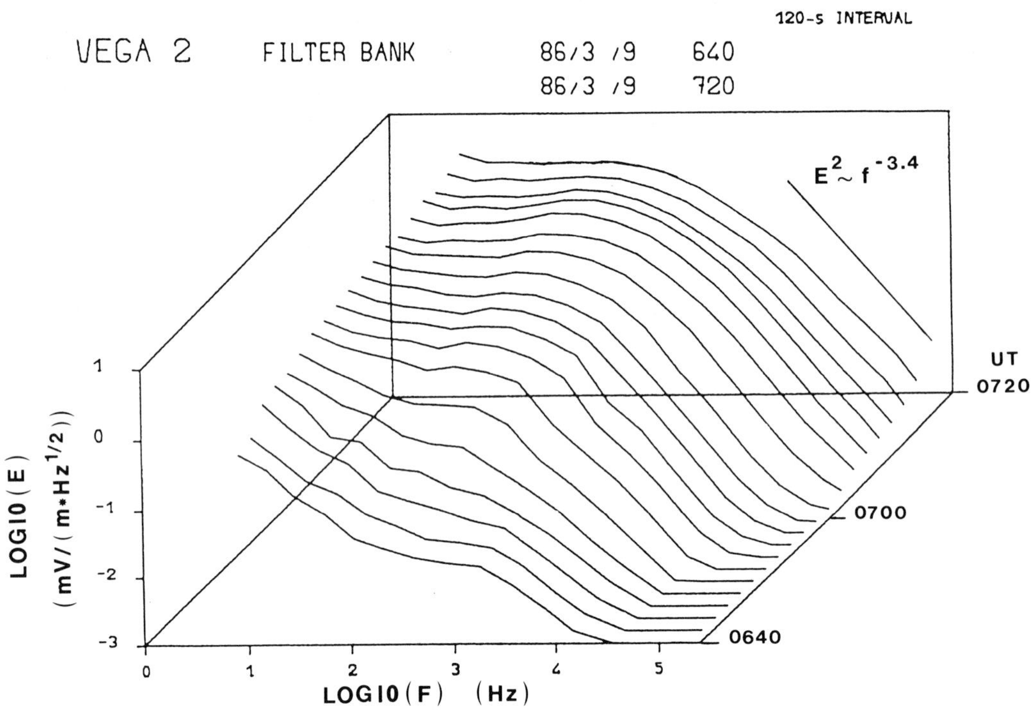

Fig. 1. Equivalent electric field spectra averaged over 2 min intervals measured by Vega 2

M_i being the mean ion/electron mass ratio. Then the highest Doppler-shifted frequency of ion-acoustic turbulence in the spacecraft frame of reference is:

$f_{max} \sim f_{LHR} + \boldsymbol{k}_{max} V_s/2\pi = f_{LHR}\ (1 + V_s/V_{Ti})$

Considering essentially protons with a temperature within the range $10^5 - 4\ 10^3$ K (Gringauz et al., 1986) and a spacecraft velocity of about 80 km s^{-1}, we get $f_{max} \sim 90$ to 350 Hz.

Could we observe electron electrostatic waves with frequencies larger than 1 kHz? In this frequency range, $f_{ce} < f < f_p$, the only undamped electron modes are Bernstein waves propagating perpendicularly to the B-field, with frequencies around the harmonics of f_{ce} and small $k_\perp$ values. Thus, such waves would be observed, as usual, around the 1st, 2nd, and possibly 3rd harmonics of f_{ce}, but not with a monotonically decreasing power law up to $\sim 100\, f_{ce}$.

Another explanation is therefore required.

2.2. *Correlation of dust flux with quasi-static and AC electric fields*

The upper part of Fig. 2 shows the quasi-static electric field recorded by Vega 2 in the bandwidth 0–0.5 Hz as a function of universal time. In the lower panel, the dust count rates simultaneously measured by the Dust Counter and Mass Analyser (DUCMA) are averaged over 10 s intervals and plotted against time relative to closest approach (The latter plot is taken from Simpson et al., 1986). The correlation between the spikes of the quasi-static signal and the flux enhancements of dust grains with mass m is much more marked for $m \geqq 1.5\ 10^{-13}$ g (curve M1) than for $m \geqq 9\ 10^{-13}$ g (curve M2); in other words, if these events are associated with dust impacts, the particles which appear to play the most significant role in out detector are those with masses $m < 10^{-12}$ g.

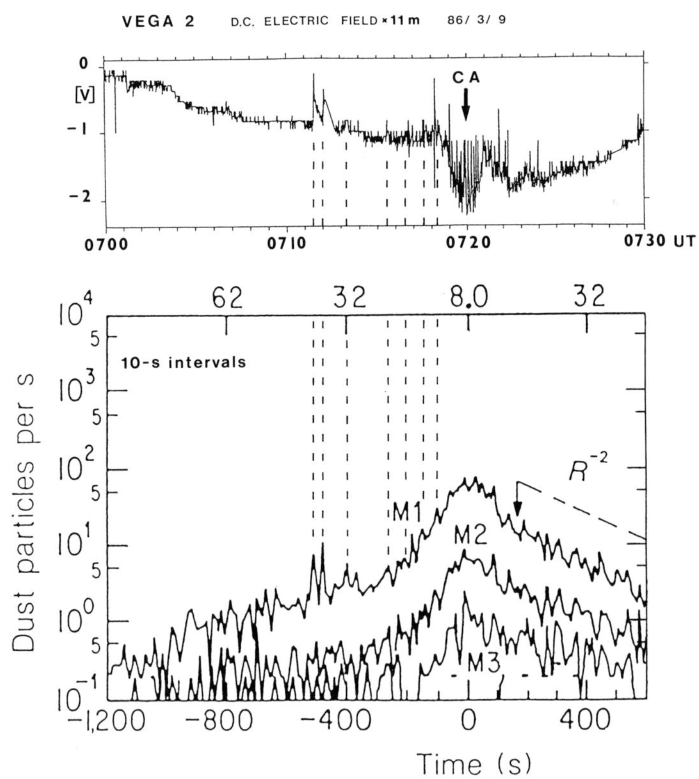

Fig. 2. The upper curve shows the quasi-static signal between 0700 and 0730 UT on March 9, 1986. The lower curve is borrowed from Simpson et al. (1986) and shows the dust count rates of the DUCMA experiment for several mass thresholds: M1, $m \geqq 1.5\ 10^{-13}$ g; M2, $m \geqq 9\ 10^{-13}$ g; M3, $m \geqq 9\ 10^{-12}$ g

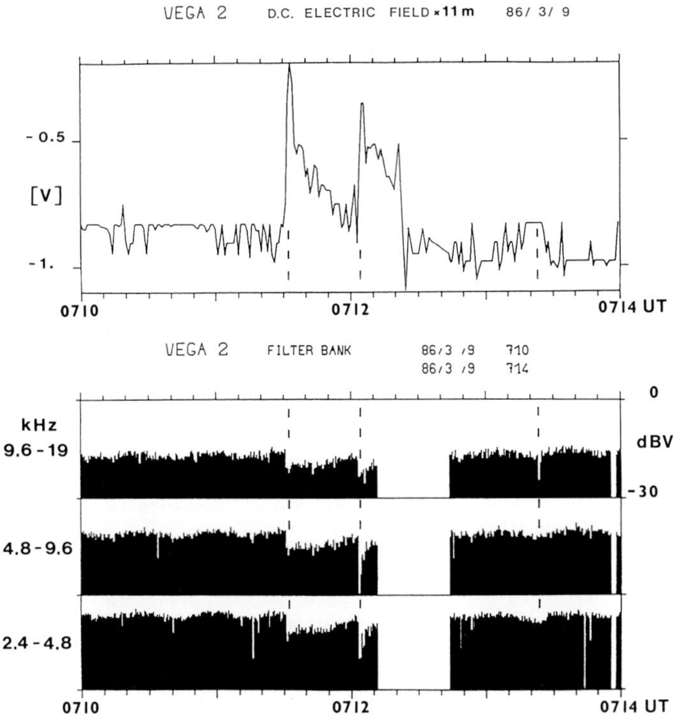

Fig. 3. Quasi-static signal (top) and noise levels within 3 frequency channels (bottom) measured by Vega 2 between 0710 and 0714 UT; 0 dB corresponds to 1 V rms. There is a data gap on the lower panel from 0712:11 to 0712:44

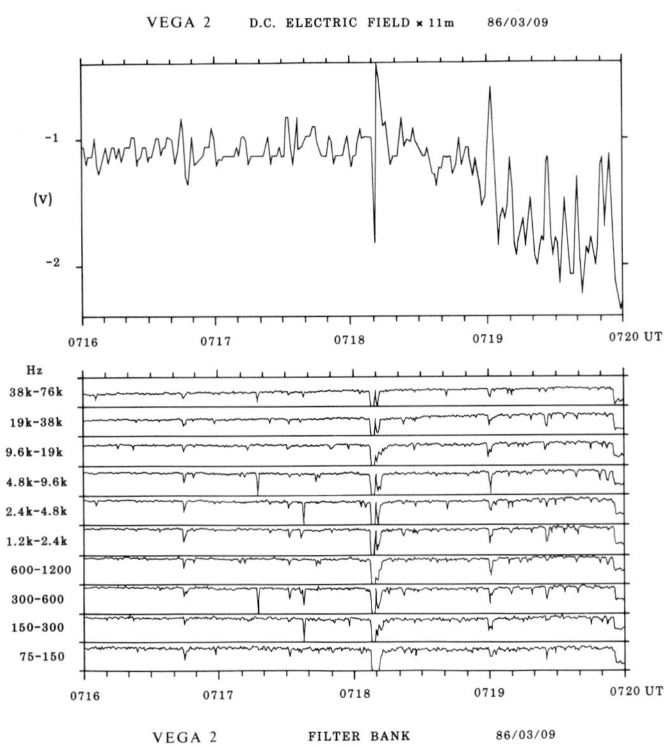

Fig. 4. Quasi-static signal (top) and noise intensities in 10 frequency channels (bottom) measured by Vega 2 within 4 min from the closest approach. In the lower panel, the ordinate axis of each plot covers the dynamic range from -50 to 0 dB

Let us now focus our attention on the spikes indicated by the first three dashed lines which have been interpreted by Galeev et al. (1986) as critical ionization velocity phenomena. These events are expanded further in Fig. 3. The quasi-static signal is plotted at the top in the time interval 0710–0714 UT (sampling rate 1 Hz), and the noise levels measured by three of the 16 filters are shown at the bottom (sampling rate 2 Hz). The frequency bandwidth of each filter is given on the left of each plot and the ordinate axis covers the dynamic range from -30 to 0 dB, 0 dB corresponding to 1 V rms. We observe that spikes on the quasi-static electric field are associated with dips in the AC electric field amplitude. Such features occur only within about 50,000 km from the nucleus, i.e. when the filter-bank signals have nearly reached their maximum levels (see Fig. 6), which suggests some saturation effect. Another example is shown in Fig. 4; note that the dips are observed over a wider frequency range, typically from 100 Hz to 100 kHz, and that they recur more frequently as the spacecraft approaches the nucleus.

2.3. Discussion of the observed correlations

As shown in Sect. 2.1 the power law index of the equivalent electric field spectrum is about -3.4 for frequencies in the range 1–300 kHz. Similar values have been reported elsewhere for random electric pulses induced by dust particles. For example, Aubier et al. (1983) and Meyer-Vernet et al. (1986b) have found a power law index of -4 in calculating the shot noise due to dust grain impacts on the monopole antennas and spacecraft when Voyager 2 went through the ring planes of Saturn and Uranus. Moreover, those authors and others (Gurnett et al., 1983; Meyer-Vernet et al., 1986a) showed that among the processes which could give rise to the intense electrostatic noise levels recorded by Voyager near the Saturnian ring plane or by ICE near comet Giacobini-Zinner, the most effective could be the impacts of small micron-sized dust particles hitting the antennas and/or the spacecraft body. Similar computations would be necessary to estimate the noise levels produced by neutral gas, taking into account both the geometry of the antennas and of the spacecraft, but they do not fall within the scope of this paper. We can however assume with a reasonable degree of confidence that a large fraction of the noise measured by the APV-V experiment is produced by dust grain impacts. The high velocity of the spacecraft V is an additional feature which supports this assumption since, the charge generated by each dust impact on the antennas and spacecraft is proportional to $V^{3.5}$ (Fechtig et al., 1978).

If we assume that the DC and AC electric field measurements are influenced by dust impacts, as suggested by the aforementioned correlations, we have to answer two important questions: (1) how can the bursts of quasi-static signal (DC component) be produced by the small-scale dust flux enhancements, with always the same polarity, and (2) why, after the dust flux has reached some threshold, is a spike in DC signal often associated with a dip in AC signal instead of an expected burst? At present, we have no definite answer to those questions and further work is needed, involving comparisons with other sources of data as well as a modelling of sensor behaviour under dust impact.

Impact ionization effects have already been studied by several authors (for a review, see Fechtig et al., 1978); a voltage pulse is induced whenever a dust particle strikes the dipole antenna with a high relative velocity, and the spectrum of the electrostatic shot noise due to grain impacts has been calculated assuming a random series of uncorrelated pulses (Aubier et al., 1983). As for conventional linear systems, the power spectrum is found to be proportional to the flux rate (for frequencies different from zero). But, it is also known that the DC part of the power spectrum for a series of randomly distributed pulses is proportional to the square of the count rate (Stern et al., 1967). Then, as the dynamic antenna response to dust impacts depends on the DC antenna voltage, the hypothesis of independant events ceases to be valid. Moreover a local enhancement of plasma density caused by dust impacts can also modify the conductivity of the spacecraft environment or perturbs the electric impedance of the sensors. Consequently, we anticipate that a nonlinear process takes place when the events come to overlap each other. This occurs as soon as the average dust count rate reaches the equivalent bandwidth, Δf, of a single impact signal, i.e. when

$$n V s \simeq \Delta f,$$

where n is the dust number density, V the relative velocity of the flow ($V \sim 80\,\mathrm{km\,s^{-1}}$) and s the area of the collecting surface ($s = 157\,\mathrm{cm}^2$ for the two spheres). For example, if we take $\Delta f = 10\,\mathrm{kHz}$ (from Fig. 1) we obtain a dust number density of about $8\,\mathrm{m}^{-3}$ which is close to the peak value measured by one of the dust experiment (Simpson et al., 1986) for grains with masses $m > 10^{-13}\,\mathrm{g}$.

We will now focus our attention on the large-scale features of the AC electric field.

3. Large-scale features associated with dust jet activity

Figure 5 shows the average equivalent electric field spectral intensity measured in three channels during successive 32-s intervals as functions of the distance of Vega 1 spacecraft from the nucleus; the Vega 1 experiment gave data up to a distance of 33,300 km from the nucleus. Two distinct regions, respectively located at ~60,000 and 120,000 km from the nucleus, exhibit gradients significantly larger than the average which corresponds roughly to an R^{-2} profile, as expected for an isotropic constant velocity radial outflow. Comparable plots for Vega 2 are shown in Fig. 6; three steep gradients are identified and are also 60,000 km apart. The two most distant ones have already been reported by Vaisberg et al. (1986) for particles with masses in the range 10^{-15}–10^{-16} g; they are indicated by the vertical lines labelled a and b in Fig. 1 of their paper (see also Mogilevsky et al., 1986). The gradient closest to the nucleus is also present in the SP-1 detector measurements (it lies between letters E and F in the same figure and is more strongly marked for lighter particles). In the AC electric field data the slope of this gradient appears to be negative rather than positive, for the threshold value of dust density estimated previously is probably reached. A good correlation is also obtained with the DUCMA experiment (Simpson et al., 1986, Fig. 2).

These gradients are believed to be associated with apacecraft crossings through regions of enhanced dust density rather than with plasma boundaries. We have plotted in Fig. 7 the normalized average electric field spectral density obtained during intervals of 32 s in four different channels, from 0625 to 0700 UT (Vega 2). The spectral densities have been multiplied by a factor $(R/R_0)^2$ to remove the R^{-2} variation; R_0 is the distance of closest approach which is equal to 8,030 km for Vega 2. It can be seen that the dust gradient, visible on the high frequency channels, differs from the cometopause discovered by Gringauz et al. (1986), which is seen only in the low frequency channels. Large fluctuations of the electric field intensity are observed below about 25–40 Hz in and

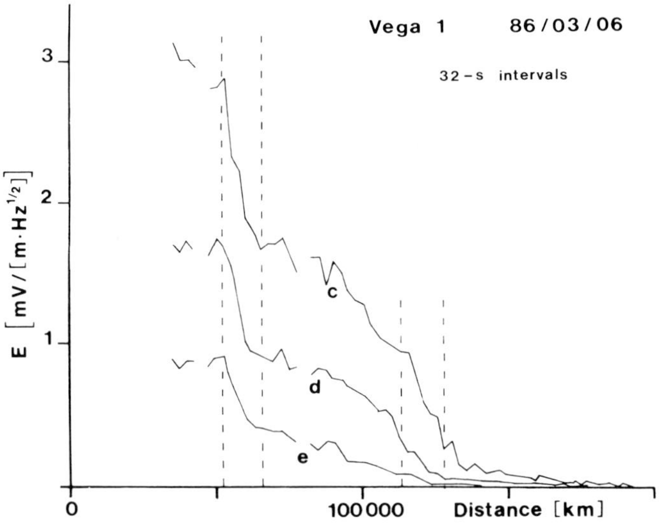

Fig. 5. Equivalent electric field spectral intensity in three channels during successive 32 s intervals against the distance of Vega 1 from nucleus: *c* 0.6–1.2 kHz; *d* 1.2–2.4 kHz; *e* 2.4–4.8 kHz. Two steep gradients are observed. The vertical dashed lines limit the boundaries of these gradients

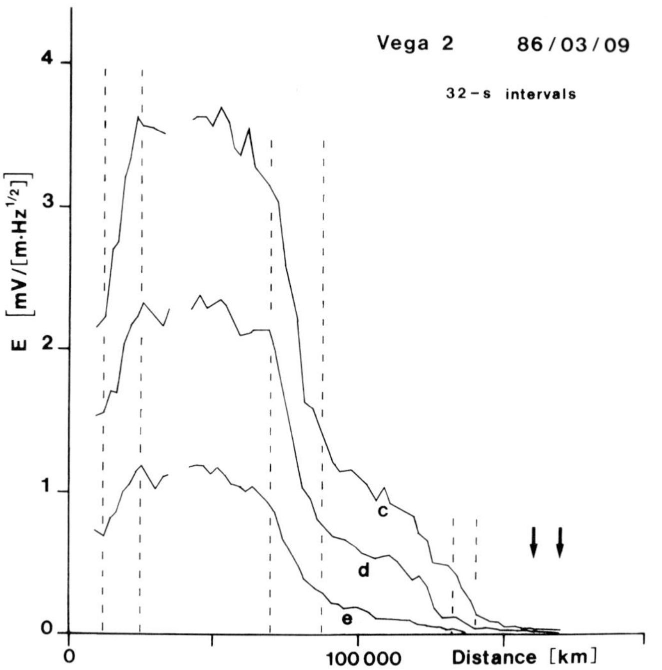

Fig. 6. Same as Fig. 5 but for Vega 2. Three steep gradients are now identified. The most distant gradient cannot be confused with the cometopause discovered by Gringauz et al. (1986) which extends between the two arrows (see also Fig. 7)

around the cometopause, while dust gradients are much more pronounced in the channels above 150–300 Hz. Note that an increase in the low frequency channel activity is already observed at 0635 UT (216,000 km) i.e. well before the cometopause.

4. Determination of the nucleus rotation period and dust particle bulk velocity

Using the location of the dust gradients, we propose a model which leads to the estimation of the apparent rotation period of

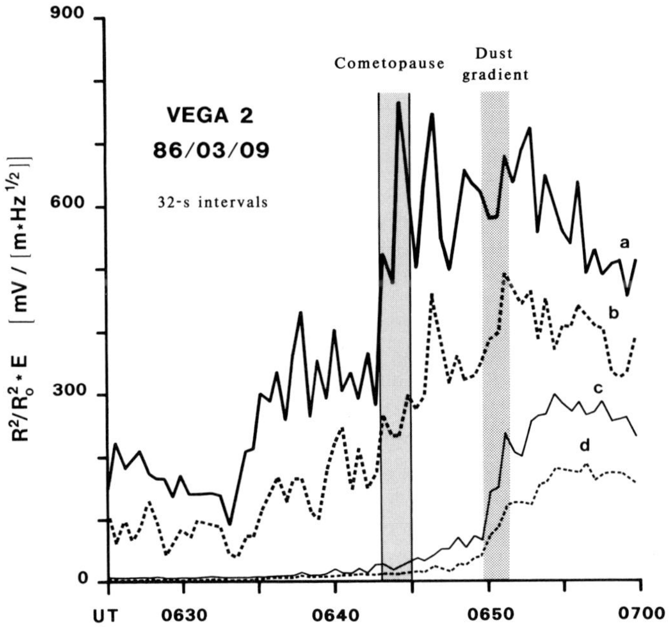

Fig. 7. Equivalent electric field spectral intensity in four channels during successive 32 s intervals measured by Vega 2 between 0625 and 0700 UT: *a* 8–14 Hz; *b* 14–25 Hz; *c* 150–300 Hz; *d* 0.6–1.2 kHz. The intensities have been multiplied by $(R/R_0)^2$ to remove the R^{-2} variation; R_0 is the distance of closest approach

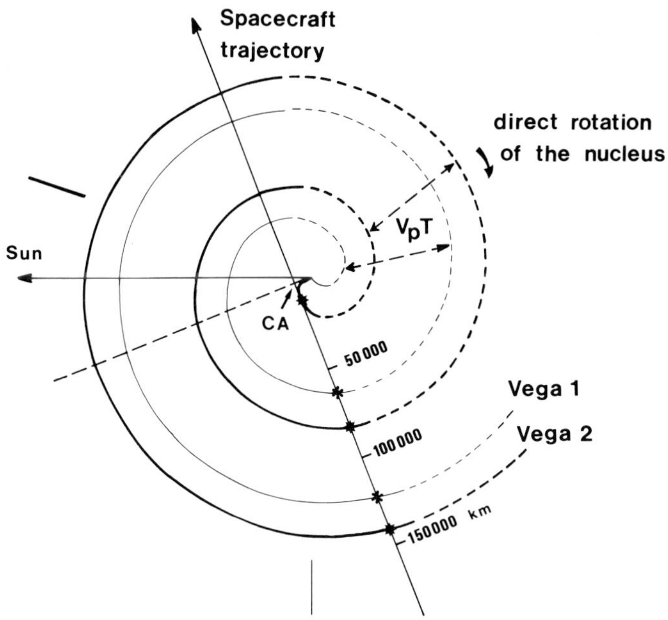

Fig. 8. Spacecraft trajectory and loci of the dust gradients crossed by Vega 1 (thin line) and Vega 2 (thick line) assuming a constant radial velocity for the dust, V_p, and only one major jet. T is the cometary nucleus rotation period. The distance from the nucleus is indicated along the spacecraft trajectory. The longitudes of the dust source at the time of the two flybys are indicated by radial pointers at the periphery of the spiral structures

the nucleus, T, and of the radial velocity, V_p, of dust particles. In this model we assume that V_p is constant and that the effective spin axis is nearly perpendicular to the plane which contains the spacecraft trajectory and the nucleus. The reality is unfortunately much more complex because the particle velocity is mass dependant; the discrete dust sources are not likely to be approximated by point sources and they are not necessarily located in the plane defined by the spacecraft trajectory and the nucleus; and the spin axis is seldom normal to this plane (see Divine et al., 1986, for a review). Nevertheless, the observations suggest that the dust particles ejected by discrete sources form well-organized structures called envelopes, These envelopes are believed to be responsible of the steep gradients observed by the APV-V experiment and, although the model we propose here is very simple, it fits the observations satisfactorily.

Let r and θ, be respectively the distance of the dust gradient from the nucleus and the angular distance with respect to the sun direction measured anticlockwise; then

$$V_p T = 2\pi \, \Delta r / \Delta \theta .$$

Taking into account the time delay of three days between Vega 1 and Vega 2 flybys, we can write

$$72 \cdot 3600 = nT/j + d/V_p ,$$

where j is the number of major jets assumed to be almost regularly distributed in longitude, n the number of complete $1/j$-rotations between Vega 1 and Vega 2 flybys. The observed distance, d, between Vega 1 and Vega 2 dust gradients is about $(19 \pm 2)\ 10^6$ m.

For a given j we can draw the locus of the dust gradient at a given time as an (r, θ) plot. Figure 8 shows the spacecraft trajectory and the loci of the dust gradients crossed by Vega 1 (thin line) and Vega 2 (thick line), assuming the existence of one single main jet ($j = 1$). The velocity of the spacecraft is much higher than

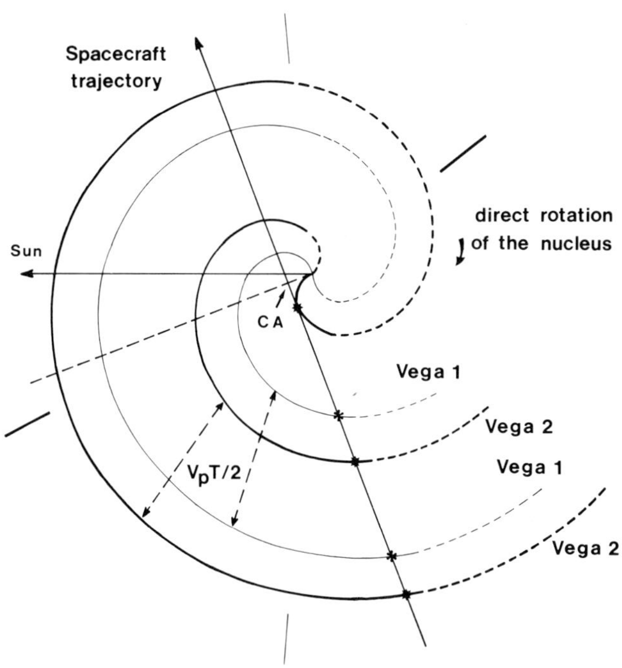

Fig. 9. Same as Fig. 8, assuming two nearly symmetrical dust jets

V_p and this sketch can be considered as a snapshot at time of closest approach. In this case, one possible solution is found to be $V_p = 300 \pm 17\ \mathrm{m\,s^{-1}}$ and $T = 54 \pm 5$ h, assuming that there is only one complete rotation between the Vega 1 and Vega 2 flybys ($n = 1$). The hypothesis of two nearly symmetrical jets per rotation ($j = 2$), as suggested by the ultraviolet imager of Suisei (Kaneda et al., 1986), leads to Fig. 9. Here a possible solution is given by $V_p = 520 \pm 30 \, m \, s^{-1}$ and $T = 61 \pm 7$ h for $n = 2$. In addition to these two solutions, other unlikely schemes are also possible. For example, three major jets distributed in longitude at approximately 120° from each other ($j = 3$) yield $T = 160 \pm 16$ h, $V_p = 296 \pm 17 \, m \, s^{-1}$ for $n = 1$.

For these three solutions, the value of V_p lies from 300 to $520 \, m \, s^{-1}$, which is compatible with the dust terminal velocity of light grains computed from a gas and dust dynamical model proposed by Gombosi et al. (1985). Indeed, in the case of P/Halley at 0.88 AU, this model predicts a dust terminal velocity between 265 and $455 \, m \, s^{-1}$ for dust size of 0.4 to 4 μm (Divine et al., 1986). The mass of such grains lies between 10^{-11} and 10^{-14} g, which is in general agreement with the value of 10^{-12} g given in Sect. 2.2, as the upper limit of dust grain masses detected by our experiment.

We have no way to choose between the different solutions for apparent roation periods and associated radial bulk velocities of dust particles. Indeed, a recent controversy concerning the rotation period of the nucleus has led Sekanina (1987) to propose a model which could reconcile most points of view: the nucleus may rotate around its major axis with a 7.4-day period but this axis precesses around a fixed direction with a period of 2.2 days. Within this uncertainty our two first solutions are compatible with the 2.2 day period, while the third one fits more the 7.4-day period. It will be necessary to make use of all possible ground and in-situ observations in order to prove such a model. The unexpected results obtained on dust impacts from an experiment, originally designed for measuring plasma-waves, may provide a significant contribution on this subject.

5. Conclusion

It appears that the plasma-wave experiment, APV-V, on board Vega 1 and Vega 2 has reacted to the impacts of dust particles. As in other space missions (Voyager and ICE) the signature of dust grain impacts is an increase of the AC electric field power over a broad range of frequencies, typically from about 100 Hz to 100 kHz, with a nearly monotonic spectrum for frequencies above 1 kHz (the power law index of this spectrum is about -3.4). It must be mentioned that the assumption about high frequency noise generation by dust impacts does not invalidate a priori the existence of instabilities observed below ~ 1 kHz which may reflect a high level of plasma wave activity (Grard et al., 1985; Mogilevsky et al., 1986). Momentary decreases of noise level can be interpreted as saturation effect of the experiment which occurs when the dust grain density reaches a threshold value of about $8 \, m^{-3}$ at 15,000 km from the nucleus. A good correlation is found between sharp increases of noise level and flux of low mass dust grains ($m < 10^{-12}$ g, Simpson et al., 1986; Vaisberg et al., 1986). These large scale gradients appear at intervals of 60,000 km, suggesting a spiral structure. The locations of the gradients seen by Vega 1 and Vega 2 can be combined within a simple model to derive, for a given distribution of the main dust sources, the apparent rotation period of the cometary nucleus and the radial bulk velocity of small dust particles.

References

Aubier, M.G., Meyer-Vernet, N., Pedersen, B.M.: 1983, *Geophys. Res. Letters* **10**, 5

Divine, N., Fechtig, H., Gombosi, T.I., Hanner, M.S., Keller, H.U., Larson, S.M., Mendis, D.A., Ray, L., Newburn, Jr., Reinhard, R., Sekanina, Z., Yeomans, D.K.: 1986, *Space Sci. Rev.* **43**,1

Fechtig, H., Grün, E., Kissel, J.: 1978, *Cosmic Dust*, ed. J.A.M. McDonnell, Wiley, New York, p. 607

Galeev, A.A., Gringauz, K.I., Klimov, S.I., Remizov, A.P., Sagdeev, R.Z., Savin, S.P., Sokolov, A.Y., Verigin, M.I., Szego, K.: 1986, *Geophys. Res. Letters* **13**, 845

Gombosi, T.I., Cravens, T.E., Nagy, A.F.: 1985, *Astrophys. J.* **293**, 328

Grard, R., Pedersen, A., Knott, K., Trotignon, J.G., Beghin, C., Formisano, V., Mogilevsky, M., Mikhaïlov, Y., Molchanov, O.: 1985, *Adv. Space Res.* **5**, 175

Grard, R., Gombosi, T.I., Sagdeev, R.Z.: 1986a, in *Space Missions to Halley's Comet*, eds. R. Reinhard, B. Battrick, ESA SP-1066, p. 49

Grard, R., Pedersen, A., Trotignon, J.G., Beghin, C., Mogilevsky, M., Mikhaïlov, Y., Molchanov, O., Formisano, V.: 1986b, *Nature* **321**, 290

Gringauz, K.I., Gombosi, T.I., Tatraallyay, M., Verigin, M.I., Remizov, A.P., Richter, A.K., Apathy, I., Szemerey, I., Dyachkov, A.V., Balakina, O.V., Nagy, A.F.: 1986, *Geophys. Res. Letters* **13**, 613

Gurnett, D.A., Grün, E., Gallagher, D., Kurth, W.S., Scarf, F.L.: 1983, *Icarus* **53**, 236

Gurnett, D.A., Averkamp, T.F., Scarf, F.L., Grün, E.: 1986, *Geophys. Res. Letters* **13**, 291

Kaneda, E., Ashihara, O., Shimizu, M., Takagi, M., Hirao, K.: 1986, *Nature* **321**, 297

Meyer-Vernet, N., Couturier, P., Hoang, S., Perche, C., Steinberg, J.L.: 1986a, *Science* **232**, 370

Meyer-Vernet, N., Aubier, M.G., Pedersen, B.M.: 1986b, *Geophys. Res. Letters* **13**, 617

Mogilevsky, M., Mikhaïlov, Y., Molchanov, O., Grard, R., Pedersen, A., Trotignon, J.G., Beghin, C., Formisano, V., Shapiro, V., Shevchenko, V.: 1986, 20th ESLAB Symp., ESA SP-250, Vol. **3**, p. 467

Pedersen, A., Grard, R., Trotignon, J.G., Beghin, C., Mikhaïlov, Y., Mogilevsky, M.: 1986, 20th ESLAB Symp., ESA SP-250, Vol. **3**, p. 425

Scarf, F.L., Coroniti, F.V., Kennel, C.F., Gurnett, D.A., Ip, W.-H., Smith, E.J.: 1986, *Science* **232**, 377

Sekanina, Z.: 1987, *Nature* **325**, 326

Simpson, J.A., Sagdeev, R.Z., Tuzzolino, A.J., Perkins, M.A., Ksanfomality, L.V., Rabinowitz, D., Lentz, G.A., Afonin, V.V., Erö, J., Keppler, E., Kosorokov, J., Petrava, E., Szabo, L., Umlauft, G.: 1986, *Nature* **321**, 278

Stern, J., de Barbeyrac, J., Poggi, R.: 1967, *Méthodes pratiques d'étude des fonctions aléatoires*, Dunod, Paris, p. 117

Vaisberg, O.L., Smirnov, V.N., Gorn, L.S., Iovlev, M.V., Balikchin, M.A., Klimov, S.I., Savin, S.P., Shapiro, V.D., Shevchenko, V.I.: 1986, *Nature* **321**, 274

Astron. Astrophys. 187, 89–93 (1987)

ASTRONOMY AND ASTROPHYSICS

Comparative study of the low-frequency waves near comet P/Halley during the Vega-1 and Vega-2 flybys

S. Savin [1], G. Avanesova [1], M. Balikhin [1], D. Wozniak [2], P. Wronowski [2], S. Klimov [1], Z. Krawczyk [3], M. Nozdrachev [1], D. Orlowski [2], A. Sokolov [1], and J. Juchniewicz [2]

[1] Space Research Institute, USSR Academy of Sciences, Profsojusnaya, SU-117810 Moscow, USSR

[2] Center for Space Research, Polish Academy of Sciences, Warsaw, Poland

[3] Aviation Institute, Warsaw, Poland

Received February 4, accepted July 10, 1987

Summary. Magnetosonic waves in the critical ionization velocity events in the inner coma of comet P/Halley are studied using the data of the low frequency wave experiments aboard Vega-1 and -2 spacecraft. Three boundaries downflow the bow shock are also found:

– "secondary wave" at the distance near 7 10^5 km from the comet;

– outer boundary of the mantle;

– cometopause.

Some common features and differences between these two flybys are discussed.

Key words: comets – plasmas – turbulence

1. Introduction

In this paper we discuss the low frequency wave data collected onboard the Vega-1 and Vega-2 spacecraft, and particularly those obtained inside of the Halley cometary bow shock. The analysis of the inner coma fine structure shows that there are some new signatures, so-called critical ionization velocity events (C IV) (Galeev et al., 1986).

2. Waves in critical ionization velocity events

The highest wave signal during flyby of comet P/Halley was found in the inner coma at a distance less than 15 10^4 km. It can be seen from Figs. 1 and 2, where 48 s averaged (middle curves), maximum (upper curves), and minimum spectrum analyzer data are shown. The common trend near the closest approach is believed to be due to the influence of dust particle impacts (Klimov et al., 1986b).

The most interesting features in the outer coma are the wave bursts at (7–8) 10^5, (3–4) 10^5, (1.5–2) 10^5 km and also at the current layer at 8 10^4 km (Vega-1).

In a picked-up solar wind plasma the ionized particles of expanding cometary atmosphere form an ion beam that relaxes under the action of plasma waves. The plasma waves in the inner coma connected with the plasma density enhancements were identified as oblique Langmuir high frequency magnetosonic waves (Galeev et al., 1986). It was also shown that these waves are the signature of the C IV phenomenon in operation in the inner coma of Halley's comet.

The C IV wave activity that may be identified only by comparison with plasma data is seen on the 24-s averaged plasma wave electric field power in the band 2–30 Hz from Vega-2 shown in Fig. 3. The occurrence of two density peaks near $(4-3)$ 10^4 km from PLAZMAG data are marked by arrows (Galeev et al., 1986).

The first peak is followed by the gap in the wave activity over the whole frequency range. The spectrum in the burst has a maximum 7 mV/m $\mathrm{Hz}^{1/2}$ near 14 Hz and in the gap it shifts to 4 Hz (Galeev et al., 1986). From the comparison of the density and wave data it can be seen that every sharp plasma density increase is connected with wave emissions in the frequency range 2–30 Hz. The mean slope of density profile is higher when strong wave activity exists.

As to the overall picture of a wave activity it is seen from Fig. 3 that the intensity of plasma waves continuously increases while the spacecraft approaches the comet nucleus. At the same time the frequency of the most intense waves grows to the distance $R = 15{,}000$ km and then suddenly drops to a fraction of Hz. Let us note that wave emissions are persisting in the whole inner coma and even where the plasma velocity value is certainly below the critical one (Galeev et al., 1986).

In the presence of the C IV related wave emissions of measured amplitudes the electrons must have superthermal tails along magnetic field lines (Galeev et al., 1986). Such anisotropic electron distribution may be an energy source for the excitation of oblique Langmuir waves with the frequencies $F = (1/5 - 1/3)\,F_{\mathrm{CE}}$ (F_{CE} is the electron cyclotron frequency). Examples of spectral peaks in the frequency band 100–400 Hz were presented in Grard et al. (1986) and Klimov et al. (1986b) using the data of APV-V (Vega-1) and APV-N (Vega-2) respectively. The overall character of the APV-N wave activity from 10 to 10^3 Hz during Vega-1 flyby is shown in Fig. 5. From 7:10 to 7:20 UT there is a prominent peak near 200 Hz. Its behaviour is similar to that of the higher frequencies. Near the closest approach after the maximum of ion density N_i (see Zastenker et al., 1986) the 200 Hz peak amplitude falls down in accordance with the proposed C IV electrons nature of the peak. At the spectrum from 2 to 10^3 Hz from the maximum

Send offprint requests to: S. Savin

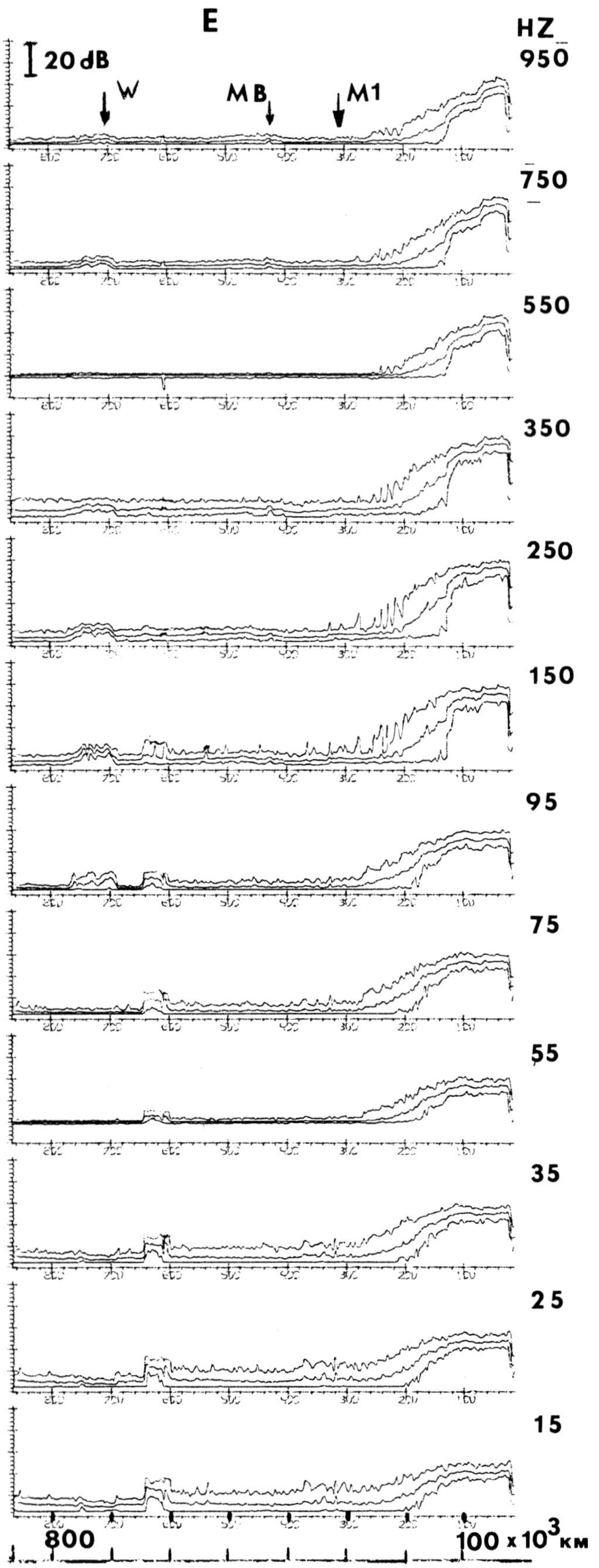

Fig. 1. Spectral components (spectral density in the different frequency ranges) of cartesian components of electric field (E) near the comet measured by Vega-1 (6.03.87) from 9 10^5 km to 8 10^3 km in arbitrary units. W – "secondary wave". M1 – the boundary from Schwingenschuh et al. (1986). Mamimum (upper curves), mean (middle curves) and minimum for 48 s

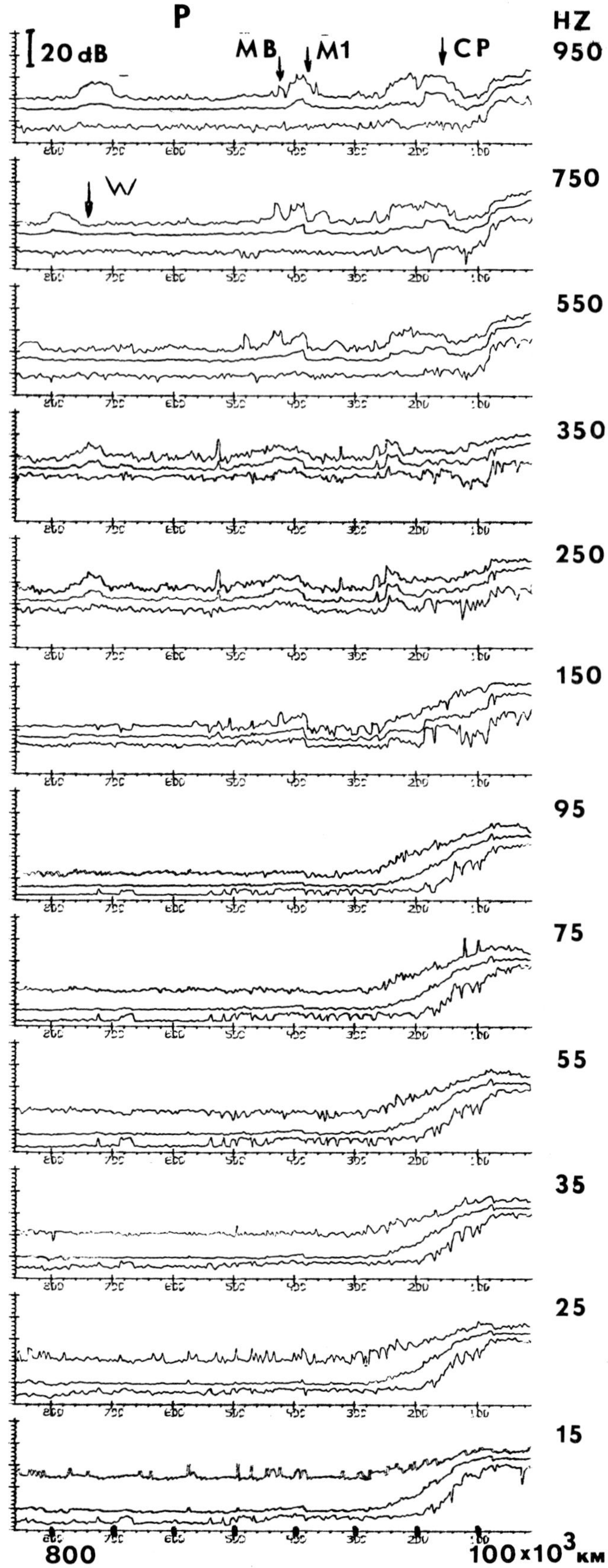

Fig. 2. Spectral components (spectral density in the different frequency ranges) of plasma flow fluctuations (P) near the comet measured by Vega-2 (9.03.87) from 9 10^5 km to 8 10^3 km in arbitrary units. W – "secondary wave". MB – outer boundary of the mantle. M1 – the boundary from Schwingenschuh et al. (1986). CP – cometopause. Maximum (upper curves), mean middle curves) and minimum for 48 s

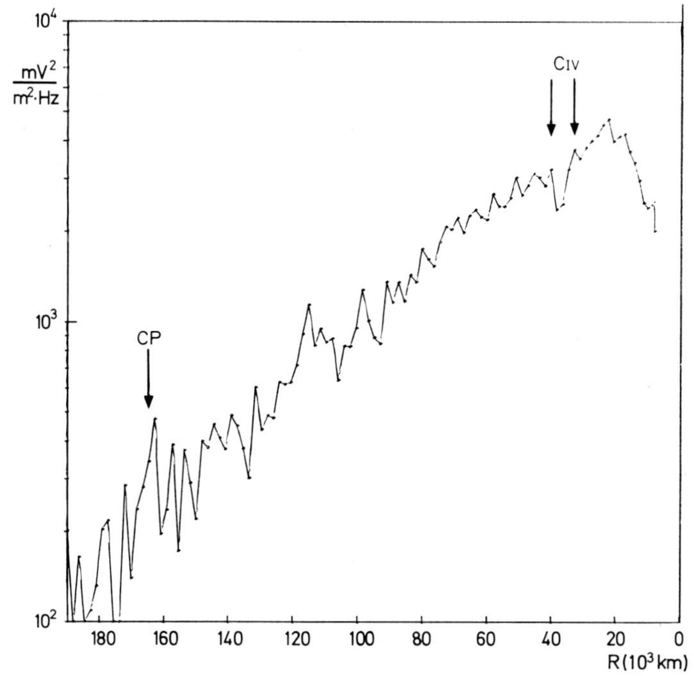

Fig. 3. The electric field power in the frequency range 2–30 Hz. The distance from the comet is in 10^3 km. The measurements from Vega-2

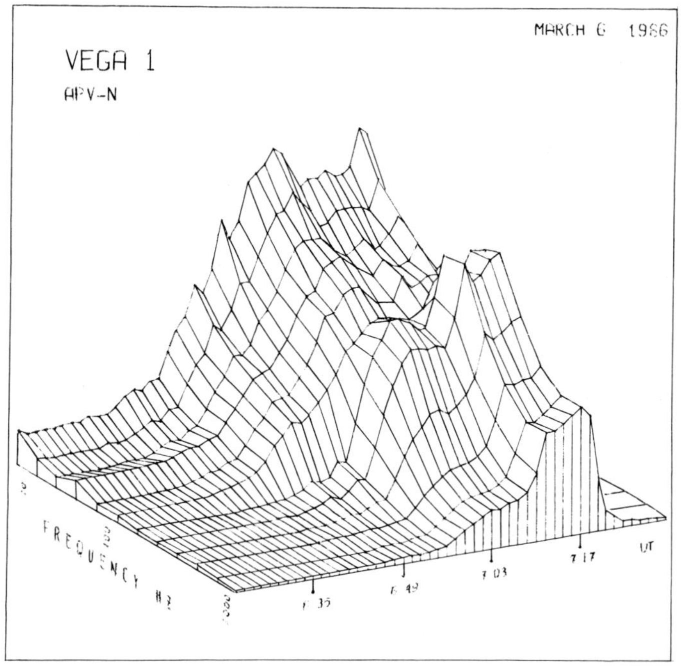

Fig. 5. Time dependence of spectra of electric field (E) fluctuations near the comet measured by Vega-1.

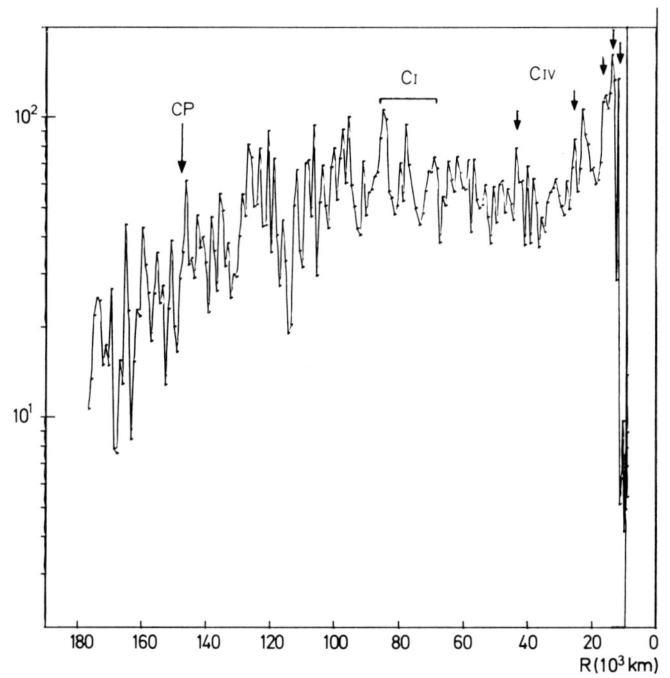

Fig. 4. The electric field power in the frequency range 1–15 Hz in arbitrary units. The distance from the comet is in 10^3 km. The measurements from Vega-1

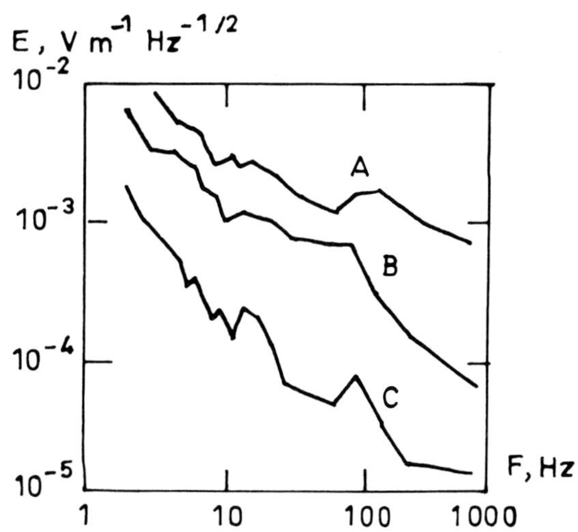

Fig. 6. Electric field spectra for 10 s intervals from Vega-1: A) 11 10^3 km (7:18 UT), B) 147 10^3 km (6:49 UT), C) 7.3 10^5 km (4:47 UT)

of N_i (Fig. 6, curve A) one can see peaks at about 15 Hz as well as at 200 Hz. From the comparison of N_i with the electric field in Fig. 4, we suggest that the first is due to C IV, because this peak exists only when there are C IV-features N_i. The second peak may be due to the instability mentioned above, but a comprehensive analysis must be done to exclude the dust impacts induced nature of it.

For Vega-1, within a distance of 60,000 km there are wave bursts in the frequency range 2–15 Hz associated with cometary ion density enhancements. This can be seen from Fig. 4 where the

12-s averaged electric field power in the band 2–15 Hz is shown. The density gradients of comet ions from BD-3 sensor of APV-N are marked by arrows (Zastenker et al., 1986). There is a good correlation between E-spikes and N_i gradients. The only prominent spike at about 25 10^3 km not marked by arrow coincides with a gap of BD-3 data. The main E-field maximum corresponds to the N_i one.

3. "Secondary wave" at the distance (7–8) 10^5 km

The first clearly seen in APV-N data boundary (denoted "w") inside the bow shock is at 7–8 10^5 km from the comet. One can see it in electric field E channels in Fig. 1 (Vega-1) and plasma (P) in Fig. 2 (Vega-2) in the band 100–1000 Hz. In the magnetic field this boundary is connected with the change of orientation, while the total magnetic field (B) is nearly constant (Schwingenschuh et al., 1986). PLAZMAG data show some extra thermalization and large plasma parameters fluctuations (Gringauz et al., 1986). The beginning of E-disturbance (Fig. 1) correlates with the end of the B_z drop, but the well developed E-signal is connected with the change in B_x, B_y orientation (Schwingenschuh et al., 1986). One can see the spectrum for the later case in Fig. 6 (curve C).

The peak near 15 Hz may be the indicator of magnetosonic or low-hybrid waves and the 100 Hz pronounced peak may be related to the "fan" instability, as in the case of cometary bow shock crossing (Klimov et al., 1986a), or to the oblique Langmuir waves (see previous section). The similarity between the shock wave and this boundary called by us "secondary wave" (W) is supported by the fact of a sharp increase in He^{2+} density (Balsiger et al., 1986). He^{2+} may be here only of the solar wind origin. But the behavior of B at W reveals the substantial difference with the cometary bow shock. One can see also a high frequency E-activity after the low frequency spike at W (Fig. 1). P-components (not shown) have similar behavior as E for Vega-1.

For Vega-2 the main disturbance at W is during the slow change of B_y sign (Fig. 2), while the sharp change in B_x, B_y and total B (Schwingenschuh et al., 1986) is less seen and there are no disturbances after W. The difference with Vega-1 may be due to the orientation of B, as in the case of bow shock (Klimov et al., 1986a). In Fig. 2 some kind of precursor of W with a raising frequency can be seen.

Whether the "secondary wave" is some kind of a bow wave in the highly supersonic (for the heavy ions) picked-up solar wind flow, a slow shock or a magnetopause – like hydrodynamic obstacle for the bow shock, or anything else, we cannot say now. But this "strange" boundary is registered by different devices aboard Vega-1, and -2, Giotto and maybe Suisei (Mukai et al., 1986), and we need either a complex analysis of plasma and wave data or a special modeling of solar wind-comet interaction to clarify the nature of this boundary.

4. Outer boundary of the mantle

The second boundary near 4 10^5 km is better seen in the P components in Fig. 2. It is associated with the outer mantle boundary (MB) and the beginning of the raising of total B (Gringauz et al., 1986; Schwingenschuh et al., 1986). For Vega-2, MB almost coincides with the boundary labelled M1 in (Schwingenschuh et al., 1986). One can see some disturbance before MB near 300 Hz and then in the entire frequency range in Fig. 2. MB is also seen in 250–350 Hz E-channels aboard Vega-2.

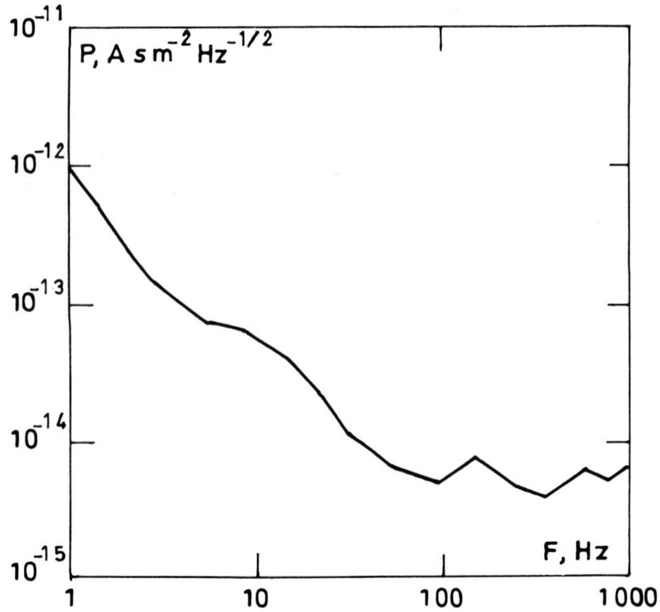

Fig. 7. Plasma flow fluctuations spectra at MB Vega-2 (see Fig. 2)

100-s averaged spectrum of P at MB is shown in Fig. 7. Besides the strong wave activity near 200 and 900 Hz a relative peak at about 10 Hz is seen. This P-spectrum is similar either to the bow shock or to the "secondary wave" E-field spectra (see curve C in Fig. 6 and Klimov et al., 1986a) besides the growth at 900 Hz. But one can see such character of P-wave activity at W (see Fig. 2).

For Vega-1 (Fig. 1) we see a wave disturbance in 100–1000 Hz band 20 min earlier than it is shown by M1 from the paper written by Schwingenschuh et al. (1986). M1 is characterized by a much less high frequency wave activity. We believe that for Vega-1 the real boundary where the solar wind proton population becomes comparable with the cometary implanted ions is MB in coincidence with the very beginning of the change of B slope. Then M1 is a shorter scale disturbance of a different nature than all other nearcomet boundaries.

5. Waves near and inside cometopause

The cometopause labelled "CP" in Figs. 2–5 (see Gringauz et al., 1986) in Vega-2 plasma wave data is characterized by a strong wave activity that is better seen in the 0.5–1 kHz P-channels in Fig. 2. Waves begin a growth at the outer side of CP firstly in the 200–400 Hz band and then in the higher one. The latter waves are seen mostly in P-channels that may be a special characteristic of ion-acoustic waves. In the 2–30 Hz band CP correlates with the beginning of the main amplitude growth, as can be seen from the electric field data in Fig. 3. There are some disturbances before CP as for 200–400 Hz, a spike just inside CP and spikes in the region of the faster growth of total magnetic field B at (10–12) 10^4 km (Schwingenschuh et al., 1986).

E-field inside 8 10^4 km in Fig. 3 was discussed in Sect. 2. In the Vega-1 2–15 Hz E-field data (Fig. 4), CP is also connected with the beginning of the substantial growth of E amplitude and there is a spike at CP. But inside CP, the E-field behaviour is quite different from Vega-2 case. Namely, there is a broad maximum with a number of spikes and then there is a minimum before the main density and wave amplitude maximum (see Sect. 2) during the Vega-1 flyby. Vega-2 data show the permanent growth of the electric field with some narrow spikes (Fig. 3).

We believe that this difference is due to the different magnetic field and corresponding plasma density signatures of the inner coma during two flybys. As it was suggested by Schwingenschuh et al. (1986) there is a current sheet (marked C1) in the inner coma that separates the stagnating and compressed inner region with frozen-in IMF having a direction opposite to that existing during the Vega-1 flyby. This moving magnetopause-like discontinuity is the source of the turbulence shown in the center of Fig. 4; the maximum peak here corresponds to the very beginning of diminution of B_x and B_y (see Schwingenschuh et al., 1986). As in the case of the Earth magnetopause the waves near C1 may be of magnetosonic nature (see Klimov et al., 1986c).

The higher frequency E-fields corresponding to Fig. 4 can be seen in Fig. 5. As in Vega-2 case, CP correlates with the beginning of the substantial growth of 10–100 Hz waves. The peak spectrum at cometopause is shown in Fig. 6, curve B. There are relative peaks in the magnetosonic region and near $(1/3-1/5)f_{ce}$. The same character with somewhat higher frequencies due to a higher B has the spectrum at 07:03 UT (Fig. 5) corresponding to the low frequency spike at C1 (see Fig. 4).

The substantial growth of high frequencies at about 07:10 UT (Fig. 5) with a spike near 300 Hz corresponds to the highest gradient of B components in the current sheet C1 (Schwingenschuh et al., 1986). The pressure and chemical gradients in the presence of currents may be a source of the waves at the current sheet (C1), cometopause (CP) and also maybe at the outer mantle boundary (MB, see Sect. 4). The spectra at these boundaries have the same features as those of the magnetopause and may have the same nature (see Klimov et al., 1986c).

Acknowledgements. The authors are grateful to Prof. A. A. Galeev for helpful discussions and evaluation of the paper.

References

Balsiger, H., Altwegg, K., Bühler, F., Geiss, J., Chielmetti, A.G., Goldstein, B.E., Goldstein, R., Huntress, W.T., Ip, W.-H., Lazarus, A.J., Meier, A., Neugebauer, M., Rettenmund, U., Rosenbauer, H., Schwenn, R., Sharp, R.D., Shelley, E.G., Ungstrup, E., Young, D.T.: 1986, *Nature* **321**, 330

Galeev, A.A., Gringauz, K.I., Klimov, S.I., Remizov, A.P., Sagdeev, R.Z., Savin, S.P., Sokolov, A.Yu., Verigin, M.I., Szegö, K.: 1986, *Geoph. Res. Lett.* **13**, 845

Grard, R., Pedersen, A., Trotignon, J.-G., Beghin, C., Mogilevsky, M., Mikhailov, Y., Molchanov, O., Formisano, V.: 1986, *Nature* **321**, 290

Gringauz, K.I., Gombosi, T.I., Remizov, A.P., Apathy, I., Szemerey, I., Verigin, M.I., Denchikova, L.I., Dyachkov, A.V., Keppler, E., Klimenko, I.N., Richter, A.K., Somogyi, A.J., Szegö, K., Szendro, S., Tatrallyay, M., Varga, A., Vladimirova, G.A.: 1986, *Nature* **321**, 282

Klimov, S., Galeev, A., Nozdrachev, M., Sagdeev, R.Z., Savin, S., Shapiro, V., Sokolov, A.: 1986, Heidelberg, ESA SP-250, 201

Klimov, S., Savin, S., Aleksevich, Ya., Avanesova, G., Balebanov, V., Balikhin, M., Galeev, A., Gribov, B., Nozdrachev, M., Smirnov, V., Sokolov, A., Vaisberg, O., Oberc, P., Krawczyk, Z., Grzedrielski, S., Juchniewcz, J., Nowak, K., Orlowski, D., Parfianovich, B., Wozniak, D., Zbyszynski, Z., Voita, Ya., Triska, P.: 1986b, *Nature* **321**, 292

Klimov, S., Nozdrachev, M., Triska, P., Voita, Ya., Galeev, A., Aleksevich, Ya., Afanasiev, Yu., Baskakov, B., Bobkov, Yu., Dunec, R., Zhdanov, A., Korepanov, V., Romanov, S., Savin, S., Sokolov, A., Schmelev, V.: 1986c, *Kosmich. Issled.* **XXIV**, 177

Mukai, T., Miyake, W., Terasawa, T., Kitayama, M., Hirao, K.: 1986, *Nature* **321**, 299

Schwingenschuh, K., Riedler, W., Schelch, G., Yeroshenko, Ye.G., Styashkin, V.A., Luhmann, J.G., Russell, C.T., Fedder, J.A.: 1986, COSPAR issue, paper 6.8.7

Zastenker, G., Vaisberg, O., Smirnov, V., Fedorov, A., Omelchenko, A.: 1986, Heidelberg, ESA SP-250, 86

Astron. Astrophys. 187, 94–96 (1987)

ASTRONOMY AND ASTROPHYSICS

Plasma flow in the cometosheath of P/Halley during the encounter of Suisei

S. Takahashi[1], T. Terasawa[1], T. Mukai[2], M. Kitayama[2], W. Miyake[3], and K. Hirao[4]

[1] Geophysical Institute, Kyoto University, Kyoto 606, Japan
[2] Institute of Space and Astronautical Science, Meguro, Tokyo 153, Japan
[3] Radio Research Laboratories, Nakaminato, Ibaragi 311-12, Japan
[4] Faculty of Engineering, Tokai University, Hiratuka, Kanagawa 259-12, Japan

Received March 4, accepted April 23, 1987

Summary. We estimate proton and alpha particle bulk parameters within the cometosheath. We observed that the density of proton was almost constant ($\sim 200\,\text{cm}^{-3}$) except near the bow shock and that of alpha particle was almost constant ($\sim 7\,\text{cm}^{-3}$) in the cometosheath. Temperature of protons seems to decrease from bow shock to about $2\,10^5$ km, and increase thereafter slightly, but that of alpha particle seems to be almost constant. This decrease is likely caused by the cooling effect accompanying with the charge exchange process as well as with the mass addition of cold pickup ions. On the contrary, the observed temperature of alpha particles stayed almost constant throughout the cometosheath. This is because that there is no effective cooling process for alpha particles within the cometosheath.

Key words: comet – cometosheath – comet-solar wind interaction

1. Introduction

One of the Japan's two Halley probes, Suisei, made a closest approach of $1.51\,10^5$ km from the Halley's nucleus at 1306 UT on 8 March 1986. During the encounter, Suisei passed through a strong interaction region, where the solar wind flow was severely perturbed by the pickup cometary ions (Mukai et al., 1986a,b,c; Terasawa et al., 1986a,b). In this paper, we shall focus on the plasma characteristics observed inside the cometosheath of Halley during the outbound leg of Suisei.

The cometosheath plasma includes multi species of ions; ions of the solar wind origin as well as those of the cometary origin. Using a specific feature in the phase space (ion pickup shells), we have succeeded in determining the densities of cometary protons and water-group ions (Mukai et al., 1986a,b,c). These shell-structured cometary ions were found superposed on the background thermal ions, which consist of the solar wind ions heated at the cometary bow shock as well as the thermalized pickup ions of cometary origin. We shall present results of least-square fitting of two-dimensional shifted isotropic Maxwellian distributions to these background thermal ions.

Send offprint requests to: T. Mukai

2. Observation

Two-dimensional distribution functions of ions in the ecliptic plane were obtained by an energy-per-charge electrostatic analyzer on Suisei (Mukai et al., 1986a,b). The field of view is $5^\circ \times 60^\circ$ with the larger dimension being perpendicular to the ecliptic plane. The geometrical factor is $2.3\,10^{-4}\,\text{cm}^2$ ster. The energy scan is made alternatively in every 16-spin period (147 s) in a high energy mode (HE mode: 248 eV/q–15.8 keV/q) or in a low energy mode (LE mode: 30 eV/q–1.92 keV/q). In Figs. 1a and 1b,

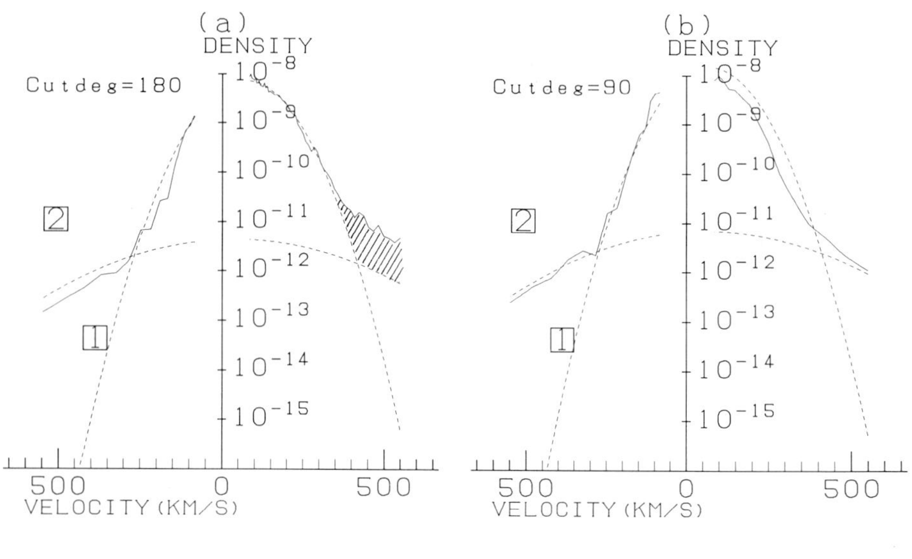

Fig. 1a and b. Cross-sectional cuts of the observed and the estimated distribution functions of ions; **a** along the line from sunward to anti-sunward, and **b** along the line perpendicular to **a.** Solid and dashed curves show the observations and the estimations, respectively. These distributions are plotted as if all ions would be protons. For the ions of the mass-to-charge ratio M/Q, the vertical and horizontal axes should be rescaled by multiplying factors $(M/Q)^2$ and $(M/Q)^{-1/2}$, respectively solid lines show cross-sectional cuts of a distribution function observed in the LE mode within the cometosheath near the closest approach ($1.51\,10^5$ km from the nucleus); (a) along a line from sunward (left) to anti-sunward (right) direction, and (b) along a line perpendicular to the above. In these panels, it is seen that the ion distribution consists of two components, a colder ('core') and a hotter ('halo') component. We have found that this 'core-halo' distribution of ions persists throughout the cometosheath. We interpret that the 'core' represents protons, and the 'halo' the alpha particles. Of course, this identification of ion species is not unique, since the plasma detector on Suisei does not distinguish between ion species. However, the similarity between the estimated velocities for 'core' and 'halo' (see below) excludes the possibility that the 'halo' consists of heavier ions. (We cannot exclude the possibility that the 'halo' consists of suprathermal protons. This interpretation, however, requires some unspecified energization process working throughout the cometosheath. This, if not impossible, seems not likely, and we shall not consider this possibility further.)

Dashed curves labeled 1 and 2 in Fig. 1 show the fitted distributions for protons and alpha particles, respectively. In the fitting procedure, we have omitted the data in the phase space region where the pickup cometary ions dominate. The hatched region in panel (a) is an example: In this region, the cometary water-group ions forming a well-defined pickup shell dominate. The estimated density of protons is $200\,\mathrm{cm}^{-3}$, which includes both protons of solar wind origin and of cometary origin. We have found that the alpha particles are much hotter than the protons: The estimated temperatures of protons and alpha particles are $9\,10^5$ K and $1.6\,10^7$ K, respectively. (Namely the thermal velocity of alpha particles is larger than that of protons by a factor of ~ 4).

The estimated bulk flow velocities for protons and alpha particles (50–$70\,\mathrm{km\,s^{-1}}$) agree within the estimated error range. The above flow velocity agrees with the previous estimation ($\sim 55\,\mathrm{km\,s^{-1}}$) based on the identification of the pickup shell (Mukai et al., 1986a,b).

The above density of protons is about a factor of 4 larger than the plasma density observed by Giotto at a similar distance from the nucleus (d'Uston et al., 1986). We attribute this difference to the fact that Suisei's encounter occurred during the passage of a stream-stream interaction region of the solar wind, while Giotto's occurred during the quiet period of the solar wind.

The pickup shell of cometary protons has been identified around the top of the background proton distribution for the same data (see Fig. 2b of Mukai et al., 1986b). The density of the shell protons ($10\,\mathrm{cm}^{-3}$) estimated previously (Terasawa et al., 1986a) is 5% of the total density of the protons. The estimated density of alpha particles is $7\,\mathrm{cm}^{-3}$, which is 3.5% of the proton number density. In the upstream solar wind, we have estimated the density of the alpha particle ranging from 3.0 to 5.0% of the proton density. We conclude that in the region near the closest approach thermalized cometary protons do not dominate over the protons of solar wind origin. (As otherwise, the fraction of alpha particle should be much less.)

Figure 2 summarizes the results of estimation: (a) number densities (cm^{-3}), (b) temperatures (K), and (c) velocity vectors ($\mathrm{km\,s^{-1}}$). Solid and dashed lines show the parameters for protons and alpha particles, respectively. Since the phase space coverage of the HE mode observation was not large enough to determine the proton parameters (the energy range was too high), they are determined only for the LE mode. Beyond $2\,10^5$ km from the nucleus we found flow turbulence with a time scale comparable to the Suisei's spin period of 9 s (Mukai et al., 1986b). The effect of this turbulence may explain the large discrepancies between the estimated velocities of protons and alpha particles beyond $2\,10^5$ km from the nucleus.

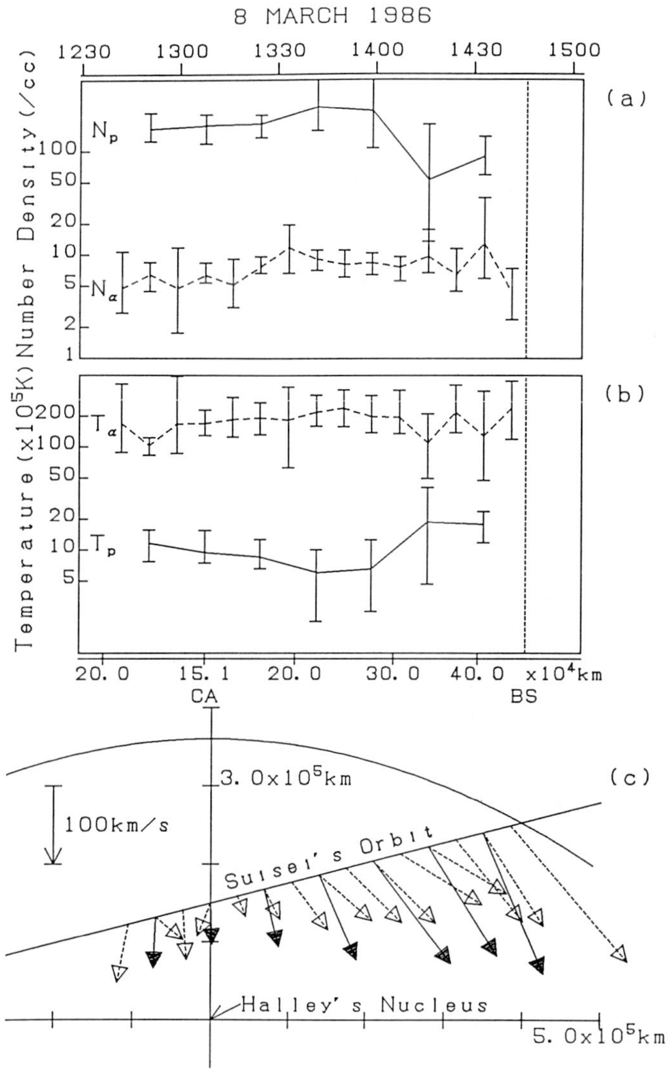

Fig. 2a–c. Estimated bulk parameters. **a** Number densities, **b** temperatures; and **c** velocities. Solid and dashed curves represent the values of the protons and the alpha particles respectively. Velocities and flow angles are represented in the rest frame of the comet. Top and bottom scales show universal time and the distance from the nucleus, respectively. 'CA' and 'BS' show the positions of the closest approach and the bow shock crossing, respectively. Y-axis of **a** is logarithmic number density (cm^{-3}), and that of **b** is logarithmic temperature (K). In **c** the downward bar corresponds to anti-sunward flow. A parabolic curve represents the cometary bow shock

In Fig. 2(a), the proton density has a plateau within $2.8\,10^5$ km from the nucleus. In Fig. 2(c), the velocity is lowest around closest approach (CA) and increases toward the bow shock (BS). Therefore the number flux nV (n density, V velocity) is not constant. This is expected since the flow structure within the cometosheath should not be one dimensional but should have a velocity divergence. Thus nVA (A; cross section of the flow region) rather than nV should be constant. The above argument is not complete, however, since these protons are lost by the charge exchange process with cometary neutral molecules. To interpret the flow structure, we need a comparison with a comprehensive hydrodynamic model of the cometosheath flow which includes losses due to charge exchange.

Since the pickup energy of the cometary protons is smaller than the thermal energy of the background protons within the subsonic cometosheath flow, both the charge exchange process between background protons with cometary hydrogen and the mass addition of cometary photoionized protons contribute to the cooling of the protons.

In Fig. 2(b), the temperature for protons reached the minimum not at the closest approach but around $2.2\,10^5$ km from the nucleus, and increased thereafter slightly. We have not yet understood the physical reason why the temperature minimum occurred at this position. (This temperature minimum is not an artifact of an inaccurate observation due to the flow turbulence. A time aliasing effect, which could be due to the turbulence, can increase an apparent temperature, but not decrease.) On the contrary, the observed temperature of alpha particles stayed almost constant throughout the cometosheath. Note that there is no effective cooling process for alpha particles within the cometosheath. Alpha particles after the charge exchange process become $H_e{}^+$, so that the charge exchange does not contribute to the cooling of alpha particles (Shelley et al.; 1986).

The spatial structure of plasma flow, especially the temperature distribution around Halley needs further study. A detailed comparison with results from other spacecrafts (e.g. Goldstein et al.; 1986, Vergin et al.; 1986) is highly desirable.

References

d'Uston, C., Reme, H., Squvaud, J.A., Cros, A., Anderson, K.A., Carlson, C.W., Curtis, D., Lin, R.P., Korth, A., Richter, A.K., Mendis, A.: 1986, 20th ESLAB Symp., ESA SP-250, p. 77

Goldstein, B.E., Neugebauer, M., Balsiger, H., Drake, J.F., Fuselier, S.A., Goldstein, R., Ip, W.-H., Rettenmund, U., Rosenbauer, H. Schwenn, R., Shelley, E.G.: 1986, 20th ESLAB Symp., ESA SP-250, p. 285

Mukai, T., Miyake, W., Terasawa, T., Kitayama, M., Hirao, K.: 1986a, *Nature* **321**, 299

Mukai, T., Miyake, W., Terasawa, T., Kitayama, M., Hirao, K.: 1986b, *Geophys. Res. Letters* **13**, 829

Mukai, T., Kitayama, M., Terasawa, T., Miyake, W., Hirao, K.: 1986c, 20th ESLAB Symp., ESA SP-250, p. 71

Shelley, E.G., Fuselier, S.A., Drake, J.F., Balsiger, H., Ip, W.-H., Goldstein, B.E., Goldstein, R., Neugebauer, M.: 1986, 20th ESLAB Symp., ESA SP-250, p. 285

Terasawa, T., Mukai, T., Miyake, W., Kitayama, M., Hirao, K.: 1986a, *Geophys. Res. Letters* **13**, 837

Terasawa, T., Takahashi, S., Mukai, T., Miyake, W., Kitayama, M., Hirao, K.: 1986b, 20th ESLAB Symp., ESA SP-250, p. 281

Vergin, M.I., Gringauz, K.I., Richter, A.K., Gombosi, T.I., Remizov, A.P., Szego, K., Apathy, I., Szemerey, I., Tatrallyay, M., Lezhen, L.A.: 1986, 20th ESLAB Symp., ESP SP-250, p. 169

Astron. Astrophys. 187, 97–102 (1987)

ASTRONOMY AND ASTROPHYSICS

MHD waves detected by ICE at distances $\geq 28\,10^6$ km from comet P/Halley: Cometary or solar wind origin?

B.T. Tsurutani[1], A.L. Brinca[1], E.J. Smith[1], R.M. Thorne[2], F.L. Scarf[3], J.T. Gosling[4], and F.M. Ipavich[5]

[1] Jet Propulsion Laboratory, California Institute of Technology, Mail Code 169–506, Pasadena, CA 91109, USA
[2] Department of Atmospheric Science, University of California, Los Angeles, CA, USA
[3] Space Sciences Department, TRW Space and Technology Group, Redondo Beach, CA, USA
[4] Los Alamos National Laboratory, Los Alamos, NM, USA
[5] Department of Physics, University of Maryland, College Park, MD, USA

Received February 13, accepted April 14, 1987

Summary. Spectral analyses of the high resolution magnetic field data are employed to determine if there is evidence of cometary heavy ion pickup when ICE was closest to Halley, $\sim 28\,10^6$ km. No evidence is found for the presence of heavy ion cyclotron waves. However, from this search, two new wave modes are discovered in the solar wind: electromagnetic waves near the proton cyclotron frequency and drift mirror mode "waves". Both modes have scales of 10–60 s (1 to 6 times the proton gyro period) in the spacecraft frame. The possibility of wave generation by cometary hydrogen pickup is explored. Theoretical arguments and further experimental evidence indicates that cometary origin is improbable. The most likely source is instabilities in high speed solar wind streams, presumably associated with stream-stream interactions. VLF electrostatic emissions are found to occur in field minima or at gradients of the drift mirror structures. Possible generation mechanisms of drift mirror mode waves, cyclotron waves and electrostatic waves are discussed.

Key words: MHD waves – Halley – ICE – electrostatic emissions – generation mechanisms

1. Introduction

It has recently been suggested by (ICE) investigators that cometary pickup ion effects may have been detected at distances beyond $28\,10^6$ km from comet Halley. Wenzel et al. (1986) have reported enhanced fluxes of energetic ions and MHD wave power at 10^{-2} Hz (the H_2O group ion cyclotron frequency) coincident with enhanced ELF/VLF plasma wave turbulence that have many characteristics similar to Giacobini-Zinner cometary waves (Scarf et al., 1986a,b).

The above articles state that for cometary neutrals to reach such large distances with sufficient flux, they would have to have velocities of 2–3 km s^{-1} or ionization time scales of $\sim 5\,10^6$ s, or both, giving effective ionization scale lengths of 10–20 10^6 km. Both sets of authors cautiously note that their observations occurred during high-speed solar wind streams and that the observations could instead, simply be due to physical processes associated with stream-stream interactions. Because of these difficulties in the interpretation of the ICE results, the above authors have encouraged further analyses to determine the source of the plasma and MHD waves and energetic particles.

In an attempt to resolve this issue, we undertake a careful examination of the ICE magnetic field data to determine if there are any distinguishing MHD wave features that could be of cometary origin. The purpose is to help resolve part of the question of the source of the enhanced MHD turbulence, energetic particles, and ELF/VLF plasma waves.

2. Observations

Fig. 1 is taken from Wenzel et al. (1986) and is repeated here to illustrate the relationship between energetic ion fluxes, solar wind velocity and density and transverse wave peak power at 10^{-2} Hz. Enhanced transverse wave power above "intermediate" solar wind conditions (Siscoe et al., 1968) and enhanced energetic ion fluxes are detected on days 82–83 and 86–87, 1986. These are illustrated in the bottom and top panels, respectively. However, these intervals occur during high-speed solar wind streams, shown in the second panel. It is thus uncertain whether the apparent enhanced wave power is a result of physical processes associated with stream-stream interactions, increased solar wind doppler shift of low frequency turbulence, or of cometary ion pickup.

Daily averaged power spectra of the transverse fluctuations of the magnetic field for the interval of interest are plotted in Fig. 2. The number of degrees of freedom are indicated in each panel. Because these latter values are large, the uncertainty is quite small. The spectra have general $\sim f^{-5/3}$ dependences consistent with Kolmogorov turbulence. The power level increases and decreases with solar wind velocity (see Fig. 1), but there is little or no indication of any pronounced wave power spectral peaks. There are no particularly remarkable features at frequencies corresponding to the cometary water ion gyrofrequency (10^{-2} Hz) or the proton gyro-frequency ($\sim 10^{-1}$ Hz). The power spectra of the field magnitude was also computed for the same days, with similar results. No obvious peaks were detected.

A careful scan of the high resolution data ($1\frac{1}{2}$ vectors s^{-1}) was made to determine if there is a possibility that H_2O^+ group (or any other ion group) waves occurred with low amplitudes or

Send offprint requests to: B.T. Tsurutani

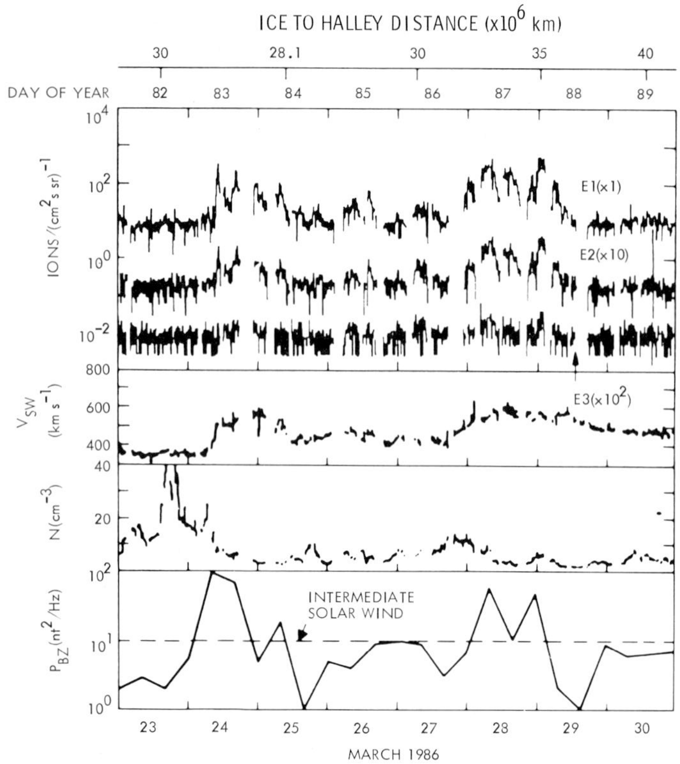

Fig. 1. From Wenzel et al. (1986). Energetic ions, plasma velocities and densities, and peak magnetic field power at 10^{-2} Hz during the ICE encounter with comet Halley

on a sporadic basis (which would not show up in a power spectral analysis). Within the limit of the natural noise present in the data, no evidence was found for such waves.

However, from the scan of the high resolution data, two other distinct modes were discovered. They are identified as non-oscillatory drift mirror mode "waves" and electromagnetic cyclotron waves. The wave amplitudes for both modes are relatively small, and the occurrence sporadic. The drift mirror waves are most apparent in the data set, while only a few definite cases of the cyclotron waves are detected.

Figure 3 gives an example of what we believe to be drift mirror mode structures. They are detected at the intersection between the end of a small stream and the beginning of a high speed stream (see Fig. 1). They are characterized by pronounced changes in field magnitude with little or no changes in field direction. This identification is based on observations of these waves in the earth's magnetosheath (Thorne and Tsurutani, 1981; Tsurutani et al., 1982, 1984). Unfortunately, because of the low data rate mode, the plasma and energetic ion data had insufficient time resolution to provide additional support for this identification.

The second panel of the figure illustrates the sine of the angle between the interplanetary magnetic field and solar wind velocity, α. For the event at 1637–1658 UT, $\sin\alpha$ is ~ 0.7. However, for the previous event occurring at 1615–1625 UT, $\sin\alpha$ was considerably smaller (~ 0.5).

The ICE plasma wave data were examined to determine if (ELF) electromagnetic lion roars were generated in the low field (high β) regions (Tsurutani et al., 1982; Thorne and Tsurutani, 1981; Moreira, 1983; Lee et al., 1987) and if (VLF) electrostatic

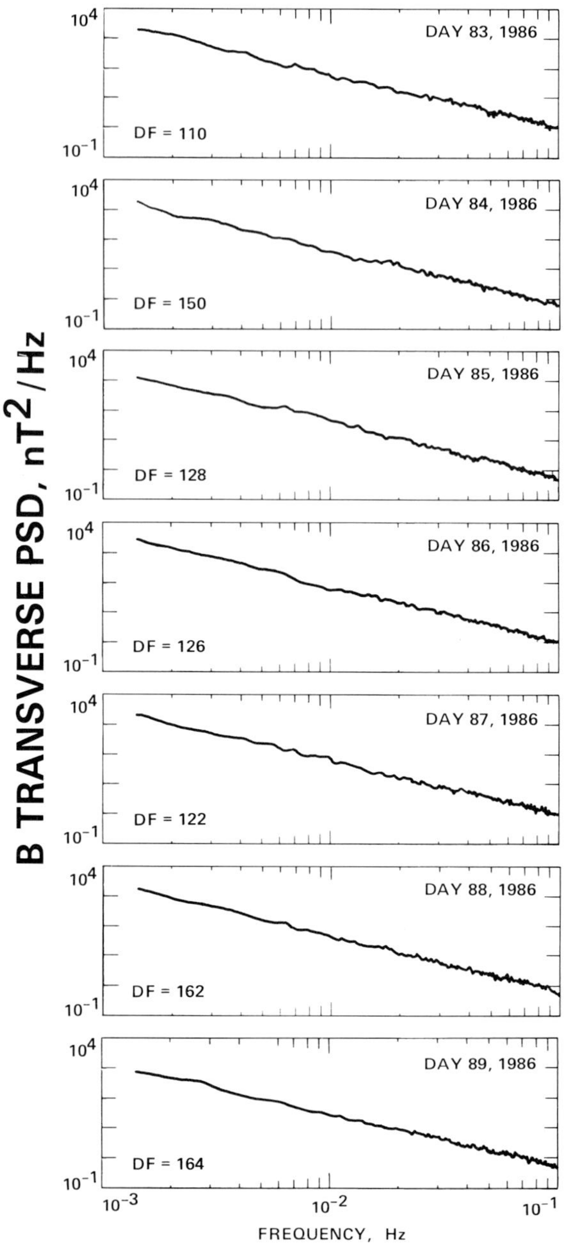

Fig. 2. Power spectral densities of the magnetic field from Days 83 to 88, 1986. The statistical uncertainties are small, indicated by the large number of degrees of freedom

ion acoustic waves are present in the high field (low β) regions (Anderson et al., 1982). Neither wave modes were found with the same correlation with drift mirror mode waves as in the Earth's magnetosheath. However, broadband electrostatic emissions were detected during the same time interval. This is illustrated in Fig. 4. Only the 1.78 KHz channel is illustrated, for brevity. Crosshatching is indicated beneath each of the electrostatic bursts. It can be seen from the figure that each burst is associated with either a dip in the magnetic field (high β region) or a magnetic field gradient.

Two easily identifiable cases of electromagnetic waves occurred on March 25 (day 84) and March 28 (day 87). The day 84 event occurred in the rising portion of a small stream and the day 87 event near the peak (on the rising portion side) of a high

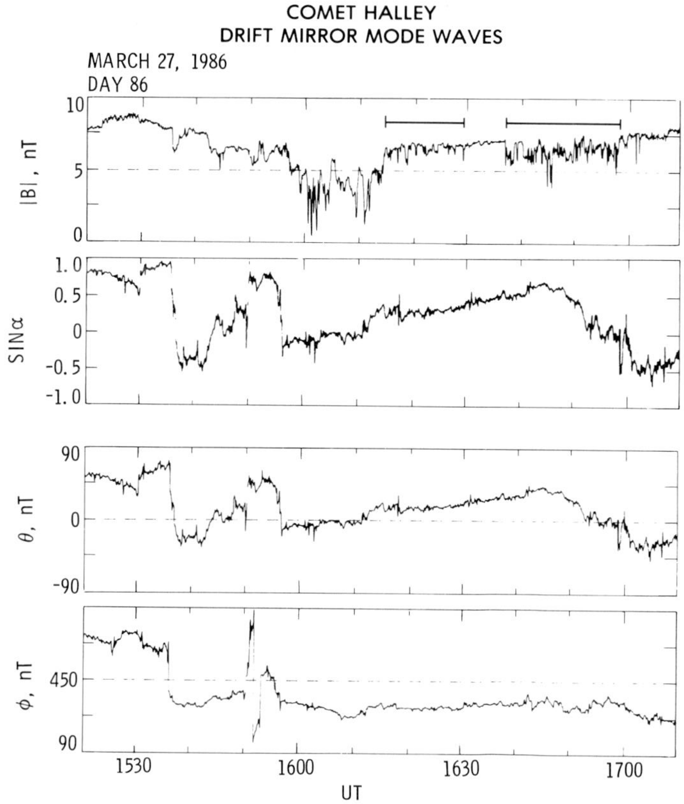

Fig. 3. Examples of events of magnetic field decreases with little changes in direction (horizontal bars). These are presumed to be drift mirror mode waves. The sine of the angle α between the solar wind velocity vector and field direction is given in the second panel. The magnetic field GSE θ and ϕ components are given in the bottom two panels

speed stream. Fig. 5 illustrates the example on day 84. The waves are almost purely transverse (little field magnitude variation). The wave period is $\sim$8 seconds. The amplitude is quite small, $\sim 1nT$ peak-to-peak with a $<0.2nT$ compressional component. The magnetic field is illustrated in the principal axis system where the ϕ variations correspond to the field rotation from B_1 to B_2 (standard notation, as in Smith and Tsurutani, 1976). The directions of the principal axes and the magnetic field are given in the figure. The waves have a left-hand sense of rotation about the ambient magnetic field (in the spacecraft frame) and are propagating nearly along the field with an average angle of 4°. The waves are elliptically polarized, with $\lambda_1/\lambda_2 \approx 4.4$.

The electromagnetic waves at 1408–1409 UT, day 87 (not shown) have periods of $\sim$16 s. The wave amplitudes are large, $\sim 6.0nT$ peak-to-peak. The magnetic field magnitude variation is $\sim 3.5nT$ peak-to-peak. The waves are elliptically polarized ($\lambda_1/\lambda_2 = 2.7$) and are propagating at relatively large angles to the ambient magnetic field, $\sim$40°. The waves are left-hand polarized in the spacecraft frame.

During both of the above cyclotron wave events, the direction of the interplanetary magnetic field is more or less along the spacecraft-sun line. $\cos^{-1}(\hat{B} \cdot \hat{X}_{SE}) = 27°$ for the day 87 event.

The detection of two new modes of waves in the solar wind, possibly associated with the pickup of cometary protons, is quite exciting. Because of the high neutral hydrogen velocities of $8\,km\,s^{-1}$ and $20\,km\,s^{-1}$ relative to the nucleus (for the dissociation of OH and H_2O, respectively), and H ionization time scales of $\sim 10^6$ s, ionization scale lengths of 8–20 10^6 km are obtained, in agreement with the values suggested as being necessary by Scarf et al. (1986a) and Wenzel et al. (1986).

To determine if these two wave modes are indeed caused by MHD instabilities associated with the pickup of comet Halley ions, below we discuss theoretical growth rates for both modes. We will also examine a prior passage of one of the corotating solar wind streams to determine if similar waves are detected. During this prior high-speed stream passage, the ICE spacecraft was considerably farther from comet Halley, $\sim$ 0.75 AU versus $\sim$0.2 AU (R. Farquhar, private communication, 1986).

3. Generation mechanisms

3.1. Drift mirror waves

Mirror-mode instability requires an anisotropic pressure distribution with $\beta_\perp/\beta_\parallel > 1 + 1/\beta_\perp$ where $\beta_\perp$ and $\beta_\parallel$ are the perpendicular and parallel plasma beta ($8\pi NkT/B^2$) values (Hasegawa, 1969; Crooker and Siscoe, 1977). The plasma will be unstable to the drift mirror mode if the pressure anisotropy, $\beta_\perp/\beta_\parallel > 1.6$ for $\beta \simeq O(1)$ or for smaller anisotropies if $\beta \gg 1$. We will calculate the effect of the addition of cometary pickup protons on the total $\beta_\perp$ at a distance of $\sim 30\,10^6$ km to determine if the pickup ions can be responsible for instability.

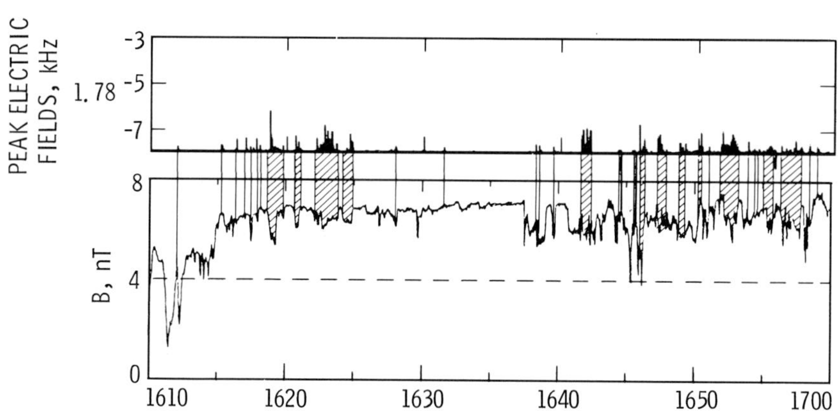

Fig. 4. The relationship between VLF electrostatic waves and drift mirror mode waves

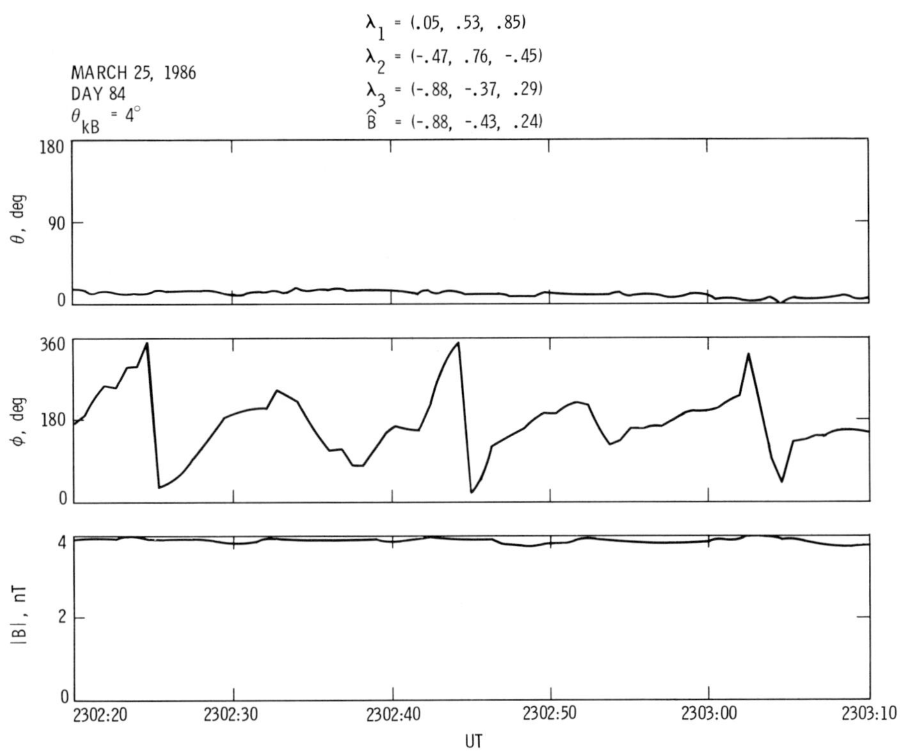

Fig. 5. An example of an electromagnetic cyclotron wave in the solar wind

The cometary neutral density at a distance r from the nucleus is $N_n = Qe^{-r/V_g\tau}/4\pi V_g r^2$, where Q is the molecular production rate, V_g the hydrogen atom velocity relative to the nucleus, and τ the ionization time. A reasonable value of Q for Halley is $4\,10^{29}\,s^{-1}$ (Reinhard, 1986 and references therein). Assuming 80% of all cometary molecules are H_2O, and using the higher value $V_g = 20\,km\,s^{-1}$, we get an upper limit of $3\,10^{-4}\,H\,cm^{-3}$ at $r = 30\,10^6$ km. By assuming all of the ions gain the maximum possible kinetic energy during pickup, $\frac{1}{2}m_i V_{SW}^2$ (where $V_{SW} = 425\,km\,s^{-1}$ at the time of drift mirror mode detection) and the ion ring formation time is long, say 10^4 s, then the beta associated with the pickup ions is $\beta_\perp = (N_i E_\perp)/(B^2/8\pi)$ or $\beta_\perp \sim 5\,10^{-3}$ (with a measured $5nT$ magnetic field). Thus, it is clear that cometary ions make a negligible contribution to the total plasma pressure anisotropy in a solar wind plasma where β is typically of order ~ 1.

3.2. Electromagnetic ion cyclotron waves

The electromagnetic cyclotron waves were detected when the IMF was nearly parallel to the solar wind velocity direction. Thus, an ion beam (Gary et al., 1981) or ring distribution (Wu and Davidson, 1972; Wu and Hartle, 1974) is the most likely source of instability. By using the previously derived number for the hydrogen atom density of $3\,10^{-4}\,cm^{-3}$ at $r = 30\,10^6$ km, an ionization time scale of 10^6 s, and a beam formation time of 10^4 s, the ion beam density can be calculated from $N_i = N_n t/\tau_i$. Substituting, we get $N_i = 3\,10^{-6}\,cm^{-3}$. For a solar wind density of $\sim 5\,cm^{-3}$, the relative ion beam density would be $N_i/N_{sw} = 6\,10^{-7}$.

This low value precludes the generation of cyclotron waves by drifting rings of newborn cometary protons: the corresponding kinetic dispersion equation does not yield growing modes.

3.3. Previous passage of the corotating event

A careful scan of days 56–64, 1986 was made for the previous appearance of the corotating stream. Similar drift mirror waves were detected. An example is shown in Fig. 6. The temporal scale of the structures is ~ 60 s. This value is somewhat larger than in the previous example, but is still comparable. Maximum amplitudes of the field decreases are $\sim 3nT$, comparable in scale to the

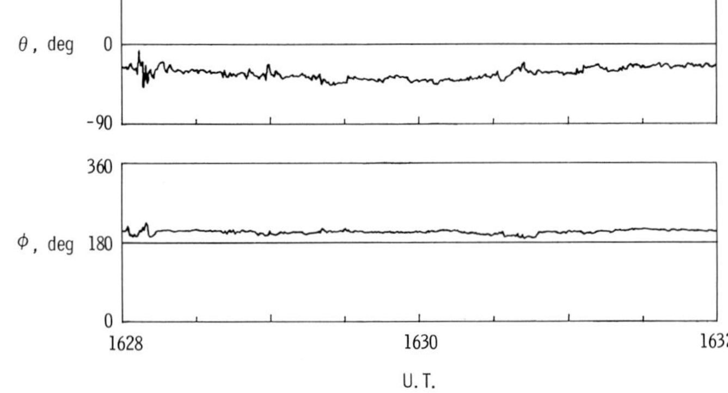

Fig. 6. Nonoscillatory drift mirror mode waves on day 61, at a distance of $110\,10^6$ km from Halley

former event. The ambient magnetic field lies at an angle of $\sim 40°$ relative to the solar wind velocity.

As stated previously, the distance of ICE from Halley at this time was ~ 0.75 AU, 3.75 times further away. Halley protons will thus be $\sim 10^3$ less dense. It is even more unlikely that these waves have any association with Halley pick-up ions.

4. Discussion

A search for waves near the cyclotron frequencies of heavy cometary ions was conducted during the interval of closest approach to comet Halley by both power spectral analyses and visual inspection. No obvious signatures were found.

Two types of electromagnetic waves have been detected for the first time in the solar wind: drift mirror mode non-oscillatory "waves" and elliptically polarized electromagnetic cyclotron waves with periods of ~ 10–60 s in the spacecraft frame. Both waves have scales near the proton gyroradius. The detection of drift mirror mode waves in a more distant solar wind stream indicates a lack of uniqueness of the Halley encounter interval. Theoretical arguments rule out the possibility of generation of either wave mode by cometary protons.

The amplitude of the waves are generally small (although some large amplitude cases were shown) and their occurrence infrequent. Their contribution to the power in solar wind fluctuations is negligible, as was indicated by the lack of any significant peaks appearing in power spectral analyses of an entire day's worth of data.

The formation of mirror mode waves at ICE must therefore be due to an anisotropic plasma pressure present in the bulk of the solar wind proton distribution. It was speculated by Tsurutani et al. (1982) that such conditions for instability could be created by the heating of solar wind plasma as it passes through an interplanetary quasiperpendicular shock. Ion and electron heating (Gosling and Robson, 1985; Feldman, 1985) can lead to a downstream β of >3. This plus the preferential ion heating in the perpendicular direction (Gosling and Robson, 1985; Kennel and Sagdeev, 1967) presumably leads to the growth of drift mirror waves in the planetary magnetosheaths of Earth, Jupiter and Saturn (Tsurutani et al., 1982).

The VLF electric noise at frequencies well above the electron cyclotron (170 Hz) but below the electron plasma (27 kHz) frequencies and the lack of any magnetic signatures, indicate that these emissions are probably electrostatic waves. The occurrence of the emissions on the gradients of the magnetic field also gives an important clue to possible generation mechanisms. Ashour-Abdalla and Okuda (1986) have suggested that cold counter-streaming ion beams, where $T_p \ll T_i \sim T_e$ and $V_{Te} > V_{ib} > C_s$, could lead to growth of electrostatic waves. Double ion beams are commonly present in the solar wind. They are most often present on the trailing edges of streams (Feldman et al., 1973), as in the case here. Adiabatic decompression of solar wind protons in the mirror mode structures will be expected to lead to the formation of anisotropic field aligned streaming and could also possibly support such a wave generation-mechanism.

The possibility that the emissions are Doppler shifted lower hybrid drift waves associated with the density gradients in the drift mirror waves (we assume magnetic plus plasma pressure is constant, or pressure isotropy) is another possible generation mechanism. Using the formulation of Huba et al. (1978) for the threshold of instability $L_n/\rho_i < (m_i/m_e)^{1/4}$ we estimate L_n by $1/2(\beta_e + \beta_i)L_B \sim L_B$ where $L_B = (1/B_0\, \delta B/\delta s)^{-1}$. L_n has a minimum value of $2\,10^3$ km at a sharp field gradient at 1658 UT day 86. We find that the threshold for instability is not met, unless higher magnetic field and plasma variations are present but escape measurement because of the temporal resolution of the instrumentation.

It is also possible to understand why lion roars are not generated in the high β regions, in contrast to the case of earth's magnetosheath. In Thorne and Tsurutani (1981) it is shown that lion roars are generated when $E_m \ll E_{th}$, where $E_m = B^2/8\pi N$ and E_{th} is the electron thermal energy. For the case illustrated here, $B \simeq 7nT$, $n \sim 3\,\mathrm{cm}^{-3}$ and thus, $E_m = 6.5\,10^{-11}$ erg. For the electromagnetic cyclotron instability, this requires an electron temperature of at least $4.7\,10^5$ K. The actual measured electron temperature is only $1.5\,10^5$ K. Thus, independent of the temperature anisotropy necessary for instability, the solar wind electron temperature is too low to lead to the generation of whistler mode lion roars.

From the above, it can be concluded that lion roars should be only rarely detected in conjuction with drift mirror waves in the interplanetary medium. Electrons have temperatures $\geq 3.2\,10^5$ only 1–2% of the time (Gosling et al., 1987) thus E_m is typically $\geq E_{th}$, leaving the solar wind electrons stable to the growth of lion roars.

5. Conclusion

We have not found any evidence of heavy ion cyclotron waves at distances $\geq 28\,10^6$ km. We have, however, detected two new solar wind electromagnetic wave modes due to this search: drift mirror mode waves and electromagnetic cyclotron waves with periods 10–60 s in the spacecraft frame. From theoretical considerations, it is most probable that these waves are generated by plasma processes within high speed streams such as stream-stream interactions, as the examples shown in this paper.

Acknowledgments. We would like to acknowledge scientific discussions with I.G. Richardson and S.P. Gary. Programming assistance was provided by J. Wolf, L. Wigglesworth and M. Burton. The Los Alamos portion of this work was performed under the auspices of the U.S. Department of Energy with support from NASA under contract S-04039-D. Part of the work described in this paper was carried out by the Jet Propulsion Laboratory, California Institute of Technology, under contract with NASA.

References

Anderson, R.R., C.C. Harvey, M.M. Hoppe, B.T. Tsurutani, T.E. Eastman, J. Echeto: 1982, *J. Geophys. Res.* **87**, 2082

Ashour-Abdalla, M., H. Okuda: 1986, *J. Geophys. Res.* **91**, 6833

Crooker, N.U., G.L. Siscoe: 1977, *J. Geophys. Res.* **82**, 185

Feldman, W.C.: 1985, In *Coll. Shocks in the Heliosphere: Revs. of Current Res.*, ed. B.T. Tsurutani, R.G. Stone, Am. Geophys. Union, Washington, DC, **35**, 195

Feldman, W.C., J.R. Asbridge, S.J. Bane, M.D. Montgomery: 1973, *J. Geophys. Res.* **78**, 2017

Gary, S.P., J.T. Gosling, D.W. Forslund: 1981, *J. Geophys. Res.* **86**, 6691

Gosling, J.T., A.E. Robson: 1985, In *Coll. Shocks in the Heliosphere: Revs. of Current Res.*, ed. B.T. Tsurutani, R.G. Stone, Am. Geophys. Union, Washington, DC, **35**, 141

Gosling, J.T., D.N. Baker, S.J. Bame, W.C. Feldman, R.D. Zwickl, E.J. Smith: 1987, *J. Geophys. Res.*

Hasegawa, A.: 1969, *Phys. Fluids* **12**, 2642

Huba, J.D., N.T. Glad, K. Papadopoulos: 1978, *J. Geophys. Res.* **83**, 5217

Kennel, C.F., R.Z. Sagdeev: 1967, *J. Geophys. Res.* **72**, 3303

Lee, L.C., C.S. Wu, C.P. Price: 1987, *J. Geophys. Res.* (to appear)

Moreira, A.: 1983, *Planet. Spa. Sci.* **31**, 1165

Price, C.P., D.W. Swift, L.C. Lee: 1986, *J. Geophys. Res.* **91**, 101,

Reinhard, R.: 1986, *Nature* **321**, 313.

Scarf, F.L., F.V. Coroniti, C.F. Kennel, D.A. Gurnett, W.-H. Ip, E.J., Smith: 1986a, *Science* **232**, 377

Scarf, F.L., F.V. Coroniti, C.F. Kennel, T.R. Sanderson, K.P. Wenzel, R.J. Hynds, E.J. Smith S.J. Bame, R.D. Zwickl: 1986b, *Geophys. Res. Letters* **13**, 857

Siscoe, G.L., L. Davis Jr., P.J. Coleman, Jr., E.J. Smith, D.E. Jones: 1968, *J. Geophys. Res.* **73**, 61

Smith E.J., B.T. Tsurutani: 1976, *J. Geophys. Res.*, **81**, 2261

Thorne, R.M., B.T. Tsurutani: 1981, *Nature* **293**, 384

Tsurutani, B.T., E.J. Smith, R.R. Anderson, K.W. Ogilvie, J.D. Scudder, D.N. Baker, S.J. Bame: 1982, *J. Geophys. Res.* **87**, 6060

Tsurutani, B.T., I.G. Richardson, R.P. Lepping, R.D. Zwickl, D.E. Jones, E.J. Smith, S.J. Bame: 1984, *Geophys. Res. Letters* **11**, 1102

Wenzel, K.P., T.R. Sanderson, I.G. Richardson, S.W.H. Cowley, R.J. Hynds, S.J. Bame, R.D. Zwickl, E.J. Smith, B.T. Tsurutani: 1986, *Geophys. Res. Letters* **13**, 861

Wu, C.S., R.C. Davidson: 1972, *J. Geophys. Res.* **77**, 5399

Wu, C.S., R.E. Hartle: 1974, *J. Geophys. Res.* **79**, 283

Astron. Astrophys. 187, 103–108 (1987)

ASTRONOMY AND ASTROPHYSICS

Plasma-tail activity at the time of the Vega encounters

M.B. Niedner, Jr.[1] **and K. Schwingenschuh**[2]

[1] Laboratory for Astronomy and Solar Physics, Code 684.1, NASA Goddard Space Flight Center, Greenbelt, MD 20771, USA
[2] Space Research Institute, Austrian Academy of Sciences, Inffeldgasse 12, A-8010 Graz, Austria

Received February 16, accepted May 11, 1987

Summary. Physical associations are sought between plasma-tail activity seen in ground-based imagery and near-comet, interplanetary magnetic field (IMF) measurements obtained by the Vega spacecraft. Emphasis is placed on Disconnection Events (DE's) and on testing the sector boundary/frontside magnetic reconnection model (Niedner and Brandt, 1978) of their origin. Strong support for the model comes from the spectacular DE whose onset, on 1986 March 7–8, is strongly correlated with a reversal of the comet's magnetic barrier observed by Vega-1 and Vega-2 at the encounters on March 6.3 and 9.3 UT, respectively, and with an IMF sector boundary observed by Vega-1 on March 7.9 UT. Additional work is needed to determine the possible role of solar-wind plasma effects in producing the DE. Relationships between other occurrences of plasma-tail activity and Halley Armada spacecraft measurements remain to be explored.

Key words: Disconnection Events (DE's) – magnetic reconnection – sector boundaries – magnetic barrier

1. Introduction

One of the earliest goals of the International Halley Watch's (IHW) Large-Scale Phenomena (L-SP) Discipline (with which author MBN is affiliated) was the assembly of an especially complete photographic record of the comet during "Encounter Week" spanned by the Vega-1, Suisei, Vega-2, Sakigake, and Giotto flyby's (Brandt et al., 1982; Niedner et al., 1982). Besides supporting the interpretation of the in situ data by providing simultaneous remote imagery of the "entire comet", it was important to take advantage of the unique chance to calibrate the *visible* solar-wind/plasma-tail interaction using near-comet solar-wind data.

In this paper, we report on an apparent physical association between a major Disconnection Event (DE) seen in remote imagery on 1986 March 8–10, and encounter/cruise-phase magnetometer data returned by the Vega-1 and Vega-2 spacecraft.

2. The sector boundary model of plasma-tail Disconnection Events

That bright, well-observed comets are seen to shed their plasma tails periodically is now generally recognized. However, the more pressing question of the physical origin of DE's is perhaps less settled. The models which have been proposed to explain DE's principally fall into two groups: the sector boundary model (Niedner and Brandt, 1978), and high-speed stream models (Ip and Mendis, 1978; Ip, 1985; Jockers, 1985; Russell et al., 1986). The element common to *some* models of both types is magnetic reconnection, which is proposed to occur on the sunward side (in the magnetic barrier) in the sector boundary model (Niedner and Brandt, 1978) and on the nightside (i.e., in the tail) in the stream model (Ip, 1985; Russell et al., 1986). Note that the stream models of Ip and Mendis (1978) and Jockers (1985) do not invoke magnetic reconnection. A more complete comparison of some of the different DE models is given by Niedner (1984).

In qualtitive form, Fig. 1 shows the essential characteristics of the sector boundary model. For more detail, the reader is referred to Niedner and Brandt (1978) (also Niedner, 1984; Niedner and Brandt, 1979). The crucial point is that the reversed interplanetary magnetic fields past the sector boundary reconnect with those fields captured into the sunward magnetic barrier from the previous sector. The end result of sustained field line cutting is the total severing of the roots of the existing plasma tail, i.e., disconnection. The available evidence suggests that the reconnection phase of a DE, that is to say the length of time for the barrier to be dissipated, is in the range $\tau_{rec} = 0.5–1.0$ days (Niedner et al., 1981; Niedner, 1982; Brosius et al., 1987).

3. The Disconnection Event of 1986 March 8–10

Figure 2 is a 27-h photographic sequence showing the most prominent plasma-tail event which occurred during the entire week of spacecraft encounters: the spectacular DE of March 8–10. The sequence shows the event 's early evolution on March 8–9; maximum visibility was achieved on March 10 (West et al., 1986). Measurements of the detached tail's linear separation from the headcenter in the last two panels yield a recession speed of 53 km s^{-1}, which at first glance seems low but is actually fairly typical of the majority of DE's (Niedner, 1981), as well as of many other measurable plasma-tail structures (Jockers et al., 1972; Jockers, 1985). Due to contaminating emission from the inner dust tail, the exact positon of the detached tail cannot be seen in panel B, only a noticeable bend in it (arrow). Panel A is noteworthy for a prominent system of long tail rays (discussed below).

Measuremants of additional images made available to the authors after the construction of Fig. 2 has made possible a better calculation of the acceleration (a), initial velocity (V_0), and time of disconnection (t_{disc}) for the DE than would have been possible

Send offprint request to: M.B. Niedner

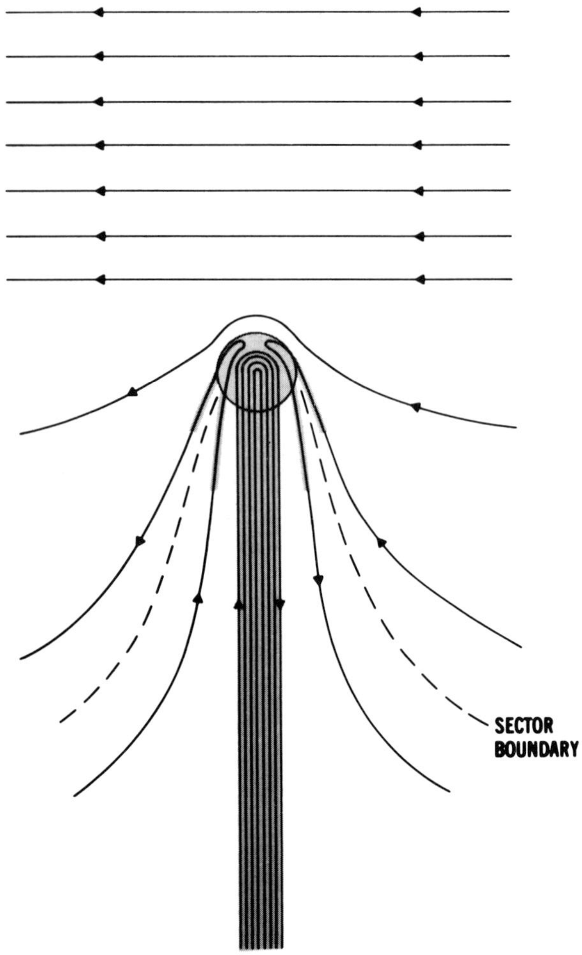

Fig. 1. Schematic diagram showing the sunward cutting of magnetic fields and the dissipation of the magnetic barrier in the sector boundary model of DE's (from Niedner and Brandt, 1978). The duration of reconnection necessary to disconnect a plasma tail is thought to be in range 0.5–1.0 days.

from the images in Fig. 2 alone. Table 1 lists the times and sources of these images, several relevant orbital parameters, and DE distances/velocities (with respect to the nucleus) obtained by assuming the DE lies approximately along the prolonged radius vector. The measurements were then taken in different combinations of three to yield exactly-determined values of a, V_0, and t_{disc}. The solutions for these parameters and the observations on which they were based are listed in Table 2. Taking an average of the three listed solutions, which represent the innermost observations of the DE, we arrive at $\langle V_0 \rangle = 24.2 \pm 3.7$ (rms) km s^{-1}, $a = 26.9 \pm 2{,}0$ cm s^{-2}, and $t_{\mathrm{disc}} =$ March 8.39 ± 0.06 UT.

The duration of activity (assumed to be magnetic reconnection) which is required to disconnent a plasma tail has been estimated (Niedner et al., 1981; Niedner, 1982) to be $\tau_{\mathrm{rec}} \sim 0.75$ d, which would place the inferred onset time of the disconnection process, t_{onset}, at $\sim$ March 7.64 UT. In other words, it is in the second half of the UT day on March 7 that the solar-wind agent responsible for the DE is inferred to have arrived at the comet.

We conclude the description of the March 8–9 images by pointing out that the first image in the Fig. 2 sequence was taken only 5 h after the derived time of disconnection, and the detached tail is presumed to be buried inside the bright inner regions of the dust tail. A strong system of ionic tail rays with maximum length $> 8\ 10^6$ km is seen at this time, and it should be noted that a well-known feature of the pre- and immediate post-disconnection phases of major DE's is a prominent system, or "burst", of tail rays (Niedner and Brandt, 1979; Jockers, 1985) which represents the plasma and fields which will build a new plasma tail.

4. The Vega magnetometer measurements

Most remarkable about the March 8 DE is what the Vega spacecraft seem to have told us about the nature of the event. Figure 3 shows Vega-1 (a) and Vega-2 (b) magnetometer data at the times of the encounters on March 6 and 9 (adapted from Riedler et al., 1986). The symbol "CA" indicates times of closest approach, $\sim$7:20 UT for both encounters. The sign convention for the field components is as follows: positive $B_{x,y,z}$ are toward the Sun, in the direction against planetary motion, and toward the north ecliptic pole, respectively. Thus, a classic Parker spiral of outward polarity (+) IMF would be composed of $B_x < 0$, $B_y > 0$ fields, and vice versa for (−) polarity IMF.

For Vega-1 on March 6, the $B_x < 0$, $B_y > 0$ orientation of the sheath fields during the 5:00–7:10 UT approach to encounter is consistent with positive (+) polarity IMF incident on the comet and being captured into the magnetic barrier. However, about 5–10 min before closest approach, $B_{x,y}$ change sign indicating that the innermost part of the barrier is composed of inward (−) polarity fields presumably captured from the IMF during an earlier time. This dual-polarity pattern is repeated in reverse sequence on the outbound trajectory.

The reversed inner barrier on March 6.3 UT has been discussed by Schwingenschuh et al. (1986) and , with emphasis on possible optical consequences, by Niedner and Schwingenschuh (1986), but it will not be further elaborated upon in this paper. A more detailed study of its probable photographic signature is in preparation (Niedner and Schwingenschuh, 1987). However, it should be noted in passing that the Vega-1 TUNDE-M experiment detected bursts of hot ions as the field reversal region was being crossed (Somogyi et al., 1986), and this feature should be considered a possible signature of magnetic reconnection. Energization of plasma and anomalous ionization effects associated with the reonnecting current layer have been proposed by Niedner (1980).

Figure 3b shows that the $B_x > 0$, $B_y < 0$ orientation of the sheath and magnetic barrier during the 6:05–7:20 UT inbound path of Vega-2 on March was consistent with inward (−) polarity IMF, although the outer sheath region, 5:00–6:05 UT, was characterized by $B_x > 0$, $B_y > 0$ fields well away from the ideal spiral orientation. Concerning the main pile-up region (6:05–7:20 UT), however, it is clear that the polarity had reversed with respect to the (+) polarity outer/newer barrier observed three days earlier by Vega-1. The obvious inference to be made is that a (+/−) magnetic sector boundary had passed the comet sometime between the encounters (Niedner, 1986; Niedner and Schwingenschuh, 1986).

We searched the Vega-1 cruise phase magnetometer data and found a sector boundary which not only would arrived at the comet between the encounters, but which "corotates" to Halley extremely close to the inferred onset time of the disconnection process ($t_{\mathrm{onset}} = t_{\mathrm{disc}} - \tau_{\mathrm{rec}} =$ March 7.64 UT) discussed earlier. The relevant data are shown in Fig. 4, a plot of Vega-1 magnetometer data spanning March 7.0–9.0 UT. Although the IMF field components show several major fluctuations during

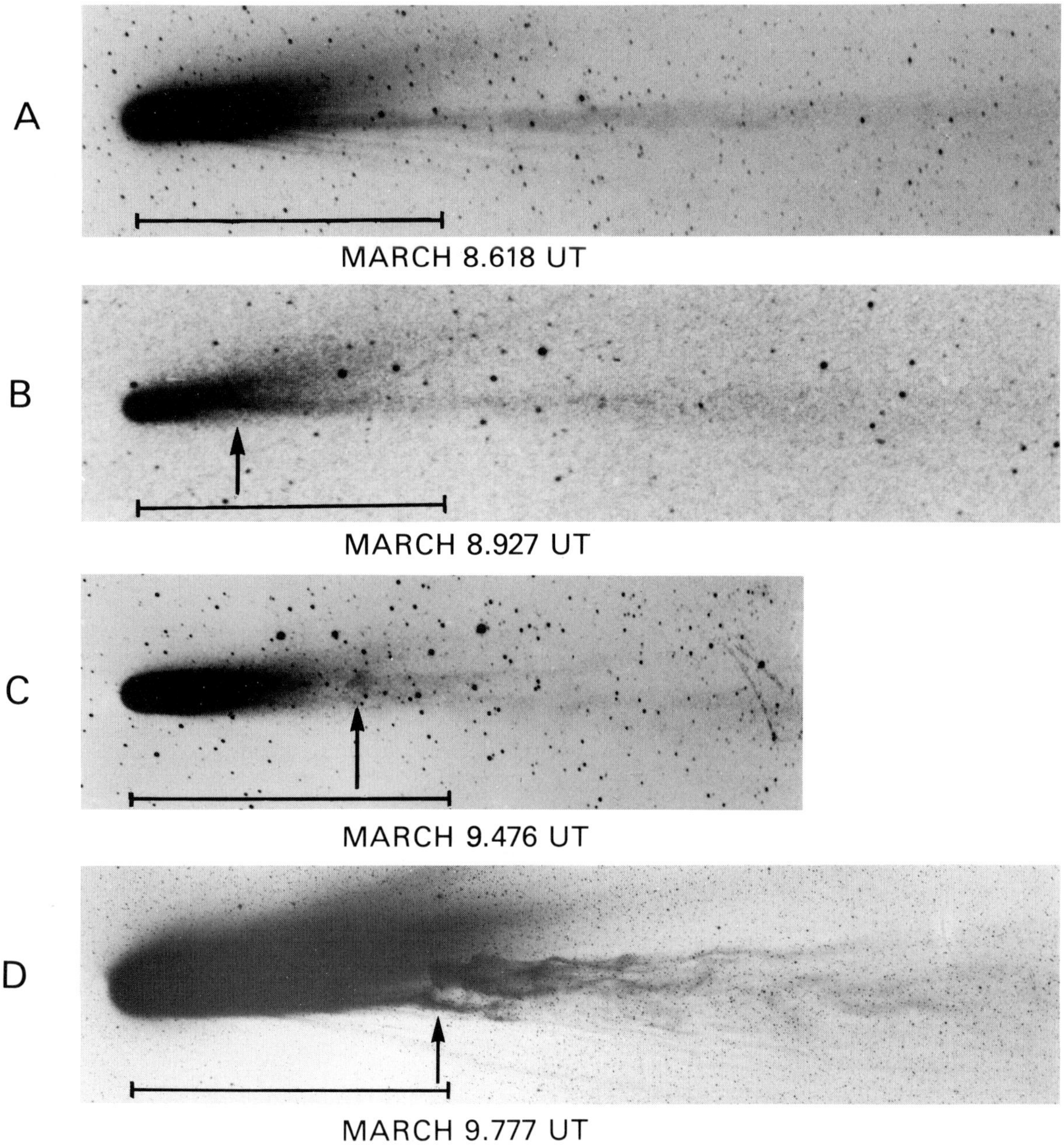

Fig. 2. March 8–9 photographic sequence showing development of a Disconnection Event (DE). Scale bar indicates a linear distance of 5 10^6 km oriented along the prolonged radius vector, and arrows in panels C and D point to the detached end of the plasma tail. The arrow in panel B denotes a bend in the detached tail; the severed end cannot be seen due to strong dust emission near the head. Photographs from top to bottom taken at: Mauna Kea Observatory, Yunnan Observatory, E.E. Barnard Observatory, and Royal Observatory (UK Schmidt, © Royal Observatory)

the interval, a shift in the $B_{x,y}$ components from predominantly (+) to predominantly (−) polarity can be seen taking place in a ~5-h transition starting on ~March 7.87 UT. As will be discussed shortly, this sector boundary was observed by several other spacecraft. Note that the observed (+/−) polarity sequence across the boundary is in agreement with the sense of the cometary magnetic barrier change observed by Vega-1 and Vega-2 (refer to previous discussion). The pair of $B_y > 0$ intervals during March 8.5–9.0 UT are small-scale reversals of 2–3 hr duration within the (−) polarity sector, and will not be further discussed here.

At the time the sector boundary swept by Vega-1, the comet-spacecraft differences in heliocentric distance, heliographic, latitude, and longitude were all very small: $\Delta r = 0.026$ AU (Halley farther from sun), $\Delta b = -1.°2$ (Halley south of Vega-1), and $\Delta l = -4.°7$ (Halley lagged in the sense of solar rotation). Working out the time delay based on Δr and Δl for an assumed 400 km s^{-1} solar-wind speed (for method, cf. Niedner et al., 1978; Brosius et al., 1987), the sector boundary should have arrived at Halley on ~March 7.67 UT, in very close agreement with the March 7.64 UT DE onset time "derived" from the imagery. We note that the spacecraft-comet latitudinal separation is so small that the corotated arrival time of the sector boundary is not significantly affected by our assumption of a north-south current sheet orientation in the corotation calculation.

Table 1. Geometric quantities for Halley and Measured positions of the March 8–10 DE

No. (1)	Date 1986 (UT) (2)	Source (3)	r (AU) (4)	Δ (AU) (5)	β (6)	DE distance (7)	DE velocity (8)
1.	Mar. 9.3444	CTIO/U. Mich. Curtis Schmidt	0.835	1.077	119°.0	2.899 E6 km	47.8 km s^{-1}
2.	Mar. 9.4760	E.E. Barnard Observatory	0.837	1.074	118°.9	3.444 E6 km	53.2 km s^{-1}
3.	Mar. 9.7765	U.K. Schmidt/Royal Observatory	0.841	1.067	118°.6	4.825 E6 km	57.9 km s^{-1}
4.	Mar. 9.9241	Yunnan Observatory	0.843	1.063	118°.5	5.564 E6 km	64.9 km s^{-1}
5.	Mar. 10.35	CTIO/U. Mich. Curtis Schmidt	0.849	1.053	118°.1	7.951 E6 km	53.6 km s^{-1}
6.	Mar. 10.4760	E.E. Barnard Observatory	0.851	1.050	118°.0	8.535 E6 km	84.0 km s^{-1}
7.	Mar. 10.7751	U.K. Schmidt/Royal Observatory	0.855	1.0425	117°.8	10.707 E6 km	

Notes:
Column (4): r is heliocentric distance of comet nucleus
Column (5): Δ is geocentric distance of comet nucleus
Column (6): β is angle between Earth-comet vector and prolonged radius vector from Sun

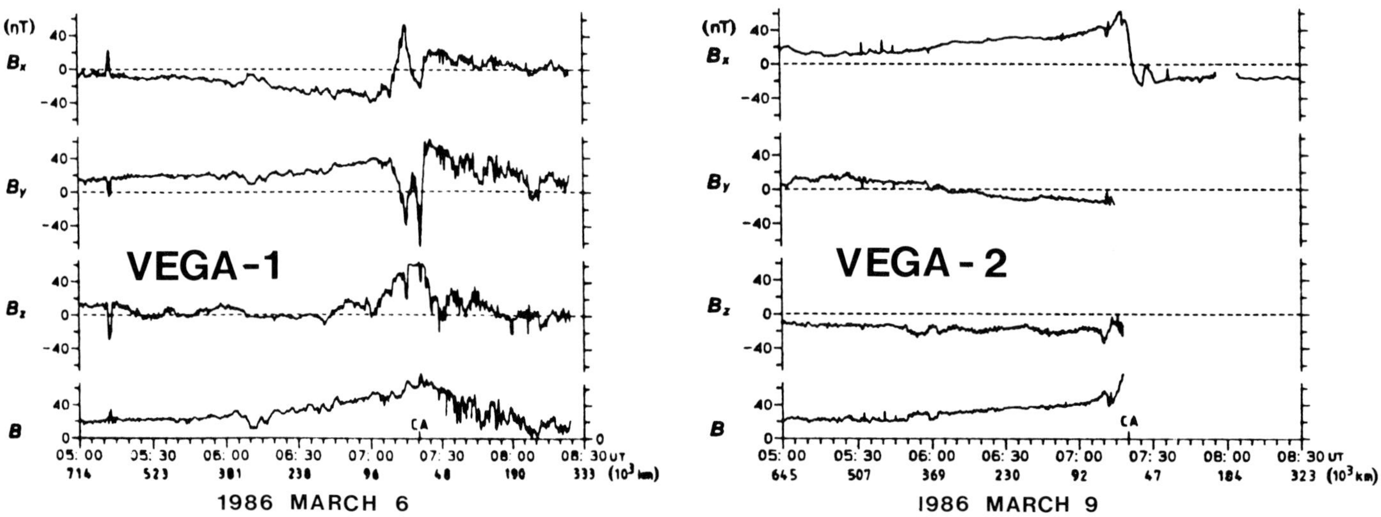

Fig. 3a and b. Magnetometer data for the Vega-1 (**a**) and Vega-2 (**b**) encounters (from Riedler et al., 1986). Refer to text for a description of the X, Y, Z coordinate system. The symbols "CA" refer to times of closest approach to Halley. Although a dual-polarity magnetic barrier is seen on March 6, the polarity of the main pile-up region changed from (+) to (−) IMF polarity between the Vega-1 and Vega-2 encounters

Table 2. Solutions of V_0, a, and t_{disc} for March 8 DE

Solution/observations used	V_0 (km s^{-1})	a (cm s^{-2})	t_{disc} (UT)
I/(1, 2, 3)	22.7	28.2	March 8.3716
II/(2, 3, 4)	28.4	24.6	March 8.4592
III/(3, 4, 5)	21.4	27.9	March 8.3374
Average:	24.2	26.9	March 8.39

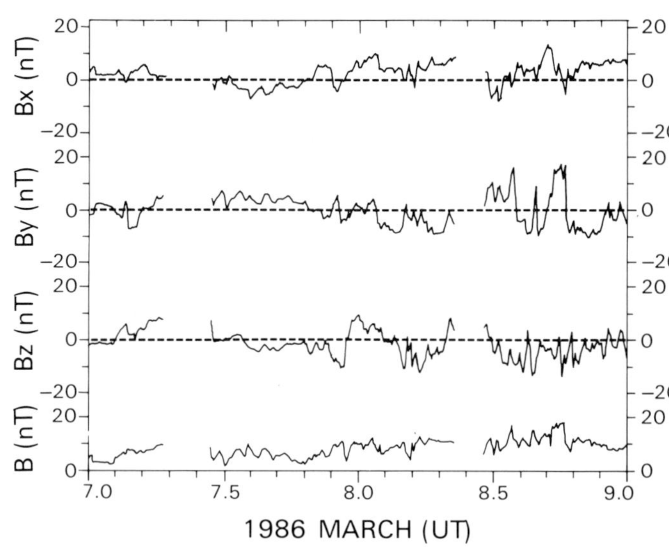

Fig. 4. Vega-1 cruise phase magnetometer data spanning March 7.0–9.0 UT; sign conventions for the field components are identical to those in Fig. 3. A shift from outward (+) to inward (−) IMF polarity begins on March 7.87 UT as B_y (B_x) acquire negative (positive) values. This (+/−) IMF sector boundary corotates almost exactly to the onset time of the disconnection process derived for the March 8–10 DE. Data have been re-drawn from original plots

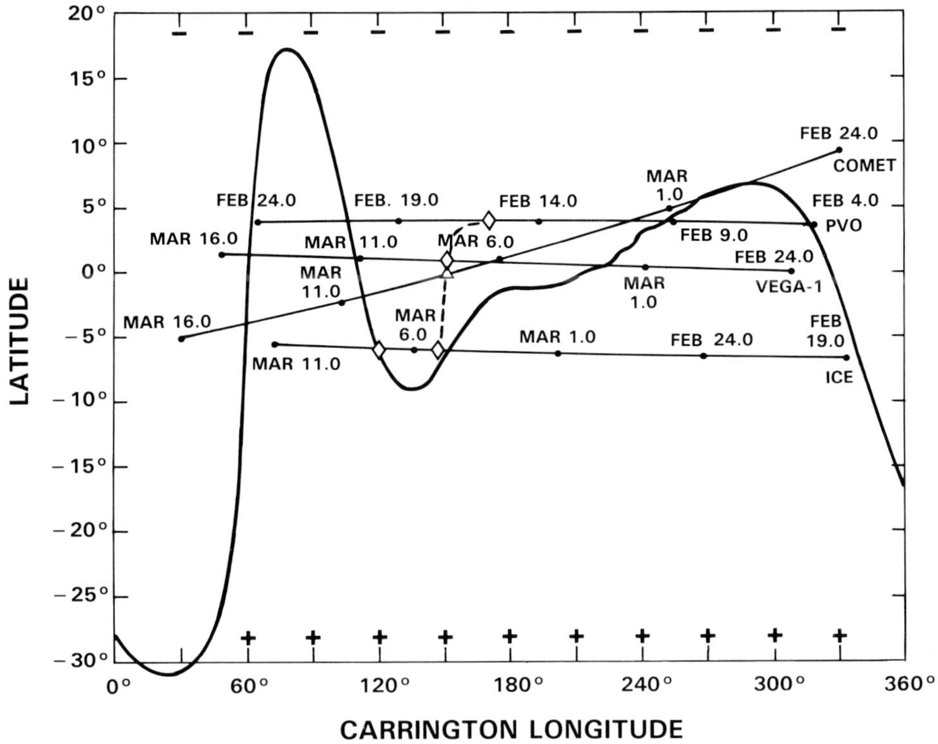

Fig. 5. The coronal neutral line (wavy) for Corrington Rotation 1772, with Vega-1, ICE, and PVO observed sector boundaries plotted as diamonds. The triangle along the comet path gives the approximate time of onset, March 7.64 UT, of the DE whose disconnection time was March 8.39 UT. Coronal longitudes and latitudes of the spacecraft and comet tracks were obtained by computing footpoint coordinates of the (constant) 400 km s^{-1} Archimedean spiral instantaneously connecting the corona and the object (i.e., spacecraft or comet). [Neutral line observations courtesy of Dr. J.T. Hoeksema and the Wilcox Solar Observatory]

Figure 5 is a 27-day synoptic diagram of Carrington Rotation 1772, showing the computed coronal neutral line (wavy), the polarity of the Sun's poles (− and + for north and south, respectively), and the coronal tracks made by the comet and several spacecraft. The neutral line observations were kindly provided by Dr. J.T. Hoeksema of Wilcox Solar Observatory, and the tracks were plotted by calculating the coordinates of the coronal footpoint of the Archimedean instantaneously connecting the Sun and the comet or spacecraft for an assumed constant solar-wind speed of 400 km s^{-1}. The diamonds on the spacecraft tracks indicate the times the various magnetometers observed a (+/−) sector boundary [exception: the second ICE-measured sector boundary was (−/+)], and the triangle on the comet track gives the inferred time of onset of the disconnection process, March 7.64 UT.

Note that although the onset time of the event is rather distant (several corotation days) from the extrapolated coronal neutral line, the Vega-1 observations of a highly-correlative sector boundary (discussed above) are supported by magnetometer data returned by other spacecraft operating in this Carrington longitude range. Specifically, the International Cometary Explorer (ICE) and Pioneer Venus Orbiter (PVO) both observed a (+/−) sector boundary near Carrington longitude 150°. If the diamonds and triangle are connected in Fig. 5 (dashed line), the interplanetary neutral surface becomes steeper than that given by the computed coronal data alone.

5. Conclusions

Our analysis of plasma-tail activity during the interval when the Vega spacecraft were in close proximity to the comet has shown that the spectacular DE of March 8–10 occurred in close temporal sequence with the arrival of an IMF sector boundary at the comet and with the reversal of the dayside magnetic barrier polarity observed by the Vega spacecraft at their encounters. The highly correlative set of events constitutes considerable support for the sector boundary/frontside magnetic reconnection model of DE's.

The next step in this effort will be to examine the Vega-1 and Vega-2 plasma data to ascertain the state of the solar-wind plasma near the sector boundary observed by Vega-1 on March 7.9 UT, and in particular, to probe the possible existence of a corotating interaction region which could perhaps have driven tailward reconnection and contributed to the March 8–10 DE. It is worth noting, however, that the observed reversal of the sunward magnetic barrier between the Vega encounters is not a prediction of high-speed stream models of DE's, but is the central feature of the sector boundary/frontside reconnection model.

Acknowledgements. It is a pleasure to thank the following individuals for their contributions to the present study: J.T. Hoeksema (coronal neutral line maps), E.J. Smith and J.A. Slavin (ICE magnetometer data), C.T. Russell (PVO magnetometer data), F.D. Miller (CTIO/Curtis Schmidt images), A. Storrs and D. Cruikshank (Mauna Kea imagery), B.W. Hadley (UK Schmidt imagery), Q. Puzhang (Yunnan Observatory imagery), and G. Emerson (E.E. Barnard Observatory imagery). The authors also thank referee K. Jockers for several helpful suggestions regarding improvement of this paper.

References

Brandt, J.C., Niedner, M.B., Rahe, J.: 1982, in *Cometary Exploration III* Budapest, Hungary, p. 47

Brosius, J., Holman, G., Niedner, M.B., Brandt, J.C., Slavin, J., Smith, E.J., Zwickl, R.D., Bame, S.J.: 1987, *Astron. Astrophys.* (in press)

Ip W.-H.: 1985, in ESA SP-235 p. 65

Ip, W.-H., Mendis, D.A.: 1978, *Astrophys. J.* **223**, 671

Jockers, K., Lust, R., Nowak, Th.: 1972, *Astron. Astrophys.* **21**, 199

Jockers, K.,: 1985, *Astron. Astrophys. Suppl.* **62**, 791

Niedner, M.B: 1980, *Astrophys. J.* **241**, 820

Niedner, M.B.: 1981, *Astrophys. J. Suppl.* **46**, 141

Niedner, M.B.: 1982, *Astrophys. J. Suppl.* **48**, 1

Niedner, M.B.: 1984, in *Magnetic Reconnection in Space and Laboratory Plasmas*, ed. E.W. Hones, AGU Geophysical Monograph No. 30, p. 79

Niedner, M.B.: 1986, *Adv. Space Res.* **6**, 315

Niedner, M.B., Brandt, J.C.: 1978, *Astrophys. J.* **223**, 655

Niedner, M.B., Brandt, J.C.: 1979, *Astrophys. J.* **234**, 723

Niedner, M.B., Ionson, J.A. Brandt, J.C.: 1981, *Astrophys. J.* **245**, 1159

Niedner, M.B., Rahe, J., Brandt, J.C.: 1982, in *Proc. ESO Workshop on The Need for Coordinated Ground-Based Observations of Halley's Comet*, Paris, p. 227

Niedner, M.B., Rothe, E.D., Brandt, J.C.: 1978, *Astrophys. J.* **221**, 1014

Niedner, M.B., Schwingenschuh, K.: 1986, *20th ESLAB Symp.*, ESA SP-250, Vol. **III**, p. 419

Niedner, M.B., Schwingenschuh, K.: 1987 (in preparation)

Riedler, W., Schwingenschuh, K., Yeroshenko, Y.G., Styashkin, V.A., Russell, C.T.: 1986, *Nature* **321**, 288

Russell, C.T., Saunders, M.A., Phillips, J.L., Fedder, J.A.: 1986, *J. Geophys. Res.* **91**, 1417

Schwingenschuh, K., Riedler, W., Schelch, G., Yeroshenko, Y.G., Styashkin, V.A., Luhmann, J.G., Russell, C.T., Fedder, J.A.: 1986, *Adv. Space Res.* **6**, 217

Somogyi, A.J., et al.: 1986, *Nature* **321**, 285

West, R.M., Pedersen, H., Monderen, P., Vio, R., Grosbol, P.: 1986, *Nature* **321**, 363

Astron. Astrophys. 187, 109–116 (1987)

ASTRONOMY AND ASTROPHYSICS

Observations of cometary plasma-wave phenomena

F.L. Scarf[1], F.V. Coroniti[1], C.F. Kennel[1], D.A. Gurnett[2], W.-H. Ip[3], and E.J. Smith[4]

[1] TRW Space and Technology Group RI/1176, Redondo Beach, CA 90278, USA
[2] University of Iowa, Iowa City, IA 52242, USA
[3] Max-Planck-Institut für Aeronomie, P.O. Box 20, D-3411 Katlenburg-Lindau, Federal Republic of Germany
[4] Jet Propulsion Laboratory, Pasadena, CA 91109, USA

Received February 24, accepted May 13, 1987

Summary. The ICE plasma wave investigation utilized very long electric antennas (90 meters tip-to-tip) and a very high sensitivity magnetic search coil to obtain: (1) significant local information on plasma physics phenomena occurring in the distant pickup regions of comet Giacobini-Zinner and comet P/Halley, and (2) information on the processes that developed in the coma and tail of Giacobini-Zinner. Since ICE traversed cometary regions that complemented those sampled by Vega and Sakigake, it is important to compare observations from the three missions that carried dedicated wave instruments. Here we summarize ICE plasma wave measurements associated with both comet encounters and relate the very high sensitivity ICE observations to corresponding measurements from the other Halley spacecraft.

Key words: comet Giacobini-Zinner – plasma waves – comet P/Halley – dust – plasma instabilities

1. Introduction

During a six-month period (September 1985 through March 1986) our understanding of comets changed drastically as six spacecraft made *in situ* measurements at comet Giacobini-Zinner (GZ) and comet Halley. These encounters provided great scientific advances in essentially every area of cometary physics, and many of the noteworthy discoveries involved plasma phenomena. The spacecraft flybys clearly demonstrated that a variety of very strong and dynamically important wave-particle interactions develop in a vast region surrounding a comet. Plasma wave instruments were included in the payloads of the International Cometary Explorer Vega-1 and -2, and Sakigake, and the comet observations show that: a) the distant ion pick-up region is characterized by the presence of significant levels of ion acoustic waves and whistler mode emissions, and; b) the closer-in region has extremely strong electromagnetic and electrostatic plasma turbulence associated with the large scale interaction of the mass-loaded solar wind and the cometary plasma. Localized wave enhancements are found during foreshock encounters, during traversals of boundaries with characteristics of collisionless bow shocks, and during entrance into the boundary of the critical ionization region.

In order to discuss the significance of the cometary wave measurements, we must take note of the differences in instrument capabilities and flyby trajectories for the various missions. Figure 1 contains a summary that covers both points for the natural comet encounters. As indicated here, short (Vega), medium (Vega and Sakigake) and long (ICE) electric field antennas were flown across the tail of GZ (ICE) and on the sunward side of Halley (Vega, Sakigake and ICE); since the instrument sensitivity is essentially proportional to the antenna length, it can be seen that the various E-field systems had great differences in the ability to detect low level wave phenomena.

Magnetic search coil wave sensors were only flown on ICE and Sakigake; however, on Sakigake the search coil was mounted directly on the spacecraft and the effective threshold was so high (H. Oya, private communication) that no significant cometary signals were detected. Thus, for all practical purposes, we can consider that, for frequencies above those covered by DC magnetometers, the ICE search coil provided the only magnetic field plasma wave information during the comet encounters.

2. Close-in observations

ICE had the most sensitive wave instrument and the ICE tracking coverage was virtually continuous for an extended interval surrounding the GZ encounter; thus it is appropriate to use the ICE plasma wave observations (Scarf et al., 1986a) as a basis for discussing all of the cometary encounters. Figure 2 contains a summary of the ICE wave measurements for a ten hour period on September 11, 1985. During this interval (in which the spacecraft-to-comet distance varied from a maximum of 375,000 km to a minimum of 7,800 km) ICE traversed the inner part of the ion pick-up region, the comet foreshock, the bow shock-strong coupling region, and the comet tail. The bottom panels in Fig. 2 show measurements from the magnetometer and the electron plasma probe, and the upper panels show the peak amplitudes measured in each of the bandpass channels of the ICE plasma wave instrument.

For each wave panel in Fig. 2, the top-to-bottom amplitude range is approximately five orders of magnitude, and it can be seen that exceptionally intense plasma waves were detected in a vast region around Giacobini-Zinner. In fact, these plasma turbulence levels are as large as any previously detected on ICE, and the low frequency B-field waves (near the lower hybrid resonance frequency, LHR) were so intense that the instrument saturated. In

---

Send offprint requests to: F.L. Scarf

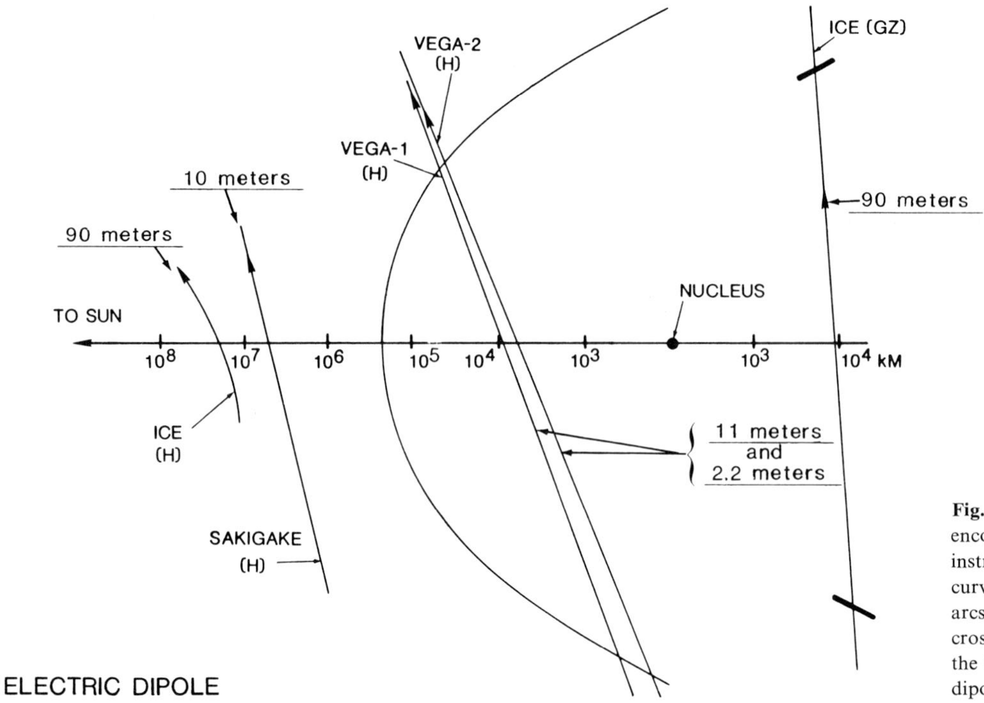

Fig. 1. Projected flyby trajectories for the comet encounter spacecraft with on-board wave instruments. The nominal Halley bow shock curve is shown; for Giacobini-Zinner the heavy arcs on the ICE(GZ) plot depict the shock crossing distances. The underlined numbers give the tip-to-tip antenna lenths for the electric dipoles

the ion acoustic wave frequency range (several hundred to several thousand Hertz), the E-field channels were nearly saturated as ICE passed through the strong coupling region (R on the order of 70,000 to 100,000 km), and we conclude that the comet-solar wind interaction is characterized by the development of exceptionally strong wave-particle interactions.

Another important aspect of the Giacobini-Zinner wave observations involves the large number of wave modes that are excited in the interaction region. Figure 3 shows calibrated peak and average E and B spectra from different regions on the inbound pass. As noted on the drawing, we were able to identify lower hybrid resonance (LHR) and ion acoustic waves, whistlers, and electron plasma oscillations (EPO). The B-field panels show a sharp jump in LHR and whistler mode amplitudes in the region near the 0911 UT discontinuity that Scarf et al. (1986a) identified as having characteristics similar to that of the bow shock, and this region was also associated with an increase in ion acoustic wave level. However, Fig. 3 clearly shows even higher intensities for all of these waves near 0930 to 1000. The latter region is generally interpreted as the "strong coupling region" (Coroniti et al., 1986), and Kennel et al. (1986) showed that here the E and B wave spectra were very similar to those measured at quasi-parallel shock transitions. In this strong coupling region it can be demonstrated that the wave-particle interactions were quite significant. Calculations by Coroniti et al. (1986) based on use of the measured wave amplitudes and plasma parameters suggest that this level of electrostatic turbulence will produce significant electron heating over time scales of minutes, while the whistlers will strongly pitch-angle scatter electrons on comparable time scales. Thus, the Giacobini-Zinner plasma wave data in the bow shock-strong coupling region clearly indicate that many wave modes are simultaneously excited by plasma instabilities, and that most of them have such large amplitudes that the associated wave-particle interactions are certain to be of great dynamical significance.

The actual identification of the Giacobini-Zinner bow shock is somewhat controversial. The lower panels in Fig. 4 show data from the energetic ion instrument along with unaveraged amplitude profiles from two E-field and two B-field filter channels. It can be seen that the wave activity was as extremely impulsive, with huge amplitude excursions from one measurement to the next, one second later. In these plots the existence of a broad strong coupling region with enhanced wave amplitudes is clearly evident (at 0930 to 1000 and at 1140 to 1220), and for the outbound case, the end of this region has a sudden shock-like transition to relatively quiet plasma conditions. The upper right panel, taken from the initial paper by Scarf et al. (1986a) shows all of E-field wave levels from this outbound boundary region, and it is clear why several of the ICE plasma wave investigators called this a bow shock.

The inbound case is more complex. At 0911, where the wave instrument detected an abrupt onset of broadband turbulence (see upper left panel), the magnetometer and electron plasma probe measured relatively smooth variations in N and B. However at this same time, it is evident that the energetic ion instrument detected large and sudden increases in the fluxes of energetic heavy ions. This suggests that the 0911 plasma wave discontinuity was associated with spacecraft penetration into the outer boundary of a region with enhanced fluxes of gyrating energetic heavy ions.

3. The comet tail traversal

The ICE encounter with comet Giacobini-Zinner provided the only traversal of a comet tail, and thus it is of great interest to analyze the plasma physics phenomena in this region. The summary plot in Fig. 2 shows that in many of the plasma wave filter channels ($f > 1$ kHz for E; $f > 18$ Hz for B), the peak wave

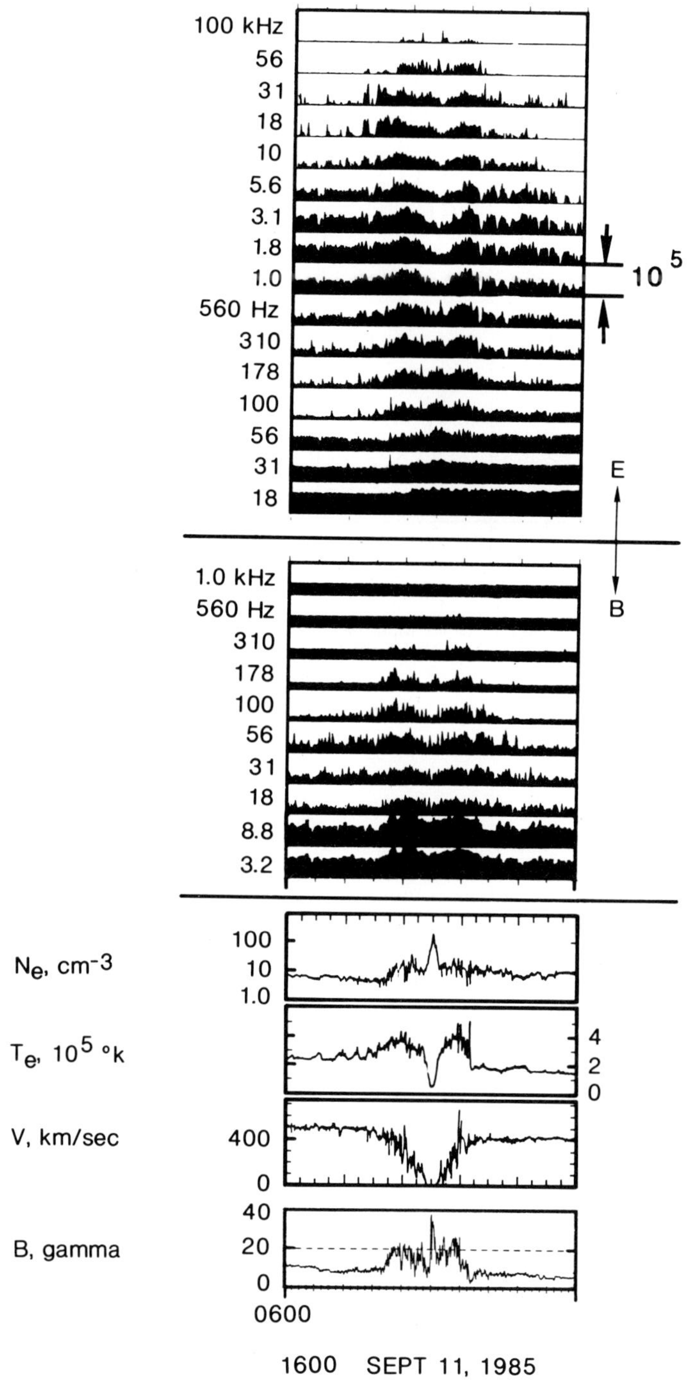

Fig. 2. Summary of plasma wave amplitude data, plasma parameters and B-field profile for the ICE encounter with comet Giacobini-Zinner

levels were relatively low near closest approach. However, for the lower frequency E and B channels, the near-comet observations included some very large amplitude readings. Indeed, the closest-approach spectra in Fig. 3 (1057–1102) show that for $f < 10$ Hz, the B-field waves were almost as intense as the peak values detected in the strong coupling region (0936–0941). Moreover, for the lower frequency E-field waves (f near 56 to 100 Hz), it turns out that the *strongest* signals were actually detected near closest approach.

Some of these tail waves can be identified readily, but in other cases, more discussion is required. It is natural to relate the low frequency magnetic field peak to the lower hybrid resonance, and the high frequency E-field enhancement to electron plasma oscillations. However, detailed analysis of the in-tail wave measurements suggests that the 56–100 Hz E-field spectral peak in Fig. 3 is actually associated with dust impacts (see Gurnett et al., 1986), and that the real E-field plasma waves had moderate amplitudes in this region.

This point is illustrated in Fig. 5, which shows wave profiles for 13 of the filter channels (again, five orders of magnitude for each panel) along with a B-field plot labeled with the magnetopause locations defined by Slavin et al. (1966). The data in the wave panels clearly have a few isolated noise spikes (dust impacts) superimposed on a low level mid-frequency spectrum that does not show any relation with the local electron cyclotron frequency. We conclude that strong whistler turbulence was not present in the comet tail and we identify these plasma waves as electrostatic. Since acoustic waves for protons and heavy ions would have (rest frame) frequencies below a few kiloHertz, these are logical candidates, and the sporadic broadband noise bursts (e.g., at 15,000 km from the tail axis) would seem to resemble the broadband electrostatic bursts detected on the boundaries of the earth's sheet (Scarf, 1986c). Thus, although the tail plasma waves only had moderate amplitudes, it is likely that they were dynamically significant.

4. Waves in the distant pick-up region

The discovery of very high intensities for cometary plasma waves was quite surprising, and another unexpected result from the *in situ* measurements is related to the huge size of the interaction region. Although it was found that Giacobini-Zinner had an extensive conventional foreshock (see, for example, Fuselier et al., 1986), Scarf et al. (1986a, b) noted that ICE measured effects of pick-up ions out to at least four million kilometers from the nucleus of Giacobini-Zinner. The full set of very remote measurements from ICE also involved direct detection of energized pick-up ions (Sanderson et al., 1986) and Scarf et al. (1986b) showed that these particle observations were correlated with simultaneous measurements of electrostatic ion acoustic waves and electromagnetic whistlers. Thus, it was concluded that these waves detected at great distances were associated with the pick-up of Giacobini-Zinner ions by the magnetized solar wind.

Late in March, 1986, ICE passed very far in front of Halley and the plasma wave search coil again detected enhanced whistler mode turbulence with characteristics very similar to those measured in the distant ion pick-up region of Comet Giacobini-Zinner (electrostatic waves were also detected here but it is likely that some of these were simply associated with the solar wind). Figure 6 compares the search coil wave profiles from Halley and Giacobini-Zinner; Scarf et al., (1986b) analyzed these observations and concluded that the March 1986 measurements of whistler mode activity were associated with the Halley ion pick-up process. Wenzel et al. (1986) reported that Halley-associated ions were also detected on ICE at this time, and a comparison of ion and wave profiles for an eight-day period in March, 1986, when ICE was near its closest approach to Halley suggested a close (but not one-to-one) correlation. Since the wave enhancements were

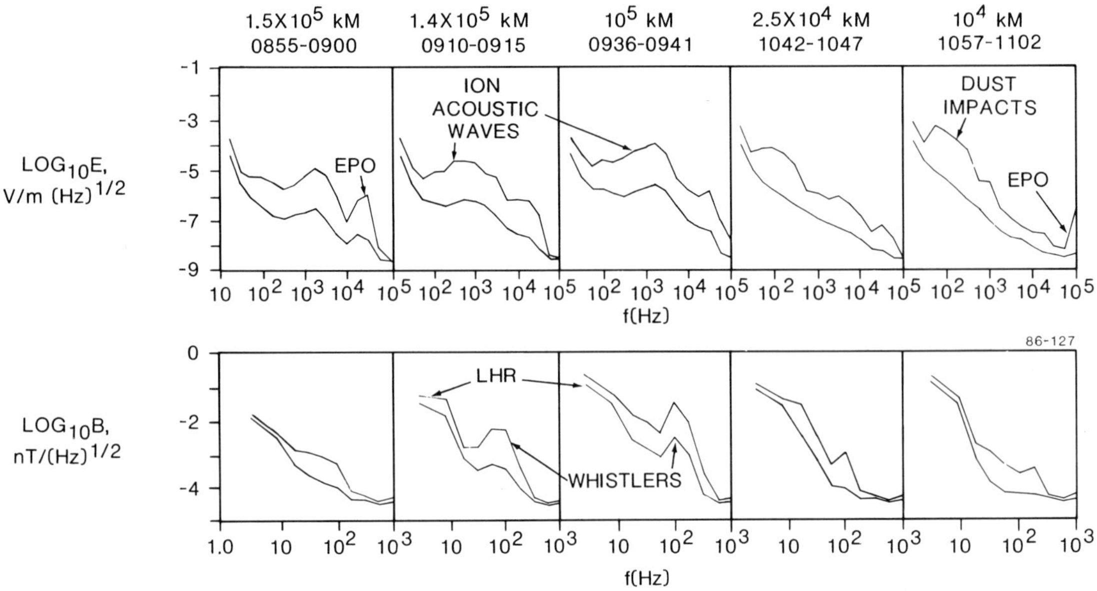

Fig. 3. Characteristic peak and average E and B wave spectra measured as ICE moved inbound from the solar wind (0855 – 0900) to Giacobini-Zinner closest approach (1102). Here LHR represents the lower hybrid resonance emission and EPO represents electron plasma oscillations

generally detected during periods when energetic ion flux enhancements were observed, this could mean that: a) the pick-up ion distribution was unstable with respect to production of plasma turbulence, and; b) wave-particle interactions associated with the pick-up process could have produced significant changes in the distribution function.

However, on the basis of particle observations from ICE, the identification of the energetic particles as heavy ions from the comet is not certain, and it is worth noting that the ICE approach to Halley occurred when the spacecraft was in the midst of a recurrent high speed solar wind stream. High solar wind velocities do make it easier to detect pick-up ions on ICE (Sanderson et al., 1986; Wenzel et al., 1986), but it is also possible that the flux enhancements represented ions of solar origin. Because of these uncertainties, we examined the ICE wave data during the previous passage of the stream. Figure 7 compares the search coil amplitudes for the first five days in March with corresponding measurements from the Halley closest approach interval. It can be seen that some low frequency electromagnetic turbulence was detected during the initial stream passage, but the near-Halley levels seem significantly higher, for longer intervals. This again suggests that ICE measured Halley-associated plasma turbulence late in March; however, the conclusion is clearly speculative and a firm decision about the size of the Halley ion pick-up region will probably have to await more definitive measurements in 2062.

5. Comet comparisons

As noted in Fig. 1, the plasma wave instruments on Sakigake and Vega used shorter dipoles, and thus they were less sensitive. The four lower plots in Figure 8 contain threshold data for Sakigake (Oya et al., 1986), Vega (Klimov et al., 1986 and Grard, private communication) and ICE; it can be seen that the long E-field antennas on ICE allowed detection of very low level waves, since the ICE threshold was as much as two orders of magnitude below those from the other spacecraft.

This difference in sensitivity did not prove to be a problem because the Vega investigators actually detected considerably stronger signals at Halley than the very intense ones previously obtained by ICE at Giacobini-Zinner. Klimov et al. (1986) and Grard et al. (1986) found wave spectra with peak intensities more than a factor of ten higher than the peaks reported during the GZ encounter. The upper curves in Fig. 8 show the most intense Halley spectra reported by the Vega teams (Grard et al., 1986; Klimov et al., 1986), and the dashed curve is the 0910–0915 peak ICE spectrum from Fig. 3, above. The Vega wave investigators also found that the inbound Halley bow shock was more clearly defined that the GZ shock (Galeev et al., 1986a), and that enhancements in lower hybrid resonance emissions appeared to be associated with critical ionization phenomena closer to the nucleus (Galeev et al., 1986b).

6. Summary

The *in situ* observations from the Halley and Giacobini-Zinner encounters conclusively demonstrate that plasma wave investigations provide first order information on the physics of the comet-solar wind interaction, the characteristics of the comet tail, mass loading and cometary pick-up processes, and phenomena associated with ionization and particle acceleration. The measured wave levels appear to produce strong wave-particle interactions over enormous distances, and the wave observations also yield extremely significant diagnostic information on the cometary plasma environment.

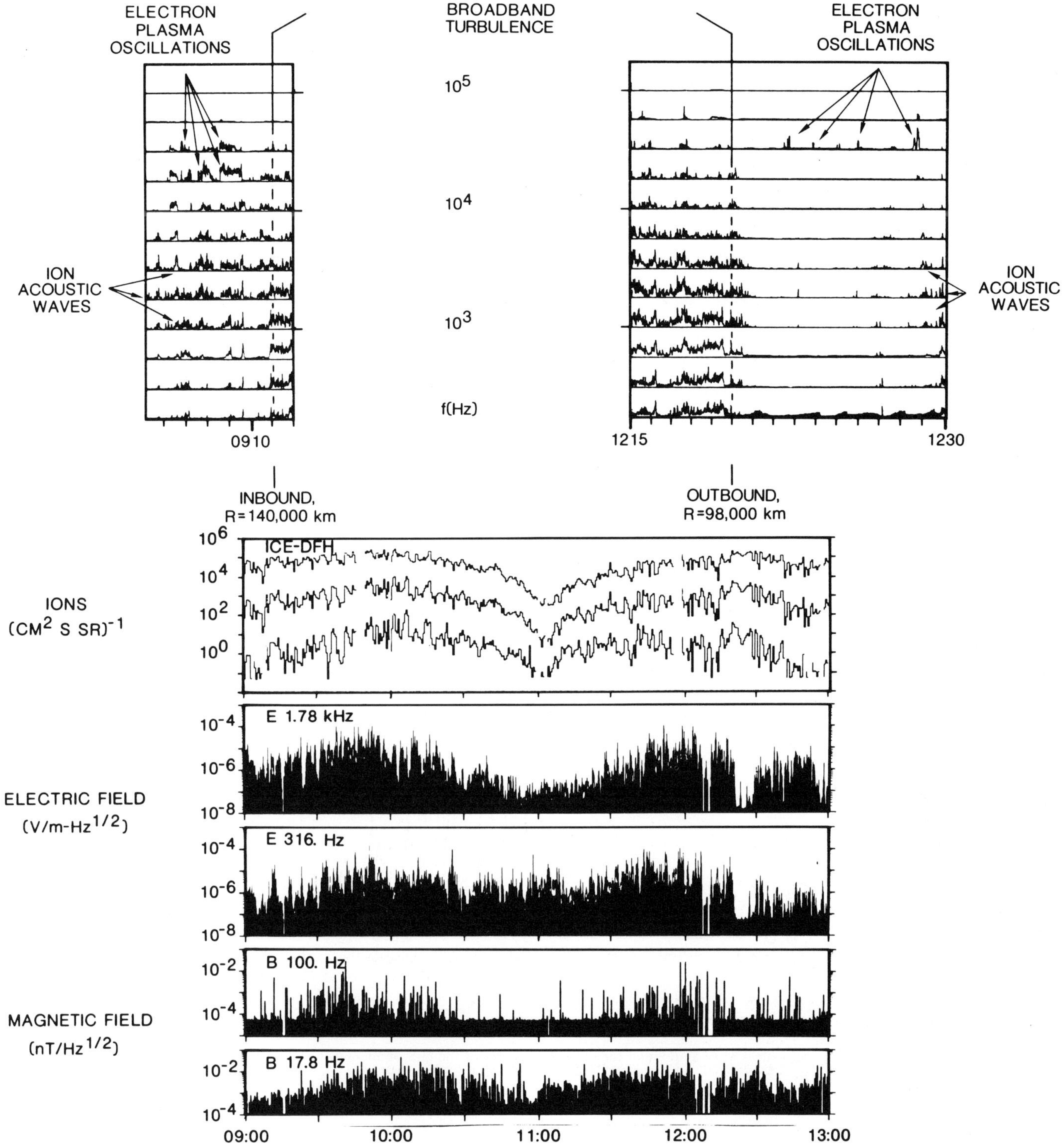

Fig. 4. Bottom: Unaveraged wave profiles in four channels, along with energetic particle fluxes. Top: Unaveraged profiles for all 16 E-field channels. The vertical dashed lines mark the events with collisionless bow shock wave characteristics

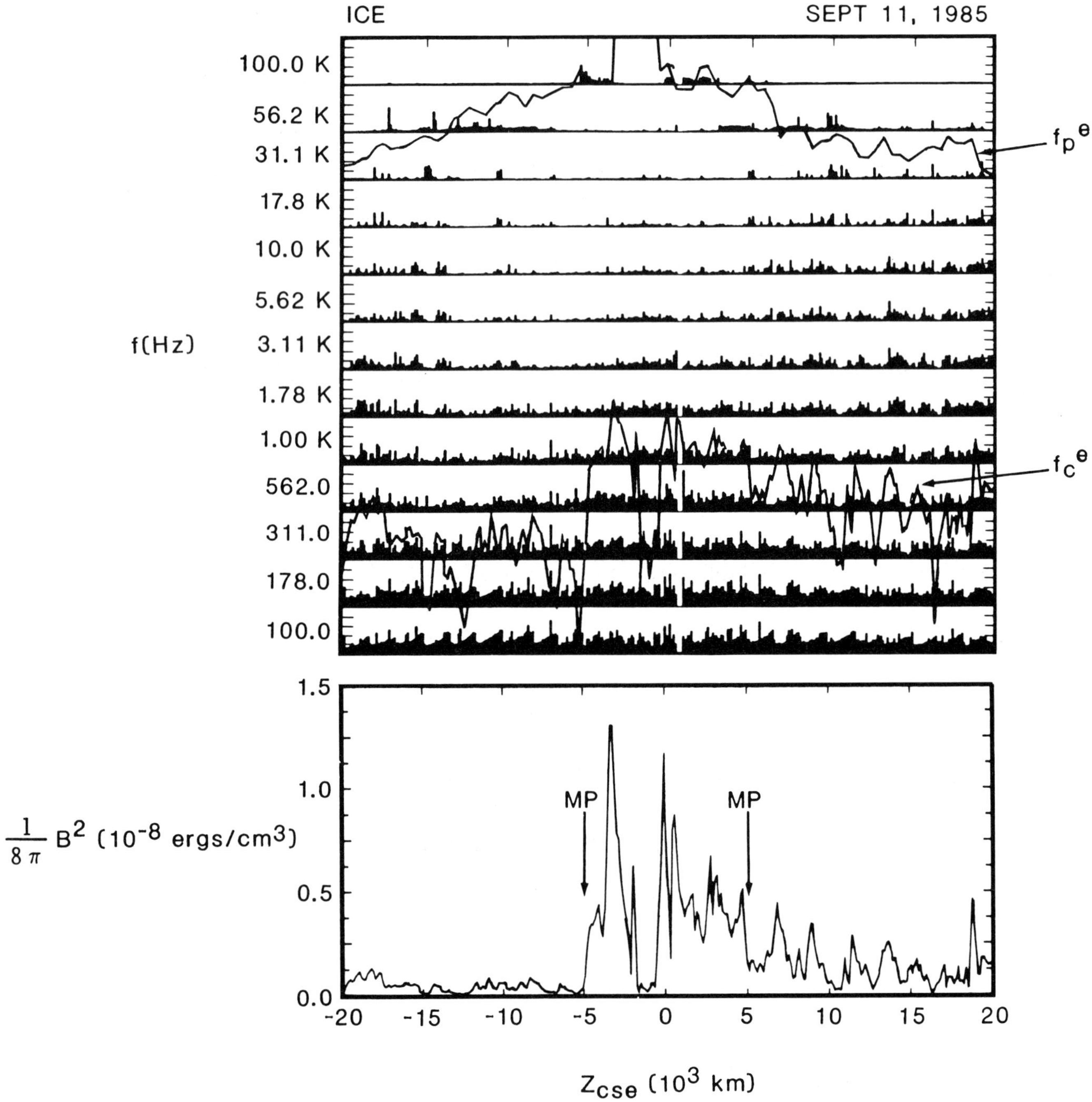

Fig. 5. B-field magnitude and 13-channel E-field wave amplitudes for the ICE crossing of the Giacobini-Zinner plasma tail

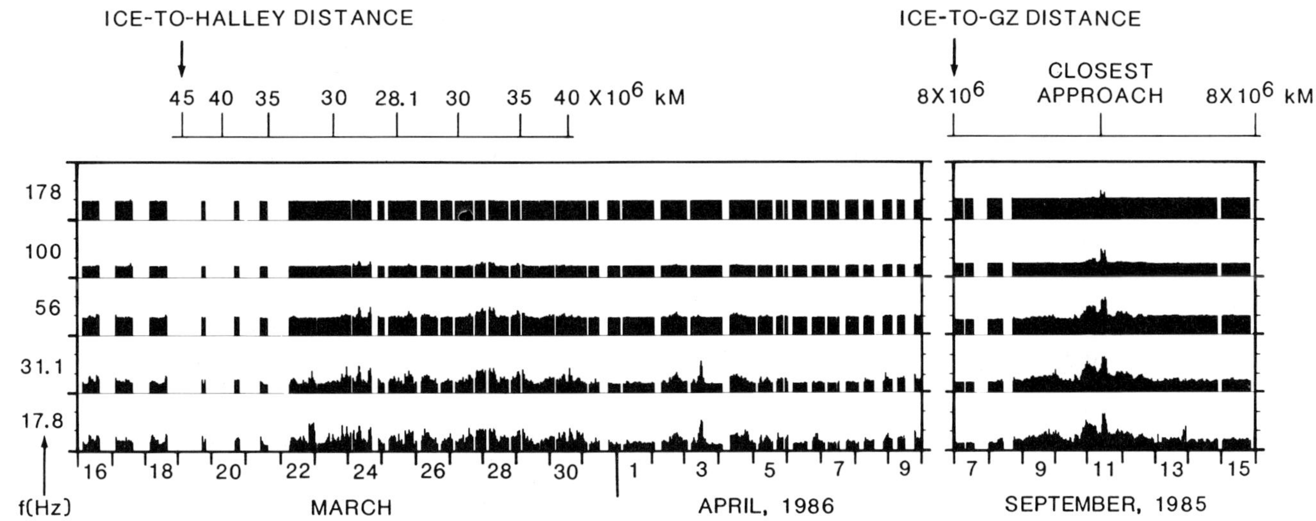

Fig. 6. Low frequency B-field waves observed during closest approach to comets Halley and Giacobini-Zinner

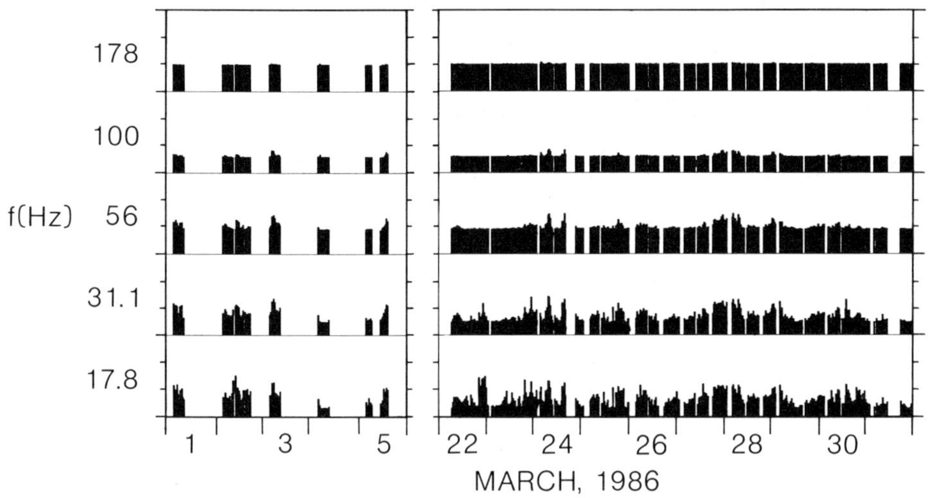

Fig. 7. B-field wave activity detected during the Halley closest approach and during the previous passage of the recurrent high-speed stream

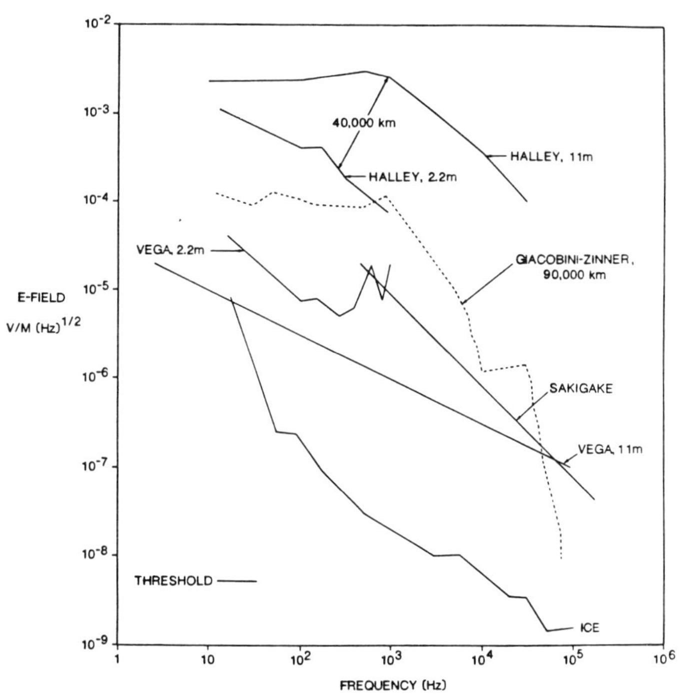

Fig. 8. Plasma wave spectra from Halley (Vega-1 and -2) and Giacobini-Zinner (ICE), together with E-field threshold curves for the comet plasma wave instruments

Acknowledgements. We are indebted to our colleagues on ICE for many helpful discussions about the Halley and Giacobini-Zinner observations, and we thank R. Grard, S. Klimov and H. Oya for allowing us to show the threshold curves and Halley results in Fig. 8. This research was supported by NASA under Contracts NAS5-28703 (TRW) and NAS5-28701 (Iowa).

References

Bame, S.J., R.C. Anderson, J.R. Asbridge, D.N. Baker et al.: 1986, *Science* **232**, 356

Coroniti, F.V., C.F. Kennel, F.L. Scarf, E.J. Smith, B.T. Tsurutani, S.J. Bame, M.F. Thomsen, R. Hynds, K.-P. Wenzel: 1986, *Geophys. Res. Lett.* **13**, 869

Fuselier, S.A., W.C. Feldman, S.J. Bame, E.J. Smith, F.L. Scarf: 1986, *Geophys. Res. Lett.* **13**, 247

Galeev, A.A., B.E. Gribov et al., T. Gombosi, K.I. Gringauz, et al.: 1986a, *Geophys. Res. Lett.* **13**, 841

Galeev, A.A., K.I. Gringauz, S.I. Klimov, A.P. Remizov, et al.: 1986b, *Geophys. Res. Lett.* **13**, 845

Grard, R., A. Pedersen, J.-G. Trotignan, C. Beghin, M. Mogilevsky, Y. Mikhailov, O. Molchanov, V. Formisano: 1986, *Nature* **321**, 290

Gurnett, D.A., T.F. Averkamp, F.L. Scarf, E. Grun: 1986, *Geophys. Res. Lett.* **13**, 291

Klimov, S., S. Savin, Ya. Aleksevich, G. Avanesova et al.: 1986, *Nature* **321**, 292

Kennel, C.F., F.V. Coroniti, F.L. Scarf, B.T. Tsurutani, E.J. Smith, S.J. Bame, J.T. Gosling: 1986, *Geophys. Res. Lett.* **13**, 921

Oya, H., A. Morioka, W. Miyake, E.J. Smith, B.T. Tsurutani: 1986, *Nature* **321**, 307

Sanderson, T.R., K.-P. Wenzel, P. Daley, S.W.H. Cowley et al.: 1986, *Geophys. Res. Lett.* **13**, 411

Scarf, F.L., F.V. Coroniti, C.F. Kennel, D.A. Gurnett, W.-H. Ip, E.J. Smith: 1986a, *Science* **232**, 377

Scarf, F.L., F.V. Coroniti, C.F. Kennel, T.R. Sanderson, K.-P. Wenzel, R.J. Hynds, E.J. Smith, S.J. Bame, R.D. Zwickl: 1986b, *Geophys. Res. Lett.* **13**, 857

Scarf, F.L.: 1986c, *Adv. Space Res.* **6**, 329, 1986c

Slavin, J.A., E.J. Smith, B.T. Tsurutani, G.L. Siscoe, D.E. Jones, D.A. Mendis: 1986, *Geophys. Res. Lett.* **13**, 283

Smith, E.J., B.T. Tsurutani, J.A. Slavin, D.E. Jones, G.L. Siscoe, D.A. Mendis: 1986 *Science* **232**, 382

Wenzel, K.-P., T.R. Sanderson, I.G. Richardson, S.W.H. Cowley et al.: 1986, *Geophys. Res. Lett.* **13**, 861

Astron. Astrophys. 187, 117–120 (1987)

ASTRONOMY AND ASTROPHYSICS

Hydromagnetic waves associated with cometary water group ions: Sakigake observation

K. Yumoto, T. Saito, and T. Nakagawa

Onagawa Magnetic Observatory and Geophysical Institute, Tohoku University, Aza-Aoba Aramaki, Sendai 980, Japan

Received February 9, accepted April 15, 1987

Summary. Long-period hydromagnetic waves near the local "water group" ion cyclotron frequency were detected at the closest approach of Sakigake to comet P/Halley. These transverse waves having $\sim 2nT$ peak-to-peak amplitude propagated approximately parallel to the ambient field. The necessary cometary pickup-ion density of "water group" ions for excitation of these waves at $7\,10^6$ km upstream from the comet can be inferred to be of the order of $10^{-3}\,\mathrm{cm}^{-3}$.

Key words: HM wave – water group ions

1. Introduction

A pickup-ion region of energetic "water group" ions of $\sim 2\,10^{-2}$ cm^{-3} was identified to extend to around $\sim 4\,10^6$ km from comet Halley by the ESP instrument on board Suisei during the encounter on 8 March 1986 (Mukai et al., 1986; Terasawa et al., 1986). Burst-like temporal enhancements of energetic heavy ion fluxes were also observed by the EPAS instrument on the International Cometary Explorer (ICE) during its closest approach ($28\,10^6$ km) to comet Halley in late March 1986 (Wenzel et al., 1986). Radio wave enhancements observed by Sakigake within a region of $7\,10^6$ km from comet Halley (Oya, 1986) and sporadic bursts of strong plasma turbulence detected near the closest approach of ICE to comet Halley ($\sim 30\,10^6$ km; Scarf et al., 1986) were interpreted in terms of plasma instabilities associated with the solar wind pickup of ions produced by heavy neutrals from Halley. These newly created cometary ions are also reported to clearly generate low-frequency hydromagnetic (HM) waves near the "water group" ion cyclotron frequency at distances of $7\,10^6$ km (Sakigake; Yumoto et al., 1986a) and of $30\,10^6$ km (ICE; Wenzel et al., 1986) from comet Halley, via a resonant instability (cf. Tsurutani and Smith, 1986a, b; Wu and Davidson, 1972; Wu et al., 1986).

In this paper, we will show the characteristics of the low-frequency HM waves observed by Sakigake in the outer turbulence region (7–$9\,10^6$ km), and then deduce the plasma parameter of the "water group" pickup ions exciting these HM waves.

2. Instrumentation

Sakigake/MS-T5 was launched on 8 January 1985 at Kagoshima Space Center of ISAS (Hirao and Itoh, 1986). A tri-axial ring-core fluxgate vector magnetometer was installed on board Sakigake at the tip of a 2 m-boom, which was deployed on 19 February 1985. Thereafter, the IMF (Yumoto et al., 1986a) data were telemetered at a high bit rate sampling a vector field measurement every $\frac{1}{8}$ s until 14 May 1985. Since then the IMF data have been successfully obtained at a low bit rate with a sampling period of 4 s, which is equivalent to a spatial resolution of ~ 120 km. The instrument had a more than adequate sensitivity ($0.03215nT$) and dynamic range (a capability to measure maximum field strengths of $\pm 50000nT$) and performed faultlessly (Saito et al., 1986). The magnetic noise level influenced by instrument performance is equivalent to $\pm 0.1nT$, rms.

Send offprint requests to: K. Yumoto

3. Observations

At the closest approach of Sakigake ($7\,10^6$ km) to comet Halley at 0417 UT on 11 March 1986, long-period HM waves were detected as shown in Fig. 1 (Yumoto et al., 1986a). These HM waves are found to be attributable to the turbulent plasma of cometary origin, which was generated by the ion-pickup phenomena in the upstream region of comet Halley. In order to identify the dominant mode of the low-frequency HM waves, the three-component IMF data in the solar ecliptic (X, Y, Z) coordinates are transformed into local magnetic field coordinates (B_1, B_2, B_3). The local IMF direction is estimated using 10-min averages of the X, Y, and Z components. The B_3 axis is taken parallel to the local IMF direction, the B_2 axis in the X, Y plane perpendicular to the B_3 axis, and the third B_1 axis is defined by $\boldsymbol{e}_{B1} = \boldsymbol{e}_{B2} \times \boldsymbol{e}_{B3}$. Figure 1 shows an amplitude-time magnetic record in the (B_1, B_2, B_3) coordinates near the closest approach of Sakigake to comet Halley ($r \sim 7\,10^6$ km) during 6 hours from 0000 to 0600 UT on 11 March 1986. In this interval, HM oscillations with 150–300 s periods were detected predominantly in the B_1 and B_2 directions, which are perpendicular to the local ambient field. Thus, these long-period HM waves can be categorized as transverse mode. The wave amplitude was as large as $\sim 2nT$ peak-to-peak. Ratios of the perturbation field to the ambient field strength ranged from ~ 0.2 to 0.5. The HM wave power spectral density appeared to peak at a frequency near the gyrofrequency of O^+ ions.

In order to examine the relation between the IMF magnitude and the predominant period of these waves observed near the closest approach during the period from 10 to 13 March 1986, dynamic spectrograms of the total field (B_{IMF}) and of its X, Y, Z components in solar ecliptic coordinates were constructed. Two

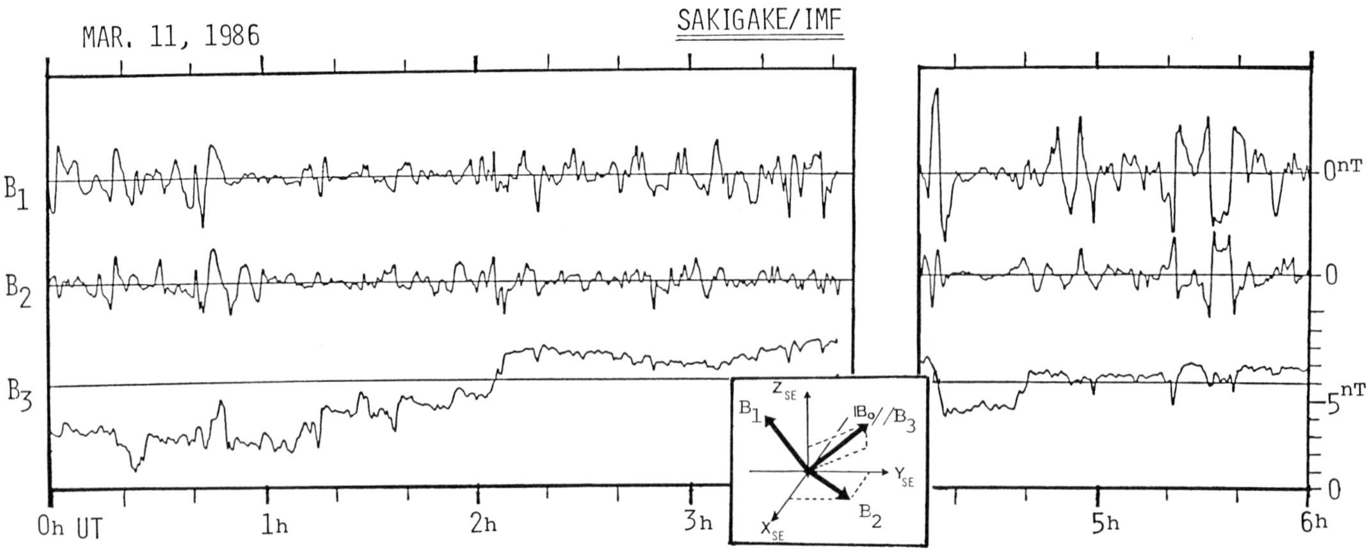

Fig. 1. Amplitude-time records of long-period HM waves in the local field-line coordinates near the closest approach of Sakigake to comet Halley

representative dynamic spectrograms associated with water-group ion cyclotron waves are shown in Fig. 2. The horizontal solid lines in each panel of the figure represent the local ion cyclotron frequencies (i.e., $\Omega_c = qB_{IMF}/2\pi m_c$) of Fe^+, CO^+, H_2O^+, O^+, CH^+, and C^+ ions, which were estimated using a 20-min average of the total field (B_{IMF}). q and m_c stand for electric charge and mass of the cometary heavy ions, respectively. Enhancements of linearly polarized oscillations at the local O^+ (or H_2O^+) ion cyclotron frequency can be seen in the Y component, with periods of ~300 s from 0000 to 0100 UT, and with periods of ~150 s from 0220 to 0350 UT on 11 March 1986. The frequencies of the enhanced oscillations are found to change as a function of the local ambient field intensity, and coincide with the local O^+ (or H_2O^+) ion cyclotron frequency. These characteristic magnetic oscillations were observed in the 6 hr interval from 0000 to 0600 UT on 11 March 1986 of the closest approach of Sakigake to comet Halley. Sporadic enhancements of HM waves near the H_2O^+–O^+ cyclotron frequency are also seen in the Z-component during the period of 0100–0210 UT on 12 March 1986 as shown in the bottom panel of Fig. 2. These characteristic waves having

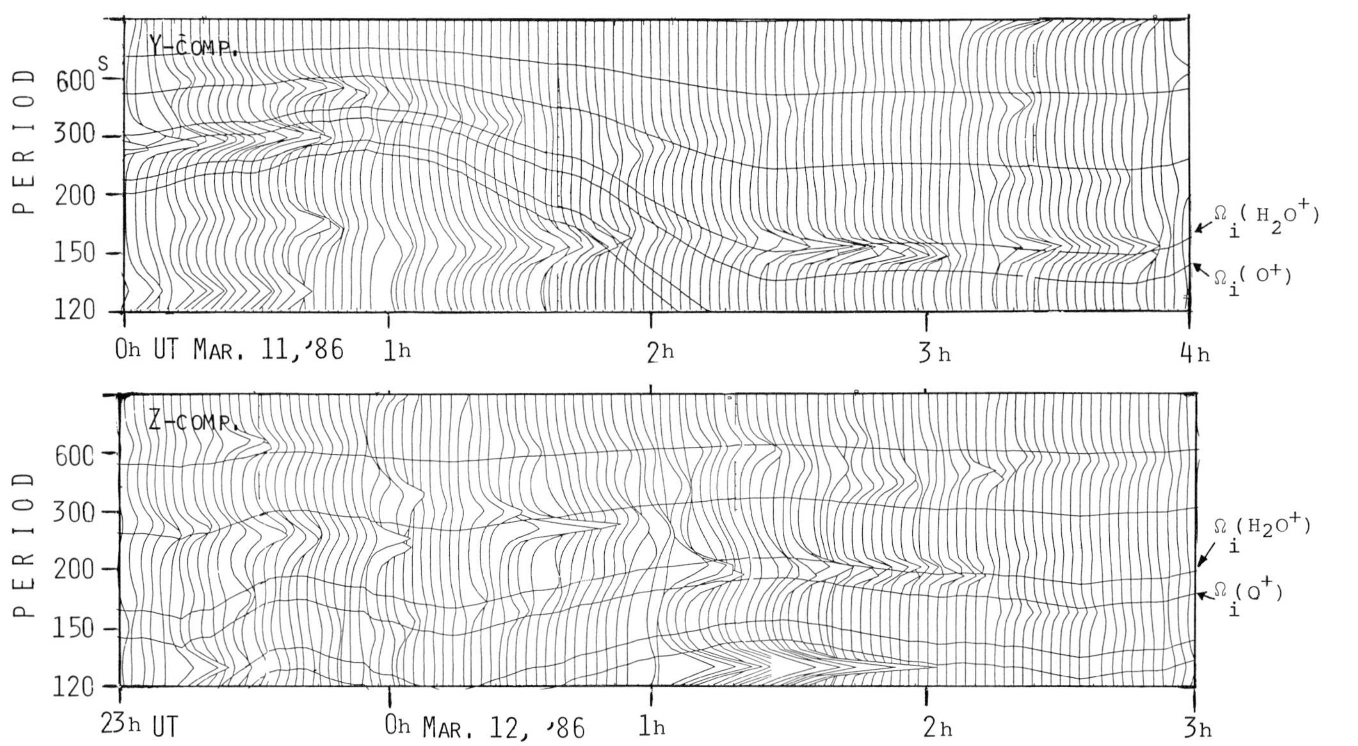

Fig. 2. Dynamic spectrograms for the Y and Z components of the IMF during the periods of 0000–0400 UT on 11 March 1986 and from 2300 UT on 11 March to 0300 UT on 12 March 1986, respectively, near the closest approach of Sakigake to comet Halley. Solid lines indicate the local ion cyclotron frequencies

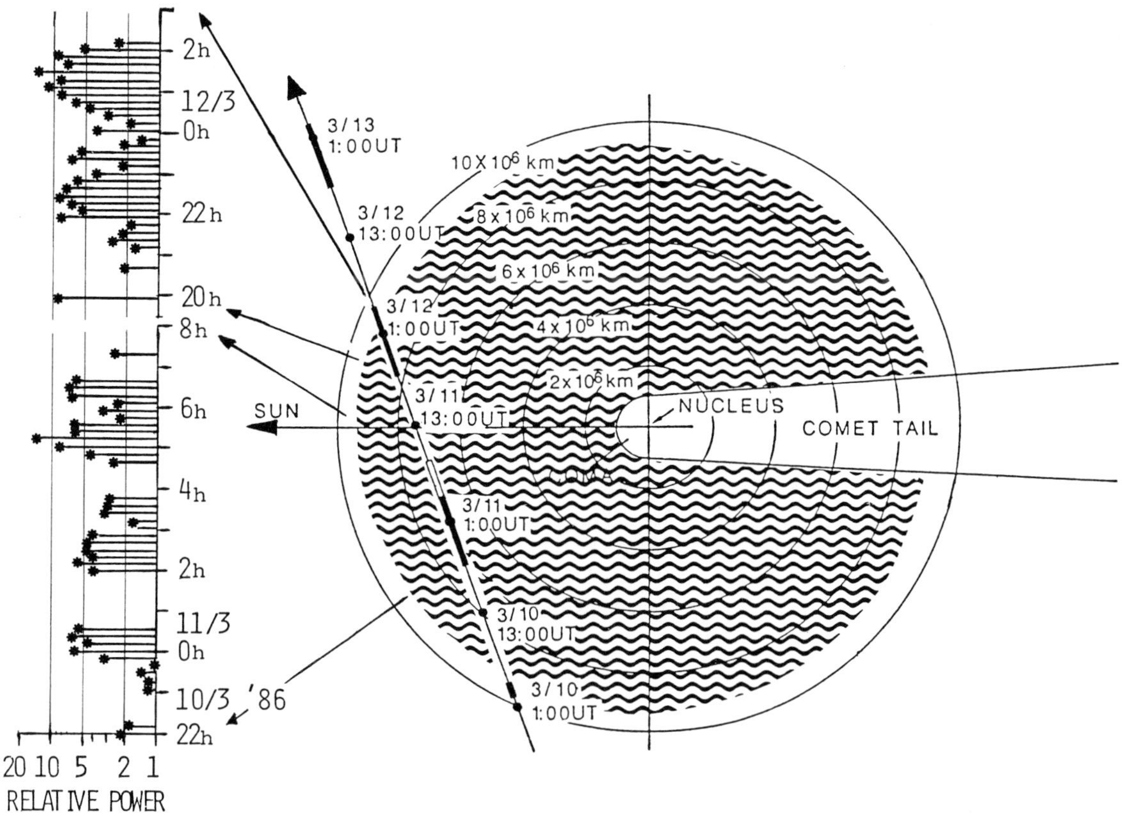

Fig. 3. Relative magnetic powers near the water group ion cyclotron period, the hydromagnetic turbulent region of $\delta B \gtrsim 2nT$, and the trajectory of Sakigake

the local "water group" ion cyclotron frequency were not detected by Sakigake in the other intervals before 0400 UT on 10 March and after 2100 UT on 12 March 1986 in the normal solar wind.

Figure 3 shows relative magnetic powers near the local O^+ ion cyclotron frequency during the period from 2150 UT on 10 March to 0300 UT on 12 March 1986. The solid line at 2 relative power marks the spectral power at an intermediate level of solar wind activity. Principal wave enhancements above this level were detected in the intervals from 0000 to 0800 UT on 11 March and from 2100 UT on 11 March to 0210 UT on 12 March 1986. These enhancements of the low-frequency HM waves are correlated with the very intense broadband emissions of electrostatic plasma waves in the 70–300 Hz frequency range (see Oya, 1986). The right panel of the figure shows the trajectory of Sakigake and HM turbulence region of $\delta B \gtrsim 2nT$ which can be identified to extend to around $\sim 9\ 10^6$ km from comet Halley by the IMF/Sakigake instrument.

4. Conclusion and discussion

In order to interpret the observational characteristics of the long-period HM waves as shown in Fig. 1, we considered generation by a resonant instability in terms of the pickup of cometary heavy ions as a possible model (Yumoto et al., 1986a). If the mean amplitude (δB) of the observed low-frequency waves is of the order of the magnetic field induced by the current of high-energy ions propagating parallel to the ambient field, i.e., $\delta B \sim \mu_0 I/(2\pi r_c) \sim \mu_0(qn_c V_{\parallel}\pi r_c^2)/2\pi r_c$ and $r_c = m_c V_{\perp}/qB_{\rm IMF}$, the lower threshold for the density n_c of resonant particles exciting these waves can be estimated as follows:

$$(n_c/n_i) = 2m_i/m_c(\delta B/B_{\rm IMF})(V_A^2/V_{\parallel}V_{\perp}), \qquad (1)$$

where n_i is the solar wind density, and V_A, $V_{\parallel}$ and $V_{\perp}$ the Alfvén speed based on the solar wind mass and density ($B_{\rm IMF}^2/\mu_0 n_i m_i$) and the resonant beam velocities of the heavy ion of mass m_c parallel and perpendicular to the magnetic field, respectively. The threshold value of Eq. (1) is consistent with that inferred by the magnetic field saturation at a value δB, which occurs when the magnetic bounce frequency $\omega_b = (qV_{\perp}\delta Bk/m_c c)^{1/2}$ for particles trapped near the bottom of the wave potential increases to a value comparable to the linear growth rate ($\gamma_k \sim \omega_{\rm in} V_{\perp}/\sqrt{2}c$, $\omega_{\rm in} = 4\pi n_c q^2/m_c$ the plasma frequency of cometary heavy ion) to saturation, i.e., $\omega_b \sim \gamma_k$ (cf., Davidson et al., 1972; Palmadesso and Schmit, 1972).

Figure 4 shows the necessary heavy ion density (n_c) as a function of wave amplitude for $m_c = 16m_p$, $m_i = m_p$ and $V_{\parallel} \sim V_{\perp}$. The moving interplanetary medium can very quickly accelerate the newly ionized particles, and the resonant beam velocity ($V_{\parallel}$) of the pickup heavy ions will become of the order of the solar wind speed ($\sim 400\,{\rm km\,s^{-1}}$). A typical Alfvén velocity in the solar wind is ~ 50–$80\,{\rm km\,s^{-1}}$ for $B_{\rm IMF} \sim 5$–$8nT$ and $n_i \sim 5\,{\rm cm^{-3}}$. Then, the resonant O^+ ion density (n_c/n_i) can be inferred to be of the order of 0.5–$1\ 10^{-3}$ for $V_{\parallel}/V_A \sim 5$–8 and $\delta B/B_{\rm IMF} \sim 0.25$. This theoretical prediction is in agreement with the observational results obtained by Suisei. Terasawa et al. (1986) estimated the density of cometary pickup ions of the water group to be $\sim 10^{-3}$ ${\rm cm^{-3}}$ at $7\ 10^6$ km from comet Halley.

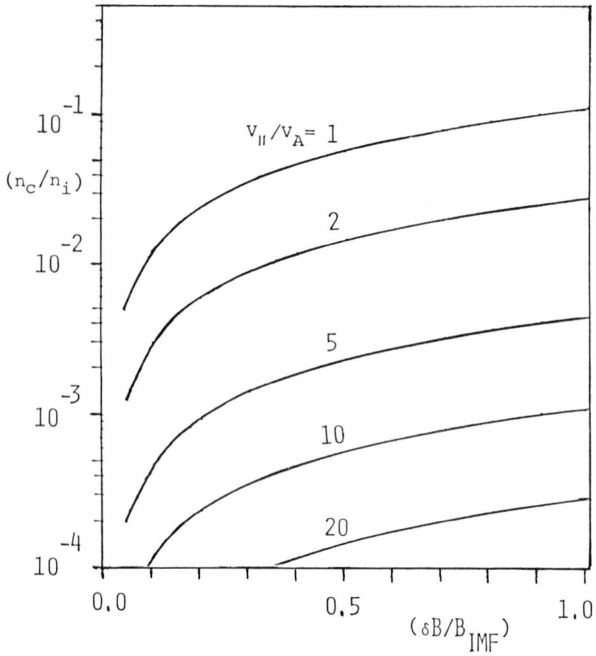

Fig. 4. Expected ion density of O^+ (or H_2O^+) normalized to the solar wind density, for excitation of the low-frequency HM waves at $7\,10^6$ km upstream from comet Halley, as a function of the relative wave amplitude for various resonant beam velocities of the cometary pickup ions normalized to the Alfvén speed

When the ratio of the density of new-born cometary ions to that of the solar wind is less than a few percent, the resonant instability, which amplifies low-frequency electromagnetic waves (Wu and Davidson, 1972; Wu and Hartle, 1974), becomes far more important than other instabilities. For example, the nonresonant firehorse instability caused by a warm beam of "water group" ions propagating parallel to the ambient magnetic field will not be viable (Goldstein and Wong, 1986). Taking O^+ ions and $B_{IMF} = 5nT$, the formation time for ring distributions of newly created ions of cometary origin, $\tau_r = 2\pi/\Omega_c$, the time scale for a shell formation via pitch-angle scattering, $\tau_s = \Omega_c^{-1}(B_{IMF}/\delta B)$, and that for "thermalization" by the diffusion process, $\tau_t \sim [0.1\,\Omega_c\,(\delta B/B_{IMF})^2]^{-1}$, (see Wu et al., 1986) are 240 s, 150 s, and 6000 s, respectively, for the turbulence level of $\delta B/B_{IMF} \sim 0.25$ in the outer turbulence region at 7–9 10^6 km (Fig. 3). Thus, we can expect to see shell distribution in this region.

The wave growth rate is theoretically given by $\gamma_k^{-1} \sim (\sqrt{2}c/\omega_{in}V_\perp)$, where ω_{in} is the plasma frequency of the newly created cometary ions $(4\pi n_c q^2/m_c)$ (see Yumoto et al., 1986a). At cometocentric distance of $r = 7\,10^6$ km the growth rate is ~ 12.8 s, which is only one-twelfth of the O^+ ion gyroperiod for $V_\perp \sim 300\,\mathrm{km\,s^{-1}}$, $\omega_{in} \sim 11$ Hz, and a heavy ion density of $n_c \sim 10^{-3}\,\mathrm{cm^{-3}}$. Therefore, the wave can grow to large amplitudes before being convected very far from the pickup region.

The low-frequency HM waves having $\sim 2nT$ peak-to-peak amplitude and the O^+ ion cyclotron frequency in the upstream region of comet Halley are concluded to be excited mainly by the resonant beam instability with high energetic "water group" pickup-ion of $\sim 10^{-3}\,\mathrm{cm^{-3}}$.

Acknowledgements. We would like to express our sincere thanks to T. Itoh, M. Oda, K. Hirao, and all members of the Sakigake team for their efforts leading to the success of this project. We are also grateful to H. Oya, K.-I. Oyama, B.T. Tsurutani and E.J. Smith for valuable comments and suggestions.

References

Davidson, R.C. et al.: 1972, *Phys. Fluids*, **15**, 317

Goldstein, M.L., Wong, H.K.: 1986, *NASA Report*, August 1986

Hirao, K., Itoh, T.: 1986, *Nature* **321**, 294

Mukai, T. et al.: 1986, *Geophys. Res. Letters* **13**, 829

Oya, H.: 1986, *20th ESLAB Symp.*, ESA SP-250, Vol. 1, 117

Palmadesso, P., Schmit, G.: 1972, *Phys. Fluids*, **14**, 1411

Saito, T. et al.: 1986, *Nature* **321**, 303

Scarf, F.L. et al.: 1986, *Geophys. Res. Letters* **13**, 857

Terasawa, T. et al.: 1986, *Geophys. Res. Letters* **13**, 837

Tsurutani, B.T., Smith, E.J.: 1986a, *Geophys. Res. Letters* **13**, 259

Tsurutani, B.T., Smith, E.J.: 1986b, *Geophys. Res. Letters* **13**, 263

Wenzel, K.-P. et al.: 1986, *Geophys. Res. Letters* **13**, 861

Wu, C.S., Davidson, R.C.: 1972, *J. Geophys. Res.* **77**, 5399

Wu, C.S., Hartle, R.E.: 1974, *J. Geophys. Res.* **79**, 283

Wu, C.S. et al.: 1986, *Geophys. Res. Letters* **13**, 865

Yumoto, K. et al.: 1986a, *Geophys. Res. Letters* **13**, 825

Yumoto, K. et al.: 1986b, *Bull. Inst. Space Astron. Sci.*, special issue No. **15**, 45

Astron. Astrophys. 187, 121–124 (1987)

ASTRONOMY AND ASTROPHYSICS

Plasma properties from the upstream region to the cometopause of comet P/Halley: Vega observations

M.I. Verigin [1], K.I. Gringauz [1], A.K. Richter [2], T.I. Gombosi [3], A.P. Remizov [1], K. Szegö [3], I. Apáthy [3], I. Szemerey [3], M. Tátrallyay [3], and L.A. Lezhen [1]

[1] Space Research Institute, USSR Academy of Sciences, Moscow, USSR
[2] Max-Planck-Institut für Aeronomie, D-3411 Katlenburg-Lindau, Federal Republic of Germany
[3] Central Research Institute for Physics, Hungarian Academy of Sciences, Budapest, Hungary

Received February 14, accepted June 4, 1987

Summary. Based on the Plasmag-1 plasma measurements on board Vega-1 and -2, evidence is provided for the deceleration upstream, for the heating at and for the thermalization and deceleration behind the bow shock of comet Halley. In the cometosheath region two separate ion populations are observed: the first one consists of cometary ions being picked up in the vicinity of the point of observation; the energy of these ions coming from the solar direction decreases much faster than the energy of the solar wind ions. The second one consists of cometary ions being picked up by the solar wind far away from the point of observation. Considerable oscillations in the plasma flow direction occur in the cometosheath region.

Key words: Vega – cometosheath plasma observations

1. Introduction

The plasma observations on board Vega-1 and Vega-2 revealed the existence of a bow shock at a distance of about 10^6 km from the nucleus of comet P/Halley and of a new boundary, called the cometopause, at a distance of about 1.6 10^5 km, respectively (Gringauz et al., 1986b–d).

The purpose of this communication is to summarize certain well-defined properties of the plasma flow and of its thermalization, deceleration and mass-loading by heavy ions originating from cometary neutrals in the region from upstream of the bow shock into the cometosheath outside of the cometopause. These properties are then discussed qualitatively.

2. Plasma observations

The plasma observations in the vicinity of comet Halley presented in this communication were obtained by the two hemispherical ion electrostatic analyzers of the Plasmag-1 experiment on board the two Vega spacecraft: the SDA and the CRA, looking into the solar direction and into the spacecraft-comet relative velocity direction, respectively. These sensors had rather wide acceptance geometries of $30° \times 38°$ and of $14° \times 32°$, and they measured ions in the range of 50–25 000 eV/Q in 60 and of 15–3500 eV/Q in 120 logarithmically spaced energy intervals. A detailed description of these sensors is given by Gringauz et al. (1983, 1986a) and by Apáthy et al. (1986).

Send offprint requests to: A.K. Richter

2.1. Up-stream deceleration and bow shock crossings

In Fig. 1 the decrease in the solar wind proton velocities with cometocentric distance R is shown, as measured by the SDA on both Vega-1 and Vega-2 during their inbound legs. At this time the SDA sensor provided seven complete ion energy-over-charge spectra per twenty minutes. The location of the bow shock, which was determined from simultaneous measurements of the plasma, the plasma waves and of the magnetic field (Galeev et al., 1986), is marked by S. As one can see, the decrease of the solar wind velocity due to mass loading by heavy cometary ions started already at a distance of 2–3 10^6 km from the nucleus, i.e. 1–2 10^6 km upstream of the bow shock.

Figure 2 presents the energy spectra measured by the SDA on board Vega-1 when crossing the bow shock on the inbound leg and on the outbound leg, respectively. Besides the gradual slowing down of protons, a gradual widening of the energy spectra can be observed when approaching the comet, i.e. the ion temperature is increasing several hours before reaching the bow shock. After a data gap of 20 min around 3.00 UT, the ion temperature is already so high that the proton distribution overlaps the α particle peak in the spectrum. The gradient of the velocity is increasing significantly when crossing the bow shock at a distance of 1.02 10^6 km from the nucleus ($\sim$ 3.46 UT). It was shown by Galeev et al. (1986) that Vega-1 crossed a quasiperpendicular and Vega-2 a quasi-parallel bow shock on their inbound legs, respectively.

The situation is more complicated when crossing the bow shock on the outbound leg of Vega-1. The highest gradient in the plasma velocity was observed between 9.00 and 9.30 UT at a distance of about 5.5 10^5 km from the nucleus. At the same time the ion temperature stayed high so that the α peak could not be distinguished from the proton distribution until 11.30 UT ($R \approx 1.2$ 10^6 km). Thus, on the outbound leg the bow shock was crossed at a distance of 5.5 ± 1 10^5 km (Galeev et al., 1986), and the high ion temperature which can be observed intil 12.20 UT for another about 8 10^5 km is associated with the high level of MHD activity in the foreshock region upstream of a quasiparallel cometary bow shock (Schwingenschuh et al., 1986).

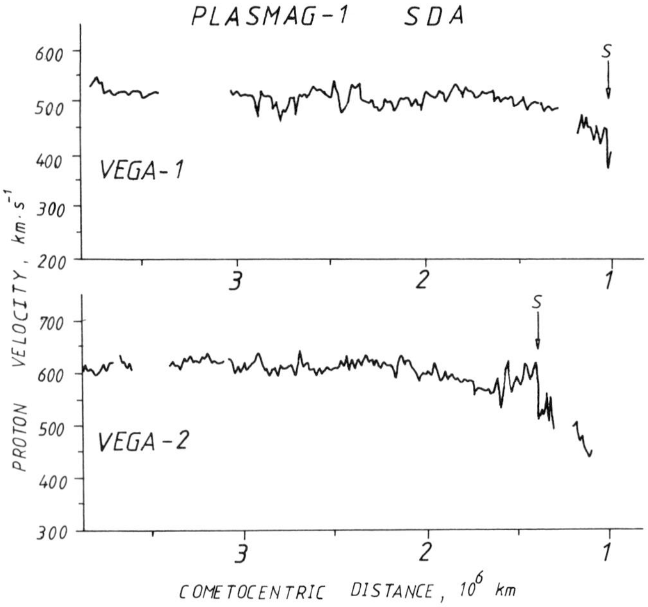

Fig. 1. Radial dependence of the solar wind bulk velocity, as observed by Vega-1 and Vega-2 on their inbound trajectories upstream of the bow shock of comet Halley (marked by S), respectively

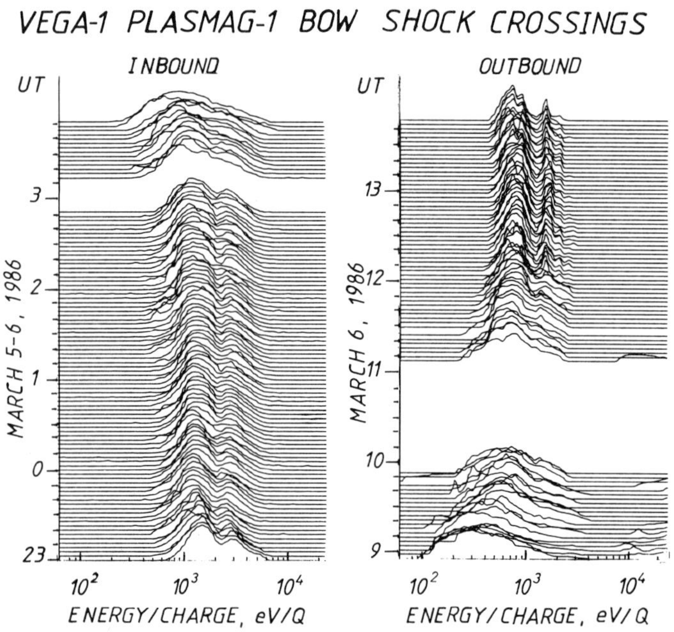

Fig. 2. Two sequences of ion energy-over-charge (E/Q) spectra measured by Vega-1 on its inbound and outbound leg in the vicinity of the bow shock of Halley's comet, respectively

2.2. *Cometosheath observations*

Downstream of the inbound bow shock both Vega spacecraft entered the region which was called cometosheath (Gringauz et al., 1986b). In Fig. 3 the ion spectrogram (intensity vs. energy-over-charge vs. time) is shown, as observed by the SDA on board Vega-1 from a cometocentric distance of $\sim 8\ 10^5$ km inbound to $\sim 3\ 10^5$ km outbound. The closest approach to comet Halley

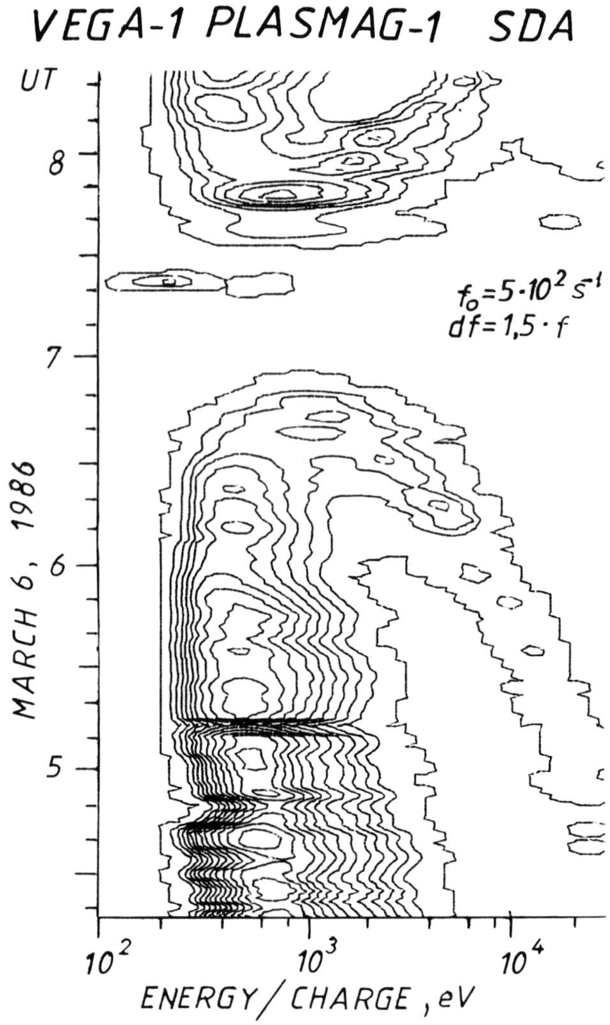

Fig. 3. Spectrogram (count rates vs. E/Q vs. time) of the ion flux throughout the cometosheath region measured by Vega-1. Closest approach at 7:20:06 UT. Normalization of iso-count rate lines: outermost line f_0; increase for any adjacent line by df

occurred at 7:20:06 UT or at a distance of 8889 km. In this spectrogram the outermost isoline corresponds to a counting rate of $f_0 = 5\ 10^2\ s^{-1}$. Each following isoline represents a counting rate which is of a factor of 1.5 higher than its adjacent outer line, i.e., $f_{n+1} = f_n + df$ with $df = 0.5 f_n$, $n = 0,1,2 \ldots$.

The main feature in this spectrogram is the occurrence of two branches in the energy distribution of ions lasting until about 6.40 UT. The left hand-side branch corresponding to the lower energy obviously describes the solar wind protons and α particles which were thermalized and slowed down at the cometary bow shock. The more energetic branch on the right hand-side must represent the cometary ions, since ions of such a high energy were never observed by the SDA in the solar wind. When extrapolating the energy peak corresponding to these cometary ions to larger distances from the nucleus, where it exceeds the upper limit of the energy range of 25 keV of the SDA, one can conclude that these cometary ions belong to the water group (e.g. O^+, OH^+, H_2O^+, or H_3O^+), since their extrapolated energy is 60–80 times higher than the proton energy. This energy then corresponds to the energy of picked-up ions of the water group moving with a

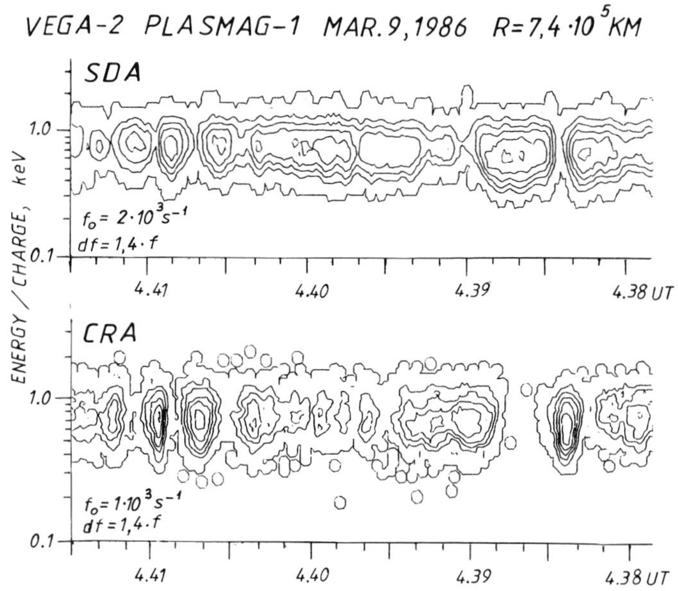

Fig. 4. Spectrograms of the ion fluxes as observed by the sun-directed (SDA) and ram-directed (CRA) ion electrostatic analyzer on board Vega-2 simultaneously in the cometosheath. Note the variations in the direction of the plasma flow of cometary ions

velocity twice the solar wind velocity. These findings have been directly verified by the JPA-IIS measurements on board Giotto (Johnstone et al., 1986).

When proceeding deeper into the cometosheath, the proton energy gradually decreases. The energy of the cometary ions observed in the SDA decreases faster, so that the ratio of the energy of heavy ions to the energy of protons is also decreasing. At distances of 3–4 10^5 km from the nucleus (6.00–6.15 UT) the velocity of cometary ions from the solar direction decreases to the value of the proton velocity due to some collective process, which is not quite understood. Also the fluxes of these two populations become comparable. Afterwards the rate of the energy decrease is further increasing for the heavy ions, while the energy of the protons practically does not change any more.

Around the cometopause ($\sim$6.45 UT), which separates the cometosheath from the cometary plasma region, solar wind protons disappear from the acceptance angle of the SDA and cometary ions produce a peak around 1 keV in the energy spectrum. When the velocity of Vega-1 relative to the comet (79.2 km s^{-1}) is taken into account, then the velocity of the heavy ions relative to the comet can be estimated to be a few times ten km s^{-1} in the vicinity of the cometopause, while the proton velocity is still around 200 km s^{-1} in this region.

After closest approach, these characteristic changes in the cometosheath plasma occur again on the outbound leg but in a reverse order.

2.3. Variations in the plasma flow

Besides these large-scale changes in the plasma flow in the cometosheath, which determine the global picture of the solar wind flowing around the comet, a high level of MHD turbulence with a wide frequency range is characteristic for this region.

In the spectrogram presented in Fig. 3, large-scale variations can be observed in the intensity of the ion flow with a period of about 15 minutes.

An example for the variations in the direction of the plasma flow within the cometosheath is presented in Fig. 4. In these spectrograms, registered by the CRA and the SDA at a distance of about $R \approx 7.4\ 10^5$ km from the nucleus, the characteristic time scale for these variations is 20–30 seconds. The variations in the flow intensity observed by both analyzers are caused by the variations in the direction of the ion flow. When the intensity of the ion flow decreases in the direction of the CRA analyzer, the counting rates observed by the SDA increase simultaneously and vice versa. The approximate value of the deviation of the proton flow from its original direction can be estimated as

$$\delta\alpha \approx \frac{2\,\mathrm{kT}}{\mathrm{mV}^2} \ln \frac{N_{\max}}{N_{\min}} \approx 5^\circ,$$

where the proton temperature is $T \approx 3\ 10^5$ K and the velocity $V \approx 350$ km s^{-1}, and where the ratio of the maximum to the minimum counting rates is $N_{\max}/N_{\min} \approx 3$–5.

3. Discussion

Already the first hydrodynamic models describing the interaction of cometary with solar wind plasma discussed the effect of mass-loading and deceleration of the solar wind by cometary ions occurring even upstream of the bow shock (e.g. Wallis, 1973). From the Vega observations we find that the deceleration length at least qualitatively corresponds to the effective ionization scale length of cometary neutrals mainly of the water-group of about 2 10^6 km (Gringauz et al., 1986b, c; Remizov et al., 1986).

The ions originating from these cometary neutrals first form a ring distribution in velocity space. The Alfvén wave turbulence being excited by the ion-cyclotron instability of such a distribution isotropizes the newly formed ions in the coordinate system moving with the solar wind (Sagdeev et al., 1986). In this way and after a few cyclotron periods the newly created ions are arriving from the solar direction with a maximum energy 4 M_i-times higher than the proton energy (M_i is the ion mass number), independently of the angle between the solar wind flow and the magnetic field direction.

In the solar wind upstream of the bow shock, the energy of these heavy ions exceeds the upper limit of the energy range of the SDA ($E/Q \leqq 25$ keV). But in the outer regions of the cometosheath, where the solar wind flow is already decelerated by the shock and by the mass-loading process, the SDA measurements can be regarded as being in accordance with the above discussed process of the implantation of cometary ions into the plasma flow at large distances from the comet.

The energy spectra of cometary ions as measured by Plasmag-1 in the inner region of the cometosheath might be interpreted in two different ways. Here the ratio of the energy of cometary ions E_i^1 to the proton energy E_p measured by the SDA is smaller than $4M_i$ ($\approx$60–80), and it gradually decreases with decreasing cometocentric distances R. The IIS sensor of the JPA instrument on board Giotto was able to measure ions up to energies of $E_i/Q \leqq 90$ keV and to estimate the mass of the ions (Johnston et al., 1986). According to the measurements of IIS there is another, more energetic branch in the energy spectrum corresponding to water group ions with an energy of $E_i^2 \approx 4\,M_i\,E_p$ beside the branch which was also observed by the SDA. If the ratio $E_i^2/E_p \approx 4\,M_i$ is constant in the cometosheath, the ions with energies E_i^2 might have been locally ionized and picked up by the process discussed above. In this case, the ions observed by the SDA with energies $E_i^1 < E_i^2$ might have been ionized further upstream of the population with energies E_i^2, and they might have lost their energy due to some collective processes afterwards (Gringauz et al., 1986c).

However, there is also another explanation for the existence of these two different cometary ion populations: the ions with the energy E_i^2 were actually created far upstream from the point of observation, but the proton energy slightly changes in the cometosheath as observed by the SDA. Hence the ratio E_i^2/E_p does not change much or stays almost constant, as in the first case. The ions registered with energies E_i^1 were created in the vicinity of the spacecraft. First these ions are only partially involved into the solar wind flow, however, when approaching the cometopause, these ions are not a minor population any more. In the beginning of this process there is no complete isotropization in the coordinate system moving with the solar wind. But when approaching the cometopause, the density of these ions (Fig. 3) is increasing and thus the energy of the solar wind flow will not be large enough to increase the velocity of all the newly created heavy ions. In the vicinity of the cometopause these ions are accelerated only to a velocity of a few times ten km s^{-1}. Since the spacecraft has a velocity of 79.2 km s^{-1} relative to the comet, the slowly moving newly created cometary ions will be observed by the SDA with an energy of about 1 keV around the cometopause. There is an additional fact in favour of the possibility that the cometary ions belonging to the less energetic branch E_i^1 were created not very far from the spacecraft: the flux of these ions is increasing with decreasing R (see Fig. 3), corresponding to the increase of the neutral gas density. The fluxes of heavy ions, however, belonging to the high-energy branch do not increase significantly when approaching the nucleus (see Thomsen et al., 1987).

The characteristic time for the variations in the direction of the plasma flow observed by Vega-2 in the cometosheath is about 20–30 s, as shown in Fig. 4. It seems that these variations are connected with the strong turbulence of the plasma flow in this region caused by the newly created heavy ions. Associated with this process, strong MHD turbulence with a maximum at 10^{-2} Hz was observed in the magnetic field (Tsurutani and Smith, 1986) and in the electron component of the plasma (Gosling et al., 1986) in the cometosheath of comet Giacobini-Zinner as measured by ICE.

In the cometosheath the variations of the plasma flow with very long periods (approximately 15 min as seen in Fig. 3) could be caused by large-scale MHD turbulence. The observed period of these variations, however, is comparable with the characteristic time of the solar wind flow in the cometosheath. As $\tau \approx R/V \approx 10^3$ s for $R \approx 3\ 10^5$ km and $V \approx 300$ km s^{-1}, their dimension along the spacecraft trajectory is of the order of $\approx 10^5$ km and thus comparable with the overall dimension of the cometosheath itself. It is therefore suggested that the plasma flow variations with a characteristic period of about 15 min observed on board Vega-2 in the cometosheath can be associated with MHD turbulence at the eigen-frequencies of the cometosheath-solar wind system.

Finally we want to propose an explanation for why the dimension (along the spacecraft trajectory) of the region of heated solar wind upstream of the bow shock on the inbound leg differs from that on the outbound leg of Vega-1. As can be seen from Fig. 2, a significant heating of protons and α particles began at about 2.30 UT ($R \approx 1.3\ 10^6$ km) as Vega-1 was approaching the comet, and hot protons and α particles (compared to the solar wind) were also registered up to almost the same cometocentric distance (to $\approx$12.20 UT) on the outbound leg. The cometary bow shock, on the other hand, was crossed at a distance of 1.02 10^6 km inbound and 5.5 ± 1.10^5 km outbound indicating that the region of heated upstream solar wind had a length of about 3 10^5 km inbound and of about 7 10^5 km outbound. One explanation for this difference could be the following one: as the inbound (outbound) bow shock is quasiperpendicular (quasiparallel), upstream, bow shock associated turbulence will extend and heat the solar wind further out outbound than inbound. Another, yet different explanation is the following one: the onset of large-amplitude turbulence is triggered by the ionization and the pick-up process of cometary neutrals exceeding some critical density value and is, therefore, symmetric as a function of cometocentric distance with respect to the inbound and outbound trajectory of Vega-1. From the neutral gas measurements on board Vega-1 (Remizov et al., 1986) we find a gradual increase in density to about 30 particles per cm^3 at a distance of about 1.3 10^6 km. This value seems to be sufficient for the onset of a turbulent heating of the solar wind ions caused by the unstable beam-like distribution of the freshly ionized cometary particles. On the other hand, the difference in the location of the bow shock is simply due to the fact that this discontinuity was crossed around the flanks on the inbound leg and much closer to the subsolar point on the outbound leg.

References

Apáthy, I. et al.: 1986, 20th ESLAB Symp., ESA SP-250, Vol. **1**, p. 65

Galeev, A.A. et al.: 1986, *Geophys. Res. Letters* **13**, 841

Gosling, J.T. et al.: 1986, *Geophys. Res. Letters* **13**, 267

Gringauz, K.I. et al.: 1983, in *Cometary Exploration III*, ed. T.I. Gombosi, Central Res. Inst. Physics Press, Budapest, p. 333

Gringauz, K.I. et al.: 1986a, in *Field, Particle and Wave Experiments on Cometary Missions*, eds. K. Schwingenschuh, W. Riedler, Verlag Österreichische Akademie der Wissenschaften, Wien, p. 203

Gringauz, K.I. et al.: 1986b, *Nature* **321**, 282

Gringauz, K.I. et al.: 1986c, *Adv. Space Res.* **5**, N. 12, 165

Gringauz, K.I. et al.: 1986d, *Geophys. Res. Letters* **13**, 613

Johnston, A. et al.: 1986, Nature **321**, 344

Remizov, A.P. et al.: 1986, 20th ESLAB Symp., ESA SP-250, Vol. **1**, p. 387

Sagdeev, R.Z. et al.: 1986, *Geophys. Res. Letters* **13**, 85

Tsurutani, B.T., Smith, E.J.: 1986, *Geophys. Res. Letters* **13**, 259

Wallis, M.K.: 1973, *Planet. Space Sci.* **21**, 1647

Astron. Astrophys. 187, 125–128 (1987)

ASTRONOMY AND ASTROPHYSICS

Observations of heavy energetic ions far upstream from comet P/Halley

T. R. Sanderson [1], K.-P. Wenzel [1], P. W. Daly [2], S. W. H. Cowley [3], R. J. Hynds [3], I. G. Richardson [3], E. J. Smith [4], S. J. Bame [5], and R. D. Zwickl [5]

[1] Space Science Department of ESA, Estec, Noordwijk, The Netherlands
[2] Max-Planck-Institut für Aeronomie, D-3411 Katlenburg-Lindau, Federal Republic of Germany
[3] Imperial College of Science and Technology, London, United Kingdom
[4] Jet Propulsion Laboratory, Calif. Inst. of Technology, Pasadena, CA, USA
[5] Los Alamos National Laboratory, Los Alamos, NM, USA

Received February 20, accepted March 29, 1987

Summary. On March 25, 1986, the ICE spacecraft came within 28 10^6 km of the nucleus of comet Halley. For several days around this time, bursts of heavy ions were observed by the ICE energetic ion experiment. These bursts were observed only during periods when the solar wind velocity was considerably higher than its nominal value. We examine the characteristics of these ions, in particular their anisotropies. Using the well known formulae for transformation of distributions from the solar wind frame of reference to the spacecraft frame, we examine the angular distributions expected from either protons, or heavy ions from the water group, and show that the measurements are consistent with heavy ions, and not with protons. We discuss other sources of heavy ions and conclude that the most likely source of these ions is comet Halley.

Key words: Halley – energetic ions

1. Introduction

The International Cometary Explorer (ICE) spacecraft passed through the tail of comet Giacobini-Zinner and then continued on to make a distant encounter with comet Halley in March 1986, passing at a distance of 28 10^6 km upstream of the nucleus. Energetic ion fluxes up to two orders of magnitude in excess of the background were observed for several days around closest approach at times when the solar wind velocity was high (Wenzel et al., 1986), the intensity being modulated by the solar wind velocity. These ions were observed to flow in the direction of the solar wind velocity vector and were highly anisotropic. This behaviour is typical for heavy energetic ions accelerated by pick-up from the solar wind. At the same time, wave activity was observed by the ICE wave experiment (Scarf et al., 1986), the wave activity being similar to that observed at Giacobini-Zinner.

In this article we examine in detail the angular characteristics of the ions, and show that the observed anisotropies are consistent with heavy ions, and not with protons accelerated in the high speed stream, and conclude that the ions we observe are most likely of cometary origin.

Send offprint requests to: T.R. Sanderson

2. Instrumentation

The ion observations were obtained with the European Space Agency/Imperial College/Utrecht energetic ion experiment (also refered to as EPAS) on the ICE spacecraft. This consists of three identical telescopes inclined at 30°, 60°, and 135° with respect to the spacecraft spin axis, the spacecraft axis being aligned normal to the plane of the ecliptic. Each telescope detects ions with unit mass in eight channels in the range 35–1600 keV, and is further subdivided into eight azimuthal sectors. Due to pulse height defects for heavy ions (Ipavich et al., 1978) the energy ranges are species dependent. For water group ions, the four lowest energy channels are: 65–95, 95–140, 140–205, and 205–310 keV. The time resolution in all energy channels during most of the encounter was 64 s. The solar wind speed measurements used here were obtained with the Los Alamos plasma instrument (Bame et al., 1978) whilst the magnetic field measurements were made with the JPL vector helium magnetometer (Frandsen et al., 1978), both of which are on the same spacecraft.

3. Observations

Figure 1 shows the ion flux measured by the energetic ion experiment on the ICE spacecraft during the Halley encounter (right hand side), and for comparison, during the Giacobini-Zinner enounter (left-hand side). In both cases the diagrams show the flux measured in sector 1 of telescope 2, the sector viewing closest to the direction of the solar wind, for the four energy channels ranging from 65–310 keV. From this it can be seen that intensities measured during the Halley encounter at around 30–35 10^6 km from Halley, are similar to those observed at Giacobini-Zinner at distances of around 3–4 10^6 km. The ions observed at Halley have characteristics similar to those observed outside of the interaction region of Giacobini-Zinner. The flux is modulated on a time scale of hours by changes in the solar wind velocity, and on a time scale of minutes by changes in the direction of the interplanetary magnetic field. The spectrum is steep, and the flow is highly anisotropic in the direction of the solar wind velocity vector (Hynds et al., 1986; Sanderson et al., 1986).

Figure 2 shows the Halley observations plotted together with the solar wind speed. Variations in the flux of ions are correlated

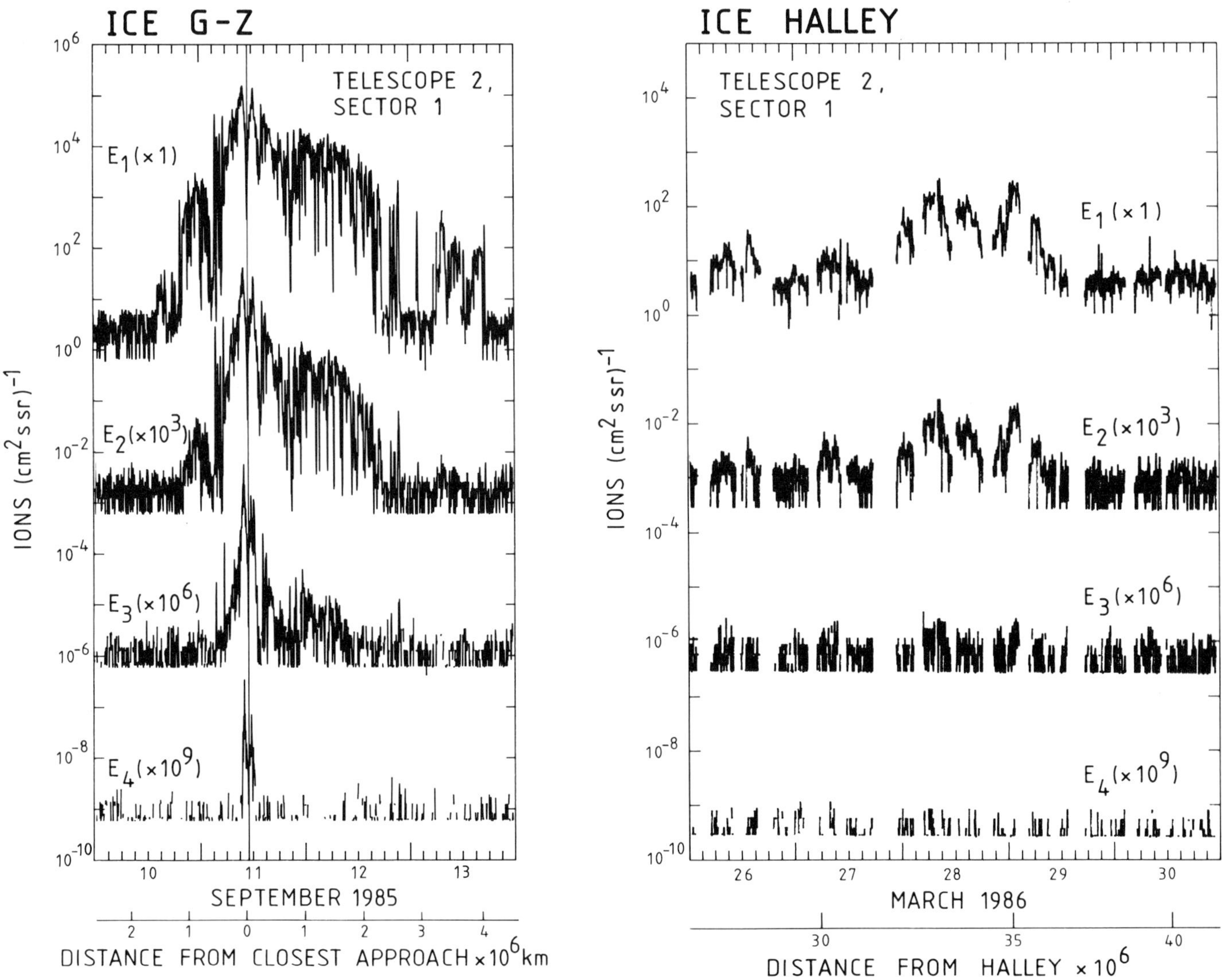

Fig. 1. Comparison of energetic ion fluxes observed during the Giacobini-Zinner and Halley encounters. The four traces show, from top to bottom, the fluxes in the 65–95, 95–110, 110–205, and 205–310 keV channels

with variations in the solar wind speed, the ions being only observed when the solar wind velocity was high. The maximum intensity was observed when the solar wind velocity was around 700 km s^{-1}.

The ions are observed flowing along the solar wind direction, as can be seen in Fig. 3. This shows anisotropies for the period 0130 to 0147 UT on 1986 March 29. This shows the viewing direction of each sector for each telescope, the sector pointing towards the sun being at the top of the figure. The number in the sector 1 of telescope 2 is the number of counts registered during the accumulation period. The plot shows the highly anisotropic nature of the angular distributions, with counts essentially only being registered in the sector pointing closest to the spacecraft-sun direction. The observed anisotropies do not change with changes in magnetic field direction.

4. Discussion

To investigate the origin of these ions, we have modeled the anisotropies expected from different ion species for the interplanetary conditions prevailing at the time of observation, and compared these models with our observations to determine the ion species responsible for the bursts. For this purpose we have used the equations of Ipavich (1974) to transform distributions from the solar wind frame of reference to the frame of reference in which the measurements were made, i.e., the spacecraft frame. We begin by assuming that in the solar wind frame of reference the distribution is isotropic, with the differential intensity following a power law in kinetic energy. Then in the spacecraft frame of reference, the observed intensity at an angle θ with respect to the solar wind direction is given by (Ipavich, 1978, Eq. (9))

$$\frac{dJ}{dE} \propto E^{-\gamma'}[1 - 2(w/v)\cos\theta + (w/v)^2]^{-(\gamma+1)} \tag{1}$$

the spectral index γ' is measured in the solar wind frame of reference, whilst the angle between the velocity vector of the ion and the solar wind velocity vector, θ, and the particle velocity, v, are measured in the spacecraft frame of reference. The relative speed of the two frames of reference is w. In the spacecraft frame of reference the measured spectral index γ_{eff} at a given angle θ with

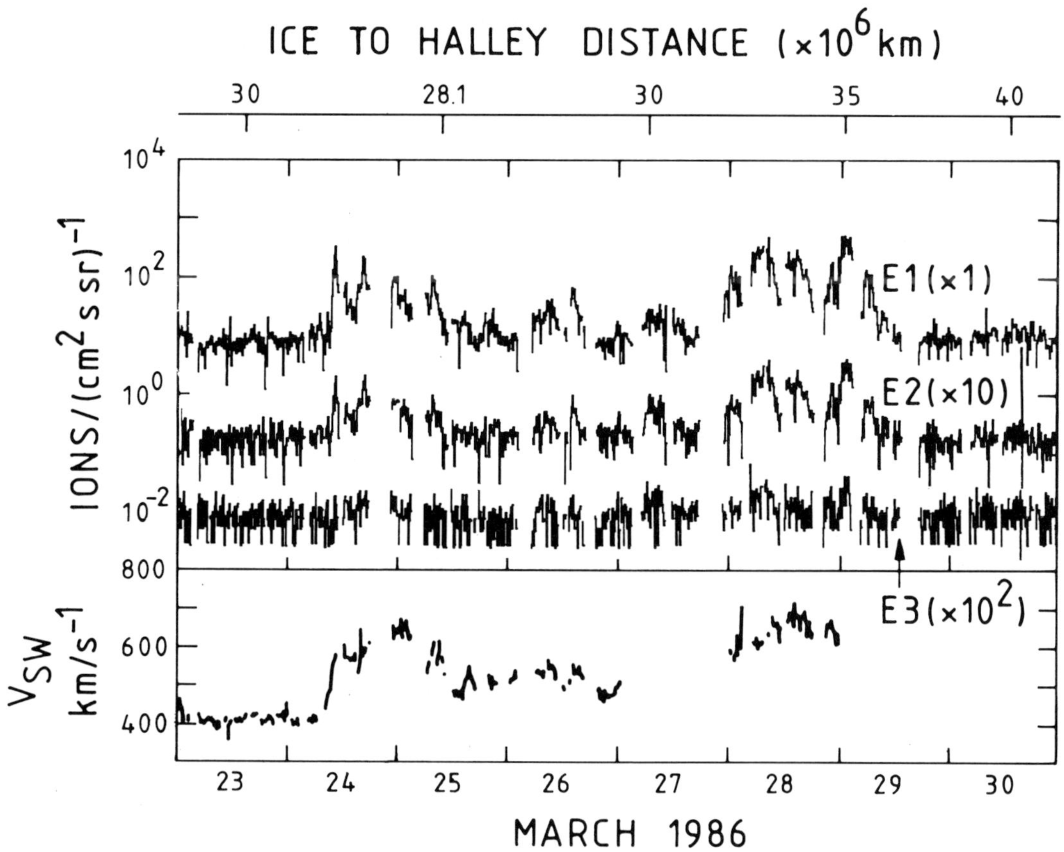

Fig. 2. Energetic ion flux in, from top to bottom, the 65–95, 95–140, and 140–205 keV channels, plotted together with solar wind velocity

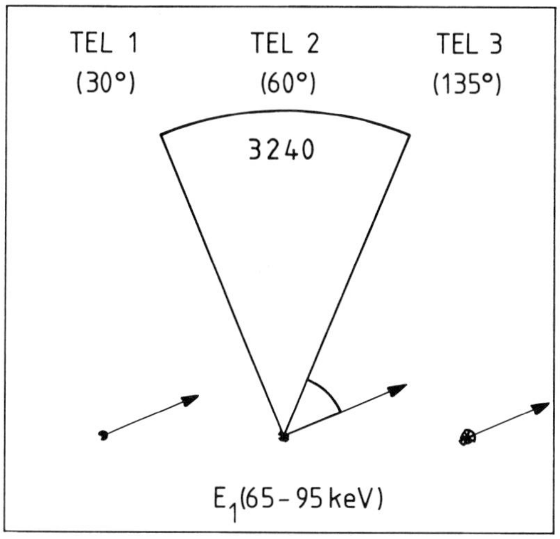

Fig. 3. Sector plots for the 65–95 keV channel for the 3 telescopes showing the counting rate in each sector. The Sun is at the top of the picture

respect to the solar wind direction, is related to the spectral index in the solar wind γ' by the relation (Ipavich, 1978, Eq. (7))

$$\frac{(\gamma_{\text{eff}}+1)}{(\gamma'+1)}=\frac{1-(w/v)\cos\theta}{1-2(w/v)\cos\theta+(w/v)^2}\,. \tag{2}$$

From our observations we have $\gamma_{\text{eff}}=7$ and $w=550\ \text{km s}^{-1}$. For protons, $v=2590\ \text{km s}^{-1}$, whilst for water group ions with mass 16, $v=880\ \text{km s}^{-1}$. Using Eq. (2) gives us γ' in the solar wind frame, hence if the ions are protons $\gamma'=2$ or if they are from the water group, $\gamma'=5.3$. Figure 4 shows a cross-section of the distribution expected in the spacecraft frame of reference for ions with mass 1 and mass 16, for a solar wind velocity w of $550\ \text{km s}^{-1}$, due to an isotropic distribution in the solar wind frame of reference which gives rise to a power law distribution with $\gamma=7$ in the spacecraft frame.

We have extended the above analysis to include the response of our instrument. Since our instrument consists of telescopes mounted at 30°, 60°, and 135° with respect to the spin axis, it is necessary to take this into account when computing the response to different ions species. To predict the response to different ion species we proceeded as follows. First, we assumed an isotropic distribution in the solar wind, with spectral index γ' calculated from Eq. (2), corresponding to a γ_{eff} of 7. We then transformed this into the spacecraft frame of reference, and integrated over the opening angle of each telescope, and over each sector. The results are shown in Fig. 5 for ion species with mass 1 and 16.

Comparing this with the data in Fig. 3, we see that the response expected for ions with mass 1 is such that, given the conditions, ($\gamma_{\text{eff}}=7$ and $w=550\ \text{km s}^{-1}$, as measured), it is impossible to explain the observations in terms of protons. This is clear, both for the forward-to-back ratio in telescope 2, and also from the ratio of counts in sector 1 of telescope 1 to sector 1 of telescope 2, and telescope 3 to telescope 2. The response is such that the observed anisotropies can only be explained in terms of heavy ions. We cannot explain our observations in terms of protons even if the protons are anisotropic in the solar wind frame of reference.

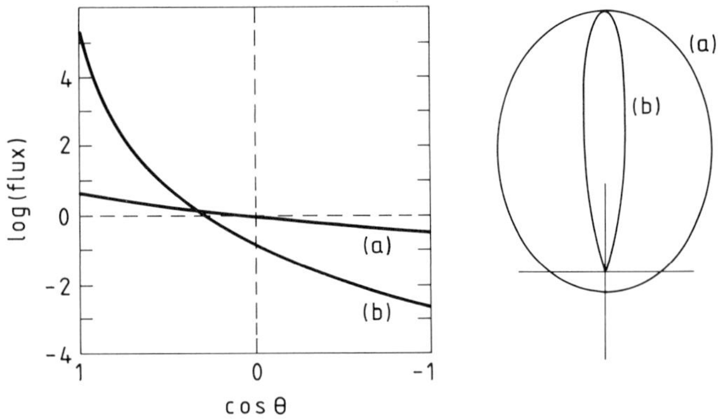

Fig. 4. Ion anisotropies expected for a solar wind velocity of 550 km s^{-1} for (a) protons and (b) oxygen ions. On the left hand side, the distribution is plotted with the logarithm of the flux versus the cosine of angle between the particle velocity vector and the Sun-comet direction, on the right hand side as a linear polar plot

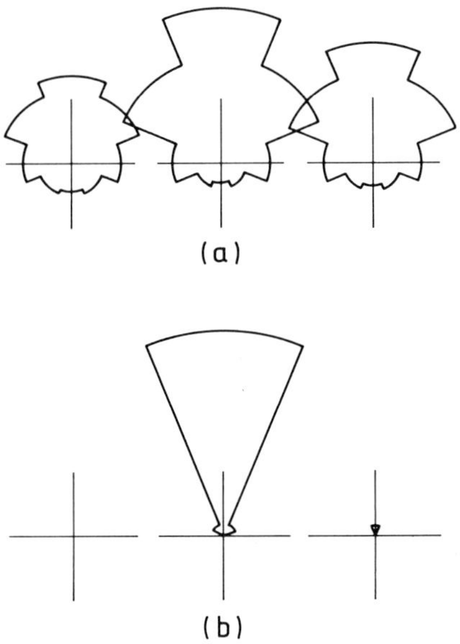

Fig. 5. Sector plots for the three telescopes for an isotropic distribution in the solar wind frame of reference for protons (top) and oxygen ions (bottom), for a solar wind velocity of 550 km s^{-1} and giving rise to a power law spectrum with index 7, observed in the lowest energy channel

Protons streaming in the solar wind frame of reference must be gyrotropic, unless gradients are present, and therefore stream along the magnetic field direction. As the field changes direction, so must the flow direction in the solar wind. The transformation of such a distribution into the spacecraft frame of reference gives rise to a distribution whose direction varies with varying field direction. This is quite unlike the distribution we observe, which is always aligned along the nominal solar wind direction.

One further possibility is that the observed ions are solar wind iron, such as observed in high speed streams by Mitchell and Roelof (1986). We believe that this is less likely than the ions being of cometary origin. First of all it is possible to explain the observed flux levels if one assumes that the cometary ions expand out from the nucleus at 2 to 3 times the nominal figure of 1 km s^{-1} and if the ionisation and charge exchange time constant is also a factor of 2 to 3 times the nominal value of 10^6 s, as has already been discussed by Wenzel et al. (1986). In addition, the efficiency for detection of the iron ions is much less than that for water group ions, for two reasons, (1) the pulse height defect is greater for iron than for oxygen (Ipavich et al., 1978), and (2) it is to be expected that the angular distribution of the iron ions will be narrower than that of the oxygens, and therefore, if the solar wind velocity vector lies in the plane of the ecliptic, the fraction of the distribution observed in telescope 2, sector 1 (which is inclined at 30° to the ecliptic) will be much less for the iron than the oxygen. Further investigations are being conducted to resolve this matter.

Acknowledgements. We thank S.T. Ho and M.J. Szumlas for assistance with the data analysis.

References

Bame, S.J., Asbridge, J.R., Felthauser, H.E., Glore, J.P., Hawk, H.L., Chavez, J.: 1978, *IEEE Trans.* **GE-16**, 160

Frandsen, A.M.A., Bonnor, B.V., van Amersfoort, J., Smith, E.J.: 1978, *IEEE Trans.* **GE-16**, 195

Hynds, R.J., Cowley, S.W.H., Sanderson, T.R., Wenzel, K.-P., van Rooijen, J.J.: 1986, *Science* **232**, 361

Ipavich, F.M., Lundgren, R.A., Lambird, B.A., Gloeckler, G.: 1978, *Nucl. Instr. Methods* **154**, 291

Ipavich, F.M.: 1974, *Geophys. Res. Letters* **1**, 149

Mitchell, D.G., Roelof, E.C., Bame, S.J.: 1986, *J. Geophys. Res.* **88**, 9059

Sanderson, T.R., Wenzel, K.-P., Daly, P., Cowley, S.W.H., Hynds, R.J., Smith, E.J., Bame, S.J., Zwickl, R.D.: 1986, *Geophys. Res. Letters* **13**, 411

Scarf, F.L., Coroniti, F.V., Kennel, C.V., Sanderson, T.R., Wenzel, K.-P., Hynds, R.J., Smith, E.J., Bame, S.J., Zwickl, R.D.: 1986, *Geophys. Res. Letters* **13**, 857

Smith, E.J., Tsurutani, B.T., Slavin, J.A., Jones, D.E., Siscoe, G.L., Mendis, D.A.: 1986, *Science* **232**, 382

Wenzel, K.-P., Sanderson, T.R., Richardson, I.G., Cowley, S.W.H., Hynds, R.J., Bame, S.J., Zwickl, R.D., Smith, E.J., Tsurutani, B.T.: 1986, *Geophys. Res. Letters* **13**, 861

Astron. Astrophys. 187, 129–131 (1987)

ASTRONOMY AND ASTROPHYSICS

Spatial distribution of water-group ions near comet P/Halley observed by Suisei

T. Mukai [1], W. Miyake [2], T. Terasawa [3], M. Kitayama [1], and K. Hirao [4]

[1] Institute of Space and Astronautical Science, Komaba, Meguro-ku, Tokyo, 153 Japan
[2] Radio Research Laboratory, Nakaminato, Ibaraki, 311–12 Japan
[3] Geophysical Institute, Kyoto University, Sakyo, Kyoto, 606 Japan
[4] Department of Astronautics and Aeronautics, Tokai University, Kitakaname, Hiratsuka, 259–12 Japan

Received March 2, accepted May 19, 1987

Summary. The spatial distribution of water-group ions observed by Suisei during the encounter with comet P/Halley is discussed in terms of the ion pickup model. The observed result is most consistently interpreted when we take the water production rate of $2\,10^{30}\,s^{-1}$. The model calculation shows the inbound-outbound asymmetry in the density of cometary ions in good agreement with the observation. It is suggested that this asymmetry is mainly caused by the difference of orbital conditions between the inbound and the outbound observations in reference to the comet.

Key words: comet Halley – ion pickup, cometary ions – water production rate – solar wind interaction

1. Introduction

The ion pickup process plays an essential role in the interaction of the solar wind with cometary gases (e.g. Biermann et al., 1967; Ip and Axford, 1982). Spatial density distribution of pickup ions provides important information on the production rate of parent neutral gas of cometary origin and on the global dynamics of the solar wind interaction with the comet. Plasma observations from Suisei clearly demonstrated the existence of the assimilation process of cometary ions inside the cometosheath of comet P/Halley as well as in the upstream region (Mukai et al., 1986a, b; Terasawa et al., 1986). Though the plasma instrument onboard Suisei is an energy per charge analyzer, we could identify the cometary ion species from the existence of the pickup shells in the phase space and estimate the density of cometary ions forming the shell. In this paper we discuss the observed density distribution of the watergroup ions in terms of a simple, but updated pickup model.

Neutral molecules in the comet are ionized through processes, such as photoionization due to solar EUV radiation, charge exchange with solar wind protons, collisional ionization by energetic electrons, and possibly Alfvén's critical ionization mechanism (see e.g., Mendis and Ip, 1977). The total ionization rate due to these processes may be varied according to the distance from the cometary nucleus, which is taken into account in the present model.

Send offprint requests to: T. Mukai

2. Observation

Figure 1 shows the orbit of Suisei from 1986 March 4 to March 13, drawn in the cometocentric solar ecliptic (CSE) coordinate system, where the *x*-axis is taken in the ecliptic plane toward the sun, and the *z*-axis perpendicular to the ecliptic plane toward the north. The *y*-axis is taken to make a right-hand coordinate system. A dotted parabola shows the approximate shape of the cometary bow shock. Shaded areas show intervals when the plasma observations were made. Bars with H^+ and O^+ respectively show the intervals when the cometary protons and water-group ions are clearly identified from the pickup shell in the phase space (Mukai et al., 1986a; Terasawa et al., 1986).

When the pickup shell is identified, we can estimate the density of cometary ions which form the shell. The spatial distribution of water-group ions (H_2O^+, OH^+, and O^+) is given in Fig. 2, where the observed density is derived by assuming that the pickup shell is filled by 50% (Mukai et al., 1986a; Terasawa et al., 1986). The vertical bar shows the range of uncertainty in the density estimation. For the intervals when the direction of the magnetic field is known from the pressure anisotropy direction of the solar wind protons (open circles), the uncertainty in the density estimation mainly comes from the uncertainty in the deter-

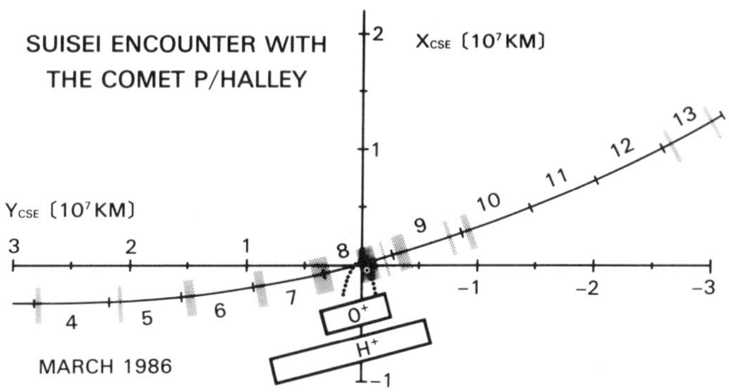

Fig. 1. Suisei's orbit near comet P/Halley. The hatched regions show when the plasma observations were made. Cometary protons and water-group ions (expected to the mainly O^+) were detected during the periods shown by bars with H^+ and O^+, respectively

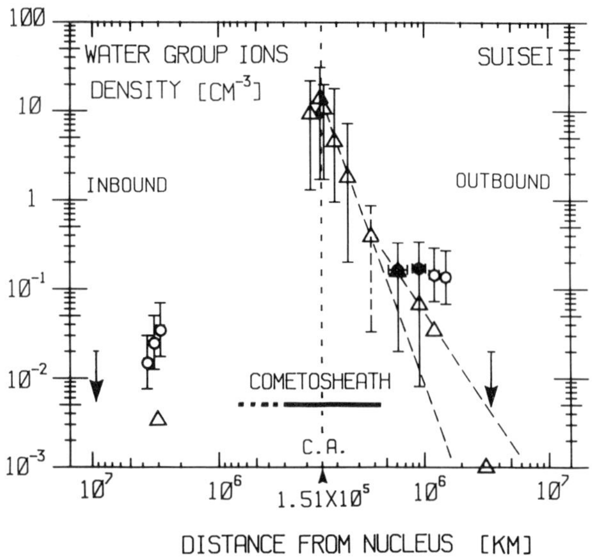

Fig. 2. Spatial distribution of cometary water-group ions. Open (solid) circles correspond to the intervals when the direction of the magnetic field can (cannot) be determined. Triangles show the calculated densities under the condition of gas production rate $Q = 2\ 10^{30}\ \mathrm{s}^{-1}$. Two dashed lines show trends of the calculated density profiles inside the cometosheath region and just outside the bow shock region, respectively

mination of the shell thickness (error $-50\% \sim +100\%$). For the intervals when the field direction is not known (solid circles), the field might be perpendicular to the ecliptic plane so that the observed "shell" would be a torus within the ecliptic plane. This possibility gives the lowest limit for the density estimation for these data. For the data obtained inside the cometosheath, only upper (100% shell) and lower (ecliptic torus) limits are given in the figure. The ion density at the closest approach ($1.5\ 10^5$ km from Halley's nucleus) is 14–20 cm^{-3}, which is consistent with the IMS observation by Giotto (Balsiger et al., 1986). The calculated densities (see next section) are also presented for comparison with the observation.

3. Discussion

We discuss the observed spatial distribution of water-group ions in terms of a simple pickup model. We assume that the parent neutral molecules escape isotropically from the cometary nucleus, since we are concerned with the neutral density far away from the collision dominated inner coma. Then the neutral density n cm^{-3} at the radial distance R is given by

$$n(R) = n_0 (R_0/R)^2 (V_0/V_n(R)) \exp\left(-\int_{R_0}^{R} \lambda(R)/V_n(R)\, dR\right), \quad (1)$$

where n_0 is the neutral density at the radial distance R_0, $\lambda(R)$ is the loss rate of neutral molecules, and V_0 and $V_n(R)$ are the escape velocities of neutral molecules at the radial distances R_0 and R, respectively. The neutral density is related to the gas production rate $Q\ \mathrm{s}^{-1}$ as follows.

$$n_0 = (Q/(4\pi R_0^2 V_0)) \exp\left(-\int_0^{R_0} \lambda(R)/V_n(R)\, dR\right). \quad (2)$$

The above formulation is essentially the same as that used previously (Mukai et al., 1986a; Ipavich et al., 1986), but can include dependence of the loss rate and the escape velocity on the cometocentric distance. During radial expansion, the neutral water molecules are lost by dissociation into H + OH and ionization including dissociative ionization. In the present model, however, we take only the ionization process into account; i.e., the loss rate due to dissociation is not taken here, since the dissociated products OH and O also belong to the parent molecules of the observed water-group ions.

The number density $N_i(P_{obs})$ of pickup ions at the observation point P_{obs} is calculated from the integral along the solar wind stream line (Mukai et al., 1986a; Ipavich et al., 1986),

$$N_i(P_{obs}) = (1/V_c) \int_{P_{obs}}^{P_{limit}} n(R)\lambda(R)\, ds, \quad (3)$$

where P_{limit} is the upstream limit and V_c is the convection velocity. Since we are concerned only with the ion density in the phase space shell, the upstream limit should be set such that the convection velocity does not change significantly over the path of integration. In the present calculation, the upstream limit is determined based on the result of the plasma flow observation (Mukai et al., 1986a; Terasawa et al., 1986). For the density calculation inside the bow shock, we take the upstream limit as $X_{limit} = 3\ 10^5$ km (a little inside the cometary bow shock). For the density calculation corresponding to observation points up to the cometocentric distance of $\sim 10^6$ km beyond the bow shock crossing, we take the integration path length of $1.2\ 10^6$ km. The upstream limit is selected as $X_{limit} = 3\ 10^6$ km for the $3\ 10^6$ km points. We neglect the loss of pickup ions due to recombination with electrons, since the transportation time from P_{limit} to P_{obs} (order of 10^3 s) is much shorter than the recombination time.

The ionization rate would vary according to variation in the plasma parameters and the neutral composition depending on the cometocentric distance. We divide the whole region of interest into several sub-regions with the cometocentric distance ranges of (a) <1.5, (b) 1.5–2.0, (c) 2.0–4.0, and (d) >4.0 10^5 km according to our plasma observation (Mukai et al., 1986b). We have calculated the ion densities corresponding to the observation points for various combinations of ionization rates in these sub-regions. First we have tried to see if a good agreement can be reached between observation and calculation for $Q = 1\ 10^{30}\ \mathrm{s}^{-1}$, but failed to find global coincidence. An enhanced ionization rate in the sub-regions (b) and (c) can give a result in good agreement with the observed values only around the closest approach ($<2\ 10^5$ km), but yields much lower ion densities than the observed values at greater cometocentric distances. In Fig. 2, one of the best fitted results (marked by triangles) is shown in comparison with the observed values. This result is obtained under the condition of $Q = 2\ 10^{30}\ \mathrm{s}^{-1}$ and the ionization rates of 1.5, 2.5, 1.5, 1.0 $10^{-6}\ \mathrm{s}^{-1}$ corresponding to the sub-regions (a)–(d), respectively. The escape velocity of neutral molecules is assumed to be $1\ \mathrm{km\,s}^{-1}$. [It is noted that the value of $Q = 1\ 10^{30}\ \mathrm{s}^{-1}$ in the caption of Fig. 5 in our previous paper (Mukai et al., 1986c) should be corrected as $Q = 2\ 10^{30}\ \mathrm{s}^{-1}$.] These values for the ionization rates are reasonably interpreted when the effective ionization mechanisms are mainly charge exchange with solar wind protons and photoionization due to solar EUV radiation. It is noted that a combination of Q and the ionization rate in the sub-region (a), which gives the identical neutral density at $R_0 = 1.5\ 10^5$ km from Eq. (2), can yield the same result as the present one, since we are concerned with the ion density in the region beyond the cometocentric distance of R_0; the observation closest to Halley's nucleus was made at the distance of

151,000 km. The escape velocity of 1 km s^{-1} is reasonable for the water-group neutrals, but some part of the parent molecules for O^+ may be atomic oxygen dissociated from CO, especially at greater cometocentric distances; in this case the escape velocity is higher than 1 km s^{-1}.

The water production rate of $Q = 2\ 10^{30}$ s^{-1} is higher when compared to the values measured by Vega-1 ($Q = 1.3\ 10^{30}$ s^{-1}; Gringauz et al., 1986) and by Giotto ($Q = 6.9\ 10^{29}$ s^{-1+} Krankowsky et al., 1986). However, the parent molecules of cometary ions observed at greater cometocentric distances would be ejected from the nucleus much earlier than the observation period; e.g., about 10 days earlier (or, around 1986 February 26) for the 10^6 km observation if the escape velocity of the neutral molecules is 1 km s^{-1}. The water production rate of Halley at that time might be higher than the values measured by Vega-1 and Giotto. Weaver et al. (1986) also reported the water production rate of 2.3 10^{30} s^{-1} on 1986 March 24, when comet P/Halley was located at the heliocentric distance of ~1 AU (post-periherion), even further away from the sun than for our observation period.

It is noted that the calculated density profile has a more gradual slope just outside the bow shock region when compared to the slope in the cometosheath region. This arises from the longer path length of the integral [Eq. (3)] outside the bow shock compared to that adopted inside the cometosheath. However, the observation shows a still more gradual slope which cannot be explained by the present calculation. The breathing of comet P/Halley (Kaneda et al., 1986) cannot account for this, either, since each observed point around 10^6 km distance represents an average density in a 2 10^5 km width of cometocentric distance. This averaging width roughly corresponds to the period of breathing (~53 hours), if the escape velocity of neutral molecules is 1 km s^{-1}.

Finally, the observation shows that the water-group ions were detected as far as 3–4 10^6 km in the inbound orbit, whereas the outbound observation only gives the upper limit. Similar asymmetry is also seen in the density profile of cometary protons (Terasawa et al., 1986). It is interesting to point out that the comparison of the calculated densities at 3 10^6 km between the inbound and outbound orbits also shows an inbound-outbound asymmetry by a factor of ~3. This asymmetry is caused mainly by the orbital difference in reference to the CSE frame. The inbound observations were made in the downstream ($X < 0$), where the cometocentric distance along the integral path in Eq. (3) decreases during negative values of X. Therefore, the integrand of Eq. (3) first increases (during $X < 0$) and then begins to decrease. On the other hand, the integrand corresponding to the outbound observation decreases monotonically along the integration path. Another cause for the symmetry arises from the flow velocity difference between the inbound and outbound orbits: The flow velocity was 430 km s^{-1} on the inbound and 570 km s^{-1} on the outbound orbit. Thus the observed asymmetry in the density distribution of cometary ions between the inbound and outbound orbits can be reasonably interpreted.

Acknowledgements. We thank all the members of the PLANET-A project team for their contributions to the success of the Suisei plasma experiment, especially T. Itoh and M. Oda. We are also grateful to T. Abe, A. Nishida, O. Ashihara, M. Shimizu, and T. Yamamoto for valuable discussions and comments.

References

Balsiger, H., Altwegg, K., Bühler, F., Fuselier, S.A., Geiss, J., Goldstein, R., Lazarus, A.J., Meier, A., Neugebauer, M., Rettenmund, U., Rosenbauer, H., Schwenn, R., Shelley, E.G., Ungstrup, E., Young, D.T.: 1986, 20th ESLAB Symp. ESA SP-250 Vol. **1**, p. 99

Biermann, L., Brosowski, B., Schmidt, H.U.: 1967, Solar Phys. **1**, 254

Gringauz, K.I., Gombosi, T.I., Remizov, A.P., Apáthy, I., Szemerey, I., Verigin, M.I., Denchikova, L.I., Dyachkov, A.V., Keppler, E., Klimenko, I.N., Richter, A.K., Somogyi, A.J., Szegö, K., Szendrö, S., Tátrallyay, M., Varga, A., Vladimirova, G.A.: 1986, *Nature* **321**, 282

Ip, W.-H., Axford, W.I.: 1982 in *Comets*, Univ. Arizona Press, Tucson, p. 588

Ipavich, F.M., Galvin, A.B., Gloeckler, G., Hovestadt, D., Klecker, B., Scholer, M.: 1986, *Science* **232**, 366

Kaneda, E., Hirao, K., Takagi, M., Ashihara, O., Itoh, T., Shimizu, M.: 1986, *Nature* **320**, 140

Krankowsky, G., Lämmerzahl, P., Herrwerth, I., Woweries, J., Eberhardt, P., Dolder, U., Herrmann, U., Schulte, W., Berthelier, J.J., Illiano, J.M., Hodges, R.R., Hoffman, J.H.: 1986, *Nature* **321**, 326

Mendis, D.A., Ip, W.-H.: 1977, *Space Sci. Rev.* **20**, 145

Mukai, T., Miyake, W., Terasawa, T., Kitayama, M., Hirao, K.: 1986a, *Nature* **321**, 299

Mukai, T., Miyake, W., Terasawa, T., Kitayama, M., Hirao, K.: 1986b, *Geophys. Res. Letters* **13**, 829

Mukai, T., Miyake, W., Terasawa, T., Kitayama, M., Hirao, K.: 1986c, 20th ESLAB-Symp., ESA SP-250 Vol. **1**, p. 71

Terasawa, T., Mukai, T., Miyake, W., Kitayama, M., Hirao, K.: 1986, *Geophys. Res. Letters* **13**, 837

Weaver, H.A., Mumma, M.J., Larson, H.P., Davis, D.S.: 1986, *Nature* **324**, 441

Astron. Astrophys. 187, 132–136 (1987)

ASTRONOMY AND ASTROPHYSICS

An interpretation of the ion pile-up region outside the ionospheric contact surface

W.-H. Ip[1], R. Schwenn[1], H. Rosenbauer[1], H. Balsiger[2], M. Neugebauer[3], and E.G. Shelley[4]

[1] Max-Planck-Institut für Aeronomie, D-3411 Katlenburg-Lindau, Federal Republic of Germany
[2] Physikalisches Institut, University of Bern, Bern, Switzerland
[3] Jet Propulsion Laboratory, Pasadena, CA, USA
[4] Lockheed Palo Alto Research Laboratory, Palo Alto, CA, USA

Received January 29, accepted May 4, 1987

Summary. An analysis interrelating some of the published results of plasma measurements by the Giotto spacecraft in the inner coma of comet P/Halley suggests that the formation of the plasma pile-up region at a distance of 10,000 km – 30,000 km may be the result of accumulation of ion density in a stagnant flow region with a sharp electron temperature transition forming a recombination front at about 10,000 km.

Key words: plasma pile-up – stagnant plasma flow – recombination loss – electron temperature

1. Introduction

One of the most interesting results from the Giotto probe to comet P/Halley concerns the finding of a dense plasma region just ahead of the magnetic field-free cavity (Balsiger et al., 1986; Krankowsky et al., 1986; Neubauer et al., 1986). In hindsight, the probable formation of layers of plasma structures had already been indicated by ground-based observations of bright comets such as comets Morehouse 1908III and Bradfield 1970II (Lüst, 1967; Delsemme and Combi, 1979). For example, plasma inhomogeneities in the form of contracting envelopes appeared on the sunward side of the optical centre of comet Morehouse 1908III had been suggested as the evidence of spatial and temporal variations in the comet-solar wind interaction (Ip and Axford, 1982). The in-situ measurements by Giotto at comet Halley perhaps have rendered this effect even more unequivocal. Because of the rapid increase of the ion density between 8,000 km and 12,000 km from the comet centre, this dense plasma region has been termed the pile-up region (Balsiger et al., 1986). In the first report of the Giotto Ion Mass Spectrometer (IMS) results, it was stated that these structures could be caused by a number of physical effects. These include:

(1) enhanced electron temperature which leads to a reduction in the electron recombination loss as well as contribution from electron impact ionization if the electron energy is large enough;

(2) rapid loss of plasma along the draped magnetic field lines in the region of minimal density;

(3) a temporal feature reflecting a change in the solar wind condition or the outgassing rate of the comet.

In this short communication, we would like to report some results of our ongoing investigation of this interesting plasma region. Because of time limitation for data analysis since the Giotto encounter much is still to be done, in particular, in the aspect of model computations of the magnetic field morphology and flow field. With the availability of experimental results from several different plasma experiments, some judicious remarks nevertheless can be made on the viability of these possibilities.

Send offprint requests to: W.-H. Ip

2. Temporal effects of neutral gas

With the reports on large variations in the surface outgassing rate either as a function of the rotational phase of the nucleus with different periodicities (i.e. 2.2 days, Kaneda et al., 1986; Sekanina and Larson, 1986 and 7.4 days, Millis and Schleicher, 1986) as well as some short term variations (Feldman et al., 1986), the presence of anisotropic and/or inhomogeneous structures in the neutral gas coma of comet P/Halley was to be expected on theoretical grounds (Kitamura, 1986; Kömle and Ip, 1987). One somewhat surprising result from the Vega and Giotto gas measurements therefore is that the published estimates of the spatial variation of neutral gas densities all followed a simple $1/r^2$ approximation (Krankowsky et al., 1986; Gringauz et al., 1986). In particular the Neutral Mass Spectrometer (NMS) experiment on Giotto showed that the H_2O gas density was quite smooth over the radial interval (5000 km – 12,000 km) where the H_2O^+ ion density had large variations. The formation of the plasma pile up region as observed by Giotto is therefore not likely caused by a large change in the local spatial distribution of the neutral coma.

3. Temporary effects of solar wind interaction

As the plasma condition in the inner coma can be significantly influenced by the solar wind interaction process in the upstream region, it is possible that the plasma pileup effect observed by the Giotto spacecraft at a distance of $1 - 2\,10^4$ km is due to a convection of the plasma structures as suggested by the formation of contracting envelopes in some comets (Lüst, 1967; Delsemme and Combi, 1979). As the distance of closest approach of the

Vega 2 spacecraft to comet Halley is about 1.5 10^4 km, the onset of the plasma density drop (if the pileup region is a stationary structure) might have just been missed by the plasma instruments onboard (see Gringauz et al., 1986).

A viable diagnostic method of the plasma structures in the inner coma is by ground-based observations so that both the spatial and temporal variations of the cometary ion structures may be monitored. Because of the problem with dust contamination, application of this technique is not straight forward in the case of the comet Halley observations. In any event, the possible role of time-variation in solar wind interaction in producing cometary plasma structures should be born in mind as it would modify the following considerations based on the absence of temporal effect.

4. Electron impact ionization

According to one theoretical model (Ip and Axford, 1982, 1987; Cravens, 1986), the cometary plasma immediately ahead of the ionospheric contact surface should be largely stagnant with field-aligned flow being the most significant transport process (see Fig. 1). The IMS experiment on Giotto also showed that the flow velocity of the cold cometary ions was very low ($\leq 0.2\,\mathrm{km\,s^{-1}}$) in the pile up region (Balsiger et al., 1986; Schwenn et al., 1986). The force balance in this broad region is achieved by an equilibrium between the $\mathbf{J} \times \mathbf{B}$ force and the frictional force between the cometary ions and the radially streaming neutral gas. A simple analytical model (Cravens, 1986; Ip and Axford, 1987) taking into account both curvature force and gradient of the magnetic pressure term has found reasonable agreement with the magnetometer observation of the formation of a field-free cavity (Neubauer et al., 1986).

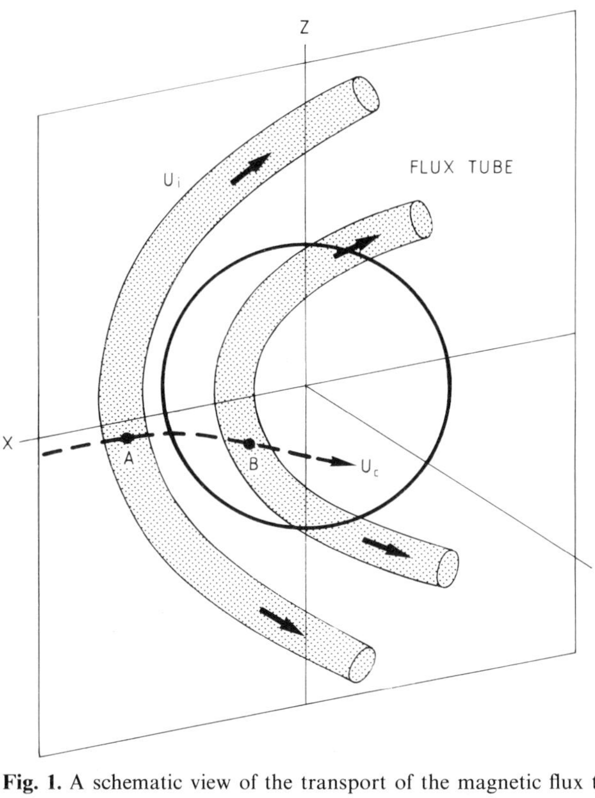

Fig. 1. A schematic view of the transport of the magnetic flux tubes in the stagnant flow region. The cometary ions are supposed to stream along the field lines as the flux tubes are being convected down-stream

One consequence of such a dynamical interaction between the draped magnetic field and the neutral gas outflow is that the larger the ion-neutral friction the more extended will be the standoff region. In other words, with increase or decrease of the ion density in the pile up region, the force balance can be adjusted by shifting the plasma density peak outward or inward. The boundary of the field-free cavity should also be moving in response of the cometary outgassing rate which determines the neutral frictional force as well as the cometary ion density in the inner coma.

While the ionization process inside the ionopause is basically governed by the photoionization effect, electron impact ionization via the so-called critical velocity ionization effect (Alfven, 1954) has been suggested to be of importance (Formisano et al., 1982; Galeev and Lipatov, 1984; Haerendel, 1986). Two events of ion density enhancements at $\sim 3.2 - 3.9\ 10^4$ km as observed by the PLASMAG experiment on Vega have been interpreted as being caused by the occurrence of the critical velocity ionization effect by Galeev et al. (1986); it is even possible that the steep increase of the plasma density in this region is related to the strong ionization effect in such transient events according to these authors.

One basic criterion for the generation of critical velocity ionization mechanism is that the relative velocity between the neutral gas and the plasma flow in the direction perpendicular to the local magnetic field must exceed the limit given by

$$\frac{mV_c^2}{2} \geq \eta \phi_i \qquad (1)$$

where ϕ_i is the ionization potential of the neutral molecules, and $\eta(\geq 1)$ an efficiency factor (Alfven, 1954; Formisano et al., 1982). In the extreme situation in which $\eta \sim 1$, the critical value of the relative velocity for H_2O^+ ions is $V_c \sim 11.2\,\mathrm{km\,s^{-1}}$. Taking advantage of the fact that the IMS experiment on Giotto can yield velocity information on the thermal cometary ions (Balsiger et al., 1986; Schwenn et al., 1986) we have examined one of the ion density enhancements similar to what had been seen in the Vega PLASMAG measurements (Galeev et al., 1986) and found that the flow velocity ($\sim 5 - 6\,\mathrm{km\,s^{-1}}$) in the ram direction during this event actually tends to be lower than the neighbouring regions (see Fig. 2). While magnetic field data must be compared before a firm conclusion can be reached, this examination of the Giotto IMS data seems to suggest that the operation of the critical velocity ionization effect was not obvious.

As the critical velocity ionization effect is basically a result of transfer of the kinetic energy of the streaming plasma flow to the ionizing electron population, the surest sign for enhanced electron impact ionization would be the detection of a significant population of suprathermal electrons in the coma. In a preliminary study of the electron number densities and velocity distributions as observed at comet Giacobini-Zinner by the International Cometary Explorer (ICE) spacecraft (Bame et al., 1986; Thomsen et al., 1986), a conclusion was drawn that the electron impact ionization effect could occasionally exceed that of the photoionization but the overall effect should not produce an order of magnitude enhancement in the ionization process (Ip, 1986a). The electron measurements on Giotto in the coma of comet Halley (Rème et al., 1986) have shown that the electron

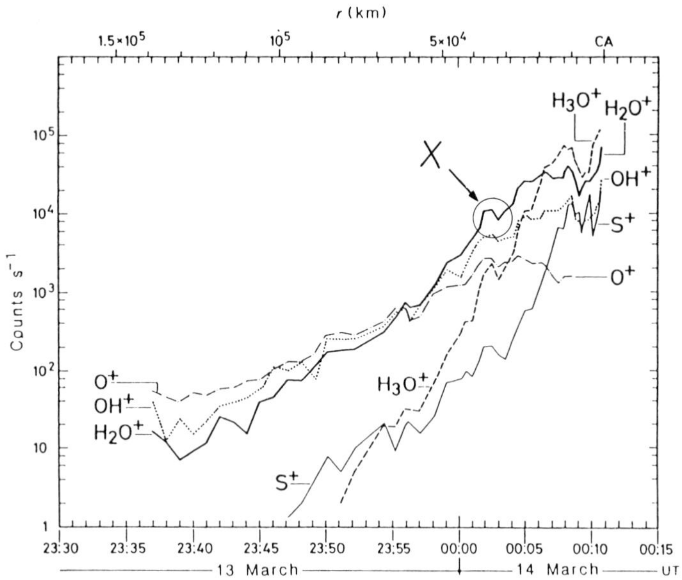

Fig. 2a. Time profiles of the IMS/HIS count rates for ion masses 16, 17, 18 and 19 as well as the mass 32–34 group. Time is ground-station received time in UT; r, distance of Giotto from the nucleus. The feature X is enlarged in Fig. 2b. From Balsiger et al. (1986)

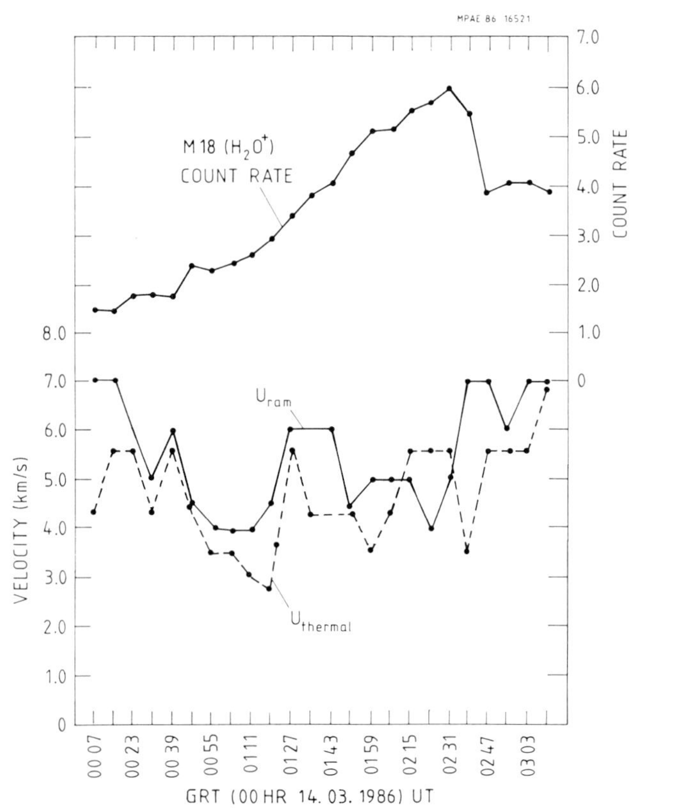

Fig. 2b. A detailed description of the ion density enhancement (feature X in Fig. 2a during 00:07 – 03:03 UT (ground station receiving time). Illustrated are the number density of the H_2O^+ ions, the ram velocity of the cometary ions relative to the spacecraft ($U_{\rm ram}$) and the thermal speed of the H_2O^+ ions ($U_{\rm thermal}$)

impact ionization rate from electrons above 10 eV is typically on the order of 10^{-7} sec or less between 12,000 km and 45,000 km along the inbound pass (R.P. Lin, priv. comm., 1986). It is thus clear that local electron impact ionization should not play an important role in the formation of the pile-up region at comet P/Halley during the Giotto encounter.

5. Transport and recombination loss

The next alternative to be investigated is the possibility of reduction of electron dissociative recombination loss of the cometary ions in the pile-up region as a result of a higher electron temperature. For electron dissociative recombination loss of the H_2O^+ ions, the rate constant can be written as (Mendis et al., 1985; D.A. Mendis, priv. comm., 1987):

$$R = 2.3\,10^{-7}(300/T_e)^{1/2}\,\mathrm{cm^3\,s^{-1}} \tag{2}$$

Hence for $T_e \geq 3000$ K, for example, the effective rate constant would be a factor of 3 or more smaller than the nominal value of $2.3\,10^{-7}\,\mathrm{cm^3\,s^{-1}}$ which was usually used in the modelling of the cometary ionospheric chemistry. This point has been noticed by Marconi and Mendis, (1984) though the focus then was the possible formation of an enhanced plasma region inside the contact surface where the plasma flow is still predominantly in the radial direction. The Giotto observations have shown that the pile-up region where the plasma flow was largely stagnant was formed outside the contact surface, however.

If the motion of the cometary ions can be approximated as being mostly in the field-aligned direction (see Fig. 1), the steady state equation of continuity for the ion number density (n_i) can be written as:

$$\frac{1}{A}\frac{d}{ds}(n_i u_i A) = q_i - l_i \tag{3}$$

where u_i is the plasma streaming speed, A the cross sectional area of the flux tube, q_i the production rate due to photoionization and l_i the loss rate from electron dissociative recombination. In a more precise computation, diffusive effect as well as momentum and energy equations should be incorporated into the calculations. These would mean a treatment of the atmospheric ion-neutral collision, Coulomb collision, and wave-particle interaction – in addition to the requirement of a self-consistent MHD model of the ion coma. To explore the first order effect involved in the interrelation between the field-aligned transport and recombination loss process, we have opted for a simplified model in which the flux tube is assumed to be circularly curved and its cross sectional area is unchanged. Furthermore, the values of T_i (ion temperature), T_e (electron temperature), and u_i are assumed to be constant.

Figure 3 illustrates the variations of the total ion number density (including OH^+, H_2O^+ and H_3O^+) at a point 90° from the subsolar point for a range of T_e and u_i. In the numerical integration of Eq. (3), the radial distance is taken to be at about 10000 km and the neutral number density is assumed to be $n_0 = 5.6\,10^5\,\mathrm{cm^{-3}}$ (i.e., an H_2O production rate of $6\,10^{29}\,\mathrm{s^{-1}}$ for an outflow velocity of about $0.9\,\mathrm{km\,s^{-1}}$). For u_i value $> 1\,\mathrm{km\,s^{-1}}$, the plasma transport time scale is sufficiently smaller than the electron dissociative recombination loss time scale such that there is little dependence on T_e for the ion number density. However, for $u_i < 1\,\mathrm{km\,s^{-1}}$, the recombination rate becomes increasingly

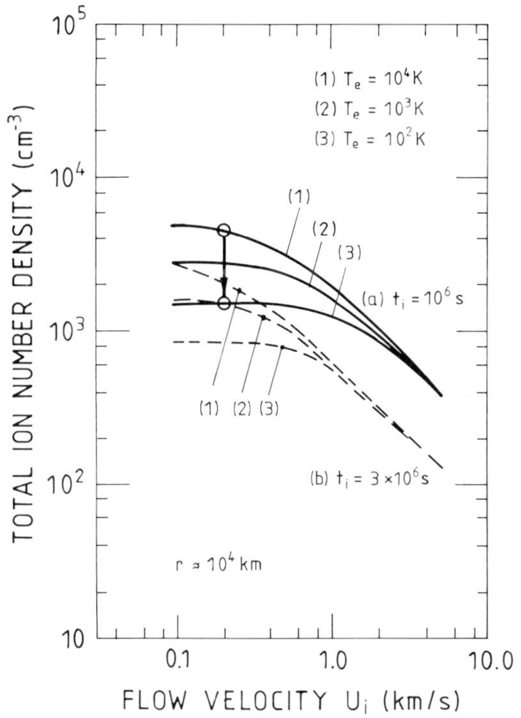

Fig. 3. A summary diagram displaying the variations of the total ion density as a function of the field-aligned streaming velocity (U_i) and the electron temperature (T_e)

important in the determination of n_i. For example, at $u \sim 0.2\,\mathrm{km\,s^{-1}}$ and an ionization time of $t_i = 10^6$ sec, $n_i \sim 4\,10^3\,\mathrm{cm^{-3}}$ for $T_e \sim 10^4$ K; and $n_i \sim 2\,10^3\,\mathrm{cm^{-3}}$ for $T_e \sim 100$ K.

Since in the coma region between 5000 km and 15000 km from the comet centre, the plasma velocity along the ram direction was indeed found to be $< 0.2\,\mathrm{km\,s^{-1}}$ (Balsiger et al., 1986; Schwenn et al., 1986), it is hence possible that the existence of a sharp change in the electron temperature from a few thousand degrees K to about one hundred degrees K at about 10,000 km could contribute to the abrupt decrease in the ion density. Inside this "electron recombination front", the electron number density will be adjusted in near-photochemical equilibrium with the low electron temperature. The total ion density thus should increase again following an inverse law of the radial distance.

One way to create a jump in the electron temperature is to invoke the thermalization of the cometary photoelectrons by collisional cooling with the H_2O molecules (Mendis et al., 1985; Ip, 1986b). Marconi and Mendis (1984) have treated this problem in a geometry of spherical symmetric outflow; for a gas production rate of $4.7\,10^{29}$ molecules/second at a solar distance of 0.9 AU ments of electron number density and electron temperature to rapidly from a value of 100 K to 10,000 K at a radial distance of about 2600 km. With a larger gas production rate ($Q \sim 6 - 7\,10^{29}$ molecules/second), anisotropic outflow pattern (Larson et al., 1986), extension of the cold electron region to a distance of 10^4 km along the inbound trajectory of the Giotto spacecraft is possible. Unfortunately, the electron analyser on Giotto (Reme et al., 1986) was not equipped to measure electron flux with T_e as low as 100 – 10,000 K, this hypothesis thus could not be tested directly. Note that the radio science experiment on the ICE spacecraft during encounter with Comet Giacobini-Zinner through the ion tail (Meyer-Vernet et al., 1986a; 1986b) provided in-situ measurements of electron number density and electron temperature to $T_e \sim 10^4$ K. A sharp rise of the electron number density at a perpendicular distance of about 8000 km was found to be coincident with a rapid drop of the electron temperature to below 10^4 K. Even though the coma environment and flow condition are very different from those of the Giotto encounter, this ICE result nevertheless is consistent with our present scenario even though a plasma pile-up region was not detected at the ion tail of comet Giacobini-Zinner.

6. Discussion

In this report we have discussed the possibility that the formation of the plasma pile-up region at comet P/Halley as observed by Giotto could be the combined result of field-aligned transport and recombination process as controlled by the electron temperature. As both data analysis and numerical modelling with respect to the spacecraft observations at comet P/Halley are still in a very early stage, the discussion above should be regarded, at best, as a guide for future lines of investigation. With some numerical efforts, a comprehensive MHD model may be constructed to shed light on the ion flow dynamics in the pile-up region so that the issue on the transport effect may be further clarified. For example, Baumgärtel and Sauer (1987) have explored some of these effects as suggested here in a one-dimensional MHD numerical model. Another aspect of the problem concerns the spatial variation of the electron temperature which eventually determines the electron dissociative recombination time scale. Laboratory study of the e – H_2O collision process, theoretical computations of the electron energy transfer in a dusty cometary coma, and observations of the H_2O gas from the ground and in space-orbiting platforms would be needed to make further progress (see Körösmezey et al., 1986). That a combination of field-aligned transport and T_e-dependence of the ion recombination rate could explain the formation of the plasma pile-up region remains a useful hypothesis in orienting our approach to the interpretation of the experimental result. Another interpretation to be pursued is the non-stationary process of plasma envelope formation as seen in some bright comets. Observational test of these alternatives may be attempted by ground-based optical observations of cometary ionspheric structures in the future.

Acknowledgements. We have benefited from useful discussions on this topic with participants, particularly Dr. A.A. Galeevin, in the WA 4 subgroup of the Lindau Workshop on Plasma, Ion Composition and Neutral Gas of comet Halley, 1986, October, 21–24. The comments by the referee, Dr. D.A. Mendis, is gratefully acknowledged. This work was supported in part by the Bundesministerium für Forschung und Technologie, the Swiss National Science Foundation, and by NASA, under contract NSAW-3729 (Lockheed), NASA Grant NGL 22-009-015 (MIT), and at the Jet Propulsion Lab, California Institute of Technology, under a contract with NASA.

References

Alfven H.: 1954, *On the Origin of the Solar System*, Oxford Clarenolon Press

Balsiger H., K. Altwegg, F. Bühler, J. Geiss, A.G. Ghielmetti, B.E. Goldstein, R. Goldstein, W.T. Huntress,W.-H. Ip, A.J.

Lazarus, A. Meier, M. Neugebauer, U. Rettenmund, H. Rosenbauer, R. Schwenn, R.D. Sharp, E.G. Shelley, E. Ungstrup, D.T. Young: 1986, *Nature* **321**, 330

Bame S.J., R.C. Anderson, J.R. Asbridge, D.N. Baker, W.C. Feldman, S.A. Fuselier, J.T. Gosling, D.J. McComas, M.F. Thomsen, D.T. Young, R.D. Zwickl: 1986, *Science* **232**, 356

Baumgärtel, K., Sauer, K.: 1987, *Astron. Astrophys.* (this issue)

Cravens T.E.: 1986, 20th ESLAB Symp., ESA SP-250, Vol. 1, p. 241

Delsemme A.H., Combi M.R.: 1979, *Astrophys. J.* **228**, 330

Feldman, P.D., Festou, M.C., A'Hearn, M.F., Arpigny, C., Butterworth, P.S., Cosmovici, C.B., Danks, A.C., Gilmozzi, R., Jackson, W.M., McFadden, L.A., Patriarchi, P., Schleicher, D.G., Tozzi, G.P., Wallis, M.K., Weaver, H.A., Woods, T.N.: 1986, 20th ESLAB Symp., ESA SP-250, Vol. 1, p. 325

Formisano V., Galeev, A.A., Sagdeev, R.Z.: 1982, *Planet Space Sci* **30**, 491

Galeev, A.A., Lipatov, A.S.: 1984, *Adv. Space Res.* **4**, 9, 229

Galeev A.A., Gringauz, K.I., Klimov, S.I., Remizov, A.P., Sagdeev., R.Z., Savin, S.P., Sokolov, A.Y., Verigin, M.I., Szego, K.: 1986, *Geophys. Res. Letters.* **13**, 845

Gringauz K.I., T.I. Gombosi, A.P. Remizov, I. Apathy, I. Szemerey, M.I. Verigin, L.I. Denchikova, A.V. Dyachkov, E. Keppler, I.N. Klimenko, A.K. Richter, A.J. Somogyi, K. Szegö, S. Szendrö, M. Tatrallyay, A. Varga, G.A. Vladimirova: 1986, *Nature* **321**, 282

Haerendel G.: 1986, *Geophys. Res. Letters.* **13**, 255

Ip W.-H.: 1986a, *Adv. Space Res.* **5**, No. 12, 47

Ip, W.-H.: 1986b, *Adv. Space Res.* **5**, No. 12, 233

Ip W.-H., Axford W.I.: 1982, in *Comets*, ed. L.L. Wilkening, p. 588

Ip W.-H, Axford W.I.: 1987, *Nature* **325**, 418

Kaneda E., O. Ashihara, M. Shimizu, M. Takagi, K. Hirao: 1986, *Nature* **321**, 297

Kitamura, Y.: 1986, *Icarus.* **66**, 241

Kömle, N., Ip, W.-H.: 1987, *Astron. Astrophys.* (this issue)

Krankowsky D., P. Lämmerzahl, I. Herrwerth, J. Woweries, P. Eberhardt, U. Dolder, U. Herrmann, W. Schulte, J.J. Berthelier, J.M. Illiano, R.R. Hodges, J.H. Hoffman: 1986, *Nature.* **321**, 326

Körösmezey, A., Cravens, T.E., Gombosi, T.I., Nagy, A.F., Mendis, D.A., Szego, K., Gribov, B.E., Sagdeev, R.Z., Shapiro, V.D., Sherchenko, V.I.: 1986, 20th ESLAB Symp. ESA SP-250, Vol. 1, p. 235

Larson, H.P., Davis, D.S., Mumma, M.J., Weaver, H.A.: 1986, 20th ESLAB Symp., ESA SP-250, Vol. 1, p. 333

Lüst R.: 1967, *Z. Astrophys.* **65**, 236

Marconi L.M., Mendis D.A.: 1984, *Astrophys. J.* **287**, 445

Mendis D.A., Houpis, H.L.F., Marconi, M.L.: 1985, *Fund. Cosmic Phys.* **10**, 1

Meyer-Vernet N., Couturier, P., Hoang, S., Perche, C., Steinberg, J.L., Fainberg, J., Meetre, C.: 1986a, *Science.* **232**, 370

Meyer-Vernet N., Couturier, P., Hoang, S., Perche, C., Steinberg, J.L.: 1986b, *Geophys. Res. Letters*, **13**, 279

Millis, R.L., Schleicher, D.G.: 1986, *Nature.* **324**, 646

Neubauer F.M., K.H. Glassmeier, M. Pohl, J. Raeder, M.H. Acuna, L.F. Burlaga, N.F. Ness, G. Musmann, F. Mariani, M.K. Wallis, E. Ungstrup, H.U. Schmidt: 1986, *Nature* **321**, 352

Reme, H., J.A. Sauvoud, C d'Uston, F. Cotin, A. Cros, K.A. Anderson, C.W. Carlson, D.W. Curtis, R.P. Lin, D.A. Mendis, A. Korth, A.K. Richter: 1986, *Nature* **321**, 349

Schwenn, R., Ip, W.-H., Rosenbauer, H., Balsiger, H., Bühler, F., Goldstein, R., Meier, A., Shelley, E.G.: 1986, 20th ESLAB Symp. ESA SP-250, Vol. 1, p. 225

Sekanina, Z., Larson, S.M.: 1986, *Astron. J.* **92**, 462

Thomsen M.F., Bame, S.J., Feldman, W.C., Gosling, J.T., McComas, D.J., Young, D.T.: 1986, *Geophys. Res. Letters* **13**, 393

Astron. Astrophys. 187, 137–140 (1987)

ASTRONOMY AND ASTROPHYSICS

Description of the main boundaries seen by the Giotto electron experiment inside comet P/Halley-solar wind interaction region

C. d'Uston [1], H. Rème [1], J.A. Sauvaud [1], A. Cros [1], K.A. Anderson [2], C.W. Carlson [2], D. Curtis [2], R.P. Lin [2], A. Korth [3], A.K. Richter [3], and A. Mendis [4]

[1] Centre d'Etude Spatiale des Rayonnements, CNRS-UPS, BP 4346, F-31029 Toulouse, France
[2] Space Sciences Laboratory, UC-Berkeley, Berkeley, CA 94720, USA
[3] Max-Planck-Institut für Aeronomie, Postfach 20, D-3411 Katlenburg-Lindau, Federal Republic of Germany
[4] EECS, UC-San Diego, La Jolla, CA 92093, USA

Received January 23, 1987, accepted May 18, 1987

Summary. The Giotto electron-plasma experiment identified several boundaries inside the comet Halley-solar wind region. Two of them are particularly interesting; they separate very different plasma regimes on quite sharp length scales and furthermore their existence was not foreseen in theoretical models. They are the limit between the transition region and the sheath detected at 550,000 km from the nucleus, and the cometopause detected at 135,000 km from the nucleus.

Characteristic features of the electron distribution functions and of the macroscopic parameters of the electron plasma through these main boundaries are described.

Key words: cometary plasma – cometary envelopes – plasma – spectrometry

1. Introduction

The particles and fields environment of comet Halley encountered by the Giotto spacecraft on March 13–14, 1986, displays a very complex structure (Balsiger et al., 1986; Johnstone et al., 1986; Rème et al., 1986a; Neubauer et al., 1986). In the midst of rapid fluctuations, one also distinguishes fairly well defined regions separated by sharp transitions. In this paper we present some characteristics of the electron plasma as observed by the Rème Plasma Analyzer (RPA) experiment at two of these plasma boundaries.

The *E*lectron *E*lectro *S*tatic *A*nalyzer (EESA) of the RPA experiment aboard Giotto was designed to measure the complete 3D distribution function of electrons between 10 eV and 30 keV. A full 4π distribution function was sampled every 2 s and the collected counts were processed on board to determine the symmetry axis of the pressure tensor. This axis coincides with the local magnetic field direction and allows one to sort the distribution function into pitch angle bins. The *P*ositive *I*on *C*luster *C*omposition *A*nalyzer (PICCA) also included in the RPA experiment consists of an electrostatic analyzer which measures the fluxes of ions with energies between 250 eV/Q and 5 keV/Q in the ram direction i.e. 73° away from the solar wind flow direction.

Send offprint request to: C. d'Uston

A complete description of the experiment is given elsewhere (Rème et al., 1986b).

From the transmitted data, several parameters were computed to characterize the various plasma regimes encountered during the inbound crossing of Halley's environment. A summary plot of this traversal is given in Fig. 1. It displays the temporal evolution of five parmeters between 18:00 UT and 24:00 UT (Ground received Time or GRT) on March 13, 1986. Several distinct plasma regions can be seen and are described in Rème et al. (1987) and Anderson et al. (1987). In chronological order there are: an upstream region, the foreshock in front of the bow shock, followed by three distinct regions, then the cometopause, and finally the cometosphere. The boundaries which will be described in this paper are (1) the transition between region 2 and region 3 ($\sim$ 22:00 UT, GRT) and (2) the cometopause ($\sim$ 23:38 UT, GRT). In Fig. 1 it can be seen that both occur an short time scales. This behavior was not foreseen by theoretical models. The first discontinuity around 22:00 UT (GRT) separates a zone containing energetic electrons (up to 2 keV) from a quiet sheath. Presently it is not understood why such energetic electrons exist in region 2. Other characteristics in this time interval make this region "mysterious" (Rème et al., 1987). The cometopause separates the quiet sheath in which the plasma flow velocity was decreasing almost down to stagnation at the cometopause, from the cometosphere where the magnetic field is piling up and reaches a maximum value just before the ionopause (Neubauer et al., 1986).

2. "Mysterious" transition at 22:00 UT (GRT)

From Fig. 1, it appears at the end of the mystery region that this transition corresponds to a sudden change in total density NTMIN calculated from a temperature estimate assuming a maxwellians shape for the electron distribution. The parmeter N 4 corresponding to 0.8–3.6 keV electron density, drops to a low value here. The flow velocity, which was decreasing very slowly, starts decreasing faster. The total flux of ram ions rises and never falls again as in the regions 1 and 2, but increases smoothly toward the nucleus. During this time the spacecraft is still far from the cold plasma region surrounding the comet nucleus: the detected ram ions are part of the compressed solar wind flow. Its large temperature causes large fluxes to be measured in the ram

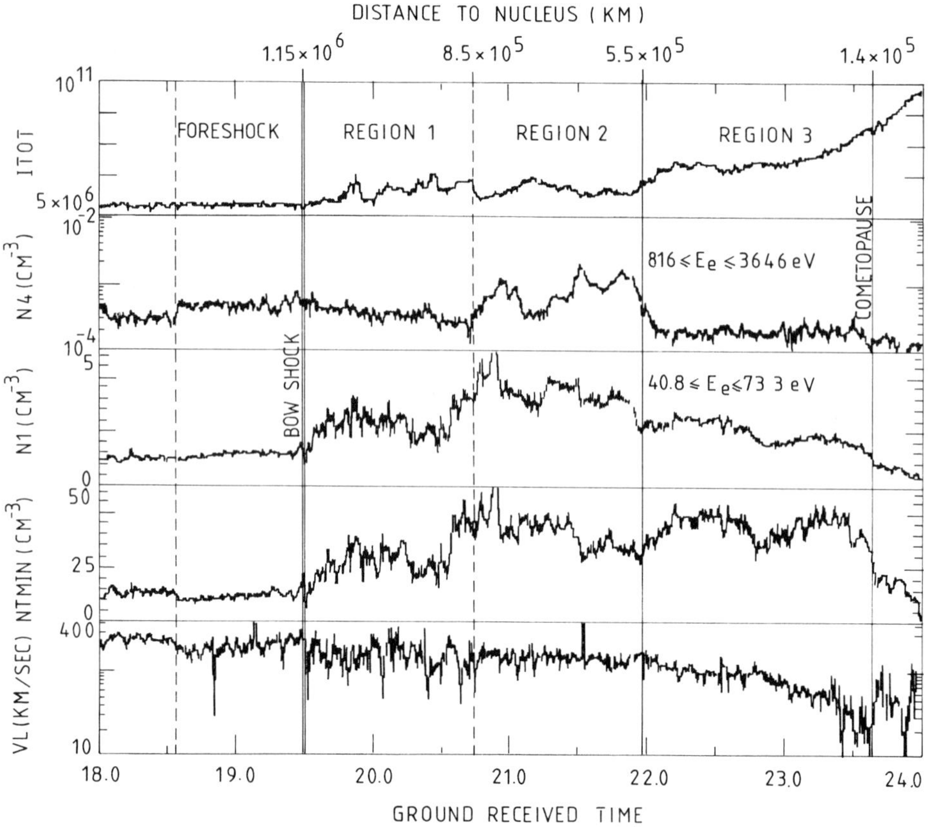

Fig. 1. Values of plasma parameters measured by the RPA experiment between 18:00 and 24:00 UT (Ground Received Time) on March 13. From bottom to top: bulk velocity of the electrons; estimate of the total electron density; density of 40 eV–73 eV electrons; density of 0.8 keV–3.6 keV electrons; total flux of ram ions with enegies between 250 eV and 5 keV in $cm^{-2} s^{-1} sr^{-1}$

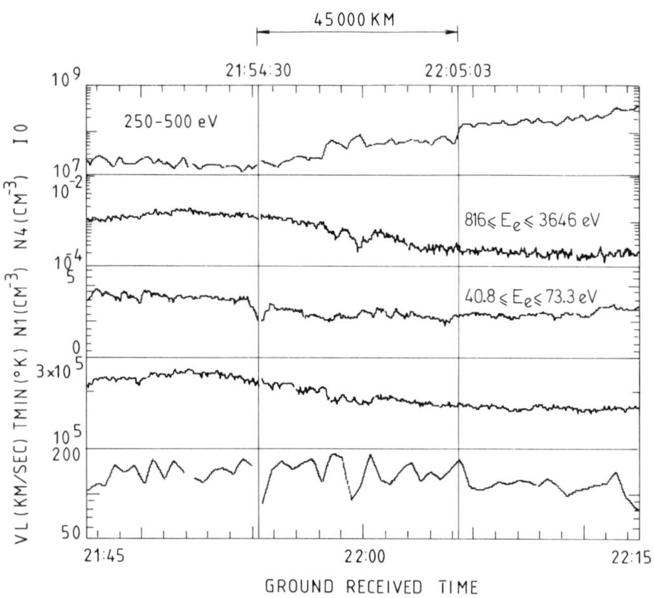

Fig. 2. Temporal variation of characteristic parameters through the "mysterious boundary" (from top to bottom: total ram ion fluxes, 0.8–3.6 keV electron density, 40–73 eV electron density, temperature of the electron distribution near 70 eV, electron bulk flow velocity). The changes in the parameters occur in ~ 10 minutes corresponding to a width of ~ 45,000 km for this boundary

direction. The transition is shown on an enlarged time scale in Fig. 2. The parameters displayed here allow a determination of the width of the transition. It takes places between 21:54:30 and 22:05:03 UT (GRT), thus giving a width of 45,000 km. Note that this transition corresponds also to a very important change in the behaviour of the solar wind and implanted ions detected by the Johnstone Plasma Analyzer (JPA) experiment. This is the time when the implanted ion distribution starts to divide into 2 components (Johnstone et al., 1986). This boundary is thus located at 5.5 10^5 km from the nucleus, roughly halfway between the shock position and the nucleus.

Bame et al. (1986) noted a similar abrupt change in the behaviour of the electron plasma inside the comet Giacobini-Zinner environment during the flyby of the International Cometary Explorer (ICE) spacecraft. In Rème et al. (1968a) it was shown that both transitions occur at the same relative position between the shock and the nucleus, thus suggesting that this structure could be a permament feature of active comets at a distance of ~1 AU from the Sun. Such a structure has not been foreseen in comet-solar wind interaction models and some theoretical work is clearly needed to interpret this behaviour. Galeev (1986) stated that the energy of the accelerated picked up ions forces the shocked solar wind to flow through the stagnation region because of their cooling by charge exchange on neutral gaz and that this effect ceases when all picked up ions are charge exchanged. He identified the mystery boundary with the moment when the cometary ions picked up in the unshocked solar wind disappear due to charge exchange. However the characteristic distance to the nucleus along the sun-nucleus line for the stagnation of the plasma flow, $R_L \sim 4\ 10^4$ km, is too short to project at the observed position of 5.5 10^5 km in the flanks. Moroever at this distance, after all the excess of thermal energy supplied by the picked up ions has been transferred to the flow, the bulk of the plasma starts to decelerate strongly and the embedded interplanetary magnetic field should start to pile up. In the electron measurements this mysterious boundary is marked by sudden drop in temperature and followed by a smooth decrease of the estimated bulk velocity but there is no magnetic field increase indicating the start of the magnetic pile up region (Neubauer et al., 1986) at this boundary. Thus it is more likely that there exists some other process which helps the shocked solar wind to maintain its flow velocity and also provides acceleration of the electrons within the first part of the cometosheath and which is overrun by the charge exchange process on a short length scale at the mysterious boundary. In this picture the plasma is actually cooled after the crossing of this boundary and the solar wind flow is still maintened though slowed until the next discontinuity which corresponds to the cometopause discussed in the following.

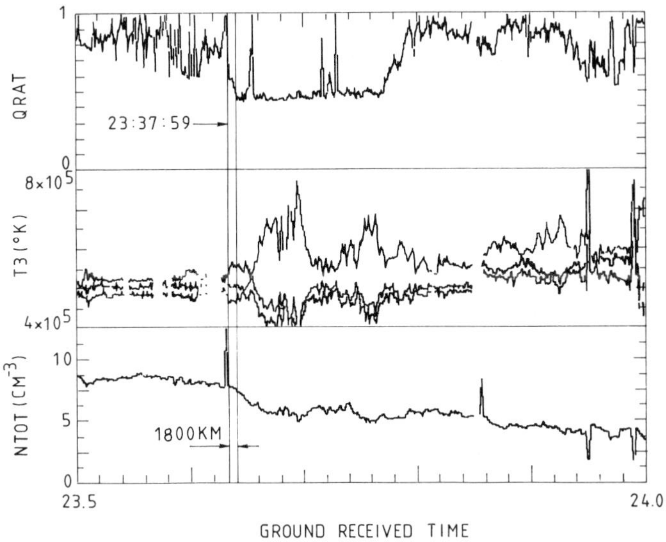

Fig. 3. Temporal variation of characteristic parameters of the electron distribution through the cometopause (note that here the time axis is labelled in decimal hours). From top to bottom are shown: heat flux ratio (see text); temperatures associated with the diagonal pressure tensor (see text); density of > 10 eV electrons. The density change starts ~ 25 seconds after the change in the spectral shape and is completed in 2–3 min

3. The cometopause at 23:38 UT (GRT)

At about 23:38 UT (GRT) and ~135,000 km from the nucleus, the RPA electron sensor detects a sudden decrease in the hot electron density ($kT_e > 10$ eV) associated with a minimum in the electron temperature, as well as in the electron bulk flow speed (Rème et al., 1986a). At the same distance, a sudden increase in the magnetic field from about 10 γ to about 30 γ is measured. This observation is consistent with the sudden slowing down of the inflowing, highly contaminated solar wind; the field lines start to pile up, giving rise to the magnetic field increase (Neubauer et al., 1986). This boundary has been called a cometopause (Gringauz et al., 1986).

RPA data pertaining to this boundary are shown in Fig. 3. NTOT represents the total density of electrons above 10 eV and shows a strong decrease starting at 23:38:25 UT (GRT) and lasting until about 23:41:30 thus giving a width of the boundary on the order of 10^4 km. In the middle panel, three curves are shown, each corresponding to one of the three temperatures given by the diagonal pressure tensor. Before the cometopause, the three temperatures are tightly linked together meaning that the electron distribution function is isotropic. After the cometopause, the temperatures evolve differently, the parallel one being very different from the perpendicular ones. Moreover the parallel temperature increases while the perpendicular ones decrease. This indicates that inside the cometosphere the electron distribution is cigar shaped with the long axis along the magnetic field direction. In the upper panel, QRAT is the ration of minimum to maximum heat flux transported by the electrons parallel and antiparallel to the magnetic field. While it was close to 1 before the cometopause, it dropped suddenly to about 0.5 at this boudary. This indicates that the electrons leak along the field lines in one preferential direction. Figure 4 illustrates the changes in the electron distribution function (D.F.) through the discontinuity. The D.F. at 23:35:00 UT (GRT) i.e. before the cometopause, shows an isotropic distribution, while the D.F. at 23:40:52 UT (GRT) shows that there exists a higher level of flux in one direction parallel to the magnetic field (x-axis). It is interesting to note the slight delay between the sudden change in the spectral shape shown by QRAT and T3 which occurs at 23:37:38 UT (GRT) and the start of the density decrease which takes place at 23:38:25 UR (GRT), ~1800 km closer to the nucleus. The first change coincides with the start of the magnetic field increase as determined by the magnetometer.

Even though there is no clear discontinuity in the total ram ion fluxes, the shape of the spectra indicates that the first cold cometary ions were detected at 23:35 (GRT), slightly before the cometopause crossing (Korth et al., 1986).

Several models predicted the formation of a magnetic barrier starting at a "collisionpause" where the strong deceleration of the solar wind is expected to commence, due to momentum transfer collisions with the outflowing neutrals (Mendis et al., 1986). But what the measurements have revealed is that this transition is very sharp, of the order of 10^4 km, and this is nor yet explained by theoretical considerations. This boundary is detected shortly after the moment when the cometary ions became the dominant species in the plasma composition (Amata et al., 1986). At this time the heavy ions have a rather small bulk velocity because they are stopped by collisions with neutrals. This leads to charge exchange processes and cools their distribution. As shown by Gringauz et al. (1986), from Vega measurements they control the plasma behavior and bring most of the energy. Then the magnetic field is progessively piling up and the field lines are drapped around the comet head toward the tail. The solar wind protons, due to their much smaller charge exchange cross section, may continue to flow further inward until they are eventually decelerated and stopped by collisions with the neutrals.

4. Conclusions

Two important boundaries detected by plasma experiment onboard Giotto in the interaction region of the solar wind with comet Halley have been presented. Their presence and characteristics are not presently understood. From electron measurements, the so called "mysterious boundary" occurs halfway between the shock and the nucleus and is spread over ~45,000 km. It is

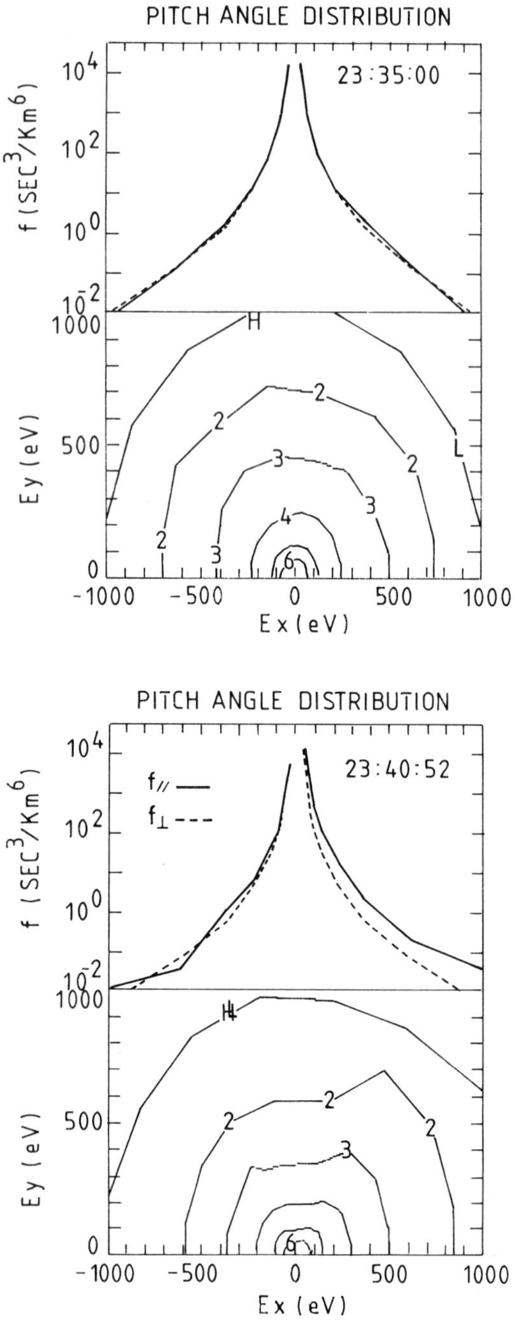

Fig. 4. Electron distribution function before (upper graph) and after (lower graph) the cometopause crossing at $\sim$ 23:38 UT (GRT). In each graph the lower panel displays isocontours of the distribution function in a frame of reference linked to the magnetic field direction (x-axis) between 30 eV and 1 keV. Isocontours are separated by a factor of 10 and isocontour 1 corresponds to $10^{-2}\,s^3/km^{-6}$. The upper panel shows cuts through the distribution functions parallel (solid line) and perpendicular (dashed line) to the magnetic field

characterized by a drop in electron density above 10 eV (and by the disappearance of the enhanced flux of energetic electrons). The cometopause detected at $\sim$135,000 km from the nucleus is very sharp ($\sim 10^4$ km width) and is characterized by a sudden change in the shape of the electron distribution functions from isotropic to field-aligned.

Acknowledegements. We are grateful to F.M. Neubauer and K.H. Glasmeier for useful discussion on the comparisons of electron and magnetic measurements.

References

Amata, E., Formisano, V., Cerulli-Irelli, R., Torrente, P., Johnstone, A., Coates, A., Wilken, B., Jockers, K., Winningham, J.D., Bryant, D., Borg, H., Thompsen, M.: 1986 20th ESLAB Symp. ESA SP-**250**, Vol. **1**, p. 312

Anderson, K.A., Carlson, C.W., Curtis, D.W., Lin, R.P., Rème, H., Sauvaud, J.A., d'Uston, C., Korth, A., Richter, A.K., Mendis, D.A.: 1987 *Astron. Astrophys.* (in press)

Balsiger, H., Altwegg, K., Bühler, F., Geiss, J., Ghielmetti, A.G., Goldstein, B.E., Goldstein, R., Huntress, W.T., Ip, W.-H., Lazarus, A.J., Meier, A., Neugebauer, M., Rettenmund, U., Rosenbauer, H., Schwenn, R., Sharp, R.D., Shelley E.G., Ungstrup, E., Young, D.T.:1986 *Nature* **321**, 330

Bame, S.J., Anderson, R.C., Asbridge, J.R., Baker, D.N., Feldman, W.C., Fuselier, S.A., Gosling, J.T., McComas, D.J., Thomsen, M.F., Young, D.T., Zwickl, R.D.: 1986, *Science* **232**, 356

Galleev, A.A.: 1986, Zoth ESLAB Symp. ESA SP-**250**, p. 3

Gringauz, K.I.,Gombosi, T.I., Tatrallyay, M., Veregin, M.I., Remizov, A.P., Richter, A.K., Apathy, I., Szemerey, I., Dyachkov, A.V., Balakina, O.V., Nagy, A.F.: 1986, *Geophys. Res. Letters*, **13**, 613

Johnstone, A., Coates, A., Kellock, S., Wilken, B., Jockers, K., Rosenbauer, H., Studemann, W., Weiss, W., Formisano, V., Amata, E., Cerulli-Irelli, R., Dobrowolny, M., Terenzi, R., Egidi, A.,Borg, H., Hultquist, B., Winningham, J., Gurgiolo, C., Byrant, D., Edwards, T., Feldman, W., Thomsen, M., Wallis, M.K., Biermann, L., Schmidt, H., Lust, R., Haerendel, G., Paschmann, G.: 1986, *Nature* **321**, 344

Korth, A., Richter, A.K., Anderson, K.A., Carlson, C.W., Curtis, D.W., Lin, R.P., Rème, H., Sauvaud, J.A., d'Uston, C., Cotin, F., Cros, A., Mendis, D.A.: 1986, *Adv. Space Res.* **5**, 221

Mendis, D.A., Smith, E.J., Tsurutani, B.T., Slavin, J.A., Jones, D.E., Siscoe, G.L.: 1986 *Geophys. Res. Letters* **13**, 239

Neubauer, F.M., Glassmeier, K.H., Pohl, M., Raeder, J., Acuna, M.H., Burlaga, L.F., Ness, N.F., Musmann, G., Mariani, F., Wallis, M.K., Ungstrup, E., Schmidt, H.U.: 186, *Nature* **321**, 352

Rème, H., Sauvaud, J.A., d'Uston, C., Cotin, F., Cros, A., Anderson, K.A., Carlson, C.W., Curtis, D.W., Lin, R.P., Mendis, D.A., Korth, A., Richter, A.K.: 1986a, *Nature* **321**, 349

Rème, H., Cotin, F., Cros, A., Médale, J.L., Sauvaud, J.A., d'Uston, C., Korth, A., Richter, A.K., Loidl, A., Anderson, K.A., Carlson, C.W., Curtis, D.W., Lin, R.P., Mendis, D.A.: 1986b, ESA SP-**1077**, p. 33

Rème, H., Savaud, J.A., d'Uston, C., Cros, A., Anderson, K.A., Carlson, C.W., Curtis, D.W., Lin, R.P., Korth, A., Richter, A.K., Mendis, D.A.: 1987, *Astron. Astrophys.* (in press)

Astron. Astrophys. 187, 141–148 (1987)

ASTRONOMY AND ASTROPHYSICS

In-situ observations of a bi-modal ion distribution in the outer coma of comet P/Halley

M. F. Thomsen[1], **W. C. Feldman**[1], **B. Wilken**[2], **K. Jockers**[2], **W. Stüdemann**[2], **A. D. Johnstone**[3], **A. Coates**[3], **V. Formisano**[4], **E. Amata**[4], **J. D. Winningham**[5], **H. Borg**[6], **D. Bryant**[7], **and M. K. Wallis**[8]

[1] Los Alamos National Laboratory, Los Alamos, NM 87545, USA
[2] Max-Planck-Institut für Aeronomie, D-3411 Katlenburg-Lindau, Federal Republic of Germany
[3] Mullard Space Science Laboratory, Holmburg St. Mary, Dorking, Surrey, RH5 6NT, UK
[4] Instituto di Fisica dello Spazio Interplanetario, I-00044 Frascati, Italy
[5] Southwest Research Institute, San Antonio, TX 78284, USA
[6] Kiruna Geophysical Institute, Umeå University, S-90187 Umeå, Sweden
[7] Rutherford Appleton Laboratory, Chilton, Didcot, Oxon, OX11 OQX, UK
[8] University College, Cardiff, Wales, CF1 1XL, UK

Received February 12, accepted May 11, 1987

Summary. Observations obtained by the Johnstone Plasma Analyzer on the Giotto fly-by of comet Halley showed a fairly sudden decrease in the count rate of energetic (~ 30 keV) water-group ions inside about 5 10^5 km from the nucleus. This decrease was accompanied by the appearance of a new water-group ion population at slightly lower energies (< 10 keV). Close inspection reveals that this lower-energy peak was also present somewhat earlier in the post-shock flow but only became prominent near the sudden transition just described. It is shown that the observed bi-modal ion distribution is well explained in terms of the velocity history of the accreting solar wind flow in the outer coma. The decline in count rate of the energetic pick-up distribution is due to a relatively sudden slowing of the bulk flow there and not to a loss of particles. Hence charge-exchange cooling of the flow is probably not important at these distances from the nucleus. Finally, the observations suggest that pitch-angle scattering is fairly efficient at least after the bow shock, but that energy diffusion is probably not very efficient.

Key words: comets – ion pick-up

1. Introduction

Ionization of neutral atoms and molecules in the extensive atmosphere of a comet causes mass-loading of the solar wind. The mass-loading results in an increase of the effective temperature of the plasma because the new-born cometary ions are picked up with gyrospeeds on the order of the bulk flow speed of the solar wind; this large gyromotion combines with the heavier masses of the pick-up particles such that their contribution to the total pressure is dominant even at relatively low cometary ion concentrations.

As the contaminated solar wind penetrates closer to the nucleus, the flow speed declines due to the increased mass-loading. Cometary ions picked up under the slowed flow conditions have lower effective thermal speeds than those picked up farther upstream in the faster flow. As a consequence, the effective temperature of the heavy ions should decline as more and more mass is accreted at the lower flow speed. Additionally, cooling of the heavy ions in the slower flow regime can be effected by charge-exchange between local cometary neutrals and the energetic ions picked up well upstream. Since charge-exchange essentially replaces a fast heavy ion with a slower one, the signature of this process should be a decline in the energetic heavy ion population concurrent with the increase in the lower energy population. Furthermore, unlike the effective "cooling" that occurs due to increased accretion at lower flow speeds, charge exchange cooling results in a reduction of the total pressure of the plasma because of a loss of the higher energy particles. The dynamical consequences of charge-exchange cooling are quite interesting and have been considered by a number of authors (e.g., Galeev et al., 1985; Wallis and Dryer, 1985).

The Implanted Ion Sensor of the Johnstone Plasma Analyzer (JPA) detected picked-up cometary ions throughout the Giotto encounter with comet Halley, from several million kilometers out until quite near the nucleus (cf., Johnstone et al., 1986b). One of the interesting features observed during the encounter was an apparent transition in the nature of the water-group ion population which occurred at ~ 5 10^5 km from the nucleus on both the inbound and outbound passes (though it was much more readily apparent in the inbound data). At this distance a relatively sudden decrease occurred in the count rate of the energetic (~ 30 keV) population of water-group ions, which had been observed since well before the bow shock crossing at $\sim 10^6$ km. This sudden decrease was accompanied by the relatively sudden appearance of a new water-group population at somewhat lower energies (< 10 keV).

This transition has been reported previously (Johnstone et al., 1986b). The purpose of the present work is to examine it in greater

Send offprint requests to: M. F. Thomsen

detail to help illuminate the physics of the outer coma. We find that the change in the appearance of the pick-up population is due primarily to a fairly sudden slowing and deflection of the contaminated solar wind, rather than to the onset of charge exchange as we had speculated earlier. We find further that the observed structure in the pick-up ion distribution is fairly well explained in terms of the velocity history of the flow in the outer coma.

2. Instrumentation

This report is based on observations obtained by the Giotto Johnstone Plasma Analyzer (Johnstone et al., 1986a). The JPA consists of two separate sensors, the Fast Ion Sensor (FIS) and the Implanted Ion Sensor (IIS). The FIS is an electrostatic analyzer designed to detect positive ions in the energy/charge (E/q) range 0.01–20 keV/q in three dimensions in velocity space. Its primary purpose was to observe the solar wind ion distribution throughout the comet encounter.

In the present study we will be primarily concerned with observations obtained with the IIS. The IIS is a set of five electrostatic analyzers followed by time-of-flight units designed to provide first-order mass resolution of the cometary ions implanted in the solar wind, as well as their energy and angular distribution (see also Wilken et al., 1986a; Weiss and Wilken, 1986). The IIS observes positively charged ions at 32 energy levels in the E/q range 0.09–90 keV/q, which it sorts into five separate mass/charge channels ($m/q = 0$–1.7, 1.8–6.4, 6.5–20, 21–48, >49). Measurements are made at five different polar angles relative to the spin axis of the spacecraft and at 16 azimuthal angles during each spin. The IIS holds each energy level for one spacecraft rotation and thus obtains a complete distribution in 32 rotations (128 s).

The viewing geometry of the IIS is of some importance in the present study and is illustrated in Fig. 1. Each of the five electrostatic analyzer/TOF units has a 10° polar field of view centered at a fixed polar angle relative to the spacecraft spin axis. As the spacecraft spins, each detector thus views a conical surface in velocity space. Figure 1 shows the intersections of the five conical fields of view with a plane containing the spin axis. During the encounter, the spin axis was parallel to the spacecraft/comet relative velocity vector so that particles at rest in the comet reference frame would have a velocity parallel to the spin axis in the spacecraft reference frame. Hence in the cold, slow flow expected near the nucleus, polar detector 5, which looks most nearly along the spin axis, should provide the best view of the ion distribution. In this study we will focus on observations made by this detector.

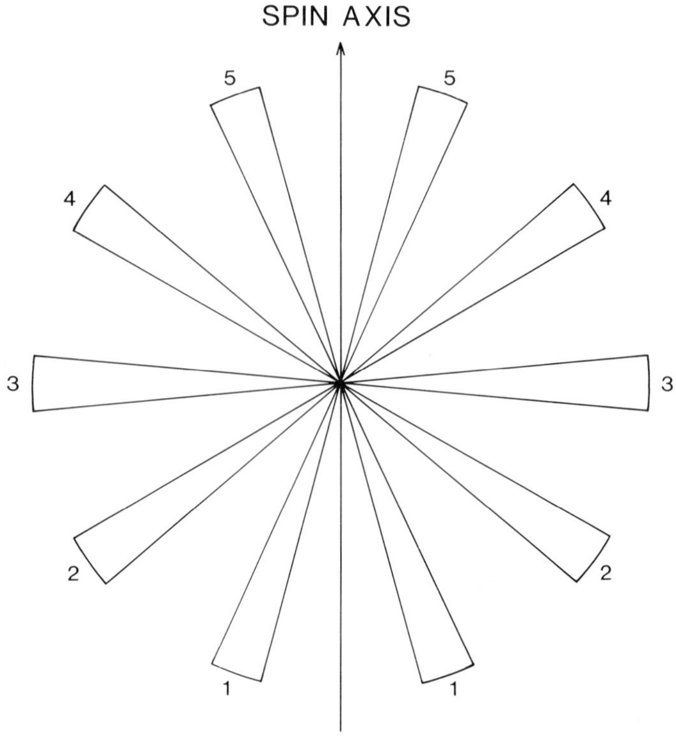

Fig. 1. Viewing geometry of the Implanted Ion Sensor of the Johnstone Plasma Analyzer. The spacecraft spin axis is parallel to the spacecraft/comet relative velocity. See text for further description

3. Observations

Throughout the encounter with the comet, ions in the mass range of $6.5 \leqq m/q \leqq 20$ dominated the cometary ion population observed by the IIS. Higher mass resolution spectra obtained by the IIS at the expense of angular resolution show that the detected ions are primarily water-group ions ($m/q \sim 16$–19). Figure 2a shows an overview of the observations summed over all 5 detectors in this mass range during the inbound pass on 1986 March 13. The figure shows the omnidirectional count rate in the water-group channel plotted as a function of energy per charge, with each successive 32-spin spectrum plotted slightly below the previous one so that the time axis runs in the downward direction. The spectral peak which appears near 400 eV is not due to real water-group ions, but rather is due to pile-up of solar wind protons in the time-of-flight unit. This proton signature can easily be distinguished from the true water-group population at higher energies.

Two transitions appear prominently in the omnidirectional water-group count rates illustrated in Fig. 2a. The first is a strong increase occurring at ~1930 GRT (~1 10^6 km from the nucleus), which corresponds to the inbound crossing of the cometary bow shock (cf., Johnstone et al., 1986b; Formisano et al., 1986). (GRT is the UT of ground receipt of the data.) The second transition in the character of the water-group population occurs at ~2200 GRT (~5 10^5 km from the nucleus). At this point there is a significant decrease in the count rate of the main heavy-ion peak as well as a shift in the location of the peak toward slightly lower energies. This decrease is fairly rapid, taking only one or two full spectra ($\lesssim 4$ minutes), and it is accompanied by the appearance of a second peak in the distribution at lower energies (~5 keV/q). This second peak is persistent and quite prominent after the 2200 transition; however, close examination of Fig. 2a reveals that a similar, though somewhat more energetic, secondary component is visible in a number of earlier spectra as well.

Figure 2a exhibits only the omnidirectionally averaged water-group count rates. In Fig. 2b we show the count rates observed during the same interval by only polar detector 5, the detector which viewed most nearly along the spacecraft/comet relative velocity (cf. Fig. 1). Comparison of Figs. 2a and 2b shows that after the bow shock crossing, counts in detector 5 accounted for a substantial fraction of the total heavy ion population observed by the IIS. In addition, the lower-energy secondary ion component

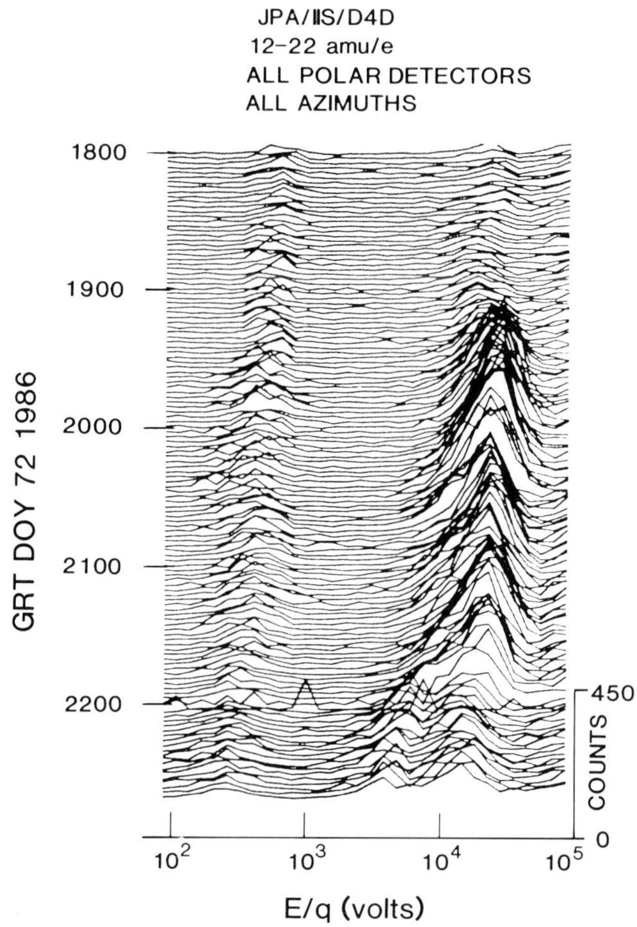

Fig. 2a. Omnidirectional count rate in the water-group mass channel of the JPA Implanted Ion Sensor during the inbound portion of the Halley encounter. The count rate is shown as a function of E/q, with each 128-second spectrum plotted slightly below the previous one so that time increases in the downward direction. The peak near 400 eV is due to pile-up of solar wind protons in the time-of-flight unit. The character of the water-group population changes at ~2200 GRT, with the emergence there of a second, lower-energy peak in the count rate and a concurrent decline in the count rate of the higher-energy peak

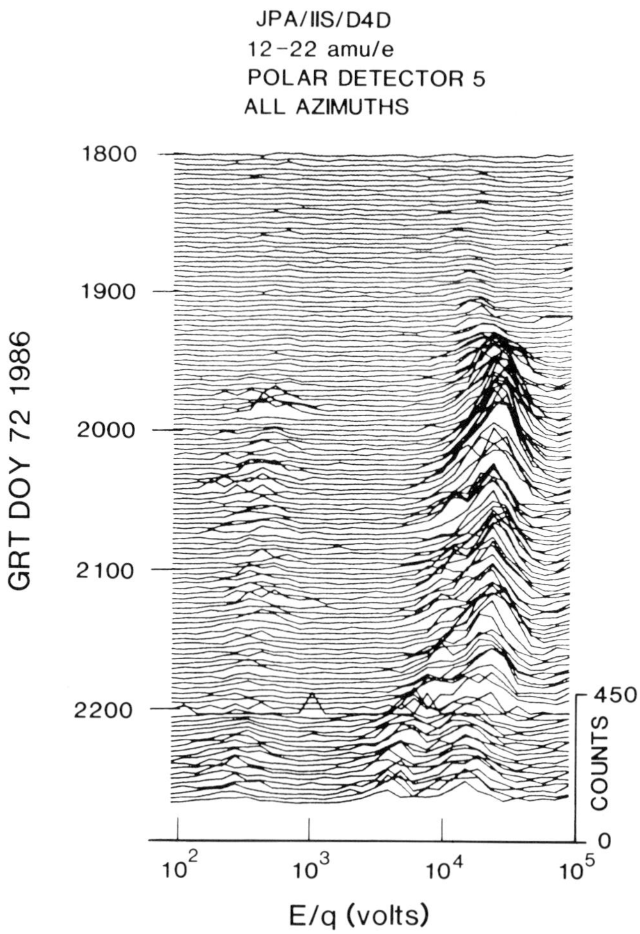

Fig. 2b. Same as Fig. 2a, except only counts observed in polar detector 5, which looks closest to the ram direction, are included

appears more prominent in Fig. 2b and better resolved from the main peak, especially at times before the 2200 transition.

The decline of the count rate of the higher-energy water-group population, more or less simultaneous with the appearance of the lower-energy heavy ions, initially suggested to us that charge-exchange between the energetic pick-up ions and local neutrals was eliminating the fast particles and replacing them with ions which were less energetic by virture of being picked up at the lower flow speed near the comet, as discussed above. However, further analysis of these data does not support this hypothesis. In particular, when the count rates shown in Fig. 2 are converted to phase space density, we find little evidence for any significant losses in the higher energy population. This fact is illustrated in Fig. 3, which shows the logarithm of the water-group phase space density (in arbitrary units) plotted as a function of particle speed in the same stacked spectrum format employed in Fig. 2, except that time now increases from the bottom to the top. The phase space densities shown here have been estimated by simply dividing the observed count rates by v^4, where v is the particle speed in the spacecraft frame. This conversion is appropriate for an electrostatic analyzer such as the IIS. The speed is calculated from the observed E/q by assuming the particles are singly charged and have a mass of 17 amu. The phase space densities have not yet been corrected for the lower instrument efficiency at lower energies. This correction would tend to raise the lower-energy part of the spectrum relative to the higher-energy part. Figure 3 corresponds most directly to Fig. 2b since only counts from polar detector 5 have been used to construct it. Only data above 1 keV/q are shown in Fig. 3 because the spurious counts at lower energies, which we attribute to proton pile-up in the TOF, are magnified to large values of the phase space density by the $1/v^4$ conversion between count rate and phase space density.

The broad, main heavy-ion pick-up component which dominates the post-shock observations is clearly evident near 500 km s^{-1} in Fig. 3, extending to 700 km s^{-1} (twice the solar wind speed, as expected for pick-up ions). The secondary peak at lower speeds shows up most clearly after 2200 but is also detectable well before that time. Figure 3 suggests that the spectral shape of the higher-energy pick-up component changes very little in the post-shock region, including the interval after the 2200 transition. What does change significantly at 2200 is the speed of the peak of this energetic population. In particular, it appears that the peak

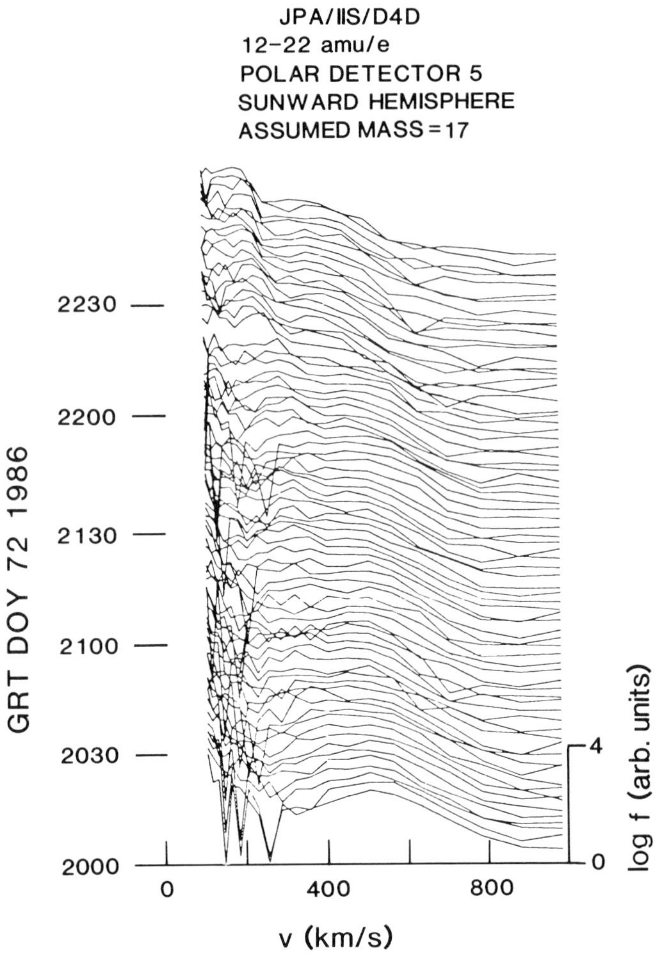

Fig. 3. Water-group phase space density (arbitrary units) calculated from observations made by polar detector 5 in the post-shock flow during the inbound portion of the encounter (see Fig. 2b). The phase space density is plotted as a function of particle speed in the spacecraft frame, in the same stacked-spectrum format used in Fig. 2, except that time now increases in the upward direction. The particle speed is based on an assumed mass of 17 amu. The phase space densities shown here are not corrected for the lower detection efficiency at lower energies

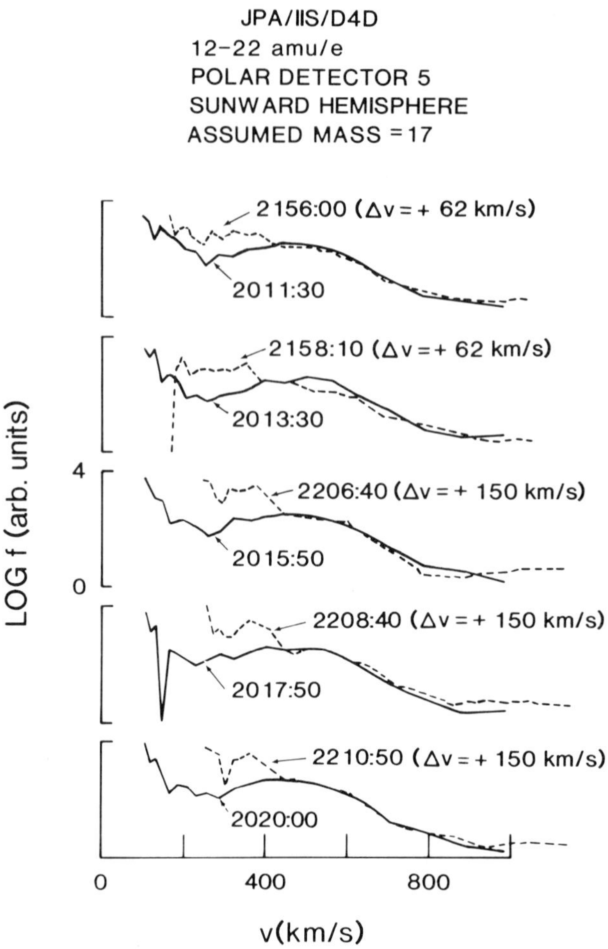

Fig. 4. Comparison of five individual water-group velocity distributions (arbitrary units) derived from polar detector 5 measurements in the immediately post-shock interval from 2011:30 to 2020:00 GRT (solid curves) with distributions obtained just before the 2200 GRT transition (top 2 dashed curves) and just after the transition (bottom 3 dashed curves). Several different spectra are illustrated for each interval in order to give a better feeling for which spectral features are persistent. The spectra all have the same normalization, but the dashed curves are offset to the right by 62 km s^{-1} and 150 km s^{-1} for the pre- and post-2200 spectra, respectively. As in Fig. 3, the particle speed is based on an assumed mass of 17 amu, and no corrections have been applied for the lower detection efficiency at lower energies

shifts downward from ~ 500 km s^{-1} before 2200 to $\lesssim 400$ km s^{-1} after 2200. This downward shift in the speed of the water-group peak is produced by an overall deceleration of the bulk flow of the plasma, which is observed in the proton mass channel of the IIS as well as in measurements by the FIS (neither of which is shown here, however). In addition to the slowing, these instruments also see a sudden increase in the deflection of the flow away from the upstream antisolar flow direction. These changes will be more fully documented elsewhere. The conclusions based on Fig. 3 are reinforced by the more detailed comparison made in Fig. 4, which shows five individual velocity distributions from the post-shock interval from 2011:30 to 2020:00 (solid lines) overlaid with two distributions from just prior to the 2200 transition (dashed line, top 2 curves) and three distributions from just after 2200 (dashed lines, bottom 3 curves). All ten curves have the same normalization, but the dashed curves have been offset to the right by 62 km s^{-1} and 150 km s^{-1} for the pre- and post-2200 spectra, respectively. These offsets were chosen to match up the higher-energy portion of the distributions. Figure 4 shows that the addition of new material at low speeds plus an overall slowing are the only notable changes in the distribution between ~ 2015 and the 2200 interval. More specifically, there does not appear to be any significant decrease in the height of the higher-energy peak across the 2200 transition. It thus appears that these higher-energy pick-up particles are not suffering significant losses at this distance from the nucleus ($\sim 5\ 10^5$ km).

4. Discussion

As noted above, Fig. 2 shows a clear decline in the count rate of the higher-energy water-group population across the 2200 transition. Yet, as we have just seen, Figs. 3 and 4 show that the phase space density of that population remains essentially constant across the transition, except that the peak shifts to lower energies. This slowing is, in fact, the key to the observed decline in the count rate: Since the count rate of an electrostatic analyzer like the IIS is proportional to the phase space density times v^4, the strong slowing across the 2200 transition produces a factor of 2–3 reduction in the peak count rate, with no reduction in peak phase space density. We thus conclude that charge exchange with local neutrals is not a strong loss mechanism for energetic pick-up ions in the neighborhood of the 2200 transition. The dynamical consequences of charge-exchange cooling must therefore be initiated closer to the nucleus. This is in agreement with the estimated charge-exchange scale radius (Wallis and Ong, 1976):

$$R_{\text{exch}} = (G\sigma/4\pi V_{\text{gas}})\, v_{\text{ion}}/u \approx 3\ 10^4\ \text{km},$$

where the gas production rate G and outflow speed V_{gas} are taken to be $7\ 10^{29}\ \text{s}^{-1}$ and $0.9\ \text{km s}^{-1}$, respectively (Krankowsky et al., 1986), the cross-section $\sigma = 2\ 10^{-15}\ \text{cm}^2$, the ion speed $v_{\text{ion}} = 500\ \text{km s}^{-1}$, and the flow velocity $u = 200\ \text{km s}^{-1}$.

While we find no clear evidence in these data for the onset of significant charge-exchange at this distance, the energy structure of the observed ions nevertheless contains interesting information regarding the ion pick-up process. In particular, it is worthwhile to compare the observed structure with that expected on the basis of a fairly simple model.

Fundamentally, the velocity space structure of the pick-up population must depend on the integrated flow history of the solar wind as it penetrates the comet. For concreteness, the discussion below assumes that all the solar wind plasma impinging on the comet experiences more or less the same flow history and that that history can be reasonably well represented by observations made by the JPA along the inbound trajectory of the Giotto spacecraft. In reality, of course, the flow history is different along different flow tubes through the system. Moreover, the Giotto trajectory was nearly perpendicular to the sun-comet line and hence did not correspond to any particular flow tube at all. And finally, newly picked-up cometary ions in general have a component of guiding center motion parallel to the magnetic field and oblique to the bulk plasma flow so that they don't even stay in the same flow tube in which they were created. The pick-up distribution at any point thus represents an integration over the flow properties within a fairly large region of the cometary coma. For these reasons, the slowing and deflection observed by Giotto, on which we base our model of ion pick-up, can only be an approximation to the integrated flow history responsible for the velocity space structure observed by the IIS.

With this important caveat in mind, in Fig. 5a we show a semi-quantitative schematic illustration of the flow history, based on JPA measurements at Halley (Johnstone et al., 1986b). The figure shows two-dimensional velocity space in the rest frame of the Giotto spacecraft. The vertical axis points in the direction of the relative velocity between the comet and the spacecraft, and the open circle on this axis shows this relative velocity, which was about $70\ \text{km s}^{-1}$. This circle indicates the velocity which cometary neutrals would have in the spacecraft frame. The heavy solid curve entering the figure from the right is the locus of the tip of the solar wind velocity vector at various distances from the comet nucleus. Point a is well upstream of the comet, where the flow is essentially in the antisolar direction at about $350\ \text{km s}^{-1}$. Point b is just before the bow shock, where the flow has slowed considerably due to upstream mass loading. The shock occurs between b and c and causes further slowing and also deflection. In the shocked plasma region between c and e, ion pick-up continues to cause slowing and

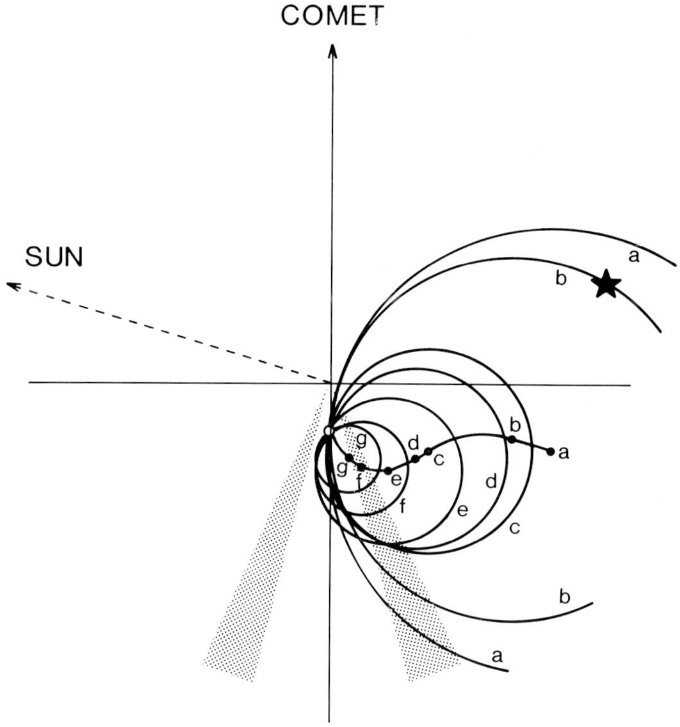

Fig. 5a. Semi-quantitative illustration of solar wind flow history and cometary ion pick-up in the vicinity of the comet. The heavy solid curve is the locus of the tip of the solar wind velocity vector in two-dimensional velocity space centered in the Giotto reference frame at various distances from the comet nucleus. The circles show the velocity space shells that would be populated at each of the indicated points by locally picked up ions, if pitch angle scattering were rapid and fairly complete. Shaded areas show the velocity space coverage of IIS detector 5. See text for further details

deflection. The transition at 2200 occurs between e and f and consists, as mentioned earlier, of a sudden slowing and deflection of the flow over a fairly short distance, possibly associated with a fairly rapid increase in mass-loading there (i.e., the new heavy-ion peak). Closer to the comet the flow slows even further, until it essentially stagnates in the comet frame, corresponding to a speed of $70\ \text{km s}^{-1}$ in the spacecraft frame.

Newly created cometary ions begin their lives with the same velocity as the neutrals, i.e., nearly at rest ($\sim 1\ \text{km s}^{-1}$ outflow) in the comet rest frame (open circle in Fig. 5a). Immediately upon ionization, however, they experience the magnetic and convective electric fields of the solar wind and begin to convect with the flow and gyrate about the magnetic field. They thus occupy a torus in velocity space which passes through the initial neutral velocity; the axis of the torus is parallel to the magnetic field and passes through the local solar wind velocity. As an example, the velocity space torus of ions created at point b (just upstream of the bow shock) would intersect the plane of Fig. 5a at the star in the upper right hand corner of the figure if the magnetic field lay in this plane at the nominal garden-hose angle of 45° relative to the solar direction. The other intersection point of the torus would, of course, be the open circle.

A toroidal velocity space distribution such as that produced by the ion pick-up process is unstable to a variety of wave modes (e.g., Winske et al., 1985). The primary effect of these instabilities is to scatter the pick-up ions in pitch angle (Galeev, 1986), particularly when the injection rate of new ions is relatively low, as it is well upstream of the comet (e.g., Gary et al., 1986). Thus, under the influence of such waves, the pick-up torus should evolve into a shell in velocity space centered on the local solar wind velocity. The circles in Fig. 5a show the intersections with the plane of the figure of the velocity space shells that could be populated by local pick-up at each of the points along the flow history, if such pitch angle scattering were rapid and fairly complete. Note that the radius of the local pick-up shells decreases as the plasma slows, especially across the shock.

The grey shaded areas on Fig. 5a show the region of velocity space viewed by detector 5 of the IIS, which views most nearly along the spacecraft/comet line (Fig. 1) and hence sees particles with velocities in the lower half of the velocity space plane of the figure. Note that for the predominant magnetic field orientations within the Halley sheath region, which were found to lie more or less in the upper left or lower right quadrants of the figure (Neubauer, 1986), unscattered pick-up particles would tend to appear primarily in the upper right portion of the illustrated pick-up shells, as does the star discussed above for pick-up at point b. The fact that detector 5 persistently and continuously saw significant fluxes of watergroup particles after the 1930 GRT bow shock crossing (Fig. 2b) thus suggests that appreciable pitch-angle scattering of the heavy pick-up ions was occurring at least at and downstream of the shock (see also the analysis by Wilken et al., 1986b). This agrees with observations made by Suisei (Mukai et al., 1986a, b). Similar conclusions have been reached for the pick-up proton population, which exhibits at least partial pitch-angle scattering even well upstream of the shock (cf. Balsiger et al., 1986; Terasawa et al., 1986; Neugebauer et al., 1986). For the rest of our discussion we therefore assume that picked-up ions scatter fairly rapidly into the shells illustrated in Fig. 5a.

Once created, these various velocity space shells are convected along with the flowing plasma, maintaining their initial radius but centered always on the local flow velocity. Hence at any given point in space, the distribution should consist of a superposition of the shells created all along the flow path. For example, the distribution observed at point f (just after the 2200 transition) should look something like the schematic distribution illustrated in Fig. 5b. Note in particular the gap between the shells created at b, just upstream of the bow shock, and at c, just downstream of the shock in the slowed flow.

We can now compare our expectations based on these considerations with the actual observations. Figure 6 shows this comparison. Three relevant quantities are presented in this figure:

The dashed lines show the speed at which a *local* pick-up shell would appear in polar detector 5 of the IIS as a function of time along the inbound trajectory of Giotto. The vertical range between the two dashed lines corresponds to the range of speeds an infinitely thin shell would cover over the finite angular width of the detector. As discussed above, the local pick-up shell radius should decline as one approaches the nucleus, with two fairly sharp drops corresponding to the rapid slowing at the shock and at the 2200 transition (e–f).

As the various pick-up shells convect along with the plasma, the speed at which they appear in the detector changes due to the overall slowing and deflection of the bulk flow (compare Figs. 5a and 5b). The solid curves in Fig. 6 show the speeds at which the detector would continue to see particles in the different shells picked up along the way. For example, the upper shaded lane shows where one would expect to see particles in the shell originally created at point b (just before the shock); and the lower shaded lane shows where detector 5 would see particles picked up just after the shock. (It is interesting to note that the deflection of

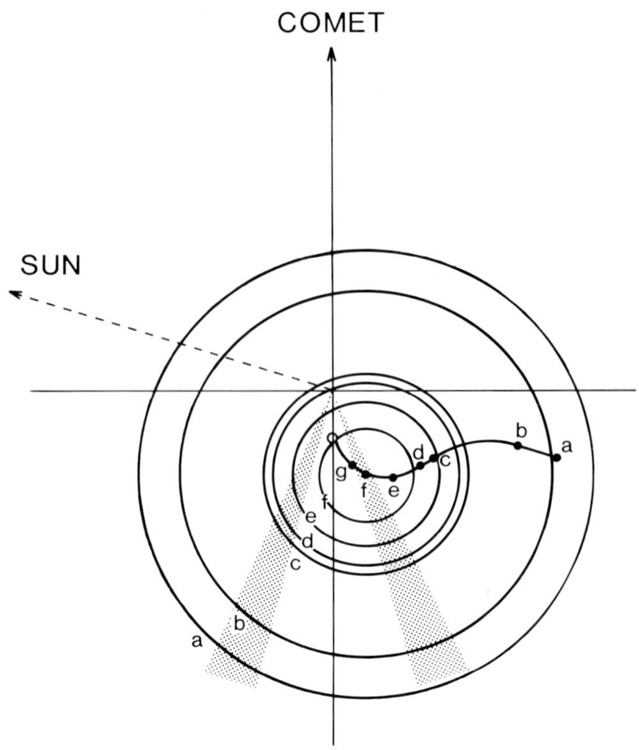

Fig. 5b. Schematic water-group velocity distribution expected at point f, just after the 2200 GRT transition, superimposed on the flow history of the solar wind as illustrated in Fig. 2a. The distribution is the superposition of shells created all along the flow path and convecting with the flowing plasma. Shaded areas again show the velocity space coverage of detector 5

the bulk flow across the shock causes the pre-shock pick-up shells to appear at higher speeds in detector 5 in the post-shock region. This effect can be seen by comparing the intersection of shell b with the detector 5 view region in Figs. 5a and 5b). As we anticipated, there is a lane between the energy regions populated before and after the shock. This lane should be populated by particles picked up during the actual shock transition. However, since the shock is relatively narrow compared to the upstream and downstream regions, there is relatively little time to populate this lane and hence it should be only sparsely filled. Such a bimodal ion distribution has been anticipated by previous authors (Wallis and Ong, 1975; Galeev et al., 1985; Cravens, 1986).

Finally, the solid dots in Fig. 6 show the locations of the peaks in the actual phase space densities calculated from polar detector 5 measurements. As in Fig. 3, the conversion from the observed E/q to the particle speed is based on the assumption that the observed ions have a mass of 17 amu, i.e., that they are water-group particles. For the higher-energy heavy ion component, this assumption is well substantiated by higher resolution mass spectra obtained by the IIS, as mentioned earlier. For the lower-energy component which appears mainly after 2200 GRT, this identification is somewhat less certain. Balsiger et al. (1986) report significant count rates in this region at $m/q = 12$, i.e., C^+. This value of m/q would also be included in the IIS mass channel examined here. However, Balsiger et al. find that the density of watergroup ions (16, 17, 18 amu) substantially exceeds that of the mass 12 ions throughout this region. Hence we assume that the lower-energy component, as well as the higher-energy component, consists primarily of water-group ions. If the particles were in fact

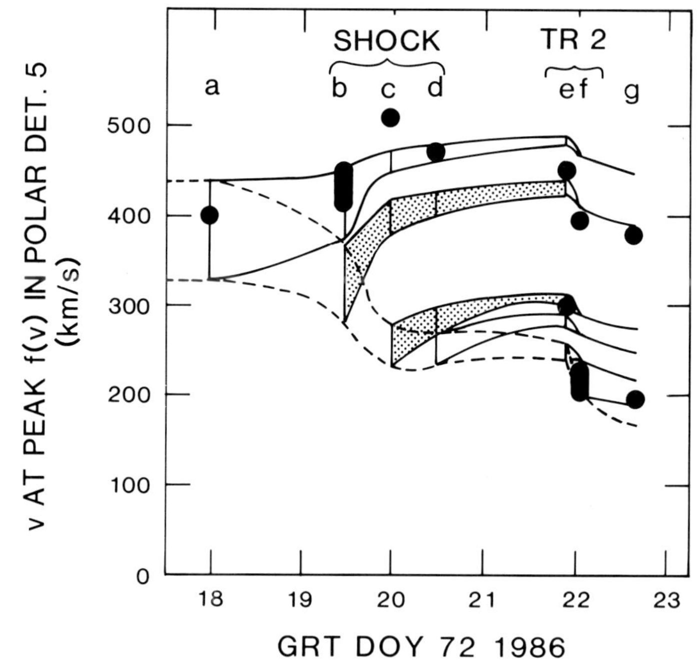

Fig. 6. Comparison of expected pick-up ion signature in IIS polar detector 5 with peaks in observed bimodal water-group population. The dashed lines indicate the range of speeds over which a local pick-up shell would appear in this detector as a function of time along the inbound Giotto trajectory. The solid curves show the speeds at which the detector would continue to see particles in the various shells picked up along the way. Note the lane between the energy regions populated before and after the shock. This lane should be sparsely filled due to the relatively short time available to populate it. The solid dots show the locations of the peaks in the observed phase space densities. See text for additional details

C^+ then the lower set of solid dots after about 2200 GRT should be raised by about 20%.

The agreement in Fig. 6 between the expected velocity space structure and that actually observed shows that in general the basic pick-up picture we have described is accurate. It confirms that the high-energy component is the signature of particles ionized upstream of the shock, while the lower-energy peak is due to pick-up after the shock. This second peak does not appear immediately after the shock because a certain amount of time is necessary to accumulate a detectable population. The decline in the energy of both components across the 2200 transition is clearly attributable to the significant decline in the bulk flow speed there. This decline is perhaps associated with a relatively sudden increase in the mass loading rate, but this needs further study.

Finally, it is worth reiterating that the fact that polar detector 5 measures such high fluxes of pick-up ions confirms that shell formation by pitch-angle scattering is relatively efficient within the cometary environment, at least inside of the bow shock, in agreement with conclusions drawn by previous authors, as noted above. And similarly, the observed persistence of the velocity-space shell structure suggests that energy diffusion of thc pick-up ions is relatively unimportant in the outer coma.

5. Conclusions

To summarize, the observations and discussion presented here suggest the following conclusions:

– The basic picture of cometary ion pick-up, as described above, is valid. The observed velocity space structure of the water-group ions can be well accounted for by the velocity history of the accreting solar wind flow in the outer coma.

– The higher-energy water-group peak observed by the IIS is composed of cometary particles ionized upstream of the bow shock.

– The lower-energy heavy-ion population is due to particles ionized in the slower post-shock flow.

– The decline in energy of both peaks at $\sim 5\ 10^5$ km from the nucleus is due to an overall slowing of the bulk flow there.

– Pitch-angle scattering of heavy ions is relatively efficient at least after the bow shock, but energy diffusion is not particularly efficient.

– Charge-exchange losses are probably not important at distances $\gtrsim 5\ 10^5$ km from the nucleus.

Acknowledgements. We would like to thank the many dedicated people associated with the Giotto project who made this historic encounter a reality. We also thank our colleagues at the various JPA institutions for their hard work to make the data readily usable. In this regard, particular thanks are due to C. Gurgiolo, R. Gowan, J. Wooten, and E. Tech. We also thank Jack Gosling for his helpful comments on this manuscript.

Work at Los Alamos was performed under the auspices of the U.S. Department of Energy and was supported by NASA through order number W-15,164. Work at Southwest Research Institute was supported by NASA contract NASW-3574.

References

Balsiger, H., Altwegg, K., Bühler, F., Geiss, J., Ghielmetti, A.G., Goldstein, B.E., Goldstein, R., Huntress, W.T., Ip, W.-H., Lazarus, A.J., Meier, A., Neugebauer, M., Rettenmund, M., Rosenbauer, H., Schwenn, R., Sharp, R.D., Shelley, E.G., Ungstrup, E., Young, D.T.: 1986, *Nature* **321**, 330

Cravens, T.E.: 1986, *Geophys. Res. Letters* **13**, 275

Galeev, A.A.: 1986, *Adv. Space Res.* **5**, 155

Galeev, A.A., Cravens, T.E., Gombosi, T.I.: 1985, *Astrophys. J.* **289**, 807

Johnstone, A.D., Bowles, J.A., Coates, A.J., Coker, A.J., Kellock, S.J., Raymont, J., Wilken, B., Stüdemann, W., Weiss, W., Cerulli-Irelli, R., Formisano, V., de Giorgi, E., Perani, P., de Bernardi, M., Borg, H., Olsen, S., Winningham, J.D., Bryant, D.A.: 1986a, *The Giotto Mission – Its Scientific Investigations*, eds. R. Reinhard, B. Battrick, ESA SP-1077, Noordwijk, The Netherlands, p. 15

Johnstone, A.D., Coates, A., Kellock, S., Wilken, B., Jockers, K., Rosenbauer, H., Stüdemann, W., Weiss, W., Formisano, V., Amata, E., Cerulli-Irelli, R., Dobrowolny, M., Terenzi, R., Egidi, A., Borg, H., Hultquist, B., Winningham, J., Gurgiolo, C., Bryant, D., Edwards, T., Feldman, W., Thomsen, M., Wallis, M.K., Biermann, L., Schmidt, H., Lüst, R., Haerendel, G., Paschmann, G.: 1986b, *Nature* **321**, 344

Krankowsky, D., Lämmerzahl, P., Herrwerth, I., Woweries, J., Eberhardt, P., Dolder, U., Herrmann, U., Schulte, W., Berthelier, J.J., Illiano, J.M., Hodges, R.R., Hoffman, J.H.: 1986, *Nature* **321**, 326

Mukai, T., Miyake, W., Terasawa, T., Kitayama, M., Hirao, K.: 1986a, *Nature* **321**, 299

Mukai, T., Miyake, W., Terasawa, T., Kitayama, M., Hirao, K.: 1986b, *Geophys. Res. Letters* **13**, 829

Neubauer, F. M.: 1986, *20th ESLAB Symp.* ESA SP-250, Vol. **I**, p. 35

Neugebauer, M., Lazarus, A.J., Altwegg, K., Balsiger, H., Goldstein, B.E., Goldstein, R., Neubauer, F.M., Rosenbauer, H., Schwenn, R., Shelley, E.G., Ungstrup, E.: 1986, *20th ESLAB Symp.*, ESA SP-250, Vol. **I**, p. 19

Terasawa, T., Mukai, T., Miyake, W., Kitayama, M., Hirao, K.: 1986, *Geophys. Res. Letters* **13**, 837

Wallis, M.K., Dryer, M.: 1985, *Nature* **318**, 646

Wallis, M.K., Ong, R.S.B.: 1975, *Planet. Space Sci.* **23**, 713

Wallis, M.K., Ong, R.S.B.: 1976, *The Study of Comets*, NASA SP-393, 856

Weiss, W., Wilken, B.: 1986, *J. Phys. E: Sci. Instr.* (submitted)

Wilken, B., Weiss, W., Stüdemann, W., Hasebe, N.: 1986a, *J. Phys. E: Sci. Instr.* (submitted)

Wilken, B., Jockers, K., Stüdemann, W., Rosenbauer, H., Thomsen, M.F., Johnstone, A., Coates, A., Formisano, V., Amata, E., Winningham, J.D., Borg, H.: 1986b, *20th ESLAB Symp.*, ESA SP-250, Vol. **I**., p. 305

Astron. Astrophys. 187, 149–152 (1987)

ASTRONOMY AND ASTROPHYSICS

The composition and radial dependence of cometary ions in the coma of comet P/Halley

A. Korth [1], A.K. Richter [1], D.A. Mendis [2], K.A. Anderson [3], C.W. Carlson [3], D.W. Curtis [3], R.P. Lin [3], D.L. Mitchell [3], H. Rème [4], J.A. Sauvaud [4], and C. d'Uston [4]

[1] Max-Planck-Institut für Aeronomie, D-3411 Katlenburg-Lindau, Federal Republic of Germany
[2] EECS, University of California, San Diego, CA, USA
[3] SSL, University of California, Berkeley, CA, USA
[4] CESR, CNRS, Toulouse, France

Received December 22, 1986; accepted March 13, 1987

Summary. The heavy ion analyzer, RPA2-PICCA, on board the Giotto spacecraft detected an increase in the densities of cometary ions within a cometocentric distance of 150000 km. The composition of cometary ions changed dramatically as the comet's nucleus was approached, but it was clearly dominated by the water group. The second and third most abundant ions identified were associated with the CO- or S-group and the CO_2-group, respectively. Ions of larger atomic mass units were also present closer to the comet, and they possibly correspond to sulphur compounds and/or various hydrocarbons. Radial profiles of various groups of heavy ions and certain abundance ratios are presented. The peak density for all mass groups was detected at a cometocentric distance of $\sim$11000 km. A distinct boundary, where the ion velocity and temperature dropped significantly, was identified at about 27000 km.

Key words: comet Halley – heavy ion composition – radial profiles – abundance ratios

1. Introduction

The heavy ion analyzer RPA2-PICCA [1] is designed to determine the composition and the energy distributions of thermal, positively charged ions in the coma of comet Halley. The instrument consists of an electrostatic analyzer and takes advantage of the large flyby velocity of the spaceprobe and the fact that the ions in the inner coma should have low temperatures and should be predominantly singly charged (Rème et al., 1986; Korth et al., 1987). Under these assumptions the energy/charge (E/Q) measurements can be directly related to the mass distribution of these ions.

Two additional mass spectrometers on board Giotto (Balsiger et al., 1986, and Krankowsky et al., 1986) studied the neutral and ionic composition up to about 60 amu (atomic mass units). The PICCA instrument is able to extend this mass range up to 200 amu and therefore to study more complex molecules.

Send offprint requests to: A. Korth

[1] RPA: Rème Plasma Analyzer,
PICCA: Positive Ion Cluster Composition Analyzer

In the following we will provide results of the composition and of the radial profiles of different ionic mass groups within a distance of 200000 km from the nucleus. We further will show temperature and flow velocity profiles in the ram direction in the vicinity of a 'cold cometary plasma boundary' at $\sim 2.7\ 10^4$ km. These profiles were obtained from a method described in a paper by Mitchell et al. (1986).

2. Composition

The PICCA instrument identified the major cometary ion groups in the inner coma (Korth et al., 1986). Figures 1 and 2 show, in a colour coded representation, the different mass groups in nominal atomic mass units (amu) versus the distance from the nucleus. This nominal mass scale is based on a flyby velocity of the Giotto spaceprobe of 68,373 km s^{-1} or 24.22 eV/amu, and thus does not take into account the flow velocity of the cometary ions themselves. The mass groups are observed sequentially from the lowest to the highest masses due to their relative abundances. After each group's initial appearance its intensity increases rather rapidly. In the lowest two mass groups the large channeltron detector is saturated as indicated by the white colour (A second small channeltron with $\sim 10^{-2}$ the effective area extends the measurements below 30 amu). The intensity maximum appears simultaneously in all mass groups at a distance of about 11000 km.

In the following we will divide the ions into several mass groups. The mass groups I to III are displayed in Fig. 1 and the mass groups IV to VII in Fig. 2, respectively. Although the PICCA instrument is capable to measure ions to about 200 amu, Fig. 2 shows only ions with masses up to 110 amu. The identification of mass peaks above 110 amu is complicated due to the presence of background which increases in intensity by approaching the nucleus; but a further investigation of these ions is planned.

With respect to the mass group I (14–24 amu) or the water group the dominant ions are likely to be O^+, OH^+, H_2O^+, and H_3O^+, but it also could include CH_2^+, CH_3^+, CH_4^+, and C_2^+. The measured peak corrected for the flow velocity in this mass group appears to occur between 18 and 19 amu.

Possible candidates for the mass group II (25–35 amu) are probably CO^+, HCO^+, H_2CO^+, and S^+. The measured peak is located between 30 and 32 amu.

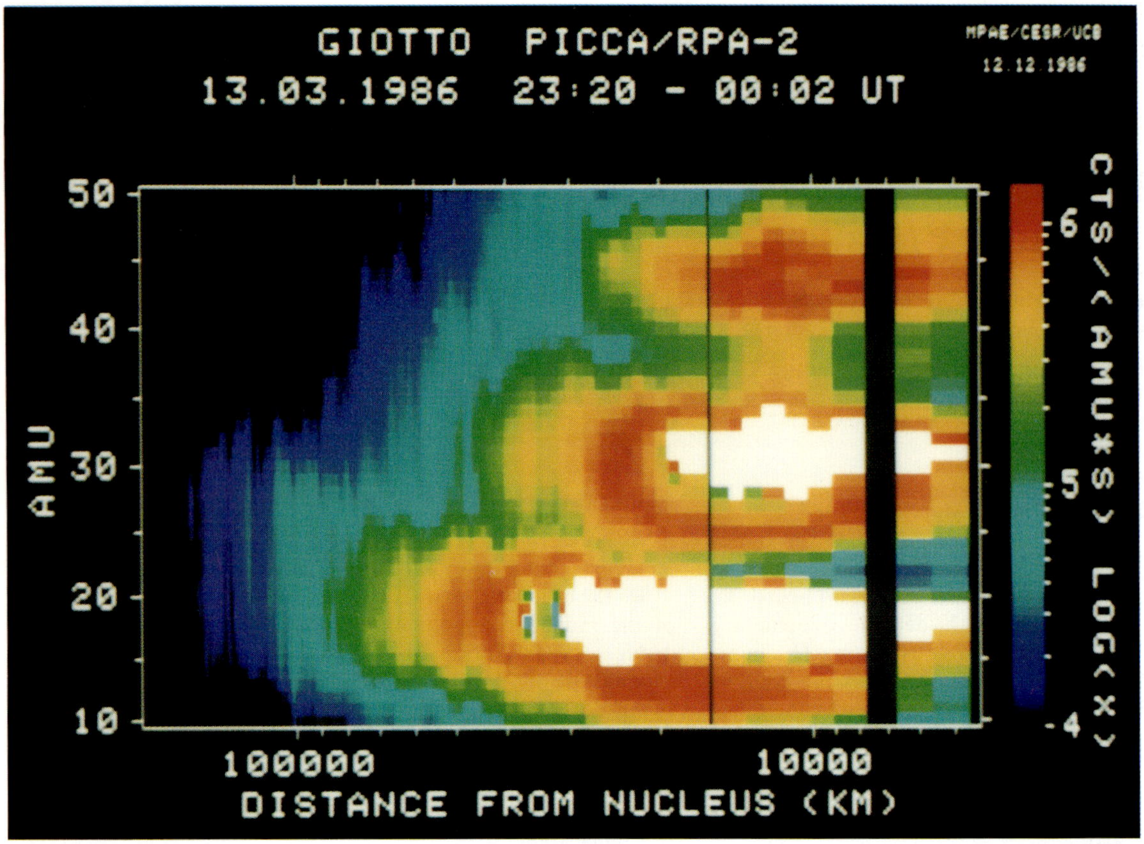

Fig. 1. Colour-coded representation of the logarithmic normalized intensity (counts $amu^{-1}\,s^{-1}$) versus distance from the nucleus (horizontal axis) and atomic mass units (vertical axis). The time interval in spacecraft event time is shown on top. Data are averaged over four spacecraft spins or 16 s

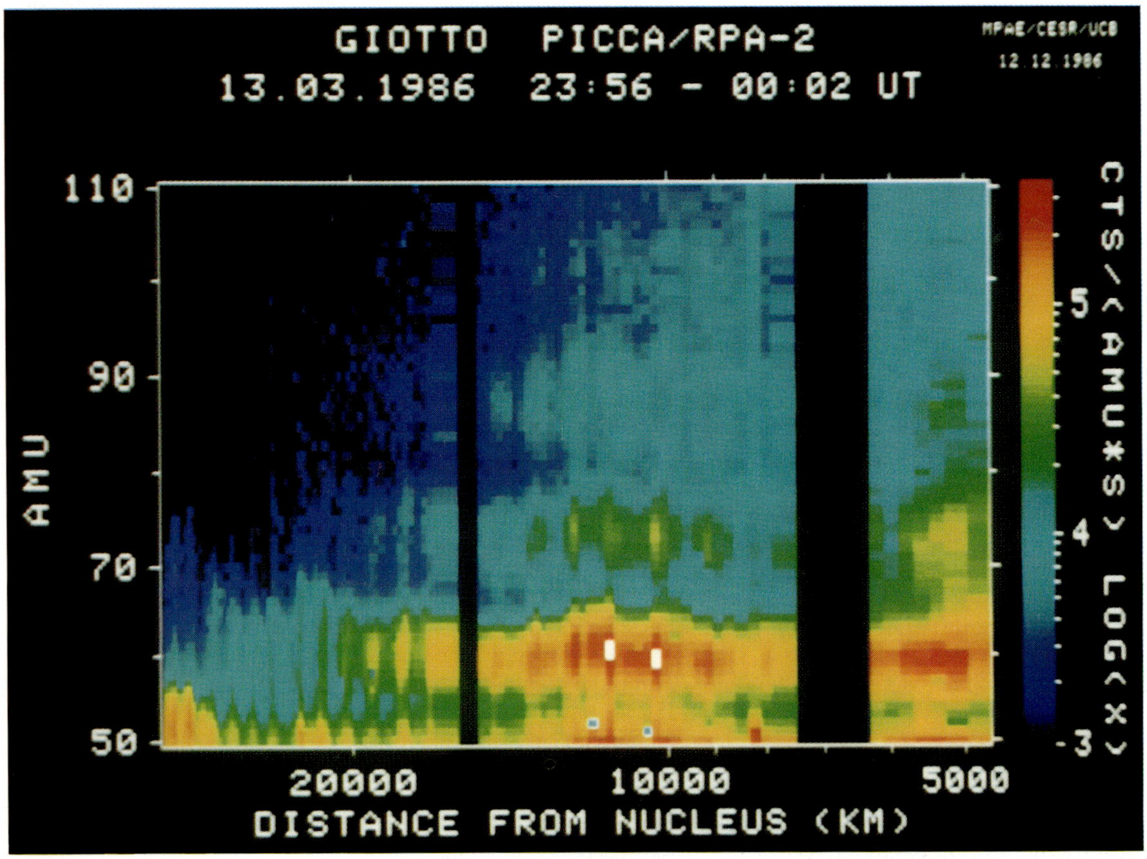

Fig. 2. Same as Fig. 1 but for higher masses and closer distances to the nucleus. Data are not averaged over four spacecraft spins but each distribution is measured every 3, 2 s. Therefore a spin modulation is observed, as the geometric factor is dependent on the direction of the incoming ions

The mass group III (39–49 amu) could include ions such as CO_2^+, CS^+, HCO_2^+, $H_2CO_2^+$, and SO^+. The measured peak occurs at mass 44. Our tentative identification is CS^+.

Further mass groups, such as 55–68 amu (IV), 68–81 amu (V), 81–97 amu (VI), and 97–110 amu (VII) can be clearly identified, and their peaks are close to the center of the mass intervals. Possible candidates for ions in mass group IV are OCS^+ and S_2^+, and for mass group V CS_2^+. The identification of higher masses in the groups IV to VII is complicated, but complex hydrocarbons or carbon rings could be possible candidates. The peaks of the different mass groups are separated by about 14–16 amu, which apparently rules out the presence of water clusters ($I^+ (H_2O)_n$). Detailed modelling has to be done in this area. Fits to the high mass peaks, done with a method given by Mitchell et al. (1986), show that each peak consists of at least two or even more separate masses.

Figure 1 also shows that far away from the comet (>30 000 km) the ion distributions are rather broad, indicating that the thermal velocity of the cometary ions is rather high. Thus, the E/Q measurements at this time cannot be directly related to a certain mass distribution. The center of the first mass group is located at about 590 eV (nominal mass of 24 amu) at a distance of about 150 000 km. However, as the comet's nucleus is approached, the ion temperature and velocity are gradually decreasing, and the peak of the first mass group is shifted towards 445 eV, which corresponds to 18 amu. At a cometocentric distance of 27 000 km the spreads in the various ion distributions and thus in the temperatures and velocities of the various ions drop rather abruptly, indicating that Giotto has now entered the cold cometary plasma region. This effect was also recorded by the two mass spectrometers (Balsiger et al., 1986, and Krankowsky et al., 1986). Inside this region the distributions of the various ions become rather narrow, C^+ can be separated quite well from the water group or mass group I, and the heavy ion mass groups show up sequentially.

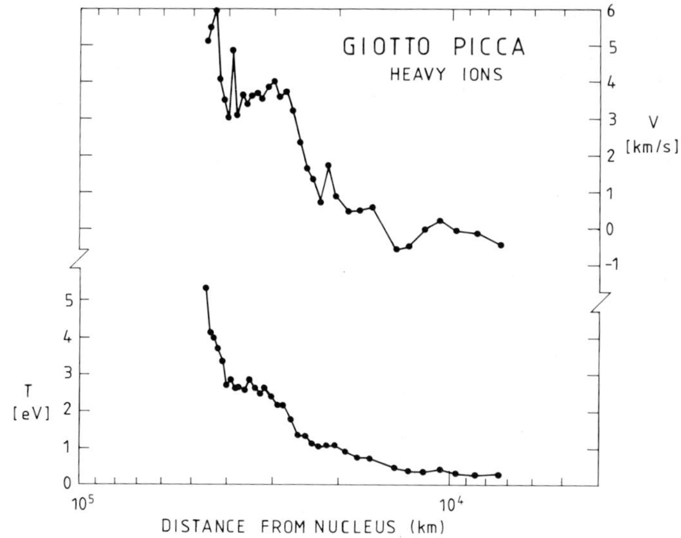

Fig. 3. Radial profiles of the flow velocity and of the temperature. A spacecraft potential of +10 volts was assumed. The data are averaged over 16 s

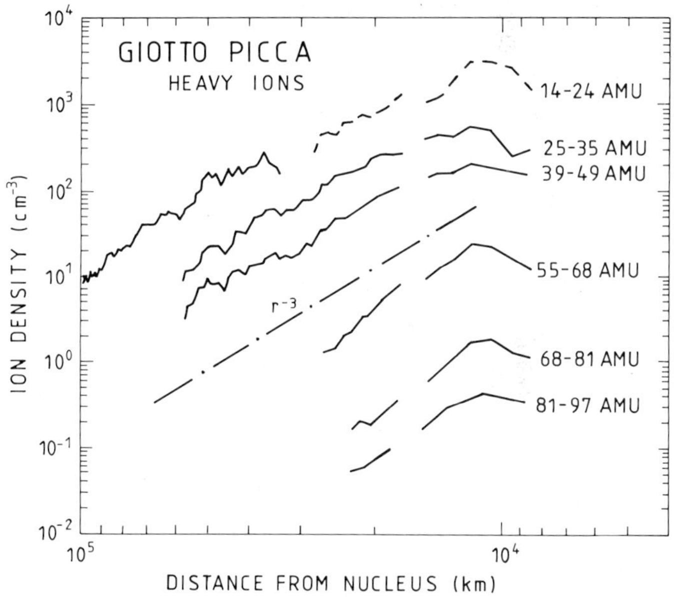

Fig. 4. Radial dependence of different ion mass groups. The dashed values for the mass group 14–24 amu are derived from the small channeltron, using preliminary cross calibrations

3. Temperature and velocity profiles

The peaks in the water group (mass group I) are fitted by a method described by Mitchell et al. (1986), Figure 3 displays the flow velocity in the ram direction and the temperature of the heavy ions versus the distance from the nucleus. For our calculations we assumed a spacecraft potential of +10 V. A potential of +16 V was estimated by the neutral mass spectrometer (Krankowsky et al., 1986), and the electron experiment (Rème et al., 1986) indicated a spacecraft potential of less than +20 V.

The velocity and temperature of the cometary ions show a general decrease with decreasing distance to the comet. A well-defined drop in temperature and in the velocity is observed at 2.7 10^4 km. Outside 4.3 10^4 km the temperature increases rapidly and the ion distributions become too broad to obtain good fits. A plateau in the velocity is observed between 2.8 10^4 km and 3.7 10^4 km.

4. Radial distributions and abundance ratios

The ion composition in the inner coma of comet Halley is clearly dominated by the water group. Figure 4 shows the ion density versus the distance from the nucleus for the first six mass groups. For reference purposes a radial dependence of r^{-3} is given as a dashed line. For all mass groups we can identify their peak density to occur at a cometocentric distance of ~11 000 km. For the water group we obtain preliminary densities of at least ~4000 ions/cm^3. Outside 11 000 km the radial dependence varies between r^{-2} and r^{-3} for masses below 50 amu. Above 50 amu the radial dependence is steeper than r^{-3}. From these radial profiles we derived the following abundance ratios at two different distances from the nucleus:

At 40 000 km:

$$\frac{(25\text{–}35)\,\text{amu}}{(14\text{–}24)\,\text{amu}} = \frac{\text{CO- or S-group}}{\text{H}_2\text{O-group}} = 0.2$$

$$\frac{(39\text{–}49)\,\text{amu}}{(14\text{–}24)\,\text{amu}} = \frac{\text{CO}_2\text{-group}}{\text{H}_2\text{O-group}} = 0.07$$

At 20 000 km:

$$\frac{(25\text{–}35)\,\text{amu}}{(14\text{–}24)\,\text{amu}} = (0.25) \qquad \frac{(39\text{–}49)\,\text{amu}}{(14\text{–}24)\,\text{amu}} = (0.1)$$

$$\frac{(55\text{–}68)\,\text{amu}}{(14\text{–}24)\,\text{amu}} = (5\ 10^{-3}) \qquad \frac{(68\text{–}81)\,\text{amu}}{(14-24)\,\text{amu}} = (3\ 10^{-4})$$

$$\frac{(55\text{–}68)\,\text{amu}}{(39\text{–}49)\,\text{amu}} = 5\ 10^{-2} \qquad \frac{(68-81)\,\text{amu}}{(39-49)\,\text{amu}} = 3\ 10^{-3}$$

The values in brackets are preleminary ratios, and they are subject to change after a further calibration is completed.

5. Conclusions

We are in the process of verifying if these ratios are consistent with our expectations that the ions of CS_2^+ and ions of its daughter products, such as S_2^+, CS^+, S^+, and C^+, are responsible for the peaks near 76, 64, 44, 32, and 12 amu.

Acknowledgements. This work was supported by the Max-Planck-Gesellschaft zur Förderung der Wissenschaften e.V., by the Bundesministerium für Forschung und Technologie under grant No 01 OF 052, by CNES under grant no 1212, and by NASA contract NASW-3375. The data reduction for the PICCA instrument was performed by H. Michels.

References

Balsiger, H., Altwegg, K., Bühler, F., Geiss, J., Ghielmetti, A.G., Goldstein, B.E., Goldstein, R., Huntress, W.T., Ip, W.H., Lazarus, A.J., Meier, A., Neugebauer, M., Rettenmund, U., Rosenbauer, H., Schwenn, R., Sharp, R.D., Shelley, E.G., Ungstrup, E., Young, D.T.: 1986, *Nature* **321**, 330

Korth, A., Richter, A.K., Loidl, A., Anderson, K.A., Carlson, C.W., Curtis, D.W., Lin, R.P., Rème, H., Sauvaud, J.A., d'Uston, C., Cotin, F., Cros, A., Mendis, D.A.: 1986, *Nature* **321**, 335

Korth, A., Richter, A.K., Loidl, A., Güttler, W., Anderson, K.A., Carlson, C.W., Curtis, D.W., Lin, R.P., Rème, H., Cotin, F., Cros, A., Médale, J.L., Sauvaud, J.A., d'Uston, C., Mendis, D.A.: 1987, *J. Physics E: Scientific Instruments* (in press)

Krankowsky, D., Lämmerzahl, P., Herrwerth, I., Woweries, J., Eberhardt, P., Dolder, U., Herrmann, U., Schulte, W., Berthelier, J.J., Illiano, J.M., Hodges, R.R., Hoffman, J.H.: 1986, *Nature* **321**, 326

Mitchell, D.L., Lin, R.P., Anderson, K.A., Carlson, C.W., Curtis, D.W., Korth, A., Richter, A.K., Rème, H., Sauvaud, J.A., d'Uston, C., Mendis, D.A.: 1986, 20th ESLAB Symp., **SP-250**, Vol. 1, p. 203

Rème, H., Cotin, F., Cros, A., Médale, J.L., Sauvaud, J.A., d'Uston, C., Korth, A., Richter, A.K., Loidl, A., Anderson, K.A., Carlson, C.W., Curtis, D.W., Lin, R.P., Mendis, D.A.: 1986, 20th ESLAB Symp., ESA **SP-1077**, p. 33

Astron. Astrophys. 187, 153–159 (1987)

ASTRONOMY AND ASTROPHYSICS

Pick-up ions at comet P/Halley's bow shock: observations with the IIS spectrometer on Giotto

B. Wilken [1], A. Johnstone [2], A. Coates [2], H. Borg [3], E. Amata [4], V. Formisano [4], K. Jockers [1], H. Rosenbauer [1], W. Stüdemann [1], M.F. Thomson [5], and J.D. Winningham [6]

[1] Max-Planck-Institut für Aeronomie, D-3411 Katlenburg-Lindau, Federal Republic of Germany
[2] Mullard Space Science Laboratory, University College London, Holmbury St. Mary, Dorking, England
[4] Instituto di Fisica dello Spazio Interplanetario, Via G. Galilei, Casella Postale 27, Frascati, Italy
[5] Los Alamos National Laboratory, POB 1663, Los Alamos, NM, USA
[6] Southwest Research Institute, San Antonio, TX 78284, USA

Received March 20, accepted April 6, 1987

Summary. Gaseous material expanding form the nucleus of comet Halley into space form the neutral coma around the comet. Ionisation in the solar UV radiation removes particles from the coma and injects them into the solar wind plasma. These freshly created ions are accelerated by the interplanetary electric field on cycloidal trajectories with gyrocenters moving with the speed of the magnetic field lines. In the solar wind frame of reference these particles move along the magnetic field lines with a fixed pitchangle. Pitchangle scattering and energy diffusion reduce quickly the initial energy anisotropy which is associated with the narrow pick-up structures. First observations of heavy cometary pick-up ions (water group ions) at the bowshock are presented. The evolution of the distribution function in the vicinity of the shock and radial density profiles are discussed.

Key words: comet – ion pick up

1. Introduction

The Implanted-Ion-Spectrometer (IIS) in the payload of the European Space Agency's spacecraft Giotto is an element of the solar wind plasma analyser JPA. The prime objective of the IIS spectrometer is the identification and analysis of cometary gases after their removal from the neutral coma by ionising processes and energisation in the solar wind. The particle identifying function is obtained from electrostatic energy per charge (E/Q) filtering in combination with a velocitiy measurement using time-of-flight techniques. The particle mass A (in amu) can be determined uniquely if the ionic charge state Q is a known quantitiy. Photoionisation in the solar radiation field and charge exchange with solar wind protons as the dominant processes lead to singly ionised cometary material making the state of ionisation indeed a well known parameter. Instrumental details of the IIS are described by Wilken et al. (1987) and Weiß and Wilken (1987). A description of IIS in the context of the JPA analyser system is given by Johnstone et al. (1987).

The Implanted-Ion-Spectrometer (IIS) employs five identical detector systems (D^1 to D^5) with viewing directions evenly spaced over 180° in the polar plane (i.e. in a plane perpendicular to the spacecraft spin plane). By dividing the spin plane into 16 angular sectors (S^0 to S^{15}) a uniform but not contiguous coverage of the unit sphere is achieved. Four-dimensional velocity distributions of cometary pick-up ions (velocity vector and ion mass) are measured with reasonable accuracy for energies ranging from 100 eV e^{-1} up to 86 keV e^{-1}. As described by Weiß and Wilken (1987) five coarse mass groups M1 to M5 are constructed in the instrument. Relevant for this paper are the hydrogen and watergroup mass classes, M1 ($1 \leqq A \leqq 3$) and M3 ($10 \leqq A \leqq 20$), respectively. The measuring geometry of IIS is shown in Fig. 1.

Cometary ionised material is subject to the electric field associated with the rapid flow of the magnetised solar wind. Freshly injected ions are accelerated on cycloidal trajectories such that the gyrocenter moves with the drift velocity of the interplanetary magnetic field lines. In the solar wind frame of reference the pick-up ions move along the magnetic field lines. In the solar wind frame of reference the pick-up ions move along the magnetic field lines with a fixed pitch angle. This injection angle is defined by the angle between the magnetic field vector and the solar wind velocity at the time of ionisation. In the solar wind frame of reference ions of cometary origin form a beam with a singularity in pitch angle. This strong pitch angle anisotropy is a source for free energy which stimulates magnetic field turbulence. Enhanced levels of wave acitivity will in turn drive pitch angle diffusion at constant energy resulting in a relaxation of the initial singular ring structures. Computer simulations by Wu et al. (1986) have shown that the formation of shell structures for pick-up ions is indeed a rather fast process requiring only a few proton gyroperiods. Observations of the three dimensional distribution function of cometary pick-up water group ions and the variation of the particle density in the vicinity of comet Halley's bow shock are the focus of this paper.

2. Observations

Figure 2 shows stacked countrate versus energy per charge profiles for water group ions observed upstream and downstream of the cometary bow shock. The distributions at around 2–2.5 keV represent attentuated 'ghosts' of solar wind protons which reflect

Send offprint request to: B. Wilken

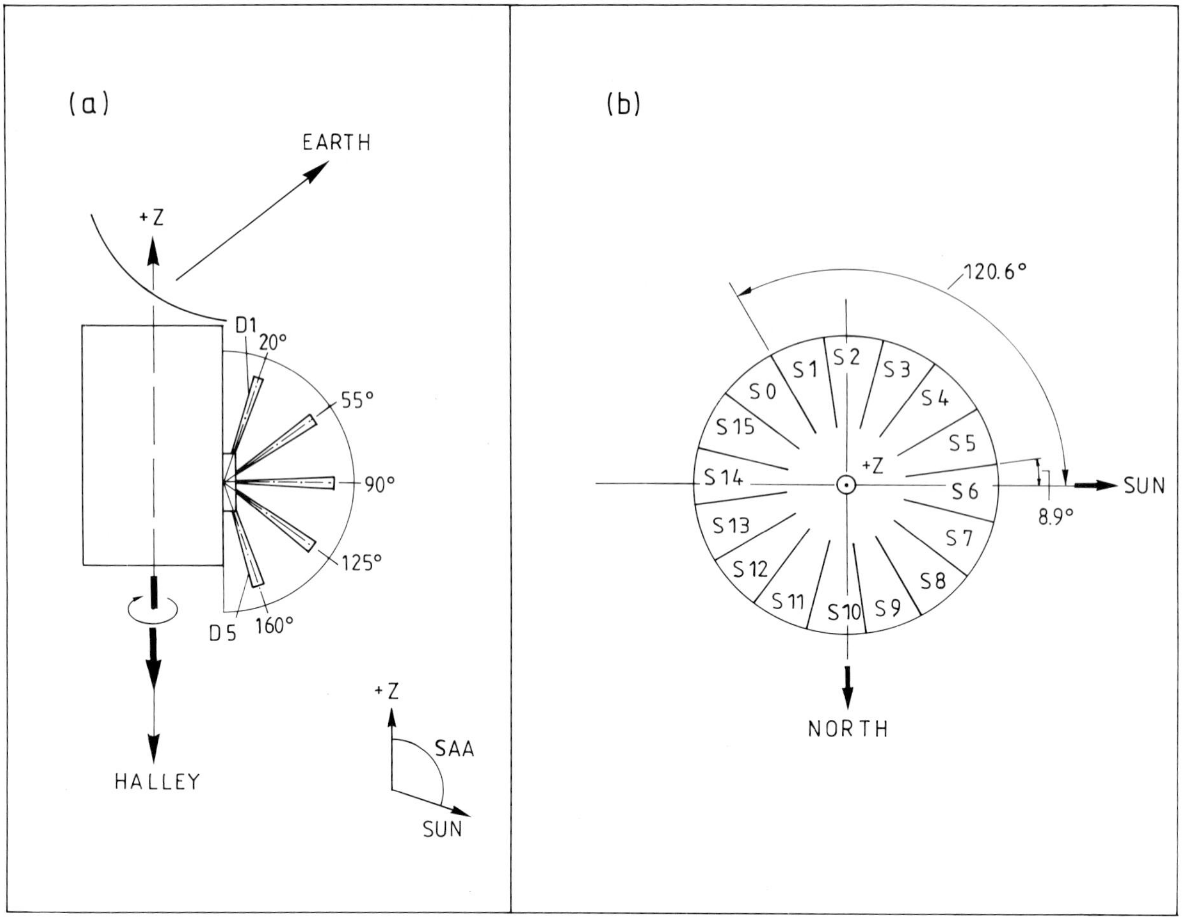

Fig. 1. Viewing directions of the five IIS sensors and the orientation of the azimuthal sectors

overload effects in the mass identification system of IIS caused by the intense solar wind beam.

Figure 2 shows clearly that energetic water group ions are already present at 1800 UT, more than 1.5 10^6 km before closest approach. At 1830 UT Giotto passed a first noticeable feature in the solar wind plasma generated by the addition of cometary material. A weak decrease in the mean energy of the water group ions associated with a similar barely visible effect in the solar wind protons is identified as the signature of a foreshock. The bow shock crossing at 1930 UT is indicated by a significant intensification of cometary pick up ions; both the cometary and the solar wind ion distributions show thermalisation effects. Downstream of the shock the solar wind speed decreases contiuously and the cometary distributions seem to respond to this evolution by and large. However, an unexpected prominent feature in the pick-up ion distribution develops at about 2100 UT when a secondary population branches off toward lower energies.

The bow shock effect in the solar wind and cometary ions are shown in the high resolution color spectogramms of Fig. 3. The two panels show for a fixed polar direction (detector D5 in Fig. 1) energy spectrogramms (counts per sample) for protons (M1-group) and water group ions (M3-group) as a function of time. The particular instrument operations require 32 spin rotations of the spacecraft or 128 s to accumulate a complete energy versus azimuthal sector distribution with a complex phase relation between energy and sector number. Rearrangement of the sequential measurements produced the spectogramm versus time representation of Fig. 3 in which a time interval of 8 s corresponds to an azimuthal sector, i.e., 128 s represent 16 sectors or 360° in azimuth.

The hydrogen spectrogramms (lower panel in Fig. 3) show clearly the presence of two different populations. The narrow solar wind beam is essentially confined to a single sector (sector S6 in Fig. 1) with the maximum intensity occurring at energies around 0.5 keV. A second broader distribution at higher energies is easily recognised as pick-up cometary hydrogen because of its characteristic relation between energy and sector. The maximum energy of approximately 1.8 keV is attained in the solar wind sector. The particle energy decreases with the azimuthal phase angle in both directions and drops below the instrument threshold when the look direction is about 4 sectors or 90° away from the solar wind direction. Such a signature in energy phase space is typical for pick-up ions. Considering that detector D5 samples actually a range of pitch angles in the solar wind frame of reference as the spacecraft rotates the implication of the relatively continuous distributions in Fig. 3 is that the ions must populate a significant fraction of a shell in velocity space rather then a narrow ring.

A first indication of the approaching bow shock is noticeabel in the azimuthal distribution centered at 1930:50 UT which shows a sudden decrease in solar wind speed and an intensity reduction in the pick-up hydrogen as a transient phenomenon. The following distribution at 1932:50 UT shows substantial azimuthal broadening in the solar wind but the speed as indicated by the intensity

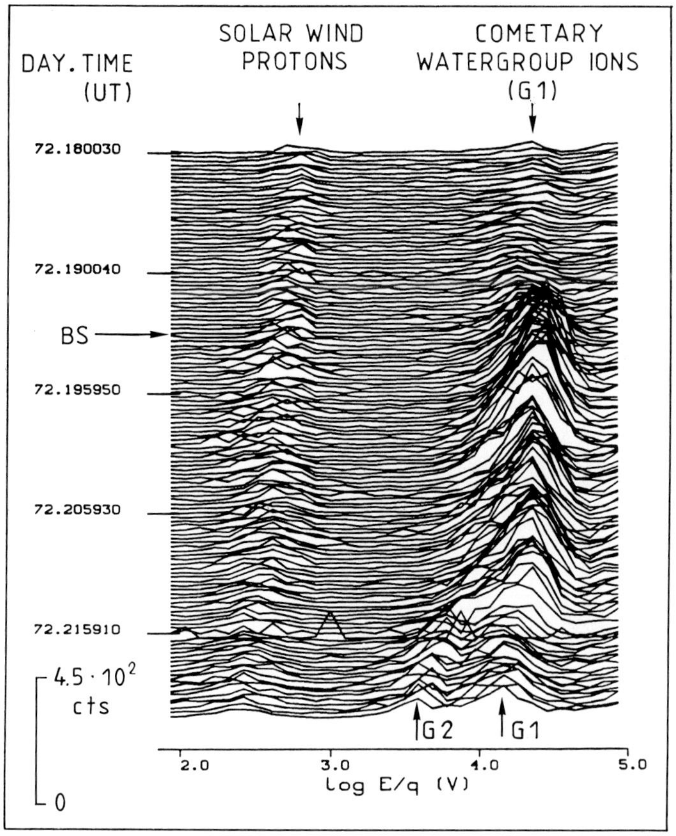

Fig. 2. Stacked counting rate versus energy profiles of comentary water group ions (summed over all angles). Overload effects create the solar wind proton distributions at low energies. Passage of the cometary bow shock (BS) is clearly indicated by the intensification and broadening of the distributions. A second water group ion population (G2) appears at 2130 UT

maximum seems to have recovered to the previous level. The following two distributions show a strong azimuthal expansion of the pick-up component filling now nearly 16 sectors (or 360° in azimuth). However, drastic changes in the solar wind bulk distribution are not observed before 1938:50 UT when strong effects of thermalisation at the bow shock become obvious. Including the precursory dip in solar wind speed the time interval needed to cross the bow shock layer as defined by the ion observations amounts to 8 ± 1 min or 33600 ± 4200 km.

A more detailed discussion of the bow shock structure including electron observations is given by Coates et al. (1987).

The upper panel in Fig. 3 shows corresponding spectograms for cometary water group ions injected into the solar wind. Prior to the shock crossing (1930–1938 UT) the distributions show the characteristic energy-azimuth signatures with peak energies in the solar wind sector of about 25 keV. This value remains below the expected maximum $E_{max} = 64\ E_p$ (with E_p denoting the solar wind proton energy) because detector D5 does not exactly sample the direction of the solar wind volocity vector. The passage of the bow shock is indicated by a sudden intensification (at 1930:50 UT) accompanied by substantial broadening in azimuth and energy. Water group ions with energies in excess of the local maximum

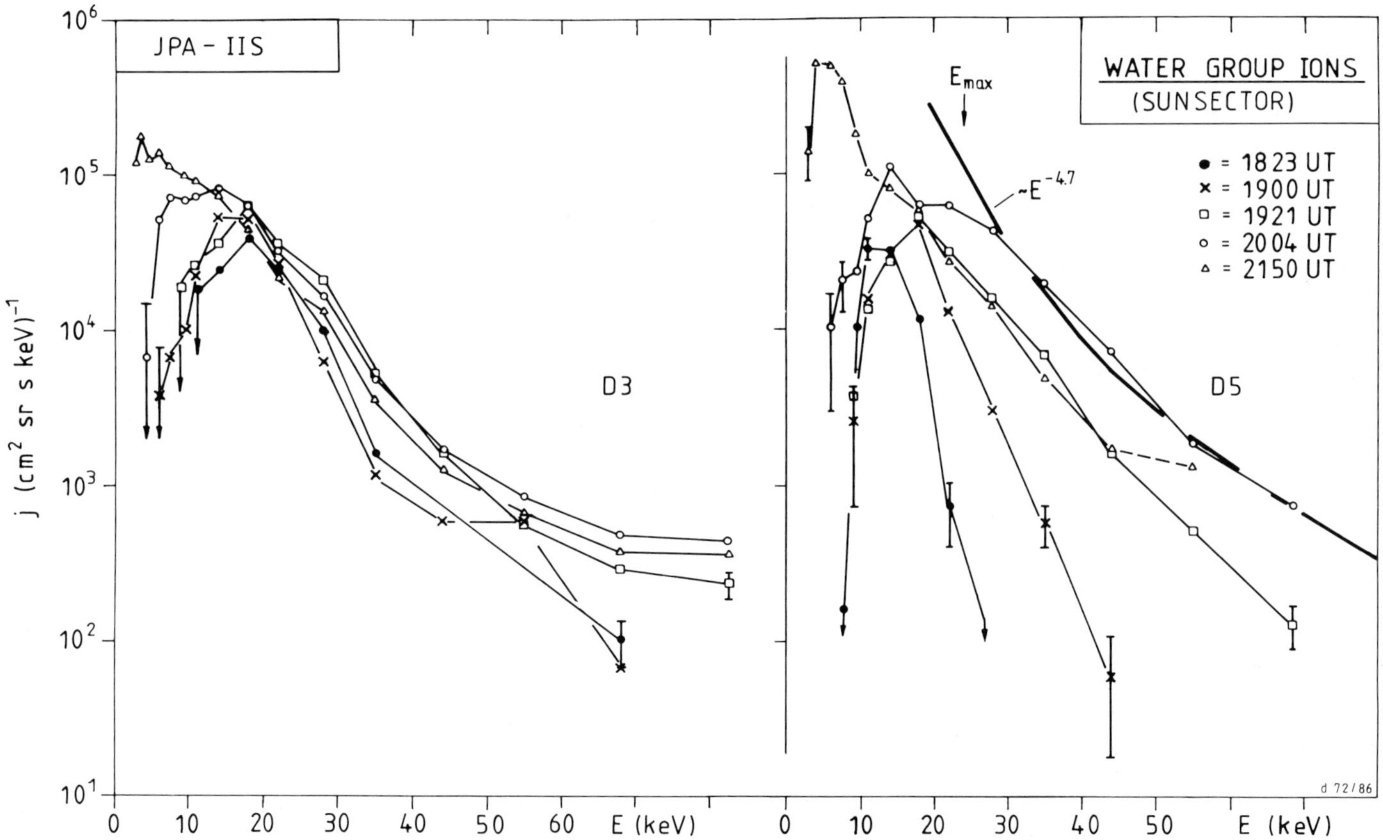

Fig. 4. Energy spectra of water group ions observed in the sunward sector S6 for two different polar directions (detector D3 and D5)

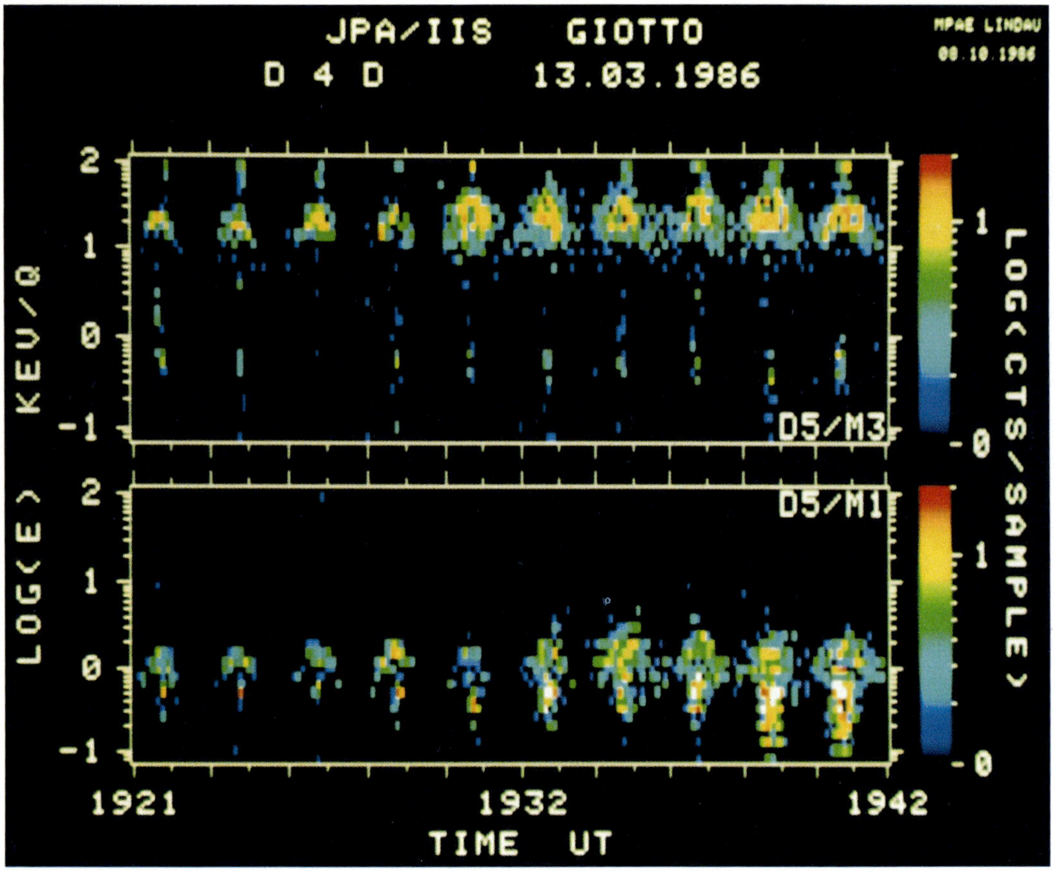

Fig. 3. High resolution spectogrammes of water group ions (upper panel) and protons (lower panel). Time intervals of 8 s correspond to an azimuthal sector

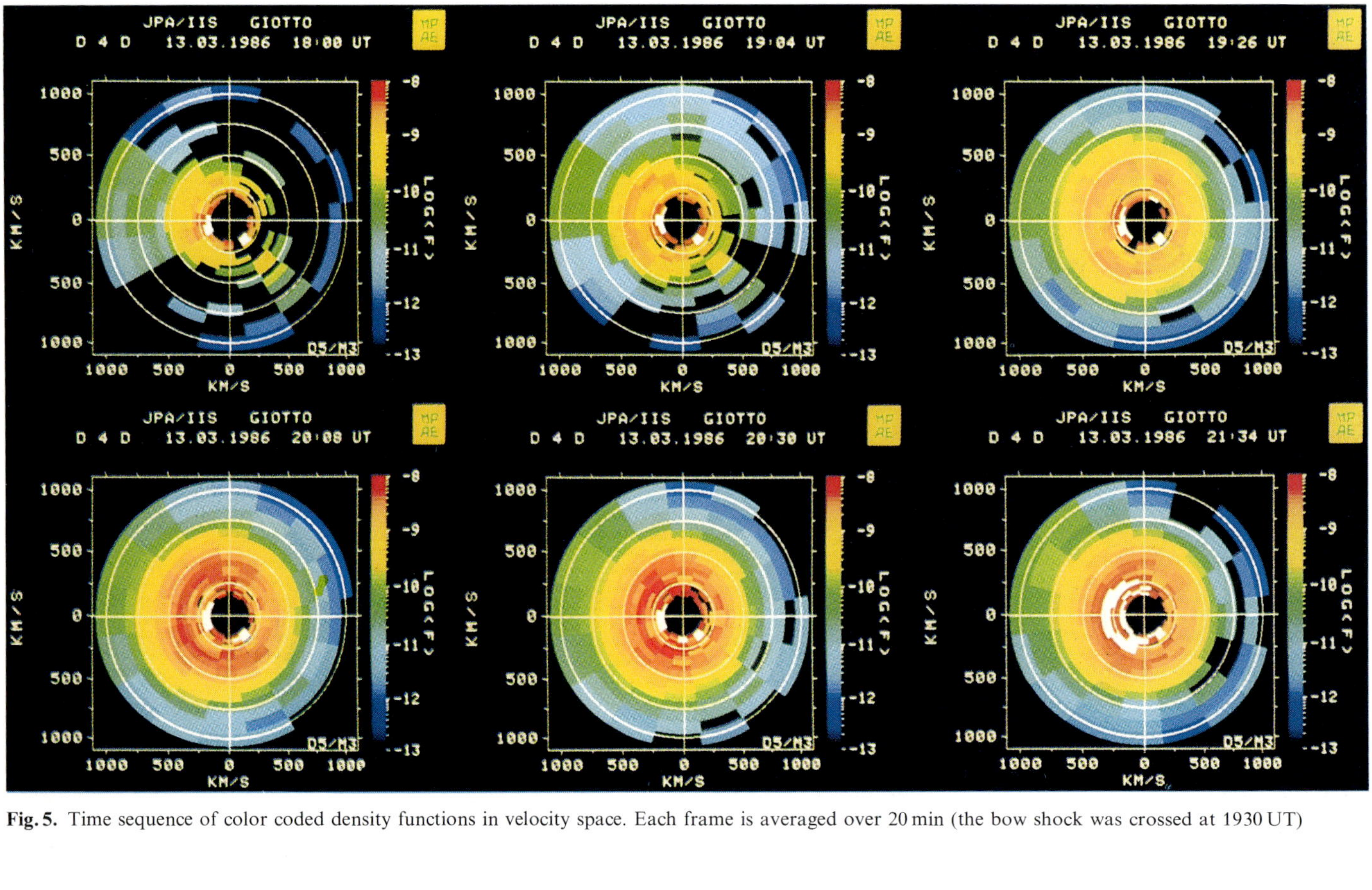

Fig. 5. Time sequence of color coded density functions in velocity space. Each frame is averaged over 20 min (the bow shock was crossed at 1930 UT)

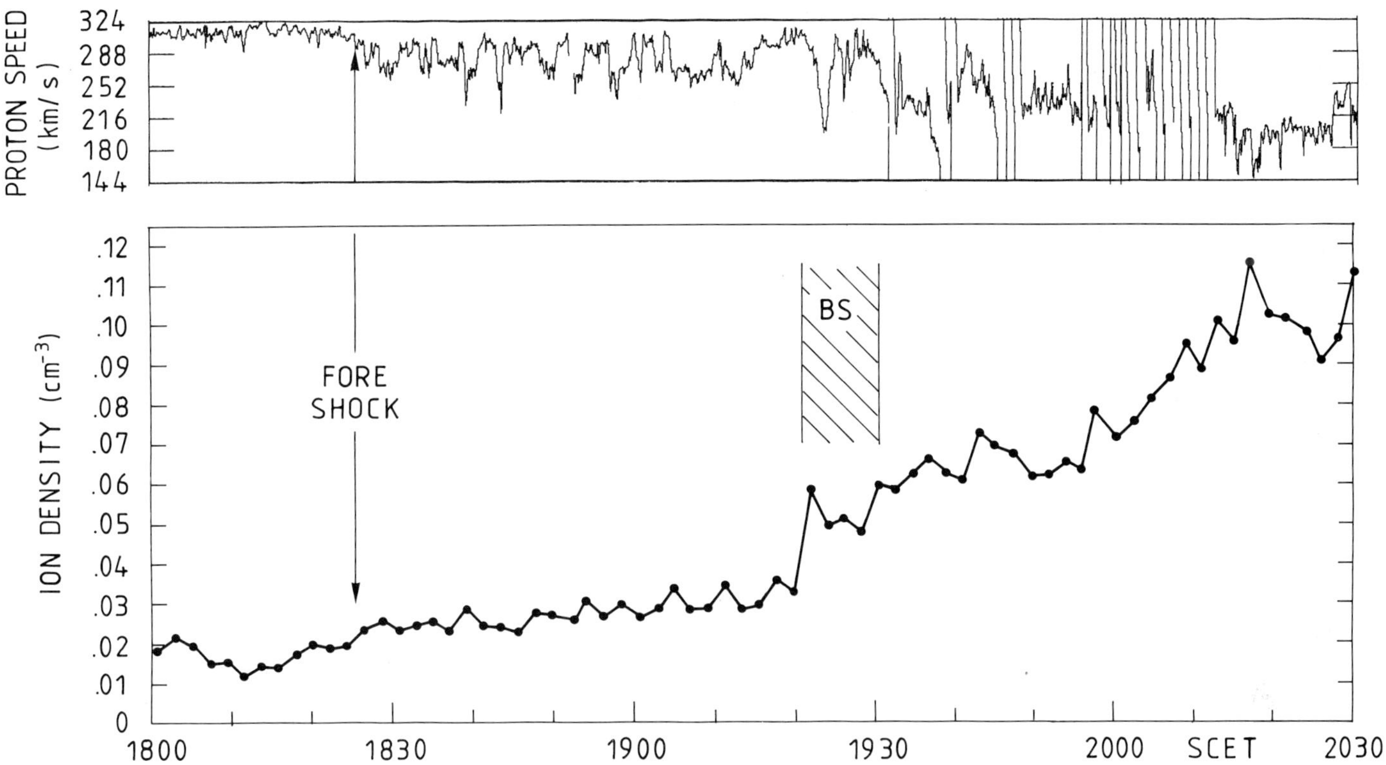

Fig. 6. Contour of the water group ion density at the cometary bow shock. The upper panel shows the solar wind speed for comparision.

pick-up energy $E_{max} = 64\ E_p = 32$ keV appear in the distributions although the corresponding hydrogen observations indicate a transient dip in solar wind speed and pick-up intensity.

A sequence of sectored energy spectra for the water group ions is shown in Fig. 4 for two different polar directions (detector D3 and D5). The spectra were observed in the sunward sector S6 (compare Fig. 1) upstream and downstream of the bow shock. Approximate values for the maximum pick-up energies calculated from average local solar wind speeds and projected into the viewing direction of the respective detector are indicated by heavy bars. For either detector the energy spectra extend to substantially higher energies than expected for local pick-up acceleration. A possible explanation is that the observed water group ions were injected into the undisturbed upstream solar wind with flow speeds higher than those in the vicinity of the bow shock. The flux increase at low energies observed at about 2000 UT or 125000 km into the cometosheath may very well be the result of local acceleration since the bulk of this new population remains below the predicted maximum pickup energy.

The functional form of the high energy tail in the spectra of Fig. 4 is not clearly defined by the measurements. An exponential dependence seems possible although deviations from a power law remain small as illustrated by the function $j \sim E^{-4.7}$. In the case of detector D3, however, a power law dependence looks generally more adequate.

Tranformation of the sectored energy spectra into phase space yields density distributions in the spacecraft frame of reference. Figure 5 shows a sequence of colour coded density functions which illustrate the evolution of pick-up water group ions as Giotto crossed the bow shock. The density functions are 20 min averages and time runs from the top left to the bottom right. The actual bow shock crossing at 1930 UT occurred in the interval starting at 1926 UT. Upstream of the shock wave (e.g. at 1904 UT) the maxima of the distribution function map out a circular structure in velocity space. A detailed analysis shows a shell with a radius of $W = 280$ km s^{-1} which is centered at $v = 300$ km s^{-1}. These values correspond quite well to pick up acceleration in the local solar wind with a speed $W_{sw} = 300$ km s^{-1} (compare e.g. Coates et al. 1987). In fact, the magnitude as well as the direction of the solar wind speed inferred from the pick-up ion distributions agree well with the direct observation of these quantities. Distribution functions of this kind are indicative for ion populations which have been injected into a reasonably steady solar wind flow. However, the observed velocity spread of 125 km s^{-1} (FWHM) and the relative uniform value at maximum density reflect a substantial degree of energy and pitchangle diffusion which has washed out the initial ring structure and transformed it into a sperical shell with noticeable width. The mean energy of the shell about the solar wind speed is found to be $E = 1.7$ keV corresponding to a $T = 2\ 10^7$ K plasma. This compares to the measured temperature of the local solar wind $W = 8\ 10^4$ K (Coates et al., 1987).

The bottom three frames in Fig. 5 show various stages of the evolution in the density distributions downstream of the shockwave. Besides substantial broadening and intensification the shell like structures appear to be more centered relative to the spacecraft system than the upstream distributions. Detailed analysis shows that the shell has maintained a radius $W = 300$ km s^{-1}

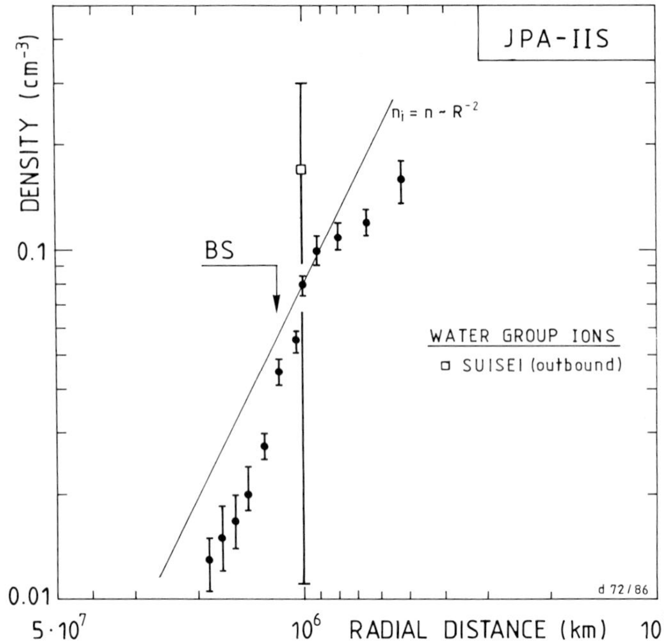

Fig. 7. Radial profile of the water group ion density in the vicinity of the bow shock. A measurement from the Japanese satellite Suisei is added for comparison. A R^{-2} dependence is also shown to guide the eye

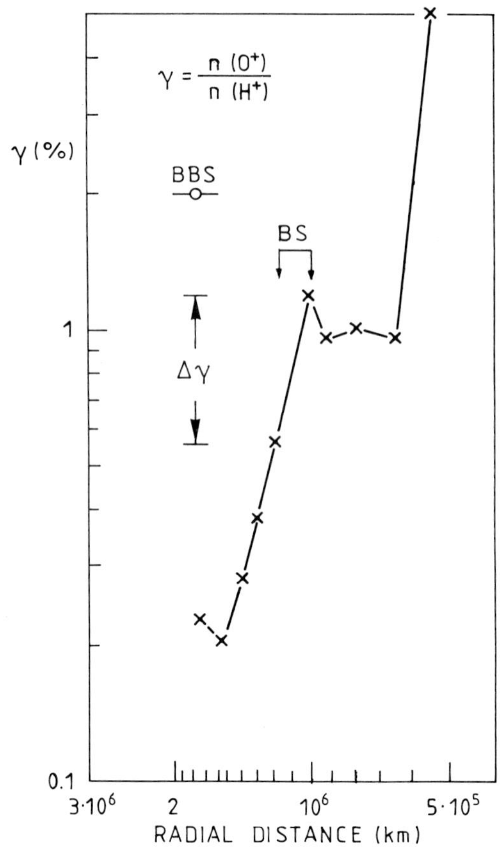

Fig. 8. The relative mass loading of the solar wind in the vicinity of the bow shock. BBS indicates the upper limit for the density ratio predicted by Biermann et al (1967)

but is now centered at an inferred local solar wind speed $W_{sw} = 235\ \mathrm{km\ s^{-1}}$. The latter value is infact quite close to the average speed of the rather unsteady solar wind at this time (Coates et al., 1987). The velocity spread (FWHM) of the shell is now $\Delta V = 240\ \mathrm{km\ s^{-1}}$ and the mean energy of the distribution increased to $E = 2.3$ keV (or $T = 2.7\ 10^{7\circ}$ K).

The shell radius $W = 300\ \mathrm{km\ s^{-1}}$ observed downstream of the shock is almost identical to the corresponding value seen upstream of the shock whereas the deceleration of the centre velocity from $V = 300\ \mathrm{km\ s^{-1}}$ to $V = 235\ \mathrm{km\ s^{-1}}$ matches the thermalisation of the solar wind. This suggests that the distributions observed to distances of at least 2.5 10^5 km downstream of the shock are dominated by cometary ions which have been created somewhere upstream of the shock. The increased velocity spread and the rather homogeneous shell structure due to enhanced levels of turbulence in the shock region are in qualitative agreement with simulations by Wu et all. (1986). However, future analysis has to show whether energy diffusion alone can account for the energisation indicated by the spectral profiles of Fig. 4.

Figure 6 shows the variation of the total water group ion density in the vicinity of the bow shock. A minute increase at 1825 SCET (spacecraft event time) indicates the crossing of the foreshock which separates the undisturbed solar wind from a region with an increased level of plasma turbulence upstream of the bow shock. Although rapid time variations are eliminated by the 2 min. averaging involved in the computation process noticeable density fluctuations are reminiscent of the unsteady solar wind conditions in this regions of space. At 1922 SCET the passage of the bow shock caused a sudden increase in the density by a factor of two. This jump in pick-up ion density is correlated with a steep transient depression in the solar wind speed as can be seen in the upper panel of Fig. 6. The downstream density gradient is obviously slightly steeper than the gradient outside the shock. Massive density excursions are conspicuous features in the downstream portion of the profile. The nature of these large scale density fluctuations and their relation to the ambient solar wind conditions has yet to be established.

The corresponding radial density profile in Fig. 7 shows that the upstream slope follows roughly a R^{-2} dependence but at the shock and further inward significant deviations are obvious. A density measurement of water group ions obtained by the Japanese Suisei satellite on the outbound path at 10^6 km radial distance is added for comparison.

The relative mass loading of the solar wind as a function of radial distance is shown in Fig. 8. The number density of water group ions $n(O^+)$ normalised to the solar wind proton density $n(H^+)$ shows a continuous steep increase as the bow shock is approached. Inside 10^6 km radial distance from the nucleus the ratio is somewhat uncertain due to strong fluctuations in the solar wind. The development of the shock wave occurred at a relative mass loading of approximately 1%. Biermann et al. (1967) pointed out that there exists an upper limit for the maximum mass loading which can be obtained in supersonic flow. Before this limit is actually reached a bow shock must be formed. For the present case this criterion predicts $\gamma = 2\%$ as a critical relative density. As has been mentioned earlier the observed value of 1% at the bow shock requires some refinement due to threshold effects in the instrument. These corrections tend to increase the density somewhat (by no more than a factor of 1.5).

Acknowledgements. This work was supported by the Bundesministerium für Forschung und Technologie under grant 01 OF

112–4, the Science and Engineering Research Council, the US Department of Energy, NASA order number W-15, 164, the Piano Spaziale Nazionale of the Consiglio Nazionale delle Ricerche, NASA contract NASW-3574, and the Swedish Board for Space Activities.

References

Biermann, L., Brosowski, B., Schmidt, H.U.: 1967, *SolarPhys.* **1**, 254

Coates, A., Johnstone, A.A., Thomson, M.F., Formasino, V., Amata, E., Wilken, B., Hockers, K., Winningham, J.D., Borg, H., Bryant, D.A.: 1987, *Astron. Astrophys.* (submitted)

Johnstone, A.D., Bowles, J.A., Coates, A.J., Coker, A.J., Kellock, S.J., Raymont, J., Wilken, B., Stüdemann, W., Weiß, W., Cerulli-Irelli, R., Formasino, V., de Giorgi, E., Perani, P., de Bernardi, M., Borg, H., Olsen, S., Winningham, J.D., Bryant, D.A.: 1986a, ESA SP-1077, 15

Weiß, W., Wilken, B.,: 1987, *J. Phys. E* (submitted)

Wilken, B., Weiß, W., Stüdemann, W., Hasebe, N.: 1987, *J. Phys. E: Sci. Instrum.* **20**, 778

Wu, C.S., Winske, D., Gaffey, J.D.Jr.: 1986, *Geophys. Res. Letters* **13**, 865

Astron. Astrophys. 187, 160–162 (1987)

ASTRONOMY AND ASTROPHYSICS

Ion temperature and flow profiles in comet P/Halley's close environment

R. Schwenn[1], **W.-H. Ip**[1], **H. Rosenbauer**[1], **H. Balsiger**[2], **F. Bühler**[2], **R. Goldstein**[3], **A. Meier**[2], **and E. G. Shelley**[4]

[1] Max-Planck-Institut für Aeronomie, P.O. Box 20, D-3411 Katlenburg-Lindau, Federal Republic of Germany
[2] Physikalisches Institut, University of Bern, Sidlerstrasse 5, CH-3012 Bern, Switzerland
[3] Jet Propulsion Laboratory, 4800 Oak Grove Drive, Pasadena, CA 91103, USA
[4] Lockheed Palo Alto Research Laboratory, 3251 Hanover Street, Palo Alto, CA 94304, USA

Received February 13, accepted May 11, 1987

Summary. The HIS instrument of the ion mass spectrometer on board the Giotto spacecraft identified the contact surface at 4800 km distance from the comet nucleus. This boundary is clearly seen by a drastic drop in the temperatures of different ion species from about 2000 K outside to values as low as 300 K inside. Also, inside the contact surface an outflow speed of $>1\ \mathrm{km\,s^{-1}}$ was measured, in contrast to a value around zero right outside. We discuss how these numbers might be affected by a potential charge-up of the spacecraft. Outside the contact surface, the ion temperature rises gradually with increasing distance. Between 9000 and 10 000 km distance the ion density increases by a factor of 4. This "ion pile-up" is not yet explained uniquely. Between 25 000 and 28 000 km distance there is again a rather abrupt jump to significantly higher temperatures and higher outflow speeds.

Key words: Halley – Giotto-IMS – ion dynamics – inner coma

The High Intensity Spectrometer (HIS) of the Giotto Ion Mass Spectrometer (IMS) was designed specifically to measure the velocity distribution functions of different ion species in the comet's innermost coma. The intrinsic field of view of HIS is $2° \times 15°$, including the direction of the spin axis, i.e. the ram direction of Giotto. Due to the spacecraft spin, the resulting field of view is conical, with a half angle of 12° and slight overlap at the center. For more instrumental details see Balsiger et al. (1986a).

A major design goal was an optimum ion mass resolution for singly ionized ions from the water group, i.e. for mass per charge ratios from 16 to 21 (in the following called M 16 to M 21). In their Fig. 5 Balsiger et al. (1986b) had presented profiles of flow speeds relative to the comet and temperatures of M 18 and M 19 in the distance range from 27 000 km to 1200 km of the nucleus. Since then the range was extended to 45 000 km, and corresponding evaluations of M 16 and M 17 were added. The results are shown in Fig. 1.

The analysis was performed with the assumption of the spacecraft potential ϕ to be zero. A potential of -20 V, e.g., would post-accelerate M 18 by $1.54\ \mathrm{km\,s^{-1}}$ and M 19 by $1.46\ \mathrm{km\,s^{-1}}$. This is to illustrate the sensitivity of the ion velocity measurements to charging of the space probe. Also, it shows that from differences of measured velocities of different ion species the actual value of ϕ can, in principle, be determined, provided the real velocities were the same. Based on the data from HIS this method could not be applied yet with sufficient accuracy. However, analogous work by Lämmerzahl et al. (1986) indicates that ϕ remained within ± 3 V and probably did not exceed ± 1 V in the inner coma. This means that the uncertainty of ion velocity measurements due to charging effects is of the order of $0.1\ \mathrm{km\,s^{-1}}$.

In order to determine the ion flow parameters as shown in Fig. 1 we assumed Maxwellian velocity distribution functions, fitted them to the measured data points, and subtracted the spacecraft's relative speed of $68.4\ \mathrm{km\,s^{-1}}$ for deriving speed values. Therefore, the speed values given in the figure refer to the component of the ion velocities in the direction opposite to the ram direction of the spacecraft. (We designate the flow as "outward" if these values are positive and "inward" if they are negative.)

Up to now only zeroth order calibration data have been taken into account. In particular, slight differences in the energy-angle interrelation for the different mass channels have been neglected so far. This may explain the apparent offset of the velocities of M 16 and M 17 relative to M 18 and M 19. Instead of absolute ion densities we still use relative units (count rates) which are proportional to the fluxes of the respective ions with reference to the spacecraft.

On going through the figure in the sense of increasing distance from the nucleus (i.e. from the right side to the left) we notice several important features:

– In the inner coma within 4000 km the ion temperatures are as low as 300 K. Whether the somewhat higher values in the innermost region are real or due to very fast S/C potential variations caused by dust impacts on the spacecraft is not yet clear.

– There is a clear-cut transition region around 4800 km from the nucleus (actually between 4400 and 5200 km). We call it the "contact surface" since it separates the very cold ions flowing away from the nucleus with a speed of $\geqq 1\ \mathrm{km\,s^{-1}}$ from much hotter ions outside which seem to be more or less at rest with the nucleus. Right at this contact surface the magnetic field starts rising sharply from essentially zero (inside the "magnetic cavity") to more than 60 nT further outside (Neubauer, 1986).

– The total ion density decreases roughly as $1/r$ out to 8500 km, without any noteworthy feature at the contact surface.

Send offprint requests to: R. Schwenn

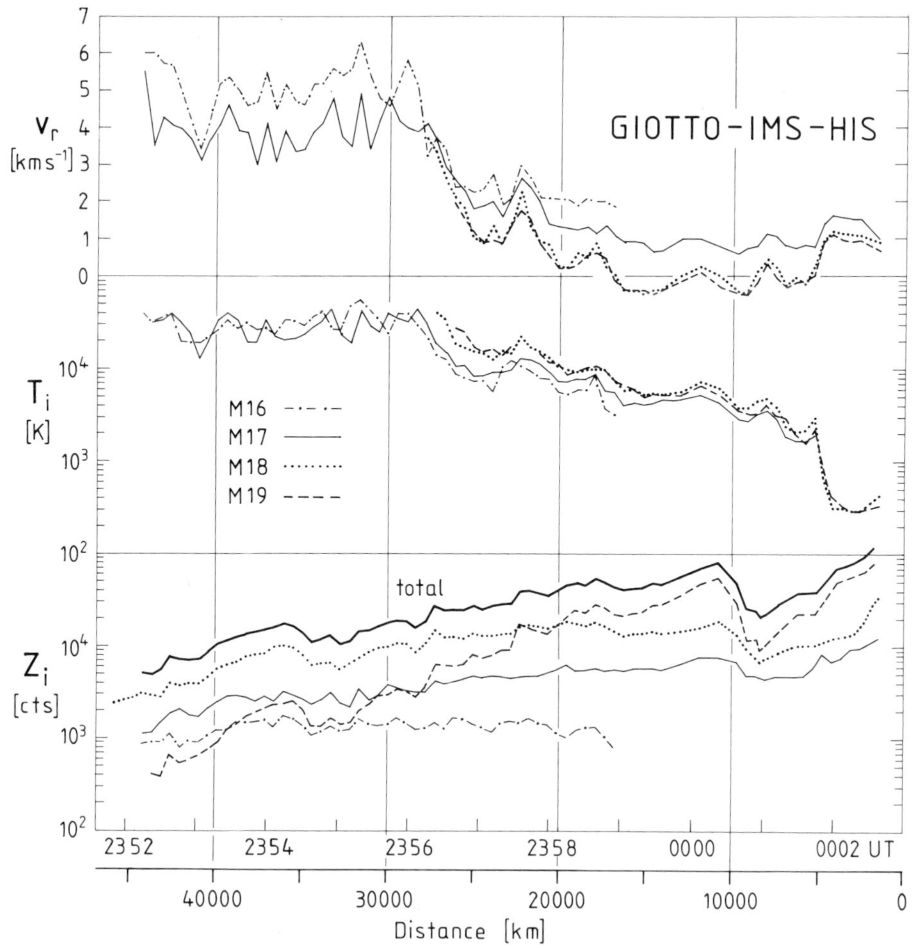

Fig. 1. Profiles of the ion velocity components (v_r) opposite to the S/C ram direction in the rest frame of the nucleus, temperatures (T_i) and the relative densitities (Z_i) for ion masses M16, M17, M18 and M19. The measurements stem from the high intensity spectrometer (HIS) instrument of the Ion Mass Spectrometer (IMS) on board the Giotto space probe

As expected (see, e.g., Balsiger et al., 1986a), the most abundant ion inside 20000 km is M19 which is mainly due to H_3O^+.

– Between 9000 and 11000 km distance the ion density suddenly rises by a factor of about 4 without any associated apparent change in flow velocity, ion temperature or magnetic field. The possible explanation for the unexpected occurrence of this "ion pile-up" were discussed by Balsiger et al. (1986b) and by Ip et al. (1986).

– There are four well-separated regions with somewhat higher outflow velocities and temperatures observed in the distance range between 8000 and 23000 km, following each other with relative distances of about 5000 km.

– Between 25000 and 28000 km the outflow velocity rises sharply to a value of around 4 km s^{-1}. Simultaneously, the temperature rises by a factor of 4. These numbers are valid out to about 45000 km distance. Further outside there is another marked increase in both radial velocity and temperature which requires more sophisticated evaluation procedures and is therefore not included in this analysis.

– There is a definite density bulge between 35000 and 40000 km. This feature is discussed in more detail by Ip et al. (1986). Similar structures had been observed from the Vega probes (Gringauz et al., 1986). One possible explanation proposed by Galeev (1986) in terms of the critical velocity effect seems to be ruled out by our measurements. Obviously differential velocities between ions and neutrals required for this effect (some 20 km s^{-1}) are well beyond the actual ion velocities.

– The average radial density profile of the water group ions shows a $1/r^2$ dependence in the range from 10000 to at least 45000 km, as would be expected for a radial outflow at constant speed with no effective local production.

– The M19 density profile shows a significantly steeper decrease with distance than all other water group ion densities. The M19/M18 density ratio varies from about 3 inside the contact surface to a value close to 1 at 9000 km, i.e. just inside the ion pile-up region. At the "center" of the pile-up the ratio reaches 3 again and only then drops continuously. It goes through 1 at 21000 km, and beyond 43000 km it is less than 0.14. Note that the absolute value of the ratio may be subject to minor adjustments once the detailed calibration data have been included. The profile of the M19/M18 ratio generally fits well with theoretical models (Balsiger et al., 1986a, b; Hübner et al., 1986). The remarkable behavior of the M19/M18 ratio around 10000 km distance may help in finding the explanation of the ion pile-up region.

Despite its preliminary character the present study has already some important implications. In the next phase we will work out further details such as angular deviations from radial flow, and the profiles of other ion species.

Acknowledgements. This study was founded in part by the German Bundesminister für Forschung und Technology under grant 010F86014.

References

Balsiger, H., Altwegg, K., Bühler, F., Fischer, J., Geiss, J., Meier, A., Rettenmund, U., Rosenbauer, H., Schwenn, R., Benson, J., Hemmerich, P., Saeger, K., Kulzer, G., Neugebauer, M., Goldstein, B.E., Shelley, E.G., Sanders, T., Simpson, D., Lazarus, A.J., Young, D.T.: 1986a, ESA SP-1077, p. 129

Balsiger, H., Altwegg, K., Bühler, F., Geiss, J., Ghielmetti, A.G., Goldstein, B.E., Goldstein, R., Huntress, W.T., Ip, W.-H., Lazarus, A.J., Meier, A., Neugebauer, M., Rettenmund, U., Rosenbauer, H., Schwenn, R., Sharp, R.D., Shelley, E.G., Ungstrup, E., Young, D.T.: 1986b, *Nature* **321**, 330

Galeev, A.A.: 1986, 20th ESLAB Symp., ESA SP-250, Vol. **I**, p. 3

Gringauz, K.J., Verigin, M.J., Richter, A.K., Gombosi, T.J., Szegö, K., Tátrallyay, M., Remizov, A.P., Apáthy, I.: 1986, 20th ESLAB Symp., ESA SP-250, Vol. **I**, p. 93

Hübner, W.F., Boice, D.C., Keady, J.J., Schmidt, H.U., Wegman, R.: 1986, 20th ESLAB Symp., ESA SP-250, Vol. **I**, p. 529

Ip, W.-H., Schwenn, R., Rosenbauer, H., Balsiger, H., Neugebauer, M., Shelley, E.G.: 1986, 20th ESLAB Symp. ESA SP-250, Vol. **I**, p. 219

Lämmerzahl, P., Krankowsky, D., Hodges, R.R., Stubbemann, U., Woweries, J., Herrwerth, I., Berthelier, J.J., Illiano, J.M., Eberhardt, P., Dolder, U., Schulte, W., Hoffman, J.H.: 1986, 20th ESLAB Symp., ESA SP-250, Vol. **I**, p. 179

Neubauer, F.M.: 1986, 20th ESLAB Symp., ESA SP-250, Vol. **I**, p. 35

Astron. Astrophys. 187, 163–168 (1987)

ASTRONOMY AND ASTROPHYSICS

The composition and dynamics of cometary ions in the outer coma of comet P/Halley

H. Balsiger [1], K. Altwegg [1], F. Bühler [1], S.A. Fuselier [2], J. Geiss [1], B.E. Goldstein [3], R. Goldstein [3], W.T. Huntress [3], W.-H. Ip [5], A.J. Lazarus [4], A. Meier [1], M. Neugebauer [3], U. Rettenmund [1], H. Rosenbauer [5], R. Schwenn [5], E.G. Shelley [2], E. Ungstrup [6], and D.T. Young [7]

[1] Physikalisches Institut, University of Bern, Sidlerstrasse 5, CH-3012 Bern, Switzerland
[2] Lockheed Palo Alto Research Laboratory, Palo Alto, CA 94304, USA
[3] Jet Propulsion Laboratory, California Institute of Technology, Pasadena, CA 91109, USA
[4] Massachusetts Institute of Technology, Cambridge, MA 02139, USA
[5] Max-Planck-Institut für Aeronomie, D-3411 Katlenburg-Lindau, Federal Republic of Germany
[6] Danish Space Research Institute, DK-2800 Lyngby, Denmark
[7] Los Alamos National Laboratory, Los Alamos, NM 87545, USA

Received February 18, accepted March 29, 1987

Summary. During its flyby at comet Halley, the Giotto spacecraft encountered high densities of cometary ions inside approximately 200 000 km from the nucleus. Their properties changed drastically as the comet was approached. We present here density profiles of solar wind alpha particles and of the major cometary ions as obtained by the IMS-HERS sensor between 340 000 km and 60 000 km from the nucleus. Typical mass spectra at various distances are presented and angular and velocity distributions of the cometary ions are discussed.

Key words: comets – Halley – Giotto-IMS – ion composition

1. Introduction

The Ion Mass Spectrometer (IMS) aboard the Giotto spacecraft measured the composition and velocity distribution of solar wind and cometary ions at distances between 7.5 10^6 and 1300 km from the nucleus of comet Halley. IMS consists of two independent sensors (Balsiger et al., 1987): HERS is optimized primarily for the interaction region of the solar wind with the cometary ions. Its field of view is a fan between 15° and 75° from the spin axis (elevation angle) which rotates with the spinning spacecraft; hence it does not include the ram direction. HIS is optimized for cold ions inside the contact surface, and its field of view is a rotating fan between −3° and 12° relative to the spin axis, thus including the ram direction. An extensive overview of the initial findings of IMS has been given in Balsiger et al. (1986). Here we present a progress report on the HERS data analysis. Whereas in Balsiger et al. (1986) only raw data (count rates) were given, we present here densities, velocities, and temperatures of the ions. This paper mainly concentrates on the region between 350 000 km and 60 000 km from the nucleus, when the cometary ions were well within the field of view of HERS. Around 60 000 km they began to be lost from the HERS field of view; hence densities given for the region between 82 000 km and 16 000 km (Fig. 4c, Table 1) are too low and can at best be used to compute abundance ratios.

Before 23.00 GRT (ground-station received time in UT) the HERS mainly measured solar wind protons and alpha particles and their interaction with the comet (Neugebauer et al., 1986; Goldstein, B. E. et al., 1986; Shelley et al., 1986). After 23:00 GRT primary cometary ions were clearly recognized: Atomic carbon (C^+), the water group ions (O^+, OH^+, H_2O^+) as well as $M/Q = 28$ and 32 were the main constituents. Around 23:50 GRT, the intensities of these ions decreased again due to the fact that HERS was losing the cometary ions out of its field of view, and by 00:06 the ion population was so stagnant and cold that HERS count rates were down to background level (except for the protons). Here, the cometary ions were analyzed by the HIS sensor (Balsiger et al., 1986; Ip et al., 1986; Schwenn et al., 1986). Shortly before closest approach (00:11 GRT) hot ions reappeared (Goldstein, R. et al., 1986) and shortly afterwards the signal was lost due to dust impact on the spacecraft. This paper will concentrate on the period when HERS measured large fluxes of cometary ions.

2. Observations

In order to investigate the nature and the properties of the heavy ions which seemed to appear rather abruptly, we have computed the densities of the major ions ($M/Q = 12, 16, 17, 18, 28$, and 32) as a function of radial distance. As a prerequisite for computing reliable physical parameters, one must ensure that the instrument is covering all or most of the phase space populated by the ions. In Fig. 1a–c examples for the angular and velocity distributions of the heavy ions in the M-mode ($12 \leqq M/Q \leqq 25$) are given. Figure 1a covers the region of increasing density of the cometary ions, Fig. 1b the peak density and Fig. 1c represents the region closer to the nucleus where fluxes of heavy ions dropped quite abruptly in the HERS sensor. In the right panel the velocity distributions of all the ions in the M-mode are shown (velocity in the spacecraft frame of reference) and in the left panel contour plots of their angular distribution. In the spacecraft frame of reference the HERS

Send offprint requests to: H. Balsiger

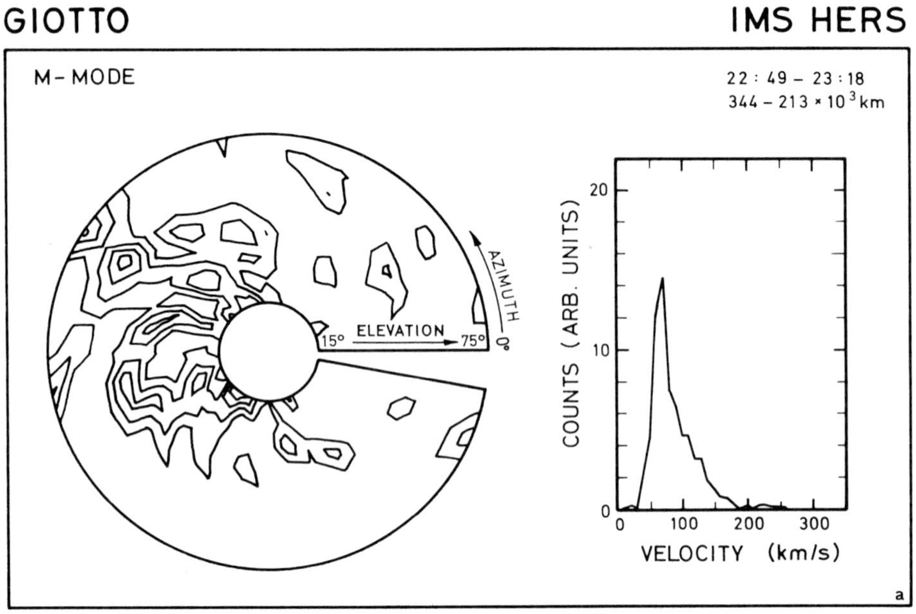

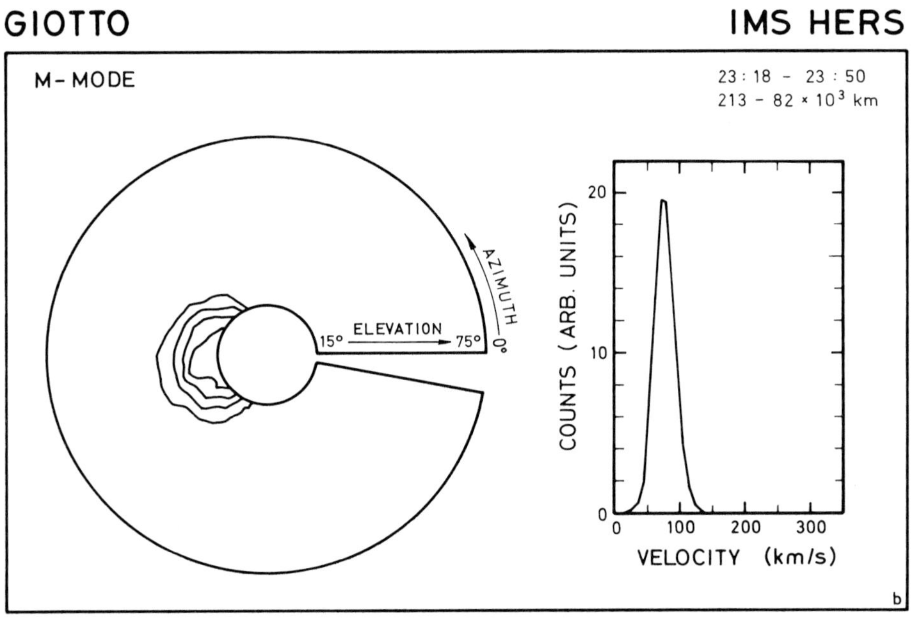

Fig. 1a–c. Contour plots for angular distributions (left) and ion velocity distribution curves (right) for the IMS-HERS for three radial intervals (times in ground-station received time, GRT). Data are summed over all masses in the M-mode ($M/Q = 12$–25). The HERS covers elevation angles 15° to 75° relative to the spin axis and 349° of spin angle (azimuth). This field of view included the free streaming solar wind (AZ ~ 180°, EL ~ 70°) but not the ram direction which is in the center of the contour plot. Contours are drawn for 80, 60, 40, and 20% of the maximum. Velocity is given in the spacecraft frame of reference

instrument covers 349° of azimuth angle and 15° to 75° of the elevation angle (plotted as radial coordinate in these plots). The free-streaming solar wind outside the bow shock was observed at an azimuth of ~180° and an elevation of ~70°, cold ions would appear in the ram direction at 0° elevation and cannot be seen by the HERS sensor. Figure 1 demonstrates how, as the spacecraft approaches the comet, the heavy ion distributions change from being quite wide in angle and energy to narrow distributions which move towards smaller elevation angles and then out of the field of view of the HERS. We have simulated the distributions in Fig. 1 with running maxwellians in the spacecraft frame of reference, and we estimate that during our main observation period (23:18 to 23:50 GRT, Fig. 1b), HERS covers 66% of the ion population. In the last period (Fig. 1c), the HERS misses about 75% of the distribution, and the computed densities can at best be used for estimating abundance ratios.

In Fig. 2 we present a radial profile of the major ions $M/Q = 2$ (He^{2+}), 12 (C^+), 16 (mainly O^+), 17 (mainly OH^+), 18 (mainly H_2O^+), 28 (mainly CO^+), and 32 (mainly S^+). For a discussion on the identification of the ions as given in parenthesis, we refer to Balsiger et al. (1986). The shaded area marks the region where cometary ions disappear out of the field of view of the HERS. As

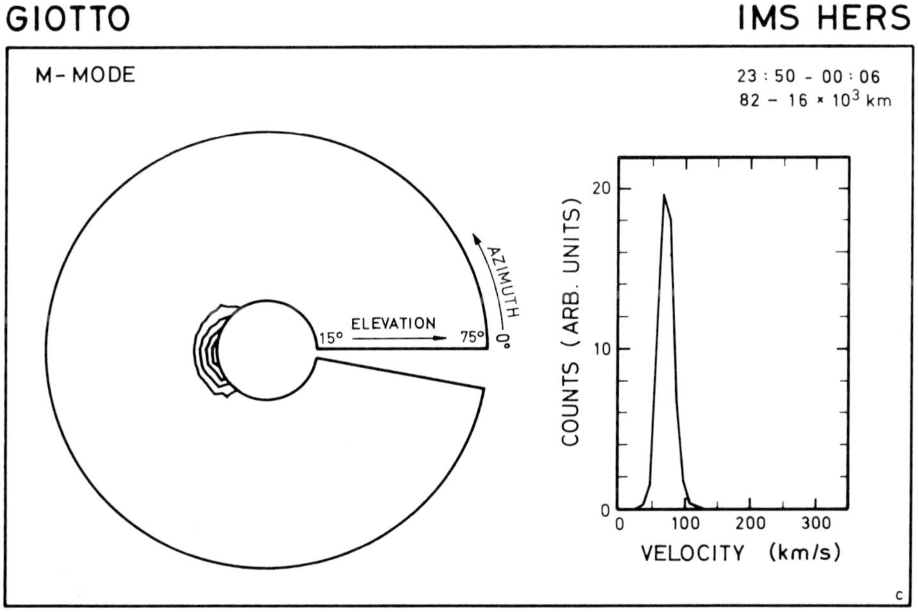

Fig. 1c

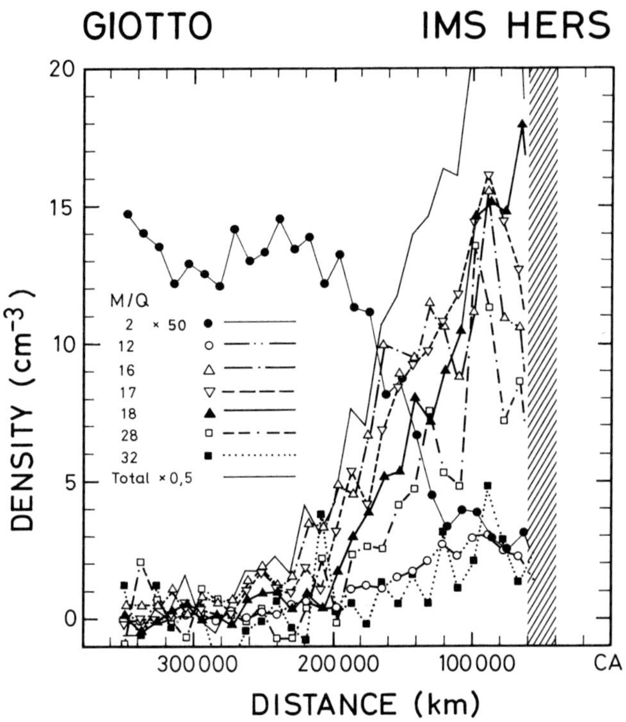

Fig. 2. Radial profiles of ion densities for the main cometary ion species. Data are averaged over 10 spins per mode, hence time resolution is 160 s. For comparison the profile of the solar wind alpha particles is also drawn. Beginning with the shaded area ions start to disappear from the field of view of the sensor (cf. Fig. 1)

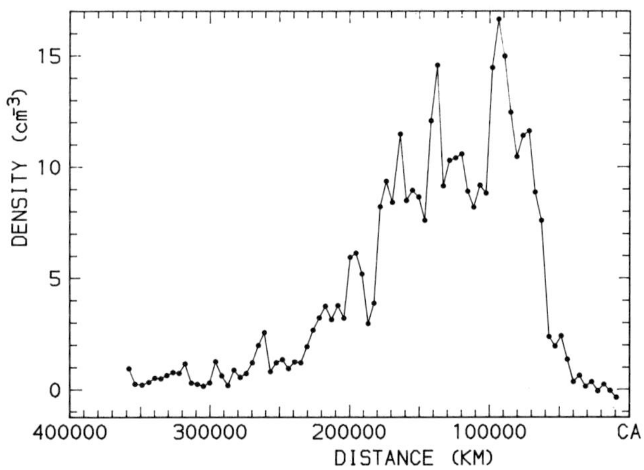

Fig. 3. Radial profile for $M/Q = 16$ with higher time resolution (64 s)

the comet is approached, density and velocity of solar wind He^{2+} decrease gradually (Balsiger et al., 1986; Goldstein, B.E. et al., 1986). Inside ~220 000 km the decrease is rapid (Fig. 2), indicating a strong interaction with cometary ions. As a matter of fact, cometary ion densities become here comparable to solar wind ion densities, and inside 200 000 km they dominate over the solar wind protons (Goldstein, B.E. et al., 1986; Shelley et al., 1986). The dependence on distance from the nucleus is similar for these major cometary ions. However, there is a tendency for the water group ions and CO^+ to increase faster (doubling of density per 30 000 km) than C^+ and S^+ (doubling of density per 50 000 km). Also, some ions (e.g. S^+) show a more structured profile than others (C^+). It would be premature to draw definite conclusions from this observation; data have been averaged over 10 spins per mode for this profile, and we shall show in the next figure that with higher time resolution the profiles tend to become more structured.

Figure 3 shows the radial profile of $M/Q = 16$ with data averaged only over 4 spins per mode. Clearly, there is more structure visible, and this picture suggests a certain layering of the heavy ions. The magnetometer shows strong variations of the B-field both in amplitude and direction in this part of the coma (Neubauer et al., 1986). In a future study, we shall attempt to correlate these two datasets more precisely (for the time being it

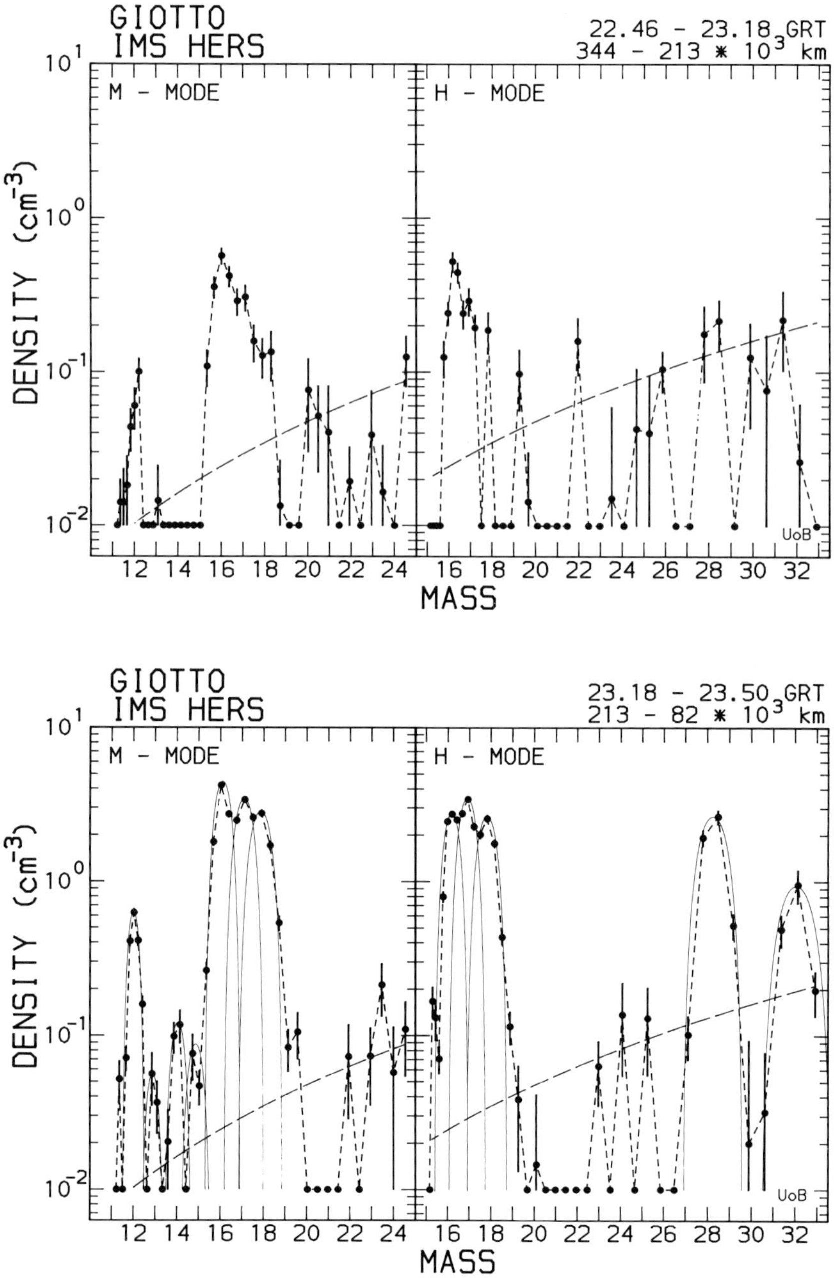

Fig. 4a–c. M/Q spectra acquired by HERS in three radial intervals in the medium and heavy ion mode. The spectra have been corrected for background (0.03 counts per second per mass channel). The dashed line is a theoretical density, calculated for a countrate of 0.01 counts/s above the background and represents a limit of confidence (see text). In **b** theoretical mass peaks have been fitted for the main masses. Errors are 1σ statistical errors. For computing absolute densities, datapoints which belong to each mass peak have to be added up (see Table 1 for absolute densities)

should be noted that these variations happen during the same period). A correlation between ion count rates and B could be caused in two ways: (a) density and B are related or (b) B-field variations move the ions out of the HERS field of view.

If one attempts to identify also minor cometary ions, it is necessary to integrate over longer time periods and to make individual background correction for each mass channel. Furthermore, we know from calibration that mass peak locations on the two dimensional detector (Balsiger et al., 1987) depend slightly on the incident direction of the ions. We present in Fig. 4a–c mass spectra of the three regions identified earlier. The average background is subtracted but the final and individual corrections for detector background and peak shifts have not been applied. Thus, only some of the minor constituents can be clearly identified at this time. We have introduced in the spectra a limit of confidence (dashed line) of 0.01 counts per second and pixel (the detector is divided up into 40 mass $\times 8$ angular sectors). Only peaks which are 2σ above this limit are discussed at this stage of the evaluation (1σ statistical errors are given in the plots). Outside 213 10^3 km (Fig. 4a), where intensities are low, only the water group ions ($M/Q = 16$–18) and C^+ (12) can be clearly identified. In the next period (Fig. 4b), where intensities are high and ions are

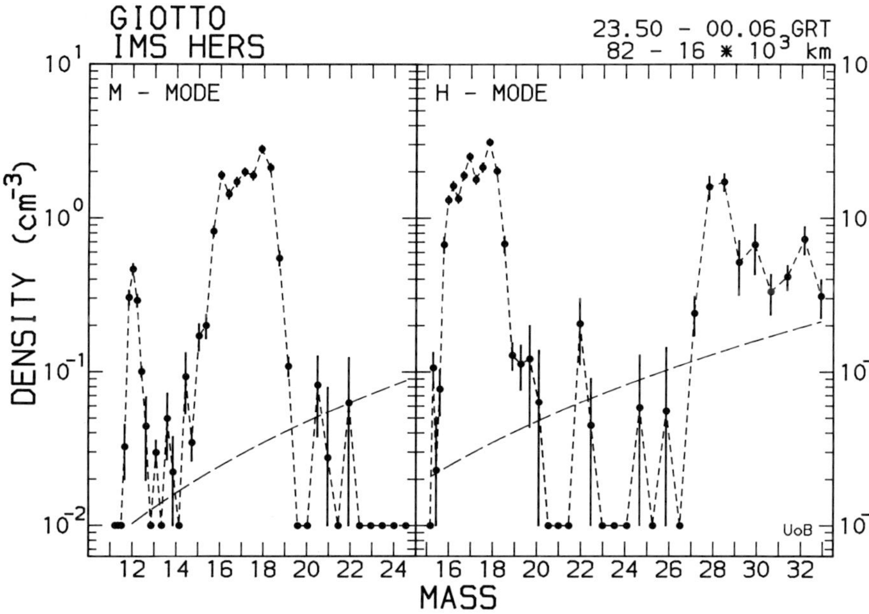

Fig. 4c

Table 1. HERS ion number densities (cm^{-3})

Range ($\times 10^3$ km)	$M/Q=2$	12	16	17	18	28	32
344 – 213	0.26 ± 0.01	0.23 ± 0.04	1.3 ± 0.2	0.7 ± 0.2	0.3 ± 0.1	0.2 ± 0.1	0.15 ± 0.1
213 – 82 [a]	0.23 ± 0.02	2.5 ± 0.4	14.5 ± 2.2	14.3 ± 2.3	10.2 ± 1.7	7.7 ± 1.4	2.5 ± 0.6
82 – 16	(0.03) ± 0.01	(1.2) ± 0.2	(5.2) ± 0.8	(7.0) ± 1.1	(7.0) ± 1.1	(3.8) ± 0.7	(1.5) ± 0.3

[a] Densities have been corrected for ions outside field of view

well within the HERS field of view, $M/Q=28$, $M/Q=32$ and also the minor ions 13 (mainly CH^+), 14 (CH_2^+ and N^+), and 15 (mainly CH_3^+) are resolved and above background. There is some evidence for ions around $M/Q=23$–24 (Na^+ and/or C_2^+), but more careful analysis is needed to establish or disprove the existence of these ions. In the innermost spectrum (Fig. 4c), $M/Q=19$ (H_3O^+) and 30 (possibly H_2CO^+ and/or NO^+) are clearly identified. Comparing the water group ions in all three figures, it is evident how the relative importance of the three peaks is changing as the comet is approached: Whereas at large distances water is strongly dissociated into H^+, OH^+, and O^+, hence making the $M/Q=16$ peak the most dominant (Fig. 4a), the H_2O^+ (18) dominates at smaller distances, H_3O^+ (19) appears as total densities go up and ion-molecule reactions become significant (Fig. 4c).

In Table 1 we give densities of the major ions averaged over the three periods of ~30 min each. For the main period (23:18–23:50 GRT) we have estimated that the HERS measures 66 ± 6% of the total distribution, and we have applied the appropriate correction to compute the densities given in Table 1. Values of the third period are given in parenthesis because they are systematically too low and should only be used for estimates of abundance ratios during this period. Errors include 2σ statistical errors as well as systematical errors (due to uncertainties in calibration and in assessing the missing ion population).

3. Discussion

A more detailed analysis of the HERS data in the region between 350 000 and 60 000 km from the nucleus has, with respect to the composition of the outer coma, generally confirmed what was reported in our first paper (Balsiger et al., 1986). The conclusions about the most likely identification of the ions and their relative abundance as deduced from Table 1 are at present unchanged. We repeat here our finding that the major elements, C, O, and S are, within the limits of error, consistent with the solar system abundance and that N may be low by an order of magnitude. We had also noted that the relatively high abundance of C^+ called for a distributed source of atomic carbon; we suggest that the fine comet dust grains release carbon in atomic form or as a highly unsaturated compound. The relatively flat radial profile of C^+ (Fig. 2) may present further evidence for this case.

In the tail of comet Giacobini-Zinner, ions of $M/Q = 23$ to 24 were reported (Ogilvie et al., 1986). In the mass spectra of Fig. 4, there is some evidence for ions in this mass region. However, the signal is too low for a positive identification at this stage of the evaluation.

Gringauz et al. (1986) have reported a new chemical boundary around 1.6 10^5 km, the cometopause, based on the sudden appearance (within 10^4 km) of cometary ions in the Vega Cometary Ram Analyzer. In the IMS-HERS instrument (which does, however, not include the ram direction), we see a more gradual increase of cometary ions (Fig. 2) over a distance of at least 10^5 km [and these ions clearly do not come from the ram direction (see Fig. 1)]. The increase of the watergroup ions would be quite consistent with a R^{-2} law, as was previously reported for the cold ions measured by the HIS sensor at smaller cometary distances (Balsiger et al., 1986). What is missing in our analysis at this time is an extension of the HIS data to 150 000 km in order to judge how much the cold ions (arriving in the ram direction) contribute to the total population. The magnetic field was quite variable during this period of the Giotto encounter (Neubauer et al., 1986), and it cannot be excluded that some of the apparent structures in the radial profile (Fig. 3) are due to these temporal or spatial variations. We note here that according to ground-based observations (cf. Lüst, 1962), well defined parabolic envelopes of ion densities are formed in the cometosheath which might be the origin of ion rays in the cometotail. A full analysis of the whole particle phase space combined with B-field data will be needed before the nature of one or several possible discontinuities can be hopefully assessed.

Acknowledgements. We acknowledge the dedicated support of the many technicians who made the Giotto mission and the IMS experiment possible. We thank J. Fischer and E. Sennhauser for graphics and I. Peter for typing. This work was supported by NASA, the German Bundesministerium für Forschung und Technologie and the Swiss National Science Foundation.

References

Balsiger, H., Altwegg, K., Bühler, F., Geiss, J., Ghielmetti, A.G., Goldstein, B.E., Goldstein, R., Huntress, W.T., Ip, W.-H., Lazarus, A.J., Meier, A., Neugebauer, M., Rettenmund, U., Rosenbauer, H., Schwenn, R., Sharp, R.D., Shelley, E.G., Ungstrup, E., Young, D.T.: 1986, *Nature* **321**, 330

Balsiger, H., Altwegg, K., Bühler, F., Fischer, J., Geiss, J., Meier, A., Rettenmund, U., Rosenbauer, H., Schwenn, R., Benson, J., Hemmerich, P., Säger, K., Kulzer, G., Neugebauer, M., Goldstein, B.E., Goldstein, R., Shelley, E.G., Sanders, T., Simpson, D., Lazarus, A.J., Young, D.T.: 1987, *J. Phys. E: Sci. Instrum.* **20**, 759

Goldstein, B.E., Neugebauer, M., Balsiger, H., Drake, J., Fuselier, S.A., Goldstein, R., Ip, W.-H., Rettenmund, U., Rosenbauer, H., Schwenn, R., Shelley, E.G.: 1986, Rosenbauer, H., Schwenn, R., Shelley, E.G.: 1986, 20th ESLAB Symp., ESA *SP-250*, Vol. **I**, p. 229

Goldstein, R., Young, D.T., Balsiger, H., Bühler, F., Goldstein, B.E., Neugebauer, M., Rosenbauer, H., Schwenn, R., Shelley, E.G.: 1986, 20th ESLAB Symp., ESA, *SP-250*, Vol. **I**, p. 207

Gringauz, K.I., Gombosi, T.I., Tátrallyay, M., Verigin, M.I., Remizov, A.P., Richter, A.K., Apáthy, I., Szemerey, I., Dyachkov, A.V., Balakina, O.V., Nagy, A.F.: 1986, *Geophys. Res. Letters* **13**, 613

Ip, W.-H., Schwenn, R., Rosenbauer, H., Balsiger, H., Neugebauer, M., Shelley, E.G.: 1986, 20th ESLAB Symp., ESA, *SP-250*, Vol. **I**, p. 219

Lüst, R.: 1962, *Z. Astrophys.* **54**, 67

Neubauer, F.M., Glassmeier, K.H., Pohl, M., Raeder, J., Acuna, M.H., Burlaga, L.F., Ness, N.F., Musmann, G., Mariani, F., Wallis, M.K., Ungstrup, E., Schmidt, H.U.: 1986, *Nature* **321**, 352

Neugebauer, M., Lazarus, A.J., Altwegg, K., Balsiger, H., Goldstein, B.E., Goldstein, R., Neubauer, F.M., Rosenbauer, H., Schwenn, R., Shelley, E.G., Ungstrup, E.: 1986, 20th ESLAB Symp., ESA, *SP-250*, Vol. **I**, p. 19

Ogilvie, K.W., Coplan, M.A., Bochsler, P., Geiss, J.: 1986, *Science*, **232**, 374

Schwenn, R., Ip, W.-H., Rosenbauer, H., Balsiger, H., Bühler, F., Goldstein, R., Meier, A., Shelley, E.G.: 1986, 20th ESLAB Symp., ESA *SP-250*, Vol. **I**, p. 225

Shelley, E.G., Fuselier, S.A., Balsiger, H., Drake, J.F., Geiss, J., Goldstein, B.E., Goldstein, R., Ip, W.-H., Lazarus, A.J., Neugebauer, M.: 1986, 20th ESLAB Symp., ESA *SP-250*, Vol. **I**, p. 285

Astron. Astrophys. 187, 169–173 (1987)

ASTRONOMY AND ASTROPHYSICS

Expansion velocity and temperatures of gas and ions measured in the coma of comet P/Halley

P. Lämmerzahl [1], D. Krankowsky [1], R. R. Hodges [2], U. Stubbemann [1], J. Woweries [1], I. Herrwerth [1,★], J. J. Berthelier [3], J. M. Illiano [3], P. Eberhardt [4], U. Dolder [4], W. Schulte [4], and J. H. Hoffman [2]

[1] Max-Planck-Institut für Kernphysik, Saupfercheckweg 1, D-6900 Heidelberg, Federal Republic of Germany
[2] University of Texas at Dallas, Richardson, TX 75083-0688, USA
[3] Centre de Recherches en Physique de l'Environnement Terrestre et Planétaire, CNET-CNRS, 4, avenue de Neptune, F-94107 Saint-Maur des Fossés, France
[4] Physikalisches Institut, University of Bern, Sidlerstrasse 5, CH-3012 Bern, Switzerland

Received February 14, accepted May 29, 1987

Summary. In situ measurements of flow velocity and temperature in the inner coma were obtained from ram energy spectra of molecules and ions observed by the NMS experiment onboard the Giotto probe to comet Halley. Radial flow speed was (800 ± 50) m s^{-1} at distances from the nucleus between 1000 km and 4000 km. Whereas ions became stagnant just outside the contact surface (located near 4500 km), the outflow of water vapour even accelerated and was above 1 km s^{-1} at 30000 km. Ion temperatures were around 200 K inside the contact surface where the plasma is collisionally coupled to the gas; at this boundary ion temperature rises by about 1000 K and then further increases to reach about 5000 K at 15000 km. Gas temperature remains below 300–500 K over this range of distances. A near zero volt spacecraft potential was inferred.

Key words: comets – outflow velocity – coma temperature – spacecraft potential

1. Introduction

The energy analyser (E-analyser), one of the two sensors of the Neutral Mass Spectrometer (NMS) experiment onboard the Giotto spacecraft, measured spectra of the ram energies of gas molecules and ions which were sampled by the instrument when moving at 68.373 km s^{-1} along the probe's inbound trajectory through the cometary atmosphere. Coma constituents are moving in the comet's frame of reference with velocities of the order of one or a few kilometres per second. The effect of superposition of the velocities on the observed ram energies is to amplify differences in kinetic energy (e.g. a flow velocity of 1 km s^{-1}, corresponding to 0.093 eV per water molecule, would shift the water peak by as much as 12.85 eV). The complementary M-analyser of the NMS, a double-focusing magnetic analyser, copes with superimposed motions and resolves masses even if the spread in energy were up to several electron volts in the comet frame. The electrostatic E-analyser, in contrast, displays all energy shifting and peak widening resulting from bulk or random motions, respectively. This allows to study the dynamic and thermodynamic properties of coma constituents, in particular, when independent information on their mass abundances is simultaneously provided by the M-analyser.

In this paper we present measurements of the flow speeds and the temperature of the gas and plasma from the inner coma of comet Halley which became accessible to our experiment down to 840 km distance from the nucleus (for calculating distances we assumed closest approach at 605 km on 14 March 1986 at 00:03:02 UT). Since the publication of first results from a preliminary NMS data analysis (Krankowsky et al., 1986b) improvements have been achieved and corrections became necessary which are reflected herein.

2. Energy spectra characteristics

In the E-analyser ions of different energies are separated in an electrostatic field between parallel plates and are then amplified on a focal plane detector consisting of microchannel plates followed by an array of 128 self-scanning anodes (pixels). A range of energies is simultaneously sampled on the detector and several energy ranges are successively scanned by varying the field strength. Modes for cometary ions and for coma gas alternate, where in the latter case the neutral molecules are ionized by electron impact in a source of fly-through geometry at the entrance of the analyser. Electron energy is either 90 eV or 17 eV. Instrumental details were described by Krankowsky et al. (1986a).

Two distinctly different plasma regimes have been identified in the inner coma (Krankowsky et al., 1986b; Balsiger et al., 1986) as characterized by a discontinuity in the ion energy distribution occurring abruptly between two ion measurements of the NMS at 4370 km and 4740 km radial distances. This boundary which is associated with a magnetic cavity (Neubauer et al., 1986) is being referred to as the "contact surface". Energy spectra of the water group ions that are typical of the populations inside and outside the contact surface are shown in Fig. 1a and b. Their different

Send offprint requests to: P. Lämmerzahl

★ Present address: Fachinformationszentrum Energie-Physik-Mathematik GmbH, D-7514 Eggenstein-Leopoldshafen, Federal Republic of Germany

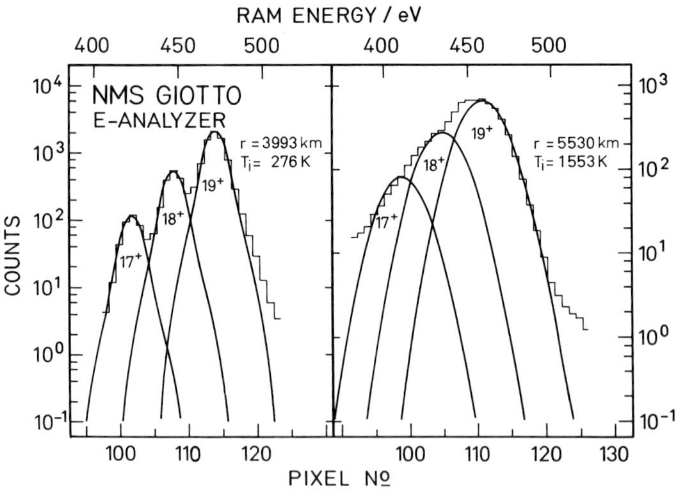

Fig. 1 a and b. Excerpts of ram energy spectra of ions typical of different ion temperature regimes. **a** Left side: an example of the cold plasma inside the contact surface. The staircase curve is the measured data corrected for background. Curved lines result from a fitting analysis and represent the relative abundances and the energy distributions of ion masses 17,18, and 19 amu/e due to their temperature ($T_i = 276$ K) and instrumental peak broadening effects; minor contributions from masses 16, 20, 21 which can be inferred with large uncertainties only are not shown. **b** Right side: an example of the hotter plasma ($T_i = 1553$ K) outside the contact surface. Relative abundances are taken from mass-analyser data

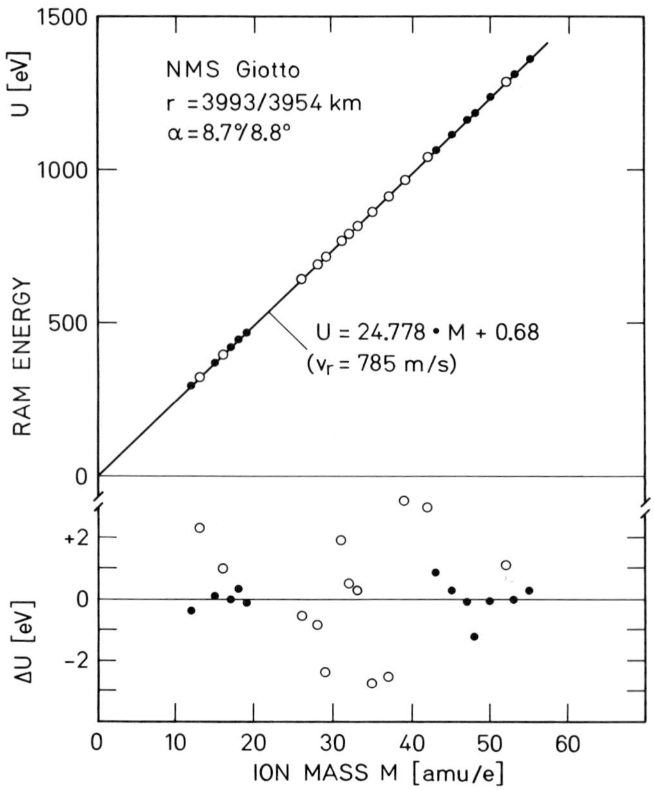

Fig. 2. Ram energies *vs.* masses of an ion spectrum. Offsets from the straight line are shown on enlarged scale (bottom). Dots are mass peaks whose energy can be determined with precision, other peaks are open circles. Radial flow velocity $v_r = 785\ \mathrm{ms}^{-1}$ and spacecraft potential $\phi = -0.68$ V are inferred from this spectrum

mass resolution is indicative of a drastic change in temperature, as will be discussed in Sect. 4.

The NMS observed at distances from 4370 km inwards ram energy spectra of a "cold" plasma. Mass-resolved peaks are found on nearly all integer mass numbers from 12 to 56 amu/e (the instrument's upper limit in the ion mode). Mass identification is unambiguous: up to 37 amu/e confirmation is given by the simultaneously measured M-analyser composition spectra and beyond that mass one can count the mass numbers peak by peak. Outside the contact surface ion spectra are composed of ram energy distributions representing groups among which individual masses overlap one another. Energy spectra of the neutral component (an example was shown by Krankowsky et al., 1986b) have similar mass resolution as those of the "cold" plasma, even at much greater distance.

3. Coma expansion

A gas expansion velocity of $(900 \pm 200)\ \mathrm{m\,s}^{-1}$ had been derived from NMS data earlier (Krankowsky et al., 1986b), and $\sim 1\ \mathrm{km\,s}^{-1}$ ion flow velocity in the inner coma region was reported from the IMS experiment (Balsiger et al., 1986) at the Giotto flyby. Our method of data anlaysis has been refined since and we are able to not only reduce the error but also to study the radial gradient.

3.1. Ion velocities

An accurate energy scale is important for velocity determination. The relationship between pixel number and ion energy which depends on analyser and detector voltages controlled by the programme logic, was calibrated over the range of energies from 60 eV to 2100 eV. A calibration facility was used that provided beams of monoenergetic ions with spread in energy of 1–2 eV only (Ghielmetti et al., 1983). Hence, absolute accuracy of the E-analyser for ion energies is between ± 2 eV and ± 4 eV (depending on the energy range) over a large part of the detector. Peak shape analysis is required to determine the centre of an energy distribution displayed on discrete anodes, since one pixel corresponds to an energy interval of between 4 eV and 10 eV for the various energy ranges. Using the three top pixels of a peak is adequate for this pupose, minimizing interference from neighbouring masses.

When plotting ram energies versus mass numbers of peaks for an ion spectrum, one obtains a straight line (Fig. 2) the slope of which (i.e. the ram energy increment per atomic mass unit) is related to the speed of the probe relative to the coma gas. Flow speeds are obtained by subtracting the probe's speed relative to the nucleus. Since the velocity field is assumed to be directed radial from the nucleus, a factor $\cos^{-1}\alpha$ is applied where α is the angle between radius and flight vectors.

Actually, a complete ion spectrum consists of two measurements (equivalent mass range split at ~ 21 amu/e) taken only 0.57 s apart. The corresponding change in radial aspect, α, would cause a change in slope which was corrected in Fig. 2 but is negligible except for one or two spectra near 1000 km. There are sections on the detector where the energy scale has irregularities, and some peaks are not sufficiently above background to determine their centres with the accuracy required. Such data

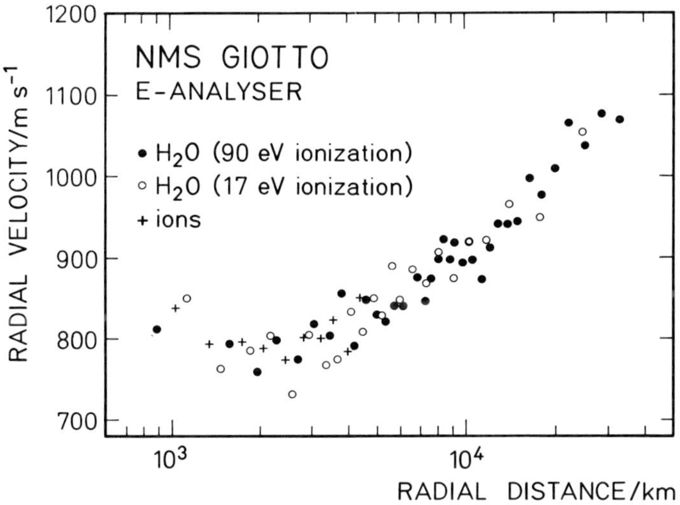

Fig. 3. Coma outflow velocities as observed from ram energies of water molecules (using two different electron energies for ionization) and of ion spectra (inside the contact surface only)

shown as open circles are not used for velocity computation. But all peaks (denoted by dots) that need not be rejected for such reason show only very small displacements of about ± 1 eV or less from a linear energy-to-mass relation within one ion spectrum. Such displacements, if interpreted as differential speeds, would amount to less than $\sim 50\,\mathrm{m\,s^{-1}}$. We conclude that within our experimental uncertainty heavy ions flow at a common speed inside the contact surface, confirming the still strong collisional coupling in this regime.

From each of the ten ion spectra obtained of the "cold" plasma both a radial flow speed, v_r (plotted in Fig. 3), and a spacecraft potential, ϕ, were derived. The average is $v_r = (806 \pm 25)\,\mathrm{m\,s^{-1}}$ and $\phi = (-0.1 \pm 0.5)$ V, errors are standard deviation ($1\,\sigma$). The total uncertainty including possible systematic errors for the energy calibration is estimated as $\pm 50\,\mathrm{m\,s^{-1}}$ for v_r.

Outside the contact surface ion peaks are less well resolved due to an increase in temperature (see Sect. 4). A preliminary analysis conducted on the identifiable peaks corresponding to dominant constituents indicates that the radial bulk velocity of ions is $0 \pm 300\,\mathrm{m\,s^{-1}}$ at distances less than a few thousand kilometres from the contact surface. Ion flow velocity is apparently close to zero for major species (masses 18 and 19) in the distance range between about 5000 km and 20000 km (Balsiger et al., 1986; Schwenn et al., 1986) while the gas outflow continues (see below). Thus, a differential speed between ions and gas builds up outside the magnetic cavity.

The best estimate of the spacecraft potential in this regime is approximately 0 V with an uncertainty of several volts. This corrects the higher value given earlier (Krankowsky et al., 1986b) and is also comparable with measurements on other spacecraft: Vega 1 was charged to less than 1 V at 33000 km (Pedersen et al., 1986), and proton energies observed on Vega 2 (Gringauz et al., 1986) between about 15000 km and 25000 km indicate low spacecraft potential.

3.2. Gas velocities

Analysis of gas outflow can be extended to larger distances since the neutral component remains cold which allows to determine ram energies of individual species. Calibrations in the laboratory with beams of molecules, however, are not available with comparable accuracy as for ions. There is an offset of the neutral energy scale due to the electric potential distribution at the electron beam of the ion source. We assumed that, inside the contact surface, both ions and neutrals have the same outflow velocity. This hypothesis should be valid in a magnetic field-free and collision-dominated regime. In a model of the thermalized plasma in the inner coma Körösmezey et al. (1986) calculated ion flow velocities which up to a transition region at the ionopause are only slightly slower than the gas outflow. Thus, velocities inferred from water which is observed at sufficient intensity to distances of ~ 30000 km, are normalized to the average ion velocity of $800\,\mathrm{m\,s^{-1}}$ measured closer than 4000 km. The H_2O flow velocities plotted in Fig. 3 are each derived from single spectra up to ~ 9000 km or ~ 6000 km, respectively, in the 90 eV or 17 eV ionization modes. At greater distances where due to low signals peak shapes are disturbed by statistical fluctuations several spectra were summed for deriving velocity from an interval of distances. This method is more suited to low signals than averaging results computed from single spectra, therefore, scatter is considerably reduced at large distances compared to our earlier paper (Lämmerzahl et al., 1986). Expansion speed apparently increases to at least 30000 km where $\sim 1080\,\mathrm{m\,s^{-1}}$ was observed. Data from much beyond that distance are not shown because OH occurring on the adjacent mass at increasing relative abundance begins to interfere with the H_2O peak. Similar results are obtained for other gaseous species, but due to lower abundances the range of possible measurements is limited to about 10000 km.

The in situ measured expansion velocity of (0.80 ± 0.05) $\mathrm{km\,s^{-1}}$ in the inner coma can be compared to results from spectral analyses of comet Halley. From observations of HCN, Despois et al. (1986) found $0.85\,\mathrm{km\,s^{-1}}$ at 1.5 AU heliocentric distance, and Schloerb et al. (1986) gave an average of $(0.87 \pm 0.12)\,\mathrm{km\,s^{-1}}$ obtained from a series of measurements extending over a six months period which did not show a significant trend with heliocentric distance. An outflow velocity of $(1.4 \pm 0.2)\,\mathrm{km\,s^{-1}}$ inferred from H_2O observations on 22 March 1986 was reported by Larson et al. (1986). From the time delay of associated dust and water vapour jets observed on the Vega 2 flyby Krasnopolsky et al. (1986) deduced a gas velocity of $(0.85 \pm 0.15)\,\mathrm{km\,s^{-1}}$. Various hydrodynamic models arrive at outflow speeds near 10^3 km cometocentric distance which are similar to our result (Crovisier, 1984; Ip, 1986) or somewhat below (Marconi and Mendis, 1984; Gombosi et al., 1986), and most of them predict a further increase by a few hundred metres per second due to photolytic heating towards the outer boundary of the collision-influenced regime at or beyond 10^4 km.

4. Temperatures

For determining temperatures of either neutral or ion species an iterative least-square-analysis is made to represent a section of the energy spectrum after appropriate background correction. The ion population observed with the E-analyser has three groups of relatively abundant masses ranging from 12 to 19, 26 to 35, and 39 to 50 amu/e with only low intensities in between. It was practical to analyse these spectral ranges separately. From the neutral constituents only water (the dominant mass 18 peak together with masses 16 and 17) has been analysed.

The "theoretical" peak shape of a single mass is assumed to be a maxwellian energy distribution convoluted with an instrument

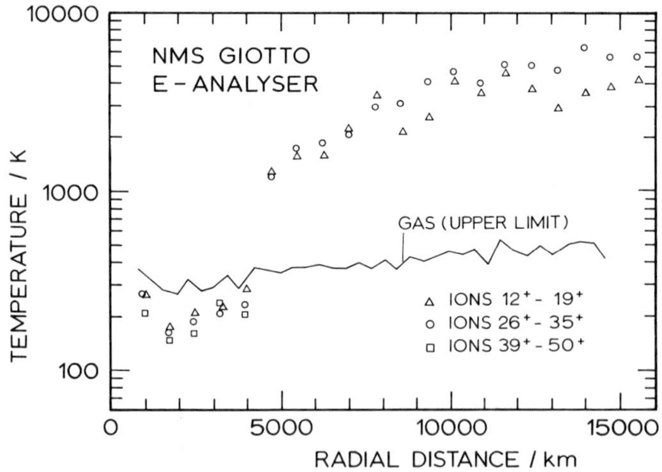

Fig. 4. Gas and ion temperatures inside and outside the contact surface. Ion temperatures were independently derived from ions grouped at different ranges of the energy spectra. Gas temperatures (from H_2O) should be considered as upper limits only, due to an experimental bias

function to account for the finite width of the entrance slit. The energy distributions of masses abundant in the respective section of the spectrum are summed. The result represents the intensity distribution on the front surface of the microchannel plate (MCP) detector. Further broadening is due to space charge spreading between the rear side of the MCP and the anodes, and to a minor extent to grid effects in front of the MCP. An empirical "detector function" is used which was calibrated with single peak laboratory spectra and which improves the fit of the peak at the 1% to 10% level. A convolution of the charge distribution at the front of the MCP with this detector function yields the charge distribution on the anodes and must be integrated over their area to represent the measured spectrum. To restrict the number of parameters all species in the spectral section are forced to have the same velocity and temperature, an assumption which is certainly justified inside the contact surface according to the observations discussed in Sect. 3. The fit provides relative mass abundances, temperature, and velocity. Due to least-square-error fitting, the derived temperature represents mostly that of the major species.

Figure 1a shows the result of such analysis for a cold plasma where ion mass peaks are resolved in the energy spectrum. Except for constituents that are below 1% of another one adjacent in mass, abundances inferred from energy spectra are generally in fair agreement with what is observed with the M-analyser. Temperature results are not very sensitive to varying background or minor masses. The same method is applied to ion energy spectra from outside the contact surface (Fig. 1b). Since individual masses are not resolved, composition is taken from the M-analyser.

Figure 4 shows the radial profile of ion temperatures which were independently derived from the three spectral ranges. Spatial resolution (distance between subsequent ion measurements) is about 380 km but we derived temperatures from every second ion spectrum only since those in between have major peaks in saturation. Markedly different temperatures characterize the plasma populations inside and outside the contact surface. From 1700 km to near the contact surface temperature increases from about 160 K to 250 K. Whether the apparent minimum is real needs further investigation; the spectrum near 1000 km was affected by high gas pressure in the analyzer. At the contact surface ion temperature rises by ~1000 K within less than 380 km (from nonsaturated minor peaks in the ion spectrum taken at 4370 km the temperature is estimated similarly low as closer to the comet), and farther out a steady increase reaching 4000–5000 K at 15000 km is observed. For the water group ions experimental uncertainty is estimated ±80 K at the cold temperatures and ±400 K at $T_i \sim 4000$ K, for heavier ions errors are larger by about 50%. Thus, results obtained from different spectral ranges are consistent within these errors for distances closer than 8000 km. At greater distances the temperature of ions 26^+–35^+ seems to be higher than that of the water group ions. Whether this offset is significant has to be further analysed. Of course, where the ion component is no longer thermalized due to less abundant collisions the assumptions made for deriving a temperature from the energy distribution of a whole group of ion species may become invalid.

The ion temperature profile in Fig. 4 shows the same radial structure as that from the IMS experiment (Balsiger et al., 1986; Schwenn et al., 1986), however, the NMS values are slightly lower. Inside the contact surface, in particular, ion temperatures are significantly below the 300 K reported by Schwenn et al. (1986) and are the lowest reported temperatures measured in situ.

In most models of the inner coma the ion and gas temperatures are assumed to be alike, except for hydrogen. Körösmezey et al. (1986) have shown that in the inner cometary ionosphere where ion-neutral coupling prevails the ion temperature can, however, exceed the temperatures of both neutrals and electrons. Measurements of the absolute gas temperature by analysing neutral energy spectra are of limited accuracy when the temperature is low, because of effects in the ion source of the instrument that cause peak broadening. When neglecting these effects temperatures of H_2O vapour as shown in Fig. 4 are obtained which apparently are high by at least 100 K when the ion temperature is considered as an upper bound to the gas temperature.

Marconi and Mendis (1986) calculated a gas temperature minimum of 64 K to occur at about 2300 km where UV photolytic heating balances expansion cooling. The lowest measured ion temperature almost agrees with this prediction within experimental errors. The measurements do not allow to tell if the gas temperature is below the ion temperature inside the contact surface. The relative variation of the gas temperature should be reliable, however, and there is an increase by about 200 K observed over the range of distances from 1000 km to 15000 km. Somewhat different rates of temperature increase have been calculated in recent coma models (Marconi and Mendis, 1986; Ip, 1986).

The abrupt rise of the ion temperature at the contact surface shows that right outside the discontinuity another energy source becomes effective. Where the magnetic field stops ion expansion, neutral-ion friction and mass loading contribute to heating the plasma. Observed ion temperatures have been used in a model calculation by Haerendel (1987) who from the temperature gradient inferred an ion inflow at velocities of $-1\ \mathrm{km\,s^{-1}}$ to $-2\ \mathrm{km\,s^{-1}}$ near 15000 km distance dropping to zero at the contact surface. Such high inflow speeds, however, were not observed (see also Schwenn et al., 1986). In order to explain the ion temperature profile, one probably must conclude that the differential speed between ions and neutrals is not the only important heat source of the plasma in this regime.

Acknowledgements. This research was supported by the German Bundesminister für Forschung und Technologie (grants 010F0519, 010F0628, and 010F85120); the Swiss National Science Foundation (grants 2.034-081 – 2.825.85); CNES (grants 81, 82/244, 83/211, and 84, 84/1211) and NASA Headquarters (contract NASW-3576).

References

Balsiger, H., Altwegg, K., Bühler, F., Geiss, J., Ghielmetti, A.G., Goldstein, B.E., Goldstein, R., Huntress, W.T., Ip, W.-H., Lazarus, A.J., Meier, A., Neugebauer, M., Rettenmund, U., Rosenbauer, H., Schwenn, R., Sharp, R.D., Shelley, E.G., Ungstrup, E., Young, D.T.: 1986, *Nature* **321**, 330

Crovisier, J.: 1984, *Astron. Astrophys.* **130**, 361

Despois, D., Crovisier, J., Bockelée-Morvan, D., Schraml, J., Forveille, T., Gérard, E.: 1986, *Astron. Astrophys.* **160**, L11

Ghielmetti, A.G., Balsiger, H., Bänninger, R., Eberhardt, P., Geiss, J., Young, D.T.: 1983, *Rev. Sci. Instrum.* **54**, 425

Gombosi, T.I., Nagy, A.F., Cravens, T.E.: 1986, *Rev. Geophys.* **24**, 667

Gringauz, K.I., Verigin, M.I., Richter, A.K., Gombosi, T.I., Szegö, K., Tátrallyay, M., Rezimov, A.P., Apáthy, I.: 1986, 20th ESLAB Symp., ESA SP-250, Vol. **I**, p. 93

Haerendel, G.: 1987, *Geophys. Res. Letters* (submitted)

Ip, W.-H.: 1986, *Adv. Space Res.* **5**, 233

Körösmezey, A., Cravens, T.E., Gombosi, T.I., Nagy, A.F., Mendis, D.A., Szegö, K., Gribov, B.E., Sagdeev, R.Z., Shapiro , V.D., Shevchenko, V.I.: 1986, 20th ESLAB Symp., ESA SP-250, Vol. **I**, p. 235

Krankowsky, D., Lämmerzahl, P., Dörflinger, D., Herrwerth, I., Stubbemann, U., Woweries, J., Eberhardt, P., Dolder, U., Fischer, J., Herrmann, U., Hofstetter, H., Jungck, M., Meier, F.O., Schulte, W., Berthelier, J.J., Illiano, J.M., Godefroy, M., Gogly, G., Thévenet, P., Hoffmann, J.H., Hodges, R.R., Wright, W.W.: 1986a, ESA SP-1077, 109

Krankowsky, D., Lämmerzahl, P., Herrwerth, I., Woweries, J., Eberhardt, P., Dolder, U., Herrmann, U., Schulte, W., Berthelier, J.J., Illiano, J.M., Hodges, R.R., Hoffman, J.H.: 1986b, *Nature* **312**, 326

Krasnopolsky, V.A., Moroz, V.I., Gogoshev, M., Gogosheva, Ts., Moreels, G., Clairemidi, J., Krysko, A.A., Jegulev, V.S., Parisot, J.P., Tkachuk, A.Y.: 1986, 20th ESLAB Symp., ESA SP-250, Vol. **I**, p. 459

Lämmerzahl, P., Krankowsky, D., Hodges, R.R., Stubbemann, U., Woweries, J., Herrwerth, I., Berthelier, J.J., Illiano, J.M., Eberhardt, P., Dolder, U., Schulte, W., Hoffman, J.H.: 1986, 20th ESLAB Symp., ESA SP-250, Vol. **I**, p. 179

Larson, H.P., Davis, D.S., Mumma, M.J., Weaver, H.A.: 1986, *Astrophys. J.* **309**, L95

Marconi, M.L., Mendis, D.A.: 1984, *Astrophys. J.* **287**, 445

Marconi, M.L., Mendis, D.A.: 1986, *Earth Moon Planets* **36**, 249

Neubauer, F.M., Glassmeier, K.H., Pohl, M., Raeder, J., Acuna, M.H., Burlaga, L.F., Ness, N.F., Musmann, G., Mariani, F., Wallis, M.K., Ungstrup, E., Schmidt, H.U.: *Nature* **321**, 352

Pedersen, A., Grard, R., Trotignon, J.G., Béghin, C., Mikhailov, M., Mogilevsky, Y.: 1986, 20th ESLAB Symp., ESA SP-250, Vol. **III**, p. 425

Schloerb, F.P., Kinzel, W.M., Swade, D.A., Irvine, W.M.: 1986, *Astrophys. J.* **310**, L55

Schwenn, R., Ip, W.-H., Rosenbauer, H., Bühler, F., Goldstein, R., Meier, A., Shelley, E.G.: 1986, 20th ESLAB Symp., ESA SP-250, Vol. **I**, p. 225

Astron. Astrophys. 187, 174–178 (1987)

ASTRONOMY
AND
ASTROPHYSICS

Giotto-IMS observations of ion-flow velocities and temperatures outside the magnetic cavity of comet P/Halley

B. E. Goldstein [1], M. Neugebauer [1], H. Balsiger [2], J. Drake [3], S. A. Fuselier [3], R. Goldstein [1], W.-H. Ip [4], U. Rettenmund [2], H. Rosenbauer [4], R. Schwenn [4], and E. G. Shelley [3]

[1] Jet Propulsion Laboratory, California Institute of Technology, Pasadena, CA 91125, USA
[2] Physikalisches Institut, University of Bern, Bern, Switzerland
[3] Lockheed Palo Alto Research Laboratory, Palo Alto, CA, USA
[4] Max-Planck-Institut für Aeronomie, Katlenburg-Lindau, Federal Republic of Germany

Received February 18, accepted June 2, 1987

Summary. Fluid parameters for He^{2+} ions obtained from the Giotto Ion Mass Spectrometer HERS sensor are presented; also investigated are proton densities and velocities and thermal speeds of protons, alpha particles, and heavy ions in the hour before closest approach. A bow shock transition lasting ten minutes is observed. A region of enhanced He^{2+} ion densities, and velocity, and decreased temperature is observed from 20:26 to 21:45. Sharp decreases in the proton density are observed at 23:30 and at 23:41; there are also sharp drops in alpha particle density and temperature at 23:30. There is a relative flow velocity between alpha particles and oxygen ions of approximately 20 km s^{-1} during a period roughly from 22:55 to 23:10; by 23:30 the difference in flow velocity is less than the experimental uncertainities.

Key words: Halley – Giotto – IMS – ion dynamics

1. Introduction

The High Energy Range Spectrometer (HERS) of the Giotto Ion Mass Spectrometer (IMS) measured positive ions for energies up to about 4 keV, depending on the mass being observed and the view direction. The instrument had a fan-like field of view that rotated about the spin axis; the inner edge (the edge towards the comet direction) of the fan was 15° from the spin axis, while the outer edge was 75° from the spin axis. The instrument has 4 modes (proton, light, medium, and heavy); each mode is measured once every four spins, so that measurements of the different species are obtained every 16 s. Further description of the experiment is provided in Balsiger et al. (1986a); first results are reported in Balsiger et al. (1986b). The spacecraft approached the comet at a speed of 68.4 km s^{-1}, with an angle between the spacecraft velocity in the comet frame and the sun-comet line of 107°, and with closest approach occurring at 00:03 UT on March 14. In the first part of this paper, the large scale features of the solar-wind interaction with the cometary plasma as determined from the HERS alpha particle observations will be discussed. The hour before closest approach will then be examined using alpha particle data, proton data, and heavy ion data. Only data obtained outside the magnetic cavity region (Neubauer et al., 1986) will be discussed.

Send offprint requests to: B. E. Goldstein

2. Ion flow parameters

Fluid parameters (density, velocity, thermal speed, elevation angle, and azimuth angle) of the flow were obtained for alpha particles by a moment method: each observation of an ion count was assumed to represent a contribution from the plasma distribution function at a location in velocity space (equivalently, energy and angular direction) corresponding to the center of the observing bin. The HERS sensor has energy windows that, for all ions other than protons, are of constant width in momentum (rather than constant width in energy as is the case for a conventional electrostatic analyzer). The consequence is that, although good energy resolution is obtained for heavy ions, the energy window for alpha particles is 64 eV full width at half maximum, and for protons is twice as wide. Two problems result when the moment method of analysis is used: 1) the thermal broadening of the distribution along the line of sight is overestimated, and 2) for low proton and alpha particle velocities, the bulk flow speed is overestimated due to the low energy cutoff of the instrument not being correctly taken into account. Therefore, for protons and alpha particles an alternative method of analyzing the data is also used which fits the observed instrument counts with the counts modelled for the instrument response to a parametrized Maxwell-Boltzmann distribution. Roughly speaking, the moment method over estimates the alpha velocity by about 20 to 30 km s^{-1} after 23:00, and before 22:30 the errors are not great. Additionally, the alpha particle estimates are subject to statistical uncertainties; this causes primarily random scatter in thermal speed estimates which is more noticeable in low density regions.

Alpha particle data are useful for examining the solar wind interaction because they are less affected by the addition of pick-up ions (H_2^+) to the flow than are protons. Tha data plotted at 16 seconds intervals in Fig. 1 are alpha particle velocity components, density, and thermal speed; velocity is referenced to the cometary rest frame. Thermal speed, w, is defined such that $w^2 = 2kT/m$, where k is the Boltzmann constant, T is the temperature, and m is the particle mass. The estimates shown

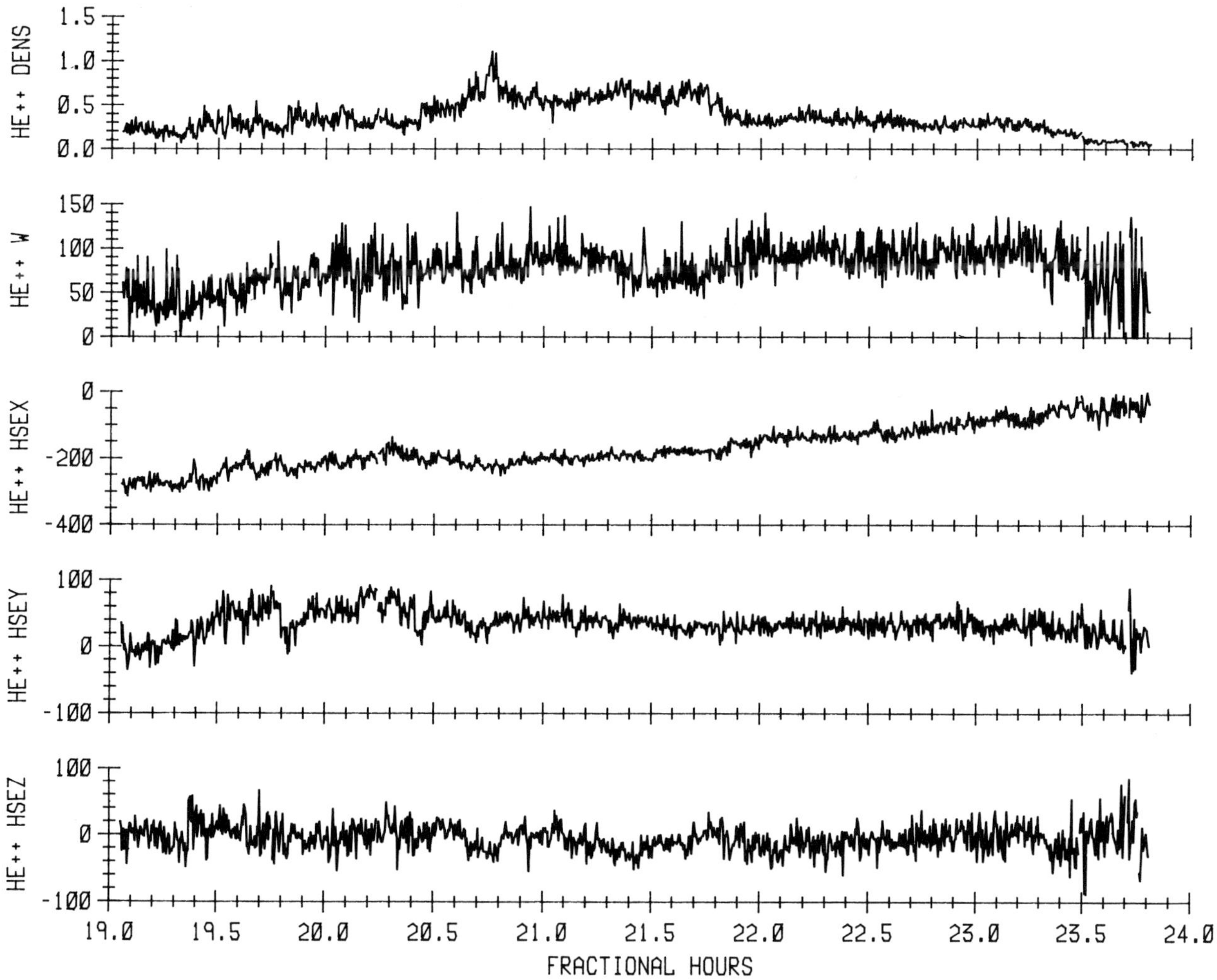

Fig. 1. Velocity components in the comet frame of reference, number densities, and thermal speeds of He^{2+} ions observed from 19:00 UT to 24:00 UT by the Giotto IMS-HERS sensor. Velocitiy components are in Halley solar-ecliptic coordinates and are measured in the km s^{-1}, number densities are per cc. Time is spacecraft event time

were obtained from the moment method. The velocities are in a Halley solar ecliptic coordinate system in which x points from the comet to the sun, z points approximately northward, and y points opposite to the direction of planetary rotation and is parallel to the ecliptic.

The bow shock is encountered from about 19:22 UT to 19:38 UT. The leading edge of the shock is coincident with the location of the shock as observed by the magnetometer group (Neubauer et al., 1986). Balsiger et al. (1986b) stated that the highest time-resolution proton data (1 spectrum every 16 sec) suggested multiple crossings of the shock. Reexamination of these data does no support this idea; it seems that fluctuations downstream of the shock were being observed. The shock structure appears to have different thicknesses in different parameters. For example, there is a sharp increase of the z velocity component from about 19:22 UT to 19:23 UT, while the x and y velocity components change gradually over the period from 19:22 UT to 19:38 UT. The scale for rise in alpha temperatures appears to be even longer, with the end of the gradient in the alpha temperature at about 19:44. The structure of the shock in both the alpha and proton velocity data is similar.

Inside the bow shock is a region of elevated alpha particle densities. The outer boundary of this region is at about 20:26 UT (about 900000 km from closest approach) where a sharp increase in alpha particle density is observed; simultaneously, the alpha particle velocity increases modestly. This increase in alpha particle density is coincident with an increase in the temperature of the solar wind electrons (d'Uston et al., 1986). Then, at about 20:37 there is another sharp increase leading to a peak of the density at about 20:45. The density declines from this peak, but maintains elevated values until about 21:45) 560000 km from closest approach, at which time the density decreases by a factor of two during a period of five minutes. The alpha particle temperature throughout this region appears generally to be anticorrelated with density. Furthermore, the density peaks in this region appear to be correlated with the electron density in the 170 to 3800 eV range (d'Uston et al., 1986); the electron and alpha temperatures are thus anticorrelated.

The region of elevated alpha particle densities is exited at about 21:50. The alpha particle thermal speed increases to values of about 100 km s^{-1}; this is higher than the post shock value of about 90 km s^{-1}. After 21:50, the lower density of the solar wind,

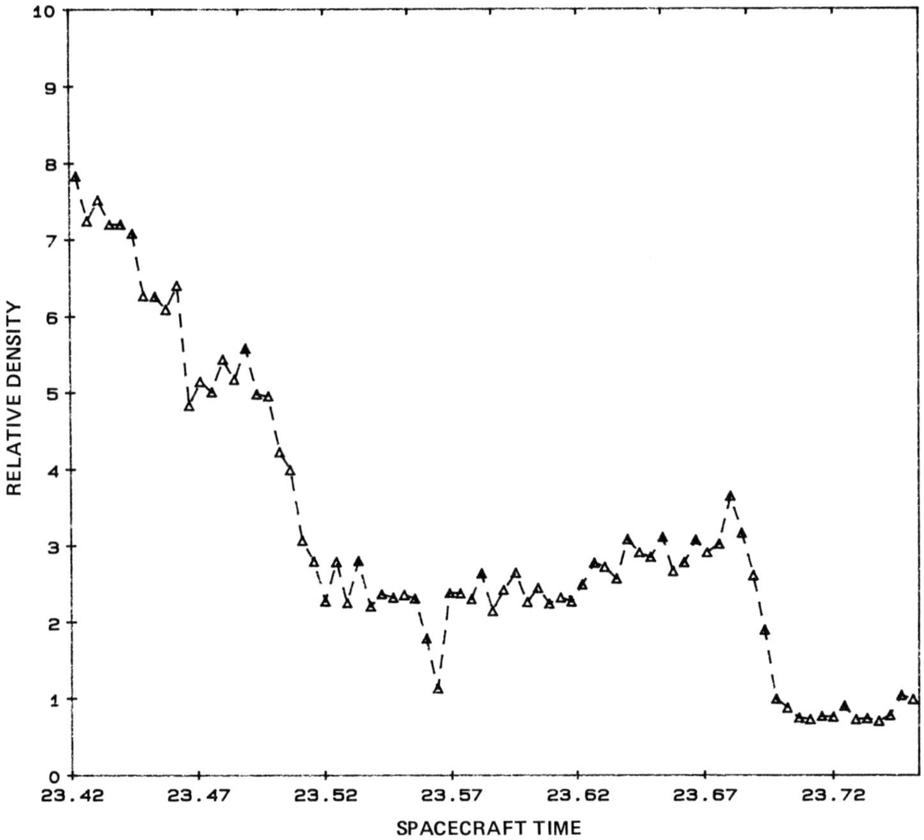

Fig. 2. Relative proton densities from 23:25 to 23:45 estimated by the moment method from data obtained by the Giotto IMS-HERS sensor

and the more rapid addition of heavy ions to the flow results in steeper gradient in the flow velocity; the alphas are no longer necessarily moving with the mass averaged flow speed, and pitch angle scattering might heat the alpha distribution. Adiabatic compression due to slowing of the flow will also heat the alpha particles. An increase in the alpha density might not occur because of charge exchange with cometary neutrals.

Further interesting structure is seen as the comet is approached. Shown in Fig. 2 is the relative proton density at the maximum time resolution (once every four spins, or 16 s per data point) for the period 23:25 to 23:45. Two features of note are apparent. First, although the proton density is generally decreasing during this time period, two periods of rapid decrease which might be characterized as discontinuities are observed. The first of these decreases is centered at about 23:30 (about 135000 km from the comet), and the second occurs at about 23:41 (about 900000 km from the comet). The outermost discontinuity (whether it is a discontinuity or a rapid decrease depends on whether the oxygen or proton gyroradius is taken as the discontinuity scale size) corresponds to the boundary of the pile-up region as observed by the Giotto magnetometer (Neubauer et al., 1986); the inner boundary occurs in a region of rapidly changing magnetic field orientation. At the outer boundary (135000 km), sharp decreases in the alpha particle density, thermal speed, and velocity components are also observed (Fig. 1). The boundary at 135000 km would apparently correspond to the "cometopause" reported by Gringauz et al. (1986). It should be noted that the discontinuity in the magnetic field upon exciting the pile-up region "was much less clearly defined" (Neubauer et al., 1986).

Ten minute averages of HERS ion speed in the spacecraft frame for masses 1, 2, 12, 16, 17, 18, and 28 are shown for the period 23:00 to 24:00 in Fig. 3. The proton and $m/q=2$ speeds have been obtained by the simulation method; other masses were obtained from the moment method. The mass ranges of the medium and heavy modes of the instrument overlap for the water group ions (Balsiger et al., 1986a). Therefore, the estimates for masses 16, 17, and 18 have been obtained by averaging the values obtained for these two modes.

The speed of each species is uncertain to the extent that a portion of each species may be outside the field of view or energy range of the HERS detector, or are not extrapolated well by the simulation method. The accuracy of the simulation method for investigating alphas and, especially, protons, is currently not well understood. Ten minute averages of thermal speeds for the various species during the last hour are shown in Fig. 4; the calculation and methods of display are the same as for Fig. 3. The apparent large speed of H_2O^+ at the beginning of the hour is due to noise and counting statistics. For the ions of mass 12 and greater, the count rates at 23:05 are dominated by masses 16 and 17, so these species provide the most statistically reliable velocity estimates. Ignoring the point at 23:05, the speeds of all species seem to be in good agreement with the exception of the $m/q=2$ particles. The $m/q=2$ particles observed before 23:22 appear to have a speed about 20 km s^{-1} greater than the other species; by 23:27 this velocity difference became much smaller or vanished. The velocity differences between the alpha particles and heavy ions before 23:20 are real; the newly added, low velocity heavy ions (see Figs. 5 and 6 in Goldstein et al., 1986) have not been pitch angle scattered into a shell, and thus have a lower velocity than the solar wind ions.

The presence of H_2^+ should not affect this conclusion; H_2^+ ions require about 30 eV less energy than He^{2+} to pass through the

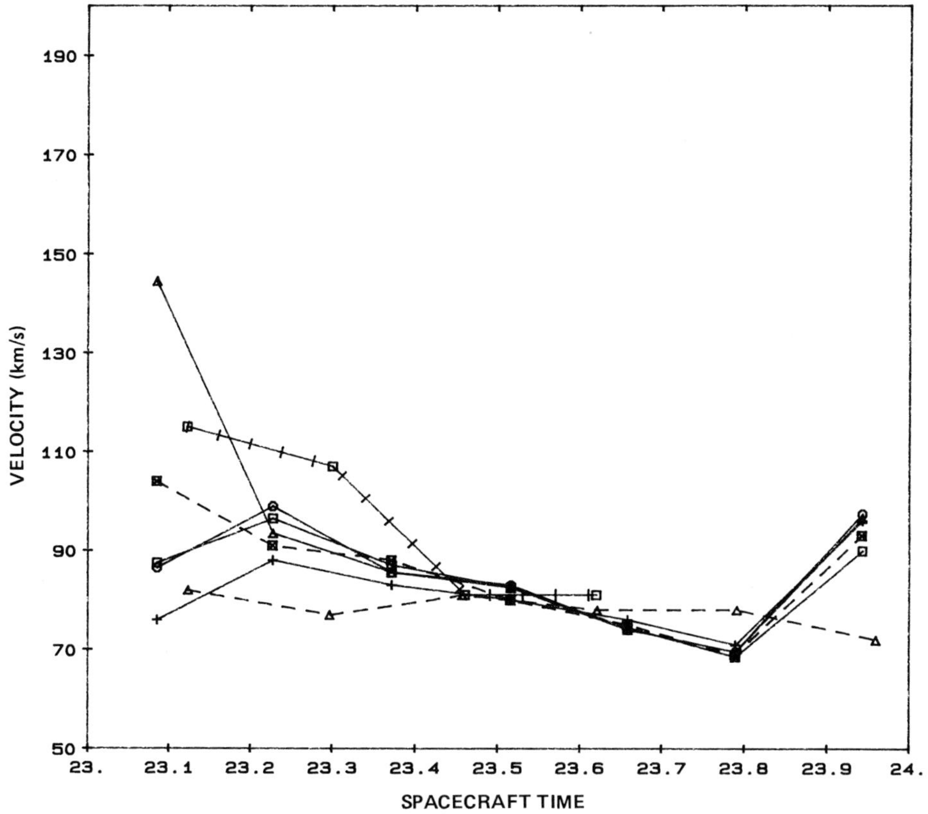

Fig. 3. Ten minute averages of bulk speeds of ions in the spacecraft frame observed from 23:00 to 24:00. The line with tic marks is $m/q = 2$, the dashed line with boxes is mass 12, the dashed line with triangles is protons, the solid line with triangles is mass 18, the solid line with circles is mass 17, the solid line with boxes is mass 16, the solid line with pluses is mass 28

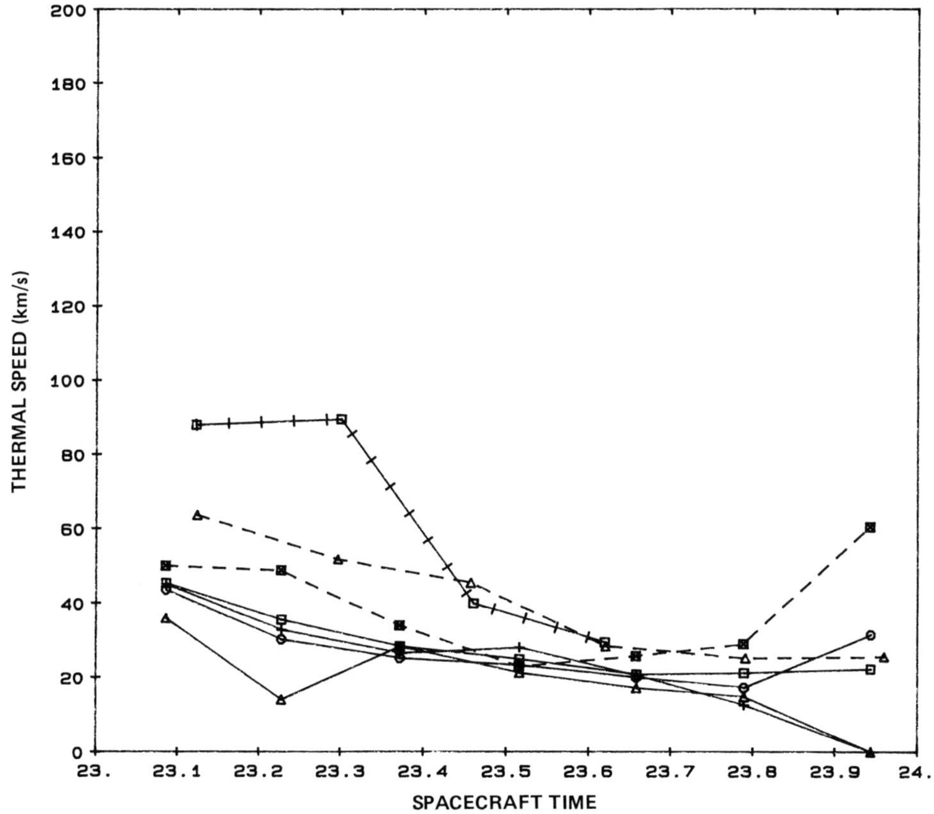

Fig. 4. Ten minute averages of thermal speeds of ions; line coding is the same as Fig. 3

HERS magnet. Balsiger et al. (1986b, see Fig. 7) find that for the period from 23:36 to 23:55 UT, roughly 30 % of the He^{2+} has charged exchanged to He^+. Hence, H_2^+ should not cause significant error in speed estimates before 23:30.

The thermal velocities of the various ion species are shown in Fig. 4. For a collisionless plasma, one might expect all pick-up ions to have similar velocity distributions in either strong-scattering or no scattering cases, and ions of pick-up origin would have different thermal speeds than solar wind ions. All species shown in Fig. 4 are rougly comparable in thermal speed and share a similar behaviour in time with the exception of the $m/q = 2$ ions. The mass 12 data (C^+) show thermal speeds slightly higher than those of the other heavy ions.

Ion-neutral collisions will be important close to the comet. If newly created ions are not pitch angle scattered into the view of the HERS sensor, the two time scales for the high energy ions are: the flow time, $\pi/2 \cdot r/V_i$, and the ion neutral collision time $1/(n_n \sigma V_i)$. Taking $\sigma = 5\ 10^{-15}\ \mathrm{cm}^2$, these processes become roughly comparable at $\sim$40000 km ($\sim$23:50 UT). Coulomb collisions with neutral H_2O would decelerate lighter ions more rapidly, but the ion that would thermalize most rapidly would be H_2O^+, with mass equal to that of the neutral water. The velocity data in Fig. 3 do not indicate a more rapid deceleration of carbon than the water group. For this reason, there is no convincing evidence for collisional effects; a more thorough investigation is needed. The last data point in Figs. 3 and 4 may be affected by instrumental noise; this needs further investigation.

Acknowledgements. The portion of this work conducted at the Jet Propulsion Laboratory, California Institute of Technology, was supported by the National Aeronautics and Space Administration, the German Bundesministerium für Forschung und Technologie, and the Swiss National Science Foundation.

References

Balsiger, H., Altwegg, K., Buhler, F., Fischer, J., Geiss, J., Meier, A., Rettenmund, U., Rosenbauer, H. Schwenn, R., Benson, J., Hemmerich, P., Sager, K., Kulzer, G., Neugebauer, M., Goldstein, B. E., Goldstein, R., Shelley, E. G., Sanders, T., Simpson, D., Lazarus, A. J. Young, D. T.: 1986a, in *The Giotto Mission – Its Scientific Investigations*, ESA SP-1077, eds. R. Reinhard, B. Battrick, p. 129

Balsiger, H., Altwegg, K., Buhler, F., Geiss, J., Ghielmetti, A. G., Goldstein, B. E., Goldstein, R., Huntress, W. T., Ip, W.-H., Lazarus, A. J., Meier, A., Neugebauer, M., Rettenmund, U., Rosenbauer, H., Schwenn, R., Sharp, R. D., Shelley, E. G., Ungstrup, E., Young, D. T.: 1986b, *Nature* **321**, 330

d'Uston, C., Rème, H., Sauvaud, J. A., Cros, A., Anderson, K. A., Carlson, C. W., Curtis, D., Lin, R. P., Korth, A., Richter, A. K., Mendis, A.: 1986, *20th ESLAB Symp.*, ESA SP-250, Vol. **I**, p. 77

Goldstein, B. E., Neugebauer, M., Balsiger, H., Drake, J., Fuselier, S. A., Goldstein, R., Ip, W.-H., Rettenmund, U., Rosenbauer, H., Schwenn, R., Shelley, E. G.: 1986, *20th ESLAB Symp.*, ESA SP-250, Vol. **I**, p. 224

Gringauz, K. I., Gombosi, T. I., Remizor, A. P., Apathy, I., Szemery, I., Verigin, M. I., Denchikova, L., I., Dyachor, A. V., Kepopler, E., Klimenko, I. N., Richter, A. K., Somogyi, A. J., Szegok, Szendro, S., Tatrallyay, M., Varga, A., Vladimova, G. A.: 1986, *Nature* **321**, 282

Neubauer, F. M., Glassmeir, K. H., Pohl, M., Raeder, J., Acuna, M. H., Burlaga, L. F., Ness, N. F., Musmann, G., Mariani, F., Wallis, M. K., Ungstrup, E., Schmidt, H. U.: 1986, *Nature* **321**, 352

Astron. Astrophys. 187, 179–182 (1987)

ASTRONOMY AND ASTROPHYSICS

The dependence of mass resolution and sensitivity of the PUMA instrument on the energy spread of ions produced by hypervelocity impacts

R. Z. Sagdeev [1], J. Kissel [2], E. N. Evlanov [1], M. N. Fomenkova [1], N. A. Inogamov [1], V. N. Khromov [1], G. G. Managadze [1], O. F. Prilutski [1], V. D. Shapiro [1], I. Y. Shutyaev [1], and B. V. Zubkov [1]

[1] Space Research Institute, Profsoyuznaya 84/32, SU-117810 Moscow GSP-7, USSR
[2] Max-Planck-Institut für Kernphysik, PO Box 103980, D-6900 Heidelberg 1, Federal Republic of Germany

Received February 9, accepted July 20, 1987

Summary. Measurements of the element composition of the dust particles in Comet Halley were made by the PUMA-1 and PUMA-2 instruments onboard the VEGA spacecraft. The time-of-flight analysis of ions produced by the hypervelocity particle impact on the target was applied with a wide ($W = 100$ eV) and a narrow ($W = 20$ eV) energy window of the instruments. The analysis of measured spectra shows that the mass resolution for "small" particles ($Q < 10^{-13}$ C) was high enough in both modes to separate individual mass lines. For large particles the resolution did not change for the narrow window, it decreased, however, for the wide one (to about 50). A relative transmission Q_w/Q_n dependence on the mass number for the wide and narrow window made has been found.

Key words: dust composition – comet Halley

1. Introduction

One of the main goals of the VEGA project was to obtain data on the element composition of the dust particle. For this purpose the PUMA instrument was built (Sagdeev et al., 1985; Kissel et al., 1986). In the PUMA instrument the time-of-flight analysis of ions produced by the hypervelocity impact of dust particle on the target was applied to measure the element composition of the dust. The schematics of the PUMA analyser are shown in Fig. 1. The hypervelocity dust particle impact initiates the ionization of the dust and target atoms. Excited atoms and ions generate a light flash measured by the photomultiplier (Fig. 1.4). The ions formed on the target (Fig. 1.1) are accelerated by an electrostatic potential between the target and accelerating grid (3), then they move along the drift tube and are reflected by the electrostatic mirror (reflector, Fig. 1.6) to the second drift tube. Finally, ions are registered by the secondary electron multiplier (SEM). To determine the ion start time from the target the method of coincidence is used for signals from the target, accelerating grid and photomultiplier. The amplitude of the signals from those electrodes depends on the total number of ions produced and characterizes the dust particle mass. For discriminator levels at the accelerating grid (AC) signal 1H, 3H, 7H the total number of ions is estimated to be 5 10^5, 5 10^6, 5 10^7, respectively.

Send offprint requests to: E. N. Evlanov

A structured target was used in PUMA-1 and a flat target in PUMA-2 (Fig. 1B). The use of both a wide and narrow energy window permitted to improve the experiment characteristics and yielded an optimized inflight functioning of the experiment over the full mass range. In the PUMA instrument the energy window is defined by the potential difference between the reflector (upper limit for the ion energy) and the retarding grid (Fig. 1.7, lower limit). The retarding grid is located near the SEM collector where the plasma effects are negligible. The interest of a narrow energy window mode for ion detection in the PUMA experiment was demonstrated in laboratory laser simulations high velocity impacts (Sagdeev et al., 1987). Plasma clouds were generated in this case with a YAG-laser (532 nm, 20 ns, 3 mJ). The dependence of the mass resolution for silver doublet (107 and 109 a.m.u.) on ion number in the cloud (an induced charge on the accelerating grid of PUMA) is shown in Fig. 2; the resolution was decreased from 300 ($N = 10^6$ ions) to 30 ($N = 10^8$). The decrease of resolution was

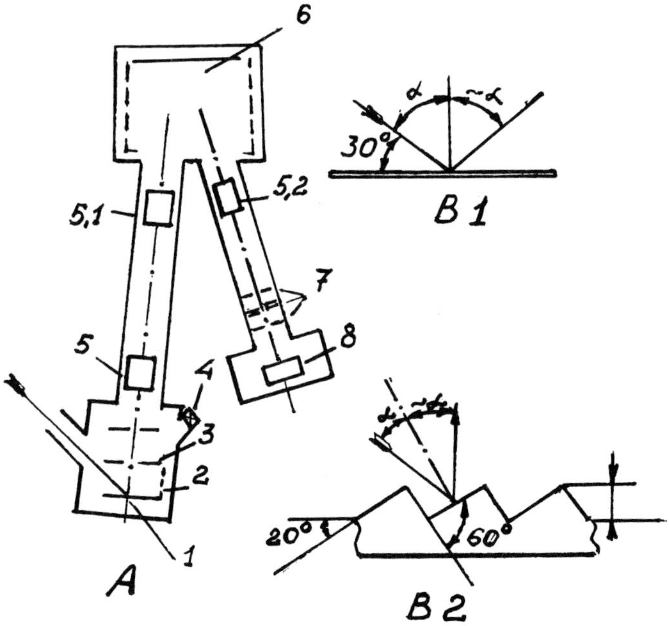

Fig. 1. PUMA mass-analyser *1* target, *2* catcher, *3* accelerating grid, *4* PM (photomultiplier), *5* ion lenses, *6* reflector, *7* retarding grid, *8* SEM (secondary electron multiplier)

mainly due to plasma screening of the accelerating electric field, which caused the spread of the ion start times.

This was confirmed by a hydrodynamical model (Sagdeev et al., 1983) of the ion cloud expansion in the external electric field ($T_e = 10$ eV, $N = 10^8$–10^{10}). An incomplete electric field penetration into the expanding plasma cloud allowed acceleration only for a small fraction of the ions in the boundary region of the cloud. When the particle concentration in the plasma cloud decreases below a critical level, the electric field penetrates into the cloud and the screening can be neglected. The main consequence of the plasma screening is an increase of the duration of the ion pulse.

Due to this effect the resolution $R = T/2DT$ is about 20 for $N = 10^{10}$. This value is approximately the same ($R = 30$ for $N = 10^8$) as that found in the laser simulations. There is a time dependence of energy of the ions in the current pulse. Ions at the front edge of the current pulse have more energy than the acceleration potential difference, while ions in the tail of the pulse have less energy. Therefore a narrow energy acceptance window reduces the spread of effective ion start times. It is necessary to choose the optimum energy and the energy spectrum width for the detected ions. The median energy of the window was selected (corresponding to the target potential) to make sure that it lies in the energy range of ions coming from the target. The energy window width was selected to optimize transmission and resolution.

From Fig. 2 (A-laser simulation) it is clear that in the narrow energy window the mass resolution is high enough ($R = 200$ for $N = 10^{10}$ ions) to separate individual mass lines over the full mass range; in the wide energy window $R = 30$. This resolution is still adequate to separate lines from the major elements H, C, N, O, Mg, Si, S, Ca, Fe, but not high enough to separate isotopes. There is a drop of sensitivity for the narrow window (about 10 times).

Based on the results of the laser simulation of plasma effects the following potential values were chosen:

a) Wide energy window.

U(target) = 1020 V
U(reflector) = 1100 V
U(retarding grid) < 1000V

b) Narrow energy window:

U(target) = 1020 V
U(reflector) = 1020 V
U(retarding grid) = 1000 V

During the fly-bys, the window was switched from narrow to wide every 30 s.

The optimum configuration of the target for PUMA-1 was defined by a computer simulation (Sagdeev et al., 1985). In the case of an inclined impact of a dust particle on the target the ion cloud is emitted at an angle close to the specular reflection angle and a major fraction of the ions cannot reach the detector. This leads to a reduction of the transmission and sensitivity (for silicon the transmission is equal to 0.01%; Sagdeev et al., 1985).

Transmission depends on the mass number A and is defined as $T = T_0/A$ (Sagdeev et al., 1985) ($T = 10\%$ for hydrogen and $T = 0.18\%$ for iron).

The implementation of a structured target can strongly increase the transmission. The optimum result (according to calculations for the transmission by Sagdeev et al., (1985) is achieved for a structured target with an angle near 30 degrees (see Fig. 1). For silicon the transmission is increased to ~10%. The geometrical parameters of the target for PUMA-1 are shown in

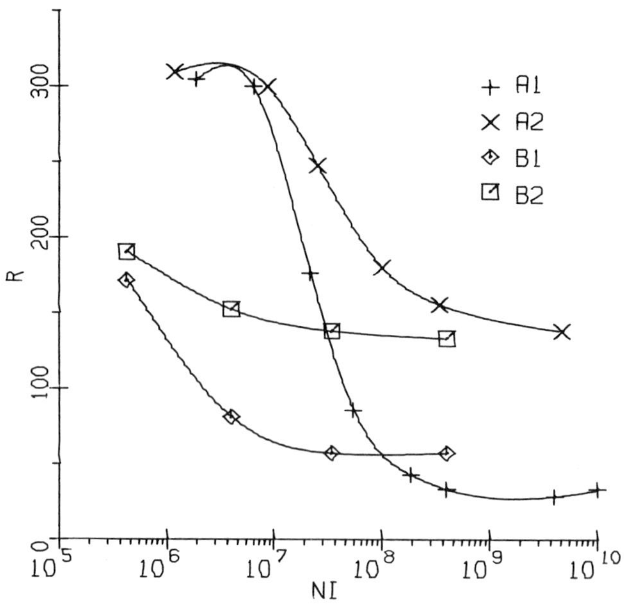

Fig. 2. Mass resolution for silver ions in calibration and cometary spectra (N: target ion yield, A: laser calibration, B: comet measurements, 1: wide and 2: narrow window)

Fig. 1 (the structure angle is close to 20 degrees and the structure height is 0.5 mm). Unfortunately a comparison of the efficiency of the PUMA-1 and PUMA-2 targets was not possible due to the decrease of the high voltage at the multiplier to 2400 V (instead of 3300V).

2. Results of in-flight measurements

During the fly-by, about 10 000 spectra of cometary dust particles for PUMA-1 and about 1000 spectra PUMA-2 were obtained. Only mode 0 spectra (ADC sampling at 15 MHz) can be used to estimate the mass resolution. The mass resolution for the silver doublet as a function of the total number of ions (charge on the acceleration grid) entering the analyser is shown in Fig. 2B. With the wide energy window, the resolution for silver ions decreases strongly for $N = 10^7$ ($R < 50$). The same effect was observed for $N = 10^8$ with the laser simulation. Resolution does not change markedly (from 180 to 140) for a range of N from 10^6 to 10^{10} (see Fig. 2B for the narrow window).

The average resolution as a function of element mass number is represented in Fig. 3. For a small dust particle ($N < 10^5$) we have an increase of resolution with the mass number in both regimes (wide and narrow energy windows). This suggests a fixed spread of ion start times in that case. For the wide energy window and $N > 10^6$ there is no actual dependence of the resolution on the mass number of the element. This results from plasma effects, since the spread of the start times will increase with mass number (Sagdeev et al., 1983).

It is difficult to make an exact comparison for the transmission of the narrow and wide windows since we have no spectra of two identical dust particles. To estimate the relative transmission of the wide and narrow energy windows, we take the ratio $j = Q_w/Q_n$ (Q_w and Q_n – average charge amplitude logarithm of mass lines for the wide and narrow windows). Values Q_w, Q_n were

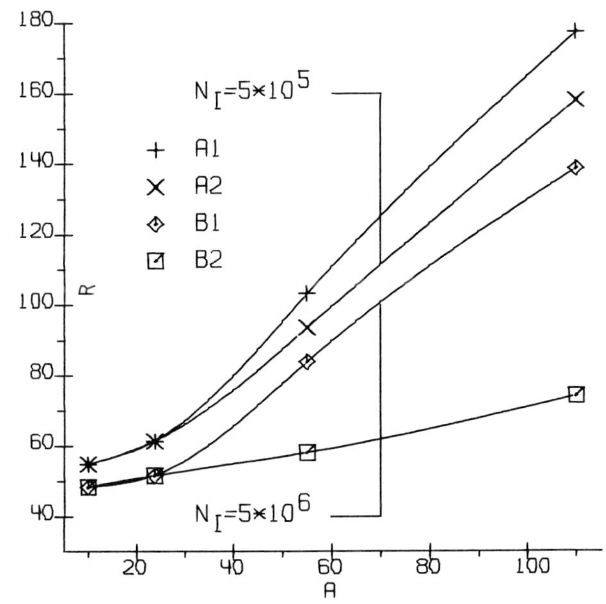

Fig. 3. Dependence of mass resolution on the mass number for the wide (A_2, B_2) and narrow (A_1, B_1) windows for two levels of the accelerating grid signal ($N = 5\ 10^5$ ions and $N = 5\ 10^6$)

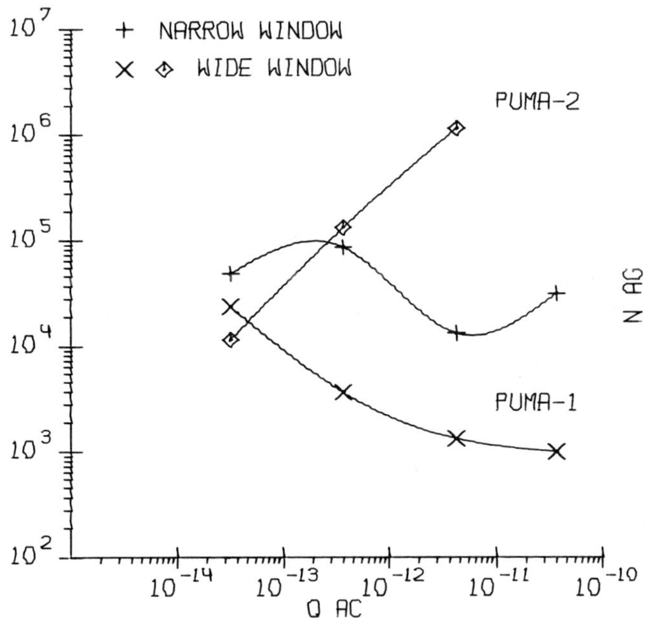

Fig. 5. Yield of silver ions as a function of accelerating grid signal for PUMA-1 and PUMA-2

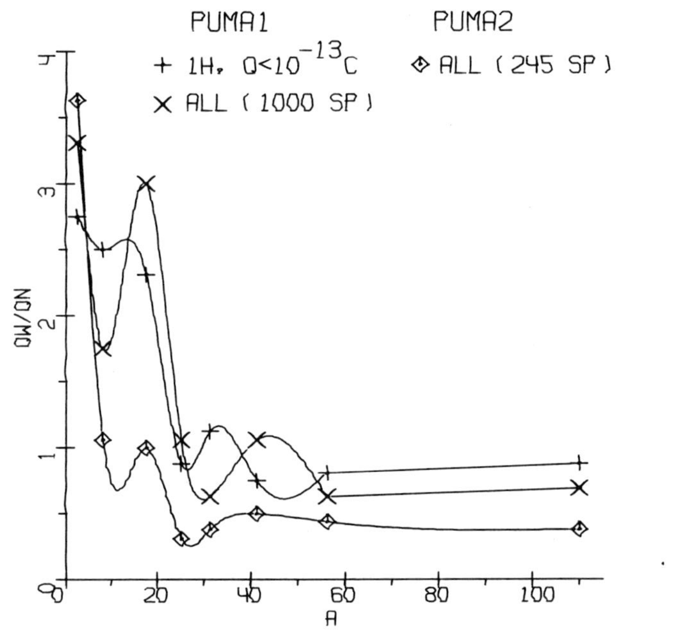

Fig. 4. Ratio of transmission for wide and narrow energy window modes as a function of mass number for small and large particles

Table 1

Window	Wide			Narrow		
AC signal	1H	3H	7H	1H	3H	7H
Number of pulses	11.1	4.0	0.9	20.8	21.7	24.4
Average amplitude	3.4	1.5	0.4	12.5	13.3	18.0

taken from the integral distribution of telemetry amplitudes over a large number of spectra:

$j = 10\, A_{tel}/25$ for PUMA-1 and
$j = 10\, A_{tel}/25$ for PUMA-2,

where A_{tel} is the difference of the average telemetry amplitude values from the integral distributions.

The dependence of j on A averaged over 1000 spectra is represented in Fig. 4. As seen from Fig. 4, for PUMA-1 (narrow window) a depletion of the mass lines with $A > 20$ are observed. The relative transmission in the mass range from 20 to 107 for small dust particles (corresponding to the signal on the accelerating grid $Q_{ac} < 10^{-13}$ C) is of the order of one. For large particles we have $j < 0.5$. The same dependence for PUMA-2 (averaged over 245 spectra) looks similar.

The lower transmission in the wide window mode for large particles compared to the narrow window mode can possibly be explained by signal saturation in the multiplier. An indication of such a saturation effect in the wide window is the dependence of the count rate of noise pulses and their average amplitude on the acceleration grid signal (see Table 1). For the narrow window the count rate and the average amplitude depend only slightly on the acceleration grid signal. In the wide window mode, a sharp decrease of the count rate and the average amplitude can be seen when the signal on the accelerating grid is increased.

Another indication of a multiplier saturation is the dependence of the number of silver ions detected by the multiplier on the accelerating grid signal (see Fig. 5). A decrease of the average number of the silver ions was observed with the increase of the AC signal in the wide window of PUMA-1 (mode 0); for the narrow window this value is constant up to $Q_{ac} = 10^{-12}$ C. For PUMA-2 the number of the silver ions is almost proportional to the AC signal (the gain coefficient is about 10^5) and the saturation of the multiplier could be negligible except for very large particles.

For PUMA-2 the saturation effects are essential only for high signals with $Q_{ac} > 10^{-10}$ C.

3. Conclusions

1. The mass resolution was increased when operating in the narrow energy window. In the narrow energy window mode the resolution achieved was high enough to separate isotopes over the full atomic mass range (0–110 a.m.u.) for large particles.

2. The relative transmission j of the wide and narrow energy windows is dependent on the element mass number and the size of the particle (for light elements $j > 1$ and $j < 1$ for elements $A > 10$). The depletion of light elements in the narrrow energy window mode could be possibly explained by a wider energy distribution due to plasma effects.

The depletion of elements with $A > 20$ in the wide energy window mode could be possibly explained by the saturation of the multiplier. The relative transmission dependence on the element mass number and the depletion of light elements must be taken into account in the calculation of relative element abundances and numbers of atoms in the dust particles.

3. The saturation effects of the multiplier for PUMA-1 are negligible in the narrow energy window when the acceleration grid signal level Q_{ac} is $< 10^{-12}$ C. In the case of the wide energy window mode the corresponding threshold is $\sim 10^{-13}$ C.

References

Kissel, J., Sagdeev, R.Z., Bertaux, J.L., Angarov, V.N., Audouze, J., Blamont, J.E., Buchler, K., Evlanov, E.N., Fechting, H., Fomenkova, M.N., von Hoerner, H., Inogamov, N.A., Khromov, V.N., Knabe, W., Krueger, F.R., Langevin, Y., Leonas, V.B., Levasseur-Regourd, A.C., Managadze, G.G., Podkolzin, S.N., Shapiro, V.D., Tabaldyev, S.R., Zubkov, B.V.: 1986, *Nature*, **321**, 280–282

Sagdeev, R.Z., Tchuraev, R.S., Shapiro, V.D., Shevtchenko, V.I.: 1983, The evaluation of plasma effects in dust time-of-flight mass spectrometer (PUMA), Space Research Institute, Moscow, Preprint No. 856

Sagdeev, R.Z. et al.: 1985 Venus-Halley Mission Louis-Jean, Gap

Sagdeev, R.Z., Anisimov, S.I., Inogamov, N.A., Konstantinov, A.B., Zhitenev, S.B.: 1985, On the transmission of time-of-flight mass-spectrometer with dust-impact ion source, Institute of Theoretical Physics, Preprint No. 19

Sagdeev, R.Z., Managadze, G.G., Evlanov, E.N., Zubkov, B.V., Khromov, V.N., Shapiro, V.D., Shutyev, I.Y.: 1987, The method to increase mass resolution of PUMA instrument (to be published)

Astron. Astrophys. 187, 183–190 (1987)

ASTRONOMY AND ASTROPHYSICS

Spatial distribution of heavy ions in comet P/Halley's coma

O. L. Vaisberg, G. Zastenker, V. Smirnov, B. Khazanov, A. Omelchenko, A. Fedorov, and D. Zakharov

Space Research Institute, USSR Academy of Sciences, Profsoyuznaya 84/32, SU-117810 Moscow, GSP-7, USSR

Received February 4, accepted June 1, 1987

Summary. Radial profiles of heavy cometary ions at large solar-zenith angles were obtained with the BD-3 plasma detector. The relatively sharp (thickness ~8000 km) boundary at ~1.5 10^5 km of the region where cometary ions dominate, may result from a violation of the frozen-in condition of solar wind protons along the field lines. The density of the heavy ions mantle varies approximately as R^{-2}. A large-scale density enhancement was observed during the inbound crossing at 65,000–25,000 km on March 6, 1986. Many small-scale (250–1,000 km) density enhancements were observed. A local maximum of ~4 10^3 cm^{-3} at ≈1.2 10^4 km was observed on inbound and on outbound trajectories of Vega-1 which shows the permanent existence of this pile-up feature. Data on the bow shock crossing and on the mantle density distribution are discussed and compared with the data of other experiments.

Key words: comets

1. Introduction

The cometary plasma environment is controlled by solar wind interaction with newly ionized neutrals outflowing from the nucleus (Alfvén, 1957; Biermann, 1967a). Hydrodynamic (Biermann et al., 1967b; Baranov et al., 1986) and MHD models (Schmidt et al., 1982) showed the main features of this flow: the initial deceleration of the solar wind due to mass-loading, low-Mach shock (see also Wallis, 1973) which is much farther than the cometary ion-rich envelope, and a monotonic increase of the ion number density towards the cometary nucleus. Interaction of the magnetized solar wind with newly created ions leads to many interesting phenomena (Biermann, 1974; Ip, 1982; Mendis et al., 1982; Galeev et al., 1984) including the physics of cometary shock (Galeev et al., 1984), possible critical ionization velocity phenomena (Galeev et al., 1984; Haerendel, 1986), formation of a ionosphere from outflowing neutrals in a region close to nucleus (Mendis et al., 1982), pile-up of ions between outflowing and mass-loaded plasma (Wallis et al., 1976), formation of an ion tail (Ip et al., 1982).

Direct measurements of the plasma properties near the comets of Giacobini-Zinner and Halley provided a large amount of data that have already been used in investigations of cometary processes (Galeev, 1986).

In addition to detailed studies of the ion and electron distribution functions, ion composition, magnetic field and plasma waves, the ion number density distribution, convective velocity and other gross features of the cometary plasma may be useful for tests of cometary models. We will consider the distribution of heavy cometary ions as obtained with the multicollector Faraday cup BD-3 on the Vega-1 and -2 spacecraft.

Send offprint requests to: O.L. Vaisberg

2. Instrumentation

The BD-3 sensor of the APV-N plasma wave instrument (Klimov et al., 1986) on Vega-1 and -2 was used for measurements of ion flux fluctuations. Collector K5 (Fig. 1) followed by an AC-amplifier provided the measurements of solar wind and cometary ions flow fluctuations due to its wide-angle aperture. BD-3 also included a small plasma spectrometer with four collectors K1–K4 and deflecting grid DG (Fig. 1). A 40 × 12 mm^2 window at the K5 collector facing encounter velocity direction was used as an entrance window of this spectrometer. The split-collector system K1, K2, K3 was used to determine the deflection of the ion flow from the encounter direction. Suppressor grids SG with potential −150 Volts prevented electrons with energy $E < 150$ eV from hitting the collectors. Four voltage steps 0, +40 V, +150 V and +500 V applied alternatively to the deflection grid DG provided limited energy-analysis capabilities of the instrument by the voltage-dependent variation of the ion current on collectors K1, K2, K3, and K4 (energy separation of ion flow on "a" and "b" beams depends on the ratio of ion energy to DG grid voltage).

The DC-amplifiers of the K1, K2, K3 collectors had a current range $-3\ 10^{-11}$ A ÷ $+6\ 10^{-9}$ A, the one of the K4 collector had a current range $-1\ 10^{-10}$ A ÷ $+2\ 10^{-8}$ A. Energy and aspect sensitivities of the plasma spectrometer were measured in the laboratory with ion and electron beams. Power supply, synchro-pulses, −150 V, and signal readouts were provided by the APV-N instrument. The energy steps changed every 12 s with a total cycle duration of 48 s. The sampling rate for all K1–K4 collectors during the fly-by was 1 Hz. The solar wind direction was screened by the spacecraft body.

3. Measurements

Data were obtained along the entire Vega-1 trajectory except near closest approach (CA) where collectors were saturated by negative current. Data on Vega-2 were obtained only along the part of inbound trajectory due to a stronger negative current and due to damage of the detector near CA. The scanning cycle became irregular at $R \lesssim 40{,}000$ km, apparently due to overload of the

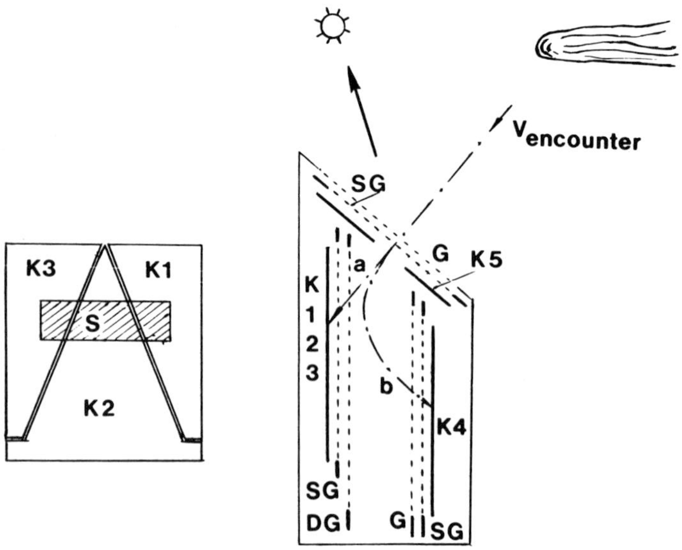

Fig. 1. The BD-3 sensor configuration: K5 – collector for plasma wave measurements with an entrance window for the plasma analyzer, K1, K2, K3 – collector system for directional measurements, K4 – collector of plane analyzer with deflecting grid DG. V_e – encounter velocity, S – shadow of entrance aperture from encounter direction on K1, K2, K3 collector system. Solar direction is indicated. "a" – undeflected ions (higher energy), "b" – deflected ions (lower energy)

power supply but data are tractable as the deflection voltage was telemetred.

4. Data analysis

Each measurement cycle provided 16 (4 collector's currents averaged for every of 4 energy steps) independent measurements. Processing of these data was made with laboratory calibration sensitivity curves and on the assumption of monoenergetic and monodirectional ion flow. This is a significant simplification as the ion distribution function has a complicated shape in velocity space (Mukai et al., 1986; Neugebauer et al., 1986). The analysis yields the direction of the ion flow and the energy in the spacecraft frame of reference and the particle flux F. The results show that at cometocentric distances $R \lesssim 200{,}000$ km the ion component entering the window from the encounter direction with energy $E/Q \approx 600$–1000 eV is easily distinguished and becomes dominant at $R \sim 150{,}000$ km. We identified it with cometary water-group ions in accordance with Mukai et al. (1986), Gringauz et al. (1986a), Krankowsky et al. (1986), Balsiger et al. (1986a), Korth et al. (1986a), Johnstone et al. (1986). In that case we can obtain an estimate of the number density of these heavy cometary ions as $n_i \approx F/V_e$ where V_e equals the encounter velocity of the spacecraft with the comet. We estimate n_i to be accurate within about a factor of 2 at $n_i \sim 10\,\mathrm{cm}^{-3}$ and within about 20% at $n_i \sim 10^3\,\mathrm{cm}^{-3}$. Model calculations with a more realistic ion distribution function will decrease the uncertainty of n_i at larger distances. Relative numbers obtained within a narrow range of distances including density variations during 12 s sub-cycles of measurements for one energy step have an accuracy of $\lesssim 20\%$.

5. Bow shock

A bow shock (or bow wave) was identified in Vega-1 inbound data by the sharp increase of the measured ion current at a distance of $\sim 1.03\ 10^6$ km (Fig. 2) (Klimov et al., 1986; Vaisberg et al., 1986a). The observed current profile was due to a significant change of flow direction as BD-3 was not sensitive to ion flow from the solar wind direction. The thickness of the transition region is $(18 \pm 6)\ 10^3$ km. Location of this Vega-1 inbound shock crossing matches well with the data of other experiments (Gringauz et al., 1986a; Riedler et al., 1986). The thickness of the shock transition agrees reasonably with most of the Vega, Suisei and Giotto results (Mukai et al., 1986; Balsiger et al., 1986a; Rème et al., 1986a; Neubauer et al., 1986; Coates et al., 1986) but is significantly less than the one given in Gringauz et al. (1986a).

Ions with energy $\approx 0{,}8$ keV/charge flowing at $15° \pm 5°$ off the solar wind direction were observed behind this boundary, corresponding to solar wind flow deflected and thermalized at the shock front.

6. Heavy ion mantle

Another population was seen in BD-3 data at $R \lesssim 200{,}000$ km as the ions with energy ~ 600 eV/Q coming from a direction close to the encounter velocity vector. This energy corresponds to water group ions (Mukai et al., 1986; Gringauz et al., 1986a; Krankowsky et al., 1986; Balsiger et al., 1986a; Korth et al., 1986a; Johnstone et al., 1986) moving with low convective velocity. A steep rise to $\sim 15\,\mathrm{cm}^{-3}$ (Fig. 3) at a distance of (150–155) 10^3 km observed both on inbound and on outbound legs of the Vega-1 trajectory marks the region where heavy ions dominate. We call this region the heavy ion mantle in agreement with what was earlier observed near Venus (Vaisberg et al., 1986b; Vaisberg et al., 1984) (see below). Its outer boundary corresponds to the cometopause (Gringauz et al., 1986a; Zastenker et al., 1986). Figure 3 shows both inbound and outbound profiles. For distances $R \gtrsim 12{,}000$ km the number density n_i in the cometary mantle varies as $\sim R^{-2}$ or somewhat steeper. Exceptions from that are the large-scale number density increase between $\approx 66{,}000$ km and $\approx 25{,}000$ km on the inbound trajectory and the steeper decrease of n_i with R closer to the external boundary of the mantle ($R \gtrsim 80{,}000$ km). Comparison with other measurements will be discussed below.

The largest number density of $\sim 5\ 10^3\,\mathrm{cm}^{-3}$ was measured at $\sim 12{,}000$ km, with subsequent decrease to $\sim 2\ 10^3\,\mathrm{cm}^{-3}$ at $\sim 9{,}000$ km. A similar behaviour was observed on the outbound leg of the Vega-1 trajectory except that the density values were lower and large-scale density bump was absent. The solar-zenith angle was $\sim 111°$ for the inbound leg and $\sim 69°$ for the outbound leg of the Vega-1 trajectory.

A local maximum at $\sim 12{,}000$ km was also observed on the inbound trajectory of the Giotto spacecraft (Balsiger et al., 1976a; Krankowsky et al., 1986), and it was found that at lower distances, $\approx 4{,}700$ km, the plasma regime changes to outflowing ionospheric plasma. Our observation of the same local maximum at inbound and outbound Vega-1 trajectories shows that this pile-up region near the interface of outflowing plasma of inner region and nearly antisunward flow of mantle plasma is a permanent feature.

The number densities of heavy ions obtained on Vega-2 in the (90–40) 10^3 km radial interval (Fig. 3), were smaller by a factor

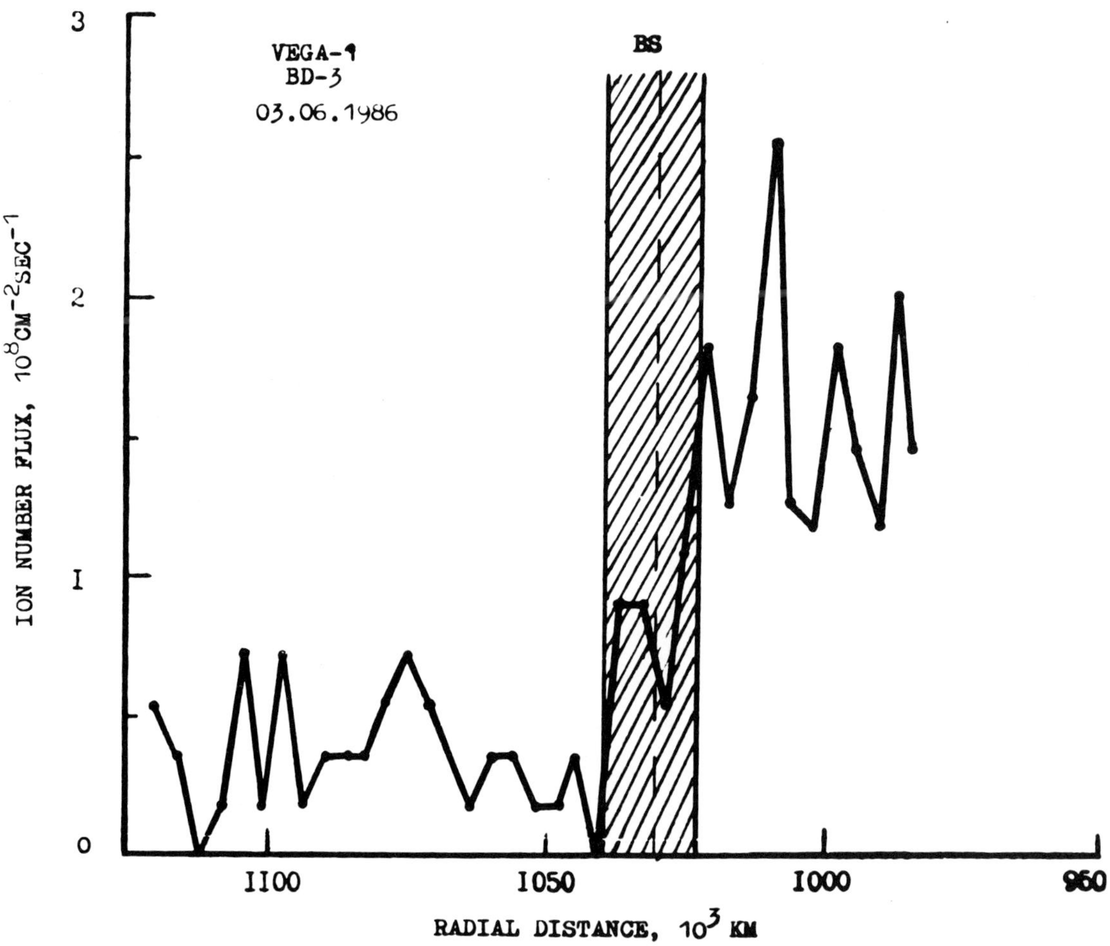

Fig. 2. Bow shock crossing (Vega-1) as seen by an increase of the ion flux in the angular aperture of BD-3. Shading indicates the approximate location and thickness of the shock

~ 3 compared to Vega-1 data at the same radial distances. The same number densities were observed on Vega-2 closer to the nucleus by a factor 1.7 compared to the Vega-1 distances.

7. Convection of heavy ions

The direction of plasma flow in the spacecraft frame of reference obtained from the ratios of currents of the K 1, K 2, K 3 collectors was different from the direction of encounter velocity V_e: the calculated angular difference $\Delta\varphi$ is directly related to the components of the plasma motion in the cometary frame of reference

$$V_\perp = (V_e \pm V_\parallel)\,\mathrm{tg}\,\Delta\varphi, \tag{1}$$

where $V_\parallel \parallel V_e$ and $V_\perp$ is transverse to V_e. The determined energy of heavy ions

$$\frac{mV^2}{2} = \frac{m(V_e \pm V_\parallel)^2}{2} + \frac{mV_\perp^2}{2} \tag{2}$$

lies in the 600 eV $\div$ 1000 eV range, that corresponds to relatively small convective velocities of water group ions, not higher than several tens of km s^{-1}, in agreement with Gringauz et al. (1986a), Balsiger et al. (1986a) Johnstone et al. (1986), and Goldstein et al. (1986). For $M \sim 16$–18 and $V_e \sim 80$ km s^{-1} it means that Eq. (1) can be approximated as

$$V_\perp \approx V_e\,\mathrm{tg}\,\Delta\varphi \tag{1'}$$

with an accuracy of about 30%. Determined $\Delta\varphi$ values were transformed in this way to $V_\perp$. They are shown in Fig. 4.

The decrease in the measured values at $R \gtrsim 100{,}000$ km is apparently caused by the limited angular diagram of the BD-3 detector. Numerical models with a ring-shaped velocity distribution of ions or with hot ions and with the aspect sensitivity of BD-3 show that for $V \gtrsim 17$ km s^{-1} the calculated angles will decrease with increasing $V_\perp$. Approximate allowance for this instrumental effect is shown in Fig. 4 by the dashed line. Another estimation of the plasma convective velocity can be obtained from inbound and outbound number density profiles (Fig. 3). Supposing that the flow is: – symmetric relative to the solar wind direction x' ($\approx 5°$ west of the Sun direction), – nearly antisunward at large cometocentric distances and at large sun-comet angles where Vega-1 crossed the outer mantle, and – incompressible, we can attribute all differences in the ion number densities between two points at the same radial distances on the inbound and outbound trajectories, $\Delta n\,(R)$, to (photo)ion pick-up (see insert in Fig. 4). This gives an estimate of the average velocity at a given R:

$$\bar{v}(R) \approx v \cdot N \cdot \Delta l(R)/\Delta n(R), \tag{3}$$

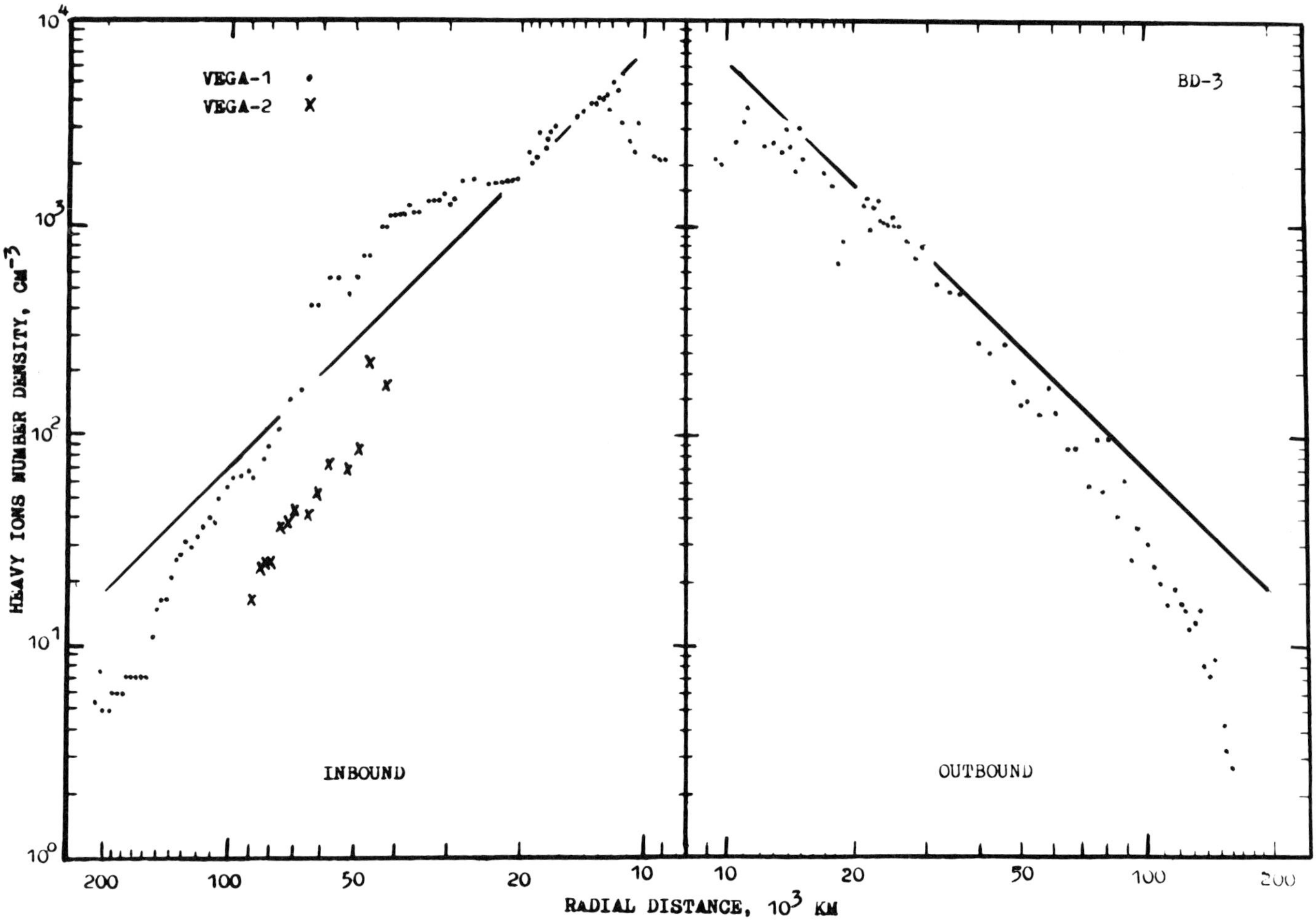

Fig. 3. Profiles of the number density of heavy ions from BD-3 measurements on Vega-1 (dots) and on Vega-2 (crosses). Straight lines show the R^{-2} profile drawn through the local inbound maximum at $\approx$12,000 km. The boundary of the heavy ions mantle or cometopause is determined by the sharp density gradient at $\approx$150,000 km

where v is the (photo) ionization rate, N is the number density of neutrals and $\Delta l(R)$ – the distance between two points on the trajectory. Values of the mean convective velocity $\bar{v}$ obtained in this way with $v = 3\ 10^{-7}\,s^{-1}$, N profile from (Gringauz et al., 1986a) and $\Delta n(R)$ from Fig. 3 is also shown (by triangles) in Fig. 4 for $R \gtrsim 70{,}000$ km (beyond large-scale inhomogeneity). Our measured values for the inner part of the mantle ($R \lesssim 70{,}000$ km) are significantly higher than the ones obtained in two Giotto experiments (Balsiger et al., 1986a; Korth et al., 1986b), but are in reasonable agreement with recent MHD calculations (Boice et al., 1987). For the outer coma ($70 \div 150\ 10^3$ km), our estimated values are about a factor of 2–3 lower than model predictions (Boice et al., 1987). This may be due to the oversimplification made in Eq (3).

8. Small-scale inhomogeneities

With twelve 1-sec measurements of currents for every energy step it was possible to study plasma fluctuations. These fluctuations were quite strong and numerous during the Vega-1 fly-by. The comparison of K1–K3 collectors currents shows that these fluctuations are due to number flux (apparently density) fluctuations rather than variations of flow direction. The large velocity of the spacecraft compared to the plasma convection velocity suggests that these variations are spatial inhomogeneities.

Examples of observed inhomogeneities and variations of their parameters (linear scale and relative variations of number density) along the Vega-1 trajectory are shown in Fig. 5. One can see that the linear dimensions of inhomogeneities vary in the range 200–2,000 km with an average value of $\sim$500 km. The variations of density amount to 10–60 %. The linear scale of inhomogeneities is not in disagreement with Vega-2 data ($\sim$800 km) (Gringauz et al., 1986a). The magnitude and number of the observed variations as well as their increase on the outbound leg of the Vega-1 trajectory suggest that they can be associated with magnetic fluctuations observed on the same spacecraft (Riedler et al., 1986).

9. Discussion

It is interesting to compare the number density measurements obtained in different experiments with theoretical calculations (Fig. 6). There are not only significant differences between the

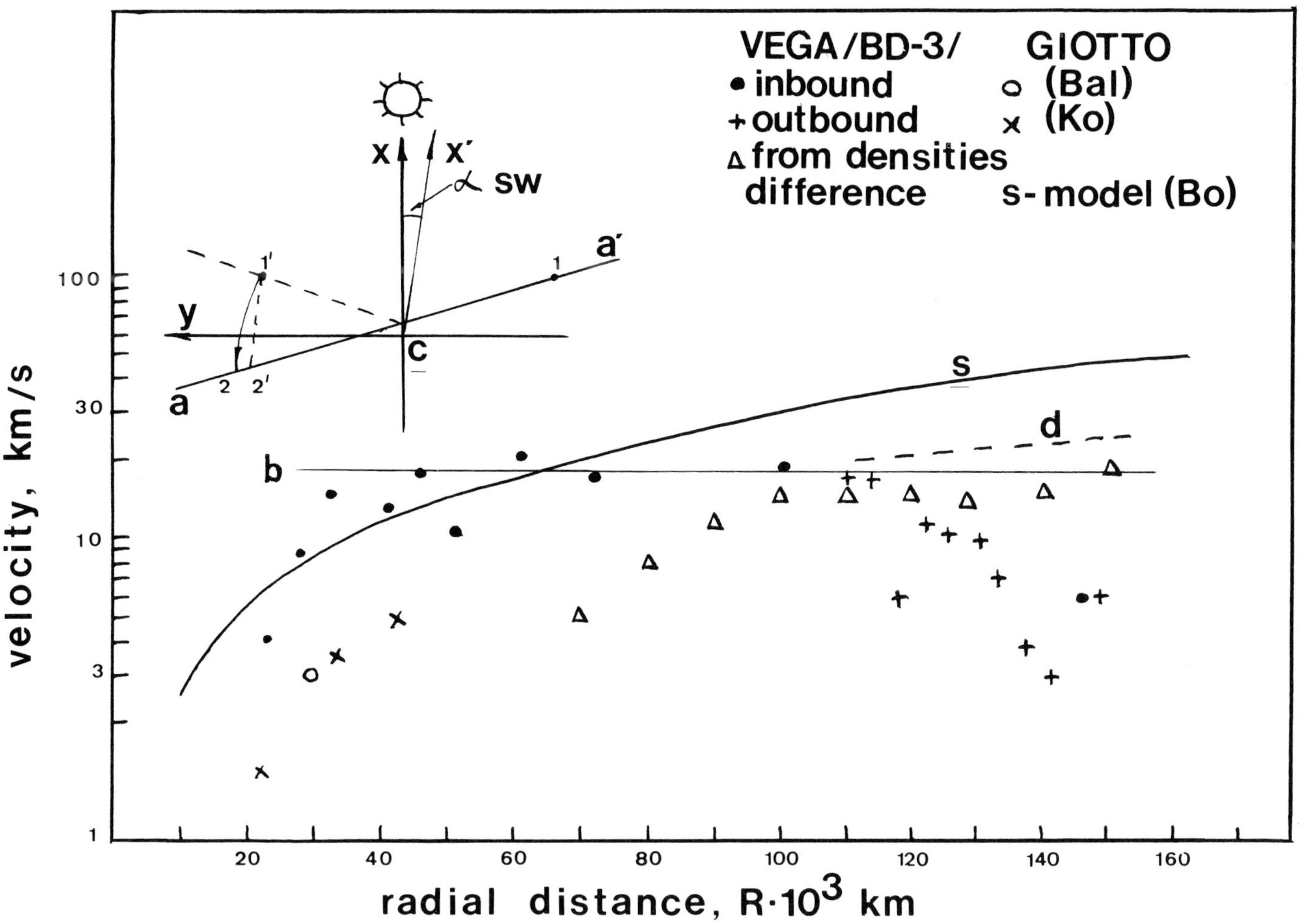

Fig. 4. Plasma convective velocity estimates from BD-3 data in comparison with some Giotto data (Balsiger et al., 1986a; Korth et al., 1986b) and with MHD model values (Boice et al., 1987) (curve S). Line b – upper limit of BD-3 values for perpendicular velocity determination, d – reevaluation of BD-3 data for $R > 100{,}000$ km (see text). Insert in upper left corner explains how the density profile was used in the velocity determination

numbers for Vega-1, Vega-2 and Giotto passes on different days but also between different experiments. This can be partly explained by the different viewing geometry of instruments that measured complicated ion distribution functions. It seems also that Langmuir probe data (Pedersen et al., 1986) are partly contaminated by photoelectrons from the spacecraft. Additional information on the absolute electron number density comes from radiooccultation measurements using Vega-1 and Vega-2 (Savich et al., 1986). The column density was $\sim 7.5\ 10^{12}\ \mathrm{cm}^{-2}$ for $R_0 \sim 8.3\ 10^3$ km (Vega-1). For R^{-2} number density distributions that corresponds roughly to a maximum value of $4.4\ 10^3\ \mathrm{cm}^{-3}$ for 1986, March 6, giving some support to our data. There are cases of reasonable agreement between data of different experiments, for example Pedersen et al. (1986) and our data on Vega-1 for $R \lesssim 60{,}000$ km, and Balsiger et al. (1986b), and Korth et al. (1986b) on Giotto at $R \approx 70{,}000$ km. The cases of significant discrepancies need more detailed analysis.

Three recent models were selected for comparison with experimental data: the gasdynamic model (Baranov et al., 1986), the magnetohydrodynamic model (Boice et al., 1987), and the MHD model allowing for CIV phenomena (Haerendel, 1986). The gas production rate Q was taken as $1\ 10^{30}$ molecules s^{-1} (Baranov et al., 1986; Haerendel, 1986), and equal to $6.9\ 10^{29}$ molecules s^{-1} in Boice et al. (1987). A shape of the MHD-profile close to R^{-2} (Boice et al., 1987) at $R \lesssim 50{,}000$ km does not contradict the observed values. At the same time, number density values from Baranov et al. (1986) are closer to Vega-1 data. Both models give smooth profiles unlike the model with CIV effects (Haerendel, 1986) that predicts a sharp rise of the number density at the outer edge of the mantle but the overall number density increase towards small distances is smaller than in other models and in the measurements. The possible importance of CIV phenomena in comet Halley was also discussed in Galeev et al. (1986) and Amata et al. (1986).

One "unexpected" boundary at 135,000–155,000 km was found in comet Halley's environment (Gringauz et al., 1986a; Vaisberg et al., 1986a; Rème et al., 1986a; Neubauer et al., 1986) and was called cometopause (Gringauz et al., 1986a), collisionopause (Rème et al., 1986a), boundary of cometary mantle (Vaisberg et al., 1986a; Zastenker et al., 1986), "chemical" boundary (Gringauz et al., 1986b) and boundary of the pile-up region (Neubauer et al., 1986).

Physical explanations of this boundary were suggested in Rème et al. (1986a), Vaisberg et al. (1986a), Zastenker et al.

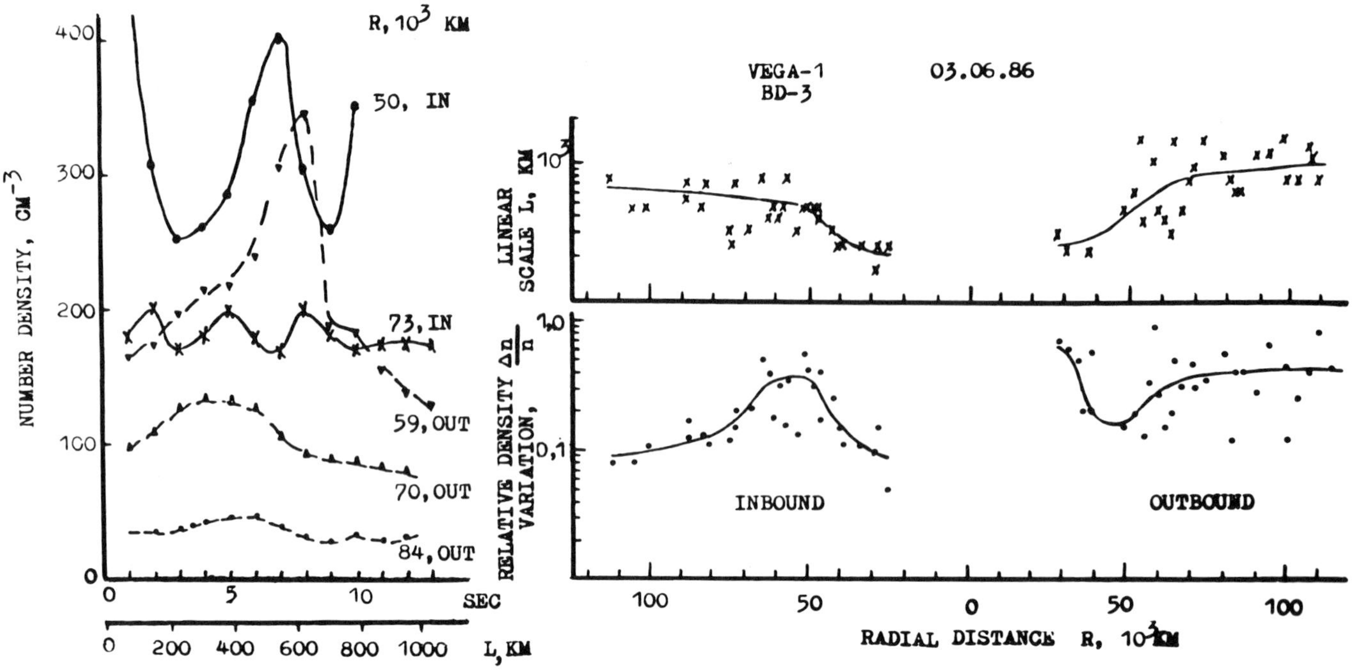

Fig. 5. Examples of observed number density inhomogeneities (left), their linear scale and modulation depth (right)

(1986), and Amata et al. (1986). The collisionpause where outflowing neutrals decelerate the solar wind by collisions at a subsolar distance of $\sim 5\ 10^4$ km (Mendis et al., 1986) was suggested to coincide with the boundary at 135,000 km observed by Giotto on the flank (Rème et al., 1986a; Rème et al., 1986b). But this model cannot explain the small thickness of the observed boundary of $\sim$8,000 km (Zastenker et al., 1986; Fig. 3 of this paper). CIV was suggested as the source of the number density discontinuity at the cometopause (Amata et al., 1986). Another mechanism of formation of the heavy ion mantle was previously developed in Vaisberg et al. (1984) as an explanation of the solar wind interaction with the upper atmosphere of Venus and it was expected to play a role near the comet (Zeleny et al., 1985). Violation of the frozen-in condition for motion of particles along the magnetic field lines combined with spatially – inhomogeneous mass-loading leads to the "sliding" of fast particles along the magnetic field lines when mass-loaded field tubes decelerate and bend. This leads to the replacement of one ion population by another one as the escape of fast ions increases mass loading rate. The radial distance R_m at which this effect becomes important can be estimated by comparing the time-of-flight of ions in external flow with the ionization (or mass-loading) time scale (Zeleny et al., 1985; Vaisberg et al., 1984, 1986a; Zastenker et al., 1986):

$$R_m/V_{\parallel} \simeq n\, m_p / \nu\, NM, \tag{4}$$

where n, m_p and $V_{\parallel}$ are the number density, mass, and parallel velocity component of protons in the external flow, N and M are the number density and mass of cometary neutrals, and ν is the (photo) ionization rate. Substituting $N = N_0 (R_0/R)^2$ and solving for R_m we obtain

$$R_m \simeq \frac{\nu\, N_0\, R_0^2\, M}{n m_p V}. \tag{5}$$

Substituting $\nu = 3\ 10^{-7}\ \mathrm{s}^{-1}$, $N_0 R_0^2 = 10^{24}\ \mathrm{cm}^{-1}$ (Gringauz et al., 1986a) $M/m = 17$, and $nV_{\parallel} = 2\ 10^8\ \mathrm{cm}^{-2}\,\mathrm{s}^{-1}$, we obtain $R_m \simeq 2.5\ 10^5$ km. Taking into account the uncertainties involved, the agreement of this value with the observed cometopause distance does not seem discouraging.

The thickness of this boundary cannot be estimated in such a simple way. At small solar-zenith angles this transition may be comparatively thick. The Vega and Giotto spacecraft crossed the cometopause on the flanks where the density profile was obtained nearly across the flow direction rather than along the plasma flow direction. Relatively weak mass-composition and velocity gradients developed closer to the subsolar point of the cometary mantle should steepen during the motion towards the flank due to velocity-dependent mass loading and that may explain the observed small thickness of the cometopause at the flank. An anisotropic 3-D MHD model will solve this problem.

It appears that some Giotto observations support this explanation. Not only solar wind protons (Gringauz et al., 1986a; Goldstein et al., 1986) and alpha-particles (Balsiger et al., 1986b) show a strong decrease at the cometopause but also fast electrons move in one direction along the magnetic field lines below the cometopause (d'Uston et al., 1986). This motion of electrons may be caused by the escape of ions of the external flow along the field lines.

It was also suggested (Neubauer, 1986) that, observed on Giotto, the magnetic field pile-up boundary coinciding with cometopause may evolve in the boundary of cometary plasma tail. This idea is essentially identical to the model of the plasma mantle that topologically connects the dayside mantle of atmospheric obstacle with the plasma (and magnetic) tail (Zeleny et al., 1985; Vaisberg, 1984).

10. Conclusions

The cometary mantle consisting mainly of heavy ions has a relatively sharp ($\sim$8,000 km) boundary at which the number density of these ions rises by a factor of $\sim$2. This boundary was

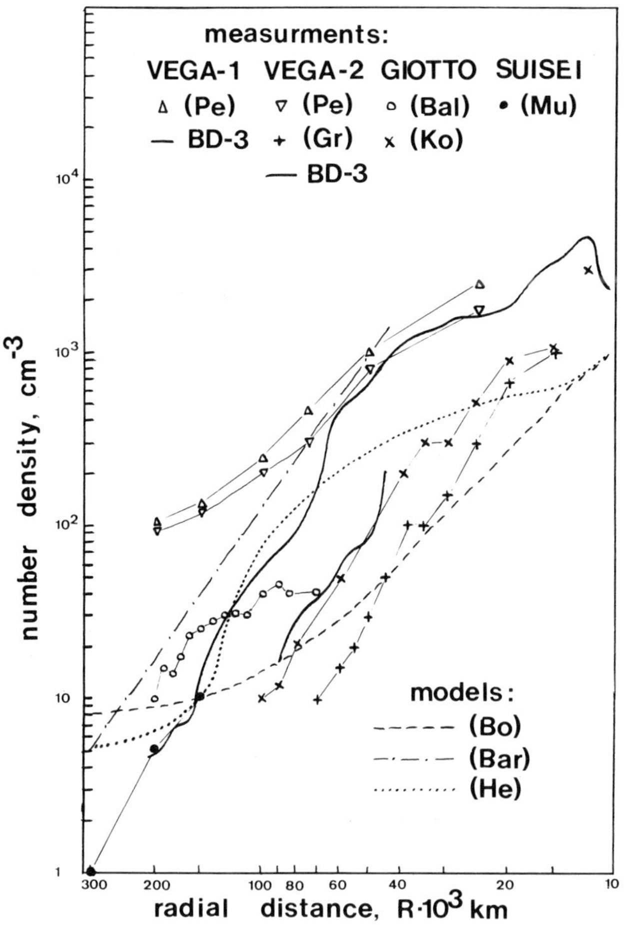

Fig. 6. Comparison of inbound profiles of the number density of heavy ions by BD-3 (Vega-1, 2) with other Vega-1, 2 (Pedersen et al., 1986; Gringauz et al., 1986a), Giotto (Korth et al., 1986b; Balsiger et al., 1986b) and Suisei (Mukai et al., 1986) data ((Pedersen et al., 1986) – electron number density). Three models are shown for comparison (Baranov et al., 1986; Haerendel, 1986; Boice et al., 1987). Data of model (Baranov et al., 1986) on mass density were converted to number density with solar wind normalizing flux $nV = 2\ 10^8$ and with the assumption that water group ions dominate. Total ion density and electron density are given in (Haerendel, 1986) and (Boice et al., 1987), respectively

located at ~150,000 km on March 6, 1986. This radial distance is in reasonable agreement with the criterion of mantle's formation near the gaseous obstacle.

The number density profile in the mantle varies approximately as R^{-2}. An increase of the ion number density by a factor of ~3 was observed on the inbound Vega-1 trajectory in the extended region between 66,000 and 25,000 km.

The local maximum of the heavy ion number density of ~4 $10^3\ cm^{-3}$ at a radial distance of ~12,000 km was observed both on the inbound and the outbound Vega-1 trajectory.

The convective plasma velocity in the outer part of the mantle is estimated as ~20 km s^{-1}.

Strong fluctuations of the ion number density on a scale of several hundred km appear to be typical for the cometary mantle.

Acknowledgements. The authors are indebted to A. Shifrin, S. Klimov, S. Savin, Z. Kravczyk, Z. Zbyshinski, H. Barthe, J.-M. Bosqued, J.-A. Sauvaud for their help in the preparation of the experiment, and to R. Isaeva, A. Dyachkov for assistance in data processing, to Prof. A. Galeev for helpful discussions, and to A. Kuznetsova for her help in the preparation of the manuscript.

References

Alfvén, H.: 1957, Tellus **9**, 192

Amata, E., Formisano, V., Cerulli-Irelli, R., Torrente, P., Johnstone, A.D., Coaters, A., Wilken, B., Jockers, K., Winningham, J.D., Bryant, D., Borg, H., Thomsen, M.: 1986, 20th ESLAB Symp., ESA SP-**250**, Vol. **I**, p. 213

Balsiger, H., Altwegg, K., Bühler, F., Geiss, J., Chielmetti, Goldstein, B.E., Goldstein, R., Huntress, W.T., Ip, W-H., Lazarus, A.J., Meier, A., Neugebauer, M., Rettenmund, U., Rosenbauer, H., Schwenn, R., Sharp, R.D., Shelley, E.G., Ungstrup, E., Young, D.T.: 1986a, *Nature* **321**, 330

Balsiger, H., Altwegg, K., Bühler, F., Fuselier, S.A., Geiss, J., Goldstein, R., Lazarus, A.J., Meier, A., Neugebauer, M., Rettenmund, U., Rosenbauer, H., Schwenn, R., Shelley, E.G., Ungstrup, E., Young D.T.: 1986b, 20th ESLAB Symp., ESA SP-**250**, Vol. **I**, p. 99

Baranov, V.B., Lebedev, M.T.: 1986, *Astron. Zh. Letters* **12**, No. 7, p. 551 (in Russian)

Biermann, L.: 1967a, *Solar Phys.* **1**, 254

Biermann, L., Brosowski, B., Schmidt, H.U.: 1967b, *Solar Phys.* **1**, 254

Biermann L.: 1974, *Solar Wind* III, ed. C.T. Russell, IGPP, UCLA Pupl., p. 396

Boice, D.C., Huebner, W.F., Schmidt, H.U., Wegmann, R.: 1987, *Astron. Astrophys.* (submitted)

Coates, A.J., Johnstone, A.D., Thomsen, M.F., Formisano, V., Amata, E., Wilken, B., Jockers, K., Winningham, J.D., Borg, H., Bryant, D.A.: 1986, 20th ESLAB Symp., ESA SP-**250**, Vol. **I**, p. 263

Galeev, A.A., Lipatov, A.S.: 1984. *Adv. Space Res.* **4**, No. 9, 229

Galeev, A.A.: 1986, 20th ESLAB Symp., ESA SP-**250**, Vol. **I**, p. 3

Galeev, A.A., Gringauz, K.I., Klimov, S.I., Remizov, A.P., Sagdeev, R.Z., Savin, S.P., Sokolov, A.Y., Verigin, M.I.: 1986, *Geophys. Res. Letters* **13**, 845

Goldstein, B.E., Neugebauer, M., Balsiger, H., Drake, J., Fuselier, S.A., Goldstein, R., Ip, W.H., Rettenmund, U., Rosenbauer, H., Schwenn, R., Shelley, E.G.: 1986, 20th ESLAB Symp., SP-**250**, Vol. **I**, p. 229

Gringauz, K.I., Verigin, M.I., Richter, A.K., Gombosi, T.I., Szegö, K., Tatrallyay, M., Remisov, A.P., Apathy, I.: 1986a, 20th ESLAB Symp., ESA SP-**250**, Vol. **I**, p. 93

Gringauz, K.I., Gombosi, T.I., Tatrallyay, M., Verigin, M.I., Remisov, A.P., Richter, A.K., Apathy, I., Szemerey, I., Dyachkov, A.V., Balakina, O.V., Nagy, A.F.: 1986b, *Geophys. Res. Letters* **13**, 613

Haerendel, G.: 1986, *Geophys. Res. Letters* **13**, 255

Ip, W.-H., Axford, W.I.: 1982, in *Comets*, ed. L.L. Wilkening, Univ. Arizona Press, p. 588

Johnstone, A., Coates, A., Kellock, S., Wilkem, B., Jocker, K., Rosenbauer, H., Studemann, W., Weiss, W., Formisano, V., Amata, E., Cerulli-Irelli, R., Dobrowolny, M., Terenzi, R., Egidi, A., Borg, H., Hultquist, B., Winningham, J., Gurgiolo, C., Bryant, D., Edwards, T., Feldman, W., Thomsen, M., Wallis, M.K., Biermann, L.: 1986, *Nature* **321**, 344

Klimov, S., Savin, S., Aleksevich, Y., Avanesova, G., Balebanov, V., Balikhin, M., Galeev, A., Gribov, B., Nozdrachev, M., Smirnov, V., Sokolov, A., Vaisberg, O., Oberc, P., Kravczyk, Z., Grzedzielski, S., Juchnewicz, J., Nowak, K., Orlowski, D., Parfianovich, B., Wozniak, D., Zbyszynski, Z., Voita, J., Triska, P., 1986, *Nature* **321**, 292

Korth, A., Richter, A.K., Loidl, A., Anderson, K.A., Carlson, C.W., Curtis, D.W., Lin, R.R., Rème, H., Sauvaud, J.A., d'Uston, C., Cotin, F., Gros, A., Mendis, D.A.: 1986a, *Nature* **321**, 335

Korth, A., Richter, A.K., Mendis, D.A., Anderson, K.A., Carlson, W., Curtis, D.W., Lin, R.P., Mitchell, D.L., Reme, H., Sauvaud, J.A., d'Uston, C.: 1986b, 20th ESLAB Symp. ESA SP-**250**, Vol. **I**, p. 199

Krankowsky, D., Lämmerzahl, P., Herrwerth, I., Woweries, J., Eberhardt, P., Dolder, U., Herrmann, U., Schulte, W., Berthelier, J.J., Illiano, J.M., Hodges, R.R., Hoffman, J.H.: 1986, *Nature* **321**, 326

Mendis, D.A., Houpis, H.L.: 1982, *Rev. Geophys. Space Phys.* **20**, 885

Mendis, D.A., Smith, E.J., Tsurutani, B.T., Slavin, J.A., Jones, D.E., Siscoe, G.L.: 1986, *Geophys. Res. Letters* **13**, 239

Mukai, T., Kitayama, M., Miyake, W., Teresawa, T., Hirao, K.: 1986, 20th ESLAB Symp., ESA SP-**250**, Vol. **I**, p. 71

Neubauer, F.M., Glassmeier, K.H., Pohl, M., Raeder, J., Acuna, M.H., Burlaga, L.F., Ness, N.F., Musmann, G., Mariani, F., Wallis, M.K., Ungstrup, E., Schmidt, H.U.: 1986, *Nature* **321**, 352

Neubauer, F.M.: 1986, 20th ESLAB Symp., ESA SP-**250**, Vol. **I**, p. 35

Neugebauer, M., Lazarus, A.J., Altwegg, K., Balsiger, H., Goldstein, B.E., Goldstein, R., Neubauer, F.M., Rosenbauer, H., Schwenn, R., Shelley, E.G., Ungstrup, E.: 1986, 20th ESLAB Symp., ESA SP-**250**, Vol. **I**, p. 19

Pedersen, A., Grard, R., Trotignon, J.G., Begin, C., Mikhailov, V., Mogilevsky, M.: 1986, 20th ESLAB Symp., ESA SP-**250**, Vol. **III**, p. 425

Rème, H., Sauvaud, J.A., d'Uston, C., Cotin, F., Cros, A., Anderson, K.A., Carlson, C.W., Curtis, D.W., Lin, K.P., Mendis, D.A., Korth, A., Richter, A.K.: 1986a, *Nature* **321**, 349

Rème, H., Sauvaud, J.A., d'Uston, C., Cros, A., Anderson, K.A., Carlson, C.W., Curtis, D.W., Lin, R.P., Korth, A., Richter, A.K., Mendis, D.A.: 1986b, 20th ESLAB Symp., ESA SP-**250**, Vol. **I**, p. 29

Riedler, W., Schwingenschuh, K., Yeroshenko, Y.G., Styashkin, V.A., Russel, C.T.: 1986, *Nature* **321**, 288

Savich, N.A., Altunin, V.I., Andreev, V.E., Bassos, Y.F., Boguslavskaia, N.-E.E., Vyshlov, A.S., Gavrik, A.L., Doroshchuk, O.N., Kruckkovich, N., Liashovskaia, V.I., Molotov, E.P., Morskoy, I.N., Nabutov, A.S., Ovsiannikova, M.A., Seleznev, V.V., Sukhanov, K.G., Sheverdiaev, A.S.: 1986, *Astron. Zh. Letters* **12**, 675 (in Russian)

Schmidt, H.U., Wegmann, R.: 1982, in *Comets*, ed. L.L. Wilkening, Univ. Arizona Press, p. 538

d'Uston, C., Rème, H., Sauvaud, J.A., Cros, A., Anderson, K.A., Carlson, C.W., Curtis, D., Lin, R.P., Korth, A., Richter, A.K., Mendis, A.: 1986, 20th ESLAB Symp. ESA SP-**250**, Vol. **I**, p. 77

Vaisberg, O., Smirnov, V., Zastenker, G., Khazanov, B., Zakharov, D., Leibov, A., Klimov, S., Savin, S., Omelchenko, A., Barthe, H.: 1986a, XXVI COSPAR Meeting, Toulouse

Vaisberg, O.L., Romanov, S.A., Smirnov, V.N., Karpinsky, I.P., Khazanov, B.I., Polenov, B.V., Bogdanov, A.V., Antonov, N.M.: 1986b, in *Physics of Solar Planetary Environments*, ed. D.J. Williams, Boulder, p. 904

Vaisberg, O., Zeleny, L., 1984, *Icarus* **58**, 412

Wallis, M.K.: 1973, *Planet. Space Sci.* **21**, 1647

Wallis, M.K., Dryer, M.: 1976, *Astrophys. J.* **205**, 895

Zastenker, G., Vaisberg, O., Smirnov, V., Fedorov, A., Omelchenko, A.: 1986, 20th ESLAB Symp., ESA SP-**250**, Vol. **I**, p. 183

Zeleny, L.M., Vaisberg, O.L.: 1985, in *Adv. Space Plasma Physics*, ed. B., Buti, World Scientific, p. 59

Astron. Astrophys. 187, 191–194 (1987)

ASTRONOMY
AND
ASTROPHYSICS

Quasi-periodic features and the radial distribution of cometary ions in the cometary plasma region of comet P/Halley

K.I. Gringauz[1], M.I. Verigin[1], A.K. Richter[2], T.I. Gombosi[3], K. Szegö[3], M. Tátrallyay[3], A.P. Remizov[1], and I. Apáthy[3]

[1] Space Research Institute, USSR Academy of Sciences, Moscow, USSR
[2] Max-Planck-Institut für Aeronomie, D-3411 Katlenburg-Lindau, Federal Republic of Germany
[3] Central Research Institute for Physics, Hungarian Academy of Sciences, Budapest, Hungary

Received February 14, accepted April 10, 1987

Summary. Based on the measurements of the ion electrostatic analyzer CRA of the Plasmag-1 instrument on board Vega-2, which was oriented along the spacecraft–comet relative velocity direction, we present observations (1) of quasi-periodic features in the intensity of cometry ions occurring inside the cometopause, (2) of the mass composition of the cometary ions in the mass range of 1 to 100 atomic mass units (amu), and (3) of the radial dependence of the density of certain, well-defined groups of cometary ions.

Key words: cometary ions – mass composition – radial dependence – Vega-2

1. Introduction

Plasma measurements performed by the Plasmag-1 experiment on board Vega-2 during its approach of Halley's comet revealed the existence of a relatively sharp, yet unexpected boundary within the cometary coma at a cometocentric distance of about $1.6\,10^5$ km (Gringauz et al., 1986a, 1986b, 1985). This boundary, the so-called cometopause, separates the cometosheath region, which is controlled by the solar wind crossing the cometary bow shock and flowing around the cometary nucleus, from the cometary plasma region in the direct vicinity of the nucleus, which is dominated by heavy cometary ions of fastly increasing densities. Based on the plasma measurements deep inside this cometary plasma region, we will present two new classes of observation: the occurrence of a quasi-periodic structure in the intensity of heavy cometary ions, and the radial dependence of certain, well-defined mass-groups of cometary ions. These observations are compared with other ones and with theoretical suggestions.

2. Observations

The data presented in this communication were obtained by the ion electrostatic analyzer of the Plasmag-1 experiment on board Vega-2, which was oriented along the spacecraft–comet relative velocity direction. This Cometary Ram Analyzer (CRA) provided a rather wide angular field-of-view of $14° \times 32°$, and it measured the energy spectra of ions in the energy-over-charge range of 15 to 3500 eV/Q in 120 logarithmically spaced energy channels every one second and with an energy resolution of $\Delta E/E = 0.055$. The CRA observations were supported by the simultaneous and independent measurements of the Ram Faraday Cup (RFC). A detailed description of these sensors can be found in the papers by Gringauz et al. (1982, 1986c) and Apáthy et al. (1986).

It was mentioned earlier (see Gringauz et al., 1985 and 1986a) that due to the enhanced fluxes of certain cometary ions, especially of the ions of the water-group, the channeltrons of the CRA started to operate in a non-linear regime in the innermost part of the cometary ion region, when significant increases in the ion flux correspond to small changes in the measured counting rates only. However, this effect will now be taken into account based on calibration data and on the measurements of the RFC sensor, and the counting rates will be transformed into density values.

As described earlier (see Gringauz et al., 1985, Fig. 3, and Gringauz et al., 1986b, Fig. 1), the CRA measurements deep inside the cometary ion region revealed the following overall scenario: the flow of heavy cometary ions was almost stagnating, i.e., their velocities were much smaller than the spacecraft relative velocity; their intensities increased towards closest approach; their temperatures were decreasing, so that from 7:10 UT or about $5\,10^4$ km onwards rather distinct peaks could be identified in their energy spectra being associated with the different populations of cometary ions; and that the proton fluxes being registered up to a cometocentric distance of about $5\,10^4$ km were significantly smaller than in the cometosheath region.

2.1. Quasi-periodic cometary ion enhancements

While penetrating deeper into the cometary plasma region, the CRA sensor observed regions of a non-steady, patchy increase in the counting rates of heavy cometary ions. This is shown in Fig. 1 exhibiting two high-resolution spectrograms, as observed during 7:00–7:04 UT and during 7:06–7:10 UT or at distances of about $8\,10^4$ km and $5\,10^4$ km, respectively (time runs from right to left). The counting rates corresponding to the outermost isolines are given by $f_0 = 4\,10^3\,\mathrm{s}^{-1}$ and $f_0 = 10^4\,\mathrm{s}^{-1}$, respectively. The increase in the counting rates is characterized in steps of df between two adjacent isolines with $df = 4\,10^2\,\mathrm{s}^{-1}$ and $df = 10^3\,\mathrm{s}^{-}$,

Send offprint request to: A.K. Richter

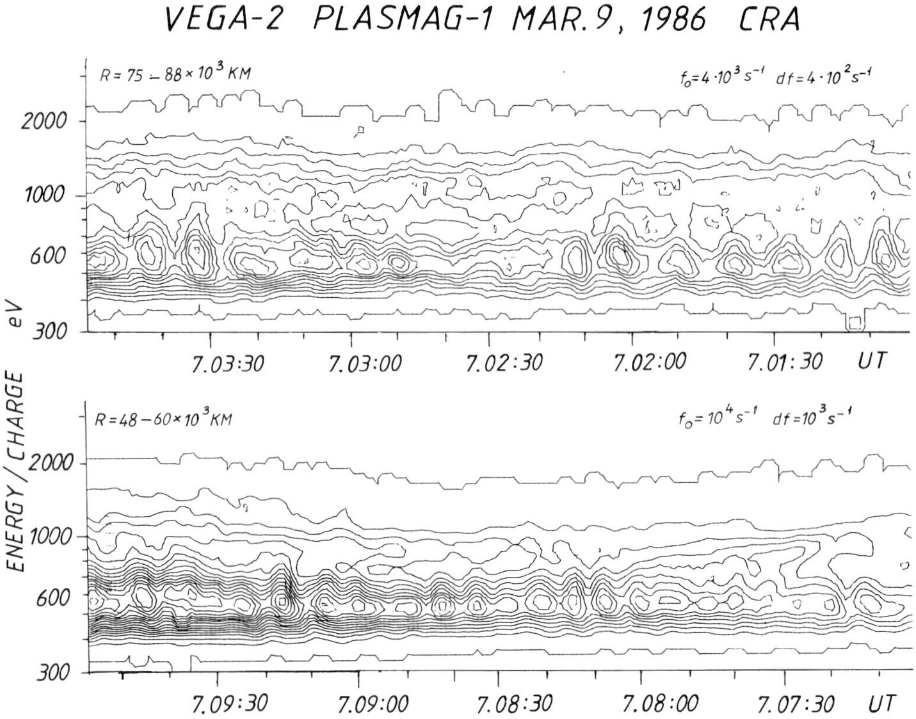

Fig. 1. Contour plots of the ion intensity vs. energy/charge and time, as measured by the Cometary Ram Analyzer (CRA) on board Vega-2 in the cometary plasma region at distances of $8\,10^4$ km and $5\,10^4$ km from the nucleus

respectively. From these figures it readily follows that a quasi-periodic modulation in the intensity of cometary ions exists. This modulation is most pronounced in the energy-over-charge range around 500–600, which (see Fig. 2) corresponds to the ions of the water group. The typical normalized amplitude in the modulation of these latter ions is about

$$A = (f_{\mathrm{Max}} - f_{\mathrm{Min}})/(f_{\mathrm{Max}} + f_{\mathrm{Min}}) \approx 0.05 - 0.1$$

Moreover, we find that the modulation period τ_M decreases when approaching the nucleus from about $\tau_M \approx 10$ s at $8\,10^4$ km to about $\tau_M \approx 8$ s at $5\,10^4$ km.

It should be noted that this type of modulation differs from the sporadic, burst-type density enhancements, which have been observed by Vega-2 at about $4\,10^4$ km and $3\,10^4$ km (Gringauz et al., 1985, Fig. 4). These bursts were associated with enhancements in the plasma wave intentity around the lower hybrid frequency, and their occurrence was therefore explained by the "critical ionization velocity effect" of Alfvén (Galeev et al., 1986).

2.2. *Radial dependence of cometary ions*

Well inside the cometary plasma region both the thermal and the bulk velocities of ions become small compared to the spacecraft relative velocity of 76.78 km s^{-1}. The energy-over-charge spectra break up in several well-defined sub-distributions with their maximum intensity being located at distinct E/Q values. This situation is depicted in Fig. 2, showing a sequence of 4 s averaged ion energy spectra as obtained by the CRA sensor at a distance of 1.4–1.7 10^4 km from the nucleus. Here the counting rates have been corrected for the non-linear effect mentioned above. From the first left-hand distribution it readily follows that the bulk velocity of the protons is close to the spacecraft relative velocity. Thus, the energy-over-charge (E/Q) spectra can be transformed into mass-over-charge (M/Q) spectra. In this way Fig. 2 suggests the presence of H^+, C^+, $CO_2{}^+$, and Fe^+ ions in the cometary plasma region. The peak at $14 \leq M/Q \leq 20$ may originate from the H_2O parent molecules, such as O^+, OH^+, H_2O^+, and H_3O^+, while the peak at $24 \leq M/Q \leq 34$ may be due to parent molecules of CO/CO_2 or to molecules containing N/S, such as CO^+, $N_2{}^+$, H_2CO^+, HCO^+, CN^+, HCN^+, or $O_2{}^+$, or to atomic ions like Mg^+, Al^+, Si^+, P^+, or S^+. Finally, the minor peaks at $M/Q = 2$, 8, 70, and 85 might stem from $H_2{}^+$ and O^{++} and from some heavier, not yet identified organic ions. We have marked these different groups of cometary ions at the top of Fig. 2.

For these groups we have determined their densities at different cometocentric radial distances by summing the respective

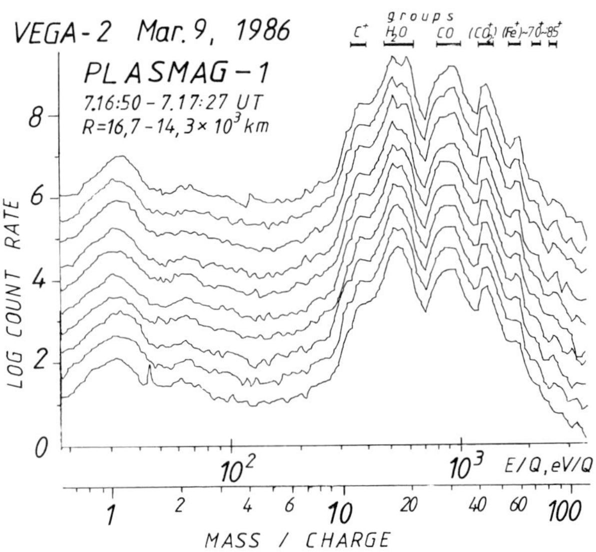

Fig. 2. Time sequence of 4 s averaged ion energy/charge (mass/charge) spectra, as observed by the CRA sensor on board Vega-2 at a cometocentric distance of (1.4–1.7) 10^4 km

count rates. The result is shown in Fig. 3 together with the total sum $\sum$ of all ions and with the anticipated R^{-2} slope. Straight lines correspond to the regions where the densities could be determined more accurately. From Fig. 3 it readily follows that, in principle, really none of the curves, except may-be the curve of mass-group 70^+, can be approximated by a pure R^{-2} dependence. For the total sum we find a dependence of $R^{-2.4}$ for $3\,10^4 \leq R \leq 1.5\,10^4$ km.

3. Discussion

At present we do not know whether the observed quasi-periodic modulation in the intensity of heavier ions in the cometary plasma region is due to density variations or to fluctuations in velocity space. However, this modulation could be explained either by a strongly non-stationary ionization process of the cometary neutral gas, mainly consisting of water at these distances, or by the development of large-scale instabilities caused by the anisotropic velocity distributions of the cometary ions. The first effect will occur as a result of a very fast ionization of the neutral gas at a characteristic time scale shorter than the gyro-period of the ions or comparable to it. If this is the case, the ions get condensed forming bunches in the phase of the cyclotron rotation. In this case the observed decrease of the fluctuation period from $\tau_M = 10$ s at $8\,10^4$ km to $\tau_M \approx 8$ s at $5\,10^4$ km could be associated with the increase of the magnetic field from $B \approx 40nT$ to $B \approx 50nT$ as observed on board Vega-2 (Schwingenschuh et al., 1986). The characteristic frequency of the fluctuations in the plasma flow, however, is a few times higher than the cyclotron frequency for

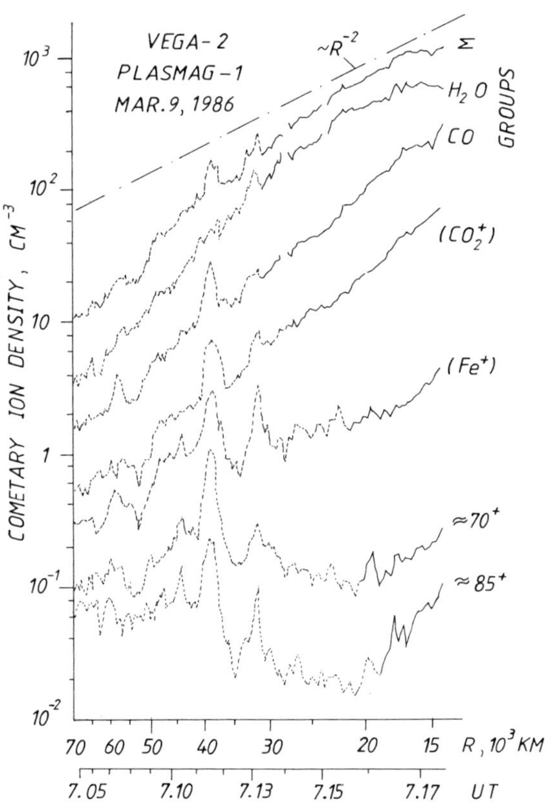

Fig. 3. Cometocentric radial dependence of the densities of certain groups of cometary ions, as defined in Fig. 2, and of the total sum $\sum$ of all ions. For comparison the R^{-2} slope is inserted

the water group ions, which dominate the plasma. Therefore it is more likely that the observed fluctuations are associated with MHD waves originating from the instabilities of the anisotropic velocity distribution of newly formed ions, in a similar way as in the solar wind upstream of the cometary bow shock. This instability is connected with the cyclotron resonance of ions by Alfvénic oscillations, and therefore the wave length of these oscillations can be estimated as $\lambda \approx v_{||}/f_{ci}$, where f_{ci} is the cyclotron frequency of the ions and $v_{||}$ their streaming velocity parallel to the magnetic field. Taking the magnetic field of $B \approx 50nT$ and the spatial dimensions of the fluctuations in the plasma flow as $\sim 8\,\text{s}\,80\,\text{km}\,\text{s}^{-1} = 640$ km into account, the parallel velocity can be estimated as $v_{||} \approx 30\,\text{km}\,\text{s}^{-1}$, which seems to be a reasonable value.

A somewhat unexpected result is the observed well defined peak at $M/Q = 56$ in the CRA ion spectra. This peak was also identified in the IMS (Balsiger et al., 1986) and NMS (Krankowsky et al., 1986) measurements on board Giotto, however, it has not been observed by the Giotto-PICCA instrument (Korth et al., 1986), although otherwise this sensor is quite similar to the CRA sensor on board Vega-2. It seems quite natural to identify this peak with the occurrence of Fe^+ ions in the innermost part of the cometary plasma region, although it should be noticed that, due to the energy resolution of the CRA sensor of 5.5%, there is an uncertainty of about 3 amu in the mass resolution. From the observed mass composition of cometary dust particles iron has been identified in most of the mass spectra analysed (Kissel et al., 1986). Thus, sputtering at dust grains with large metallic cores and no layer of ice could account for the release of metallic ions in the cometary coma (Ip and Axford, 1986), although also other processes discussed by Geiss et al. (1986) could account for the occurrence of metal ions.

As indicated in Fig. 3, the density of cometary ions decreases somewhat faster than R^{-2}. This result is consistent with the observations of the IMS (Balsiger et al., 1986) and the PICCA instrument (Korth et al., 1987) on board Giotto. The neutral gas density, however, decreases as R^{-2}, as observed both by Vega (Gringauz et al., 1986a; Gringauz et al., 1985; Remizov et al., 1986; and Grard et al., 1986) and by Giotto (Krankowsky et al., 1986). This somewhat faster decrease in the ion density could be associated with the fact, that the loss rate of ions increases with increasing distances from the nucleus due to an increasing convective outflow of the cometary ions. Vega-2 observed a neutral gas density of $N_n \approx 1.5\,10^5\,\text{cm}^{-3}$ at $2\,10^4$ km (Remizov et al., 1986). For a characteristic time of ionization of $\tau_i \approx 10^6$ s, the local production rate for ions is $Q_i \approx N_n/\tau_i \approx 0.15\,\text{cm}^{-3}\,\text{s}^{-1}$. In order to obtain an ion density of $N_i \approx 10^3\,\text{cm}^{-3}$ at these distances, a value for the velocity of the convective outflow of $V_i \approx Q_i R/N_i \approx 3\,\text{km}\,\text{s}^{-1}$ would be sufficient. This value for V_i is in reasonable agreement with the velocity determined for the water group ions from the IMS measurement on board Giotto (Balsiger et al., 1986).

On the other hand, if the loss of cometary ions is mainly caused by the convective outflow, then the ratios in the densities of the ions and of their parent molecules must be close to each other. Krankowsky et al. (1986) estimated a value of 5% for the CO_2 content relative to the content of H_2O. From our CRA measurements we find that 70–80% of the total ion content belongs to the water group ions and 2–5% to the ions with $M/Q \approx 44$ (CO_2^+). In turn, as we find 15–20% for the content of the CO/CO_2 group ions, we can estimate the content of the parent molecules H_2O and CO/CO_2 based on our CRA data to be 70–80% and 15–20%, respectively. This agrees with the production rates of H_2O and CO, as determined from the UV measurements on board the IUE spacecraft (Festou et al., 1986).

Acknowledgement. The authors were indebted to R.Z. Sagdeev and A.A. Galeev for useful discussions.

References

Apàthy, I., et al.: 1986, 20th ESLAB Symp. ESA **SP-250**, Vol. 1, 65
Balsiger, H., et al.: 1986, *Nature* **321**, 330
Festou, M.C., et al.: 1986, *Nature* **321**, 361
Galeev, A.A., et al.: 1986, *Geophys. Res. Letters* **13**, 845
Geiss, J., et al.: 1986, *Astron. Astrophys.* **166**, L1
Grard, R., et al.: 1985, *Adv. Space Res.* **5**, 149
Gringauz, K.I., et al.: 1982, in *Cometary Exploration III*, ed. T.I. Gombosi, Hungarian Academy of Sciences Press, Budapest, p. 333
Gringauz, K.I., et al.: 1985, *Adv. Space Res.* **5**, Vol. 12, 165
Gringauz, K.I., et al.: 1986a, *Nature* **321**, 282
Gringauz, K.I., et al.: 1986b, *Geophys. Res. Letters* **13**, 613
Gringauz, K.I., et al.: 1986c, in *Field, Particle and Wave Experiments on Cometary Missions*, eds. K. Schwingenschuh, W. Riedler, Verlag der Österreichischen Akademie der Wissenschaft, Wien, p. 203
Ip. W.-H., Axford, W.I.: 1986, *Nature* **321**, 682
Kissel, J., et al.: 1986, *Nature* **321**, 280
Korth, A., et al.: 1986, *Nature* **321**, 335
Korth, A., et al.: 1987, *Astron. Astrophys.* (this volume)
Krankowsky, D., et al.: 1986, *Nature* **321**, 326
Remizov, A.P., et al.: 1986, 20th ESLAB Symp. ESA **SP-250**, Vol. 1, 387
Schwingenschuh, K., et al.: 1986, *Adv. Space Res.* **6**, 217

Astron. Astrophys. 187, 195–200 (1987)

ASTRONOMY AND ASTROPHYSICS

Electronic spectroscopy and relaxation of some molecular cations of cometary interest

S. Leach[1,2]

[1] DAMAP★, Observatoire de Paris–Meudon, F-92190 Meudon, France
[2] Laboratoire de Photophysique Moléculaire du CNRS★★, Université de Paris-Sud, F-91405 Orsay, France

Received February 13, accepted March 11, 1987

Summary. Recent laboratory studies of the electronic spectra and nonradiative relaxation properties of a number of singly-charged molecular cations are discussed. Their interest for cometary physics is emphasized. Spectroscopic and relevant cometary information is given on CS^+, SH^+, $CO_2{}^+$, $CS_2{}^+$ and H_2S^+. Coincidence techniques have been used for studying radiationless transitions in $CO_2{}^+$, $NH_3{}^+$ and other polyatomic ions. The results have been amplified by the use of laser excitation methods in the case of $CO_2{}^+$. Some discussion is given on recent cometary observations of $CO_2{}^+$ emission. The fluorescence quantum yields are measured for several molecular ions which could be formed from parent molecules likely to exist as frozen gases in the nucleus. The possibility of spectral observation of these cations in comets is considered.

Key words: cometary molecular ions – electronic spectra – radiationless transitions – fluorescence quantum yields

1. Introduction

The present review brings up to date part of an earlier account of studies of some molecular ions of interest in astrophysics (Leach, 1985a), but here emphasizing singly-charged cations in their cometary context. Among the species investigated by optical spectroscopy and/or by coincidence techniques at Orsay are the following: diatomics: OH^+, $N_2{}^+$, $O_2{}^+$, CS^+, SH^+, SO^+; triatomics: $CO_2{}^+$, COS^+, $CS_2{}^+$, N_2O^+, H_2S^+, $SO_2{}^+$; polyatomics: $C_2H_2{}^+$, $C_2N_2{}^+$, $NH_3{}^+$, CH_3CN^+.

Some of these species could be formed by nondissociative ionization of parent molecules assumed to be stored, as frozen gases, in comet nuclei (Huebner, 1985)). Others could result from fragmentation processes, e.g. OH^+, CS^+, SH^+. The possibility of optical detection of cometary ions depends largely on whether they emit photons, i.e. on the magnitude of their fluorescence quantum yields.

The present study first introduces the subject of radiative and nonradiative transitions in molecular ions and discusses some techniques used for their investigation. Mainly spectroscopic results are presented on CS^+, $CS_2{}^+$, SH^+, H_2S^+ and $CO_2{}^+$. Interpretation of cometary spectra is shown in some cases to require higher spectral resolution observations than at present available. Fluorescence quantum yield measurements on $CO_2{}^+$, $SO_2{}^+$, CH_3CN^+, $NH_3{}^+$, $C_2H_2{}^+$ and $C_2N_2{}^+$ are discussed and their significance for cometary studies is considered.

Send offprint request to: S. Leach[1]

★ Unité associée au CNRS (UA N°812).

★★ Laboratoire associé à l'Université Paris-Sud.

2. Radiative and nonradiative transitions in molecular ions: techniques

A number of coincidence techniques were used to measure molecular ion fluorescence quantum yields Φ_F and lifetimes τ_m (Leach, 1985b). These quantities are related to the nonradiative, k_{nr}, and radiative, k_r, rates by the expressions: $k_r = \Phi_F \tau_m^{-1}$ and $k_{nr} = (1 - \Phi_F)\tau_m^{-1}$. In these equations, τ_m is assumed to correspond to a single monoexponential decay process.

The measured lifetime and quantum yield can be used to obtain an approximate value of the electronic transition oscillator strength $f = 1.5\ \Phi_F/\tilde{\nu}^2\tau_m$ (Calvert and Pitts, 1966), (where $\tilde{\nu}$ is an average transition frequency in cm^{-1} and the measured lifetime τ_m is in seconds) and so have many applications in modelling excitation and de-excitation processes in astrophysics.

A number of special ion sources have been developed at Orsay to obtain high resolution emission spectra of molecular ions that fluoresce (Leach, 1985a).

Coincidence techniques used in studies of intramolecular relaxation of molecular ions are tabulated in Table 1 where acronyms are made explicit. The PIFCO technique is used to measure the average fluorescence lifetime τ_m and quantum yield Φ_F, the average being over occupied vibrational levels of a particular emitting electronic state of the ion. The other techniques in Table 1 allow measurements to be made at specific vibronic excitation energy levels within a resolution of 30–100 meV.

In the PIFCO, PEFCO and PEPICO techniques, information is obtained on processes resulting mainly from direct ionization whereas in T-PEFCO and T-PEPICO the data obtained relate not only to direct ionization processes but also to those of superexcited autoionizing states.

3. Results

3.1. CS^+ and $CS_2{}^+$

Gauyacq and Horani (1978) carried out rotational analysis of the $A^2\Pi \rightarrow X^2\Sigma^+$ emission bands of CS^+ in the 6000–7000 Å region.

Table 1. Coincidence techniques used for the study of intramolecular relaxation processes in singly-charged molecular cations. (RGL: rare gas lamp, MSR: monochromatized synchrotron radiation.)

Acronym	Coincidence between	Excitation source	Principal ionization mechanism: direct (D), autoionization (A)	Channel studied
PIFCO	*P*hoto*i*on-*f*luorescence Photon	RGL	D	Ion-selected parent or fragment-ion fluorescence
PEFCO	Photo*E*lectron-*F*luorescence Photon	RGL	D	Energy-selected parent or fragment-ion fluorescence
T-PEFCO	*T*hreshold-*p*hoto-*E*lectron-*F*luorescence Photon	MSR	D,A	Energy-selected parent or fragment-ion fluorescence
	*P*hoto*e*lectron *P*hoto*i*on	RGL	D	Energy-selected parent or fragment-ion formation
T-PEPICO	*T*hreshold *p*hoto-*E*lectron-*p*hoto*i*on	MSR	D,A	Energy-selected parent or fragment-ion formation

The spectrum consists of a doubled-headed progression of five red-degraded bands assigned to $A^2\Pi(v') \rightarrow X^2\Sigma^+(v''=0)$ where $v'=1$ to 5. Each band exhibits two double-headed subbands whose interval of 300 cm^{-1} is approximately equal to the spin-orbit coupling constant of $A^2\Pi$. This is the only transition of CS^+ to have been rotationally resolved. $A^2\Pi$–$X^2\Sigma^+$ transition bands involving higher v' and v'' levels have been observed by Coxon et al. (1976) and by Mizukami et al. (1985), using Penning and charge transfer dissociative ionization techniques respectively. Gauyacq and Horani, who used a low-pressure hot cathode discharge, also observed a group of bands around 4000 Å which were tentatively assigned to the $B^2\Sigma^+$–$A^2\Pi$ transition of CS^+ (Leach, 1972). In particular, two strong red-degraded features at 4059.65 and 4110.57 Å were assigned to the 0, 0 band of this transition. The assignment has been confirmed by a vibrational analysis of the $B^2\Sigma^+$–$A^2\Pi$ spectrum produced by dissociative charge transfer reactions of He^+ on CS_2 or OCS (Tsuji et al., 1980a).

The $B^2\Sigma^+$–$X^2\Pi$ emission spectrum has also been observed as a relatively weak transition in the 2400–3400 Å region, and analysed vibrationally (Tsuji et al., 1980b). Erman and Larsson (1981) have measured the lifetime of the $B^2\Sigma^+$, $v=0$ level as 425 ± 25 ns, using a high frequency deflection technique, whereas Obase et al. (1981) have derived the variation with internuclear distance of the electronic transition moment of the $B^2\Sigma^+$–$X^2\Sigma^+$ and $B^2\Sigma^+$–$A^2\Pi$ transitions from emission data obtained via a metastable reaction. No experimental data exist on the strength of the $A^2\Pi$–$X^2\Sigma^+$ transition, but calculations have been made of the A–X oscillator strength as well as the B–X and B–A transition moments and radiative lifetimes of the $B^2\Sigma^+$ and $A^2\Pi$ states (Larsson, 1985). The $A^2\Pi$ is predicted to have a long lifetime [$\tau(v=0) = 22.3\,\mu$s], whereas the calculated lifetime of the $B^2\Sigma^+$ state is 222 ns, i.e. shorter than the measured lifetime, which may indicate that CS^+ interelectronic state interactions exist which lead to lifetime lengthening of the $B^2\Sigma^+$ state.

Cometary emission of neutral CS was first observed in comet West 1976 VI (Smith et al., 1980), and has since been observed in many comets with the International Ultraviolet Explorer (IUE) satellite (Weaver et al., 1981), including IRAS-Araki-Alcock 1983d. In interpreting the relative production rates of S atoms and CS in Comet West, Smith et al. (1980) have suggested that photoionization of CS could occur, and propose searching for $B^2\Sigma$–$X^2\Sigma^+$ emission bands of CS^+. Since this transition lies in a spectral region heavily overlaid with strong neutral CS $A^1\Pi$–$X^1\Sigma^+$ bands, it may be difficult to identify the ion in this way. Search should also be made for $B^2\Sigma^+$–$A^2\Pi$ and $A^2\Pi$–$X^2\Sigma^+$ bands in the visible region.

It has been proposed that CS is formed by photodissociation of CS_2 released from the nucleus (Jackson et al., 1986; Jackson et al., 1982; A'Hearn et al., 1983). CS_2 would be as difficult to detect directly in comets as is its isovalent relative CO_2. A search should therefore be made for cometary emissions of the $CS_2{}^+$ ion. Emission spectra of this ion are well known in the ultraviolet, transition $\tilde{B}^2\Sigma_u^+$–$\tilde{X}^2\Pi_g$ bands between 2800 and 3100 Å, and in the visible, transition $\tilde{A}^2\Pi_u$–$\tilde{X}^2\Pi_g$ between 4300 and 5200 Å (Callomon, 1958; Horani and Leach, 1959; Leach, 1964: Leach, 1970; Balfour, 1976), and have been rotationally and vibrationally analyzed. The fluorescence quantum yields and lifetimes of the $\tilde{B}^2\Sigma_u^+$ and $\tilde{A}^2\Pi_u$ states have been measured by PIFCO (Eland et al., 1976) PEFCO (Maier and Thommen, 1980) and other (Ibuki and Sugita, 1984) techniques so providing useful information for eventual cometary modelling.

3.2. SH^+

The laboratory emission spectrum of SH^+ was first observed, between 3170 and 4000 Å, and vibrationally analyzed, by Horani et al. (1963); rotational analysis of the 0, 0 band of the $A^3\Pi$–$X^3\Sigma^-$ transition of SH^+ and SD^+ was also carried out (Horani et al., 1967). More recently, the (1 – 0), (0 – 0), (0 – 1) and (0 – 2) bands of both isotopic species have been rotationally analyzed and precise molecular constants obtained for the $A^3\Pi$ and $X^3\Sigma^-$ states (Rostas et al., 1984). Predissociation of the $A^3\Pi$ state for $v' > 1$ allows only about 15 lines of the $(1-0)$ band of SH^+ to be detected in emission. Further information on predissociation has been obtained by Edwards et al., (1984) who studied the SH^+ photofragmentation spectrum by laser exciting the $A^3\Pi$–$X^3\Sigma^-$ $(1-3)$ band in an SH^+ ion beam. They also made an analysis of proton hyperfine structure in both electronic states. The radiative and nonradiative properties of the $A^3\Pi$ state have been discussed in detail (Rostas et al., 1984; Edwards et al., 1984). Relevant spectroscopic properties in the UV, IR and microwave domains have been determined or estimated, and assessed, in order to aid detection of SH^+ in astrophysical objects, including comet tails (Horani et al., 1985). Predictions of the ultraviolet absorption and emission spectra have been made for various temperature conditions. To aid in the detection of SH^+ in the mm, near and far infrared regions, a set of rotational term values has been determined for the $X^3\Sigma^-$, $v = 0$, 1 and 2 levels. The rotational lines involving $N = 0$ i.e. $1_0 - 0_1$ and $1_2 - 0_2$, are predicted around 340 and 350 GHz respectively and are expected to be split into doublets by hyperfine interactions.

A likely domain of astrophysical occurrence of SH^+ is comet tails (Huebner, 1985), especially in view of reports of sulphur species observations such as S_2 (A'Hearn et al., 1983) and H_2S^+ (Cosmovici et al., 1984) in comet IRAS-Araki-Alcock 1983d.

3.3. H_2S^+

The emission spectrum of H_2S^+ occurs between 4000 and 5000 Å; it was first observed (Horani et al., 1963) by electron excitation, using a crossed beam technique (Horani and Leach, 1959). Vibrational (Dixon et al., 1971) and rotational (Duxbury et al., 1972) analyses of the $\tilde{A}^2A_1$–$\tilde{X}^2B_1$ transition have been carried out. The main band system consists of a progression in the excited state bending vibration v_2. Subsidiary progressions in v_2, with one and two quanta of v_2 excited are also observed. Both ground and upper states are bent, the barrier to linearity in the $\tilde{A}^2A_1$ state being 4612 cm^{-1} (Duxbury et al., 1983). In the longer wavelength bands of H_2S^+, corresponding to $v_2 < 4$, the most prominent features are two violet degraded heads, assigned to $^RQ_{0,N}$ and $^PQ_{1,N}$ branches. Bands with $v_2 > 4$ have only one head, assigned to $^PQ_{1,N}$ branches.

In recent years there have been *ab initio* studies of the electronic states of H_2S^+, including calculation of the potential energy curves (Bruna et al., 1980). Predissociation of the ion has been further investigated, by measurement of the emission lifetimes of selected rovibronic levels in the $\tilde{A}^2A_1$ state (Mohlmann and de Heer, 1975) and via ion beam photodissociation spectroscopy (Edwards et al., 1982). Further analysis of the optical spectra, with emphasis on Renner-Teller couplings in this ion, has also been carried out (Duxbury et al., 1983). In a review of cometary spectra, following the observation of H_2O^+ in Comet Kohoutek (Wehinger et al., 1982), Herbig (1976) suggested that a cometary search be made for the isosteric species H_2S^+. A tentative assignment of a few emission bands of H_2S^+ in spectra of Comet Bradfield 1980t was made by Cosmovici et al. (1982). Very recently Cosmovici and Ortolani (1984) have assigned nine emission bands in the visible spectrum of Comet IRAS-Araki-Alcock 1983d to H_2S^+. The published spectra, taken at a resolution of 6 Å, exhibit a low signal-to-noise ratio for the H_2S^+ bands assigned on the microdensitometer tracing. Spectra of higher resolution and contrast are desirable in order to confirm the assignment.

3.4. CO_2^+

Spectral emission of the CO_2^+ ion has been assigned to two electronic transitions, $\tilde{B}^2\Sigma_u^+ \rightarrow \tilde{X}^2\Pi_g$ around 2900 Å and $\tilde{A}^2\Pi_u \rightarrow \tilde{X}^2\Pi_g$ between 2800 and 5000 Å (Gauyacq et al., 1975; 1979).

Bands corresponding to these transitions are among the important features observed by Mariners 6, 7 and 9 in the atmosphere of Mars (Barth, 1985). They are also observed in cometary spectra (Huebner, 1985). Interpretation of CO_2^+ emissions in astrophysical situations requires a knowledge of the $\tilde{B}^2\Sigma_u^+$ and $\tilde{A}^2\Pi_u$ state excitation and subsequent emission processes. In the original work on the Martian airglow (Dalgarno et al., 1970; Dalgarno and Degges, 1971; McConnell, 1973), the three excitation processes considered were photoionization, resonance fluorescence and photoelectron impact. The following explicit or implicit assumptions were made:
(i) the electronic state branching ratios are independent of excitation wavelengths, (ii) the $\tilde{C}^2\Sigma_g^+$ state of CO_2^+ decays to the $\tilde{A}^2\Pi_u$ and $\tilde{B}^2\Sigma_u^+$ states with a cascade branching ratio of 4:1, (iii) collisional deactivation of $\tilde{A}^2\Pi_u$ and $\tilde{B}^2\Sigma_u^+$ states is negligible, (iv) no intramolecular coupling between $\tilde{B}^2\Sigma_u^+$ and $\tilde{A}^2\Pi_u$ states.

It was shown elsewhere (Leach, 1977) that assumptions (i), (ii) and (iv) are invalid. Furthermore, insufficiently accurate data on vibrational distributions was used in the modelling studies. A critical examination of earlier modelling assumptions and data, taking also into account new information available from the Viking I probe (Nier et al., 1976), led to revised values for the excited state production rate (Leach, 1977). In particular, the emission rates are notably affected by intramolecular coupling which was shown to exist between the $\tilde{B}^2\Sigma_u^+$ and $\tilde{A}^2\Pi_u$ states of CO_2^+ (Leach et al., 1978a; 1978b). It was found that emission from $\tilde{B}^2\Pi_u^+$ rotational levels occurs not only in the "normal" 2900 Å region but also at $\lambda > 3000$ Å. Thus the Mariner observations must include considerable emission redshifted to the $\tilde{A}$–$\tilde{X}$ spectral region from $\tilde{B}^2\Sigma_u^+$ rovibronic levels populated in the original photoionization, fluorescent scattering or photoelectron impact excitation. PIFCO measurements on CO_2^+ have shown that the $\tilde{B}^2\Sigma_u^+ \rightarrow \tilde{A}^2\Pi_u$ radiationless yield is 0.41 ± 0.08 (Leach et al., 1978a). This value has recently been confirmed in a laser excitation experiment on timegated electron impact created CO_2^+ (Johnson et al., 1984) in which $\Phi_{RSF}(J')$, the $\tilde{B} \rightarrow A$ radiationless yield for individual rotational levels J' of the $\tilde{B}^2\Sigma_u{}^+$ state, was measured and integrated over a rotational distribution corresponding to that resulting in a photoionization experiment.

Fox and Dalgarno (1979) have also published a revised model of the upper atmosphere of Mars, again based on Viking I data and taking into account the changes in basic parameter values resulting from the absence of cascade processes from the $\tilde{C}^2\Sigma_g^+$ state as well our results on intramolecular coupling between the $\tilde{B}$ and $\tilde{A}$ states.

Thus analysis of the Martian dayglow observations has depended critically on (i) the results of detailed laboratory high resolution spectroscopic studies on CO_2^+ to determine the relevant energy levels and transition probabilities, (ii) knowl-

Table 2. Fluorescence yield Φ_F of some molecular ions of cometary interest (low or zero Φ_F cases). (F = Fluorescence, D = Dissociation, I.C. = Internal Conversion)

Ion	Excited state energy (eV)	Dissociation energy (eV)	$\bar{\Phi}_F$	Relaxation process
$SO_2{}^+$	3.69	3.63	$6\,10^{-5}$	F; D
CH_3CN^+	2.92	1.80	$< 10^{-4}$	D
$NH_3{}^+$	4.83	5.58	$< 5\,10^{-5}$	I.C.
$C_2H_2{}^+$	4.97	5.82	$< 10^{-4}$	I.C.
$C_2N_2{}^+$	2.11	4.10	$< 10^{-4}$	I.C.

edge of $CO_2{}^+$ spectral perturbations and (iii) radiationless transition yields, the latter being determined from coincidence studies and from laser excitation experiments. However, the fact that the "red-shifted fluorescence" yield Φ_{RSF} is a strong function of the rotational level of the $\tilde{B}^2\Sigma_u^+$ state implies that detailed astrophysical interpretation of $CO_2{}^+$ emission requires a knowledge of the rotational population. It thus necessitates higher spectral resolution than has hitherto been available in either planetary or cometary observations on $CO_2{}^+$. Present interpretations of $CO_2{}^+$ emissions in planetary or cometary atmospheres have not taken into account this rotational dependence and have used a global Φ_{RSF} value corresponding to a rotational temperature of ≈ 350 K.

Interaction between electronic states of $CO_2{}^+$ has recently been studied by other techniques. Fluorescence polarization studies (Guest et al., 1984) have confirmed the global Φ_{RSF} yield due to $\tilde{B}$, $\tilde{A}$ state interactions. Fluorescence lifetime measurements have added new information on $\tilde{B}$, $\tilde{A}$ state mutual perturbations and vibronic coupling effects in $CO_2{}^+$ (Larson, 1985), as well as revealing the existence of coupling between low lying vibrational levels of the $\tilde{A}^2\Pi_u$ state and high vibrational levels of the $\tilde{X}^2\Pi_g$ ground state (Erman et al., 1983). More detailed spectroscophic information on $\tilde{B}$, $\tilde{A}$ couplings and perturbations have resulted from optical-optical double resonance (Johnson et al., 1982) and electron impact (Rostas and Tuckett, 1982) experiments.

Observations of $CO_2{}^+$ emission in a number of cometary spectra have been interpreted using the Fox and Dalgarno excitation model to determine $CO_2{}^+$ fluorescence efficiencies and surface brightness (Festou et al., 1982; A'Hearn et al., 1986). The results indicate that $CO_2{}^+$ is produced mainly by photoionization and is excited through resonance fluorescence of solar radiation. $CO_2{}^+$ emission has been observed by IUE observations in Comet P/Halley (1982i) (Feldman et al., 1986a) and recently has been discussed in connection with outbursts of comet Halley (Feldman et al., 1986b); the nature of the ionization source near the nucleus remains to be clarified.

Festou et al. (1982), in their study of $CO_2{}^+$ emission bands in three comets observed by the IUE satellite, have noted that the 3/2–3/2 subbands of the $\tilde{A}^2\Pi_u$–$\tilde{X}^2\Pi_g$ transition are consistently weaker than the 1/2–1/2 subbands. The origin for this apparent "population reversal" effect is unclear. Again, higher spectral resolution in cometary observations are called for in order to clarify details of the $CO_2{}^+$ excitation and fluorescence mechanisms. The strongest bands of the $\tilde{A}^2\Pi_u$–$\tilde{X}^2\Pi_g$ transition have been detailed by ground-based observations. Some previously non-identified bands in the 3900–4600 Å region of Halley's comet tall spectra (Wyckoff et al., 1986; Arpigny et al., 1986) have been shown to be due to $CO_2{}^+$ by direct comparison with laboratory spectra obtained by photon or electron excitation of CO_2 (Leach, 1986, see also Wyckoff et al., 1986). Among these bands are some "red-shifted fluorescence" features; thus information on $\tilde{B}^2\Sigma_u^+$ state excitation can in principle be obtained by observations in a spectral region accessible to ground based observations, whereas the direct $\tilde{B}^2\Sigma_u^+ \rightarrow \tilde{X}^2\Pi_g$ emission at ≈ 2890 Å lies beyond the ozone limit.

3.5. Molecular ions of low fluorescence quantum yield

The ions listed in Table 2 are all present or of potential cometary interest, in particular as indicators of parent molecules in the nucleus. It is therefore important to know whether these ions fluoresce, in oıder to be able to observe them directly.

PIFCO coincidence measurements have been carried out to determine the fluorescence yields of these ions (Table 2). Only in $SO_2{}^+$ was fluorescence detected positively (Dujardin and Leach, 1981), albeit with a very low yield. In all other cases in Table 2 no fluorescence was observed above the experimental noise level and only an upper limit to the yield could be given. *These results indicate the extreme improbability of observing cometary emission spectra of the ions listed in Table 2.*

The low or zero fluorescence yield is readily understandable for $SO_2{}^+$ and CH_3CN^+ where the excited state energy lies above the lowest dissociation limit, so that the dominant relaxation process is fragmentation. However, for the other ions the excited state lies *below* the first dissociation limit. In these cases, interpretation of effective intramolecular fluorescence quenching involves postulating that internal conversion processes occur, in which the excited state couples with high vibrational levels of lower-lying (generally the ground) states. Such processes have been shown to be highly probable for $NH_3{}^+$ (Dujardin and Leach, 1985; Krier et al., 1985). The result is to effectively lengthen the lifetime of the excited state so that, it is eventually deactivated by a collisional process or by vibrational radiative relaxation in the infrared (Dujardin et al., 1980). Radiative relaxation processes of such long lived states might be observable in certain astrophysical situations where low particle densities occur. However, the low effective fluorescence quantum yields render exceedingly difficult the observation in the laboratory of dispersed fluorescence spectra of these ions. Thus the emission band near 2338 Å observed by Herzberg and Lew in an electron impact experiment on ammonia and tentatively assigned to $NH_3{}^+$ (Herzberg, 1980a, b), is unlikely to be due to this ion. The observation of $NH_3{}^+$ in comets is therefore improbable by emission spectroscopy.

4. Conclusion

Molecular ion observation in comets has generally concerned the identification of these species and only more rarely the study of their various energy states. However, it is important to study such energy states, and as a function of time, in order to understand the internal dynamics of these molecular ions and the chemistry in which they are involved. Thus optical spectroscopy at the highest possible resolution is required. Both emission and absorption studies should be made if possible since, as we have seen, intramolecular radiationless transitions and dissociative processes may effectively quench spectral emission of many molecular ions.

The emission spectra of most polyatomic molecular ions are expected to lie at $\lambda > 3300$ Å (Leach et al., 1980). Molecular ions in comets can also be observed by infrared and microwave spectroscopy and, along with optical spectroscopic studies, can be used for investigating excitation mechanisms. The experimental data discussed here have mainly been obtained by the various use of spectroscopic, laser and coincidence techniques. A considerable extension of laboratory work on molecular ions is timely in relation to future observations by satellites and space probes and the extension of ground-based spectroscopic astrophysical observations. New, and it is hoped higher resolution, information on astrophysical molecular ions is certain to result, with consequent benefit for cometary science.

References

A'Hearn, M.F. Feldman, P.D. Schleicher, D.G.: 1983, *Astrophys. J.* **274**, L99–L103

A'Hearn, M.F. McFadden, L.A. Feldman, P.D. Böhnhardt, H. Rahe, J. Festou, M. Brandt, J.C. Maran, S.P. Niedner, M.B. Smith, A.M. Schleicher, D.G.: 1986, *New Insights in Astrophysics, Proc. Joint NASA/ESA/SERC Conf.*, ESA SP-263, 35–38

Arpigny, C. Dossin, F. Zucconi, J.M. Andrillat, Y. Manfroid, J. Hutsemékers, D.: 1986, *20th ESLAB Symp. ESA SP-250*, Vol. I, p. 471

Balfour, W.J.: 1976, *Can. J. Phys.* **54**, 1969

Barth, C.A.: 1985, in *The Photochemistry of Atmospheres*, J.S. Levine, ed. London, Academic Press, p. 337

Bruna, P.J. Hirsch, G. Peric, M. Peyerimhoff, S.D. Buenker, R.J.: 1980, *Molec. Phys.* **40**, 521

Callomon, J.H.: 1958, *Proc. Roy. Soc.* **244A**, 220

Calvert, J.G. Pitts Jr, J.N.: 1966, *Photochemistry*, New York, Wiley

Cosmovici, C.B. Barbieri, C. Bonoli, C. Bortoletto, F. Hamzaoglu, E.: 1982, *Astron. Astrophys.* **114**, 373

Cosmovici, C.B. Ortolani, S.: 1984, *Nature* **310**, 122

Coxon, J.A. Marcoux, P.J. Setser, D.W.: 1976, *Chem. Phys.* **17**, 403

Dalgarno, A. Degges, T.C.: 1971 in *Planetary Atmospheres*, C. Sagan, et al., eds. IAU Symp No. **40**, New York, Springer Verlag, p. 337

Dalgarno, A. Degges, T.C. Stewart, A.I.: 1970, *Science* **167**, 1490

Dixon, R.N. Duxbury, G. Horani, M. Rostas, J.: 1971, *Molec. Phys.* **22**, 977

Dujardin, G. Leach, S.: 1981, *J. Chem. Phys.* **75**, 2521

Dujardin, G. Leach, S.: 1985, *Can. J. Chem.* **63**, 1386

Dujardin, G. Leach S. Taïeb, G. Maier, J.P. Gelbart, W.M.: 1980, *J. Chem. Phys.* **73**, 4987

Duxbury, G. Horani, M. Rostas, J.: 1972, *Proc. Roy. Soc. Lond.* **A331**, 109

Duxbury, G. Jungen, Ch. Rostas, J.: 1983, *Molec. Phys.* **48**, 719

Edwards, C.P. MacLean, C.S. Sarre, P.J.: 1982, *Chem. Phys. Letters* **87**, 11

Edwards, C.P. McLean, C.S. Sarre, P.J.: 1984, *Molec. Phys.* **52**, 1453

Eland, J.H.D. Devoret, M. Leach, S.: 1976, *Chem. Phys. Letters* **43**, 97

Erman, P. Larsson, M.: 1981, *Phys. Scripta* **23**, 1052

Erman, P. Grimm, F.A. Gustafsson, O. Larsson, M.: 1983, *Phys. Scripta* **28**, 611

Feldman, P.D. Festou, M.C. A'Hearn, M.F. Arpigny, C. Butterworth, P.S. Cosmovici, C.B. Danks, A.C. Gilmozzi, R. Jackson, W.M. McFadden, L.A. Patriarchi, P. Schleicher, D.G. Tozzi, G.P. Wallis, M.K. Weaver, H.A. Woods, T.N.: 1986a, *New Insights in Astrophysics, Proc. Joint NASA/ESA/SERC Conf.*, ESA SP-263, 39

Feldman, P.D. A'Hearn, M.F. Festou, M.C. McFadden, L.A. Weaver, H.A. Woods, T.N.: 1986b, *Nature* **324**, 433

Festou, M.C. Feldman, P.D. Weaver, H.A.: 1982, *Astrophys. J.* **256**, 331

Fox, J.L. Dalgarno, A.: 1979, *J. Geophys. Res.* **84**, 7315

Gauyacq, D. Horani, M.: 1978, *Can. J. Phys.* **56**, 587

Gauyacq, D. Horani, M. Leach, S. Rostas, J.: 1975, *Can. J. Phys.* **53**, 2040

Gauyacq, D. Larcher, C. Rostas, J.: 1979, *Can. J. Phys.* **57**, 1634

Guest, J.A. O'Halloran, M.A. Zare, R.N.: 1984, *J. Chem. Phys.* **81**, 2689

Herbig, G.H.: 1976, in *The Study of Comets*, B. Donn, et al., eds. IAU Colloq. No. **25**, NASA SP-393, 136

Herzberg, G.: 1980a, in *Interstellar Molecules*, B.H. Andrew, ed. IAU Symp. **87**, 231

Herzberg, G.: 1980b, *J. Phys. Chem.* **84**, 2095

Horani, M. Leach, S.: 1959, *C.R. Acad. Sci. (Paris)* **248**, 2196

Horani, M. Leach, S. Rostas, J.: 1963, *VI Conf. Int. Pheno. Ionisation dans les Gaz*, Paris, SERMA Vol. **1**, 45

Horani, M. Leach, S. Rostas, J.: 1967, *J. Molec. Spectrosc.* **23**, 115

Horani, M. Rostas, J. Roueff, E.: 1985, *Astron. Astrophys.* **142**, 346

Huebner, W.F.: 1985, in *The Photochemistry of Atmospheres* ed Levine, J.S. London, Academic Press, 437

Ibuki, T. Sugita, N.: 1984, *J. Chem. Phys.* **80**, 4625

Jackson, W.M. Halpern, J. Feldman, P.D. Rahe, J.: 1982, *Astron. Astrophys.* **107**, 385

Jackson, W.M. Butterworth, P.S. Ballard, D.: 1986, *Astrophys. J.* **304**, 515

Johnson, M.A. Rostas, J. Zare, R.N.: 1982, *Chem. Phys. Letters* **92**, 225

Johnson, M.A. Zare, R.N. Rostas, J. Leach, S.: 1984, *J. Chem. Phys.* **80**, 2407

Krier, C. Praet, M. Th. Lorquet, J.C.: 1985, *J. Chem. Phys.* **82**, 4073

Larsson, M.: 1985a, *Chem. Phys. Letters* **117**, 331

Larsson, M.: 1985b, *Phys. Scripta* **31**, 262

Leach, S.: 1964, *J. Chim. Phys.* **61**, 1493

Leach, S.: 1970, *J. Chim. Phys.* **67** (supplément), 74

Leach, S.: 1972, *Faraday Disc. Chem. Soc.* **54**, 68

Leach, S.: 1977, in *Les Spectres des Molécules Simples au*

Laboratoire et en Astrophysique, Colloque Int. d'Astrophys. 21, Liège, Soc. Roy. Sci., 403

Leach, S.: 1985a in *Photophysics and Photochemistry above 6 eV*, ed Lahmani, F. Amsterdam, Elsevier, 113

Leach, S.: 1985b, in *Photophysics and Photochemistry in the Vacuum Ultraviolet*, eds. McGlynn, S. et al. Dordrecht, Reidel, 297

Leach, S.: 1986, (unpublished results)

Leach, S. Devoret, M. Eland, J.H.D.: 1978a, *Chem. Phys.* **33**, 113

Leach, S. Stannard, P.H. Gelbart, W.M.: 1978b, *Molec. Phys.* **36**, 1119

Leach, S. Dujardin, G. Taïeb, G.: 1980, *J. Chim. Phys* **77**, 705

Maier, J.P. Thommen, F.: 1980, *Chem. Phys.* **51**, 319

McConnell, J.C.: 1973, in *Physics and Chemistry of Upper Atmospheres* ed. McCormac, B.M. Dordrecht, Reidel, 309

Mizukami, K. Obase, H. Tsuji, M. Nishimura, Y.: 1985, *Chem. Phys. Letters* **116**, 510

Mohlmann, G.R. De Heer, F.J.: 1975, *Chem. Phys. Letters* **36**, 353

Nier, A.O. Hanson, W.B. Seiff, A. McElroy, M.B. Spencer, N.W. Duckett, R.J. Knight, T.C.D. Cook, W.S.: 1976, *Science* **193**, 786

Obase, H. Tsuji, M. Nishimura, Y.: 1981, *Chem. Phys.* **57**, 89

Rostas, J. Tuckett, R.P.: 1982, *J. Molec. Spectrosc.* **96**, 77

Rostas, J. Horani, M. Brion, J. Daumont, D. Malicet, J.: 1984, *Molec. Phys.* **52**, 1431

Smith, A.M. Stecher, T.P. Casswell, L.: 1980, *Astrophys. J.* **242**, 402

Tsuji, M. Obase, H. Matsuo, M. Endoh, M. Nishimura, Y.: 1980a, *Chem. Phys.* **50**, 195

Tsuji, M. Obase, H. Nishimura, Y.: 1980b, *J. Chem. Phys.* **73**, 2575

Weaver, H.A. Feldman, P.D. Festou, M.C. A'Hearn, M.F. Keller, H.U.: 1981, *lcarus* **47**, 449

Wehinger, P.A. Wyckoff, S. Herbig, G.H. Herzberg, G. Lew, H.: 1974, *Astrophys. J.* **190**, L43

Wyckoff, S. Wehinger, P.A. Spinrad, H. Belton, M.J.S.: 1986, *20th ESLAB Symp.* ESA SP-250 Vol. I, p. 311

Astron. Astrophys. 187, 201–208 (1987)

ASTRONOMY
AND
ASTROPHYSICS

Possible models on disturbances of the plasma tail of comet Halley during the 1985–1986 apparition

T. Saito [1,★], **K. Saito** [2], **T. Aoki** [3], **and K. Yumoto** [4]

[1] Institute for Space and Aeronautical Science, Komaba, Meguro-ku, Tokyo, 153 Japan
[2] Hosei University, Fujimi 2-17-1, Chiyoda-ku, Tokyo, 102 Japan
[3] Tokyo Astronomical Observatory, The University of Tokyo, Mitaka, Tokyo, 181 Japan
[4] Onagawa Magnetic Observatory and Geophysical Institute, Tohoku University, Sendai, 980 Japan

Received March 6, accepted June 29, 1987

Summary. More than 500 photographs of the plasma tail of comet P/Halley taken from the ground during this apparition are surveyed to study the solar wind-plasma tail interaction. The main disturbances of the plasma tail are tabulated, classifying them as outstanding streamers, rays, condensations helices, arcades, kinks, or disconnection events (DE). Based on the photographs, a standard three-dimensional model of the cometary magnetosphere is proposed. The classified plasma tail disturbances are explained further by possible interaction models taking into account of the IMF observation by Sakigake.

Key words: comets – hydromagnetics – interplanetary medium – plasmas – the Sun: solar wind

1. Introduction

A study of the plasma tail of comets is very important to understand not only cometary physics, but also comparative magnetospheres. We cannot observe overall configuration and dynamics of our own earth's magnetosphere, because our magnetosphere is too close to us and because our magnetosphere is not brightly visualized. However, we can infer the earth's magnetosphere by observing the plasma tail taking into account of an essential difference in the magnetospheres of magnetized bodies like the earth and of unmagnetized bodies like a cometary nucleus. Among numerous stars, comets are the only single specimens whose entire magnetosphere can be visually observed from the earth. In this sense, the study of the plasma tail disturbances must be really important.

A single event or a few events of the plasma tail disturbances of a comet or of comet Halley have ever been studied in relation to the mechanism of the solar wind-cometary magnetosphere interaction. This way of studying is sometimes risky to derive controversies among one result to the others. Hence, the purpose of the present paper is to survey all the main disturbances in the plasma tail of comet Halley throughout the present apparition and to find possible causes for the classified disturbances by comparing each of them with possible models.

Send offprint requests to: T. Saito (Sendai address)

★ Present address: Onagawa Magnetic Observatory and Geophysical Institute, Tohoku University, Aoba, Sendai, 980 Japan

2. Data

During this apparition, many photographs of comet Halley were sent to one of the present authors (T.S.) to compare them with the solar wind data obtained by Sakigake. Another author (K.S.) compiled a large amount of photographs of the comet taken by many professional and amateur astronomers who sent these to the Astronomical Society of Japan. The 105 cm Schmidt telescope was efficiently used by the other author (T.A.) and his colleagues in the Kiso Astronomical Observatory to take many high-resolution photographs of the comet.

Surveying carefully more than 500 sheets of these photographs, we classified only the main signatures in the plasma tail structure into outstanding streamers (S), outstanding rays (R), condensation (C), helix (H), arcade (A), kink (K), and disconnection event (DE).

These signatures are listed on a daily basis as shown in Table 1, where the data sources are restricted to the photographs taken by Japanese astronomers. In order to understand the main day-to-day signatures of the plasma tail during this apparition, 64 representative photographs out of the >500 are displayed on a daily basis in Fig. 1. The date, time (UT), place where the photograph was taken, name of the person who provided it kindly, and the data book from which the photograph was reproduced, are listed in Table 2.

3. Models

3.1. Streamer, ray, condensation, helix, and arcade

The plasma tail of a comet was well explained by a draping process of the interplanetary magnetic field (IMF) lines by which the dense plasmas of the cometary origin are picked up (Alfvén, 1957). The Alfvén model was well supported by comet Kobayashi-Berger-Milon (Brandt, 1985), comet Giacobini-Zinner (Slavin et al., 1986), comet Halley (Neubauer et al., 1986), etc. The plasma density around the nucleus is not always homogeneous, as multi-

Table 1. Catalogue of comet Halley disturbances

Date		S	R	C	H	A	K	D	Date		S	R	C	H	A	K	D
Dec.	3		*						Feb.	25	?						
	5		*							28	*						
	8		*						Mar.	3	*						
	10		*	*				*		4	*	*					
	12		*			*		*		5	*						
	13		*			*				6	*					?	
	14		*							7	*						
	15		*							8	*	*	*	*	*		*
	16		*							9		*	*	*		*	
	17		*							10		*		*		?	*
	24		*							11	*	*		*			
	29									12		*		*			
	31	*	*	*	*		?	?		13		*		*		*	
Jan.	2		*	*	*					14		*	*			*	?
	4		*			*				15		*					
	5		*				*			16		*		*		?	
	6		*	*						17		*				*	
	7		*	*	*	*				19	*						
	8		*	?	?		?			20		*	*			*	?
	9		*		?					21		*	?				
	10		*		?		*			25							
	11		*	*	*				Apr.	9		*	*				
	12	*	*							13		*					
	13		*	*	?					15		*	*				
	14	*								16		*				*	
	15	*								17		*	*				
	16	*								29	*						
	17	*							May	6	*						
	20	?								7	*						
Feb.	22		*		*		*			8	*						

layered density structure (Schlosser et al., 1986; Chakaveh et al., 1986) and bursts (A'Hearn et al., 1986; Larson et al., 1986) were detected in the coma of comet Halley for many days during this apparition. The inhomogeneous distribution of the particles is possible to form the ray structure or condensation after these ionized process. The solar wind irregularities such as tangential and rotational discontinuities (Nakagawa et al., 1986) would give rise to the ray structure or condensation when these are pressed into the comet head. When the plasma density distribution is not so inhomogeneous and the solar wind does not contain much irregularities, the plasma tail forms a streamer without significant ray structure (e.g. 1986 Mar. 2).

The draped ray structure near the ionopause is not well visually observed in the coma, because the coma is much brighter than the structured rays. If the IMF are draped monotonously by the ionopause, a bundle of straight rays in the tail must be seemingly convergent into a point in the upstream region (cf. Fig. 8 A of Russell et al., 1985). However, the rays tend to converge observationally into the downstream region of the coma as seen in many of the photographs in Fig. 1.

An actual analysis of observed data revealed that even 90 % of the converged points are distributed in the downstream region of the coma of comet Halley (Saito et al., 1987a). This fact implies that the IMF must be overdraped toward the downstream region as in Fig. 8 B of Russell et al. (1985), or as shown in Fig. 2. As for the three other observed facts which could be explained further by the overdraping process, the reader may refer to the previous paper (Saito et al., 1986f).

Various kinds of hydromagnetic waves observed in the comets Giacobini-Zinner and Halley may give rise to the helical type structure in the tail. It is noteworthy that in many cases helical rays coexist with the straight rays (e.g. 1986 Mar. 12). The helical structure may correspond to Alfvén waves, while a series of bright spots with regular spacing (e.g. 1986 Mar. 10) may correspond to compressional HM waves. Sometimes, a 3-dimensional field-aligned current system like that during the earth's auroral substorm (e.g. Fig. 18 of Alfvén, 1977) could be set up in the cometary magnetosphere. When the current system is suddenly set up at the onset of the expansion phase of a substorm in the case of the earth's magnetosphere, the field lines make a helical motion that is observed as a geomagnetic Pi 2 pulsation (Saito, 1987). Approximately more than ten Pi 2 pulsations are observed generally on the night side of the earth within 24 h in sunspot minimum (Saito, 1969). The current-induced helices may be expected more in comet, because the magnetic fields are weaker in the comet than in the geotail. Nevertheless, the occurrence of the helices is observationally not so frequent than our expectation from the earth's substorm. This difference between comet and the earth may partly be caused by the following difference in the field-line configuration: In the case of the earth, the northern tail-to-earth field-aligned current is connected via the northern nightside ionospheric current to the northern earth-to-tail field-aligned

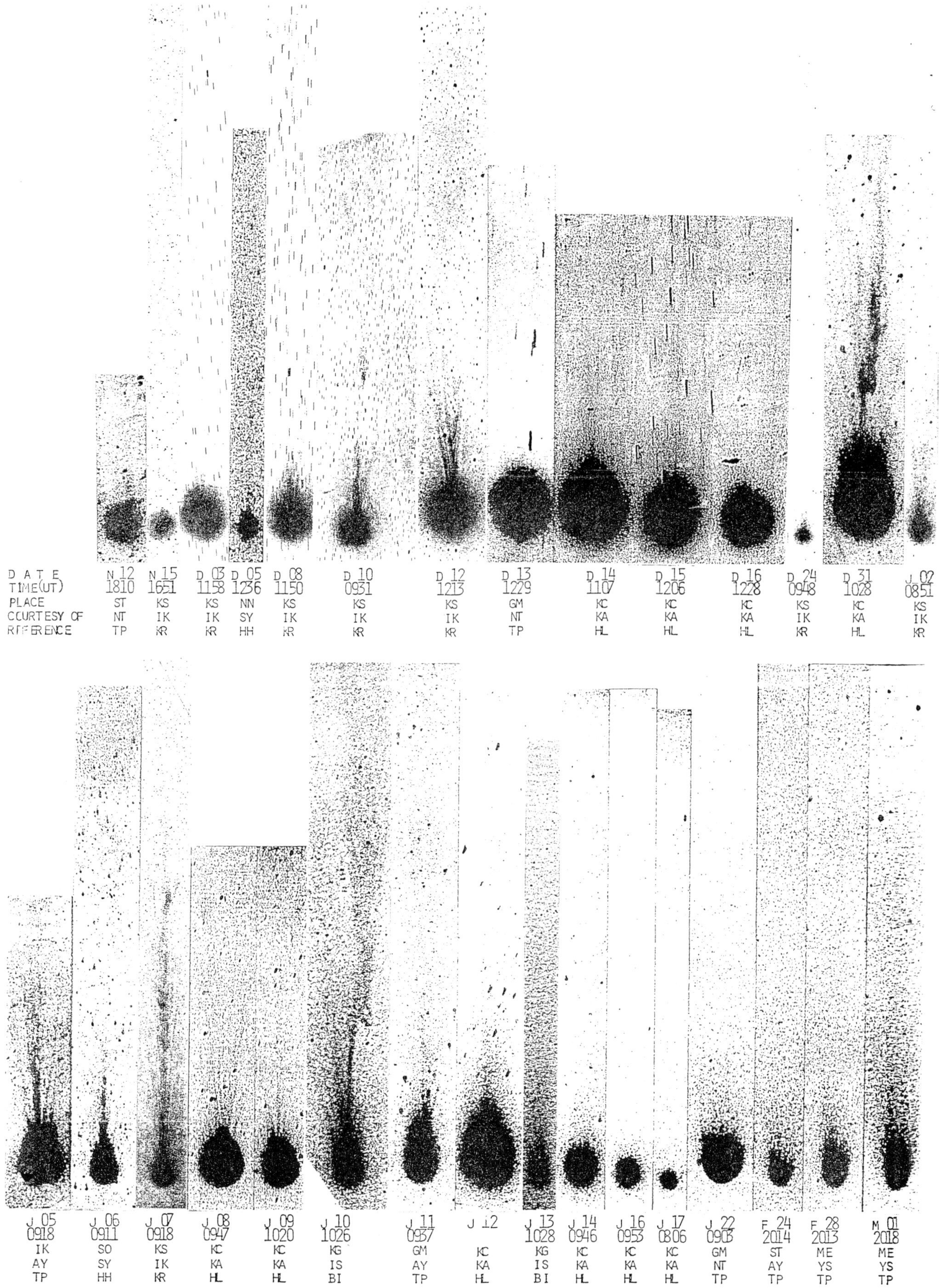

Fig. 1. Daily representative photographs of comet Halley showing main features of the plasma tail disturbances during this apparition. These are restricted to the photographs taken by Japanese astronomers. Types of the disturbances are catalogued in Table 1. See also Table 2 for the abbreviation of place, reference, etc.

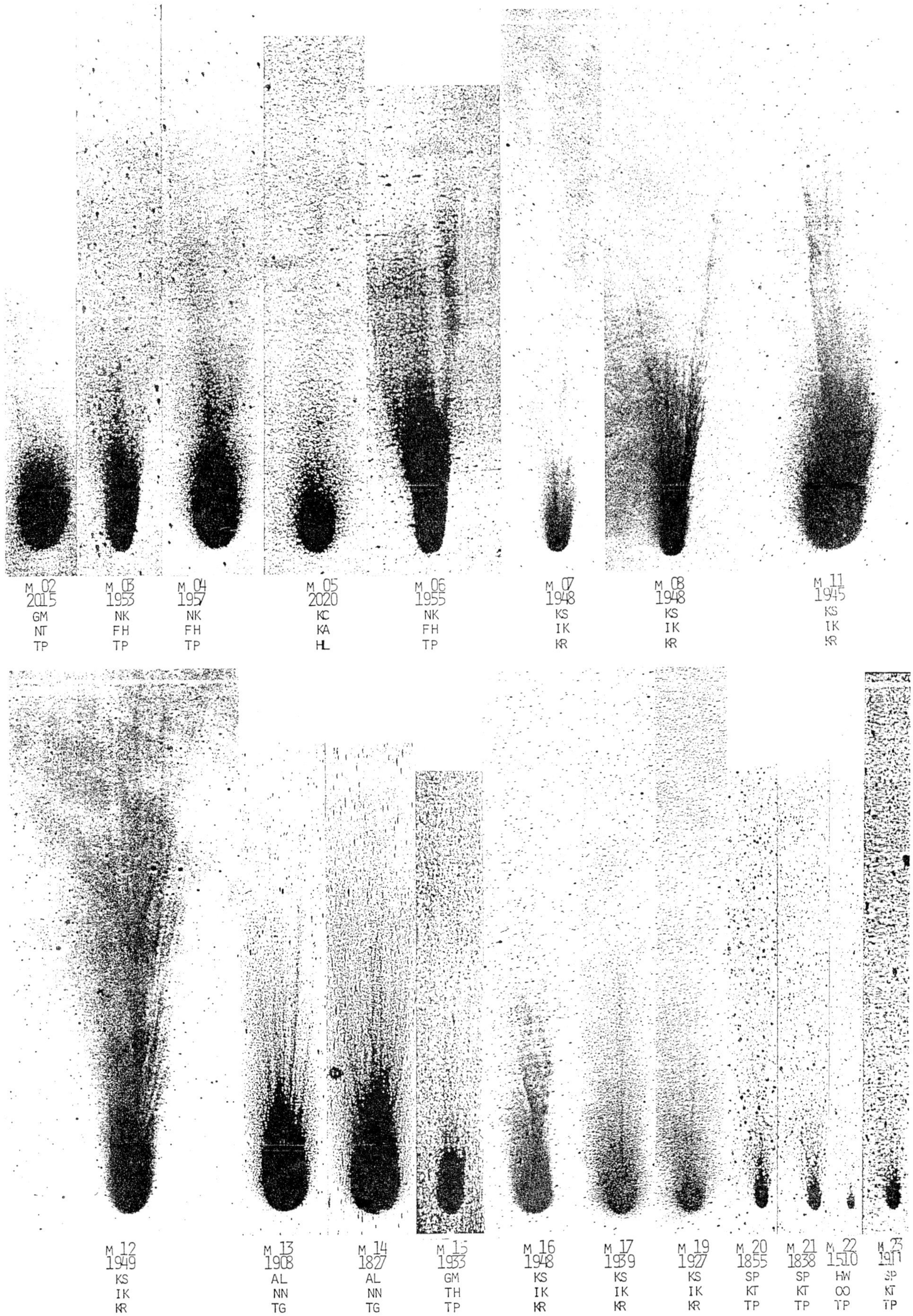

Fig. 1 (continued)

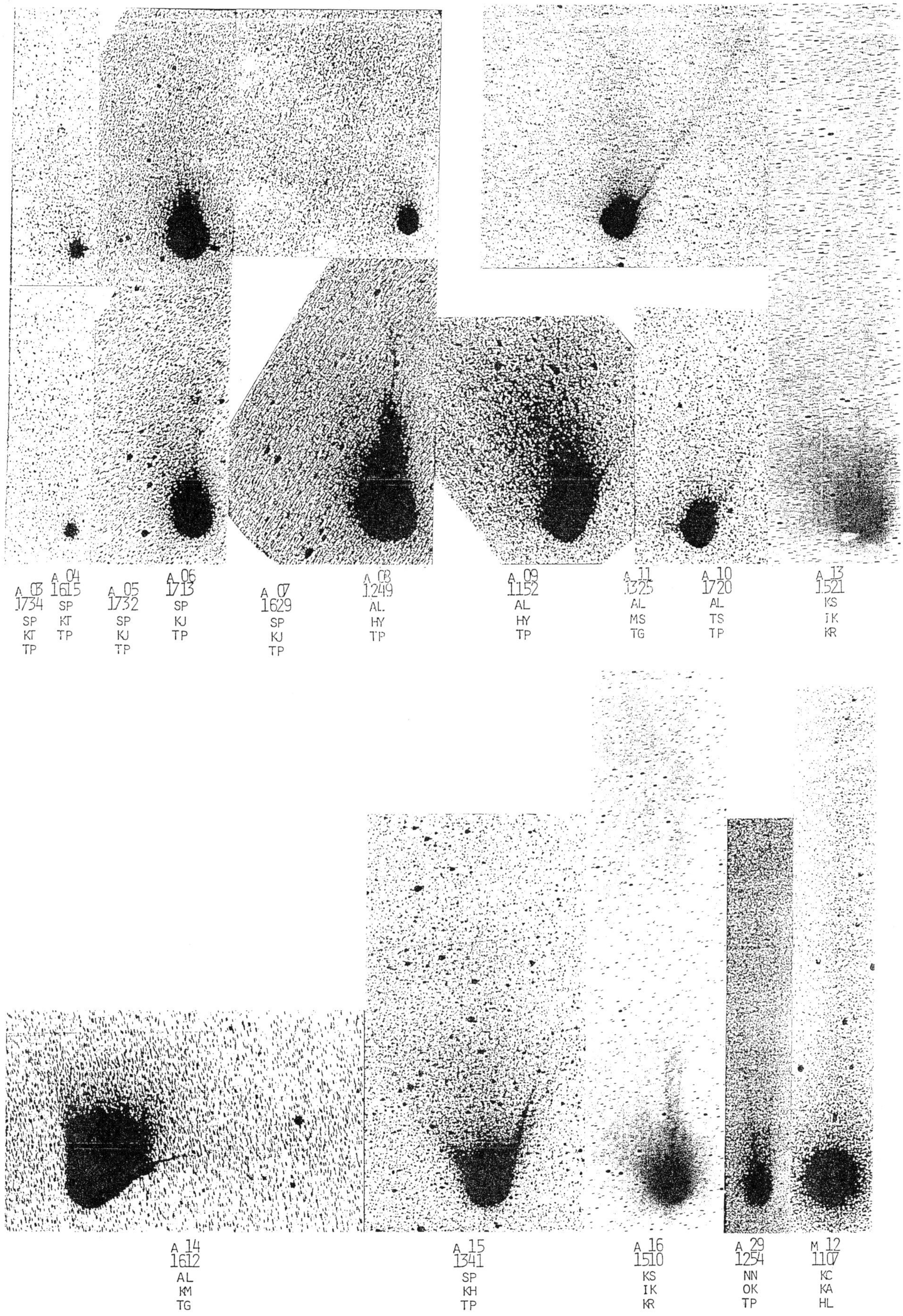

Fig. 1 (continued)

Table 2. Place, name, and reference (cf. Fig. 1)

	Place
KS	Kiso
ST	Saitama
KC	Kochi
IK	Ibaraki
GM	Gumma
KG	Kagoshima
ME	Mie
NN	Nagano
AL	Australia
SP	Saipan
HW	Hawaii
SO	Shizuoka
NK	Norikura

	Reference
TP	Tokyo Univ. Press
TG	Tenmon Gaido
TB	Tokyo Astron. Obs. Bull.
HL	Comet Halley 85/86
HH	Hoshi to Hito
BI	Bull. ISAS

	Courtesy of
NT	T. Niijima
IK	K. Ishida
KA	A. Kawazoe
AY	Y. Arai
YS	S. Yuasa
NO	Norikura
KN	N. Kimura
ST	T. Sudo
TH	Tokai Univ. Comet Halley Obs. Team
KT	T. Koseki
OO	O. Okoshi
TJ	J. Kojima
SH	H. Suzuki
HY	Y. Hirose
TS	S. Takahashi
KM	M. Koishikawa
KH	H. Kitamura
OK	K. Ogiwara
IS	ISAS
SY	Y. Sakai
MS	S. Murotani
NN	N. Nishitani
FH	H. Fukushima

current forming a separate circuit from the southern current system. If we denote conventionally the northern and the southern tail field lines in the case of the comet in the equatorial view in Fig. 2, the southern tail-to-coma field line is connected via the dayside coma to the northern coma-to-tail field line. Since the northern field lines are basically connected to the southern field lines owing to the draping process, the substorm current as in the case of the earth can hardly be expected in the cometary magnetosphere.

The draped IMF has to be swept away by the solar wind from a via b, c, d, to e in Fig. 2. The arcade structure (Fig. 3 of Saito et al., 1986f, for example) in the plasma tail is possible to be the state e. An isolated arcade is sometimes possible to be misunderstood as DE.

As for the kink structure, the reader may refer to Saito et al. (1986d) and Tomita et al. (1987) on the four possible processes: Wind socks, Vns, direction changing, and stream intrusion.

3.2. Disconnection event (DE)

The famous model on disconnection event (DE) was given by Niedner et al. (1978) that an old plasma tail is disconnected from the coma from which a new plasma tail grows out when the comet passes relatively through an interplanetary sector boundary, in other words, the heliospheric neutral sheet. According to a model on solar cycle variation of the three-dimensional heliosphere (Saito et al., 1986a), the heliospheric neutral sheet is always essentially one throughout a solar cycle. The model advocates further that the polarity change of the magnetic field in the heliographic polar cap accompanies a turning over (Saito et al.,

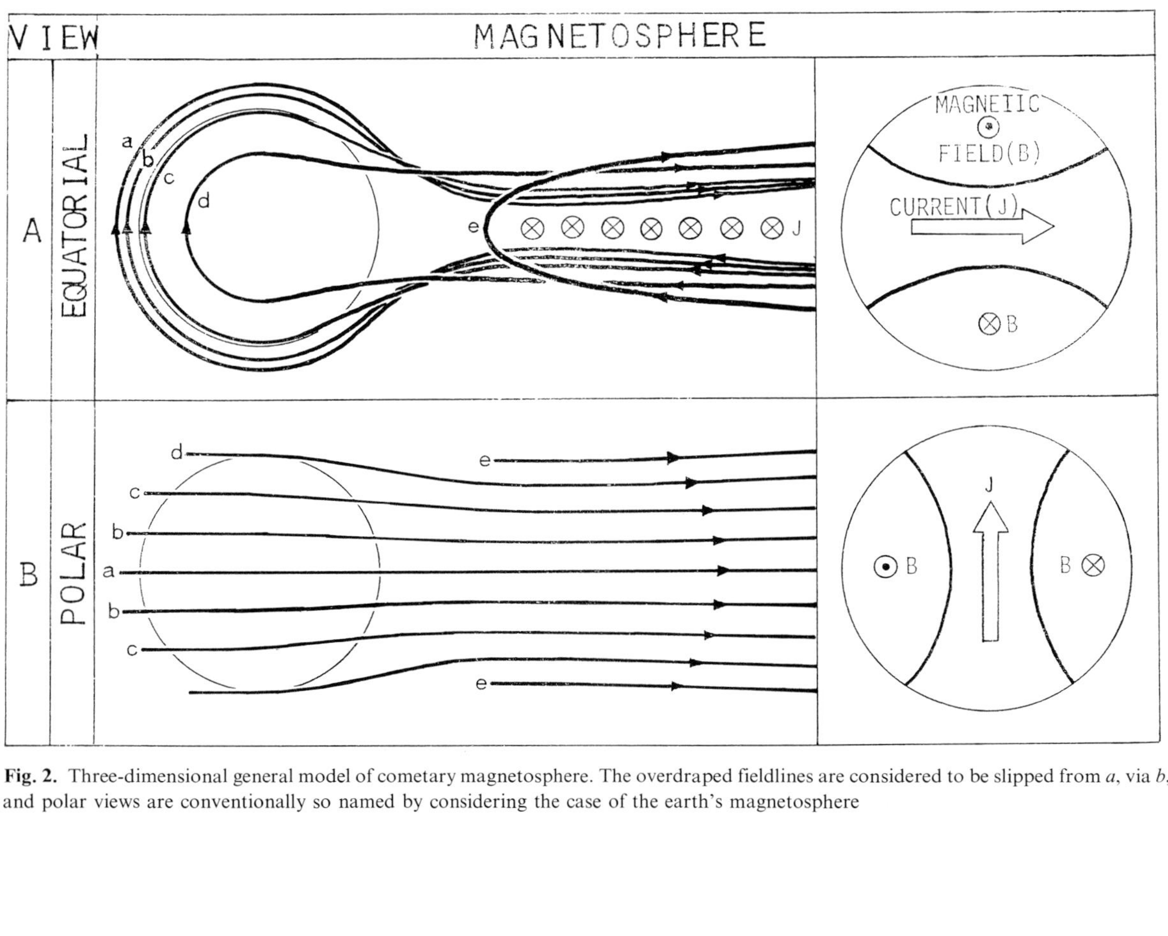

Fig. 2. Three-dimensional general model of cometary magnetosphere. The overdraped fieldlines are considered to be slipped from *a*, via *b*, *c*, *d*, to *e*. The equatorial and polar views are conventionally so named by considering the case of the earth's magnetosphere

1987c) of the heliospheric neutral sheet during sunspot maximum (see Fig. 1 of Saito et al., 1986d, or Fig. 4 of Tomita et al., 1987). The model implies that the sheet is nearly vertical in sunspot maximum years, while nearly parallel in minimum years. Hence, a comet with an orbit having a large inclination, *i*, has chances to cross the heliospheric neutral sheet generally more frequently in reversing phase than in aligned phase (e.g. see the outstanding kind of comet Bennett ($i \sim 90°$) in 1970 in Fig. 7 of Saito et al., 1986d). On the contrary, a comet having a very small inclination has chances of crossing it more in aligned phase than in reversing phase, because the aligned heliospheric neutral sheet has multiple undulations with small amplitudes (e.g. see Fig. 2 of Saito et al., 1986c). It is noteworthy that the apparition period of comet Halley is nearly seven times solar cycle, in other words, the comet tends to appear near sunspot minimum phase.

When we infer the sector boundary crossing of comet from some data, we have to take into account of the following possible flattening effect of the heliospheric neutral sheet. Generally we have four methods (scanning, midline, minimum velocity, and potential, respectively, see Saito et al., 1987b) to construct observationally the heliospheric neutral sheet. The undulation amplitude of the sheet obtained from the former three methods tend to be smaller than that obtained from the latter one, potential method, especially in excursion and aligned phases (e.g. see Fig. 8 of Saito, 1986). So, it is quite likely that during the solar wind flows from the solar source surface to the interplanetary space, the heliolatitudinal range of the undulating neutral sheet becomes smaller owing to the following process. Let us take two points, P ($B_P = 20°$, $L_P = 70°$, $S_P =$ away) and Q ($B_Q = 20°$, $L_Q = 40°$, $S_Q =$ toward), on the solar source surface, where *B*, *L*, and *S* are heliolatitude, heliolongitude, and sector polarity, respectively, as shown in Fig. 3. When the solar winds flow out from P with $V_P = 300$ km s^{-1} and from Q with $V_Q = 500$ km s^{-1}, respectively, the solar wind from Q arrives at 1 AU earlier than the wind from P. The Sakigake data, solar magnetic field data (Hoeksema, 1986), and IPS data (Kojima et al., 1987) suggest that this situation occurred actually in solar rotation No. 1770. Since the expected away-polarity from P was to be replaced by the toward-polarity from Q, it would give rise to the flattening of the undulating heliospheric neutral sheet. This flattening effect must be important when we consider the mechanisms of DE's with the sector boundary crossing process for comet Halley during this apparition.

Many observers reported many DE's of comet Halley during this apparition. Besides the dayside reconnection process due to the sector boundary crossing, there must be the following possibilities that some of the DE's were reported when the following processes took place:

(1) Dynamic pressure process (Saito et al., 1986b, e, 1987d) when the dynamic pressure of the solar wind increased suddenly (e.g. 1985, Dec. 31, see also Russell et al., 1986).

(2) The slipping IMF process when high-density plasmas along an aracade-shaped field line were swept away by the solar wind (e.g. 1985, Dec. 13).

(3) Rotating IMF process when the IMF direction rotates gradually as shown in Fig. 4 (e.g. 1986 Mar. 8; 1986 Mar. 16; 1910 May 4).

(4) Hones' process. A southward turning of the solar wind magnetic field is known to generate an auroral substorm on the earth. Based on a data obtained by ISEE-3, Hones proposed that the substorm is associated with a plasmoid that is conveyed down stream. According to him the DE that occurred in comet Morehouse in 1908 Sept. 30 – Oct. 2, is an evidence supporting this Hones' process. The southward turning of the northward IMF for the magnetized earth may correspond to the sector boundary crossing for the nonmagnetized cometary nucleus. It is probable

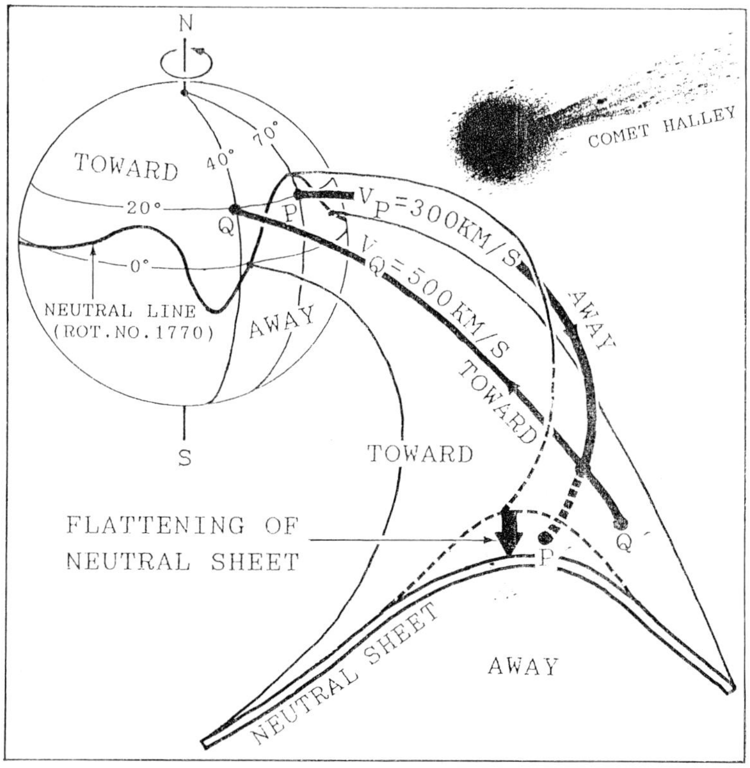

Fig. 3. Flattening mechanism of the undulating heliomagnetospheric neutral sheet explaining the observed difference of the undulation amplitudes between the solar sector boundary and the interplanetary sector boundary. The large heliolongitudinal velocity gradient (e.g. between P and Q) is shown in Fig. 4 Dg of Tomita et al. (1987), which is schematically illustrated from the Sakigake and the IPS data

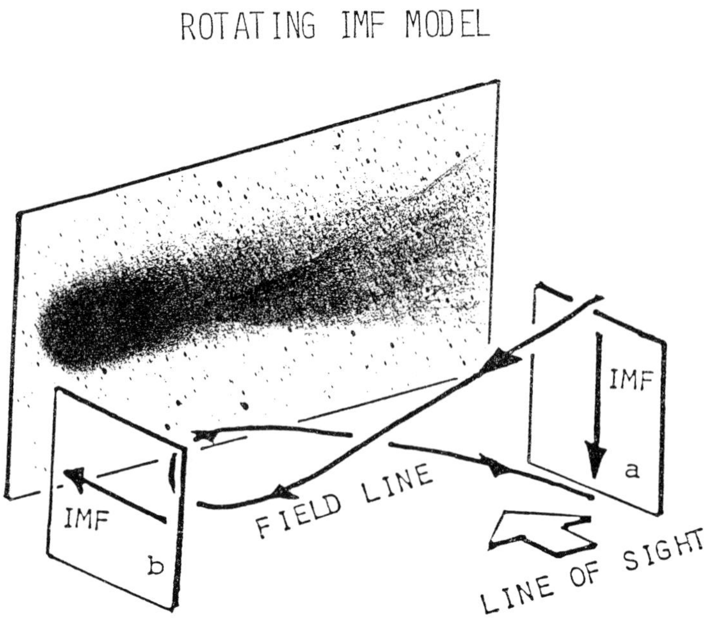

Fig. 4. Rotating IMF model for the fan-shaped plasma tail. The photograph of comet Halley taken at the Kiso Astronomical Observatory on March 16, 1986 is used here as an example. The comet is considered to be changing from the equatorial view to the polar view (cf. Fig. 2) owing to the change of the IMF direction from *a* to *b*

that a DE occurs in the cometary plasma tail through the Hones' process.

(5) Perspective effect on kink. If a sharp kink occurs in the plasma tail, this is sometimes possible to be reported as DE, when the perspective is appropriate.

Acknowledgements. We would like to express our sincere thanks to many astronomers who provided us kindly their valuable photographs of comet Halley. Our special thanks go to the astronomers and publishers in Table 2. Thanks are also due to Dr. T. Oki for his valuable discussions.

References

A'Hearn, M.F., Hoban, S., Birch, P.V., Bowers, C., Martin, R., Klinglesmith, III: 1986, 20th ESLAB Symp. ESA SP-250, Vol. I, 483
Alfvén, H.: 1957, *Tellus* **9**, 92
Alfvén, H.: 1977, *Rev. Geophys. Space Phys.* **15**, 271
Brandt, J.C.: 1985, *Comets*, Univ. Arizona Press, Tucson, p. 519
Chakaveh, S.H., Green, S.F., Ridley, J.K., M.McDonnell, J.A., Hughes, D.W.: 1986, 20th ESLAB Symp. ESA SP-250, Vol. II, 163
Hoeksema, J.T.: 1986 (private communication)
Kojima, M., Kakinuma, T.: 1987, *J. Geophys. Res.* **92**, 7269
Larson, S., Sekanina, Z., Levy, D., Tapia, S., Senay, N.: 1986, 20th ESLAB Symp. ESA SP-250, Vol. II, p. 145
Nakagawa, T., Saito, T., Yumoto, K., Hirao, K.: 1986, *ISAS Research Note, No.* **349**
Neubauer, F.M.: 1986, 20th ESLAB Symp. ESA SP-250, Vol. I, p. 35
Niedner, M.B., Jr., Brandt, J.C.: 1978, *Astrophys. J.* **223**, 655
Russell, C.T., Luhmann, J.G., Elphic, R.C., Neugebauer, M.: 1985, in *Comets*, Univ. Arizona Press, Tucson, p. 561
Russell, C.T., Saunders, M.A., Phillips, J.L., Fedder, J.A.: 1986, *J. Geophys. Res.* **91**, A2, 1417
Schlosser, W., Schulz, R., Koczet, P.: 1986, 20th ESLAB Symp. ESA SP-250, Vol. III, p. 495
Saito, T.: 1969, *Space Sci. Rev.* **10**, 319
Saito, T.: 1986, *ISAS Bull. No.* **15**, p. 69
Saito, T., Yumoto, K., Hirao, K., Aoyama, I., Smith, E.J.: 1986a, in *The Sun and the Heliosphere in Three Dimensions.*, ed. R.G. Marsden, Reidel, Dordrecht, p. 281
Saito, T., Yumoto, K., Hirao, K., Saito, K., Nakagawa, T., Smith, E.J.: 1986b, *Geophys. Res. Letters* **13**, No. 8, 821
Saito, T., Yumoto, K., Hirao, K., Nakagawa, T., Saito, K.: 1986c, *Nature* **321**, 303
Saito, T., Saito, K.: 1986d, 20th ESLAB Symp. ESA SP-250, Vol. I, p. 135
Saito, T., Yumoto, K., Hirao, K., Saito, K., Nakagawa, T., Smith, E.J.: 1986e, 20th ESLAB Symp. ESA SP-250, Vol. III, p. 149
Saito, K., Saito, T., Aoki, T., Yumoto, K.: 1986f, 20th ESLAB Symp. ESA SP-250, Vol. III, p. 155
Saito, T.: *Aurora, Comet, and Heliosphere*, Kyôritsu Publ. Co. (in press)
Saito, T., Saito, K., Aoki, T.: 1987a, *8th Symp. Solar System Sci.*, p. 38
Saito, T., Hirao, K., Oyama, K.-I., Itoh, T., Nakagawa, T., Kojima, M.: 1987b, *ISAS Bull.* No. **46**, 1
Saito, T., Akasofu, S.-I.: 1987c, *J. Geophys. Res.* **92**, A2, 1255
Saito, T., Yumoto, K., Hirao, K., Minami, S., Saito, K., Smith, E.J.: 1987d, *Astron. Astrophys.* (this issue)
Slavin, J.A., Smith, E.J., Tsurutani, B.T., Siscoe, G.L., Jones, D., Mendis, D.A.: 1986, *Geophys. Res. Letters* **13** (3), 283
Tomita, K., Saito, T., Minami, S.: 1987, *Astron. Astrophys.* **187**, 215

Note added in proof: Some details and related topics of this work are to be appeared in review papers "Interactions between the heliomagnetosphere and cometary magnetosphere" by Saito and Oki in *Laboratory and Space Plasmas*, ed. by H. Kikuchi, Springer Verlag (in press), and "Interactions between comet Halley and the solar wind magnetic field" by Saito in *Comet Halley 1986*, Ellis Horwood Ltd.

Astron. Astrophys. 187, 209–214 (1987)

ASTRONOMY AND ASTROPHYSICS

Structure and dynamics of the plasma tail of comet P/Halley

I. Knot event on December 31, 1985

T. Saito [1,★], K. Yumoto [2], K. Hirao [3], S. Minami [4], K. Saito [5], and E. Smith [6]

[1] Institute for Space and Aeronautical Science, Komaba, Meguro-ku, Tokyo, 153 Japan
[2] Onagawa Magnetic Observatory and Geophysical Institute, Tohoku University, Sendai, 980 Japan
[3] Department of Aeronautics and Astronautics, Tokai University, Hiratsuka, Kanagawa, 259-12 Japan
[4] Department of Electrical Engineering, Osaka City University, Sumiyoshi, Osaka, 558 Japan
[5] Hosei University, Fujimi 2-17-1, Chiyoda-ku, Tokyo, 102 Japan
[6] Jet Propulsion Laboratory, California Institute of Technology, Pasadena, CA 91109, USA

Received March 6, accepted August 3, 1987

Summary. Many photographs of an outstanding disconnection event (DE)-like knot in the plasma tail of comet Halley were taken by many Japanese astronomers on December 31, 1985. According to the solar wind magnetic field observed by Sakigake, comet Halley encountered not a sector boundary, but a high-speed stream on that day, demanding an explanation for the DE-like event, different from the sector boundary crossing model. The flow of the knot relative to the nucleus is obtained from the 20 photographs taken for 2.6 h having an averaged velocity of $V \sim 58\ \mathrm{km\,s^{-1}}$ for the distance range from the nucleus, $X = 4 \sim 9\ 10^5$ km. The high and almost constant velocity in the large distance range near the nucleus suggests that some impulsive force was applied to the tail plasma at $X < 4\ 10^5$ km. This evidence implies the previously proposed dynamic pressure process that a sudden increase in the dynamic pressure of the solar wind gave rise to a reconnection of the tail field-lines causing a plasmoid that was impulsively accelerated tailwards. The observed values of X for the observed time, t, coincide quite well with the theoretical $X-t$ relation expected from the dynamic pressure model: $t - t_\mathrm{R} = \int_0^{X-X_\mathrm{R}} \cosh(X/X_0)/[V_0\{\cosh(X/X_0)-1\}+V_\mathrm{R}]\,dX$, where t_R, X_R X_0, V_0, and V_R mean the time of reconnection, distance of the reconnection point from the nucleus, characteristic distance, solar wind velocity, and accelerated velocity by the reconnection, respectively. Since the model was already substantiated also by a computer simulation, all the evidence are provided from observed characteristics, computer simulation, and model calculation, to conclude that the December 31 knot event is well explained by the dynamic pressure model.

Key words: comets – hydromagnetics – interplanetary medium – plasmas – the Sun: solar wind

1. Introduction

Many scientists watched comet Halley expecting occurrences of a disconnection event [abbreviated by Niedner and Brandt (1978) as DE]. According to them, "DE is a major condensation that has become detached from the region of the comet's head, where condensations and knots are sometimes used interchangeably" (Brandt, 1985). They hypothesized that a DE occurs when a comet encounters the sector boundary (that is, the heliomagnetospheric neutral sheet). The world-first chance of an in situ observation of the heliospheric neutral sheet that was just going to cross a comet came on the very day of the closest approach of Sakigake to comet Halley. However, the comet did not show any distinct DE. Then we proposed in a previous paper a "quasi-parallel" model for the observational fact that the plasma tail of a comet was not always drastically disconnected when a comet penetrated through the heliospheric neutral sheet, provided the sheet is quasi-parallel to its plasma tail axis (Saito et al., 1986b). In the present paper, we further study the solar wind-plasma tail interaction by using the Sakigake data for another interval during which one of the most outstanding disturbances of the 1985–1986 apparition of comet Halley was observed.

Hundreds of photographs of comet Halley have very kindly been sent to our Sakigake/IMF team and to the Japanese Astronomical Society by many observers. Surveying over 500 photographs from more than 80 observers, Saito et al. (1986c) tabulated disturbances of the plasma tail throughout the comet's apparition, classifying them as outstanding rays, streamer, helix, kink, arcade, and DE (see the classification by Brandt, 1985). The December 31, 1985 event simultaneoulsly includes various disturbances as well as an outstanding DE-like knot.

The purpose of the present paper is to analyze 20 photographs taken by Japanese astronomers to examine the dynamic pressure model that was proposed in our previous paper (Saito et al., 1986d) to explain the mechanism of the knot event appeared in the plasma tail of the comet in New Year Eve.

2. Observation of comet Halley and the heliosphere

The photographs of the whole feature of comet Halley on this day with $\sim 1.3\ 10^7$ km length of the plasma tail are displayed in Fig. 2 of Saito et al. (1986d). The photographs show that in the plasma tail there are many rays, condensations, helices, knot, etc., which are especially recognized in the two original plates obtained by the

Send offprint requests to: T. Saito (Sendai address)

★ Present address: Onagawa Magnetic Observatory and Geophysical Institute, Tohoku University, Aoba, Sendai, 980 Japan

Table 1. Twenty-two instances at which the DE-like knot event was pictured on December 31, 1985. The pictures on Nos. 18 and 22 were not used in the analysis, because these were sent to us after the submission of our paper by the courtesy of Dr. Liu Zong-li

No.	Time (UT)	By the courtesy of
1	$08^{\mathrm{h}}36^{\mathrm{m}}.0$	M. Koishikawa
2	08 56.0	M. Koishikawa
3	09 16.5	M. Koishikawa
4	09 20.5	K. Ishida
5	09 36.0	M. Honda
6	09 45.3	K. Ishida
7	09 45.5	K. Okamura
8	09 45.8	T. Takenouchi
9	09 52.5	K. Koishikawa
10	09 58.5	K. Okamura
11	10 00.5	A. Fujii
12	10 04.0	M. Honda
13	10 06.5	M. Koishikawa
14	10 11.0	A. Kawazoe
15	10 20.5	M. Koishikawa
16	10 24.5	A. Kawazoe
17	10 33.0	M. Honda
18	10 38.5	Liu Zong-li
19	10 43.5	M. Honda
20	10 57.5	M. Honda
21	11 12.0	M. Honda
22	11 33.5	Liu Zong-li

⌀105 cm Schmidt telescope at $09^{\mathrm{h}}21^{\mathrm{m}}.5$ and $09^{\mathrm{h}}45^{\mathrm{m}}.3$ UT, respectively. We find from the comparison that each of the structure has a downstream flow speed, V, and that V increases from $\sim 60\ \mathrm{km\,s^{-1}}$ to $\sim 240\ \mathrm{km\,s^{-1}}$ with increasing of the distance, X, from the nucleus to the structure. The precise V–X curve for the whole microstructure will be obtained observationally and be discussed theoretically together with a new digital unsharp masking (DUM) method of analysis (Aoki et al., 1987) in the near future. The focus of the present paper will be put on both flow velocity and formation mechanism of the most peculiar knot event that was observed in Japan.

An especially distinct structure of the DE-like knot was pictured at 20 different instances at six different places in Japan, at least, as listed in Table 1. Out of the 20 images, Fig. 1 displays 15 representative photographs taken at 15 different instances at the six different places (Kiso, ⌀105 cm Schmidt telescope; Kohchi, ⌀60 cm reflector; Kagoshima Space Center (KSC), ⌀50 cm Schmidt; Shirakawa, ⌀30 cm reflector; Saitama, ⌀30 cm re-

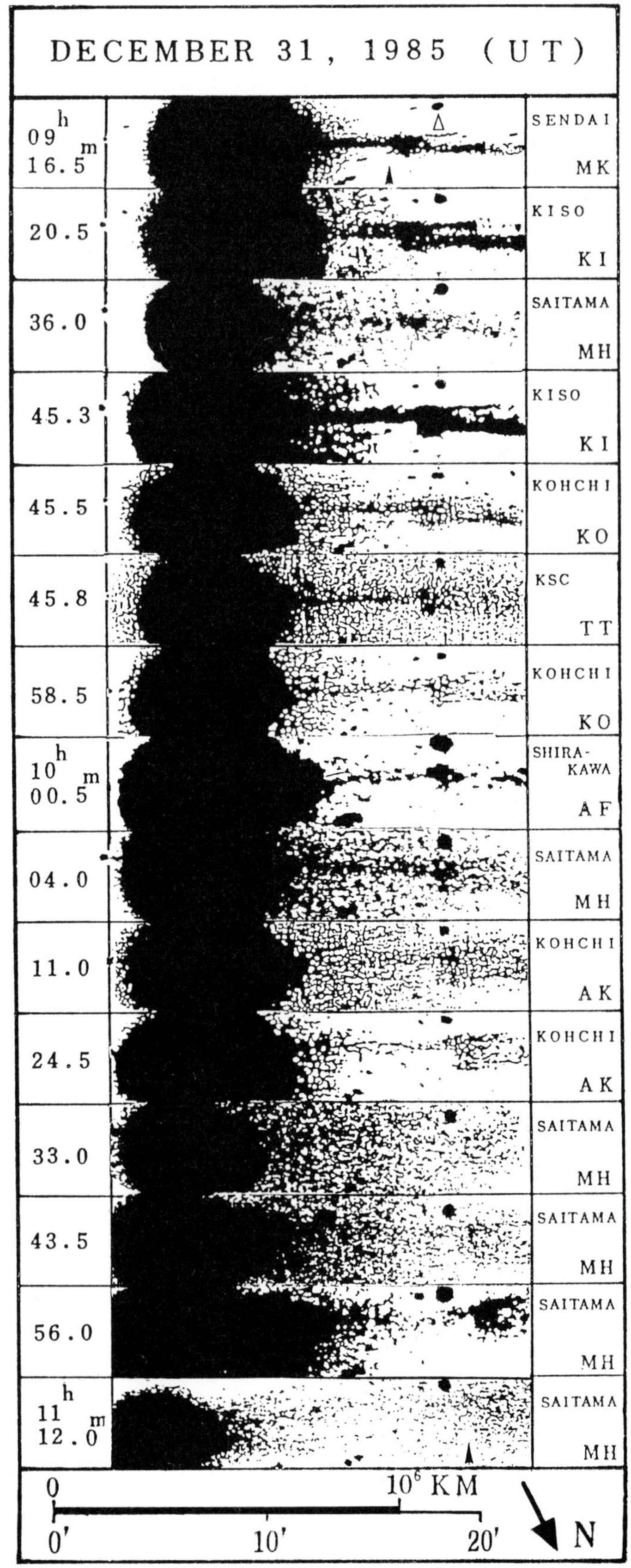

Fig. 1. Knot event of comet Halley observed on December 31, 1986. Representative fifteen photographs among the 22 that were kindly sent to be authors are displayed here in the order of time. Each panels represent the identical portion of the sky. The fixed star indicated by the hollow arrow is BD $-2°5734$ (magnitude: 9.5). Leftward orbital motion of the coma and the rightward flow of the knot (indicated by the solid arrows) are seen clearly. The solid arrows are put on the left side of the knot, because it has the more distinct boundary owing to the pinch effect of the reconnection, while the right side is obscured owing to a diffusion along the open field

flector; and Sendai, ⌀20 cm reflector). In the figure, the 15 panels show the identical portion of the sky, and the knot is indicated by the solid arrows. As is clearly shown in the figure, the coma's orbital motion was upstream of the solar wind, while the knot moved downstream. The knot changed gradually to an arcade and was going to disconnect seemingly from the head at about $10^{h}50^{m}$ UT. (Based on the DUM analysis, the detailed change of the microstructure and its relation to DE will be described later in a separate paper.)

As was reported in our previous paper (Saito et al., 1986d), comet Halley was far from the sector boundary of that day. Let us consider that the solar wind started from the source point P_H on the solar source surface ($r=2.5\,R_\odot$) and arrived at comet Halley on December 31.4 UT, 1985. All the following three conditions played positive roles in shifting the source point P_H toward a higher northern heliomagnetic latitude. (1) The north pole of the solar rotational axis tilted toward comet Halley by 7°.2. (2) The north pole of the equivalent centered dipole for the source-surface magnetic field tilted toward comet Halley by 15°.0. (3) Comet Halley was 8°.7 above the ecliptic plane. Hence, the source point P_H was about 18° north to the warped neutral sheet in heliographic latitude. [As for the detailed description, see Saito et al. (1986d, e).]

According to the two-hemisphere model (Saito, 1975) on solar-cycle variation of the heliomagnetosphere, the heliospheric neutral sheet is almost always one throughout a solar cycle, and is less warped in sunspot minimum phase (see Fig. 4 of Tomita et al., 1987). The less-warped sheet was substantiated by the Sakigake/IMF observation (Saito et al., 1986f; Suzuki et al., 1987), and by the IPS (interplanetary scintillation) observation (Kojima et al., 1987a, b).

These evidences imply that comet Halley was far from the heliospheric neutral sheet when the DE-like knot event took place in that night. Therefore, the evidences would demand for the formation of the DE-like knot some mechanism other than the sector boundary crossing model.

3. Observed flow of the knot

The observed distance X of the knot from the nucleus is read from the 20 photographs of the knot. The distance means here the component along the sun-nucleus line, namely,

$$X = \Delta \sin\alpha \cdot \operatorname{cosec}(\beta-\alpha)\,, \qquad (1)$$

where Δ, α, and β mean the earth-nucleus distance, the nucleus-earth-knot angle, and the earth-nucleus-sun angle, respectively. The observed values of X are plotted against the time, t, by the hollow circles in Fig. 2.

The nearly straightly arranged distribution of the 20 circles suggests a constant velocity of about 58 km s^{-1} as was calculated by Koishikawa and Kasahara (1986). The straight distribution at such small distances from 4 to 9 10^5 km implies as if a plasma cloud was ejected from the nucleus at about 6^{h}.6 UT of the same day with the constant velocity of 58 km s^{-1}. However, cometary particles are generally regarded to leave its nucleus never with a speed of ~60 km s^{-1}, but with the initial thermal speed of ~1 km s^{-1} and to be accelerated very gradually by the ordinary $\boldsymbol{J}\times\boldsymbol{B}$ force as will be explained by the following section. If we call this process an ordinary flow, the observed flow velocities of many microstructures in the plasma tails of comet Bennett and comet Halley were very well explained by this ordinary flow (Minami and White, 1986; Celnik, 1986; Tomita et al., 1987).

In the case of the outstanding kink event of comet Halley on January 10–11, 1986 (Tomita et al., 1987), for example, the velocity of the ordinary flow showed much smaller value (<1.5 km s^{-1}) than in our case of the December 31 knot event (~58 km s^{-1}) in the same range of distance (4~9 10^5 km).

Consequently, we have to explain for the knot event three observational peculiarities; (1) the nearly constant and (2) higher velocities (~58 km s^{-1}) in comparison with the ordinary flow (3) in the wide range at small distances (4~9 10^5 km). All of these three will be explained in Sect. 5 by the dynamic pressure model which was proposed in the earlier papers (Saito et al., 1986a, b). Before the explanation in Sect. 5, let us compare quantitatively our observation with the ordinary flow.

4. Explanation of the ordinary flow

The plasma in the plasma tail with the externally applied magnetic field perpendicular to the solar wind velocity, V_0, is assumed to be accelerated by the $\boldsymbol{J}\times\boldsymbol{B}$ force (Minami and White, 1986), where $\boldsymbol{J}$ is the current flow across the plasma tail.

The kinetic equation is expressed as

$$\frac{m_i N_c V_T}{V}\frac{dv}{dt} = \frac{\sigma_1\sigma_2(V_0-V)B^2}{\sigma_1+\sigma_2}\,, \qquad (2)$$

using a molecular mass, m_i, comet plasma density in the coma, N_c, the initial thermal velocity, V_T, the velocity of the cometary

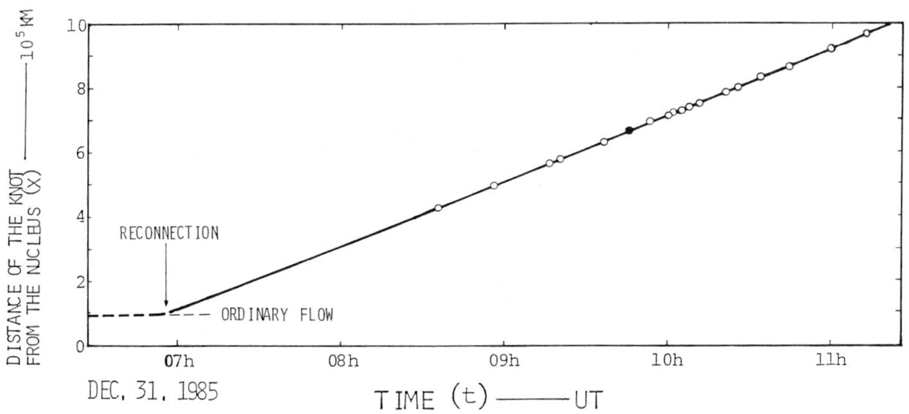

Fig. 2. Flow of the knot as expressed by the nucleus-knot distance (X) against time (t) in UT. The hollow circles represent the observed values (see Fig. 1). The solid circle represent the three observed values that cannot be separately distinguished in the diagram. The dashed curve expresses the ordinary flow [see Eq. (6)], while the solid one expresses the best-fit curve of the accelerated flow [see Eq. (9)]. The acceleration is considered to be generated by the tail field reconnection

plasma, V, the solar wind velocity, V_0, conductivity of the solar wind plasma, σ_1, and conductivity of the cometary plasma, σ_2, respectively. By assuming the magnetic field distribution in the tail as

$$B = B_0\ [\tanh(X/X_0)]^{1/2}, \tag{3}$$

the equation (2) becomes

$$V = [V_0\{\cosh(X/X_0) - 1\} + V_T]/\cosh(X/X_0), \tag{4}$$

where X_0 is the characteristic length for the acceleration as

$$X_0 = \frac{m_i N_c V_T}{B_0^2} \cdot \frac{\sigma_1 + \sigma_2}{\sigma_1 \cdot \sigma_2}. \tag{5}$$

Then the X–t relation becomes

$$t = \int_0^X \frac{\cosh(X/X_0)\,dX}{V_0\{\cosh(X/X_0) - 1\} + V_T}. \tag{6}$$

The dashed line in Fig. 2 indicates an example of the V–t relation when the plasmas left the nucleus with $V_T = 1\ \mathrm{km\,s^{-1}}$ at 0150 UT on December 30, 1986. As seen in the figure, the velocity of the ordinarily accelerated flow (dashed line) is quite smaller than the velocity expected from the observation (hollow circles).

Then, let us explain the observed values by the dynamic pressure model in the next section.

5. Dynamic pressure model for formation of the knot

Cometary rays in the plasma tails were noticed by Saito et al. (1986d) to converge seemingly in most cases not towards the upstream region of the nucleus as are expected from the ordinary draping process, but frequently towards the downstream region as if a DE is starting in the coma. In some cases 90% of the assumed convergent points distributed in the downstream region (Saito et al., 1987a), even though any DE did not occur afterwards (e.g. see the detailed rays on March 12, 1986 that was photographically processed by Marine). This fact implies that the overdraping model (Saito et al., 1986d) must be the general case especially when the dynamic pressure of the solar wind is high as shown in Fig. 3. They proposed the dynamic pressure model that, when the pressure increases suddenly, a field-line reconnection takes place in the plasma tail, causing a plasmoid like the knot as was observed on December 31, 1986. It could be quite similar with the plasmoid in the earth's magnetotail during the so-called SC-triggered auroral substorm.

The magnetic field lines around the plasmoid are considered to be closed like the letter O in the case of the earth's magnetotail, while opened like the letter U in this knot event as inferred by the arcade-shaped rays, which are evident at $09^h20^m.5$ UT or $10^h56^m.0$ UT, for example, in Fig. 1. (The opened structure is seen more significantly by the DUM analysis, though the result will be described in much more detail in a separate paper.) The very difference between the O-type and the U-type characterizes an essential difference between the closed magnetosphere of planets having intrinsic magnetic field (like the earth, for example) and the open magnetosphere of planets having no intrinsic magnetic field (like the cometary nucleus, for example). Hence, this knot event must be important also from the viewpoint of comparative magnetospheres.

Ogino et al. (1986) carried out various computer simulations, some of which were noticed to confirm evidently the dynamic pressure model as seen in Fig. 5 of Saito et al. (1986c, d). Russell et al. (1986) also proposed that the plasmoid is generated in general cometary magnetosphere by the sudden pressure increase as well as an encounter of a shock front to a comet. Since the dynamic

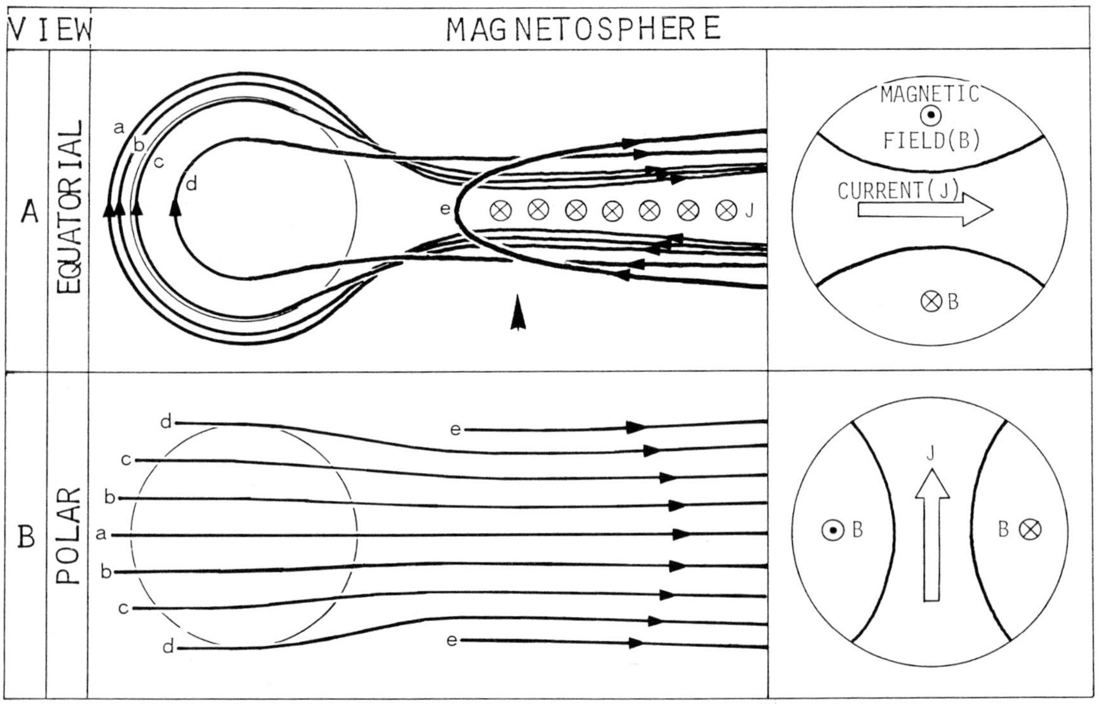

Fig. 3. Three-dimensional general cometary magnetosphere. The equatorial and polar views are so named conventionally by considering the case of the earth's magnetosphere. The over-draped fieldline a is considered to be slipped via b, c, d, to e. When the dynamic pressure of the solar wind increases suddenly, a reconnection of the field lines is considered to take place on the tail axis near the region as indicated by the arrow

pressure model has been confirmed from the viewpoints of both observed characteristics and computer simulation in these ways, then let us compare the model quantitatively with the observed flow data in Fig. 2.

If the reconnection of the tail field lines takes place as is expected from the dynamic pressure model, the tail field lines through the reconnection point at $X = X_R$ move rapidly so that the sharp angle between the two reconnected lines changes to the more and more blunt angle. The rapid movement of the field lines is associated with a rapid movement of the ambient plasma, forming the sweepings, which must be observed as a plasmoid or a knot.

Then the V–X relation in Eq. (4) and the X–t relation in (6) must be expressed by changing V_T in these equations to V_R as

$$V = [V_0\{\cosh(X/X_0) - 1\} + V_R]/\cosh(X/X_0), \tag{7}$$

and

$$t = \int_0^{X-X_R} \frac{\cosh(X/X_0)\,dX}{V_0\{\cosh(X/X_0) - 1\} + V_R} + t_R, \tag{8}$$

where V_R is the plasma velocity secondarily gained by the reconnection, X_R is the distance of the reconnection point from the nucleus, and t_R is the time of the reconnection as expressed by

$$t_R = \int_0^{X_R} \frac{\cosh(X/X_0)\,dX}{V_0\{\cosh(X/X_0) - 1\} + V_T}. \tag{9}$$

Hence, the Eqs. (4) and (6) must be used for the range, $0 < X < X_R$, and $0 < t < t_R$, while the Eqs. (7) and (8) for the range, $X > X_R$, and $t > t_R$, respectively.

The solid line in Fig. 2 indicates the best-fit curve of (8) for the observation. Note that the solid line is not an exactly straight line, but is faintly curved, and that the expected curve coincides quite well with the observed points.

Hence, we may consider in Fig. 2 from the dynamic pressure model that initially the cometary plasma moved slowly as the ordinary flow with the $\boldsymbol{J} \times \boldsymbol{B}$ acceleration along the dashed curve, and secondarily moved rapidly along the solid curve owing to the field line reconnection. The reconnection point must be located at $X_R \leqq 3\ 10^5$ km, because a distinct plasmoid (knot) was already formed at $X \sim 4\ 10^5$ km, the point must also be located at $X_R \geqq 10^5$ km, because the reconnection seems not to occur at so small distance when the result from the computer simulation is taken into consideration. Therefore the turning point from some dashed curve of the ordinary flow to the solid curve of the accelerated flow by the reconnection would exist in the range of the distance $10^5 \leqq X \leqq 3\ 10^5$ km.

Figure 4 displays the V–t curve for the Eqs. (4) and (7) expected from Fig. 2. The impulsive acceleration due to the reconnection is expressed in Fig. 4 with the sudden increase of V at $X = 1\ 10^5$ km. The three observed peculiarities that were pointed out in Sect. 3 can be explained by the dynamic pressure model as expressed in Fig. 4.

6. Discussion and conclusion

The observed knot was explained in terms of the dynamic pressure model based on the 20 photographs. While the present paper was revised, two photographs taken at the Beijing Astronomical Observatory at $10^{\mathrm{h}}38^{\mathrm{m}}.5$ UT and $11^{\mathrm{h}}03^{\mathrm{m}}.5$ UT were kindly sent to us by Dr. Liu Zong-li. The positions of the knot on the two photographs are quite in agreement with our result.

We discussed on the knot event pictured at $4 \sim 9\ 10^5$ km. The overall V–X relation of microstructures in the plasma tail on the same day from $9\ 10^5$ to $1.2\ 10^7$ km was obtained based on the DUM analysis of the two plates (see Fig. 2 of Saito et al., 1986d).

One may have a question that a large-scale sector boundary is not required to detach a tail in the frontside reconnection model; a local, $0.5 - 1.0$ day magnetic polarity reversal in the IMF can also a job, and such small-scale reversals certainly can occur well away from the neutral sheet. Then, let us describe our opinion to this question on the local polarity reversal.

Generally we have four main methods (scanning, midline, minimum-velocity, and potential; see Saito et al., 1987b for the detail) to obtain the heliospheric neutral sheet (or the sector boundary). The undulation amplitudes of the neutral sheet obtained by the first three methods tend to be generally smaller than those obtained by the last, the potential method. The difference may be caused by the three possible reasons; (A) difficulty in estimating the heliopolar field, (B) flattening of the heliomagnetospheric neutral sheet from zero to one AU by high-velocity streams with large heliolongitudinal velocity gradient (see Fig. 3 of Saito et al., 1987c), and (C) with large heliolatitudinal velocity gradient, respectively. Apart from these possible reasons on the difference (Saito et al., 1987b), let us put a focus on the local polarity reversals by using the potential data, since the potential data is very frequently used by many scientists.

The local reversal of the sector polarity on the source surface forms in the interplanetary space a sub-neutral sheet, which has been studied by Saito (1983) with the name of island (on the solar synoptic chart) or subcone neutral sheet (in the interplanetary space, see Fig. 4c, d of Tomita et al., 1987). Such small islands appeared actually on the solar source surface as a result of the potential field calculation, e.g., once in 1978, seven times in 1979, five times in 1980, and once in 1981 (see Hoeksema et al., 1983). It is evident from the data that the islands temporally appear only in sunspot maximum phase when the tilt angle of the main neutral sheet is large, and never appear near around minimum phase when the tilt angle is small and the aligned dipole component is large as in the case of the 1985–1986 apparition of comet Halley.

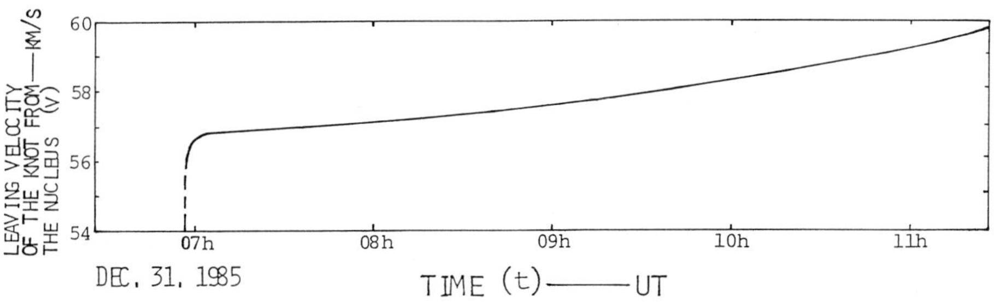

Fig. 4. Relation between the time (t) and the flow velocity (V) obtained by differentiate the curve in Fig. 2. abruptly increased velocity at about 0700 which is considered to be due to the tail field reconnection

The position of the source point, P_H, of the solar wind on the solar source surface to comet Halley on December 31.4, 1987, was estimated to be $B_C = 14°.0$ and $L_C = 159°.0$, where B_C and L_C are heliolatitude and Carrington longitude of P_H, respectively (Saito et al., 1987b). A comparison of this point with the synoptic chart of the solar wind velocity obtained by the IPS observation (Kojima et al., 1987b) revealed a good agreement with the dynamic pressure model.

All of the 20 photographs of the knot event, the overall V–X relation, the effect of the longitudinal velocity gradient, and the comparison of P_H with the IPS observation will be presented and discussed in more detail in separate papers (Saito, 1987, etc.) that are being prepared.

Finally, our conclusion is that the previous dynamic pressure model on the December 31 knot event of comet Halley is substantiated by observed characteristics, computer simulation, and model calculation.

Acknowledgements. The authors would like to express their sincere thanks to ISAS, especially to Prof. M. Oda, Director, and Prof. T. Itoh, Project Manager of the Planet-A mission. Thanks are also due to Prof. Liu Zong-li at the Beijing Astronomical Observatory, Prof. K. Ishida and Dr. T. Aoki at the Kiso Astronomical Observatory, Ex-prof. T. Takenouchi and Mr. M. Seo at KSC, ISAS, Dr. J. Ossaka and Mr. M. Koishikawa at the Sendai Astronomical Observatory, Mr. A. Kawazoe and Mr. K. Okamura at the Geisei Astronomical Observatory, Mr. A. Fujii at the Shirakawa Astronomical Observatory, and Mr. M. Honda for providing valuable photographs of comet Halley. We also thank Dr. M. Niedner of NASA, Prof. T. Kakinuma, Dr. T. Ogino, and Dr. M. Kojima, Nagoya University, and Dr. T. Oki for their valuable discussions.

References

Aoki, T., Saito, T., Okamura, S.: 1987, Analysis of comet Halley by means of the digital unsharp masking (DUM) method and the slipping IMF model, Abstracts, 62th Semiannual Meeting of Japanese Astronomical Society, p. 16

Brandt, J.C.: 1985, Observations and dynamics of plasma tails, in *Comets*, ed. L.L. Wilkening, Univ. Arizona Press, pp. 519–537

Celnik, W.E.: 1986, The acceleration within the plasma tail, the rotational period of the nucleus and the observation of the plasma tail of comet P/Halley 1986, *Proc. 20th ESLAB Symp. on the Exploration of Halley's Comet*, Heidelberg, 27–31 October 1986, ESA SP-250, Vol. **I**, p. 53

Hoeksema, J.T., Wilcox, J.M., Scherrer, P.H.: 1983, *J. Geophys. Res.* **88**, 9910

Koishikawa, M., Kasahara, S.: 1986, Analysis of a peculiar cloudy structure in the plasma tail of comet Halley observed on December 31, 1985 (in Japanese), *Hoshino Techô*, Vol. **3**, 110

Kojima, M., Kakinuma, T.: 1987a, *J. Geophys. Res.* (in press)

Kojima, M., Kakinuma, T., Oyama, K., Mukai, T., Hirao, T., Miyake, W.: 1987b, Solar wind structure observed by interplanetary scintinllation and spacecraft in 1985 and 1986, *Proc. Res. Inst. Atmosph. Nagoya Univ.* (in press)

Minami, S., White, R.S.: 1986, *Geophys. Res. Letters* **13**, 849

Niedner, M.B., Jr., Brandt, J.C.: 1978, *Astrophys. J.* **223**, 655

Ogino, T., Walker, R.J., Ashour-Abdalla, M.: 1986, *Geophys. Res. Letters*, **13**, 929

Russell, C.T., Saunders, M.A., Phillips, J.L., Fedder, J.A.: 1986, *J. Geophys. Res.* **91**, 1417

Saito, T.: 1975, Two-hemisphere model of the three-dimensional magnetic structure of the interplanetary space. Sci. Rep. Tohoku Univ., Ser. 5, **23**, 37–54

Saito, T.: 1983, *Proc. 16th ISAS Lunar and Planetary Symp.*, held at ISAS on July 4–5, 1983, 32–35

Saito, T., Takenouchi, T., Saito, K., Yumoto, K.: 1986a, Halley's plasma tail disturbances and solar wind observed by Sakigake, *Proc. 7th Symp. on Solar System Sci.* held in ISAS on 16–17, January 1986 at ISAS, Publ. by ISAS, Tokyo, pp. 6

Saito, T., Yumoto, K., Hirao, K., Nakagawa, T., Saito, K.: 1986b, *Nature* **321**, 6067, 303

Saito, K., Saito, T., Aoki, T., Yumoto, K.: 1986c, *Proc. 20th ESLAB Symp. on the Exploration of Halley's Comet*, Heidelberg, 27–31 October 1986, ESA SP-250, Vol. **III**, p. 155

Saito, T., Yumoto, K., Hirao, K., Saito, K., Nakagawa, T., Smith, E.J.: 1986d, A disturbance of the ion tail of comet Halley and the heliospheric structure as observed by Sakigake, *Geophys. Res. Letters* **13**, No. 8, 821

Saito, T., Yumoto, K., Hirao, K., Saito, K., Nakagawa, T., Smith, E.J.: 1986e, *Proc. ESLAB Symp. on the Exploration of Halley's Comet*, Heidelberg, 27–31 October 1986, ESA SP-250, Vol. **III**

Saito, T., Yumoto, K., Hirao, K., Aoyama, I., Smith, E.J.: 1986f, *The Sun and the Heliosphere in Three Dimensions*. ed. R.G. Marsden, Reidel, Dordrecht, pp. 281

Saito, T., Saito, K., Aoki, T.: 1987a, *Proc. 8th Solar System Sci. Symp.*, 8–9 Jan. 1987, 38

Saito, T., Yumoto, K., Hirao, K., Oyama, K.-I., Itoh, T., Nakagawa, T., Kojima, M.: 1987b, Ion tail disturbances of comet Halley and the solar wind structure as observed by SAKIGAKE, (I), ISAS Bulletin, No. 46, 1–15

Saito, T., Saito, K., Aoki, T., Yumoto, K.: 1987c, *Astron. Astrophys.* **187**, 201

Saito, T.: 1987d, Interactions between the plasma tail of comet Halley and the solar wind magnetic field. Comet Halley 1986, Publ. by Ellis Horwood Publ. Co. (in press)

Suzuki, Y., Saito, T., Yumoto, K., Nakagawa, T , Oyama, K.: 1987, *Proc. 8th Symp. on Solar System Sci.* (in press)

Tomita, K., Saito, T., Minami, S.: 1987, *Astron. Astrophys.* **187**, 215

Note added in proof: Some details and related topics of this work are to be appeared in review papers "Interactions between the heliomagnetosphere and cometary magnetosphere" by Saito and Oki in *Laboratory and Space Plasmas*, ed. by H. Kikuchi, Springer Verlag (in press).

Astron. Astrophys. 187, 215–219 (1987)

ASTRONOMY AND ASTROPHYSICS

Structure and dynamics of the plasma tail of comet P/Halley

II. Kink event on January 10–11, 1986

K. Tomita [1], **T. Saito** [2,*] **and S. Minami** [3]

[1] Advanced Engineering Services Co. 2027 Uenomuro, Sakura-mura, Niihari-gun, Ibaraki, 305 Japan
[2] Institute for Space and Aeronautical Science, Komaba, Meguro-ku, Tokyo, 153 Japan
[3] Department of Electrical Engineering, Osaka City University, Sumiyoshi, Osaka, 558 Japan

Receive March 6, accepted July 9, 1987

Summary. An outstanding DE-like kink event occurred in the plasma tail of comet P/Halley on January 10–11, 1986. The kink was traced for 33 h in these days by surveying 24 photographs of the comet. The images of the kink on eleven representative photographs out of the 24 were transcribed on the Palomar Sky Survey Atlas to investigate its motion and formation. In order to explain the kink event, the following two problems are discussed: one is how the observed relation of the nucleus-kink distance, X, versus time, t, is explained, while the other is how the kink is formed. As for the observed X-t relation, a theoretical function, $t = \int \cosh(X/X_0)/[V_0\{\cosh(X/X_0) - 1\} + V_\mathrm{T}]\,dX$, is applied being based on the $\boldsymbol{J} \times \boldsymbol{B}$ acceleration process, where X_0, V_0, and V_T are characteristic length, final solar wind velocity, and initial (thermal) velocity, respectively. The expected curve coincides very well with the observation. As for the mechanism of the kink formation, an intrusion of an intermediate-velocity flow into a low-velocity flow along the tail axis is presented to be responsible taking into account a result of MHD computer simulation.

Key words: comets – hydromagnetics – interplanetary medium – plasmas – the Sun: solar wind

1. Introduction

Besides an interest in cometary physics itself, the authors present two kinds of approaches for the study of a comet. One is comparative magnetosphere; among a large amount of stars, a comet is the only celestial body whose entire magnetosphere can be visually observed from the earth. The other is three-dimensional heliosphere; as a sensitive moving wind vane, a comet, having generally a larger orbit inclination than other planets and almost all interplanetary spacecraft, provides us with important information on high-latitude heliospheric conditions which can hardly be detected by other kinds of observations. From these viewpoints Saito et al. (1987a) compiled a catalogue of the main plasma tail disturbances of comet Halley during its 1985/1986 apparition.

By surveying the catalogue, we concluded that the most outstanding kink event occurred in the comet Halley's plasma tail on 10–11 January 1986. (The relation between the kink event and the disconnection event (DE) that was reported by Brandt et al. (1986) and others will be commented upon in the next section.) The purpose of the present paper is to trace the flow of the kink for more than 33 hours based on many photographs and to propose a mechanism of the kink formation.

2. Data and display

Twenty-four photographs of this kink event were collected by one of the present authors [K.T.] from the world-wide places through various ways. The best eleven photographs out of the 24 were exhibited during the Heidelberg Conference (Saito et al., 1986b). Analyses of the same kink event were published also by many researchers in Heidelberg (Kubacek et al., 1986; Shkodrov et al., 1986; Liu, 1986). Brandt and Niedner (1986), and Caron et al. (1986) reported that a spectacular disconnection event (DE) occurred during the same days. The time range reported as DE seems to be a little different from that when the kink was observed. In the present paper, the plasma tail disturbance on the quoted days will be studied from the viewpoint of the kink event in the following way. The plasma tail with the kink on the eleven photographs are transcribed on the copy of the Palomar Sky Survey Atlas. Since the faintest of the stars pictured on all the photographs is registered on the Atlas, the configuration of the plasma tail can be exactly transcribed on the Atlas. This method is regarded to be accurate enough to locate such a faint and diffuse phenomenon as the plasma tail structure without using a measuring machine. The necessary accuracy is fully satisfied by this method, because one arc minute corresponds to one mm length on the Atlas.

In the order of the observation time, Fig. 1 exhibits the kink transcribed in this way on the eleven panels with the identical portion of the sky. The place and the time of the observation are noted on both sides of the panels. Figure 2 displays the movement of the kink in a time-versus-distance diagram, where the latter means the distance of the kink from the nucleus which is located by the Yeomans' calculation. The distance, X, is expressed along the sun-comet line taking into account of the relative position among the comet, the sun, and the earth [see Eq. (1) of Saito et al.

Send offprint requests to: T. Saito

* Present address: Onagawa Magnetic Observatory and Geophysical Institute, Tohoku University, Aoba, Sendai, 980 Japan

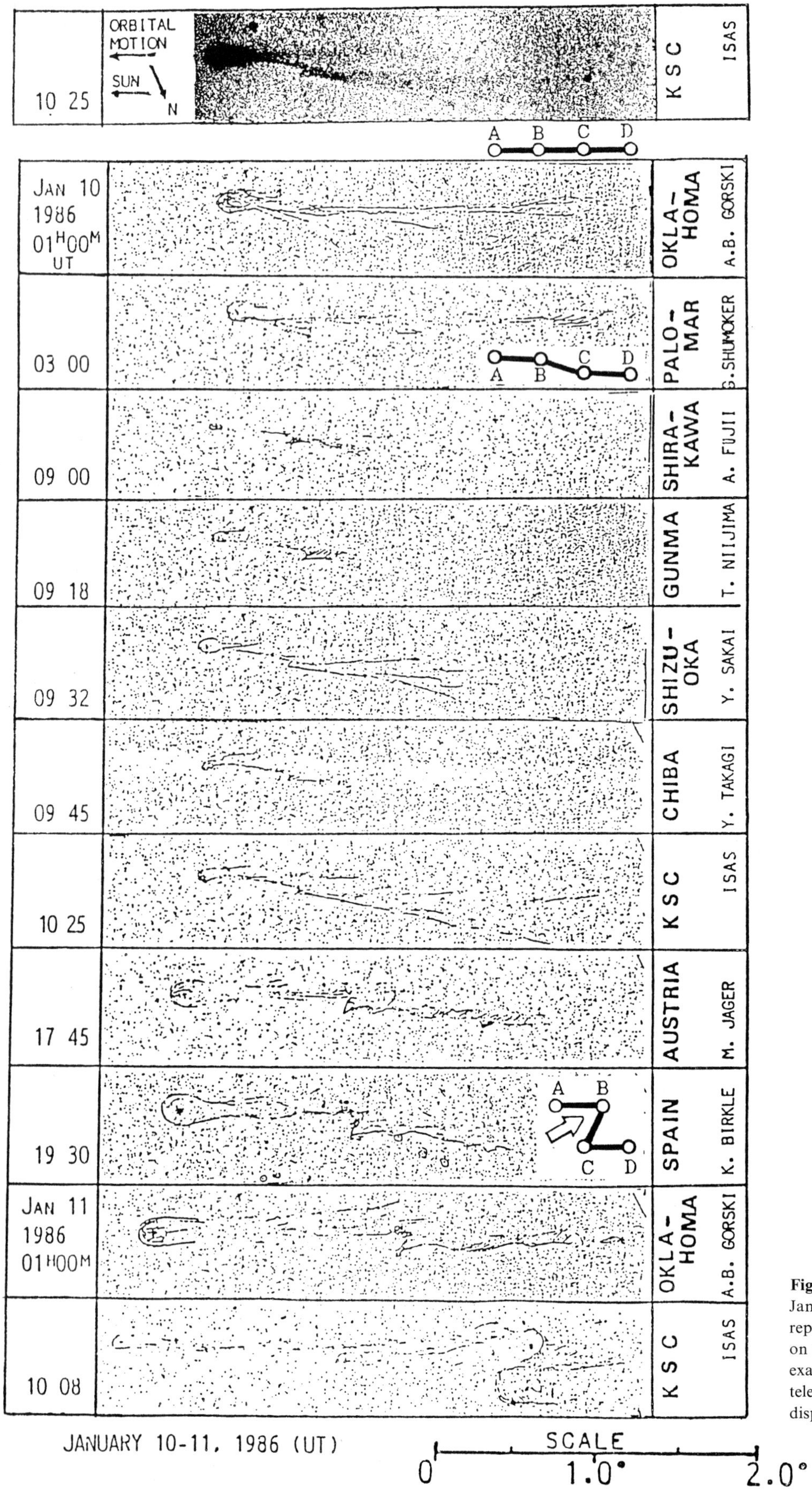

Fig. 1. Kink event of comet Halley observed on January 10–11, 1986. Comet Halley on representative eleven photographs are transcribed on the same area of Palomar Sky Atlas. An example photograph taken by the 50-cm Schmidt telescope at the Kagoshima Space Center is displayed at the top

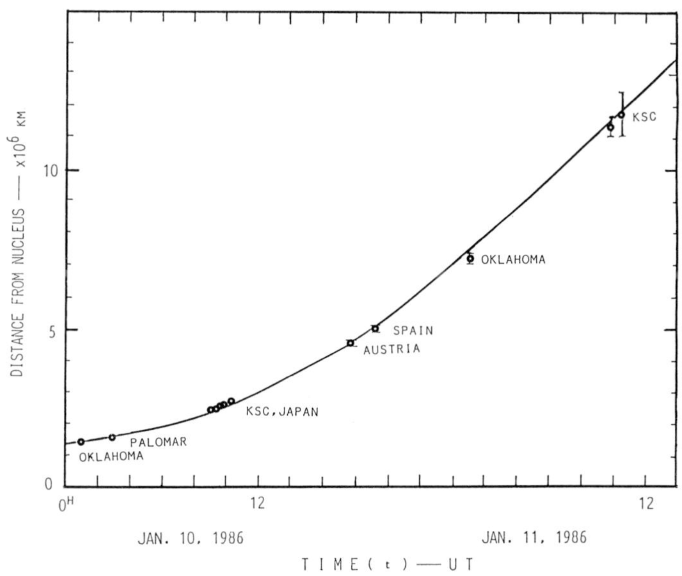

Fig. 2. Flow of the kink as expressed by the nucleus-kink distance (X) against time (t) in UT. The observed values (hollow circles) are expressed well with the best fit curve from the Eq. (1)

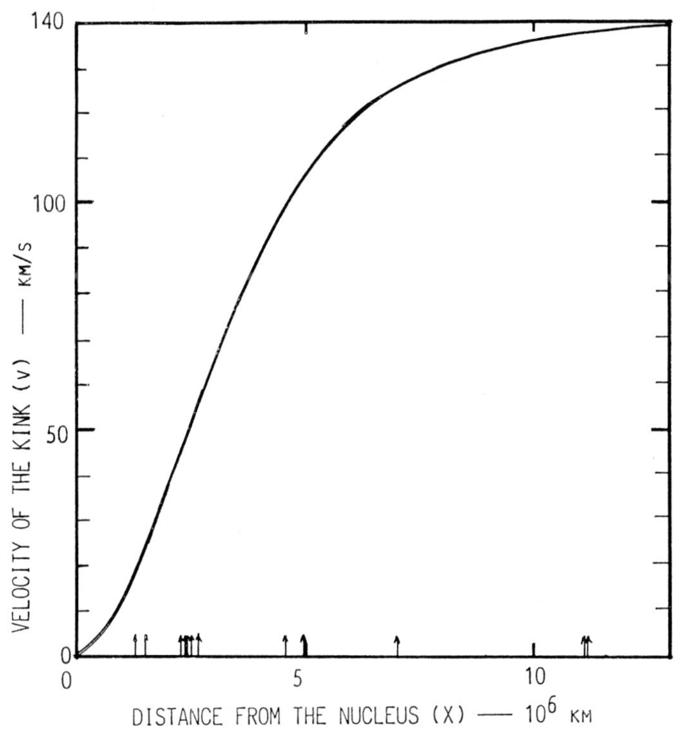

Fig. 3. Relation between the distance (X) and the flow velocity (V) obtained by differentiating the curve in Fig. 2. See the ordinary flow without any abrupt change like that in Fig. 4 of Saito et al. (1987b)

(1987b)]. The observed distance, X, of the kink against the time, t, in UT is expressed by the hollow circles in Fig. 2.

3. Flow velocity of the kink

In order to explain the observed X-t relation, we considered first the ordinary $\boldsymbol{J} \times \boldsymbol{B}$ force to the cometary plasma. Since the additional acceleration due to the tail field reconnection is not taken into consideration here, let us call this case the ordinary flow. The equation for the ordinary flow is given as

$$t = \int_0^X \frac{\cosh(X/X_0)\, dX}{V_0\{\cosh(X/X_0) - 1\} + V_T}, \tag{1}$$

where X_0 is characteristic length, V_0 is the solar wind velocity, and V_T is the thermal velocity (see the detail in Minami and White, 1986; or Saito et al., 1987b).

The best-fit X-t curve of Eq. (1) for the observed values is given by the solid curve in Fig. 2. The figure indicates that the observed values are very well expressed by the ordinary flow. Fig. 3 indicates the V-t relation that was obtained by differentiating Eq. (1). The name of the ordinary flow for this case may be appropriate, because there is no abrupt increase of V in the curve in Fig. 3, in contrast to the impulsive acceleration due to the tail field reconnection in Fig. 2B of Saito et al. (1987b).

4. Formation of the kink

Let us put the four points, A, B, C, and D from left to right on a line string AD as shown in the inserted illustration in Fig. 1. When we move the part AB to the upper part of the fixed segment CD, the four points form the letter "Z". If we denote ∢ ABC (= ∢ BCD) the kink angle, it decreases generally from 180° to ~40° during the movement of AB.

In the case of the January 10–11 kink event, the kink angle decreased from ~160° at 00^h00^m on January 10 gradually to ~50° at 01^h00^m on January 11. Associated with the decrease of the kink angle, the length of BC increased from ~5 arcmin at 09^h00^m on 10th to ~15′ at 01^h00^m. About nine hours later, the Z-shape turned at last to the large inverted S-shape. These changes of the kink were as if a higher velocity stream intruded towards the corner B of the letter Z (see the arrow in the insert in Fig. 1).

If all the flow around the comet is stable and laminar, the ordinary flow along the tail axis is slowest, because it starts from the nucleus only with $V_T \sim 1\,\mathrm{km\,s^{-1}}$ to V_0, while the external (solar wind) flow with the constant velocity of V_0 is fastest. The flow velocity in the space between the tail axis to the constant solar wind region must be intermediate so that the velocity increases with increasing of the radial distance from the tail axis. There must be a large velocity shear between the axial flow and the external flow especially near the coma.

If the velocity of the external flow on the northern side, for example, of the tail increases suddenly, the increased velocity shear makes an intrusion of the solar wind from the northern side into the intermediate region of the tail. The sudden deformation of the northern boundary of the cometary magnetosphere would not be visualized, because the density of the cometary plasma is extremely low there. However, the intrusion on the boundary would give rise to a sudden intrusion of the intermediate flow from the northern side to the axial ordinary flow. The intrusion to the ordinary flow with high plasma density must be visually observed as a kinked deformation of the plasma tail as the case of the January 10–11 event.

In this case, the external solar wind does not need a large time lag (as in the case of the dayside draping around the cometary ionosphere) to make the intrusion of the intermediate flow to the ordinary one. If we assume $450\,\mathrm{km\,s^{-1}}$ for the high-velocity

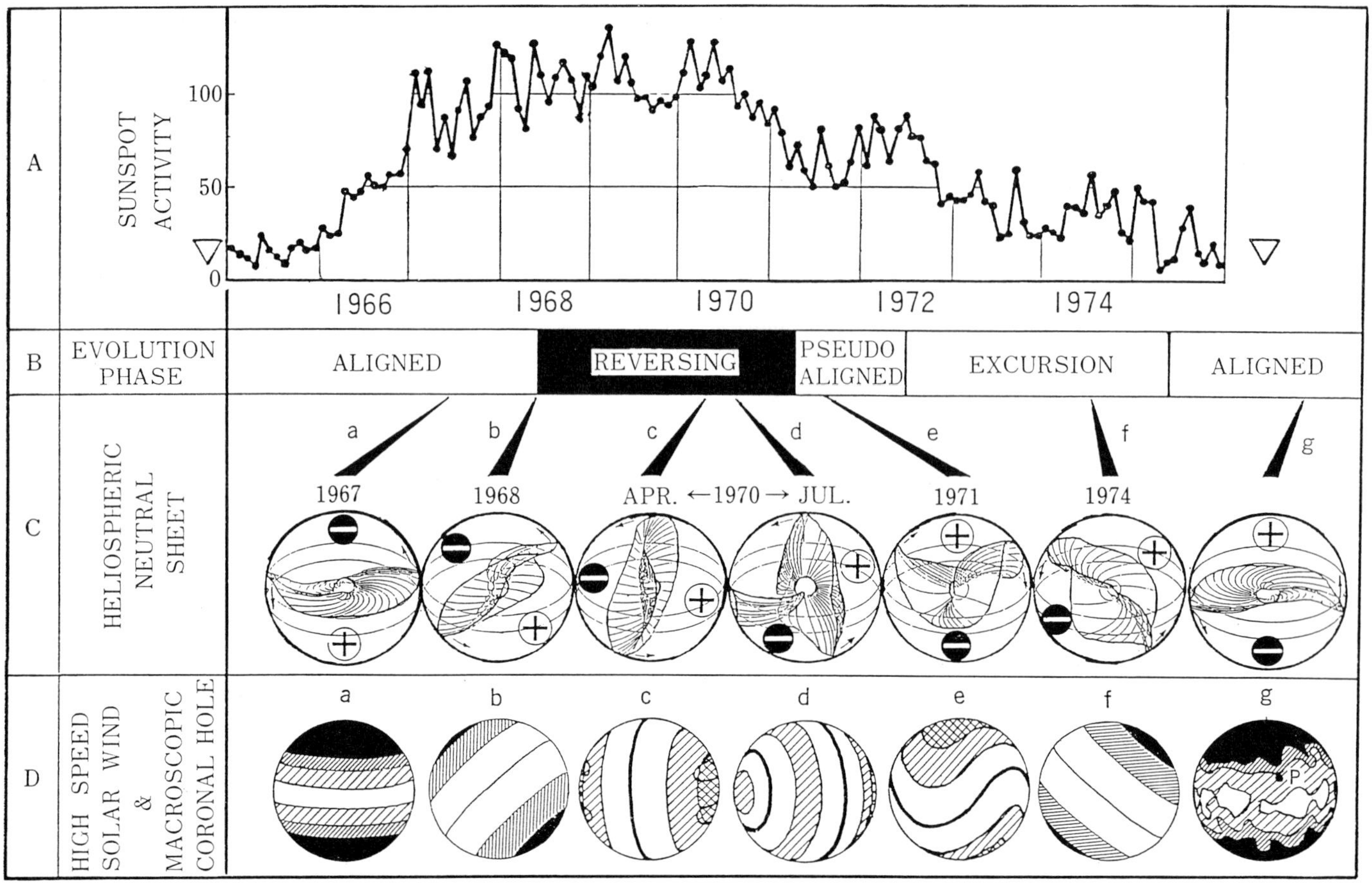

Fig. 4. Two-hemisphere model on solar cycle variation of the heliosphere. Solar cycle phases are expressed from the viewpoints of (A) sunspot, (B) heliomagnetosphere (C) heliospheric neutral sheet, and (D) distribution of the solar wind and coronal hole. The solar wind stream that gave rise to the kink event is considered to flow out from the point P in Dg of the figure. The wind speed distribution on the solar source surface as viewed from $B = +7°0$, $L = 0°0$ is expressed being based on the IPS observation. See the difference of the distribution before average (Dg) and after average (Da ~ Df)

stream, the source of the stream can be located at $B = 17°7$ and $L = 10°6$ on the solar source surface, where B and L are heliographic latitude and Carrington longitude, respectively. The source location seems to be very appropriate, because a source region from which a high velocity stream with 450 km s^{-1} flows out can be identified there (see the point P in Fig. 4Dg) on the synoptic chart of the solar wind velocity distribution that was obtained by the interplanetary scintillation (IPS) observation (Kojima et al., 1987).

Finally we may write the following possible scenario on the formation of the January 10–11 kink event. A high-velocity stream flew out from the region at $B = 17°7$ and $L = 10°6$ on the source surface and attacked the cometary magnetosphere on January 10.

The attack of the external wind gave rise inside the cometary magnetosphere to the secondary effect of a southward intrusion of the intermediate-velocity flow into the low-velocity ordinary flow along the tail axis, forming the January 10–11 kink event.

5. Discussion and conclusion

A possible interpretation was given for the observed flow in Sect. 3 and the observed configuration of the kink in Sect. 4. Since the attack of the high-velocity stream was assumed to be moderate, the movement of the kink was explained by a small modification of the ordinary flow with considering neither an extremely large acceleration nor a drastic magnetic field reconnection as in the case of the December 31 knot event (Saito et al., 1987b).

The 1986 apparition of comet Halley was predicted to correspond to an aligned phase in the heliomagnetospheric cycle (see Fig. 1 of Saito et al., 1986a; or Fig. 2 of Saito et al., 1987c), when a low velocity region near the flat solar sector boundary is generally sandwiched by two high velocity regions in higher heliographic latitudes as seen in the coarse axisymmetric distribution of the solar wind velocity on the source surface in Fig. 4Da. Besides this general latitudinal distribution, there are actually longitudinal inhomogeneity of the velocities, forming several tongues of high velocity wind source that extend from the higher latitude regions towards the solar equation (Kojima et al., 1987). Our conception on the fine distribution of the solar wind velocity is expressed schematically on the source surface as viewed from the Carrington longitude of 0°0 in Fig. 4Dg, which is illustrated being based on the IPS observation in Fig. 4 of Kojima et al. (1987). The source region at $B = 17°7$ and $L = 10°6$ was one of the several tongues of the coronal hole extending from north to the equator as shown in Fig. 4Dg. This situation agrees with our model, because the kink was observed on the northern side of the plasma tail as seen in Fig. 1 (Note that the solar wind is expressed to flow from left to right in all our figures as is accustomed to be so in the case of the earth's magnetosphere).

The change of the configuration of the kink suggested the intrusion process in Sect. 4. In Sect. 3, we read the middle point between B and C in each panel of the kink in Fig. 1 as the first approximation and expressed by the hollow circles in Fig. 2. The ordinary flow of the point C and the slightly modified flow of point B will be separately expressed on an X-t diagram in our next paper to substantiate the intrusion process.

Acknowledgements. The authors wish to express their sincere thanks to the observers, whose names are cordially cited in Fig. 1, which was based on valuable photographs taken by them. Our hearty thanks are to Dr. J.C. Brandt of Colorado University, Dr. M.B. Niedner of NASA, and Dr. T. Oki for their very valuable discussion to this study. Thanks are also to Dr. K. Hurukawa of Tokyo Astronomical Observatory for providing us with his valuable data of subcomet position on the Sun.

References

Brandt, J.C., Niedner, M.B., Jr.: 1986, 20th ESLAB Symp., ESA SP-250, Vol. **I**, p. 47

Caron, R., Guerin, P., Koutchmy, S., Sarrazin, M., Zimmermann, J.P.: 1986, 20th ESLAB Symp., ESA SP-250, Vol. **III**, p. 143

Kojima, M., Kakinuma, T.: 1987, *J. Geophys. Res.* (in press)

Kubacek, D., Pittich, E.M., Zvolankova, J.: 1986, 20th ESLAB Symp., ESA SP-250, Vol. **III**, p. 277

Liu Zong-li: 1986, 20th ESLAB Symp., ESA SP-250, Vol. **I**, p. 145

Minami, S., White, R.S.: 1986, *Geophys. Res. Letters* **13**, 849

Niedner, M.B., Jr., Brandt, J.C.: 1978, *Astrophys. J.* **223**, 655

Saito, T., Yumoto, K., Hirao, K., Aoyama, I., Smith, E.J.: 1986a in *The Sun and the Heliosphere in Three Dimensions.*, ed. R.G. Marsden, Reidel, Dordrecht, p. 28

Saito, K., Saito, T., Aoki, T., Yumoto, K.: 1986b, 20th ESLAB Symp., ESA SP-250, Vol. **III**, p. 155

Saito, T., Saito, K., Aoki, T., Yumoto, K.: 1987a, *Astron. Astrophys.* (this issue)

Saito, T., Yumoto, K., Hirao, K., Minami, S., Saito, K., Smith, E.: 1987, *Astron. Astrophys.* **187**, 209

Saito, T., Akasofu, S.-I.: 1987c, *J. Geophys. Res.* **92**, 1255

Shkodrov, V., Ivanova, V., Bonev, T.: 1986, 20th ESLAB Symp., ESA SP-250, Vol. **III**, p. 195

Note added in proof: Some details and related topics of this work are to be appeared in review papers "Interactions between the heliomagnetosphere and cometary magnetosphere" by Saito and Oki in *Laboratory and Space Plasmas*, ed. by H. Kikuchi, Springer Verlag (in press), and "Interactions between comet Halley and the solar wind magnetic field" by Saito in *Comet Halley 1986*, Ellis Horwood Ltd.

Astron. Astrophys. 187, 220–224 (1987)

ASTRONOMY
AND
ASTROPHYSICS

Hot ions observed by the Giotto ion mass spectrometer at the comet P/Halley contact surface

R. Goldstein [1], D.T. Young [2], H. Balsiger [3], F. Buehler [3], B.E. Goldstein [1], M. Neugebauer [1], H. Rosenbauer [5], R. Schwenn [5], and E.G. Shelley [4]

[1] Jet Propulsion Laboratory, California Institute of Technology, Pasadena, CA 91109, USA
[2] Los Alamos National Laboratory, Los Alamos, NM 87545, USA
[3] Physikalisches Institut, University of Bern, CH-3012, Bern, Switzerland
[4] Lockheed Palo Alto Research Laboratory, Palo Alto, CA 94304, USA
[5] Max-Planck-Institut für Aeronomie, D-3411 Katlenburg-Lindau, Federal Republic of Germany

Received February 18, accepted June 19, 1987

Summary. Just inside the contact surface (~ 4700 km) the High Energy Range Spectrometer (HERS) sensor of the Giotto Ion Mass Spectrometer (IMS) detected a sudden, intense burst of ions that lasted until the HERS sensor ceased transmitting data at a distance of ~ 3000 km from comet P/Hallley. During this brief interval ions with $M/Q = 1, 2, 12, 14, 16–19, 24$, and 28 were observed. The heavier ions appear in two populations (in the S/C frame): 1) a very low energy, almost omnidirectional distribution, and 2) a more energetic ($\sim$ ram speed) population coming from the ram direction. In this paper we investigate the possible origin(s) of these two populations. In particular, we discuss whether the low energy ions belong to the natural Halley enviroment or are generated at the spacecraft by dust and gas bombardment.

Key words: comet P/Halley – Giotto – cometary ion composition and energy

1. Introduction

The HERS sensor of the Giotto Ion Mass Spectrometer measured three dimensional, mass and velocity analyzed ion distributions during the comet Halley flyby (Balsiger et al., 1986). The viewing geometry of the instrument in relation to the spacecraft trajectory is shown in Fig. 1. Ions are observed (in the spacecraft frame) in a fan 60° wide in elevation (or polar) angle from 15° to 75° of the ram direction. This fan, about 3° thick, sweeps out an annular field of view in azimuth angle as the spacecraft spins. The azimuth angle is divided into 32 sectors (sector 32 is blanked while the electronics switches modes); the diagram shows that sectors 1 and 32 are in the comet direction, while sectors 16 and 17 correspond to pointing in the solar direction. The HERS sensor measures cyclically in each of four mass ranges: light, from $M/Q = 2$ to 4, medium, from 12 to 26, heavy, from 16 to 34, and protons only. In

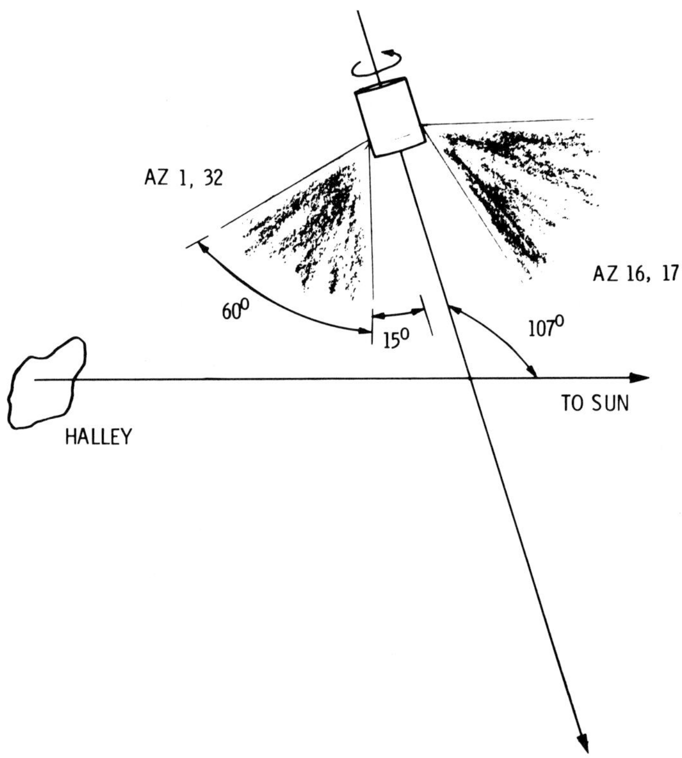

Fig. 1. IMS HERS viewing geometry. The spin axis is closely aligned with the spacecraft velocity vector

the encounter mode data from two spins of each range are summed before being telemetered down.

As described in Balsiger et al. (1986) inside the bow shock, at a distance of 100,000 km, heavy ions flowing toward the comet were observed by the HERS sensor. At about 62,000 km from the comet, the plasma became colder and was barely observable in the instrument field of view. At several thousand kilometers from the comet the HERS sensor measured a sudden burst of ions in all

Send offprint requests to: R. Goldstein

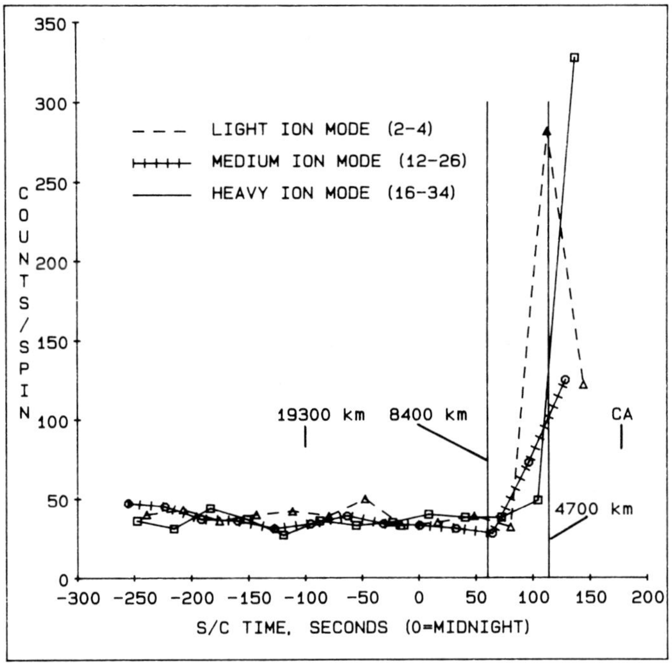

Fig. 2. Total HERS counts/spin measured in three mass ranges shortly before closest approach (CA). The vertical line labeled 8400 km corresponds to the point at which the magnetometer reported the beginning of a sharp drop in the magnetic field intensity, which eventually fell to zero at the point labelled 4700 km

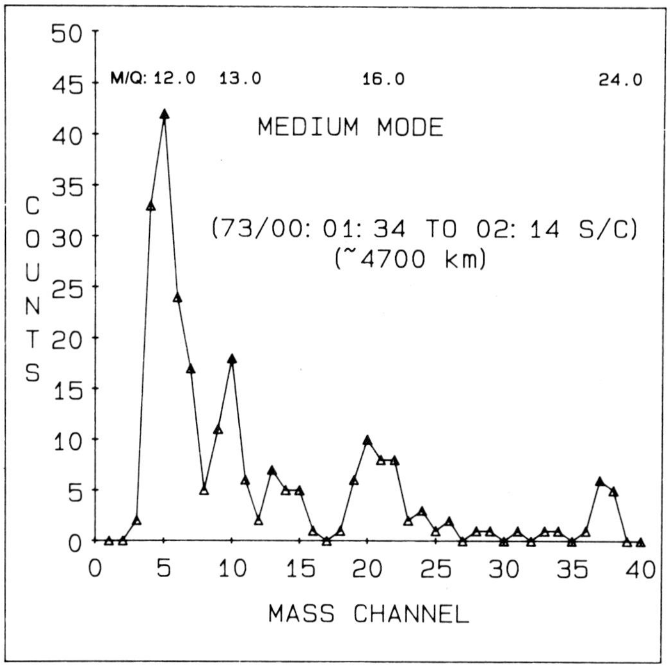

Fig. 3. HERS medium ion mass spectrum in contact surface region (Distance for the comet was from 6019 to 3019 km)

mass ranges, and then, as a result of an electrical failure, ceased transmission of data.

Figure 2 shows the total ion counts per spin in three of the mass ranges during this period. (The proton countrates will not be discussed in this paper.) Note that the background countrate at this time is estimated to by about 32 counts per spin. At a distance of about 8400 km, marked in Fig. 2, the magnetometer reported (Neubauer et al., 1986) the beginning of a rapid drop in the field strength magnitude, which eventually fell essentially to zero at about 4700 km, also marked in the figure. This feature has been identified as passage of the spacecraft through the contact surface. From this diagram it is seen that the sudden burst measured by the HERS sensor occurred during passage into this region. Tantalizingly, the data abruptly end here. The time of closest approach (CA) is also marked in the figure. In this paper we discuss the results of anlyzing the data in this "burst" region, and their interpretation.

2. Mass spectra

The counts measured by the HERS sensor during the "burst" are real ions and not noise generated by arcing or other disturbance within the instrument just before failure; this can be seen by examining the mass spectra which these data represent. The mass spectra, of course, also help describe the character of the plasma. Figure 3 shows the mass spectrum obtained for the last four spins of the medium ion mode. (Background counts have been partially removed for this figure.) The approximate spacecraft time and distance to the comet for these data are given in the Figure.

The predominant peaks are identified by their M/Q values. The spectrum is similar to those observed earlier in this mass range. The spectra for the other mass ranges are also similar to the corresponding ones observed at other distances from the nucleus.

In the remainder of the paper we will discuss details only of the ions in the $M/Q = 12$ peak, although preliminary analysis indicates that at least the water group ions behave in a similar way.

3. Azimuth angle distributions

In order to understand the character of the plasma in the contact surface region, we have compared the measured azimuth angle distribution with that at two other times, as shown in Fig. 4. Because of low countrates, we have summed pairs of azimuth bins for this figure. The solid curve shows the data summed over 26 spins relatively far from the comet. The prominent peak centered at about the solar direction (see Fig. 1) indicates a flow in the antisolar direction, and the width shows that the plasma was relatively warm.

The dashed curve, for data from 40 spins at a closer distance when the HERS sensor measured a very low count rate, also indicates flow away from the sun and the narrower peak shows that the plasma here was much cooler than it was farther out.

The ticked curve corresponds to the last four spins, while passing through the contact region, and shows an entirely different kind of plasma. Although there is a very small peak in the antisolar direction, the distribution is essentially isotropic in azimuth.

4. Velocity distributions

The energy and elevation angle distribution of the ions in the contact surface are also different from those at greater distances

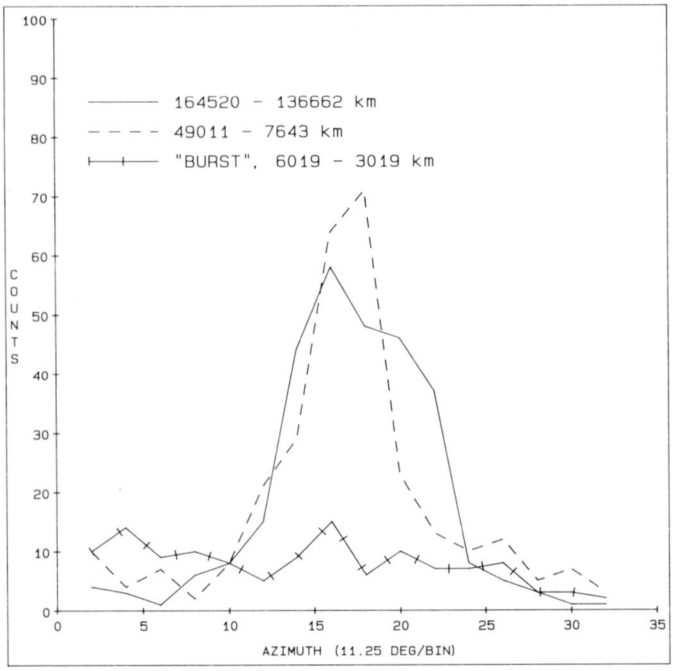

Fig. 4. HERS $M/Q = 12$ azimuth angle distribution at three spacecraft to comet distance ranges

from the comet. Figures 5–7 show scatter plots of the data as function of speed and elevation angle in the comet frame of reference, for the same three spacecraft to comet distances discussed above. Ideally, contour plots would be desired here but because of poor counting statistics for four spins, only these scatter plots are shown.

The field of view of the sensor transformed into this space is shown by the dashed lines. This field of view corresponds to the elevation angle fan sketched in Fig. 1, and would also sweep out an annulus as the spacecraft spins about the spacecraft velocity direction. The 10 eV lower energy cutoff of the sensor results in the truncation near 60 km s^{-1}. The data are plotted to correspond to the mean angle of each of the 8 elevation angle anodes, although there is, in fact, continuous elevation angle coverage over the whole field of view.

In Figs. 5 and 6 the heavy straight lines indicate the range of speed and angle over which the $M/Q = 12$ ions were observed at those distances indicated. Figure 5 represents the data from about 136662 to 164520 km. They show a relatively broad angle and velocity distribution, consistent with the observation above that in this region the plasma was relatively warm. Note that because of the azimuth distribution (Fig. 4), these ions appear only in a relatively narrow portion of the annulus swept out by the spacecraft rotation, flowing anti-sunward.

Figure 6 represents the data from about 49011 to 7643 km, the region where the plasma had become colder. Thus, as expected, the velocity and angle distributions are narrower. Again, as indicated by the azimuth distribution (Fig. 4) these ions appear only in a restricted portion of the field of view swept out by the spacecraft rotation, flowing anti-sunward.

Finally, Fig. 7 shows the distribution (in more detail than in Figs. 5 and 6) of the data from the contact surface region. Although the ions have a nearly isotropic azimuth distribution (Fig. 4), for convenience, the sum over all azimuths is plotted here in the same sector as in Fig. 5 and 6. Here we see a distinctly new

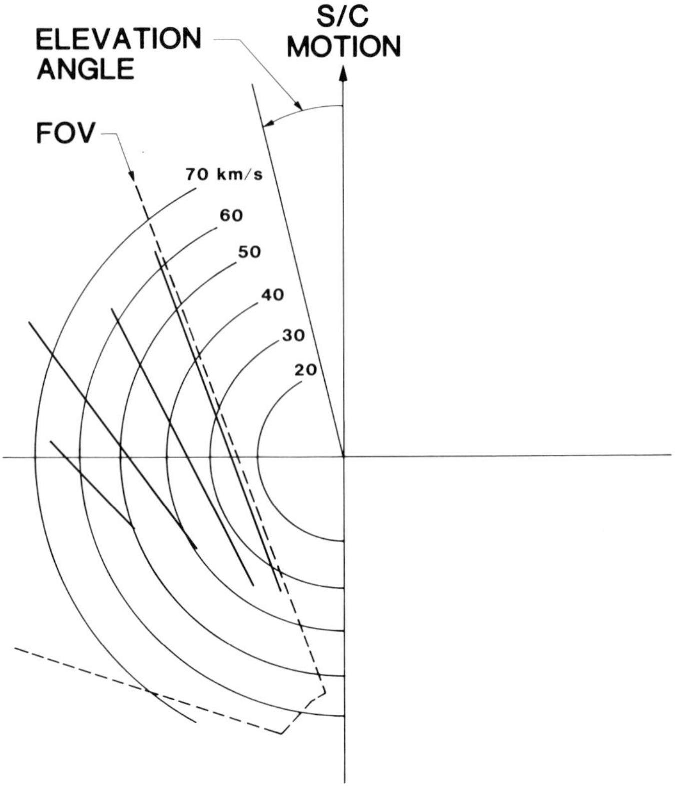

Fig. 5. HERS $M/Q = 12$ velocity-elevation angle distribution in the comet frame, distance range 16,4520 km to 13,6662 km to the comet

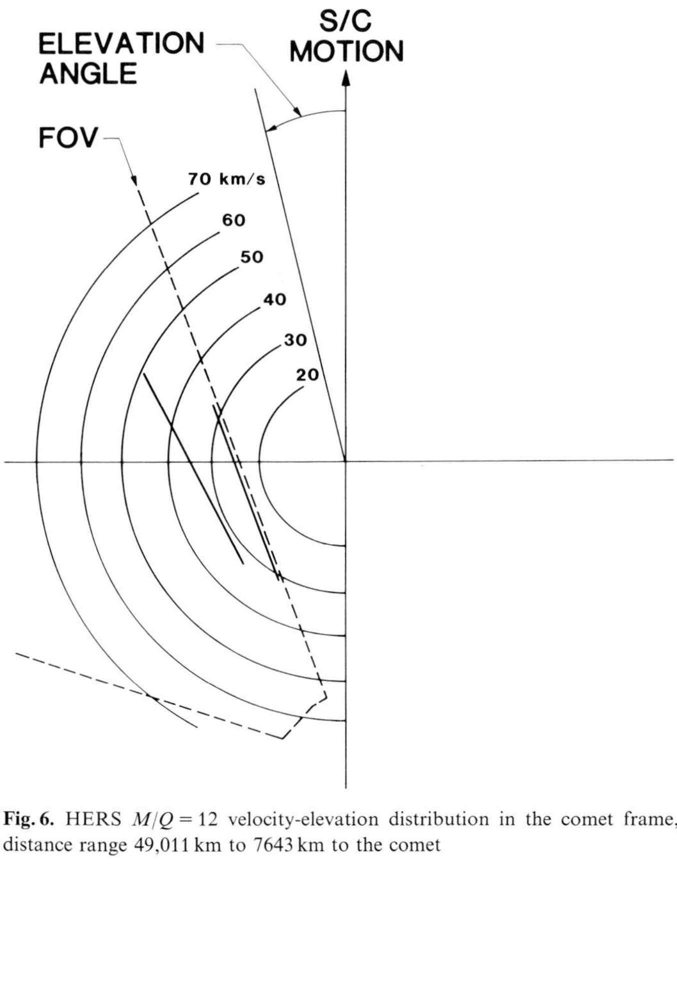

Fig. 6. HERS $M/Q = 12$ velocity-elevation distribution in the comet frame, distance range 49,011 km to 7643 km to the comet

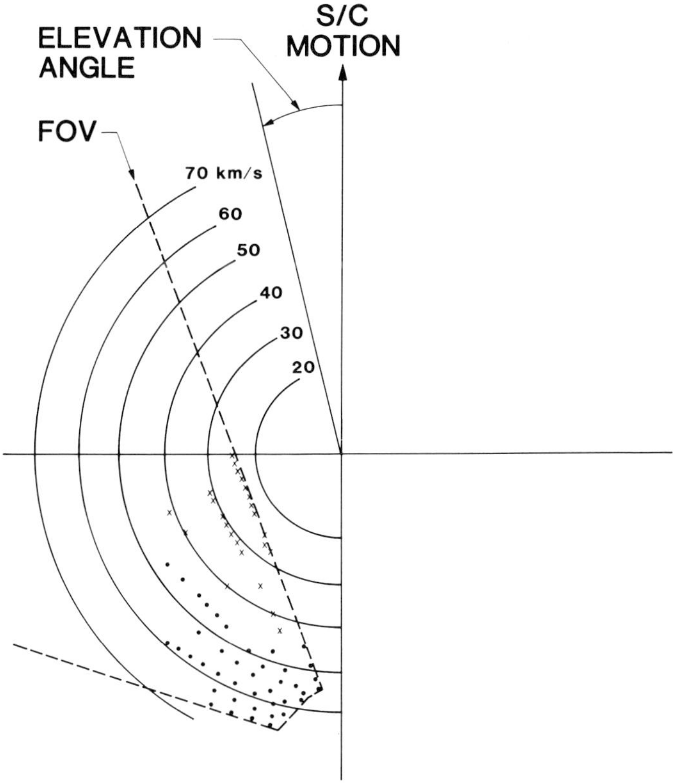

Fig. 7. HERS $M/Q = 12$ velocity-elevation angle distribution in the comet frame, distance range 6019 km to 3019 km to the comet

population not previously seen during the encounter, in addition to the ram direction ions seen previously. These new ions have a broad elevation angle distribution and appear down to the lowest energy measurable by the sensor (10 eV). Inspection of Fig. 7 indicates that the ions from the ram direction can be interpreted as corresponding to a distribution with near-zero flow speed relative to the comet; this is also consistent with their broad azimuth distribution. If we assume that the ion velocity distribution represents a Maxwellian at rest relative to the comet, we obtain about 100 eV ($\sim 10^6$ K) as a rough estimate of the ion temperature. (The ions marked by X's were used for this calculation.) This is considerably higher than the $\sim 10^3$ K temperature reported by Schwenn et al. (1987) for the ions inside the contact surface.

The second population of ions (the dots in Fig. 7) appears to have a velocity of the order of 60 km s^{-1} in the comet frame coming from the solar hemisphere. Such a flow was unexpected at the contact surface. Since these ions are at low energy relative to the spacecraft (down to the 10 eV cutoff of the instrument) they could also be interpreted as ions generated at and travelling with the spacecraft.

5. Discussion

The ions stationary or slowly moving relative to the comet are to be expected inside the contact surface, where primarily uncontaminated cometary ions are moving out from the nucleus at about 1 km s^{-1} (Schween et al., 1987). However, other measurements of these ions show them to be cold, the order of 10^3 K or less (Balsiger et al., 1986; Schwenn et al., 1987; Lämmerzahl et al., 1986), so the origin of the hotter ions observed by the HERS sensor is unclear at the present. There may, however, be heating processes resulting from the second population discussed below.

The second population observed is broadly distributed in elevation angle, but shows a predominance of flux flowing from the solar direction and at energies down to the low energy cutoff of the instrument. The combination of angular field of view and energy cutoff prevented observation of more of this distribution. These ions can be interpreted as either moving along with the spacecraft or flowing roughly towards the comet. The former case could be explained by secondary or sputtered ions created at the spacecraft. The mass spectrum shows no indication of spacecraft material, however (e.g., Al^+ or Si^+). The viewing geometry of the HERS sensor requires the secondary ions to spread out and flow down the side of the spacecraft in order for them to be detected in the sensor. Modelling of secondary ion production (Parker, 1984), especially if there is even a slight positive charging of the pacecraft, indicated that the ions would spread out but nor flow near the Giotto sides. In addition, there was no strong increase measured at the contact surface of either the neutral gas density or dust flux. Conceivably, there may have been a jet of dust particles too fine to be measured by the dust detectors on board Giotto. However, it does not appear that these are spurious ions produced at the spacecraft.

We have also considered the possibility of spacecraft charging in interpreting our data. In the results for the Giotto Neutral Mass Spectrometer (Lämmerzahl, 1986) it is reported that the spacecraft potential was zero around the time of contact surface crossing. Also, a spacecraft potential of more than a few volts (positive or negative) would produce a measureable M/Q dependence on ion velocity, which we did nor observe. Hence, we believe that charging effects are unimportant for interpreting our data.

Our current working hypothesis to explain these ions is that they are the result of charge exchange of fast ions flowing inward in the outer coma, producing a flow of fast neutrals. These neutrals are unaffected by the magnetic field and hence can penetrate deeply into the inner coma. These neutrals, however, eventually encounter ambient density high enough to re-convert them to ions by charge exchange or neutral-neutral collisions.

Since the energy and direction of these ions are not appreciably affected by such processes, they represent an "image", so to speak, of the ion population in some region farther out in the coma, where, in fact, the spacecraft has not actually passed. Such processes have been observed in the case of the earth's magnetosphere (Scholler et al., 1975), and has also been discussed for Jupiter, Saturn, and comets (Barbosa and Eviatar, 1986).

We are presently working on a detailed model along these lines to explain our measurements, which will be reported in a future publication. Additional detailed analysis of the ions of other M/Q obeserved in the contact surface region is also planned.

Acknowledgements. The work described in this paper was carried out in part by the Jet Propulsion Laboratory, California Institute of Technology, under a contract with the National Aeronautics and Space Administration. Funding was also provided by the German Bundesministerium für Forschung und Technologie and the Swiss National Science Foundation.

References

Balsiger, H., Altwegg, K., Buhler, F., Geiss, J., Ghielmetti, A.G., Goldstein, B.E., Goldstein, R., Huntress, W.T., Ip, W.-H., Lazarus, A.J., Meier, A., Neugebauer, M., Rettenmund, U., Rosenbauer, H., Schwenn, R., Sharp, R.D., Shelley, E.G., Ungstrup, E., Young, D.T.: 1986, *Nature* **321**, 330

Barbosa, D.D., Eviatar, A.: 1986, *Astrophys. J.* **310**, 927

Lämmerzahl, P., Kankowsky, D., Hodges, R.R., Stubbemann, U., Woweries, J., Herrwerth, I., Berthelier, J.J., Illiano, J.M., Eberhardt, P., Dolder, U., Schulte, W., Hoffman, J.H.: 1986, 20th ESLAB Symp., ESA SP-**250**, Vol. **I**, p. 179

Neubauer, F.M., Glassmeier, K.H., Pohl, M., Raeder, J., Acuna, M.H., Burlaga, L.F., Ness, N.F., Musman, G., Mariani, F., Wallis, M.K., Ungstrup, E., Schmidt, H.U.: 1986, *Nature* **321**, 352

Parker, LW.: 1984, *Proc. Giotto PEWG Meeting*, ESA SP-**224**, 65

Schwenn, R., Ip, W.-H., Rosenbauer, H., Balsiger, H., Buhler, F., Goldstein, R., Meier, A., Shelley, e.G.: 1987, *Astron. Astrophys.* (this issue).

Astron. Astrophys. 187, 225–228 (1987)

ASTRONOMY AND ASTROPHYSICS

Photographic observations of tail-formation activities of comet P/Halley in November 1985

Z. L. Liu

Beijing Astronomical Observatory, Academia Sinica, Beijing, People's Republic of China

Received January 30, accepted April 25, 1987

Summary. On November 12, 1985 the plasma tail of comet P/Halley made its first appearance during the latest apparition. Subsequently the comet displayed a great variety of tail activities. These included structures of different shapes in different bands even on the same night. The activities in the *U* and *B* bands usually preceded those in the *V* and *R* bands and were more prominent. According to the morphologies and lengths of the structures which appeared in the *B* and *U* bands they were plasma tails rather than dust jets. The deviation of the tail from the antisolar direction seemed to show a general tendency of decrease as the comet got closer and closer to the sun.

Key words: comet P/Halley – tail formation – plasma tail

1. Introduction

1985 November marked an important stage for the formation of the tail of comet Halley. Some outstanding events took place during this period, which aroused deep interest of many astronomers. In their papers Grün et al. (1986) thought they had observed jets on 1985 November 12/13, while Jockers (1985) and Watanabe et al. (1986) considered the twisted structure of November 12/13 to be an ion tail. In view of the importance in clarifying the physical conditions and the mechanism of the tail formation the preliminary results of our photographic observations are presented in this paper. And the nature of the activities of the tail of comet Halley in November 1985 is briefly discussed.

2. Observations

On 1985 November 9–20 comet Halley was observed with the 60/90-cm Schmidt telescope (f/3,5) at the Beijing Observatory. The scale was 115″/mm and the plate size was 16 cm × 16 cm. The plate/filter combinations were Kodak IIaO, IIaO+UG2, IIaO+GG13, IIaD+GG14 and IIIaF(103aF)+RG1. On 9, 10, 11, 12 November 1985 the plates were hypersensitized with hydrogen. All plates were calibrated using a spot sensitometer with 9 steps. Some details are given in Table 1.

3. The tail activities of comet Halley

During this period a lot of tail activities were observed, some of them were very conspicuous and showing up clearly even on the prints. On November 12 a prominent extended tail was observed in the *U* band. It was ejected from the south of the nucleus. Later it was curved, pointing to the SSW (about 220°) in the plane of the plate. It formed an angle (about 37°) with the projected antisolar direction (257°). It covered a length of about 4 10^5 km in the projected plane of the sky. Two kinks could be seen in the tail. It is shown in Fig. 1. On November 13 a long thin tail spouted from the NW part of the head straight towards the SW (∼225°) in the *B* band. It formed an angle of about 30° with the antisolar direction (255°). Its projected distance in the sky was about 6 10^5 km. It is shown in Fig. 2. Two short tails in the *U* band were seen, pointing to the SW (∼230°). On November 16 there was a tail pointing to the SW (∼210°) in the *B* band. The projected distance of the tail in the sky was 4 10^5 km. It formed an angle of about 26° with the antisolar direction (236°) in the projected plane of the sky. Besides this tail there was a secondary tail in the west, pointing to the SW (∼220°), as shown in Fig. 3. A long tail pointed to the SW (∼220°) in the *U* band on the same night. Its projected 6 10^5 km. On November 17 there was a tail ejected from the south of the head in the *B* and *U* bands. It formed an angle of about 22° with the antisolar direction (202°) (shown in Fig. 4). On November 20 there was a prominent long tail pointing to the east (90°) on the unfiltered plate. Its projected distance in the sky was as long as 7 10^5 km. Besides this tail there were two secondary tails. As shown in Fig. 5, they pointed to the SEE (∼100°) and NE (∼50°). The morphology of the tail in the *B* band was virtually the same as on the unfiltered plate, although it was weaker. Besides the tail activities in the *U* and *B* bands a short tail was observed in *V* band on November 19 and in *R* band on November 20 separately. Some details of tail activities are given in Table 1.

4. Some preliminary results and a discussion

The tail activities of comet Halley became apparent earlier in the *U* and *B* than in the *V* and *R* bands, and they were also more prominent. Because most of the emission lines of gases and ions appear in the shorter wavelength regions (*U* and *B* bands) the activities should be attributed primarily to gases and ions, such as CN, CO^+, CO_2^+ etc., at the early stage of tail formation.

As was mentioned above, there were deviations between the tails and the antisolar direction during this period. They varied between 2°–52°. The larger deviations can have resulted from two effects. One was a perspective effect as the comet was near the opposition. Another was that the tail activities were related to explosions on the comet. At the early stage they were like a sort of

Table 1. Tail activity observations of comet P/Halley (1985 November 9–20)

UT in November	Exposure time (min)	Plate/filter combination	Tail activity	Position angle of tail	Position angle of antisolar direction	Deviation	Length of tail (km)
9.6691	15	IIaO	Obscure				
9.7375	20	IIaO+GG13	Obscure				
9.7635	25	IIaD+GG14					
9.8156	47	IIaO+UG2	Obscure				
10.6444	20	IIaO+GG13					
10.6799	20	IIaD+GG14					
10.7139	50	IIaO+UG2	Obscure				
10.7778	50	IIIaF+RG1					
11.6375	20	IIaO+GG13					
11.6903	20	IIaO+GG13					
11.7750	50	IIIaF+RG1					
11.8243	20	IIaD+GG14					
12.6347	20	IIaO+GG13	Less conspicuous	SW			
12.6837	17	IIaD+GG14					
12.7138	40	IIIaF+RG1					
12.7538	45	IIaO+UG2	Conspicuous	220°	257°	37°	4 10^5
13.6556	30	IIaO+GG13	Conspicuous	225°	255°	30°	6 10^5
13.6840	10	IIaO+GG13	Less conspicuous				
13.7406	61	IIaO+UG2	Conspicuous	230°	255°	25°	
13.7972	30	IIaD+GG14					
14.5931	30	IIaO+GG13					
14.6198	5	IIaO+GG13					
14.6354	8	IIaO+UG2					
16.6007	30	IIaO+GG13	Conspicuous	210°, 220°	236°	26°, 16°	4 10^5
16.6316	5	IIaO+GG13					
16.6792	80	IIaO+UG2	Conspicuous	220°	235°	15°	6 10^5
16.7358	15	IIaO+UG2					
16.7764	50	103aF+RG1					
16.8186	30	IIaD+GG14					
17.6132	30	IIaO+GG13	Conspicuous	180°	202°	22°	2 10^5
17.6340	6	IIaO+GG13					
17.6660	50	103aF+RG1					
17.6986	10	103aF+RG1					
17.7424	80	IIaO+UG2	Conspicuous	180°	193°	13°	
17.7844	15	IIaO+UG2					
17.8118	30	IIaD+GG14	Obscure				
17.8358	5	IIaD+GG14					
18.9080	11	IIaO+GG13					
19.7000	30	IIaO+GG13	Obscure				
19.7556	6	IIaO+GG13					
19.7917	80	IIaO+UG2					
19.8330	15	IIaO+UG2					
19.8580	25	IIaD+GG14	Conspicuous	102°	100°	2°	
19.8781	5	IIaD+GG14	Less conspicuous				
20.6972	30	IIaO	Conspicuous	90°, 100°, 50°	102°	12°, 2°, 52°	7 10^5
20.7264	30	IIaO+GG13	Conspicuous	90°, 100°, 50°	102°	12°, 2°, 52°	
20.7451	6	IIaO+GG13	Less conspicuous				
20.7826	80	IIaO+UG2	Conspicuous	50°	101°	51°	
20.8240	15	IIaO+UG2	Obscure				
20.8542	48	103aF+RG1	Conspicuous	102°	100°	2°	

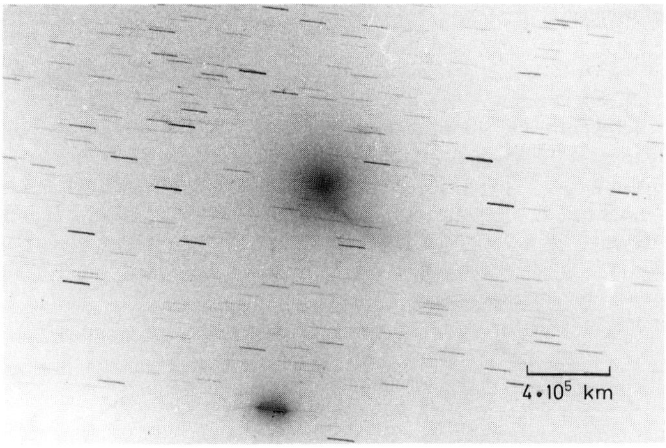

Fig. 1. 1985 November 12, 17^h43^m–18^h28^m (UT), IIaO + UG2

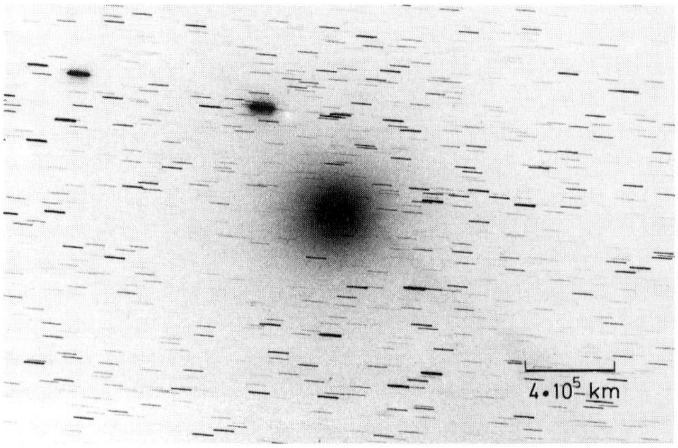

Fig. 2. 1985 November 13, 15^h29^m–15^h59^m (UT), IIaO + GG13

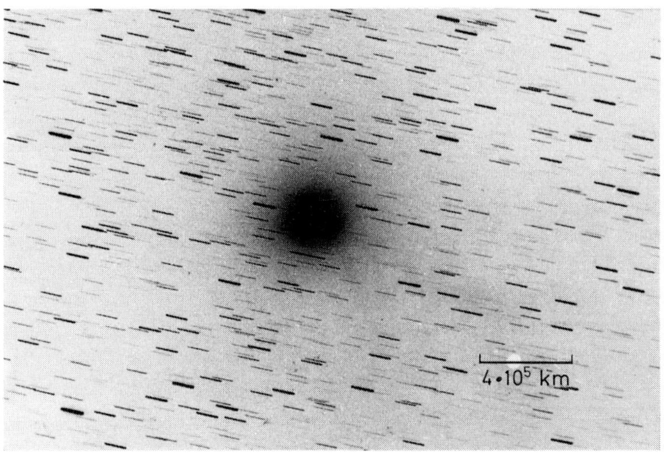

Fig. 3. 1985 November 16, 14^h10^m–14^h40^m (UT), IIaO + GG13

ejections which consisted of gases and ions. Afterwards they gradually became conventional plasma tails which coincided with the antisolar direction. From November 12 to 20 the deviation seemed to show a general tendency of decrease. It seemed to be related to the enhancement of the solar wind as the comet got closer and closer to the sun.

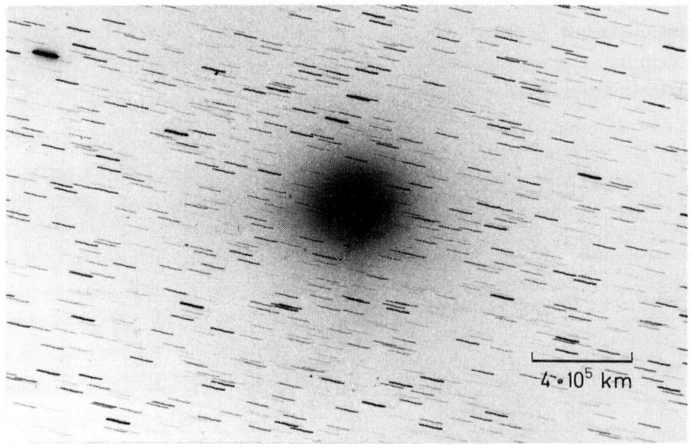

Fig. 4. 1985 November 17, 14^h28^m–14^h58^m (UT), IIaO + GG13

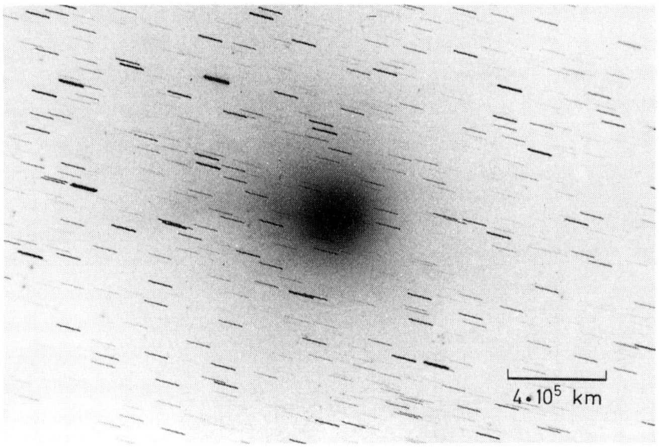

Fig. 5. 1985 November 20, 16^h29^m–16^h59^m (UT), IIaO

The tail activities in different bands were different in shape and pointed to different directions even on the same night. So we see that from the beginning to tail formation the activities of cometary gas and dust were not synchronous. And the activities of different gases and ions were different as well. Some images obtained at the same time in different bands are expected to determine why the tail activity was different in *U* and *B* bands on the same night.

From the morphologies and lengths of the structures which appeared in the *U* and *B* bands, as indicated above, there is no doubt that they were plasma tails rather than dust jets. And the plasma tail appeared for the first time on November 12 during the 1985–1986 apparition. It probably consisted of CO_2^+ ions. The tail which appeared in the *V* band on November 19 and in the *R* band on November 20 would be the dust tail.

The tails observed in the same band but on different dates were of different shapes and intensities. There might be two reasons resulting in this situation. One was that the previous shape of tail showed rapid changes. Another was that the old tail disappeared and a new one appeared. For example, there was one tail in the *B* band on November 17. There was no tail in the *B* band on November 19. And there was one main tail and two secondary tails in the *B* band on November 20. This might indicate that the ratio between gas and dust, and the ratio of different kinds of gases and ions changed with time.

Before the prominent extended tail was observed in *U* band on 12.7538 (UT) November 1985 some tail activities could already be seen on the plates, e.g. in the *B* band on 12.6347 (UT) November 1985, which seemed to point to the SW. After the image enhancement will be made, more details about tail formation activities will be discussed.

Acknowledgements. It is a pleasure to express my gratitude to Ke Da-rong, Zhao Li-min, Wang Ke-min, and Zheng Jia-qing for their help with the observations and calculations of solar direction.

References

Grün, E., Graser, U., Kohoutek, L., Thiele, U., Massonne, L., Schwehm, G.: 1986, *Nature* **321**, 144

Jockers, K., Geyer, E.H.: 1985, ESO Messenger **42**, 9

Watanabe, J., Takatou, N., Kawakami, H., Tomita, T., Kinoshita, H., Nakamura, T., Kozai, Y.: 1986, 20th ESLAB Symp. ESA-250, Vol. **III**, 119

Astron. Astrophys. 187, 229–232 (1987)

ASTRONOMY AND ASTROPHYSICS

The outburst of comet P/Halley on December 12, 1985

J. Watanabe[1], H. Kawakami[1], K. Tomita[2], H. Kinoshita[2], T. Nakamura[2], and Y. Kozai[2]

[1] Department of Astronomy, Faculty of Science, University of Tokyo, Bunkyo-ku, Tokyo, 113, Japan
[2] Tokyo Astronomical Observatory, University of Tokyo, Mitaka, Tokyo, 181, Japan

Received January 28, accepted April 25, 1987

Summary. A strong outburst with a southward jet was found during 9^h07^m to 12^h29^m UT on December 12, 1985 while we were observing the near nucleus region of comet P/Halley. The projected expansion velocity is estimated to be $0.3 \pm 0.1\,\mathrm{km\,s^{-1}}$ and the dust number density is 9 times as large as that in the coma background, assuming the jet cone angle is 30°. As the width of the jet in the C_2 emission band is 10% broader than that in white light, it can be concluded that the C_2 molecules are created from some parent body in the ejected matter.

Key words: comets – P/Halley – near nucleus phenomena – outburst – jet

1. Introduction

Extensive studies and observations of several comets made it clear that inhomogeneous masses are frequently ejected from the cometary nucleus. The mass ejection rate and the brightness increase suddenly on such occasions and such phenomena are called outbursts. On the other hand, Sekanina (1981) proposed the "active region model" of the cometary nucleus. According to this model, a few active regions on the surface show violent ejection of material intermittently, which is observed as a "jet". The interrelation between active regions and the outburst phenomena has been described by Whipple (1980) in the case of comet P/Schwassmann-Wachmann 1. Sekanina and Larson (1984) identified some jet phenomena of comet P/Halley during its 1910 apparition. Giotto and Vega have succeeded in observing similar jets in this apparition (Keller et al., 1986; Sagdeev et al., 1986).

We observed near nucleus features of comet P/Halley with a CCD camera, and found an outburst with a southward jet on December 12, 1985. This paper reports the characteristics of this phenomenon.

2. Observations

Observations were carried out during December 10–12, following the IHW-day, with the 188-cm telescope at the Okayama Astrophysical Observatory. The total number of images taken at the Newton focus is 86, of which 60 are with no filter (10 s integration) and 2 with a C_2 band pass filter (60 s integration). It was on December 12 from 9^h07^m to 12^h29^m UT that we found a strong outburst. The heliocentric and geocentric distances were 1.32 and 0.77 AU, respectively. On December 12 clouds often prevented long exposures. The C_2 band pass filter centered at 514 nm ($\Delta\lambda = 10$ nm) is the one supplied by the IHW headquarters. The pixel number is 410 × 490, and the pixel size is 13.5 μm × 23 μm for our NEC CCD (type μ PD 3511D). As electrons in the neighbouring two pixels were combined within the CCD under the control of vertical clock in order to reduce the readout noise, the effective pixel size is 27 μm × 23 μm, which corresponds to 0″.60 × 0″.52 on the sky.

All frames were flat-fielded and the dark current was removed, without absolute intensity calibration.

Send offprint requests to: J. Watanabe

3. Analysis

Figures 1 and 2 show near-nucleus images of comet P/Halley on December 12 and 11, respectively. Figure 2 has a fairly round shape, while figure 1 has an elongated jet feature in the central region. Our observation was performed continuously for 212 minutes on December 12. We applied a second moment method proposed by Stobie (1980) to find the morphological change during this time interval. Ellipses are fitted to the contours of different intensity levels.

We estimate the position angle of the jet direction from the major axis of an inner ellipse. Figure 3 shows the variation of the position angle with the mean radius of contours, where the position angle is measured counterclockwise from north (in equatorial coordinates). Although the direction of the elongation of the innermost contour is uncertain because of seeing effects, the position angles of the ellipses within 15″ are nearly 180°. This indicates that the jet did not change the direction of the ejection. It also suggests that the active region, if any, will be near the pole of the nucleus, or the rotational velocity of the nucleus will be slow. The position angle of the north pole determined by Sekanina and Larson (1984) is 175° for the date of our observations, which is close to the observed direction of the jet. This direction of the jet is not the same as that of the bright jet observed by the Giotto spacecraft (Keller et al., 1986), however, is close to that of the jet observed on December 20 by Takagishi et al. (1985).

Although the major axes of the inner elliptical contours did not rotate during the observation, those of the outer ellipses rotated clockwise possibly due to the expansion of the jet. The portion where a steep decline takes place in the intensity profile

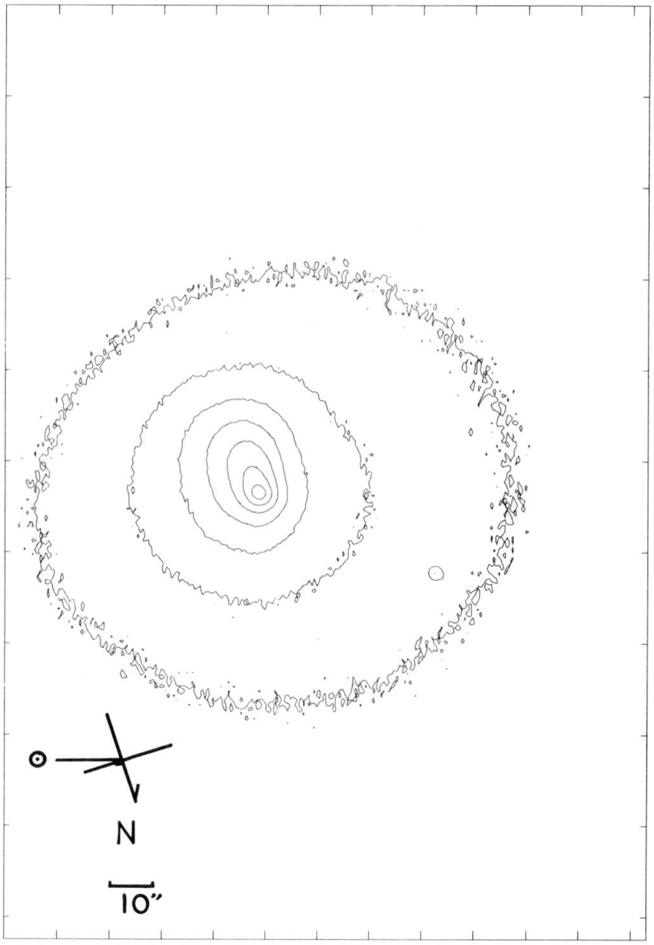

Fig. 1. The image of the central region of comet P/Halley taken at 12^h28^m UT on December 12, 1985, using the 1.8 m telescope at Okayama Astrophysical Observatory. The field of view is 2.′5 × 3.′5; north is on the right, 68° below the horizontal line; the apparent motion of the comet is leftward. The sun is on the left, 1° above the horizontal line and 42° above the image plane. ⊙ and N express the sunward and north pole directions. Exposure time is 10 sec.

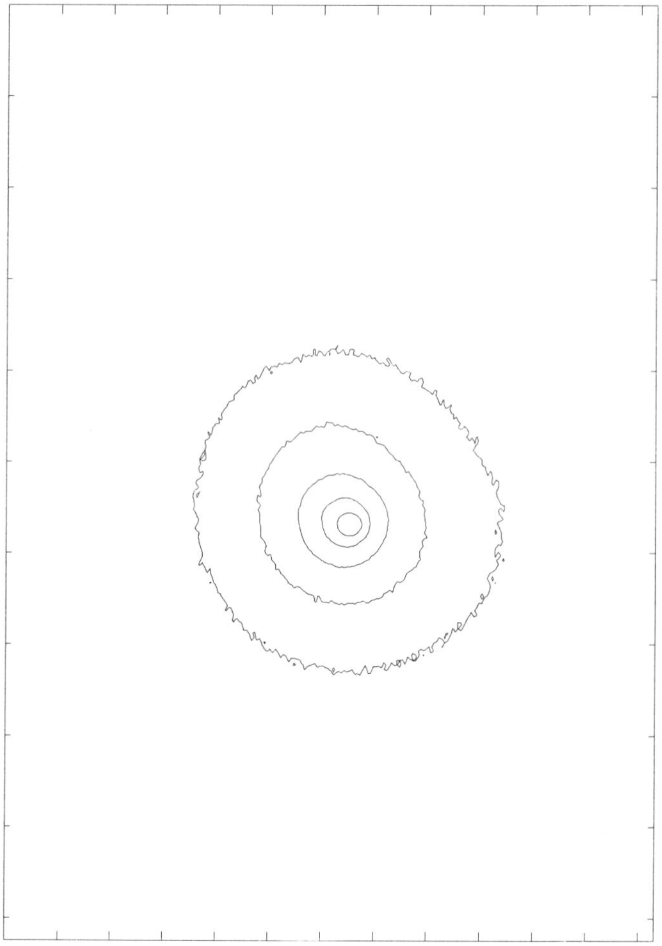

Fig. 2. Comet P/Halley at 9^h56^m UT on December 11. Exposure time is 30 secs. Other descriptions are the same as in Fig. 1

along the jet, that is the front of the jet, shifts with time. This steep decline is direct evidence for the expansion of the jet. From its moving speed, it is estimated that the jet began at about 2^h UT of December 12. Kaminski and Griep (1985) also reported this jet phenomenon as about 2″ northsouth elongation of the central condensation at 12.25 UT. A Lyman-α outburst was also observed by the Japanese interplanetary spacecraft Suisei (Kaneda et al., 1986a) at 12.2 UT.

Analysis of the front motion also gives 0.3 ± 0.1 km s^{-1} as the expansion speed, if the jet is assumed to be transverse to the line of sight. Although this value is consistent with the terminal speed of dust particles of comet Arend-Roland obtained by Finson and Probstein (1968), it is 20–30% less than that predicted by Whipple (1978, 1980, 1982).

Figure 4 shows the contour map of a C_2 image. This map is very similar to the white light image shown in Fig. 1 except for the width of the jet. The result of ellipse fitting indicates that the width of the jet in the C_2 band is about 10% broader than that in white light. The contribution from the continuum has not been subtracted from our C_2 images, it will, however, be negligible

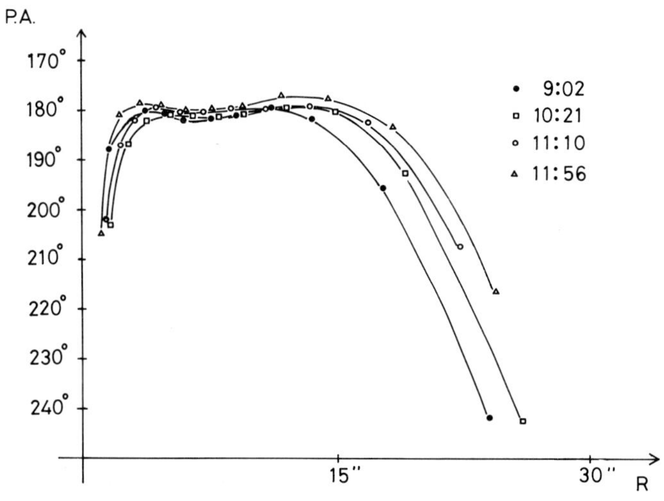

Fig. 3. The relation between the position angle of the major axis and the mean radius of each contour. The expansion of the jet affects the position angle of the outer contours

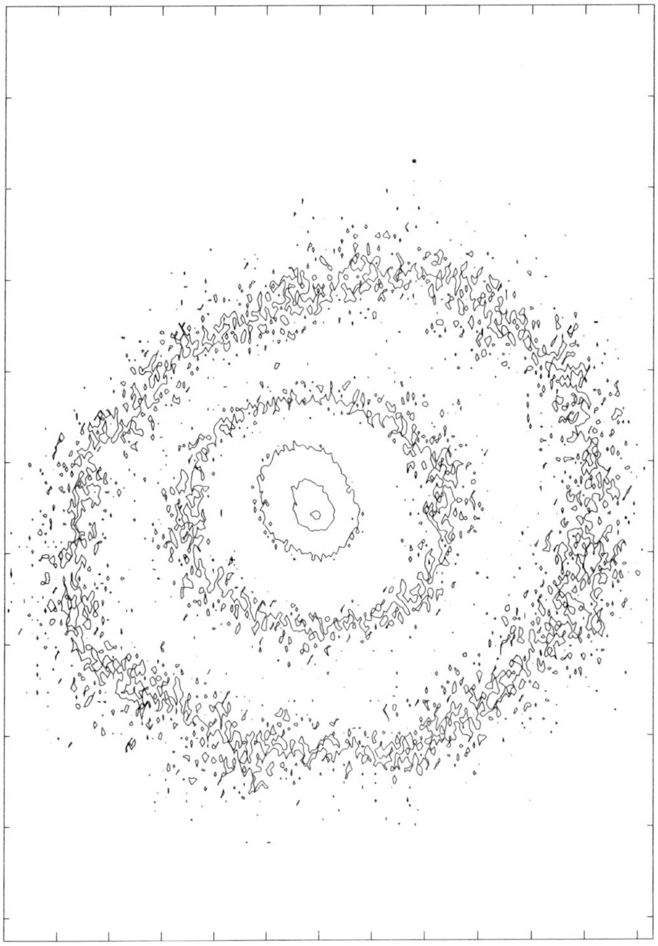

Fig. 4. C_2 image of comet P/Halley taken at 12^h30^m UT on December 12. The step of the contour is 0.5 mag. Other descriptions are the same as in Fig. 1

except within a few arcseconds from the 'nucleus' (A'Hearn et al., 1986).

The above mentioned fact shows that the C_2 molecules expand more easily in the direction perpendicular to the jet and dissipate more rapidly. This can be explained in the following ways. The C_2 molecules may be created from larger molecules near the nucleus. The jet profile seen in the C_2 band is distorted. Another possible explanation is that the C_2 molecules come from small dust particles in the jet. A'Hearn et al. (1986) suggest that the CN molecules originate from CHON particles found by spacecraft (Kissel et al., 1986a, 1986b). Similar processes may be applied to our broad C_2 jet.

We roughly evaluate the ratio of the number density of dust particles in the jet to that in the coma background. We adopt the method of Larson and Sekanina (1984).

We made the following assumption for the dust density distribution.

(a) The ejected dust moves along straight lines with a constant velocity.

(b) The dust ejection rate from the quiet surface changes neither with place nor with time.

(c) The dust neither splits nor evaporates in the inner coma.

From assumptions (a)–(c), the dust distribution in the coma back-

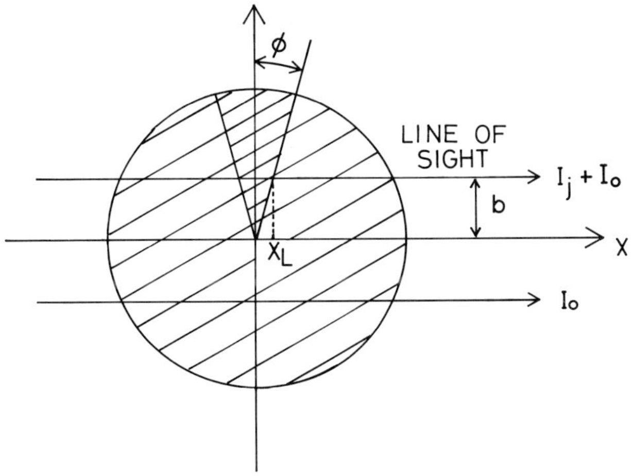

Fig. 5. Schematic diagram of the jet model

ground $n(r)$ is expressed by

$$n(r) = \alpha/r^2\,, \tag{1}$$

where r denotes the distance from the nucleus, α is an arbitrary constant. We describe the dust density in the jet $N(r)$ as

$$N(r) = k(r) \cdot n(r)\,, \tag{2}$$

where $k(r)$ is the ratio of the dust density in the jet to that in the background region. We make three more assumptions to simplify matters:

(d) The direction of the jet is transverse to the line of sight.

(e) The near nucleus region is optically thin.

(f) The white light image reflects the dust distribution.

Then, the intensity ratio of the jet region to the coma background $R(b)$ is written as follows;

$$R(b) = [I_j + I_0]/I_0\,, \tag{3}$$

$$I_j = 2 \int_0^{x_L} k(r) \cdot n(r)\, dx\,, \tag{4}$$

$$x_L = b \cdot \tan(\phi)\,, \tag{5}$$

where I_j and I_0 denote the observed intensity at each position defined in Fig. 5, b the projected distance from the nucleus, x_L the geometrical path length of the jet, dx the infinitely small path length and ϕ half the vertex angle of the jet cone (see Fig. 5). Equations (1)–(5) give the following relation,

$$R(b) = 1 + 2\phi(k(b) - 1)/\pi\,, \tag{6}$$

where $k(r)$ can be approximated by $k(b)$ when ϕ is small. This equation permits us to obtain the dust density ratio $k(r)$ from the observed intensity ratio $R(b)$ at each apparent distance from the nucleus. The result is shown in Fig. 6. From the width of the jet the angle ϕ is estimated to be smaller than 15°. Hence the dust density in this jet is found to be at least 9 times larger than that in the coma background.

4. Discussion

Generally, a jet blows nearly in the sunward direction (Keller et al., 1986; Sekanina and Larson, 1984). However, in our case, the angle between the observed jet and the direction of the sun is about 70°, or even larger because of the projection effect. This

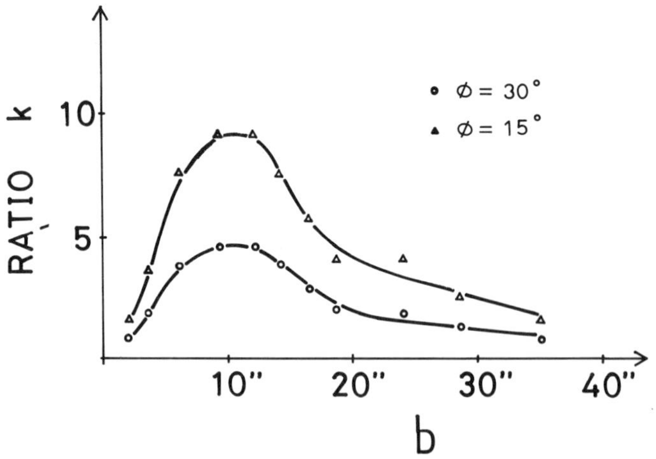

Fig. 6. The relation between the projected distance b and the jet-to-background ratio of the dust density. Triangles show the results for the case $\phi = 15°$ and circles those for $\phi = 30°$

means that the altitude of the sun at the active region of this jet on the surface is lower than 20°.

In spite of this the jet phenomenon which occurred on December 12 is one of the strongest outbursts before the perihelion passage of comet P/Halley. The dust density in this jet is at least 9 times larger than that in the coma background. In this case it should be noted that our observed white light image includes the contribution from C_2 and CN emissions. Therefore, the dust density of the jet can even be larger than this estimated value. This outburst caused the discontinuous arcade structure of the ion tail reported by Saito et al. (1987).

All these facts may indicate that this outburst is one of the 'turning on' events of the active region. The outburst on December 20 observed by Takagishi et al. (1985) may be the recurrent one of the same active region. We notice that there is an interval of about 7 days between these two southward jet phenomena. On December 27 we found a faint southward distortion in the contours of CCD images taken by Takagishi. This fact may be related to the 7.4-day periodicity of activities proposed by Millis and Schleicher (1986). On the other hand, Kaneda et al. (1986b) determined the outburst period of 2.2 ± 0.1 day from Lyman α emission observation. They named the strongest source of the outburst S_1. The activity on December 12 may be caused by S_1. These two conflicting periods need to be studied further. This was done recently by Sekanina (1987) with a dynamical theory of a torque-free rigid rotator, and by Julian (1987) who modelled the nucleus as a rigid homogeneous ellipsoid with three unequal principal axes in torque-free rotation.

Acknowledgements. The authors would like to thank Mr. Y. Norimoto for setting up the CCD system and observation. They are also thankful to Messrs. M. Shimizu and K. Okita who were helpful. Special thanks are due to Professor S. Nishimura who made a CCD data acquisition programs, and the staff of NEC corporation who kindly supplied CCD chips (type μPD 3511D). Data acquisition was performed under the control of FACOM S-3300 computer of the Okayama Astrophysical Observatory. HITAC M-280H at the Computer Center, University of Tokyo was also used for the data reduction.

References

A'Hearn, M.F., Hoban, S., Birch, P.V., Bowers, C., Martin, R., Klinglesmith III, D.A.: 1986, *Nature* **324**, 649

Finson, M.L., Probstein, R.F.: 1968, *Astron. J.* **154**, 353

Julian, W.H.: 1987, *Nature* **326**, 57

Kaminski, C., Griep, D.: 1985, *IAU Circular* **4154**

Kaneda, E., Hirao, K., Takagi, M., Ashihara, O., Itoh, T., Shimizu, M.: 1986a, *Nature* **320**, 140

Kaneda, E., Ashihara, O., Shimizu, M., Takagi, M., Hirao, K.: 1986b, *Nature* **321**, 297

Keller, H.U., Arpigny, C., Barbieri, C., Bonnt, R.M., Cazes, S., Coradini, M., Cosmovici, C.B., Delamere, W.A., Huebner, W.F., Hughes, D.W., Jamar, C., Malaise, D., Reitsema, H.J., Schmidt, H.U., Schmidt, W.K.H., Seige, P., Whipple, F.L., Whihelm, K.: 1986, *Nature* **321**, 320

Kissel, J., Sagdeev, R.Z., Bertaux, J.L., Angarov, V.N., Audouze, J., Blamont, J.E., Büchler, K., Evlanov, E.N., Fechtig, H., Fomenkova, M.N., von Hoerner, H., Inogamov, N.A., Khromov, V.N., Knabe, W., Krueger, F.R., Langevin, Y., Leonas, V.B., Levasseur-Regourd, A.C., Managadze, G.G., Podkolzin, S.N., Shapiro, V.D., Tabaldyev, S.R., Zubkov, B.V.: 1986a, *Nature* **321**, 280

Kissel, J., Brownlee, D.E., Büchler, K., Clark, B.C., Fechtig, H., Grün, E., Hornung, K., Igenbergs, E.B., Jessberger, E.K., Krueger, F.R., Kuczera, H., McDonnell, J.A.M., Morfill, G.M. Rahe, J., Schwehm, G.H., Sekanina, Z., Utterback, N.G., Völk, H.G., Zook, H.A.: 1986b, *Nature* **321**, 336

Larson, S.M., Sekanina, Z.: 1984, *Astron. J.* **89**, 571

Millis, R.L., Schleicher, D.G.: 1986, *Nature* **324**, 646

Sagdeev, R.Z., Szabó, F., Avanesov, G.A., Cruvellier, P., Szabó, L., Szegö, K., Abergel, A., Balazs, A., Barinov, I.V., Bertaux, J.L., Blamont, J., Detaille, M., Demarelis, E., Dul'nev, G.N., Endröczy, G., Gardos, M., Kanyo, M., Kostenko, V.I., Krasikov, V.A., Nguyen-Trong, T., Nyitrai, Z., Reny, I., Rusznyak, P., Shamis, V.A., Smith, B., Sukhanov, K.G., Szabó, F., Szalai, S., Tarnopolsky, V.I., Toth, I., Tsukanova, G., Valníček, B.I., Varhalmi, L., Zaiko, Yu.K., Zatsepin, S.I., Ziman, Ya.L., Zsenei, M., Zhukov, B.S.: 1986, *Nature* **321**, 262

Saito, K., Saito, T., Aoki, T., Yumoto, K.: 1986, 20th ESLAB Symp., ESA SP-250, vol. **III**, 155

Sekanina, Z.: 1981, *Ann. Rev. Earth Planet. Sci.* **9**, 113

Sekanina, Z., Larson, S.M.: 1984, *Astron. J.* **89**, 1408

Sekanina, Z.: 1987, *Nature* **325**, 326

Stobie, R.S.: 1980, *J. British Interplanet. Soc.* **33**, 323

Takagishi, K., Eiraku, M., Tomita, K., Matsuoka, M.: 1985, *IAU Circular* **4154**

Whipple, F.L.: 1978, *Nature* **273**, 134

Whipple, F.L.: 1980, *Astron. J.* **85**, 305

Whipple, F.L.: 1982, in Comets, ed. L.L., Wilkening, Univ. Arizona, Tucson, p. 277

Astron. Astrophys. 187, 233–248 (1987)

ASTRONOMY AND ASTROPHYSICS

Structure and dynamics of plasma-tail condensations of comet P/Halley 1986 and inferences on the structure and activity of the cometary nucleus*

W.E. Celnik and Th. Schmidt-Kaler**

Astronomisches Institut, Ruhr-Universität, Postfach 10 21 48, D-4630 Bochum 1, Federal Republic of Germany

Received February 12, accepted April 29, 1987

Summary. Comet P/Halley has been observed photographically by Bochum observers from 1986 Feb. 17, to Apr. 17. Coma, dust, and plasma tail were recorded with high resolution and wide-field cameras. From a detailed analysis of the plasma tail dynamics we derived the distributions of velocity and acceleration along the tail up to distances of $50\,10^6$ km from the nucleus while the tail length detected reached $80\,10^6$ km. The mean value of the acceleration within the plasma tail is $29 \pm 1\,\mathrm{cm\,s^{-2}}$ while the initial velocity of the condensations is in most cases $0.54 \pm 0.08\,\mathrm{km\,s^{-1}}$ corresponding to the outflow velocities of gas and dust jets. Some larger values observed may be real, due to strong interaction with solar wind particles. The mean outflow velocity from all determinations up to now is $0.61 \pm 0.05\,\mathrm{km\,s^{-1}}$. We establish a consistent model of mass ejection activity from the surface of the nucleus and the distribution of the sources along the equator on the nucleus. The extension of the strong sources is less than 0.7 km. Main activity occurs at the long ends of the deformed ellipsoidal nucleus, the activity contrast between long ends and broad sides is 5:1. The sidereal rotation period of the nucleus is 50.5 ± 0.6 hours. The mean of all determinations up to now is 52.3 ± 0.4 hours. We explain the appearance of a 7.4-days period observed by other authors and suggested also by our data. As shown by the comparison of tail and coma dynamics plasma and neutral gas clouds are probably ejected at the same time from the same source on the nucleus. We discuss the variation of coma shell expansion with heliocentric distance. The variation of tail aberration seems to correlate more to the interplanetary magnetic field than to solar wind parameters. Tail disconnection events appear correlated to the magnetic field neutral sheets.

Key words: comets – comet P/Halley – plasma tail – nucleus – interplanetary medium

1. Introduction

Comet P/Halley (1982i) was observed photographically from 1986 February 17 to April 17, at the European Southern Observatory (ESO), La Silla, by a group of the Astronomisches Institut der Ruhr-Universität Bochum. Collaborators were W.E. Celnik, P. Koczet, W. Schlosser, R. Schulz and K. Weißbauer. The special equipment described later was used. Photoelectric photometry of the cometary coma was obtained from February 24 to April 17, 1986, using the 61 cm-Bochum telescope.

The observations were aimed to study the structure, dynamics, and physical properties of coma, dust and plasma tails in full spatial extent, and to search for correlations in different parts of the comet with the solar wind.

Until now a few aspects only out of the rich spectrum of research possibilities opened up by our data material have been investigated: the structure and dynamics within the plasma tail and the neutral gas coma. First results on these matters were published by Celnik et al. (1986a, b), Celnik (1986), and Schlosser et al. (1986).

The present paper presents results from a deeper and more refined analysis of the plasma tail dynamics during the observation period. In Sect. (2) the "Bochum Halley Project" is described in more detail, the instrumentation and the full data base of the whole photographic campaign are summarized. Section (3) describes the method of extracting useful data from the photographic images. In Sect. (4) the results concerning plasma tail, neutral coma, and cometary nucleus are presented while in the last section some physical conclusions are drawn. In a subsequent paper (Paper II) information will be given on instrumentation, data base, and results, more detailed than possible here.

2. Photographic work

2.1. Photographic instrumentation

Special equipment was set up under the open sky: A very stable Alt 6AD parallactic mounting with programmable step motors in both right ascension and declination in order to follow the comet's motion automatically. Guiding on the central coma was controlled using a small achromatic refractor $f/10$ with $f = 1000$ mm providing a wide field of view (25′) and a visual enlargement of more than 150x. A Lichtenknecker Flat-Field-Camera (FFC) $f/4$ with a focal length of 760 mm was used in combination with a Canon FTB camera housing for imaging the head of P/Halley on 35 mm film resulting in a field of view of 1°.8 × 2°.7 and a best angular resolution of 5″. Sensitized 103a-E, 103a-F and Technical Pan 2415 emulsions were used. In order to detect the tail in full extent four wide-field cameras (WFC) of the format

Send offprint requests to: W.E. Celnik

* Based on observations made at the European Southern Observatory, La Silla, Chile

** Present address: Wilhelm-Foerster-Sternwarte, Munsterdamm 90, D-1000 Berlin 41

6 × 6 cm were installed (Hasselblad 2000 FC/M housings fitted with Zeiss Planar F 2.0/110 mm T lenses) delivering a field of view of 28°.6 × 28°.6 and a resolution of approximately 30″. Sensitized IIIa-F and IIa-O emulsions on glass plates were used for the 6 × 6 cameras avoiding focus problems due to motion of thin film sheets. Besides this main observational instruments a Pentacon $f/4$, 300 mm tele lens was fitted to a Pentax MX 35 mm housing, with a field of view of 4°.6 × 6°.9 and a resolution of 12″. Here 103a-F 35 mm film was used. In addition some images were taken on 103a-F 35 mm film through the optical system of the 61-cm telescope photometer resulting in a focal ratio of $f/14$ and a focal length of approximately 8.5 m. Their angular resolution is seeing-limited.

Concerning optical filters we followed the recommendation of the International Halley Watch and chose filters with characteristics listed in Paper II which gives all the combinations of optics, emulsions, filters and maximum exposure times used. To obtain high contrast photographs of plasma tail and neutral coma we used interference filters for CO^+ (λ 4260) and CN (λ 3880) both for the FFC and WFC. The dust tail was recorded using Schott OG 530 filters for FFC and WFC and RG 645 for the 300 mm tele lens, while no filter was used for the 61 cm telescope photographs. To allow shorter exposure times for the plasma tail we used a broad-band filter combination of a long and short wave pass filter for the fourth WFC resulting in a better time resolution but giving a lower density contrast. For the same reason we took many pictures with the FFC in combination with a long wave pass filter GG 420, cutting off the ultra-bright CN coma at 3880 Å. Additionally, some photographs were taken with the WFC using interference filters for H_2O^+, N_2^+ and CO_2^+ molecular spectral bands with the N_2^+ and CO_2^+ wavelengths lying near the transmission limit of the planar lens (cf. Paper II). When using the FFC the glass filters were set into the converging $f/4$ beam between primary mirror and camera housing (cf. Paper II) slightly altering the transmission characteristics of the filters. Concerning the other cameras the filters were set in front of the optics. In order to detect structures both in the outer and innermost regions of the coma, also very short exposure times down to 1 second were applied to the FFC and WFC.

2.2. *Photographic data base*

Due to the excellent weather conditions at La Silla (Feb. 17 to Apr. 17, 1986) it was possible to obtain photographic images of P/Halley during 57 of a total of 60 nights. More than 1200 images can be used to study structures within tail and coma. Most of the nights were of photometric quality. In order to allow photometric analysis of the photographic images densitometer spots were applied every night for FFC, WFC, and the 300 mm tele lens. For calibration purposes the galaxy M83 was photographed every night with nearly all cameras. A table giving a complete listing of all cometary images taken, dates, exposure times, filtering will be presented in Paper II. All WFC-plates and FFC-films have been sensitized by ESO staff in the usual manner (IIIa-, IIa-O, 103a-). TP 2415 film was baked in forming gas at 65° C for 5 hrs. using a pressure of 500 kg cm^{-2}. The hypersensitizing processes remained identical during the whole campaign. All sky and calibration images of a single night were developed together in unthinned Ilford Microphen at 19° C for 25 min., the development conditions remaining constant throughout the whole observational run. In Paper II the total number of images obtained with different instruments and optical filters is given, where the distribution of images used for this investigation across the various instruments is indicated. Nearly the complete sample of CO^+ wide field images and a large part of the blue broad-band images were investigated, a total amount of 124 images spread approximately continuously over the whole observational run. All these pictures are presented as a full sample in Paper II.

3. Data reductions

As the first step enlargements of the 124 pictures were made on photographic paper of gradation 3 and 5. The scale of these prints was chosen to 6′ mm^{-1} for the WFC and 0′.6 mm^{-1} for the FFC pictures. The maximum length of the CO^+ tail was measured on the high contrast prints and is given in Fig. 1 as a function of date and of heliocentric distance. On the photographic prints the angular distance of the visible end of the

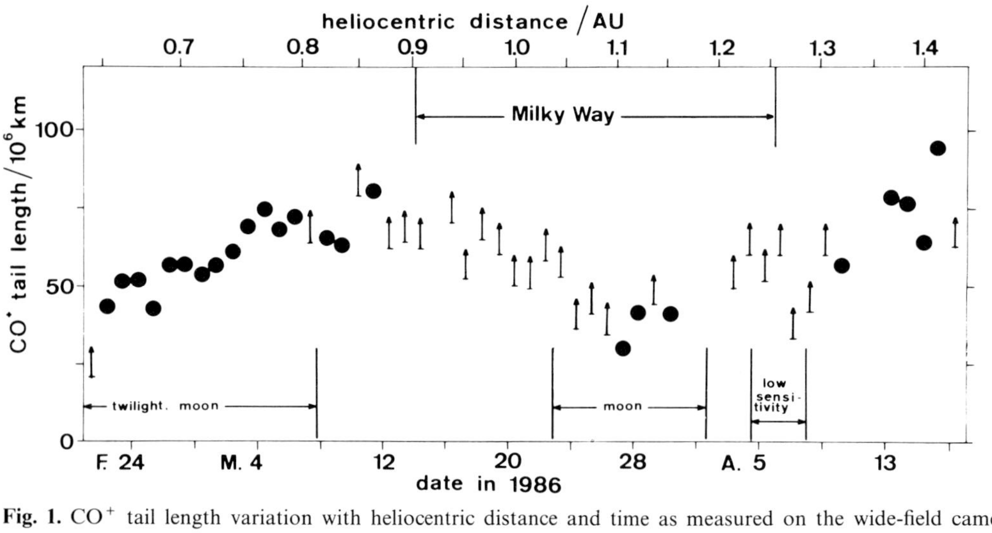

Fig. 1. CO^+ tail length variation with heliocentric distance and time as measured on the wide-field camera high contrast prints. Arrows mark lower limits. The tail length varies only apparently due to disturbing twilight and moon, and to the Milky Way. The observations can be represented by a constant tail length of 75 10^6 km

plasma tail was converted into true linear distance using the ephemeris of Yeomans (1986a) and assuming the tail pointing into the antisolar direction.

The plasma tail structure on all medium and high contrast prints was checked extensively and the tail appearances on different pictures were compared in detail. Altogether 134 tail structures like plasma condensations, tail disconnections, and kinks were found which can be followed day by day on different plates drifting along the tail off the coma. Table 1 summarizes the dates when very prominent structures were visible in the plasma tail, the most famous being the disconnection event best visible on March 10. The angular distance of tail structures from the nucleus was measured in the same way as the visible tail end and each structure observed was provided with a symbol for identi-

Table 1. Occurrence of prominent tail structures[a]

(1)	(2)	(3)	(4)	(5)	(6)	(7)	(8)	(9)
Feb. 18					a			18
19								19
20								20
21			+				+	21
22		+						22
23		+						23
24								24
25								25
26	k, o							26
27							+	27
28	p							28
Mar. 1				+				1
2				+				2
3								3
4		x, z			x!			4
5			+					5
6			+					6
7	e1, h1	i1			i1!!		+	7
8		j1		+				8
9				+				9
10	v1	v1		+	v1!			10
11						+		11
12				+				12
13				+				13
14								14
15		d2, f2			f2!, g2			15
16	h2		+	+				16
17		e2				+		17
18	p2	p2			p2!!			18
19			+					19
20								20
21								21
22				+				22
23			+					23
24							+	24
25								25
26						+		26
27						+		27

[a] *Columns:* 1) date in 1986, 2) calculated outflow of condensation marked with symbol, 3) outflow of a kink, 4) observation of tail waves, 5) observation of tail rays, 6) formation of a disconnection event, 7) observation of high initial velocities ($v_0 > 3\,km\,s^{-1}$), 8) observation of high acceleration ($\langle a \rangle > 40\,cm\,s^{-2}$), 9) date in 1986.
Structure marked with a letter means date of formation.
Structure marked with a + means date of observation.
!: rather prominent disconnection event (DE)
!!: very prominent DE

Table 1 (continued)

(1)	(2)	(3)	(4)	(5)	(6)	(7)	(8)	(9)
28						+		28
29								29
30	w3				w3!		+	30
31								31
Apr. 1		e4			e4!	+		1
2		i4		+		+		2
3								3
4								4
5							+	5
6	d5	d5		+	d5!!	+	+	6
7	v4			+	v4!!	+	+	7
8								8
9				+				9
10				+			+	10
11	n5	n5			n5	+	+	11
12	y5				y5!		+	12
13	z5, c6			+			+	13
14			+	+				14
15			+	+				15
16				+				16
17								17

fication. The best visible structures within the tail are marked with their identification symbols in Paper II. For each structure the true linear distance and the number of plates on which it was measured is listed in a table of Paper II. For these, from each adjacent pair of time and distance, velocities are calculated and given therein. In order to determine the mean acceleration (a) in the tail, the initial velocity (v_0) of each plasma cloud, and the time of zero distance from the nucleus (t_0), the data of a single cloud or event were fitted with

$$s = \frac{1}{2} a \cdot (t - t_0)^2 + v_0(t - t_0) \tag{1}$$

where t is the UT of the picture and s the distance from the nucleus measured. A similar fit was made by Celnik et al. (1986a) for the prominent plasma clouds visible on March 9 and the following days, and by Celnik (1986) and Celnik et al. (1986b) for the sample of structures but assuming the initial velocity to be zero.

4. Results

4.1. Velocities in the plasma tail

The distribution of cloud velocities averaged over the whole observation run in suitable distance intervals along the tail is given in Fig. 2. The intervals were chosen so that each interval covers approximately equal numbers of data pairs and the resulting mean values are weighted with the uncertainty of the velocity

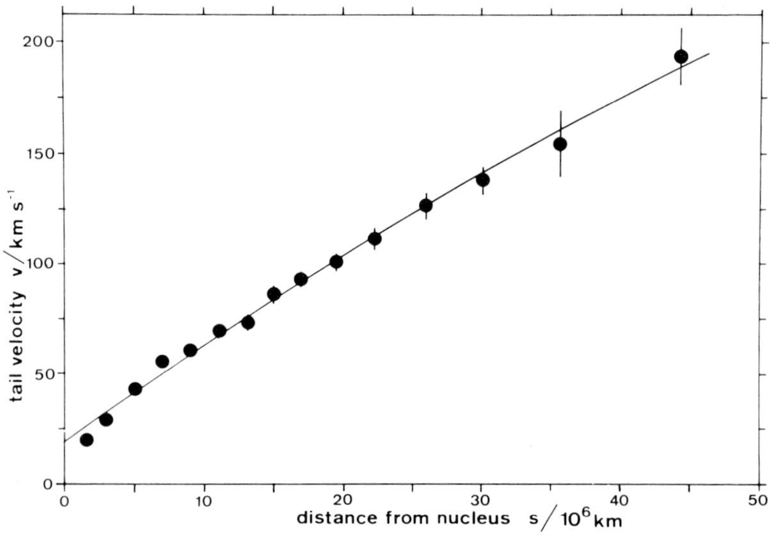

Fig. 2. Mean velocity distribution of condensations along the plasma tail; the fitted line is a second degree polynomial (see text). Errors of mean values are indicated

data. Figure 2 shows also the best fit with the function

$$v(s) = m \cdot s^2 + n \cdot s + v_0 \quad (2)$$

where the resulting parameters are $m = -0.0013 \pm 0.0003/(10^6\,\text{km/day})$, $n = 0.39 \pm 0.01/\text{day}$, and $v_0 = 19 \pm 2\,\text{km s}^{-1}$. The velocity v is measured in km s^{-1} and the distance s in 10^6 km from the nucleus. The distribution of velocities along the tail is highly non-linear. Even the second degree polynomial (2) does not fit the data well; the true initial velocity is certainly much smaller than $19\,\text{km s}^{-1}$ as obtained by the fit.

The deduced velocity distribution even at distances of $50\,10^6$ km from the coma is still steeply increasing to values far beyond $200\,\text{km s}^{-1}$. Because the solar wind is responsible for the acceleration it is clear that its velocity must be much larger than $200\,\text{km s}^{-1}$ and even larger than $320\,\text{km s}^{-1}$, the extreme value measured in the tail. As a matter of fact the mean value of the solar wind velocity during our observations was $480\,\text{km s}^{-1}$ with large temporal variations (Oyama et al., 1986). Thus it appears that even at the largest distances of recognizable structures in the tail the full solar wind effect has not yet been reached.

4.2. Accelerations in the plasma tail

4.2.1. Accelerations as a function of distance from the nucleus

There are different ways to determine the mean acceleration in the plasma tail. One of these was presented by Celnik (1986), calculating the acceleration from time/velocity data and averaging the results in distance intervals. Because the velocity data are widely scattering (due to real scatter and measuring errors) these accelerations obtained as velocity gradients seem not to be very reliable.

For this reason the distribution of accelerations along the tail was calculated from the *averaged* velocity distribution (cf. Fig. 2) using

$$a = 0.5\,(v_2^2 - v_1^2)/(s_2 - s_1) \quad (3)$$

where a is the acceleration and v_i is the velocity at distance s_i. The result is shown in Fig. 3. Although the accelerations are somewhat scattering there is a clear trend to higher values at larger distances. This may be a sign for the increasing efficiency of the acceleration mechanism due to the expansion of the gas clouds while drifting along the tail. Possibly dense plasma clouds are screened from the solar wind when they are still near the tail axis in the first part of the tail. The acceleration is increasing by $1.24 \pm 0.02\,\text{cm s}^{-2}/10^6$ km, the initial value is $14.5 \pm 0.3\,\text{cm s}^{-2}$. Similar to the velocity distribution of Fig. 2 the behaviour of the accelerations at distances smaller than $2\,10^6$ km is uncertain. The mean value of acceleration in the tail obtained here is $27.0 \pm 3.4\,\text{cm s}^{-2}$ in good agreement with the results derived in the next chapter.

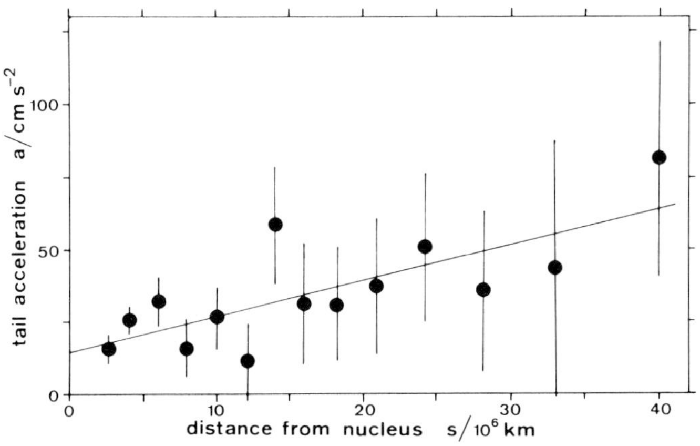

Fig. 3. Mean acceleration distribution of condensations along the plasma tail; a straight line was fitted, acceleration values are determined from Fig. 2. Errors of mean values are indicated. No effect whatsoever of the varying distance from the Sun has been noted

4.2.2. Assuming constant acceleration in the tail

Looking at the scatter and the uncertainties of the acceleration values (Fig. 3) a fit with constant acceleration to the velocity data of Fig. 2 appears reasonable. The result is $31.3 \pm 1.0\,\text{cm s}^{-2}$ and an initial velocity of $0 \pm 9\,\text{km s}^{-1}$, quite similar to the values obtained by Celnik et al. (1986a) for a single prominent structure. In Sect. 3 we fitted a constant acceleration and an initial velocity to the observed motion of single tail structures. From these we obtain mean values averaged over the whole observational run: The mean acceleration is $27.7 \pm 1.0\,\text{cm s}^{-2}$ and the mean initial velocity $4.5 \pm 1.7\,\text{km s}^{-1}$. However, it is obvious that the data split up into two different groups. The first group comprises with less than 14% of the data, the initial velocities larger than $3\,\text{km s}^{-1}$ (with a mean of $30 \pm 4\,\text{km s}^{-1}$) and rather low accelerations (mean $19 \pm 3\,\text{cm s}^{-2}$). The second group covers more than 86% of the data. It shows significantly smaller velocities (mean $0.54 \pm 0.08\,\text{km s}^{-1}$) and relatively large accelerations (mean $29.1 \pm 1.3\,\text{cm s}^{-2}$) in good agreement with the results obtained above. This value of the initial velocity corresponds quite well to the outflow velocities of gas and dust jets ($0.63\,\text{km s}^{-1}$), for which results were obtained by several authors as summarized in Table 2. While the outflow velocities of dust, CN, and plasma are quite similar ($0.47 \ldots 0.57\,\text{km s}^{-1}$) the gas velocity tends to significant larger values ($0.79\,\text{km s}^{-1}$). The first conclusion following from these observational fact is that CN molecules are possibly originating from dust grains, what was already mentioned by A'Hearn et al. (1986). Secondly, because of the significant difference between the near-nucleus streaming velocities of neutral gas and ions, slow molecules seem to undergo ionization much more efficiently than high velocity molecules, what may be important for the theory of interaction between cometary and solar wind matter. Expansion velocities within the coma are discussed in Sect. 4.5.

In view of the large scatter within the high velocity group, the low percentage of the complete sample and the comparatively lower accuracy of these data (velocity and acceleration are determined from significantly fewer plates than the second group values, the structures are in most cases very faint, some plates are underexposed, the tail is crossing the Milky Way) it seemed likely that the high initial velocities might result from misidentifications of plasma tail structures on different, adjacent plates. However, there are some hints that at least some of the high initial velocities are real. Firstly more than 83% of the large v_0 are connected to the occurrence of tail rays (cf. Table 1). Secondly structure "v_4" is a very prominent plasma cloud and the result of a tail disconnection event which happened on April 7.06. The fit to this structure dynamics gives $18.1 \pm 0.2\,\text{km s}^{-1}$ for the initial velocity and an acceleration of $27.6 \pm 0.9\,\text{cm s}^{-2}$ which

Table 2. Outflow velocities observed by different authors. Errors in parenthesis are assumed values. As measured by Lämmerzahl et al. (1986) plasma and neutral gas are collisionally coupled below the contact surface at 4500 km distance from the nucleus

Authors	Aim	Velocity ($km\,s^{-1}$)
A'Hearn et al. (1986)	CN	0.67 ± 0.33
Balsiger et al. (1986)	Ions	$1.0 \pm (0.5)$
Bockelèe-Morvan et al. (1986)	HCN	<1.35
Despois et al. (1986)	HCN	<0.85
Edenhofer et al. (1986)	Dust	$0.4 \pm (0.2)$
Körösmezey et al. (1986)	Ions	$0.6 \pm (0.2)$
Krankowsky et al. (1986)	Gas	0.9 ± 0.2
Krasnopolsky et al. (1986)	Dust	0.57 ± 0.05
	Gas	0.85 ± 0.15
Krasnopolsky, Tkachuk (1986)	OH	$0.32 < v < 1.85$
Lämmerzahl et al. (1986)	H_2O, ions	<0.81
Larson et al. (1986a)	H_2O	0.9 ± 0.2
		1.4 ± 0.2
Opal et al. (1986)	Gas, UV spectra	$1.0 \pm (0.5)$
Samarasinha et al. (1986)	Dust	0.5 ± 0.25
Schloerb et al. (1986a)	HCN	0.87 ± 0.12
Schloerb et al. (1986b)	HCN	$1.0 \pm (0.2)$
	CN	$0.59 \pm (0.2)$
Schloerb et al. (1986c)	OH parents	0.58 ± 0.03
Schlosser (1966)	Ions	0.44 ± 0.05
Schlosser et al. (1986)	CN	<0.8
Schwenn et al. (1986)	Ions	$1.0 \pm (0.5)$
Simpson et al. (1986)	Dust	$0.30 < v < 0.42$
Vaisberg et al. (1986)	Gas	0.61 ± 0.04
This paper	CO^+	0.54 ± 0.08
	CN	$0.23 < v < 0.7$
Mean values	Total	0.626 ± 0.047
(weighted with ε^{-1})	Ions	0.534 ± 0.080
	Gas	0.724 ± 0.062
	Gas without CN	0.785 ± 0.076
	CN	0.566 ± 0.056
	Dust	0.469 ± 0.056

is close to the mean value of the high initial velocity sample ($19 \pm 3\,km\,s^{-1}$). Any misidentification can be excluded in this case. Thirdly all high velocity events observed seem to be correlated with extreme densities occurring in the solar wind as measured by the Japanese space probe Sakigake (Oyama et al., 1986). Common observational time is from Feb. 18 to Apr. 5. In Table 3 the t_0 values of all high initial velocities found are opposed to the time of relative maxima within the solar wind density ($t_{\varrho max}$). The mean interval between these dates is 1.76 days. The last column in Table 3 quotes for each structure the shortest distance (s_1) at which the structure was observed, giving a mean of $4.14\,10^6$ km. If we assume a dense solar wind plasma cloud possibly intruding into the coma and causing a high velocity of a cometary plasma cloud by strong interaction with the cometary gas on the distance scale s_1, we can derive a formal velocity of this cloud of

$$v_p = \langle s_1 \rangle / \langle (t_0 - t_{\varrho max}) \rangle = 27.2\,km\,s^{-1}, \tag{4}$$

a value corresponding quite well to averaged initial velocity of the cometary plasma clouds of $28.6\,km\,s^{-1}$ during the common observation time, a fact which supports this hypothesis. The most important verification was given by e.g. Goldstein et al. (1986), Korth et al. (1986), Schwenn et al. (1986), and Zastenker et al. (1986) who observed tailward high velocity plasma streamers showing velocities up to $60\,km\,s^{-1}$ at relatively low nuclear distances (≈ 45000 km). Goldstein et al. corroborating our hypothesis suggested that the plasma observed is a mixture of cometary and solar wind ions. Thus it must be accepted that the higher initial velocities obtained are not necessarily wrong measurements but are at least partly real. Disconnection events are discussed later in Sect. 4.7.

4.2.3. The variation of acceleration with heliocentric distance

Figure 4 shows the variation of acceleration in the plasma tail with time and heliocentric distance as averaged over one day.

Table 3. Extreme initial velocities of plasma clouds and maximum density of the solar wind (further explanations see text)

Structure	t_0 (UT)	$t_{\varrho max}$ (UT)	$t_0 - t_{\varrho max}$ (days)	s_1 (10^6 km)
q1	M.11.57	M.10.9	0.67	9.0
e2	17.89	10.9	6.99	13.9
g3	26.09	24.31	1.78	7.3
o3	26.52	26.38	0.14	0.7
n3	27.42	27.32	0.10	2.7
m3	27.46	27.32	0.14	3.6
s3	28.81	27.32	1.49	0.6
b4	A. 1.68	31.33	1.35	2.0
c4	2.09	31.33	1.76	3.9
u4	6.85	A. 4.40	2.45	0.9
t4	6.87	4.40	2.47	0.9
Mean			1.76	4.14

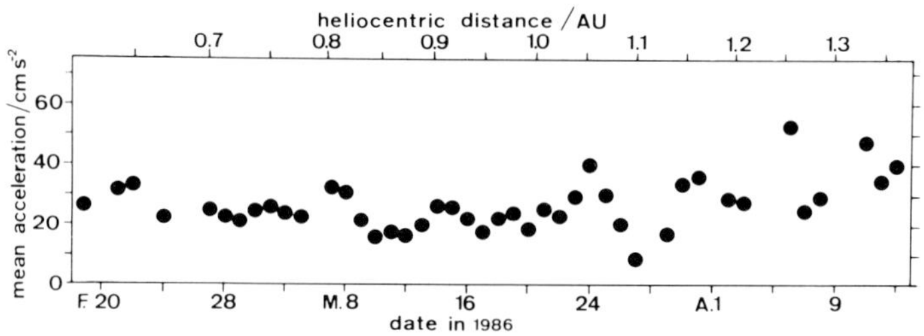

Fig. 4. Variation of acceleration (assumed to be constant along the tail) with heliocentric distance and time. Mean values as averaged over one day are plotted. The mean uncertainty is $\pm 3\,\mathrm{cm\,s^{-2}}$

Our observations cover the distance range from 0.61 to 1.42 AU. Single measurements of acceleration (assumed to be constant along the tail) vary by a factor of 10 between 9 and 89 $\mathrm{cm\,s^{-2}}$ excluding the high initial velocity points. We have found two possible effects in our data. Firstly the variations are not erratical but seem to show a 7-days-period. Secondly maxima in the acceleration may be correlated to dates of high density within the solar wind and to disconnection events. These effects will be dealt with in a forthcoming paper. Long-time observations of interplanetary magnetic field parameters as measured by space probes during our observations have not been compared to our results on the plasma tail. But there are a lot of hints that magnetic field sector boundaries passing by the comet have great influence on the plasma tail structure. This aspect will be discussed later in Sects. 4.4 and 4.7.

4.3. *Jet activity of the nucleus*

4.3.1. Rotational period of the nucleus

By fitting a constant acceleration to the time-distance data of a certain tail condensation, simultaneously the time of zero distance from the nucleus (ejection time t_o) of the material considered could be derived with a mean error of 0.06 days. These ejection dates are not scattered arbitrarily, but are ordered in groups around some "central dates" with a mean scatter of 0.5 days. The

Table 4. Starting dates t_0 and rotational period

Central t_0 (UT)	t_0 (UT)	Δt_0 (days)	Nucleus longitude
F.18.82	F.18.82		253°
		1.79	
20.60	20.60		198
		1.30	
21.90	21.71		27
	22.09		93
		1.91	
23.81	23.81		27
		2 × 1.82	
	26.77		174
	26.91		198
	27.01		215
27.44	27.21		249
	27.71		334
	27.78		345
	28.67		138
		2.33	
M. 1.77	M. 1.77		327
		2.39	
	3.11		195
	3.45		254
	4.15		14
4.16	4.41		58
	4.50		73
	4.56		84
	4.92		154
		2 × 2.02	
	7.16		167
	7.70		261
	7.93		299
	8.02		315
8.19	8.19		344
	8.25		355
	9.14		147
	9.16		149
		2.26	
	10.28		342
10.45	10.38		359
	10.69		51
		1.60	
	11.57		202
12.05	11.95		267
	12.62		22
		1.54	
13.59	13.59		187
		1.95	
	14.43		331
	14.85		42
	14.91		554
	15.27		115
15.53	15.52		158
	15.68		186
	16.15		265
	16.32		294
	16.77		12

Table 4 (continued)

Central t_0 (UT)	t_0 (UT)	Δt_0 (days)	Nucleus longitude
		2.81	
	17.38		115
	17.60		153
	17.89		203
	17.92		207
	18.04		229
18.35	18.14		246
	18.75		351
	19.03		38
	19.20		67
	19.54		125
		2.47	
20.82	20.55		298
	21.10		31
		1.65	
22.47	22.07		197
	22.87		335
		2.27	
	24.15		194
	24.27		215
24.74	24.60		271
	25.18		10
	25.47		60
		2.02	
	26.09		166
	26.16		178
	26.52		238
26.76	26.90		304
	27.42		34
	27.46		39
		2.06	
28.81	28.81		271
		1.50	
	30.10		131
	30.15		139
	30.26		159
30.02	30.39		181
	30.44		189
	30.57		211
		2.92	
	A. 1.58		195
	1.68		213
	1.94		257

Central t_0 (UT)	t_0 (UT)	Δt_0 (days)	Nucleus longitude
	2.09		283
A.2.23	2.23		307
	2.36		330
	2.68		24
	3.27		125
		2 × 2.26	
	5.55		154
	5.72		183
	5.89		213
	5.94		221
	6.06		241
	6.29		282
	6.61		336
	6.69		349
	6.71		352
6.75	6.85		17
	6.87		20
	6.93		30
	7.06		52
	7.10		59
	7.39		109
	7.77		174
	7.96		206
	8.16		241
		2 × 2.35	
	10.57		293
	10.68		311
	11.24		47
	11.36		68
11.44	11.72		129
	11.80		143
	12.01		179
	12.16		205
		1.82	
	12.81		315
	12.81		315
	13.39		55
13.26	13.41		59
	13.42		60
	13.48		70
	13.52		78°
± 0.23	± 0.06		± 3°
Rotation period (hours)			Syn. 50.1 ± 0.6 Sid. 50.5 ± 0.6

mean time interval between these central dates is 50.1 ± 0.6 hours (Table 4), which is interpreted as the synodic rotation period of the nucleus. Taking into account the varying aspect of the nucleus to the sun during our observations of 72°, we derive a sidereal rotation period of 50.5 ± 0.6 hours (=2.105 days). This value is in good agreement with the results of other authors (cf. Table 5) giving a mean of 52.3 ± 0.4 hours. The possible appearance of a 7.4 days period is discussed later in Sect. 4.3.5.

4.3.2. Distribution of activity along longitude at the surface of the nucleus

With the sidereal rotation period derived above, each of the ejection dates was converted into longitude λ at the surface of the nucleus; the origin was chosen arbitrarily, λ is increasing with time (Col. 4 of Table 4). The mean error of λ is 10° corresponding to 0.06 days. The frequency of ejection events per 20° interval

Table 5. Rotation period of the nucleus, errors in parenthesis are assumed values

Authors	Period (hours)
Belton et al. (1986)	$\{50.3 \pm 0.1$[a] 54.0 ± 0.1
Celnik (1986)	53.5 ± 1.7
Kaneda et al. (1986a)	52.8 ± 2.4
Kaneda et al. (1986b)	52.9 ± 1.0
Leibowitz, Brosch (1986)	52 ± 4
Sagdeev et al. (1986a)	53 ± 4
Sagdeev et al. (1986b)	53.5 ± 1.0
Sagdeev et al. (1986c)	$52 \pm (2)$
Schlosser et al. (1986)	53.4 ± 0.9
Sekanina (1985)	$52 \pm (2)$
Sekanina (1986a)	52.5 ± 1.6
Sekanina (1986b)	$53 \pm (2)$
Vaisberg et al. (1986)	$50 \pm (2)$
Wilhelm et al. (1986)	54 ± 1
this paper	50.5 ± 0.6
Mean (weighted with ε^{-1})	52.3 ± 0.4

[a] Belton et al. quote only a period of about 54 hours, but it seems from their Fig. 3 that 50.3 hours are possible too with a similar probability

in longitude is shown in Fig. 5. A maximum of activity occurs at longitudes 190° . . . 210° and 30° . . . 50°, while sources at 90° . . . 110° and 230° show less activity. The activity contrast between maximum and minimum frequency values is 5.3 ± 1.0 in good agreement with results obtained by Sekanina and Larson (1986). If the 2^d-period indeed describes the rotation of the nucleus, the sources of activity at the nucleus' surface must be arranged near the equator in order to explain the sharp activity contrast observed. This hypothesis is supported by e.g. Samarasinha et al. (1986), Sekanina et al. (1986), and Pankiewicz and McDonnell (1986). No results could be obtained from our data concerning the spin axis orientation. But it seems not unlikely that it is orientated approximately perpendicular to the orbit plane

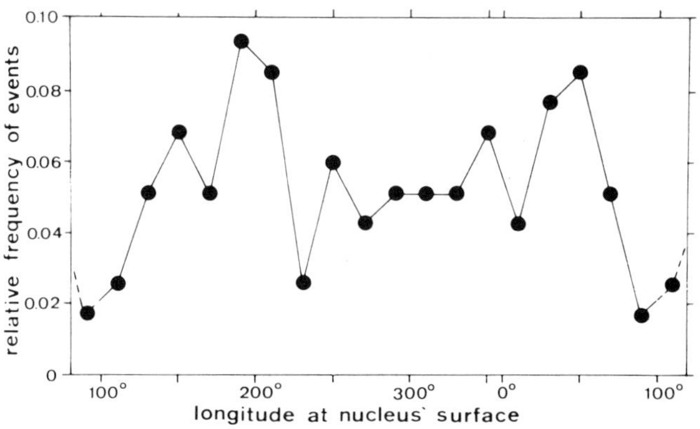

Fig. 5. Relative frequency of source activity observed as a function of longitude on the surface of the nucleus (using a rotation period of 2.105 days). Maximum activity occurs at the long ends of the ellipsoidal nucleus (cf. Fig. 7). The origin of the longitude scale is chosen arbitrarily

(Samarasinha et al. (1986), Sekanina et al. (1986), Sagdeev et al. (1986b,c), Wilhelm et al. (1986), Yeomans (1986b)) although the results obtained by different authors are strongly varying. The distance intervals in between the active ejection sources derived here and obtained by Kaneda et al. (1986c) are almost identical and in good agreement with the Giotto, Vega-2, and Suisei results (Pankiewicz et al., 1986; Kaneda et al., 1986) (Fig. 6). Edenhofer et al. (1986) studied the dust jets crossed by Giotto and mapped them back to the nucleus, but without considering the curvature. The location of their four features were therefore not taken over into Fig. 6. Their results are, however, qualitatively compatible with ours. The mean scatter in Fig. 6 is 7°8 corresponding to 0.7 km on the nucleus, adopting a circumference of 34.6 km. Thus we estimate that the diameter of sources on the nucleus is smaller than 0.7 km. This corresponds to the results of Edenhofer et al. and Huebner et al. (1986) who obtained 300 ± 50 m and 0.5 km for the extension of activity sources.

The alternative period $2^d\!.18 = 52^h\!.3$ seems less preferable as it produces a more erratic source distribution on the nucleus (Fig. 5) with an activity contrast of only 3.7. The $7^d\!.4$-period which is further discussed below produces again a sensible pattern of activity sources with a contrast of less than 3.6.

4.3.3. Variation of activity with time

An investigation of the activity distribution in time for sources at different longitudes showed that there are two very active sources occurring at longitudes 190° . . . 210° and 30° . . . 50° remaining active steadily for a long time, while there are three longitude zones showing activity only sporadically and for short time: 110° (Mar. 15 . . . Apr. 7), 230° (Mar. 18 . . . Apr. 6), and 270° (Mar. 8 . . . 29). These results are contradictory to those of Sekanina (1986c) who found from pictures obtained 1910 that most sources were active only once.

4.3.4. Distribution of activity sources with respect to long and short axis of the nucleus

Sagdeev et al. (1986a) found that the nucleus of P/Halley has the shape of a deformed rotational ellipsoid of 7.5 × 7.5 × 14.0 km, showing a thicker and a sharper end at the long axis. Other authors found slightly different values (cf. e.g. Whipple, 1986; Möhlmann et al., 1986; Reitsema et al., 1986; Wilhelm et al., 1986).

Further they quote that the space probe Vega-2 probably has seen the nucleus broadside 40° after closest approach while

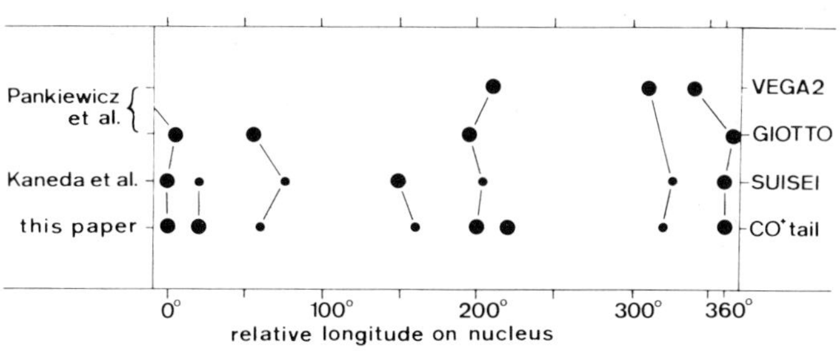

Fig. 6. Distribution of strong (●) and weak (•) sources on the nucleus with relative longitude (arbitrary origin). The difference between the longitude scale origins of Fig. 5 (this paper) and Fig. 1 of Pankiewcz et al. is assumed to 50°. Note the various definitions of source strengths by different authors: Kaneda et al. and Pankiewicz et al.: intensity, this paper: frequency

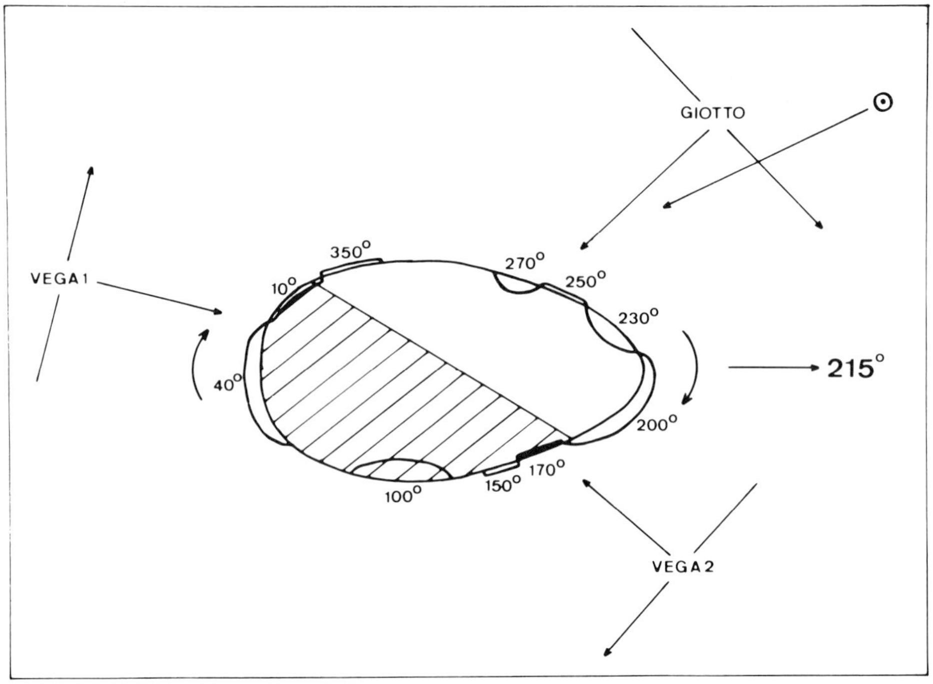

Fig. 7. Our model of source activity distribution on the nucleus, compared to the view from different space probes during their flyby as seen from above the ecliptical plane. The daylight configuration for the Giotto flyby is indicated, when the longitude interval from 180° to 360° was illuminated by the Sun (which is in the paper plane). Thick and thin ends of the nucleus are indicated as obtained from the rotation period and the Vega1/2 views at encounter. The thin end points into $\lambda = 215°$ of our coordinate system on the nucleus. The sense of rotation is marked and the activity of sources is represented by bulges on the periphery of the nucleus. The uncertainty of source locations is approximately $\pm 3°$

Vega-1 has seen the thick end of the long axis approximately 5° before closest approach. From these results and from our data follows that at the times of nearest distance the longitude 21° (in units of Fig. 5) was facing Vega-1 and 174° Vega-2, respectively. The conclusion is illustrated in Fig. 7: The two main sources of activity at the surface must be located near the *ends* of the long axis of the nucleus while low-activity zones are following at distances 30° and 60° on the *broad sides* of the surface. In order to compare our model with the images of the European space probe Giotto we have indicated the longitude facing Giotto during its closest approach and the direction to the Sun. There is excellent agreement between our model and the image obtained from 18270 km distance presented by Keller et al. (1986a). For an even better image of the nucleus cf. Fig. 2 of Reitsema et al., which is a composite of 60 images taken by Giotto. Firstly the orientation of the nucleus concerning long/short axes and narrow/thick ends corresponds to our result in Fig. 7. The four jet sources visible in the composite image are equivalent to our sources in Fig. 5 at 200°, 210°, 250° and approximately 300°.

We conclude that it was possible to establish a consistent model of mass ejection activity distribution across the surface of the nucleus from the time of ejection, rotation period and the Vega space probe observations. More than 37% of the events observed occurred on less than 8% of the surface. This result corresponds well to the observations of other authors who found that actually only 10 ... 15% of the surface show jet activity (cf. e.g. Whipple, 1986; Huebner et al., 1986; Ferrin and Gil, 1986; Keller et al., 1986a). Thus the long ends of the nucleus seem to be privileged in undergoing mass ejection while the broad sides show only sporadic activity. This is also supported by the activity ratio of 5:1 for long end to broad side activity. Plausible causes for this observational fact are manifold. The nucleus seems to be envelopped by a dense dust atmosphere of relatively small scale height as observed by Sagdeev et al. (1986d) and Keller et al. (1986b). In this case, the long ends of the nucleus will be heated more effectively than the broad sides due to smaller extinction of sunlight supporting the activity of the long ends. The temperature gradient between cometary day and night then should be larger at the long ends than on the broad sides again furthering activity at the long ends by increased thermal stresses influencing the structure of the surface. The same effect is caused even more efficiently by the crusty dust layer on the surface of the nucleus. This dust layer was predicted by many authors (cf. the review by Whipple, 1986) and now observed directly by the space probes Vega 1/2 and Giotto. Kührt et al. (1986) modelled thermal stresses on the dust mantle on the surface and the dust kinematics near the surface. They found that the mantle is supported by dust grains falling back from the jets to the terrain surrounding the active source. Kührt et al. also quoted that this dust layer reduces the activity of the covered surface. If this is the case the dust layer cannot be uniform across the whole surface due to the activity of distinct sources. This is supported by the Giotto composite image of the nucleus made by Reitsema et al.: Although the spatial resolution is varying across the image it is clear that the bright source at the broad side of the nucleus looks quite different compared to the long end sources. Reitsema et al. suggested that there are different surface optical properties. If we assume the dust layer on the broad sides of the nucleus being thicker than on the long ends, activity at the long ends is favoured. Dust grains falling back from jets at the long ends should first be visible as a dust atmosphere which prefers layering down at the broad sides strengthening the dust mantle and hindering jet emission in these regions. The result of this self-reinforcing process is just what we observe: a strong activity at the long ends and less activity at the broad sides of the nucleus. Our observations also support the "icy-glue model" of Houpis and Gombosi (1986) who predict long lifetimes for the main sources.

4.3.5. Is there a 7.4-day period?

A 7.4-day period was proposed e.g. by Festou (1986) from the variability of photometric light curves and by Millis and

Schleicher (1986) (7.4 ± 0.1 days) from production rate variations. It is probably visible also in a long-time light curve established from naked eye observations by amateur astronomers (Kammerer, 1987). A'Hearn et al. (1986) state that both periods (7.4 and 2.1 days) are possible from their near-nucleus imaging data.

From our observations we find as mean time interval between two events at the same place on the surface of the nucleus 7.7 ± 0.3 days, a value mainly determined by the sporadic sources at the broad side of the nucleus (8.8 ± 0.4 days) rather than by the sources at the long ends (5.7 ± 0.3 days). It is a striking fact that all 7.4-day period observers had used light curves while observers investigating the nucleus and structures within tails and coma only find a 2.1 ... 2.2 day period. A 7.4 day period would be visible in light curves consistently besides the 2.1 day period (what was observed therein by Belton et al., 1986) if the sporadically active sources at the broad side of the nucleus were much more intensive than those at the long ends. Broad-side sources once breaking through the dust mantle should be much "fresher" compared to the long-end sources, if we follow the model suggested in the previous section. Occasionally (that is every 7.4 d.) appearing very bright sources will show up in light curves but hardly in statistical data like ours. Nevertheless we searched our data for 7.4-day time intervals in the activity of a single source. We did not find any interval of 7.4 days, but three times a *14.8 day* interval occurred, all of these appearing in broad-side sources (70°, 250°) 180° apart. The latter result agrees with Sekanina's (1987) model of a 7.4-day rotation around the long axis and a shorter (2.2 days) precession period of this axis around an angle of 77°. If the model is correct a certain broad-side source would have the same aspect to the Sun every 14.8 days. If there are sources of such kind at opposite broad-sides a 7.4-day period as observed can be understood also from our data. Belton (1987) suggests that the 54 hours and 7.4 days period observed in light curves are beat periods from the combination of a true 49 hours rotation and a true 106 hours nutation period. Our 50.5 hours period seems to correspond more to the 49 hours because it is determined geometrically rather than by photometry.

4.4. *Aberration angle of plasma tail*

The aberration angle of the plasma tail was measured on 103 wide-field images (blue-broad-band, CO^+) from Feb. 21 to Apr. 17, 1986. Aberration values were measured as a tangent to the tail near the coma and reach from $-5\overset{\circ}{.}3$ to $+13\overset{\circ}{.}4$ with a mean uncertainty of approximately $1\overset{\circ}{.}0$ which is caused by the relatively low angular resolution of the wide-field images. In Fig. 8 the aberration angle averaged over all plates obtained during one night is shown as a function of date. The mean value averaged over the whole observing run is $+1\overset{\circ}{.}8 \pm 0\overset{\circ}{.}3$. Surely the slow variation of aberration angle ε as indicated in Fig. 8 by the fitted polynomial line is due to projection effects rather than the obvious sudden changes on short time scales. The strongest deviations ($\Delta\varepsilon$) from the smooth curve appear on Feb. 21 ($-9\overset{\circ}{.}3$), Feb. 22 ($-6\overset{\circ}{.}8$), Feb. 26 ($+3\overset{\circ}{.}0$), Mar. 11/12 ($+5\overset{\circ}{.}6$), Mar. 25 ($+3\overset{\circ}{.}5$), Mar. 26 ($-3\overset{\circ}{.}7$), Mar. 28 ($-4\overset{\circ}{.}3$), Apr. 9 ($+2\overset{\circ}{.}7$), and Apr. 15/16 ($+6\overset{\circ}{.}4$) (cf. Tab. 6).

Niedner (1986), Saito et al. (1986), and Riedler et al. (1986) calculated or observed sector boundaries of the interplanetary magnetic field passing by the coma. In Table 6 such events are indicated and compared with aberration angles $\Delta\varepsilon$ measured

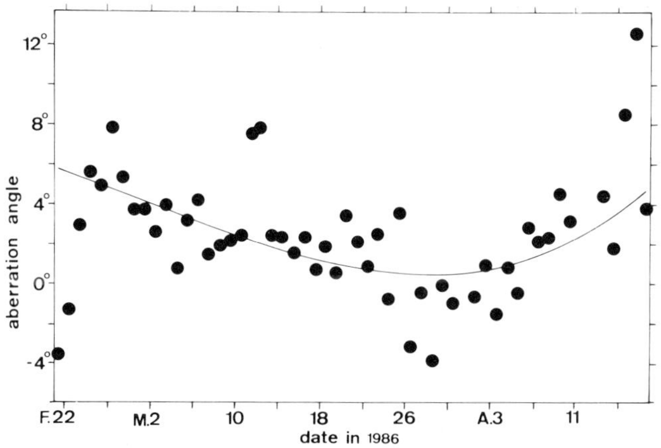

Fig. 8. Variation of the aberration angle of the plasma tail as a function of time as measured on the wide-field camera prints as a tangent to the tail near the nucleus. The fitted polynomial line indicates the smooth drift of aberration probably due to projection effects. Uncertainty of a single value is approximately $1\overset{\circ}{.}0$

during the time of passing boundaries. The comparison shows that within an uncertainty of 1^d nearly 80% of boundaries passing the coma are correlated with aberration angles $|\Delta\varepsilon| > 1.5$ degrees. The portion decreases for larger $|\Delta\varepsilon|$. Conversely more than 70% out of $|\Delta\varepsilon| > 1\overset{\circ}{.}5$ are correlated to sector boundaries passing within 1^d, but the portion decreases only to 66% for larger $|\Delta\varepsilon|$. Thus it is possible that sector boundaries of the interplanetary magnetic field passing the comet are responsible for the deviations of the plasma tail aberration angle. A comparison of the aberration angle with solar wind parameters like velocity, density, and temperature (Oyama et al., 1986) indicated no correlation. A possible connection of the aberration angle with solar wind direction could not be checked.

4.5. *CN-coma expansion*

Schlosser et al. (1986) investigated the 42 FFC-images in the light of CN λ 3880 obtained from 1986 March 5 to April 17, as described above in Sect. 2. They showed that there exist shell-like structures within an expanding CN coma. The expansion velocity is decreasing with increasing heliocentric distance.

We have taken the radii of CN shells measured by Schlosser et al. (1986, cf. Fig. 1 therein) and re-interpreted the CN-shells from these data (Fig. 9). Fitting a linear function to each shell given by more than two measured radii resulted in a starting time and an expansion velocity (which was assumed to be constant). Any variable expansion velocity cannot be verified from these data. Table 7 gives the results obtained from the fit for each shell. In Fig. 10 the shell expansion velocity is plotted as a function of heliocentric distance. It is clearly decreasing with increasing distance although the scatter is very large. The scatter may be due to incomplete data sets of shell radii identified by inspection on the original plates which might result in wrong shell identifications. The data are most complete in shells no. b, g, l, m, q, s, and y. In Fig. 10 all these shells are within a narrow band with very low scattering, except shell No. q which is possibly composed of two different shells (cf. Fig. 9). The best fit to the data points in Fig. 10 is given by

$$\frac{v}{\mathrm{km\,s^{-1}}} = 0.32\left(\frac{r_{\mathrm{hel}}}{\mathrm{AU}}\right)^{-1/4} + 0.40\left(\frac{r_{\mathrm{hel}}}{\mathrm{AU}}\right)^{-2} + 0.23 \tag{5}$$

Table 6. Tail disconnection events and correlation with aberration angle and magnetic field sector boundaries (connected signs mark identical events, ! and !! mark prominentest events)

Date in 1986	Disconn. event		Aberration angle $\Delta\varepsilon$	Sector boundaries		
	Here	Niedner		Niedner	Saito et al.	Riedler et al.
Feb. 18	+		–			
21			−9.3			
22		+	−6.8	+		
23			−2.4			
26			+3.0			
Mar. 1/2		+	−0.9	+		
4	+!	+	−2.8	+		
6			+1.1			+
7	+!!		−1.5			
8		+	−0.8	+		
9			−0.4			+
10	+!		+0.1		+	
11			+5.4		+	
12			+5.8			
13/14		+	+0.6	+		
15	+!		0.0			
18	+!!		+0.7		+	
19			−0.5		+	
20		+	+3.4	+		
23			+2.1			
25			+3.5			
26			−3.7			
28			−4.3			
30	+!		−1.4			
Apr. 1	+!		−1.3			
3			−2.3			
6	+!!		+1.6			
7	+!!		+0.7			
8		+	+0.7	+		
9			+2.7			
11	+	+	–	+		
12	+		–			
14			−1.7			
15/16		+	+6.4	+		

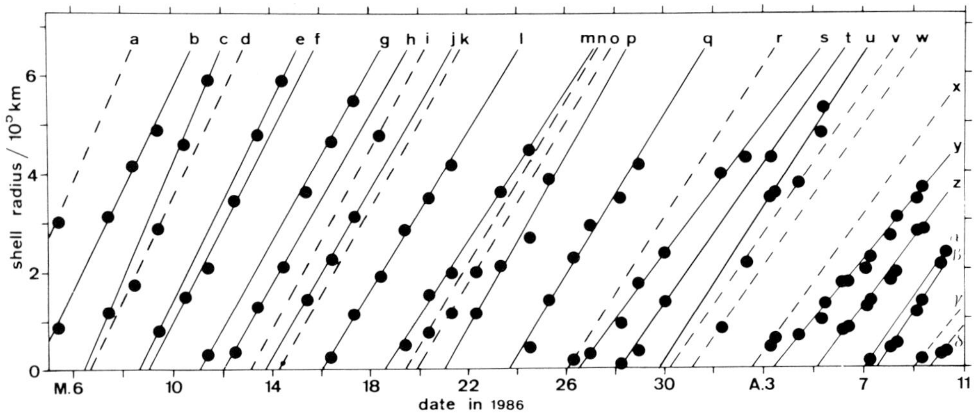

Fig. 9. Alternatively interpreted CN shells using the radii measured by Schlosser et al. (1986). Straight lines were fitted to the expanding radii of each shell. Broken lines result from the fit in Fig. 10. Mean error of a single point is apprroximately 0.3 10^5 km

Table 7. CN shell expansion (values in parenthesis are taken from the fit to the data in Fig. 11)

Shell	Points	t_0 (UT)	v_{exp} ($km\,s^{-1}$)
a	1	M. (2.70)	(1.29)
b	4	4.46	1.18
c	3	6.44	1.36
d	2	(6.69)	(1.20)
e	3	8.63	1.16
f	3	9.02	1.12
g	4	11.12	1.01
h	3	12.01	0.99
i	1	(13.18)	(1.06)
j	3	13.76	0.99
k	1	(14.31)	(1.04)
l	6	16.03	0.93
m	5	18.62	0.88
n	1	(19.48)	(0.96)
o	2	(19.95)	(0.96)
p	4	21.04	1.00
q	6	23.68	0.94
r	1	(26.04)	0.88
s	6	26.55	0.77
t	4	28.23	0.83
u	4	29.85	0.91
v	2	(30.27)	(0.84)
w	1	(31.12)	(0.83)
x	2	A. (2.59)	(0.81)
y	11	3.43	0.71
z	8	5.17	0.78
α	3	7.08	0.85
β	4	7.56	0.89
γ	1	(9.03)	(0.77)
δ	2	(9.71)	(0.76)
Errors		±0.06	±0.02

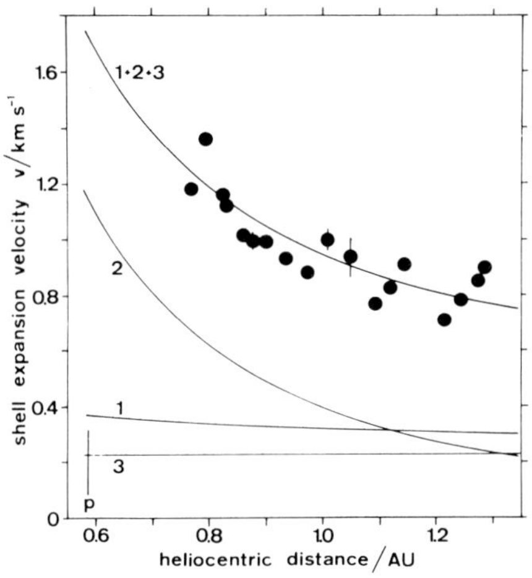

Fig. 10. Variation of CN shell expansion velocity with heliocentric distance. The fitted line is composed of three components indicated by the numbers 1 . . . 3 and described in the text by Eq. (5)

where the first term describes the thermal part of expansion, assuming the gas temperature being equal to the surface temperature of the sunlit nucleus. Cooling processes and shadowing effects have been neglected, therefore the constant factor of 0.32 obtained using the formulae of Lang (1980) is an upper limit. The albedo of the nucleus' surface was assumed to 0.03 (cf. e.g. Reinhard, 1986), and the temperature was estimated for a slowly rotating body. Because the shell expansion is faster than thermal there must be additional mechanisms causing larger velocities. The second term in Eq. (5) describes an effect depending quadratically on the heliocentric distance. Possibly this is due to an excess velocity resulting from photodissociation of the mother molecules of CN, of which at least two species are known, HCN and CH_3CN (Krasnopolsky et al., 1986a). The third, constant term in Eq. (5) turned out to be necessary for a reasonable fit. Probably it is produced by the initial (jet) velocity of mother molecules or other material from which CN is originating.

Taking into account cooling losses within the first term of Eq. (5) the first factor should decrease. If it is smaller than 0.2 the last term should increase to values beyond 0.4 to fit the data. In this case the initial velocity (third term) is reaching the outflow velocity of dust jets from the nucleus as determined by several authors (cf. Table 2). A'Hearn et al. (1986) discuss the possibility of CN molecules being directly generated by grains in dust jets what is also suggested by the close mean values of the outflow velocities of dust and CN (cf. Table 2). If indeed the third term is equal to the outflow velocity this would verify the above interpretation of that term in Eq. (5). In any case we can give a minimum value for the initial (outflow) velocity of 0.23 and a maximum value of 0.7 $km\,s^{-1}$ from our CN shell expansion data.

4.6. *Mass ejection times from tail and coma observations*

A comparison of ejection times t_0 obtained from plasma tail structure observations (Table 4) and from coma shell observations (Table 7), respectively, shows that 36% of the CN dates coincide with CO^+ dates within 0.1 days, 54% within 0.2 days, and 86% within 0.5 days, i.e. within a quarter of one rotation period. This agreement of results obtained by quite different kinds of observation is astonishingly good, especially in view of the mean error of 0.06 days each, of the intrinsic uncertainties like structure and shell identification problems, and of the completeness of both the coma and tail data samples. It follows that it is highly probable that the plasma and neutral gas clouds observed have been ejected at the same time and approximately from the same regions on the surface of the nucleus.

We made another correlation of our ejection times with the "onset" times given by Schleicher et al. (1986) from production rate variations. The t_0 dates from our tail observations (Table 4) are in the average 0.35 ± 0.24 days earlier than the onset dates from gas production. From the CN shell expansion (Table 7) the difference is 0.05 ± 0.27 days. The correlation between photometric and geometric data is remarkable. Thus it seems that a brightening in CN is observed in coincidence with the release of a new shell, while it is delayed with respect to "plasma" cloud formation by 0.4 days. During this time interval the potential "plasma" is drifting from the nucleus to radii of 16000 ± 11000 km, assuming an outflow velocity of $0.54 \pm 0.08\,km\,s^{-1}$. Nevertheless, within the uncertainties brightening, CN, and potential plasma release occur at the same time.

4.7. *Tail disconnection events*

The understanding of disconnection events (DE) occurring in the plasma tail of comets involves many problems. It seems therefore worthwhile to correlate the observation of the DE's with other parameters of the comet or in the interplanetary neighbourhood. Niedner (1986) e.g. has investigated the coincidence of DE's and sector boundaries within the interplanetary magnetic field passing the comet. Looking at our Tables 1 and 6 there are a lot of parameters to be compared with the occurrence of DE's (altogether 17 in number). More than 69% of the DE's observed are connected to the occurrence of tail rays, 73% to magnetic field sector boundaries passing by, 63% to high accelerations in the tail following the event, 57% to large aberration angle deviations, 57% to high initial velocities of detaching plasma clouds, but only 35% to high densities in the solar wind stream. (cf. Oyama et al., 1986). It follows that independently of the strength of DE's they are less connected to variations of solar wind properties than to those of the interplanetary magnetic field passing by. That the interplanetary magnetic field has also some influence on the tail dynamics is verified by the strong correlation between tail rays and high initial velocities of plasma clouds. Nevertheless in some cases sector boundaries passing by seem not to be sufficient for a DE to occur. On March 18.75 a dense plasma cloud "p_2" had detached from the coma which was observed by a lot of observers. A strong intensity outburst beginning on March 18.81 was detected by the IUE ultraviolet satellite (Feldman et al., 1986). On March 18.2 the space probe Sakigake measured the highest density within the solar wind ever observed between Nov. 1, 1985 and Apr. 6, 1986. Thus it seems that DE's may possibly be triggered also by extremely dense clouds in the solar wind, at least the event of March 18. However, an important role doubtlessly is due to the activity of the nucleus. Coincidence of high activity, sector boundaries and high solar wind density may cause spectacular DE's in the plasma tail.

5. Conclusions

We have shown up the possibility to extract rich information on cometary nuclei by observing the structure of plasma tails photographically over long time periods. It is highly important to dispose of data collected continuously over a long time lapse in order to obtain reliable results of statistical kind.

Thus it was possible from a detailed investigation of CO^+ tail wide-field images to establish a consistent model of source activity distribution around the periphery of the nucleus. Main activity occurs at the long ends of the deformed ellipsoidal nucleus. The ratio of activity at the long ends and at the broad sides is approximately 5:1 which seems to be caused by increased heating at the long ends and differences in the dust layer covering the surface of the nucleus. The sidereal rotation period of the nucleus is 50.5 ± 0.6 hours; a 7.4-day period is just suggested. The obviously longer 7.4 day period found by some authors from light curves could be explained by the theory that the less active broad-side sources are brighter than those at the long ends; this is supported by our results. From a comparison of dynamics of tail and coma it turned out that plasma and neutral gas clouds are probably produced approximately at the same time and ejected from the same regions on the surface of the nucleus. A three-component fit to the CN coma expansion with heliocentric distance resulted in the conclusion that the expansion velocities observed are due to heating of the nucleus' surface, possibly to photodissociation of mother molecules, and to their initial velocity which equals the velocity of dust (not gas) jets erasing from the nucleus.

The mean acceleration within the plasma tail is $30\,\mathrm{cm\,s^{-2}}$ showing the tendency to increase backwards. The mean initial velocity of plasma clouds observed is mostly $0.5\,\mathrm{km\,s^{-1}}$ corresponding to the jet outflow velocities. 14% of plasma clouds show larger velocities ($\approx 30\,\mathrm{km\,s^{-1}}$) which cannot be explained as mere measuring errors. Tail acceleration and plasma initial velocities are usually not correlated to solar wind parameters but seem to be connected to the interplanetary magnetic field sector boundaries.

Large deviations from the mean aberration angle of the plasma tail are not correlated to solar wind parameters, and their correlation to the passing of sector boundaries is only weak. On the other hand most plasma tail disconnection events are clearly connected to discontinuities in the interplanetary magnetic field and mostly not to the solar wind. But there must exist additional processes advancing disconnecting events, e.g. the jet activity of the nucleus itself and in some cases the interaction of cometary plasma with extremely dense solar wind plasma clouds.

Acknowledgements. We thank the Director General of the European Southern Observatory for the allowance to set up our special equipment on ESO ground, and the ESO staff for valuable help during the observational run. Thanks are due to Dr. K. Jockers for lending us some optical filters and Mr. W. Hünecke for the lot of work he had with the photographic prints. The observations were supported by the Deutsche Forschungs-gemeinschaft under grant Ce 21/1-1 what is gratefully acknowledged.

References

A'Hearn, M.F., Hoban, S., Birch, P.V., Bowers, C., Martin, R., Klinglesmith III., D.A.: 1986, *Nature* **324**, 649

Balsiger, H., Altwegg, K., Bühler, F., Geiss, J., Ghielmetti, A.G., Goldstein, B.E., Goldstein, R., Huntress, W.T., Ip, W.H., Lazarus, A.J., Meier, A., Neugebauer, M., et al.: 1986, *Nature* **321**, 330

Belton, M.J.S., Wehinger, P., Wyckoff, S., Spinrad, H.: 1986, 20th ESLAB Symp. ESA SP-250, Vol. **I**, 599

Belton, M.J.S.: 1987, (preprint, to be published in Planet. Rep.)

Bockelèe-Morvan, D., Crovisier, J., Despois, D., Forveille, T., Gèrard, E., Schraml, J., Thum, C.: 1986, 20th ESLAB Symp. ESA SP-250, Vol. **I**, 365

Celnik, W.E.: 1986, 20th ESLAB Symp. ESA SP-250, Vol. **I**, 53

Celnik, W.E., Schulz, R., Weissbauer, K.: 1986a, *Astron. Astrophys.* **163**, L7

Celnik, W.E., Koczet, P., Schulz, R., Weissbauer, K.: 1986b, *Mitt. Astron. Ges.* **67**, 271

Despois, D., Crovisier, J., Bockelèe-Morvan, D., Schraml, J., Forveille, T., Gèrard, E.: 1986, *Astron. Astrophys.* **160**, L11

Edenhofer, P., Buschert, H., Bird, M.K., Volland, H., Porsche, H., Brenkle, J.P., Kursinsky, E.R., Mottinger, N.A., Stelzried, C.T.: 1986, 20th ESLAB Symp. ESA SP-250, Vol. **II**, 215

Feldman, P.D., A'Hearn, M.F.A., Festou, M.C., McFadden, L.A., Weaver, H.A., Woods, T.N.: 1986, *Nature* **324**, 433
Ferrin, I., Gil, C.: 1986, 20th ESLAB Symp. ESA SP-250, Vol. **II**, 427
Festou, M.C.: 1986, Contrib. 20th ESLAB Symp. Heidelberg, Oct. 27-31
Goldstein, R., Young, D.T., Balsigner, H., Buehler, F., Goldstein, B.E., Neugebauer, M., Rosenbauer, H., Schwenn, R., Shelley, E.G.: 1986, 20th ESLAB symp. ESA SP-250, Vol. **I**, 207
Houpis, H.L.F., Gombosi, T.I.: 1986, 20th ESLAB Symp. ESA SP-250, Vol. **II**, 397
Huebner, W.F., Keller, H.U., Wilhelm, K., Whipple, F.L., Delamere, W.A., Reitsema, H.J., Schmidt, H.U.: 1986, 20th ESLAB Symp. ESA SP-250, Vol. **II**, 363
Kammerer, A.: 1987, (private communication)
Kaneda, E., Hirao, K., Takagi, M., Ashihara, O., Itoh, T., Shimizu, M.: 1986a, *Nature* **320**, 140
Kaneda, E., Takagi, M., Hirao, K., Shimizu, M., Ashihara, O.: 1986b, 20th ESLAB Symp. ESA SP-250, Vol. **I**, 397
Kaneda, E., Ashihara, O., Shimizu, M., Takagi, M., Hirao, K.: 1986c, *Nature* **321**, 297
Keller, H.U., Arpigny, C., Barbieri, C., Bonnet, R.M., Cazes, S., Coradini, M., Cosmovici, C.B., Delamere, W.A., Huebner, W.F., Hughes, D.W., Jamar, C., Malaise, D., et al.: 1986a, *Nature* **321**, 320
Keller, H.U., Delamere, W.A., Huebner, W.F., Reitsema, H., Schmidt, H.U., Schmidt, W.K.H., Whipple, F.L., Wilhelm, K.: 1986b, 20th ESLAB Symp. ESA SP-250, Vol. **II**, 359
Körösmezey, A., Cravens, T.E., Gombosi, T.I., Nagy, A.F., Mendis, D.A., Szegö, K., Gribov, B.E., Sagdeev, R.Z., Shapiro, V.D., Shevchenko, V.I.: 1986, 20th ESLAB Symp. ESA SP-250, Vol. **I**, 235
Korth, A., Richter, A.K., Mendis, D.A., Anderson, K.A., Carlson, W., Curtis, D.W., Lin, R.P., Mitchell, D.L., Rème, H., Sauvaud, J.A., d'Uston, C.: 1986, 20th ESLAB Symp. ESA SP-250, Vol. **I**, 199
Krankowsky, D., Lämmerzahl, P., Herrwerth, I., Woweries, J., Eberhardt, P., Dolder, U., Herrmann, U., Schulte, W., Berthelier, J.J., Iniano, J.M., Hodges, R.R., Hoffman, J.H.: 1986, *Nature* **321**, 326
Krasnopolsky, V.A., Moroz, V.I., Hogoshev, M., Gogosheva, Ts., Moreels, G., Clairemidi, J., Parisot, J.P., Krysko, A.A., Jegulev, V.S., Tkachuk, A.Y.: 1986a, 20th ESLAB Symp. ESA SP-250, Vol. **I**, 459
Krasnopolsky, V.A., Tkachuk, A.Y.: 1986b, 20th ESLAB Symp. ESA SP-250, Vol. **I**, 465
Kührt, E., Möhlmann, D., Giese, B., Tauber, F.: 1986, 20th ESLAB Symp. ESA SP-250, Vol. II, 385
Lämmerzahl, P., Krankowsky, D., Hodges, R.R., Stubbemann, U., Woweries, J., Herrwerth, I., Berthelier, J.J., Illiano, J.M., Eberhardt, P., Dolder, U., Schulte, W., Hoffman, J.H.: 1986, 20th ESLAB Symp. ESA SP-250, Vol. **I**, 179
Lang, K.R.: 1980, Astrophysical Formulae, Springer Berlin Heidelberg New York p. 23
Larson, H.P., Davis, D.S., Mumma, M.J., Weaver, H.A.: 1986a, *Astrophys. J.* **309**, L 95
Leibowitz, E.M., Brosch, N.: 1986, 20th ESLAB Symp. ESA SP-250, Vol. **I**, 605
Millis, R.L., Schleicher, D.G.: 1986, *Nature* **324**, 646
Möhlmann, D., Börner, H., Danz, M., Elter, G., Mangoldt, T., Rubbert, B., Weidlich, U.: 1986, 20th ESLAB Symp., ESA SP-250, Vol. **II**, 339
Niedner, M.B.: 1986, (preprint, to be published in Adv. Space Res.)
Opal, C.B., McCoy, R.P., Carruthers, G.R.: 1986, 20th ESLAB Symp., ESA SP-250, Vol. **I**, 425
Oyama, K., Hirao, K., Hirano, T., Yumato, K., Saito, T.: 1986, *Nature* **321**, 310
Pankiewicz, G.S., McDonnell, J.A.M.: 1986, 20th ESLAB Symp., ESA SP-250, Vol. **II**, 201
Reinhard, R.: 1986, *Nature* **321**, 313
Reitsema, H.J., Delamere, W.A., Huebner, W.F., Keller, H.U., Schmidt, W.K.H., Wilhelm, K., Schmidt, H.U., Whipple, F.L.: 1986, 20th ESLAB Symp., ESA SP-250, Vol. **II**, 351
Riedler, W., Schwingenschuh, K., Yeroshenko, Y.G., Styashkin, V.A., Russell, C.T.: 1986, *Nature* **321**, 288
Sagdeev, R.Z., Szabò, F., Avanesov, G.A., Cruvellier, P., Szabò, L., Szegö, K., Abergel, A., Balazs, A., Barinov, I.V., Bertaux, J.L., Blamont, J., Detaille, M., et al.: 1986a, *Nature* **321**, 262
Sagdeev, R.Z., Krasikov, V.A., Szegö, K., Tòth, I., Smith, B., Larson, S., Merènyi, E., Shamis, V.A., Tarnopolski, V.I.: 1986b, 20th ESLAB Symp., ESA SP-250, Vol. **II**, 335
Sagdeev, R.Z., Ziman, Y.L., Avanesov, G.A., Tarnopolsky, V.I., Krasikov, V.A., Usikov, D.A., Debabov, A.S., Sukhanov, K.G., Kogan, A.Y., Kheifets, V.N.: 1986c, 20th ESLAB Symp., ESA SP-250, Vol. **II**, 307
Sagdeev, R.Z., Avanesov, G.A., Ziman, Y.L., Moroz, V.I., Tarnopolsky, V.I., Zhukov, B.S., Shamis, V.A., Smith, B., Toth, I.: 1986d, 20th ESLAB Symp., ESA SP-250, Vol. **II**, 317
Saito, T., Yumoto, K., Hirao, K., Nakagawa, T., Saito, K.: 1986, *Nature* **321**, 303
Samarasinha, N.H., A'Hearn, M.F., Hoban, S., Klinglesmith, D.A.: 1986, 20th ESLAB Symp., ESA SP-250, Vol. **I**, 487
Schleicher, D.G., Millis, R.L., Tholen, D., Lark, N., Birch, P.V., Martin, R., A'Hearn, M.F.: 1986, 20th ESLAB Symp., ESA SP-250, Vol. **I**, 565
Schloerb, F.P., Kinzel, W.M., Swade, D.A., Irvine, W.M.: 1986a, *Astrophys. J.* **310**, L55
Schloerb, F.P., Kinzel, W.M., Swade, D.A., Irvine, W.M.: 1986b, 20th ESLAB Symp., ESA SP-250, Vol. **I**, 577
Schloerb, F.P., Claussen, M.J., Tacconi-Garman, L.: 1986c, 20th ESLAB Symp., ESA SP-250, Vol. **I**, 583
Schlosser, W.: 1966, Thesis, Univ. Hamburg, Federal Republic of Germany
Schlosser, W., Schulz, R., Koczet, P.: 1986, 20th ESLAB Symp., ESA SP-250, Vol. **III**, 495
Schwenn, R., Ip, W.H., Rosenbauer, H., Balsiger, H., Bühler, F., Goldstein, R., Meier, A., Shelley, E.G.: 1986, 20th ESLAB Symp., ESA SP-250, Vol. **I**, 225
Sekanina, Z.: 1985, *IAU Circular* **4151**
Sekanina, Z.: 1986a, *IAU Circular* **4187**
Sekanina, Z.: 1986b, *Adv. Space Res.* **5**, 307
Sekanina, Z.: 1986c, 20th ESLAB Symp., ESA SP-250, Vol. **II**, 131
Sekanina, Z.: 1987, *Nature* **325**, 326
Sekanina, Z., Larson, S.M., Emerson, G., Helin, E.F., Schmidt, R.E.: 1986, 20th ESLAB Symp., ESA SP-250, Vol. **II**, 177
Sekanina, Z., Larson, S.M.: 1986, *Nature* **321**, 357
Simpson, J.A., Rabinowitz, D., Ksanfomality, L.V., Sagdeev,

R.Z., Tuzzolino, A.J.: 1986, 20th ESLAB Symp., ESA SP-250, Vol. **II**, 11

Vaisberg, O.L., Smirnov, V.N., Gorn, L.S., Jovlev, M.V., Balikchin, M.A., Klimov, S.I., Savin, S.P., Shapiro, V.D., Shevchenko, V.I.: 1986, *Nature* **321**, 274

Whipple, F.L.: 1986, 20th ESLAB Symp., ESA SP-250, Vol. **II**, 281

Wilhelm, K., Cosmovici, C.B., Delamere, W.A., Huebner, W.F., Keller, H.U., Reitsema, H., Schmidt, H.U., Whipple, F.L.: 1986, 20th ESLAB Symp., ESA SP-250, Vol. **II**, 367

Yeomans, D.K.: 1986a, *IHW Newsletter* **8**, 51

Yeomans, D.K.: 1986b, 20th ESLAB Symp., ESA SP-250, Vol. **II**, 419

Zastenker, G., Vaisberg, O., Smirnov, V., Fedorov, A., Omelchenko, A.: 1986, 20th ESLAB Symp., ESA SP-250, Vol. **I**, 183

Astron. Astrophys. 187, 249–255 (1987)

ASTRONOMY AND ASTROPHYSICS

Observations of the coma of comet P/Halley and the outburst of 1986 March 24–25 (UT)

T.W. Rettig [1], J.R. Kern [1, 2], R. Ruchti [1], B. Baumbaugh [1], A.E. Baumbaugh [3], K.L. Knickerbocker [3], and J. Dawe [4]

[1] University of Notre Dame, Notre Dame, ID 46556, USA
[2] Rochester Institute of Technology, Rochester, NY, USA
[3] Fermi National Accelerator Laboratory, Chicago, IL, USA
[4] Siding Spring Observatory, Australia

Received February 17, accepted April 3, 1987

Summary. Observations using a fast CCD and custom built Video Data Acquisition System at the Nasmyth focus of the 2.3 meter telescope at Siding Spring Observatory, Australia produced digitized spectra and images of comet Halley. On March 25 (UT), an overall brightening and anti-sunward jet was recorded continuously over several hours and is shown to be consistent with the enhanced spectral activity recorded on March 24 (UT). The outburst and is consistent with the Giotto measurements. Preliminary analysis of the spectral enhancement of March 24 is presented. The time resolution system allowed a spectral range from 3800Å to 7000Å to be analyzed every $^1/_2$ s. Indications of unexplained rapid variations in several molecular emmission band persist and are noted.

Key words: coma variation – spectra – images– core outburst – rotation

1. Introduction

The typical variations in cometary spectra, image shape and image size with distance from the sun is fairly well studied phenomena, and outbursts have been noted most recently by Lutz and Wagner (1986), Larson et al. (1986), Cochran and Barker (1986), Russell et al. (1986), and Mumma et al. (1986). Usually, cometary images and spectra are sampled over intervals separated by minutes or hours but this investigation has made systematic observations (Rettig et al., 1986) to search for very rapid variations in the inner one arc minute of the coma of Comet Halley by using a fast Fairchild CCD and recording individual digitized frames at the rate of 30 per second.

Electronic imaging devices such as CCD cameras have been used extensively for astronomical observation (Timothy, 1983). These detectors provide high quantum efficiency over a considerable range. In a typical application, the CCD is operated at LN_2 temperature to eliminate thermal noise, and the image accumulated over a period of minutes. These devices are then read slowly to avoid additional noise introduced during the read-out cycle. While excellent for long time exposures, such devices do not provide temporal information for physical systems in which the characteristic time constants are substantially shorter than the image integration and readout time. The fast CCD and Video Data Acquisition System (VDAS) provides excellent time resolvability in a fraction of a second as well as good spatial resolution and quantum efficiency (Ruchti et al., 1986)

Send offprint requests to: W. Rettig

2. Equipment and observations

2.1. Data acquisistion system

For the observations discussed in this paper the equipment consisted of microchannel plate image intensifiers, a fast CCD camera and the VDAS attached to the Nasmyth focus of the Australian 2.3 meter Advanced Technology Telescope (ATT) for narrow field imaging and the Cassegrain focus of a 14-inch telescope for a wider field. The equipment schematic shown in Fig. 1 includes a satellite video uplink since the comet observations were provided to Australian, European and American stations.

The data were digitized and recorded at 30 CCD frames/s (5.5 Mbytes of pixel information/s). With the VDAS (First-In-First-Out memory) recently developed at Fermi National Accelerator Laboratory and the University of Notre Dame, we were capable of integrating, compacting and storing a large number of CCD images (Ruchti et al., 1985). The effective exposure time could be selected from $^1/_{30}$ to many seconds as required. For the observations at Siding Spring the system typically digitized and averaged 16 frames every 12 s.

The CCD electronic camera utilized is a Fairchild CCD 3000 F device capable of producing a complete video frame (two interlaced fields of 244×388 elements) in 30 ms. The video data recording system consistent of a storage and processing system which digitizes the CCD information at 30 MHz using a 6 bit flash ADC, compacts or sparsifies the data and eliminates data below a user selected threshold, provides temporary storage of the data in a FI/FO memory of 8 M depth, operates on the data with numerous image processing functions including hardware frame averaging, pseudo-color look up tables, etc., writes processed data to video or to a higher bandwidth system such as magnetic tape or disk and has a modular memory capable of expansion to 356 Mbytes (Baumbaugh et al., 1986).

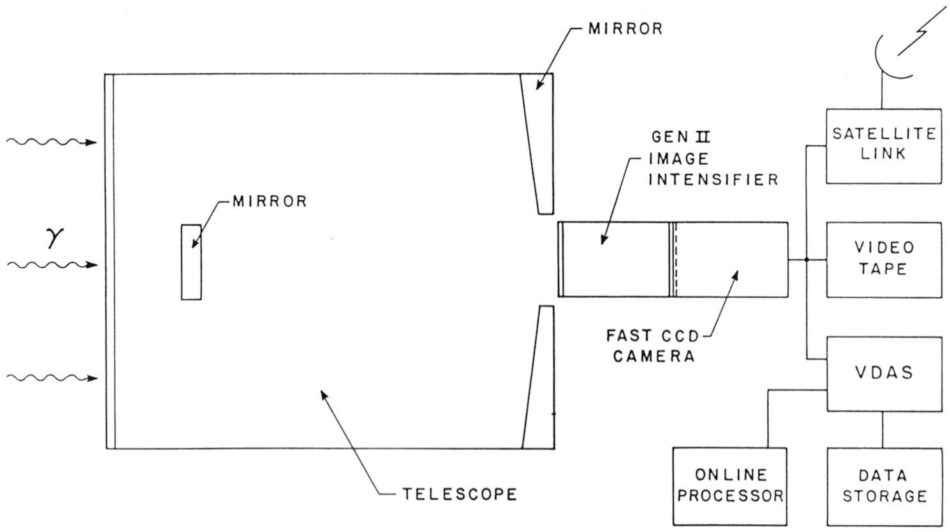

Fig. 1. Schematic of the Astronomical Fast Imaging System

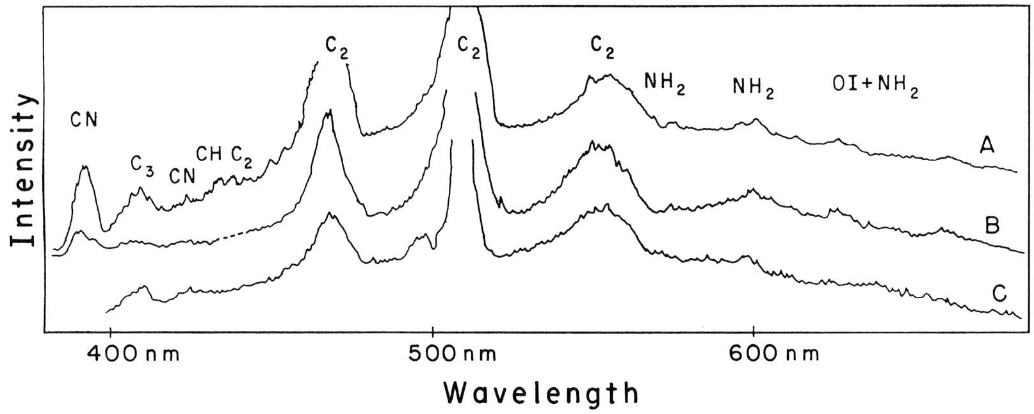

Fig. 2. Spectra (390nm to 700nm) from two nights at the 2.3m ATT. Total integration time for each region is $^1/_2$ s.

2.2. Observations using the 2.3 meter Advanced Technology Telescope.

During an observing period from March 20 to April 9, 1986 at Siding Spring Observatory, Australia, we obtained approximately fifty hours of digitized video and spectral images of comet Halley to search for variations on different time scales. The goal was to obtain images as close as possible to perihelion passage while cometary activity was maximal.

To search for rapid spectral fluctuations in the inner core of the coma, a grating spectrograph with 20Å resolution was attached to the imaging system. The spectral scans covered the range of 3800Å to 7000Å. Each spectral frame covered an interval of 600Å and was digitized at the 30 MHz rate. With the 2.3 meter *f*/18 focal arrangement, the spectral output gives spatial intensity measurements from the core and continuum out to $\pm 30''$. Figure 2 shows spectra and continuum with prominent molecular lines identified for March 24 (A and B) and March 23 (C). Scan B is offset 2′ from the core. The spectra in Fig. 2 are integrated spatially over $\pm 30''$. All spectral illustrations are $^1/_2$ s averages and no corrections for background or continuum have been applied. Figure 3 shows a spectral interval containing the CN(0,0) and C_3 molecular bands integrated spatially over $\pm 5''$ centered on the coma. By offsetting the telescope, the spectral information was extended to a maximum of 6′ at which point the intensifier and CCD dark current dominated observable signal. Changes in the inner coma of the prominent molecular bands of CN, C_3, C_2, NH_2, and CH were monitored.

3. Outburst of 1986 March 24–25

3.1. Optical images

The 2.3 meter ATT images from the nights of March 22, 24, and 25 (all dates and times in this paper are expressed in UT) presented in Fig. 4 show the inner 1′ coma with prominent variations in the size and contour of the bow shock from night to night. On March 25.72 an obvious overall brightening and prominent antisunward jet was evident. The jet extended a minimum of 28″ west of the core. Using the integrated intensity from an artificial diaphragm of $66'' \times 53''$, the coma flux increased during this outburst by a

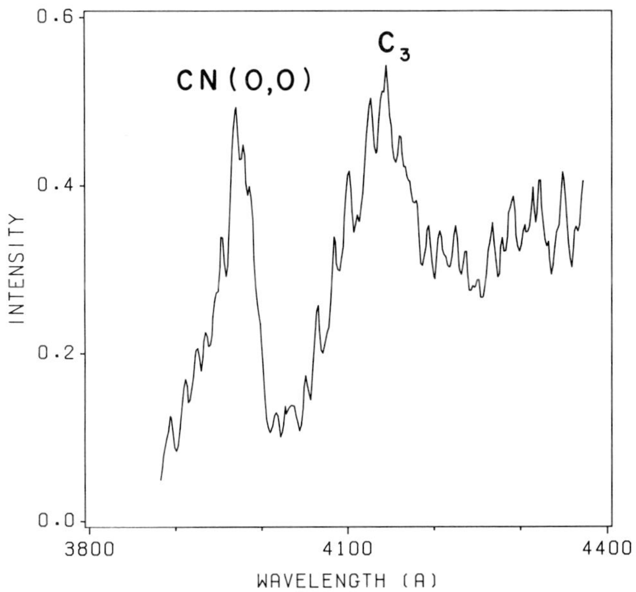

Fig. 3. Spectral interval showing the CN(0,0) and C_3 bands. The data are integrated over 10″ centered on the coma. The integration time is $^1/_2$ s

factor of 3.6 and by a factor of 2.6 within a 35″ diameter diaphragm.

The brightness profiles of the images in Fig. 5 show the comet on three different nights. All profiles are along a direction aligned approximately with the jet. The two inner profiles (A, B) are for the observations of March 22 and March 24. A slight increase in brightness in noted toward the sunward bow shock front of profile (B) but is, in general, consistent with the previous image. The image profiles (C) and (D) from March 25.6 and 25.7 show the dramatic increase from the previous night and show a continued increase in the overall brightness in the coma over the observation interval. The asymmetry left of center is due to the antisunward jet shown in the previous contours of Fig. 4. The intensity profile indicates the integrated intensity increase is not entirely due the antisunward outburst. The profiles on the sunward side show a $r^{-1.2}$ dependence prior to the outburst and following the outburst changes to $r^{-0.7}$. The antisunward region immediately behind the core shows approximately r^{-2} dependence and the jet shows a $r^{-0.1}$ dependence.

The several hours of digitized images from the 2.3 meter telescope provide an opportunity for a detailed analysis of the progress and rate of the outburst. Analysis of the secondary jet to the lower right in Fig. 4 has not been included. Velocities were determined by following an intensity contour of the jet as it moved radially antisunward. The images provide projected ejection velocities of 0.11 km s^{-1} ± 0.07 km s^{-1} at a distance of 7.70 10^3 km (15″), a velocity of 0.32 (± 0.05) km s^{-1} at a distance of 1.07 10^4 km (20″) and a velocity of 0.47 (±0.05) km s^{-1} at a distance of 1.47 10^4 km s^{-1} (28″).

Nominal outflow velocities range from 0.615 km s^{-1} from the Delsemme relation, to the Giotto inner coma outflow velocitiy of 0.9 km s^{-1} given by Weaver et al. (1986) and Larson et al. (1986). The generally accepted dust ejection rate from Vega II data is 0.55 km s^{-1} (Krasnopolsky et al., 1986). Using the 0.47 km s^{-1} velocity, a linear extrapolation of the 1.47 10^4 km region of the outburst to the core results in an estimate of −8.7 h (March 25.3)

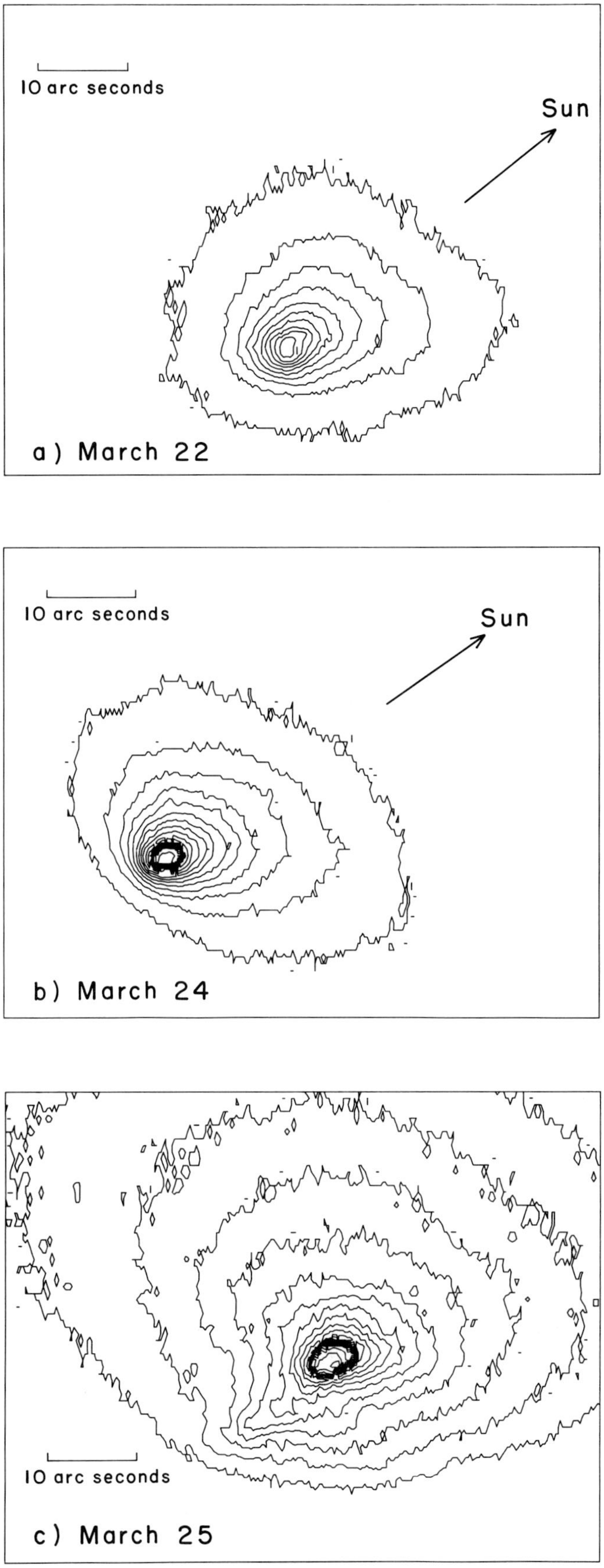

Fig. 4a–c. Contoured images of March 22, March 24 and March 25 showing the changes in the bow front and development of an antisunward jet

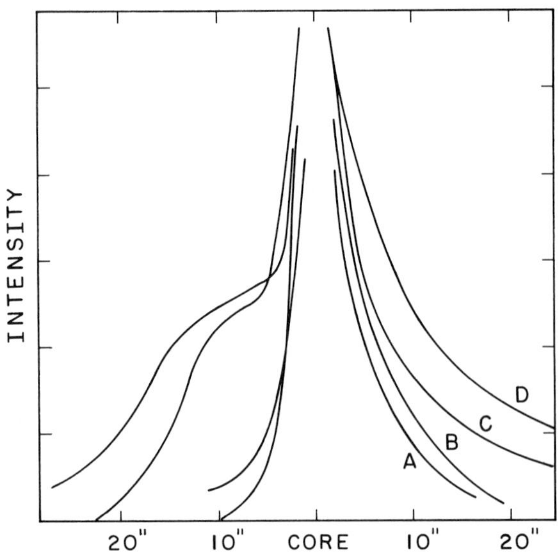

Fig. 5. Brightness profiles for the three images shown in Fig. 4. Profile A is for March 22 and profile B is for March 24 showing only minor variations. Profiles C and D for times separated by 103 min. on March 25 and the profile directions are aligned with the antisunward jet

as the time the outer jet material left the core. For the material projected to be nearest the core. The resulting time of initial outburst is -19.4 (± 5) hours or March 24.9 (± 0.2). If the velocity of $0.47\,\mathrm{km\,s^{-1}}$ in this relatively narrow jet is perpendicular to the line of sight, the implication is for dust and ice grains as major components of the antisunward outburst.

We note also the nearly linear nature of this jet is in contrast to the helical jets analyzed by A'Hearn et al. (1986) and that it does not project back to the central core brightness but intersects with the secondary jet to the lower right approximately 3″–5″ from the core. If the material is assumed to originate from the same antisunward outburst we might presume that the outburst occurs over a minimum time interval from -19.4 h to -8.7 h (10.7 h) and that the dust particles are ejected at various velocities with respect to mass or the inner material has been slowed through interaction with the inner coma material. More probable is that the lowest inner velocity measurement may indicate an initial ejection direction nearer the line of sight and the variation in velocities from the core is a result of sight and the variation in velocity from the core is a result of core rotation as described by Larson and Sekanina (1984). Assuming rotation as the explanation for the variation in transverse velocities as a function of distance from the core, a lower limit to the rotation period of the core is found to be 2.1 ($\pm$ 0.4) days depending on the maximum unprojected velocity. This is very near the limits determined by Belton et al. (1986). From the offset of the apparent jet origin and assuming the feature is on the equatorial region of the rotating core, the spin axis is tilted at approximately 3°–5° to the line of sight.

The KAO observations of H_2O at 2.65 μm by Weaver et al. (1986) indicate a sharp increase in flux over a maximum 1.9 h interval just prior to 24.8 March 1986, approximately 22 h before our observations of the jet. The projected transverse velocity, $0.11\,\mathrm{km\,s^{-1}}$ noted previously, predicts nearly the same initial outburst time. The IR observations by Hanner et al. (1987) show a large thermal brightness increase on March 24.77 and the IUE observations by McFadden et al. (1986) show a rise in the UV flux to a peak at March 25.0. This increase occurs over a 1.2 day time interval more indicative of the typical development time of the daughter species (OH). Our spectral results (see Sec. 3.3) show a similar dramatic rise in the line profile brightnesses especially near the core and continuum between March 24.72 and March 24.77. From the coincidence of times for the spectral outburst reported in this paper and the above authors, we suggest these events are associated. If the antisunward jet we project to begin prior to March 25.3–24.9 is associated with the same outburst reported by others, the conclusion will require the comet to rotate (approximately 1 day from sunward to tailward) to correlate the sunward brightening noted by Weaver (1986), and provide additional indirect support for the rotation period observed by Giotto (Sagdeev et al., 1986).

3.2. Implication of the outburst

The magnitude of this narrow antisunward jet implies a rather confined outburst region and must result from an eruption of volatiles degassing through the dust-rich and non-volatile surface layers. Various steady state processes continually deplete the subsurface volatile ices and concentrates the low thermally conductive nonvolatiles in the surface layers. This dusty, porous regolith as described by Hartmann et al. (1982) traps upward outgassing material and slowly releases the volatiles unless the gas pressure increases beyond the surface strenth of the non-volatiles and an eruption occurs. With low thermally conductive surface materials such as dust and hydrogen cyanide polymers suggested by Matthews and Ludicky (1986) the outburst can be delayed and is not necessarily confined to the sunward region. The initial explosive process (Mumma et al., 1986) from the spectral observations of Sec. 3.3, the explosive event may have an initial rise time of less than one hour and, if the subsequent jet as suggested in Sec. 3.1 is a later product of the outburst, the event must continue over a period of at least 0.8 days to account for the continuous nature of the feature.

Further evidence for this assumption occurs in the IR observations of the 10 μm silicate emission feature by Gehrz and Ney (1986). In the March 25.5 observations the emission feature was completely missing and reappears in the subsequent observations on March 28.6. This silicate emission feature is not expected to disappear completely until the comet recedes beyond a heliocentric distance of 1 AU. During this instance the dust albedo fell from 10–15% to 5% implying the darker carbonaceous material was dominating the continuum. The excess of this dark material suggests the outburst ejecta dominates the reflected light after the surface is drastically disturbed as in the instance of March 24.7. This effect could be expected to persist in the coma for several days.

3.3. Spectral outburst of March 24

Observations obtained during the March 24.72 to March 24.77 interval covering approximately 72 min show a dramatic increase in the spectral line intensities. This is concurrent with the time frame of the IR observations by Hanner et al. (1987), the KAO observations and IUE observations reported by Weaver et al. (1986) and McFadden et al. (1986) respectively. For the spectral observations reported in this papaer the slit direction was oriented aproximately (N-S) perpendicular to the tail direction and extends from the core out to approximately $\pm 20''$. Note the images obtained immediately after the initial spectral enhancement show only minor brightening of the sunward bow region (see profile B in Fig. 5).

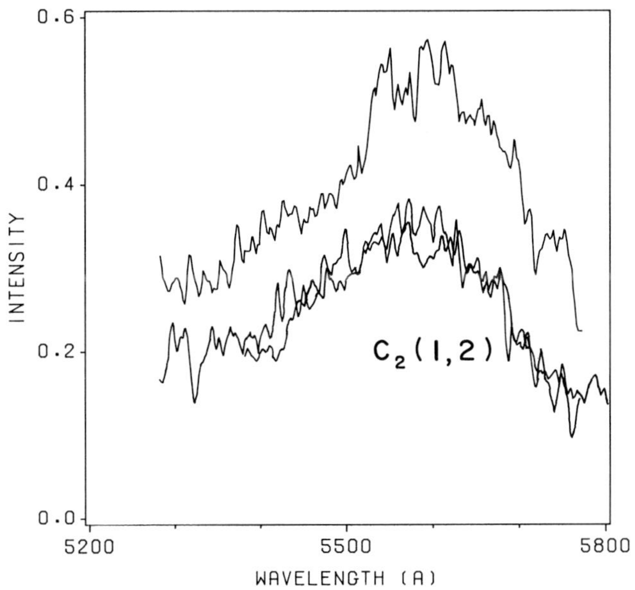

Fig. 6. $C_2(1,2)$ band profiles for March 24.72 and the previous night (bottom curves) and for March 24.77 (upper curve) showing a dramatic enhancement of the spectra in a 72 min. period on March 24

Table 1. Integration limits used in producing spatial profiles along the slit direction

Feature	Integration limits (Å)
CN (0, 0)	3883–4000
C_3	4040–4190
C_2 (1, 2)	5500–5650
NH_2 (0, 9, 0)	5980–6045
Continuum	5310–5370

Figure 6 shows the $C_2(1,2)$ band profiles for the 72 minute interval and a comparison profile. The lower profiles consist of the March 24.72 data prior to enhancement and an identical region from the March 23 data. This night-to-night similarity extends to other spectral species. The upper profils shows the obvious brightening for March 24.77. These profiles employ the same spatial integration scheme used in Fig. 3.

The sudden brightening over the 72 min. period on March 24 is also apparent in other molecular bands. The spectral profile for each feature was measured along the N/S slit and integrated over the wavelength ranges given in Table 1. Figure 7a–e presents these spatial profiles for CN(0,0), C_3, $C_2(1,2)$, $NH_2(0.9,0)$, and the continuum at 5340 Å, respectively, for March 24.72 (lower curves). Table 2 shows the N-fold increase in intensity as a function of distance from the core out to 20″ on either side of the core and the last column gives a typical value for the intensity increases, excluding the core values, for each spectral feature. Continuum contributions near the core have not been subtracted for either the data of parts a-d of Fig. 7 or from the values given a Table 2. The zero position in the slit is chosen by the location of the pixel row passing through the center of the continuum across the entire CCD frame. We note that the molecular emission profiles peak just south (−) of the positon where the continuum peaks.

Table 2. N-fold increases in molecular band intensities during the March 24 outburst as a function of distance from the core

Feature	Distance from core					
	South			North		
	20″	10″	0″	10″	20″	
CN (0, 0)	2.7	3.2	3.5	3.2	3.0	<3.1>
C_3	2.0	3.0	2.9	3.0	2.3	<2.6>
C_2 (1, 2)	0.9	1.2	1.3	1.4	1.2	<1.2>
NH_2 (0, 9, 0)	–	1.4	1.3	1.1	1.3	<1.3>
Continuum	1.0	1.0	2.4	2.7	1.0	<2.4>

These intensity enhancements appear to show the same general rise as in the IUE observations of comet Halley by McFadden et al. (1987) which result in a value of 2.5 for the rise in OH over a period of approximately one day and an increase of 4 in the value of the dust contribution. Photometric data by Schleicher et al. (1986) clearly show variation of a factor of about 2.5 in the production rates of all observed species over a time scale of several days. Similar more rapid temporal variations have been noted by Lutz and Wagner (1986) in comet IRAS-Araki-Alcok 1983d and in comet Halley by Cochran and Barker (1986) and may be associated with the rapid changes noted in the Giotto flyby.

4. Possible rapid spectral variations

The rapid data collection technique has allowed us to look for very short interval fluctuations in the intensity profiles of the various molecular band. We have noted these rapid variations (Rettig et al., 1986). Similar variations (Moreels, 1986, private communication have been noted in the Vega 2 spectrophotometry (Moreels et al., 1986) of the $C_2(0,0)$ band but remain unconfirmed. If these short interval variations are confirmed, the molecular band fluctuations could provide physical constraints for gas production rates and gas-to-dust ratios of the inner coma. We continue to investigate possible sources of non-cometary time variations including gain changes, variable seeing and transparency to resolve the observations but none provide a source for selective molecular emission variation. Further analysis is needed before a conclusion can be reached.

Acknowledgements. Special thanks are due to the Australian astronomers and staff. In particular we would like to thank Dr. Mike Bessel, Dr. Don Mathewson, Paul Harding, Patrick Keogh, Hilton Lewis and Dennis Andriolo for their assistance with observing time and advice. Also, special thanks to Gary Seiber and Phil Kurz for their initial assistance. Research has been funded in part by the U.S. National Science Foundation, the University of Notre Dame, the Planetary Society, the National Geographic Society and United Airlines.

References

A'Hearn, M., Hoban, S., Birch, P.V., Bowers, C., Martin, R., Klinglesmith, D.A.: 1986, 20th ESLAB, ESA SP-250, Vol. **1**, p. 483

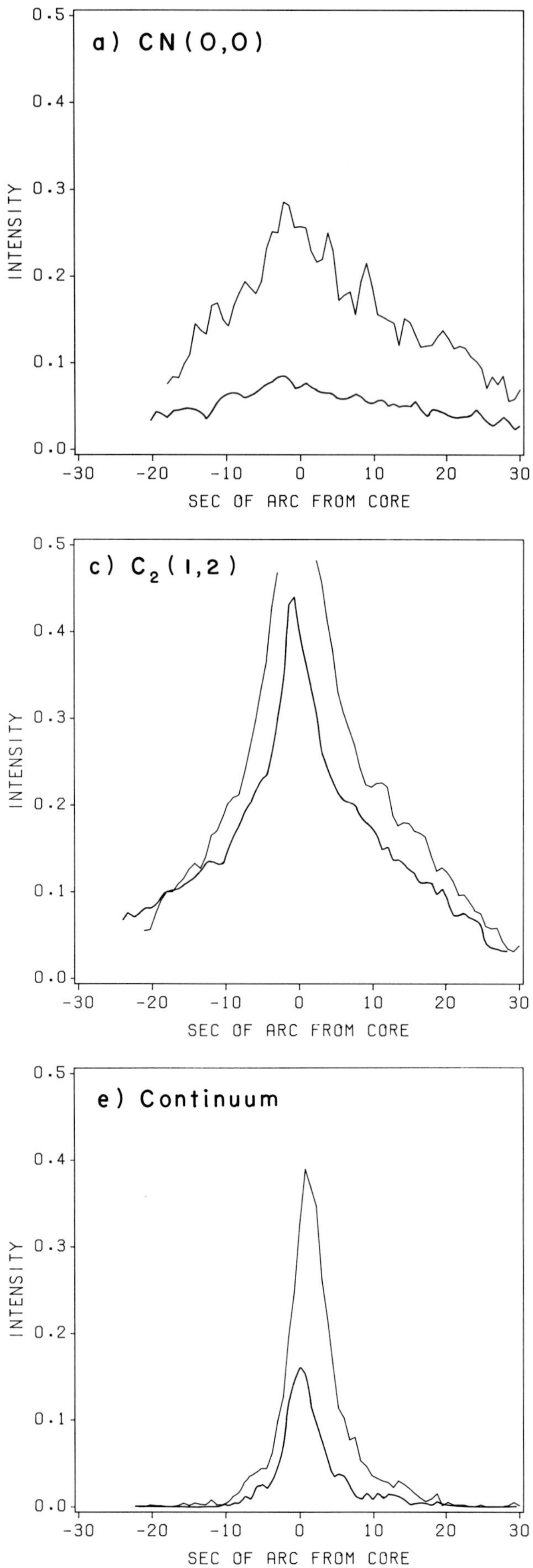

Fig. 7a–e. Intensity as a function of position along the slit for the spectra of several species (**a**)–(**d**) with the continuum contribution included and (**e**) the continuum at 5340 Å. Negative values on the horizontal axis indicate regions south of the core

Baumbaugh, A.E., Knickerbocker, K.L., Baumbaugh, B., Retting, T., Ruchti, R.: 1986, *IEEE Trans. Nuclear Sc.*,**NS-33**, 903

Belton, M.J., Wehinger, P., Wyckoff, S., Spinrad, H.: 1986, 20th ESLAB Symp., ESA SP-250, Vol. **I**, p. 599

Cochran, A.L., Barker, E.S.: 1986, 20th ESLAB Symp., ESA SP-250, Vol. **I**, p. 439

Gehrz, R.D., Ney, E.P.: 1986, 20th ESLAB Symp., ESA SP-250, Vol. **II**, p. 101

Hanner, M.S., Tokunaga, A.T., Golisch, W.F., Griep, D.M., Kaminski, C.D.: 1987, *Astron. Astrophys.* (preprint)

Hartmann, W.K., Cruishank, D.P., Degewij, J.: 1982, *Icarus* **52**, 377

Krasnopolsky, V.A., Gogoshev, M., Moreels, G., Moroz, V.I., Krysko, A.A., Gogosheva, T., Palazov, K., Sargoichev, S., Clairemidi, J., Vincent, M., Bertaux, J.L., Blammont, J.E., Troshin, V.S., Valnicek, B.: 1986, *Nature* **321**, 269

Larson H., Davis, D., Mumma, M., Weaver, H.: 1986, *Astrophys. Matthews, C.N., Ludicky, R.: 1986, 20th ESLAB Symp., ESA, SP-250, Vol.* **II**, p. 273

McFadden, L., A'Hearn, M.F., Feldman, P.D., Butterworth, P.S., Weaver, H.A., Woods, T.N.: 1986, Abstr. 20th ESLAB Symp., 223, and private communication

Moreels, G., Vincent, M., Clairemidi, J., Mougin, B., Parisot, J.P., Zucconi, J.M., Bertaux, J.-L., Blamont, J.E., Herse, M., chalon, G., Lepage, J.P., Runavot, J., Krasnopsky, V.A., Moroz, V.I., Krysko, A.A., Troshin, V.S., Barke, V.V., Jegulev, V.S., Sanko, N.F., Tomashova, G.V., Parashev, V.A., Tkachuk, A.Y., Novikov, B.S., Perminov, V.G., Sulakov, I.I.et al.: 1985, *Adv. Space Res.* **5**, 12, 137

Moreels, G.: 1986 (private communication)

Mumma, M.J., Weaver, H.A., Larson, H.P.: 1986, 20th Eslab Symp., ESA SP-250, Vol. **I**, p. 341

Rettig, T.W., Ruchti, R., Baumbaugh, B., Baumbaugh, A., Knickerbocker, K.: 1986, 20th ESLAB Symp., ESA SP-250, Vol. **III**

Ruchti, R., Rettig, T., Baumbaugh, B., Baumbaugh, A., Knickerbocker, K.: 1985, *IEEE Trans. Nuclear Sci.*, **NS-33**, 151

Russell, R.W., Lynch, D.K., Rudy, R.J., Rossano, G.S., Hackwell, J.A., Campins, H.C.: 1986, 20th ESLAB Symp., ESA SP-250, Vol. *II*, p. 125

Sagdeev, R.Z., Szabo, F., Avanesov, G.A., Cruvellier, P., Szabo, L., Szego, K., Abergel, A., Balzs, A., Barinov, I.V., Bertaux, J.-L., Blamont, J., Betaille, M., Demarelis, E., Dul'nev, G.N., Endroczy, G., Gardos, M., Kanyo, M., Kostenko, V.I., Drasikov, V.A., Nguyen-Trong, T., Nyitrai, Z., Reny, I., Rusznyak, P., Shamis, V.A., Smith, B. et al.: 1986, *Nature* **321**, 262

Schleicher, D.G., Millis, R.L., Tholen, D., Lark, N., Birch, P.V., Martin, R., A'Hearn, M.F.: 1986 20th ESLAB Symp., ESA SP-250, Vol. **I**, p. 565

Timothy, J.G.: 1983, *Publ. Astron. Soc. Pacific* **95**, 810

Weaver, H.A., Mumma, M.J. Larson, H.P., Davis, D.S.: 1986, *Nature* **324**, 441

Astron. Astrophys. 187, 256–260 (1987)

ASTRONOMY AND ASTROPHYSICS

Observations of ions in comet P/Halley with a focal reducer *

K. Jockers [1], E.H. Geyer [2], H. Rosenbauer [1], and A. Hänel [2]

[1] Max-Planck-Institut für Aeronomie, D-3411 Katlenburg-Lindau, Federal Republic of Germany
[2] Observatorium Hoher List, D-5568 Daun, Federal Republic of Germany

Received February 11, accepted March 3, 1987

Summary. Comet Halley was observed with the ESO 1 m telescope and a dedicated focal reducer instrumentation. Narrow-band images of the comet in the light of the CO_2^+ ion at 367.4 nm were obtained for the first time. On these images, the cometary plasma can be observed as close as 10^4 km to the nucleus. No plasma envelopes are visible. Multislit spectra show emissions of OH^+, CO_2^+, CO^+, N_2^+ and CH^+. They indicate the presence of CO^+ ions in a large halo around the comet, not only in the tail streamers.

Key words: comet Halley – observational methods – molecular ions

1. Introduction

For several reasons it is important to understand the behaviour of cometary ions in the cometary coma.

a) The cometary ions H_2O^+, CO_2^+ and CO^+ relate directly to cometary mother substances H_2O, CO_2 and possibly CO.

b) Because of their large cross-sections ion molecular reactions are most important for the chemistry of the inner coma.

c) The distribution of the ion sources determines the nature of the solar wind interaction.

To study the behaviour of ions in the coma of comet P/Halley a special instrumentation was put together at the Observatory Hoher List and the Max-Planck-Institute for Aeronomy and employed at the ESO 1 m telescope between March 10–16 and April 4–11, 1986. We wanted to explore the following questions:

1) What is the kinematics of ions in the inner coma? How do ion rays form in the inner coma?

2) What are the column densities of individual ion species and how are they distributed within the comet?

Even though our observations will eventually allow to address both questions this paper will be mainly concerned with the first one. During the course of our observations we obtained for the first time images in the light of the CO_2^+ ion at 367.4 nm. These images show very little continuum radiation from the dust coma and virtually no emission of neutral molecules. With our instrumentation, which will be described in the following section, it is possible to obtain either narrow band images or multislit spectra. In Sect. 3 we will present an example of a multislit spectrum and in Sect. 4a sequence of narrow-band images. We will discuss the implications of these observations regarding the spatial ion distribution and its temporal changes.

2. The Instrumentation

For the observations the focal reducer of the Observatory Hoher List (Geyer et al., 1979) was used. The Max-Planck-Institut for Aeronomy provided a lens camera with an image intensifier, a UV grating prism ("grism") and several interference filters. The optical scheme of the focal reducer is shown in Fig. 1. Behind the Cassegrain focus the optical beam is recollimated by a dioptric collimator (exit pupil diameter 50 nm) and an image of reduced size is formed by a compound objective lens of focal ratio f/2.8. This image is detected by a two-stage image intensifier with bialkali photocathode and recorded on hypersensitized IIIa-F plates pressed against the fiberglass exit window of the image intensifier. At the ESO 1 m telescope the resulting image scale is about 1 arcmin/mm and the field size is about 22′. The instrument can be operated in an imaging and in a spectrographic mode. To change between the two configurations takes about 5 minutes.

In the imaging mode a narrow-gap tunable Fabry-Perot filter (Atherton and Reay, 1981) with an interference prefilter is mounted in the parallel beam. This etalon is equipped with capacitor pads and piezoelectric spacers which allow precise control of parallelism and gap setting. In the wavelength range from 353 to 436 nm the Fabry-Perot mirrors have a reflectivity between 83 and 86 %, i.e. a reflection finesse of about 19. The total finesse is about 12. In the mode used during the comet observations the free spectral range was between 13 and 17 nm. Four interference prefilters (369 ± 5 nm, 390 ± 6 nm, 404 ± 5 nm, 425 ± 4 nm) were available to suppress adjacent orders. This setup allowed to obtain monochromatic images with about 1 nm bandwidth at most of the interesting emissions of ions and neutrals in the blue and near-UV wavelength range.

In the spectrographic mode a multislit diaphragm was put into the Cassegrain focus and the Fabry-Perot filter was replaced by a grating prism. The slit plate comprises of a total of 69 slits arranged in 5 columns in such a way that the individual spectra do not overlap for a certain wavelength range and reciprocal angular dispersion. Projected onto the sky the slit columns have an angular distance of about 4.′6 and the distance of the individual slits (size 2.″8 × 30″) within the slit column is 1.′4. Two grating prisms with 300 and 600 lines per mm were used and gave a dispersion of 200 and 100 Å mm^{-1}, respectively. To avoid the

---

Send offprint requests to: K. Jockers

* Paper based on observations collected at the European Southern Observatory, La Silla, Chile

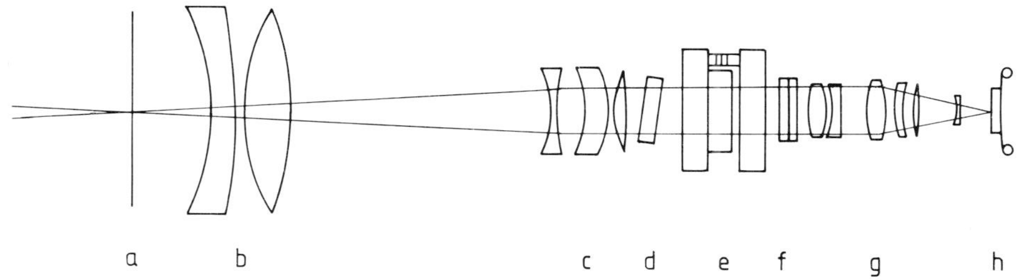

Fig. 1. Optical arrangement of the focal reducer in the imaging mode. a: Cassegrain focus; b: field lens; c: collimator lens; d: coloured glas prefilter; e: tunable Fabry-Perot-filter; f: interference filter; g: camera lens f/2.8; h: two-stage image intensifier. In the spectrograph mode the Fabry-Perot filter is exchanges by a grating and a slit mask is positioned at Cassegrain focus

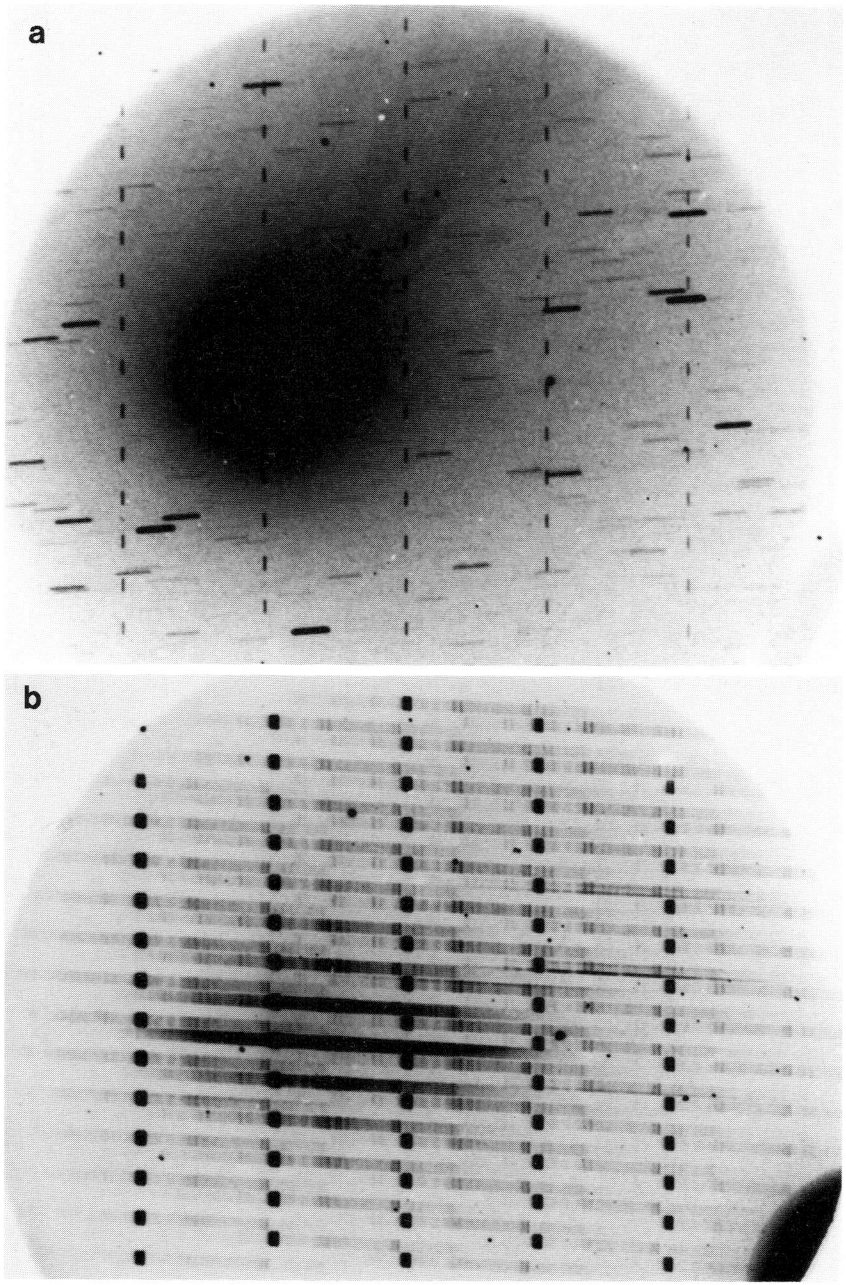

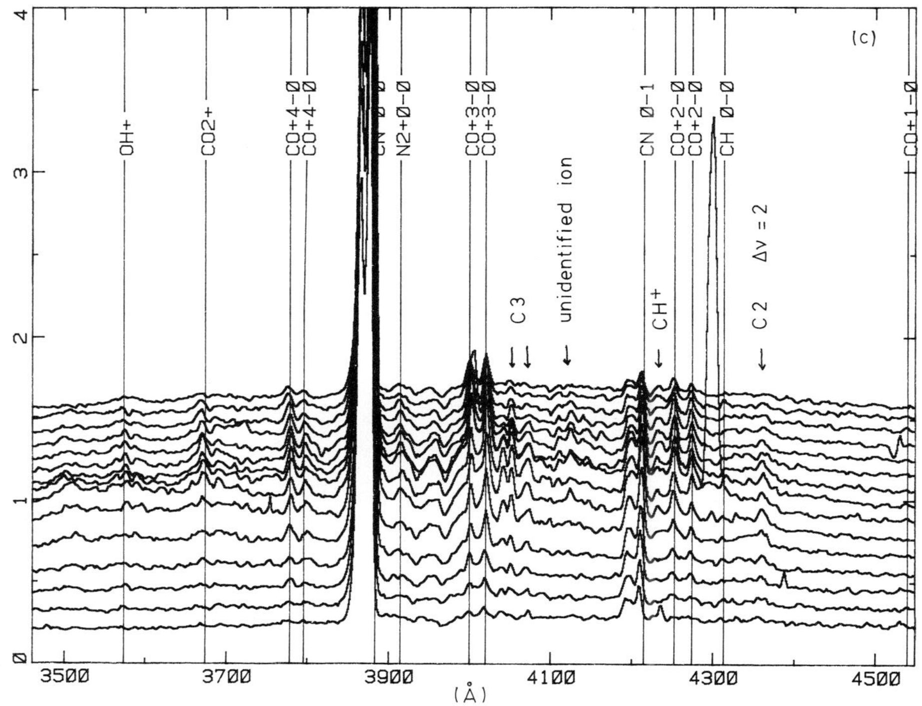

Fig. 2a–c. A multislit spectrum of Comet Halley from April 10. **a** Reference image to show location of slits with respect to the comet. **b** Reproduction of spectrum plate. **c** Intensity tracings of the spectra of the central column of slits

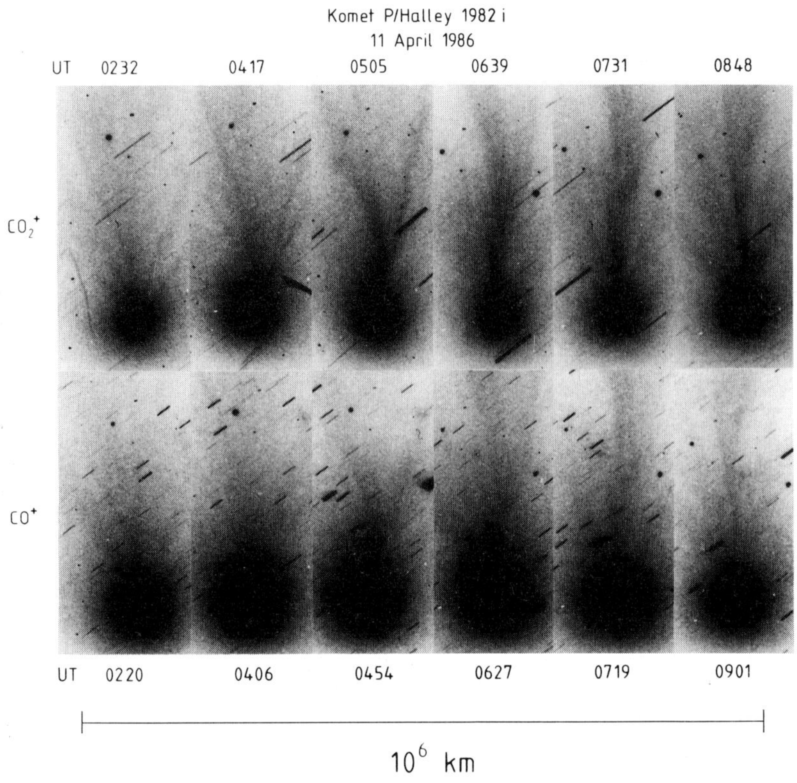

Fig. 3. Motions in the inner plasma tail of Comet Halley recorded in the light of the CO_2^+ and CO^+ ions

overlapping of the spectra we had to limit the spectral range to $350 \leqq \lambda \leqq 430$ nm for the high resolution and to $350 \leqq \lambda \leqq 500$ nm for the low resolution grating by putting a colour filter in front of the grism. To obtain knowledge of the location of the comet with respect to the slits we exposed an image of the comet through the interference filter 369 ± 5 nm (CO_2^+). Then we would insert the slit plate at the Cassegrain focus, remove the interference filter and expose the slit pattern on the same plate.

The instrumentation worked satisfactorily but two major problems occurred which affect the quality of the data.

a. Despite extremely careful calibration of the etalon setting it seems that the etalon was not tuned properly to the wavelength of 425.2 nm employed to record the CO^+ ions. We attribute this to a response of the etalon coatings to humidity which is most apparent at the long-wave cutoff of the mirror coating. This effect needs to be further investigated during controlled laboratory tests.

b. Even though the proximity focused image intensifier had a very good response in the near-UV and blue spectral regions the photometric quality of the intensifier plate combination is unfortunately rather poor. Bright spots in the image of the intensifier are surrounded by an extended halo which creates a crosstalk between individual multislit spectra. Hypersensitation of the plates varied from plate to plate. This introduces large photometric errors when different plates are compared.

3. A multislit spectrum obtained on 1986, April 10

In Fig. 2 we present a multislit spectrum obtained on April 10 between 0505-0535 UT. Figure 2a shows the slit pattern superimposed on the comet image, which was obtained with the CO_2^+ prefilter (without etalon). Note the five columns with the interlaced slits. In Fig. 2b the corresponding spectrum plate is reproduced. To reduce overlap of individual spectra the dispersion direction has been set to an angle not exactly perpendicular to the slits. The strongest emission is the CN 0–0 band which is blurred because of blooming in the image intensifier. The wavelengths increase to the right. The tracings of the spectra of the central column are presented in Fig. 2c. From Fig. 2a it is seen that the first 7 slits of the central column are in the comet tail and slits 8–10 are closest to the inner coma. Correspondingly, the ion emissions show up more strongly in the upper tracings and the neutral emissions of C_3 and C_2 more in the central tracings. The ion emissions of OH^+, CO_2^+, CO^+, N_2^+ and CH^+ are present. In addition an unidentified ion emission is visible around 413 nm. It has also been observed by Wyckoff et al. (1986) and tentatively identified as another band of CO_2^+. The spectra are disturbed by solar absorption lines (in particular H and K) which appear in the dust continuum of the coma and dust tail. Therefore a proper evaluation of scalelengths has to await a subtraction of the solar spectrum.

A study of the ion lines in Fig. 2b reveals that the cometary plasma is not confined to the ion streamers but is as well present in other parts of the coma, in particular on the sunward side of the comet. The ion rays appear to be superimposed on a smooth background distribution. Consequently, a lot of diffuse background in the direct image (Fig. 2a) must be caused by CO_2^+ emission. For CO^+ and H_2O^+ Hoban et al. (1986) arrive at a similar conclusion from their interference filter observations.

4. A sequence of direct images in the CO_2^+ and CO^+ radiation obtained 1986, April 11

On 1986, April 11, between 0045 and 1008 UT comet images were taken alternatively in the light of the CO^+ and CO_2^+ ions. In total

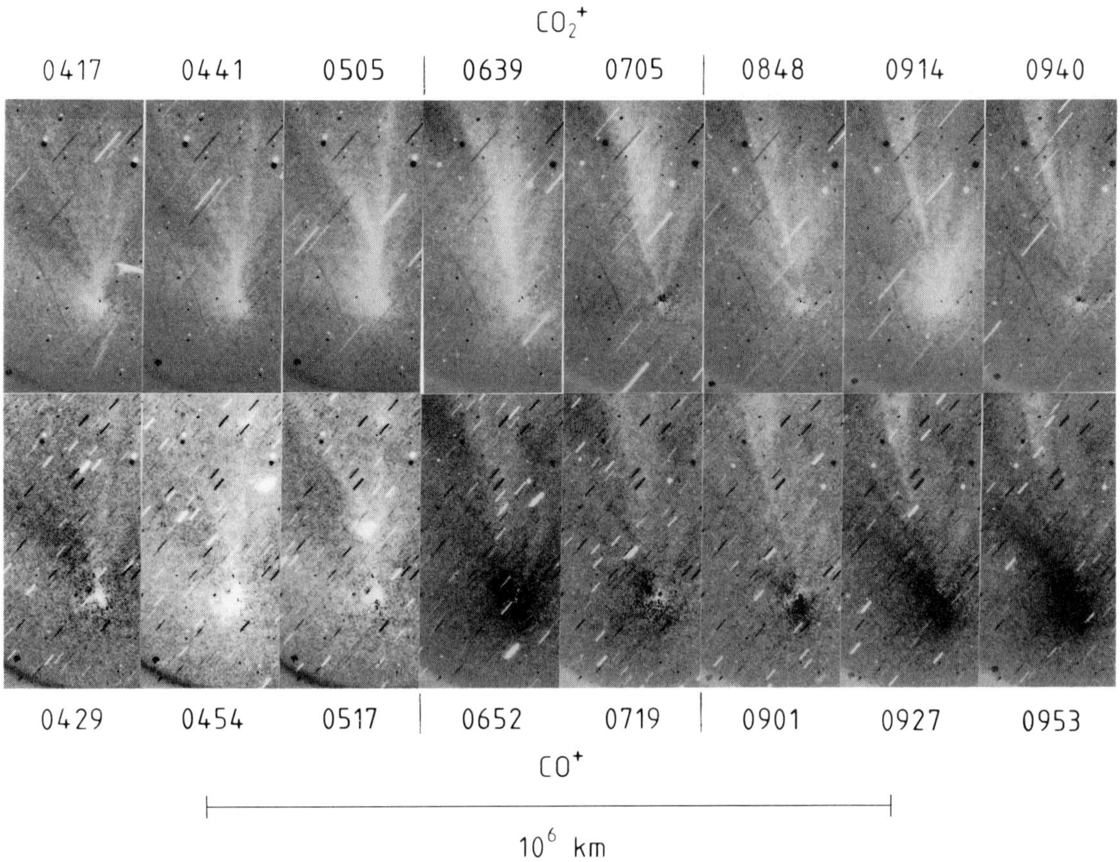

Fig. 4. Difference images of the motions in the inner plasma tail. The leftmost images of Fig. 3 have been subtracted from the other images to enhance changes in ion content

18 CO^+ images (exposure time 5 min, centre wavelength 425.2 nm) and 17 CO_2^+ images (exposure time 15 min, centre wavelength 367.4 nm) were obtained. A series of selected images is shown in Fig. 3. Despite of the small bandpass of the filter the images in Fig. 3 still show a coma, the extent of which is, however, much reduced. In an unfiltered exposure the CN coma would fill almost the whole field of the frames of Fig. 3. Moreover, the size of the coma appears exaggerated in Fig. 3 by the reduced contrast of the printed picture as compared to the original plate.

To our opinion the coma shown in the frames of Fig. 3 is a dust coma. In case of CO_2^+ there is no significant neutral emission present in the whole passband of the prefilter of 369 ± 5 nm (see Fig. 2b). In case of CO^+ there is a slight possibility of contamination by the CN 0–1 band at 421.5 nm but the suppression of this band as compared to CO^+ should be about a factor of 30. A contamination with CH can be safely excluded. If the response function of the instrumentation and the extinction is taken into account the relative brightness of the coma in the two passbands is consistent with a dust coma of solar colour. On the original CO_2^+ plates a sunward extension of the coma as typical for dust is clearly seen.

Apart from the fact that the ion structures can be discerned more closely to the nucleus the ion features of Fig. 3 behave similar to those visible on large-scale tail photographs as, e.g., described by Jockers (1985). At the beginning of the sequence a tail ray grows to the upper right corner of the frames and at the same time the main tail is pushed to the right, in correspondence with a special rule established in the above mentioned paper. Later on a pair of tail rays develops and a tail disconnection occurs (0823 UT).

In order to eliminate the coma from the CO_2^+ and CO^+ images they were scanned with a microdensitometer and transferred to relative intensities. Even though the plates were exposed equally long and processed under identical conditions it turned out that the background intensities were different on the different plates. Most of the effect seems to come from the hypersensitation which varies from plate to plate. To be able to compare the images the background was determined in the outer part of the image and was normalized to unity. Then, in analogy to the process of eliminating the F corona from coronal images by forming differences between images obtained at different times (see Sheeley et al., 1980) the normalized image taken at 0232 UT (CO_2^+) was subtracted from the other CO_2^+ images and the normalized image taken at 0220 UT (CO^+) from the other CO^+ images. This way the stationary dust coma is eliminated and the temporal changes are enhanced. A stationary ion component, necessarily, would also be cancelled out. The result is shown in Fig. 4.

Let us first discuss the CO_2^+ images. They show ejection of a plasma cloud which has its largest column density around 0639 UT. On images where the intensities close to the nucleus are cancelled out by the subtraction, like e.g. the frames at 0705, 0848 and 0940 UT plasma structures radiate out from the nucleus and there is no indication of a diffuse plasma cloud or plasma envelopes as observed in comet Morehouse 1908 III (see

Eddington, 1910; Lüst, 1967). On the frames at 0505 and 0914 diffuse clouds surround the nucleus but it is not clear if they are real or caused by insufficient cancellation. The CO^+ images are much more noisy. This is partly caused by the fact that the etalon bandpass covers only one subband of the CO^+ double bands. In addition, because of the much stronger continuum coma the exposure time had to be kept short. The high constrast between dust coma and plasma features makes it difficult to achieve a good coma cancellation. The most striking feature is that the ejected plasma cloud is well visible in the CO^+ frames only at the edge of the frame. This may, however, be an instrumental effect, since with etalous and interference filters the bandpass is always shifted slightly to the blue in the outer parts of the field. If the etalon, because of reasons discussed in Sect. 2, was turned slightly to longer wavelengths the transmission at the field edge is shifted to the line center and the tail will be better visible at the field edge.

5. Conclusion

The focal reducer instrumentation employed to observe comet Halley has prove to be sensitive enough and of sufficient spectral purity to allow registration of the individual cometary ions in an extended field around the nucleus. The methods of multislit spectra and direct imaging through a Fabry-Perot filter complement each other nicely. The observations have suffered from photometric inaccuracies caused by the image intensifier-plate combination used as a detector. It is planned to replace this detector in future observations with a CCD.

References

Atherton, P.D., Reay, N.K.: 1981, *Monthly Notices Roy. Astron. Soc.* **197**, 507

Eddington, A.S.: 1910, *Monthly Notices Roy. Astron. Soc.* **70**, 442

Geyer, E.H., Hoffmann, M., Nelles, B.: 1979, *Astron. Astrophys.* **80**, 248

Hoban, S., A'Hearn, M.F., Birch, P.V., Candy, M.P., Martin, R., Klinglesmith, D.A.: 1986, 20th ESLAB Symp. ESA SP-250, Vol. 1, 505

Jockers, K.: 1985, *Astron. Astrophys. Suppl.* **62**, 791

Lüst, Rh.: 1967, *Z. Astrophys.* **65**, 236

Sheeley, N.R. Jr., Michels, D.J., Howard, R.A., Koomen, M.J.: 1980, *Astrophys. J.* **237**, L99

Wyckoff, S., Wehinger, P.A., Spinrad, H., Belton, M.J.S.: 1986, 20th ESLAB Symp., ESA SP-250, Vol. 1, 311

Astron. Astrophys. 187, 261–263 (1987)

ASTRONOMY
AND
ASTROPHYSICS

Two disconnection events in comet P/Halley and possible solar causes

H. Lundstedt [1] **and P. Magnusson** [2]

[1] Lund Observatory, Box 43, S-221 00 Lund, Sweden
[2] Uppsala Astronomical Observatory, Box 515, S-751 20 Uppsala, Sweden

Received February 16, accepted May 18, 1987

Summary. Disconnection events in comet P/Halley in April 1986 are discussed. Observations were made both from La Palma, Canary Islands and Siding Spring, Australia. Different solar phenomena, such as solar flares, coronal mass ejections, coronal holes and the coronal neutral line, are examined as possible causes of the disconnection events. The solar wind counterpart of the coronal holes and the coronal neutral lines, the so called high speed plasma streams and the heliospheric current sheet, are the most often suggested causes. In our two cases we found the heliospheric current sheet to be the most likely cause, but also a high speed plasma stream might have caused the DE of April 12.

Key words: comet P/Halley – disconnetion events

1. Introduction

Alfvén (1957) was the first to suggest that the solar wind magnetic field gets hung up on the cometary plasma head. The magnetic field lines are draped around the head and organized into two lobes of opposite magnetic polarity. The International Cometary Explorer's fly-through of Giacobini-Zinner (Slavin et al., 1986) showed this picture to be basically correct.

A major cometary phenomenon is a so called disconnection event (DE), i.e. a detachment of the plasma tail from the head. The major DE's generally occur every week (Pedersen, 1986). 12–24 h before the tail comes off, ray bursts, innertail narrowing and brightening are seen. DE's have been explained through the interaction between the comet and the heliospheric current sheet (Niedner and Brandt, 1978; Lundstedt, 1983) or between the comet and high speed solar wind streams (Ip and Mendis, 1978; Jockers, 1985; Russell et al., 1986). In this article we will discuss two major DE's observed in April 1986.

2. Observations

Comet Halley was observed both from La Palma, Canary Islands, and from Siding Spring, Australia. Our site of observations, on La Palma, (long = 15° west, lat = 29° north) was close to the Swedish solar telescope with excellent seeing at an altitude of 2400 m. Observations on La Palma were made during two weeks centered on March 10 and two weeks centered on April 11. Observations at the Siding Spring Observatory (long = 149° east, lat = 31° south) were made during an even longer period, between March 6 and May 15.

On La Palma we used a Hasselblad camera 200 FC/M, with a F 2.0, 100 mm lens. The fields of view was 28° × 28°. The images were then PDS-scanned and some image processing was performed, mainly histogram equalization. The observations at Siding Spring were made with the Uppsala Schmidt telescope with $f = 173$ cm. The field of view was 3°.98 × 3°.98.

Figure 1 and Fig. 2 illustrate two DE's. The first one was observed from La Palma on April 12 at 01^h46^m UT with an 8 min exposure time. We used a Agfacrome 1000 ASA. The photograph was measured with an aperture size of 47 µm × 47 µm. Fig. 1 is a black and white copy of the false color processed original. 1° i.e. 1/8 th of the image length corresponds to 3.5 10^6 km. The same DE of April 12, but about 10 hours later, is illustrated in Brandt and Niedner (1986). In that image the plasma tail is now clearly detached from the head. The second DE was observed from Siding Spring on April 14 at 15^h35^m UT with a 30 min exposure time. We used an emulsion-filter combination IIa-OGG 385. 1° i.e. 1/28 th of the image length corresponds to 3.6 10^6 km.

3. Possible solar causes of the two DE's

Let us first examine if any transient solar phenomena took place. The solar activity was, as expected, very low. No solar flares were reported (Solar Geophysical Data, 1986) in the period 1986 April 6 to 16. Several active regions were present during the above period. Solar active region 4721 was located at (S 02°, W 85°) on April 8, region 4722 at (S 09°, W 15°) on April 8, region 4723 at (S 12°, E 21°) on April 16 and finally region 4724 at (S 07°, E 43°) on April 23. Coronal mass ejections (CME) occur at a rate of about 0.2–0.3/day during solar minimum (Hildner, 1986). A slow streamer (Hildner, private communication) disrupted on April 5 at a position angle 262°, 8° south of the solar equator. The corona showed no changes from 6 through 12 April, so we may disregard CMEs as causes of the two DE's.

Let us now examine what kind of long-lived solar phenomena were present during the period of interest. Polar coronal holes (Geophysical Solar Data, 1986) were present during the whole period. No equatorial holes were present. The polar holes could have caused high speed solar wind streams. Actually, a compression of the magnetic field in the solar wind was observed by IMP-8 (Lepping, private communication) on April 12 indicating a high speed plasma stream. The coronal neutral line for Carrington rotation number 1774 (1986 April 6 – May 3) at 2.5 source surface, was calculated (T. Hoeksema, private communication) from observations of the large-scale magnetic field made at Wilcox

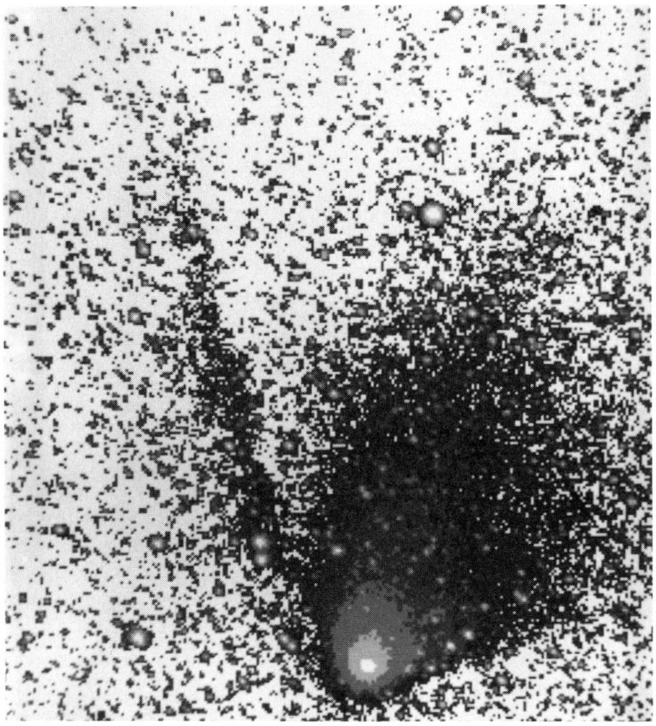

Fig. 1. A disconnection event in comet Halley on April 12, 1986 at 01^h46^m UT. The photograph was taken on La Palma, Canary Islands and image processed at Lund Observatory

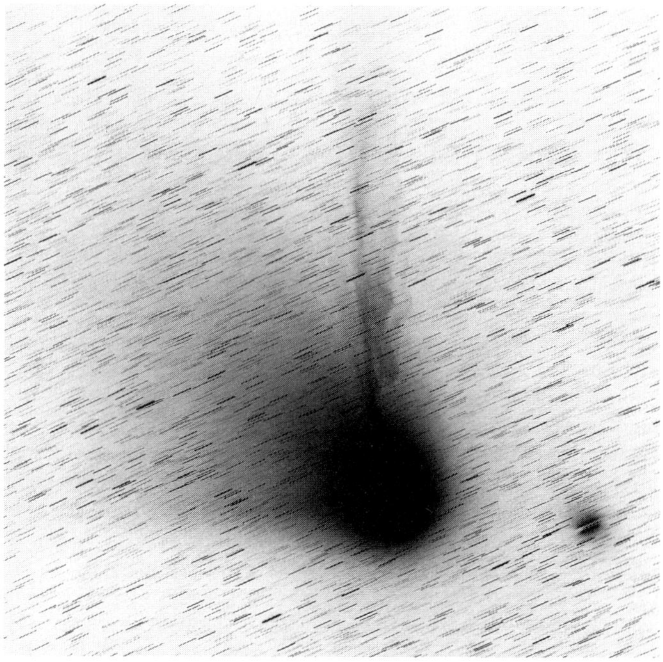

Fig. 2. A disconnection event in comet Halley on April 14, 1986 at 15^h35^m UT. The plate was taken from Siding Spring Australia

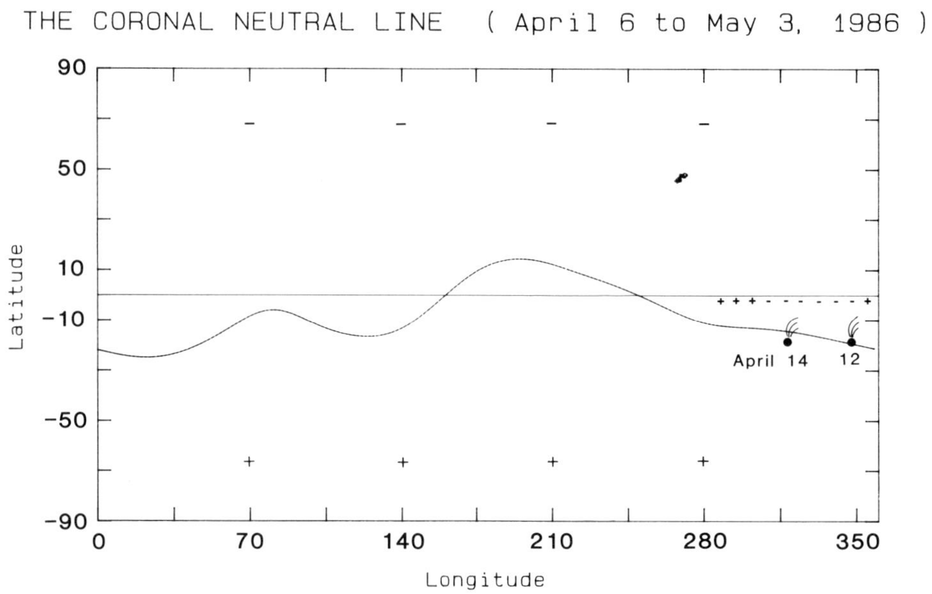

Fig. 3. Synoptic map the coronal source surface for Carrington Rotation 1774. The wave line is the coronal neutral line across which the magnetic polarity reverses. The position of Comet Halley relative the coronal neutral line for dates April 12 and 14, are marked. The magnetic polarity, measured by IMP-8 at Earth, for April 11 through April 15 is also marked twice a day (longitude and latitude in degrees)

Solar Observatory, Stanford. In our plot (Fig. 3) the wave line outline the coronal neutral line. The coronal neutral line is convected out into interplanetary space, giving rise to the magnetic sectors and the heliospheric current sheet. The position of comet Halley on April 12 and on April 14 on the source surface, was obtained by computing the footpoint longitude and latitude of the Archimedean spiral connecting the sun and the comet for a constant solar wind speed of $400\,\mathrm{km\,s^{-1}}$. We also marked the polarity, measured by IMP-8 at Earth, in Fig. 3. From Fig. 3 it seems likely that comet Halley has crossed a $-/+$ sector boundary on April 13/14. As the IMP-8 Data indicate, we may also have had small-scale ripples superimposed on the large-scale warped current sheet causing repeated sector boundary crossings. That might have caused the April 12 DE.

4. Conclusions

We examined two disconnection events in comet Halley which took place around 12 April and around 14 April 1986. We found that a −/+ sector boundary crossing seems likely for the 14 April event. The other event might also have been caused by a crossing of a high speed solar wind stream, however a current sheet with a small waves could also have caused that DE.

References

Alfvén, H.:1957, *Tellus* **IX**, 92

Brandt, J.C., Niedner, M.B.: 1986, *20th ESLAB Symp.* ESA SP-250, Vol. **1**, p. 47

Hildner, E.: 1986, *XXVI COSPAR*, June 30–July 11 1986, Toulouse, France

Ip, W.H., Mendis, D.A.: 1978, *Astrophys. J.* **223**, 671

Jockers, K.: 1985, *Astron. Astrophys. Suppl.* **62**, 791

Lundstedt, H.: 1983, in *Asteriods, Comets and Meteors* eds C.-I. Lagerkvist, H. Rickman, Uppsala, p. 311

Niedner, M.B., Brandt, J.C.: 1978, *Astrophys. J.* **223**, 655

Russell, C.T., Saunders, M.A., Phillips, J.L.: 1986, *J. Geophys. Res.* **91**, 1417

Slavin, J.A., Smith, E.J., Tsurutani, B.T., Siscoe, G.L., Mendis, D.A., Jones, D.E.: 1986, *XXVI COSPAR*, June 30–July 11 1986, Toulouse, France

Solar Geophysical Data, prompt reports 1986, Nos. 501 and 502, Part 1

Astron. Astrophys. 187, 264–266 (1987)

ASTRONOMY AND ASTROPHYSICS

Activity of the plasma tail of comet P/Halley in March, 1986

M. C. Wu and P. Z. Qiu

Yunnan Observatory, Academia Sinica, P.O. Box 110, Kunming, Yunnan, China

Received January 29, accepted April 22, 1987

Summary. A photographic sequence of Comet Halley was taken at the Simao Observation Station of the Yunnan Observatory from March 4 to 22, recording the large-scale activity of the plasma tail. The various phenomena are described in the present article, and it is suggested from the analyses that at least two disconnection events occurred on March 8–9 and 21, respectively. The observed results are in accordance with those obtained by the observers in other countries. Finally, the process of the activity of the plasma tail and the correlation between the activity of the plasma tail and the solar activity are preliminarily discussed.

Key words: plasma tail – disconnection event

1. Introduction

The large scale phenomena of the plasma tail of Halley's Comet after its passage through the perihelion have been observed at the Simao Observation Station (100°58′18″E, 22°55′16″N; with an elevation of 2004 m) of the Yunnan Observatory by using a tracking camera with an aperture of 10 cm, a focal length of 25 cm and a field of view of 24°.2 × 17°.6, and plates of Kodak IIa–o. The observations from March 4 to 22 (except for March 7) have been obtained, especially during the encounter of Halley's Comet with the five spacecraft such as Vega, Giotto, etc. In view of the importance of the data obtained in this time interval, we communicate them here.

2. Observed results

March 4. There are two plasma tails at p.a. = 268° and 275°, respectively (where p.a. is the position angle, calculated from north to east; the same below). The former is short and the latter is longer (main tail). Because of the strong moon-light a deeper exposure could not be made. The visible comet tails are all shorter and their details are not clear.

March 5. At p.a. = 268° lies a slender and straight plasma tail (which is often simply called comet tail; similarly hereinafter), and at p.a. = 274° is the main tail in the shape of a wave, showing a disturbance.

March 6. The section of the main tail between the cometary nucleus and the place about 2°.2 away from the nucleus (the distance is calculated from the nucleus along the tail axis; the same below) is slender and straight, and then the main tail fluctuates like a wave, which is the continuation of the disturbance of March 5.

March 7. It is cloudy and no data can be obtained.

March 8. It can be seen from the plates of 8.93 and 8.94 UT (centre of exposure time in UT in fractions of a day, similarly hereinafter) that the main tail, which is situated at p.a. = 280°, has a Z-shaped twist at the place about 0°.2 away from the nucleus, which shows that a new disturbance occurs in the near-coma region. The twisted tail is narrow and straight from 0°.2 to 2°.2, and then expands and has a wavy structure, which is the remains of the previous disturbance. In addition, there is a shorter tail at p.a. = 267° (see Fig. 1).

March 9. On the plates of 9.90 and 9.92 UT, one can see that there appear three broader plasma tails at p.a. = 262°, 270° and 279°, respectively, showing up a symmetrical distribution. There is also a filamentary subtail at p.a. = 287°. It can easily be seen that the Z-shaped twist in the near-coma region of March 8 has receded away from the Sun to the place 1°.3 away from the nucleus and separated into two branches, parallel to each other, with irregular fluctuation. In addition, there is a disconnection between the near-cometary head region and the two branches, and the twisted tail has an obvious wavy structure. It can be inferred that an important disconnection event (which is abbreviated to DE) occurred in the comet's tail (see Fig. 2).

March 10. On the plates of 10.91 and 10.93 UT, the trace of the DE still exists and recedes to the place 3°.3 away from the nucleus. The distance between the two branches increases and a broader tail forms, while the main branch expands outward from the Z-shaped twist. Besides, two shorter tails are at p.a. = 259° and 270°, respectively, and at p.a. = 277°, between the two twisted branches, lies a slender and straight tail (see Fig. 4).

March 11. On the plates of 11.91 and 11.93 UT, the main tail, which is situated at p.a. = 279°, is 9° long. At the position 7°.3 away from the nucleus one can still see the remnants of the DE. There appear three quite slender plasma tails at p.a. = 260°, 272° and 290°, respectively, with the middle one being as thin as a needle. They may be a tail ray system.

March 12. On the plate of 12.91 UT, those slender and straight tails draw close to the main tail and merge into two, and the main tail has a twist at the position 1°.1 away from the nucleus.

March 13. The twist recedes to the position 1°.5 away from the nucleus, the tail blurs and thus has no details.

March 14. The twist recedes to the position 2°.8 away from the nucleus, with the former section of it being narrow and straight and the latter expanding like a wave. It seems that there is a gap at the joint between the two sections. And there exists a short and straight tail at p.a. = 271°.

Send offprint requests to: M. C. Wu

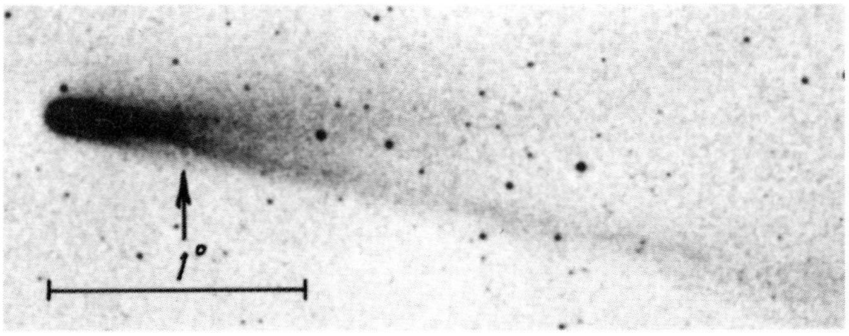

Fig. 1. March 8.8 UT

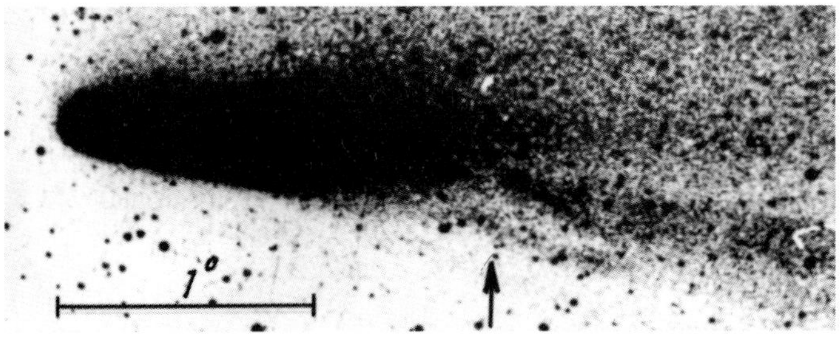

Fig. 2. March 9.9 UT

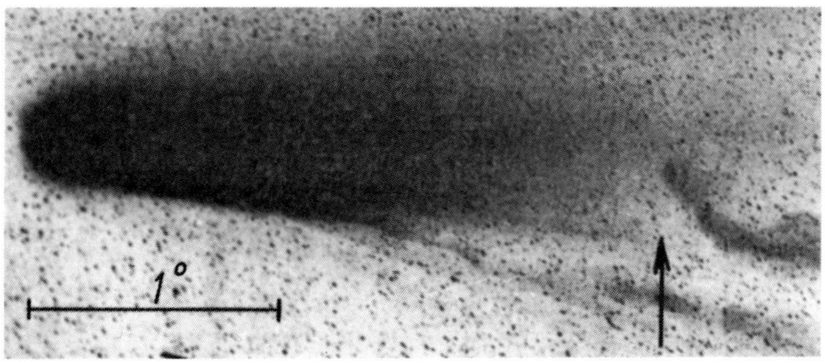

Fig. 3. March 10.4 UT, reproduced from West et al. (1986)

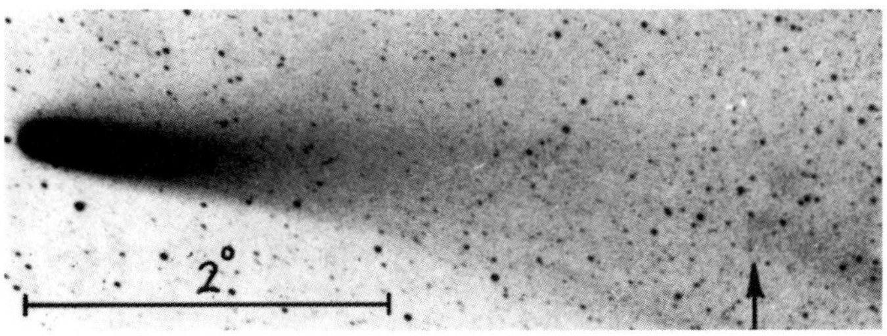

Fig. 4. March 10.9 UT

March 15. The main tail extends to the position 13°.3 away from the nucleus. A thick and short tail and a filamentary tail are at p.a. = 268° and 288°, respectively.

March 16. The main tail is 18° in length (equivalent to 58.9 10^6 km) and is the longest tail observed in March. At the position 4°.6 away from the nucleus the tail has a wind, and in the nearcoma region, at 0°.2 away from the nucleus, a Z-shaped twist appears again. The subtail grows 3°.3 in length.

March 17. On the plates of 17.89 and 17.92 UT, the main tail has twists at positions 1°.8, 2°.7, 4°.2 and 6°.2 away from the nucleus, showing up a large scale wavy structure which has symmetrical tail axes and gradually expands outwards. It exhibits a strong activity. A thin subtail about 2°.6 in length at p.a. = 278° intersects the main tail at the place 1°.8 away from the nucleus and then extends about 0°.2 in length. A disconnection may appear here. In addition, at p.a. = 266° lies a thick and short tail.

March 18. On the plate of 18.91 UT, the thin subtail extends to the positions 4°.4 away from the nucleus, intersects the twisted main tail at the position 3°.8 away from the nucleus, with an overlap of about 0°.6, and then stretches out. The first twist of the main tail recedes to the position 3°.8 away from the nucleus, the second twist can still be identified and the distance between the two twists increases obviously. There appears a bright spraying along the tail axis in the near-coma region, and a thick and short tail lies at p.a. = 271°.

March 19. The phenomena which occurred on March 17 and 18 disappear and only a plasma tail, thin and straight like a needle, exists and may be the main tail rays.

March 20. The plasma tail is narrow and straight down to 1°.5 away from the nucleus and then one side of it opens northwards, making an angle of about 10° with the other side.

March 21. The opening tail recedes to the place 4° away from the nucleus and is disconnected here. It trends northwards, making an angle of about 8° with the main tail axis. This is another DE.

March 22. Three tail rays occur at p.a. = 269°, 272° and 277°, respectively, and gradually become three tails.

3. Discussion

The observations show that in March when the five spacecraft intercepted Halley's Comet, the plasma tail of the comet Halley exhibited rich and colourful tail phenomena. For example, multiple plasma tails were a common sight, and wavy structures, rays, even disconnection events appeared frequently. On March 9, just when the spacecraft Vega-2 encountered Halley's Comet, an important DE happened. Besides, on March 17 an important DE might happen; on March 13 a smaller DE might appear; the DE which occurred from March 20 to 21 may undoubtedly be a small one.

The DE's have their common characteristics: After the disconnections happen, the near-coma region has sprayed backwards nearly simultaneously, forming tail rays or tail streams; the wavy fluctuations at the comet's tail reveal the unstability in the plasma. Especially, the large scale wavy structure of March 17 is exactly like the phenomenon of unstability of the electric current pinch (Hasegawa, 1981). Then the tail rays gradually draw close, the wavy structure disappears and the details are blurred; and therefore a homogeneous plasma tail forms. The whole process takes about two or three days. It is worth noticing that the twist which happens in the near-coma region may be the omen of a DE, or even indicates that a DE already happened.

The above-mentioned results are independently put forward by the authors themselves according to their own observations. The two important DE's were reported as soon as the observation was finished. Having completed this paper, the authors read the article published by R. M. West et al. (1986), in which the observed results of March 5 to 14 are given. The results are in accordance with ours. In particular, the judgement on the DE of March 9 proposed by West et al. fully tallies with that given by us. From the arrangement of our plates of 8.9, 9.9 and 10.9 UT and the plate of

10.4 UT of West et al. in sequence of date (see Fig. 1 to 4), one may clearly see how the activity progresses.

The recessive velocity of the DE calculated from our own plates is 60 km s^{-1}, while that given by West et al. is 70 ± 20 km s^{-1}. Both accord well with each other. According to our calculations, the time when the DE happens is 8.8 UT, but according to West et al., it is 9.4 UT. However, from the plate of 8.93 UT one can see the twist in the near-coma region, which is morphologically consistent with the start-point of the DE shown on the plate of 9.92 UT. From this, it may be much more reasonable that the DE should happen before 8.93 UT. But at that time it was daylight at the European Southern Observatory and the phenomenon could not be observed there.

Our assessment of the above-mentioned phenomena also tallies with the result published by IHW (International Halley Watch) Large-Scale Phenomena Network (1986).

The activity of a cometary plasma tail has close correlation to solar wind events (Mendis et al., 1982). According to the relevant data (SESC, 1986), a series of flares had erupted on the solar disc before and after March 5. If it is considered that the DE which began on March 8 is caused by the encounter of the comet Halley with the tangential disconnection plane of the interplanetary shock wave produced by flares, taking the average velocity of the shock wave to be 730 km s^{-1} and letting the distance between the comet and the Sun be 0.81 AU, one can see from the calculations that the shock wave should be emitted from the solar surface 46 hours ealier, i.e. at 14^h UT on March 6. This is in accordance with a flare of Importance 1F accompanied by a proton event of 10 MeV which erupted on the solar disc at $16^h 37^m$ UT on March 6. As for other DE's, the solar activities corresponding to them have not been found. Whether they have relation to the passage of the sector boundary of solar wind awaits being verified when more data are available.

Acknowledgements. We extend our heartfelt thanks to the other members of this group: Dr. Zhang Ying, Dr. Liu Zhizhong, Dr. Yu Jianming, Dr. Wang Jinsuo, Mr. Wang Chaodong and Mr. Lou Ke, for their useful help.

References

Hasegawa, A.: 1981, *Plasma Instabilities and Nonlinear Effects*, Beijing, Science Press (in Chinese)

IHW Large-Scale Phenomena Network: 1986, Circular No. **5**, April

Mendis, D. A., Houpis, L. F.: 1982, *Rev. Geophys. Space Phys.* **20**, 885

SESC: 1986, Preliminary Report and Forecast of Solar Geophysical Data, PRF. **548**, **549**, **550**

West, R. M., Pederson, H., Monderen, R. V., Grosbol, P.: 1986, *Nature* **321**, 363

Astron. Astrophys. 187, 267–275 (1987)

ASTRONOMY AND ASTROPHYSICS

The cause of two plasma-tail disconnection events in comet P/Halley during the ICE-Halley radial period

J.W. Brosius [1,2], G.D. Holman [2], M.B. Niedner [2], J.C. Brandt [2,*], J.A. Slavin [3,], E.J. Smith [3], R.D. Zwickl [4], and S.J. Bame [4]**

[1] ST Systems Corporation, Lanham, MD 20706, USA
[2] Code 682, Laboratory for Astronomy and Solar Physics, NASA/Goddard Space Flight Center, Greenbelt, MD 20771, USA
[3] 169-506, Jet Propulsion Laboratory, 4800 Oak Grove Drive, Pasadena, CA 91109, USA
[4] Space Physics Group, Los Alamos National Laboratory, Los Alamos, NM 87545, USA

Received January 19, accepted March 19, 1987

Summary. We have examined ICE magnetometer and electron plasma data for possible causes of the plasma tail disconnection events (DE's) which were observed in Halley's comet on 1986 March 20–22 and April 11–12. We attribute the DE of March 20–22 to an interplanetary magnetic field polarity reversal, and the DE of April 11–12 to either a compression region in the solar wind, an interplanetary magnetic field polarity reversal, or a combination of the two. The measured average speeds of tail recession determined from pairs of successive photographs for both events vary between about 40 and 90 km s^{-1}, and the average acceleration determined from successive pairs of average velocity measurements is quite variable. The calculated times of disconnection for the two events are, in UT decimal dates, March 19.55 ± 0.16 and April 10.91$\pm$0.36. Assuming that the two disconnection events are due to frontside reconnection after an interplanetary magnetic field reversal (the most likely explanation for the March 20–22 event, and one possible explanation for the April 11–12 event), we estimate that the time period between the onset of reconnection and the final disconnection of the tail is $0.1 \leqq \tau_{rec} \leqq 0.6$ day. This suggests that the average speed at which reconnection proceeds through the cometary magnetic field pile-up region is $1 \leqq v_{rec} \leqq 6$ km s^{-1}, or several tenths the local Alfvén speed.

Key words: comets – interplanetary medium

1. Introduction

For nearly a century, cometary plasma tails have been known to exhibit a variety of phenomena, including disconnection events, helical structures, ray folding, side rays, and kinks. Easily the most impressive of these is the disconnection event, in which the plasma tail is uprooted from the head and recedes from the head approximately along the sun-comet line. For an observational review of these structures and their probable solar wind causes, the reader is referred to Brandt and Mendis (1979), Ershkovich (1980), Brandt (1982), Jockers (1985), and Niedner (1986).

The solar wind was postulated based on observations of the gross behavior of cometary tails (Biermann, 1951), and the very existence of the tails themselves was suggested to depend upon the solar wind plasma dragging "frozen in" solar magnetic fields into the interplanetary medium and getting "hung up" in the cometary ionosphere (Alfvén, 1957). Wind-sock theory was used to infer the solar wind speed (and changes therein) as manifested by cometary plasma tail dynamical aberration angles (Brandt, 1961, 1967; Brandt and Rothe, 1976; Mendis and Ip, 1977), but not until the era of interplanetary spacecraft were these velocity changes, as well as changes in other parameters of the interplanetary medium (e.g., density, temperature, magnetic field strength and polarity) fully appreciated. Indeed, not until the encounter of the International Cometary Explorer (ICE) with comet Giacobini-Zinner on 1985 September 11 was Alfvén's (1957) draped magnetotail theory confirmed (Smith et al., 1986; Slavin et al., 1986a, b). We will refer to the dynamic parameters of the interplanetary medium (velocity and density) by the term "solar wind", and to the magnetic field parameters of the interplanetary medium (field strength and polarity) by the term "IMF" (interplanetary magnetic field). It is to the observed changes in the solar wind/IMF that numerous authors ascribe the observed cometary plasma tail phenomena (Niedner et al., 1978; Ip and Mendis, 1978; Niedner and Brandt, 1978, 1979; Ip, 1980; Ershkovich, 1980; Niedner, 1980, 1982, 1986; Ip and Axford, 1982; Saito et al., 1986; Russell et al., 1986).

In this paper we treat two disconnection events which occurred during the ICE – Halley radial period, 1986 March–April. As the photographic data-base of the International Halley Watch (IHW) network increases, additional phenomena will also be trated in some detail.

Several theories have been put forth to explain the occurrence of disconnection events. Among them are (i) the brute force dynamic pressure theory of Ip (1980), in which an enhancement in solar wind dynamic pressure simply compresses the cometary ionosphere out of existence, thus freeing the captured field lines which presumably confine the tail plasma to the head (Alfvén, 1957), allowing the tail to escape; (ii) the flute instability theory of

Send offprint requests to: J.W. Brosius [2]

* Present address: Laboratory for Atmospheric and Space Physics, Box 392, University of Colorado at Boulder, Boulder, CO 80309, USA
** Present address: Code 696, Laboratory for Extraterrestrial Physics, NASA/Goddard Space Flight Center, Greenbelt, MD 20771, USA

Ip and Mendis (1978), in which an enhancement in solar wind dynamic pressure lowers the growth time of the flute instability (to which the cometary ionosphere is marginally stable under quiescent conditions), temporarily enhancing the ionization of neutrals and ultimately depleting the ionosphere at a later time; (iii) the backside (near-tail) reconnection theory of Ip (1985) and Russell et al. (1986), in which either interplanetary shocks or variations in the Alfvén Mach number trigger reconnection on the tail side of the nucleus, in analogy to reconnection in the earth's magnetotail; and (iv) the frontside reconnection theory of Niedner and Brandt (1978, 1979), in which field lines of one predominant polarity are compressed against field lines of opposite polarity after an interplanetary sector boundary crossing, forcing the field lines which are draped around the comet head to undergo reconnection, severing the field lines from the comet head at the point of reconnection. The first three theories require enhanced solar wind ram pressure (in the compression region of a "high speed" stream, for example) in order to operate, while the latter theory requires a reversal of the IMF polarity (e.g., traversal of an interplanetary sector boundary). Russell et al. (1986), using the analogy of reconnection in the earth's magnetotail, also suggest that backside reconnection may be triggered by sudden changes in the direction of the IMF.

Ultimately, the question becomes the following: Do disconnection events correlate more closely with "streams" or with magnetic field polarity reversals (sector boundary crossings)? What makes the question so vexing is that sector boundaries frequently closely precede the compression regions (high density, low speed) of streams by $\leqq 1$ day, which in turn precede the high speed (low density) regions by $\leqq 1$ day, leaving some uncertainty concerning what actually does the work in bringing about a DE. There is not a one-to-one correspondence between sectors and streams (Niedner and Brandt, 1979), however, a fact which may be helpful in distinguishing which of the possible mechanisms actually cause DE's. In this paper, we make use of the photographs currently available through the IHW network, as well as in situ solar wind/IMF monitoring with the ICE spacecraft, in order to determine the mechanism responsible for two plasma tail disconnection events during 1986 March–April.

2. The data

Through the IHW Large-Scale Phenomena network, we currently have five good-quality photographs available for DE's of both 1986 March 20–22 and April 11–12 (see Figs. 1 and 2). The UT times and sources of these imagery are given in Table 1, and the relevant geometric parameters are given in Table 2.

Measurements of important quantities in the interplanetary medium – flow speed, density, magnetic field strength, and magnetic field polarity – were obtained with the ICE spacecraft. These are the primary data used to correlate solar wind/IMF features with DE's. These data are, to some extent, supplemented by Vega-1 magnetometer data (Schwingenschuh, 1986) and interplanetary solar neutral sheet maps (Hoeksema, 1986).

We seek to correlate the DE's of March 20 and April 11 with "high-speed" streams and/or IMF polarity reversals (sector boundary crossings). A sector boundary is defined as the boundary between a region in which the field lines are directed primarily toward the sun (in the ecliptic plane) and a region in which the field lines are directed primarily away from the sun (in the ecliptic plane). The angle of the field projected onto the ecliptic plane, or longitude angle φ_{xy}, is the traditional probe of such crossings. To first order any frontside reconnection process should not be sensitive to the sun-comet x-component of the field, however, and in this paper we use the two-dimensional "clock angle" φ_{yz} ($\phi_{yz} = \arctan(B_z/B_y)$) which includes the component of the magnetic field perpendicular to the ecliptic plane (z), and the component perpendicular to the sun-comet line in the plane of the ecliptic (y). A polarity reversal is encountered when the clock angle undergoes a 180° change, and such changes usually correspond to the traditional sector boundary crossing, in which ϕ_{xy} changes by $\sim 180°$. We look for correlations between DE's and 180° changes in ϕ_{yz}. Typically, the dynamic pressure of the solar wind is greatest in the compression region of a stream, where the density is elevated above average values; it is to such compression regions that we also look for correlations with disconnection events.

3. Data reduction and analysis

3.1. *Derivation of times of disconnection*

Because in most instances we lacked the original glass plates or good contact film copies of the originals, the measurements described below were made on enlargement photographic prints. For each photograph, a pair of SAO reference stars was selected, and the position of the detached end of the tail was measured with respect to the reference line connecting the stars by means of a finely-scaled ruler and a protractor. These measurements and the measured length of the reference line were entered into a computer program which calculates the coordinates (RA and Dec) of the severed tail, the coordinates of the nucleus at the time of observation, and the linear distance (km) of the detached tail from the nucleus on the assumption that the tail is oriented along the prolonged radius vector. The astrometry performed on the tail is considered accurate to $\leqq 20''$.

The average velocity v_{av} of the front edge of the disconnected tail during a given time interval is obtained by dividing the distance traveled by the change in mid-times of exposure between the corresponding photographs. Under the assumption of uniformly accelerated motion, the average velocity was taken to occur at the center of the time interval over which the average velocity was calculated. The average acceleration a_{av} was obtained by dividing the average velocity change by the corresponding time interval. In Table 3 we list results for the two disconnection events studied in this paper. The decimal date is given in Universal Time (UT), the distance of the front edge of the disconnected tail from the nucleus (x) is given in km, v_{av} is in km s^{-1}, and a_{av} is in cm s^{-2}.

From the equations of uniformly accelerated rectilinear motion, we obtain the time of disconnection t_D as

$$t_D = t_i - \left(\frac{v_i - v_0}{a}\right), \tag{1}$$

where v_i, the instantaneous speed inferred at time t_i under the assumption of uniform acceleration, is given by

$$v_i = v_{av(i,\, i+1)} - a\left(\frac{t_{i+1} - t_i}{2}\right), \tag{2}$$

and v_0, the "initial speed" when the tail first disconnects from the head, is given by

$$v_0 = (v_i^2 - 2ax_i)^{1/2}, \tag{3}$$

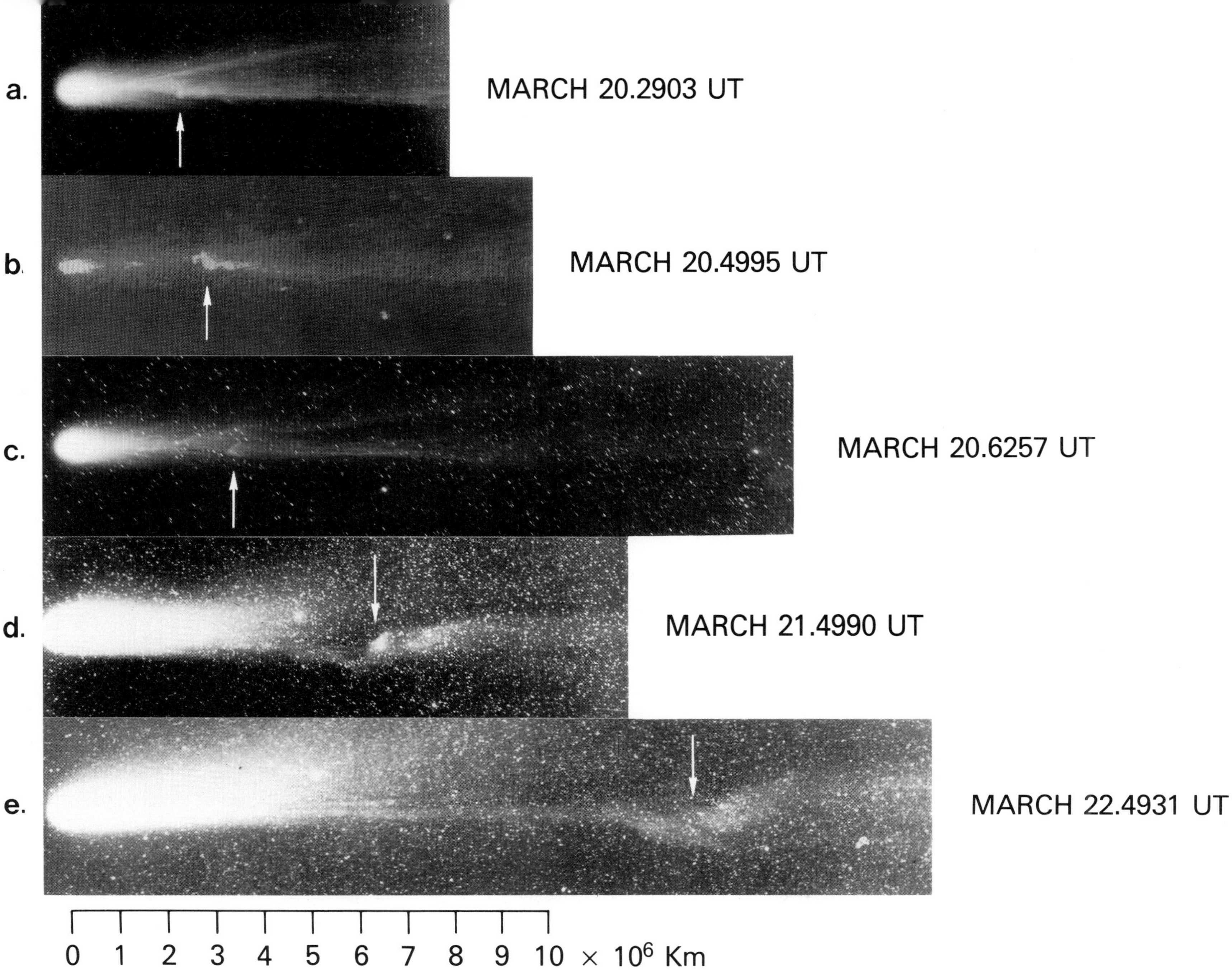

Fig. 1a–e. 1986 March 20–22 photographic sequence of Halley's comet showing the development of the first disconnection event (DE) studied in this paper. The images were obtained: **a** March 20.2903 UT at Cerro Tololo Interamerican Observatory (University of Michigan Curtis Schmidt), **b** March 20.4995 UT at Table Mountain Observatory (University College London detector), **c** March 20.6257 UT at Mauna Kea Observatory, **d** March 21.4990 UT at Easter Island (part of International Halley Watch's "Island Network"), and **e** March 22.4931 UT at the Joint Observatory for Cometary Research. For additional observational details concerning the images, see Table 1. The arrows locate the near end of the rejected tail, whose linear separations from the nucleus are listed in Table 3. All photographs are printed at the same angular scale; the linear scale at the bottom of the figure yields distances along the prolonged radius vector at the mid-time of the sequence

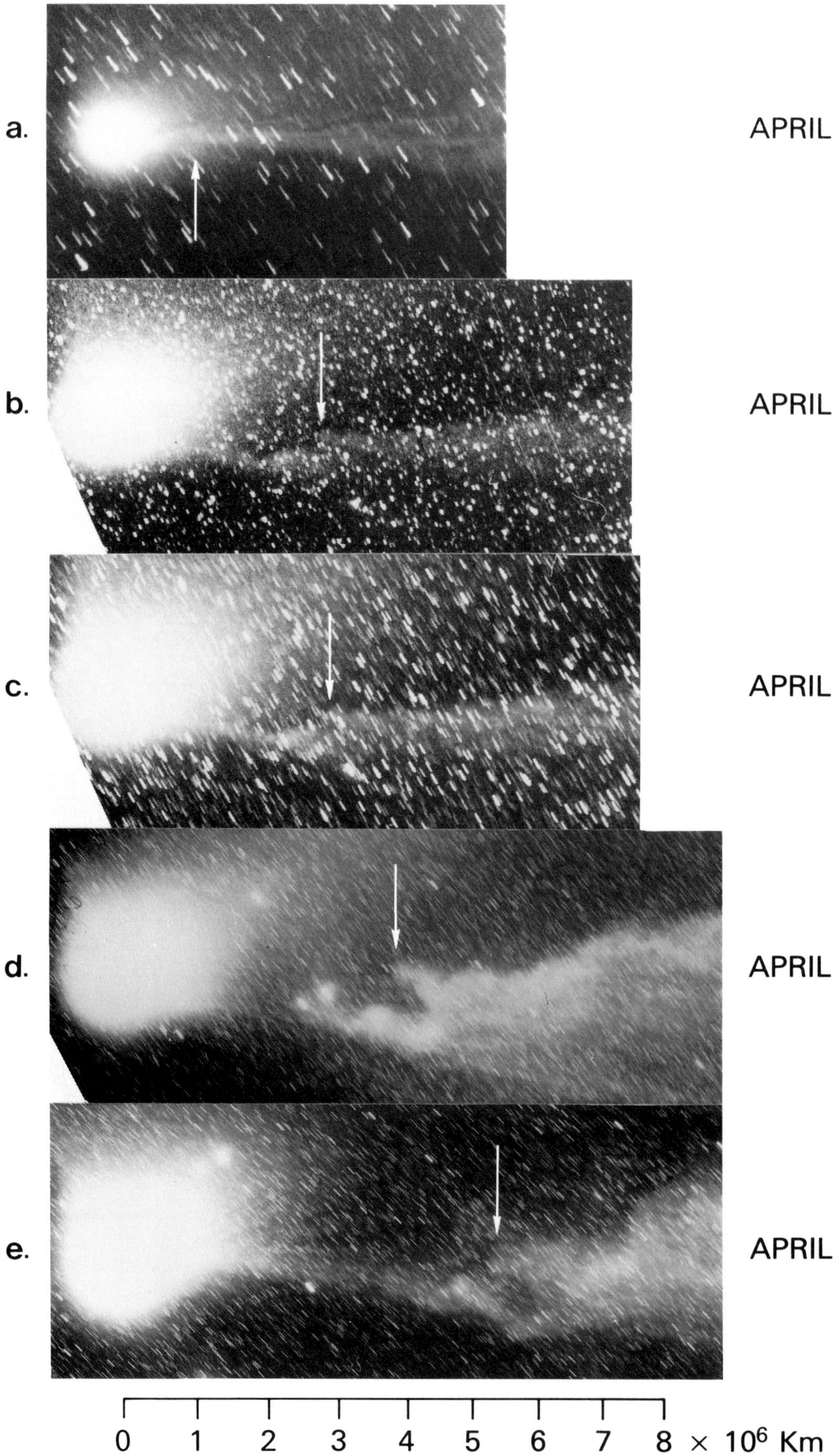
a.
APRIL 11.3306 UT
b.
APRIL 11.8712 UT
c.
APRIL 11.8875 UT
d.
APRIL 12.1042 UT
e.
APRIL 12.3090 UT
0 1 2 3 4 5 6 7 8 × 10^6 Km

Table 1. Photographic observations used in the analysis of the disconnection events of march 20–22 and april 11–12

Obs. no. (this paper)	Mid-exposure (UT)	Source	Plate no.	Exp. length (min)	Emulsion	Filter
March 20–22 event						
1.	March 20.2903	Univ. Mich. Curtis Schmidt (CTIO)	29953	20	IIa-O	–
2.	March 20.4995	Univ. Col. London (Table Mtn Obs.)	–	20	Imaging photon det.	CO^+
3.	March 20.6257	Mauna Kea Observatory	–	30	IIIa-J	–
4.	March 21.4990	IHW "Island Net" – Easter Island	S-134	16	2415TP	–
5.	March 22.4931	Joint Obs. for Cometary Res. (JOCR)	S-2049	10	IIa-O	–
April 11–12 event						
1.	April 11.3306	IHW "Island Net" – Tahiti	–	30	2415TP	W47B
2.	April 11.8712	IHW "Island Net" – Reunion I.	–	9	2415TP	–
3.	April 11.8875	IHW "Island Net" – Reunion I.	–	20	2415TP	–
4.	April 12.1042	Univ. Mich. Curtis Schmidt (CTIO)	30065	20	103a-O	–
5.	April 12.3090	Univ. Mich. Curtis Schmidt (CTIO)	30072	20	103a-O	–

Table 2. Geometric parameters used in the analysis of the disconnection events of march 20–22 and april 11–12

Obs. no.	Mid-exposure	Δ [a]	β [b]	R_H [c]	R_I [d]	θ_{HE} [e]	θ_{IH} [f]
March 20–22 event							
1.	March 20.2903	0.8032	113.9	0.9972	0.9303	47.93	−15.59
2.	March 20.4995	0.7979	113.9	1.0004	0.9304	47.51	−15.14
3.	March 20.6257	0.7948	113.9	1.0023	0.9305	47.26	−14.87
4.	March 21.4990	0.7727	113.9	1.0157	0.9308	45.55	−13.05
5.	March 22.4931	0.7477	114.1	1.0309	0.9312	43.63	−11.00
April 11–12 event							
1.	April 11.3306	0.4174	149.1	1.3371	0.9426	9.56	25.51
2.	April 11.8712	0.4184	150.6	1.3454	0.9430	8.72	26.42
3.	April 11.8875	0.4185	150.6	1.3456	0.9430	8.70	26.44
4.	April 12.1042	0.4191	151.2	1.3489	0.9432	8.36	26.80
5.	April 12.3090	0.4198	151.7	1.3521	0.9433	8.04	27.14

[a] Δ = geocentric distance of Halley in AU
[b] β = tail foreshortening angle in degrees
[c] R_H = heliocentric distance of Halley in AU
[d] R_I = heliocentric distance of ICE in AU
[e] θ_{HE} = Halley-Earth longitude difference in degrees (positive indicates Halley leads)
[f] θ_{IH} = ICE-Halley longitude difference in degrees (positive indicates ICE leads)

and where $i = 1, 2, 3, 4, 5$ correspond to the five observations of x, t for each of the two events. The velocities and accelerations in Tables 3 and 4 are well within the range of values quoted by previous authors (Mendis and Ip, 1977; Niedner, 1981; Guerin et al., 1986). The observed variability in acceleration may be a manifestation of the expectation that different forces (e.g., Lorentz force and solar wind dynamic pressure force) dominate the disconnected tail motions at different distances from the nucleus, as well as the expectation that these individual forces vary with position and/or time. Some combinations of data points yield an imaginary initial speed, probably due to the non-uniform acceleration. These imaginary values are listed along with the real values in Table 4.

Combining the results for the various trios of x, t listed in Table 4 for each of the two disconnection events, we obtain t_D = March 19.55 ± 0.16 for the event of March 20, and t_D = April

Fig. 2a–e. 1986 April 11–12 photographic sequence showing the development of the second DE studied. The images were obtained **a** April 11.3306 UT at Tahiti (part of the IHW's "Island Network"), **b** April 11.8712 UT and **c** April 11.8875 UT at Reunion Island (part of the IHW's "Island Network"), **d** April 12.1042 UT and **e** April 12.3090 UT at CTIO/University of Michigan Curtis Schmidt. For additional details concerning the photographs, refer to Table 1. The arrows locate the near end of the rejected tail, whose linear separations from the nucleus are listed in Table 3. All photographs are printed at the same angular scale; the linear scale at the bottom of the figure yields distances along the prolonged radius vector at the mid-time of the sequence

Table 3. Distance, average velocity, and average acceleration of the front edge of the disconnected tail for the disconnection events of March 20 and April 11

Obs. no.	t = date (UT)	x (km)	v_{av} (km s^{-1})	a_{av} (cm s^{-2})
1	3/20.2903	2.43 10^6		
			40.4	
2	3/20.4995	3.16 10^6		− 23.0
			37.0	
3	3/20.6257	3.56 10^6		3.1
			38.4	
4	3/21.4990	6.46 10^6		50.9
			79.5	
5	3/22.4931	1.33 10^7		
1	4/11.3306	1.08 10^6		
			42.7	
2	4/11.8712	3.07 10^6		46.9
			54.0	
3	4/11.8875	3.15 10^6		− 18.8
			52.1	
4	4/12.1042	4.12 10^6		215.6
			91.3	
5	4/12.3090	5.74 10^6		

Table 4. The measured average acceleration, and the calculated initial speed and time of disconnection for various trios of data points for the disconnection events of March 20 and April 11. The symbol "j" denotes imaginary initial speeds

Obs. trio	a_{av} (cm s^{-2})	t_D (date)	v_0 (km s^{-1})
March 20–22 event			
1, 2, 3	− 23.0	19.71	54.0
2, 3, 4	3.1	19.47	34.1
3, 4, 5	50.9	–	57.1j
1, 2, 4	− 4.1	19.62	43.2
1, 3, 4	− 1.4	19.58	40.2
1, 2, 5	19.3	19.39	23.6
April 11–12 event			
1, 2, 3	46.9	10.55	2.1j (∼0)
2, 3, 4	− 18.8	11.27	63.9
3, 4, 5	215.6	–	112.1j
1, 2, 4	28.5	10.93	26.2
1, 3, 4	27.1	10.94	27.3
1, 2, 5	65.8	–	26.0j

10.91 ± 0.36 for the event of April 11 (all dates are given as UT decimal dates). These ranges in disconnection time reflect the observed variability in acceleration, and the corresponding error associated with the assumption of uniform acceleration. A better knowledge of disconnected plasma tail motions for distances less than 10^6 km from the nucleus are needed in order to reduce the uncertainties. Recently, Jockers et al. (1986) have obtained a sequence of photographs of the event of April 11 when the front edge of the disconnected tail was less than 10^6 km from the nucleus; our disconnection time of April 10.91 agrees with their observations. Concerning the March event, a visible/UV outburst in comet Halley, beginning March 18.8, has been reported by Feldman et al. (1986). Such outbursts may be attributed to sector boundary crossings and the ensuing reconnection process (Niedner, 1980). It is therefore possible that the outburst reported by Feldman et al. (1986) marks the arrival of an IMF polarity reversal and the onset of reconnection. We will pursue this further in the Discussion section.

3.2. The solar wind and IMF measured by ICE

The corotation time from Halley to ICE is given, to first order, by

$$\Delta t = \frac{\theta_2 - \theta_1}{\Omega} + \frac{R_2 - R_1}{V_{sw}}, \tag{4}$$

where $\theta_2 - \theta_1$ is the angular separation between ICE and Halley in the plane of the solar equator (where angles are taken to increase in the direction of solar rotation), Ω is the sideral equatorial angular rotation velocity of the sun ($14^\circ\!.4\,\mathrm{day}^{-1}$), R_2 and R_1 are the heliocentric distances of ICE and Halley, and V_{sw} is the solar wind speed. High-order corrections take into account ICE's motion during the first order corotation interval and are incorporated by an iterative procedure in a computer code. The average solar wind speed at 1 AU is 400 km s^{-1}, but since we do not know a priori what the solar wind speed is at Halley at the time of the DE, we calculate a range in Δt based on the observation that the solar wind speed typically ranges between 300 and 700 km s^{-1}. The geometrical parameters used in the corotation procedure are listed in Table 2.

For the DE of March 20, we obtain corotation times of −1.63 days, −1.54 days, and −1.43 days for solar wind speeds of 300, 400, and 700 km s^{-1}, respectively; for the event of April 11, we obtain corotation times of −0.58 day, 0.04 day, and 0.82 days for the same solar wind speeds. A negative sign means that the feature arrived at ICE before it arrived at Halley. Combining these with the times of disconnection obtained earlier, we obtain the times of disconnection mapped to ICE, which to first order indicates at what times the responsible solar wind feature(s) would have swept by the spacecraft. The corotation time for the March 20 event is relatively insensitive to the solar wind speed because the radial separation between ICE and Halley is small (∼0.04 AU). The corotation time for the April 11 event has a fairly strong dependence on solar wind speed due to a significant radial separation (∼0.37 AU); however, an examination of ICE electron plasma data during the time interval obtained by combining the disconnection and corotation times shows that 300 km s^{-1} $< V_{sw} <$ 400 km s^{-1} (see Fig. 4), suggesting that the more appropriate corotation time range is −0.58 day to 0.04 day. The "disconnection windows" at ICE for the two events are therefore

$$t_{ICE}\,(\text{March 20}) = 18.02 \pm 0.26,$$

and

$$t_{ICE}\,(\text{April 11}) = 10.64 \pm 0.67.$$

These windows are marked as solid lines on the ICE electron plasma and magnetometer data shown in Figs. 3 and 4. We now consider additional possible sources of uncertainty before examining the ICE data during these time intervals.

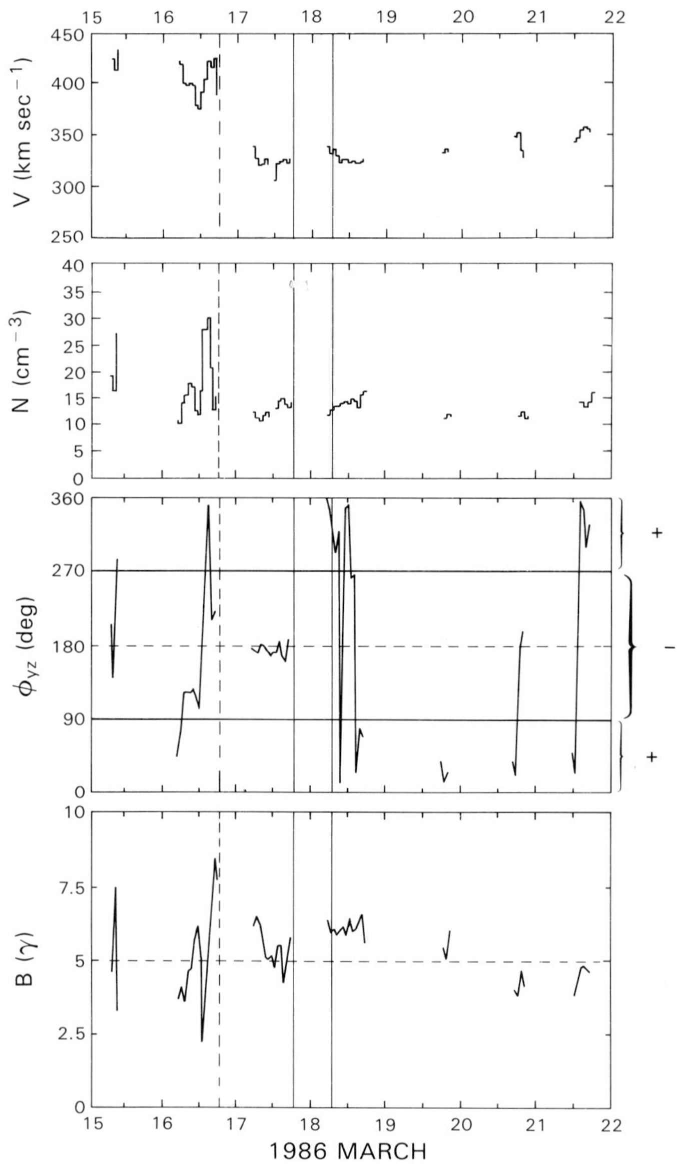

Fig. 3. The solar wind flow speed (a) in km s^{-1} and the solar wind electron number density (b) in cm^{-3} from the ICE electron plasma experiment around the time of the March 20 DE. The interplanetary magnetic field clock angle ϕ_{yz} (c) and the magnetic field strength (d) in units of 10^{-5} Gauss ($=1\gamma$) from the ICE magnetometer around the time of the March 20 DE. The solid lines indicate the arrival window as calculated in the text, and the dashed line allows for an additional 1 day response time. The IMF is directed predominantly toward the sun for $90° \leqq \phi_{yz} \leqq 270°$

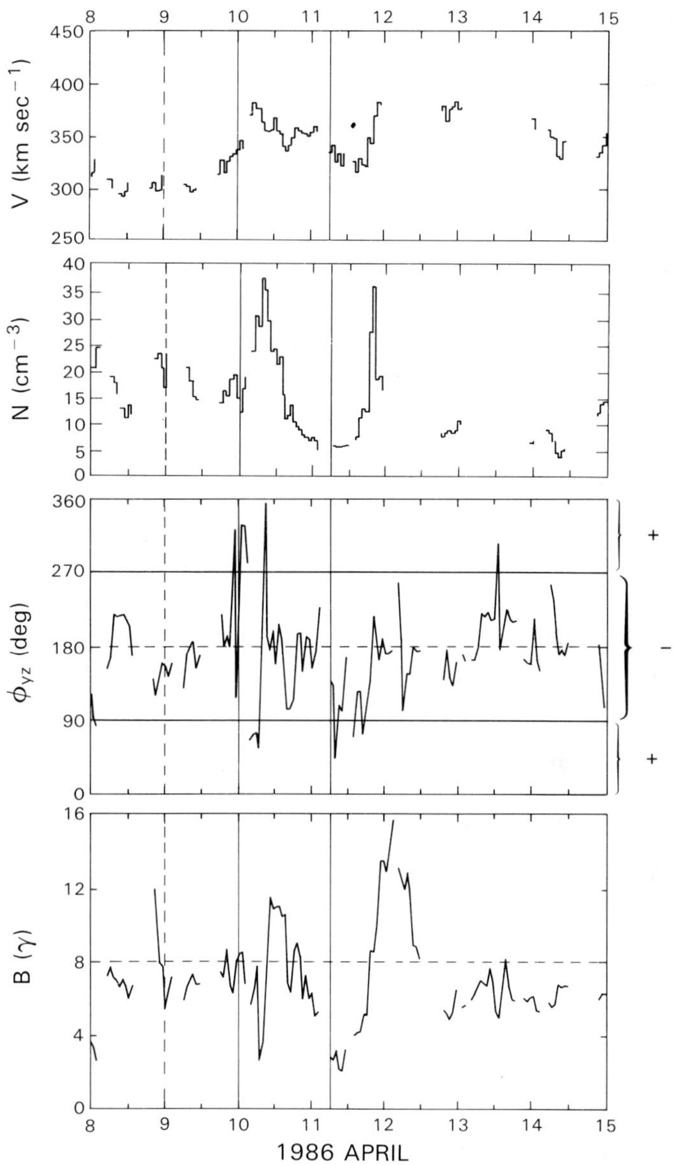

Fig. 4. The solar wind flow speed (a) in km s^{-1} and the solar wind electron number density (b) in cm^{-3} from the ICE electron plasma experiment around the time of the April 11 DE. The interplanetary magnetic field clock angle ϕ_{yz} (c) and the magnetic field strength (d) in units of 10^{-5} Gauss ($=1\gamma$) from the ICE magnetometer around the time of the April 11 DE. The solid lines indicate the arrival window as calculated in the text, and the dashed line allows for an additional 1 day response time

The response time between the arrival of a solar wind feature and the onset of an observable cometary phenomenon is not a well known parameter. Niedner (1980, 1982) and Niedner and Brandt (1978) estimate a time period $\tau_{\rm rec} \sim 0.5$–1.0 day from the onset of frontside reconnection to a final disconnected tail. Ip and Mendis (1978) estimate the growth time for the flute instability to be $\tau_f \sim 1$ h, based on pre-Halley Encounter estimates of the relevant parameters. We compensate for the uncertain response time for any agent causing a disconnection event by expanding the front edge of the ICE arrival windows listed above by 1.0 day, obtaining March 16.76–18.28 and April 8.97–11.31. It is these windows which we examine (see Figs. 3 and 4) for solar wind features responsible for the observed disconnection events. The inclusion of a response time in the ICE arrival window does not affect the arguments concerning the observed solar wind speed in the arrival window which were used to narrow the arrival window for the April 11 event (i.e., no solar wind speeds outside the range $300 < V_{\rm sw} < 400$ km s^{-1} were found between April 8.97 and 9.97).

An additional possible source of uncertainty which we mention, but at present do not incorporate in a detailed way, is the heliographic latitude difference between ICE and Halley for the March 20 (−1°.6) and April 11 (−11°.5) events. It is possible that the solar wind and IMF properties at ICE are different from those at Halley, especially for the April 11 event. For example, Rhodes and Smith (1975, 1976) found a dependence of solar wind speed on heliographic latitude $\sim 13\ \mathrm{km\,s^{-1}\,deg^{-1}}$, while Schwenn et al. (1978) found that a high speed stream which was observed at one satellite was not observed at another, separated by only 10° in latitude from the first. There is no way to be certain that the solar wind speed, density, field strength, or polarity measured near the ecliptic plane (with ICE) are representative of those same quantities at Halley. Independent analysis of the orientation and latitude extent of the solar neutral sheet presumed to be responsible for the interplanetary sector structure (Rosenberg and Coleman, 1969; Smith et al., 1978) may be done by extrapolating the observed solar photospheric magnetic field out to a "source surface" via multipole expansion (Svalgaard and Wilcox, 1978). The extension of the resulting coronal neutral line into the interplanetary medium beyond the source surface is assumed to be radial. Such calculations have been provided by Hoeksema (1986), and the resulting neutral sheet maps agree reasonably well with spacecraft measurements in that the extrapolated sector crossings in the ecliptic plane usually occur where ICE and Pioneer Venus Orbiter (PVO) spacecraft measure them to occur (Niedner, 1986). There is, however, sometimes substantial disagreement (of several days) between the location of the crossing as determined by spacecraft and the location of the crossing as determined by neutral sheet maps (Niedner, 1986). We therefore feel that using the neutral sheet maps to incorporate further corrections to the corotation times between ICE and Halley due to neutral sheet tilt angles and ICE/Halley latitude differences would not be justified. Nevertheless, for the sake of completeness, we note that the neutral sheet tilt angles derived from the coronal neutral line maps yield corrections of $+0.15$ day for the March 20 DE and -0.61 day for the April 11 DE. These corrections do not change the arguments in what follows.

The disconnection windows at ICE are shown as solid lines in Figs. 3 and 4; the dashed lines indicate the early edges of the arrival windows at ICE if a response time of 1 day is assumed. We give below the salient features of the solar wind/IMF in the arrival windows, and discuss their possible relation to the observed DE's.

For the March 20 event, the arrival window extends from March 16.76 to 18.28. The flow speed remains between 300 and $430\ \mathrm{km\,s^{-1}}$ throughout the entire window. The magnetic field is of mixed polarity on the 16th; the predominant polarity prior to the 16th cannot be determined from the ICE data because there is essentially no data for the preceding ~4 days, but Vega-1 data currently available to the authors (Schwingenschuh, 1986) suggest a sector (of predominantly inward polarity) of ~2–3 days duration. A 12-h data gap extends from March 16.69 to March 17.19. After the gap, the field strength, polarity, and electron density remain fairly constant for 12 hours. After another 12-h data gap, the electron density and field strength maintain their pre-gap values, but the magnetic field polarity is rotated by 180°. Although the data on the following days are sparse, it appears that the new (predominantly outward) polarity persists for at least 6 days.

Clearly, there is one feature observed in the ICE arrival window for the March 20 DE to which the DE may be attributed: a magnetic field polarity reversal occurring during the data gap between March 17.67 and March 18.17. We conclude that this magnetic field polarity reversal which occurred well withinn the ICE arrival window, without assumption of response time, is the most likely cause of the event. Note, however, the compression region of several hours duration which occurred around the middle of the 16th (the density maintained a peak value of $\sim 30\ \mathrm{cm^{-3}}$ for ~1.7 h). It is possible that a similar short-lived compression region, with solar wind and IMF signatures hidden in the data gaps, may also be responsible for the DE. (The possibility that a short-lived compression region occurred in the same data gap within which the polarity reversal occurred will be settled after further acquisition of Vega-1 data).

For the April 11 event, the arrival window extends from April 8.97 to 11.31. The flow speed remains between 300 and $400\ \mathrm{km\,s^{-1}}$ throughout the entire window. The clock angle and field strength remain constant through most of the 9th. A magnetic field polarity reversal of ~0.5 day duration occurs between April 9.88 and April 10.42. Near the end of this polarity reversal the field strength and the electron plasma density undergo steep, short-term enhancements. The field strength and electron density then drop to their lowest values in about a 7-day interval, before increasing to their highest respective values in the same 7-day interval. The density remains at about $38\ \mathrm{cm^{-3}}$ between 1900 UT and 2100 UT on the 11th, outside of the arrival window. The field polarity is turbulent, but remains fairly steadily at 180° after the 0.5 day reversal on the 9th–10th.

There are two features in the ICE arrival window to which the DE of April 11 may be attributed: a polarity reversal, and a compression region which occurs essentially simultaneously with the polarity reversal. It is impossible to determine whether one or the other, or both in concert, brought about this DE. Indeed, with such a large heliographic latitude separation between ICE and Halley for this event (11°.5), we cannot rule out the possibility that the compression region outside of the arrival window, or even some solar wind/IMF feature not measured by ICE, brought about this DE.

4. Discussion

Our primary conclusions are that the DE of March 20–22 was most likely caused by an IMF polarity reversal, and the DE of April 11–12 could have been caused by a compression region, an IMF polarity reversal, or a combination of the two. We discuss some implications of these conclusions in this section.

For the March 20 event, the earliest and latest possible times at which the IMF polarity reversal could have occurred at ICE are March 17.67 and 18.17; these times correspond to the beginning and the end of the data gap within which the polarity reversed (see Fig. 3). Using the corotation time of $+1.53 \pm 0.10$ days from ICE to Halley, the earliest and latest possible times at which the polarity reversal could have arrived at Halley are March 19.10 and 19.80. The latest possible disconnection time, March 19.71, is 0.61 day after the earliest possible arrival of the polarity reversal. Since the polarity reversal must have arrived at Halley before the onset of disconnection (assuming that frontside reconnection is the appropriate mechanism), we find that $0.0\ \text{day} \leqq \tau_{rec} \leqq 0.61\ \text{day}$. This is not inconsistent with the results of Niedner (1980, 1982) and Niedner and Brandt (1978), but places a tight restriction on the reconnection timescale. A similar calculation for the April 11 event, again assuming frontside reconnection, yields $0.09\ \text{day} \leqq \tau_{rec} \leqq 1.43$ day. Since, however, the polarity reversal itself lasts only 0.54 day, and since the timescale required for reconnection to operate cannot exceed the lifetime of the IMF feature driving reconnection if a totally detached tail is to result, we obtain $0.09\ \text{day} \leqq \tau_{rec} \leqq 0.54$ day. This result is consistent with that for the March 20 DE.

Assuming that the reconnection time scale τ_{rec} corresponds to the travel time of the reconnection diffusion region across the magnetic field pile-up region (Neubauer et al., 1986) between the "collisionopause" (Réme et al., 1986) and the contact surface

(Balsiger et al., 1986; Reinhard, 1986), we can calculate the average speed v_{rec} at which reconnection occurs in the pile-up region. With an average distance of 5.5 10^4 km between the collisionopause and the nucleus along the sun-comet line (Rème et al., 1986), and an average distance of 4.7 10^3 km between the contact surface and the nucleus along the sun-comet line (Balsiger et al., 1986; Reinhard, 1986), we obtain $1.0 \leqq v_{rec} \leqq 6.4$ km s^{-1} for the reconnection time scales quoted above. For the magnetic field strength of $30\gamma \leqq B \leqq 70\gamma$ (Neubauer et al., 1986; Riedler et al., 1986) and the electron number density of 100 cm$^{-3} \leqq n_e \leqq 4000$ cm^{-3} (Grard et al., 1986) measured in the pile-up region (where 4000 cm^{-3} was measured just before instrument failure), we estimate the Alfvén speed ~ 5.7 km s$^{-1} \leqq V_A \leqq \sim 15$ km s^{-1}. It is interesting that v_{rec} is comparable to the Petschek reconnection rate, ~ 0.1 to $1.0\, V_A$ (cf. Sonnerup, 1985).

Diffusion region properties (dimensions and resistivity) have been studied by Niedner et al. (1981) based on $\tau_{rec} \sim 0.75$ days along with estimates of other cometary parameters. With actual measurements (not estimates) of cometary parameters from the Halley and Giacobini-Zinner Encounters, along with the reconnection time scales obtained in this paper, an improved study of the reconnection process in cometary atmospheres can now be made. This topic will be pursued in a later paper.

It is interesting that the outburst of March 18.8 reported by Feldman et al. (1986) occurred so close in time to the commencement of a spectacular disconnection event. Either the visibility of the DE was enhanced by the outburst (loading CO_2^+ ions into the tail), or the outburst itself was a result of the solar wind/IMF feature (presumably IMF polarity reversal in this case) giving rise to the DE. As discussed earlier in this section, the earliest and latest possible arrivals of the IMF polarity reversal at Halley were March 19.10 and March 19.80. Thus the polarity reversal arrived 0.3–1.0 day after the onset of the outburst. It thus seems unlikely that the outburst may be attributed to frontside reconnection during an IMF polarity reversal.

Acknowledgements. This work was supported in part by the ISEE/ICE Guest Investigator Program. The research performed at the Jet Propulsion Laboratory of the California Institute of Technology was carried out under contract to the National Aeronautics and Space Administration. The work done at Los Alamos National Laboratory was under the auspices of the United States Department of Energy with support from NASA under S-04039-D.

References

Alfvén, H.: 1957, *Tellus* **9**, 92

Balsiger, H., Altwegg, K., Buhler, F., Geiss, J., Ghielmetti, A.G., Goldstein, B.E., Goldstein, R., Huntress, W.T., Ip, W.-H., Lazarus, A.J., Meier, A., Neugebauer, M., Rettenmund, U., Rosenbauer, H., Schwenn, R., Sharp, R.D., Shelley, E.G., Ungstrup, E., Young, D.T.: 1986, *Nature* **321**, 330

Biermann, L.: 1951, *Z. Astrophys* **29**, 274

Brandt, J.C.: 1961, *Astrophys. J.* **133**, 1091

Brandt, J.C.: 1967, *Astrophys. J.* **147**, 201

Brandt, J.C.: 1982, in *Comets*, ed. L.L. Wilkening, University of Arizona Press, p. 519

Brandt, J.C., Mendis, D.A.: 1979, in *Solar System Plasma Physics*, eds. C.F. Kennel, L.J. Lanzerotti, E.N. Parker, North-Holland, Amsterdam, p. 255

Brandt, J.C., Rothe, E.D.: 1976, in *The Study of Comets*, ed. B. Donn et al., NASA SP-393, p. 878

Ershkovich, A.I.: 1980, *Space Sci. Rev.* **25**, 3

Feldman, P.D., A'Hearn, M.F., Festou, M.C., McFadden, L.A., Weaver, H.A., Woods, T.N.: 1986, *Nature* **324**, 433

Grard, R., Pedersen, A., Trotignon, J.-G., Beghin, C., Mogilevsky, M., Mikhailov, Y., Molchanov, O., Formisano, V.: 1986, *Nature* **321**, 290

Guerin, P., Koutchmy, S., Vial, J.C.: 1986, *Astron. Astrophys.* **167**, 395

Hoeksema, T.: 1986 (private communication)

Ip, W.-H.: 1980, *Astrophys. J.* **238**, 388

Ip, W.-H.: 1985, ESA SP-235, Proc. ESA Workshop on *Future Missions in Solar, Heliospheric, and Space Plasma Physics*, p. 65

Ip, W.-H., Axford, W.I.: 1982, in *Comets*, ed. L.L. Wilkening, University of Arizona Press, Tucson, p. 588

Ip, W.-H., Mendis, D.A.: 1978, *Astrophys. J.* **223**, 671

Jockers, K.: 1985, *Astron. Astrophys. Suppl.* **62**, 791

Jockers, K., Geyer, E.H., Rosenbauer, H., Hanel, A.: 1986, 20th ESLAB Symp., ESA SP-250, Vol. **I**, p. 51

Mendis, D.A., Ip, W.-H.: 1977, *Space Sci. Rev.* **20**, 145

Neubauer, F.M., Glassmeier, K.H., Pohl, M., Raeder, J., Acuna, M.H., Burlaga, L.F., Ness, N.F., Musmann, G., Mariani, F., Wallis, M.K., Ungstrup, E., Schmidt, H.U.: 1986, *Nature* **321**, 352

Niedner, M.B.: 1980, *Astrophys. J.* **241**, 820

Niedner, M.B.: 1981, *Astrophys. J. Suppl.* **46**, 141

Niedner, M.B.: 1982, *Astrophys. J. Suppl.* **48**, 1

Niedner, M.B.: 1986, *Adv. Space Res.* **6**, 315

Niedner, M.B., Brandt, J.C.: 1978, *Astrophys. J.* **223**, 655

Niedner, M.B., Brandt, J.C.: 1979, *Astrophys. J.* **234**, 723

Niedner, M.B., Ionson, J.A., Brandt, J.C.: 1981, *Astrophys. J.* **245**, 1159

Niedner, M.B., Rothe, E.D., Brandt, J.C.: 1978, *Astrophys. J.* **221**, 1014

Reinhard, R.: 1986, *Nature* **321**, 313

Reme, H., Sauvaud, J.A., d'Uston, C., Cotin, F., Cros, A., Anderson, K.A., Carlson, C.W., Curtis, D.W., Lin, R.P., Mendis, D.A., Korth, A., Richter, A.K.: 1986, *Nature* **321**, 349

Rhodes, E.J., Smith, E.J.: 1975, *J. Geophys. Res.* **80**, 917

Rhodes, E.J., Smith, E.J.: 1976, *J. Geophys. Res.* **81**, 2123

Riedler, W., Schwingenschuh, K., Yeroshenko, Y.G., Styashkin, V.A., Russell, C.T.: 1986, *Nature* **321**, 288

Rosenberg, R.L., Coleman, P.J.: 1969, *J. Geophys. Res.* **74**, 5611

Russell, C.T., Saunders, M.A., Phillips, J.L., Fedder, J.A.: 1986, *J. Geophys. Res.* **91**, 1417

Saito, T., Yumoto, K., Hirao, K., Saito, K., Nakagawa, T., Smith, E.J.: 1986, *Geophys. Res. Letters* **13**, 821

Schwenn, R., Montgomery, M.D., Rosenbauer, H., Miggenrieder, H., Muhlhauser, K.H., Bame, S.J., Feldman, W.C., Hansen, R.T.: 1978, *J. Geophys. Res.* **83**, 1011

Schwingenschuh, K.: 1986 (private communication)

Slavin, J.A., Goldberg, B.A., Smith, E.J., McComas, D.J., Bame, S.J., Strauss, M.A., Spinrad, H.: 1986a, *Geophys. Res. Letters* **13**, 1085

Slavin, J.A., Smith, E.J., Tsurutani, B.T., Siscoe, G.L., Jones, D.E., Mendis, D.A.: 1986b, *Geophys. Res. Letters* **13**, 283

Smith, E.J., Tsurutani, B.T., Rosenberg, R.L.: 1978, *J. Geophys. Res.* **83**, 717

Smith, E.J., Tsurutani, B.T., Slavin, J.A., Jones, D.E., Siscoe, G.L., Mendis, D.A.: 1986, *Science* **232**, 382

Sonnerup, B.U.O.: 1985, in *Unstable Current Systems and Plasma Instabilities in Astrophysics*, eds. M.R. Kundu, G.D. Holman, Reidel, Dordrecht, p. 5

Svalgaard, L., Wilcox, J.M.: 1978, *Ann. Rev. Astron. Astrophys.* **16**, 429

Astron. Astrophys. 187, 276–280 (1987)

ASTRONOMY AND ASTROPHYSICS

Energy spectra of energetic ions in the vicinity of comet P/Giacobini-Zinner

I.G. Richardson[1], S.W.H. Cowley[1], V. Moore[1], K. Staines[1], R.J. Hynds[1], T.R. Sanderson[2], K.-P. Wenzel[2], and P.W. Daly[3]

[1] Blackett Laboratory, Imperial College, London SW7 2BZ, United Kingdom
[2] Space Science Department of ESA, ESTEC, NL-2200 AG Noordwijk, The Netherlands
[3] Max-Planck-Institut für Aeronomie, Postfach 20, D-3411, Katlenburg-Lindau, Federal Republic of Germany

Received February 13, accepted April 21, 1987

Summary. The energy spectrum of water-group pick-up ions with energies >35 keV observed during the encounter of comet Giacobini-Zinner by the ICE spacecraft is examined using data from the EPAS experiment, combined with published data from the ULECA sensor. In the plasma frame the combined spectra consist mainly of a steeply-falling tail extending to energies of a few hundred keV, but there is also evidence for flattening at speeds somewhat below the local pick-up speed. The latter feature is most evident in the solar wind outside the mass-loaded region of slowed flow, where the local pick-up speed is sufficiently fast that it falls within the range of the combined EPAS and ULECA data. Above the local ion pick-up speed the ion distribution function is well approximated by a decreasing exponential of ion speed. The characteristic speed of this distribution increases (i.e. the spectrum hardens) as the comet is approached. It is demonstrated that EPAS data can give detailed information on this portion of the spectrum with 32 s time resolution.

Key words: comets – Giacobini-Zinner – energetic ion spectra

1. Introduction

During the encounter of the International Cometary Explorer (ICE) spacecraft with comet Giacobini-Zinner in September 1985, large fluxes of energetic water-group ions were detected over a region several million kilometers in extent surrounding the comet (Hynds et al., 1986; Ipavich et al., 1986). These particles originate from the extensive coma of neutral atoms and molecules formed by emission from the nucleus, which, upon being ionised by solar UV radiation or by charge-exchange with solar wind ions, are picked up and accelerated by the solar wind flow to energies of several tens of keV. More precisely, since the coma neutrals flow outwards from the nucleus at speeds of only $\sim 1\,\mathrm{km\,s^{-1}}$, a newly formed ion is created essentially at rest in the comet's (~spacecraft's) rest frame. Consequently, in the solar wind rest frame ions are created with a speed equal to the solar wind speed V_{sw} in the comet's rest frame. Subsequent scattering of the ions by waves and irregularities in the solar wind magnetic field will tend to isotropize their distribution in the rest frame of the flow, forming a spherical shell in velocity space centred on V_{sw} and having a radius equal to V_{sw}. The ions are also expected to be diffused in energy, forming a flattened distribution function at speeds below V_{sw}, and a well-developed high energy tail at speeds above (Ip and Axford, 1986, and references therein). Thus the pick-up process is expected to result in a characteristic form for the ion spectrum, with a "break" in slope at the local pick-up speed (V_{sw}) in the solar wind rest frame.

In this paper we examine energetic ion data from the Imperial College/Space Science Department of ESA/Space Research Laboratory, Utrecht EPAS experiment (Balogh et al., 1978), together with data from the ULECA sensor on the ICE spacecraft (Hovestadt et al., 1978) published by Gloeckler et al. (1986), in order to determine the pick-up ion spectrum in the vicinity of comet Giacobini-Zinner. Much of the data obtained by these two instruments relate to water-group ion speeds well above the local pick-up speed in the flow rest frame. In this regime it is shown that the ion distribution function is well approximated by a single decreasing exponential in ion speed. However, in regions of higher speed flow outside the bow shock, the measured range encompasses the local pick-up speed, and it is then shown that evidence exists for a flattening of the spectrum below this speed, as expected theoretically.

2. Observations

During the Giacobini-Zinner encounter the EPAS experiment measured the fluxes of energetic water-group ions in 24 different directions (3 independent telescopes binned into eight azimuthal sectors) in eight logarithmically-spaced energy channels every 32 s, thus providing detailed three-dimensional information on the ion velocity distribution. An overview of the direction-averaged ion fluxes from the lowest 5 energy channels for the 10 hours around closest approach ($\pm 4\,10^5$ km) is shown in Fig. 1 at the 32 s sampling rate. For water-group ions these channels (E1 to E5) correspond to the energy ranges 65–95, 95–140, 140–205, 205–310 and 310–480 keV (Hynds et al., 1986). The bow shock crossings centred at ~0927 and ~1220 UT are indicated (Tranquille et al., 1986; Smith et al., 1986), and ICE was within the cold plasma tail between 1050 and 1115 UT (Bame et al., 1986; Meyer-Vernet et al., 1986). The ion fluxes increase up to and beyond the bow shock and then decline within $\sim 10^5$ km of the comet, with a steep fall around the tail crossing.

Send offprint requests to: I.G. Richardson

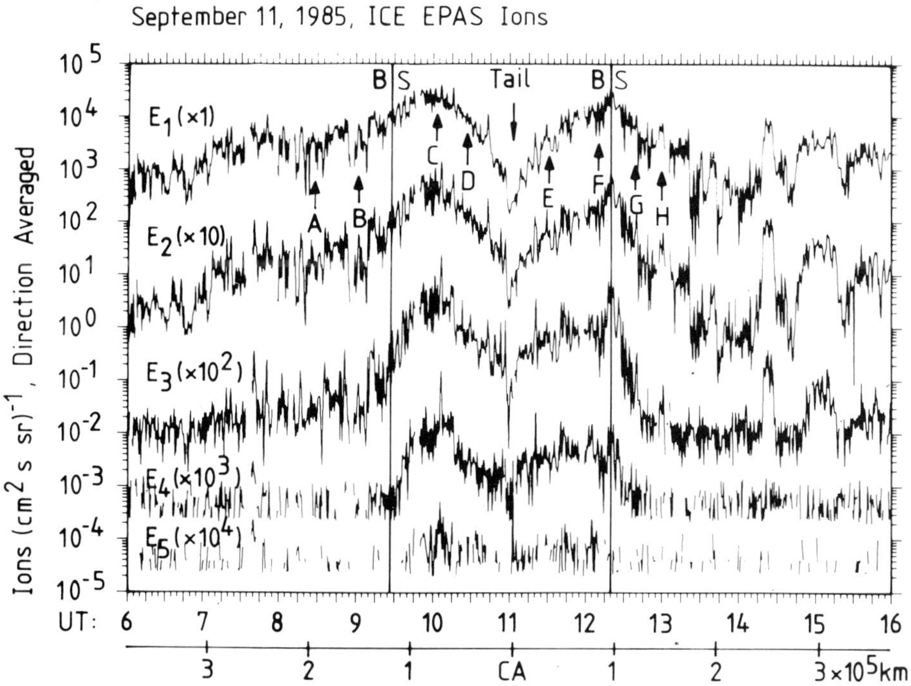

Fig. 1. The direction-averaged ion fluxes in EPAS energy channels 1 to 5 (E1 to E5) at 32 s resolution for a ten hour interval centred on closest approach to comet Giacobini-Zinner. The flux scale refers to E1, while fluxes in the other energy channels should be multiplied by the factors indicated. The vertical lines indicate the bow shock crossings (BS), while the letters and arrows indicate the positions of the distribution functions in the ion rest frame shown in Fig. 2

To investigate the ion spectrum, the directional fluxes are transformed into the ion bulk rest frame, the bulk velocity of the ions being determined from the ion data itself using the method of Gloeckler et al. (1984). It is assumed that the ions are isotropic in their rest frame, and a transformation speed is then sought such that the measurements in different directions and at different energies form a continuous smooth spectrum. Richardson et al. (1986) showed that in the mass-loaded region and its environs such a transformation could indeed be made, and that the ion bulk speed so determined assuming water-group ions is similar to that obtained from thermal electron observations (Bame et al., 1986). This result indicates that in this region the bulk of the ions have indeed been scattered into near isotropy in the solar wind frame. In Fig. 2 we show 8 examples of ion rest-frame spectra obtained from the 32 s EPAS data at the times indicated in Fig. 1. These spectra show that EPAS provides detailed information on the ion spectrum at the highest available time resolution; previously published examples from EPAS and ULECA have represented averages over 10 min or more (Richardson et al., 1986; Ipavich et al., 1986; Gloeckler et al., 1986). The straight lines are obtained from fitting a function of the form $f \sim \exp(-v/v_0)$ to each spectrum, and the characteristic speed v_0 is noted in each case. The ion bulk speed (V_c, $\sim V_{sw}$) is determined by finding the transformation velocity which gives the least scatter in a fit to a function of this form, as described by Richardson et al. (1986). It can be seen that this function gives a good fit to the data, and that there is no evidence of features in the spectrum over the range sampled. The exponential spectra simply become harder (v_0 increases) as the comet is approached, notably in the vicinity of the bow shock (Richardson et al., 1986; Tranquille et al., 1986). The values of the ion bulk speed determined from the data (given in the figure) also indicate the speed with which newly-created pick-up ions are produced in the ion rest frame as discussed above. In each case the lowest measured transformed ion speed lies well above the local pick-up speed, so that these EPAS data cannot be used to investigate the ion spectrum in its vicinity. The data are, however, consistent with the expectation that energy diffusion will result in the formation of an extensive high-energy tail to the ion distribution, extending to energies of a few hundred keV in the ion rest frame.

To investigate the spectrum close to the local ion pick-up speed measurements at lower energies are required, two sources of which are available. First, EPAS energy "channel zero" (E0) records the fluxes of water-group ions in the energy range 45 to 65 keV, in 24 directions, as for the other energy channels, but with a measurement every 256 s. This channel usually monitors instrument noise, but a significant ion flux well above the background level was seen during the comet encounter. Second, data from the ULECA sensor cover the energy range ~35 to 150 keV, and four rest-frame spectra from this instrument were presented by Gloeckler et al. (1986). In Fig. 3 we have combined these ULECA spectra with EPAS channel E0 to E5 data, averaged over the same time intervals and transformed using the same velocities as in Gloeckler et al. (1986). (Note that these velocities were determined from thermal electron observations and may not represent the best fit for EPAS data alone, thus accounting for some of the scatter at higher energies). The ULECA results have also been multiplied by a previously determined instrumental cross-calibration factor of 2.1 (Daly et al., 1985). As in Fig. 1, the solid lines show the best-fit exponentials determined from the EPAS data (E0 to E5), while the vertical arrows indicate the transformation speeds (V_c) employed in each case (the local pick-up speed in the ion rest frame).

Figure 3 illustrates two important points. First, in the region where they overlap, the ULECA and EPAS spectra are in good agreement. Second, and more importantly, it should be noted that spectra (a) and (d) were obtained in the solar wind upstream from the bow shock (inbound and outbound respectively) where the flow speed is relatively high, such that the local pick-up speed lies just within the lower range of measurements. Spectra (b) and (c) were obtained in the post-shock mass-loaded region of slowed flow on either side of closest approach, where the measured range

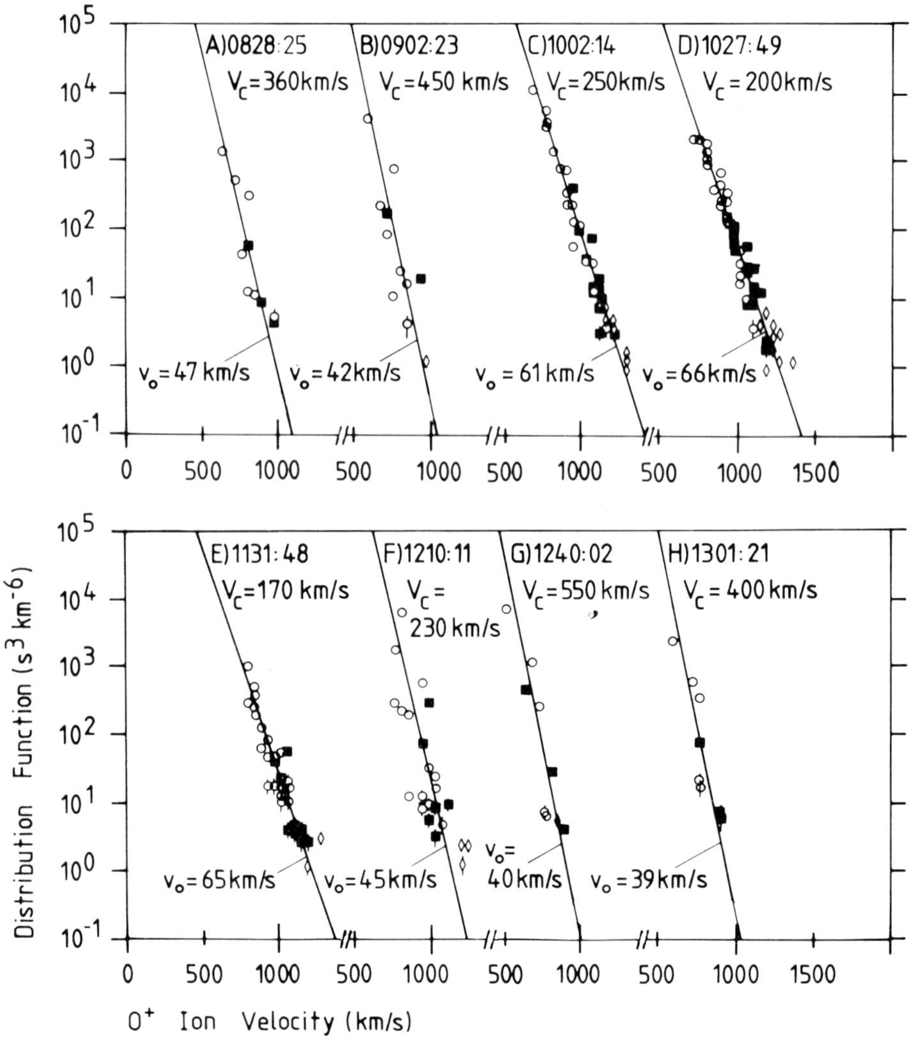

Fig. 2. Distribution functions in the ion rest frame at the positions indicated by arrows in Fig. 1. Each distribution function is obtained from a 32 s interval following the times indicated. Open circles, solid squares and open diamonds show points determined from EPAS channels E1 to E3 respectively (as in Fig. (3)), with statistical uncertainties less than 30%. The straight lines show the best-fit exponentials of the form $f \sim \exp(-v/v_0)$. The ion bulk velocity (V_c) is indicated for each interval

does not encompass the local pick-up speed (though only just so for spectrum (b)). It can be seen that spectra (a) and (d) contain evidence for flattening of the spectrum at speeds just below the local pick-up speed, in that the lowest energy point in each case falls below the fitted line, particularly in case (a), in agreement with the general conclusions of Gloeckler et al. (1986). In case (b) the lowest four energy points also show moderate depressions below the fitted line suggestive of a reduction in spectral slope near to the local pick-up speed. However, in case (c) where the local pick-up speed lies well below the lowest point there is little evidence for flattening over the measured range.

3. Summary and conclusions

In this paper we have examined the spectrum of water-group ions observed during the encounter of comet Giacobini-Zinner by the ICE spacecraft, using data from the EPAS and ULECA experiments (the latter published by Gloeckler et al. (1986)). It has been shown that in the energy range where they overlap the EPAS and ULECA data are in good agreement. The combined spectra consist of a steeply falling tail extending to energies of a few hundred keV, together with a somewhat flattened portion at speeds somewhat below the local pick-up speed. These results are in conformity with general theoretical expectations (Ip and Axford, 1986). However, the low energy flattening of the ion spectrum is only clearly evident in the solar wind region upstream from the bow shock where the relatively high flow speeds result in the local ion pick-up speed being within the range of the ULECA data. In the mass-loaded region the local pick-up speed lies below the ULECA/EPAS energy range and only modest evidence is then found for features in the spectrum over the measured interval. Above the local pick-up speed the ion distribution function is well approximated by a single decreasing exponential in ion speed, whose characteristic speed v_0 increases (the spectrum hardens) with decreasing distance to the comet. The EPAS data have been shown to provide information on this portion of the ion spectrum with 32 s time resolution and this will enable EPAS data to be combined with data from the other ICE instruments to provide a detailed examination of the plasma and energetic particle properties during the Giacobini-Zinner encounter.

Acknowledgements. The EPAS is a joint project of the Blackett Laboratory, Imperial College, the Space Science Department of ESA, ESTEC and the Space Research Laboratory, Utrecht. IGR

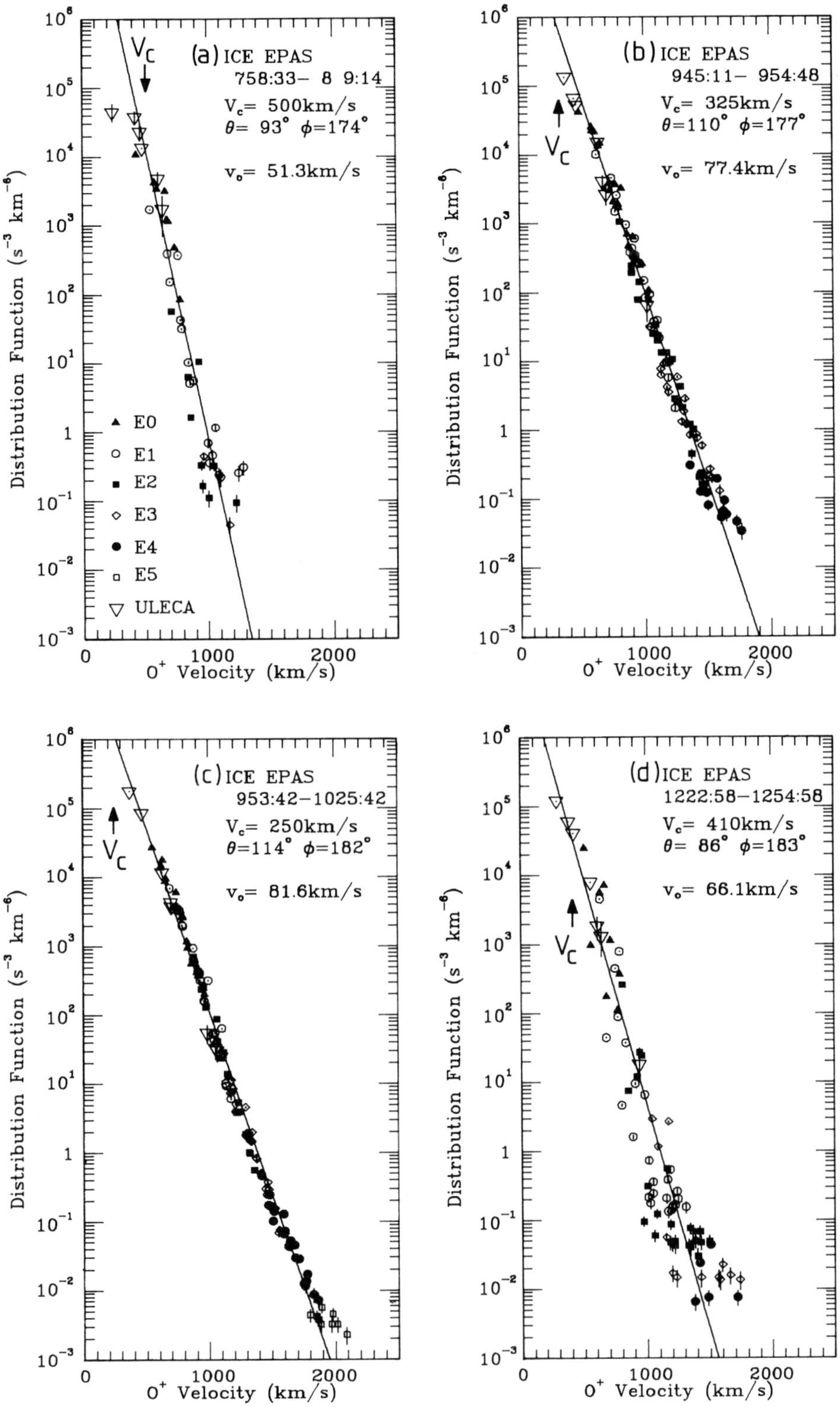

Fig. 3. Combined ULECA (triangles) and EPAS (channels as indicated in the key) bulk rest frame ion distribution functions, in a similar format to Fig. (2). The ion bulk speed and direction used to transform the EPAS data are indicated. The bulk speeds are those given by Gloeckler et al. (1986) while the directions (not given by Gloecker et al.) were determined from a spherical harmonic fit to the EPAS E1 data. The straight lines show the best fit exponential to the EPAS E0 to E5 data, and the characteristic speed v_0 is also indicated. The vertical arrows indicate the local pick-up ion speed in the ion bulk frame, equal to the transformation speed V_c

acknowledges a SERC Research Assistantship and KS a SERC Research Studentship.

References

Balogh, A., G. van Dijen, J. van Genechten, J. Henrion, R. Hynds, G. Korfmann, T. Iversen, J. van Rooijen, T. Sanderson, G. Stevens, K.-P. Wenzel: 1978, *IEEE Trans. Geosci Electron* **GE-16**, 176

Bame, S.J., R.C. Anderson, J.R. Asbridge, D.N. Baker, W.C. Feldman, S.A. Fuselier, J.T. Gosling, D.J. McComas, M.F. Thomsen, D.T. Young, R.D. Zwickl: 1986, *Science* **232**, 356

Daly, P.W., T.R. Sanderson, K.-P. Wenzel: 1985, *J. Geophys. Res.* **90**, 1499

Gloeckler, G., M. Scholer, F.M. Ipavich, D. Hovestadt, B. Klecker, and A.B. Galvin: 1984, *Geophys. Res. Letters* **11**, 603

Gloeckler, G., D. Hovestadt, F.M. Ipavich, M. Scholer, B. Klecker, and A.B. Galvin: 1986, *Geophys. Res. Letters* **13**, 251

Hovestadt, D., G. Gloeckler, C.Y. Fan, L.A. Fisk, F.M. Ipavich, B. Klecker, J.J. O'Gallagher, M. Scholer, H. Arbinger, J. Cain, H. Hofner, E. Kunneth, P. Laeverenz, E. Tums: 1978, *IEEE Trans. Geosci Electron* **GE-16**, 166

Hynds, R.J., S.W.H. Cowley, T.R. Sanderson, K.-P. Wenzel, J.J. van Rooijen: 1986, *Science* **232**, 361

Ip, W.-H., W.I. Axford, 1986, *Planet. Space Sci.* **34**, 1061

Ipavich, F.M., A.B. Galvin, G. Gloeckler, D. Hovestadt, B. Klecker, M. Scholer: 1986, *Science* **232**, 366

Meyer-Vernet, P. Couturier, S. Hoang, C. Perche, J.L. Steinberg, J. Fainberg, C. Meetre: 1986, *Science* **232**, 370

Richardson, I.G., S.W.H. Cowley, R.J. Hynds, T.R. Sanderson, K.-P.Wenzel, P.W. Daly: 1986, *Geophys. Res. Letters* **13**, 415

Smith, E.J., J.A. Slavin, S.J. Bame, M.F. Thomsen, S.W.H. Cowley, I.G. Richardson, D. Hovestadt, F.M. Ipavich, K.W. Ogilvie, M.A. Coplan, T.R. Sanderson, K.-P. Wenzel, F.L. Scarf, A.F. Vinar, J.D. Scudder: 1986, 20th ESLAB Symp, ESA SP-250, Vol. III, p. 461

Tranquille, C., I.G. Richardson, S.W.H. Cowley, T.R. Sanderson, K.-P. Wenzel, R.J. Hynds: 1986, *Geophys. Res. Letters* **13**, 853

Astron. Astrophys. 187, 281–286 (1987)

ASTRONOMY AND ASTROPHYSICS

Plasma structures in comets P/Halley and Giacobini-Zinner

J.C. Brandt [1] **and M.B. Niedner, Jr.** [2]

[1] Laboratory for Atmospheric and Space Sciences, University of Colorado, Boulder, CO 80309, USA
[2] Laboratory for Astronomy and Solar Physics, NASA-Goddard Space Flight Center, Greenbelt, MD 20771, USA

Received February 16, accepted April 27, 1987

Summary. An overview of large-scale plasma phenomena is presented based on results of spacecraft probing of comets Halley and Giacobini-Zinner and on worldwide submissions to the Large-Scale Phenomena Discipline Specialist Team of the International Halley Watch. Examples of tail phenomena and science are presented with emphasis on observed disconnection events. The archive of this material will clearly be very valuable for studying the comet/solar-wind interaction during the 1985–1986 apparition of Halley's comet.

Key words: plasma tails – disconnection events – solar wind interaction

1. Introduction

Our pre-encounter view of plasma structures in comets was based on the ideas of Biermann (1951) as extended by Alfvén (1957) to include the effects of the solar-wind magnetic field. The solar wind interacts with bright comets through its magnetic field and the cometary ions which are produced in an extended region around the comet. The field lines away from the comet are unimpeded and thus wrap around the comet to create a magnetic field in a hairpin configuration, i.e., a configuration with lobes of opposite polarity separated by a current sheet. Some part of this total magnetic structure includes the visible plasma tail. Because the comet is an obstacle in a supersonic and super-Alfvénic solar wind, a bow shock was also expected. Despite the sketchiness of many of the details, this picture presented specific tests of the data returned by the flyby spacecraft sent to comets Giacobini-Zinner and Halley.

2. In situ results

Direct exploration of comets began with the International Cometary Explorer (ICE) intercept of comet Giacobini-Zinner on 1985, September 11. The intercept was 7800 km from the nucleus on the tailward side.

Draping of field lines was observed, confirming Alfvén's (1957) basic picture, and the current sheet in the center of the tail was clearly detected. Extensive plasma wave activity was recorded, and the plasma wave experiment also served as a dust impact detector. The composition of the plasma was dominated by water-group ions. As the spacecraft traversed the comet, the flow speed of the plasma decreased from solar-wind values to low speeds on the order of a few tens of km s^{-1} while the density

Fig. 1. Several images of Halley's comet on November 15, 1985, showing the beginnings of plasma tail activity (Bulgarian National Observatory photographs)

Send offprint requests to: J.C. Brandt

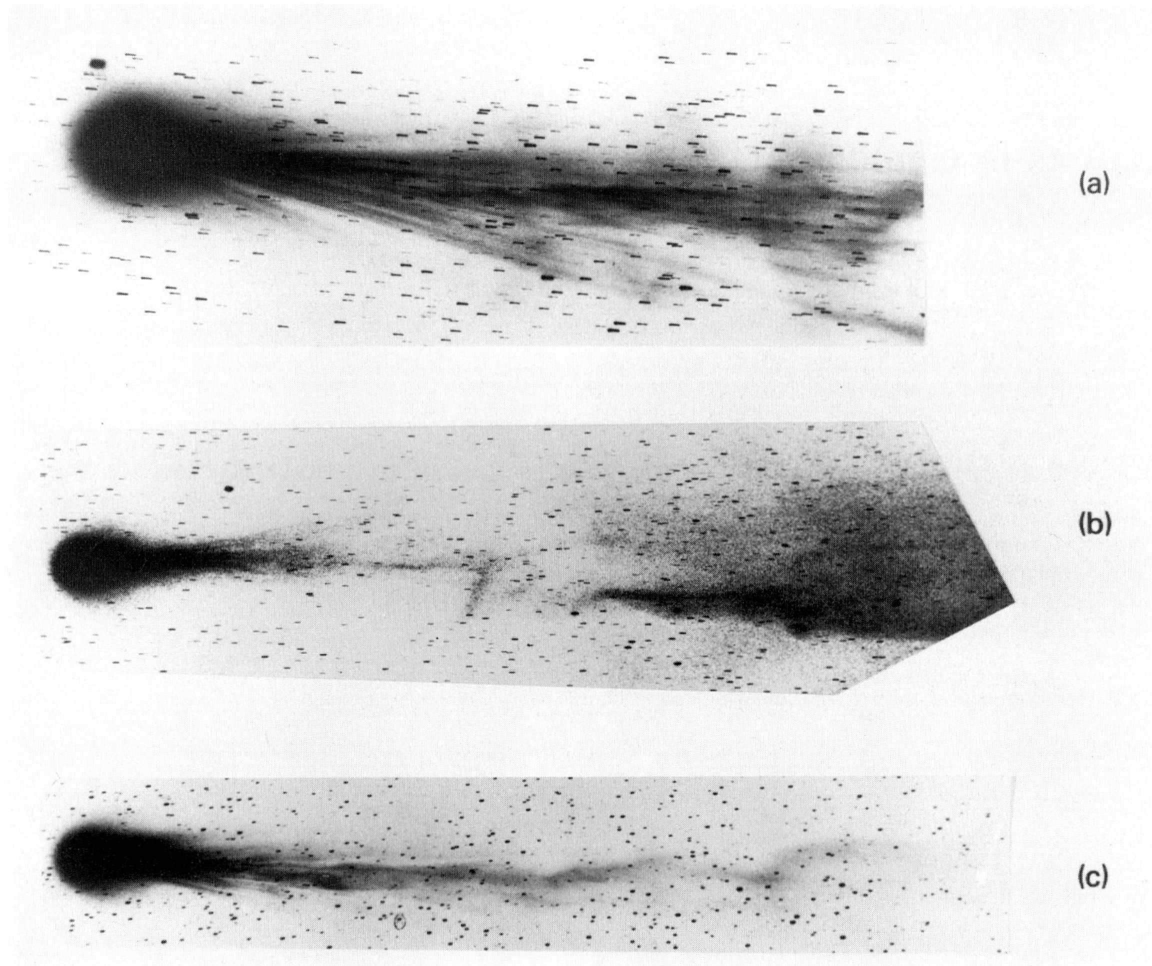

Fig. 2. Images of Halley's comet obtained from the Calar Alto Observatory of the Max-Planck-Institut für Astronomie on 1986 January 9 and 10, (a) and (b); and from the Haute-Provence Observatory on January 11, (c). A Disconnection Event (DE) is clearly shown

increased to a maximum of approximately 600 cm^{-3} in the center of the tail. The dense plasma regions were observed to be cold, ~20,000 K. Away from the central tail region, large fluctuations in most parameters were observed. Energetic ions were observed by two experiments. A substantial fraction of the ions are probably produced by the "pick-up" process involved in the solar-wind interaction, but the existence of ions in the 500 keV energy range requires an additional acceleration process.

The "bow shock" in comet Giacobini-Zinner was not as most investigators expected. It was an extended deceleration region which achieved the same purpose as a true shock, namely, the deceleration of the incident solar wind. Note that this region was sampled only on the flanks, and thus, there may have been a classical bow shock at the nose region. In broad outline, the pre-encounter view of the plasma structure was confirmed, along with a few surprises and many quantitative details. For a summary of ICE results, see the 1986 March and April issues of *Geophysical Research Letters*, and the 18 April 1986 issue of *Science*.

Very roughly, the dimensions of plasma structures in Halley are the comet Giacobini-Zinner values scaled upward by a factor of seven. Two very important differences are as follows. First, the close approach of Giotto enabled it to penetrate the contact surface and establish the existence of a magnetic field free region (radius ~5000 km) around the nucleus; the size of Giacobini-Zinner's ionosphere is unknown because ICE did not traverse it. Second, the bow wave exhibited the characteristics of a shock. Bear in the mind that all spacecraft to comet Halley passed on the sunward side. The large dimensions of the plasma interaction can be illustrated by the ICE measurements at Halley. The spacecraft passed Halley by 0.2 AU sunward in late-March 1986 and clearly detected plasma waves and energetic ions associated with the comet. Preliminary results from the Halley missions were published in the 1986 May 15, issue of *Nature*.

3. Physical framework for plasma morphology

The large-scale plasma morphology and solar-wind interaction for a medium-to-large water ice-dominated comet is dominated by two physical developments: (1) the turn-on of plasma phenomena as the comet appraches the sun (and the turn-off as the comet recedes), and specifically, the development of an ionosphere; and (2) the cyclic building and discarding of plasma structures and tails associated with Disconnection Events, or DE's.

For the turn-on description, we follow Mendis and Flammer (1984), who extended the earlier work of Biermann et al. (1967). The key parameter in the discussion is the mean molecular weight of the mass-loaded solar wind as the flow approaches the nucleus.

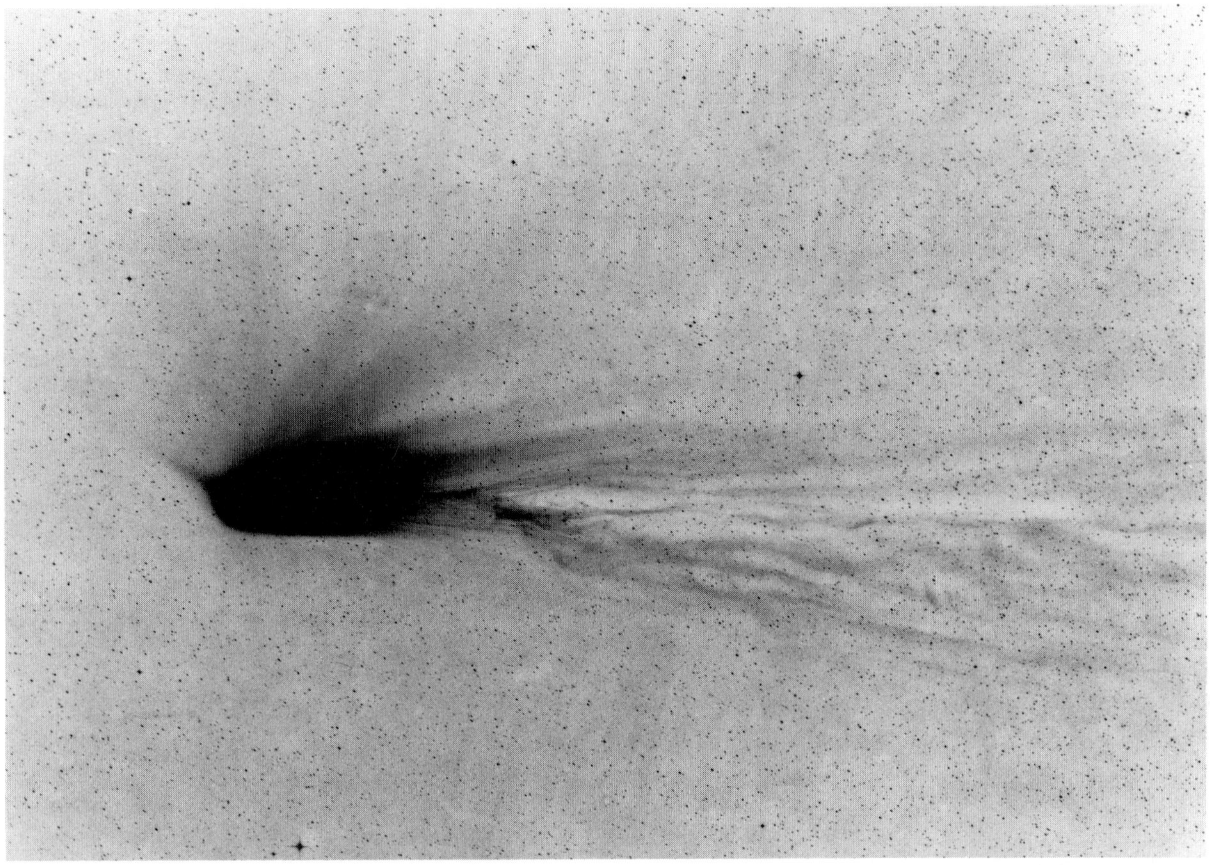

Fig. 3. The comet as photographed by the U.K. Schmidt telescope in Australia on 1986 February 22.78 UT, showing an anti-tail (left), dust structures (above), and the plasma tail (right, below). (photography by B.W. Hadley, © Royal Observatory, Edinburgh)

According to standard gas dynamic theory, the mean molecular weight, normalized to the value far from the comet, cannot exceed 4/3 without the creation of a shock wave upstream of the nucleus.

The cometary atmosphere is sufficiently tenuous that the solar wind can penetrate to the nuclear surface without violating the critical value (4/3) for comet Halley outside heliocentric distances of about 3 AU. Specifically, this is the distance at which the critical value is reached one ion (solar-wind proton) Larmor radius from the surface. Inside the 3 AU distance, a shock wave forms upstream to divert the flow around the comet, thereby ensuring that the critical value is not exceeded in the supersonic regime. The collisionless shock is expected to have a thickness of an ion Larmor radius, thus leading to the condition described above.

To form an ionopause (separating the mass-loaded solar-wind flow from the pure cometary ions by a tangential discontinuity) requires sufficient momentum in the cometary outflow, which is generally supplied by the neutrals where they are collisionally-coupled to the ions. Houpis and Mendis (1981) showed that the solar wind would penetrate the cometary atmosphere until the proton-cometary neutral momentum transfer mean free path is equal to the radial distance from the nucleus. This is the ionophause distance and the ionopause is well defined when this distance exceeds the Larmor radius. This occurs at a heliocentric of about 2.5 AU or less.

When the ionopause is developed and stable (cf. Ershkovich and Mendis, 1983), a visible tail is formed and is subject to the evolution caused by the solar-wind magnetic sector structure. We follow the description of Niedner and Brandt (1978), who extended the basic physical description of the comet/solar-wind interaction given by Alfvén (1957) to include reversals in the interplanetary magnetic field. The capture of magnetic fields from the solar wind is changed when a sector boundary is traversed and fields of opposite polarity are pressed into the old field rooted into the head region. Magnetic reconnection occurs on the sunward side and ultimately severs the flux looped around the head. Then the old plasma/magnetic structure disconnects from the head to form a DE and the comet forms a new plasma/magnetic tail with the new polarity. Niedner and Brandt (1980) have assembled observational evidence that the DE is the key part of a cyclic evolution with a characteristic morphology before and after the DE. This model of DE's is discussed in the context of Halley imagery in Sect. 5.

In the sections below, we compare the observations of comet Halley with the physical framework.

4. Summary of imaging results

Plasma tail activity for comet Halley was observed to commence in mid-November 1985 (Fig. 1), when $r = 1.8$ AU; such a "turn on" distance is in basic agreement with the theory of Mendis and Flammer (1984) when the uncertainties in the observations and the theory are considered. The plasma tail was fully developed by

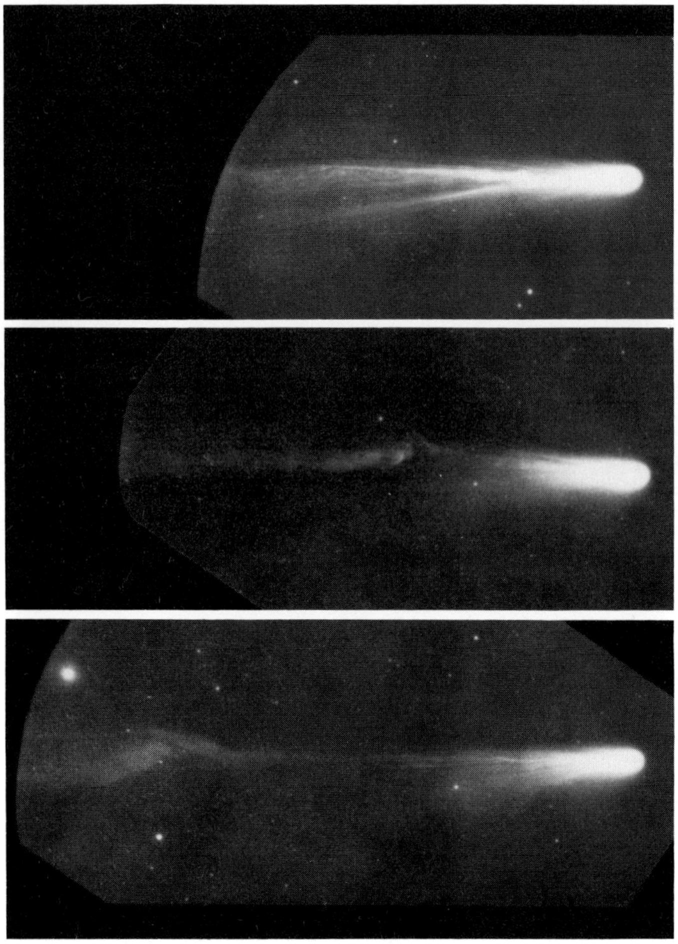

Fig. 4. A prominent DE developed in the plasma tail of Halley on March 20, as shown in this 3-day sequence spanning, top to bottom, March 20–22. All photographs were obtained at about 12 h UT with the 18″ Palomar Schmidt Camera

early December 1985. Perhaps the first *dramatic* DE for comet Halley was centered on 1986 January 10, (Fig. 2).

Early post-perihelion images from Schmidt telescopes in the southern hemisphere were spectacular. Multiple dust dails and a probable DE were seen on February 22 (Fig. 3) as was a DE on 1986, March 10. The DE of 1986 March 20–22, is shown in Fig. 4, and an even more spectacular event on April 12 is shown in Fig. 5. Activity continued at least to June 14 and the photograph taken on June 14 (Fig. 6) is currently one of the last ones available to us.

5. Disconnection events

Niedner and Brandt (1978, 1979) rediscovered DE's (known in the early-1900's) and proposed the sector boundary/magnetic reconnection model discussed above. We have extensively advertised this apparition as a potentially rich source of DE's and as a test bed for this hypothesis; Niedner (1986) has presented a list of 16 prominent DE's occurring from December to April. Thus, our first expectation has been amply fulfilled. Verification of the physical model is more complex although initial results (Niedner, 1986; Niedner and Schwingenschuh, 1987) from this apparition of Halley's Comet indicate that the sector boundary/magnetic reconnection model of DE's (Niedner and Brandt, 1978) is consistent with the times of known DE's in Halley and with the magnetic polarity data available from interplanetary and near-earth satellites, and from solar data.

The DE of March 8, 1986 illustrates the exceptional power of both *in situ* data and wide-field images; here we follow Niedner (1986) and Niedner and Schwingenschuh (1987). Figure 7 shows the sector boundary/current sheet location during the week of spacecraft encounters. The DE clearly visible on March 10 (see imagery of West et al., 1986) probably detached from the head close to March 8.4 UT. As Fig. 7 shows, the current sheet was nearly a parallel of solar latitude during this period. Thus, circumstances would appear to be very unfavorable for determining a crossing. However, the time of detachment falls between two times of "Space Truth" as determined by Vega's 1 and 2. The polarity of the draped magnetic field around Halley's Comet had reversed between March 6 and March 9. Therefore, a reasonable scenario involves the passage of a reversal in the polarity of the solar-wind magnetic field circa March 8, ultimately resulting in the field reversal measured by Vega's 1 and 2 and the spectacular DE of March 8.4 UT.

The DE's of March 20–22 (shown in Fig. 4) and April 11–12 (shown in Fig. 5) have been investigated by Brosius et al. (1987), who used ICE plasma and magnetometer data taken at a time when the comet and spacecraft were relatively close together, to examine possible solar-wind causes of these two DE's. The results are that: (1) the March 20–22 event occurred at a time of sector boundary crossing and uncharacteristically low solar-wind proton densities; (2) the April 11–12 event was associated both with magnetic reversals in the interplanetary magnetic field and with compression regions in the solar-wind plasma. The results are hence consistent with the sector boundary model (Niedner and Brandt 1978) of DE's, but compression effects and tailside reconnection (Ip, 1985; Russell et al. 1986) probably cannot be completely ruled our for the April 11–12 event.

6. Conclusion

The activities associated with the 1985–1986 return of Halley's Comet have provided us with good imaging coverage of rich tail phenomena and extensive *in situ* data. When fully archived, these will be an invaluable data base for studies of large-scale and related phenomena.

To date, the data show that: (1) the basic mechanism for the comet's interaction with the solar wind (Alfvén, 1957) is correct; (2) that the turn-on and turn-off of plasma activity follows the description of Mendis and Flammer (1984); and (3) that disconnection events are associated with sector boundary crossings and are an essential element in the morphology of developed plasma tails (Niedner and Brandt 1978).

Acknowledgements. The authors, along with Dr. J. Rahe of Remeis Observatory, Bamberg, and NASA Headquarters, Washington, DC, are Discipline Specialists for Large-Scale Phenomena of the International Halley Watch. For a description of this Network see Brandt et al. (1982) and Niedner et al. (1982).

Fig. 5. Halley's comet as photographed by F. D. Miller on April 12.1042 UT with the University of Michigan Curtis Schmidt, La Serena, Chile. Note the spectacular DE

Fig. 6. Halley's comet on June 14, 1986 (photograph taken by E. P. Moore, Joint Observatory for Cometary Research, NASA-Goddard Space Flight Center, and New Mexico Institute of Mining and Technology)

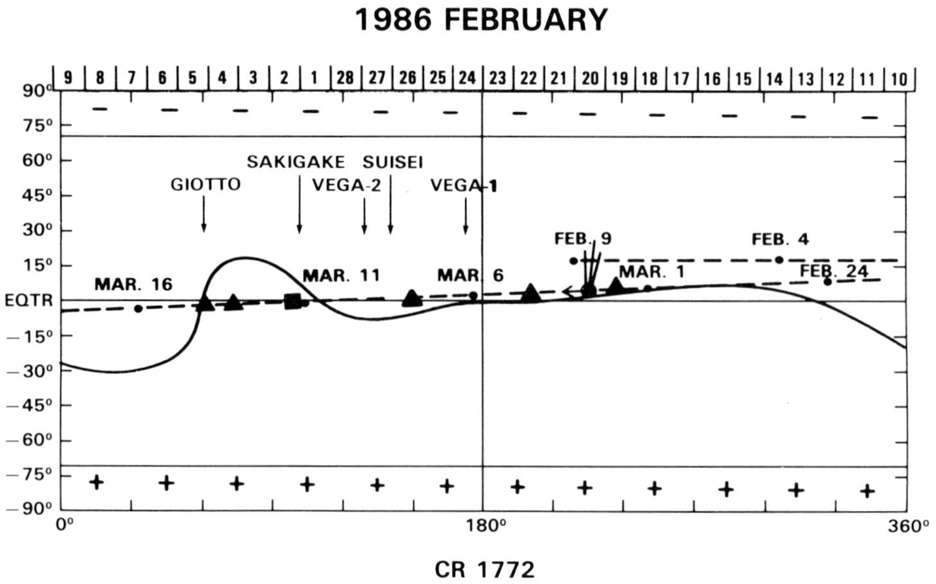

Fig. 7. Synoptic map of the coronal "source surface" of the solar wind and interplanetary magnetic field for Carrington Rotation 1772 (after Niedner, 1986). The heavy wavy line is the coronal neutral line across which magnetic polarity reverses and which is convected out into interplanetary space, giving rise to magnetic sectors. The track of the comet across the corona was obtained by computing the foot point longitude and latitude of the Archmidean spiral connecting the sun and comet for a constant solar-wind speed of 400 km s^{-1}. The times of the Halley spacecraft encounters are indicated, and the triangles and squares along the comet track denote the times of prominent and suspected DE's, respectively. The dates along the top of the figure are the times of central meridian passage at Earth (the coronal neutral line coordinates were kindly provided by Dr. J.T. Hoeksema, Stanford University)

References

Alfvén, H.: 1957, *Tellus* **9**, 192

Biermann, L.: 1951, *Z. Astrophys.* **29**, 274

Biermann, L., Brosowski, B., Schmidt, H.U.: 1967, *Solar Phys.* **1**, 254

Brandt, J.C., Niedner, M.B., Rahe, J.: 1982, in *Cometary Exploration III*, p. 47

Brosius, J., Holman, G., Niedner, M.B., Brandt, J.C., Slavin, J., Smith, E.J., Zwickl, R.D., Bame, S.J.: 1987, *Astron. Astrophys.* (in press)

Ershkovich, A.I., Mendis, D.A.: 1983, *Astrophys. J.* **269**, 743

Houpis, H.L.F., Mendis, D.A.: 1981, *Astrophys. J.* **243**, 1088

Mendis, D.A., Flammer, K.R.: 1984, *Earth, Moon, Planets* **31**, 301

Niedner, M.B.: 1986, *Adv. Space Res.* **6**, 315

Niedner, M.B., Brandt, J.C.: 1978, *Astrophys. J.* **223**, 655

Niedner, M.B., Brandt, J.C.: 1979, *Astrophys. J.* **234**, 723

Niedner, M.B., Brandt, J.C.: 1980, *Icarus* **42**, 257

Niedner, M.B., Rahe, J., Brandt, J.C.: 1982, in *Proc. ESO Workshop on the Need for Coordinated Ground-Based Observations of Halley's Comet*, Paris, p. 227

Niedner, M.B., Schwingenschuh, K.: 1987, *Astron. Astrophys.* (in press)

West, R.M., Pedersen, H., Monderen, P., Vio, R., Grosbol, P.: 1986, *Nature* **321**, 363

Astron. Astrophys. 187, 287–289 (1987)

ASTRONOMY AND ASTROPHYSICS

Analysis of the electron measurements from the Plasmag-1 experiment on board Vega 2 in the vicinity of comet P/Halley

K.I. Gringauz[1], **A.P. Remizov**[1], **M.I. Verigin**[1], **A.K. Richter**[2], **M. Tàtrallyay**[3], **K. Szegö**[3], **I.N. Klimenko**[1], **I. Apàthy**[3], **T.I. Gombosi**[3], **and T. Szemerey**[3]

[1] Space Research Institute, USSR Academy of Sciences, Moscow, USSR

[2] Max-Planck-Institut für Aeronomie, Postfach 20, D-3411 Katlenburg-Lindau, Federal Republic of Germany

[3] Central Research Institute for Physics, Hungarian Academy of Sciences, Budapest, Hungary

Received February 14, accepted March 27, 1987

Summary. Measurements of electron spectra, as obtained by the Plasmag-1 experiment on board Vega 2 in the vicinity of comet Halley, are presented. It is shown that the temperature for thermal electrons gradually decreases when the comet is approached from about $4\,10^5$ K behind the cometary bow shock to about $2\,10^5$ K at the cometopause. In the region inside the cometopause a fast increase in the flux of energetic electrons of about 1 keV energy is observed. Various possibilities are discussed regarding the differences in the electron spectra measured by Vega 2 and Giotto, respectively.

Key words: electron measurements – comet Halley

1. Introduction

The electron electrostatic analyzer EA was part of the plasma instrument Plasmag-1 on board the Vega 2 spaceprobe, which encountered comet Halley in March 1986. This sensor was designed to measure the energy spectra of electrons in the interplanetary and in the cometary plasma. It was oriented perpendicular to the ecliptic plane, and it observed the flux of electrons in the energy range of 3 eV to 10 keV in 30 logarithmically spaced energy intervals. A detailed description of this experiment can be found in Gringauz et al. (1986a) and Apàthy et al. (1986), and first measurements have been published by Gringauz et al. (1985 and 1986b). In the present paper we will present the Vega 2 electron measurements in the vicinity of Halley's comet, and we will compare these observations with those obtained on board the Giotto spaceprobe.

2. Observations

Figure 1 shows the sequence of the two minute averaged electron spectra as observed during the inbound leg of Vega 2 between app. $8\,10^5$ and $1.5\,10^4$ km from the cometary nucleus. The intensity is presented on a logarithmic scale (see insert in the upper right-hand corner) where the baseline corresponds to $10^2\,s^{-1}$. The first spectra in Fig. 1 were measured downstream of the cometary bow shock, which was crossed at about 2.20 UT at a cometocentric distance of about $1.3\,10^6$ km (Galeev et al., 1986), and the sequence includes the crossing of the cometopause (Gringauz et al., 1986c), which was located at a distance of about $1.6\,10^5$ km or at about 6.45 UT, respectively.

A first, well defined maximum in the electron intensity is identified at an energy E_m of a few times ten eV, corresponding to the thermal electrons. While passing through the cometosheath region towards the cometopause, we find a gradual decrease of the energy E_m. Assuming an isotropic Maxwellian distribution for these thermal electrons, the counting rate is proportional to $E^2 \exp(-E/kT_e)$, with T_e being the electron temperature and k

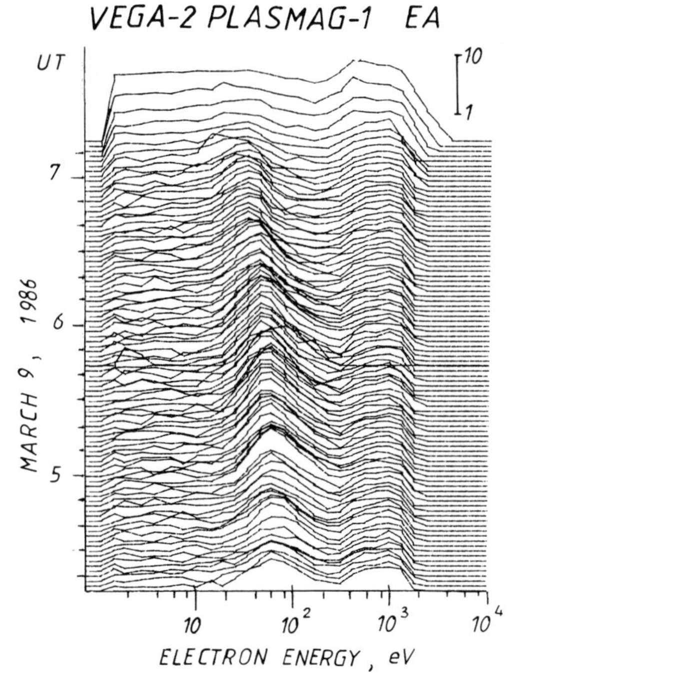

Fig. 1. Time sequence of 2 min averaged energy spectra, i.e. counting rate vs. energy, of electrons measured by the electron electrostatic analyzer EA on board Vega 2 between $8\,10^5$ and $1.5\,10^4$ km from the nucleus of comet Halley

Send offprint requests to: A.K. Richter

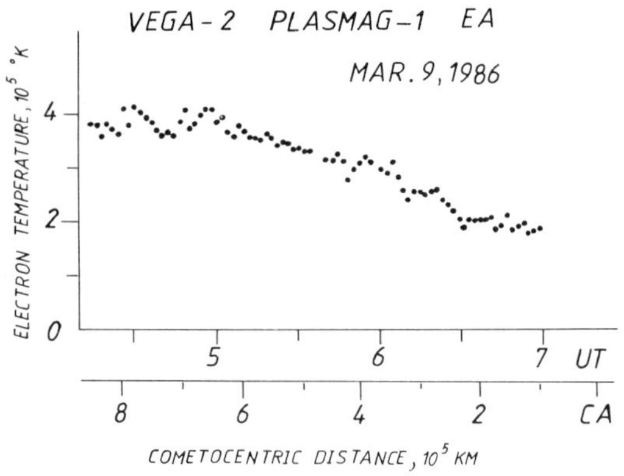

Fig. 2. Distribution of the electron temperature in the cometosheath of comet Halley between 8 10^5 and 1.6 10^5 km (cometopause). CA = closest approach

the Boltzmann constant. Thus, $T_e = 2E_m/k$. In Fig. 2 the radial dependence of T_e is shown in the distance range from about 8 10^5 to 1.6 10^5 km, respectively.

From this Figure it readily follows (1) that there is an overall decrease of T_e by a factor of two from about 4 10^5 K to about 2 10^5 K within this region, (2) that T_e decreases fastest within the region of about 5.5–6 10^5 km, but stays app. constant outside, and (3) that there are significant fluctuations in T_e present in the region outside of about 6 10^5 km. It should be mentioned that at a cometocentric distance of about 5–6 10^6 km, i.e. in the undisturbed solar wind, the electron temperature was also about 2 10^5 K, and that it increased to its maximum value of about 3.5–4 10^5 K already at a distance of about 3 10^5 km upstream of the cometary bow shock.

Reconsidering Fig. 1 we find that during the time of closest approach to the cometary nucleus, i.e., after about 7.10 UT, the counting rates are increasing for the keV electrons. This situation is depicted in Fig. 3, representing the sequence of 10 s

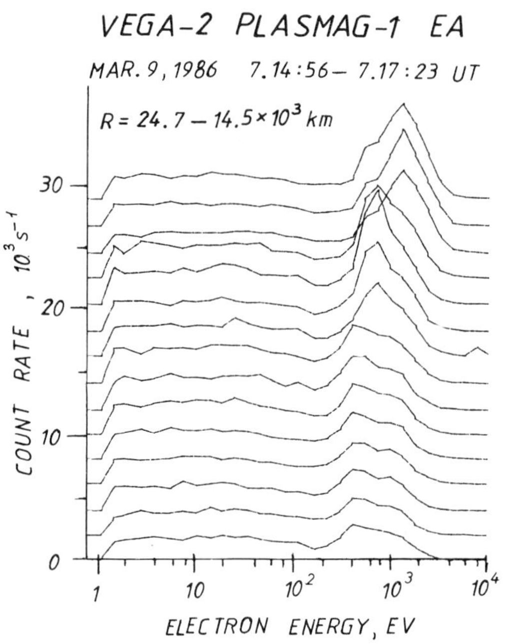

Fig. 3. Time sequence of 10 s averaged energy spectra of electrons measured by Vega 2 between 2.5 10^4 and 1.5 10^4 km from the nucleus of comet Halley

averaged electron energy spectra within the distance range of 2.5–1.5 10^4 km. The maximum counting rates are now observed in the energy interval of 0.3–3 keV with a fast increase of the maximum intensity and with a change in the distribution. At the end of this sequence the Plasmag 1 experiment temporarily ceased operating, presumably because it was hit by a cometary dust particle.

3. Comparison with the Giotto observations

During the encounter of the Giotto spacecraft with comet Halley, electrons were measured by the electron electrostatic analyser EESA of the RPA experiment (Rème et al., 1986). In the undisturbed solar wind the electron temperature was about 2.5 10^5 K, and it increased to about 3.5 10^5 K a few times 10^5 km upstream of the bow shock. Downstream, in the region from 1.15 10^6 km to 5.5 10^5 km T_e stayed practically constant (3.5 10^5 K) but with large fluctuations being superimposed, while afterwards, in approaching the cometopause, T_e decreased to about 2 10^5 K. Inside the cometopause and during closest approach, however, the RPA experiment did not observe any electrons in the keV energy range.

Thus, outside the cometopause both the Vega 2 and Giotto electron measurements are in excellent agreement with each other, not only with respect to the actual values of T_e but also with respect to its overall radial dependence. The only, yet essential difference between both observations is the observation of keV electrons of Vega 2 during its closest approach to comet Halley.

4. Discussion

We propose that the decrease in the electron temperature from about 4 10^5 K near the bow shock to about 2 10^5 K at the cometopause, as observed by Vega 2 and Giotto independently, might be due at least to some extent to inelastic collisions between the thermal electrons and the cometary neutral gas. Near the cometopause at $R \approx 1.6\ 10^5$ km the neutral density is about $N_n \approx 5\ 10^3\ \mathrm{cm}^{-3}$ (Remizov et al., 1986). The energy loss of electrons caused by inelastic collisions with neutrals of the water group is $N_n L_e \approx 2\ 10^{-11}\ \mathrm{eV\ cm}^{-1}$, where $L_e \approx 4\ 10^{-15}\ \mathrm{cm^2\ eV}$ is the energy loss function for electrons with an energy of about 40 eV (Olivero et al., 1972). The plasma velocity near the cometopause is $V \approx 200\ \mathrm{km\ s}^{-1}$ (Gringauz et al., 1986c). The characteristic time for the plasma flow through the cometopause is therefore $\tau = R/V \approx 800$ s. During this time the thermal electrons at an average velocity of about 4000 km s^{-1} have traversed a distance of about 3 10^6 km. Thus, according to the above equation their energy loss is about 6 eV.

The observation of electrons at an energy of about 1 keV by Vega 2 at a distance of 1.5–2 10^4 km from the nucleus could either be due to a purely instrumental effect of the EA analyser, or to the sporadic occurence of these particles in the sense that they were observed by Vega 2 on March 9 but not by Giotto on March 13/14, 1986. Let us first discuss the question of the instrumental effect: After crossing the cometopause, the EA sensor observed a significant increase in the overall background counting rate inside the cometary plasma region (see also Fig. 1). Besides in its normal mode of operation, the EA sensor also provided the possibility to measure electrons in the first 15 energy intervals

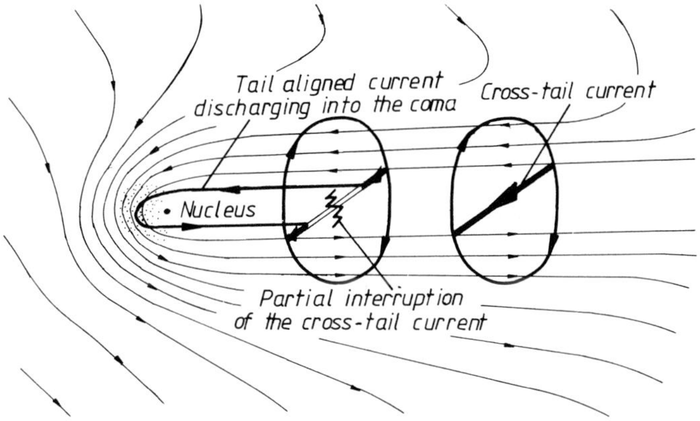

Fig. 4. Schematics of a cometary cross-tail disruptive event, which could have produced the keV energetic electrons observed by Vega 2 during closest approach (see Fig. 3)

($E_e < 150$ eV) at an about 100 times lower sensitivity for every forth spectrum. Inside the cometary plasma region the counting rates for electrons <150 eV did practically not change when switching from the normal mode to this lower sensitivity mode. Thus, the fast increase in the intensity of these thermal electrons at the crossing of the cometopause (Fig. 1) seems to be a result of an increase in the instrumental background rather than a real increase of the electron flux. This may also hold for the energy range $E_e > 150$ eV, although there was no decreased sensitivity mode for this energy range. Yet, according to Fig. 3 the maximum intensity at about 1 keV increased much faster as compared to the lower energies from a distance of about $2\ 10^4$ km onwards, and it was even shifted to higher energies further inside the cometary plasma region. At this time the maximum counting rate at about 1 keV was 5 to 7 times higher than the background rates at about 200 eV. Moreover, several laboratory tests have been performed in the meantime with an electron analyzer similar to the flight sensor, but an effect like this observed one has not yet been identified. Thus, the observation of keV electrons by Vega 2 cannot be explained by an instrumental effect only. On the other hand, if the occurence of such energetic electrons is real, what sporadic acceleration mechanism could have caused their appearance? The onset of a magnetospheric substorm similar to the events occuring in the terrestrial magnetosphere could provide the explanation. The occurence of such events in the cometary magnetotail has been proposed by Ip and Mendis (1976) and by Ip (1986). A schematic representation of a cometary substorm, as suggested by Ip and Axford (1982) and by Mendis et al. (1985), is shown in Fig. 4. In the stationary case the flow system of the tail electric current adopts the usual θ shape configuration. If the cross-tail current gets partially disrupted, induced tail-aligned currents can discharge through the cometary coma. By such a process electrons can be accelerated to some few keV energies, as in case of auroral electrons in the terrestrial upper atmosphere. Vega 2 might have observed such substorm-generated energetic electrons, while Giotto did not.

Finally, we would like to summarize another observation pointing also towards the possibility that energetic electrons may occur sporadically in the coma of comet Halley. In the paper by Feldman et al. (1986) two UV spectra measured by IUE on March 18–19, 1986, on the tailward side of the nucleus at a distance of about 40,000 km and about 30 min apart were compared. It was found that the CO_2^+ emission had decreased by more than a factor of four, while the OH^+ brightness had remained nearly constant. The authors proposed that this difference could be explained by the existence of localized currents of energetic electrons close to the nucleus.

Acknowledgements. The authors are indebted to D.A. Mendis for useful discussions.

References

Apáthy, I. et al.: 1986, 20th ESLAB Symp., ESA **SP-250**, Vol. 1 65

Feldman, P.D. et al.: 1986, *Nature* **324**, 433

Galeev, A.A. et al.: 1986, *Geophys. Res. Letters* **13**, 841

Gringauz, K.I. et al.: 1985, *Adv. Space Res.* **5**, Vol. 12, 165

Gringauz, K.I. et al.: 1986a, in *Field, Particle and Wave Experiments on Cometary Missions*, eds. K. Schwingenschuh, W. Riedler, Verlag der Österreichischen Akademie der Wissenschaften, Wien, p. 203

Gringauz, K.I. et al.: 1986b, *Nature* **321**, 282

Gringauz, K.I. et al.: 1986c, *Geophys. Res. Letters* **13**, 613

Ip, W.-H., Mendis, D.A.: 1976, *Icarus* **39**, 147

Ip, W.-H., Axford, W.I.: 1982, in *Comets*, ed. L.L. Wilkening, University of Arizona Press, Tucson, p. 588

Ip, W.-H.: 1986, *Planet. Space Sci.* **27**, 121

Mendis, D.A., Houpis, H.L., Marconi, M.L.: 1985, in *Fund. Cosmic Phys.* **10**, 1

Olivero, I.I., Stagat, R.W., Green, A.E.S.: 1972, *J. Geophys. Res.* **77**, 4797

Rème; H. et al.: 1986, *Nature* **321**, 349

Remizov, A.P. et al.: 1986, 20th ESLAB Symp., ESA **SP-250**, Vol. 1, 387

Astron. Astrophys. 187, 290–292 (1987)

ASTRONOMY AND ASTROPHYSICS

The upstream region, foreshock and bow shock wave at comet P/Halley from plasma electron measurements

K.A. Anderson[1,2], C.W. Carlson[1], D.W. Curtis[1], R.P. Lin[1], H. Rème[3], J.A. Sauvaud[3], C. d'Uston[3], A. Korth[4], A.K. Richter[4], and D.A. Mendis[5]

[1] Space Sciences Laboratory, University of California, Berkeley, CA 94720, USA
[2] Physics Department, University of California, Berkeley, CA 94720, USA
[3] Centre d'Etude Spatiale des Rayonnements, 9, avenue Colonel-Roche, B.P. 4346, F-31029 Toulouse Cedex, France
[4] Max-Planck-Institut für Aeronomie, Postfach 20, D-3411 Katlenburg-Lindau 3, Federal Republic of Germany
[5] Department of Electrical Engineering and Computer Sciences, University of California at San Diego, La Jolla, CA 92093, USA

Received January 29, accepted March 19, 1987

Summary. We survey the plasma electron parameters from a distance 2.7 10^6 km from the comet nucleus to the bow shock wave at about 1.1 10^6 km and somewhat beyond. We describe the features of the electron foreshock lying outside the shock to a distance of about 230,000 km. It is a region of intense solar wind-comet plasma interaction in which energetic electrons are prominent. Several spikes of electrons whose energies extend to about 2.5 keV appear in front of the shock. These energetic electrons may have been accelerated in the same way electrons are accelerated at the Earth's bow shock to energies of 1 to 10 keV. The direction of the electron bulk flow direction changes abruptly between 1920 and 1922 UT, and the flow speed begins a sharp decline at the same time. We believe the spacecraft has entered the bow shock wave sometime between 1920 and 1922 UT. We find that the electron density variations at P/Halley are very much smaller than those reported at Giacobini-Zinner.

Key words: comets – plasmas – shock waves – Sun (the): solar wind

1. Introduction

The purpose of this paper is to further describe and interpret results from the plasma electron analyzer carried by the Giotto spacecraft to Comet Halley in March 1986. The emphasis here is on the interaction of the solar wind with the cometary plasma from distance of about 2.7 10^6 km from the comet nucleus to the crossing of the bow shock wave, a distance of about 10^6 km from it. We have already published some results describing the solar wind-cometary plasma interaction (Rème et al., 1986). Experiments done at Comet Giacobini-Zinner by the International Comet Explorer were the first to directly demonstrate that the solar wind interacts strongly with cometary plasma (Bame, 1986). Among the manifestations of the interaction reported by Bame et al. are: (1) counterstreaming hot electrons (reverse heat flow) observed outside the bow wave; (2) a "transition region" in which the plasma density and temperature increased while the flow velocity decreased (however, they found no evidence in the plasma electrons for a conventional bow shock at the edge of the transition region or inside it); and (3) very large fluctuations in the electron density ascribed to strong plasma turbulence. We will show in this paper that while some of the interaction features are similar at the two comets there are also striking differences.

Send offprint requests to: K.A. Anderson[1]

2. Observations

Figure 1 plots several plasma electron parameters beginning about 11 hours before closest approach to the comet. During this interval solar wind-comet interaction is continuously evident. Earlier during the same day backstreaming electrons were occasionally seen at distances up to 4 10^6 km from the nucleus. We attribute these effects to the presence of the comet. Two important features of the solar wind-cometary plasma interaction as seen by the plasma electron measurements are:

2.1. The slowing of the solar wind

Beginning about 0900 Spacecraft Event Time (SCET) the bulk flow speed of the solar wind determined from the plasma electron distribution function begins to decrease, no doubt due to mass loading resulting from the ionization of cometary molecules. The speed decrease is generally systematic over the several hour interval but there are significant shorter term variations. A large decrease of about 100 km s^{-1} occurs about 1826 SCET and this is followed by a nearly complete recovery within one hour. A second drop in speed of about 100 km s^{-1} occurs at 1922. The first decrease signals entry into the electron foreshock, the second decrease marks the passage through the bow shock structure.

2.2. Appearance of energetic electrons

From 1400 SCET energetic electrons—defined to be those with energy greater than 165 eV—become prominent. We attribute the appearance of these electrons to the solar wind-comet interaction. At the times of magnetic field connection to the cometosheath increases in the energetic electron flux occur. These

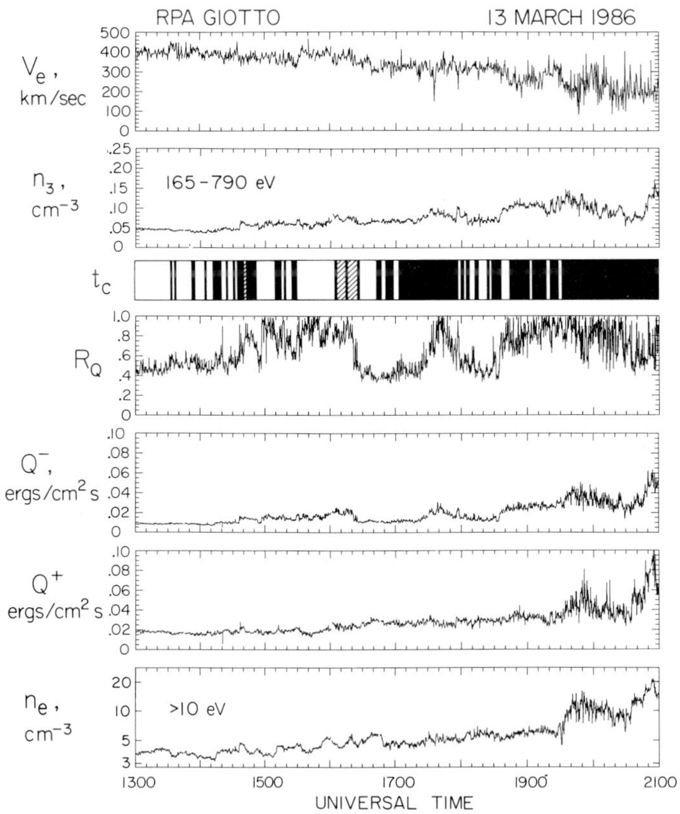

Fig. 1. A survey of several plasma electron parameters covering an 8-hour interval beginning a distance of 2.7 10^6 km from the comet nucleus. The solar wind-comet interaction is evident throughout this interval. V_e is the bulk speed of the electrons, n_e their density above 10 eV, and n_3 their density between 165 and 790 eV. Times when the spacecraft is connected to the cometosheath by the magnetic field are denoted by dark bars on the panel labelled t_c. The electron heat flux away from the Sun and toward the Sun are denoted by Q^+ and Q^- respectively. R_Q is the ratio Q^-/Q^+

electrons are backstreaming from the cometosheath. There is also a component of energetic electrons that systematically increases from 1400 onward and is particularly prominent in the electron foreshock.

A striking feature of the solar wind-comet interaction appears at 1834 SCET and extends to the bow shock. The corresponding spatial distance is approximately 230,000 km along the spacecraft trajectory. We refer to this feature as the electron foreshock. It is apparent at large increases in the energetic electron density and reverse heat flow, Q^-, but much less so in the total density. The plasma flow speed drops suddenly and significantly at the leading edge of the foreshock but recovers to its previous value just before encounter with the bow shock. The electron temperature in the foreshock reaches a value of roughly 2.6 10^5 K and the electron distributions are not strictly Maxwellian at these times. This temperature was determined by finding the maximum slope of the electron distribution function. The energy at which maximum slope occurs is usually in the range 50 to 100 eV. In the foreshock and at other times when energetic electrons are present the slope of the distribution function decreases above 100 eV.

Beginning about 15 minutes before encounter with what we believe to be the bow shock wave at 1922, spikes of energetic electron appear. Two of them, marked B and C, can be seen in the middle panel of Fig. 2. Electrons of energy up to about 2.5 keV were detected in these spikes. We have used the direction of the symmetry axis of the electron angular distribution to determine the direction of the magnetic field at the time of the energetic electron spikes. Although it is not possible to prove that the spacecraft field line becomes tangent to the bow shock surface at the precise times of the three spikes, the field line must have passed through the tangent condition at a time at least close to the time of the spikes. We therefore suggest that the energetic electrons in these spikes may have been accelerated in essentially the same way electron acceleration occurs at Earth's bow shock (Anderson et al., 1979).

From Fig. 2 we see that between 1920 and 1922 SCET large changes in the flow speed and direction occur. The electron temperature reaches values as high as 3 10^5 K in that time interval. The density increases beginning at 1920 but then falls to a very low value at 1924. This increase in density is associated with the rapidly falling portion of the largest of the hard electron spikes. Both these effects may be associated with the direction of the magnetic field, which is changing rapidly at this time. The density variations from 1920 to about 1950 suggest large amplitude wave disturbances.

One of the striking differences between the solar wind-comet interactions at Giacobini-Zinner and Halley is the very much

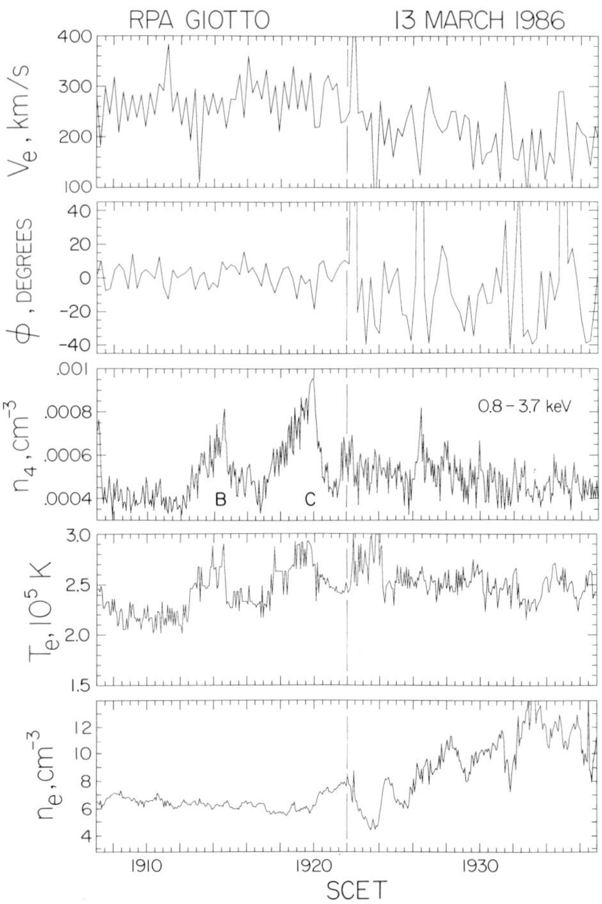

Fig. 2. The appearance of energetic electron spikes from 1906 to 1920 Spacecraft Event Time (SCET) is taken to be evidence that the spacecraft is close to a shock wave. The changes in flow speed and direction lead us to believe that the shock wave was encountered sometime between 1920 and 1922 SCET

smaller density fluctuations at Halley. The largest electron density changes outside the bow shock occur from 1300 SCET to the time of the shock crossing which begins about 1922. The changes are never more than 40% from valley to peak and except from 1826 to 1922 they are rather slow changes. The great contrast between the two comets at the times of the spacecraft encounters is illustrated in Fig. 3. In that figure the density and time scales have been made the same. Due to the greatly different spacecraft-comet relative velocities the spatial scales are quite different, as indicated in the figure. Note also that the density changes at Halley are much more suggestive of a bow shock wave.

From 1924 to 1933 the density rises to a value of about twice its value in the electron foreshock. The flow speed drops from $300\,\mathrm{km\,s^{-1}}$ to $200\,\mathrm{km\,s^{-1}}$ between 1922 and 1924. Around 1950 the speed briefly returns to a value of around $300\,\mathrm{km\,s^{-1}}$, then falls again to around $200\,\mathrm{km\,s^{-1}}$ (Fig. 1). Despite such somewhat anomalous behavior, the general features of many plasma electron parameters lead us to believe a shock wave was encountered sometime between 1920 and 1922 SCET.

The plasma electron analyzer sometimes is able to detect the presence of wave disturbances most clearly via changes in the direction of the symmetry axis of the electron angular distribution which usually approximates the magnetic field direction. The electron symmetry direction shows fluctuations not characteristic of the undisturbed solar wind throughout March 13 from the time the instrument is turned on at about 0500 SCET. From 1200 to 1500 the peak spacings are usually in the range 150 to 300 s. An interval when the quasi-periodic character of these fluctuations is particularly clear is shown in Fig. 4. Here the peaks are rather uniformly spaced at an interval of close to 200 s, close to the gyroperiod of oxygen or water-group ions, suggesting that implantation of heavy ions in the vicinity of the spacecraft is producing these plasma disturbances (Tsurutani and Smith, 1986).

The changes in electron density during the wave disturbance shown in Fig. 4 are small. We believe that at this time the interaction is not as strong as at Giacobini-Zinner just outside its shock. There the waves are grown to be highly non-linear. This difference may be due to the larger production rate of ions in the vicinity of Giacobini-Zinner's bow wave compared to that rate near Halley's bow shock. Although the emission rate of molecules is greater at Halley the shock wave is 10 times further away than it is at Giacobini-Zinner.

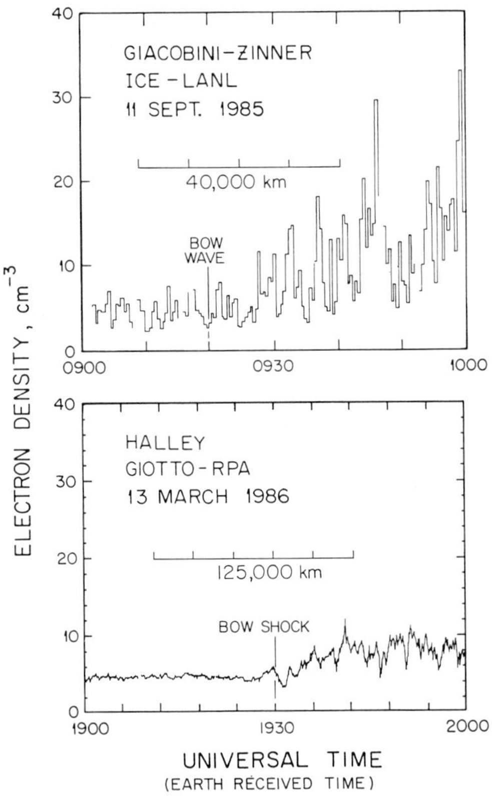

Fig. 3. Comparison of the plasma electron density fluctuations at comets Giacobini-Zinner and Halley. At Halley the density changes are more clearly indicative of a bow shock wave and their amplitude is markedly less than at Giacobini-Zinner

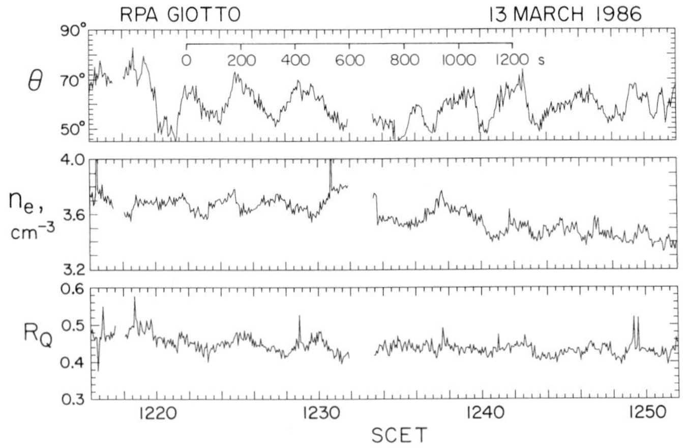

Fig. 4. Large, irregular fluctuations in the direction of the electron angular distribution symmetry axis are characteristic during approach to the comet on 13 March. This example is one for which characteristic peak spacings of about 200 s are well defined. Changes in the heat flux ratio are loosely correlated with the angle changes but the density fluctuations seem to be largely incoherent

Acknowledgements. This work was supported in the U.S. by NASA contract NASW-3575, in France by CNES under grant no. 1212, and in W. Germany by the Max-Planck-Gesellschaft zur Förderung der Wissenchaften and the Bundesministerium für Forschung und Technologie under grant no. 01 OF 052.

References

Anderson, K.A., Lin, R.P., Martel, F., Lin, C.S., Parks, G.K., Rème, H.: 1979, *Geophys. Res. Letters* **6**, 401

Bame, S.J.: 1986, *Science* **232**, 356

Rème, H., Sauvaud, J.A., d'Uston, C., Cotin, F., Cros, A., Anderson, K.A., Carlson, C.W., Curtis, D.W., Lin, R.P., Mendis, D.A., Korth, A., Richter, A.K.: 1986, *Nature* **321**, 349

Tsurutani, B.T., Smith, E.J.: 1986, *Geophys. Res. Letters* **13**, 263

Astron. Astrophys. 187, 293–296 (1987)

ASTRONOMY AND ASTROPHYSICS

Stochastic Fermi acceleration of ions in the pre-shock region of comet P/Halley

B.E. Gribov[1], K. Kecskeméty[2], R.Z. Sagdeev[1], V.D. Shapiro[1], V.I. Shevchenko[1], A.J. Somogyi[2], K. Szegő[2], G. Erdős[2], E.G. Eroshenko[3], K.I. Gringauz[1], E. Keppler[4], R.G. Marsden[5], A.P. Remizov[1], A.K. Richter[4], W. Riedler[6], K. Schwingenschuh[6], and K.-P. Wenzel[5]

[1] Space Research Institute, Profsoyuznaya 84/32, 117810 Moscow, USSR
[2] Central Research Inst. of Physics, H-1525 Budapest POB 49, Hungary
[3] IZMIRAN, USSR Academy of Sciences, 142092 Troitsk, USSR
[4] Max-Planck-Institut für Aeronomie, D-3411 Katlenburg-Lindau, Federal Republic of Germany
[5] Space Science Department of ESA, ESTEC, 2200 AG Noordwijk, The Netherlands
[6] Space Research Institute, Inffeldgasse 12, A-8010 Graz, Austria

Received February 13, accepted April 21, 1987

Summary. Energetic cometary ion fluxes measured between $2.5\,10^6$ km and 10^6 km from the cometary nucleus along the inbound trajectory of s/c VEGA-1 are used to derive the temperatures of the ion distributions in the solar wind frame. The increase of the temperature is modelled by the temperature change derived from a Fokker-Planck type equation with a source term and a stochastic acceleration term. The temperature increase predicted by theory is about 3 keV, higher than the observed one ($\simeq 1.4$ keV). The difference may be due to the approximations applied. The second order Fermi mechanism is thus capable of producing the temperature increase observed.

Key words: comets – ion-acceleration mechanisms

1. Introduction

The TUNDE-M instrument on board the spacecraft VEGA-1 recorded ions of cometary origin in the energy range of 40 to 630 keV (nominal values, Somogyi et al., 1986). Upstream of the bow shock the lowest four energy channels were effective, while in the region inside the shock, up to the phase of closest approach all the 52 channels were operational. As discussed in Kecskeméty et al. (1986), most of the ions observed in the upstream region belong most probably to the water group and are mainly O^+ and OH^+. Considering this the energies of the ions detected in the lowest four channels are between 78 and 132 keV. Time profiles of the ion fluxes as measured in the region $r = 12\,10^6$ km to 10^6 km from the cometary nucleus along the inbound leg of the s/c trajectory (this part of the trajectory will be referred to as the pre-bow-shock (PBSh) region in what follows, since the bow shock crossing in the inbound leg occurred at about $r = 10^6$ km) are shown in Fig. 2 of Kecskeméty et al. (1986).

More instrumental details (including observational geometry) have also been given in Kecskeméty et al. (1986) in which correlations with magnetic field directions and energy distributions of ions in the solar wind reference frame (SWRF) were discussed.

The purpose of this paper is to study the effectiveness of the stochastic second order Fermi acceleration in the PBSh region. Large, quasi-periodic enhancements of the ionic intensities were observed between $r = 12\,10^6$ and $2.5\,10^6$ km in the PBSh region (see Fig. 2 in Somogyi et al., 1986) which must be of very complex origin. We thus confine our investigations to the relatively quiet region between 2.5 million km to 1 million km from the nucleus.

The possibility of stochastic Fermi acceleration in the solar wind- cometary interaction have been considered previously by Amata and Formisano (1985) and by Ip and Axford (1986). In the latter paper, various other mechanisms for particle acceleration (adiabatic compression, magnetic field reconnection, first order Fermi acceleration in a diffuse shock) have also been taken into account and the conclusion was reached that second order Fermi acceleration was "likely to be of particularly importance" in cometary environments such as P/Giacaboni-Zinner and P/Halley. The time evolution of a $\delta(t)\,\delta(T - T_0)$ injection of the density of ions with kinetic energy T_0 at the time t was also given in that paper. Here we use a different approach: instead of solving the first momentum equation (for ionic density) we derive and solve an equation for the time dependence of the ionic temperature (Sect 2 of this paper) for subsequent comparison with experiment (Sect 3).

2. Time-dependence of ionic temperature in a second order Fermi acceleration

2.1. Injection mechanism

It is well known that cometary neutral molecules moving away of the cometary nucleus with relative velocities $1\,\mathrm{km\,s^{-1}}$ get ionized by the UV radiation of the sun and also by charge-exchange with solar wind ions. The freshly born cometary ions are picked up by the solar wind in which they gyrate around a centre moving parallel to the magnetic field with a velocity $u\cos\theta$ (where u is the solar wind velocity and θ is the angle with the magnetic field direction, Ip and Axford, 1986). The resulting ion

Send offprint requests to: A.J. Somogyi

cyclotron instability gives rise to Alfvén waves and makes the magnetic field turbulent. The solar wind loses speed as a consequence of getting loaded with the pick-up ions. Scattering in the turbulent field makes the ion distribution isotropic, this process ends up after several gyration periods with a uniformly distributed $|u|$ in the SWRF. Essential features of this injection mechanism are: a) the ionization process and b) the scattering on the MHD turbulence excited by the cometary ions moving parallel to the magnetic field.

The processes of ionization were dealt in many papers (cf. Ip and Axford, 1982), while the theory of the build-up of magnetic turbulence was first given recently (Galeev et al., 1985; Sagdeev et al., 1986). Comparison with the values of the level of magnetic turbulence as measured near P/G–Z (Tsurutani and Smith, 1986) and P/Halley (Riedler et al., 1986) showed an overall agreement with values predicted by the theory (Galeev et al., 1986).

2.2. *The basic equation for resonant stochastic acceleration*

As it is known, this acceleration can be described by the equation

$$u\frac{\partial f}{\partial x} = \frac{Q\delta(v-u)}{4\pi r^2 \lambda v^2}\exp(-r/\lambda) + \frac{1}{v^2}\frac{\partial}{\partial v}\left(v^2 D\frac{\partial f}{\partial v}\right) \tag{1}$$

where

$$D = \frac{e^2 v_A^2}{4m_i^2 c^2}\cdot\frac{1}{v}\int_0^{\pi/2} \frac{d\theta \sin^3\theta}{\cos\theta}[B_k^2]_{k=\omega_{Bi}/v\cos\theta} \tag{2}$$

with the following notations: the x axis is directed along the solar wind flow; f is the six dimensional density function (assumed to be independent of y and z); Q is the total cometary gas flow per second ($\simeq 1.2\,10^{30}$ molecules/s for P/Halley); r is the radial distance from the nucleus; λ is the characteristic length of ionization for cometary neutral molecules ($\simeq 2\,10^6$ km, see Gringauz et al., 1986); v is the particle velocity, v_A the Alfvén velocity of solar wind flow, $\omega_{Bi} = eB/m_i c$ is the ion cyclotron frequency (m_i being the ionic mass) and B_k^2 is the spectral density of magnetic fluctuations i.e. energy density contained at the wave number interval of $(k, k+dk)$. Values of B measured in the interval selected for this study ($2.5\,10^6\text{ km} > r > 10^6$ km in the PBSh region) show that with a reasonable approximation,

$$B_k^2 = \tilde{B}^2/k \tag{3}$$

Figure 1 shows an example of the magnetic field power spectra as measured in the region envisaged.

The second term on the r.h.s. of Eq. (1) is the quasi-linear term describing the diffusive acceleration of particles by MHD turbulence, while the first term on the r.h.s. is the source term which describes the production of new ions due to photoionization and their subsequent isotropization by Alfvén turbulence. If we neglect the diffusion term, the solution of the Eq. (1) has the form

$$f_0(v,x) = \frac{3}{8\pi m_i}\frac{\rho_\infty u_\infty}{uv^3}\left(1-\frac{u_\infty^{1/2}}{2v^{1/2}}\right)\sigma(v-u)\sigma(u_\infty - v)\,, \tag{4}$$

where ρ_∞ and u_∞ are the density and the bulk velocity of the undisturbed solar wind, respectively; $\sigma(x)$ is the step function: $\sigma = 1$ for $x > 0$, $\sigma = 0$ for $x < 0$. Due to the second term on the r.h.s. of Eq. (1), a high energy tail of the distribution appears. For the case under consideration, when $\tilde{B}^2 \sim 1/k$, the high energy tail has the form

$$f(v) \propto \exp(-m_i v^2/2T), \text{ for } v > u_\infty\,. \tag{5}$$

2.3. *The temperature variation of ions during the 2nd order Fermi process*

The value of T, the ionic temperature, can be obtained from Eq. (1) by multiplying it by $m_i v^2$ and integrating over the three components of v. Taking into account Eqs. (2) and (3) this procedure yields

$$u\frac{dT}{dx} = \frac{1}{3}m_i v_A^2 \omega_{Bi}\frac{\tilde{B}^2}{B^2} \tag{6}$$

In the case of $\Lambda/R = 2$ (Λ is the scattering mean free path, R is the ion gyroradius) Eq. (6) coincides with the result obtained by Ip and Axford (1986) under the assumption of approaching the strong turbulence limit:

$$\frac{dT}{dt} = \frac{3R}{4\Lambda}ZeBv_A^2 \tag{7}$$

On the r.h.s. of Eq. (6), v_A, ω_{Bi}, B^2, and $\tilde{B}^2$ are functions of x. Note that there is no contribution to the value of T from the source term, since T only determines the shape of the high energy tail of the distribution.

Equation (6) makes it possible to compare the observed T_{obs} values with calculated ones:

$$T_c = \frac{1}{u(x)}\int dx\,\frac{1}{3}m_i v_A^2 \omega_{Bi}\frac{\tilde{B}^2}{B^2} \tag{8}$$

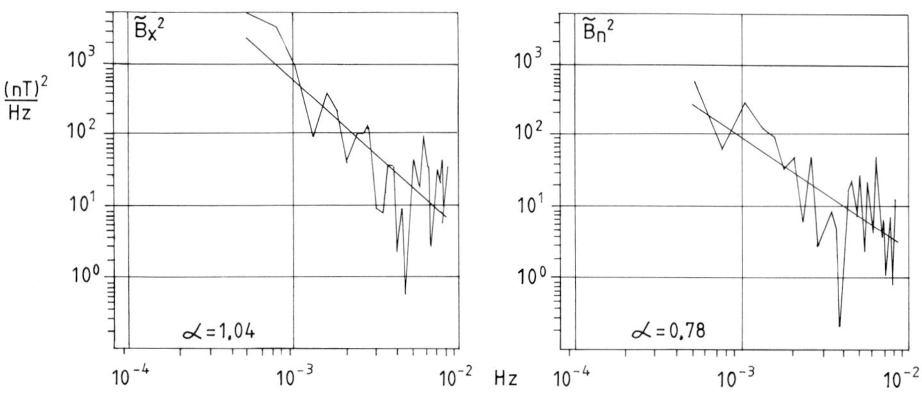

Fig. 1. Power spectrum density $[(nT)^2/\text{Hz}]$ of magnetic fluctuations between 10^{-4} and 10^{-2} Hz. Values of α are spectral indices of power function fits. Left panel refers to B_x, the sun-spacecraft component of B; right panel: same for B_n (normal component)

3. Comparison of observed and calculated ion temperatures

To calculate T_c on the basis of Eq. (8) the functions $u(x)$, $B(x)$, $\rho(x)$, and $\tilde{B}^2(x)$ must be known. Values of $B(x)$ and $\tilde{B}^2(x)$ were taken from magnetic field measurements (Riedler et al., 1986), those of $u(x)$ and $\rho(x)$ from plasma measurements (Gringauz et al., 1986) made onboard VEGA-1 in the PBSh region between $r = 2.5\,10^6$ and 10^6 km. The calculated values of T_c are shown in the lower part of Fig. 2 as a function of time and the cometary distance.

To determine the values of T_{obs} on the basis of the intensity profiles of cometary ions by the instrument TUNDE-M onboard VEGA-1 (simultaneously with values of u, B, ρ, $\tilde{B}^2$), the procedure described in Sect. 4 of Kecskeméty et al. (1986) was used. This is an iterative process to transform the time profiles $I(E, t)$ observed in the spacecraft frame of reference in a fixed direction, into a velocity distribution in the SWRF which is assumed to be isotropic. Due account is taken of the geometric sensitivity of the measuring device which, because of the specific conditions mentioned, is energy dependent. Up to now, the process has only been used to calculate six T_{obs} values spread over the entire range of the trajectory part selected. The resulting T_{obs} values are shown in the upper panel of Fig. 2 as a function of time and distance r from the nucleus. For the sake of comparison, the observed $I(E, t)$ time profiles are shown for the four operational energy channels together with the measured $\tilde{B}^2(t)$ values in Fig 3.

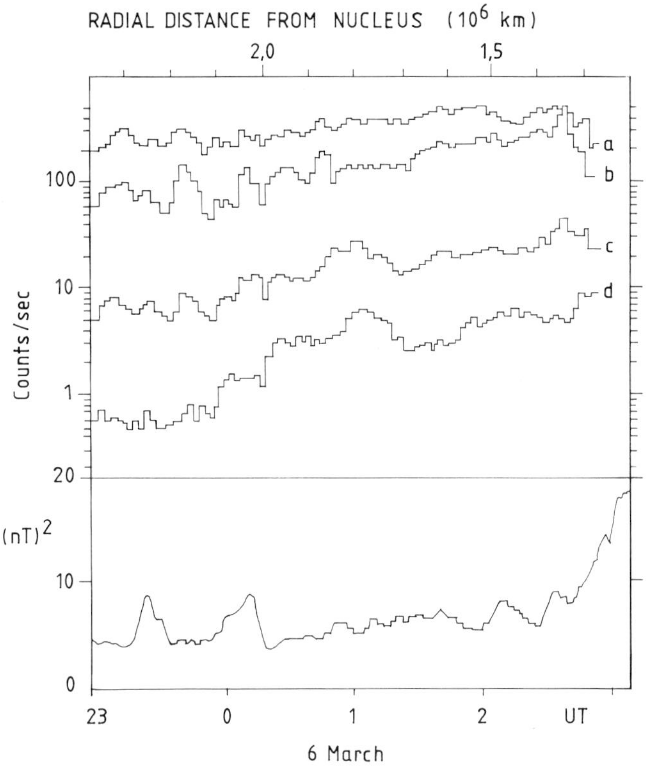

Fig. 3. Main input values used to calculate T_{obs} and T_c. Upper panel: counting rates of ions in the 4 operational channels (a: 78–88, b: 88–107, c: 107–119, d: 119–132 keV); lower panel: variation of $\tilde{B}^2$ as calculated from MISCHA measurements

4. Discussion

First of all, it should be noted that the pick-up mechanism alone cannot produce the ions observed because the viewing direction of the telescope (Somogyi et al., 1986 and Kecskeméty et al., 1986). The mere presence of ions observed proves the operation of an additional acceleration mechanism capable of producing abundant fluxes of ions: several thousands ions per ($cm^2 \cdot sr \cdot s$) of energies 78–88 keV and several tens of ions per ($cm^2 \cdot sr \cdot s$) of energies 119–132 keV (assuming O^+ ions, Fig. 3). The geometry factor of the telescope for isotropic distributions is 0.25 cm^2 sr.

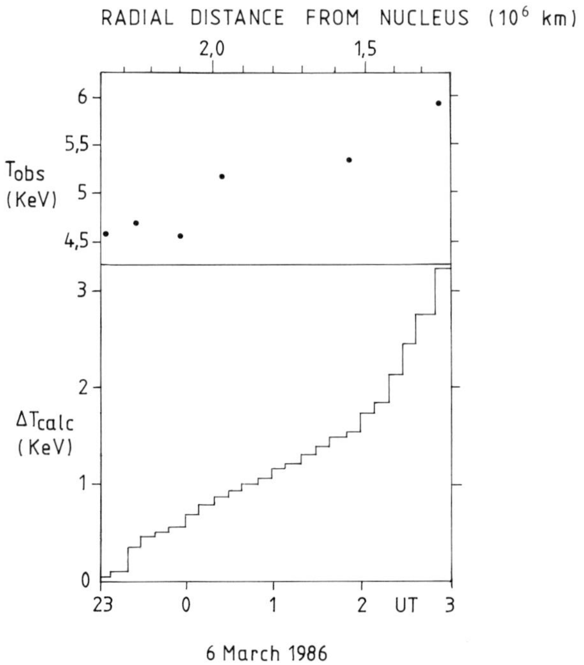

Fig. 2. Upper panel: observed ion temperatures from TUNDE-M using magnetic field (MISCHA) and solar wind (PLASMAG-1) data; lower panel: calculated ion temperatures based on stochastic Fermi acceleration

The change of ionic temperatures predicted on the basis of the theory of the second order Fermi mechanism is shown in the lower panel of Fig. 2, the actually measured temperatures are shown in the upper panel. The theoretical values are averages of subsequent 10 minute intervals, the experimental values are also based on 10 min flux averages spaced at about 40 min distances from each other.

The experimental curve exhibits a rather uniform increase of about 1.4 keV during the period studied (from 2300 UT on March 5 to 0300 UT on March 6). The decrease of the solar wind speed was about 6% during this period (from 500 km s^{-1} to 470 km s^{-1}) resulting in a compression factor of $F \simeq 1.06$. Adiabatic compression with this factor gives rise to a temperature rise of $F^{2/3} \simeq 1.04$ which is negligible. The adiabatic compression is thus not efficient enough to produce the temperature increase observed. The second order Fermi mechanism would produce a temperature increase of about 3 keV, i.e. larger than the observed one.

The difference may, however, be explained by the approximations applied both in the theoretical part, both in the data reduction. In the theoretical part, the magnetic power spectrum was assumed to be proportional to k^{-1}. The main sources of error in the observed T_{obs} values come from the uncertainties of the background flux in the lowest energy channel and the deviations from exponential variation in the velocity distributions observed. At the present stage, no better quantitative agreement could have been expected.

In conclusion: The increase of ionic temperature between distances of 2.5 million and 1 million km from the comet nucleus is established with certainty and its most probable source is the second order Fermi mechanism.

Acknowledgement. The authors are greatly indebted to Dr. T.E. Cravens and Mr. L. Gan for carrying out the calculations necessary to determine the T_{obs} values displayed in Fig. 2.

References

Amata, E., V. Formisano: 1985, *Planet. Space Sci.* **11**, 1243

Galeev, A.A., B.E. Gribov, T.I. Gombosi, K.I. Gringauz, S.I. Klimov, P. Oberc, A.P. Remizov, W. Riedler, R.Z. Sagdeev, S.P. Savin, A.Y. Sokolov, V.D. Shapiro, V.I. Shevchenko, K. Szegő, M.I. Verigin, E.G. Eroshenko: 1986, *Geophys. Res. Letters* **13**, 841

Gringauz, K.I., T.I. Gombosi, A.P. Remizov, I. Apáthy, I. Szemerey, M.I. Verigin, L.I. Denschikova, A.V. Dyachkov, E. Keppler, I.N. Klimenko, A.K. Richter, A.J. Somogyi, K. Szegő, S. Szendrő, M. Tátrallyay, A. Varga, G.A. Vladimirova: 1986, *Nature* **321**, 282–285

Ip, W.-H., W.I. Axford: 1982, in: *Comets*, ed. L.L. Wilkening, Tucson, Arizona, Univ. Arizona Press, p. 588

Ip, W.-H., W.I. Axford: 1986, *Planet. Space Sci.* **34**, 1061

Kecskeméty, K., T.E. Cravens, V.V. Afonin, G. Erdős, E.G. Eroshenko, L. Gan, T.I. Gombosi, K.I. Gringauz, E. Keppler, I.N. Klimenko, R.G. Marsden, A.F. Nagy, A.P. Remizov, A.K. Richter, W. Riedler, K. Schwingenschuh, A.J. Somogyi, K. Szegő, M. Tátrallyay, A. Varga, M.I. Verigin, K.-P. Wenzel: 1986, 20th ESLAB Symp., ESA-SP-250 Vol. 1, p. 109

Riedler, W., K. Schwingenschuh, E.G. Yeroshenko, V.A. Styashkin, C.T. Russell: 1986, *Nature* **321**, 288

Sagdeev, R.Z., V.D. Shapiro, V.I. Shevchenko, K. Szegő: 1986, *Geophys. Res. Letters* **13**, 85–88

Somogyi, A.J., K.I. Gringauz, K. Szegő, L. Szabó, Gy. Kőzma, A.P. Remizov, J. Erő jr., I.N. Klimenko, I.T.-Szűcs, M.I. Verigin, J. Windberg, T.E. Cravens, A. Dyachkov, G. Erdős, M. Faragó, T.I. Gombosi, K. Kecskeméty, E. Keppler, T. Kovács jr., A. Kondor, Yu. I. Logachev, L. Lohonyai, R. Marsden, R. Redl, A.K. Richter, V.G. Stolpovskii, J. Szabó, I. Szentpétery, A. Szepesváry, M. Tátrallyay, A. Varga, G.A. Vladimirova, K.P. Wenzel, A. Zarándy: 1986, *Nature* **321**, 285–287

Tsurutani, B.T., E.J. Smith: 1986, *Geophys. Res. Letters* **13**, 259

Astron. Astrophys. 187, 297–303 (1987)

ASTRONOMY AND ASTROPHYSICS

Measurements of low energy electrons and spacecraft potentials near comet P/Halley

A. Pedersen[1], R. Grard[1], J.G. Trotignon[2], C. Beghin[2], Y. Mikhailov[3], and M. Mogilevsky[3]

[1] Space Science Department, ESA/ESTEC, Keplerlaan, 2200 AG Noordwijk, The Netherlands
[2] Laboratorie de Physique et de Chimie de l'Environnement, CNRS, F-45045 Orléans, France
[3] IZMIRAN, 142092 Troitsk, USSR

Received February 18, accepted April 9, 1987

Summary. Data collected with the Langmuir probes and the electric field sensors on Vega-1 and Vega-2 show that the spacecraft potentials were a few volts positive in the outer coma and less positive near the closest approach. This leads to the conclusion that the energy of the secondary electrons generated by gas and dust impacts was lower than assumed in previous model work. The electron density in the outer coma were of the order $50\,cm^{-3}$ for Vega-1 and $25\,cm^{-3}$ for Vega-2 with wavelike fluctuations over a time interval of 10 min, or a distance of 500,000 km. Steep increases of the electron density were observed at a distance of 150,000 km from the nucleus, in the vicinity of the cometopause. Langmuir probe data are judged to suffer little influence from gas and dust impacts at distances larger than 50,000 km where the electron densities for both flights were of the order of $10^3\,cm^{-3}$. The electron mean kinetic energy was typically 0.5 eV for both flights, a slight increase was observed by Vega-1 within 150,000 km. Asymmetries between the potentials of two electric field probes, measured relative to the spacecraft, are thought to result from small dust impacts. Such asymmetries are characterized by a gradual increase (Vega-1) or a clear onset (Vega-2) at distances of about 80,000 km; a strong enhancement of this effect was also seen on Vega-2 at distances less than 10,000 km from the nucleus.

Key words: comet Halley – Vega – Langmuir probe – plasma – electron density – spacecraft charging

1. Introduction

The two Vega spacecraft carried Langmuir probes to perform plasma diagnostics, and a double probe antenna to measure electric fields from dc to several hundred kHz.

This paper presents the results obtained with the Langmuir probe which, in addition to electron density and energy, gives the spacecraft potential relative to the ambient plasma.

The electric antenna yields complementary information about spacecraft potential and secondary electron emission due to dust impacts on the probes and the spacecraft. The high frequency wave measurements are dealt with in a separate paper (Trotignon et al., this issue).

Send offprint requests to: A. Pedersen

Relatively infrequent large dust particle impacts on the spacecraft and on the probes are observed as individual, short spikes in the Langmuir probe current, and to a lesser degree in the electric field measurements. The analysis of these discrete phenomena is not the object of this paper but is left for a future publication.

This study is based on information collected within a distance of 850,000 km, which therefore does not cover the bow shock. The Vega-1 experiment performed satisfactorily up to a distance of 28,600 km from the nucleus, when the data flow was suddenly interrupted, probably by the impact of a large dust particle. One of the Vega-2 Langmuir probes suffered a partial damage possibly caused by a direct hit on a connecting cable, at a distance of 70,000 km.

2. Electric field and Langmuir probes

2.1. *Instrument description*

Figure 1 is a schematic drawing showing the positions of the electric field probes, P1 and P2, and Langmuir probes, LP1 and LP2. These probes are mounted on two booms extending from the outer solar panels; P1 and P2 are located at 1.5 m from the edge of the panels and are separated from each other by a distance of 11 m; LP1 and LP2 are at approximately 0.60 m from the edge of the panels. The conductive surface which is used as

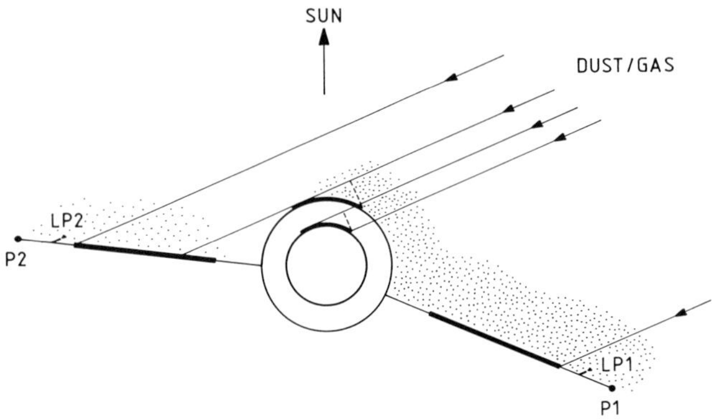

Fig. 1. Schematic representation of the Vega spacecraft showing the arrangement of the electric field sensors (P1, P2) and Langmuir probes (LP1, LP2). Also shown are the conductive surface elements oriented towards the Sun, the ram direction for dust and gas and the secondary electron cloud induced by impacts

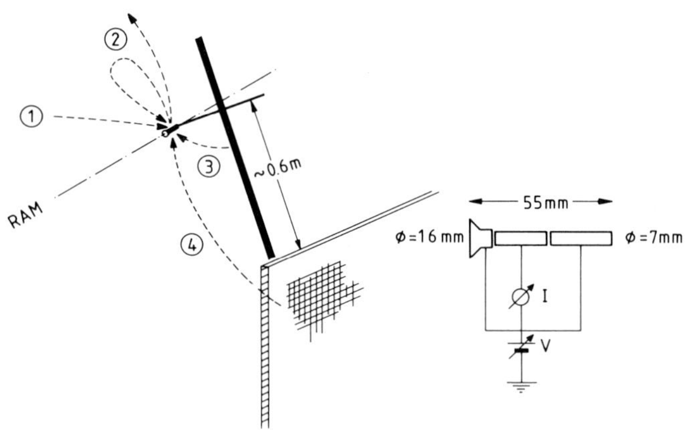

Fig. 2. The axis of the Langmuir probe shown on the right is aligned with the ram direction. The current components to be considered are those of ambient electrons (1), escaping photoelectrons (2), photoelectrons emitted from the boom supports (3) and from the solar panels (4)

a reference for potential measurements has an area of 1–2 m^2, is facing the Sun and is electrically connected to the spacecraft structure. The ram direction for gas and dust is indicated in Fig. 1.

The electric field probes are two spheres, 10 cm in diameter, painted with a conductive carbon material. The potentials of these sensors are biased towards that of space by means of 50 nA currents which are injected from the probes into the plasma. Each sphere is connected with a coaxial cable to a preamplifier; the shield of the signal wire is polarized at the same potential as the probe in order to minimize the capacitance at the input of the electronic circuitry. The potential differences between the two probes, $V(\mathrm{P1}) - V(\mathrm{P2})$, and between one probe and the spacecraft structure $V(\mathrm{P2}) - V(\mathrm{S})$ are measured, the former up to several kHz. Here we will discuss only information derived from the quasistatic channels (0–0.5 Hz).

One Langmuir probe is shown in some detail in Fig. 2. Each probe, of cylindrical shape, is aligned within a few degrees from the ram direction (i.e. the orientation of the spacecraft motion with respect to the comet) during the encounter. The Langmuir probe consists of three elements biased to the same voltage relative to the spacecraft structure. The "trumpet" element facing the ram direction is designed for picking up cometary gas and dust; it shields the middle element connected to the current amplifier from secondary emission caused by impacting gas and dust. The third element serves as a guard to the titanium rod supporting the Langmuir probe. All probe elements are covered with a conductive carbon paint. The leading probe was swept sinusoidally with a periodicity of 32 s during the cometary approach. A slow sweep was selected in order to avoid any influence on the wave experiments and also because of bit rate limitation. The other probe was kept at a fixed bias of +5 V in order to detect fast variations of the electron flux.

2.2. *Interpretation of Langmuir probe measurements*

A Langmuir probe is best suited for measurements in a plasma where the Debye length is small compared to its dimensions. This situation was approached in the inner part of the coma, but not in the solar wind; a discussion of the probe performance in the solar wind and the outer coma is therefore required.

Under the assumption of isotropic electron distributions, and in the absence of secondary or photo-electron emissions a probe at local plasma potential collects an electron current I_0 given by:

$$I_0 = \sum_i I_{0i} = \sum_i N_i e v_{ei} A/4 \tag{1}$$

where A is the probe area, and N_i and v_{ei} are the densities and mean thermal speeds of the different populations, namely the ambient electrons and the photoelectrons emitted from the solar panel and the supporting boom. For a probe biased negatively at a potential V relative to the local plasma potential, the electron current is

$$I = \sum_i I_{0i} \exp(-V/V_{ei}), \tag{2}$$

where $V_{ei} = kT_{ei}/e$. This equation is valid for Maxwellian distributions and for negligible bulk speeds. Maxwellian distributions are approximations of the real conditions whereas the bulk speeds in the coma are negligible compared to the electron speeds.

For a probe at a positive potential V, we assume that the electron collection process is approximately the same as for a sphere when the Debye length is larger than its radius. This approach is justified on the basis that the equipotential surfaces around any probe have a nearly spherical shape at large distances when the bias potential equals a few times V_{ei}. The electron current for a positive potential V can consequently be approximated by the formula for a sphere at attractive potentials in a tenuous plasma;

$$I = \sum_i I_{0i}(1 + V/V_{ei}). \tag{3}$$

Figure 2 also shows a few trajectories of photoelectrons emitted from the probe itself, some escaping and some orbiting back to its surface. The flux of returning photoelectrons increases when the probe potential becomes more positive until there is no net escape. The space charge due to photo-electrons is important near large planar sunlit surfaces. However photo-electrons emitted and returning to a small probe have a negligible effect in terms of modifying Eq. (3). Typical model current-voltage characteristics are illustrated in Fig. 3, for solar wind conditions (10 cm^{-3}, 10 eV) in the top panel, and for an additional cometary electron population (50 cm^{-3}, 0.5 eV) in the lower panel. The spacecraft potential, which is used as reference, is a few volts more positive than that of the ambient plasma, $V(\mathrm{PL})$. The current measurements performed during one voltage sweep yield a characteristic which can be decomposed into parts 1, 2, 3 and 4, with reference to Figs. 2 and 3. The first component represents the current of cometary electrons (Fig. 3b). This is the parameter sought and which is competing with other electron currents. The solar wind ion and electron currents are labelled p(sw) and e(sw) respectively for typical solar wind conditions. Ion currents can be neglected compared to electron currents. The relative importance of ions is further reduced in the coma where the cometary ions have a smaller bulk speed. The current of photoelectrons escaping from the probe surface is represented by component 2 and the contributions due to photoemission from the solar panel, the boom and the probe support are combined in the component marked 3 + 4.

It is possible to estimate from a careful analysis of solar wind current voltage curves that the density of the photoelectrons emitted from surfaces other than the Langmuir probes is of the order of 20 cm^{-3} in the vicinity of these sensors. Consequently

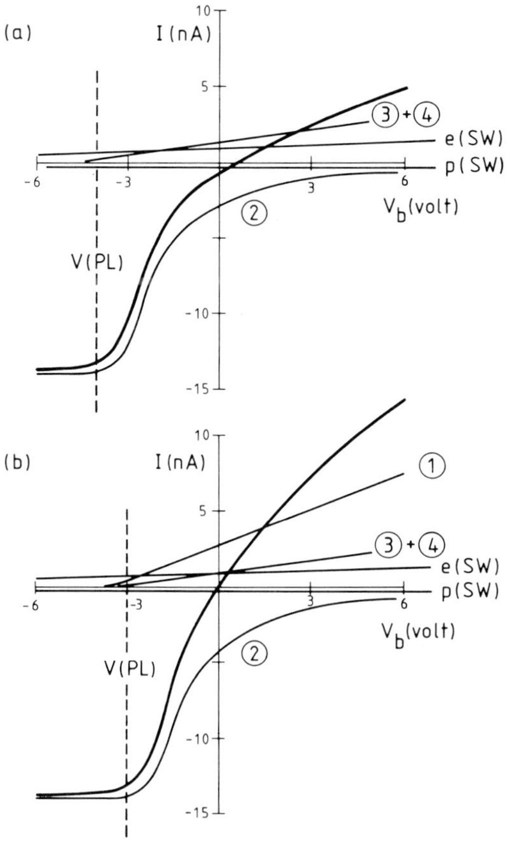

Fig. 3. Model current-voltage characteristics for solar wind in upper panel ($N_e = 10\ \text{cm}^{-3}, kTe = 10\,\text{eV}$) and coma environment in lower panel ($N_e = 50\,\text{cm}^{-3}$ and $kTe = 0.5\,\text{eV}$. The current I is positive when electrons are collected by the probe; V_b is the bias voltage relative to the spacecraft structure and $V(\text{PL})$ is the plasma potential. The components 1–4 refer to Fig. 2; *e*(sw) and *p*(sw) represent contributions from solar wind electrons and ions for conditions specified

it is not possible to measure the electron density and temperature of the solar wind under average conditions. These measurements can however prove extremely useful to characterize the level of contamination associated with photoemission (Fig. 3a). For example, the saturation photoelectron current density emitted by the probe itself (component 2) is $9.5\,\text{nA}\,\text{cm}^{-2}$ for Vega-1 at 0.792 AU and $8.6\,\text{nA}\,\text{cm}^{-2}$ for Vega-2 at 0.834 AU; this corresponds to a level of $6\,\text{nA}\,\text{cm}^{-2}$ at 1 AU, which is a typical value for this carbon paint near Earth.

The procedure to determine the cometary electron density and characteristic energy is the following. From the solar wind data it is possible to establish the components 2 and 3 + 4. When measuring a real current-voltage curve in a cometary environment these components can be subtracted to yield the cometary electron component provided it is of sufficient magnitude.

It was feared that the Langmuir probes would be influenced by secondary electron emission due to gas and dust impacts. In fact, no increase was observed during the flyby above the saturation photo-electron current measured in the solar wind, which indicates that the conical shield insured an efficient protection for the probe surface. With regard to secondary emission from the solar panels and booms it was found experimentally (Grard and Gringauz, 1986) that this production rate is less important than photoemission at distances larger than 40,000 – 50,000 km but may have to be taken into account at smaller distances from the nucleus.

In the laboratory or in planetary ionospheres, the electron density and temperature are usually evaluated with Eq. (2) in the retarding potential regime ($V < V(\text{PL})$). In the present case, the same approach is feasible provided the ambient electron density is larger than approximately $500\,\text{cm}^{-3}$, which gives rise to a current in the retarding potential regime large enough to be differentiated from the photoelectron background. However, for a situation such as that pictured in Fig. 3b information about the ambient plasma can only be derived by modelling the different current components over the retarding as well as the accelerating parts of the I-V curve. The electron density is defined with a better accuracy than the temperature, the reason being that Eq. (3) is rather insensitive to the latter parameter. For densities less than about $50\,\text{cm}^{-3}$ it is therefore necessary to assume the value of the temperature; for electron densities above $50\,\text{cm}^{-3}$, it is possible to define the combination of these two parameters which fits the observations best.

3. Probe and spacecraft potentials

The floating potential of a probe or a spacecraft is reached, in first approximation, when the flow of photoelectrons and secondary electrons which escape from its surface exactly balance the incoming flow of ambient electrons. In the case of a probe, the ambient electrons include not only particles of cometary origin but also secondary electrons and photo-electrons emitted from the spacecraft itself.

Figure 4 gives the equilibrium potentials of the Vega-1 and Vega-2 spacecraft, $V(\text{S})$, measured relative to Langmuir probe LP1 when the latter is biased at the local plasma potential. The spacecraft potential can be derived from the measured current-voltage curves, which have the same shape as the model curves in Fig. 3. The plasma potential $V(\text{PL})$ can be detected from the photo-electron saturation current, and the spacecraft floating potential for $I = 0$. The floating potentials are of the order of $+3.0$ V for both spacecraft in the outer coma ($8\ 10^5$ km); those values are close to those expected at that distance from the Sun and in such a plasma environment (see next section). The drops

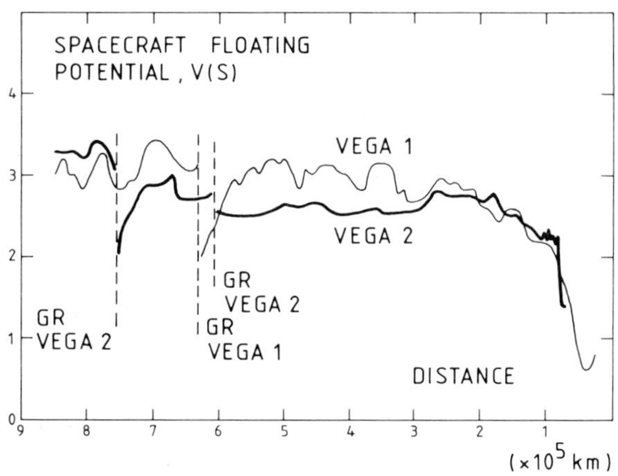

Fig. 4. Vega-1 and Vega-2 spacecraft potentials derived from the Langmuir probe characteristics. The events marked GR are nitrogen gas releases from the cooling system of an infrared instrument. The ionized nitrogen causes a reduction of the spacecraft potentials

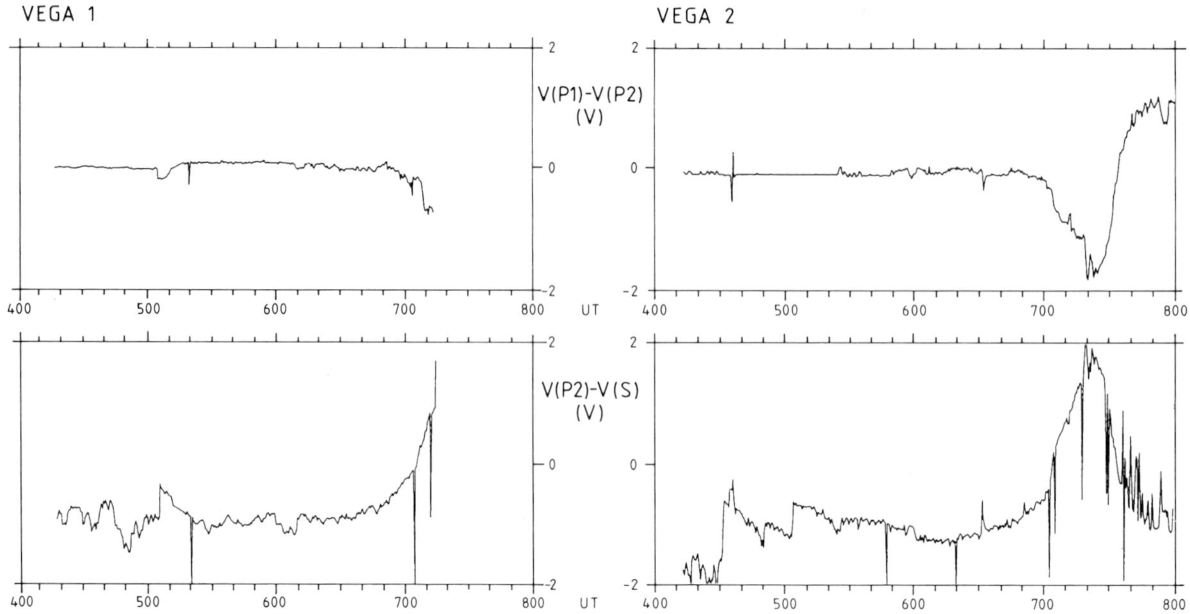

Fig. 5. The quasistatic signals V(p1) − V(P2) and V(P2) − V(S) for the two encounters (30 s averages)

in spacecraft potential labelled GR are associated with nitrogen gas releases by the cooling system of another instrument. Ionization of the nitrogen gas gives rise to an additional cold plasma component in the spacecraft environment, which entails a reduction of the floating potential.

The quasistatic signals measured with the spherical probes V(P1) − V(P2) and V(P2) − V(S) are shown in Fig. 5. The level of V(P1) − V(P2) is very close to zero in the outer coma as expected with two symmetric probes. It appears that V(P2) − V(S) is of the order of -1.0 V at $8\,10^5$ km; V(S) is indeed larger than V(P2) for two reasons: (a) the ratio of the area of conductive surface in contact with the plasma over the cross-section of the same surface illuminated by the Sun is smaller for the spacecraft than for the spherical sensor as shown in Fig. 1 and (b) the sensor is biased by a current of 50 nA which reduces its potential. The change of polarity of V(P2) − V(S) and the sudden decrease of V(P1) − V(P2) in the inner coma are more surprising, but we shall attempt to give a qualitative interpretation of these phenomena.

Figure 6 shows model current-voltage curves which are meant to explain the above potential measurements. The current I (normalized to the photo-emitting area) is positive for collected ambient electrons and negative for emitted electrons. In the inner coma these two components dominate. The voltage scale is referred to the ambient plasma.

The floating potentials of probes as well as the spacecraft are, in the absence of secondary electron emission from gas and dust, determined by the two currents mentioned above. The curves on the left demonstrate that in the absence of gas and dust impacts the spacecraft is ~1 V more positive than the probes due to a smaller ratio of total area to sunlit area (sunlit conductive area in Fig. 1).

Secondary electron emission from gas and dust will at some stage dominate over photo-electron emission, and the probes will have a larger secondary electron emission than the spacecraft due to spallation products from the spacecraft. This is the likely explanation why the spacecraft and probe 1 are at the same potential as illustrated on the right side of Fig. 6. The relatively small value of the spacecraft potentials is confirmed by the measurement of the ram energy of H^+ which shows very little retardation due to surface charging (Gringauz et al., this issue). This potential is much less than expected from preflight calculations (Young, 1986) because the mean kinetic energy of the secondary electrons is probably smaller than anticipated.

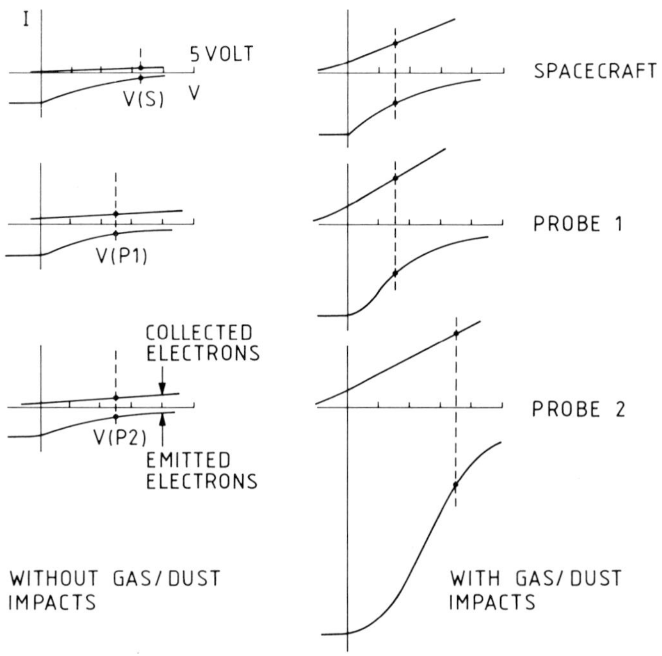

Fig. 6. Left side: schematic current-voltage characteristics illustrating the potential differences V(P1) − V(P2) observed in the outer coma (Fig. 5). Right side: representation of characteristics in the inner coma under the influence of gas and dust impacts (see text for explanations)

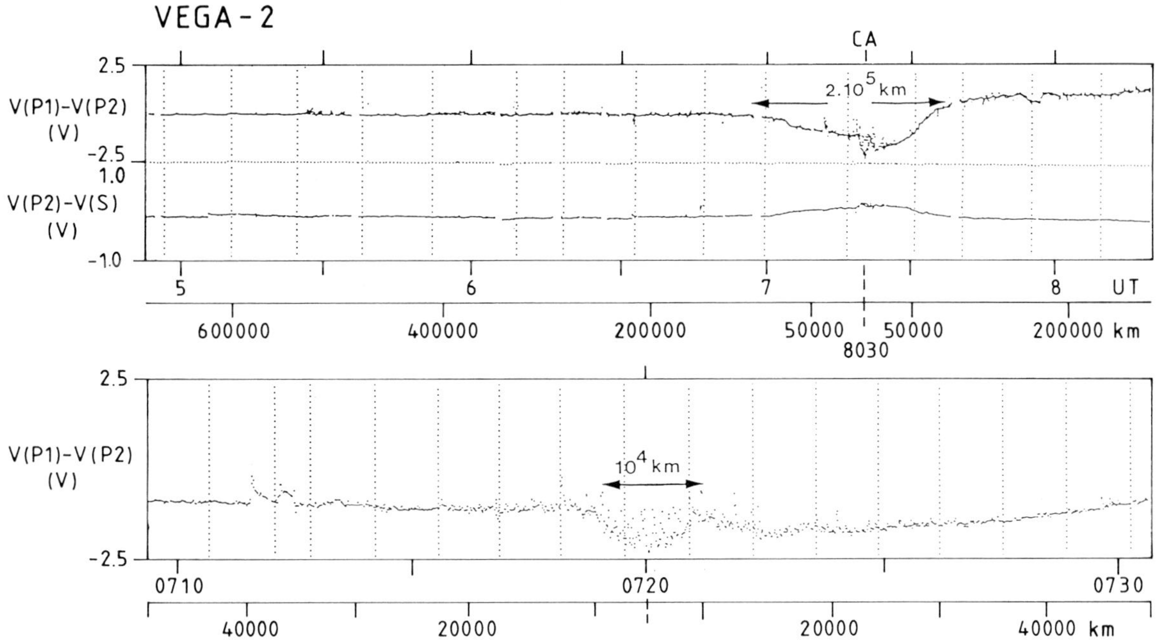

Fig. 7. High resolution plots of $V(P1) - V(P2)$ on Vega-2 showing fluctuations of this parameter supposedly associated with variations of the dust count-rate

Why is the potential of P2 approximately two volts more positive than those of P1 and the spacecraft? The only plausible explanation is that secondary electron emission from P2 is relatively larger than from P1 and the spacecraft because of the dust and gas flow direction. Spallation products from dust impacts increase in number by orders of magnitude for small incidence angles and are strongly focussed along the forward direction of the impacting particle (Zook et al., 1984). The lagging probe, P2, is certainly more exposed to spallation products than the leading probe, P1 (Figure 1); a higher production rate of sputtered material and secondary electrons from its surface may therefore be expected, as illustrated in Fig. 6 (right side).

A decrease of $V(P1) - V(P2)$ or an increase of $V(P2) - V(S)$ should therefore be detected whenever the dust impact rate increases; this trend can indeed be found in the Vega-2 data. At approximately 0705 UT (80,000 km) it can be seen in Fig. 5, and with better time resolution in Fig. 7, that $V(P1) - V(P2)$ goes more negative and $V(P2) - V(S)$ more positive. This is also the time when the impact rate of dust particles in the mass range 10^{-16}–10^{-10} g increases, as observed by the SP-1 experiment (Vaisberg et al. 1986). There is a further marked decrease of $V(P1) - V(P2)$ in a 10,000 km wide region around the closest approach. The lowest panel of Fig. 7 has sufficient time resolution to show this. Again this corresponds well with dust data, mainly with a maximum of the counts of small dust particles in the mass range 10^{-15}–10^{-16} g (Vaisberg et al. 1986).

4. Electron density and temperature

The electron density, N_e, and mean kinetic energy, kT_e measured by Vega-1 are plotted in Fig. 8 for cometary distances less than $8.5\,10^5$ km. The total density is measured with an estimated accuracy of $\pm 20\%$ and is represented by the full line. The shaded area gives the density range of cometary electrons after substraction of the solar wind component and removal of the spacecraft photoelectron environment.

The density shows wavelike variations between 40 and $80\,\text{cm}^{-3}$ at distances between 9.5 and $6.35\,10^5$ km, well within the bow shock. At a distance of $6.35\,10^5$ km the nitrogen gas release from the cooling system of an infrared optical instrument gives rise to an additional plasma component, whose onset appears quite clearly in the data sequence. For the electrons of cometary origin as well as for those generated by photoionization of nitrogen, a value of 0.5 eV for kT_e fits the data reasonably well. The nitrogen gas flow decreases steadily, but possibly

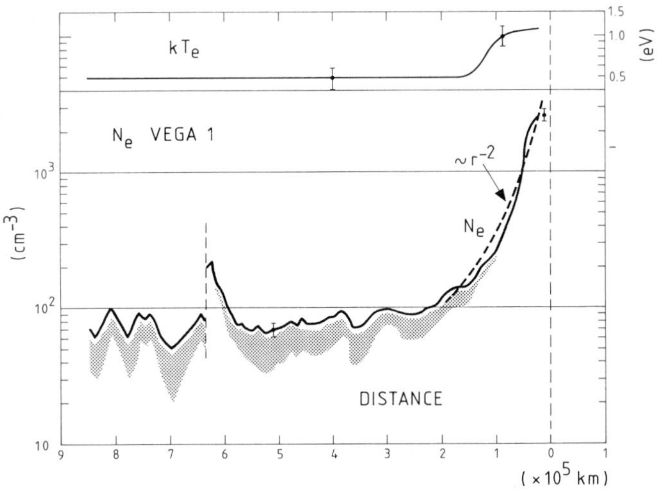

Fig. 8. Electron density, N_e, and mean kinetic energy, kT_e, versus distance from nucleus. The peak at 6.3×10^5 km is due to a nitrogen gas release. The uncertainty in the determination of the cometary electron density is given by the shaded area. The full line corresponds to the sum of the ambient cometary electrons and spacecraft secondary and photoelecrons. The two sets of results do not differ significantly within a cometocentric distance of 100,000 km when the flux of electrons emitted from the spacecraft surface can be neglected compared to that of ambient cometary electrons. The curve $1/r^2$ is given for reference

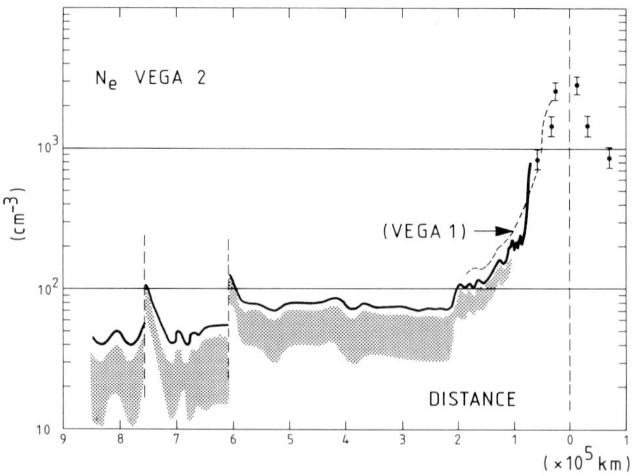

Fig. 9. Display similar to that of Fig. 8 but for the electron density measured by Vega-2. The voltage bias sweep system failed at a distance of about 10^5 km but the density profile can be extrapolated with a reduced accuracy (vertical bars). Two nitrogen gas releases were initiated between 8 and 6 × 10^5 km. The Vega-1 results are represented by a thin dashed line for comparison

contaminates the plasma measurements down to a distance of 3 10^5 km. However, the increase in electron density observed at 3.8 10^5 km reflects a local property of the cometary plasma. The density measurements at distances less than 2 10^5 km are probably not significantly influenced by nitrogen releases or photoemission from the spacecraft surface. The density reaches a value of nearly 3000 cm^{-3} and the electron mean kinetic energy is of the order of 1 eV at a distance of about 30,000 km when the instrument is probably damaged by dust impact.

The Vega-2 results shown in Fig. 9 display density fluctuations at distances of the order of 8 10^5 km similar to those already observed with Vega-1; the average density, however, appears to be lower by a factor approximately 2. The measurements are believed to be influenced by the openings of two nitrogen tanks probably down to a cometocentric distance of 2 10^5 km (the two gas releases were simultaneous on Vega-1). At 75,000 km from the nucleus, the voltage sweep system of the probe fails; the electron density can still be estimated (vertical bars) assuming that the mean kinetic energy is of the order of 1 eV. In the close vicinity of the nucleus, the instrument is saturated and no data is available.

5. Discussion

We shall now review the results which have been obtained from this study, progressing radially outwards from the inner regions towards the outer coma up to a cometocentric distance r = 850,000 km.

It has been established with Giotto that the plasma density falls off as $1/r$ within the contact surface and $1/r^2$ outside this boundary which was crossed at 4,500 km (Balsiger et al., 1986); this finding is in agreement with the model of Körösmezey et al., (1986), for example.

It is also seen from Fig 8 that the electron density measured on Vega-1 varies approximately like $1/r^2$ between 50,000 km and 200,000 km; i.e. this would agree with a model where the plasma flow outside the contact surface is nearly radial and recombinations can be neglected. There are however some significant deviations from a simple $1/r^2$ dependence near 40,000 km and a tendancy for the density to level off close to the nucleus; this feature is probably caused by a plasma enhancement coinciding with a dust and gas jet. The Vega-2 results also show small scale fluctuations associated with a jet near 83,000 km.

The cometopause was observed by Gringauz et al. (1986) at 150,000 km; this boundary which marks the transition from the solar wind to the cometary plasma has no clear signature in the electron density profile and can be related at the utmost with the increase observed at 200,000 km on Vega-2.

The maximum electron density measured with the Langmuir probe on Vega-1 is 2–3 10^3 cm^{-3}, in good agreement with other results; Zastenker et al. (this issue) reported a density of 2 10^3 cm^{-3} at 30,000 km during the same flyby.

The electron mean kinetic energy observed on Vega-1 is about 1 eV inside the cometopause, in fair agreement with the model of Körösmezey et al. (1986). It is at present not understood why the electron energy should be lower outside the cometopause. Most electrons are born near the contact surface and relatively few neutrals are ionized further away. In first approximation, the electron mean kinetic energy should remain unchanged outside the cometopause where the mean free path is large.

Simultaneous dips in electron density and magnetic field are observed on Vega-1 at a distance of 350,000 km. It is possible that a change of the solar wind magnetic field gives rise to a reduction in the pile-up of magnetic flux, providing an easier escape for the plasma.

The contamination caused by the nitrogen gas releases is clearly visible in Figs. 8 and 9, information about electron density and temperature is consequently lost for 1–2 10^5 km following the beginning of the releases. It is interesting to note that the electron mean kinetic energy in the ionized nitrogen cloud which surrounds the spacecraft is the same as in the outer coma (0.5 eV).

The electron density at 800,000 km is higher than it would follow from a simple $1/r^2$ dependence. This effect is particularly pronounced for Vega-1 and is possibly associated with a compression of the solar wind plasma within the bow shock. Rème et al. (1986) have indeed observed on Giotto an increase of the electron density from 5 to 8 cm^{-3} at the bow shock, in the energy range above 10 eV. This compression is also acting on the low energy electron population which is sampled by the Langmuir probe on VEGA.

The cold electron densities of 50 cm^{-3} measured on Vega-1 are much too large to be solely explained in terms of contamination by photoelectrons. There are two possibilities for the origin of this electron population:

1. It is formed in the inner coma and after a nearly radial transport piles up inside the bow shock in a way which is controlled by the connection between the flux tubes and the solar wind.

2. It originates from the ionization of neutrals in the outer coma. The ionization rate is low, but gives rise to significant electron densities if the outward plasma motion is impeded.

The real situation may result from a combination of both effects. The analysis of the Vega-1 data across the bow shock, which has not been performed so far, will hopefully provide more information about the creation and transport of this cold plasma.

The Vega-2 data show a much lower cold electron density in the outer coma, close to the limit of detectability, which is a possible consequence of a reduced gas production of Halley's comet combined with a different condition for the solar wind environment.

The floating potential of the conductive surface elements connected to the spacecraft structure have been obtained by measuring the plasma potential with the Langmuir probe. This potential is of the order +3.0 V in the outer coma and drops to less than 1.0 V near 50,000 km; these results are compatible with the determination of the H^+ ram energy of 30 eV by Gringauz et al. (this issue).

Differences between the potentials of the electric sensors are explained in terms of asymmetry in the generation of spallation products caused by dust impacts on the spacecraft surface; the potential of the trailing probe P2 is more positive than that of the leading one P1 and its variation is correlated with that of the dust impact rate measured by Vaisberg et al. (1986).

References

Balsiger, H., Altwegg, K., Bühler, F., Geis, J., Ghielmetti, A., Goldstein, B., Goldstein, R., Huntress, W., Ip, W-H., Lazarus, A., Meier, A., Neugebauer, M., Rettenmund, W., Rosenbauer, H., Schwenn, R., Sharp, R., Shelley, E., Ungstrup, E., Young D.: 1986, *Nature*, **321**, 330

Grard, R., Gringauz, K.: 1986, *Geophy. Res. Lett.* **13**, 877

Gringauz, K., Gombosi, T., Remizov, A., Apathy, I., Szemerey, I., Verigin, M., Denchikova, L., Dyachkov, A., Keppler, E., Klimenko, I., Richter, A., Somogyi, A., Szegö, K., Szendrös, S., Tatrallyay, M., Varga, A., Vladimirova, G.: 1986, *Nature* **321**, 282

Gringauz, K., Veregin, M., Richter, A., Gombosi, T., Szegö. K., Tatrallyay, M., Remizov, A., Apathy I.: 1987, *Astron. Astrophys.*

K̈orösmezey, A., Cravens, T., Gombosi, T., Nagy, A., Mendis, D., Szegö, K., Gribov, B., Sagdeev, R., Shapiro, V., Shevchenko, W.: 1986, *J. Geophys. Res.* (submitted)

Rème, H., Sauvaud, J., d'Uston, C., Cotin, F., Cros, A., Anderson, K., Carlson, C., Curtis, D., Lin, R., Mendis, D., Korth, A., Richter, A.: 1986, *Nature*, **321**, 349

Trotignon, J.G., Beghin, C., Grard, R., Pedersen, A., Formisano, V., Mogilevsky, M., Mikhaïlov, Y.: 1987 *Astron. Astrophys.*

Vaisberg, O., Smirnov, V., Gorn, L., Iovlev, M., Balikchin, M., Klimov, S., Savin, S., Shapiro, V., Shevchenko V.: 1986, *Nature* **321**, 274

Young, D.: 1986, *ESA SP-1066*, 85

Zastenker, G., Vaisberg, O., Smirnov, V., Fedorov, A., Omelchenko, A.: 987, *Astron. Astrophys.*

Zook, H., Lange, G., Grün, E., Fechtig, H.: 1984, *Lunar Planet Sci 15*, NASA/LPS USRA Report

Astron. Astrophys. 187, 304–306 (1987)

ASTRONOMY AND ASTROPHYSICS

Charge exchange of solar wind ions in the coma of comet P/Halley

E.G. Shelley [1], S.A. Fuselier [1], H. Balsiger [2], J.F. Drake [1], J. Geiss [2], B.E. Goldstein [3], R. Goldstein [3], W.-H. Ip [4], A.J. Lazarus [5], and M. Neugebauer [3]

[1] Lockheed Palo Alto Research Laboratory, Dept. 91-20, Bldg. 255, 3251 Hanover St., Palo Alto, CA 94304, USA
[2] Physikalisches Institut, University of Bern, Bern, Switzerland
[3] Jet Propulsion Laboratory, California Institute of Technology, Pasadena, CA, 91109, USA
[4] Max-Planck-Institut für Aeronomie, D-3411 Katlenburg-Lindau, Federal Republic of Germany
[5] Massachusetts Institute of Technology, Cambridge, MA 02139, USA

Received January 29, accepted July 10, 1987

Summary. In addition to solar wind pick-up of cometary ions, charge exchange between solar wind ions and cometary neutrals is, potentially, an important comet-solar wind interaction closer to the nucleus. One monitor of the charge exchange rate is the relative abundances of solar wind He^{2+} and He^{+}. Radial profiles of these two ions measured by the Giotto IMS/HERS instrument on the inbound trajectory to comet Halley are presented. Results indicate that charge exchange may be an important solar wind loss process in the comet Halley coma at distances $\sim 10^5$ km from the nucleus.

Key words: charge exchange – comet-solar wind interactions – cometary modeling

1. Introduction

The process of comet-solar wind interaction is to a large degree determined by the solar wind pick-up of cometary ions in the extensive coma of neutral gas surrounding the nucleus (Ip and Axford, 1982). However, at small distances ($\sim 10^5$ km from the nucleus), charge exchange between the solar wind ions and the cometary neutral gas can become an important solar wind loss process (Ip and Hsieh, 1982). Experimentally it is difficult to uniquely determine the solar wind hydrogen charge exchange loss in the comet coma because there are competing cometary processes which can affect the ion density and because the energetic neutral hydrogen atoms are not easily detectable. However, the relative concentrations of He^{2+} and He^{+} provides an excellent monitor of the charge exchange process since no measurable He^{+} is expected from the comet. If the solar wind He^{2+} charge exchanges to He^{+} much more rapidly than the He^{+} charge exchanges to He, the ratio of the two charge states is a direct indicator of the absolute charge exchange process. Including a significant loss of He^{+} by charge exchange, the relative concentrations of He^{2+} and He^{+} provides at least a lower bound on the charge exchange process in the comet Halley coma.

The Giotto IMS/HERS instrument observed a significant increase in the relative concentrations of He^{2+} and He^{+} in the Halley coma at a distance of about 10^5 km from the nucleus (Balsiger et al., 1986b). The IMS/HERS instrument (Balsiger et al., 1986a) measured the three dimensional velocity distributions of both He^{+} and He^{2+} within a 60 degree wide annular cone extending from 15 to 75 degrees relative to the spacecraft spin axis (which was also the spacecraft velocity vector relative to the comet). This cone included most of the deflected solar wind velocity distribution and a significant fraction of the cometary pick-up ion velocity distributions at distances greater than about 10^5 km.

In this paper we present radial profiles of both charge states of helium ions measured by the IMS/HERS instrument. Predicted concentrations of He^{2+} and He^{+} based on a fluid model of incompressible flow (Ip and Hsieh, 1982) are compared to observed concentrations. Although both the predicted and observed concentrations of He^{+} relative to He^{2+} increase with decreasing distance from the comet, the predicted increase is much smaller than the observed increase.

2. Observations

Figure 1 shows the density for Mass per Charge (M/Q) = 1, 2, and 4 ions versus time for March 13, 14 from 2000-0010 UT. (All times in this paper are ground received times. To obtain spacecraft event times, subtract 8.06 min) The time interval shown covers the region after the cometary bow shock crossing at about 1930 UT up to the contact surface at 0010 UT. All densities are based on ion fluxes observed within the finite field of view of the IMS/HERS instrument (Balsiger et al., 1986a). The $M/Q = 1$ ion densities are averaged over ten minutes and arbitrarily scaled to 25 times the $M/Q = 2$ ion density at 2200 UT. The $M/Q = 2$ densities are averaged over five minutes and, to improve statistics, the $M/Q = 4$ densities are averaged over 30 min from 2000 to 2300 UT and 10 min from 2300 to 0010 UT. Error bars shown are one sigma based on counting statistics only.

The $M/Q = 1$ density, which consists primarily of solar wind protons, remains approximately constant from 2200 to 2320 UT after which it drops rapidly by more than an order of magnitude. Similarly, the $M/Q = 2$ density, which is dominated by solar wind helium (but also contains cometary H_2^+), is nearly constant from 2200 to 2330 UT after which is also drops rapidly, being down by several orders of magnitude by 0010 UT. In contrast, the $M/Q = 4$

Send offprint requests to: E.G. Shelley

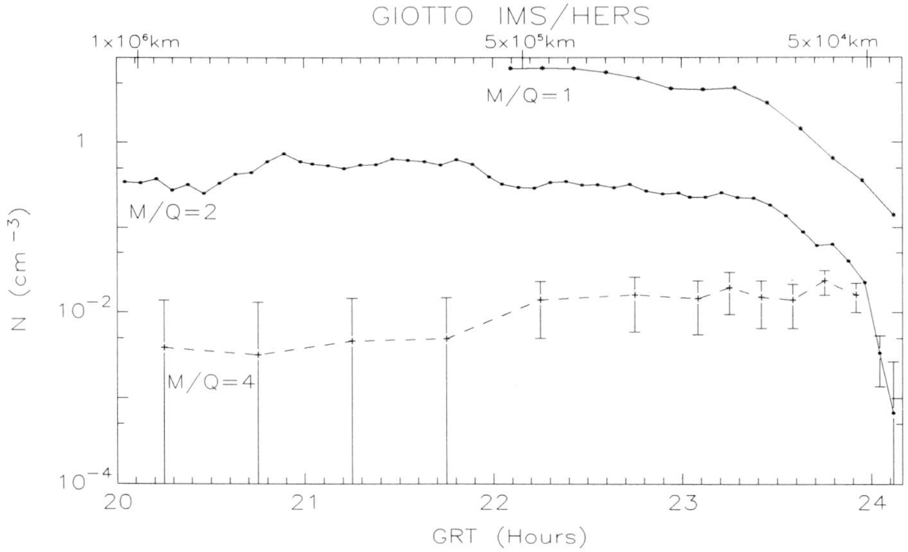

Fig. 1. Density versus time for $M/Q=1$ ions (predominantly solar wind protons), $M/Q=2$ ions (primarily solar wind helium), and $M/Q=$ ions (predominantly He^+). This plot covers the region between the inbound bow shock crossing (at 1930 UT) and the contact surface crossing (at about 0010 UT). The proton density is arbitrarily normalized to 25 times the $M/Q=2$ density at 2200 UT

density increases from 2000 to 2400 UT before dropping abruptly at 0010 UT.

In the absence of sources or losses, the solar wind ion components would be expected to increase with decreasing distance to the comet, conserving total flux as the flow speed is reduced. Because the cometary neutral molecular density also increases as the spacecraft approaches the comet, the observed decrease in $M/Q=1$ density in Fig. 1 is qualitatively consistent with increasing conversion of solar wind protons to neutral hydrogen by charge exchange with cometary neutrals. The decrease in $M/Q=2$ density in Fig. 1 coupled with the increase in $M/Q=4$ density is also qualitatively consistent with charge exchange of solar wind He^{2+} ($M/Q=2$) to He^+ ($M/Q=4$).

The above qualitative discussion does not take into consideration the finite field of view of the IMS/HERS instrument. As the spacecraft approaches the comet, the solar wind is slowed due to mass loading and the flow in the spacecraft frame moves more into the ram direction. Since the field of view of the IMS/HERS instrument does not include the 15 degree half-angle cone centered on the ram direction, this movement results in an apparent loss of ions. Phase space contour plots (not shown here) indicate that about 1/2 of the $M/Q=2$ ion distribution is outside the field of view of the IMS/HERS instrument after 2330 UT and nearly all of the distribution is outside the field of view of the instrument by 0010 UT. From 2330 to 0010 UT, the IMS/HIS instrument, which looks primarily in the ram direction, shows a continued increase in cometary ion flux ($M/Q=12$ to 44) (Balsiger et al., 1986b). This increase can be attributed in part to a movement of these ion distributions into the ram direction.

Since charge exchange from He^{2+} to He^+ should not result in significant momentum transfer, the He^{2+} and He^+ ions should have the same velocity distributions. Therefore, the relative densities of these ions should be unaffected by the finite field of view of the IMS/HERS instrument. This is equally true for relative proton and helium densities since both solar wind components have nearly the same velocity distributions. This argument is valid only up to about 2400 UT in Fig. 1. After 2400 UT, the distributions are sufficiently out of the field of view of the IMS/HERS instrument that small differences in velocity distributions and poor counting statistics result in significant uncertainties even in the relative density.

Figure 2 shows the ratio of $M/Q=4$ to $M/Q=2+4$ densities from Fig. 1 versus distance from the comet nucleus. In analogy with He^{2+} and He^+ charge exchange in neutral Hydrogen, (Smith and Bewtra, 1978), the charge exchange cross section for He^{2+} to He^+ in H_2O is expected to be significantly greater than the cross section for He^+ to He. Therefore, Fig. 2 is expected to reasonably represent the fraction of charge exchanged solar wind helium as a function of distance from the comet nucleus along the spacecraft trajectory. If there is a significant charge exchange loss of He^+, then Fig. 2 represents a lower bound on the charge exchange process. This assumes all the $M/Q=4$ ions are charge exchanged solar wind helium and all the $M/Q=2$ ions are solar wind helium.

In fact, not all the $M/Q=4$ density in Fig. 1 is due to charge exchanged solar wind helium. Some of the density is intrinsic $M/Q=4$ ions in the solar wind (mainly Si^{7+}). However, this amount is small ($\ll 1\%$ of the solar wind helium density) (Boschsler, 1986) and therefore is negligible relative to the fraction in Fig. 2 for distances less than 6 10^5 km. Also, some of the $M/Q=4$ density may be multiply charge exchanged O^{4+} from solar wind O^{7+} and O^{6+}. Again, this contribution is expected to be negligible. The ratio of oxygen to helium in the solar wind is typically about one part in 70 (Boschsler, 1986). After multiple charge exchanges the oxygen will be distributed over several charge states, thus the O^{4+} is expected to be $\ll 1\%$ of the total $M/Q=4$ density.

Contamination of the $M/Q=2$ density by cometary H_2^+ presents a potentially more serious problem. Cometary H_2^+, produced by dissociation and ionization of H_2O and CH_4, is itself dissociated into atomic hydrogen far from the comet, however it may exist in significant concentrations at distances $\leqq 10^5$ km (W.T. Huntress, personal communication, 1986). While the small difference in mass per charge between H_2^+ and He^{2+} ($\sim 0.7\%$) is smaller than the IMS/HERS mass resolution, it has been possible to exploit the small difference to estimate the relative concentrations of the two species. Initial estimates indicate that H_2^+ may account for about half of the total $M/Q=2$ density for distances $\leqq 10^5$ km (Balsiger et al., 1986b). For distances greater than 10^5 km, there is no clear indication of H_2^+ and it is estimated to contribute much less than 50% to total $M/Q=2$ density. Further study is needed before these estimates can be improved.

Because of the cometary H_2^+ contamination and the fact that some fraction of the He^+ is lost by charge exchange to neutral He,

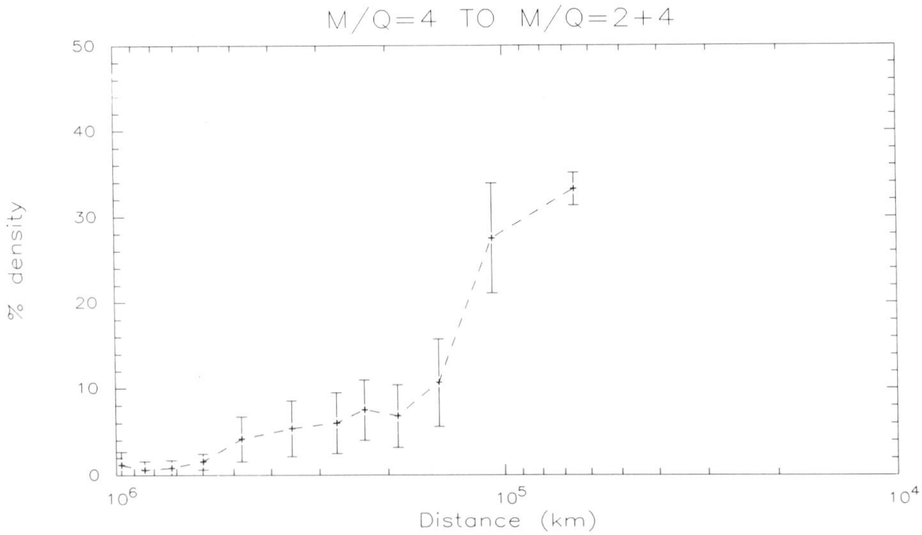

Fig. 2. Ratio of the density of $M/Q=4$ ions to the combined densities of $M/Q=2$ and 4 ions versus distance to the nucleus along the inbound trajectory

the plot in Fig. 2 represents a lower bound of the fraction of charge exchanged solar wind helium. If the H_2^+ accounts for half of the total $M/Q=2$ density inside 10^5 km, then the fraction increases to a maximum of about 50% at 7 10^4 km.

3. Discussion

Since the comet-solar wind interaction is largely determined by the pick up of cometary ions in the neutral gas surrounding the comet, little attention has been given to the charge exchange of solar wind ions in the cometary coma (see Ip and Axford, 1982 for a review). Results from 1-dimensional incompressible flow modeling of the charge exchange process (Ip and Hsieh, 1982, Fig. 2) show that charge exchange becomes important inside $\sim 10^4$ km. At distances on the order of 10^5, Ip and Hsieh predict only a few percent fractional charge exchange of solar wind helium.

These predictions are not in agreement with the observations in Fig. 2, which shows that the lower bound on the observed fraction of solar wind helium is greater than 30% at a distance of $\sim 10^5$ km. The experimental errors in the measurements of the relative densities of He^+ and He^{2+} are discussed in the previous section and are thought to be small compared with the discrepancy between the measured and predicted values. Possible source of error in the model used by Ip and Hsieh (1982) include the neutral gas production rate, plasma flow field assumptions and the charge exchange cross sections.

The neutral gas column density along the ion path is linearly dependent on the neutral gas production rate, therefore the fraction of charge exchanged helium is approximately linearly dependent on the gas production rate. Ip and Hsieh assume a gas production rate about a factor of three smaller than the observed rate of $\sim 7\ 10^{29}\,s^{-1}$ (Krankowsky et al., 1986). Using the observed gas production rate would thus raise the predicted fractional charge exchanged helium to at most about 10%.

Since the Giotto spacecraft encountered comet Halley nearly perpendicular to the sun-comet line, the spacecraft trajectory did not follow any single solar wind flux tube. Because of this, it is difficult to assess how the incompressible flow assumption along a single flux tube affects the fraction of charge exchanged helium predicted by Ip and Hsieh. Even if a new model for the solar wind charge exchange process incorporating a different flow field is devised, comparison with experimental results is difficult because of one important unknown factor. While the charge exchange cross sections for He^+ and H^+ in H_2O and CO_2 are known (Koopman, 1968), to our knowledge, no cross sections for He^{2+} in H_2O or CO_2 at low energies (~ 100 eV) have been measured in the laboratory. Ip and Hsieh assumed that the He^{2+} charge exchange cross section in water was three times that of He^+ or 3 $10^{-16}\,cm^2$. A significantly greater cross section for He^{2+} in H_2O could account for the observed large fraction of charge exchanged solar wind helium in Fig. 2. Further modeling of the charge exchange profile of solar wind helium awaits a laboratory measurement of the charge exchange cross section for He^{2+} in H_2O and CO_2.

Acknowledgements. This work was supported by NASA through the Jet Propulsion Laboratory, California Institute of Technology and by NASA under the contract NASW-3729 at Lockheed. Also support was received from the German Bundesministerium für Forschung und Technologie, and the Swiss National Science Foundation.

References

Balsiger, H., Altwegg, K., Bühler, F., Fischer, J., Geiss, J., Meier, A., Rettenmund, U., Rosenbauer, H., Schwenn, R., Benson, J., Hemmerich, P., Säger, K., Kulzer, G., Neugebauer, M., Goldstein, B.E., Goldstein, R., Shelley, E.G., Sanders, T., Simpson, A.J., Lazarus, A.J., Young, D.T.: 1986a, 20th ESLAB Symp., ESA SP-1077, p. 129

Balsiger, H., Altwegg, K., Bühler, F., Geiss, J., Ghielmetti, A.G., Goldstein, B.E., Goldstein, R., Huntress, W.T., Ip, W.-H., Lazarus, A.J., Meier, A., Neugebauer, M., Rettenmund, U., Rosenbauer, H., Young, D.T.: 1986b, *Nature* **321**, 330

Boschsler, P.: 1986, *Physica Scripta* (submitted)

Ip, W.-H.: 1986, *Adv. Space Res.* **5**, 233

Ip, W.-H., Hsieh, K.C.: 1982, in *Cometary Exploration I*, ed. T.I. Gombosi, Hungarian Academy of Sciences, p. 265

Ip, W.-H., Axford, W.I.: 1982, in *Comets*, ed. L.L. Wilkening, The University of Arizona Press, Tucson, p. 588

Krankowsky, D., Lammerzahl, P., Herrwerth, I., Woweries, J., Eberhardt, P., Dolder, U., Herrmann, U., Schulte, W., Berthelier, J.J., Illiano, J.M., Hodges, R.R., Hoffman, J.H.: 1986, *Nature* **321**, 326

Koopman, D.W.: 1968, *Phys. Rev.* **166**, 57

Smith, P.H., Bewtra, N.K.: 1978, *Space Sci. Rev.* **22**, 301

Astron. Astrophys. 187, 307–310 (1987)

ASTRONOMY AND ASTROPHYSICS

Fluid simulation of comet P/Halley's ionosphere

K. Baumgärtel and K. Sauer

Institute of Cosmical Research, Academy of Sciences of the GDR, 1199 Berlin, German Democratic Republic

Received February 25, accepted April 15, 1987

Summary. The formation of a magnetic cavity around the cometary nucleus is studied in a non-stationary quasi one-dimensional fluid treatment. Characteristic features of spatial density, velocity and magnetic field profiles observed on comet P/Halley by the Giotto spacecraft can be reproduced. The observed relative density maximum ahead of the cavity (pile-up region) may be interpreted as a consequence of reduced recombination.

Key words: comets – ionospheres – fluid simulation – magnetic field – plasmas

1. Introduction

One of the most interesting results of the Giotto mission was the detection of a magnetic field-free region around the cometary nucleus of P/Halley with an extension of approximately 5,000 km in sunward direction. Its boundary is commonly identified with the contact discontinuity which separates purely cometary plasma from the magnetized strongly contaminated solar wind plasma.

From a theoretical point of view, cometary ionospheres have been discussed in the review papers of Ip and Axford (1982), Mendis et al. (1985) and recently by Ip et al. (1987) and Cravens (1986, 1987). In the latter papers, a balance between the magnetic pressure gradient force and the ion-neutral drag force has been assumed to control the flow in the vicinity of the contact surface.

This paper is an improved version of our earlier contribution (Sauer and Baumgärtel, 1987). The formation of the magnetic cavity and the contact discontinuity is described in a self-consistent way on the basis of an initial value problem for a set of fluid equations.

2. The model

The comet is considered as a point-like source of a radial neutral gas flow of constant velocity. It is subject to ionization by solar uv radiation and thus produces a purely cometary plasma flow which encounters the solar wind controlled plasma flow carrying the compressed interplanetary magnetic field. The plasma is treated as a cold fluid and the frozen-in approximation is used to describe the magnetic field.

We follow the philosophy developed in the early steady-state approach of Biermann et al. (1967) and use an axisymmetric version of the fluid equations which describe the plasma flow along the symmetry axis sun-comet in a stream tube of variable cross-sectional area. This area is allowed here to vary also in time and it obeys a separate equation which can be derived in a similar way as its analog in the paper of Biermann et al. (1967). The basic equations read

$$\frac{\partial n}{\partial t} + \frac{\partial}{\partial x}(nv) = \sigma n_0 - 2n\varphi - \alpha n^2, \tag{1}$$

$$\frac{\partial \rho}{\partial t} + \frac{\partial}{\partial x}(\rho v) = \sigma m_0 n_0 - 2\rho\varphi - \alpha\rho n, \tag{2}$$

$$\frac{\partial}{\partial t}(\rho v) + \frac{\partial}{\partial x}(\rho v^2 + p_m) = \sigma m_0 n_0 v_0 - 2\rho v\varphi + \beta n n_0 \frac{m m_0}{m + m_0}(v - v_0), \tag{3}$$

$$\frac{\partial}{\partial t}(\rho\varphi) + \frac{\partial}{\partial x}(\rho v\varphi) - \frac{1}{R}\frac{d}{dx}p_m = \sigma m_0 n_0 \frac{v_0}{x} - 3\rho\varphi^2 + \beta(v_0/x - \varphi), \tag{4}$$

$$\frac{\partial B}{\partial t} + \frac{\partial}{\partial x}(vB) = -B\varphi. \tag{5}$$

n_0, v_0 are, respectively, the density and the velocity of the cometary neutral gas; m_0 is the neutral mass. n, ρ, v denote the particle number density, the mass density and the velocity of the plasma ($m = \rho/n$). B is the frozen-in magnetic field and $p_m = B^2/2\mu$ is the magnetic pressure. The quantity φ accounts for lateral flow effects. It is related to the local stream tube cross section A by $\varphi = (v/2A) \cdot dA/dx \cdot \sigma$ is the photoionization rate, α is the recombination coefficient and β is the coefficient for momentum exchange between gas und plasma. Equation (4) contains the curvature radius R of the surfaces of constant magnetic pressure at their cross points with the x-axis. It is set here equal to twice the distance from the nucleurs. Magnetic diffusion is ignored.

The initial configuration consists simply of a uniform cold magnetized plasma flow with magnetic field perpendicular to the flow direction, penetrated by the neutral gas flow emanating from the comet (sketched in Fig. 1a). Photoionization is "switched on" at zero time creating cometary ions moving in

Send offprint requests to: K. Baumgärtel

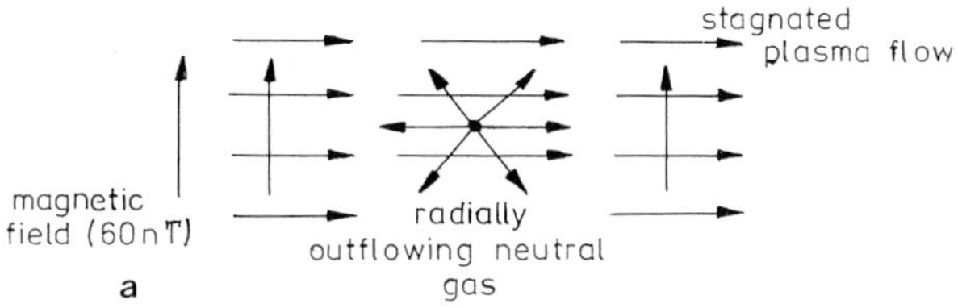

Fig. 1a. Sketch of the initial configuration: radially outflowing cometary neutral gas, imbedded in strongly mass-loaded solar wind, carrying a compressed frozen-in magnetic field perpendicular to the flow direction

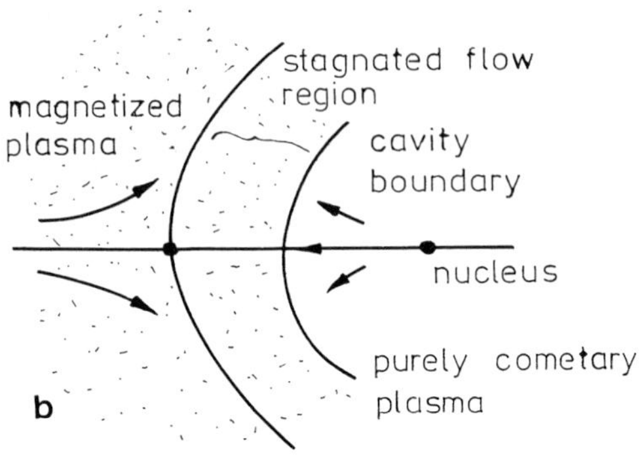

Fig. 1b. Scheme of the near-axis flow after formation of a steady-state ionosphere (nearly stagnant plasma flow, contact discontinuity, magnetic cavity)

radial direction. This causes the magnetic field to be frozen out of the vicinity of the nucleus. This freezing out proceeds until a configuration of relative stability is established which consists of a magnetic cavity filled with radially outflowing purely cometary plasma, separated from the magnetized, strongly mass-loaded solar wind plasma through a contact discontinuity as indicated in Fig. 1b.

In a strictly one-dimensional treatment, a steady-state contact discontinuity cannot exist without particle/mass sinks. The system (1)–(5) is capable of describing stagnation because of the presence of sinks due to both lateral flow and recombination. The lateral loss rate in Eq. (5) for the magnetic field is reduced by a factor two (compared, for example, with that in the density equation). This accounts for the fact that only lateral flow perpendicular to B contributes to a freezing out of the magnetic field.

For the solution of the initial value problem for Eqs. (1)–(5), the flux-corrected transport scheme ETBFCT (Boris et al., 1981) on an Eulerian equally spaced grid is employed.

3. Results

Velocity, magnetic field and density of the "artificial" uniform initial plasma flow have been taken to be $v = 1\,\mathrm{km\,s^{-1}}$, $B = 60n\,T$, and $n = 500\,\mathrm{cm^{-3}}$. This parameter set is thought to be representative for the flow conditions at a distance of some 10^4 km upstream from the nucleus and its choice is oriented at measurements (Balsinger et al., 1986; Neubauer et al., 1986). The neutral gas density is determined by a production rate of $Q = 10^{30}\,\mathrm{s^{-1}}$ and a flow velocity of $v_0 = 1\,\mathrm{km\,s^{-1}}$. The neutral mass is assumed to be $m_0 = 20 m_p$ (m_p proton mass). We adopt further a photoionization rate $\sigma = 10^{-6}\,\mathrm{s^{-1}}$, a recombination rate $\alpha = 3\,10^{-7}\,\mathrm{cm^3\,s^{-1}}$ and a neutral gas-plasma friction rate coefficient $\beta = 10^{-9}\,\mathrm{cm^3\,s^{-1}}$. The calculation box covers a region which extends from the nucleus $2\,10^4$ km toward the sun and $5\,10^3$ km away from the sun. At the left-hand boundary, density, momentum and magnetic field are kept at their initial level while free-streaming conditions are realized at the other boundary.

Figure 2 shows spatial profiles of particle number density, velocity and magnetic field after an integration time of $1.5\,10^4$ s which represents a quasi steady-state configuration. The two plateaus in the velocity profile left and right the center, in which the plasma velocity coincides with the neutral gas outflow velocity, reflect the purely cometary plasma flow produced by ionization of the neutrals. In these regions, therefore, gas-plasma friction does not operate. While the velocity keeps its plateau value in downstream direction, the encounter with the oppositely flowing magnetized solar wind controlled plasma at the sunward side leads to the formation of a stagnation point ($v = 0$) which marks the position of the contacct discontinuity at the x-axis.

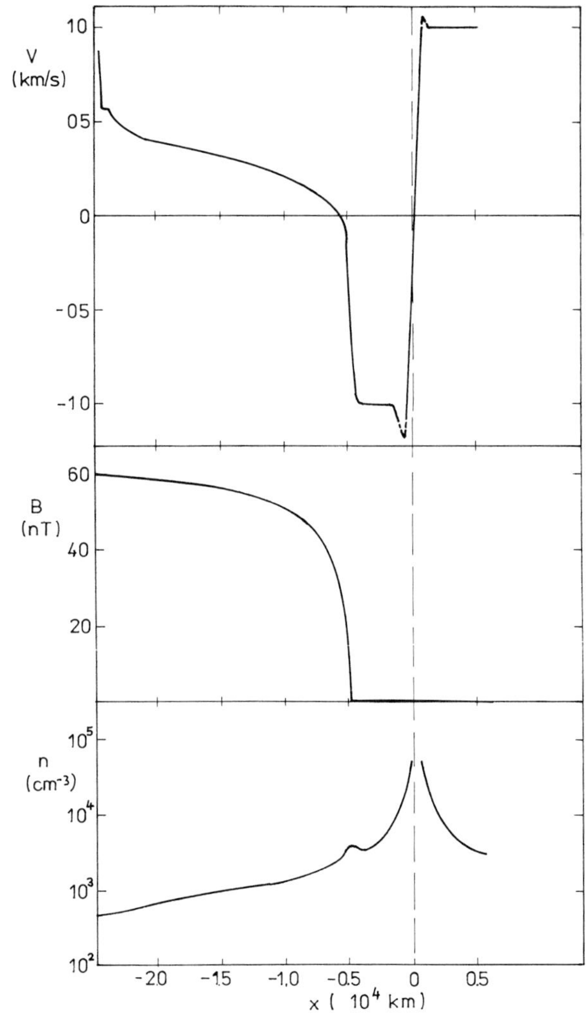

Fig. 2. Spatial profiles of plasma velocity, magnetic field and plasma density for the stationary state (after $t = 1.5\,10^4$ s). Recombination ($\alpha = 3\,10^{-7}\,\mathrm{cm^3\,s^{-1}}$) and ion-neutral drag ($\beta = 10^{-9}\,\mathrm{cm^3\,s^{-1}}$) are included

The magnetic cavity is seen to extend from the nucleus to a distance of about 5,000 km toward the sun, where a strong drop in the magnetic field indicates its boundary. In the opposite direction, the cavity is continously stretched in time. Stagnation point and location of the steepest magnetic field gradient are close together. In addition, the stagnation point is indicated in the density profile by a weak relative maximum. Both magnetic pressure gradient forces and plasma-neutral gas friction are responsible for the plasma velocity not to exceed the $0.5\,\mathrm{km\,s^{-1}}$-level ahead of the magnetic cavity.

Figure 3 shows results obtained for the same parameter combination with the only modification that recombination is ignored outside a fixed region beyond the cavity. This is achieved by setting the recombination rate α equal to zero ahead of the arbitrarily chosen position $x = -10^4$ km. Such a drastic reduction of the recombination may be caused by an increase in the electron temperature. Although there are no electron temperature measurements, this behaviour is suggested by in situ measurements of the ion temperature (Balsinger et al., 1986) which was shown to increase by an order of magnitude outside the cavity. (See also the discussion in the paper of Ip et al., 1987). While the profiles of velocity and magnetic field do not alter seriously, a significant relative maximum in the plasma density is produced. When penetrating the inner region of enhanced recombination, the plasma stream rarefies drastically and the density is reduced by a factor of about 3.5.

When plasma-neutral gas friction is neglected, the size of the cavity shrinks to half the value shown in Fig. 3. This indicates the importance of momentum exchange between plasma and neutral gas outside the cavity for maintaining the proper cavity diameter. When lateral flow effects are ignored, i.e. $\varphi \equiv 0$ in Eqs. (1)–(5), no steady state will be reached in a comparable integration time. The plasma velocity remains negative at the sunward side and no stagnation occurs. When neither recombination nor friction is taken into consideration, the solution of the initial value problem again merges into a steady-state configuration with a magnetic cavity of nearly the same size as in Fig. 3. Here, the magnetic pressure is balanced by the dynamic pressure of the cometary plasma flow.

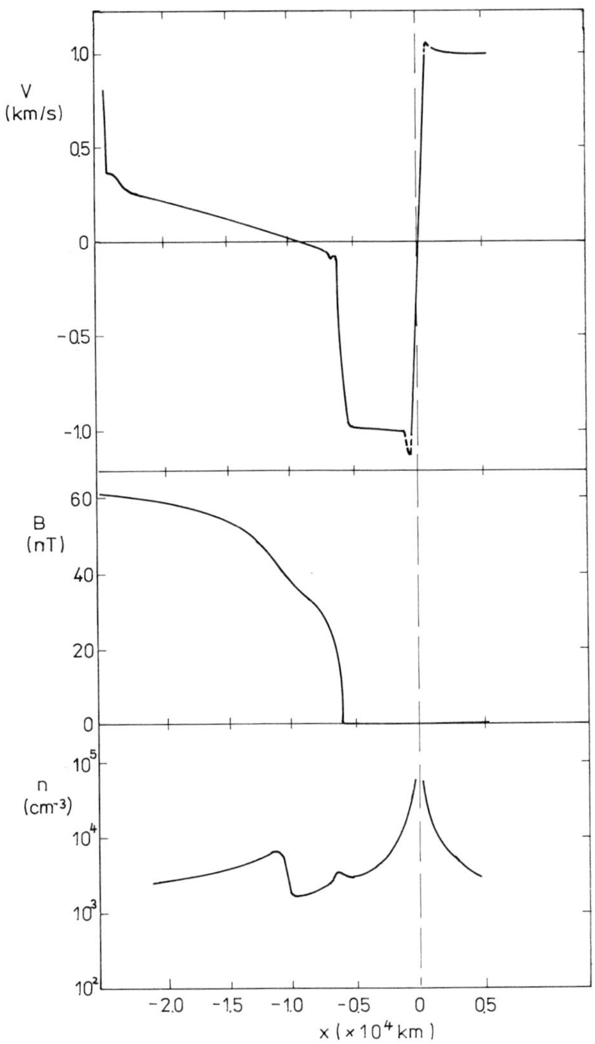

Fig. 3. The same as Fig. 2 but the recombination rate beyond $x = -10^4$ km is assumed to be zero

4. Comparison with Giotto observations

The Giotto spacecraft registered a magnetic field-free region around P/Halley with an extension of about 8500 km along the trajectory (Neubauer et al., 1986). This corresponds to a cometocentric cavity radius of approximately 4500 km at the sunward side. This is almost in accordance with the calculated cavity size shown in Figs. 2, 3 ($r \simeq 5000$ km). The results shown in Fig. 3 are of particular interest for a comparison with the Giotto plasma measurements of Balsinger et al. (1986). There, a spatial density profile is reported which exhibits a relative maximum ahead of the magnetic cavity ($n_{\max} \simeq 5\,10^3\,\mathrm{cm}^{-3}$) which is shifted from the cavity boundary by about 7000 km. In accordance with the discussion of Ip et al. (1987), we suggest that a cut in the recombination rate is the reason for this extremum to occur. We point further to the observed velocity profiles which reveal a broad region of nearly stagnant flow outside the cavity up to a distance of $1.5\,10^4$ km, in agreement with the calculated flow profile.

5. Conclusions

The formation of a magnetic cavity and an associated contact surface surrounding the cometary nucleus of P/Halley has been studied in an initial value problem for quasi one-dimensional fluid equations. The influence of lateral flow, recombination and ion-neutral drag on the spatial profiles of plasma velocity, plasma density and magnetic field has been demonstrated. There is agreement with respect to several features of Giotto measurements.

References

Balsinger, H., Altweg, K., Bühler, F., Geiss, J., Ghielmetti, A.G., Goldstein, B.E., Goldstein, R., Huntress, W.T., Ip, W.H., Lazarus, A.J., Meier, A., Neugebauer, M., Rettenmund, U., Rosenbauer, H., Schwenn, R., Sharp, R.D., Shelley, E.G., Ungstrup, E., Young, D.T.: 1986, Nature **321**, 330

Biermann, L., Brosowski, B., Schmidt, H-U.: 1967, Solar Phys. **1**, 254

Book, D.L., Boris, J.P., Zalesak, S.T.: 1981, Springer Series Comp. Phys., Springer-Verlag, Heidelberg, Berlin, New York, p. 29

Cravens, T.E.: 1986, Proc. XXVI COSPAR Meeting, Pergamon Press, p. 198

Cravens, T.E.: 1987, 20th ESLAB Symp. ESA SP-250, Vol. **I**, p. 241

Ip, W.H., Axford, W.I.: 1982, in *Comets*, ed. L.L. Wilkening, Tuscon, Arizona, Univ. Arizona Press, p. 588

Ip, W.H., Schwenn, R., Rosenbauer, H., Balsinger, H., Neugebauer, M., Shelley, E.G.: 1987, 20th ESLAB Symp., ESA SP-250, Vol. **I**, p. 219

Mendis, D.A., Houpis, H.L.F., Marconi, H.L.: 1985, The Physics of Comets, ed. V.M. Canuto, Bruce G. Elmegreen, Gordon and Breach Science Publishers, New York, p. 1

Sauer, K., Baumgärtel, K.: 1987, 20th ESLAB Symp., ESA SP-250, Vol. **III**, p. 401

Astron. Astrophys. 187, 311–319 (1987)

ASTRONOMY AND ASTROPHYSICS

Unusual characteristics of electromagnetic waves excited by cometary newborn ions with large perpendicular energies

A.L. Brinca★ and B.T. Tsurutani

Jet Propulsion Laboratory, California Institute of Technology, Mail Code 169-506, Pasadena, California 91109, USA

Received January 26, accepted April 8, 1987

Summary. Solution of the linear kinetic dispersion equation shows that cometary new-born ions with large perpendicular energies (large initial angles between the solar wind velocity and the IMF) can excite a wave mode with rest frame frequencies of the order of the heavy ion cyclotron frequency which exhibits maximum growth slightly away (6°–15°) from parallel propagation, together with high mass density compression ratios and almost linear polarization. The association of these properties is relevant to space observations of MHD-like waves at comets and in planetary foreshocks and has not been reported in the vast literature on possible instability mechanisms. Dilute drifting oxygen ion and photoelectron rings with finite thermal spreads model the free energy source in the solar wind frame. For parallel propagation, the mode of interest is left-hand circularly polarized, moves in the direction of the newborn particles, and is predominantly fed by the perpendicular energy of the heavy ions; as obliquity increases, its polarization and compression change rapidly. Modification of the free energy source parameters (ring density, drift velocity, perpendicular speed and thermal spread) determines their relative influence, defines the domain of existence of the mode, and shows that it might evolve into a nonoscillatory, purely growing structure distinct from the mirror wave. The results are also applicable to non-cometary environments.

Key words: comets – waves – instabilities – polarization – compression

1. Introduction

Electromagnetic wave activity below a few Ω_p (proton cyclotron frequency) is observed in distinct space environments. Apart from inner magnetosphere pulsations where finite geometry effects play a major role (Southwood and Hughes, 1983), hydromagnetic modes have been detected in the distant magnetotail (e.g. Tsurutani and Smith, 1984), in the magnetosheath (Kaufmann et al., 1970; Tsurutani et al., 1982; Anderson et al., 1982), upstream of both the earth's (Hoppe et al., 1981) and other planet's (Smith et al., 1976; Goldstein et al., 1985; Smith and Lee, 1986) bow shocks, upstream of interplanetary shocks (Tsurutani et al., 1983), and near comets (Tsurutani and Smith, 1986 a, b; Yumoto et al., 1986; Neubauer et al., 1986; Tsurutani et al., 1987).

Theoretical analyses of possible excitation mechanisms have been conducted, with emphasis on (proton or heavier) ion beams with Maxwellian (e.g. Gary et al., 1985) or drifting ring (e.g. Wu and Hartle, 1974; Winske et al., 1985) velocity distributions, and parallel propagation, albeit oblique behavior has also been studied (e.g. Gary et al., 1984; Sharma and Patel, 1986). The vast parameter space accessible to the background and active media precludes the presentation of a complete instability survey.

In general, a cometary environment contains newborn hydrogen and heavier ions, with relative densities depending on the distance from the nucleus. Each isolated ion beam is able to excite instabilities but the coexistence of different beams warrants the assessment of their eventual interaction and relative efficiency in wave excitation. Preliminary results to be presented elsewhere on coexisting hydrogen and oxygen ion beams suggest that (i) the mutual influence is negligible in most domains of the Brillouin plane for the resonant modes, and (ii) comparable (in physical nature and beam density) instabilities are more efficiently generated by the newborn protons.

Here we concentrate on a model (solar wind permeated by dilute drifting ring distributions of electrons and oxygen ions with finite thermal spreads) with parameters compatible with the ICE (Interstellar Cometary Explorer) observations at the Giacobini-Zinner (G–Z) comet, though the results might also prove helpful in other environments. Although four free energy sources (perpendicular and streaming energies of each ring) are available, the modes excited in the hydromagnetic frequency range (below a few Ω_p) are predominantly fed by the heavy ions. The free energy depends on the relative orientation of the solar wind velocity and the IMF, defined through the angle α. Changes on this orientation are created by discontinuities (Burlaga, 1971; Smith, 1973; Tsurutani and Smith, 1979; Neugebauer et al., 1984, 1986; Lepping and Behannon, 1986) and Alfvén waves (Coleman, 1966; Belcher and Davis, 1971; Roberts, et al., 1986; Tsurutani and Gonzalez, 1987) in the solar wind. As the angle α increases from zero, the situation evolves from purely streaming beams with almost no perpendicular energy ($\sin\alpha \sim 0$) to rings with almost no drift ($\cos\alpha \sim 0$). Near $\sin\alpha \sim 0$ the free energy destabilizes nonresonant (firehose instability in the hydromagnetic approximation) and resonant (high- and low-frequency) modes (Winske et al., 1985), whereas the anisotropy associated with the $\cos\alpha \sim 0$ regime favors the excitation of ion cyclotron waves (Kennel and Petschek, 1966) and the mirror mode (Hasegawa, 1969; Hada and Ashour-Abdalla, 1986; Price et al., 1986). Winske

Send offprint requests to: B.T. Tsurutani

★ Permanent address: Centro de Electrodinamica (INIC), Instituto Superior Tecnico, P-1096 Lisboa Codex, Portugal et al. (1985) analyze the linear properties of three (nonresonant, low- and high-frequency resonant) electromagnetic instabilities that can couple newborn ions to the solar wind in the direction parallel to the magnetic field, and follow their nonlinear behavior by means of computer simulation; they find that the high-frequency resonant instability saturates at low levels and thus is not important for the ion pickup process. Hada and Ashour-Abdalla (1986) study the effect of a heavy ion species on the mirror and electromagnetic ion cyclotron waves driven by ion temperature anisotropies (with $T_\perp > T_\parallel$); the linear and simulation results show the occurrence of a frequency band where the mirror growth exceeds the ion cyclotron growth.

Our attention will be mainly focussed on the region of small $\cos\alpha$ (almost a pure ring distribution), and one of the unstable modes to be identified below. For parallel propagation this wave is left hand (LH) circularly polarized, exhibits growth in a frequency range containing the oxygen cyclotron frequency, Ω_0, and propagates in the direction of the ring drift along the stationary magnetic field (Note, however, that for low perpendicular energies the mode is stable for parallel propagation and only grows in a given range of oblique propagation). The wave has maximum growth at small obliquity (as low as 6°) in association with large mass density compressions and almost linear polarization. These co-existing features, to the best of our knowledge, have not yet been reported in the literature, and may be relevant to hydromagnetic wave space observations, namely the interpretation of wave steepening and polarization. This new mode arises from the coupling of the heavy (oxygen) ion beam to the left-hand mode dispersion modified by the two ion species; a newborn proton population would not be able to feed the instability of interest here.

A necessary condition for the onset of steepening and eventual formation of a shock is the occurrence of (mass density) compression. Oblique electromagnetic modes yield density fluctuations, but there are instances (Vinãs et al., 1984; Tsurutani and Smith, 1986a, 1987) of quasi-parallel electromagnetic propagation associated with compression; for large β plasmas the fast MHD mode is compressional but, being purely electrostatic for parallel propagation, is an unlikely source of the electromagnetic quasi-parallel compression. Nonlinear (parametric instability) mechanisms (Wong and Goldstein, 1986) or parallel growth followed by refraction into oblique propagation (Hada, 1985) have been invoked to explain these features. The mode to be studied in this paper has considerably different initial conditions ($\cos\alpha \sim 0$) and yields maximum growth and large density fluctuations at small obliquities. This mechanism could provide an alternative means of compressive quasi-parallel propagation waves under different solar wind conditions and should be thought of as being complementary to the ideas above.

Although the three MHD modes are linearly polarized at oblique propagation, one should recall that the MHD approximation is only valid for very small frequencies (with respect to the smallest cyclotron frequency). As the frequency increases, they rapidly acquire sizeable elliptic polarizations which, as parallel propagation is approached, become (left and right) circular and electrostatic (fast or slow mode in large or small β plasmas). Space observations of quasi-parallel hydromagnetic waves frequently show almost circular polarizations (e.g. Smith et al., 1985) but recent cometary experiments (Tsurutani and Smith, 1986a; Neubauer et al., 1986; Riedler et al., 1986; Yumoto et al., 1986) have detected linearity polarized modes near parallel propagation. Again, the characteristics of the waves to be studied could prove helpful in the interpretation of these observations. Clearly, it is necessary to determine the initial solar wind conditions (α) to determine the instability and plasma processes responsible for the linearly polarized parallel propagating waves.

The search conducted in the free energy parameter space has shown that the wave under analysis may evolve into a nonoscillatory, purely growing mode both at parallel and oblique propagation. These structures are not analyzed in detail, but they warrant further study; existing at parallel propagation, they are not identical to the mirror waves, which only occur for oblique propagation, but they might bring about observable features.

Upon definition of the model adopted for the background solar wind and the dilute newborn particles, we describe the approach to be used in the study of the wave characteristics. The presentation of the results and concluding discussion is preceded by the identification of the mode of interest here.

2. Model

We perform the analysis in the solar wind frame. This background is permeated by an interplanetary magnetic field (IMF) $\boldsymbol{B}_o = 8\,nT$, and is composed of isotropic Maxwellian electron and proton populations with a common number density of $N_p = N_e = 4.95\,\mathrm{cm}^{-3}$, temperatures of $T_e = 2\,10^5$ K and $T_p = 8\,10^4$ K. In the ICE spacecraft frame the solar wind has a velocity of magnitude $V_{sw} = 400\,\mathrm{km\,s}^{-1}$, forming an angle α with the direction of $\boldsymbol{B}_o$; the newborn (water group) oxygen ions and corresponding photoelectrons are assumed to have zero initial velocities in this frame. Table 1 displays the values of these and other related physical parameters.

The free energy source lies in the newborn particles. Their initial speed in the solar wind frame is V_{sw}. In a quiet background (energy density of turbulence small compared to the initial energy density of the newborn particles), and assuming that instabilities excited by the initial beam-like distributions have growth rates smaller than Ω_0, the ionization of the oxygen atoms brings about drifting ring distributions with pitch angle α; they drift with a speed of $|V_{sw}\cos\alpha|$ along the IMF and their radius (perpendicular speed) is $|V_{sw}\sin\alpha|$. We shall assess the influence of different ring densities, since they depend on the distance to the comet. Our nominal model given in Table 2 assumes a number density two

Table 1. Background (solar wind) magnetoplasma

	Electrons (s = e)	Protons (s = p)
Number density, N (cm^{-3})	4.95	4.95
Temperature, T_s (K/eV)	$2\,10^5$/17.2	$8\,10^4$/6.9
Plasma frequency, f_{ps} (kHz)	20	0.466
Thermal speeds, v_{ts} (km s^{-1})	2460	36.4
Cyclotron frequencies, $\Omega_s/2\pi$ (Hz)	224	0.122
Gyration radii, ρ_s (km)	1.75	47.4
Plasma beta, β_s	0.54	0.22
IMF	$\boldsymbol{B}_o = 8\,nT$	
Solar wind speed	$V_{sw} = 400\,\mathrm{km\,s}^{-1}$	
Alfvén speed	$v_A = 77.8\,\mathrm{km\,s}^{-1}$	

Table 2. Newborn particles

	Oxygen ions ($s = 0$)	Photoelectrons ($s =$ ei)
Number density, N_s (cm^{-3})	0.05	0.05
Parallel thermal speeds, v_{ts} (km s^{-1}/eV)	1.095/0.1	187.5/0.1
Drift speed, V_s	$V_{sw} \cos\alpha$	$V_{sw} \cos\alpha$
Parameters, Eq. (2), $a_s = 1.5 b_s$	$1.64\ 10^5 \sin^2\alpha$	$5.61 \sin^2\alpha$

orders of magnitude below the background density ($N_0 = N_{ei} = 0.05\,\mathrm{cm}^{-3}$). This value corresponds to the maximum possible density upstream of the bow wave, assuming no scattering (Ipavich et al., 1986). Thus this and lower numbers are representations of the comet-solar wind interactions.

The drifting ring distributions are separable in velocity space, $f_s(v_{\parallel}, v_{\perp}) = f_{\parallel s}(v_{\parallel})\, f_{\perp s}(v_{\perp})$; the parallel component is modelled by a drifting Maxwellian,

$$f_{\parallel s}(v_{\parallel}) = \frac{1}{\sqrt{\pi}\, v_{ts}} \exp\left[-\left(\frac{v_{\parallel} - V_{sw}\cos\alpha}{v_{ts}}\right)^2\right] \tag{1}$$

and the perpendicular ring is simulated with a loss-cone type distribution obtained through the subtraction of two Maxwellians with different temperatures,

$$f_{\perp s}(v_{\perp}) = \frac{1}{\pi(a_s - b_s)v_{ts}^2}\left[\exp\left(-\frac{v_{\perp}^2}{a_s v_{ts}^2}\right) - \exp\left(-\frac{v_{\perp}^2}{b_s v_{ts}^2}\right)\right],\ a_s > b_s \tag{2}$$

The maximum of $f_{\perp s}$ ($=0$ at $v_{\perp} = 0, \infty$) occurs for

$$v_{\perp}^2 = v_{\perp ms}^2 = a_s v_{ts}^2 \frac{\ln r_s}{r_s - 1},\ r_s = a_s/b_s > 1$$

and the mean square value is

$$\langle v_{\perp}^2 \rangle_s = v_{\perp ms}^2 \frac{r_s^2 - 1}{r_s \ln r_s} = (a_s + b_s)v_{ts}^2$$

The parameters of Table 2 yield $v_{\perp ms}^2 = V_{sw}^2 \sin^2\alpha$ and $\langle v_{\perp}^2 \rangle_s = 2.055\, v_{\perp ms}^2$. The (negligible) effect of larger parallel thermal spreads on the wave characteristics shall be discussed below.

3. Approach

Solution of the linearized Maxwell-Vlasov equations for harmonic waves with complex field amplitudes varying as $\exp j(\boldsymbol{k} \cdot \boldsymbol{r} - \omega t)$ yields the wave equation

$$\boldsymbol{k} \times (\boldsymbol{k} \times \boldsymbol{E}) + \frac{\omega^2}{c^2}\, \boldsymbol{\varepsilon} \cdot \boldsymbol{E} = \mathbf{D}(\omega, \boldsymbol{k}) \cdot \boldsymbol{E} = 0 \tag{3}$$

where $\boldsymbol{E}$ is the wave electric field complex amplitude, c the vacuum light speed, and

$$\boldsymbol{\varepsilon} = \mathbf{U} + \sum_s \boldsymbol{\varepsilon}_s \tag{4}$$

is the dielectric tensor, with contributions arising from free space ($\mathbf{U}$, unit dyadic) and the four species (solar wind electrons and protons, oxygen ions, photoelectrons; $s =$ e, p, O, ei).

The frequency, $\omega = \omega_r + j\gamma$, and wave number vector, $\boldsymbol{k}$, of the eigen modes satisfy the dispersion equation det $(\mathbf{D}) = 0$. In the initial value problem to be analyzed, the wave number $\boldsymbol{k}$ is real, and unstable modes have $\gamma > 0$. Expressions for the components of $\mathbf{D}$, or $\boldsymbol{\varepsilon}$, can be found in the plasma wave literature (e.g. Krall and Trivelpiece, 1973).

The wave modes shall be characterized by their real, $\omega_r(\boldsymbol{k})$, and imaginary, $\gamma(\boldsymbol{k})$, dispersion, mass density compression, $\delta\rho$, and polarization. Using the continuity equation, and expressing the average harmonic species velocities in terms of $\boldsymbol{E}$ and $\boldsymbol{\varepsilon}_s$, we obtain for the mass density perturbation

$$\delta\rho = -j\,\omega\varepsilon_0 \sum \frac{m_s}{q_s(\omega - \boldsymbol{k} \cdot \boldsymbol{V}_s)}\, \boldsymbol{k} \cdot (\boldsymbol{\varepsilon}_s \cdot \boldsymbol{E}) \tag{5}$$

where ε_0 is the vacuum dielectric constant, and m_s, q_s, and $\boldsymbol{V}_s$ are the mass, electric charge, and drift velocity of species s. The results will show the compression ratio $|\delta\rho/\rho_o|$, with $\rho_o = \sum m_s N_s$, when $|\boldsymbol{E}| = 1\,\mathrm{mV\,m}^{-1}$.

The polarization is determined by Eq. (3). Its state is obtained using two orthogonal components of the wave magnetic field in a plane perpendicular to $\boldsymbol{k}$. Adopting the Poincare's sphere representation (Kraus, 1966), we compute the sine of the latitude of the polarization state, $\sin 2\varepsilon$. Linear polarization lies in the equator ($\sin 2\varepsilon = 0$), and circular polarization at the poles ($\sin 2\varepsilon = \pm 1$). With our sign conventions, the southern (northern) hemisphere, $\sin 2\varepsilon < 0$ (>0), corresponds to left (right) hand waves in the plasma physics sense. The ratio of the major to minor axes in the general elliptic polarization is $AR = \cot|\varepsilon|$, $|\varepsilon| < \pi/4$, so that polarizations with $|\sin 2\varepsilon| < 0.4$ have $AR > 4.8$ and are, therefore, approximately linear.

We use program WHAMP (Rönmark, 1982) to solve numerically the exact linear dispersion equation, for parallel and oblique propagation, with the given equilibrium velocity distributions of the four particle species; the infinite sums of modified Bessel functions which appear in the dielectric tensor components are reduced into a summable form with recourse to Padé approximations for the plasma dispersion function (Rönmark, 1983). In configuration space the z-axis lies along $\boldsymbol{B}_o$, and the $\boldsymbol{k}$ vector, forming an angle θ with $\boldsymbol{B}_o$, is contained in the x-z plane. The proton cyclotron frequency, Ω_p, and gyroradius, ρ_p, are used to normalize the complex wave dispersion ($\omega_r/\Omega_p, \gamma/\Omega_p, k\rho_p$). We take ω_r and k to be positive, so that the sign of the drift velocity indicates whether the mode moves with ($V_s > 0$), or against ($V_s < 0$) the ring beams.

Having adopted separable velocity distributions for all species, $f_s(v_{\parallel}, v_{\perp}) = f_{\parallel s}(v_{\parallel})\, f_{\perp s}(v_{\perp})$, the characteristics of parallel wave propagation are not affected by the detailed shape of $f_{\perp s}(v_{\perp})$, depending only on the mean perpendicular energy, $\langle v_{\perp}^2 \rangle_s$, of the species (e.g. Lee and Crawford, 1970).

4. Wave mode identification

The addition of a heavy ion (viz. oxygen) population to a hydrogen magnetoplasma drastically modifies the dispersion of the hydromagnetic modes. The original parallel LH wave below Ω_p splits into two branches, the lower resonating at Ω_0, and the upper evolving from a cutoff frequency ω_{cf} ($>\Omega_0$) to a resonance at Ω_p. The parallel RH mode is not affected by the heavy ions but, for oblique propagation, it becomes coupled to the upper

LH mode near the crossover frequency ω_{cr}, where polarization conversion occurs. We have for oxygen ions (Smith and Brice, 1964):

$$\omega_{cr} = \Omega_0[1 + r_n(r_m^2 - 1)]^{1/2}$$

and, for $\omega^2 \ll \Omega_e^2$,

$$\omega_{cf} = \Omega_0[1 + r_n(r_m - 1)],$$

with $r_n = N_0/N_e$ and $r_m = m_0/m_p$.

Figure 1, adapted from Young et al. (1981), sketches the cold dispersion below Ω_p for parallel and oblique propagation. Consideration of finite temperatures for the isotropic particle distributions does not change significantly the real dispersion of these modes. However, attenuation sets in and the LH branches become heavily cyclotron damped as their resonances are approached.

Centering the attention on the LH modes, and using for the heavy ions a nondrifting ring ($\alpha = 90°$) distribution of oxygen with $v_{\perp mo} \gg v_{t\|0}$, we find that the lower LH branch is destabilized below Ω_0, as sketched in Fig. 2 for parallel propagation. The coupling between the ring ($\omega = \Omega_0$) and the LH mode brings about a different real dispersion in the small wavenumber (stable) region. The ring parallel velocity spread is responsible for the upper limit in the unstable wavenumber band, and the perpendicular energy of the oxygen ions is the clear free energy source of the instability.

Allowing the oxygen ring to drift along $\boldsymbol{B}_o$ with velocity $\boldsymbol{V}_0$ modifies the instability. At parallel propagation, as schematically depicted in Fig. 3, two unstable frequency bands arise, one containing Ω_o and another above Ω_p. The latter development is

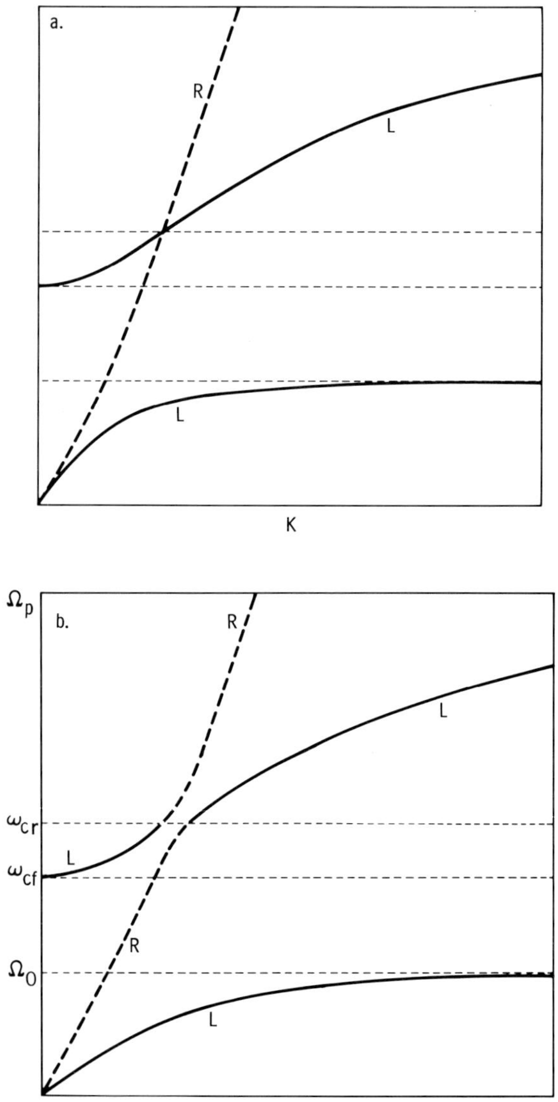

Fig. 1a and b. Sketch (not to scale) of wave dispersion in a multispecies (hydrogen plasma with oxygen ions) cold magnetoplasma (adapted from Young et al., 1981); *R* and *L* identify the wave polarization. **a** Parallel ($\theta = 0$) propagation. **b** Oblique ($\theta = 45°$) propagation

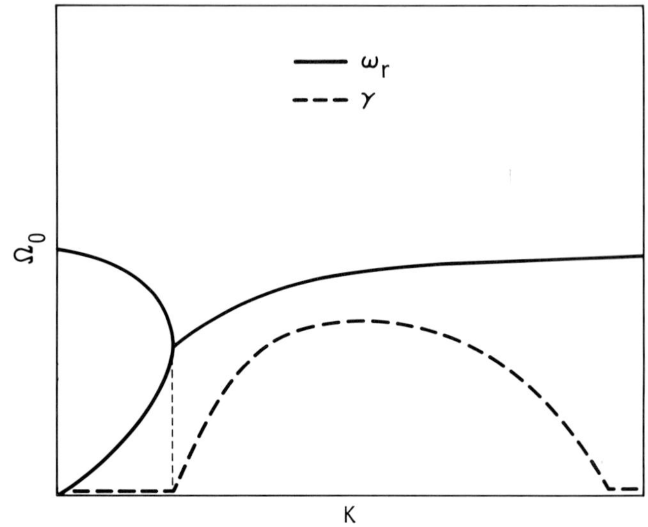

Fig. 2. Sketch (not to scale) of the complex dispersion ($\omega = \omega_r + j\gamma$) of the parallel lower LH mode in a hydrogen magnetoplasma permeated by a nondrifting ring of oxygen ions

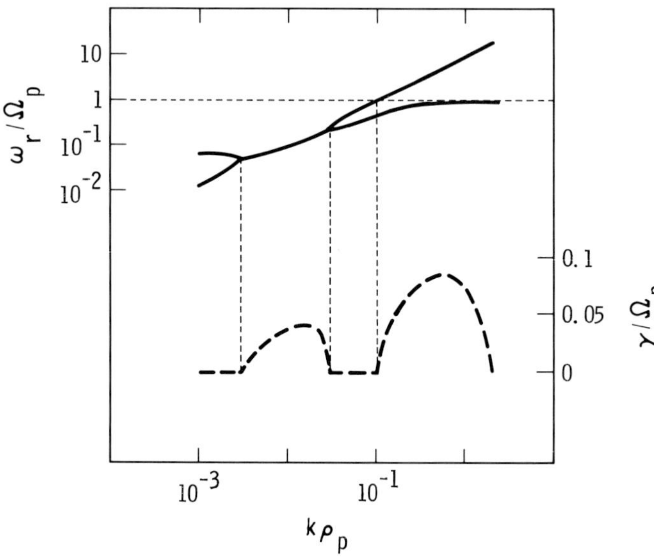

Fig. 3. Sketch of the complex dispersion ($\omega = \omega_r + j\gamma$) of the parallel LH mode in a hydrogen magnetoplasma permeated by a drifting ring of oxygen ions. The characteristics of the two unstable frequency bands depend on the ring drift velocity: see discussion in the text

better understood recalling that the beam mode satisfies $\omega = kV_0 + \Omega_0$. The instability starting at Ω_p is basically of this beam type, whereas the lower unstable band arises from the combined effects of the intrinsic ring anisotropy (already responsible for the instability below Ω_0 when $V_0 = 0$) and the coupling of the beam dispersion to the lower portion of the upper LH mode; when the two dispersion curves intersect, we speak of cyclotron resonance although finite velocity spreads may allow for strong destabilizing coupling even without curve intersection. Resonance arises because the drifting positive charged ions feel a wave Doppler-shifted frequency equal to their own cyclotron frequency, with the particles and the wave fields gyrating in the same LH sense.

Clearly, the efficiency and dispersion of these couplings are strongly dependent on the oxygen ring drift velocity (for the modes and parameter range to be analyzed, the role of the electron ring is negligible.) As V_0 (>0) increases, the beam dispersion moves away from the unperturbed LH modes, and the coupling becomes ineffective; conversely, lower drift velocities can strongly couple the beam to the upper unperturbed LH branch, and eventually bring about the merging of the two unstable bands depicted in Fig. 3, creating a continuous unstable band that contains the interval $\Omega_0 - \Omega_p$. Negative drift velocities (oxygen beam moving against the LH waves) are not our main concern here: the beam dispersion intersects the lower LH branch and the ensuing coupling brings about a resonant instability which has been widely discussed in the literature for hydrogen plasmas (proton beams and LH hydromagnetic modes, or the dual problem of electron beams and whistler modes).

We identify the mode of interest for the present study, as the unstable wave in Fig. 2, which evolves into the lower frequency unstable wave of Fig. 3. Its characteristics shall be analyzed for parallel and oblique propagation, as the free energy source parameters are varied. Other unstable modes coexist with this instability and shall be discussed elsewhere; their relative importance cannot be estimated simply by comparing the various linear growth rates and would require an assessment of their nonlinear evolution which is outside the scope of this work. We refrain from naming the mode of interest here. Although the frequency range is hydromagnetic, utilization of the classical MHD terminology (fast, intermediate and slow waves) in our medium (multispecies magnetoplasma with particle beams), with its inherent dispersion, compression and polarization characteristics, is clearly unwarranted.

5. Results

5.1. *MHD behavior*

The main results presented below deal with the mass compression and polarization of oblique waves excited by newborn particles. In order to assess quantitatively their importance, we provide the corresponding characteristics of the MHD-like modes supported by the background solar wind plasma, without the newborn populations (Table 1). This medium has $\beta < 1$, so that the slow mode is electrostatic for parallel propagation. We focus attention on the fast (RH circularly polarized at $\theta = 0$) and intermediate (LH circularly polarized at $\theta = 0$) modes. Figure 4 depicts their dependence on the obliquity θ for $k\rho_p = 0.02$. Both waves show reasonably rapid changes of polarization: at $\theta \sim 30°$, the intermediate mode is purely linear, and the fast mode has $\sin 2\varepsilon \sim 0.24$ ($AR \sim 8.2$). For waves with electric field amplitudes

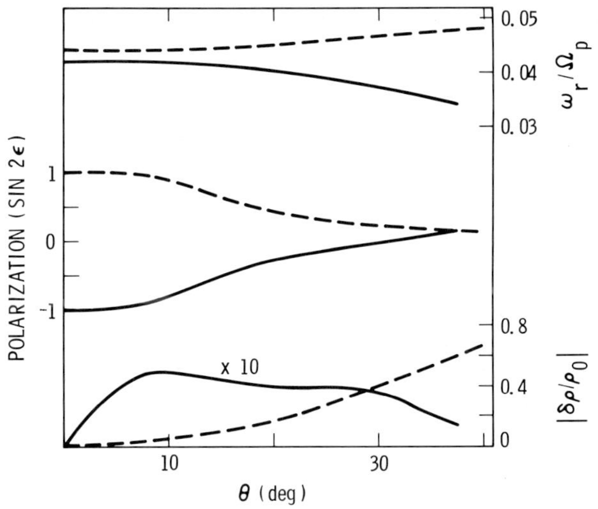

Fig. 4. Dependence on the obliquity angle θ of the real frequency (ω_r), polarization ($\sin 2\varepsilon$), and mass density compression ratio ($|\delta\rho/\rho_o|$) of the fast (dashed) and intermediate (solid) MHD-like modes. Magnetoplasma of Table 1 and $k\rho_p = 0.02$

of $1\,\mathrm{mV\,m^{-1}}$, the intermediate mode compression ratio, $|\delta\rho/\rho_o|$ maximizes around 0.05, at $\theta \sim 10°$, whereas the fast wave compression increases monotonically (~ 0.41 at $\theta \sim 30°$).

5.2. *Parallel propagation*

We now consider the newborn particles with drifting ring velocity distributions in the solar wind. At parallel propagation the electromagnetic mode of interest has LH circular polarization and is noncompressional. For the parameters shown in Tables 1 and 2 with $\sin^2\alpha = 0.9$ ($\alpha = 71.57°$), the complex dispersion of our unstable mode is displayed in Fig. 5. The unstable frequencies range from $0.772\,\Omega_0$ to $3.67\,\Omega_0$ with maximum growth of $\gamma/\Omega_0 \sim 0.67$ when $k\rho_p = k_m\rho_p \sim 0.038$. The wave moves with the drifting rings ($V_0 = V_{ei} = 1.6\,v_A$), with phase velocities within $(1.34 - 3.13)\,v_A$.

The separation between the real dispersion curves of the mode, $\omega_r(k)$, and the oxygen beam, $\omega_r = \Omega_0 + kV_0$, together with the small parallel thermal spread of the newborn ions, suggests

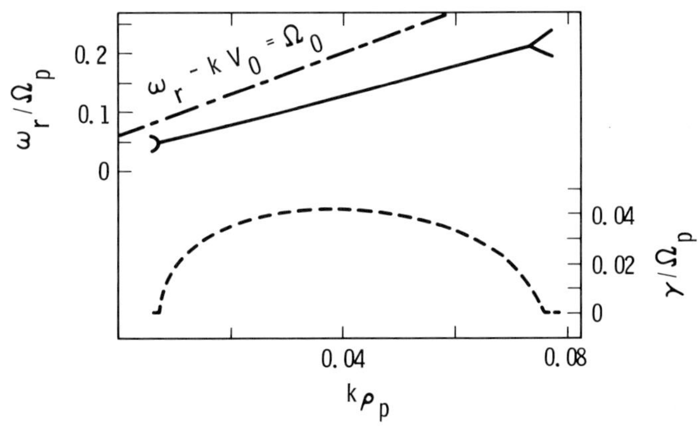

Fig. 5. Complex ($\omega = \omega_r + j\gamma$) parallel dispersion for the model given in Tables 1 and 2, with $\sin^2\alpha = 0.9$. The dispersion of the oxygen beam, $\omega_r - kV_0 = \Omega_0$, is also shown

that the instability is nonresonant and arises from the large ion ring reservoir of perpendicular energy. The ring photoelectrons are unable to couple to this mode ($\omega_r \neq -\Omega_e + kV_{ei}$); they could be removed from the medium without significantly affecting the characteristics to be discussed. Modification of the parallel ring temperatures also brings about negligible effects: a fifty fold increase to $T_{\|0} = T_{\|ei} = 5$ eV yields dispersion changes of less than 1%.

Changes in the angle α between $\boldsymbol{V}_{sw}$ and $\boldsymbol{B}_o$ influence the ring characteristics, as shown in Table 2: the drift velocity and the perpendicular energy vary as $\cos\alpha$ and $\sin^2\alpha$, respectively. An increase in α pushes the oxygen beam closer to resonance and adds to the perpendicular energy reservoir, thus enhancing the instability. Conversely, the instability is quenched when the ring drift velocity exceeds $\sim 1.91\, v_A$, i.e., $\alpha < 68°$. Figure 6 depicts the dependence of the complex frequency, $\omega = \omega_r + j\gamma$, on α, through the corresponding variation of the ring drift velocity (since ω_r and k are taken as positive, negative drift velocities imply counter-streaming beams). Consideration of the evolution of the beam dispersion curve in association with the dispersion curves of the unperturbed modes (Fig. 1) explains the reduction of the real frequency with the drift velocity. At drifts around $-0.79\, v_A$ the mode becomes nonoscillatory ($\omega_r = 0$), purely growing with $\gamma/\Omega_0 \sim 1.36$. (Further reduction of the drift velocity excites a RH circularly polarized wave propagating with the beams.) These nonoscillatory structures also occur at oblique propagation, although they do not originate in the mirror instability which is unable to excite parallel modes.

5.3. Oblique propagation

We start with the wavenumber that maximizes parallel growth, $k_m\rho_p = 0.038$, when $\sin^2\alpha = 0.9$, and vary the obliquity θ. As shown in Fig. 7, the growth rate maximizes off-axis ($\theta \sim 9°$), around $\gamma/\Omega_0 \sim 0.88$, when $\omega_r/\Omega_0 \sim 1.39$. At this obliquity, the polarization is already close to linear ($\sin 2\varepsilon \sim -0.38$, $AR \sim 5.1$) and the wave is strongly compressional, $|\delta\rho/\rho_o| \sim 2.4$. We emphasize that the oxygen ions make a dominant contribution to the mass density perturbation: their number density perturbation and larger mass compensate for the smaller equilibrium number density.

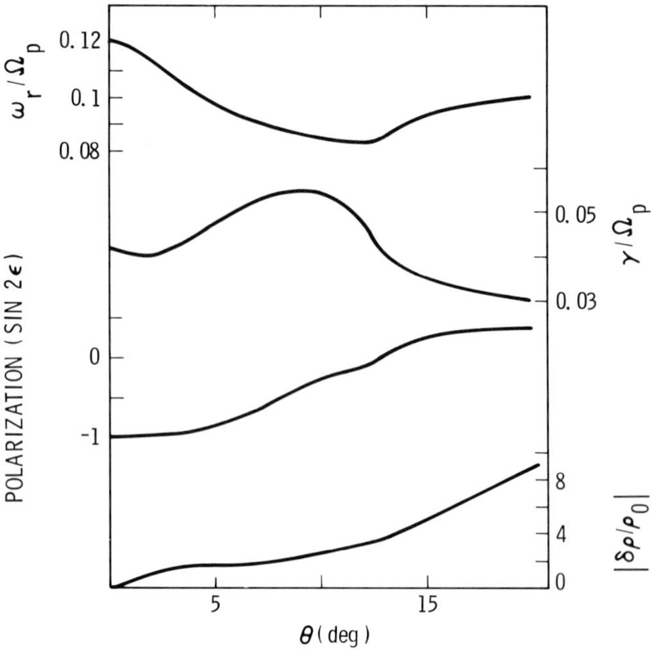

Fig. 7. Dependence on the obliquity angle θ of the real frequency (ω_r), growth rate (γ), polarization ($\sin 2\varepsilon$), and mass density compression ratio. Model of Tables 1 and 2, with $\sin^2\alpha = 0.9$ and $k\rho_p = 0.038$ (maximum growth in Fig. 5)

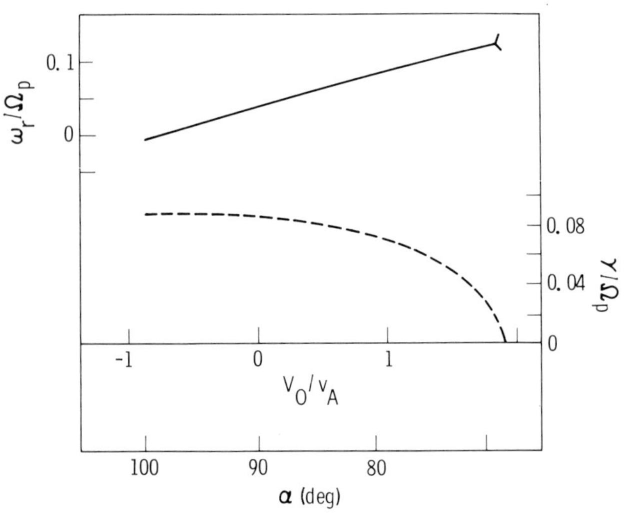

Fig. 6. Dependence of the complex ($\omega = \omega_r + j\gamma$) parallel dispersion on the angle α (or ring drift velocity, $V_0 = V_{sw}\cos\alpha$). Model of Tables 1 and 2 with $k\rho_p = 0.038$ (maximum growth in Fig. 5)

Variation of the wavenumber at a constant obliquity of $\theta = 9°$ shows two broad maximums of the growth rate, $\gamma/\Omega_0 \sim 0.88$, centered at $k\rho_p \sim 0.04$ and 0.063. A repeated exploration of the dispersion dependence on θ, this time with $k\rho_p = 0.063$, yields maximum growth of $\gamma/\Omega_0 \sim 0.91$ at $\theta \sim 6°$ and $\omega_r/\Omega_0 \sim 2.33$, when the polarization is almost linear ($\sin 2\varepsilon \sim -0.29$, $AR \sim 6.75$) and the compression ratio large, $|\delta\rho/\rho_0| \sim 2.2$; the maximum is fairly broad and, at $\theta \sim 10°$, we still find $\gamma/\Omega_0 \sim 0.86$.

These searches were conducted for the parameter values given in Tables 1 and 2, using $\sin^2\alpha = 0.9$. The question thus arises as to what extent the characteristics of our mode depend on the model adopted to describe the solar wind and the newborn particles. The influence of the parameters in Table 1 shall not be discussed: they are typical G–Z ICE background values whose variation within realistic bounds would not significantly affect the wave properties. In contrast, the angle α and some of the free energy source parameters of Table 2 strongly dominate the wave stability and dispersion, thus warranting further analysis.

Figure 8 displays the wave properties as a function of the ring drift velocity, depicting their dependence on α ($V_0 = V_{sw}\cos\alpha$), for $k\rho_p = 0.063$ and $\theta = 6°$. The qualitative evolution of both the real frequency and the growth rate is similar to that found in parallel propagation: ω_r decreases with V_0, whereas γ shows an opposite behavior. We find a nonoscillatory ($\omega_r = 0$), purely growing mode ($\gamma/\Omega_0 \sim 1.28$) when the rings are nondrifting. The reduction in V_0 (>0) tends to increase the compression ratio and to approximate the wave to linear polarization.

We have already indicated that the photoelectrons and the parallel ring temperatures do not significantly affect the wave characteristics in our parameter domain. We now assess the influence of the remaining factors (ring density and perpendicular

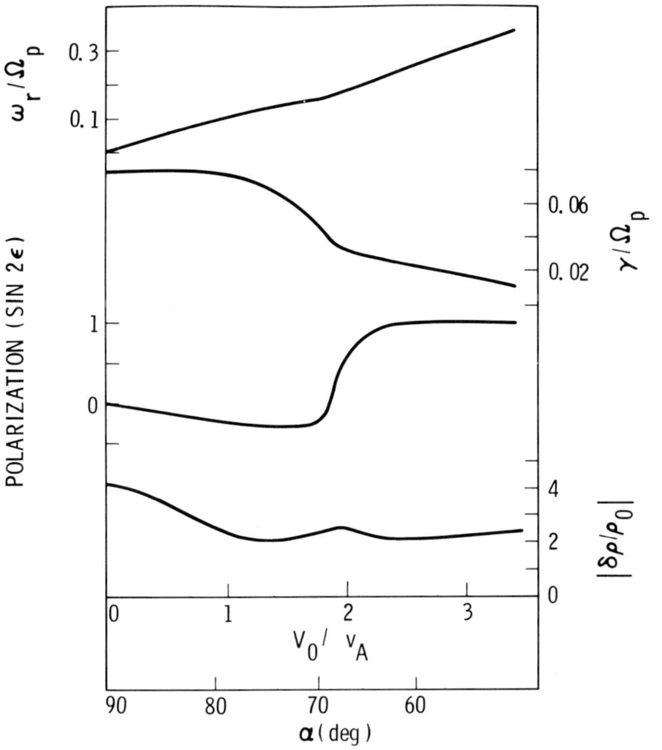

Fig. 8. Dependence on the angle α (or ring drift velocity, $V_0 = V_{sw} \cos \alpha$) of the real frequency (ω_r), growth rate (γ), polarization ($\sin 2\varepsilon$), and mass density compression ratio. Model of Tables 1 and 2, with $\theta = 6°$ and $k\rho_p = 0.063$ (see text)

energy), using $k\rho_p = 0.063$, $\theta = 6°$ and $\sin^2 \alpha = 0.9$. The perpendicular energy of the rings is reduced with respect to its nominal value in Table 2 through corresponding reductions in the parameters a_s and b_b ($a_s/b_s = 1.5$; $s = 0,\mathrm{ei}$): we recall that $\langle v_\perp^2 \rangle_s = (a_s + b_s)v_{ts}^2$. Figure 9 shows the effect of perpendicular energy reductions down to $\sim 20\%$ of the nominal value: the polarization becomes almost LH circular, and the growth rates rapidly decay. However, as evidenced in Fig. 10, the mode recovers its basic characteristics (almost linear polarization and large compressions associated with maximum growth at oblique propagation) at somewhat larger θ: with a perpendicular ring energy equal to 25% of its nominal (Table 2) value, the wave is stable and LH circularly polarized at parallel propagation but, as the obliquity increases, exhibits maximum growth at $\theta \sim 12.5°$ ($\gamma/\Omega_0 \sim 0.42$, $\omega_r/\Omega_0 \sim 2.3$, $|\delta\rho/\rho_o| \sim 6.1$) and is already linearly polarized at $\theta \sim 11.5°$. Clearly, the mode might exist only for oblique propagation if the ring perpendicular energy is insufficient to destabilize the parallel wave.

The modifications brought about by a reduction of the newborn particle density are given in Fig. 11. The polarization gradually evolves into RH circular, and the growth rate is slowly reduced; when the ring density is four orders of magnitude smaller ($5\,10^{-4}\,\mathrm{cm}^{-3}$) than the solar wind density, the growth rate is still significant ($\gamma/\Omega_0 \sim 0.16$, $\omega_r/\Omega_0 \sim 2.3$). The decrease in the compression ratio is associated with the ring density reduction (the ring ions made a major contribution to $\delta\rho$), and the gradual conversion of the mode's nature, as evidenced in the polarization change. Indeed, as the ring density diminishes, the wave evolves into the wellknown RH mode (e.g. Gary et al., 1984) and, in the parameter domain considered here, loses its unusual characteristics for oblique propagation; in particular, the growth rate now maximizes at parallel propagation. Note that a decrease in density translates into an increase in the distance to the comet, so

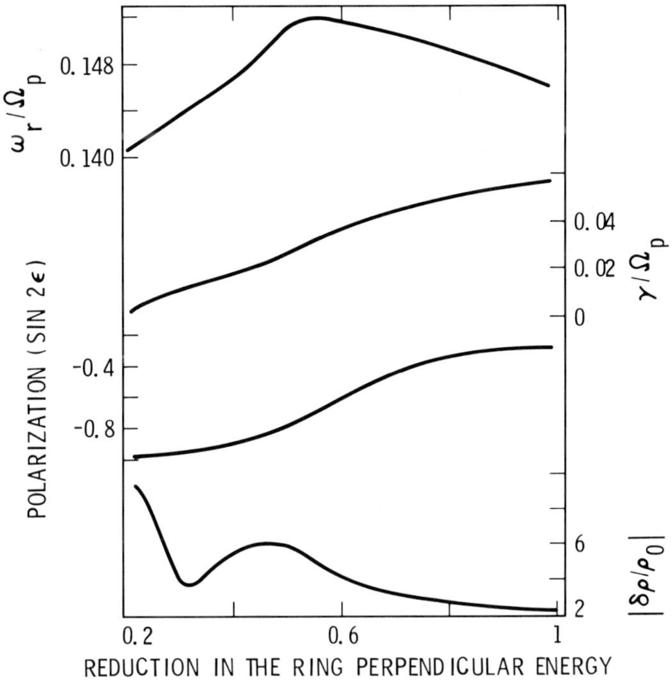

Fig. 9. Dependence on the perpendicular energy of the rings, which is reduced from the nominal value in Table 2. Remaining parameters as in Tables 1 and 2, with $\sin^2 \alpha = 0.9$, $\theta = 6°$ and $k\rho_p = 0.063$ (see text)

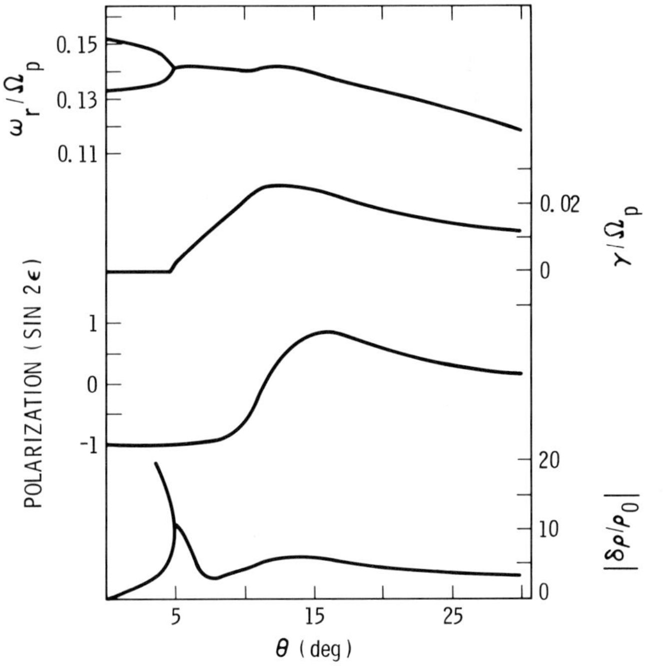

Fig. 10. Dependence on the obliquity angle θ of the real frequency (ω_r), growth rate (γ), polarization ($\sin 2\varepsilon$), and mass density compression ratio. Model of Tables 1 and 2, with $\sin^2 \alpha = 0.9$ and $k\rho_p = 0.063$ (see text), and the perpendicular energy of the rings reduced to one fourth of its nominal value in Table 2

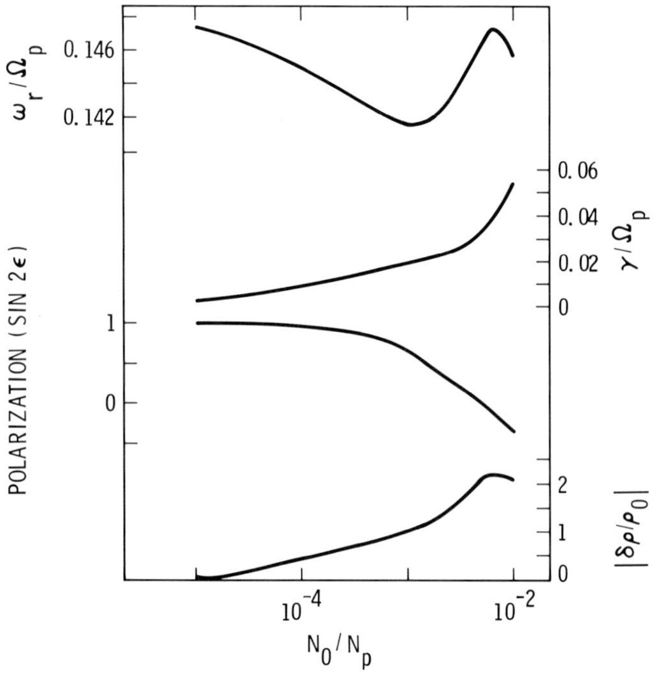

Fig. 11. Dependence on the relative density of the newborn particles of the real frequency (ω_r), growth rate (γ), polarization ($\sin 2\varepsilon$), and mass density compression ratio. Remaining parameters as in Tables 1 and 2, with $\sin^2 \alpha = 0.9$, $\theta = 6°$ and $k\rho_p = 0.063$ (see text)

that this mode conversion might have to do with observed differences in wave properties as spacecrafts approach comets.

6. Discussion

Newborn ions with large perpendicular energies (high α) can excite electromagnetic waves with frequencies of the order of the heavy ion cyclotron frequency, Ω_i, and unusual propagation characteristics at small obliquity angles. Because $\alpha \sim 90°$ and the waves are quasi-parallel, the solar wind Doppler shift is small and the perturbations should be detected near Ω_i in the spacecraft frame. For parallel propagation the mode is LH circularly polarized, might be unstable in a frequency range containing Ω_i, and moves in the direction of the newborn ion drift along the static magnetic field. Slight deviations from parallel propagation bring about drastic changes in the wave polarization, which becomes almost linear, and in the compressional behavior, with sharp increases in the mass density perturbations.

These properties are of interest in themselves; they provide further insight into the interpretation of space observations of almost parallel MHD-like waves with respect to polarization and nonlinear behavior (wave steepening and shock formation). However, the distinguishing feature of these waves lies in the coexisting association of maximum growth with almost linear polarization and large compressions: the growth rates maximize at small obliquities.

Although four free energy sources are available (the two rings radii and parallel streaming), the mode is predominantly excited by the perpendicular energy of the newborn ions. Within realistic parameter domains, the properties of the mode do not significantly depend on the details of the perpendicular velocity distribution of the newborn ions (the dependence is nonexistent for parallel propagation and separable velocity distributions). Therefore, albeit cast for cometary environments with ring distributions, we expect this study to be relevant in other space regions (e.g. planetary foreshocks) with different free energy sources, provided that the perpendicular energy of the heavy ions is sufficiently large.

Acknowledgments. This work was done while one of the authors (A.L.B.) held a National Research Council – NASA Research Associateship. The research described in this paper was performed by the Jet Propulsion Laboratory, California Institute of Technology, under contract with the National Aeronautics and Space Administration.

References

Anderson, R.R., C.C. Harvey, M.M. Hoppe, B.T. Tsurutani, T.E. Eastman, J. Etcheto: 1982, *J. Geophys. Res.* **87**, 2082

Belcher, J.W., L. Davis, Jr.: 1971, *J. Geophys. Res.* **76**, 3534

Burlaga, L.F.: 1971, *J. Geophys. Res.* **76**, 4360

Coleman, P.J., Jr.: 1966, *J. Geophys. Res.* **71**, 5509

Gary, S.P., C.W. Smith, M.A. Lee, M.L. Goldstein, D.W. Forslund: 1984, *Phys. Fluids* **27**, 1852

Gary, S.P., C.D. Madland, B.T. Tsurutani.: 1985, *Phys. Fluids* **28**, 3691

Goldstein, M., W. Wong, A. Viñas, C. Smith: 1985, *J. Geophys. Res.* **90**, 302

Hada, T.: 1985, *Ph.D. Thesis*, UCLA

Hada, T., M. Ashour-Abdalla.: 1986, *EOS* **67**, 1167

Hasegawa, A.: 1969, *Phys. Fluids* **12**, 2642

Hoppe, M., C.T. Russel, L.A. Frank, T.E. Eastman, E.W. Greenstadt: 1981, *J. Geophys. Res.* **86**, 4471

Ipavich, F.M., A.B. Galvin, G. Gloeckler, D. Hovestadt, B. Klecker, M. Scholer: 1986, *Science* **232**, 366

Kaufmann, R.L., J.T. Horng, A. Wolfe: 1970, *J. Geophys. Res.* **75**, 4666

Kennel, C.F., H.E. Petschek: 1966, *J. Geophys. Res.* **71**, 1

Krall, N.A., and A.W. Trievelpiece: 1973, *Principles of Plasma Physics*, McGraw-Hill, New York, p. 405

Kraus, J.D.: 1966, *Radio Astronomy*, Chap. 4, McGraw-Hill, New York

Lanzerotti, L.J., A. Hasegawa, C.G. Maclennan: 1969, *J. Geophys. Res.* **74**, 5565

Lee, J.C., F.W. Crawford.: 1970, *J. Geophys. Res.* **75**, 85

Lepping, R.P., K.W. Behannon: 1986, *J. Geophys. Res.* **91**, 8725

Neubauer, F., K. Glessmeier, M. Pohl, J. Raeder, M. Acuña, L. Burlaga, N. Ness, G. Musmann, F. Mariani, M. Wallis, E. Ungstrup, H. Schmidt: 1986, *Nature* **321**, 352

Neugebauer, M., D.R. Clay, B.E. Goldstein, B.T. Tsurutani, R.D. Zwickl: 1984, *J. Geophys. Res.* **89**, 5395

Neugebauer, M., C.J. Alexander, R. Schwenn, A.K. Richter: 1986, *J. Geophys. Res.* **91**, 13694

Price, C.P., D.W. Swift, L.C. Lee: 1986, *J. Geophys. Res.* **91**, 101

Riedler, W., K. Schwingenschuk, Ye. G. Yeroshenko, V.A. Styashkin, C. T. Russel: 1986, *Nature* **321**, 288

Roberts, D.A., L.W. Klein, M.L. Goldstein: 1986, *EOS* **67**, 1149

Rönnmark, K.: 1982, *Kiruna Geophysical Institute Report No. 179*

Rönnmark, K.: 1983, *Plasma Phys.* **25**, 699

Sharma, O., V. Patel: 1986, *J. Geophys. Res.* **91**, 1529

Smith, E.J.: 1973, *J. Geophys. Res.* **78**, 2054

Smith, R.L., N. Brice: 1964, *J. Geophys. Res.* **69**, 5029
Smith, C.W., M.A. Lee: 1986, *J. Geophys. Res.* **91**, 81
Smith, E.J., B.T. Tsurutani, D.L. Chenette, T.F. Conlon, J.A. Simpson: 1976, *J. Geophys. Res.* **81**, 65
Smith, C.W., M.L. Goldstein, S.P. Gary, C.T. Russel: 1985, *J. Geophys. Res.* **90**, 1429
Southwood, D.J., W.J. Hughes: 1983, *Space Sci. Rev.* **35**, 301
Tsurutani, B.T., W.D. Gonzalez: 1987, *Planet. Space Sci.* (in press)
Tsurutani, B.T., E.J. Smith: 1979, *J. Geophys. Res.* **84**, 2773
Tsurutani, B.T., E.J. Smith: 1984, *Geophys. Res. Letters* **11**, 331
Tsurutani, B.T., E.J. Smith: 1986a, *Geophys. Res. Letters* **13**, 263
Tsurutani, B.T., E.J. Smith: 1986b, *Geophys. Res. Letters* **13**, 259
Tsurutani, B.T., E.J. Smith, R.R. Anderson, K.W. Ogilvie, J.D. Scudder, D.N. Baker, S.J. Bame: 1982, *J. Geophys. Res.* **87**, 6060
Tsurutani, B.T., E.J. Smith, D.E. Jones: 1983, *J. Geophys. Res.* **88**, 5645
Tsurutani, B.T., R.M. Thorne, E.J. Smith, J.T. Gosling, H. Matsumoto: 1987, *J. Geophys. Res.* (submitted)
Viñas, A.I., M.L. Goldstein, M.H. Acuña: 1984, *J. Geophys. Res.* **89**, 6813
Winske, D., C.S. Wu, Y.Y. Li, Z.Z. Mou, S.Y. Guo: 1985, *J. Geophys. Res.* **90**, 2713
Wong, H.K., M.L. Goldstein: 1986, *J. Geophys. Res.* **91**, 5617
Wu, C.S., R.E. Hartle: 1974, *J. Geophys. Res.* **79**, 283
Young, D.T., S. Perrault, A. Roux, C. de Villedary, R. Gendrin, A. Korth, G. Kremser, D. Jones: 1981, *J. Geophys. Res.* **86**, 6755
Yumoto, K., T. Saito, T. Nakagawa: 1986, *Geophys. Res. Letters* **13**, 825

Astron. Astrophys. 187, 320–324 (1987)

ASTRONOMY AND ASTROPHYSICS

Far-ultraviolet objective spectra of comet P/Halley from sounding rockets

C.B. Opal [1], **R.P. McCoy** [2], **and G.R. Carruthers** [2]

[1] McDonald Observatory, University of Texas, Austin, TX 78712, USA

[2] E.O. Hulburt Center for Space Research, U.S. Naval Research Laboratory, Washington, DC 20375, USA

Received February 23, accepted May 19, 1987

Summary. A sounding rocket payload to obtain far ultraviolet spectral images of comet P/Halley was launched on 1986 February 24 and again on 1986 March 13. The second flight occurred 13 hours prior to the Giotto spacecraft flyby of the comet. The payload included an objective grating spectrograph, consisting of an $F/1.5$ microchannel plate intensified electrographic Schmidt camera pointed at a plane grating. The instrument field of view was 12° with a spatial resolution of 1′ and a point source wavelength resolution of 0.1 nm. High quality images of the oxygen coma at 130.4 nm, the carbon coma at 156.1 nm and 165.7 nm, and the sulphur coma at 181.4 nm were obtained. The images have been analyzed using a radial outflow model to derive production rates for these species.

Key words: gaseous coma – elemental abundances – ultraviolet spectra

1. Introduction

Wide field imagery of comets in the ultraviolet provides a useful means of determining such fundamental parameters as the production rates, lifetimes, and outflow velocities of daughter photodissociation fragments in the coma. Such measurements can yield the abundance ratios of the common volatile elements, providing important information on the conditions under which comets formed. Far-ultraviolet imagery has been used to measure the production rates and outflow characteristics of hydrogen for comets Bennett (Code et al., 1972), Kohoutek (Opal et al. 1974; Meier et al., 1976), and West (Opal and Carruthers, 1977a), for C and O in comet Kohoutek (Opal and Carruthers, 1977b), and for C, S in comet West (Smith et al., 1980). More recently these elements have been observed spectroscopically in a number of other comets from the International Ultraviolet Explorer (IUE) spacecraft (Feldman, 1982).

We present here results from two rocket flights of an ultraviolet spectroscopic imaging experiment to observe comet Halley. The rockets were launched from White Sands Missile Range, New Mexico on February 24 at 11:53:30 UT (at which time R, the sun-comet distance, was 0.672 AU) and 1986 March 13, at 10:54 UT (when R was 0.895 AU). The first rocket (NASA 36.010 DL) was launched as soon after perihelion as practical in order to observe the comet at high luminosity, while the second rocket (NASA 36.017 DL) was launched as close in time as possible to the Giotto spacecraft flyby of comet Halley. Both experiments were coordinated with rocket observations of the comet by Johns Hopkins University experimenters, who used a far-ultraviolet long-slit spectrograph (Woods et al., 1986).

Our observations yielded direct electrographic images of the hydrogen coma of the comet at Lyman-α (121.6 nm) and objective spectra containing images of the coma at the oxygen (130.4 nm), carbon (156.1 nm and 165.7 nm) and sulfur (181.4 nm) resonance multiplets. The Lyman-α images from the two flights are described by McCoy et al., 1986. The images of oxygen, carbon and sulfur emissions obtained with the objective grating spectrograph are presented below and production rates for these elements are derived.

2. Observations

2.1. Instrumentation

The rocket payload instruments were similar to those used previously in sounding rocket observations of comet Kohoutek (Opal et al., 1974). The payload contained a direct imaging camera covering the wavelength range 105–160 nm (including the hydrogen Lyman-α line at 121.6 nm), and an objective grating spectrograph covering the 123–200 nm range (Fig. 1). The spectrograph was based on an electrographic Schmidt camera having a CaF_2 corrector plate and CsI photocathode, with a focal length of 150 mm and a $f/1.5$ focal ratio. It used a 1200 line/mm objective grating, yielding a spectral dispersion at the photocathode of approximately 3 nm/mm. It had a 12° square field of view and angular resolution of about 1′. The corresponding point-source spectral resolution was about 0.1 nm. This camera was much more sensitive than that used in the previous objective spectroscopy of comet Kohoutek, because it incorporated a microchannel plate (MCP) intensification stage (Carruthers, 1979). The MCP was operated at a gain of about 100, yielding a corresponding increase in effective diffuse-source speed.

2.2. Calibration and data reduction

Laboratory calibrations consisted of measurements of relative wavelength sensitivity combined with an absolute measurement of a monochromatic point source. The reduction of the laboratory

Send offprint requests to: C.B. Opal

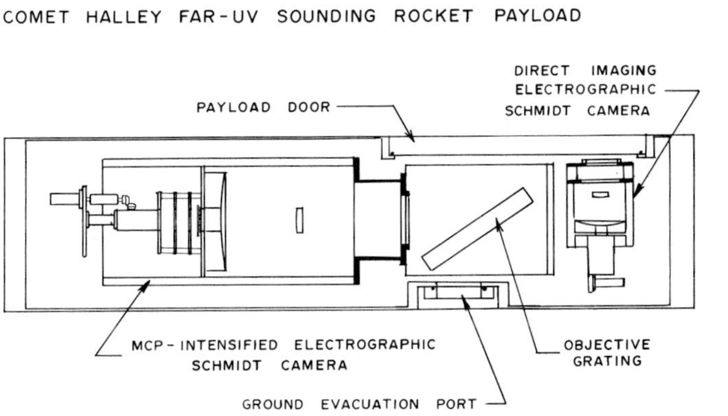

Fig. 1. Schematic layout drawing of the rocket payload, showing the objective spectrograph and Lyman-α camera. The portion shown is approximately 0.4 1.5 m in size. The payload was flown to an altitude of 312 km aboard a Terrier boosted Black Brant V rocket. A gyro-stabilized attitude control system pointed the field of view at the comet to an accuracy of about 1°. Approximately 6 minutes of data at altitudes above 200 km were obtained on each flight

calibration data is complex and has not yet been completed; consequently we used an in-flight calibration based on the spectrum of the star τ Capricornis, which appeared next to the comet spectrum on the first flight. In addition, objective spectra of a star field centered near δ Scorpii were obtained in each flight; by comparing stars in this field an accurate relative calibration between the two flights was achieved (the sensitivities on the two flights were nearly identical).

The flight and calibration exposures were scanned and digitized using a PDS microdensitometer at the University of Texas and a second PDS recently put into service at NRL; the two instruments were cross calibrated using sensitometry spots. The images were scanned using a 16×16 micron aperture on a 2048×1536 array; the measurements were then summed together on a 2×2 basis to form a 1024×768 array of 32 micron pixels for final analysis (the resultant sampling was 2–3 pixels/resolution element). Nonlinearities are small in the electrographic process below a film density of about 4.0, and were avoided by using data from shorter exposures for the brighter features.

In the first flight, the comet was only 0°.6 away from the star τ Cap ($V = 5.22$, spectral type B6 III). As a result, high-quality spectrograms of this star were obtained which could be compared directly with those of the comet. This star has been observed by instruments on the OAO-2 and TD-1 satellites. Unfortunately, the shortest-wavelength OAO-2 photometry of τ Cap (Code et al., 1980) was in a band centered at 191 nm. This measurement was used ot normalize the OAO-2 spectrum of ζ Dra (Code et al., 1979), which is of the same spectral type and color ($(B-V) = -0.12$). TD-1 spectrophotometry of τ Cap is available down to 138 nm, but is about 15% below the OAO-2 measurement at 191 nm (Jamar, 1976). The microdensitometry of the τ Cap spectrum was summed in the direction perpendicular to the dispersion, and the background in an equal width of the image outside the spectrum was subtracted. This was then smoothed to the OAO-2 resolution of 2.0 nm, and divided by the derived τ Cap flux distribution to yield the instrument sensitivity vs. wavelength. There was no convenient stellar spectrum near the comet on the second flight; consequently we normalized the observations on the second flight to those of the first using stellar spectra observed in the Scorpius field. We believe the absolute accuracy of the spectral data is about ±35%, including the uncertainty in the comparison star spectrophotometry, but the relative wavelength variation in the calibration is about ±10%. The accuracy of the ratios of the fluxes between the two flights is also good (better than 10%).

3. Results

3.1. *Objective spectral images of O, C, and S*

In both rocket flights objective spectral images of the comet were made with exposure times of 2.5, 9.5, and 29.5 s. Grayscale reproductions of images from the first flight for a 9.5 s exposure, a 29.5 s exposure, and the sum of nine 29.5 s exposures are shown in Fig. 2. Contours of the surface brightness of the O 130.4 nm, C 156.1 nm and 165.7 nm, and S 181.4 nm multiplets are shown in Fig. 3. Since the innermost regions of the 130.4 nm and 165.7 nm images were saturated on the film for exposure times longer than 2.5 s, the isophote contours shown for these features are from a combination of images with long and short exposure times. Contours for isophotes greater than 100 R were taken from 2.5 s exposures for both features. The outer isophotes of the 130.4 nm image were taken from a 9.5 s exposure; those of the C 165.7 nm, C 156.1 nm, and S 181.4 nm were taken from the summed 29.5 s images. Note that the lines in the sulphur multiplet are separated enough to be resolved into two groups at this resolution. The spatial scales along and perpendicular to the dispersion are different in an objective spectrograph, as can be seen by differentiating the grating equation. For the present configuration the scale slong the dispersion was a factor of 1.22 higher at 130 nm and 1.39 higher at 181 nm. This distortion was corrected for in plotting the isophotes. Some additional distortion was observed on the long wavelength side of the 130.4 nm contours, which we attribute to emission from C^+ at 133.5 nm as identified by Woods et al. (1986). Other than these effects, no azimuthal asymmetries were apparent in the contours.

Because of the azimuthal symmetry of the coma images, it was possible to generate radial brightness profiles by integrating the data in circular annuli to improve the signal-to-noise ratio at large radii. The resulting radial brightness profiles are shown in Fig. 4. The profiles have been synthesized from data for different exposure times, as discussed above for the contours. Also, to avoid zero-order star images and spectra which cross the coma images, profiles were determined separately for each quadrant. The final profiles were obtained by choosing data from quadrants least affected by such contamination. Unfortunately, the sulphur emission was too faint on the second flight for us to measure the profile. The center of the sulphur emission was taken to be midway between the two peaks, so the profile near the center is not realistic. We have not yet reduced the C 156.1 nm data since it provides the same information as the C 165.7 nm emission but at lower signal-to-noise ratio.

3.2. *Atomic production rates*

If the lifetime of the species is known, the production rate, Q, in atoms/s, can be determined from

$$4\pi F = Q(\tau_0 R^2)(g_0/R^2) = Q\tau_0 g_0, \tag{1}$$

where $4\pi F$ is the total luminosity of the comet in photons/s, g_0 is the number of solar photons scattered per s per atom at 1 AU and τ_0 is the lifetime of the species at 1 AU. The difficulty with this approach is that most of the emission comes from the outer part of

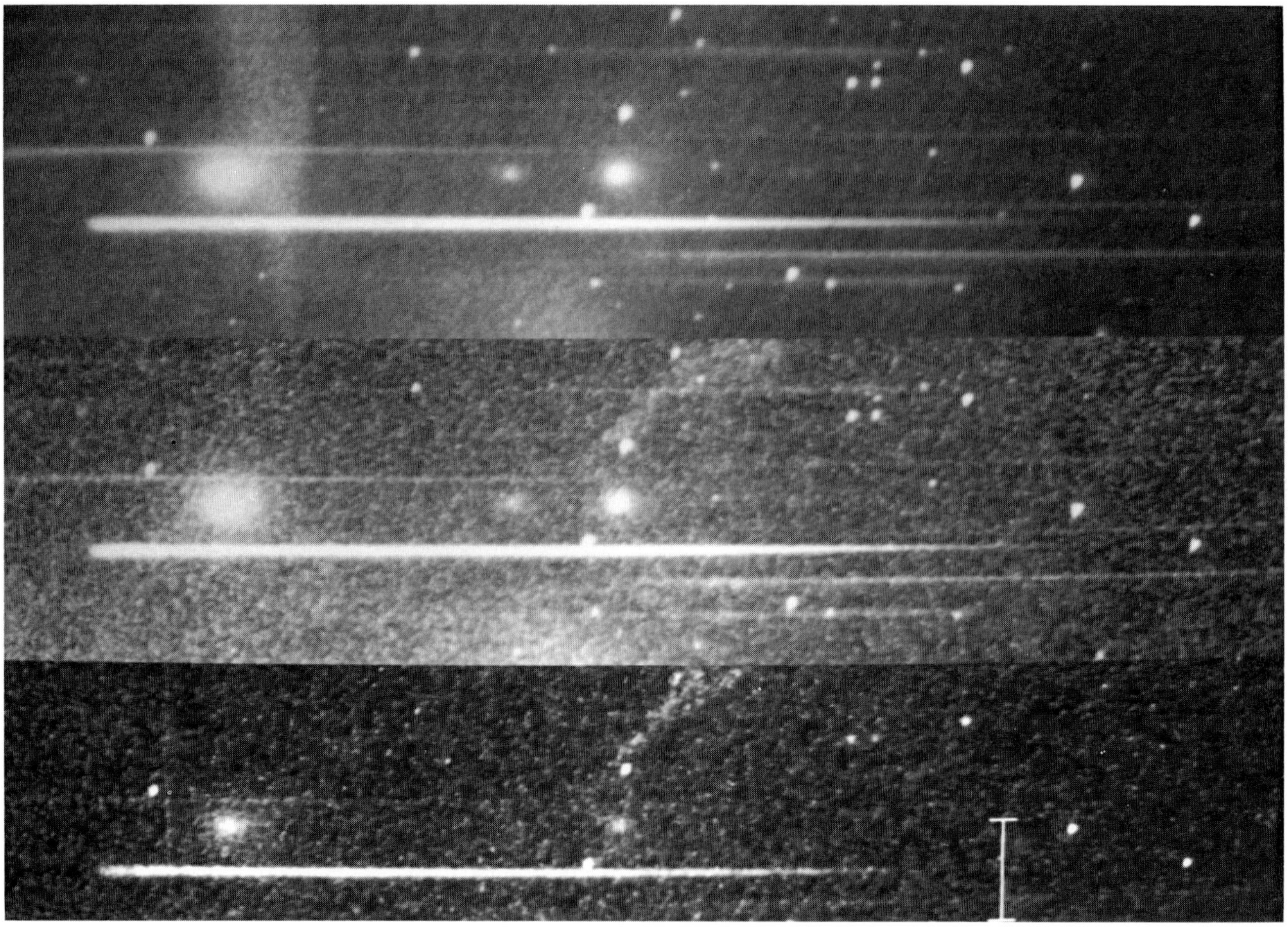

Fig. 2. Objective spectra from the February 24 flight. The lower spectrum is a 9.5 s exposure, the central one 30 s, and the top spectrum is the sum of 9 × 30 s images. The spectrum of the comet appears as a fuzzy, moderately elliptical images at each emission wavelength. The spectrum of τ Cap lies immediately below that of the comet. The leftmost emission is that from O at 130.4 nm, while the second strongest feature is due to C at 165.7 nm near center. The C 156.1 nm image lies to the left of C 165.7 nm. Faintly evident in the topmost spectrum is a double feature due to S at 181.4 nm, and several bands of CO between the C and O features. The size marker indicates an angular extent of 1°

the coma and the total is very sensitive to errors in the determination of the background to subtract from the image.

Because of the background subtraction problem, the production rates of C, O and S were determined from the radial profiles by fitting them to a simple radial-outflow model. In this model atoms are assumed to expand outward isotropically from a negligibly small source region. The production can be expressed as:

$$Q = \varrho B(\varrho) \{4\ 10^6\, \bar{v}\, (R^2/g_0)\} \tag{2}$$

where ϱ is the projected distance from the nucleus, $B(\varrho)$ is the observed brightness at that distance measured in Rayleighs, $\bar{v}$ is the mean outflow velocity. A radial outflow model which includes a finite source region and loss processes was given by Haser (1957). A more accurate model which takes into account that the dissociation products flow radially outward from the point of generation, rather than radially from the nucleus (the vectorial model), has been studied by Festou (1981). Additional effects, such as bulk motions, collisions, and the anisotropy of the photo-dissociation process, can also modify the profiles. And, at least in the case of comet Halley, there are additional complications due to the anisotropic and highly time-variable production of the parent molecules, as observed by the flybys and from earth.

The simple radial outflow model will only be valid in the region where ϱ is large compared to the source region, so that anisotropies and time variations will average out, but small compared to the scale length, γ, associated with loss processes. Opal and Carruthers (1977b) give γ_0 (the scale length at 1 AU) as 1.4 10^6 for O and 1.6 10^6 km for C, while Smith et al. (1980) derive a lifetime for S of 1.6 10^6 km, which gives a γ_0 of 1.6 10^6 km for the species assuming a typical radial outflow velocity to 1 km s^{-1}. At the time of our February 24 observation, the corresponding scale lengths are 6.3 10^5 km for 0 and 7.2 10^5 km for S and C.

Production rates derived from the simple radial outflow model for the three species on February 24 and for O and C on March 13 are summarized in Table 1. In each case the brightness observations were taken at a point on the curves where the brightness falloff followed a $1/\varrho$ dependence, with as small a value of ϱ as possible in order to minimize lifetime effects. For consistency with previous calculations we have adopted an outflow velocity of 1 km s^{-1}, although this may require revision in the light of recent in situ measurements. The sulphur brightness listed is the sum

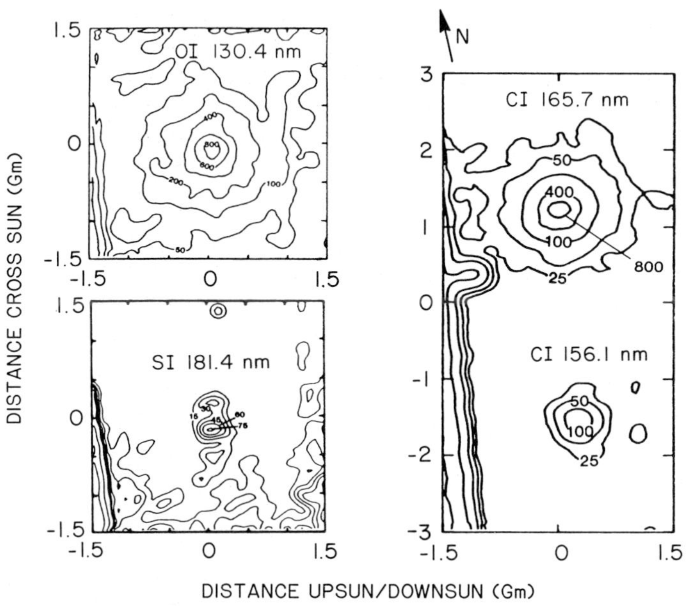

Fig. 3. Brightness contours (in Rayleighs) of the objective spectral images of C, O, and S resonance multiplets observed during the flight of February 24. The images have been compressed in the spectral direction to correct for the non-isotropic plate scale of an objective spectrograph. The contours have been oriented so that the sun is positioned to the left of the plots and the spatial scales in both dimensions are in Gm. A background was subtracted from all four images which was taken from an average in the vicinity of each feature. The vertical streaks on the left of the images are from the spectrum of τ Cap which was used for calibration. The right side of the images show a small amount of contamination from another stellar spectrum, and contours from zero order star images are seen

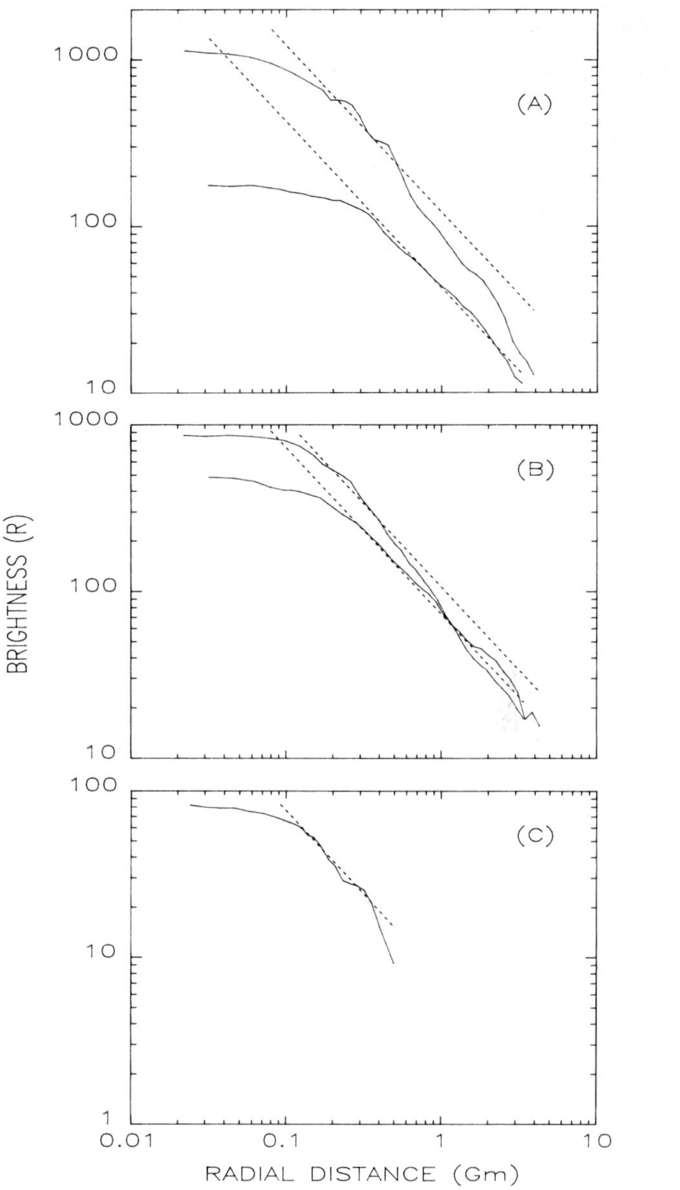

Fig. 4A–C. Radial brightness profiles (in Rayleighs) for **A** O at 130.4 nm, **B** C at 165.7 nm, and **C** S at 181.4 nm. The lower trace on the C and O profiles are the March 13 observations; the other traces are from February 24. The dashed lines are $1/\varrho$ fits to the data in the outer regions

from the two bright lines, and is the least reliable because of the poor signal-to-noise ratio for this multiplet and because of the overlapping brightness distributions from the resolved multiplet. Note that even though the O coma was much brighter on February 24, the production rate was actually lower than on March 13, when the Doppler shift was unfavorable for exciting this transition.

Lifetime effects can be quite significant even at rather small values of ϱ. For example, at $\varrho = 0.3\,\gamma$ (approximately the distance used for the O and C calculations in Table 1), the Haser (1957) model gives a brightness almost a factor of two lower than given by a $1/\varrho$ dependence. However, there are indications that the scale lengths are much longer than given above, at least for C and O. For example, at $\varrho = 2\,\gamma$, the departure from the $1/\varrho$ dependence should be a factor of 16, whereas for both C and O on February 24 the brightness at this distance (1.5 Gm) is down only about a factor of 2 from $1/\varrho$. There are several possible explanations for this effect. The lifetimes of these species may be longer than previously believed, or the outflow velocities may be substantially higher than 1 km s^{-1}. Alternatively, because the atoms at 1.5 Gm are on the order of two weeks old, the greater brightness at this distance may be the result of a much higher production rate at earlier times. Because of the uncertainties in the scale lengths, we did not apply corrections for lifetime effects.

4. Discussion

Our February 24 results can be compared with those of Woods et al. (1986) measured on February 26, and our March 13 results can

Table 1. Production rates

Species	ϱ (Gm)	$B(\varrho)$ (R)	g_0 (1 AU) (s^{-1})	Q (atoms/s)
O (24 Feb.)	0.2	610	5.3 (−6)	4.2 (29)
C	0.2	532	2.3 (−5)	8.4 (28)
S	0.15	51	5.9 (−5)	2.4 (27)
O (13 Mar.)	0.4	142	1.8 (−6)	1.0 (30)
C	0.2	365	2.2 (−5)	1.0 (29)

Notes: Numbers in parenthesis are powers of 10 by which numbers should be multiplied. Values for g_0 for C and O from Feldman et al. (1976); g_0 for S from Smith et al. (1980)

be compared with their simultaneous measurements. Their spatial resolution was about a factor of five higher than ours, so the data cannot be compared directly in the inner regions. Our observations farther out, near 10^5 km, are in reasonable agreement (within better than a factor of two). They derive a C production rate of 7 10^{28} on February 26, about 20% less than our value of February 24, which we consider excellent agreement (they do not give a rate for March 13). The O values (derived from their H_2O production rates) are 1.2 10^{30} and 6 10^{29}, the latter is similar to ours but the former is almost a factor of 3 higher. We attribute the discrepancy to the models used because to the extent that the data can be compared they are in reasonable agreement. Woods et al. used a photochemical model for the inner region of the coma, whereas we used a radial outflow model for the outer region.

In any case, we agree with Woods et al. (1986) that carbon is cosmically underabundant relative to oxygen in the volatile component of this comet. However, Giotto measurements indicate a large carbon content in the dust (Kissel et al., 1986), which may explain the deficiency of gaseous carbon. Our C/S ratio of 35 is not very different from the value of 22 given in Smith et al. (1980) (the cosmic ratio is 21); however, this result is tentative because of the split multiplet and it will be necessary to model the response of the instrument to this coma to obtain an accurate value.

In this paper we have concentrated on determining production rates for these species from the brightness observed in the radial outflow region. However, additional information can be obtained from these data, including loss scale lengths and possibly some characteristics of the source region. This requires modelling expected brightness profiles, convolving them with the instrument response, and comparing the results with the data. The results of these comparisons will be described in a future paper.

References

Carruthers, G.R.: 1979, *Proc. SPIE* **172**, 304

Code, A.D., Houck, T.E., Lillie, C.F.: 1972, *The Scientific Results from Orbiting Astronomical Observatory* (*OAO-2*), ed. A.D. Code, NASA SP-310, Washington DC, p. 109

Code, A.D., Hom, A.V., Bottemiller, R.L.: 1980, *Astrophys. J. Suppl.* **43**, 501

Code, A.D., Meade, M.R.: 1979, *Astrophys. J. Suppl.* **39**, 195

Feldman, P.D.: 1982, in *Comets*, ed. L.L. Wilkening, Tucson, Univ Arizona Press, p. 461

Feldman, P.D., Opal, C.B., Meier, R.R., Nicolas, K.R.: 1976, in *The Study of Comets*, eds. B. Donn et al., NASA SP-393, Washington DC, p. 773

Festou, M.: 1981, *Astron. Astrophys.* **95**, 69

Haser, L.: 1957, *Bull. Acad. Roy. Belgique*, Classe de Science, **43**, 740

Jamar, C.: 1976, *An Ultraviolet Bright-star Spectrophotometric Catalog (ESA Sr-27)*, November 1976, European Space Agency

Kissel, J. et al.: 1986, *Nature* **321**, 336

McCoy, R.P., Opal, C.B., Carruthers, G.R.: 1986, *Nature* **324**, 439

Meier, R.R., Opal, C.B., Keller, H.U., Page, T.L., Carruthers, G.R.: 1976, *Astron. Astrophys.* **52**, 272

Opal, C.B., Carruthers, G.R., Prinz, D.K., Meier, R.R.: 1974, *Science* **185**, 702

Opal, C.B., Carruthers, G.R.: 1977a, *Icarus* **31**, 503

Opal, C.B., Carruthers, G.R.: 1977b, *Astrophys. J.* **211**, 294

Smith, A.M., Stecher, T.P., Casswell, L.: 1980, *Astrophys. J.* **242**, 402

Woods, T.N., Feldman, P.D., Dymond, K.F., Sahnow, D.J.: 1986, *Nature* **324**, 436

Astron. Astrophys. 187, 325–328 (1987)

ASTRONOMY AND ASTROPHYSICS

IUE observations of comet P/Halley: evolution of the ultraviolet spectrum between September 1985 and July 1986

P.D. Feldman[1]★, M.C. Festou[2]★, M.F. A'Hearn[3]★, C. Arpigny[4]★, P.S. Butterworth[5]★, C.B. Cosmovici[6]★, A.C. Danks[7]★, R. Gilmozzi[8]★, W.M. Jackson[9]★, L.A. McFadden[3], P. Patriarchi[10]★, D.G. Schleicher[11], G.P. Tozzi[10]★, M.K. Wallis[12]★, H.A. Weaver[13]★, and T.N. Woods[1]

[1] Johns Hopkins University, Baltimore, MD 21218, USA
[2] Observatoire de Besançon, F-25000 Besançon, France
[3] University of Maryland, College Park, MD 20742, USA
[4] Institut d'Astrophysique de Liège, B-4200 Liège, Belgium
[5] NASA/Goddard Space Flight Center, Greenbelt, MD 20771, USA
[6] IFSI-CNR, I-00044 Frascati, Italy
[7] Florida International University, Miami, FL 33199, USA
[8] IUE Tracking Station, ESA/Vilspa, E-28080 Madrid, Spain
[9] University of California, Davis, CA 95616, USA
[10] Osservatorio Astrofisico di Arcetri, I-50125 Firenze, Italy
[11] Lowell Observatory, Flagstaff, AZ 86001, USA
[12] University College, Cardiff CF1 1XL, UK
[13] Space Telescope Science Institute, Baltimore, MD 21218, USA

Received January 27, accepted March 19, 1987

Summary. The ultraviolet spectrum of comet P/Halley (1982 i) was monitored with the International Ultraviolet Explorer (IUE) satellite observatory between 1985 September 12 and 1986 July 8 ($r < 2.6$ AU pre- and post-perihelion) at regular time intervals except for a two-month period around the time of perihelion. A complete characterization of the UV spectrum of the comet was obtained which allows us to derive coma abundances and to study the light emission mechanisms of the observed species. IUE observations at the time of the Giotto encounter provide a unique opportunity to compare the *in situ* measurements with remote observations of the principal coma species. The Fine Error Sensor (FES) camera of the IUE was used to photometrically investigate the coma brightness variation on time scales of the order of hours. Spectroscopic observations as well as FES measurements show that the activity of the nucleus was highly variable, particularly at the end of December 1985 and during March and April 1986. The production rates of OH, CS and dust are derived for the entire period of the observations. The total water loss rate for this period is estimated to be 3 10^8 metric tons.

Key words: comet P/Halley – spectroscopy – ultraviolet

Send offprint requests to: P. D. Feldman

★ Guest Observer with the *International Ultraviolet Explorer* satellite observatory.

1. Introduction

The IUE was used in an extensive joint US-European program to monitor the spectral characteristics and evolution of comet Halley between September 1985 and July 1986, corresponding to heliocentric distance $r < 2.6$ AU both pre- and post-perihelion. Observations, from both Goddard Space Flight Center and Vilspa, were made at regular intervals during this period except between January 1 and March 8, 1986 when the solar elongation angle of the comet was too small. A special observing technique, developed by the GSFC IUE staff, made possible the almost daily observations during the week of the Vega-2 and Giotto encounters.

An unexpected benefit of the IUE program was the use of the Fine Error Sensor (FES) as a photometric monitor over relatively long time periods of short-term fluctuations in cometary activity. Large amplitude variations in brightness (in the 18" square central region of the coma) were particularly evident the last week of December 1985 and throughout March and April 1986. These variations are attributed to a combination of nucleus rotation and outburst of individual jets or active areas on the surface of the nucleus. Contiguous observing shifts have made it possible to study the behavior of the UV emissions during several periods of outburst activity.

This report presents a summary of the observations made through July 1986 and a preliminary analysis of the long-term evolution of the gaseous coma of P/Halley. The highlights of the observations made during the spacecraft encounters have been presented by Festou et al. (1986). In this regard it is worth noting the excellent agreement obtained between the water production rate based on IUE measurements of OH emission and the vectorial model of Festou (1981) and that based on the *in situ* neutral mass spectrometer data from Giotto (Krankowsky et al., 1986). The *in situ* measurements also showed the validity of the inverse square distribution and radial outflow velocity (0.9 ± 0.2 km s^{-1}) of the H_2O parent molecules used in the coma models, even though some question remains concerning the value of the photodestruction lifetime of H_2O.

2. Spectroscopy

The first spectrum of comet Halley ever recorded from space, a long wavelength three-hour exposure taken on 1985 September 12.0, is shown in Fig. 1. The OH (0, 0) band at 3085 Å is readily apparent, indicating a significant H_2O production rate already at 2.59 AU. Typical of the period following perihelion is the exposure from 1986 March 11, shown in Fig. 2, which displays all of the known neutral cometary emission features in this spectral range as well as a very strong component of sunlight reflected by particulate grains in the coma. Comparison with the measured solar spectrum (Mount and Rottman, 1981), shown as the dotted curve in Figure 2, indicates that the reflected spectrum is more reddened (7% per 100 Å) than that of more distant, dusty comets observed previously by the IUE (Feldman and A'Hearn, 1985). The source of the reddening remains uncertain and is contrary to the behavior expected from the large population of small grains detected by the *in situ* dust experiments. The strong solar component tends to mask the emission from the CO_2^+ doublet near 2890 Å. However, when a properly reddened solar spectrum is subtracted from a spectrum such as in Fig. 2, the resultant CO_2^+ emission suggests an abundance relative to OH comparable to that found in other comets (Festou et al., 1982). The absence of detectable CO^+ First Negative band emission near 2200 Å probably results from a combination of lower fluorescence efficiency for these transitions (relative to the blue Comet Tail bands) (Magnani and A'Hearn, 1986) and the poor sensitivity of the LWP camera in this spectral range.

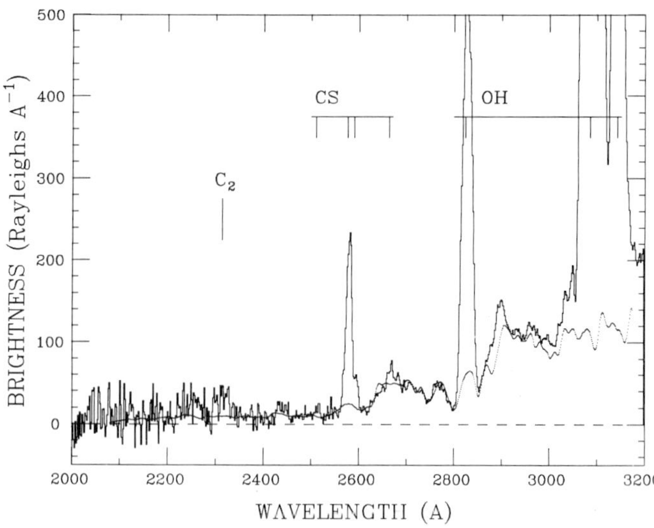

Fig. 2. A long wavelength 15 minute exposure of P/Halley (LWP 7773) obtained on 1986 March 11

As with the long-wavelength spectra, the short-wavelength spectrum (Fig. 3) does not show any previously undetected cometary emission features. In fact, the apparent weakness of O I, C I and S I, relative to those emissions in the spectrum of comet Bradfield (1979 X) (Weaver et al., 1981), may reflect lower fluorescence efficiencies and photodissociation rates due to the lower solar UV fluxes characteristic of solar minimum as compared with 1980 which was the year of solar maximum.

3. Variability

Visible light photometric monitoring of P/Halley was accomplished by using the Fine Error Sensor (FES) camera of the IUE (normally used for target acquisition and tracking) as a broadband (4000–6500 Å) photometer. In this mode, the FES samples a region approximately 18 arc-seconds square around the center of brightness of the comet. These observations are made auto-

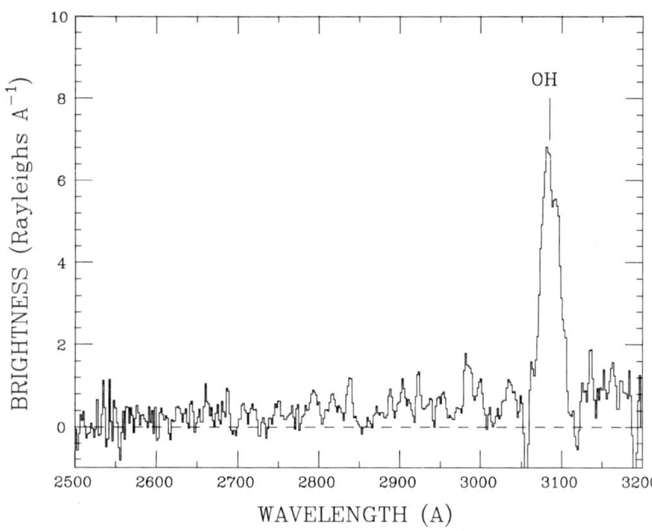

Fig. 1. Long-wavelength spectrum of P/Halley (LWP 6720) obtained on 1985 September 12. The exposure time was 180 minutes

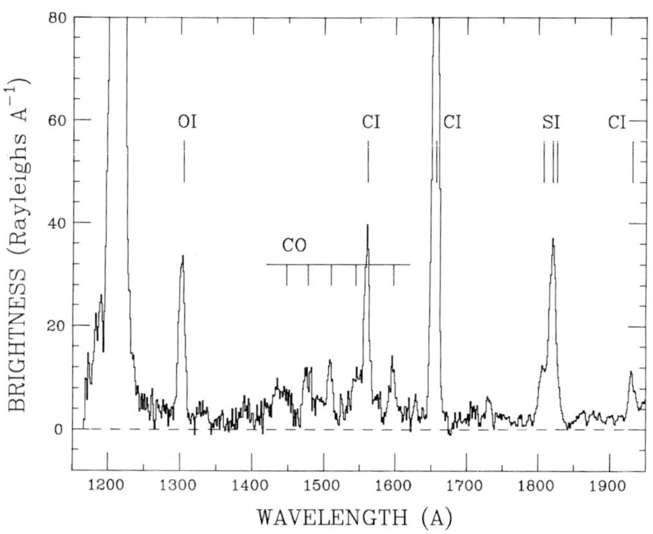

Fig. 3. Short-wavelength spectrum of P/Halley (SWP 27884) obtained on 1986 March 9. The exposure time was 45 minutes

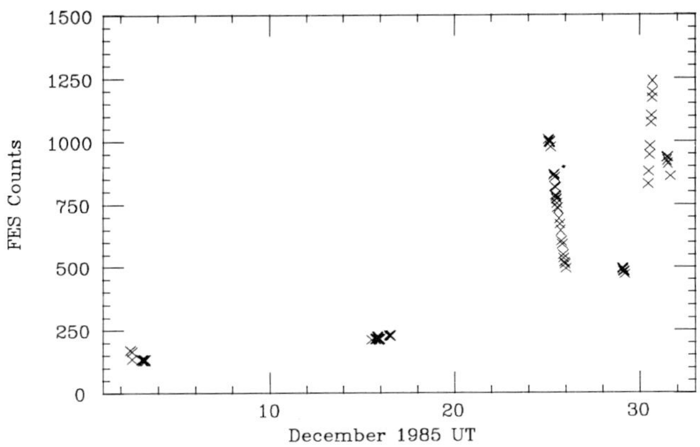

Fig. 4. FES photometry for December 1985

matically as the IUE tracking of the moving comet is verified at intervals of ~ 30 to 60 minutes. The short-term variability of comet Halley can be seen in Figs. 4 and 5 which show the FES data for December 1985 and March-April 1986, respectively. The FES responds to a mixture of gas and dust emission and a comparison with the ultraviolet spectra (taken at much less frequent intervals) indicates that both the UV continuum and the CS emission follow the changes in cometary activity more rapidly than does the OH emission, which arises from a parent molecule whose lifetime is about 1 day (McFadden et al., 1987). Amongst many apparent transient events seen in Fig. 5, a sharp outburst on March 18-19 was found to be correlated with CO_2^+ emission observed by IUE 150,000 km tailward from the nucleus during the outburst and with ground-based observations of a large detached CO^+ cloud on 20 March 1986 (Feldman et al., 1986). Otherwise, the variations in Fig. 5 are found to closely follow the 7.4-day period light curve derived from ground-based photometry by Millis and Schleicher (1986). However, it should be noted that IUE observations made after April did not show any significant variation in the FES count rate during an 8-hour observing shift.

4. Production rates

One of the principal goals of the IUE Halley program was the determination of the variation of gas production rate with heliocentric distance both before and after perihelion. The derivation of this rate from the observed surface brightness is generally achieved with the use of steady-state, spherically symmetric models. Although the validity of these assumptions is clearly questionable in light of both *in situ* and remote observations of P/Halley, they provide the only means of intercomparing observations made over a large range of heliocentric and geocentric distances. However, as noted above, the preliminary results from the Giotto neutral mass spectrometer experiment (Krankowsky et al., 1986) confirmed two of the basic assumptions of the models, the radial outflow velocity and the inverse square density distribution of the parent molecules. For the present, only the production rates of the parents of OH and CS are considered. For the grains, the production rate can be related to the quantity $Af\rho$, the product of albedo, filling factor and aperture radius (A'Hearn et al., 1984). The vectorial model (Festou, 1981) and OH fluorescence efficiencies of Schleicher (1983) are used to derive Q_{H_2O}, while Q_{CS}, assumed to be equal to the production rate of its short-lived parent, CS_2 (Jackson et al., 1986), is derived from a Haser model and the assumption of heliocentric velocity independent fluorescence efficiency. For the parent molecules, the outflow velocity was taken to be 1 km s^{-1}, while the photodissociation lifetimes of both H_2O and OH were taken from Festou (1981). The use of a shorter H_2O scale length, as suggested by Krankowsky et al. (1986) would yield values of Q about 60% of those shown in Fig. 6. In all three cases, the emissions are assumed to be optically thin. While this is not exactly so for the strongest lines of the OH (0, 0) band used in the analysis, the effect of resonance re-radiation on the derived production rates is estimated to be begligible.

The results, shown in Figs. 6-8, are somewhat surprising. Both the CS and the dust show larger short-term fluctuations with comet variability than does OH, but also exhibit a uniform decrease in production rate to larger heliocentric distance. The H_2O production rate, on the other hand, while increasing dramatically between 2 and 3 AU pre-perihelion, and then more gradually to 1 AU, did not begin to decrease appreciably until the comet was approaching 2 AU post perihelion. This behavior suggests that in P/Halley the CS and dust production are not totally controlled by the water vaporization, a conclusion already reached for two other comets previously observed by IUE, IRAS-Araki-Alcock (1983 VII) (Feldman et al., 1984) and P/Encke (A'Hearn et al., 1985).

Finally, we note that the water production rates shown in Fig. 6 can be integrated over the entire period from September 1985

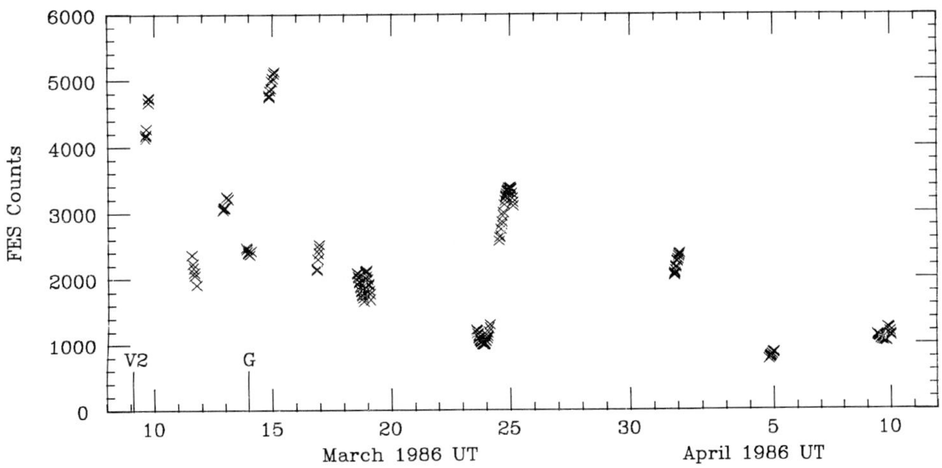

Fig. 5. FES photometry for March and April 1986. The times of the Vega-2 and Giotto encounters are indicated

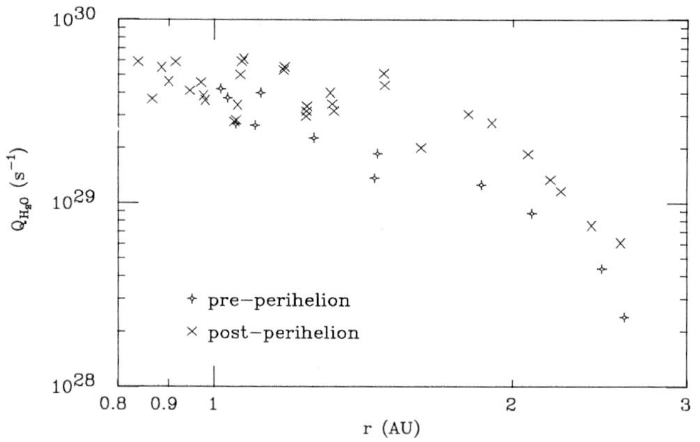

Fig. 6. Water production rate for comet Halley as a function of heliocentric distance as derived from IUE observations of OH

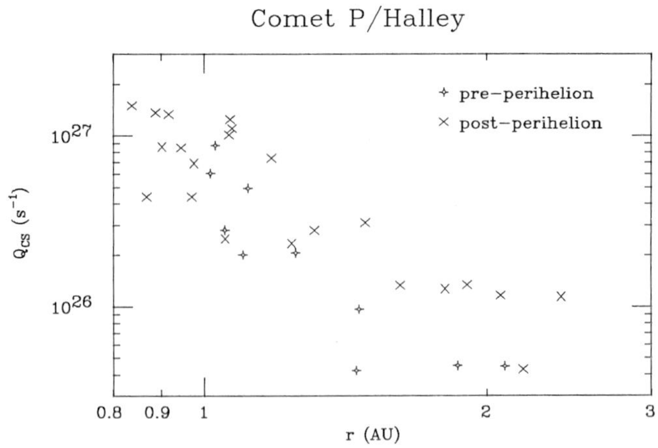

Fig. 8. Production rate of CS as a function of heliocentric distance

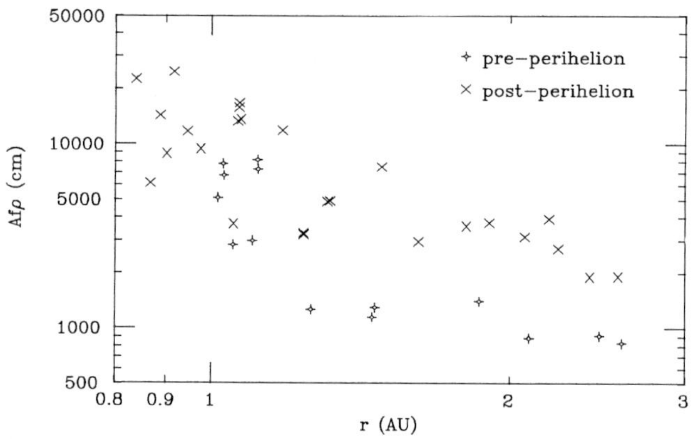

Fig. 7. Dust production rate variability as a function of heliocentric distance

through the beginning of July 1986 to estimate the total amount of water lost by P/Halley during this apparition. For the period around perihelion when there are no IUE observations, a constant value of $Q = 1\,10^{30}$ molecules s^{-1} is used. This value is consistent with the H I Lyman-α observations made by the ultraviolet spectrometer on Pioneer Venus Orbiter during this period (Stewart, 1987). The total H_2O lost, estimated this way, is $3\,10^8$ tons.

5. Conclusion

The IUE observations of comet Halley provide a unique source of information about the composition, evolution, and short-term behavior of the cometary coma, and establish a context for the inter-comparison of the *in situ* encounter snapshots of Halley and the large body of Earth-based observations of the comet.

Acknowledgements. We thank the IUE staffs at Goddard Space Flight Center and Vilspa for their assistance in the operations and aquisition of data. E. Roettger assisted in the data analysis. This work was supported by NASA grants NSG-5393 to the Johns Hopkins University and NAG 5-252 to the University of Maryland.

References

A'Hearn, M.F., Schleicher, D.G., Feldman, P.D., Millis, R.L., Thompson, D.T.: 1984, *Astron. J.* **89**, 579

A'Hearn, M.F., Birch, P.V., Feldman, P.D., Millis, R.L.: 1985, *Icarus* **64**, 1

Feldman, P.D., A'Hearn, M.F., Millis, R.L.: 1984, *Astrophys, J.* **282**, 799

Feldman, P.D., A'Hearn, M.F.: 1985, in *Ices in the Solar System*, (eds. J. Klinger et al.) D. Reidel Publ. Co., Dordrecht, p. 453

Feldman, P.D., A'Hearn, M.F., Festou, M.C., McFadden, L.A., Weaver, H.A., Woods, T.N.: 1986, *Nature* **324**, 433

Festou, M.C.: 1981, *Astron. Astrophys.* **95**, 69

Festou, M.C., Feldman, P.D., Weaver, H.A.: 1982, *Astrophys. J.* **256**, 331

Festou, M.C., Feldman, P.D., A'Hearn, M.F., Arpigny, C., Cosmovici, C.B., Danks, A.C., McFadden, L.A., Gilmozzi, R., Patriarchi, P., Tozzi, G.P., Wallis, M.K., Weaver, H.A.: 1986, *Nature* **321**, 361

Jackson, W.M., Butterworth, P.S., Ballard, D.: 1986, *Astrophys. J.* **304**, 515

Krankowsky, D., Lämmerzahl, P., Herrwerth, I., Woweries, J., Eberhardt, P., Dolder, U., Herrmann, U., Schulte, W., Berthelier, J.J., Illiano, J.M., Hodges, R.R., Hoffman, J.H.: 1986, *Nature* **321**, 326

Magnani, L., A'Hearn, M.F.: 1986, *Astrophys. J.* **302**, 477

McFadden, L.A., A'Hearn, M.F., Feldman, P.D., Roettger, E.E., Edsall, D.M., Weaver, H.A., Butterworth, P.S.: 1987, *Astron. Astrophys.* (*in press*)

Millis, R.L., Schleicher, D. G.: 1986, *Nature* **324**, 646

Mount, G.H., Rottman, G. J.: 1981, *J. Geophys. Res.* **86**, 9193

Schleicher, D.G.: 1983, Ph.D. Dissertation, University of Maryland

Stewart, A.I.F.: 1987, Astron. Astrophys. (in press)

Weaver, H.A., Feldman, P.D., Festou, M.C., A'Hearn, M.F., Keller, H.U.: 1981, *Icarus* **47**, 449

Astron. Astrophys. 187, 329–332 (1987)

ASTRONOMY AND ASTROPHYSICS

Some diatomic molecules from comet P/Halley's UV spectra near spacecraft flybys

Max K. Wallis [1] **and K.S. Krishna Swamy** [2,⋆]

[1] Department of Applied Mathematics and Astronomy, University College, Cardiff CF1 1XL, Wales, UK
[2] NASA Ames Research Center, Moffett Field, CA 94035, USA

Received February 13, accepted May 22, 1987

Summary. The abundances of diatomic molecules seen in the UV both give clues to the parent compounds and help unravel the gas and ion mass spectrometry. From IUE spectra of comet Halley, we find upper limits on SH and CS^+ column densities, and estimates of probable NO, S_2, and SO. In particular, judged from the 226 nm γ-band, NO was relatively abundant at 2–8 10^{13} cm^{-2} on March 9–14. The production rate of S_2 was around 1 10^{27} molecules s^{-1} at that time, but both showed day-to-day variability by 2–3 times.

Key words: gas coma – coma molecules – UV spectroscopy

1. Introduction

While cometary spectroscopy can potentially reveal a wide range of molecules in the coma and help resolve the mass coincidences in the Halley probes' ion and gas mass spectrometry, the capability of earth-based observations is limited by a number of factors. The signal: noise ratio is often low, while the molecular peaks were also low compared with dust continuum in comet P/Halley's spectra. Secondly, short-lived parents concentrated within a few hundred km of the nucleus appear relatively weak in apertures of several arc sec. Thirdly, very few ions or molecules with 3 or more atoms have been spectrally characterized. Some progress was achieved with comet IRAS-Araki-Alcock which approached unusually close to the Earth; HCO and possibly other new molecules were tentatively identified in the visible region (Cosmovici and Ortolani, 1984) while S_2 was identified in the UV (A'Hearn et al., 1983).

The UV spectra secured with the International Ultraviolet Explorer (IUE) orbiting telescope present particular problems. The noise is far from random "white noise" and the effective spectral resolution for extended objects is poor at $\sim$1 nm. However, 2-D spectral images are registered, so cosmic ray tracks can be recognized and filtered out. Corrections for variable pixel sensitivity are now well-determined for the long-used SWP and LWR cameras (though not for the LWP camera in the 200–320 nm range, used generally only since late 1985).

It is becoming clear that some of the structures in cometary IUE spectra previously neglected as noise are really molecular features. Examples now confirmed are the S_2 bands in 280–310 nm, CS in 250–280 nm (Krishna Swamy and Wallis, 1987) and CO in 140–180 nm (Woods et al., 1986). Others in the rich array of features, that are evident in several comets, are undoubtedly molecules not yet identified. One technique for improving signal: noise that has so far been little exploited is to combine information from several bands of a molecule – bands that individually are only at the "noise" level. Recently, we have made some progress (Krishna Swamy and Wallis, 1986, 1987) by comparing several spectra and using the multiplicity of wavelength correspondances and the general patterns of molecular emissions. This approach sidesteps the uncertainties in relative band strengths due to unknown Swings effects (intensity variations due to Doppler shifts relative to Fraunhofer lines in the exciting solar spectrum). We do give some weight to molecular Franck-Condon factors when they are known (for diatomics). We tentatively identified the SO radical in IUE spectra of several of the fainter and generally more distant comets compiled by Weaver et al. (1981). The observed pattern for a total of 11 bands of the (B–X) system, whose wavelengths lie around 220–240 nm, agreed satisfactorily with the laboratory wavelengths. The S_2 dimer, whose main bands were displayed in comet IRAS-Araki-Alcock, has similarly been detected in a number of comets in which individual bands appeared marginal. In this paper we report searches for these and some other diatomics in comet Halley spectra, available to us as members of the European team granted observing time at ESA's VILSPA station.

Above 220 nm, the continuum of dust-scattered solar radiation can be significant. No broad feature analogous to the 220 nm interstellar feature has been seen, just a gradual reddening relative to the solar spectrum (Feldman and A'Hearn, 1984) appropriate to a wide range of dust particle sizes, shapes and compositions. Since the dust emission is an extended source, with intensity $\sim R^{-1}$ from the comet nucleus, a solar spectrum degraded to about 1 nm resolution used by Feldman and A'Hearn (1984) is appropriate. Its magnitude has to be judged ("educated guess") to lie below the total spectrum to within the noise level. With high signal: noise ratio, as 13 MAR spectrum of Fig. 2, this procedure is quite precise. In cases with relatively weaker signals and where one suspects unidentified molecular emissions, as 9 MAR spectrum of Fig. 1, the continuum level is somewhat subjective. In setting the level, one takes into account the expectation that real emissions have FWHM of a few Å, and judges the detector-noise by the magnitude of changes in adjacent pixels. To the 1 nm resolution, the solar spectrum above 220 nm

---

Send offprint requests to: M.K. Wallis

⋆ Permanent address: Tata Institute for Fundamental Research, Kolaba 5, India

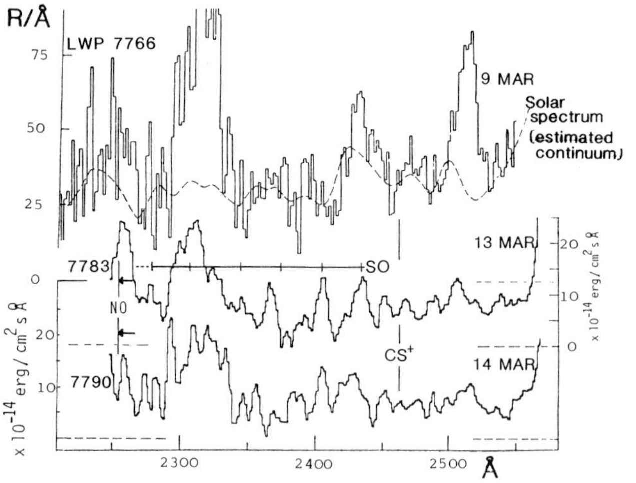

Fig. 1. Three IUE spectra in the wavelength region of the SO bands. The resolution for such an extended emission ~10 Å FWHM (Krishna Swamy and Wallis, 1986). The solar spectrum (upper example) shows that features near the noise level can be obscured by variations in the dust-scattered continuum. The 2320 Å feature is probably C_2 and 2430 Å may be second-order H Lα. From the other marked bands, SO is marginal. The NO γ-band head at 2260 Å corresponds to the shortward limit of a strong feature in all three spectra. The CS^+ (5,0) band indicated at 2462 Å does not seem to be present

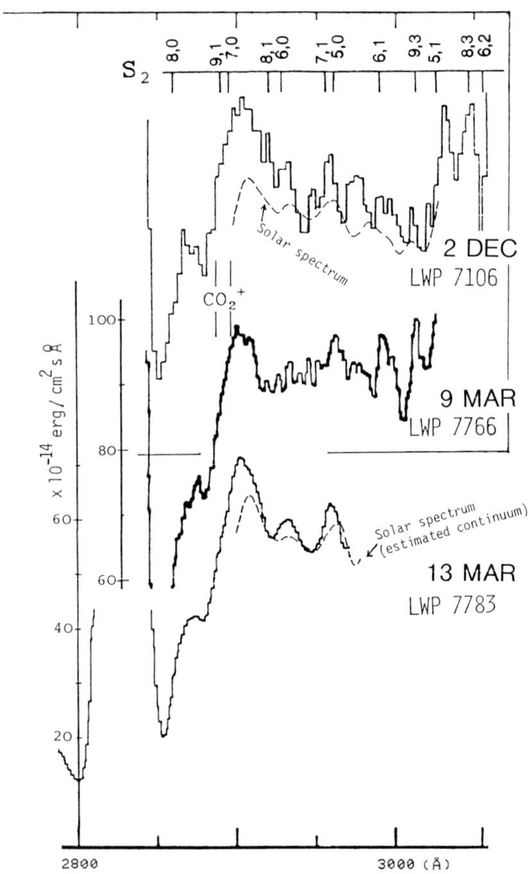

Fig. 2. IUE spectra near the S_2 bands. The dust-scattered continuum is strong, and simulated by a solar spectrum of adjustable intensity. CO_2^+ intrudes around 2890 Å, but all indicated S_2 bands can be present. The source of the 2970–2980 Å feature is not likely to be S_2 and remains uncertain. For calculating S_2 strengths (Table 1), the (7,1), (5,0) blend is used

varies by only a few % with the sun's rotation; while absolute uncertainties may be 50% as judged by various rocket measurements, the relative uncertainties over 10–20 nm ranges are judged as the 10% difference between the different data sets (Simon, 1981; Brasseur and Simon, 1981).

2. Sulphur monoxide, SO

The laboratory wavelengths of the $B^3\Sigma$–$X^3\Sigma$ system range over 215–245 nm (Rosen, 1970), but since the LWP camera sensitivity falls off at the shorter end we concentrate on the region longwards of 225 nm. In Halley (Fig. 1), many features are evident and vary relatively from spectrum to spectrum of different dates. Features that could be attributed to SO should be of comparable intensity, being composed of a series of over-lapping bands having similar Franck-Condon factors (Krishna Swamy and Wallis, 1987). This is confirmed by preliminary results of fluorescence calculations predicting the relative intensities now underway (Krishna Swamy and Wallis, in preparation), which should improve the qualitative and quantitative accuracy.

The upper 9 MAR spectrum of Fig. 1 has shorter exposure so higher noise level: the continuum has been set at the highest level compatible with the spectrum smoothed to 1 nm resolution, when the minor features virtually disappear, apart from those at 226 and 243 nm. Retention of the higher resolution shown is justified only to show correspondances with the other two spectra with their lower noise, a combination of marginal identifications having higher significance than the individual ones. Various stronger emissions in the Halley spectra make it more difficult to identify SO than in comets previously examined (Krishna Swamy and Wallis, 1987). C_2 is strong at 232 nm, second order H Lα probably contributes at 243 nm while there are unknown emissions at 237, 241 and 242 nm among others. Only the emissions at 228 and 234.5 nm may be clean: our fluorescence calculations give their intensity ratio as 1.5:1 (231 nm is 1.4 and the others $\lesssim 1$). The observed ratio of about 2:1 is considered to agree well within present uncertainties.

For the fluorescence efficiency rate, we adopt the mean lifetime of 17 ns (Smith, 1969) and take solar fluxes appropriate to solar minimum conditions (Brasseur and Simon, 1981). Having identified the bands where not overlaid by other emissions (an example shown in Fig. 1), we measure the brightness of the 234.3 nm peak above estimated dust continuum and derive the column density via the formula (Feldman, 1982) $N = r^2\,[\mathrm{AU}] \times 10^6\, B/g$, where B is the brightness in Rayleighs and g is the excitation rate at 1 AU. The results of particular dates are given in Table 1. In comparison, one of the "faint" comets (Weaver et al., 1981 – comet Meier at 1.5 AU) had 20 R intensity and rather less SO in absolute terms at $3\ 10^{10}\ \mathrm{cm}^{-2}$.

3. Molecular sulphur, S_2

We have recently studied the identification of S_2 in several comets from the multiple weak bands in 290–320 nm. The bands prominant in the exceptionally close comet IRAS-Araki-Alcock (A'Hearn et al., 1983) are in fact combinations; at lower intensities they appear divided and somewhat variable comet to comet (a Swings effect). Another difficulty is that the mean solar spectrum mimics some of the S_2 features.

The molecule was evident in comet Halley from early in December, when gas activity was increasing and the ion tail was changing to the long classical form. The upper spectrum of Fig. 2

Table 1. Intensities of selected bands in Halley spectra (upper rows, units of Rayleighs) and derived molecular column densities, with solar fluxes appropriate to solar minimum from Brasseur and Simon (1981). For CS^+ and SH, the values are upper limits

Band nomenclature		NO 226 nm A–X(0,0)	SO 234.5 nm B–X(4,1)	CS^+ 246.2 nm B–X(5,0)	S_2 296 nm B–X(0,0)	SH 167.7 nm B–X(0,0)
Mar. 9	LWP 7766	300 R	80	$\lesssim 40$	100	
Mar. 11	SWP 27898					5
Mar. 13	LWP 7783	800	310	<20	120	
Mar. 14	LWP 7790	360	300	40	$\lesssim 50$	
f_λ		4.0 10^{-4}	4.7 10^{-2}	9.1 10^{-4}	–	10^{-3}
Solar flux ($\times 10^{11}$ phot/cm^2 s Å)		4.68	4.65	6.4	81.7	0.33
g (s^{-1}) at 1 AU		0.83 10^{-5}	1.0 10^{-3}	3.9 10^{-5}	4.0 10^{-4}	8 10^{-7}
Col. density N(cm^{-2})		2–8 10^{13}	5–24 10^{10}	<7 10^{11}	2 ± 0.3 10^{11}	5 10^{12}

shows the possible presence of S_2 from the band pattern in 290–300 nm when the comet was at 1.46 AU. In comparison we show a solar spectrum of 1.1 nm FWHM (Feldman, 1982) of possible intensity, that might be expected from dust scattering of the solar radiation. Residual intensities attributable to S_2 are unavoidable.

Later Halley spectra from the encounter period are also shown in Fig. 2: at that time the solar contribution was relatively stronger, but still cannot explain the whole structure. A CO_2^+ contribution is evident at 289 nm, and there are others at 290 and 297-8 nm that do not correspond to the indicated S_2 bands. The feature at 298 might be identified with S_2(8,2), but that band is expected to be weaker by a factor 20 according to fluorescence calculations (smooth solar spectrum, omitting Fraunhofer lines – Krishna Swamy and Wallis, in preparation). There is sufficient correspondance between the bands in the three spectra, even in the short section available for 14 March, to allow us to confirm S_2, and determine its abundance from the 297 nm peak. This blend of the (5,0) and (7,1) bands is considered the least affected by uncertainty in the scattered solar continuum (as fitted to the lowest spectrum) and the effects of predissociation (A'Hearn et al., 1983). We derive the intensities given in Table 1, using the excitation rate of A'Hearn et al. adjusted for solar flux differences.

In the Vega-2 spectrum of 9 March (Krasnopolsky et al., 1986), the S_2 bands around 290 nm are marginally indicated. But a more convincing identification comes from correlating features in the 350–380 nm region with higher S_2 bands (Krishna Swamy and Wallis, in preparation).

4. Nitrogen monoxide radical, NO

The radical has not previously been identified in cometary spectra. Its spectral structure has been well studied in the laboratory, and has generally strong γ-bands below 230 nm. There are numerous bands in 150–190 nm (Okabe, 1978) but the solar flux is weak there. The (0,0) γ-band is expected to be the most evident. We can see a broad structure extending from the band-head at 226.3 nm in several comets as well as Halley; cf. comets IRAS-Araki-Alcock and the "weaker" comets Meier, Stephan-Oterma and Borrelly (Krishna Swamy and Wallis 1986), in which it lies just longwards of a possible (weaker) SO band.

In the Halley spectra (Fig. 1), the variability of the 227 nm feature is striking (March 13 greater than the March 9 and 14). While it could be noise, in all three spectra it stands out above the solar continuum and is stronger than the mean noise. Taking this tentative identification and adopting the f-value from Okabe (1978), we obtain the column density cited in Table 1. By comparison, we find some ten times less at <4 10^{-12} cm^{-2} in comet Meier (spectrum of Weaver et al., 1981).

The molecule CS^+ with several bands in 240–280 nm (Tsuji et al., 1980) has been suspected in some comets (Singh, 1982), partly because CS is generally prominant in cometary UV spectra. But its excitation rate is much lower and ions being daughter products with relatively high speeds are difficult to detect with the small IUE apertures. We have looked for the 246.2 nm (5,0) band in a relatively clean part of the spectrum (Fig. 1), but do not find any consistent emission. For setting the upper limits on column density in Table 1, we have made use of the oscillator strength given by Singh (1982) to calculate the excitation rate.

5. Hydrogen sulphide radical, SH

As OH is strong and sulphur is a few % of oxygen, SH may be expected as another decomposition product in the coma. It possesses several bands in the 160–190 nm region, as indicated in Fig. 3. Features compatible with the 167.7 [(0,0) and (1,1)], 168.2 [(2,2) and (3,3)] and 192 nm (3,0) bands of the (B–X) system can be seen, while expected intermediate bands at 175 and 183 nm are blended with CO and submerged under S, resp. Our identification is therefore again tentative. Shortwards of 180 nm, the main features (apart from C I 165.7 nm) and apparent "noise" are known to be CO, from the rocket observations of Woods et al. (1986). The relative strengths as calculated by these authors are indicated: the relatively poor correspondance at e.g. 173-4 nm may be a "Swings" effect giving stronger higher-order band elements (as in their spectrum), but CO cannot contribute at 168 nm. In their spectrum, the 167.7, 168.2 feature is also evident as a shoulder on the intense C I line. While the feature may be real, it could be other than SH. The f-value of SH is unknown, so we estimate it at 10^{-3} s^{-1} for the (0,0) band of such a diatomic and derive an upper limit to the column density presented in Table 1.

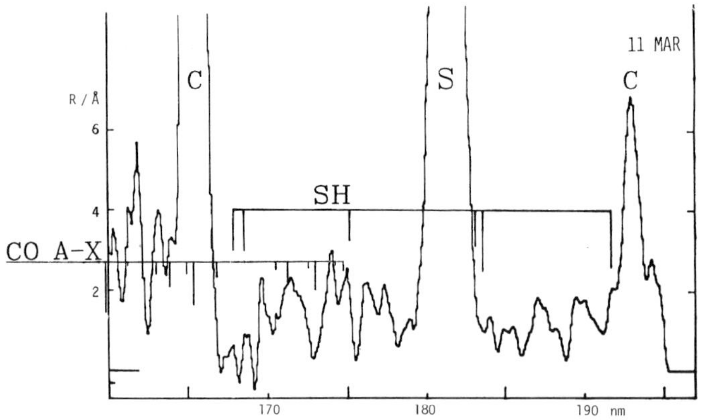

Fig. 3. IUE spectrum of March 11 in the short wavelength region (SWP 27898), indicating CO-band strengths (Woods et al., 1986) and SH-band positions. Only the 168 nm complex of SH features is potentially free from blending problems, so an identification with SH here and in the Woods et al., rocket spectrum can only be tentative

6. Conclusions

The abundance of S_2 is the only well-established number among our results; that of SO is quite low, having a few times the column densities of minor components like CS and S in the fainter comets (Weaver et al., 1981) and less relative to OH etc. The NO comes out substantially stronger, and would have to be released rather close to the nucleus for such a value to be plausible – the required parent scale of $\leq 10^4$ km and long lifetime have to be similar to the values measured in situ for CO (Eberhardt et al., 1986). The intermediate tentative value for SH is plausible, as its dissociation properties and lifetime are comparable to OH (Okabe, 1978). For the CS^+ ion, it is not clear whether our rather high limit significantly constrains its contribution to the mass 44 ion peak (Balsiger et al., 1986). Protonated derivatives of our molecules or simply their ions appear as plausible contributors to the mass spectra sampled by Giotto (Balsiger et al., 1986) at mass numbers 30–31 (NO), 33–34 (SH), and 48–49 (SO).

To derive production rates, we need to know length scales of both the molecule and its parent. For S_2 the derivation is straightforward, as the molecule is produced near the nucleus and destroyed (200 s lifetime) while still within the IUE aperture. From the $N = 1.7\ 10^{11}\ \mathrm{cm}^{-2}$ on March 9, we calculate a total of 1.8 10^{19} within the $10'' \times 20''$ aperture; adopting a lifetime at 0.83 AU of 175 s (De Almeida and Singh, 1986) we find the S_2 source strength of 1.0 10^{27} molecules s^{-1}, 0.1% of the H_2O production at the Vega encounter time. For the Giotto encounter, the comparable figure from the higher intensity and longer lifetime at 0.9 AU is 1.25 10^{27} S_2 molecules s^{-1}, perhaps 0.2% of the total gas production (Festou et al., 1986) and very similar to its fraction in comet IRAS-Araki-Alcock (A'Hearn et al., 1983, De Almeida and Singh, 1986).

Acknowledgements. We obtained early copies of the Halley spectra as members of the European team granted time at ESA's VILSPA station for IUE observations of the comet. A Visiting Fellowship from the UK's SERC for KSKS and hospitality at University College Cardiff are also acknowledged.

References

A'Hearn, M.F., Feldman, P.D., Schleicher, D.G.: 1983, *Astrophys. J.* **274**, L99
Balsiger, H. et al.: 1986, *Nature* **321**, 330
Brasseur, G., Simon, P.C.: 1981, *J. Geophys. Res.* **86**, 7343
Cosmovici, C.B., Ortolani, S.: 1984, *Nature* **310**, 122
De Almeida, A.S., Singh, P.D.: 1986, *Earth, Moon, Planets* **36**, 117
Eberhardt, P. et al.: 1986, 20th ESLAB Symp., ESA SP-250, Vol. **I**, p. 383
Feldman, P.D.: 1982, in *Comets*, ed. L.L. Wilkening, University of Arizona Press, Tucson, p. 461
Feldman, P.D., A'Hearn, M.F.: 1984, in *Ices in the Solar System*, eds. J. Klinger et al., Reidel, Dordrecht, p. 453
Festou, M.C. et al.: 1986, *Nature* **321**, 361
Krasnopolsky, V.A. et al.: 1986, *Nature* **321**, 269
Krishna Swamy, K.S., Wallis, M.K.: 1986, in *New Insights in Astrophysics*, ESA SP-**243**, p. 133
Krishna Swamy, K.S., Wallis, M.K.: 1987, *Monthly Notices Roy. Astron Soc.* (in press)
Okabe, H.: 1978, *Photochemistry of Small Molecules*, Wiley, New York, Chap. 5
Rosen, B.: 1970, *Selected Constants-Spectroscopic Data Relative to Diatomic Molecules*, Pergamon, Oxford
Simon, P.C.: 1981, *Solar Phys.* **74**, 273
Singh, P.D.: 1982, *Astron. Astrophys.* **108**, 369
Smith, W.H.: 1969, *J. Quant. Spectr. Rad. Transfer.* **9**, 1191
Tsuji, M., Obase, H., Nishimura, Y.: 1980, *J. Chem. Phys.* **73**, 2575
Weaver, H.A., Feldman, P.D., Festou, M.C., A'Hearn, M.F., Keller, H.U.: 1981, *Icarus* **47**, 449
Woods, T.N., Feldman, P.D., Dymond, K., Sahnow, D.J.: 1986, *Nature* **324**, 436

Astron. Astrophys. 187, 333–338 (1987)

ASTRONOMY AND ASTROPHYSICS

Activity of comet P/Halley 23–25 March, 1986: IUE observations

L.A. McFadden [1], **M.F. A'Hearn** [1,★], **P.D. Feldman** [2,★], **E.E. Roettger** [2], **D.M. Edsall** [1], and **P.S. Butterworth** [3,★]

[1] Astronomy Program, University of Maryland, College Park, MD 20742, USA
[2] Department of Physics and Astronomy, Johns Hopkins University, Baltimore, MD 21212, USA
[3] Science Applications Research, Inc. Code 633, Goddard Space Flight Center, Greenbelt, MD 20771, USA

Received February 23, accepted June 2, 1987

Summary. A large but gradual increase in the brightness of the emission bands of comet Halley and thus in the coma abundance of OH, CS, CO_2^+ and dust was observed with the International Ultraviolet Explorer (IUE) satellite from 23 to 25 March 1986. This brightness change was also monitored by the Fine Error Sensor (FES) tracking and acquisition camera at a higher temporal resolution than that of the spectrophotometric measurements. The amplitude of the increase varied among the species depending on the lifetime of the parent molecule. By comparing the FES lightcurve with the optical lightcurve measured in the light of C_2 by Millis and Schleicher (1986) we can show that this increase in brightness corresponds to part of the periodic lightcurve of the comet. None of the IUE observations show evidence of the large increase in H_2O reported to have occurred on March 24.7 by Weaver et al. (1986). The activity recorded between March 23 and 25 is different from the smaller and more rapid increase in brightness observed on March 18–19 also with the IUE (Feldman et al., 1986). We conclude that the activity reported here is controlled primarily by photodissociation and ionization of molecules ejected from an active area of the nucleus that rotates into the field of view of the sun approximately every 7.4 days.

Key words: comets – spectrophotometry – UV radiation

1. Introduction

The ultraviolet spectrum of comet P/Halley was monitored extensively during the comet's 1985–1986 apparition by the International Ultraviolet Explorer (IUE) satellite (Feldman et al., 1987). During our observations to study the abundance of UV-active species and their emission mechanisms in the coma, we observed variations in the visual brightness of the comet and the flux of the UV emission features.

We have previously reported (Feldman et al., 1986) a very short-lived outburst on March 18–19 which showed a major increase in the abundance of ions without a corresponding increase in neutrals. We suggested that this outburst could be explained by the sudden release of a pocket of CO_2 which was then very rapidly ionized.

Send offprint requests to: M.F. A'Hearn

★ Guest Observer with the International Ultraviolet Explorer Observatory

Other investigators have also reported very sudden outbursts observed with a variety of techniques. For example, Weaver et al. (1987) have reported sudden increases in the amount of H_2O as observed in the infrared while Rettig et al. (1986) have reported sudden changes in the abundance of NH_2. Slower variations are also seen in various sets of data. Radio observations by Gerard et al. (1986), Schloerb et al. (1986), and Mirabel (1986) exhibit day-to-day variations in the abundances of OH and HCN. Images in the light of CN and other radicals exhibit significant day-to-day variation (Schlosser et al., 1986; A'Hearn et al., 1986a, b). Many of these day-to-day fluctuations are thought to be correlated with the periodic variations in activity dramatically documented by Millis and Schleicher (1986). It has been argued that these variations are due to the rotation of the nucleus with active areas on the surface rotating in and out of sunlight.

In this paper we consider the variations observed with IUE in the period March 23–25. We will show that the variation is correlated with the periodic variations observed by Millis and Schleicher (1986), and is most probably explained as a gradual increase in outgassing as an active area rotates into sunlight. This activity was of a quite different character than that observed on March 18–19.

2. Observations

We measured a change in brightness by a factor of 3.2 over a 24-h interval with the FES (Fig. 1) which monitored the comet periodically for tracking purposes. The FES is used in a photometric mode to track moving targets such as comets. From these data we get a measure of the brightness of the comet over ~ 432 arcsec^2 at $\sim 1/2$ h intervals in a spectral bandpass with 1/2-power points at 3800 and 6500 Å. Temporal variations of the ultraviolet emission bands were observed with the long wavelength primary (LWP) cameea in low dispersion mode ($\delta\lambda = 18$ Å). Short exposures of less than 60 s were made with the LWP spectrograph to measure strong features while longer exposures, typically 10–15 min, were made to record the weaker features. A few very long exposures were made to observe the very weak C_2 Mulliken bands. A representative LWP spectrum can be found in Feldman et al. (1987). The integrated flux of each emission band is directly proportional to the column abundance of these species assuming a negligible optical depth and provided the geometrical parameters ($r = 1.047$–1.070 AU, $\Delta = 0.722$–0.684 AU) and the comet's heliocentric velocity (26.65–26.72 km s^{-1}) remain nearly constant as they did for the time period discussed in this paper.

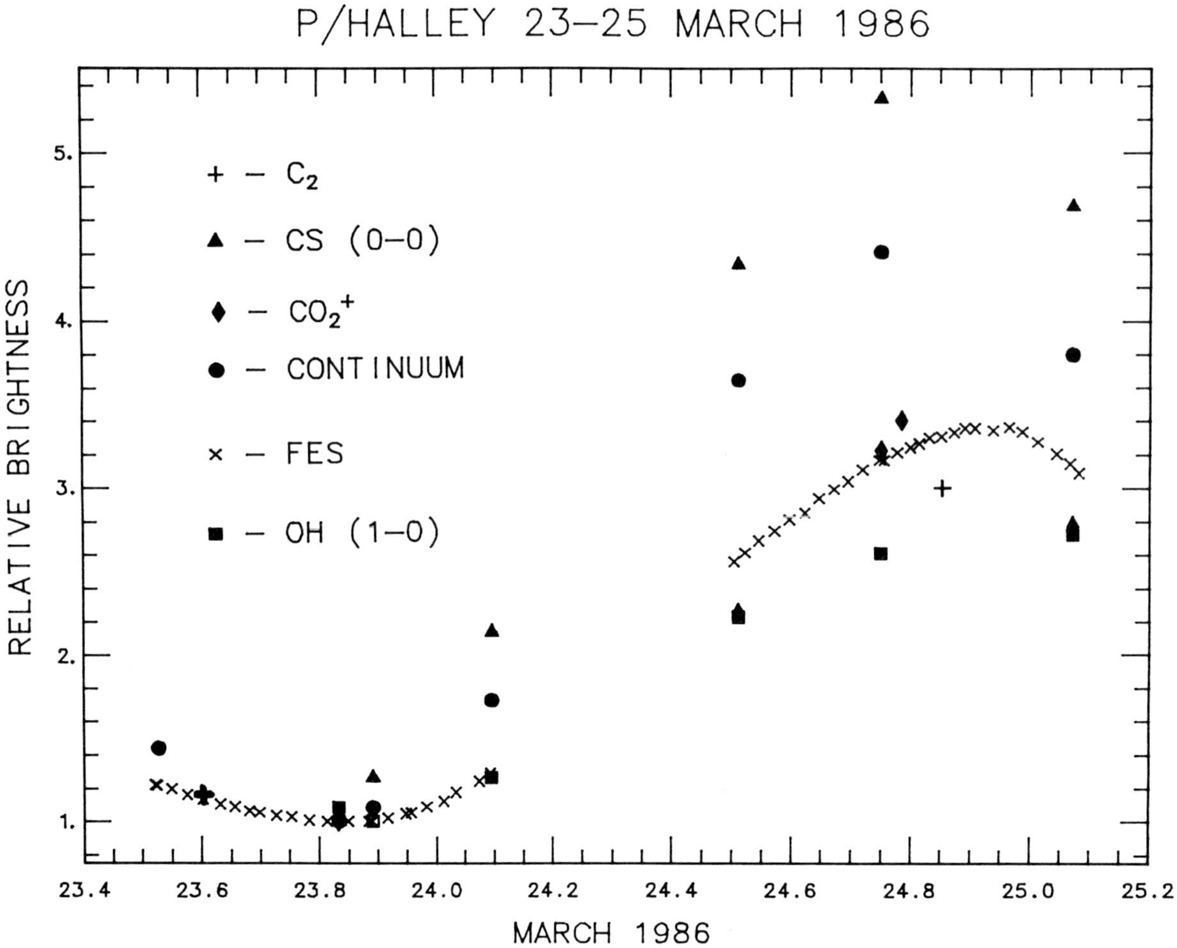

Fig. 1. The relative brightness of neutral and ionic species, the continuum and FES fluxes vs. time. Values are scaled to the brightness at minimum light, March 23.83. The C_2 brightness is scaled to the FES flux at March 23.60

Our initial data reduction procedure is the same as that used in previous analyses of IUE spectra of comets (e.g., Weaver et al., 1981a). The spectra were extracted using a slit size of 10″ × 15″ (5100 × 7600 km). The flux between 2940 and 2980 Å, a region free of emission bands and with a strong signal on the short exposures, was integrated to measure the continuum flux and thus the abundance of dust. The cometary continuum was fit with the solar spectrum, allowing for a linear slope, using six regions between 2550 and 3050 Å which were free of emission bands. The solar spectrum was obtained from the average of six solar analogs previously observed with IUE. The fitted solar spectrum was subtracted from the cometary spectrum to obtain the net emission spectrum. This was integrated over standard wavelength regions to derive the integrated band fluxes listed in Table 1.

3. Results

The fluxes from Table 1 are shown graphically in Fig. 1, together with the brightness measured with the FES. We have normalized all species to the flux measured at March 23.83, nearly at the minimum of the variation. Emission by C_2 was measureable only on two spectra, not including the one on March 23.83. These two data are plotted by normalizing the datum at March 23.60 to the value of the FES lightcurve at that time. The fluxes of CS, CO_2^+, and the continuum increase gradually until approximately March 24.8, after which they decrease. The OH, on the other hand, appears to continue to increase until March 25.07. It is clear that the various species exhibit different amplitudes, ranging from a factor of 6 for CS to roughly 2.5 for OH. Since the maximum may not have been seen in OH, the amplitude could be larger. There are large uncertainties in the results for CO_2^+ because the emission is weak and the underlying reflected continuum is strong and shows significant spectral structure.

On March 24 and 25, offset exposures were made pointing 30″ both tailward and sunward from the nucleus. On March 24.0, near the minimum of the lightcurve, the neutral and ionic species were enhanced in the tailward direction relative to the sunward direction by about 12% in OH and by a factor of 2.5 in CS and CO_2^+ (Table 2). The continuum flux, however, was enhanced in the opposite sense, the flux of the sunward pointing being about twice as bright as that of the tailward pointing (Table 2). On March 25.0 near the maximum of the lightcurve, the OH (0, 0) and (1, 0) bands were saturated, but the CS and CO_2^+ showed enhancements of the opposite sense compared to the previous day (Table 2). The flux in the sunward direction was approximately 2 times higher than that in the tailward direction for both CS and CO_2^+. The OH (1, 1) band did not show a sunward enhancement at this time. The continuum flux had also changed its sense of

Table 1. Flux of UV emission bands in comet Halley

Date 1986 March	OH (1 – 0)	CS (0 – 0)	CO_2^+	C_2	Continuum (10^{-13} erg $cm^{-2}s^{-1}Å^{-1}$)	Image	Exposure time (min:sec)
	R						
23.53	–	–	–	–	1.66	LWP 7848	00:40
23.60	–	–	–	310 ± 55	–	LWP 7849	140:00
23.83	4850 ± 90	1780 ± 120	290 ± 80	–	1.15	LWP 7851	15:00
23.89	4490 ± 220	2250 ± 640	–	–	1.25	LWP 7852	02:00
24.10	5680 ± 340	3810 ± 810	–	–	1.99	LWP 7856	01:10
24.51	10000 ± 170	7710 ± 190	660 ± 190	–	4.70	LWP 7863	10:00
24.72	–	–	–	–	4.88	LWP 7865	00:30
24.75	11700 ± 190	9470 ± 330	944 ± 220	–	5.08	LWP 7866	04:00
24.78	–	–	996 ± 180	–	–	LWP 7867	15:00
24.85	–	–	–	835 ± 190	–	LWP 7868	45:00
25.04	–	–	–	–	4.63	LWP 7871	00:40
25.07	12200 ± 250	8330 ± 200	811 ± 200	–	4.38	LWP 7872	07:00

Table 2. Flux of offset pointings at comet Halley

Date 1986 March	Offset[a] (arcsec)	OH (1 – 0)	CS (0 – 0)	CO_2[a]	Continuum (10^{-13} erg $cm^{-2}s^{-1}Å^{-1}$)	Image	Exposure time (min:sec)
		R					
23.94	30 T	2530 ± 90	300 ± 130	300 ± 70	0.199	LWP 7853	30:00
23.99	30 S	2180 ± 100	70 ± 140	180 ± 100	0.398	LWP 7854	30:00
24.05	30 T	2380 ± 20	200 ± 40	220 ± 20	0.181	LWP 7855	48:00
24.95	30 S	–	1190 ± 130	650 ± 90	0.199	LWP 7869	30:00
25.00	30 T	–	700 ± 100	350 ± 80	0.370	LWP 7870	30:00

[a] S = Sunward, T = Tailward

enhancement and was higher in the tailward direction by almost a factor of 2 (Table 2).

The variation in the coma from March 24.0 to March 25.0 is also well illustrated by considering the offset brightness relative to the nuclear brightness interpolated to the same time as the offset exposures. Relative to the nucleus, the sunward coma brightened by a factor of 6 in CS and a factor of 1.5 in CO_2^+. It faded by a factor 6 in the continuum. The tailward coma was unchanged in CS, and faded by a factor of 1.5 in CO_2^+ and in the continuum. Thus the neutral and ionic species are ejected primarily in the sunward direction. The fading of the dust on the sunward side is presumably due to outflow of grains from the previous maximum of activity, those grains not yet being replaced by the newly released grains from the nucleus which travel much more slowly than does the gas.

4. Discussion

The change in the continuum follows the same trend as the neutral and ionic species (Fig. 1) suggesting that the dust was ejected simultaneously with the volatiles. This flux increases both as the grains flow away from the nucleus and as the area of activity rotates into the sunlight. The OH increases in abundance slowly because of the long dissociation lifetime of its parent molecule. In the 8.2 10^4 seconds (or even the 4.3 10^4 s as suggested by Krankowsky et al., 1986) during which the H_2O molecules dissociate, many of them flow out of the field of view. Thus the amplitude of the increase in OH is less than for other species whose parents have shorter lifetimes.

The parent molecule of CS has a short lifetime (500 s; Jackson et al., 1986) relative to that of the parents of OH and C_2 (8.2 10^4 s and 1.7 10^4 s respectively). Because of this difference in the lifetime of the species, during the time of increasing activity at the comet, the daughter molecule, CS is created faster than are OH and C_2, and before the parent leaves the field of view of the telescope. This explains the observed difference in amplitudes between CS and OH. The CS should also peak in time before the OH flux for this reason but we do not have sufficient temporal resolution to observe and measure this phase lag.

Because the time resolution of the IUE exposures is low, we place only approximate limits on the ionization lifetime of CO_2. It appears to be shorter than the photodissociation lifetime of H_2O (8.2 10^4 s) but longer than 500 s, the lifetime of CS_2 (the presumed parent of CS). If we assume the difference in amplitude of CS and OH is due to the difference in the fraction of the parent before it leaves the aperture, then we can infer that the lifetime of the parent of CO_2^+ is somewhere between the lifetime of the parent of CS and OH.

Since the photoionization lifetime of CO_2 is substantially larger than the lifetime of H_2O against dissociation, an additional source of ionization must be present in the inner coma as suggested by Feldman et al. (1986).

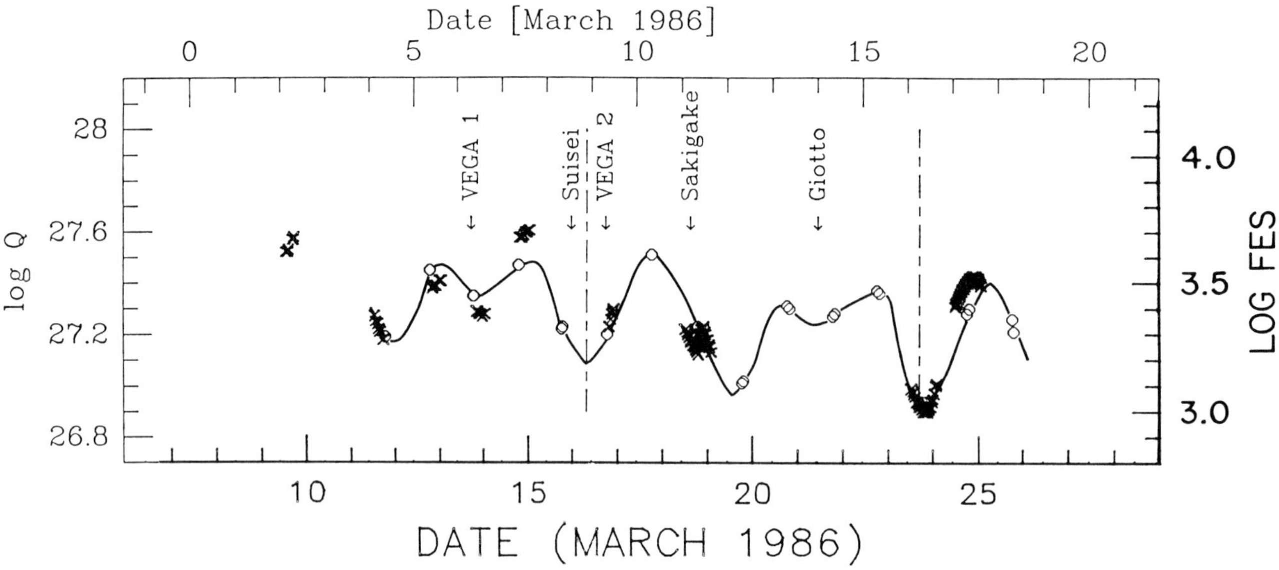

Fig. 2. The FES lightcurve (x's) and the lightcurve measured by the C_2 production rate (o's) (Millis and Schleicher, 1986) scaled to the log of the FES flux. The axes for IUE data are on the bottom and right side of the figure. The axes for Millis and Schleicher data are on the top and left of the figure. Note the phase shift of one data set relative to the other by 7.4 days almost exactly one period

The change in the sense of the sunward-tailward asymmetry observed in the offset exposures between minimum and maximum light supports the hypothesis that the source of the increase in flux is vaporization on the sunlit side of the comet. This activity is probably controlled by incident sunlight on a readily vaporized region of the nucleus. The neutral and ionic species are seen in the sunward direction for a longer time period than the dust flux because of the time lag that is necessary before the daughter molecules are observed. Thus we observe the dust flux increase in the sunward direction before the increase in the flux of the emission bands is observed. The relative increases in the flux of the neutral and ionic species as seen at the two tailward pointings taken a few hours apart (Table 2) also demonstrates the relative lifetimes of the three parent molecules, OH being the longest, followed by CO_2^+ and CS.

4.1. Relationship to the optical lightcurve

Millis and Schleicher (1986) observed photometric variations of the comet that repeat approximately every 7.4 days. The lightcurve had multiple maxima presumably corresponding to different ective regions of the nucleus. Figure 2 shows the FES lightcurve of measurements acquired between 1986 March 10–25 plotted with the C_2 production rate measured by Millis and Schleicher between 1986 March 4–19. Whereas there are both a phase shift and amplitude variations between these two data sets, the FES lightcurve supports the 7.4 day period derived by Millis and Schleicher (1986). We do not expect an exact match to the optical lightcurve because the bandpass of the FES is much wider than the C_2 filter and different sized apertures were used for the two data sets.

4.2. Relationship to reported activity of H_2O

Weaver et al. (1986, 1987) used the University of Arizona's Fourier Transform Spectrometer on the Kuiper Airborne Observatory (KAO) to directly measure the brightness of water in a 41″ aperture on March 24. A series of measurements over a week or so with this instrument consistently yields production rates for water which are a factor of 2 to 3 higher than those we deduce from the IUE observations of OH. Weaver et al. (1986) reported separate observations on March 24.65 and 24.73 which implied an increase by 60% (revised to 40% in Weaver et al., 1987) in the number of water molecules in their beam. This corresponds to an excess of nearly 10^{34} molecules appearing in their beam in less than 2 h. As can be seen in Fig. 1, this reported activity took place near the peak of the FES light curve when the production of OH appeared to be rising slowly but nearing its peak. Clearly there is no evidence for an outburst in the data for OH. According to IUE data, the OH appears to be rising smoothly from March 23.9 until March 25.1.

As Weaver et al. (1986) point out, they were observing a parent molecule whereas we were observing a dissociation product and they should therefore better monitor the actual variability of the nucleus. On the other hand, an outburst of that size should have effects easily detected in the dissociation products that we observe. Since the parent of CS (presumed to be CS_2) has a very short lifetime (500 s at 1 AU), the activity of CS also closely monitors the nuclear activity. There is no obvious outburst in the production of CS. The absence of a detectable outburst of CS requires that the relative abundance of CS_2 in the outburst was at least a factor 5 less than typical.

If the outburst of H_2O was a delta function, then we should have observed a measureable deviation of the OH production above its smooth variation unless the outburst took place no earlier than about ten minutes prior to the second KAO observation, a rather tight constraint on the outburst. If, on the other hand, the abundance of H_2O increased steadily between the time of the two observations, then the smooth behavior of OH requires that the linear increase had also been in progress since March 23.9. Given the slope of the variation in H_2O between March 24.65 and March 24.73 observed by Weaver et al., the total amplitude of variation must have been orders of magnitude, again a severe constraint on any activity.

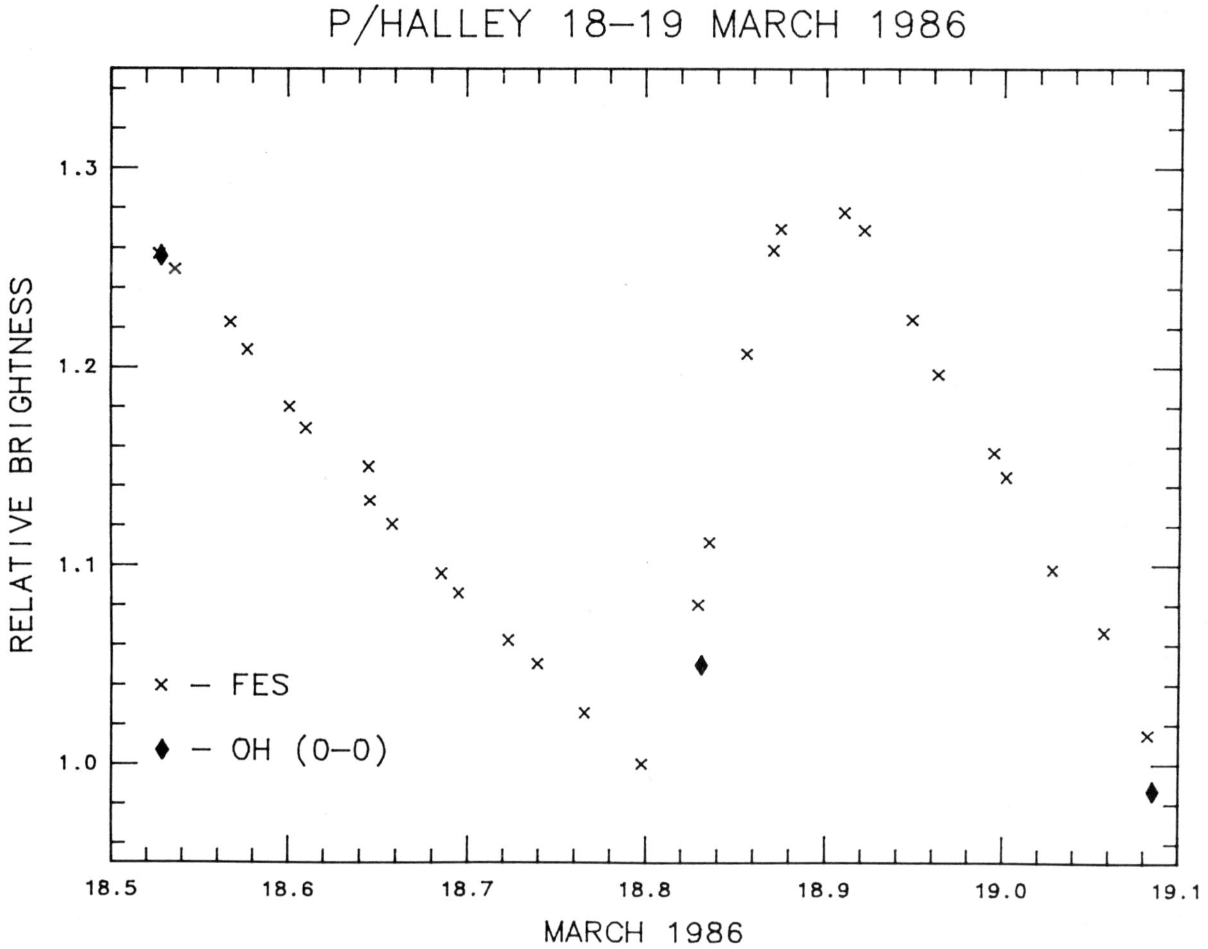

Fig. 3. March 18–19 FES flux and the OH (0–0) brightness. The OH brightness is 5.55 10^5 Rayleighs at 18.5 March UT

The limited temporal resolution achieved in the IUE spectra, combined with the 1-h transit time of the IUE aperture, allows some short outbursts to go undetected. However, the FES lightcurve has much better temporal resolution. All other activity in Halley has been correlated with variations observed in the FES and the absence of such variation in this instance indicates that we were not observing an outburst. Further detailed modeling to test the relationship of the IUE and the KAO results is postponed to a later paper.

4.3. Comparison with the Outburst on 18–19 March

On 1986 March 18–19 an event with a totally different character was observed with the IUE (Feldman et al., 1986). That event was cleary a transient event (Fig. 3), much shorter (0.3 days total) than the activity reported here and of much lower amplitude (approximately 30% increase in the white-light measurements with the FES). The OH abundance decreased monotonically through this interval as shown in Fig. 3 (Feldman et al., 1986). The lightcurve of Millis and Schleicher predicts a minimum of activity at about the end of that outburst. Thus the outburst on March 18–19 is not correlated with the periodic variations observed by Millis and Schleicher although that lightcurve naturally explains the observed decreasing trend of the OH.

Feldman et al. (1986) reported a dramatic increase in the tailward abundance of CO_2^+ during the March 18–19 outburst that was not seen in any of the neutral species. In the activity on March 23–25, however, the evidence suggests that the neutral species increase dramatically on the sunward side of the coma, a totally different behavior from that observed on March 18–19. On March 23–25 we do not see any dramatic rise in the relative abundance of CO_2^+ in either the nuclear or the offset pointings; the ions do not increase any more than the neutrals, contrary to the behavior of the ions on March 18–19.

The strong correlation between the activity observed with IUE on March 23–25 and the lightcurve of Millis and Schleicher, the enhancement of abundances on the sunward side of the coma, and the gradual nature of the increasing activity of the comet, argue for an interpretation of this activity as due to rotation of an active area or jet into sunlight. The activity of March 18–19, on the other hand was more likely due to sudden eruption of a pocket of CO_2 or a sudden and very dramatic enhancement of ionization as discussed in Feldman et al. (1986).

5. Conclusions

We have shown that all species: neutrals, ions, and dust, increase gradually over the period 1986 March 23–25 exhibiting a pattern that is well correlated with the lightcurve measured both by the FES and by the various molecular production rates measured by Millis and Schleicher (1986). The gradual changes, the correlation with the 7.4 day periodicity, and the spatial asymmetries argue strongly for control of the event by rotation of an active area or jet into sunlight.

The amplitudes of the various species are correlated with the lifetimes of the parent molecules. Since the amplitude of the variation of CO_2^+ is intermediate between those of CS and of dust on the one hand and of OH on the other, this suggests that the ionization time of CO_2^+ is intermediate between 500 s and 8 10^4 s, the lifetimes of the respective parents of CS and OH. The ionization mechanism is not known since photoionization is too slow but the mechanism may be collisional as suggested by Woods et al. (1987). A more detailed modeling of the temporal variation will be required to determine the actual value of the ionization time.

We find no activity in the record from IUE which can be related to the outburst reported using the KAO (Weaver et al., 1986, 1987). Preliminary modeling suggests that the only plausible model for that outburst would involve very abrupt activity nearly at the time of one of their observations. Further temporal modeling will be required in order to determine whether plausible models of the outburst can be reconciled with the lack of observed effects in the data from IUE.

During its apparition, comet Halley exhibited several types of variability of which at least two very different types have now been documented with data from IUE. The activity observed on March 23 – 25 was of a totally different character than that observed on March 18–19.

Acknowledgements. We thank the staff of the IUE for their usual superb assistance with the observations. We thank Hal Weaver for numerous discussions of the results from the KAO and we thank Dave Schleicher for providing the lightcurve for use in Fig. 2. Computer time for data reduction was provided by the University of Maryland Computer Science Center. This work was supported by NASA grants NAG 5-252 (U MD) and NSG-5393 (JHU).

References

A'Hearn, M.F., Hoban, S., Birch, P.V., Bowers, C., Martin, R., Klinglesmith, D.A.: 1986a, *Nature* **324**, 649

A'Hearn, M.F., Hoban, S., Birch, P.V., Bowers, C., Martin, R., Klinglesmith, D.A.: 1986b, 20th ESLAB Symp., ESA SP-250, Vol. **I**, p. 483

Feldman, P.D., Festou, M.C., A'Hearn, M.F., Arpigny, C., Butterworth, P.S., Cosmovici, C.B., Danks, A.C., Gilmozzi, R., Jackson, W.M., McFadden, L.A., Patriarchi P., Schleicher, D.G., Tozzi, G.P., Wallis, M.K., Weaver, H.A., Woods, T.N.: 1987, *Astron. Astrophys.* (submitted)

Feldman, P.D., A'Hearn, M.F., Festou, M.C., McFadden, L.A., Weaver, H.A., Woods, T.N.: 1986, *Nature* **324**, 433

Gerard, E., Bockelée-Morvan, D., Bourgois, G., Colom, P., Crovisier, J.: 1986, 20th ESLAB Symp., ESA SP-250, Vol. **I**, p. 589

Jackson, W.M., Butterworth, P.S., Ballard, D.: 1986, *Astrophys. J.* **304**, 515

Krankowsky, D., Lämmerzahl, P., Herrwerth, I., Woweries, J., Eberhardt, P., Dolder, U., Hermann, U., Schulte, W., Berthelier, J.J., Illiano, J.M., Hodges, R.R., Hoffman, J.H.: 1986, *Nature* **321**, 326

Millis, R.L., Schleicher, D.G.: 1986, *Nature* **324**, 646

Mirabel, I.F., Bajaja, E., Arnal, E.M., Cersosimo, J.C., Colomb, F.R., Martin, M.C., Mazzaro, J., Morras, R., Poppell, W.G.L., Silva, A.M., Boriakoff, V.: 1986, 20th ESLAB Symp., ESA SP-250, Vol. **I**, p. 595

Rettig, T., Ruchti, R., Baumbaugh, B., Baumbaugh, A., Knickerbocker, K., Dawe, J.: 1986, 20th ESLAB Symp., ESA SP-250, Vol. **III**, p. 93

Schloerb, F.P., Kinzel, W.M., Swade, D.A., Irvine, W.M.: 1986, 20th ESLAB Symp., ESA SP-250, Vol. **I**, p. 577

Schlosser, W., Schulz, R., Koczet, P.: 1986, 20th ESLAB Symp., ESA SP-250, Vol. **III**, p. 495

Weaver, H.A., Feldman, P.D., Festou, M.C., A'Hearn, M.F.: 1981a, *Astrophys. J.* **251**, 809

Weaver, H.A., Feldman, P.D., Festou, M.C., A'Hearn, M.F., Keller, H.U.: 1981b, *Icarus* **47**, 449

Weaver, H.A., Mumma, M.J., Larson, H.P., Davis, D.S.: 1986, *Nature* **324**, 441

Weaver, H.A., Mumma, M.J., Larson, H.P.: 1987, *Astron. Astrophys.* (submitted)

Woods, T.N., Feldman, P.D., Dymond, K.F.: 1987, *Astron. Astrophys.* (submitted)

Astron. Astrophys. 187, 339–350 (1987)

ASTRONOMY AND ASTROPHYSICS

Cometary MHD and chemistry

R. Wegmann [1], H.U. Schmidt [1], W.F. Huebner [2], and D.C. Boice [2]

[1] Max-Planck-Institut für Astrophysik, Karl-Schwarzschild-Strasse 1, D-8046 Garching bei München, Federal Republic of Germany
[2] Southwest Research Institute, P.O. Drawer 28510, San Antonio, TX 78284, USA

Received February 19, accepted May 21, 1987

Summary. Our magneto-hydrodynamical (MHD) and chemical comet-coma model has been applied to describe and analyze the plasma flow, the magnetic field and the ion abundances in comet P/Halley in a consistent manner. We assume the volatile composition to consist of 80% water and 20% carbon-, nitrogen-, oxygen-, and sulfur-compounds. The radius of the nucleus is 3.36 km. With hemispherical illumination, this is equivalent to the active area on the nucleus during the Giotto encounter. The physics and chemistry of the coma are modeled in great detail, including photoprocesses, gas-phase chemical kinetics, energy balance with a separate electron temperature, multifluid hydrodynamics with a transition to free molecular flow, fast-streaming atomic and molecular hydrogen, counter and cross-streaming of the ionized species relative to the neutral species in the coma-solar wind interaction region with momentum exchange by elastic collisions, mass-loading through ion pick-up, and Lorentz-forces of the advected magnetic field. A comparison of the results is made with the data from the Giotto mission, especially from the HIS ion mass spectrometer. We resolve the contact surface (magnetopause); its position is in agreement with observations. We also find the enhancement of the ion density just outside of the contact surface and agreement for the three groups in the ion mass spectra, particularly for the first group up to 21 amu.

Key words: comets – hydromagnetics – interplanetary medium – magnetic field – plasmas

1. Introduction

In order to achieve a consistent description for the ionized coma of a comet and its interaction with the solar wind we combine and improve, in iterative steps, our existing computer programs for the detailed chemical evolution of a spherically expanding coma (Huebner, 1985) and for the magneto-hydrodynamical (MHD) flow of plasma and magnetic field in a comet (Schmidt and Wegmann, 1982; Wegmann 1987). We apply this model to the data from the Giotto mission to comet Halley, especially to the ion abundances as measured with the HIS ion mass spectrometer (IMS) (Balsiger et al., 1986). One important goal of this exercise is to adjust the free parameters in the model, such as the abundances of the minor species on the nuclear surface until the calculations agree with the relevant observations made during the Giotto encounter. The ion species abundances in the coma and the ion density enhancement just outside of the contact surface are parts of this goal and therefore are a major topic of this report.

Morphologically, the neutral coma gas, produced by sublimation from the illuminated hemisphere, spreads over the entire nucleus to establish pressure balance with the unlit hemisphere and expands uniformly. The ions produced in the inner coma expand uniformly only up to the region where they begin to interact with the solar wind. In the coma-solar wind interaction region, the plasma flow is axisymmetric about the sun-comet line in a first approximation, with the comet acting as an obstacle in the fluid flow. Transverse magnetic fields imbedded in the solar wind cause only mild deviations from the axial symmetry which partially cancel when averaged over fluctuating field orientations. Therefore, our model consists of two sections that are coupled by the continuity of physics and chemistry: (1) the radially symmetric inner coma and (2) the solar wind interaction region, which is iteratively improved over various levels of complexity. The radially symmetric inner coma part of the model simulates photoprocesses with wavelength-dependent solar flux attenuation, attenuation-dependent coma energy balance including a separate determination of the electron temperature which enhances electron-collisional dissociation and ionization, gasphase chemical kinetics for H-, C-, N-, O-, and S-compounds, multifluid hydrodynamics for fast-streaming atomic and molecular hydrogen, and a transition to free molecular flow (Huebner, 1985). The radially symmetric part of the model for the neutral gas also extends throughout the coma-solar wind interaction region.

The ionized coma interacts with the solar wind and therefore deviates from spherical symmetry. The embedded magnetic field destroys even axial symmetry. Therefore, a detailed description of the region would require a three-dimensional MHD model with full chemistry. For economy, we approximate this by an iterative procedure in which we first calculate the full chemistry axisymmetrically and then use the averaged chemical quantities in a three-dimensional MHD calculation with detailed magnetic field.

An outline of an earlier, less advanced version of this model with applications to the data from the ICE mission to comet Giacobini-Zinner has been given by Wegmann (1987).

In Sect. 2 we list the model parameters for the comet and the solar wind which we use throughout this paper. In Sect. 3 we describe the axisymmetric part of the model with angle-averaged magnetic field. The results of the axisymmetric calculation of Sect. 3 are averaged over species for each kind of reaction to

Send offprint requests to: H.U. Schmidt

obtain mean rate coefficients for use in the three-dimensional MHD calculation with detailed magnetic field which is presented in Sect. 4. In Sect. 5 we unfold the total values of the ions and neutrals by reemploying the full chemistry and using the flow field obtained in Sect. 4. We discuss the results both inside and outside of the contact surface and compare them to relevant Giotto observations but specifically to the HIS ion mass spectra (Balsiger et al., 1986).

2. Model parameters

To apply this general coma model to Halley we have chosen the following parameters. During encounter, Halley had a heliocentric distance of $r = 0.89$ AU. We assumed an albedo of 0.03 (Keller et al., 1986) and an infrared emissivity of 0.97. A nucleus with a radius $R_0 = 3.36$ km, when hemispherically illuminated, gives an equivalent area about equal to the active area during encounter.

Using the above model parameters, the gas production rate from the hemispherically illuminated nucleus is $6.9\ 10^{29}$ mol s^{-1}, in agreement with the value measured by the neutral mass spectrometer (NMS) (Krankowsky et al., 1986a).

The assumed composition of the nucleus is given in Table 1. Further modifications of this composition will be made as the analysis progresses. We have made use of molecular identifications in the coma of Halley. This very limited list was supplimented with other molecules that we expect in the nucleus at a level of a few percent or less. Among the identified species are H_2O, with an abundance of $\sim 80\%$ (Krankowsky et al., 1986a), CO_2 with an abundance of 1.5% (Combes et al., 1986) to 3.5% (Krankowsky et al., 1986a) relative to water, HCN at a level of $<1\%$ (Bockelée-Morvan et al., 1986), and possibly CO (Combes et al., 1986) and H_2CO (Combes et al., 1986; Synder et al., 1987).

In the assumed mixture, the abundance ratios of elements are approximately H:C:N:O:S = 100:10:2:53:1. These values are typical for the elemental composition of a waterdominated comet (Delsemme, 1982) except N which is deficient. To account for minor abundances in the observed mass spectra we assumed a solar distribution for the isotopes; specifically, $^{12}C:^{13}C = 98.9:1.1$, $^{14}N:^{15}N = 99.6:0.4$, $^{16}O:^{18}O = 99.8:0.2$, and $^{32}S^+:^{33}S:^{34}S = 95.0:0.8:4.2$. For comparison, Krankowsky et al. (1986a) report $^{32}S^+:^{34}S^+ = 100:4.5$.

Table 1. Assumed chemical composition of volatiles in the nucleus

Species	Abundance	Comment
H_2O	80.0%	Identified
CO	8.0	Possibly identified
CO_2	3.0	Identified
CH_4	2.0	
NH_3	2.0	
H_2CO	2.0	Possibly identified
CS_2	1.0	
N_2	0.68	
C_2H_2	0.67	
$H_2C_3H_2$	0.05	
H_2CO_2	0.30	
CH_3CN	0.16	
NH_2CH_3	0.08	
HCN	0.06	Identified

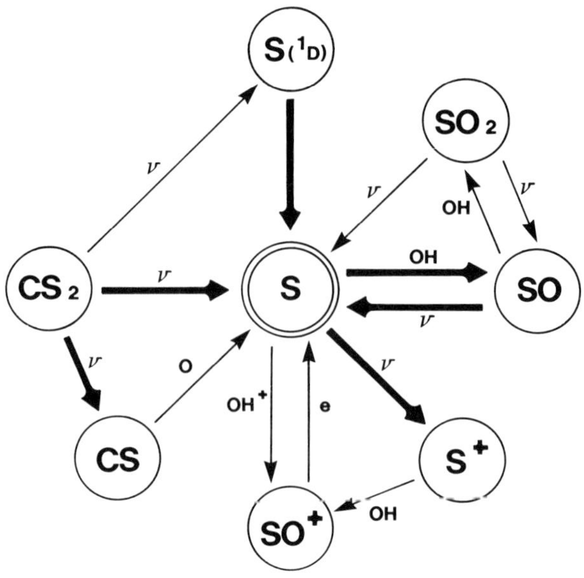

Fig. 1. Major sulfur reactions at cometocentric distance $R = 1500$ km assuming CS_2 as the parent of sulfur-bearing species. Heavy arrows indicate the most important reaction paths. Symbols next to the arrows are the reactants; ν indicates photodissociation or photoionization

New in our model is the sulfur chemistry. We have added 40 sulfur-containing species and about 200 sulfur reactions to the chemical reaction network which brings the total to 135 relevant species and about 950 reactions. CS_2 is the only sulfur compound assumed to be in the nucleus. We have chosen CS_2 as the parent molecule as observations of CS and S in many comets show similar spatial distributions which could be explained by photodissociation of CS_2. However, the actual parent molecules of sulfur-bearing species remain unknown and could be CS_2, H_2S, S_2, or other compounds.

The primary reactions that form sulfur are photoprocesses, the most important being the photodissociation of CS_2. Figure 1 illustrates the major sulfur reactions at a cometocentric distance of $R = 1500$ km. A major destruction path of sulfur is the neutral rearrangement with OH to form SO. OH appears to be an important component in the sulfur network as it plays a major role in the production of SO_2 and SO^+, which is also formed by the reaction of OH^+ with S. These identifications are tentative as there exist uncertainties in the rates of many sulfur reactions. Additional laboratory work is needed to determine accurate rates for many gas-phase sulfur reactions.

The unperturbed solar wind was characterized by a density of 7 protons cm^{-3}, a velocity of 380 km s^{-1}, an electron temperature of $2.5\ 10^5$ K, an ion temperature of 10^5 K, and a magnetic field of 7 nT.

3. Axisymmetric model with magnetic field averaged over angles

The ionized coma interacts with the solar wind and therefore deviates from spherical symmetry. In the first approximation, used throughout this section, we describe the solar wind interaction with a stationary, axisymmetric MHD-model in which axisymmetry is achieved by averaging the Lorentz forces of the transverse magnetic field over all azimuthal orientations (Schmidt and Wegmann, 1982). The neutrals are hardly influenced by the solar wind and therefore are accurately described by the spherically symmetric part of the model. In the interaction region the ions are embedded into this neutral background. The computational grid for the interaction region starts at a distance of 1500 km from the nucleus. Deviating from Schmidt and Wegmann (1982), at the inner boundary all variables match to the ionized component of the spherically symmetric model. The ionized coma is treated as a multicomponent fluid with 60 chemical species and with separate temperatures for electrons and ions. For each ion species we use a separate continuity equation with a common bulk velocity vector field and source terms determined by the chemical reactions of the electrons, ions, and the neutral background. The photo-processes and the chemical kinetics in the inner coma are extended consistently into the axisymmetric region where counter and cross-streaming of neutrals and ions, momentum exchange by elastic collisions, mass-loading of the solar wind, the bow shock, the contact surface, and angularly averaged magnetic field effects are modeled. With the exception of the aforementioned interaction processes and the inner boundary condition, the plasma, and the neutral flow are decoupled. This model presents a further improvement of the model discribed by Huebner et al. (1987a).

At a large distance from the nucleus the solar wind flow is decelerated by mass-loading and momentum exchange but is not yet influenced by dynamic effects of the imbedded magnetic field. A bow shock is formed and an axisymmetric hydrodynamic description still suffices, but in this flow the magnetic field is piled up and bent around the nucleus with Lorentz forces gradually becoming important. During a travel time of at least 4 hours from bow shock to stagnation, the interplanetary magnetic field can suddenly change its direction by about 1 radian several times. Assuming that the field changes its transverse direction at random and so frequently that the flow cannot adjust to the azimuthal forces, we can describe the average flow pattern of all variables, including the field strength, as axisymmetric with proper account of the magnetic advection and of the effects of the average Lorentz forces (Schmidt and Wegmann, 1982).

Because this axisymmetric model totally neglects the relaxation of magnetic stresses by small non-axisymmetric azimuthal velocity components, it must produce a divergent pile-up of magnetic field near stagnation, as was already shown by Lees (1964). Therefore the flow in the stagnation region never reaches a fully stationary state but instead oscillates. The profiles of the plasma variables on the sun – comet axis, as calculated in this first approximation, are given in Fig. 2. The comparison with the same profiles as obtained in the three-dimensional model described in the following section, shows that this simple approximation is valid throughout most of the interaction region. The exaggerated magnetic pile-up is obvious in the total field strength near stagnation. It pushes the contact surface too far inward by a factor of two. The difference in the shock position is due to changes of the model parameters, namely in the ratio γ of specific heats and the assumed cross-sections σ for elastic collisions: ($\gamma = 2$, $\sigma = 6\ 10^{-15}\ \mathrm{cm}^2$ in 2D, $\gamma = 5/3$, $\sigma = 2.4\ 10^{-16} - 3.5\ 10^{-15}\ \mathrm{cm}^2$ in 3D).

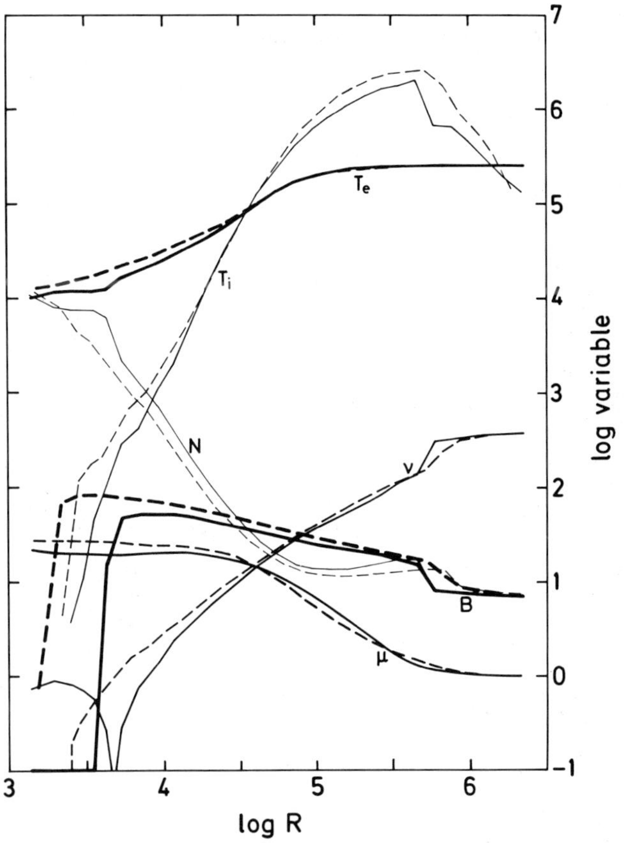

Fig. 2. Flow variables along the axis for the axisymmetric model of Sect. 3 (dashed lines) and the three-dimensional model (solid lines): T_i = ion temperature [K], T_e = electron temperature [K], v = velocity [km s^{-1}], N = number density [cm^{-3}], B = magnetic field strength [nT], μ = mean molecular weight, R = distance from nucleus [km]

4. Three-dimensional model with stationary magnetic field

In the next step we can now construct the full three-dimensional flow field including its azimuthal components which necessarily arise from the Lorentz forces of the transverse magnetic field.

The plasma is treated by the MHD conservation laws with source terms for particle numbers, mass, momentum, electron energy, and ion energy. For this model we assume an inclined, non-fluctuating interplanetary magnetic field. We derive source terms which describe the effects of all photo-processes, electron impact reactions, recombination, charge exchange, and elastic collisions in the following manner. We start with the source terms as they were calculated in the axisymmetric model described in the last section. For each type of reaction we divide the respective contribution by the appropriate total density (neutrals, ions, or electrons) from the axisymmetric model. In this way we obtain effective rate coefficients (some of them multiplied by suitable molecular weights, particle momenta, and energies). The source terms are obtained by multiplication of these effective rate coefficients by the actual densities in the three-dimensional model. For all general dependences on molecular weights, ordered velocities, and temperatures, the actual values are used. In this procedure we consider the remaining variances as explicit functions of the distance from the nucleus.

The large scale features of the model comet can be seen in Fig. 3, which shows the situation in a meridional plane parallel to the interplanetary field. The nucleus is in the center with the sun above. Each plot extends one million kilometers on each side of the nucleus. The variables are: electron density, mean molecular weight, velocity, magnetic field strength, electron temperature, and ion temperature. The color coding is such that blue

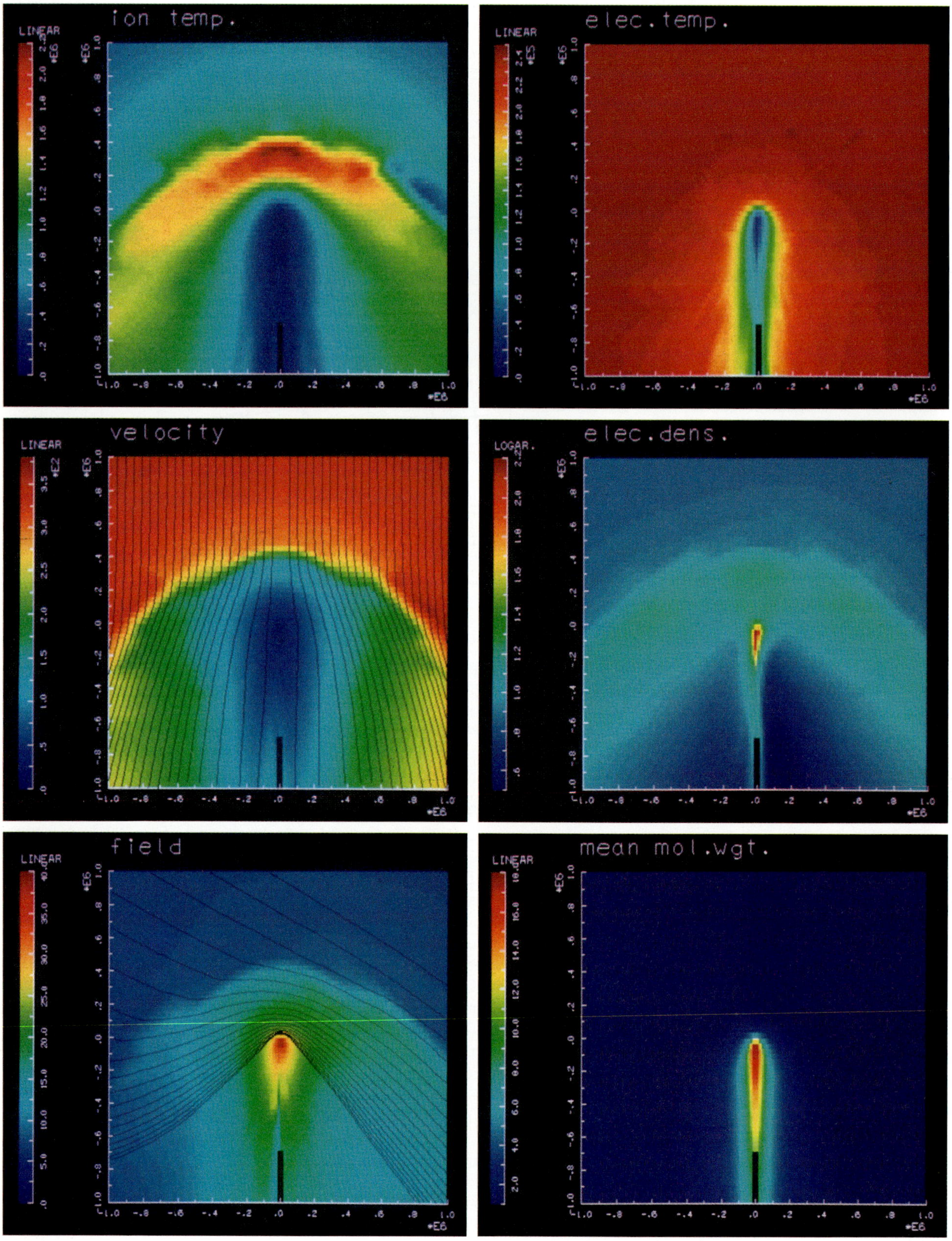

Fig. 3a–f. The bow shock: The color contour maps extend to 10^6 km on each side of the nucleus. **a** Ion temperature [K], **b** velocity [km s^{-1}] with streamlines, **c** The intensity of the magnetic field [nT] with fieldlines. The fieldlines follow each other in time intervals of 600 s. **d** Electron temperature [K], **e** Number density of the electrons [cm^{-3}] on a logarithmic scale, **f** the mean molecular weight of the ions (including solar wind protons)

corresponds to a low value of the variable, red to a high value. A paraboloid-shaped region around the tail axis is not covered by the computational grid and is indicated by the black spike in the tail direction.

The shock front is characterized by discontinuities in the density, velocity, magnetic field and ion temperature. We assume that the interplanetary magnetic field is inclined with respect to the velocity vector by an angle of 56 degrees. Since only the component parallel to the shock is compressed, the field strength is asymmetric behind the shock. The magnetic field lines are broken in the shock transition and then are draped around the nucleus. Since the field is frozen-in, the field lines are also isochrones. The field lines in Fig. 3 follow each other in time intervals of ten minutes. The 19 field lines cover a total time interval of 3 h.

The tail region consists of dense, slow plasma with low ion and electron temperatures. This region consists mainly of cometary ions. It coincides essentially with the visible plasma tail. The inclination of the interplanetary magnetic field imposes a slight aberration of the tail.

Figure 2 shows the flow variables along the sun-comet axis. The solar wind flow is gradually decelerated and heated by mass-loading with slow cometary ions and collisions with neutrals. It remains a parallel flow until, at a distance of 500000 km in front of the nucleus, a shock occurs which has the effect that the flow becomes divergent. Therefore, the number density which was increased by the shock, decreases again slightly behind the shock. The mass density increases, however, since heavy cometary ions are continually added. This is indicated by the increasing mean molecular weight. The ions are heated by the shock to 2.2 10^6 K but then are cooled very efficiently by collisions and charge exchange. The temperature of the electrons is unaffected by the shock. The electrons loose their energy in electron impact reactions.

Up to a distance of 30000 km the velocity decrease is roughly proportional to the cometocentric distance R as the flow approaches the comet. A stagnation region follows which is characterized by a steep increase of the ion density with an R^{-2} dependence and a steep decrease of the velocity until the flow comes to a stop at 4800 km in front of the nucleus. A contact surface is formed which separates the contaminated solar wind from the pure cometary plasma inside. On the inner side of the contact surface there is also a stagnation region. Further inward the flow matches to the spherically symmetric purely cometary flow with a plasma density proportional to R^{-1}. At the contact surface, two plasmas of different origin converge without penetration creating a discontinuity in several flow variables.

The magnetic field is compressed by the shock and then rises steadily until it reaches a maximum of 53 nT in the stagnation region. Then it drops off very rapidly near the contact surface.

This flow field is the result of a delicate balance of a number of different effects, which are illustrated in Fig. 5.

a) The mass-loading is described by a source term in the continuity equation. In the solar wind the main contribution comes from charge exchange. In the inner region photoionization is by far the most efficient process. In the stagnation region the high ion density and the low electron temperature favor recombination which at the stagnation point, removes slightly more ions, than are produced by photoreactions. In this way a stationary stagnation flow becomes possible.

b) The electrons are heated by the surplus energy from the ionizing photons in photoprocesses. This energy is consumed, to a large extent, by electron impact reactions.

c) Elastic collisions and charge exchange cause a volume force on the flow which varies roughly like R^{-2}. This force is balanced by pressure forces. At the stagnation region the reservoir of thermal energy which was built up in the shock, is exhausted and Lorentz forces take over.

d) Elastic collisions and charge exchange cool the plasma with an efficiency roughly proportional to R^{-1}. This is partly balanced by the work which is done by the pressure forces and in the stagnation region by the Lorentz-forces. This cooling maintains the pressure gradient which drives the flow inward against the forces exerted by collisions.

Figure 4 shows the innermost region of the flow in three perpendicular planes: the meridional plane which contains the interplanetary magnetic field, the meridional plane perpendicular to the first plane, and a plane perpendicular to the velocity vector of the solar wind. The nucleus is in the center and the graphs extend 30000 km on either side of the nucleus. The figures on the right hand side show the magnetic field strength. There is an extended dark blue zone, which marks a region with zero field. This magnetic activity with almost circular cross-section extends about 5000 km to the side and 15000 to 20000 km into the tail. In the stagnation region in front of this cavity the field is piled up to its maximum value of 53 nT. The original inclination of the field has almost completely disappeared in this central region. The field is compressed to a bundle of field lines, which forms two lobes of opposite polarity adjacent to the cavity. The field lines in Fig. 4 follow each other in time intervals of five minutes. The region covered by these field lines is transversed by the plasma in 1.5 h.

The graphs on the left hand side show the velocity, which ranges from 0 to 18 $\mathrm{km\,s^{-1}}$. The streamlines flow around the contact surface. The flow accumulates near the contact surface which acts as a sink due to recombination processes.

The extreme magnetic pile-up in axisymmetric stagnation is reduced in a fully threedimensional flow. The flow becomes obviously divergent in the direction perpendicular to the field but since the mass-loading is reduced, this is compensated by a convergent flow along the field lines.

The profiles of the flow variables along the Giotto trajectory as predicted by our model are shown in Fig. 6. The discontinuities on the left and right hand sides are associated with the contact surface and the shock front, respectively. Since the Giotto trajectory is inclined by an angle of 102 degrees with respect to the solar wind, the distances of the shock is smaller on the outbound than on the inbound trajectory. The model calculation predicts the bow shock to be in the range 1.–2. 10^6 km from closest approach. The bow shock position measured by instruments on the Giotto spacecraft is at $\sim$1.1 10^6 km inbound (Balsiger et al., 1986; Johnstone et al., 1986; Rème et al., 1986; Neubauer et al., 1986).

The model predicts the positions of the contact surface to be in the range 5000–6000 km from closest approach, it has been measured to be at 4600 km inbound (Krankowsky et al., 1986; Balsiger et al., 1986; Neubauer et al., 1986).

The total ion density (equal to the electron density) as a function of cometocentric distances, R, generally varies as R^{-1} inside the contact surface and steepens to an R^{-2} dependence outside. This is in qualitative agreement with the observations reported by Balsiger et al. (1986).

5. Improved axisymmetric model: ion abundances

For economy we calculate detailed ion abundances only for axisymmetric flow fields. This is justified since the total density

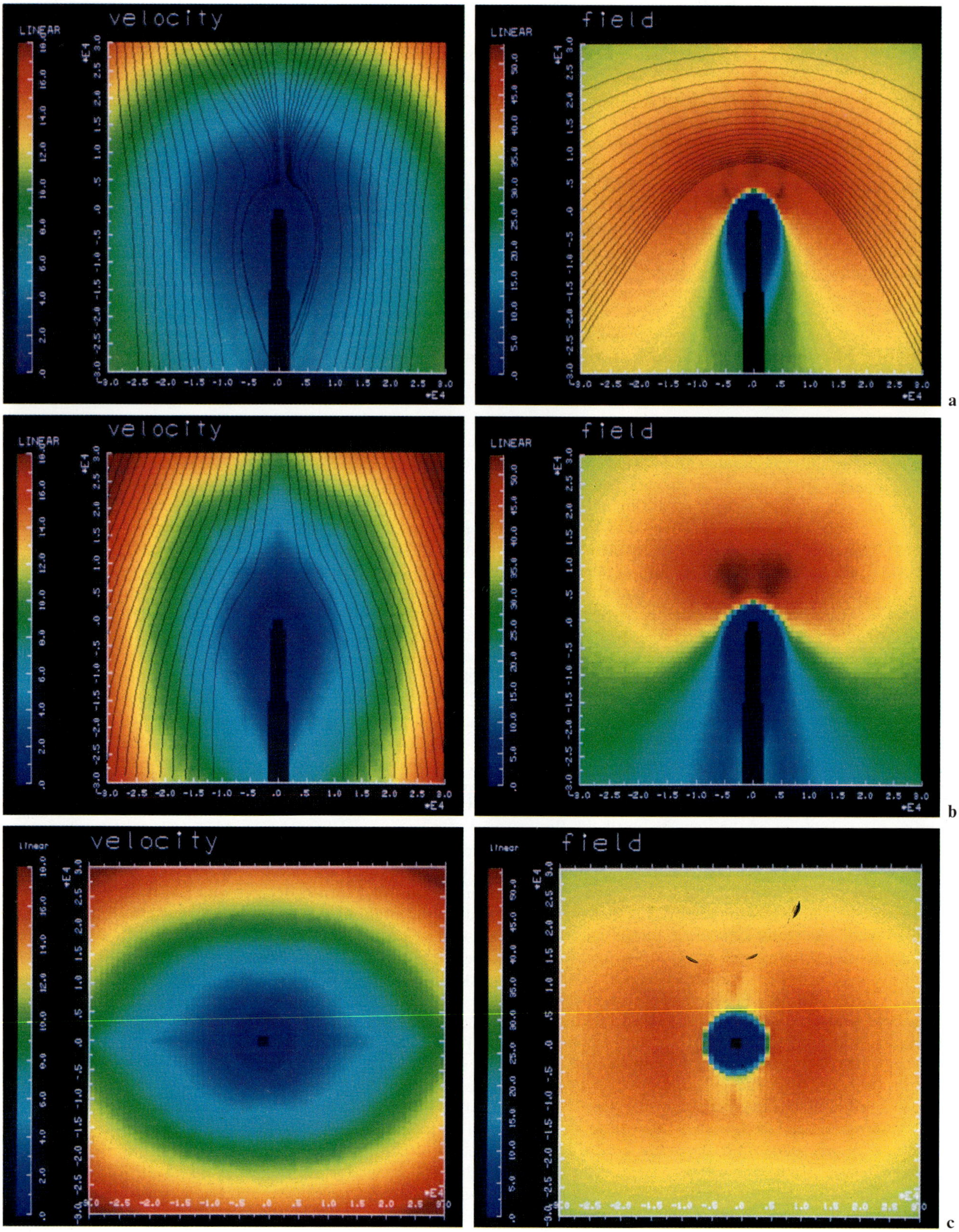

Fig. 4a–c. The contact surface: Velocity and magnetic field in three perpendicular planes: **a** the meridional plane parallel to the IMF, **b** the meridional plane perpendicular to **a**, **c** plane perpendicular to the unperturbed solar wind velocity vector. The magnetic field lines follow each other in time intervals of 300 s

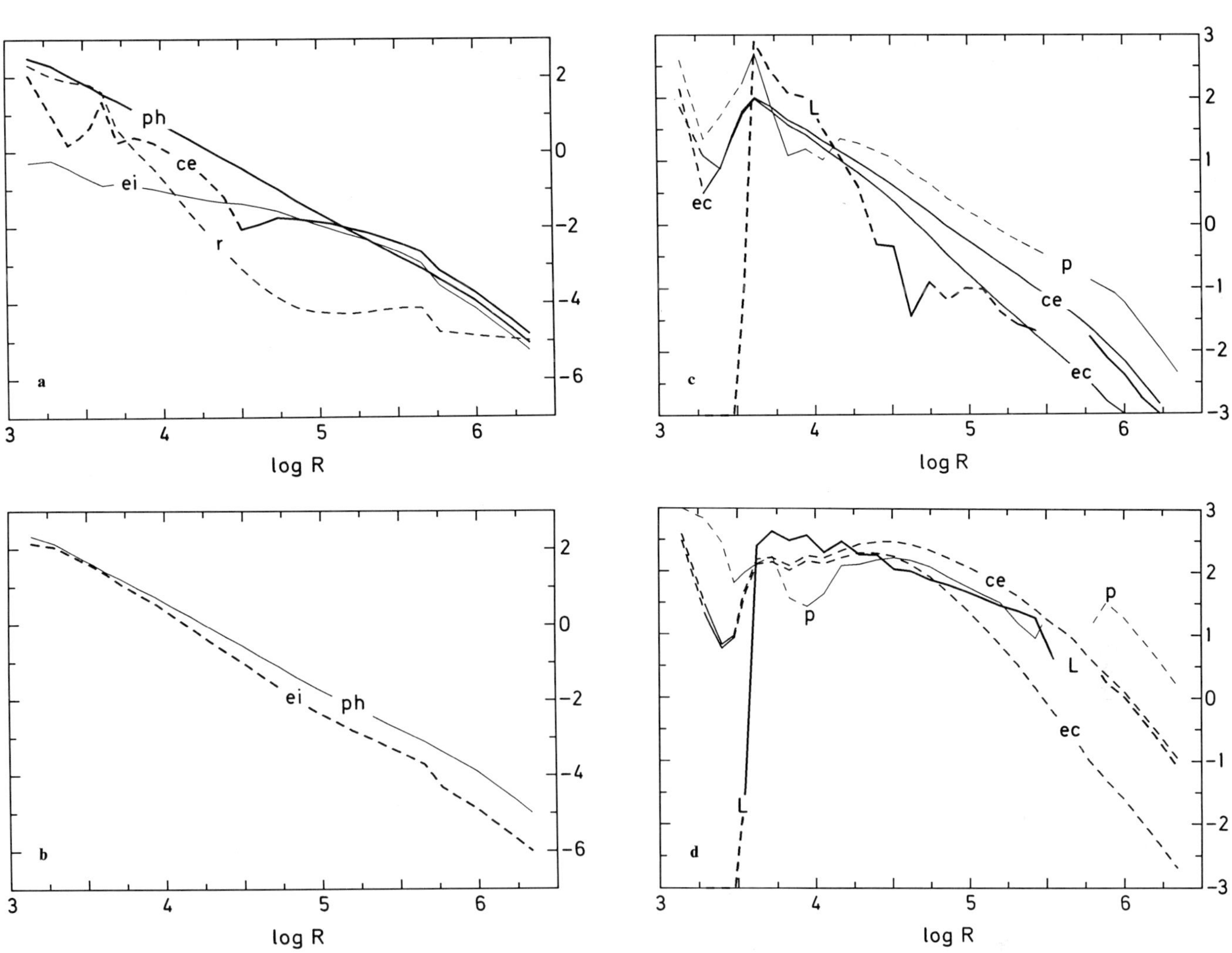

Fig. 5a–d. Major contributions to source terms along the axis for the three-dimensional model: **a** Sources and sinks of ion mass, **b** Heating and cooling of electrons, **c** Forces on the plasma, **d** Total energy of plasma. The symbols in the figure have the following meaning: photoprocesses (ph), electron impact processes (ei), charge exchange (ce), recombination (r), elastic collisions (ec), Lorentz forces (L), pressure forces (p). Negative contributions are indicated by dashed lines, terms due to lateral outflow or inflow of momentum or energy are not shown

distribution in our three-dimensional part of the model is nearly axisymmetric except in the far tail region (Wegmann, 1987). In order to use the best possible axisymmetric flow field, we average the threedimensional flow field of the model in Sect. 4 over the azimuthal angle. The chemistry in this flow field is calculated in the same manner as outlined in Sect. 3.

For comparison with ground-based observations we evaluate model column densities for the ions CO^+, H_2O^+, CO_2^+, and OH^+ (Fig. 7). Figure 8 shows the density profiles for 15 ions along the inbound Giotto trajectory. The sample includes those ions which are the most abundant either in the inner or outer coma. Figure 9 shows density profiles of 5 ions from this sample on a linear length scale for direct comparison with data reported in Balsiger et al. (1986).

In Fig. 10 the model ion densities are compared with uncorrected peak count rates from the high-intensity spectrometer (HIS) of the Giotto ion mass spectrometer at $R = 1500$ km. Balsiger et al. (1986) note that the mass channels, as presented from the HIS sensor, are not necessarily identical with the mass to charge ratio, m/q. Also, no clear interpretation is given for the count rates in the half-channels below $m/q = 26$. In our comparison we assume that the HIS mass channels are identical with m/q and that the peaks of adjacent half-channels contain the meaningful count rates. These assumptions are strengthened by reference to the NMS operating in the ion mode for $m/q = 17$–21 (Eberhardt et al., 1986) and $m/q = 44$–47 (Krankowsky et al., 1986b). However, we note, that a shift by $\Delta m/q = 1$ for the high mass channels $m/q > 40$ gives a better agreement with the abundance pattern. This may be fortuitous since the abundances in these high mass channels are still quite uncertain in the model.

Since related, but different quantities are compared, the measured peak count rates are normalized to the model number density of H_3O^+ ($m/q = 19$). Table 2 contains prominent ions in the m/q bins that are included in our model. There is very general agreement in the pattern of three high abundance mass groupings from $m/q = 6$ to 21, 26 to 35, and about 40 to 50. In the first of these groups agreement is very good for $m/q = 12$, 15, 17, 18, 19, and 21. Channel 19 contains only H_3O^+ depleted of $H_3{}^{18}O^+$ which is the only contributor in channel 21. Deuterium H_2DO^+ depletion has a minor effect on channel 19. According to our model, channel 18 is composed of 55% NH_4^+ and 45% H_2O^+ depleted of $H_2{}^{18}O^+$. At this distance, the NH_4^+ is produced mainly by the reaction

$$NH_3 + H_3O^+ \rightarrow NH_4^+ + H_2O, \quad k = 2.2\,10^{-9}\,\mathrm{cm^3\,s^{-1}}, \qquad (1)$$

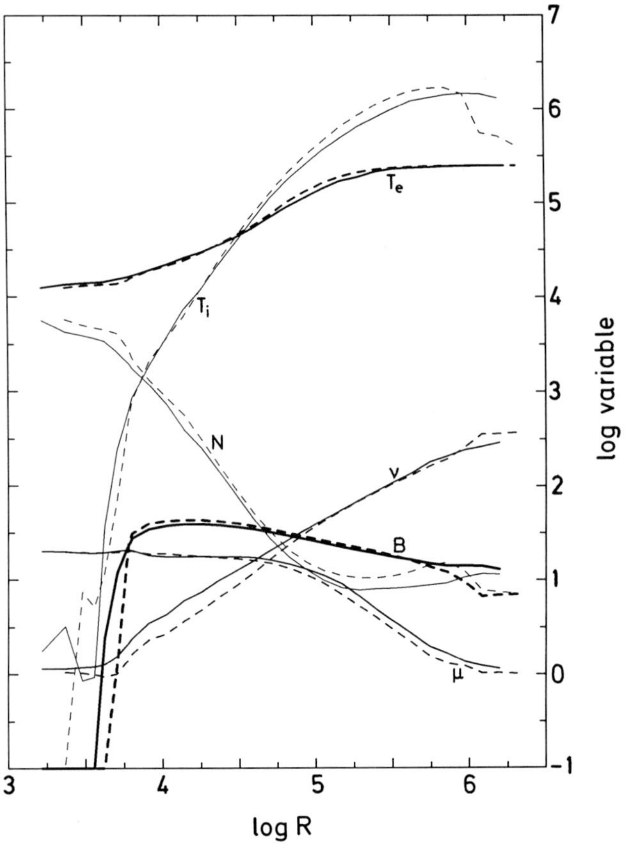

Fig. 6. Flow variables along the Giotto trajectory for the three-dimensional model: inbound (solid), outbound (dashed). R = distance from closest approach [km], all other variables as in Fig. 2

from the assumed 2% of NH_3 in the frozen mixture in the nucleus. Using less NH_3 reduces the abundance in channel 18 relative to 19. On the other hand the amount of $H_2{}^{18}O^+$ (about 0.2% of the H_2O^+ in channel 18) may be too low when compared to the observed count rate. This would suggest that there is not enough H_2O^+ in channel 18 and too much NH_4^+. However, the count rate in channel 20 is on the shoulder of the very high count rate in channel 19 which may contaminate channel 20 and prevent the assignment of an approximate value for the amount of H_2O^+ in channel 18. The inclusion of deuterated species will improve the agreement with the HIS mass spectra, particularly channel 20, when an accurate D/H ratio is determined for Halley. Only more detailed data analyses can resolve the possible discrepancy.

Channel 17 contains about 80% $NH_3{}^+$ and 20% OH^+, while channel 16 has about 45% $CH_4{}^+$, 40% O^+, and 15% NH_2^+. The group of channels 16, 17, 18, and 19 has been called the water group channels. According to this analysis this seems hardly justifiable at this distance, but at larger distances beyond about 5000 km, H_2O^+ and O^+ dominate in channels 16 and 18. However, it is not until about 2 10^4 km that OH^+ becomes more abundant than NH_3^+ in channel 17.

Channel 16 contains $CH_4{}^+$, O^+, and NH_2^+. Because of the steep abundance change in the half-channels, it is difficult to compare the model with the measurements. A shift by a half-channel or a spill-over from an adjacent half-channel can alter the measured abundance drastically. More detailed analyses of the data are needed. Channel 15, which contains CH_3^+ and NH^+ is in reasonable agreement with measurements. Channel 14 contains CH_2^+ and N^+, and channel 13 contains CH^+. Model abundances for both of these channels are low. It is possible that there is another source for CH_2^+ and CH^+ that has not been considered in the model. Kissel et al. (1986) have reported that most of the dust particles are rich on H, C, N, and O, and Korth et al. (1986) and

Table 2. Dominant ions in mass/charge bins

m/q	Ion species	m/q	Ion species
12	C^+	34	$^{34}S^+$, H_2S^+
13	CH^+	35	H_3S^+
14	$CH_2{}^+$, N^+	36	$C_3{}^+$
15	$CH_3{}^+$, NH^+	37	C_3H^+, H_3O^+. H_2O
16	$CH_4{}^+$, O^+, NH_2^+	38	C_2N^+, $C_3H_2^+$
17	$NH_3{}^+$, OH^+, CH_5^+	39	$C_3H_3^+$
18	$NH_4{}^+$, H_2O^+	40	$C_3H_4^+$, CH_2CN^+
19	H_3O^+	41	$C_3H_5^+$
20	$H_2{}^{18}O^+$	42	
21	$H_3{}^{18}O^+$	43	CH_3CO^+
22		44	CO_2^+, CS^+, $C_3H_8^+$
23		45	HCS^+, $HCO_2{}^+$
24	$C_2{}^+$	46	$H_2CO_2^+$, NS^+, H_2CS^+
25	C_2H^+	47	H_3CS^+, HNS^+
26	$C_2H_2^+$	48	SO^+
27	$C_2H_3^+$, HCN^+	49	HSO^+
28	H_2CN^+, CO^+, N_2^+, $C_2H_4^+$	50	$C_4H_2^+$
29	HCO^+, $C_2H_5^+$, N_2H^+	51	$C_4H_3^+$
30	H_2CO^+, CH_4N^+, NO^+, $C_2H_6^+$	52	$C_3H_2N^+$
31	CH_2OH^+, HNO^+	53	$C_4H_5{}^+$
32	S^+, O_2^+	54	
33	HS^+, $^{33}S^+$, O_2H^+	55	$C_3H_3O^+$, $C_4H_7^+$

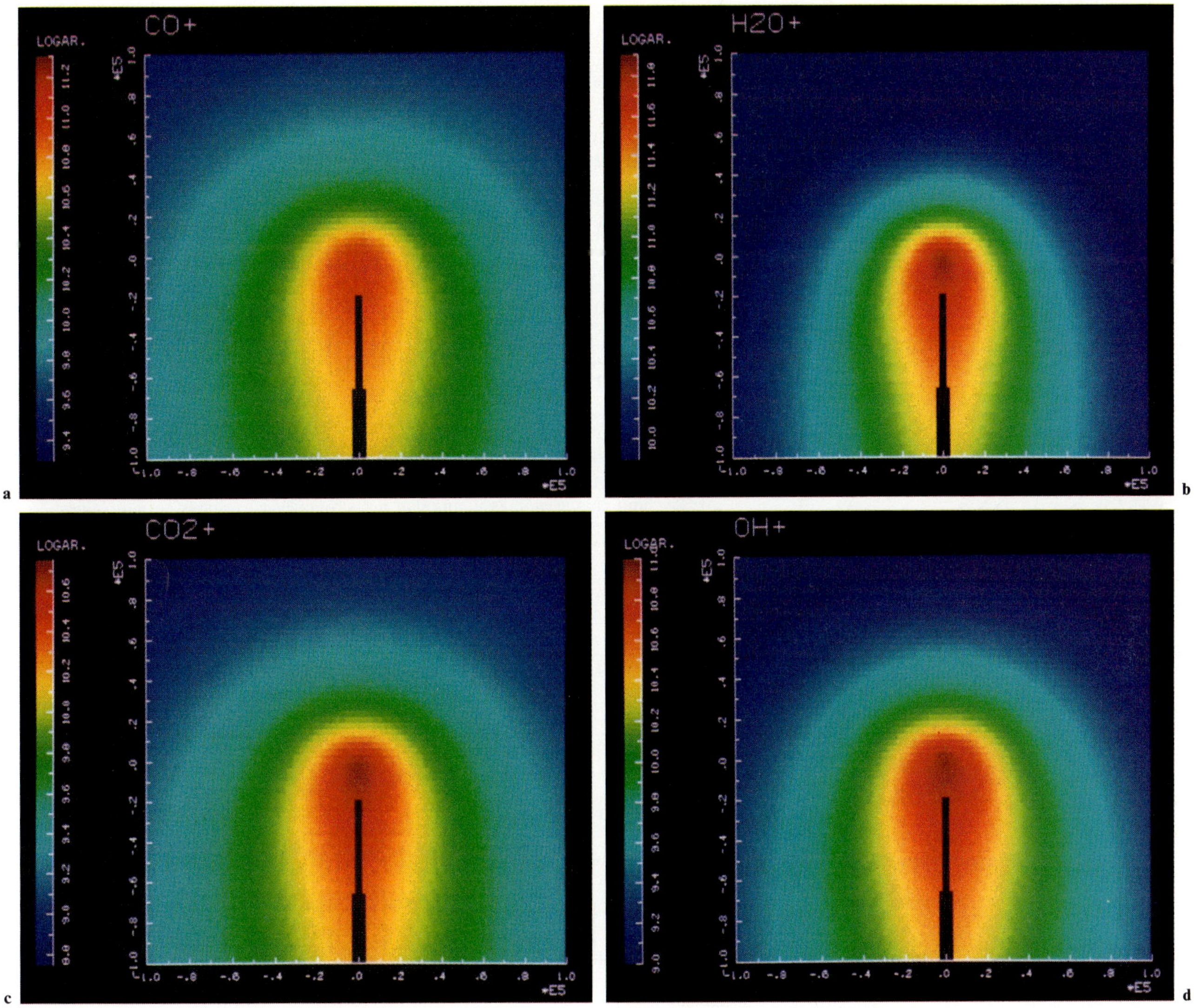

Fig. 7. The visible ion coma from the improved axisymmetric calculation. The contour maps extend to 10^5 km on either side of the nucleus. The four contour maps show the column densities [cm^{-2}] for the ions CO^+, H_2O^+, CO_2^+, and OH^+ on a logarithmic scale which covers two orders of magnitude in each case

Mitchell et al. (1986) have reported heavy ion mass spectra suggesting hydrocarbons.

It is likely that the suggested hydrocarboons, possibly polyoxymethylene, are on dust grains that disintegrate and are a source for additional CH_2 and CH and their ions. It is also possible that the lifetime for dissociation of CH in our model is too small, but that would not explain the model underabundance of CH_2^+. Other possibilities are incorrect rate coefficients and greater uncertainty because of the low count rates in the HIS. The model carbon ion abundance in channel 12 agrees well with the measurements.

In the second dominant group of ions we generally see more discrepancies in the model and the HIS data. The model yields an abundance too low for C_2^+ (channel 24), but in very good agreement for channels 25 and 26 which contain C_2H^+ and $C_2H_2^+$, respectively.

The model abundance for channel 27 ($C_2H_3{}^+$ and HCN^+) is too low by about a factor of 10. The total abundance in channel 28, which contains H_2CN^+, CO^+, N_2^+, and $C_2H_4^+$, agrees well. The model abundances in channel 29 (HCO^+, $C_2H_5^+$, and N_2H^+) and 30 (H_2CO^+, CH_4N^+, NO^+, and $C_2H_6^+$) are low. The abundences of channels 31 (CH_2OH^+ and HNO^+) and 32 (S^+ and O_2^+) agree to within a factor of two, the model predicting too much for channel 31 and too little for channel 32.

The predictions for channels 33 through 43 are much too low, except channel 39 which is low by only a factor of three. In the third group, channels 44, 46, 48, and 49 are in good agreement. The model abundances for all other channels in the third group are too low. If all the m/q values observed in the spectrum of group three are higher by one than the mass channel numbers, then the abundance pattern of the model would correlate better with the observations, but would still be too low. However, we note that the NMS data in the ion mode for $m/q = 44$, 45, and 47 (Krankowsky et al., 1986b) corroborate the mass channel calibration of the HIS. In particular, the ions at mass 45 are the most abundant species in this mass range and may be possibly $CH_3OCH_2^+$, resulting from photodestruction of polyoxymethylene (POM) (Huebner et al., 1987b). As we have not incorporated the POM chemistry into the coma chemistry network, this identification is tentative.

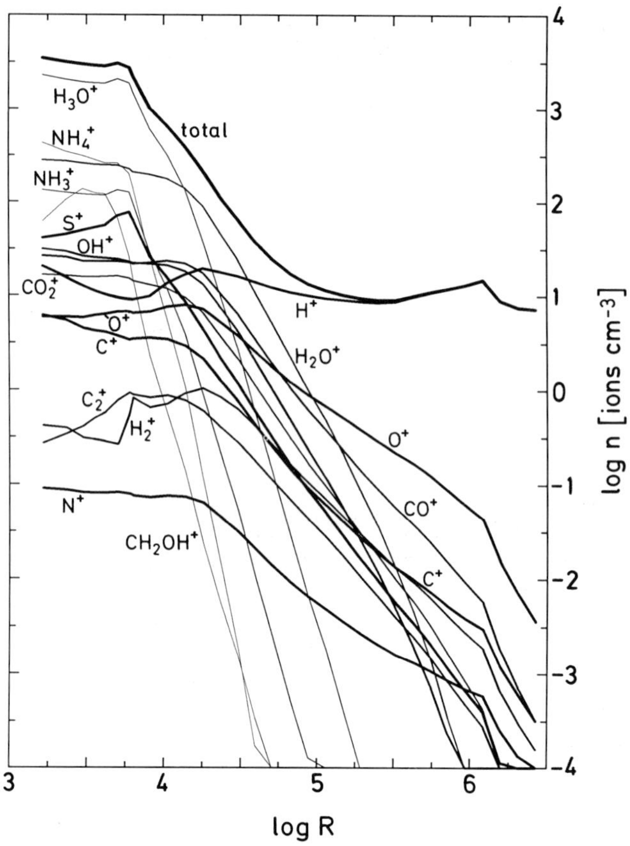

Fig. 8. Ion densities [cm^{-3}] along the inbound Giotto trajectory

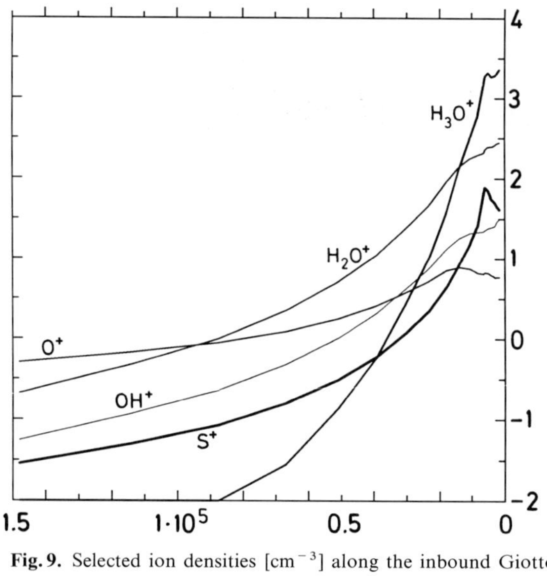

Fig. 9. Selected ion densities [cm^{-3}] along the inbound Giotto trajectory for comparison to Fig. 3 in Balsiger et al. (1986)

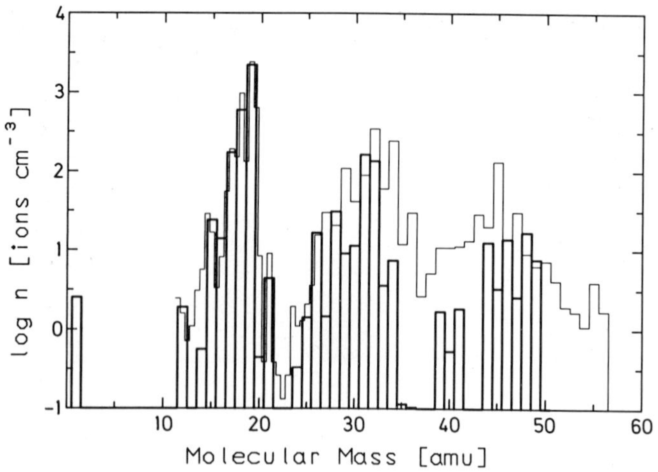

Fig. 10. Comparison of model ion densities (solid line) with HIS peak count rates (thin line) at $R = 1500$ km (within contact surface) as a function of mass per unit charge m/q, assuming that mass channels correspond to m/q values

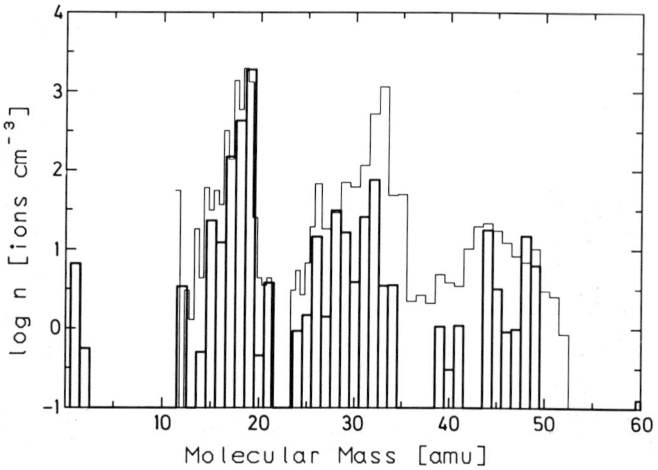

Fig. 11. Comparison of model ion densities (solid line) with HIS peak count rates (thin line) at $R = 6000$ km. At this distance (outside the contact surface) the model calculations include the solar wind interaction

The ion abundances at 6000 km from the nucleus are compared to the HIS data in Fig. 11. Again we see qualitative agreement in the three groups of ions but generally poorer than at 1500 km (Fig. 10). In the first group, the model abundances are low by about a factor of two in channels 16 through 18 and low by about an order of magnitude in channels 12 through 14. At this distance, H_2O^+ is slightly more abundant (53%) than NH_4^+ in channel 18. Channel 17 consists of 85% $NH_3{}^+$ and 15% OH^+. In channel 16 O^+ (55%) is the dominant ion. If the first half-channel at 12 is real, then the amount of C^+ in the model is very underabundant. Channel 21, which contains only $H_3{}^{18}O^+$, is in excellent agreement, assuming the solar isotopic ratio of ^{18}O.

In the second group the model abundances are low by about a factor of 2 in channels 24 (C_2^+), 25 (C_2H^+), 26 ($C_2H_2^+$), 29 (HCO^+ and $C_2H_5^+$), 31 (CH_2OH^+), and 32 (S^+, O_2^+), but agree very well in channel 28 (H_2CN^+, CO^+, $N_2{}^+$, and $C_2H_4{}^+$). Again channel 27 (HCN^+ and $C_2H_3^+$) is underabundant by a factor of ten, the same applies to channel 30. All other channels are low by an order of magnitude or more.

The model abundances in channels 44 (CO_2^+ and CS^+), and 49 (HSO^+) in the third group agree reasonably with the HIS data. The channels 45 (HCS^+), 46 ($H_2CO_2^+$, NS^+, H_2CS^+), and 47 (HNS^+, H_3CS^+) are low by a factor of three or more. The same applies to the masses 39 ($C_3H_3^+$), and 41 ($C_3H_5^+$), while SO^+ with mass 48 is overabundant. Other channels are low by an order of magnitude or more. Clearly, more effort is needed to resolve these discrepancies among the heavier ions.

Some of the more complex ions are essentially confined to the region inside the contact surface (see Fig. 8). Therefore, their abundances depend very sensitive on the flow field. The same holds true for some of the mass channels, e.g. for channel 46 which is mainly filled by $H_2CO_2^+$.

6. Conclusions

By alternating iterations between axisymmetric models with detailed chemistry and three-dimensional models with rigorous account of the Lorentz forces we derived a consistent description of the plasma flow and chemical evolution in the ionized cometary coma. We have constructed such models appropriate to comet Halley in which the magnetized plasma flow, the bow shock, the magnetic cavity of size 5000 km, and the model ion abundances at 1500 km and 6000 km from the nucleus are in qualitative agreement with the Giotto data. The improved axisymmetric model correctly shows the three groupings of ions at each distance. In particular, the model abundances for the light ions, up to 21 amu, are in very good agreement with the 1500 km observations. The comparison generally becomes worse as one considers higher molecular masses and greater distances from the nucleus. But even for $H_2CO_2^+$ (46 amu) the iteration gives a dramatic improvement. We find that small amounts of CS_2 as a parent molecule have negligible effect on the surface sublimation rate or dynamics, but play an important role in the chemistry and production of some ions of higher molecular mass. Isotopes of C, N, O, and S have been considered and are good candidates for species in mass channels 20 ($H_2{}^{18}O^+$), 21 ($H_3{}^{18}O^+$), 33 ($^{33}S^+$), and 34 ($^{34}S^+$). Other heavier species still need to be considered for populating mass channels that are in poor agreement. Additional analysis of the HIS data is needed to resolve issues such as, the physical meaning of the half-channels below 26 and possible contamination by adjacent half-channels, an accurate calibration of the mass channels to m/q and peak count rates to number densities, and the noise (background) level of the detector.

The inclusion of additional physics, e.g., radiation pressure on the neutrals, nonthermal distribution of pick-up ions, multicomponent electron temperature, POM chemistry, repeated charge exchange, and time-dependent three-dimensional magnetic field calculations for appropriately fluctuating IMF orientations, will improve this model even further.

As we continue to revise and adapt such models to additional constraints from the many observations of comet Halley we can hope to determine a set of model parameters, such as initial chemical composition of the nuclear surface layer, that uniquely describe comet Halley.

Acknowledgement. In part, this research was supported by funds from the NASA Planetary Atmospheres Program.

References

Balsiger, H., Altwegg, K., Buehler, F., Geiss, J., Ghielmetti, A.G., Goldstein, B.E., Goldstein, R., Huntress, W.T., Ip, W.-H., Lazarus, A.J., Meier, A., Neugebauer, M., Rettenmund, U., Rosenbauer, H., Schwenn, R., Sharp, R.D., Shelley, E.G., Ungstrup, E., Young, D.T.: 1986, *Nature* **321**, 330

Bockelee-Morvan, D., Crovisier, J., Despois, D., Forveille, T., Gerard, E., Schraml, J., Thum, C.: 1986, Abstracts 20th ESLAB Symp., p. 73

Combes, M., Moroz, V., Crifo, J.F., Bibring, J.P., Coron, N., Crovisier, J., Encrenaz, T., Sanko, N., Grigoriev, A., Bockelée-Morvan, D., Gispert, R., Emerich, C., Lamarre, J.M., Rocard, F., Krasnopolsky, V., Owen, T.: 1986, Abstracts 20th ESLAB Symp., p. 70

Delsemme, A.H.: 1982, in *Comets*, ed. L.L. Wilkening, Univ. Arizona Press, Tucson, AZ, p. 85

Eberhardt, P., Dolder, U., Schult, W., Krankowsky, D., Lämmerzahl, P., Hoffman, J.H., Hodges, R.R., Berthelier, J.J., Illiano, J.M.: 1986, 20th ESLAB Symp., ESA SP-250, Vol. **I**, p. 529

Huebner, W.F.: 1985, in *The Photochemistry of Atmospheres: Earth, the Other Planets, and Comets*, ed., J.S. Levine, Academic Press, New York, p. 437

Huebner, W.F., Keady, J.J., Boice, D.C., Schmidt, H.U., Wegmann, R.: 1987a, in *Astrochemistry*, eds. M.S. Vardya, S.P. Tarafdar, *IAU Coll.* **120**, Reidel, Dordrecht, p. 431

Huebner, W.F., Boice, D.C., Sharp, C.M.: 1987b, Symp. *Diversity and Similarity of Comets*, p. 65

Johnstone, A., Coates, A., Kellock, S., Wilken, B., Jockers, K., Rosenbauer, H., Studemann, W., Weiss, W., Formisano, V., Amata, E., Cerulli-Irelli, R., Dobrovolny, M., Terenzi, R., Egidi, A., Borg, H., Hultquist, B., Winningham, J., Gurgiolo, C., Bryant, D., Edwards, T., Feldman, W., Thomsen, M., Wallis, M.K., Biermann, L., Schmidt, H.U., Luest, R., Haerendel, G., Paschmann, G.: 1986, *Nature* **321**, 344

Keller, H.U., Arpigny, C., Barbieri, C., Bonnet, R.M., Cazes, S., Coradini, M., Cosmovici, C.B., Delamere, W.A., Huebner, W.F., Hughes, D.W., Jamar, C., Malaise, D., Reitsema, H.J., Schmidt, H.U., Schmidt, W.K.H., Seige, P., Whipple, F. L., Wilhelm, K.: 1986, *Nature* **321**, 320

Kissel, J., Brownlee, D.E., Buechler, K., Clark, B.C., Fechtig, H., Gruen, E., Hornung, K., Igenbergs, E.B., Jessberger, E.K., Krueger, F.R., Kuczera, H., McDonnell, J.A.M., Morfill, G.M., Rahe, J., Schwehm, G.H., Sekanina, Z., Utterback, N.G., Völk, H.J., Zook, H.A.: 1986, *Nature* **321**, 336

Korth, A., Richter, A., Loidl, A., Anderson, K.A., Carlson, C.W., Curtis, D.W., Lin, R.P., Rème, H., Sauvaud, J.A., d'Uston, C., Cotin, F., Cros, A., Mendis, D.A.: 1986, *Nature* **321**, 335

Krankowsky, D., Lämmerzahl, P., Herrwerth, I., Woweries, J., Eberhardt, P., Dolder, U., Herrmann, U., Schulte, W., Berthelier, J.J., Illiano, J.M., Hodges, R.R. Hoffman, J.H.: 1986a, *Nature* **321**, 326

Krankowsky, D., Eberhardt, P., Bethelier, J.J., Dolder, U., Hodges, R.R., Hoffman, J.H., Illiano, J.M., Lämmerzahl, P., Schulte, W., Stubbeman, U., Woweries, J.: 1986b, 20th ESLAB Symp., ESA SP-250, Vol. **I**, p. 381

Lees, L.: 1964, *Amer. Inst. Aero. Astronaut. J.* **2**, 1576

Mitchell, D.L., Lin, R.P., Anderson, K.A., Cartlson, C.W., Curtis, D.W., Korth, A., Richter, A.K., Rème, H., Sauvaud, J.A., d'Ustın, C., Mendis, D.A., 1986: 20th ESLAB Symp. SP-250, Vol. **I**, p. 203

Neubauer, F. M., Glassmeier, K. H., Pohl, M., Raeder, J., Acuna, M. H., Burlaga, L. F., Ness, N. F., Musmann, G., Mariani, F., Wallis, M. K., Ungstrup, E., Schmidt, H. U.: 1986, *Nature* **321**, 352

Rème, H., Sauvaud, J. A., d'Uston, C., Cotin, F., Cros, A., Anderson, K. A., Carlson, C. W., Curtis, D. W., Lin, R. P., Mendis, D. A., Korth, A., Richter, A. K.: 1986, *Nature* **321**, 349

Schmidt, H. U., Wegmann, R.: 1982, in *Comets*, ed. L. L. Wilkening, Univ. Arizona Press, Tucson, p. 552

Synder, L. E., Palmer, P., DePater, I.: 1987, *Astron. J.* (submitted)

Wegmann, R.: 1987, Proc. Conf. *Physical Processes in Comets, Stars and Active Galaxies*, Springer, p. 34

Astron. Astrophys. 187, 351–356 (1987)

ASTRONOMY AND ASTROPHYSICS

Atomic hydrogen production rates for comet P/Halley from observations with Dynamics Explorer 1

J. D. Craven and L. A. Frank

Department of Physics and Astronomy, The University of Iowa, Iowa City, IA 52242-1479, USA

Received February 17, accepted July 15, 1987

Summary. The distribution of atomic hydrogen surrounding comet Halley is observed in resonantly scattered solar Lyman-α radiation with the imaging photometer for vacuum-ultraviolet wavelengths on the Earth-orbiting spacecraft Dynamics Explorer 1. Measurements are made of the total Lyman-α flux at Earth due to the cometary neutral hydrogen distribution and the hydrogen production rate determined as a function of heliocentric distance, r. Corrections are made for the finite field-of-view of the photometer and for the $\sim$700-R background at the spacecraft arising from the presence of geocoronal, interplanetary and galactic hydrogen. The distribution of hydrogen surrounding the comet nucleus at time t is dominated by atoms released from the comet in parent molecules at about time $t-\tau$, where the hydrogen lifetime is τ. Corrections are made for this time difference. For distances 1.5 to 0.68 AU before perihelion passage the hydrogen production rate varies as $Q_0 r^{-n}$, where $n=2.30$ and $Q_0 = 9.1\ 10^{29}$ atoms/s. Using the same functional dependence in the post-perihelion period at distances 0.63 to 1.2 AU, preliminary values based on a limited data set are $n=1.62$ and $Q_0 = 1.1\ 10^{30}$ atoms/s. These production rates are consistent with *in situ* measurements from the Giotto and Vega spacecraft, and with remote observations using two sounding rockets and with the Pioneer-Venus and IUE spacecraft.

Key words: comets – neutral hydrogen – Lyman-alpha radiation

1. Introduction

In this report we extend the initial work of Craven et al. (1986) using newly analyzed observations to determine the hydrogen production rate for comet Halley at heliocentric distances, r, less than $\sim$1.5 astronomical units (AU). The observations with the Earth-orbiting spacecraft Dynamics Explorer 1 comprise fully two-dimensional imaging of the comet at the 121.6-nm wavelength of resonantly scattered solar Lyman-α radiation. In the absence of absorption by hydrogen near the cometary nucleus and of background emissions in the imaging field-of-view, the method by which this production rate is determined is independent of the actual hydrogen distribution, except for small corrections due to the finite field-of-view of the photometer. Model-dependent corrections are applied in order to determine the hydrogen production rate in the presence of the background emissions. These background emissions are due to scattering of solar Lyman-α radiation into the image field-of-view by geocoronal, interplanetary and interstellar hydrogen.

The spacecraft Dynamics Explorer 1 (DE 1) was launched on 3 August 1981 into an eccentric, polar orbit with initial perigee and apogee altitudes of $\sim$570 km and 3.65 Earth radii, respectively. The orbital period is $\sim$6.83 h and the line of apsides advances at $\sim$0.33°/day. The spacecraft spin axis is normally oriented perpendicular to the orbit plane, initially directed toward right ascencion $\alpha = 250°$ and declination $\delta = 0°$, and is adjusted by means of magnetic torquing coils controlled by ground command. Observations of comet Halley in the post-perihelion period are available due to significant reorientations of the spacecraft spin vector. Additional information concerning the spacecraft is provided by Hoffman et al. (1981). For the period of observations discussed here, apogee of DE 1 is located at high southern latitudes which provide a nearly ideal viewing geometry for imaging comet Halley. A detailed description of this instrumentation is given by Frank et al. (1981), and many of the details relevant to imaging of comet Halley are discussed by Craven et al. (1986).

2. Analysis of observations

For an equilibrium distribution of steady-state hydrogen production at a rate Q_H and loss after a mean lifetime τ,

$$Q_H = \Phi / g\tau \ \textit{atoms/s}. \tag{1}$$

The quantity g is the scattering rate for Lyman-α radiation in units of photons (atom s)$^{-1}$ and Φ the total rate at which the incident solar Lyman-α radiation is scattered by the cometary hydrogen distribution (e.g., Krishna Swamy, 1986). Isotropic single scattering with no absorption in an optically thin medium is assumed. The quantity Φ is obtained empirically by measuring the Lyman-α flux, L photons (cm^2 s)$^{-1}$, scattered by the cometary hydrogen at a geocentric distance Δ and observed at Earth: $\Phi = 4\pi\Delta^2 L$. An example of the projected brightness distribution observed with DE 1 on 29 January 1986 is shown in Fig. 1.

A *lower limit* for the Lyman-α flux L is determined for the DE-1 observations by a summation over the field of image pixels,

$$L = \frac{10^9\, \Sigma\Omega}{4\pi S} \text{ photons } (\text{cm}^2\,\text{s})^{-1}, \tag{2}$$

Send offprint requests to: J. D. Craven

Fig. 1. False-color image of Comet Halley at the 121.6-nm wavelength of resonantly scattered solar Lyman-α radiation. Brightness at the center of the coma is ~ 17 kilorayleighs (kR), and the average brightness at the outer edge is 1.3 kR. This threshold level is established in image processing, and does not identify the outer edge of the hydrogen distribution. The diameter of the hydrogen coma for a brightness greater than 1.3 kR is $\sim 9\ 10^6$ km. From Earth, the angular separation between the center of the hydrogen coma and the star 15 Aquarii is $\sim 3.7°$. Approximate diameter of the Sun is denoted by the two parallel lines below the comet. Field-of-view of the imaging photometer is ~ 0.6 of the solar diameter. Heliocentric distance to the comet is 0.64 AU. Geocentric distance is 1.55 AU

where

$$\Sigma = \frac{1}{I}\left[\sum_{i=1}^{n} C_i - \frac{n}{m}\sum_{j=1}^{m} c_j\right] \text{counts.} \tag{3}$$

The quantity Σ is the total number of photometer counts in an image, such as that shown in Fig. 1, due to the Lyman-α flux scattered towards Earth by the cometary hydrogen. The summation is over a composite image (comprising I superposed images) in which there are n pixels of response C_i counts/pixel centered on the comet, and m background pixels of response c_j counts/pixel from parts of the image adjacent to the n-pixel array. When the background contribution is small, L is a good measure of the cometary Lyman-α flux at Earth. However, as the cometary Lyman-α flux decreases with increasing heliocentric distance of the comet, the relative contribution of the background to the total photometer response increases and L is underestimated. Responses due to VUV-bright stars are removed prior to calculating Σ (see also the discussion by Craven et al., 1986). Photometer sensitivity averaged over the pixel array is S counts/kR, or $10^9/4\pi S$ photons (cm^2 s sr count)$^{-1}$. Uncertainty in instrument sensitivity for the January-to-March 1986 period is estimated at $<50\%$ for this analysis (R. L. Rairden, private communication). Effective solid angle of each pixel is $\Omega = 2.0\ 10^{-5}$ sr. For Δ in astronomical units, a *lower limit* Q_s for the production rate Q_H is inferred from the summation over the image field by

$$Q_s = 2.25\ 10^{35}\frac{\Sigma\Omega\Delta^2}{Sg\tau}\ \text{atoms/s.} \tag{4}$$

The line-center Lyman-α flux at Earth is estimated to be $F = 1.8\ 10^{12}$ photons (cm^2 s nm)$^{-1}$ as determined from measurements of the total solar Lyman-α flux with the Solar Mesospheric Explorer ($\sim 2.2\ 10^{11}$ photons (cm^2 s)$^{-1}$). The correction to line-center value is given by Vidal-Madjar and Phissamay (1980). However, this correction appears to be uncertain by as much as a factor of two for periods of minimum solar activity. The value of g for this line-center flux is 10^{-3} photons (atom s)$^{-1}$ at a heliocentric distance $r = 1$ AU, and varies as r^{-2}. Additional uncertainties arise as the Comet-Sun-Earth angle can be quite large and the SME spacecraft only determines the total Lyman-α flux from the Earth-facing sector of the Sun.

The lifetime of neutral hydrogen is limited by charge exchange with solar-wind protons, electron impact ionization and photoionization by solar EUV radiation. At $r = 1$ AU the dominant contribution to hydrogen loss is charge exchange, with typical lifetimes of $1-3\ 10^6$ s for a 42-day period in late 1985 (Combi et al., 1986). In the absence of *in situ* measurements of the solar-wind parameters for proper determination of τ, it is generally assumed that τ is proportional to r^2, and hence that the product $g\tau$ is independent of r. While this assumption might be reasonable for a steady, symmetric model of the coronal expansion (see Hundhausen (1972) for the model parameters), the large variations in lifetime demonstrated by Combi et al. for typical variations in solar-wind proton fluxes (due to shocks, high-speed streams, etc.) are ample warning of the difficulties associated with this problem. However, for the determination of Q_s we assume that $\tau = 1.0\ 10^6$ s at 1 AU and that $g\tau$ is 10^3 photons/atom.

Utilizing the method outlined here and new software for semiautomatic computer analysis of the composite images, the original 17 images discussed by Craven et al. (1986) and an additional 36 images are analyzed to determine the variations of Q_s with heliocentric radial distance, r. These data are displayed in Fig. 2 as a function of the logarithm of r for the pre-perihelion

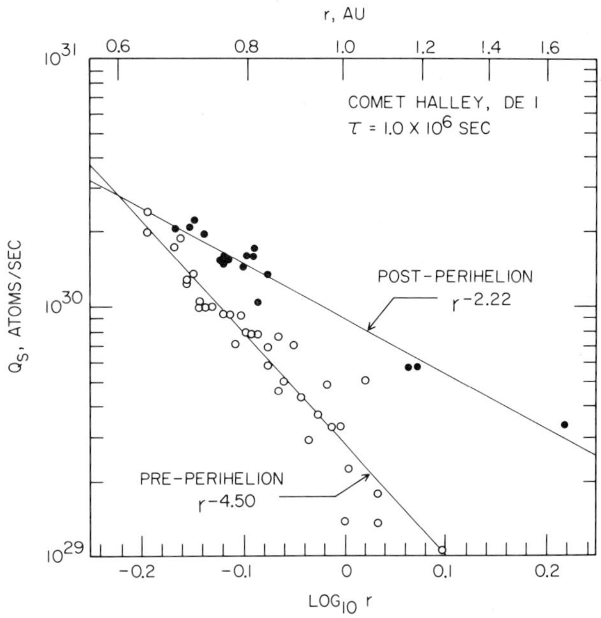

Fig. 2. Initial estimates of the production rate for neutral atomic hydrogen by comet Halley from observations with Dynamics Explorer 1. The rate Q_s is determined by measurement of the two-dimensional projected brightness distribution (e.g., Plate 1) in the presence of a $\sim$700-R background. For increasing heliocentric distance r, Q_s increasingly underestimates the true production rate. Solid lines identify least-squares fits of the form $Q_0 r^{-n}$ in the pre- and post-perihelion periods

(open circles) and post-perihelion (solid dots) periods. The 36 measurements in the pre-perihelion period were gained between 16 December 1985 ($r = 1.25$ AU) and 29 January 1986 ($r = 0.64$ AU). For the post-perihelion period 17 imaging sequences have been analyzed to date for observations from 25 February ($r = 0.68$) through 2 May 1986 ($r = 1.65$). Least-squares fits to the function $Q_s = Q_0 r^{-n}$ yield $n = 4.50$ and 2.22, respectively, for the two periods.

In the next section of this paper a method is developed to obtain the hydrogen production rate Q_H from the observed lower limit Q_s and the scale height H_s for the radial distribution of the luminosity. To do this it will be convenient to reduce the number of measurements for correction by averaging in increments of r. This has the added benefit of smoothing short-term variations on the order of several days believed to be associated with a possible $\sim$2.2-day rotation rate of the nucleus (e.g., Kaneda et al., 1986) and rapid variations in solar-wind parameters that determine charge-exchange lifetimes.

Values of Q_s are averaged in intervals $\Delta x_i = |\log_{10}(r_{i+1}/r_i)| = 0.04$ centered at $x = \log_{10} r = 0$, ± 0.04, ± 0.08, ... These averages, $\bar{Q}_s$, and unweighted least-squares fits are presented in Fig. 3a for the pre- and post-perihelion periods. Values of the index n do not change appreciably from the values derived using the individual samples (see Fig. 2).

Assuming axial symmetry about the bright central emission region at the nucleus, radial brightness distributions are computed by Craven et al. (1986) for 17 composite images in the pre-perihelion period, and it is shown that the observed portions of the radial profiles are reasonably well represented by exponential functions with scale height H_s. Typical values for observations in the period 1–29 January 1986 are $H_s \simeq 0.7$ to 1.5 million kilometers. For observations in the post-perihelion period,

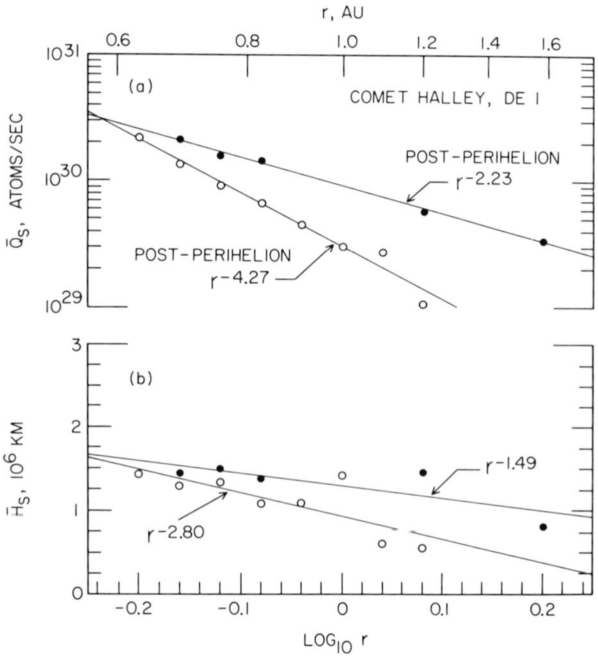

Fig. 3. **a** Averages of the measured values Q_s given in Fig. 2, in increments of $\log r = 0.04$. **b** Averages of H_s in the same intervals of $\log_{10} r$ as for **a**, above. $H_s(\varrho)$ is the least-squares scale height for an exponential decrease in observed brightness with increasing projected distance ϱ from the nucleus

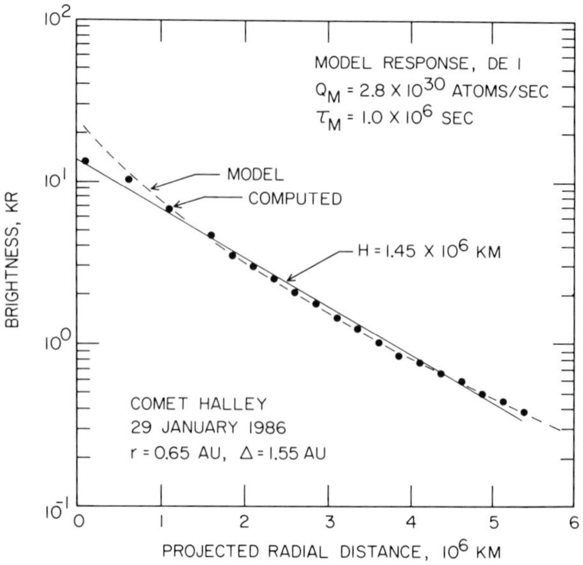

Fig. 4. Model response for the DE-1 imager when viewing comet Halley on 29 January 1986. The model brightness distribution (dashed line) is obtained with a simple radial outflow model for the hydrogen distribution using a radial outflow speed of 8 km s^{-1}. A correction is applied for small values of ϱ as the optical depth exceeds unity. The brightness distribution observed with DE 1 (solid dots) is computed by numerical integration using the model hydrogen distribution and instrument response function. The solid line identifies the least-squares fit through the computed brightness distribution; the scale height for the model fit is H

elongation of the hydrogen distribution in the antisolar direction is more readily observed due to a more favorable Comet-Sun-Earth angle. However, for the preliminary analysis of the post-perihelion observations, axial symmetry is assumed. Values of H_s for the 53 composite images analyzed here are averaged in the same intervals of $x = \log_{10} r$ as noted above and are shown in Fig. 3b for the pre- and post-perihelion periods. Simple, unweighted least-squares of the form $\bar{H}_s = H_0 - n \log_{10} r$ yield $n = 2.80$ and 1.49, respectively, for the two periods.

3. Determination of the hydrogen production rate

To determine the hydrogen production rate as a function of r, a quantitative model for the hydrogen distribution is established and computer codes developed to calculate the photometer response. A simple radial outflow model (e.g., Krishna Swamy, 1986) with a single hydrogen outflow speed of 8 km s^{-1} (e.g., McCoy et al., 1986) is used for this estimate of $Q_H(r)$. Free parameters are the hydrogen production rate Q_m and lifetime τ_m. To correct for the optically thick region at projected distances $\varrho \lesssim 10^6$ km from the nucleus, radial profiles for the model brightness distributions $B(\varrho)$ are modified at small ϱ by a simple shielding function $(1 + N(\varrho)/N_0)^{-1}$, where $N(\varrho)$ is the column density of hydrogen at a projected distance ϱ from the nucleus. A value of $N_0 = 10^{13}$ atoms/cm^2 is selected empirically so that the computed radial profile for 29 January 1986 approaches that observed with DE 1 for small values of ϱ. For a Maxwellian hydrogen distribution at 300 K the line-center cross section is $3.5\ 10^{-13}$ cm^2, so that a column of 10^{13} hydrogen atoms corresponds to ~3 optical depths. Hence, the empirical fit to the observed radial profile correctly indicates that important deviations in the luminosity profiles from those predicted assuming an optically thin medium are present for column densities in excess of several optical depths.

A model brightness profile is shown as the dashed line in Fig. 4 for the DE-1 viewing geometry of 29 January 1986 and parameters $Q_m = 2.8\ 10^{30}$ atoms/s and $\tau_m = 10^6$ s. Numerical integrations using the model brightness distribution, the known instrument response function and the methods discussed in Sect. 2 of this paper determine the computed values for the instrument responses and the brightnesses identified by solid dots in Fig. 4. A least-squares fit through these computed values yields a scale height of $1.45\ 10^6$ km and an intercept of 15 kR at the nucleus. The computed values compare favorably with the observed radial brightness profile presented in Fig. 3 of Craven et al. (1986). It is suggested that the observed and computed radial brightness profiles for DE 1 differ from those of Meier and Keller (1985) because of the finite field-of-view of the instrument and, at smaller heliocentric distances, because of the optically thick hydrogen distribution at projected radial distances $\lesssim 10^6$ km not taken into account in their model.

Determination of the best-fit hydrogen production rate is achieved by searching through $Q_m - \tau_m$ parameter space for the values $Q_H(r)$ and τ which yield each of the observed pairs of values $\bar{Q}_s$ and $\bar{H}_s$ given in Fig. 3. The results of this search are presented in Fig. 5. Hydrogen lifetimes at 1 AU derived for the assumed radial outflow speed of 8 km s^{-1} are $\sim 10^6$ s. Variations in the hydrogen production rate with heliocentric distance ($Q_H = Q_0 r^{-n}$) are described by $Q_0 = 6.31\ 10^{29}$ atoms/sec and $n = 2.71$ in the pre-perihelion period. Estimates for the post-perihelion period from

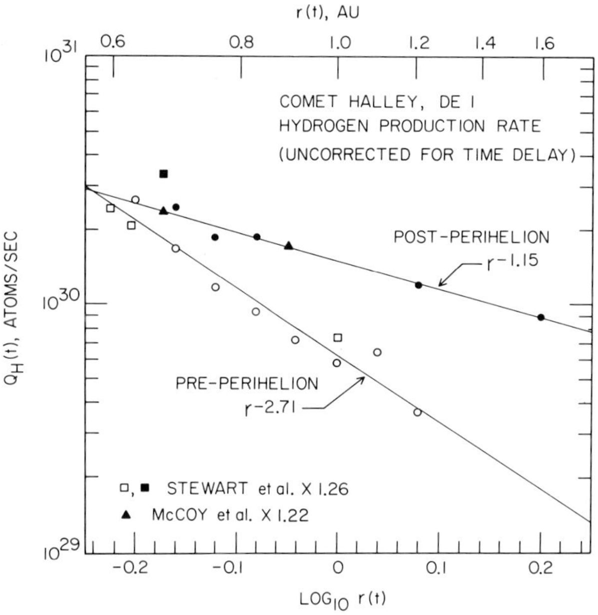

Fig. 5. Production rate for neutral atomic hydrogen by comet Halley after model-dependent corrections for background emissions are applied to data shown in Fig. 3. Values of Q_H inferred from observations of the VUV brightness distribution during two rocket flights are also indicated (by solid triangles) for comparison of absolute magnitudes. Production rates derived from Pioneer-Venus observations for three selected heliocentric distances in the pre-perihelion period (open squares) and one in the post-perihelion period (solid square) are given

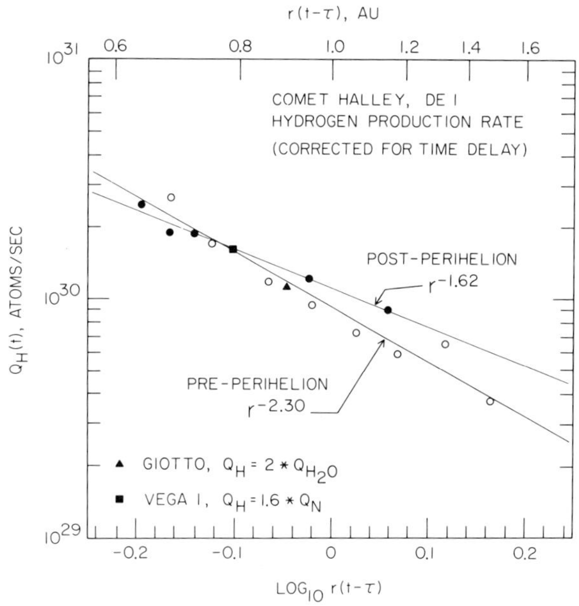

Fig. 6. Production rate for neutral atomic hydrogen by comet Halley at heliocentric distance $r(t-\tau)$ for observations at time t. The hydrogen lifetime is τ. Hydrogen production rates are estimated from in situ measurements at Giotto (solid triangle) and at Vega 1 (solid square), as discussed in the text

which fewer images are available to date for processing yield $Q_0 = 1.51\ 10^{30}$ atoms/s and $n = 1.15$.

Rocket observations of the Lyman-α brightness distribution by McCoy et al. (1986) on 24 February and 13 March 1986 with well-calibrated instrumentation provide complementary, though not simultaneous observations of Q_H. These hydrogen production rates are determined not by integrating over the brightness distributions, but by comparing the observed brightnesses with values calculated using a simple radial-outflow model and a mean outflow velocity of 8 km s^{-1}. The comparisons are made at $\varrho = 2$ and 1 million kilometers, respectively, for the two rocket flights. These production rates are increased here by a factor of 1.22 to correct for a different choice of line-center solar Lyman-α flux, and are shown as solid triangles in Fig. 5. These rocket measurements are not used when determining the least-squares fit. When it is recalled that the DE-1 instrumentation has been in orbit for more than five years with a significant accumulated radiation dosage, the agreement is quite remarkable.

Observations of Lyman-α emissions from Comet Halley are also available with the Pioneer-Venus spacecraft (Stewart, 1987). Representative hydrogen production rates for Comet Halley at three heliocentric distances from the pre-perihelion period are indicated in Fig. 5 (open squares), and a single value (solid square) is given from the post-perihelion period (data courtesy of A.I.F. Stewart). The production rates are increased here by a factor of 1.26 to correct for a different choice of line-center solar Lyman-α flux. Agreement between the DE-1 and Pioneer-Venus observations is good.

The distribution of hydrogen surrounding the comet nucleus at time t is dominated by atoms released earlier from the nucleus in parent molecules. The release time is $t-\tau$ for simple radial-outflow models for which the lifetimes of the parent molecules are short compared to τ. Similarly, the instantaneous brightness of the cometary hydrogen coma at $r(t)$ is associated with the gas production rate at $r(t-\tau)$. The hydrogen production rates corrected for the time delay are shown in Fig. 6, with $\tau = 10^6\ r^2$ s (r in AU). Variations in the hydrogen production rate with heliocentric distance ($Q_H = Q_0\, r^{-n}$) are described by $Q_0 = 9.1\ 10^{29}$ atoms/s and $n = 2.30$ in the pre-perihelion period. Preliminary estimates for the post-perihelion period yield $Q_0 = 1.1\ 10^{30}$ atoms/s and $n = 1.62$. The radial profiles are more nearly symmetric about perihelion, with significant differences observed only at $r \gtrsim 1$ AU.

In situ observations at Comet Halley with Giotto and the Vega-1 spacecraft provide further conformation of absolute values shown in Fig. 6. A measurement of the water production rate $Q_{H_2O} = 5.5\ 10^{29}$ mol/s obtained with Giotto (Krankowsky et al., 1986) is doubled ($Q_H \simeq 2\, Q_{H_2O}$) to estimate the hydrogen production rate (solid triangle). A measurement of the total gas production rate $Q_N = 6.9\ 10^{29}$ mol/s with Vega 1 (Curtis et al., 1987) is used to estimate the hydrogen production rate. First, the water production rate is estimate as $Q_{H_2O} = 0.8\, Q_N$ (Krankowsky et al., 1986). The hydrogen production rate is then again assumed equal to twice the hydrogen production rate, or $Q_H \simeq 1.6\ 10^{30}$ atoms/s. This value is identified in Fig. 6 by a solid square.

Observations of the (0,0)-band fluorescent emissions of OH with the International Ultraviolet Explorer (IUE) spacecraft provide additional, independent determinations of the water production rate (Feldman et al., 1987) which can also be compared with hydrogen production rates $Q_H(t-\tau)$ obtained here at identical heliocentric distances. (The lifetimes of H_2O and

OH are short relative to that for H). For $r \sim 1$ to 1.2 AU, observations with the two spacecraft are in complete agreement both before and after perihelion. Extrapolation of the DE-1 pre-perihelion radial profile to $r \sim 2.5$ AU continues to yield good agreement. IUE values for the production rate are systematically lower that the DE-1 production rates (by $\sim 25\%$) for $r \lesssim 1$ AU.

Acknowledgements. We thank Ms. M.R. Dvorsky, Dr. R.L. Rairden and Mr. G.T. Parmentier at the University of Iowa for their significant contributions during the processing of the data. The timely distribution of SME solar Lyman-α flux measurements by Dr. C.A. Barth at the University of Colorado's Laboratory for Atmospheric and Space Physics is gratefully acknowledged. We are also indebted to Dr. A.I.F. Stewart and his colleagues, also at the University of Colorado, for permission to publish hydrogen production rates from the Pioneer-Venus mission prior to publication of the complete body of observations. This research was supported in part by the National Aeronautics and Space Administration under grants NAG 5–483 and NGL-16-001-002 and the Office of Naval Research under grant N 00014-85-K-0404.

References

Combi, M.R., Stewart, A.I.F., Smyth, W.H.: 1986, *Geophys. Res. Letters* **13**, 385

Craven, J.D., Frank, L.A., Rairden, R.L., Dvorsky, M.R.: 1986, *Geophys. Res. Letters* **13**, 873

Curtis, C.C., et al.: 1987, *Astron. Astrophys.* (this issue)

Feldman, P.D., et al.: 1987 *Astron. Astrophys.* (this issue)

Frank, L.A., Craven, J.D., Ackerson, K.L., English, M.R., Eather, R.H., Carovillano, R.L.: 1981, *Sp. Sci. Instr.* **5**, 369

Hoffman, R.A., Hogan, G.D., Maehl, R.C.: 1981, *Sp. Sci. Instr.* **5**, 349

Hundhausen, A.J.: 1972, *Coronal Expansion and Solar Wind*, Springer, Berlin, Heidelberg, New York

Kaneda, E., Ashihara, O., Shimizu, M., Takagi, M., Hirao, K.: 1986, *Nature* **321**, 297

Krankowsky, D., Lämmerzahl, P., Herrwerth, I., Woweries, J., Eberhardt, P., Dolder, U., Herrmann, U., Schulte, W., Berthelier, J.J., Illiano, J.M., Hodges, R.R., Hoffman, J.H.: 1986, *Nature* **321**, 326

Krishna Swamy, K.S.: 1986, *Physics of Comets*, World Scientific Publ. Co., Singapore

McCoy, R.P., Opal, C.B., Carruthers, G.R.: 1986, *Nature* **324**, 439

Meier, R.R., Keller, H.U.: 1985, *Icarus* **62**, 521

Stewart, A.I.F.: 1987, *Astron. Astrophys.* (this issue)

Vidal-Madjar, A., Phissamay, B.: 1980, *Solar Phys.* **66**, 259

Astron. Astrophys. 187, 357–359 (1987)

ASTRONOMY
AND
ASTROPHYSICS

The spectrum of P/Halley's coma obtained with an objective-prism

A. Florsch [1], **J. Marcout** [1], **G. Traversa** [2]

[1] Observatoire de Strasbourg, 11 rue de l'Université, F-67000 Strasbourg, France
[2] Observatoire de Haute-Provence, Saint-Michel l'Observatoire, F-04380 Forcalquier, France

Received February 12, accepted April 17, 1987

Summary. Spectra of comet P/Halley were taken with an objective prism at the Haute-Provence Observatory in November, December 85 and January 86. They lead to approximate values of the size of the coma for C2, C3 and CN and to relative surface brightness for these molecules.

Key words: comet Halley – objective prism

1. Instrumentation and observations

A set of twenty-three spectra of the comet was taken at the Haute-Provence Observatory with one of Fehrenbach's Objective prism ($D = 40$ cm, $f = 400$ cm, dispersion = 110 Å mm^{-1}) in November, December 85 and early January 86. We used IIaO plates and the spectral range from 3800 Å to 5000 Å, which includes the CN, C3, CH, C2 emissions. The advantage of the objective prism is to furnish monochromatic images of the comet. Unfortunately, these images are enlarged in that case because molecules emit bands rather than narrow lines. CN gives a better image of the coma than C2 and C3, because the images in the latter case are broadened in the dispersion direction that is the direction of declination (Refer to slit-spectra given by C. Arpigny et al., 1986). The main drawback is indeed the overlap of the image due to the size of the coma. For this reason, objective prism spectra may be useful only as long as the coma is small, and at the same time the comet is relatively faint.

Figure 1 shows three spectra which give an outline of their evolution over the period of three months. We notice on them the growing importance of images overlapping of C2 (0,0) 5165 Å; C2 (1,0) 4737 Å; C2 (2,0) 4382 Å; CH 4315 Å; CN (0,1) 4216 Å and C3 4000–4074 Å. We notice also the relative purity of the image given by CN (0,0) 3883 Å.

All plates were calibrated for their response in density as a function of illumination. The final response, including all absorption and transmitting effects as a function of wavelength, was estimated from a stellar A0 spectra. All spectra were scanned along their axis in a narrow strip in the direction of the dispersion and in a perpendicular direction through CN (0,0) C3 and C2 (0,0).

2. Size of the coma

Objective prism images of a comet cannot lead to a fine spectral analysis, but can provide values for coma size for different molecules, at least in a direction perpendicular to the dispersion. Let R be the projected distance between a point of the coma and the nucleus. We measured R for different values of the surface brightness only in the direction of right ascension opposite the tail. We limited these measurements to three main features, C2 (1,0) C3 and CN (0,0). Since no absolute calibration was available for these plates, we referred each value to the maximal value at the same wavelength near the center of the coma, taken as unity. Figure 2 shows a part of the results for the values $R10$ and $R20$ of R at points where the brightnesses are respectively one tenth and one twentieth of the maximum. If the mean of these values varies smoothly over the period of two months, then one can explain the

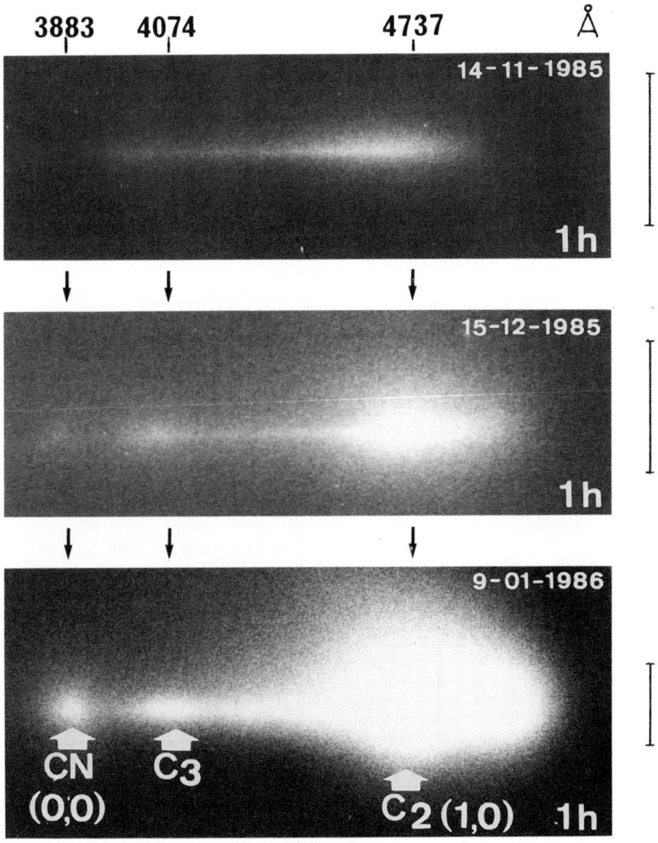

Fig. 1. The figure shows one hour exposure spectra of each observing period. A scale corresponding to 100,000 km is given on the right of each exposure

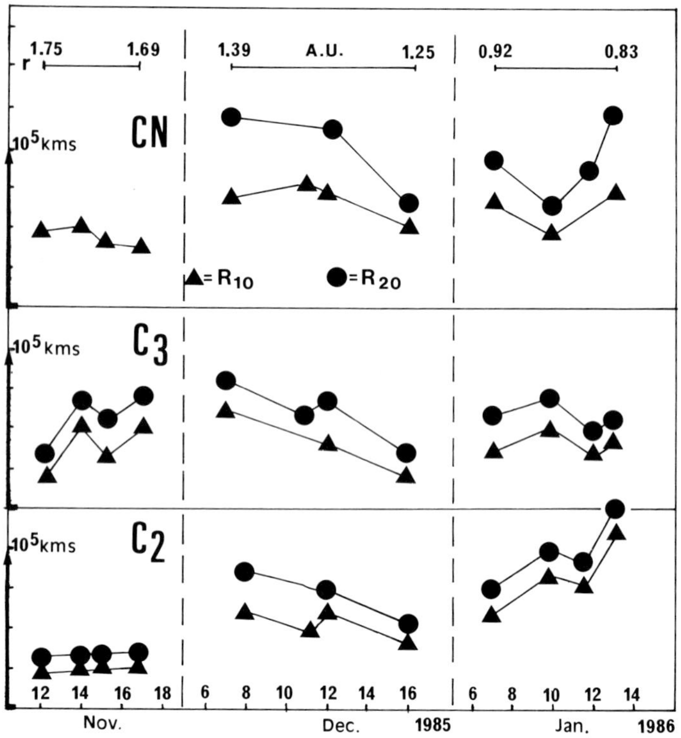

Fig. 2. The figure shows the evolution of the values $R10$ and $R20$ for CN, C3 and C2 of the projected distance R between the center of the coma and the points where the surface-brightness are one tenth and one twentieth of the central brightness. The dates are given at the bottom and the corresponding values of heliocentric distance r at the top

rapid change occurring in a few days over each period by sudden brightness increase or decrease in the center of the coma, around the nucleus. In fact, these values have to be compared to changes in magnitude of the comet, thus the variation in size of the coma, to which the changes in $R10$ and $R20$ are related, follow the comet's activity.

Table 1

		mean date			mean value
		Nov. 14	Dec. 11	Jan. 10	
B C_3/C_2	Center	1,51 ± 0,03	1,51 ± 0,02	1,48 ± 0,06	1,51 ± 0,03
	R25000	1,54 ± 0,01	1,50 ± 0,04	1,44 ± 0,11	1,50 ± 0,06
B CN/C_2	Center	1,67 ± 0,04	1,71 ± 0,08	1,78 ± 0,08	1,71 ± 0,05
	R25000	1,72 ± 0,03	1,70 ± 0,03	1,73 ± 0,10	1,71 ± 0,07

3. Surface brightness

We compared the surface brightness of the C3 and CN images at their centers (around the nucleus) taking the C2 image as unity. Since the results could be affected by the continuum, we measured also these values aside the spectrum axis, at a distance "a" corresponding to 25 000 km, opposite to the tail (see Fig. 3). The results are given in Table 1. We notice that they are not significantly different.

The surface brightness measurement was made on each spectrum for each molecule C2, C3 and CN at distances where one fifth, one tenth and one twentieth of the top value occur. The set of these values lead of a surface brightness law of the form kR^x (as previously we consider only the half side toward the sun). We calculated the values of x from the values of R comprised between $R5$ and $R20$. For this reason they may not be valuable out of this range. The results are given in Table 2. The values for each period of the observations (December, November and January) do not vary very much so that it is possible to give mean values of x for the whole. For comparison we estimated x using the surface

Table 2

Values of x, R_5, R_{20} (km)			mean date			mean value
			Nov. 14	Dec. 11	Jan. 10	
	C2	x	−1,95 ± 0,15	−2,38 ± 0,15	−3,40 ± 0,60	−2,25 ± 0,20
		R_5	20 000	50 000	75 000	
		R_{20}	40 000	80 000	100 000	
	C3	x	−1,45 ± 0,15	−2,02 ± 0,25	−2,03 ± 0,15	− 1,74 ± 0,15
		R_5	25 000	25 000	30 000	
		R_{20}	50 000	80 000	60 000	
	CN	x	−1,91 ± 0,35	−2,14 ± 0,30	−1,84 ± 0,30	−2,01 ± 0,20
		R_5	35 000	55 000	50 000	
		R_{20}	75 000	100 000	95 000	

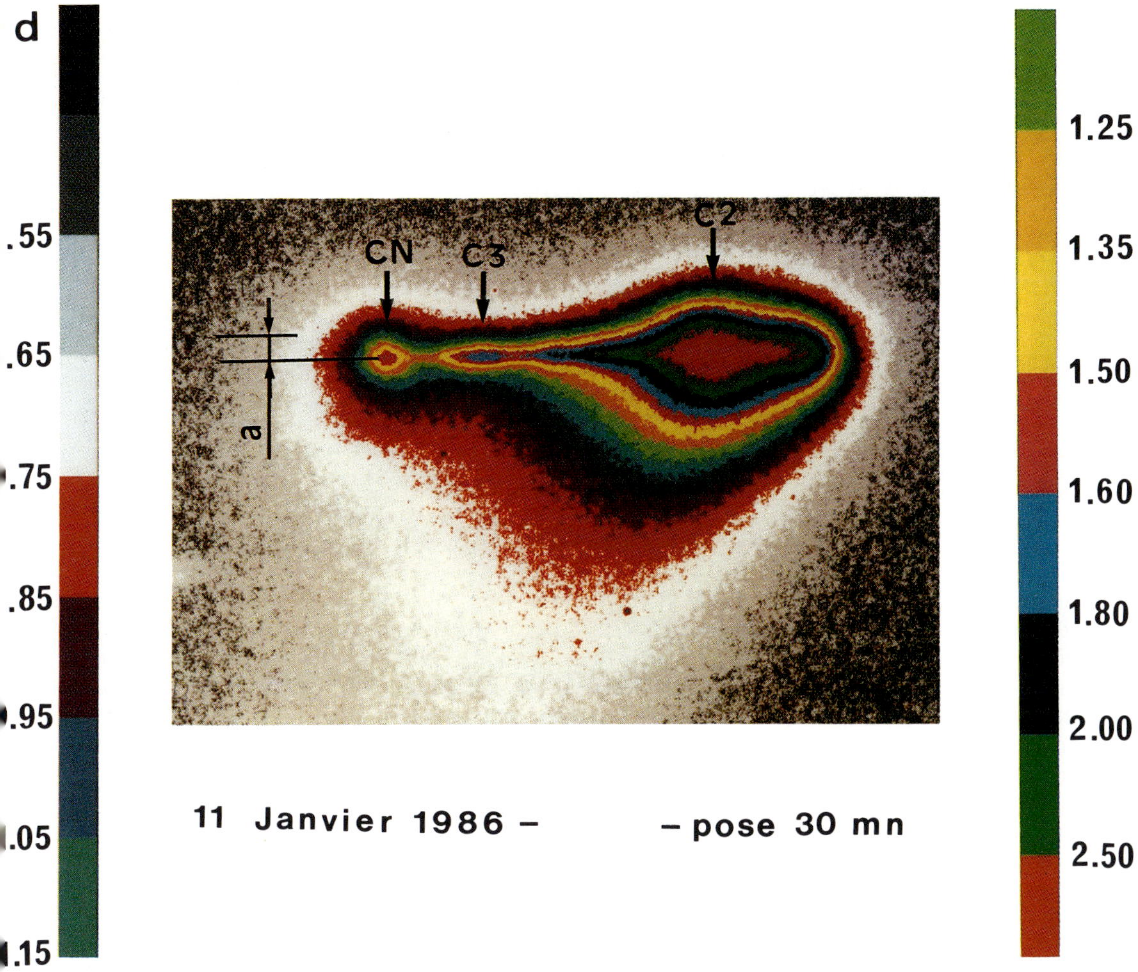

Fig. 3. This figure is a scan of the spectrum on january the 11th. The coma in CN appears as separated in the spectrum. The values are photographic densities. The distance "a" corresponds to 25,000 km

brightness values for different projected distances R given by Rees et al. (1986) in one of their diagrams, we found for the region between 2 10^5 km and 10^6 km a value $x = -1.94$. Attention must be drawn to the low values of x for C3 which is rather unusual. These values are in contradiction with those we found for two other comets: Honda 1968e and Hartley-Good 1985l for which the measurements give values of x larger than 3. Nevertheless the records that we made of the different profiles show that the emission of C3 is steeper than C2 and CN emissions at small distances from the nucleus, where we did make no measurements estimating that they could be of poor quality because of photographic overexposure. Further investigations are necessary to find the reason for these small values of x which may lie in contamination by other emissions.

Acknowledgements. We thank Ph. Véron, Director of the Haute-Provence Observatory and the local staff, especially R. Burnage for their support, J.L. Halbwachs (Strasbourg), A. Laugier (Haute-Provence Observatory) who took part in the observations and J. Florsch (Strasbourg) for his help in the measurements. We are grateful also to C. Arpigny for his useful comments.

References

Arpigny, C., Dossin, F., Zucconi, J.M., Andrillat, Y., Manfroid, J., Hudsemekers, D.: 1986, 20th ESLAB Symp. ESA SP-250, Vol. **I**, 471

Rees, D., Meredith, N.P., Wallis, M.K.: 1986, 20th ESLAB Symp. ESA SP-250, Vol. **I**, 493

Astron. Astrophys. 187, 360-362 (1987)

ASTRONOMY AND ASTROPHYSICS

Comet P/Halley neutral gas density profile along the Vega-1 trajectory measured by the Neutral Gas Experiment

C.C. Curtis[1], C.Y. Fan[1], K.C. Hsieh[1], D.M. Hunten[2], W.-H. Ip[3], E. Keppler[3], A.K. Richter[3], G. Umlauft[3], V.V. Afonin[4], A.V. Dyachkov[4], J. Erö, Jr.[5], and A.J. Somogyi[5]

[1] Department of Physics, University of Arizona, Tucson, AZ 85721, USA
[2] Department of Planetary Sciences, University of Arizona, Tucson, AZ 85721, USA
[3] Max-Planck-Institut für Aeronomie, D-3411 Katlenburg-Lindau 3, Federal Republic of Germany
[4] Space Research Institute, Profsoyuznaya 84/32, 117810 Moscow, USSR
[5] Central Research Institute for Physics, P.O. Box 49, H-1525, Budapest, Hungary

Received January 27, accepted April 27, 1987

Summary. Two complementary gas analyzers comprised the Neutral Gas Experiment (NGE) on the two Vega spacecraft sent to comet Halley. Although no significant mass spectra were obtained, data returned from the Vega-1 experiment have permitted the determination of the total neutral gas density profile along the spacecraft trajectory during the comet encounter. Discounting small fluctuations, the FIS instrument measured a density profile which varied approximately as the inverse radial distance squared. Data from the EIS instrument yielded a series of calibration points; e.g. the neutral density at a radial distance of 10^5 km was about $10^4\,\mathrm{cm}^{-3}$. The combined data provided a calibrated total density profile, and implied a neutral production rate of $1\,10^{30}$ molecules/second (assuming an outflow velocity of $1\,\mathrm{km\,s}^{-1}$) during the hours prior to the spacecraft encounter.

Key words: comets

1. The gas analyzers

Two complementary instruments comprised the Neutral Gas Experiment (NGE) on the Vega spacecraft (Keppler et al., 1986). One unit (FIS) employed a Field Ionization Source which largely preserved the parent molecules in the cometary gas during the ionization process, but had an undesired sensitivity to background gas originating at the spacecraft. The second unit (EIS) used an Electron Impact ionization Source (Curtis and Hsieh, 1986) and an analyzer which discriminated against thermal background gas, but this ion source fragmented the incoming molecules to a greater extent than did FIS.

The FIS ion source produced field ions near the surfaces of 40 needle tips of about 0.1 micron radius, biased at about +30 kV. The largely monoenergetic ions with energies near 30 keV were directed through a 0.3 microgram/cm^2 carbon foil, along trajectories nearly perpendicular to the paths of the incoming neutrals. Electrons ejected from the foil by a penetrating ion triggered a two-stage microchannel plate (MCP) detector to produce a start signal. Some nanoseconds later, the ion itself struck a similar detector 7.4 cm downstream from the carbon foil to generate a stop signal. Time of flight (TOF) circuitry converted the start/stop signal pair into a voltage pulse proportional to the square root of the ion mass. FIS produced, over a one second interval, a complete mass spectrum of particles having flight times within the range of 2 to 255 ns. It generated such a spectrum every two seconds.

The EIS employed two cold-cathode electron sources (one of them redundant) to ionize a fraction of the incoming gas. Cometary ions were excluded by repelling electrodes at the gas inlet and outlet. Like the FIS, this instrument was of "flythrough" design, with internal structure which presented a minimum cross section to entering dust particles and neutrals not ionized. The ion source was designed to preserve the initial velocity of the incoming molecules, which was a nearly uniform $79\,\mathrm{km\,s}^{-1}$ for the cometary gas. Ions from this source having $E/e < 30\,\mathrm{eV}/e$ were rejected by the EIS hyperbolic electrostatic analyzer, which dispersed the ions along a 5 cm-long MCP detector. Position sensing was accomplished by measuring the relative amounts of charge collected at the two ends of a resistive anode at the MCP output. EIS was designed to accumulate a spectrum covering the mass interval from 1 to 14 amu over a one-second period, and then switch ranges to accumulate a spectrum covering masses 14 to 28 amu during a later one-second interval.

2. Flight results

Both the FIS and EIS on the Vega-1 spacecraft responded to the passage of comet Halley on March 6, 1986, although neither instrument matched the performance of the lab prototypes. The FIS produced data leading to a total neutral gas density profile near the comet, but gave little mass information because of a combination of instrument problems and excessive counting rates.

The lower trace in Fig. 1 shows a spectrum produced by the FIS over a 20-second interval nearly one hour before the closest approach ($\sim$275,000 km from the nucleus), which exhibits a peak at channel 40 (mass 28) probably due to molecular nitrogen released from another experiment on the spacecraft. This peak

Send offprint requests to: C.C. Curtis

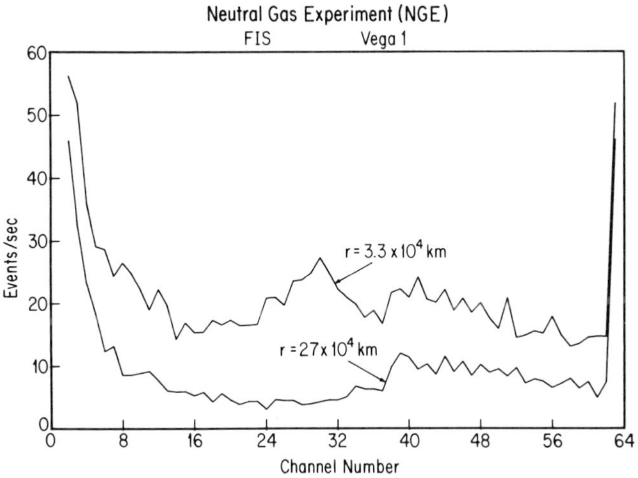

Fig. 1. (Lower trace): FIS spectrum ~275,000 km from the comet nucleus, showing a peak near channel 40 (mass 28) due to N_2 produced at the spacecraft. (Upper trace): FIS spectrum ~33,000 km from the comet nucleus, showing peaks near channel 30 (mass 18) and channel 40 (mass 28) served as a calibration point. Additional peaks at higher mass channels may be largely caused by straggling in the carbon foil. At this distance, spacecraft outgassing masked any gases produced by the comet.

At some intermediate distances, particularly in the outbound leg, a mass peak probably associated with cometary water appears in the spectra. The upper trace in Fig. 1 shows a 20-second spectrum accumulated at a distance of about 33,000 km from the nucleus, with a mass peak occurring at channel 30 (mass 18), in addition to the previously observed signals at channel 40 and above. This peak position shifts between channels 29 and 30, and represents molecules in the mass 16–18 range. The mass resolution is not sufficient for us to clearly identify these events as H_2O rather than O or OH resulting from the photodissociation of water molecules. At closer distances, events at shorter flight times are predominant, and the FIS spectra become relatively featureless. The counting rates were too high for the TOF electronics to resolve particle masses, although the overall rates did not exhibit saturation.

Because of the limited mass resolution of the FIS spectra, only a total neutral density profile was generated by summing the counting rates of channels 8 through 62 (masses 2 to 80 at 30 kV tip bias) as a function of radial distance from the nucleus. (The omitted channels exhibit continuously high counting rates.) The upper traces in Fig. 2 display the radial dependence of the total counting rate for these 55 channels, averaged over 20-second intervals, converted to number densities. The gap near 10^5 km results from a data loss. The density profiles in both legs display an r^{-2} dependency, with a small exponential correction, but the inbound density reaches a maximum about one minute before the closest approach. From its calculated sensitivity, the uncorrected FIS counting rate was several times higher than expected for the number densities deduced from the EIS data. Uncertainties in the ionization or detection efficiencies, or other instrument effects, could account for this discrepancy.

The EIS data complement the FIS data by providing closer estimates of the cometary neutral gas density at distances between 40,000 and 150,000 km from the comet nucleus. Although the EIS counting rate in all channels was inflated by secondary ions produced by electrons leaking from the ion source into the analyzer region, the EIS should not have been strongly sensitive to thermalized gas in the vicinity of the spacecraft. As happened with the FIS instrument, mass resolution was severely compromised by electronics problems. The EIS density profile for neutrals in the mass range from about 6 to 18 is given by the lower traces in Fig. 2. For the inbound leg, power law behavior of the counting rate was observed in the distance range between 40,000 and 150,000 km. An apparent saturation of the instrument caused a deviation from a constant slope of about −2 at distances less than 40,000 km, and a turnover in the total counting rate at a distance of about 18,000 km from the nucleus. As the spacecraft reached the closest approach point, the EIS counting rate dropped precipitously. While it recovered somewhat during the outbound leg, the counting rate remained below the inbound rate at comparable distances, in agreement with the FIS results. Note that at radial distance greater than ~40,000 km, there is general agreement between FIS and EIS on both inbound and outbound legs. At closer distances, the EIS saturation results in an incorrect profile from that instrument.

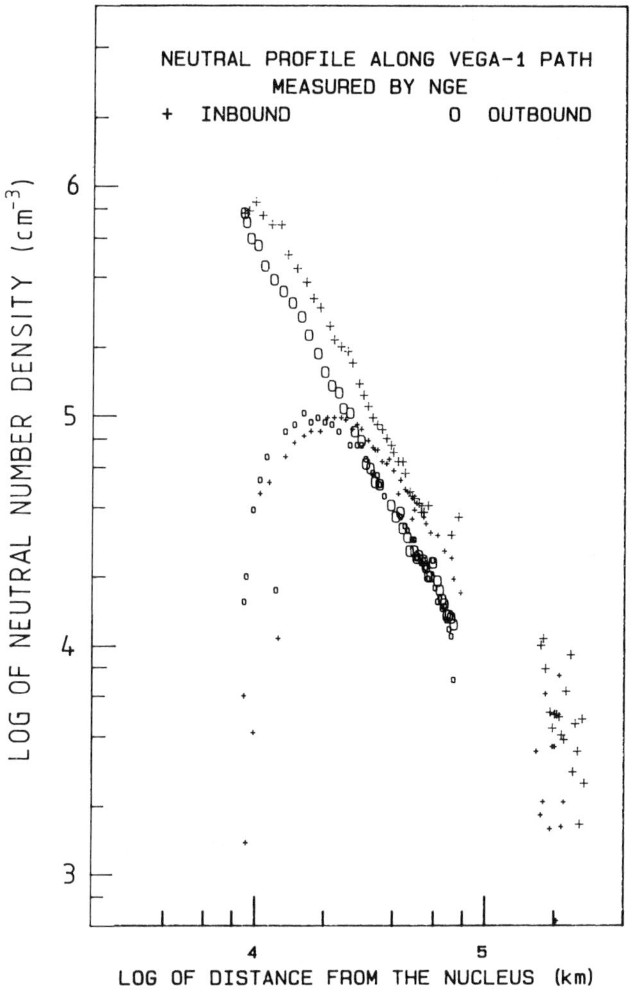

Fig. 2. Total neutral density profiles deduced from the FIS (large + and 0 symbols) and EIS (small symbols) counting rates. The EIS profiles show instrument saturation at distances less than ~40,000 km, and are inaccurate within that distance range

3. The neutral density

From the calculated EIS sensitivity, we estimate the total neutral density at a distance of 10^5 km on the inbound leg to have been about 10^4 cm^{-3}, to within a factor of 4, allowing for uncertainties in the emission current of the electron source and the overall efficiency of the detection system. This figure is in basic agreement with that of Gringauz et al. (1986) on Vega 1, but is more than an order of magnitude higher than the density for water obtained by Krankowsky et al. (1986) on the Giotto spacecraft. There are two factors which may account for this. First, it appears that comet P/Halley was considerably more active (was producing more gas) just prior to the Vega 1 encounter on March 6 than it was for the later Giotto encounter. Gringauz et al. inferred a total gas production rate of about $1.3\,10^{30}\,s^{-1}$, while Krankowsky et al. calculated a total production rate only half as great during their later encounter. Second, the water density profile data of Krankowsky et al. show a substantial deviation from a simple power law dependence, owing to a large photodestruction effect. Our analysis of our data shows only a small exponential correction to a simple r^{-2} dependency over the distance range shown. While water was presumably the major constituent of gas released at the nucleus, it was probably only a minor neutral species at 10^5 km.

It is also possible that a higher rate of dust production during the Vega 1 encounter resulted in larger amounts of water being transported as subliming ice to distances beyond 10^4 km, providing fresh sources of neutral gas molecules at distances between 10^4 and 10^5 km. Further, when rotation of the comet nucleus is taken into account, azimuthal variations in gas production at the nucleus have a substantial effect on the observed radial dependence of the neutral gas density (Hsieh et al., 1987).

From our estimate of the neutral gas density at 10^5 km, we infer a total neutral gas production rate of about $1\,10^{30}\,s^{-1}$ at the comet surface about one day prior to the encounter, assuming a uniform flux with a $1\,km\,s^{-1}$ constant velocity. (This production rate is insensitive to the inclusion of a photoionization length of about $2\,10^6$ km in the calculation.) The number is in basic agreement with the projections of both Gringauz et al. and Krankowsky et al., since their measurements of the neutral density at distances of 10^4 km and less (where we have no good measurement) differ by only a factor of 2.

Acknowledgements. We are indebted to many individuals at our respective institutions for their major contributions to this project. In particular, we wish to thank A.M. Hudor, W. Axford, H. Rosenbauer, V. Vasyliunas, C. Römer, K. Fischer, W. Weiss, H. Lauche, H. Sjuts, A. Loidl, W. Boogaerts, I. Schrader, K. Szegö, and V. Valentina. Participation of the University of Arizona was made possible by NASA Contract NASW3627 and NSF grant INT8205325.

References

Curtis, C.C., Hsieh, K.C.: 1984, *Rev. Sci. Instrum.* **57**(5), 989

Hsieh, K.C., Curtis, C.C., Fan, C.Y., Hunten, D.M., Ip, W.-H., Keppler, E., Richter, A.K., Umlauft, G., Afonin, V.V., Erö, J., Jr., Somogyi, J.A.: 1987, *Astron. Astrophys.* (this issue)

Gringauz, K.I., Gombosi, T.I., Renizov, A.P., Apáthy, I., Szemerey, I., Verigin, M.I., Denchikova, L.I., Dyachkov, A.V., Keppler, E., Klimemko, I.N., Richter, A.K., Somogyi, A.J., Szegö, K., Szendrö, S., Tátrallyay, M., Varga, A., Vladimirova, G.A.: 1986, *Nature* **321**, 282

Keppler, E., Afonin, V.V., Curtis, C.C., Dyachkov, A.V., Erö, J., Jr., Fan, C.Y., Hsieh, K.C., Hunten, D.M., Ip, W.-H., Richter, A.K., Somogyi, A.J., Umlauft, G.: 1986, *Nature* **321**, 273

Krankowsky, D., Lämmerzahl, P., Herrwerth, I., Wowries, J., Eberhardt, P., Dolder, U., Herrmann, U., Schulte, W., Berthelier, J.J., Illiano, J.M., Hodges, R.R., Hoffman, J.H.: 1986, *Nature* **321**, 326

Astron. Astrophys. 187, 363–368 (1987)

ASTRONOMY AND ASTROPHYSICS

Low-resolution maps of comet P/Halley in principal atomic and molecular species

M.R. Kidger[1,2], J.A. Acosta[1,3], F. Garzón[1], M. Prieto[1], and R. Gómez[1]

[1] Instituto de Astrofísica de Canarias, Universidad de La Laguna, Camino de La Hornera, Tenerife, Spain
[2] Department of Physics, Queen Mary College, Mile End Road, London. E1 4NS, United Kingdom
[3] SISSA, Trieste, Italy

Received February 9, accepted May 4, 1987

Summary. We present maps of a 3′ × 2′ region of the inner coma of comet P/Halley, at 10″ resolution, in CN (3883 Å); C_2 (5100 Å); C_3 (4040 Å); CN (4216 Å); NH_2 (6107 Å); [O I] (6300 Å) and CH (4322 Å). The maps were prepared from multi-position absolute spectrophotometry of comet Halley, with 3.6 Å resolution, in the wavelength range 3700–7598 Å, using the 2.5-m "Isaac Newton Telescope" in La Palma, Spain. This data represents a sub-set of a larger database of mapping at a range of heliocentric distances.

Key words: comet – spectrophotometry – observational methods

1. Introduction

Very little two dimensional photometry has been performed of the distribution of any of the major species seen in cometary spectra, mainly due to the lack of a reliable method of calibration of spectra recorded on photographic plates. For discussion of available information on bi-dimensional photometry and its importance in cometary physics, we refer the reader to an excellent review article by A'Hearn (1982) and to our previously published work eg: Prieto et al. (1986), Acosta et al. (1986).

In this paper, we report multipositional spectrophotometry using an Image Photon Counting System (IPCS), the observations being part of a sub-set of a larger collection obtained on a number of nights, using the same equipment. By the date of observation (1986 May 20–21), the comet had faded below naked eye visibility, having a total magnitude of around 6.1 and showing a diffuse coma and short tail in small telescopes (Keitch, private communication, 1986).

In presenting these data, we hope to demonstrate the potential of multi-position long-slit spectrophotometry, when coupled to a large telescope. We note some of the pitfalls and their possible solution, as a guide to observers who wish to use this technique to observe future comets.

2. The observations

Our observations were carried out using the Royal Greenwich Observatory (RGO) Image Photon Counting System (IPCS) attached to the 235 camera of the 100 inch (2.5 m) Isaac Newton Telescope (INT), in "El Observatorio del Roque de los Muchachos" (hencefore "el Roque", for brevity), on the island of La Palma, Spain, as part of a coordinated program of observation of comet P/Halley in the infrared and visible (see Prieto et al., 1985; Prieto et al., 1986; Acosta et al., 1986; Lázaro et al., 1986; Garzón et al., 1987a, b, for further details). A 300 line per mm, 138.5 Å mm^{-1} grating was used, giving a spectral resolution of 3.6 Å, and a range from 3430 Å to 7598 Å, although a WG 360 colour filter, used to restrict interference at the red end of the spectrum, from second order, greatly restricted the throughput at wavelengths shortward of 3700 Å. This filter, which has greater than 99% internal transmittance redward of its cut-off wavelength, was used in preference to the GG395 filter used in some previous published work within this program (e.g. Prieto et al., 1986), due to its preferable transmittance characteristics (Pettini, 1983). The maximum possible slit length allowed by the system, 3.′9, was used for all the observations (although this resulted in some vignetting at the edges of the field and hence, in this work, we have, for simplicity of reduction, used only the central three minutes of the slit), giving a format of 2044 pixels in the wavelength axis and 90 sections (lines) along the slit and hence a spatial resolution of about 2.″7.

Eleven exposures of the comet have been used, each of them taken with an exposure of 500 s and the slit aligned W-E. Exposures were taken at 10″ intervals, starting at and centered on the nucleus, out to 40″ north and south, with a further exposure at 60″ north and south. Unfortunately, the short observational window (the INT has a hardware limit at a zenith distance of 70°) meant that no observations could be taken at 50″, hence the gap in the data in these positions. Sky subtraction was carried out using identical exposures taken 3° and 8° to the north of the comet, in a region free of extended cometary emission.

3. Reduction

Reduction of the observations was carried out at the Instituto de Astrofísica de Canarias (IAC) using PANDORA, a suite of computer programs originally designed in Gröningen for use with the telescopes at Mount Stromlo and Siding Springs Observatories and specially modified for use at the IAC.

Calibration in wavelength was carried out using 3 copper-argon and one copper-argon/copper-neon arc spectra, which give

Send offprint requests to: M. Kidger[1]

internal errors no greater than ±0.4 Å over the range 3700–7598 Å. The absence of usable lines in the calibration arcs, shortward of 3700 Å, means that the wavelength calibration here is rather more doubtful. Comparison of theoretical and observed wavelengths for a number of the most prominent lines in the cometary spectrum, within the wavelength range 3700–7598 Å reveals errors consistent with the known resolution of the data.

Photometric calibration was carried out by taking two spectra of the subdwarf Ross 627 (Oke, 1974), one before and one after observations of the comet. Comparison with later observations of the standard star GD190 shows that there was a change of less than 1% in the transparency of the atmosphere during the night. Information on dust extinction was provided by the astronomers on the Carlsberg Automatic Transit Circle which is also based at "el Roque". This was combined with the standard atmospheric model for La Palma, to give the final calibration.

Since the sky in La Palma has a very clean spectrum over the range 4000–7000 Å, with only a single prominent line in emission, the 5577 Å line of neutral oxygen, which is not observed in comets, an exact sky subtraction can ensured by scaling the height above the continuum, of the principal night sky line in the sky spectra, to the height of this feature above the continuum in the cometary spectra. By definition, so long as there is no emission feature in the cometary spectrum at precisely this wavelength, the subtraction can be made exactly, independent of the level of the cometary continuum. The region of each cometary spectrum close to the night sky line was very carefully examined and the subtraction was made iteratively. This allowed us to compensate for changes in the intensity of the sky brightness during the observations caused by the changing altitude of the Moon, although a small amount of under or over-subtraction is still possible.

A major problem encountered in subtracting the sky was that of an apparent colour change in it during the observations, in addition to its changing brightness. At the start of the observing window, the comet was comparatively high in the sky and the Moon (phase 88%), comparatively low. This situation was reversed by the end of observations, at which point the sky exposures were taken. In the first two or three cometary exposures, we find that an exact subtraction of the principal night sky line encountered in "el Roque", the 5577 Å line of oxygen, results in over-subtraction in the blue. This problem had to be solved empirically, by using a different correction in the blue of these spectra. We thus caution that there may be a zero point error in integrated (that is, line + continuum) fluxes, in this region, but since the sky spectrum shows a very steep fall in flux after about 4000 Å, the problem diminishes rapidly longward of this wavelength. Exact sky subtraction for an extended object which is larger than the slit length is not trivial because of such problems as this. In a time critical situation, with a comparatively short observing window, a compromise has to be found between the taking of many sky exposures, thus assuring a perfect sky subtraction, but few cometary exposures and taking very few sky exposures to make time for greater coverage of the comet, with its concurrent risk of a resultant poor sky subtraction. This problem is exacerbated by the need for sky exposures to be as close as possible in length to those of the object, so that the subtraction is a good one. Our experience suggests that, in conditions where the Moon's altitude (and thus the sky brightness and color) are changing, the best procedure is to take sky exposures at the beginning, during and at the end of observations, although this may not always be possible (as in this case), for technical, or meteorological reasons.

All spectra have been smoothed to their theoretical resolution (1.8 pixels on the IPCS, equivalent to 3.6 Å) using a gaussian filter.

4. The maps

Because of the gap in the data at 50″ north and south of the nucleus and the need to use equally spaced data in producing the maps, we prepared two data sets from the reduced spectra. The first set used data spaced at 10″ intervals, from 40″ north, to the same distance south of the nucleus. This gave us an acceptable spatial resolution in the declination axis, but did not use all of the data, hence a second data set was produced, using spectra spaced at 20″ intervals, over the range 1′ in north, to 1′ in south. This gave us information over an area of sky 50% larger, but at the cost of losing resolution in the declination axis. We do though, have the opportunity to compare the maps produced by two, partially overlapping, data sets and thus checking the extent of differences between them. As a result, we feel that we have more confidence as to the reliability, or otherwise, of any structure encountered in the maps.

The lines and bands to be used in the preparation of the maps were identified in spectra of the nucleus. A wavelength range for the line was defined by careful inspection of the spectrum and this range used to extract fluxes in all the spectra that formed the data set. Due to the large number of transitions and the comparatively low resolution of the spectra, it is very difficult to define, with confidence, a continuum. This is especially true in the red end of the spectrum. The method which we adopted was to use a polynomial routine to fit a curve through the data, which was then forced to pass through control points defined as continuum. This is an alternative method to that of subtracting the solar spectrum directly. In both cases it is necessary to make some assumptions as the absolute level of the continuum. Since most of the lines extracted are fairly narrow, we believe that there is little difference between the two methods in terms of resultant extracted line fluxes.

Once line fluxes were extracted from the reduced spectra, they were stored in a matrix and a standard contour plotting routine using spline interpolation, was used to fit isophotes. In practice, it was found that the data had to be smoothed, to give a similar resolution in the Right Ascension (RA) axis to the data spacing in Declination (DEC), partly to reduce noise in the data and partly to avoid a spurious appearance to the isophotes caused by having data which was much more highly sampled in one axis, than in the other (we note though, that contours very close to the nucleus, will still be distorted due to undersampling). The maps are thus produced from matrices smoothed with a five-point running mean (giving an effective resolution of 13″ in RA), this type of mean being selected for its ability to reveal the trend in otherwise noisy data. An odd number of points must be chosen for the mean, to avoid shifts.

Maps are presented here in the following transitions: CN (3883 Å); C_3 (4040 Å); C_2 (5100 Å); NH_2 (6107 Å); CH (4322 Å); [O I] (6300 Å) and CN (4216 Å). The projected size of the maps within the coma is of some 188,000 km × 125,000 km (at 20″ separation of spectra) and 188,000 km × 84,000 km (for maps composed of spectra at 10″ separation), with the nucleus in the centre. Our resolution corresponds to 13,400 km at the comet in the E-W direction and either 10,500 km, or 21,000 km (dependent on the data set used, in the N-S axis).

Of the species studied, the expected spatial distribution varied widely. From the scale lengths accepted by A'Hearn (1982), we would, at the epoch of observation, with the heliocentric distance, r, of 1.92 AU, expect a radius for the CN coma of $1.1\ 10^6$ km, which would give a radius, in the sky, of 18′. For C_2, the values are expected to be $6.0\ 10^5$ km and 10′ respectively and for C_3, $2.2\ 10^5$ km and 3.′4. These three species would be expected to fill the entire area of sky which was mapped and in the case of CN especially, we are only examining the central part of the emission. The available information about other species is highly limited. In the cases of CH and NH_2, emission is known to be highly concentrated close to the nucleus, which is thought to be the case for [O I] 6300 Å, too.

The maps are presented in Fig. 1(a–g), at 10″ resolution in declination and Fig. 2(a–g), at 20″ resolution. All plots are on a logarithmic scale.

As expected, the emission in CN and C_2 is considerably more extended than the area of sky mapped. We note also, that the nucleus in CN is off-set slightly from the centre of the CN coma, an effect that we have noted previously during observations of comet P/Halley. This we believe to be an effect of a preferred direction of outgassing from the nucleus and was present in our earliest reported data (Prieto et al., 1986). There is a suggestion that the emission is more extended to the north-west as the isophotes are more widely separated in this direction than to the south-east. This effect is visible in both maps of CN (3883 Å) and also in the 20″ map of C_2 although not in maps of other species

Table 1. Parameters of the contour maps plotted in Fig. 1. Units are erg cm^{-2} $Å^{-1}$ s^{-1}

Line or band	Maximum contour	Minimum contour	Interval
CN (3883 Å)	$3.96\ 10^{-13}$	$8.84\ 10^{-14}$	$2.05\ 10^{-14}$
CN (4216 Å)	$3.94\ 10^{-14}$	$3.55\ 10^{-15}$	$4.80\ 10^{-17}$
C_3	$5.11\ 10^{-13}$	$7.46\ 10^{-14}$	$3.35\ 10^{-14}$
C_2	$6.99\ 10^{-13}$	$1.19\ 10^{-13}$	$3.86\ 10^{-14}$
CH	$2.56\ 10^{-14}$	$8.51\ 10^{-15}$	$1.90\ 10^{-15}$
NH_2	$9.17\ 10^{-15}$	$5.23\ 10^{-15}$	$7.87\ 10^{-16}$
[O I]	$4.93\ 10^{-14}$	$1.76\ 10^{-14}$	$3.16\ 10^{-15}$

Table 2. Parameters of the contour maps plotted in Fig. 2. Units are erg cm^{-2} $Å^{-1}$ s^{-1}

Line or band	Maximum contour	Minimum contour	Interval
CN (3883 Å)	$3.96\ 10^{-13}$	$7.86\ 10^{-14}$	$2.11\ 10^{-14}$
CN (4216 Å)	$3.94\ 10^{-14}$	$1.93\ 10^{-14}$	$2.51\ 10^{-15}$
C_3	$5.11\ 10^{-13}$	$4.49\ 10^{-14}$	$3.33\ 10^{-14}$
C_2	$6.99\ 10^{-13}$	$1.11\ 10^{-13}$	$3.92\ 10^{-14}$
NH_2	$9.18\ 10^{-15}$	$5.15\ 10^{-15}$	$8.06\ 10^{-16}$
CH	$2.56\ 10^{-14}$	$9.44\ 10^{-15}$	$1.79\ 10^{-15}$
[O I]	$4.93\ 10^{-14}$	$1.79\ 10^{-14}$	$3.14\ 10^{-15}$

and thus it may well be real. The E-W elongation visible in several of the maps at 10″ resolution though, is almost certainly due to a small zero point error in the sky subtraction of one of the central exposures and almost certainly is not real.

Emission in C_3 is also extended to the edge of the sampled area. In contrast to CN and C_2, the emission is highly symmetrical in both maps and there is little suggestion of any preferred direction. NH_2 and CH though are found to be highly concentrated close to the nucleus, in the case of NH_2, exceptionally so, indeed, the diameter of the coma in NH_2 appears to be of the order of only fifteen arcseconds, as compared to about 30–40 for CH, although this may be dominated by the relative weakness of the NH_2 (6107 Å) line compared to other transitions presented here. To study the spatial distribution of such weak lines in real detail we would advise future observers to use much longer exposures than would be thought to be necessary normally as, even when the CN, C_2 and C_3 emissions are heavily exposed, lines such as NH_2 will still have relatively few counts. This may well be the actual limiting factor in using this method for studying many transitions within the same spectrum.

In the case of [O I] we see something that appears to be small-scale structure and of a strikingly different form from that suspected in CN and C_2, the maps showing an almost triangular form for the outer contours. Again, the "tail" visible to the west is probably due to a small zero point error in sky subtraction. CCD images obtained by one of the authors with a narrow band [O I] filter during late April, also show this characteristic form, hence we believe it to be real, however the reason for this precise shape and the difference in the direction of elongation compared to CN and C_2 remains obscure.

One surprise is the different form of the coma in the 4216 Å line of CN compared to its appearance in the 3883 Å line. We would expect the two maps to be very similar, since the only difference is in the transition (1 – 0) instead of (0 – 0) and the line is more extended spatially than C_3. We find though, that the emission is not detected over a diameter greater than about one arcminute. Again, this is probably chiefly due to the comparative weakness of the CN (4216 Å) line compared to some of the others that were studied.

5. Conclusions

We find that the method of multipositional absolute spectrophotometry is capable of extracting useful spatial information on the distribution of a number of principal cometary atomic and molecular species. Such spatial information should provide a useful database for the calculation of scale lengths and the study of physical processes and chemistry active within the inner coma. The principal advantages of this method over conventional photometry are that: 1.) Many transitions may be observed at once; 2.) The spectral band-pass can be precisely defined by the observer in the reduction of the data and is better able to isolate emission in only the required transition than photometry with special filters; 3.) Whilst subtraction of the continuum is still not a trivial exercise, it may be done more accurately using this method than in standard photometry. Particular difficulties are sky subtraction and giving sufficiently long exposures to measure accurately the weakest lines. A further difficulty is photometric calibration if the observing conditions are less than perfect.

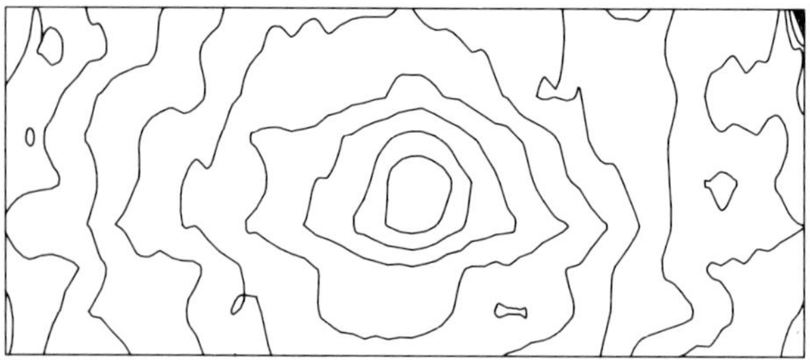

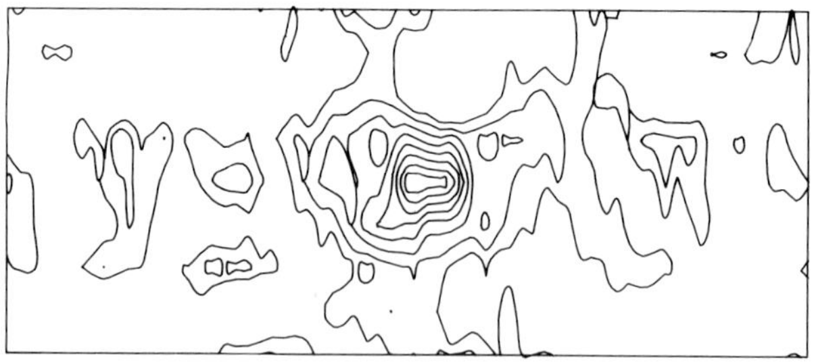

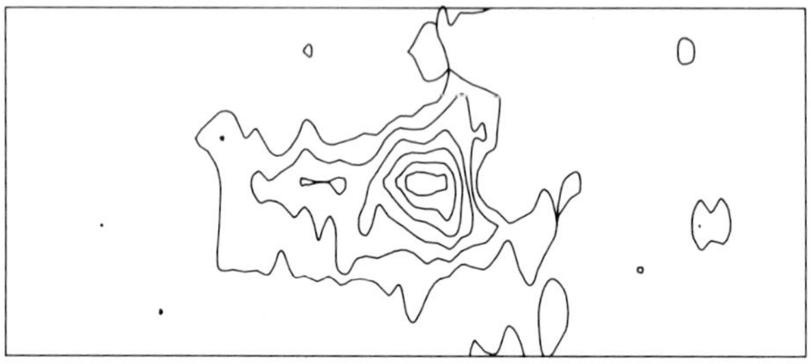

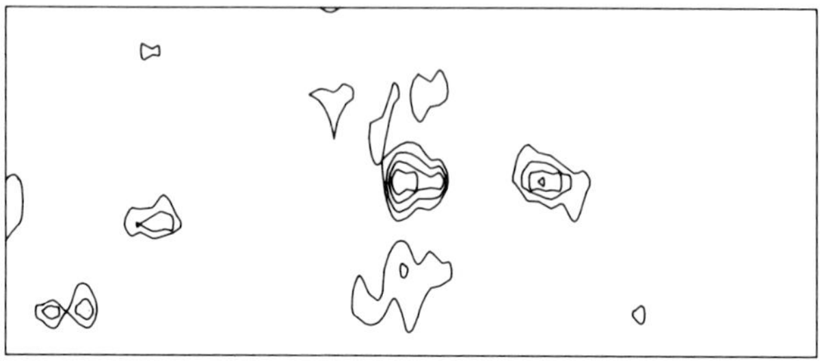

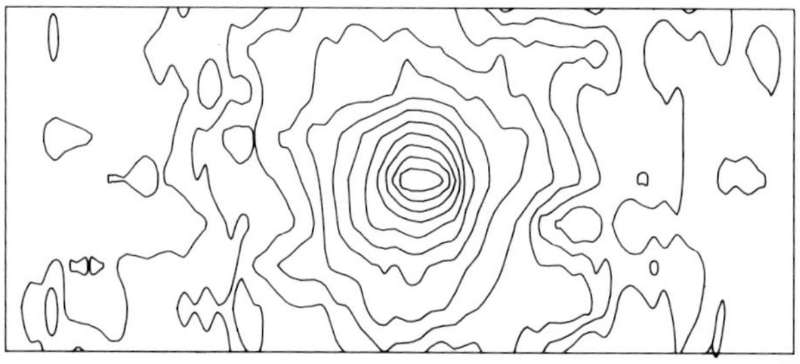

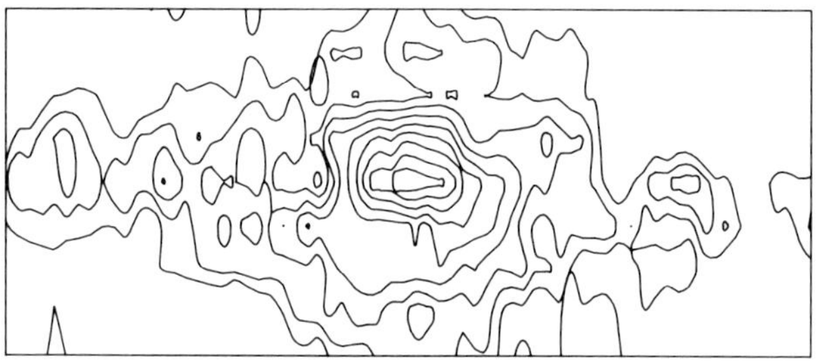

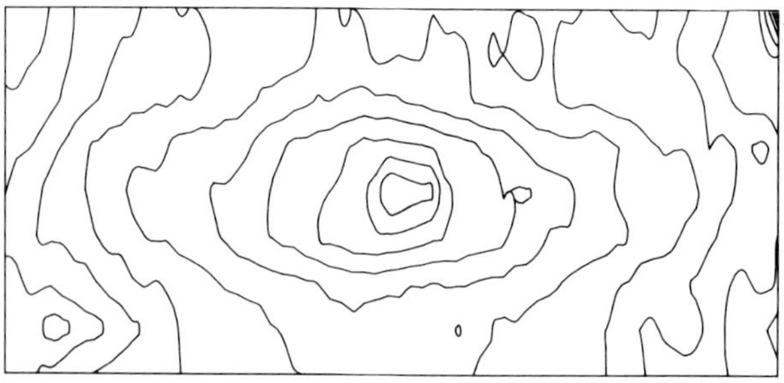

Fig. 1a–g. Maps of a 180″ × 80″ region of the inner coma of comet Halley, at 10″ resolution in declination, produced by the method described in the text. The characteristics of each map are given in Table 1

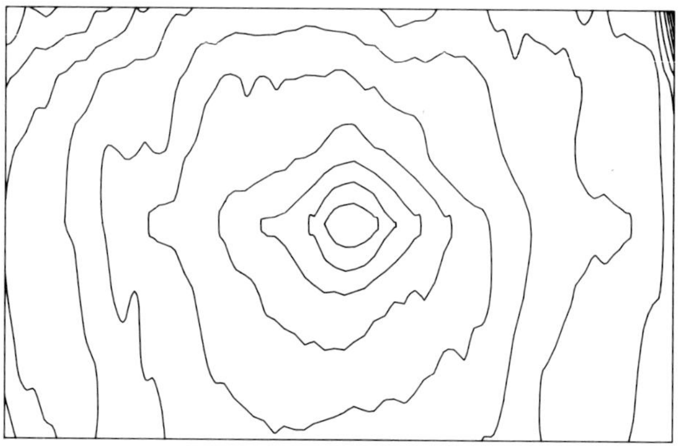

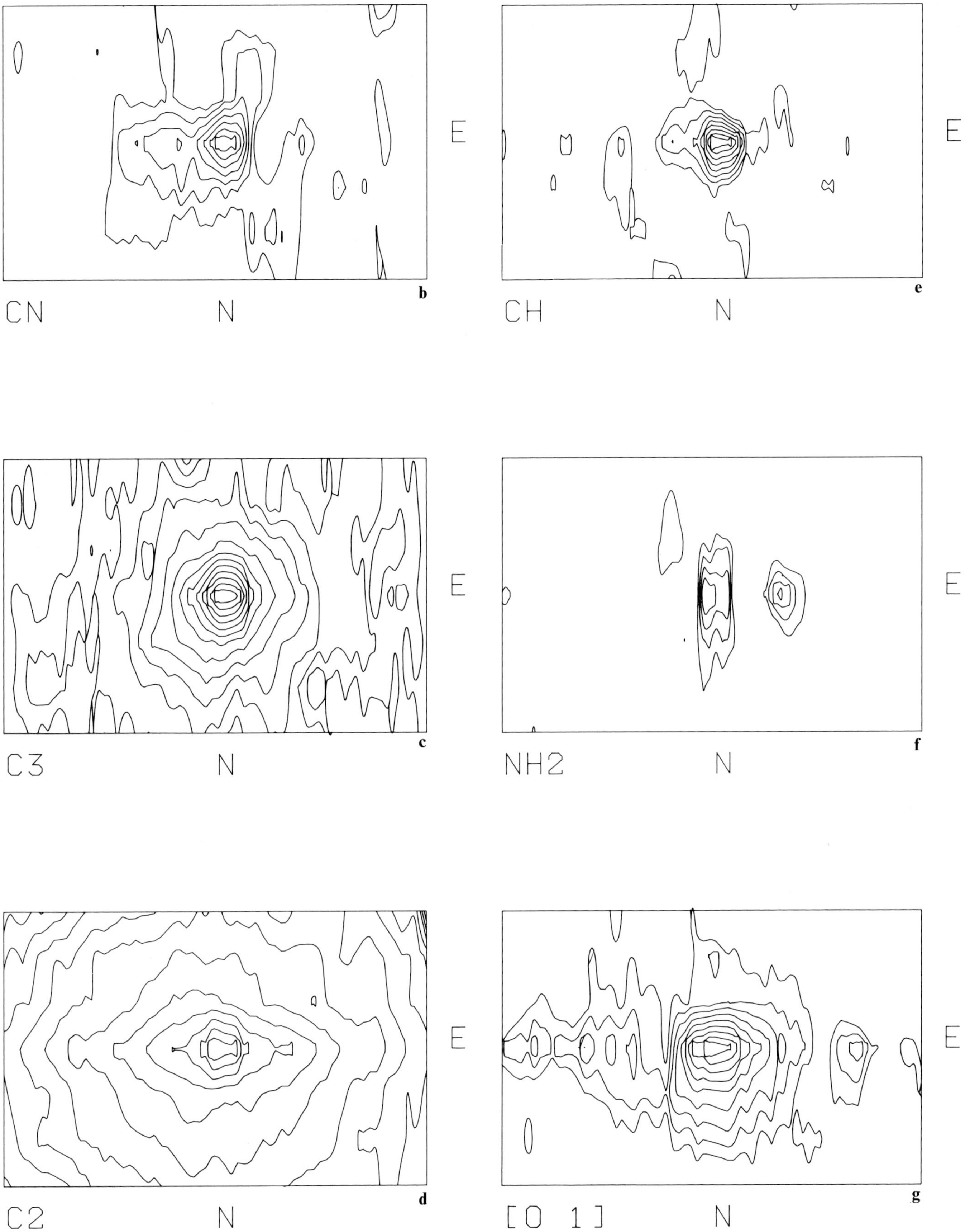

Fig. 2a–g. Maps of a 180″ × 120″ region of the inner coma of comet Halley, at 20″ resolution in declination, produced by the method described in the text. The characteristics of each map are given in Table 2

Recently, various groups have presented some extremely interesting results as to small-scale structure in the gas coma, particularly in the case of CN. Several papers have reported the observation of CN jets in imaging of the cometary coma with the IHW filter set (e.g. A'Hearn et al., 1986; Samarasinha et al., 1986; Cosmovici et al., 1987 and Ip et al., 1987), although at least one group has looked for such jets using very narrow band-pass interference filters and failed to find evidence for them (Rees et al., 1986). The data presented here are, in their present state, unable to settle this conflict, although a careful re-reduction, especially of the CN, C_2 and C_3 maps should be able to confirm, or deny the presence of jets on the night of observation. Whilst comparison of such results with simultaneous, or near-simultaneous direct imaging is planned, it has not been possible as yet. We note though, that a number of results suggest strongly that most, if not all molecules observed in Comet Halley have a dispersed source (e.g. profiles and calculations of scale lengths suggest deviations from the Haser model: Wyckoff, 1987, private communication; Acosta et al., 1986; Rees et al., 1986; Wallis et al., 1987; whilst infrared observations of a colour gradient in the inner coma suggest that there is fragmentation of dust grains which could, in the process, be releasing volatiles and thus serving as the dispersed source: see Garzon et al., 1987b; Rieke and Campins, 1987). We also note that Schlosser et al. (1987) observed CN shells expanding from the nucleus, although these were reported to be generally spherically symmetrical and thus are not a likely cause of the distortion seen here.

Some general comments are still possible. The best (cleanest) maps presented here are those in CN 3883 Å and C_3 4040 Å. In the case of C_3, there is little evidence of any deviation from spherical symmetry, however, in the case of CN, we note that there is a clear asymmetry, with the contours in the north-west being much more widely spaced than in other directions. This is consistent with the suggestion that there may be jets underlying the data in this direction, although we reserve judgement on this until the data have been further processed. Since it has been pointed out by Ip et al. (1987) and by Cosmovici et al. (1987) that features observed in CN generally do not correspond with features observed in other filters, it is not inconsistent that we do not see the same features in C_3 as in CN.

One interesting inconsistency between these data and other published work occurs in the case of the [O I] 6300 Å line. Roesler et al. (1986, 1987) have presented results that show no significant asymmetry in the spatial distribution of the flux from this line. In contrast, even when the outermost, noisy contours are removed, there is some evidence of strongly asymmetric emission. The reason for this discrepancy is unclear at present, although the method used by Roesler et al. is considerably more sophisticated than ours, mainly due to their high spectral resolution which allows them to separate out cometary [O I] 6300 Å from airglow within the Earth's atmosphere and also from cometary NH_2.

Acknowledgements. We would like to thank our support astronomer, Neil Reid and Sr. Carlos Garcia, for their assistance during these observations and the IAC for a summer studentship granted one of us (JAA) to reduce these data. MRK gratefully acknowledges an SERC "Instant Award" and the kind hospitality of the Instituto de Astrofisica de Canarias.

The Isaac Newton Telescope, on the island of La Palma is operated by the RGO in collaboration with the IAC.

References

Acosta, J.A., Kidger, M.R., Prieto, M., Cepa, J., Muñoz, C.: 1986, Revista Mexicana de Astronomia y Astrofisica (in press)

A'Hearn, M.F.; 1982, in *Comets*, ed. L. Wilkening, University of Arizona Press, p. 433

A'Hearn, M.F., Hoban, S., Birch, P.V., Bowers, C., Martin, R., Klinglesmith, D.A.; 1986, 20th ESLAB Symp. ESA SP-250, Vol. **I**, 483

Cosmovici, C.B., Mack, P., Ip, W.H., Craubner, H., Schwarz, G.; 1987, Proc. Symp. *Diversity and Similarity of Comets* (in press)

Garzón, F., Kidger, M.R., Prieto, M., Cepa, J.; 1987, *Astron. Astrophys.* (submitted)

Garzón, F., Lázaro, C., Arevalo, M.; 1987, *Astron. Astrophys.* (this volume)

Ip, W.-H., Cosmovici, C.B., Mack, P., Craubner, H., Schwarz, G.; 1987, *Proc. Symp. Diversity and Similarity of Comets* (in press)

Lázaro, C., Garzón, F., Arevalo, M.; 1986, 20th ESLAB Symp. ESA SP-250, Vol. **III**, 389

Oke, J.B.; 1974, *Astrophys. J. Suppl.* **27**, 21

Pettini, M.: 1983, UK Optical Telescopes, La Palma, Observers Guide

Prieto, M., Kidger, M.R., Beckman, J.E.; 1985, *IAU Circular* **4115**

Prieto, M., Kidger, M.R., Beckman, J.E., Rosa, F.; 1986, *Astron. Astrophys. Letters* **163**, L1

Rees, D., Meredith, N.P., Wallis, M.K.; 1986, 20th ESLAB Symp., ESA SP-250, Vol. **I**, 493

Rieke, M.J., Campins, H.; 1987, *Proc. Symp. Diversity and Similarity of Comets*, Book of Abstracts, p. 161

Roesler, F.L., Scherb, F., Magee-Sauer, K., Harlander, J., Reynolds, R.J. Oliverson, R.J., Yelle, R.V., Broadfoot, A.L.; 1986, 20th ESLAB Symp. ESA SP-250, Vol. **I**, 483

Roesler, F.L., Scherb, F., Magee-Sauer, K., Harlander, J., Reynolds, R.J., Oliverson, R.J.; 1987, *Proc. Symp. Diversity and Similarity of Comets* (in press)

Samarasinha, N.H., A'Hearn, M.F., Hoban, S., Klinglesmith, D.A.; 1986, 20th ESLAB Symp. ESA SP-250, Vol. **I**, 487

Schlosser, W., Schulz, R., Koczet, P.; 1986, 20th ESLAB Symp. ESA SP-250, Vol. **III**, 495

Wallis, M.K., Meredith, N.P., Rees, D.; 1987, Proc. Symp. *Diversity and Similarity of Comets* (in press)

Wyckoff, S.; 1987, (private communication)

Astron. Astrophys. 187, 369–374 (1987)

ASTRONOMY
AND
ASTROPHYSICS

Pioneer Venus measurements of H, O, and C production in comet P/Halley near perihelion

A. I. F. Stewart

Laboratory for Atmospheric and Space Physics, and Department of Astrophysical, Planetary and Atmospheric Sciences, University of Colorado, Boulder, CO 80309, USA

Received February 17, accepted May 11, 1987

Summary. The Pioneer Venus Orbiter Ultraviolet Spectrometer measured the production of H, O, and C in comet Halley from late December 1985 to early March 1986. Water production rates obtained from the H data rose from 3.3 $10^{29}\,s^{-1}$ at 1 AU inbound to 1.2 $10^{30}\,s^{-1}$ at perihelion; there was a further increase to 1.6 $10^{30}\,s^{-1}$ followed by a slow decline to 1.4 $10^{30}\,s^{-1}$ at the time of the Vega-1 encounter. The H measurements show a periodic behavior that imply variations in the water production rate having structure within a 7.4-day sidereal period, in phase with similar variations seen in radical and dust production in March and April. Average H:O:C ratios of 1.0:0.7:.07 were measured from 10 to 26 days after perihelion.

Key words: comets – ultraviolet – photometry – spectroscopy

1. Introduction

Comet P/Halley passed within 0.27 AU of Venus on 1986 Feb. 3.8, and was 0.40 AU from Venus at perihelion on 1986 Feb. 9.5. The Pioneer Venus orbiter (Colin and Hall, 1977; Colin, 1979, 1980) was thus well placed to observe the comet during perihelion passage, when observation from Earth was difficult. The Pioneer Venus Orbiter Ultraviolet Spectrometer (PVOUVS, Stewart, 1980) made systematic measurements of the comet from 1985 Dec. 28 through 1986 Jan. 6, and again from 1986 Jan. 31 through Mar. 7. These observations in large part span the gaps in measurements made from Earth or Earth orbit due to the comet's superior conjunction.

PVOUVS measurements of the resonance UV emission from atomic hydrogen, oxygen, and carbon in Halley's coma are reported here. They are interpreted in terms of the water production rate and the O/H and C/H ratios in the gases subliming from the nucleus.

2. Observations

2.1. Orbital considerations

Figure 1 shows portions of the orbits of Halley, Venus, and Earth projected into the ecliptic plane. The PVOUVS observations occurred between the superior conjunctions of Halley as seen from Venus which occurred on 1985 Dec. 5.8 and 1986 Mar. 25.3. A spacecraft constraint required the comet to be 26°.5 or more from the Sun for observations to be possible; the orbital arcs begin and end on the first and last actual observing days, which were 1985 Dec. 28 and 1986 Mar. 7. Because Halley's perihelion lies considerably above the ecliptic plane, the comet remained at an acceptable angular distance from the sun throughout this period, as shown in Fig. 2a. Figure 2b shows the comet's distance from the Sun and from Venus for this period. No data were obtained during the period 1987 Jan. 7 – Jan. 30, because of Venus' angular proximity to the Sun (superior conjunction was on 1986 Jan. 19.1).

2.2. Spacecraft maneuvers

The PVOUVS has no articulation in its telescope; its optic axis is fixed at 59°.95 from the spacecraft spin axis. Therefore, to observe the comet the spacecraft must be appropriately maneuvered. In addition to the requirement of maintaining the spin axis about 60° away from the comet, it is necessary to keep the angle between the spin axis and the Sun in the range 86°.5 – 100°, for power and thermal control reasons, and the angle between the spin axis and the earth in the range 70° – 107°, to permit full gain communication. Usually, these latter requirements constrain the spin axis to within a few degrees of one of the ecliptic poles; the south pole is the normal location for observations of Venus. However, during Halley observations, the angular proximity of the Earth and Sun allowed the spacecraft to be rotated freely around the Venus-Sun direction. For operational reasons it was decided to maneuver the spacecraft spin axis once per day. More information on spacecraft operations while observing Halley is given by Jackson et al. (1987).

2.3. Instrument operation

The PVOUVS instrument has a field of view of 24 × 2.5 mrad, spectral resolution of about 1.4 nm, and sensitivities of 130, 240, and 100 counts/s/kR at 122, 130, and 166 nm respectively. Throughout the observations of Halley it was operated in its fixed-grating mode, in which its diffraction grating is held at the position appropriate for the emission being measured. The instrument then operates on a cycle synchronized with the spacecraft spin period (about 13 s): during each spin, data is collected from a swath of sky, stored in an internal buffer, and read out by the spacecraft telemetry system. Using the spacecraft sun-sensor pulse as a spatial reference, commands to the instrument were constructed so that this data arc was centered on the comet's position at

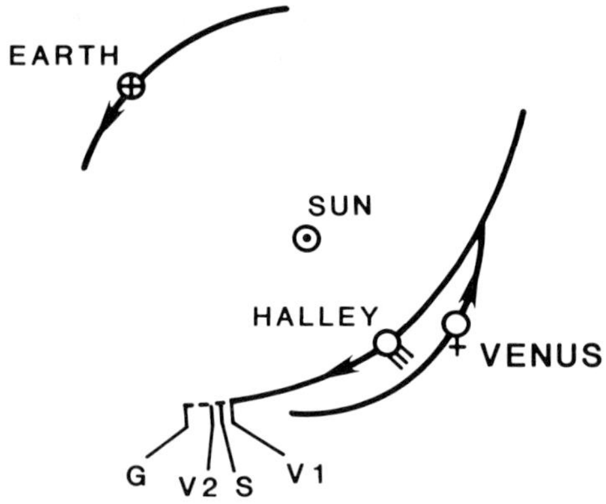

Fig. 1. Partial orbits of Halley, Venus, and Earth, for the period of Pioneer Venus observations of Halley. The symbols show the positions of the bodies at perihelion on February 9. The orbit of Halley is extended beyond the last Pioneer Venus observation to show its position at the encounters of Vega-1 (V1), Suisei (S), Vega 2 (V2), and Giotto (G)

0600 UT each day. The actual angular length of the arc depended on the downlink telemetry rate; it varied from 20° to 80°. For measurements of hydrogen, command parameters were varied to ensure an arc length of 40° or more, commensurate with the scale lengths of atomic hydrogen in the coma. Observations covering the entire spacecraft spin were also made so that the interplanetary Ly α signal could be measured and substracted from the cometary signal. The angular resolution along the data swath was determined by the instrument integration period, since the rotational smear was greater than the instrument field of view. It varied from 3.3 to 13.1 mrad.

2.4. Data charcteristics

The characteristics of the cometary signal recorded by PVOUVS depended on the activity level of the comet, its distance from the sun (r), its distance from Venus (δ), and on the downlink telemetry rate and quality. All of these factors varied substantially during Halley observations. Concerning the downlink, only the degradation and interruption of the telemetry during the major solar storm of February 3 and 4 need be mentioned here. Except for this period, the signal-to-noise in the Ly α data was uniformly satisfactory. Figure 3a shows the Ly α signal plotted against distance from the nucleus on 1986 Jan. 1, Feb. 9, and Mar. 7; on these dates the values of r and δ in AU were 1.00, 0.59, 0.80 and 1.49, 0.40, 1.38 respectively. It can be seen that the cometary signal substantially exceeded background at all times (the background itself increased from 170 R to 400 R between January 1 and March 7, as the comet moved across the sky). The effect of solar radiation pressure on the hydrogen coma is clearly seen on February 9, when the PVOUVS swaths crossed the coma in a nearly antisunward-sunward direction. It is also evident that the comet was intrinsically brighter on March 7 than on January 1, by a factor greater than can be accounted for by the differences in r and δ.

Figure 3b and c show the signals at 130 nm (atomic oxygen) and 166 nm (atomic carbon). These signals are much weaker, due to the smaller production rates and photon scattering rates of these species. The background for these measurements is set by the instrument dark count rate, which was $1-2\,s^{-1}$ during these observations. The noise-equivalent brightness is greater at 166 nm than at 130 nm because of the lesser sensitivity. The apparent asymmetry in the oxygen signal on March 7 is not statistically significant.

3. Data analysis

Typically, data were collected at a given wavelength for a period of an hour; before further analysis, all of these data were averaged together to improve the signal-to-noise. For most of the Halley observing period, these one-hour blocks contained the average of between 230 and 270 swaths across the coma; examples are shown in Fig. 3. The signal from regions near the nucleus were further averaged over a 39 mrad arc, to reduce the sensitivity of the analysis to the precise location of the nucleus with respect to the instrument line-of-sight, which is known only to $\pm 0.25°$. The quantity to be analyzed therefore represented the average brightness within an effective field of view measuring 39×24 mrad, whose projected size varied from 1.6×1.0 Gm at closest approach (February 3.8) to 8×5 Gm on March 7. These dimensions were always smaller than the scale lengths of atomic hydrogen, and always larger than those for atomic oxygen and carbon. Thus, the hydrogen analysis retained some spatial

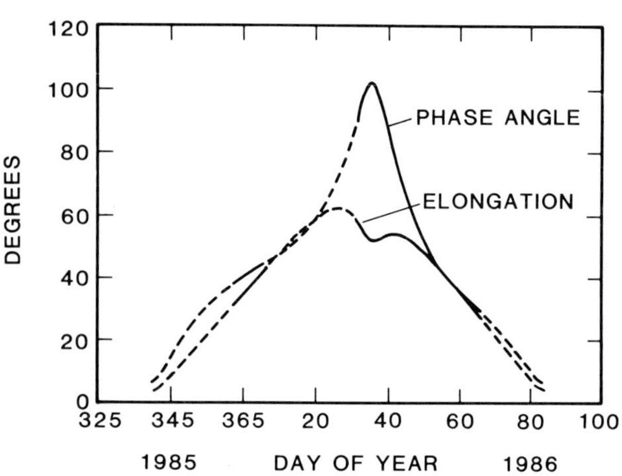

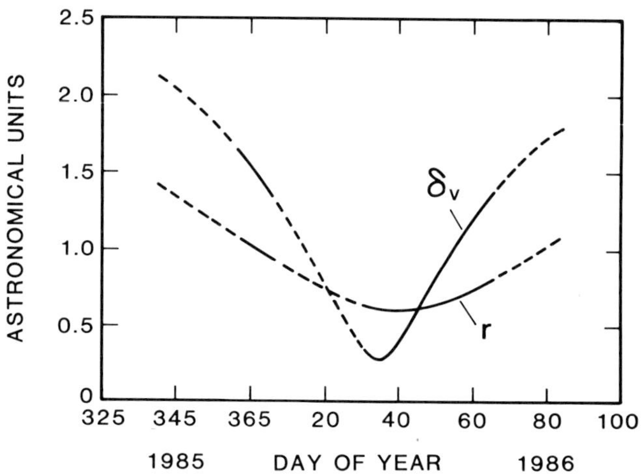

Fig. 2. Variation of observational parameters during the period December 1985–March 1986. The curves are solid during the times of Pioneer Venus observations. **a** Phase angle and elongation from the Sun. **b** Distances from the Sun (r) and Venus (δ_v), in AU

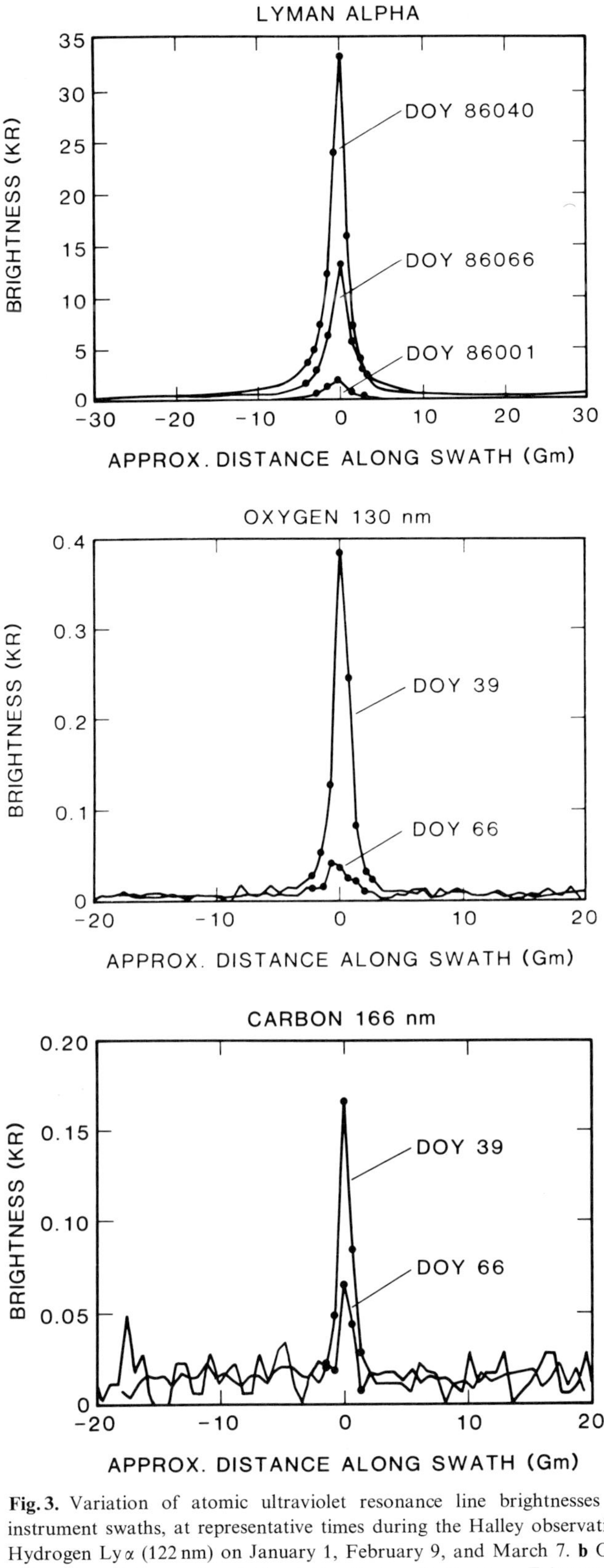

Fig. 3. Variation of atomic ultraviolet resonance line brightnesses along instrument swaths, at representative times during the Halley observations. **a** Hydrogen Ly α (122 nm) on January 1, February 9, and March 7. **b** Oxygen (130 nm) on February 8 and March 7. **c** Carbon (166 nm) on February 8 and March 7

Table 1. Parameters used in the radial-outflow coma model

Atom	g-factor[a] (s^{-1})	Lifetime[a] (s)	Velocity ($km\,s^{-1}$)	Scale length[a] (km)
H (fast)	1.26 10^{-3}	1.5 10^6	20	3.0 10^7
H (slow)	1.26 10^{-3}	1.5 10^6	8	1.2 10^7
O	4.0 10^{-6}	1.4 10^6	1.4	2.0 10^6
C	1.2 10^{-5}	2.4 10^6	1.0	2.4 10^6

[a] Values are given for a heliocentric distance of 1 AU

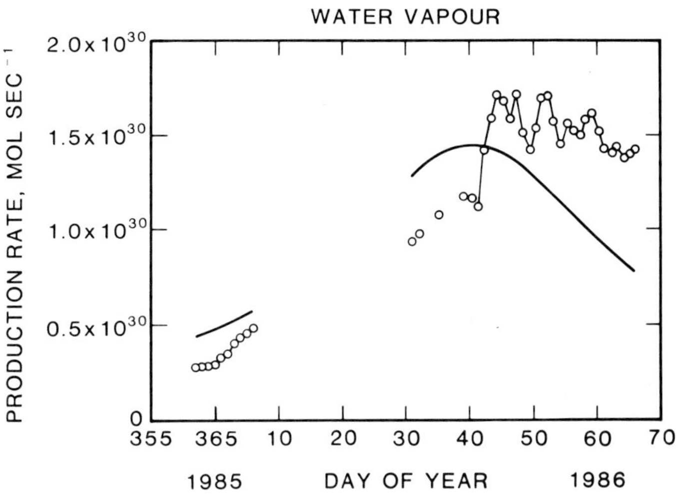

Fig. 4. Apparent water production rates obtained from Ly α measurements using a static radial-outflow coma model. The solid curve represents a rate equal to 5 $10^{29}\ r^{-2} s^{-1}$, for comparison

resolution, while the oxygen and carbon analyses were effectively of whole-coma measurements. The effective field-of-view dimensions were also smaller than the distances over which asymmetries in the H coma produced by solar radiation pressure were apparent.

The analysis was carried out by modeling the averaged brightness using a radial-outflow coma model. The model assumed that hydrogen atoms were produced in equal numbers at velocities of 8 and 20 $km\,s^{-1}$. Other model parameters are shown in Table 1, in which lifetimes, photon scattering rates (g-factors) and scale lengths are given at 1 AU. Lifetimes and scale lengths were assumed to vary as r^2; for the g-factors the Swings effect was also taken into account, as were the variations in solar line intensities measured daily by the SME satellite (G.J. Rottman, private communication). The water production rate was obtained from the H atom production rate by assuming that each water molecule on average yielded 1.8 hydrogen atoms.

4. Results

4.1. Water production rate

The water production rates obtained by the above modeling process are shown in Fig. 4. The data yielded daily values from 1985 Dec. 28, to 1986 Jan. 6, and again from 1986 Jan. 31 to Mar.

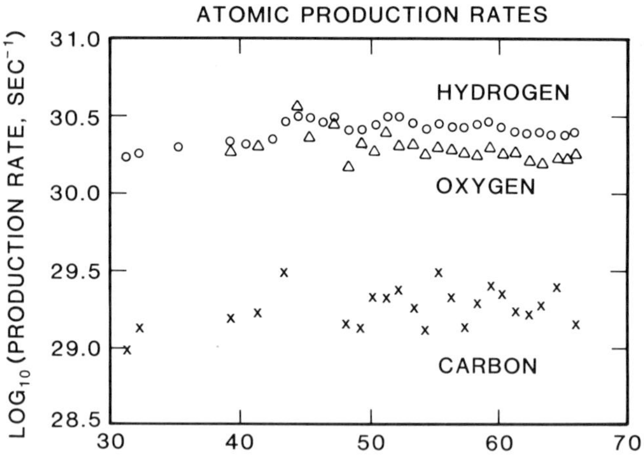

Fig. 5. Apparent production rates of hydrogen (○), oxygen (△) and carbon (×). The average rates for the period 8 February–7 March 1986 are: hydrogen, 2.7 10^{30} s^{-1}; oxygen, 1.9 10^{30} s^{-1}; carbon, 2.0 10^{29} s^{-1}; yielding H:O:C ratios of 1.0:0.7:0.07

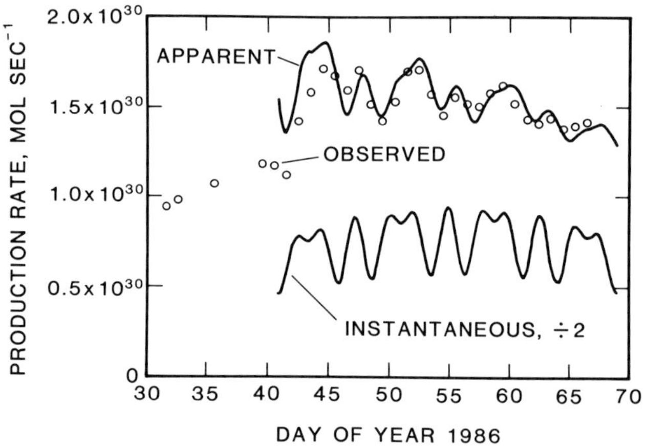

Fig. 6. Comparison between a periodic radial-outflow coma model and the water production rates from Fig. 4. Circles, apparent rates obtained from Ly α measurements; lower solid curve, assumed actual instantaneous rate, reduced by a factor of 2 for clarity; upper solid curve, modeled apparent rate. The assumed instantaneous rate has a synodic period of 7.7 days. The coefficients of the constant and first three harmonic terms are 1.0, −0.10, 0.15, and 0.15, giving a peak-to-trough ratio of 1.7 and an amplitude of 0.26

7, with the exceptions of February 2–3 and 5–7. Figure 4 compares the PVOUVS measurements with a model in which $Q(H_2O) = 5\ 10^{29}\, r^{-2}$ molecules s^{-1}; that is, a model in which all the sunlight falling on a spherical nucleus of 3 km radius (a nominal pre-1986 model) is used to sublimate water. The asymmetry about perihelion is immediately obvious; the ratio of the measured points to the symmetric model increases from about 0.6 in late December 1985 to about 1.7 in early March 1986.

4.2. Atomic oxygen and carbon

Figure 5 shows the production rates of H, O, and C during Febuary and March 1986. Measurements of O and C were planned only for the times when the nucleus was near the center of the instrument swath. Because of errors in spacecraft maneuvers during the first part of February, some of these measurements were executed with the nucleus near the edge of, or outside, the swath, and these measurements have large uncertainties. This problem disappeared about ten days after perihelion, and the subsequent scatter in the carbon and oxygen measurements are close to what would be expected from the counting statistics. During the last 17 days of observations, the O/H and C/H ratios remained essentially constant at 0.70 and 0.07 respectively.

4.3. Periodic behavior

A Fourier analysis applied to the water production rates in Fig. 4 showed no components with periods near 2.2 days, a value reported from 1910 telescopic data (Sekanina and Larson, 1986) and from early Suisei observations (Kaneda et al., 1985). However, in the post-perihelion data a structured component with a period near 8 days is apparent even to the eye. The structure consists of a broad (3–4 days) peak followed by a narrow (1-day) peak. The amplitude of the fluctuation decreases from 0.14 to 0.08 to .05 over the three cycles contained in the data. The dip near Dec. 31 may be identified with the minima on Feb. 9–10, Feb. 17–18, Feb. 26, and Mar. 5–6 if a synodic period of 8.1 days is adopted. If it is to be identified with the minima on Feb. 15, Feb. 23, and Mar. 3, a synodic period of 7.7. days is required.

Because the observed period is comparable to both the lifetime of H atoms and to the time required for H atoms to traverse the effective field-of-view used here, the amplitude of the variations in the actual production rate of water is likely to be much larger than the amplitude of the variations in the observed signal. A simple periodic source was used which reproduced the structure seen in CN, C_2, and dust by Millis and Schleicher (1986). This was fed into the radial-outflow model, and the resulting apparent production rates are shown in Fig. 6, together with the observed apparent rates from Fig. 4. It was found that the PVOUVS data in the second half of February and early March could be satisfactorily reproduced if the assumed source had a peak-to-trough ratio of about 1.7. It was also found that the phase lag in early March was 1 day, making it possible to identify the March 4 trough found by Millis and Schleicher with the March 5–6 trough in the PVOUVS data. The PVOUVS data from February and early March thus are in good agreement with the March and April observations of Millis and Schleicher, although they found a larger peak-to-trough ratio (about 2.5) than is required here.

5. Discussion

5.1. Water production rates

The water production rates derived here are of course sensitive to the parameters of the model. Specifically, the rates are proportional to v/g, the ratio of atom velocity to photon scattering rate. They also depend on the assumed scale length $v\tau$, where τ is the atoms' lifetime. Some experimentation with the model showed that reducing the lifetime from 1.5 10^6 to 1.0 10^6 s increased the rates by about 20%, while adopting an infinite lifetime decreased them by about 25%.

The rates obtained at the end of December 1985 agree very well with those obtained from the Internationel Ultraviolet Explorer (IUE) measurements of OH by Feldman et al. (1986). On February 24 McCoy et al. (1986) obtained a rocket image of the comet in Ly α, from which they deduced an H production rate of $1.9\ 10^{30}\ s^{-1}$, compared to the present value of $2.8\ 10^{30}\ s^{-1}$ on the same date. The difference is entirely accounted for by the different model parameters used. The Vega-1 values on March 6 were $1\ 10^{30}\ s^{-1}$ (Curtis et al., 1986), and $1.3\ 10^{30}\ s^{-1}$ (Remizov et al., 1986), somewhat less than the present value of $1.4\ 10^{30}\ s^{-1}$. Thus, the PVOUVS measurements are in good to excellent agreement with simultaneous measurements by other techniques.

The integrated loss of water during the times when Halley was observed by the PVOUVS totals $1.8\ 10^{11}$ kg, if the loss rate is assumed to vary as r^{-2} during the superior conjunction gap in January. Combined with the IUE measurements (Feldman et al., 1986), the orbit-integrated value is of order $2.7\ 10^{11}$ kg, very close to the estimate given by Whipple (1986).

5.2. *O/H and C/H ratios*

Since the production rate of O and C shown in Fig. 5 were obtained from essentially whole-coma measurements, for which the luminosity $L = Q\ g\tau$, they are relatively insensitive to v: for carbon, increasing v from 1 to 5 km $^{-1}$ results in an increase in the derived atomic production rate of only about 50%. They are, however, sensitive to the quantity $g\tau$. The uncertainties in g and τ are substantial, especially for O for which the Swings effect is large in early March (Feldman et al., 1976). The nominal H:O:C ratios of 1.00:0.70:0.07 are consistent with an H_2O:CO_2 ratio of 7:1 or an H_2O:CO ratio of 4:1. The large difference is because approximately half of any CO produced from the nucleus is converted to CO^+ (M.R. Combi, private communication) and removed by the solar wind, without producing a detectable C atom. Since direct measurements of CO_2 from Giotto (Krankowsky et al., 1986) suggest an H_2O/CO_2 ratio of nearly 30, the present results suggest that CO is more abundant as a parent molecule than CO_2. Indeed, the present results are consistent with the results of Eberhardt et al. (1986) from Giotto, who find that $CO/CO_2 = 1.4$–4.2. They are also in good accord with the rocket measurements of CO by Woods et al. (1986).

5.3. *Periodic behavior*

The periodic variations in the PVOUVS Ly α measurements are in general agreement with the variations seen in radical emissions and dust continuum by other workers (Millis and Schleicher, 1986; Samarasinha et al., 1986) as regards their period, phase, and amplitude. A puzzling question is why the Ly α images from Suisei (Kaneda et al., 1986) apparently show the 2.2-day period reported by Sekanina and Larson (1986) and subsequently found in other data (Belton et al., 1986), while the PVOUVS sees the 7.4-day (sidereal) period. Because the 2.2-day period is seen most strongly in its first harmonic, it would be difficult to detect with the analysis techniques used in this paper, because of the 1-day observing cycle and the subsequent daily averaging. There is unfortunately almost no overlap between Suisei and Pioneer Venus measurements; however, the dim Suisei image of February 25.6 and the bright image of February 28.4 are matched by a trough and a peak, respectively, in Fig. 4.

6. Conclusions

Pioneer Venus OUVS Ly α measurements yield water production rates rising from about $3.3\ 10^{29}\ s^{-1}$ at 1 AU inbound to $1.2\ 10^{30}\ s^{-1}$ at perihelion, followed by a rapid increase to $1.6\ 10^{30}\ s^{-1}$ and a slower decline to $1.4\ 10^{30}\ s^{-1}$ by 0.8 AU outbound. These measurements are in good to excellent accord with simultaneous measurements by other techniques.

The total water loss for the 1986 apparition, based on Pioneer Venus near-perihelion measurements and IUE measurements further out, is about $2.7\ 10^{11}$ kg.

The H:O:C ratios measured by Pioneer Venus are consistent with compositions satisfying the relationship $4\,Q_{CO} + 7\,Q_{CO_2} = Q_{H_2O}$.

The Pioneer Venus Ly α measurements show variations consistent with a structured, 7.4-day (sidereal) periodicity in the production of H_2O from the nucleus. The period, phase, and to a lesser extent the amplitude of these variations are consistent with similar periodicities detected by other techniques.

Acknowledgements. This work was performed under NASA contract NAS2-9477 and NASA grant NAGW-389. I thank my colleagues on the Pioneer Venus Science Steering Group for their support and for foregoing Venus observations while these observations of Halley were made. It is a pleasure to thank the Pioneer Project Office for its enthusiasm, ingenuity, and hard work in making the observations possible, and D.K. Chantala of the University of Colorado for her efficiency and dedication in providing computational and operational support.

References

Belton, M.J.S., Wehinger, P., Wyckoff, S., Spinrad, H.: 1986, 20th ESLAB Symp., ESA SP-250, Vol. **I**, p. 599

Colin, L.: 1979, *Science* **203**, 743

Colin, L.: 1986, *J. Geophys. Res.* **85**, 7575

Colin, L., Hall, C.F.: 1977, *Space Sci. Rev.* **20**, 283

Curtis, C.C., Fan, C.Y., Hsieh, K.C., Hunten, D.M., Ip, W.-H., Keppler, E., Richter, A.K., Umlauft, G., Afonin, V.V., Dyachkov, A.V., Erö, J., Jr., Somogyi, A.J.: 1986, 20th ESLAB Symp., ESA SP-250, Vol. **I**, p. 391

Eberhardt, P., Krankowsky, D., Schulte, W., Dolder, U., Lämmerzahl, P., Berthelier, J.J., Woweries, J., Stubbemann, U., Hodges, R.R., Hoffman, J.H., Illiano, J.M.: 1986, 20th ESLAB Symp., ESA SP-250, Vol. **I**, p. 383

Feldman, P.D., Opal, C.B., Meier, R.R., Nicolas, K.R.: 1976, in *The Study of Comets* NASA SP-393, Vol. **2**, p. 773

Feldman, P.D., Festou, M.C., A'Hearn, M.F., Arpigny, C., Butterworth, P.S., Cosmovici, C.B., Danks, A.C., Gilmozzi, R., Jackson, W.M., McFadden, L.A., Patriarchi, P., Schleicher, D.G., Tozzi, G.P., Wallis, M.K., Weaver, H.A., Woods, T.N.: 1986, 20th ESLAB Symp., ESA SP-250, Vol. **I**, p. 325

Jackson, R.W., Lozier, D.W., Phillips, J.R., Smith, M.A.: 1987, paper no. AIAA-87-0500 presented at 25th AIAA Aerospace Sciences Meeting, Reno, January 1987

Kaneda, E., Hirao, K., Takagi, M., Ashihara, O., Itoh, T., Shimizu, M.: 1986, *Nature* **320**, 140

Kaneda, E., Takagi, M., Hirao, K., Shimizu, M., Ashihara, O.: 1986 20th ESLAB Symp., ESA SP-250, Vol. **I**, p. 397

Krankowsky, D., Lämmerzahl, P., Herrwerth, I., Woweries, J., Eberhardt, P., Dolder, U., Herrmann, U., Schulte, W., Berthelier, J.J., Illiano, J.M., Hodges, R.R., Hofman, J.H.: 1986, *Nature* **321**, 326

McCoy, R. P., Opal, C. B., Carruthers, G.R.: 1986, 20th ESLAB Symp., ESA SP-250, Vol. **I**, p. 403

Millis, R. L., Schleicher, D. G.: 1986, *Nature* **324**, 646

Remizov, A. P., Verigin, M.I., Gringauz, K.I., Apáthy, I., Szemerey, I., Gombosi, T.I., Richter, A.K.: 1986, 20th ESLAB Symp., ESA SP-250, Vol. **I**, p. 387

Samarasinha, N.H., A'Hearn, M.F., Hoban, S., Klinglesmith, D. A.: 1986, 20th ESLAB Symp., ESA SP-250, Vol. **I**, p. 487

Sekanina, Z., Larson, S. M.: 1986, *Astron. J.* **92**, 462

Stewart, A.I.F.: 1980, *IEEE Trans. Geosci. Remote Sensing* GE-18, 65

Whipple, F. L.: 1986, 20th ESLAB Symp., ESA SP-250, Vol. **II**, p. 281

Woods, T.N., Feldman, P.D., Dymond, K.F., Sahnow, D.J.: 1986, *Nature* **324**, 436

Astron. Astrophys. 187, 375–379 (1987)

ASTRONOMY AND ASTROPHYSICS

Anisotropy of the neutral gas distribution of comet P/Halley deduced from NGE/Vega 1 measurements

K.C. Hsieh [1], C.C. Curtis [1], C.Y. Fan [1], D.M. Hunten [2], W.-H. Ip [3], E. Keppler [3], A.K. Richter [3], G. Umlauft [3], V.V. Afonin [4], J. Erö, Jr. [5], and A.J. Somogyi [5]

[1] Department of Physics, University of Arizona, Tucson, AZ 85721, USA
[2] Department of Planetary Sciences, University of Arizona, Tucson, AZ 85721, USA
[3] Max-Planck-Institut für Aeronomie, D-3411 Katlenburg-Lindau 3, Federal Republic of Germany
[4] Space Research Institute, Profsoyuznaya 84/32, SU-117810 Moscow, USSR
[5] Central Research Institute for Physics, P.O. Box 49, H-1525 Budapest, Hungary

Received January 27, accepted May 12, 1987

Summary. The density of neutral gas near comet Halley was measured by the Neutral Gas Experiment on the Vega 1 spacecraft. Plots of gas density vs. distance to the comet nucleus show an asymmetry between the inbound and the outbound legs of the spacecraft flyby. A multidimensional analysis is presented. It permits gas densities detected at distances between 8.9 10^3 and 7.2 10^4 km to be mapped back to source regions on or near the comet nucleus, thus providing information on the spatio-temporal properties of the source of neutral gas in comet Halley. Sample results of the mapping-back process using an outflow velocity of 0.9 km s^{-1}, scale lengths of 2.9 10^5 km and 7.0 10^4 km, and the faster rotation period of 2.09 days are presented.

Key words: comets

1. Introduction

The preliminary report of the Neutral Gas Experiment (NGE) on Vega 1 showed an asymmetry in the counting rates, with respect to the distance R to the nucleus, between the inbound and the outbound legs of the comet Halley fly-by on 6 March 1986 (Keppler et al., 1987). This asymmetry, confirmed by Curtis et al. (1987), and the consequent deviation from a model of spherical expansion with losses for the neutral gas, must be caused by variations in the gas production rate in specific regions on or near the surface of the comet nucleus. To map these anisotropies back to the nucleus, one needs to know the outflow velocity V of the gas, the scale length L of the effective loss mechanisms, and the rotation rate Ω of the comet nucleus about a known axis. Using available information and one simplifying approximation regarding the spin axis, we show a qualitative view of the consequences of such an analysis. Examples of our results are given. Comparison of our measurements with ion, dust, and optical observations from Vega 1, UV observations from Suisei, and ground-based observations could yield the true values of L, V, and Ω; conversely, using confirmed values of L, V, and Ω, our analysis could lead to a better understanding of the surface features and gas dynamics of comet Halley.

Send offprint requests to: K.C. Hsieh

2. Data

Curtis et al. (1987) showed that after subtracting constant background rates from the total counting rates of FIS (Field Ionization Spectrometer) and EIS (Electron Impact Spectrometer) of NGE on Vega 1, the instruments agreed well in both the inbound and the outbound legs when EIS was not saturated, i.e. when the counting rate decreased with decreasing R (see their Fig. 2). We shall concentrate on the data collected at $R < 10^5$ km. For completeness, the five EIS data points between 72,095 and 57,030 km on the inbound leg are added to the 79 FIS data points between 55,865 km on the inbound leg and 72,173 km on the outbound leg to form the database. The two FIS data points at about 7 10^4 km on the inbound leg are excluded due to their large deviation from the smooth curve.

3. Analysis

The measured neutral gas density clearly deviates from spherical symmetry. The fact that the gas leaves the comet at speeds much less than the speed of the spacecraft relative to the comet makes it clear that gas sampled comes from different portions of the rotating nucleus at various times. Ground observations also show that the gas production varies from region to region and from time to time (Whipple, 1978; Green and Morris, 1986, and references therein). Therefore, it is reasonable that in situ neutral gas data be analyzed not only in terms of R, but with consideration of the other spatial variables and time.

Figure 1 shows the geometry of the Halley/Vega 1 system near the closest approach. The Vega 1 trajectory forms an angle $\eta = 9\overset{\circ}{.}4$ with the orbital plane of comet Halley. The location of the spacecraft at time t is given by (R, θ, ϕ). At closest approach t_0 ($07^h20^m06^s$ UT on March 6), the spacecraft is at $b = 8889$ km, $\theta_0 = 22\overset{\circ}{.}6$, $\phi_0 = 72\overset{\circ}{.}1$.

The controversy over the rotation periods, whether it is 2.2 days (Sekanina and Larson, 1986) or 7.4 days (Millis and Schleicher, 1986), seems to have settled to the understanding that the nucleus has two rotational motions, one having a longer period, 7.4 or possibly 14.6 days, about the long axis of the nucleus, and the other having a shorter period of 2.09 days about an axis nearly perpendicular to the comet's orbital plane (Festou

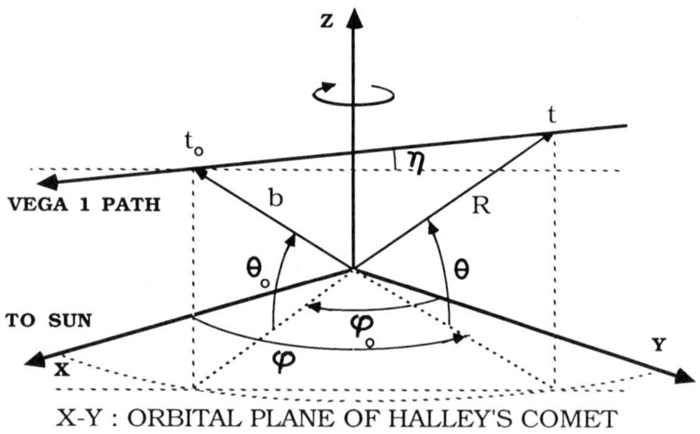

Fig. 1. The trajectory of Vega 1 is shown in the Halley-Sun frame. At closest approach, $t_0 = 07^h20^m06^s$ UT on 1986 March 6, $b = 8889$ km, $\theta_0 = 22.6°$, $\phi_0 = 72.1°$

et al., 1987). It is the motion with the shorter period that affects most significantly the spatio-temporal distributions of gas and dust expanding from the nucleus (Festou, 1987; Larson, 1987). This is especially true for observations such as ours during the fly-by, since the sampling duration is much shorter than 2.09 days. To demonstrate the method of analysis and to gain a qualitative view of the possible distributions at or near the source regions on the nucleus, we consider only the faster rotation and set the axis of this 2.09 day rotation normal to the comet's orbital plane, as shown in Fig. 1. This approximation will provide us a rough idea on the relative gas emission rates at the source regions directly connected to Vega 1, assuming a radially expanding gas.

For a radially expanding neutral gas that is subject to losses, the density at any point in the coma can be written in the form

$$n(R, \theta, \phi, t) = a_0 f R^{-2} \exp(-R/L), \quad (1)$$

where a_0 is a normalization constant, $n_0 R_0^2$. The function f is the measure of deviation from the simple R-dependence at (R, θ, ϕ, t), and also represents the relative gas density or production rate at some source point on or near the nucleus at an earlier time $(R_0, \theta, \Sigma, t')$. The angle θ does not change in the process, because of our choice of the axis of rotation. The angle Σ, fixed to the coordinate system that corotates with the nucleus about the axis normal to the orbital plane, and measured from the Halley-sun line at t_0, is given by

$$\Sigma = \phi - \Omega R/V + \Omega(t - t_0). \quad (2)$$

One can now deduce from the measured density $n(R, \theta, \phi, t)$, for a given L, an anisotropy function f which can be mapped back to a source region $(R_0, \theta, \Sigma, t')$ on or near the surface of the nucleus to represent the relative gas production rate from that region at a time $t' = t - R/V$. To obtain a qualitative picture of (1) the observed anisotropy along the Vega 1 path during the fly-by, and (2) the relative gas emission rates at possible locations of the sources of the measured anisotropy, we choose $\Omega = 1.99\ 10^{-3}$ deg s^{-1} for a rotation period $P = 2.09$ days of Festou et al. (1987) which is very close to that given by Sekanina and Larson (1986); and an outflow velocity of $V = 0.9$ km s^{-1} (Festou, 1987), which is not quite in agreement with Larson et al. (1986). For L, the scale length for the neutral gas, we use two values, 2.9 10^5 km and 7.0 10^4 km. The shorter L is that for the photodissociation of H_2O at 1 AU (Festou, 1981), while the longer L is obtained from fitting Eq. (1) (without f) to our outbound data. A similar process applied to our inbound data yields $L = 1.7\ 10^5$ km. We choose the two more divergent values to demonstrate the effect of L in the analysis. A discussion on the significance of L will be given in Sect. 4 of this paper.

The normalization factor a_0 in Eq. (1) is set at 5.6 10^{13} cm^{-3} km^2, based on the least-squares fit of Eq. (1), again without f, to our outbound data. This value corresponds to a density of 5.6 10^{11} mol cm^{-3} at a cometocentric distance of 10 km, or a production rate of 6.3 10^{29} mol s^{-1} for an outflow speed of 0.9 km s^{-1}. If the inbound data were used, the corresponding values would be 8.9 10^{13} cm^{-3} km^2, 8.9 10^{11} mol cm^{-3}, and 1.0 10^{30} mol s^{-1}. The lower values from the outbound data are chosen so that the resulting f will be >1, an arbitrary choice for convenience. Independent of the value of a_0, the resulting f gives only the relative anisotropy, but $a_0 f$ gives the absolute production rate at a given source region.

Using the selected values for the parameters stated above, we expect a relative error in $f < 20\%$; the largest error comes from L, especially for the case $L = (7.0 \pm 1.0)\ 10^4$ km, i.e. 14%.

4. Results and discussion

The anisotropy f as a function of ϕ is shown in Fig. 2 for the two values of L. For both values of L, f shows a minimum near the Halley-Sun line ($\phi = 0°$), and higher values at $\phi > 60°$ and $\phi < -40°$. In these ranges of ϕ, which correspond to large cometocentric distances R, the effect of L is clearly manifest. For a smaller L, say, 7.0 10^4 km, a larger f is required to compensate the more rapid exponential decrease at larger R; similarly, a larger L, say, 2.9 10^5 km, needs a smaller f at greater R. We shall come back to the discussion on L later.

Figure 2 describes only the deviations from $a_0 R^{-2} \exp(-R/L)$ along the Vega 1 path. How one compares these results directly with other observations is not immediately obvious. However, the results of directly tracing these anisotropies back to locations on or near the surface of the nucleus may be more readily interpreted, and thus more useful in correlative studies.

Using the values of V and Ω stated earlier, Eq. (2) yields, as shown in Fig. 3, a track of gas emitting regions on the nucleus seen by NGE/Vega 1. The angle $\theta = 0°$ represents the cut through the nucleus by the orbital plane at $t = t_0 = 07^h20^m06^s$ UT on March 6, the time of closest approach. The angle $\Sigma = 0°$ signifies the Halley-Sun direction at t_0. According to the television observations of both Vegas and Giotto (Sagdeev et al., 1986, and Wilhelm et al., 1986), the Halley-Sun line at t_0 nearly coincided with the long axis of the nucleus with the bigger end facing the sun. The angle Σ lies in the orbital plane and is defined with respect to this particular orientation. Figures 3 shows that NGE/Vega 1 observations did not cover the full range of Σ and even a narrower range in θ.

The relative anisotropies along the Vega 1 trajectory, which are also the relative emission rates at the source regions, shown in Fig. 3, are displayed in Figs. 4 and 5 as functions of Σ, for the respective values of L, 7.0 10^4 km and 2.9 10^5 km. The effect of L shown in Fig. 2 is carried over into Figs. 4 and 5.

Not shown in these figures are the times of emission, t'. More complete information is provided in Table 1, in which the times are given in UT (hr:min:sec); the minus signs designate March 5 and plus signs March 6. Also listed in Table 1 are the two relative anisotropies or emission rates f_1 and f_2. It is now easier to see that the density observed by NGE/Vega 1 at (R, ϕ, θ, t) is from a source region on the nucleus emitted at an earlier time $(R_0, \Sigma, \theta, t')$.

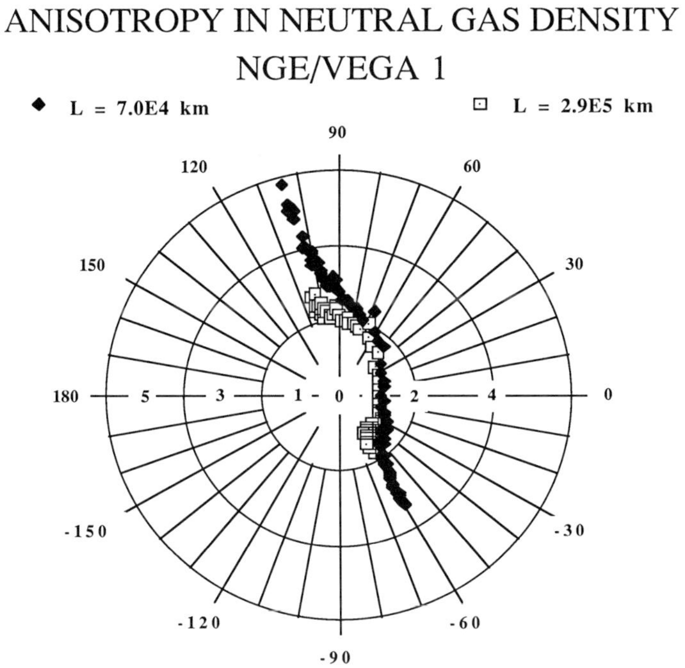

Fig. 2. The function f vs. angle ϕ with $a_0 = 5.6\ 10^{13}\ \mathrm{cm}^{-3}\,\mathrm{km}^2$ for $L = 2.9\ 10^5$ km and $7.0\ 10^4$ km. The angle 0° points to the Sun. The smaller L yields higher anisotropy. It appears that there is more gas in the morning sector

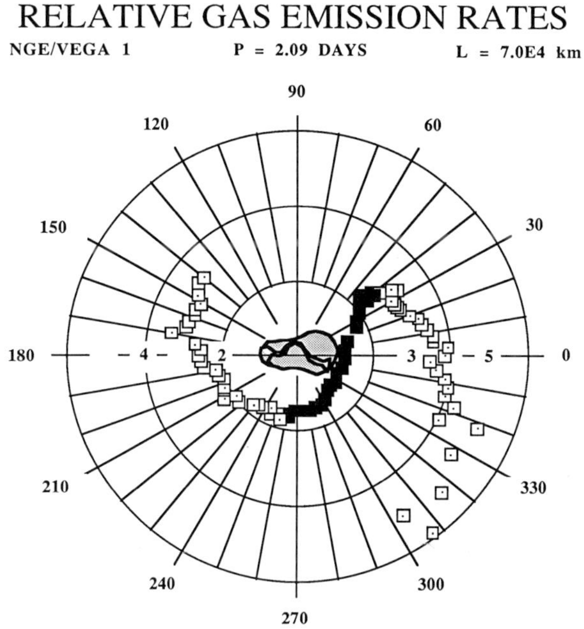

Fig. 4. The function $f(L = 7.0\ 10^4$ km and $P = 2.09$ days), f_1 in Table 1, mapped back to the nucleus. The angles are Σ's. The 0°–180° line is the long axis of the nucleus with the bigger end at 0°. The θ- and t'-dependence are given in Table 1

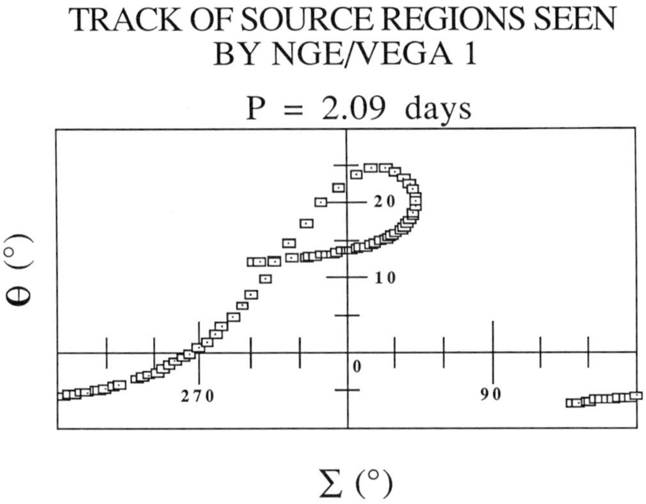

Fig. 3. Locations (θ, Σ) of the source regions connected to Vega 1 are shown on a reference sphere that rotates about an axis normal to the orbital plane with a period of 2.09 days. The meridian ($\Sigma = 0°$) is defined by the Halley-Sun line at the closest approach, t_0. Samples from the same Σ are from different θ's. The emission time at each site and the relative emission rate are given in Table 1. The rotation about the axis $\Sigma = 0°$ is ignored here

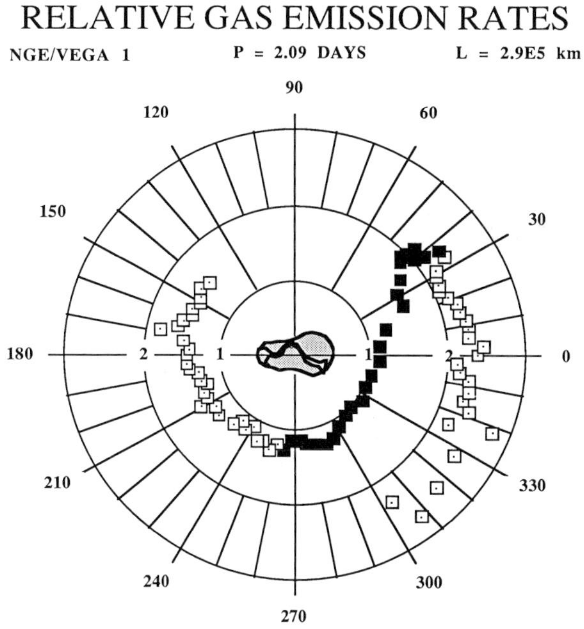

Fig. 5. Same as Fig. 4, except the scale length $L = 2.9\ 10^5$ km. (See f_2 in Table 1)

This mapping-back process depends explicitly on the outflow velocity V and the model of rotation. The deduced relative gas production rates at the specific sources, however, depend also on the scale length L. We considered two values of L in this report for good reasons. First, there is not yet a definitive value of L that we can use. For example, although the L for H_2O at 1 AU is supposed to be $(7.0 \pm 1.0)\ 10^4$ km (Festou, 1987), Krankowsky et al., (1986) and Weaver et al., (1986) reported only about half of that. Second, due to loss of mass resolution (Curtis et al., 1987), what NGE/Vega 1 measured may well include the photodissociation products of H_2O, i.e., OH and O, whose scale length is of the order of $2\ 10^5$ km at 1 AU (Weaver et al., 1979; Krankowsky et al., 1986; Festou, 1987). Third, there may be other mechanisms that extend the range of most abundant neutral molecules,

Table 1. NGE/Vega 1 neutral density data and deduced source parameters

R(km)	θ (°)	ϕ (°)	log n	f_1	f_2	Σ (°)	t′ (UT)	t (UT)	R(km)	θ (°)	ϕ (°)	log n	f_1	f_2	Σ (°)	t′ (UT)	t (UT)
72095	12.12	105.05	4.29	5.1	2.3	303.4	-08:49:57	07:05:02	9548	17.34	- 4.16	5.77	1.1	1.0	334.8	+04:24:01	07:20:50
70523	12.18	104.91	4.38	5.8	2.7	306.8	-09:19:24	07:05:22	10234	14.64	-12.30	5.74	1.2	1.1	325.1	+04:11:38	07:21:10
66437	12.34	104.50	4.41	5.2	2.6	315.6	-10:35:56	07:06:14	11104	12.10	-19.14	5.65	1.2	1.0	316.4	+03:55:52	07:12:29
61103	12.59	103.86	4.48	4.8	2.5	326.9	-12:15:50	07:07:22	12121	9.82	-24.80	5.59	1.2	1.1	308.5	+03:37:22	07:21:49
57030	12.81	103.30	4.49	4.1	2.2	335.5	-13:32:08	07:08:14	13249	7.84	-29.48	5.54	1.3	1.1	301.4	+03:16:48	07:22:09
55856	12.88	103.13	4.61	5.0	2.8	337.9	-13:54:07	07:08:30	14463	6.15	-33.36	5.49	1.4	1.2	294.9	+02:54:40	07:22:30
54292	12.97	102.87	4.58	4.3	2.4	341.2	-14:23:25	07:08:49	15743	4.70	-36.61	5.43	1.5	1.3	288.8	+02:31:17	07:22:50
52730	13.07	102.60	4.58	4.0	2.3	344.4	-14:52:41	07:09:09	17075	3.46	-39.34	5.35	1.5	1.2	283.2	+02:06:57	07:23:10
51168	13.18	102.33	4.62	4.1	2.3	347.6	-15:21:56	07:09:29	18446	2.39	-41.68	5.27	1.5	1.2	277.8	+01:41:54	07:23:30
49609	13.30	102.02	4.64	3.9	2.3	350.8	-15:51:08	07:09:50	19849	1.47	-43.68	5.19	1.5	1.2	272.8	+01:16:15	07:23:49
48050	13.42	101.71	4.65	3.7	2.2	354.0	-16:20:21	07:10:10	21278	0.68	-45.41	5.13	1.5	1.2	267.9	+00:50:07	07:24:09
46494	13.55	101.36	4.67	3.5	2.1	357.1	-16:49:30	07:10:30	22727	-0.02	-46.93	5.10	1.6	1.3	263.2	+00:23:37	07:24:29
44940	13.68	100.99	4.75	3.9	2.4	0.3	-17:18:36	07:10:49	24194	-0.63	-48.25	5.03	1.6	1.2	258.7	-23:56:47	07:24:50
43388	13.83	100.60	4.80	3.9	2.5	3.3	-17:47:41	07:11:09	25748	-1.20	-49.49	5.01	1.8	1.3	254.0	-23:28:22	07:25:10
41838	13.99	100.17	4.80	3.6	2.3	6.4	-18:16:43	07:11:29	27166	-1.66	-50.48	4.93	1.7	1.2	249.9	-23:02:25	07:25:30
40291	14.15	99.72	4.84	3.6	2.3	9.4	-18:45:42	07:11:50	28668	-2.09	-51.42	4.89	1.7	1.3	245.7	-22:34:56	07:25:49
38747	14.33	99.22	4.87	3.5	2.3	12.4	-19:14:37	07:12:10	30179	-2.48	-52.26	4.79	1.5	1.1	241.6	-22:07:17	07:26:09
37206	14.53	98.68	4.90	3.3	2.2	15.3	-19:43:30	07:12:30	31696	-2.83	-53.03	4.77	1.7	1.2	237.5	-21:39:32	07:26:29
35670	14.74	98.09	4.94	3.3	2.2	18.2	-20:12:16	07:12:50	33220	-3.15	-53.72	4.71	1.6	1.1	233.4	-21:11:38	07:26:50
34137	14.96	97.46	4.96	3.1	2.1	21.0	-20:40:59	07:13:09	34750	-3.44	-54.35	4.70	1.8	1.2	229.5	-20:43:38	07:27:10
32610	15.21	96.75	4.99	3.0	2.1	23.7	-21:09:36	07:13:29	39364	-4.17	-55.96	4.61	2.0	1.3	217.7	-19:19:12	07:28:09
31088	15.48	95.98	5.04	3.0	2.1	26.3	-21:38:07	07:13:49	40909	-4.38	-56.41	4.56	2.0	1.3	213.9	-18:50:55	07:28:29
29573	15.77	95.13	5.09	2.9	2.1	28.9	-22:06:31	07:14:10	42457	-4.57	-56.83	4.58	2.2	1.4	210.1	-18:22:35	07:28:49
28066	16.09	94.17	5.14	2.9	2.1	31.3	-22:34:45	07:14:30	44008	-4.75	-57.21	4.51	2.1	1.3	206.3	-17:54:12	07:29:10
26568	16.44	93.11	5.23	3.1	2.4	33.6	-23:02:49	07:14:50	45561	-4.91	-57.59	4.47	2.1	1.3	202.5	-17:25:46	07:29:30
25080	16.82	91.92	5.28	3.1	2.3	35.7	-23:30:43	07:15:09	47116	-5.07	-57.93	4.41	2.0	1.2	198.8	-16:57:18	07:29:50
23605	17.25	90.57	5.30	2.8	2.2	37.7	-23:58:22	07:15:29	48674	-5.21	-58.24	4.41	2.2	1.3	195.1	-16:28:47	07:30:09
22145	17.72	89.03	5.33	2.6	2.0	39.4	+00:25:44	07:15.49	50232	-5.34	-58.54	4.38	2.2	1.3	191.3	-16:00:16	07:30:29
20704	18.24	87.27	5.39	2.5	2.0	40.9	+00:52:45	07:16:10	51793	-5.47	-58.82	4.39	2.5	1.4	187.7	-15:31:42	07:30:49
19285	18.81	85.22	5.47	2.6	2.1	42.0	+01:19:22	07:16:30	53354	-5.59	-59.09	4.37	2.6	1.4	184.0	-15:03:07	07:31:10
17893	19.46	82.85	5.51	2.4	2.0	42.8	+01:45:28	07:16:50	54918	-5.70	-59.33	4.34	2.6	1.4	180.3	-14:34:29	07:31:30
16537	20.16	80.05	5.58	2.4	2.0	43.0	+02:10:55	07:17:09	56482	-5.81	-59.57	4.30	2.6	1.4	176.6	-14:05:52	07:31:50
15224	20.94	76.73	5.64	2.2	1.9	42.7	+02:35:34	07:17:29	58047	-5.91	-59.79	4.30	2.8	1.5	173.0	-13:37:13	07:32:09
13969	21.77	72.75	5.70	2.1	1.8	41.5	+02:59:08	07:17:49	59614	-6.00	-60.00	4.36	3.4	1.8	169.3	-13:08:32	07:32:29
12786	22.64	67.97	5.83	2.4	2.1	39.4	+03:21:23	07:18:09	61181	-6.09	-60.20	4.28	3.1	1.6	165.7	-12:39:51	07:32:49
11699	23.49	62.17	5.83	2.0	1.7	36.0	+03:41:51	07:18:30	62749	-6.18	-60.39	4.24	3.0	1.5	162.1	-12:11:08	07:33:09
10737	24.21	55.17	5.87	1.8	1.6	31.2	+04:00:00	07:18:50	64318	-6.26	-60.57	4.20	2.9	1.5	158.5	-11:42:25	07:33:30
9935	24.65	46.86	5.93	1.7	1.6	24.7	+04:15:11	07:19:10	65888	-6.34	-60.74	4.18	3.0	1.5	154.8	-11:13:41	07:33:50
9336	24.58	37.25	5.89	1.4	1.3	16.5	+04:26:36	07:19:29	67458	-6.41	-60.90	4.13	2.9	1.4	151.2	-10:44:56	07:34:10
8979	23.81	26.68	5.88	1.2	1.1	6.7	+04:33:33	07:19:49	69029	-6.48	-61.06	4.13	3.1	1.5	147.6	-10:16:11	07:34:29
8895	22.23	15.78	5.88	1.2	1.1	356.1	+04:35:26	07:20:09	70601	-6.55	-61.20	4.12	3.2	1.5	144.0	-09:47:24	07:34:49
9090	19.98	5.30	5.84	1.2	1.1	345.2	+04:32:09	07:20:30	72173	-6.61	-61.35	4.09	3.2	1.5	140.5	-09:18:37	07:35:09

H_2O, into large distances from the comet. For example, H_2O may sublime from icy grains (Curtis et al., 1987), which could possibly explain the observation of Gringauz et al. (1986), giving $L = 2\ 10^6$ km. Fourth, the values of L determined by the observations mentioned above were all based on a steady state spherically symmetric model of the expanding gas, which we now know is not the case. Therefore, we did not use only one L value; and in doing so, we also demonstrated the effect of L on the deduced relative production rates at the sources.

We have ignored any rotation about the long axis of the nucleus, which may have a period of 7.6 or 14.6 days. From Table 1 we see that the gas sampled covers a 19.7 hour span in t'. During this time interval, the nucleus could have rotated about the long axis at most 12°.3 or as little as 6°.2. This sets the possible error in θ.

The results of our analysis show that there is a decrease in gas emission between 08^h50^m UT March 5 and 04^h35^m UT March 6. This is qualitatively consistent with ground based observations (Schleicher et al., 1986).

5. Conclusion

The observed asymmetry in the neutral density along the path of Vega 1 with respect to the point of closest approach to comet Halley (Fig. 2 of Curtis et al., 1987) led us to interpret our observation in terms of a simple but realistic model of the nucleus and the expanding gas. We considered only a single rotation axis for the nucleus, which is normal to the orbital plane of the comet and has a rotation period is 2.09 days. We also assumed that the outflow velocity is radial at 0.9 km s^{-1}. Two values for the scale length, 7.0 10^4 km and 2.9 10^5 km, were used to determine the relative emission rates at the mapped-back sources. We intend to refine this analysis to incorporate a better determined rotational model of the nucleus so that the results on the source regions and their relative emission rates can be compared with other observations, such as dust and photometric, to improve our understanding of the dust and gas production at specific region of the nucleus.

Acknowledgements. Participation of the University of Arizona in the Vega Mission and this investigation is partially supported by NASA contract NASW 3627 and NSF grant INT 8205325. We also thank the many dedicated and capable personnel working with us at the several institutions with which we are affiliated. We are grateful to A.J. Tuzzolino and D. Rabinowitz for their critical reading of our earlier draft, and to M.C. Festou for helpful comments to improve the final draft.

References

Curtis, C.C., Hsieh, K.C., Fan, C.Y., Hunten, D.M., Ip, W.-H., Keppler, E., Richter, A.K., Umlauft, G., Afonin, V.V., Erö, J.,Jr., Somogyi, A.J.: 1987, *Astron. Astrophys.* (in press)

Festou, M.C.: 1987 (private communication)

Festou, M.C., Drossart, P., Lecacheux, J., Encrenaz, T., Puel, F., Kohl-Moreira, J.L.: 1987, *Astron. Astrophys.* (in press)

Green, D.W.E., Morris, C.S.: 1986, 20th ESLAB Symp., ESA SP-250, Vol. **I**, p. 613

Gringauz, K.I., Gombosi, T.I., A.P., Apáthy, I., Szemerey, I., Verigin, M.I., Denchikova, L.I., Dyachkov, A.V., Keppler, E., Lklimemko, I.N., Richter, A.K., Somogyi, A.J., Szegö, K., Szendrö, S., Tátrallyay, M., Varga, A., Vladimirova, G.A.: 1986, *Nature* **321**, 282

Keppler, E., Afonin, V.V., Curtic, C.C., Dyachkov, A.V., Erö, J., Jr., Fan, C.Y., Hsieh, K.C., Hunten, D.M., Ip, W.-H., Richter, A.K., Somogyi, A.J., Umlauft, G.: 1986, *Nature* **321**, 326

Krankowsky, D., Lämmerzahl, P., Herrwerth, I., Wowries, J., Eberhardt, P., Dolder, U., Herrmann, U., Schulte, W., Berthelier, J.J., Illiano, J.M., Hodges, R.R., Hoffman, J.H.: 1986, *Nature* **321**, 326

Larson, H.P., Davis, D.S., Mumma, M.J., Weaver, H.A.: 1986, Steward Obs., Preprint No. **676**, University of Arizona

Larson, S.M.: 1987 (private communication)

Millis, R.L., Schleicher, D.G.: 1986, *Nature* **324**, 646

Schleicher, D.G., Mllis, R.L., Tholen, D., Lark, N., Birch, P.V., Martin, R., A'Hearn, M.F.: 1986, 20th ESLAB Symp., ESA SP-250, Vol. **I**, p. 565

Sekanina, Z., Larson, S.M.: 1986, *Nature* **321**, 357

Sagdeev, R.Z., Krasikov, V.A., Shamis, V.A., Tarnopolski, V.I., Szegö, K., Tóth, I., Smith, B., Larson, S., Merényi, E.: 1986, 20th ESLAB Symp., ESA SP-250, Vol. **II**, p. 335

Weaver, H.A., Feldman, P.D., Festou, M.C., A'Hearn, M.F.: 1979, *Astrophys. J.* **251**, 809

Weaver, H.A., Mumma, M.J., Larson, H.P., Davis, D.S.: 1986, *Nature* **324**, 441

Whipple, F.L.: 1978, *Moon Planets* **18**, 343

Wilhelm, K., Cosmovici, C.B., Delamere, W.A., Huebner, W.F., Keller, H.U., Reitsema, H., Schmidt, H.U., Whipple, F.L.: 1986, Max-Planck-Institut für Aeronomie Preprint MPAE-W-55-86-31

Astron. Astrophys. 187, 380–384 (1987)

ASTRONOMY AND ASTROPHYSICS

The atomic carbon distribution in the coma of comet P/Halley

T.N. Woods, P.D. Feldman, and K.F. Dymond

Department of Physics and Astronomy, Johns Hopkins University, Baltimore, MD 21218, USA

Received January 27, accepted April 27, 1987

Summary. The radial distributions of CO, O I, C I, and C II emissions in the coma of comet Halley were measured by a long-slit far ultraviolet spectrograph aboard a sounding rocket on 26 February and 13 March 1986. While radial outflow models of CO can match the observed CO profiles at cometocentric distances $> 10^4$ km, the observed carbon distribution is not consistent with the model, suggesting an additional source of atomic carbon in the inner coma. The *in situ* plasma measurements from the Vega and Giotto spacecraft suggest a possible additional source from the dissociative recombination of electron-impact ionized CO in the inner coma.

Key words: comet P/Halley – UV radiation – C and CO abundances – electron impact excitation

1. Introduction

The derived abundances of atomic carbon produced by photodestruction of carbon-bearing parents, such as CO, have been found to be dissimilar for different comets (Festou, 1984; A'Hearn and Feldman, 1980). For example, the derived carbon abundance in the coma of comet West (1976 VI) is consistent with a model of carbon as a daughter of CO (Feldman, 1978), while the data from comet Bradfield (1979 X) are not consistent with models of carbon either as the daughter of CO or the granddaughter of CO_2 (Weaver, 1981). While this suggests that the cometary ice may have differing composition, the overall similarities in the ultraviolet spectra of comets observed with the International Ultraviolet Explorer (IUE) (Weaver et al., 1981) cannot be neglected. Therefore, the assumptions of the mechanisms for producing atomic carbon from the principal carbon-bearing parents in the coma (usually CO, CO_2, and CH_4) need to be examined more closely before strongly concluding that the molecular abundances in different comets are intrinsically different.

This paper presents a summary of the results from two observations of carbon emissions in comet Halley made with far ultraviolet instrumentation aboard a sounding rocket together with a discussion of the apparent excess atomic carbon abundance (relative to the observed CO) and a possible mechanism to account for the observed carbon distribution.

Send offprint requests to: T.N. Woods

2. Sounding rocket observations

The sounding rocket experiment, which observed comet Halley on both 26 February and 13 March, has been described by Woods et al. (1986a). Briefly, long-slit (7′.7) spectra were obtained in the far ultraviolet range of 1200 to 1750 Å at 12.5 Å resolution. The effective spatial resolution was 20″ due to the tracking drift and pointing jitter. Spatial profiles of CO, O I, C I, and C II emissions were obtained along the sun-comet line. The observation conditions and the derived brightnesses of the observed emissions from the central coma are listed in Table 1, while the observed radial profiles of CO, O, and C are shown in Fig. 1. The two observations were made 15 days apart at almost identical points, two cycles apart, on the 7.4 day period light curve obtained by Millis and Schleicher (1986).

The initial analysis of these data (Woods et al., 1986a) showed that the observed CO profile could be modelled by a radial outflow model (Haser, 1957; Festou, 1981) for a parent molecule and that the atomic oxygen profile could be modelled by a radial outflow model with its dominant parent being H_2O. Comparing the observed CO and O I emission profiles to the profiles derived from a radial outflow model, which includes both photodissociation and charge exchange processes, the derived production rates were $1.2\,10^{30}\,s^{-1}$ for H_2O and $2.4\,10^{29}\,s^{-1}$ for CO on 26 February 1986, and $6\,10^{29}\,s^{-1}$ for H_2O and $1.0\,10^{29}\,s^{-1}$ for CO on 13 March 1986. The derived H_2O production rates for both observations are consistent with those derived from other measurements made at almost the same times by either Pioneer Venus Orbiter UVS (Stewart, 1987), the sounding rocket experiment of McCoy et al. (1986), IUE (Festou et al., 1986), or Giotto (Krankowsky et al., 1986).

A preliminary analysis of the Giotto neutral mass spectrometer (NMS) measurements of the CO density (Eberhardt et al., 1986) showed that the CO has a distribution similar to that of a parent molecule (which varies as ρ^{-2}, ρ = cometocentric distance) outside a distance of $2\,10^4$ km from the nucleus, but their CO profile, being less steep inside 10^4 km, implies a source of CO in the inner coma in addition to the CO sublimated directly from the nucleus. The observed flattening of the CO emission profile in Fig. la is indicative of this additional distributed source of CO. The CO density profile from the NMS data is consistent with the sounding rocket measurement (13 March 1986) which has an effective spatial resolution of $2\,10^4$ km. The identity of this additional source of CO is not known, but it must have a photodestruction lifetime shorter than 10^4 s or be more active within the inner coma region where collisions and chemical reactions may play an important role.

Table 1. Summary of the sounding rocket observations

		FLIGHTS	
		NASA 21.093 UG	NASA 21.095 UG
Launch Date		26 Feb. 1986	13 Mar. 1986
Launch Time (UT)		12:02	11:20
Heliocentric r (AU)		0.695	0.895
Heliocentric v ($km\ s^{-1}$)		19.50	25.56
Geocentric Δ (AU)		1.32	0.976
Geocentric v ($km\ s^{-1}$)		-34.16	-43.30
Brightness†	OI λ1304	1460	280
	CII λ1335	130	25
	OI λ1356	60	< 10
	CO λ1478	175	65
	CO λ1510	140	60
	CI λ1561	480	150
	CI λ1657	1600	520
Production Rate*	H_2O	12.	6.0
	CO	2.4	1.0
	C	0.7	-

†Brightness of the central coma (10 x 16 arc s) is in units of Rayleighs (10^6 photons s^{-1} cm^{-2}).

*Production rate as derived from an observed emission profile is in units of 10^{29} s^{-1}.

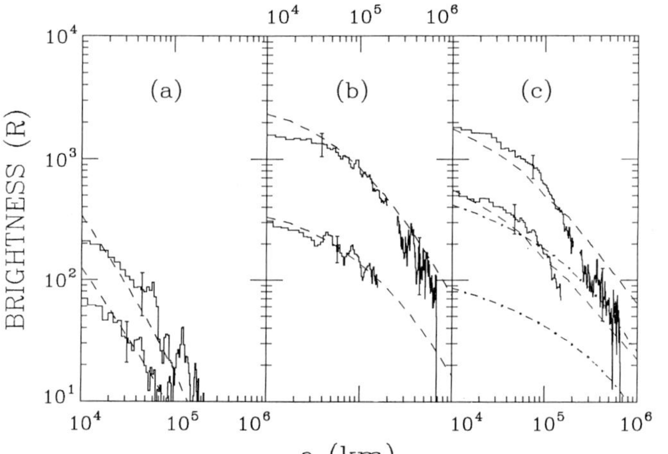

Fig. 1. The observed radial distribution of (a) the 1478 Å CO band, (b) O I λ 1304, and (c) C I λ 1657 emissions for both sounding rocket flights with the data from the first flight being the upper curve in each panel. The dot-dashed lines in Fig. 1c are the carbon emission profiles from an outflow model that includes CO, CO_2, and CH_4 at the rate of 20%, 3.5%, and 7% of the water production rate, respectively. The dashed lines are emission profiles from a preliminary outflow model which includes electron impact ionization and dissociative recombination based on the *in situ* plasma measurements from the Vega and Giotto spacecraft

3. The carbon puzzle

The analysis of the ultraviolet spectra of several comets has shown that the atomic carbon abundance in the inner coma cannot be satisfactorily accounted for by the photodissociation of carbon-bearing molecules such as CO, CO_2, and CH_4 (Festou, 1984). The present experiment was specifically designed to address this problem. The brightness of the carbon emission given by a radial outflow model with CO as a parent molecule lies about an order of magnitude below the observed carbon emission brightness near the center of the coma ($\rho < 10^5$ km) while it reproduces reasonably well the data in the outer coma ($\rho > 3\ 10^5$ km) (Woods et al., 1986a). An *ad hoc* additional source of carbon at the nucleus at a rate of ~2% of the water production rate was shown to give a better fit to the observed carbon distribution (Woods et al., 1986b). An improved fit is also possible with an additional distributed source of carbon within the inner coma, which is considered here to mean within the collision sphere ($\rho \leqq 10^4$ km). The recent *in situ* measurements at comet Halley have provided new information bearing on this problem, and suggest possible mechanisms for the production of atomic carbon in the inner coma. These are discussed below.

4. Other carbon-bearing parents

First, we consider a possible solution proposed by Festou (1984), namely the photodissociation of undetected abundant carbon-bearing parent molecules. The most abundant carbon-bearing molecules that were measured by the Giotto mass spectrometers were CO, CO_2, and CH_4. The upper limits for these species, from the preliminary analysis of the Giotto data, are 20% for CO, 3.5% for CO_2, and 7% for CH_4, all relative to H_2O (Krankowsky et al., 1986; Balsiger et al., 1986). The CO upper limit is consistent with the CO abundance derived from our data as given above. A detailed analysis of the Giotto ion mass spectrometer (IMS) data derived an abundance of CH_4 of 2% relative to H_2O (Allen et al., 1987). The inclusion of CO_2 and CH_4 into the CO and H_2O radial outflow model, used for interpretation of the sounding rocket data, provides a minor additional source of carbon but is insufficient to match the observed carbon profiles (see the dot-dashed lines in Fig. 1c). While this model, which includes both photolytic and charge exchange processes, excludes the known carbon-bearing parents as the additional source of carbon, there could still be unidentified heavier molecules with a short lifetime ($\tau \leq 10^4$ s) so that most of the carbon from these molecules would be produced in the inner coma.

Because the preliminary analysis of the CO density measured by the Giotto mass spectrometer revealed that there could be an additional source of CO within the inner coma (Eberhardt et al., 1986), the yet unidentified source of CO might also produce atomic carbon in the inner coma. Eberhardt et al. (1986) have suggested that the vaporization of dust particles may be the source of CO in the inner coma, and Clark et al. (1986) report that some of the CHON particles (rich with C, H, O and N) detected by the Giotto PIA experiment may be volatile icy grains. Further analysis of the production of CO and atomic carbon from dust and CHON particles in the inner coma is clearly warranted when additional PIA data becomes available.

5. Electron impact processes

It is clear that a different approach to the problem is needed. Assuming, as the NMS data indicate, that CO and CO_2 are the principal carbon-bearing species in the coma, the problem is to identify a locally strong source of dissociation (other than photodissociation) in the inner coma. The presence of the O I λ1356 emission in the UV spectrum of comet Halley provides spectroscopic evidence for electron impact excitation in the inner coma; hence, electron impact ionization and dissociative recombination were investigated as possible sources of carbon in this region. Photodissociation and photoionization by solar radiation are considered the dominant destruction processes for water in most of the coma. There is also a fast ion-molecule charge transfer/dissociation process which is limited to the inner coma (Mendis et al., 1985). However, the dominant destruction process for CO_2 and CO can be electron impact ionization, if the electron flux is as large as that measured by the Vega PLASMAG experiment at $1.5\,10^4$ km from the nucleus of comet Halley (Gringauz et al., 1986; Cravens et al., 1986). The frequency of destruction of a molecule in the coma by electron impact ionization or dissociation can be calculated using the available laboratory cross sections for the molecule (Massey, 1969; Orient and Srivastava, 1985) and the *in situ* measurements of the electron flux. The results of this calculation are given in Table 2 along with the photodestruction frequency (Huebner and Carpenter, 1979) scaled to 0.83 AU, the heliocentric distance at the time of the Vega 2 encounter. It should be noted that the destruction of a molecule by electron impact results primarily in the ionization of the molecule. The electron impact destruction frequency of CO is about an order of magnitude larger than its photolytic frequency, while that for CO_2 is about the same. For comparison, the electron impact destruction frequency of H_2O is about an order of magnitude smaller than its photolytic frequency. This implies that the CO chemistry in the inner coma, unlike that for H_2O, is dominated more by electron impact mechanisms than the photolytic processes. Cravens et al. (1986) have also investigated the electron impact ionization process for other molecules and atoms that are typically found in the coma, and their results also indicate that the electron impact process could be as important as the photolytic process.

The second most abundant product, after CO^+, from the electron impact on CO is C^+ (Orient and Srivastava, 1985). The calculated frequency of the destruction of CO into C^+ using the *in situ* electron flux at $\rho = 1.5\,10^4$ km is $4.6\,10^{-7}\,s^{-1}$. Because this frequency is about the same magnitude as the photoionization and charge exchange reaction frequency of C and the density of CO is ~100 times larger than the density of C for $\rho < 10^4$ km, the majority of C^+ is probably from electron impact dissociation of CO. This process may explain the unexpectedly large abundance of C^+ in the coma of comet Halley measured by the Giotto IMS (Balsiger et al., 1986) and observed (C II λ1335 emissions) by the sounding rocket experiment.

The photolytic lifetimes vary as the square of the heliocentric distance *r*; however, the variation of the electron impact destruction frequencies with *r* is not clearly known. The electron flux,

Table 2. Destruction frequencies at $r = 0.83$ AU and $\rho = 1.5\,10^4$ km

Species	Photolytic[a] (s^{-1})	Electron Impact[b] (s^{-1})	Dissociative Recombination of its ion[c] (s^{-1})
H_2O	1.7 (-5)†	2.7 (-6)	1.4 (-3)
CO_2	2.9 (-6)	5.2 (-6)	8.5 (-4)
CO	9.4 (-7)	4.1 (-6)	3.8 (-4)
C	6.0 (-7)	2.0 (-6)	9.0 (-9)

† Numbers are represented as d.d (m) for d.d × 10^m

[a] The rates from Huebner and Carpenter (1979) are scaled to 0.83 AU.

[b] Calculated with the in situ electron flux from Cravens et al. (1986) and the cross sections from Massey (1969) and Orient and Srivastava (1985). The frequency for C is estimated as the frequency for O.

[c] Calculated from the recombination rates taken from Mendis et al. (1985) and the in situ electron measurements from Grard et al. (1986).

as modelled by Ashihara (1978) for a cometary atmosphere with an electron energy range up to 50 eV, varies approximately as r^{-4} for 20 eV electrons, but only about r^{-3} for 40 eV electrons. For electron impact excitation the cross sections typically peak near 20 eV, so electron impact excitation would probably vary as r^{-4} in the coma. The electron impact ionization cross sections typically peak near 100 eV with a slow decline towards higher energies; therefore, it is difficult to predict the variation with r for the electron impact destruction frequencies without knowing the behavior of the $50 - 10^4$ eV electrons with heliocentric distance.

Some of the comet models, such as the radial outflow model and the vectorial model (Haser, 1957; Festou, 1981), which have previously assumed a lifetime of a molecule to be independent of the cometocentric distance, can be modified to include the electron impact process if the variation of the electron impact destruction frequency with ρ is known. The effective lifetime of a molecule due to the electron impact process is the inverse of the electron impact destruction frequency as a function of ρ (i.e. the lifetime of a molecule can be represented as a lifetime that varies with ρ). The 40 eV electron flux modelled by Ashihara (1978) varied approximately as ρ^{-1} for $\rho > 10^4$ km, while Korosmezey et al. (1986) also derive a ρ^{-1} variation of the electron flux in the outer coma and a flat profile of the electron flux in the inner coma from their comet model. Further analysis of the *in situ* measurements of the electron flux should provide a clearer understanding of the variation of electron flux with the cometocentric distance.

To examine the importance of recombination of CO^+ to form C and O, the electron density and temperature are needed as a function of the cometocentric distance ρ. In Table 2, the dissociative recombination frequencies of CO^+, CO_2^+, and H_2O^+ at $\rho = 1.5\,10^4$ km have been estimated using the *in situ* measurements of electron density and temperature by the Vega APV-V experiment (Grard et al., 1986). The extrapolated value of the *in situ* electron density at $\rho = 1.5\,10^4$ km is $1.6\,10^4\ \mathrm{cm}^{-3}$ assuming the ρ^{-2} variation of the measured ion density from the Giotto IMS data (Balsiger et al., 1986). Grard et al. (1986) estimated an electron temperature of $1.7\,10^4$ K at $\rho = 3\,10^4$ km. Because the dissociative recombination frequency for CO^+ is large, this mechanism may be an important additional source of C in the inner coma.

The electron density measured by the Vega APV-V experiment (Grard et al., 1986) and the total ion density measured by the Giotto mass spectrometers (Krankowsky et al., 1986; Balsiger et al., 1986) can be used to determine the variation of the electron density as a function of ρ. Inside the "contact surface" near $5\,10^3$ km, the ion density (and hence the electron density) was found to vary as ρ^{-1}. Directly outside the contact surface, the ion density varied as ρ^{-2}; however, the ion density seems to fall more rapidly with larger ρ. Until there are more *in situ* data available on the electron temperature, we have assumed an electron temperature profile similar to that from the comprehensive cometary atmosphere model of Mendis et al. (1985).

6. Electron impact modelling

The ability of the electron impact process to produce the observed emissions from comet Halley are investigated with the *in situ* electron measurements. As discussed in Sect. 5, the results from comet models (Mendis et al., 1985; Ashihara, 1978; Korosmezey et al., 1986) are used to estimate the variations of the plasma parameters with ρ and r, while the *in situ* plasma measurements are used to normalize the functions of the electron flux, the electron density, and the electron temperature.

Electron impact excitation can be modelled if the density distribution of the species is measured or derived from a comet model. The observed O I λ 1304 emission profiles along the cometocentric distance ρ (see Fig. 1b) allow the derivation of the oxygen density, $n_O(\rho)$, from a radial outflow model of H_2O and CO. The O I λ 1356 brightness in Rayleighs is expected to be:

$$B(\rho) = 10^{-6} \int_{-\infty}^{\infty} n_O(R) g_e(R)\, dz \tag{1}$$

where

$$g_e(R) = \left(\frac{0.83\ \mathrm{AU}}{r}\right)^4 \left(\frac{1.5\,10^9\ \mathrm{cm}}{R}\right) \int_0^{\infty} \sigma(E) F_e(E)\, dE\,, \tag{2}$$

$R^2 = \rho^2 + z^2$, F_e is the measured electron flux, and σ is the electron impact excitation cross section. The excitation frequency, g_e, is calculated to be $8.9\,10^{-8}\ \mathrm{s}^{-1}$ with the O I λ 1356 cross section (60% of that given by Stone and Zipf, 1974) for $r = 0.83$ AU and $R = 1.5\,10^9$ cm. The calculated O I λ 1356 brightness from Eq. 1 is $50R$ on 26 February and $5R$ on 13 March. This is consistent with the observed O I λ 1356 brightness of $60 \pm 25R$ on 26 February and an upper limit of $10R$ on 13 March.

The electron impact process should not substantially affect the distribution of a parent molecule. The revision of the radial outflow model for a lifetime that is a function of the cometocentric distance ρ yields the density of the parent species to be:

$$n(\rho) = \frac{Q}{4\pi\rho^2 v} e^{-D(\rho,\rho_n)} \tag{3}$$

where

$$D(\rho_2, \rho_1) = \int_{\rho_1}^{\rho_2} \beta(\rho')\, d\rho'\,, \tag{4}$$

and

$$\beta(\rho) = \frac{1}{\tau(\rho) v}\,. \tag{5}$$

Here, v is the constant outflow velocity and ρ_n is the radius of the nucleus. From this simple model, the electron impact process would only affect the distribution of the parent at distances comparable to the scalelength β, which is $\sim 10^5$ km for CO when the electron impact destruction frequency is included and the electron flux is assumed relatively flat in the inner coma. Because the sounding rocket observations only detected the CO emissions out to $\sim 10^5$ km, the derived production rates of CO, using the outflow model that only considered photodestruction, do not need to be revised to be consistent with the model that includes the electron impact destruction of CO. A sensitive parameter in the derivation of the production rate is the outflow velocity, chosen as $0.9\ \mathrm{km\,s^{-1}}$ for the H_2O and CO molecules in these models.

7. Discussion

A preliminary investigation of the electron impact ionization and recombination in the coma suggests that the sounding rocket observations may be properly modelled by including ionization of CO enhanced by electron impact in the inner coma, together with the rapid recombination of CO^+ into C and O. Using a similar recombination model where the CO^+ was produced only from the photoionization of CO, Feldman (1978) satisfactorily modelled the CO and C emissions (particularly C I λ 1931) observed from comet West (1976 VI). However, that earlier experiment only measured a total flux and did not provide spatial profiles. Because the electron impact process is limited mostly to the inner coma ($\rho \leqq 10^4$ km), the C from the recombination of CO^+ may be a sufficient additional source of C to match the observed C I emissions as shown in Fig. 1c (dashed lines). While this preliminary model does not perfectly predict the observed emissions, it does demonstrate the plausibility and importance of electron impact processes in producing atomic carbon from CO. The electron impact process may also resolve the discrepancy in the observed Q_{CO}/Q_C ratio (Woods et al., 1986a) by providing a different ratio of ionization and dissociation of CO than that due to photolytic processes alone. A preliminary examination of the $C(^1D)$ emissions at 1931 Å observed by IUE has shown that only a small percentage of the dissociative recombination of CO^+ into $C(^1D)$ is required to provide the $\sim 50R$ of $C(^1D)$ λ 1931 emission as measured on 1986 March 11–14. The increased production of O from the recombination of CO^+ remains a minor contribution compared to its production from the photodestruction of H_2O. The preliminary model requires the water production rates, as derived with a standard outflow model, to be lower by a small amount. From the preliminary analysis presented here, it appears that electron impact ionization and dissociative recombination are important mechanisms in the inner coma and require further investigation as the majority of the electron parameters needed are derived from the *in situ* plasma measurements.

Acknowledgements. We acknowledge helpful conversations with M.C. Festou, T.I. Gombosi, and L.J. Paxton. This work was supported by NASA grant NAG 5-619.

References

A'Hearn, M.F., Feldman, P.D.: 1980, *Astrophys. J.* **242**, L187

Allen, M., Schwenn, R., Balsiger, H., Geiss, J.: 1987, "Evidence for Methane and Ammonia in the coma of comet Halley" (preprint)

Ashihara, O.: 1978, *Icarus* **35**, 369

Balsiger, H., Altwegg, K., Buhler, F., Geiss, J., Ghielmetti, A.G., Goldstein, B.E., Goldstein, R., Huntress, W.T., Ip, W.-H., Lazarus, A.J., Meier, A., Neugebauer, M., Rettenmund, U., Rosenbauer, H., Schwenn, R., Sharp, R.D., Shelley, E.G., Ungstrup, E., Young, D.T.: 1986, *Nature* **321**, 330

Clark, B., Mason, L.W., Kissel, J.: 1986, *20th ESLAB Symp. ESA SP-250 Vol III*, 353

Cravens, T.E., Kozyra, J.U., Nagy, A.F., Gombosi, T.I., Kurtz, M.: 1986, *J. Geophys. Res.* **92** (A7), 7341

Eberhardt, P., Krankowsky, D., Schulte, W., Dolder, U., Lämmerzahl, P., Berthelier, J.J., Woweries, J., Stubbemann, U., Hodges, R.R., Hoffman, J.H., Illiano, J.M.: 1986, *20th ESLAB Symp. ESA SP-250 Vol I*, 383

Feldman, P.D.: 1978, *Astron. Astrophys.* **70**, 547

Festou, M.C.: 1981, *Astron. Astrophys.* **95**, 69

Festou, M.C.: 1984, *Adv. Space Res.* **4** (9), 165

Festou, M.C., Feldman, P.D., A'Hearn, M.F., Arpigny, C., Cosmovici, C.B., Danks, A.C., McFadden, L.A., Gilmozzi, R., Patriarchi, P., Tozzi, G.P., Wallis, M.K., Weaver, H.A.: 1986, *Nature* **321**, 361

Grard, R., Pedersen, A., Trotignon, J.-G., Beghin, C., Mogilevsky, M., Mikhailov, Y., Molchanov, O., Formisano, V.: 1986, *Nature* **321**, 290

Gringauz, K.I., Gombosi, T.I., Remizov, A.P., Apáthy, I., Szemerey, I., Verigin, M.I., Denchikova, L.I., Dyachkov, A.V., Keppler, E., Klimenko, I.N., Richter, A.K., Somogyi, A.J., Szegö, K., Szendrö, S., Tátrallyay, M., Varga, A., Vladimirova, G.A.: 1986, *Nature* **321**, 282

Haser, L.: 1957, *Bull. Acad. Roy. Belgique* **43**, 740

Huebner, W.F., Carpenter, C.W.: 1979, *Solar Photo Rate Coefficients*, Los Alamos Scientific Laboratory Report LA-8085-MS

Korosmezey, A., Cravens, T.E., Gombosi, T.I., Nagy, A.F., Mendis, D.A., Szego, K., Gribov, B.E., Sagdeev, R.Z., Shapiro, V.D., Shevchenko, V.I.: 1986, *20th ESLAB Symp. ESA SP-250 Vol I*, 235

Krankowsky, D., Lämmerzahl, P., Herrwerth, I., Woweries, J., Eberhardt, P., Dolder, U., Herrmann, U., Schulte, W., Berthelier, J.J., Illiano, J.M., Hodges, R.R., Hoffman, J.H.: 1986, *Nature* **321**, 326

McCoy, R.P., Opal, C.B., Carruthers, G.R.: 1986, *Nature* **324**, 439

Massey, H.S.W.: 1969, *Electronic and Ionic Impact Phenomena*, London, Oxford Univ. Press, 1021–1048

Mendis, D.A., Houpis, H.L.F., Marconi, M.L.: 1985, *Fundamentals of Cosmic Phys.* **10**, 1

Millis, R.L., Schleicher, D.G.: 1986, *Nature* **324**, 646

Orient, O.J., Srivastava, S.K.: 1985, *Conf. Phys. Electronic and Atomic Collisions*, Palo Alto 24–30 July 1985, 273

Stewart, A.I.F.: 1987, *Astro. and Astrophys.* (in press)

Stone, E.J., Zipf, E.C.: 1974, *J. Chem. Phys.* **60**, 4237

Weaver, H.A.: 1981, Ph.D. dissertation, The Johns Hopkins University

Weaver, H.A., Feldman, P.D., Festou, M.C., A'Hearn, M.F., Keller, H.U.: 1981, *Icarus* **47**, 449

Woods, T.N., Feldman, P.D., Dymond, K.F., Sahnow, D.J.: 1986a, *Nature* **324**, 436

Woods, T.N., Feldman, P.D., Dymond, K.F., Sahnow, D.J.: 1986b, *Adv. Space Res.* **5** (12), 289

Astron. Astrophys. 187, 385–387 (1987)

ASTRONOMY AND ASTROPHYSICS

Carbon-isotope ratio in PUMA 1 spectra of P/Halley dust

M. Šolc [1,*], V. Vanýsek [2], and J. Kissel [1]

[1] Max-Planck-Institut für Kernphysik, P.O. Box 103980, D-6900 Heidelberg, Federal Republic of Germany

[2] Charles University Prague, Dept. of Astronomy and Astrophysics, Švédská 8, CS-15000 Praha 5, Czechoslovakia

Received February 18, accepted May 11, 1987

Summary. Carbon isotopes ^{12}C and ^{13}C have yet been identified both in cometary gas and dust. Although the determination from optical C_2 and CN emissions from ground-based observations is not precise, the ratio $^{12}C/^{13}C$ is roughly the same as the mean ratio derived from PUMA 1 mass spectra of cometary grains. This value both in volatile and refractory material does not differ much from the nominal solar system value 90. However, the collection of 38 PUMA 1 mode 0 spectra shows surprisingly variable 12/13 amu lines ratio within 3 decades, increasing with carbon abundance in the grains. Possible instrumental effects are discussed.

Key words: comets – isotopic ratio

1. Introduction

Isotopic ratios in comets may contain a key information on physical conditions in the environment where the cometary nuclei or pre-cometary dust particles were formed. The only data concerning the isotopes in about five comets before P/Halley are those obtained for stable carbon isotopes ^{12}C and ^{13}C by the groundbased methods in optical spectra from isotopically shifted emission of $^{12}C^{12}C$ $(1-0)$ and $^{12}C^{13}C$ $(1-0)$ P branch bands (Encrenaz, 1984; Vanýsek and Rahe, 1978; Vanýsek, 1986). The average ratio $^{12}C/^{13}C$ varies from 60 to 135, and although the error bars of the particular estimates are large, the data indicate slight enrichment in ^{12}C relatively to the terrestrial ratio 89.

Similarly, the $^{12}C/^{13}C$ ratios derived from ground based spectroscopy of comet Halley approach the terrestrial value, being in the range 80 ± 20 (Festou, 1986; Wehinger, 1986). These results have been obtained from the ^{13}CN and $^{12}C^{13}C$ lines in the violet band of CN and C_2 $(0-0)$ band of the Swan system in case of high spectral resolution. However, as follows from the theoretical comparison with the synthetic spectra, the determination of carbon isotopes ratio from the profile analysis of $(1-0)$ Swan band, which extends over a large spectral portion, may be better than that derived from the $(0-0)$ band which is limited to a very narrow region (Krishna Swamy, 1986). Some systematic differencies in the results obtained from different bands may therefore be expected. Nevertheless, an acceptable range of the stable carbon isotopic ratio found from gaseous emission in comets is 70–120.

It will be very important and interesting to compare these results with isotopic ratio in the cometary solid particles. Mass spectrometers flown on-board spacecrafts to P/Halley enable us to investigate for the first time the elemental and isotopical composition of the solid cometary grains (Jessberger et al., 1986). The comprehensive description of these dust impact analyzers is given elsewhere (e.g. Sagdeev et al., 1985, Kissel, 1986). About 50 mass spectra collected by the experiment PUMA 1 and transferred to the Earth in the less compressed mode (mode 0) are of sufficient quality for distinguishing typical patterns of some stable isotopes. The aim of this paper is to discuss the occurence of 12 and 13 amu signals in PUMA 1 mode 0 mass spectra in terms of the possible abundances of stable carbon isotopes.

2. Criteria for the experimental data selection

One of the essential problems concerning the evaluation of the mass spectra measured by the particulate impact analyzer is the determination of the ion yields for individual elements. However, the estimation of the relative isotopic abundances can be done directly because it can be assumed that ionization yields are of the same value for all isotopes of the given element. Consequently it can be assumed that the ion rate of the isotopic species is proportional to the true isotopic ratio in solid particles.

The identification of the isotopic patterns can be successfully performed either in spectra where (a) the less abundant isotope is represented by a signal definitely distinguishable from the noise or (b) in case of absence of such a signal, if there are peaks that can be assigned to the other isotopes of the same element.

The inspection of about 300 mode 0 spectra obtained by PUMA 1 provides a set of 38 best spectra selected by the following criteria:

i) The mass resolution better than 1 amu in the region between 10 and 60 amu is required;

ii) Spectra with doubled, tripled and broadened structures enter the set only if the peak definition allows to separate clearly the structures at 12 and 13 amu;

iii) The height of the 12 amu signal exceeds the noise level (i.e. by about 2 decades) in most cases.

The first two requirements are reasonably fulfilled even in the spectra of less massive grains, while high peaks that obey item iii) are characteristic mainly for the massive particles. Thus a selection effect is involved and the resulting set of the spectra is somewhat biased to more massive grains.

Send offprint requests to: M. Šolc

* On leave from Charles University Prague, Dept. of Astronomy and Astrophysics, Czechoslovakia

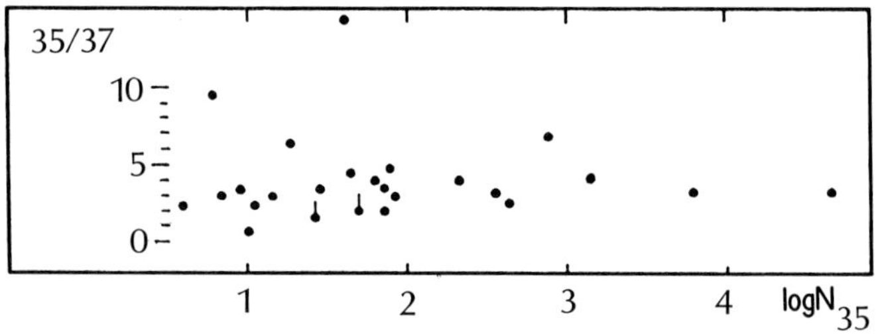

Fig. 1. The ratio of ion numbers at 35 and 37 amu in 25 PUMA 1 mode 0 spectra vs. ion number at 35 amu. (The vertical scale is linear)

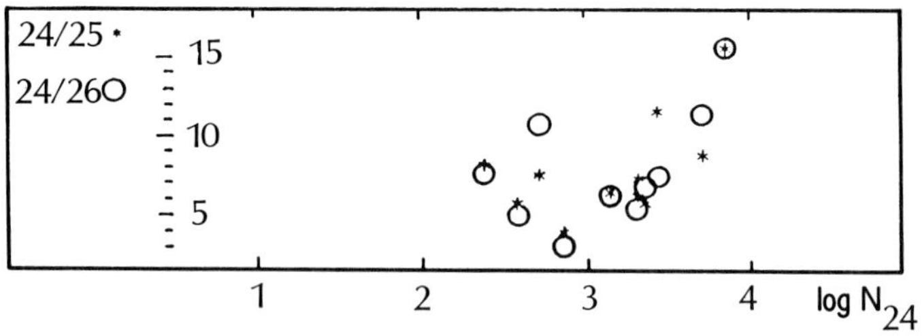

Fig. 2. The ratio of ion numbers at 24, 25, and 26 amu in 10 PUMA 1 mode 0 spectra vs. ion number at 24 amu. (The vertical scale is linear)

3. Results

The examination of all 38 spectra selected according to the criteria mentioned above reveals that the ratio of peak heights at 12 and 13 amu scales from 0.5 to about 3000. This surprisingly large variability corresponds neither to the range of carbon isotopic ratio derived from interstellar gas nor to somewhat higher values found in solar system bodies including comets. In contrary to this, other elements in the sample of these 38 carbon-rich PUMA 1 spectra do not give isotopic ratio with such a large differences.

For instance, the peaks at 35 and 37 amu do not show such a variability. If interpreted as chlorine isotopes then the ratio of 35/37 amu signals approaches well the natural isotopic ratio $^{35}Cl/^{37}Cl$ which is about 3. However, the line intensity suggests a strikingly high abundance of chlorine, more than 3000 times greater than the one in carbonaceous chondrites C1 (when compared to the bulk elemental composition; Jessberger et al., 1986). This makes the interpretation in terms of cometary chlorine unlikely and a contamination should be supposed as more probable because high 35 and 37 signals are not seen in spectra obtained by analogic experiments on-board other two spacecrafts. But anyway the deviations of the 35/37 amu signal ratio from the ratural ratio of possible corresponding isotopes are less than in other isotopic patterns in all PUMA 1 mode 0 spectra. The mean value of it is 3:1, and variations are between 2:1 to 14:1 (Fig. 1). In one case only (No. 52823) the peak at 37 amu is higher than that at 35 amu.

Another example is represented by the magnesium isotopes with masses 24, 25, and 26 amu. The terrestrial ratio 24/25 is about 8 and 24/26 about 7. As can be seen in 10 spectra from those of 38 carbon-rich particles, the ratios vary from 3.8 to 11.5 and most of them fall between 7 and 8 (Fig. 2).

The unusually enormous variability of 12/13 amu signals makes the direct interpretation in terms of carbon stable isotopes somewhat problematic. Some effects may be suggested to affect the measured results, and discussed as follows:

a) According to Kissel and Krüger (1987) the scarcely formed double ionized atoms such as Mg^{2+} cannot substantially contribute to 12 amu line. It can be shown in spectrum No. 53265 where the 24 amu line exceeds 4 decades while 12 amu line is hidden in the noise.

b) The contribution of C_2 clusters to 24 amu line may be expected when the signal at 12 amu is strong. However, there are numerous spectra dominated by a very strong 12 amu line but with only moderate 24 amu peak, so that the formation of biatomic carbon clusters does not likely alter the 12/13 line ratio.

c) The substantial source of errors remains in interpreting the splited or very broad peaks by computer model simulated line profiles. However, in all of 8 spectra with very high 12 amu line the profiles are sharp enough to maintain the estimated error of ion numbers well below the factor of 2.

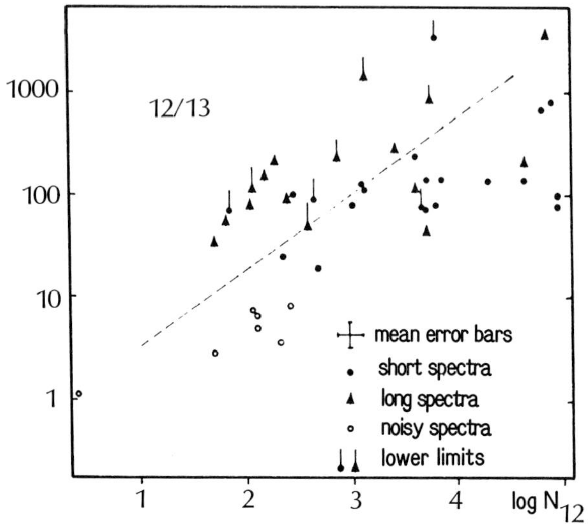

Fig. 3. The ratio of ion numbers at 12 and 13 amu in 38 PUMA 1 mode 0 spectra vs. ion number at 12 amu. (The vertical scale is logarithmic)

d) The very high ratio of 12/13 amu lines may be due to an instrumental artifact. If the 12 amu signal is too strong, the 13 amu feature may be suppressed due to some saturation processes in the ion-counting multiplier. On the other hand, there are spectra (e. g. Nos. 54825, 52823, 51341) with moderate 12 amu line where the 12/13 ratio is about 130. This value seems to be still an acceptable measure of the true isotopic ratio as derived from cometary gas.

e) The spectra showing low 12/13 ratio belong frequently to those of lower quality, with broad or split lines. The characteristic height of 12 amu line is 2 decades while the 13 amu line, even if with well defined profile, is comparable to the noise at a level of 1 decade. Of course, one can admit the possibility of contaminating it by CH clusters. However, in such a case an unrealistic high amount must be formed during the impact process on the silver target. Another argument against it follows from Fig. 3; for both long and short PUMA 1 spectra the plotted 12/13 amu values vs. the intensity of 12 amu lines does not depend on hydrogen abundance.

4. Conclusions

The average ratio of signals at 12 and 13 amu which if interpreted as carbon isotopes ratio tend to exceed 100 in massive and carbon rich particles from the PUMA 1 mode 0 collection, while the moderately sized and small particles exhibit slightly lower ratio of about 60–70. The mean value of this ratio is consistent with that determined from optical emission spectra of C_2 and CN. The mean value indicates that there is probably no significant difference between the gross abundances of stable carbon isotopes in volatile and refractory materials. The very high 12/13 amu signal ratio found in about 5 spectra that does not match the solar system value is probably due to instrumental artifact. Low values of the 12/13 amu signal ratio are characteristic for spectra with moderate or small lines in the region between 11 and 16 amu so that the contamination of 13 amu line by other species than ^{13}C is probable.

References

Encrenaz, T.: 1984, in *Isotopic Ratios in the Solar System*, C.N.E.S. Paris, p. 175

Festou, M.C.: 1986 (private communication)

Jessberger, K.E., Kissel, J., Fechtig, H., Krueger, F.: 1986, in *Comet Nucleus Sample Return*, ESA SP-240, p. 27

Kissel, J.: 1986, The Giotto Mission, ESA SP-1077, p. 67

Kissel, J., Krueger, F.: 1987, *Nature* **326**, 755

Krishna Swamy, K.S.: 1986, 20the ESLAB Symp., ESA SP-250, Vol. **III**, p. 3

Sagdeev, R.Z. et al.: 1985, Venus-Halley Mission, ed. Louis-Jean, Gap

Vanýsek, V., Rahe, J.: 1978, *Moon Planets* **18**, 441

Vanýsek, V.: 1986, in *Astrochemistry*, *IAU Symp.* **120**, eds. M.S. Vardya, S.P. Tarafdar, Dordrecht, Reidel, Dordrecht, p. 461

Wehinger, P.A.: 1986, 20th ESLAB Symp., ESA SP-240, Vol. **III**, p. 473

Wehinger, P.A.: 1986, 20th ESLAB Symp., ESA SP-250, Vol. **III**,p. 475

Astron. Astrophys. 187, 388–390 (1987)

ASTRONOMY AND ASTROPHYSICS

Study of the isotopic features of Swan bands in comets

K.S. Krishna Swamy★

Ames Research Center, Moffett Field, CA, 94035, USA

Received March 9, accepted April 1, 1987

Summary. It is shown from a detailed statistical equilibrium calculation of the $^{12}C^{13}C$ molecule that the interpretation of the observed intensities of Swan bands of the normal and the isotopic molecule of C_2 in terms of the abundance ratio of ^{12}C and ^{13}C is a reasonable one. The synthetic profile of some isotopic features in the (0, 0) Swan band is compared with the observed profiles for comet West.

Key words: Swan bands – isotopic features – comets.

1. Introduction

The determination of the isotopic ratio of elements of cometary material is of prime interest, as it is intimately related to the origin of comets. However, at the present time the results are very limited in extent. Until very recently, the only isotopic ratio known was the $^{12}C/^{13}C$ ratio (Stawikowski and Greenstein, 1964; Danks et al., 1974; Lambert and Danks, 1983). But the recent space missions to comet Halley have made it possible to determine other isotopic ratios (Krankowsky et al., 1986).

At the present time, the study of the isotopic ratio $^{12}C/^{13}C$ is based mainly on the analysis of the Swan band intensities of the C_2 molecule. In these studies the implicit assumption is that the ratio of the band intensities of the normal and the isotopic molecule give directly the abundance ratio, $^{12}C/^{13}C$. This means that the population distribution in various levels of $^{12}C^{12}C$ and $^{12}C^{13}C$ molecules are the same. We would like to investigate this assumption in some detail based on the statistical equilibrium calculations of the $^{12}C^{13}C$ molecule. We will also show that the profile analysis of the isotopic feature of the (1, 0) Swan band should give a good estimate of the $^{12}C/^{13}C$ ratio as the blending of NH_2 lines occurring at 4744.6 and 4744.3Å can be avoided. Finally, we compare the synthetic profile of the isotopic feature of the (0, 0) Swan band with the observed profile of Lambert and Danks (1983) for comet West.

2. Results and analysis

2.1. $^{12}C^{13}C$ molecule

The major difference between the normal and the isotopic molecule is that certain transitions which are forbidden in the $^{12}C^{12}C$ molecule are allowed in the $^{12}C^{13}C$ molecule. In particular, the vibrational transitions in the lower electronic states which are forbidden in a homonuclear molecule are permitted for the $^{12}C^{13}C$ molecule. These transitions should be taken into account, as they can influence the overall level population distribution and hence have an effect on the determined $^{12}C/^{13}C$ ratio. However, the calculation of the vibrational transition probabilities involves a knowledge of the value of the relevant transition moment. Unfortunately, this moment is not known for $^{12}C^{13}C$ at the present time. In the absence of this knowledge, we have made an estimate and it is about $6.3\,10^{-7}$ debye (see Appendix). Here we have considered ten electronic states with 14 vibrational levels in each of them (Krishna Swamy and O'Dell, 1987). For the singlet-triplet transitions, we have considered (*a-x*), (*b-x*) and (*c-x*) transitions (Lambert and Danks 1983; Johnson et al., 1983). The Franck-Condon factors for the transitions of the isotopic molecule have been calculated using the program of Albritton (1975), which uses an RKR potential. The wavelengths of the lines have been calculated using the molecular constants of Philips and Davis (1968) and Phillips (1968). The Höln-London factors are those of Schadee (1964). We have used a smoothed radiation field obtained from various sources (Krishna Swamy, 1981). The statistical equilibrium equations have been solved as discussed earlier (Krishna Swamy and O'Dell, 1977, 1979) with resonance fluorescence as the excitation process. The resulting population distribution in various levels turns out to be very similar to that of the $^{12}C^{12}C$ molecule. This implies that the ratio of the band intensities of Swan bands of $^{12}C^{12}C$ and $^{12}C^{13}C$ molecules should directly reflect their abundance ratios, as anticipated by Danks et al. (1974).

2.2. Isotopic features

The observed cometary Swan band of the C_2 molecule arise out of $^3\pi_g$, $^3\pi_u$ transitions. These in turn have three sub-states corresponding to $^3\pi_0$, $^3\pi_1$, and $^3\pi_2$ states. We have considered 80 rotational levels in each of the lower and upper substates in the solution of statistical equilibrium equations. The wavelength of the transitions for the normal molecule are calculated as before. For the isotopic molecule, the value of T_e is adjusted to make the calculated wavelength of the lines agree with those of tabular values (Stawikowski and Greenstein, 1964; Danks et al, 1974). For a heteronuclear molecule like $^{12}C^{13}C$, the rotational transitions are allowed and should be considered. The importance of these transitions depend upon the value of the dipole moment. For the estimated value of the dipole moment, the typical lifetime of rotational levels is of the order of 10^5 years (Appendix). Therefore these transitions can be neglected, again in agreement with the assumption made by Danks et al. (1974). Knowing the popu-

★ NAS-NRC-NASA-Senior Research Associate on leave from Tata Institute of Fundamental Research, Bombay

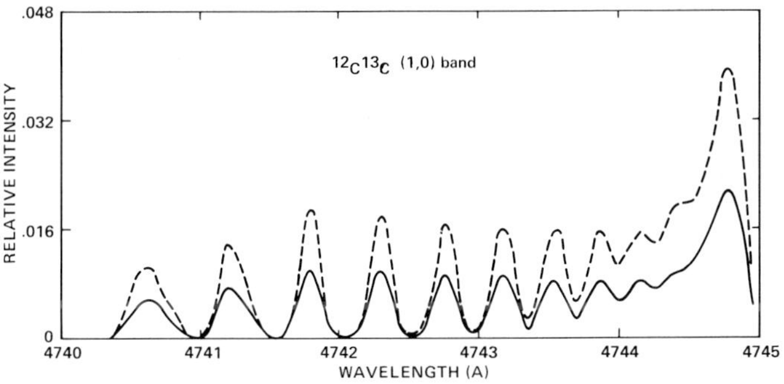

Fig. 1. Calculated profile of (1,0) Swan band of $^{12}C^{13}C$ feature for $\Delta\lambda = 0.15$Å. Continuous and dashed lines refer to $^{12}C/^{13}C$ of 90 and 50 respectively. The profiles are normalized to the peak of the $^{12}C^{12}C$ feature

lation distribution in various rotational levels, the total intensity can be calculated by summing over all the lines. We have assumed equal weight for the three sub-states. The synthetic profile is finally convolved with the gaussian profile (Krishna Swamy, 1986). The results of these calculations for the (1,0) Swan band of the $^{12}C^{13}C$ molecule are shown in Fig. 1. The curves are normalized to the peak of the $^{12}C^{12}C$ feature. It can be seen from the figure that the rotational structure is well resolved for resolutions ≤ 0.15Å. Therefore from a profile fit to the observed features, in which one can avoid the blending of NH_2 lines, it is possible to get a good estimate of the $^{12}C/^{13}C$ ratio. This type of observations could possibly be carried out with existing techniques on a bright comet with weak continuum.

The profile of (0,0) Swan band can also be calculated similar to that of (1,0) band. The wavelength of the transitions for the isotopic molecule are checked against laboratory values (Dhumwad et al., 1981; Amiot, 1983). The calculated profile of the isotopic feature of the (0,0) Swan band is compared with those of observed profiles of Lambert and Danks for comet West in Fig. 2. Because of low signal to noise ratio, the observed isotopic features should be taken with caution. Indeed, as Lambert and Danks themselves pointed out, no definite conclusion can be drawn as to the $^{12}C/^{13}C$ ratio in this comet.

Finally, it may be noted that the (1,0) band has certain advantages over the (0,0) band in the determination of $^{12}C/^{13}C$ ratio, in that it requires less resolution and in addition it provides many resolved lines.

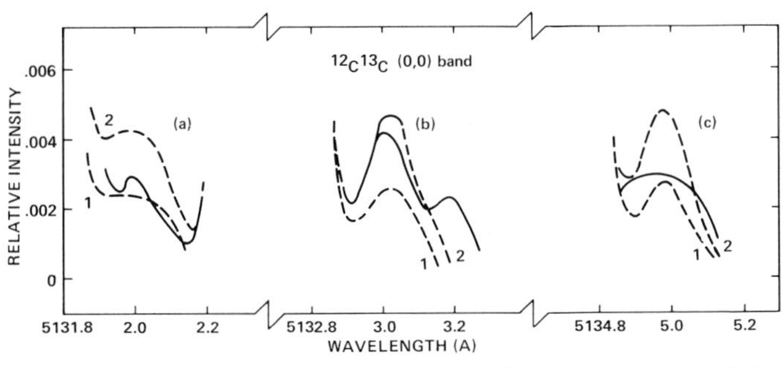

Fig. 2. Calculated profiles (dashed, $\Delta\lambda = 0.15$Å) are compared with the observed profiles (continuous) of Lambert and Danks for comet West. Curves marked 1 and 2 correspond to $^{12}C/^{13}C$ of 90 and 50 respectively. The calculated and the observed profiles are normalized to the peak of the corresponding P-lines

Acknowledgement. It is a pleasure to thank Dr. Jeremy T. Tatum for kindly allowing me to include the material on the estimate of the dipole moment of the $^{12}C^{13}C$ molecule. I would also like to thank Dr. C. Arpigny for valuable comments on the paper.

Appendix

(Dipole moment of $^{12}C^{13}C$-contributed by Jeremy B. Tatum).
According to Bunker (1973), the dipole moment of HD is about $5.9\,10^{-4}$ debye, which is about $2.32\,10^{-4}$ a_0e or $1.97\,10^{-33}$ coulomb-meter.
If we assume that the dipole moment is very roughly proportional to $\frac{(m_1 - m_2)}{m_1 m_2}$ (Bunker, 1973), we estimate that for $^{12}C^{13}C$ it is $2.5\,10^{-35}$ coulomb meter.
The Einstein A value between the levels N and $N-1$ is given by (Tatum and Gillespie, 1977; Arpigny, 1964)

$$A_{N,N-1} = \frac{64\pi^3\mu^2 B_0^3}{3\varepsilon_0 h} \frac{N^4}{N+\frac{1}{2}}$$

From the molecular constants of B_e and α_e of $^{12}C_2$ from Huber and Herzberg (1979), we can deduce that, for the ground state of $^{12}C^{13}C$

$$B_0 = 174\,\mathrm{m}^{-1}.$$

Therefore, we find that

$$A_{N,N-1} = 3.78\,10^{-16} \frac{N^4}{N+\frac{1}{2}}\,\mathrm{s}^{-1}$$

For a "typical" level, say $N = 10$, then

$$\tau(10,9) = 2.78\,10^{12}\mathrm{s} = 8.8\,10^4\,\text{years}.$$

Similarly, the typical lifetime for vibrational levels (see Krishna Swamy, 1979) for $V = 1$ to 0 is of the order of $3\,10^9$ s.

Therefore, one can argue that the dipole moment of $^{12}C^{13}C$ is indeed negligible and that $^{12}C^{13}C$ may after all be treated as a homonuclear molecule.

References

Albritton, D.L.: 1975, (private communication)
Amiot, C.: 1983, *Astrophys. J. Suppl.* **52**, 329
Arpigny, C.: 1964, *Ann. Astrophys.* **27**, 393
Bunker, P.R.: 1973, *J. Mol. Spectrosc.* **46**, 119
Danks, A.C., Lambert, D.L., Arpigny, C.: 1974, *Astrophys. J.* **194**, 745
Dhumwad, R.K., Patwardhan, A.B., Kulkarni, V.T.: 1981, *J. Mol. Spectrosc.* **85**, 177
Huber, K.P., Herzberg, C.: 1979, *Molecular Spectra and Molecular Structure Vol. IV*, Constants of the Diatomic Molecules, Van Nostrand Reinhold, New York
Johnson, J.R., Fink, Uwe., Larson, H.P.: 1983, *Astrophys. J.* **270**, 769
Krankowsky, D., et al., 1986, *Nature* **321**, 326
Krishna Swamy, K.S.: 1979, *Astrophys. J.* **227**, 1082
Krishna Swamy, K.S.: 1981, *Astron. Astrophys.* **97**, 110

Krishna Swamy, K.S.: 1986, *Publ. Astron. Soc. Pacific* **98**, 252
Krishna Swamy, K.S., O'Dell, C.R.: 1977, *Astrophys. J.* **216**, 158
Krishna Swamy, K.S., O'Dell, C.R.: 1987, *Astrophys. J.* **317**, 543
Lambert, D.L., Danks, A.C.: 1983, *Astrophys. J.* **268**, 428
Phillips, J.G.: 1968, *J. Mol. Spectrosc.* **28**, 233
Phillips, J.G., Davis, S.P.: 1968, *Berkeley Analysis of Molecular Spectra, Vol.* **21**, University of California Press, Berkeley
Schadee, A.: 1964, *Bull. Astron. Inst. Netherlands* **17**, 311
Stawikowski, A., Greenstein, J.L.: 1964, *Astrophys. J.* **140**, 1280
Tatum, J.B., Gillespie, M.I.: 1977, *Astrophys. J.* **218**, 569

Astron. Astrophys. 187, 391–397 (1987)

ASTRONOMY AND ASTROPHYSICS

Kinematic properties of the neutral gas outflow from comet P/Halley

H.P. Larson[1], **M.J. Mumma**[2], **and H.A. Weaver**[3]

[1] Lunar and Planetary Laboratory, University of Arizona, Tucson, AZ 85721, USA
[2] Planetary Systems Branch, Code 693, Goddard Space Flight Center, Greenbelt, MD 20771, USA
[3] Space Telescope Science Institute, Baltimore, MD 21218, USA

Received February 6, accepted April 27, 1987

Summary. High resolution airborne infrared spectra of H_2O in comet Halley were used to characterize the velocity field in the neutral gas outflow from the nuclear surface. The observed H_2O emission appeared to be the partially isotropized sum of gas production from numerous active areas primarily on the Sunlit side of the nucleus. The spatial distribution and expansion velocity of the H_2O molecules were deduced from the positions, width, and shape of the H_2O lines. The line profiles demonstrated that the H_2O distribution was anisotropic in the circular field-of-view of approximately 10^4 km radius centered on the nucleus. The blue-shifted nucleocentric line positions indicated that the outflow was predominantly into the Sun-facing hemisphere. Direct observations of the H_2O spatial brightness distribution confirmed that the outflow was concentrated in the Sunward direction, although some H_2O emission was detected in the tail. The measured expansion velocities were 0.9 ± 0.2 and $1.4 \pm 0.2\,\mathrm{km\,s^{-1}}$ pre- and postperihelion, respectively. These values are in general agreement with thermodynamic predictions and other measurements, but the observed H_2O velocities are apparently not a simple function of heliocentric distance. Outbursts and longer term temporal variations in the H_2O production rate may influence the velocity field in the coma.

Key words: Comets – Comet Halley – H_2O molecule – infrared spectroscopy

1. Introduction

Studies of the gas and dust escaping from cometary nuclei offer complementary insight into the composition of cometary matter and physical and chemical processes in the coma. The continuum spectrum of the dust component is readily observed with spectrophotometric techniques at ultraviolet and visible wavelengths by reflected solar radiation and at infrared wavelengths by its thermal emission. The accompanying neutral gas component is not very well characterized, however, because opportunities to observe the spectral signatures of parent molecules are very limited. The requirements on instrumentation are more demanding because higher resolution is necessary to study discrete spectral features, and the spectra of many plausible parent molecules are blocked by the same species in Earth's atmosphere. This obscuration is especially severe for H_2O, one of the most important parent molecule candidates. Many Earth-based and spaceborne experiments searched for H_2O in Comet Halley in order to resolve fundamental questions concerning the role of H_2O in cometary matter. The first definitive detection was made from the Kuiper Airborne Observatory (KAO) using the techniques of high resolution infrared spectroscopy (Mumma et al., 1986). This detection demonstrated conclusively that H_2O is an abundant constituent of cometary matter, as had long been suspected from multiple lines of circumstantial evidence. In addition, the high information content of these airborne spectra allow the H_2O molecule to be used as a sensitive remote probe of physical conditions in the coma. In this paper we relate our high resolution H_2O line profiles to the dynamics of the neutral gas outflow from Comet Halley.

2. Observations

We acquired numerous emission spectra of the ν_3 band of H_2O at 2.65 μm during our pre- and postperihelion airborne studies of comet P/Halley; a complete log of observations is included in Weaver et al. (1987). The resolving power in two of these spectra was sufficiently high ($\geq 10^5$) that the H_2O line profiles reveal properties of the velocity field in the neutral gas outflow. A preperihelion spectrum was recorded on UT 24.11 December 1985 ($R = 1.13$ AU, $\Delta = 1.00$ AU, $\dot{\Delta} = +35.0\,\mathrm{km\,s^{-1}}$) at a velocity resolution of $3.2\,\mathrm{km\,s^{-1}}$. Postperihelion observations on UT 22.72 March 1986 ($R = 1.03$ AU, $\Delta = 0.74$ AU, $\dot{\Delta} = -43.4\,\mathrm{km\,s^{-1}}$) had a velocity resolution of $1.1\,\mathrm{km\,s^{-1}}$. We also recorded lunar and stellar comparison spectra in order to determine instrumental transmission factors and to calibrate residual telluric absorptions.

We used a Fourier transform spectrometer (FTS) (Davis et al., 1980) for all observations. The angular field-of-view was 41″ dia and the spectral bandpass was established by cold filters in the detector systems. This type of spectrometer has several properties that are advantageous for cometary spectroscopy, including intrinsic absolute wavenumber calibration, continuously variable spectral resolution, and high optical throughput. The spectra presented here represent combinations of resolving power, field-of-view, and spectral bandwidth that could not have been achieved with other types of infrared spectrometers.

Our two highest resolution spectra of comet P/Halley are presented in Fig. 1, in addition to a lunar comparison spectrum

Send offprint requests to: H.P. Larson

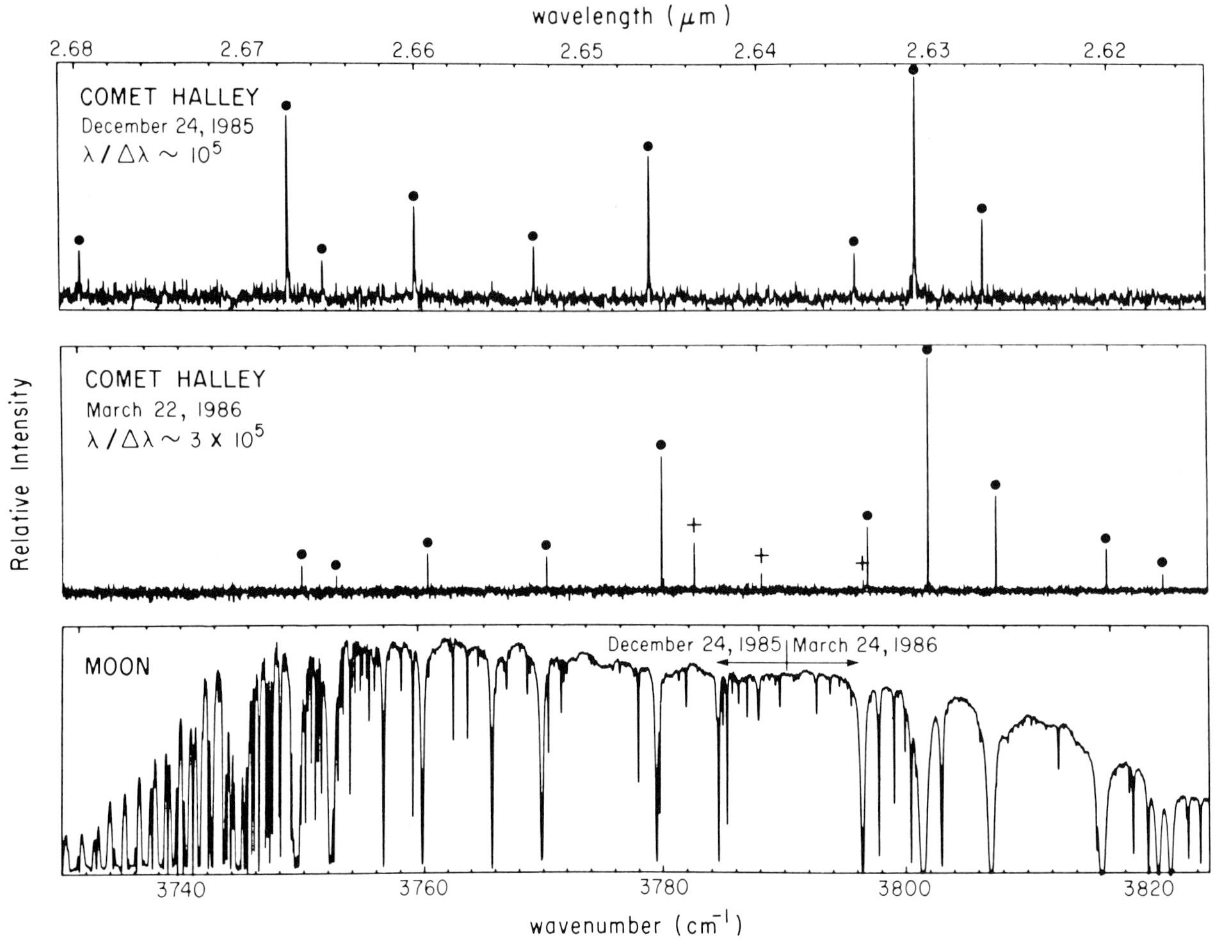

Fig. 1. High resolution airborne observations of H_2O in comet P/Halley. The line intensities are uncorrected for instrumental and atmospheric transmission factors, which are derived from the accompanying lunar comparison spectrum. Residual telluric absorptions are H_2O and, below 3750 cm^{-1}, CO_2. Slightly different interference filters were used to limit the spectral bandpass in the pre- and postperihelion observations. Both filters had a cutoff wavelength near 2.61 μm. The long wavelength cutoff was 2.67 and 2.71 μm for the pre- and postperihelion spectra, respectively. The lunar spectrum is a composite assembled from individual spectra acquired with the two different filters. Differences between the two cometary spectra which are not simply due to filter transmission are discussed in the text. The filled circles designate transitions in the ν_3 H_2O band; the plus signs identify transitions in the "hot" ν_3 band. All assignments are listed in Table 1. The transition near 3779.5 cm^{-1} is shown on an expanded scale in Fig. 2. The small number of cometary H_2O lines, relative to those in Earth's atmosphere, indicates non-LTE excitation in the coma

indicating atmospheric and instrumental transmission factors. One line from each spectrum is displayed on an expanded wavenumber scale in Fig. 2 to illustrate the observed H_2O line profiles. By scheduling observations near the comet's maximum pre- and postperihelion geocentric velocities, each cometary emission line was Doppler-shifted to the far wings of the absorption line profile of the same transition in Earth's atmosphere. This strategy was essential because telluric absorption is still severe in the centers of the low-J lines of the ν_3 H_2O band at aircraft altitude (12.4 km).

A total of 14 H_2O lines appear in the combined spectra of comet P/Halley in Fig. 1. These are listed in Table 1. Eleven are assigned to the 001-000 H_2O band; three belong to the "hot" 011-010 H_2O band. All positions in Table 1 are expressed as absolute wavenumbers in vacuum. The calibration procedure, described in Larson et al. (1986a), was modified here to include improved H_2O laboratory positions (Toth, 1986). As a consequence, the uncertainties in our observed H_2O line positions in Table 1 are lower and our calculated velocities in Sect. 3 slightly different than previously reported (Larson et al., 1986a,b; Mumma et al., 1986).

The two spectra of comet P/Halley in Fig. 1 differ in the number of observed H_2O lines and, especially, in their observed relative intensities. These differences are due to pre- and postperihelion changes in (1) the achieved signal-to-noise ratio (SNR), (2) the spectral bandpass of the FTS, (3) atmospheric transmission, and (4) excitation conditions in the coma. The SNR was about 2 times higher and the resolving power about 3 times higher in the March, 1986 spectrum. This combination provided higher spectroscopic sensitivity, thus contributing to the postperihelion detection of the weakest H_2O lines in Table 1 (e.g. 3787.7, 3796.1, and 3820.7 cm^{-1}). The interference filters used pre- and postperihelion had similar transmission from approximately 2.64 μm to their short wavelength cutoff at 2.61 μm. The two spectra in Fig. 1 should therefore appear similar over this interval, everything else being equal. The filters differed, however, in their red responses. The long wavelength cutoff was 2.71 μm preperihelion and 2.67 μm postperihelion. Thus, postperihelion H_2O

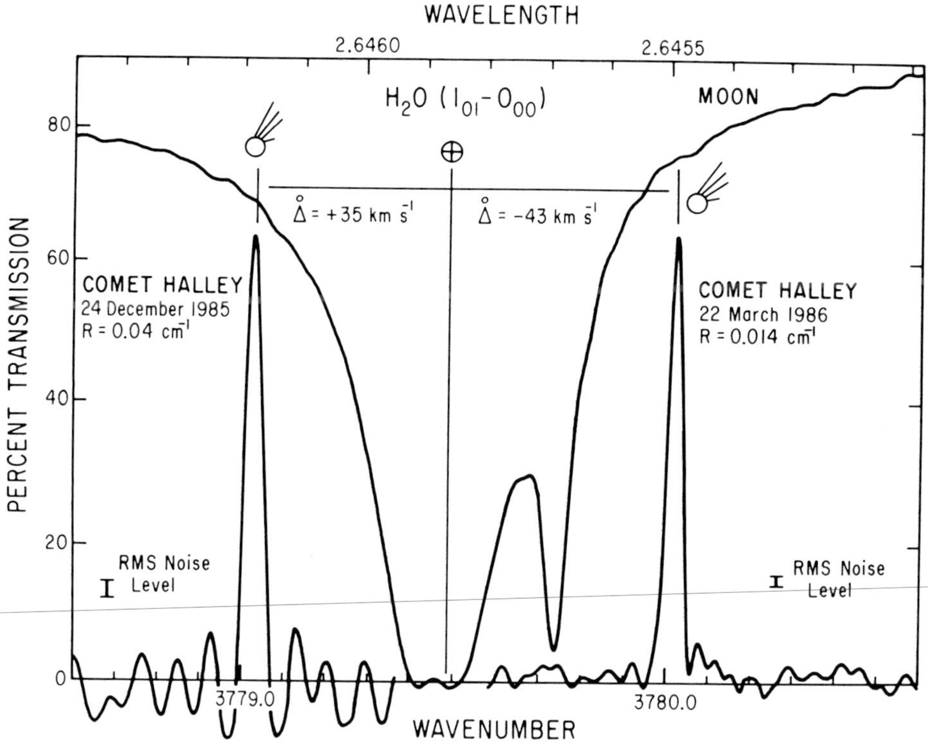

Fig. 2. The same transition in the ν_3 band of H_2O appears three times: in absorption at its rest frequency (3779.49 cm^{-1}) in Earth's atmosphere, and in emission in comet P/Halley at Doppler-shifted positions (3779.05, 3780.04 cm^{-1}) due to pre- and postperihelion geocentric velocities. The ordinate scale is relative transmission in the lunar spectrum. The cometary lines are arbitrarily scaled to the same height for direct comparison of their line profiles. The preperihelion profile is unresolved. The structure at the base of this profile is due to the side lobes of the unapodized instrumental line shape (a sinc function) of a Fourier transform spectrometer. The higher resolution postperihelion line profile lacks these side lobes because the cometary line is at least partially resolved. (After Larson et al., 1986a)

Table 1. H_2O emission lines observed in comet Halley

Transition[1] Vibration	Transition[1] Rotation	Observed Line Position Dec 1985 (cm^{-1})	Observed Line Position Mar 1986 (cm^{-1})	Nucleocentric Line Position Dec 1985 (cm^{-1})	Nucleocentric Line Position Mar 1986 (cm^{-1})	Rest Line Position[2] (cm^{-1})
001-000	0_{00}-1_{01}	3731.6978±0.0074	---	3732.1368	---	3732.1344
001-000	1_{11}-1_{10}	3748.8907±0.0018	3749.8775±0.0039	3749.3317	3749.3333	3749.3292
001-000	2_{20}-2_{21}	3751.7706±0.0090	3752.7576±0.0071	3752.2119	3752.2133	3752.2126
001-000	1_{10}-1_{11}	3759.4051±0.0036	3760.3931±0.0034	3759.8476	3759.8474	3759.8446
001-000	2_{11}-2_{12}	3769.4541±0.0064	3770.4395±0.0037	3769.8976	3769.8924	3769.8890
001-000	1_{01}-0_{00}	3779.0534±0.0023	3780.0449±0.0020	3779.4978	3779.4964	3779.4931
011-010	2_{02}-1_{01}	---	3782.7328±0.0029	---	3782.1840	3782.1791
011-010	2_{11}-1_{10}	---	3788.2221±0.0069	---	3787.6725	3787.6660
011-000	3_{13}-2_{12}	---	3796.6342±0.0134	---	3796.0834	3796.0828
001-000	2_{12}-1_{11}	3795.9962±0.0091	3796.9942±0.0024	3796.4427	3796.4432	3796.4396
001-000	2_{02}-1_{01}	3800.9769±0.0016	3801.9731±0.0020	3801.4242	3801.4215	3801.4187
001-000	2_{11}-1_{10}	3806.5717±0.0045	3807.5690±0.0021	3807.0196	3807.0164	3807.0137
001-000	3_{13}-2_{12}	---	3816.6492±0.0030	---	3816.0956	3816.0916
001-000	3_{03}-2_{02}	---	3821.2953±0.0076	---	3820.7409	3820.7385

[1]See Herzberg (1945) for an explanation of the quantum assignments

[2]Data from Toth (1986). Uncertainties are ±0.00025 cm^{-1}.

lines at $\lambda > 2.64\,\mu m$ in Fig. 1 are increasingly attenuated relative to the preperihelion lines, and one (e.g. $3732.1\,cm^{-1}$) could not even be detected. The large geocentric velocities guaranteed that cometary H_2O lines would be Doppler-shifted away from the same transitions in Earth's atmosphere. Coincidences with neighboring telluric lines could not always be avoided, however, as demonstrated by the cometary H_2O line at $3816.1\,cm^{-1}$. This line was apparently not present preperihelion because its Doppler-shifted position accidentally coincided with another strong telluric H_2O line. This interfering line appears as a partially resolved component to the saturated $3816.1\,cm^{-1}$ telluric line in the lunar spectrum in Fig. 1.

After allowances are made for the three effects described above, there still remain differences between the comet's spectra in Fig. 1. These differences may be due to physical changes in the coma itself. The postperihelion appearance of three lines in the "hot" ν_3 band of H_2O is one example. At least the strongest of these lines would have been obvious at its preperihelion Doppler-shifted position of $3781.7\,cm^{-1}$ if the relative intensities observed in March, 1986 also applied to the December, 1985 data. Other indications of changes in the pre- and postperihelion comae at $R \approx 1$ AU include the measured H_2O expansion velocities (see Section 3.2) and the calculated H_2O production rates (Weaver et al., 1986a,b; Weaver, Mumma and Larson, 1987). In general, these airborne spectra have clearly demonstrated that excitation of H_2O in comets is more complex than predicted by fluorescence equilibrium (Crovisier, 1984; Weaver and Mumma, 1984). More realistic simulations of the airborne spectra result when the effects of collisions and optical depth are included (Bockelée-Morvan and Crovisier, 1986).

3. The velocity field in the coma

The observed H_2O line profile is the superposition of Doppler-shifted emission from H_2O molecules moving away from the nucleus with a characteristic expansion velocity V_{H_2O}. The velocity components may range from $-V_{H_2O}$ to $+V_{H_2O}$ along the comet-Earth line-of-sight, and they will be weighted according to the spatial distribution of H_2O molecules around the nucleus and the phase angle at which the coma is viewed. Uniform outflow into a sphere, for example, would generate a symmetric profile (FWHM $\approx 2V_{H_2O}$) centered at the nucleocentric rest frequency. The spectroscopic signatures of anisotropic outflow include shifts from rest positions and asymmetric line shapes, both of which are present in our data. As discussed in detail in the sections below, our observations are consistent with an anisotropic distribution in which the outflow is concentrated into the Sun-facing hemisphere.

3.1. The nucleocentric velocity of H_2O

The observed positions of the H_2O lines in Table 1 are shifted from laboratory rest values because of three velocity components: (1) the comet's geocentric velocity; (2) the observatory's motion caused by Earth's rotation and the flight track of the KAO (topocentric component); and (3) the radial outflow velocity V_{H_2O} of H_2O molecules from the nucleus (nucleocentric component). The first two velocity components are known, and their wavenumber shifts were removed from each observed line position. The resulting nucleocentric line positions (see Table 1) are still slightly higher than laboratory rest values. We attribute this difference to the net outflow of H_2O molecules toward Earth (a "blue" shift). Each H_2O line is an independent measure of this velocity. Thus, from the entries in Table 1 we obtain two weighted velocity averages: $-0.33 \pm 0.08\,km\,s^{-1}$ preperihelion and $-0.26 \pm 0.07\,km\,s^{-1}$ postperihelion. These averages are not V_{H_2O} itself, which is expected to be about $1\,km\,s^{-1}$ (e.g. Crovisier, 1984). Rather, they represent both V_{H_2O} and its spatial distribution averaged over the spectrometer's field-of-view. One plausible model is H_2O emission into just the Sun-facing hemisphere. This geometry is certainly valid for the dust emission, as demonstrated by visible images from spacecraft (e.g. Keller et al., 1986). It is also supported by our measured asymmetry in the H_2O spatial brightness distribution (Weaver et al., 1986a). The Earth-comet-Sun phase angles were 54.4° and 66.1° for our pre- and postperihelion viewing geometries, respectively. The superposition of velocity-shifted H_2O emission lines for uniform outflow at $V_{H_2O} \approx 1\,km\,s^{-1}$ into the Sun-facing hemisphere would introduce small blue shifts into the spectral line profiles, just as we observed. This result implies that the observed H_2O distribution is the isotropized sum of the gas production from numerous active areas located primarily on the Sunlit side of the nuclear surface. The observations are not indicative of major outbursts or a few strong jets. Physical processes in the coma have not completely smoothed out surface phenomena, however, since asymmetry in the neutral gas outflow has been detected at distances up to 10^5 km from the nucleus (Hsieh et al., 1986; Weaver et al., 1986a).

3.2. The expansion velocity of H_2O

The width of a fully resolved cometary H_2O line profile gives V_{H_2O} if the velocity distribution function (VDF) is known. As discussed above, uniform outflow into the Sun-facing hemisphere appears to be a useful approximation to the VDF. However, our observed cometary H_2O line profiles, examples of which are displayed in Fig. 2, are influenced by the finite resolving power of the spectrometer so V_{H_2O} cannot be read directly from them. These profiles are the convolution of the cometary line shape with the instrumental line shape (ILS). The ILS for an FTS is a sinc function [$\mathrm{sinc}(x) \equiv (\sin x)/x$], characterized by prominent side lobes. The resolution of an FTS is equal to FWHM/1.206, where FWHM is the full width at the half intensity points of the sinc function. For the high resolution spectra in Fig. 1, the velocity resolution was $3.2\,km\,s^{-1}$ ($3.9\,km\,s^{-1}$ FWHM) preperihelion and $1.1\,km\,s^{-1}$ ($1.3\,km\,s^{-1}$ FWHM) postperihelion. The preperihelion line profile in Fig. 2 is essentially a sinc function, which means that the width of the cometary line must be much less than $3.9\,km\,s^{-1}$ FWHM. The postperihelion line profile is at least partially resolved, however, and most of its width (about $2.9\,km\,s^{-1}$ FWHM) is due to the VDF of the gas field in the coma. To measure V_{H_2O} we convolved the appropriate ILS with simple velocity distribution functions (e.g. rectangular, triangular, Gaussian, etc.) and compared the calculated line profiles with the observations. The results are discussed below.

3.2.1. Preperihelion V_{H_2O}.

A maximum width of $1.8 \pm 0.4\,km\,s^{-1}$ is consistent with the unresolved preperihelion H_2O line profile, from which $V_{H_2O} \leq 0.9 \pm 0.2\,km\,s^{-1}$. However, we noted above that V_{H_2O} must have been approximately $1\,km\,s^{-1}$ in order to produce the blue shift of $-0.33 \pm 0.08\,km\,s^{-1}$ in the preperihelion H_2O line positions.

Since this independent estimate establishes a lower bound to V_{H_2O}, we therefore adopt $V_{H_2O} = 0.9 \pm 0.2\,km\,s^{-1}$ as our measured preperihelion value.

3.2.2. Postperihelion V_{H_2O}.

A width of $2.8 \pm 0.4\,km\,s^{-1}$ is necessary to fit the resolved postperihelion H_2O line profile, from which $V_{H_2O} = 1.4 \pm 0.2\,km\,s^{-1}$. This deconvolved width is nearly equal to the observed width which means that the postperihelion H_2O line profile in Fig. 2 is almost completely resolved. The observed blue shift of $-0.26 \pm 0.07\,km\,s^{-1}$ also requires that V_{H_2O} be approximately $1\,km\,s^{-1}$. These postperihelion V_{H_2O} measurements were sensitive to the shape of the assumed VDF because the ILS was narrower than the H_2O line. We found, for example, that a rectangular VDF did not reproduce the observed line profile, upon convolution with the ILS, as well as other trial functions (e.g. Gaussian or triangular functions). The actual VDF was asymmetric, however, since the observed postperihelion line profile in Fig. 2 is generally wider on the red side of line center. This asymmetry is probably due to several factors, including anisotropy in the outflow and optical depth effects in the coma. A comprehensive excitation/outflow model is therefore required to interpret details of the perihelion line profiles; we will present this analysis in a future paper.

Calculations for a dust-free coma by Crovisier (1984) predict an expansion velocity $V_{H_2O} \approx 0.9\,km\,s^{-1}$ (for $R = 1\,AU$ and $Q_{H_2O} = 1.1\,10^{29}\,mol\,s^{-1}$), in excellent agreement with our preperihelion measurement of $V_{H_2O} = 0.9 \pm 0.2\,km\,s^{-1}$ (at $Q_{H_2O} \approx 2\,10^{29}\,mol\,s^{-1}$ and $R = 1.13\,AU$). Other direct measurements of the expansion velocity of neutral species in the coma of comet P/Halley include $V_{H_2O} = 0.8$–$1.1\,km\,s^{-1}$ from *Giotto* (Lämmerzahl et al., 1986), $V_{HCN} = 0.85 \pm 0.15\,km\,s^{-1}$ ($R = 1.5\,AU$) (Despois et al., 1986), and $V_{HCN} = 0.8\,km\,s^{-1}$ ($R = 1.5\,AU$) and $1.3\,km\,s^{-1}$ ($R = 0.6\,AU$) (Bockelée-Morvan et al., 1986). Schloerb et al. (1986b) found that their measurements of V_{HCN} were consistent with the power law $V_{HCN} = 1.0\,R^{-1/2}$, from which $V_{HCN} = 1.0\,km\,s^{-1}$ at $R = 1\,AU$. These measurements are generally consistent with thermodynamic models of the coma, although there are some indications of pre- and postperihelion asymmetry. Our postperihelion result ($V_{H_2O} = 1.4 \pm 0.2\,km\,s^{-1}$ at $R = 1.03\,AU$), for example, was definitely larger than our preperihelion value. Wider postperihelion lines were also noted in OH (Schloerb, et al., 1986) and possibly in HCN (Schloerb et al., 1986a; but see discussion in Schloerb et al., 1986b). The observed expansion velocities in the coma, therefore, may not be simple functions of heliocentric distance. Other influences, such as temporal changes in the gas and dust production rates, need to be investigated in order to interpret these measurements.

4. Discussion

The diagrams in Fig. 3 characterize the gas outflow from comet P/Halley according to analyses of our infrared spectra. The basic picture shows that H_2O was widely dispersed about the nucleus, but with a strong Sunward-tailward asymmetry. We established this geometry along the comet-Sun axis through direct measurements of the H_2O spatial brightness distribution (Weaver et al., 1986a). Complementary information along the comet-Earth axis was provided by kinematic analyses of the high resolution H_2O spectral line profiles. The observed widths, blue shifts, and shapes of the H_2O lines are consistent with outflow into the Sun-facing hemisphere. The H_2O emission that we observed in the tailward direction suggests that some areas may have remained active at reduced rates on the dark side of the nucleus. Alternatively, collisions near the nucleus are expected to cause some isotropization even if surface gas production were limited to the Sunlit hemisphere. More comprehensive analyses of our airborne spectra may eventually clarify some of these details.

An obvious property of the gas outflow from comet P/Halley was its variability. The specific intensities of the H_2O lines demonstrated that the gas production from comet P/Halley varied significantly on many time scales: pre/postperihelion asymmetries, day-to-day variations, and outbursts. Our measured H_2O production rates differed by almost an order of magnitude in pre- and postperihelion observations at $R \approx 1\,AU$. Asymmetries in the gas production are also apparent in observations of HCN (Schloerb et al., 1986b), OH (see discussion in Weaver et al., 1987), H (Kaneda et al., 1986), and O (Stewart, 1987). We also observed large daily variations in Q_{H_2O}. Between UT December 22.1 and 24.1, 1985, for example, Q_{H_2O} increased by a factor of 3 (Mumma et al., 1986); radio OH measurements covering the same period displayed comparable variation (Gérard et al., 1986). Similar temporal variations in gas and dust production throughout most of comet P/Halley's apparition were reported by many observers (e.g. Festou et al., 1986; Feldman et al., 1986). We observed several H_2O outbursts during our airborne observations. On UT 24 March, 1986, for example, the flux in the H_2O lines increased by 40% in 100 min. On this day Fink, Schultz and DiSanti (1987) recorded a pronounced jet oriented approximately 115° from the comet-Sun axis in CCD images of H_2O^+ and continuum emission. Changes in molecular band strengths on even shorter time scales were noted by Rettig et al. (1986).

In conclusion, our spectra of H_2O in comet P/Halley have preserved a record of kinematic conditions in the gas outflow that only a parent molecule could reveal. Our analyses indicate that the neutral gas distribution is anisotropic to at least $10^4\,km$ from the nucleus and the expansion velocities may not be a simple function of heliocentric distance. More realistic outflow models need to be developed in order to clarify the physical significance of these measurements and their possible correlation with outbursts and our observed asymmetry in Q_{H_2O}.

Acknowledgements. We express our appreciation to L. Haughney and the staff of the Kuiper Airborne Observatory for their enthusiastic support of the airborne comet P/Halley program. The successful operation of the spectrometer throughout the flight program must be attributed to the many years of dedicated effort by D.S. Davis and M. Williams. We thank R. Toth for allowing us to use his latest H_2O laboratory measurements in advance of their publication. H.P.L. appreciated the assistance of R.E.L. in the preparation of this manuscript. We gratefully acknowledge support from the following sources: NASA Grant NAG 2-206 to the University of Arizona; NASA Grant NAG 2-356 to the Johns Hopkins University; and RTOP's 188-41-55, 196-41-54, and 352-02-03 to the Goddard Space Flight Center.

References

Bockelée-Morvan, D., Crovisier, J.: 1986, *20th ESLAB Symp.*, ESA SP-250, Vol. III, 11

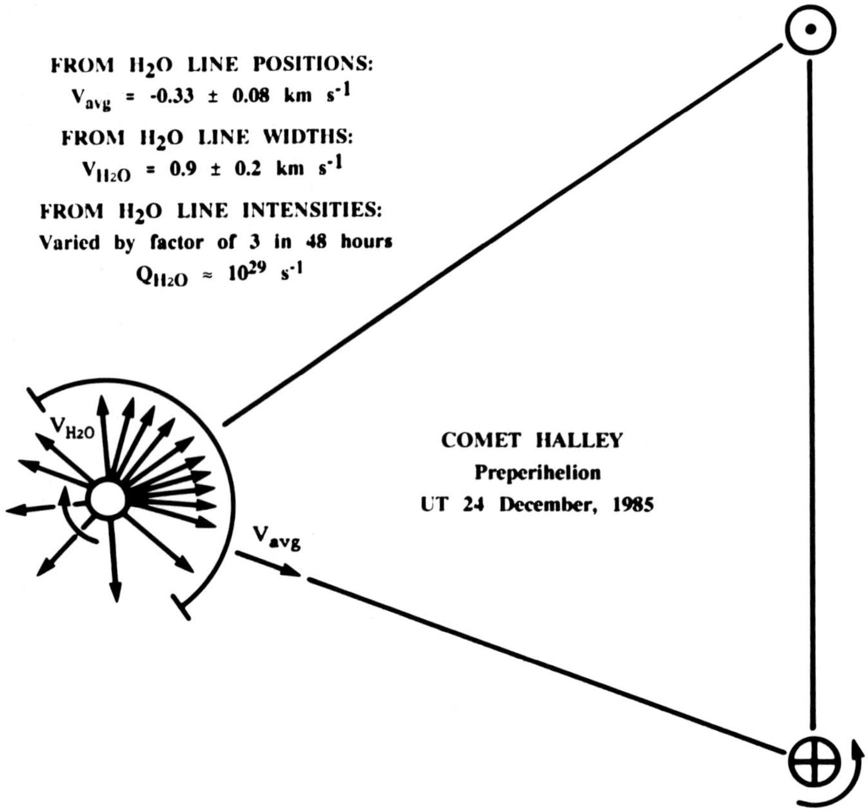

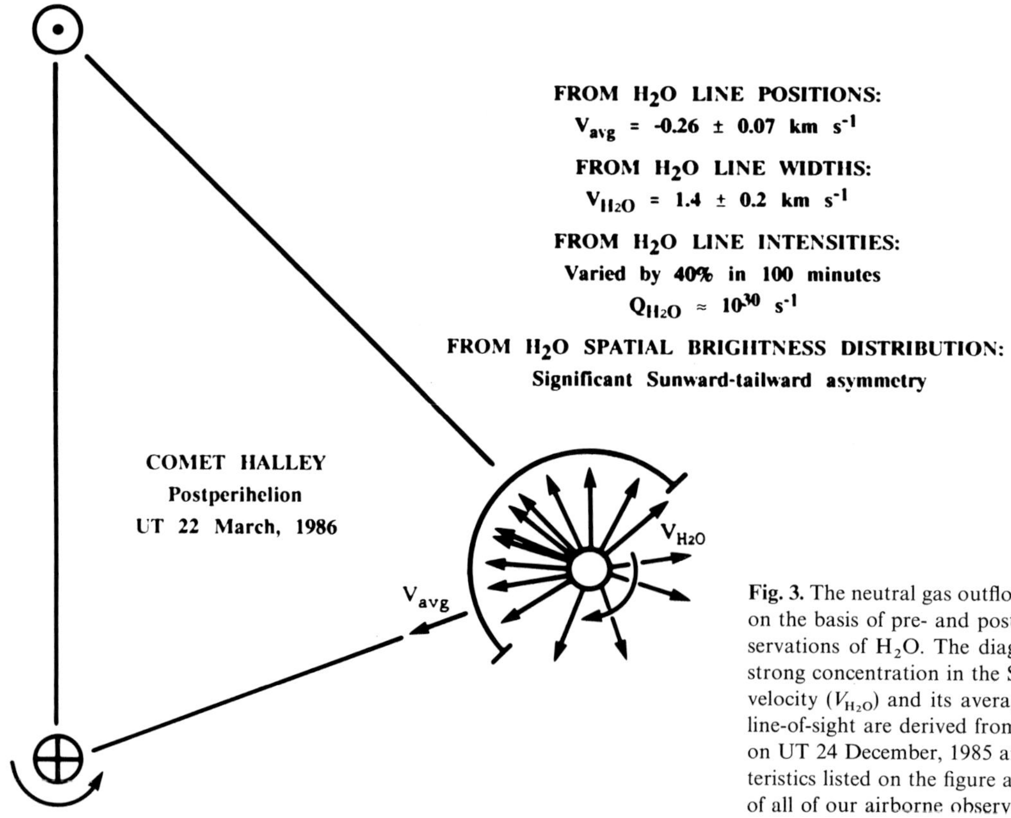

Fig. 3. The neutral gas outflow from comet P/Halley is summarized on the basis of pre- and postperihelion airborne spectroscopic observations of H_2O. The diagrams indicate spherical outflow with strong concentration in the Sun-facing hemisphere. The expansion velocity (V_{H_2O}) and its average value (V_{avg}) along the comet-Earth line-of-sight are derived from the high resolution spectra acquired on UT 24 December, 1985 and 22 March, 1986. The other characteristics listed on the figure are representative values from analyses of all of our airborne observations of H_2O in comet Halley

Bockelée-Morvan, D., Crovisier, J., Despois, D., Forveille, T., Gérard, E., Schraml, J., Thum, C.: 1986, *20th ESLAB Symp.*, ESA SP-250, Vol. I, 365

Crovisier, J.: 1984, *Astron. Astrophys.* **130**, 361

Davis, D.S., Larson, H.P., Williams, M., Michel, G., Connes, P.: 1980, *Appl. Optics* **19**, 4138

Despois, D., Crovisier, J., Bockelée-Morvan, D., Schraml, J., Forveille, T., Gérard, E.: 1986, *Astron. Astrophys.* **160**, L11

Feldman, P.D., Festou, M.C., A'Hearn, M.F., Arpigny, C., Butterworth, P.S., Cosmovici, C.B., Danks, A.C., Gilmozzi, R., Jackson, W.M., McFadden, L.A., Patriarchi, P., Schleicher, D.G., Tozzi, G.P., Wallis, M.K., Weaver, H.A., Woods, T.N.: 1986, *20th ESLAB Symp.*, ESA SP-250, Vol. I, 325

Festou, M.C., Feldman, P.D., A'Hearn, M.F., Arpigny, C., Cosmovici, C.B., Danks, A.G., McFadden, L.A., Gilmozzi, R., Patriarchi, P., Tozzi, G.P., Wallis, M.K., Weaver, H.A.: 1986, *Nature* **321**, 361

Fink, U., Schultz, A., DiSanti, M.: 1987, private communication of unpublished data

Gérard, E., Bockelée-Morvan, D., Bourgois, G., Colom, P., Crovisier, J.: 1986, *20th ESLAB Symp.*, ESA SP-250, Vol. I, 589

Herzberg, G.: 1945, *Molecular Spectra and Molecular Structure*, Vol. 2, *Infrared and Raman Spectra of Polyatomic Molecules* (New York: D. Van Nostrand)

Hsieh, K.C., Curtis, C.C., Fan, C.Y., Hunten, D.M., Ip, W.-H., Keppler, E., Richter, A.K., Umlauft, G., Afonin, V.V., Erö, J., Somogyi, A.J.: 1986, *20th ESLAB Symp.*, ESA SP-250, Vol. I, 417

Kaneda, E., Hirao, K., Shimizu, M., Ashihara, O.: 1986, *Geophys. Res. Letters* **13**, 833

Keller, H.U., Arpigny, C., Barbieri, C., Bonnet, R.M., Cazes, S., Coradini, M., Cosmovici, C.B., Delamere, W.A., Huebner, W.F., Hughes, D.W., Jamar, C., Malaise, D., Reitsema, H.J., Schmidt, H.U., Schmidt, W.K.H., Seige, P., Whipple, F.L., Wilhelm, K.: 1986, *Nature* **321**, 320

Lämmerzahl, P., Krankowsky, D., Hodges, R.R., Stubbemann, U., Woweries, J., Herrwerth, I., Berthelier, J.J., Illiano, J.M., Eberhardt, P., Dolder, U., Schulte, W., Hoffman, J.H.: 1986, *20th ESLAB Symp.*, ESA SP-250, Vol. I, 179

Larson, H.P., Davis, D.S., Mumma, M.J., Weaver, H.A.: 1986a, *Astrophys. J. Letters* **309**, L95

Larson, H.P., Davis, D.S., Mumma, M.J., Weaver, H.A.: 1986b, *20th ESLAB Symp.*, ESA SP-250, Vol. III, 223

Mumma, M.J., Weaver, H.A., Larson, H.P., Davis, D.S., Williams, M.: 1986, *Science* **232**, 1523

Rettig, T., Ruchti, R., Baumbaugh, B., Baumbaugh, A., Knickerbocker, K., Dawe, J.: 1986, *20th ESLAB Symp.*, ESA SP-250, Vol. III, 93

Schloerb, F.P., Claussen, M.J., Tacconi-Garman, L.: 1986, *20th ESLAB Symp.*, ESA SP-250, Vol. I, 583

Schloerb, F.P., Kinzel, W.M., Swade, D.A., Irvine, W.M.: 1986a, *Astrophys. J. Letters* **310**, L55

Schloerb, F.P., Kinzel, W.M., Swade, D.A., Irvine, W.M.: 1986b, *20th ESLAB Symp.*, ESA SP-250, Vol. I, 577

Stewart, A.I.F.: 1987, *Astron. Astrophys.* **187**, 369

Toth, R.: 1986, personal communication of unpublished data

Weaver, H.A., Mumma, M.J.: 1984, *Astrophys. J.* **276**, 782. Also see *erratum: Astrophys. J.* **285**, 872

Weaver, H.A., Mumma, M.J., Larson, H.P., Davis, D.S.: 1986a, *Nature* **324**, 441

Weaver, H.A., Mumma, M.J., Larson, H.P., Davis, D.S.: 1986b, *20th ESLAB Symp.*, ESA SP-250, Vol. I, 329

Weaver, H.A., Mumma, M.J., Larson, H.P.: 1987, *Astron. Astrophys.*, this volume

Astron. Astrophys. 187, 398–404 (1987)

ASTRONOMY AND ASTROPHYSICS

The spectrum of comet P/Halley between 0.9 and 2.5 microns

J.P. Maillard[1], **J. Crovisier**[2], **T. Encrenaz**[2], **and M. Combes**[2]

[1] Institut d'Astrophysique de Paris, F-75014 Paris, France
[2] Observatoire de Paris, Section de Meudon, F-92195 Meudon-Cedex, France

Received February 3, accepted March 20, 1987

Summary. High resolution Fourier-transform spectra of comet P/Halley have been recorded at the CFH 3.60 m telescope (Hawaii) on 1985 December 20–23, and 1986 April 28–30. The spectral interval ranges from 3800 to 12000 cm^{-1} (0.83 to 2.6 μm) and the spectral resolution is equal to 10 cm^{-1}. Dust temperatures of 300 ± 40 K and 240 ± 40 K are derived for 1985 December 20–23, and 1986 April 28–30, respectively, from the slope of the thermal component. Bands of the CN red system are observed at 1.46 μm (1–2), 1.10 μm (0–0), 0.94 μm (2–1), and 0.92 μm (1–0). In addition, an emission feature centered at 2.44 μm (4100 cm^{-1}) is detected. We tentatively identify this feature as the $\nu_1 + \nu_3 - 2\nu_2$ band of H_2O. Ground-based monitoring of this feature on future bright comets might be a direct method for measuring their H_2O production rate.

Key words: comet – comet P/Halley – molecules – H_2O – infrared spectroscopy

1. Introduction

Infrared spectrophotometry of comets is a powerful tool for studying their dust properties and for detecting possible features due to ices and parent molecules. Indeed, in contrast with the UV and visible ranges which show most of the emissions due to cometary radicals and ions, the IR range is especially favourable for the detection of molecular vibrational bands.

The highest possible resolving power is desirable, since these vibrational bands consist of a discrete number of Doppler broadened emission lines. Possible candidates, in the near infrared range, are H_2O, NH_3, CO_2, CH_4, CO, HCN..., which are expected to be excited by fluorescence from the solar infrared radiation field. Theoretical estimates of these fluorescence emissions have been calculated by Encrenaz et al. (1983), Crovisier and Encrenaz (1983), Crovisier and Le Bourlot (1983), Crovisier (1984; 1987), and Weaver and Mumma (1984).

However, high resolution spectroscopy of a comet in the near infrared is difficult, and requires a bright object. A high quality Fourier-transform spectrum was recorded by Johnson et al. (1983) on the very bright, non periodic comet West (1976 VI).

In this paper we present high-resolution spectra of comet P/Halley recorded in December 1985 and April 1986, between 0.9 μm and 2.6 μm. In this spectral range where the flux mostly comes from scattered sunlight, the following features can be searched for: broad signatures due to H_2O ice (at 1.50, 1.65 and 2.0 μm) as well as ices of NH_3, CH_4, CO_2, etc (Fink and Sill, 1982), emission signatures of possible parent molecules (H_2O, NH_3, CH_4, CO) or radicals (C_2, CN, OH). Our observations are described in Sect. 2, and the results are presented in Sect. 3. The CN red system is analyzed in Sect. 4. In Sect. 5, we discuss the identification of an emission feature at 2.44 μm, interpreted as a hot band of gaseous H_2O. Our conclusions are summarized in Sect. 6.

Send offprint requests to: T. Encrenaz

2. Observations

Comet P/Halley was observed at the 3.60 m CFH telescope of Mauna Kea Observatory (Hawaii) with a Fourier-transform interferometer on 1985 December 20–23, and 1986 April 28–30 i.e. before and after perihelion. The exact UT dates and observing conditions are listed in Table 1.

The Fourier-transform spectrometer is a Cassegrain focus double input-double output Michelson interferometer with a maximum path difference of 60 cm (Maillard and Michel, 1982). The instrument is equipped with two InSb detectors. A 6 mm thick piece of red glass in front of each detector was used as a broad band filter. The time needed for a raw interferogram was approximately 7 min. We used beam-switching, i.e. the comet was centered alternatively on one aperture and on the other, and each pair of interferograms was co-added in order to remove background signal, which was limiting the signal-to-noise ratio. Various solar-type stars were used for calibration (see Table 1).

In order to improve the signal-to-noise ratio, the spectra transformed from each pair of interferograms were averaged with appropriately assigned weights. Spectra recorded over several nights were co-added. The April spectrum was more difficult to secure, only one detector being available.

3. Results

Figures 1a and 1b show the observed spectra of comet P/Halley, in 1985 December and 1986 April respectively. These spectra are not corrected for the atmospheric transmission and the instrumental transmission. For comparison, Fig. 1c shows the spectrum of a solar-type calibration star (BS 1856) recorded during the 1985 December observing run.

3.1. Absolute calibration of the spectra

For a quantitative analysis of the cometary emissions, an absolute calibration is required. For this purpose, we used the

Table 1. Observations of comet P/Halley

UT Date	1985 Dec 20.2 & and Dec 23.2	1986 April 28.2, 29.2, 30.2
Heliocentric distance	1.19 to 1.15 AU	1.57 to 1.62 AU
Geocentric distance	0.92 to 0.98 AU	0.69 to 0.77 AU
Aperture beam	8″	12″
Aperture separation	53″	53″
Aperture diameter at the coma	5700 km	6570 km
Spectral range	3900–12000 cm^{-1} (0.83–2.56 μm)	3900–9600 cm^{-1} (1.04–2.56 μm)
Spectral resolution	10 cm^{-1}	10 cm^{-1}
Air mass range	1.2 to 3.2	1.3 to 2.6
Integration time	2^h	5^h 40
Calibration stars	BS 1856 (G3) β Lep (G5) ζ Pup (G3)	BS 5089 (G8) BS 5019 (G6) BS 5072 (G5)
IR magnitude[a]		
J	10.2	9.1
H	9.8	8.7
K	9.9	8.5

[a] This magnitude is estimated for the aperture used in our observation, assuming that the flux is proportional to the aperture diameter

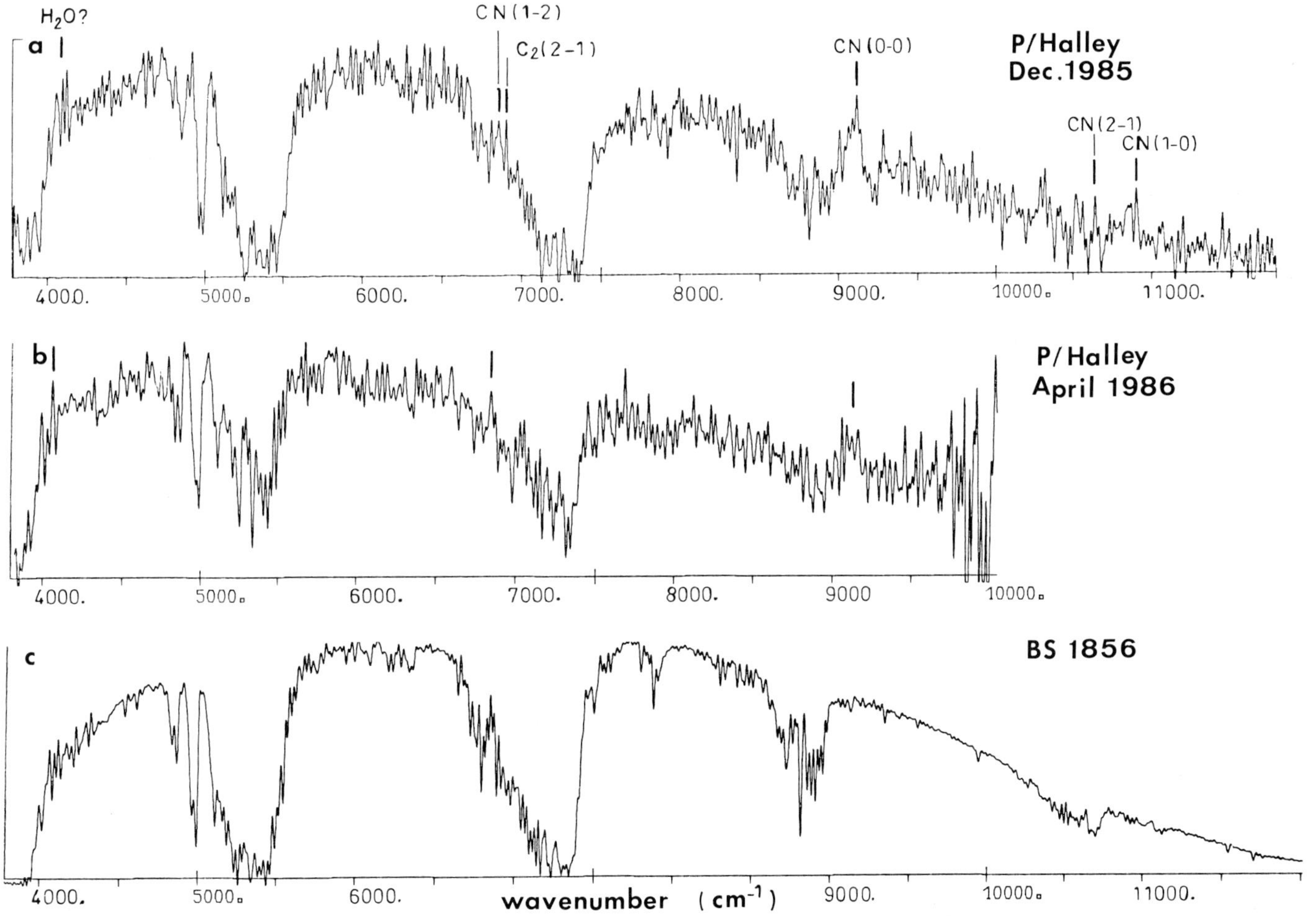

Fig. 1. The spectrum of comet P/Halley between 3900 cm^{-1} and 12,000 cm^{-1}. (a) 1985 December 20.2 and 23.2; (b): 1986 April 28.2, 29.2 and 30.2, (c) Solar-type calibration spectrum (BS 1856), 1985 Dec.

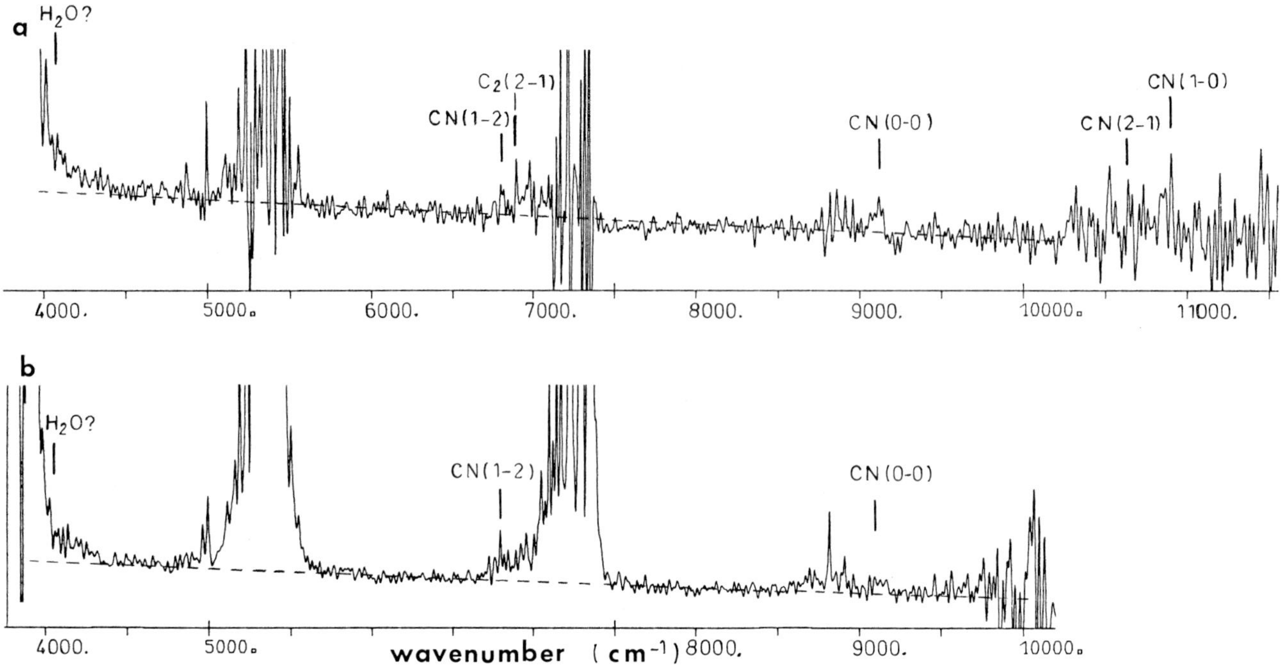

Fig. 2. Ratio of the P/Halley spectrum to the BS 1856 spectrum. (a): 1985 December data; (b): 1986 April data

comet's photometry performed at ESO (Chile) before and after perihelion in the *J*, *H* and *K* filters (Bouchet et al., 1987; Danks et al., 1987). In addition, we used *JHK* photometry recorded at Haute-Provence Observatory on 1985 December 18–21 (Brockmann et al., 1985). Unfortunately, we have no data corresponding to December 22 and 23. However, there is some evidence that the comet's activity was strongly varying from day to day at that time, by as much as one magnitude (Bouchet et al., 1987). We thus consider the uncertainty to be as much as a factor of 2 in the fluxes derived from the *J*, *H*, *K* magnitudes used in this paper (see Table 1).

In order to estimate the integrated fluxes of the cometary emissions, we computed the ratio of the cometary spectrum by a solar-type star spectrum (Fig. 2). It can be seen that the telluric absorption bands are not completely removed: this is because the cometary spectra were recorded over a large range of air mass (1.2 to 3.2), which was difficult to fit with one short record of a calibration star. It can be noticed also that, between the telluric bands, the slope of the continuum is not exactly zero. A possible explanation is that the star used for calibration (BS 1856, spectral type G3) has a general spectrum shape which slightly differs from the slope of the solar spectrum; another reason is probably that the comet's albedo is not exactly constant, as also observed in the case of comet West (Johnson et al., 1983). At 4500 cm^{-1}, the continuum shows a departure from a straight line: this is due to the thermal emission of cometary dust, as will be discussed below.

For calibrating the cometary emissions, the absolute flux density for the continuum at 4545 cm^{-1}, 6060 cm^{-1} and 8064 cm^{-1} were taken from the *K*, *H* and *J* magnitudes given in Table 1. The resulting integrated fluxes of cometary emissions are listed in Table 2.

3.2. *The thermal component*

It can be seen in Fig. 2a and b that the general shape of the cometary spectrum is well fitted by the shape of the solar spectrum, at wavenumbers higher than 4500 cm^{-1}. Below this value,

Table 2. Emission features in the 0.9–2.5 μm spectrum of comet P/Halley

Observed wavelength (μm)	Emission wavenumber (cm^{-1})	Identified species	Laboratory wavenumber (cm^{-1})	Type of transition	Integrated flux ($W/m^{-2}sr^{-1}$)	
					Dec. 85	Apr. 86
2.44	4100	H_2O?	4098	$\nu_1 + \nu_3 - 2\nu_2$	$1.7\ 10^{-7}$	$1.3\ 10^{-7}$
1.46	6850	CN	6846	$A \rightarrow X\ (1 \rightarrow 2)$	$2.5\ 10^{-7}$	$5.7\ 10^{-7}$
1.45	6900	C_2	6889	$b \rightarrow a\ (2 \rightarrow 1)$		
1.10	9110	CN	9117	$A \rightarrow X\ (0 \rightarrow 0)$	$3.0\ 10^{-6}$	$2.5\ 10^{-6}$
0.973	10280	OH?	10230	$X^2\Pi\ (3 \rightarrow 0)$	$2.5\ 10^{-6}$	
0.952	10500	?				
0.941	10630	CN	10624	$A \rightarrow X\ (2 \rightarrow 1)$	$1.3\ 10^{-6}$	
0.917	10900	CN	10904	$A \rightarrow X\ (1 \rightarrow 0)$	$5.1\ 10^{-6}$	

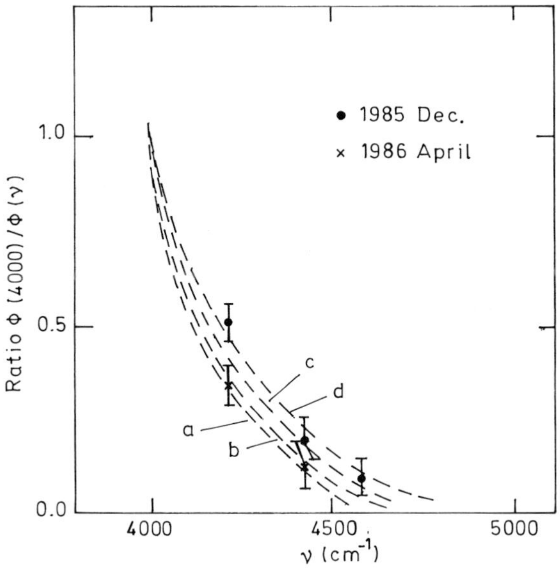

Fig. 3. Determination of the dust temperature, from the slope of the thermal emission. The ordinate is the thermal flux, normalized to its value at $4000\,cm^{-1}$. Curves a, b, c and d are blackbody spectra, corresponding respectively to $T = 230\,K$, 250 K, 280 K and 320 K

the continuum shows a second component in the cometary spectrum which increases toward shorter wavenumbers, due to the cometary dust emission. An estimate of its temperature can be derived from the slope of this component. For this purpose we measured the thermal component $\phi(v)$ at $v = 4000$, 4200, 4400 and $4600\,cm^{-1}$ respectively and we plotted the quantities $\phi(v)/\phi(4000\,cm^{-1})$ versus a set of blackbody curves (Fig. 3). Taking into account the uncertainty introduced by the poor atmospheric cancellation, we derive from this comparison the following results:

$T = 300 \pm 40\,K$ on December 20–23, 1985
$T = 240 \pm 40\,K$ on April 28–30, 1986.

These results are in very good agreement with other estimates obtained from infrared photometry, if we assume $T \propto r_H^{-1/2}$, r_H being the heliocentric distance. Indeed, a temperature of about 300 K is found for $r_H = 1.1$ AU and a temperature of 350–400 K is derived for $r_H = 0.67$ (Bouchet et al., 1987; Danks et al., 1987). From the $r_H^{-1/2}$ dependence, we derive $T = 300\,K$ and $T = 249\,K$ for 1985 December 20–23 and 1986 April 28–30, respectively, which is fully consistent with the present results.

3.3. Cometary emission features

Some emission features of cometary origin are easily observable in Fig. 1 (see Table 2). First, the CN red system ($A^2\Pi \rightarrow X^2\Sigma_g^+$) is observable in the $(0-0)$ band at $9117\,cm^{-1}$ and the $(1-0)$ band at $10904\,cm^{-1}$; the $(1-2)$ band at $6846\,cm^{-1}$ is also present, possibly blended with the C_2 $b \rightarrow a$ $(2-1)$ band at $6889\,cm^{-1}$; a marginal feature might also be attributed to the CN $(2-1)$ band at $10624\,cm^{-1}$ in the December spectrum. The CN emissions are analyzed in Sect. 4. There is a marginal emission at about $10280\,cm^{-1}$ which might be attributed to OH $(3-0)$ in the December spectrum; but since there is no clear evidence for OH $(2-0)$ around $6970\,cm^{-1}$, this identification is dubious. An emission could be present at $10500\,cm^{-1}$, presently unidentified. Another unexpected feature, which appears in emission at about $4100\,cm^{-1}$ (i.e. 2.44 μm), will be discussed in detail in Sect. 5.

In Fig. 1, we see no emission due to the fundamental bands of H_2O, NH_3, CH_4 or CO. In the case of H_2O, the v_3 band centered at $3755\,cm^{-1}$ is known to be present, as shown on the high resolution KAO spectra recorded in December 85 by Mumma et al. (1986); but on our ground-based spectra, this emission is hidden by the telluric absorption. There is no signature of any kind of ice in our spectrum. This can probably be explained by the rather large field ($\sim$ 6000 km) of our observations.

It is also interesting to compare our data with the spectrum of Comet West (1976 VI) recorded by Johnson et al. (1983). For the latter, the spectral resolution and the signal-to-noise ratio are approximately twice better, due to the higher intrinsic brightness of the source, by a factor of 100. All emission lines recorded in our data are present in Johnson et al's spectrum, except possibly the $10280\,cm^{-1}$ feature, but the latter emission is marginal in our spectrum. In particular, the $4100\,cm^{-1}$ is also clearly visible in Comet West's spectrum, but unassigned by the authors. Weaker bands of C_2 and CN, not visible in Halley's spectrum, are also visible on the West spectrum.

4. The CN red system

The fluorescence of the CN radical is one of the best modelled cometary fluorescence, as discussed in the recent work of Zucconi and Festou (1985). The expected relative intensities of the $(1-0)$, $(2-1)$ and $(1-2)$ bands of the red system, with respect to the $(0-0)$ band, are respectively 0.78, 0.19 and 0.13, according to Zucconi and Festou (1985). In December, we observed relative intensities of 1.7, 0.4 and 0.08. The discrepancies with the model values may be attributed to the low signal-to-noise ratio, and to absolute calibration uncertainties.

The $(0-0)$ CN band, observed here with a good signal-to-noise ratio, may be used to derive a CN abundance. Assuming a lifetime of 8 μs for the $A^2\Pi v = 0$ state of CN, we obtain a column density N(CN) of $6.4\,10^{11}\,cm^{-2}$ in our 5700 km field of view in 1985 December, and $1.0\,10^{12}\,cm^{-2}$ in the 6600 km field of view in 1986 April. The December value is to be compared with the value of $1.6\,10^{10}\,cm^{-2}$ obtained from the CN violet system by Catalano et al. (1986), on 1985 December 17–18, in a 90 000 km diameter field. The difference cannot be completely explained by the CN spatial distribution (one would suspect N(CN) to vary less steeply than the inverse of the diameter), and must be partly due to calibration uncertainties, CN red band strength uncertainty, and gas production variations.

Our estimate of the CN column density can also be compared with the result obtained by Johnson et al. (1983) for comet West. The CN $(0-0)$ signal observed on comet West was 100 times stronger than the Halley signal in 1985 December. Taking into account the increase of the excitation rate due to the smaller heliocentric distance, Johnson et al. (1983) derived for comet West a CN column density of $1.04\,10^{13}\,cm^{-2}$, in a field of view similar to ours.

5. The $4100\,cm^{-1}$ emission feature

The $4100\,cm^{-1}$ feature was detected in both spectra of P/Halley and was previously observed in the spectrum of comet West; these three independent detections give credit to its reality.

5.1. Possible identification of the 4100 cm^{-1} feature

This emission feature is relatively narrow (extending from 3980 cm^{-1} to 4150 cm^{-1}) with possible structure. This suggests that it comes from gas emission rather than from dust emission. As shown by Crovisier and Encrenaz (1983), the most efficient radiation mechanism for molecules in the infrared is fluorescence of vibrational bands induced by the solar radiation field. For our feature, we may invoke strong bands of minor species or weaker bands of abundant species. We will see that the possibilities are very limited around 4100 cm^{-1}.

5.1.1. Fundamental vibrational bands

These bands have the strongest emission rates according to Crovisier and Encrenaz (1983). However, no such band is known to exist near 4100 cm^{-1}. One might think that the present feature is due to some lines of the wing of the strong ν_3 band of H_2O. Such lines, however, can only originate from high rotational levels ($J \gtrsim 6$) whereas both observations (Mumma et al., 1986) and theoretical models (Bockelée-Morvan and Crovisier, 1986; Bockelée-Morvan, 1987) show that the cometary water molecule is relaxed to the lowest rotational levels.

5.1.2. Harmonic and combination bands

These bands are not expected to have high emission rates, because molecules preferentially decay from high vibrational levels by steps of $\Delta v = 1$ jumps rather than directly to the ground state. Indeed, the 4100 cm^{-1} feature coincides with the $2\nu_3$ band of OCS at 4101 cm^{-1} and the 2-0 band of CN ($X^2\Sigma^+$) at 4060 cm^{-1}. However, the observed intensity of the band (Sect. 5.4) would correspond to unreasonably high production rates of these molecules.

5.1.3. Hot bands

The strongest hot bands are those with $\Delta v = 1$ jumps; they coincide in frequency with the corresponding bands, and none exist at the frequency of our feature. Other hot bands correspond to transfer of energy between different modes of vibration and have therefore a weak strength. If the 4100 cm^{-1} feature is indeed a hot band, it should be that of an abundant species. For reasonably low vibrational levels, the only possibility is the $\nu_1 + \nu_3 - 2\nu_2$ [or (1, 0, 1)–(0, 2, 0)] band of water, centered at 4098.2 cm^{-1}, according to the water energy levels published by Flaud et al. (1981).

5.2. The $\nu_1 + \nu_3 - 2\nu_2$ water band and its excitation

This band has never been, to our knowledge, observed in the laboratory. Of course, this hot band cannot be observed in absorption. The (1, 0, 1) vibrational level, which has an energy corresponding to a temperature of ~ 10500 K, is only marginally populated during laboratory measurements of spectral emissivity of hot water, which are conducted at less than 3000 K (Ferriso and Ludwig, 1964; Coppalle and Verrisch, 1986). In these measurements, the $\nu_1 + \nu_3 - 2\nu_2$ band does not show up in the wing of the ν_3 band.

The $\nu_1 + \nu_3 - 2\nu_2$ band which transfers energy between stretching and bending modes, is forbidden in the pure harmonic oscillator approximation, but not on the basis of symmetry selection rules. In fact, Fermi-resonance is expected to occur between the (1, 0, 1) vibrational level at 7249.8 cm^{-1} and (0, 2, 1) at 6871.5 cm^{-1}. It can be noted that (0, 2, 1) – (0, 2, 0) has a strength comparable to that of the strong ν_3 band.

In cometary conditions, the (1, 0, 1) vibrational level is expected to be excited by the solar near infrared field with a rate $g_{000-101} = 2.7\,10^{-5}\,s^{-1}$ at $r_h = 1$ AU (Crovisier, 1984). The molecules may decay following different paths (see Fig. 4), with branching ratios proportional to the band Einstein coefficients A. $A_{101-020}$ is unknown, which will hamper the continuation of our analysis.

A conservative upper limit for $A_{101-020}$ is that of $A_{021-020} = 54\,s^{-1}$, because of the Fermi resonance noted above. The emission rate of the $\nu_1 + \nu_3 - 2\nu_2$ band may be expressed:

$$g_{101-020} = 2.7\,10^{-5} \frac{A_{101-020}}{104 + A_{101-020}} \, [s^{-1}]$$

and is at most $9\,10^{-6}\,s^{-1}$. In fact, the Fermi resonance is expected to be of the order of a few percent, and $g_{101-020}$ might not reasonably exceed $10^{-6}\,s^{-1}$.

5.3. The structure of the $\nu_1 + \nu_3 - 2\nu_2$ band

The structure of the band (e.g. distribution within the ro-vibrational lines) is determined by the rotational population distribution of the ground vibrational state. Since the $\nu_1 + \nu_3 - 2\nu_2$ band has the same symmetry as the ν_3 band, the two bands should roughly have the same structure, and differ only by a wavelength shift and a global intensity scaling factor. This is not quite true because:

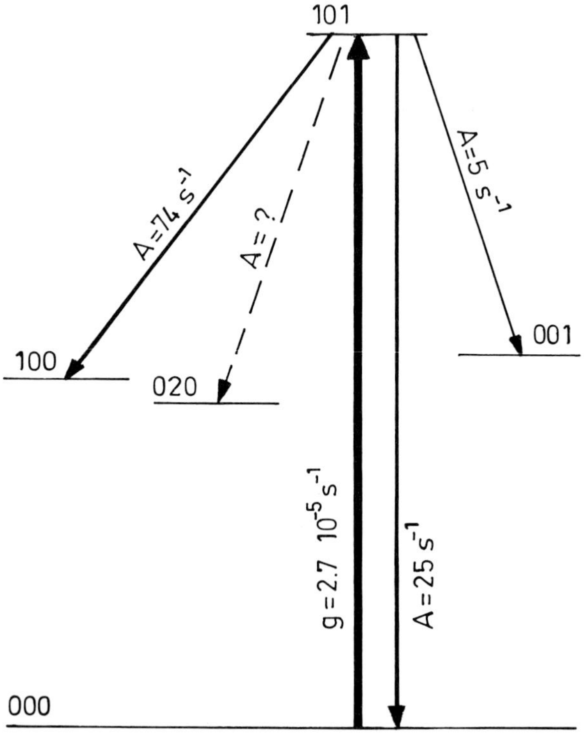

Fig. 4. Excitation of the (1, 0, 1) vibrational level of H_2O and subsequent fluorescence cascades

– perturbations affect the rotational levels of high vibrational states. We have computed the accurate wavenumbers of the first ro-vibrational lines of the band from the (1, 0, 1) and (0, 2, 0) rotational level energies listed in the GEISA data bank (Husson et al., 1982).

– the relative intensities of the ro-vibrational Einstein coefficients are not the same for this band and for ν_3, because Fermi resonance between (1, 0, 1) and (0, 2, 1) depends on the rotational level.

In the outer coma, the water molecule is at fluorescence equilibrium and is relaxed on the lowest rotational levels 0_{00} and 1_{01}. Fig. 5 shows, in these conditions, the structure of the $\nu_1 + \nu_3 - 2\nu_2$ band, deduced from that of the ν_3 band (Crovisier, 1984; Weaver and Mumma, 1984) and assuming the same relative Einstein coefficients.

In the actual case, we observed the $r < 6000$ km region of the coma. In this region, the water molecule is not yet relaxed to the lower rotational levels (Bockelée-Morvan, 1987), and more ro-vibrational lines are expected. This is indeed what was observed for the ν_3 band of water in the $r < 15000$ km region on Dec. 24 (Mumma et al., 1986). Therefore, it is not surprising that the lines indicated for fluorescence equilibrium at the bottom of Fig. 5 are not clearly present in the observed spectra. The observation of the individual lines might be difficult, taking into account the signal-to-noise ratio and the spectral resolution of our data.

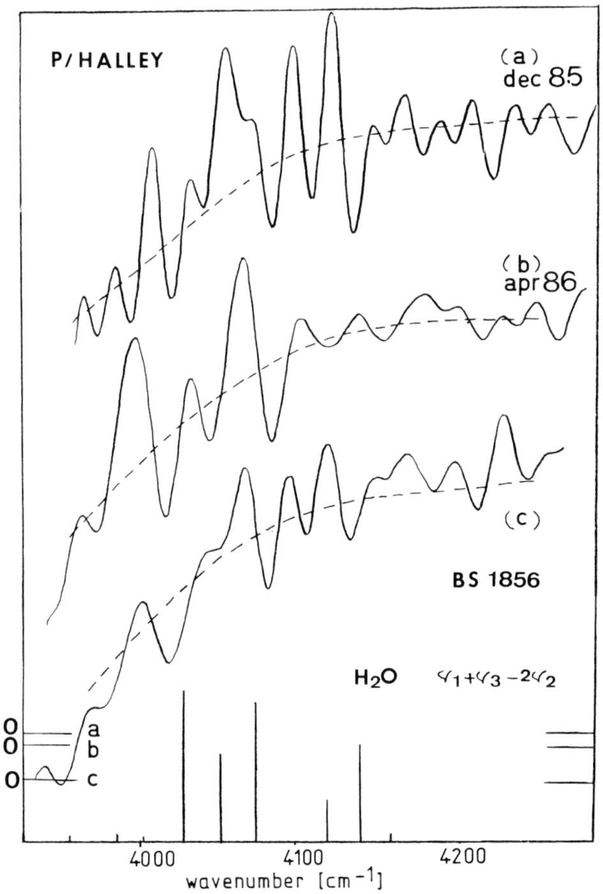

Fig. 5. Relative intensities of the ro-vibrational lines of H_2O ($\nu_1 + \nu_3 - 2\nu_2$) at fluorescence equilibrium. Spectra of P/Halley and BS 1856 are shown for comparison

5.4. Intensity of the 4100 cm^{-1} band and water production rate

Estimates of the intensity of the 2.44 μm band are given in Table 2 for the two cometary spectra. As discussed above, the uncertainty of these numbers is about a factor of 2.

The integrated flux W of the whole water band does not depend on its rotational structure. For a resolved water coma, it is related to the water production rate Q through:

$$W = h\nu g \frac{1}{4\Pi v_{\text{exp}}} \cdot \frac{Q}{\Delta\theta}$$

Assuming a gas expansion velocity $v_{\text{exp}} = 0.8\,\text{km}\,\text{s}^{-1}$ and $g = 10^{-6}\,\text{s}^{-1}$ (Sect. 5.2), we obtain $Q = 1.1\ 10^{29}\,\text{s}^{-1}$ for December and $1.7\,10^{29}\,\text{s}^{-1}$ for April. These rates are comparable to the water rates measured at the same periods by other means: $Q\,[H_2O] = 4\ 10^{28}\,\text{s}^{-1}$ on Dec. 22.1, $1.7\ 10^{29}\ \text{s}^{-1}$ on Dec 24.1 (Mumma et al., 1986), $2\,10^{29}\,\text{s}^{-1}$ on Apr 29.8 (Feldman et al., 1986).

Thus, the identification of the 4100 cm^{-1} emission feature with the $\nu_1 + \nu_3 - {}^2\nu_2$ water band seems reasonable, and its emission intensity can be perfectly understood, if the Einstein A-coefficient of this band is indeed of the order of $5\,\text{s}^{-1}$.

6. Conclusion

The present paper illustrates the scientific interest of high resolution infrared spectroscopy for cometary studies. Although the signal recorded on P/Halley in December 1985 and April 1986 was about 100 times lower than the signal of comet West in March 1976, high resolution (10^3) spectroscopy of comet Halley has still been possible, thanks to the developments of infrared technology.

Our data have allowed a determination of the dust temperature, an estimate of the CN column density, and the detection of an emission band at 2.44 μm, identified for the first time as a hot band of water vapor. Theoretical work to ascertain the strength of this band is obviously needed.

If our identification of this feature is confirmed by future cometary observations, the ground-based monitoring of the 2.44 μm emission on bright comets could become a direct and powerful method for measuring their H_2O production rate. Indeed, the various ground-based methods used to derive the H_2O production rate (UV and radio measurements of OH) have often led to conflicting results; moreover, they have shown strong fluctuations of the production rate from day to day. A direct measurement of H_2O which could be monitored from the ground would then provide a very important contribution.

Acknowledgements. We are grateful to the CFHT staff members for their technical support. We thank J.M. Flaud and C. Camy-Peyret for very helpful discussions concerning this paper.

References

Bockelée-Morvan, D.: 1987, *Astron. Astrophys.* (in press)

Bockelée-Morvan, D., Crovisier, J.: 1986, in *Asteroids, Comets, Meteors II,* eds C.-I. Lagerkvist et al., Uppsala University Press, p. 279

Bouchet, P., Chalabaev, A., Danks, A., Encrenaz, T., Epchtein, N., Le Bertre, T.: 1987, *Astron. Astrophys.* **174**, 288

Brockmann, B., Encrenaz, T., Epchtein, N.: 1985, IHW Hotline, P. Wehinger, ed., Dec. 23

Catalano, F.A., Baratta, G., Lo Presti, C., Strazzula, G.: 1986, *Astron. Astrophys.* **168**, 341
Coppalle, A., Verrisch, P.: 1986, *J. Quant. Spectr. Rad. Transfer* **35**, 121
Crovisier, J.: 1984, *Astron. Astrophys.* **130**, 361
Crovisier, J.: 1987, *Astron. Astrophys. Suppl.* **68**, 223
Crovisier, J., Encrenaz, T.: 1983, *Astron. Astrophys.* **126**, 170
Crovisier, J., Le Bourlot, J.: 1983, *Astron. Astrophys.* **123**, 61
Danks, A., Bouchet, P., Le Bertre, T. Chalabaev, A., Encrenaz, T.: 1987, *Astron Astrophys.* (in press).
Encrenaz, T., Crovisier, J., Combes, M. Le Bourlot, J.: 1983, in *Asteroids, Comets, Meteors,* C.I. Lagerkvist, H. Rickman, eds., Univ. of Uppsala, p. 291
Feldman, P.D., Festou, M.C., A'Hearn, M.F., Arpigny C., Butterworth, C., Coslovici, B., Danks, A.C., Gilmozzi R., Jackson, W.M., McFaden, L.A., Patriarchi, P., Schleicher, D.G., Tozzi, G.P., Wallis, M.K., Weaver, H.A., Woods, T.N.: 1986, in *New Insights in Astrophysics*, ESA-SP 263, p. 39
Ferriso, C.C., Ludwig, C.B.: 1964, *J. Quant Spectr. Rad. Transfer* **4**, 215
Fink, U., Sill G.T.: 1982, in *Comets*, L. Wilkening, ed. Univ. of Arizona Press, p. 164
Flaud, J.M., Camy-Peyret, C., Toth, R.A.: 1981, *Int Tables Selected Constants* **19**, Pergamon Press
Husson, N., Chedin, A., Scott, N.A., Cohen-Hallaleh, I., Berroir, A.: 1982, La banque de données GEISA, mise à jour N°3, *Lab. Météorol. Dyn. CNRS,* Internal Note **116**
Johnson, J.R., Fink, U. Larson, H.P.: 1983, *Astrophys. J.* **270**, 769
Maillard, J.P., Michel, G.: 1982 in *Instrumentation for Astronomy with Large Optical Telescopes,* ed. C.M. Humphries, Reidel, p. 213
Mumma, M.J., Weaver, H.A., Larson, H.P., Davis, D.S., Williams, M.: 1986, *Science* **232**, 1523
Weaver, H.A., Mumma, M.J.: 1984, *Astrophys. J.* **276**, 782
Zucconi, J.M., Festou, M.: 1985, *Astron. Astrophys.* **150**, 180

Astron. Astrophys. 187, 405–410 (1987)

ASTRONOMY AND ASTROPHYSICS

Anisotropic non-stationary gas flow dynamics in the coma of comet P/Halley

N.I. Kömle* and W.-H. Ip

Max-Planck-Institut für Aeronomie, D-3411 Katlenburg-Lindau, Federal Republic of Germany

Received February 22, accepted March 24, 1987

Summary. One of the major results from the Giotto and Vega observations was that the gas outflow from the nucleus should be far from being spherically symmetric. Rather, the areas of active outgassing are restricted to a small part of the nucleus surface on the sunlit side and – due to the rotation of the nucleus – sources should be turned on and off periodically. In this paper we present results from numerical model calculations which address two particular outflow patterns that may occur at comet Halley: (a) steady state outflow dominated by a strong gas source at the subsolar point and (b) steady state gas flow caused by a ring-like outgassing region. In both cases we find that the anisotropies introduced at the comet's surface are maintained out to rather large distances, i.e. several thousand kilometers. In addition to the steady state solutions we discuss the transient stages of the gas flow immediately after onset or offset of the strong outgassing at the subsolar point.

Key words: comet P/Halley – anisotropic gas jets – time variation of gas emission

1. Introduction

The observational results obtained by the 'comet Halley armada' have for the first time given some evidence about the nature of the gas- and dust environment in the near-nucleus region of comet Halley (Sagdeev et al., 1986a; Keller et al., 1986; and Ip, 1986; Sekanina, 1986 for reviews). One of the most outstanding findings was the fact that the main sources of gas and dust emission are apparently restricted to a relatively small area of the nucleus surface; moreover, on the nightside hemisphere no significant outgassing activity could be detected. Due to the rotation of the nucleus the local sources become active as soon as they are exposed to the sunlight. The duration of their activity could be highly dependent on the local surface topology and can be restricted to a few hours near the subsolar meridian. In addition a background component seems to be present originating from the sunlit hemisphere (Ip and Rickman, 1986). In addition to the dust jets clearly seen by the Halley Multicolour Camera (HMC) and the Television System (TVS) on Giotto and Vega, respectively (Keller et al., 1986; Sagdeev et al., 1986), the Three-Channel-Spectrometer (TKS) on Vega2 found evidence for gas jets extending over several thousand kilometers on the sunward side of the nucleus (Moreels et al., 1986). The global behavior of the time-dependent gas flow in response to the rotation of the nucleus was most clearly seen in the Lyman α observations from the Susei spacecraft (Kaneda et al., 1986).

The observations summarized above imply a highly anisotropic, time-variable, and inhomogeneous flow pattern. They indicate the need for new models of the cometary atmosphere, which could overcome the restrictions of most of the existing cometary atmosphere models, with the assumptions of a spherically symmetric, steady outflow from the nucleus. An important work in this direction has been contributed by Kitamura (1986), who calculated steady state solutions for a dusty gas jet assuming axial symmetry. In the present paper we discuss some results from similar numerical model calculations explaining the consequences of different possible outgassing patterns on the gas flow. While some of Kitamura's results are confirmed by our calculations, several new aspects are also discovered. We will show numerical results for four plausible situations which may occur at comet Halley:

(a) quasi-steady state gas flow with a strong source active at the subsolar point superimposed on a much weaker background flow;

(b) quasi-steady state gas flow with a 'ring-like' activity region around the subsolar point to examine the effect of an inactive region inside an active region;

(c) transient flow when a strong source at the subsolar point starts to be active;

(d) transient flow when a strong source at the subsolar point stops to be active.

While we are aware that these examples still contain a high degree of idealization as compared to the real flow, they are nevertheless useful to underline a number of important aspects of the gas phenomena at comet Halley. In particular the lateral expansion of the flow (case(a)) and the formation of convergent 'jets' (case (b)) could be applied to the spacecraft observations. The following results correspond to the dust-free case. Gas-dust-coupling will be included in a later step of this investigation.

Send offprint requests to: N. Kömle

* Permanent Address: Space Research Institute, Austrian Academy of Sciences, Lustbühel Observatory, A-8042 Graz, Austria (after June 7, 1987)

2. Basic equations and numerical method

Keeping in mind that our computations are restricted to axisymmetric conditions (with the origin at the comet's centre and the symmetry axis along the sun-comet line) the conservation equations (as discussed in Kitamura, 1986) for the gas can be written as follows:

continuity equation:

$$\frac{\partial}{\partial t}(r^3 \rho e^{t/\tau}) = -\text{div}(\rho \mathbf{v}) e^{t/\tau} \tag{1}$$

momentum equations:

$$\frac{\partial}{\partial t}(r^3 \rho v_r e^{t/\tau}) = \left\{ -\text{div}(\rho v_r \mathbf{v}) - \frac{\partial}{\partial x}(r^2 p) + r^2 \rho v_\theta^2 + 2r^2 p \right\} e^{t/\tau} \tag{2a}$$

$$\frac{\partial}{\partial t}(r^3 \rho v_\theta e^{t/\tau}) = \left\{ -\text{div}(\rho v_\theta \mathbf{v}) - \frac{\partial}{\partial \theta}(r^2 p) - r^2 \rho v_r v_\theta \right\} e^{t/\tau} \tag{2b}$$

energy equation:

$$\frac{\partial}{\partial t}(r^3 \varepsilon e^{t/\tau}) = \{ -\text{div}(\varepsilon \mathbf{v}) - p \,\text{div}\, \mathbf{v} + r^3 (Q_{\text{heat}} - Q_{\text{cool}}) \} e^{t/\tau} \tag{3}$$

Equations (1)–(3) are formulated in spherical geometry, thus the operator 'div' is defined as

$$\text{div}\, \mathbf{A} = \frac{\partial}{\partial x}(r^2 A_r) + \frac{r^2}{\sin\theta} \frac{\partial}{\partial \theta}(\sin\theta A_\theta),$$

where r is the distance from the nucleus centre and θ the polar angle ($\theta = 0°$ along subsolar line). x is a logarithmic space variable defined by $x = \ln(r/r_{\text{nuc}})$. These equations are supplemented by an equation of state given as

$$\varepsilon = \frac{p}{\gamma - 1}, \tag{4}$$

where ε is the internal energy per unit volume. For the adiabatic index γ the value 4/3 (corresponding to water vapor) is used and the dissociation time scale τ for the H_2O-molecules is taken to be 10^5 seconds. The heating and cooling terms in Eq. (3) are defined as

$$Q_{\text{heat}} = 2.63\, 10^{-7} J_{\text{UV}} \rho \tag{5}$$

$$Q_{\text{cool}} = \frac{8.5\, 10^{-19}}{m_{H_2O}} \frac{\rho^2 T^2}{\rho + 2.7\, 10^7 m_{H_2O} T} \tag{6}$$

with $J_{\text{UV}} = 3.5\, 10^{12}$ photons $\text{cm}^{-2}\,\text{s}^{-1}$. The chemical heating term is chosen to be the same as in the spherically symmetric gas-dust interaction model of Gombosi et al. (1985). The cooling term Q_{cool} (caused by IR-radiation of H_2O-molecules) is given by Shimizu (1976).

We have performed computations in two different spatial domains which must be matched together to get a composite picture. In the region between the nucleus surface ($r = 6$ km) and $r = 100$ km a high resolution grid consisting of 31×55 grid points in $r - \theta$ space was used. In the r-direction uniform spacing in the logarithmic variable x is chosen. This choice allows high resolution close to the nucleus. In the θ-direction the grid has high resolution ($\Delta\theta = 1°$) in the domain $0° \leq \theta \leq 12°$ while for $12° < \theta < 180°$ the cell size is $\Delta\theta = 4°$. For the calculations between $r = 100$ km and $r = 10000$ km we used a coarser grid which was uniform in θ-direction with $\Delta\theta = 5°$.

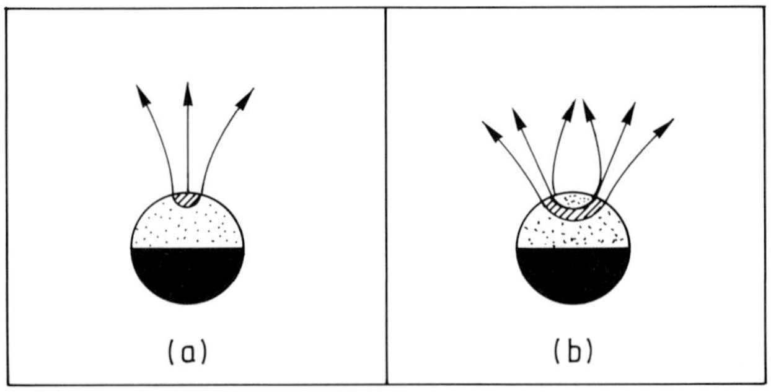

Fig. 1. Possible anisotropic outgassing patterns from the cometary surface: (a) strong source with a half width of 5° centered at the subsolar point; (b) ring-like source distribution with a hole at the subsolar point

3. Boundary conditions on comet surface

We will assume that the gas leaves the comet's surface at the local sound speed. In regions of strong outgassing we set the gas temperature equal to the free sublimation temperature of water ice, $T = 208$ K. In regions where the outgassing rate is reduced, say, due to the existence of a dust crust, $T = 350$ K is considered to be an appropriate choice while on the night hemisphere we set $T = 150$ K. Figure 1 shows schematically the different outgassing patterns on the comet's surface, corresponding to case (a) and (b). The background outgassing rate on the dayside is estimated to be $Z_{\text{day}} = 2\, 10^{16}\,\text{cm}^{-2}\,\text{s}^{-1}$. For the nightside the outgassing rate is assumed to be reduced by a factor 10, i.e. $Z_{\text{night}} = 2\, 10^{15}\,\text{cm}^{-2}\,\text{s}^{-1}$. For case (a) we take a maximum outgassing rate (at the subsolar point) of $Z_{\text{peak}} = 1.5\, 10^{18}\,\text{cm}^{-2}\,\text{s}^{-1}$ according to the estimate given by Ip and Rickman (1986). This peak value falls rapidly off to the background value in a gaussian distribution with a half-width of 5° in θ. In case (b) a flux of $Z_{\text{peak}}/2$ is adopted at $\theta = 25°$ and thus forms a ring-like density distribution around the symmetry line. This situation allows to simulate the formation of a strong compression region near the subsolar line.

4. Numerical results

4.1. Near-nucleus region

Figures 2 and 3 illustrate the quasi-steady state gas flow caused by a strong source at the subsolar point. The half width of the source density profile is chosen to be 5°. Figure 2 shows the velocity field within $r \leq 20$ km. It can be seen that in the region around the strong source the flow is laterally diverted as it may be expected from such a density spike; an interaction region between the 'jet flow' around the subsolar line and the background flow originating at larger θ has formed. Deviations from radial flow are restricted to a cone $\leq 45°$, i.e. no gas from the jet will be able to reach the nightside.

Figure 3 shows a more quantitative representation of the flow variables. The density profile is 'narrow' (i.e. comparable to the 5° half width of the source) only in the immediate vicinity of the nucleus, out to ≈ 4 km above the nucleus. Within this region the maximum of ρ lies close to the symmetry axis and ρ decreases to the background value within $\theta < 12°$. Further outward in r a secondary maximum begins to form, which becomes more and more pronounced, while the first maximum disappears. At $r = 15$ km the secondary maximum (i.e. the expansion cone) begins to

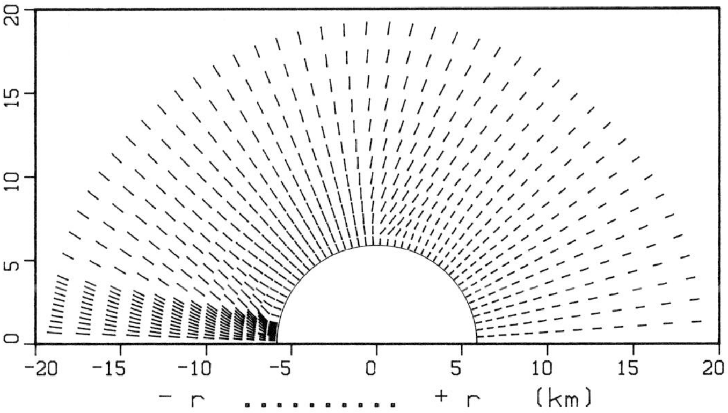

Fig. 2. Steady state velocity field resulting from a strong gas outflow at the subsolar point

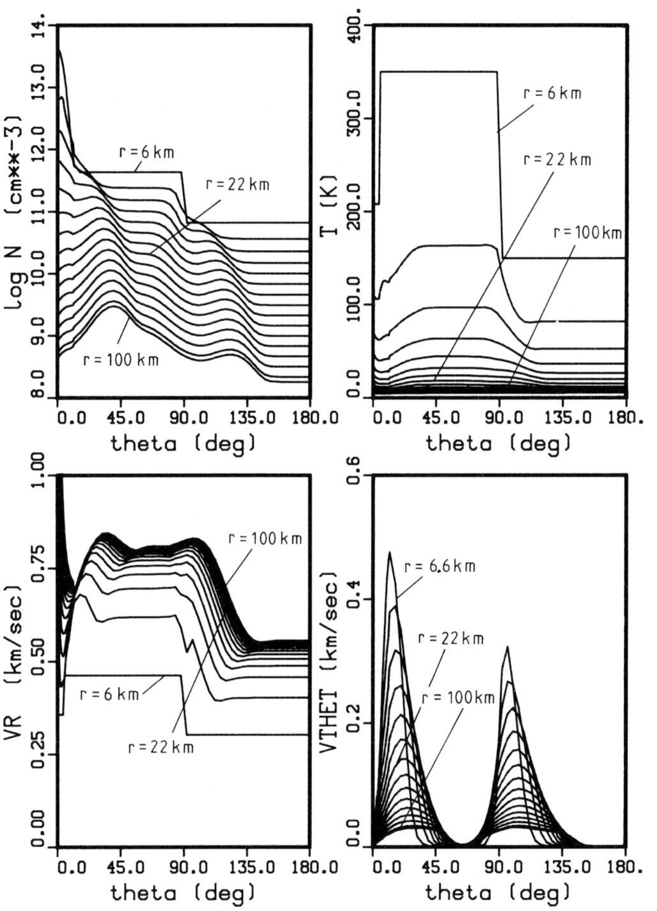

Fig. 3. Steady state solution with a strong outgassing source at the subsolar point ($\theta = 0°$). This corresponds to our case (a)

dominate. It shifts towards larger θ, as r increases. At $r = 100$ km the maximum of ρ appears at $\theta \approx 40°$. It may be of interest to compare the maximum and minimum number density on the comet's dayside for different radial distances: at the surface, $r = 6$ km, one get $N_{max}/N_{min} \approx 100$; at $r = 100$ km this ratio is still a factor of ≈ 7 according to Fig. 3. This shows that at $r = 100$ km one still has a rather strong anisotropy of density and pressure although it is smeared out over a broader region. On the other hand, the anisotropy of the temperature has completely disappeared at this distance, as is obvious from Fig. 3.

A few more words should be said about the behaviour of the azimuthal and radial velocity components. v_θ shows two peaks: The smaller one at $\theta > 90°$ is due to the lateral flow at the terminator towards the nightside because of the sharp change in the surface sublimation rates assumed. The other one, with an amplitude of $v_{\theta max} \approx 0.55\,\mathrm{km\,s^{-1}}$ at 0.5 km above the surface, is due to the density and pressure spike at the subsolar point. The maximum of v_θ shifts from $\theta \simeq 10°$ close to the nucleus to $\theta \simeq 30°$ at $r = 100$ km.

The radial velocity component v_r shows an interesting behaviour close to the subsolar line. While it goes through a minimum a few degrees off the axis, one gets a strong increase towards the subsolar line. Even at large distances ($r = 100$ km) the width of this region of increasing v_r remains comparable to the half width of the original density distribution at the surface. Thus one gets a narrow beam close to the subsolar line, where the radial mass flux is increased compared to the flux at $\theta > 10°$. The formation of this high-speed beam is most likely caused by the divergent geometry maximized along the central axis. We note that this particular effect was not reported in the work of Kitamura (1986) and it appears possible that the height of this central maximum in v_r is over-estimated in our computations. Numerical experiments revealed further that this feature depends also rather sensitively on the adopted cooling profile. If the radiative cooling term is switched off in the calculation this velocity peak tends to isotropize. However, the gradual distortion of the central density peak near the nucleus surface to a hollow cone with density peaking at a few ten's of degrees off the central axis at large radial distances is in good agreement with Kitamura's published results.

Figure 4 shows the flow caused by a ring-like outgassing region with a width of $\pm 5°$ centered at $\theta = 25°$ (case (b)). It shows a number of qualitative differences compared to case (a): Because of the 'hole' around the subsolar point the gas can also expand laterally towards the subsolar line, where it is diverted back to the radial direction. While close to the surface the maximum of density and pressure lies at $\theta \simeq 25°$, this maximum disappears and a new maximum forms close to the subsolar line. At $r = r_{nuc}$ the ratio N_{max}/N_{min} is ≈ 40, at $r = 100$ km this ratio is still a factor of ≈ 10. Expansion towards the other direction also leads to a local maximum which has shifted to $\theta \approx 60°$ at $r = 100$ km.

Looking at the velocity components, one can distinguish several pronounced maxima and minima in v_θ: at $\theta < 25°$ a flow towards the axis develops with peak velocities of $0.25\,\mathrm{km\,s^{-1}}$. In the tailward direction a v_θ flow with an amplitude of $\approx 0.4\,\mathrm{km\,s^{-1}}$ close to the surface develops. Beyond $r \approx 50$ km the maxima caused by the 'ringlike' outflow and by the terminator-flow merge and form one broad maximum of low amplitude ($v_\theta \simeq 0.05\,\mathrm{km\,s^{-1}}$) extending from 30° to 135°.

The radial component v_r has a slight peak at $\theta = 25°$ indicating the radial acceleration of the gas due to lateral expansion off from the $\theta = 25°$ line. In the region very close to the subsolar line one will have a very strong gas-gas interaction. The boundary condition $v_\theta = 0$ at the axis means in a sense that the expanding gas encounters a 'rigid wall' where it cannot flow through. Therefore, after having reached a steady state, a 'stand off shock' may have formed close to the symmetry axis which diverts the gas back to the radial direction. In addition to this stand off shock a contact surface separating streamlines originating in the 'hole' from the diverting streamlines originating in the 'ring' must have

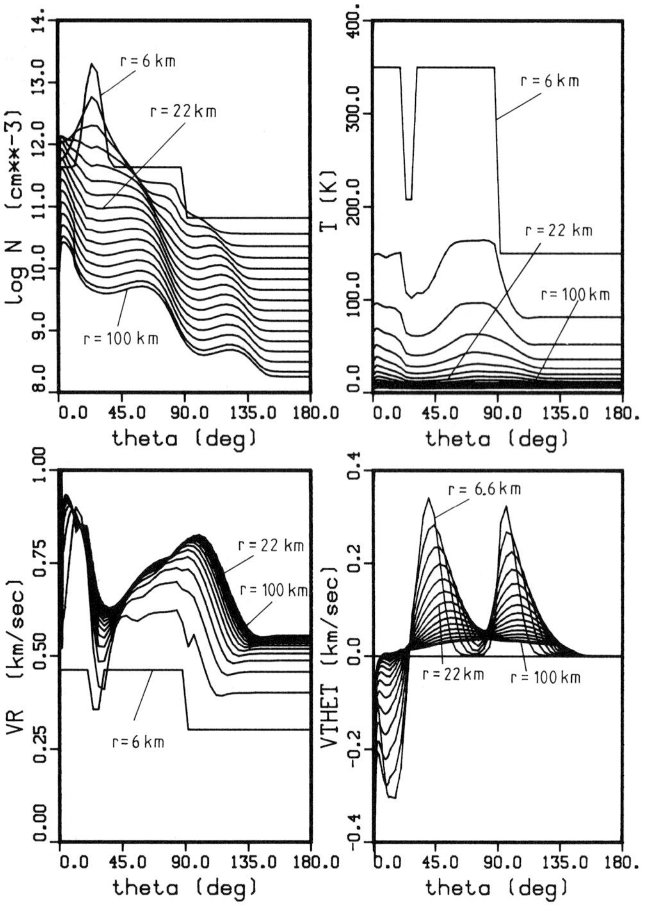

Fig. 4. Steady state solution with a 'ring-like' outgassing source at $\theta = 25°$. This corresponds to our case (b)

formed. Despite the increased resolution close to the axis those discontinuities are not well resolved in our numerical calculation; the sharp spike in v_r and v_θ close to $\theta = 0$ may, however, be interpreted as an indication of such a discontinuity.

4.2. Time-dependent behaviour

Figure 5 shows the initial stages of the transient flow after the onset of strong gas outflow at the subsolar point (case (c)). The density at different times after 'turn on' of the source is shown. As start solution an outgassing pattern with hemispherical asymmetry is taken. At $t = 0$ s the strong source described in case (a) becomes active. The density variations after 10 s, 30 s and 120 s are shown. Note that the situation at $t = 120$ s is already close to the steady state solution. Figure 6 shows the reverse process, namely the development of the time-dependent solution after the strong source has stopped outgassing (case (d)). Again the density after 10 s, 30 s, 60 s and 120 s is shown. If the interpretation that 'expanding halos' frequently observed in cometary comae are caused by spin-modulation of the outgassing rate (Ip, 1981; Gombosi et al., 1985) is correct, the gas flow shown in Figs. 5 and 6 may represent the very early stages of halo formation: The compression front seen in Fig. 5 would develop into the leading edge of an outward propagating shell, while the rarefaction wave shown in Fig. 6 might later form the wake of the halo. It should be noted that during the post-perihelion phase of comet Halley 1986 very distinct 'expanding halo' structures have been found in monochromatic CN-images (Schlosser et al., 1986).

4.3. Outer coma behaviour

In order to study the behaviour of the solution at larger distances from the nucleus, we have extended the calculations out to $r \approx 10000$ km for case (a). For this purpose the solutions obtained from the inner coma calculation at $r = 87$ km were used as new inner boundary conditions. Because in these outer regions a coarser grid and, hence, a much larger timestep may be used, is possible to integrate over several hours until a new steady state is reached in the whole domain.

Two main features can be seen in Fig. 7 which shows density-, temperature- and velocity-profiles between 87 km and 10000 km: (a) The temperature begins to increase again because the chemical heating terms become important and (b) the density anisotropy caused by the density spike at the subsolar point is maintained out to $r \approx 1000$ km, then it begins to flatten gradually. At $r = 6000$ km the anisotropy is still recognizeable, while at $r = 10000$ km it has almost flattened out. The reason for the flattening of the profile is the increase of the temperature due to chemical heating, which tends to isotropize density inhomogeneities. The temperature increase due to chemical heating may be somewhat over-estimated in our model as the extension of the collision-

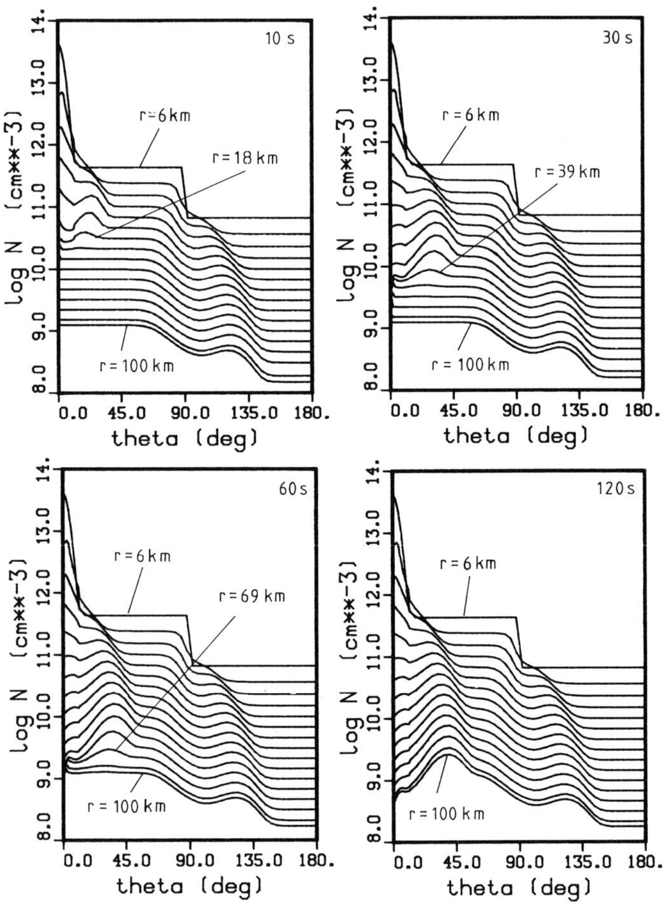

Fig. 5. Transient stages of the gas flow after onset of the strong source at the subsolar point. The density after 10 s, 30 s, 60 and 120 s is shown (case (c))

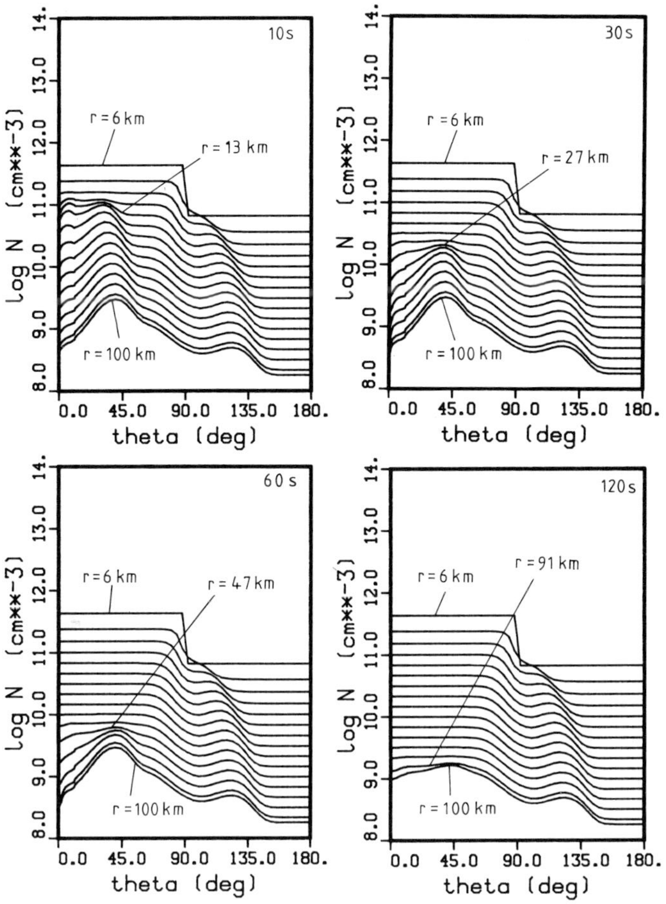

Fig. 6. Transient stages of gas flow after the strong source at the subsolar point has ceased outgassing. Density profiles after 10 s, 30 s, 60 s and 120 s are shown (case (d))

dominated coma could be smaller than 10000 km. In this case the density anisotropy in the outer coma would remain even larger than predicted by our model. However, the neutral mass spectrometer experiment (NMS) aboard Giotto yielded a number density of H_2O-molecules $n_{H_2O} \approx 4.6\,10^7\,cm^{-3}$ at $r = 1000$ km (Krankowsky et al., 1986). From this result the extension of the collision-dominated coma can be estimated to $r_{coll} \approx 2\,10^4$ km. Thus our model assumption that there is continuum flow out to $r \approx 10000$ km is consistent with the measured density profile. Moreover, at distances >500 km negative v_θ-components occur which reach about $-0.06\,km\,s^{-1}$ at $r = 10000$ km and, due to the onset of the chemical heating, the radial outflow velocity increases again.

5. Discussion and conclusions

Two basically different patterns of outgassing have been studied in the present paper: Case (a) – a strong source at the subsolar point; this would correspond to a limited region covered with H_2O-ice from which the main outgassing comes, while outside this region the gas flow is strongly reduced by an insulating dust crust (see Brin and Mendis, 1979; Houpis et al., 1985; Ip and Rickman, 1986). In this case one finds that the central source develops into a cone with maximum density at $\theta \approx 40°$ and a hole in the center (= subsolar line).

Case (b) – an outgassing source placed along a ring at $\theta = 25°$ from the subsolar point; this may be a reasonable model for the flow resulting from outgassing along an ensemble of irregularly shaped cracks connected along a polygon. In this case, due to the lateral expansion off from the center of the ring, one gets a streamerlike interaction with the background gas both in the tailward direction and in the direction towards the subsolar line. While close to the surface the maximum density is found directly above the ring-like outgassing region, at $r = 100$ km the density maximum occurs close to the subsolar line. Thus, in contrast to case (a), where one gets a 'cone' of increased density, in case (b) the gas is concentrated along a line and may look more similar to a collimated 'jet' when observed from outside. The density anisotropy introduced by anisotropic outgassing from the comet's surface is maintained out to several thousand kilometers above the nucleus where it gradually smoothens out due to the temperature increase caused by chemical heating of the gas. A comparison of the numerical models as discussed here with the observations of the H_2O vapor anisotropy from comet Halley (Weaver et al., 1986) should be very interesting.

Although the present calculations have not yet included equations for the dust, some conclusions regarding the expected distribution of dust particles may be drawn from the calculated velocity fields for the gas. Above the strong gas-dust coupling region (which is restricted to a few hundred meters above the surface) the dust has no more dynamical influence on the gas, it

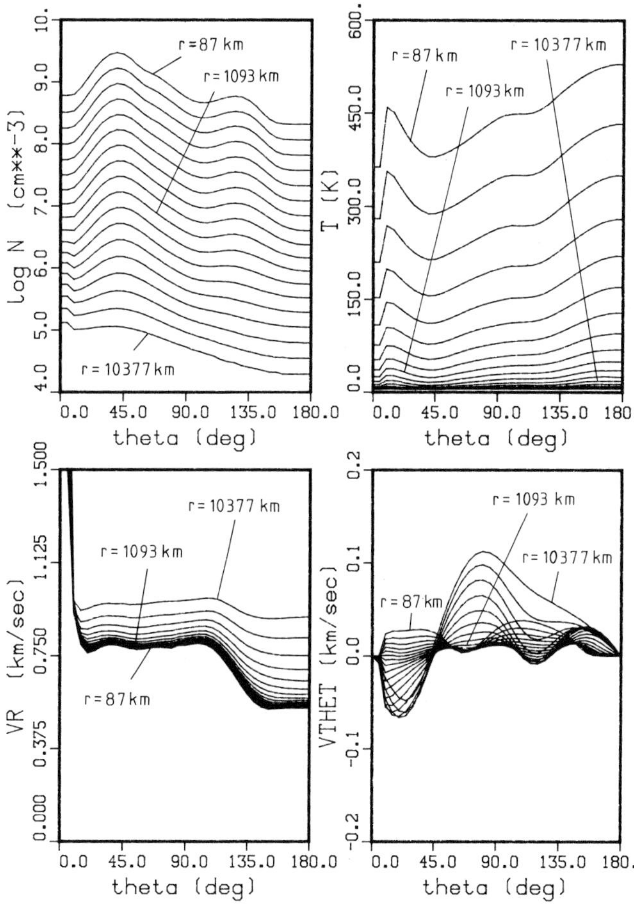

Fig. 7. Density-, temperature- and velocity profiles in the region between 87 km and 10300 km

just follows the motion of the gas until it decouples completely from the gas at $r \approx 100$ km; therefore formation of dust streamers may be favoured in regions with a strong negative gradient in v_θ while regions with positive v_θ-gradient should become depleted from dust. This means that for our case (a) the dust should concentrate near the outer border of the cone (see also Kitamura, 1986). In our case (b), however, the dust originating from $\theta < 25°$ would be swept towards the subsolar line, while dust originating from $\theta > 25°$ would accumulate in a cone around $\theta \approx 60°$. Furthermore, the divergent flow pattern established by the strong difference in the outgassing rates between the dayside and the nightside could also lead to a deflection of dust particles across the terminator – as might have been suggested by the Vega observations (Sagdeev et al., 1986). We plan to pursue a more realistic modeling of the dust-gas interaction for these various situations in the next step of this investigation.

References

Brin, G.D., Mendis, D.A.: 1979, *Astrophys. J.* **229**, 402

Gombosi, T.I., Cravens, T.E., Nagy, A.F.: 1985, *Astrophys. J.* **293**, 328

Houpis, H.L., Ip, W.-H., Mendis, D.A.: 1985, *Astrophys. J.* **295**, 654

Ip, W.-H.: 1986, *Adv. Space Res.* **5**, 233

Ip, W.-H.: 1981, *Nature* **289**, 269

Ip, W.-H., Rickman, H.: 1986, ESA-SP-249, 181

Kaneda, E., Ashihara, O., Shimizu, M., Takagi, M., Hirao, K.: 1986, *Nature* **321**, 297

Keller, H.U., Arpigny, C., Barbieri, C., Bonnet, R.M., Cazes, S., Coradini, M., Cosmovici, C.B., Delamere, W.A., Huebner, W.F., Hughes, D.W., Jamar, C., Malaise, D., Reitsema, H.J., Schmidt, H.U., Schmidt, W.K.H., Seige, P., Whipple, F.L., Wilhelm, K.: 1986, *Nature* **321**, 320

Kitamura, Y.: 1986, *Icarus* **66**, 241

Krankowsky, D., Lämmerzahl, P., Herrwerth, I., Woweries, J., Eberhardt, P., Dolder, U., Hermann, U., Schulte, W., Berthelier, J.J., Illiano, J.M., Hodges, R.R., Hoffmann, J.H.: 1986, *Nature* **321**, 326

Moreels, G., Vincent, M., Clairemidi, J., Mougin, B., Parisot, J.P., Zucconi, J.M., Bertaux, J.-L., Blamont, J.E., Hersè, M., Chalon, G., Lepage, J.P., Runavot, J., Krasnopolsky, V.A., Moroz, V.I., Krysko, A.A., Troshin, V.S., Barke, V.V., Jegulev, V.S., Sanko, N.F., Tomashova, G.V., Parshev, V.A., Tkachuk, A.Yu., Novikov, B.S., Perminov, V.G., Sulakov, I.I., Fedorov, O.S., Gogoshev, M., Gogosheva, Ts., Sargoichev, S., Palasov, K., Georgiev, A., Kostadinov, I., Kanev, K., Spasov, S., Werner, R., Mendev, D., Valnicek, B., Vanysek, V.: 1986, *Adv. Space Res.* **5**, 137

Sagdeev, R.Z., Szabó, F., Avanesov, G.A., Cruvellier, P., Szabó, L., Szegö, K., Abergel, A., Balazs, A., Barinov, I.V., Bertaux, J.-L., Blamont, J., Detaille, M., Demarelis, E., Dul'nev, G.N., Endröczy, G., Gardos, M., Kanyo, M., Kostenko, V.I., Krasikov, V.A., Nguyen-Trong, T., Nyitrai, Z., Reny, I., Rusznyak, P., Shamis, V.A., Smith, B., Sukhanov, K.G., Szabó, F., Szalai, S., Tarnopolsky, V.I., Toth, I., Tsukanova, G., Valniček, B.I., Varhalmi, L., Zaiko, Yu.K., Zatsepin, S.I., Ziman, Ya.L., Zsenei, M., Zhukov, B.S.: 1986a, *Nature* **321**, 262

Sekanina, Z.: 1986, *Adv. Space Res.* **5**(12), 307

Schlosser, W., Schulz, R., Koczet, P.: 1986, in 20th ESLAB Symp., ESA-SP-250 Vol. III, 495

Shimizu, M.: 1976, *Astrophys. Space Sci.* **40**, 149

Smith, B., Szegö, K., Larson, S., Merenyi, E., Toth, I., Sagdeev, R.Z., Avanesov, G.A., Krasikov, V.A., Shamis, V.A., Tarnapolski, V.I.: 1986, in 20th ESLAB Symp., ESA-SP-250 Vol. II, 327

Weaver, H.A., Mumma, M.J., Larson, H.P., Davies, D.S.: 1986, in 20th ESLAB Symp., ESA-SP-250 Vol. I, 329

Astron. Astrophys. 187, 411–418 (1987)

ASTRONOMY
AND
ASTROPHYSICS

Infrared investigation of water in comet P/Halley

H. A. Weaver [1], **M. J. Mumma** [2], **and H. P. Larson** [3]

[1] Space Telescope Science Institute, 3700 San Martin Drive, Baltimore, MD 21218, USA
[2] Planetary Systems Branch, NASA Goddard Space Flight Center, Greenbelt, MD 20771, USA
[3] Lunar and Planetary Laboratory, The University of Arizona, Tucson, AZ 85721, USA

Received February 16, accepted May 29, 1987

Summary. Gaseous, neutral water was detected in the coma of comet P/Halley during infrared (IR) observations from the NASA Kuiper Airborne Observatory (KAO). High resolution spectra revealed intense, solar-pumped IR fluorescent emission in the (001-000) band of H_2O near $\lambda \sim 2.65\,\mu m$. Three lines of the (011-010) "hot" band were also observed. In this paper, we summarize the results of our KAO H_2O investigation. Besides detecting H_2O and measuring its absolute production rate, we measured its spatial distribution, temporal variability, outflow velocity, and have determined the H_2O ortho/para ratio. The measured line intensities imply that water in the coma is rotationally relaxed, as predicted by recent nonthermal equilibrium models. However, significant differences in excitation are detected between a fluorescence equilibrium model and the observed spectra, and these differences may yield important information on the coma kinetic temperature. We find that the aperture-averaged excitation temperatures for the low-lying rotational transitions are ~ 40 K, indicating that the coma kinetic temperature is extremely cold. Excitation differences are also detected between the pre- and post-perihelion data, and between nucleus-centered and offset exposures. Swings effects may explain some of the differences. We have detected significant short-term temporal variations in the H_2O line intensities, and find a strong pre- to post-perihelion asymmetry (a factor of ~ 7) in the water production rate at a heliocentric distance of ~ 1 AU. Finally, we comment on possible future directions for high resolution IR spectroscopy of comets.

Key words: comets – molecules: H_2O – infrared radiation – lines: identification

1. Introduction

Prior to the recent apparition of comet P/Halley, the evidence for water as the primary volatile constituent of cometary nuclei was based solely on circumstantial arguments. The discovery of H_2O^+ emission from comet Kohoutek (Herzberg and Lew, 1974), and subsequent H_2O^+ detections in other comets (cf., Wyckoff, 1982), argued that H_2O was present in cometary comae. The strength of the observed H_2O^+ emission suggested that H_2O was probably abundant in the coma, but the physics relating ionic emissions to absolute neutral abundances is notoriously difficult to model owing to many poorly known parameters, and quantitative information was never derived. Ultraviolet (UV) observations of many comets (cf., Feldman, 1983) provided strong evidence that the abundant H, O, and OH seen in the coma were produced by the photodissociation of H_2O. These observations provided the first quantitative support for H_2O as the *dominant* volatile constituent in comets. However, the direct observation of neutral H_2O has proven to be an elusive goal which has generated considerable controversy.

A broad, near-infrared (IR) absorption feature attributed to water ice has been reported in two comets (Campins et al., 1983; Hanner, 1984) observed at large heliocentric distance (R), but abundance estimates could not be established, and the identification itself has been challenged (A'Hearn et al., 1984). Marginal detections of the H_2O rotational line near 22 GHz ($6_{16} - 5_{23}$) were reported for two comets (Jackson et al., 1976; Altenhoff et al., 1983) but, again, no quantitative information could be derived since there was no known mechanism for exciting this line in comets.

The first indisputable, direct detection of H_2O in any comet was made during IR observations of comet Halley from the NASA Kuiper Airborne Observatory (KAO) in December 1985 (Mumma et al., 1986). We used the University of Arizona Fourier transform spectrometer (FTS; Davis et al., 1980) to obtain high resolution spectra of the (001-000) and (011-010) bands of H_2O near $\lambda \sim 2.65\,\mu m$. This paper summarizes the results of this investigation. We describe in detail how the observed H_2O excitation differs from our original expectations, and analyze the differences between the pre- and post-perihelion data. New discussion is presented on the observed short-term temporal variations, and further evidence is introduced which supports the reality of the previously reported large pre- to post-perihelion asymmetry in water production at $R \sim 1$ AU. Regarding the latter point, we have compared our findings to those from UV and radio observations of OH, and to Suisei and Pioneer Venus observations of H. Finally, we also comment on possible future directions for high resolution IR spectroscopy of comets.

2. Observations

Spectra of comet Halley were obtained both pre- and post-perihelion. Airplane altitudes (~ 12 km) and large geocentric radial velocities ($\dot{\Delta} \sim 35\,\mathrm{km\,s^{-1}}$ for the pre- and $\dot{\Delta} \sim -43\,\mathrm{km\,s^{-1}}$ for the post-perihelion observations) were required to obtain good

Send offprint requests to: H. A. Weaver

atmospheric transmittance at the positions of cometary H_2O emission lines. Table 1 summarizes other relevant observational information from our H_2O investigation. The instrumental parameters (e.g., filter bandwidth and spectral resolving power; see Table 1) were varied according to specific objectives, since the multiple scientific goals of this investigation were incompatible with a single instrument configuration.

Our pre-observation prediction of the (001-000) band relative line intensities (Weaver and Mumma, 1984) are compared with the transmittance properties of three filters in Fig. 1. Table 2 summarizes the advantages of each filter. A fourth filter which had full transmittance over the entire range shown in Fig. 1 was used during one observation pre-perihelion, but the poor signal-to-noise ratio achieved in this spectrum rendered it unsuitable for quantitative analysis.

Only three filters can be accomodated by the FTS at one time. For each of our two observing runs (pre- and post-perihelion), the chosen filters consisted of: one relatively narrow H_2O filter (e.g., F1 or F2), one broad H_2O filter (e.g., F3), and a third filter chosen for a different application (for example, in searching for OH and CH_4).

Unfortunately, absorption by CO_2 in the terrestrial atmosphere remains strong even at airplane altitudes making observations between 3600 and 3750 cm^{-1} difficult, at best. The filter F1 in Fig. 1 has a low wavenumber cut-on which roughly matches the region where CO_2 absorption ceases to be a problem.

Table 1. Log of KAO IR observations of comet Halley

Date [a] (UT)	Helio-centric distance (AU)	Geo-centric distance (AU)	Region observed	Resolving power	Spectral [b] interval (cm^{-1})	Band [c] intensity (10^{-8} W cm^{-2} sr^{-1})	Q_{H_2O} [d] (10^{30} mol s^{-1})
22.08 Dec 85	1.16	0.96	Nucleus	9.3 10^4	3685–3839	0.025	0.070
23.14 Dec 85	1.14	0.98	Nucleus	8.8 10^4	3413–3521	*	–
24.07 Dec 85	1.13	1.00	Nucleus	2.5 10^3	3135–4132	–	–
24.11 Dec 85	1.13	1.00	Nucleus	9.3 10^4	3685–3839	0.079	0.22
20.70 Mar 86	1.00	0.79	Nucleus	1.1 10^4	2959–3049	*	–
20.74 Mar 86	1.00	0.79	Nucleus	2.9 10^3	3736–3839	0.83	1.4
22.68 Mar 86	1.03	0.74	Nucleus	2.9 10^3	3736–3839	0.49	0.82
22.72 Mar 86	1.03	0.74	Nucleus	2.5 10^5	3736–3839	0.48	0.81
24.65 Mar 86	1.06	0.69	Nucleus	2.9 10^3	3736–3839	0.71	1.2
24.68 Mar 86	1.06	0.69	Sunward (65″)	2.9 10^3	3736–3839	0.036	–
24.71 Mar 86	1.07	0.69	Tailward (65″)	2.9 10^3	3736–3839	0.019	–
24.73 Mar 86	1.07	0.69	Nucleus	2.9 10^3	3736–3839	0.99	1.6
26.64 Mar 86	1.09	0.65	Nucleus	2.9 10^3	3736–3839	0.57	0.91
26.66 Mar 86	1.10	0.65	Sunward (97″)	2.9 10^3	3736–3839	0.013	–
26.70 Mar 86	1.10	0.65	Nucleus	5.4 10^3	3704–5000	–	–

[a] Midpoint of observation
[b] Bandwidth at 10% relative transmittance points
[c] Average for the H_2O ν_3 band in a 41″ diameter field of view; calculated assuming that the intensity in the $2_{02}-1_{01}$ line is 11% of the total band intensity for the pre-perihelion period and for the offset pointings, and 7.8% for the post-perihelion nuclear pointings
[d] $v_{H_2O}=1$ km s^{-1} and $\tau_{H_2O}=8.2\ 10^4$ s at 1 AU from the sun; Water production rates reported here differ from previously reported values (Mumma et al., 1986; Weaver et al., 1986) due to slightly revised absolute band intensities and the use of different Haser model input parameters
* The observation on 23.14 Dec 1985 was to search for the OH (1, 0) band. The observation on 20.70 March 1986 was to search for the ν_3 band of CH_4

Table 2. Filter bandwidths and applications

	F1	F2	F3
Bandwidth [a]	3736-3839	3685-3839	3704-5000
Observing period	Post-perihelion	Pre-perihelion	Post-perihelion
Advantages	Lowest noise spatial mapping temporal variability high spectral resolving power	Low noise high spectral resolving power ortho/para	High and low *J* lines ortho/para

[a] Bandwidth at 10% relative transmittance points in cm^{-1}

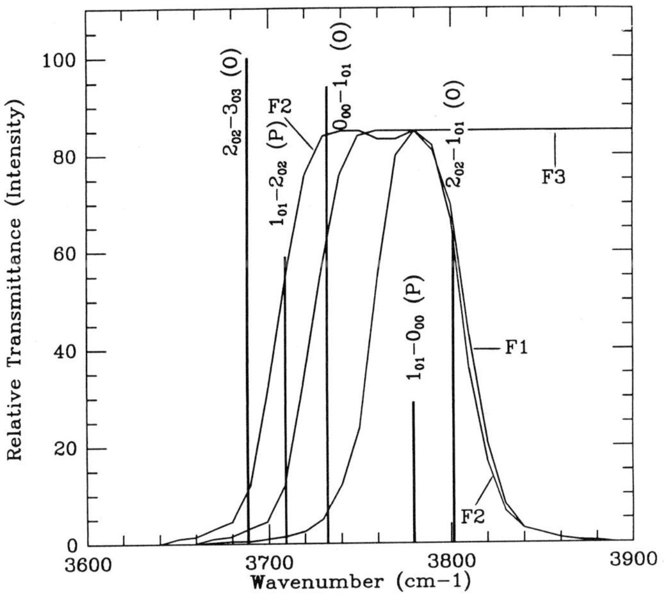

Fig. 1. Relative transmittance curves are shown for three of the filters used in the KAO Halley H_2O investigation. They are marked F1, F2, and F3. Also shown is a stick spectrum giving the predicted intensity distribution within the (001-000) band for fluorescence equilibrium. The five emission lines are marked with their state designations (J_{K_a,K_c}) and whether they belong to the ortho (o) or para (p) species of H_2O

F1 also has good transmittance in the relatively unobscured region containing the $1_{01}-0_{00}$ and $2_{02}-1_{01}$ lines. Since background noise decreases with bandwidth, F1 yielded the highest sensitivity for these two lines, and our highest resolution spectrum (see Larson et al., 1987) was taken with it.

F2 (Fig. 1) is also a relatively narrow filter which can be used for high resolution observations. In addition, it extends into the region of CO_2 absorption and includes the very important $0_{00}-1_{01}$ line, which is needed for determining the ortho/para ratio. The observability of this line depends critically on $\dot{\Delta}$, which determines where the H_2O line lies relative to the terrestrial CO_2 lines. Figure 2a shows typical spectra taken through F2.

High-J ($J \equiv$ total angular momentum quantum number) H_2O lines occur mainly at wavenumbers $\geqq 3800\,\text{cm}^{-1}$. F3 was used (see Fig. 2b) to verify that no significant high-J H_2O emission was present and that reliable estimates of the total (001-000) band emission could be obtained from the individual line intensities measured with the narrow filters, F1 and F2.

3. Scientific results

3.1. General summary

The KAO Halley H_2O investigation addressed a variety of scientific goals. The achieved objectives include:
1. direct detection of water,
2. determination of absolute water production rates,
3. retrieval of detailed water excitation information,
4. mapping of the water spatial brightness distribution,
5. measurement of water outflow velocities,
6. monitoring of short-term temporal variability,
7. measurement of the pre- to post-perihelion asymmetry in water production, and
8. determination of the water ortho/para ratio.

Many of these results have been described in three recently published papers (Mumma et al., 1986; Larson et al., 1986; Weaver et al., 1986), and will only be summarized here.

The absolute water production rates are large ($\sim 10^{29}$–$10^{30}\,\text{mol s}^{-1}$) confirming that H_2O is the dominant volatile constituent of cometary nuclei. The observed H_2O emission is due to solar pumped IR fluorescence, and the H_2O molecules have a strongly relaxed rotational population distribution, thereby supporting recent nonthermal equilibrium models. The H_2O spatial brightness profile can be fit by that of a parent molecule. However, short-term cometary temporal activity can cause large brightness variations ($\sim 40\%$) on time scales of 2 hrs, and this complicates the derivation of an H_2O scalelength from the brightness profile. A significant sunward-tailward brightness asymmetry was observed which suggests ejection primarily into the sunlit hemisphere. We find that the post-perihelion water production rates are systematically about a factor of 7 larger than the pre-perihelion values (both at $R \sim 1$ AU), although short-term temporal variability complicates this analysis as well.

The derived ratio of ortho/para H_2O is consistent with the statistical equilibrium value of 3 for both the pre- and post-perihelion data. The derived H_2O outflow velocity was 0.9 (± 0.2) km s^{-1} pre-perihelion and 1.4 (± 0.2) km s^{-1} post-perihelion. In depth discussion of the ortho/para ratio and the kinematics of outflow is given in two other articles in this volume (Mumma et al., 1987; Larson et al., 1987).

3.2. Water excitation

3.2.1. Interpreting the spectra

Weaver and Mumma (1984) predicted that solar-pumped IR fluorescence in the vibrational fundamental bands of cometary "parent" molecules would produce emissions exceeding that from cometary grains by orders of magnitude, and should be readily detectable for abundant species observed with high spectral resolution instruments. In the case of H_2O, they showed that radiative rotational transitions in the ground vibrational level are so rapid that H_2O molecules in cometary comae should exhibit a very "cold" rotational distribution (see also: Crovisier, 1984). For simplicity, they assumed that the H_2O molecules were completely relaxed so that only the lowest ortho and para states were populated. For this *fluorescence equilibrium* case, there are only five strong emission lines in the (001-000) band (Fig. 1).

The measured relative intensities for both the pre- and post-perihelion observations are shown in Fig. 3. Due to absorption by CO_2 in the earth's atmosphere, only three of the predicted lines could be accurately measured and these are labeled. The observed spectrum is obviously more complex than the fluorescense equilibrium spectrum. At least 12 (001-000) band lines are observed while only 3 were predicted. Yet, the lack of agreement between theory and observation is not nearly as severe as a first impression indicates. The two unobserved lines $2_{02}-3_{03}$ and $1_{01}-2_{02}$ have the same upper states as two of the observed lines, and since the branching ratios in the IR transitions are known to an accuracy of $\sim 1\%$, the intensities of these lines can be calculated. Upon doing this, we find that these two lines are the strongest ortho and para lines, respectively, in the comet's spectrum. Optical depth effects will change the line ratios from the values expected from branching ratios, but in each of these two cases the observed lines are more strongly affected by saturation than the unobserved lines so that our calculations give *lower* limits to the intensities. There can be little doubt that the lines predicted

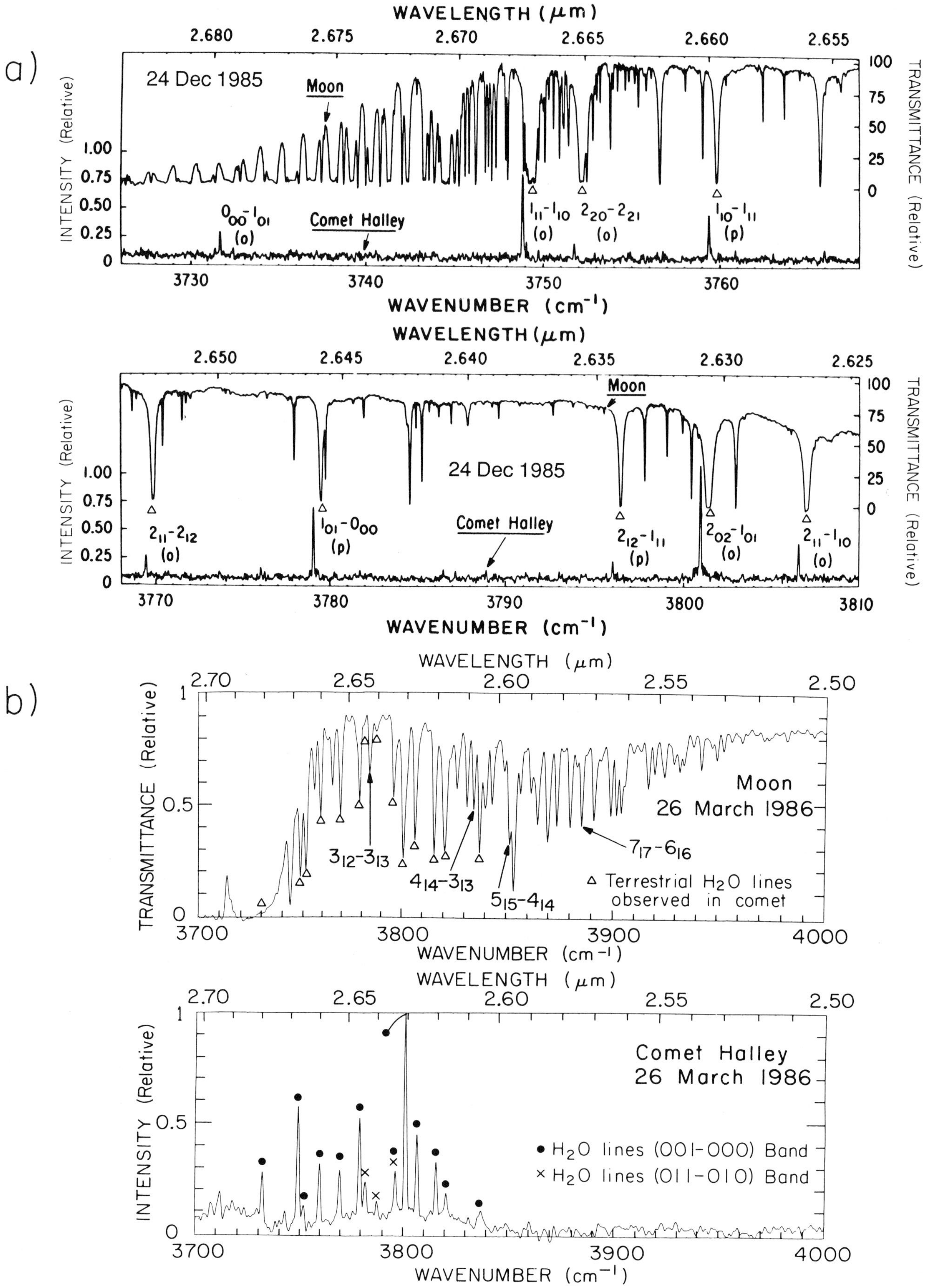
a)
WAVELENGTH (μm)
2.680 2.675 2.670 2.665 2.660 2.655
24 Dec 1985
Moon
INTENSITY (Relative)
TRANSMITTANCE (Relative)
1.00 0.75 0.50 0.25 0
100 75 50 25 0
$0_{00}-1_{01}$ (o)
Comet Halley
$1_{11}-1_{10}$ (o)
$2_{20}-2_{21}$ (o)
$1_{10}-1_{11}$ (p)
3730 3740 3750 3760
WAVENUMBER (cm^{-1})
WAVELENGTH (μm)
2.650 2.645 2.640 2.635 2.630 2.625
Moon
24 Dec 1985
$2_{11}-2_{12}$ (o)
$1_{01}-0_{00}$ (p)
Comet Halley
$2_{12}-1_{11}$ (p)
$2_{02}-1_{01}$ (o)
$2_{11}-1_{10}$ (o)
3770 3780 3790 3800 3810
WAVENUMBER (cm^{-1})
b)
WAVELENGTH (μm)
2.70 2.65 2.60 2.55 2.50
TRANSMITTANCE (Relative)
1 0.5 0
Moon
26 March 1986
$3_{12}-3_{13}$
$4_{14}-3_{13}$
$5_{15}-4_{14}$
$7_{17}-6_{16}$
△ Terrestrial H_2O lines observed in comet
3700 3800 3900 4000
WAVENUMBER (cm^{-1})
WAVELENGTH (μm)
2.70 2.65 2.60 2.55 2.50
INTENSITY (Relative)
Comet Halley
26 March 1986
● H_2O lines (001-000) Band
× H_2O lines (011-010) Band
3700 3800 3900 4000
WAVENUMBER (cm^{-1})

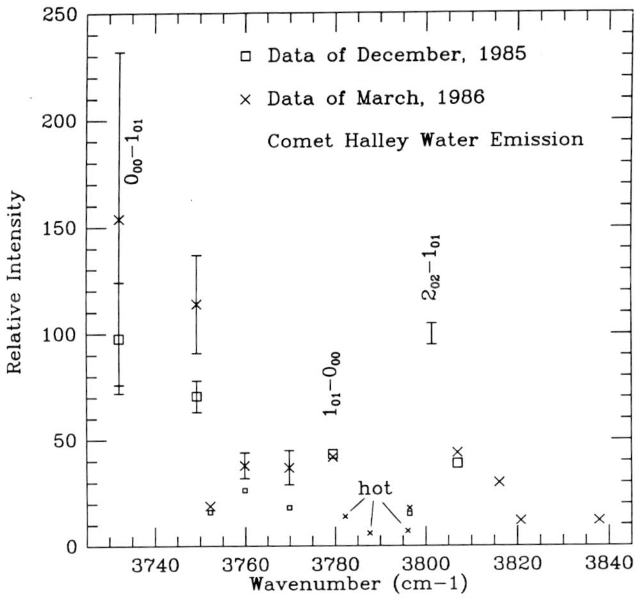

Fig. 3. A comparsion of the pre-perihelion (□), and post-perihelion (×), relative intensities for the comet Halley H_2O emission. The pre-perihelion data are derived from observations on 1985 Dec 24, while the post-perihelion data are from observations on 1986 March 22 and 24. All lines belong to the (001-000) band, except for the three lines marked "hot" which belong to the (011-010) band. All line intensities are relative to the $2_{02}-1_{01}$ line at 3801 cm^{-1}. For the lines without error bars, the size of the symbol gives the approximate error. State designations are given for the three lines predicted to be strong in fluorescence equilibrium

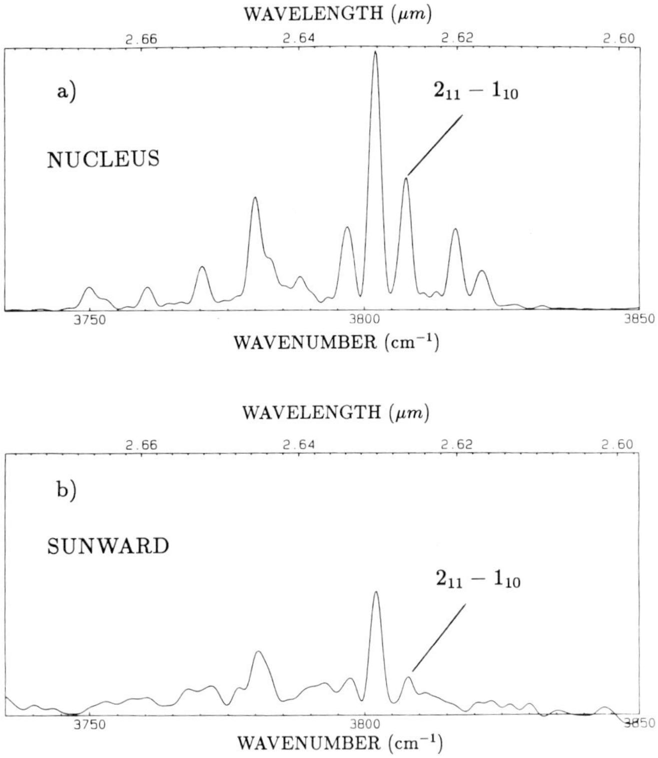

Fig. 4a and b. Moderate resolution spectra (resolving power ~1300) taken through filter F1 on 1986 March 24. **a** is an exposure centered on the nucleus and **b** is offset sunward by ~3.3 10^4 km. The intensity scale in **b** is magnified by a factor of 10 compared to **a**. The $2_{11}-1_{10}$ line has become relatively weaker in the offset spectrum, indicating relaxation towards fluorscence equilibrium

to be strongest in fluorescence equilibrium are also the strongest in Halley's H_2O spectrum. This, coupled with an examination of Fig. 2b which shows no transitions having a lower state $J>3$, demonstrates that the fundamental theoretical picture of an extremely "cold" H_2O rotational distribution has been verified.

Significant excitation differences are observed between the pre- and post-perihelion data and between nucleus-centered and offset exposures. Figure 4 compares the spectrum of a pointing on the nucleus to one which is offset sunward by ~3.3 10^4 km. Although the achieved signal-to-noise ratio is smaller in the offset data due to the drop in absolute intensities as one moves off the nucleus, the *relative* brightness of the $2_{11}-1_{10}$ line has clearly decreased for the offset spectrum. At least qualitatively, the offset spectrum more closely resembles the fluorescence equilibrium case. This is expected for two reasons: (1) the aperture-averaged column densities are significantly smaller for the offset exposure implying that optical depth effects should be less important, and (2) the aperture for the offset pointing is well outside the collision zone implying that the molecules should be relaxing towards fluorescence equilibrium. As discussed by Bockelée-Morvan and Crovisier (1986), complete fluorescence equilibrium will not be achieved until the molecules are ~10^5 km from the nucleus.

Another indication of excitation changes within our data can be obtained by calculating the fraction of total band intensity emerging in the $2_{02}-1_{01}$ line. Our pre-perihelion data give 11% (±1%) while our post-perihelion nucleus-centered data give 7.8% (±1%). The fraction predicted by fluorescence equilibrium is ~18%.

The "weaker" lines in the spectrum have become *relatively* stronger post-perihelion, while the relative intensities of the fluorescence equilibrium lines have remained constant. Since water production rates were much larger post-perihelion (see Sect. 3.3), a combination of collisional and optical depth effects is probably producing this effect. Indeed, the pre- to post-perihelion relative intensity changes shown in Fig. 3 provide direct evidence of optical depth effects in the ro-vibrational transitions. The lines at 3770 cm^{-1} ($2_{11}-2_{12}$) and 3807 cm^{-1} ($2_{11}-1_{10}$) have the same upper level, so their observed intensity ratio can be compared to the value calculated from branching ratios. The predicted ratio (3807/3770) is 1.74 (±0.03), while the pre- and post-perihelion values are 2.14 (±0.24) and 1.19 (±0.27), respectively. The decrease in the ratio is significant and indicates that the $2_{11}-1_{10}$ line is becoming saturated post-perihelion. This conclusion is also consistent with our previous finding (Weaver et al., 1986) that the

Fig. 2a and b. Examples of **a** pre- and **b** post-perihelion KAO spectra of comet Halley taken when $R\sim1.1$ AU (**a** is adapted from Mumma et al., 1986 and **b** is adapted from Weaver et al., 1986). The spectrum in **a** was taken at a resolving power of ~10^5 through filter F2 and shows 9 lines in the (001-000) band of H_2O. The lower resolution (resolving power ~2500) spectrum in **b** was taken through filter F3 and shows 12 lines in the (001-000) band, as well as 3 lines in the (011-010) "hot" band. Four H_2O lines whose lower states are the upper states for rotational transitions which have been suggested as probes of H_2O in comets are identified, but are not seen in the cometary spectrum. Both cometary spectra are uncorrected for atmospheric transmission and instrumental response, but the lunar spectra (also shown) can be used to calibrate the relative intensities *aperture-averaged* optical depths for the strongest lines are ~ 1 post-perihelion.

The pure rotational transitions between low-lying levels of the ground vibrational state become optically thick at H_2O column densities which are at least a factor of 100 smaller than our aperture-averaged values post-perihelion. When the rotational line optical depths become this large, collisions can maintain LTE at lower densities than that required in the optically thin domain. In this case, the lower level populations are related to the coma kinetic temperature, and the H_2O spectrum will show departures from fluorescence equilibrium. Thus, the intensities of the "weak" lines in the observed spectra may prove invaluable in determining the neutral coma kinetic temperature, which has previously had no experimental determination. The lowest-lying rotational transitions in each of the two H_2O species (ortho and para) are expected to be the last to depart from LTE as the molecules expand from the inner to the outer coma. From our pre-perihelion data we estimate excitation temperatures of ~ 27 K $(+14, -27)$, ~ 47 K $(+16, -11)$, and ~ 40 K (± 3) for the $2_{12}-1_{01}$, $1_{10}-1_{01}$, and $1_{11}-0_{00}$ transitions, respectively. The errors in the temperatures do not include possible large systematic contributions, but all three values are consistent with an extremely low coma kinetic temperature, lending at least qualitative support to recent hydrodynamic models (cf. Crovisier, 1984). Similar calculations have not yet been completed for the post-perihelion data.

Bockelée-Morvan (1986) has developed a sophisticated model which treats collisional and optical depth effects. She has applied her model to our post-perihelion observations (Bockelée-Morvan and Crovisier, 1986) and predicts relative intensities which compare quite well with the observed values, lending support to the picture described above. Inclusion of Swings effects (see Sect. 3.2.2), non-spherical geometries, and fine-tuning of various model parameters can be expected to improve the agreement and yield further valuable insights into coma physics.

3.2.2. Missing lines

Six lines were discovered post-perihelion which were not seen pre-perihelion. The three highest wavenumber lines in Fig. 3 are relatively weak (001-000) band lines. The line at 3838 cm^{-1} (possibly a blend of two lines) could not be observed pre-perihelion because the filter did not transmit this line. Since water production rates were much higher for the post-perihelion observations, the absence of the line at 3821 cm^{-1} is expected based on signal to noise considerations alone. However, the strongest of the group cannot be so easily dismissed. We have found that the cometary emission pre-perihelion was Doppler-shifted into "accidental" coincidence with a terrestrial absorption line, thereby explaining its absence.

The three weak lines marked "hot" in Fig. 3 belong to the (011-010) band of H_2O, and are seen only in the post-perihelion data. We have calculated g-factors for this band and find that the calculated and measured post-perihelion intensities are nearly identical. The absence of two of the lines pre-perihelion is expected based on signal to noise considerations. However, the strongest line (at 3783 cm^{-1}) should have been observable pre-perihelion with a signal to noise of ~ 10, if the pre- and post-perihelion excitation conditions were not significantly different. There is no evidence in the data for terrestrial absorption causing the discrepancy.

The "hot" band line in question is the $2_{02}-1_{01}$ transition, and the intensity in this line can be compared to the intensity in the "equivalent" fundamental transition [i.e., $2_{02}-1_{01}$ in the (001-000) band]. Since both lines are pumped from the same lower level, neither collisions nor optical depth effects in the rotational transitions can alter the ratio of line intensities (these processes simply redistribute population among the rotational levels of the ground vibrational state). However, optical depth effects in the vibrational transitions can make the ratio ("hot"/fundamental) larger than the optically thin prediction. For the fundamental line, more of the emission from the de-excitation of 2_{02} will emerge in the $2_{02}-3_{03}$ line (which is the preferred de-excitation route anyway) as the optical depth in the $2_{02}-1_{01}$ line increases. Since the "hot" band lines remain optically thin, this "robbing" of flux from $2_{02}-1_{01}$ does not occur for this band and the ratio of "hot" line intensity to fundamental line intensity for $2_{02}-1_{01}$ will increase. Since optical depths were almost an order of magnitude larger post-perihelion compared to pre-perihelion, this effect is in the right direction to explain the non-detection of the "hot" line pre-perihelion. On the other hand, since the observed post-perihelion ratios of "hot" to fundamental line intensities are almost identical to the optically thin values, this effect may not be large enough to completely account for the missing line.

One possible source of the discrepancy is the Swings effect, in which the line excitation rate is a function of the comet's heliocentric radial velocity. During the pre-perihelion observations this velocity was ~ -27 km s^{-1}, while the post-perihelion value was ~ 27 km s^{-1}. Using a recently obtained high resolution solar spectrum (Farmer, private communication), we can verify directly that no important variation in the solar spectrum occurs over the wavenumber range relevant for the excitation of the fundamental line $2_{02}-1_{01}$. Unfortunately, we do not yet have a similar spectrum for the region of the (011-000) band ($\lambda \sim 1.9$ μm). Examination of the solar spectrum in the region of the fundamental band has revealed that the excitation of some fundamental lines (although *not* $2_{02}-1_{01}$) can be changed by up to 50% due to the presence of solar absorption lines. Since the solar spectrum probably contains more structure near 1.9 μm than near 2.7 μm, Swings effects are expected to be important in explaining the excitation of "hot" band lines. Until a high resolution solar spectrum for this region can be obtained, the Swings effect cannot be ruled out as the reason for the missing "hot" band line pre-perihelion.

3.3. Temporal variability and Q_{H_2O}

3.3.1. Short-term variability

Comet Halley exhibited pronounced short-term temporal variations in its UV (cf., Feldman et al., 1986), visible (cf., Rettig et al., 1986), IR (cf., Weaver et al., 1986), and radio (cf., Gérard et al., 1986) emissions. Since parent species sublimate directly from the nucleus, while dissociation fragments have a "global" source region (i.e., the entire coma), one would expect the parents to be higher fidelity indicators of nucleus activity. Comparing variations of H_2O and OH seems to confirm this point; during a 108 minute observing period on 1986 March 24 UT, the H_2O line brightnesses measured from the KAO increased by a factor of ~ 1.4 (± 0.2) while the OH brightness measured by the International Ultraviolet Explorer (IUE) apparently increased by less than 6% (Weaver et al., 1986; our earlier reported value of a 64% change in Q_{H_2O} has been reduced after consideration of air mass effects). Brightness variations observed in several UV species (McFadden et al., 1987), lend further support to this hypothesis. For instance, the "parent-like" molecule, CS, showed a temporal variation over one 38 hr period (1986 March 23.5–25.1 UT) which was ~3 times larger than the OH variation.

The latter examples refer to aperture-averaged variations. We have used the KAO data to estimate the magnitude of the corresponding activity at the nucleus. Assuming that the increased H_2O brightness in the KAO aperture was due to a "step" increase which begins at the beginning of our observing interval and ends 108 minutes later, we calculate that the average Q_{H_2O} during this time must have been 1.9 times the "steady-state" value. If, instead, there was a linear rise over the interval, then the final Q_{H_2O} must be 3.7 times its initial value. Such variations are large but not necessarily unrealistic. Similar changes were seen in visible spectra and images taken during this same period (Rettig et al., 1986). Also, pictures of Halley taken by the flyby spacecraft (Keller et al., 1986; Saagdeev et al., 1986) show dramatic jet activity suggesting that large, short-term temporal variations are possible.

McFadden et al. (1987) have addressed the question of whether or not the large rise in H_2O brightness is consistent with the small increase seen in OH. Their preliminary analysis suggests that a larger rise in OH should have been observed, if our nominal value for the H_2O rise (40%) is correct. More sophisticated temporal modeling is needed to compare the relationship between H_2O and OH in detail.

3.3.2. Pre- to post-perihelion asymmetry

Figure 5 shows Q_{H_2O} vs. R for the pre- and post-perihelion KAO observations. An isotropic, radial outflow model with $v_{H_2O} = 1$ km s^{-1} and $\tau_{H_2O} = 8\ 10^4$ s at $R = 1$ AU was used in calculating H_2O column densities. Water production rates are derived from measured band brightnesses assuming that the column densities are directly proportional to column brightnesses. As discussed earlier, this latter assumption of small optical depths is not strictly correct. However, Bockelée-Morvan and Crovisier (1986) have demonstrated that the optically thin assumption produces only small errors in deriving Q_{H_2O}.

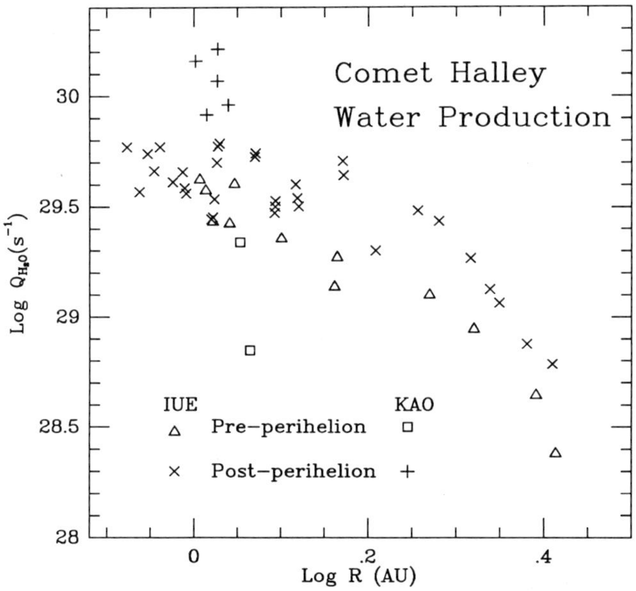

Fig. 5. Water production rates (Q_{H_2O}) derived from KAO observations of H_2O (+ and □) and from IUE observations of OH (△ and × Feldman et al., 1986) are plotted as a function of heliocentric distance (R). For the IUE data, the relative measurement uncertainties are significantly smaller than the symbol sizes. Within a given epoch (i.e., either pre- or post-perihelion), the KAO relative uncertainties are approximately the size of the symbols. See the text for further discussion of the measurement uncertainties

A striking feature of the KAO data in Fig. 5 is the apparent large increase in Q_{H_2O} from pre- to post-perihelion. In light of the above discussion on short-term temporal variability, one must be careful to ensure that a true systematic pre- to post-perihelion asymmetry is being observed, and not an effect due to sampling the cometary activity at different "phases". Based on simultaneous, or nearly simultaneous, IUE monitoring of Halley's visible light curve, the KAO observations on 1985 Dec 24.11 UT and 1986 March 24.73 UT were certainly made near peaks in cometary activity. The ratio of post- to pre-perihelion water production dates for these dates is ~7.

The KAO absolute water production rates have a fairly large systematic error (~±50%) due to absolute calibration uncertainties. However, a relative comparison of the two pre-perihelion rates, or of one post-perihelion rate to another, should be more accurate since common calibration stars are used for all observations within a single epoch. A possible source of systematic error in the relative measurements is the change in atmospheric transmittance from one observation to another. Unlike our previously reported intensities, an attempt has been made to correct the values reported here for this effect. Assuming that the transmittance correction has been done properly, the remaining dominant source of relative error is pointing uncertainty. We estimate that our pointing error ($\leqq 15''$) translates into a ±10% uncertainty in the intensity measurements.

There is also independent evidence that the ratios of post- to pre-perihelion rates are more accurate than indicated by the formal absolute calibration errors. High resolution (resolving power $\geqq 10^5$) KAO spectra were taken on 1986 March 22.72 UT and on 1985 Dec 24.11 UT and we have compared the signal-to-noise ratios (S/N) on the brightest line ($2_{02} - 1_{01}$) in each spectrum. Even if Q_{H_2O} was identical on each date, differences in the achieved S/N would be expected due to differences in Δ, R, integration time, filter bandwidth, spectral resolution, H_2O excitation, and terrestrial transmittance. However, one can estimate the magnitude of each of these effects; we find that if Q_{H_2O} were identical on both dates, then the post-perihelion S/N would have been 0.74 times the pre-perihelion value. Since the measured post-perihelion S/N, was 1.98 times the pre-perihelion value, Q_{H_2O} must have have been larger post-perihelion by a factor of 2.7, which is in good agreement with the value calculated using the absolute production rates given in Table 1 (3.7). Applying this information to a comparison of the peak pre- and post-perihelion water production rates yields a ratio of 5.3 for the pre- to post-perihelion asymmetry. This value is guite similar to the number quoted above and argues that, indeed, a large asymmetry in Q_{H_2O} was detected for $R \sim 1$ AU.

Prior to the direct detection H_2O, observations of OH and H were used to monitor H_2O production. We have examined the UV data on OH and H, and the radio data on OH, to determine if these species show a pre- to post-perihelion asymmetry. Fig. 4 shows Q_{H_2O} as derived from IUE OH observations over the entire Halley observing period (Feldman et al., 1986). Model parameters for the H_2O parent are identical to the values used in deriving the KAO production rates, while the OH density is calculated using the vectorial method (Festou, 1981). The IUE data show systematically higher production rates post-perihelion for large R, but this asymmetry becomes quite small for $R \leqq 1.1$ AU. For $R \sim 1$ AU, the ratio (post/pre) of peak production rates is ~1.5. Doing the same thing for the radio OH data (Gérard et al., 1986) also yields a ratio of ~1.5, although these data show variations up to factor of ~6 even within the pre-perihelion data for $R \sim 1$ AU. Only one measurement at $R \sim 1$ AU has been reported by the H Lyman-α imager on Suisei (Kaneda et al., 1986), but the data at $R \sim 1.3$ AU suggest that Halley's water production rates were a factor of ~5 larger post-perihelion. The H Lyman-α measurements from Pioneer Venus (Stewart, 1987) included $R = 1$ AU pre-perihelion, but stopped at $R = 0.8$ AU post-perihelion. Nevertheless, a pronounced pre- to post-perihelion asymmetry was observed and extrapolation of the post-perihelion data to ~1 AU suggests that Q_{H_2O} was ~3–4 times larger post-perihelion. Perhaps future modeling efforts can resolve these discrepancies, but that remains to be demonstrated.

4. Future IR investigations

The comet Halley KAO H_2O investigation represents only the first step in what we hope develops into a method for determining the detailed molecular inventory of cometary ices in many comets. Good candidate molecules for future IR detection include CO, CO_2, CH_4, H_2CO, and NH_3, among others. The Halley spectrum taken by the Vega IKS experiment (Combes et al., 1986) already presents an intriguing view into what the near-IR spectral region promises for future investigations. Besides detecting H_2O, the IKS team observed CO_2, a broad emission feature near 3.4 µm, and has tentatively identified H_2CO and CO. However, higher resolving power ($\geqq 10^3$) is required to characterize the molecular excitation and also, in some cases, to identify the source of the emission. It should be particularly interesting to examine with high resolution the broad feature near 3.4 µm described by the IKS team as possibly arising from a fundamental vibration band of a C-H bearing species. At least some observers (Baas et al., 1986; Knacke et al., 1986) have suggested that this feature contains spectral structure; if so, then this would be an important clue in identifying the source of the feature and understanding the excitation mechanism.

High resolution IR spectrometers placed on flyby spacecraft, in earth orbit, and on airplanes and balloons, can potentially provide important, new data leading to a deeper understanding of comets.

Acknowledgements. We would like to thank Lou Haughney and the staff of the Kuiper Airborne Observatory for providing excellent support of our investigation. The help of Scott Davis and Mike Williams was crucial for the operation of the spectrometer. We thank Barney Farmer for sending us *Atmos* solar spectra prior to their publication, Gordy Bjoraker for assistance in correcting the data for atmospheric transmittance effects, and A.R. Molloy for his continuing insistence that the Swings effect may be important in interpreting the data. Financial support from NASA Grants NAG 2-356 to The Johns Hopkins University and NAG 2-206 to the Lunar and Planetary Laboratory (The University of Arizona), and from NASA RTOP's 188-41-55, 196-41-54, and 352-02-03 to the Goddard Space Flight Center, is gratefully acknowledged.

References

A'Hearn, M.F., Dwek, E., Tokunaga, A.T.: 1984, *Astrophys. J.* **282**, 803

Altenhoff, W.J., Batrla, W., Huchtmeier, W.K., Schmidt, J., Stumpff, P., Walmsley, M.: 1983, *Astron. Astrophys.* **125**, L 19

Baas, F., Geballe, T.R., Walther, D.M.: 1986, *Astrophys. J. Letters* **311**, L 97

Bockelée-Morvan, D., Crovisier, J.: 1986, in *20th ESLAB Symp.* ESA SP-250, Vol. **III**, p. 11

Bockelée-Morvan, D.: 1987, *Astron. Astrophys.* (in press)

Campins, H., Rieke, G.H., Lebofsky, M.J.: 1983, *Nature* **301**, 405

Combes, M., Moroz, V., Crifo, J.F., Bibring, J.P., Coron, N., Crovisier, J., Encrenaz, T., Sanko, N., Grigoriev, A., Bockelée-Morvan, D., Gispert, R., Emerich, C., Lamarre, J.M., Rocard, F., Krasnopolsky, V., Owen, T.: 1986, in *20th ESLAB Symp.*, ESA SP-250, Vol. **I**, p. 353

Crovisier, J.: 1984, *Astron. Astrophys.* **130**, 361

Davis, D.S., Larson, H.P., Williams, M., Michel, G., Connes, P.: 1980, *Appl. Optics* **19**, 4138

Feldman, P.D.: 1983, *Science* **219**, 347

Feldman, P.D., Festou, M.C., A'Hearn, M.F., Arpigny, C., Butterworth, P.S., Cosmovici, C.B., Danks, A.C., Gilmozzi, R., Jackson, W.M., McFadden, L.A., Patriarchi, P., Schleicher, D.G., Tozzi, G.P., Wallis, M.K., Weaver, H.A., Woods, T.N.: 1986, in *20th ESLAB Symp.*, ESA SP-250, Vol. **I**, p. 325

Festou, M.C.: 1981, *Astron. Astrophys.* **95**, 69

Gérard, E., Bockelée-Morvan, D., Bourgois, G., Colom, P., Crovisier, J.: 1986, in *20th ESLAB Symp.*, ESA SP-250, Vol. **I**, p. 589

Hanner, M.: 1984, *Astrophys. J. Letters* **277**, L 75

Herzberg, G., Lew, H.: 1974, *Astron. Astrophys.* **31**, 123

Jackson, W.M., Clark, T., Donn, B.: 1976, in *The Study of Comets*, NASA SP-393, Part I, p. 272

Kaneda, E., Hirao, K., Shimizu, M., Ashihara, O.: 1986, *Geophys. Res. Letters* **13**, 833

Keller, H.U., Arpigny, C., Barbieri, C., Bonnet, R.M., Cazes, S., Coradini, M., Cosmovici, C.B., Delamere, W.A., Huebner, W.F., Hughes, D.W., Jamar, C., Malaise, D., Reitsema, H.J., Schmidt, H.U., Schmidt, W.K.H., Seige, P., Whipple, F.L., Wilhelm, K.: 1986, *Nature* **321**, 320

Knacke, R.F., Brooke, T.Y., Joyce, R.R.: 1986, *Astrophys. J. Letters* **310**, L 49

Larson, H.P., Davis, D.S., Mumma, M.J., Weaver, H.A.: 1986, *Astrophys. J. Letters* **309**, 195

Larson, H.P., Mumma, M.J., Weaver, H.A.: 1987, *Astron. Astrophys.* (in press)

McFadden, L.A., A'Hearn, M.F., Feldman, P.D., Roettger, E.E., Edsall, D.M., Butterworth, P.S.: 1987, *Astron. Astrophys.* (in press)

Mumma, M.J., Weaver, H.A., Larson, H.P., Davis, D.S.: 1986, *Science* **232**, 1523

Mumma, M.J., Weaver, H.A., Larson, H.P.: 1987, *Astron. Astrophys.* (in press)

Rettig, T., Ruchti, R., Baumbaugh, B., Baumbaugh, A., Knickerbocker, K., Daws, J.: 1986, in *20th ESLAB Symp.*, ESA SP-250, Vol. **III**, p. 93

Sagdeev, R.Z., et al.: 1986, *Nature* **321**, 262

Stewart, A.I.F.: 1987, *Astron. Astrophys.* (submitted)

Weaver, H.A., Mumma, M.J.: 1984, *Astrophys. J.* **276**, 782; also see erratum: *Astrophys. J.* **285**, 872

Weaver, H.A., Mumma, M.J., Larson, H.P., Davis, D.S.: 1986, *Nature* **324**, 441

Wyckoff, S.: 1982, in *Comets*, ed. L. Wilkening, Univ. Arizona Press, Tucson, p. 3

Astron. Astrophys. 187, 419–424 (1987)

ASTRONOMY AND ASTROPHYSICS

The ortho-para ratio of water vapor in comet P/Halley

M.J. Mumma [1], **H.A. Weaver** [2], **and H.P. Larson** [3]

[1] NASA/Goddard Space Flight Center, Planetary Systems Branch, Greenbelt, MD 20771, USA
[2] Space Telescope Science Institute, 3700 San Martin Drive, Baltimore, MD 21218, USA
[3] University of Arizona, Lunar and Planetary Laboratory, Tucson, AZ 85721, USA

Received February 16, accepted May 11, 1987

Summary. The abundance ratio of ortho-H_2O to para-H_2O is shown to be an invariant in the cometary coma, its dependence on temperature in thermal equilibrium is given, and the nuclear-spin-temperature is defined. The relation between the physical temperature of the cometary ice and the nuclear-spin-temperature is discussed, and the prospects for using the observed ortho-para ratio to infer properties of the cometary nucleus are explored. The ortho-para ratio in Halley's comet, derived from high resolution infrared spectra of the ν_3-band near 2.7 μm wavelength, is given. On UT 1985 December 24.1, it was 2.73 ± 0.17, and on UT 1986 March 22.7/26.7, it was 3.23 ± 0.37. The error estimates contain a sizeable subjective component for atmospheric transmittance, and may be reduced pending further modelling. The nuclear-spin-temperature was 35 K (+9 K, −5 K) pre-perihelion, and >40 K post-perihelion, at the 67% confidence limit. Both numbers are consistent with modelled values of the equilibrium temperature of Halley's nucleus at aphelion (~47 K). However, at the 95% confidence limit they are also fully consistent with temperatures exceeding 50 K, corresponding to an ortho-para ratio of ~3.0.

Key words: atomic and molecular processes – comets – infrared radiation – solar system: general – spectroscopy.

1. Introduction

Water vapor was definitively detected for the first time in any comet on December 22.1, 1985 by high resolution infrared spectroscopic observations of the ν_3-band (2.7 μm) in Halley's comet from NASA's Kuiper Airborne Observatory (KAO, Mumma et al., 1986). Its presence was confirmed by additional observations with the KAO on 1985 Dec. 24.1, and by the IKS-infrared spectrometer on the Vega-1 spacecraft on 1986 Mar 6, (Combes et al., 1986). An in-depth study was conducted from the KAO on 1986 Mar 20–26, and analyses of the KAO data set provided many insights into the physics of H_2O in the cometary coma (Mumma et al., 1986; Weaver et al., 1986, 1987; Larson et al., 1986, 1987).

Two high resolution spectra were acquired, on 1985 UT Dec. 24.1, and 1986 UT Mar 22.7, respectively (Figs. 1 and 2), and subsequent analysis of these two spectra, together with a survey spectrum taken on 1986 UT Mar 26.7, provided line-by-line intensities with sufficient accuracy to permit determination of the ortho-para ratio for H_2O in the comet (Mumma et al., 1986; Weaver et al., 1986). In the present paper, we discuss the determination of the ortho-para ratio for H_2O in Halley's comet and its use as a probe of the temperature of the icy nucleus.

1.1. *The quantum states of ortho-H_2O and para-H_2O*

In addition to their usual quantized properties (vibration, rotation, electron spin angular momentum, orbital angular momentum) molecular energy levels are also quantized according to the total nuclear spin angular momentum. For isotopically symmetric molecules, the requirement of invariance of the wave function under exchange of identical nuclei, leads to a segregation of the rotational states according to the value of the total nuclear spin. Indeed, in 1927, Heisenberg (1927) and Hund (1927) independently postulated the existence of two species of molecular hydrogen (ortho-H_2 and para-H_2), based on the then new quantum theory, and the experimental confirmation of that prediction by Dennison in 1927 constituted one of the real triumphs of the day. Subsequent work demonstrated that all even-J rotational levels belong to para-H_2 ($I=0$, ↑↓), and all odd-J belong to ortho-H_2 ($I=1$, ↑↑). Transitions between the two species ($J \rightarrow J \pm 1$) are strictly forbidden both by radiative [$A \sim 10^{-20}\,s^{-1}$, see Raich and Good (1964), and Dodelson (1986)] and collision-induced processes, the latter having a halflife of ~3 years for the ortho-para conversion at atmospheric pressure (for an early review, see Farkas, 1935).

Other volatiles of astrophysical interest also exhibit nuclear-spin-species (e.g., CH_4, CO_2, C_2H_2, H_2O, ...) and it is interesting to consider the role of the nuclear-spin-temperature as a tracer of physical processes in low-temperature media. Like H_2, H_2O is organized into ortho ($I=1$) and para ($I=0$) species, according to the hydrogen nuclear spins being parallel or anti-parallel, with statistical weights $2I+1$ (Herzberg, 1945). The lowest ortho-level lies $\sim 24\,cm^{-1}$ (or ~34 K) above the lowest para-level, so the ortho-para ratio will be temperature-dependent. It will, however, require an exceedingly long time to re-establish equilibrium spin populations after a change in the physical temperature, and so the measured spin temperature may be used to infer the physical temperature at which the nuclear spins were last equilibrated.

In thermal equilibrium, the ortho-para ratio (OPR) may be defined by

$$\mathrm{OPR} = \frac{(2I_o+1)\,\Sigma\,(2J+1)\exp\left(-\dfrac{E_o(J,K_a,K_c)}{kT}\right)}{(2I_p+1)\,\Sigma\,(2J+1)\exp\left(-\dfrac{E_p(J,K_a,K_c)}{kT}\right)}. \quad (1)$$

Send offprint requests to: M.J. Mumma

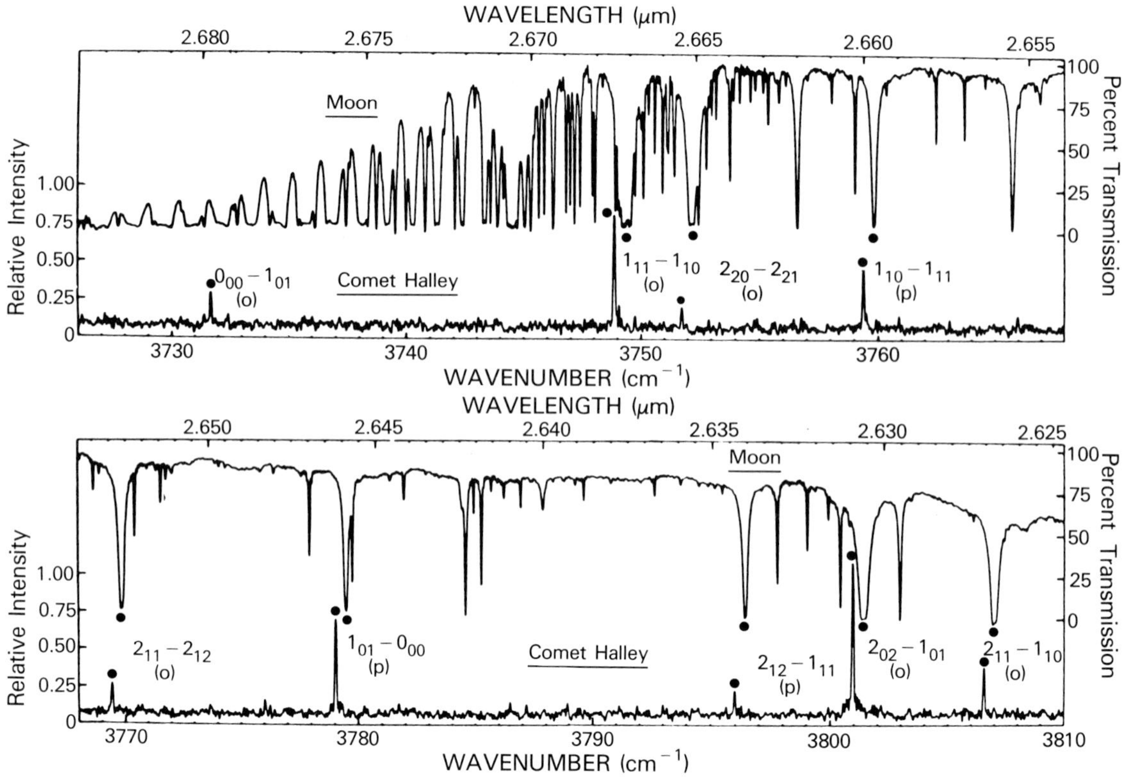

Fig. 1. High Resolution Infrared Spectra of Halley's Comet and of the Moon, measured on 1985 UT Dec. 24.1, (Mumma et al., 1986). Lines originating from ortho-H_2O (o) and from para-H_2O (p) are fully resolved, permitting determination of the ortho-para ratio after correction for atmospheric transmittance. Intensities are given in Mumma et al. (1986)

The sum in the numerator (denominator) extends over all ortho (para) levels (J, K_a, K_c). Part of the energy level diagram is given in Fig. 3.

Equation 1 has been evaluated for temperatures of interest, using the energy levels given in Hall and Dowling (1967). The results are shown in Fig. 4. The comet's nuclear-spin-temperature may be determined by locating the observed ortho-para ratio on this curve.

2. The nuclear-spin-temperature as a probe of the cometary nucleus

Figure 4 demonstrates (solid curve) that statistical equilibrium (OPR = 3.0) is achieved at temperatures exceeding ~60 K, but that for lower temperatures the ortho-para ratio will differ significantly from the statistical equilibrium value. Then it may be a useful probe of the temperature at which the water was last equilibrated, and under favorable conditions may demonstrate that cometary water vapor originated as a low-temperature ice. Ultimately, measurements obtained throughout a perihelion passage may permit the determination of thermal conductivity for the active layers of the cometary nucleus. In order to establish the potential application of nuclear-spin-temperatures for cometary studies, we must demonstrate three conditions: 1) that interior temperatures are low enough to produce measureably relaxed ortho-para ratios, 2) that those ratios do not change significantly during the time available for warming the ice to sublimation, and 3) that subsequent processes in the coma do not modify the ortho-para ratio significantly.

2.1. The interior temperatures of cometary nuclei

The equilibrium temperature of a spherical, rapidly rotating body at Halley's aphelion distance (~35 AU) would be ~47 K, if its albedo were independent of wavelength (grey-body). However, the temperature of Halley's nucleus at aphelion is expected to differ from the grey-body value, owing to several effects: 1) the contribution of libration- and mineral-bands (for example) will make the nucleus non-grey, 2) Halley's nucleus is aspherical and compositionally heterogeneous, consequently its exact geometry, surface topology, and orientation will all affect the equilibrium temperature, and 3) some portion of the insolation received near perihelion is effective in heating the entire nucleus. Herman and Weissman (1986) estimate that this latter effect could increase the equilibrium temperature to ~68 K, but inclusion of the effects of nuclear rotation reduces the estimated temperature to ~65 K because of enhanced radiative cooling (Weissman, private communication). Also, their model requires the thermal conductivity and heat capacity of the nucleus, but these are poorly known, leading to corresponding uncertainty in their temperature estimates. For example, for a thermal conductivity ~20% of water ice, Herman and Weissman found a reduction in temperature of ~10 K. The very low apparent densities (0.2–0.4 gm cm^{-3}) deduced for the nucleus (this volume) imply a very porous mechanical structure, with concomitantly reduced thermal conductivity and reduced equilibrium temperature. Thus, all three effects work to modify Halley's aphelion temperature, from the value of 47 K, possibly placing it in the range for which the ortho-para ratios are measurably relaxed (Fig. 4). Dynamically new

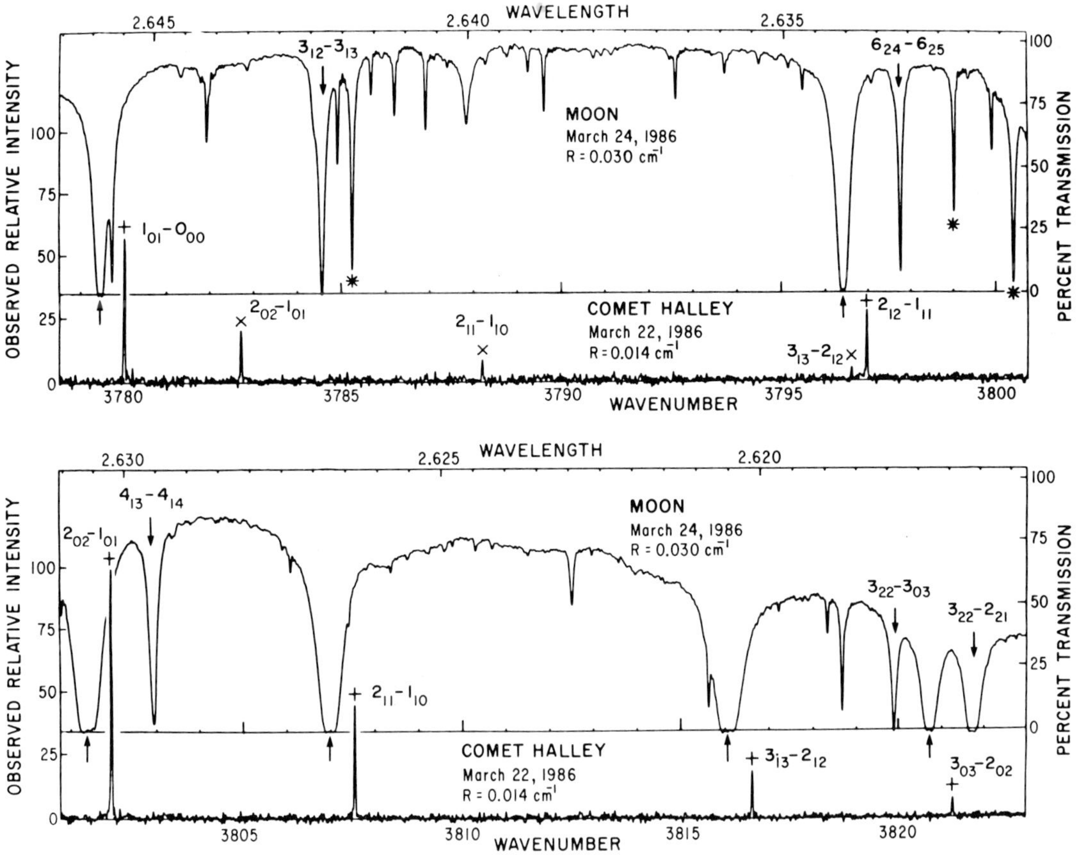

Fig. 2. Ultra-high Resolution Infrared Spectra of Halley's comet and of the Moon, measured on 1986 UT Mar 22.7, and UT Mar 24, respectively. Lines of the ν_3-band of H_2O are marked (+) and those of the (011-010) band are marked (x), see Weaver et al. (1986) and Larson et al. (1986). Individual line identifications are the same as in Fig. 1 for ortho-H_2O and para-H_2O, and the intensities are given in Weaver et al. (1986)

comets are expected to have very low internal temperatures (~ 10 K) since they originate far from the sun, and thus they may be particularly suitable for investigation by this approach.

2.2. *Ortho-para invariance in the cometary nucleus*

As a comet enters the inner solar system, a strong thermal gradient is established near the surface, ultimately producing an active zone of sublimation. Owing to low thermal conductivity, the transition region from the warmer surface to the very cold internal temperature is expected to be quite thin. A typical model (Kuhrt, 1984) predicts a thickness $(D) \sim 10$ m for amorphous ice. The low density found for Halley's comet (this volume) suggests correspondingly reduced thermal conductivity, hence an even thinner transition region – perhaps as thin as one meter. The actively sublimating zone is undoubtedly still thinner, and migrates into the interior of the nucleus at some rate $\dot{R}$ (cm s^{-1}), thus the nuclear ice is warmed from its cold interior temperature to sublimation in a characteristic time $D/\dot{R}$. If the ortho-para conversion time in the solid is τ, then we require

$$\tau > D/\dot{R} \tag{2}$$

in order that the ortho-para ratio not be modified significantly.

The mean local rate of surface depletion is estimated from the mass loss rate ($\dot{M}$), and other parameters:

$$\dot{R} \sim \frac{\dot{M}}{4\pi R_s^2 F_a \varrho} \sim \frac{QMF_T}{4\pi R_s^2 F_a \varrho} \tag{3}$$

Evaluating this rate for our KAO observations of H_2O in March 1986 (Weaver et al., 1986), and using various other parameters presented for Halley's comet (this volume), we take:

$Q = 1(30)$ molecules s^{-1}, the water production rate,
$M = 3(-23)$ g molecule^{-1}, the water molecular mass,
$F_T = 1.5 \sim$ ratio of total mass loss to H_2O loss,
$R_s \sim 5(5)$ cm, the effective nuclear radius,
$F_a \sim 0.1$, the fraction of surface active at any time,
$\varrho \sim 0.3$ gm cm^{-3}, the mean density of the cometary nucleus.

Hence

$$\dot{R} \sim 5(-4)\ \text{cm s}^{-1},\ \text{or}\ \sim 1.8\ \text{cm h}^{-1},\ \text{or}\ \sim 43\ \text{cm day}^{-1}. \tag{4}$$

Thus, very cold ice would seem to be warmed from its aphelion temperature and vaporized in less than a few days time, at least during this very active phase.

In contrast to this steady state rate, the characteristic time can be very much shorter during outbursts. We observed the H_2O content of our beam to increase by $\sim 40\%$ in 108 minutes during an outburst on UT March 24.6 (Weaver et al., 1987). About 114,000 metric tons of material were ejected during this outburst, corresponding to a cylindrical volume of depth 58 m and diameter 116 m if $\varrho = 0.3$ gm cm^{-3}. For this (assumed) geometry

$$\dot{R} \sim 0.9\ \text{cm s}^{-1} \tag{5}$$

i.e., the local erosional rate was several thousand times larger than the steady-state rate.

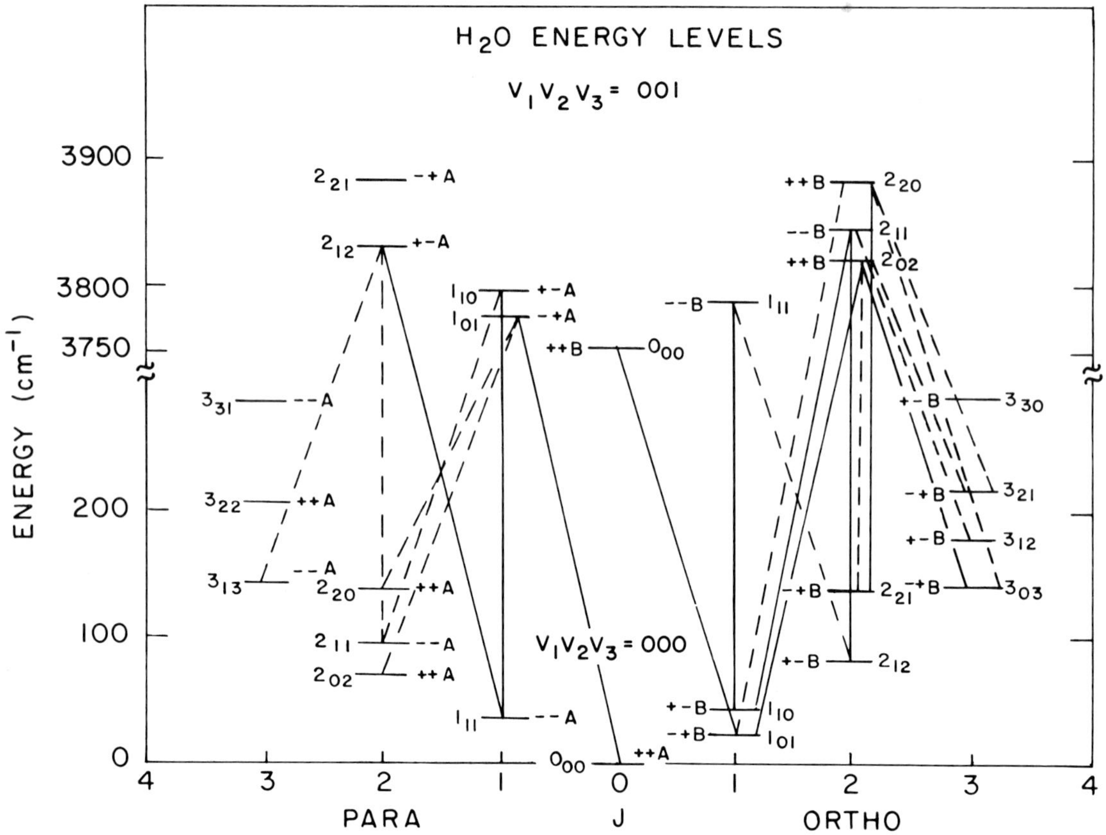

Fig. 3. Partial energy level diagram for H_2O, showing quantum states of the ortho-H_2O and para H_2O species in the (001) and (000) vibrational states. Transitions observed on 1985 Dec. 24.1 (Fig. 1) are drawn as solid lines, and those which must have been present, but which were obscured by terrestrial CO_2, are drawn as dashed lines. The ortho-para ratio is derived from the individual line intensities as described in the text

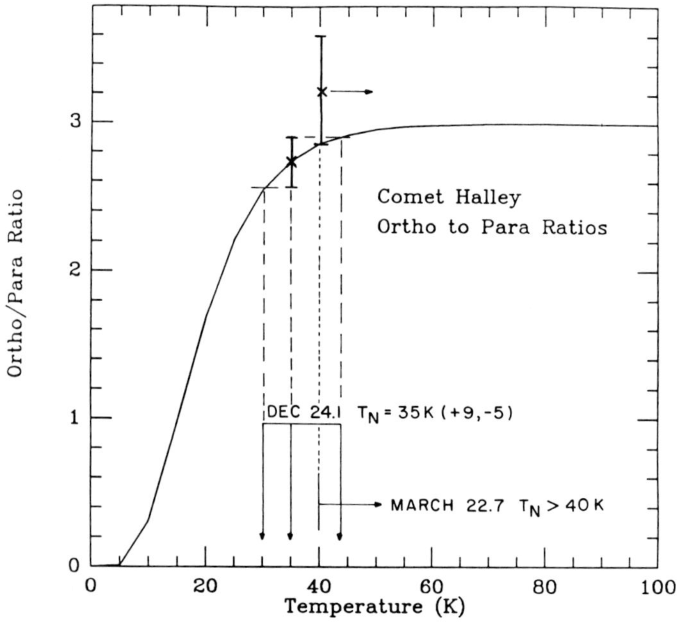

Fig. 4. The ortho-para ratio for low-temperature gaseous H_2O in thermal equilibrium is given by the solid line, calculated according to Eq. (1). The ortho-para ratios determined for Halley's comet are also shown (2.73 ± 0.17 for 1985 UT Dec. 24.1; and 3.23 ± 0.37 for 1986 March). At the 95% confidence limit, both measurements are consistent with a nuclear-spin-temperature $\gtrsim 50$ K

The time scales for re-equilibration of the ortho-para ratio upon temperature changes in the solid phase are very poorly studied. Some laboratory work has been done on H_2O trapped in rare gas matrices, within which it is free to rotate. Relaxation times then are typically hours, except when catalysed by paramagnetic materials ($\tau \sim 10$ min) (Hopkins et al., 1968). Regrettably, these data are not applicable to the present problem. No data exist for dirty H_2O ice, and the sole investigation of pure H_2O ice was based upon neutron moderation cross-section measurements which suggested ortho-para relaxation times of order hours at 4 K, and at 77 K (Borst et al., 1961). No information exists on the effect of lattice bonding (hindered rotation) on ortho-para quenching, nor is it known whether the sublimated vapor retains the ortho-para ratio of its natal ice. However, no physical process is known that might change it. Clearly, significant laboratory investigations are needed to establish a rigorous logical chain connecting observed ortho-para ratios in the coma, with the internal temperature of the nucleus. The only data available for pure water ice would seem to indicate that the water vapor has insufficient time to re-equilibrate its ortho-para ratio during outbursts, and probably during steady state sublimation as well.

2.3. Invariance of the ortho-para ratio in the coma

Using even a crude coma model, it is easily demonstrated that the ortho-para ratio is a "frozen" parameter throughout the coma. Collisions (ions, neutrals, electrons, dust) are far too few to modify the OPR, and radiative processes are also too slow by many orders of magnitude.

The total number of neutral-neutral collisions experienced by an individual H_2O molecule is easily estimated. The molecule may be characterized by a thermal velocity (v_{th}) in the local center-of-mass frame, which moves at outflow velocity (v_0). It experiences a local density (n, cm^{-3}) which decreases with time as the material flows outward. The density is taken to be approximately constant within a range of one collision length, leading to an over-estimate of the number of collisions in the outer coma. The instantaneous collision rate (α, s^{-1}) is then

$$\alpha = n\sigma v_{th}, \tag{6}$$

where σ is the momentum transfer cross-section ($\sim 10^{-14}\,cm^2$). Spherically symmetric, uniform outflow with $v_{th}/v_0 = 0.3$ is assumed. Then, near the nucleus the collision rate (α_N^s) is

$$\alpha_N^s \sim \frac{Q\sigma}{4\pi R_s^2}\frac{v_{th}}{v_0} \sim \frac{10^{30}(10^{-14})}{4\pi(2.5\ 10^{11})}\,0.3 \sim 1000\,s^{-1}. \tag{7}$$

The total number of collisions (S) experienced during outflow is

$$S \sim \int_0^\infty \alpha\, dt = \frac{Q\sigma}{4\pi}\frac{v_{th}}{v_0}\int_{R_s}^\infty \frac{\exp(-R/L)}{R^2}\frac{dR}{v_0}. \tag{8}$$

Neglecting the exponential factor and hydrodynamic effects

$$S \sim \frac{\sigma v_{th}}{4\pi v_0}\frac{Q}{v_0 R_s} = \alpha_N^s \frac{R_s}{v_0} \sim 10^3\,\frac{5\ 10^5}{1\ 10^5} \sim 5000.$$

Proper treatment of hydrodynamic outflow will increase the numbers by perhaps ten-fold, an unimportant amount for the present problem. Since $\sim 2.4\ 10^{17}$ collisions are needed for ortho-para species conversion by collisions in H_2, and since conversion is not observed in the lab for gas-phase H_2O, we can safely neglect collisions with neutrals in the coma. In passing, we note that exchange collisions with hydrogen atoms are too infrequent to be significant in the coma.

Collisional conversion by ions and electrons is also negligible. For thermal ions

$$\alpha_{ions}^s \sim 10^6\,cm^{-3}\ 10^{-11}\,cm^3\,s^{-1} = 10^{-5}\,s^{-1}$$

and

$$S_{ions} \sim 5\ 10^{-5}. \tag{9}$$

The probability that a particular H_2O molecule experiences an ion-molecule reaction is less than 1/10,000. For electrons, the elastic scattering rate (much larger than the spin-conversion rate) is

$$\alpha_{electrons}^s = 10^6\,cm^{-3}\ 10^{-15}\,cm^2\ 6\ 10^7\,\sqrt{E}\,cm\,s^{-1}.$$

Assuming $E \sim 0.025\,eV$ near the nucleus, then

$$\alpha_{electrons}^s \sim 10^{-2}\,s^{-1}$$

and

$$S_{electron} \sim 5\ 10^{-2}. \tag{10}$$

Even allowing higher energies in the outer coma, collisions with electrons are also negligible.

Collisions with dust grains are estimated as follows: The grain diameter is taken to be 0.1 µm, and the mass loss rate ($\dot{M}_d$) to be $0.2\,\dot{M}_{H_2O}$ (this volume). The surface density (n_g^s) of grains is

$$n_g^s = \frac{\dot{M}_D}{4\pi R_N^2 v_S}\frac{6}{\pi D^3 \varrho_g} \tag{11}$$

we take

$\dot{M}_D \sim 6(6)\,gm\,s^{-1}$, $v_S \sim 3(4)\,cm\,s^{-1}$, $D \sim 1(-5)\,cm$, and $\varrho_g \sim 0.3\,gm\,cm^{-3}$.

Then $n_g^s \sim 4(5)\,cm^{-3}$, and the cross-sectional area (σ_g) of a typical grain is $\sim 7.8(-11)\,cm^2$. The instantaneous collision rate for a given H_2O molecule with grains is then

$$\alpha_{dust}^s \sim n_g^s \sigma_g v_{th} \sim 1\,s^{-1}. \tag{12}$$

The total number of collisions (S_d) a molecule experiences with dust in traversing the coma is then

$$S_d \sim \alpha_{dust}^s \frac{R_s}{v_0} \sim 5. \tag{13}$$

It would thus require an accomodation coefficient near unity in order for collisions with dust grains to affect the OPR significantly. Even in the case of paramagnetic substances (platinum, activated charcoal) tens of minutes are required to convert ortho-H_2 to para-H_2 at a pressure of one atmosphere (Farkas, 1935). It therefore seems unlikely that collisions with grains (even if paramagnetic) could change the ortho-para ratio in the coma.

Radiative coupling of the ortho-para species is strictly forbidden; thus known processes are ineffective in changing the ortho-para ratio in the coma.

2.4. *Water vapor as a daughter molecule*

There remains the possibility that H_2O is itself a daughter of some short-lived parent with scale-length <1000 km. In that event, we would expect the H_2O to be produced in the ratio of the number of final quantum states available, i.e., ortho-para = 3.0. At present the method cannot distinguish between H_2O which was equilibrated in the ice phase at $T_N \gtrsim 60\,K$, and that which might have been produced in the coma with a ratio of 3.0. This distinction is only possible if $T_N \leqq 60\,K$ is observed.

3. The observed ortho-para ratio

The ortho-para ratio of water vapor in Halley's comet may be determined from high-resolution infrared spectra in the following way. The total intensity emitted by ortho-H_2O (I_o) and by para-H_2O (I_p) is proportional to the fluorescence efficiency (g_o, g_p), where $g_o = g_p$ to a good approximation. (Differences do occur owing to the effects of weak solar lines. These have recently been measured by space infrared spectroscopy (C.B. Farmer, private communication) and will produce a small Swings effect in the Halley data.)

$$I_o = g_o N_o = \sum_i I_o(i) \qquad o = \text{ortho}$$

$$I_p = g_p N_p = \sum_j I_p(j) \qquad p = \text{para} \tag{14}$$

where i, j designate the individual lines and the sum proceeds over all lines for each species.

Some spectral lines are obscured by terrestrial CO_2 (dashed lines, Fig. 3), but these can be included by using known branching ratios (B_o, B_p) to derive the total multiplet intensity from the observed individual line intensities. This is a good assumption if all lines are optically thin. Then,

$$\text{OPR} = \frac{N_o}{N_p} = \frac{\sum_i I_o(i)/B_o(i)}{\sum_j I_p(j)/B_p(j)}. \tag{15}$$

In the event that more than one observed line originates from a given upper state, it is understood that I/B refers to the sum of their intensities divided by the sum of their individual branching ratios.

The ortho-para ratios derived from our observed spectra are

$$\mathrm{OPR} = 2.73 \pm 0.17, \text{ 1985 UT Dec. 24.1,}$$
$$\mathrm{OPR} = 3.23 \pm 0.37, \text{ 1986 UT Mar 22.7.} \quad (16)$$

These have been related to the nuclear-spin-temperatures by locating them in Fig. 4. On December 24.1 the nuclear-spin-temperature (T_N) was 35 K (+9, −5) and in March it was greater than 40 K, at the 67% confidence limit. The internal temperature of a spherical, grey-body nucleus at aphelion would be ~47 K.

Each error estimate is composed of two components. The first is the statistical error associated with the individual spectral line intensities, including instrumental sensitivity, while the second is a subjective estimate of the error associated with the atmospheric transmittance correction. The lunar comparison spectra do not provide an unambiguous measure of atmospheric transmittance during the cometary observations, owing to variability in the atmospheric water content along the flight path and from day-to-day.

Our earlier result for 1986 UT Dec. 24.1, (2.66 ± 0.13, Mumma et al., 1986) has changed owing to revised processing of the spectra, and upward revision of the error estimate. However, the transmittance errors are correlated from line-to-line and further treatment may result in reduced error estimates for the OPR. Measurements of the column burden of H_2O are available for most flights and will be used together with lunar and stellar spectra, to improve the accuracy of atmospheric transmittance correction.

Pending careful modelling of the atmospheric transmittances in the wings of the H_2O lines, and a correlated error propagation analysis, we must accept the present error estimates. Then we find that, at the 95% confidence limit, the ortho-para ratio is consistent with the statistical equilibrium value ($T_N \gtrsim 50$ K) of 3/1.

Acknowledgements. We gratefully acknowledge support from the following sources: NASA Grant NAG 2-206 to the University of Arizona; NASA Grant NAG 2-356 to the John Hopkins University; and NASA RTOP's 188-41-55, 196-41-54, and 352-02-03 to the Goddard Space Flight Center.

References

Borst, L.B., Borst, S.L., Koysooko, L., Patel, H., Stusnick, E.: 1961, *Phys. Rev. Letters* **7**, 343
Combes et al.: 1986, *Nature* **321**, 266
Dennison, D.M.: 1927, *Proc. Roy. Soc. A*, **115**, 483
Dodelson, S.: 1986, *J. Phys. B. At. Mol. Phys.* **19**, 2871
Farkas, A.: 1935, *Orthohydrogen, Parahydrogen, and Heavy Hydrogen*, Cambridge University Press, Cambridge
Hall, R.T., Dowling, J.M.: 1967, *J. Chem. Phys.* **47**, 2454
Heisenberg, W.Z.: 1927, *Z Physik* **41**, 239
Herman, G., Weissman, P.R.: 1986, *Icarus* (in press)
Herzberg, G.: 1945, *Molecular Spectra and Molecular Structure, Vol. 2, Infrared and Raman Spectra of Polyatomic Molecules*, Van Nostrand, New York
Hopkins, H.P.Jr., Curl, R.F., Pitzer, K.R.: 1968, *J. Chem. Phys.* **48**, 2959
Hund, F.: 1927, *Z. Physik* **42**, 93
Larson, H.P., Davis, D.S., Mumma, M.J., Weaver, H.A.: 1986, *Astrophys. J. Letters* **309**, L95
Larson, H.P., Mumma, M.J., Weaver, H.A.: 1987, *Astron. Astrophys.* (in press)
Kuhrt, E.: 1984, *Icarus* **60**, 512
Mumma, M.J., Weaver, H.A., Larson, H.P., Davis, D.S.: 1986, *Science* **232**, 1523
Raich, J.C., Good, Jr. R.H.: 1964, *Astrophys. J.* **139**, 1004
Weaver, H.A., Mumma, M.J., Larson, H.P., Davis, D.S.: 1986, *Nature* **324**, 441
Weaver, H.A., Mumma, M.J., Larson, H.P., Davis, D.S.: 1987, *Astron. Astrophys.* (in press)

Astron. Astrophys. 187, 425–430 (1987)

ASTRONOMY
AND
ASTROPHYSICS

The 2.7 μm water band of comet P/Halley: interpretation of observations by an excitation model

D. Bockelée–Morvan and J. Crovisier

Observatoire de Paris, Section de Meudon, F-92195 Meudon, France

Received January 30, accepted April 16, 1987

Summary. In order to interpret infrared observations of parent molecules in comets, we have developed an excitation model which takes into account excitation by collisions, pumping of the fundamental bands of vibration by the infrared solar radiation field, and radiation trapping in the rotational and rovibrational lines. This model allows us to derive infrared synthetic spectra for various species. It is applied here to interpret recent infrared observations of water in comet P/Halley. The saturation of the 2.7 μm water band observed by the IKS instrument on Vega is evaluated. The relative water line intensities observed by the KAO in March 1986 are well fitted by the model, when a coma kinetic temperature of ~60 K is assumed in the $r \gtrsim 1000$ km region.

Key words: comets – comet P/Halley – parent molecules – infrared spectroscopy – models

1. Introduction

Among the highlights of comet P/Halley observations are the detections of several vibrational bands of parent molecules in the infrared. A detailed evaluation of the excitation of these molecular bands is now needed to interpret the observations. First-order models were developed formerly to study the feasibility of such observations (Yamamoto, 1982; Crovisier and Encrenaz, 1983; Weaver and Mumma, 1984). More sophisticated models were also elaborated for specific parent molecules (Crovisier and Le Bourlot, 1983; Chin and Weaver, 1984; Bockelée-Morvan et al., 1984; Crovisier, 1984; Bockelée-Morvan and Crovisier, 1985). We have elaborated a generalization of these models, which follows the evolution of molecular excitation as a function of distance to nucleus, from collision-dominated excitation in the inner coma to radiation-dominated excitation in the outer coma, and which takes into account, when necessary, optical depths of rotational and ro-vibrational transitions. Such a model was applied to the water molecule (Bockelée-Morvan and Crovisier, 1986; Bockelée-Morvan, 1987) and to linear molecules (CO, CO_2, HCN, HC_3N, C_2N_2, C_2H_2, OCS) for which spectroscopic properties may be easily parametrized (Crovisier, 1987). This model allows us to compute synthetic infrared spectra as well as rotational line intensities. Applications of such a model are two-fold: a) Reliable molecular column densities and production rates may be derived from the observed spectra. b) Excitation conditions, and therefore coma physical parameters, may be retrieved from observed band shapes or by comparing relative intensities of different lines.

Since detailed descriptions of this model are published elsewhere (Bockelée-Morvan, 1987; Crovisier, 1987), only a summary is given here (Sect. 2). We present in Sect. 3 applications of our model to recent infrared spectroscopic observations of P/Halley: the medium-infrared spectrum obtained by Vega/IKS (Infra Krasnoi Spectrometer, Combes et al., 1986; Moroz et al., 1987) and the high-resolution observation of the ν_3 band of H_2O with the KAO (Kuiper Airborne Observatory, Mumma et al., 1986a; Weaver et al., 1986).

Send offprint requests to: D. Bockelée–Morvan

2. The excitation model

The prevailing excitation mechanisms of parent molecules in comets are radiative excitation of the fundamental bands of vibration by the solar infrared field and thermal excitation by collisions in the inner coma (Crovisier and Encrenaz, 1983). For abundant species, large optical depths may be encountered in both rotational and ro-vibrational transitions, so that line transfer must be included.

Collisional excitation is difficult to figure out accurately because little is known on the cross-sections and the kinetic temperature of the coma. The main collision partner in the inner coma is the water molecule. In our nominal model, we assume a total collisional cross-section for rotational relaxation of $\sigma = 5 \times 10^{-14}\,\mathrm{cm}^2$ and $10^{-14}\,\mathrm{cm}^2$ for respectively H_2O—H_2O collisions and collisions between H_2O and other polar molecules. These values are consistent with laboratory line-broadening measurements. Since almost nothing is known on the selection rules between rotational levels, we assume that each collision redistributes the molecules on the rotational levels according to the Boltzmann distribution. We provisionally use a kinetic temperature law similar to that derived by Crovisier (1984) with a minimum of ~16 K at ~80 km from the nucleus. Collisional coupling of the vibrational levels is negligible.

In the outer coma, low densities preclude any collisional or opacity effects. Balance between infrared excitation by the solar field and spontaneous decay tends to establish "fluorescence equilibrium" of the population distribution among rotational levels. It is generally sufficient to consider excitation of one or two fundamental bands of vibration. Excitation rates of the strongest vibrational bands are typically of the order of a few $10^{-4}\,s^{-1}$ at 1 AU.

Our numerical method consists in following the time evolution of the rotational population distributions as the molecules expand from the nucleus, by solving the statistical equilibrium differential equations. Steady-state equations are used for the vibrational levels. The molecule density is computed for a gas expansion velocity of $0.8\,km\,s^{-1}$ and taking into account photodissociation by solar UV (the H_2O life-time is assumed to be 10^5 s at 1 AU from the Sun).

Most possible cometary parent molecules have small column densities and their rotational lines are thin. A particular case is that of the H_2O molecule for which rotational lines are optically thick in a large part of the coma. This strongly affects the rotational population distribution and complicates the line transfer treatment. Radiation trapping in the ro-vibrational lines, which modifies the excitation and fluorescence of the vibrational bands, must also be considered. These optical depth effects are modelled using the escape probability method, assuming spherical geometry and constant radial velocity.

Once the rotational population distribution as a function of nucleus distance is determined, the rotational and ro-vibrational line intensities may be computed by volume integration within the instrumental beam, taking into account, when necessary, optical depth effects.

3. Applications to P/Halley observations

Examples of synthetic infrared spectra for various species and typical observational conditions, as well as rotational line intensity evaluations, may be found in Bockelée-Morvan (1987) and Crovisier (1987). Our model has also been used to interpret molecular observations of P/Halley at millimetre wavelengths (Bockelée-Morvan et al., 1987). We shall rather present and discuss here synthetic spectra pertaining to the ν_3 water band at 2.7 μm observed in P/Halley.

3.1. The medium-infrared spectrum from IKS

The 2.5 to 5 μm spectra of P/Halley taken by IKS (Combes et al., 1986; Moroz et al., 1987) show the spectral signatures of H_2O (ν_3 band), CO_2 (ν_3 band), CH-bearing molecules, and possibly other species. The water production rate cannot be directly retrieved from the band integrated flux by a simple method such as that of Crovisier and Encrenaz (1983) or Crovisier (1984): the excitation and emission rates are affected by optical thickness within the band and the signal is not proportional to the water production rate. On the other hand, the signal is no longer proportional to $1/D$ (D = distance of the spacecraft to the comet) when the spacecraft is approaching the comet, as could be expected from a resolved coma with a r^{-2} density law. An evaluation of these saturation effects needs the knowledge of the rotational population distribution, and therefore the use of our model.

We have computed several synthetic spectra of the ν_3 water band, pertaining to the IKS observing conditions. Examples are shown in Fig. 1. They correspond to $Q[H_2O] = 10^{30}\,s^{-1}$ or $3.1\,10^{29}\,s^{-1}$, and comet-spacecraft distances $D = 10^4$ or 10^5 km. For all these spectra, the instrument beam diameter is 1°, and the model parameters are those described in Sect. 2. The effect of line saturation, as can be clearly seen in the figure, is to spread

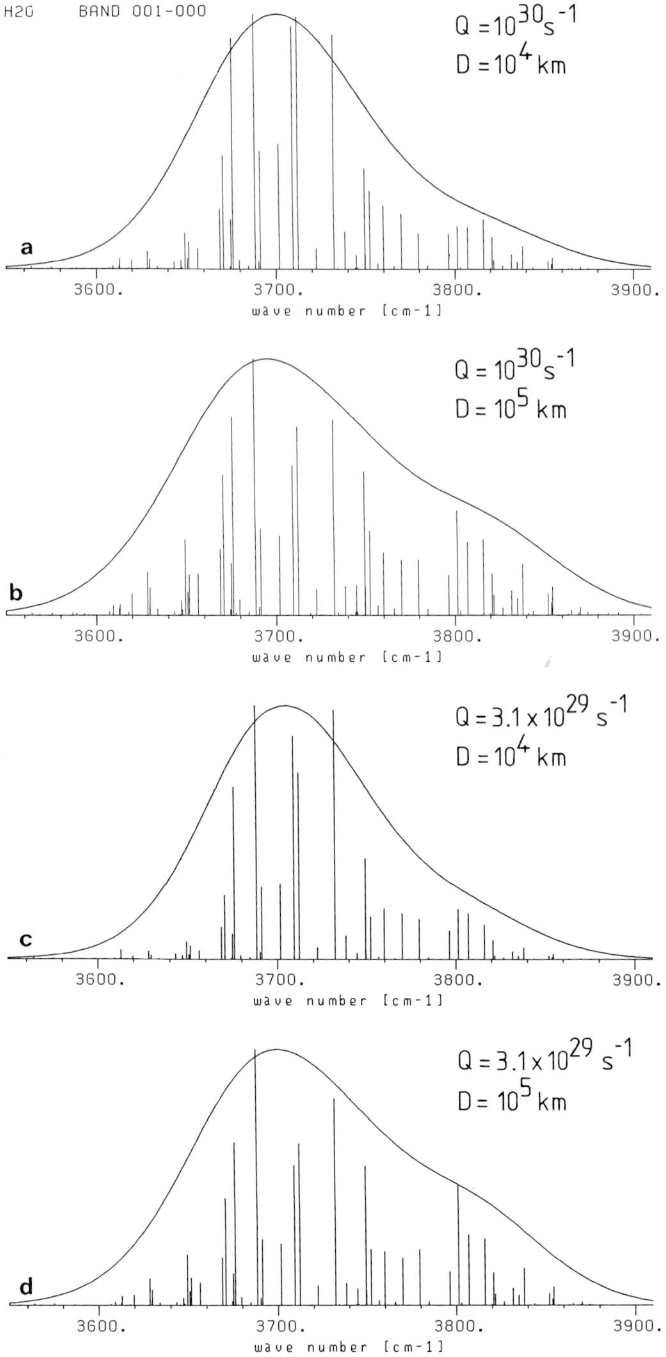

Fig. 1a–d. Synthetic ν_3 band profiles of H_2O computed for IKS observing conditions. The assumed water production rates Q and distances of instrument to comet D are given for each spectrum, and the other model parameters are given in the text. The figure shows the relative intensities of each ro-vibrational transition as well as the expected band profile for a $\Delta\lambda/\lambda = 1/50$ spectral resolution

the band energy over more ro-vibrational transitions as the production rate is increased. One can also see that the band is broader when the comet is observed at a larger distance. This is due to the coma kinetic temperature law we have assumed: at $D = 10^4$ km, the instrument is sampling a colder region than at $D = 10^5$ km.

The band shape critically depends upon the assumptions

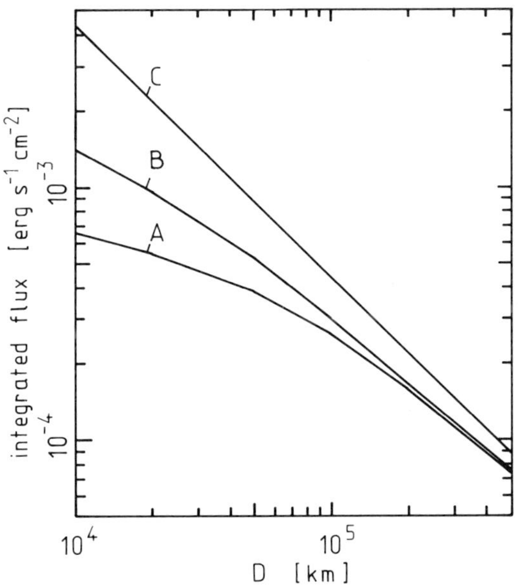

Fig. 2. Integrated flux of the H_2O ν_3 band as a function of distance D of instrument to comet, for IKS observing conditions. $Q[H_2O]$ is fixed to 10^{30} s^{-1}. (**A**) for the "cold" kinetic temperature law given in Section 2.1; (**B**) for a uniform kinetic coma temperature of 100 K and (**C**) for the limit case of an optically thin coma

made on the kinetic temperature of the inner coma (this effect has also been demonstrated for the ν_2 water band at 6.3 μm, for similar observing conditions, by Bisikalo et al., 1986). So one can in principle retrieve temperature information from the observed band shape. This is unfortunately not possible for the IKS spectrum, because of its moderate spectral resolution ($\Delta\lambda/\lambda \sim 1/50$) and of the uncertainty on the observed band profile.

The expected band integrated flux as a function of D is shown in Fig. 2 as curve A (for $Q[H_2O] = 10^{30}$ s^{-1} and the same model parameters as above). For comparison, the integrated flux variation for the limit case of an optically thin coma is also indicated (curve C): one can see that optical depth effects decrease the observed signal by a factor of $\sim$1.4 at $D_1 = 200000$ km, a factor of $\sim$2.6 at $D_2 = 40000$ km. Also shown in Fig. 2 (curve B) is the flux variation expected when the coma kinetic temperature is constant and equal to 100 K. In this case, the temperature is larger, for the coma region sampled by the instrument, than that assumed above, and since in this region the population distribution is imposed by collisions, the rotational population is distributed over more levels. As a consequence, the line optical depth effects are less important, and curve B is closer to the $1/D$ law expected in the thin case. The dependence with distance of the water signal observed by the IKS instrument is not yet analysed with sufficient accuracy (Fig. 2a of Moroz et al., 1987) to allow us to discriminate between these curves.

Saturation curves were also obtained by Krasnopolsky and Tkachuk (1986), using single-scattering approximation and a LTE population distribution for the water rotational levels. They are in agreement with those of Fig. 2.

In fact, in addition to the ν_3 band, the 2.7 μm emission feature observed by IKS also includes the ν_1 water band as well as combination bands ($\nu_2 + \nu_3 - \nu_2$ and $\nu_1 + \nu_3 - \nu_1$). However, these bands have small strengths, or their lower vibrational levels are not significantly populated. Therefore, they are much less affected by optical depth effects, and their emissions can simply be evaluated using the global fluorescence rates of Crovisier (1984). In the most representative IKS spectrum (which is the difference of signals taken at distances $D_1 = 200000$ km and $D_2 = 40000$ km), the ν_3 band was attenuated by a factor of 3.4. The other bands, which would have, in the optically thin case, a contribution of 0.29 that of ν_3, are expected to have a total intensity comparable to that of ν_3. This result was used to determine $Q[H_2O] = 10^{30}$ s^{-1} for the IKS spectra (Combes et al., 1986; Moroz et al., 1987).

Although $Q[CO_2]$ is much smaller than $Q[H_2O]$, the ν_3 band of CO_2 is also affected by optical depth effects, because its strength is 10 times that of the H_2O ν_3 band. For the IKS observing conditions at $D = 40000$ km, the signal decrease due to saturation is estimated to be a factor of 1.5. The resulting $Q[CO_2]$ is $\sim 2\,10^{28}$ s^{-1} (Moroz et al., 1987).

3.2. The high resolution spectra of the ν_3 band of water from the KAO

Mumma et al. (1986a) and Weaver et al. (1986) performed high resolution observations of the ν_3 band of water in P/Halley with a Fourier-transform spectrometer on the NASA-Kuiper Airborne Observatory. Nine individual lines of the band were detected at the end of December 1985 and resulted in the first definite detection of water vapor in comets. During post-perihelion runs in March 1986, twelve lines of the ν_3 band were detected. The line-by-line intensities (Fig. 3a) may be used for testing excitation conditions of cometary water molecules.

The identification and relative intensities of observed spectral lines demonstrate that rotational states are maintained in a cold distribution in the coma region (r $\leq$ 15000 km) sampled by the beam (41″ diametre). Fig. 3b shows a synthetic spectrum where rotational LTE at $T = 150$ K is assumed. The lines observed on March 24 (Weaver et al., 1986) are plotted for comparison. Clearly, they are not fitted by the thermal spectrum. However, the observed spectrum is roughly consistent with that expected from LTE at a much lower temperature ($T \sim 50$ K). The absence of lines originating from high ground state rotational levels shows that these states are not significantly excited in the coma. The rotational populations of the ground vibrational state can be derived from the observed relative intensities, neglecting optical depth effects: Mumma et al. (1986a) have shown that, for their December observation, they deviate strongly from LTE conditions, which confirms the need of nonthermal-equilibrium models.

The shape of the ν_3 band, when the rotational population distribution is at fluorescence equilibrium, is plotted in Fig. 3c. Since the infrared excitation rates are small with respect to vibrational and rotational spontaneous emission rates, only the ortho 1_{01} and para 0_{00} ground state levels are significantly populated. This results in a limited number of ν_3 fluorescent lines. The observed band definitely shows strong emission lines not predicted by fluorescence equilibrium, which means that this equilibrium is not yet achieved in the region $r \leq 15000$ km. Indeed, our excitation model predicts that it sets up at distances typically $\sim 10^5$ km from the nucleus.

The synthetic spectra obtained with our model (Bockelée-Morvan, 1987) typically lie between thermal and fluorescence equilibrium spectra. It is thus tempting to adjust the model parameters in order to try to match the observed spectra. In contrast with the IKS observations, saturation effects are not

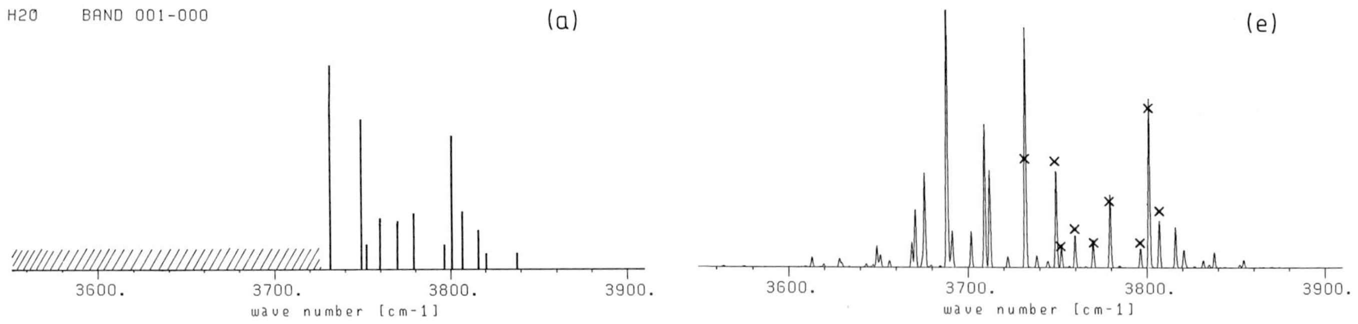

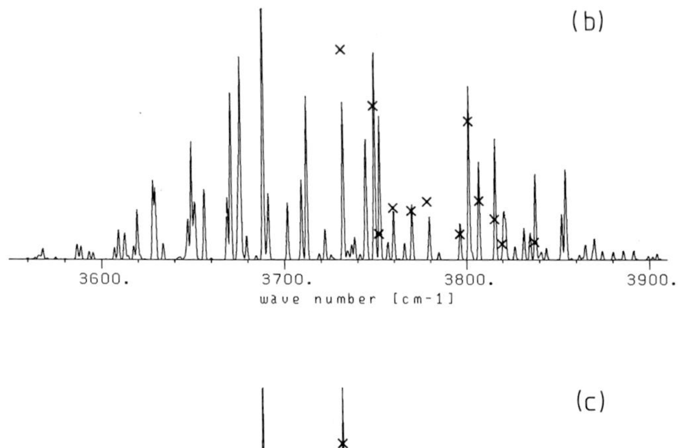

Fig. 3. a Relative intensities of the lines of the ν_3 band of water observed by the KAO on March 24, 1986 (Mumma et al., 1986). Weak lines from the $\nu_2 + \nu_3 - \nu_2$ band are omitted. The hatched spectral region was not accessible to the instrument. **b–d** Synthetic spectra of the ν_3 band of water. Crosses indicate the line positions and relative intensities of the KAO March 24 observation for comparison. **b** For rotational LTE at $T = 150$ K. **c** For fluorescence equilibrium. **d** For our model (KAO observing conditions with $Q[H_2O] = 2\,10^{30}\,s^{-1}$, $\sigma = 5\,10^{-14}\,cm^2$ and $T = C^{te} = 60$ K). **e** Synthetic spectrum for our model (KAO observing conditions of December 24, 1985, $Q[H_2O] = 5\,10^{29}\,s^{-1}$, $\sigma = 5\,10^{-14}\,cm^2$ and $T = C^{te} = 100$ K). Crosses indicate the line positions and relative intensities of the KAO December 24 observation

important here for the whole ν_3 band ($\sim 10\%$), since the KAO beam is sampling the coma region where most H_2O molecules have thin ro-vibrational lines (Bockelée-Morvan, 1987). Therefore, the H_2O production rate can be derived confidently from the measured band intensity assuming no optical depth effects (see Mumma et al., 1986a), and does not depend on the choice of the model collisional parameters. The model parameters remaining free to be adjusted are thus the collision and temperature laws.

The relative intensities of the ro-vibrational lines reflect the rotational population distribution inside the region sampled by the beam, and thus are expected to depend critically on the size of the region where rotational LTE is achieved and on the kinetic temperature. First we used the nominal collisional cross-section of $5\,10^{-14}\,cm^2$ and a production rate of $2\,10^{30}\,s^{-1}$ to explore which range of temperature can fit the data. The energy distribution within the band was found to be very sensitive to the kinetic temperature in the $r \geq 1000$ km region. The best fit was obtained for temperature laws with $T = 60$ K in the $r > 1000$ km region (Fig. 3d). From the comparison of Figs. 3a and 3d, we see that not only the number of lines predicted to be present in the band of the spectrometer is consistent with the observed spectrum, but their intensities are in excellent agreement. The size of the LTE region is determined by both the collisional cross-section and optical depth effects. Since molecules relax to an extremely cold distribution when rotational LTE breaks down, larger temperatures are needed to fit the data when smaller total collisional cross-sections are used: for $\sigma = 10^{-14}\,cm^2$, we need $T = 70$ K. Radiation trapping in the rotational lines delays radiative decay and therefore maintains LTE at lower densities than required in optically thin conditions. For the KAO observing conditions in March, a large fraction of the water molecules sampled by the beam were in the LTE region. An underestimation of the optical depth effects should overestimate the derived temperature and inversely. The uncertainties due to the approximations in our line transfer treatment are however difficult to figure out.

KAO observations were also performed on March 22, 1986, and late December 1985 with similar geometrical conditions ($r_h \sim 1$ AU, $\Delta = 0.8$–1.0 AU, phase angle $\sim 60°$). However, the H_2O production rates measured by the KAO were only $\sim 10^{30}\,s^{-1}$ on March 22, $\sim 2\,10^{29}\,s^{-1}$ on December 24. Therefore, we should expect the lines originating from pumping of levels other than the lowest ortho and para states (1_{01} and 0_{00}) to be relatively smaller. The behaviour of the ν_3 energy distribution from December 1985 to March 1986 is qualitatively consistent with this prediction. However, we failed to reproduce the spectrum of December 24,

1985 with $\sigma = 5\,10^{-14}\,\mathrm{cm}^2$ and a H_2O production rate of $2\,10^{29}\,\mathrm{s}^{-1}$, whatever is the temperature law. In fact, in contrast with the March observations, the beam sampled the coma region where most H_2O molecules have relaxed to the lower states, and the energy distribution is not very sensitive to the kinetic temperature. Various explanations may be proposed:

1) The KAO data have large absolute calibration uncertainties, and the H_2O production rate could have been underestimated. Using a H_2O production rate of $5\,10^{29}\,\mathrm{s}^{-1}$, it is possible to reasonably fit the December 24 spectrum with a kinetic temperature of $\sim 100\,\mathrm{K}$ in the $r = 1000$–$10000\,\mathrm{km}$ region. This fit is shown in Fig. 3e (the 0_{00}–1_{01} line at $3732.14\,\mathrm{cm}^{-1}$ is in fact ill-fitted: one should note that, according to Mumma et al., 1986a, the calibration uncertainty on this line is very large). This explanation has the advantage to reconcile the KAO H_2O observations with the IUE observations of OH (Feldman et al., 1986), from which a higher production rate was derived for the same period.

2) Radiation trapping effects could have been underestimated in our line transfer treatment, leading to an underestimation of the size of the LTE region. Then, they were a fortiori underestimated in the model fitting of March 24 data, and the derived kinetic temperature was overestimated.

3) Our model assumes steady and isotropic gas production, whereas the comet activity is notoriously variable, and the gas may be expected to be partly distributed in jets. Therefore, for the same total gas content, the gas real density may be higher in restricted regions, where the optical depth effects are enhanced. Note that on March 22, the 1_{11}–1_{10}, 2_{20}–2_{21} and 1_{10}–1_{11} lines were unexpectedly relatively stronger than on March 24, whereas the production rate was weaker: this might be due to this effect.

We assumed in the model an ortho/para population ratio of 3 for water. The ratio might be different if water was formed at very low temperatures. Mumma et al. (1986b) did not find any significant departure from this ratio in their analysis of the KAO data. This result is confirmed by the good fit of the spectra by our model.

4. Conclusion

The confrontation of models and observations of spectra of cometary parent molecules is of primary importance to investigate the physical conditions within cometary atmospheres. Reliable excitation models are also a requisite for interpreting spectroscopic observations of parent molecules and deriving molecular abundances. The recent observations of the ν_3 band of water in P/Halley with the KAO have clearly shown that water is not at LTE, and that its emission spectrum cannot be explained by simple fluorescence equilibrium. Our excitation model, which takes into account the evolution between LTE and fluorescence equilibrium as well as radiation trapping, seems to be adequate to interpret such observations. The spectrum observed on March 24, 1986, can be nicely reproduced by the model when the kinetic temperature is assumed to be $\sim 60\,\mathrm{K}$ at a few 1000 km from the nucleus. The agreement is excellent, despite the preliminary nature of some of the model ingredients, and the calibration uncertainties of the KAO pioneering observations. This indicates that cometary water excitation is fairly well understood, and that no important phenomenon has been overlooked.

No direct measurement of the temperature of cometary molecules is available up to now. Our results suggest a cool temperature, at least at $r \sim 1000\,\mathrm{km}$. This is cooler than the temperature ($\sim 200\,\mathrm{K}$) measured in situ for ions by the NMS instrument aboard Giotto (Lämmerzahl et al., 1986) in the $r < 5000\,\mathrm{km}$ region. In this region, the ions are believed to be thermalized by collisions with neutrals, and should partake their temperature.

Many cometary thermodynamical models have been elaborated in order to predict the gas kinetic temperature (see Mendis et al., 1985, for a review). Most models predict that, starting from the nuclear region at a value close to sublimation equilibrium temperature ($\sim 180\,\mathrm{K}$), the gas temperature rapidly decreases in the first 100 km, due to nearly adiabatic expansion. Afterwards, the temperature may increase due to photolytic heating, but radiative cooling by water rotational emission may also play a role. After the molecules escape the collisional region, their temperature is frozen. Bockelée-Morvan and Crovisier (1987) have recently proposed a cometary thermodynamical model for pure water which includes the processes mentioned above and which solves water excitation and thermal balance in a consistent way. Their model predicts gas temperatures in full agreement with that suggested by the present analysis.

Acknowledgements. We are grateful to Drs D.S. Davis, H.P. Larson, M.J. Mumma and H.A. Weaver for discussions and for communicating their results prior to publication.

References

Bisikalo, D.V., Repin, S.V., Strelnitskij, V.S.: 1986, *Astron. Tsirk. No* **1450**

Bockelée-Morvan, D.: 1987, *Astron. Astrophys.* **181**, 169

Bockelée-Morvan, D., Crovisier, J.: 1985, *Astron. Astrophys.* **151**, 90

Bockelée-Morvan, D., Crovisier, J.: 1986, in *Asteroids, Comets, Meteors II*, edts. Cl. Lagerqvist et al., Uppsala University Press, p. 279

Bockelée-Morvan, D., Crovisier, J.: 1987, in *Symposium on the Diversity and Similarity of Comets*, ESA SP-278 (in press)

Bockelée-Morvan, D., Crovisier, J., Baudry, A., Despois, D., Perault, M., Irvine, W.M., Schloerb, F.P., Swade, D.: 1984, *Astron. Astrophys.* **141**, 411

Bockelée-Morvan, D., Crovisier, J., Despois, D., Forveille, T., Gérard, E., Schraml, J., Thum, C.: 1987, *Astron. Astrophys.* **180**, 253

Chin, G., Weaver, H.A.: 1984, *Astrophys. J.* **285**, 858

Combes, M., Moroz, V.I., Crifo, J.F., Lamarre, J.M., Charra, J., Sanko, N.F., Soufflot, A., Bibring, J.P., Cazes, S., Coron, N., Crovisier, J., Emerich, C., Encrenaz, T., Gispert, R., Grigoryev, A.V., Guyot, G., Krasnopolsky, V.A., Nikolsky, Yu. V., Rocard, F.: 1986, *Nature* **321**, 266

Crovisier, J.: 1984, *Astron. Astrophys.* **130**, 361

Crovisier, J.: 1987, *Astron. Astrophys. Suppl.* **68**, 223

Crovisier, J., Encrenaz, T.: 1983, *Astron. Astrophys.* **126**, 170

Crovisier, J., Le Bourlot, J.: 1983, *Astron. Astrophys.* **123**, 61

Feldman, P.D., Festou, M.C., A'Hearn, M.F., Arpigny, C., Butterworth, C., Cosmovici, B., Danks, A.C., Gilmozzi, R., Jackson, W.M., McFadden, L.A., Patriarchi, P., Schleicher, D.G., Tozzi, G.P., Wallis, M.K., Weaver, H.A., Woods, T.N.: 1986, in *New Insights in Astrophysics*, ESA SP-263, p. 39

Krasnopolsky, V.A., Tkachuk, A. Yu.: 1986, in 20th ESLAB

Symp., ESA SP-250, Vol. **1**, 465

Lämmerzahl, P., Krankowsky, D., Hodges, R.R., Stubbemann, U., Woweries, J., Herrwerth, I., Berthelier, J.J., Illiano, J.M., Eberhart, P., Dolder, U., Schulte, W., Hoffman, J.H.: 1986, in 20th ESLAB Symp., ESA SP-250, Vol. **1**, 179

Mendis, D.A., Houpis, H.L.F., Marconi, M.L.: 1985, *Fund. Cosmic Physics* **10**, 1

Moroz, V.I., Combes, M., Bibring, J.P., Coron, N., Crovisier, J., Encrenaz, T., Crifo, J.F., Sanko, N.F., Grigoryev, A.V., Bockelée-Morvan, D., Gispert, R., Emerich, C., Lamarre, J.M., Rocard, F., Krasnopolsky, V.A., Owen, T.: 1987, *Astron. Astrophys.* (in press)

Mumma, M.J., Weaver, H.A., Larson, H.P.: 1986b, in 20th ESLAB Symp., ESA SP-250, Vol. **1**, 341

Mumma, M.J., Weaver, H.A., Larson, H.P., Davis, D.S., Williams, M.: 1986a, *Science* **232**, 1523

Weaver, H.A., Mumma, M.J.: 1984, *Astrophys. J.* **276**, 782

Weaver, H.A., Mumma, M.J., Larson, H.P., Davis, D.S.: 1986, *Nature* **324**, 441

Yamamoto, T.: 1982, *Astron. Astrophys.* **109**, 326

Astron. Astrophys. 187, 431–434 (1987)

ASTRONOMY AND ASTROPHYSICS

Curves of growth of emission lines in cometary spectra. Implications for H_2O and OH bands of comet P/Halley

V.A. Krasnopolsky and A.Y. Tkachuk

Space Research Institute, Academy of Sciences, Profsoyuznaya 84/32, SU-117810 Moscow, USSR

Received February 12, accepted May 13, 1987

Summary. Curves of growth of emission lines are calculated for both parent and secondary species of comets assuming a Haser model distribution. Variations of these curves when varying input parameters are discussed. The technique developed is applied to the measurements of H_2O 2.7 and 1.38 μm bands by the Vega spacecraft and to OH 3090 Å band measurements by IUE. The evaluation of $10^{30}\,s^{-1}$ for the H_2O total production rate on 1986 March 9 is obtained by this technique.

Key words: comets – spectroscopy – gas

1. Introduction

Fluorescence as a dominant process in spectra of comets implies direct conversion of a measured intensity $4\pi I$ to column density N of a radiating component: $4\pi I = gN$, where the emission rate factor g is calculated with the known transition probability and solar radiation intensity. However, this simple relation is valid only for optically thin conditions $\tau_0 < 1$ in the line center. The aim of this paper is to calculate the curves of growth, i.e. the relationship between the abundance of a species and the intensity of its emission line for optically thick condition. A curve of growth depends on the process which defines the line shape. It is Doppler broadening in the case of cometary lines, and spatial distributions of number density and velocity of the species considered should be known. We adopt the Haser model for the former while the latter is given as the sum of a constant radial component V_0 and a constant component V_T assumed to be distributed isotropically and giving a rectangular local line shape. V_T is equivalent to the thermal velocity and formed by collisions and dissociation events.

2. Geometry and result of calculations

We use the cylindrical system of coordinates with the center at the nucleus and the z-axis directed to the sun (Fig. 1). Position of the optical axis is given by angles ϕ and α and distance ϱ (see Fig. 1); ϕ is the phase angle. The main variable quantity is the angle θ which determines a point B on the optical axis. Transmission of the coma to the light scattered by point B is assumed equal to

$$T = \exp(-\tau_z)\,[\omega + (1-\omega)\exp(-\tau_a)],$$

where τ_z is the optical thickness of the coma for the incident solar radiation, τ_a is that along the optical axis, $\omega = A_i/\Sigma A_i$ is the single scattering albedo equal to the ratio of transition probabilities. If $\omega \approx 0$ then $T \approx \exp(-\tau_z - \tau_a)$; if $\omega \approx 1$ then $T \approx \exp(-\tau_z)$, i.e. we ignore scattering along the optical axis which changes directions of photons but not their number. Therefore a fraction ω is re-emitted and not absorbed while a fraction $(1-\omega)$ is re-emitted in other lines and lost when considering the given line. If these non-resonant lines are of interest also, the relationship for T should be changed. This approach is a slightly modified version of the single scattering approximation.

We do not consider details of the determination of τ_z, τ_a and some other quantities but give final relationships. Both τ_z and τ_a depend on the direction of the thermal velocity of a molecule at point B which is given by the angles x and y (Fig. 2).

2.1. Parent molecules

For ϱ much less than the scale length

$$\tau_z(\theta, x, y) = \frac{QSc}{8\pi\varrho v V_0 V_T}\cos\theta\left\{\arcsin\left[\sin\alpha(\theta) + \frac{V_T}{V_0}(1+\sin x\cos y\sin\phi - \cos x\cos\phi)\right] - \alpha(\theta)\right\}$$

$$\tau_a(\theta, x) = \frac{QSc}{8\pi\varrho v V_0 V_T}\,\frac{1}{\sin\phi}\int_{\theta}^{\theta(\psi_{max} = \arccos[\cos\psi - (V_T/V_0)(1-\cos x)])}\cos^2\alpha\, d\theta$$

If the expression in the square brackets exceeds unity, it is assumed equal to unity.

Here Q is the total production rate of the species, S is the line strength, v is the wavenumber, c is the light velocity:

$$\alpha(\theta) = \mathrm{arctg}\,[\cos\theta\,\mathrm{tg}\,\alpha_0 - \sin\theta\,\mathrm{ctg}\,\phi]$$

$$\psi(\theta) = 2\,\mathrm{arctg}\sqrt{\frac{(p-b)(p-d)}{p(p-a)}}$$

with

$$a = \frac{1}{\cos\alpha_0}, \quad b = \frac{1}{\cos\alpha\cos\theta},$$

$$d = \frac{\mathrm{tg}\,\theta}{\sin\phi}, \quad p = \frac{1}{2}(a+b+d)$$

Send offprint requests to: V.A. Krasnopolsky

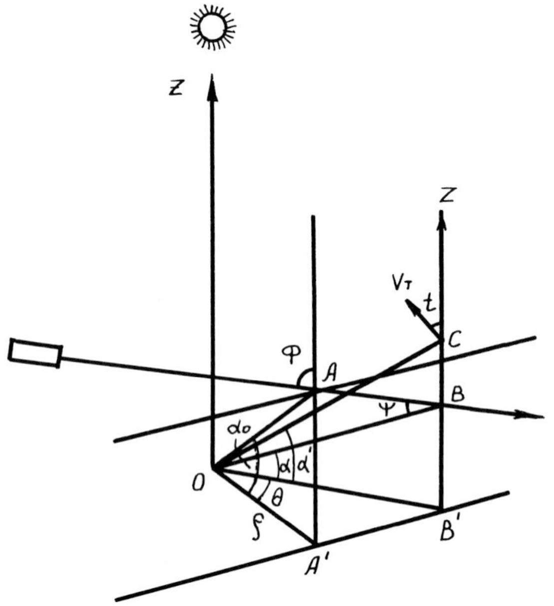

Fig. 1. Geometry of calculations

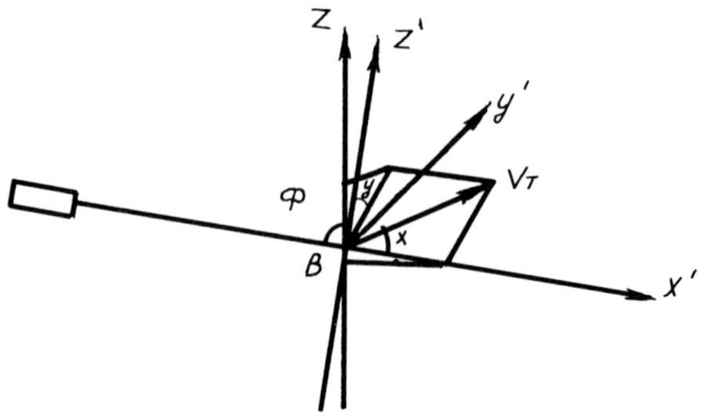

Fig. 2. Velocity projections at the point on the optical axis

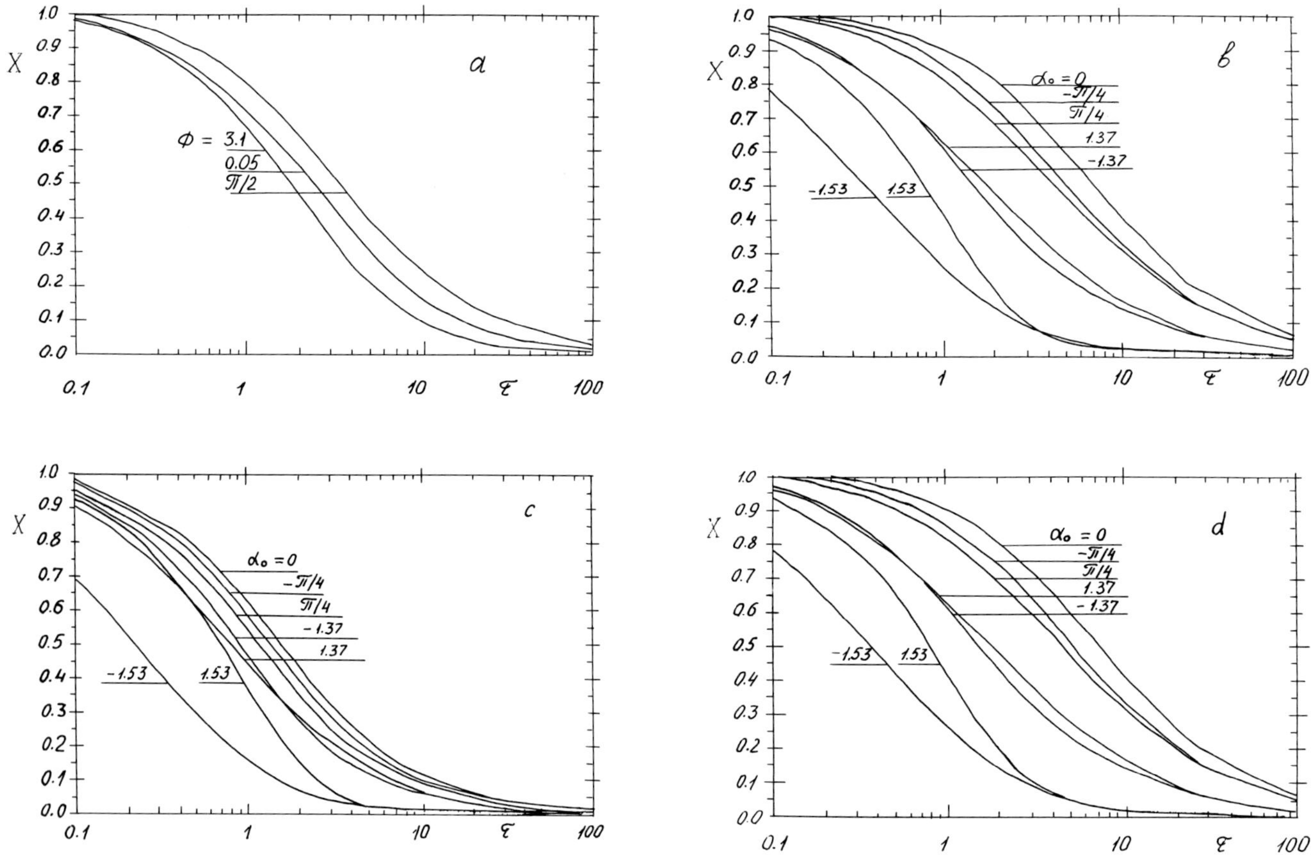

Fig. 3a–d. Curves of growth of emission lines for parent molecules: **a** variations with phase angle ϕ for $\alpha_0 = 0$, $V_T/V_0 = 0.2$, $\omega = 0.25$; **b** variations with position angle α_0 for $\phi = \pi/2$, $V_T/V_0 = 0.2$, $\omega = 0.25$; **c** the same for $\phi = \pi/2$, $V_T/V_0 = 0.4$, $\omega = 0.25$; **d** the same for $\phi = \pi/2$, $V_T/V_0 = 0.2$, $\omega = 1$

The effect of saturation may be described by the ratio of the calculated intensity $4\pi I$ to that in the absence of absorption, $4\pi I_0$:

$$X=\frac{4\pi I}{4\pi I_0}=\frac{\int\limits_{-\pi/2}^{\pi/2}\cos^2\alpha\, d\theta \int\limits_0^{\pi}[\omega+(1-\omega)\exp(-\tau_a(\theta,x))]\sin x\, dx \int\limits_0^{2\pi}\exp(-\tau_z)\, dy}{4\pi\int\limits_{-\pi/2}^{\pi/2}\cos^2\alpha\, d\theta}$$

The curves of growth represent this ratio as a function of

$$\tau=\frac{QSc}{8\pi\, rv\, V_0\, V_T}$$

where $r=\varrho/\cos\alpha_0$. They depend also on ϕ, α_0, V_T/V_0, ω (Fig. 3). Variations of X with the phase angle ϕ are low (Fig. 3a) and all other curves are calculated with $\phi=\pi/2$. The dimensionless parameter τ is the optical thickness of the coma along the instrument axis at the line center.

2.2. Secondary molecules

Relationships obtained for the secondary molecules are the following (l_2 is their scale length and l_1 that of their parent):

$$K(\alpha)=\exp\left(-\frac{r\cos\alpha_0}{l_1\cos\theta\cos\alpha}\right)-\exp\left(-\frac{r\cos\alpha_0}{l_2\cos\theta\cos\alpha}\right)$$

$$C=\frac{QSc}{8\pi rv\, V_0\, V_T(l_1/l_2-1)}$$

$$\tau_z(\theta,x,y)=C\,\frac{\cos\theta}{\cos\alpha_0}\int\limits_{\alpha(\theta)}^{\arcsin[\sin\alpha(\theta)+(V_T/V_0)(1+\sin x\cos y\cos\phi-\cos x\cos\phi)]}K(\alpha')\, d\alpha'$$

$$\tau_a(\theta,x,y)=C\,\frac{1}{\cos\alpha_0\,\sin\phi}\int\limits_{\theta}^{\theta(\psi_{max}=\arccos[\cos\psi(\theta)-(V_T/V_0)(1-\cos x)])}K(\alpha)\cos^2\alpha\, d\alpha$$

$$X=\frac{4\pi I}{4\pi I_0}=\frac{\int\limits_{-\pi/2}^{\pi/2}K(\alpha)\cos^2\alpha\, d\theta\int\limits_0^{\pi}[\omega+(1-\omega)\exp(-\tau_a(\theta,x))]\sin x\, dx\int\limits_0^{2\pi}\exp(-\tau_z)\, dy}{4\pi\int\limits_{-\pi/2}^{\pi/2}K(\alpha)\cos^2\alpha\, d\alpha}.$$

3. Implication for observations of comet Halley

3.1. Water vapor band at 2.7 μm

This band was measured by the Vega 1 infrared spectrometer (Combes et al., 1986a, b) with an aperture of 1°. The difference between band intensities measured at the spacecraft – nucleus distance D of $4\ 10^4$ km and $2\ 10^5$ km equals $1.6\ 10^{-7}$ W cm^{-2} sr^{-1}. Evaluation of the total production rate of water vapor by means of these data needs the saturation effect.

Table 1 summarizes wavenumbers, upper and lower vibrational levels and band strengths (Rothman, 1981) of absorption bands which result in excitation of the appropriate levels, and vibrational levels and emission rate factors of the emission bands which form the spectral feature at 2.7 μm. Contribution of hot bands is about 20% for optically thin conditions and becomes dominant with increasing optical thickness.

With the gas temperature profile calculated by Mendis et al. (1985) the mean gas temperature averaged along the spectrometer line of sight and the field of view is 40 K. A simple evaluation shows that collisional relaxation of water vapor is more effective than rotational, and less effective than vibrational radiative relaxation, at distances from 30 to 5000 km from the nucleus for $Q\approx 10^{30}$ s^{-1}. Therefore the Boltzmann distribution of rotational levels is assumed for $v''=000$. Every absorbed photon of the v_3 and v_1 bands is re-emitted in the same band ($\omega=1$) which simplifies calculations. The hot bahnds have $\omega\approx 0.8$ which is also close to the conservative scattering: the saturation effect is not very prominent for them and the mean ω is a rather good approximation inplace of the line-by-line calculations. Summing intensities of the separate lines we obtain intensity of the 2.7 μm feature as a function of distance D from the spacecraft to the nucleus with different total production rates of water vapor given in Fig. 4. The measured difference of intensities at $4\ 10^4$ and $2\ 10^5$ km corresponds to $Q_{H_2O}=1.1\ 10^{30}$ s^{-1}.

3.2. Water vapor band at 1.38 μm

Measurements of this band by the Vega 2 three-channel spectrometer (Krasnopolsky et al., 1986b) demonstrate ejection of water vapor by comet Halley in a cone with solid angle less than π steradian. That is why the curve of growth of this band calculated for the isotropic outflow model can be used to derive Q only with

Table 1. Wavenumbers, vibrational levels and strengths of absorption bands which result in excitation of the emission feature at 2.7 μm and vibrational levels and emission rate factors of emission bands which contribute to this feature

Absorption			Emission	
v (cm^{-1})	V'–V''	S (cm)	V'–V''	g (s^{-1})
3756	001–000	6.92–18 [a]	001–000	2.7–4
3657	100–000	4.86–19	100–000	1.9–5
1595	010–000	1.04–17	–	–
5331	011–000	8.04–19	011–010	2.5–5
7250	101–000	7.47–19	101–100	2.5–5
			101–001	

[a] $6.92-18=6.92\ 10^{-18}$

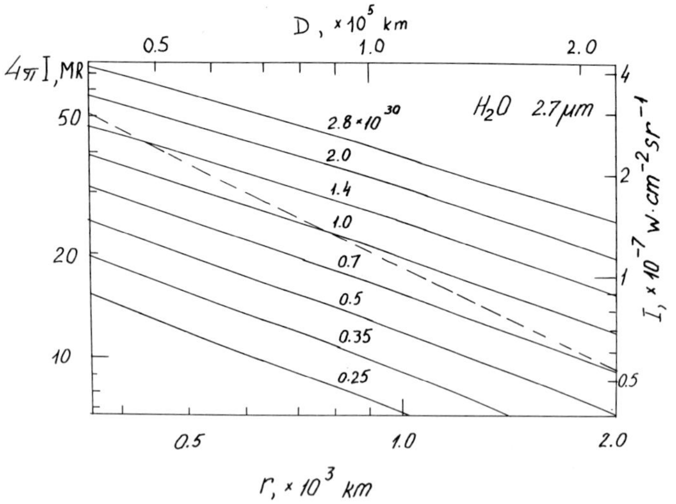

Fig. 4. Intensity of the H_2O band at 2.7 μm averaged on the IKS field of view as a function of the instrument – comet distance D or radius of the field of view r calculated for a set of Q_{H_2O} values. Dashes are the same for $Q_{H_2O} = 0.5\ 10^{30}\ s^{-1}$ with optically thin conditions

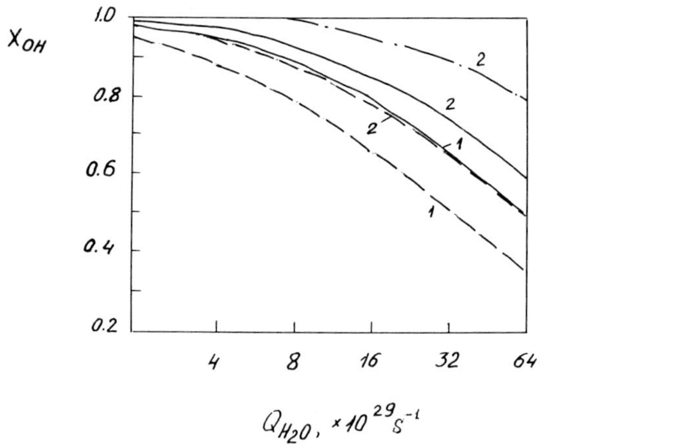

Fig. 5. Curves of growth of the OH band at 3090 Å for $\phi = \pi/2$, $\alpha_0 = \pi/4$ and optical axis to nucleus distance of $3\ 10^3$ km (dashes), 10^4 km (full lines), $3\ 10^4$ km (dash-dot lines): $l_{H_2O} = 3.1\ 10^4$ km (1) and $6.2\ 10^4$ km (2) at 0.834 AU

substantial corrections for the real geometry of the observation and the outflow.

3.3. Hydroxyl ($A^2\Sigma^+ - X^2\Pi$) 3090 A band

According to Schleicher and A'Hearn (1982) almost all excitation of this band is due to transitions at 3082, 3078 and 3072 Å from the lowest ground state rotational level $J'' = 3/2$, $N'' = 1$ to the three lowest levels in the $A^2\Sigma^+\ v' = 0$ state. The second and the third lines consist of two components when the parity rule is taken into account. Intervals between these components exceed the OH velocities in comet assumed equal to 1 km s^{-1}, therefore the band may be considered as five separate lines of similar strength with $\omega = 1$ to calculate the saturation effect. Calculation are made with the H_2O scale length measured by the Giotto mass spectrometer (Krankowsky et al., 1986) and that given in Krasnopolsky (1984) for solar minimum, with the OH lifetime of $1.3\ 10^5$ s (Van Dishoek and Dalgarno, 1984). The results are shown in Fig. 5. The intensity of the OH band of 0.9 MR measured by IUE on 1986 March 9 (Festou et al., 1986) results in a total production rate of $0.65\ 10^{30}$ and $1.15\ 10^{30}\ s^{-1}$ for the H_2O scale lengths discussed above.

4. Conclusion

Thus corrections caused by the saturation effect for parent molecules are determined mainly by a value of $\tau = QSc/8\pi rv\, V_0\, V_T$, which is the optical thickness of the coma along the instrument axis at the line center. These corrections are of a factor of 2 for $\tau \approx 2$ and depend also on the single scattering albedo ω, position angle of the optical axis α_0 and ratio of thermal to radial velocities V_T/V_0. These corrections for Vega and IUE measurements of comet Halley result in the total production rate of water vapor equal to $\approx 10^{30}\ s^{-1}$ with the isotropic outflow model.

References

Combes, M., Moroz, V.I., Crifo, J.F., Lamarre, J.M., Charra, J., Sanko, N.F., Soufflot, A., Bibring, J.P., Cazes, S., Coron, N., Crovisier, J., Emerich, C., Encrenaz, T., Gispert, R., Grigoryev, A.V., Guyot, G., Krasnopolsky, V.A., Nicolsky, Y.V., Rocard, F.: 1986a, *Nature* **321**, 266

Combes, M., Moroz, V.I., Crifo, J.F., Bibring, J.P., Coron, N., Crovisier, J., Encrenaz, T., Sanko, N.F., Grigoriev, A.V., Bockelee-Morvan, D., Gispert, R., Emerich, C., Lamarre, J.M., Rocard, F., Krasnopolsky, V.A., Owen, T.: 1986b, *Adv. Space Res.* **5**, No. 12, 127

Festou, M.C., Feldman, P.D., A'Hearn, M.F., Arpigny, C., Cosmovici, C.B., Danks, A.C., McFadden, L.A., Gilmozzi, R., Patriarchi, P., Tozzi, G.P., Wallis, M.K., Weaver, H.A.: 1986, *Nature* **321**, 269

Krankowsky, D., Lämmerzahl, P., Herwerth, I., Woweries, G., Eberhardt, P., Dolder, U., Hermann, U., Schulte, W., Berthelier, J.J., Illiano, J.M., Hodges, R.R., Hoffman, J.H.: 1986, *Nature* **321**, 326

Krasnopolsky, V.A.: 1984, *Cosmic Res.* **22**, 292

Krasnopolsky, V.A., Gogoshev, M., Moreels, G., Moroz, V.I., Krysko, A.A., Gogosheva, T., Palazov, K., Sargoichev, S., Clairemidi, J., Vincent, M., Bertaux, J.L., Blamont, J.L., Troshin, V.S., Valnichek, B.: 1986a, *Nature* **321**, 269

Krasnopolsky, V.A., Gogoshev, M., Moreels, G., Moroz, V.I., Krysko, A.A., Barke, V.V., Jegulev, V.S., Parshev, V.A., Sanko, N.F., Tomashova, G.V., Tkachuk, A.Y., Troshin, V.S., Novikov, B.S., Perminov, V.G., Sulakov, I.I., Fedorov, O.S., Gogosheva, T., Sargoichev, S., Palasov, K., Georgiev, A., Nedkov, I., Kanev, K., Clairemidi, J., Vincent, M. et al.: 1986b, *Adv. Space Res.* **5**, No. 12, 143

Mendis, D.A., Houpis, H.L., Marconi, M.L.: 1985, *Fund. Cosmic Phys.* **10**, 1

Rothman, L.S.: 1981, *Appl. Optics* **20**, 791

Schleicher, D.C., A'Hearn, M.F.: 1982, *Astrophys. J.* **268**, 864

Van Dishoek, E.F., Dalgarno, A.: 1984, *Icarus* **59**, 305

Astron. Astrophys. 187, 435–437 (1987)

ASTRONOMY AND ASTROPHYSICS

The D/H ratio in water from comet P/Halley

P. Eberhardt[1], U. Dolder[1], W. Schulte[1], D. Krankowsky[2], P. Lämmerzahl[2], J.H. Hoffman[3], R.R. Hodges[3], J.J. Berthelier[4], and J.M. Illiano[4]

[1] Physikalisches Institut, University of Bern, Sidlerstrasse 5, CH-3012 Bern, Switzerland
[2] Max-Planck-Institut für Kernphysik, Postfach 103980, D-6900 Heidelberg, Federal Republic of Germany
[3] University of Texas at Dallas, M.S. FO 2.2, P.O. Box 688, Richardson, TX 75080, USA
[4] Centre de Recherches en Physique de l'Environnement Terrestre et Planétaire, CNET-CNSR, 4, avenue de Neptune, F-94107 Saint-Maur des Fossés, France

Received February 14, accepted March 27, 1987

Summary. The neutral gas mass spectrometer (NMS) on the Giotto spacecraft made neutral and ion composition measurements with a high mass resolution. From a preliminary evaluation of the ion data within the contact surface we obtain for the water in Halley an ^{18}O abundance of $^{18}O/^{16}O = 0.0023 \pm 0.0006$ and for the D/H ratio $0.6\,10^{-4} \leqq D/H \leqq 4.8\,10^{-4}$. The cometary ^{18}O abundance is within error limits identical to the terrestrial one. The D/H ratio in Halley is definitely not compatible with the D/H in gaseous hydrogen of the protosolar nebula or the Jovian and Saturnian atmospheres. It is in the range observed for hydrogen in solar system objects which acquired their hydrogen as part of volatile molecules, e.g. as ices.

Key words: deuterium – hydrogen – oxygen – isotopes

Of all stable isotopes, the pair deuterium-hydrogen has the largest relative mass difference. Therefore, hydrogen is the element for which the largest non-nuclear isotopic effects can be expected in natural systems. This is borne out by the actual observations of D/H ratios in such varied objects as interstellar molecules, meteorites and planetary atmospheres. Enrichment factors larger than 10^3 have been observed (Geiss and Reeves, 1981). The D/H ratio in cometary volatiles can thus serve as a tracer for the origin of the cometary volatiles and the formation of comets.

The in situ measurement of the isotopic composition of the volatile light elements (H, C, N, O) in a cometary coma with a mass spectrometer is difficult, because the low abundance isotopes (D, ^{13}C, ^{15}N, ^{17}O, ^{18}O) are always heavier than the abundant ones (^1H, ^{12}C, ^{14}N, ^{16}O) and can thus be masked by hydrids (H_2, CH, NH, OH, H_2O). Only for the ion H_3O^+ is the interference for the isotopic H_2DO^+ limited to a few possible species and corrections can be applied for these with sufficient accuracy. In this paper we report a preliminary evaluation of the $^{18}O/^{16}O$ and the D/H ratios in water of comet Halley (cf. also Eberhardt et al., 1986). The results are based on the ion measurements of the NMS within the contact surface.

A description of the NMS has been published (Krankowsky et al., 1986). The instrument had two independent analyzers: the M-analyzer and the E-analyzer. The data in this paper are essentially based on the M-analyzer, a double focussing mass spectrometer covering the mass range from 1 amu/e to 37 amu/e. The half peak width at the 1% level is ~0.5 amu essentially independent of the mass, and the abundance sensitivity for adjacent integer mass numbers is $>10^3$. In the ion mode of the instrument the electron beam used in the neutral mode for the ionization of the gas was suppressed and the ion source potentials were reconfigured to allow the cometary ions access to the analyzer region. A focal plane detector allowed the simultaneous measurement of all masses over the whole mass range of the instrument. The dynamic range of the detector at a fixed voltage of the microchannelplates (MCP) was $>10^3$. To increase this dynamic range every second ion spectrum was recorded with an MCP gain increased by approximately a factor of 30.

Figure 1 shows the mass 17 amu/e to 21 amu/e region of an ion spectrum recorded inside the contact surface. The mass peaks 20^+ and 21^+ are clearly resolved despite the fact that their abundance is only of the order of 10^{-3} of the dominant 19^+ peak. An evaluation of their intensities is thus possible without exact knowledge of the actual peak shape. The different ion species which can contribute to the masses 18^+ to 21^+ are compiled in Table 1. For 18^+ and 19^+ only the main contributing ions are shown, for 20^+ and 21^+ the list is comprehensive, including all volatiles and their isotopes. Doubly charged ions are not considered. The solar system abundance of F is low ($F/S = 1.64\,10^{-3}$, Anders and Ebihara, 1982) and the ionization potentials of F and HF are high (17.4 eV and 16.0 eV, respectively). We thus do not expect any appreciable contribution of F^+ and HF^+ ions to masses 19 and 20 amu/e. Neon is a very volatile element (condensation temperature at 10^{-6} bar is 8 K, for H_2 it is 7 K) and expected to be highly depleted in comets. Furthermore, Ne has a very high ionization potential of 21.5 eV and we do not expect any significant Ne^+ contribution. The contribution from ions containing three or more isotopes with low abundance is always negligible and they are not listed in Table 1. The same holds for ions with two low abundance isotopes, except if e.g. D would be

Send offprint requests to: P. Eberhardt

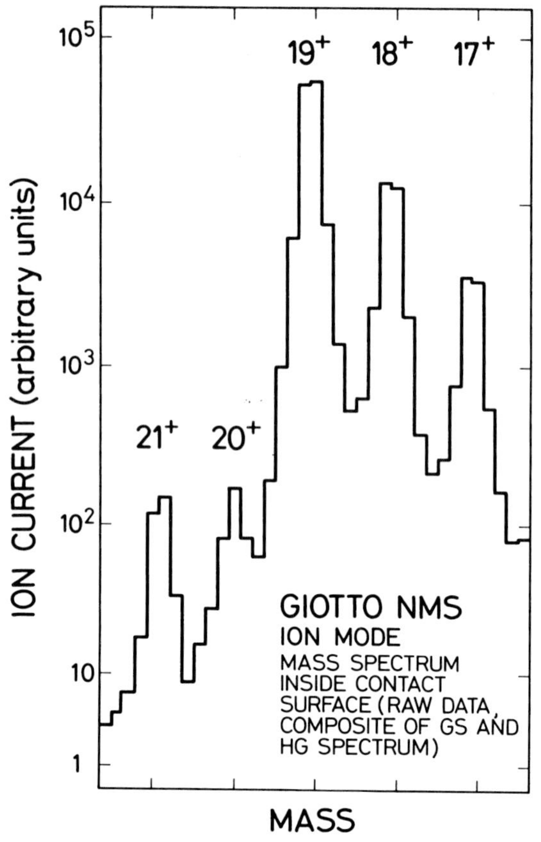

Fig. 1. Ion mass spectrum obtained with the M-analyzer of the NMS inside the contact surface. The 21.5 to 19.5 amu/e mass range is from a spectrum recorded in the high gain (HG) mode, the remainder from a gain step (GS) spectrum (Krankowsky et al., 1986). Corrections for gain variations between individual pixels and other instrumental discrimination have not been applied

extremely enriched (enrichment factor >1000) in a species such as NH_3. As we have no evidence in our data for such extreme enrichments we can also neglect the contributions from ions containing two minor isotopes (e.g. $D_2{}^{16}O$).

The $21^+/19^+$ ratio gives directly the $^{18}O/^{16}O$ ratio in the water. From the preliminary evaluation of the ion spectra obtained within the contact surface we obtain

$$^{18}O/^{16}O = 0.0023 \pm 0.0006 . \quad (1)$$

Within the experimental errors this ratio is identical with the terrestrial $^{18}O/^{16}O$ ratio of 0.00205. The ^{16}O abundance variations observed in meteorites are less than 6% (Clayton et al., 1977) and no meaningful comparison of the meteoritic with the cometary $^{18}O/^{16}O$ is yet possible. For determining the $H_3{}^{17}O^+$ contribution on 20^+ we assumed therefore terrestrial ^{16}O, ^{17}O, and ^{18}O abundances. The correction for $H_3{}^{17}O^+$ was calculated both from $H_3{}^{18}O^+$ and $H_3{}^{16}O^+$ to take the uncertainty in the cometary $^{18}O/^{16}O$ ratio into account.

A difficulty in making the $H_2{}^{18}O^+$ correction is the unknown mixing ratio of $H_2{}^{16}O^+$ and $^{14}NH_4^+$ ions on mass 18^+. Depending on the NH_3 abundance either $^{14}NH_4^+$ or H_2O^+ could be the dominant ion on mass 18 amu/e in the distance range from 1500 to 4500 km (Huebner and Giguere, 1980; Ip, 1986). For this preliminary evaluation we have assumed as limiting cases that the $^{14}NH_4^+/H_2{}^{16}O^+$ ratio is either zero or infinity and included the resulting uncertainty in the error limits for the D/H ratio. With these assumptions we obtain for the D/H in water from comet Halley

$$0.6\,10^{-4} \leq D/H \leq 4.8\,10^{-4} . \quad (2)$$

This corresponds to an enrichment factor V of 3 to 24 relative to the protosolar $D/H = (2 \pm 1)\,10^{-5}$ (Geiss and Bochsler, 1981). Our upper limit of $4.8\,10^{-4}$ is a factor of 8 lower than the limit obtained by A'Hearn et al. (1985) for the OD/OH ratio in comet Austin 1982VI from UV spectra taken with the IUE satellite. For comet Halley Schleicher et al. (1986) report an upper limit of $OD/OH \leq 4\,10^{-4}$ from IUE observations. This limit is essentially in agreement with our upper limit. Of course, these authors could not report a lower limit for the deuterium abundance in comet Halley. For the further discussion we will therefore use the D/H range obtained from our in-situ mass spectrometric measurements.

Table 1. Ions contributing to masses 18 to 21 amu/e. For 18^+ and 19^+ the ions containing low abundance isotopes are not listed

Mass	Number of low abundance isotopes in molecule		
	0	1	2
18^+	$H_2{}^{16}O^+$		
	$^{14}NH_4{}^+$		
19^+	$H_3{}^{16}O^+$		
	F^+		
20^+	HF^+	$H_2{}^{18}O^+$	$HD^{17}O^+$
	$^{20}Ne^+$	$H_3{}^{17}O^+$	$D_2{}^{16}O^+$
		$H_2D^{16}O^+$	$^{15}NH_3D^+$
21^+	$^{21}Ne^+$	$H_3{}^{18}O^+$	$HD^{18}O^+$
		DF^+	$H_2D^{17}O^+$
			$HD_2{}^{16}O^+$

Figure 2 gives an overview of the deuterium abundance in different solar system and galactic reservoirs. The very high D enrichments observed in different molecules in dark interstellar clouds are believed to be the result of fractionation in ion-molecule reactions (Watson, 1974). It is to be expected that D is thus enriched in icy interstellar grains or in the proposed volatile mantles of the grains (Greenberg, 1982). Additional D enrichment might also have occurred in the protosolar nebula, either by ion-molecule reactions or in low temperature chemical reactions. For a detailed discussion cf. Geiss and Reeves (1981).

Due to incomplete data evaluation, the uncertainty in the cometary D/H is, at present, quite substantial, Nevertheless, some conclusions are quite evident (cf. Fig. 2):

– Deuterium is distinctly enriched in Halley water ice compared with protosolar hydrogen and with hydrogen acccreted in gaseous form from protosolar nebula (i.e. Jupiter and Saturn).

– The D/H ratio in Halley water ice is compatible with other solar system hydrogen that was accreted as part of volatile molecules, e.g. H_2O, CH_4 ices etc. as in Titan, Meteorites and the Earth.

– The Earth, Meteorites, Titan and comet Halley could have acquired their hydrogen from the same region of the protosolar nebula, that is they could have formed in the same heliocentric distance range.

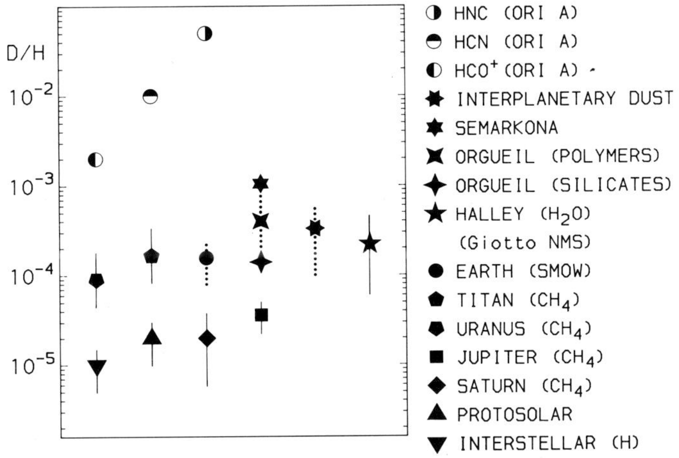

Fig. 2. Deuterium abundances in comet Halley compared to different solar system and galactic reservoirs. The dotted lines indicate the range of D/H ratios observed in separated components of meteorites, in different terrestrial reservoirs, and in interplanetary particles. The data are from different sources, for details see Penzias (1979); Snell and Wootten (1979); Kolodny et al. (1980); Geiss and Reeves (1981); Drossart et al. (1982); Knacke et al. (1982); Kunde et al. (1982); Vidal-Madjar et al. (1983); Yang and Epstein (1983); Pillinger (1984); Bjoraker (1985); McKeegan et al. (1985); Owen et al. (1986)

– The D/H ratio in Halley water ice does not contradict theories, as discussed in Anders (1968), Anders and Owen (1977), Sill and Wilkening (1978), that the earth and other terrestrial planets acquired their volatiles from a veneer of volatile rich materials of e.g. cometary origin.

– A cometary origin of carbonaceous chondrites is not excluded by the observed D/H ratio.

– The D/H ratios observed in interplanetary dust particles (McKeegan et al., 1985) are compatible with a cometary origin, but cannot be considered a proof.

We expect that the more detailed data analysis taking the measurements outside the contact surface into account will result in a much more precise value for the cometary D/H. This should allow us to draw more detailed conclusions on the origin of comets and their relationship with the other solid bodies in the solar system.

Acknowledgements. We would like to thank R. Wright for graphics and B. Grose for typing. This research was supported by the Swiss National Science Foundation (grants 2.034–0.81 – 2.825.85); the German Bundesminister für Forschung und Technologie (grants 010F0519, 010F0628 and 010F85120); CNES (grants 81, 82/244 – 83/211 – 84, 85/1211); and NASA headquarters (contract NASW-3576).

References

A'Hearn, M.F., Schleicher, D.G., West, R.A.: 1985, *Astrophys. J.* **297**, 826

Anders, E.: 1968, *Acc. Chem. Res.* **1**, 289

Anders, E., Ebihara, M.: 1982, *Geochim. Cosmochim. Acta* **46**, 2363

Anders, E., Owen, T.: 1977, *Science* **198**, 453

Bjoraker, G.L.: 1985, thesis, Univ. Arizona (Tucson)

Clayton, R.N., Onuma, N., Grossman, L., Mayeda, T.K.: 1977, *Earth Planet. Sci. Letters* **34**, 209

Drossart, P., Encrenaz, T., Kunde, V., Hanel, R., Combes, M.: 1982, *Icarus* **49**, 416

Eberhardt, P., Dolder, U., Schulte, W., Krankowsky, D., Lämmerzahl, P., Hoffman, J.H., Hodges, R.R., Berthelier, J.J., Illiano, J.M.: 1986, 20th ESLAB Symp *ESA SP-250*, Vol. I, 539

Geiss, J., Bochsler, P.: 1981, in *Solar Wind Four*, ed. H. Rosenbauer, Proc. 4th Solar Wind Conf., Burghausen, 1978, Max-Planck-Institut für Aeronomie, Katlenburg-Lindau, FRG, Rep. No. MPAE-W-100-81-31, 403

Geiss, J., Reeves, H.: 1981, *Astron. Astrophys.* **93**, 189

Greenberg, J.M.: 1982, in *Comets*, ed. L.L. Wilkening, Univ. Arizona Press, Tucson, p. 131

Huebner, W.F., Giguere, P.T.: 1980, *Astrophys. J.* **238**, 753

Ip, W.H.: 1986, *Adv. Space Res.* **5**, 233

Knacke, R.F., Kim, S.J., Ridgway, S.T., Tokunaga, A.T.: 1982, *Astrophys. J.* **262**, 388

Kolodny, Y., Kerridge, J.F., Kaplan, I.R.: 1980, *Earth Planet. Sci. Letters* **46**, 149

Krankowsky, D., Lämmerzahl, P., Dörflinger, D., Herrwerth, I., Stubbemann, U., Woweries, J., Eberhardt, P., Dolder, U., Fischer, J., Herrmann, U., Jungck, M., Meier, F.O., Schulte, W., Berthelier, J.J., Illiano, J.M., Godefroy, M., Gogly, G., Thévenet, P., Hoffman, J.H., Hodges, R.R., Wright, W.W.: 1986, *ESA SP-1077*, 109

Kunde, V.G., Hanel, R., Maguire, W., Gautier, D., Baluteau, J.P., Marten, A., Chedin, A., Husson, N., Scott, N.: 1982, *Astrophys. J.* **263**, 443

McKeegan, K.D., Walker, R.M., Zinner, E.: 1985, *Geochim. Cosmochim. Acta* **49**, 1971

Owen, T., Lutz, B.L., de Bergh, C.: 1986, *Nature* **320**, 244

Penzias, A.A.: 1979, *Astrophys. J.* **228**, 430

Pillinger, C.T.: 1984, *Geochim. Cosmochim. Acta* **48**, 2739

Schleicher, D.G., A'Hearn, M.F., and the NASA and ESA IUE teams for comet Halley: 1986, *Proc. IUE conference* (London)

Sill, G.T., Wilkening, L.L.: 1978, *Icarus* **33**, 13

Snell, R.L., Wootten, A.: 1979, *Astrophys. J.* **228**, 748

Vidal-Madjar, A., Laurent, C., Gry, C., Bruston, P., Ferlet, R., York, D.G.: 1983, *Astron. Astrophys.* **120**, 58

Watson, W.D.: 1974, *Astrophys. J.* **188**, 35

Yang, J., Epstein, S.: 1983, *Geochim. Cosmochim. Acta* **47**, 2199

Astron. Astrophys. 187, 438–450 (1987)

ASTRONOMY
AND
ASTROPHYSICS

Improved gas-kinetic treatment of cometary water sublimation and recondensation: application to comet P/Halley

J. F. Crifo

L.P.S.P.–C.N.R.S., B.P. No 10, F-91371 Verrières le Buisson Cedex, France

Received February 2, accepted May 11, 1987

Summary. Two effects associated with the solid-gas phase transition assumed to control cometary activity, i.e. nucleus ice sublimation, and homogeneous recondensation in the fluid expansion region are reconsidered using appropriate gasdynamic methods. (1) Physically consistent expressions are used for the initial boundary conditions at the base of the fluid expansion region. It is found that this reduces the gas velocities by 5–10% as compared with currently accepted assumptions, and the terminal dust velocities by about the same amount, the reduction increasing with the grain size. The nucleus surface temperature is affected by less than 1%. Finally, the theory leads to a rogorous formula for the momentum loss coefficient needed in the computations of the non-gravitational forces exerted on the nucleus. (2) The classical homogeneous nucleation theory is applied taking into account the heating of the gas by dust and by photo-dissociation products, and using rigorous boundary conditions at the nucleus surface. It is found that water vapour expands in an initially supersaturated state, contrary to previous assumption. However, due to the gas heating effects, the nucleation rate is smaller than in the case of pure adiabatic expansion. The magnitude of the perturbation of the gas parameters resulting from nucleation is found to depend not only upon the total gas production rate, but also upon the total dust production rate and upon the dust grain size and chemical nature. In all cases, the optical emission from the halo of nucleated ice grains is computed to be several orders of magnitude fainter than the emission from the dust, making the detectability of this halo from Earth very implausible.

The limitations of these results are discussed and suggestions for future improvements are given.

Key words: comets – nucleus sublimation – water recondensation

1. Introduction

Following the initial suggestion of Whipple (1950), solar controlled sublimation of ice has been accepted as the key process governing cometary activity. Elaborate models aimed at predicting the thermal budget of the nucleus (e.g. Weissman and Kieffer, 1984; Squyres et al., 1985; Herman and Podolak, 1985), the magnitude of the non-gravitational forces (e.g. Fanale and Salvail, 1986; Rickman and Froeschlé, 1983), and the hydrodynamic expansion of the sublimating gas-dust mixtures (Hellmich, 1979; Marconi and Mendis, 1983; Crovisier, 1984; Gombosi et al., 1985; Crifo, 1986a, b; Kitamura, 1986) have been developed, which in general incorporate as a key building block formulae which relate together the nucleus surface temperature, the local rate of gas production, and the initial gas parameters.

The works cited above can be classified in two groups according to the way in which they assume that the ice sublimation occurs. One assumption is that bare ice with inclusions of grains is directly exposed to solar light; the other assumes that the sublimating level is buried below a mantle of brittle outgassed grains (see the recent review by Gombosi et al., 1986). We consider here the first instance. This is not meant to argue against the existence of the second process, but simply to advance the modelling of the first one, to permit convincing comparative analysis of experimental data.

Coming back to the works based on "direct" or "surface" sublimation, one can summarize the assumptions generally accepted as follows: it is assumed (1) that the local gas sublimation rate is given by the Hertz-Knudsen formula, as if the coma was collisionless, (2) that the gas temperature and pressure are equal to the ice temperature and equilibrium sublimation pressure, as if the vapour was not in vacuum expansion, (3) that the vapour expands with an initial velocity which is sometimes derived by hydrodynamic methods, and sometimes guessed: see Delsemme and Miller (1971) whom many authors cite in support for the above practices. These assumptions do not comply with the gas dynamic theory of fast sublimation, which is in use in the other fields of physics where the problem is encountered (e.g. Anisimov, 1958; Luikov, 1971; Ytrehus, 1975; Cercignani, 1981). One noticeable exception to the previous practices is the work of Shul'man (1970, 1975), but it is significant that it has not been given attention in the cometary community. The assumptions mentioned above certainly give correct orders of magnitude, and it is understandable that they remain in use when measurements are not available that would permit detailed verifications of model predictions. However, the post Halley era offers a more favorable context. For instance, the development of high resolution spectroscopy of primary molecules (Larson et al., 1986) opens the possibility of checking the gas temperature throughout the coma (Crovisier, 1984; Bockelée Morvan, 1987). It will also be possible to test models of Comet Halley non-gravitational forces with unprecedented accuracy, based on our knowledge of its nucleus and coma and of its orbit.

We devote the first part of this paper to an application of the gas kinetic theory to the sublimation of cometary ice. We thus update the work of Shul'man, and apply our formalism to the case of Comet Halley at the time of the flyby missions.

We devote the second part of this study to a different problem, which is closely related to the first one since it is the reverse process of gas phase recondensation into molecular clusters (very tiny ice grains). This phenomenon was only recently considered in the cometary context, first by Yamamoto and Ashihara (1985), then by Ip (1986). The importance of accurately assessing this effect is two-fold. First, one can ask whether the nucleated ice grains could be observed more easily than their parent water molecules. Second, the release of latent heat in the process may alter severely the gas temperature which, as already mentioned, will become a measurable parameter in the future. Recently, Kitamura and Yamamoto (1986) gave a very comprehensive treatment of the nucleation, growth, and ultimate sublimation of the clusters, based on an hydrodynamic model including their interaction with the gas. Yamamoto and Ashihara (1985) computed the spectrum of the cluster emission and suggested that it could be detectable. These authors assumed an adiabatic water atmosphere and initial undersaturation of the vapour over the ice. In fact, comets and comet Halley in particular, are heated by convective and radiative exchanges with dust, and by absorption of energy from fast photodissociation products. Besides, as we will see, a correct gaskinetic treatment of the solid/gas interface leads to the conclusion that the vapour is initially supersaturated. Here, we compute the hydrodynamic expansion of a dusty and photodissociating water coma with correct initial boundary conditions, and taking into account homogeneous nucleation. We compare the emission of the nucleated ice clusters with the emission of the dust. We show that convective exchanges between the dust and the gas as well as the gas heating by photodissociation products reduce the effects of nucleation in proportions which depend upon the properties of the dust. As to the optical emission of the clusters, it is found to be several orders of magnitude below the intensity of the dust emission, so that its detectability will probably not be easier than that of water vapour itself, or than that of an eventual halo of "primary" ice grains.

A preliminary presentation of part of the results obtained in this study was made in Crifo (1986b) and Crifo (1986c).

2. Determination of the water vapour parameters at the base of the fluid expansion region

In this section, we assume that the nucleus surface in an active region of the nucleus of a comet consists in pure ice.

2.1. *Assumptions in use in the cometary literature*

To compute the heat balance, the rate of gas release, of momentum loss, etc. of a nucleus surface element, one needs to specify the sublimating gas parameters above the surface. One can choose the initial gas pressure, p_g^0, temperature, T_g^0, and expansion velocity V_g^0. Or, alternatively, the dimensionless numbers:

$$M^0 = V_g^0 / \sqrt{\gamma k T_g^0 / m_g}, \tag{1}$$

$$T = T_g^0 / T_n, \tag{2}$$

$$\Pi = p_g^0 / p_n, \tag{3}$$

where T_n is the nucleus temperature and p_n a convenient reference pressure. Here, where we consider the surface to consist in pure ice, p_n will be taken as the equilibrium sublimation pressure of the ice, $p_s(T_n)$. The remaining quantities γ and m_g are, respectively, the gas ratio of specific heats and the gas molecule mass, and k is the Boltzman constant.

Using the perfect gas law, the local surface sublimation rate Z (molecule cm^{-2} s^{-1}), is given by:

$$Z = \frac{p_g^0 V_g^0}{k T_g^0} = \frac{\Pi M_0 \sqrt{\gamma}}{\sqrt{T}} \frac{p_s(T_n)}{\sqrt{m_g k T_n}} = \left(\Pi M_0 \sqrt{\frac{2\pi\gamma}{T}}\right) Z_{HK}(T_n), \tag{4}$$

where the Hertz-Knudsen sublimation rate

$$Z_{HK} = \frac{p_s(T_n)}{\sqrt{2\pi m_g k T_n}} \tag{5}$$

has been introduced. Similarly, the rate of momentum loss per unit surface element is:

$$\frac{dF}{dA} = Z m_g V_g^0 = (\Pi M_0 \gamma)\, p_s(T_n). \tag{6}$$

We have already mentioned and will not repeat further that Shul'man (1970, 1975) has computed Π and T (as a function of M_0) by a method slightly different from the one that will be used here, but essentially correct. We now review the choices made by the other authors.

2.1.1. Gas temperature

$T = 1$ is always assumed. In fact it is given as an evidence, not as an assumption.

2.1.2. Gas velocity

Here, one must distinguish between authors who solve the hydrodynamic equations governing the gas expansion, and the others. The first group, after having set Π and T, compute M_0 from the requirement of having a "well behaved" solution. This is of course a quite correct procedure, and the resulting initial velocity would be correct if Π and T were given their gas-dynamically correct value (see next section). The second group makes an assumption for V_g^0, after having assumed $T = 1$. This assumption has the form

$$V_g^0 = \eta \left(\frac{8 k T_n}{\pi m_g}\right)^{1/2}, \tag{7}$$

where η is set equal to 0.6 (Delsemme and Miller, 1971), or is left unspecified (Rickman and Froeschlé, 1983 with $0 \leqq \eta \leqq 1$). This of course conceals an assumption for M_0, which can be computed from (7) and (1). Note that η is a local parameter, and therefore should in general vary over the surface of the nucleus, as T_n and M_0.

2.1.3. Gas pressure

In general, rather than specifying Π, cometary authors assume $Z/Z_{HK} = 1$ (Delsemme and Miller, 1971; Marconi and Mendis, 1983; Weissman and Kieffer, 1984; Herman and Podolak, 1985; Squyres et al., 1985; Fanale and Salvail, 1986; Rickman and Froeschlé, 1983; Kitamura and Yamamoto, 1986). From relation (4), the implied assumption for Π can be found.

For the sake of clarity, we must note that some works do not make a complete set of assumptions. For instance those interested in the nucleus thermal history, neglecting coma effects, do not

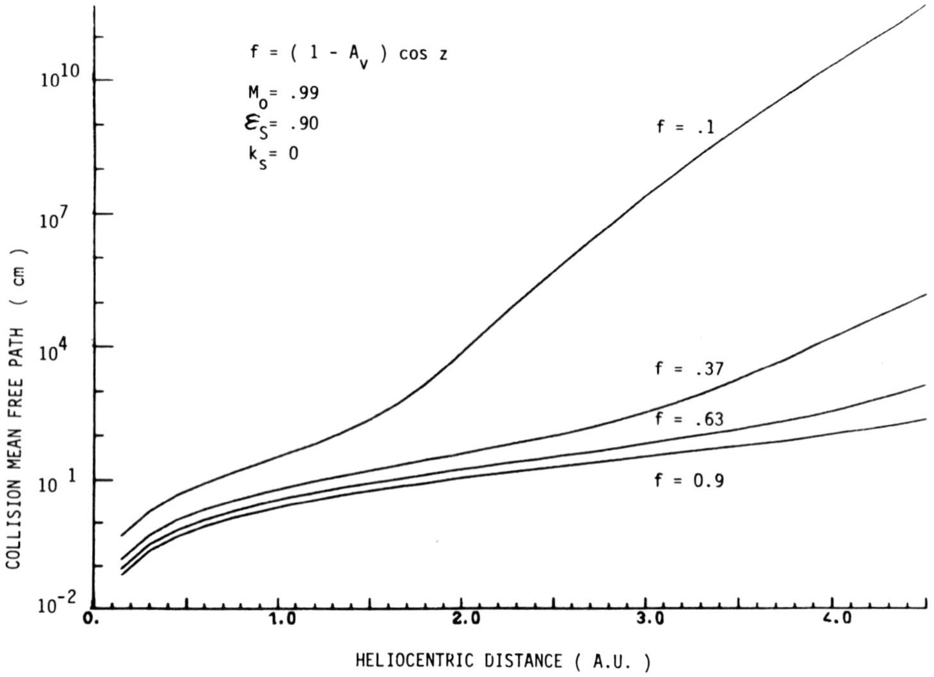

Fig. 1. Mean free path in the vicinity of a sublimating ice surface. The water vapour free path is plotted as a function of the ice heliocentric distance. The computation takes into account the difference in gas and surface temperature and the dependence of the gas collision section on temperature. Note that these results are applicable only if the ice transverse dimensions are much greater than the computed free paths. A_v is the surface albedo in the visible, z the solar zenith angle, ε_s the infrared emissivity, k_s the thermal conductivity, M_0 the vapour initial Mach number

need assumptions on V_g^0 (e.g. Herman and Podolak, 1985), while those strictly interested in coma hydrodynamics without looking for consistency at the nucleus surface do not need a relation between Z and T_g, and specify them independently (e.g. Crovisier, 1984).

2.2. *Validity of the preceding assumptions*

The assumption that Z equals Z_{HK} is valid to the extent that no molecule emitted by the ice returns to it. This obviously requires that the molecule mean free path against collisions, Λ_{gg}, be much larger than the characteristic size of the nucleus, R_n otherwise a large fraction of the molecules backscattered during their first collisions will return to the ice and recondense.

Using the collision cross section compiled by Crifo, 1986b, and the exact theory of sublimation presented below in Sect. 2.3, one obtains the variation of Λ_{gg} with distance to the Sun shown in Fig. 1. If one requires, for instance, that $R_n \leqq 10\,\Lambda_{gg}$, one sees that the free molecular formulae are applicable to a kilometer size nucleus only beyond 2.7 AU.

When this condition is not met, one has to write:

$$Z = Z_{\mathrm{HK}} - Z_{\mathrm{RTN}} = Z_{\mathrm{HK}}(1-\alpha), \tag{8}$$

where Z_{RTN} is a return flux to the ice resulting from the collisions, and α a parameter introduced here for later convenience. α can be expressed as a function of M_0, Π and T using Eq. (4).

It is less evident to convince oneself that the assumption $T = 1$ is incorrect. First, one must realize that the vapour cannot be in microscopic equilibrium close to the ice, in other words its distribution function f is not a Maxwell-Boltzman distribution. Consider a point close to the ice (with respect to Λ_{gg}): the molecules leaving the ice are represented by a "half-space" Maxwellian with temperature T_n, and with density $p_s(T_n)/kT_n$, according to the Hertz-Knudsen theory. The molecules returning to the ice are represented by a "half space" function f_{RTN} which depends upon the gas conditions at large distances (compared to Λ_{gg}). Thus we have, at the ice surface:

$$f \equiv f_s = f_{\mathrm{HK}}(\mathrm{T_n}) + f_{\mathrm{RTN}}(M_0, \Pi, T, T_n). \tag{9}$$

This form for f is obviously not a maxwellian.

As we move away from the ice, the form of f changes under the effect of collisions until finally, at large distance (compared to Λ_{gg}) a local (drifting) Maxwellian f_0 is reached. We can now define rigorously the "initial gas parameters" T_g^0, p_g^0, V_g^0 as those prevailing at the distance where $f = f_0$.

If Λ_{gg} is much larger than R_n, one has $f_{\mathrm{RTN}} = 0$, while at sublimation equilibrium ($V_g^0 = 0$) f_{RTN} is just the "mirror symmetric" of f_{HK} (in velocity space). The sublimation flux is the first (velocity) moment of f, thus (9) shows why it is given by (8). The nucleus momentum loss is the second moment of f and therefore (9) shows that it is not given by the simple expression $Z_{\mathrm{HK}}\, m_g V_g^0$, but by (6). And finally the gas temperature is the second moment of f and thus will not in general be equal to T_n since f is not the required Maxwellian function at T_n.

2.3. *Computation of Π and T from an elementary gaskinetic theory*

A fully comprehensive approach of the problem requires solving a Boltzman equation which f satisfies, using boundary conditions over the ice and at infinity (see e.g. Ytrehus, 1975; Cercignani, 1981). However, if one is interested only in Π and T (not in the gas properties within the disequilibrium zone), it is possible to avoid the integration by using a "moment method", as follows. One postulates for f_{RTN} a simple but physically acceptable form at the ice surface:

$$f_{\mathrm{RTN}}(v_g^{\perp}, v_g^{\parallel}) = \begin{cases} 0 & \text{if } v_g^{\perp} \geqq 0 \\ \beta^- f_0(v_g^{\perp}, v_g^{\parallel}) & \text{if } v_g^{\perp} \leqq 0 \end{cases}, \tag{10}$$

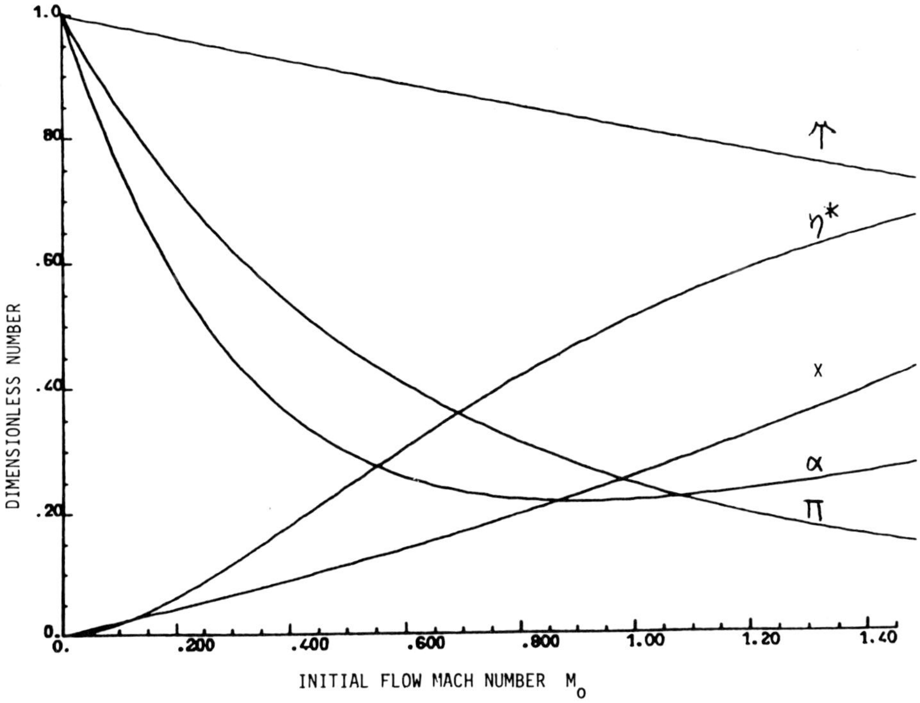

Fig. 2. Boundary layer parameters. The ice surface is assumed at 200 K. The curves show the variation with vapour Mach number M_0 of the following dimensionless parameters: Π and T defined by Eqs. (2) and (3), α defined by (8), η^* defined by (18) and $X = 0.1 \log_{10}(S_S)$ with S_S defined by (14)

where $v_g^\perp, v_g^\parallel$ are the components of the molecule velocity perpendicular to, and parallel to the surface of the ice, and f_0 the drifting maxwellian prevailing at the base of the expanding fluid:

$$f_0 = \frac{(p_g^0/kT_g^0)}{(2\pi kT_g^0/m_g)^{3/2}} \exp\left\{-\left[\frac{(v_g^\perp - V_g^0)^2 + v_g^{\parallel 2}}{2kT_g^0/m_g}\right]\right\}. \tag{11}$$

β^- is a dimensionless parameter which is to be determined. There are now four unknown parameters (Π, T, M_0 and β^-). Three relations between them and T_n can be obtained by writing that mass, momentum and energy ("MME") are preserved accross the disequilibrium region (since they are collisional invariants):

$$\{\mathrm{MME}\}_S = \{\mathrm{MME}\}_{HK} - \{\mathrm{MME}\}_{RTN} = \{\mathrm{MME}\}_0, \tag{12}$$

where the subscripts have the same meaning as when labelling the distribution functions. One is thus left with one free parameter, for which we choose M_0 since it has to be found by solving the hydrodynamic expansion of the vapour in the coma. Physically, the fact that there is one free parameter means that there can be an infinity of different sublimation regimes associated with a given T_n, a fact that current experience tells us indeed.

We do not reproduce the explicit form of Eqs. (12): they can be found in many places, for instance in Cercignani (1981) or Ytrehus (1975). One obtains after elimination of β^-:

$$\sqrt{T} = -\frac{\sqrt{\pi}}{2}\left(S_M\frac{\gamma-1}{\gamma+1}\right) + \sqrt{1+\frac{1}{4}\left(S_M\frac{\gamma-1}{\gamma+1}\right)^2}, \tag{13a}$$

$$\Pi = \frac{1}{2} - S_M\sqrt{\frac{T}{\pi}} + \left[\left(S_M^2+\frac{1}{2}\right)\sqrt{T} - \frac{S_M}{2}\sqrt{\pi}\right] \cdot \operatorname{erfc}(S_M)\exp(S_M^2) \tag{13b}$$

in which

$$S_M = M_0\sqrt{\frac{\gamma}{2}} \tag{13c}$$

Formulae (13a)–(13c) provide a simple and physically consistant means of predicting Π and T once M_0 is known. Since the determination of M_0 from the downstream hydrodynamic equations requires prior knowledge of Π and T, the global problem can only be solved by an iteration procedure.

2.4. *Results*

Figure 2 shows the variation with M_0 of Π and T computed from (13a) to (13c) for vibrationally relaxed H_2O ($\gamma = 4/3$, see Crovisier, 1984). Also shown is the ratio α of the backscattered flux to the direct Hertz-Knudsen flux, and the gas supersaturation ratio:

$$S_S = p_g/p_s(T_g). \tag{14}$$

Integration of the hydrodynamic equations for comets with conditions more or less similar to comet Halley at the time of the flyby missions indicate that initial flow Mach number lie in the range $0.75 \leqq M_0 \leqq 1$ (Probstein, 1968; Hellmich, 1979; Marconi and Mendis, 1983; Gombosi, 1985; Crifo, 1986a, b). Therefore, to comply with correct gaskinetic considerations, we recommend to adopt the values:

$$\alpha \simeq 0.25, \quad \Pi \simeq 0.30, \quad T \simeq 0.85. \tag{15}$$

We now evaluate the consequence of Eq. (15).

2.4.1. Nucleus temperature

The sublimation energy loss E_{vap} which enters into the surface energy budget equation must be written:

$$E_{vap} = ZL_V = (1-\alpha)\,Z_{HK}\,L_V = (1-\alpha)\,(E_{vap})_{HK} \tag{16}$$

(where L_V is the molecular latent heat of sublimation). We have solved surface energy budget equations for an heliocentric distance

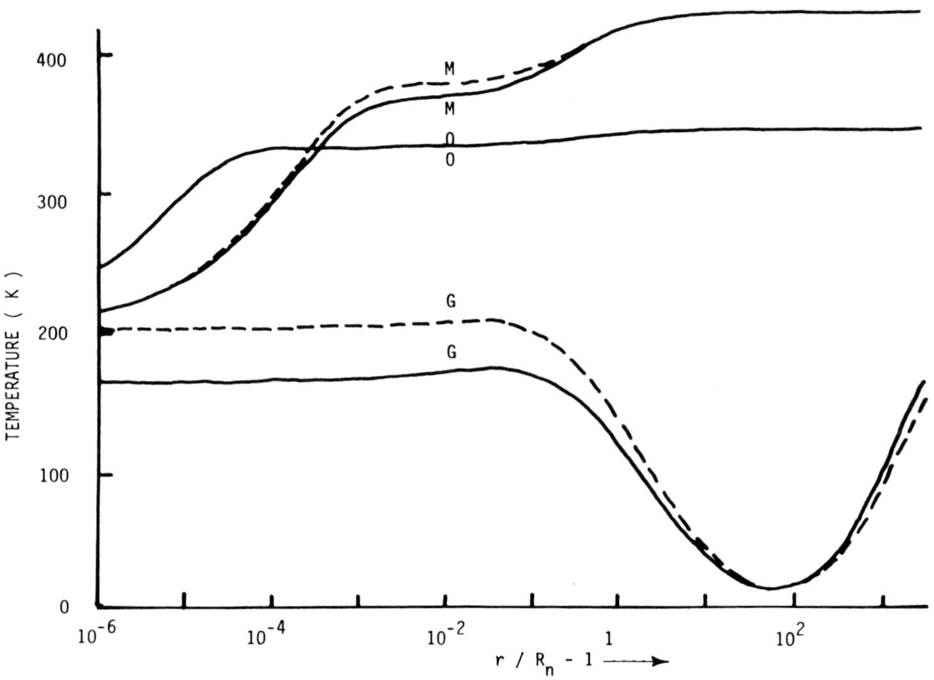

Fig. 3. Effect of the ice vapour boundary layer on the temperatures in the coma. Plotted as a function of relative altitude above the comet nucleus are: the water vapour temperature (label G), the olivine grain temperature (label O) and the magnetite grain temperature) label M). The dashed curves are computed ignoring the boundary layer, and the full curves with correct boundary layer parameters. The model parameters are given in Sect. 2.4.3

equal to 0.8 AU and found that the shift in T_n is $\Delta T_n \simeq +1.5$ K (near $T_n \simeq 200$ K). This is perhaps not very significant for nucleus studies (the heat input into the nucleus will be increased by less than 1 %).

2.4.2. Non-gravitational forces

Following the practices cited in Sects. 2.1.2 and 2.1.3, we write the rate of momentum loss per unit surface element under the form:

$$\frac{dF}{dA} = \eta^* \left(\frac{8}{\pi} m_g k T_n\right)^{1/2} \tag{17}$$

and we take this equation as a definition of η^*. Using (6), (4) and (1) one obtains easily the value of η^*:

$$\eta^* = \left(\frac{\pi}{2}\right) \Pi M_0^2 \gamma \tag{18}$$

Figure 2 shows the variation of η^* with M_0. The range of possible values corresponding to the interval $0.75 \leqq M_0 \leqq 1$ is:

$$0.39 \leqq \eta^* \leqq 0.50 . \tag{19}$$

We note that this result does not depend upon the "quality" of the ice surface: any accidents with spatial dimensions less than Λ_{gg} ($\simeq 5$ cm, near 1 AU) are immaterial, and accidents greater than Λ_{gg} will be taken into account in the surface integration of (17): the hydrodynamic equations are local and provide η^* locally. Therefore it seems difficult to imagine configurations giving $\eta^* \simeq 1$ as Rickman and Froeschlé (1983) seemed once tempted to postulate. The value $\eta^* = 0.6$ suggested by Delsemme and Miller (1971) and adopted by several authors is somewhat too large and authors looking for accurate values of η^* should take them within the above indicated range. In any case, formulae (13b) and (18) will give the correct value once M_0 has been determined.

2.4.3. Gas and dust temperatures in the coma

For illustration purposes, we have solved a set of three-component hydrodynamic equations with, and without, Eqs. (13a) – (13c). In the second case, we used the usual (and incorrect) assumptions $T = 1$, $Z = Z_{HK}$. The local water production rate was assumed to be 9.95 10^{17} molecule cm^{-2} s^{-1}, the heliocentric distance 0.8 AU, the dust-to-gas mass ratio was 0.1 for olivine grains of 0.2 μm radius, and 0.2 for magnetite grains of 0.9 μm radius. Figure 3 shows the results: no significant difference is found in the radial dust temperature profiles, even very close to the nucleus. The radial gas temperature profiles are very different, in the immediate vicinity of the surface; higher in the coma, the two profiles come closer, under the effect of expansion cooling and photo-dissociative heating. The main consequence of the difference in low altitude profiles is the fact that the gas leaves in a supersaturated state ($S_S \simeq 10^3$) when conditions 13a–c are included, and undersaturated otherwise (as in Kitamura and Yamamoto, 1986). This paradoxical result, apparently first pointed out in Crifo (1986b) has been independently discussed by Hermans and Beenakker (1986).

2.4.4. Gas and dust velocities in the coma

Table 1 indicates for the two solutions the terminal dust velocities and the gas velocities at several altitudes. If we recall that the extension of the dust coma is proportional to the square of the grain ejection velocity, we realize that the effect is significant. The differences in gas velocities are appreciable at low altitudes, but are subsequently reduced by photodissociative heating. However, if minority molecules with short lifetimes are considered, or secondary molecules reminiscent of the inner coma velocities, then the difference of Table 1 may lead to observable effects.

2.5. Discussion

The simple assumptions (10–11) represent the minimum consistent approach to the problem. Evidently, the real gas distribution function cannot be as simple as that. But examination of the results obtained by authors who use different assumptions for f indicates that the numerical results for Π and T are quite

Table 1. Coma velocities. The coma composition is given in Sect. 2.4.4. The subsript (0) indicates values with boundary layer ignored, and the subscript (1) the values when the boundary layer is taken into account. The velocities are given in km s^{-1}

Altitude above nucleus $r/R_n - 1$	10^{-2}	10^{-1}	1	10	10^2	10^3
V_0 magnetite	0.045	0.12	0.26	0.336	0.336	0.336
V_1 magnetite	0.040	0.11	0.24	0.324	0.324	0.324
V_0 olivine	0.10	0.26	0.47	0.600	0.600	0.600
V_1 olivine	0.098	0.23	0.44	0.570	0.570	0.570
V_0 gas	0.357	0.43	0.65	0.86	0.93	0.99
V_1 gas	0.325	0.37	0.60	0.80	0.87	0.95

similar to those given by formulae 13a–c. This is for example the case for Shul'man's work. On the other hand, the conservation requirements made here [Eq. (12)] are necessary, but not sufficient. A more comprehensive treatment consists in deriving f as a function of distance to the ice surface, using the Boltzman equation (Anisimov, 1958; Luikov, 1971; Ytrehus, 1975; Cercignani, 1981). One then can assess the distance over which f differs markedly from its fluid region form f_0. The results is that $f=f_0$ beyond about 10 mean free paths above the ice as long as $M_0 \leqq 1$. If $M_0 \to 1$ or exceeds 1, the disequilibrium region where $f \neq f_0$ seems to extend without bounds! This means that there can probably be no steady state initially supersonic sublimation of pure ice in free space. In reality, the cometary ice is dusty, and this seems to imply $0.75 \leqq M_0 \leqq 1$ as already mentioned.

Since the dust acceleration occurs in much more than ten gas free paths (see e.g. Crifo, 1986b), the dust number density is high just above the ice, so that the treatment of Sect 2.3 must be generalized in the future to account for the MME exchanges between gas and dust in this region. This problem, i.e. correct kinetic treatment of the vacuum surface sublimation of dusty ice, should not be confused with the quite different approach to the "operation" of comets which postulates "sublimation" of the ice below a mantle across which the vapour diffuses into space (see the recent review by Gombosi et al., 1986). Authors which work out such models also obtain formulae for Z/Z_{HK}, T_g^0, and V_g^0. There is no reason why their formulae should coincide with (13a–c) since the physics involved is quite different. Nonetheless, Gombosi et al. (1986) state that their formulae proposed for an "ice-under-mantle" situation are applicable as well to a "bald nucleus" because the sublimation must take place "below a microscopic layer of evacuated dust, else this dust cannot be dragged away". One can show that this statement is incorrect by computing numerical values. Assume that a comet loses 5 ton/s of dust grains with radius $a_d = 0.25\,\mu\mathrm{m}$, from an active area of $10^{11}\,\mathrm{cm}^2$ at a temperature of 200 K. One computes easily that the grains, which are liberated with Brownian speeds (since they are bombarded by gas molecules from all directions – not only from below) will have, above the surface, a mean spacing $\delta \simeq 0.1$ mm $\gg a_d$. This is a worst case, as δ increase as $a_d^{3/2}$. But it already indicates that we deal with a dilute suspension of grains in the gas, not with gaseous diffusion across a dust blanket. Furthermore, δ is increased by a factor 4 within the first millimeter above the surface!

3. Water recondensation in the lower coma

The pressure p_g of a gas which undergoes strong vacuum expansion generally becomes quickly smaller than its equilibrium vaporization or sublimation pressure $p_s(T_g)$: the gas becomes unstable with respect to recondensation (see e.g. Wegener and Wu, 1977; Hagena, 1981). Molecular clusters form as a consequence of density fluctuations. According to the classical nucleation theory, those which have a certain "critical radius" a_* do not preserve this size, but grow by accretion of molecules. The latent heat released in the process is transferred by convection to the gas, and consequently S_S decreases. This evolution of the gas towards equilibrium ($S_S = 1$) is controled by the rate of formation J^* of the "critical" clusters (of radius a_*). The classical theory (see e.g. Wegener and Wu, 1977) gives:

$$J_* = \frac{1}{\varrho_c}\left(\frac{p_g}{kT_g}\right)^2 \sqrt{\frac{2\sigma m_g}{\pi}} \exp\left(-\frac{4\pi\sigma a_*^2}{3kT_g}\right) \mathrm{cm}^{-3}\,\mathrm{s}^{-1}, \tag{20a}$$

in which ϱ_c is the ice specific mass and σ the ice surface tension constant, and

$$a_* = \frac{2\sigma m_g}{\varrho_c kT_g \ln(S_S)}. \tag{20b}$$

Although this effect has been occasionally mentioned in the cometary litterature, the first comprehensive treatment has been published by the ISAS group (Yamamoto and Ashihara, 1985; Kitamura and Yamamoto, 1986) using parameters appropriate for a dust-free comet Halley. Their results indicate a strong heating of the gas (see also Ip, 1986) and a moderate gas depletion not exceeding 10% by number at $\simeq 50$ km above the nucleus.

Since S_S is an extremely fast varying function of T_g, however, it is not obvious that these results apply in a heated atmosphere (as the cited authors themselves recognize): hydrodynamic models of the expansion of cometary dust-gas mixtures show clearly an increase in T_g in the immediate vicinity of the nucleus, resulting from convective heat exchanges with the grains (Marconi and Mendis, 1983; Crifo, 1986a, b and this study Figs. 3 and 5). Another source of heat is the hot atoms produced by photodissociation of the gas. This source is most efficient at high altitude, but its effectiveness extends at lower and lower levels with increasing total gas production rate, Q. As a consequence of heating, the water supersaturation is reduced. On the other hand, because $T \leqq 1$, water leaves the ice with $S_S \geqq 1$ rather than $S_S \leqq 1$ as assumed by the ISAS group. Finally, we have discovered an omission in the work of Kitamura and Yamamoto (1986), also present in our preliminary publication (Crifo, 1986c): computing the evolution of ice clusters, one must allow for the variation of the cluster sublimation pressure with the cluster radius a_c. This variation is given by the Kelvin equation:

$$\frac{p_s^*(T_c, a_c)}{p_s(T_c)} = \exp\left(\frac{2\, m_g \sigma}{kT\, \varrho_c a_c}\right) \simeq (S_S)^{a_*/a_c}, \tag{20c}$$

where p_s^* is the sublimation pressure effective at radius a_c.

This effect is indeed the source of the existence of the critical radius a_* given by (20b). It is negligible for ice grains with radius in excess of $\simeq 300$ Å, but is a strong effect otherwise, as S_S is a large number. The effect of Eq. (20c) is to reduce the maximum size of the nucleated grains as well as the maximum altitude that they can reach.

In the following, the problem of nucleation is reexamined to take the presence of dust into account, and using gaskinetically correct boundary conditions over the ice and a correct size-dependant cluster sublimation pressure.

3.1. Hydrodynamic equations

We choose to represent the nucleated icy grains by one fluid component of a global multifluid model. In doing so, we have to neglect the size distribution of the icy grains: we assume that *all* icy grains have at an altitude r a common radius $a_c(r)$, temperature $T_c(r)$ and radial velocity $V_c(r)$, even though they are formed at different altitudes. For this set $a_c(r)$, $T_c(r)$, $V_c(r)$, we choose the a_c, T_c and V_c of those grains which are produced at the altitude r_* where J_* has its maximum. To compute these a_c, T_c, V_c, we assume that these grains are fully accommodated to the gas at $r = r_*$:

$$a_c(r_*) = a_*(r_*), \tag{21a}$$

$$T_c(r_*) = T_g(r_*), \tag{21b}$$

$$V_c(r_*) = V_g(r_*), \tag{21c}$$

and subsequently we follow their evolution using the equations which express MME interchanges between gas and grains under hypersonic free molecular conditions, and taking into account absorption of solar light and thermal reradiation (Kitamura and Yamamoto, 1986; Crifo, 1986a, b). The total number density of icy grains is given by:

$$n_c(r) = 0 \qquad \text{for } r \leqq r_*, \tag{22a}$$

$$\frac{1}{r^2}\frac{d}{dr}(r^2 V_c n_c) = J_*(r) \qquad \text{for } r \geqq r_*. \tag{22b}$$

As a consequence of cluster formation, the gas hydrodynamic equations must be modified to include gas mass and momentum loss terms $\dot{m}_{cg}$ and $\dot{p}_{cg}$ and a gas energy source $\dot{q}_{cg}$, given by:

$$\dot{m}_{cg}(r) = -J_*(r)\, 4\pi/3\, [a_*(r)]^3 \varrho_c, \tag{23a}$$

$$\dot{p}_{cg}(r) = -\dot{m}_{cg}(r)\, V_g, \tag{23b}$$

$$\dot{q}_{cg}(r) = -(\dot{m}_{cg}/m_g)\, L_v, \tag{23c}$$

where L_v is the molecular latent heat of formation of bulk ice.

Finally, we include in these gas equations terms representing the already mentioned MME exchanges between gas and evolving ice grains (Crifo, 1986a, b; Kitamura and Yamamoto, 1986). The resulting system of coupled differential equations is explicitly given in Appendix A.

3.2. The numerical code

We have modified along the lines described in the preceding section a multicomponent hydrodynamic computer code developed precedingly and which did incorporate as one of its components volatile grains (Crifo, 1986a). This model assumes that water vapour expands with spherical symmetry, dragging with it spherical grains of magnetite, olivine, and ice. The transport equations for mass, momentum, energy of each constituent are written down, taking into account solar illumination effects and radiative losses. The model includes attenuation of solar light by the dust and the gas, and heating of water by its photodissociation products. This latter effect is modelled according to the Monte Carlo simulation of Bockelée-Morvan and Crovisier (1987).

The water molecule thermodynamic properties are taken from (Crovisier, 1984). The optical properties of the grains and clusters are computed from Rayleigh scattering theory, Mie theory, or the Complex Angular Momentum theory, depending upon the ratio of particle radius to wavelength; the particle optical constants are taken from reliable experimental data (see references in Crifo, 1982). The resulting system of coupled differential equations (see Appendix A) is solved using a marching method, and the initial Mach number is found by a shooting procedure. The initial boundary conditions for the gas are derived self-consistently from Eqs. (13a–c). The initial conditions for the dust grains assume Brownian motion at temperature T_n.

3.3. Results

For the sake of simplicity, we assumed the cometary atmosphere to be either pure H_2O, or H_2O loaded with 30% by mass of dust. Two different dust compositions were assumed; "composition I" consists in 15% by mass of dirty amorphous olivine ($k_v = 0.03$) plus 15% by mass of magnetite, all grains having a common radius $a_d = 0.2$ µm; "composition II' assumes 10% by mass of impure amorphous olivine, plus 20% by mass of magnetite, all grains having a common radius equal to 1 µm. The production of water was assumed spherically distributed around a nucleus of radius 2 km at 0.8 AU from the Sun. Runs were performed with total water production rates Q ranging from $2\ 10^{28}$ up to $4\ 10^{30}$ molecule/s, and with nucleation effective or nucleation suppressed.

3.3.1. Cluster distribution and perturbation of the gas parameters

The results indicate that the nucleation has the general behaviour found by the ISAS group: the clusters form almost instantaneously at $r = r_*$ (of the order of a few R_n), grow up to radii

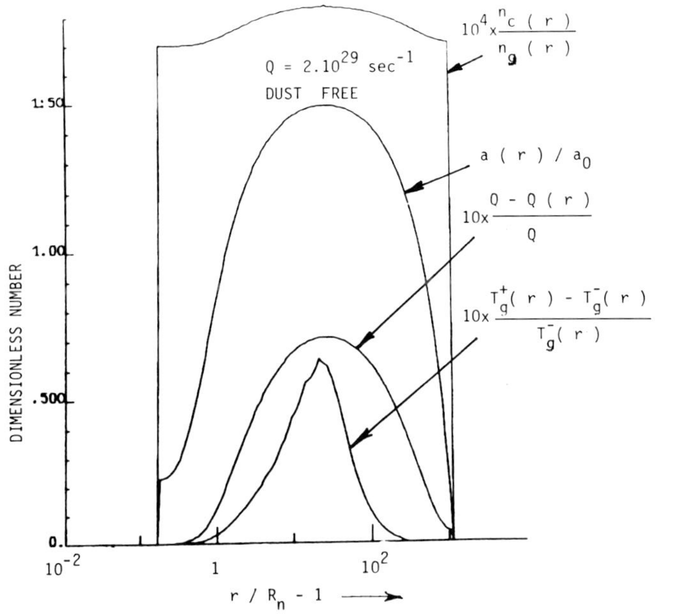

Fig. 4. Dust-free nucleation. The symbols not defined in the text are: n_g, water number density, a_0 a normalizing radius equal to 10 Å, $Q(r)$ the total flux of water molecules across the sphere of radius r, T_g^+ and T_g^- the gas temperature computed, respectively, with and without nucleation

of 5–15Å reached between 20 and 90 R_n, and subsequently decay by solar sublimation, reaching however altitudes eventually as high as 1000 R_n. Figure 4 shows an example of nucleation in a dust-free coma (still non adiabatic as photodissociative heating is at play). By comparison with Fig. 2 of Crifo (1986c), one sees the reduction in cluster size and maximum altitude resulting from the adoption of the correct formula (20c). The figure also shows the difference between the gas parameters computed with and without nucleation.

Figure 5 compares the gas temperature profiles obtained in a dust-free coma and in the dusty comae defined precedingly, at different total water production rates Q. Also indicated on the figures is the cluster temperature in the range over which they exist. One sees how, in a dust free atmosphere, the perturbation of the gas temperature first varies from a negligible up to a strong effect, as Q increases from $2\ 10^{28}\ s^{-1}$ up to $5\ 10^{29}\ s^{-1}$; it is a surprise to see that, above this value it decreases again, as photodissociative heating is now sufficiently effective at low altitudes to quench the nucleation.

The addition of dust distributed in fine grains, as in composition I, eliminates almost completely the perturbations of the gas by nucleation. However, the same mass load of dust distributed in larger grains (composition II) leaves an appreciable perturbation. The variation of this residual perturbation with Q has the same pattern as in the case of dust-free nucleation. This dependence of nucleation quenching effectiveness on dust size had not been noticed in our preliminary study (Crifo, 1986c).

Table 2 allows a detailed comparison between some of the solutions obtained in this study, and between these solutions and the results of the ISAS group.

3.3.2. Detectability of the halo of ice clusters

Since the hydrodynamic code described in Sect. 3.2 computes the dust and ice grain temperatures at all altitudes by balancing the grain convective and radiative exchanges, the grain and ice emission spectrum is easily obtained as a sub-product. The grain emission rate at altitude r in direction θ at wavelength λ is (in Watt per steradian):

$$g^i_\lambda(r,\theta) = \pi a_d^{i^2}(r)\cdot[f_\odot(\lambda)\, Q_S(a^i_d,\lambda)\, p^i(a^i_d,\lambda,\theta) + Q_A(a^i_d,\lambda)\, B_\lambda(T^i(r))] \tag{24}$$

where Q_A and Q_s are the absorption and scattering efficiencies, $f_\odot$ the incident solar flux, B_λ Plank's function, and p the scattering phase function at the scattering angle θ. The superscript i labels the grain kind, i.e., olivine, magnetite, or ice.

We have used (24) to compute the intensity from the coma of magnetite, olivine and ice clusters under opposition conditions, i.e.:

$$j^i(\lambda) = \int_{r=R_n}^{r=r^i_M} g^i_\lambda(r,\pi)\, n^i_d(r)\, dr \tag{25}$$

This corresponds approximately to the observing conditions of the Vega spacecraft during closest approach of comet Halley. In (25), r^i_M is the maximum altitude reached by the grains on the subsolar axis, due to sublimation (ice grains) or radiation pressure (magnetite and olivine grains); $n^i_d(r)$ is the number density of grains of kind i, given by Eq. (22b) for ice grains and by similar ones (with rhs = 0) for magnetite and olivine.

We first ran the program with no dust in the coma, in order to compare the cluster emission with its value computed by Yamamoto and Ashihara (1985). We find an emission spectrum similar to their spectrum, and absolute intensities of the same order of magnitude as theirs (see Table 2). This last agreement is more or less fortuitous, however, since their computation was made with a primitive version of their model. We expect the intensities corresponding to the model described by Kitamura and Yamamoto (1986) to be higher than ours, due to their neglect of formula (20c).

We then have run the program with the two compositions defined in Sect. 3.3. We find that composition II provides an emission spectrum closer to the one observed from Comet Halley (Tokunaga et al., 1986). It is also found to be less efficient in quenching the nucleation than composition I.

Figure 6 compares the cluster intensity with the magnetite plus olivine grain intensity. From these results, most observers will probably conclude that the emission from the clusters is hopelessly concealed by the emission from the dust. The situation is probably worse than when trying to detect haloes of large ice grains hypothetically fragmented from the nucleus (see Crifo and Emerich, 1985). It is also not easier to look for clusters than it is to look for fluorescence emissions of water itself, as manifested by the fact that these emissions have been unquestionably detected from the Vega spacecraft (Crovisier et al., 1987) while possible signatures from icy grains or icy clusters remain to be looked for in the data.

The detectability of the clusters from the Earth should be even more difficult for two reasons: (1) if the observations is not made at opposition, then the ratio of scattered light from the clusters to scattered light from the dust will decrease since the former are in Rayleigh regime and the latter in Mie regime. Thus the conspicuous icy cluster scattering signatures below $\lambda = 4\ \mu m$ will sink further below the dust continuum; (2) these observations in general integrate the emissions over most of the coma, and therefore the ratio of the number of icy clusters to the number of dust grains within the field of view will be an order of magnitude smaller than in the case of the one-dimensional integration (25), due to the difference in r^i_M.

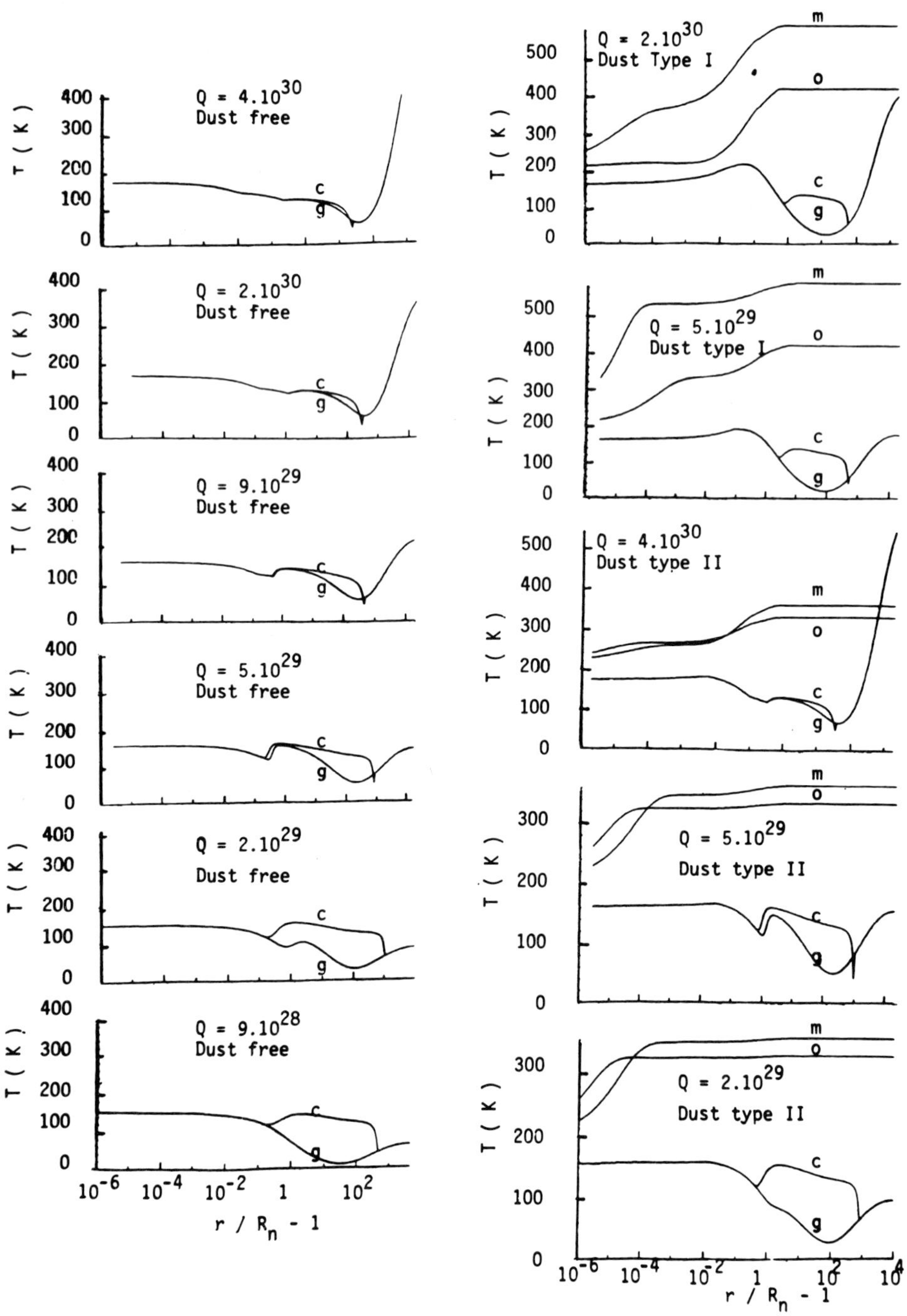

Fig. 5. Radial temperature profiles of water vapour (G) and of the water clusters (C). Curves on the left are for a dust-free coma and those to the right are for a coma with dust composition I or II given in Sect. 3.3. The total water production rate increases from bottom to top as indicated. Curves labeled (M) and (O) show the temperatures of the magnetite and olivine grains

3.4. Discussion

The results presented here indicate that homogeneous nucleation is strongly affected by non-adiabatic effects in the coma. The magnitude of the effect depends upon the dust composition and upon the total water production rate. For the dust and gas production rates which had been defined as "nominal Halley flyby conditions", the effects of nucleation on the gas are negligible if the dust is concentrated in submicron grains ("composition I"). At higher production rates, or if the dust is spread over a large range of sizes, nucleation will introduce limited perturbations in the gas. The presence of even traces of dust also makes the detectability of a halo of ice clusters from Earth extremely difficult.

One effect which could further reduce the effect of nucleation, especially at high water production rates, is the IR heating of the molecules by the dust radiation. This effect results from the fact, first established by Crovisier (1984) that the water rotational lines are optically thick. This effect has been treated by Marconi and Mendis (1986), who predict a strong heating of the gas in the inner coma. However it is not obvious that their simple approach provides satisfactory results. Comparison with a detailed level by level budget of the rotational state of the water molecule (as developed by Bockelée Morvan, 1987) is not included in their

Table 2. Nucleation parameters. The two first columns correspond to very similar initial gas conditions: the difference in grain evolution is due to differences in the optical model of the grains and in the grain sublimation rate used in the ISAS study and here

	ISAS MODEL	THIS STUDY			THIS STUDY		
	$Q = 5.10^{29}$ $R_n = 3$ km	$Q = 2.10^{29}$ $R_n = 2$ km			$Q = 5.10^{29}$ $R_n = 2$ km		
	ADIABATIC	DUST FREE	DUST COMPOSITION I	DUST COMPOSITION II	DUST FREE	DUST COMPOSITION I	DUST COMPOSITION II
$r_* \ (R_n)$	1.2	1.16	2.40	1.46	1.20	3.48	1.50
$10^4\ n_c/n_g$ at $r = r_*$	17	1.7	4.4	1.6	9.6	5.6	6.
Maximum grain radius (Å)	6.4	14.9	5.53	11.7	9.87	3.48	11.0
Altitude of maximum gas depletion (R_n)	17	25.6	25.5	25.5	38	38	38
Maximum gas density depletion (%)	10	7.1	0.08	3.28	13	0.25	10.
Altitude of terminal grain sublimation (km)	1500	1100	368	844	869	528	969
Optical intensity at $\lambda = 12.3$ µm (in 10^{-8} watt cm^{-2} $ster^{-1}$ μm^{-1})	> 2.	1.0	0.0045	0.28	9.6	0.015	3.34

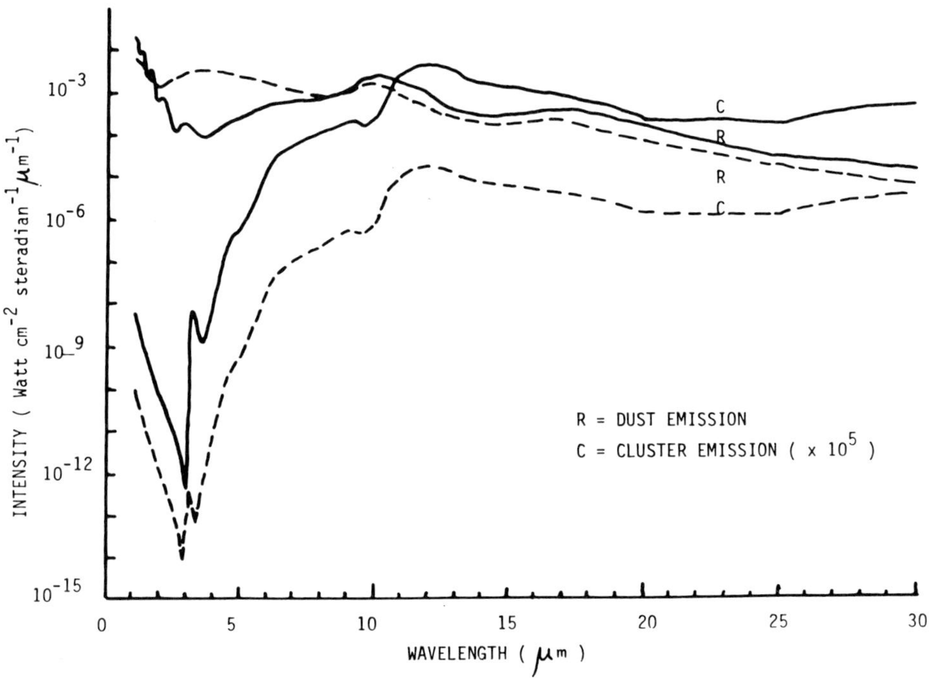

Fig. 6. Clusters and refractory dust emission. The curves represent integrated intensities radiated in the direction of the Sun by grains located on the Comet-to-Sun-axis. Dashed curves correspond to a "composition I" coma, and full line curves to a "composition II". (R) labels the total emission of the magnetite and olivine grains, and (C) the emission from the clusters, after multiplication by 10^5. For all curve, $Q = 5\ 10^{29}$ molecules second

publication. Therefore we prefer to leave to a future study the evaluation of this effect on nucleation.

The results presented here were obtained assuming that dust is a mixture of magnetite and olivine grains. There are now indices that the absorbing component may consist in carbonaceous material. We performed several computer runs with amorphous carbon substituted to magnetite. No significant difference in the nucleation properties were found. This is due to the fact that both materials have similar hydrodynamic and optical properties.

Our model uses spherical geometry for the coma (Appendix A). The Halley flyby data demonstrate clearly that the near-nucleus outflow of comet Halley dust is anything but spherically symmetric. Most of the dust issues in discrete jets, and presumably so does the gas (Huebner et al., 1987). However, we believe that the present results represent adequately the physics of the most significant regions of the coma, i.e. the jets: it has been shown long since (Ashkenas and Sherman, 1966) that the flow properties of an axially symmetric jet near its axis are well represented by the solution of spherically symmetric equations.

The credibility of the present results (and of those of the ISAS Group as well), and the usefulness of developing more sophisticated treatments (for instance by incorporation if the ice grain size distribution) can be questioned on fundamental grounds: the classical nucleation theory is generally unable to provide quantitatively reliable predictions (Hagena, 1981). The onset of nucleation is predicted correctly, but not the nucleation rate nor the critical radius a_*. H_2O clustering, in particular, has been the subject of considerable investigations and it appears that so far no successful theory is available (Kassner et al., 1980). The difficulties of the theory are here at their maximum, because of the small values of the critical cluster radius a_* which are predicted. These values (a few Å) correspond to clusters of a few molecules only, which cannot be ascribed macroscopic parameters (ϱ_d, σ, T_c, and even a_* itself!).

An additional difficulty is set by the need to perform optical computations to derive the cluster temperature and spectrum: clusters of a few molecules cannot be treated as solid state grains (which requires $\simeq 10^3$ atoms at minimum), but should be treated as Van der Waals molecules. The uncertainty in the computed intensity is high at wavelengths where the thermal emission is dominant, i.e. beyond 4 µm in wavelength. However, our conclusion that the detectability is very difficult is safe: even if all the coma water had condensed into small clusters, the resulting increase in intensity would be only an order of magnitude above the value computed here, and this is still several oders of magnitude below the emission from the refractory grains. The absence of ice signatures in ground based spectra of comets cannot be argued against the effectiveness of nucleation, exactly as it cannot be used to exclude the presence of an icy halo (Crifo, 1982 and references therein).

It appears that a definitive assessment of nucleation phenomena in comets (and elsewhere) must await future progresses at the basic physics level.

Acknowledgements. I am indebted to Dr. M.K. Wallis for indicating useful references, to Dr. T. Ytrehus for communication of his work on sublimation. Thanks are due to Miss F. Marchand for assistance in literature reviews, to Mr. F. Bougnet for assistance in computer operation and to Mrs. G. Bargot for edition of the manuscript.

Appendix A

The following equations govern the vacuum expansion of a mixture of condensible vapor and "small" spherical dust grains ($a_d^i \ll \Lambda_{gg}$), in the presence of solar illumination. They assume spherical symmetry, except that the solar illumination comes from a given direction. Therefore, they are strictly speaking applicable only in the vicinity of the comet-to-Sun axis.

Neglecting Brownian motion, the conservation equations for dust grains of kind i are:

$$\frac{d}{dr}(r^2 n_d^i V_d^i) = 0\,, \tag{A1}$$

$$\frac{d}{dr} a_d^i = 0\,, \tag{A2}$$

$$\frac{4}{3}\pi (a_d^i)^3 \varrho_d^i V_d^i \frac{dV_d^i}{dr} = -G\frac{M_n}{r^2}\frac{4}{3}\pi (a_d^i)^3 \varrho_d^i + \dot{P}_{gd}^i\,, \tag{A3}$$

$$\frac{4}{3}\pi (a_d^i)^3 \varrho_d^i C_d^i V_d^i \frac{dT_d^i}{dr} = \dot{E}_d^{abs,i} - \dot{E}_d^{rad,i} + \dot{E}_{gd}^i\,, \tag{A4}$$

where G is the gravitational constant, M_n the nucleus mass, $\dot{P}_{gd}$ and $\dot{E}_{gd}$ the rates of momentum and energy exchange with the gas, $\dot{E}_d^{abs}$ and $\dot{E}_d^{rad}$ the rates of light absorption and emission, C_d the specific heat constant, ϱ_d the specific mass.

One has:

$$\dot{P}_{gd}^i = \frac{1}{2}\pi (a_d^i)^2 n_g m_g (V_g - V_d^i)^2 \left[\frac{\exp(-S_i^2)}{\sqrt{\pi}}\left(\frac{1}{S_i^3} + \frac{2}{S_i}\right) + 2\,\mathrm{erf}(S_i)\left(1 + \frac{1}{S_i^2} - \frac{1}{4S_i^4}\right) + \frac{2}{3S_i}\sqrt{\pi\frac{T_d^i}{T_g}}\right],$$

$$\dot{E}_{gd}^i = \pi (a_d^i)^2 n_g k T_g \sqrt{2kT_g/m_g} \cdot \left\{\Phi(S_i)\left[\frac{\gamma(T_g)}{\gamma(T_g)-1} + S_i^2 - \frac{\gamma(T_d)+1}{2\gamma(T_d)-2}\,\frac{T_d}{T_g}\right] - \frac{\mathrm{erf}(S_i)}{2S_i}\right\},$$

$$\dot{E}_d^{abs,i} = \pi (a_d^i)^2 \int Q_A(a_d^i,\lambda)\left\{f_\odot(\lambda)\, e^{-\tau_\lambda} + \left[\varepsilon_n B_\lambda(T_n) + \frac{A_n}{\pi}\right] 2\pi\left(1 - \frac{R_n^2}{r^2}\right)^{1/2} \exp[\tau_\lambda - \tau_\lambda(R_n)]\right\} d\lambda\,,$$

$$\dot{E}_d^{rad,i} = 4\pi^2 (a_d^i)^2 \int Q_A(a_d^i,\lambda)\, B_\lambda(T_d^i)\, d\lambda\,,$$

where

$$S_i \equiv (V_g - V_d^i)/(2kT_g/m_g)^{1/2}\,,$$

$$\Phi(S) \equiv \left(1 + \frac{1}{2S^2}\right)\mathrm{erf}(S) + \frac{1}{S^2}\,\frac{\exp(-S^2)}{\sqrt{\pi}}\,.$$

A_n and ε_n are the nucleus albedo and emissivity, and τ_λ is the coma optical thickness given by:

$$\frac{d\tau_\lambda}{dr} = -\sum_i n_d^i\, \pi (a_d^i)^2 \left[Q_S(a_d^i,\lambda) + Q_A(a_d^i,\lambda)\right] - n_g \sigma_{H_2O}(\lambda)\,,$$

where σ_{H_2O} is the water molecule absorption cross section.

The equations governing the formation and evolution of the water clusters are:

$$\frac{1}{r^2}\frac{d}{dr}(r^2 n_c V_c) = J_*, \quad \text{(A5)}$$

$$4\pi a_c^2 \varrho_c V_c \frac{da_c}{dr} = -\dot{M}_{cg}, \quad \text{(A6)}$$

$$\frac{4}{3}\pi a_c^3 \varrho_c V_c \frac{dV_c}{dr} = \dot{P}_{gc}, \quad \text{(A7)}$$

$$\frac{4}{3}\pi a_c^3 \varrho_c C_c V_c \frac{dT_c}{dr} = \dot{E}_c^{abs}(a_c, r) - \dot{E}_c^{rad}(a_c, T_c) + \dot{E}_{gc}, \quad \text{(A8)}$$

where

$$\dot{M}_{cg} = 4\pi a_c^2 m_g \left\{ \frac{p_s(T_c)}{\sqrt{2\pi m_c k T_c}} \exp\left(\frac{2 m_g \sigma}{k T_c \varrho_c a_c}\right) - n_g \frac{V_g - V_c}{4} \Phi(S_c) \right\},$$

$$\dot{P}_{gc} = \frac{1}{2}\pi a_c^2 n_g m_g (V_g - V_c)^2 \cdot \left[\frac{\exp(-S_c^2)}{\sqrt{\pi}} \left(\frac{1}{S_c^3} + \frac{2}{S_c}\right) + 2\,\mathrm{erf}(S_c)\left(1 + \frac{1}{S_c^2} - \frac{1}{4 S_c^4}\right) \right],$$

$$\dot{E}_{gc} = n_g k T_g \sqrt{2 k T_g / m_g} \cdot \left\{ \Phi(S_c) \left[\frac{L_V}{k T_g} + \frac{\gamma(T_g)}{\gamma(T_g) - 1} + S_c^2 - \frac{\gamma(T_c) + 1}{2\gamma(T_c) - 2} \frac{T_c}{T_g} \right] - \frac{\mathrm{erf}(S_c)}{2 S_c} \right\} - \frac{L_V p_s(T_c)}{\sqrt{\pi m k T_c / 8}} \exp\left(\frac{2 m_g \sigma}{k T_c \varrho_c a_c}\right).$$

S_c, $\dot{E}_c^{abs}$, and $\dot{E}_c^{rad}$ are given by expressions similar to those given above for the dust grains.

Finally, the equations governing the evolution of the gas are:

$$\frac{1}{r^2}\frac{d}{dr}(r^2 n_g m_g V_g) = \dot{M}_{cg} n_c + \dot{m}_{cg} + \dot{m}_{ph}, \quad \text{(A9)}$$

$$\frac{1}{r^2}\frac{d}{dr}(r^2 n_g m_g V_g^2) = -\frac{d}{dr}(n_g k T_g) + n_c(\dot{M}_{cg} V_c + \dot{P}_{cg}) + \dot{p}_{cg} + \sum_i n_d^i \dot{P}_{dg}^i + \dot{P}_{ph} + \dot{p}_{ph}, \quad \text{(A10)}$$

$$\frac{1}{r^2}\frac{d}{dr}\left[r^2 n_g V_g \left(m_g \frac{V_g^2}{2} + \frac{\gamma(T_g)}{\gamma(T_g) - 1} k T_g \right) \right] = n_c(\dot{E}_{cg} + V_c \dot{P}_{cg}) + \tfrac{1}{2} n_c \dot{M}_{cg} V_c^2 + \dot{e}_{cg} + \sum_i n_d^i (\dot{E}_{dg}^i + V_d^i \dot{P}_{dg}^i) + \dot{E}_{ph} + \dot{e}_{ph}, \quad \text{(A11)}$$

where $\dot{m}_{cg}$, $\dot{p}_{cg}$ and $\dot{e}_{cg}$ have been defined in the text, $\dot{m}_{ph}$, $\dot{p}_{ph}$ and $\dot{e}_{ph}$ are loss terms corresponding to the removal of molecules by photodissociation:

$$\dot{m}_{ph} = -n_g m_g \int f_\odot(\lambda)\, e^{-\tau_\lambda} \sigma_{H_2O}(\lambda)\, d\lambda,$$

$$\dot{p}_{ph} = \dot{m}_{ph} V_g,$$

$$\dot{e}_{ph} = \dot{m}_{ph} \left[\frac{V_g^2}{2} + \frac{\gamma(T_g)}{\gamma(T_g) - 1} \frac{k T_g}{m_g} \right],$$

and were $\dot{E}_{ph}$ and $\dot{P}_{ph}$ are source terms resulting from collisions with the fast photodissociation products of the H_2O molecule. For these terms, the analytic expressions derived by Bockelée-Morvan and Crovisier (1987) were adopted.

Finally, one has:

$$\dot{P}_{cg} = -\dot{P}_{gc}$$
$$\dot{P}_{dg} = -\dot{P}_{gd}$$
$$\dot{E}_{dg}^i = -\dot{E}_{gd}$$

but:

$$\dot{E}_{cg} = \pi a_c^2 n_c k T_g \sqrt{2 k T_g / m_g} \left\{ \frac{\mathrm{erf}(S_c)}{2 S_c} - \left[\frac{\gamma(T_g)}{\gamma(T_g) - 1} + S_c^2 \right] \Phi(S_c) \right\} + \pi a^2 \frac{\gamma(T_c) + 1}{\gamma(T_c) - 1} \left(\frac{2\pi k T_c}{m_g}\right)^{1/2} p_s(T_c) \exp\left(\frac{2 m_g \sigma}{k T_c \varrho_c a_c}\right).$$

A detailed justification of these equations will be presented elsewhere.

References

Anisimov, S.I.: 1958, *Soviet Phys. JETP* **27**, 321

Ashkenas, H., Sherman, F.S.: 1966, *IVth Rarefield Gas Dynamics Symposium*, ed. J.H. de Leeuw, Academic Press, NY, Vol. **2**, p. 84

Bockelée-Morvan, D.: 1987, *Astron. Astrophys.* **181**, 169

Bockelée-Morvan, D., Crovisier, J.: 1987, in: Proceedings of the International Colloquium on the Similarity and Diversity of Comets, ESA SP-278 (in press)

Cercignani, C.: 1981, Progr. Astron. Aeron. **74**, Vol. 1, p. 305

Crifo, J.F.: 1982, in *Cometary Exploration*, ed. T. Gombosi, Hungarian Academy of Sciences Press, Vol. II, p. 167

Crifo, J.F., Emerich, C.: 1985, in *Ices in the Solar System*, ed. J. Klinger et al., Reidel, Dordrecht, p. 429

Crifo, J.F.: 1986a, in *Asteroids, Comets, Meteors II*, eds. C.I. Lagerkvist et al., Uppsala University Press, p. 389

Crifo, J.F.: 1986b, in *Rarefied Gas Dynamics XV*, ed. V. Boffi, C. Cercignani, Teubner, p. 229

Crifo, J.F.: 1986c, in 20th ESLAB Symp., ESA SP 250, Vol. **I**, p. 533

Crifo, J.F.: 1987, in: Proceedings of the Intenational Colloquium on the Similarity and Diversity of Comets, ESA SP-278 (in press)

Crovisier, J.: 1984, *Astron. Astrophys.* **130**, 361

Crovisier, J., Moroz, V., Encrenaz, T., Combes, M., Grigoriev, A.V., Crifo, J.F., Bibring, J.P., Coron, N., Sanyko, N.F., Bockelée-Morvan, D., Gispert, R., Emerich, C., Lamarre, J.M., Rocard, F., Krasnopolski, V.A., Owen, T.: 1987, in: Proceedings of the International Colloquium on the Similarity and Diversity of Comets, ESA SP-278 (in press)

Delsemme, A.H., Miller, D.C.: 1971, *Planet Space Sci.* **19**, 1229

Fanale, F.P., Salvail, J.R.: 1986, *Icarus* **66**, 154

Gombosi, T.I., Cravens, T.E., Nagy, A.F.: 1985, *Astrophys. J.* **293**, 328

Gombosi, T.I., Nagy, A.F., Cravens, T.E.: 1986, *Rev. Geophys.* **24**, 3, 667

Hagena, O.F.: 1981, *Surface Sci.* **106**, 101

Hellmich, R.: 1979, Ph. D. Thesis, Universität Göttingen

Herman, G., Podolak, M.: 1985, *Icarus* **61**, 252

Hermans, L.J.F., Beenakker, J.J.M.: 1986, *Phys. Fluids* **29**, 4231

Huebner, W.F., Keller, H.U., Wilhelm, K., Whipple, F.L., Delamere, W.A., Reitsema, H.J., Schmidt, H.U.: 1987, 20th ESLAB Symp., ESA SP-250, Vol. **II**, p. 363

Ip, W.H.: 1986, *Astrophys. J.* **300**, 456

Kassner, J.L. et al.: 1980, in *Atmospheric Water Vapor*, ed. A. Deepak et al., Academic Press, New York, p. 613
Kitamura, Y., Yamamoto, T.: 1986, *Icarus* **68**, 266
Kitamura, Y.: 1986, *Icarus* **66**, 241
Larson, H.P., Scott Davis, D., Mumma, M.J., Weaver, H.A.: 1986, *Astrophys. J.* **309**, L95
Luikov, A.V.: 1971, *Int. J. Heat Mass Transfer* **14**, 177
Marconi, M.L., Mendis, D.A.: 1983, *Astrophys. J.* **273**, 381
Marconi, M.L., Mendis, D.A.: 1986, *Moon Planets* **36**, 249
Probstein, R.F.: 1968, in *Problems of Hydrodynamics and Continuum Mechanics*, Philadelphia Soc. Ind. Appl. Math., p. 568
Rickman, H., Froeschle, C.: 1983, in *Cometary Exploration*, Vol. 3, ed. T.I. Gombosi, Hungarian Acad. Sci., p. 109
Shul'man, L.M.: 1970, *Astrometr. Astrofizs.* **4** (translated as NASA-N70-30776, 100)
Shul'man, L.M.: 1975, *Astrometr. Astrofiz.* **32**, 197 (in Russian)
Squyres, S.W., McKay, C.P., Reynolds, R.T.: 1985, *J. Geophys. Res.* **90**, B14, 12381
Tokunaga, A.T., Golish, W.F., Griep, D.M., Kaminski, C.D., Hanner, M.S.: 1986, *Astron. J.* **92**, 1183
Weissman, P.R., Kieffer, H.H.: 1984, *J. Geophys. Res.* **89**, Suppl. C358
Whipple, F.: 1950, *Astrophys. J.* **111**, 375
Wegener, P.P., Wu, B.J.C.: 1977, *Adv. Colloid Interface Sci.* **7**, 325
Yamamoto, T., Ashihara, O.: 1985, *Astron. Astrophys.* **152,** L17
Ytrehus, T.: 1975, Von Karman Institute for Fluid Dynamics (Belgium), TN **112**

Note added in proof. Based on data from the "Giotto" and "Vega" flyby spacecrafts and from ground-based infrared instrumentation, a realistic multifluid hydrodynamic model of comet P/Halley has been developed (Crifo, 1987). It confirms that the "type II" model presented here more closely approximates nucleation in the comet than the "type I" model, which was based on estimates made prior to the return of the comet.

Astron. Astrophys. 187, 451–454 (1987)

ASTRONOMY AND ASTROPHYSICS

Detection of OH rotational emission from comet P/Halley in the far-infrared

G.J. Stacey[1], J.B. Lugten[1], and R. Genzel[1,2]

[1] Department of Physics, University of California, Berkeley, California 94720, USA
[2] Max Planck Institut für Physik und Astrophysik, D-8046 Garching, Federal Republic of Germany

Received January 30, accepted March 27, 1987

Summary. We have detected the $^2\Pi_{3/2}$ ($J = \frac{5}{2} \rightarrow \frac{3}{2}$; $+ \rightarrow -$ parity) transition of OH in comet Halley at 119.44 μm and placed an upper limit to the line intensity of the $- \rightarrow +$ parity transition at 119.23 μm. The ratio of these lines is consistent with radiative pumping and inconsistent with photochemical pumping of the OH rotational levels. These far-infrared lines probe the inner regions of the comet coma where OH is produced through photodissociation of a parent (presumably H_2O) molecule. Hence, these lines complement the 18 cm radio measurements which are sensitive to the outer regions of the comet coma. The OH production rate on May 3, 1986 was $\sim 3.6\,10^{29}$ molecules s^{-1}.

Key words: comet Halley – far-infrared spectra – OH production rates

1. Introduction

Water is believed to be one of the primary constituents of comet mantles. However, observations of cometary water lines are hampered to a large degree by the presence of deep telluric water vapor absorption features. The problem is particularly severe for the far-infrared (FIR) rotational transitions. Consequently, water vapor in comet comas has eluded detection until quite recently (Mumma et al., 1986). In contrast, the OH radical, which is believed to be formed by the photodissociation of H_2O has been extensively studied both in its ultraviolet bands at 3090 Å and in its 18 cm ground state Λ doublet line. Comet Halley has been studied extensively through both of these transitions from July, 1985 through July, 1986 (cf. Feldman et al., 1986; Gérard et al., 1986).

We report the first detection of a FIR rotational transition of OH from a comet. The FIR lines arise from rotational levels of the ground vibrational state which are radiatively pumped through absorptions of solar ultraviolet photons. The FIR observations complement the 18 cm observations as the small ($\sim 50''$) beam size of the FIR observations allow one to probe the inner regions of the comet coma–regions in which the optical pumping schemes for the 18 cm lines fail due to collisional redistribution of the ground state Λ doublet population (Elitzur, 1981). The FIR emitting levels, having a much shorter lifetime, are not affected by collisional processes.

Send offprint requests to: G.J. Stacey

2. Observations

The lowest rotational levels of the ground vibrational state of OH are indicated in Fig. 1. We observed the $^2\Pi_{3/2}$ ($J = \frac{5}{2} \rightarrow \frac{3}{2}$) rotational lines in emission from comet Halley on May 3, 1986 in a flight from Christchurch, New Zealand with the 91.4 cm telescope on board NASA's Kuiper Airborne Observatory (KAO). The instrument used was the U.C. Berkeley tandem Fabry-Perot spectrometer described in detail by Lugten et al. (1987). In brief, the instrument is a high resolution ($\lambda/\Delta\lambda \leqq 3\,10^4$), far-infrared ($45\,\mu m \leqq \lambda \leqq 200\,\mu m$) tandem Fabry-Perot spectrometer. At the transition wavelength, 119 μm, the detective element used is a Ge : Sb photoconductor (Watson et al., 1980). The system noise equivalent power referred to the sky was $4\,10^{-15}$ W Hz$^{-1/2}$. The telescope secondary was chopped at a frequency of 24 Hz with an amplitude of 4′. Standard beam switching techniques were employed which ensures an accurate determination of the zero level of the spectrum. The spectrum of comet Halley was calibrated by division by a spectrum of Saturn obtained on the same flight. This both provided an absolute flux calibration and ensured a featureless baseline for the Halley spectrum. Saturn was

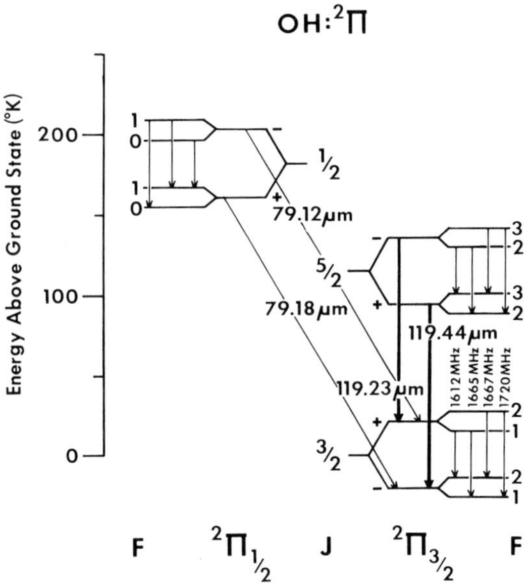

Fig. 1. Schematic representation of the low lying rotational levels of the OH radical. The Λ doubling and hyperfine splittings are not drawn to scale

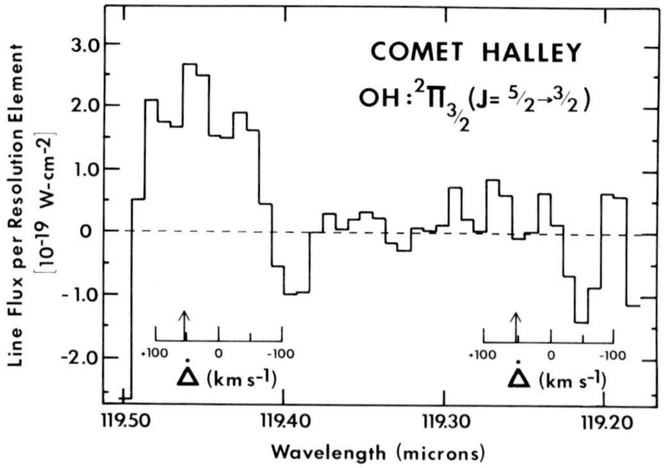

Fig. 2. Spectrum of comet Halley at 119 μm obtained on May 3, 1986. The spectrum has been divided by one of Saturn obtained on the same flight for flux calibration and to ensure a flat baseline

assumed to emit like a 100 K blackbody at 119 μm (Hildebrand et al., 1985). Our beam size was 53″ full-width-half-maximum ($8.0\,10^{-8}$ sr) which is equivalent to a disk of radius $2.1\,10^4$ km at the earth-comet distance ($\Delta = 0.88$ AU) at the time of observation. The spectrum (Fig. 2) was obtained in fourty minutes of integration time on comet Halley. Arrows on the velocity scale indicate the geocentric velocity of the comet, $\dot{\Delta} = 52.8\,\mathrm{km\,s^{-1}}$. The 119.44 μm ($+ \rightarrow -$ parity) line is clearly detected with an integrated line flux of $2.2 \pm 0.4\,10^{-19}\,\mathrm{W\,cm^{-2}}$. The 119.23 μm ($- \rightarrow +$ parity) transition is not detected. A 3σ upper limit to the flux is $1.2\,10^{-19}\,\mathrm{W\,cm^{-2}}$, which means the line flux ratio $F_{119.44}/F_{119.23}$ is greater than 1.8. The lines are not resolved at our velocity resolution of $\sim 75\,\mathrm{km\,s^{-1}}$ (0.03 μm). Absolute line intensities are believed to be accurate to $\sim 30\%$.

3. Results

Andresen (1986) has suggested that the level populations of cometary OH molecules may be largely determined by the formation process of photodissociation of H_2O. His calculations predict these processes could lead to the observed inversion of the ground state Λ doublet manifest as maser emission in the 18 cm lines. However, the formation process would lead to a 119.44 μm to 119.23 μm line flux ratio less than unity, in contrast to our measured line ratio. Thus, photochemical processes do not play a large role in the determination of cometary OH level populations.

The rotational level populations are also unaffected by collisional processes. For densities and velocities typical of the inner coma, ($n_{H_2O} \sim 10^4\,\mathrm{cm^{-3}}$; $v_{OH} \sim 1\,\mathrm{km\,s^{-1}}$ – see below), collisional excitation (or de-excitation) timescales are of the order of 10^6 seconds. An OH molecule will typically absorb a solar UV photon every 10^3 seconds at a solar distance of 1 AU (Mies, 1974). Hence, UV pumping of the excited levels of OH dominates collisional processes. Since the lifetime of the rotational levels is ~ 6 seconds, collisional depopulation of excited rotational levels may also be safely ignored.

The relative level populations of OH in fluorescence equilibrium with the solar spectrum has been extensively studied by several authors (cf. Snyder, 1986 and references therein). The physical mechanism is that OH initially in the ground $^2\Pi_{3/2}$ ($J = \frac{3}{2}$) level absorbs solar UV photons in transitions to the excited $^2\Sigma^+$ level. The subsequent radiative relaxation of the excited levels leads to small equilibrium populations in rotationally excited levels. By far the most populated excited levels are the $J = \frac{5}{2}$ level of the $^2\Pi_{3/2}$ rotational ladder and the $J = \frac{1}{2}$ and $J = \frac{3}{2}$ levels of the $^2\Pi_{1/2}$ rotational ladder. For comets with heliocentric distance, r_H, of the order of 0.7 AU, these levels will have relative populations $0.1 \rightarrow 1\%$ (Schleicher and A'Hearn, 1982). Changes in the relative populations of the Λ doublet states of each rotational level occur as the pumping frequencies are Doppler shifted into and out of solar Fraunhofer lines causing selective pumping of one or the other Λ doublet state. This is the familiar "Swings" effect (Swings, 1941) which often results in inversions of the ground state Λ doublet and consequent masering of the 18 cm line.

Schleicher (1983) (see also Schleicher and A'Hearn, 1982, and Piehler and Kegel, 1986) predicts the fluorescence equilibrium populations of the lower rotational levels of the first three vibrational bands of the $^2\Sigma^+$ and $^2\Pi$ electronic states of OH including the effects of nine UV pumping bands. For $r_H = 0.713$ AU and a heliocentric velocity, $\dot{r}_H$, of $24\,\mathrm{km\,s^{-1}}$ the populations in the $^2\Pi_{3/2}$ ($J = \frac{5}{2}$) rotational levels relative to the total OH population are $4.09\,10^{-3}$ and $2.11\,10^{-3}$ respectively for the positive and negative parity states. The fluorescence efficiency scales approximately linearly with the incident solar flux for $r_H \geqq 0.5$ AU. At the time of the observations, $r_H = 1.66$ AU, so we expect the level populations to be down by a factor of 5.4.

Since the Einstein A coefficients for the 119.44 μm and the 119.23 μm lines are identical, the model of Schleicher (1983) predicts a line flux ratio $F_{119.44}/F_{119.23} = 1.94$. The fluorescence equilibrium model of Piehler and Kegel (1986) predict a similar line ratio (1.75). This is consistent with our lower limit derived above, confirming the validity of the UV pumping model together with the Swings' effect on the adopted solar spectrum.
The flux we detect in our beam is given by:

$$\mathrm{F}_{119.44\,\mu m} = \frac{n_+ (J = \frac{5}{2})}{n_{OH}} A_{119} \frac{hc}{\lambda} \frac{N_{OH}}{4\pi\Delta^2} \,\mathrm{erg\,cm^{-2}s^{-1}} \tag{1}$$

where

$\frac{n_+ (J = \frac{5}{2})}{n_{OH}}$ = relative population of the positive parity state of the $J = \frac{5}{2}$ rotational level ($7.6\,10^{-4}$).

A_{119} = Einstein A coefficient for spontaneous emission of a 119 μm photon ($0.14\,\mathrm{s^{-1}}$).

Δ = Earth comet distance (0.86 AU = $1.29\,10^8$ km)

N_{OH} = total number of OH molecules in the beam.

Our measured 119.44 μm flux ($2.2\,10^{-19}\,\mathrm{W\,cm^{-2}}$) requires then $N_{OH} = 2.6\,10^{33}$ molecules.

4. Discussion

OH is believed to be the primary daughter product of the photodissociation of H_2O. Using a Haser model (uniform isotropic radial outflow) the density of OH molecules, $n_{OH}(r)$, in the comet coma at radius r is given by (Despois et al., 1981):

$$n_{OH}(r) = \frac{Q_P}{4\pi r^2 v_{OH}} \left(\frac{L_{OH}}{L_{OH} - L_P}\right) \{e^{-r/L_{OH}} - e^{-r/L_P}\} \tag{2}$$

where Q_P is the production rate of the parent molecule of scale length L_P, and v_{OH} is the velocity of the OH molecules of scale length L_{OH}. The total number of OH molecules sampled by our beam includes contributions from both the inner coma ($r < r_{beam} = 2.1\,10^4$ km) and the overlying OH along the same column. Integrating Eq. 2 over the inner coma, we have (assuming $L_{OH} = 4\,10^5$ km, $L_P = 5\,10^4$ km, $v_{OH} = 1.0\,\mathrm{km\,s^{-1}}$ at 1 AU):

$$N_{\mathrm{inner\ coma}} = 2.0\,10^3\,Q_P$$

We have assumed that the scale lengths scale as $r_H^{1.75}$ and v_{OH} scales as $r_H^{-0.25}$. Over the remaining column along the line of sight there is a comparable amount of OH. Thus using this simple Haser model we calculate a total parent (hence OH) production rate of: $Q_P \sim 6.5\,10^{29}$ molecules $\mathrm{s^{-1}}$.

The FIR measurements sample gas within the region where OH is being created and therefore are sensitive to parameters in this regime. In particular, our model is sensitive to the parent scalelength, L_P. Decreasing L_P to $3.9\,10^4$ km as suggested by the *in situ* measurements of Krankowski et al. (1986), decreases the required outgassing rate by ~20%. Also, the energy imparted to the daughter OH molecule following photodissociation is comparable to the kinetic energy of the parent H_2O molecule. Hence, the velocity field of the daughter molecules is probably far from that of uniform radial expansion. Some molecules may be expected to travel outwards with twice the initial velocity while others may have near zero, or even net inward velocity after photodissociation. Since collisional timescales are at the order of 10^6 s, many of these OH molecules may remain in the inner coma unaffected by the general flow for their lifetimes ($\sim 4\,10^5$ s). The net result is considerable enhancement of OH lifetimes, hence densities, within the inner coma. The vectorial model of Festou (1981) addresses this problem. This model predicts OH lifetimes to be enhanced by a factor of 1.8 relative to the Haser model, within the inner ($r \leqq 2 \times 10^4$ km) coma. Using this model, our calculated OH production rates would be:

$$Q_{OH} \sim 3.6\,10^{29}\ \mathrm{molecules\ s^{-1}}.$$

Feldman et al. (1986) have determined OH production rates from comet Halley through IUE measurements of the ultraviolet emission lines. The IUE beam samples the inner 10″ × 20″ regions of the comet coma. Unfortunately, no data were taken on May 3, 1986. However, data were taken on April 29 and May 12 resulting in calculated OH production rates of 2.0 and 3.0×10^{29} molecules $\mathrm{s^{-1}}$ respectively. Considering that the UV determined OH production rates were variable by factors of two over timescales of days (the rate was as large as $5\,10^{29}$ molecules $\mathrm{s^{-1}}$ on April 21), the agreement between the production rates calculated from the IUE data and the FIR data is very good. One should expect good agreement as the FIR beam is not much larger than the IUE beam and neither the ultraviolet nor the FIR transitions are heavily affected by collisional processes or optical depth effects.

Gérard et al. (1986) have estimated the OH production rates from comet Halley through measurements of the 18 cm line. They used the Nançay radio telescope which has a 3.5′ × 18′ beam at this wavelength. On May 3 they estimate an OH production rate of $\sim 1.6 \times 10^{29}$ molecules $\mathrm{s^{-1}}$, significantly smaller than the number derived from our FIR measurements. The radio production rates are also systematically lower than those derived from IUE observations. This difference is probably due to several different effects largely related to the substantially larger beam size used in the radio measurements relative to beams used for the FIR and IUE measurements.

First the radio lines are probably significantly affected by quenching (collisional depopulation) in the inner comet coma. This lowers the measured 18 cm flux and consequently lowers the calculated OH production rates. Gérard et al. (1986) point out that if the 18 cm lines are collisionally depopulated to a distance of 10^5 km from the comet nucleus (as suggested by some theoretical arguments, Elitzur (1981)) rather than the 4×10^4 km taken by the authors, this would increase their outgassing rates by 50%.

Secondly, the large beam sizes make the radio estimates particularly sensitive to OH lifetimes. Gérard et al. (1986) note that decreasing the assumed OH lifetime from the assumed value of 4×10^5 s to 1.6×10^5 s (as predicted theoretically by the calculations of van Dishoeck and Dalgarno (1984)) will also increase their calculated OH production rates a factor of two. Production rates calculated from the FIR or IUE measurements are largely unaffected by the OH lifetimes due to their much smaller beams.

5. Conclusions

We have made the first detection of a rotational transition from OH in a comet. The ratio of the opposite parity rotational lines is consistent with radiative pumping schemes and exclude photochemical pumping models. The OH production rate calculated from these far-infrared lines is similar to the rates calculated from IUE measurements. The production rate of OH on May 3, 1986 from comet Halley was $\sim 3.6\,10^{29}$ molecules $\mathrm{s^{-1}}$.

Acknowledgements. We thank Dr. Michael J. Mumma for several useful discussions. Special thanks are due the team of the Kuiper Airborne Observatory whose efforts in Christchurch, New Zealand made this experiment possible. This work was supported by NASA grant NAG2-208.

References

Andresen, P.: 1986, *Astron. Astrophys.* **154**, 42

Despois, D., Gérard, E., Crovisier, J., Kazès, I.: 1981, *Astron. Astrophys.* **99**, 320

van Dishoeck, E.F., Dalgarno, A.: 1984, *Icarus* **59**, 305

Elitzur, M.: 1981, *Astrophys. J.* **246**, 354

Feldman, P.D., Festou, M.C., A'Hearn, M.F., Arpigny, C., Butterworth, P.S., Cosmovici, C.B., Danks, A.C., Gilmozzi, R., Jackson, W.M., McFadden, L.A., Patriarchi, P., Schleicher, D.G., Tozzi, G.P., Wallis, M.K., Weaver, H.A., Woods, T.N.: 1986, 20th ESLAB Symp., Heidelberg

Festou, M.C.: 1981, *Astron. Astrophys.* **95**, 69

Gérard, E. Bocklée-Morvan, D., Bourgois, G., Colom P., Crovisier, J.: 1986, in *Workshop on Cometary Radio Astronomy*, held at Green Bank, West Virginia, USA, Sept. 24-25, 1986

Hildebrand, R.H., Loewenstein, R.F., Harper, D.A., Orton, G.S., Keene, J., Whitcomb, S.E.: 1985, *Icarus* **64**, 64

Krankowsky, D., Lämmerzahl, P., Herrwerth, I., Woweries, J., Eberhardt, P., Dolder, U., Herrmann, U., Schulte, W., Berthelier, J.J., Illiano, J.M., Hodges, R.R., Hoffman, J.H.: 1986, *Nature* **321**, 326

Lugten, J.B., Crawford, M.K. Genzel, R.: 1987, (in preparation)
Mies, F.H.: 1974, *Astrophys. J. Letters* **191**, L145
Mumma, M.J., Weaver, H.A., Larson, H.P., Davis, D.S., Williams, M.: 1986, *Science* **232**, 1523
Piehler, G., Kegel, W.H.: 1986, *Astron. Astrophys.* **155**, L13
Schleicher, D.G.: 1983, Ph.D. Dissertation, University of Maryland
Schleicher, D.G., A'Hearn, M.F.: 1982, *Astrophys. J.* **258**, 864
Snyder, L.E.: 1986, *Astron. J.* **91**, 163
Swings, P.: 1941, *Lick Obs. Bull. No.* 508, **19**, 131
Watson, D.M., Storey, J.W.V., Townes, C.H., Haller, E.E., Hansen, W.L.: 1980, *Astrophys. J. Letters* **239**, L129

Astron. Astrophys. 187, 455–461 (1987)

ASTRONOMY AND ASTROPHYSICS

18-cm wavelength radio monitoring of the OH radical in comet P/Halley 1982i

E. Gérard, D. Bockelée-Morvan, G. Bourgois, P. Colom, and J. Crovisier

Observatoire de Paris, Section de Meudon, F-92195 Meudon, France

Received January 30, accepted March 11, 1987

Summary. The 1667 MHz and 1665 MHz transitions of the OH radical were detected and monitored nearly daily in comet Halley 1982i with the Nançay radio telescope. The observations cover the period July 1985 to July 1986 and provide a unique test for the excitation mechanisms of the OH molecule and the kinematic models of the OH coma. The heliocentric velocity range was fully observed between -26.8 and 26.8 km s^{-1} and 6 line reversals were covered. The OH-parent production rate of P/Halley was measured between 3.4 AU (pre-perihelion) and 2.8 AU from the Sun (post-perihelion). We find new effects like significant short-term variations of the OH-parent production rate in a few days, large variations of the half power profile width and anomalous line shapes.

Key words: comets – comet Halley – OH radical – radio astronomy

1. Introduction

The 1667 MHz and 1665 MHz transitions of the OH radical were recently monitored in several comets with the Nançay radio telescope: P/Halley 1982i (July 1985 to July 1986), P/Giacobini-Zinner 1985 XIII (June to November, 1985), Hartley-Good 1985 XVII (September 1985 to January 1986), Thiele 1985 XIX (November to December 1985), and Wilson 1986l (from August 1986 on). This represents the latest outcome of our continuous effort to observe the 18 cm OH lines in bright comets (Biraud et al., 1974; Despois et al., 1981; Bockelée-Morvan et al., 1981; Gérard et al., 1987). The monitoring of the first two comets is the contribution of the Nançay radio observatory to the International Halley Watch (IHW). A preliminary report of the observations of comet P/Halley is given here. Further analysis of the data is continuing and more detailed results will be published in forthcoming papers.

2. The observations

The observations were made at the Nançay radio telescope (at 18 cm: RA × dec beam = 3.′5 × 19′, aperture efficiency = 0.35 or 0.87 K Jy^{-1}, beam efficiency = 0.48) with a 45 K system temperature receiver. The back-end consists of four banks analysed by a 1024-channel autocorrelator split into 4 × 256 channels with 782 Hz separation (0.14 km s^{-1}). We chose to observe simultaneously the 1667 MHz and 1665 MHz transitions of the OH Λ-doublet with both left and right circular polarizations. The observation mode was frequency switching with a frequency shift of 100 kHz. The daily tracking time with the Nançay meridian telescope is a little over one hour. The radio source NGC 2024 (W12) was observed daily during the comet observing runs in order to check the calibration and stability of the local oscillators.

Comet Halley was tracked using the successive updates of the orbital elements of the IHW astrometry network. The first observing runs started in late January 1985. The 1667 MHz OH line was unambiguously detected for the first time in late August 1985 when the heliocentric distance was ~2.8 AU (Bockelée-Morvan et al., 1985). But the observations from 1 to 10 July 1985, at 3.4 AU from the Sun, already suggested a detection at a 2-σ level. The 1667 and 1665 MHz OH lines were daily monitored whenever the comet was above $-38°$ declination from the end of July 1985 until the end of July 1986. The log of the observations is given in the IHW Radio Science Net Logbook (Schloerb and Irvine, 1986). Most observations were performed with the beam aimed at the nucleus, but from 11 November to 23 December 1985, we spent two thirds of the daily integration time tracking 3.5′ East or West of the nucleus in order to measure the OH extension. On 29–30 October 1985, we attempted to detect the occultation of the strong radio source Taurus A.

During this full year of observations, 6 line reversals were observed in the heliocentric velocity range -26.8 to 26.8 km s^{-1}. Figure 1 shows the evolution of the line area (mJy km s^{-1}/beam) of centre exposures at 1667 MHz as a function of time. The flux density reached ~300 mJy/beam on 23 January 1986 and the r.m.s. noise was ~30 mJy/beam in a 782 Hz channel for most daily observations. Some representative line profiles averaged over several consecutive observing days are shown in Fig. 2.

3. Discussion

3.1. The excitation model

The excitation mechanism of the cometary OH molecule is mostly solar UV pumping followed by fluorescence, which determines the populations n_u and n_l of the upper and lower Λ sublevels of the $^2\Pi_{3/2}$ ground state, and depends upon the comet radial velocity (Swings effect) and the differential motions of the radicals inside the coma (Greenstein effect). Depending on the sign of the

Send offprint requests to: E. Gérard

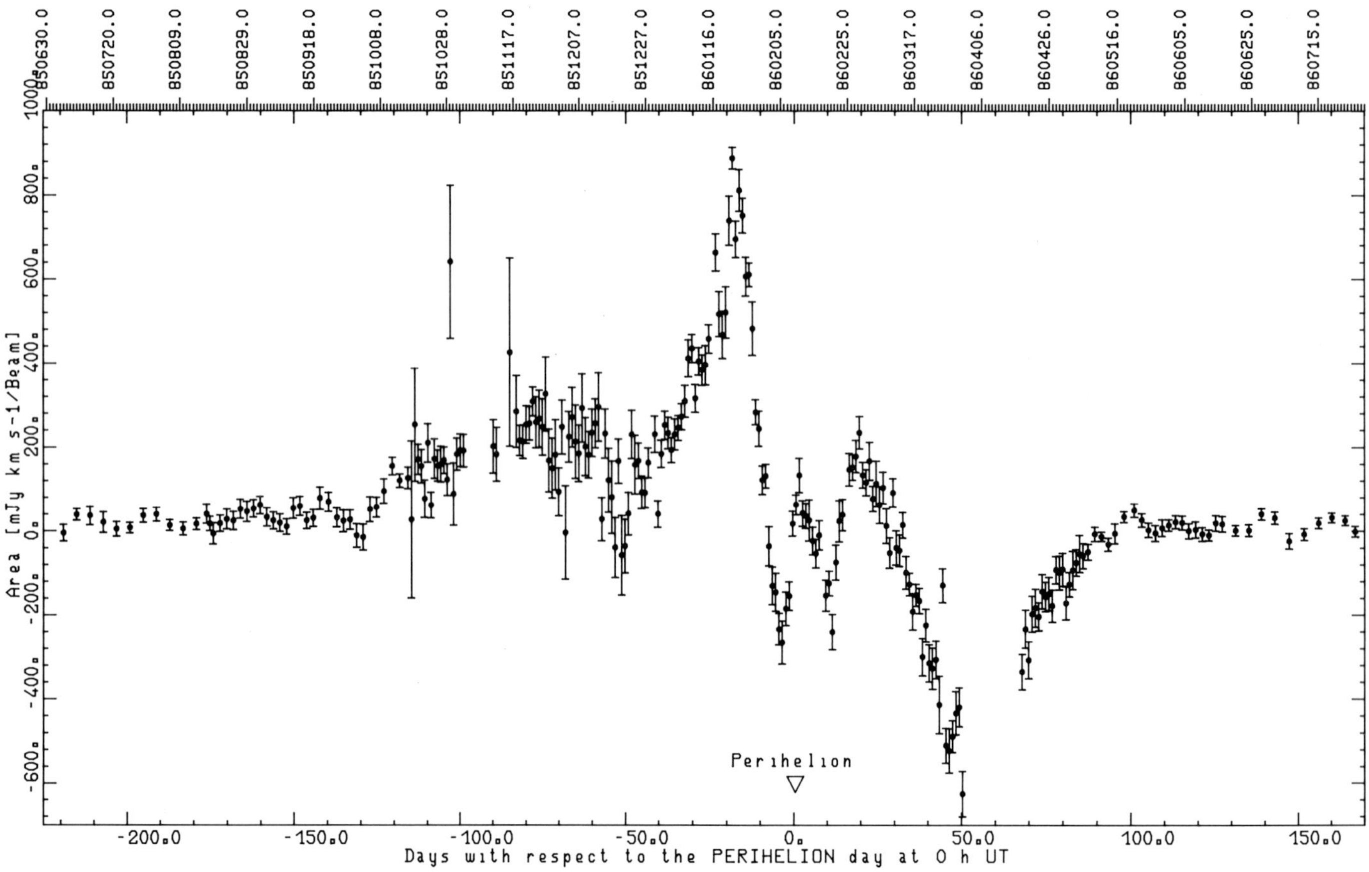

Fig. 1. Line area, with ± 1-σ error bars, of the 1667 MHz OH line in comet P/Halley 1982i as a function of time. Running means calculated over 8 days (1 July to 21 August 1985, 16 June to 29 July 1986), 4 days (18 August to 18 October 1985, 5 May to 17 June 1986) and daily values (17 October 1985 to 5 May 1986) are displayed

inversion $i = (n_u - n_l)/(n_u + n_l)$, the signal appears either in emission or in absorption. In the inner coma, however, collisions may thermalize the Λ doublet levels and quench the signal. The observations of P/Halley spanned a large range of heliocentric velocities, with part of it explored for the first time. Therefore these observations can be expected to provide a check of two available pumping models (Despois et al., 1981, and Schleicher, 1983), which, indeed, do not differ very much.

The behaviour of the OH emission with heliocentric velocities is, in general, consistent with the two excitation models mentioned above. Some remarks can be already given, however. In the velocity range 0 to $\sim 6\,\mathrm{km\,s^{-1}}$, covered by P/Halley in the 9 to 14 February 1986 period, the two models differ. In particular, while Schleicher's (1983) calculations predict a peak inversion of 0.08 and a change from emission to absorption on February 15 for P/Halley, the model of Despois et al. (1981) predicts a peak of only 0.01 and a turn-over on February 13 (Schloerb and Gerard, 1985). The evolution of the signal during this period (Fig. 1) is in favour of Schleicher's model. On the contrary, the observations from March 5 to 8, 1986 (heliocentric velocity $\sim 23.5\,\mathrm{km\,s^{-1}}$) and from May 31 to mid-June, 1986 are more consistent with the pumping model of Despois et al. Thus, the preliminary estimates of the OH-parent production rate widely differ in some heliocentric velocity ranges, depending on which model was used: this is particularly true when the inversion is small.

The 1667 MHz/1665 MHz hyperfine line ratio provides another test of the validity of the UV pumping model. The strongest rotational transitions and the 1667 MHz radio transition get optically thick for column densities of the order of $3\,10^{13}\,\mathrm{mol\,cm^{-2}}$ (Despois et al., 1981): thus the 1667/1665 MHz line area ratio could deviate from the LTE ratio (1.8). In January 1986, the signal-to-noise ratio was high and Fig. 3 shows the 1667/1665 MHz ratio for 4-day averages. There is good general agreement with LTE value, except around 12 January. On the other hand, the 4-day averaging procedure may smooth out significant departures from LTE on isolated days.

3.2. Gaseous production

Once the excitation mechanism is fairly well understood, it is possible to convert the average radio brightness over the beam into an average OH column density. With a radio telescope of small beam size like Arecibo (3′), the average OH column density (hence the signal strength) is large, but the antenna beam only contains a small fraction of the total number of OH molecules: the determination of the OH-parent production rate heavily relies on the density model assumed. For a telescope of large beam size like the Dominion Radio Observatory 30-m antenna, the situation is reversed.

With its 3′.5 × 19′ beam size, the Nançay radio telescope lies in between. Although the angular size of the OH coma – which

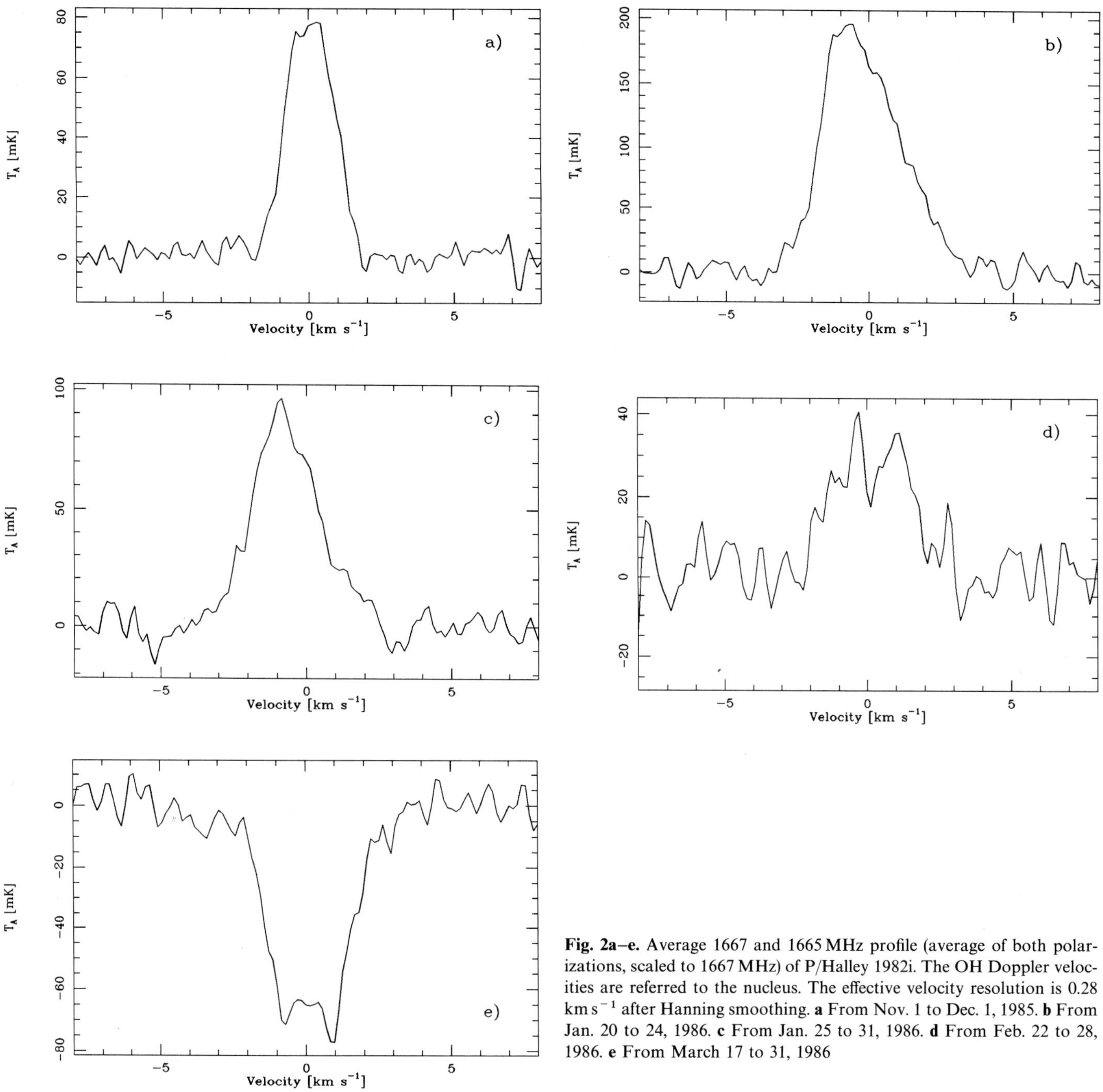

Fig. 2a–e. Average 1667 and 1665 MHz profile (average of both polarizations, scaled to 1667 MHz) of P/Halley 1982i. The OH Doppler velocities are referred to the nucleus. The effective velocity resolution is 0.28 km s^{-1} after Hanning smoothing. **a** From Nov. 1 to Dec. 1, 1985. **b** From Jan. 20 to 24, 1986. **c** From Jan. 25 to 31, 1986. **d** From Feb. 22 to 28, 1986. **e** From March 17 to 31, 1986

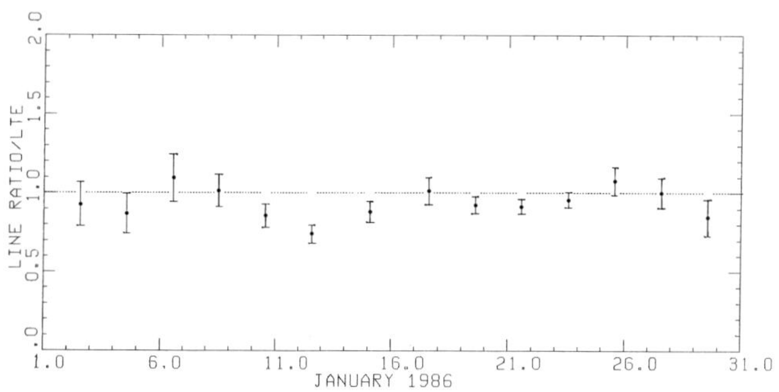

Fig. 3. P/Halley 1667 MHz/1665 MHz line area ratio during January 1986, normalized to the LTE value of 1.8

scales roughly as $r_h{}^2/\Delta$ – varied by a factor of 18 during the observing period, it was always resolved by the beam. Indeed, a preliminary comparison with the DRO 30-m (beam 30′) observations (Galt, 1987) and the Green Bank 43-m (beam 18′) observations (Schloerb et al., 1986) shows that the Nancay point source flux density was always smaller than the other two. The mapping made at Nançay in November and December 1985 was obviously not sufficient to settle the problem of density models, and it is clear that we will greatly benefit from intercomparisons of observations from different radio telescopes throughout 1985–1986. In the mean time, we will rely on existing OH density models to derive OH-parent production rate Q_p. Such models are subject to revision. For instance, recent evaluations of the H_2O scalelength (Krankowsky et al., 1986; Weaver et al., 1986) suggest smaller values than expected from the standard photochemical model: if the parent scalelength is reduced, the OH scalelength should be increased in order to fit the OH UV brightness profiles (Festou, 1981; Bockelée-Morvan, 1982). Clearly, one cannot dispense from mapping the OH coma in order to improve the determination of the model parameters and of Q_p itself.

The determination of Q_p from UV measurements is even more model dependent, since the field of view of many instruments is typically 1′ or less. But the total number of OH molecules can also be computed by integrations of the 309 nm isophotes (Festou, 1981b). Since the OH radio results give systematically lower production rates than the UV results, partly because of the shorter OH scalelength used by UV observers, a new set of parameters (now referred to as 1986a; see Table 1) was proposed during the Green Bank Workshop on Cometary Radio Astronomy (1986 September 24–26), in an effort to reconcile both determinations. This set assumes a parent expansion velocity of $0.9\,\mathrm{km\,s^{-1}}$, in agreement with recent direct measurements of the gas expansion velocity (Despois et al., 1986, Krankowsky et al., 1986), an OH ejection velocity of $1.15\,\mathrm{km\,s^{-1}}$ (Festou, 1986), a parent lifetime of $8\,10^4$ s and an OH lifetime of 10^5 s at $r_h = 1$ AU. OH and OH-parent Haser-equivalent scalelengths are then calculated following the geometrical approximation of Combi and Delsemme (1980). The quenching radius r_q, inside of which the Λ doublet maser is quenched by collisions, is assumed to be nil. We initially presented the OH production rates of P/Halley as a function of time using model 1986a (Gérard et al., 1986a). However, we feel that it is premature to adopt this model for the following reasons: (i) If, as suggested above, the H_2O lifetime is shorter than $8\,10^4$ s at 1 AU and close to $5\,10^4$ s, the vectorial model predicts an OH lifetime of about $3\,10^5$ s at 1 AU. (ii) The Arecibo mapping of the OH coma in late 1985 (Cordes et al., 1986) does not support the model 1986a parameters, and a preliminary analysis of the Nançay fluxes measured on- and off-nucleus leads to the same conclusion: the OH scalelength is 3 to $4\,10^5$ km. (iii) Model 1986a was a useful step in comparing radio and UV OH data, but the radio production rates are still a factor of 2 lower than the UV OH production rates (Schloerb et al., 1986). For these reasons, we will rather present here Q_p determinations using the model parameters of Despois et al. (1981), listed in Table 1.

Figure 4 plots the evolution of the OH production rate of P/Halley as a function of time. 1667 MHz spectra, averaged over 8 days, 4 days, or daily spectra are used (see Fig. 1), depending on the signal-to-noise ratio. The peak production rate reached $\sim 3\,10^{29}\,\mathrm{mol\,s^{-1}}$ in early 1986, when one discards some observations of February 1986 leading to suspiciously high production rates, presumably due to the uncertainties of the excitation model, the inversion being very low at these heliocentric velocities. Figure 4 shows an asymmetry with respect to perihelion, the OH production rate being higher after perihelion for comparable heliocentric distances. Note that this behaviour is observed for the comet heliocentric magnitude and for other gas tracers (Feldman et al., 1986). It is more clearly seen in Fig. 5, which plots the production rate as function of the heliocentric distance.

The OH production rates derived using either the set of parameters 1986a or that of Despois et al. (1981) are still systematically smaller than those derived from UV observations (Feldman et al., 1986) or from other data. Figure 6 shows, for comparison, the radio OH production rates for both sets of parameters for January 1986. Choosing the 1986a set does increase the derived Q_p, but not sufficiently to match the UV values. Setting a quenching radius of the order of $4\,10^4$ km still does not match the data. Adjusting these parameters to reconcile the results is beyond the scope of the present paper. We note, however, that the relative variations of the observed OH production rate practically do not depend on the model parameters, and are reliable.

The OH production rate exhibited also large short-term variations, as evidenced by the January data shown in Fig. 6. During the slow increase from January 1 to 27, there are maxima around 9 and 16, but also minima around 4, 11, and 18. The intensity ratio between maximum and minimum is typically 1.5. In March, Schleicher et al. (1986) reported variations of about 2.5 in the production of OH, NH, CN, C_2 and C_3. Furthermore, Millis and Schleicher (1986) found a 7.4-day periodicity in the light curve of all species during March and April. Such a periodicity could also be present in our data and a detailed analysis is in progress. For such day-to-day variations, the comparison of radio data from different observatories will be invaluable. Large spectral asymmetries were also observed during the January period (Fig. 2b and Sect. 3.3), demonstrating that the nucleus outgassing was anisotropic.

Finally, the OH parent production rate could be compared with the "heliocentric magnitude", m_1^h, deduced from the total visual magnitude, m_1, according to the expression $m_1^h = m_1 - 5\log(\Delta)$. On Fig. 7, we plot $\log(Q_p)$ versus m_1^h, the magnitudes being compiled from the IAU Circulars, over the same time interval as Q_p. A least-squares fit yields:

$$\log(Q_p) = 29.55 - 0.12\,m_1^h\,,$$

Table 1. The set of Haser-equivalent parameters and excitation parameters used to derive the OH-parent production rates

	Model 1986a	Adapted from Despois et al. (1981)
OH-parent scalelength [km][a]	$6\,10^4$	$8\,10^4$
OH scalelength [km][a]	$1.4\,10^5$	$6\,10^5$
OH radial velocity [$\mathrm{km\,s^{-1}}$]	1.4	1.5
Inversion model	Despois et al. (1981)	Despois et al. (1981)
Quenching radius [km]	0	0

[a] at $r_h = 1$ AU

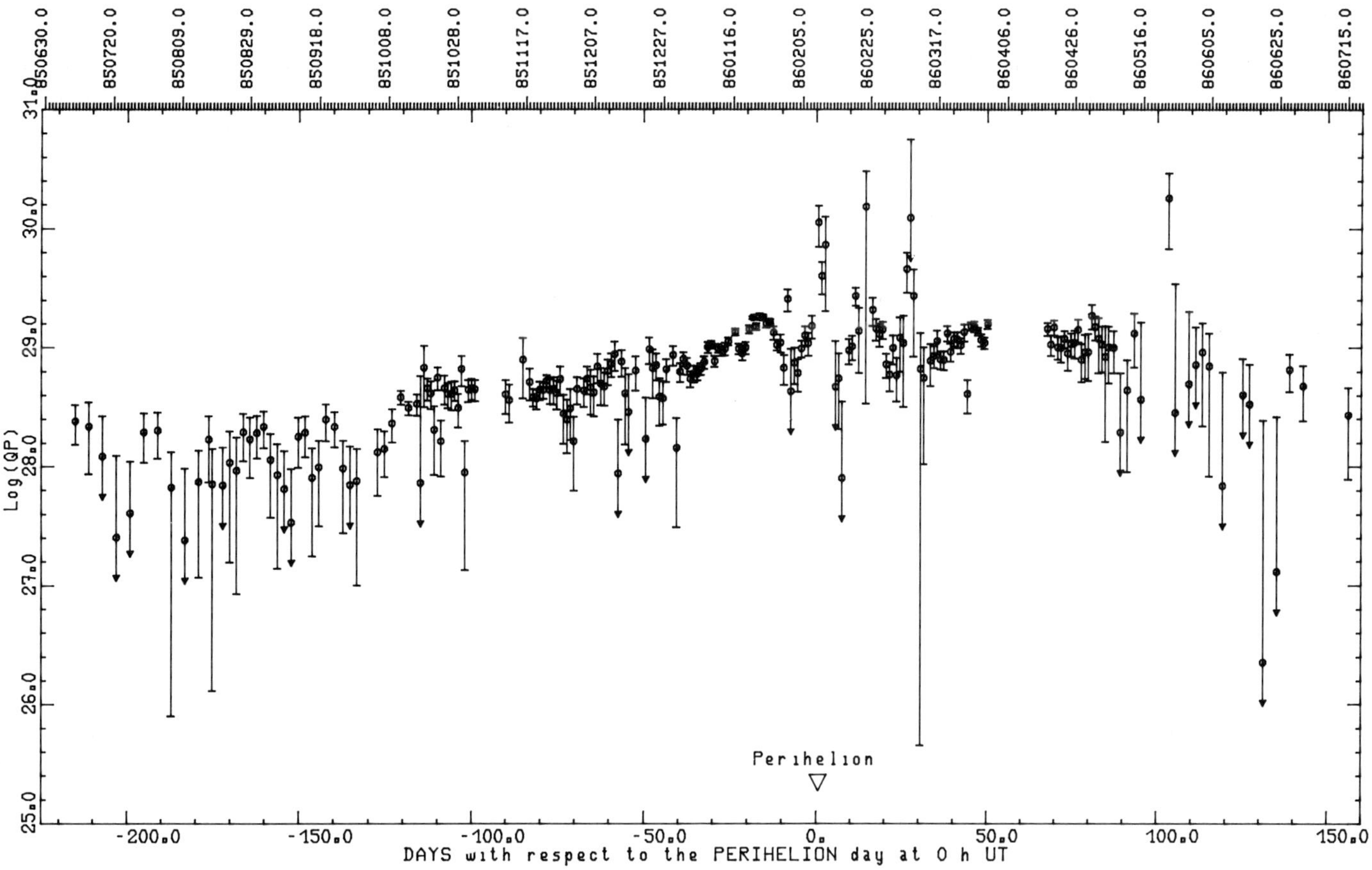

Fig. 4. OH production rate of P/Halley 1982i as a function of time, with ±1-σ error bars (a triangle indicates a 1-σ error larger than the calculated production rate). Calculations are performed with the second set of parameters given in Table 1, adapted from Despois et al. (1981)

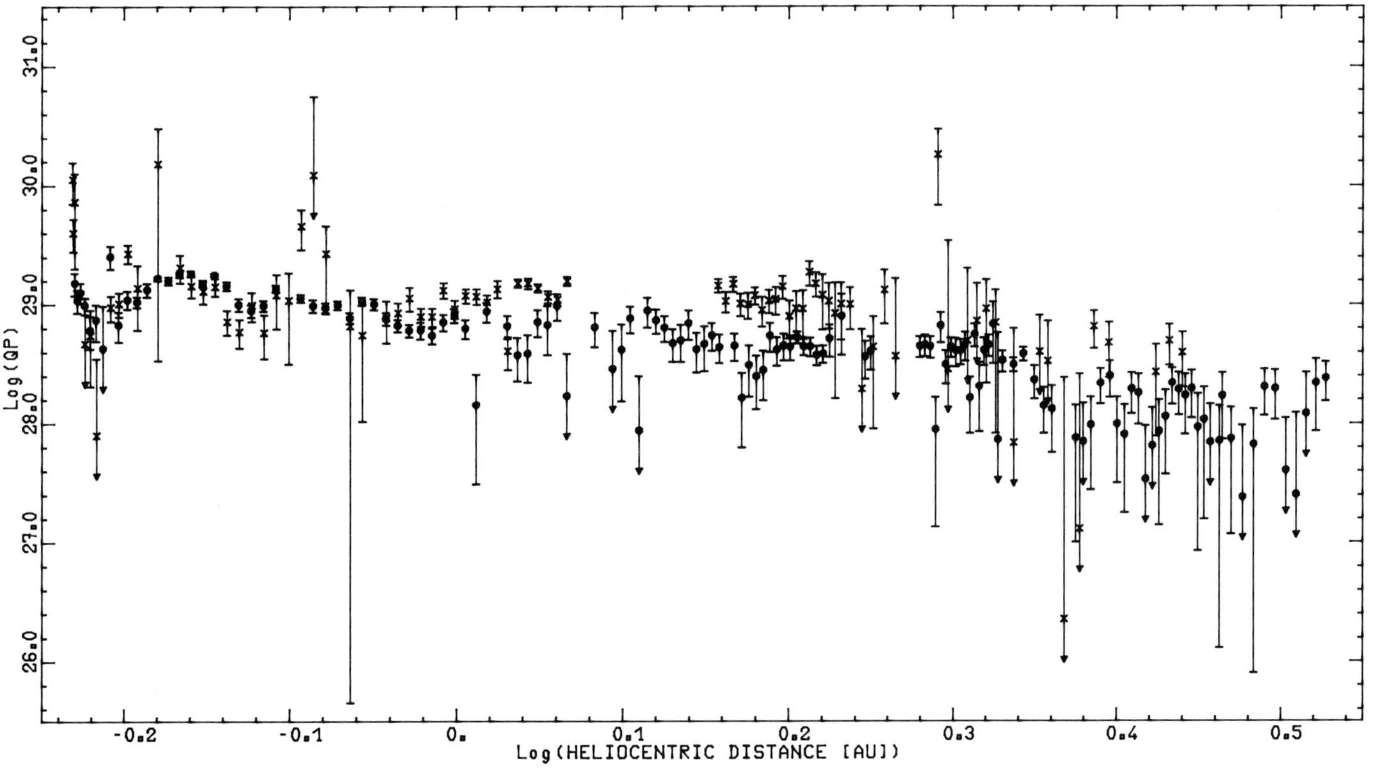

Fig. 5. OH production rate of P/Halley as function of heliocentric distance, calculated from the second set of parameters of Table 1. Circles: pre-perihelion. Crosses: post-perihelion

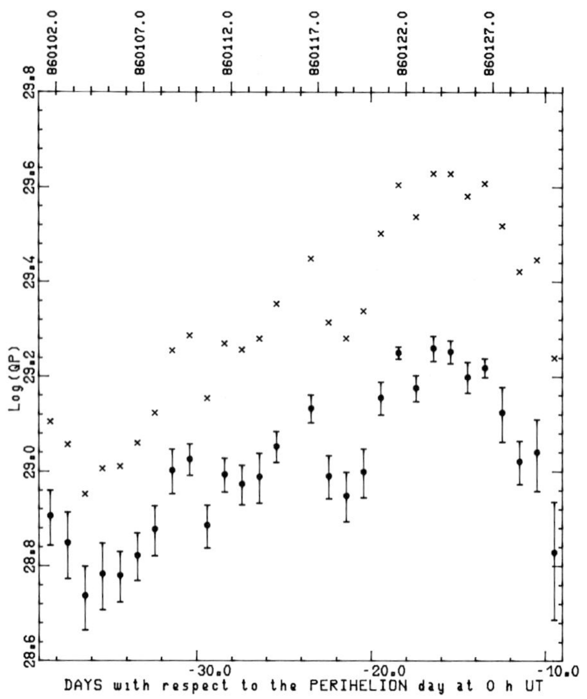

Fig. 6. OH production rate of P/Halley in January 1986. Circles and error bars are from the second set of parameters of Table 1, the crosses are from the first set (1986a)

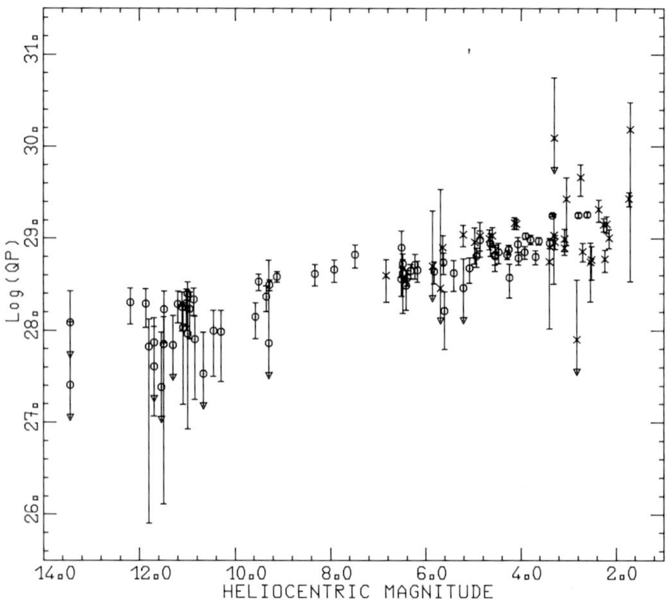

Fig. 7. Variation of the OH production rate Q_p (still computed from the second set of parameters of Table 1) with heliocentric magnitude m_1^h. Circles: pre-perihelion. Crosses: post-perihelion

with a correlation coefficient of 0.86. The variation of $\log(Q_p)$ with m_1^h is nearly the same pre and post-perihelion, although the range is smaller post-perihelion.

3.3. Kinematics of the OH coma

The high velocity resolution of radio spectroscopy allows measurements of the velocity distribution of the OH molecules within the coma. In all spectra, the maximum OH radial velocity with respect to the nucleus does not exceed $3\,\mathrm{km\,s^{-1}}$, which constraints the velocity of possible jets. Many spectra are reasonably symmetrical with respect to the rest velocity of the nucleus, or show an asymmetry which can be analysed in terms of the Greenstein effect (Figs. 2a, c, d, e). Some observations, however, present evidence of strong spectral asymmetries, which could be attributed to anisotropic outgassing (Bockelée-Morvan and Gérard, 1984): the spectrum of comet Halley on January 20 to 24 (Fig. 2b), where no Greenstein effect is expected, has an asymmetrical shape. Note that it does not resemble that shown in Fig. 2c, where a large Greenstein effect is present. The spectrum of Fig. 2d shows a lack of OH radicals with zero radial velocities and does not resemble that generally observed in comets; a similar shape was observed in P/Crommelin 1983n and was interpreted by a peculiar outgassing pattern (Bockelée-Morvan et al., 1985).

We also find considerable variations in the linewidths of the spectra. For example, the linewidth of the average spectrum of P/Halley from Nov. 1 to Dec. 1 (Fig. 2a) is only $2\,\mathrm{km\,s^{-1}}$ while that measured in March 1986 (Fig. 2e) reaches $3\,\mathrm{km\,s^{-1}}$. Note that a similar behaviour was found for the radio lines of HCN between November 1985 and February 1986 (Bockelée-Morvan et al., 1986). Heliocentric distance is likely a factor determining the outflow gas velocity. A quick analysis of the observations of P/Halley suggests that the linewidth decreases as the comet recedes from the Sun. However, the small linewidth of $\sim 2\,\mathrm{km\,s^{-1}}$ observed for comet Hartley-Good 1985 XVII (Gérard et al., 1986b), at $r_h \simeq 0.8$ AU, shows that other factors like gaseous production, anisotropic outgassing... must be considered to explain the observed linewidths.

References

Biraud, F., Bourgois, G., Crovisier, J., Fillit, R., Gérard, E., Kazès, I.: 1974, *Astron. Astrophys.* **34**, 163

Bockelée-Morvan, D.: 1982, Thèse de troisième cycle, Université Paris 6

Bockelée-Morvan, D., Bourgois, G., Crovisier, J., Gérard, E.: 1985, *IAU Circular No* **4106**

Bockelée-Morvan, D., Crovisier, J., Gérard, E., Henkel, C., Jewell, P.R., Snyder, L.E., Clemens, C.A., Molloy, A.R., Schloerb, F.P.: 1985, *Astron. J.* **90**, 2586

Bockelée-Morvan, D., Crovisier, J., Gérard, E., Kazès, I.: 1981, *Icarus* **47**, 464

Bockelée-Morvan, D., Gérard, E.: 1984, *Astron. Astrophys.* **131**, 111

Combi, M.R., Delsemme, A.H.: 1980, *Astrophys. J.* **237**, 633

Cordes, J., Falchi, A., Tofani, G., Lewis, B.M., Terzian, Y.: 1986, 20th ESLAB Symp., ESA SP-250, Vol. **III**, p. 113

Despois, D., Crovisier, J., Bockelée-Morvan, D., Schraml, J., Forveille, T., Gérard, E.: 1986, *Astron. Astrophys.* **160**, L11

Despois, D., Gérard, E., Crovisier, J., Kazès, I.: 1981, *Astron. Astrophys.* **99**, 320

Feldman, P.D., Festou, M.C., A'Hearn, M.F., Arpigny, C., Butterworth, C., Cosmovici, B., Danks, A.C., Gilmozzi, R., Jackson, W.M., McFadden, L.A., Patriarchi, P., Schleicher, D.G., Tozzi, G.P., Wallis, M.K., Weaver, H.A., Woods, T.N.: 1986, in *New Insights in Astrophysics*, ESA SP-263, p. 39

Festou, M.C.: 1981, *Astron. Astrophys.* **96**, 52

Galt, J.: 1987, *Astron. J.* **93**, 747

Gérard, E., Bockelée-Morvan, D., Bourgois, G., Colom, P., Crovisier, J.: 1986a, 20th ESLAB Symp., ESA SP-250, Vol. **I**, p. 589

Gérard, E., Bockelée-Morvan, D., Bourgois, G., Colom, P., Crovisier, J.: 1986b, in "Workshop on Cometary Radio Astronomy", eds. W.M. Irvine, F.P. Schloerb, L. Tacconi-Garman, NRAO, Green Bank, WV, USA, p. 125

Gérard, E., Bockelée-Morvan, D., Crovisier, J.: 1987, (in preparation)

Krankowsky, D., Lämmerzahl, P., Herrwerth, I., Woweries, J., Eberhardt, P., Dolder, U., Herrmann, U., Schulte, W., Berthelier, J.J., Illiano, J.M., Hodges, R.R., Hoffman, J.H.: 1986, *Nature* **321**, 326

Millis, R.L., Schleicher, D.G.: 1986, *Nature* **324**, 646

Schleicher, D.G.: 1983, Ph. D. dissertation, University of Maryland

Schleicher, D.G., Millis, R.L., Tholen, D., Lark, N., Birch, P.V., Martin, R., A'Hearn, M.F.: 1986, 20th ESLAB Symp., ESA SP-250, Vol. **I**, p. 565

Schloerb, F.P., Claussen, M.J., Tacconi-Garman, L.: 1986, 20th ESLAB Symp., ESA SP-250, Vol. **I**, p. 583

Schloerb, F.P., Gérard, E.: 1985, *Astron. J.* **90**, 1117

Schloerb, F.P., Irvine, W.M.: 1986, *Log of Radio Observations of Halley's comet and comet P/Giacobini-Zinner*", International Halley Watch Radio Science Net Logbook

Weaver, H.A., Mumma, M.J., Larson, H.P., Davis, D.S.: 1986, *Nature* **324**, 441

Astron. Astrophys. 187, 462–464 (1987)

ASTRONOMY
AND
ASTROPHYSICS

10.7 GHz continuum observations of comet P/Halley

A. Falchi, L. Gagliardi, F. Palagi, G. Tofani, and G. Comoretto

Osservatorio Astrofisico di Arcetri, Largo Enrico Fermi 5, I-50125 Florence, Italy

Received February 9, accepted April 21, 1987

Summary. A search for continuum emission at 10.7 GHz from Halley's comet has been performed with the Bologna 32 m radiotelescope near the first comet perigee. Owing to the expected low fluxes the beam switching ON-OFF technique was adopted for the observations. Sky background level was determined by subtraction of instrumental and sky temperature long term drifts using a least square fit method. Long time integrations (>3000 s) were obtained for 7 days. The results indicate 2 detections greater than 3σ level with fluxes of $\sim$14 mJy.

Key words: comets – radio continuum

1. Introduction

Radio continuum emission has been successfully detected from few comets: Kohoutek 1973 XII (Hobbs et al., 1975), West 1976 VI (Hobbs et al., 1977) and Iras-Araki-Alcock (IAA) 1983d (Altenhoff, 1983). Strong differences in flux density were observed for Kohoutek and West on time scale of a day, suggesting that the continuum may be a transient phenomenon.

Gibson and Hobbs (1981) interpreted the cm wavelength emission of Kohoutek and West in terms of an Icy Grain Halo (ICH) surrounding the nucleus. More recent observations of comet Austin at 6.1 cm (Snyder et al., 1983) established a lower limit to the flux density which is far below the level predicted by the IGH model.

The single dish emission of IAA at 1.3 cm (Altenhoff, 1983) is consistent with a nucleus of 7 km diameter at 200 K. The nearly simultaneous observation of IAA with VLA at 2 cm (Irvine et al., 1984) gives a lower limit for the flux density significantly below the detection at 1.3 cm. The discrepancy may be explained if the emission is due to an extended (100–400 km) component of grains.

The emission of Kohoutek and West corresponds to a monochromatic luminosity order of magnitude larger than that found in Iras-Araki-Alcock. Probably different mechanisms should be invoked for these types of emission.

In the case of the Halley comet the effort of the astronomical community may contribute to better interpret this continuum emission. For this reason a program of observations at 2.8 cm was performed with the new radiotelescope in Bologna. The method of observations and the results are discussed in the following.

Send offprint requests to: A. Falchi

2. Instrument and observations

The Bologna antenna is a 32 m Cassegrain operating at the frequencies of 1.6, 5, 10.7 and 22 GHz. The sensitivity is determined by cryogenic FET amplifiers, quality of the antenna surface and pointing accuracy. The search for comet emission was done at 10.7 GHz as a compromise between sensitivity, spectrum of emission and weather conditions at the epoch of observations. At small zenith angles the system temperature is about 100 K (50 MHz bandwidth) and the antenna gain is 0.1 K/Jy. The Half Power Beam Width is 2.′5, the absolute pointing accuracy, tested on a set of point sources, is about 15″; first sidelobe levels are below 18 dB. The linear response of the instrument to calibration sources with flux lower than 1 Jy was tested (Fig. 1). Fluxes, where necessary, are corrected for zenith angle gain variations.

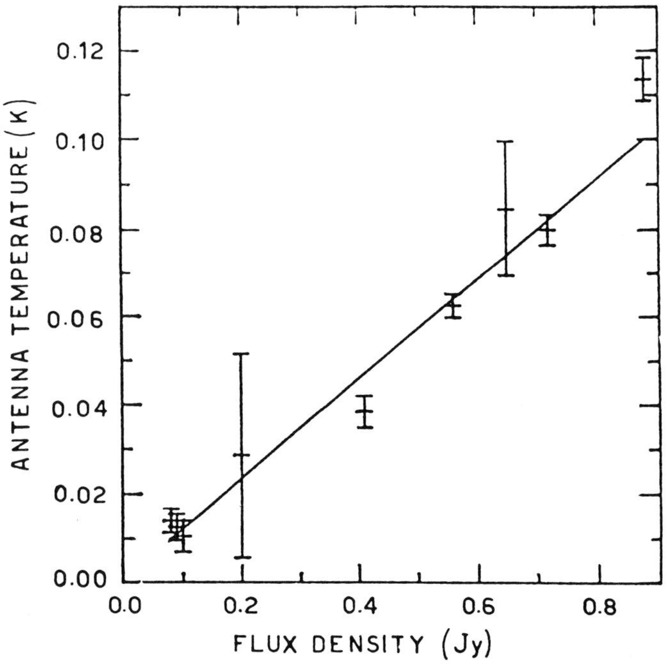

Fig. 1. Antenna temperature of a set of standard calibration sources. The mean slope represents the antenna gain

The comet nucleus was tracked at sidereal rate by updating the nominal position every 60 s according to the IHW Ephemeris program. Observations were performed with the total power ON-OFF technique with a basic cycle of 10 s ON source and 10 s OFF source (15′ Eastward in Azimuth). Every 300 s the signal from a noise source was recorded. Hereafter we call observation cycle the period between two noise calibration.

In Fig. 2 the ON-OFF antenna response for 3 different sources is shown. The integration time for each acquisition ON or OFF source is 10 s.

Tracking has been done for several hours every day during the period from 6 November to 2 December 1985. Weather conditions and receiver stability limited the available data to 8 days.

3. Data reduction

An integration time of 3000 s was required in order to reach one sigma sensitivity less than 5 mJy. During this time interval the observation cycles may be both affected by fast fluctuations due to interferences and long drifts due to the atmosphere and receiver gain changes. Both effects were removed before the integration.

From the analysis of each observation cycle all signals with an r.m.s. exceeding 5σ level were rejected (1σ is the accepted error level corresponding to the integration time of one cycle). However the excluded cycles do not exceed the 20% of the total.

We assume that sky temperature long term drifts may be fitted by linear or square functions of time using a least square method.

In each observation cycle there are 15 ON-OFF acquisitions and for each of them we can write the following equations:

$$T_{\mathrm{sys,ON}} = T_{\mathrm{rec}} + \alpha t + \beta t^2 + T_{a(\mathrm{source})}$$

$$T_{\mathrm{sys,OFF}} = T_{\mathrm{rec}} + \alpha t + \beta t^2$$

where T_{sys} (K) is the system temperature, T_{rec} (K) the receiver noise temperature, t is the time for each acquisition in ON or OFF position, α and β the coefficients of the linear and square term respectively. $T_{a(\mathrm{source})}$ is the source temperature contribution. The least square solution of this equation system is derived for ON and OFF cycles together. It supplies the value of $T_{a(\mathrm{source})}$ and its standard error. The square term is used only if the data cannot be fitted with the linear one. We check further that the obtained value of $T_{a(\mathrm{source})}$ does not have a residuals systematic drift with time. The line fitted to $T_{a(\mathrm{source})}$ has a slope of about $10^{-4}\,\mathrm{K\,s^{-1}}$.

The antenna temperature for each cycle is averaged with weight of its standard error to obtain the observed antenna temperature for each day. The observation period, the integration time on the source, the measured flux density, the r.m.s. are given in Table 1.

The encounters of Halley with background sources during these periods of observation were checked on the list provided by IHW. Only on 21 and 29 November two sources, OD + 181 and OC + 132, lie in the sidelobe structure. Their fluxes at 10.7 GHz were derived from the quoted 1.4 GHz values assuming a spectral index of −0.7. The resulting possible contribution is 1.5 mJy at maximum.

The time spent OFF-source may be used to check the internal consistency of our data. For each observation cycle we write a system of equations, as the former ones, where the ON and OFF data are formed by the sequence of the real OFF data only. The expected value of $T_{a(\mathrm{source})}$ should be of the order of zero, within the error, if OFF contributions during one observing session are always the same. In Table 2 integration time, measured

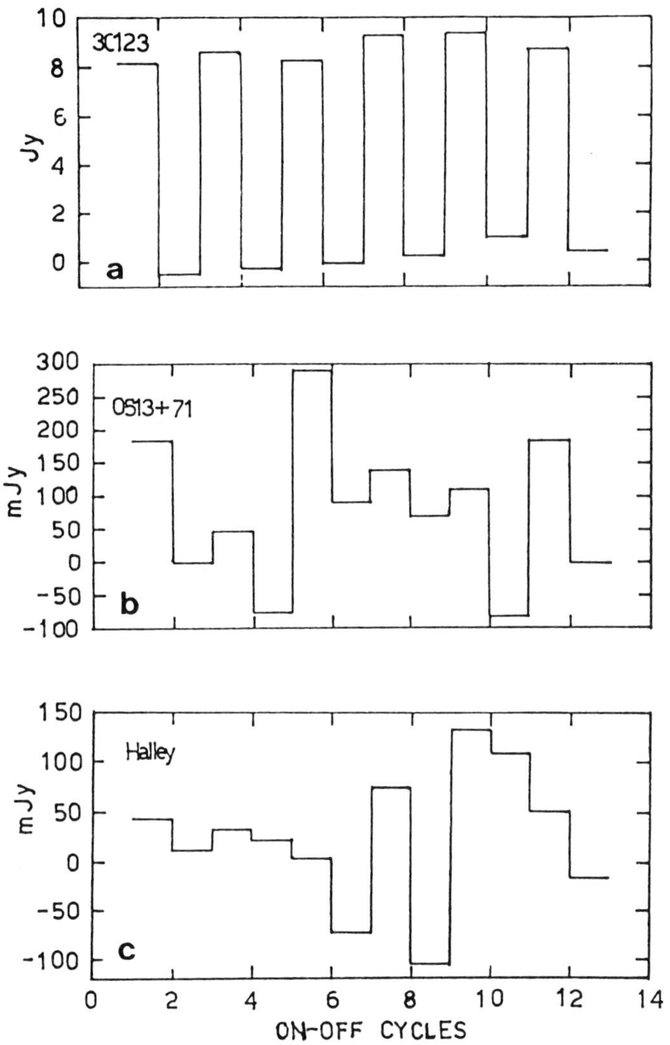

Fig. 2a–c. Examples of the ON-OFF cycles for sources of different intensity. **a** 3C 123, flux 8.23 Jy; **b** 0513 + 71, flux 0.09 Jy; **c** Halley comet. The integration time for each ON or OFF position is 10 s

Table 1. Observed flux density and its error for different days

Date (UT) 1985	Int. time (s)	Flux (mJy)	r.m.s. (mJy)
Nov. 7.78– 7.99	4800	15	3.3
15.79–16.06	4800	8.6	2.5
21.67–21.86	4200	14.3	4.1
26.82–27.06	5250	5.0	2.4
27.64–27.14	1500	−1.	6.8
28.66–29.04	7200	3.5	1.7
29.67–30.04	6180	13.7	4.5
Dec. 1.59– 1.75	3500	−1.6	4.1

Table 2. Flux density and its error for the time spent OFF – source on different days

Date 1985	Int. time (s)	Flux (mJy)	r.m.s. (mJy)
Nov. 7	2200	−1.8	6.9
Nov. 15	2180	9.4	4.3
Nov. 21	1890	−3.5	7.5
Nov. 26	2270	−6.8	4.0
Nov. 27	750	4.5	13.6
Nov. 28	3600	−7.1	3.5
Nov. 29	2060	7.9	6.2
Dec. 1	1750	−18.0	7.0

OFF contribution and r.m.s. are given. It must be noticed that the errors of Table 2 are respectively higher than the ones in Table 1 due to the reduction by a factor of two in the integration time.

From the analysis of Table 1, possible detections at 3σ level occur on 7, 15, 21, 29 November. The comparison between Table 1 and 2 indicates: **a)** a good consistency between noise level and off contribution on November 7 and 21, **b)** an off contribution higher than the noise level but lower than the signal on November 29, **c)** an off contribution higher than the noise level and of the order of the signal on November 15.

We think that the observations referring to 7 and 21 may be considered real detections, while the ones referring to 15 and 29 have probably some contribution from other sources and cannot be considered detections.

4. Conclusion

Radio continuum emission from Halley has been observed in the same period by Altenhoff et al. (1986) at 86 and 226 GHz with the 30 m MRT on the Pico Veleta. Their day by day results gave only upper limits and a comparison between our and their data may be tentative.

The real overlap between our and their observations is only on 3 days: November 21 and 28 and December 1. On November 28 and December 1 no emission is detected at any frequency. On November 21 a flux of 14 mJy ± 4 is detected at 10.7 GHz, an upper limit of 36 mJy is given at 86 GHz and no observation is quoted at 226 GHz.

The expected flux density of a nucleus with a radius of 5 km and a temperature of 200–400 K is some order of magnitude lower than the 10.7 GHz measured fluxes or than average values given by Altenhoff et al. at both frequencies.

The alternative model of an Icy Grain Halo cannot explain at the same time our and Altenhoff observations because it refers to a thermal emission from an optically thick sphere. If with a proper choice of the parameters such as the size or the emissivity of grains, the IGH model may be matched to the observed fluxes at one frequency, there are no ways to use the same parameters for all the frequencies. The existence of different mechanisms for the emission at cm and mm wavelength should be investigated. The detection of a signal on two days only, out of seven, may suggest that continuum radiation is due to sporadic outbursts with expulsion of mm-cm sized dust grains.

Acknowledgements. We would like to acknowledge the International Halley Watch team for providing us the Ephemeris program.

This research was carried out with financial support of the Italian Consiglio Nazionale delle Ricerche.

References

Altenhoff, W.J.: 1983, *Astron. Astrophys.* **125**, L19
Altenhoff, W.J., Huchtmeier, W.K., Schmidt, J., Schraml, J.B., Stumpff, P., Thum, C.: 1986, *Astron. Astrophys.* **164**, 227
Gibson, D.M., Hobbs, R.W.: 1981, *Astrophys. J.* **248**, 863
Hobbs, R.W., Maran, S.P., Brandt, J.C., Webster, W.J., Swamy, K.S.K.: 1975, *Astrophys. J.* **201**, 749
Hobbs, R.W., Brandt, J.C., Maran, S.P.: 1977, *Astrophys. J.* **218**, 573
Irvine, W.M. et al.: 1984, *Icarus* **60**, 215
Snyder, L.E.: 1982, *Icarus* **51**, 1
Snyder, L.E., Palmer, P., Wade, C.M.: 1983, *Astrophys. J.* **269**, L21
Walmsley, C.M.: 1985, *Astron. Astrophys.* **142**, 437

Astron. Astrophys. 187, 465–468 (1987)

ASTRONOMY AND ASTROPHYSICS

Rotational structure of the (2, 0) Phillips band of C_2 in comet P/Halley ★

I. Appenzeller [1] and G. Münch [2]

[1] Landessternwarte, Königstuhl, D-6900 Heidelberg 1, Federal Republic of Germany
[2] Max-Planck-Institut für Astronomie, Königstuhl 17, D-6900 Heidelberg 1, Federal Republic of Germany

Received October 24, 1986; accepted March 3, 1987

Summary. The results of the study of a digital echellogram of the nuclear region of comet Halley, containing the (2,0) Phillips band of C_2 fully resolved, are presented. The intensities of individual lines, measured from the R-branch head up to $R36$, $Q30$, and $P24$, indicate a very high degree of rotational excitation. The population of the levels from where the lines arise, when approximated by a Boltzmann distribution, is found to be characterized by a temperature higher than that obtained from the Swan bands in other comets. An absolute calibration of the line fluxes has been obtained from narrow band CCD-imagery, to make feasible their comparison with measurements in the Swan system, probably carried out by other observers of comet Halley.

Key words: comets – molecules – spectrophotometry

1. Introduction

The interpretation of the C_2 Swan bands, $d^3\Sigma_g - a^3\Pi_u$, in cometary spectra, dominant emissions in the visible range, raises problems not yet completely solved, consequent to the homopolar nature of the molecule. A molecule brought by resonance fluorescence of sunlight to an excited vibrational level $v'' \geqq 4$ of $a^3\Pi_u$ does not decay to $v''=0$ by a radiative cascade within the system, because it would involve quadrupole transitions of very small probability. Instead, the cascade takes place mostly by the emission of two infrared photons in the Ballik-Ramsay system, $b^3\Sigma_g^- - a^3\Pi_u$. The v'' levels of $a^3\Pi_u$ can alternatively decay by spontaneous transitions to the $v''=0$ levels of the ground state $X^1\Sigma_g^+$ of the Phillips and Mullikan systems. In the absence of intersystem transitions, the vibrational temperature of the Swan bands would be expected to be close to the color temperature of the sun (Stockhausen and Osterbrock, 1965), independently of heliocentric distance. Observationally such is not found to be the case. The vibrational temperatures of Swan bands predicted by Krishna-Swamy and O'Dell (1977, 1979; referred to below as KSO), taking into account mixing of $a^3\Pi_u$ with $b^3\Sigma_g^-$ and $X^1\Sigma_g^+$, depend on heliocentric distance and are considerably lower than 6000 K. The transition moment $|R_e|^2$ for the intersystem transitions $a^3\Pi_u - X^1\Sigma_g^+$, however, is not accurately known. The value $T_{v'} = 3500 \pm 400$ K, determined by Lambert and Danks (1983) for the $1 \leqq v' \leqq 5$ levels of $d^3\Pi_g$ in comet West at 0.75 AU, agrees with theory (KSO) if $|R_e|^2 \simeq 5\ 10^{-5}$, but much smaller values have been derived theoretically and from measured cometary flux ratios between triplet and singlet bands, so far carried out for the latter with very low spectral resolution. The populations of the states corresponding to the various band systems are thus uncertain, shortcoming with implications for the understanding of the formation process of C_2.

The flux ratio between Phillips and Swan bands has been predicted (KSO) to discriminate $|R_e|^2$ more sensitively than the ratio between Swan bands (vibrational temperatures). Unfortunately, the Phillips bands, because of their small f-value, are weaker than the Swan bands by two orders of magnitude and fall in an observationally difficult range (0.76–1.2 μm). Most of the reputed measurements of unresolved Phillips bands are really no more than simple detections and, to our knowledge, not a single observation of the rotational structure within a Phillips band has been recorded in the literature. This communication contains the results of the study of a single digital spectrogram of comet Halley showing the rotational structure of the (2,0) Phillips band distinctly resolved. Because it was obtained during twilight time, as a fill-in to a program concerning extragalactic objects, comparative observations of the Swan bands and a spectrophotometric calibration were not secured. However, through narrow band CCD-images of Halley at 985 nm, obtained simultaneously to the spectral observations with another telescope, we have determined an absolute scale for the line fluxes. It is thus believed that the data here presented provides a basis for a comparison with Swan band fluxes that other Halley observers may have obtained, and hence for a determination of $|R_e|^2$ more accurate than those heretofore attempted.

2. Observations

2.1. Spectroscopy

The spectrograms of comet Halley here discussed were obtained on March 3.39 and 5.39 (UT), 1986, at the 3.6 m telescope of the European Southern Observatory, using the CASPEC spectrograph (D'Odorico et al., 1983) configured with a 31.6 grooves/mm echelle grating, a 300 grooves/mm cross disperser, and an $f/1.5$ camera. The detector was a CCD-array (RCA Type SID 501 DX). Both spectrograms have a resolution of about 0.45 Å (FWHM)

Send offprint requests to: I. Appenzeller

★ Based on observations collected at the European Southern Observatory, La Silla, Chile and cover the range 810–910 nm. On March 3 a 2″ × 22″ slit centered on the nucleus and oriented East-West was used. The frame, exposed for 5 min, shows a strong Fraunhofer spectrum reflected by the nuclear region, together with C_2 line emission weakly recorded beyond the nuclear region. Since the C_2 line intensities are decreasing with distance to the nucleus less steeply than that of the solar continuum, on March 5 a longer slit (2″ × 28″) was used, centered 6″ West of the nucleus and also in p.a. 90°, with an exposure increased to 10 min. In order to account for airglow contamination, a 5 min exposure of the "night sky" 2′ West from the nucleus was secured immediately after the cometary exposure.

The spectrograms have been reduced and corrected for instrumental effects using image processing software existing at the Landessternwarte Heidelberg. Since the March 5 spectrogram was obtained with the maximum slit length compatible with order separation, a reliable flat field correction could not be established. Moreover, the limited time available for the cometary observations did not allow the obtention of exposures of standard stars to establish an absolute spectrophotometric calibration. The flux distribution in the spectrograms, therefore, has been normalized to a flat arbitrary continuum level. Notwithstanding the flat-field correction problem referred to above, which could introduce errors up to 15% in the relative flux scale, the March 5 spectrogram contains information about the C_2(2,0) Phillips band more extensive and accurate than the March 3 does. Our discussion below is thus restricted to the March 5 results.

The spectra extracted from the echelle orders 65 and 64, covering the range 8660–8935 Å, which includes the C_2(2,0) band from the *R*-branch head up to the *R*36, *P*24, and *Q*30 lines, are shown in Fig. 1. Cometary lines, some probably due to C_2, can be seen beyond 8935 Å, but their measurement is practically precluded by the onset of heavy terrestrial H_2O absorption (air mass 2.1) and at $\lambda > 9148$ Å by the presence of the very strong emission in the CN(1,0) red system cometary band. The lower tracings of Fig. 1 show the spectral flux integrated over the whole 2″ × 28″ slit, while the upper tracings show the spectrum through a "software slit" of 2″ × 8″ centered 11″ West of the nucleus. The C_2 emission in the latter is enhanced with respect to the continuum and hence is less affected by reflected Fraunhofer lines. No attempt to subtract the OH airglow lines of the (7,3) band, with an *R*-branch head unfortunately near that of C_2(2,0), has been undertaken. The most prominent OH lines, however, are indicated in Fig. 1 at the wavelengths calculated from the term values given by Coxon (1980) and Coxon and Foster (1982). The identification of a number of strong Fraunhofer lines has also been entered in Fig. 1. The wavelength scale is in the observer's reference frame and hence the cometary lines exhibit a shift of $-39\,\mathrm{km\,s^{-1}}$, while the solar absorption lines show an additional shift due to the $+23\,\mathrm{km\,s^{-1}}$ comet velocity relative to the sun. The identification of the C_2(2,0) lines, as given in Fig. 1 and Table 1, has been carried out on the bais of the laboratory measurements of Chauville et al. (1977). In general, the agreement between observed and laboratory wavelengths is within 1 pixel ($10\,\mathrm{km\,s^{-1}}$). The peak

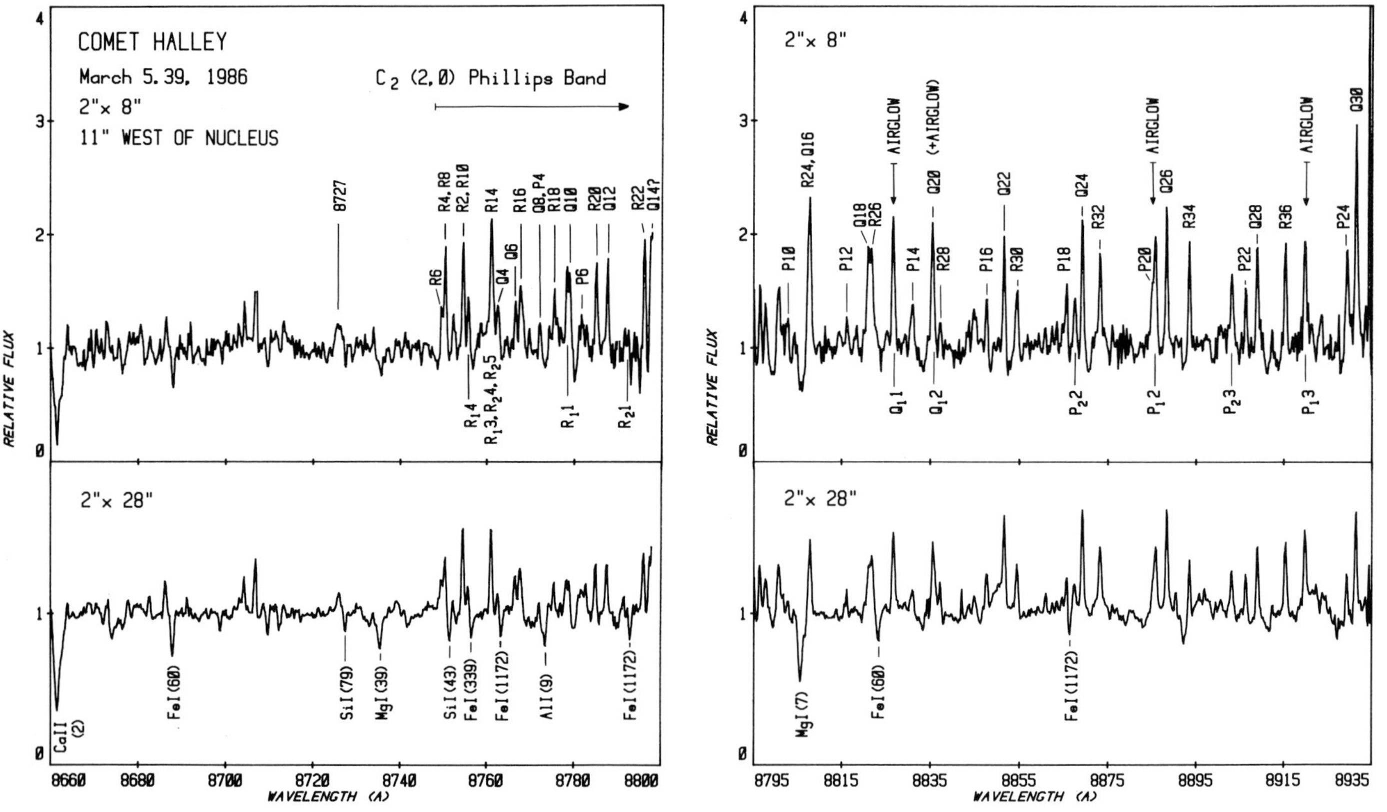

Fig. 1. Sections of the comet Halley spectrum containing the (2, 0) Phillips band of C_2, from the *R*-branch head up to the *Q*30 line. The upper tracings refer to a 2″ × 8″ area of the coma 11″ West of the nucleus. The lower tracings refer to the whole 2″ × 28″ slit and include the nucleus. In the upper tracings the identification of C_2 lines has been entered at the top, while that of OH (7, 3) airglow lines is given at the bottom. Some of the scattered solar absorption lines are identified in the lower tracings

Table 1. Identified lines of the C_2 (2,0) Phillips band

λ_{lab} (Å)	Identi-fication	Relative peak int.	W_λ (Å)	Remarks
8750.85	*R*6	0.36:		
8751.49	*R*8			Blend (*R*4)
8751.69	*R*4			Blend (*R*8)
8753.58	*R*10			Blend (*R*2)
8753.95	*R*2			Blend (*R*10)
8762.15	*R*14	1.12:	0.41:	Blend (airglow)
8763.75	*Q*4	0.37		
8767.76	*Q*6	0.40		
8768.63	*R*16	0.53		
8773.22	*Q*8	0.21:		Blend (*P*4)
8773.43	*P*4			Blend (*Q*8)
8776.61	*R*18	0.53		
8780.14	*Q*10	<0.64		Blend (airglow)
8782.31	*P*6	0.29		
8786.04	*R*20	0.72	0.24	
8788.56	*Q*12	0.77	0.23	
8796.97	*R*22	0.93:		Near edge of frame
8798.46	*Q*14	0.93:		Edge of frame
8804.50	*P*10	0.27		
8809.40	*R*24			Blend (*Q*16)
8809.84	*Q*16	<1.31		Blend (*R*24)
8817.83	*P*12	0.27		
8822.73	*Q*18	≦0.88		Blend (*R*26)
8823.37	*R*26	≦0.85		Blend (*Q*18)
8832.68	*P*14	0.37	0.14	
8837.12	*Q*20	≦1.09		Blend? (airglow)
8838.84	*R*28	>0.21		Weakened by CCD defect
8849.07	*P*16	0.43	0.15	
8853.04	*Q*22	0.97	0.42	
8855.87	*R*30	0.51		
8866.99	*P*18	0.56		
8870.51	*Q*24	1.12	0.45	
8874.46	*R*32	0.82	0.23	Abs.?
8886.48	*P*20	0.56:		Blend (airglow)
8889.53	*Q*26	1.23	0.42	
8894.62	*R*34	0.91		Abs. in blue wing
8907.54	*P*22	0.51	0.28	
8910.13	*Q*28	0.85	0.36	
8916.40	*R*36	0.91	0.43	
8930.17	*P*24	0.83:	0.24	Near edge of frame
8932.33	*Q*30	<1.92		Near edge of frame

intensities of the C_2 lines identified in the upper tracings, in units of the continuum level in the same tracings, is given in column 3 of Table 1. The equivalent widths, referred to the nuclear continuum, of a few apparently unblended strong C_2 lines seen in the lower tracings, is also given in column 4 of Table 1.

2.2. Photometric calibration

On 1986 March 5.39, CCD-images of comet Halley were obtained by Trauger et al. (1987) at the 2.2 m telescope at ESO, with an *f*/1.0 focal reducer through an interference filter (FWHM = 8.3 Å) peaked at 9850 Å in parallel light. Three exposures, each of 5 minute duration, were obtained with the filter tilted at different angles, in order to ascertain that the cometary C I 9850 Å emission did not contribute significantly to the recorded signal. The processing of these photometrically calibrated frames, carried out by Dr. H. Hippelein at our request, provides for the unit continuum of the lower tracings in Fig. 1, a mean intensity 2.2 10^{-11} erg cm^{-2} s^{-1} sr^{-1} Å^{-1}. The slight extrapolation of the cometary continuum from 9850 to 8900 Å, carried out with a solar color gradient, is not critical.

3. Results

3.1. Degree of excitation

The presence in our spectra of a strong $R36$ line, arising from a level with rotational energy $E(J') = 3860$ K, indicates a very high degree of excitation within the $v' = 2$ state. The near impossibility to measure intensities of lines with upper rotational quantum number $J' \geqq 38$, leads us to characterize approximately the population of the $v' = 2$ state as a whole by a Boltzmann distribution at temperature T_r. This procedure has been applied in the past to many comets (Arpigny, 1966; Danylewich et al., 1978; A'Hearn, 1978; Lambert and Danks, 1982). The measured intensities $I(J'J'')$ of lines at frequencies $\nu(J'J'')$ are then expressed as

$$I(J'J'') \propto (2J'+1)\exp\{-E(J')/kT_r\}\, S_{J'}\, \nu^4(J'J''),$$

where the transition probabilities have been written in terms of the Hönl-London factors normalized to $2J'+1$ (Herzberg, 1950). The slope best fitting the logarithmic plot of the intensities shown in Fig. 2 then determines T_r^{-1}. Disregarding obviously unreliable measurements, from a formal unweighted least squares fit we find

$$10^4\, T_r^{-1} = 0.97 \pm 0.40 \quad (T_r \text{ in K})$$

for the mean slope and its standard deviation ($\pm\sigma$). The large σ affecting this determination of T_r^{-1} arises from the weak dependence of T_r on the intensities of lines with $E(J') < kT_r$, as those contained in our sample. Nevertheless, our T_r would appear to be in accord with the 5000 K rotational temperature given by Arpigny (1966) for the (0,0) and (1,0) Swan bands in various comets, since it would imply an error in our determination between 2σ and 3σ. In order to make our T_r compatible with values in the range 4000–3000 K obtained for the same bands in more recent and presumably more accurate determinations (Danylewich et al., 1978; A'Hearn, 1978; Lambert and Danks, 1982), our error would have to be between 4σ and 6σ. On the whole, our data would thus seem to indicate for the $v' = 2$ state of the Phillips system a rotational temperature not smaller than that characterizing the Swan bands.

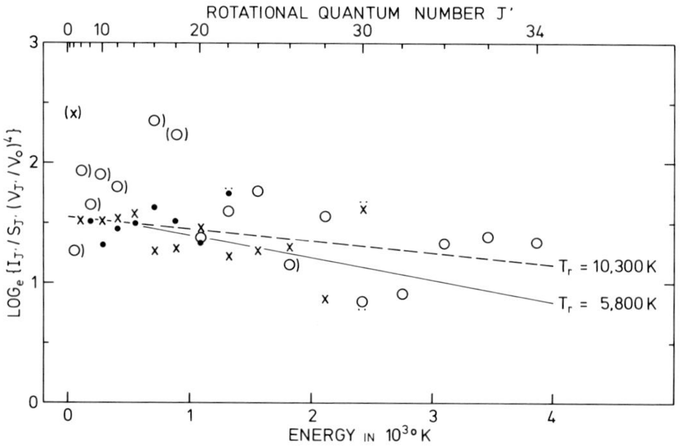

Fig. 2. Fit of a Boltzmann distribution at temperature T_r to the populations of the upper rotational levels J' from where the observed lines originate. Empty circles, product signs, and filled small circles, refer to R, Q, and P lines, respectively. Symbols within parenthesis indicate lines blended with OH lines, while those followed by a closing parenthesis denote blends of C_2 lines. Lines with uncertain intensities due to underlying solar absorption lines or afected by edge effects are marked with two dots

3.2. Emission feature at 8727 Å

The emission line appearing in Fig. 1 near the wavelength 8727.13 Å of the [C I] $^1S \rightarrow {}^1D$ transition deserves some comments, because cometary emission in the C I 1931 Å ($3\,^1P^0 \rightarrow {}^1D$) permitted line has been observed in comets West and Bradfield (A'Hearn and Feldman 1980) with emission rates of 150 and 50 R (Rayleigh), respectively. Since the $3\,^1P^0$ state can decay either to the ^1D or the ^1S metastable levels, with relative probability 10:1 (Nussbaumer and Storey, 1984), it may be readily seen that the emission rate in 8727 Å should be only 1/10 of that in 1931 Å. The true emission rate of the 8727 Å feature cannot be determined accurately, because of the underlying Si I 8728.02 Å Fraunhoffer line, but probably it is not less than about 30 R. It is thus too strong to be due to [C I] alone, even in the unlikely case that the column density of metastable carbon in comet Halley, at 0.77 AU heliocentric distance, were as high as that determined in comet West at 0.35 AU. Actually, the rather broad appearance of the 8727 Å feature suggests that it is blend, with the NH_2 lines at 8726.32 and 8727.76 Å listed by Johns et al. (1976) as likely components.

Acknowledgements. We are indebted to Dr. J.T. Trauger and H. Hippelein for communicating to us data in advance of publication. Our thanks are also due to Dr. G. Klare for his help in the preparation of the observations.

References

Arpigny, C.: 1966, *Proc. Liège Coll.* **13**, 165

A'Hearn, M.F.: 1978, *Astrophys. J.* **219**, 768

A'Hearn, M.F., Feldman, P.D.: 1980, *Astrophys. J. Letters* **242**, L187

Chauville, J., Maillard, J.P., Mantz, A.W.: 1977, *J. Molec. Spectr.* **68**, 399

Coxon, J.A.: 1980, *Canadian J. Phys.* **58**, 933

Coxon, J.A., Foster, S.C.: 1982, *Canadian J. Phys.* **60**, 94

Danylewych, L.L., Nicholls, R.W., Neff, J.S., Tatum, J.B.: 1978, *Icarus* **35**, 112

D'Odorico, S., Enard, D., Lizon, J.L., Ljung, B., Nees, B., Ponz, D., Raffi, G., Tanne, J.F.: 1983, *The Messenger* **33**, 2

Herzberg, G.: 1950, *Molecular Spectra and Molecular Structure I. Spectra of Diatomic Molecules*, 2nd ed. Van Nostrand, London. p. 208

Johns, J.W.C., Ramsay, D.A., Ross, S.C.: 1978, *Canadian J. Phys.* **54**, 1804

Krishna-Swamy, D.S., O'Dell, C.R.: 1977, *Astrophys. J.* **216**, 158

Krishna-Swamy, D.S., O'Dell, C.R.: 1979, *Astrophys. J.* **231**, 624

Lambert, D.F., Danks, A.C.: 1983, *Astrophys. J.* **268**, 428

Nussbaumer, H., Storey, P.J.: 1984, *Astron. Astrophys.* **140**, 383

Stockhausen, R.E., Osterbrock, D.E.: 1965, *Astrophys. J.* **141**, 287

Trauger, J.T., Hippelein, H., Münch, G., Roesler, R.: (in preparation)

Astron. Astrophys. 187, 469–474 (1987)

ASTRONOMY AND ASTROPHYSICS

OH radio observations of comet P/Halley

F.P. Schloerb, M.J. Claussen, and L. Tacconi-Garman

Five College Radio Astronomy Observatory, 617 Lederle GRC, University of Massachusetts, Amherst, MA 01003, USA

Received February 2, accepted April 16, 1987

Summary. We present the first results of an extensive program of OH observations of P/Halley obtained with the NRAO 43 m antenna between July 1985 and May 1986. Using Radio Model 1986a, we find that the OH production rate derived from these data follows an r_h^{-2} law for heliocentric distances in the range $r_h < 2$ AU. Beyond 2 AU, the production rate decreases in a manner which is qualitatively consistent with nominal models of the sublimation of water ice, although quantitative agreement would require some modification of the nominal physical properties of the nucleus. Comparison of radio OH production rates to those estimated by ultraviolet techniques continues to show that the radio technique finds systematically lower production rates. Observations of the OH line shape show that the line width increases as the comet approaches perihelion, and that the broader line widths persist after perihelion to at least $r_h = 1$ AU. Detailed modeling of the profiles suggests that this may be accounted for through an increase in the velocity of the OH parent. However, the OH parent velocities which are required to produce the observed line widths are not quantitatively consistent with other estimates of the parent velocity. Finally, the mean velocity of the OH emission (after correction for the OH Greenstein effect) is slightly blue shifted compared to the nucleus velocity and suggests that the outgassing from the nucleus may occur preferentially on its sunward side.

Key words: comets – comet P/Halley – OH – coma – radio

1. Introduction

Radio observations of the OH radical via its $^2\Pi_{3/2}$ $J=3/2$ Λ-doublet transitions at 18-cm wavelength can contribute greatly to the study of comets since they provide direct information on both the OH production and the kinematics of the cometary coma (*cf.* Schloerb and Gérard, 1985). In the study of Halley's comet, both of these capabilities have made radio observations fruitful. The ability of radio telescopes to observe the comet close to perihelion has filled a gap in the OH production curve that would have existed if only ultraviolet/optical techniques were employed, and perhaps more importantly, the high spectral resolution of radio receivers has permitted direct measurements of the velocity distribution of OH molecules in the coma.

In this paper, we describe the first results of our group's efforts to monitor the 18-cm OH transitions from comet Halley with the US National Radio Astronomy Observatory (NRAO) 43 m telescope. Since this telescope is equipped with the world's most sensitive 18-cm receiver, our effort was largely directed toward achieving several high signal-to-noise ratio spectra for the purpose of modeling the kinematics of OH in the coma. In this initial work, we shall emphasize both the long-term average behavior of the OH production rate, as monitored by the OH radio lines, and the OH kinematics derived from model fitting of the observed line profiles.

Send offprint requests to: M.J. Claussen

2. Observations

The observations were carried out at the NRAO 43 m telescope in Green Bank, West Virginia. Eleven observing sessions between July 1985 and May 1986 were used to monitor the strength and line shape of the OH transitions. Typically, observations of the $^2\Pi_{3/2}$ $J=3/2$ $F=2$-2 transition at 1667 MHz and the $^2\Pi_{3/2}$ $J=3/2$ $F=1$-1 transition at 1665 MHz were carried out simultaneously in both right and left circular polarizations. In some instances, when the emission from the comet was strong, observations of the $^2\Pi_{3/2}$ $J=3/2$ $F=2$-1 at 1720 MHz were also carried out to study the relative intensities of the hyperfine lines. However, in this paper, we shall be concerned only with the behavior of the 1667 MHz line, which is the strongest of the four 18-cm OH transitions.

The receiver was a dual-channel, cooled FET amplifier, which received orthogonal linear polarizations. The linear polarizations were mixed in the IF to produce right- and left-circular polarizations; the system temperature with this receiver was typically 25 K. The spectrometer was a 1024 channel auto-correlator which was usually divided into four 256-channel backends for the simultaneous observations of two polarizations for each of the 1667 and 1665 MHz transitions. The spectral resolution of this system was 0.22 km s^{-1}, and the beam size of the 43 m telescope at 18-cm wavelength was measured to be 18′. Calibration was based on observations of the standard radio source 3C286, which was assumed to have a flux of 13.6 Jy at the frequency of the OH transitions (Baars et al., 1977). The sensitivity of the system was 0.32 K Jy^{-1}, and checks of the calibration were also made on the International Halley Watch standard source W12. Sample OH 1667 MHz spectra of comet Halley are presented in Fig. 1 for three representative epochs: 1985 August 21.9 (the first detection of OH by our group), 1985 November 16.3, and 1986 February 19.8.

3. OH Observational results

3.1. OH production rate

In Table 1, we present the OH production rate derived from observations of the 1667 MHz line. To compute the OH

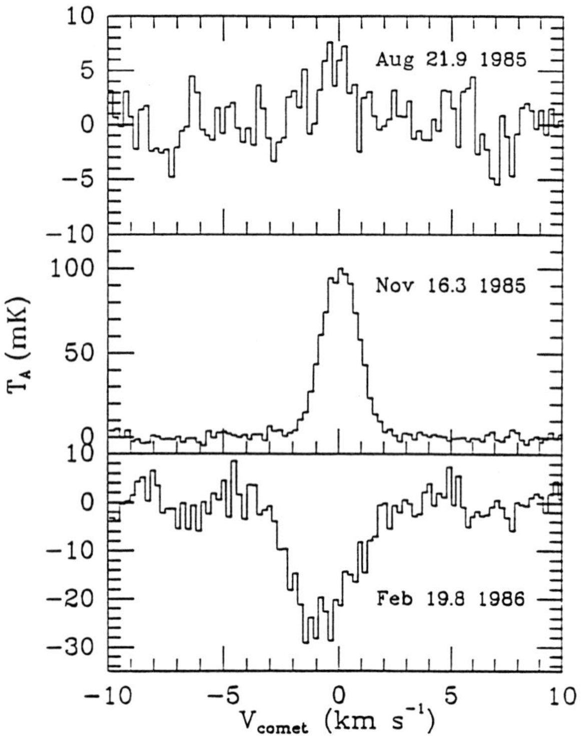

Fig. 1. Sample observations of the OH 1667 MHz transition in Halley's comet obtained at the NRAO 43 m antenna. The August 1985 spectrum represents our first detection of Halley's comet at radio wavelengths. The other spectra represent data obtained near comet perigee in November and near perihelion in February 1986. The appearance of the line in emission or absorption is produced by the UV excitation of the lambda doublet level that is observed. Note that the width of the OH line near perihelion in February is much greater than in November

production rate, we have followed the formalism of Schloerb and Gérard (1985), using the inversion curve of Schleicher (1983) to specify the excitation of the groundstate Λ-doublet. To determine the total production rate of OH from the observed beam-averaged column density, it is also necessary to adopt a model for the distribution of molecules in the coma. We adopt a vectorial model

Table 2. Parameters for radio models of the OH coma

Parameter	Radio Model 1986a Green Bank Workshop	STD Radio Model Despois *et al.* (1981)
Type	Vectorial	Haser
τ_{OH} (s)	$1\times10^5 r_h^2$	$4\times10^5 r_h^2$
τ_P (s)	$8\times10^4 r_h^2$	$8\times10^4 r_h^2$
V_{out} (km s^{-1})	$V_{OH}=1.15$ $V_P=0.9$	$V=1.5$

in which the OH and OH parent velocities and lifetimes are in accord with recent theoretical work and comparable to those typically adopted by UV observers of OH in comets (cf. Feldman et al., 1986). This new model, known as Radio Model 1986a, has been motivated primarily by a longstanding, and not understood, difference between UV and radio estimates of the production rate. It was recently adopted as a standard for the interpretation of radio OH observations to facilitate their intercomparison; the parameters of this model are summarized in Table 2, along with the previous standard radio model of Despois et al. (1981). In Fig. 2, we display the production rates derived from both models as a function of heliocentric distance, r_h, and compare them to the OH parent production rates derived from UV observations using the IUE satellite (Feldman et al., 1986).

The OH parent production rates, Q_P(OH), shown in Fig. 2 demonstrate that even though the new model 1986a removes some of the discrepancy between the radio and UV production rates, there is still a factor of 2 difference which remains to be understood. Previous authors (Schloerb and Gérard 1985; Schleicher 1983; Despois et al. 1981) have speculated that some of this difference may be accounted for by the quenching of the OH Λ-doublet by collisions in the inner coma. However, to account for a full factor of 2 by this process requires that the size of the quenching region be comparable in size to the OH scale length, which is unlikely to be the case. Thus, the difference continues to be a challenging problem for cometary astronomers.

Table 1. Radio OH observations of P/Halley

Mean Date			Δ	r_h	∫SdV	Q_{OH}	<V>	ΔV
Month	Day	Year	(AU)	(AU)	mJy-km/s	10^{27} mol/s	km/s	km/s
Jul	11.4	1985	4.24	3.34	8.6± 7.1	<8	---	---
Aug	21.9	1985	3.18	2.81	31.7± 8.3	8± 2	-0.31±0.18	1.78±0.62
Sep	22.0	1985	2.33	2.46	64.0± 7.3	16± 2	-0.39±0.14	2.00±0.29
Oct	20.2	1985	1.43	2.09	291.1±11.9	49± 2	-0.05±0.04	2.04±0.12
Nov	16.3	1985	0.72	1.71	639.3± 7.5	80± 1	+0.05±0.01	1.92±0.03
Dec	11.0	1985	0.75	1.33	456.9±13.9	122± 4	+0.24±0.04	1.98±0.10
Jan	4.9	1986	1.24	0.95	398.8±15.2	329±11	+0.24±0.04	2.31±0.12
Feb	4.8	1986	1.56	0.60	365.7±21.6	575±34	-0.03±0.11	3.65±0.27
Feb	19.8	1986	1.43	0.63	239.7±10.7	563±26	-0.52±0.07	2.94±0.18
Mar	17.5	1986	0.87	0.96	426.2±15.8	194± 8	+0.14±0.09	3.39±0.22
May	4.0	1986	0.89	1.68	179.9± 7.4	61± 2	+0.51±0.04	1.73±0.11

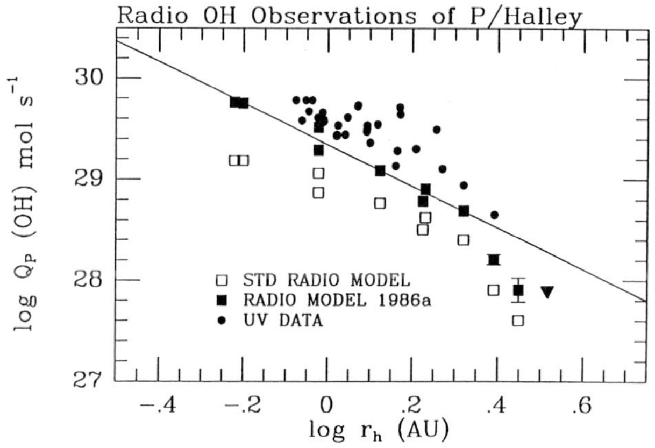

Fig. 2. Production rate of the OH parent derived from the NRAO observations of the 18-cm OH lines. The filled squares are the results for Radio Model 1986a, whose parameters are given in Table 2. We note that the production rate beyond about 2 AU falls below an r_h^{-2} extrapolation of the production rates inside of 2 AU, which is consistent with the expected behavior of water ice sublimation. Results for the Standard Radio Model of Despois et al. (1981) and from UV observations (Feldman et al., 1986) are also shown

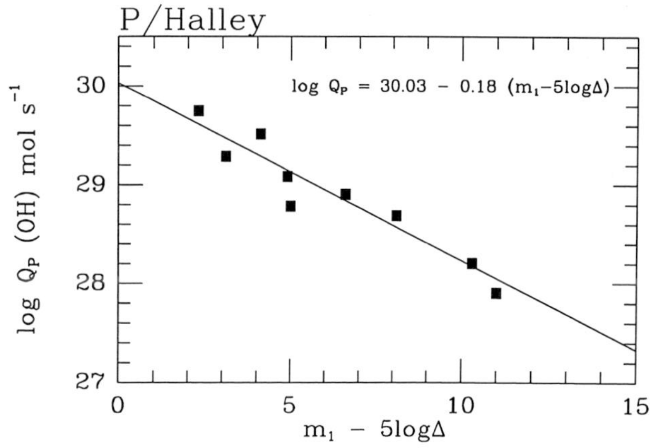

Fig. 4. Correlation of OH production with the visual heliocentric magnitude of the comet. The slope of the correlation suggests that the visual magnitude varies as Q^2, which would be consistent with most of the visible light arising from the C_2 radical

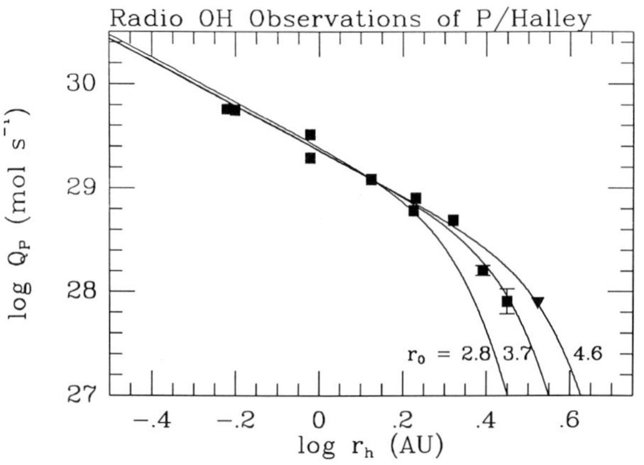

Fig. 3. OH production rates compared to models of the sublimation of water ice. The functional form of the models is derived from the work of Marsden et al. (1973); the parameter r_0 that distinguishes the models is indicated in the figure

The behavior of the production rate derived from our data is interesting because there does not appear to be a strong asymmetry in the pre- and post-perihelion production rates, as found in the UV shown above and by other observers. Indeed, all our radio OH observations are well fit by an r_h^{-2} law for $r_h < 2$ AU. Beyond 2 AU, the OH production rates fall below this trend in a manner that is similar to sublimation curves for water ice (cf. Delsemme and Miller 1971), and it is, therefore, interesting to attempt a quantitative comparison to such models. For simplicity, we have adopted the approximation to the water ice sublimation law that was developed by Marsden et al. (1973) for computation of the non-gravitational forces. This law uses the function $g(r)$ to describe the heliocentric behavior of ice sublimation, where

$$g(r) = \alpha \left(\frac{r}{r_0}\right)^{-m} \left[\left(1 + \left(\frac{r}{r_0}\right)^{n}\right]^{-k},$$

the factor α is chosen to normalize the function, and m, k, and n are adjusted to give the best fit to sublimation models. For our work, we have adopted the standard values of m, k, and n derived by Marsden et al. (1973), and computed production rate curves for a range of values of r_0. The results are shown in Fig. 3, where we note that the behavior is best fit by a single value of r_0 of approximately 3.7. This result is somewhat different from the nominal value of r_0 adopted by Marsden et al. (1973) which is 2.8, corresponding to the value obtained from fits to sublimation models with low visiual albedos (0.1) and high infrared emissivities (0.9). This discrepancy, taken at face value, suggests that some adjustment of these nominal parameters is required by the data. Since the visual albedo of the nucleus is known to be very low, the IR emissivity would have to be decreased to about 0.6 in order to achieve a fit to our data. Alternatively, since OH has been observed at even greater heliocentric distances (cf. A'Hearn et al., 1984), it is also possible that this difference implies that simple vaporization of H_2O ice may not be the only mechanism that controls OH production.

Finally, in order to address the question of the validity of radio OH production rates as a tracer of the gas production, it is interesting to compare our OH production rates to the total visual magnitude of the comet. Such a comparison was done originally by Despois et al. (1981), and in Fig. 4, we show the relationship we have derived for P/Halley between Q_P(OH) and the heliocentric magnitude, m_H, where $m_H = m_1 - 5\log\Delta$. The existence of this relation attests to the validity of radio OH measurements, and comparison to similar relations derived by other radio groups (Gérard et al., 1986) has shown a remarkable consistency between radio telescopes using the Radio Model 1986a. The slope of this relation suggests that the visual magnitude of the comet is proportional to Q_P^2(OH), which would be the case if C_2 emission dominates the continuum in the visible passband (cf. Divine et al., 1986).

3.2. *OH kinematics*

The spectral resolution (0.22 km s^{-1}) and high signal-to-noise ratio of our OH spectra make them ideal for the study of the kinematics of OH in the coma. Previous work by Bockelée-

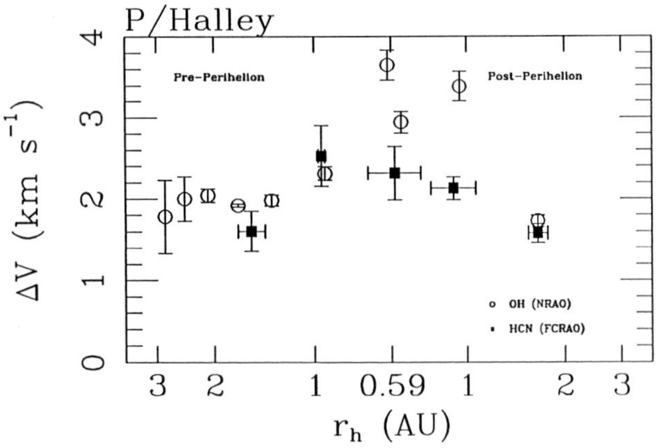

Fig. 5. OH and HCN linewidth during the 1985–1986 apparition of Halley's comet. Both lines increase in width towards perihelion due to the increase in outflow velocity that occurs are the comet approaches the sun. HCN linewidths are always less than those of OH, since daughter molecules are given additional kinetic energy as a result of photodissociation of the parent. Thus, the line width behavior of HCN and OH are consistent with their parent and daughter status

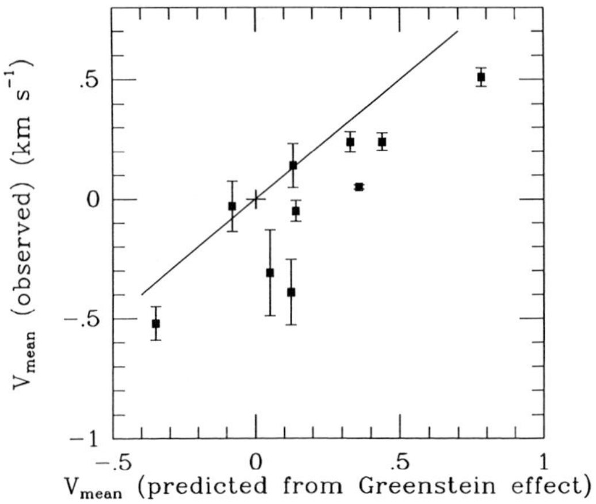

Fig. 6. Mean velocity of the OH 1667 MHz transitions with respect to the nucleus. Due to the radiative pumping of the OH radio transition via its UV electronic transitions, a small offset from the nuclear velocity is expected due to the Greenstein effect. The line in the figure shows the expected behavior if the Greenstein effect alone were responsible for shifts from the nuclear velocity, and a small residual blueshift from this prediction is nearly always present. The magnitude of the blueshift (about $0.2\,\mathrm{km\,s^{-1}}$) is consistent with that derived from HCN data and it probably results from anisotropic outgassing from the nucleus

Morvan and Gérard (1984) has shown that the OH line shape cannot be explained by the simple Haser model of pure radial outflow. Rather, the full vectorial model (cf. Festou, 1981) is required to provide a match to the data. In this model, the linewidth is provided by the vectorial combination of the outflow velocity of the OH parent, v_p, and the velocity taken up by the OH radical itself upon photodissociation of the parent, v_{OH}.

In general, we find that the shapes of the OH lines are well fit by gaussian profiles, and in Table 1 we have summarized the results of gaussian fits to the 1667 MHz lines obtained in this study. The line width data (full width at half maximum) show a marked trend toward greater values as the comet approaches the sun. In Fig. 5, we present a plot of the OH linewidth versus R_h; HCN line widths (Schloerb et al., 1986a) are also presented in the same figure in order to facilitate their intercomparison. Several trends are worth noting in this figure. The first is that the HCN linewidth is, within the errors, always less than the OH linewidth, as would be expected from their roles as parent and daughter molecules, respectively. The second is that both the HCN and OH linewidths increase with decreasing r_h, although the HCN increase, which should trace the parent velocities, does not seem to be sufficient to explain the full increase observed in the OH line width. Finally, we note that the OH data show an asymmetry in the linewidth between pre-perihelion and post-perihelion at the same heliocentric distances. The post-perihelion OH linewidth at 1 AU is about 40% greater than that observed pre-perihelion. Since Larson et al. (1986) have made a similar finding in velocity resolved observations of H_2O, it appears that heliocentric distance alone does not determine the outflow velocity of the coma.

The measurement of the mean velocity of the OH emission provides an excellent oportunity to directly determine whether the outflow of gas from the nucleus is homogeneous and isotropic, or more realistically, anisotropic with a strong perference towards production on the sunward-facing side of the nucleus. However, before the mean OH velocity may be properly interpreted, one must account for the OH Greenstein effect, in which the shape of the line is distorted by a difference in the excitation of OH molecules flowing towards and away from the sun. In Fig. 6, we illustrate this effect with a plot of the observed mean velocity, from Table 1, versus the mean velocity that would be expected from the Greenstein effect alone. It is clear from the figure that there is a strong correlation between the observed and predicted mean velocities, but it is also apparent that the observed mean velocity is generally blueshifted by approximately $0.2\,\mathrm{km\,s^{-1}}$ from the prediction of the Greenstein effect. We interpret this residual blueshift of the OH emission in terms of the anisotropic outgassing from the sunward side of the nucleus. Since the phase angle was never greater than about 60° for our observations, the sunward side of the nucleus was typically pointed towards the observer, and the preferential production of gas on that side of the nucleus would cause a net flow towards the observer. A similar effect has also been observed in HCN spectra (Schloerb et al., 1986b), H_2O spectra (Larson et al., 1986) and OH spectra (Cordes et al., 1986) and has been attributed to the same cause.

4. Vectorial models of the 18-cm line profiles

In order to fully interpret the line profiles obtained in this study, we have begun to construct synthetic line profiles, based upon nominal models of the OH kinematics, for comparison to the data. The models are vectorial models, and they have been calculated using the Monte Carlo techniques outlined by Combi and Delsemme (1980). The Monte Carlo technique allows us to easily include a variety of effects, including the OH Greenstein effect, anisotropic outgassing from the nucleus, and the effects of beam resolution of the coma by the telescope. In this initial work, we present only the first results obtained using models in which the parents flow outward homogeneously and isotropically.

There are many parameters required to construct a model of the OH coma. For this initial work, we have specified all parameters based upon recent theoretical work except for v_p, the outflow velocity of the OH parent, which we allow to vary in order to achieve a fit to our observed line profiles. Thus, for the OH inversion, we adopt the model of Schleicher (1983). For the

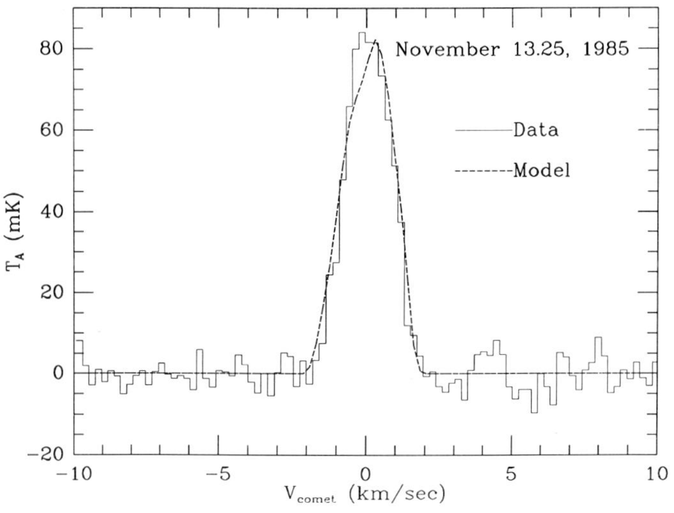

Fig. 7. Vectorial model fit to the OH spectrum of November 13.25 1985. The best fitting parent outflow velocity is 0.54 ± 0.04 km s^{-1}

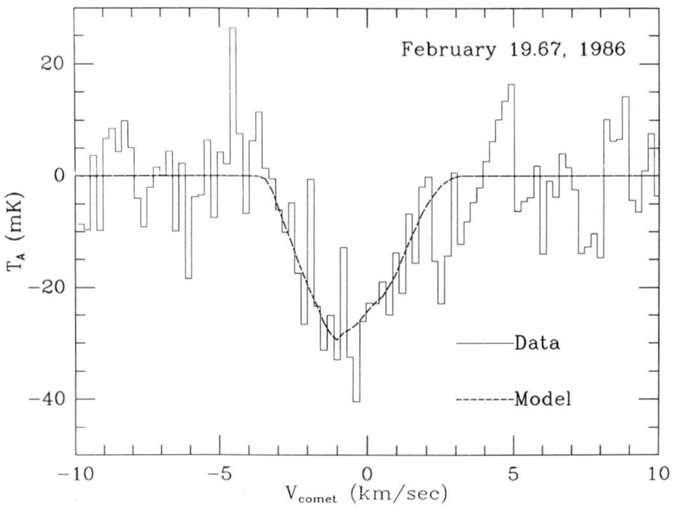

Fig. 8. Vectorial model fit to the OH spectrum of February 19.67 1986. The best fitting parent outflow velocity is 1.97 ± 0.16 km s^{-1}

velocity distribution of OH radicals upon photodissociation, we have adopted values for the principal channel for OH production from H_2O dissociation derived by Festou (1981). Lifetimes of OH and its parent are adopted from Van Dishoeck and Dalgarno (OH: 1984) and Festou (H_2O: 1981), and we subject the OH radical to an acceleration due to solar radiation pressure. For an isotropic outflow model, the only remaining parameter to specify is the velocity distribution about v_P. We adopt a model which specifies the distribution about v_P as maxwellian with an effective temperature of 10 K to simulate the low temperatures which may characterize the region of the coma where the molecules become collisionally decoupled (cf. Crovisier, 1984).

Upon specification of a coma model, it is possible to fit the OH line profiles to determine the parent outflow velocity. We caution that the v_P determined in this way is model dependent both because of the assumption of isotropic outgassing from the nucleus and because of the necessity to adopt other parameters in the model at this stage of the analysis. Nevertheless, the initial results are instructive and illustrative of the power of the modeling approach. We have chosen to initially fit spectra taken on two days, 1985 November 13.25 and 1986 February 19.67, since each day's integration resulted in an acceptable signal-to-noise ratio and since these periods are representative of different physical conditions at different heliocentric distances (see e.g. Fig. 4).

The model fitting results are shown in Figs. 7 and 8 for the November and February data, respectively. Both models provide an acceptable fit to the data, although the fit to the November data leaves a small residual spike near 0 km s^{-1} which may be significant. The parent outflow velocities that are required by the model are 0.53 ± 0.04 km s^{-1} for the November observations and 1.97 ± 0.16 km s^{-1} for the observations made in February. It is interesting that the model fits give results which are *qualitatively* consistent with the observations of HCN velocities (Schloerb et al., 1986), and show that a reasonable increase in the OH parent velocity can account for the increased OH linewidth near perihelion. However, we note that the *quantitative* estimates of parent velocities from HCN differ significantly (by 0.2 km s^{-1} or more) from those required by the OH models. Thus, as discussed in the previous section, further work is required to reconcile the OH kinematic models with all the data that are capable of constraining them.

5. Summary and future work

In this paper, we have discussed the OH production and kinematics of comet Halley derived from a set of radio observations of the 1667 MHz OH transition. In general, we find that the behavior of P/Halley that is derived from these observations is qualitatively consistent with theory and the observations of other using different techniques. However, quantitative agreement continues to be as difficult a goal for this set of cometary observations as it is for most others.

The production rates follow trends which are similar to those seen in UV OH observations, but even with the adoption of a coma model that is similar to the UV standard models, there continues to be a discrepancy between UV and radio derived production rates. The production rates derived in our analysis behave similarly to models of the sublimation of water ice, yet here again a quantitative difference exists between the models and the data. The OH line shape has been used to provide first order information on the coma kinematics through measurements of the OH line width and mean velocity with respect to the nucleus. The line width increases with decreasing heliocentric distance, and in addition, we find that the line widths are typically greater after perihelion than before perihelion at heliocentric distances less than 1 AU. The mean velocity of the line is determined in part by the OH Greenstein effect, but even allowing for this effect, we find that there is a significant tendency for the OH molecules to be blueshifted with respect to the nucleus. This effect is likely to be due to preferential outgassing from the sunward face of the nucleus, as discussed by other observers (Schloerb et al., 1986b; Larson et al., 1986; Cordes et al., 1986). Preliminary attempts to construct detailed models of the line profiles have resulted in acceptable fits to the data. However, again, quantitative agreement between the derived OH parent outflow velocities and the parent outflow velocities derived from HCN observations has not been obtained.

Finally, although this paper presents many more questions than answers, we note that the data which have been presented represent only a part of the full data set that was obtained during the Halley apparition and further analysis and model fitting of this data set is in progress. Notably, in addition to observation of the 1665 MHz transition, which may be used to test the excitation and optical depth of OH, we have also obtained maps of the OH emission. The 18′ beam of the 43 m telescope was capable of resolving the coma at many times during the apparition, and such direct observations will provide constraints on the OH scale length that bear on many of the important questions that have been discussed above. It is to be hoped that when all data are fully analyzed and compared with results from other observers, there will be a sufficient number of constaints on models of the OH coma to provide solutions to the paradoxes that now exist.

Acknowledgements. It is a pleasure to acknowledge the cheerful and diligent assistance of the staff of the National Radio Astronomy Observatory in Greenbank W. Va. who gave us trouble free observing with the world's most sensitive cometary OH system. We also appreciate helpful referee comments from E. Gérard. Planetary Astronomy at the University of Massachusetts is funded by NASA grant NGL 22-010-023, and some aspects of cometary research also are supported by the NSF US-France Cooperative Science Program. FPS also acknowledges support by the International Halley Watch for his role as a Discipline Specialist in Radio Science.

References

A'Hearn, M.F., Schleicher, D.G., Feldman, P.D., Millis, R.L., Thompson, D.T.: 1984, *Astron. J.* **89**, 579

Baars, J.W.M., Genzel, R., Pauliny-Toth, I.I.K., Witzel, A.: 1977, *Astron. Astrophys.* **61**, 99

Bockelée-Morvan, D., Gérard, E.: 1984, *Astron. Astrophys.* **131**, 111

Combi, M.R., Delsemme, A.H.: 1980, *Astrophys. J.* **237**, 633

Crovisier, J.: 1984, *Astron. Astrophys.* **130**, 360

Delsemme, A.H., Miller, D.C.: 1971, *Planet. Space Sci.* **19**, 1229

Despois, D., Gérard, E., Crovisier, J., Kazes, I.: 1981, *Astron. Astrophys.* **99**, 320

Divine, N., et al.: 1986, *Space Sci. Rev.* (in press)

Feldman, P.D., Festou, M.C., A'Hearn, M.F., Arpigny, C., Butterworth, P.S., Cosmovici, B., Danks, A.C., Gilmozzi, R., Jackson, W.M., McFadden, L.A., Patriarchi, P., Schleicher, D.G., Tozzi, G.P., Wallis, M.K., Weaver, H.A., Woods, T.N.: 1986, in *New Insights in Astrophysics, 8 Years of UV Astronomy with IUE*, London

Festou, M.C.: 1981, *Astron. Astrophys.* **95**, 69

Gérard, E., Bockelée-Morvan, D., Bourgois, G., Colom, P., Crovisier, J.: 1986, *20th ESLAB Symp.*, ESA Sp-250, Vol **I**, 589

Larson, H.P., Davis, D.S., Mumma, M.J., Weaver, H.A.: 1986, *Astrophys. J.* **309**, L95

Marsden, B.G., Sekanina, Z., Yeomans, D.K.: 1973, *Astron. J.* **78**, 211

Schleicher, D.G.: 1983, Ph.D. dissertation, University of Maryland

Schloerb, F.P., Gérard, E.: 1985, *Astron. J.* **90**, 1117

Schloerb, F.P., Kinzel, W.M., Swade, D.A., Irvine, W.M.: 1986a, *20th ESLAB Symp.*, ESA SP-250, Vol **I**, 577

Schloerb, F.P., Kinzel, W.M., Swade, D.A., Irvine, W.M.: 1986b, Astrophys. J. Letters **310**, L55

Van Dishoeck, E.F., Dalgarno, A.: 1984, *Icarus* **59**, 305

Astron. Astrophys. 187, 475–480 (1987)

ASTRONOMY AND ASTROPHYSICS

Observations of HCN in comet P/Halley

F. P. Schloerb, W. M. Kinzel, D. A. Swade, and W. M. Irvine

Five College Radio Astronomy Observatory, Lederle GTWR B, University of Massachusetts, Amherst, MA 01003, USA

Received February 2, accepted April 16, 1987

Summary. We present observations of the HCN $J=1-0$ rotational transition at 3.4 mm wavelength in comet P/Halley. The data were obtained during a total of 56 individual observing sessions between November 1985 and May 1986 and represent the first time that a cometary parent molecule has been so extensively monitored. The HCN production rate is well correlated with the total visual magnitude of the comet, and comparison of the HCN production to the total gas production of the comet indicates that it is a relatively minor constituent with 0.1% the abundance of H_2O. Comparison of HCN and CN production suggests that HCN is a major parent molecule of CN, but probably not the sole parent. HCN spectra obtained by binning the data with heliocentric distance show that the line width, and thus the parent outflow velocity, increases with decreasing heliocentric distance, and that there is a tendency for the lines to be blue shifted due to anisotropic outgassing from the nucleus. Finally, there is evidence of day-to-day time variability in the total HCN emission and in the hyperfine ratios. The time variation of the total emission is consistent with the known time variable behavior of the comet, and detailed comparisons to optical data, where possible, confirm this interpretation. However, non-LTE values of the hyperfine ratios are not consistent with theoretical modeling of the excitation of these transitions.

Key words: comets – comet Halley – coma – molecules

1. Introduction

The optical spectral lines of the CN radical are among the strongest features in the spectrum of a comet. It has long been realized, however, that CN owes its existence to the presence of one or more complex parent molecules. HCN now appears to be at least one of the parent molecules of the CN radical based upon its initial detection in comet Kohoutek (Huebner et al., 1973) and its recent detection in comet Halley by three independent groups (Schloerb et al., 1986a; Despois et al., 1986; Winnberg et al., 1987). In this paper, we discuss our group's observations of HCN in comet Halley, which trace the behavior of this parent molecule throughout most of the 1985–1986 apparition. Some aspects of this work have also been discussed in Schloerb et al. (1986a; hereafter Paper I).

Send offprint requests to: F. P. Schloerb

2. Observations

The HCN observations of comet Halley were carried out at the Five College Radio Astronomy Observatory using the 13.7 m antenna during a total of 56 individual observing sessions between 1985 November 18 and 1986 May 11. At the frequency of the HCN $J=1-0$ transition (88.6 GHz) the antenna has a half power beam width of 60″ and a main beam efficiency of 0.55. The receiver was a cooled Schottky diode mixer and was calibrated by the standard "chopper wheel" method to correct for atmospheric attenuation and some telescope losses. Chopper wheel system temperatures obtained during the Halley observations varied from approximately 300 K–600 K single sideband, depending on the weather at the time of observations. The spectrometer was a 256-channel 100 kHz filterbank which yielded a spectral resolution of 0.34 km s^{-1}; other details of the observing procedure may be found in Paper I. An average spectrum of the data obtained close in time to the Giotto encounter with Halley is shown in Fig. 1.

3. Results

3.1. HCN production rate

In Paper I, we presented HCN production rates derived from several sub-averages of the data during the comet Halley apparition. Each sub-average was derived from data obtained over a maximum time span of 4 days to improve the signal-to-noise ratio of the spectra and study the average behavior of the comet. As will be seen later in this paper, this averaging procedure is somewhat dangerous because significant day-to-day variations in the strength of the line were observed. Nevertheless, because of the weakness of the HCN lines, time binning of the data is usually necessary to achieve useful detections.

The average HCN production rates are summarized in Table 1. Calculation of the HCN production rate is based upon the Haser model assuming a radial outflow velocity of 0.8 km s^{-1} for all periods and an HCN lifetime of 8 10^4 s at 1 AU; the excitation of the HCN molecule in the coma is specified by the IR fluorescence equilibrium model of Bockelée-Morvan and Crovisier (1985). A comparison of these production rates to the visual magnitude, derived from the IAU circulars, is shown in Fig. 2. The similar behavior of HCN and visual magnitude suggests that the HCN production follows the overall gas and dust production of the comet.

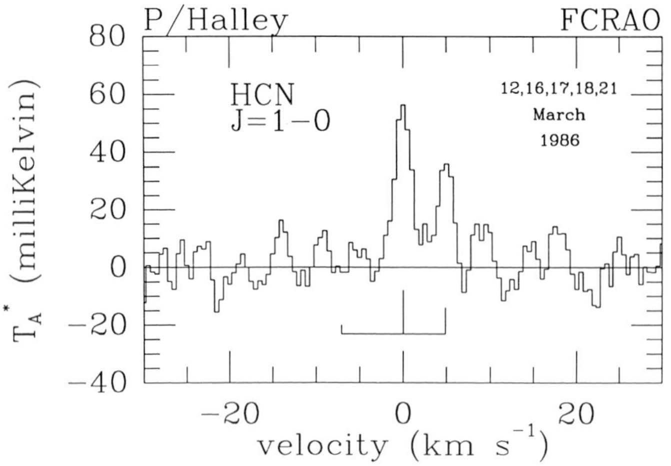

Fig. 1. Average spectrum of HCN observations obtained for the days indicated during March, 1986, near the time of the Giotto encounter. The figure beneath the spectrum indicates the locations and relative intensities of the hyperfine components

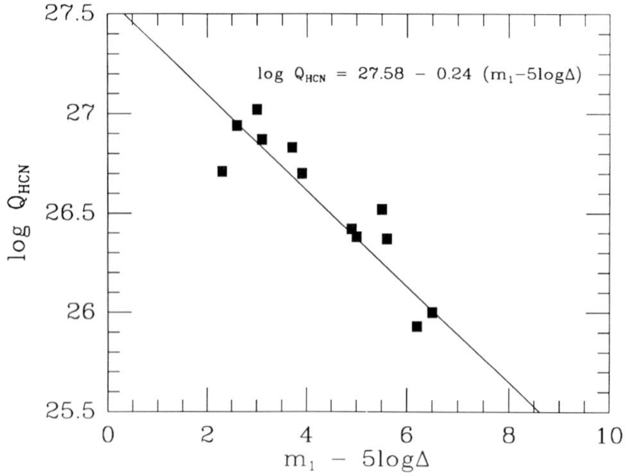

Fig. 2. Comparison of HCN production rates to heliocentric visual magnitude (visual magnitude, m_1, corrested for geocentric distance)

Table 1. HCN production rates for comet Halley

Average Date	Days	Δ (AU)	r_h (AU)	$\int T_A^*(F=2-1)dV$ (mK km s^{-1})	Q_{HCN} (mol s^{-1})
11/20.0/1985	2	0.66	1.65	42± 10	1.0×10^{26}
12/ 1.4/1985	4	0.63	1.48	35± 9	8.5×10^{25}
1/ 4.2/1986	3	1.22	0.96	54± 10	5.0×10^{26}
1/27.2/1986	3	1.54	0.66	37± 12	1.1×10^{27}
2/ 9.1/1986	4	1.55	0.59	32± 12	$<1.3\times10^{27}$
2/13.1/1986	4	1.52	0.59	51± 15	1.9×10^{27}
2/28.0/1986	4	1.29	0.71	29± 9	5.1×10^{26}
3/ 6.0/1986	4	1.16	0.79	70± 12	8.6×10^{26}
3/10.1/1986	4	1.06	0.84	19± 21	$<6.1\times10^{26}$
3/14.1/1986	2	0.96	0.90	101± 31	7.4×10^{26}
3/17.5/1986	3	0.87	0.96	127± 14	7.4×10^{26}
3/23.0/1986	2	0.74	1.04	154± 21	6.7×10^{26}
3/28.0/1986	2	0.61	1.12	113± 50	$<4.5\times10^{26}$
4/25.0/1986	3	0.64	1.54	109± 16	2.6×10^{26}
4/28.0/1986	3	0.72	1.59	88± 17	2.3×10^{26}
5/ 1.0/1986	3	0.80	1.63	29± 19	$<1.6\times10^{26}$
5/ 4.0/1986	3	0.89	1.68	72± 10	2.4×10^{26}
5/10.0/1986	3	1.08	1.76	80± 15	3.3×10^{26}

Comparison of the HCN production rate to that of H_2O and the CN radical is interesting because it demonstrates the importance of HCN as a cometary parent molecule. During the time of the Vega and Giotto flybys, the gas production rate from P/Halley was approximately 1 10^{30} molecules s^{-1} (cf. Reinhard, 1986; Sagdeev et al., 1986), while the HCN production rate in this interval was approximately 9 10^{26} molecules s^{-1}. Thus, the HCN production rate was approximately 10^{-3} that of the total gas production, and HCN is only a minor constituent of the cometary nucleus. Since HCN is a probable parent of the CN radical, it is also interesting to compare its production rate with that of CN. During this time, the CN production rate is estimated to be approximately 2 10^{27} molecules s^{-1} (Cochran and Barker, 1986; Schleicher et al., 1986), which exceeds the HCN production rate at this time by factors of 2–4. Thus, taken at face value, the results seem to indicate that HCN cannot be the only parent molecule of the CN radical. However, since Millis and Schleicher (1986) and Schleicher et al. (1986) have shown that the CN production rate varies by a factor of 3 from day-to-day in this time interval, and since both the HCN and CN results are model dependent, it seems prudent to defer final judgement on this question until more complete comparisons can be made.

3.2. *Kinematics of HCN*

One of the main advantages of radio spectroscopic observations of the cometary coma is that they allow a direct measurement of the line of sight velocity distribution of molecules in the coma. In Paper I, we presented the kinematic properties of the averaged spectra used to obtain the production rates shown in Table 1. These data showed that the mean line velocity was typically blue shifted, as would be expected for anisotropic outflow with more gas production from the sunward facing side of the nucleus. Interestingly enough, however, the line width in these spectra did not significantly increase as the comet approached perihelion, although there was marginal evidence for an increase in the line width after perihelion. Since both previous determinations of the coma outflow velocity (Bobrovnikov, 1954; Whipple, 1980; Delsemme, 1982) and the P/Halley HCN data of Bockelée-Morvan et al. (1986) show a clear dependence on heliocentric distance, we have reexamined this question by constructing new average spectra within different bins in heliocentric distance. These new averages typically include many more spectra than the "four day" averages of Table 1, and therefore, their signal-to-noise ratio is typically 10–20 rather than 3–5 as in the spectra of

Table 2. Kinematics of HCN in comet Halley

Date range (MM/DD/YY)	r_h range AU	Days	Peak T_A^* mK	Mean V km s^{-1}	ΔV km s^{-1}
11/18/85-12/03/85	1.68-1.45	6	20±3	-0.09±0.13	1.60±0.25
01/03/86-01/06/86	0.98-0.94	3	23±3	-0.40±0.22	2.53±0.37
01/25/86-03/02/86	0.73-0.59	14	13±2	-0.16±0.22	2.32±0.33
03/04/86-03/25/86	1.06-0.77	11	37±2	+0.04±0.08	2.13±0.14
04/24/86-05/11/86	1.78-1.53	15	23±3	-0.16±0.06	1.58±0.12

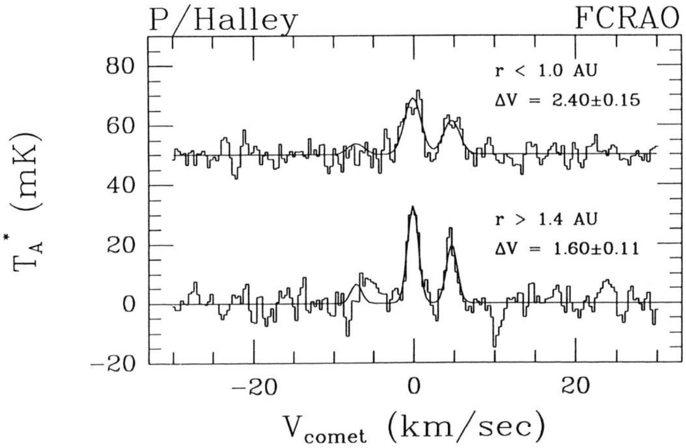

Fig. 3. Comparison of average spectra for heliocentric distance >1.4 AU (bottom) and <1.0 AU (top). The line width of the HCN emission increases with decreasing heliocentric distance

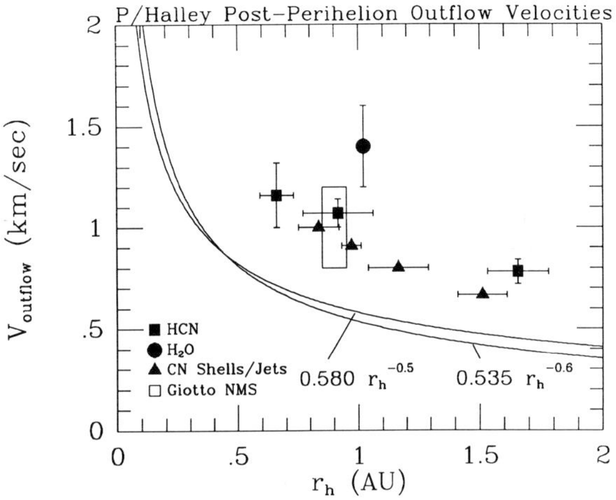

Fig. 4. Outflow velocities of Helley's comet derived from various techniques. The solid curves represent the velocity laws derived by Whipple (1980) and Delsemme (1982) from observations of several comets. The box labeled Giotto NMS represents the range of outflow velocities detected by that instrument at different points within the coma

Paper I. The results of fitting a gaussian model including all three HCN hyperfine components to these new averages is shown in Table 2, and a strong tendency for the line width to increase with decreasing heliocentric distance is revealed. In Fig. 3, we illustrate this effect with average spectra which include, respectively, all observations with $r > 1.4$ AU and all observations with $r < 1.0$ AU.

Although detailed comparisons of the HCN linewidth to other measures of the expansion velocity await a theoretical treatment of the HCN line profiles in the context of a realistic coma model, a reasonable approximation to the outflow velocity is made by adopting 1/2 of the gaussian full width at half maximum (FWHM) line width. This approximation is valid to an accuracy of about 10% for the observing conditions that pertained to comet Halley under the assumption that the HCN line profiles may be described by the Haser model of isotropic radial outflow. In Fig. 4, we have plotted the outflow velocity derived in this way for the post-perihelion observations. Also shown in Fig. 4 are previously determined curves of outflow velocity derived by Whipple (1980) and Delsemme (1982) from observations of several comets. We note that the heliocentric behavior of the HCN derived velocities is similar to the previous model fits, but that the magnitude is typically 50–100% greater. The latter difference is supported by several other independent measures of the expansion velocity that were obtained during the same time interval that is represented in the figure (CN Shells: Schlosser et al., 1986; CN Jets: A'Hearn et al., 1986; H_2O: Larson et al., 1986; Giotto NMS: Lämmerzahl et al., 1986). Thus, the gas velocities observed in Halley's comet are not consistent with the previously derived velocity laws.

Another feature of the line fits in Table 2 is the tendency of the central velocity of the gaussian profile to be blue shifted form the nuclear velocity. This blue shift, which was also found in the "four day" averages of Paper I, is thus verified by the higher signal to noise ratio spectra. The tendency for the lines to be blue shifted is in good agreement with spacecraft observations which show that the activity of the comet was largely restricted to its sunlit hemisphere (cf. Sagdeev et al., 1986; Keller et al., 1986). Since the phase angle of the comet was never above about 70 degrees during our observations, gas from the sunlit side of the nucleus moves primarily towards the observer in all cases, and the result is a general blue shift of the emission. Finally, we note that an average blue shift has also been observed and similarly interpreted by other parent molecule observations (H_2O by Larson et al., 1986) and observations of daughters (OH by Schloerb et al., 1986b).

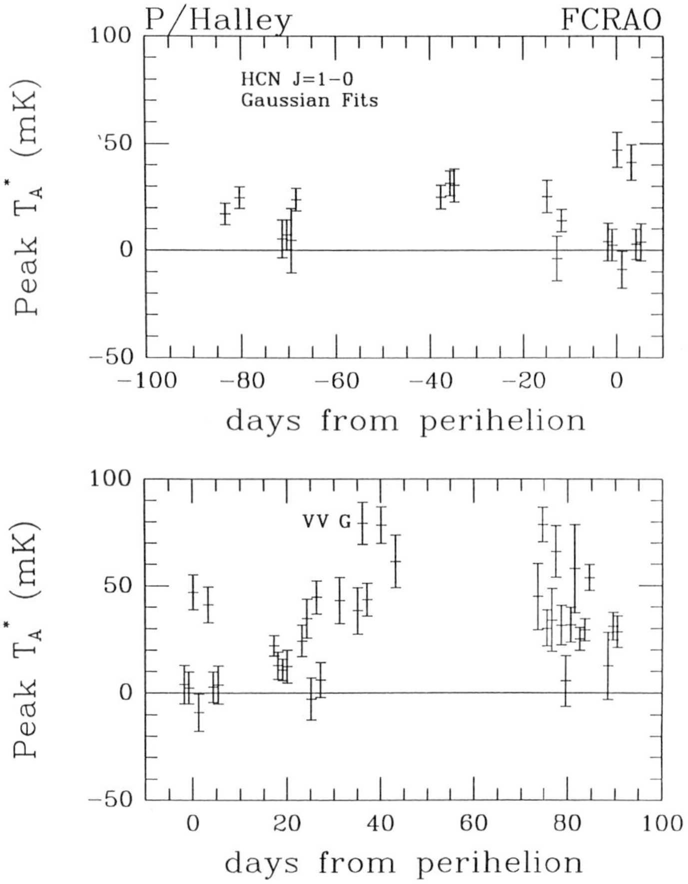

Fig. 5. Peak intensity of the HCN $J=1-0/F=2-1$ line derived from gaussian fits to individual days. The times of the two Vega encounters (V) and the Giotto encounter (G) are indicated

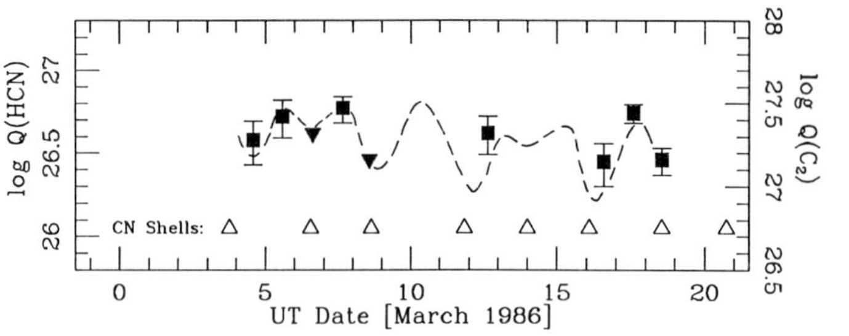

Fig. 6. Day-to-day variability of HCN production compared to the C_2 production rate curve (dashed) of Millis and Schleicher (1986). The origin times of CN shells observed by Schlosser et al. (1986) are also shown as triangles

3.3. Time variability

There is now good evidence, on a long term basis, for a correspondence between the visual magnitude and HCN production rate. On a day-to-day basis, however, much work needs to be done in order to establish whether a similar correspondence exists between HCN production and the frequent optical outbursts that were observed in comet Halley. Due to the general weakness of the HCN emission, it has been difficult to make a strong case for day-to-day HCN variations based on the radio data alone. However, as other data become available for comparison, the case for day-to-day variability of the HCN emission has been strengthened. Thus, it is to be hoped that when the temporal behavior of Halley is better documented, a detailed comparison between optical and HCN data will demonstrate the reality of variations in the latter and elucidate their cause.

In Fig. 5, we present our evidence for day-to-day HCN variability based upon gaussian fits to the radio data alone. The gaussian models include all three hyperfine components in their nominal 5:3:1 ratio. Each component has the same fixed linewidth of 2.0 km s^{-1}, and the line center is fixed at 0 km s^{-1} with respect to the nucleus. The gaussian fitting technique improves the signal-to-noise ratio of a typical day's data by a factor of about 2 over the use of the $F=2-1$ line observations alone, due to the use of all molecules in the $J=1$ level, and it may be viewed as a filter which is optimized for the detection of HCN in our spectra. Tests of this fitting process on portions of the spectra which do not include HCN lines do not yield spurious results and demonstrate that the model fitting errors are consistent with those expected from the rms of the spectrum.

The data in Fig. 5 show many interesting variations which are probably related to activity in the comet. For example, we note that outbursts were apparently detected near perihelion on February 9.6 and February 12.6, and that these times agree well with the times of outbursts that Sekanina (1986) has derived for this time period from straie observed in the tail during late February. We also find that, where direct comparisons to published optical data are possible, there is an excellent correlation between the HCN and total gas production rate on a day-to-day basis. In Fig. 6, we show a comparison of our HCN data and the C_2 production rates of Millis and Schleicher (1986) during the time period surrounding the spacecraft encounters. Since the C_2 production rate is well correlated to the comet's total gas production rate (cf. Millis and Schleicher, 1986; A'Hearn and Millis, 1980), the correlation of these species indicates that HCN production is well correlated to the activity of the comet. Increases in the HCN production rate are also well correlated to the appearance of the CN shells observed by Schlosser et al. (1986), and in Fig. 6, we have noted the origin times of the shells during this time interval.

3.4. Hyperfine ratio anomalies

In previous papers on HCN in comet Halley (Schloerb et al., 1986; Despois et al., 1986; Winnberg et al., 1987), there has been discussion of the possibility of deviations of the HCN hyperfine ratios from those expected based upon the statistical weights of the hyperfine levels. The possibility of anomalous values for the HCN hyperfine ratios is worthy of further consideration since it is totally unexpected theoretically and might eventually prove to be a useful clue about the physical conditions in the coma. One important clue that the hyperfine ratios may be anomalous arises from an average spectrum of all HCN observations obtained in this study. This average spectrum is shown in Fig. 7, along with a gaussian model which includes all three hyperfine lines with their nominal strengths. It is clear from this model prediction that the gaussian fit with nominal hyperfine ratios predicts a larger $F=0-1$ line than is observed, and a gaussian model fit which permits variations in the hyperfine ratios from their LTE values supports this conclusion. We find that the $F=1-1/F=2-1$ ratio is consistent with the LTE ratio (0.62 ± 0.05), while the $F=0-1/F=2-1$ ratio appears to be too low (0.04 ± 0.05) compared to the LTE ratio of 0.2.

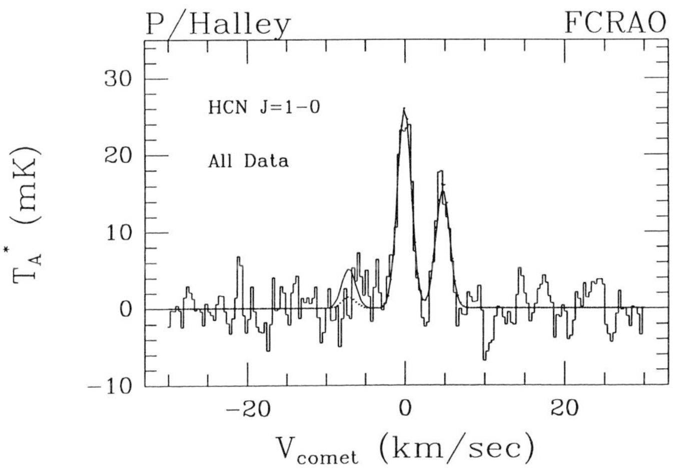

Fig. 7. Average spectrum of all HCN data shown to illustrate the weakness of the $F=0-1$ hyperfine component. The solid curve is a gaussian model fit assuming the nominal hyperfine ratios. The dashed curve gives the gaussian model with the best fitting hyperfine ratios

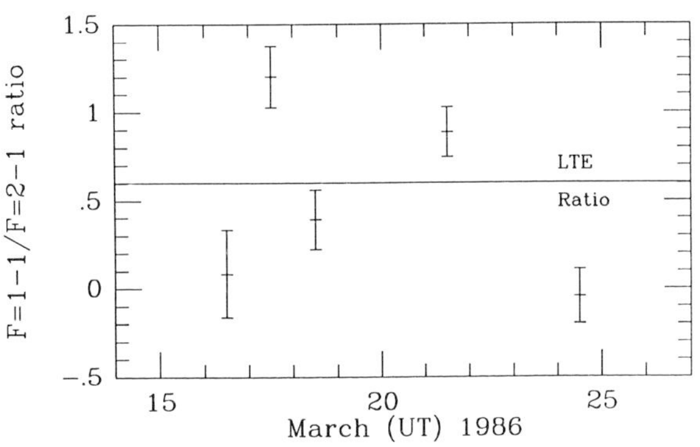

Fig. 8. Determination of the $F=1-1/F=2-1$ hyperfine ratio for several days in 1986 March. The ratio is determined from gaussian fits to the line peak intensities

The long term average spectrum shown above indicates that the $F=0-1$ component of the HCN $J=1-0$ transition may be anomalously weak while, on the average, the $F=1-1$ component appears to have approximately the correct strength relative to the $F=2-1$ component. In Fig. 8, we present evidence which suggests that the $F=1-1/F=2-1$ ratio may be variable on shorter time scales. This figure presents the results of fitting for the $F=1-1/F=2-1$ line ratio on several individual days in March, 1986, when inspection of the spectra suggested that large variations were occuring. We find that, indeed, three of the points shown in this time interval lie more than 2 standard errors away from the LTE value of 0.6, and chi-square tests on the data suggest that this has less than a 1% chance of being due solely to statistical errors. The estimation of systematic errors which might reproduce this effect is much less straightforward, but we note that most possible systematic effects such as calibration and pointing errors tend to cancel in the measurement of a line ratio. Thus, we believe that the HCN hyperfine ratios may be variable, even though, as discussed by Despois et al. (1986), existing theories for the excitation of HCN in comets cannot explain deviations from the 5:3:1 ratio expected from the statistical weights. Further theoretical and data analysis work on this problem are in progress by our group.

4. Discussion

Our observations of HCN in the coma of comet Halley have reaffirmed its reality as a cometary parent molecule and served to brighten the prospects for the study of cometary parent molecules via their millimeter-wave rotational transitions. Such data are extremely valuable since they offer a direct means to monitor both the production of parent molecules and their kinematics. The HCN production rates appear to follow the overall gas and dust production of the comet closely, on both a long term and day-to-day basis. HCN accounts for only 0.1% of the estimated gas production rate from the nucleus and is evidently a minor constituent. However, comparisons of HCN and CN production suggest that HCN could be a major parent of the CN radical even though the best information available suggests that it is unlikely to be the only CN parent.

The kinematic behavior of gas in the coma is well characterized through the direct HCN observations of parent outflow velocities. The HCN lines are typically blue shifted due to the anisotropic outgassing of the nucleus. Moreover, the HCN line width increases with decreasing heliocentric distance and increasing gas production rate as predicted by theoretical models (cf. Crovisier, 1984) and previous observations of outflow velocities in comets (e.g. Bobrovnikov, 1954; Whipple, 1980; Delsemme, 1982). It is unfortunate that the heliocentric distance and the geocentric distance were anticorrelated during the observations, since this means that it is difficult to definitively distinguish between a true velocity increase with decreasing heliocentric distance and any change in line width that might occur as a result of different resolution on the coma. However, we do not believe that resolution effects will bias the linewidth determination to a great extent, and in any case, it is quite clear that the velocities determined by HCN and by other techniques in comet Halley exceed those of the previous empirical velocity laws. Conceivably, this difference reflects either anomalous behavior by P/Halley or the indirectness of the previous techniques for velocity measurement.

In conclusion, we note that the comparison of HCN observations to other types of observations should yield much new information when data can be intercompared over the full Halley apparition. To the extent that the comparisons that are possible at the present time reflect the average behavior of the comet, there are already conclusions which may be drawn from the comparison of our data to the CN and C_2 data that exist. For example, comparison of the HCN and CN velocities in Fig. 4 suggests that the CN velocity is likely to be less than the HCN velocity. Such a case would be difficult to explain if the observed CN were produced directly from HCN, since the photodissociation process should impart an additional 1 km s^{-1} of velocity of the daughter. Thus, the bulk of the CN must either arise from a lower velocity parent or the CN must be "thermalized" as discussed by Delsemme (1982). Clearly, the former case would be consistent with a grain origin for some CN in the coma as proposed by A'Hearn et al. (1986), since the grain velocities are expected to be less than those of the gas. Another interesting point may be made by comparing the C_2 production rate curve to the HCN time variations in Fig. 6. Here we note that the HCN and C_2 production rates appear to follow one another closely with no large time lag in spite of their parent and daughter status. Since the time variations are long compared to the HCN lifetime during this period, it follows that "outbursts" may last over a time period of days, rather than being purely impulsive events.

Acknowledgements. We acknowledge and appreciate the cooperation of our FCRAO colleagues who lost observing time during this "once in a lifetime" opportunity to observe Halley's comet. We also thank J. Crovisier for his helpful referee comments. The Five College Radio Astronomy Observatory is supported by NSF grant AST 85-12903 and by the Commonwealth of Massachusetts, and it is operated with the permission of the Metropoliton District Commission. Planetary Astronomy at the FCRAO is funded by NASA grant NGL 22-010-023. Partial funding for cometary research is also provided by the NSF US-France Cooperative science program. FPS and WMI acknowledge support by the International Halley Watch in their role as Radio Science Discipline specialists.

References

A'Hearn, M.F., Hoban, S., Birch, P.V., Bowers, C., Martin, R., Klinglesmith III, D.A.: 1986, *Nature* **324**, 649

A'Hearn, M.F., Millis, R.L.: 1980, *Astron. J.* **85**, 1528

Bobrovnikov, N.: 1954, *Astron. J.* **59**, 537

Bockelée-Morvan, D., Crovisier, J.: 1985, *Astron. Astrophys.* **151**, 90

Bockelée-Morvan, D., Crovisier, J., Despois, D., Forveille, T., Gérard, E., Schraml, J., Thum, C.: 1986, *20th ESLAB Symp.*, ESA SP-250, Vol. **I**, p. 365

Cochran, A.L., Barker, E.S.: 1986, *20th ESLAB Symp.*, ESA SP-250, Vol. *I*, p. 439

Crovisier, J.: 1984, *Astron. Astrophys.* **130**, 361

Delsemme, A.: 1982, in *Comets*, ed. L.L. Wilkening, Univ. Arizona Press, Tuscon, AZ

Despois, D., Crovisier, J., Bockelée-Morvan, D., Schraml, J., Forveille, T., Gérard, E.: 1986, *Astron. Astrophys.* **160**, L11

Huebner, W.F., Snyder, L.E., Buhl, D.: 1974, *Icarus* **23**, 580

Keller, H.U., Arpigny, C., Barbieri, C., Bonnet, R.M., Cazes, S., Coradini, M., Cosmovici, C.B., Delamere, W.A., Huebner, W.F., Hughes, D.W., Jamar, J., Malaise, D., Reitsema, H.J., Schmidt, H.U., Schmidt, W.K.H., Seige, P., Whipple, F.L., Wilhelm, K.: 1986, *Nature* **321**, 320

Lämmerzahl, P., Krankowsky, D., Hodges, R.R., Stubbemann, U., Woweries, J., Herrwerth, I., Berthelier, J., Illiano, J.M., Eberhardt, P., Dolder, U., Schulte, W., Hoffman, J.H.: 1986, *20th ESLAB Symp.*, ESA SP-250, Vol. **I**, p. 179

Larson, H.P., Davis, D.S., Mumma, M.J., Weaver, H.A.: 1986, *Astrophys. J.* **309**, L95

Millis, R.L., Schleicher, D.G.: 1986, *Nature* **324**, 646

Reinhard, R.: 1986, *Nature* **231**, 313

Sagdeev, R.Z., Blamont, J., Galeev, A.A., Moroz, V.I., Shapiro, V.D., Shevchenko, V.I., Szego, K.: 1986, *Nature* **321**, 259

Schleicher, D.G., Millis, R.L., Tholen, D., Lark, N., P.V., Martin, R., A'Hearn, M.F.: 1986, *20th ESLAB Symp.*, ESA SP-250, Vol. **I**, p. 565

Schloerb, F.P., Kinzel, W.M., Swade, D.A., Irvine, W.M.: 1986a, *Astrophys. J.* **310**, L55

Schloerb, F.P., Claussen, M.J., Tacconi-Garman, L.: 1986b, *20th ESLAB Symp.*, ESA SP-250, Vol. **I**, p. 583

Schlosser, W., Schultz, R., Koczet, P.: 1986, *20th ESLAB Symp.*, ESA SP-250, Vol. **III**, p. 495

Sekanina, Z.: 1986, *IAU Circular* No. **4187**

Whipple, F.L.: 1980, *Astron. J.* **85**, 385

Winnberg, A., Ekelund, L., Ekelund, A.: 1987, *Astron. Astrophys.* (in press)

Astron. Astrophys. 187, 481–484 (1987)

ASTRONOMY AND ASTROPHYSICS

The CO and N_2 abundance in comet P/Halley

P. Eberhardt[1], D. Krankowsky[2], W. Schulte[1], U. Dolder[1], P. Lämmerzahl[2], J.J. Berthelier[3], J. Woweries[2], U. Stubbemann[2], R.R. Hodges[4], J.H. Hoffman[4], and J.M. Illiano[3]

[1] Physikalisches Institut, University of Bern, Sidlerstrasse 5, CH-3012 Bern, Switzerland
[2] Max-Planck-Institut für Kernphysik, Postfach 103980, D-6900 Heidelberg 1, Federal Republic of Germany
[3] Centre de Recherches en Physique de l'Environnement Terrestre et Planétaire, CNET-CNRS, 4, avenue de Neptune, F-94107 Saint-Maur des Fossés, France
[4] University of Texas at Dallas, M.S. FO 2.2, P.O. Box 688, Richardson, TX 75080, USA

Received February 14, accepted May 11, 1987

Summary. A preliminary evaluation of the mass 28 amu/e signal observed in the neutral mode of the Giotto neutral gas mass spectrometer (NMS) is presented. At 1000 km from the nucleus we obtain for the CO number density $n(\mathrm{CO})/n(\mathrm{H_2O}) \leqq 0.07$. The production rate of CO as a parent molecule directly from the nucleus is thus less than 7% of the H_2O production rate. However, CO is also produced from an extended source in the inner coma ($R < 20\,000$ km) and at 20 000 km from the nucleus we obtain for the total equivalent CO production rate $0.05 \leqq Q(\mathrm{CO})/Q(\mathrm{H_2O}) \leqq 0.15$. For N_2 an upper limit $Q(\mathrm{N_2})/Q(\mathrm{H_2O}) \leqq 0.1$ is derived. No parent molecule for the CO could be identified which would be in agreement with the NMS measurements. It is proposed that CO or a very short-lived parent is relased in the coma from cometary dust grains, such as the "CHON" particles.

Key words: carbon monoxide – nitrogen – dust

The Neutral Gas Mass Spectrometer (NMS) on the Giotto spacecraft observed the first peaks due to neutral gas species at distances from the nucleus $>150\,000$ km. The last neutral mass spectrum before the NMS was damaged, probably by a large-amplitude electrical transient, was obtained in a distance of ~ 850 km. The radial profile of the H_2O density and upper limits of the CO_2, NH_3, and CH_4 abundances have been published (Krankowsky et al., 1986b). In this paper we give a preliminary evaluation of the CO and N_2 densities and related parameters (cf. also Eberhardt et al., 1986).

A description of the NMS has been published (Krankowsky et al., 1986a). The instrument had two independent analyzers: the M-analyzer and the E-analyzer. The data reported in this paper are based essentially on the M-analyzer, which is a double focussing mass spectrometer covering the mass range from 1 amu/e to 37 amu/e. The half peak width at the 1% level is ~ 0.5 amu essentially independent of mass and the abundance sensitivity for adjacent integer mass number is $>10^3$. An electron impact ion source with fly-through geometry and two electron energies of ~ 90 eV (HeV mode) and ~ 17 eV (LeV mode) was used. Cometary ions were suppressed by electrostatic deflection plates and a potential barrier. A focal plane detector allowed simultaneous measurement of all masses over the whole mass range of the instrument.

The evaluation of mass spectra obtained by an electron impact source is complicated by the fact that electron impact not only ionizes the neutral gas molecules but also produces charged fragments. Thus a complex gas mixture such as a cometary atmosphere close to the nucleus produces a complicated mass spectrum in a neutral gas mass spectrometer. In fact at distances $<11\,000$ km we observed a signal on virtually every integer mass number in the mass range 12 to 37 amu/e. For electrons with energies a few eV above the ionization potential as in the LeV mode the fragmentation process is negligible and identification of the parent molecule is facilitated. However, ionization cross sections are then one to two orders of magnitude lower compared to the HeV mode and hence the efficiency of the ion source is correspondingly smaller. A detailed interpretation of the neutral gas mass spectra obtained by the NMS requires thus the complete reduction of all spectra recorded in the HeV and LeV modes, a task which needs considerably more time.

The two major contributors to the mass 28 amu/e peak in a cometary atmosphere are CO and N_2. However, C_2H_4 and fragments from heavier hydrocarbons, CO_2 and other CO or N_2 containing molecules can also contribute. Assuming that all these contributions are negligible and that CO is the only molecule responsible for the 28 amu/e peak, an upper limit for the CO density can be deduced. At a distance of 1000 km from the nucleus we obtain for the number densities n:

$$n(\mathrm{CO}, 1000)/n(\mathrm{H_2O}, 1000) \leqq 0.07\,. \tag{1}$$

The in situ measured photodestruction length for H_2O is $\sim 39\,000$ km and H_2O is undoubtedly a parent molecule (Krankowsky et al., 1986b, constant expansion velocity assumed). Using the photodestruction rates published for H_2O and CO for 1 AU and solar minimum UV flux (Appendix I of Levine, 1985) we calculate for CO a photodestruction scale length of $\sim 7\;10^5$ km for the conditions prevailing during the encounter. Hence, the density ratio at 1000 km is also an upper limit for the production rate of CO as parent molecule:

$$Q^*(\mathrm{CO})/Q(\mathrm{H_2O}) \leqq 0.07\,. \tag{2}$$

(We denote in this paper the direct production rate of a molecule as parent molecule with Q^*; the production rate as daughter from

Send offprint requests to: P. Eberhardt

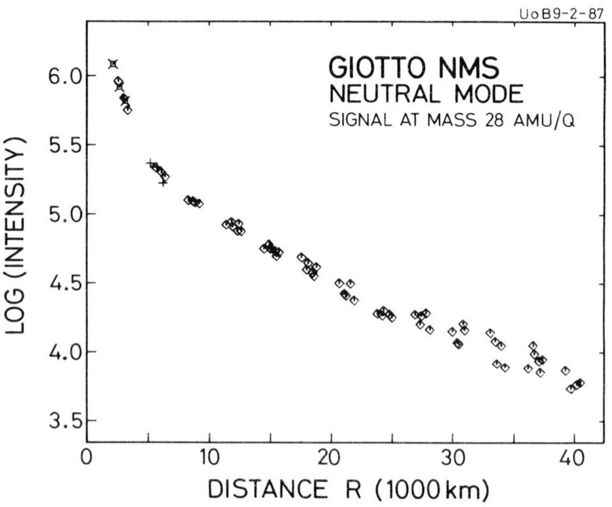

Fig. 1. Ion current in M-analyzer at mass 28 amu/e (neutral mode, HeV) as function of radial distance from nucleus. Different symbols refer to different detector gains. Preliminary data reduction

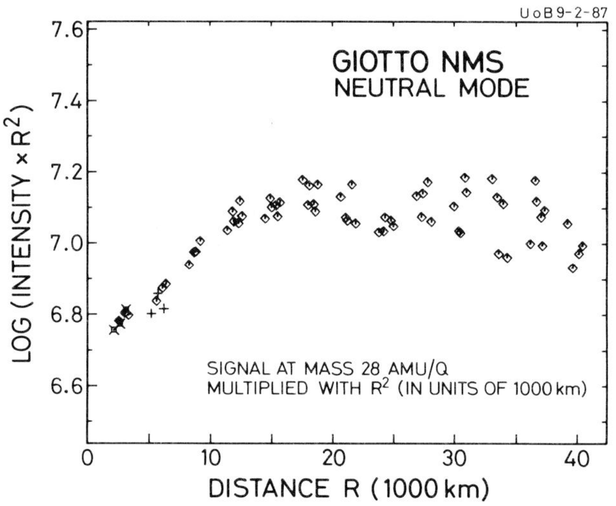

Fig. 2. Ion current in M-analyzer at mass 28 amu/e (neutral mode, HeV) multiplied with the square of the distance. Different symbols refer to different detector gains. Preliminary data reduction. If no sources or sinks are present, then this value should be independent of the distance from the nucleus provided the expansion velocity is constant. The increase observed for distances <20000 km shows that a parent molecule or dust grains must be decaying into fragments contributing to the mass 28 amu/e signal

a parent molecule by Q' and the total production rate of a molecule as Q.) This limit is a factor of 3 lower than the production ratio of 0.17 derived by Woods et al. (1986) from UV observations on a sounding rocket flight on March 13, 1986, 11:20 UT, approximately 13 hours before the Giotto encounter.

The intensity of the 28 amu/e peak shows a non $1/R^2$ dependence (R = distance from nucleus) (Fig. 1). This is especially evident when the mass 28 amu/e signal strength multiplied with the square of the distance from the nucleus is plotted (Fig. 2). If the number density $n(R)$ of a species decreases only because of the expansion then $n(R) \cdot R^2$ should be constant provided the expansion velocity is constant. It is evident from Fig. 2 that the mass 28 amu/e signal shows an approximately exponential growth doubling in intensity within 10000 km. At distances >20000 km the signal becomes essentially constant. The scatter of the data points obtained in the preliminary data reduction is too large to determine a reliable scale length for distances >20000 km.

From the intensity and radial behaviour of the signals at masses 27 and 30 amu/e we can exclude ethane (C_2H_6) and ethene (C_2H_4) as molecules responsible for the increase at mass 28 amu/e. The E-analyzer shows no heavy species of sufficient abundance in the mass range 37–86 amu/e which could by fragmentation in the ion source explain the increase. Hence, we conclude that the observed increase in the signal on mass 28 amu/e is due to a corresponding increase in either the CO or N_2 relative abundance in the cometary coma, indicating an extended source.

If we assume that at 20000 km only CO contributes to the mass 28 amu/e signal we obtain for the number densities

$$n(\mathrm{CO}, 20000)/n(\mathrm{H_2O}, 20000) \leqq 0.24 \tag{3}$$

and – after correction for the different scale lengths for photodestruction – for the total production rates of CO and H_2O

$$Q(\mathrm{CO})/Q(\mathrm{H_2O}) \leqq 0.15 \tag{4}$$

Essentially the same production rate ratio results if the data in the 45000 km to 75000 km distance range are used. Within error limits this ratio is compatible with the production ratio of 0.17 $\pm$0.04 given by Woods et al. (1986). However, these authors did not recognize the fact that a substantial fraction of the CO must, if their production ratio holds, be due to a CO containing parent molecule.

The LeV mode data can be used to set an upper limit on the N_2/CO ratio and hence a lower limit for the CO density. The ionization potential of N_2 is ~15.6 eV, that of CO ~14 eV. Thus, CO is preferentially ionized in the LeV mode relative to N_2. From the observed mass 28 amu/e signal strength ratio of the two modes in the 20000 to 75000 km range we conclude that in this distance range

$$n(\mathrm{N_2})/n(\mathrm{CO}) < 2 \tag{5}$$

and hence

$$0.05 \leqq Q(\mathrm{CO})/Q(\mathrm{H_2O}) \leqq 0.15\,. \tag{6}$$

The lower limit for Q(CO) corresponds to an upper limit for a possible N_2 production of

$$Q(\mathrm{N_2})/Q(\mathrm{H_2O}) \leqq 0.1\,. \tag{7}$$

A model calculation which takes the radial dependence of the expansion velocity into account (Lämmerzahl et al., 1986) shows that the parent molecule responsible for the observed increase on mass 28 amu/e must have a 3 to 4 times higher photodestruction rate than H_2O. The molecule HNCO photodissociates as follows (rate coefficients c for 1 AU):

$$\mathrm{HNCO} + h\nu \rightarrow \mathrm{NH} + \mathrm{CO} \qquad c = 15\ 10^{-6}\,\mathrm{s}^{-1}$$
$$\mathrm{HNCO} + h\nu \rightarrow \mathrm{H} + \mathrm{NCO} \qquad c = 14\ 10^{-6}\,\mathrm{s}^{-1} \tag{8}$$

The total rate coefficient of 29 $10^{-6}\,\mathrm{s}^{-1}$ is approximately 3 times larger than the one for H_2O (12 $10^{-6}\,\mathrm{s}^{-1}$). However, the upper limit on the HNCO abundance which can be derived from the signal observed at mass position 43 amu/e in the E-analyzer is much too low to account for the observed increase and mass 15 amu/e (NH) does not show any increase as would be expected from the dissociation of HNCO into NH and CO. These observations rule HNCO out as a possible parent moecule of CO.

At present we have not yet identified any other possible parent molecule which could explain the observed increase in the CO or N_2 density and is compatible with the measurements made by the NMS on Giotto.

One possible source could be the cometary dust particles. Adsorbed or otherwise bound CO could be released as the dust particles streaming away from the nucleus are heated by the sunlight. Depending on the optical properties particles smaller than ~ 1 µm can reach temperatures several hundred degrees above the blackbody equilibrium temperature (Hanner, 1982) and fairly refractory organic compounds might decompose and evaporate. Very close to the nucleus gas-dust reactions could also be a non-negligible source for certain gas species. However, in the 5000 km to 10 000 km distance range the gas density is already so low that an unreasonably high dust concentration would be required to explain by gas-dust reactions the observed 50% increase in the mass 28 amu/e equivalent production rate (cf. Fig. 2). The contribution from sputtering by ions is probably negligible (Geiss et al., 1986). It is tempting to speculate that the grains rich in the light volatile elements H, C, N, and O ("CHON"-particles; Kissel et al., 1986) are this additional CO source. To explain the whole increase by CO release from dust would require $Q'(CO)/Q(H_2O) \sim 0.08$. The dust to water mass production ratio in Halley is ~ 0.2 (Mazets et al., 1986; McDonnel et al., 1986). A very substantial fraction of the grains would have to evaporate if this explanation is true (the evaporated grains are not necessarily included in the dust production rate as this production rate is partly based on dust densities measured at distances >8000 km).

There is additional evidence that dust particles are a source of neutral gas in Halley's coma. A'Hearn et al. (1986) observed CN jets reaching as far out as 50 000 km. The only tenable explanation seems that CN is directly released from the surfaces of grains, probably the "CHON" particles. It is interesting to note that some of the "CHON" particles show extremely variable compositions (Langevin, 1986).

The idea that dust supplies CO on an extended scale in the coma is also supported by direct dust observations. Vaisberg et al. (1986) observed a distinct change of the grain size distribution with distance from the comet. Fine grains become overabundant and the relative abundance of medium sized grains seems to decrease with increasing distance. The depletion starts closest to the nucleus ($R \lesssim 10\,000$ km) with grains in the 10^{-14} to 10^{-13} g range. At greater distances also larger dust grains show a depletion. This and other evidence is interpreted by Vaisberg et al. (1986) as the result of splitting of larger grains during their outward motion in the coma. Aggregate particles seem to lose their interstitial material by evaporation and to eventually break up into very small grains. This process should proceed faster with smaller grains and the lifetime of the aggregates thus increases with their size. Once the break-up begins, additional surface area coated with volatile material will be exposed to space and the sun, leading to an enhancement of the gas production rate. Figure 3 shows the radial dependence of the CO production rate required to explain the observed increase in the measured 28 amu/e signal (Fig. 2). The quantity $\Delta Q'(R)$ is the CO production rate in a shell of 2000 km thickness in a distance R from the nucleus. The shape of the curve, showing a maximum at ~ 9000 km is in qualitative agreement with the gas production rate expected from disintegrating aggregates.

The exact chemical nature of the material evaporating from the grains either as CO and CN or as a short lived parent of these two molecules is at present uncertain. The proposed photolyzed

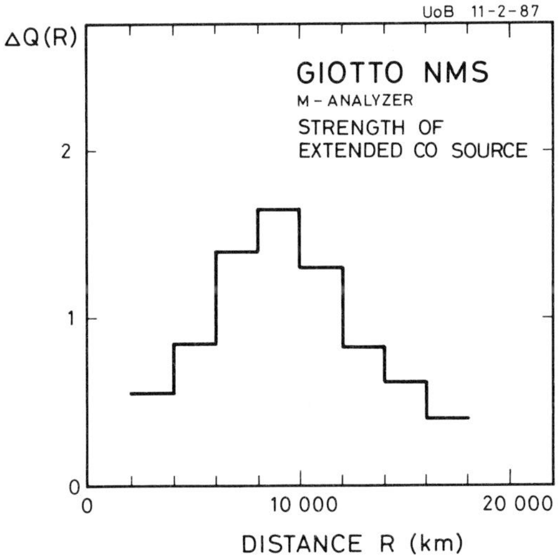

Fig. 3. Radial dependence of the strength of the CO source required to explain the observed radial increase in the 28 amu/e signal (multiplied by R^2). The quantity $\Delta Q'(R)$ (in arbitrary units) is the total amount of gas released in concentric shells with 2000 km thickness

outer mantle of interstellar grains (Greenberg, 1982) could be a source of evaporating organic molecules, including CO or a parent of CO. In interstellar clouds CO is condensing directly on the grain surfaces as solid CO (Lacy et al., 1984) which requires temperatures as low as 17 K (Léger, 1983). Irradiating CO, NH_3 and other ice mixtures in the laboratory with UV has shown that molecules containing cyano ($C \equiv N$) groups are formed and that these reaction products have much lower vapor pressure and are not lost even at temperatures as high as 150 K (Lacy et al., 1984). From IR absorption spectra Lacy et al. conclude that XCN molecule formation also occurs on interstellar grains. The observation of the extended CO and CN sources in the coma of comet Halley may thus be an indication that relatively unaltered interstellar grains are released from the nucleus.

Acknowledgements. We would like to thank R. Wright for graphics and B. Grose for typing. This research was supported by the Swiss National Science Foundation (grants 2.034-0.81–2.825.85); the German Bundesminister für Forschung und Technologie (grants 010F0519, 010F0628 and 010F85120; CNES (grants 81, 82/244, 83/211 and 84, 84/1211); and NASA headquarters (contract NASW-3576).

References

A'Hearn, M.F., Hoban, S., Birch, P.V., Bowers, C., Martin, R., Klinglesmith, D.A.: 1986, *Nature* **324**, 649

Eberhardt, P., Krankowsky, D., Schulte, W., Dolder, U., Lämmerzahl, P., Berthelier, J.J., Woweries, J., Stubbemann, U., Hodges, R.R., Hoffmann, J.H., Illiano, J.M.: 1986, 20th ESLAB Symp., *ESA SP*-**250**, Vol. I, p. 383

Geiss, J., Bochsler, P., Ogilvie, K.W., Coplan, M.A.: 1986, *Astron. Astrophys.* **166**, L1

Greenberg, J.M.: 1982, in *Cometary Exploration II*, ed. T.I. Gombosi, Hungarian Academy of Sciences, p. 23

Hanner, M.S.: 1982, in *Cometary Exploration II*, ed. T.I. Gombosi, Hungarian Academy of Sciences, p. 1

Kissel, J., Brownlee, D. E., Büchler, K., Clark, B. C., Fechtig, H., Grün, E., Hornung, K., Igenbergs, E. B., Jessberger, E. K., Krueger, F. R., Kuczera, H., McDonnell, J. A. M., Morfill, G. M., Rahe, J., Schwehm, G. H., Sekanina, Z., Utterback, N. G., Völk, H. J., Zook, H. A.: 1986, *Nature* **321**, 336

Krankowsky, D., Lämmerzahl, P., Dörflinger, D., Herrwerth, I., Stubbemann, U., Woweries, J., Eberhardt, P., Dolder, U., Fischer, J., Herrmann, U., Jungck, M., Meier, F. O., Schulte, W., Berthelier, J. J., Illiano, J. M., Godefroy, M., Gogly, G., Thévenet, P., Hoffman, J. H., Hodges, R. R., Wright, W. W.: 1986a, *ESA SP*-**1077**, p. 109

Krankowsky, D., Lämmerzahl, P., Herrwerth, I., Woweries, J., Eberhardt, P., Dolder, U., Herrmann, U., Schulte, W., Berthelier, J. J., Illiano, J. M., Hodges, R. R., Hoffman, J. H.: 1986b, *Nature* **321**, 326

Lacy, J. H., Baas, F., Allamandola, L. J., Persson, S. E., McGregor, P. J., Lonsdale, C. J., Geballe, T. R., van de Bult, C. E. P.: 1984, *Astrophys. J.* **276**, 533

Lämmerzahl, P., Krankowsky, D., Hodges, R. R., Stubbemann, U., Woweries, J., Herrwerth, I., Berthelier, J. J., Illiano, J. M., Eberhardt, P., Dolder, U., Schulte, W., Hoffman, J. H.: 1986, 20th ESLAB Symp., *ESA SP*-**250**, Vol. I, p. 179

Langevin, Y.: 1986 (private communication)

Léger, A.: 1983, *Astron. Astrophys.* **123**, 271

Levine, J. S.: 1985, *The photochemistry of atmospheres*, Academic Press, New York

Mazets, E. P., Aptekar, R. L., Golenetskii, S. V., Guryan, Y. A., Dyachkov, A. V., Ilyinskii, V. N., Panov, V. N., Petrov, G. G., Savvin, A. V., Sagdeev, R. Z., Sokolov, I. A., Khavenson, N. G., Shapiro, V. D., Shevchenko, V. I.: 1986, *Nature* **321**, 276

McDonnel, J. A. M., Alexander, W. M., Burton, W. M., Bussoletti, E., Clark, D. H., Grard, R. J. L., Grün, E., Hanner, M. S., Hughes, D. W., Igenbergs, E., Kuczera, H., Lindblad, B. A., Mandeville, J.-C., Minafra, A., Schwehm, G. H., Sekanina, Z., Wallis, M. K., Zarnecki, J. C., Chakaveh, S. C., Evans, G. C., Evans, S. T., Firth, J. G., Littler, A. N., Massonne, L., Olearczyk, R. E., Pankiewicz, G. S., Stevenson, T. J., Turner, R. F.: 1986, *Nature* **321**, 338

Vaisberg, O., Smirnov, V., Omelchenko, A.: 1986, 20th ESLAB Symp., *ESA SP*-**250**, Vol. II, p. 17

Woods, T. N., Feldman, P. D., Dymond, K. F., Sahnow, D. J.: 1986, *Nature* **324**, 436

Astron. Astrophys. 187, 485–488 (1987)

ASTRONOMY AND ASTROPHYSICS

Resolution of the [O I] + NH_2 blend in comet P/Halley*

C. Arpigny [1], P. Magain [2], J. Manfroid [1,**], F. Dossin [1], A.C. Danks [3,***], and D.L. Lambert [4]

[1] Institut d'Astrophysique, Université de Liège, B-4200 Ougrée-Liège, Belgium
[2] European Southern Observatory, Casilla 19001, Santiago 19, Chile
[3] Department of Physics and Astronomy, Michigan State University, East Lansing, MI 48224, USA
[4] McDonald Observatory and Department of Astronomy, University of Texas, Austin, TX 78712, USA

Received February 20, accepted May 11, 1987

Summary. Spectra of comet P/Halley have been taken at very high resolution (0.015 nm) with a view to evaluating the contamination of the forbidden oxygen doublet at 630 nm by NH_2 features at low resolution. Comparison is made with a few other bright comets.

Appreciable variations are found not only in the NH_2/[O I] ratio but also in the relative intensities of the various NH_2 emissions themselves. We comment upon the accuracy obtained on the oxygen abundance derived from the [O I] lines. Mapping of these emissions over the coma is required in order to correctly remove any important NH_2 contribution at low resolution. This should in addition provide information bearing upon the origin of the oxygen atoms in the ^1D level.

Key words: forbidden oxygen lines – coma – abundance

1. Introduction

The determination of the abundance and of the origin of the oxygen atoms present in the gas produced by comets represents one of the important items in our endeavour to unravel the problem of the chemical composition of the cometary nuclei. Now, the forbidden lines of oxygen at 630.0 and 636.4 nm, which contrary to the other cometary emissions cannot be excited by fluorescence on account of their very low transition probabilities, are emitted by atoms directly produced in their upper ^1D level and decaying radiatively at a well-known rate. As a result, the measurement of their intensity in principle provides a straightforward means for evaluating the abundance or production rate of those oxygen atoms, as well as of their parents provided the relevant O (^1D) yield is known. That the latter oxygen-bearing molecules are important cometary constituents is indicated by the strength of the red doublet observed in many spectra and this was already pointed out by Biermann and Trefftz (1964). The situation is not so simple, however, because aside from the superposition of the telluric [O I] emission which can be taken care of readily, the oxygen emission is also rather severely blended with cometary NH_2 features to a degree that should be estimated in each case. Excellent detailed discussions of this problem have been given by Festou and Feldman (1981), Spinrad (1982), and by Fink and Johnson (1984).

Halley's comet has offered the opportunity to use a very high resolution to separate all the components of the [O I] + NH_2 blend and we report briefly here on our observations of this spectral region. We wanted, in particular, to assess the validity of the assumption made by Spinrad and others that the NH_2 contribution to the intensity of the 630.0 feature observed at low resolution is equal to the intensity of another, purely NH_2 feature at 633.4. Comparison has also been made with spectra of four other comets observed at fairly high resolution.

2. Observations

Our own observations of comet P/Halley were carried out at the European Southern Observatory (ESO) with the Coudé Echelle Spectrometer equipped with a Reticon and fed by the 1.4 m Coudé Auxiliary Telescope on 6, 27, and 28 April 1986. The heliocentric distances of the comet were 1.26, 1.58 and 1.60 a.u., respectively. The resolving power was about 50000 (resolution $\sim$0.015 nm) and the slit height was 20″, corresponding to approximately 6500 km and 10000 km projected on the comet on the first date and on the other two dates, respectively. In most cases the comet's image was centred on the slit. At this very high resolution the Doppler shift is more than enough to separate the cometary and telluric [O I] emissions and the problem of sky contamination is eliminated ipso facto. An example of such a spectrum appears in Fig. 1. On the other hand, the usual reduction procedure was followed and the spectrum of Mars and of the star HR 6165 were used for photometric calibration.

In order to compare P/Halley with other bright comets we were fortunate to have a series of high quality spectra obtained by various astronomers at the Lick Observatory using the 120-inch and 24-inch auxiliary aperture together with image intensifiers, and kindly made available to some of us by Dr. G. Herbig. These spectra of comets Kohoutek (1973 XII), Bradfield (1974 III), Kobayashi-Berger-Milon (1975 IX), and West (1976 VI), have resolutions of about 0.08 to about 0.2 nm; they were microphotometered using a slit covering 5000 to 9000 km centred on the comet.

Send offprint requests to: C. Arpigny

* Based on observations made at the European Southern Observatory (Chile) and at the Lick Observatory (USA)

** Senior research associate, Belgian National Fund for Scientific Research

*** Present address: Dept. of Physics, Florida International University, Miami, FL 33199, USA

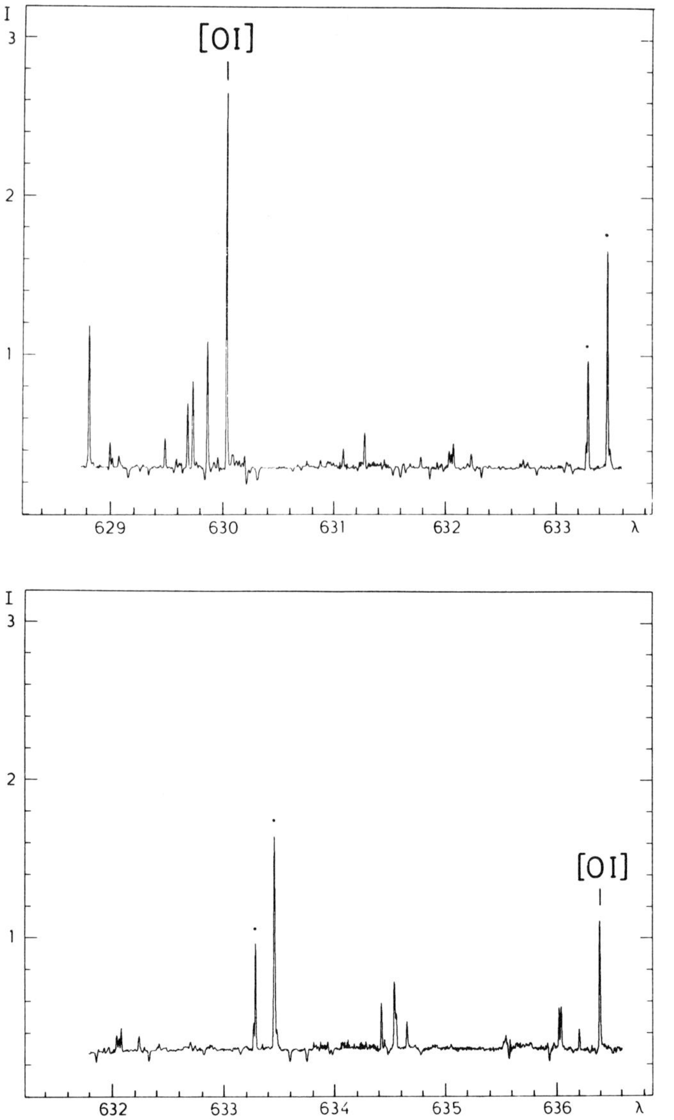

Fig. 1. High-resolution (0.015 nm) spectrum of the central region of comet P/Halley showing the two components of the [O I] red doublet and the neighbouring emissions due to the (0, 8, 0) – (0, 0, 0) band of NH_2. The intensity is plotted on an arbitrary scale. The two NH_2 lines marked by dots merge at low resolution into a single feature designated by c in the text and used as a reference in previous work

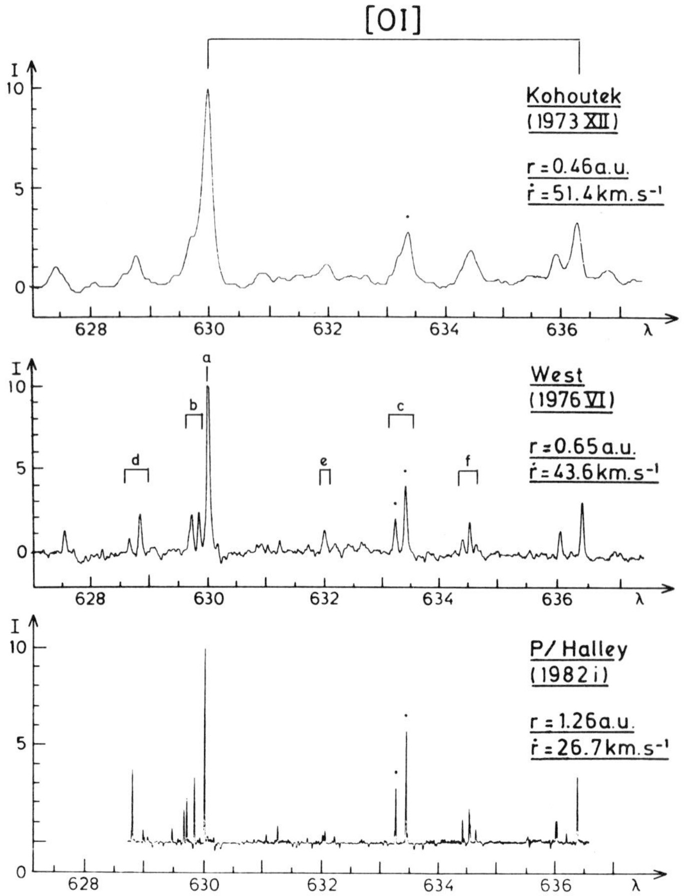

Fig. 2. The region of the [O I] red doublet at progressively higher resolution (0.2, 0.08, and 0.015 nm, respectively)

3. Results

The complete separation of the various NH_2 lines (which all belong to the 8–0 band) and of the forbidden oxygen lines in comet P/Halley allows the measurement of the relative intensities of the individual components of the blend they would form at lower resolution and hence the evaluation of their contributions to these blends. Let us refer to the middle panel in Fig. 2, which defines various groupings of the lines; the letters will also designate the sum of intensities of the lines in each group, except for a which represents the intensity of the [O I] 630.023 line by itself.

In order to estimate the effect of the blending we consider two types of resolutions, say (1) 0.4–0.6 nm and (2) 1.0–1.5 nm:

– in Case (1), we would observe the two blends $(a + b)$ and c separately from the other features,

– in Case (2), d would be included in the first blend, e and f in the second; let us further call $B = b + d$ and $C = e + c + f$ in this case.

Thus the "feature" at 633.4 (dots on the spectra shown here), used as a reference in the "decontamination" procedure mentioned above, actually represents either c or C in our examples.

We have accordingly calculated the b/c and B/C ratios and found values of 0.5 – 0.65 and 0.85 respectively from the spectra of Halley's comet.

As for the other comets, Figs. 2–4 show some definite variations of the various NH_2 emissions either among different comets or in a given comet at different distances from the Sun. In the 0.2 nm resolution spectra the b contribution to the $(a + b)$ blend has been extracted approximately by very simple curve fitting using gaussians. The above ratios there range from 0.4 to 1.3 and from 0.6 to 1.0, respectively. Thus, appreciable departures from the assumed value of 1.0 exist. A detailed analysis would be required to examine whether the fluorescence excitation of the NH_2 emissions could explain these variations.

It is very difficult to carry out a detailed evaluation of the uncertainties of intensity measurements derived from photographic spectra, particularly from spectra obtained with image tubes: many factors would have to be taken into account, most of them unknown. Fortunately, however, a significant indicator of the overall precision achieved in the reduction process is provided by the [O I] 636.4/630.0 intensity *ratio*, since this should be equal

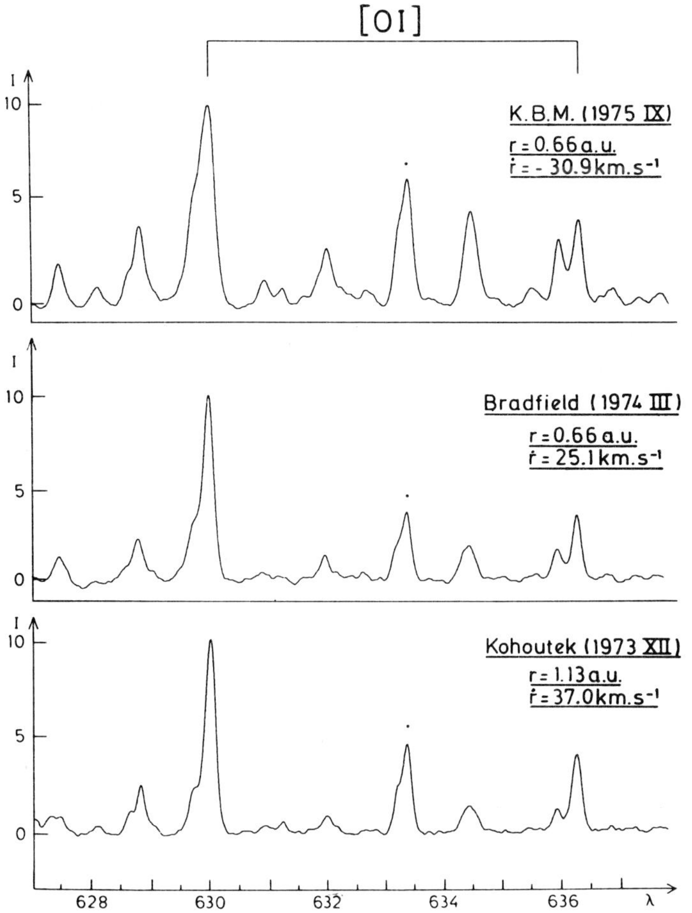

Fig. 3. Comparison of the region of the [O I] red doublet in three comets (resolution 0.2 nm)

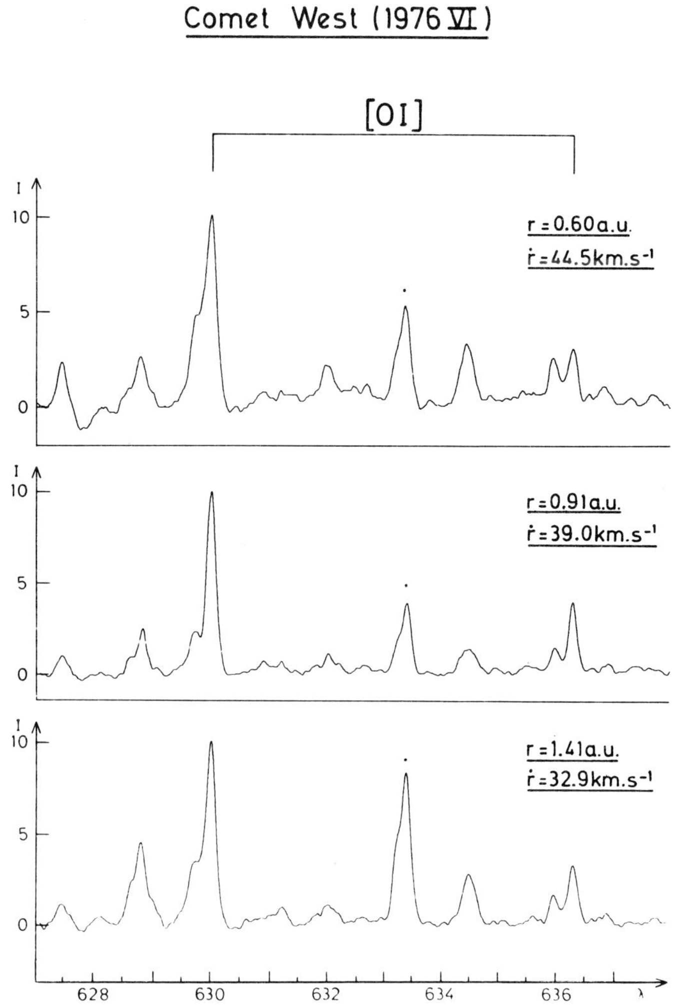

Fig. 4. Evolution with heliocentric distance of the region of the [O I] red doublet in comet West (resolution 0.2 nm)

to the ratio of the corresponding transition probabilities. This quantity is known both from experimental measurements, 0.32 ± 0.03 (Kernahan and Pang, 1975), and from theoretical calculations, 0.325, with an estimated error not exceeding 10% (Garstang, 1956; Mendoza, 1983). The value obtained from our high-resolution spectrum of comet P/Halley comes out very close to these estimates: 0.34, with an uncertainty placed at a few percent. On the other hand, as can already be seen from Figs. 2–4, this oxygen line ratio remains rather constant, as it should, from one cometary spectrum to the next. This relative constancy and the agreement with the above numbers of the average value derived from measurements on the eight photographic spectra, 0.33 ± 0.04 (r.m.s. deviation), seem quite satisfactory. This rather good accuracy no doubt results from the conjunction of several favourable circumstances (*relative* intensities of lines separated by a small wavelength interval, low and uniform plate background, unsaturated exposures, low geometrical distortion).

However, the important question is to know how these results affect the estimate of the production rate of the O (^1D) atoms. To investigate this we evaluate a/a^*, where a^* represents the value of the [O I] 630.0 intensity (or flux) which would be derived under the assumption that $b = c$ or $B = C$, i.e. $a^* = (a + b) - c$ or $(a + B) - C$ in Case (1) or (2), respectively. This ratio turns out to vary as well, as shown in Table 1.

Table 1. The a/a^* ratio

Comet	Case (1)	Case (2)
P/Halley	1.7	1.3
Other comets	0.9–2.3	1.0–2.2

a is the true intensity of the 630.0 component of the red doublet, a^* the value derived for this intensity from low-resolution spectra by simply subtracting the intensity of the 633.4 feature from that of the blend observed at 630.0

We have examined the incidence of the errors of measurements, estimated as explained above, on the a/a^* ratio and found that the uncertainty on the latter ratio is less than 10% for the spectra of comet P/Halley, while it lies between 15 and 30% for the photographic spectra of the other comets. Thus, we see, for instance, that observations at low resolution of the comets in our sample would, in some cases, have led to underestimate the [O I] flux or luminosity and the corresponding production rate Q by a factor of the order of 2, had the above hypothesis been adopted, even if this hypothesis sometimes happens to be appropriate.

4. Conclusion

In view of the importance of the determination of $Q(\mathrm{O}\ ^1D)$ for identifying the origin of the excited oxygen atoms, it appears desirable, when the NH_2 emission is appreciable, to ascertain its contribution to the blends observed at low resolution by means of an appropriate method. This can be achieved by taking advantage of the different spatial distributions of the NH_2 and [O I] emissions. Such a procedure has already been applied on a few comets (Delsemme and Combi, 1983; Fink and Johnson, 1984). Furthermore, mapping the [O I] radiation over the coma provides additional data related to the production mechanism(s) of the corresponding atoms and to the nature of their parents. Although there are indications that the photodissociation of H_2O plays a major rôle, it would be interesting to study the possible existence of other sources such as the photodissociation of CO_2 or the dissociative recombination of ions like H_2O^+ or CO_2^+.

Acknowledgements. We are very grateful to G. Herbig for lending us a number of excellent cometary spectrogrammes secured at the Lick Observatory. We also thank D. Baade for taking a spectrum of P/Halley on his own run and H. Lindgren for help in some of our observations at the ESO Coudé Auxiliary Telescope.

References

Biermann, L., Trefftz, E.: 1964, *Z. Astrophys.* **59**, 1
Delsemme, A.H., Combi, M.R.: 1983, *Astrophys. J.* **271**, 388
Festou, M.C., Feldman, P.D.: 1981, *Astron. Astrophys.* **103**, 154
Fink, U., Johnson, F.R.: 1984, *Astron. J.* **89**, 1565
Garstang, R.H.: 1956, in *The Airglow and the Aurorae*, eds. E.B. Armstrong, A. Dalgarno, New York, Pergamon Press, p. 324
Kernahan, J.A., Pang, P.H.-L.: 1975, *Can. J. Phys.* **53**, 455
Mendoza, C.: 1983, in *Planetary Nebulae, IAU Symp.* **103**, ed. D. Flower, Reidel, Dordrecht, p. 143
Spinrad, H.: 1982, *Publ. Astron. Soc. Pacif.* **94**, 1008

Astron. Astrophys. 187, 489–496 (1987)

ASTRONOMY AND ASTROPHYSICS

A rotational-state population analysis of the high-resolution IUE observation of CS emission in comet P/Halley

M.G. Prisant [1] **and W.M. Jackson** [2,*]

[1] Department of Chemistry, University of California, Berkeley, CA 94720, USA
[2] Department of Chemistry, University of California, Davis, CA 95616, USA

Received February 21, accepted May 13, 1987

Summary. The high resolution spectrum of comet P/Halley has been measured in the region of the CS emission. CS in comet Halley is produced by photodissociation of CS_2 and subsequently re-excited by broad band solar radiation. An inversion procedure is introduced to extract the rotational population distribution from the dispersed fluorescence spectrum. The fluorescence profile from the *R* branch of the $v' = 0$ to $v'' = 0$ is analyzed using this method. The extracted rotational distribution shows a peak at $J'' = 1$ superimposed on a background extending to $J'' = 29$. The distribution is interpreted by considering radiative and collisional relaxation processes in the comet. Intensity anomalies in the wavelength region corresponding to *P* and *Q* branch emission are considered.

Key words: comets – atomic and molecular processes

1. Introduction

The CS molecule holds the potential of playing an important role in our understanding of comet chemistry (Swamy, 1986). The CS radical is produced by photodissociation of CS_2 in the comet. The scale length of the CS daughter has been shown to be greater than 10^5 km: this corresponds to a sufficiently long lifetime for the detection of molecular fluorescence due to subsequent reexcitation by solar radiation (Swamy, 1986; Jackson et al., 1986). The scale length of CS_2 parent, on the other hand, has been shown to be less than 10^3 km (Swamy, 1986): this means that CS radicals are formed in the inner region of the comet coma where collisions could conceivably affect the rotational populations of newly formed product (Swamy, 1986; Jackson, 1986). CS is the only molecule which is consistently observed in the spectra of comets that originates from a parent of such short scale-length. The CS species is thus uniquely positioned to monitor local conditions in the inner region of the comet coma.

The rotational distribution of CS radical reflects conditions in the inner coma. In the case of complete thermalization of CS product, we would expect a rotational distribution well described by a Boltzmann functional form (Jackson et al., 1982); the temperature of this distribution would then be the temperature of the inner coma. In the case of radiative equilibrium (in some sense the opposite extreme) we would anticipate a distribution of a highly non-Boltzmann character in which most molecules were relaxed to very low *J* levels; a "temperature" determined from such a distribution would bear no relation to the temperature of the inner coma.

The form of the CS rotational distribution as determined from the fluorescence contour thus weights heavily in evaluating the character and content of information which may be learned on the inner coma conditions. Unfortunately the CS spectrum, like many other molecular systems of astrophysical interest, is sufficiently complex and overlapped to prohibit a straightforward determination of individual rotational populations. The procedure generally used – spectral simulation – is not only inappropriate in this situation because it begins by presuming the *form* of the distribution which is of special interest, but also inadequate because it provides no measure of goodness of fit or parameter sensitivity.

The problem of extracting populations from congested spectra has recently been treated by Prisant and Zare (1985) and refined by Prisant and Levine (1986). These authors define a linear transformation connecting spectral intensities and state populations. Definition of this transformation permits direct extraction of population estimates through use of linear regression. In this work, the method of Prisant and Zare will be adapted to analyze high resolution spectra obtained in an IUE observation of comet Halley on 1985 December 25 and 26. As this technique bears on population analysis of astrophysical spectra in a completely general fashion, extensive discussion will be given to the modalities of its application. The results of the analysis are prerequisite to distinguishing between radiative equilibrium and collisional thermalization mechanisms for the production of the CS rotational distribution in Halley's comet.

2. Observation of CS radical in comet fluorescence

CS has been seen in many comets since its first observation in comet West 1976 VI (Prisant and Levine, 1986). Spatial resolution of CS emission in other comet sightings has shown that CS originates not from the comet nucleus but rather from the decay of another molecule at a height of less than 1000 km in the cometary atmosphere. Analysis of emission from comet IRAS-Araki-Alcock 1983d using the model of Haser (Haser, 1957; Festou, 1981) established a scale length of 300 km for the parent of CS (Jackson et al., 1982).

Send offprint requests to: W.M. Jackson

* Guest observer with the International Ultraviolet Explorer Satellite

It has been argued that CS radical is formed in comet atmospheres by the photodissociation of CS_2 (Jackson et al. 1985):

$$CS_2 + h\nu(\text{solar radiation: } 190\,\text{nm} < \lambda < 210\,\text{nm}) \rightarrow CS + S\,. \quad (1)$$

Identification of CS_2 as the parent species is based on the photochemical unsuitability of other species containing a CS bond found in comet atmospheres as parents of CS and the photochemical suitability of CS_2 as the only simple molecule containing CS with a large absorption coefficient at wavelengths that are reasonably intense in solar radiation (Jackson et al., 1985). We note that CS_2 photodissociates efficiently throughout the 100–210 nm region. Okabe identifies the ground state product as the major channel in the region from 190 to 210 nm and production of $A^1\Pi$ as a major primary process in 120 to 140 nm region: the former occurring through predissociation of the $^1B_2$ state and the latter through direct dissociation of Rydberg states. Three lines of reasoning argue for the predissociation mechanism followed by solar excitation as the source of CS emission in the comet. First, the CS parent lifetime deduced from scale length measurements in the comet (500 s) is consistent with the calculated predissociation lifetime from the $^1B_2$ state (590 s) (Jackson et al., 1985). Second, as will be seen, the observed rotational and vibrational distributions in emission are highly relaxed; if CS A state directly produced by photodissociation were a major contributing source of emission intensity, we would expect an emission pattern characteristic of highly excited product. Third, solar intensities are much reduced in 120–140 nm region relative to those in the 190–210 region. In our analysis it will be assumed that the CS observed in cometary fluorescence is produced by photodissociation of CS_2 yielding ground state product and subsequently reexcited by solar radiation.

Photodissociation of CS_2 in laboratory experiments is known to lead to the production of highly vibrationally and rotationally excited CS radicals (McCrary et al., 1985). The radicals appear to be formed with energy up to the thermodynamic limit, so that in excitation at 193 nm vibrational levels up to $v''=12$ and substantial amounts of rotational energy are populated. It is unlikely that CS radicals formed in the cometary environment will survive in the highly excited vibrational levels of the nascent product state distribution. The natural radiative lifetime for the a CS radiacal radiating in the infrared region via:

$$CS(v'') \rightarrow CS(v''-1) + h\nu \quad (2)$$

is on the order of milliseconds. We therefore expect that all radicals formed in vibrationally excited states will cascade down to the ground state on a time scale of 10^{-2} s. This is short compared to the solar excitation rate (on the order of 10^3 s at 1 AU) and the collision rate (on the order of seconds at 300 km from the nucleus). It may therefore be concluded that most of the detected CS radicals are in the $v''=0$ vibrational level.

In contrast to vibration, full radiative relaxation of rotational excitation in the CS radical in the cometary atmosphere is not expected. The B value for the CS radical is $\sim 0.8\,\text{cm}^{-1}$ and approximate spacing between J levels of $2BJ$ suggest that the radiative lifetime of all J'' levels $< J = 60$ will be greater than 1 s. This is the largest number of J levels which must be considered in any rotational analysis and defines the dimensionality of the problem.

We now describe the observation of CS in Halley's comet. CS produced in the comet coma is excited by solar radiation; resulting fluorescence is detected by the IUE telescope. The spectrum of the exciting solar radiation is shown in Fig. 1: this data presented by A'Hearn (1983) has been averaged over the solar disk. The solar radiation is unpolarized. Boggess et al. (1978) has previously described the IUE instrumentation: only the observational details and the results will be given in this section. Figure 2 shows the high resolution spectrum obtained with the LWP camera during observation of comet Halley between 257.5 and 257.9 nm. This portion of the spectrum has been assigned to emission from $CS(^1\Pi\,(v=0) - X^1\Sigma^+\,(v=0))$. The spectrum was observed using a 720 minute exposure on 1985, 25 and 26 December (LWP 7383). The CS spectrum was observed in two different orders but in this paper we present only the order that occurs near the center of the camera.

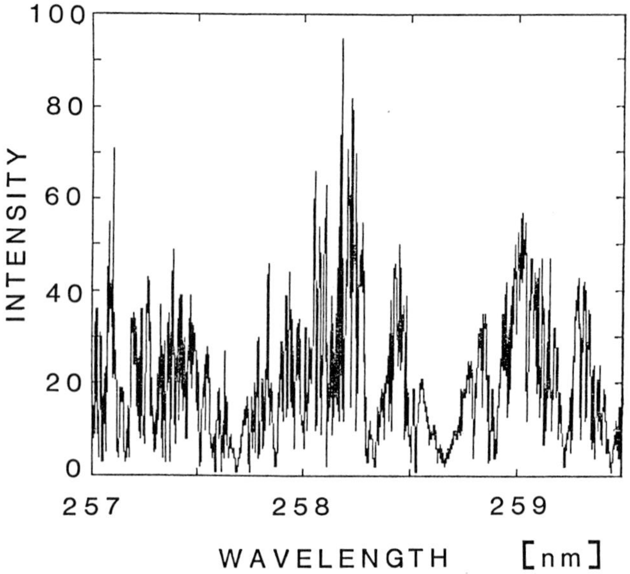

Fig. 1. The solar emission spectrum between 257 and 259 nm. An air wavelength scale is used. The comet sees this spectrum Doppler shifted ~ 0.02 nm to the blue

3. Theory

3.1. An inversion procedure for extraction of rotational populations from the fluorescence contours of broad-band excitation spectra

The fluorescence contour viewed by the observation telescope is the result of broad band excitation by solar radiation followed by emission. In the discussion which follows we first develop the linear transformation which relates the vector of emission intensities to that of rotational populations. We then go on to develop a parameterization scheme for the rotational populations.

The emission intensity at wavelength λ_l, $I_l(\lambda_l)$ is a linear function of the ground state populations and is given by:

$$I_l(\lambda_l) = K \sum_i \left\{ N_i \left[\sum_{jk} s_{ij} s_{jk} v_{ij} v_{jk}{}^3 O(v_{ij}) \phi(\lambda_{jk} - \lambda_l) \right] \right\}, \quad (3)$$

where K is a proportionality constant. The first summation over i is over the populated ground state levels, the second summation over j is over the intermediate levels excited by solar radiation, and the third summation over k is over all ground state levels optically coupled in fluorescence. The terms relevant to the absorption process are: N_i, the population flux of ground state absorbers is in

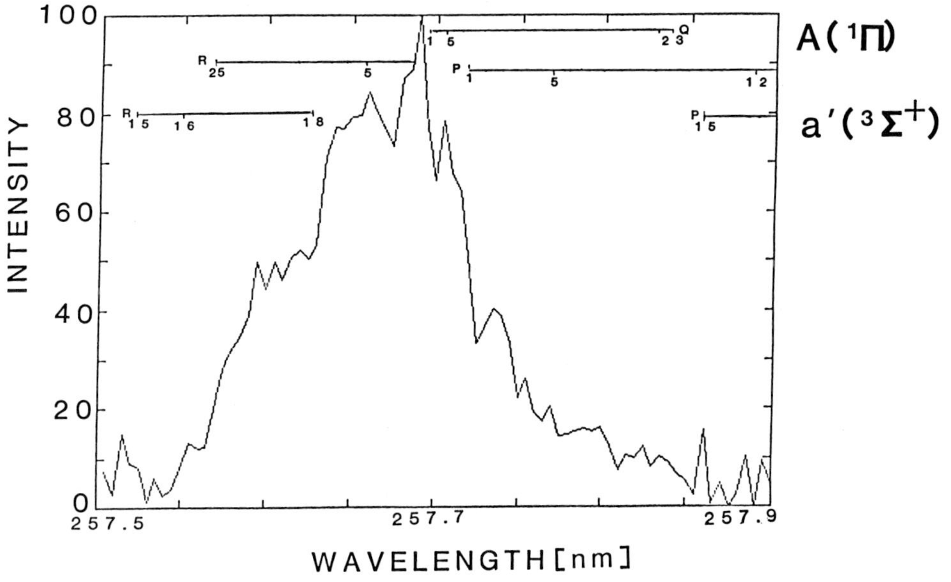

Fig. 2. The portion of the emission spectrum of Halley's comet captured by the IUE satellite assigned to CS emission. An air wavelength scale is used. The motion of the satellite relative to the comet shifts the emission lines ~0.02 nm to the red of their values for a stationary observer. The intensity scale is relative and does not indicate the level of background scattered light not due to CS fluorescence. Positions of the assigned transitions are shown: the *R* branch line occurs to the blue of 257.68 nm and the *P* and *Q* branch lines to the red

state i, v_{ij}, the Doppler shifted absorption frequency, $O(v_{ij})$ the solar radiation flux at v_{ij}, and s_{ij}, the natural line strength of the absorption transition. The terms relevant to the emission process are: $\phi(\lambda_{jk}-\lambda_l)$, the instrument function of the telescope, s_{jk}, the natural line strength of the emission transition, v_{jk}, the Doppler shifted emission frequency, and λ_{jk}, the Doppler shifted emission wavelength. We also make the following notes. Doppler shifts in absorption originate from the motion of the comet from the sun; Doppler shifts in emission originate from the motion of the observer telescope relative to that of the comet. The natural line strength is given by:

$$s_{jk}=q^2(v',v'')\,\mu_e{}^2 S_{jk}/(2J+1)\,, \tag{4}$$

where $q(v'v'')$ is the Franck-Condon factor between levels v' and v'', μ_e is the electronic transition moment, $2J+1$ is the degeneracy of the populated state in emission or absorption with J the total angular momentum quantum number, and S_{jk} is the Hönl-London line strength factor for the transition. Finally the telescope instrument function is taken to be a grating function or triangle with FWHM equal to the instrument resolution.

The original equation for emission intensity may be recast in matrix form such that:

$$\boldsymbol{I}=\mathbf{Q}\boldsymbol{N}\,, \tag{5}$$

where $\boldsymbol{I}$ is an L-dimensional column vector with L being the number of measured points, $\boldsymbol{N}$ is an I-dimensional vector with I being the number of ground state rovibrational levels, and $\mathbf{Q}$ is an L by I matrix where each element corresponds to the contribution made by a specific ground state level i with unit population to the emission intensity at wavelength l.

If the telescope instrument function is such that all emission lines are resolved then the population of all the initial levels is overdetermined due to the presence of *P*, *Q*, and *R* branches originating from each level. This is generally not the case and we must consider a parameterization of the distribution in order to reduce the number of fitted parameters and define a model matrix which is not rank deficient. Let each population N_i be given by the expansion:

$$N_i(vJ)=\sum_q a_q F_q(N_i)/f_q\,, \tag{6}$$

where $\{F_q\}$ is a Q-dimensional set of orthogonal basis functions (Prisant and Levine, 1986). The factor f_q is the normalization constant of basis function F_q determined by:

$$f_q=\sum_i F_q(N_i)\,F_q(N_i)\,. \tag{7}$$

The $\{a_q\}$ are the independent moments (Prisant and Levine, 1986) of the population distribution $N(vJ)$ defined by:

$$a_q=\langle N(vJ)\,|\,F_q\rangle \tag{8}$$

with

$$\langle F_p|F_q\rangle=f_q\,\delta_{pq} \quad \text{and} \quad \langle N(vJ)|F_q\rangle=\sum_i N_i F_q(N_i)\,. \tag{9}$$

A number of different orthogonal basis sets may be chosen for the parameterization (Prisant and Levine, 1986); in this work we use the spline functions defined by:

$$F_q(N_i)=\begin{cases}0 & \text{if} \quad i\notin[n(q),n(q+1))\\ 1 & \text{if} \quad i\in[n(q),n(q+1))\,,\end{cases} \tag{10}$$

where the $n(q)$ are indices which define the level limits of the spline. The normalization factor is given by:

$$f_q=n(q+1)-n(q)\,. \tag{11}$$

For a spline function basis, the orthogonal moments are proportional to the integrated or average populations over the limits of the spline.

We may now expand the population vector such that $\boldsymbol{N}=\mathbf{F}\boldsymbol{a}$ where $\mathbf{F}$ is a matrix whose elements $F_{iq}=F_q(N_i)$ and $\boldsymbol{a}$ is the vector of moments. Substituting into the matrix equation for the emission intensities we find that:

$$\boldsymbol{I}=\mathbf{M}\boldsymbol{a}\,, \quad \mathbf{M}=\mathbf{QF}\,, \tag{12}$$

where $\mathbf{M}$ is the model matrix; each element M_{lq} corresponds to the contribution of basis function q with unit weight to the emission intensity at wavelength l. A set of estimated parameter values $\tilde{A}$ and associated errors which minimize the least squares measure:

$$(\boldsymbol{I}-\boldsymbol{y})^2=(\boldsymbol{I}-\mathbf{M}\boldsymbol{a})^2 \tag{13}$$

Table 1. Spectroscopic constants of CS electronic states (taken from Bergeman and Cossart, 1981)

State	T_{e0}	ω_e	$\omega_e x_e$	B_e	$\alpha_e\,[10^{-2}\,\mathrm{cm}^{-1}]$	R_e [Å]
	$[\mathrm{cm}^{-1}]$					
$X^1\Sigma^+$	0.0	1285.154	6.502	0.8200449	0.5921	1.53496
$a^3\Pi$	27030.37	1135.41	7.747	0.78478	0.692	1.5691
$a'^3\Sigma^+$	30695.90	829.24	4.943	0.64727	0.570	1.728
$d^3\Delta$	35041.28	796.17	4.966	0.63685	0.614	1.742
$e^3\Sigma^-$	38040.75	752.93	4.955	0.62251	0.627	1.762
$A^1\Pi$	38255.13	1077.23	10.639	0.78760	0.83	1.566

may be determined using linear regression. Here $\boldsymbol{y}$ is the vector of measured experimental emission intensities corrected for detector response and emission background not due to molecular fluorescence. We remark that the error estimate associated with each parameter a_q determine the variance in the estimated value of the *integrated* populations.

Properly, the inversion procedure determines the best-fit parameters and estimates their associated errors rather than determining the exact form of the distribution itself. The procedure extracts, in the parameterization scheme chosen, the best-fit independent spline moments. Due to the congestion of the spectrum and consequent loss of information, the extracted moments necessarily represent a partial set. The inversion procedure does not distinguish between distributions characterized by this partial set of moments nor does it necessarily determine the best choice of orthogonal basis.

The extracted best-fit moments thus constrain but do not determine the functional form of the actual distribution. The constraints on the form of the reconstructed distribution, $P(vJ)$ are given by:

$$\tilde{a}_q = \langle P(vJ) | F_q \rangle, \tag{14}$$

where the $\tilde{a}_q$ are the fitted values of the moments and

$$P(vJ) > 0 \tag{15}$$

for all vJ contributing to the spectrum.

Any ansatz for reconstruction of the form of the distribution must select one member of the functional family meeting these constraints. This choice is somewhat arbitrary but may be guided by quantitative criterion of the overall variance and qualitative criteria concerning the expected form of the distribution. Acceptable methods for reconstruction of the distribution include generation of all members of the population vector from the partially determined set of moments and truncated fitting basis set or optimization of a second functional form to the values of the determined moments.

3.2. Spectroscopy of the CS molecule

Emission detected from comet Halley in the 257–259 nm region corresponds to the $A^1\Pi - X^1\Sigma^+$ electronic band system of CS (Swamy, 1986; Jackson and Prisant, 1986). The lines observed originate from the $v' = 0$ to $v'' = 0$ transition. This band system has been extensively studied and is a model system for the study of spin-orbit perturbations.

Molecular constants and term values for the $X^1\Sigma^+$ have been derived from analysis of microwave and IR spectra (Todd, 1977; Yamada and Hizota, 1979). The Λ doubled energy levels in the $A^1\Pi$ state are perturbed by near-lying levels in the $a^3\Pi$, $a'^3\Sigma^+$, $d^3\Delta$, and $e^3\Sigma^-$ states. A set of deperturbed molecular constants has been derived by Cossart and Bergeman (Bergeman and Cossart, 1981) for this system from the collected emission and absorption data for CS: these are given in Table 1. The deperturbation analysis and term values are fully described in the definitive works of Cossart and Bergeman (Bergeman and Cossart, 1984; Cossart, 1974). It was determined that the most important influence on the $A^1\Pi\,(v=0)$ originates from the $a'^3\Sigma^+\,(v=10)$ level with the $d^3\Delta\,(v=2$ and $4)$ making more minor contributions (Field and Bergeman, 1971). The rotational "resonance" with the a' state reaches a maximum at $J=16$ ($\approx 10\%$ mixing) and thus directly concerns us in this study. The introduction of a' character into the A state has the consequence of altering the energy levels in the A state and diminishing the optical coupling to the X state. Conversely, the introduction of A character into the rotational levels of the a' state means that those levels may be optically coupled to the ground state.

Lefebvre-Brion and Field (1986) point out that the natural line strengths of a perturbed transition depend on "which of the two possible electronic transitions has oscillator strength in the spectral region being monitored; the vibrational and rotational line-strength factors which are applicable for both main and extra lines are those associated with the electronic-vibrational band with nonzero-oscillator strength." Even for levels which emit due to small admixtures of the optically coupled state, the electronic and rovibrational wavefunctions of the non-optically coupled state are completely "irrelevant" to observed linestrengths except insofar as they determine the fractional character of optically coupled state (Lefebvre-Brion and Field, 1986). The CS band system in question is thus to be distinguished from the familiar example of intensity interference seen in NO where the eigenstates are formed from the admixture of two strongly coupled bases, both of which have allowed optical transitions to the ground state.

Accordingly, the form of rotational line strength for the observed CS transitions follows that of a $^1\Pi - {}^1\Sigma^+$ transition. Each J level of the Π state has two lambda doublet components labelled e and f. The inversion parity of the rotational wave functions associated with the e or Π^+ levels is given by $(-1)^J$ and that of those associated with the f or Π^+ levels by $-(-1)^J$. The inversion parity of the rotational wave functions associated with the $^1\Sigma^+$ state is given by $(-1)^J$. The parity selection rule forces all P and R branch transitions to originate from e levels and all Q

branch transitions to originate from f levels of the Π state (Lefebvre-Brior and Field, 1986; Gauyacq et al., 1986). The modified natural line strength for the perturbed CS $^1\Pi - {}^1\Sigma$ transition incorporates Hönl-London factors weighted by the fractional character of optically coupled A state (Lefebvre-Brion and Field, 1986). These may be written:

$$S'_{jk} = p_j S_{jk} \quad \text{with} \quad S_{jk} = \begin{cases} (J''-1)/4 & P(J''=J'-1) \\ (2J''+1)/4 & Q(J''=J') \\ (J''+2)/4 & R(J''=J'+1), \end{cases} \tag{16}$$

where p_j is the fractional A character of the A state rotational level and J'' the quantum number in the X state. We note that the Q branch line strengths are approximately twice of the P and R branches.

4. Analysis and results

We begin by making the following comments on the experimental data presented in Fig. 2. The wavelength scale from the IUE guest observer tape is in air Ångstroms and has not been corrected for the relative motion of the comet and the telescope. The Doppler corrections due to these motions are significant in the analysis of the spectrum. At the time of observation, the comet was moving with a velocity of $\sim -26\,\mathrm{km\,s^{-1}}$ toward the sun and a velocity of $\sim +35\,\mathrm{km\,s^{-1}}$ away from the IUE observer telescope. This corresponds to a blue shift of $3.3\,\mathrm{cm^{-1}}$ or -0.02 nm of solar emission as seen by the comet and a red shift of $-4.5\,\mathrm{cm^{-1}}$ or $+0.02$ nm of comet fluorescence as seen by the telescope camera. Correcting for the Doppler shift in emission, we find that emission to the blue of 257.68 nm originates from R branch transitions and that to the red from P and Q branch transitions. The width of the CS feature suggests rotational states populated up to $J \sim 30$.

The intensity data taken from the IUE tape is reported at 92 irregularly spaced points between 257.5 and 257.9 nm. We have made a projection, using linear interpolation, of this data onto a regular grid of 80 points spaced at 0.005 nm: this projection produces no visible change in the spectrum. The intensity scale in the figure is relative and uncorrected for background light scattering not due to CS fluorescence. The IUE instrumental bandwidth is quoted to be 0.08 nm for an extended source and 0.02 nm for a point source: we find that it is the later which best duplicates the experimental spectrum. The signal to noise ratio may be estimated from the negative excursion peaks in the spectrum – this is found to be $\sim 33/1$ in the Halley spectrum. The shape of the intensity profile – peaking near the Doppler shifted band origin – suggests highly rotationally relaxed CS. However, given that the Hönl-London factors for the Q branch are approximately two fold greater than those of the P and R branches and that the P and Q branches overlap, we would expect a spectrum whose intensity fell off more rapidly in the blue with shouldering in the red. The observed spectrum is almost symmetric with slightly more integrated intensity in the wavelength region corresponding to the R branch emission than that in the region corresponding to the Q and P branches.

A computer program has been written which simulates fluorescence spectra excited by broad band solar radiation and extracts estimates of population parameters through direct inversion of experimental intensities. The term values for the $X(v=0)$ state and the term values and fractional A character for the $a'(v=10)$, and $A(v=0)$ states have been tabulated by Cossart and Bergeman (1981). Transition frequencies for the band system are

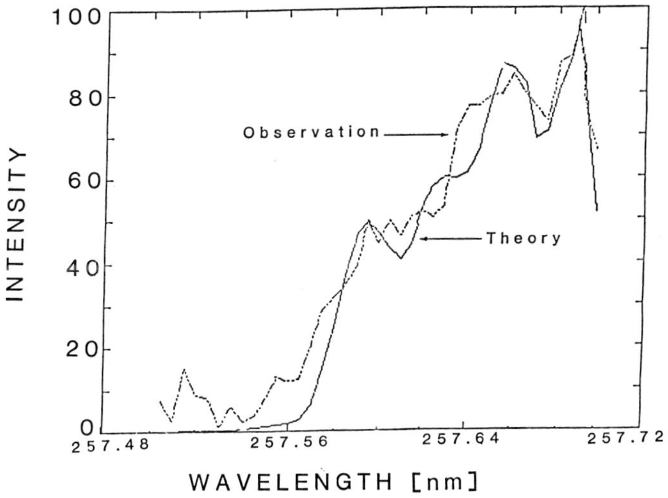

Fig. 3. The fitted R branch of the emission spectrum. The dotted line represents the experimental data; the solid line shows the spectrum simulated according to the parameters of the rotational population distribution extracted in the inversion procedure

calculated through use of these tables. Natural line strengths are calculated through use of Eqs. (4) and (16) subject to the selection rules described in Sect. 3.2. Transitions in absorption are weighted according to the Doppler shifted intensity of solar radiation as given by A'Hearn. For each upper state populated in absorption, all allowed transitions in emission are calculated. For simulation, the Doppler shifted air wavelength is sorted into a histogram bin of regular spaced observation points and weighted by population and a grating instrument function depending on its displacement from the center of the bin. Alternatively a model matrix for inversion may be tabulated by cross-sorting with a parameter index and unit population weighting. Given the tabulated model matrix, a linear regression routine is used to determine the best-fit parameters and associated estimated errors.

Inversion of the R branch portion of the spectrum yields good agreement with reported intensities. Figure 3 compares the spectrum simulated according to extracted population parameters with the experimentally obtained spectrum. Figure 4 shows the population distribution reconstructed from the fitted parameters. The estimated variance for the fitted parameters was sensitive to the choice of orthogonal spline basis set though the overall form of the distribution was not. The $J=1$ and $J=0$ level parameters are the most highly populated; a background extends out to $J \approx 30$ which is at ≈ 0.2 of this maximum level. Attempts to fit the entire spectrum and the P and Q branch portion separately yielded consistently inferior results. If the red portion of the spectrum corresponding to the P and Q branch emission was enhanced by a factor of 2.5 relative to the blue portion corresponding to R branch emission, then a reasonable quality fit was obtained and the population distribution resembles that shown in Fig. 4.

5. Discussion

Initial attempts to model the observed spectrum with a Boltzmann distribution have been described in a previous publication (Jackson and Prisant, 1986); these efforts showed that a Boltzmann distribution cannot account for the observed form of

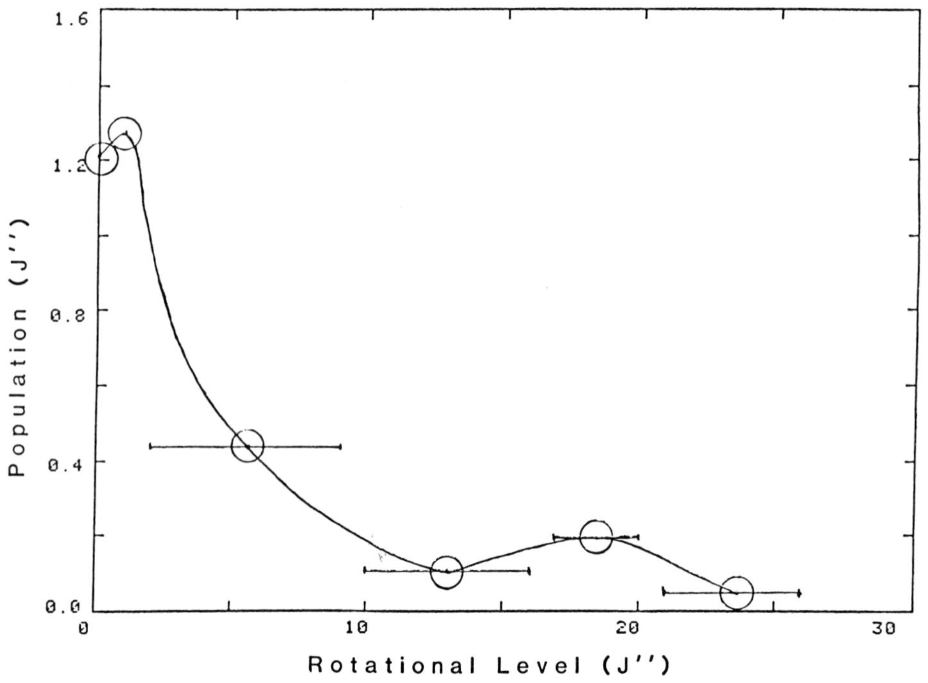

Fig. 4. The rotational population distribution reconstructed from linear regression best fit parameters. The bars represent the bin width and the circles represent the average population per J level over the interval. The nature of the errors are described in the text

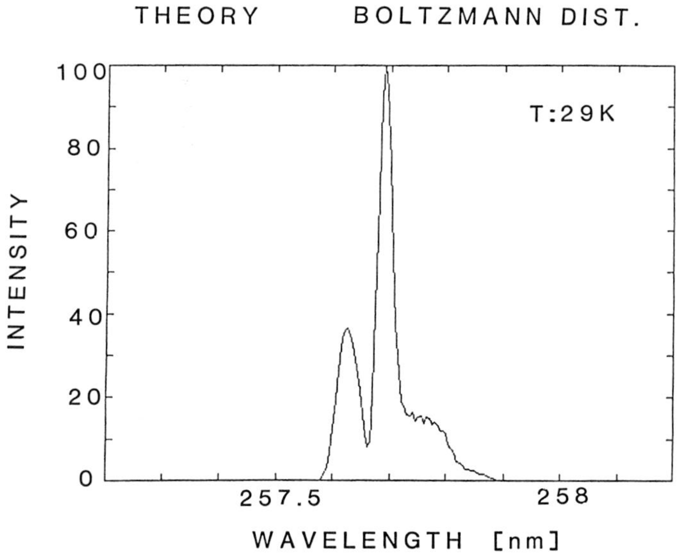

Fig. 5. The $A^1\Pi - X^1\Sigma^+ (v=0)(v=0)$ emission simulated for a 29 K Boltzmann distribution of absorbers

the CS fluorescence contour. If a low rotational temperature of the order of 7 K is used, the spectral bandshape is roughly triangular but much narrower than the observed spectrum. Increasing the temperature to broaden the band profile results in the development of a characteristic dip separating the R and P, Q halves of the simulated spectrum. Figure 5 shows a plot of the spectrum resulting from simulation of a 29 K or best-fit Boltzmann rotational distribution. While accounting for the breadth of the fluorescence contour, the overall shape of the spectrum and that of constituent branches are clearly incorrect. No reasonable adjustment of the IUE instrument bandwidth was able to eliminate this disagreement. In earlier work on CS in comet Bradfield (Jackson et al., 1982), a Boltzmann distribution appeared to give a reasonable fit to the observed data. Subsequent consideration suggests that this earlier agreement was fortuitous and due to the lower signal to noise ratio in the Bradfield spectrum and that required excessive smoothing of the original data. Since a Boltzmann distribution does not fit the observed spectra it is unlikely that collisions are responsible for *most* of the CS rotational distribution.

Also examined in a previous publication (Jackson and Prisant, 1986) were simulations resulting from several different functional forms of the rotational distribution. These included a linear distribution peaking at $J''=0$ and monotonically decreasing to $J''=10$, a bimodal distribution peaking at $J''=0$ and at $J''=20$, and uniform distributions that cut off at some upper J'' level. Though none of these distributions provided a satisfactory fit to the overall observed spectrum, the bimodal distributions were superior and provided a better accounting of the general shape of the R branch half of the spectrum.

Use of the inversion procedure showed that the fit variance could be dramatically improved by artificially suppressing the P and Q branches of the spectrum relative to the R branch. In other words, the general form of the spectrum is incompatible with the known rotational line strengths of CS. Because of this incompatibility we have chosen to use the inversion program to only fit the R-branch and derive populations from this branch.

The spectrum simulated with the population distribution reconstructed according to the fitted parameters and the reconstructed distribution itself are shown in Figs. 3 and 4 respectively. The results show reasonable agreement with the experimental observations. To reduce the overall error and improve the fit it is necessary to limit the number of parameters that are determined in the least squares fitting procedure. We have chosen to to this integrating the population in four bins for $J''=2$ through 27. The bin sizes indicated in the figure by the horizontal lines the error on the integrated populations is very low, with a two

Table 2. Some characteristic times for elementary processes involving CS in comets at 1 AU

Process	Time	Ref.
1. Electronic excitation	1400	Jackson (1982)
2. Rotational radiative relaxation	$1.9\ 10^5/(J'')^3$	Winnewisser and
for $J''=5$	1500	Cook (1968)
for $J''=4$	2970	Guillory (1977)
3. Time between collisions	$5.7\ 10^{-15}(r)^2$	Jackson and
for $r=$ 500 km	14	Donn (1966)
for $r=1900$ km	206	
for $r=5000$ km	1425	
4. Photochemical production	500	Jackson (1986)

sigma value indicated by the size of the circles. As would be expected, the estimated errors on the individually fitted populations of the $J''=0$ and 1 levels are much larger, with one sigma values of 0.1 and 0.6, respectively. The errors on the $J''=0$ and 1 populations can be improved with a different choice of basis; the overall form of the reconstructed distribution is not, however, terribly sensitive to this choice. After experimentation with a number of different bases, we found that the distribution reconstructed with the basis shown produced the optimal variance and yielded positive values for all populations even in the high J domain.

The populations derived from fitting the R branch are generally satisfactory if we can find a reasonable explanation for the inability of the inversion procedure to fit the Q and P branches. Several explanations come to mind but as further discussion will indicate most are unlikely. We first consider an optical depth effect. To be suitable, the absorber must have an optical thickness 0.4 units greater in the P and Q branch region of the spectrum than that in the R branch region. Calculations using the oscillator strength and measured production rates to compute column density for CS indicate that its optical thickness is only 0.02; this rules out any explanation invoking self-absorption. The presence of a strong absorber other than CS itself that overlaps just the P and Q branches may be ruled out because 1) there is no indication in the spectrum for such a species and 2) given the large oscillator strength of CS and its production close to the nucleus, it is improbable that another unknown absorber could exist whose optical thickness would exceed that of CS by an order of magnitude. A second possibility is that a polarization effect is being observed in the emission spectrum. This seems unlikely since 1) the exciting radiation is unpolarized and 2) for a given anisotropic ensemble of emitters, the Q branch would be expected to polarized opposite to the P and R branches whereas the observed anomalous relative intensities are for P and Q with respect to R.

A plausible explanation for the observed discrepancy is the contamination of the R branch measured intensity by a higher contribution of scattered sunlight than the P and Q branches. This explanation is suggested by examination of the photowrite images in the region of OH spectrum where there is clear indication that a continuum is present and contributes to the measured emission intensity. Consistent with this hypothesis, the spectral distribution of scattered sunlight overlapping the CS band system (shown in Fig. 1) exhibits a dip in the P and Q branch region.

Halley was 1.08 AU from from the earth at the time of the present measurements. The observed column of emitting gas had a length of 10^5 km and a radius of 4000 km with the telescope observation axis lying along the column axis. Prerequisite to understanding the bimodal rotational distribution are the characteristic times for the various processes involving CS in comets presented in Table 2. The 500 s reciprocal rate of CS photochemical production and the observed 300 km locus of CS production above the comet nucleus are consistent with an outward gas flow velocity of $0.6\ \mathrm{km\,s^{-1}}$. The reciprocal rate or characteristic time for CS electronic excitation is 1400 s; this means that multiple excitation of individual molecules will contribute to a spreading of the distribution. As the carrier gas expands outward from the comet nucleus diminishing in density, the time between collisions varies from 5 s at 300 km to 74 s at 1140 km. At 300 km, the characteristic collision time is greater than the characteristic radiative relaxation time for all levels above $J''=34$ while at 1140 km, the comparable value is $J''=14$. The rotational population distribution at a given altitude will reflect competition between collisional thermalization and radiative relaxation with radiative relaxation increasing and collisional thermalisation decreasing in importance with increasing distance from the nucleus. The extracted distribution averages over the entire observation volume.

The results reported in the figure are clearly not at odds with the qualitative ideas presented above. The fitted populations are dominated by a low J component that probably originates from those radicals that have radiatively relaxed to the lowest J'' levels. This larger component is superimposed on a weaker component which represents those radicals that sample the inner collisional region. To obtain more quantitative information from the observed distribution a detailed quantitative model is needed which incorporates the qualitative ideas given above.

Acknowledgements. WMJ wishes to thank G. Basir for the ANA program and his help in its use and P. Butterworth for his aid in this project. MGP wishes to acknowledge fertile conversations with J. Rostas, D. Cossart, R.W. Field, and T. Bergeman on the spectroscopy of CS. He also wishes to thank S. Fredin for pointing out the possibility of parity selectivity in consecutive absorption and emission processes. This work was supported by NASA under grants NAGW-446 and NAGW-903.

References

A'Hearn, M.F., Ohlmacher, J.T., Schleicher, D.G.: 1983, *A High Resolution Solar Atlas for Fluorescence Calculations*, University of Maryland Technical Report, TR Ap 83-044

Barrow, R.F., Dixon, R.N., Lagerqvist, A., Wright, C.V.: 1960, *Ark. Fys.* **18**, 543

Bergeman, T., Cossart, D.: 1981, *J. Mol. Spec.* **87**, 119

Boggess, A. et al.: 1978, *Nature* **275**, 377

Cossart, D.: 1974, *Etude a Haute Resolution des Spectres et des Perturbations Electroniques du Radical CS*, These d'Etat, Universite de Paris XI

Cossart, D., Horani, M., Rostas, J.: 1977, *J. Mol. Spec.* **67**, 283

Cossart, D.: 1980, *J. Phys.* **41**, 489

Crawford, F.H., Shurcliff, W.A.: 1934, *Phys. Rev.* **45**, 860

Feldman, P.D., Weaver, H.A., Festou, M.C., A'Hearn, M.F., Jackson, W.M., Donn, B., Rahe, J., Smith, A.M., Benvenutti, P.: 1980, *Nature* **286**, 132

Festou, M.: 1981, *Astron. Astrophys.* **95**, 69

Field, R.W., Bergeman, T.: 1971, *J. Chem. Phys.* **54**, 2936

Gauyacq, D., Fredin, S., Jungen, Ch.: 1986, *Molecular Physics* (submitted)

Guillory, W.A.: 1977, *An Introduction to Molecular Structure and Spectroscopy*, Allyn and Bacon, Boston, p. 33

Haser, L.: 1957, *Bull. Acad. Roy. Belgique, Classe des Sciences* **43**, 740

Haser, L.: 1966, *Mém. Soc. Roy. Sci. Liège. 5e Ser.* **12**, 233

Jackson, W.M., Donn, B.: 1966, *Mém. Soc. Roy. Sci. Liège. 5e Ser.* **12**, 133

Jackson, W.M., Rahe, J., Donn, B., Smith, A.M., Keller, H.U., Benvenutti, P., Delsemme, A.H., Owen, T.: 1978, *Astron. Astrophys.* **73**, L7

Jackson, W.M., Halpern, J.B., Feldman, P.D., Rahe, J.: 1982, *Astron. Astrophys.* **107**, 385

Jackson, W.M., Butterworth, P.S., Ballard, D.: 1986, *Astrophys. J.* **304**, 515

Jackson, W.M., Prisant, M.G.: 1986, *20th ESLAB Symp.* ESA SP-250, 545

Lagerqvist, A., Westerlund, H., Wright, C.V., Barrow, R.F.: 1958, *Ark. Fys.* **14**, 387

Lefebvre-Brion, H., Field, R.W.: 1986, *Perturbations in the Spectra of Diatomic Molecules*, Academic Press, New York, Chap. 5

McCrary, V.R., Lu, R., Zakheim, D., Russel, J.A., Halpern, J.B., Jackson, W.M.: 1985, *J. Phys. Chem.* **83**, 3481

Prisant, M.G., Zare, R.N.: 1985, in *Gas Phase Chemiluminescence and Chemi-Ionization*, ed. Fontijn, A., North-Holland, Amsterdam, p. 189

Prisant, M.G., Zare, R.N.: 1985, *J. Chem. Phys.* **83**, 5458

Prisant, M.G., Levine, R.D.: 1986 (unpublished)

Robbe, J.-M., Schamps, J.: 1972, *Chem. Phys. Lett.* **15**, 596

Smith, A.M., Stecher, T.P., Caswell, L.: 1980, *Astrophys. J.* **242**, 402

Swamy, K.S.K.: 1986, *The Physics of Comets*, World Scientific, Philadelphia

Todd, T.R.: 1977, *J. Mol. Spec.* **66**, 162

Todd, T.R., Olson, W.B.: 1979, *J. Mol. Spec.* **74**, 190

Winnewisser, G., Cook, R.L.: 1968, *J. Mol. Spec.* **28**, 266

Yamada, C., Hirota, E.: 1979, *J. Mol. Spec.* **74**, 203

Astron. Astrophys. 187, 497–501 (1987)

ASTRONOMY AND ASTROPHYSICS

Search for methane in comet P/Halley

S. Drapatz[1], **H.P. Larson**[2], **and D.S. Davis**[2]

[1] Max-Planck-Institut für extraterrestrische Physik, D-8046 Garching, Federal Republic of Germany
[2] The University of Arizona, Tucson, AZ 85721, USA

Received February 14, accepted April 29, 1987

Summary. A search for gaseous neutral CH_4 has been conducted in the coma of comet Halley on March 20, 1986 using a Fourier Transform Spectrometer on the NASA Kuiper Airborne Observatory. In the field-of-view (41″ diam. ≙ 24,000 km) the CH_4 rotational temperature is equal to the kinetic temperature of the coma molecules due to collisions in most of the region and is frozen in at the outer edge. The spectrum of the ν_3 band is due to fluorescence with the R(3) to R(6) lines being most suitable for Earth-based observations. Measured upper (3σ) limits of $\sim 4\ 10^{-19}$ W cm^{-2} for several lines lead to a (5σ) upper limit of $4\ 10^{28}$ CH_4 molecules per second. This value corresponds to $\sim 4\%$ of the water vapor production rate at the time of observation.

Key words: comets – molecules – abundances – infrared observations

1. Introduction

The search for methane in comet Halley was motivated by the fact that CH_4 was an important parent molecule candidate for comets (Whipple, 1950) with abundance estimates ranging up to 10% and higher (e.g. Hübner, 1985). However, the uncertainties were very high: while CH_4 is expected to be the dominant carbon compound at equilibrium conditions in the outer regions of a cold presolar nebula, the temperature and degree of LTE in cometary nuclei during their condensation from the interstellar medium is unknown, and subsequent fractionation processes could have altered the primordial value. Also, CH_4 might be depleted with respect to other species due to its high vapor pressure. Observations of methane in comet Halley and other comets could therefore provide important clues to conditions during the formation of cometary nuclei and to their subsequent thermal evolution.

Another important reason to search for methane is its usefulness in testing cometary thermodynamical models. Due to its molecular structure the population distribution of rotational levels is determined by collisional rather than fluorescence excitation. Thus, CH_4 spectral lines provide a tool to probe coma kinetic temperatures. This information cannot be obtained from line width measurements of cometary species, since they are due to coma kinematics rather than kinetic temperature. The observation of gaseous methane could also shed some light on the molecule production process, i.e. whether the gas sublimates directly from the nucleus surface or from ice grains released from the nucleus and whether partial recondensation of molecules into ice particles takes place.

Methane cannot be detected in the ultraviolet spectral region because its spectrum is continuous (Herzberg, 1960), nor can it be observed in the radio spectral region because it does not possess a dipole moment (except for some small charge fluctuations). However, CH_4 does have two strong infrared vibration bands, ν_3 at 3.3 μm and ν_4 at 7.7 μm. The former is more favorable for cometary observations because of reduced radiation from cometary dust (emission as well as scattering) and higher efficiency of near-infrared line excitation by sunlight. However, terrestrial H_2O and CH_4 absorption lines complicate the search for cometary CH_4 from ground-based telescopes. This limitation can be significantly relaxed through the use of high resolution spectroscopic techniques at airborne observatories. At flight altitudes of 12 km and higher, telluric H_2O lines effectively disappear in the region of the CH_4 ν_3 band and the telluric CH_4 features, although still prominent, are narrower. The comet's geocentric velocity can then be used to shift the positions of cometary CH_4 features to the far wings of the terrestrial CH_4 lines. This strategy was used effectively in airborne studies of H_2O in comet Halley (see Larson et al., 1986). The first attempt to detect the CH_4 ν_3 band in a comet (Kohoutek) from an airborne observatory was not successful (Roche et al., 1975). We present here the results of a search for CH_4 in airborne observations of comet Halley. Although CH_4 was not detected, the upper limit is sufficiently low that the predictions of certain models can definitely be excluded.

2. Observations

A spectrum of comet Halley in the region of the CH_4 ν_3 band was acquired on UT 20 March 1986 from the NASA Kuiper Airborne Observatory (KAO). The University of Arizona Fourier Transform Spectrometer (FTS; Davis et al., 1980) was used at a spectral resolution of 0.27 cm^{-1} ($\lambda/\Delta\lambda \sim 1.1\ 10^4$). The integration time was about 60 min. A cold filter limited the spectral bandpass from 2950 to 3100 cm^{-1}. This region includes the CH_4 *Q*-brach and low-*J* lines in the *P*- and *R*-branches. The angular field-of-view was $\theta = 41''$, equivalent to a spatial field of 2.4 10^4 km at the comet's geocentric distance $\Delta = 0.8$ AU. The comet's geocentric velocity was $\dot{\Delta} = -43.5$ km s^{-1}, and the heliocentric distance was $R = 1.0$ AU. The lunar spectrum in Fig. 1 was used to evaluate

Send offprint requests to: S. Drapatz

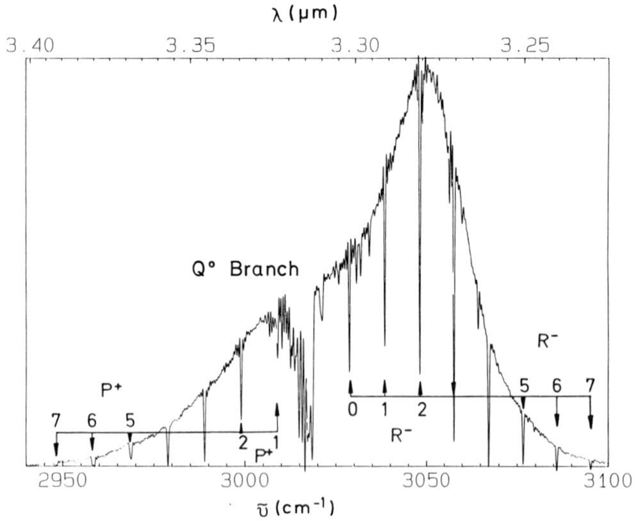

Fig. 1. Absorption spectrum of telluric CH_4 with the Moon as a background source. The shape of the continuum is determined by the filter transmission curve

instrumental transmission factors. The minimum detectable flux (1σ limit per spectral element) in the comet's spectrum (see Fig. 5) was determined to be $1.25\ 10^{-19}\ \mathrm{W\ cm^{-2}}$. This was checked against independently calibrated spectra of cometary H_2O that were acquired during the same flight program. The ordinate scales were internally consistent to 15%, although the absolute calibration of the KAO data may have higher systematic error (see discussion in Weaver et al., 1986).

3. Model of cometary CH_4 emission

3.1. The outflow model

In a simple spherically symmetric model the number density of molecules is given by

$$n_{\mathrm{mol}} = Q_{\mathrm{mol}}/4\pi r^2 v \cdot \exp(-r/v\tau_{\mathrm{mol}}), \tag{1}$$

where Q_{mol} is the molecular production rate (molecules/sec), r is the radial distance to the center of the nucleus (km), v is the radial outflow velocity of the molecules ($\mathrm{km\ s^{-1}}$), and τ_{mol} is the molecular lifetime against photolysis (seconds). The dominant component in the outflow is most likely H_2O. At the time of our CH_4 observations the water production rate was about $10^{30}\ \mathrm{mol\ s^{-1}}$ (Weaver et al., 1986). For the outflow velocity we adopt $v \sim 1\ \mathrm{km\ s^{-1}}$ and for the photochemical lifetime we use $\tau_{H_2O} = 8.3\ 10^4$ s (Hübner and Carpenter, 1979). For CH_4 we adopt $\tau_{CH_4} = 1.3\ 10^5$ s (Hübner, 1986). According to preliminary analyses of comet Halley H_2O data, the postperihelion outflow velocity was somewhat higher (Larson et al., 1986) and the H_2O lifetime could have been significantly shorter (factor 2–3, Weaver et al., 1986; Krankowsky et al., 1986) due to other processes (e.g. collisional ionization). Since additional analysis is needed to clarify the physical significance of these measurements and since their influence on our model is small, we will not consider their effect on our CH_4 upper limit determination.

The number of CH_4 molecules in the field-of-view is

$$N(CH_4) = f\left(\frac{r_0}{v\tau}\right) \int_0^{r_0} n(CH_4)\, 4\pi r^2\, dr \sim \frac{\pi r_0}{2v}\, Q(CH_4). \tag{2}$$

The integral includes the molecules in the sphere around the nucleus with radius $r_0 = \theta\Delta = 1.2\ 10^4$ km corresponding to the field-of-view. The correction factor $f(r_0/v\tau) \sim \pi/2$ (Yamamoto, 1982) takes molecules into account which are in the beam but outside the sphere. The characteristic expansion time of the coma is defined as

$$\tau_{\exp} = \frac{n}{v}\left(\frac{dn}{dr}\right)^{-1} = \frac{1}{v}\left(\frac{2}{r} + \frac{1}{v\tau}\right)^{-1} \approx \frac{r}{2v}. \tag{3}$$

The temperature profile of the dominant component (H_2O) for the heliocentric distance $R = 1$ AU and $Q = 10^{30}\ \mathrm{s^{-1}}$ is taken from Crovisier (1986), who investigated the thermodynamics of the coma. A simple fit gives

$$T_{\mathrm{kin}} = 0.5\, r^{0.7}. \tag{4}$$

3.2. The role of collisions

Whether the energy levels of CH_4 are in vibrational or rotational equilibrium at that temperature depends on the relative importance of collisional and radiative transitions. Following Weaver and Mumma (1984) we define disequilibrium distances $D_\uparrow$ for excitation and $D_\downarrow$ for relaxation. These distances, measured from the nucleus, indicate zones in the coma beyond which collisions are negligible factors in molecular excitation:

$$D_\uparrow = (Q(H_2O)\, \sigma_\uparrow / 4\pi \langle G \rangle)^{1/2}$$
$$D_\downarrow = (Q(H_2O)\, \sigma_\downarrow / 4\pi \langle A \rangle)^{1/2}. \tag{5}$$

Here σ is the corresponding collision cross section, $\langle G \rangle$ is the radiative excitation in the whole band, and $\langle A \rangle$ is an average Einstein transition probability. We will demonstrate that the prevailing excitation mechanism for the vibrational transitions is fluorescence induced by solar pumping, while the population of the ground state rotational levels is due to thermal excitation through collisions in the inner coma.

Vibrational relaxation of the v_3 band occurs in two steps: energy transfer from the v_3 mode to other vibrational modes with eventual transfer to the v_4 mode; and relaxation of the v_4 mode. The cross section for the first process is $\sim 10\%$ of the rotational relaxation cross section, i.e. $\sim 3\ 10^{-16}\ \mathrm{cm^2}$ (Lambert, 1977), and that for the second process is $\sim 10^3$ times smaller. The cross section $\sigma_\uparrow$ follows from microscopic reversibility. One obtains disequilibrium distances below 10 km if $\langle G \rangle$ and $\langle A \rangle$ values from the following paragraph are used.

The mean number of collisions needed for rotational energy transfer in pure CH_4 gas at 300 K is $Z \sim 12$ (e.g. Karbach and Hess, 1985) and decreases to $Z \sim 3$ at 77 K (Lambert, 1977). Collisions with a polar molecule like H_2O are generally more efficient and rotational transitions are expected to occur for every gas kinetic collision. Therefore, the collision cross section for rotational downward transitions is assumed to be $\sim 3\ 10^{-15}\ \mathrm{cm^2}$, i.e. close to the gas kinetic cross section of H_2O and CH_4 molecules (Lambert, 1977).

There exist no allowed rotational transitions in CH_4. Centrifugal distortion effects produce a small permanent dipole moment (Rosenberg et al., 1972) leading to Einstein transition probabilities of the order of $10^{-9}\ \mathrm{s^{-1}}$. Therefore one has $D_\downarrow \gg r_0$ and $D_\uparrow \gg r_0$. This means that the rotational modes of the CH_4 molecules in the field-of-view are coupled to the translational modes of the molecular gas, i.e. $T_{\mathrm{rot}} = T_{\mathrm{kin}}$, as long as the collision

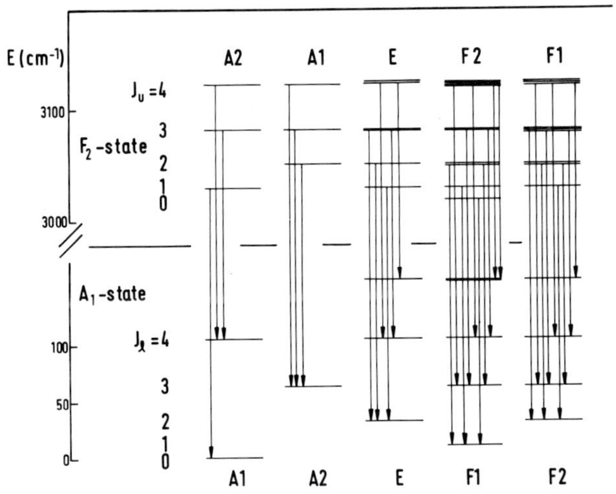

Fig. 2. Lower rotational states of the A_1 ground state and F_2 excited state of the ν_3 vibrational band of CH_4. Rotational levels of one species (A, E or F) can only combine with levels of the same species

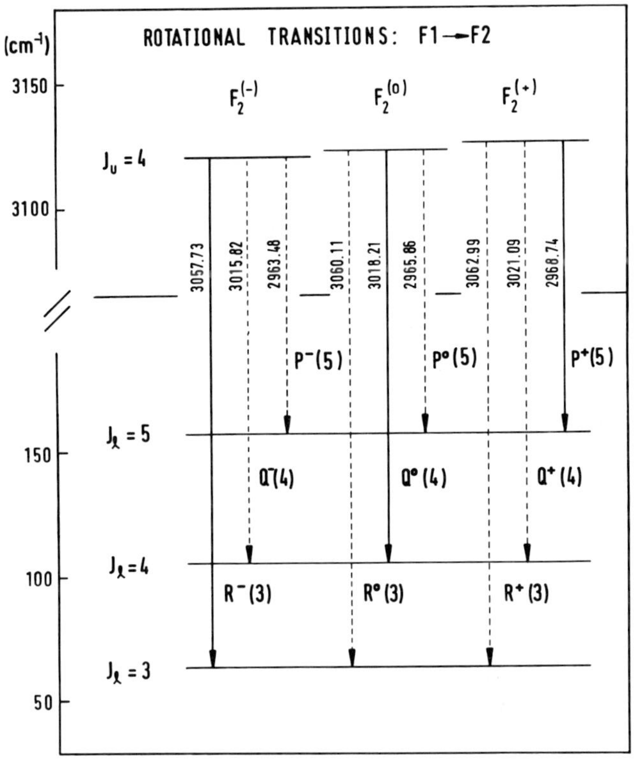

Fig. 3. More detailed presentation of a part of the F_1–F_2 transition scheme (right-hand side in Fig. 2). Allowed transitions (full lines) occur only for the combinations $F^{(-)}$ and $\Delta J=1$, $F^{(0)}$ and $\Delta J=0$, $F^{(+)}$ and $\Delta J=-1$

time is smaller than the characteristic expansion time. Beyond that regime the rotational temperature will be frozen in. From Eq. (3)

$$\frac{r_f}{2v} = (n(H_2O)\,\langle v\rangle\,\sigma)^{-1}\,. \tag{6}$$

The corresponding radius is $r_f \sim 10^4$ km if the thermal and turbulent velocity is assumed to be $\langle v\rangle \sim 1\ \mathrm{km\,s^{-1}}$.

3.3. *The methane spectrum*

In the asymmetric stretching mode ν_3 of CH_4 the vibrational ground state A_1 can only combine with vibrational states of species F_2 (Fig. 2 illustrates the situation for the lower-J transitions). There are three non-combining modifications (analogous to ortho- and para-H_2) with the selection rules $A_1 \leftrightarrow A_2$, $E \leftrightarrow E$, $F_1 \leftrightarrow F_2$ (Hougen, 1976) yielding a group of sub-lines for each R-branch ($J_u = J_l + 1$), Q-branch ($J_u = J_l$) and P-branch ($J_u = J_l - 1$) transition (indices u and l refer to upper and lower states respectively). The rotational levels of the triply degenerate vibrational state F_2 split into $F^{(+)}$, $F^{(0)}$, and $F^{(-)}$ sub-levels due to Coriolis interaction (see Fig. 3, where part of the $F_1 - F_2$ transitions of Fig. 2 are shown in detail). The upper states are $F^{(-)}$ for R-branch, $F^{(0)}$ for Q-branch and $F^{(+)}$ for P-branch allowed transitions, i.e. in the primarily allowed transitions no upper state or ground state appears twice. Hence fluorescence in vibrational lines only changes the population of the ground state rotational level if secondarily allowed or forbidden transitions are taken into account. The transition probabilities of these lines are generally smaller by more than one order of magnitude (Barnes et al., 1972) and their influence on the level population is less than 10% until the CH_4 molecules are destroyed by photolysis.

3.4. *Interference of telluric lines*

The profiles of terrestrial CH_4 lines are broad in ground-based observations due to saturation effects and spectral sub-structure. Under such conditions the search for cometary CH_4 is very inefficient and may not even be possible, and the interpretation of suspected cometary features may not be quantitatively reliable. At aircraft altitudes the terrestrial CH_4 line profiles are narrower (HWHM $< 0.2\ \mathrm{cm^{-1}}$) and the comet's geocentric velocity may then be sufficient to separate some features. In this study the comet's geocentric velocity produced a wavenumber shift of $+0.43\ \mathrm{cm^{-1}}$. This would not help much in the Q-branch where the separation of individual lines is very small ($\Delta E \lesssim 0.3\ \mathrm{cm^{-1}}$); cometary CH_4 lines would just be Doppler-shifted into neighboring telluric CH_4 lines. The P- and R-branch lines, however, are widely spaced (by $2B \approx 10.5\ \mathrm{cm^{-1}}$, where B is the rotational constant for CH_4) and the sub-structure of the low-J line groups is contained within narrow limits (see Table 1).

The situation has been investigated quantitatively by using high resolution ($0.089\ \mathrm{cm^{-1}}$) spectra of telluric methane. In Fig. 4 the $R(3)$ line group is shown for illustration. Although the air mass for the cometary observation was higher by a factor 1.48 (different elevation angles of the airborne telescope for the observation of the atmosphere and the comet), the transmission for the cometary CH_4 lines is still in the 90% region. This small influence of atmospheric absorption is neglected in the calculation of line intensities.

3.5. *Line intensities*

Since only a single allowed line connects an upper state u with a lower state l, the fluorescence intensity is given by

$$I_{ul} = G_{lu}\,N_l\,h\,\tilde{\nu}\,c\,(4\pi\,\Delta^2)^{-1}\,, \tag{7}$$

where N_l is the number of molecules in the lower level. The excitation rate G_{lu} ($\mathrm{s^{-1}}$) is the product of the radiation field density ϱ ($\mathrm{erg\,cm^{-3}/cm^{-1}}$) and the Einstein coefficient B_{lu}. Since scattered solar radiation and thermal emission of the cometary

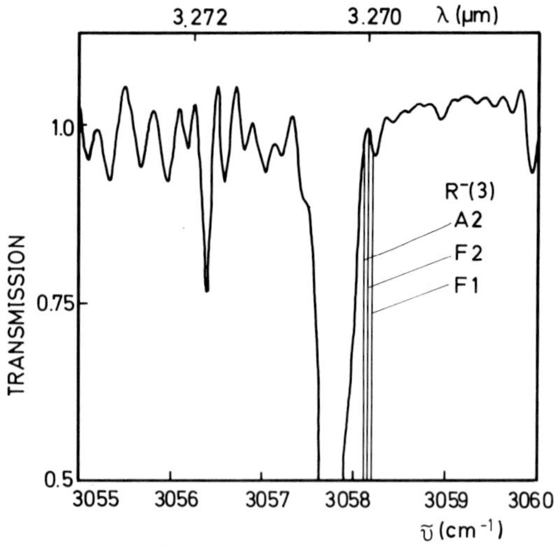

Fig. 4. Measured profile of the $R(3)$ line group of telluric methane and location of the corresponding cometary lines (note that the ordinate does not start at zero transmission)

nucleus and dust are negligible in this wavelength range (Crovisier and Encrenaz, 1983), ϱ is given just by the direct solar radiation field ($T_0 = 5500$ K)

$$G_{\mathrm{lu}} = \left(\frac{R_\odot}{2R}\right)^2 (\exp(h\tilde{\nu}_{\mathrm{lu}}\, c/kT_\odot) - 1)^{-1} \frac{\omega_{\mathrm{u}}}{\omega_{\mathrm{l}}} A_{\mathrm{ul}}. \tag{8}$$

Only the geometrical dilution factor applies, since the optical depth in the CH_4 lines is small (see below). The product of Einstein coefficient A_{ul} and statistical weight ω is related to the peak line intensity $\hat{S}_{\mathrm{lu}}$ of laboratory measurements (at $T_{\mathrm{L}} = 294$ K, Gray and Robiette, 1979) by (see Drapatz et al., 1986)

$$\omega_{\mathrm{u}} A_{\mathrm{ul}} = 7.5\ 10^{-11}\ \tilde{\nu}_{\mathrm{ul}}^3 \exp(h\tilde{\nu}_1\, c/kT_{\mathrm{L}})\ Z(T_{\mathrm{L}})\ \hat{S}_{\mathrm{lu}}(T_{\mathrm{L}}). \tag{9}$$

$Z(T)$ is the rotation vibration partition function (Herzberg, 1960; Fox, 1970). The number of molecules in the lower state is given by

$$N_{\mathrm{l}} = \frac{\omega_{\mathrm{l}} N}{Z(T)} \exp(-h\tilde{\nu}_{\mathrm{l}}\, c/kT). \tag{10}$$

From Eq. (9) the mean Einstein probability $\langle A\rangle$ averaged over the level population is obtained

$$\langle A\rangle = \sum_{\mathrm{ul}} \frac{\omega_{\mathrm{u}}}{\omega_{\mathrm{l}}} A_{\mathrm{ul}} N_{\mathrm{l}} \Big/ \sum_{\mathrm{l}} N_{\mathrm{l}} \approx 2 \sum_{\mathrm{ul}} \hat{S}_{\mathrm{ul}} \approx 75\ \mathrm{s}^{-1}. \tag{11}$$

Using this result and Eq. (8) the corresponding mean band excitation rate is

$$\langle G\rangle = \sum_{\mathrm{ul}} G_{\mathrm{ul}} = 3.4\ 10^{-4}\ \mathrm{s}^{-1}. \tag{12}$$

These numbers were used in Sect. 3.2 to determine the relative importance of radiative and collisional transitions.

The optical depth of cometary CH_4 lines is given by (population of upper vibrational states is negligible)

$$\tau_{\mathrm{opt}} = \frac{\omega_{\mathrm{u}} A_{\mathrm{ul}}}{\omega_{\mathrm{l}}\, 8\pi^{3/2} \langle v\rangle\, \tilde{\nu}_{\mathrm{ul}}^3} \int_{\infty}^{r} n_l(\mathrm{CH_4})\, dr. \tag{13}$$

For a mean $T = 300$ K and neglecting the exponential term in Eq. (1) one obtains

$$\tau_{\mathrm{opt}} < 1.35\ 10^{-23}\, \hat{S}_{\mathrm{ul}}\, r\, Q(\mathrm{CH_4}). \tag{14}$$

Thus, for $Q(\mathrm{CH_4}) = 10^{29}\ \mathrm{s}^{-1}$ cometary CH_4 is transparent ($\tau_{\mathrm{opt}} < 1$) to within $r = 10$ km of the nucleus, i.e. even for a more realistic outflow model with an asymmetric flow pattern and jets, optical depth effects can be neglected for intensity line calculations.

4. Results and discussion

We use the density model Eq. (1) with the rotational temperature given by Eq. (4) for $r < r_f$ and $T = T(r_f)$ for $r > r_f$. The intensity of a line can then be obtained from Eq. (7) by use of Eqs. (8)–(10) and integrating over all molecules in the beam (we neglect those within $r = 100$ km)

$$I_{\mathrm{ul}} = 3.7\ 10^{-66}\ \tilde{\nu}_{\mathrm{ul}}^4\, \hat{S}_{\mathrm{lu}}(T_{\mathrm{L}}) \int \frac{\exp(h\tilde{\nu}_{\mathrm{l}}\, c/kT_{\mathrm{L}})\, Z(T_{\mathrm{L}})}{\exp(h\tilde{\nu}_{\mathrm{l}}\, c/kT)\, Z(T)}\, n(r)\, dr. \tag{15}$$

The corresponding intensities of the prominent line groups within our filter are given in Table 1 together with their intensity-weighted mean wavenumber $\langle\tilde{\nu}\rangle$ and the spacing between the two outer lines $\tilde{\nu}_> - \tilde{\nu}_<$. The calculations assumed $Q_{\mathrm{CH}} = 4\ 10^{28}\ \mathrm{s}^{-1}$, in order to produce a $\sim 3\sigma$ intensity in the strongest lines. The key lines were those CH_4 line groups having predicted high intensities and narrow envelopes, as measured by $\tilde{\nu}_> - \tilde{\nu}_<$ in Table 1. For the envelope width being less than about $0.2\ \mathrm{cm}^{-1}$, the cometary emission line group is shifted away from the corresponding terrestrial group in absorption, thus minimizing telluric obscuration, and all components of the cometary group would be contained within one resolution element ($0.27\ \mathrm{cm}^{-1}$), thus maximizing the spectral contrast. The line groups $R^-(2)$ to $R^-(5)$ met all of these criteria. However, as can be seen in Fig. 5, no cometary CH_4 lines were observed. The CH_4 production rate that yielded $\sim 3\sigma$ intensities in the important line groups (corresponding to a 5σ total limit) was $4\ 10^{28}\ \mathrm{s}^{-1}$, or about 4% of the H_2O production at the time of our observations (Weaver et al., 1986). This result is comparable to the upper limit of about 7% obtained directly with the neutral mass spectrometer on Giotto (Krankowsky et al., 1986). If the CH_4 modification ratio were frozen in at very low temperatures ($T_{\mathrm{m}} \sim 10$ K), the $R(0)$ line would have to be extremely strong. However, for somewhat elevated temperatures ($T_{\mathrm{m}} \sim 30$ K) the line intensities would not be strongly influenced, since the distance between the lowest A, E, F levels is very small (about half the value for ortho/para levels in H_2O).

The observed upper limit of CH_4 is unexpectedly low. One possible reason is that CH-bearing molecules at the surface of the nucleus may have been processed by irradiation. If the broadband emission features detected in the 3.2 to 3.7 μm region by the Vega IKS experiment (Combes et al., 1986) and by ground-based observations (Knacke et al., 1986; Danks et al., 1986) are due to such processed materials, the corresponding production rates for CH molecules would be 10^{28}–$10^{29}\ \mathrm{s}^{-1}$, i.e. of the order expected for CH_4. Studies at higher spectral resolution ($\sim 7\ \mathrm{cm}^{-1}$, Baas et al., 1986) indicate some substructure in the emission feature. As the source of the radiation, the authors propose ultraviolet-pumped infrared fluorescence of a small number of complex hydrocarbons, which are either free-floating or occur in very small grains. The present investigation, however, does not show any prominent line structure at high spectral resolution (Fig. 5). A spectrum obtained at reduced resolution ($2.7\ \mathrm{cm}^{-1}$) shows some excess radiation (at a low confidence level) around 3000 and $3050\ \mathrm{cm}^{-1}$, i.e. in the range indicative of resonance radiation due to saturated as well as unsaturated hydrocarbons.

Table 1. CH_4 cometary line intensities

	Line group	$\langle\tilde{\nu}\rangle_{rest}$ [a]	$\tilde{\nu}_{>}-\tilde{\nu}_{<}$	W cm^{-2} line^{-1} [b]
$P^+(4)$	A_1 A_2 E F_1 F_2	2978.88	0.362	2.6 10^{-19}
$P^+(3)$	A_2 F_1 F_2	2988.90	0.239	2.4 10^{-19}
$P^+(2)$	E F_2	2999.02	0.066	1.0 10^{-19}
$R^-(0)$	A_1	3028.75	{ one	2.2 10^{-19}
$R^-(1)$	F_1	3038.50	line	1.6 10^{-19}
$R^-(2)$	E F_2	3048.16	0.016	2.6 10^{-19}
$R^-(3)$	A_2 F_1 F_2	3057.72	0.073	4.9 10^{-19}
$R^-(4)$	A_1 E F_1 F_2	3067.25	0.136	5.0 10^{-19}
$R^-(5)$	E F_1 F_1 F_2	3076.64	0.177	3.5 10^{-19}
$R^-(6)$	A_1 A_2 E F_2 F_2	3085.95	0.253	5.2 10^{-19}

[a] $\langle\tilde{\nu}\rangle_{comet}=\langle\tilde{\nu}\rangle_{rest}+0.43\ \mathrm{cm}^{-1}$, $\langle\tilde{\nu}\rangle_{rest}$ from Gray and Robiette (1979)
[b] For $Q(CH_4)=4\ 10^{28}\ \mathrm{s}^{-1}$

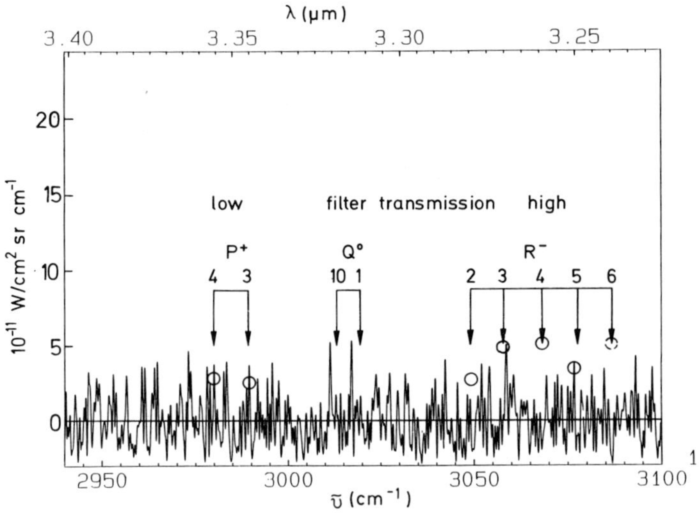

Fig. 5. Spectrum of comet Halley in the CH_4 window. Expected intensities for certain lines (see Table 1) are indicated by circles

If a radiation-resistant crust of the cometary nucleus protects the underlying primitive matter, then the coma could contain two CH components: neutral gaseous CH_4 subliming from the deep nuclear ice at lower than expected abundance; and CH-bearing molecules at higher than anticipated abundance that either boil off the crust or are sputtered from particles in the coma by the local plasma or are evaporated off the grains. It is therefore important to conduct even more sensitive searches with high spectral resolution in the 3000 cm^{-1} region in order to clarify the role of CH species in the origin and evolution of cometary matter.

Acknowledgements. We express our appreciation to Mr. Lou Haughney and the staff of the NASA Kuiper Airborne Observatory for their support during our Southern Hemisphere program of airborne studies of comet Halley. This research was supported by NASA Grant NAG 2-206.

References

Baas, F., Geballe, T.R., Walther, D.M.: 1986, *Astrophys. J.* **311**, L97
Barnes, W.L., Süsskind, R.H., Hunt, R.H., Plyler, E.K.: 1972, *J. Chem. Phys.* **56**, 5160
Combes, M. et al.: 1986, 20th ESLAB Symp., ESA SP-250, Vol. **I**, p. 353
Crovisier, J., Encrenaz, T.: 1983, *Astron. Astrophys.* **126**, 170
Crovisier, J.: 1986, Proc. Workshop *Cometary Radio Astronomy*, Green Bank, West Virginia, USA
Danks, A., Encrenaz, T., Bauchet, P., LeBerte, T., Chalabaev, A., Epchtein, N.: 1986, 20th ESLAB Symp., ESA SP-250, Vol. **III**, p. 103
Davis, D.S., Larson, H.P., Williams, M., Michel, G., Connes, P.: 1980, *Appl. Optics* **19**, 4138
Drapatz, S., Larson, H.P., Davis, D.S.: 1986, 20th ESLAB Symp., ESA SP-250, Vol. **I**, p. 347
Gray, D.L., Robiette, A.G.: 1979, *J. Mol. Spectr.* **77**, 440
Fox, K.: 1970, *J. Quant. Spectr. Radiat. Transfer* **10**, 1335
Herzberg, G.: 1960, *Molecular Spectra and Molecular Structure II*, D. v. Nostrand Comp., Princeton, NJ
Hougen, J.T.: 1976, MTP Int. Rev. Science, Phys. Chem. Series II, Vol. **3**, *Spectroscopy*, Ramsay, D.A., ed. Butterworths, London
Hübner, W.F., Carpenter, C.W.: 1979, Los Alamos Sci. Lab. Rept., LA-8085-MS
Hübner, W.F.: 1985, in *Molecular Astrophysics*, eds. G.H.F. Diercksen, W.F. Hübner, P.W. Langhoff, Reidel, Dordrecht, p. 311
Hübner, W.F.: 1986 (private communication)
Karbach, A., Hess, P.: 1985, *J. Chem. Phys.* **83**, 1075
Knacke, R., Brooke, T.Y., Joyce, R.R.: 1986, *Astrophys. J.* **310**, L49
Krankowsky, D. et al.: 1986, *Nature* **321**, 326
Lambert, J.D.: 1977, *Vibrational and Rotational Relaxation in Gases*, Clarendon Press, Oxford
Larson, H.P., Davis, D.S., Mumma, M.J., Weaver, H.A.: 1986, *Astrophys. J.* **309**, L95
Roche, A.E., Cosmovici, C.B., Drapatz, S., Michel, K.W., Wells, W.C.: 1975, *Icarus* **24**, 120
Rosenberg, A., Ozier, I., Kudian, A.K.: 1972, *J. Chem. Phys.* **57**, 568
Weaver, H.A., Mumma, M.J.: 1984, *Astrophys. J.* **276**, 782
Weaver, H.A., Mumma, M.J., Larson, H.P., Davis, D.S.: 1986, *Nature* **324**, 441
Whipple, F.L.: 1950, *Astrophys. J.* **111**, 375
Yamamoto, T.: 1982, *Astron. Astrophys.* **109**, 326

Astron. Astrophys. 187, 502–512 (1987)

ASTRONOMY
AND
ASTROPHYSICS

Evidence for methane and ammonia in the coma of comet P/Halley

M. Allen[1,2], M. Delitsky[1], W. Huntress[1], Y. Yung[2], W.-H. Ip[3], R. Schwenn[3], H. Rosenbauer[3], E. Shelley[4], H. Balsiger[5], and J. Geiss[5]

[1] Jet Propulsion Laboratory, California Institute of Technology, 4800 Oak Grove Drive, Pasadena, CA 91109, USA
[2] Division of Geological and Planetary Sciences, California Institute of Technology, Pasadena, CA 91125, USA
[3] Max Planck Institut fur Aeronomie, D-3411 Katlenburg-Lindau, Federal Republic of Germany
[4] Lockheed Palo Alto Research Laboratory, Palo Alto, CA 94304, USA
[5] Physikalishes Institut, University of Bern, 3012 Bern, Switzerland

Received February 10, accepted April 30, 1987

Summary. Methane and ammonia abundances in the coma of Halley are derived from Giotto IMS data using an Eulerian model of chemical and physical processes inside the contact surface to simulate Giotto HIS ion mass spectral data for mass-to-charge ratios (m/q) from 15 to 19. The ratio $m/q = 19/18$ as a function of distance from the nucleus is not reproduced by a model for a pure water coma. It is necessary to include the presence of NH_3, and uniquely NH_3, in coma gases in order to explain the data. A ratio of production rates $Q(NH_3)/Q(H_2O) = 0.01–0.02$ results in model values approximating the Giotto data. Methane is identified as the most probable source of the distinct peak at $m/q = 15$. The observations are fit best with $Q(CH_4)/Q(H_2O) = 0.02$. The chemical composition of the comet nucleus implied by these production rate ratios is unlike that of the outer planets. On the other hand, there are also significant differences from observations of gas phase interstellar material.

Key words: atomic and molecular processes – comets

1. Introduction

The ion mass spectra of the coma of comet Halley are richly complex, as illustrated by the early results published by Balsiger et al. (1986). In this paper we concentrate on the most abundant group of ions in Halley ion mass spectra; the group with masses between 12 and 19 amu. This group of ions contains information on the abundances of water (H_2O), methane (CH_4), and ammonia (NH_3); three of the simplest potential parent molecules containing C, N, and O atoms.

We focus on an analysis of ion processes inside the contact surface (along the Giotto trajectory, ~4600 km from the nucleus; Balsiger et al. 1986) where the ion chemistry is relatively straightforward, and where photoionization is the dominant ionization process. Also, ion flow is predominantly radial in this region, which simplifies the model analysis. This investigation uses the data from the high-intensity spectrometer (HIS) of the Giotto ion mass spectrometer (IMS) experiment. The HIS instrument was optimized specifically for making ion measurements inside the contact surface.

Table 1 shows the ions that potentially could contribute to the peaks in IMS spectra at mass-to-charge ratios (m/q) from 12 to 19. In a search for evidence of methane and ammonia in coma gases, we concentrate on the peaks at $m/q = 15–19$ in the HIS ion mass spectra (cf. Fig. 8 in Balsiger et al., 1986). The peak at $m/q = 15$ is diagnostic for CH_3^+ ions, but with a potential contribution from NH^+. Balsiger et al. (1986) have reported that the dominance of the 15 peak relative to 16 over the entire inner coma identifies this peak as mainly CH_3^+ rather than NH^+. This is because of all ions with $m/q \leq 16$ [i.e., CH_n^+ ($n = 0–4$), NH_n^+ ($n = 0–2$), and O^+], only CH_3^+ does not react with water; the main constituent of the coma. Thus, in a water-dominated coma, the unreactive CH_3^+ ions at $m/q = 15$ become prominant at this position in the mass spectrum. This fortunate circumstance allows the identification of the methyl cation in the presence of dominating quantities of water, and allows the identification to be distinguished from NH^+ ions. There is no equivalent circumstance in neutral mass spectra. In this paper, we confirm the identification of $m/q = 15$ as CH_3^+ and derive a relative production rate of its parent by comparison with chemical modeling of ion-molecule reactions in the coma.

Since water dominates the composition of Halley coma gases, any peaks which might be due to NH_2^+ at $m/q = 16$ and NH_3^+ at $m/q = 17$ will have significant contributions from O^+ and OH^+. Contributions from CH_2^+ and CH_3^+ at $m/q = 14$ and 15

Table 1. Ion masses and contributing species

m/q	Possible species
12	C^+
13	CH^+
14	CH_2^+, N^+
15	CH_3^+, NH^+
16	CH_4^+, NH_2^+, O^+
17	CH_5^+, NH_3^+, OH^+
18	NH_4^+, H_2O^+
19	H_3O^+

Send offprint requests to: M. Allen

can also mask the identification of N^+ and NH^+. This makes the identification of small quantities of ammonia exceedingly difficult in both ion and neutral mass spectra of the coma. Ion-molecule chemistry can be used to diagnose the presence of ammonia in this case, but no well-defined peak results as in the case for CH_3^+. Instead, the effect of adding ammonia is to significantly alter the ratio of $m/q = 19$ relative to $m/q = 18$. This is because the H_3O^+ ion reacts rapidly with NH_3 to give NH_4^+:

$$H_3O^+ + NH_3 \rightarrow NH_4^+ + H_2O$$

The effect of this reaction is to decrease the 19/18 ratio with increasing ammonia concentration as H_3O^+ at $m/q = 19$ is converted to NH_4^+ at $m/q = 18$.

We shall show that the 19/18 ion ratio computed for a pure water Halley atmosphere is always too high compared to the HIS data within the contact surface. This suggests that some species is required which preferentially reacts with H_3O^+ at $m/q = 19$ relative to H_2O^+ at $m/q = 18$. However, any species that reacts with H_3O^+ most likely will react with H_2O^+ also. This is because the affinity of OH for a proton is less than that of H_2O, so that H_2O^+ will transfer a proton to a larger suite of chemical species than will H_3O^+. Also, H_2O^+ has a higher ionization potential than H_3O^+ so that it will transfer charge to a larger number of species. H_2O^+ is an odd-electron ion and is in general more reactive than H_3O^+. Therefore, the effect of adding reactive neutrals to water is to depress the abundances of ions at both $m/q = 18$ and 19, so that the ratio does not change dramatically. The only species so far found to have the desired effect is ammonia, because the product of the reactions of NH_3 with both H_2O^+ and H_3O^+ is NH_4^+ at $m/q = 18$. Thus, unless some other species can eventually be found to have this effect, the low 19/18 ion ratio in the coma is best explained by the presence of ammonia. In this way we derive a relative production rate for NH_3.

2. The coupled transport/chemistry model

The computer model utilized in these calculations is a version of a general one-dimensional coupled transport/chemistry program previously used for studies of the atmospheres of the earth and other planets (e.g., Allen et al., 1981). The model is an Eulerian formulation that solves the continuity equation for each species i self-consistently at a set of fixed distances from the center of the comet nucleus:

$$\frac{\partial n_i}{\partial t} + \frac{1}{r^2}\frac{\partial}{\partial r}(r^2 \Phi_i) = P_i + L_i \qquad (1)$$

where n_i, Φ_i, P_i, and L_i are the concentration, flux, production, and loss rates of i at a distance r and time t. These quantities are also functions of r. The flux Φ_i is purely the result of outflowing (or inflowing) material:

$$\Phi_i(r) = n_i(r) v_i(r) \qquad (2)$$

where the velocity v can vary with distance and be different for different species. To seek a numerical solution of (1), with boundary condition (2), we evaluate the spatial variables at a discrete set of points and use time marching for time evolution. The resulting coupled finite difference equations are solved using an implicit method adopted from Richtmeyer (1957). The results of test runs in which a constant outflow velocity and photodestruction scale length were prescribed are found to agree with values derived from the analytical formulation of Haser (1957) to within a few percent. In this paper, the results of steady-state calculations ($\partial n_i/\partial t = 0$) will be presented.

The grid points at which the calculations are performed extend from the surface of the nucleus (assumed to be 7.5 km from the nucleus center) out to a distance of 10^6 km, with fourteen points per decade of distance. At each grid point, a temperature T and velocity v were prescribed. We adopted a temperature for neutrals, ions, and electrons of 340 K (Balsiger et al., 1986) constant with distance, assuming that the neutral, ion, and electron temperatures are closely coupled inside the contact surface.

Lämmerzahl et al. (1986) report smoothly increasing radial flow speeds from $\sim$1000 km out to $\sim$30,000 km from the nucleus. Their data inside the contact surface comes from measurements of ion species, while outside the contact surface measurements of neutral species were utilized. Based on their work, we adopted a simple function for the outflow velocity for both ions and neutrals varying with distance: 0.8 km s^{-1} within 1000 km of the nucleus, increasing at a rate of 1.13 cm s^{-1} km^{-1} out to 3 10^4 km, and constant at 1.13 km s^{-1} at larger distances. This assumes that a strong coupling between neutral and ion motions exists in the inner coma allowing the neutral species to be assigned the same outflow velocity as the ions.

The lower boundary (nucleus surface) conditions are prescribed for each species in the calculation in the form of a concentration or flux consistent with a desired surface-integrated production rate $Q(i)$. For other than the "parent" species released from the nucleus, the lower boundary condition was zero flux. The upper boundary condition for all species is an outflow velocity of 1.13 km s^{-1}.

The solar flux used in the photolysis calculations comes from results of satellite and/or rocket measurements at solar minimum reported by Torr and Torr (1985) for $\lambda \leq 1035$ Å and Mount and Rottman (1983) for $\lambda \geq 1190$ Å. At Lyman alpha, $\lambda = 1215.7$ Å, the solar flux at 1 AU distance from the sun was set to a value of 2.22 10^{11} photons cm^{-2} s^{-1} (R. Clancy, private communication, 1986) as observed by the Solar Mesosphere Explorer satellite (SME; Rottman et al., 1982) on the day of the Giotto encounter with Halley. The solar flux is scaled to the distance of the Giotto encounter. Opacity due to absorption by H_2O in the coma is included in the internal radiation field computation and is allowed to vary as the distribution of H_2O changes. This is a minor effect as the radiation field is attenuated by more than 10% only within a few hundred kilometers of the nucleus surface. In general, it is a good assumption that the photolytic rate constants are constant throughout the coma. Opacity due to the presence of dust in the coma has not been taken into account; we await further reports from the *in situ* and remote observations of Halley.

The discussion in this paper is focussed upon the abundances of light ions – H_nO^+, NH_n^+, and CH_n^+ – within the contact surface. For the purposes of this paper, we calculate the radial distributions of only the "parent" species H_2O, NH_3, CH_4, the neutral "child" species H, O, OH, and the ions H^+, $H_{n=0-3}O^+$, $NH_{n=1-4}^+$, and $CH_{n=2-5}^+$. The interrelated chemistry of other species was considered and found not to be important in this spatial region. The key reactions involving these species within the contact surface are listed in Table 2. Total charge neutrality is assumed to hold at all distances; the electron profile is calculated iteratively by summing over the computed positive ion abundances. This is an accurate procedure because the ions with

Table 2. Key reactions

Reaction	Rate constant[a]	Reference
1. $H_2O + h\nu \rightarrow$ all products	$1.12\ 10^{-5}$[b] $\lambda \leq 1975$ Å	(1)
2. $H_2O + h\nu \rightarrow H + OH$	$9.37\ 10^{-6}$ 610 Å $\leq \lambda \leq 1975$ Å	(1)
3. $H_2O + h\nu \rightarrow 2H + O$	$6.49\ 10^{-7}$ 810 Å $\leq \lambda \leq 1275$ Å	(1)
4. $H_2O + h\nu \rightarrow H_2O^+ + e^-$	$4.35\ 10^{-7}$ $\lambda \leq 983$ Å	(1)
5. $H_2O + h\nu \rightarrow H + OH^+ + e^-$	$8.32\ 10^{-8}$ $\lambda \leq 687$ Å	(1)
6. $H_2O + h\nu \rightarrow H^+ + OH + e^-$	$3.58\ 10^{-8}$ $\lambda \leq 663$ Å	(1)
7. $H_2O + h\nu \rightarrow O^+$ + products $+ e^-$	$4.01\ 10^{-9}$ $\lambda \leq 662$ Å	(1)
8. $NH_3 + h\nu \rightarrow$ all products	$1.49\ 10^{-4}$ $\lambda \leq 2225$ Å	(1)
9. $NH_3 + h\nu \rightarrow NH_3^+ + e^-$	$8.52\ 10^{-7}$ $\lambda \leq 1231$ Å	(1)
10. $NH_3 + h\nu \rightarrow NH_2^+ + H + e^-$	$2.00\ 10^{-7}$ $\lambda \leq 786$ Å	(1)
11. $NH_3 + h\nu \rightarrow NH^+$ + products $+ e^-$	$6.07\ 10^{-9}$ $\lambda \leq 775$ Å	(1)
12. $NH_3 + h\nu \rightarrow H^+ + 2H$ + products $+ e^-$	$4.76\ 10^{-9}$ $\lambda \leq 681$ Å	(1)
13. $CH_4 + h\nu \rightarrow$ all products	$9.16\ 10^{-6}$ $\lambda \leq 1625$ Å	(1)
14. $CH_4 + h\nu \rightarrow CH_4^+ + e^-$	$4.65\ 10^{-7}$ $\lambda \leq 945$ Å	(1)
15. $CH_4 + h\nu \rightarrow CH_3^+ + H + e^-$	$2.52\ 10^{-7}$ $\lambda \leq 866$ Å	(1)
16. $CH_4 + h\nu \rightarrow CH_2^+$ + products $+ e^-$	$2.63\ 10^{-8}$ $\lambda \leq 822$ Å	(1)
17. $H + h\nu \rightarrow H^+ + e^-$	$8.09\ 10^{-8}$ $\lambda \leq 910$ Å	(1)
18. $O + h\nu \rightarrow O^+ + e^-$	$2.55\ 10^{-7}$ $\lambda \leq 910$ Å	(1)
19. $OH + h\nu \rightarrow O + H$	$1.50\ 10^{-5}$ $\lambda \leq 1925$ Å	(1)
20. $H_2O^+ + e^- \rightarrow H + OH$	$9.1\ 10^{-6}\ T_e^{-0.5}$	(2)
21. $H_2O^+ + e^- \rightarrow O$ + products	$3.9\ 10^{-6}\ T_e^{-0.5}$	(2)
22. $H_3O^+ + e^- \rightarrow H_2O^+ + H$		
23. $H_3O + e^- \rightarrow OH + 2H$	$3.39\ 10^{-4}\ T_e^{-0.9}$	(3)
24. $H_3O^+ + e^- \rightarrow OH$ + other products		
25. $NH_3^+ + e^- \rightarrow H$ + products	$4.1\ 10^{-5}\ T_e^{-0.5}$	estimate
26. $NH_4^+ + e^- \rightarrow NH_3 + H$	$4.1\ 10^{-5}\ T_e^{-0.6}$	(4)
27. $NH_4^+ + e^- \rightarrow 2H$ + products		
28. $CH_2^+ + e^- \rightarrow$ products	$8.7\ 10^{-6}\ T_e^{-0.5}$	(5)
29. $CH_3^+ + e^- \rightarrow$ products	$1.2\ 10^{-5}\ T_e^{-0.5}$	(5)
30. $CH_4^+ + e^- \rightarrow H$ + products	$1.2\ 10^{-5}\ T_e^{-0.5}$	(5)
31. $CH_4^+ + e^- \rightarrow 2H$ + other products		
32. $H^+ + H_2O \rightarrow H_2O^+ + H$	$8.2\ 10^{-9}$	(6)

[a] Photolytic rate constants are in units of s^{-1}, two body rate constants in units of $cm^3\ s^{-1}$.

[b] Values for 0.9 AU with no opacity. Indicated also is the wavelength range in which the cross-sections are significant.

References: (1) Allen et al. (1987, in preparation); (2) McGowan et al. (1979); (3) Heppner et al. (1976); (4) Alge et al. (1983); (5) McGowan and Mitchell (1984); (6) Anicich and Huntress (1986).

Table 2 (continued)

Reaction	Rate constant[a]	Reference
33. $H^+ + CH_4 \rightarrow CH_4^+ + H$	$1.6\ 10^{-9}$	(6)
34. $H^+ + CH_4 \rightarrow CH_3^+ +$ products	$2.4\ 10^{-9}$	(6)
35. $O^+ + H_2O \rightarrow H_2O^+ + O$	$3.2\ 10^{-9}$	(6)
36. $OH^+ + H_2O \rightarrow H_2O^+ + OH$	$1.5\ 10^{-9}$	(6)
37. $OH^+ + H_2O \rightarrow H_3O^+ + O$	$1.4\ 10^{-9}$	(6)
38. $OH^+ + O \rightarrow H +$ products	$1.0\ 10^{-9}$	Estimate
39. $OH^+ + NH_3 \rightarrow NH_4^+ + O$	$1.2\ 10^{-9}$	(6)
40. $OH^+ + NH_3 \rightarrow NH_3^+ + OH$	$1.2\ 10^{-9}$	(6)
41. $H_2O^+ + H_2O \rightarrow H_3O^+ + OH$	$2.0\ 10^{-9}$	(6)
42. $H_2O^+ + NH_3 \rightarrow NH_3^+ + H_2O$	$2.2\ 10^{-9}$	(6)
43. $H_2O^+ + NH_3 \rightarrow NH_4^+ + OH$	$1.0\ 10^{-9}$	(6)
44. $H_3O^+ + NH_3 \rightarrow NH_4^+ + H_2O$	$2.2\ 10^{-9}$	(6)
45. $NH^+ + H_2O \rightarrow H_3O^+ +$ products	$1.1\ 10^{-9}$	(6)
46. $NH^+ + H_2O \rightarrow H_2O^+ +$ products	$1.1\ 10^{-9}$	(6)
47. $NH^+ + H_2O \rightarrow NH_2^+ + OH$	$8.75\ 10^{-10}$	(6)
48. $NH^+ + H_2O \rightarrow NH_3^+ + O$	$1.75\ 10^{-10}$	(6)
49. $NH^+ + H_2O \rightarrow$ other products	$3.5\ 10^{-10}$	(6)
50. $NH_2^+ + H_2O \rightarrow H_3O^+ +$ products	$2.75\ 10^{-9}$	(6)
51. $NH_2^+ + H_2O \rightarrow NH_4^+ + O$	$1.45\ 10^{-10}$	(6)
52. $NH_3^+ + OH \rightarrow NH_4^+ + O$	$1.0\ 10^{-9}$	Estimate
53. $NH_3^+ + O \rightarrow$ products	$1.0\ 10^{-9}$	Estimate
54. $NH_3^+ + NH_3 \rightarrow NH_4^+ +$ products	$2.2\ 10^{-9}$	(6)
55. $NH_3^+ + CH_4 \rightarrow NH_4^+ +$ products	$4.8\ 10^{-10}$	(6)
56. $CH_2^+ + H_2O \rightarrow H +$ products	$1.2\ 10^{-9}$	(6)
57. $CH_2^+ + OH \rightarrow H +$ products	$1.0\ 10^{-9}$	Estimate
58. $CH_2^+ + NH_3 \rightarrow H +$ products	$2.5\ 10^{-9}$	(6)
59. $CH_2^+ + CH_4 \rightarrow$ products	$1.2\ 10^{-9}$	(6)
60. $CH_3^+ + OH \rightarrow$ products	$1.0\ 10^{-9}$	Estimate
61. $CH_3^+ + O \rightarrow$ products	$4.4\ 10^{-10}$	(6)
62. $CH_3^+ + NH_3 \rightarrow NH_4^+ +$ products	$3.0\ 10^{-10}$	(6)
63. $CH_3^+ + NH_3 \rightarrow$ other products	$1.29\ 10^{-9}$	(6)
64. $CH_3^+ + CH_4 \rightarrow$ products	$1.2\ 10^{-9}$	(6)
65. $CH_4^+ + H_2O \rightarrow H_3O^+ +$ products	$2.5\ 10^{-9}$	(6)
66. $CH_4^+ + OH \rightarrow$ products	$1.0\ 10^{-9}$	Estimate
67. $CH_4^+ + NH_3 \rightarrow NH_3^+ + CH_4$	$1.6\ 10^{-9}$	(6)
68. $CH_4^+ + NH_3 \rightarrow NH_4^+ +$ products	$1.5\ 10^{-9}$	(6)
69. $CH_4^+ + NH_3 \rightarrow CH_5^+ +$ products	$6.32\ 10^{-11}$	(6)
70. $CH_5^+ + H_2O \rightarrow H_3O^+ + CH_4$	$3.7\ 10^{-9}$	(6)

$m/q \leq 19$ account for almost all of the observed ionization in the inner coma (Balsiger et al., 1986).

3. The radial distribution of parent species

The radial profiles of coma ions are directly related to the distribution within the coma of the neutral molecules (usually "parent" species directly released from the nucleus) from which the ions are derived. Few observations of the spatial profiles of parent species have been published for comet Halley with which our model for the neutral precursor species can be verified. However, the distribution of H_2O within $\sim$ 40,000 km of the nucleus is available from the Giotto NMS experiment (Krankowsky et al., 1986).

In a note added in proof, Krankowsky et al. indicate that their published data includes an error in the distance scale, 230 km farther from the nucleus than in actuality. Therefore, we chose to refit the published NMS experimental results, correcting the distances, using the Haser (1957) formulation. Averaging results from each NMS analyzer, we find a value for $Q(H_2O)/v = 4.9\ 10^{14}\ cm^{-3}\ km^2$ and, assuming $v = 0.9\ km\ s^{-1}$, $Q(H_2O) = 4.4\ 10^{29}$ molec s^{-1}. If v is closer to $0.8\ km\ s^{-1}$, then $Q(H_2O)$ is proportionately smaller. This $Q(H_2O)$ is $\sim 20\%$ lower than that reported by Krankowsky et al. and is what is used to derive the model boundary condition for H_2O at the nucleus. A comparable analysis for the NMS CO_2 data yields $Q(CO_2)/Q(H_2O) = 0.042$, which is to be compared with the value of 0.035 reported by Krankowsky et al.

The radial profile for H_2O calculated by the model is compared with our representation of the Giotto NMS results in Fig. 1a. Only a small difference between these two curves is noticeable. A more detailed comparison between model results and measurements is shown in Fig. 1b where both H_2O profiles are

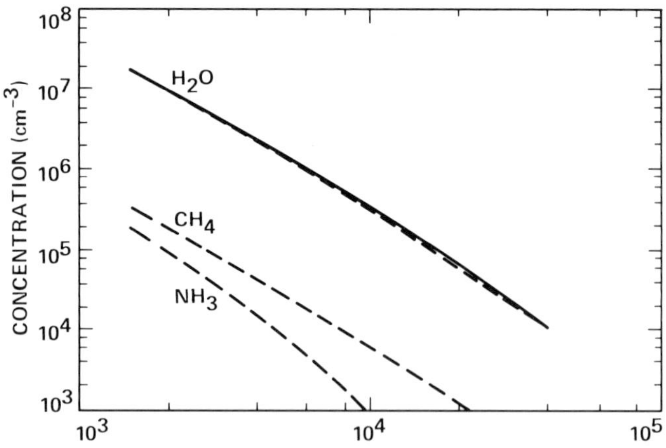

Fig. 1a. Comparison of model calculations (dashed lines) of the radial distribution of H_2O, NH_3, and CH_4 with the Giotto neutral mass spectrometer data for H_2O (solid line: Krankowsky et al., 1986).

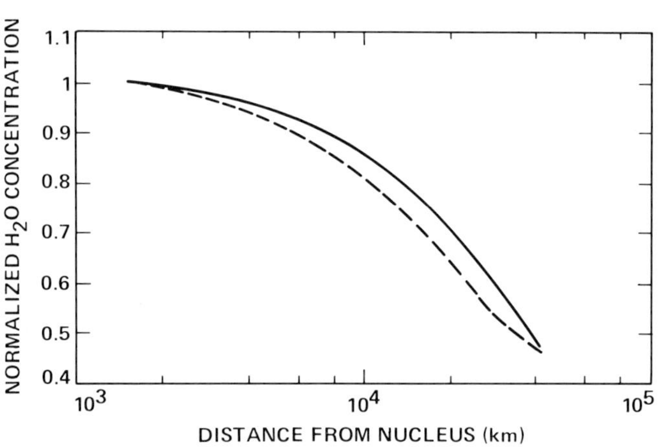

Fig. 1b. Giotto NMS results (solid line) and model values (dashed line) for H_2O normalized by the value at the smallest distance and scaled by r^2 to correct for spatial dilution from spherical outflow

replotted, scaled by r^2, and normalized by the value at the smallest distance. The differences are now more visible, but still the largest discrepancy is only $\sim 7\%$. This manner of presenting the spatial distribution of a species is of interest because the effect of transport and chemistry can be separated. If H_2O were inert, a distribution varying as r^{-2} would be the expected result of the spatial dilution resulting from the radial outflow. This would appear in Fig. 1b as a constant profile with the value of unity. The observed departure from unity reflects the effect of chemical loss.

Krankowsky et al. calculate from their data a value for $v/J(H_2O)$ of $3.9\,10^4$ km, where J is the total photodestruction rate constant for H_2O. With $v = 0.9\,\mathrm{km\,s^{-1}}$, $J(H_2O) = 2.3\,10^{-5}\,\mathrm{s^{-1}}$. This deduced value for $J(H_2O)$ is much larger than that usually calculated by photochemical models ($\sim 1.1\,10^{-5}\,\mathrm{s^{-1}}$) as noted by Krankowsky et al. However, the $J(H_2O)$ derived from a Haser model analysis of the NMS data is reduced significantly after correcting the NMS distance scale and adopting the new estimate of radial velocity. However such an analysis assumes a constant outflow velocity, which is shown to be invalid by the more recent Lämmerzahl et al. (1986) work. Figure 1b shows that the model chemistry summarized in Table 2 reproduces the observed H_2O profile.

As the agreement between observations and model results for one parent species, H_2O, is excellent, we feel that other parent species distributions calculated by our model are fairly accurate and have reasonable confidence in our model ion source terms. For later reference we also plot in Fig. 1a the calculated profiles for NH_3 and CH_4, using values for $Q(NH_3)/Q(H_2O)$ and $Q(CH_4)/Q(H_2O)$ derived below.

4. The ratio m/q = 19/18 and the inferred ratio $Q(NH_3)/Q(H_2O)$

Within the contact surface, H_nO^+ ions are formed directly from H_2O by ionization and indirectly by charge exchange or by dissociation followed by ionization. These ions react subsequently with H_2O in one or more steps leading to production of H_3O^+. This is illustrated in Fig. 2. In a pure H_2O coma, the only loss process for H_3O^+ is relatively slow recombination with electrons. As a consequence, H_3O^+ at $m/q = 19$ is expected to be the most abundant ion. Figure 3 shows the spatial variation of the ratio $m/q = 19/18$ from HIS spectra compared with the results of model calculations for the pure H_2O coma case. The model shows a different variation with distance than is seen in the HIS data, and the model ratios are significantly higher at the smaller distances.

Unlike the case for methane, it is not possible to identify a specific ion in the HIS spectra as derived from ammonia. The presence of ammonia is inferred from the behavior of the ratio $m/q = 19/18$ with distance from the comet. The important reactions inside the contact surface involving $NH_n{}^+$ ions are illustrated in Fig. 4. The direct products of NH_3 ionization react rapidly with H_2O to form $NH_4{}^+$ at $m/q = 18$. The proton exchange between H_3O^+ and NH_3 is a significant loss process for H_3O^+ and a major source for $NH_4{}^+$. The presence of ammonia introduces a mechanism for transferring ion density from mass 19 to mass 18 with a consequent reduction in the ratio $m/q = 19/18$. The results of model runs with varying amounts of ammonia are shown in Fig. 3. It is clear that ammonia is necessary in order to account for the very low 19/18 ratios observed at distances closest to the comet. Values of $Q(NH_3)/Q(H_2O) = 0.01$–0.02 result in the best fit to the Giotto data.

Among other likely agents for extracting a proton from H_3O^+ is HCN, which has been detected in Halley's coma by its millimeter line emission (Despois et al., 1986; Schloerb et al., 1986; Winnberg et al., 1987). However, the $Q(HCN)/Q(H_2O)$ is only $\sim 10^{-3}$, which is too small to have an appreciable effect on H_3O^+. Moreover, if proton transfer to some species other than NH_3 were to dominate the loss of H_3O^+, then a peak at higher mass comparable to the intensity at $m/q = 19$ would be expected. This is not observed.

Another possible process to reduce the 19/18 ratio is photodissociation of H_3O^+. To be important, a photodissociation rate constant of $10^{-3} - 10^{-2}\,\mathrm{s^{-1}}$ is necessary. Laboratory cross-sections for this process are unavailable, but a value this large is unlikely. This is illustrated in Table 2 by comparison with the isoelectronic species NH_3, whose photodissociation rate constant is much smaller than the required values.

Recently, Murad and Bochsler (1987) proposed a source for H_3O^+ not included in our model chemistry: photoionization of the water dimer directly vaporizing from the nucleus and/or sputtered from icy coma grains. If their chemistry were to be included

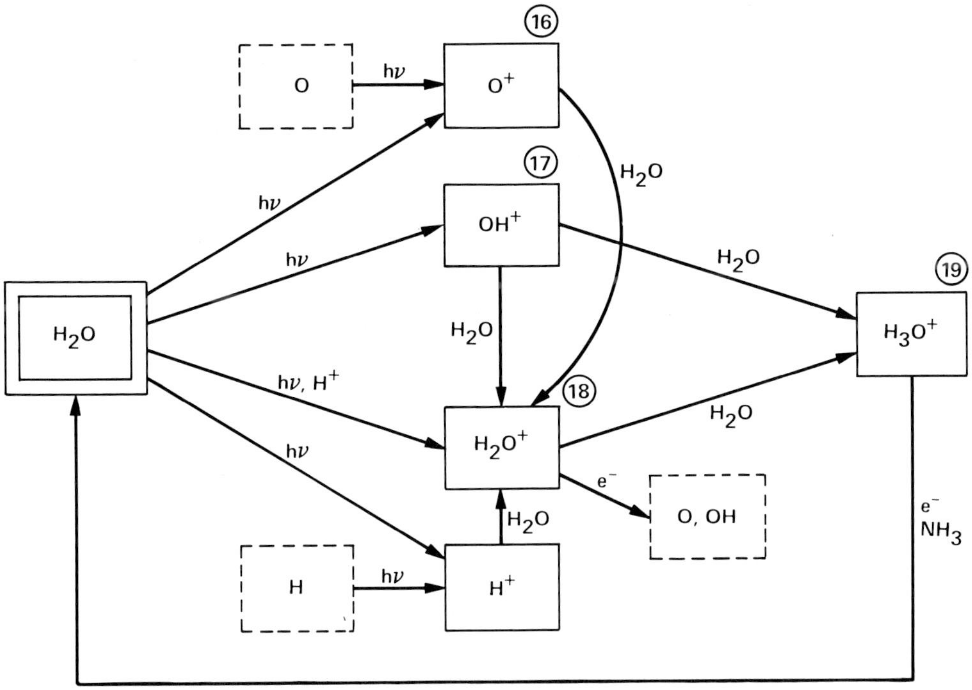

Fig. 2. Important reactions for H_nO^+ inside of the contact surface. Masses of ions indicated in circles

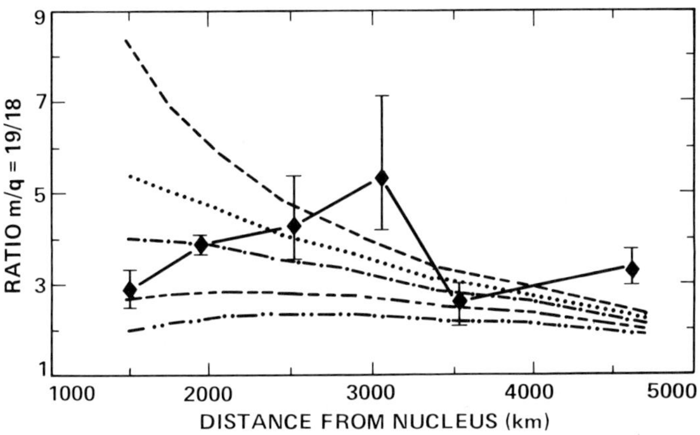

Fig. 3. The ratio $m/q = 19/18$ from Giotto HIS measurements (diamonds) compared with model calculations (pure H_2O coma: dashed line; $Q(NH_3)/Q(H_2O) = 0.005$: dotted line; $Q(NH_3)/Q(H_2O) = 0.01$: dash-dot line; $Q(NH_3)/Q(H_2O) = 0.02$: long dash-short dash line; $Q(NH_3)/Q(H_2O) = 0.03$: dash-double dot line). Uncertainties in our estimation of the HIS ratio values are indicated. Giotto values connected by solid lines for visualization purposes

in our model calculations, the model values for the 19/18 ratio would increase, requiring the presence of even more ammonia to fit the Giotto results. Without the availability of absolute ion concentrations, there appears to be no need to add this speculative source of H_3O^+ at this time.

The major feature in the data that leads to the necessity to include ammonia is the relatively constant value for the 19/18 ratio with cometary distance, while the prediction for the case without ammonia is significantly different. There are large excursions in the data which are not predicted by the model. We have chosen to weight heaviest in our model/observation comparison the points closest to the nucleus and closest to the contact surface. The reality of the large "bulge" in the 19/18 data will be tested in future analysis of the HIS data. Moreover, reducing the amount of ammonia to fit the central points in Fig. 3 results in a poorer fit at the other m/q ratios to be discussed below.

5. The ratio m/q = 15/18 and the inferred ratio $Q(CH_4)/Q(H_2O)$

There are a number of potential sources for the CH_3^+ ion identified in the HIS ion mass spectrum. These include the dissociative ionization of methane, hydrocarbons, and other organic compounds containing the methyl (CH_3) group: such as methyl acetylene (CH_3CCH), methanol (CH_3OH), methyl amine (CH_3NH_2), and methyl cyanide (CH_3CN). Methane is the simplest possible and most efficient per-molecule source of CH_3^+, and is a likely candidate for cometary volatiles since it is common in the atmospheres of the outer planets and is the principal carbon-containing molecule in equilibrium solar nebula models. There is also additional evidence in the high-energy-range spectrometer (HERS) ion spectra that support methane as the major source of CH_3^+. Balsiger et al. (1986) have reported that the relative abundances of $m/q = 13–15$ in HERS data at large distance follow well the predicted behavior for these ions if they were due entirely to methane photolysis and subsequent photoionization.

The important reactions for CH_n^+ ions in the inner coma are illustrated in Fig. 5. All of the ions directly produced by photoionization of CH_4 react with H_2O except for CH_3^+. The dominant loss processes for CH_3^+ are electron recombination and reaction with NH_3. Because of the latter process, it is necessary

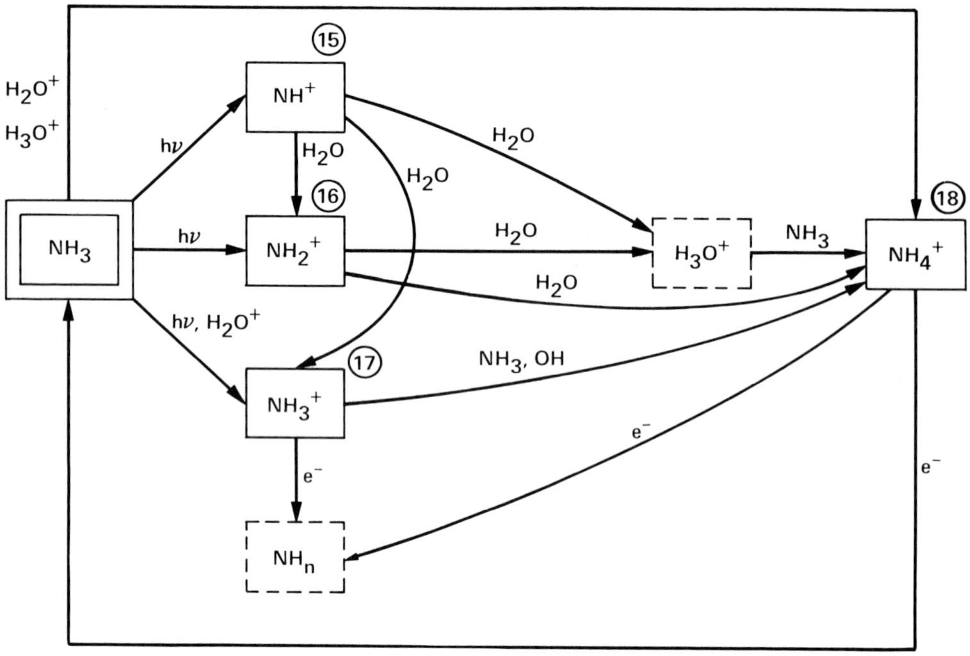

Fig. 4. Important reactions for NH_n^+ inside of the contact surface. Masses of ions indicated in circles

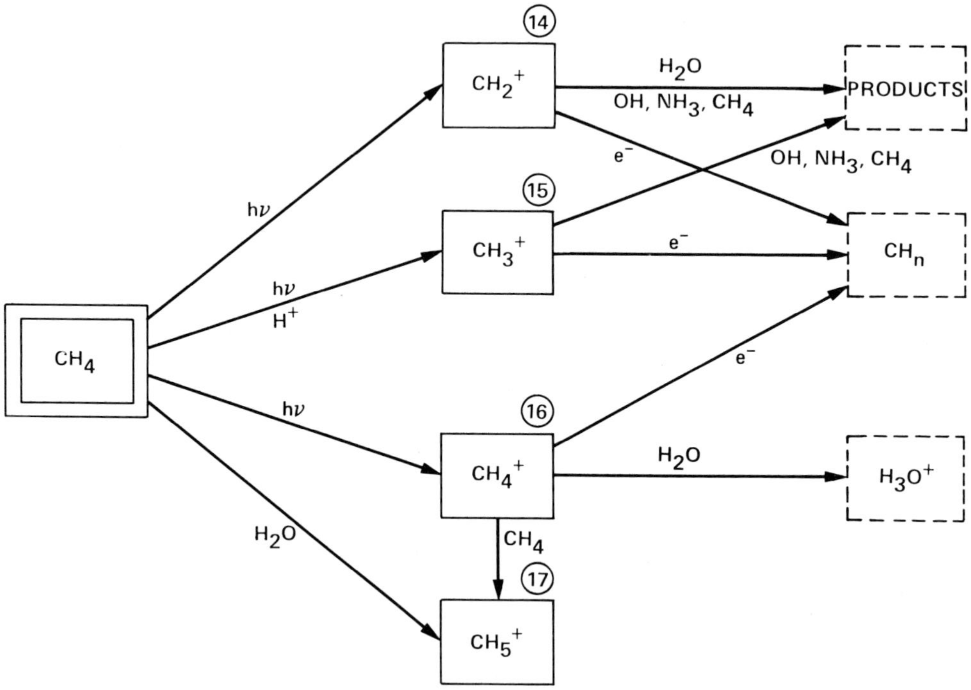

Fig. 5. Important reactions for CH_n^+ inside of the contact surface. Masses of ions indicated in circles

to determine the ratio $Q(NH_3)/Q(H_2O)$ before modelling the CH_n^+ ion chemistry. Figure 6 shows the intensities at $m/q = 15$ relative to 18 in HIS ion spectra inside the contact surface. Shown on the same figure are the results of model runs in which the values for both $Q(CH_4)/Q(H_2O)$ and $Q(NH_3)/Q(H_2O)$ are varied. A comparison of the profiles with and without ammonia shows that NH_3 is necessary in the model in order to produce the correct functional form for the abundance of CH_3^+ with distance. While the abundance of NH_3 is the controlling factor in the shape of the curve, the abundance of CH_4 is the controlling factor in the amplitude of the curve. A model with $Q(CH_4)/Q(H_2O) = 0.02$ and $Q(NH_3)/Q(H_2O) = 0.015$ fits the HIS data in Fig. 6 fairly well. Larger amounts of CH_4 require larger amounts of NH_3 for a reasonable fit, but the NH_3 value is already constrained by the 19/18 data. Smaller amounts of CH_4 do not produce sufficient $m/q = 15$ at the larger distances.

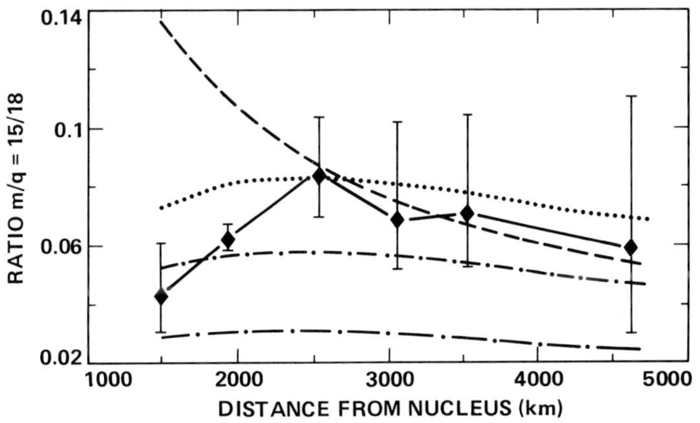

Fig. 6. The ratio $m/q = 15/18$ from Giotto HIS measurements (diamonds) compared with model calculations ($Q(CH_4)/Q(H_2O) = 0.02$, $Q(NH_3)/Q(H_2O) = 0$: dashed line; $Q(CH_4)/Q(H_2O) = 0.03$, $Q(NH_3)/Q(H_2O) = 0.015$: dotted line; $Q(CH_4)/Q(H_2O) = 0.02$, $Q(NH_3)/Q(H_2O) = 0.015$: dash-dot line; $Q(CH_4)/Q(H_2O) = 0.01$, $Q(NH_3)/Q(H_2O) = 0.015$: long dash-short dash line). Uncertainties in our estimation of the HIS ratio values are indicated. Giotto values connected by solid lines for visualization purposes

6. Further tests of the derived relative production rates for NH_3 and CH_4

The remaining major distinguishable peak in the mass range $m/q = 13$–19 spectra inside the contact surface is $m/q = 17$. The $m/q = 16$ ion appears as a shoulder between $m/q = 15$ and 17 in HIS ion spectra and is contaminated by overlap counts from 15 and 17. Figure 7 shows the 17/18 ratio versus cometary distance compared to a pure H_2O model ($CH_5{}^+$ abundances are insignificant) and a model with $Q(NH_3)/Q(H_2O)$ and $Q(CH_4)/Q(H_2O)$ as inferred above. The model with ammonia fits the data much better than that without ammonia. Ammonia adds $NH_3{}^+$ ions to the existing OH^+ ions to make up needed ion density at $m/q = 17$. It is significant that ammonia appears necessary to fit the ion density versus distance at every mass where ammonia would have the appropriate effect.

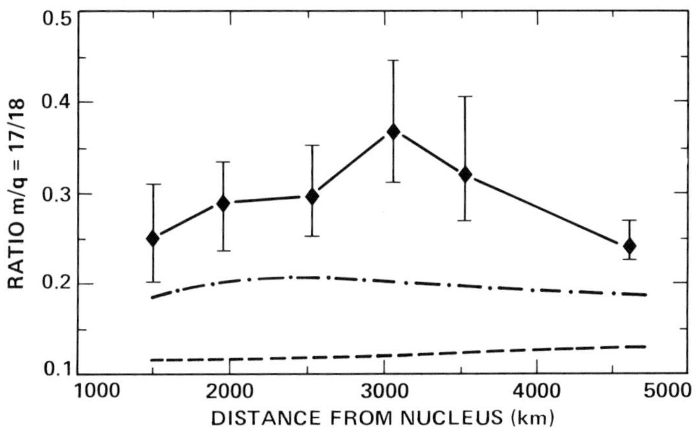

Fig. 7. The ratio $m/q = 17/18$ from Giotto HIS measurements (diamonds) compared with model calculations (pure H_2O coma: dashed line; $Q(CH_4)/Q(H_2O) = 0.02$, $Q(NH_3)/Q(H_2O) = 0.015$: dash-dot line). Uncertainties in our estimation of the HIS ratio values are indicated. Giotto values connected by solid lines for visualization purposes

One further test was applied to examine the consistency between HIS and HERS data regarding the presence of ammonia. HERS spectra were taken primarily at large distances beyond the photodissociation scale length for ammonia. The ultimate photodissociation product from NH_3 is the N atom. We examined HERS ion mass spectra at ranges from 100,000–200,000 km from the nucleus to determine an upper limit for the amount of N^+ ($m/q = 14$) relative to O^+ ($m/q = 16$) and hence an upper limit for the amount of N from NH_3 photolysis relative to O from H_2O photolysis. At these ranges, water is largely dissociated, but carbon monoxide is not, so that almost all O^+ ions in the ion mass spectrum originate from photolysis of H_2O to O atoms via several steps, followed by ionization. For the purpose of this exercise we assumed photoionization to be the major source of ionization with rates of $3\,10^{-7}\,s^{-1}$ for O and $2\,10^{-7}\,s^{-1}$ for N. All $m/q = 14$ ions are assumed to be N^+ from NH_3 photoprocessing, and the loss rate for all ions is assumed equal via transport. The result is a maximum NH_3/H_2O ratio of 2%. Considering the assumptions involved and the errors in measurement, this is consistent with the ratio determined from HIS ion spectra in the inner coma.

Carbon, nitrogen, and oxygen in the form of both ions and radical compounds have been detected by other experiments on Halley spacecraft and from terrestrial platforms. A detailed comparison with these data awaits further information about the observing geometries. While the evidence for the presence of ammonia and methane seems fairly conclusive from HIS ion mass spectra, the value derived for the relative production rates appears large compared to related results from Vega on the column abundances for NH and NH_2. Moreels et al. (1986) report column abundance ratios of $NH/OH = 1.4\,10^{-4}$ and $NH_2/OH = 7\,10^{-4}$. Although not directly related to $Q(NH_3)/Q(H_2O)$, these values are quite small compared to our result of 1–$2\,10^{-2}$. On the other hand, Moreels et al. (1986) report for CH a ratio $CH/OH = 0.05$, and Wyckoff et al. (1986) find from ground-based measurements in the visible spectrum $CH^+/H_2O^+ = 0.04$. Although possibly fortuitous, these results are close to our result, $Q(CH_4)/Q(H_2O) = 0.02$. Drapatz et al. (1986) have attempted a direct measurement of CH_4 using high resolution infrared spectroscopic measurements at the ν_3 line of CH_4 from the Kuiper Airborne Observatory. Their upper limit of $CH_4/H_2O < 0.04$ is consistent with our measurement from HIS data.

7. The radial distribution of ion concentrations

We have derived above the relative production rates $Q(NH_3)/Q(H_2O) = 0.01$–0.02 and $Q(CH_4)/Q(H_2O) = 0.02$. We shall refer to this as the "best fit" case and for the purposes of calculations use $Q(NH_3)/Q(H_2O) = 0.015$. In Figs. 8–10, we show model results for the distribution of the light oxygen, nitrogen, and carbon ions inside the contact surface using the rate constants in Table 2 and the value for $Q(H_2O)$ derived from the Giotto NMS data. As seen in Fig. 1a, the NH_3/H_2O abundance ratio drops from a value of 0.01 near 1500 km to a value of 0.005 at 5000 km (due to differential photodissociation of NH_3 relative to H_2O). As the source of ions is decreasing with increasing distance, the abundance of $NH_n{}^+$ relative to H_nO^+ also decreases with distance as shown in Figs. 8 and 9. In fact the abundance of ammonia drops sufficiently that H_nO^+ is not controlled by the H_3O^+/NH_3 reaction at larger distances, with the result that $m/q = 19/18$ ap-

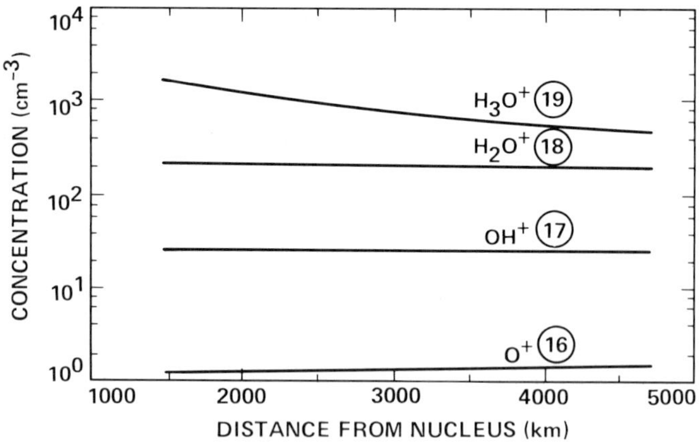

Fig. 8. Spatial distribution of the abundances of H_nO^+ from the "best fit" case. Masses of ions indicated in circles

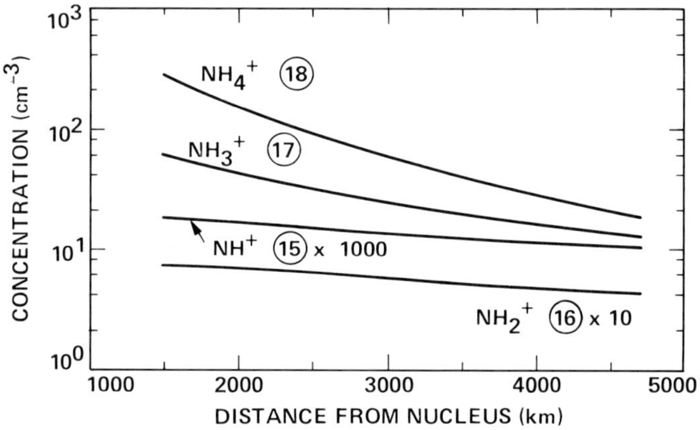

Fig. 9. Spatial distribution of the abundances of $NH_n{}^+$ from the "best fit" case. Masses of ions indicated in circles

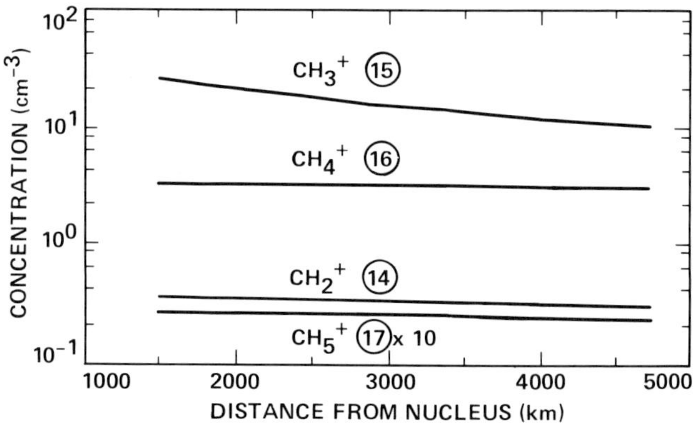

Fig. 10. Spatial distribution of the abundances of $CH_n{}^+$ from the "best fit" case. Masses of ions indicated in circles

proaches the value for the pure water coma (see Fig. 3). At the smallest distances, the contribution to $m/q = 18$ is almost equally from $NH_4{}^+$ and H_2O^+, but it is almost all H_2O^+ at the larger distances. While $NH_3{}^+$ is the main component of $m/q = 17$ within 2500 km, beyond this distance OH^+ dominates. At all distances, $CH_3{}^+$ dominates at $m/q = 15$. These results are of importance in trying to derive isotope ratios from ion mass spectra (see, for example, Eberhardt et al., 1986b).

With the chemistry illustrated in Figs. 2, 4, and 5 in mind, the model results in Figs. 8–10 illustrate two further points. (1) The most abundant hydride ion species are the ones that do not react with H_2O. (2) The photolytic source terms for all the ions scale with the spatial profile of the parent species (roughly r^{-2} in the inner coma). The concentrations of those ions destroyed predominantly through reactions with parent species vary very little with distance in the inner coma since the chemical loss factors also have a r^{-2} dependence. However the concentrations of those ions destroyed predominantly by electron recombination decrease with increasing distance from the nucleus more like r^{-1} as would be expected from the analysis of Ip (1986).

8. Model uncertainties

The results presented so far have been derived from calculations in which a simple physical model of the inner coma of comet Halley has been assumed: a constant temperature and a variable outflow velocity, which is the same for every species. Our calculations are quite sensitive to the chosen values for the electron temperature T_e as the abundances of the "terminal ions" are controlled by temperature-sensitive electron recombination. The value of T_e we adopted assumed that the ion temperatures measured by Giotto experiments inside the contact surface would be equivalent to the electron temperatures. We have used the temperature measured by the IMS experiment. A somewhat lower value for the ion temperature of 200 K was measured by the NMS experiment (Lämmerzahl et al., 1986). Model calculations using this temperature show a decrease in the ratio $m/q = 19/18$ of 6–20%, the largest change occurring near the contact surface. On the other hand, the results for 15/18 and 17/18 are reduced by $\sim 10\%$ at 1500 km and show minimal differences at the largest distances. These differences would not change our choice of the "best fit" values.

On the other hand, Marconi and Mendis (1984) calculate theoretically an increase in T_e from 4 K to 4000 K within 5000 km from the nucleus. In these computations the electron and ion temperatures are decoupled quite close to the nucleus. We were unable to match the Giotto 19/18 data between 1500 km and 5000 km using the variation in T_e predicted by Marconi and Mendis (1984). High values of T_e yield divergent results at larger distances near the contact surface, and low values give poor agreement at all distances. High values of T_e result in low model values for $m/q = 15/18$ and 17/18, and vice versa. Large variations in T_e with distance will not produce model results that are as constant with distance as is suggested by the Giotto HIS data. Limits on an inferred electron temperature might be derived from our model calculations. Our model results suggest that the electron temperature is approximately a few hundred degrees K and varies by no more than a few hundred degrees K throughout the inner coma.

The model ion chemistry inside the contact surface is close to being in local photochemical equilibrium so that the ion concentrations are relatively insensitive to the adopted velocities for the ions. The velocity profile we have used does affect the distribution of the neutral parent species and, therefore, indirectly the ion profiles. A comparison with the available data discussed earlier in this paper shows that the H_2O distribution is in good agreement with in situ measurements.

Our results are sensitive to uncertainties in several other parameters adopted for these calculations. The accuracy of the photodissociation and photoionization rates which drive the coma chemistry directly depends on the accuracy of the solar flux and cross-sections utilized. Except at Lyman alpha, the solar flux is not derived from the epoch of the Giotto encounter. Solar variability is possible in the extreme ultraviolet where photoionization occurs. Absorption cross-sections have been measured in the laboratory for various parent molecules, and the branching ratios for the various photoionization channels are becoming more available, but these laboratory data are still incomplete – especially for radicals.

Ion-molecule reactions control the ion densities in the inner coma and our model is very sensitive to this chemistry. Rate constants for these reactions have been measured by a variety of different techniques and good agreement generally exists among the various experimental groups for the reactions relevant to this work (Anicich and Huntress, 1986). Our results are also very sensitive to the values adopted for positive ion-electron recombination coefficients. Values derived from the merged beam experiments (McGowan and Mitchell, 1984) often differ significantly from the results of other techniques. The choice of values used for H_3O^+ and $NH_4{}^+$ recombination coefficients strongly impacts the derived NH_3 and CH_4 abundances. We have chosen to use the results from microwave experiments since these values result in model CH_4 and NH_3 abundances which are more consistent with constraints from the HERS experiment.

Calibration of the Giotto HIS data set is not yet completed, and the assignment of data channels to a particular m/q value is not yet completely resolved. Estimates of possible uncertainties are indicated in the figures of the relative m/q ratios.

Taken all together, the cumulative uncertainties in the input parameters and in the Giotto data set can increase the inferred relative production rates derived in this paper by as much as a factor of four, but might decrease the relative production rates by a similar amount. We have not yet pursued an exhaustive error analysis.

9. Discussion

The abundance of CH_4 and NH_3 in the coma are important indicators of the origin of comet Halley. Clues to the origin of comets can be found in the abundances of the elements and in the chemical distribution of each element (J. Geiss, this issue). Comparison of cometary bulk atomic ratios, isotope ratios and chemical composition with these same properties in other solar system and galactic objects may reveal commonalities that suggest possible scenarios for the origin of comets.

The distribution of cometary carbon atoms into CO, CO_2, CH_4, other hydrocarbons and organic species, and as amorphous solid, graphitic material or trapped radical species, is a critical indicator of the origin of comets. Carbon in the original solar nebula in equilibrium models is mainly in the form of CH_4 in the cooler parts of the nebula where the outer planets and comets are believed to have condensed. Lewis and Prinn (1980), however, have shown that this prediction must be modified if the cooling solar nebula gases did not reach thermodynamic equilibrium, and have shown that considerable CO can be retained in the gas phase relative to CH_4 under non-equilibrium conditions.

In the interstellar medium, CO is the major form of carbon observed. It is now well established that more that 10% of the whole carbon inventory may be in the form of gas-phase CO in some regions (Black and Willner, 1984). In some cool clouds, atomic carbon can be present in abundance approaching that of CO (Phillips et al., 1980). Knacke et al. (1985) have reported an upper limit to the gas-phase interstellar CH_4/CO abundance ratio of 10^{-2}. It is generally assumed that any missing elemental carbon is contained in interstellar grains. The detection of the appropriate infrared absorption and/or emission bands has revealed the existence of species in the condensed phase containing CH and/or CO bonds (Leger and Puget, 1984; Lacy et al., 1984; Allamandola et al., 1985). The amount of carbon appears to be less than the amount of gas-phase CO.

The CO abundance in Halley has been reported as 17–20% of water in rocket observations (Woods et al., 1986). The NMS experiment (Eberhardt et al., 1986a) yields a production rate for "native" CO relative to that for H_2O of ~ 0.07 with an equivalent production rate for a species easily decomposing to CO. The CO/H_2O value from the Giotto IMS (HERS) experiment is about 0.2 (Balsiger et al., 1986). As reported earlier in this paper, our reanalysis of the NMS data yields a value of 0.042 for $Q(CO_2)/Q(H_2O)$ so that the CO_2/CO ratio is ~ 0.5 or less. As pointed out by Prinn (private communication, 1987), the use of coma abundances to infer the composition of the nucleus volatiles implicity assumes that there is no differentiation among these species in the course of being released from the nucleus. Having made that assumption, we note that the rather low ratio of CH_4 to CO observed in Halley's coma, $CH_4/CO \sim 0.1$–0.3, would tend to argue for an interstellar origin, unless the solar nebula did not reach thermodynamic equilibrium and retained considerable amounts of CO from the pre-nebula cloud phase. On the other hand, the fact that the CH_4/CO ratio in Halley is higher than observed interstellar gas-phase values suggests that Halley may not contain pristine (i.e., totally unprocessed) interstellar material.

Just as for carbon, the distribution of nitrogen into NH_3, N_2 and various other compounds (such as HCN and organics) provides clues to comet origin and primordial nebula composition. Equilibrium models of the solar nebula yield NH_3 as the major form of nitrogen, whereas non-equilibrium models have significant amounts of N_2 (Lewis and Prinn, 1980). Models of the chemistry in dense interstellar clouds show N_2 and atomic N as the principal forms of nitrogen with smaller amounts of NH_3 and HCN (Prasad and Huntress, 1980). The Giotto IMS results show a low abundance for nitrogen in general, and if the present interpretation of the HIS results is correct, then NH_3 is the major form of Halley gas phase nitrogen. The upper limit for N_2 in Halley's coma is less than 2% of water (from $N_2/CO < 0.1$, Balsiger et al., 1986), so that $N_2/NH_3 < 1$. The ground-based results for HCN show its abundance to be less than 0.1% of water. It is important in any origin hypothesis to explain the low bulk N/O and N_2/NH_3 ratios. A potential scenario is that N_2 was a major form of nitrogen in the region of cometary origin, but was not retained in cometary condensates due to its volatility. This requires CO to be trapped preferentially relative to N_2 in amorphous ice, clathrates, or other solid phases at low temperature.

10. Conclusion

We have inferred the production rates for NH_3 and CH_4 relative to H_2O from the variation with distance of Giotto HIS ion mass spectra data. The profile for $m/q = 19/18$ is not well simulated by models with a pure water coma. A reasonable fit results for the ratio of production rates $Q(NH_3)/Q(H_2O) = 0.01-0.02$. The presence of a distinct peak at $m/q = 15$, taken together with HERS data at $m/q = 13-15$ at larger distances, indicates the presence of CH_4 in the coma gas. The Giotto HIS data for $m/q = 15/18$ is best reproduced with a model production rate ratio $Q(CH_4)/Q(H_2O) = 0.02$. The fit for methane also requires the presence of ammonia in coma gases so that the derived relative production rates are correlated. If a lower value of NH_3 is adopted, then less CH_4 is needed to fit the $m/q = 15/18$ data.

A number of consequences follow from these results. The partitioning of carbon between CH_4 and CO in Comet Halley is unlike speculated conditions in the primitive outer solar nebula or what is seen today in the atmospheres of the giant outer planets. On the other hand, the dissimilarity to interstellar observations suggests that Halley may not be composed of pristine interstellar material. Similarly, the inferred N/O abundance ratio implies a unique origin for Halley.

The work presented in this paper represents the beginning of an analysis procedure that will hopefully reveal the details of the composition of the coma of comet Halley. In this paper we have concentrated on only a limited portion of the Giotto IMS data, and on only H_2O, NH_3, and CH_4. There is a rich harvest yet to be made of the Giotto ion composition measurements.

Acknowledgements. This work represents one phase of research carried out at the Jet Propulsion Laboratory, California Institute of Technology, under contract to the National Aeronautics and Space Administration. Partial support also was received from NASA grant NSG-7376 to the California Institute of Technology. The Giotto IMS experiment was supported by the National Aeronautics and Space Administration, the German Bundesministerium für Forschung und Technologie, and the Swiss National Science Foundation.

References

Allamandola, L.J., Tielens, A.G.G.M., Barker, J.R.: 1985, *Astrophys. J.* **290**, L25

Anicich, V.G., Huntress, W.T.: 1986, *Astrophys. J. Suppl.* **62**, 553

Allen, M., Yung, Y.L., Waters, J.W.: 1981, *J. Geophys. Res.* **86**, 3617

Alge, E., Adams, N.G., Smith, D.: 1983, *J. Phys. B: At. Mol. Phys.* **16**, 1433

Balsiger, H., Altwegg, K., Bühler, F., Geiss, J., Ghielmetti, A.G., Goldstein, B.E., Goldstein, R., Huntress, W.T., Ip, W.-H., Lazarus, A.J., Meier, A., Neugebauer, M., Rettenmund, U., Rosenbauer, H., Schwenn, R., Sharp, R. D., Shelley, E.G., Ungstrup, E., Young, D.T.: 1986, *Nature* **321**, 330

Black, J.H., Willner, S.P.: 1984, *Astrophys. J.* **279**, 673

Despois, D., Crovisier, J., Bockelée-Morvan, D., Schraml, J., Forveille, T., Gérard, E.: 1986, *Astron. Astrophys.* **160**, L11

Drapatz, S., Larson, H.P., Davis, D.S.: 1986, *20th ESLAB Symp* ESA SP-250, Vol. 1, p. 347

Eberhardt, P., Krankowsky, D., Schulte, W., Dolder, U., Lämmerzahl, P., Berthelier, J.J., Woweries, J., Stubbemann, U., Hodges, R.R., Hoffman, J.H., Illiano, J.M.: 1986a, *20th ESLAB Symp* ESA SP-250, Vol. 1, p. 383

Eberhardt, P., Dolder, U., Schulte, W., Krankowsky, D., Lämmerzahl, P., Hoffman, J.H., Hodges, R.R., Berthelier, J.J., Illiano, J.M.: 1986b, *20th ESLAB Symp* ESA SP-250, Vol. 1, p. 539

Haser, L.: 1957, *Bull. Acad. Roy. Belgique* (*Cl. Sci.*) **18**, 740

Heppner, R.A., Walls, F.L., Armstrong, W.T., Dunn, G.H.: 1976, *Phys. Rev. A.* **13**, 1000

Ip, W.-H.: 1986, *Adv. Space Res.* **5**, 233

Knacke, R.F., Geballe, T.R., Noll, K.S., Tokunaga, A.T.: 1985, *Astrophys. J.* **298** L67

Krankowsky, D., Lämmerzahl, P., Herrwerth, I., Woweries, J., Eberhardt, P., Dolder, U., Herrmann, U., Schulte, W., Berthelier, J.J., Illiano, J.M., Hodges, R.R., Hoffman, J.H.: 1986, *Nature* **321**, 326

Lacy, J.H., Baas, F., Allamandola, L.J., Persson, S.E., McGregor, P.J., Lonsdale, C.J., Geballe, T.R., Van de Bult, C.E.P.: 1984, *Astrophys. J.* **276**, 533

Lämmerzahl, P., Krankowsky, D., Hodges, R.R., Stubbemann, U., Woweries, J., Herrwerth, I., Berthelier, J.J., Illiano, J.M., Eberhardt, P., Dolder, U., Schulter, W., Hoffman, J.-H.: 1986, *20th ESLAB Symp* ESA SP-250, Vol. 1, p. 179

Leger, A., Puget, J.L.: 1984, *Astron. Astrophys.* **137**, L5

Lewis, J.S., Prinn, R.G.: 1980, *Astrophys. J.* **238**, 357.

McGowan, J.W., Mitchell, J.B.A.: 1984, *Electron-Molecule Interactions and their Applications, Vol. 2*, ed. L.G. Christophorou, Academic Press, Orlando, p. 65

McGowan, J.W., Mul, P.M., D'Angelo, V.S., Mitchell, J.B.A., Defrance, P., Froelich, H.R.: 1979, *Phys. Rev. Lett.* **42**, 373; erratum, 1186

Moreels, G., Clairemidi, J., Parisot, J.P., Zucconi, J.M., Festou, M., Bertaux, J.L., Blamont, J.E., Herse, M., Krasnopolsky, V.A., Moroz, V.I., Parshev, V.A., Tkachuk, A. Yu., Gogoshev, M., Gogosheva, Ts., Werner, R., Spasov, S.: 1986, *20th ESLAB Symp* ESA SP-250, Vol. p. 451

Mount, G.H., Rottman, G.J.: 1983, *J. Geophys. Res.* **88**, 5403

Murad, E., Bochsler, P.: 1987, *Nature* **326**, 366

Phillips, T.G., Huggins, P.J., Kuiper, T.B.H., Miller, R.E.: 1980, *Astrophys. J.* **238**, L103

Prasad, S.S., Huntress, W.T.: 1980, *Astrophys. J. Suppl.* **43**, 1

Richtmeyer, R.D.: 1957, *Difference Methods for Initial Value Problems*, Interscience Publ., New York

Rottman, G.J., Barth, C.A., Thomas, R.J., Mount, G.H., Lawrence, G.M., Rusch, D.W., Sanders, R.W., Thomas, G.E., London, J.: 1982, *Geophys. Res. Lett.* **9**, 587

Schloerb, F.P., Kinzel, W.M., Swade, D.A., Irvine, W.M.: 1986, *Astrophys. J.* **310**, L55

Torr, M.R., Torr, D.G.: 1985, *J. Geophys. Res.* **90**, 6675

Winnberg, A., Ekelund, L., Ekelund, A.: 1987, *Astron. Astrophys.* **172**, 335

Woods, T.N., Feldman, P.D., Dymond, K.F.: 1986, *20th ESLAB Symp* ESA SP-250, Vol. 1, p. 431

Wyckoff, S., Wehinger, P.A., Spinrad, H., Belton, M.J.S.: 1986, *20th ESLAB Symp* ESA SP-250, Vol. 1, p. 311

Astron. Astrophys. 187, 513–518 (1987)

ASTRONOMY AND ASTROPHYSICS

Detection of parent molecules in comet P/Halley from the IKS-Vega experiment

V.I. Moroz[1], M. Combes[2], J.P. Bibring[3], N. Coron[4], J. Crovisier[2], T. Encrenaz[2], J.F. Crifo[4], N. Sanko[1], A.V. Grigoryev[1], D. Bockelée-Morvan[2], R. Gispert[4], Y.V. Nikolsky, C. Emerich[4], J.M. Lamarre[4], F. Rocard[3], V.A. Krasnopolsky[1], and T. Owen[5]

[1] Space Research Institute, Profsoyuznaya, 84/32, 117810 Moscow GSP-7, USSR
[2] Observatoire de Paris, F-92195 Meudon-Cedex, France
[3] Laboratoire René Bernas, BP No 1, F-91406 Orsay, France
[4] Laboratoire de Physique Stellaire et Planétaire, BP No 10, F-91370 Verrières-le-Buisson, France
[5] ESS Department, State University of New York, Stony Brook, NY 11794, USA

Received Feburary 3, accepted March 17, 1987

Summary. Several parent molecules have been detected in comet P/Halley using the 2.5–5 μm channel of the IKS experiment aboard Vega 1. The spectral resolving power was about 40 at 2.5 μm and 70 at 5 μm. The ν_3 band of H_2O has been observed at 2.7 μm, with an intensity corresponding to a production rate of about $10^{30}\,s^{-1}$. The ν_3 band of CO_2 has been detected for the first time at 4.3 μm; the derived CO_2/H_2O ratio is $2\,10^{-2}$. In addition, there are emission features between 3.3 μm and 3.7 μm, also observed for the first time in a comet, which are attributed to CH-bearing molecules. Their mixing ratio relative to H_2O is on the order of a few percent. The ν_1 band of H_2CO might be present at 3.6 μm, but this interpretation is tentative. A weaker feature also appears at 2.8 μm; it could be tentatively identified as the OH (1 – 0) band. The CO/H_2O ratio is lower than 0.2.

Key words: comets – comet P/Halley – infrared spectroscopy

1. Introduction

As pointed out by several authors over the past few years, the medium infrared range appears especially suitable for the study of parent molecules in cometary atmospheres. Indeed, most of the possible parent molecules exhibit strong vibration-rotation bands between about 2 and 15 μm. First calculations of synthetic spectra were performed by Encrenaz et al. (1982), Yamamoto (1982) and Crovisier and Encrenaz (1983). The main emission process is expected to be resonant fluorescence of the fundamental vibrational bands induced by the solar radiation field. Some molecules have been studied in detail: H_2O (Crovisier, 1984; Weaver and Mumma, 1984; Bockelée-Morvan and Crovisier, 1986) and CO (Crovisier and Le Bourlot, 1983). A review of synthetic spectra of a large number of molecules can be found in Crovisier (1987).

The infrared instrument IKS ("InfraKrasnoi Spectrometre", or infrared spectrometer) flown on-board the Vega space-probe was designed for the detection of emission bands of parent molecules and for a measurement of the size of the thermal emitting nuclear region and of its temperature. For this purpose, the instrument was equipped with three channels: two spectroscopic channels, operating at 2.5–5 μm and 6–12 μm respectively, and one "imaging channel" designed to modulate the infrared signal of the nucleus in two perpendicular directions. A description of the instrument can be found in Crifo (1981), Arduini et al. (1983) and Bibring et al. (1984).

In this paper, we present a summary of the results which have been obtained in the 2.5–5 μm spectroscopic channel. Preliminary results have been published in Combes et al. (1986a, b, c). A complete discussion of the data reduction and analysis will be found in Combes et al. (1987).

2. The experiment

The IKS telescope has a 13 cm diameter and a 1° field of view. The spectroscopic separation is achieved by two circular variable filters located on two tracks of an encoding wheel. The resolving power is about 40 at 2.5 μm and about 70 at 5 μm. The integrating time of each spectrum is 18 seconds, which corresponds to 128 rotations of the filter wheel. An internal black-body is used for in-flight calibration. The minimum detectable intensity in an 18s spectrum is about $5\,10^{-8}\,W\,\mu m^{-1}\,cm^{-2}\,sr^{-1}$ at 3 μm. The detectors are cooled by an active cryogenic system based on the expansion of pressurized gaseous nitrogen.

On Vega 1, the IKS spectroscopic experiment operated successfully before the flyby, as the distance of the probe to the nucleus ranged between about 250,000 km and 40,000 km. Near closest approach, the data flux was interrupted for 30 min, due to an erroneous command received by the instrument. Spectra were again recorded after the flyby, and were used for calibration. Before the flyby, 165 cometary spectra of 18 s integration each, have been recorded. On Vega 2, no result could be obtained, due to a failure of the cryogenic system.

Since the instrument was not cooled, the internal background signal caused by the instrument itself was, as expected, very strong (about 10^4 times the cometary signal). Because of the large apparent size of the source during the flyby, it was not possible to use sky chopping in order to eliminate the internal signal. Thus, the signal was only spectrally modulated by the rotation of the wheel, and the cometary signal was superimposed on the

Send offprint request to: T. Encrenaz

internal background signal. To first order, this internal signal was removed by taking the difference of two spectra, between a current spectrum and a reference "background" spectrum with no (or very little) cometary signal, obtained at the very beginning of the observing sequence. However, due to the temperature variation inside the instrument between the acquisition of the two spectra, there is still an internal signal component after this subtraction. This component has been removed by using the difference of spectra recorded at 200,000 km from the nucleus, before and after the flyby; this difference corresponds to the maximum instrumental temperature difference (about 1.5 °K) and contains no cometary signal.

Another effect had to be corrected also: a spectral modulation appeared in some individual spectra at wavelengths lower than 3 μm. This modulation is apparently correlated to small variations of the monitored detector temperature. In the preliminary reduction presented in Combes et al. (1986a, b, c), the spectra showing this modulation were simply removed. In the present analysis, we have been able to correct this effect by making appropriate differences between these cometary spectra, and other spectra recorded after flyby which showed a similar effect due to similar detector temperature variations. By this method we have been able to use almost all the cometary spectra which have been recorded during the observation sequence. A complete description of the procedure used for these corrections will be given in Combes et al. (1987).

Once the two corrections described above are applied, the resulting spectrum clearly shows cometary emissions, but it still contains a continuous component which is not likely of cometary origin.

We decided to remove the continuous component, even of cometary origin. Several procedures have been used: a Fourier-transform filtering method (Combes et al. 1986a), a subtraction of polynomials (Combes et al., 1986b, c). In the present study, we have used a polynomial of 4th order, which fits the spectrum at various wavelengths, where we assume there is no emission line from the comet. These wavelengths are: 2.32 μm, 2.34 μm, 2.50 μm, 2.51 μm, 2.53 μm, 2.87 μm, 3.04 μm, 3.18 μm, 3.53 μm, 3.67 μm, 4.06 μm, 4.39 μm and 4.86 μm. We have checked that a slight change in the choice of these numbers does not significantly affect the result. A more complete analysis of the subtraction procedure will be found in Combes et al. (1987).

3. Results and discussion

Figure 1 shows the cometary spectrum retrieved by the method described above. Cometary emissions are clearly visible, varying with the distance to the nucleus. Figure 1a is the difference of spectra recorded at nuclear distances of about 90,000 and 200,000 km respectively. The spectrum is the sum of 20 individual scans of 18 s integration each Fig. 1b is the difference of spectra recorded at nuclear distances of 42,000 km and 200,000 km respectively. The spectrum is the sum of 5 individual scans.

A strong emission is present at 2.7 μm and can be attributed to the H_2O ν_3 band. Another strong emission is visible at 4.3 μm, and can be identified as the CO_2 ν_3 band. Between 3.2 and 3.7 μm there is a series of emission features: we attribute them to a mixture of CH-bearing molecules which all exhibit a signature characteristic of the C—H bond, around 3.4 μm. The 3.6 μm emission feature might be attributed to H_2CO; but this identification is tentative, as will be discussed in Sect. 3.4.

In order to study the variation of the cometary signal as a function of the nuclear distance, and to derive production rates, the following procedure has been used. The series of spectra has been divided sequentially in 5 groups, each group corresponding to a mean nuclear distance D. The mean spectrum was calculated for each group. The first group corresponds to $D = 90{,}000$ km (Fig. 1a) and the last group corresponds to $D = 42{,}000$ km (Fig. 1b). The integrated flux (W cm^{-2} sr^{-1}) of each cometary feature was calculated in each case, and plotted as a function of $x = \left[\frac{1}{D} - \frac{1}{D_0}\right]$. Here, $D_0 = 200{,}000$ km, which corresponds to the beginning of the observing sequence.

If the cometary features are signatures of parent molecules, which have a constant production rate over the observation time, and if the lines are optically thin, we expect the signal to be proportional to D^{-1}; thus, the above curve is expected to be a straight line with a slope proportional to the production rate Q of the molecule. For each molecular species, we tried to find a

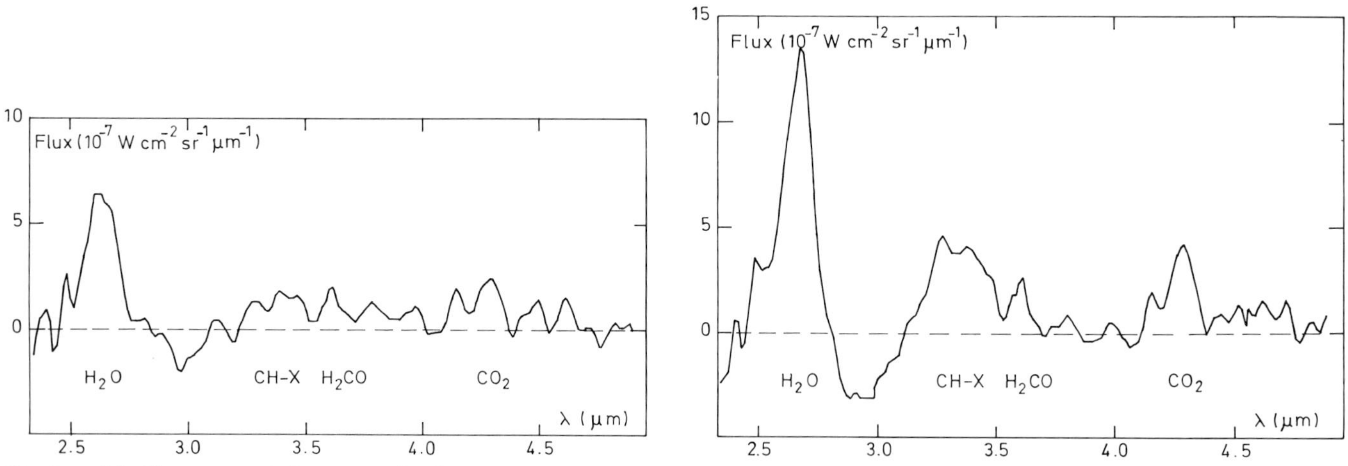

Fig. 1a and b. The spectrum of comet Halley between 2.5 and 5 μm, obtained from the difference of 2 spectra recorded for 2 different values of the probe-nucleus distance, D and D_0. $D_0 = 200{,}000$ km, which corresponds to the beginning of the observing sequence. **a** $D = 90{,}000$ km; **b** $D = 42{,}000$ km law

$$W = a + bx$$

a and b being defined by least mean squares. If the observed emission feature is entirely of cometary origin, we expect $a = 0$. When a value of a different from zero is found, we interpret it as a residual of the instrumental signal, and we take as the final value of the cometary integrated flux:

$$W_0 = W - a = bx$$

The production rate Q is then derived from the b coefficient, using the following formula:

$$Q = \frac{4\pi V\theta}{h\bar{v}g} \cdot b$$

where θ is the field of view angular diameter, V the gas expansion velocity and g the fluorescence rate. The fluorescence induced by the solar radiation field is the dominant process (Crovisier and Encrenaz, 1983). Thus the g factor is given by:

$$g = \frac{\Omega_\odot}{4} 3.08\,10^{-8}\,\bar{v}^2 S \frac{1}{\exp\left(\dfrac{hc\bar{v}}{kT_\odot}\right)} [\mathrm{s}^{-1}]$$

where $\Omega_\odot$ is the solar solid angle, $\bar{v}$ the central frequency of the band, S the strength of the band in $\mathrm{cm}^{-2}\,\mathrm{atm}^{-1}$ and $T_\odot$ the solar temperature.

Following recent determinations (e.g. Despois et al., 1986; Lämmerzahl et al., 1986; Krasnopolsky et al., 1986b) we assumed $V = 0.8\,\mathrm{km\,s}^{-1}$.

The variation of the cometary signal versus the distance to the nucleus is shown in Fig. 2 for H_2O, CO_2 and the CH—X molecules. The values of the production rates are given in Table 1.

3.1. H_2O

As mentioned previously, the excitation of cometary H_2O molecules was modelled before the observations by several authors (Crovisier and Encrenaz, 1983; Crovisier, 1984; Weaver and Mumma 1984; Bockelée-Morvan and Crovisier, 1986; Bockelée-Morvan, 1987).

As shown in Fig. 2, the linearity of the H_2O signal versus x is remarkable. This implies that the scalelength of H_2O is significantly larger than the largest observed coma diameter (5000 km). This is fully consistent with the estimate of $3.9\,10^4$ km for the H_2O scalelength derived from the NMS Giotto experiment (Krankowsky et al., 1986).

The H_2O lines are not optically thin, as shown by Bockelée-Morvan (1987), Krasnopolsky and Tkachuk (1986), and Bockelée-Morvan and Crovisier (1986, 1987). Using models where the radiative transfer is considered, it can be found that the attenuation of the ν_3 band due to saturation effects is by a factor as high as 3.4 in Fig. 1b. On the other hand, the 2.7 μm water band consists not only of the ν_3 band ($g = 2.57\,10^{-4}\,\mathrm{s}^{-1}$), but also includes the ν_1 band and other combination bands with smaller, but not entirely negligible, g-factors, given here for $r_h = 1$ AU (Crovisier, 1984):

$$\nu_1\colon g = 3.19\,10^{-5}\,\mathrm{s}^{-1}$$
$$(\nu_2 + \nu_3 - \nu_2)\colon g = 2.32\,10^{-5}\,\mathrm{s}^{-1}$$
$$(\nu_1 + \nu_3 - \nu_1)\colon g = 1.94\,10^{-5}\,\mathrm{s}^{-1}$$

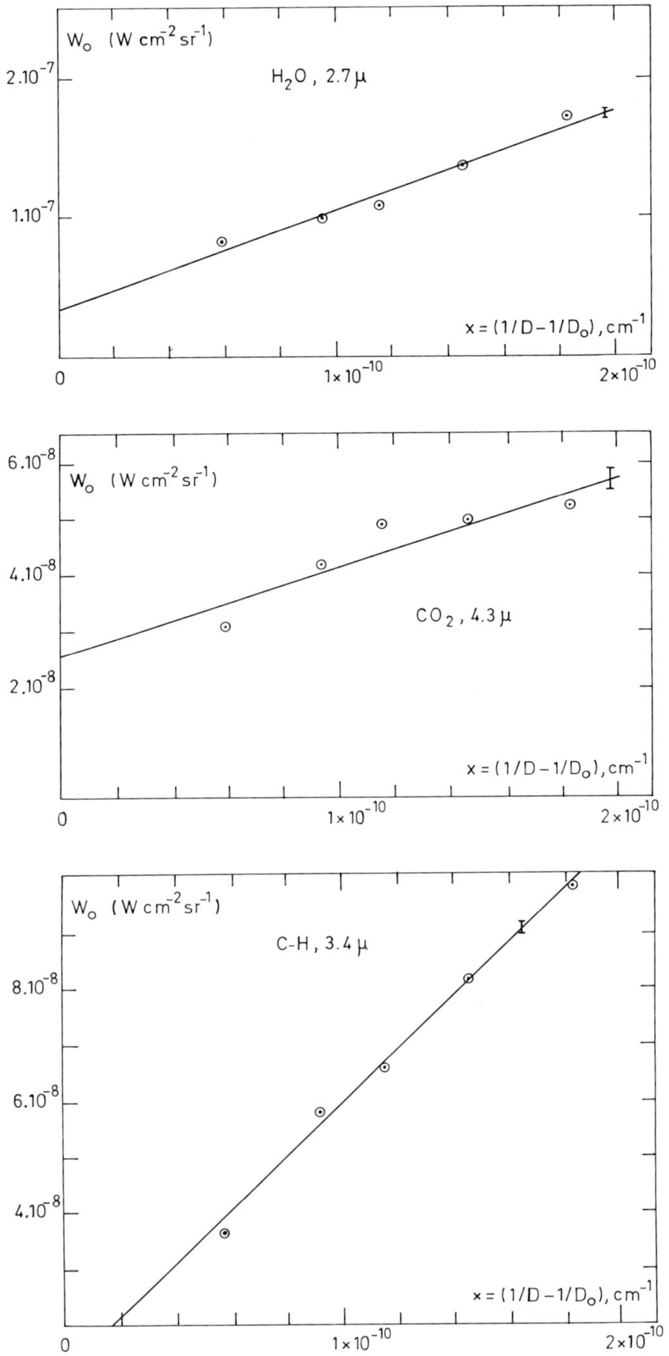

Fig. 2a–c. Variation of the integrated fluxes of the cometary emissions as a function of the parameter $x = \frac{1}{D} - \frac{1}{D_0}$ ($D_0 = 200{,}000$ km). **a** H_2O; **b** CO_2; **c** CH—X. A straight line is expected if the density of the molecule follows a R^{-2} law (R distance to the nucleus). The regression lines are drawn, as well as error bars corresponding to the standard deviations to the regression lines. Integration limits are 2.58–2.84 μm (H_2O), 4.17–4.37 μm (CO_2) and 3.23–3.50 μm (CH—X)

In the optically thin case, these bands are weaker than ν_3 by a factor 0.29, and are not, or not significantly, saturated. As a consequence, while ν_3 is attenuated by a factor 3.4, these weak bands have an intensity comparable to ν_3. Accounting for this effect, the derived production rate for water is $Q_{H_2O} \sim 10^{30}\,\mathrm{s}^{-1}$.

Table 1

Molecule	Wave-number (cm^{-1})	S ($cm^{-2}\,atm^{-1}$)	g^{a} (s^{-1})	$G = gh\gamma^{b}$ (W/mol)	b ($W\,sr^{-1}\,cm^{-1}$)	Q (s^{-1})
H_2O	3755	190	$4.5\,10^{-4}$ [c]	$3.4\,10^{-23}$ [c]	700	10^{30} [d]
[C—H]	2940	1000	$2.1\,10^{-3}$	$1.2\,10^{-22}$	500	$7\,10^{28}$
CO_2	2349	2500	$4.5\,10^{-3}$	$2.1\,10^{-22}$	160	$2\,10^{28}$ [d]

[a] At 1 AU
[b] At 0.8 AU
[c] For γ_3 only, optically thin case;
[d] Corrected for optical depth effects

This value can be compared with other results of the Vega and Giotto flyby missions. From the three-channel spectrometer TKS on Vega 2, Krasnopolsky et al. (1986a, b) have found, three days after our measurement, a similar result from both the 1.38 μm H_2O band and from the OH emission at 1.6–1.9 μm and in the near UV, when assuming that all OH comes from H_2O dissociation.

On Giotto, 7 days later, at a time corresponding to a fainter activity, from the neutral mass spectrometer NMS, the derived value $Q_{(H_2O)}$ is $5.5\,10^{29}\,s^{-1}$ (Krankowsky et al., 1986).

Previous ground-based infrared observations of comet Halley have also led to the identification of H_2O. In particular, emission lines of the H_2O ν_3 band have been observed from the Kuiper Airborne Observatory before and after perihelion (Mumma et al., 1986; Weaver et al., 1986). The derived production rates seem to be in good overall agreement with the present results, taking into account the variations of the heliocentric distance, and the outgassing variability of the comet.

3.2. CO_2

The CO_2 emission feature shown in Fig. 1 is the first detection of neutral carbon dioxide in a comet.

Although not as accurately defined as for H_2O, the linear dependence of the CO_2 signal versus x is clear. It is interesting to note that the small emission feature which appears at 4.13 μm in the wing of the CO_2 band, shows no variation with the distance to the nucleus. For this reason we conclude that this feature is probably not of cometary origin and we remove its contribution in the calculation of the CO_2 integrated flux.

By taking into account the opacity effect in a radiative transfer calculation, we derive a production rate of $2\,10^{28}$ for CO_2 (Bockelée-Morvan and Crovisier, 1986), i.e. 2% of the H_2O production rate.

The presence of CO_2 was expected in comets before the Vega flyby, especially because of the presence of CO_2^+, but this parent molecule had never been directly identified before the present experiment. A CO_2 production rate of $2\,10^{28}$ is in good agreement with the value of $2\,10^{28}$ reported by Krankowsky et al. (1986) from the M = 44 signal of the Giotto mass spectrometer experiment. But, as pointed out by these authors, this latter value is rather an upper limit, if part of the M = 44 signal in the Giotto mass spectrometer comes from another molecule of same mass (CS, C_3H_8 ...).

3.3. *Hydrocarbons*

Figure 2c shows the variation of the 3.2–3.5 μm signal as a function of x. As in the case of H_2O, the variation of the integrated flux is very well represented by a linear law. We thus conclude that the substance responsible for this emission is directly outgassed from the nucleus.

The emission feature shown in Fig. 1 between 3.2 and 3.6 μm could actually extend up to 4 microns, raising the continuum under the CO_2 emission. This feature, observed for the first time in a comet, is too broad to be attributed to a single molecule. As already pointed out Combes et al, 1986a) the whole band could be attributed to a mixture of various hydrocarbons, which all exhibit around 3.3 μm–3.4 μm a spectral signature characteristic of the C—H bond. Our spectra show several peaks at 3.28 μm, 3.40 μm, and 3.63 μm (see Fig. 1). The 3.4 μm signature is characteristic of the CH stretching frequencies (CH_3, CH_2) of saturated hydrocarbons (alkanes). In contrast, the 3.3 μm signature is characteristic of the CH stretching of unsaturated hydrocarbons, in particular alkenes (—CH=CH—, —CH=CH_2) and aromatic compounds (=C—H) (Bellamy, 1975). It is not possible to discriminate between the gaseous or solid nature of the emitter; if we assume that the emission is due to a gas, we can reasonably state that both saturated and unsaturated hydrocarbons seem to be present as parent molecules in Halley's Comet. Their production rate can be estimated by taking a "mean" band strength of about $1000\,cm^{-2}\,atm^{-1}$ (within a factor of 3). By comparison with H_2O and CO_2, we derive a "mean" production rate of about $7\,10^{28}\,s^{-1}$, for all the hydrocarbons (unsaturated and saturated). If we notice (Finkel, 1966) that the strength of the 3.4 μm band increases linearly with the number of CH_2 groups in the molecule ($S = 1000\,cm^{-2}\,atm^{-1}$ corresponding to 3 CH_2 groups) we derive a total number of carbon atoms of about $2\,10^{29}\,s^{-1}$; this can be compared with the value of $10^{29}\,s^{-1}$ found for the CH production rate (Krasnopolsky et al., 1986a).

There is additional evidence for the presence of carbonaceous compounds in the immediate surroundings of the nucleus: the high temperature of the nuclear region measured by IKS (Combes et al., 1986a; Lamarre et al., 1986), the low surface albedo (Sagdeev et al., 1986), the indication of a source of atomic carbon given by the IMS experiment (Balsiger et al., 1986) and the PUMA dust analyses (Kissel et al., 1986) which demonstrate the presence of a high proportion of grains containing predominantly H, C, O and N atoms, up to more than 100,000 km from the nucleus. Their origin might be the following: mixtures of ices, such as

$H_2O + CO_2 + NH_3$, are transformed under laboratory irradiation into molecular complexes with spectral signatures in the 3.2–3.4 μm region. These species appear after irradiation either by UV photons (Greenberg, 1982) or high energy particles (Rocard, 1986). In particular, it has been shown that dark residues with polymeric structure are produced by ion bombardment of frozen volatiles containing H and C atoms. In order to account for the observed spectral features, such organic matter could come from submicron grains or mantles around icy or refractory cores. In that case, the mechanism required would operate in a way similar to that invoked for interpreting infrared absorption features observed in the interstellar medium (Butchart et al., 1986).

It is interesting to note that the 3–4 μm range is also observable from the ground. An attempt was made in December 1985 and no emission feature was detected (Danks et al., 1986). After the Vega fly-by new attempts were made, late March and April 1986. An emission feature centered at 3.4 μm was reported by Allen and Wickramasinghe (1986), Knacke et al. (1986), Baas et al. (1986) and Danks et al. (1986). All authors observed a broad emission centered at 3.36–3.37 μm. Additional emissions were reported at 3.5 μm and 3.28 μm. All these data are in good agreement with the IKS results.

3.4. Other emission features

At 3.6 μm there is an emission feature which seems to be well above the noise level. However, its dependence as a function of x is not clearly represented by a linear law, so we cannot definitely conclude on its cometary origin. If this feature is actually due to the comet, it could be interpreted as H_2CO which has its ν_1 band centered at 3.59 microns. The intensity of the band would correspond to a production rate of about $5\,10^{28}$, i.e. 5 percent of the H_2O production rate. But again, this identification is tentative.

At 2.8 μm there is a weak emission, especially in Fig. 1b, superimposed on the broader absorption feature tentatively attributed to H_2O ice. This emission could be due to OH. Its intensity appears to be too strong to be due to fluorescence, and we would have to admit that the emission comes from the relaxation of OH molecules produced in vibrationally excited states from H_2O photolysis (as also suggested by the presence of $\Delta v = 2$ OH bands in the TKS near-IR spectra; Krasnopolsky et al., 1986a).

In preliminary analyses of our data (Combes et al., 1986b, c) we mentioned that CO could be marginally present at 4.7 μm. However, the improved data analysis used in the present study does not result in a definite identification of CO. We just derive from our spectra an upper limit of $2\,10^{29}$ for the CO production rate, which corresponds to an upper limit of 20% for the CO mixing ratio. This upper limit is consistent with the mixing ratio of 10–20% estimated from the IMS Giotto spectrometer (Balsiger et al., 1986) and from the IUE observations (Festou et al., 1986).

4. Conclusion

The main results of the present analysis can be summarized as follows:

1) New parent molecules are detected for the first time in a comet: carbon dioxide and saturated hydrocarbons. Other possible candidates are unsaturated hydrocarbons. In addition, the strong signature of the H_2O ν_3 band is observed at 2.7 μm. H_2CO and OH are tentatively identified at 3.6 μm and 2.8 μm respectively.

2) The production rates derived for the parent molecules are $10^{30}\,s^{-1}$ for H_2O, $2\,10^{28}\,s^{-1}$ for CO_2. An estimate of $7\,10^{28}\,s^{-1}$ is derived for hydrocarbons. The production rate of CO is lower than $2\,10^{29}\,s^{-1}$.

3) The column densities of H_2O, CO_2 and the hydrocarbons as a function of nuclear distance are in good agreement with the D^{-1} law. This is consistent with the distribution in R^{-2} expected for parent molecules in the vicinity of the nucleus.

The evidence for carbonaceous material is one of the major results of comet Halley's exploration. The present study provides part of this evidence with the detection of the signature of carbon species in the 3.4 μm region and of CO_2. This carbon material could be either hydrocarbons in the gaseous phase or submicron grains. If it is in the gaseous phase, the total amount of carbon atoms in hydrocarbons and CO_2 should be close to $2\,10^{29}\,s^{-1}$, or about 20% of the H_2O production rate.

Acknowledgements. We are grateful to the French and Soviet technical team members who have made this experiment possible, and who have developed a highly efficient cooperative effort. We thank A.V. Diachkov and H.G. Havenson for their help. Financial support was provided by the Centre National d'Etudes Spatiales.

References

Arduini, M., Bibring, J.P., Cazes, S., Combes, M., Coron, N., Crifo, J.F., Encrenaz, T., Gispert, R., Harduin, D., Lamarre, J.M.: 1983. *Adv. Space Res.* **2**, 113

Baas, F., Geballe, T., Walther, D.: 1986, *Astrophys. J.* **311**, L97

Balsiger, H., Altwegg, K., Bühler, F., Geiss, J., Ghielmetti, A.G., Goldstein, B.E., Goldstein, R., Huntress, W.T., Ip., W.H., Lazarus, A.J., Meier, A., Neugebauer, M., Rettenmund, U., Rosenbauer, H., Schwenn, R., Sharp, R.D., Shelley, E.G., Ungstrup, E., Young, D.T.: 1986, *Nature* **321**, 330

Bellamy, L.J.: 1975, *The Infrared Spectra of Complex Molecules*, Chapman and Hall, London

Bibring, J.P., Cazes, S., Charra, J., Combes, M., Coron, N., Cougrand, B., Crifo, J.F., Crovisier, J., Emerich, C., Encrenaz, T., Gispert, R., Gondet, B., Guyot, G., Harduin, D., Lamarre, J.M., Levanti, G., Maurel, C., Parisot, D., Rocard, F., Salvetat, P., Soufflot, A.: 1984, *Adv. Space Res.* **4**, 273

Bockelée-Morvan, D., Crovisier, J.: 1986, *in "Asteroids, Comets, Meteors II"*, eds. C.I. Lagerkvist et al., Uppsala University, p. 279

Bockelée-Morvan, D., Crovisier, J.: 1987, *Astron. Astrophys.* (in press)

Bockelée-Morvan, D: 1987, *Astron. Astrophys.* (in press).

Butchart, I., McFadzean, A.D., Whittet, D.C.B., Geballe, T.R., Greenberg, J.M.: 1986, *Astron. Astrophys.* **154**, L5

Combes, M., Moroz, V.I., Crifo, J.F., Lamarre, J.M., Charra, J., Sanko, N.F., Soufflot, A., Bibring, J.P., Cazes, S., Coron, N., Crovisier, J., Emerich, C., Encrenaz, T., Gispert, R., Grigoryev, A.V., Guyot, G., Krasnopolsky, V.A., Nilkolsky, Y.V., Rocard, F.: 1986a, *Nature* **321**, 266

Combes, M., Moroz, V., Crifo, J.F., Bibring, J.P., Coron, N., Crovisier, J., Encrenaz, T., Sanko, N., Grigoriev, A., Bockelée-Morvan, D., Gispert, R., Emerich, E., Lamarre, J.M., Rocard, F., Krasnopolsky, V., Owen, T.: 1986b, *Adv. Space Res.* **5**, 127

Combes, M., Moroz, V., Crifo, J.F., Bibring, J.P., Coron, N., Crovisier, J., Encrenaz, T., Sanko, N., Grigoriev, A., Bockelée-Morvan, D., Gispert, R., Emerich, C., Lamarre, J.M., Rocard, F., Krasnopolsky, V., Owen, T.: 1986c, 20th ESLAB Symp., ESA SP-250, Vol. I, p. 353

Combes, M., Moroz, V.I., Crovisier, J., Encrenaz, T., Bibring, J.P., Grigoryev, A.V., Sanko, N.F., Coron, N., Crifo, J.F., Gispert, R., Bockelée-Morvan, D., Nikolsky, Y.V., Krasnopolsky, V.A., Owen, T., Emerich, C., Lamarre, J.M., Rocard, F.: 1987, (in preparation)

Crifo, J.F.: 1981, in *The Solar System and its Exploration*, ESA SP-164, 229

Crovisier, J.: 1984, *Astron. Astrophys.* **130**,

Crovisier, J.: 1987, *Astron. Astrophys. Suppl.* **68**, 223

Crovisier, J., Le Bourlot: 1983, *Astron. Astrophys.* **130**, 61

Danks, A., Encrenaz, T., Bouchet, P., Le Bertre, T., Chalabaev, A., Epchtein, N.: 1986. 20th ESLAB Symp., ESA SP-250 Vol. 3, p. 103

Despois, D., Forveille, T., Schraml, J., Bockelée-Morvan, D., Crovisier, J., Gérard, E.: 1986, *Astron. Astrophys.* **160**, L11

Encrenaz, T., Crovisier, J., Combes, M., Crifo, J.F.: 1982, *Icarus* **51**, 660

Festou, M.C., Feldman, P.D., A'Hearn, M.E., Arpigny, C., Cosmovici, C.B., Danks, A.C., McFadden, L.A., Gilmozzi, R., Patriarchi, P., Tozzi, G.P., Wallis, M.K., Weaver, H.A.: 1986, *Nature* **321**, 361

Finkel, A.G.: 1966, *Opt. Spectr.* **20**, 432

Greenberg, M.: 1982, Proc. *International Conference on Cometary Exploration II*, Budapest, T.I. Gombosi, ed., p. 23

Keller, H.U., Arpigny, C., Barbieri, C., Bonnet, R.M., Cazes, S., Coradini, M., Cosmovici, C.B., Delamere, W.A., Huebner, W.F., Hughes, D.W., Jamar, C., Malaise, D., Reitsema, H.J., Schmidt, H.V., Schmidt, W.K.H., Seige, P., Whipple, F.L., Wilheim, K.: 1986, *Nature* **321**, 320

Kissel, J., Sagdeev, R.Z., Bertaux, J.L., Angarov, V.N., Audouze, J., Blamont, J.E., Büchler, K., Evlanov, E.N., Fechtig, H., Fomenkova, M.N., von Hoerner, H., Inogamov, N.A., Khromov, V.N., Knabe, W., Krueger, F.R., Langevin, Y., Leonas, V.B., Levasseur-Regourd, A.C., Managadze, G.G., Podkolzin, S.N., Shapiro, V.P., Tabaldyev, S.R., Zubkov, B.: 1986, *Nature* **321**, 280

Knacke, R.F., Brooke, T.Y, Joyce, R.R.: 1986, *Astrophys. J.* **310**, L49

Krankowsky, D., Lammerzahl, P., Herrwerth, I., Woweries, J., Eberhardt, P., Dolder, U., Herrmann, U., Schulte, W., Berthelier, J.J., Illiano, J.M., Hodges, R.R., Hoffman, J.H.: 1986, *Nature* **321**, 326

Krasnopolsky, V.A., Gogoshev, M., Moreels, G., Moroz, V.I., Kryoko, A.A., Barke, V.V., Segalev, V.S., Parshev, V.A., Sanko, N.F., Tomashova, G.V., Tkachuk, A.Yu. Troshin, V.S., Novikov, B.S., Perminov, V.G., Sulakov, I.I., Fedorov, O.S., Gogosheva, Ts., Sargoichev, S., Palasov, K., Geogiev, A., Nedkov, I., Kanev, K., Clairemidi, J., Vincent, M., et al.: 1986b, *Adv. Space Res.* **5**, 143

Krasnopolsky, V.A., Gogoshev, M., Moreels, G., Moroz, V.I., Krysko, A.A., Gogosheva, Ts. Palasov, K., Sargoichev, S., Clairemidi, J., Vincent, M., Bertaux, J.L., Blamont, J.E., Troshin, V.S., Valnicek, B.: 1986a, *Nature* **321**, 269

Krasnopolsky, V.A., Tkachuk, A.Yu.: 1986, *Cosmic Research* (in press)

Lamarre, J.M., Emerich, C., Moroz, V., Combes, M., Crifo, J.F., Sanko, N.F., Rocard, F., Gispert, R., Nikolsky, Yu. V., Bibring, J.P., Coron, N., Crovisier, J., Encrenaz, T.: 1986. *Adv. Space Res.* **5**, 123

Lämmerzahl, P., Krankowsky, D., Hodges, R.R., Stubbemann, U., Woweries, J., Herrwerth, I., Berthelier, J.J., Illiano, J.M., Eberhardt, P., Dolder, U., Schulte, W., Hoffmann, J.H.: 1986, 20th ESLAB Symp. ESA SP-250, Vol. 1, p. 179

Mumma, M.J., Weaver, H.A., Larson, H.P., Davis, D.S., Williams, M.: 1986, *Science* **232**, 1523

Rocard, F.: 1986, PhD Thesis, Université Paris XI-Orsay

Sagdeev, R.Z., Szabo, F., Avanesov, G., Cruvellier, P., Szabo, L., Szego, K., Abergel, A., Balazs, A., Barinov, I.V., Bertaux, J.L., Blamont, J., Detaille, M., Demarelis, E., Dul'nev, G.N., Endroczy, G., Gardos, M., Kanyo, M., Kostenko, V.I., Krasikov, V.A., Nguyen-Trong, T., Nyitrai, Z., Reny, I., Rusznyak, P., Shamis, V.A., Smith, B. et al.: 1986, *Nature* **321**, 262

Weaver, H.A., Mumma, M.J.: 1984, *Astrophys. J.* **276**, 782

Weaver, H.A., Mumma, M.J., Larson, H.P., Davis, D.S.: 1986, *Nature* **324**, 441

Wickramasinghe, D., Allen, D.: 1986, *Nature* **323**, 44

Yamamoto, T.: 1982, *Astron. Astrophys.* **109**, 326

Astron. Astrophys. 187, 519–522 (1987)

ASTRONOMY AND ASTROPHYSICS

Detection of a new emission band at 2.8 μm in comet P/Halley

A. T. Tokunaga*, T. Nagata**, and R. G. Smith**,***

Institute for Astronomy, University of Hawaii, 2680 Woodlawn Drive, Honolulu, HI 96822, USA

Received February 18, accepted May 18, 1987

Summary. A 2.8- to 3.7-μm spectrum of comet Halley taken in May 1986 is presented. The spectrum shows (1) the 3.36-μm emission feature that most likely arises from carbonaceous material, (2) a previously unknown emission band at 2.8–2.9 μm, and (3) no evidence for the 3-μm water-ice band either in reflection or absorption. The 2.8- to 2.9-μm emission feature does not arise from infrared fluorescence from water or OH. Since LTE emission from water is unlikely, the nature of this feature is uncertain.

Key words: comets – infrared radiation – infrared spectroscopy

1. Introduction

Infrared observations of comet P/Halley at thermal wavelengths (wavelengths longer than 2.5 μm) permit the study of neutral "parent" molecules such as methane and the solid state components of the nucleus such as water ice (Yamamoto, 1982; Crovisier and Encrenaz, 1983; Weaver and Mumma, 1984; Hanner, 1981). During the recent apparition of comet Halley, some of these cometary constituents, such as water (Mumma et al., 1986) and CO_2 (Combes et al., 1986a, b), were observed for the first time. One of the most important but unexpected discoveries was that of a new emission band at 3.4 μm, which most likely arises from hydrocarbon material (Combes et al., 1986a, b; Wickramasinghe and Allen, 1986; Knacke et al., 1986; Baas et al., 1986; Danks et al., 1986).

Our original motivation for obtaining a 3-μm spectrum of Halley was to search for the water-ice band in absorption against the scattered light while the comet was at a large heliocentric distance. The detection of water ice in comets remains doubtful at the present time, since current observations are sparse and have a low signal-to-noise ratio. The evidence for water ice is summarized by Tokunaga et al. (1986b). The existence of the 3.36 μm emission feature also complicates the interpretation of previous data upon which the evidence for water ice rests, since such a feature might cause the appearance of a spectral minimum at 3 μm that is coincident with the water-ice band. Although we failed to detect the water-ice band, we found a new emission band in comet Halley at a wavelength <2.9 μm. The 3.4-μm band was also seen in emission at a heliocentric distance of 1.9 AU, in agreement with previous work (Baas et al., 1986).

Send offprint requests to: A. T. Tokunaga

* Staff Astronomer, NASA Infrared Telescope Facility

** Visiting Astronomer, Infrared Telescope Facility, which is operated by the University of Hawaii, under contract with the National Aeronautics and Space Administration

*** Current address: Max-Planck-Institut für Extraterrestrische Physik, D-8046 Garching, Federal Republic of Germany

2. Observations

The observations were made with the Cooled-Grating Array Spectrometer (CGAS) at the NASA Infrared Telescope Facility at Mauna Kea, Hawaii. Comet Halley was observed from 19–21 May 1986 UT. During this time the comet moved from 1.89 to 1.92 AU. A 75 line mm^{-1} grating blazed at 2.5 μm was used to achieve a resolving power of 160 to 210 at 2.85 to 3.70 μm. The entrance aperture was 2.7 arcsec in diameter. The detector array employed has 32 detectors and was constructed by Cincinnati Electronics (model IMS-3201).

To remove the background emission from the sky and telescope, positions 30″ north and south from the nucleus were subtracted. The total integration time, comet and sky combined, was 2464, 3600, and 4032 seconds on 19, 20, and 21 May, respectively. The standard star used for atmospheric correction was SAO 137575 (spectral type A 3) and was located near the comet. This star was measured often enough (alternating between the star and the comet) that the airmass difference between it and comet Halley was <0.1 airmass. For such small airmass differences, the ratio of the comet by the standard star with correction for atmospheric extinction was sufficient to remove the telluric absorption bands completely. To demonstrate this, we show in Fig. 1 the ratio of individual standard star scans. These ratios do not show any correlation with atmospheric bands except for the strong 3.3-μm telluric methane band where the noise is somewhat higher.

The available time allowed observations in only one grating position each night, and the spectra for each grating position are shown separately in Fig. 2. A combined spectrum for 19 and 20 May is also shown. These spectra were averaged in the region of spectral overlap. The spectrum for 21 May was not added to the others because on this day the 3.4-μm feature does not appear to be present as it was on 19 and 20 May. However, the addition of spectra from all three nights still shows that 3.4-μm feature. The absolute flux density scale was established from observations of Alpha Leo and has an uncertainty of ~15%.

3. Discussion

The combined May 19 and 20 spectra show two emission features superimposed on the scattered light continuum: the 3.36-μm emission feature observed previously and a new emission feature at wavelengths <2.9 μm. Other apparent features, such as those at

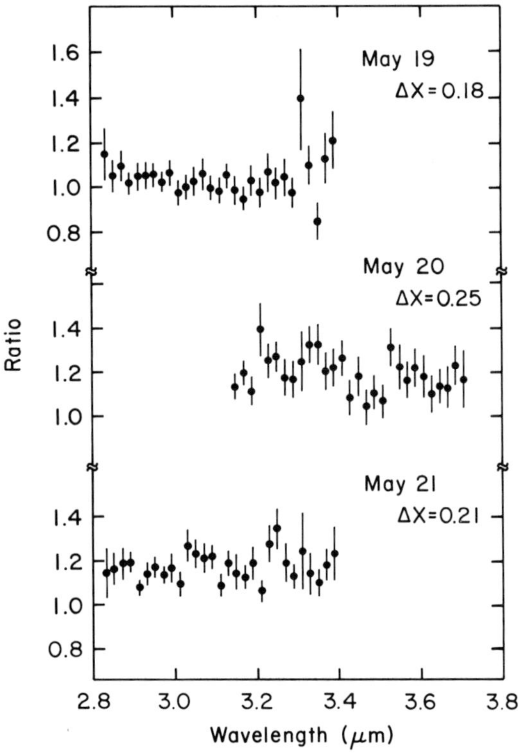

Fig. 1. Ratio of SAO 137575 by itself at different airmasses (the difference denoted by "ΔX"). The increased noise at 3.3 μm on May 19 is a result of the strong telluric methane band. Ratios of single scans corrected for the airmass difference are shown

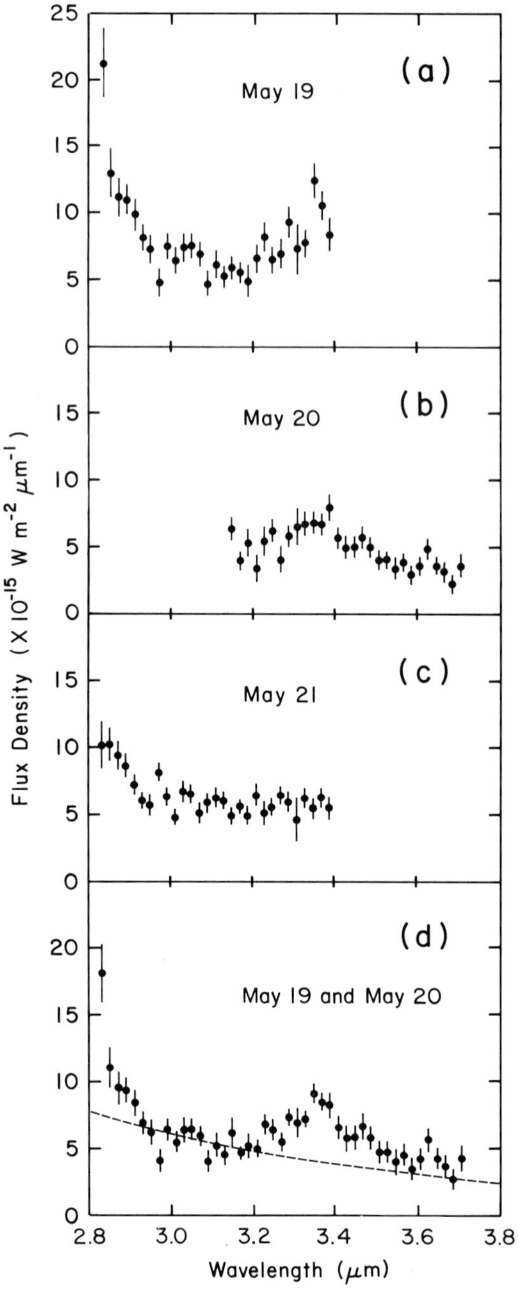

Fig. 2. a–c Spectra of comet Halley at single grating positions for May 19, 20, and 21. **d** Summed spectra from 19 and 20 May. The dashed line indicates the spectral shape of the reflected solar spectrum

2.95 and 3.63 μm, are not significant, as they do not appear consistently on individual spectra. In any case they are not significant by more than 2 σ.

Figure 3 shows the overall scattered light and thermal emission spectrum of the comet near the dates of our observation. This data is part of an ongoing comet Halley monitoring program described by Tokunaga et al. (1986a), and the data shown in the figure are taken from Tokunaga (1986). From this figure, we infer that the infrared continuum at 3 μm is primarily scattered solar radiation.

3.1. The 3.36-μm band

The possible nature of this band has been discussed by Combes et al. (1986a, b), Wickramasinghe and Allen (1986), Knacke et al. (1986), Baas et al. (1986), and Danks et al. (1986). As discussed by these authors, this band most likely originates from a hydrocarbon or carbonaceous material, but the existing data set is insufficient to identify the material.

We find the peak of the band to be 3.36 ± 0.02 μm, in agreement with previous authors. The intensity of the band relative to the continuum is consistent with the Baas et al. spectrum obtained only three days later than ours. Also, when corrected for aperture, the absolute level of the emission band is consistent within the uncertainties. The signal-to-noise ratio in our spectrum is too low to confirm any of the weaker features reported by Combes et al. (1986a, b), Knacke et al. (1986), Baas et al. (1986), or Danks et al. (1986).

As noted above, the 3.36-μm feature seems to have decreased in intensity substantially on May 21 (compare Fig. 2a to that of 2c). Since there is no indication from the standard star data of an instrumental or observational problem and the emission at 2.8–2.9 μm is present on both May 19 and 21, we conclude that this variation is real. Some evidence for intensity variations in the 3.36-μm feature can be seen in the series of spectra from 30 March to 1 April 1986 obtained by Wickramasinghe and Allen (1986), although no variation was seen in the 26–29 March 1986 spectra of Knacke et al. (1986).

Variations in the silicate emission band at 10 μm were observed by Gehrz and Ney (1986). They find little sign of the silicate band on 25 March 1986, although it is strong on 28 March. From the concurrent variations in the overall grain albedo, they surmise that the darker grains, perhaps the carbonaceous grains, are more

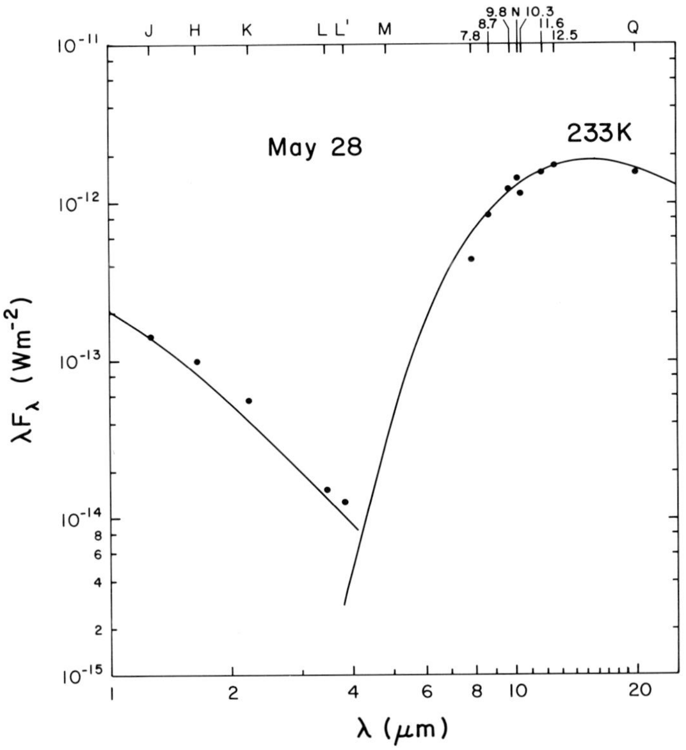

Fig. 3. 1–20 μm spectrum of comet Halley. The solid line from 1–4 μm is the normalized spectral shape of the Sun. The 3 μm spectral region is primarily scattered light from the Sun. The solid line from 4–20 μm is the estimated thermal emission spectrum obtained by fitting a blackbody to the 10-μm observations. The blackbody color temperature is 233 K

abundant when the silicate band is weaker. Such a process may give rise to variations in the 3.36-μm emission band from day to day and might be anticorrelated with the strength of the silicate emission. Whatever the cause, there is evidence in our data for fluctuating abundance of the material giving rise to the 3.36-μm feature.

Figure 3 shows that the characteristic dust temperature in the coma was ~230 K in late May, and Baas et al. (1986) argue that it is not likely that the 3.36-μm emission feature arises from hot dust. If so, infrared fluorescence is the most likely mechanism giving rise to the 3.36-μm emission band. However, the calculation for the hot dust emission depends on the grain temperature, the band strength, and the grain abundance – all of which are highly uncertain as discussed by Knacke et al. (1986).

Of the comets for which 3-μm spectra exist, comet Halley is thus far the only comet for which the 3.36-μm feature has been seen. Comet West (Oishi et al., 1978) and comet IRAS-Araki-Alcock (Hanner et al., 1985) both show featureless spectra at 3 μm. Is the 3.36-μm feature common or rare in comets? Observations of more comets are clearly necessary.

3.2. The short-wavelength emission feature

Part of an emission feature was observed from 2.8 to 2.9 μm. We are confident about the reality of this feature because this feature was seen on two separate days and we can rule out poor atmospheric cancellation as demonstrated by Fig. 1. In addition, Weaver et al. (1986) detected what appears to be broad emission features at 2.6–2.7 μm in comet Halley, and the long-wavelength feature may join the feature we see at 2.8–2.9-μm. The bands observed by Weaver et al. cannot be attributed to instrumental or observational effects and therefore are most likely real (H. P. Larson and M. J. Mumma, private communication). There is uncertainty about the identification of the 2.8- to 2.9-μm feature, since only part of it was observed; indeed we do not know what the width and central wavelength of this feature is. There is also some indication of this feature on the long-wavelength wing of the 2.7-μm H_2O emission band observed by the IKS-Vega 1 experiment (Combes et al., 1986a).

We can rule out water and OH fluorescence emission as the source of this feature. Water emission lines were observed in comet Halley by Mumma et al. (1986) and Weaver et al. (1986), but the emission arose from infrared fluorescence and was confined to the 2.63 to 2.77 μm spectral region. Strong lines from OH are expected at 2.81, 2.87, and 2.90 μm (Weaver and Mumma, 1984), but these are not observed in our spectra.

An alternate possibility is the emission from a solid-state band such as hydrated silicates. Chlorite and serpentine, for example, have strong absorption bands just shortward of 3 μm, as shown by Knacke et al. (1985). Depending on the type of mineral, the 3-μm band could arise either from bound water or OH. This possibility can be ruled out because the even stronger 10-μm band in hydrated silicates was not present (see Fig. 3 and Knacke et al., 1985).

Significant LTE water emission is not expected from the coma of comets, since the region from which collisions could populate the upper levels of the water molecule is expected to be small (Weaver and Mumma, 1984; Mumma et al., 1986, Crovisier and Encrenaz, 1983). In addition, calculations by Bockelée-Morvan and Crovisier (1987) indicate that LTE emission cannot explain the observations by Mumma et al. (1986). Thus LTE water emission seems unlikely to explain the 2.8–2.9 μm emission feature.

On May 19, there is a spectral point at 2.833 μm that is significantly higher than the other points. This point does not appear enhanced on May 21. It does seem to be real, however, since our standard star ratios are flat. As we have no information on the width of the 2.8 to 2.9 μm emission band, it is impossible to ascertain whether this spectral point is a single point or part of a broad emission structure.

We obtained a 3-μm spectrum of comet Halley in December 1985 (Tokunaga et al., 1986a). Unfortunately, the spectral range is much smaller since it consisted of a single grating position and is inadequate for us to determine conclusively whether or not the 2.8- to 2.9-μm and 3.36-μm emission features existed in December 1985.

3.3 Search for water ice

The rationale for continuing the search for water ice was stated previously by Tokunaga et al. (1986a). At a heliocentric distance of 1.5 AU or larger, one may expect to see water ice in absorption against the scattered light from the Sun at 3 μm if there is a large enough ice column density. In this situation, thermal emission at 3 μm is negligible compared to the scattered light. In addition, the lifetime of the ice grains increases dramatically with increasing heliocentric distances (Hanner, 1981; Crifo, 1982).

A definitive detection of water ice in any comet has not yet been achieved. Previous spectroscopic searches for the 3-μm water-ice band in comets with large heliocentric distances are limited to comet Cernis (Hanner, 1984) and, recently, comet

Halley (Bregman et al., 1986). In both cases, the results seemed positive, but only at the 2–3 σ level. The infrared spectrum of comet Halley obtained with the Vega 1 spacecraft at a heliocentric distance of 0.79 AU showed a flux minimum at 2.9 µm that was also tentatively attributed to water ice (Combes et al., 1986a).

The minimum in the scattering cross-section for water ice is at 2.85 µm. Unfortunately, the short-wavelength emission band in our spectrum makes it impossible to see the water-ice absorption band against the continuum. There is no evidence for a reduction in flux at the position of the water-ice band, so that only an upper limit can be placed on the amount of ice in the coma.

The spectral shape of the continuum seems to be different in comets Cernis and Halley. In comet Cernis, the continuum is lower at 2.88 µm compared to 3.16 µm (Hanner, 1984), but in Halley the continuum is flat or slightly rising. This indicates that some comets might show clearly the water-ice band so long sought. Only further observations of future comets can settle this point as the sample of comets observed spectroscopically at 3 µm and at a heliocentric distance >1.5 AU is only two thus far.

4. Summary

1. The 3.36-µm emission feature was observed in comet Halley at a heliocentric distance of 1.9 AU. The peak wavelength and intensity were consistent with other published spectra. There is evidence, however, that the intensity of feature changes on a time scale of days.

2. An emission feature at 2.8–2.9 µm was observed; however, we can rule out fluorescent emission from water and OH, LTE emission from water, and also emission from a solid-state band of hydrated silicates.

3. No 3-µm absorption band from water ice was observed. Unfortunately, the emission band at 2.8–2.9 µm may have masked the presence of the ice band. Because of the limited number of 3-µm spectra of comets at large heliocentric distances, more spectra of comets are needed.

Acknowledgements. A. Tokunaga acknowledges discussions with H. P. Larson and M. J. Mumma that led to the initial recognition that the 2.8–2.9 µm emission band was unusual. A. Tokunaga and R. G. Smith were supported by NASA contract NASW 3159 during the time this work was completed, and T. Nagata by NSF grant AST 84-20775. The CGAS was constructed in part by NSF grant AST 82-16217.

References

Baas, F., Geballe, T.R., Walther, D.M.: 1986, *Astrophys. J.* **311**, L 97

Bregman, J.D., Witteborn, F.C., Rank, D.M., Wooden, D.: 1986, *Bull. Am. Astron. Soc.* **18**, 634

Bockelée-Morvan, D., Crovisier, J.: 1987, *Astron. Astrophys.* in press

Combes, M., Moroz, V., Crifo, J.F., Bibring, J.P., Coron, N., Crovisier, J., Encrenaz, T., Sanko, N., Grigoriev, A., Bockelée-Morvan, D., Gispert, R., Emerich, C., Lamarre, J.M., Rocard, F., Krasnopolsky, V., Owen, T.: 1986a, *20th ESLAB Symp.*, ESA SP-250, Vol. **1**, p.353

Combes, M., Moroz, V.I., Crifo, J.F., Lamarre, J.M., Charra, J., Sanko, N.F., Soufflot, A., Bibring, J.P., Cazes, S., Coron, N., Crovisier, J., Emerich, C., Encrenaz, T., Gispert, R., Grigoryev, A.V., Guyot, G., Kransnopolsky, V.A., Nikolsky, Y.V., Rocard, F.: 1986b, *Nature* **321**, 266

Crifo, J.F.: 1982, in Proc. Intern. Conf. *Cometary Exploration*, p. 167

Crovisier, J.: 1984, *Astron. Astrophys.* **130**, 361

Crovisier, J., Encrenaz, T.: 1983, *Astron. Astrophys.* **126**, 170

Danks, A., Encrenaz, T., Bouchet, P., Le Bertre, T., Chalabaev, A., Epchtein, N.: 1986, in *20th ESLAB Symp.*, ESA SP-250, Vol. **3**, p. 103

Gehrz, R.D., Ney, E.P.: 1986, *20th ESLAB Symp.*, ESA SP-250, Vol. **2**, p. 101

Hanner, M.S.: 1981, *Icarus* **47**, 342

Hanner, M.S.: 1984, *Astrophys. J.***277**, L 75

Hanner, M.S., Aitken, D.K., Knacke, R., McCorkle, S., Roche, P.F., Tokunaga, A.T.: 1985, *Icarus* **62**, 97

Knacke, R.F., Brooke, T.Y., Joyce, R.R.: 1986, *Astrophys. J.* **310**, L 49

Knacke, R.F., Puetter, R.C., Erickson, E., McCorkle, S.: 1985, *Astron. J.* **90**, 1828

Mumma, M.J., Weaver, H.A., Larson, H.P., Davis, D.S., Williams, M.: 1986, *Science* **232**, 1523

Oishi, M., Kawara, K., Kobayashi, Y., Maihara, T., Noguchi, K., Okuda, H., Sato, S., Iijima, T., Ono, T.: 1978, *Publ. Astron. Soc. Japan* **30**, 149

Tokunaga, A.: 1986, *IAU Circ.* **4226**

Tokunaga, A.T., Golisch, W.F., Griep, D.M., Kaminski, C.D., Hanner, M.S.: 1986a, *Astron. J.* **92**, 1183

Tokunaga, A.T., Smith, R.G., Nagata, T., DePoy, D.L., Sellgren, K.: 1986b, *Astrophys. J.* **310**, L 45

Weaver, H.A., Mumma, M.J.: 1984, *Astrophys. J.* **276**, 782

Weaver, H.A., Mumma, M.J., Larson, H.P., Davis, D.S.: 1986, *Nature* **324**, 441

Wickramasinghe, D.T., Allen, D.A.: 1986, *Nature* **323**, 44

Yamamoto, T.: 1982, *Astron. Astrophys.* **109**, 326

Astron. Astrophys. 187, 523–525 (1987)

ASTRONOMY AND ASTROPHYSICS

Photometry of P/Halley (1982i)*

C. Sterken [1,**], J. Manfroid [2,***], and C. Arpigny [2]

[1] Astrofysisch Instituut, Vrije Universiteit Brussel, Pleinlaan 2, B-1050 Brussels, Belgium
[2] Institut d'Astrophysique, Université de Liège, Avenue de Cointe, B-4200 Cointe-Ougrée, Belgium

Received February 7, accepted May 25, 1987

Summary. We report the results of photoelectric observations carried out during several weeks between 1985 September 12, and 1986 June 24. All measurements were made using the standard IAU/IHW cometary filters. Large amplitude night-to-night changes are superimposed on the general trend of slow brightness variation and support a 7^{d}.4 period. The changes are seen as smooth variations with a range up to 0.2 mag on time scales of several hours.

Key words: coma filter photometry – continuum – neutrals and ions – lightcurves – P/Halley

1. Introduction

Two of the authors participated in the IHW photometry net and carried out photoelectric observations of comet P/Halley between 1985 September 12, and 1986 June 24 at the European Southern Observatory, Chile, and at Mt. John Observatory, New Zealand.

2. The Observations

2.1. General aspects

All observations were carried out with two standard IAU/IHW filtersets which were used in photoelectric photometers mounted on telescopes with apertures of 50 cm and 60 cm at both sites. Table1 gives the journal of observations an particular details on the equipment used.

The filtersets were in principle identical, but for the sake of completeness the sets are labeled. Set number 1 is the filterset belonging to the Universities of Liège/Brussels, the second set is the one which was available at Mount John Observatory.

During each night several standard stars (solar analogs and early type flux standards) were observed at different airmasses. Frequent measurements of sky background were made at distances of 30′ up to one degree off the nucleus.

Send offprint requests to: C. Sterken

* Based on observations obtained at the European Southern Observatory (Chile) and at Mount John Observatory (New Zealand)

** Research Associate NFWO Belgium

*** Senior Research Associate, FNRS Belgium

Integration times of single measurements in one filter were chosen in accordance with the count rate in the particular band. In those cases where the speed of the comet across the sky was too high, and where no adequately adjustable telescope tracking speeds were available, multiple short integrations (separated by careful recentering) in each filter were taken. The results of these multiple short integrations were averaged afterwards.

Measurements through several diaphragms of different sizes were made whenever possible (see Table 1 for the sizes of the diaphragms). One problem, especially during the time when the comet was at its brightest, consisted of the limited choice of diaphragm sizes which were available at the different telescopes.

2.2. Specific details

2.2.1. European Southern Observatory (March 1986)

The weather was good, and the exceptional quality of the site allowed us to observe without difficulties in the OH λ 3085 filter. The largest available diaphragm was 40″. During several hours in different nights we also alternately monitored the comet and a nearby comparison star in order to detect the presence of any variations on a timescale of hours or minutes.

2.2.2. Mount John Observatory

The telescope was the Optical Craftsman 60 cm telescope equipped with the Mount John photometer. Since the photomultiplier tube was not sensitive in the red, no measurements in the H_2O^+ filter were attempted. The weather was poor during the first four weeks of the observing run.

2.2.3. ESO (June 1986)

The same equipment as in March–April was used. Due to extremely bad weather, this two week observing run only yielded one night of good quality, which resulted in one series of excellent measurements on June 24.

3. Data reduction procedure

The reductions were done with a programme based on the method explained by Manfroid and Heck (1983). The main features of this programme are the multinight technique and the optimal use of every measurement of every standard and constant star for the calculation of the parameters of the reduction. The reduction was

Table 1. Journal of observations

Site	ESO	Mt John	ESO
Country	Chile	New Zealand	Chile
Date	March 27-April 8	March 4-April 8 April 20-May 8	June 9-June 24
Observer	J. Manfroid	C. Sterken	C. Sterken
Telescope	50 cm	60 cm	50 cm
PM Tube	RCA 34032	EMI 6094 B	RCA 34032
Filterset	1	2	1
Diaphragm sizes	10″, 15″, 21″ 30″ and 40″	70″	80″

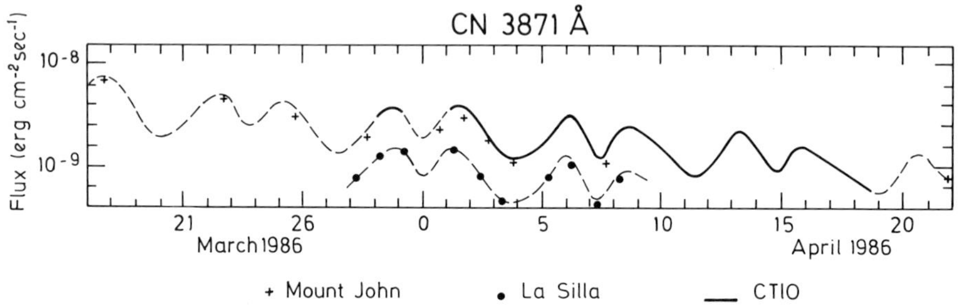

Fig. 1. Variation of CN λ 3871 emission band flux in March–April 1986. The data by Millis and Schleicher are represented by a solid line. For the Mt John and La Silla observations repectively a 70″ and a 40″ focal plane diaphragm was used. CTIO data are based on a diaphragm of 78″

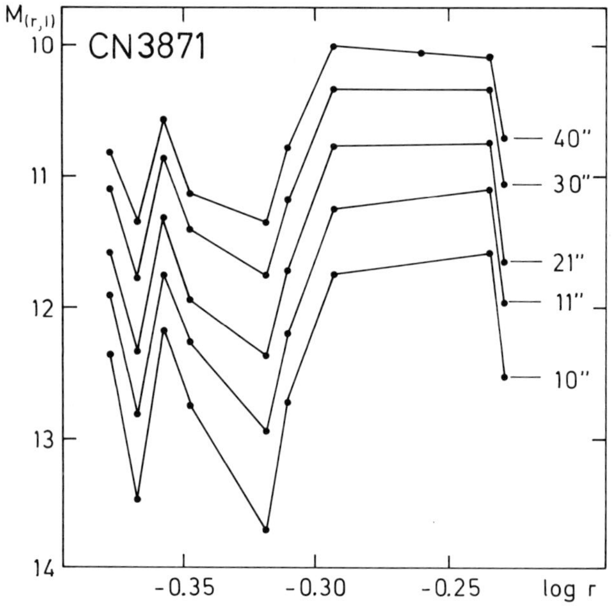

Fig. 2. Variation of CN-magnitude with heliocentric distance (in astronomical units) for each diaphragm

made to the natural system corresponding to each individual site; the zero points being derived from standard photometry published by the Photometry Section of the International Halley Watch.

The resulting magnitudes were transformed to flux values with the equations provided by A'Hearn (1986). Because of the relatively large errors which affect the observations and also the flux derivations, we made no attempt to include any corrections for radial velocity effects.

4. Results

4.1. Night to night variations

Figure 1 illustrates the variation of the comet's flux in March and April 1986, in the CN λ 3871 emission band. The data obtained by Millis and Schleicher (1986) are included for comparison. The agreement between the results (considering the differences in diaphragm size and in frequency of good weather conditions) is fairly good.

The simultaneous measurements clearly show that the cometary head exhibited large amplitude flare-like irregular variations with a range of about 2 mag. over a time interval of one or two days. Similar results were also obtained by Cochran and Barker (1986). When these bursts occur, they show up in all emission lines and also in the continuum band. The combined data agree with the period of $7^{\mathrm{d}}\!.4$ found by Millis and Schleicher (1986) and is likely related to the rotation period.

Figure 2 illustrates the behaviour of the CN λ 3871 emission line strength in function of the heliocentric distance for all diaphragm sizes used during the observing run at ESO.

4.2. Short term fluctuations

Figure 3 shows the light curve of comet Halley obtained at higher time resolution during three nights in 1986. The data shown here

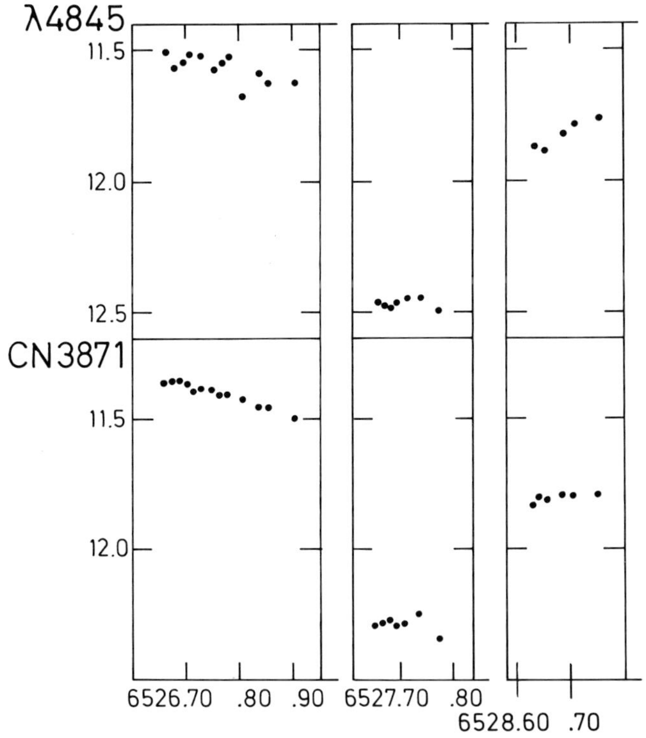

Fig. 3. Lightcurve of comet Halley obtained at higher time resolution during three nights in 1986. A diaphragm of 20″ was used.

represent the CN λ 3871 emission and the λ 4845 continuum bands, and result from additional sequences of differential measurements between the comet and nearby comparison stars. The accuracy of each measurement is of the order of 0.01 mag.

The data collected during the last two nights indicate little activity, while on the first night we observe a very strong monotonic decline in the intensity in the CN λ 3871 band over about 0.2 mag. in a time interval of about 6 h. The variation in the continuum shows the same trend, but has a more erratic character. The slope of the light curve is in agreement with the general trend of the strong night to night variations (see Fig. 1). Also do the nightly mean magnitudes confirm the trend visible in Fig. 3. The occurrence of strong bursts supports the assumption that they are provoked by sudden releases of material from the comet's head during the rotation of the nucleus.

Those observations agree with those of Millis and Schleicher (1986) which favor nucleus rotation with a period of $7^d.4$. The dust and molecules ejected from the sunward hemisphere of the nucleus provoke a two-peaked general light curve.

5. Conclusions

Ground-based post-perihelion observations of comet Halley confirm the existence of a $7^d.4$ photometric periodicity characterized by a double peaked lightcurve. This periodicity was originally found by Millis and Schleicher (1986). Our data extend the time interval over which the periodicity is recognized. Current rotation models of comet Halley's nucleus point towards a rotation period of $53^h.5$ (Smith et al., 1987) and a torque-free precession with a $7^d.4$ period about an axis approximately perpendicular to the long axis of the nucleus. The modulation of the solar flux on the active region of the comet results in the observed periodic gas and dust production rates.

Acknowledgements. Financial support for this research was provided by the Belgian National Fund for Scientific Research (NFWO/FNRS), the Vrije Universiteit Brussel, and the Université de l'Etat à Liège. The authors gratefully acknowledge observing time granted by the European Southern Observatory, and by Mount John Observatory.

The work benefitted very much from the efficient way in which the IHW photometry net functioned. C.S. is particularly indebted to Mr. John Baker (Mt. John Obs.) for technical assistance during the test period which preceeded the observations.

References

A'Hearn, M.F.: 1986, Photometric Transformations, International Halley Watch Communication, January 26, 1986

Cochran, A.L., Barker, E.S.: 1986, 20th ESLAB Symp., ESA SP-250, Vol. **I**, p. 439

Manfroid, J., Heck, A.: 1983, *Astron. Astrohys.* **120**, 302

Millis, R.L., Schleicher, D.G.: 1986, *Nature* **324**, 646

Smith, B.A., Larson , S.M., Szegö, K., Sagdeev, R.Z.: 1987, *Nature* **326**, 573

Astron. Astrophys. 187, 526–530 (1987)

ASTRONOMY AND ASTROPHYSICS

Polarimetry of comet P/Halley: continuum versus molecular bands

J.F. Le Borgne, J.L. Leroy, and J. Arnaud

Observatoires du Pic du Midi et de Toulouse, 14 Avenue Edouard Belin, F-31400 Toulouse, France

Received January 26, accepted May 4, 1987

Summary. Measurements of the linear polarization of molecular bands in comets P/Halley and Hartley-Good are presented. The measured bands are the C2 Swan (5140 Å), the CN (3880 Å), the C3 (4060 Å) and the OH (3090 Å). The phase angle dependence of the polarization of C2, C3 and CN is in good agreement with theory. However the polarization of OH is found 5 times smaller than expected. The direction of polarization of C2, C3 and OH is usually perpendicular to the scattering plane, though some measurements are parallel to it. For CN half the measures show significant deviations from the perpendicular to the scattering plane.

Key words: comets – polarization – comet P/Halley

1. Introduction

Polarimetry of molecular bands in comets in the visible and UV spectral ranges is a field which has been seldom explored, either observationally or theoretically. The theory of polarization of molecular bands was studied a long time ago by Mrozowski (1936) although his work does not deal with molecular bands observed in comets. No theoretical work was done since then on that subject. As far as observations are concerned, the most significative work still dates from the 1940's (Ohman, 1939, 1941). Some measurements have been published in between: Blackwell and Willstrop (1957), Bappu et al. (1967), Kharitonov and Rebristyi (1974) (see the review by Dobrovolsky et al., 1986). However, in most cases it is not clear whether continuum light was correctly subtracted from measurements.

Here we report new measurements of the linear polarization of the main ground observable molecular bands obtained in comets P/Halley and Hartley-Good. A first analysis is published in Le Borgne et al. (1987) (Paper I).

2. Observations

The photopolarimeter STERENN was set on the 1-m telescope at Pic du Midi Observatory. It was equipped with interference filters to isolate the C2 Swan band at 5140 Å, the CN band at 3880 Å, the C3 band at 4060 Å and the OH band at 3090 Å. Furthermore, measurements in two continuum filters were done (4845 Å and 3650 Å). The observation procedure is described in Paper I. We have observed P/Halley for 17 nights from 1985 October 10 to 1986 May 9 and Hartley-Good for 3 nights on 1985 October 15 to 17.

Table 1 gives the diameter in arcsec of the diaphragms used and their projected dimension on comets Halley and Hartley-Good.

Tables 2 to 8 display the measurements: the degree of polarization is given in column 4 and equatorial direction of polarization in column 5. Column 6 gives the phase angle and 7 the deviation from the direction perpendicular to the plane of scattering. The data concerning the molecular bands (Tables 2 to 5 for comet Halley and 8 for Hartley-Good) are obtained after careful subtraction of the continuum light (auxiliary measurements of the continuum light transmitted through the filters were obtained on the solar-type star HD 30455).

3. Phase angle dependence of polarization

As shown in Paper I for C2 and CN, our measurements of the polarization of the emission bands (except OH) are in good agreement with the theoretical dependence of band polarization with phase angle (Fig. 1). This dependence was proposed by Ohman (1941):

$$p_{\text{th}} = \frac{p_{\max} \sin^2\alpha}{1 + p_{\max}\cos^2\alpha}$$

where $p_{\max}$ is the polarization for phase angle $\alpha = 90°$.

Table 1. Linear dimension of the diaphragms projected on comets P/Halley and Hartley-Good

Date	Diaphragm ($''$)	Dimension on comet Halley (km)
1985 Oct	32	35 to 40 10^3
1985 Dec	65	33 to 37 10^3
1985 Dec	10	5 to 6 10^3
1986 May	32	19 to 25 10^3
Date	**Diaphragm ($''$)**	**Dimension on comet Hartley-Good (km)**
1985 Oct	32	13 10^3

Send offprint requests to: J.F. Le Borgne

Table 2. Polarization of the 5140 Å C2 Swan Band in comet P/Halley. Continuum light is subtracted.

Date	Diaphragm (″)	p(%)	θ (°)	Phase angle (°)	$\theta - \theta_\perp$ (°)
1985 Oct 10 and 13[a]	32	3.1 ± 2.5	155 ± 25	25.7	153
1985 Oct 16 and 17[a]	32	2.0 ± 1.3	90 ± 20	25.5	88
1985 Dec 9–10	65	3.5 ± 0.2	157 ± 1.5	45.7	−1
1985 Dec 12–13	65	3.6 ± 0.2	157 ± 1.6	49.1	0
1985 Dec 12–13	10	4 ± 3	173 ± 18	49.1	16
1985 Dec 13–14	65	4.2 ± 0.2	156 ± 1.4	50.0	−1
1985 Dec 14–15	65	4.5 ± 0.2	158 ± 1.2	50.8	1
1985 Dec 15–16	65	4.4 ± 0.2	156 ± 1.1	51.6	0
1986 May 1–2	32	1.5 ± 0.5	177 ± 9	29.3	−9
1986 May 6–7	32	2.3 ± 0.8	16 ± 8	30.8	5
1986 May 7–8	32	2.5 ± 1.4	19 ± 14	31.0	7
1986 May 9–10	32	1.7 ± 0.7	13 ± 9	31.2	0

[a] Average over several nights

Table 3. Polarization of the 4060 Å C3 Band in comet P/Halley. Continuum light is subtracted.

Date	Diaphragm (″)	p(%)	θ (°)	Phase angle (°)	$\theta - \theta_\perp$ (°)
1985 Oct 13 to 17[a]	32	1.5 ± 1.1	120 ± 25	25.5	−62
1985 Dec 9–10	65	3.7 ± 0.3	160 ± 2	45.7	2
1985 Dec 12–13	65	3.3 ± 0.5	156 ± 4	49.1	−1
1985 Dec 13–14	65	3.7 ± 0.6	161 ± 4	50.0	4
1985 Dec 14–15	65	4.5 ± 0.4	158 ± 2	50.8	1
1985 Dec 15–16	65	3.5 ± 0.5	156 ± 3	51.6	0
1986 May 1–2	32	1.2 ± 1.2	—	29.3	
1986 May 9–10	32	2.5 ± 0.8	18 ± 8	31.2	5

[a] Average over several nights

Table 4. Polarization of the 3880 Å CN Band in comet P/Halley. Continuum light is subtracted. Significant deviations from perpendicular to the scattering plane are underlined

Date	Diaphragm (″)	p(%)	θ (°)	Phase angle (°)	$\theta - \theta_\perp$ (°)
1985 Oct 11 and 16[a]	32	2.0 ± 0.6	0 ± 10	25.7	−2
1985 Dec 9–10	65	3.3 ± 0.2	159 ± 2	45.7	1
1985 Dec 12–13	65	4.6 ± 0.2	160 ± 1.0	49.1	<u>3</u>
1985 Dec 13–14	65	4.0 ± 0.2	154 ± 1.2	50.0	−3
1985 Dec 14–15	65	3.9 ± 0.2	153 ± 1.2	50.8	<u>−4</u>
1985 Dec 15–16	65	3.8 ± 0.2	156 ± 1.5	51.6	0
1986 May 1–2	32	0.7 ± 0.5	12 ± 16	29.3	6
1986 May 6–7	32	0.9 ± 0.7	23 ± 20	30.8	12
1986 May 7–8	32	2.8 ± 0.6	168 ± 5	31.0	<u>−24</u>
1986 May 9–10	32	2.3 ± 0.5	0 ± 5	31.2	−13

[a] Average over several nights

Table 5. Polarization of the 3090 Å OH Band in comet P/Halley. Continuum light is subtracted

Date	Diaphragm (″)	p(%)	θ (°)	Phase angle (°)	$\theta - \theta_\perp$ (°)
1985 Dec 9–10	65	2.6 ± 2.0	78 ± 19	45.7°	−80°
1985 Dec 12–13	65	1.7 ± 0.9	170 ± 13	49.1°	13°
1985 Dec 13–14	65	2.5 ± 1.1	146 ± 11	50.0°	−11°
1985 Dec 14–15	65	2.5 ± 1.1	152 ± 11	50.8°	−5°

Table 6. Polarization of continuum light at 4845 Å in comet P/Halley. Significant deviations from perpendicular to the scatterring plane are underlined

Date	Diaphragm (″)	p(%)	θ (°)	Phase angle (°)	$\theta - \theta_\perp$ (°)
1985 Oct 11–12	32	1.7 ± 0.9	178 ± 15	25.7	−4
1985 Oct 15–16	32	1.4 ± 0.6	172 ± 11	25.5	−10
1985 Oct 17–18	32	2.7 ± 0.7	12 ± 7	25.3	10
1985 Dec 9–10	65	7.3 ± 0.2	157.1 ± 0.6	45.7	−1
1985 Dec 12–13	65	9.6 ± 0.3	157.9 ± 0.5	49.1	1
1985 Dec 12–13	10	11.2 ± 1.4	148 ± 3	49.1	−9
1985 Dec 13–14	65	9.5 ± 0.3	156.8 ± 0.7	50.0	0
1985 Dec 13–14	10	11.5 ± 1.4	165 ± 3	50.0	8
1985 Dec 14–15	65	9.2 ± 0.2	158.2 ± 1.2	50.8	1
1985 Dec 15–16	65	9.5 ± 0.2	157.3 ± 0.5	51.6	1
1986 May 1–2	32	2.5 ± 0.2	7 ± 2	29.3	1
1986 May 6–7	32	3.0 ± 0.4	13 ± 3	30.8	2
1986 May 7–8	32	3.0 ± 0.4	19 ± 3	31.0	7
1986 May 9–10	32	2.9 ± 0.2	15 ± 2	31.2	2

Table 7. Polarization of continuum light at 3650 Å in comet P/Halley. Significant deviations from perpendicular to the scattering plane are underlined

Date	Diaphragm (″)	p(%)	θ(°)	Phase angle (°)	$\theta - \theta_\perp$ (°)
1985 Oct 12–13	32	1.8 ± 2.2	170 ± 31	25.7	−12
1985 Dec 9–10	65	5.5 ± 0.9	160 ± 4	45.7	2
1985 Dec 12–13	65	7.8 ± 0.4	157 ± 1	49.1	0
1985 Dec 13–14	65	8.1 ± 0.4	162 ± 1	50.0	5
1985 Dec 14–15	65	7.8 ± 0.4	157 ± 1.5	50.8	0
1986 May 1–2	32	2.5 ± 1.7	22 ± 16	29.3	16
1986 May 6–7	32	4.3 ± 1.7	10 ± 10	30.8	−1
1986 May 9–10	32	9.5 ± 2.0	10 ± 5	31.2	−3

Table 8. Polarization of comet Hartley-Good through a 32″ diaphragm. Continuum light is subtracted from molecular band data. Significant deviations from perpendicular to the scattering plane are underlined

Date	Filter (Å)	p(%)	θ (°)	Phase angle (°)	$\theta - \theta_\perp$ (°)
1985 Oct 15–17	C2 5140	4.8 ± 0.8	170 ± 5	51	2
1985 Oct 15	C3 4060	4 ± 2	159 ± 11	51	−9
1985 Oct 17	C3 4060	11 ± 7	159 ± 5	51	−9
1985 Oct 16–17	CN 3880	5.9 ± 1.1	149 ± 5	51	<u>−19</u>
1985 Oct 15–17	continuum 4845	6.5 ± 0.8	173 ± 3	51	5

* Average over several nights

Following Mrozowski's computations (1936), Ohman proposed that $p_{max} = 7.7\%$. However, our measurements show significant departures from the theoretical curve, especially for CN, which do not seem to be due to data reduction or instrumental effects.

For the OH band the situation seems to be quite different. Swings (1956), also following Mrozowski, announced that p_{max} could be as high as 19% for this band. Our measurements, the mean value of which is $1.8 \pm 0.7\%$ at $\alpha = 50°$, are the first published. With $p_{max} = 19\%$ we should have found $p(\alpha = 50°) = 10\%$. There is no reason why Mrozowski's theory could not be applied to OH as for C2 or CN, but, in comets, there is a difference between OH and C2 or CN as suggested by Festou (private communication): the theory assumes that fluorescence equilibrium is reached, which may not be true for OH.

In Fig. 1, we also plot our measurements of the polarization of the continuum light at 4845 Å for comparison. The curve, fitting the continuum measurements, is (Kiselev and Chernova, 1978):

$$p = \frac{p_{max} \sin\alpha (\sin\alpha - \sin\alpha_0)}{1 - \sin\alpha_0 + p_{max}\cos^2\alpha}$$

with $p_{max} = 20\%$ and $\alpha_0 = 15°$ (α_0 is the phase angle, $\neq 0$, where $p = 0\%$). It is the best fit we can obtain with the few measurements we got.

4. Direction of polarization

As in the case of Thomson and Rayleigh scattering, it is expected that the polarization of molecular emission by fluorescence scattering is perpendicular to the plane of scattering (that is to say perpendicular to the direction to the Sun projected on the celestial sphere). As it can be seen in Tables 2 to 8, our measurements show that it is usually the case, with, however, some significant exceptions.

For the CN band, some measurements deviate significantly from the perpendicular of the scattering plane, from 3° to 24°. The polarization was, however, never found in the scattering plane. In Paper I, we show that there exists a possible correlation between the deviations of direction and degree of polarization. There is no explanation for these abnormal direction of polarization, though we suggest in Paper I that the magnetic field could play a prominent role (Crosley et al., 1968; Landolfi and Landi Degl'Innocenti, 1986).

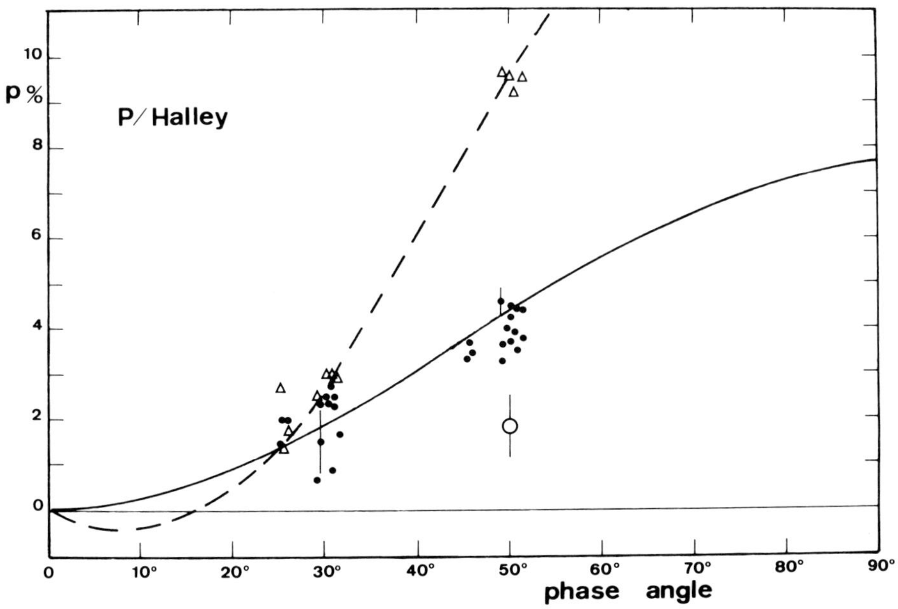

Fig. 1. Phase angle dependence of the degree of polarization of comet P/Halley for continuum at 4845Å (triangles) and molecular bands (full circles: C2, C3 and CN; open circle: OH). The full line is the theoretical dependence for molecular bands ($p_{max} = 7.7\%$). The dashed line is a fit of theoretical dependence of polarization on continuum measurements

The polarization of the continuum bands is always found close to the perpendicular of the scattering plane. However, some measurements deviate significantly from it.

5. Conclusion

Our measurements in comets P/Halley and Hartley-Good concerning the linear polarization of molecular bands show that:

1) the polarization of the bands of C2 at 5140 Å and C3 at 4060 Å is in good agreement with classical fluorescence theory, both in degree and direction.

2) the polarization of the OH band at 3090 Å (for which these measurements are the first ones ever published) is 5 times smaller than expected. However, a complete theory of OH polarization in comets remains to be done, since the calculations made by Swings (1956) are approximative.

3) we observed significant deviations in the direction of polarization from the perpendicular of the scattering plane for the CN band at 3880 Å. We have no quantitative explanation of this phenomenon, though we suggest that the magnetic field can play a prominent role.

Postscript

The theoretical polarization of the molecular bands quoted in the present paper has recently been studied (J. F. Le Borgne and J. Crovisier: 1987, "Symposium on the diversity and similarity of comets", Brussels, 6–9 April 1987). It is shown that the degree of polarization of OH cannot exceed 2% (as observed) and greatly depends on the heliocentric velocity of the comet (Swings effect). The polarization could even be negative. The observed polarization of C_2 and of CN is still in agreement with the theory. However, the theory predicts that the polarization percentage of C_3 should be much greater than observed.

References

Bappu, M.K.V., Sivaraman, K.R., Bhatnagar, A., Natarajan, V.: 1967, *Monthly Notices Roy. Astron. Soc.* **136**, 19

Blackwell, D.E., Willstrop, R.V.: 1957, *Monthly Notices Roy. Astron. Soc.* **117**, 590

Crosley, D.R., Tango, W.J., Zare, R.N.: 1968, *Astrophys. J.* **154**, L153

Dobrovolsky, O.V., Kiselev, N.N., Chernova, G.P.: 1986, *Earth, Moon Planets* **34**, 189

Kharitonov, A.V., Rebristyi, V.T.: 1974, *Sov. Astron.* **17**, 672

Kiselev, N.N., Chernova, G.P.: 1978, *Sov. Astron.* **22**, 607

Landolfi, M., Landi Degl'Innocenti, E.: 1986, *Astron. Astrophys.* **167**, 200

Le Borgne, J.F., Leroy, J.L., Arnaud, J.: 1987, *Astron. Astrophys.* **173**, 180 (Paper I)

Mrozowski, S.: 1936, *Acta Phys. Pol.* **5**, 85

Ohman, Y.: 1939, *Monthly Notices Roy. Astron. Soc.* **99**, 631

Ohman, Y.: 1941, *Stockholm Obs. Ann.* **13**, No. 11, 3

Swings, P.: 1956, *Vistas Astron* **2**, 958

Astron. Astrophys. 187, 531–538 (1987)

ASTRONOMY
AND
ASTROPHYSICS

Photometric observations of comet P/Giacobini-Zinner

D. G. Schleicher [1], R. L. Millis [1], and P. V. Birch [2]

[1] Lowell Observatory, 1400 W. Mars Hill Rd., Flagstaff, AZ 86001, USA
[2] Perth Observatory, Walnut Rd., Bickley, Western Australia 6076

Received February 3, accepted March 16, 1987

Summary. Narrowband filter photometry of comet P/Giacobini-Zinner was obtained on 15 nights at Lowell Observatory in the interval from 15 June to 20 November 1985. The emission bands of OH, NH, CN, C_3, and C_2 were isolated, along with continuum points at 3650 Å and 4845 Å. All observed species showed pronounced asymmetry about perihelion, with pre-perihelion abundances being greater by more than a factor of 2 than post-perihelion abundances over the heliocentric distance range of 1.0–1.5 AU. The characteristics of the asymmetry were significantly different than observed in previous apparitions, with the peak production rates occurring in late July/early August. The unusually low C_2/CN abundance ratio seen in previous apparitions is confirmed by the present observations. Furthermore, our OH data demonstrate that the C_2 in Giacobini-Zinner (along with C_3 and NH) is depleted when compared to water (the dominant species), rather than CN being unusually over-abundant.

Key words: cometary coma – filter photometry – molecular abundances

1. Introduction

In September 1985, comet P/Giacobini-Zinner (G-Z) became the first comet to have measurements made *in situ*. These measurements were obtained by the International Comet Explorer (ICE) as it passed through the tail of G-Z to within 7800 km of the nucleus on 11 September. Originally known as the International Sun-Earth Explorer Number Three (ISEE-3), the ICE spacecraft was not designed for cometary observations, but was redirected for the encounter after fulfilling its original mission of observing the interaction of Earth's magnetosphere and the solar wind (von Rosenvinge et al., 1986). For this reason, instrumentation was unavailable for determining certain critical parameters, such as the composition and rate of production of the various neutral species. Accordingly, we conducted a program of narrowband filter photometry of comet Giacobini-Zinner during the interval from mid-June to mid-November 1985 using several telescopes at Lowell Observatory. Production rates of the gaseous species OH, NH, CN, C_3, and C_2, as well as estimates of dust production are presented here, so that the spacecraft data and other ground-based data obtained at the time of the ICE encounter can be put into proper context, and compared with the comet's activity throughout the 1985 apparition. We also compare this apparition to results obtained in previous ones, since the light curve of G-Z is known to have visually varied from apparition to apparition (Sekanina, 1985). Finally, because it has been known for some time that the relative molecular abundances in Giacobini-Zinner are anomalous (cf. Bobrovnikoff, 1927; Herbig, 1976), we also compare G-Z with other comets observed in our on-going filter photometry program. These anomalous abundances can influence the application of the ICE results to other comets.

Send offprint requests to: D. G. Schleicher

2. Observations and reductions

The ground-based, narrowband filter photometry observations presented here are a continuation of a 10-year-old comet program initiated by A'Hearn and Millis (see A'Hearn et al., 1977). The same general observing and reduction techniques are again used as before (e.g. A'Hearn et al., 1979, 1981), but many of the numeric parameters have changed over the years due to the use of new filters and revised molecular parameters. We use the standard IAU filters to isolate the emission bands of CN, C_3, C_2, and CO^+, as well as two continuum points (λ 3650 and λ 4845). In addition, a new filter for OH was adopted to replace our original OH filter, while we continue to use the same NH filter as has been used since 1982. The procedure used for flux calibrations for similar filters can be found in two papers by A'Hearn et al. (1979, 1981). Details of the OH extinction correction for objects having differing spectral distributions, and coefficients of the magnitude-to-flux transformations for all of the current filters are given by A'Hearn (1985, 1986a). We note, however, that the extinction correction for the OH filter has been subsequently revised based on empirical data, as have the transformation coefficients for the λ 4845 continuum filter, for which our recent observations show larger C_2 contamination than theoretically calculated (A'Hearn, 1986b). Even after this correction, residual contamination remains in one or both of the continuum filters, resulting in spurious CO^+ measurements. Consequently, CO^+ will not be discussed further in this paper. This residual contamination, apparently at the few percent level of the continuum flux, also gives rise to slightly low (by up to 20%) band fluxes for the weaker molecular species. These errors will be discussed in more detail with the data sets (Sect. 3).

After subtraction of the appropriate underlying continuum for each molecular species, the resulting fluxes were reduced to column densities (M) by using the fluorescence efficiency per molecule (L/N, also known as the g-factor). For the C_3 band we use the value for L/N from A'Hearn (1982), while for C_2 we use the revised value from A'Hearn et al. (1985) (see Table 1). For OH and CN, where there is a heliocentric velocity ($\dot{r}_H$) dependence in L/N due to the Swings effect, we have interpolated the table of values given by Schleicher (1983), multiplying the CN 0-0 band value by 1.08 to approximate the additional contribution of the CN 1-1 band. Finally, in this paper we discuss the NH molecule for the first time, because of its importance in understanding the chemistry behind the anomalous molecular abundances found in Giacobini-Zinner. While NH should have a quite strong Swings effect, the fluorescence equilibrium calculations have not yet been performed adequately, and so for the time being we have adopted an L/N value of 3.1 10^{-13} erg s^{-1} molecule^{-1}, based on the work of Litvak and Rodriguez-Kuiper (1982). Uncertainty in the absolute magnitude of this value will not affect relative abundances as compared to those observed in other comets, as these will directly scale. However, we estimate that the Swings effect may yield variations by up to a factor of 3 in the relative fluorescence efficiency for a particular velocity, based on the structure of the solar spectrum, the number of rotational levels populated and the corresponding results calculated for CN and OH. The fluorescence efficiencies for all species have been scaled by r_H^{-2} during the actual reductions.

To obtain an aperture-independent quantity, we continue our practice of converting the molecular column densities to production rates, Q, by using the Haser model, and assuming an outflow velocity, v, of 1 km s^{-1}. Other values of v can be accommodated by dividing the quoted production rates by v with units of km s^{-1}. We continue to use the scale lengths for OH and C_3 from A'Hearn (1982), as well as the values given by the same source for NH (originally from Combi, 1978). The values for NH given in A'Hearn's Table II are in fact for a heliocentric distance of 0.644 AU, *not* 1 AU. We have scaled these values to 1 AU by r_H^2. In the cases of CN and C_2, we have used the scale lengths recently determined by Combi and Delsemme (1986), even though other values have recently been put forth (Newburn and Spinrad, 1984; Cochran, 1985a) and considerable debate continues as to which set of values are best. A summary of the scale lengths used in this paper is given in Table 1. The effect of altering the scale lengths is most readily observed when greatly differing apertures are utilized or when a comet's heliocentric distance varies greatly. Neither is the case in the present data for Giacobini-Zinner or in the data for the comets we use for comparison. Therefore, the conclusions we reach are relatively independent of the scale lengths used. We again acknowledge the physical limitations of the Haser model, but it is quite adequate for our purposes as long as one keeps in mind that the Haser model scale lengths are not construed to be true molecular scale lengths, but rather values which best fit the observed spatial distribution when using the Haser model formulation. One can usually directly scale the resulting production rates to obtain a more physically relevant production rate such as is obtained by using the vectorial model (Festou, 1981), although this procedure is not necessary if one seeks only to make comparisons among various species for a variety of comets.

For the continuum measurements, we have again employed the quantity $Af\varrho$, the product of the albedo of the grains, the fraction of the aperture filled by grains, and the aperture radius, as a measure of the production of grains (see A'Hearn et al., 1984). The advantage of using $Af\varrho$ to quantify the continuum measurements is two-fold. First, it makes no apriori assumptions as to the size distribution or albedo of the grains. Second, $Af\varrho$ is independent of aperture size if the grains are flowing radially out from the nucleus, their physical characteristics remain unchanged

Table 1. Parameters used in reductions

Species	L/N[a] (erg s^{-1} molecule^{-1})	Haser Scale Length[b] Parent (X_P) (km)	Daughter (X_D) (km)
OH (0-0)	c	4.1 $\times 10^4$	1.16$\times 10^5$
NH (Δv=0)	3.1$\times 10^{-13}$	5.8 $\times 10^3$	4.34$\times 10^5$
CN (Δv=0)	c	2.19$\times 10^4$	3.0 $\times 10^5$
C_3 (λ4050)	1.0$\times 10^{-12}$	1.0 $\times 10^3$	6.0 $\times 10^4$
C_2 (Δv=0)	4.5$\times 10^{-13}$	1.6 $\times 10^4$	1.1 $\times 10^5$

Continuum	$q(F_\odot)$
λ3650	2.244$\times 10^{17}$
λ4845	1.301$\times 10^{17}$
λ6840	1.701$\times 10^{17}$

Notes:

a all fluorescence efficiencies (L/N; for $r_H = 1$ AU) are scaled by r_H^{-2}.

b all scale lengths (for $r_H = 1$ AU) are scaled by r_H^2, except for the CN parent which is scaled by $r_H^{1.5}$.

c L/N for OH (0-0) and CN (0-0) bands are function of $\dot{r}_H$ (tabulated in Schleicher 1983). The CN (0-0) L/N values are multiplied by 1.08 to approximate the contribution of the CN (1-1) band. The resulting values are listed in Table 2.

as they move outward, and if the column of grains is optically thin. One can compute $Af\varrho$ in units of cm from the equation $Af\varrho = q(F_\odot)[r_H^2\Delta/\theta]F_c$, where $q(F_\odot)$ contains the various unit conversions and the solar flux at 1 AU for the appropriate wavelength range and is tabulated in Table 1, r_H and Δ – the heliocentric and geocentric distances, respectively – have units of AU, θ is the aperture diameter in arcsec, and F_c is the observed continuum flux in erg cm^{-2} s^{-1} Å^{-1}. We note that, for completeness, we have also included in Table 1 the value of $q(F_\odot)$ for the IHW λ 6840 continuum filter, even though no observations of G-Z were made with this filter.

Photometric observations of Giacobini-Zinner were obtained on 15 nights at Lowell Observatory in the interval from 15 June to 20 November 1985 using a conventional single-channel photometer having pulse-counting electronics. Observing circumstances are summarized in Table 2. In Table 3 we list the molecular and continuum fluxes, while in Table 4 we present the resulting column densities and production rates for the molecular species and $Af\varrho$ for the two continuum points. Each observational set typically consisted of three 30-sec integrations for each filter along with sky measurements more than one-half degree from the comet. On many of the nights, several sets of data were obtained, often using differing apertures. These are included as separate entries in Tables 3 and 4. The aperture radius projected at the comet ranged from 2.7 10^3 to 16.4 10^3 km for these observations.

3. Discussion

In Fig. 1 we have plotted the production rates of each molecular species observed, as well as $Af\varrho$ for the continuum filters. A formal error, including photon statistics, inaccuracy in centering the

Table 2. Observational circumstances for P/Giacobini-Zinner

1985 UT Date	Telescope	r_H (AU)	Δ (AU)	$\dot{r}_H$ (km s^{-1})	log L/N[a] (erg s^{-1} molecule^{-1}) OH	CN
Jun 15.4	LO 72-in	1.507	1.001	-15.50	-14.741	-12.459
Jun 16.4	LO 72-in	1.498	0.991	-15.40	-14.732	-12.460
Aug 11.4	LO 72-in	1.087	0.546	-7.80	-14.779	-12.409
Aug 12.4	LO 72-in	1.082	0.541	-7.60	-14.785	-12.411
Aug 13.4	LO 72-in	1.078	0.535	-7.30	-14.789	-12.414
Aug 20.4	LO 72-in	1.053	0.502	-5.20	-14.761	-12.451
Sep 09.4	LO 42-in	1.030	0.469	1.40	-14.788	-12.530
Sep 12.4	LO 31-in	1.033	0.473	2.40	-14.729	-12.458
Sep 13.5	LO 31-in	1.035	0.475	2.80	-14.703	-12.430
Sep 16.4	LO 72-in	1.041	0.482	3.80	-14.644	-12.372
Sep 17.5	LO 72-in	1.043	0.485	4.10	-14.628	-12.359
Oct 13.5	LO 72-in	1.159	0.613	10.90	-14.434	-12.329
Oct 15.5	LO 72-in	1.173	0.626	11.30	-14.403	-12.336
Nov 07.5	LO 42-in	1.346	0.768	14.40	-14.233	-12.365
Nov 20.5	LO 42-in	1.457	0.845	15.20	-14.230	-12.360

Notes:
a all fluorescence efficiencies (L/N) are for r_H = 1 AU, and are scaled by r_H^{-2} in the reductions.

Table 3. Photometric data of comet P/Giacobini-Zinner

1985 UT Date	Aperture diam. (θ) (arcsec)	radius (ρ) (10^4 km)	log Emission Band Flux (erg cm^{-2} s^{-1}) OH	NH	CN	C_3	C_2	log Continuum Flux (erg cm^{-2} s^{-1} Å^{-1}) λ3650	λ4845
Jun 15.311	19.5	0.71	-11.42	-12.95	-11.21	-12.33	-12.10	-14.22	-13.91
Jun 15.342	19.5	0.71	-11.43	-13.11	-11.21	-12.40	-12.09	-14.18	-13.90
Jun 15.431	13.7	0.50	–	–	-11.47	-12.62	-12.46	-14.32	-14.01
Jun 16.295	19.5	0.70	-11.42	-12.89	-11.21	-12.31	-12.17	-14.21	-13.89
Jun 16.367	19.5	0.70	-11.38	-13.02	-11.19	-12.34	-12.11	-14.20	-13.89
Jun 16.433	13.7	0.49	–	–	-11.44	-12.56	-12.41	-14.29	-13.98
Aug 11.362	27.6	0.55	-10.12	-11.57	-10.25	-11.39	-11.02	-13.10	-12.84
Aug 11.434	19.5	0.39	-10.38	-11.86	-10.50	-11.56	-11.28	-13.27	-13.00
Aug 12.372	27.6	0.54	-10.13	-11.63	-10.24	-11.34	-11.06	-13.17	-12.87
Aug 12.431	38.9	0.76	-9.89	-11.38	-10.00	-11.15	-10.78	-13.03	-12.74
Aug 12.468	13.7	0.27	-10.65	–	-10.73	-11.73	-11.57	-13.48	-13.16
Aug 13.423	27.6	0.54	-10.15	-11.59	-10.24	-11.34	-11.00	-13.14	-12.86
Aug 20.378	27.6	0.50	-10.15	-11.50	-10.29	-11.32	-11.05	-13.20	-12.89
Sep 09.375	37.8	0.64	-10.17	-11.56	-10.36	-11.39	-11.04	-13.27	-12.96
Sep 09.424	37.8	0.64	-10.16	-11.61	-10.35	-11.37	-11.01	-13.25	-12.98
Sep 09.459	26.5	0.45	-10.41	-11.70	-10.58	-11.54	-11.26	-13.41	-13.12
Sep 09.481	53.4	0.91	-9.92	-11.23	-10.12	-11.14	-10.78	-13.14	-12.84
Sep 12.421	55.1	0.95	-9.85	-11.23	-10.03	-11.42	-10.77	-13.11	-12.85
Sep 12.444	55.1	0.95	-9.85	-11.15	-10.03	-11.25	-10.78	-13.13	-12.85
Sep 12.468	38.7	0.66	-10.11	-11.54	-10.26	-11.45	-11.06	-13.29	-12.99
Sep 13.457	55.1	0.95	-9.86	-11.21	-10.02	-11.35	-10.76	-13.07	-12.87
Sep 16.392	27.6	0.48	-10.37	-11.78	-10.48	-11.63	-11.29	-13.46	-13.16
Sep 17.476	27.6	0.49	-10.35	-11.72	-10.48	-11.67	-11.33	-13.43	-13.15
Oct 13.509	27.6	0.61	-10.50	-12.03	-10.75	-11.79	-11.66	-13.72	-13.33
Oct 15.476	19.5	0.44	-10.70	-12.54	-11.01	-12.19	-11.83	-13.78	-13.47
Nov 07.470	37.8	1.05	-10.33	-13.04	-10.80	-12.09	-11.60	-13.64	-13.38
Nov 20.456	53.4	1.64	-10.21	-12.13	-10.71	-11.88	-11.56	-13.72	-13.37

Table 4. Column densities and production rates of comet P/Giacobini-Zinner

1985	r_H	ρ	log Column Density Integrated Over ρ (molecule)					log Q (molecule s^{-1})					log $Af\rho$ (cm)	
UT Date	(AU)	(10^4 km)	OH	NH	CN	C_3	C_2	OH	NH	CN	C_3	C_2	λ3650	λ4845
Jun 15.311	1.507	0.71	31.13	27.36	29.05	27.48	28.05	28.14	23.74	25.76	23.56	24.75	2.20	2.27
Jun 15.342	1.507	0.71	31.12	27.20	29.06	27.40	28.07	28.13	23.58	25.76	23.49	24.76	2.24	2.28
Jun 15.431	1.507	0.50	–	–	28.79	27.19	27.69	–	–	25.76	23.46	24.64	2.25	2.33
Jun 16.295	1.498	0.70	31.10	27.41	29.04	27.48	27.96	28.11	23.79	25.75	23.57	24.66	2.19	2.29
Jun 16.367	1.498	0.70	31.14	27.28	29.06	27.45	28.03	28.16	23.66	25.77	23.54	24.72	2.21	2.29
Jun 16.433	1.498	0.49	–	–	28.81	27.23	27.73	–	–	25.78	23.50	24.69	2.27	2.35
Aug 11.362	1.087	0.55	31.65	27.94	29.15	27.60	28.32	28.65	24.34	25.90	23.78	25.02	2.62	2.64
Aug 11.434	1.087	0.39	31.39	27.65	28.90	27.43	28.06	28.65	24.27	25.91	23.78	25.01	2.60	2.63
Aug 12.372	1.082	0.54	31.64	27.86	29.15	27.65	28.27	28.64	24.27	25.91	23.83	24.97	2.55	2.61
Aug 12.431	1.082	0.76	31.88	28.12	29.39	27.84	28.55	28.63	24.30	25.90	23.86	25.02	2.53	2.59
Aug 12.468	1.082	0.27	31.12	–	28.66	27.25	27.76	28.65	–	25.94	23.79	24.96	2.54	2.62
Aug 13.423	1.078	0.54	31.61	27.88	29.15	27.63	28.31	28.63	24.29	25.91	23.81	25.02	2.56	2.61
Aug 20.378	1.053	0.50	31.51	27.90	29.06	27.57	28.19	28.55	24.34	25.86	23.79	24.93	2.46	2.53
Sep 09.375	1.030	0.64	31.44	27.76	28.99	27.43	28.12	28.29	24.04	25.60	23.53	24.68	2.20	2.27
Sep 09.424	1.030	0.64	31.44	27.72	28.99	27.45	28.15	28.29	23.99	25.61	23.55	24.71	2.22	2.25
Sep 09.459	1.030	0.45	31.19	27.63	28.77	27.27	27.90	28.30	24.13	25.63	23.54	24.70	2.22	2.27
Sep 09.481	1.030	0.91	31.69	28.10	29.23	27.68	28.38	28.29	24.16	25.60	23.63	24.70	2.18	2.24
Sep 12.421	1.033	0.95	31.70	28.10	29.25	27.41	28.40	28.28	24.14	25.60	23.35	24.70	2.20	2.23
Sep 12.444	1.033	0.95	31.70	28.18	29.26	27.58	28.39	28.28	24.22	25.60	23.51	24.69	2.18	2.23
Sep 12.468	1.033	0.66	31.45	27.79	29.02	27.37	28.11	28.28	24.05	25.61	23.46	24.65	2.18	2.24
Sep 13.457	1.035	0.95	31.67	28.13	29.24	27.48	28.42	28.25	24.17	25.59	23.42	24.72	2.24	2.21
Sep 16.392	1.041	0.48	31.12	27.58	28.74	27.22	27.91	28.19	24.04	25.56	23.45	24.67	2.17	2.23
Sep 17.476	1.043	0.49	31.13	27.65	28.74	27.18	27.87	28.20	24.10	25.56	23.41	24.63	2.20	2.24
Oct 13.509	1.159	0.61	31.09	27.63	28.73	27.36	27.84	28.04	23.98	25.42	23.49	24.49	2.10	2.26
Oct 15.476	1.173	0.44	30.88	27.15	28.50	26.99	27.70	28.08	23.72	25.44	23.27	24.59	2.22	2.29
Nov 07.470	1.346	1.05	31.38	26.94	29.04	27.38	28.23	28.03	23.01	25.41	23.27	24.58	2.28	2.30
Nov 20.456	1.457	1.64	31.65	28.01	29.28	27.75	28.42	28.03	23.84	25.38	23.44	24.51	2.16	2.27

comet in the aperture, contamination of continuum filters, and intrinsic activity of the comet or other violated model assumptions, cannot accurately be determined. The most practical estimate of the observational errors associated with the data (not including the contribution due to contamination) is obtained from the spead in the results for each species on a single night (if multiple observations were obtained). As can be seen, the CN and OH measurements are clearly the best determined, followed by C_2, and then C_3 and NH. The direct effect of contamination of the continuum filters is greatest for the weakest emission bands – C_3 and NH – resulting in production rates which may be as much as 20% underdetermined. The ground-based OH observations shown in Fig. 1 have been overlain by the results obtained with the International Ultraviolet Explorer (IUE) spacecraft (McFadden et al., 1987), after first adjusting the IUE values for the difference in model parameters used. The agreement to within ~15% in the absolute calibration between the two sets of data is similar to that obtained previously for several other comets (e.g. A'Hearn et al., 1983; Schleicher et al., 1986), even though the IUE aperture ($10'' \times 15''$) is much smaller than the apertures we have used, and the two observing techniques are quite different. To obtain H_2O production rates consistent with the Festou vectorial model, these Haser model production rates for OH should be multiplied by 1.5.

The asymmetric behavior about perihelion is present for all species and will be discussed later in this section. First we will look at the relative abundances among the various species. Listed in Table 5 are ratios of production rates for G-Z, along with corresponding "normal" values obtained from a sample of 6 comets previously measured. The quoted G-Z abundance ratios are the mean values for all of the data sets obtained throughout the apparition (i.e. from Table 4). The comparison values (hereafter defined as "normal") were deduced from observations obtained in our comet program since spring 1984 when we switched to the same IHW filters used for the G-Z observations. Similarly, only comets whose heliocentric distance at the time of the observation was less than 1.7 AU were included to minimize the effect of errors in the adopted scale lengths. The comets in the "normal" sample – P/Boethin, P/Crommelin, P/Encke, Hartley-Good, P/Halley, and Thiele – were reduced using the same model parameters as discussed in Sect. 2 and mean abundance ratios were computed for each comet individually. Unweighted mean production rate ratios and formal sigmas of the mean were then calculated for the six comets, and these are listed in Table 5. While these sigmas may be somewhat artificial due to the small sample size, they are indicitive of the similarity usually seen in abundance ratios among comets. We have also compared our normal, relative production

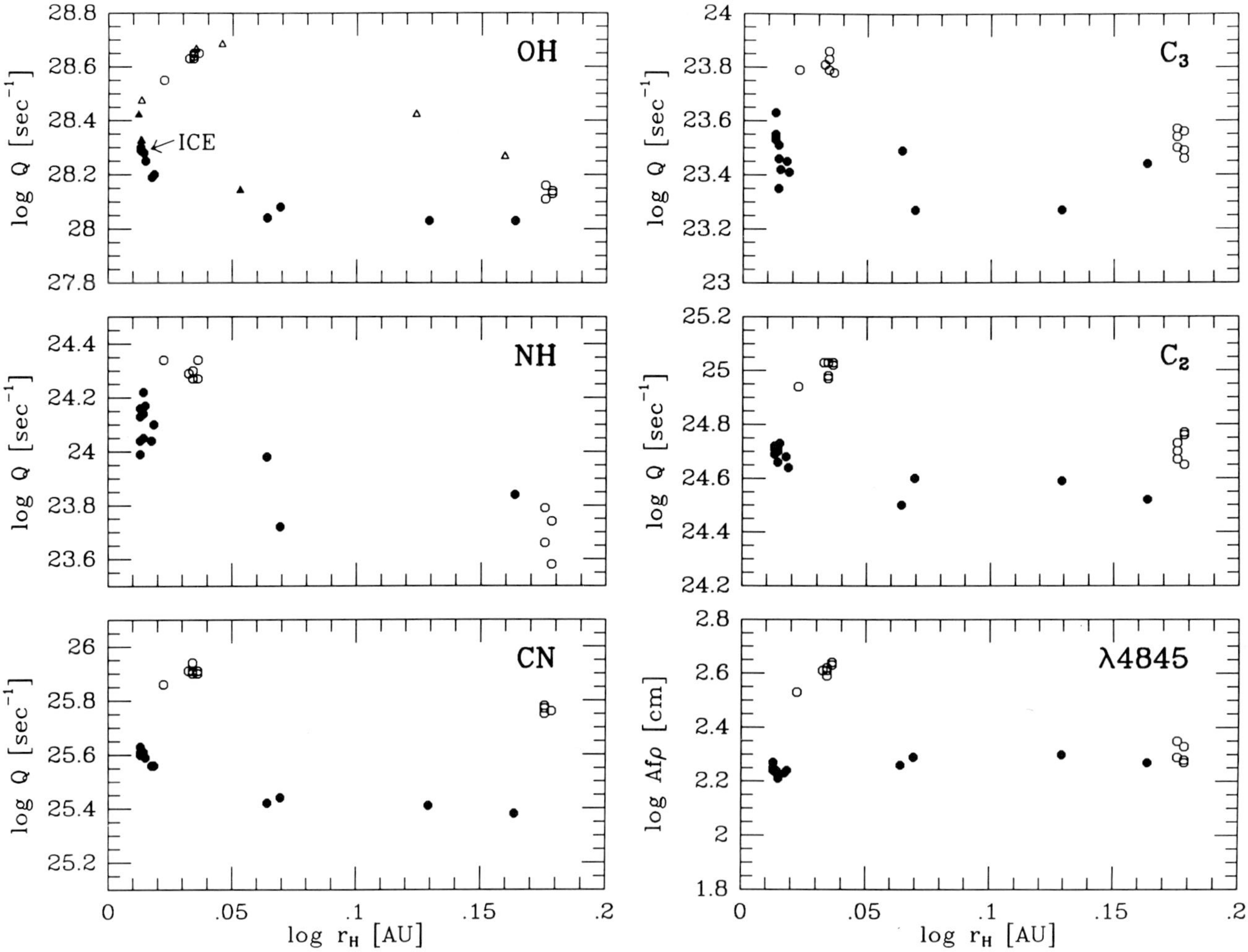

Fig. 1. Production rates as a function of heliocentric distance (r_H) for each of the observed molecular species and $Af\varrho$ (defined in the text) for the continuum at λ4845. Pre-perihelion photometry data are given by open circles, while post-perihelion data are given by filled circles. In the plot of OH production, IUE data (McFadden et al., 1986) are shown plotted as triangles after adjustment for differing model parameters. The time of the ICE spacecraft encounter is indicated by the arrow in the plot of the OH data

Table 5. Relative molecular production rates

Ratio	Giacobini-Zinner	"Normal"
Q_{OH}/Q_{CN}	430	330±160
Q_{OH}/Q_{C_2}	3500	480±240
Q_{OH}/Q_{C_3}	56000	5100±2000
Q_{OH}/Q_{NH}	18000	3800±1500
Q_{CN}/Q_{C_2}	8.7	1.5±0.3
Q_{CN}/Q_{C_3}	135	17±5
Q_{CN}/Q_{NH}	43	13±4
Q_{C_2}/Q_{C_3}	15	11±2
Q_{C_2}/Q_{NH}	5	9±3

rates to those recently determined by Cochran (1987a), who used a much larger, but less homogeneous, data set. While at first it appears that Cochran's production rate ratio for CN-to-C_2 of 0.68 is differs significantly from our value of 1.5, we have confirmed that this difference is due to the use of differing fluorescence efficiencies and scale lengths. Our 'normal' production rate ratios are in fact completely consistent with Cochran's. It is readily apparent from Table 5 that the abundances of C_2, C_3, and NH relative to CN and OH are all quite low in G-Z as compared with those normally seen in comets.

As mentioned in the introduction, C_2 emission in G-Z had been noted to be unusually weak as compared to that of CN already more than 50 years ago (Bobrovnikoff, 1927) and both C_2 and C_3 were shown to be low with respect to CN by Herbig (1976). These results were first confirmed spectrophotometrically early in the current (1985) apparition by Schleicher (1985) and Cochran (1985). By also observing the OH bands in the near-ultraviolet, Schleicher was able to show that it was in fact the pure carbon species which were anomalous when compared to the dominant species, H_2O, rather than CN having an unusually high abundance. Our data confirm that conclusion. While the C_2 production rate mimics the total gas production rate fairly well for G-Z (at least over the range of heliocentric distances reported here), it is clear that CN is a better monitor of the absolute gas production rate if none of the water dissociation products are observable. In order of increasing depletion when compared to OH, are CN (normal), NH (one-fifth normal), C_2 (one-seventh normal), and C_3 (one-eleventh normal). Based on the quoted sigmas of the normal ratios, the depletions of NH, C_2, and C_3 in G-Z are all significant at about the 10-sigma level or better. When comparing abundances among the minor constituents directly, one sees that the C_2/C_3 ratio is high at the 2-sigma level compared to what is typical for other comets, while the C_2/NH ratio is low by an equivalent amount. While the probable contamination of the continuum filters, resulting in artificially low molecular abundance determinations, is important, it is not sufficient to modifiy these results to a large degree. Even if contamination exists at the 10 % level of that of the continuum (more than twice what we believe to be the case), it would translate to under determinations by approximately 15 % of the abundances of C_2 and NH, and by approximately 30 % for the abundance of C_3. The fractional depletions of NH, C_2, and C_3 relative to OH would at this level of continuum contamination be one-fourth, one-sixth, and one-eighth, respectively. We stress that these depletions are for a quite large contamination, and that the true values are very probably between the formal results and these values. Insofar as incorrect model parameters are concerned, we reiterate that the normal ratios quoted are for comets having similar viewing circumstances, resulting in relatively small variations in the ratios even if scale lengths are drastically in error. Errors of this nature could indeed account for much of the dispersion within the normal results, but are insufficient to explain the G-Z results. In addition, while the Haser model assumes isotropic, quasi-static equilibrium, the manifestations which would result if this were not the case would be more apparent in several of the comets in our normal sample which showed signs of temporal variations than for G-Z. While temporal variations could explain particular data points, it is inconsistent with the total set of data. We therefore conclude that the observed depletions of nearly an order of magnitude in C_2 and C_3 are real, as is the smaller depletion in NH (by a factor of 5). It remains uncertain whether the deviations of the relative abundances among C_2, C_3, and NH from normal are significant. The depletions of C_2 and C_3 reported here are similar to those observed by Cochran (1987b), who gives production rates for both pure carbon species as being one-fifth normal (using CN, rather than OH as the base), compared to our formal values of one-sixth and one-eighth, respectively. A relative determination of the gas-to-dust ratio for G-Z can be made by computing the ratio of the OH production rate to $A f \varrho$, resulting in a value of 1 10^{26}. This value is nearly identical with that for three comets defining normal, while the remaining three comets have widely differing values. Therefore, while the gas-to-dust ratio is known to vary considerably among comets as a whole, it is clear that G-Z was not at all unusual (i.e. an extreme case) in this regard. Finally, the color of the grains in G-Z, characterized by the difference in $A f \varrho$ determined from the two continuum filters, is red by 15 %. While slightly redder than typical, it falls within the range observed in our sample set of comets. More importantly, the differences among the sample set may be due entirely to contamination in the continuum bands, and so no significance has been attached to these color determinations at this time.

While we are not yet ready to put forth a model to explain the chemistry in detail, a few key items which must be accounted for can be stated. First, G-Z has a normal ratio of CN to OH, while C_2 and C_3 are nearly equally depleted by a very large amount. Perhaps more important is the depletion of NH, since it is not a carbon compound at all. Unfortunately, no data has been published as to whether NH_2 is depleted in G-Z, and if so, by how much. We suspect that the abundances of NH and NH_2 in G-Z may prove to be the one of the more useful clues obtained at this apparition for use by chemical modelers. It is difficult to imagine any mechanism acting in the coma of G-Z which could give rise to anomolous abundances in the coma from normal abundances at the source. Hence, we conclude that the nucleus itself must have, at least in localized regions if not throughout, abnormal parent abundances. The most obvious explanation is that G-Z formed in a different portion of the solar nebula than most other comets which thus far have been measured. However, we cannot exclude the possibility that an isolated pocket of highly volatile material having unusual abundances is the primary source of the coma. For instance, many of the recent results for Halley independently showed that perhaps as much as 90 % of Halley's coma is produced from localized regions of activity, and that most of its surface may be virtually inert (e.g. Keller et al., 1986). A model of G-Z having a rapidly rotating nucleus (as proposed by Sekanina, 1985) with one active area having abnormal abundances cannot be excluded by our data. One possible scenario would picture the nucleus of Giacobini-Zinner as being originally composed of sub-entities, or "cometesimals", at least one of which has very different abundances and is currently exposed at the surface of the nucleus. This cometesimal would need to be relatively large to survive over the many apparitions for which abnormal abundances have been noted. Based on the observed production rate of OH, we estimate that at least 1 10^{14} gm of water has been released during the time that anomalous abundances have been reported. This amount of water represents more than 1 % of the expected mass of a typical comet nucleus. This hypothesis, therefore, requires relatively large cometesimals accreting in the final stages of a comet's formation. Such a line of reasoning would further imply that G-Z may not have formed in a different part of the solar nebula from most other comets, but that it may actually be quite normal in its total chemical and physical make-up, and that the fraction of comets showing these characteristics would be indicitive of the percentage of cometesimals in all comets having unusual abundances. In fact, G-Z is not completely unique in its relative abundances. Swings and Haser (1956) listed several comets apparently having low C_2 to CN ratios, while A'Hearn et al. (1979) showed that P/Grigg-Skjellerup had low C_2 and C_3 abundances when compared to CN. It is not known, though, whether in these cases the CN abundance is normal (compared to OH), making these comets similar to G-Z, or instead that the CN abundance is in fact abnormally high. A major difficulty with this scenario is that even if only 10 % of the coma comes from the remaining, relatively inert nucleus, then the abundances of the active cometesimal must be extremely abnormal to compensate for dilution. One must also explain why this cometisimal is active while the remainder of the surface is not. A third hypothesis is that G-Z has evolved differently from most comets, resulting in different surface properties (e.g. Cochran, 1987b). This seems unlikely, however, due to the large quantity of material which would have to have been modified differently than for most other comets. In summary, if the majority of G-Z coma can be shown to come from the surface as a whole, rather than from a discrete active region on a rapidly rotating body, then it is most likely that the nucleus of G-Z is homogeneous throughout in its unusual abundances.

We now return to the characteristics of the over-all light curve for this apparition. All observed species showed similar, pronounced asymmetry about perihelion, with pre-perihelion abundances typically being greater by more than a factor of 2 than post-perihelion abundances over the heliocentric distance range of 1.0 to 1.5 AU. The peak production rate occurred in late July/early August, four-to-six weeks prior to perihelion and the ICE encounter. Because of the shape of the production rate curves and

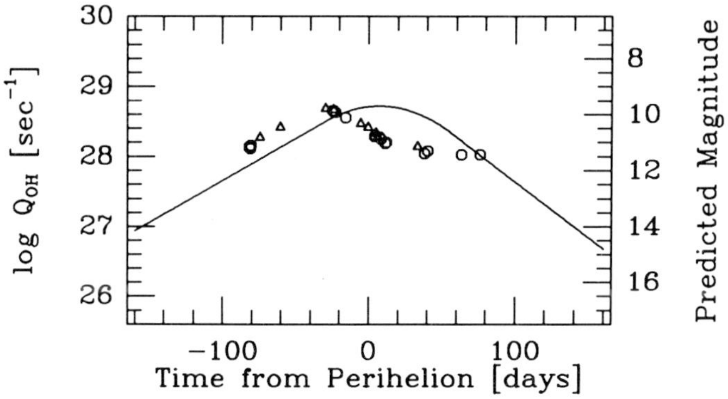

Fig. 2. OH production rate of comet Giacobini-Zinner as a function of time from perihelion. The open circles are observations from this paper. IUE data are represented by triangles. The curve represents the normalized total magnitude prediction by Sekanina (1985)

the relatively small range of heliocentric distances over which G-Z was observed, we have not determined the slopes of Q for each species as a function of r_H. These slopes clearly would not be indicative of Q at larger distances, as can be seen in the case of OH post-perihelion when the slope is nearly zero (Fig. 1). The behaviour of each individual molecular species varies somewhat with r_H, although this may be partially attributed to incorrect scale lengths. Incorrect scale lengths cannot explain, however, the differences in pre- and post-perihelion behaviour of CN when compared to the behaviour of the other species. All other species, continuum included, had roughly the same production rate (within 50%) near 1.5 AU both pre- and post-perihelion, while CN varied by a factor of 2.5. Because the geocentric distance (Δ) was similar pre- and post-perihelion, this difference in CN as compared to other species cannot be due to inaccurate scale lengths coupled with aperture effects. Since the Swings effect is already accounted for in the reductions for CN, we must conclude that the behaviour of CN is tracing real differences in the mechanism of CN production as distinguished from those of the other observed species. While NH does not have widely differing production rates pre- and post-perihelion near 1.5 AU, it is the only species which has a lower value pre-perihelion. If the difference is significant (the sigmas are relatively large), quite possibly it is simply a reflection of the Swings effect, the correction for which, as we noted in Sect. 2, is not known for NH. Our estimated variation of the fluorescence efficiency by up to a factor of 3 with heliocentric velocity is consistent with the variation in NH as compared to OH and C_2, so we will not attempt to attribute to these data any physical significance at this time.

A pronounced asymmetry of the light curve about perihelion has been observed on previous apparitions, and indeed these asymmetries have been modelled by Sekanina (1985) to determine the rotational axis, rotation rate, and precession of the nucleus on the assumption that the asymmetry was due to seasonal effects. His best model fit to visual light curves obtained since 1900 implies a very fast rotation period (less than 2 h), and precession which resulted in the nucleus tipping over during the 1959 apparition, thereby explaining the large change in the visual light curve which occurred at that time. For comparison to the visual light curves observed at previous apparitions as shown by Sekanini (his Figs. 1–6), we have plotted in Fig. 2 the production rate of OH as a function of time from our 1985 observations. One characteristic, the relatively flat "fall-off" following perihelion, is similar to that

Table 6. Estimated production rates for P/Giacobini-Zinner at the ICE encounter

log Q (molecule s^{-1})					
OH	NH	CN	C_3	C_2	$Af\rho$ (cm)
28.3	24.1	25.6	23.5	24.7	2.2

witnessed in 1972, but otherwise the 1985 apparition appears quite different from previous ones. Sekanina also predicted the values of the coefficients for the visual light curve for the 1985 apparition, and, using these, we have computed the normalized total magnitude light curve. This curve is also shown in Fig. 2. The observations are clearly dissimilar to the prediction, with the time of maximum brightness differing by more than one and a half months. Not having the visual light curve for the 1985 apparition available, we do not known with certainty whether the visual light curve mimics the gas production rate. It has been shown, in fact, in the case of P/Encke (A'Hearn et al. 1985) that the two do not always follow one another. However, in the case of G-Z the shape of the OH production rate curve is quite similar to the shapes of the curves for C_2 and $Af\varrho$, the two quantities which should best match what is seen visually. Consequently, we assume that the OH production does match the visual brightness. The difference in the Sekanina prediction and our values seems to imply that further modeling in necessary before the light curves of G-Z can be considered fully understood.

Finally, our estimates of the production rates of the various molecular species and dust at the time of the ICE spacecraft's fly-by of G-Z are summarized in Table 6. While we have no data taken on the night of the encounter itself (11 Sept.), we do have data a few days before and immediately after the encounter. Data obtained during the encounter by IUE and our own data from August and September indicate that G-Z was essentially quiescent, and so interpolation of our data to the time of the encounter should be quite satisfactory. We again remind the reader that the values given in Table 6 are based on the Haser model parameters given in Table 1, and that a water production rate (presumably equal to the total gas production rate) using Festou's vectorial model would be approximately 50% greater than our OH production rate of 2.0 10^{28} molecules s^{-1}.

4. Conclusion

Observationally we have shown that the coma of comet P/Giacobini-Zinner is depleted in C_2, C_3, and to a lesser degree NH, when compared to the dominant species OH (and therefore H_2O). The CN abundance is normal. These observed relative abundances imply unusual parent molecular relative abundances at the nucleus. This fact can most simply be explained by a model of G-Z in which the comet has a homogeneous nucleus which formed in a different portion of the solar nebula than those of most other comets. It is possible, however, that G-Z is heterogeneous, that the coma is primarily produced by one active region, and that this region is the remnant of a cometesimal having abnormal abundances. In this scenario, it is only the atypical cometestimal which would have formed in a different part of the solar nebula. Cometesimals such as this could in fact be incorporated in most comets, and the percentage of comets showing the characteristics of G-Z may be indictive of the fraction of this material in most comet nuclei. The characteristic asymmetry about perihelion seen in the total visual magnitude of comet P/Giacobini-Zinner in previous apparitions is observed in the production rates of all species. However, the detailed nature of the asymmetry we observed is markedly different. In particular, pre-perihelion abundances were more than a factor of 2 greater than post-perihelion abundances over the heliocentric distance range of 1.0 to 1.5 AU, and the peak production rate occurred in late July/early August – more than a month before perihelion. These results are inconsistent with predictions by Sekanina (1985), indicating that further modeling in required.

Acknowledgements. We thank M. A'Hearn for numerous helpful discussions. S.J. Bus assisted in many of the reductions. This research was supported by NASA Grant NGR 03-003-001 and by the government of Western Australia.

References

A'Hearn, M.F.: 1982, in *Comets*, ed. L. Wilkenning, Univ. Arizona Press, p. 433

A'Hearn, M.F.: 1985, *IHW Photometry Polarimetry Net Letter*, 30 Aug. 1985

A'Hearn, M.F.: 1986a, *IHW Photometry Polarimetry Net Letter*, 3 Feb. 1986

A'Hearn, M.F.: 1986b (private communication)

A'Hearn, M.F., Birch, P.V., Feldman, P.D., Millis, R.L.: 1985, *Icarus* **64**, 1

A'Hearn, M.F., Millis, R.L., Birch, P.V.: 1979, *Astron. J.* **84**, 570

A'Hearn, M.F., Millis, R.L., Birch, P.V.: 1981, *Astron. J.* **86**, 1559

A'Hearn, M.F., Millis, R.L., Thompson, D.T.: 1983, *Icarus* **55**, 250

A'Hearn, M.F., Thurber, C.H., Millis, R.L.: 1977, *Astron. J.* **82**, 518

A'Hearn, M.F., Schleicher, D.G., Feldman, P.D., Millis, R.L., Thompson, D.T.: 1984, *Astron. J.* **89**, 579

Bobrovnikoff, N.T.: 1927, *Astrophys. J.* **66**, 439

Cochran, A.L.: 1985a, *Astron. J.* **90**, 2609

Cochran, A.L.: 1985b, *Bull. Am. Astron. Soc.* **17**, 920

Cochran, A.L.: 1987a, *Astron. J.* **92**, 231

Cochran, A.L.: 1987b, *Astron. J.* **92**, 239

Combi, M.R.: 1978, *Astron. J.* **83**, 1459

Combi, M.R., Delsemme, A.H.: 1986, *Astrophys. J.* **308**, 472

Festou, M.C.: 1981, *Astron. Astrophys.* **95**, 69

Herbig, G.H.: 1976, in *The Study of Comets*, eds. B. Donn, M. Mumma, W. Jackson, M. A'Hearn, NASA SP-393, p. 136

Keller, H.U. Arpigny, C., Barbieri, C., Bonnet, R.M., Cazes, S., Coradini, M., Cosmovici, C.B., Delamere, W.A., Huebner, W.F., Hughes, D.W., Jamar, C., Malaise, D., Reitsema, H.J., Schmidt, H.U., Schmidt, W.K.H., Seige, P., Whipple, F.L., Wilhelm, K.: 1986, *Nature* **321**, 320

Litvak, M.M., Rodriguez-Kuiper, E.N.: 1982, *Astrophys. J.* **253**, 622

McFadden, L.A., A'Hearn, M.F., Feldman, P.D., Bohnhardt, H., Rahe, J., Festou, M.C., Brandt, J.C., Maran, S.P., Niedner, M.B., Smith, A.M., Schleicher, D.G.: 1987, *Icarus* **69**, 329

Newburn, R.L., Spinrad, H.: 1984, *Astron. J.* **89**, 289

Schleicher, D.G.: 1983, Ph. D. Dissertation, Univ. of Maryland

Schleicher, D.G.: 1985, *Publ. Astron. Soc. Pacific* **97**, 903

Schleicher, D.G., Millis, R.L., Tholen, D., Lark, N., Birch, P.V., Martin, R., A'Hearn, M.F.: 1986, *20th ESLAB Symp.*, ESA SP-250, Vol. **I**, p. 565

Swings, P., Haser, L.: 1956, *Atlas of Representative Cometary Spectra*, University of Liège Astrophysical Institute

Sekanina, Z.: 1985, *Astron. J.* **90**, 827

von Rosenvinge, T.T., Brandt, J.C., Farquhar, R.W.: 1986, *Science* **232**, 353

Astron. Astrophys. 187, 539–542 (1987)

ASTRONOMY AND ASTROPHYSICS

Circular polarization near the nucleus of comet P/Halley*

K. Metz and R. Haefner

Universitäts-Sternwarte München, Scheinerstraße 1, D-8000 München 80, Federal Republic of Germany

Received December 18, 1986; accepted March 25, 1987

Summary. Circular polarization measurements at different positions near the nucleus of comet P/Halley have been carried out for phase angles normally producing a high linear polarization. The observed circular polarization degree up to 2% was highly variable but never changed its sign.

Key words: circular polarimetry – comets

1. Introduction

Originally it was intended to perform simultaneous polarimetric (linear and circular), photometric and spectroscopic observations of comet P/Halley near the greatest possible Sun-Comet-Earth phase angle of 66°1 at 1986, March 20, when in the cometary coma a fairly strong linear polarization could be expected to be caused by Mie-scattering. However, since at that time no suitable polarimeter was available at the ESO site we decided to convert the standard one-channel photometer of the ESO 0.5 m telescope into an auxiliary polarimeter. Since at that time Halley was visible only in the last part of the night this conversion had to be done in a simple way and should not affect the preceding photometric observations of other objects. For reasons given in the next section only circular polarization measurements were performed. Contrary to theoretical predictions of circular polarization degrees up to 4% in comets (Dolginov and Mitrofanov, 1975) no hint for circular effects could be found in either comet Kohoutek or Bradfield or West (Michalsky, 1981) or Tago-Sato-Kosaka (Wolf, 1972). This negative result could be caused by using too large apertures (diameter up to 42″) centered on the core of the comet thus averaging over too extended areas. This is supported by the fact, that the extremely dustrich comet Bennett showed up to 5% and even once 18% circular polarization measured for a phase angle of about 90° with a 14″ aperture placed on different areas within the coma (Metz, 1970).

2. Structure of the auxiliary polarimeter

The determination of polarization requires that the intensity of the light will subsequently be measured in at least two (determination of the Stokes parameters I, V for circular polarization) or in three (determination of the Stokes parameters I, Q, U for linear polarization) selected angle positions of a suitable polarization modulator. To reduce the error of a single polarization measurement this sequence must be repeated several times.

The photometer installed at the 0.5 m telescope allows to select a certain filter sequence and to change quickly the different filter positions within this sequence. Therefore it was nearly obvious to insert a set of appropriately oriented polarization modulators into the filter wheel thus allowing for a rapid change from one modulator position to the desired next one.

Though moonlight, often essentially limiting the efficiency of linear polarimetric measurements, was not present during our observing runs, we restricted ourselves to a determination of the circular polarization for two reasons: The time during which useful polarimetric measurements could be performed was restricted to about 50 min at the beginning and 120 min at the end of our observing time. The determination of linear polarization requires, however, much more time than circular polarimetry due to the necessary determination of three Stokes parameters and more frequent sky measurements. Moreover, there is still a need for circular polarization measurements of comets. Therefore such observations were strongly recommended by the International Halley Watch (IHW).

To measure the Stokes parameter V (describing circular polarized light) the right or left handed circular polarization has to be converted into a linear one by means of two quarter-wave plates placed in front of the polarizers and rotated with their fast axes exactly by $-45°$ (modulator A) and $+45°$ (modulator B) versus the corresponding transmission axes of the polarizers (the designation of the fast axis is in contradiction to the description given by E. Käsmemann Ltd). Thus each modulator set was composed of one polarizing – and two precisely oriented retarding – foils.

An economic and above all simple method of modulating the signal was to use Polarex polarizers (typ P–W 64 with a transmission of 32% for unpolarized white light) and quarter-wave sheets (typ P–V 140 with retardation of $\pi/2$ for $\lambda = 560$ nm) as well. Both modulators were fabricated by E. Käsemann Ltd./Germany. To avoid instrumental problems that can arise from highly linear polarized light each Polaroid behind each Polarex sheet was combined with a quarter-wave foil in order to convert the linear polarization into a circular one. Therefore these foils were placed with their fast axes rotated by 45° versus the polarization axis of the polarizers. Laboratory tests showed that the problem of cross-talking between linear and circular Stokes parameters, which may arise from the presence of a strong cometary linear polarization, is not relevant within the error limits of 0.5% (see Sect. 3).

Send offprint requests to: K. Metz

* Based on observations collected at the European Southern Observatory

3. Observations and reduction

The modulators were carefully aligned and fixed in two adjacent holes of the filter wheel of the pulse counting one-channel photometer equipped with a thermoelectric cooled RCA 31034 Quantacon. Comet Halley showed a pronounced central condensation with a diameter of a few arcsec as seen through the view finder of the photometer. The cross wire and concentric rings of this view finder allowed an accurate positioning of selected areas near the core in the S–N and E–W directions to be measured. The tracking speed of the telescope was adjusted to the actual motion of the comet. However, the observations of Halley had to be performed through high air mass. Therefore differential refraction could cause a very small uncompensated motion. This could influence the result of the two subsequent measurements necessary for one polarization determination since due to photon noise an integration time of 20 s per modulator had to be chosen. To overcome this the modulator sequence ABBA was taken for one single polarization measurement. Mostly a circular diaphragm opening of 10″ was used. To study the suspected correlation between aperture diameter and measured degree of polarization several larger apertures were applied for the central condensation and a selected area nearby. The unpolarized standard stars of solar type HD 98281, HD 102438 and HD 125184 (recommended by the IHW) were measured every night before and, if possible, after the Halley observations in order to derive the instrumental effects. Also sky measurements were obtained mostly at the beginning and the end of the Halley monitoring. During all these observations perfect meteorological conditions prevailed except on March 19, when some measurements were disturbed by cirri. From March 16 to 20 altogether about 140 polarimetric observations for different areas in Halley's coma could be obtained.

The wavelength dependent deviation from the quarter-wave condition of the retarders as well as the cometary emission lines caused a depolarization. To derive roughly a quantitative value for this depolarization the wavelength dependence of the resulting signal influenced e.g. by the spectral distribution of the coma, the transmission through the terrestrial atmosphere and the modulators as well as the characteristics of the multiplier was determined performing *UBVRI* measurements of a sunlike unpolarized star through the Polarex polarizers which mainly define the transmission curve of the modulators. This allowed to calculate the wavelength dependent influence of the departure from the quarter-wave condition of the retarders on the depolarization. It should be noted that due to this deviation the component of the unknown linear polarization parallel to the transmission direction of the polarizers can also influence the measured polarization. A calculation of the resulting Mueller matrices of both retarders shows that the countrates are proportional to:

$$I + Q\cos\tau\cos 2\delta_A + U\cos\tau\sin 2\delta_A + V\sin\tau$$

for modulator A and

$$I + Q\cos\tau\cos 2\delta_B + U\cos\tau\sin 2\delta_B - V\sin\tau$$

for modulator B

The Stokes parameter (Q, U) describe the unknown linear polarization and τ near 90° is the retardation of the phase plates. δ_A, δ_B are the angle orientations of the retarders A, B within the filter wheel. In order to exclude falsifications of the measured circular polarization by the linear one the two modulators were equally oriented within the filter wheel with $\delta_A = \delta_B \approx 0°$. Only once modulator B was rotated against modulator A by exactly 90° for a check of cross-talking. Assuming a different localisation for the origin of continuum and emission lines and assuming further no circular polarization for these lines which then only increase the total intensity, the influence of the emissions could be inferred from intensities of IDS spectra obtained a few days prior to the polarimetric observations. Summarizing the influence of all these effects a depolarization factor of $D = 1.34$ was derived.

The (sky subtracted) photon count rates obtained through the modulators A and B are

$$N_A = \alpha I (1 + (P'_c + P'_i))/2 \quad \text{modulator A}$$
$$N_B = \beta I (1 - (P'_c + P'_i))/2 \quad \text{modulator B}$$

where α, β refer to different absorptions within the modulators. I is the intensity of light, P'_c and P'_i are the measured cometary and instrumental polarizations respectively.

It proved convenient to use the following approximation for the reduction which is valid for polarization degrees up to 10%:

$$P'_c = (\ln (N_A/N_B)_c - R)/2$$

where $R = \ln(\alpha/\beta) + 2P_i \equiv \ln(N_A/N_B)_{St}$ is derived from the measurements of the unpolarized standard stars. The intrinsic circular polarization of the comet is then

$$P_c = P'_c/D$$

Table 1 shows the mean values of R for each night. From the standard deviations and from earlier experience the polarimetric error of the device is estimated to be about 0.5%. For the sake of clearness only polarization values P_c with $P'_c \geq 0.5\%$ are therefore presented in Table 2. Subsequent measurements at the same position obtained through the same diaphragm opening were averaged. The standard deviations are, except a few cases, within a limit of 0.5%. Theses larger deviations are not caused by bad measurements rather than by larger differences in the polarization degree which were derived from measurements showing normal photon statistics. Measurements with abnormal photon statistics which mainly were obtained on March 17 (especially in East-direction) were rejected. Red (dust) images of the near-nucleus region observed by West et al. (1986) with the ESO 40 cm double astrograph show that the nucleus was extremely active during this night as compared with the remaining ones here in question. Thus brightness variations on short time scales caused by jets emanating from the sun-heated side of the core influenced

Table 1. Mean values for R as determined for different nights using the comparison stars. N gives the number of measurements where one measurement consists of the sequence ABBA. For the first night the modulators were fixed in a position different by 90° with respect to that used for the following nights

Date 1986 March	R	s.d.	N
16	+0.039	0.003	6
17	−0.020	0.003	10
18	−0.021	0.002	8
19	−0.021	0.005	6
20	−0.020	0.003	8

Table 2. Journal of circular polarization measurements of comet Halley

DATE 1986 March	Time (UT)	Direction*)	Distance from core (′)	Diaphragm**) (″)	Number of measure-ments	Start and end of ob-servation (UT)	P (%)	s.d.
16	8^h 57^m	N	0.5	10	2		-2.2	0.1
	9 01	S	0.5	10	2		-0.9	0.4
	9 05	W	0.5	10	2	8^h 36^m	-1.5	1.2
	9 10	N	1.0	10	2	–	-1.8	1.1
	9 14	W	1.0	10	2	9^h 29^m	-1.4	0.3
	9 17	W	3.0	10	1		-1.2	–
	9 21	E	0.5	10	1		-1.5	–
	9 24	CC	0	15	1		-0.9	–
17	8 44	CC	0	21	2	8^h 21^m	-0.8	0.2
	8 58	E	1.0	10	1	–	-2.0	–
	9 12	W	1.0	10	2	9^h 31^m	-0.7	0.2
18	8 25	CC	0	10	2	7^h 58^m	-1.1	0.4
	8 30	E	0.5	10	2	–	-0.7	0.4
	8 34	E	0.25	10	2		-1.0	0.4
	8 49	E	0.25	15	1	9^h 30^m	-1.0	–
19***)	8 14	CC	0	15	2		-1.2	0.1
	8 29	E	0.25	10	1	7^h 47^m	-1.8	–
	8 33	E	0.5	10	2	–	-2.3	0.4
	9 12	N	0.25	10	1	8^h 35^m	-0.7	–
	9 16	N	0.5	10	2		-0.7	0.0
	9 20	S	0.25	10	2		-1.4	0.3
	9 23	S	0.5	10	2	9^h 12^m	-0.7	0.1
	9 27	W	0.25	10	2	–	-1.2	0.5
	9 35	E	0.25	10	1	9^h 47^m	-1.7	–
	9 42	E	0.5	10	1		-1.1	–
20	8 13	CC	0	15	2		-1.0	0.1
	8 32	E	0.5	10	3	7^h 33^m	-1.0	0.4
	8 36	E	1.0	10	2		-0.7	0.0
	8 53	N	1.0	10	2	–	-1.1	0.7
	8 59	S	0.25	10	3		-0.9	0.4
	9 14	W	0.5	10	2	9^h 35^m	-0.8	0.2
	9 17	W	1.0	10	2		-0.7	0.3
	9 31	E	0.5	10	2		-1.5	0.2

*) CC = Central Condensation
E = EAST
W = WEST
N = NORTH
S = SOUTH

**) 10" corresponds to a length of roughly 9000 km on Halley during these observations

***) = interrupted due to cirri

the quality of these measurements. Most of the observations on March 18 exhibit normal photon statistics and revealed almost no polarization probably as a consequence of the fact that the core was still active, however, on a lower level. Comparing the results for the remaining three nights showing almost no activity it appears that the higher polarization degrees were measured in East-direction, i.e. sunwards. These areas are certainly the most dust-rich regions in the coma. Nearly all the regions given in Table 2 were probed each night partly several times. It turned out that the polarization degree was variable for the same region on short as well as on long time scales. Further, it is interesting to note, that three subsequent measurements on March 20 using 10″, 21″ and again 10″ openings for position E (0.′5) revealed a circular polarization of −1.0%, −0.2% and −1.3% respectively. The central condensation was measured using also different apertures (10″, 15″, 21″). Mostly the 15″ opening yielded the highest polarization degree.

4. Discussion

A generally accepted mechanism for producing circular polarization in comets is single scattering from aligned non-spherical dielectric particles giving rise to circular effects up to 4% polarization degree (Dolginov and Mitrofanov, 1975), whereas multiple scattering is thought to be unlikely except near the nucleus of very dusty comets. From the photographs of the coma taken during our observing runs it can be seen that the nucleus of Halley exhibited a pronounced activity strongly varying from night to night. This activity could also be noticed by our polarimetric measurements: If the nucleus was active very often a high and extremely variable polarization could be detected. However, if the standard deviation of the recorded photon counts was also high the conclusion was obvious that the polarization was only pretended by dust jets disturbing the rates of photon counts determining a complete polarization measurement. On the other hand sometimes during successive polarization determinations at the same area a high variability of P_c' was found though the standard deviations of the registered photon counts remained normal. Since due to the ln-approximation any continuous changes can be eliminated the conclusion still remains that such observed variations in the polarization are intrinsic.

If one assumes that single scattering by aligned non-spherical particles is responsible for the measured circular polarization, then the observed influence of jet streamers on the results forces the conclusion that the disturbed particles at the observed areas must sufficiently fast be realigned everywhere in the coma in the same way to produce, as observed, all the time the same sign of the circular Stokes parameter. This entire condition can hardly be understood since after the perturbations caused by a jet the succeeding realignment most likely has to be caused by radiation pressure. By that no real alignment can be caused, but however a separation of particles of different orientation relatively to the Sun. To get an approximate idea about the time scales necessary for such an alignment a simple idealizing calculation was made assuming only one disturbing jet. Further, the particles to be separated are thought to have a constant diameter of 0.5 μm and an extremely low density of 0.04 g cm^{-3} (Schwehm et al., 1986; Kissel et al., 1986). The particles were assumed to be very flattened circular cylinders with a height of one quarter of the diameter. For calculating the separation by collisions with photons coming from the Sun only particles oriented either with disk surface or rim against the Sun were considered, thus introducing an unrealistic high difference of the geometrical cross sections. Even with these extremely favourable conditions the time necessary to separate different oriented particles over a distance of 4500 km (corresponding to roughly half the 10″ aperture) amounts to 2 10^5 s. Due to the favourable conditions this time is only a lower limit. Therefore, a homogeneous alignment of non-spherical particles within the coma has to be excluded at least at the time of our observations.

Since linear polarized light scattered by arbitrary particles (including also spheres) results generally in elliptically polarized light (Van de Hulst, 1957) and since the linear polarization of Halley during our observations was reported to be of the order of 18% (Mukai et al., 1986), the circular polarization may also have been caused by multiple scattering. For phase angles, where only a low linear polarization is prevailing, a reduced or vanishing circular polarization is then expected. This is supported by measurements performed by Dollfus et al. (1986) about four weeks after our observations at phase angles between about 40° and 22°. At that time the linear polarization was smaller than 6% and the circular polarization smaller than 0.4%. A similar behaviour was also suggested by additional observations of comet Bennett performed 1.5 months after the detection of an extremely high linear and also circular polarization in the coma: whereas the linear polarization had decreased to 1% no circular polarization could be detected (Metz, 1970).

One hint for the validity of a multiple scattering mechanism to produce circular polarization may also be derived from the fact that the observed polarization degree normally was decreasing with increasing aperture opening. If the cometary nucleus is not regarded to be relevant for these considerations since too large errors may be caused by irregular evaporations from the central core then any model has to explain the dependence of the observed circular polarization upon the aperture diameter. Single scattering does not accomplish these conditions if small scale effects very near the nucleus are excluded. Therefore the circular polarization was simply determined by Monte Carlo calculations assuming a linear polarization of 18% in the coma and assuming further only double scattering where the first scatterer was regarded as a Rayleigh – instead of a more realistic Mie – scatterer. The second scattering, caused by particles of very high albedo, partly transforms the linear light produced by the first scattering process into a circular one. The high albedo was introduced only to simulate higher particle densities and has no real physical meaning. The basic idea for these simplified calculations was to show that, due to an increased contribution of particles with unfavourable high or small scattering angles in greater areas, the linear and therefore also the circular polarization must decrease if greater aperture openings are selected. The results show that the circular polarization decreases by a factor of 3 if an aperture opening of 10″ is replaced by one of 21″. Within the accuracy limits this is in accordance with the observations and may also confirm that multiple scattering was prevailing in the coma of Halley.

Acknowledgement. We thank A. Urquieta for his technical support on La Silla, Dr. R. West for providing unpublished GPO plates, Drs. W. Schlosser, W. Celnik and R. Schulz for copies of their narrow band images and M. Roth for performing the *UBVRI* photometry. The paper has benefited from comments given by Dr. A. Dollfus.

References

Dolginov, A.Z., Mitrofanov, I.G.: 1975, *Russ. Astron. J.* **52**, 1268

Dollfus, A., Suchail, J.-L., Curssaire, D., Killinger, R.: 1986, (preprint)

Kissel, I., et al.: 1986, *Nature* **321**, 336

Metz, K.: 1970, Thesis, University of Munich

Michalsky, J.J.: 1981, *Icarus* **47**, 388

Mukai, T., Mukai, S., Kikuchi, S.: 1986, 20th ESLAB Symp., ESA SP-250, Vol. 2, 59

Schwehm, G.H. et al.: 1986, 20th ESLAB Symp., Abstracts, 282

Van de Hulst, H.C.: 1957, Light scattering by small particles, J. Wiley and Sons, New York

Wolf, G.W.: 1972, *Astron. J.* **77**, 576

Astron. Astrophys. 187, 543–550 (1987)

ASTRONOMY AND ASTROPHYSICS

Spectrophotometry of comet P/Halley

I. Flux, column density and emission gradients within the coma in the emission bands and the continuum

K.R. Sivaraman, G.S.D. Babu, B.S. Shylaja, and R. Rajamohan

Indian Institute of Astrophysics, Bangalore 560034, India

Received March 4, accepted July 15, 1987

Summary. We have made spectral scans of comet P/Halley using the scanner at the Cassegrain focus (*f*/13) of the 102 cm reflector at Kavalur. The scans cover generally the wavelength region from 3900 to 6200 Å and are at 40 Å resolution and on a few occasions in the blue region at 20 Å resolution. The *f*/13 beam provided an image of the coma with a scale of 15″.5 mm^{-1}. We have obtained scans on several consecutive nights in March and April, 1986 with a 25″.9 entrance aperture of the scanner at discrete and systematically displaced locations within the coma in the anti-sunward direction as well as normal to it about the position of the nucleus. We have derived the brightness profiles of the neutrals and dust within the coma and have discussed the variations of the spatial gradients of the profiles.

Key words: spectrophotometry of comet P/Halley – spatial distribution of gas and dust

1. Introduction

Spectrophotometric measurements of fluxes in the emission bands of the different radicals in the neutral atmospheres of comets constitute the principal diagnostics for understanding the origin, behaviour and the chemical composition of these species in the environment of cometary comae. The law governing the heliocentric dependence of the flux in the emission bands provides clues to the volatility of the snows. Furthermore, if emission profiles of the various radicals are secured at independent locations within the coma from reasonably high spatial resolution observations, emission flux gradients can be evaluated and these would directly lead to an improved understanding of the structure of the inner coma in terms of the neutrals and the dust. One of the earlier attempts in this direction was on Comet West (Bappu et al., 1980).

An extensive programme of observations of Comet Halley consisting of spectrophotometry, imaging photometry and astrometry were carried out in 1985 and 1986 using the telescopes at the Vainu Bappu Observatory, Kavalur of the Indian Institute of Astrophysics. Here, we report the results of our spectrophotometric observations viz., the fluxes in the emission bands and in the continuum derived therefrom as well as the distribution of fluxes as a function of the radial distance (ϱ) from the nucleus within the coma.

Send offprint requests to: K.R. Sivaraman

2. Observations

We observed the comet on 6 nights in November and December 1985 and on 11 nights in March, April, and May 1986. Our observations consisted of spectral scans with an automated spectrum scanner that used a thermoelectrically cooled RCA C31034 tube in the pulse counting mode at the *f*/13 cassegrain focus of the 102 cm reflector. The entrance aperture to the scanner was a circular diaphragm of 25″.9 in diameter and the exit slot had a band pass of 40 Å in the first order. With this, we obtained scans on the comet covering the range 4500–7200 Å on all the 3 nights of November and on Dec 20.570 and from 3900 to 6200 Å on the remaining nights except on April 17.694 where the scans extend upto 6700 Å. The telescope has an image scale of 15″.5 mm^{-1} at the cassegrain focus and this large scale coupled with the 25″.9 diaphragm enabled us to sample different regions within the coma. We located this diaphragm on every night of observation at the brightest central part of the coma, namely the nucleus. In addition on 6 nights we systematically displaced the diaphragm to locations in the antisunward direction and on 3 nights to locations approximately normal to the sun-comet radius vector. Since one of the positions of the diaphragm every night was centred on the nucleus, all measurements of each night could be referred to this position. We have been able to observe regions out to as much as $\simeq 2'$ from the nucleus. These are schematically shown in Figs. 3 and 4 as insets. For each such position of the diaphragm, we scanned the neighbouring sky either before or after each scan on the comet and used this to eliminate the contribution of the background sky. We have then derived the energy distribution curves for all the scans on the comet following the reduction procedure described by Sivaraman et al. (1979) with the help of the observations of the standard stars θ Crt, 58 Aql, η Hya, and 109 Vir. In Fig. 1 we show a sample of these curves for March 31.904 where the emission bands of C_3(4050), CH + CN(0,1), C_2(1,0), NH_2(0,13,0), C_2(0,0), C_2(0,1), and C_2(2,4) are prominently seen.

We notice the presence of NH_2(0,7,0) emission in our scans of November which extends till 7200 Å. NH_2(0,13,0) emission can be marginally seen at 4900 Å in the November scans, but is definitely seen in the scans on Dec 8.813 and 19.576 although their flux is not significant. NH_2(0,13,0) emission became unusually intense by March and is seen conspicuously in the scans of all the nights of March both at the nucleus as well as at locations away from the nucleus (Fig. 1). Since on these occasions our scans end at 6200 Å we have no way of checking the presence of NH_2(0,7,0)

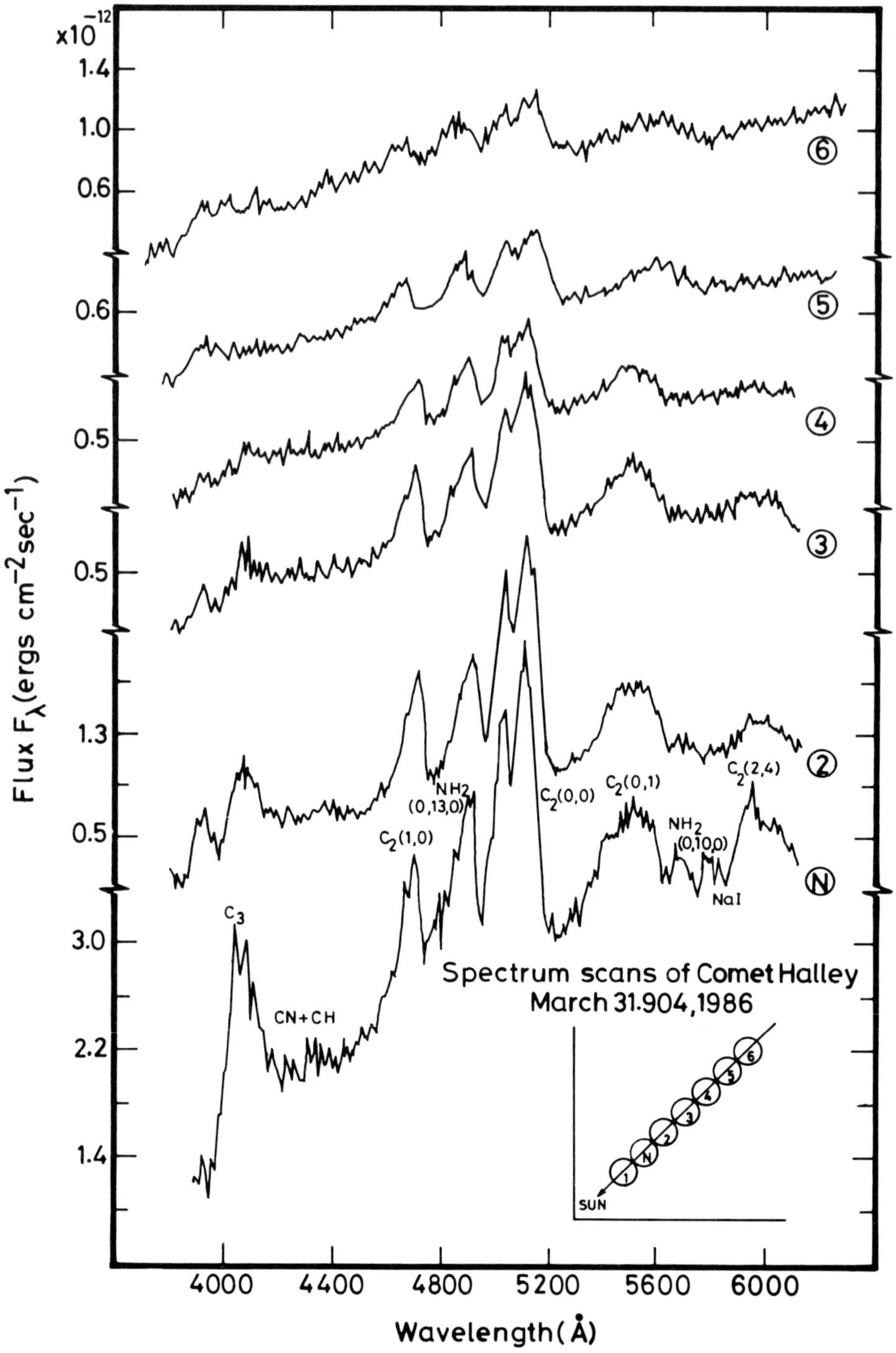

Fig. 1. Plot of flux F_λ vs. wavelength corresponding to 6 locations (designated N, 2, 3...6) of the 25″.9 diaphragm within the coma, N being the position of the nucleus. The 5 successive positions of the diaphragm (~28″ apart from each other) in the anti-sunward direction are schematically shown in the inset on the lower right corner. Notice the fast decrease in emission in the bands as the diaphragm is displaced systematically away from the nucleus

from our records. Again on April 17.694 when our scans extend upto 6700 Å, $NH_2(0,7,0)$ is seen well, but the flux in $NH_2(0,13,0)$ had decreased to the December level. Similarly $NH_2(0,10,0)$ emission is seen in all our records and on the nights of March is present significantly. Spectra of Arpigny et al. (1987) show all these bands of NH_2 clearly. Na I emission, although present in measurable quantities on the nights of March, is most likely to be contaminated by the bigger contribution from $NH_2(0,10,0)$. Since it may not be possible to separate the NH_2 emission from that of Na I reliably at the spectral resolutions we have employed, we have not attempted to calculate the fluxes of emission of Na I or $NH_2(0,10,0)$.

3. Results and discussion

3.1. Emission band fluxes

We have evaluated from these spectral energy curves, the fluxes in the emission bands of C_3, CN, CH, Swan bands, NH_2 and H_2O^+

Table 1. Observed emission band fluxes for the nucleus of comet P/Halley

Date U.T. 1985-86	r AU	Δ AU	Log F										
			CN (λ3883)	C_3 (λ4050)	CN (λ4200)	CH (λ4350)	$C_2(1,0)$ (λ4737)	$NH_2(0,13,0)$ (λ4900)	$C_2(0,0)$ (λ5165)	$C_2(0,1)$ (λ5635)	$C_2(2,4)$ (λ6050)	$NH_2(0,7,0)$ (λ6600)	H_2O^+ (λ7050)
Nov. 16.904	1.71	0.72	-	-	-	-	-10.91	-	-10.66	-11.02	-11.01	-11.31	-10.62
Nov. 17.748	1.69	0.71	-	-	-	-	-10.93	-	-10.64	-11.04	-11.30	-11.01	-10.70
Nov. 18.762	1.68	0.69	-	-	-	-	-10.90	-	-10.65	-11.03	-11.33	-11.06	-10.74
Dec. 8.813	1.38	0.70	-	-	-	-	-10.83	-	-10.66	-10.97	-11.27	-	-
Dec. 19.576	1.21	0.90	-10.92	-10.39	-11.21	-11.32	-10.30	-	- 9.89	-10.08	-10.65	-	-
Dec. 20.590	1.20	0.92	-10.95	-10.53	-11.54	-11.40	-10.37	-	- 9.84	-10.07	-10.51	-10.85	-11.04
Mar. 21.918	1.00	0.79	-	-10.03	-10.64	-10.68	-10.09	-10.12	- 9.64	- 9.80	-10.34	-	-
Mar. 23.950	1.04	0.74	-	-10.21	-11.05	-11.02	-10.21	- 9.99	- 9.68	- 9.86	-10.30	-	-
Mar. 28.883	1.11	0.62	-	- 9.69	-10.47	-10.95	- 9.96	- 9.82	- 9.55	- 9.72	-10.15	-	-
Mar. 29.896	1.13	0.59	-	- 9.75	-10.39	-10.41	- 9.70	- 9.82	- 9.27	- 9.59	-10.15	-	-
Mar. 30.903	1.14	0.57	-	-10.06	-10.81	-11.08	-10.20	-10.03	- 9.73	-10.03	-10.38	-	-
Mar. 31.904	1.16	0.55	-	- 9.64	-10.70	-10.70	- 9.90	- 9.78	- 9.39	- 9.73	-10.11	-	-
Apr. 5.848	1.24	0.46	-	- 9.70	-10.35	-10.40	- 9.79	-	- 9.54	- 9.89	-10.40	-	-
Apr. 12.850	1.34	0.42	-	- 9.59	-10.29	-10.35	-10.00	-	- 9.76	- 9.87	-10.22	-	-
Apr. 17.694	1.42	0.47	-	- 9.91	-10.65	-10.88	-10.25	-	-10.03	-10.18	-10.41	-10.40	-
Apr. 30.638	1.62	0.77	-	-10.63	-11.19	-11.34	-10.60	-	-10.35	-10.45	-10.44	-	-
May 7.622	1.72	0.99	-	-10.83	-11.06	-11.02	-10.73	-	-10.60	-10.82	-11.16	-	-

r = heliocentric distance of comet in AU.

Δ = geocentric distance of comet in AU.

Log F = log of total flux in the emission band in ergs cm^{-2} s^{-1}.

above the smoothed continuum level and also the flux in the continuum at 4860 Å. These are given in Table 1. We have then evaluated the total number of molecules (N) of each of these species contained in a cylinder of diameter 25″.9 in the line of sight and extending through the comet using the following relation which is valid for the resonance fluorescence mechanism (A'Hearn and Cowan, 1975).

$$N = L_i \frac{m_e \lambda_i}{\pi e^2 f \lambda_a p \varrho_v} = \frac{L_i}{g_i} \quad (1)$$

where

$L_i = 4\pi \Delta^2 F_\lambda$

Δ comet – earth distance in cm

F_λ – band flux observed in the 26″ diaphragm

λ_i – the mean wavelength of the emission band

f – the oscillator strength

p – the spontaneous transition probability from the upper to the lower level and

ϱ_v – the solar radiation density at the site of the comet at the mean wavelength (λ_a) of absorption to the upper level.

The reciprocal of the coefficient of L_i of Eq. (1) is represented by g_i which is the fluorescence efficiency of the species for the emission band concerned. The values of g_i, adopted by us for the various bands are presented in Table 2. We have derived the efficiencies for the $C_2(0,0)$ and $C_2(0,1)$ sequences using the $C_2(1,0)$ efficiency given by Oliversen et al. (1985) and the C_2 band sequence flux ratios given by A'Hearn (1978). We have used the g's that are averaged over all the radial velocities of the comet. The values of the g's when inserted in Eq. (1) lead to the following relations

Table 2. Fluorescence efficiency used for reduction of fluxes to column densities

Species	Efficiency g* ergs sec^{-1}	Source
C_3	1.5×10^{-12}	[1] KSK
$C_2(1,0)$	2.4×10^{-13}	[2] Oliversen et al.
$C_2(0,0)$	4.42×10^{-13}	[3] A'Hearn
$C_2(0,1)$	2.08×10^{-13}	[3] A'Hearn
$NH_2(0,7,0)$	1.75×10^{-14}	[4] A'Hearn
H_2O^+	2.98×10^{-12}	[5] Wycoff & Wehinger

* Fluorescence efficiency per molecule at r = 1 AU.

1. Krishna Swamy, K.S. 1986, Personal Communication.
2. Oliversen, R.F., Hollis, J.M. and Brown, L.W., 1985, ICARUS, 63, 339.
3. A'Hearn, M.F. 1978, Astrophys. J. **219**, 768.
4. A'Hearn, M.F., 1983, In 'Comets' ed. by L.Wilkening, Universit of Arizona Press, P.431.
5. Wycoff, S. and Wehinger, P.A., 1976, Astrophys. J. **204**, 604.

Table 3. Column densities of molecules at the nucleus integrated over the radius R, i.e. total number of molecules in the observed column. R is the projected radius of the 25.″9 diaphragm on the comet at the geocentric distances of observations

Date U.T. 1985–86	R in 10^4 Km	Log N					
		C_3	$C_2(1,0)$	$C_2(0,0)$	$C_2(0,1)$	$NH_2(0,7,0)$	H_2O^+
Nov. 16.904	0.67		29.34	29.31	29.28	30.07	28.54
Nov. 17.748	0.66		29.29	29.32	29.25	30.35	28.43
Nov. 18.762	0.65		29.29	29.27	29.23	30.27	28.36
Dec. 8.813	0.65		29.20	29.11	28.13		
Dec. 19.576	0.84	28.94	29.83	29.98	30.12		
Dec. 20.590	0.86	28.82	29.78	30.05	30.14	30.43	28.02
Mar. 21.918	0.74	29.02	29.77	29.95	30.12		
Mar. 23.950	0.69	28.83	29.63	29.89	30.04		
Mar. 28.883	0.58	29.25	29.78	29.92	30.08		
Mar. 29.896	0.55	29.16	30.01	30.18	30.19		
Mar. 30.903	0.53	28.82	29.49	29.70	29.72		
Mar. 31.904	0.51	29.23	29.77	30.02	30.00		
Apr. 5.848	0.43	29.07	29.79	29.77	29.75		
Apr. 12.850	0.39	29.17	29.56	29.54	29.76		
Apr. 17.694	0.44	29.00	29.47	29.42	29.59	30.45	
Apr. 30.638	0.72	28.83	29.65	29.64	29.87		
May 7.622	0.93	29.39	29.80	29.66	29.77		

which we have used to derive the total number of molecules (N) of each species in the field of view.

$$\log N_{C_3} = \log F_\lambda + 2\log r\Delta + 39.262$$
$$\log N_{C_2(1,0)} = \log F_\lambda + 2\log r\Delta + 40.066$$
$$\log N_{C_2(0,0)} = \log F_\lambda + 2\log r\Delta + 39.80$$
$$\log N_{C_2(0,1)} = \log F_\lambda + 2\log r\Delta + 40.127$$
$$\log N_{NH_2(0,7,0)} = \log F_\lambda + 2\log r\Delta + 41.203$$
$$\log N_{H_2O^+} = \log F_\lambda + 2\log r\Delta + 38.972.$$

In Table 3 we have given the calculated values of the number of molecules (N) of each species for the position of the nucleus.

3.2. *Swan band sequence flux ratios*

The classical problem of the C_2 molecule has been the gross mismatch between the observed and theoretical band sequence flux ratios for the Swan bands based on resonance fluorescence mechanism (Stockhausen and Osterbrock, 1965; Mayer and O'Dell, 1968). Also, the vibrational temperature for the population distribution in the triplet levels needed to explain the observed intensities of the Swan bands were found to be discordant with the conditions prevailing in cometary atmospheres.

Krishna Swamy and O'Dell (1977, 1986) have constructed a model based on detailed fluorescence equilibrium calculations for C_2 including the Ballik-Ramsay bands, the Fox-Herzberg bands in addition to the Swan bands from which they predict the band sequence flux ratios and their dependence on the heliocentric distance of the comet. By comparing these theoretical values with the observed flux ratios, the efficacy of the model can be assessed. Recently O'Dell and Tegler (1987) have brought to light the discrepancy that exists between their observations on two

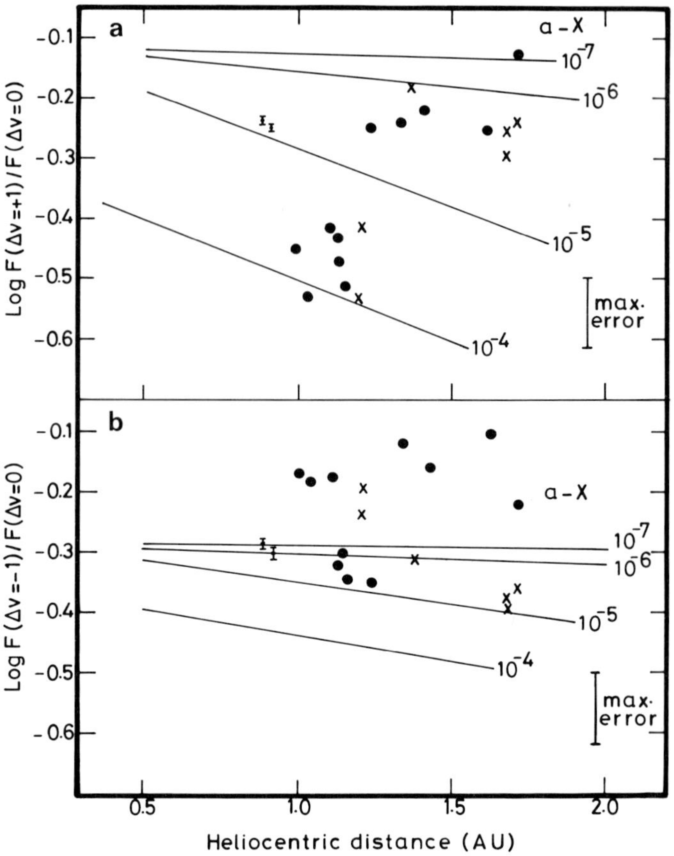

Fig. 2. **a** Swan band sequence flux ratio $\log F(\Delta V=+1)/F(\Delta V=0)$ as a function of heliocentric distance. Crosses: observations of pre-perihelion period; filled circles: observations of post-perihelion period. The solid lines are theoretical predictions reproduced here from Fig. 1 of O'Dell and Tegler (1987). **b** Same as Fig. 2a except for the $\log F(\Delta V=-1)/F(\Delta V=0)$ ratios. The solid lines are reproduced from Fig. 2 of O'Dell and Tegler (1987)

occasions on Comet Halley and the theoretical predictions of the flux ratios based on this model. They find that, increasing the data points by pooling of similar observations from the literature for earlier comets does not bring down the discrepancy. Although, this would enhance the number of data points, the diversity existing in the different observations may not permit an impartial comparison, whereas it would be desirable to have a large number of observations on a single comet for this purpose. In this connection our flux ratios constitute a homogeneous set of observations obtained on a single comet using the same instrumentation. We have plotted the flux ratios $F(\Delta V=+1)/F(\Delta V=0)$ and $F(\Delta V=-1)/F(\Delta V=0)$ for all the nights of Table 1 onto Figs. 1 and 2 of O'Dell and Tegler (1987). These are shown in Figs. 2a and 2b. The discrepancies are obvious. It is surprising that even the sophisticated model for C_2 of Krishna Swamy and O'Dell is unable to resolve the discrepancy, thus leaving the problem of C_2 as enigmatic as ever.

3.3. *Emission profiles within the coma*

On 9 nights we obtained spectral scans at many locations within the coma, along and normal to the anti-sunward direction; both the directions passing through the nucleus. We have shown in Fig. 1 a set of traces of one night's observations. For all other similar sets of observations, we have followed the same reduction

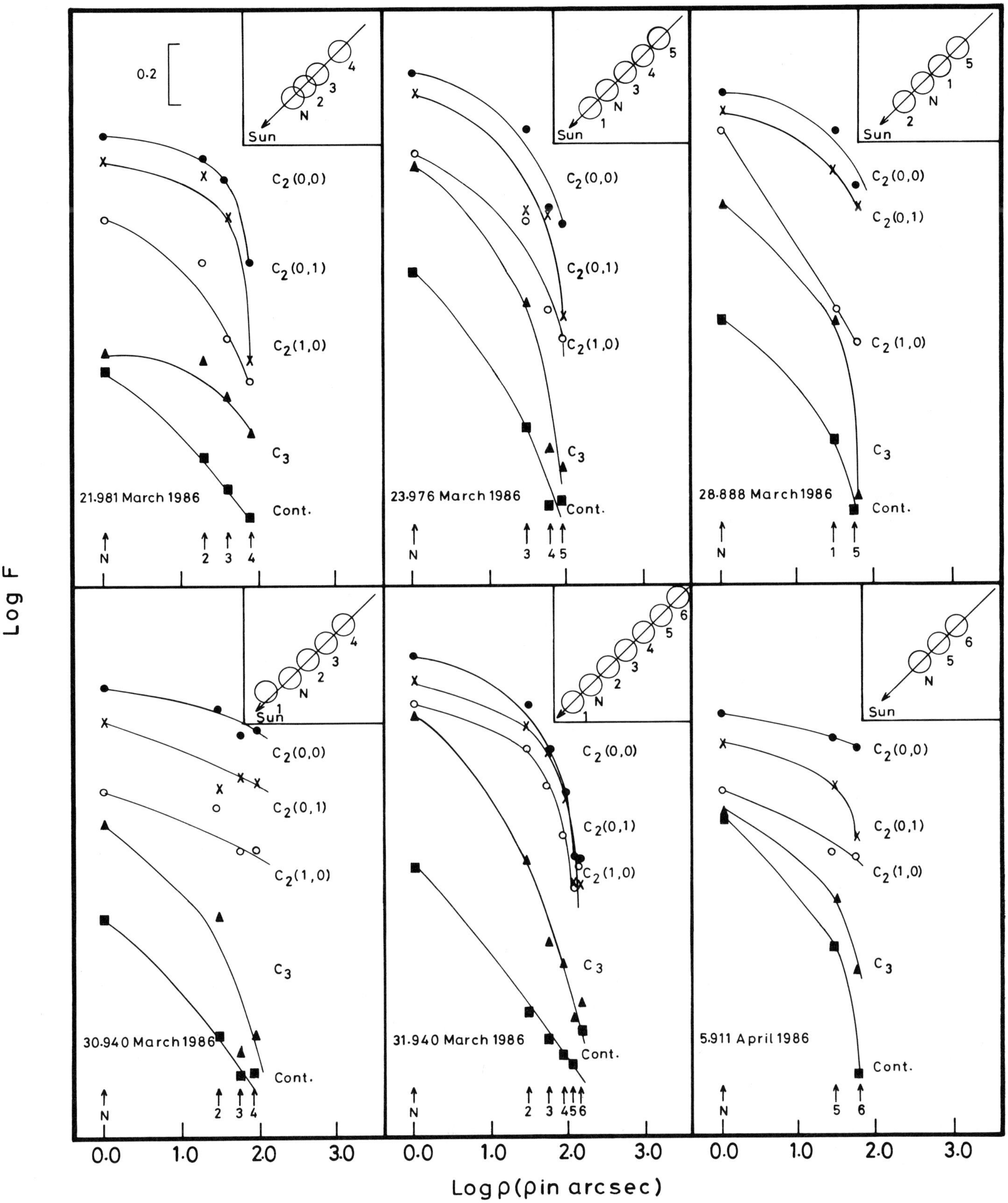

Fig. 3. Emission profiles for C_3, Swan bands and continuum as a function of the distance (ϱ) in arcsec of the centre of the diaphragm in the anti-sunward direction from the nucleus *N*. The positions of the diaphragm are schematically shown in the inset on the top right corner of each box and bearing designations *N*, 1, 2, 3 ... etc. as the case may be. These correspond to nucleus, positions 1, 2, 3 ... etc. and the same designations are reproduced on the *x*-axis in addition to the arcsec markings for convenience. Log of flux (log *F*) is along the *y*-axis. Within each box the profiles are shifted vertically to avoid overlap. The bar on the top left box represents 0.2 units in log *F*

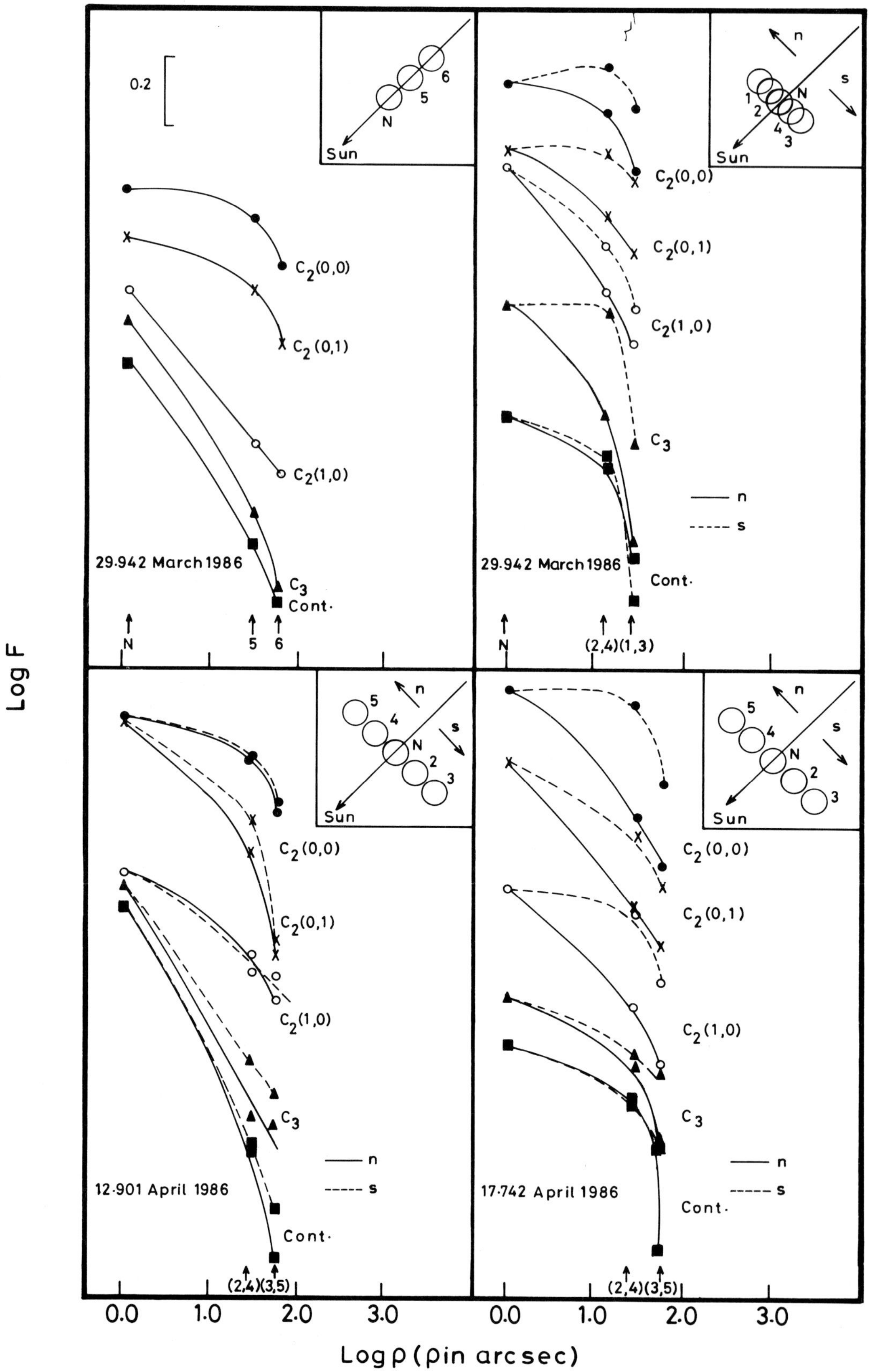

Fig. 4. Same as Fig. 3 except that the diaphragm positions are normal to the radius vector through the nucleus *N*. Notice that the emission profiles in the directions *n* (——) and *s* (– – – – –) are well separated bringing out the asymmetries. See text for details

Table 4. Dust-gas ratio for comet P/Halley for the nucleus

Date U.T. 1985-86	r AU	Phase angle in degrees	F_{cont} (λ4860) at Δ = 1 AU (erg s^{-1} cm^{-2} $Å^{-1}$)	F_{cont} (4000-6000) / $F_{Swan\ bands}$	F_{cont} (4000-6000) / ($F_{C_3(4050)}$ + $F_{Swan\ bands}$)
Nov. 16.904	1.71	4.0	1.99×10^{-13}	18.81	-
Nov. 17.748	1.69	2.4	3.03	27.49	-
Nov. 18.762	1.68	1.4	2.43	23.23	-
Dec. 8.813	1.38	42.9	1.52	16.62	-
Dec. 19.576	1.21	53.3	2.51	2.61	2.26
Dec. 20.590	1.20	53.7	3.68	4.11	3.71
Mar. 21.918	1.00	66.1	6.29	4.45	3.72
Mar. 23.950	1.04	65.8	4.98	4.73	4.11
Mar. 28.883	1.11	62.8	7.60	6.96	5.14
Mar. 29.896	1.13	61.8	7.43	6.35	5.39
Mar. 30.903	1.14	60.5	3.94	7.06	5.64
Mar. 31.904	1.16	59.1	7.57	7.42	5.62
Apr. 5.848	1.24	49.1	5.60	9.14	6.79
Apr. 12.850	1.34	29.1	4.45	12.26	7.56
Apr. 17.694	1.42	21.1	4.42	19.97	12.27
Apr. 30.638	1.62	28.4	1.90	8.39	6.85
May 7.622	1.72	30.8	0.40	28.35	15.87

procedure as for the scan on the nucleus and have obtained the flux (F_λ) in the emission bands. Using these, we have constructed the emission profiles for C_3, Swan bands of C_2 and the continuum at 4860 Å, which are plots of $\log F_\lambda$ vs. $\log \varrho$, ϱ being the distance of the centre of the diaphragm in arcsecs from the nucleus. These emission profiles are shown in Figs. 3 and 4.

The emission profiles along the radius vector show that, on the same night the three bands of C_2 show the same gradients while the profiles of C_3 and continuum are generally steeper. All the profiles of March 23.976 and 31.948 are steeper than their counterparts of the remaining nights showing that, there are nightly variations. These are possibly due to minor explosions in the nucleus resulting in changes in the gradients from an average value. The dust profiles also show similar variations and this cannot be completely attributed to changes in phase angle, which are small in this short span of time.

Of greater interest is the behaviour of the gradients normal to the radius vector. The diaphragm positions lying above the radius vector are designated as direction n and those below this line as direction s for the sake of the present discussion. On March 29.942 we have profiles along the radius vector as well as normal to it. These show that the gradients are steeper in the normal direction than along the radius vector for C_2, while for C_3 and continuum the reverse is true. Further, the gradients along the directions n and s also widely differ from each other. These asymmetries are most prominent in the C_2 profiles of March 29.942 and April 17.742. The continuum is symmetrical in n and s directions. The asymmetries in the distribution of the neutrals and dust within the coma of Halley seem to be also time dependent. We propose to discuss these in greater detail in terms of the production rates in our next paper.

3.4. *Dust-gas ratio*

We have computed the ratios of the fluxes in the continuum over the range 4000–6000 Å to the sum of the fluxes in the Swan bands on all nights. This ratio may serve as an indicator of the dust-gas ratio in comets. But in the case of Comet Halley C_3 (4050 Å) emission was quite strong in the spectrum and we have added its contribution to those of the Swan bands while calculating the ratio. Ideally, the CN(0,0) band flux and those of CH, NH_2, H_2O^+ should also have been included. Our spectra do not reach the CN(0,0) band and the other bands are rather weak and may not alter the ratios significantly. We have presented these ratios in Table 4, where we have included the heliocentric distances (r) and the phase angle values (ϕ). The ratio decreased fast from November 16.904 ($r = 1.71$ AU) towards December 20.590 ($r = 1.20$ AU). After perihelion, the ratio increased steadily from March 21.918 ($r = 1.0$ AU) to May 7.622 ($r = 1.72$ AU). This behaviour of the ratio is consistent with the scenario that, as the comet approaches the Sun, the grains undergo sputtering by UV radiation as well as solar wind and are broken up into fine particles whose net contribution to the continuum flux can be expected to go down. In the vicinity of the Sun, as the grains evaporate or are destroyed, the gaseous component is likely to increase and so also the emission in the bands. Thus the ratio of the flux in the continuum to that in the emission bands will be further reduced. After perihelion passage, the grains will start growing and therefore the trend will be reversed as shown in our Table 4. The influence due to changes in phase angle appears to be secondary, when compared to that of the heliocentric distance parameter.

4. Conclusions

1. The fluxes in the emission bands derived from the energy distribution curves for 17 nights (Table 1) show the time dependent evolution in the pre and post perihelion periods.

2. The Swan band-sequence flux ratios $F(\Delta V = +1)/F(\Delta V = 0)$ and $F(\Delta V = -1)/F(\Delta V = 0)$ obtained from our observations do

not match with the ratios predicted by the model for C_2 emission by Krishna Swamy and O'Dell (1977) (Figs. 2a and b).

3. The emission profiles in C_3, Swan bands and the continuum obtained from the spatial scans show varying gradients for the different species. These seem to change with time and also exhibit spatial asymmetries (Figs. 3 and 4).

Acknowledgements. The authors are thankful to Prof. J.C. Bhattacharyya, Director, Indian Institute of Astrophysics for the kind encouragement and for providing graciously telescope time and to the Department of Space, for partial fincancial support. The valuable help by K. Kuppuswamy, M.J. Rozario, and K. Jayakumar at the telescope and by R. Kariyappa and J.S. Nathan in the computations are gratefully acknowledged. We are thankful to G.A. Shah for reading through the manuscript and for his suggestions.

References

A'Hearn, M.F., Cowan, J.J.: 1975, *Astron. J.* **80**, 852

A'Hearn, M.F.: 1978, *Astrophys. J.* **219**, 768

Arpigny, C., Dossin, F., Zucconi, J.M., Andrillat, Y., Manfroid, F., Hutsemekers, D.: 1987, Proc. 20th ESLAB Symposium on *Exploration of Halley's Comet*, ESA SP-250, p. 471

Bappu, M.K.V., Parthasarathy, M., Sivaraman, K.R., Babu, G.S.D.: 1980, *Monthly Notices Roy. Astron. Soc.* **192**, 641

Krishna Swamy, K.S., O'Dell, C.R.: 1977, *Astrophys. J.* **216**, 158

Krishna Swamy, K.S., O'Dell, C.R.: 1986, *Astrophys. J.* (submitted) (personal communication)

Mayer, P., O'Dell, C.R.: 1968, *Astrophys. J.* **153**, 951

O'Dell, C.R., Tegler, S.C.: 1987, Proc. 20th ESLAB Symposium on *Exploration of Halley's Comet*, ESA SP-250, p. 561

Oliversen, R.J., Hollis, J.M., Brown, L.W., 1985, *Icarus*, **63**, 339

Sivaraman, K.R., Babu, G.S.D., Bappu, M.K.V., Parthasarathy, M.: 1980, *Monthly Notices Roy. Astron. Soc.* **189**, 897

Stockhausen, R.E., Osterbrock, D.E., 1965, *Astrophys. J.* **141**, 287

Astron. Astrophys. 187, 551–559 (1987)

ASTRONOMY AND ASTROPHYSICS

Spectrophotometry of comet P/Halley at wavelengths 275–710 nm from Vega 2

G. Moreels[1], J. Clairemidi[1], J.P. Parisot[1], J.M. Zucconi[1], J.L. Bertaux[2], J.E. Blamont[2], M. Hersé[2], V.A. Krasnopolsky[3], V.I. Moroz[3], A.A. Krysko[3], A.Y. Tkachuk[3], M. Gogoshev[4], T. Gogosheva[4], R. Werner[4], and S. Spasov[4]

[1] Observatoire de Besançon, 41 bis Avenue de l'Observatoire, F-25000 Besançon, France
[2] Service d'Aeronomie, CNRS, B.P. 3, F-91370 Verrières le Buisson, France
[3] Institute of Space Research, Profsoyuznaya 84/32, SU-117810 Moscow GSP-7, USSR
[4] Astronomical Observatory of Stara Zagora, BG-6000 Stara Zagora, Bulgaria

Received February 16, accepted May 19, 1987

Summary. Spectrophotometric observations of the inner coma of comet Halley in the range 275–710 nm were conducted from Vega 2 on March 8, 9, 10, 11, 1986. During the encounter session, when the impact parameter was minimum, $p = 120$ km, the gaseous emissions were still present, superimposed on an intense continuum. In the inner coma, C_2 and C_3 emissions steadily increase when the impact parameter decreases. The emissions of OH at 309 nm and CN in its band head at 388 nm show a plateau between 1500 and 600 km. The decrease in OH emission when the impact parameter is less than 500 km is interpreted in terms of high optical thickness of water vapor at Lyman-alpha. Pixel-to-pixel ratios of spectra show that the color of dust continuum presents small fluctuations which might be produced by a discontinuous emission of icy grains.

Key words: comet Halley – spectrophotometry – inner coma

1. Introduction

The inner coma of comets has long been considered as a region of major interest for the study of mechanisms leading to production of daughter molecules. Mother molecules are ejected from the nucleus or may be released from the icy matrix of grains when they vaporize. Daughter molecules are produced by photolysis of mother molecules at rates fixed by these species' lifetimes but they might also be stored in the icy grains and behave like parent molecules. From an observational point of view, the inner coma may be defined from spectrophotometry in the visible and near UV as the sphere centered on the nucleus where the continuum is more intense than the three major gaseous emissions: OH 309 nm, CN 388 nm, and C_2 516 nm ($\Delta v = 0$). The radius of this sphere is of the order of several thousand kilometers and is comparable to the radius of the thermalization sphere. Given the high resolution necessary to conduct spatially-resolved spectrophotometry of the circumnuclear region from ground-based telescopes (Jockers et al., 1986; Falciani et al., 1986) the near ultraviolet and visible (NUVI) channel of the Vega three-channel spectrometer appears as a very useful tool to obtain a spectroscopic description of this region. This channel was designed to reach the following scientific objectives: to study in detail the spectroscopic components of the coma, to obtain monochromatic charts in the different optical emissions and to study the processes leading from parent to daughter molecules. Another objective was to take advantage of the presence on the movable platform of the TVS camera and IKS infra-red spectrometer aligned along the same optical axis to proceed to comparative studies.

The coma spectrum presents two components: dust continuum with a given wavelength distribution and gaseous emissions superimposed on it (Levasseur-Regourd et al., 1986). The NUVI channel resolution was chosen to allow discrimination of gaseous emissions against dust. Combined with the zoom effect resulting from the relative movement of spacecraft and comet, these instrumental capacities made it possible to obtain a detailed description of the inner parts of the coma.

Send offprint requests to: G. Moreels

2. Instrumentation

The NUVI channel is completely static and consists of a 2.5×0.050 mm^2 slit, a concave holographic grating of aperture $f/3.5$, a plane mirror and an intensified detector. The concave grating works as the combination of three elements: collimator, plane grating and objective lens (Gogoshev et al., 1985). The intensified detector is built in cementing a linear array of 1024 2.5×0.025 mm^2 photodiodes equipped with a fiber optic window to the output screen of an 18 mm proximity focused image intensifier tube which is operated under a variable MCP voltage. The photometric gain is selected by the microprocessor code from the intensity of the 516 nm C_2 band in order to have a dynamic range larger than 10^4.

The instrument sensitivity was measured in Toulouse in September 1984 using the lines of a mercury lamp at 313.0, 404.7, 435.8, 546.1 nm and of a zinc lamp at 636.2 nm. The lowest detectable signal was of the order of 200 Rayleighs. This calibration was verified in Moscow in October 1984 by the Soviet team using a tungsten lamp. An agreement was found for wavelengths longer than 480 nm but the sensitivity at shorter wavelengths was found approximately 20% higher during this

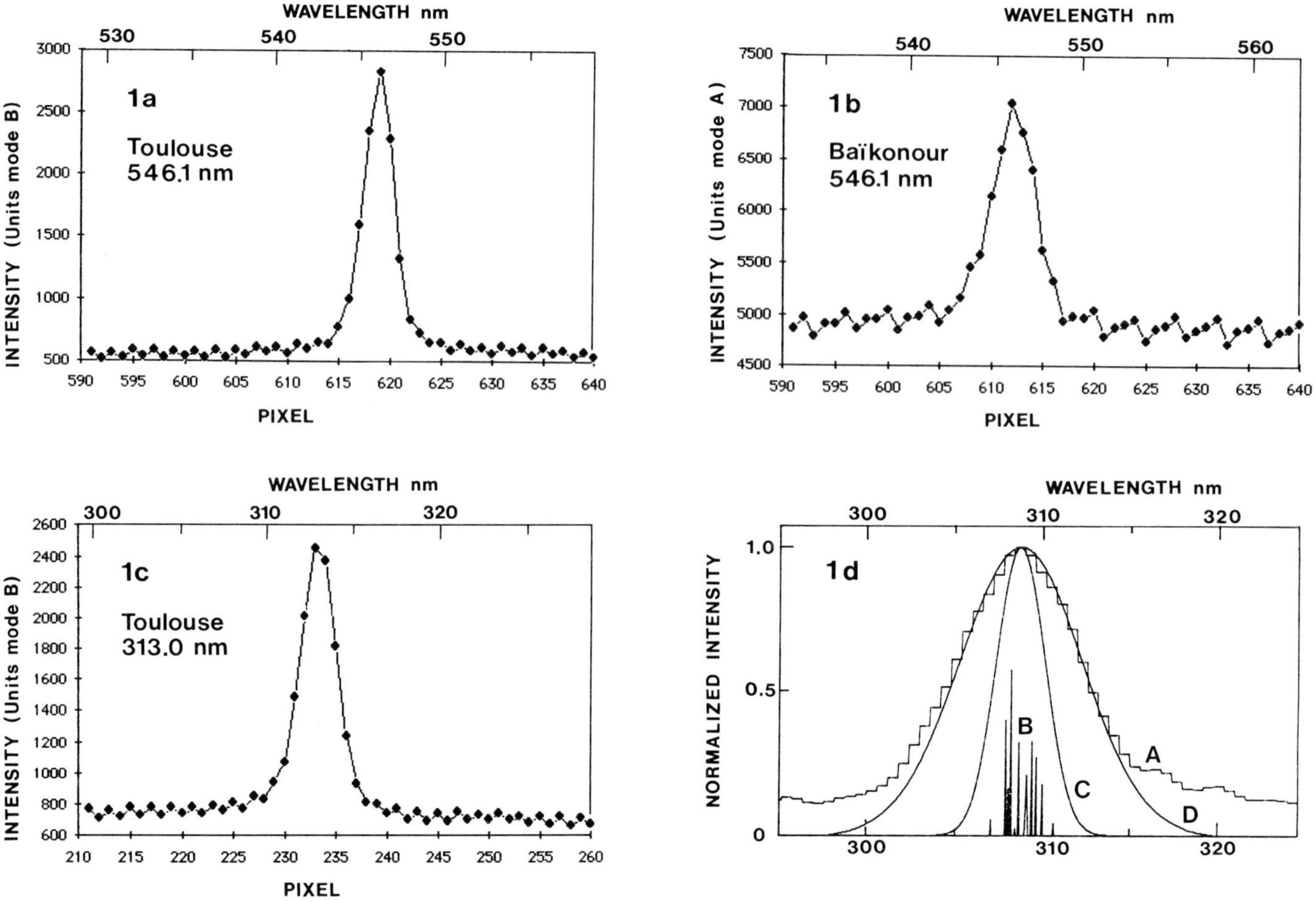

Fig. 1 a–d. Spectral response function of the NUVI channel of TKS spectrometer. **a** and **b** spectral responses to the 546.1 nm mercury line measured in Toulouse and in Baïkonour. **c** spectral response to the 313.0 nm mercury line measured in Toulouse. **d**, *A* measured spectrum of the OH 309 nm band; *B* OH fluorescence spectrum; *C* and *D* convolution of this spectrum with a 2.7 nm and a 7.0 nm bandpasses

second test session. A small reference tungsten lamp placed inside the telescope and used to check the instrument sensitivity during flight showed that the resulting test signal decreased by a factor 1.6 in June 1985 when the distance to the Sun was 0.7 A.U. during the Venus flyby. The spectral response function measured in Toulouse and Baïkonour with a mercury lamp was equal to 4.5 pixels or 2.7 nm at half-height (Fig. 1a–c). The examination of flight spectra shows that this bandpass during the cometary observation sessions was broader. In Fig. 1d, a synthetic fluorescence spectrum of the OH (0, 0) band has been convolved with a spectral gaussian response function having an adjustable width at half-height. The resulting spectrum is compared to an in-flight spectrum. The best fit is found for a spectral width of 7 nm or 12 pixels.

The changes in the instrument characteristics have been put in evidence by the test sessions performed during the interplanetary flight. Since the signal produced by the reference lamp was attenuated by the same factor at all wavelengths and showed no sign of any variation in the tungsten lamp temperature, the instrument sensitivity decreased by a factor 1.6. The reason for these changes may probably be found in the relatively high temperature, of the order of 60–70 °C, suffered by the image tube during the Venus flyby. As suggested by the image tube manufacturer, some degassing material probably was released by the entrance window cement, producing some spoiling of the order-sorting filter placed in front of the window.

The instrument worked under either a chart mode, named A, or a line mode, named B. Under mode A the field of view was rectangular $2^\circ \times 1.5^\circ$ and was scanned in an array of 7 lines of 15 positions numbered 1 to F in hexadecimal. Under mode B only the central line was scanned in a see-saw motion. This mode was used only on March 9 during the encounter period from 06:46:30 to 07:21:30 U.T.

3. Observations during encounter day 1986 March 9

During the encounter day, 1200 spectra were recorded during a two-hour session which finished at 07:21:30 U.T., i.e. 90 s after the closest approach. As the distance between the spacecraft and nucleus decreases, spectra gradually exhibit variations in intensity, in gas/dust emission ratios and in continuum color. Two spectra obtained in mode B at 06:47 U.T. and one minute later when the distance to nucleus was about 150 000 km and impact parameters were 5240 and 2400 km are presented in Fig. 2. The following features can be identified in the spectra: OH, bands (1,0), (0,0), (1,1); NH; CN, bands (1,0), (0,0), (0,1); C_3; CH; C_2, sequences $\Delta v = 2, 1, 0, -1$; NH_2; H_2O^+. The plotted sensitivity

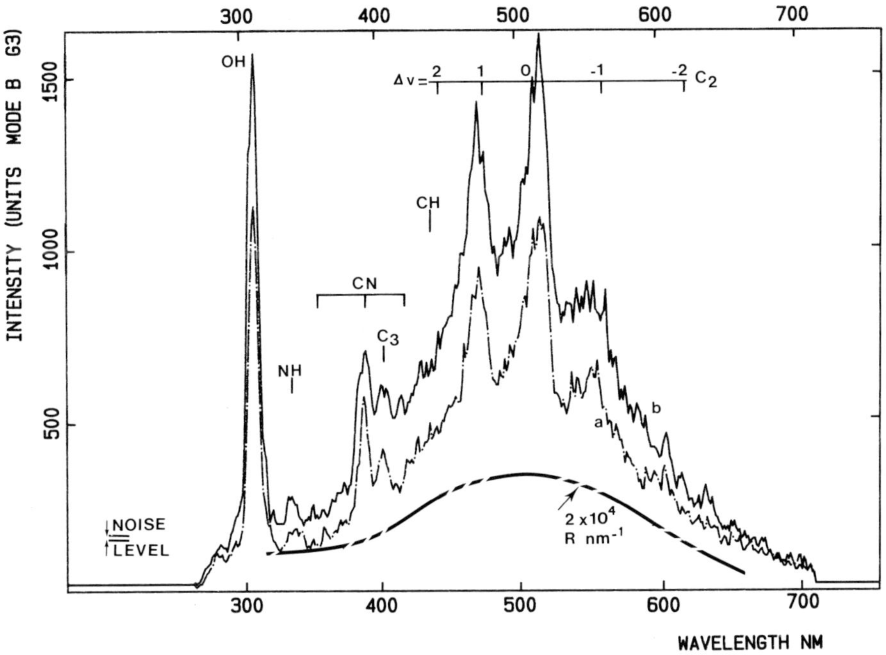

Fig. 2. Spectra of comet Halley on March 9, 1986 at 06:47 U.T. (**a** and **b** impact parameters: 5240 and 2400 km). Intensities are given in units of mode B, image tube gain G3. Dashed line is the instrument response to an intensity of 20000 Rayleighs nm^{-1}. Noise level is indicated

curve is obtained as follows. The instrument response to a monochromatic line having an intensity expressed in Rayleighs is converted into the response to a continuum radiation by using the calibration data obtained in Toulouse with a measured $\Delta\lambda$ of 2.7 nm (Chalon, 1984). Then, in order to take into account the spectral broadening during the flight, it is assumed that the total response to a monochromatic line did not change, i.e. had a smaller maximum amplitude but was displayed on a greater number of pixels. Finally the sensitivity is divided by 1.6 to take into account the measured decrease of the reference lamp signal during the Venus flyby. The gaseous emissions have an intensity comparable to the dust continuum. The emission of CO_2^+ at 289 nm is present in a number of spectra but its apparition is not regular. Since the exposure time was 0.4 s and a spectrum was transmitted every 5 s, the instrument was able to follow rapid signal variations. In fact, fluctuations from one spectrum to the following one 5 s apart, were apparent on the C_2 bands. The same kind of rapid variation has been reported by Rettig et al. (1986) and apparently shows that, at a time scale of 5 s, steady state is not achieved for the emissive species.

Three spectra taken at 07:17:20, 07:20:00 and 07:21:10 U.T., i.e. in a period of 4 minutes including the time of closest approach, are presented in Fig. 3. Impact parameters are for spectra a: 490 km, b: 330 km and c: 145 km. At these distances, gaseous emissions can still be distinguished against the dust continuum. Three features are clearly apparent on these spectra: (i) the OH

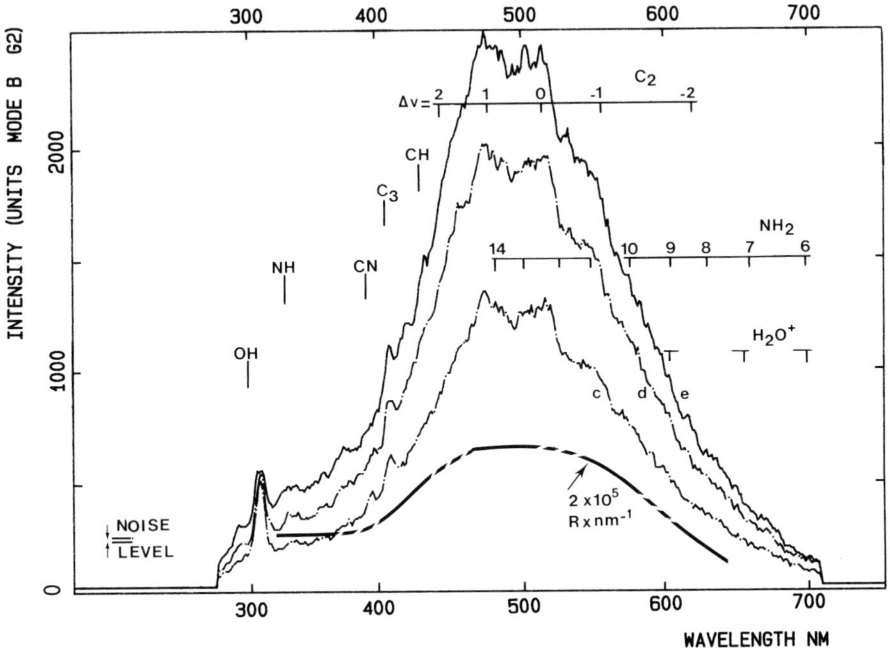

Fig. 3. Spectra of comet Halley on March 9 during closest approach around 07:20 U.T. (2c, 2d, and 2e: impact parameters: 490, 330 and 145 km). Intensities are given in units of mode B, image tube gain G2, i.e. 5 times less sensitive than gain G3. Dashed line is the instrument response to an intensity of 200 000 Rayleighs nm^{-1}. Noise level is indicated. It is smaller than in Fig. 2 because the signal is higher

intensity decreases whereas the dust continuum increases, (ii) The C_3 band becomes very intense when compared to CN at 388 nm, (iii) the variation of the continuum with wavelength between 300 and 450 nm is very different from that measured in Fig. 2. This reflects either an additional diffuse emission in the 350–400 nm range or a difference in the color of dust. Spectra obtained during this period need to be studied in more detail because, during the platform rotation, the optical axis scanned regions of pronounced anisotropy, as shown by the TVS camera (Sagdeev et al., 1986) and by dust detector measurements (Mazets et al., 1986).

During 34 min before the closest approach, the instrument worked in the line B mode. The scanned line was located above the nucleus, the angular distance between the center of the line and nucleus being equal to 1°.4 (see Fig. 8). During this period the distance to the nucleus decreased from 154000 to 8030 km and the impact parameter varied from 5800 to 120 km. The 390 spectra were aligned in order to follow the intensity variation with time. A sequence of spectra relative to the 15 positions of a line is given in Fig. 4. This representation is analogous to a prism objective spectrum; it shows, at any wavelength, the intensity variation along the line which is equivalent to a long slit. In order to monitor the evolution with impact parameter of the main emissions and dust continuum, six series of 10 consecutive pixels are selected and averaged. In the case of the dust continuum, results are given in Fig. 5 where the intensities at wavelengths 296, 327, 376, 410, 526, and 610 nm are plotted as a function of time. Each line scan motion of the secondary mirror tilting system produces an amplitude oscillation of the signal which increases gradually as the distance between the spacecraft and nucleus decreases. The curve follows the $1/p$ dependence, where p is the impact parameter, generally predicted by dust models (Divine, 1981; Moroz, 1985) and measured from the ground using large telescopes (Green and Hughes, 1986).

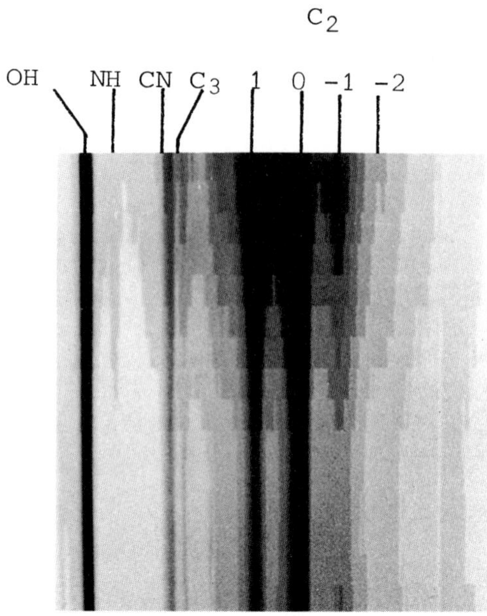

Fig. 4. Objective spectrum of comet Halley in the 290–700 nm range. The 15 consecutive spectra obtained at 06:41 UT are superimposed and show the variation of intensity at different wavelengths along a radius of length 6000 km. The nucleus is located slightly above the picture. The representation is analogous to a photographic negative. The main bands from left to right are: OH (0,0), CN ($\Delta v = 0$), C_2 ($\Delta v = 1, 0, -1$)

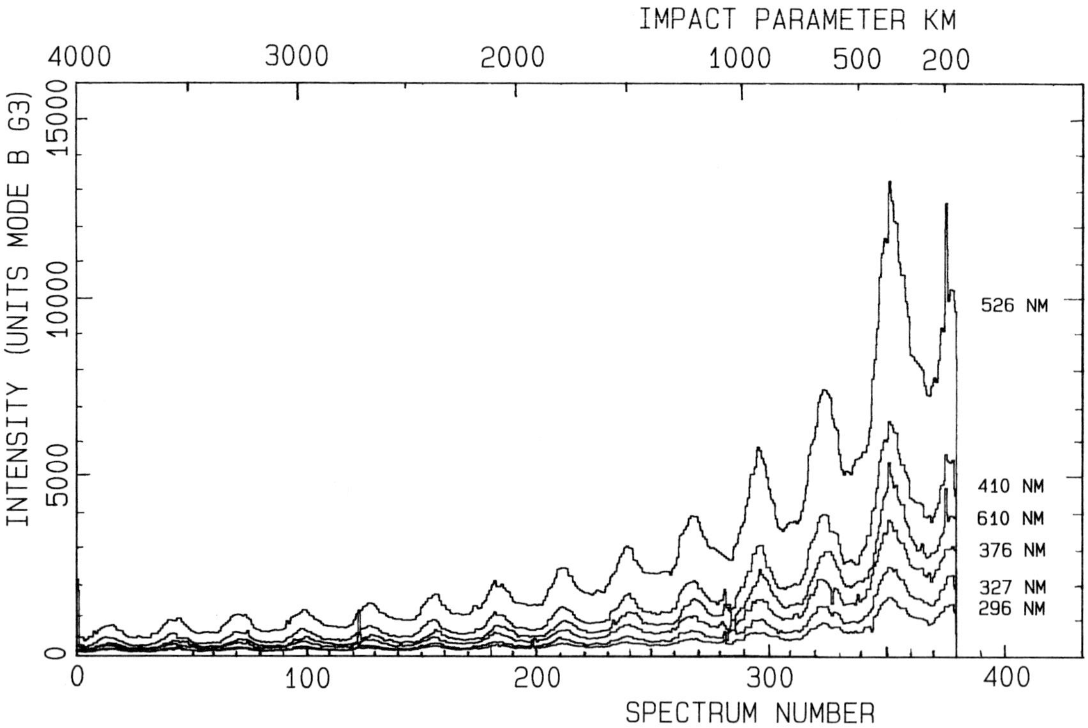

Fig. 5. Relative variation of dust continuum with time and decreasing impact parameter from 06:45:30 to 07:21:30 UT on March 9, 1986

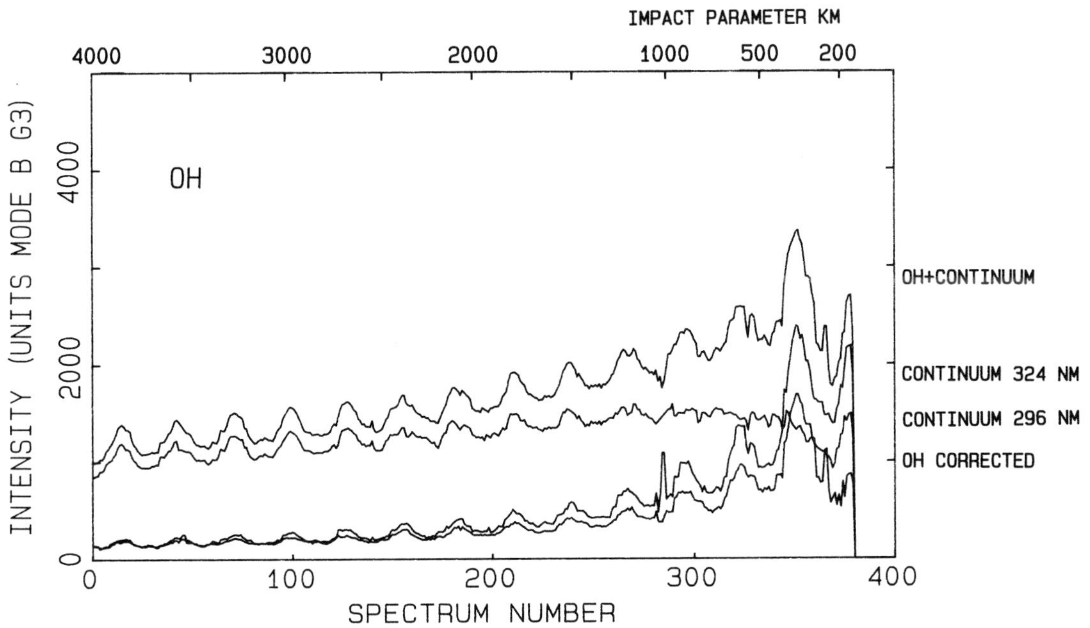

Fig. 6. Relative variation of the OH 309 nm emission with time. Horizontal scale: same as in Fig. 5

In the case of gaseous emissions, the same curves can be drawn, but the continuum must be subtracted. This subtraction may be ambiguous during the last 5 mn due to the large amplitude of the continuum and due to the fact that its distribution as a function of wavelength shows strong variations as noted in the comparison of Figs. 2 and 3. The problems in subtracting the continuum are pointed out, for instance in the case of H_2O^+, by Arpigny et al. (1986). Figure 6 presents the results for the OH 309 nm band and the continuum intensities at 296 and 324 nm. The result is that the corrected OH emission is relatively flat in the inner coma and, as shown in Fig. 3, it decreases when the impact parameter is less than 500 km.

The variations of OH, CN, C_3, and C_2 emissions corrected from the continuum are presented in Fig. 7 as a function of impact parameter p. Both C_2 and C_3 show a comparable increase when p becomes smaller. CN does not increase at the same rate as C_2 and C_3.

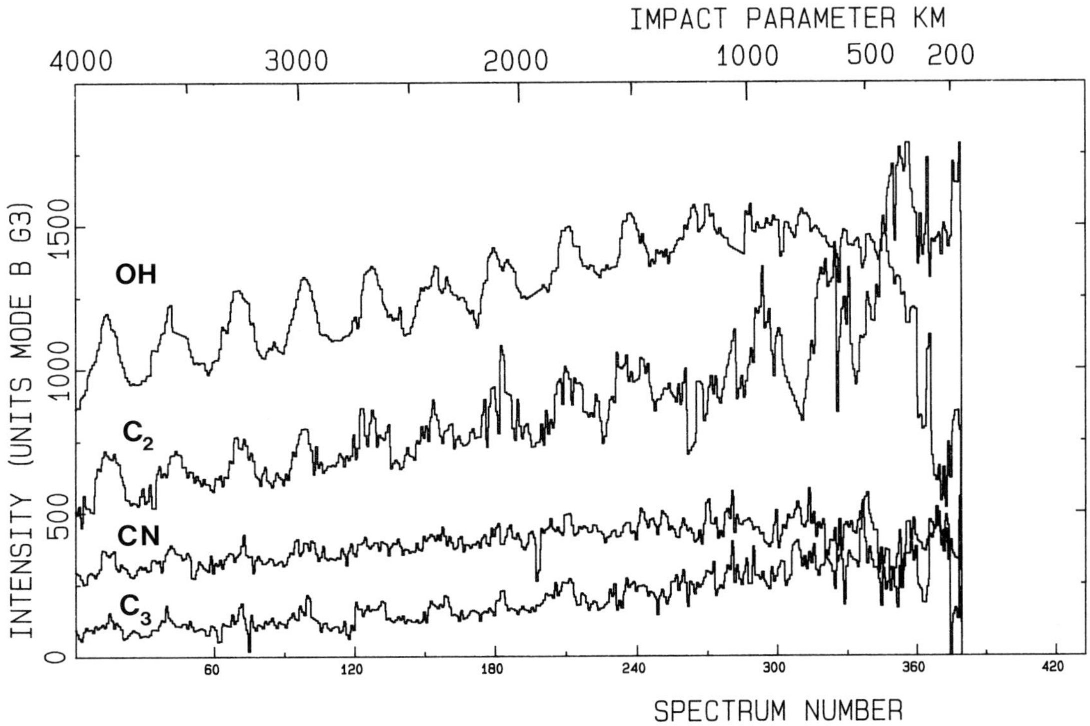

Fig. 7. Relative variation with time of the gaseous emissions averaged over 10 pixels and corrected from dust continuum. Horizontal scale: same as in Figs. 5 and 6

4. Discussion

When the Vega spacecraft penetrates into the inner coma, the continuum radiation measured by the NUVI channel of TKS becomes fairly bright. The albedo of dust may be determined by dividing the comet spectrum by the spectrum of the Sun. This operation requires a precise knowledge of the absolute sensitivity curve of the instrument. In order to overcome this point, the following differential method provides useful information about the evolution of the albedo of dust with impact parameter in the inner coma. In this method, advantage is taken of the fact that, in mode B a line of an angular width of 2° is scanned in 15 positions with a distance of 8.′6 from each other. This see-saw motion allowed the instrument to explore, during the 40 min before closest approach, a triangular field of view with a summit at the nucleus. In Fig. 8, two groups of 6 adjacent positions with two different impact parameters: 3550 km and 360 km, have been selected. Inside each group, four pixel-to-pixel ratios of adjacent spectra are depicted. Each ratio is independent of sensitivity and is proportional to the ratio of albedos in the spectral regions where no gaseous emission is superimposed on the dust continuum. In the second group of ratios (Fig. 8b), OH and CN gaseous emissions are depicted by depressions which clearly show that they increase at a much smaller rate than the continuum in the inner coma. Between adjacent positions, distant of about 150 km, the ratios of albedos may present an increase in the near UV region or at the other end of the spectrum. As a result, the color of the dust continuum presents fluctuations in the inner coma. At some locations, the ratios of spectra show an increase in the 300–400 nm which is a sign of an albedo slightly higher in the UV. This observation may be correlated with colorimetric and spectro-photometric studies of outer solar system materials (Hartmann et al., 1985) or of comet Halley at large radial distances (Belton et al., 1986; Meech et al., 1986). It might provide some evidence of a discontinuous emission by icy grains into the dusty environment of the inner coma.

If observations conducted during 1986 March 8, 9, and 10 are put together, a complete distribution with an impact parameter in the range of 200–200 000 km may be drawn. Figure 9 presents this distribution for the OH (0,0) band. In the case of comet Halley, the intensity of the OH (0,0) band is constant and equal to 1.8 ± 0.4 MegaRayleighs at impact parameters shorter than 1500 km. From this intensity, Krasnopolsky and Tkachuk (1986) have calculated, using a curve of growth method, that the OH gas production rate on March 9, 1986 was $1.4 \pm 0.5 \times 10^{30}$ mol s^{-1}. They assume that the scalelengths of H_2O and OH are 4.6 10^4 km and 9 10^4 km and the velocity of the OH radical is 1 km s^{-1}. With the same OH emission value, the vectorial model predicts an OH production rate slightly lower: 9 10^{29} mol s^{-1} (Moreels et al., 1986). In fact, the activity of comet Halley is highly variable and on March 10 an intensity approximately two times higher than on March 8 was measured. This variation is in good agreement with the curves given by Schleicher et al. (1986). Close to the nucleus, the intensity does not increase. The observation is in apparent contradiction

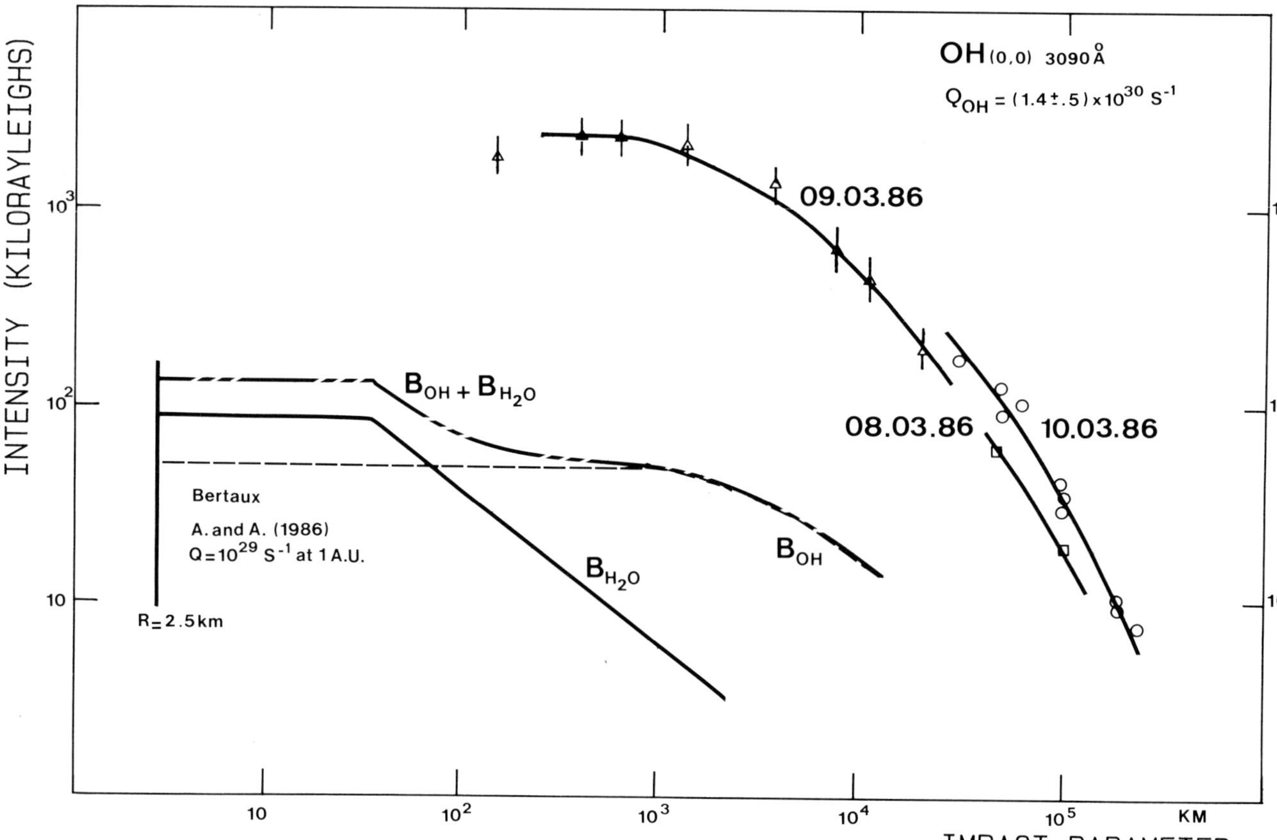

Fig. 9. Variation of the OH intensity with impact parameter p. Three sets of data are drawn on this graph: March 8, 9, and 10. Comparison with a model calculation by Bertaux (1986). B_{OH} is the amount of OH emission produced by fluorescence. B_{H_2O} is the emission due to the direct production on excited OH radicals in the photolysis of water vapor

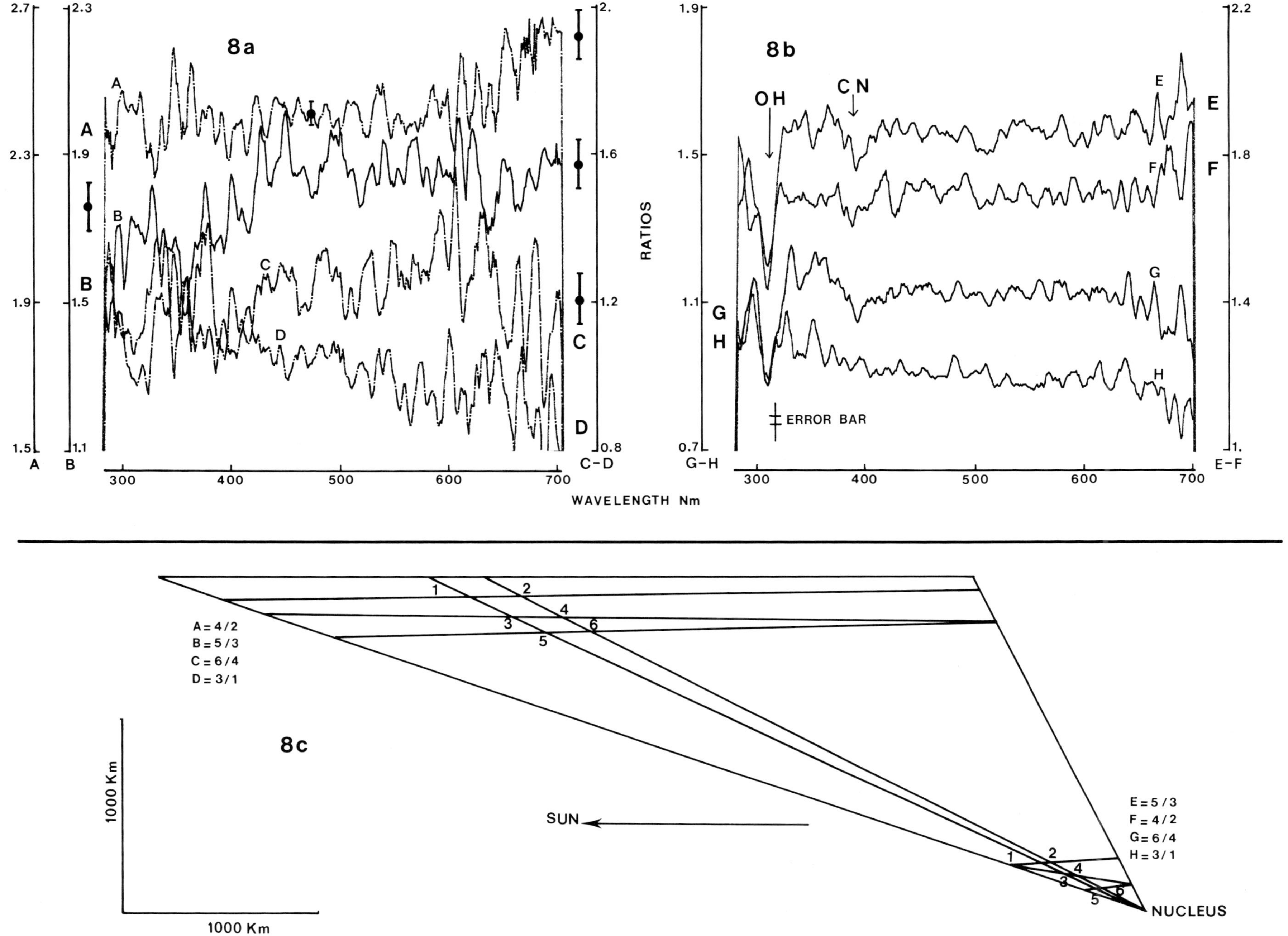

Fig. 8a–c. Pixel-to-pixel ratios of spectra. Two groups of 6 adjacent positions at radial distances of 3550 km (**a**) and 360 km (**b**) are chosen along a radius. **a** and **b** shows pixel-to-pixel ratios of adjacent spectra. Close to the nucleus, the OH and CN emission ratios appear as depressions. Errors bars due to noise in the signal are shown. **c** shows the field of view scanned by the optical axis of the instrument in mode B. This field of view is a triangle with a summit at the nucleus

with the intensity predicted by Bertaux (1986). The OH intensity should increase in the few tens of kilometers around the nucleus due to the existence of a second source of OH ($A^2\Sigma^+$) radicals directly produced by H_2O photolysis. It was stated that the measurement of this second intensity component should provide a semi-direct way to evaluate water vapor and detect local inhomogeneities. In present case this increase was not observed. The reason is probably the following. At the end of the mode B session on March 9, the impact parameter was 120 km. The optical thickness of water vapor at Ly α may be calculated assuming an absorption cross section of 1.45 10^{-17} cm^2 (Watanabe and Zelikoff, 1953), an ejection velocity of 1 km s^{-1} and a production rate of 1.4 10^{30} s^{-1}. The column density may be evaluated using the simplified expression $N = Q/4pV$. The calculated optical thickness is then $\tau = 2$ at $p = 120$ km. A fraction of 24 to 39% of the water vapor, depending upon solar activity, is dissociated at 121.6 nm according to the following processes (Festou, 1981):

$$H_2O + h\nu \xrightarrow[\lambda=121.6\,nm]{} \begin{array}{ll} H + OH(X^2\Pi) & 68\% \\ H + OH(A^2\Sigma^+) & 7\% \\ H_2 + O(^1D) & 25\%. \end{array}$$

At longer wavelengths, the absorption cross-section is much smaller. The coma remains optically thin for impact parameters as small as 50 km and OH ($X^2\Pi$) radicals are still produced at this distance. Consequently, the high optical thickness of water vapor at Ly α is probably the cause of the observed OH emission decrease by about 25% at $p = 120$ km. At this wavelength, the two mechanisms responsible for the OH 309 nm emission: fluorescence of OH ($X^2\Pi$) or direct production of OH ($A^2\Sigma^+$) become both inefficient. This observation provides an additional argument in favor of the assumption that water vapor is the major source of OH radicals.

5. Conclusion

During the four observation sessions conducted with the NUVI channel of TKS instrument on March 8 to 11, 1986, the following results were obtained: (i) spectra in the 275–710 nm wavelength range showing progressive increase of the dust continuum when the impact parameter decreases from 4000 to 120 km; (ii) variation with impact parameter of gaseous emissions and the continuum at selected wavelengths in bandpasses of 6 nm (10 pixels); (iii) decrease of the OH 309 nm emission when the impact parameter is less than 300 km. The following features are present in the spectra: OH, (1,0), (0,0), and (1,1) bands; CO_2^+ at 289 nm in some spectra only; NH; CN; C_3; CH; C_2, $\Delta v = 1, 0, -1$; NH_2. The variation of gaseous emissions with decreasing impact parameter p shows that C_2 and C_3 increase until the end of the encounter session when $p = 120$ km. The intensity of the CN band head at 388 nm shows a plateau for p values in the 1500 – 600 km range. The intensity of the OH 309 nm band is flat in the same distance range and decreases when the impact parameter is less than 300 km. This observation is probably due to the fact that water vapor becomes optically thick at Ly α inside a sphere of radius 150 km if the assumed production rate is 1.4 10^{30} s^{-1}. Pixel-to-pixel ratios of spectra taken at adjacent positions distant of 150 km give evidence that the color of dust continuum shows slight fluctuations which might result from a discontinuous emission of icy grains into the dusty environment of the inner coma. Strong emphasis has been put on the variability of comet Halley by a number of observers. Present measurements of the OH 309 nm band show that, for p values in the range 2 10^4 – 2 10^5 km, the intensity was higher by a factor 2 on March 10 than on March 8 and 9. This is in agreement with IUE and ground-based observations of P/Halley during the same period.

Acknowledgements. We would like to thank P. Angebault, M. Athanase, G. Chalon, H. Choux, J. P. Lepage, P. Mérat, B. Mougin, B. Taillade and A. Vincent for their hand work on the TKS instrument. Special thanks are due to J. Runavot, CNES Vega Project Manager and M. Vincent, Technical Manager of the part of TKS devolved to the French team. Support was provided by the Centre National d'Etudes Spatiales under a contract with the Université de Besançon – Franche-Comté.

References

A'Hearn, M.F., Hoban, S., Birch, P.V., Bowers, C., Martin, R., Klingelsmith III, D.A.: 1986, *Nature* **324**, 649

Arpigny, G., Dossin, F., Manfroid, J., Magain, P., Danks, A.C., Lambert, D.L., Sterken, C.: 1986, *The Messenger* **45**, 8

Belton, M.J.S., Wehinger, P., Wyckoff, S., Spinrad, H.: 1986, 20th ESLAB Symp. SP-250, Vol. **I**, p. 599

Bertaux, J.L.: 1986, *Astron. Astrophys.* **160**, L7

Chalon, G.: 1984, Thèse de Docteur-Ingénieur, Université de Besancon

Divine, N.: 1981, Proc. Joint NASA/ESA Working Group Meeting, ESA, SP-174, 47

Falciani, R., Festou, M.C., Smaldone, L.A., Tozzi, G.P.: 1986, 20th ESLAB Symp. ESA SP-250, Vol. **III**, p. 75

Feldman, P.D., Festou, M.C., A'Hearn, M.F., Arpigny, C., Butterworth, P.S., Cosmovici, C.B., Danks, A.C., Gilmozzi, R., Jackson, W.M., McFadden, L.A., Patriarchi, P., Schleicher, D.E., Tozzi, G.P., Wallis, M.K., Weaver, H.A., Woods, T.N.: 1986, 20th ESLAB Symp. ESA SP-250, Vol. **I**, p. 325

Festou, M.C.: 1981, *Astron. Astrophys.* **96**, 52

Gogoshev, M., Gogosheva, T., Sargoichev, S., Palasov, K., Georgiev, A., Kostadinov, I., Kanev, K., Spasov, S., Werner, R., Mendev, I., Moreels, G., Clairemidi, J., Vincent, M., Mougin, B., Parisot, J.-P., Zucconi, J.-M., Festou, M., Lepage, J.P., Runavot, J., Bertaux, J.-L., Blamont, J.E., Hersé, M., Krasnopolsky, V.A., Moroz, V.I., Krysko, A.A., Troshin, V.S., Barke, V.V., Jegulev, V.S., Sanko, N.F., Tomashova, G.V., Parshev, V.A., Tkachuk, A.Y., Novikov, B.S., Perminov, V.G., Sulakov, I.I., Fedorov, O.S.: 1985, *Adv. Space Res.* **5**, 12, 133

Green, S.F., Hughes, D.W.: 1986, 20th ESLAB Symp. ESA SP-250, Vol. **II**, p. 157

Hartmann, W.K., Cruikshank, D.P., Tholen, D.J.: 1985, in *Ices in the Solar System*, (eds. J. Klinger et al.) Reidel, Dordrecht, p. 169

Jockers, K., Houpis, H.L.F., Geffert, M.: 1986, in *Asteroids, Comets, Meteors* **II**, p. 331

Krasnopolsky, V.A., Moroz, V.I., Krysko, A.A., Jegulev, V.S., Tkachuk, A.Y., Gogoshev, M., Gogosheva, T., Moreels, G., Clairemidi, J., Parisot, J.P.: 1986, 20th ESLAB Symp. ESA, SP-250, Vol. **I**, p. 459

Krasnopolsky, V.A., Tkachuk, A.Y.: 1986, 20th ESLAB Symp. ESA, SP-250, Vol. **I**, p. 465

Levasseur-Regourd, A.C., Bertaux, J.L., Dumont, R., Festou, M., Giese, R.H., Giovane, F., Lamy, P., LeBlanc, J.M., Llebaria, A., Weinberg, J.L.: 1986, *Nature* **321**, 341

Mazets, E.P., Aptekar, R.L., Golenetskii, S.V., Guryan, Y.A., Dyachkov, A.V., Ilyinskii, V.N., Panov, V.N., Petrov, G.G., Savvin, A.V., Sagdeev, R.Z., Sokolov, I.A., Khavenson, N.G., Shapiro, V.D., Shevchenko, V.I.: 1986, *Nature* **321**, 276

Meech, K.J., Jewitt, D.C., Ricker, G.R.: 1986, *Icarus* **86**, 561

Moreels, G., Gogoshev, M., Krasnopolsky, V.A., Clairemidi, J., Vincent, M., Parisot, J.-P., Bertaux, J.-L., Blamont, J.E., Festou, M.C., Gogosheva, T., Sargoichev, S., Palasov, K., Moroz, V.I., Krysko, A.A., Vanysek, V., 1986: *Nature* **321**, 271

Moroz, V.I.: 1985, *Adv. Space Res.* **5**, 12, 113

Rettig, T., Ruchti, R., Baumbaugh, B., Baumbaugh, A., Knickerbocker, K., Dawe, J.: 1986, 20th ESLAB Symp. ESA, SP-250, Vol. **III**, p. 93

Sagdeev, R.Z., Avanesov, G.A., Ziman, Y.L., Moroz, V.I., Tarnopolsky, V.I., Zhukov, B.S., Shamis, V.A., Smith, B., Toth, I.: 1986, 20th ESLAB Symp. ESA, SP-250, Vol. **II**, p. 317

Schleicher, D.G., Millis, R.L., Tholen, D., Lark, N., Birch, P.V., Martin, R., A'Hearn, M.F.: 1986, 20th ESLAB Symp. ESA, SP-250, Vol. **I**, p. 565

Watanabe, K., Zelikoff, M.: 1953, *J. Opt. Soc. Am.* **43**, 753

Astron. Astrophys. 187, 560–568 (1987)

ASTRONOMY
AND
ASTROPHYSICS

The visual brightness behavior of P/Halley during 1981–1987

D.W.E. Green [1] **and C.S. Morris** [2]

[1] Harvard-Smithsonian Center for Astrophysics, 60 Garden St., Cambridge, MA 02138, USA
[2] Jet Propulsion Laboratory, California Institute of Technology, 4800 Oak Grove Dr., Pasadena, CA 91109, USA

Received February 18, accepted May 18, 1987

Summary. The overall behavior of the total visual coma brightness of P/Halley during 1981–1987 is discussed. Observations used in this study include ~75 published "V" magnitudes obtained with large telescopes (usually with CCDs) prior to July 1985 and ~3500 total magnitudes (published in the *International Comet Quarterly*) estimated visually with smaller telescopes during the period July 1985–February 1987. We discuss the observed macroscopic light curve and provide preliminary least-square fits to the traditional power-law formula. While the heliocentric brightness (H) variation with respect to heliocentric distance (r) did not follow expectations closely, P/Halley's maximum brightness during 1986 Feb.–Apr. fell somewhere between the most pessimistic and the most optimistic forecasts. We also review brightness variations on smaller timescales (hours, days), noting that several definite outbursts in total visual magnitude occurred during Oct. 1985–May 1986 which can be correlated with outbursts observed at other wavelengths.

Key words: total visual magnitude – photometry – light curve

1. Introduction

The most continuous set of physical data for comets available for study of cometary activity, except for astrometric data used for improving orbital elements and for studying nongravitational forces, is the visual photometric base of observations describing the total magnitude (m_1) of the coma. No comet has been as well observed, or has as thorough and continuous a set of observations at all wavelengths, as P/Halley at its current return to the inner solar system. The visual base of brightness data describes cometary activity associated with the resonance-fluorescence of diatomic carbon (C_2) and with the release from the cometary nucleus of dust, spectroscopically referred to as the continuum (Newburn, 1981). More than 4000 total visual magnitudes of periodic comet Halley 1982*i* (hereafter, P/Halley) obtained by ~100 observers during 1985–1987 have been published (Green, 1985a, 1986a, 1987) in the *International Comet Quarterly* (*ICQ*), and an additional 1000 or so magnitudes have been recently received for publication in the *ICQ*. In addition, ~75 photometric V m_1 measures are available, for $r > 5$ AU, which should be fairly compatible with the visual data obtained. Using these data, this paper describes some of the major features of P/Halley's light curve during its current return, from observations made through early 1987 February. P/Halley's light curve will serve as a useful tool in the study of cometary physics, with such a complete set of data available.

Send offprint requests to: D.W.E. Green

2. Observations

There still is no simple, pragmatic method for obtaining the total, integrated brightness of a comet's coma by electronic means, except where the coma diameter is quite small ($\lesssim 20''$), due to telescope-diaphragm sizes and sky/star background-subtraction constraints. It now appears hopeful that CCDs can provide complementary integrated visual magnitudes for cometary comae when comets are at large r and quite faint. Provided the CCD data are true V magnitudes, they may be used with the visual data, the latter having a scatter of ± 0.4 magnitude among experienced observers most of the time. For stars, V magnitudes differ from visual by ~0.1–0.2 mag, depending upon the peak wavelength of the radiation (Green and Morris, 1982), so the correction of CCD V data to visual magnitudes can be assumed negligible once other variables such as instrumentation are properly adjusted. Several sets of V magnitude data (~75 observations) are available from CCD observations (Schild, 1985; Green, 1986b; Festou et al., 1986; and references therein) made while the comet's coma was quite small and $r > 6$ AU. The collective data set of m_1 photometry obtained at $r > 5$ AU (prior to mid-1985) is scattered into several different photometric systems, leading to uncertainties (due to transformations to a common system) which are even greater than that quoted (up to ± 0.5 mag) by the observers. Several sets of V magnitudes were not measured in the V bandpass and had to be corrected to V magnitudes; this was usually done according to recommendations supplied by the original observers/authors.

Carefully-measured visual m_1 data should be useful in combination with properly-obtained charge-coupled device (CCD) V magnitudes obtained with larger telescopes when comets are at $m_1 > 15$. Both the so-called "nuclear" and total magnitudes (cf. Green et al., 1986) of comets vary considerably at different wavelengths (Roemer, 1976), and photographic (blue) and visual (yellow) data should never be thrown together without appropriate correction. Observed magnitudes are also dependent upon instrument type, aperture, and focal length (Roemer, 1976; Meisel and Morris, 1976, and Sect. 6 of this paper), but the degree of dependency differs from comet to comet. It appears that the degree of central condensation plays a significant role in visual m_1

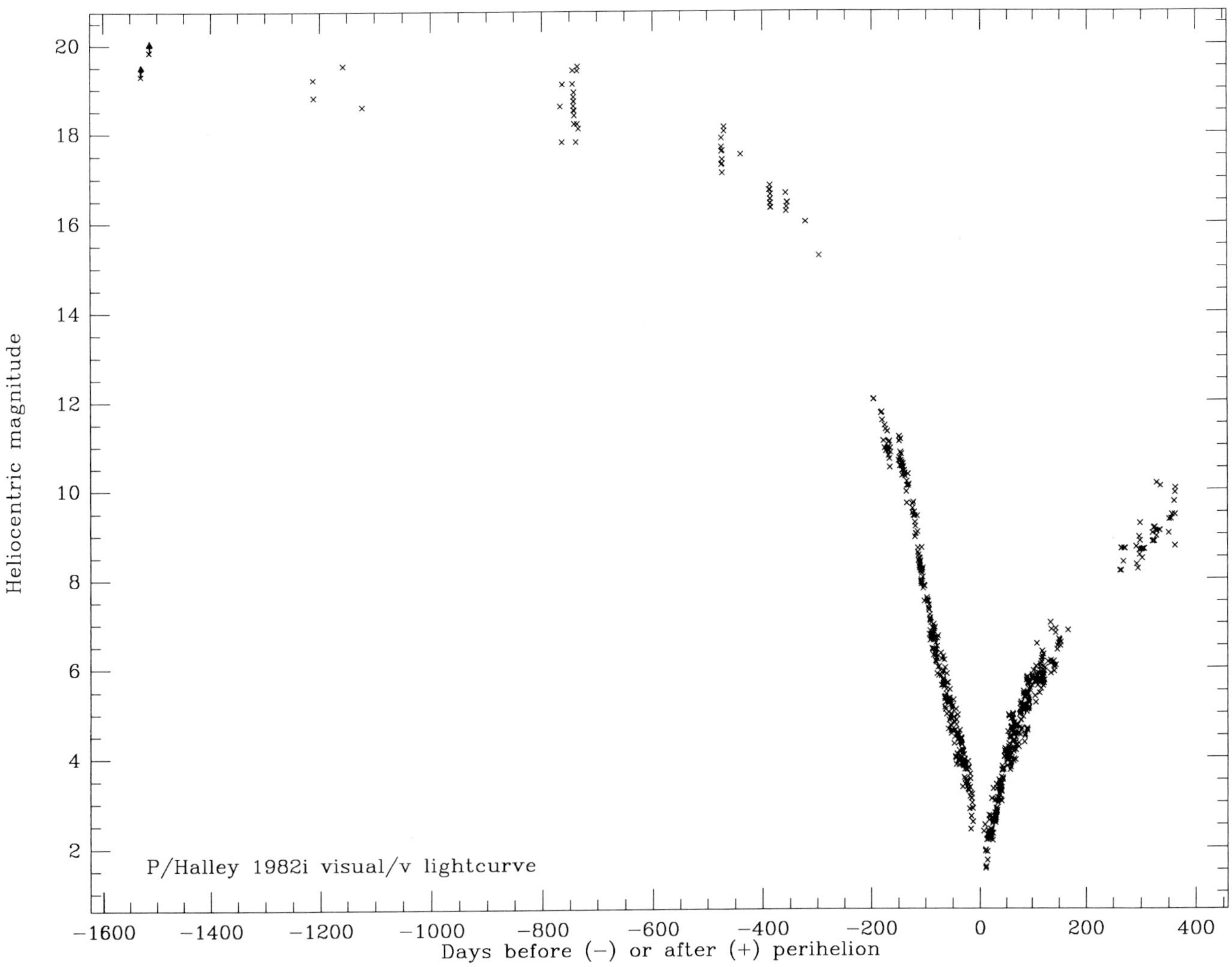

Fig. 1. The entire range of visual/V m_1 data for P/Halley from 1981–1987, plotted as a function of time versus heliocentric magnitude

estimation; for example, a comet with a strong central condensation and a large, much fainter coma shows more scatter in visual magnitude data, a problem which occurred with P/Halley even among experienced observers especially during 1985 Dec.–1986 June.

Visual m_1 data of P/Halley published thus far in the ICQ (as of February 1987; cf. Green, 1985a, 1986a) cover the period 1985 Jan. 23–1986 Aug. 9; additional data through early February 1987 have been received (and will be published in the April 1987 ICQ) and were included in this study. The ICQ archive contains visual m_1 estimates for every UT date from 1985 Oct. 6 to 1986 Jan. 28 (except Oct. 31) and from 1986 Feb. 15 to June 14. There are ~4200 such estimates of P/Halley now (April 1987) in the archive (with ~3500 published by the end of 1986), and it is anticipated that ~5500 P/Halley m_1 values may be in the ICQ database by early 1988 as more data are reported. Green et al. (1986) have recently reviewed the ICQ archive and announced the availability of the archive on magnetic tape to researchers. We have selected 650 observations from six experienced observers (J.E. Bortle, R. Bouma, D.W.E. Green, R. Keen, C.S. Morris, and D.A.J. Seargent) for the least-squares power-law solutions (see Sect. 6). All six observers have observed many comets visually over an average of >12 yr per observer, and all made visual m_1 estimates of P/Halley on >80 nights spanning at least from 1985 September through 1986 June, and all six observers had the majority of their data published in the ICQ archive at the time of this study. The selection of these very experienced observers gives a good representation of the comet's brightness behavior. One visual m_1 estimate per observer per night was selected; when an observer made > one m_1 estimate per night, the selected m_1 estimate was generally that made with the smallest instrument to *easily* see the comet. Figure 1 shows the selected visual and CCD V m_1 estimates reduced to "heliocentric magnitude" ($H = m_1 - 5 \log \Delta$, where Δ is the comet's geocentric distance in AU) and plotted *versus* time; also, 46 additional m_1 estimates made between 1986 Oct. and 1987 Feb. are depicted in all figures with this article, including 13 by A. Hale, all to be published in the April 1987 ICQ.

The pre-perihelion data as of this writing cover ~6 times the range of post-perihelion observations, in terms of r. Note that, due largely to small elongations from the sun, there are no B- or V-magnitude photometric data for P/Halley for the following periods: 1983 Feb. 14–Dec. 30, 1984 Mar. 5–Sept. 21, 1985

Apr. 10–July 18, 1986 Jan. 29–Feb. 14, and 1986 Aug. 11–Oct. 26. The two magnitudes at $\Delta T \sim -1500^d$ represent upper limits (i.e., P/Halley was not detected) to m_1.

3. P/Halley at $5 < r < 13$ AU

When P/Halley was recovered (Jewitt et al., 1982a) at $r = 11.04$ AU in 1982 October, it was at $V \simeq 24.2$ ($H \sim 19.0$) the faintest object ever detected and known to be in the solar system. The comet slowly brightened to $V \simeq 20$ (corresponding to $H \simeq 17$) by late 1984, although there are indications (Lecacheux et al., 1984; West and Pedersen, 1984) of large-amplitude (up to 1.5 mag) variation in brightness in timescales as short as 1–2 days and even 1–2 h. It seems that such short-term variations in brightness may be due to rotation of the cometary nucleus (see also Sect. 5), and though various early attempts to fit regular rotational periods to the magnitude data proved fruitless (e.g., Sekanina, 1985), this may not be so enigmatic in view of the irregular shape of the nucleus (Reitsema et al., 1986).

The first definite appearance (Spinrad et al., 1984) of a coma did not occur until 1984 September (at $r = 6.14$ AU). Wyckoff et al. (1985) speculated that P/Halley's sublimation began at $r \sim 6$ AU. While it does seem obvious that activity began increasing noticeably at that time, caution should be used in making assumptions based on spectra and photometry alone, as variations in brightness of the nuclear condensation, apparently not attributable to nucleus rotation (Sekanina, 1985), and the presence of coma activity in comets at large heliocentric distances ($r > 6$ AU) implies that some sublimation is present while comets are much further than 6 AU from the sun (Roemer, 1976; Sekanina, 1976; Green et al., 1981). In particular, Flammer et al. (1986) show that many of P/Halley's brightness flares at large r can be correlated with high-speed solar wind streams near the comet, and they suggest that fine, charged dust is blown off the nucleus by electrostatic levitation. Assuming that there was negligible coma contribution to the observed magnitudes of P/Halley when at $r > 7$ AU, an absolute "nuclear" V magnitude (H_N) of $\simeq 14.0$ can be derived (Green, 1984; Sekanina, 1985; Wyckoff et al., 1985; Meech et al., 1986). However, the comet could not be detected in late 1981 with the 5.1-m Palomar reflector (Jewitt et al., 1982b), implying (Jewitt and Danielson, 1984) that $H_N > 14.3$.

Indeed, Festou et al. (1986) have found $H_N \simeq 14.8$, noting that when $m_1 > 22.5$, a dust coma $\leqq 2''$ in diameter will likely not be detectable due to low surface brightness and thus might be contaminating the nucleus. Weissman (1985) notes that the Infrared Astronomical Satellite (IRAS) found P/Schwassmann-Wachmann 1 "very bright in the infrared even during quiescent periods" at $r \sim 6$ AU (during which the comet appears essentially stellar at visual wavelengths), indicating that there may be some coma contaminating the observed light on a regular basis; it is possible, though, that for some time after an outburst there may linger dust (visible in infrared but not in visible light) in the comet's vicinity. Comet Bowell 1982 I has been observed recently as a diffuse object (coma size as much as 25″) as far out as $r \sim 14$ AU (Meech and Jewitt, 1987; Scotti, 1987).

When the pre-perihelion m_1 data are plotted as a function of H vs. r (see Fig. 2 of Green and Morris, 1986), it can be seen that, while a really clear departure from an r^{-2} curve does not occur until $r \sim 6$–7 AU, a departure may have begun as far out as 11 AU; the data are not conclusive on this matter, and better-calibrated post-perihelion V data will be needed to help resolve the problem. Few, if any, observers doing photometry of P/Halley at $r > 6$ AU (pre-perihelion) used comparison stars within a few magnitudes of the comet. Several observers used comparison stars that were 10^4–10^6 times brighter than the comet; while CCDs are known to be very linear in response, and only ~ 0.1–0.2 mag of error may be introduced due to nonlinearity, observers are encouraged to try using comparison stars closer to the brightness of the comet. Of more serious concern are uncertainties in the transformation coefficients between the various photometric systems employed by various observers (see Sect. 2), and observers are urged to obtain Johnson V magnitudes of P/Halley as much as possible at $r > 7$ AU post-perihelion.

4. P/Halley at $r < 6$ AU

Comet 1982i continued its slow brightness increase during the 1984–1985 observing season, and was still at $V \sim 19$ when lost in the solar glare in 1985 April. The first visual detection of P/Halley at this return was apparently made by O'Meara (1985; see also Green, 1985b) on 1985 January 23–24 with a 24-inch (61-cm) Cassegrain reflector at Mauna Kea ($m_v = 19.6$, $H \simeq 16.4$); his magnitude estimate was consistent with V magnitudes derived via CCD photometry (though obtained with much larger telescopes) at the same time. The comet was not seen visually again until it was located after conjunction in 1985 July, when the comet had brightened to $m_v \sim 15$ ($H \sim 12$). In 1985 August, the comet appeared to begin a more rapid brightening, only to face again; while the moon was near full (late August), the comet faded noticeably in brightness. This "false start" in brightening (at $r \sim 3$ AU) may have also occurred at a similar pre-perihelion distance during the 1909–1910 apparition (cf. Morris and Green, 1982). By 1985 September 15, P/Halley was at $m_v \simeq 13.0$ ($H \simeq 11.0$), increasing more rapidly than expected to $m_v \sim 10$ ($H \sim 9$) by mid-October and to naked-eye brightness ($H \sim 7$) in the first half of November. The comet then rose slowly to $m_v \simeq 3.5$ ($H \simeq 2.6$) by late 1986 January, when it was lost in evening twilight as it neared perihelion on February 9 (see Fig. 2).

When first picked up a week after perihelion, P/Halley was at $m_v \simeq 2.5$ ($H \simeq 1.7$), and it stayed at $2.0 \leqq m_v \leqq 3.5$ until late April (when $m_v \sim 4.0$ and $H \simeq 4.2$). The comet gradually faded to $m_v \sim 7$ by early June, to ~ 9 by early July, and to ~ 10 ($H \sim 7.2$) by early August. After conjunction with the sun in mid-September 1986, P/Halley was again picked up by visual observers in late October at $m_v \simeq 11.5$ ($H \sim 8.2$), and the comet had faded by no less than a full apparent magnitude by early February 1987, when $H \sim 9.4$. Interestingly, only about eight or ten observers are known to have made accurate m_1 estimates of P/Halley since July 1986.

H was brighter at a given r after perihelion than at the same heliocentric distance before perihelion (see Fig. 1), a phenomenon known to be associated with P/Halley at previous returns (Morris and Green, 1982; Sekanina, 1983; Weissman, 1986). In February 1987, P/Halley was a full six magnitudes brighter than at an equivalent r during its outbound leg. This asymmetry is noted by most observers who made intensity measurements both before and after perihelion (cf. references in Battrick et al., 1986). The coma size of P/Halley became too large for accurate CCD m_1 determination around 1985 Aug.–Sept., and visual m_1 estimates are the best data available from then through early 1987.

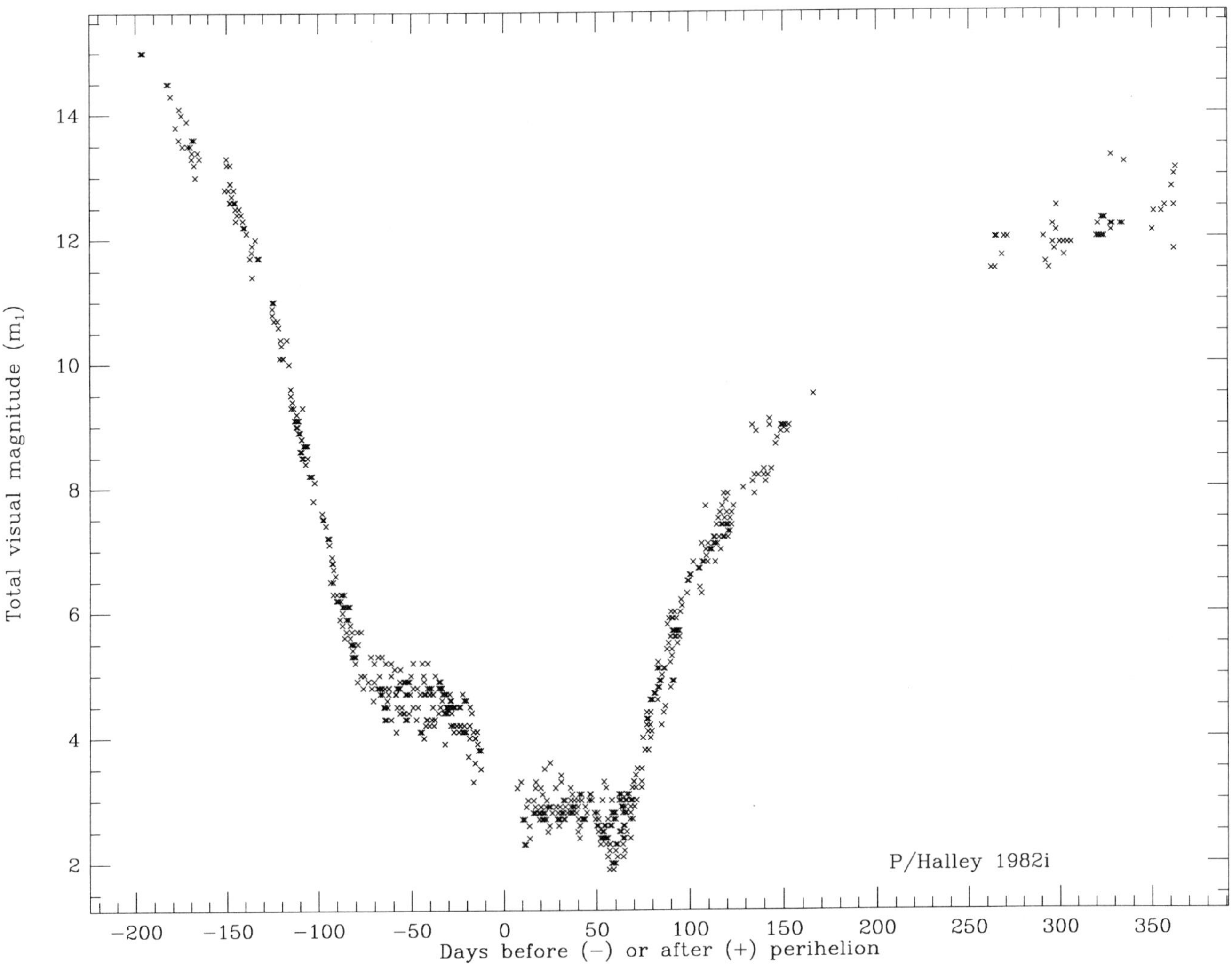

Fig. 2. The total visual magnitude data plotted versus time for P/Halley from July 1986 to February 1987

5. Short-term brightness variations

P/Halley has relatively stable, slightly increasing, nongravitational effects appearing in its orbital motion (Marsden, 1985). Unlike some comets which exhibit significantly increasing or wild nongravitational effects, a pattern also sometimes associated with occasional marked outbursts in cometary brightness, P/Halley does not have an obvious recurring history of large (>2 mag), sporadic outbursts in brightness. Sekanina (1983) has noted an apparent 2- to 3-magnitude outburst in 1836 some 70 days after perihelion, and Bortle and Morris (1984) also looked at possible historical outbursts. No such outburst has been observed at the current apparition, although outbursts of 0.5- to 1.5-mag amplitude (in ultraviolet, *B*, *V*, and infrared magnitudes) have been noted (Lecacheux et al., 1984; West and Pedersen, 1984; Spinrad et al., 1984; le Bertre, 1986; Ney and Knutson, 1986; West, 1986; Hale and Green, 1986; Feldman et al., 1986; and many references in Battrick et al., 1986) at the current return, including an infrared change of two magnitudes in 3 days ($\Delta m_L = -2.0$ during 1986 Feb. 9.8–12.7 UT). Outbursts detected in narrow-band photometry usually refer to brightness changes in the near-nuclear region; such outbursts of a magnitude or more, if visible in the visual wavelengths, would be expected to increase the integrated coma brightness by only a few tenths of a magnitude. During 1986 April 5.5–7.5, there was a noticeable visual brightening of perhaps as much as $\simeq 0.6$ mag in $\sim 1^d$. The frequency of historical outbursts is not well established; the absence of exceptional brightness outbursts at this return may be partly due to the fact that solar activity is now at the minimum point of its 11-year cycle (Coffey, 1986).

Though there may be some evidence that P/Halley's 2.2- and 7.4-day rotation rates (Kaneda et al., 1986; Millis and Schleicher, 1986; see also Sect. 3) are visible in the ICQ archival data, such "regular" brightness variations will not be investigated here. Indeed, some of the fluctuations discussed here probably *are* related to nucleus rotation. Irregular brightness variations on the order of $\lesssim 1$–2 days are quite noticeable in the ICQ archive; with m_1 estimates internally consistent (among experienced observers) to ± 0.1–0.2 mag, we can detect with varying degrees of confidence several rises and drops in brightness on the order of 0.3 mag or more over such timescales.

We have searched all of the ICQ-published m_1 estimates of P/Halley (~ 3500 by Feb. 1987; see Sect. 2) by observer for significant increases and decreases in brightness on timescales of a

Table 1. Significant short-term brightness variations in ICQ archival data. All magnitudes in column 2 refer to total visual magnitudes; a negative variation indicates a brightening and a positive value indicates a fading. Rating scale is authors' estimation of reliability (4 = definite, 1 = questionable; see text)

Approximate Date (UT) of maximum or minimum brightness	Average observed variation (mag/day)	Rating	Note	Remarks
1985 Oct. 19.5–21.4	−0.5	2		
1985 Nov. 5–9	(−0.5)	4	a	(1), (2)
1985 Nov. 14–18	−0.5	2		(1), (3), (4)
1985 Nov. 19	+0.2	1		(5)
1985 Dec. 3.6–4.8	−0.3	2		(6)
1985 Dec. 9.8	−0.4	1		(7)
1985 Dec. 12.4–14.8	−0.3	4		(4), (8)
1986 Jan. 7.5–9.0	−0.3	4		(9)
1986 Jan. 12.7–13.1	−0.3	4		(10)
1986 Mar. 1.5	+0.3	1		(11)
1986 Mar. 3.4–4.1	+0.3	2		(12), (13)
1986 Mar. 22.1–23.7	−0.3	3		(13), (14)
1986 Mar. 31.2–Apr. 3.1	−0.3	2	b	(15)
1986 Apr. 3.5–4.7	+0.3	2		(13)
1986 Apr. 6.5–7.9	−0.35	4		(13), (16)
1986 Apr. 8.2–8.7	+0.2	1		(13), (17)
1986 Apr. 13.9–14.9	−0.3	1		(13), (18)
1986 Apr. 29.9	−0.3	1		(19)
1986 Apr. 30.4–May 5.8	+0.3	1	c	
1987 Apr. 21–24	−0.7	4		(20)

Notes:

(a) From Oct. 29 to Nov. 9, there was an unusually rapid increase in brightness (from mag 8.5 to 6.1, ~1 mag more than expected from a normal power-law formula). The moon was likely a factor during Oct. 29–Nov. 4.

(b) The moon was possibly a factor (last quarter on Mar. 30)

(c) From Apr. 30–May 5, several observers noted an average decrease of −0.3 mag/day, whereas only −0.1 mag/day was expected from a normal power-law formula.

Remarks (all references are in Battrick et al., 1986, unless otherwise noted by an asterisk, *, in which case they can be found in the Reference listing at the end of this paper):

(1) Photometry and photography showed "sharp brightness enhancements evidently associated with outbursts of cometary material" on Nov. 5.9 and 15.8 (Dobrovolsky et al., Vol. III, 31).

(2) The CN brightness was greater on Nov. 9.0 than on Nov. 12.0 (Svoreň, Vol. III, 39). On Nov. 7, there was a large increase in OH 1667-MHz peak line flux density (Cordes et al., Vol. III, 113). Significant infrared flux increase on Nov. 4.0 (Stanga et al., Vol. III, 365). Significant infrared outburst on Nov. 8 (Lynch et al., Vol. III, 479).

(3) On Nov. 19.9, the HCN production rate was ~3 times higher than on Nov. 20.9 (Bockelée-Morvan et al., Vol. I, 365). The CN brightness on Nov. 16.0 was greater than on Nov. 18.0 (Svoreň, Vol. III, 39). Also, continuum outburst observed by Meech and Jewitt (Vol. I, 553). On Nov. 15, "spectrum exhibits an unusually high continuum intensity" (Fink et al., Vol. III, 485).

(4) C_2, C_3, CN production higher on Nov. 15 and Dec. 13 than on surrounding dates (Cochran and Barker, Vol. I, 439).

(5) On Nov. 20, Cordes et al. (Vol. III, 113) observed a decrease in OH 1667-MHz peak line flux density from the previous day.

(6) Two outbursts on Dec. 3–4 are suggested from observations in visual-light gas emission (Falciani et al., Vol. III, 75). Outstanding rays in the tail structure on photos taken Dec. 3 and 5 (Saito et al., Vol. III, 155).

(7) Significant CN brightness increase between Dec. 3 and 12 (Svoreň, Vol. III, 39). Fitzsimmons et al. (Vol. III, 63) noted a large increase in OH luminosity and in flux from Dec. 8.9 to 9.9 in 365- and 484-nm filters. "DE" and outstanding rays in tail structure on photos taken Dec. 10 (Saito et al., Vol. III, 155).

(8) Strong outburst on Dec. 12.5 (Watanabe et al., Abstracts, 438, and Vol. III, 267). Also outburst at infrared wavelengths (J, H, K, L) on Dec. 12.2 detected by Tokunaga et al. (1986*). "DE" in photos taken Dec. 12 (Saito et al., Vol. III, 155).

(9) Outburst in infrared brightness (K, L) on Jan. 8.1 (Tokunaga et al., 1986*). Significant tail disturbances in photographs obtained Jan. 7 and 8 (Saito et al., Vol. III, 155). "DE" reported on Jan. 8.7 by Kubáček et al. (Vol. III, 277).

Table 1 (continued)

(10) IR brightness more than doubled on Jan. 10 (Lynch et al., Abstracts, 71; also Vol. III, 479). Dust outburst noted by Sekanina et al. (Vol. II, 77). "Dramatic" "DE" on Jan. 10 (Brandt and Niedner, Vol. I, 47; also Moore, Abstracts, 410). Outstanding tail streamers and rays in photos taken Jan. 12 (Saito et al., Vol. III, 155).

(11) On March 1–2, in increase of 1.25 mag in infrared band L was observed by le Bertre (1986*). The nucleus was much brighter on photo taken Mar. 2.4 than on one taken Mar. 1.4 (West, 1986*).

(12) Decrease observed in CN and OH during Mar. 3–4 (Schleicher et al., Vol. I, 565). Central coma was ~0.5 mag fainter on Mar. 3.8 than on Mar. 4.8 (Green and Hughes, Vol. II, 157). Outstanding tail streamers and rays in photos taken Jan. 12 (Saito et al., Vol. III, 155). On March 5, Na I emission very much stronger than on other dates (Fink et al., Vol. III, 485).

(13) Miller and Liller (Vol. III, 137) report tail disturbances photographed with Schmidt telescopes on the following dates: 1986 March 2–3 amd 20–23, and April 3–5, 7–11, and 14–18.

(14) IUE outburst over Mar. 23–25 (McFadden et al., Abstracts, 223); also "violent" outburst noted by Chen et al. (Abstracts, 421) during this time. Sharp IUE outburst on Mar. 18/19 (Feldman et al., Vol. I, 325). Kuiper Observatory found increase in H_2O line brightness on Mar. 24 (Weaver et al., Vol. I, 329). Rettig et al. (Vol. III, 93) reported that on Mar. 24 there was an "anti-sunward core outburst; the comet brightness was estimated to have increased by a factor of 2–3". Large outbursts in brightness on Mar. 22 and 23 at 365, 406, 426, and 484.5 nm (Gong, Vol. III, 261). Significant infrared flux increase on Mar. 21.3 (Stanga et al., Vol. III, 365).

(15) The 10.8-μm flux was up 50% on Mar. 29.7 over Mar. 28.7 (Campins et al., Vol. II, 91); IR continuum and 3.36-μm fluxes increased by almost 3 times over same time (Knacke et al., Vol. II, 95).

(16) 5- to 9-μm brightness was ~2 times higher on Apr. 8.6 than Apr. 10.5 (Campins et al., Vol. II, 121); also 10-μm flux rose ~3 times from Apr. 7.6 to Apr. 8.6, peaking on Apr. 9.6 (Russell et al., Vol. II, 125). Large "DE" on Apr. 8 (Guérin et al., Abstracts, 409).

(17) "DE" observed on Apr. 8.5 by Caron et al. (Vol. III, 143).

(18) "DE" on Apr. 13/14 (Lundstedt and Magnusson, Vol. I, 141); outbursts on Apr. 13 and 15 (Wang et al., Abstracts, 439).

(19) Outstanding tail streamers appear on photos taken Apr. 29 (Saito et al., Vol. III, 155).

(20) An increase of 1.38 mag (red filter, 16″ diaphragm aperture, 1.5-m reflector) was observed from 1987 Apr. 20.17 to 23.19 (Wisniewski et al., 1987*)

---

few days or less. The results are presented in Table 1. As the heaviest observational coverage occurred within 3 months of perihelion, most of the cases listed in Table 1 lie in the period Nov. 1985–early May 1986. There were many instances where individual observers reported lone significant night-to-night changes in brightness (±0.3–0.4 mag or more), in which there were no contradictory observations, but we do not list such brightness variations unless we feel there is a fairly strong case with 3 or more observers detecting a change with no (or very few) conflicting observations. Table 1 includes a column titled "Rating", which contains a number from 1 to 4 which is the authors' evaluation of each apparent brightness increase/decrease, as follows: 4 = definite outburst (steep brightness gradient observed by numerous observers), 3 = probable outburst, 2 = possible outburst or rapid fading, 1 = questionable outburst or rapid fading. The column titled "Remarks" lists significant brightness variations reported over a wide range of wavelengths. Six visual cases are strong enough to term "definite". The most significant outbursts in the ICQ archival data occurred during 1985 Dec. 12.4–14.8 and 1986 Apr. 6.5–7.9. (During the referee process, a large outburst was reported in late 1987 April, and we have added that to Table 1.) And there is not a definite correlation between brightness outbursts and disconnection events (DEs) in the tail, although several DEs are shown to occur near times of noticeable brightness variations.

6. Representing the light curve

A comet's apparent brightness has traditionally been represented in magnitudes by the equation

$$m_1 = H_0 + 5 \log \Delta + 2.5n \log r\,, \qquad (1)$$

where H_0 is called the absolute magnitude and n is the heliocentric power-law exponent. This equation does not usually provide (with one set of parameters) a continuous fit through perihelion. However, the formula is still widely used to describe roughly and predict the brightness of comets, as it is the simplest available formula for producing a quick idea of a comet's activity. It is useful especially for giving observers an estimate of what a newly-discovered comet will do in the following few months, to permit planning of observing programs. It would be difficult to produce *any* formula to fit closely the observed brightness of P/Halley throughout its passage in the inner solar system, due to the odd shape of the nucleus, to the extremely non-uniform ejection of gas and dust from the nucleus (Reitsema et al., 1986), and to the comet's being significantly brighter after perihelion than before. Pre- and post-perihelion least-square solutions yield different sets of (H_0, n), and the concept of H_0 has no real physical meaning as represented in this formula (Roemer, 1963). It appears that H_0, and possibly n, are functions of r in some manner, since the coma size changes with r; because the total reflecting area (gas and dust molecules in the coma) is changing with r, there is nothing really "absolute" about the reflecting/fluorescing area or cross-section. We have derived (H_0, n) values for P/Halley at its current apparition from the m_1 data mentioned in Sect. 2 of this paper.

To account for instrument effects (e.g., Meisel and Morris, 1976), visual magnitude estimates are often empirically corrected to a "standard" aperture of 6.78 cm. The observations selected in our analysis, except those made after conjunction in September 1986, require little or no aperture correction and none was applied. This does not mean that aperture/instrument effects were absent through most of the apparition, but rather that we primarily chose estimates made using binoculars, which require minimal correction, during those periods when these effects were significant. When the comet was no longer a binocular object, and certainly after conjunction in Sept. 1986, the comet was very diffuse and the magnitude estimates probably should be corrected for aperture/instrument effects. However, the magnitude of this correction is indeterminate (although probably ~0.5 mag). We also have chosen to display the observations in the figures without corrections; however, none of the post-conjunction observations were used in the calculation of the power-law parameters.

Additional information on instrument effects for P/Halley observations is given by Morris (1986).

A continuous pre-perihelion fit using Eq. (1) is not possible, and we divided the pre-perihelion light curve for $r < 5$ AU into two separate segments. The resulting pre-perihelion (H_0, n) parameters are as follows: $5.2 > r > 1.5$ (1985 Jan.–1985 Dec. 2), (1.8, 8.4); $1.5 > r > 0.6$ (1985 Dec. 3–1986 Jan. 29), (4.3, 3.2). Figure 4 of Green and Morris (1986) shows the (1.8, 8.4)-fit drawn through all of the pre-perihelion data for $r < 5.8$, and it can be seen how the data leave the curve at $r \simeq 1.5$ ($\log r \simeq 0.16$) and at $r \simeq 5.6$ ($\log r \simeq 0.75$). For post-perihelion, we derive (3.4, 3.0) for $0.6 < r < 2.9$, and Fig. 3 shows this least-squares fit as a line with respect to the individual data points through early Feb. 1987; the observations at $r > 4$ AU are beginning to fade away from this line. For comparison, R. Millis (private communication, Oct. 1986) and colleagues found the following production variations from spectrophotometry: $Q(\mathrm{OH}) \propto r^{-2.2}$ pre-perihelion (pre-T) and $\propto r^{-1.8}$ post-T; $Q(C_2) \propto r^{-3.6}$ pre-T and $\propto r^{-2.5}$ post-T; and $Q(\mathrm{CN}) \propto r^{-3.0}$ pre-T and $\propto r^{-2.2}$ post-T.

Just as Marsden (1985 and references therein) and colleagues obtained a semi-empirical formulation for nongravitational forces based upon the vaporization flux of H_2O snow with r, attempts have been made to correlate visual magnitude variation with gas and dust production rates (Newburn, 1984). The complex physical theory connecting total visual magnitude variation to dust and gas production rates (as functions of r), and also to nuclear precession and rotation, is still being researched (Newburn, 1984; Sekanina, 1984; Rickman et al., 1985; Festou, 1986; Whipple and Sekanina, 1979). However, the visible coma is a result of not only H_2O gas production (and dissociation into daughter molecules) but also CO_2 gas production and dissociation, the dust production resulting from the vaporization of the ices, and some contribution from other volatile ices in the nucleus. Water vapor seems to account for $> 80\%$ of the non-radical gas in P/Halley's inner coma, and CO_2 is perhaps second in population to H_2O (Krankowsky et al., 1986; Combes et al., 1986). Vega 1 found a production rate (at $r \sim 0.8$ AU) for carbon dioxide, $Q(CO_2)$, around $0.015\, Q(H_2O)$, and suggested $Q(H_2CO) \sim 0.01\, Q(H_2O)$ (Combes et al., 1986). Early Giotto results (Eberhardt et al., 1986) yield $Q(\mathrm{CO}) \leqq 0.15\, Q(H_2O)$ and $Q(N_2) \leqq 0.1\, Q(H_2O)$. Ground-based observations have yielded the apparent detections of HCN ($\nu = 88.6$ GHz), H_2O ($\lambda = 1.4$, 1.9, 2.63–2.67 μm), and H_2CO ($\lambda = 404.4$, 469.5 nm) (references cited in Green and Morris, 1986). Vaporization curves for H_2O show (Delsemme,

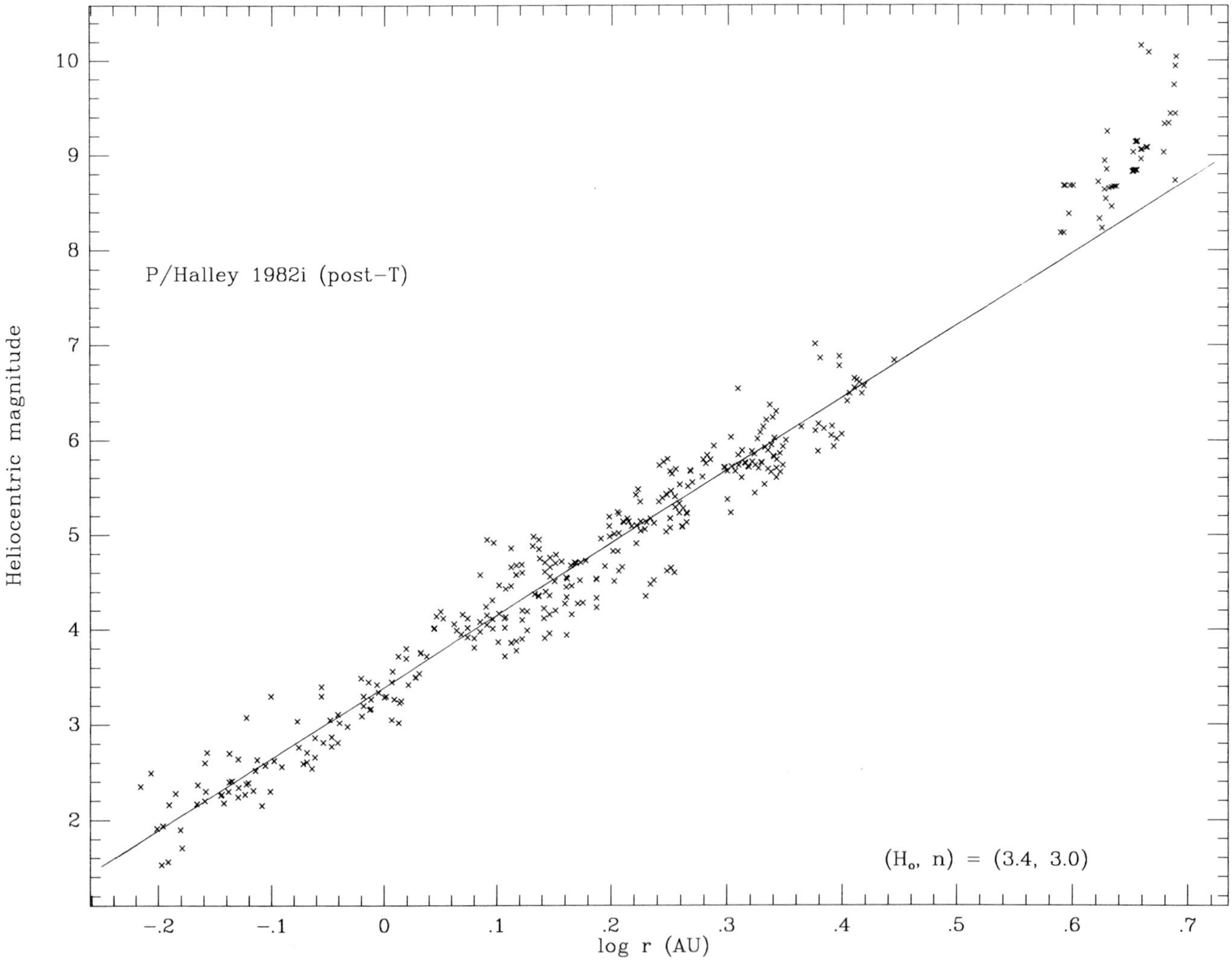

Fig. 3. The post-perihelion visual m_1 data for P/Halley 1982i, with the straight line representing the least-squares fit to the standard power-law formula for $H_0 = 3.4$ and $n = 3.0$. Note that following conjunction in Sept. 1986, the comet's brightness begins to deviate from the power law

1982) a significant drop after $r = 2.5$ AU, and H_2O outgassing appears negligible certainly by $r = 3.5$ AU if not by $r = 2.8$ AU. The active comae present in many comets well outside $r = 3.5$ AU, including P/Halley, indicate that the contribution of these ices more volatile than H_2O ultimately must be fully accounted for. Cochran (1985) suggests that the main contributor to the visual brightness, C_2, chiefly comes from the parent molecule C_2H_2 (via photodissociation into C_2H); Huebner et al. (1986) suggest that C_2H_2 accounts for some 0.8% of the total composition.

The rate of decrease of H_2O production as observed by the *International Ultraviolet Explorer* is very similar to the rate of decrease in H when plotted with respect to r over the post-perihelion interval $0.8\,\mathrm{AU} < r < 2.6\,\mathrm{AU}$. Schloerb et al. (1986) find good correlation between visual m_1 data and HCN production rates. Feldman et al. (1986) note that CS and dust production apparently are not totally controlled by water vaporization for $r < 2$ AU. Delsemme (1983) has stated that only about 1 comet in 4 has a smooth-enough light curve such that the curve can be fitted to water vaporization; he did obtain (Delsemme, 1985) a preperihelion fit to P/Halley's 1909–1910 light curve using the H_2O sublimation model employed by Marsden et al. for representing nongravitational forces, though no good post-perihelion fit was found. The problem is indeed complex, with the fractional composition of volatile ices and dust still unknown. Analysis of spectrophotometric data should yield information concerning how much the continuum contributes to m_1 brightness in the scotopic passband, but the dust contribution does seem to be small. The Vega instruments found a dust-to-gas ratio $d \sim 0.1 - 0.25$ in P/Halley's inner coma (Mazets et al., 1986). This ratio does not appear to be constant throughout the coma, however, and varies with r; Newburn (1981) found that $d \propto r^2$ for high gas flux and $d \propto r^3$ for low gas flux.

7. Conclusions

While the heliocentric brightness (H) variation with respect to heliocentric distance (r) did not follow expectations closely (especially during pre-perihelion; cf. Morris, 1986, and references listed therein; Marcus, 1986), P/Halley's maximum brightness during 1986 Feb.–Apr. fell somewhere between the most pessimistic and the most optimistic forecasts. It has been demonstrated (e.g. A'Hearn et al., 1979; Festou, 1986; Schloerb et al., 1986; Newburn, 1984) that carefully-made visual magnitude estimates, using standard methods and comparison-star references, by experienced observers can yield meaningful data for use in modelling the gas- and dust-production rates of comets. The P/Halley m_1 archival data describe the overall coma activity throughout the apparition, and definite instances of significant brightness variations are present on timescales of ~1–2 days.

At this writing (February 1987), P/Halley is still observable visually in small (0.20-m) telescopes, and we should have a good extension to the arc of available post-perihelion photometry for P/Halley 1982*i*. With great potential value in a very long arc of visual/V broadband m_1 photometry, observers with access to large telescopes and CCDs (which have much better linearity than other detectors) are strongly encouraged to obtain such V data of P/Halley beginning in 1987 and continuing as long as the comet's brightness will permit.

Acknowledgements. The research carried out at the Jet Propulsion Laboratory, California Institute of Technology, was performed under contract with the National Aeronautics and Space Administration. Alain C. Porter, Caltech, provided useful discussions regarding CCD photometry of faint, extended objects. Paul D. Feldman, Johns Hopkins University, kindly provided IUE H_2O production rates, and Robert Millis, Lowell Observatory, provided information from his spectrophotometric data. We thank Brian G. Marsden, Center for Astrophysics, for a critical reading of an earlier draft of this paper. Thanks are extended also to the many observers contributing photometric data to the ICQ.

References

A'Hearn, M.F., Millis, R.L., Birch, P.V.: 1979, *Astron. J.* **84**, 570

Battrick, B., Rolfe, E.J., Reinhard, R., eds.: 1986, 20th ESLAB Symp., ESA SP-250, Abstracts and Vols. I–III

Bortle, J.E., Morris, C.S.: 1984, *Sky Telesc.* **67**, 9

Cochran, A.L.: 1985, *Astrophys. J.* **289**, 388

Coffey, H., ed.: 1986, *Solar-Geophysical Data – Part I*, No. 502, 10

Combes, M., Moroz, V., Crifo, J.F., Bibring, J.P., Coron, N., Crovisier, J., Encrenaz, T., Sanko, N., Grigoriev, A., Bockelée-Morvan, D., Gispert, R., Emerich, C., Lamarre, J.M., Rocard, F., Krasnopolsky, V., Owen, T.: 1986, 20th ESLAB Symp., ESA SP-250, Vol. **I**, p. 353

Delsemme, A.H.: 1982, in *Comets*, L.L. Wilkening, ed., University of Arizona Press, Tucson, p. 85

Delsemme, A.H.: 1983, *J. Phys. Chem.* **87**, 4214

Delsemme, A.H.: 1985, in *Ices in the Solar System*, eds. J. Klinger et al., Reidel, Dordrecht, p. 367

Eberhardt, P., Krankowsky, D., Schulte, W., Dolder, U., Lämmerzahl, P., Berthelier, J.J., Woweries, J., Stubbemann, U., Hodges, R.R., Hoffmann, J.H., Illiano, J.M.: 1986, 20th ESLAB Symp., ESA SP-250, Vol. **I**, p. 383

Feldman, P.D., Festou, M.C., A'Hearn, M.F., Arpigny, C., Butterworth, P.S., Cosmovici, C.B., Danks, A.C., Gilmozzi, R., Jackson, W.M., McFadden, L.A., Patriarchi, P., Schleicher, D.G., Tozzi, G.P., Wallis, M.K., Weaver, H.A., Woods, T.N.: 1986, 20th ESLAB Symp., ESA SP-250, Vol. **I**, p. 325

Festou, M.C.: 1986, in *Asteroids, Comets, Meteors II*, eds. C.-I. Lagerkvist et al., Uppsala universitet Reprocentralen HSC, p. 299

Festou, M.C., Lecacheux, J., Kohl, J.L., Encrenaz, T., Baudrand, J., Combes, M., Despiau, R., Laques, P., Lefèvre, O., Lemonnier, J.P., Lelièvre, G., Mathez, G., Pierre, M., Vidal, J.L.: 1986, *Astron. Astrophys.* **169**, 336

Flammer, K.R., Jackson, B., Mendis, D.A.: 1986, *Earth, Moon, Planets* **35**, 203

Green, D.W.E.: 1984, *Int. Comet Q.* **6**, 98

Green, D.W.E., ed.: 1985a, *Int. Comet Q.* **7**, 65, 81, 105

Green, D.W.E.: 1985b, *Int. Comet Q.* **7**, 40

Green, D.W.E., ed.: 1986a, *Int. Comet Q.* **8**, 25, 47, 93, 135

Green, D.W.E.: 1986b, *Int. Comet Q.* **8**, 75

Green, D.W.E., ed.: 1987, *Int. Comet Q.* **9** (in press)

Green, D.W.E., Marsden, B.G., Whipple, F.L., Meisel, D.D., Morris, C.S., Roemer, E., Sekanina, Z., Spinrad, H.: 1981, *Sky Telesc.* **62**, 552

Green, D.W.E., Rokoske, T.L., Morris, C.S.: 1986, 20th ESLAB Symp., ESA SP-250, Abstracts, 431, and Vol. **III**, p. 249

Green, D.W.E., Morris, C.S.: 1982, *Int. Comet Q.* **4**, 5

Green, D.W.E., Morris, C.S.: 1986, 20th ESLAB Symp., ESA SP-250, Vol. **I**, p. 613

Hale, A., Green, D.W.E.: 1986, *Int. Comet Q.* **8**, 82
Huebner, W.F., Boice, D.C., Keady, J.J., Schmidt, H.U., Wegman, R.: 1986, 20th ESLAB Symp., ESA SP-250, Vol. **I**, p. 529
Jewitt, D.C., Danielson, G.E., Gunn, J.E., Westphal, J.A., Schneider, D.P., Dressler, A., Schmidt, M., Zimmerman, B.A.: 1982a, *IAU Circ.* **3737**
Jewitt, D.C., Danielson, G.E., Terrile, R.J.: 1982b, *IAU Circ.* **3688**
Jewitt, D.C., G.E. Danielson: 1984, *Icarus* **60**, 435
Kaneda, E., Takagi, M., Hirao, K., Shimizu, M., Ashihara, O.: 1986: 20th ESLAB Symp., ESA SP-250, Vol. **I**, p. 397
Krankowsky, D., Lämmerzahl, P., Herrwerth, I., Woweries, J., Eberhardt, P., Dolder, U., Herrmann, U., Schulte, W., Berthelier, J.J., Illiano, J.M., Hodges, R.R., Hoffman, J.H.: 1986, *Nature* **321**, 326
le Bertre, T.: 1986: *IAU Circ.* **4189**
Lecacheux, J., Lefèvre, O., Baudrand, J., Lemonnier, J.P., Mathez, G., Lelièvc, G.: 1984, *IAU Circ.* **3928**
Marcus, J.N.: 1986, 20th ESLAB Symp., ESA SP-250, Vol. **III**, p.307
Marsden, B.G.: 1985, in *Dynamics of Comets: Their Origin and Evolution*, eds. A. Carusi, G.B. Valsecchi, Reidel, Dordrecht, p. 343
Mazets, E.P., Aptekar, R.L., Golenetskii, S.V., Guryan, Y.A., Dyachkov, A.V., Ilyinskii, V.N., Panov, V.N., Petrov, G.G., Savvin, A.V., Sagdeev, R.Z., Sokolov, I.A., Khavenson, N.G., Shapiro, V.D., Shevchenko, V.I.: 1986, *Nature* **321**, 276
Meech, K.J., Jewitt, D., Ricker, G.R.: 1986, *Icarus* **66**, 561
Meech, K.J., Jewitt, D.: 1987, "Comet Bowell at Record Heliocentric Distance", *Nature* submitted; see also *Minor Planet Circ.* **11466**
Meisel, D.D., Morris, C.S.: 1976, in *The Study of Comets*, eds. B. Donn et al., NASA SP-393, Washington: U.S. Government Printing Office, p. 410
Millis, R.L., Schleicher, D.G.: 1986, *Nature* **324**, 646
Morris, C.S.: 1986, *Int. Comet Q.* **8**, 7
Morris, C.S., Green, D.W.E.: 1982, *Astron. J.* **87**, 918
Newburn, R.L., Jr.: 1981, in *The Comet Halley Dust and Gas Environment*, ESA SP-174, ESTEC, Noordwijk, The Netherlands, p. 3
Newburn, R.L., Jr.: 1984, *Adv. Space Res.* **4**, 185
Ney, E.P., Knutson, A.G.: 1986, *IAU Circ.* **4178**
O'Meara, S.J.: 1985, *Sky Telesc.* **69**, 376
Reitsema, H.J., Delamere, W.A., Huebner, W.F., Keller, H.U., Schmidt, W.K.H., Wilhelm, K., Schmidt, H.U., Whipple, F.L.: 1986, 20th ESLAB Symp., ESA SP-250, Vol. **II**, p. 351
Rickman, H., Froeschlé, C., Klinger, J.: 1985, in *Ices in the Solar System*, eds. J. Klinger et al., Reidel, Dordrecht, p. 419
Roemer, E.: 1963, in *The Moon, Meteorites, and Comets*, eds. B.M. Middlehurst, G.P. Kuiper, Univ. of Chicago Press, p. 527
Roemer, E.: 1976, in *The Study of Comets*, eds. B. Donn et al., NASA SP-393, Washington: U.S. Government Printing Office, p. 380
Schild, R.: 1985, *Publ. Astron. Soc. Pacific* **97**, 1011
Schloerb, F.P., Kinzel, W.M., Swade, D.A., Irvine, W.M.: 1986, *Astrophys. J.* **310**, L55
Scotti, J.V.: 1987, *Minor Planet Circ.* **11466, 11563**
Sekanina, Z.: 1976, in *The Study of Comets*, eds. B. Donn et al., NASA SP-393, Washington: U.S. Government Printing Office, p. 537
Sekanina, Z.: 1983, *Adv. Space Res.* **2**, 121
Sekanina, Z.: 1984, *Astron. J.* **89**, 1573
Sekanina, Z.: 1985, *Astron. Astrophys.* **148**, 299
Spinrad, H., Djorgovski, S., Belton, M.J.S.: 1984, *IAU Circ.* **3996**
Tokunaga, A.T., Golisch, W.F., Griep, D.M., Kaminski, C.D., Hanner, M.S.: 1986, *Astron. J.* **92**, 1183
Weissman, P.R.: 1985, in *Dynamics of Comets: Their Origin and Evolution*, eds. A. Carusi, G.B. Valsecchi, Reidel, Dordrecht, p. 165
Weissman, P.R.: 1986, 20th ESLAB Symp., ESA SP-250, Vol. **III**, 517
West, R.M.: 1986, *IAU Circ.* **4196**
West, R.M., Pedersen, H.: 1984, *IAU Circ.* **3943**
Whipple, F.L., Sekanina, Z.: 1979, *Astron. J.* **84**, 1894
Wisniewski, W., Levy, D., Fink, U.: 1987, *IAU Circ.* **4372**
Wyckoff, S., Wagner, R.M., Wehinger, P.A., Schleicher, D.G., Festou, M.C.: 1985, *Nature* **316**, 241

Astron. Astrophys. 187, 569–574 (1987)

ASTRONOMY AND ASTROPHYSICS

The spectral behavior of P/Halley at large heliocentric distance in light of the Giotto/Vega results

M.J.S. Belton [1], **H. Spinrad** [2], **P.A. Wehinger** [3], **S. Wyckoff** [3] **and D.K. Yeomans** [4]

[1] National Optical Astronomy Observatories, Tucson, AZ 85726, USA
[2] Astronomy Department, University of California, Berkeley, CA 94720, USA
[3] Physics Department, Arizona State University, Tempe, AZ 85287, USA
[4] Jet Propulsion Laboratory, Pasadena, CA 91109, USA

Received January 30, accepted June 19, 1987

Summary. We present results on five spectra of P/Halley obtained when it was between 8.8 and 5.6 AU from the sun. V magnitudes and $(g-r)$ and $(V-R)$ colours are derived which show a trend to increasing redness with decreasing heliocentric distance. The $(V-R)$ colour of the comet while near 8 AU is unusual in that it is significantly bluer than the Sun. The spectra show no signs of emission lines or bands. Two possible interpretations are given: The first relies on our previously reported detection of periodicity in the comet's lightcurve while it was still at large heliocentric distance and which we associate with rotational motion of the nucleus. With this interpretation the light in all of our spectra has a significant component reflected from the nucleus. In spectra taken in Oct./Nov. 1985 roughly half of the total light is attributable to the nucleus; while for earlier spectra the nucleus dominates. The unusual $(V-R)$ colour seen near 8 AU contrasts with the neutral colour for the nucleus reported by Vega experimenters and implies that the spectral reflectivity of the surface beyond 6000 A changed as it approached the sun. In the alternate explanation we recognize that our spectra could have been influenced by cometary outbursts. In this case the $(V-R)$ colour near 8 AU could be due to small coma particulates similar to those found by Moreels et al. (1987) close to the nucleus of the comet at the time of the Vega-2 encounter.

Key words: comets – solar system – spectroscopy – photometry

1. Introduction

From the point of view of ground-based observers faced with the limitations of atmospheric seeing, the light reflected from the nucleus of an active comet near the sun is usually completely overwhelmed by light scattered from the surrounding coma. Only with the spatial resolution afforded by spacecraft can the nucleus be seen against the extended coma. From the ground it is necessary to observe the comet at distances beyond about 5 AU (cf. Belton, 1985) if there is to be a chance of obtaining direct information about the nucleus. At such distances a comet nucleus is usually very faint ($V > 21$) and large telescopes and special observing techniques are needed to make spectroscopic observations; moreover, although water may have ceased to be an effective agent in causing "activity" at these distances, other materials, e.g. CO_2, with higher vapor pressure, may be present in sufficient quantities to support comatic activity that at times can produce enough scattering cross-section to overwhelm the light from the nucleus (cf. Festou et al., 1986). Thus while observations of cometary light at large heliocentric distances may contain the signature of the nucleus, this is by no means guaranteed.

In Sect. 2 we describe preperihelion spectroscopic observations of P/Halley while it was beyond 5.6 AU, the reduction procedures we followed, and derive V magnitudes and $(g-r)$ and $(V-R)$ colour indices.

In Sect. 3 we outline two possible interpretations of these observations. The first, which ascribes part of the light in the spectra to reflection from the nucleus and part to reflection from grains in the developing coma, depends on our earlier interpretation of a 54-h periodicity in the comet's brightness in terms of nuclear rotation (Belton et al., 1987). Uncertainties in this interpretation due to the discovery of a second photometric periodicity near 7.4 days (Millis and Schleicher, 1987) and evidence of sporadic activity at large heliocentric distances lead us to consider a second possible interpretation in which coma particulates maybe entirely responsible for the spectrophotometric behavior. Our general conclusions are summarized in Sect. 4.

2. Spectroscopy beyond 5 AU

2.1. Observations

The observations reported here consist of five spectra taken with the Mayall 4 m telescope/CRYOCAM combination on Kitt Peak when the comet was between 8.8 and 5.6 AU from the sun. CRYOCAM is a fast, cryogenic, camera-spectrograph with a *Texas Instruments* 800 × 800 CCD detector (DeVeny, 1985). The first of these spectra, taken by Djorgovski and Spinrad and in which the comet is barely registered, is only discussed in a qualitative way. We have included some information about it since it represents a rough measure of "the state of the art" for detection of spectra of this type.

Send offprint requests to: M.J.S. Belton

Table 1. Log of spectrograms of P/Halley ($r = 8.8$ to 5.6 AU)

No.	UT date	UT mid-exposure	Exposure (s)	Sky condition	Zenith distance (deg)	FWHM seeing (arcsec)	Standard stars
1 [a]	83 Oct 10	11:28.3	4000	clear	29.3	2	L98–313
2	84 Feb 4	03:46.5	3600	Partly Cloudy	21.9	2	Feige 15; L92–433
3	84 Oct 30	08:37.0	1800	clear	41.6	1.5	L98–313; G99-37
4	84 Oct 30	09:11.0	3600	clear	32.1	1.5	L98–313; G99-37
5	84 Nov 26	10:25.0	3000	Thin Cirrus	24.2	4	Feige 25; L98-313

[a] H. Spinrad and S. Djorgovski made this observation assisted by W. Halbedel, based on an ephemeris, charts, and astrometric offsets supplied by D.K. Yeomans and M.J.S. Belton

Table 2. Cometary parameters and instrumental setup

No.	Heliocentric distance (AU)	Earth-comet distance (AU)	Phase angle (deg)	Tracking rate (arcsec/h)	PA of slit (deg.)	Grism	Slit width (arcsec)
1	8.8	8.8	6.5	3.52	35.5	770 [a]	3.2
2	8.0	7.2	4.9	19.23	100.7	770	3.2
3	5.8	5.3	8.8	16.17	81.1	770	3.2
4	5.8	5.3	8.8	16.17	81.1	770	3.2
5	5.6	4.8	5.6	31.05	87.8	770	3.2

[a] 300 lines/mm; gives 4.3 Å/pixel with spectral resolution of ~10 Å and covers the region from 4800 to 7900 Å. Used with a GG-420 blocking filter.

Table 1 contains a log of observations and Table 2 the parameters which describe the configuration of the comet and the instrumental setup.

Acquisition of these spectra required the use of blind offset techniques onto a moving object in a crowded star field because Halley at $V \sim 20$–23 mag was much too faint to be seen in the 4m-telescope field-acquisition instrumentation. The procedure (cf. Wyckoff et al., 1986) relied on an accurate ephemeris for the comet generated by DKY, the generation of accurate astrometric positions for several faint stars near the track of the comet, and the excellent quality of the pointing control of the Mayall telescope. Since the comet's path was against the background of the Milky Way, requested nights and specific exposure intervals were preselected by examining the track of the comet on the relevant Palomar Sky Survey (POSS) glass plates. This procedure eliminated the chance that background stars down to $V \sim 20$ would interfere with our exposures. Although the KPNO Telescope Allocation Committee largely supported our requests, nature was not always so kind. Several of our spectra, including some discussed in this paper, were contaminated by light from faint stars ($V \sim 21$–22) at or below the POSS limiting magnitude.

The spectrograph slit was oriented so that it was parallel to the predicted motion of the comet on the sky and was kept wide enough (3.″2) so that the probability the comet would end up in the slit – given potential errors in the timing of the observation, the ephemeris, and the offset maneuver – was maximized. The observation themselves indicate that the above strategy was successful since the spectra in fact appeared around one pixel (0.″84) off the predicted position on the detector and their cross-dispersion sharpness (cf. Fig. 2) corresponded well with the seeing properties of the particular night.

Standard KPNO/CRYOCAM setup procedures (DeVeny, 1985) were followed; calibration was accomplished using flat-field exposures on the inside of the illuminated dome, and observations of the G-type stars in Landolt's (1973, 1983) catalogs and flux standards in the KPNO/IIDS catalog (Strom, 1977). Final reduction of the data was done using the NOAO/IRAF Longslit Spectrogaphic Data software package (Anderson, 1986) at NOAO; preliminary reductions were carried out independently by HS and by PW with other software and gave results which were effectively identical.

2.2. *Properties of the spectra*

The best example of the spectra that we obtained is illustrated in Fig. 1 and clearly shows how night-sky lines dominate the signal. The average signal-to-noise ratio (snr) is about 1.7 in each independent element of the comet spectrum. In November 1984 the spectrum was characterized by snr = 0.8 per element; and in Febuary, 1985 by snr = 0.1 making it very hard to see even in

Fig. 1. Spectrum of P/Halley on 30 Oct 1984. The spectrum is dominated by night sky lines and covers the region from 5000 (bottom) to 7300 Å. Halley is the narrow spectrum to the right; a trailed (due to tracking the comet during the exposure) spectrum of a faint star on the slit is to the left

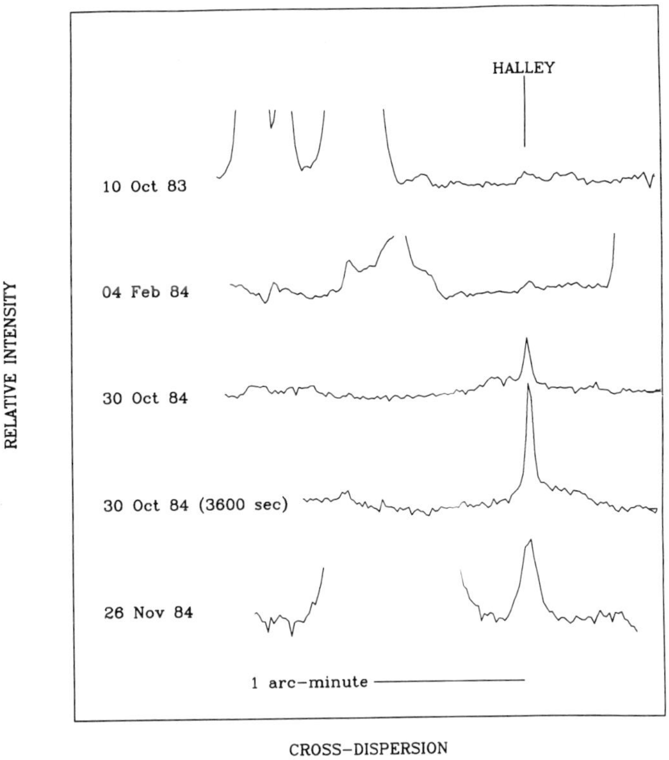

Fig. 2. Cross-dispersion distribution of intensity in the spectra averaged over wavelength in order to maximize the visibility of the Halley signal. The effects of a faint star passing behind the comet can be seen in the Oct 1984 spectra. The colors and magnitude given in Table 3 have been corrected for this

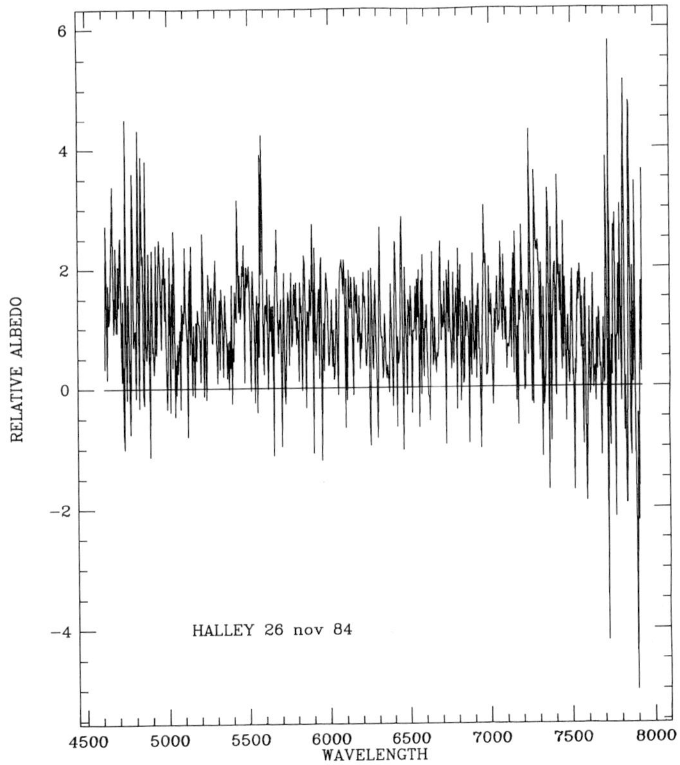

Fig. 3. Relative spectral reflectivity for P/Halley on 26 Nov 84. The spectrum is unfiltered to illustrate the signal-to-noise available at the pixel level in the data

highly enhanced, processed images. Cross-dispersion cuts across the spectra, which are summed over a broad spectral band to greatly increase the snr in the comet signal and which are used to identify Halley in the data, are illustrated in Fig. 2. A fully reduced, normalized, reflectivity spectrum (i.e. the wavelength dependence of the solar continuum has been removed) is illustrated in Fig. 3, and in Table 3 we list magnitudes and relative colors that we have derived.

These photometric quantities are clearly subject to considerable uncertainty. However, due to the limited number of spectra that we were able to obtain and the intrinsic variability in the comet, it is difficult for us to quantify the magnitude of the possible errors. The *formal* errors in the broad-band magnitudes, calculated from the spectra, are all well below 1% but experience (see below) suggests that they are unlikely to be a reliable guide. On October 30, 1984, when we were able to obtain two spectra on a single night, the differences between independent measurements of the same photometric quantities are at the 8% level. We believe

Table 3. Colors and magnitudes of P/Halley from distant spectra (A discussion of uncertainties is given in the text)

Date	V	$\Delta V(54)$ [a]	$\Delta V(53)$ [b]	$(g-r)$ [c]	$(V-R)$ [c]
83 Oct 10	23.5:	+0.2	+0.8	?	−0.1:
84 Feb 04	23.3	+0.1	+0.3	+0.11 ± 0.1	−0.15 ± 0.1
84 Oct 30	21.1	−1.2	−0.7	+0.11 ± 0.05	+0.05 ± 0.05
84 Nov 26	20.6	−1.9	−0.8	+0.22 ± 0.05	+0.09 ± 0.05

[a] $\Delta V(53) = V - V_{\text{(predicted for nucleus)}}$ based on 53.96 h period
[b] $\Delta V(54) = V - V_{\text{(predicted for nucleus)}}$ based on 54.125 h period; see text
[c] Relative to the sun. In $(g-r)$ the sun ~0.18 (Jewitt and Danielson, 1984); and in $(V-R)$ the Sun is ~0.52 (Allen, 1973). The (g, r) system is described in Thuan and Gunn (1976). A colon denotes great uncertainty

that this level of uncertainty is reasonable for our derived colours (which are relative measurements) but provides little insight into the errors that might be latent in the magnitudes themselves. Experience with direct CCD broad-band filter photomery of faint stars near the sky limit (Carney and Seitzer, 1986) shows that mean errors of ± 0.1 mag in $B-V$ and slightly less than ± 0.1 in V can be achieved on objects between 23 and 23.5 mag with the 4m telescope under good photometric conditions and with integration times trough the V filter of approx 4000 seconds. For brighter objects ($V < 20.8$ but > 18) their uncertainties reduce to ± 0.054 mag in colour (per object) and ± 0.031 in magnitude. Our effective integration times are roughly the same as theirs (cf. Table 1) and we have the advantage (for colour measurements) that light in all bands is measured simultaneously. We therefore expect our broad-band colours to show similar uncertainties as theirs. For our February, 1984, spectrum (V = 23.3) this implies an uncertainty of ± 0.1 mag. In the case of the later, brighter, spectra the uncertainties should be roughly ± 0.05.

The uncertainty in the V magnitudes is difficult to quantify since they are more susceptible to systematic errors due to seeing and atmospheric extinction. They will be clearly larger than the uncertainties found by Carney and Seitzer since our observing conditions were far from ideal (cf. Table 1): Our best guess is that the uncertainties are ± several tenths of a magnitude but probably less than ± 0.5 mag based on our previous experience with this type of spectrophotometry.

The spectra are continuous and show no signs of cometary emission features, e.g. C_2, [O I], and NH_2 which might just be "turning-on" at these distances. Between 1984 Nov. 26 and 1985 Feb. 17, when Spinrad (cf. Belton, 1985) and Wyckoff et al. (1985) independently detected the first emission features, [O I] 6300 Å and CN (0–0) 3883 Å, respectively, we did not have any clear weather. However, R. Ellis (private communication) obtained a spectrum on 1984 Dec. 22, when the comet was at 5.4 AU, with the 2.5 m Isaac Newton Telescope on La Palma which showed no emission features either.

The derived colours (cf. Table 3), taken when the comet was between 5 and 6 AU, show that the comet's reflectivity is within the estimated uncertainty of the measurements is effectively neutral but possibly slightly redder than the Sun, not unlike the colours seen in the continua of many, dusty, active comets (Newburn and Spinrad, 1985; Jewitt and Meech, 1986). The colours may also show a trend to increased redness with decreasing heliocentric distance similar to that which has been seen earlier in other comets by Hartmann and Cruikshank (1984). At 8 AU the $(V-R)$ color is anomalous in that it is significantly bluer than the Sun. Inspection of the 1984 Feb. 4, spectrum indicates that the albedo decreases to the red of 6000 Å. The blue continuum ($\lambda\lambda \sim 5000$–6000 Å), as measured by the $(g-r)$ index, nevertheless maintains a red slope.

Independent checks on these data are few. In January 1984, when the comet was near 8.2 AU and at a phase angle of 1.9 deg, Jewitt and Danielson (1984) obtained a series of magnitudes from which a mean $(g-r)$ index of $+0.38 \pm 0.10$ can be derived; the associated error estimate is a formal error and may be too small since it does not take into account any intrinsic changes that may have occurred in the magnitude of colour of the comet as the individual magnitudes were acquired at different times. All that can be said is that the results are in rough agreement. Similarly, our derived visual magnitudes are in general accordance with the more accurate photometry of West and Pedersen (1984), LeFevre et al. (1984), and Meech et al. (1986). We do not recommend that the V mags. derived here be used in hight precision studies of the photometric variabilty of the comet, but they should be reliable indicators of the general trend of the brightness variation and can be used (see below) to estimate the relative contributions of light reflected from the nucleus and from the inner coma.

3. Interpretation of the spectra

A primary issue is to what degree these data refer to light reflected for the nucleus. The simplest view might be that the earlier, more distant, spectra reflect the properties of the nucleus, while the spectra taken later in October and November, 1984, reflect the scattering properties of coma particles. This is supported by the first obvious appearance of comatic activity in the middle of the time interval covered by our spectra (late September, 1984), when a substantial coma (±6″ in extent) was imaged by Djorgovski and Spinrad (1984). Crotts (unpublished) also took an image in November, 1984, which showed evidence of coma (cf. Belton, 1985).

However, there are immediate problems with such a simple interpretation: first the spectral reflectivity of the surface of the nucleus, as determined by Vega experimenters (Sagdeev et al., 1986), was grey, of effectively colorless, and not coloured as in our spectra. Secondly analysis of photometric observations obtained between October, 1984, and January, 1985, by Meech et al. (1986) indicates that a double-peaked 54 h periodicity was present in the light variations at that time (Belton et al., 1987). Because the amplitude and phase of these variations are in good agreement with geometric expectations based on Giotto and Vega results, this variability is almost certainly the signature of the elongated shape of the rotating nucleus. It follows therefore that our Oct./Nov., 1984, spectra should contain an appreciable amount of light reflected directly off the surface of the nucleus. We can make a very rough estimate of the relative contributions by comparing the observed brightness with the predicted nuclear brightness based on a simple "asteroidal" model and on photometric parameters yielded by the periodicity analysis analysis mentioned above. In our model we have used the parameters given in Belton et al. (1987) for periods of 53.96 and 54.125 h and assumed an asteroidal phase effect (Bowell and Lumme, 1979) like that of a C-type asteroid. Differences between the predicted magnitudes for the nucleus and the observed magnitudes are included in Table 3 and show that in Oct./Nov., 1984, the measured brightness is two or three times the predicted brightness. For the earlier spectra the predicted brightness are about 40% higher than the observed brightnesses however such a small difference is probably not significant for data of this quality. We conclude that the Oct./Nov., 1984, spectra are probably composed of roughly equal proportions of light scattered from the nucleus an the coma while the earlier spectra primarily refer to the nucleus.

The fact that the Vega experimenters (Sagdeev et al., 1986) found the nucleus to be colourless implies that our peculiarly coloured Feb. 4 spectrum must reflect a major change in the spectral reflectance of the nuclear surface. The change would have to be such that the albedo beyond 6000 Å was depressed at 8 AU (relative to what was seen by the Vega experimenters) for the visual magnitudes are consistent with the predicted behavior of a simple nuclear reflection model.

At the Heidelberg Symposium two results were announced which provide the basis for an alternate interpretation of these spectra. These were: (i) the announcement of 7.4 day periodicty (Millis and Schleicher, 1987) in the brightness of the comet while it was active an close to the sun; and, (ii) the discovery of a class of coma particles which have a very blue spectrum and which apparently only existed within 500 km of the nucleus (Moreels et al., 1987).

The work of Millis and Schleicher has caused us to reexamine our earlier work on the photometric variabilty of the comet at large heliocentric distance and search for the 7.4 day period. If both periodicities are present then substantial changes can be expected in the results of the simple model described above. Only a summary of work in progress can be given but we find that there maybe evidence for periodicities at both 7.37 $\pm$0.03 days as well as at 54-h. However the two variations have different characteristics: The 54-h periodicity is primarily in the 2nd harmonic (ie. two maxima or minima per cycle) while that at 7.37 days is primarily 1 st harmonic. Both periodicities have about the same amplitude.

As we suggested at the Heidelberg Symposium, the basic reason for the multiplicity of periods (we now suspect there may be three!) is probably a generalized nutational motion of the nucleus. Details of our model (based on the motion of a asymmetric top) will be published elsewhere, so that data from the current observing semester can be included in its definition. We agree with Sekanina (1987) that the 7.37 day period is probably related to an apparent component of rotation about the long axis of Halley's nucleus and that the 54-h period is a reflection of the apparent nutational motion of the other components of the total spin vector about the total angular mometum vector of the nucleus whose direction in effectively fixed in space. What is important for the interpretation of our spectra is that the 7.37 day periodicity seems to be primarily in the first harmonic, i.e. one maximum in brightness per rotation period. We interpret this to mean (cf. Ostro and Connelly, 1984) that, unlike the case for the 54-h period, this periodicity does not reflect the geometric shape of the nucleus, but rather follows a variation of albedo over the surface or, as with Millis and Schleicher's data, variability in the level of coma activity.

There is little data on which to make a choice between these last two possibilities. If the 7.37 day variability reflects the effect of a spot of contrasting reflectivity on the surface of the nucleus then the implication for our spectra would be the same as for the case of the 54-h periodicity. i.e. the light reflected from the nucleus is a significant component in our spetra. If, however, the variability represents activity in the coma then the case for a large nuclear contribution is weakened and this latter possibility provides the basis for an alternate explanation of the blue ($V-R$) colour seen at large heliocentric distance. In this case the interpretation would be based on an enhancement of the blue end of the spectrum as a result of selective scattering by very small ($\sim$ 0.1 micron radius) grains in the coma. Jewitt and Danielson (1984) have shown that a substantial coma could be present in faint, sky-limited, observations even though the image is stellar in appearance.

Our February 4 spectrum was obtained within a few hours of a major flare in the brightness of Halley in the *B* passband (Lecachaux et al., 1984). The rapidity of the reported brightening argues for the release of a large number of small grains from the nucleus at this time. Since this flare occurred hours after our spectrum was taken, the possibilty that grains associated with it could have been directly responsible for the enhanced blue reflectivity in our spectrum in excluded. However, the level of activity does indicate that coma particles, perhaps from an unrecorded outburst 54-h – or 7.4 days – earlier, could have been present and contributed to the spectrum. If so the stimulus for this acitvity would be unlikely to be H_2O since the temperature is too low. A more likely stimulus would be a pocket of highly volatile material such as CO of CO_2 as suggested most recently by Feldman et al. (1986). If we suppose that the particulates which are liberated from a center of activity are largely composed of the same volatile material that drives the activity then the coma lifetime would be very short and help explain why there was no observed build-up of a substantial coma until the comet moved within 5 AU of the sun. The "blue" color and the short life-times of the particulates found by Moreels et al. (1986) suggests that there may be a direct relationship between these particles and, should they exist, the grains responsilble for the enhanced "blue" spectrum observed on 1984 Feb. 4.

4. Conclusions

The interpretation of the five spectra that we have obtained of comet Halley when it was well beyond 5 AU for the sun is complex and lacks uniqueness. An improvement in this situation will have to wait until the nutational motion of the nucleus and its contribution to the photometric variability are more clearly understood.

Tentatively we believe that there is ample evidence for sporadic activity on the comet to as great as 8 AU from the sun, probably due to localized pockets of highly volatile materials. However, even with this activity, light reflected directly from the surface of the nucleus may still be a substantial contributor to the total light for most of the time that the comet was beyond 5 AU preperihelion. The anomalous ($V-R$) colour in our 1984 Feb. 4 data may be due to the release of an abundance of very small coma particulates, perhaps composed of the same material that drives the outbursts, and perhaps identical in nature to the short-lived "blue" particles found in the innermost region of the coma during the Vega encounter.

The flat, slightly red coloured spectra obtained in October and November, 1984, appear to have roughly equal contributions of light from the nucleus and from the more usual variety of cometary dust.

References

Allen, C.W.: 1973, *Astrophysical Quantities*, The Athlone Press, University of London

Anderson, E.: 1986, "Reduction of Longslit Spectrographic Data using IRAF", CCS, National Optical Astronomy Observatories, Tucson, Arizona

Belton, M.J.S.: 1985, *Science* **230**, 1229

Belton, M.J.S., Wehinger, P., Wyckoff, S., Spinrad, H.: 1987, *20th ESLAB Symp.*, ESA SP-**250**, Vol. **I**, p. 599

Bowell, E., Lumme, K.: 1979, *in Asteroids* ed. T. Gehrels, University of Arizona Press, p. 132

Carney, B.W., Seitzer, P.: 1986, *Astron. J.* **92**, 23

De Veny, J.: 1985, *An Observer's Manual for the Cryogenic Camera*, KPNO, National Optical Astronomy Observatories, Tucson, Arizona

Djorgovski, S., Spinrad, H.: 1984, *IAU Circ.* **3996**

Feldman, P.D., A'Hearn, M.F., Festou, M.C., McFadden, L.A., Weaver, H.A., Woods, T.N., 1986, *Nature* **324**, 433

Festou, M.C., Lecacheux, J., Kohl, J.L., Encrenaz, T., Baudrand, J., Combes, M., Despiau, R., Laques, P., LeFevre, O., Lemmonier, J.P., Lelième, G., Mathez, G., Pierre, M., Vidal, J.L.: 1986, *Astron. Astrophys.* **169**, 336

Hartmann, W.K., Cruikshank, D.P.: 1984, *Icarus* **57**, 55

Jewitt, D., Danielson, G.E.: 1984, *Icarus* **60**, 435

Jewitt, D., Meech, K.J.: 1986, *Astrophy. J.* **310**, 937

Landolt, A.U.: 1973, *Astron. J.* **78**, 959

Landolt, A.U.: 1983, *Astron. J.* **88**, 439

Lecachaux, J., Le Fevre, O., Baudrand, J., Lemonnier, J.P., Mathez, G.: 1984, *IAU Circ* **3928**

LeFevre, O., Lecacheux, J., Mathez, G., Lelièvre, G., Baudrand, J., Lemmonier, J.P.: 1984, *Astron. Astrophys.* **138**, L1

Meech, K.J., Jewitt, D., Ricker, G.R.: 1986, *Icarus* **66**, 561

Millis, R., Schleicher, D.: 1987, *Nature* **324**, 646

Moreels, G., Clairmidi, J., Parisot, J.P., Zucconi, J.M., Krasnopolsky, V.A., Moroz, V.I., Gogoshev, M., Gogosheva, T., Krysko, A.A., Tkachuk, A.Y., Werner, R., Spasov, S., Bertaux, J.L., Blamont, J.E., Herse, M.: 1987, *20th ESLAB Symp.*, ESA SP-250, Vol. I, p. 451

Newburn, R.L., Spinrad, H.: 1985, *Astron. J.* **90**, 2591

Ostro, S.J., Connally, R.: 1984, *Icarus* **57**, 443

Sagdeev, R.Z., Avanesov, G.A., Ziman, Y.L., Moroz, V.I., Tarnopolsky, V.I., Zhukov, B.S., Shamis, V.A., Smith, B., Toth, I.: 1986, *20th ESLAB Symp.* ESA SP-250, Vol. **I**, p. 317

Sekanina, Z.: 1987, *Nature* **325**, 326

Strom, K.M.: 1977, Standard Stars for IIDS Observations, KPNO, National Optical Astronomy Observatories, Tucson, Arizona

Thuan, T.X., Gunn, J.E.: 1976, *Publ. Astron. Soc. Pacific* **88**, 543

West, R.M., Pedersen, H.: 1984, *Astron. Astrophys.* **138**, L9

Wyckoff, S., Wehinger, P.A., Belton, M.J.S.: 1986, in *Asteroids, Comets, Meteors, II*, eds. C.-I. Lagerkvist, H. Rickman, Uppsala, Sweden, p. 487

Astron. Astrophys. 187, 575–580 (1987)

ASTRONOMY AND ASTROPHYSICS

Periodicities in the light curve of P/Halley and the rotation of its nucleus

M.C. Festou [1], P. Drossart [2], J. Lecacheux [2], T. Encrenaz [2], F. Puel [1], and J.L. Kohl-Moreira [2]

[1] Observatoire de Besançon, F-25044 Besançon Cedex, France
[2] Observatoire de Meudon, F-92195 Meudon Cedex, France

Received February 16, accepted April 27, 1987

Summary. Early photometric measurements of comet P/Halley exhibit a photometric periodicity of 7.307 + −0.005 days. Brightness variations occur together with the periodic release of solid particles during bursts of activity or activity increases of the nucleus. The mechanical behaviour of the comet nucleus can be described as that of an axisymmetric ellipsoid. In the torque free approximation it is found that the nucleus rotates about its long axis in 14.614 days. A free precession about an axis tilted by 84° to the rotation axis occurs in 2.088 days. Due to this peculiar geometry, the true photometric period is 14.6 days with two halves nearly identical, thus mimicking a periodicity of 7.3 days. The permanence of the rotation during at least half the apparition of the comet indicates that no internal damping of the nucleus rotation exists and that the nucleus must then be rigid. Since from the examination of nucleus images it is found that the precession axis lies close to the perpendicular to the orbital plane, then the photometric behaviour of the comet and that of the nucleus gaseous output during the entire apparition can be explained if most of the activity takes place in one main active (or one group of) spot(s) located not too close from the poles of the nucleus. The mean albedo of the dark areas of the nucleus is found with that model to be about 3%. The icy patches have an albedo of 8–12% and are thus very dusty.

Key words: photometry – nucleus rotation – comet Halley

1. Introduction

The scientific objective of the observations of P/Halley made before 1985 was threefold: 1) determination of the size of the nucleus (actually the product of the geometric albedo p_v by the nucleus cross section S from which, after assuming a value for p_v, the value of S is deduced), 2) determination of the spin period of the nucleus and 3) determination of the activity onset of the nucleus. The three goals require a reliable light curve be recorded over a large range of heliocentric distances. The first objective was reached after combining accurate photometric measurements of the reduced magnitude of the nucleus with information provided by the Vega and Giotto images on the geometrical cross section S of the nucleus. It was found that the mean geometric albedo was very low, of order 0.04 (Keller et al., 1986; Festou et al., 1986a). The third objective was partially attained since, while it was clearly established that the gaseous emission had started already in February 1985 (Wyckoff et al., 1985), there were many signs that the activity could have begun when the comet was farther away from the sun, at least at 6.9 AU from it pre-perihelion (Meech et al., 1986) and more probably as fas as 8 AU pre-perihelion (Festou et al., 1986a).

Despite the constant efforts of many teams devoted to the evaluation of the spin period from photometric observations, the data set gathered on the reduced magnitude of the comet was surprisingly found to be inappropriate to provide an unambiguous value of the spin period of the nucleus. Numerous values were proposed between about a few hours and about 50 h (see Wilhelm et al., 1986 for a review on the subject). After the claim by Sekanina (1985) that structures in the head of P/Halley observed in 1910 were explained if a rotation of about $52\overset{h}{.}5$ was assumed, many additional indirect evidences favouring that value appeared, e. g. studies on the photometric behaviour of the comet (Leibowitz and Brosch, 1986), the breathing of the hydrogen coma (Kaneda et al., 1986) or the formation of cyanogen shells (Schlosser et al., 1986). Finally, direct images obtained by cameras placed on board the Vega probes were fully compatible with any value close to 53 h (Sagdeev et al., 1986a).

The discovery by Millis and Schleicher (1986) that a periodicity of about 7.37 + −0.007 days existed in the production curve of the C_2 radical in March and April 1986 led to a new series of investigations of the problem either to establish the facts (Festou et al., 1986b) or to explain the apparent discrepancy between the two most likely to exist periodicities, ≈2.2 days and ≈7.4 days (Sekanina, 1986a; Wilhelm, 1987; Julian, 1987).

In this study, we present a re-investigation of the determination of the presence of recurrences in the value of the reduced magnitude of comet P/Halley when this latter was at large heliocentric distances, pre-perihelion. We shall see that a periodicity of 7.307 days can be found. It will be then shown that using complementary information provided by observations obtained during the encounter of the comet with the spacecraft, the true rotation period of the nucleus can be determined. Finally, the evolution of the light curve of the comet (whichever is the phenomenon used to define it, the reduced magnitude, the gas production rate of any species or the radius of neutral or dust shells) can be undestood and/or predicted.

2. Period determination

Most previous studies on the subject have used limited data sets for which it was considered that either the light curve was not sufficiently sampled (Le Fèvre et al., 1984) or that observations were spoiled by measurements errors (Meech et al., 1986). Festou et al. (1986a) used the most complete set of photometric data and

Send offprint requests to: M. Festou

Table 1. Corrected magnitudes used in our period search. Fluxes are given in units of $B = 25$, adjusted for the heliocentric and geocentric distance of the comet on 1984 January 3

Date[1]	Flux	Uncertainty	Date[1]	Flux	Uncertainty	Date[1]	Flux	Uncertainty	Date[1]	Flux	Uncertainty
003.2538	06.210	0.86	029.1669	08.036	1.64	332.1187	14.000	7.00	385.1749	12.706	1.17
003.2597	08.730	1.54	029.1776	07.962	1.62	333.5677	15.276	3.55	385.2208	12.246	1.70
003.2690	08.338	1.14	032.3500	15.172	2.10	333.7178	10.965	5.48	385.2458	11.066	1.02
003.2818	08.110	0.98	034.2493	03.508	0.98	354.0458	18.365	5.14	385.2666	10.666	2.48
003.2917	08.492	1.16	034.2823	04.218	1.60	354.0763	14.588	4.08	385.2902	09.817	0.91
003.3032	06.934	0.96	034.3182	10.594	1.96	356.1556	06.668	1.87	386.1131	12.589	1.74
003.3128	06.040	1.80	034.3873	15.312	2.84	354.0240	15.000	7.50	389.3332	04.365	0.60
006.2391	16.636	3.38	035.2404	03.200	0.88	355.0276	15.000	7.50	389.3484	03.020	0.42
006.2963	09.750	2.26	035.2747	05.560	2.10	354.4999	07.656	2.14	389.4079	08.710	1.21
006.3930	15.596	2.16	035.3118	08.036	3.02	362.0989	10.666	1.98	389.4148	14.454	2.68
006.4015	15.596	2.60	035.3631	07.328	2.40	362.1288	14.723	2.73	389.4464	10.965	2.03
007.2509	06.622	0.92	036.3546	11.616	2.16	362.1600	13.062	3.66	390.5755	13.183	6.59
007.2582	04.754	0.94	083.2543	16.182	5.30	362.9545	06.252	0.87	391.5633	13.183	6.59
007.2653	04.064	0.94	085.2458	16.182	2.24	362.9823	06.607	1.22	391.5942	13.183	6.59
007.4200	04.256	1.04	265.7662	17.378	8.69	363.1191	03.873	0.72	392.4310	04.018	1.32
026.1011	03.412	0.70	265.7930	17.378	8.69	363.8608	03.105	0.87	393.2873	09.204	1.71
026.1335	04.498	0.92	268.4868	04.920	1.64	363.8962	11.169	1.03	393.3222	14.588	2.02
027.0630	05.984	1.22	269.7807	15.849	7.93	363.9490	14.588	1.35	393.3310	11.066	1.02
027.0972	06.684	1.36	270.4802	12.134	2.43	363.9740	03.281	1.24	393.3422	14.588	2.70
027.1356	07.962	1.62	295.4019	07.586	2.12	364.0538	01.368	1.30	393.3588	15.996	1.48
027.1498	08.186	1.66	295.4505	05.495	0.76	364.9073	13.677	3.18	393.3699	14.588	4.08
027.1640	07.130	1.44	295.4720	05.702	1.60	364.9372	11.169	2.07	394.3449	14.588	2.70
027.1783	07.816	1.58	295.5137	04.406	1.23	364.9837	07.586	1.41	394.3543	08.790	1.22
028.0675	06.684	1.36	297.3403	06.546	3.12	365.0289	03.698	1.39	394.4080	11.066	1.53
028.0839	06.684	1.36	297.3556	06.792	1.26	365.0747	06.546	1.52	409.8263	16.000	8.00
028.0982	05.310	1.08	297.3695	07.586	1.41	383.1353	08.954	0.83	411.8602	15.000	7.50
028.1792	05.508	1.12	297.3806	07.379	1.02	384.1228	11.169	1.03	413.1198	20.000	7.09
029.0620	08.570	1.74	297.3917	05.702	1.06	384.1458	11.482	3.21	413.2656	20.000	7.09
029.0793	08.650	1.76	297.4112	08.710	1.21	384.1944	10.375	1.44	413.2997	08.630	2.00
029.0936	10.790	2.20	297.5042	07.379	1.37	384.2173	08.472	0.78	415.8302	18.000	9.00
029.1078	08.892	1.82	300.3782	03.499	0.65	384.3083	07.943	0.73	415.9198	18.000	9.00
029.1220	08.974	1.82	300.3893	03.873	0.72	384.3326	08.790	1.63	438.1786	18.000	9.00
029.1384	10.594	2.16	303.5148	01.690	0.56	385.1083	09.462	0.87	438.8610	04.000	3.00
029.1526	08.036	1.64	330.3980	05.058	1.42	385.1527	09.550	2.22			

[1] **Origin is 0.0 January 1984**

failed in their attempt too. Belton et al. (1986) by using a selected set of 70 measurements believed to be of uniform quality found two values for the period (53^h58^m and 54^h07^m) between which they are unable to choose instead of an expected unique value.

After publication of the results obtained by Millis and Schleicher (1986), we first looked at similar data obtained with the IUE spacecraft (Festou et al., 1986b) and we observed that the result was independent of the species used in the determination of the periodicity (the emissions of OH, CS and the light reflected by the dust were tested; a periodicity of about 7.4 days was found in each case, although the gas production curves were not found to be exactly in phase, a result which will be explained later). It was then clear that if the variation of the gas production rates was related to the apparition of emissive areas because of the rotation of the nucleus, such a periodicity should be found in older photometric data. We consequently re-examined the available 135 accurate photometric measurements of the magnitude of the comet obtained between 1984 January and 1985 February (very early data, not photometrically accurate, were eliminated) because our preceding lack of success could well have been due to the fact we had limited our search to the period interval 0^h–60^h chosen in the hope of finding the then currently accepted value of about 53^h.

Standard search methods (spectral analysis) do not lead to an unambiguous period determination. Many peaks of similar amplitude appear in the power spectrum, the two strongests of which are those near 49 hours and 29.5 days (the synodic month). Note that 2.2 days is a beat frequency for those two latter values. Once the data are folded according to the peak frequencies, none of the resulting light curves is acceptable to the eye: they reflect too much the negative information contained in the data set, i.e. the lack of data at some times. In addition, the existence of recurring outbursts (Festou et al., 1986a) possibly masks the presence of a periodicity. To overcome that difficulty, we corrected the original data, the $B(r, \Delta, 0)$ magnitudes, by a scaling factor (s) calculated as follows. It was assumed that the brightness variations observed in early 1983 were solely due to the rotation of the bare nucleus, whence $s = 1$ for that period. Since it was established in the above mentioned study that the amplitude of the brightness variations were proportional to the incident solar flux, the indication that the recorded light was mostly light reflected by solid particles released when the comet was active, we applied to the data a correction factor which is the inverse of coefficient k appearing in Table 4 of Festou et al. (1986a). As a consequence, the reduced magnitude of the nucleus $V(1, 1, 0)$ is contained between two fixed values, $V = 14.9$ and $V = 13.7$. The final data set used in this study is given in Table 1. Since a periodicity had been found by other observers near specific values, we explored in great detail the period intervals mentioned by Belton et al. (1986) and the 7.25–7.45 day region and judged *visually* the quality of the fit, the most simple means to take into account the errors which affect the measurements. The period search led to the unique solution of 7.307 ± 0.005 days (the

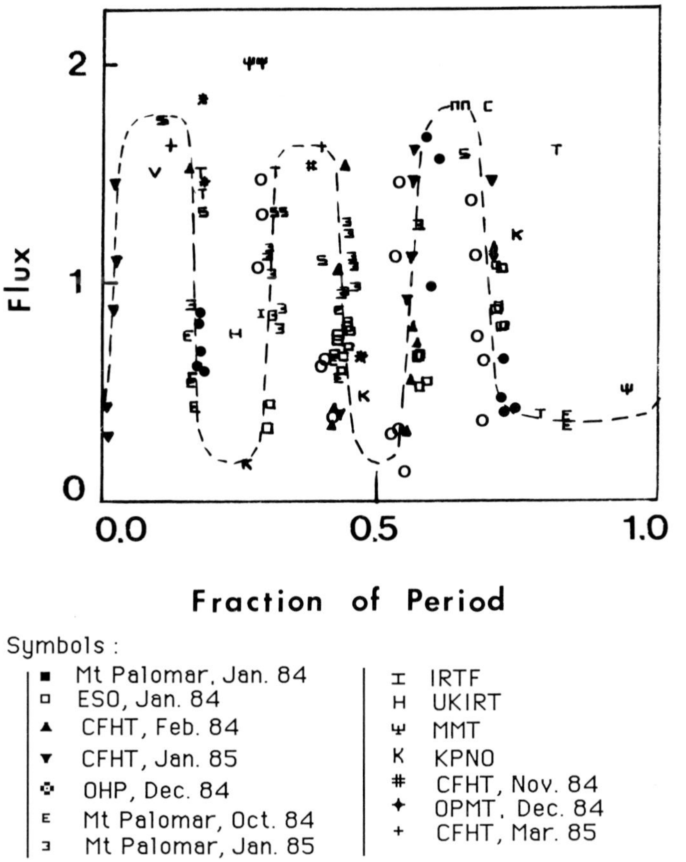

Fig. 1. Light curve of P/Halley for an assumed periodicity in the recurrence of the values of the corrected reduced magnitude (see text for more details) of 7.307 days. The three consecutive maxima are separated by about 2.05 days while the two maxima separated by a deep minimum occur about 3.2 days apart from each other. Data taken into account here were obtained between 1984 January and 1985 February. (Flux in arbitrary units)

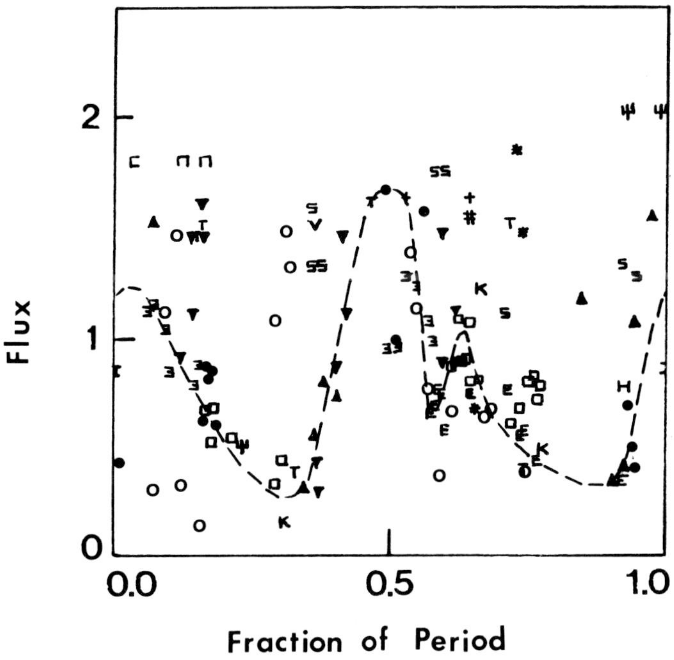

Fig. 2. Curve similar to that presented in Fig. 1 for one of the two frequency intervals indicated by Belton et al. (1986) as giving the solution of $54^{\rm h}11$. Symbols are the same as those in Fig. 1. (Flux in arbitrary units)

fit disappears at distances larger than ± 0.01 day from the critical value of 7.307 days).

No correction for the co-rotation effect due the displacement of the comet during the 13 month period considered was applied. It can be checked that the change of the true anomaly of the comet at that time has an effect which is much smaller than the few hour mean dispersion which appears in the phased data (the shorter the period, the better this holds).

Results are shown in Fig. 1. Note that the modification of the original experimental values does not create a spurious fit since the periodicity is proven by the existence of the phase of the rapid brightness changes. This way, only the formation of "clouds" around the maxima of brightness is avoided and the appearance of a regular pattern is favoured. Conversely, the exact shape of the light curve near the brightness maxima is lost. Figure 2 represents the result of a similar period search in our data set for one of the two intervals ($54^{\rm h}11 \pm 0^{\rm h}03$) indicated by Belton et al. (1986) as containing a periodic solution: no fit is found. A similar negative result is found for the second window mentioned by these authors.

The direct consequence of the assumption made above about the decrease of the minimum value of the reduced V magnitude is that changes of the comet brightness are due to the apparition (or disparition) or bright emissive areas in sunlight. This is in agreement with the fact it is nearly impossible to find a reasonable nucleus shape which produces the rapid observed changes while the albedo of the surface would be rather uniform. Only the gross structure of the light curve as shown in Fig. 1 can be investigated. One can see that they are *at most* three important active areas which have similar surfaces (otherwise one would not expect after we applied our correction to find a constant value of the brightness maxima). The spot(s) is (are) exposed in front of the sun during about 1/7 of the period. Maxima of brightness occur after time intervals of about 2/7, 2/7 and 1/2 the period, respectively. Note that a simple rotation about one axis would imply an apparition of the spot(s) every period and would imply a transit time close to one half of that period in most cases.

3. Rotation and precession of the nucleus of comet P/Halley

Whatever is the periodicity found in the data, what is truly remarkable is the fact that this periodicity exists during very long time intervals (13 months in our case, probably all the time since a periodicity is always present in observations of the gaseous output of the comet, Leibowitz and Brosch, 1986; Millis and Schleicher, 1986). Without that constancy, no fit would have been found by our search method. In this analysis, we consider that the only possibility to explain the observations is to relate them to spinning motions of the nucleus. This does not exclude the activity to be sometimes of explosive nature: the amplitude of the observed phenomena would be changed, not the phase. As a matter of fact, it is because we have separated the effect of the outbursts from that of the pure rotation (by applying the "s" corrective factor to the original magnitudes) that Fig. 1 reveals so clearly the 7.3 day periodicity.

A periodicity which is sustained during a very long time indicates that the motion of the nucleus is not chaotic (tumbling), i.e. since we know from the spacecraft images that the nucleus shape can be approximated by that of an ellipsoid, that the mechanical behaviour of Halley's comet nucleus is very close to that of an axisymmetric ellipsoid. Far from the sun, the torque free approximation must be valid.

Let us first examine the case in which the nucleus rotates about a single axis. This latter must necessarily be one of the two extreme axis of the ellipsoid. Whichever is the axis of rotation, the short or the long nucleus axis, the quasi constancy of the brightness maxima does not allow two important spots to be facing the sun at the same time while only one or none would be present at other times. The observations are thus compatible only with the existence of three similar spots located not too far from one pole; the rotation axis would be tilted in such a way that only one spot would be lit at a given time. Because of the motion of the comet around the sun, a seasonal effect would imply that sooner or later the three spots would be simultaneously lit and that the periodicity would disappear. A complementary argument against the existence of three main active areas might be simply the fact the general structure of the light curve (values of the extrema, width of the peaks, same rates of brightness decrease and brightness increase) corresponds much better to the appearance/disparition of a single spot in a more complicated, but natural, motion that will be described now.

A second possibility is a free rotation coupled with a free precession. The two axis of rotation are fixed by Euler's equations, as described in Sekanina (1986a). As will be seen below, the main difference between our solution and that of Sekanina is that the second one does not provide a light curve which presents a periodicity of 7.4 days but rather one of 14.8 days. In addition, Sekanina started his calculations in accepting the existence of the two periods of 2.2 and 7.4 days, even though evidences in favour of the 2.2 day value are weak. In our case we consider that only the existence of the 7.3 day period is firmly documented. We add as a complementary working assumption (that we believed proven by our investigation and those of the kind performed by Millis and Schleicher) that the periodic pattern is maintained during the entire apparition of the comet. It is not possible to determine a unique solution to our problem with these sole initial conditions. Our additional working assumption implies that the two periods we are searching are in a simple ratio (otherwise, the photometric period which is the smallest common multiple of the two spin periods would be much larger and the periodic pattern would be more complicated and would get undetected).

The theoretical light curve cannot be computed without additional information. We first assumed that most of the brightness of the nucleus was due to a single active area (nucleus feature descriptions given by Keller et al., 1986, Reitsema et al., 1986 and Wilhelm et al., 1986 indicate that active areas are close to each other and, as far as the photometric behaviour of the nucleus is concerned, can be considered as forming a single spot). Second, we assumed that the fastest spin axis was nearly perpendicular to the orbital plane (and as a consequence of Euler's equations, solved for the same ellipsoidal nucleus shape than assumed by Sekanina, the slowest spin axis is directed along the long axis of the nucleus, i.e. this axis is the rotation axis), as suggested by the analysis of direct images of the nucleus (Sagdeev et al., 1986b). Thirdly, we assumed that this emissive area was not located near the long axis ends, otherwise the light curve would exhibit only the fastest rotation period, an assumption fully validated by the inspection of nucleus images (Keller et al., 1986) if an active area lifetime is of the order or larger than a few years.

Now, we can test various simple ratios of the two periods by comparison of the observed light curve to our photometric model

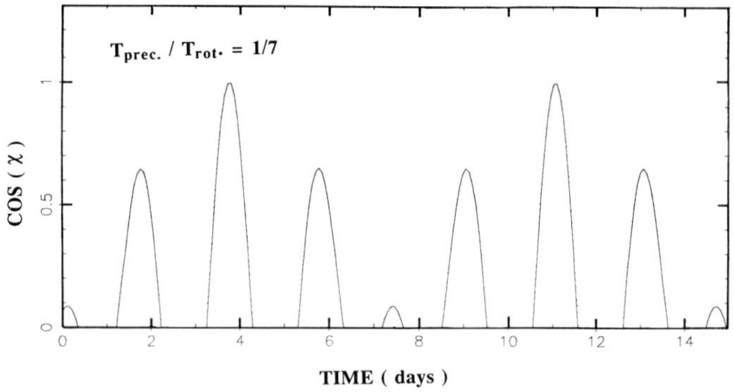

Fig. 3. Surface of the lit part of an active area projected on a plane perpendicular to the sunn-comet line as a function of time when the ratio $T_{prec}/T_{rot} = 1/7$

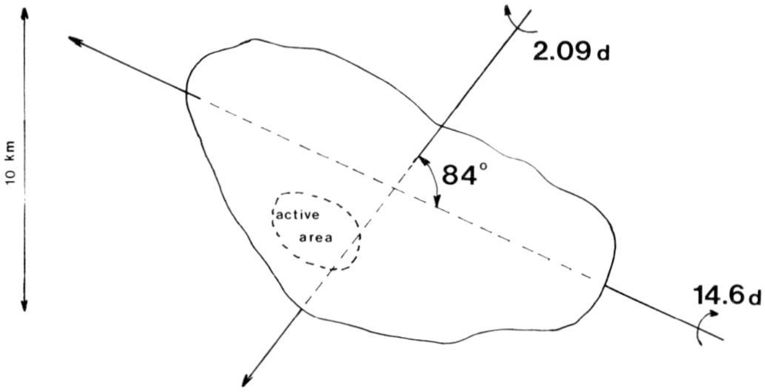

Fig. 4. Approximative shape of the nucleus, position of the rotation and precession axis and approximative position of the main active area as seen from the sun at the time of the Giotto spacecraft encounter with comet Halley

of the nucleus. No sub-multiple of 7.3 days provides an adequate light curve. The most simple ratio which provides a light curve resembling that observed is seven when the rotation period is taken as twice 7.3 days. The theoretical light curve which results from such assumptions is shown in Fig. 3. We can observe that the real photometric period is 14.6 days and that the light curve presents an axis of symmetry only because the active area is on the equator of the nucleus. Even in the case the active area would not be located exactly on the nucleus equator, because of observational uncertainties, the true light curve would still appear as having a 7.3 days periodicity. Note that the distance between two consecutive maxima is either 2.05 days or 3.20 days.

We consider that this is the most probable solution since the two resulting periods are 14.614 days for the rotation period and 2.088 days (or 50 hours) for the precession period. The angle between the two spin axis is 84°. Figure 4 is a sketch representing our solution. Figure 5 shows the comparison between the calculated light curve and the observed one: one should remember that the correction applied to the initial data points flattens the real light curve, so that only the general trend should be considered to estimate the quality of the fit. As a matter of fact, it is very important to observe here that the solution which is obtained crucially depends on the position of the bright spot at the beginning of the calculations. In other words, for a given period ratio, many light curves can be obtained. All of them are solutions of our problem. However, they are valid on a limited portion of

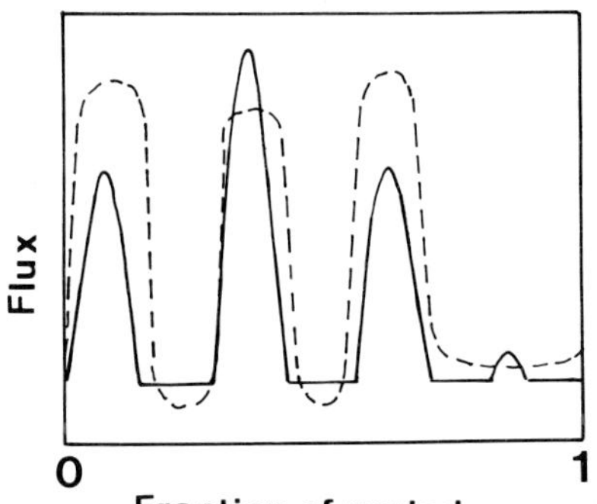

Fig. 5. Comparison between the observations and a theoretical photometric light curve produced by one main active area. A constant vealue has been added to the periodic function described in the text to allow for a minimum value of the light flux reflected by the nucleus. The shape of the observed curve near the maxima is probably due to our reduction procedure and cannot be considered as real. (Flux in arbitrary units)

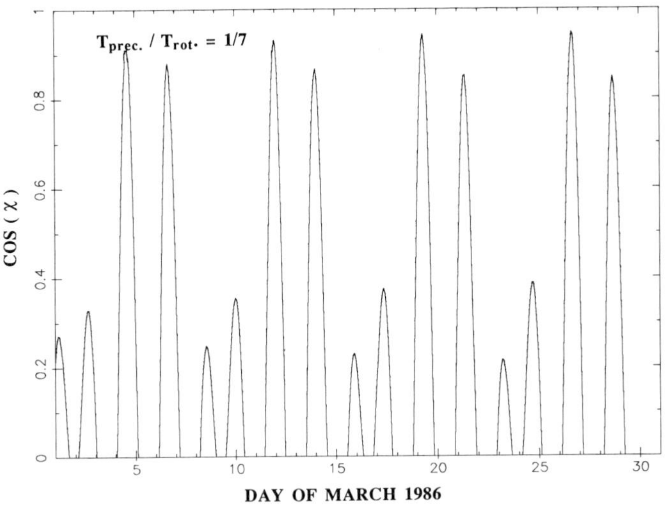

Fig. 6. Theoretical light curve which is produced by our model for the month of March 1986. Note the double maximum of cos (χ) followed by two minima

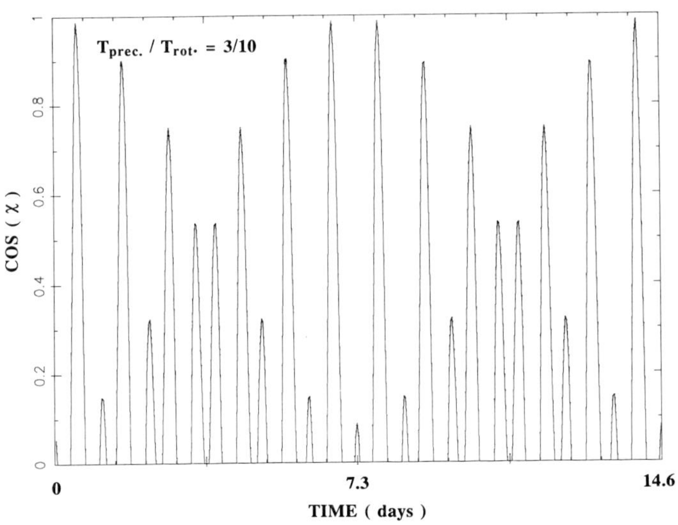

Fig. 7. Theoretical photometric curve produced by one main active area in a configuration close to that of Sekanina ($P_{\text{rot}} = 7.3$ days; $P_{\text{prec}} = 2.09$ days). The resulting photometric period is 14.6 days, not 7.3 days

the orbit. As an example, it can be predicted that using the above mentioned parameters and assumptions, the light curve of the nucleus of comet Halley (or the gaseous output of its nucleus) will present in March 1986 two maxima of similar amplitude followed by two deep minima (Fig. 6). Figure 7 provides another illustration of what was just said: it applies to conditions close to those considered by Sekanina (1987); 13 extrema are found in a period of 7.3 days and no initial conditions produce three maxima per 7.3 day interval.

4. Discussion

The above analysis does not present "the" solution of the rotation of the nucleus of comet Halley problem. It is only one acceptable solution, in agreement with a very large body of observations. The arguments presented here to justify our approach can be used to demonstrate that the solutions proposed by Sekanina (1986a) and Wilhelm (1987) are not in agreement with photometric observations of the comet, both at large and small heliocentric distances. Sekanina's solution is correct within the basic assumptions made by the author on the point of view of mechanics but it does not produce a photometric periodicity of 7.3 days. Wilhelm has an original and interesting approach which produces a solution very different to ours, even though the rotation and precession periods are alike. Here, a photometric periodicity of 7.4 days can be predicted if one neglects the effect over a few day time interval of the forced oscillation. Unfortunately, the amplitude of the nucleus magnitude or of the gaseous output will never be detectable because of the small value of the angle between the two spin axis. We have another objection to Wilhelm's solution: it assumes a rotation of the nucleus without precession far from the sun, an unacceptable solution for the light curve at large heliocentric distances.

Our analysis indicates that it is essential that a few assumptions be made because it is clear that no observations obtained until now allow to give an unambiguous answer to the problem treated here. We will briefly discuss the most important of them. It is expected that a high quality light curve will be obtained when the comet will be far from the sun on the outbound leg of its orbit. However, we already know that this will not bring "the" solution but only put stronger constraints on models of the nucleus.

We first assumed that the brightness variations observed when the comet was far from the sun were mostly due to the rotation of the nucleus. If one assumes that the small cross section of the nucleus ($\approx 50\,\text{km}^2$) corresponds to the observed maximum magnitude $V(1,1,0)$ of 14.8, the mean albedo of that cross section is 3%. Conversely, the largest cross section corresponds to a mean albedo ($V(1,1,0) \approx 13.5$) of no more than 5.5%. The gas production rate measured near the sun ($\approx$5–10 $10^{29}\,\text{s}^{-1}$) implies from water vaporization equilibrium calculations a total outgassing area of 30 to 50 km^2, slightly above figures deduced from HMC images (Keller et al., 1986; Huebner et al., 1986). This indicates that the "ice" of Halley's comet nucleus is vers dark with an albedo of 8–12%. This is consistent with the fact dust is observed leaving the active areas region. Our analysis does not demand the totality of the emission occur from one spot but rather that one or an ensemble of nearby spots dominantly contributes to the gas production. It is evident that our resolution is not good enough to separate the effect of two not too distant sources.

Secondly, the observational evidence is that the maximum of brightness lasts as long as the spot is exposed to sunlight (sharp brightness variations). We cannot speculate here about the duration and the occurrence frequency of those outbursts. We simply know that they exist (Festou et al., 1986a; Feldman et al., 1986) and the present data analysis implies that they occur often enought when an active area is exposed to the sun light to simulate the existence of a plateau in the light curve.

Third, we must comment on the fact Belton et al.'s (1986) analysis results to two values for the photometric period. It is possible to explain that the search leads to two values insteadt of the expected unique solution because the number of data points is significantly smaller than ours; it may also simply mean that no solution is found. Remind that all the early analysis (Wilhelm et al., 1986) led to values smaller than 54^h. It seems to us that the limited search interval those authors selected is responsible for their results.

Fourth, we assumed that no force was acting on the nucleus. The calculations performed by Wilhelm (1987) show clearly that outgassing rates as small as those encountered far from the sun can only induce a forced precession of very small amplitude and cone angle. The permanence of the periodicity during our 13 month observing period complemented by the fact the periodicity observed by Millis and Schleicher had the same value than that reported here shows that the nucleus is a rigid rotator and that no significant damping of the rotation occurs between two consecutive apparitions of the comet. Wilhelm's assumption that the nucleus had forgotten any previous excitation is thus questionable if one admits that our light curve shows a real periodicity. The constancy of the rotation and precession periods is supported by Sekanina and Larson's claim (1984) that structures in the head of comet Healley recorded in 1910 were showing a periodicity of 1.73 $\pm$ 0.4 days (a value of 2.2 days was reported later by Sekanina, 1986a), close to the period of 2.05 days found above. In our model the intervals separating four consecutive apparitions of the active area during one 7.3 day period is 2.05 days, 2.05 days and 3.20 days, respectively.

Fifth, we did not consider a short axis mode for the rotation of the nucleus. As mentioned by Julian (1987), this mode is dynamically possible for a tri-axis ellipsoid; but the angle between the rotation and the precession axis deduced from Julian (11°) seems to be too small to allow large variations of the photometric curve of the nucleus. We did not consider other multiples of the 7.3 day period in our model because the resulting photometric periods were found different (and much larger) than the value of 7.3 days.

It seems worth mentioning that the formation of cyanogen shells or jets of neutral gas associated with dust structures should neither show the period of about 2.088 days nor those of 7.3 and 14.6 days because what governs those come structures are the rapid changes in the dust-gaseous output of the nucleus that occur due to the apparent motion of active areas (approximately mearured by the distance between two consecutive peaks in Fig. 1). One can thus expect to observe the apparition of consecutive identical phenomena at intervals of about 2.05 and 3.2 days, the exact value depending on the magnitude of the co-rotation effect (+0.3 day on March 1, 1986 and less than +0.1 day in April 1986) and on the exact latitude of the spot. Dust structures periodicities detected by Vaisberg et al. (1986) and Sekanina (1986b) are likely to be completely compatible with a value in this interval. As an illustration of what we said, let us mention that the times of formation of the cyanogen shells indicated by Schlosser et al. (1986) follow a regular pattern in which two shells are formed about 2.1 to 2.3 days apart while the third one appears 3 to 3.2 days later.

Finally, let us mention that the phase differences (Festou et al., 1986b, c) observed between the various gas production curves (C_2, OH, CS, dust) structures are perfectly explained by the difference which exist between the various lifetimes of the coma species (parent molecules and radicals). The CS gas production exhibits a time response to gaseous output changes which are of the same order than that shown by the continuum emission. On the other hand, the OH coma slowly responds to similar variations. As a consequence, one may expect that for a species which is not rapidly photodestroyed the signal will recorded a smoothed variation of the gaseous output and short time scale structures of the order or smaller than one day will not be found. The light curves of the CN and C_2 radicals recorded by Millis and Schleicher (1986) must be analyzed using our model. We qualitatively checked that the structure of the light curve will rapidly change and that the co-rotation should decrease the photometric period from 7.5–7.6 days in March 1986 to 7.4 days in April 1986. One can predict that any frequency search will favour the apparent period of 7.3 days because of the tendency for rather long lived radicals to smooth out double peak structures that should appear at that time.

References

Belton, M.J.S., Wehinger, P., Wyckoff, S., Spinrad, H.: 1986, 20th ESLAB Symp. ESA SP-250, Vol. **I**, p. 599

Feldman, P.D., A'Hearn, M.F., Festou, M.C., McFadden, L.A., Weaver, H.A., Woods, T.N.: 1986, *Nature* **324**, 433

Festou, M.C., et al.: 1986a, *Astron. Astrophys.* **169**, 336

Festou, M.C., Lecacheux, J., Kohl, J.L., Encrenaz, T.: 1986b, unpublished communication presented at the 20th ESLAB Symp.

Festou, M.C., et al.: 1986c, *Nature* **321**, 361

Huebner, W.F., Keller, H.U., Wilhelm, K., Whipple, F.L., Delamere W.A., Reitsema, H.J., Schmidt, H.U.: 1986, 20th ESLAB Symp. ESA SP-250, Vol. **II**, p. 363

Julian, W.J.: 1987, *Nature* **326**, 57

Kaneda, E. et al.: 1986, *Nature* **320**, 140

Keller, H.U. et al.: 1986, *Nature* **321**, 320

Le Fēvre, O., Lecacheux, J., Mathez, G., Lelièvre, G., Baudrand, J., Lemonnier, J.P.: 1984, *Astron. Astrophys.* **138**, L1

Leibowitz, E.M., Brosch, N.: 1986, *Icarus* **68**, 418

Meech, K.J., Jewitt, D., Ricker, G.R.: 1986, *Icarus* **66**, 561

Millis, R.L., Schleicher, D.G.: 1986, *Nature* **324**, 646

Reitsema, H.J., Delamere, W.A., Huebner, W.F., Keller, H.U., Schmidt, W.K.H., Wilhelm, K., Schmidt, H.U., Whipple, F.L.: 1986, 20th ESLAB Symp. ESA SP-250, Vol. **II**, p. 351

Sagdeev, R.Z. et al.: 1986a, *Nature* **321**, 62

Sagdeev, R.Z. et al.: 1986b, 20th ESLAB Symp. ESA SP-250, Vol. **II**, p. 335

Sekanina, Z.: 1985, *IAU Circ.* **4151**

Sekanina, Z.: 1986a, *Nature* **325**, 326

Sekanina, Z.: 1986b, *IAU Circ.* **4187**

Sekanina, Z., Larson, S.M.: 1984, *Astron. J.* **89**, 1408

Schlosser, W., Schultz, R., Koczet, P.: 1986, 20th ESLAB Symp. ESA SP-250, Vol. **III**, p. 495

Vaisberg et al.: 1986, *Nature* **321**, 274

Wilhelm, K.: 1987, *Nature* **327**, 27

Wilhelm, K., Cosmovici, C.B., Delamere, W.A., Huebner, W.F., Keller, H.U., Reitsema, H., Schmidt, H.U., Whipple, F.L.: 1986, 20th ESLAB Symp. ESA SP-250, Vol. **II**, p. 367

Wyckoff, S., Wagner, M., Wehinger, P., Schleicher, D.G., Festou, M.C.: 1985, *Nature* **316**, 241

Astron. Astrophys. 187, 581–584 (1987)

ASTRONOMY AND ASTROPHYSICS

Photometry of comet P/Halley at near post-perihelion phases

T. Neckel and G. Münch

Max-Planck-Institut für Astronomie, Königstuhl 17, D-6900 Heidelberg 1, Federal Republic of Germany

Received April 16, accepted June 20, 1987

Summary. Photometric measurements of Halley's comet at phases around its recent perigee are presented. Narrow band filters isolating C_2 Swan band emission and a neighboring continuum, besides the standard *UBV*, have been used with 4 nucleus centered field stops subtending angles between 41″ and 1050″. The brightness fluctuations detected, especially conspicuous in C_2 and the nearby continuum, indicate a period of 7.3 days. Some evidence for the double peaked nature of the periodic variation is marginally present.

Key words: comets – photometry

1. Introduction

A rotational period of 2.2 days has been deduced by Sekanina and Larson (1986) for the rotation of Halley's comet nucleus from the study of recurring events observed during the 1910 apparition. The expected occurrence of brightness fluctuations in the comet's inner coma, related to akin events, led us to undertake an extended photometric program during the recent perihelion passage. This paper reports in detail the results of measurements secured during 26 nights in March-April 1986, which show periodic light variations consistent with the period and phase values determined by Millis and Schleicher (1986). Incomplete sampling of the light curve precludes the recognition of any harmonic variation due to the 2.2 day period of Sekanina and Larson (1986), recently ascribed (Julian, 1987) to a free precessional mode of the spinning nucleus.

2. Observations and reductions

The measurements were carried out by one of us (TN) with a 50 cm reflector at the Gamsberg station of the Max-Planck-Institute for Astronomy in South West Africa/Namibia over the period March 10 to April 19, 1986. Although the "rainy" season at the location occurs around that time, measurements could be made during 24 nights. A single channel photometer, fitted with a thermoelectrically cooled EMI 6256 photomultiplier and integrating pulses over fixed periods of 10 s, was employed. The filters used were the standard *UBV* and two narrow pass-band ones, peaking at 514 and 525 nm. The former narrow filter (hereafter referred to as *L*) is the standard IAU/IHW filter for isolating strong C_2 emission in the $\Delta v = 0$ sequence of the Swan bands. The other narrow filter (*C*) has widths of 96, 118 and 136 Å, respectively at the 50, 20 and 10 % transmission levels and an effective equivalent width of 100.0 Å. It is thus essentially devoid of significant cometary emission lines. Four circular field stops, with diameters of 41″, 317″, 575″, and 1051″ (referred to below as D1, D2, D3, and D4, respectively) were used. A measurement for each filter consisted of four 10 s successive integrations. Since the telescope has only a sidereal drive, the centering of the comet had to be verified on the smallest stop after every two 10 s integrations. The instrumental magnitudes *u*, *b*, *v* have been reduced to the Johnson *UBV*-system by relating them to the set of 18 standard stars (Gutiérrez-Moreno et al., 1966) given in Table 1, through the procedure described in detail by Klare and Neckel (1977). Since the passbands of the *L* and *C* narrow filters are very close to the effective wavelength of the *V*-band, we have fixed the corresponding *L* and *C* magnitudes of the standard stars (given in Table 1) as closely as possible to *V*.

Table 1. Magnitudes of standard stars

Star	V	$B-V$	$U-B$	$V-C$	$V-L$
HR 4468	4.694	−0.067	−0.181	+0.032	+0.080
4779	5.590	+0.864	+0.482	−0.033	−0.076
4781	5.487	−0.043	−0.102	+0.036	+0.075
5019	4.729	+0.717	+0.269	−0.018	−0.057
5570	4.476	+0.317	0.044	+0.007	+0.031
HR 5794	3.596	+1.377	+1.584	−0.090	−0.155
5928	3.884	−0.200	−0.817	+0.045	+0.094
6147	4.261	+0.934	+0.715	−0.034	−0.072
6243	4.638	0.476	+0.080	−0.004	+0.002
6567	4.627	+0.118	−0.196	+0.028	+0.049
HR 6700	4.735	−0.031	−0.024	+0.032	+0.075
7063	4.208	+1.101	+0.872	−0.057	−0.118
7340	3.925	+0.219	+0.146	+0.013	+0.049
7623	4.369	−0.150	−0.682	+0.037	+0.083
HR 166197	6.146	−0.141	−0.821	+0.033	+0.080
191639	6.481	−0.168	−0.873	+0.040	+0.086
193901	8.654	+0.532	−0.127	−0.010	−0.021
198188	8.132	+0.582	+0.088	−0.012	−0.026

Send offprint requests to: T. Neckel

Table 2. Magnitudes of comet Halley

T	D	V	B-V	U-B	C-V	V-L	T	D	V	B-V	U-B	C-V	V-L
70.13	2	4.096	.854	.442	.736	0.900	97.04	1	6.916	.783	.052	.411	0.565
.14	3	3.342	.696	.682	.669	0.809	.05	2	5.000	.882	.226	.677	0.846
.15	4	3.137	.628	.699	.540	0.713	.06	3	3.428	.967	.424	.949	1.012
71.13	2	4.650	.853	.350	.350	0.770	.06	4	2.892	.952	.488	.908	0.979
.14	3	3.788	.683	.601	.455	0.644	.09	1	6.959	.796	.057	.349	0.553
72.13	2	4.269	.918	.413	.862	0.962	.92	1	7.011	.696	.146	.401	0.632
.14	3	3.605	.770	.531	.642	0.803	.92	2	5.248	.807	.221	.562	0.795
.15	4	3.382	.694	.589	.528	0.706	.93	3	3.825	.911	.354	.692	0.875
75.12	2	4.591	.857	.357	.637	0.814	.94	4	3.192	.910	.473	.756	0.904
.13	3	3.634	.731	.679	.679	0.833	.99	1	6.927	.700	.149	.412	0.625
.14	4	3.375	.644	.730	.614	0.739	98.00	2	5.213	.799	.255	.598	0.824
.15	1	6.225	.823	.140	.487	0.687	99.07	1	7.195	.727	.101	.088	1.151
76.11	1	5.891	.830	.202	.648	0.865	.08	2	4.870	.840	.265	.705	0.892
.12	2	4.515	.889	.316	.766	0.920	.08	3	3.413	.935	.408	.911	1.000
.13	3	3.816	.776	.507	.592	0.812	99.09	4	2.962	.916	.466	.872	0.973
.14	4	3.515	.696	.590	.528	0.694	100.05	1	6.999	.737	.079	.346	0.540
77.12	2	3.857	.936	.393	.940	0.993	.06	2	5.168	.837	.227	.595	0.803
.12	3	3.092	.863	.572	.893	0.954	.07	3	3.617	.926	.381	.798	0.932
.13	4	2.922	.799	.604	.794	0.841	.07	4	3.034	.921	.486	.839	0.952
79.09	1	6.499	.815	.233	.689	0.840	103.80	1	6.797	.722	.096	.350	0.579
.10	2	4.849	.874	.289	.692	0.836	.81	2	5.027	.824	.282	.667	0.876
.11	3	3.899	.779	.537	.617	0.767	.86	1	6.863	.715	.114	.334	0.594
.12	4	3.588	.682	.648	.545	0.691	.87	2	5.053	.828	.290	.665	0.864
82.12	2	4.488	.921	.343	.850	0.973	.87	3	3.598	.937	.404	.853	0.970
.13	3	3.413	.892	.566	.935	1.011	103.88	4	3.127	.925	.435	.801	0.937
.14	4	3.165	.832	.626	.838	0.958	104.87	1	7.425	.758	.020	.260	0.432
.16	1	6.204	.812	.114	.507	0.732	.88	2	5.514	.845	.186	.464	0.690
84.11	2	4.304	.901	.314	.977	1.021	.88	3	3.964	.923	.378	.720	0.884
.12	3	3.407	.890	.464	.998	0.991	104.89	4	3.215	.917	.481	.765	0.908
.13	4	3.157	.805	.536	.924	0.902	106.04	1	6.960	.712	.083	.331	0.549
85.11	1	6.518	.821	.180	.543	0.741	.05	2	5.248	.792	.266	.632	0.848
.12	2	4.649	.918	.318	.750	0.904	.09	1	6.985	.706	.059	.336	0.554
.13	3	3.446	.937	.580	.951	0.992	.10	2	5.248	.802	.247	.633	0.845
.13	4	3.135	.871	.654	.911	0.949	.11	3	3.878	.889	.341	.752	0.914
86.10	1	7.179	.789	.185	.479	0.696	.12	4	3.414	.409	.725	.725	0.894
.11	2	5.193	.881	.252	.575	0.772	.15	1	6.990	.715	.033	.331	0.535
.12	3	3.884	.867	.522	.697	0.846	106.76	1	7.244	.734	.044	.289	0.508
.12	4	3.447	.787	.649	.696	0.822	.77	2	5.445	.822	.218	.541	0.786
87.10	1	7.014	.765	.215	.591	0.787	.78	3	3.884	.916	.374	.768	0.928
.11	2	5.146	.887	.301	.737	0.889	.79	4	3.351	.913	.438	.763	0.911
.11	3	3.962	.872	.436	.680	0.840	107.80	1	7.467	.711	.035	.258	0.478
.12	4	3.576	.797	.549	.624	0.775	.80	2	5.663	.810	.197	.508	0.731
88.09	1	6.631	.730	.238	.550	0.777	.83	1	7.460	.723	.030	.279	0.460
.10	2	4.995	.883	.343	.826	0.953	.84	2	5.670	.810	.203	.513	0.734
.11	3	3.804	.915	.458	.793	0.912	.89	1	7.463	.727	.051	.269	0.454
.13	4	3.423	.860	.522	.709	0.339	.94	1	7.504	.723	.039	.279	0.453
.14	1	6.613	.734	.224	.580	0.787	108.04	1	7.529	.735	.020	.291	0.446
.15	1	6.598	.735	.222	.581	0.794	.04	2	5.730	.816	.180	.494	0.726
89.11	1	6.217	.777	.147	.493	0.725	.05	3	4.201	.909	.304	.683	0.844
.12	2	4.535	.909	.363	.900	1.001	.06	4	3.630	.927	.413	.729	0.861
.12	3	3.348	.960	.524	1.007	1.032	.09	1	7.562	.735	.022	.275	0.435
.13	4	3.019	.923	.576	.915	0.965	108.80	1	7.846	.750	.010	.228	0.381
93.02	1	7.154	.790	.109	.430	0.646	.81	2	5.989	.824	.133	.400	0.614
.02	2	5.118	.884	.229	.654	0.839	.83	1	7.863	.744	.012	.223	0.360
.03	3	3.655	.946	.440	.823	0.953	.98	1	7.876	.744	.005	.228	0.364
.07	4	3.143	.915	.593	.868	0.950	.99	2	6.033	.817	.124	.387	0.594
94.04	1	7.481	.739	.131	.424	0.620	109.00	3	4.461	.893	.293	.566	0.760
.05	2	5.484	.834	.217	.564	0.771	.01	4	3.874	.892	.405	.625	0.799
.05	3	4.026	.873	.374	.621	0.810	.04	1	7.731	.727	.029	.177	0.339
94.06	4	3.484	.839	.520	.635	0.816	109.09	1	7.708	.738	.052	.217	0.343

The standards' colors $V-L$ and $V-C$ then differ from the instrumental ones $v-l$ and $v-c$ only by constants which have r.m.s. deviations amounting to 0.005 mag. Differential extinction between the V, L, and C colors has been found negligible. The internal accuracies (r.m.s. deviations) of the system are 0.01 mag in V and $B-V$ and 0.035 mag in $U-B$.

3. Results

The magnitudes of Halley's comet given in Table 2 on simple inspection reveal fluctuations far in excess of the measuring accuracy. Accounting for the brightness dependence on heliocentric distance $R\,(0.86 < R < 1.45)$ by a R^{-2} law, the magnitudes at $R=1.0$ for the stops D1 and D2, L_1 and C_1, behave in time as shown in Fig. 1. A dependence of C_1 on geocentric distance $\Delta\,(0.86 < \Delta < 0.52)$ is not apparent, while L_1 shows an increasing trend. The simple appearance of the light curves immediately suggests their periodic nature. During the interval $106 < T < 109$ (T stands for day of the year with decimal fraction in GMT), best covered by closely spaced observations, C_1 steadily increased by about 0.8 mag and L_1 by 1.0 mag in D1. This single measurement sequence rules out a period near 2 days. We have therefore numerically fitted to the light curves sinusoids of period P and amplitude A, superimposed on a baseline of constant slope $m_0' = \Delta m/\Delta T$. The parameter values which minimize the sum of the squared differences between the sine curves and the measurements have been searched by a computer program, initialized by 4 pairs (m, T) approximately describing with a cursor in a graphic display the range of variation. The goodness of fit can be estimated by comparing the parameter values obtained with different inputs. The program also provides the value of T_0, the epoch of a maximum, and the magnitude m_0 of the baseline at that epoch. The best fit curves thus obtained have been drawn in Fig. 1 and the corresponding parameters are given in Table 3.

The magnitudes measured with the larger field stops D3 and D4 have a strong dependence on Δ, but still suggest periodic fluctuations. In order to conveniently show the periodic behaviour of the light curves in D3 we have plotted the values of $C_1 - 0.75 \log \Delta$ and $L_1 - 0.75 \log \Delta$ in Fig. 1, together with the best fit sine curves, which have parameters given in Table 3.

Table 3. Parameters of fitted light cures

Stop	Color	m_0 (mag)	T_0 (day)	m_0 (m/d)	A (mag)	P (day)
D1	C	6.90	105.4	0.0	0.32	7.5
41″	L	6.10	105.4	+0.024	0.34	7.7
D2	C	5.23	105.2	+0.004	0.22	7.1
317″	L	4.00	105.2	+0.009	0.32	7.3
D3	C	4.21	105.3	−0.009	0.15	7.1
575″	L	2.52	105.3	−0.012	0.28	7.5

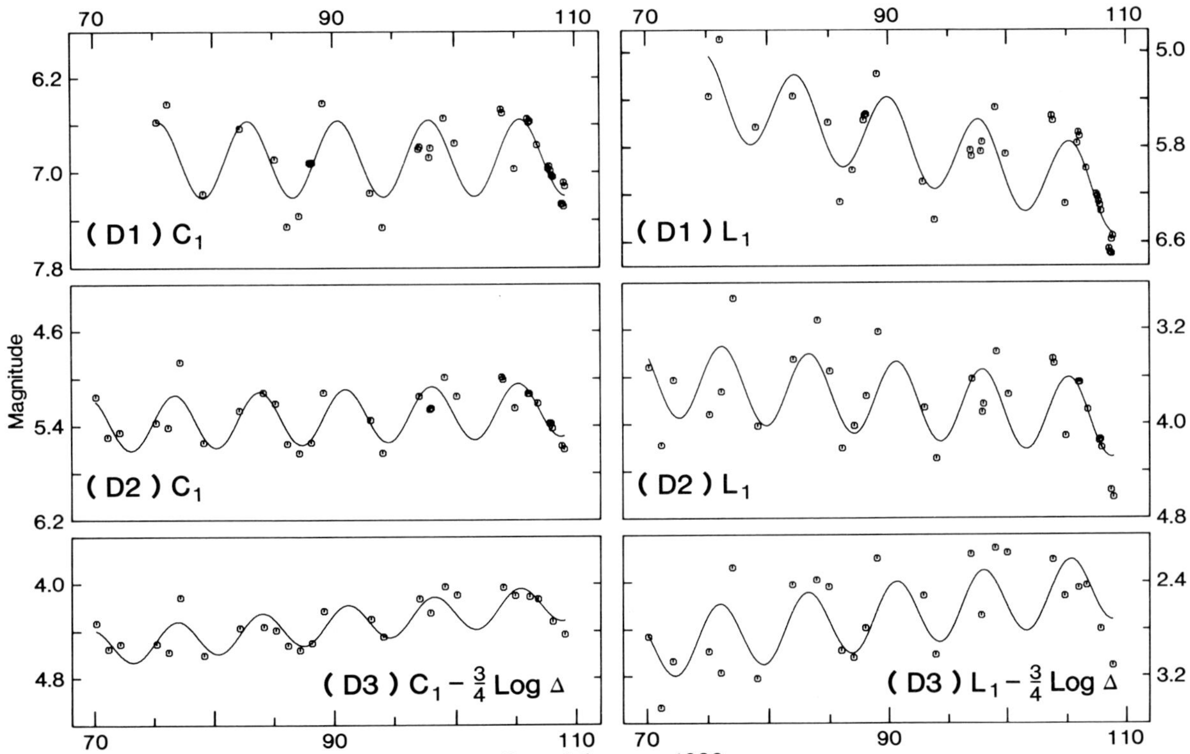

Fig. 1. Time behaviour of the magnitudes L_1 and C_1 for the coma of Halley's comet through the filters isolating the C_2 (0, 0) Swan band (right panels) and its neighbouring continuum (left panels), respectively. For the D1, D2, and D3 diaphragms, the magnitudes have been reduced to unit heliocentric distance R by a R^{-2} law. An allowance for the varying geocentric distance Δ has been introduced arbitrarily only for D3 (lower pannels). The sine curves (solid lines) are best fits in a least squares sense, unconstrained in period, amplitude, phase and baseline

Regarding the parametrization of the light curves by means of simple sinusoids, the following comments are in order:

a) The observed light curves were first fitted by sinusoids with periods around 7 days in September 1986, without knowledge of the results of Millis and Schleicher (1986), announced at the ESLAB Symposium on Halley's comet held in Heidelberg October 27–31, 1986 (Schleicher et al., 1986).

b) The close agreement between the periods determined for the six sets of independent measurements contained in Fig. 1 is a manifestation of the "stability" of the solutions provided by the numerical procedure followed.

c) The deviations of individual measurement from the best fit curves are appreciable and are in L systematically larger than in C. In particular, the L magnitude difference between points near the maxima and minima often exceeds 1.0 mag in D1, while the full amplitude of the best fit sine curve is $2A = 0.64$ mag. This means that the representation of the periodic variations by a single sine function is not a good one.

d) The light variations produced by sources within the solid angle subtended by D1 contribute, of course, to the fluctuations measured with other diaphragms. The light curve corresponding to the difference between the fluxes measured through D1 and D2, however, definitely shows a periodic behaviour, especially in the L-band, indicating periodic variability in the number of C_2 molecules contained within the cone defined by the annulus between D2 and D1. Simple arithmetic of the parameters characterizing the sinusoidal representation of the light curves also readily shows that the amplitudes determined for D2 and D3 are too large to be accounted for by light fluctuations occurring only within D1 and D2, respectively.

e) The steep dependence on Δ of the magnitudes corresponding to the outermost annulus, between D4 and D3, together with our incomplete sampling of data, does not allow to recognize any definite pattern in the light variations.

4. Conclusions

The values of the periods, 7.1 to 7.4 days, given in Table 3 for the light curves of Fig. 1, are in essential agreement with the 7.37 day value obtained by Millis and Schleicher (1986). The coverage secured by these authors over $91 < T < 109$, much denser than our's, allowed them to establish that the light curve is double-peaked. Our data suggest such a double peak wave only when folded into a single period as shown in Fig. 2. The two maxima present in such composite light curve occur nearly at the phases of the peaks directly observed by Millis and Schleicher (1986, Fig. 1).

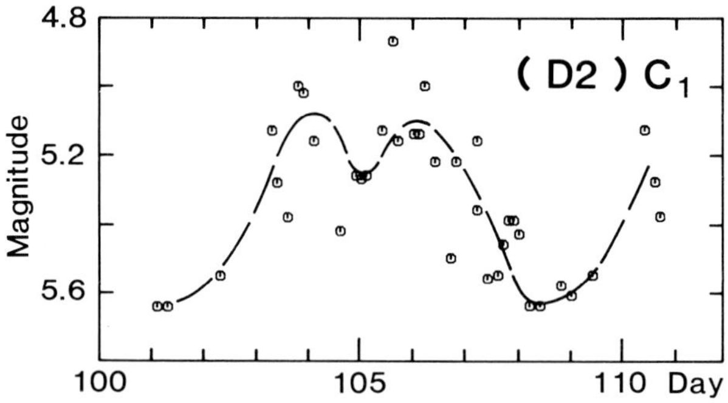

Fig. 2. Composite light curve of Halley's comet obtained by folding the continuum magnitudes C_1 through D2 with a 7.1 day period around the epoch $T_0 = 105.2$ day. The freehand drawn curve represents an estimate of the double peaked single periodic function best representing the observations

The full amplitude $2A = 0{,}7$ mag of our fitted (D1) L sine light curve is only one half of that characterizing the composite light curve of Millis and Schleicher (1986) for the same color and a diaphragm twice as large as D1. For the reasons explained above (Sect. 3c) we do not consider this difference to be significant.

Significant departures of the cometary brightness from a strictly periodic trend are clearly revealed by our measurements, indicating "activity" in the nucleus on time scales of one day. Their true nature will emerge only when all the data that may have been obtained at various sites are jointly analyzed.

Acknowledgements. We are greatly indebted to Hans Hippelein for providing the software basis of our analysis. Our thanks are also due to Michael Sarcander for his contribution to the reduction and preliminary study of the observations.

References

Gutiérrez-Moreno, A., Moreno, H., Stock, J., Torres, C., Wroblewski, L.: 1966, *Publ. Dept. Astron. Universidad Chile* **1**, 1

Julian, H.: 1987, *Nature* **326**, 57

Klare, G., Neckel, Th.: 1977, *Astron. Astrophys. Suppl.* **27**, 215

Millis, R.L., Schleicher, D.G.: 1986, *Nature* **324**, 646

Schleicher, D.G., Millis, R.L., Tholen, D., Lark, N., Birch, P.V., Martin, R., A'Hearn, M.F.: 1986, 20th Proc. ESLAB Symp., ESA SP-250, Vol. I, p. 565

Sekanina, Z., Larson, S.M.: 1986, *Astron. J.* **92**, 462

Astron. Astrophys. 187, 585–593 (1987)

ASTRONOMY AND ASTROPHYSICS

Observations of comet P/Halley at minimum phase angle*

K.J. Meech** and **D.C. Jewitt**

Department of Earth, Atmospheric and Planetary Sciences, Massachusetts Institute of Technology, 54–416, Cambridge, MA 02139, USA

Received March 5, accepted May 27, 1987

Summary. We present time-series spectrophotometric observations of comet P/Halley obtained over 8 nights in 1985 November in near backscattering geometry (phase angle $1° \leqq \alpha \leqq 10°$). We have used the spectra to search for enhanced backscattering of the type reported earlier in comets P/Ashbrook-Jackson, Bowell (1982I), and P/Stephan-Oterma and to study the time dependence of the molecular band and continuum emissions. No opposition surge in brightness greater than ≈20% was observed in P/Halley; instead, the brightening is consistent with a small linear phase coefficient $\beta = 0.02 \pm 0.01$ mag/deg in the observed phase angle range. The phase functions for P/Halley and the other comets are steeper than the phase function of the interplanetary dust. The comet phase functions are also unlike those of the low albedo asteroids, which typically show an opposition surge. Two continuum brightness outbursts observed on UT 1985 November 15 and 21 are discussed. Temporal variations in the molecular emission bands of the type seen by Millis and Schleicher (1986) have a timescale of variation which is consistent with the period of 7.37 days found in their data; however, the shape of the lightcurve is different.

Key words: backscattering – spectrophotometry – grains – nucleus activity – phase function

1. Introduction

Measurement of the scattering properties of particles as a function of phase angle is an important means of learning about the composition and physical properties of the scatterers. The phase functions of cometary grains have been measured in the range $30° \leqq \alpha \leqq 150°$ (Ney and Merrill, 1976; Ney, 1982) and are consistent with scattering from μm-sized dielectric spheres. At small phase angles ($\alpha \leqq 7°$), atmosphereless solar system bodies exhibit a surge in brightness (the so called "opposition effect") caused by the disappearance of interparticle shadows. The exact height and shape of the surge are dependent upon the roughness and porosity of the surface and upon the single scattering albedo of the particles (Lumme and Bowell, 1981). In comets, however, the nucleus is rarely directly visible. Generally, the scattering cross section in active comets lies in small solid particles in the extended coma. An opposition surge may or may not be produced in the coma particles, depending on their shape, composition and size. Some cometary dust grains may be similar to the Brownlee particles (Brownlee, 1978) collected in the Earth's upper atmosphere: aggregates of silicate grains bound together by an amorphous carbon-rich material (Bradley et al., 1984). Hanner et al. (1981) and Mukai et al. (1982) have shown that enhanced backscattering is possible for slightly absorbing fluffy particles such as these. Resonant backscattering ("glory") may also be produced (Oetking, 1966) if the particles are both spherical and transparent. The Gegenschein may be caused by enhanced backscattering from the zodiacal light particles (Dumont and Sánchez, 1975) which may themselves be cometary debris (Whipple, 1967). It therefore seems reasonable to search for an opposition surge or enhanced backscattering in the grain comae of active comets. Although the scattering function can certainly not be used as a unique diagnostic of the properties of cometary grains, it may be possible to use it to place useful constraints on the nature of the grains.

There have been several attempts to observe backscattering phase effects in comets. Kiselev and Chernova (1981) obtained broadband photometric measurements as a function of phase angle ($2° \leqq \alpha \leqq 25°$) for the comae of comets Meier (1978f) and P/Ashbrook-Jackson. Based on additional polarization measurements and Mie theory calculations they interpreted the enhanced brightening at small phase angles to be caused by backscattering from large (radius $a \geqq 1$ μm) dielectric particles. It is likely that the broadband filters used by Kiselev and Chernova took in considerable gas band emission in the case of comet Meier, so that a simple interpretation of their data may not be possible. Millis et al. (1982) made *narrowband* observations of the dust continuum of comet P/Stephan-Oterma as a function of phase angle ($3° \leqq \alpha < 30°$) and reported a factor of two increase in brightness at small phase angles which they interpreted as single particle enhanced backscattering. Finally, similar observations have been obtained for comet Bowell (1982I), by A'Hearn et al. (1984), who argued that the backscattering brightness (near $\alpha = 0°$) is approximately a factor of 3 greater than at moderate phase angles.

We have taken advantage of the fortuitous passage of P/Halley through small phase angles in late 1985 to examine its brightness as a function of phase, and to study temporal changes in the gas bands and continuum. The major problem with any attempted measurements of the phase functions of comets is that the heliocentric distance, R, and the phase angle, α, vary

Send offprint requests to: K.J. Meech

* Observations taken at the McGraw-Hill Observatory, which is operated by a consortium including the University of Michigan, Dartmouth College and the Massachusetts Institute of Technology

** Now at: Institute for Astronomy, 2680 Woodlawn Drive, Honolulu, HI 96822, USA

Table 1

UT date 1985	JD[a] −2440000	R (AU)[b]	Δ (AU)[c]	α[d] (deg)
15 Nov	6384.5	1.72	0.74	5.68
16 Nov	6385.5	1.71	0.72	4.02
17 Nov	6386.5	1.69	0.70	2.38
18 Nov	6387.5	1.68	0.69	1.37
19 Nov	6388.5	1.66	0.67	2.41
20 Nov	6389.5	1.65	0.66	4.31
21 Nov	6390.5	1.63	0.65	6.42
22 Nov	6391.5	1.62	0.64	8.63

[a] Julian Day Number for 0^hUT of date
[b] Heliocentric Distance
[c] Geocentric Distance
[d] Phase angle. All positions were computed with the authors' two-body ephemeris program using elements from MPC 10155

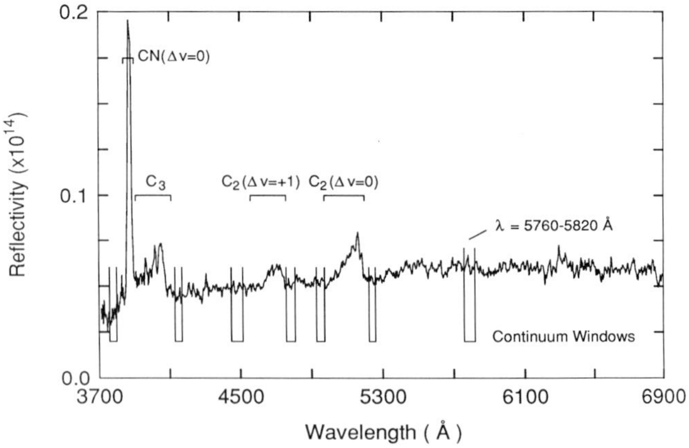

Fig. 1. Average of 9 P/Halley reflectivity spectra from UT 1985 November 22. Gas emission bands of CN ($\Delta v = 0$), C_3, C_2 ($\Delta v = +1$) and C_2 ($\Delta v = 0$) are marked as are their surrounding continuum regions. An additional continuum window at $\lambda_c = 5790$ Å is also marked

simultaneously, so that a given observed brightness change may be attributed to either variable. Temporal variations intrinsic to the comet may also be present. However, the geometry of P/Halley was particularly suitable for this project since the heliocentric distance remained virtually constant ($1.62 \text{ AU} < R < 1.72 \text{ AU}$) while the phase changed from $1^\circ.3 \leqq \alpha \leqq 8^\circ.6$. In the remainder of this paper, we describe our observations and give a brief review of the known backscattering measurements of comets. This paper replaces a less detailed report of our work given at the Heidelberg ESLAB Conference (Meech and Jewitt, 1986).

2. Observations

Observations were obtained during UT 1985 November 15–22 at the McGraw-Hill Observatory 1.3 m telescope at Kitt Peak. The MkII spectrograph and a 2048 channel intensified Reticon detector (Shectman and Hiltner, 1976) were used at the $f/7.5$ Cassegrain focus. All observations were taken with a 300 line/mm grating, giving a dispersion of 2.3 Å per channel. The effective spectrograph resolution over the range of wavelengths $3700 \text{ Å} \leqq \lambda \leqq 6900 \text{ Å}$ was $\Delta\lambda \approx 15$ Å. All observations were taken through an 8″ diameter circular diaphragm which was chopped to a nominal sky position 40″ west of the nucleus. The chop position always contained coma, however, and was used as a second data position. True sky subtraction was achieved by periodically observing the sky 3° north of the nucleus. Guiding on the comet was achieved by using an intensified CCD camera to monitor the image of the comet reflected from the spectrograph slit plate. We estimate that the guiding was accurate to better than ±1″. Exposure durations for both the central and 40″ west positions were between 300–600 s. Table 1 lists the geometry of the comet on each of the 8 nights during which we obtained data.

The Reticon counting rate was kept below 4000 photon s^{-1} via the use of neutral density filters, in order to avoid coincidence detection problems. The same filters used to observe the comet were also used for the standard stars since the transmission curves of the supposedly neutral filters are known to be slightly wavelength dependent. A single neutral density filter was used on all nights except for UT 1985 November 21 during which the comet brightened significantly. Ironically, as a consequence of the additional neutral density used on this night, the spectra of the comet at its brightest have the smallest signal to noise ratio.

Spectrophotometric calibration was obtained from observations of the standard stars HZ 15, Feige 15 and Feige 25 (Stone, 1977). These standards were chosen to be at similar airmass to the comet so as to minimize possible errors in assuming the Kitt Peak mean extinction coefficients. Wavelength comparison Hg and Kr spectra were taken before and after each observation in order to correct for instrumental flexure. Third order polynomial fits to 16 spectral lines provided good dispersion solutions, with residuals typically less than 1 channel (2.3 Å). No significant differences were found in the instrumental response curves computed from each standard star, hence a mean curve was formed from all of the standards observed each night. The standard star magnitudes were reproducible to within $\approx \pm 0.1$ mag over the 8 nights of observation, providing an empirical measure of the photometric quality of the data.

3. Analysis and discussion

In order to study the scattering properties of the dust, the signature of the solar spectrum must be removed from the data. The solar spectrum (Arvesen et al., 1969) was smoothed to match the MkII spectral resolution by convolving it with the instrumental spectral line profile, which was approximated by a gaussian of full width at half maximum (FWHM) equal to $\Delta\lambda = 15$ Å. Reflectivity spectra were computed by dividing each of the spectra of P/Halley by the convolved solar spectrum. The effectiveness of the division can be seen in Figs. 1 and 2 by the removal of the solar H and K calcium and the hydrogen-alpha lines from the spectra.

The reflectivity spectra show molecular emission bands (due to CN ($\Delta v = 0$), C_3, C_2 ($\Delta v = +1$) and C_2 ($\Delta v = 0$) as well as weaker bands due to CH ($\Delta v = 0$) and [O I]) superimposed on a strong dust continuum. In order to separate the continuum brightness from the molecular emissions, the continuum brightness was measured in a narrow emission-free region between $5760 \text{ Å} \leqq \lambda \leqq 5820$ Å. The location of this "continuum window" is shown in Fig. 1, which is an average of 9 reflectivity spectra obtained on UT 1985 November 22. Several of the gas bands are indicated in Fig. 1, together with their corresponding continuum windows. The wavelengths used for the gas bands and continuum windows

Table 2a

Species	Emission[a]	Continuum[a]
CN ($\Delta v = 0$)	3840–3900	3760–3800, 4130–4170
C_3	3910–4110	3760–3800, 4130–4170
C_2 ($\Delta v = +1$)	4560–4760	4450–4515, 4760–4810
C_2 ($\Delta v = 0$)	4975–5200	4930–4975, 5225–5260

[a] Wavelength range in Å

Table 2b

Start	End
(wavelength in Å)	
4390	4450
4750	4825
5200	5250
5760	5820
6235	6265
6380	6475
6780	6820

are listed in Table 2a. Nightly averages of the reflectivity spectra are presented in Fig. 2a–e. The spectra are plotted at a fixed absolute scale for ease of comparison. It is evident from the figure that substantial changes occurred in the dust continuum and molecular band components of the spectrum of P/Halley. For example, note the fading of the continuum from UT 1985 November 15 to 16 and from UT 1985 November 21 to 22 and the changes in the strength of the CN band from night to night.

Although there are substantial changes in the strength of the dust continuum, the continuum color does not change significantly from night to night. Measurements of the brightness in 7 continuum windows between 4390 Å $\leqq \lambda \leqq$ 6820 Å (see Table 2b) were used to compute the normalized reflectivity gradient, $S'(4390\,\text{Å}, 6820\,\text{Å}) = [dS/d\lambda]/S_{\text{mean}}$, expressed in percent per 10^3 Å, where $dS/d\lambda$ is the slope of the reflectivity spectrum and S_{mean} is the mean reflectivity in the wavelength interval. The normalized reflectivity gradient is a measure of the grain color; positive values indicating grain reddening, or increased scattering efficiency at longer wavelengths. Uncertainties in the color of the sun can introduce systematic errors in the color gradient of about $\pm 2\%/10^3$ Å, which is roughly equal to the reflectivity gradient measurement error (Jewitt and Meech, 1986). Emission band contamination of the continuum windows is minimal as all spectra (including those of other active comets) were examined to determine the emission free regions. The measurements show that the continuum of P/Halley is reddened with respect to the sun by $S'(4390\,\text{Å}, 6820\,\text{Å}) = (9 \pm 2)\%/10^3$ Å. This value is consistent with the reddening $[5 \pm 2 < S'(\%/10^3\,\text{Å}) < 18 \pm 2]$ seen for a sample of nine other comets (Jewitt and Meech, 1986).

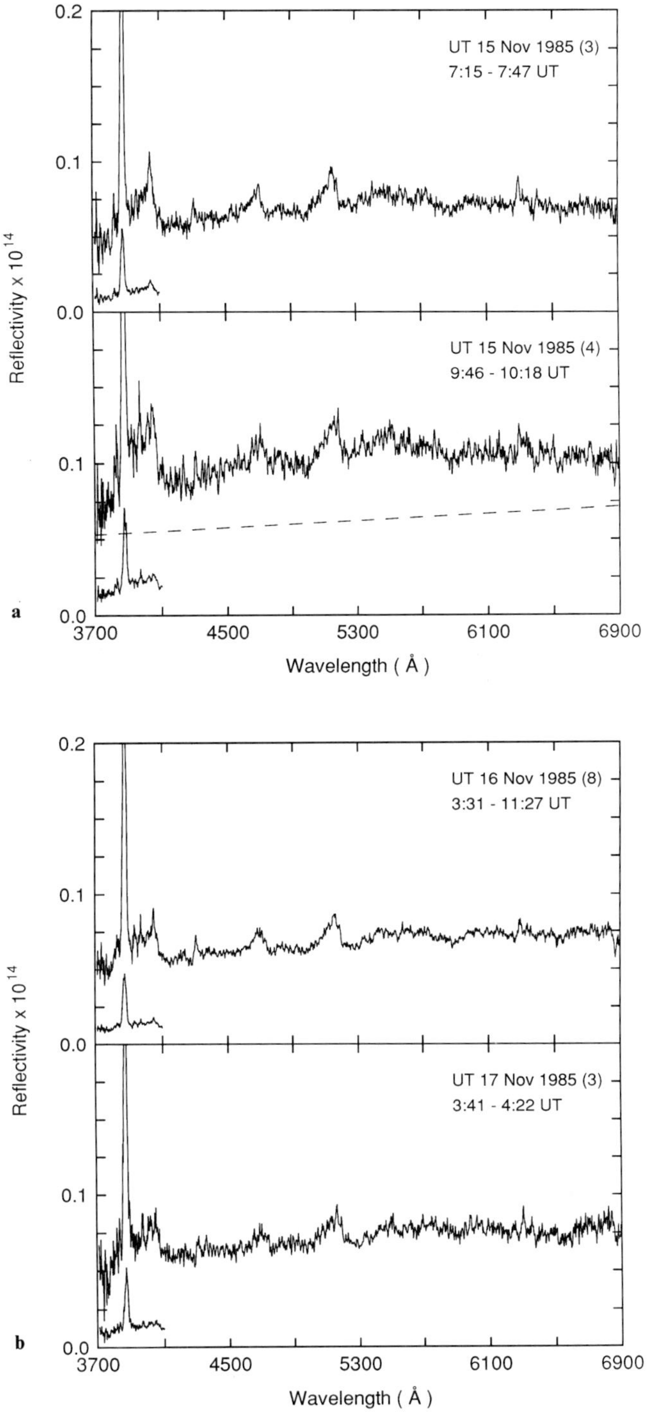

Fig. 2a–e. Nightly averages of P/Halley reflectivity spectra versus wavelength. Note that all spectra are plotted at the same absolute scale for ease in comparing nightly changes in both the line strength and continuum brightness. The CN band is also plotted at one tenth scale. The UT date and range of times for the spectra used to compute the average reflectivities are indicated in the figure. The number in parentheses following the date is the number of spectra used in the average. Outbursts in the continuum brightness are evident on UT 1985 November 15 and 21. For comparison, the mean continuum brightness in the 5760 Å $\leqq \lambda \leqq$ 5820 Å continuum window for all other nights is indicated by the dashed lines. The increased noise in the reflectivity on UT November 21 is due to the large neutral density filter employed (see text)

3.1. Phase function

3.1.1. Comet P/Halley

Figure 3 shows a plot of the mean continuum flux density, $F_{\lambda=5790}$ (J s^{-1} m^{-2} Å^{-1}), at the central and 40″ West positions

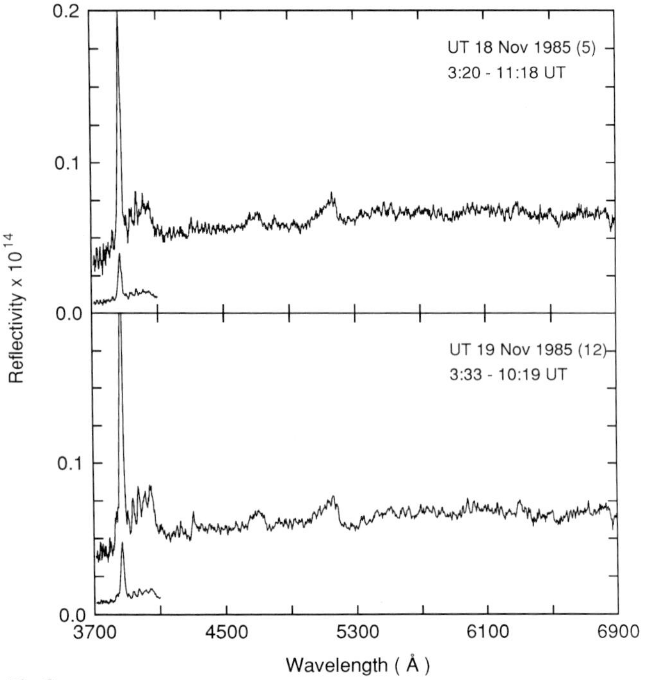

Fig. 2c

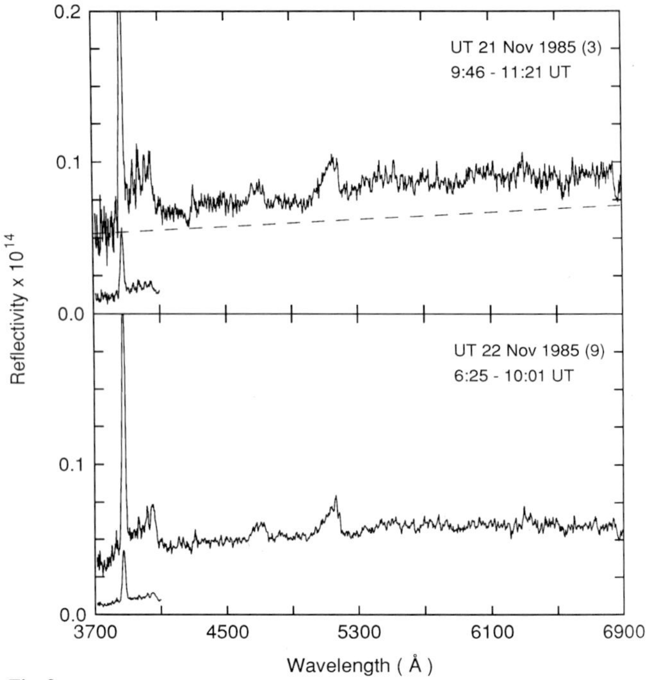

Fig. 2e

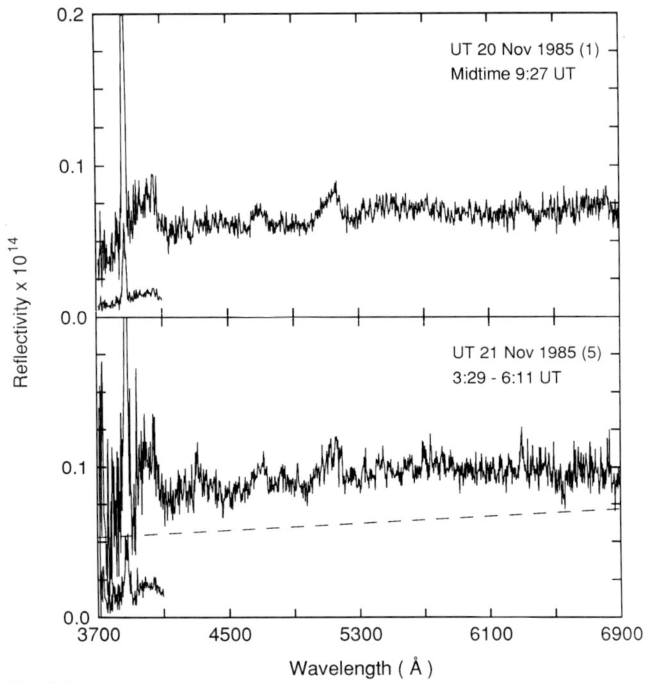

Fig. 2d

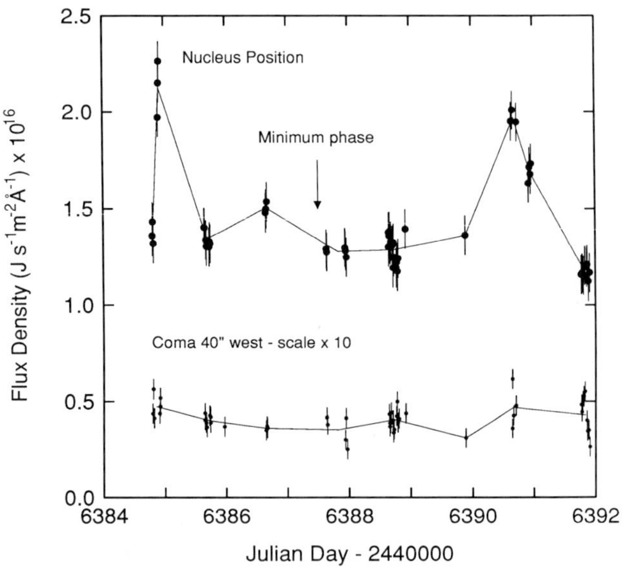

Fig. 3. Plot of the P/Halley continuum flux density ($5760\,\text{Å} \leqq \lambda \leqq 5820\,\text{Å}$) in the central and coma positions as a function of Julian Day (JD 2446384.5 = 1985 November 15 at 0^{h}UT). Minimum phase angle ($\alpha = 1\overset{\circ}{.}37$) occurred near JD = 2446387.5; no phase related brightening is apparent. However, two brightness outbursts in the dust coma are evident near JD = 2446385.0 and JD = 2446390.7, (UT 1985 November 15 and 21) with brightness increases of 0.6 and 0.4 mag, respectively. The continuum flux density versus JD for the coma position (40″ west of the nucleus) shows no delayed brightness surge. The flux density measured at the coma position has been multiplied by 10 in the figure

as a function of the time of observation. It is evident from Fig. 3 that the continuum flux density varies among nights by a factor of order 2. There are two distinct outbursts in the continuum centered near JD 2446385.0 and JD 2446390.7, each with a duration of about 1 day. In between the outbursts, the continuum is a relatively smooth function of time. The substantial maxima apparent in the central ("nucleus") diaphragm are muted or absent at the 40″ West position.

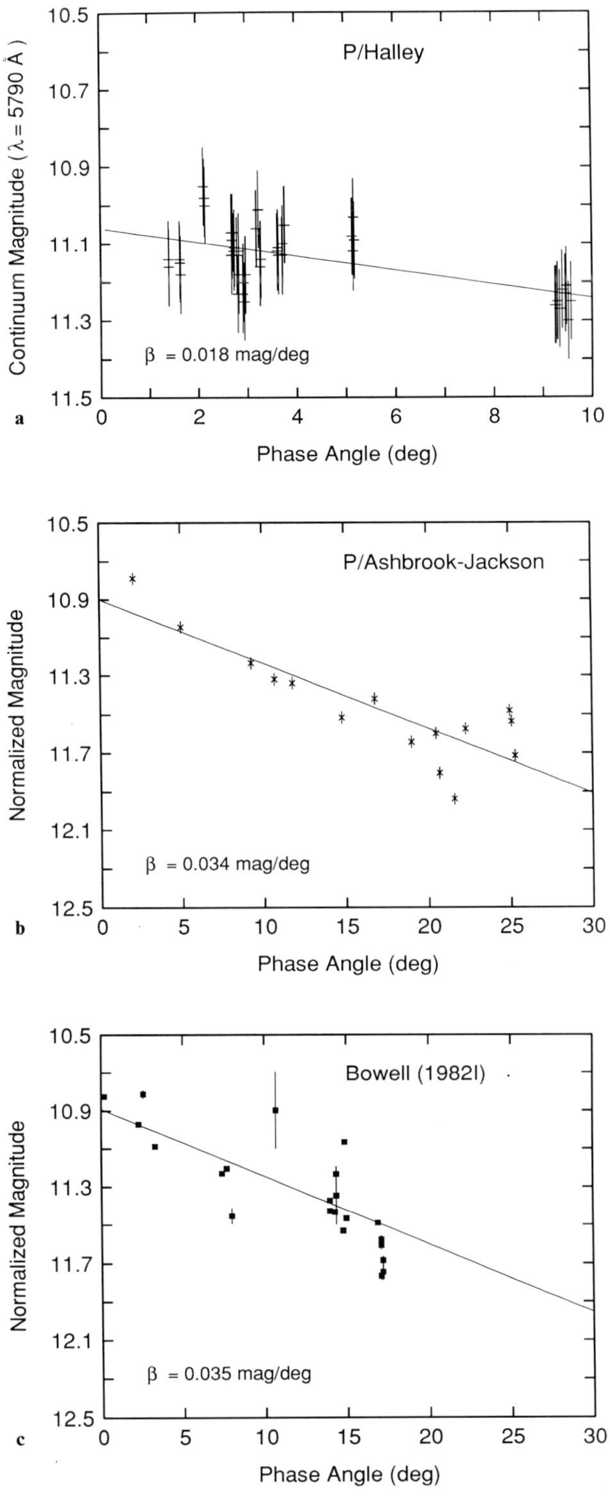

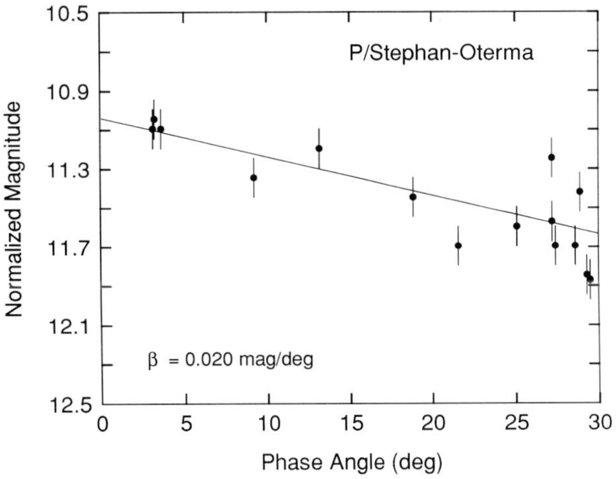

Fig. 4d

Fig. 4.a–d. Normalized magnitude versus phase angle for **a** P/Halley (data from the present work), where the magnitude is that computed from the continuum window at $\lambda = 5790\,\text{Å}$ ($m_{5790\,\text{Å}} \approx V$), and only magnitudes measured outside the two "outbursts" have been plotted for clarity; **b** P/Ashbrook-Jackson (data from Kiselev and Chernova, 1981), where the errors of 0.03 mag are as reported in Kiselev and Chernova; **c** Bowell (data from A'Hearn et al., 1984); the errors, where shown, are the errors on log (F_λ) as taken from the original paper; and **d** Stephan-Oterma (data from Millis et al., 1982) where arbitary 0.1 mag errors are plotted. A linear least squares fit for the linear phase coefficient, β, is shown for each comet. All comets are consistent with no opposition effect $\geqq 0.2$ mag for $0° < \alpha < 10°$

The data from the central diaphragm are shown in Fig. 4a, but this time the continuum magnitude

$$m_{5790} = -2.5 \log(F_\lambda) - 28.58 \tag{1}$$

is plotted against the phase angle, α. (For practical purposes, m_{5790} is close to the V magnitude of the continuum.) Data judged to be from the two outbursts in Fig. 3 have been omitted from Fig. 4a to attempt to emphasize any phase variations which may be present. The data are consistent with a small linear phase coefficient $\beta = 0.02 \pm 0.01$ mag/deg. There is no evidence for an opposition surge $\geqq 20\%$. Most importantly, the factor of 2 to 3 ($\Delta m \approx 1$ mag) backscattering enhancements reported for other comets are clearly not seen in comet Halley over the phase angle range $1 \leqq \alpha \leqq 10°$. It is possible, although unlikely, that intrinsic activity acted to cancel phase related variations in P/Halley.

To set the P/Halley data in proper context, we now make a critical assessment of the brightness versus phase data for three other comets.

3.1.2. Comet P/Ashbrook-Jackson

Data from Kiselev and Chernova (1981) for comet P/Ashbrook-Jackson have been re-analyzed and are shown in Fig. 4b. The magnitudes have been reduced to unit R and Δ, where Δ is the geocentric distance in AU, by subtracting $2.5 \log(R^2 \Delta)$ from the data. (Note that Kiselev and Chernova assumed a Δ^2 dependence in their analysis which is appropriate for a point source but not for an extended comet). Additionally, the magnitudes have been normalized so that the brightness at $\alpha = 10°$ is equal to the brightness of P/Halley at this phase angle ($V \approx 11.24$ mag). Although the P/Ashbrook-Jackson observations were made with the broadband V filter ($\lambda_{\text{eff}} = 5480\,\text{Å}$, $\Delta\lambda = 860\,\text{Å}$), Larson (1978) found that the comet spectrum was pure continuum, so that gas contamination should not be a problem. Furthermore, the change in R during the interval of observation (4.5 months) was only $\Delta R \approx 0.1$ AU, so that variations with R may also be neglected. Even though Kiselev and Chernova have fitted an asteroid-like phase coefficient (plus a non-linear term), we find that the data can be equally well fitted with a simple linear phase coefficient $\beta = 0.034 \pm 0.006$ mag/deg. In our estimation any more complicated fit to the Kiselev and Chernova data is unjustified given that the data were sparsely sampled during the 4.5 month observation

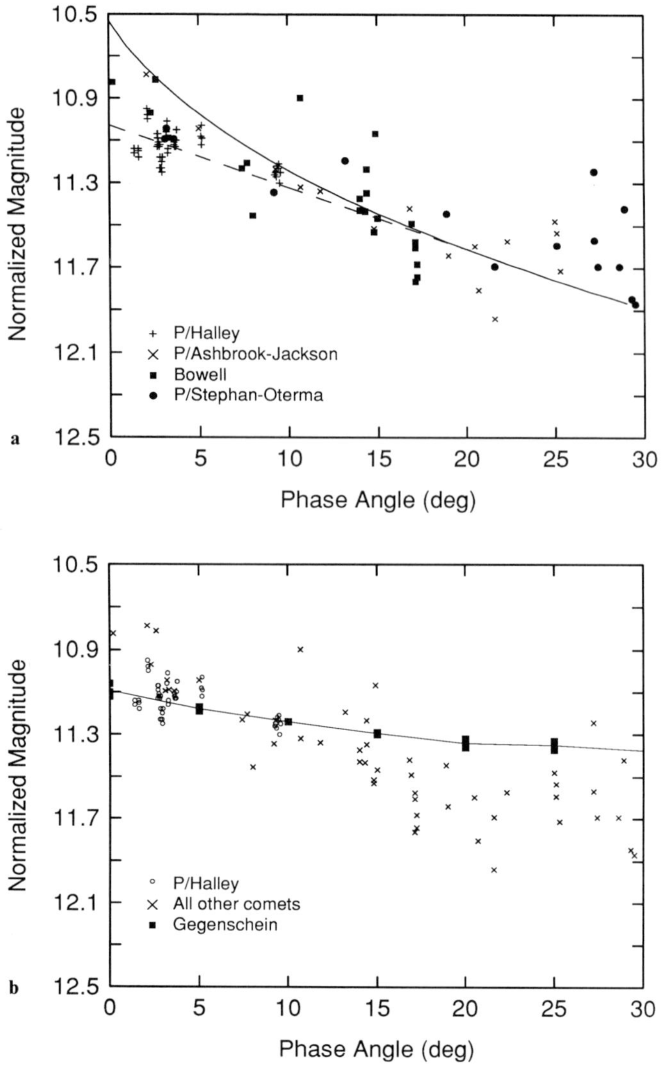

Fig. 5. **a** Normalized V magnitude versus phase angle for all of the comets shown in Fig. 4. A Bowell and Lumme (1979) phase function for a dark asteroid ($p_v \approx 0.07$, $\beta \approx 0.037$ mag/deg) is plotted as the solid line. The dashed line indicates a linear phase coefficient of 0.03 mag/deg. The comet data do not show the backscattering enhancement of $\Delta V \approx 0.4$ mag over the linear portion of the phase curve which is typically seen for asteroids. **b** Comparison of the normalized V magnitude versus phase angle for all of the comets with the phase dependence of the zodiacal light (data from Dumont and Sánchez, 1975; Leinert, 1975). The data on P/Halley are plotted with open circles, whereas all other comets are plotted with crosses. The zodiacal dust data are plotted as filled boxes

interval, thus precluding any estimate of the intrinsic cometary activity.

3.1.3. Comet Bowell

Data from A'Hearn et al. (1984) for comet Bowell (1982I) have been used to create the plot in Fig. 4c. The continuum magnitudes were derived from mean flux densities, F_λ ($\lambda_{\mathrm{eff}} = 5246$ Å, $\Delta\lambda = 72$ Å), found in Table 1 of their paper. The normalized V magnitude plotted in Fig. 4c was computed via:

$$V = -2.5\left[\log\left(\frac{\phi_0 R^2 \Delta}{\phi}\right) + \log(F_\lambda)\right] + \mathrm{C}_0, \qquad (2)$$

where ϕ is the observing diaphragm diameter, $\phi_0 = 20''$ is a standard diaphragm diameter and C_0 is a normalization constant. The observations of comet Bowell were taken over a large range of R (3.7 AU $\leq R \leq$ 5.6 AU) so that temporal and heliocentric variations may be suspected. However, the cross section of the coma of comet Bowell was nearly constant during the interval of observation (e.g. Jewitt, 1984), suggesting that the trend in brightness was due to the changing phase. We find a best fitting linear phase coefficient $\beta = 0.035 \pm 0.005$ mag/deg, and we find no evidence for an opposition surge.

3.1.4. Comet P/Stephan-Oterma

Finally, the normalized magnitudes of P/Stephan-Oterma (Millis et al. 1982) are presented as a function of phase angle in Fig. 4d. Unlike the previous authors, Millis et al. did attempt to remove the effect of intrinsic activity from their data by examining the ratio of the dust production rate to the C_2 gas production rate as a function of phase angle. The success of their method demands that the ratio of the dust to gas production rates did not change as a function of R. The normalized magnitudes shown in Fig. 4d were computed from the data found in Table 2 of Millis et al. by assuming that the coma brightness is proportional to the amount of dust in the coma and so to the production rate. A change in the dust/gas production ratio is thus equivalent to a change in the dust brightness, which can be converted to a normalized V magnitude using:

$$V = -2.5\log(Q_{\mathrm{R}}) + \mathrm{C}_1, \qquad (3)$$

where Q_{R} is the ratio of the dust production rate to the C_2 gas production rate, and C_1 is the normalization constant to the P/Halley data. As with the previous comets, we find no evidence for an opposition surge. Instead, a linear phase coefficient represents the data well; for P/Stephan-Oterma we find $\beta = 0.020 \pm 0.004$.

3.1.5. Comparison with other solar system bodies

In all three references discussed above, the comets were described as showing either a "backscattering peak" or "enhancement", suggesting an asteroid-like opposition surge in some cases. We find no convincing evidence for an asteroid-like opposition surge, neither in our data on P/Halley nor in the other comets for which phase curve measurements exist. The present spectrophotometric data set for P/Halley is probably the best suited for searching for the opposition surge, since the data were taken over a small interval in time and R. Frequent sampling allowed gross temporal variations to be easily identified.

In Fig. 5a all of the normalized phase curves, including the present data for P/Halley, are plotted versus phase angle. For comparison, a theoretical phase curve for a low albedo asteroid (after Bowell and Lumme, 1979) has been plotted as a solid line. The dashed line shows a linear phase coefficient of 0.03 mag/deg, normalized to the theoretical curve for $\alpha > 15°$. The theoretical phase function has been normalized to the P/Halley magnitude at $\alpha = 10°$. It is evident from Fig. 5a that none of the comet phase curves exhibit the 0.4 mag brightness enhancement over the linear portion of the curve as seen for the model. Models for surfaces with higher albedos have smaller opposition surges because multiple scattering destroys interparticle shadows; however, these models do not fit the large phase angle portion of the comet phase data as well as the low albedo models. The fact that the asteroid regolith scattering models do not fit the data well may suggest that the phase dependence is not caused by interparticle shadowing and that effects due to the single scattering albedos of the cometary particles are more important. We conclude that the comets show a linear phase coefficient $\beta = 0.02$–0.04 mag/deg for phase angles $1° \leqq \alpha \leqq 30°$, but that there is no evidence for an opposition surge greater then about 0.2 mag. In this sense, the backscattering from comets is different from that seen in typical asteroids.

There are occasions when a regolith opposition effect can be extremely narrow and therefore go unobserved unless the measurements extend to very small phase angles $0° < \alpha < 2°$. Such is the case for the phase curves of the Uranian satellites Oberon, Ariel and Titania where the sharp rise in brightness does not occur until the phase angle is less than $\alpha \approx 1°$ (cf. Brown and Cruikshank, 1983; Hapke, 1986). According to Hapke (1986), the width of the opposition peak is a function of both the filling factor and the particle size distribution. For power-law particle size distributions with indices $1 \leqq k \leqq 4$ and a large ratio of the largest to smallest particle sizes such as has been measured for P/Halley (McDonnell et al., 1986), the opposition surge width can be extremely narrow. Furthermore, the surge can also be narrow when the particle volume filling factor is low. It is possible that an opposition surge in P/Halley was unobserved by us because it was extremely narrow.

At the other extreme, an opposition surge would be difficult to detect if it were broad and of low amplitude. For example, the Gegenschein is a broad backscattering enhancement from interplanetary particles which are themselves believed to originate in comets. The Gegenschein is an antisolar brightness excess of about 20% with a diameter approximately 20 deg between the points of half intensity (Dumont and Sánchez, 1975). Lamy and Perrin (1986) have utilized extensive zodiacal light brightness measurements in order to obtain the scattering phase function per unit volume of the interplanetary dust. They concluded that the backscattering enhancement was only a factor of 2 over the minimum value near phase angle $\alpha = 80°$; furthermore, the models do not suggest a strong surge at small phase angles ($\alpha \leqq 10°$). Leinert (1975) found similar brightness gradients in the Gegenschein of about 1–2% per deg, within approximately $\alpha \leqq 10°$ of the center. This corresponds to about a 0.1–0.2 mag brightness enhancement within the phase angle range observed for P/Halley.

In Fig. 5b we compare the phase function of the zodiacal dust (Dumont and Sánchez, 1975; Leinert, 1975) with the brightness versus phase data for the comets. As in Fig. 5a, all data are normalized to $V = 11.24$ at $\alpha = 10°$. The very small amplitude of the Gegenschein is apparent from the figure. Considerable scatter among the phase curves of the comets makes a comparison with the zodiacal dust rather difficult. However, it appears that the phase curves of the comets are steeper than the phase curve of the interplanetary dust, especially for phase angles $\alpha \geqq 10°$.

In summary, the backscattering phase functions of the comets are different from a majority of asteroid and satellite phase functions (including that of the moon, see Gehrels et al., 1964) which show a pronounced opposition surge (see Fig. 5a). Furthermore, there appear to be differences between the phase functions of the interplanetary dust and the cometary dust (see Fig. 5b). On the other hand, a similarity to surfaces with very narrow backscattering enhancements such as Oberon, Ariel and Titania cannot be ruled out. The slopes of the phase curves shown in Fig. 5a are compatible with a simple extrapolation to small phase angles of the phase curves presented for angles $30° \leqq \alpha \leqq 150°$ by Ney (1982) for comet West. It should also be noted that there is no evidence for glory-like backscatter in the present data. Further measurements of scattering from the comae of comets at small to moderate phase angles, while taking account of the intrinsic variability of the coma brightness, will allow more detailed comparison of the scattering properties of comet dust, the interplanetary dust and asteroid surfaces.

3.2. *Outbursts*

The two prominent peaks in the continuum brightness (near JD ≈ 2446385.0, $\Delta V \approx 0.6$ mag; and near JD ≈ 2446390.7, $\Delta V \approx 0.4$ mag) are probably caused by a sudden increase in the dust production from the nucleus (see Fig. 3). The brightening seen on these two dates (UT 1985 November 15 and 21) is quite clearly seen at all continuum wavelengths (see Fig. 2). Dashed lines in Fig. 2 indicate the mean level of the continuum during UT 1985 November 16–20 and 22. Note the roughly factor of two increase in the mean continuum level during an outburst.

A considerable mass of grains was ejected into the coma on a short time-scale during each outburst. During quiescent periods, the continuum magnitude was near $V \approx 11.1$–11.2 mag. During the first outburst on UT 1985 November 15, the peak magnitude was $V \approx 10.5$–10.6 and during the second $V \approx 10.7$. The product of the cross section (C) times the geometric albedo of the dust (as a function of R, Δ, and solar magnitude, $V_0 = -26.74$) is given by:

$$p_v C = 2.24\ 10^{22}\ \pi R^2 \Delta\ 10^{0.4(V_0 - V)}\ (\mathrm{m}^2)\,. \tag{4}$$

Adopting the albedo $p_v \approx 0.04$ (Keller et al., 1986) we find that a total dust cross section

$$C \approx 2\ 10^3\ \mathrm{km}^2 \tag{5}$$

was ejected in the first outburst. The outburst rise time, from Fig. 3, was $\Delta t \approx 3$ hours (10^4 s). Assuming µm-sized grains with density $\varrho = 10^3\ \mathrm{kg\,m^{-3}}$, this corresponds to a mass $M = 3\ 10^6$ kg, or a mass loss rate $dM/dt \approx 3\ 10^2\ \mathrm{kg\,s^{-1}}$ averaged over the rise time of the outburst. This is a significant fraction (≈ 15%) of the total mass loss rate from P/Halley at this R, $dM/dt \approx 2\ 10^3\ \mathrm{kg\,s^{-1}}$ (Tokunaga et al., 1986).

The outbursts are not seen at the 40″ west coma position, delayed in time as would be expected if the ejected material subsequently passed into the coma. This suggests that either the material was in the form of icy grains which sublimated over a distance shorter than the projected separation of the observing positions (40″ $\approx 2\ 10^7$ m), or that the ejected material did not achieve escape velocity and fell back on the nucleus. It is also possible that the ejected material was confined to a narrow jet which did not cross the 40″ west diaphragm.

Figure 2 shows that there were not only brightness outbursts in the continuum, but that there were also variations in the gas band strengths, most noticeably in the CN emission. The total gas band fluxes of CN ($\Delta v = 0$), C_3, C_2 ($\Delta v = 0$) (see Fig. 1) were computed using:

$$F_T = \Delta\lambda \sum_{i=\lambda_1}^{\lambda_2} (S(\lambda_i) - S_c(\lambda_i))\, F_{sun}(\lambda_i)\,, \tag{6}$$

where λ_1 and λ_2 are the end wavelengths of the gas band (see Table 2a), and $\Delta\lambda = \lambda_1 - \lambda_2$. A linear least squares fit to the local continuum reflectivity near each gas band (see Table 2a), $S_c(\lambda)$, was subtracted from the reflectivity in each band, $S(\lambda)$, and the result was multiplied channel by channel with the solar flux density, $F_{sun}(\lambda_i)$. Figure 6 shows the ratio of the continuum flux at $\lambda = 5790$ Å to the CN ($\Delta v = 0$) and C_2 ($\Delta v = 0$) band fluxes as a function of time. The plotted points represent nightly averages

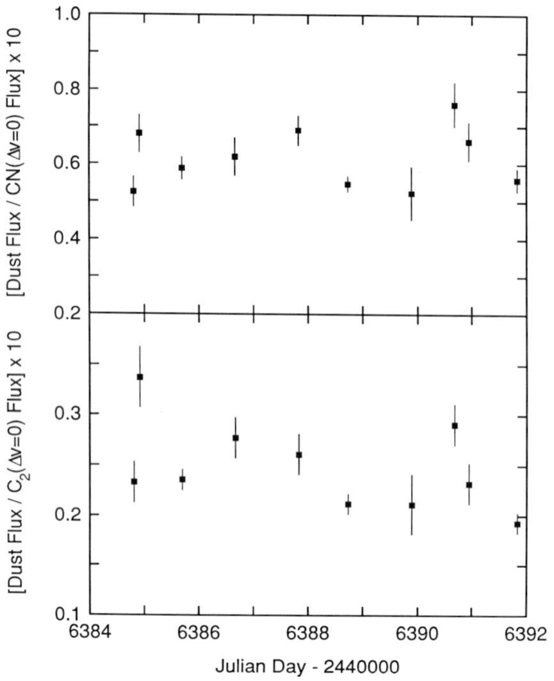

Fig. 6. Flux in the 5760–5820 Å continuum window divided by the flux in the CN ($\Delta v = 0$) and C_2 ($\Delta v = 0$) emission bands. Note the dust outbursts near JD = 2446385.0 and 6390.7

Table 3

Variables	r_{corr} [a]	N [b]	$P(r > r_{corr})$ [c]
Dust, CN	0.62	53	$< 10^{-5}$
Dust, C_2	0.21	53	0.13

[a] Linear correlation coefficient
[b] Number of data pairs in the calculation
[c] Probability that a correlation larger than r_{corr} would be found in an uncorrelated parent population

(except during an outburst when ≈2 points per night are plotted). The error bars show the standard error on the mean. At the time of the first outburst the dust/CN ratio changed only slightly, implying that the increase in dust was approximately coincident with an increase in the CN column density, however for C_2, there is a peak in the ratio, implying that the C_2 release was not as well correlated with the dust. For both molecules at time of the second outburst, there is a peak in the dust/gas ratio. This suggests that the second outburst was primarily in the dust.

The release of dust is normally assumed to be the result of ice sublimation, hence a correlated gas enhancement is expected. To search for correlations between the dust and gas, linear correlation coefficients were computed between the dust and CN ($\Delta v = 0$) fluxes and between the dust and C_2 ($\Delta v = 0$) fluxes. The correlation coefficients and the probabilities that correlation coefficients larger than these could come from an uncorrelated parent population are presented in Table 3. A probability of $P < 0.01$ indicates that there may be a significant correlation. It is apparent from Table 3 that there is a strong correlation between the dust and CN fluxes (even though the second outburst was primarily dust) whereas the dust and C_2 ($\Delta v = 0$) fluxes do not appear to be as significantly correlated.

Two possible explanations for the dust bursts come to mind: (1) the CN and other optical species may not be good tracers of the total gas production rate (cf. A'Hearn et al., 1985), alternatively, (2) the dust expulsion may have been the result of a non-thermal surface event (e. g. local failure of a dust crust). Ground-based and spacecraft measurements showing highly anisotropic ejection support the existence of such a crust (Keller et al., 1986). The considerable variations in the apparent continuum/molecular band ratio (Fig. 6) indicate that instantaneous measurements of this ratio may not always fully characterize the dust/gas in a comet. Similar observations of comet IRAS-Araki-Alcock by Lutz and Wagner (1986), showed not only a variation in the gas/dust ratios, but also variations in the gas to gas ratios, all of which varied as a function of position in the coma.

3.3. Temporal variations

There is another aspect of brightness variations which is more subtle than the outbursts or the phase dependence of the brightness, yet which is extremely interesting in light of the recent paper by Millis and Schleicher (1986). These authors found a periodic variation in the continuum brightness and molecular band strengths of P/Halley which they suggested can be attributed to modulation by the rotation of the nucleus. From their data Millis and Schleicher obtained a rotation period near 7.37 days. The present spectrophotometry shows a similar slow day-to day brightness variation which is best seen in the molecular emission bands. Figure 7 shows the variation of the CN ($\Delta v = 0$) and C_2 ($\Delta v = 0$) fluxes as a function of JD. The variation is remarkably similar in the two bands. Our observations cover a timebase too short to independently determine the period; however, our data are consistent with a period of 7.37 days. Unlike the Millis and Schleicher data which show similar variations in both the gas and dust, our data show considerable shape differences between the gas and dust variations.

4. Conclusions

1. No opposition effect ≧20% is detected in the continuum of P/Halley in the phase angle range $1° < \alpha < 10°$. Instead, the backscattered continuum is adequately represented by a linear phase coefficient, $\beta = 0.02 \pm 0.01$ mag/deg.

2. We find no evidence for opposition effects greater than 20% in a re-examination of published data on comets P/Ashbrook-Jackson, Bowell and P/Stephan-Oterma. The back-scattering phase functions of these comets are best described by small (but significant) linear phase coefficients, $\beta = 0.02$–0.04 mag/deg.

3. The cometary phase coefficients are similar to those of low albedo asteroids. However, the comets differ from low albedo asteroids in that they lack observable opposition surges. The back scattering phase functions of comets are steeper than the phase function of the zodiacal light in the range $0° < \alpha < 30°$.

4. Reflectivity spectra of comet P/Halley show a strong dust continuum with bands of CN ($\Delta v = 0$), C_3, C_2 ($\Delta v = +1$) and C_2 ($\Delta v = 0$) superimposed. The dust continuum of P/Halley is reddened with respect to the sun by $S' = (9 \pm 2)$ %/10^3 Å in the wavelength interval 4390 Å $\leqq \lambda \leqq$ 6820 Å. This value is typical of other comets.

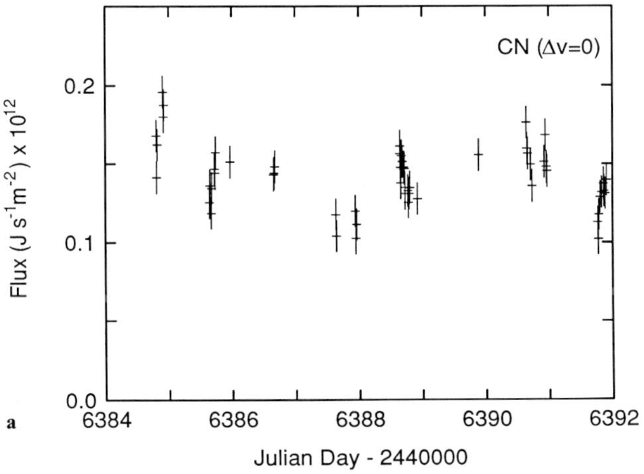

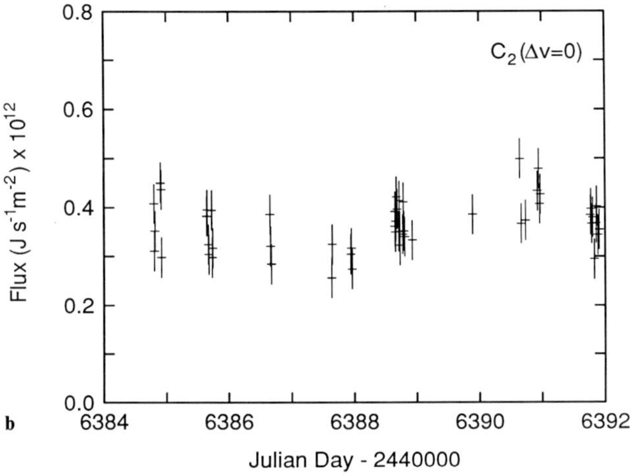

Fig. 7a and b. Temporal variation in the **a** CN ($\Delta v=0$) and **b** C_2 ($\Delta v=0$) gas bands

5. The apparent dust/gas ratio varied during the period of observation by a factor of order 2. The largest variations were associated with two discrete outbursts seen on UT 1985 November 15 and 21. The outbursts were not seen in the 40″ west aperture even though the observations covered the time during which the time delayed surge should have been visible. Both of these were primarily dust events. Possible implications are that either these molecules are not good indicators of the total gas production – or that a dust crust on the surface of the nucleus underwent local fragmentation.

6. The molecular emission bands show a temporal variation, qualitatively similar to that reported by Millis and Schleicher (1986). No period has been determined for the variation, although it is consistent with the 7.37 ± 0.07 day period found by Millis and Schleicher.

Acknowledgements. We thank Ron Remillard for invaluable assistance with the MkII and with its spectral reduction software. Some of the calculations were done on the MIT Planetary Astronomy VAX. This work was supported in part, by a grant from the NASA Planetary Astronomy Program (NAGW-770).

References

A'Hearn, M.F., Birch, P.V., Feldman, P.D., Millis, R.L.: 1985, *Icarus* **64**, 1
A'Hearn, M.F., Schleicher, D.G., Feldman, P.D., Millis, R.L., Thompson, D.T.: 1984, *Astron. J.* **89**, 579
Arvesen, J.C., Griffin, R.N., Pearson, B.D.: 1969, *Appl. Optics* **8**, 2215
Bowell, E., Lumme, K.: 1979, in *Asteroids*, ed. T. Gehrels, Univ. Arizona Press, Tucson, p. 132
Bradley, J.P., Brownlee, D.E., Fraundorf, P.: 1984, *Science* **223**, 56
Brown, R.H., Cruikshank, D.P.: 1983, *Icarus* **55**, 83
Brownlee, D.E.: 1978, in *Cosmic Dust*, ed. J.A.M. McDonnell, Wiley, New York, p. 295
Dumont, R., Sánchez, F.: 1975, *Astron. Astrophys.* **38**, 405
Gehrels, T., Coffeen, T., Owings, D.: 1964, *Astron. J.* **69**, 826
Hanner, M.S., Giese, R.H., Weiss, K., Zerull, R.: 1981, *Astron. Astrophys.* **104**, 42
Hapke, B.: 1986, *Icarus* **67**, 264
Jewitt, D.: 1984, *Icarus* **60**, 373
Jewitt, D., Meech, K.: 1986, *Astrophys. J.* **310**, 937
Keller, H.U., Arpigny, C., Barbieri, C., Bonnet, R.M., Cazes, S. Coradini, M., Cosmovici, C.B., Delamere, W.A., Huebner, W.F., Hughes, D.W., Jamar, C., Malaise, D., Reitsema, H.J., Schmidt, H.U., Schmidt, W.K.H., Seige, P., Whipple, F.L., Wilhelm, K.: 1986, *Nature* **321**, 320
Kiselev, N.N., Chernova, G.P.: 1981, *Icarus* **48**, 473
Lamy, P.L., Perrin, J.-M.: 1986, *Astron. Astrophys.* **163**, 269
Larson, S.: 1978, *IAU Circ. 3293*
Leinert, C.: 1975, *Space Sci. Rev.*, **18**, 281
Lumme, K., Bowell, E.: 1981, *Astron. J.* **86**, 1705
Lutz, B.L., Wagner, R.M.: 1986, *Astrophys. J.* **308**, 993
McDonnell, J.A.M., Alexander, W.M., Burton, W.M., Bussoletti, E., Clark, D.H., Grard, R.J.L., Grün, E., Hanner, M.S., Hughes, D.W., Igenbergs, E., Kuczera, H., Lindblad, B.A., Mandeville, J.-C., Minafra, A., Schwehm, G.H., Sekanina, Z., Wallis, M.K., Zarnecki, J.C., Chakaveh, S.C., Evans, G.C., Evans, S.T., Firth, J.G., Littler, A.N., Massonne, L., et al.: 1986, *Nature* **321**, 338
Meech, K.J., Jewitt, D.: 1986, in *20th ESLAB Symp.* ESA SP-250, Vol. **I**, p. 553
Millis, R.L., A'Hearn, M.F., Thompson, D.T.: 1982, *Astron. J.* **87**, 1310
Millis, R.L., Schleicher, D.G.: 1986, *Nature* **324**, 646
Mukai, S., Mukai, T., Giese, R.H., Weiss, K., Zerull, R.H.: 1982, *Moon Planets* **26**, 197
Ney, E.P.: 1982, in *Comets*, ed. L.L. Wilkening, Univ. Arizona Press, Tucson, p. 323
Ney, E.P., Merrill, K.M.: 1976, *Science* **194**, 1051
Oetking, P.: 1966, *J. Geophys. Res.* **71**, 2505
Shectman, S., Hiltner, W.A.: 1976, *Publ. Astron. Soc. Pacific* **88**, 960
Stone, R.P.S.: 1977, *Astrophys. J.* **218**, 767
Tokunaga, A.T., Golisch, W.F., Griep, D.M., Kaminski, C.D., Hanner, M.S.: 1986, *Astron. J.* **92**, 1183
Whipple, F.L.: 1967, in *The Zodiacal Light and the Interplanetary Medium*, ed. J.L. Weinberg, NASA SP-150, p. 409

Astron. Astrophys. 187, 594–600 (1987)

ASTRONOMY AND ASTROPHYSICS

Chinese observations of comet P/Halley in China and abroad

S.M. Gong (Kung) [1], **G.J. Wu** [2], **P.S. Chen** [2], **X.F. Zhang** [3], **and S.S. Sun** [4]

[1] Purple Mountain Observatory, Nanjing, China
[2] Yunnan Observatory, Kunming, China
[3] Shanghai Museum of Natural History. Shanghai, China
[4] Qingdao Station, Oceanographic Institute, Qingdao, China

Received February 2, accepted June 22, 1987

Summary. A brief account of Chinese observations of Halley's comet in China and abroad together with some preliminary results is given.

Key words: Halley's comet – observations – general account – preliminary results

1. Introduction

After comet P/Halley was first observed, as shown in Fig. 1, in China in the early morning of November 5, 1984 with the 1 m reflector plus CCD at Yunnan Observatory (Yeomans et al., 1985) investigators from Purple Mountain Observatory (PMO), Shanghai and Beijing Observatories have participated in using the equipment for Near-Nucleus Studies (NNS) and photometry of the comet. The different types of observations carried out at various sites are shown in Table 1.

PMO began to engage in its astrometry work in August 1985 and telexed the comet's positions to the JPL, as did the Qingdao Station and Shanghai Observatory. Most of the telexed positions, especially those of PMO, ranked high in the observatories all over the world, and were good enough to be used to improve the determination of the orbit of Halley's comet (Yeomans et al., 1985).

From mid-November 1985 Beijing Observatory devoted its twin 40 cm astrograph and 60–90 Schmidt to the Large Scale Phenomenon (LSP) and obtained a series of clear photos with an intensity calibration which will be shown elsewhere in these Proceedings. PMO and Yunnan Observatory did the same type of work. A team was despatched by Shanghai Observatory to Ke-Shi in the west, to study LSP. Due to the fine weather and less extinction in the high altitude, good photos of comet Halley were obtained.

In using its newly built radio telescope with a 25 m dish at Mei-xian, Shanghai Observatory, with the collaboration of the Geophysics Department of Beijing University and the 39th Institute of the Electronic Industry Ministry, successfully observed the continuous emission of comet Halley at 2.8 cm in December 1985 and then the OH 18 cm absorption line, with a signal-to-noise ratio greater than 4 in April 1986.

Send offprint requests to: S.M. Gong (Kung)

Good photometric, infrared and spectroscopic observations of comet Halley were carried out with 60 cm and 1 m reflectors at Yunnan Observatory from November 1985. An outburst of Halley's comet was observed between March 22 and 23, 1986 (Chen et al., 1987).

To observe Halley's comet after its perihelion passage, we sent several teams to Hainan Island and one each to Paracel-Inseln Island and Si-mao. These teams made observations for the LSP and NNS disciplines of the International Halley Watch (IHW). In addition, a team of 4 people flew to New Zealand and a team of 2 people went to Australia to make the same type of studies. All of the teams worked successfully except for the teams at Hainan Island which experienced bad weather.

An outburst phenomenon was observed by the Chinese oversea team on March 25, 1986 at John Observatory, New Zealand. At first glance, it seemed to them to be a small-scale splitting of the nucleus. However, its true nature is most probably some kind of jet stream or streams. At Mount Stromlo Observatory, Australia, the Chinese astronomers observed the rapid increases of the brightness of comet Halley by means of an image intensifier plus a CCD and a television camera attached to the 76 cm reflector on both April 13 and 15, 1986. They interpreted them as two outbursts of Halley's comet. The center of the observing band was 6880 Å with a half width of 850 Å.

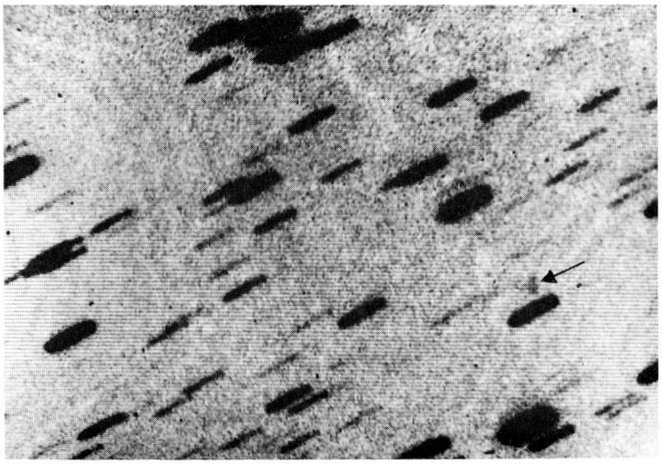

Fig. 1. Comet Halley first observed in China on Nov. 5, 1984 with 1 m reflector plus CCD at Yunnan Observatory. An arrow is added to indicate comet Halley

Table 1. Log of observations at various sites

Site	Latitude	Longitude	Date	Instrumentation	Type of work
Yunnan Obs.	25°02′ N	102°45′ E	Mar 20–Apr 26, 1986 Nov 4–Nov 22, 1985 Mar 25–Apr 2, 1986 Mar 21–Mar 26, 1986	1 M Refl. F/13 1 M Refl. F/13, CCD 1 M Refl. F/13 60 cm Refl.	Spectrophotometry NNS, Astrometry Photoelectric Photometry, Infrared Photoelectric Photometry, Optical
Shanghai Obs.	31°06′ N	121°11′ E	Oct 7, 1985–May 15, 1986	40 cm Twin Astrograph, F/7	Astrometry
Purple Mountain Obs.	32°04′ N	118°49′ E	Aug 27–Dec 25, 1985	40 cm Twin Astrograph, F/5	Astrometry
Beijing Obs.	40°06′ N	116°20′ E	Sep 20, 1985–Jan 11, 1986 Dec 8, 1985–Jan 9, 1986	60/90 Schmidt, F/3 40 cm Twin Astrograph, F/4	LSP, NNS LSP
Qingdao Station	36°04′ N	120°19′ E	Sep 18, 1985–Jan 1, 1986	32 cm Astrograph, F/11	Astrometry, LSP
San-ya (Hainan Island)	18°12′ N	109°32′ E	Mar 6–Apr 12, 1986	16/32 Schmidt, F/1.23	Photometry of Coma
Paracel Inseln Island	16°50′ N	112°20′ E	Mar 8–Apr 15, 1986	15 cm Astrograph F/5	LSP
Mei-xian	34°10′ N	107°48′ E	Nov 29–Dec 7, 1985 Mar 31–Apr 17, 1986	25 M Dish at 2.8, Cont., 18 cm Line	Radio Science
Ke-shi	39°24′ N	76°42′ E	Nov 10, 1985–Jan 5, 1986	15 cm Astrograph F/5, 10 cm HaΦa-3 Camera, F/2.5	LSP
Si-mao	22°55′ N	100°58′ E	Mar 4–Apr 16, 1986	10 cm HaΦa-3 Camera, F/2.5	LSP
Mount John Univ. Obs.	43°59′ S	189°32′ E	Mar 16–Apr 22, 1986 Mar 25–Apr 27, 1986	61 cm Refl., F/14 12.5 cm Ross Camera, F/7	NNS LSP
Mount Stromlo Obs.	35°19′ S	149°0.5′ E	Apr 9–Apr 17, 1986	76 cm Refl. F/18 + TV Camera (Image Intesifier + CCD) 66 cm Astrograph F/15	NNS Astrometry

2. Preliminary results

2.1. *Spectroscopy and spectrophotometry*

In March-April of 1986, while comet Halley had a heliocentric distance of about 1 AU and 1.5 AU, respectively, Wu et al. (1987) carried out some long-slit spectral observations using a Cassegrain spectrograph attached to the 1 m RCC telescope of Yunnan Observatory. This work is only concerned with the 6 nitrogen-baked Kodak 103a–O plates which have a spectral range of 3000–5100 Å with a dispersion of 165 Å mm^{-1} and a resolution of no less than 7 Å. The dispersion curve of every plate was obtained by a polynomial fit of wavelengths versus positions of a sequence of strong cometary emission lines. At first, the spectra were scanned with a PDS at PMO. The intensities of spectral lines were derived from densities with a 7-point gaussian smoothing process.

The tracing used for the identification of spectral lines has a scale 20–40 times that of the spectra. The estimated accuracy of the measured wavelength is about 3 Å. The spectra and the tracing are shown, respectively, in Figs. 2 and 3. The identification of spectral lines was carried out mainly by means of referring to the works by Dobrowolsky (1966) and Allen (1976) and a wavelength table. Out of 69 emission lines, 59 have been identified; all the 29 absorption lines are identified as well. However, as the plates shown here are actually not very high signal-to-noise, many of the emission and absorption features are most likely not real but noise. The result of the identification of spectral lines makes it clear that there are molecular emission lines of CN, C_2, C_3, CH, NH, and OH; perhaps emission lines of NH_2 and CH^+ exist as well. The existence of a number of strong absorption lines, strong continuum and very few ion lines, suggest that comet Halley is dust-rich. OH (3090 Å) emission was discovered, although the emission is weak and appears only on the spectrum taken on April

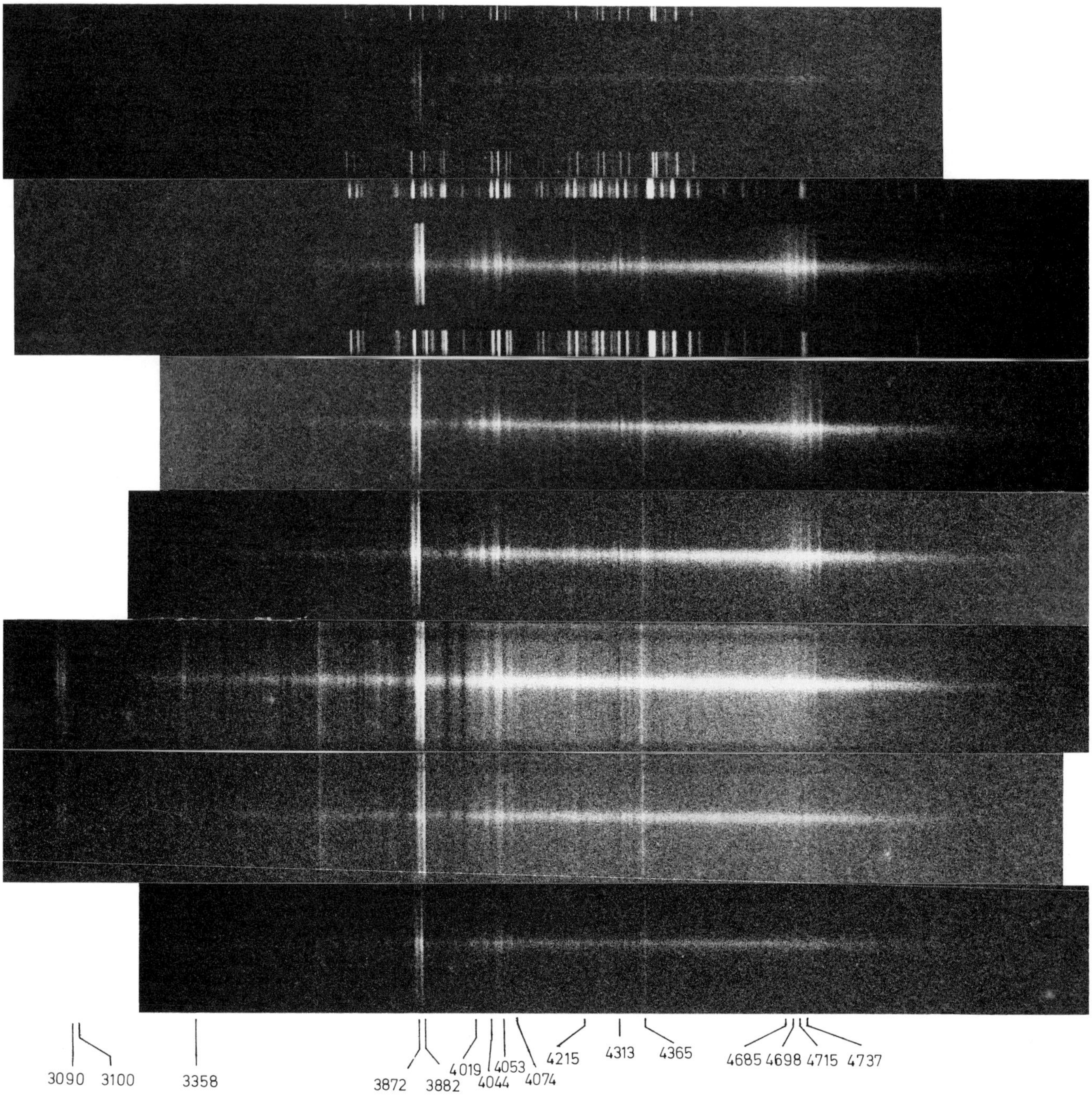

Fig. 2. Spectra of Halley's comet in March–April, 1986, taken at Yunnan Observatory with 1 M RCC telescope at its Cassegrain focus at a dispersion of 165 A/mm on baked 103a-O plates. From top to bottom, the exposure times are (UT): Mar 20, 21^h03^m–22^h29^m; Mar 21, 21^h10^m–22^h28^m; Mar 22, 20^h50^m–22^h24^m; Mar 24, 20^h25^m–22^h28^m; Apr 24, 12^h56^m–17^h05^m; Apr 25, 12^h28^m–17^h28^m; Apr 26, 12^h21^m–16^h03^m

24. This emission is a clue in the research on the development of the hydrogen cloud around this comet. Its discovery indicates that the conditions for ultraviolet observations at Yunnan Observatory, at about 2000 km above sea level, are very good.

2.2. *Optical and infrared photoelectric photometry*

During March 20 April 3, 1986, Chen et al. (1987) made a photoelectric photometry for Halley's comet by using the 60 cm telescope at Yunnan Observatory. The diaphragm used was 2 mm corresponding to a field of view about 33″. Six standard narrow filters were used in the observation, their specification being shown in Table 2.

Observations were made after perihelion when the comet's heliocentric distance increased from 1.0065 AU to 1.1917 AU and its geocentric distance decreased from 0,788 AU to 0.511 AU. The standard star was 96 Her as recommended by the IHW and defined to be $5^m_.0$ at all the 6 wavelengths used in the

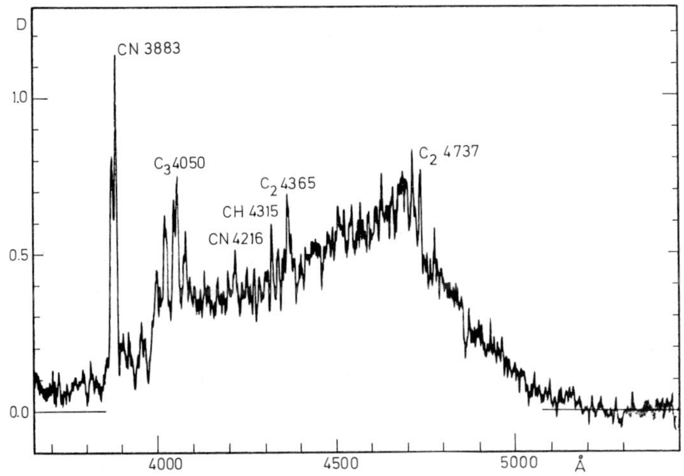

Fig. 3. The tracing of the spectra in Fig. 3

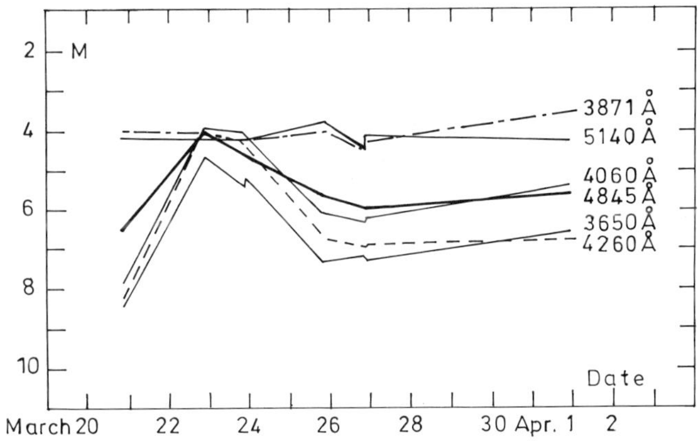

Fig. 4. Light curves of Halley's comet in optical wavelength range

Table 2. Filter specification

No.	Wave band	Center (Å)	Maximum transmissivity	Half-width
1	Continuum 1	3650	>0.25	80
2	CN	3871	>0.20	50
3	C_3	4060	>0.40	70
4	$CO + N_2$	4260	>0.40	70
5	Continuum	4845	>0.50	65
6	C_2	5140	>0.50	90

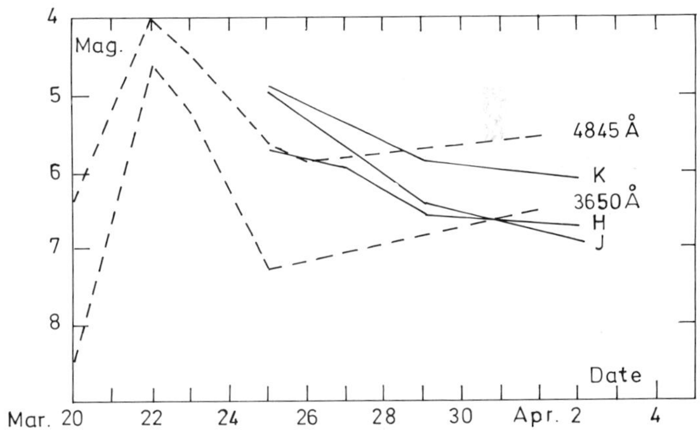

Fig. 5. Light curves of Halley's comet in optical and near-infrared ranges

measurements (A'Hearn, 1984). The calibration stars nearby were 52 Sgr. and τ Sgr.

For the convenience of comparison, magnitudes were all reduced to "the reduced absolute magnitude" M, i.e., the heliocentric and geocentric distance were all set equal to 1 AU.

The results of observation are shown in Fig. 4, from which one readily observes an outburst of comet Halley. From Mar. 21 to 22 the brightness of comet Halley increased greatly in bands 1, 3, 4, and 5, their magnitudes being 2^m–4^m brighter. In the band 4 (4260 Å of $CO^+ + N2^+$), it brightened the most, up to $4\overset{m}{.}27$. The whole process of outburst lasted several days until Mar 26. However, it is worthy to notice that no obvious change of brightness happens at band 2 (CN, 3871 Å) and 6 (C_2, 5140 Å).

The conspicuous increase of brightness at bands 1 and 5, as interpreted by Chen et al., was due to the reflection of the solar radiation at these two continuums. Canuto et al. (1985) pointed out that the number density of the dust was almost inversely proportional to its distance from the nucleus. It follows that the continuum should brighter earlier in the response to the outburst. Ions CO^+ and N_2^+, being produced by photoionization of neutral gases CO_2 and N_2, formed readily from some point nearby the nucleus outward during the expulsion of these molecules in the outburst and thus the number density of these ions could reach a maximum at about several hundred kilometers from the nucleus, as shown by Canuto et al. This explains the great increase of brightness at the band 4 ($CO^+ + N_2^+$, 4260 Å).

For the infrared photometry, Chen et al. made the observations of comet Halley at J, H, and K bands and in the range 1.3–2.6 μm with the 1 m telescope of Yunnan Observatory during Mar 25 to Apr 2, 1986. The stars α Lyr and τ Sgr were selected, respectively, as the fundamental standard star for calibration and the standard star for photoelectric photometry. As above, "the reduced absolute magnitude" was adopted.

Figure 5 shows the reduced absolute magnitudes M of comet Halley at J, H, and K bands during the interval of the observation. All the M's had a conspicuous increase. In Fig. 5 are also shown the variations of the M's for the continuums at $\lambda = 3650$ Å and $\lambda = 4845$ Å for comparison. It shows that the decrease of infrared brightness in the interval Mar 25–Apr 2 is the aftereffect of the aforementioned outburst on Mar 22.

Chen et al. also used a circular variable filter (CVF) to measure the energy spectrum in the near infrared. The filter covering 1.3–2.6 μm has a resolution $\Delta\lambda/\lambda$ equal to 2%. After assigning the magnitudes of the standard star α Lyr at all wavelengths to be 0^m, i.e., its radiation flux being defined to be 1 unit, the results of observations over 3 days are illustrated in Fig. 6. Evidently, comet Halley had 2 emission peaks at 1.4 and 1.9 μm and a third one outside 2.16 μm. The authors agree with Knacke's view (Knacke, 1986) that the first 2 peaks are produced by H_2O. As to the emission peak outside 2.6 μm, it has not been identified yet, probably due to OH (Canuto et al., 1985).

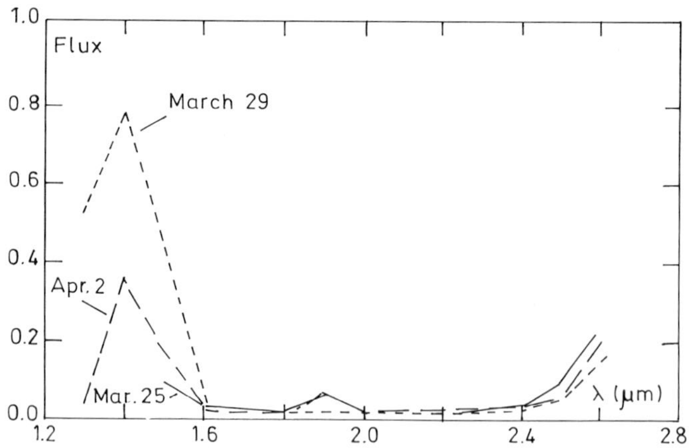

Fig. 6. The normalized flux for Halley's comet

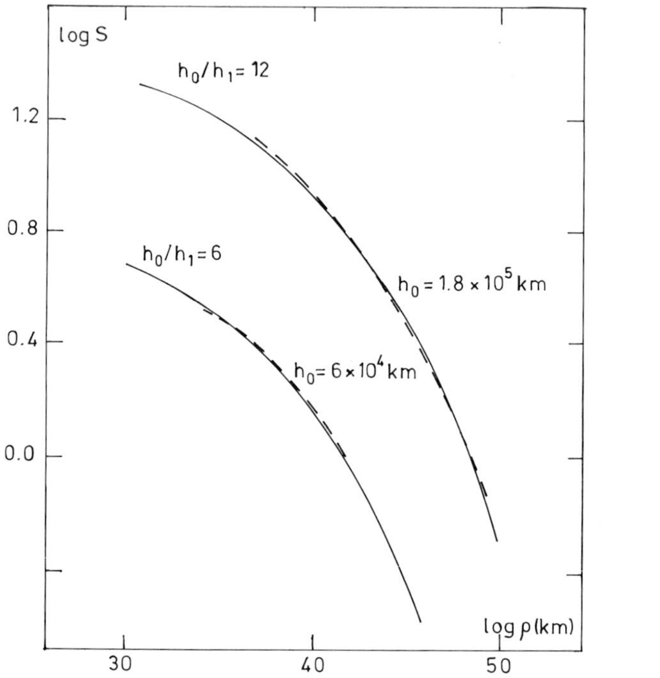

Fig. 7. The fitting of the theoretical curve (———) to the observed curve (------)

Fig. 8a. The photo of comet Halley taken on 1986, Mar 17

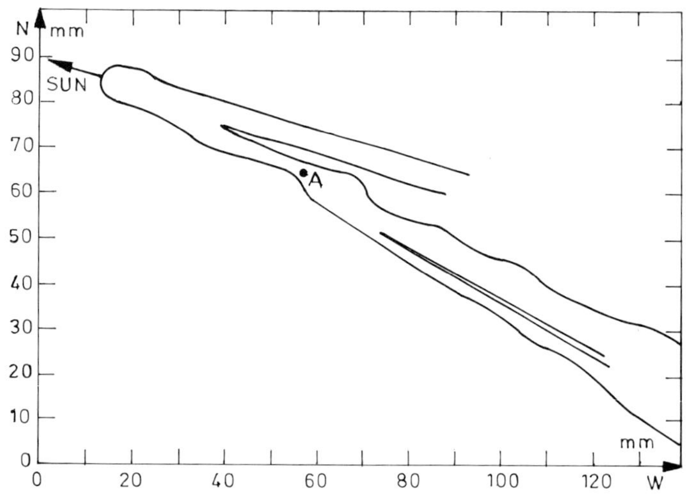

Fig. 8b. The sketch of the photo in Fig. 8a. Scale: 1 mm = 1$'$.83 = 69,000 km

2.3. The photographic photometry of the coma of comet Halley

Zhang et al. made the observation of comet Halley with a Schmidt camera of 16/23 cm at San-ya on Hainan Island in April 1986. The plate used was Ilford (HP5) and the filter had the center wavelength at 4235 Å with a halfwidth of 100 Å. The work is concerned with 2 plates taken on Apr 4 and 6. They were calibrated and measured with a PDS at PMO.

Taking the coma to be spherically symmetric (as a convenient working hypothesis) the densities at 4 points of equal projectional distance ϱ from the nucleus at $\pi/2$ apart from each other were measured. By this procedure $\bar{D}(\varrho)$, the average density at ϱ and then $\bar{S}(\varrho)$, the average surface brightness at ϱ can be obtained in the density-intensity transformation. With Haser's theoretical coma model (Brandt and Chapman, 1981), the following formula in the thin outer part of coma can be derived,

$$\log S(\varrho) - \log S_0 = -\log(\varrho/h_0) + \log\left[\int_{\varrho/h_0}^{\varrho/h_1} K_0(y)\,dy\right], \tag{1}$$

where h_0 and h_1 are, respectively, the scale length of daughter and parent molecules; S_0, a parameter related to h_0 and h_1 and K_0, the zero order Bessel function of the second kind. By suitable choice of h_0 and h_1, $\bar{S}(\varrho)$, calculated from Eq. (1), is able to fit to the observed curve $\bar{S}(\varrho)$, as shown in Fig. 7. From the 2 curves, the mean values of h_0 and h_1 are obtained as follows:

$$h_0 = 1.2\ 10^5\ \text{km}, \quad \text{and} \quad h_1 = 1.25\ 10^4\ \text{km}.$$

The numerical results obtained here agree in order of magnitude with that derived from CN (Newburn and Spinrad, 1984; Cochran, 1985). Assuming the outflow velocities of the daughter and parent molecules, respectively, to be $v_0 = 0.9\ \text{km s}^{-1}$ and $v_1 = 0.6\ \text{km s}^{-1}$ with the definition of scale length, $h_0 = v_0\tau_0$, $h_1 = v_1\tau_1$, the lifetimes of the daughter molecule CN and its parent molecule are,

$$\tau_0 = 1.3\ 10^5\ \text{s}, \quad \text{and} \quad \tau_1 = 2.08\ 10^5\ \text{s}.$$

Fig. 9a. The photo of comet Halley taken on 1986, Apr 13

Fig. 10a. The photo of comet Halley taken on 1986, Apr 15

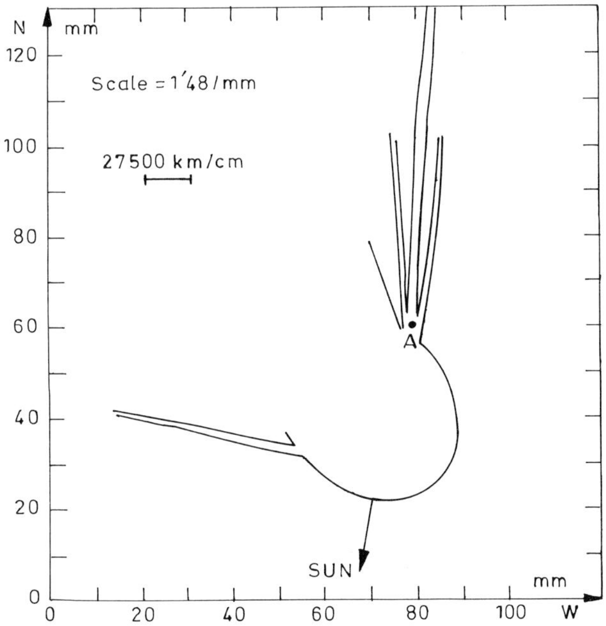

Fig. 9b. The sketch of the photo in Fig. 9a. Scale: 1 mm = 1′.48 = 27,500 km

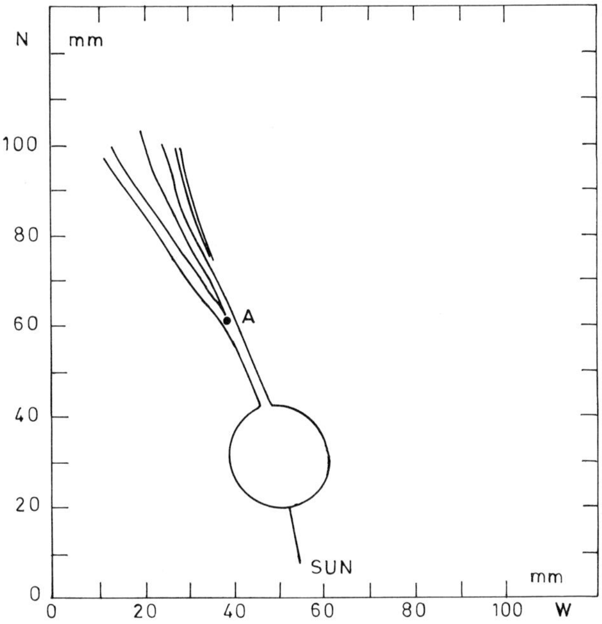

Fig. 10b. The sketch of the photo in Fig. 10a. Scale: 1 mm = 1′.39 = 27,000 km

2.4. Large-scale phenomena (LSP)

Many groups in China have worked on LSP as mentioned in the introduction. Most of them are still in the stage of reduction and analysis. The work reported in this section is obtained from the photos taken on Kodak IIa–O plates without filter with a 15 cm astrograph of 75 cm focal length at the Paracel-Inseln Island by Sun et al. in March and April 1986.

The photo HC 16, shown in Fig. 8a and its sketch in Fig. 8b, was taken on 1986 March 17, at 20^h44^m UT with an exposure time of 60 min. It showed the distortion of the plasma tail of comet Halley when the position angle (PA) of the sun was 77°.7 and that of the tail was 253°.8. One can see from Fig. 8a and 8b that at a distance of $4.26\ 10^6$ km from the nucleus, indicated by point A in Fig. 8b, the tail suddenly changed its direction and turned counterclockwise by 33°; then, after extending $7.7\ 10^5$ km, returned approximately to its original direction. After the distortion, the plasma tail formed several streamers, extending and departing to two sides.

Figures 9a and 9b show, respectively, the photo HC 66 and its sketch of comet Halley taken on 1986 Apr 13, at 15^h55^m UT with an exposure time of 90 min. It is the second of 3 photos taken on the same day. The first was taken at 14^h32^m UT with an exposure time of 60 min. From the disturbing point, indicated by point A in Fig. 9b, at about 1.7 10^5 km from the nucleus, 2 rays radiated and could be clearly seen on the second photo, as shown in Figs. 9a and 9b. The third photo was taken at 19^h31^m UT. From these photos one could see the distrubing point moving away from the nucleus with a decreasing velocity. The position angle of the sun was 170° and that of the tail was 355°.

The photo HC 77, shown in Fig. 10a and its sketch Fig. 10b, was taken on Apr 15, at 16^h23^m UT with an exposure time of 40 min. This is the second of 3 photos obtained on the same day. On the first one, taken at 15^h07^m, at the position, indicated by point A, of about 9.7 10^5 km from the nucleus, the plasma tail separated itself into 3 branches, the main branch of which had a PA of 24°, approximately the same as that of the main tail before its breaking. After 1^h16^m the west branch broke away as shown on the second picture (10a and 10b). After 2.5 hr, the main branch also broke away, as seen from the third photo, taken at 18^h54^m GMT with an exposure time of 40 min, leaving only the east branch in connection with the main tail. The measurement showed that the main tail turned clockwise. After separating from the main tail, both the west and the main branches were contracting due to the gradual disappearance of their farthest parts.

Acknowledgements. We are indebted for the referee's valuable comments and suggestions which led us to made the proper modifications. We thank Y.L. Li, J.K. Li, and W.M. Shi of Yunnan Observatory for their collaboration in the spectroscopic work; Y.K. Chen, L. Lu, H. Kao, and Y. Zhang of Yunnan Observatory and J. Yang and M. Mei of Beijing Normal University for their participation in the photometry and infrared works; J. Xue and L. Wang of Shanghai Museum of Natural History for their work in the coma photometry; Y.J. Shao, and C.Z. Dong of Beijing Observatory and Y.Q. Hui and B.L. Zhang of Beijing Planetarium for their taking part in the work of Large Scale Phenomenon. The financial support of Academia Sinica is gratefully acknowledged.

References

Allen, C.W.: 1976, *Astrophysical Quantities*, p. 79
A'Hearn, M.F.: 1984, Letter to the Members of IHW, Nov. 8
Bobrovnikoff, N.T.: 1931, *Lick Obs. Publ.* **17**, 305
Brandt, J.C., Chapman, R.D.: 1981, *Introduction to Comets*, Chap. 4
Canuto, V.M.: 1985, *Fund. Cosmic Phys.* **10**
Chen, P.S., Chen, Y.K., Lu, L., Kao, H., Yang, J., Mei, M.: 1987, *Acta Astron. Sinica* **28**, No. 1
Cochran, A.L.: 1985, *Astron. J.* **90**, 2609
Combi, M.R., Delsemme, A.H.: 1980, *Astrophys. J.* **237**, 633
Dobrowolsky, O.V.: 1966, *Comets*, Moscow, p. 67
Greenstein, J.L., Arpigny, C.: 1962, *Astrophys. J.* **135**, 892
Knacke, R.: 1986, *IHW Newsletter* **8**, 45
Mo, J.E., Fan, C.Y.: 1979, *Acta Astron. Sinica* **20**, 43
Moore, C.E.: 1966, *The Solar Spectrum* from 2935 Å to 8770 Å, Washington, D.C.
Newburn, R.L., Jr., Spinrad, H.: 1984, *Astron. J.* **89**, 289
Swings, P., Haser, L.: 1956, *Atlas of Representative Cometary Spectra*, Liège
Wu, G.J., Li, Y.L., Li, J.K., Shi, W.M.: 1987, *Acta Astrophys Sinica* (in press)
Yeomans, D.K.: 1985, *IHW Newsletter* **6**, 31
Yeomans, D.K.: 1985, *IHW Astrometry Newsletter*, Oct; ibid., 1986 Jan

Astron. Astrophys. 187, 601–604 (1987)

ASTRONOMY AND ASTROPHYSICS

Thermal infrared imaging of comet P/Halley

H. Campins[1,★], C.M. Telesco[2,★], R. Decher[2,★], and B.D. Ramsey[2,★]

[1] Planetary Science Institute, SAIC, 2030 E. Speedway, Tucson, AZ 85719, USA
[2] Space Science Laboratory. NASA-Marshall Space Flight Center, ES-63, Huntsville, AL 35812, USA

Received February 16, accepted April 23, 1987

Summary. Thermal infrared images ($\lambda = 10.8$, 12.8 and 19.2 μm) of comet P/Halley were obtained on 1986 March 28.7, 29.7 and 30.7 UT using the NASA-MSFC 20-pixel bolometer array at the NASA Infrared Telescope Facility. A preliminary analysis of the observations shows the following results: (1) the presence of spatial structure indicative of dust ejection occurring mainly on the sunward side of the nucleus; (2) a dust coma deficient in large particles (radius $\gtrsim 100\,\mu$m) compared with that of comet P/Giacobini-Zinner; (3) temporal variability which is consistent with an extrapolation of a quasi-sinusoidal lightcurve observed by IUE on 1986 March 23, 24, and 25; (4) a maximum, at the nuclear condensation, of the color temperature and the strength of the silicate emission feature.

Key words: comets – infrared-images

1. Introduction

Thermal infrared imaging ($\lambda \gtrsim 3\,\mu$m) is a new and powerful technique for the study of the solid particles in the comae of comets which combines the benefits of infrared photometry and image analysis. The first groundbased thermal infrared images of any comet were of P/Giacobini-Zinner (hereafter GZ) and have been reported by Campins et al. (1986) and Telesco et al. (1986). We have used this technique to observe comet Halley before and after perihelion. The preperihelion thermal images were taken simultaneously with visible images which isolate the scattered continuum. Those observations have been combined to produce albedo maps of the coma and are discussed by Hammel et al. (1987). In this work we present a preliminary analysis of thermal images obtained on 3 consecutive days in March of 1986. A more detailed discussion of the observations and their implications will be presented by Telesco et al. (in preparation).

2. Observations

The observations were made at the 3.0 m NASA Infrared Telescope Facility on 1986 March 28.7, 29.7, and 30.7 UT, using the 20-pixel bolometer spatial array developed at the NASA Marshall Space Flight Center (Telesco et al., 1986). The heliocentric and geocentric distances and the phase angles corresponding to the observing times are given in Table 1. Each square pixel of the array had side dimensions of 4.″3 (~1800 km) and the pixels were configured in a 5 × 4 (RA × DEC) arrangement. The center-to-center pixel separation was 4.″5. The chopper throw was approximately 100″ in declination. Based on a ρ^{-1} decrease of surface brightness with projected distance ρ from the nucleus we estimate that the contribution from the coma to the reference beams does not exceed 10% of the signal in the inner 20″. Further out, the contribution of the coma to the reference beams can be larger if the ρ^{-1} dependence holds; this would make the interpretation of any detailed structure in the outer coma very difficult. In reality, however, the ρ^{-1} dependence of the surface brightness holds only for a distance of approximately 10″ from the nucleus in comets at heliocentric and geocentric distances comparable to those of P/Halley (Jewitt and Meech, 1987). Beyond 10″ or so from the nucleus the surface brightness decreases as ρ^{-2} or faster for almost all comets observed. In such cases the contribution of the outermost coma to the reference beams is acceptably low. The three bandpasses used were centered at 10.8, 12.8, and 19.2 μm and had full widths at half maximum of 5.3, 1.2 and 5.2 μm, respectively. The star α Her was used as a standard, and the magnitudes adopted were −3.94, −3.99, and −4.17. The absolute calibration given by Tokunaga (1984) was used for the transformation from magnitudes to flux densities. There is an uncertainty of about ±7% associated with the absolute calibration.

On 28.7 and 29.7 March, 10.8 μm measurements were made on a grid of 25 × 20 points (112.″3 × 90″, RA × DEC) centered on the nuclear condensation. The resultant maps are shown in Fig. 1. On 30.7 March poor weather restricted our observations to a single field of 5 × 4 points. Observations at 12.8 and 19.2 μm were obtained on the second day in three fields of 5 × 4 points each. One field was approximately centered on the nuclear con-

Table 1

UT date (March 1986)	R (AU)	Δ (AU)	Phase angle (degrees)
28.7	1.13	0.60	62
29.7	1.14	0.58	61
30.7	1.16	0.56	60

Send offprint requests to: H. Campins

★ Visiting Astronomer at the Infrared Telescope Facility, operated by the University of Hawaii under contract from the National Aeronautics and Space Administration

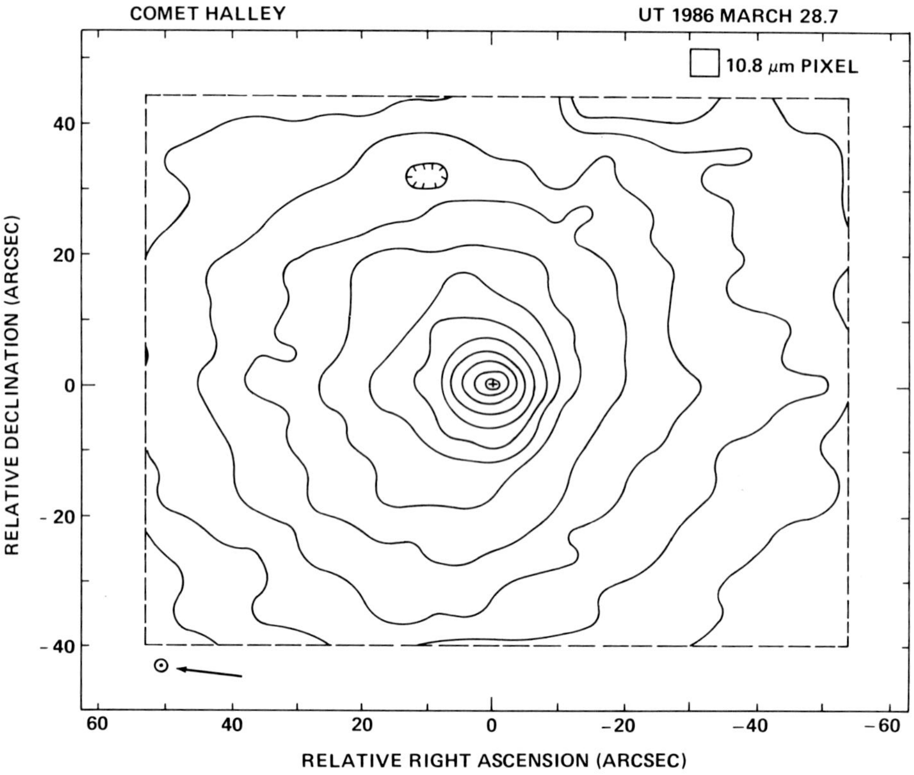

a

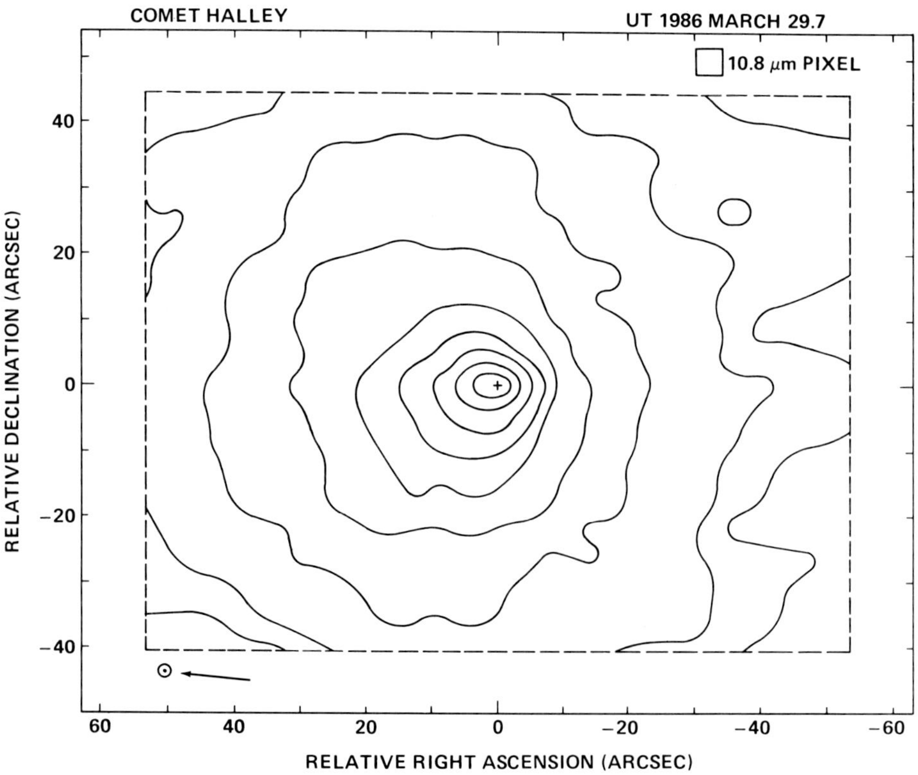

b

Fig. 1. Contour maps of the 10.8 μm flux density of comet Halley on **a** 1986, March 28.7 and **b** 29.7. The contours are logarithmically spaced. The peak flux densities (at the cross) are 80 ± 0.7 and 191 ± 1.7 Janskys (Jy) on March 28.7 and 29.7, respectively. The values of the contours are (in Jy/pixel): 56.2, 39.8, 28.2, 20.0, 14.1, 10.0, 7.1, 5.0, 3.5, 2.5, 1.8, 1.3 for March 28.7; and 107.2, 60.3, 33.9, 19.1, 10.7, 6.0, 3.4, 1.9, 1.1 for March 29.7. Typical measurement noises are 0.2 and 0.4 Jy for the first and second day, respectively. The size of each pixel is 4.″3 and is indicated in the upper right

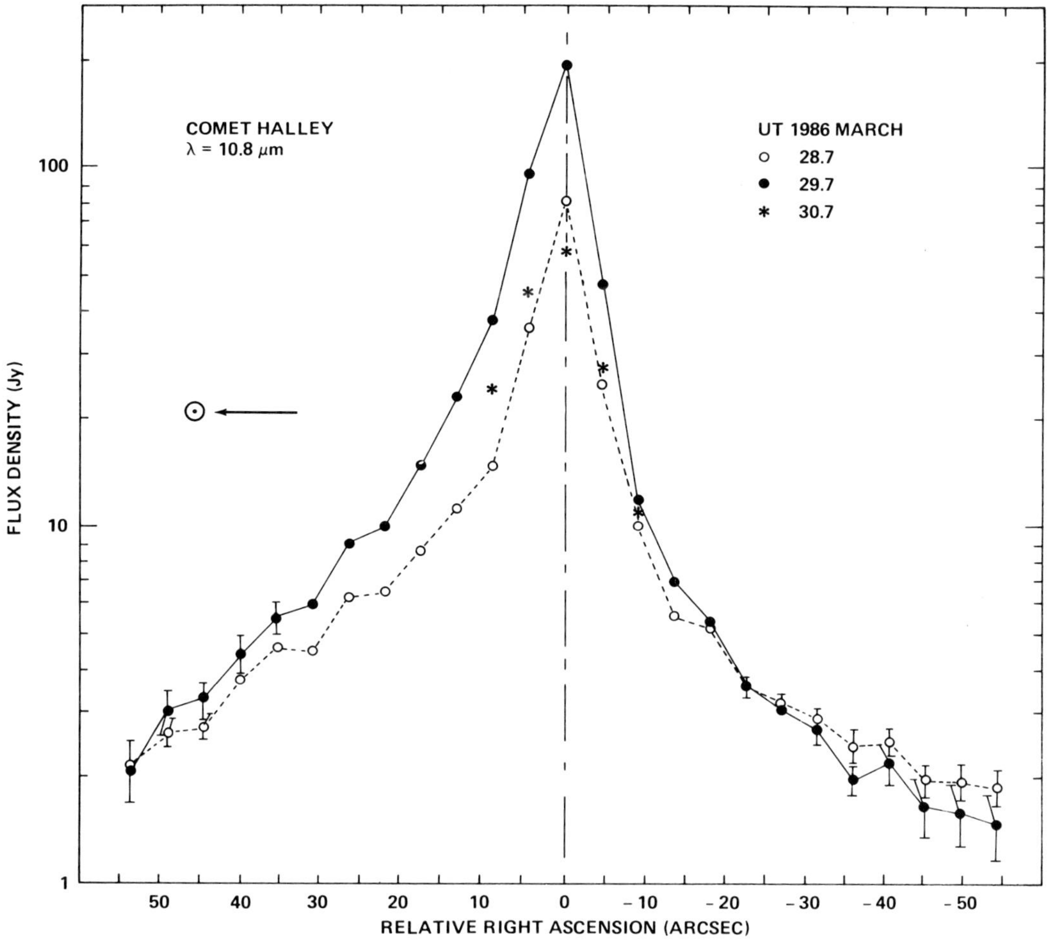

Fig. 2. The east-west flux density scans of comet Halley at 10.8 μm on three successive dates. Note the dramatic day-to-day brightness variations and the well defined asymmetry about the nuclear condensation

densation, with the other two fields displaced from the first one by 15″ along position angles 128° and 308°.

Figure 2 shows a plot of the 10.8 μm flux density along the east-west direction for each of the 3 days. This direction is nearly the same as the projected sun-tail direction indicated by the arrow in Figs. 1 and 2. The sun-tail direction was about 30° from the plane of the sky (i.e., the plane perpendicular to the line of sight. The heliocentric and geocentric distances and the observing geometry of comet Halley on these dates are almost identical to those of comet GZ when we observed it in 1985 (Telesco et al., 1986). The relevance of this similarity will be discussed in Sect. 3.

3. Results

The main results of these observations can be divided into three categories: morphology, time variability, and surface photometry. The morphology of the coma has two main features, the presence of a sunward fan and the absence of a well defined tail. The sunward fan was present on all three days, as is evident in Figs. 1 and 2. A comparison of the observed infrared brightness distributions with the models discussed by Wallace and Miller (1958) indicates that most of the dust ejection occurs on the sunward side of the nucleus, in agreement with the spacecraft observations of the nucleus itself. The absence of a well defined tail is notable since the observing geometry and field coverage of Comet Halley is almost identical to that of comet GZ when it was observed by us. The GZ map shows a well defined tail which has been explained (Telesco et al., 1986) by the presence of a significant population of large particles (radius $\gtrsim 100\,\mu$m). The absence of a similar tail in the observations of comet Halley under such similar conditions indicates that the dust coma of this comet is deficient in large particles compared with that of comet GZ. This conclusion is supported by the analysis of albedo maps of comets Halley and GZ discussed by Hammel et al. (1987). More detailed modelling of the P/Halley images is under way in order to constrain further the particle size distribution.

The temporal variability of the nuclear condensation is consistent with an extrapolation of the quasi-sinusoidal lightcurve observed in blue light by the International Ultraviolet Explorer (IUE) on 1986 March 23, 24 and 25 (Feldman et al., 1986). A period of 2.2 days was assumed by us when extrapolating the IUE data. However, more recently an additional period of 7.4 days has been proposed (Millis and Schleicher, 1987). The compatibility of our observations with this new period is being investigated. Further information about the behavior of the dust production rate during the periods preceding our observations is contained in the radial structure of the dust coma. A simple model[1], which assumes the temporal variability of the dust production rate to be equal to the IUE lightcurve and a mean dust expansion velocity with respect to the nucleus of about 0.1 km s^{-1} yields brightness profiles similar to those observed in Fig. 2. Although this result is not unique, the apparent agreement between the temporal behavior at the two wavelengths suggests an intimate relationship between the dust responsible for scattering the blue continuum and that emitting the infrared radiation.

The observations obtained at 12.8 and 19.2 μm have been combined with the 10.8 μm maps in order to study the spatial behavior of the shape of the thermal spectrum. The ratio of the broadband 10.8 μm and the narrowband 12.8 μm flux densities is used as an indicator of the strength of the silicate emission feature observed to peak near 10 μm (see, e.g., Bregman et al., 1987). This ratio indicates the silicate feature to be strongest at the nuclear condensation and decreasing in all directions, but more slowly towards the sun. This result of a silicate emission stronger in the sunward fan than elsewhere in the coma (except for the nuclear condensation) is in agreement with other observations (e.g., Campins et al., 1987) which indicate that the sunward fan has a higher abundance of small grains (radius $\lesssim 1$ μm) than the rest of the coma. The ratio of the 12.8 to 19.2 μm fluxes is related to the color temperature of the particles and indicates a maximum color temperature at the nuclear condensation. An estimate of the actual values of the color temperatures requires further modelling. This is because the 19.2 μm filter includes much of another emission feature which is also attributed to silicates and is possibly present in this comet (Herter, Campins, and Gull, 1987). The strength of this feature must be estimated and subtracted in order to calculate a true color temperature which refers to the continuum.

[1] This model considers exclusively the simple case of single-velocity and single-sized particles, and it is intended only to illustrate the overall morphological similarities between the IUE and the infrared observations.

Acknowledgements. We would like to thank D. Griep of the IRTF for his assistance with the observations, and C. Sisk and C. Benson at NASA-MSFC for assistance with the data reduction and image processing. HC acknowledges the support of the National Science Foundation Grant No. AST-8414737. The Planetary Science Institute is a Division of Science Applications International Corporation. This is PSI contribution No. 230.

References

Bregman J.D., Campins H., Witteborn F.C., Wooden D.H., Rank D.M., Allamandola L.J., Cohen M., Tielens A.G.G.M.: 1987, (this issue).

Campins H., Joy M., Harvey P.M., Lester D.F., Ellis H.B.: 1987, (this issue).

Campins H., Telesco, C.M., Decher R., Mozurkewich D., and Thronson, H.A. Jr.: 1986; *Geophys. Res. Letters* **13**, 295–298.

Feldman P.D., et al.: 1986: *New Insights in Astrophysics*, London, UK, 14–16 July 1986

Hammel H.B., Telesco C.M., Decher R., Campins H., Storrs A.D., Cruikshank D.P.: 1987 this issue.

Herter. T., Campins, H., Gull, G.E.: 1987, Airborne: 1987, (this issue).

Jewitt, D.C., Meech K.J.: 1987, *Astrophys. J.* (in press).

Millis R.L., Schleicher D.G.: 1987, *Nature* **324**, 646–649.

Telesco C.M. et al.: 1986, *Astrophys. J. Letters* **310**, L61

Tokunuga A.T.: 1984, *Astron. J.* **89**, 172–175.

Wallace L.V., Miller F.D.: 1958, *Astron. J.* **63**, 213–220.

Astron. Astrophys. 187, 605–608 (1987)

ASTRONOMY AND ASTROPHYSICS

Low resolution mapping of comet P/Halley in the near-infrared

C. Lázaro, F. Garzón, and M.J. Arévalo.

Instituto de Astrofisica de Canarias, E-38200 La Laguna (Tenerife), Spain

Received February 16, accepted April 7, 1987

Summary. We present low resolution maps, of a 4′ × 0′.5 region, of comet Halley, in photometric bands J and K, a central scan in the H band and $JHKL$ photometry taken with the 1.5 m Sánchez-Magro Telescope of the Instituto de Astrofisica de Canarias. We obtain central and integrated magnitudes and colours and discuss its variation with distance from the nucleus. The black-body colour temperature of the grains is found to be ~ 545 K, significantly greater than the black-body equilibrium temperature (228 K). The profile of the scans seem to follow an r^{-2} density law in the dust distribution, although the central core has a much steeper density law.

Key words: comets – dust – infrared radiation

1. Introduction

$JHKL$ photometry centered on the visual condensation of comet Halley was performed on the night of 1986 April 22. From these data can be obtained an estimation of the colour temperature of the emitting dust and a comparison with previous estimations made by other authors at different heliocentric distances of Halley.

Low resolution near-infrared mapping in photometric bands J and K of the inner coma of comet Halley was made in the night of 1986 April 23. This was combined with scans in photometric band H, made on the same night, with the aim of studying possible colour changes with distance from the nucleus, which would be indicative of spatial variation in the dust to gas ratio within the coma. These observations were made less than two weeks after the comet's closest approach to the Earth, with its geocentric and heliocentric distances 0.61 AU and 1.53 AU, respectively. The low geocentric distance permits us a comparatively high spatial resolution, with 1″ in the sky corresponding to a linear distance of 440 km at the comet; hence our measurements have an effective resolution at the comet of about 7000 km. The integrated visual magnitude at the epoch of observation was 4.5.

Whilst a great deal of effort has gone into photometry of the nuclear condensation of comets in general and of comet Halley in particular, with a highly active infrared observing net providing extensive photometry in J, H, K, L and M (e.g. Tokunaga et al., 1986; Bouchet et al., 1986), there is hardly any published information on the two-dimensional distribution of infrared emission. The current authors are aware of only four studies on this problem. Ney (1982) presented two-dimensional photometry of comet Kohoutek (1973f), which revealed the extent of its (then invisible) anti-tail. Walker et al. (1984) presented isophotal mapping of comet IRAS-Iraki-Alcock (1983d) by the Infrared Astronomical Satellite. Their results revealed the presence of a hitherto unknown and extensive dust tail and a gradual change in both the size and the temperature of the dust grains within it. Most recently, Philips et al. (1986) made scans of comet Halley in J, H, K and L, although only in one position, finding the emission from the nuclear condensation to be essentially from a barely resolved central core, with the extended coma most strongly present in J. Finally, Garzon et al. (1986) presented low resolution maps in J and K, finding, in contrast, the extended coma to be significantly larger in K and encountering a possible change in the J-K colour of the outer coma compared to that of the nuclear condensation.

The importance of two-dimensional photometry lies in the spatial information that may be obtained on the extension of dust emission. Despite in situ measurements by the Giotto, Vega and Planet-A missions (see, for example Viasberg et al., 1986; Matzets et al., 1986; McDonnell et al., 1986), a powerful method of studying dust distribution and production rates is by near-infrared photometry. It is a well established fact that, whilst the visible emission of comets is almost totally due to solar-reflected light, or to gas fluorescence, dust emission begins to "cut-on" slightly shortward of 2 μm and in bands from K longward, the total emission is almost entirely dust emission (see, for example, Ney, 1982). The results from the Vegas and Giotto suggested that the comet's dust is highly concentrated near the nucleus of the comet. Whilst it obeys a r^{-2} distribution up to 5000 km from the nucleus, the distribution is a very much steeper one closer in (McDonnell et al., 1986; Levasseur-Regourd et al., 1986). The diameter of this inner distribution of dust, which dominates the dust emission of the comet, corresponds to roughly 6″ at 1 AU and hence, for infrared observations with our aperture diameter of 20″, is effectively a point source.

2. Observational procedure

Observations were taken with the 1.5m Sánchez-Magro at the Observatorio del Teide, using the multi-filter, single detector photometer at the $f/13.5$ Cassegrain focus, in two-beam chopping mode. For the scans a circular aperture of nominal diameter

Send offprint requests to: C. Lázaro

Table 1. Scan positions refer to their position relative to the nucleus, thus 10″ S implies that the aperture was placed 10″ to the south of the centre of the visible nuclear condensation

Night (1986)	Scan position	Band	Number of averaged scans
April 23/24	Central	*K*	4
23/24	5″ S	*K*	2
23/24	10″ S	*K*	1
23/24	15″ S	*K*	1
23/24	5″ N	*K*	2
23/24	Central	*H*	2
23/24	Central	*J*	7
23/24	5″ S	*J*	1
23/24	10″ S	*J*	1
23/24	15″ S	*J*	2
23/24	20″ S	*J*	1
23/24	5″ N	*J*	2
23/24	10″ N	*J*	1
23/24	15″ N	*J*	1

20″; a tracking rate of $2'' s^{-1}$; a scan length of 4′; a chopper throw of 50″ and a rate of data-taking of 3 measures per second, were used throughout. In all cases, scans were taken in right ascension. In each position, scans were averaged until a satisfactory signal to noise was obtained. We note here, that the very slow rate of data collection, allowed a high signal to noise to be obtained in averaging even very few scans. Table 1 shows the observing log for the observations presented here. The photometry was taken using the same equipment, also with a circular aperture of 20″ of nominal diameter and a chopper throw of 50″.

3. Data reduction

The data from the scans were reduced using a suite of standard programs developed for the reduction of surface photometry of galaxies in the infrared. These programs align the separate scans in each position by finding the peak emission in each one and then sliding the data, to place the peak in the same relative position in all scans. Due to the slow rate of data collection and subsequent high signal to noise of the individual scans, this technique worked well even in off-nucleus positions. The average positional error in placing the nuclear condensation in the start position for each scan is estimated to be approximately 3″–4″.

The data were then calibrated photometrically using measures of the standard star BS4030 taken in each filter, both before and after observation of the comet, and a standard model of the photometric extinction at the observatory. Both the photometric magnitudes of the standard star and the absolute flux calibration were taken from Koornneef (1983a, 1983b). The comparison star was close enough to the comet to have an almost identical airmass during the observations. A splines routine was used to interpolate and plot contours through the reduced data.

Two standard stars were used for comparison in the photometry, BS3982 and BS4030, both with air masses close to that of the comet. Also, the magnitudes for these stars and the absolute flux calibration were taken from Koornneef (1983a, 1983b).

4. Discussion

As can be seen from Fig. 1, the shape of the scans is remarkably wider than the point source response shape. This feature is even more evident when comparing the width of the scans at one tenth of the peak. In Table 2 we summarise the width of the central scans for the comet and the standard star at half and one tenth of the maximum. Nevertheless, due to the low signal to noise ratio at one tenth of the peak, the differences between the values quoted for *K* and both *J* and *H* bands must be taken with caution.

It must be noticed that, given the amplitude of the chopper of 50″, the measured signal is probably contaminated by the emission of the outer coma. From Eq. (3) in Tokunaga et al. (1986), this contamination can be estimated to be around 10% of the central maximum. As the reference beam was located at the south of the nucleus, the contaminating effect will increase when moving to the north from the nucleus. As the central scans were formed averaging the two beams, the contamination for these scans will be different to that in the off-nucleus scans, which were formed averaging only one beam.

In Table 3 the central and total integrated magnitude in *J* filter and colours are shown, together with the photometric results. Central magnitude and colours are derived with the nucleus of the comet centered on the aperture of 20″ of nominal diameter. Integrated values are obtained from the central scans, circular symmetry assumed, on a pseudo-aperture with diameter 160″.

Our central (*J-H*) colour is redder than the solar one, but is within the range of the values quoted by different authors. Also, as can be seen from Table 3, the integrated colours are slightly redder than the central ones. This seems to indicate an excess of emission in the longer wavelengths of the outer coma with respect to the central region. Apart from the central core, no appreciable changes in colour were found with the distance from the nucleus. This fact was already noticed by others authors in different epochs (Phillips et al., 1986; Garzón et al., 1986), and seems to be due to a broad invariancy of the general properties of the dust distribution.

One of the most interesting results which has arisen from the space missions devoted to the study of comet Halley is the richer knowledge we have, nowadays, of the spatial distribution of dust around the nucleus. Both the TKS experiment aboard the Vega spacecraft and OPE optical photopolarimeter aboard Giotto, show a rapid increase in dust density towards the nucleus. At the time of the encounters with the comet, a dust distribution following a r^{-2} law with the distance from the nucleus is derived for distances ranging from $5\,10^3$ to 10^5 km, but becomes more peaked closer to the nucleus.

We have compared the radial trend of the central scans fluxes in *JHK* with those expected from a dust distribution $\sim r^{-2}$, assuming, for simplicity, spherical symmetry. We find a good agreement for the outer regions, roughly 20″ in radius apart from the central core, but the central fluxes are considerably higher than the corresponding values for that distribution. It may be attributed to the existence of a central core of dust, with a steeper increase in density, or to a considerable difference in grains properties with respect to the outer surroundings, in accordance to the results of the spatial probes (Encrenaz, 1986).

The colours (*J-H*) from photometry and scans are in very good agreement. This is not the case for the colours (*J-K*), which

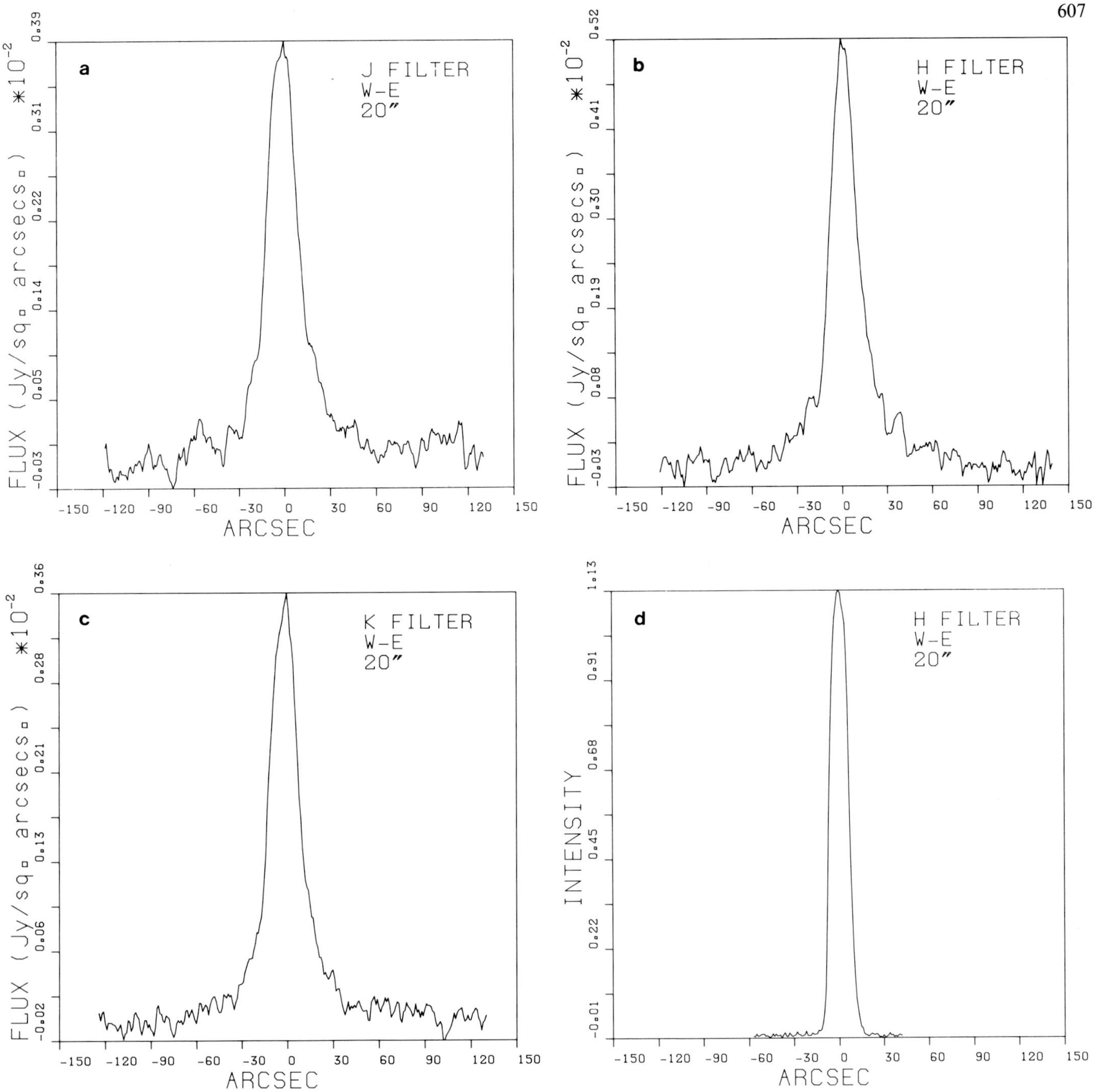

Fig. 1a–d. Central scans of comet Halley in each filter (7a–c). In figure d is shown the point source response in the H filter and in arbitrary units. The response of the standard star in J and K is of the same slope as in Fig. 1d

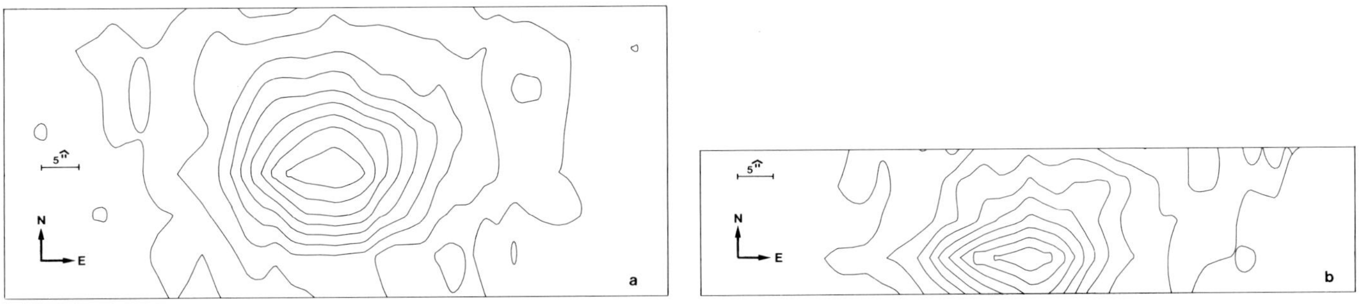

Fig. 2a and b. Maps in J and K filters. There are 9 levels, the maximum being $3.87\,10^{-3}$ in J and $3.53\,10^{-3}$ in K, both in Jy arcsec^{-2}. The interval between the contours is 0.1 Jr arcsec^{-2} in all cases

Table 2. Width of the central scans in each filter and of the standard star at half and one tenth of the maximum

	FWHM	1/10 peak
J	24″	90″
H	22.5″	90″
K	22.5″	72″
Standard	16″	24″

Table 3. Central and total integrated magnitudes and colours from the scans with photometry in each filter

	J	(*J-H*)	(*J-K*)	(*J-L*)
Central	8.08	0.47	0.79	
Integrated	6.39	0.59	0.85	
Photometry	7.97	0.43	0.55	1.84

show an appreciable change from one day to another, in the sense that the K magnitude in the scans is greater with respect to that of the photometry. One could hardly explain this difference as a real change in the emission from the comet, although part of this discrepancy may be due to a changes in the atmospheric transmission or errors in centering the visual condensation of the comet in the aperture. Despite these uncertainties, we think that the photometric data are yet valuable for a rough estimation of the dust temperature.

In calculating the dust temperature, we have assumed that the H fluxes are strongly dominated by scattering of the solar radiance. Then we have subtracted from the measured fluxes in K and L, the fraction corresponding to the scattering of solar radiation, taking into account the solar colours (H-K) and (H-L) from Koornneef (1983b). With these assumptions, we derive a black-body colour temperature of ~ 545 K, which is well in excess with respect to that for grains in thermal equilibrium at this heliocentric distance, $T \sim 228$ K. A similar excess has been previously reported (Phillips et al., 1986). For comparison, we note that assuming an emissivity $\varepsilon \sim \lambda^{-m}$ this colour temperature should decrease to ~ 445 K for $m = 2$. To derive the observed colours with dust at the equilibrium temperature would imply an unreasonable value for m. The high colour temperature observed is perhaps indicating the existence of another population of dust with small size, as has been suggested by Ney (1974) from IR observations of comet Kohoutek.

However, if the reflectivity of the grains increases with wavelength, as previously suggested by A'Hearn et al. (1981), the flux excess in the K and L bands, which we have attributed to the thermal emission of the grains, will decrease, and hence, so will the derived temperature for the dust. In that case, only in the L band will remain an appreciable excess of flux, which could be attributed to thermal emission of the grains.

Acknowledgements. The authors are much indebted to Mark Kidger for his valuable comments and suggestions in preparing this work.

The 1.5m Sánchez-Magro Telescope, on the island of Tenerife, is operated by the Instituto de Astrofisica de Canarias at its Spanish Observatorio del Teide.

References

A'Hearn, M.F., Dwek, E., Tokunaga, A.T.: 1981, *Astrophys. J. Letters* **248**, L147

Bouchet, P., Chalabbev, A., Danks, A., Encrenaz, T., Epchtein, N., LeBertre, T.: 1986, *Astron. Astrophys.* (in press)

Encrenaz, T.: 1986, (preprint.)

Hanner, M.S., Tokunaga, A.T., Veeder, G.J., A'Hearn, M.F.: 1984, *Astron. J.* **89**, 162

Jewitt, D.: 1984, *Icarus* **60**, 373

Kaneda, E., Hirao, K., Takagi, M., Ashihara, O., Itoh, T., Shimizu, M.: 1986, *Nature* **320**, 140

Koornneef, J.: 1983a, *Astron. Astrophys. Suppl.* **51**, 489

Koornneef, J.: 1983b *Astron. Astrophys.* **128**, 84

Garzón, F., Kidger, M.R., Prieto M., Cepa, J.: 1986, *Astron Astrophys.* (in press)

Levasseur-Regourd, A.C., Bertaux, J.L., Dumont, R., Festou, M., Geise, R.H., Giovane, F., Lamy, P., Le Blanc, J.M., Llebaria, A., Weinberg, J.L.: 1986, *Nature* **321**, 341

Matzets, E.P., Aptekar, R.L., Golenetskii, S.V., Guryan, Yu.A., Dyachkov, A.V., Ilyinskii, V.N., Panov, V.N., Petrov, G.G., Savvin, A.V., Sagdeev, R.Z., Sokolov, I.A., Khavenson, N.G., Shapiro, V.D. Shevchenko, V.I.: 1986, *Nature* **321**, 276

McDonnell, J.A.M., Alexander, W.M., Burton, W.M., Bussoletti, E., Clark, D.H., Grard, R.J.L., Grün, E., Hanner, M.S., Hughes, D.W., Igenbergs, E.B., Kuczera, H., Lindblad, B.A., Mandeville, J-C., Minafra, A., Schwehm, G.H., Sekanina, Z., Wallis, M.K., Zarnecki, J.C., Chakaveh, S.V., Evans, G.C., Evans, S.T., Firth, J.G., Littler, A.N., Massonne, L., Olearczyk, R.E., Pankiewicz, G.S., Stevenson, T.J.. Turner, R.F.: 1986, *Nature* **312**, 338

Ney, E.P.: 1974, *Astrophys. J. Letters* **189**, L141.

Ney, E.P.: 1982, In *Comets*, ed. L.L. Wilkening, University of Arizona Press, p. 323

Philips, P., Mampaso, A., Garzón, F.: 1986, *Astron. Astrophys. Letters* **161**, L17

Prieto, M., Kidger, M.R., Beckman, J.E., Rosa, F.: 1986, *Astron. Astrophys Letters* **163**, L1

Tokunaga, A.T., Golisch, W.F., Griep, D.M., Kaminsky, C.D., Hanner, M.S.: 1986, *Astron. J.* **92**, 1183

Vaisberg, O.L., Smirnov, V.N., Gorn, L.S., Iovlev, M.V., Balikchin, M.A., Klimov, S.I., Savin, S.P., Shapiro, V.D., Shevchenko, V.I.: 1986, *Nature* **321**, 274

Walker, R.G., Aumann, H.H., Davis, J., Green, S., DeJong, T., Houck, J.R., Soifer, B.T.: 1984, *Astrophys. J. Letters* **278**, L11

Astron. Astrophys. 187, 609–615 (1987)

ASTRONOMY AND ASTROPHYSICS

Infrared monitoring of comet P/Halley *

D. Lorenzetti [1], **A. Moneti** [2,3], **R. Stanga** [4,3], **and F. Strafella** [5]

[1] Istituto di Fisica dello Spazio Interplanetario – CNR, C. P. 27, I-00044 Frascati, Italy
[2] Osservatorio Astrofisico Arcetri, Largo Enrico Fermi 5, I-50125 Firenze, Italy
[3] European Southern Observatory, Karl-Schwarzschild-Strasse 2, D-8046 Garching bei München, Federal Republic of Germany
[4] Istituto di Astronomia, Universitá di Firenze, Largo Enrico Fermi 5, I-50125 Firenze, Italy
[5] Istituto di Fisica, Universitá di Lecce, I-73100 Lecce, Italy

Received February 13, accepted June 22, 1987

Summary. We present the results of an infrared monitoring program of comet P/Halley. Multiaperture photometry between 1 and 5 μm was obtained during 12 nights in the preperihelion phase from the Gornergrat observatory, and during 4 nights in the postperihelion phase from the ESO 1-m telescope at La Silla. In addition, 10 and 20 μm photometry was obtained on November 4 and 7, when the comet was at 1.88 AU from the Sun. An analysis of the *JHK* data indicates that Halley was typically somewhat redder than the Sun ($\langle J-H\rangle = 0.48$, $\langle H-K\rangle = 0.18$) and that its colors did not change throughout our monitoring period, implying that the properties of the dust in the coma remained constant. Also, with the exception of November 4 and March 21, when Halley was presumably undergoing an outburst, the multiaperture data are consistent with an r^{-2} density distribution of scattering dust in the coma. Furthermore, the flux from Halley in its quiescent phase increased as $R^{-4.5}$, where R is the heliocentric distance. The 10 μm silicate feature in emission was detected at about the 2σ level on 4 and 7 November, and from the spectral energy distribution at that time we deduced an albedo of 0.05 for the scattering dust in the coma.

Key words: Halley's comet – infrared continuum – dust: silicates

1. Introduction

The return of comet Halley has stimulated a large amount of activity among astronomers, both in terms of space programs and of ground based observations. A program of coordinated UV, optical, infrared and radio (OH) observations has been undertaken at the Arcetri Observatory in collaboration with other research institutions. These observations were in turn coordinated with the International Halley Watch (IHW). We present here the results of an infrared monitoring program that was carried out within this framework. These results form a uniform photometric record of comet Halley during the period when it was visible from the Gornergrat Observatory which, due to the high latitude of that observatory (+46°) lasted from October to December 1985. Most of the observations were obtained in the near infrared and with aperture sizes ranging from 10″ to 30″. Large apertures are important for the study of the density structure of the coma, and they are not usually available on larger telescopes. In addition to the photometry we attempted to obtain spectrophotometry in the 3 μm region when Halley was at ~2 AU in order to look for the 3.1 μm ice feature; unfortunately, however, the comet was too faint at that time for detection in that region of the spectrum with the TIRGO telescope. Our preperihelion observations also include some 10–20 μm observations that were obtained when the comet was at $R = 1.88$ AU.

We will also present some 1–5 μm observations obtained with the European Southern Observatory (ESO) 1-m telescope at La Silla after perihelion when the comet was at 1 AU, and we will compare them to the preperihelion observations.

Several investigators have observed the infrared spectral energy distribution of comets (see Ney, 1982, and references therein), and they generally concluded that the infrared emission is due mostly to scattering and thermal emission from dust in the coma. More specifically, the infrared spectral energy distribution of a comet is determined by several factors which can be subdivided in two groups: the *extrinsic* factors, such as heliocentric distance, R, and the Sun-comet-earth angle, β, which depend on the position of the comet in its orbit, and the *intrinsic* factors, which are the physical and chemical properties of the dust as well as the amount of dust present. Some of the intrinsic factors may depend on extrinsic factors: for instance, the amount of dust present in the coma depends on the dust production rate, which in turn is determined by the amount of sunlight incident on the comet, and hence on the heliocentric distance. It follows that an understanding of the nature of dust in a comet requires that both the temporal and the spectral behaviour of the infrared emission be studied.

2. Observations

2.1. Preperihelion observations

All the preperihelion observations were carried out with the Gornergrat Infrared Observatory (TIRGO) 1.5-m telescope between 14 October and 12 December 1985. Most of the observing

Send offprint requests to: A. Moneti, ESO

* Based on data collected at the TIRGO Observatory, and at the European Southern Observatory, La Silla, Chile

Table 1. Journal of the observations

Date	Time (UT)	Δ (AU)	R (AU)	β (°)	km in 1″	Filters	Qual[a]
1985 Oct 14	04:00	1.62	2.17	25.7	1179	*JHK*	c
Oct 15	01:50	1.59	2.16	25.6	1157	*JHK*	b
Oct 16	03:45	1.56	2.14	25.5	1135	*JHK*	a
Oct 17	02:10	1.53	2.13	25.5	1113	*JHK*	a
Oct 18	02:10	1.50	2.11	25.3	1091	*JHK*	b
Oct 20	02:30	1.43	2.09	25.0	1040	*JHKL*	b
Nov 04	01:00	0.99	1.88	18.7	720	*JHKL*	a
	05:00					*N-Q*	a
Nov 07	02:30	0.91	1.84	16.1	662	*N-Q*	b
Nov 14	23:45	0.76	1.73	7.3	553	*JHKL*	a
Nov 16	00:30	0.74	1.72	5.7	538	*JHKLM*	b
Nov 16	23:40	0.72	1.71	4.0	524	*JHKLM*	c
Dec 01	20:00	0.63	1.48	29.7	458	*JHKL*	a
Dec 11	19:00	0.75	1.33	46.9	446	*JHKLM*	c
1986 Mar 19	08:50	0.98	0.83	66.1	715	*JHKLM*	c
Mar 20	08:30	1.00	0.80	66.2	726	*JHKLM*	a
Mar 21	09:00	1.01	0.79	66.1	738	*JHKLM*	a
Mar 22	08:20	1.03	0.75	66.0	749	*JHK*	b

[a] Night quality code: see text for details

nights were during three periods selected by the IHW. During this time P/Halley moved from 2.17 to 1.33 AU. In Table 1 we present a journal of the observations: for each night we give the heliocentric and geocentric distances, R and Δ, the Sun-comet-earth angle, β, the projected distance corresponding to 1″, the filters used, and a code that denotes the quality of the night: *a* indicates a photometric night, *b* indicates a clear night but with variable exctinction, and *c* indicates that there were some clouds, but the sky around the comet was clear.

The 1–5 μm measurements were made with the Arcetri InSb photometer operating at solid nitrogen temperature using the standard TIRGO *JHKLM* filter set (see Table 2 for filter specifications). Standard infrared techniques of chopping and beam switching were used for data acquisition. Chop throws from about one to several arcmin were used depending on the size of the coma; Tokunaga et al. (1986) have shown that the contribution from the coma to the reference beam for a 1′ beam separation is 3%, and that it decreases linearly with beam separation. East-west chopping was used at all times. The telescope was set to track the comet, and centering was checked every 3–5 min. Standard stars from the TIRGO standard star list (Hunt, 1986) were used for photometric calibration. The TIRGO photometric system is based on the the faint infrared standard stars of Elias et al. (1982), plus other stars that were calibrated with respect to Elias's standards. The photometric system is thus ultimately referred to α Lyrae which is assumed to have 0 mag in all bandpasses. The uncertainties in the zero points are <0.01 mag.

The comet and the photometric standards were always observed at less than 1.5 air masses, and always within 0.2 air masses of each other. An attempt at determining the air mass correction showed that it was smaller than the uncertainties, and thus no air mass correction was applied to the data. The data are presented in Table 3. When more than one measurement was obtained for a given filter-aperture combination during the same night (typically within 2–3 h), the data were combined and their average is presented in the table. From the spread of these data we also determined the uncertainties in the measurements. The typical uncertainties are 0.05 mag in the *JHK* filters for apertures >10″, 0.10 mag for *JHK* in a 10″ aperture, and 0.2 mag in the *L* and *M* filters at all apertures. The data taken with the 10″ aperture are subject to larger uncertainties arising mostly from inaccuracies in the motion of the chopping secondary and in the nodding of the telescope.

From the way in which the uncertainties were determined, it follows that they are observational uncertainties; possible systematic errors in the zero point and in the magnitudes of the

Table 2. TIRGO and ESO filter sets

Filter	TIRGO		ESO	
	λ[a]	$\Delta\lambda$[a]	λ[a]	$\Delta\lambda$[a]
J	1.24	0.28	1.24	0.32
H	1.65	0.30	1.63	0.28
K	2.22	0.36	2.19	0.39
L	3.87	0.64	3.79	0.68
M	4.74	0.62	4.64	0.63
N1	8.65	1.30		
N2	9.97	1.60		
N	10.99	5.80		
N3	11.55	2.10		
Q1	19.5	11.0		
Q	22.75	8.5		

[a] In μm

Table 3. Pre-perihelion data

Date (1985)	Time (UT)	Apt (arcsec)	J	H	K	L	M
			(magnitude)				
Oct 14	4:00	10	12.61	12.04	11.92		
		17	11.90	11.51	11.42		
		27	11.81	11.23	11.01		
Oct 15	1:50	17	12.11	11.44	11.23		
		27	11.57	10.94	10.76		
Oct 16	3:45	10	12.01	11.51	11.29	9.72	
		17	11.46	11.06	10.84		
Oct 17	2:10	10	12.07	11.57	11.35		
		17	11.50	11.04	10.82		
		27	11.10	10.55	10.40		
Oct 18	2:10	10	11.87	11.36	11.64		
		17	11.42	10.97	10.79		
		24	11.16	10.64	10.42		
		27	11.16	10.55	10.39		
Oct 20	2:30	10	11.76	11.24	11.24	9.86	
		17	11.29	10.84	10.59		
Nov 4	1:00	10	12.15	12.09	11.95	10.24	
		17	10.89	10.44	10.43	9.06	
		27	10.12	9.59	9.37	8.82	
Nov 14	23:45	10	10.28	10.34	10.26	8.83	
		17	9.52	9.12	8.94	8.12	
		27	9.13	8.57	8.44		
Nov 16	0:30	17	9.12	8.79	8.48	8.00	5.40
		24	8.91	8.32	8.16	7.18	
Nov 16	23:40	17	8.97	8.55	8.44	7.67	8.17
		24	8.68	8.14	8.04	7.21	
Dec 1	10:00	10	9.94	9.45	9.31	8.19	
		17	9.31	8.84	8.69		
		27	8.87	8.27	8.14		
Dec 11	19:00	14	8.80	8.44	8.25	6.23	4.72
		20	8.63	8.15	7.98	6.05	

standards stars are much smaller than the observational ones and hence they have not been considered explicitly.

Some of the data may also be affected by contamination from field stars falling in either the source or the reference beam; this problem could affect the data of the October observing period, when the comet was relatively faint and when it was crossing the galactic plane. Naturally, no observations were obtained when a visible star appeared on the TV monitor to fall within either beam, but we estimate that the monitor reached to 15 mag (visual), and the presence of stars bright enough in the infrared to contaminate the observations cannot be excluded. During the other observing periods the comet was so bright that any contribution from field stars is expected to be negligible.

During the nights of 4 and 7 November we also obtained 8–20 μm photometry with the IFSI-CNR (Frascati) Ge bolometer (Saraceno et al., 1982). That instrument uses a fixed 12″ aperture and it contains wideband N and Q filters and narrow band filters selected to investigate the silicate feature (see Table 2). Chop throws of 20″ and 40″ were used on November 4 and 7, respectively, and α Tau was used as a photometric standard. The data are presented in Table 4. A single observation for each filter was obtained on November 4, and the uncertainties presented for those data are derived from the counting statistics. On November 7 three separate observations were obtained for each filter (only two for Q), and the uncertainties presented reflect the spread in the individual measurements.

The observed magnitudes should be corrected to account for the differences in the spectral energy distribution of the source compared to the standards as described in Hanner et al. (1984). For our M and Q filters these corrections amount to $\lesssim 0.05$ mag which is much less than our uncertainties, and hence they have not been applied there. A larger correction could be necessary for the L filter data; no correction was applied, however, since significant scattered sunlight is present in the L filter in addition to thermal radiation from the comet. No correction is necessary for narrowband filters.

2.2. Postperihelion observations

The postperihelion observations were obtained at ESO with the 1-m telescope between 1986 March 19 and 22 with the ESO InSb photometer, using the standard ESO *JHKLM* filter set (see Table 2). Three apertures between 10″ and 30″ were used, and the chop throw was 60″ EW. The telescope was set to track the comet and

Table 4. Mid-infrared results

Date (1985)	Time (UT)	$N1$	$N2$	$N3$	$Q1$	Q
		(magnitude[a])				
Nov 4	5:00	2.5 (0.1)	1.7 (0.1)	1.3 (0.5)	−0.6 (0.1)	−0.8 (0.2)
Nov 7	2:30	2.6 (0.1)	1.6 (0.1)	1.4 (0.3)	−0.6 (0.1)	−1.0 (0.1)

[a] The numbers in parenthesis are the uncertainties

Table 5. Post-perihelion near-IR observations

Date (1986)	Time (UT)	Apt (arcsec)	J	H	K	L	M
			(magnitude)				
Mar 19	9:33	15.0	7.25	6.78	6.57	4.13	2.42
		22.5	6.72	6.35	6.14	3.68	1.97
		30.0	6.47	6.07	5.85	3.40	1.71
Mar 20	8:23	15.0	7.31	6.84	6.64	4.28	2.50
		22.5	6.82	6.39	6.19	3.84	2.13
		30.0	6.51	6.08	5.88	3.50	1.78
Mar 21	8:54	15.0	6.26	5.84	5.61	3.15	1.61
		22.5	5.86	5.43	5.23	2.71	1.18
		30.0	5.62	5.19	4.98	2.48	0.95
Mar 22	8:17	15.0	7.36	6.90	6.70		
		22.5	6.73	6.24	6.06		
		30.0	6.31	5.85	5.66		

the pointing was checked every 5 minutes. The observations were always at an air mass <1.5. Photometric standard stars were chosen from the ESO standard star list (Koornneef 1983). The results are collected in Table 5. When compared to the preperihelion results obtained at similar heliocentric distance, our limited data show that Halley was roughly two magnitudes brighter after perihelion.

3. Discussion

3.1. Near-IR multiaperture data: coma density distribution

In Fig. 1 we present the multiaperture J fluxes as a function of the physical dimension of the observed region. Similar results are found at H and K. If $F_\lambda(d)$ is the flux density at wavelength λ within an aperture that corresponds to a diameter d at the distance of Halley, and assuming that $F_\lambda(d) \propto d^\alpha$, we find for the preperihelion data that $\alpha_J = 0.8 \pm 0.2$, $\alpha_H = 0.9 \pm 0.3$, and $\alpha_K = 1.0 \pm 0.3$, where the errors are the standard deviations. We have excluded from the average the data of November 4 for which $\alpha_J = 1.9$, $\alpha_H = 2.3$, and $\alpha_K = 2.4$ which are significantly different from the mean values.

The above result is consistent with $\alpha = 1$ independent of wavelength, which is indicative of a dust space density distribution of the form $\varrho(r) \propto r^{-2}$ in the coma, where r is the distance from the nucleus. Such a dust distribution would arise if the radial velocity of the escaping dust were constant and if the mass loss rate and the dust size distribution did not vary significantly over the time-scale needed to cross the coma (about 1 day). Furthermore, with the exceptions noted above, α did not vary over our monitoring period, which indicates that the same density distribution occurred for $2000 < r < 15000$ km.

For the postperihelion observations we find $\alpha \approx 1.0$ on March 19 and 20, but on March 21 $\alpha \approx 0.85$, and on March 22 $\alpha \approx 1.40$, and again α was independent of wavelength. The variation in α may be related to the luminosity increase observed on March 21: on that date an increase in flux by a factor slightly larger than 2 was observed in all wavebands; the flux returned to normal on the following day (see Table 5). The luminosity of the coma is related to the number of grains there, and Fulle et al. (1987) found that luminosity variations resulted primarily from variations of the mean grain size and not from variations in the mass loss rate. In other words, sudden luminosity increases occurred when the same amount of mass was ejected in the form of many small grains rather than fewer large grains. Accordingly, we will henceforth use the term *outburst* to mean an increase in the *number* of dust grains emitted per unit time, $\dot{N}$, rather an increase in the mass loss rate.

Just after an outburst there would be an anomalously large number of dust grains in the inner regions of the coma; this would be detected as increased flux in the smaller apertures and would result in a lower α. This dust would then move outward and eventually result in an enhancement in the flux detected in the larger apertures compared to the smaller ones, and result in α becoming higher than its nominal value of 1. If this were the case, however, the flux observed on March 22 in the largest aperture

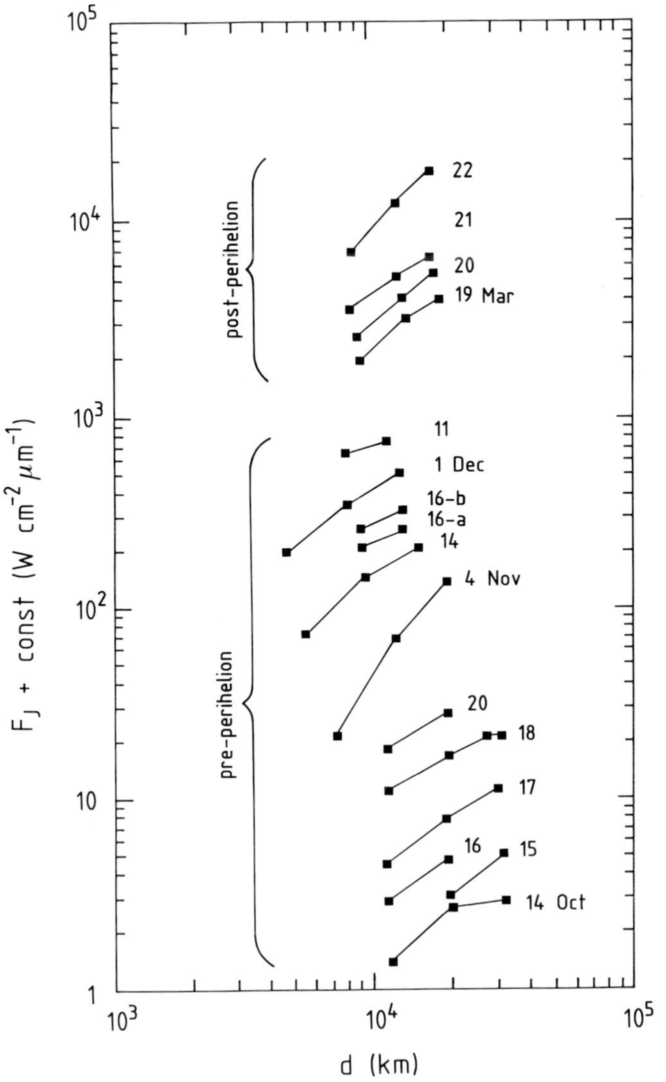

Fig. 1. Observed J fluxes plotted against the physical size of the comet corresponding to the observing aperture. Each line is identified by the observing date, and is shifted by an arbitray amont to avoid superpositions

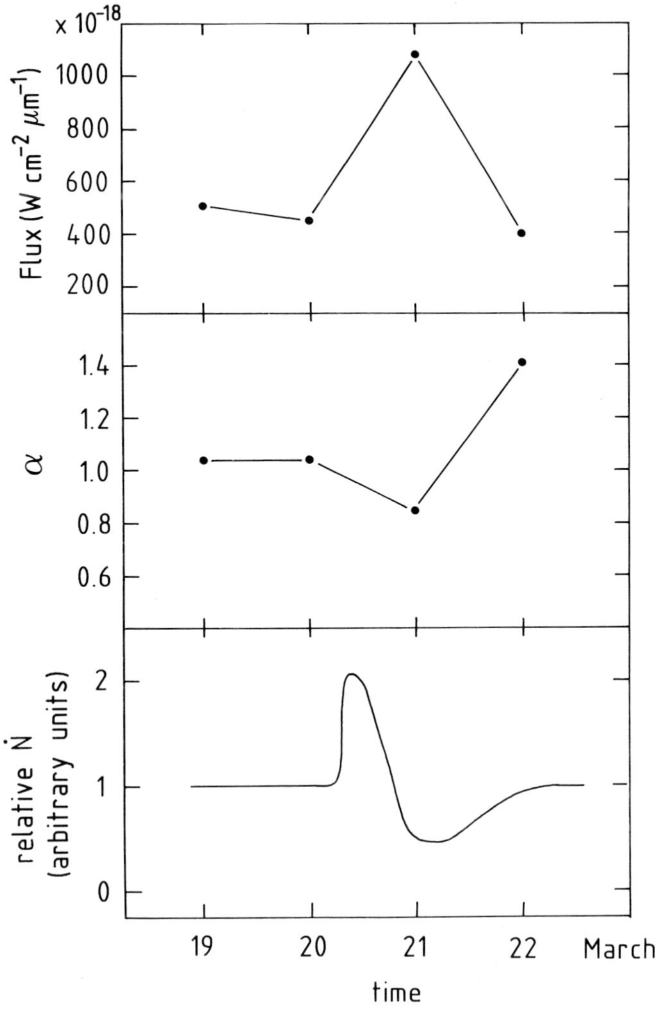

Fig. 2. Halley's behaviour during the period of 1986 March 19–21. Top panel: flux density at J showing the outburst on 21 March; an increase by a factor of about 2 was noted also in the H, K, and L filters. Middle panel: α. Bottom panel: our interpretation of how the rate of dust grain production, $\dot{N}$, varied during the outburst

should be similar to that observed on March 21, since the dust released during the outburst would still be within that aperture. The fact that the flux decreased considerably suggests that $\dot{N}$ decreased temporarily to below the "quiescent" level before returning to normal. This interpretation is shown graphically in Fig. 2, where the flux and α are plotted vs. time. The third panel of Fig. 2 shows how $\dot{N}$ may have varied during that period.

From this interpretation we can derive an estimate of the dust outflow velocity: assuming that the dust from the outburst moved outward by $15'' = 11000$ km (from our smallest to our largest aperture) between March 21 and 22, we derive $v_{\text{dust}} = 0.13\ \text{km s}^{-1}$, which we expect to be correct to within a factor of 2. This result is in excellent agreement with the results of Fulle et al. (1987) who determined the dust ejection velocity using different techniques. An outburst may also have occurred just prior to November 4 and have caused the unusually high α at that time, but our temporal coverage is inadequate to discuss in detail the observations of that date.

3.2. Brightness variation with heliocentric distance

The observed flux density, F_{obs}, of a comet as it travels through the solar system can be expressed as

$$F_{\text{obs}} = F_0 \left(\frac{R}{R_0}\right)^{-n} \left(\frac{\Delta}{\Delta_0}\right)^{-2} \Phi(\beta) \left(\frac{d}{d_0}\right)^{\alpha}, \tag{1}$$

where F_0 is the absolute flux density, $\Phi(\beta)$ is the phase function, d and α are the same as in Sect. 3.1, and R_0, Δ_0, and d_0 are normalization parameters. In the following we will adopt $R_0 = \Delta_0 = 1$ AU. At wavelengths shortward of 2.5 µm, comets are seen mostly by sunlight scattered by dust in the coma, and a value of n different from 2 reflects the change in the amount of scattering dust in the coma. In the following discussion, we will assume that the phase function is constant over the values of β for which we obtained observations (see Table 1); this is justified by the results of Ney (1982, see his Fig. 5).

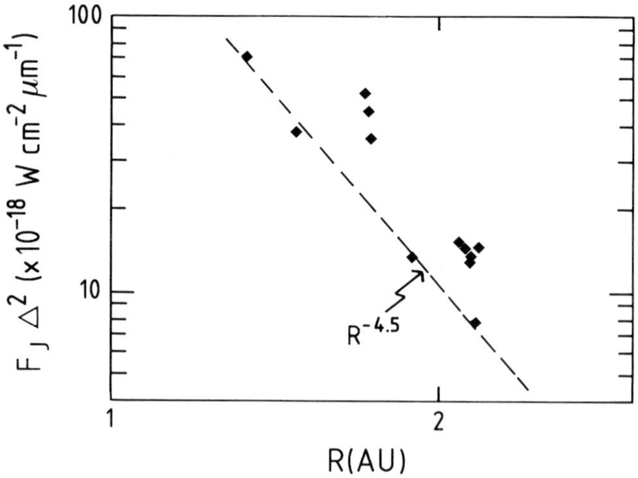

Fig. 3. Fluxes in the J band normalized to 1 AU vs. R. The points fitted by the line represent the comet during its quiescent phase, while the remaining points are interpreted as observations obtained during outbursts

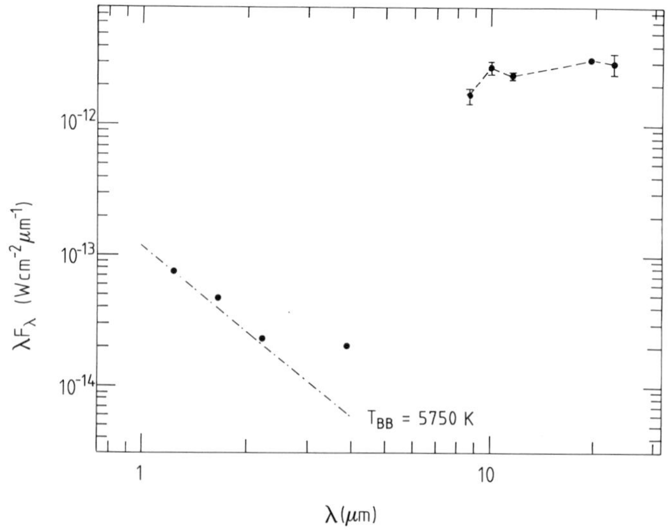

Fig. 4. Observed infrared spectral energy distribution of comet Halley. The $JHKL$ data are from November 4, and the 10–20 μm data are the November 4 and 7 data averaged together (see text for details). The shape of the solar continuum ($T_{BB} = 5750$ K) is shown normalized to the J datum; a dashed line is drawn through the 10–20 μm data to guide the eye

In order to look for brightness variations with R, we first determined the expected flux density within an aperture that corresponds to a diameter of $d = 13000$ km at the distance of the comet; this diameter was chosen because it was almost always within the largest and the smallest apertures used. These flux densities were determined from the fits discussed in Sect. 3.1, and they were then normalized to $\Delta = 1$ AU using Eq. (1). Figure 3 is a plot of $F_J(\Delta/1 \text{ AU})^2$ vs. R. The plot shows four points that fall very well along a straight line, together with other points above that line. We interpret the points on the line as observations of the comet in its quiescent phase and the other points as observations during outbursts. The former points are fit by a line going as $R^{-4.5}$. A similar result was found for comet Kohoutek (Ney, 1982) and it is generally more typical of non-periodic comets than of periodic ones (Mendis, et al., 1985).

3.3 $J-H$ and $H-K$ colors

The $J-H$ and $H-K$ colors of Halley were computed from each night's data excluding those data obtained with the 10 arcsec aperture. The latter were rejected because of the uncertainties discussed in Sect. 2. The $J-H$ and $H-K$ colors remained essentially constant in time, and no appreciable difference is noted between the pre- and the postperihelion data. The mean colors were $\langle J-H \rangle = 0.48 \pm 0.06$ and $\langle H-K \rangle = 0.18 \pm 0.04$ where the uncertainties are the standard deviations. Our results are consistent with the $J-K$ color of Halley at $R \approx 5$ AU (Cruikshank et al., 1985) and with the colors measured by Tokunaga et al. (1986). The constancy of the colors suggests that the composition of the scattering dust did not change throughout the period in which Halley travelled from 5 AU before perihelion to 1 AU after. Furthermore, Halley's colors remained consistently somewhat redder than the Sun's, a property that has been noted in other comets (Ney, 1982). Veeder and Hanner (1981) find that the observed colors could be due to silicates with some absorbing impurities in their matrix. The presence of silicate grains is also consistent with our possible detection of the 10 μm feature on 4 and 7 November (see below).

3.4 The silicate feature

The 10–20 μm photometry of November 4 and 7, when Halley was at $R = 1.88$ AU shows evidence for the emission feature at 10 μm which is typically ascribed to silicates, and which is detected at about the 2σ level. The spectral energy distribution of Halley is shown in Fig. 4. The $JHKL$ data are from 4 November, and they have been scaled to an equivalent 12″ aperture using Eq. 1; the 10–20 μm data are the mean of the November 4 and 7 observations, since they were the same, within the uncertainties. Our results are similar to those of Green et al. (1986, see their Fig. 1) who observed the Halley on November 6. The presence of the silicate feature in a comet at a heliocentric distance of 1.88 AU is unusual (Rieke and Lee, 1974; Campins and Hanner, 1982), and in fact, the feature did not become clear until about a month after our observations (Tokunaga et al., 1986).

3.5. Albedo

From the spectral energy distribution we can determine the albedo of the dust in the coma. Following the method of Ney (1982), we evaluate the total scattered energy as the energy from a blackbody a 5750 K normalized to the comet's flux density at J, and the total energy reradiated was determined from a blackbody fit to the thermal infrared data. The latter fit yields a blackbody temperature of $T_{BB} = 240 \pm 20$ K; which is consistent with the results of Green et al. (1986) who observed Halley on November 6, and with Tokunaga et al. (1986) who observed it on November 10. From our fit we determine an albedo of $\gamma = 0.04 \pm 0.01$, where the uncertainty was determined by varying T_{BB} by 20 K and the blackbody normalization by 10%. This albedo is similar to that measured in the visible (Dollfus et al., 1986), and it is consistent with the model in Giese et al. (1986). This albedo is rather low compared to that of other comets, both periodic and nonperiodic, which typically have albedos of 0.1 to 0.2 (Ney, 1974, 1982).

4. Conclusions

We have presented the results of an infrared monitoring program of Halley's comet conducted at the TIRGO observatory and covering the period from October to December 1985, together with some additional observations obtained in March 1986 from La Silla. Our main results are:

a) Near-infrared multiaperture observations are consistent with a space density distribution of the dust in the coma varying as r^{-2}. This is interpreted in terms of a constant dust outflow velocity. Steeper dust density distributions were observed on November 4 and on March 22; in the latter instance we show that this phenomenon may be related to an outburst that occurred on the previous day. Further analysis of the latter outburst indicates a dust outflow velocity of $\sim 0.13 \mathrm{~km} \mathrm{~s}^{-1}$.

b) The luminosity of the comet in its quiescent phase has been found to vary as $R^{-4.5}$ during the preperihelion observations.

c) The $J-H$ and $H-K$ colors are slighlty redder that those of the Sun, and they remained essentially constant in time. This is taken as an indication that the composition of the dust did not change appreciably as Halley approached the Sun.

d) The albedo of the scattering dust in the coma was ~ 0.04 in early November. This indicates that Halley's dust is quite dark compared to that of most other comets.

e) There is some evidence for the presence of the silicate feature when the Halley was at a heliocentric distance $R=1.88$ AU. The feature, however, is only detected at about the 2σ level.

A more complete understanding of the dust component of Halley's comet must await the compilation and analysis of all of the data gathered by the IHW.

Acknowledgements. We wish to thank G. Calamai for his help during some of the observing runs, and the TIRGO time allocation committee for granting extensive time to the project.

References

Campins, H., Hanner, M.S.: 1982, in *Comets*, ed. L. Wilkening, Univ. Arizona Press, Tucson, p. 341

Cruikshank, D. P., Hartmann, W.K., Tholen, D.J.: 1985, *Nature* **315**, 122

Dollfus, A., Suchail, J.-L., Crussaire, D., Killinger, R., 1986, in *Exploration of Halley's Comet*, eds. B. Battrick, E.J. Rolfe, R. Reinhard, ESA, Noordvijk, p. 41

Elias, J.H., Frogel, J.A., Matthews, K., Neugebauer, G.: 1982, *Astron. J.* **87**, 1029

Fulle, M., Barbieri, C., Cremonese, G.: 1987 (preprint)

Giese, R.H., Killinger, R.T., Kneissel, B., Zerull, R.H.: 1986, in *Exploration of Halley's Comet*, eds. B. Battrick, E.J. Rolfe, R. Reinhard, ESA, Noordvijk, p. 131

Green, S.F., McDonnell, J.A.M., Pankiewicz, G.S.A., Zarnecki, J.C.: 1986, in *Exploration of Halley's Comet*, eds. B. Battrick, E.J. Rolfe, R. Reinhard, ESA, Noordvijk, p. 131

Hanner, M.S., Tokunaga, A.T., Veeder, G.J., A'Hearn, M.F.: 1984, *Astron. J.* **89**, 162

Hunt, L.: 1986, *Oss. Astrofis. Arcetri.* (internal note)

Koornneef, J.: 1983, *Astron. Astroph. Suppl.* **51**, 489

Mendis, D.A., Houpis, H.L.F., Marconi, M.L.: 1985, *Fund. Cosm. Phys.* **10**, 1

Ney, E.P.: 1974, *Icarus* **23**, 551

Ney, E.P.: 1982, in *Comets*, eds. L.Wilkening, Univ. Arizona Press, Tucson, p. 323

Rieke, G.H., Lee, T.A.: 1974, *Nature* **248**, 737

Saraceno, P., Lorenzetti, D., Messi, R., Orfei, R.: 1982, in *Proc. 2nd ESO Infrared Workshop*, eds. A.F.M. Moorwood, K. Kjaer, p. 93

Tokunaga, A.T., Golish, W.F., Griep, D.M., Kaminski, C.D., Hanner, M.S.: 1986, *Astron. J.* **92**, 1183

Veeder, G.J., Hanner, M., S.: 1981, *Icarus* **47**, 381

Astron. Astrophys. 187, 616–620 (1987)

ASTRONOMY AND ASTROPHYSICS

Airborne and groundbased spectrophotometry of comet P/Halley from 5-13 micrometers

J.D. Bregman[1], H. Campins[2], F.C. Witteborn[1], D.H. Wooden[1,3], D.M. Rank[3], L.J. Allamandola[1], M. Cohen[4], and A.G.G.M. Tielens[1,5]

[1] Space Sciences Division, MS 245-6, NASA Ames Research Center, Moffett Field, CA 94035, USA
[2] Planetary Science Institute, SAIC, 2030 E. Speedway. Suite 201, Tucson, AZ 86719, USA
[3] Board of Studies University of California at Santa Cruz and Lick Observatory, UCSC, Santa Cruz, CA 95064, USA
[4] Radio Astronomy Laboratory, University of California at Berkeley, Berkeley, CA 94720, USA
[5] Space Sciences Laboratory, University of California at Berkeley, Berkeley, CA 94720, USA

Received January 29, accepted April 15, 1987

Summary. Spectrophotometry from 5-10 μm ($\Delta\lambda/\lambda \approx 0.02$) of comet Halley was obtained from the Kuiper Airborne Observatory on 1985 December 12.1 and 1986 April 8.6 and 10.5, UT. 8–13 μm data were obtained on 17.2 December 1985 from the Nickel Telescope at Lick Observatory. The spectra show a strong broad emission band at 10 μm and a weak feature at 6.8 μm. We do not confirm the strong 7.5 μm emission feature observed by the Vega 1 spacecraft.

The 10 μm band, identified with silicate materials, has substructure indicative of crystalline material. The band can be fitted by combining spectral data from a sample of interplanetary dust particles. The primary component of the silicate emission is due to olivine. The 6.8 μm emission feature can be due either to carbonates or the C—H deformation mode in organic molecules. The lack of other emission bands is used to place limits on the types of organic molecules responsible for the emission observed by others at 3.4 μm.

Color temperatures significantly higher than the equilibrium blackbody temperature indicate that small particles are abundant in the coma. Significant spatial and temporal variations in the spectrum have been observed and show trends similar to those observed by the spacecraft and from the ground. Temporal variability of the silicate emission relative to the 5–8 μm continuum suggests that there are at least two physically separate components of the dust.

Key words: comets – infrared radiation

1. Introduction

The spectral region from 5.5–8 μm is unobservable from the ground and no previous data on a comet exists at these wavelengths. It is an important region to observe for several reasons. First, the 5–8 μm region contains the OH, NH, and CH deformation modes in alcohols, amines, and hydrocarbons, the CO stretching mode in carbonyl containing molecules (e.g., formaldehyde) and the important C—C stretching modes in aromatic molecules (e.g., polycyclic aromatic hydrocarbons, hereafter PAH's). Together with data from other infrared wavelength regions, this can be used as a "fingerprint" to identify the molecular subgroup in the absorbing molecules (Allamandola, 1984). The recent discovery of a 3.4 μm emission band in the spectrum of comet Halley (Combes et al., 1986; Baas et al., 1986; Wickramasinghe and Allen, 1986; Knacke et al., 1986; Danks et al., 1986) makes these observations even more important. Additionally, interplanetary dust particle (IDP) spectra sometimes show a band due to carbonates centered at 6.8 μm. If IDP's come from comets, then we might observe this band in the coma of comet Halley. Secondly, this is a good region in which to define the dust temperature as the expected peak of the thermal emission from dust in comets near 1AU from the sun occurs near this wavelength region. The continuum can also be used to constrain the particle size distribution of the dust.

In this work we present the results of a preliminary analysis of airborne and groundbased spectrophotometry of comet Halley in the 5 to 13 micrometer region.

Send offprint requests to: J. Bregman

2. Observations

Observations were obtained on 1985 Dec 12.1 UT from 5.2–9.0 μm, and 1986 April 8.6 and 10.5 UT from 5.2–10.0 μm from the Kuiper Airborne Observatory (KOA). Ground-based data were obtained on Dec 17.2 from 8.2–13.2 μm from the Lick Observatory Nickel telescope. All data were obtained with the NASA-Ames FOGS 24-channel array ($\Delta\lambda/\lambda \approx 0.02$), liquid helium cooled grating spectrometer (Witteborn and Bregman, 1984). The aperture was 21″ (FWHM), and a 100″ chop aligned approximately perpendicular to the tail was used. Spectra were obtained at various positions as shown in Table 1. All the data were divided by standard stars and multiplied by the intrinsic fluxes of the standards. The start α Tau and β Gem were used in December, while α Boo and α Hya were used in April (Gezari et al., 1984; Thomas et al., 1973). An air mass correction was computed for the April data from a model of the earth's atmosphere and applied to the data to correct for the difference since the standards were obtained at significantly different elevation angles than the comet. This procedure was not necessary for the December data. The airborne and ground-based data taken in December were combined by normalizing the ground-based data to the airborne data using the overlapping channels from 8–9 μm.

The spectra were divided by blackbody curves of different temperatures until the 5–8 μm section of the spectra were flat.

Table 1

Date	R(AU)	Δ(AU)	Position observed	Color temp. (K ± 5)
85/12/12	1.32	0.77	Nucleus	320
85/17/12	1.32	0.77	Nucleus	—
86/08/4	1.30	0.42	Nucleus	325
			Sunward (21″ SE)	340
			Tailward (21″ NW)	335
86/10/4	1.33	0.42	Nucleus	325
			Sunward (21″ SE)	325
			Tailward (21″ NW)	325
			Up (21″ NE)	320
			Down (21″ SW)	315

This gave the color temperatures listed in Table 1. Heliocentric and geocentric distances are also given in Table 1.

Figures 1–3 show the spectra of the central condensation on the three dates while Figures 4–6 show the same spectra after division by the appropriate blackbody.

3. Results

After division of our data by blackbodies, there are residual features left in the spectra, the most obvious one being the silicate emission in the 10 μm region. Additionally, there is a weak emission feature at 6.8 μm (especially obvious in the December data) possibly due to carbonates or the C—H deformation mode of organic molecules. Also, there is an unidentified upturn (relative to a blackbody) at the shortest wavelengths.

There is no evidence for an emission feature in the 7–8 μm region as reported by Combes et al. (1986) from the IKS spectrograph on the Vega 1 spacecraft. The IKS instrument observed a much smaller central part of the coma than our instrument, and it is possible that the emission close to the nucleus is different

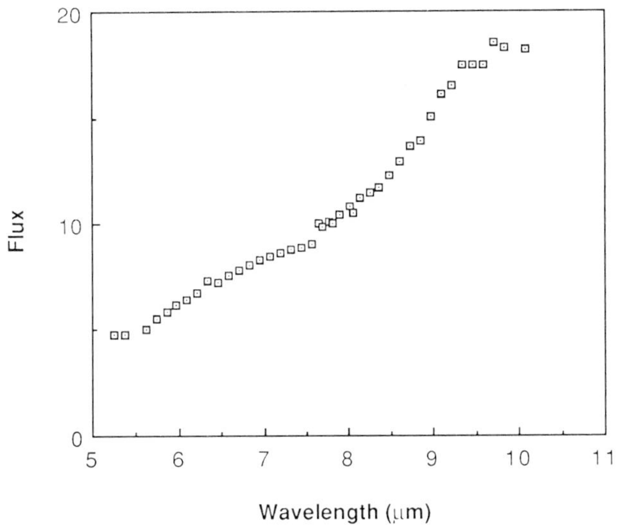

Fig. 2. The 5–9 μm spectrum of the central condensation of comet Halley taken on 1986 April 8.6, UT. Error bars representing the statistical errors are shown when the errors exceed the size of the data point. Units for the flux scale are $10^{-16}\,\mathrm{W\,cm^{-2}\,\mu m^{-1}}$

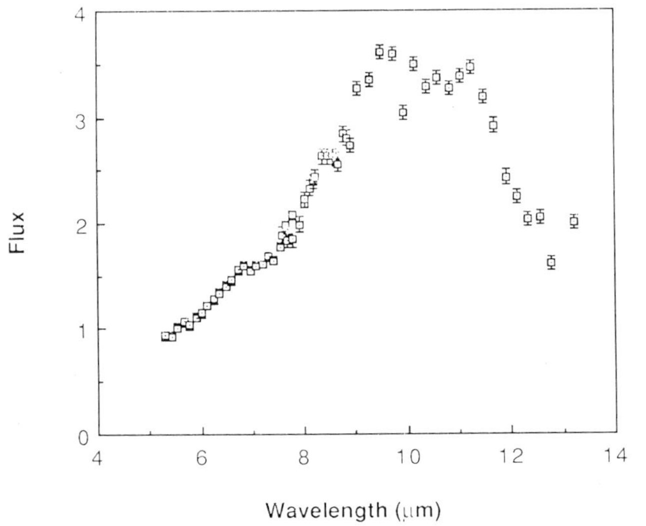

Fig. 1. The 5–13 μm spectrum of the central condensation of comet Halley taken on 1985 December 12.1 and 17.2, UT. Error bars representing the statistical errors are shown when the errors exceed the size of the data point. Units for the flux scale are $10^{-16}\,\mathrm{W\,cm^{-2}\,\mu m^{-1}}$. The low data point at 9.7 μm is due to a miscorrection for terrestrial ozone

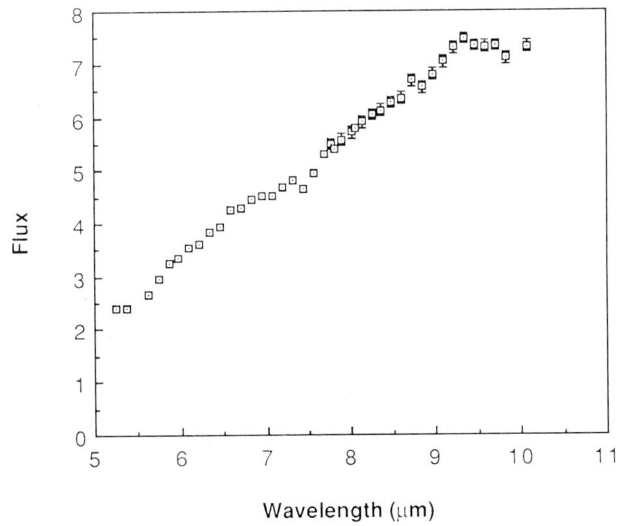

Fig. 3. The 5–9 μm spectrum of the central condensation of comet Halley taken on 1986 April 10.5, UT. Error bars representing the statistical errors are shown when the errors exceed the size of the data point. Units for the flux scale are $10^{-16}\,\mathrm{W\,cm^{-2}\,\mu m^{-1}}$

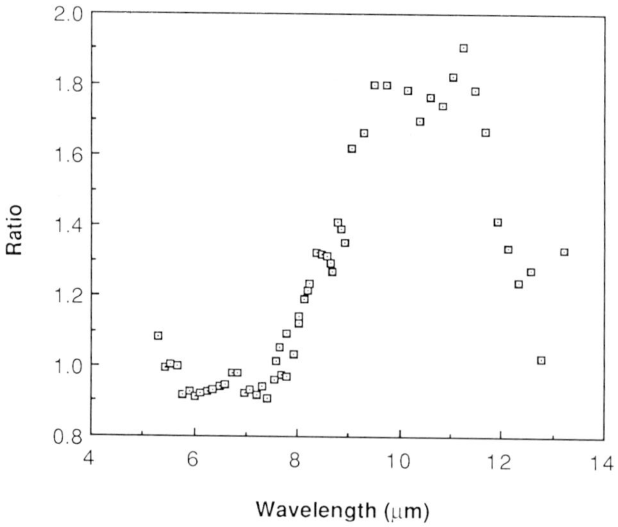

Fig. 4. The ratio of the 1985 December spectrum to a 320 K blackbody normalized to 1.0 at 6 μm. Note the deviations from a blackbody shortward at 5.6 μm and longward of 8 μm, and the return to the blackbody line at 13 μm

than farther out. However, the published IKS results were preliminary and had difficulties with calibration and instrument response variations (T. Encrenaz, private communication). Further analysis of the IKS data may resolve the discrepancy.

3.1. *The continuum emission and the temperature of the dust*

The shape of the continuum can yield information on the temperature and size distribution of the particles. If the thermal emission was produced by particles of only one size, fits to single blackbody curves would characterize the temperature of the particles. However, particles of a broad range of sizes are responsible for the thermal emission. At the shorter wavelengths, such as the 5 to 8 μm region, the emission will be dominated by small particles (small particles will be warmer than large ones) while at longer wavelengths, the larger grains will contribute most of the flux. For this reason, the color temperatures obtained from fits at shorter wavelengths are higher than those obtained from fits at the longer wavelengths. Hence, modeling of the thermal emission yields information on the sizes of the particles which contribute most of the light in this spectral region. We have not modeled the particle size distribution in this paper, but just taken a single blackbody that reasonably fits the 5.6–8 μm range. This wavelength range is short enough to be approximated with a single temperature. The excess above a blackbody shortward of 5.5 μm rises sharply, and may not be explained by emission from a population of small hot dust grains.

The presence of a 10 μm emission feature implies that particles small compared to the wavelength (radius less than about 1 μm) dominate the emission from 8–13 μm. Large particles become optically thick at wavelengths outside the emission features, so if large particles dominated, the spectrum would be a featureless blackbody. Thus, the presence of such a strong feature implies the dominance of small particles. The temperatures measured from the 5–8 μm continuum on the nucleus and in the vicinity of the nucleus are all higher than the equilibrium blackbody temperature expected for large particles ($r > 1$ μm), indicating that small dust particles ($r < 1$ μm) are also dominant at these wavelengths. Note that the color temperature is not a good measure of the dust temperature. The Table shows that on all three dates, the nuclear temperature was the same, while on 1986, April 8 the sunward and tailward temperatures were significantly higher. It is difficult to see how the nuclear temperatures on the two dates could be the same while the dust temperatures 20″ (6600 km) from the nucleus differ by 10–15 K unless there was an ejection of small particles from the nucleus a few hours before the observations of April 8. An increase of 10–15 K requires a decrease of the dominant particle size by about 20%. There was an outburst observed from the nucleus on April 8 that might have provided the increased abundance of small particles. The outburst was also observed in the IR brightness, as the flux from the comet was twice as bright on the 8th as on the 10th. Also, the silicate emission was 80% higher than the 5–8 μm continuum on the 8th as compared to 40% higher on the 10th. This variation of the

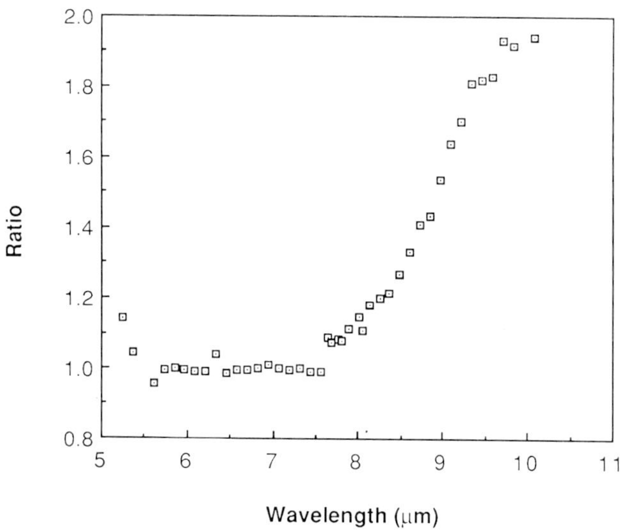

Fig. 5. The ratio of the 1986 April 8.6 data divided by a 325 K blackbody normalized to 1.0 at 6 μm

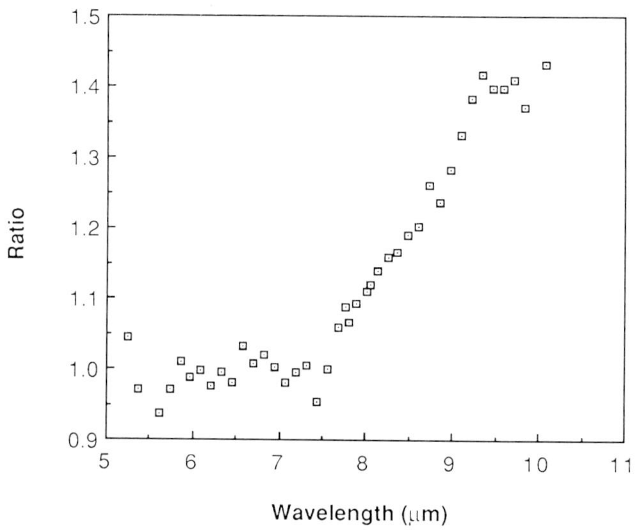

Fig. 6. The ratio of the 1986 April 10.5 data divided by a 325 K blackbody normalized to 1.0 at 6 μm

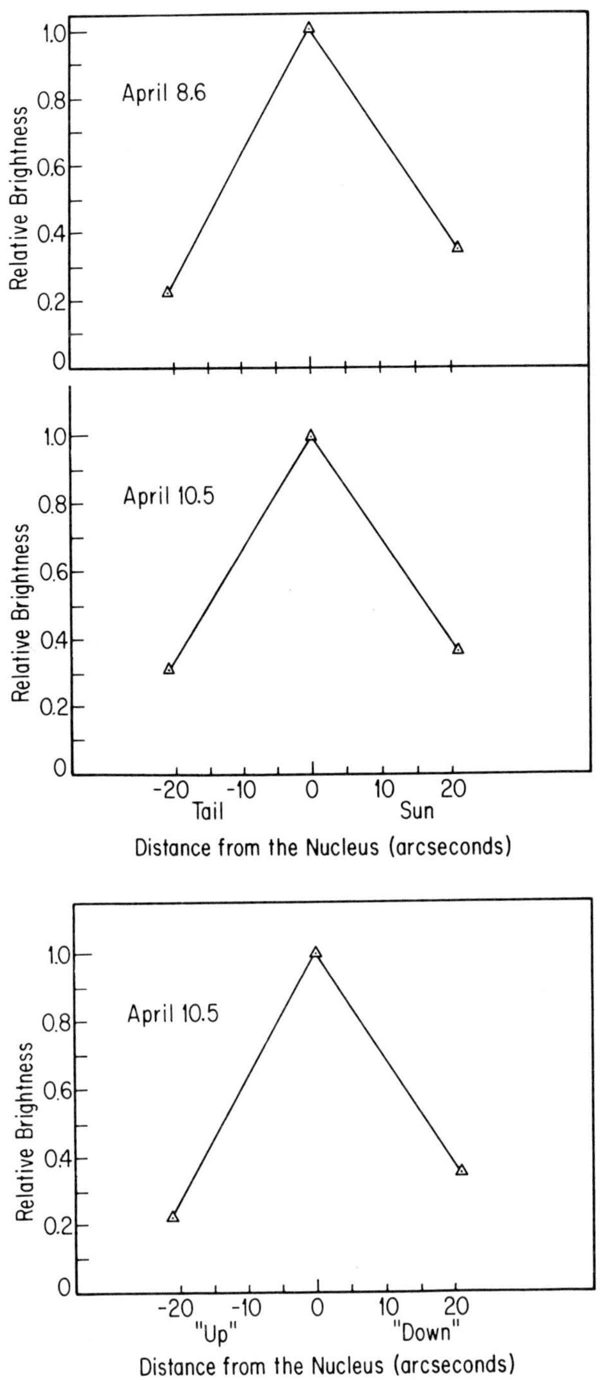

Fig. 7. Comparison of the brightness of the coma at various positions. Note the increased brightness in the sunward direction relative to the anti-sunward direction

strength of the silicate feature suggests that there are two components to the dust, the silicates producing the 10 μm emission and a featureless component (such as graphite or amorphous carbon) producing the continuum from 5–8 μm. Large silicate particles are ruled out as the source of the continuum since the continuum temperature is higher than the equilibrium black-body temperature (and because the 10 μm silicate feature is so prominent). The timescale for observing variability in the dust emission is hours since it only takes about 9 hours for a dust particle to travel half a beam diameter (3300 km) at a velocity of 0.1 km s^{-1}. These observations also imply that the small particles move away from the nucleus faster than the large particles, which is expected if they are all ejected from the nucleus with either the same force or the same pressure.

Figure 7 shows the relative intensities measured in the 5.5–8 μm region of different positions in the coma measured on April 8 and 10. The sunward fluxes on both days are brighter than the tailward fluxes. The asymmetry is greater on the 8th than the 10th which is when the comet was brightest, in agreement with spacecraft and groundbased observations. Measurements made above ("up") and below ("down") the nucleus on April 10 also show an asymmetry. These observations indicate that most of the dust ejection occurs on the sunward side of the nucleus.

3.2. *Emission features*

In Fig. 8 we show a fit to the silicate emission by using a sample of interplanetary dust particle (IDP) spectra. The IDP spectra have been reproduced by using the data of Sandford and Walker (1985). The fit is quite good, and consists of a mixture of the three different mineral classes of IDP's observed, olivines, pyroxenes, and layer lattice silicates. A reasonable fit requires including either pyroxenes or layer lattice silicates (but not necessarily both) combined with olivines. Olivines are essential for fitting the peak at 11.2 μm. The presence of structure in the data means that the silicates are crystalline, unlike the generally featureless silicates observed in the interstellar medium and around stars.

The 5–8 μm spectral results are surprising for their paucity of emission features. There is a small emission feature centered at 6.8 μm, which is strongest in the December data, that can be attributed either to the C—H deformation mode in organic compounds (see below), or emission from carbonates. Carbonates comprise about 1% of IDP's, and the observed 6.8 μm feature is

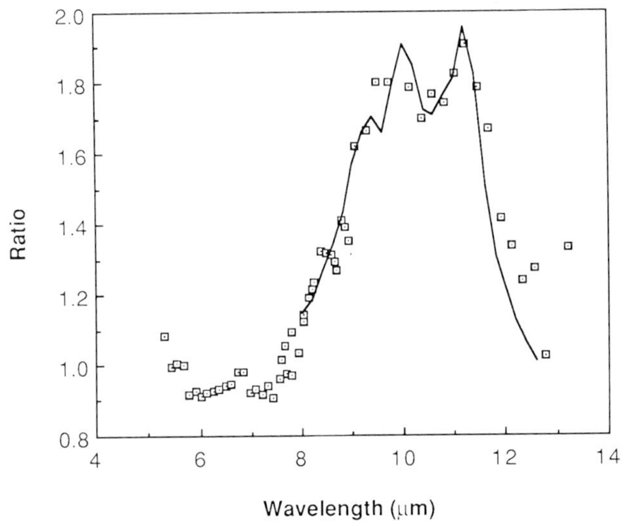

Fig. 8. The 1985 December data compared to a spectrum (the solid line) generated by combining spectra from a variety of interplanetary dust particles (Sandford and Walker, 1985). The dominant component is olivine, with some pyroxenes and a small amount of layer lattice silicate. The small feature at 6.8 μm could be due to carbonates associated with the layer lattice component in interplanetary dust particles or to simple saturated aliphatic hydrocarbons

about the same strength relative to the 10 μm silicate emission as it would be in a mixture of IDP's.

The strongest band observed in the 3–4 μm region is at 3.4 μm, which is due to simple saturated aliphatic hydrocarbons such as alkanes or alcohols. Either of these classes is allowed since their corresponding CH deformation modes near 6.8 μm are relatively weaker, and could in fact account for the weak emission observed at 6.8 μm. While the position of the 3.54 μm band is suggestive of aldehydes, it may well be a higher transition in the molecules responsible for the 3.4 μm peak, and is observed at 3.54 μm due to the anharmonic term in the energy levels. Since this implies a considerable amount of excitation energy per vibrational mode. it implies that molecular sized species are responsible (Barker, et al., 1987).

An intriguing result is the lack of other features in the 5.5–8 μm region, especially in light of the observed hydrocarbon emission features near 3.4 μm (Combes et al., 1986; Baas et al., 1986). Our results also place strong constraints on the types of hydrocarbons responsible for the 3.4 μm emission by eliminating classes of molecules. The spectrum of comet Halley obtained by Baas et al. (1986) shows a strong band at 3.34 μm. Two weaker bands at 3.28 and 3.54 μm may also be present. A 3.28 μm emission feature is observed in many interstellar nebulae and is generally attributed to IR fluorescence from PAH's (Leger and Puget, 1984; Allamandola et al., 1985). However, if aromatics were present, there should be corresponding bands at 6.2 and 7.7 μm; both of these are absent at levels well below what is expected, ruling out aromatics similar to the PAH's observed in interstellar nebulae as the origin of the 3.28 μm band. In a similar fashion we can rule out the presence of molecules such as aldehydes and ketones as these would have a strong carbonyl feature in the 5.5–5.9 μm range. Olefinic groups, which could account for the 3.28 μm band, cannot be ruled out since their corresponding CH deformation modes, which fall in the 10 μm region, would not be evident above the strong silicate emission. Gas phase formaldehyde is possible since its 5.8 μm band would be about the same strength as the ones at 3.54 μm and 3.60 μm. Bands at 3.54 and 3.60 μm may have been observed by the IKS spectrometer (Combes et al., 1986). Associated formaldehyde or formaldehyde mixed in an ice is not possible as these two bands shift to shorter wavelengths (van der Zwet et al., 1985). In general, large molecules of these types are ruled out since they would show the strong carbonyl (C=O) band in the 5–8 μm region.

4. Summary

Spectrophotometry of comet Halley was obtained from 5–13 μm in December, 1985 and twice from 5–10 μm in April, 1986. The results show that:

1. A strong silicate emission band extending from 8–13 μm which varies in strength relative to the 5–8 μm continuum, suggests that there are at least two physically separate dust components.
2. The infrared brightness towards the sun is higher than in the anti-sun direction, consistent with the dust ejection from the nucleus being predominantly in the sunward direction.
3. There is a weak emission feature at 6.8 μm which could be due either to carbonates or the C—H deformation mode in saturated aliphatic hydrocarbons. The only organic molecules that could simultaneously account for the 3.4 μm emission band observed by other authors and our data are alkanes and alcohols. The lack of other emission features in the 5–8 μm wavelength region also rules out significant amounts of free aromatic molecules.
4. The strong emission feature reported by Combes et al. (1986) at 7.5 μm is not present in any of our data. Either the species responsible for the feature is confined to the inner coma observed by the IKS instrument, or the preliminary calibration of the IKS data is not correct and subsequent recalibration will resolve the discrepancy.
5. The 8–13 μm silicate emission band shows two peaks, indicating that the silicates are crystalline. A good fit to the spectrum can be made by using a mixture of interplanetary dust particle spectra. Olivines are required to fit the 11.2 μm peak, while either layer lattice silicates or pyroxenes could account for the remaining emission.

Acknowledgements. We want to thank the crew and support staff of the Kuiper Airborne Observatory. Without their dedication and hard work these observations would not have been possible. Harold Crean played a key role in preparing the equipment for use on the KAO. We also want to thank the staff of Lick Observatory for making telescope time available on such short notice, and working to get our equipment mounted on the telescope in time to make the observations. Part of this research was supported by NASA Contract NAS2-12323 to the Planetary Science Institute, a division of Science Application International Corporation. This is PSI Contribution No. 227. We wish to thank Scott Sandford for suggesting the comparison of the 10 μm emission feature with a composite of IDP spectra, and for assisting with the comparison.

References

Allamandola, L.J.: 1984, in *Galactic and Extragalactic IR Spectroscopy*, eds. Kessler and Phillips, Reidel, Holland

Allamandola, L.J., Tielens, A.G.G.M., Barker, J.R.: 1985, *Astrophys. J. Letters*, **290**, L25

Baas, F., Geballe, T.R., Walther, D.M.: 1986, *Astrophys. J. Letters*, **311**, L97

Barker, J.R., Allamandola, L.J., Tielens, A.G.G.M.: 1987, *Astrophys. J. Letters* (in press)

Combes, M., Moroz, V.I., Crifo, J.F., Lamarre, J.M., Charra, J., Snako, N.F., Soufflot, A., Bibring, J.P., Cazes, S., Coron, N., Crovisier, J., Emerich, C., Encrenaz, T., Gispert, R., Grigoryev, A.V., Guyot, G., Krasnopolsky, V.A., Nikolsky, Y.V., Rocard, F.: 1986, *Nature* **321**, 266

Danks, A., Encrenaz, T., Bouchet, P., Le Bertre, T., Chalabaev, A., Epchtein, N.: 1986, in *20th ESLAB Symp.* **ESA SP-250**, 103

Gezari, D.Y., Schmitz, M., Mead, J.M.: 1984, *Catalog of Infrared Observations*, NASA Reference Publication 1118

Knacke, R.F., Brooke, T.Y., Joyce, R.R.: 1986, *Astrophys. J. Letters*, **310**, L49

Leger, A., Puget, J.L.: 1984, *Astron. Astrophys. Letters*, **142**, L19

Sandford, S.A., Walker, R.M.: 1985, *Astrophys. J.* **291**, 838

Thomas, J.A., Hyland, A.R., Robinson, G.: 1973, *Monthly Notices Roy. Astron. Soc.* **165**, 201

Wickramasinghe, D.T., Allen, D.A.: 1986, *Nature* **323**, 44

Witteborn, F.C., Bregman, J.D.: 1984, *Proc. SPIE* **509**, 123

Note added in proof: Further reduction of the *IKS* Vega data had resulted in a new spectrum that agrees with our data, showing no 7 μm emission, but instead a doubly peaked silicate emission (Encrenaz, private communication).

Astron. Astrophys. 187, 621–624 (1987)

ASTRONOMY AND ASTROPHYSICS

The near-infrared polarization and color of comet P/Halley

T.Y. Brooke[1,★,★★,★★★], **R.F. Knacke**[1,★,★★], **and R.R. Joyce**[2,★]

[1] Astronomy Program, Department of Earth and Space Sciences, State University of New York at Stony Brook, Stony Brook, NY 11794 USA

[2] Kitt Peak National Observatory, National Optical Astronomy Observatories, Tucson, AZ 85726 USA

Received February 2, accepted April 23, 1987

Summary. The linear polarization and color of comet P/Halley in broadband filters $J(1.25\,\mu m)$, $H(1.65\,\mu m)$, and $K(2.2\,\mu m)$ were measured at phase angles $9^\circ\!.4 \le \beta \le 65^\circ\!.3$. Comet Halley exhibited a red color characteristic of comet dust at these wavelengths with no clear variation with β. The JHK polarizations reach 25% at the largest phase angles with a reversal at $\beta = 20^\circ$ and negative polarizations of a few per cent at smaller β. The polarization increases with wavelength at the larger phase angles. Mie theory calculations of the polarization from a grain size distribution based on measurements by the Giotto spacecraft indicate that no single component fits the data; a two-component model is needed. The two components that give the best fit correspond loosely to a "dirty" silicate and a moderately absorbing material. The model gives fair agreement with both the polarization and color of the Halley dust. The average albedo of the grains in the model is 0.02. Calculated phase functions show no increase toward backscattering angles; a model incorporating rough particles may be needed.

Key words: comets – infrared radiation – polarization

1. Introduction

The linear polarization and color of light scattered by the dust grains in comets depend, in general, on the composition, size, and structure of the grains. To help determine these grain properties, we measured the linear polarization and color of comet Halley at near-infrared wavelengths over a wide range of phase angles and heliocentric distances. The data are presented here, along with an attempt to model the variation of polarization with phase angle and wavelength, and the color of the scattered light with Mie theory calculations. The grain size distribution was based on measurements of the dust impact detectors on the Giotto spacecraft.

Mie theory calculations are strictly valid only for homogeneous spheres, while comet grains are more likely to be irregularly shaped aggregates with rough surfaces, similar to the interplanetary particles collected by Brownlee (1978). A more detailed study of the effects of scattering by rough particles on the conclusions presented here is planned.

2. Observations

Comet Halley was observed from 1985 September 21 UT to 1986 May 28 UT as it passed through phase angles $9^\circ\!.4 \le \beta \le 65^\circ\!.3$ and heliocentric distances $1.06\,\mathrm{AU} \le r \le 2.47\,\mathrm{AU}$. The standard $J(1.25\,\mu m)$, $H(1.65\,\mu m)$, and $K(2.2\,\mu m)$ filters were used at three telescopes: the Infrared Telescope Facility on Mauna Kea, the 1.3-m on Kitt Peak, and the 1.5-m on Cerro Tololo, Chile. The aperture size varied from 8″ to 33″ corresponding to 5000 to 21,000 km at the comet. The aperture was centered on the position of the peak infrared signal before each integration.

The linear polarization of the light in each filter was measured by rotating a wire grid polarizer which was mounted in front of the InSb dewar. Observations of unpolarized stars on each night were used to determine the instrumental polarization, typically a few per cent. Polarizer efficiencies were measured in the laboratory; these were found to be 90% at J, 95% at H, and 97% at K. The position angle of the polarization was calibrated with observations of standard sources, e.g., the BN source. Light polarized perpendicular to the scattering plane of the comet is defined to be positive; that polarized parallel to the scattering plan is negative. The position angle of the light from comet Halley was always either perpendicular or parallel to the scattering plane within the uncertainties; we found no evidence for aligned grains.

3. Results

The polarizations of Halley and two other comets are plotted *vs.* scattering angle, i.e., (180°-phase angle), in Fig. 1. The uncertainties are one standard deviation of the mean of several measurements on each night. The data vary smoothly both before and after perihelion. This suggests, but does not guarantee, that

Send offprint requests to: T.Y. Brooke

★ Visiting Astronomer, Cerro-Tololo Inter-American Observatory, National Optical Astronomy Observatories, which is operated by AURA, Inc. under contract with the National Science Foundation

★★ Visiting Astronomer at the Infrared Telescope Facility which is operated by the University of Hawaii under contract to the National Aeronautics and Space Administration

★★★ Visiting Astronomer, Kitt Peak National Observatory, National Optical Astronomy Observatories, which is operated by AURA, Inc. under contract with the National Science Foundation

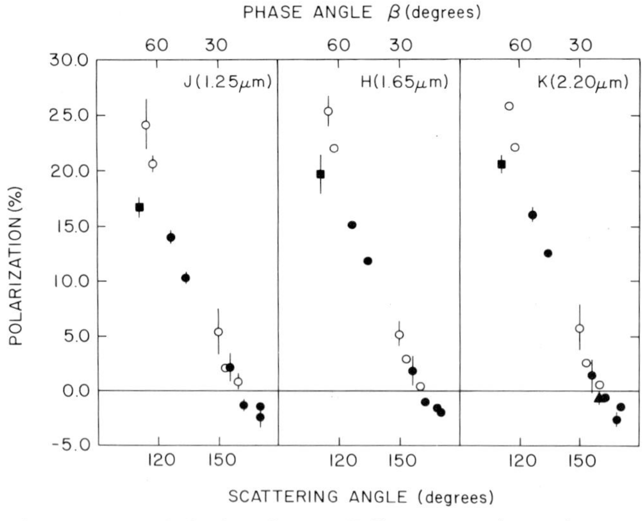

Fig. 1. Linear polarization of comet Halley *vs.* scattering angle pre-perihelion (filled circles) and post-perihelion (open circles). Filled squares are comet Hartley-Good (1985 l). Filled triangle is comet Thiele (1985 m)

the scattering properties of the dust did not change drastically over the course of the observations. The *JHK* polarizations vary approximately linearly with phase angle from negative values of a few percent at $\beta < 20°$ to positive values near 25% at $\beta = 65°$. Comet Hartley-Good shows a significantly lower polarization than Halley at a similar phase angle. The overall shape of the curves is similar to the visible polarization of the dust continuum in P/Halley (Bastien et al., 1986; Dollfus et al., 1986; Kiselev et al., 1986; Le Borgne et al., 1986; Mukai et al., 1986) and other comets (Kiselev and Chernova, 1978; Myers and Nordsieck, 1984; Myers, 1985), and the data merge smoothly with the near-infrared polarimetry of comet West at higher phase angles (Oishi et al., 1978a).

One new result of this investigation is the change in the wavelength dependence of the polarization as the comet passed through different phase angles. There is a definite trend for the polarization to increase with wavelength at large phase angles, while there is no clear trend at smaller β. This is evident in Fig. 2a, where the data are plotted as a function of wavelength. While the uncertainties in some of the data are high, polarization increasing with wavelength, primarily between 1.25 μm and 1.65 μm, is seen in four of the Halley observations and in observations of comet Hartley-Good. The near-infrared data of comet West show no clear wavelength dependence (Oishi et al., 1978a). A flat wavelength dependence in the visible has been reported for several comets (Michalsky, 1981; Myers and Nordsieck, 1984; Myers, 1985). However, comparison with visible polarimetry of the dust continuum in comet Halley (Mukai et al., 1986) suggests that the wavelength dependence observed in the near-infrared continues into the visible.

Plotted in Fig. 2b are the colors of comet Halley divided by the solar color and normalized at 1.65 μm. Comet Halley exhibited a red color typical of many comets at these wavelengths (Jewitt and Meech, 1986), with no clear variation with phase angle or heliocentric distance. This is further evidence that the properties of the grains did not change significantly over time. The average colors are $J - H = 0.47 \pm 0.02$ and $H - K = 0.18 \pm 0.02$, in agreement with the results of Tokunaga et al. (1986).

4. Models

Previous models of the near-infrared polarization of comet West at phase angles near 90° (Oishi et al., 1978b) fail to match the Halley data at smaller phase angles. These authors used the Mie theory to calculate the polarization from power-law size distributions of absorbing and silicate spheres. While their models do exhibit a negative branch at small β, the magnitude of the negative peak is too high to fit the Halley observations.

We calculated the polarization from a size distribution of grains in comet Halley inferred from a revised Giotto DIDSY cumulative mass distribution (McDonnell et al., 1986a; McDonnell et al., 1986b, Fig. 11), assuming a density of 1 g cm^{-3}. The lower and upper limits to the grain radius were taken to be 0.1 μm and 140 μm; particles with radii outside this range do not contribute significantly to the scattered light for the range of refractive indices and scattering angles considered here. The effective radius of this distribution, defined as the geometric cross-section weighted average (Jewitt and Meech, 1986), is 7 μm.

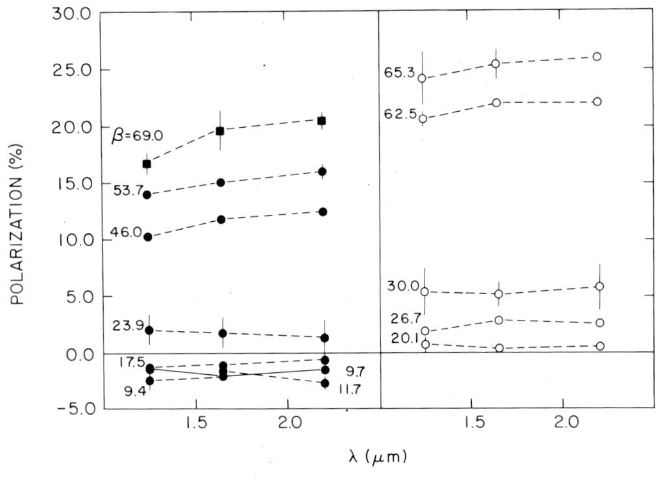

Fig. 2a. Polarization as a function of wavelength. β is the phase angle. Symbols are the same as Figure 1

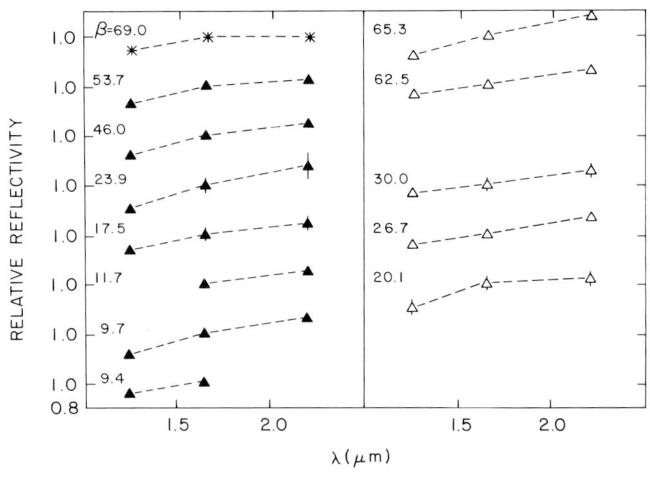

Fig. 2b. Relative reflectivity normalized at 1.65 μm pre-perihelion (filled triangles) and post-perihelion (open triangles). Crossed symbols are comet Hartley-Good

A more recent interpretation of the Giotto spacecraft data suggests that a slope change in the mass distribution could occur for masses greater than 10^{-5} g (McDonnell et al., 1986b, Fig. 16). If this interpretation is confirmed, grains larger than 140 μm could contribute significantly to the near-infrared scattered light, though they will not necessarily be dominant. Mie theory calculations are impractical for very large grains. We plan to consider the effect of these large grains within the context of scattering by rough particles in the future. Mie theory fits using the results of the Vega SP-2 detector (Mazets et al., 1986) were found to be qualitatively similar to those of the Giotto distribution with an upper radius cutoff of 140 μm.

Single-component models were tested first. A grid in the real and imaginary parts of the complex refractive index, $n = n' + in''$, was constructed. Values of the real part tested were 1.3(0.1)2.0, 3.0; values of the imaginary part were 10^{-4}, 10^{-3}, 0.01, 0.05, 0.1(0.1)1.0, 2.0, 3.0. This range includes the indices of most of the materials suggested as being present in cometary dust. Calculated polarization *vs.* scattering angle curves were compared to the data at 2.2 μm.

None of the refractive indices tested provided a good fit to the observations. Models that successfully reproduced the negative branch of the polarization had too steep a slope at larger phase angles. Similar behavior was seen using power-law size distributions (Brooke et al., 1986). Myers (1985) found that no single component Mie theory model fit the visible polarization *vs.* phase angle curve.

A two-component model provided a better fit to the data. For each combination of refractive indices, the relative abundance of the two components was varied for the best fit to the polarization at 2.2 μm. The best fit was for refractive indices $n_1 = 1.7 + i0.3$ for the first component and $n_2 = 1.4 + i0.05$ for the second with a relative abundance ($n_1 : n_2$) of 3.5 by number. The two components correspond to a moderately absorbing material similar to the "composite" material ($n = 1.66 + i0.4$) proposed by Hanner (1983) to explain the thermal emission of several comets, and a "dirty" silicate. The fit is shown in Fig. 3. Scattering by the silicate gives the negative branch, while the absorbing material dominates the polarization at larger phase angles, as in the models of Oishi et al. (1978b). Two-component fits with water ice, terrestrial silicates, organic tholins (Khare et al., 1984), or strongly absorbing materials like graphite or magnetite did not match the data well, though these or similar materials could be present in a multi-component model.

The dust mass spectrometers aboard the Vega and Giotto spacecraft found that many of the particles were rich in hydrogen and carbon (Kissel et al., 1986a; 1986b). These may be the absorbing particles that appear to be needed to fit the polarization *vs.* phase angle data.

The identification of the second component as a silicate should be taken with some caution, though silicates were present in Halley (Tokunaga et al., 1986). Terrestrial silicates have real refractive indices in the range 1.44–1.54 and imaginary indices $8.4\,10^{-5} - 3.9\,10^{-3}$ at these wavelengths (Pollack et al., 1973). Hanner (1983) found that raising the imaginary part of the silicate in her model to $n'' = 0.01$ improved the fit to the scattered light from comets. Higher imaginary indices may be indicative of additional absorbing material mixed with the silicate in these grains.

With the relative abundance fixed by the fit at 2.2 μm, the refractive indices of the two components were allowed to vary in order to fit the polarization and color at the other wavelengths. Red colors in this wavelength region are typical of the scattering by distributions of particles larger than a few microns in size for a wide range of materials, even for constant refractive index (Jewitt and Meech, 1986). However, models with constant refractive index failed to give the correct wavelength dependence of the polarization. Refractive indices increasing with wavelength were needed to fit the polarization and color at J and H. The best fits were for $n_1 = 1.64 + i0.19$, $n_2 = 1.38 + i0.04$ at 1.25 μm and $n_1 = 1.70 + i0.26$, $n_2 = 1.40 + i0.05$ at 1.65 μm (Fig. 3). The calculated relative reflectivities averaged over angle and normalized at 1.65 μm are 0.88 at J and 1.06 at K, in good agreement with the observations (Fig. 2b). Refractive indices increasing with wave-

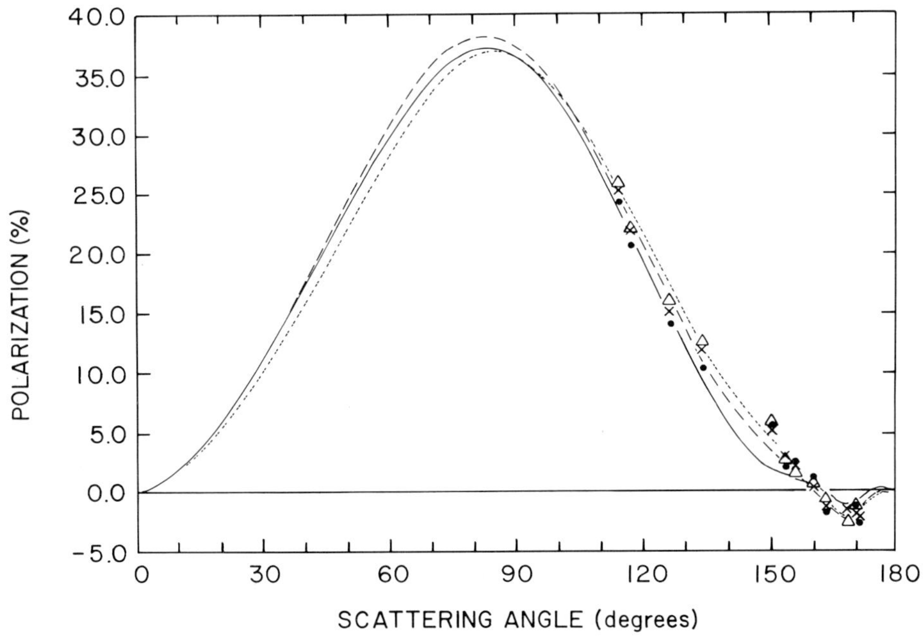

Fig. 3. Model fits to the polarization of comet Halley (see text). Filled circles and solid line are the data and model, respectively, at J(1.25 μm); crossed symbols and long dashed line are for H(1.65 μm); open triangles and short dashed line are for K(2.2 μm). Error bars in the data have been omitted for clarity. Note that the points at a scattering angle of 150° have low weight

length are common to absorbing materials such as carbon and magnetite.

The grains in the best fit model have a low albedo. The angle-averaged albedo of the size distribution defined as in Hanner et al. (1981) is $A_p = 0.02$ at 1.25 μm, lower than the value of 0.03 derived for comet Halley by Tokunaga et al. (1986), though it is within the range of other comets in their survey. Scattering by rough particles may explain some of this discrepancy. Hanner et al. (1981) noted that measurements of the angular scattering by irregular absorbing particles yield a higher albedo than spheres of equal geometric cross-section at phase angles $\beta < 90°$.

One difficulty with the present Mie theory models is that the calculated phase functions at J, H, and K show no increase toward backscattering angles as is observed in many comets (Ney and Merrill, 1976; Millis et al., 1982; Hanner et al., 1985), including Halley (Tokunaga et al., 1986). This may be additional evidence for the presence of rough particles, as both laboratory measurements (e.g., Zerull, 1985) and theoretical studies (Mukai et al., 1982; Perrin and Lamy, 1983) of rough particles show enhanced backscattering.

An alternative interpretation of comet polarimetry relies upon the similarity between the polarization *vs.* phase angle curves of comets and those of asteroid and satellite surfaces (Kiselev and Chernova, 1978; Myers and Nordsieck, 1984; Myers, 1985). A theoretical study (Wolff, 1975) indicates that the negative branch of the polarization in these bodies is due to double reflections and shadowing by rough surfaces. The slope of the polarization *vs.* phase angle curves for asteroids is correlated with albedo (Dollfus and Zellner, 1979). If comet grains scatter like asteroid surfaces, an approximate albedo may be derived. Though there has been little near-infrared polarimetry of asteroids, comparison of our data with visible polarimetry of asteroids suggests that the albedo of the dust in Halley is low, less than 0.05 (cf. Bastien et al., 1986; Dollfus et al., 1986). Regardless of whether the comet polarization is interpreted as due to scattering by spherical grains or by rough surfaces, a low albedo is indicated.

We conclude that Mie theory models incorporating two grain components can match the near-infrared polarization and color of comet Halley, and provide useful constraints on the composition of the grains. Models incorporating scattering by rough particles may be needed to complete our understanding of the optical properties of cometary dust.

Acknowledgements. This research was supported at Stony Brook by National Science Foundation grant AST 8414825.

References

Bastien, P., Ménard, F., Nadeau, R.: 1986, *Monthly Notices Roy. Astron. Soc.* **223**, 827

Brooke, T.Y., Knacke, R.F., Joyce, R.R.: 1986, *20th ESLAB Symp. ESA SP-250, Vol. II*, p. 87

Brownlee, D.E.: 1978, in *Cosmic Dust*, ed. J.A.M. McDonnell, John Wiley and Sons, New York, p. 295

Dollfus, A., Suchail, J.-L., Crussaire, D., Killinger, R.: 1986, *20th ESLAB Symp. ESA SP-250, Vol. II*, p. 41

Dollfus, A., Zellner, B.: 1979, in *Asteroids*, ed. T. Gehrels, Univ. Arizona Press, Tucson, p. 170

Hanner, M.S.: 1983, in *Cometary Exploration*, ed. T.I. Gombosi, Hungarian Academy of Sciences, Budapest, 2, p. 1

Hanner, M.S., Giese, R.H., Weiss, K., Zerull, R.: 1981, *Astron. Astrophys.* **104**, 42

Hanner, M.S., Tedesco, E., Tokunaga, A.T., Veeder, G.J., Lester, D.F., Witteborn, F.C. Bregman, J.D., Gradie, J., Lebofsky, L.: 1985, *Icarus* **64**, 11

Jewitt, D., Meech, K.J.: 1986, *Astrophys. J.* **310**, 937

Khare, B.N., Sagan, C., Arakawa, E.T., Suits, F., Callcott, T.A., Williams, M.W.: 1984, *Icarus* **60**, 127

Kiselev, N.N., Chernova, G.P.: 1978, *Sov. Astron.* **22**, 607

Kiselev, N.N., Pushnin, P.A., Siklitsky, V.I., Chernova, G.P.: 1986, *20th ESLAB Symp. ESA SP-250, Vol. III*, p. 29

Kissel, J., Sagdeev, R.Z., Bertaux, J.L., Angarov, V.N., Audouze, J., Blamont, J.E., Büchler, K., Evlanov, E.N., Fechtig, H., Fomenkova, M.N., von Hoerner, H., Inogamov, N.A., Khromov, V.N., Knabe, W., Krueger, F.R., Langevin, Y., Leonas, V.B., Levasseur-Regourd, A.C., Managadze, G.G., Podkolzin, S.N., Shapiro, V.D., Tabaldyev, S.R., Zubkov, B.V.: 1986a, *Nature* **321**, 280

Kissel, J., Brownlee, D.E., Büchler, K., Clark, B.C., Fechtig, H., Grün, E., Hornung, K., Igenbergs, E.B., Jessberger, E.K., Krueger, F.R., Kuczera, H., McDonnell, J.A.M., Morfill, G.M., Rahe, J., Schwehm, G.H., Sekanina, Z., Utterback, N.G., Völk, H.J., Zook, H.A.: 1986b, *Nature* **321**, 336

Le Borgne, J.F., Leroy, J.L., Arnaud, J.: 1986, *20th ESLAB Symp. ESA SP-250, Vol. I*, p. 571

Mazets, E.P., Sagdeev, R.Z., Aptekar, R.L., Golenetskii, S.V., Guryan, Yu.A., Dyatchkov, A.V., Ilyinskii, V.N., Panov, V.N., Petrov, G.G., Savvin, A.V., Sokolov, I.A., Frederiks, D.D., Khavenson, N.G., Shapiro, V.D., Shevchenko, V.I.: 1986, *20th ESLAB Symp. ESA SP-250, Vol. II*, p. 3

McDonnell, J.A.M., Kissel, J., Grun, E., Grard, R.J.L.: 1986a, in *20th ESLAB Symp.*

McDonnell, J.A.M., Kissel, J., Grun, E., Grard, R.J.L., Langevin, Y., Olearczyk, R.E., Perry, C.H., Zarnecki, J.C.: 1986b, *20th ESLAB Symp. ESA SP-250, Vol II*, p. 25

Michalsky, J.J.: 1981, *Icarus* **47**, 388

Millis, R.L., A'Hearn, M.F., Thompson, D.T.: 1982, *Astron. J.* **87**, 1310

Mukai, S., Mukai, T., Giese, R.H., Weiss, K., Zerull, R.H.: 1982, *Moon Planets* **26**, 197

Mukai, T., Mukai, S., Kikuchi, S.: 1986, *20th ESLAB Symp. ESA SP-250, Vol. II*, p. 59

Myers, R.V.: 1985, *Icarus* **63**, 206

Myers, R.V., Nordsieck, K.H.: 1984, *Icarus* **58**, 431

Ney, E.P., Merrill, K.M.: 1976, *Science* **194**, 1051

Oishi, M., Kawara, K., Kobayashi, Y., Maihara, T., Noguchi, K., Okuda, H., Sato, S., Iijima, T., Ono, T.: 1978a, *Publ. Astron. Soc. Japan* **30**, 149

Oishi, M., Okuda, H., Wickramasinghe, N.C.: 1978b, *Publ. Astron. Soc. Japan* **30**, 161

Perrin, J.M., Lamy, P.L.: 1983, *Opt. Acta* **30**, 1223

Pollack, J.B., Toon, O.B., Khare, B.N.: 1973, *Icarus* **19**, 372

Tokunaga, A.T., Golisch, W.F., Griep, D.M., Kaminski, C.D., Hanner, M.S.: 1986, *Astron. J.* **92**, 1183

Wolff, M.: 1975, *Appl. Opt.* **14**, 1395

Zerull, R.H.: 1985, in *Properties and Interactions of Interplanetary Dust, IAU Coll.* **85**, eds. R.H. Giese, P.L. Lamy, Reidel, Dordrecht, p. 197

Astron. Astrophys. 187, 625–628 (1987)

ASTRONOMY AND ASTROPHYSICS

The 3.2–3.6 μm emission features in comet P/Halley: spectral identifications and similarities

R.F. Knacke[1,★,★★], T.Y. Brooke[1,★,★★], and R.R. Joyce[2,★,★★]

[1] Astronomy Program, Department of Earth and Space Sciences, State University of New York, Stony Brook, NY 11794, USA
[2] Kitt Peak National Observatory, P. O. Box 26732, Tucson, AZ 85726, USA

Received January 29, accepted March 30, 1987

Summary. Ground-based spectroscopic observations of emission features near 3.3 μm in comet Halley are reported. At a resolution of $\Delta\lambda/\lambda = 0.012$ the emissions consist of several components between 3.2 μm and 3.6 μm, probably superpositions of C—H group bands. The grains responsible for the 3–20 μm thermal emission do not appear to be the site of the band emitter, but either molecular or small particle emission is possible. The band shape differs from those of interstellar emitting substances in having strong features longward of 3.3 μm bands. This extra structure may be due to alkyl side-chains or to compounds characteristic of low formation temperatures. Emission at 3.52 ± .02 μm could be evidence for oxygen containing molecules, possibly formaldehyde.

Key words: dust – comets – infrared radiation – interplanetary medium: dust

1. Introduction

The infrared spectrometer aboard the Vega spacecraft detected emissions between 3.2–3.4 μm in comet Halley that were not identified with a specific compound, although the shorter wavelengths are highly suggestive of C—H stretch vibrations in carbonaceous material (Combes et al., 1986a). Because of the evident importance of a carbonaceous substance in or near the nucleus of a comet, we initiated ground-based observations to confirm the existence of the 3.2–3.4 μm feature and to study it further. Ground-based observations were also completed recently by Wickramasinghe and Allen (1986), Baas et al. (1986), and Danks et al. (1987).

2. Observations

Comet Halley was observed on the nights of 26, 27, and 29 March 1986 UT at the Cerro Tololo (CTIO) 1.5m telescope and on 25 and 27 April at the Infrared Telescope Facility (IRTF) on Mauna Kea. Spectra centered on the brightest spot in the coma of comet Halley are shown in Fig. 1. The observing procedures and analysis are contained in Knacke et al. (1986). Some relevant aspects of the observations are summarized here.

The infrared flux of comet Halley changed rapidly during the CTIO observations. On 29 March both the continuum and the 3.36 μm band flux increased by a factor of almost 3 over the 27 March values. The same event was observed by Telesco et al. (1986).

The data of 25 April show that the 3.36 and 3.52 μm features had the same shape as in March, but the continuum slope and the line to continuum ratio changed. The 3–4 μm spectral region is in the intersection of the scattered solar radiation and the thermal emission continuum. The dust cooled from 26 March to 25 April as comet Halley moved from a heliocentric distance of 1.09 to 1.55 AU. This shifted the exponential side of the blackbody curve so that thermal emission dominated the 3–4 μm continuum in March and scattering dominated at the end of April.

In April, the emission feature was stronger relative to the scattered continuum than it was to the thermal continuum of

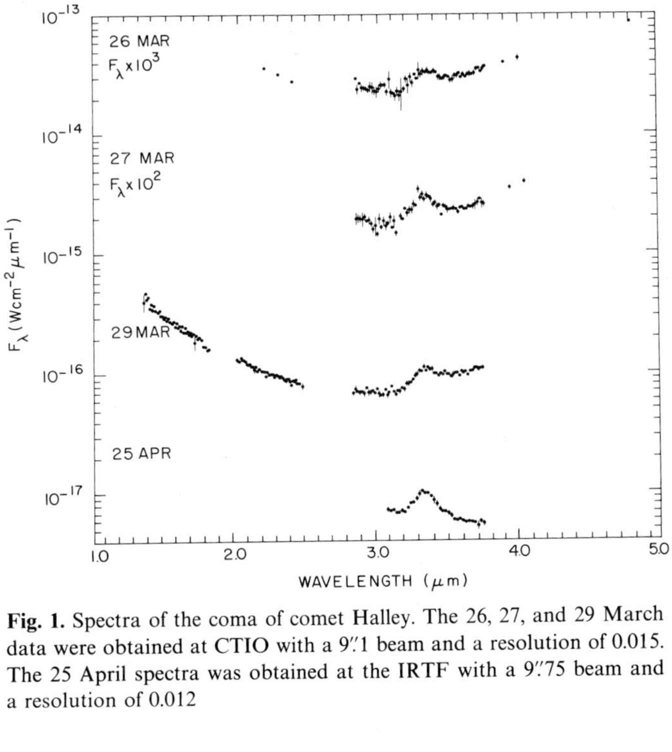

Fig. 1. Spectra of the coma of comet Halley. The 26, 27, and 29 March data were obtained at CTIO with a 9″.1 beam and a resolution of 0.015. The 25 April spectra was obtained at the IRTF with a 9″.75 beam and a resolution of 0.012

Send offprint requests to: R.F. Knacke

★ Visiting astronomer, Cerro Tololo Inter-American Observatory, supported by the National Science Foundation.

★★ Visiting Astronomer at the Infrared Telescope Facility which is operated by the University of Hawaii under contract to the National Aeronautics and Space Administration almost a month earlier. Therefore the 3.36 μm emission cannot be coupled to the thermal continuum emission that dominates the 4 to 20 μm region and that includes the emissions by silicate grains. It must be from a separate emitter – either another dust component or a gaseous source.

3. Radiation mechanisms and abundance

Inspection of the Vega IKS spectra suggests that the loss rate of the 3.36 μm emitter is of order that of the water if the radiation mechanism and band strengths are similar. Estimates of molecular scattering to give the 3.36 μm feature assuming reasonable g-values confirm that a high production rate of the C—H emitter would be required, and this has led to suggestions that other emission mechanisms might be involved such as small particle emission or ultraviolet pumped infrared fluorescence (Knacke et al., 1986; Baas et al., 1986).

4. Towards a spectral identification

At this point we have not identified the emission mechanism nor determined whether the source is a gas or small particles. Nevertheless, it is useful to see what can be inferred about the chemical composition of the emitter, based on comparison of the infrared bands with laboratory and astronomical spectra of pure compounds, meteorites, and natural or synthetic substances.

The comet emission bands show structure with at least three, and possibly more, features between 3.2 and 3.6 μm. There is a suggestion of a band or shoulder at 3.28 μm in Fig. 1 that shows up clearly in the higher resolution spectrum obtained by Baas et al. (1986). A strong feature is centered at 3.36 μm; observations are needed to see if it breaks up into components at higher resolution. There is an emission peak at 3.52 μm in all the ground-based observations. Baas et al. suggest that it might be a doublet. There may also be a weak absorption at 3.22 μm.

The structure and complexity of the band would be expected from C—H vibrations. The 3 μm spectra of complicated organic molecules or mixtures consist of overlapping bands of C—H groups with shifts in the fundamental frequencies on the order of tenths of microns. Unfortunately the 3 μm absorptions are so strong and ubiquitous that their interpretation is often difficult. Despite this, we believe that some preliminary proposals can be made towards identification of the comet emitters.

The 3.28 μm (3040 cm^{-1}) feature is near the position of an interstellar emission usually centered near 3030 cm^{-1} (Baas et al., 1986; Geballe et al., 1985). In the interstellar case, this feature has been proposed to originate in polycyclic aromatic hydrocarbons ("PAH"), multiple ring compounds (Duley and Williams, 1981; Léger and Puget, 1984; Allamandola et al., 1985). Possibly related compounds ("QCC") were suggested by the laboratory syntheses of Sakata et al. (1984). While the PAH or QCC identifications are not yet secure, these compounds are plausible organic components in the regions near ultraviolet sources where the emission is found.

A distinct difference between the interstellar emitter and the comet emitter is the presence of only weak emissions longward of 3.3 μm in the former (Geballe et al., 1985), and the dominant emission centered at 3.36 μm in the Halley spectrum. The 3030 cm^{-1} absorption in the laboratory spectra of aromatic hydrocarbons is the vibration of the =C—H bond of the ring carbons. In

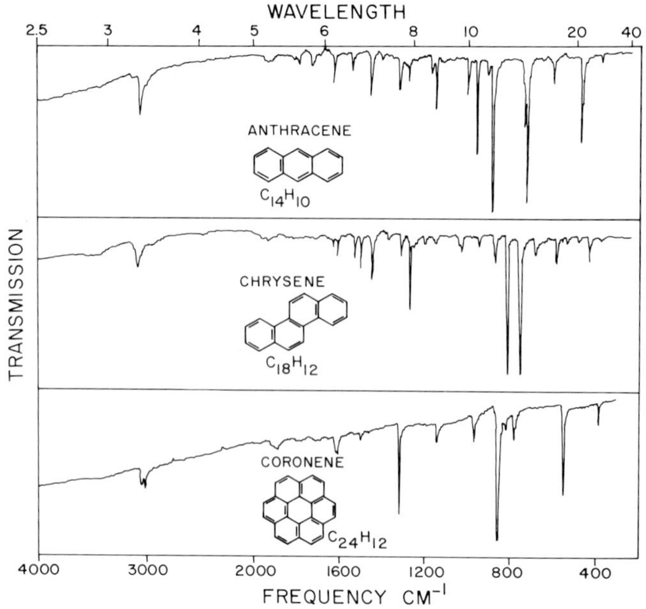

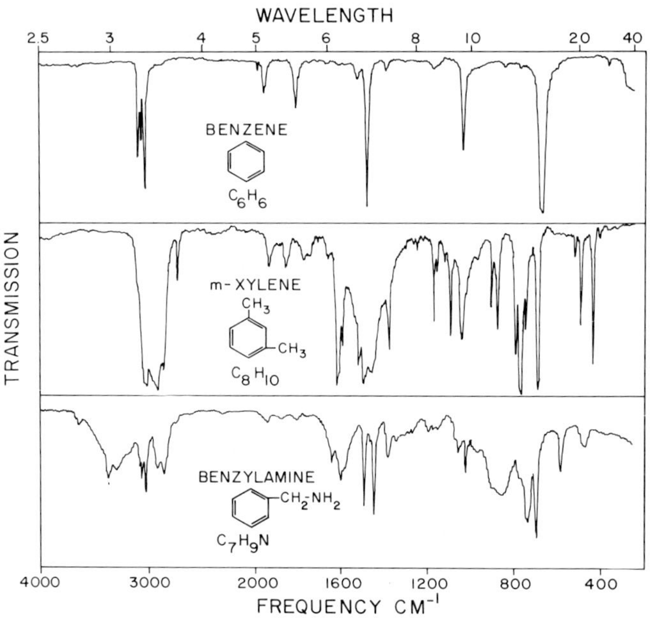

Fig. 2. (Top) Spectra of typical aromatic hydrocarbons without side chains. (Bottom) Spectra of aromatic hydrocarbons with side chains or substitutions (Stadler Standard Spectra, 1969)

aromatic hydrocarbons without alkyl chains (—CH_3 and other groups), this absorption is dominant in the 3 μm region as shown in Fig. 2. In this figure one can also see that the spectra tend to become simpler as the molecules become more complicated. The largest one shown, coronene (Léger and Puget, 1984), corresponds most closely to the interstellar absorptions, although the agreement is far from exact.

Aromatics with alkyl chains have more complicated 3 μm spectra which include the C—H vibrations of the chain hydrocarbons. Noteworthy is that the chain groups have *stronger* band strengths than do the ring groups. This suggests that if the PAH identification is indeed correct, then the interstellar rings only rarely have alkyl side chains. The small features observed by Geballe et al., (1985) at lower frequencies than the 3030 cm^{-1} band may originate from such groups, but they are evidently few.

The comet spectra could, in this interpretation, contain features of PAHs with abundant side chains to give the broad 3.36 μm feature. The spectra could equally well consist of spectral bands of C—H's *not* bound to ring molecules, but to some other structure, since absorptions in this wavelength region are generally observed from $—CH_2$ and $—CH_3$ groups. In any event, the Halley emitter cannot be dominated by aromatics that lack side chains, as the interstellar emitter may be.

The interstellar 3.4 μm features (Allen and Wickramasinghe, 1981; Smith et al., 1986) probably originate from C—H groups also, but the wavelength precludes that they are mainly PAHs. These features are seen in absorption in dense molecular clouds. The ground-based data show that the band shapes in molecular clouds and comet Halley are clearly different, although the differences are in detail; C—H groups in different structures could explain them both.

The third clearly separated structure is at 3.52 μm (the spectrum obtained by Baas et al., (1986) is suggestive of a doublet feature at 3.51 and 3.53 μm). Oxygen containing molecules such as formaldehyde or methanol have bands near these wavelengths (Knacke et al., 1986). Further evidence for formaldehyde in comet Halley's spectra is given by Danks et al., (1987) and Combes et al., (1986b).

This interpretation of the spectra has implications for the chemistry and particularly for the thermal history of the carbon compounds. In aromatics the rings are more stable than the attached alkyl chains. Indeed, the number and complexity of alkyl chains in complex substances is correlated with the temperature history of the material. The number of alkyl chains increases with decreasing formation temperature, shown in Fig. 3 (Blumer, 1976).

Since the interstellar emission feature at 3030 cm^{-1} has the fewest side chains (weakest features longward of 3.3 μm), we infer that these are the highest temperature compounds. The comet emitter, and the interstellar 3.4 μm absorber, contain evidence for *lower temperature compounds* (features longward of 3.3 μm). This spectroscopic evidence is consistent with what we know about environments in which the substances formed. The 3.3 μm emitter is located in or near H II regions and planetary nebulae where high temperature processes occur and there is a strong ultraviolet flux. Indeed, there is strong evidence that the emitting particles reach temperatures on the order of 1000 K (Sellgren, 1984). Temperatures between 10 and 100 K characterize comet formation and molecular clouds.

5. Comparison with analogue substances

Another approach to gain insight into the 3.36 μm emitter is to compare the spectra with those of natural or synthetic materials that are believed to resemble the comet substance. That organic substances would be found as parent molecules of the observed gaseous carbon compounds has been proposed many times. Summary discussion are given in Wilkening (1982).

The carbonaceous component in the carbonaceous chondrites is a candidate, or suspect, for the comet emitter. Approximately 70% of the carbonaceous material consists of an insoluble polymer that is difficult to characterize but is apparently composed of aromatic chains (Hayatsu et al., 1977). The soluble material is rich in aromatics and alkanes, particularly chains of length 11 to 18. Structure in the 3.5 μm spectra region shows up in the wings of the very strong —OH band centered near 2.8 μm. We have not found detailed meteoritic spectra needed to make more careful comparisons with organic features. Such spectra would be useful and interesting in view of the comet results.

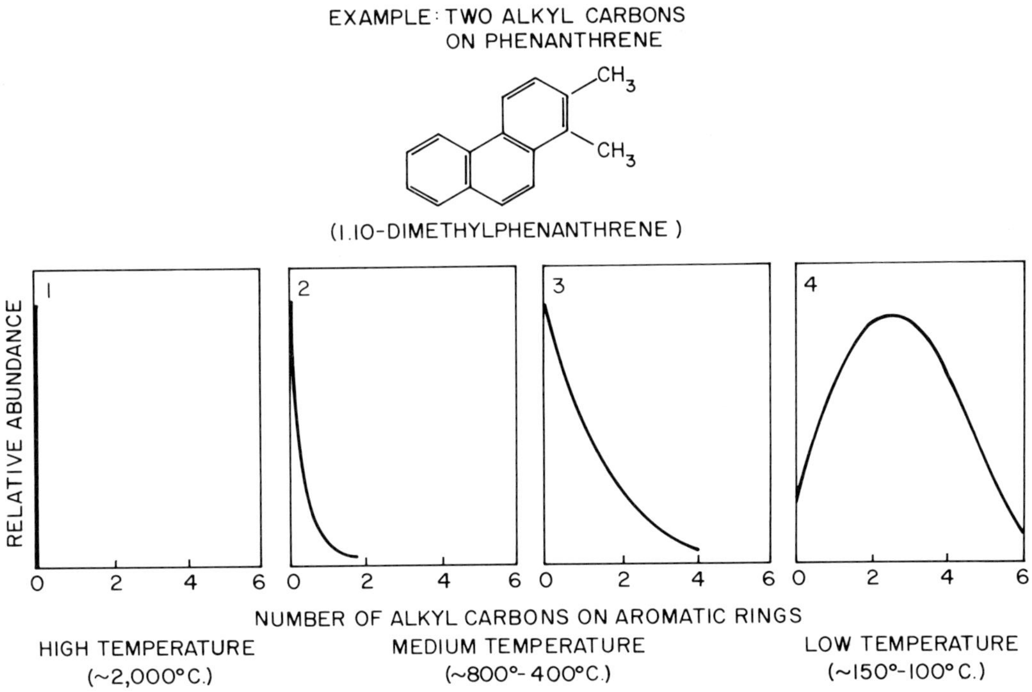

Fig. 3. Number of side chains as a function of temperature (adapted from Blumer, 1976)

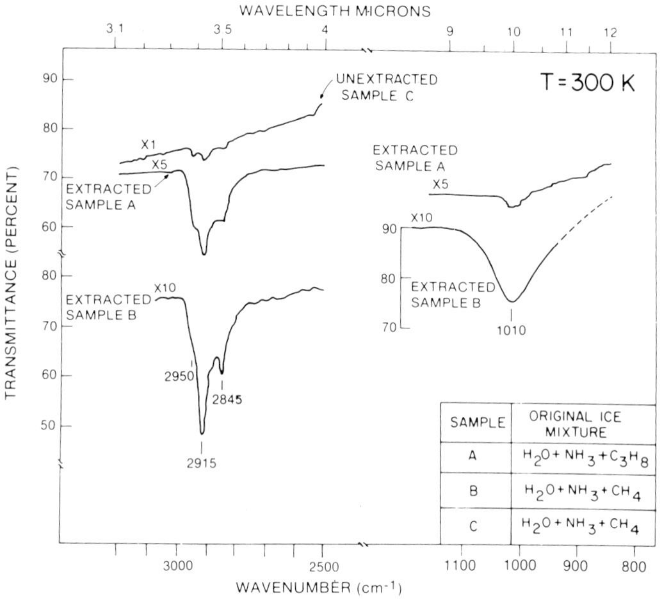

Fig. 4. Spectra of proton irradiated ices (Moore and Donn, 1982)

Numerous attempts to replicate astrophysical conditions in the laboratory have been made in recent years. Many of these materials are synthesized by irradiation of ice mixtures by ultraviolet (Greenberg, 1982), by protons (Moore and Donn, 1982), and atomic species (Bibring and Rocard, 1984). These materials show structure in the 3–4 μm region, some of it similar to comet spectra. A spectrum of the proton irradiated residue (Fig. 4) looks strikingly like the comet spectrum in having a 3.52 μm feature. A similar spectrum was recently obtained by Strazzula (1985). Further investigations are needed to assess the significance of these similarities.

These comparisons with natural and synthetic materials suggest new lines of spectral investigation. Their study has also provided valuable insights into the chemistry of interstellar and solar system processes. However, the materials are a difficult problem for spectral analysis at this stage because they are complex mixtures of organic compounds that do not readily yield chemical identifications.

6. Discussion

We have not found exact spectra correspondence with any of the materials that seem plausible candidates for the comet emitter. The suggestions made in this paper are preliminary ones based on the available data; the real situation may turn out to be different. The comet emitter probably consists of a complex mixture of organic compounds, which judging by laboratory and astrophysical experience, will be difficult to characterize exactly. Studies of this band, particularly at higher resolution, are needed in other comets to make progress in the identification of the features. It is not likely that the comet substance is precisely identical to materials in other environments, but this should not obscure the significance of the presence of carbonaceous material in many "primitive" objects.

Acknowledgement. This research was supported at Stony Brook by National Science Foundation Grant No. AST 8414825.

References

Allamandola, L.J., Tielens, A.G.G.M., Barker, J.R.: 1985, *Astrophys. J. Letters* **290**, L25

Allen, D.A., Wickramasinghe, D.T.: 1981, *Nature* **294**, 239

Baas, F., Geballe, T.R., Walther, D.M.: 1986, *Astrophys. J. Letters* **311**, L97

Bibring, J.P., Rocard, F.: 1984, *Adv. Space Res.* **4**, 12, 103

Blumer, M.: 1976, *Scientific Am.* **234**, 34

Combes, M., Moroz, V.I., Crifo, J.F., Lamarre, J.M., Charra, J., Sanko, N.F., Soufflot, A., Bibring, J.P., Cazes, S., Coron, N., Crovisier, J., Emerich, C., Encrenaz, T., Gispert, R., Grigoryev, A.V., Guyot, G., Krasnopolsky, V.A., Nikolsky, Yu.V., Rocard, F.: 1986a, *Nature* **321**, 266

Combes, M., Moroz, V., Crifo, J.F., Bibring, J.P., Coron, N., Crovisier, J., Encrenaz, T., Sanko, N., Grigoriev, A., Bockelée-Morvan, D., Gispert, R., Emerich, C., Lamarre, J.M., Rocard, F., Krasnopolsky, V., Owen, T.: 1986b, *Bull. Amer. Astron. Soc.* **18**, 794

Danks, A., Encrenaz, T., Bouchet, P., LeBertre, T., Chalabev, A., Epchtein, N.: 1987, *Astron. Astrophys.* (in press)

Duley, W.W., Williams, D.A.: 1981, *Monthly Notices Roy. Astron. Soc.* **196**, 269

Geballe, T.R., Lacy, J.H., Persson, S.E., McGregor, P.J., Soifer, B.T.: 1985, *Astrophys. J.* **292**, 500

Greenberg, J.M.: 1982, in *Comets*, ed. L.L. Wilkening, Univ. Arizona Press, Tucson, p. 131

Hayatsu, R., Matsuoka, S., Scott, R.G., Studier, M.H., Anders, E.: 1977, *Geochim. Cosmochim. Acta* **41**, 1325

Knacke, R.F., Brooke, T.Y., Joyce, R.R.: 1986, *Astrophys. J. Letters* **310**, L49

Léger, A., Puget, J.L.: 1984, *Astron. Astrophys.* **137**, L5

Moore, M.H., Donn, B.: 1982, *Astrophys. J. Letters* **257**, L47

Sakata, A., Wada, S., Tanabé, T., Onaka, T.: 1984, *Astrophys. J. Letters* **287**, L51

Sellgren, K.: 1984, *Astrophys. J.* **277**, 623

Smith, R.G., Sellgren, K., Tokunaga, A.T.: 1987 (preprint)

Stadler Standard Spectra: 1969, Stadler Research Laboratories, Philadelphia

Strazzula, G.: 1985, *Icarus* **61**, 48

Telesco, C.M., Decher, R., Ramsey, B.D., Campins, H.: 1986 (private communication)

Wickramasinghe, D.T., Allen, D.A.: 1986, *Nature* **323**, 44

Wilkening, L.L.: 1982, ed., *Comets*, Univ. Arizona Press, Tucson

Astron. Astrophys. 187, 629–631 (1987)

ASTRONOMY AND ASTROPHYSICS

Airborne spectrophotometry of P/Halley from 16 to 30 microns

T. Herter[1], **Humberto Campins**[2], **and G.E. Gull**[1]

[1] Space Sciences Building, Cornell University, Ithaca, NY 14853, USA

[2] Planetary Science Institute, 2030 E. Speedway, No. 201, Tucson, AZ 85719, USA

Received February 17, accepted April 23, 1987

Summary. We report on the first medium resolution spectrophotometric observations of a comet ever obtained in the 16- to 30-μm region. Comet P/Halley was observed using the Cornell University 7-channel spectrometer ($\Delta\lambda/\lambda \sim 0.02$ on board the Kuiper Airborne Observatory on 1985 December 14.2. Measurements were made centered on the nuclear condensation. These observations show a complex spectrum which could not be fit with simple models. The 20 μm silicate feature, if present, is very weak; and a relatively narrow ($\Delta\lambda \sim 0.5\,\mu$m), strong feature centered at approximately 28.4 μm is observed and attributed to dust emission.

Key words: airborne – infrared – dust

1. Introduction

The 16- to 30-μm region of the spectrum is mostly inaccessible to ground-based telescopes because of atmospheric absorption. For this reason the only observations of comets in this region which were available until now are broad and narrowband photometry near 20 μm (e.g., Ney, 1974; Rieke and Lee, 1974), which show the presence of an emission band. This band, along with another emission band observed near 10 μm, has been attributed to silicates. The longer wavelength band when combined with the 10 μm one is useful in determining the type of silicate present, and also whether the cometary silicates resemble more closely interstellar silicates or the silicates found in meteorites (Zailkowski and Knacke, 1975) and in Interplanetary Dust Particles (Sandford and Walker, 1985). Except for photometric observations like those just mentioned, this spectral region has been virtually unexplored. In addition to the compositional information available in the spectral features, the shape of the thermal continuum can be used to determine the color temperature of the emitting particles and constrain their size distribution. The observations presented in this work represent the first 16- to 30-μm spectrum of a comet ever obtained.

2. Observations

Comet P/Halley was observed on 14 December 1985 using the 91-cm telescope of the Kuiper Airborne Observatory (KAO) on a flight out of NASA-Ames, Moffett Field, California. The comet was located at a distance of 0.80 AU from the Earth and had a heliocentric distance of 1.28 AU at the time of the observations. The flight altitude for the Halley leg of the flight was in excess of 11.7 km. The Cornell 7-channel, helium-cooled spectrometer was used. The Si:Sb array system has a resolution of about 0.2 μm for the 16–30 μm range over which Halley was observed. The beam size, when this instrument is used on the KAO, is approximately 30″ in diameter (Full Width Half Maximum). The stars μ Cep and α Ori were used as calibrators and were observed before and after the comet observations respectively. α Ori served as the primary flux calibrator. The α Ori spectrum is taken from Forrest et al. (1979, hereafter FHM) except for wavelengths shortward of 20 μm where a cross-calibration with a lunar spectrum obtained in August 1985 indicates that the FHM spectrum may be in error. The lunar calibration indicates the spectrum of α Ori rises to shorter wavelengths with the same slope (in a semi-log plot) as at nearby longer wavelengths and does not turn over as in FHM.

Although the on-board radiometers were not working properly, subsequent examination of radiosonde and satellite data indicates that the amount of precipitable water vapor along the line of sight was about 15 μm and was relatively constant over the duration of the flight (McMahon and Kuhn, private communication). The reliability of these estimates are difficult to determine, and unfortunately, telluric water vapor can significantly effect the observed spectrum, particularly near 28.4 μm where a possible feature has been found (see Sect. 4).

Spectra at one point per resolution element from 16–30 μm were obtained by performing four ten-second integrations per grating position and stepping the grating through ten different positions. Seven spectral positions are sampled per grating position. Nine complete spectral scans were obtained over a period of approximately 1.75 hours. The spectrum obtained by averaging the complete data set and calibrating by α Ori is shown in Fig. 1.

3. The model

In order to define the strength and shape of any spectral features which may be present, a continuum level must be established. To establish this continuum level we combine the current spectrum with that of Tokunaga et al. (1986) and Campins et al. (1986). The Tokunaga et al. spectrum consists of 8–13 μm photometry and a 20-μm photometric measurement. Because of differing beam sizes, the Tokunaga et al. spectrum had been scaled

Send offprint requests to: T. Herter

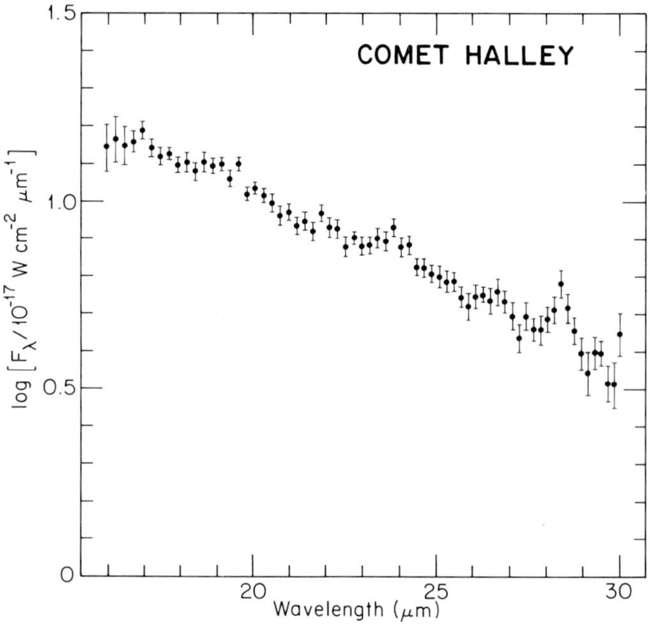

Fig. 1. 16–30 μm spectrum of comet P/Halley obtained with a 30″ beam size and a spectral resolution of $\Delta\lambda \sim 0.2\,\mu$m obtained on 14.2 December 1985. See Sect. 4 for a discussion of the reality of the features

to agree with the current spectrum at 20 μm. This combined data is shown in Fig. 2.

The best blackbody fit to the data is 285 K, shown in Fig. 2 also. This fit demonstrates the complex structure of the spectrum. In practice the observed emission is due to a mixture of grains of varying composition, size and temperature. As demonstrated by the blackbody fit, the whole spectrum cannot be fit by thermal emission from dark absorbing grains (with emissivities which vary smoothly with wavelength) at any temperature, although the spectrum for $\lambda > 16\,\mu$m seems well matched by a blackbody, probably indicating the presence of relatively large dust grains. These grains are believed to be responsible for most of the continuum emission from the coma.

For simplicity a preliminary analysis has been carried out using a two-component model for the dust particles responsible for the thermal emission in the coma of the comet. Based on the evidence in favor of the presence of silicates in comets, we have chosen to use silicates as the one component. The emissivity of the silicate material is taken from FHM who fit the observed emission circumstellar sources, although the possible problems with the FHM calibration discussed in Sect. 2 could invalidate the appropriateness of applying their deduced emissivity function. These emissivities are taken as representative of carbonaceous chondrite materials (see discussion of FHM), and is used only in a first attempt to match the overall shape of the spectrum; further work will be necessary to determine the actual composition of the cometary silicates. It must be pointed out that silicates alone are also not sufficient to match our spectrum.

For our second component we elected to try an emissivity law which varies smoothly with wavelength, such as is the case for graphite. Our finding, however, is that a graphite component, e.g., the emissivity law proportional to $\lambda^{-1.5}$, cannot be used along or in conjunction with the silicate component to fit the spectrum. As such, we use as our second component greybody emission which has an emissivity law independent of wavelength.

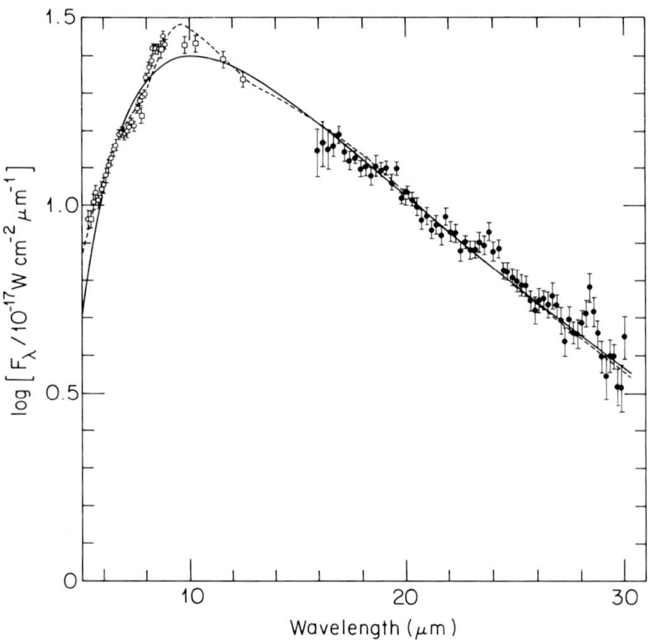

Fig. 2. Combined spectra from the present work (filled circles), data from 1985 December 14; Campins et al. (open circles), data from 1985 December 12; and Tokunaga et al. (open boxes), data from 1985 December 13. The data of Tokunaga et al. have been scaled to match the current observations at 20 μm. The solid curve is a blackbody fit, while the dashed line is a combined blackbody and silicate fit (see Table 1). The emission features near 23.5 and 28.4 μm have been ignored in the fitting process

4. Results

Table 1 gives the results of our simple models which attempt to fit the data, the first being the blackbody fit. The other model has a silicate component and a blackbody component. As can be seen from Fig. 2, neither of the models adequately reproduce the spectrum, but do indicate that a mixture of types of grains need to be combined to understand the spectrum. More detailed models which perform simultaneous fits to the 10- and 20-μm silicate features and to the 5- to 8-μm and 30- to 65-μm continuum (observed on other flights just before and after our flights; Campins et al., 1986 and Glaccum, et al., 1986) may be necessary to constrain significantly the composition of the particles and their size distribution and the relative abundance of silicates and dark absorbing grains. Such fits can be attempted, but the temporal variability of the comet and the corrections

Table 1[a]. Fits to 5–30 μm spectrum

Model number	Silicate component		Blackbody component	
	a	T(K)	a	T(K)
1		–	13.3	285
2	1.7	422	12.0	272

[a] Silicate component uses emissivity of Forrest et al. Component amplitudes, a, are the flux contributions to each fit (in units of $10^{-17}\,\mathrm{W\,cm^{-2}\,\mu m^{-1}}$) at 18 μm

necessary for the effects of different aperture sizes used by the different instruments complicate matters further.

In addition to the silicate feature, other features are evident in the spectrum. The strongest, a sharp feature at approximately 28.4 μm, which is about 50–60% above the continuum and has a full width at half maximum of 0.5 μm, has never before been observed in an astronomical source. This feature falls in a region where there are water vapor transitions. We have carried out modelling of the atmospheric water vapor absorption to see if the observed feature is a spurious result produced by an improper atmospheric correction.

The change of water content necessary to artificially generate the 28.4-μm feature requires that the comet and the calibrator were observed through 5.0 and 40.0 μm of precipitable water vapor respectively. Both the height and width of the feature are mimicked by this amount of change in line-of-sight water vapor. This change seems extreme when compared with the evidence for rather uniform water vapor content cited earlier and to the stability of the water vapor during the standard star and flight legs.

We note that there seemed to be little change in the line-of-sight water vapor within the comet P/Halley leg (approximately three hours long) and within the calibrator leg. Dividing the Halley data into three segments (beginning, middle and end), no difference is found between the spectra. The feature appears in all three spectra with the same strength. Comparison of data taken at the beginning and end of the α Ori leg indicates that there is no water vapor variation over the calibration leg. Thus a quantum jump from 5 to 40 μm of precipitable water vapor within 15 minutes would have been necessary to mimic the observed feature.

Since the spectral resolution of a grating spectrometer can vary between point and extended sources, dividing comet P/Halley, an extended source, by α Ori, a point-like source, could spuriously produce a feature. We checked this by calibrating a lunar spectrum obtained in August 1985 spectrometer with α Ori. The calibrated lunar spectrum is smooth, showing no evidence of a feature at 28 μm.

In summary, although we have no absolute proof that it is not artifically generated by improper correction for telluric water vapor absorption, the indirect evidence strongly suggests that it is real. As a cautionary note, however, Glaccum et al. observed comet P/Halley two days following our observations covering the 28 μm spectral region with a spectral resolution of ~1 μm and did not see the feature.

If we assume the 28 μm feature is real, then a possible identification must be sought. Taking the emission to be due to a gaseous component, then the flux observed in the feature which is approximately $10^{-17}\,\mathrm{W\,cm^{-2}}$ is given by $F = fNA/d^2\, h\nu/4\pi$ where N is the total number of emitting molecules within the beam, $h\nu$ is the photon energy at 28.4 μm, d is the distance to the comet, A is the transition probability, and f is the fraction of emitters in the level from which the emission originates. We also have the production rate of a given species as $Q \sim Nv/R$ where v is the outflow velocity ($\sim 1\,\mathrm{km\,s^{-1}}$) and R is the projected beam radius on the comet (~9000 km). Combining these equations and taking Q to be the water production rate at the time of the observation (cf. Mumma et al.) yields $f = [4\,10^{-3}/A][v/(1\,\mathrm{km\,s^{-1}})][Q/10^{29}\,\mathrm{s^{-1}}]^{-1}$. Assuming a typical A-value for a rotational transition (which is the likely type of transition for this wavelength) of ~0.01, and assuming water to be the emitting species (i.e., $Q = 10^{29}\,\mathrm{s^{-1}}$), we obtain a population fraction of 4% in the emitting level. Since direct observations and a detailed model of the H_2O transition of H_2O in comet Halley already exist (Mumma et al., 1986) it is possible to compare the f-value calculated above with that predicted by the model and rule out H_2O molecules as the emitters responsible for the feature. Other possible gaseous sources including H_2 have also been considered for this feature, but they all appear to be too weak by several orders of magnitude (by the above analysis). Emission due to a solid state feature produced by the dust seems more likely.

Two weaker features are observed at approximately 23.8 and 26.7 μm. It is possible that these weaker features and even the 28.4 μm one are due to structure in the spectrum of the silicate particles which produce a broad feature near 19 μm. In fact, the laboratory spectra of different silicates (Zailkowski et al., 1975) show complex structure in the 16- to 30-μm region. Further work to match these features must be explored to refine our preliminary interpretation.

Acknowledgements. This work was supported in part by NASA Contract NAS2-12323 to the Planetary Science Institute (PSI) which is a division of Science Applications International Corporation, and by NASA Grant NAG2-207 to Cornell University. We would like to thank the staff of the KAO for their excellent support which made these observations possible. We thank the referee for valuable comments. This is PSI contribution number 233.

Note added in proof: Glaccum et al. (private communication) report that a recent reduction of Halley data from December 20 does show the presence of a 28.4 μm feature. This feature is not seen in data taken on December 17, April 15, or April 17.

References

Campins, H., Bregman, J.D., Witteborn, F.C., Wooden, D.H., Rank, D.M., Allamandola, L.J., Cohen, M., Tielens, A.G. G.M.: 1986, 20th ESLAB Symp. ESA SP-250 Vol. II, p. 121

Forrest, W.J., McCarthy, J.F., Houck, J.R.: 1978, *Astrophys. J.* **233**, 611

Glaccum, W., Moseley, S.H., Campins, H., Loewenstein, R.F.: 1986, 20th ESLAB Symp. ESA SP-250, Vol. II, p. 111

Huffman, D.R.: 1977, *Adv. Phys.* **20**, 129

Mumma, M.J., Weaver, H.A., Larson, H.P., Davis, D.S., Williams, M.: 1986, *Science* **232**, 1523

Ney, E.P.: 1974, *Icarus* **23**, 551

Rieke, G.H., Lee, T.A.: 1974, *Nature* **248**, 737

Sandford, S.A., Walker, R.M.: 1985, *Astrophys. J.* **291**, 838

Tokunaga, A.T., Golisch, W.F., Griep, D.M., Kaminiski, C., Hanner, M.S.: 1986, *Astron. J.* (submitted)

Zaikowski, A., Knacke, R.F.: 1975, *Astrophys. Space Sci.* **35**, 97

Zaikowski, A., Knacke, R.F., Povoco, C.C.: 1975, *Astrophys. Space Sci.* **35**, 97

Astron. Astrophys. 187, 632–634 (1987)

ASTRONOMY AND ASTROPHYSICS

Photometry of comet P/Halley from 40 to 160 microns

H. Campins[1], M. Joy[2], P.M. Harvey[2], D.F. Lester[2], and H.B. Ellis, Jr.[2]

[1] Planetary Science Institute, SAIC, 2030 E. Speedway, Tucson, AZ 85719 USA
[2] University of Texas at Austin, Dept. of Astronomy, R.L. Moore Hall, Austin TX 78712 USA

Received February 17, accepted April 23, 1987

Summary. Far infrared photometry of comet P/Halley was obtained from NASA's Kuiper Airborne Observatory. A preliminary analysis of the observations shows the following results: (1) The brightness of the comet decreased by nearly a factor of two on the second day, in phase with the variability observed in blue light by IUE during the same time period. (2) The energy distribution varied marginally with a shallower (cooler) spectrum on the second day. (3) Observations centered 30″ and 60″ away from the nuclear condensation show lower 50–100 μm color temperatures on both dates. This is consistent with spacecraft observations which indicate a higher proportion of large particles away from the nucleus.

Key words: comets – infrared-photometry

1. Introduction

Far infrared observations of comets can provide information on the size distribution and emissivity of particles in the coma which are larger and cooler than the particles sampled by groundbased observations. The absolute level of the far-infrared flux depends on the total amount of these large particles present in the coma; the shape of the spectrum depends on the emissivity of the particles, which in turn is a function of composition and particle size.

The only previously published far-infrared photometry of a comet are the Infrared Astronomical Satellite (IRAS) observations of comet IRAS-Araki-Alcock (1983d) by Walker et al. (1984). In this paper we discuss the results from a preliminary analysis of the only far-infrared photometry obtained of comet Halley.

2. Observations

Our observations were obtained during a recent deployment of the Kuiper Airborne Observatory (KAO) (Cameron, 1976) to Christchurch, New Zealand. The comet was observed on two successive nights, 15 March and 16 March, 1986, at about 17 hours UT on both nights. The heliocentric distances were 0.92 and 0.93 AU respectively. We used the University of Texas 6-channel far-infrared photometer described by Wilking et al. (1984). On both nights photometry was obtained centered on the nuclear condensation at four far-infrared wavelengths. The effective wavelengths of the four bands were 40, 47, 95, and 160 μm with $\Delta\lambda/\lambda \sim 0.5$ at the two shorter wavelengths and 0.7 at the two larger wavelengths. Several additional photometric measurements were obtained at positions 30″–60″ away from the nucleus at 47 and 95 μm. The measured half-power-widths of the beams were 25″, 26″, 26″, and 36″ at the four wavelengths. The chopping secondary mirror of the KAO was oriented in azimuth with an amplitude of 3.′5. At this time the comet's tail was pointing about 20° away from the vertical direction.

The observations were calibrated relative to Mars for which we assumed brightness temperatures of 230, 228, 227, and 218 K respectively at 40, 47, 95, and 160 μm (Wright, 1976). The uncertainty in the absolute flux which arises from the Mars model is estimated to be about 15% at 40 and 47 μm; and 25% at 100 and 160 μm. Because of the different shape and size of the beams at the different wavelengths due to diffraction effects we have corrected all the flux densities (Fig. 1) to values which would have been observed with beam shapes equal to that of the 50 μm. This correction is only significant in the case of the 160 μm point and was estimated based on the spatial distribution of the thermal emission from this comet observed from the ground (Campins et al., 1987).

3. Results

The main results from these observations are shown in Figs. 1 and 2. Figure 1 illustrates the spectral energy distribution measured on the two successive days in the far-infrared together with that expected from a 290 K blackbody, the equilibrium temperature of a black body at the comet's heliocentric distance. Figure 2 shows the observed profiles measured along the sun-tail direction.

Figure 1 shows that the far-infrared flux density of comet Halley decreases only slightly faster with wavelength than does a blackbody. If the thermal emission was produced by particles of only one size, fits to a blackbody function modified for the wavelength dependence of the emissivity would characterize the temperature of the particles. This function would have the following form: $F(\lambda) \propto B(\lambda)\lambda^{-N}$, where $F(\lambda)$ is the thermal flux observed at a given wavelength; $B(\lambda)$ is the Planck function; λ is the wavelength and the quantity N is the spectral index for the dust emissivity. For particles which are much smaller than the wavelength "N" has values between 1 and 2. In comets, however, particles of a broad range of sizes are responsible for the thermal emission;

Send offprint requests to: H. Campins

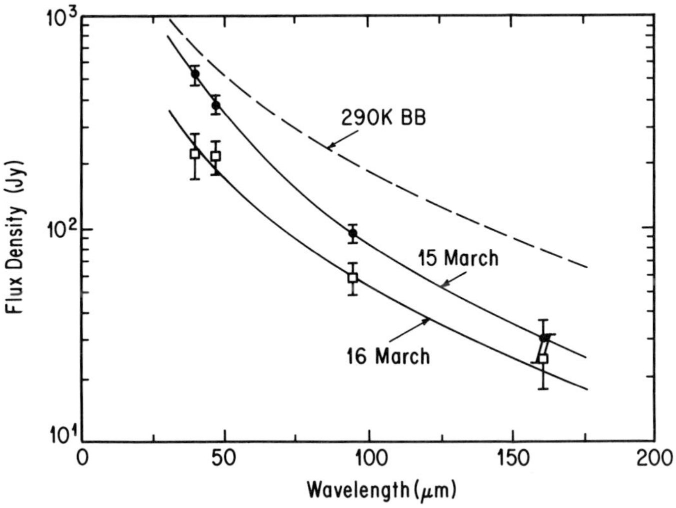

Fig. 1. The spectral energy distribution for comet Halley as measured on two successive days. The fluxes shown are for a 26″ aperture. In addition to the uncertainties shown, there is an absolute calibration uncertainty of about 15% at 40 and 47 μm and of 25% at 100 and 160 μm

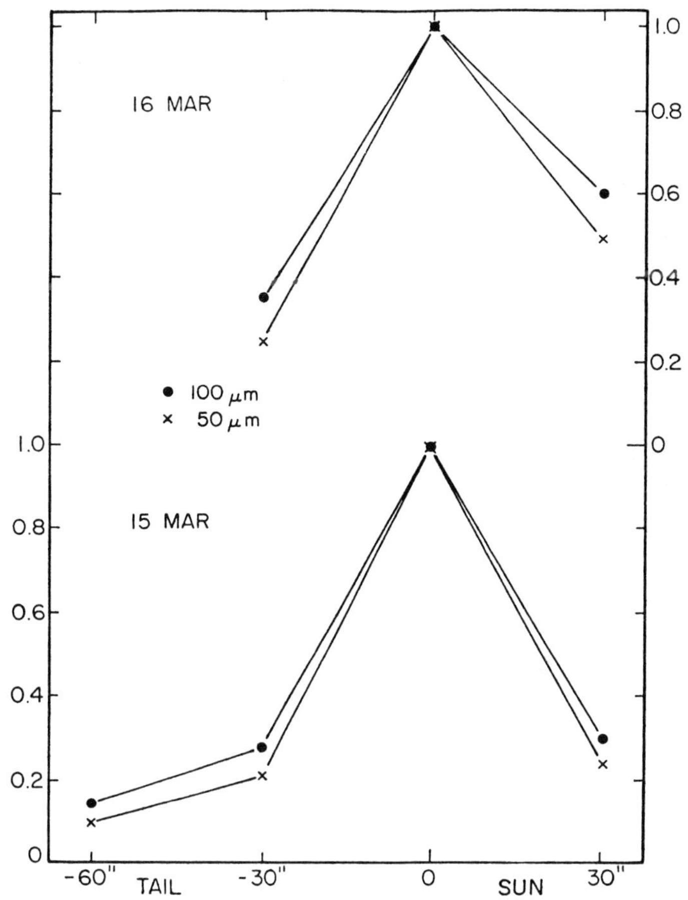

Fig. 2. The spatial distribution of the far-infrared flux of comet Halley normalized to the peak central surface brightness

in such a case, fits to this function can be used to obtain a first order estimate of the shape of the particle size distribution.

We find values of $N = 0.45 \pm 0.17$ on 15 March and $N = 0.15 \pm 0.17$ on March 16. In both cases the value is significantly less than the values of 1–2 expected for micron-sized particles and smaller. Therefore, a significant population of particles with sizes of order a few tenths of microns must exist in the coma of comet Halley. A more refined fit to comet Halley's thermal spectrum based on several particle size distributions, including that observed by the Vega spacecraft (Mazets, 1986), is given by Glaccum et al. (1987).

It is interesting to compare our far-infrared photometry with that obtained by Walker et al. (1984) of comet IRAS-Araki-Alcock. They found a wavelength dependence for the dust emissivity of $\lambda^{-0.45}$ for regions close to the nucleus and derived a mean particle radius of 30 μm. Our observations suggest that Halley had a comparable particle size distribution on 15 March. Considering the uncertainties, the value of N is only marginally lower on March 16; if real this would indicate a decrease in the relative abundance of small particles (radius $\lesssim 10\,\mu m$) on that day. A similar effect, of a shallower (cooler) energy distribution when the comet was fainter, was observed in the 5 to 10 μm region by Bregman et al. (1987).

One major difference between comet Halley and comet IRAS-Araki-Alcock is the far-infrared surface brightness. Walker et al. (1984) found, for instance, a peak 100 μm flux of $2.3\,10^8$ Jy sr^{-1} at a spatial resolution of a few thousands km. Our 95 μm flux density of 15 March implies a surface brightness of $F \sim 6\,10^9$ Jy sr^{-1} at a substantially lower spatial resolution of 18,000 km. This implies a dust column density at least 30 times larger in comet Halley.

The behavior of the spectrum with distance from the central condensation is illustrated in Fig. 2 which shows that the ratio of 50 to 100 μm flux density decreases away from the central position. We constructed a complete set of Halley source profiles which are consistent with the data (which include the "canonical" ρ^{-1} surface brightness dependence, where ρ is the radial distance from the nucleus), and convolved these with the measured beam profiles. From a comparison of these model profiles and the observed distribution, we infer that *at most* half of the radial decrease in the 50 to 100 μm flux ratio can be attributed to the slightly wider 100 μm beam profile. Thus, on the two observed days in March, there is evidence for a cooler dust color temperature and/or larger average dust particle size 30–60″ away from the nucleus of comet Halley. This is consistent with the Giotto (McDonnell, et al., 1986) and Vega (Mazets, et al., 1986) observations of a larger proportion of large particles away from the nucleus; and also with the groundbased thermal infrared images (Campins, et al., 1987) which show a temperature maximum at the central condensation. This same effect, of lower temperatures away from the central condensation was observed in comet IRAS-Araki-Alcock by IRAS. Walker et al. (1984) attribute the lower temperatures in that comet to a greater abundance of *smaller* particles away from the nucleus. As stated above, the opposite (more large particles away from the nucleus) is true in comet Halley. A re-examination of the IRAS observations and their interpretation may help resolve this apparent discrepancy.

There are peculiarities in the spatial distribution of the brightness and the color temperature which deserve attention. The asymmetry along the sun-tail direction (sunward fan) is most pronounced on March 16 which is when the comet was faintest (this is contrary to what was observed in groundbased thermal images 11 days later, Campins et al. 1987). The absolute brightness and the color temperature of the sunward observations are the same on both dates; on the other hand, the central condensation and the 30″ tailward observations became fainter and cooler on the second day. We have found no simple model to account for this behavior. More detailed modelling is in progress.

Our observations of the temporal variability appear to be in phase with those made by the International Ultraviolet Explorer (IUE) (Festou, et al, 1986) in blue light. On March 16 we observed almost simultaneously with IUE, while on 15 March they observed roughly 12 hours earlier. Our results are consistent that a maximum brightness near 8 UT on 15 March and minimum brightness at the time of our observations on 16 March. An agreement between more detailed groundbased thermal observations and IUE observations later in March (Campins et al., 1987) also indicates that the dust responsible for scattering the blue continuum exhibits the same temporal behavior as the dust responsible for the thermal emission.

Acknowledgements. We thank the ground and flight crew of the KAO for their heroic efforts in operating the observatory during this deployment. This work was supported by NASA grant NAG 2-67 to the University of Texas and NASA Contract NAS2-12323 to the Planetary Science Institute. This is PSI contribution No. 229.

References

Bregman J.D., Campins H., Witteborn F.C., Wooden D.H., Rank D.M., Allamandola L.J., Cohen M., Tielens A.G.G.M.: 1987, (this issue)

Cameron, R.M.: 1976, Sky Telescope: **52**, 327

Campins H., Telesco, C.M., Decher, R., Ramsey, B.O.: 1987 (this issue)

Festou, M.C. et al.: 1986, *Nature* **321**, 361

Glaccum, W., Moseley, S.H., Campins, H., and Loewenstein, R.F.: 1987, (this issue)

Mazets, E.P. et al.: 1986, *Nature* **321**, 276

McDonnell, J.A.M. et al.: 1986, *Nature* **321**, 338

Walker, R.G. et al.: 1984, *Astrophys. J. (Letters)* **278**, L11

Wilking, B.A., Harvey P.M., Lada C.J., Joy, M., Doering, C.R.: 1984, *Astrophys. J.* **279**, 291

Wright, E.L.: 1976, *Ap. J.* **210**, 250

Astron. Astrophys. 187, 635–638 (1987)

ASTRONOMY AND ASTROPHYSICS

Airborne spectrophotometry of P/Halley from 20 to 65 microns

W. Glaccum [1,2], **S.H. Moseley** [1], **H. Campins** [3], **and R.F. Loewenstein** [2]

[1] Laboratory for Astronomy and Solar Physics, Code 685, NASA Goddard Space Flight Center, Greenbelt, MD 20771, USA
[2] Yerkes Observatory, University of Chicago, Box 258, Williams Bay, WI 53191, USA
[3] Planetary Science Institute, SAIC, 2030 E. Speedway No. 201, Tuscon, AZ 85719, USA

Received February 16, accepted May 11, 1987

Summary. We report on simultaneous 20–65 μm spectrometry and 100 μm photometry of P/Halley obtained on board the Kuiper Airborne Observatory (KAO) in 1985 December and 1986 April. Spectra with resolution $\lambda/\Delta\lambda \simeq 20$–$40$ were obtained with the NASA/Goddard 24-channel grating spectrometer. Measurements were made on the nucleus as well as 5 points along and perpendicular to the sun-tail direction. The comet spectrum is smooth and contains no strong (>6%) spectral features. The color temperature of the dust is higher than that expected for a rapidly rotating blackbody and varies over time scales as short as 2 days. The coma is brighter at these wavelengths on the suward side than on the tailward side, but the color temperature does not vary within one arcmin of the nucleus. The spectrum is fitted with a power law distribution of dark, featureless, spherical grains.

Key words: infrared – comets

1. Introduction

The 20–65 micron region in the spectra of comets has been mostly unexplored because telluric absorption makes these wavelengths inaccessible to groundbased telescopes. The only published observations of comets in this region are the photometric measurements at 25 and 60 μm made by the Infrared Astronomical Satellite (IRAS) (Walker et al., 1984; Walker and Aumann, 1984; Matson, 1986). Nevertheless, observations of comets in this spectral range are of great interest because they can yield information about the composition, emissivity, and size distribution of the solid particles in the coma which cannot be obtained with observations at shorter wavelengths. In this work we present the first 20 to 65 μm spectra of a comet ever obtained and we discuss the results of a preliminary analysis of the observations.

2. Observations

We observed comet Halley from the 0.9 m Kuiper Airborne Observatory (KAO) on two nights in December and two nights in April (Table 1). The measurements were made using a 24-channel spectrophotometer made at NASA/Goddard Space Flight Center (GSFC) and the University of Chicago 100–300 μm photometer. The spectrometer (Moseley and Glaccum, 1987) has 24 detectors

Send offprint requests to: H. Campins

Table 1. Observations

Date	Time UT	Geocentric distance (AU)	Heliocentric distance (AU)	Phase angle
85/12/17	3:20	0.88	1.23	52.5
85/12/20	3:30	0.92	1.20	53.1
86/4/15	15:20	0.45	1.40	22.9
86/4/17	16:20	0.48	1.43	22.2

and 2 gratings on an Ebert-Fastie mount. The detectors are composite bolometers operated at 0.36 K. In one configuration the 20–35 μm range is covered with 0.9 μm resolution and in the other configuration the 35–68 μm range is covered with 1.8 μm resolution. The photometer 100 μm passband has $\lambda/\Delta\lambda = 5$. Beams were defined by 42″ field stops inside both instruments. All measurements were made with a wobbling secondary with a 2.′2 throw, perpendicular to the tail direction. We estimate the cometary flux in the reference beam to be <5% of the peak flux, so little photometric error arises from chopping.

On all nights except Dec. 20, we observed the comet over the entire 20–68 μm range; on Dec. 20, we observed only the 20–35 μm range. The Dec. 1985 measurements were calibrated with respect to α Ori, which we had calibrated to Mars on Dec. 15. The apparent structure in the 40–65 μm spectrum of the comet on Dec. 15 (see Fig. 1) is spurious, since it was seen in other reference objects on this flight as well. The Dec. 20 observations were noisier and are not reported here.

On both flights in Apr. 1986, we were able to calibrate the comet directly against Mars, our primary calibration standard. On Apr. 17 we mapped our beams in both grating settings on Mars, which had a 10″ diameter disk. On Apr. 17 we also measured the comet photometrically at 100 μm.

3. Data reduction and calibration

The measured ratio of the unknown object to the calibration object must be used to derive the flux of the unknown object. This requires that we have a model for the emission of the standard object and that we can correct for any changes in atmospheric extinction between the observations of the two objects. The flux of

Mars which was the primary calibrator was taken from the model of Wright (1976, also S. Odenwald private communication). The model for atmospheric extinction was computed for a 12.5 km altitude using the program of Traub and Stier (1975). The spectral response for each channel was measured at GSFC before the December observations and both before and after the April observations.

Flux densities are corrected for H_2O absorption by adjusting the H_2O column density for both the object and the calibrator until the telluric H_2O lines are removed from the spectrum of the object. We do not use values from the KAO water vapor radiometers, as we have found that they are poorly correlated with values derived from the spectra themselves. Flux densities are corrected for diffraction into and out of the beam by using theoretical beam profiles which are integrated over a spatial model of the object. The "flux per beam" we report is the flux within a 21″ radius of the beam center.

Several stars expected to have smooth spectra were observed and calibrated using the same procedure as on comet Halley. On Dec. 20, we observed α Tau, a star with no dust emission. On Apr. 15 we observed α Sco, which has weak emission from silicate dust, and on Apr. 17 we observed γ Cru, which has no thermal dust emission. The dustless stars have smooth Rayleigh-Jeans spectra, as expected, while α Sco has a smooth spectrum with some enhancement near 20 μm. Measurements of these and other stars to 65 μm with high signal-to-noise (Glaccum, 1987) on these flights and a subsequent flight are in agreement with ground-based 20 μm measurements found in the literature and the IRAS 25 μm fluxes, to within the errors quoted by the authors. Therefore the model for Mars has the correct shape from 20–35 μm and the actual spectrum of Mars is featureless from 20–65 μm. Furthermore, the reduction of stars (point sources) to Mars (Observed at diameters of 6″, 10″, and 22″) using theoretical beam models does not generate spurious spectral features.

The absolute uncertainty of the flux density determination for the comet is about 8 % at 20 μm, which arises from the uncertainty of the Mars model. Near 50 μm the uncertainty in the Mars model is 15 % and there is an additional error of 2–3 % (worst case: 5 %) from imperfect water vapor corrections and $\simeq$5 % arising from imperfect models for the comet's spatial distribution and the instrumental beam profile. However, on April 15, we measured Uranus, for which there exists a Voyager (IRIS) spectrum, absolutely calibrated to 1 %. Conrath (private communication) has produced a full-disk spectrum based on the Voyager data which allows for the polar orientation of Uranus. Our measurements differ from the Voyager-based model by less than 3 % at 50 μm. Thus our total flux uncertainty near 50 μm is $\simeq$8 % for our Apr. 15 measurements. We estimate the uncertainty in the ratio of the flux of the comet at 20 and 50 μm to be 8 %.

The use of "flux per beam" minimizes the error due to the uncertainty in the distribution of surface brightness of the comet. Since the beam is much larger than the spreading from diffraction and the disk of the calibrator, there is negligible error if the surface brightness falls off monotonically from the nucleus and is azimuthally symmetric. Overall, the measured beam patterns match the theoretical beams, so there is no systematic error in flux between 20 and 65 μm unless the surface brightness distribution of the comet is strongly wavelength-dependent. The reduction will generate spurious spectral features if the beam patterns differ among channels. The measured beam patterns do have very small differences on small spatial scales. Pending more complete analalysis, one should be wary of attributing significance to features smaller than $\simeq$5 %.

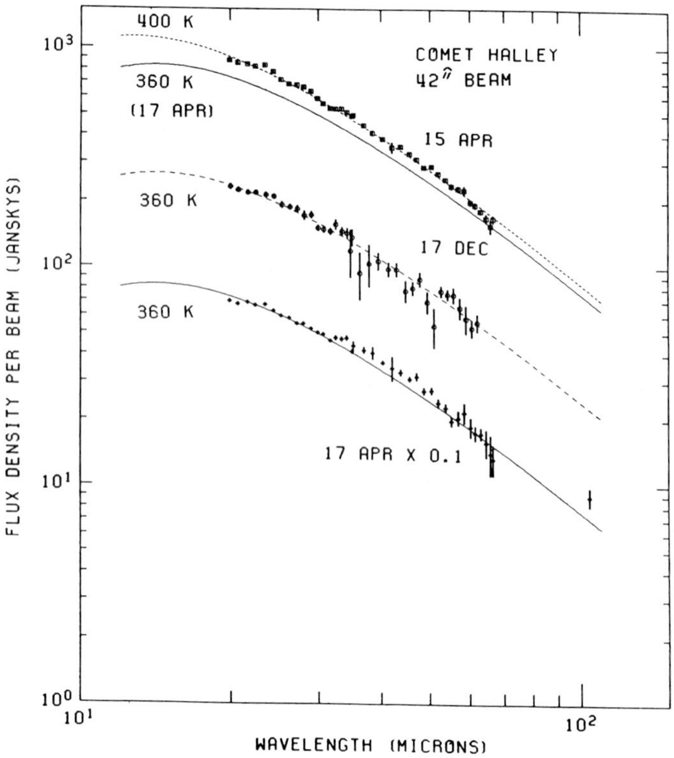

Fig. 1. Temporal dependence of the spectrum. The spectrum of P/Halley on 17 December 1986, 15 April 1986, and 17 April 1986 from 20 to 65 microns. The spectrum for 17 April is displaced by a factor of 10. The curves are for blackbodies

4. Discussion

The comet Halley nuclear fluxes are shown in Fig. 1. Only statistical errors are shown. The spectra are smooth with no feature as much as 6 % above a smooth continuum. The comet brightened by a factor of $\simeq$3.5 from December to April, which is 70 % more than expected from the change in geocentric distance, assuming a $1/r$ surface brightness distribution. There is a significant flattening of the spectrum from Apr. 15 to Apr. 17, where it returned to the same shape as December, as indicated by the lower color temperature. The curves drawn through the data are blackbodies fit by eye by adjusting the temperature and intensity. They are not meant to represent any physical model, but are merely presented as a visual aid.

In Fig. 2 we show the spatial dependence of the 20–35 μm spectrum. The off-nucleus positions are one beam FWHM away from the nucleus in the direction along and perpendicular to the tail. The sunward side is nearly twice as bright as the corresponding position opposite the nucleus, and the comet is brighter on one side of the sun-comet line than the other. There is also a very steep gradient in flux between 40″ sunward and 80″ sunward; it is 50 % steeper than an inverse distance law would predict. The overall shapes of the off-nucleus spectra are the same as that of the nucleus. However they are much less smooth than the spectrum of the nucleus. This is probably due to differences in spatial response with wavelength. We have data to address this point, but the work is not complete at this time. This does suggest that one should not attribute significance to the weak bumps in the nuclear spectrum pending the completion of this work.

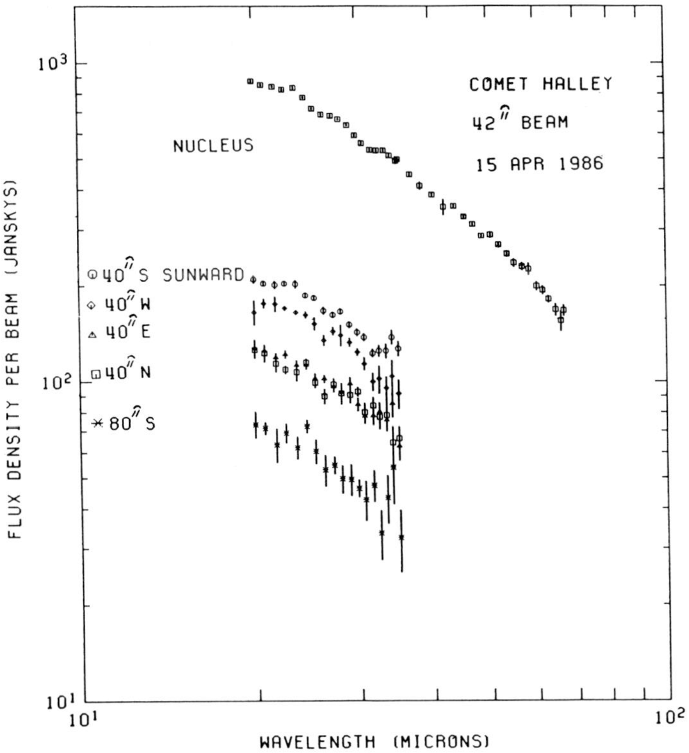

Fig. 2. Spatial dependence of the spectrum of 15 April 1986

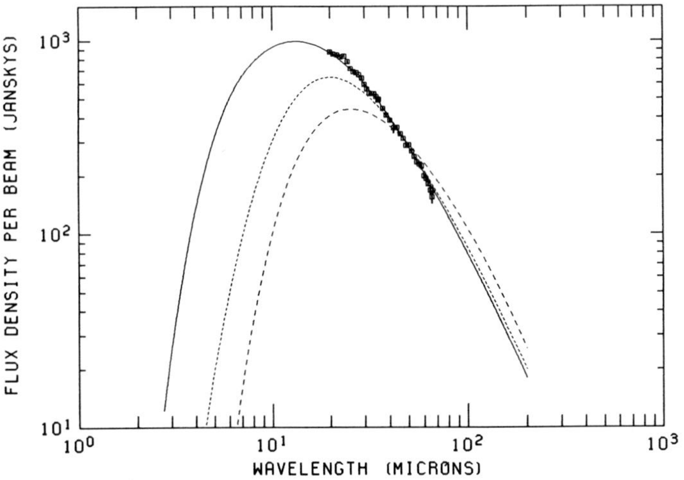

Fig. 3. Model fits to the 15 April spectrum. Models fits for 3 size distributions: (———) Power law, index −3.3. (------) Extrapolation from Vega SP-2 experiment. (— — —) Extrapolation from Giotto DIDSY experiment

Herter et al. (1986) observed a strong emission feature at 28.5 μm. Such a feature would appear in our spectra as a $>10\%$ emission feature in a single channel at 28 μm. We observed no excess at 28 μm on any flight. On our first December flight, 3 days after Herter's observations, the feature must have weakened by more than a factor of 3 relative to the continuum.

We have fit our observations with a very simple model of thermal emission from a distribution of grains. Since the 20–65 μm spectrum is smooth and because infrared spectra of the comet at shorter wavelengths indicate that most of the emission is from featureless grains, our model uses grains with no features. We were unable to find any optical constants for amorphous carbon at long wavelengths in the literature, so we assumed the grains were made of material with index $m = 1.6 + 0.7i$, independent of wavelength. We then used Mie theory for homogeneous spheres to derive Q_{abs}. Regardless of whether such an index is physically possible, the Q_{abs} has the appropriate behavior: the grains are highly absorbing in the visible and for $\lambda \gg a$, where a is the grain radius, $Q_{abs} \propto 1/\lambda$, as does amorphous carbon. We compute the emission from an ensemble of grains heated by the sun. Three emission spectra are shown in Fig. 3 along with the data from Apr. 15. The curve that fits the data well is a power law for the size distribution with $n(a) \propto a^{-3.3}$ for $0.01\ \mu m \leqq a \leqq 1000\ \mu m$. In this model, grains with radii between 1 and 50 μm contribute most of the flux, with the larger grains contributing an increasing fraction of the flux at longer wavelengths.

We also modeled the spectrum using grain size distributions from the Vega Sp-2 experiment (Mazets et al., 1986) and the Giotto DIDSY experiment (McDonnell et al., 1986). The distributions were taken from the cumulative mass distributions of both experiments for the closest distance to the nucleus, which is the appropriate distance for our 42″ beam. We assumed a density of $1\ g\,cm^{-3}$ for the grains. In both cases the grain size distribution required an extrapolation from the spacecraft data for large grains. For Vega we continued the distribution with $n(a) \propto a^{-3.3}$ for $100\ \mu m \leqq a \leqq 1000\ \mu m$, where $n(a)\,da$ is the number of grains in the range $(a, +da)$. The resulting spectrum appears colder because there are fewer of the warm, small grains ($a < 2\ \mu m$) than in the power law distribution. To reconcile the observations with the Vega mass distribution we require some or all of the grains to have a steeper wavelength dependence of emissivity. Merely changing the density requires $\varrho > 50\ g\,cm^{-3}$. For the fit to the Giotto size distribution, we assumed that there were no grains with $a > 140\ \mu m$. This model is dominated by the 10–60 μm grains and fits poorly because there are not enough larger or smaller particles to flatten the spectrum. We are aware of the problem with the mass distribution observed by Giotto and we plan to use a revised distribution in our models as soon as one is published.

From our data alone we cannot rule out the presence of a large number of dirty, amorphous silicate grains, which could contribute significantly to the flux between 20 and 35 μm, since most circumstellar silicate grains are featureless in this wavelength range (Glaccum, 1987). Silicates are seen in emission at 10 μm (e.g., Tokunaga et al., 1987) and spacecraft mass spectrometers (Kissel et al., 1986a, b) indicate that silicates account for 10–20% of the total number of grains observed. However, the 16–30 μm spectrum of Herter et al. (1986) shows no evidence of the 18 μm silicate feature, and in their fits to the 8–30 μm spectrum with silicate and amorphous carbon grains, the silicates contribute negligibly to the emission for $\lambda > 16\ \mu m$.

Acknowledgements. We thank the staff of the KAO for their usual excellent support of our observations. We thank D. A. Harper for the use of the Univ. Chicago photometer. We acknowledge support for this research from the NASA Airbone Astronomy Program. H.C. was supported by NASA under contract NAS2-12323 to the Planetary Science Institute, a division of Science Applications International Corporation. This is PSI contribution No. 232. W.G. acknowledges the support of the NASA Graduate Student Researchers Program. An anonymous referee pointed out an error in the models.

References

Glaccum, W.: 1987, Ph.D. Thesis (in preparation)

Herter, T., Campins, H., Gull, G.E.: 1986, 20th ESLAB Symp., ESA SP-250, Vol. **II**, p. 117

Kissel, J., Sagdeev, R.Z., Bertaux, J.L., Angarov, V.N., Audouze, J., Blamont, J.E., Büchler, K., Evlanov, E.N., Fechtig, H., Fomenkova, M.N., von Hoerner, H., Inogamov, N.A., Khromov, V.N., Knabe, W., Krueger, F.R., Langevin, Y., Leonas, V.B., Levasseur-Regourd, A.C., Managadze, G.G., Podkolzin, S.N., Shapiro, V.D., Tabaldyev, S.R., Zubkov, B.V.: 1986a, *Nature* **321**, 280

Kissel, J., Brownlee, D.E., Büchler, K., Clark, B.C., Fechtig, H., Grün, E., Hornung, K., Igenbergs, E.B., Jessberger, E.K., Kreuger, F.R., Kuczera, H., McDonnell, J.A.M., Morfill, G.M., Rahe, J., Schwehm, G.H., Sekanina, Z., Utterback, N.G., Völk, H.J., Zook, H.A.: 1986b, *Nature* **321**, 336

Matson, D.L., ed.: 1986, *IRAS Asteroid and Comet Survey*, Preprint Version No. 1, Infrared Processing and Analysis Center, Jet Propulsion Laboratory, California Institute of Technology, Pasadena, California, USA

Mazets, E.P., Aptekar, R.L., Golenetskii, S.V., Guryan, Y.A., Dyachtov, A.V., Ilyinskii, V.N., Panov, V.N., Petrov, G.G., Savvin, A.V., Sagdeev, R.Z., Sokolov, I.A., Khavenson, N.G., Shapiro, V.D., Shevchenko, V.I.: 1986, *Nature* **321**, 276

McDonnell, J.A.M., Alexander, W.M., Burton, W.M., Bussoletti, E., Clark, D.H., Grard, R.J.L., Grün, E., Hanner, M.S., Hughes, D.W., Igenbergs, E., Kuczera, H., Lindblad, B.A., Mandeville, J.-C., Minafra, A., Schwehm, G.H., Sekanina, Z., Wallis, M.K., Zarnecki, J.C., Chakaveh, S.C., Evans, G.C., Evans, S.T., Firth, J.G., Littler, A.N., Massonne, L., Olearczyk, R.E., Pankiewicz, G.S., Stevenson, T.J., Turner, R.F.: 1986, *Nature* **321**, 338

Moseley, S.H., Glaccum, W.: 1987 (in preparation)

Tokunaga, A.T., Golisch, W.F., Griep, D.M., Kaminski, C.M., Hanner, M.S.: 1986, *Astron. J.* **92**, 1183

Traub, W.A., Stier, M.T.: 1975, *Appl. Optics* **15**, 364

Walker, R.G., Aumann, H.H.: 1984, *Adv. Space Res.* **4**, No. 9, 197

Walker, R.G., Aumann, H.H., Davies, J., Green, S., de Jong, T., Houck, J.R., Soifer, B.T.: 1984, *Astrophys. J. Letters* **278**, L 11

Wright, E.L.: 1976, *Astrophys. J.* **210**, 250

Note added in proof: We have re-reduced the data from Dec. 20 and the noise is about the same as on Dec. 17. The spectrum is the same as Dec. 17 in shape and absolute level. These data can be found in the archive of comet P/Halley observations compiled by the International Halley Watch.

Astron. Astrophys. 187, 639–644 (1987)

ASTRONOMY AND ASTROPHYSICS

Comet P/Halley near-nucleus phenomena in 1986

S. Larson [1], **Z. Sekanina** [2], **D. Levy** [1], **S. Tapia** [3], **and M. Senay** [4]

[1] Lunar and Planetary Laboratory, University of Arizona, Tucson, AZ 85721, USA
[2] Jet Propulsion Laboratory, California Institute of Technology, 4800 Oak Grove Drive, Pasadena, CA 91109, USA
[3] Steward Observatory, University of Arizona, Tucson, AZ 86721, USA
[4] Boyden Observatory, University of the Orange Free State, P.O. Box 339, Bloemfontein 9300, Republic of South Africa

Received February 6, accepted April 3, 1987

Summary. Computer enhanced high resolution groundbased images taken during Halley's most active phases in 1985–86 exhibit all of the coma structures seen in 1910 photographs. Improved detector technology compensated for the poor observing circumstances of the current apparition yielding a wealth of complex coma structure. This structure resulted mostly from discrete dust jet activity producing spirals, arcs, and jagged-edged envelopes. Several episodes of the "multiple nucleus" and antisunward jet phenomenon were also observed. The variation in dust jet activity during the Vega and Giotto flybys is easily seen on the groundbased images. Our images of jets in the 388 nm CN (0,0) band show that they usually do not correlate very well with visible dust jets. As in 1910, the observed expansion velocities and jet curvature indicate a rotation period near 2.2 days. Because the short period dominates the shape of the jets, detailed jet modeling will be required to detect longer term rotation components such as the proposed 7.4 day period.

Key words: comet Halley – dust jets – CN jets – imaging

1. Introduction

A large number of high resolution groundbased images of comet Halley has been obtained for near-nucleus studies coordinated by the International Halley Watch. We report here a small sample of one CCD data set to qualitatively compare the appearance and behavior of coma features in 1986 with those in 1910. Complete reduction and enhancement of these images, in combination with other image sets will be the first step in a quantitative analysis of the spatial and temporal distribution of dust to determine jet source locations, their evolution and the particle size distribution. Since coma feature contrast is typically much less than 20% at groundbased resolution, computer enhancement is necessary to overcome the strong radial intensity gradient. We have applied the radial and rotational shift-difference algorithm introduced by Larson and Sekanina (1984) to recent CCD images to locate the edges of jet and envelope features in the coma. Figure 1 shows how shift-difference processing improves coma feature visibility. It must be noted that the result is a map of intensity discontinuities and should not be confused with the visual appearance. We found that the broad linear response of the CCD requires displaying the base ten logarithm of the enhanced image to improve visibility of the features and for comparison with the enhanced 1910 photographic images. Pertinent data on all images can be found in Table 1. The near-nucleus structure showed significant daily change (Figs. 2, 3) resulting from jet emission episodes as the jet sources rotated into sunlight. Because of the uneven distribution of sources on the nucleus, a two day modulation of activity was evident. The increased post-perihelion dust production mirrored a general increase in jet activity.

Send offprint requests to: S. M. Larson

2. 1910–1986 comparison

Qualitatively, all of the coma features seen in 1910 (Larson and Sekanina 1984, 1985; Sekanina and Larson, 1984, 1986b) were present in 1986 (Fig. 4). The most frequent coma pattern was spiral arcs produced by continuous emission of dust from sources in the sunlit hemisphere of the nucleus (Sekanina and Larson, 1986a). The directions of the spirals were consistent with the prograde rotation, and a small obliquity rotation pole, but detailed modeling will be required to refine the spin vector. The stronger spiral patterns evolved two or three days later into envelopes which can be seen on lower resolution, wider field Schmidt photographs. In many cases, feature identification is complicated by the superposition of multiple jets. During the post perihelion period in late March, a complex of jets from the more active hemisphere produced a strong counterclockwise spiral pattern every other day. As in 1910, envelopes sometimes showed jagged edges due to short duration variations in dust production, and/or closely spaced jet vents. When the earth was near the nucleus equatorial plane in April, the jets were linear near the nucleus and became curved by solar radiation pressure in the outer coma. Several outburst episodes producing halos and antisunward jets were seen in 1986, and as in 1910, occurred when the comet was 0.8 to 1.1 AU from the Sun before and after perihelion. Typically, these outbursts first appear as a brightening of the nucleus condensation followed by an elongation in the tailward direction that sometimes gives the appearance of a multiple nucleus. Usually, after a few hours a halo, concentric with the nucleus and containing the emerging "antisunward jet" can be seen (Fig. 5, March 19). These outbursts, as well as the onset of strong spiral jets, correlate well with reported photometric outbursts.

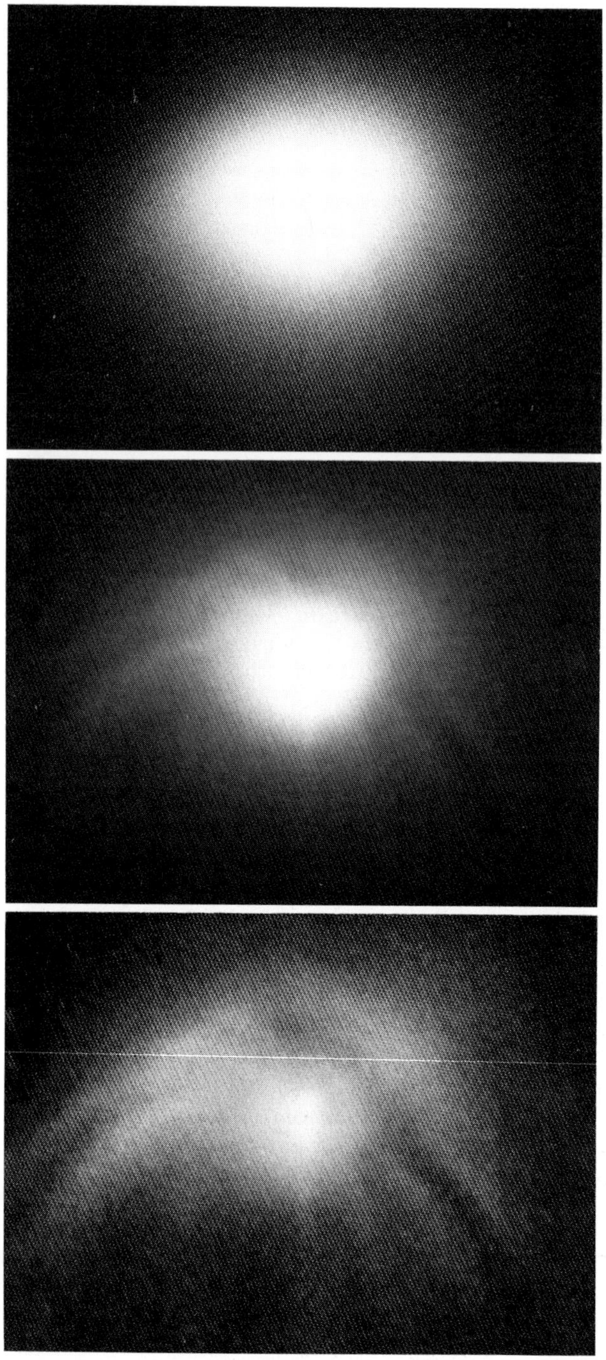

Fig. 1. Four co-added broadband red (650–950 nm) CCD images taken with the 1.5 m Catalina reflector on 1986 January 3.1 UT (top). The same image after a 10 degree rotational, and 5 pixel radial shift-difference (center), and displayed as the logarithm (bottom) reveals many intensity discontinuities due to anisotropic dust distribution in the coma

3. Activity during the spacecraft encounters

The degree of complexity of dust coma anisotropy observed from the ground at the times of the spacecraft encounters correlates well with the spacecraft imaging experiment results (Fig. 6). Dust jet activity had been intense prior to the Vega 1 encounter, and might have contributed to the attenuated contrast of the nucleus surface. Vega 2, on the other hand, imaged the nucleus during minimal jet activity. Fairly intense jet activity was observed at the time of the Giotto encounter. Correlation of the jets seen in spacecraft and groundbased images will require quantitative modeling.

4. Jets seen in gas emission bands

The discovery of jet structure in the strong CN (0,0) and C2 (0,0) bands by A'Hearn et al. (1986) introduces a new type of near-nucleus phenomenon. In many cases these jets do not coincide with continuum band jets (Fig. 7) and show less curvature indicative of higher ejection velocities. A'Hearn et al. conclude that CN must be produced in situ from dust grains in the coma, and Sekanina (1986) argues that very small (optically invisible) particles are the source. More work is needed to understand the CN jet mechanism, but it is more evidence of inhomogenous nucleus composition. In addition to the late April 1986 epoch reported by A'Hearn et al. we observed CN jets in early January 1986. We have not found any CN jets in our November or December 1985 images, and have not yet looked for jets in other emission bands.

5. Nucleus rotation

The March-April 1986 photometric light curve of comet Halley obtained by Millis and Schleicher (1986) shows a 7.4 day double-peaked period which has been interpreted as the nucleus rotation period. Analysis is underway to see if this period is reflected in jet activity. We have noted that, like in 1910, there was a general two day cycle of jet activity in January and March of 1986, and that episodes of the "antisunward jet" occurs at 2.2 day intervals. Preliminary measurements of the projected expansion velocities of the dust feature edges in 1986 are consistent with those obtained in 1910 which, with the observed curvature, indicats a 2.2 day rotation period. For ejection velocities in the 200–600 ms^{-1} range, a 7.4 day period produces less curvature of the jets than that produced by solar radiation pressure. Until the April 1986 images are processed and modeled, we can only state that the data taken together indicates a complex rotational motion that may take some time to understand.

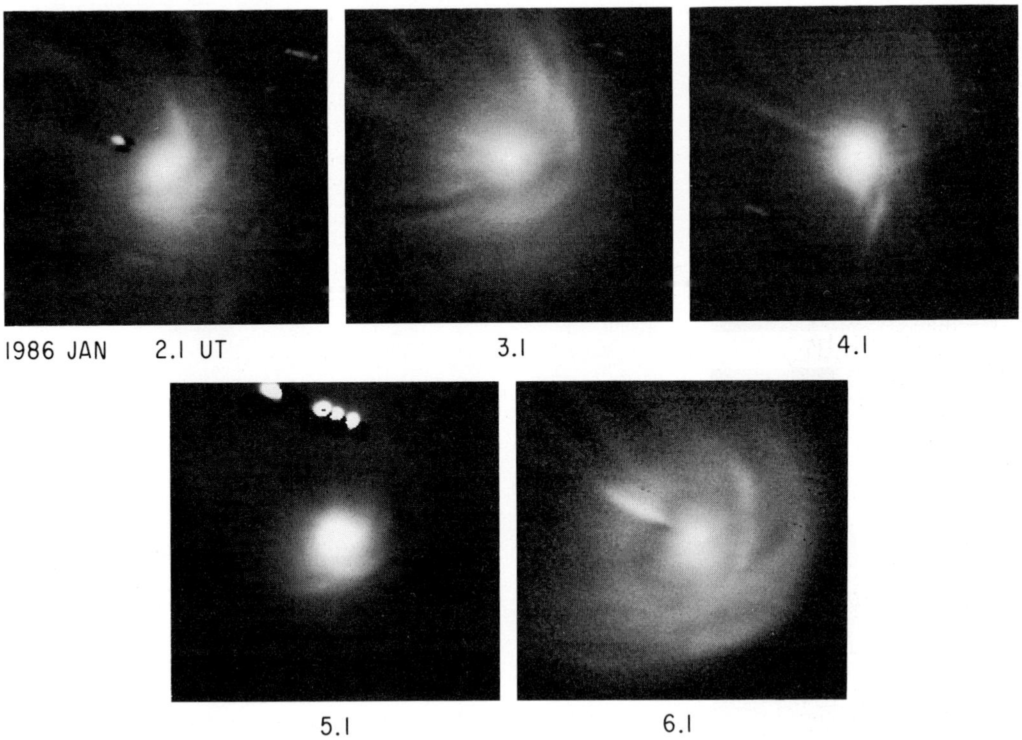

Fig. 2. Enhanced broadband red CCD images showing daily changes in the preperihelion dust jet distribution for 1986 January 2–6 UT. Orientation is north up and east to the left and each edge is 150,000 km at the comet. The direction to the Sun is towards the lower right

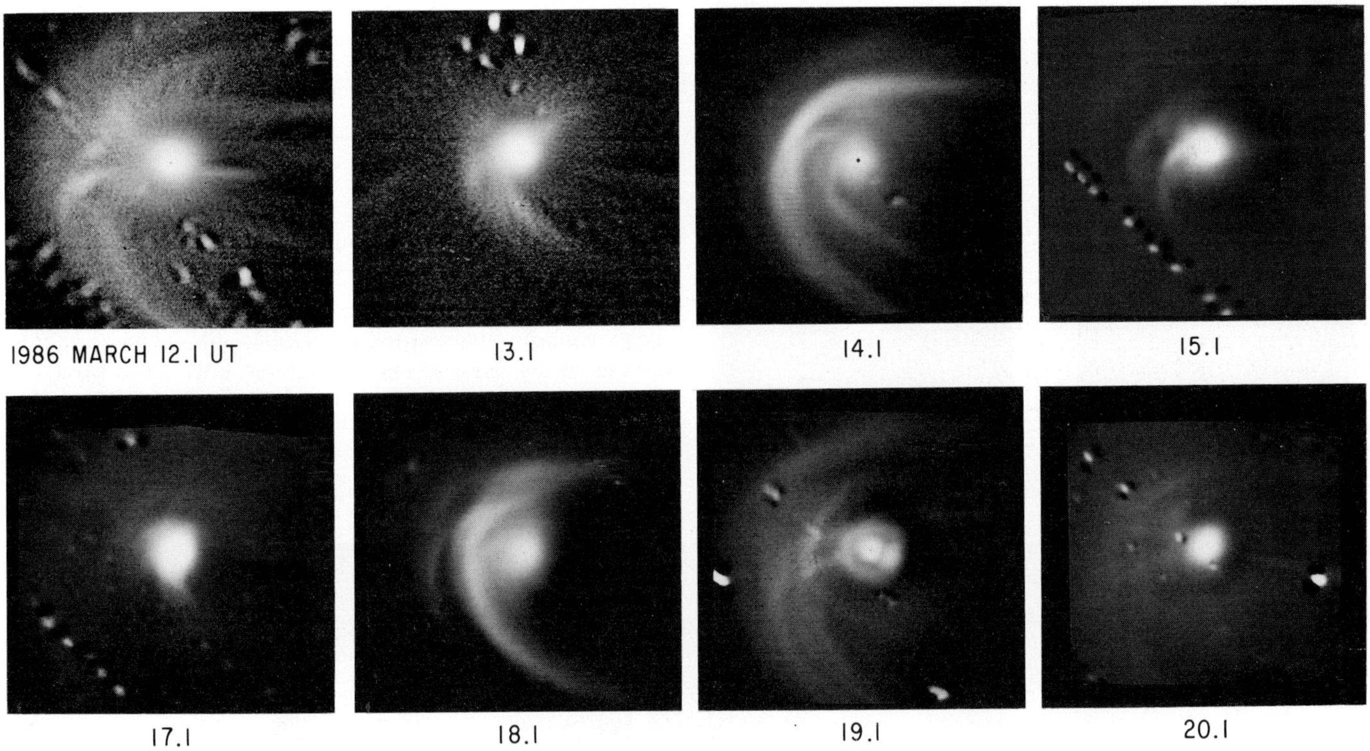

Fig. 3. Same as Fig. 2 for 1986 March 12–15, 17–20 UT. The Sun is to the left

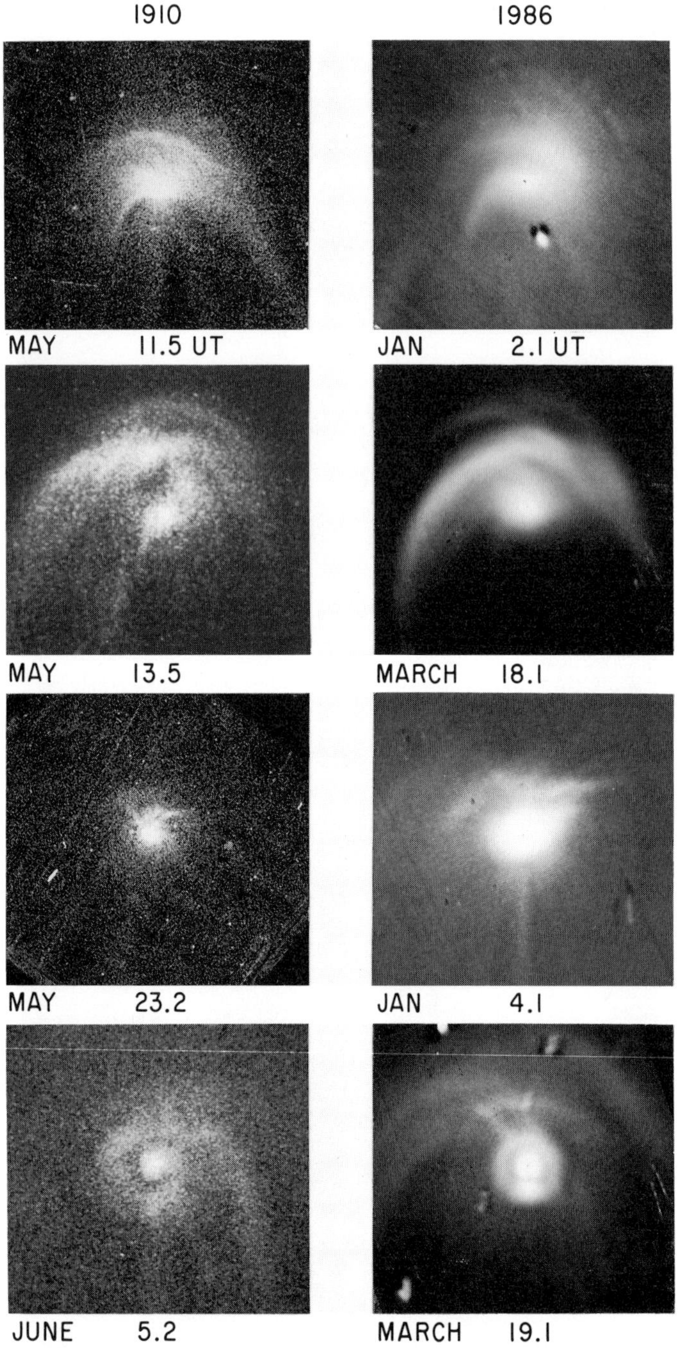

Fig. 4. Comparison between 1910 (left) and 1986 (right) coma morphology. The sun is at the top, and each edge is 120,000 km at the comet. See text for discussion. The 1910 images were taken on (from top to bottom) May 11, May 13, May 23 and June 5. The 1986 images were taken on (from top to bottom) Jan. 2, March 18, Jan. 4 and March 19

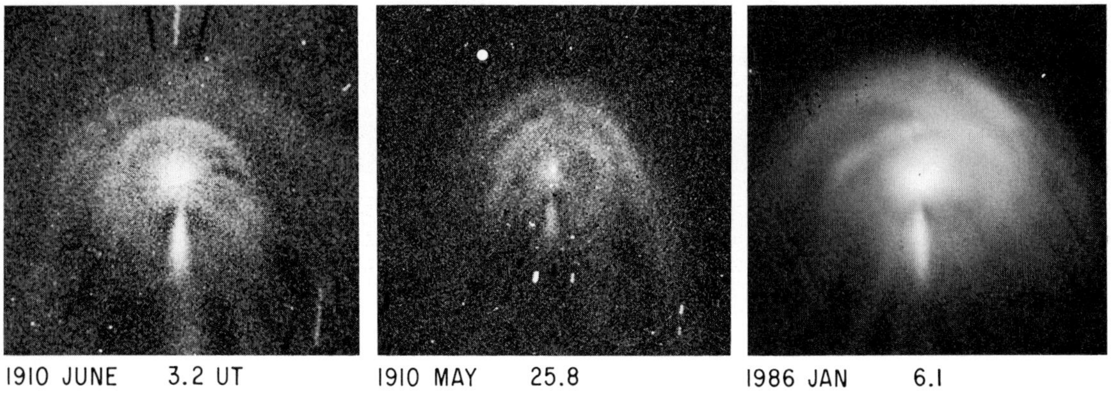

Fig. 5. 1910 June 3 (left), May 25 (middle) and 1986 Jan. 6 (right) images of the antisunward jet phenomenon. Sun is at top, each edge is 160,000 km at the comet

Table 1. Image data[a]

UT Date		Detector	Observer	Telescope	Fig(s).
1910					
May	11.49	Seed 23	Ritchey	Mt. Wilson 1.5 m	4
May	13.48	Inst. Iso.	Curtis	Lick 0.9 m	4
May	23.23	Inst. Iso.	Curtis	Lick 0.9 m	4
May	25.76	Monarch	Shaw	Helwan 0.8 m	5
June	3.23	Seed 23	Ritchey	Mt. Wilson 1.5 m	5
June	5.24	Seed 27	Ritchey	Mt. Wilson 1.5 m	4
1986					
Jan	2.08	CCD, red	Larson, Levy	Catalina 1.5 m	2, 4, 7
Jan	2.13	CCD, CN	Larson, Levy	Catalina 1.5 m	7
Jan	3.07	CCD, red	Larson, Levy	Catalina 1.5 m	1, 2
Jan	4.08	CCD, red	Larson, Levy	Catalina 1.5 m	2, 4
Jan	5.08	CCD, red	Larson, Levy	Catalina 1.5 m	2
Jan	6.08	CCD, red	Larson, Levy	Catalina 1.5 m	2, 5
Mar	7.13	CCD, red	Tapia, Senay	Boyden 1.5 m	6
Mar	7.13	CCD, red	Tapia, Senay	Boyden 1.5 m	6
Mar	10.10	CCD, red	Tapia, Senay	Boyden 1.5 m	6
Mar	12.10	CCD, red	Tapia, Senay	Boyden 1.5 m	3
Mar	13.10	CCD, red	Tapia, Senay	Boyden 1.5 m	3
Mar	14.10	CCD, red	Tapia, Senay	Boyden 1.5 m	3, 6
Mar	15.11	CCD, red	Tapia, Senay	Boyden 1.5 m	3
Mar	17.10	CCD, red	Tapia, Senay	Boyden 1.5 m	3
Mar	18.10	CCD, red	Tapia, Senay	Boyden 1.5 m	3, 4
Mar	19.11	CCD, red	Tapia, Senay	Boyden 1.5 m	3, 4
Mar	20.10	CCD, red	Tapia, Senay	Boyden 1.5 m	3
Apr	6.06	CCD, red	Tapia, Senay	Boyden 1.5 m	7
Apr	6.09	CCD, CN	Tapia, Senay	Boyden 1.5 m	7
Apr	7.00	CCD, red	Tapia, Senay	Boyden 1.5 m	7
Apr	7.02	CCD, CN	Tapia, Senay	Boyden 1.5 m	7
Apr	12.00	CCD, red	Tapia, Senay	Boyden 1.5 m	7
Apr	12.08	CCD, CN	Tapia, Senay	Boyden 1.5 m	7
Apr	30.80	CCD, red	Tapia, Senay	Boyden 1.5 m	7
Apr	30.83	CCD, CN	Tapia, Senay	Boyden 1.5 m	7

[a] A fuller explanation of the 1910 images may be found in Larson and Sekanina (1984 and 1985)

Fig. 6. Enhanced broadband red CCD images taken shortly after the Vega 1 (1986 March 7, left), and the Vega 2 (March 10, middle) encounters, and at the time of the Giotto encounter (March 14, right). Orientation and scale the same as in Fig. 3

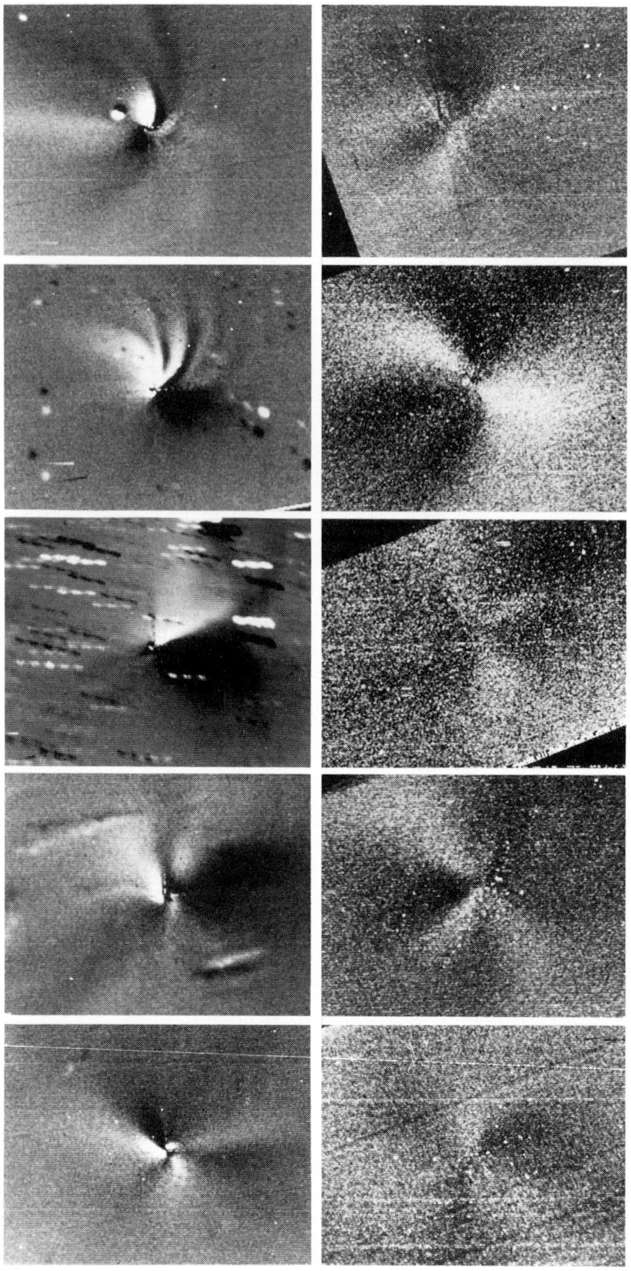

Fig. 7. Comparison broadband red dust distribution (left) and 388 nm CN distribution (right) for (top to bottom) 1986 January 2, April 6, 7, 12 and 30 UT. Orientation is north up, and east to the left. We show here the same angular field of 4.'3 by 3.'4. These images are single direction rotational shift-differences of 15° (left) and 30° (right)

Note added in proof: Comparison of the relative brightness of the inner coma of Comet Halley recorded in the Boyden Observatory red CCD images confirm the light curve of Millis and Schleicher (1986) during the same time interval in April 1986. The brightness peaks coincide with the production of strong dust jets, however these same jets show curvature and expansion velocities more consistent with a 2.2 day period. Publication of a detailed study of the light curve and changing jet morphology is in preparation.

Acknowledgements. We thank Dr. A.H. Jarrett, Director of the Boyden Observatory, University of the Orange Free State, for use of the 1.5 m telescope and logistical support. This work is supported in part by NASA grant NAGW 247. The research described in this paper was carried out jointly by the Lunar and Planetary Laboratory, University of Arizona, and the Jet Propulsion Laboratory, California Institute of Technology, under contract with the National Aeronautics and Space Administration.

References

A'Hearn, M.F., Hoban, S., Birch, P.V., Bowers, C., Martin, R., Klinglesmith, D.A.: 1986, Nature **324**, 649

Larson, S.M., Sekanina, Z.: 1984, Astron. J. **89**, 571, 600

Larson, S.M., Sekanina, Z.: 1985, Astron. J. **90**, 823, 917, 923

Millis, R.L., Schleicher, D.G.: 1986, Nature **324**, 646–649

Sekanina, Z.: 1986 Adv. Space Res. (in press)

Sekanina, Z., Larson, S.M.: 1984, Astron. J. **89**, 1408, 1446

Sekanina, Z., Larson, S.M.: 1986a, Nature **321**, 357

Sekanina, Z., Larson, S.M.: 1986b, Astron. J. **92**, 426

Astron. Astrophys. 187, 645–649 (1987)

ASTRONOMY AND ASTROPHYSICS

The sunward spike of Halley's comet

Z. Sekanina [1], S.M. Larson [2], G. Emerson [3], E.F. Helin [1], and R.E. Schmidt [4]

[1] Jet Propulsion Laboratory, California Institute of Technology, Pasadena, CA 91109, USA
[2] Lunar and Planetary Laboratory, University of Arizona, Tucson, AZ 85721, USA
[3] E.E. Barnard Observatory, Boulder, CO 80307, USA
[4] US Naval Observatory, Washington, DC 20390, USA

Received February 13, accepted April 3, 1987

Summary. On wide-field photographs from late April till early June 1986, comet Halley is seen to display a spike-like extension in the general direction of the Sun, projecting to distances of about 700,000 km from the nucleus. The spike was composed of dust and its enormous sunward extent (compared to other dust features) suggests an anomalously high ratio of particle ejection velocity to solar radiation pressure. The grains are either dielectric or slightly absorbing, $\ll 0.1$ μm in size, and undetected optically from Earth except when it is located in or very near a plane of their concentration. The only plane to which these grains ejected from Halley's wobbling nucleus can possibly be confined for long is the plane normal to the comet's angular momentum vector. This concept is applied to interpret the spike observations.

Key words: comet Halley – comet dust – sunward spike – rotation – light scattering

1. Introduction

Reports of the photographic detection, in late April and early May 1986, of a spike extending from comet Halley's head to within 20° of the direction to the Sun were entirely unexpected. The feature had nothing in common with the antitail that had been photographed extensively two months earlier. The interesting aspects of the spike include its projected length – up to some 700,000 km or more (almost ten times the distance typically reached by dust features on the sunward side of the nucleus) – and its appearance shortly after A'Hearn et al. (1986a) had observed prominent cyano jets and within two weeks of the predicted time of the Earth's transit across what was called Halley's equatorial plane (Sekanina and Larson, 1986).

2. Observations

We have collected a total of 34 photographic observations that show the spike, with a varying degree of clarity, from 1986 April 28 till June 7 (Table 1). Unfortunately, the distribution of the observations in time is less than satisfactory, with a wide gap between 12 May and 1 June.

Send offprint requests to: Z. Sekanina

As a rule, the spike is more pronounced on the panchromatic 2415 emulsion than on the blue sensitive IIa-O. This is regarded as evidence that the spike's light is continuous and that it therefore is made of dust. The fact that the IIa-O emulsion is much more grainy than 2415 must also contribute to the difference.

3. Image processing and positional reduction

Because of the spike's low surface brightness, reliable measurement of its direction and length requires computer processing. The digitization, image enhancement, and positional reduction have now been completed for 27 observations. The best results have been achieved with a technique based on double linear-shift differencing in the direction perpendicular to the spike. This approach permitted measurement of points along the spike that indicates a precision to $\pm 2°$ in position angle. Each photograph has been oriented with the aid of several SAO field stars. The length of the spike is determined with lesser precision.

Nine examples of the digitally processed images are displayed in Fig. 1. The spike points to the west-northwest of the nucleus. The long trail to the east-southeast is a streamer in the dust tail, which makes an angle of 10–15° with the somewhat wavy ion tail to the north of it (Sect. 5). The positional data on both the spike and the streamer determined from the 27 photographs are listed in Table 2. Columns 1–4 and 6–7 are self-explanatory. The reader is referred to Sects. 4 and 5 for remarks on the contents of the other columns.

4. Results and interpretation

The spike's position angle was changing very slowly, if at all, during the 40-day period of observation, even though the Sun-Halley vector rotated by 24°, from 89° on April 28 to 113° on June 7. The spike's projected length of $\sim$700,000 km implies characteristic flight times of two weeks or more for dust particles ejected at initial velocities of 0.5–0.7 km s^{-1}. A simple fountain model gives a crude upper limit of $\beta \simeq 0.1$–0.2 the solar attraction for the particle accelerations by solar radiation pressure. Based on theoretical calculations of the scattering efficiency for radiation pressure, the particles must be extremely small ($\ll 0.1$ μm in size) and dielectric or slightly absorbing, in order to satisfy the constraints imposed on both the ejection velocity and the repulsive

Table 1. List of photographic observations

Mid-exposure 1986 (UT)		Instr. [a]	Emulsion	Exp. (min)	Observer	Mid-exposure 1986 (UT)		Instr. [a]	Emulsion	Exp. (min)	Observer
April	28.204	122	IIIa-J [b]	15	Helin	May	4.130	V20	2415	15	Schmidt
	29.189	122	IIa-O [b]	15	Helin		4.145	V14	2415	25	Chester
	29.225	122	IIIa-J [b]	45	Helin		4.147	V20	2415	7	Schmidt
	30.180	P46	IIa-D	5	Helin		4.168	V14	2415	20	Chester
	30.194	P46	IIa-D	10	Helin		4.198	B30	2415	30	Emerson
	30.201	B30	2415	20	Emerson		10.185	B30	IIa-O	8	Emerson
	30.301	122	IIIa-J [b]	14	Helin		11.089	V14	2415	25	Chester
May	1.165	B30	2415	32	Emerson		11.113	V14	2415	25	Chester
	1.185	B30	IIa-O	10	Emerson		12.082	V20	2415	25	Schmidt
	2.238	P46	IIa-D	3	Helin		12.100	V20	2415	10	Schmidt
	2.256	P46	IIa-D	8	Helin		12.115	V20	2415	15	Schmidt
	3.093	V20	2415	8	Schmidt	June	1.184	B30	2415	30	Emerson
	3.139	V20	2415	10	Schmidt		4.105	V14	2415	30	Chester
	3.147	V20	2415	7	Schmidt		4.120	V14	2415	30	Chester
	4.094	V14	2415	25	Chester		6.194	122	IIa-O [b]	20	Helin
	4.116	V20	2415	15	Schmidt		6.198	P46	4415	8	Helin
	4.124	V14	2415	30	Chester		7.208	122	IIa-D [b]	20	Helin

[a] Instrument abbreviations:
122 = 122-cm f/2.5 Schmidt, Palomar Mountain Observatory
P46 = 46-cm f/2 Schmidt, Palomar Mountain Observatory
B30 = 30-cm f/1.8 Schmidt, Barnard Observatory
V20 = 20-cm f/1.5 Schmidt, near Paris, VA
V14 = 14-cm f/3.6 Schmidt/Newtonian, near Rixeyville, VA
[b] With a GG385 filter.

force. These grains scatter essentially according to the Rayleigh law and their optical detection is exceedingly difficult, except possibly at very high optical depths.

The existence of inefficiently scattering, subfemtogram grains in the comet's atmosphere is documented by the various dust experiments on the Halley missions; their obvious presence about the time of the spike's detection is inferred from prominent cyano and C_2 jets invisible in continuum bandpasses (A'Hearn et al., 1986a, b).

The appearance of the sunward spike is suggestive of a thin sheet of dust observed edge-on and is indeed reminiscent of an antitail during the Earth's transit across a comet's orbital plane. However, the situation is different in the sense that the time scale is shorter and, primarily, because it is the ejection velocity (much higher than that of the antitail particles) and not the radiation pressure that determines the ejecta's motions. If the subfemtogram grains are found to be concentrated in a particular plane, it is so because they were expelled from the comet that way. For a rotating nucleus, continuous emissions from discrete vents could remain confined to a plane for an appreciable period of time only if the sources are located in or near an invariable plane and ejecta are released parallel to it. If Halley's nucleus were in pure rotation, this plane would coincide with the equatorial plane. Since this apparently is not so (Sekanina, 1986a, 1987), the invariable plane is normal to the comet's angular momentum vector, about which the spin axis must freely precess, unless of course outgassing exerts torques on the nucleus. Reasonable estimates of these torques suggest that Halley's angular momentum vector is likely to be essentially constant on time scales of at least a few weeks. To determine the position of the angular momentum vector, we have compared the measured position angles of the spike with the time-history models for the projected invariable plane. This plane does not project in the sky as a straight line except at the time of transit and we have assumed that, in general, the spike's direction is perpendicular to the plane of maximum projection foreshortening. We have then optimized the position of the angular momentum vector to achieve the best fit to the observations. The residuals of the solution are presented in column 5 of Table 2. The mean residual is $\pm 1\overset{\circ}{.}9$, which is comparable to the error of measurement (Sect. 3). If defined, similarly to the rotation-pole position (Sekanina, 1981), by two Eulerian parameters, the angular momentum vector is determined by the "obliquity" $I_0 = 12°$ and by the "argument" $\Phi_0 = 300°$, corresponding to right ascension 17° and declination $-63°$ (equinox 1950.0). These coordinates are almost identical with those established previously for what was termed the "rotation pole" (Sekanina, 1986b; Sekanina et al., 1986).

5. Discussion and conclusions

Our measurements are in excellent agreement with those made recently by M. Jäger on his unprocessed images of Halley obtained on five nights between 1986 May 3 and 25 (Fischer, 1987, private communication). He employed his 20-cm f/1.5 Schmidt camera, using the 2415 emulsion. He finds the spike's position angle to be constant at 287° during the 22-day interval and its projected length between 570,000 and 840,000 km.

Although it is our intention to continue collecting more information on Halley's sunward spike, the presented photo-

SUNWARD SPIKE OF COMET HALLEY

LATE APRIL-EARLY JUNE 1986

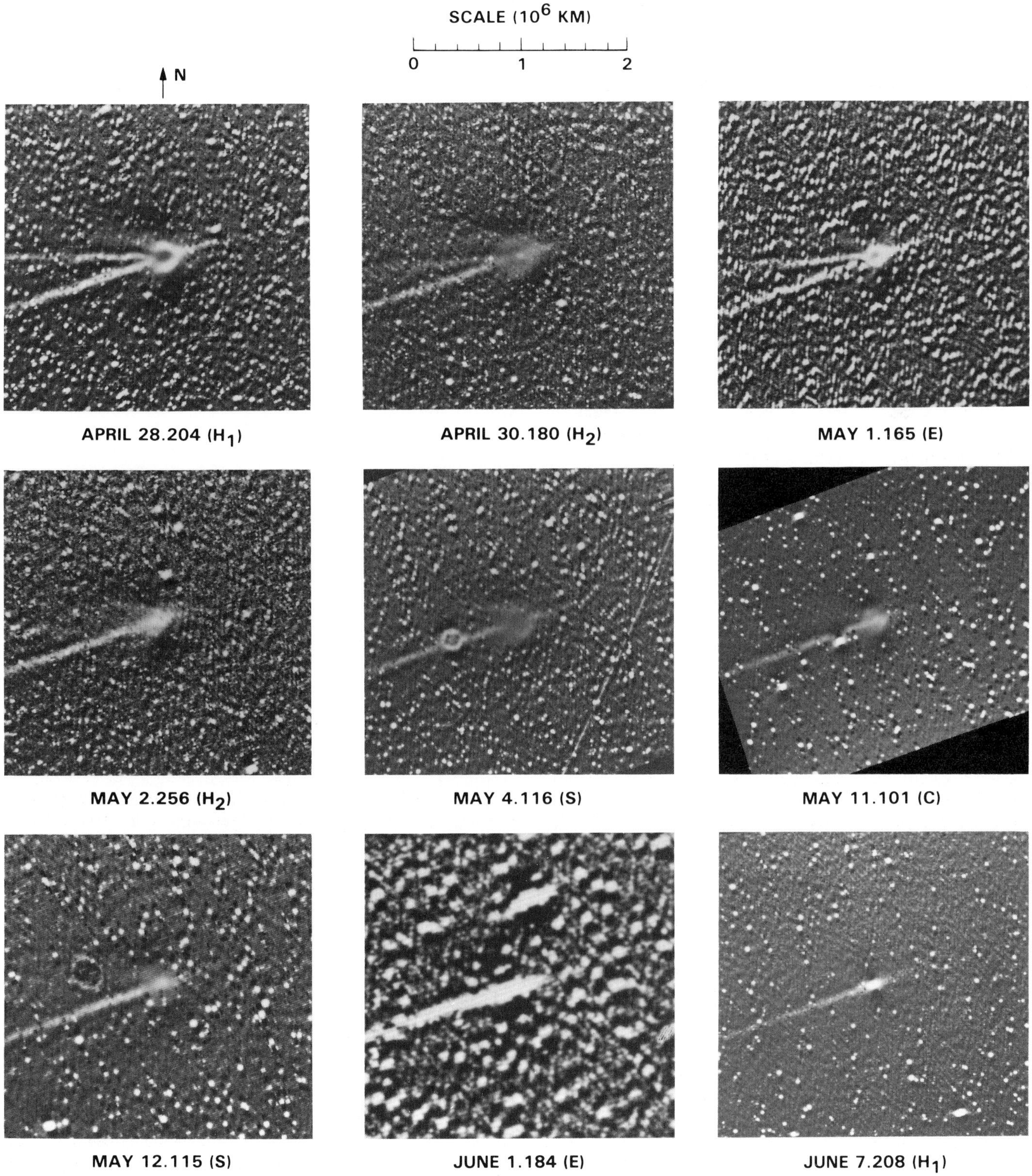

Fig. 1. Examples of digitally processed photographs that show the sunward spike of comet Halley. The prints have a common scale and are oriented with the north up and the east to the left. The times are the mid-exposures in UT and the parenthesized letters identify the observers: C = G. Chester; E = G. Emerson; H_1, H_2 = E. F. Helin, using, respectively, the 122-cm and 46-cm telescopes; and S = R. E. Schmidt

Table 2. Positional measurements of the spike and streamer

Time of mid-exposure 1986 (UT)		Sunward spike: Length apparent (arcmin)	Sunward spike: Length projected (10^6 km)	Sunward spike: Position angle obs. [a] (deg)	Sunward spike: Position angle o-c (deg)	Streamer: Apparent length (deg)	Streamer: Position angle obs. (deg)	Streamer: Position angle o-c (deg)	Position angle of streamer, minus position angle of spike, plus 180° (deg)
April	28.204	16.0	0.50	283.7	+0.5	2.9	108.7	0.0	+5.0
	29.225	23.5	0.77	281.0	-2.6	4.2	109.7	+0.8	+8.7
	30.180	12.6	0.43	282.6	-1.5	5.3	108.0	-0.9	+5.4
	30.201	19.2	0.65	283.8	-0.3	>1.6	107.9	-1.0	+4.1
	30.301	19.5	0.66	283.3	-0.8	2.8	109.0	+0.1	+5.7
May	1.165	14.8	0.52	(285.)	(+0.7)	>1.2	109.5	+0.5	(+4.5)
	2.238	12.8	0.47	287.4	+2.7	4.7	109.1	0.0	+1.7
	3.093	13.5	0.51	287.5	+2.6	>3.0	108.8	-0.3	+1.3
	3.139	11.1	0.42	285.7	+0.8	>3.0	108.4	-0.7	+2.7
	3.147	17.5	0.66	284.8	-0.1	>3.0	109.4	+0.3	+4.6
	4.094	6.7	0.26	(286.)	(+0.8)	2.6	108.9	-0.2	(+2.9)
	4.116	15.1	0.59	287.4	+2.2	4.4	109.4	+0.3	+2.0
	4.124	<16.3	<0.64	287.9	+2.7	2.6	109.3	+0.2	+1.4
	4.130	13.3	0.52	287.6	+2.4	4.6	109.0	-0.1	+1.4
	4.145	8.4	0.33	283.7	-1.5	>2.1	108.6	-0.5	+4.9
	4.147	15.0	0.59	285.8	+0.6	4.2	109.4	+0.3	+3.6
	4.168	6.8	0.27	288.9	+3.7	>2.1	109.2	+0.1	+0.3
	4.198	13.0	0.51	287.4	+2.2	>1.8	109.1	0.0	+1.7
	10.185	10.7	0.51	286.2	-0.2	>1.4	108.7	-0.2	+2.5
	11.101 [b]	9.3	0.45	290.4	+3.9	>1.8	109.3	+0.5	-1.1
	12.082	11.1	0.55	288.4	+1.7	4.4	108.8	0.0	+0.4
	12.100	12.9	0.64	290.1	+3.4	4.4	108.7	-0.1	-1.4
	12.115	14.5	0.73	288.6	+1.9	4.4	108.6	-0.2	0.0
June	1.184	4.9	0.39	287.3	-0.4	>1.3	107.8	-0.2	+0.5
	4.112 [b]	3.9	0.32	(287.)	(-0.7)	>1.4	108.1	+0.2	(+1.1)
	6.194	10.1	0.87	287.4	-0.3	2.8	108.2	+0.3	+0.8
	7.208	5.3	0.46	285.9	-1.8	2.6	108.2	+0.3	+2.3

[a] Uncertain data are parenthesized.
[b] Measurement from two consecutive, digitally co-added images (mid-exposure times in Table 1).

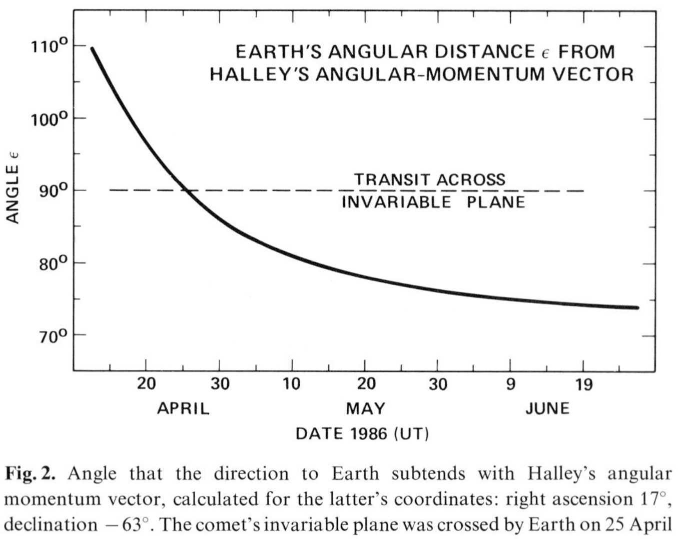

Fig. 2. Angle that the direction to Earth subtends with Halley's angular momentum vector, calculated for the latter's coordinates: right ascension 17°, declination −63°. The comet's invariable plane was crossed by Earth on 25 April 1986. Note that Earth was moving almost parallel to the plane in June

graphic evidence is already sufficient to conclude that the feature pointed in essentially the same direction for at least 40 days, apparently concentrating in the plane normal to the comet's angular momentum vector. If Halley's nucleus does indeed wobble, as it now appears to be the case, the position of the angular momentum vector is not expected to coincide on the Giotto images with the normal to the long, almost straight limb on the night side of the nucleus, which Keller et al. (1986) identified with the presumed "spin axis".

The comparative persistence of the spike until June is understood from Fig. 2, in which the Earth's angular distance from the comet's angular momentum vector is plotted as a function of time. While this angle was changing rapidly in late April, it leveled off during June, when Earth was staying within 16° of Halley's invariable plane. This qualitatively explains why the spike showed up rather suddenly but was fading very gradually. Figure 2 also suggests that the spike could have been detected on photographs taken as early as mid-April. Unfortunately, the moon began to interfere about that time.

Also examined was a possible relationship between the spike and the long streamer that appeared totally straight and pointed in the direction opposite to that of the spike. The last column of

Table 2 shows that the angle the streamer made with the spike varied with time and was not exactly 180°, implying the absence of any obvious relationship between the two phenomena. On the other hand, the streamer's position angles can be fitted with a mean residual of $\pm 0\overset{\circ}{.}4$ on the assumption that the feature represented a signature of an outburst of dust that took place some 27 ± 7 days before perihelion or within a week of 1986 January 13. The large uncertainty in the time of ejection is due to the well-known "pileup" effect that drastically limits the time resolution of old dust emissions. The maximum measured lengths of the streamer (Table 2) suggest, rather consistently, peak radiation-pressure accelerations between 0.02 and 0.026 the solar attraction, implying the presence of submillimeter-sized and larger particles. On a number of photographs the streamer can be followed all the way to the edge. The position-angle residuals for the individual observations are listed in column 8 of Table 2. No attempt has as yet been made to verify the outburst on photographs taken before the end to April, but attention is called to Lamy's (1986) identification of a dust event centered 13 days before perihelion and extending over some 12 days and to Gérard et al.'s (1986) detection of two OH flare-ups on 9 and 16 January 1986.

Potential effects of electrodynamic forces on the spike particles have not been investigated. Although such forces are believed by some to be significant for grains whose dimensions are smaller than ~0.1 μm, this was shown by Finson and Probstein (1968) to depend on the particles' typical separation distance with respect to the Debye length. Besides, the forces acting on charged grains are strongly particle-size dependent and should cause them to disperse in space. A major component of these forces is expected to be in the direction normal to the orbital plane (Horanyi and Mendis, 1985), an effect not observed with the spike.

Acknowledgements. We thank D. H. Levy for his assistance during image processing of the data, G. Chester for permission to use his observations, and D. Fischer for communicating the observations by M. Jäger of Austria. S. Singer-Brewster assisted E.F.H. during her observing runs. The research described in this paper was carried out in part by the Jet Propulsion Laboratory, California Institute of Technology, and by the Lunar and Planetary Laboratory, University of Arizona, under contract with the National Aeronautics and Space Administration (Z.S., S.M.L., E.F.H.). E.F.H. also acknowledges support from The Planetary Society and from the World Space Foundation.

References

A'Hearn, M. F., Hoban, S., Birch, P. V., Bowers, C., Martin, R., Klinglesmith, D. A.: 1986a, *Nature* **324**, 649

A'Hearn, M. F., Hoban, S., Birch, P. V., Bowers, C., Martin, R., Klinglesmith, D. A.: 1986b, 20th ESLAB Symp., ESA SP-250, Vol. I, p. 483

Finson, M. L., Probstein, R. F.: 1968, *Astrophys. J.* **154**, 327

Gérard, E., Bockelée-Morvan, D., Bourgois, G., Colom, P., Crovisier, J.: 1986, 20th ESLAB Symp., ESA SP-250, Vol. I, p. 589

Horanyi, M., Mendis, D. A.: 1985, *Astrophys. J.* **294**, 357

Keller, H. U., Arpigny, C., Barbieri, C., Bonnet, R. M., Casez, S., Cordini, M., Cosmovici, C. B., Delamare, W. A., Huebner, W. G., Hughes, D. W., Jamar, C., Malaise, D., Reitsema, H. J., Schmidt, H. U., Schmidt, W. K. H., Seige, P., Whipple, F. L., Wilhelm, K.: 1986, *Nature* **321**, 320

Lamy, P.: 1986, *Adv. Space Res.* **5**, No. 12, p. 317

Sekanina, Z.: 1981, *Ann. Rev. Earth Planet. Sci.* **9**, 113

Sekanina, Z.: 1986a, *IAU Circ. Nol* **4273**

Sekanina, Z.: 1986b, *Adv. Space Res.* **5**, No. 12, p. 307

Sekanina, Z.: 1987, *Nature* **325**, 326

Sekanina, Z., Larson, S. M.: 1986, *Astron. J.* **92**, 462

Sekanina, Z., Larson, S. M., Emerson, G., Helin, E. F., Schmidt, R. E.: 1986, 20th ESLAB Symp., ESA SP-250, Vol. II, p. 177

Astron. Astrophys. 187, 650–652 (1987)

ASTRONOMY AND ASTROPHYSICS

Complex refractive index of grain material deduced from the visible polarimetry of comet P/Halley

T. Mukai[1], **S. Mukai**[1], **and S. Kikuchi**[2]

[1] Kanazawa Institute of Technology, Nonoichi, Ishikawa 921, Japan
[2] Tokyo Astronomical Observatory, Mitaka, Tokyo 181, Japan

Received January 26, accepted March 3, 1987

Summary. Visible polarimetry of comet Halley over the phase angle α (sun-comet-earth angle) from 1°.6 to 64°.4 has shown negative values of the polarization p in $\alpha \leq 20°$ and linearly increase of p with $\alpha > 20°$. In addition, a slight increase of p with wavelength λ is likely to exist. By applying the Mie calculations to explain both α- and λ-dependences of p, we have found that it's likely for the cometary dust to consist of the material with the average complex refractive index ($m = n - i \cdot k$), i.e. $(n, k) = (1.392, 0.024)$ at $\lambda = 0.365\,\mu$m, $(1.387, 0.031)$ at $\lambda = 0.484\,\mu$m, $(1.385, 0.035)$ at $\lambda = 0.62\,\mu$m and $(1.383, 0.038)$ at $\lambda = 0.73\,\mu$m, for grains assumed to have the size spectrum reported by the Vega missions.

Key words: complex refractive index of grain material – cometary dust – comet Halley

1. Introduction

The polarimetric observations of comets over wide wavelength λ/phase angle α (sun-comet-earth angle) ranges provide fruitful data for studying the nature of cometary dust grain because a single scattering of sunlight by individual grain plays a major role in cometary atmosphere. The great increase in our observational knowledge of the dust properties in comet Halley allows us to make more definitive statements about the optical properties of cometary dust.

For polarimetry of comet Halley, Kikuchi et al. (1987) have found in $0.36\,\mu\text{m} \leqq \lambda \leqq 0.90\,\mu\text{m}$, (i) negative polarization in $\alpha \leqq 20°$, (ii) a quasi-linear increase of polarization with α in $\alpha > 20°$ and (iii) a slight increase of polarization with wavelength λ. Le Borgne et al. (1986), Kiselev et al. (1986), Dollfus et al. (1986), and Bastien et al. (1986) have also found a similar dependence of polarization p on α. Furthermore, Brooke et al. (1986) have revealed the increase of polarization with λ in the near infrared wavelengths, i.e. J, H and K bands.

In this paper, we shall estimate the optical properties of cometary grains based on the observed phase angle-/wavelength-dependences of the polarization, by using Mie calculations.

Send offprint requests to: T. Mukai

2. Dust model

We shall try to estimate the average complex refractive index of cometary grains by fitting the computed α- and λ-dependences of p to those observed in several comets including comet Halley. The grain model with multi-components was applied to explain the infrared behaviour of comets (e.g. Campins and Hanner, 1982). Some of the features observed in the scattered light might be explained by the grain model assuming the surface roughness/irregular shape of the particles (e.g. Giese, 1980). In addition, it is natural to assume that a variation of cometary dust properties with a heliocentric distance r of the comet exists. The local change of polarization in the coma was also reported in the high-spatial-resolution polarization maps of comet West (1976 VI) by Isobe et al. (1978).

Therefore, it should be kept in mind that our trial is devoted to deduce only the average complex refractive index of grain materials over a wide range of r and over the observed area of the cometary coma. Of course, we do not reject the multi-components dust model for cometary grains. Furthermore, we assume that the grains are spheres with radius s, and their size spectrum is approximated by that reported in in-situ measurements of comet Halley by the Vega missions (Mazets et al., 1986), i.e.

$$\begin{aligned} &n(s) \sim s^{-2} && \text{in } s \leqq 0.62\,\mu\text{m}, \\ &n(s) \sim s^{-2.75} && \text{in } 0.62\,\mu\text{m} < s \leqq 6.2\,\mu\text{m and} \\ &n(s) \sim s^{-3.4} && \text{in } 6.2\,\mu\text{m} < s, \end{aligned} \tag{1}$$

where s is in unit of μm, and the mass density of a grain is assumed to be $1\,\text{g}\,\text{cm}^{-3}$.

3. Phase-angle dependence of the polarization

Based on the observed α-dependence of p in several comets (e.g. Dobrovolsky et al., 1986 and Lamy, 1986) including comet Halley (e,g, Kikuchi et al., 1987), we restrict ourselves to the computed results of α-dependence of p, i.e. at least $p < 0$ in $\alpha \leqq 20°$. In addition, the conditions of $p_{\min} \geqq -10\%$ at $\alpha \sim 10°$ and $p_{\max} \leqq 30\%$ at $\alpha \sim 90°$ were adopted for limiting the optical constant $m = n - i \cdot k$ of grain material.

A large number of values of (n, k) were employed to compute the α-dependence of p in terms of Mie calculations. Figure 1 shows a part of our calculations by illustrations of p as a function of α for each of values of n and k at $\lambda = 0.62\,\mu$m. By using these results, we can estimate a preferable region of (n, k), whose values

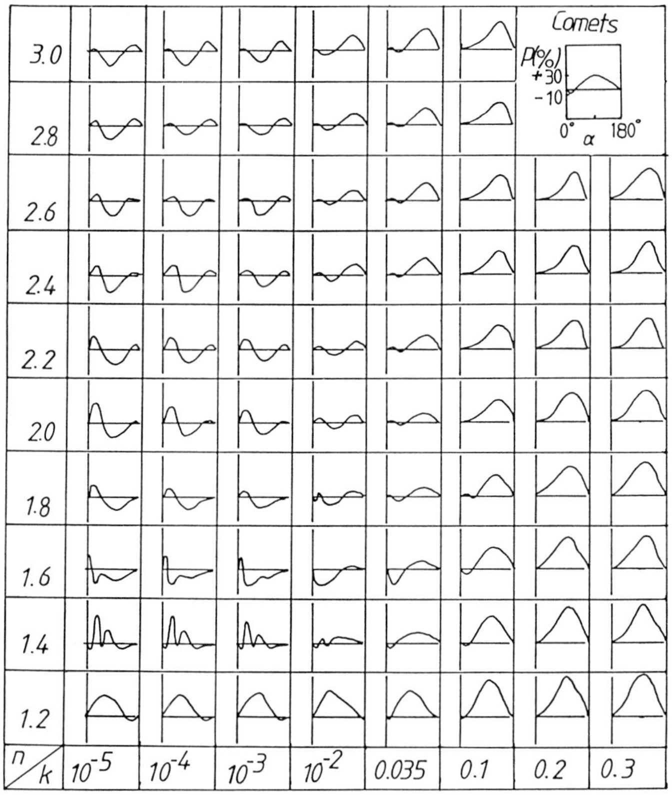

Fig. 1. Computed phase angle α-dependence of the polarization p at a wavelength $\lambda = 0.62\,\mu$m for the grain model with the complex refractive index $m = n - i \cdot k$, assuming the size spectrum of grains reported in Mazets et al. (1986). "Comets" in the upper-right corner denotes the general feature of the observed α-dependence of p derived from several comets

give a behaviour of phase function of p similar to that derived from several comets (as shown by "comets" in the top-right corner of Fig. 1). Strong condition on p, i.e. $-10\% \leqq p \leqq 30\%$ leads a narrow region of (n, k), e.g. $n = 1.387$ and $k = 0.031$ are included there at $\lambda = 0.484\,\mu$m (see Figs. 2 and 3).

4. Wavelength dependence of the polarization

As noted in Dobrovolsky et al. (1986) and Lamy (1986), the previous reports of the λ-dependence of p conflict to each other. Neutral polarization was reported in the visible in comets West, Austin and P/Churyumov-Gerasimenko, in contrast with an increase of p with λ in comets Bennett and Kohoutek (see references cited in Lamy 1986).

Our results for comet Halley (Kikuchi et al., 1987) are likely to suggest an increase of p with λ. It should be mentioned, however, that at $\alpha = 24^\circ\!.1$ where the values of p are small, rather neutral polarization was recorded in our measurements. Therefore, we have the feeling that the λ-dependence of p varies with α. In the near infrared region (i.e., J, H and K bands), Brooke et al. (1986) have reported the similar increase of p with λ in comet Halley. In addition, they have also mentioned the dependence of p with λ weakened with decreasing α.

From our observed results of p at four continuum wavelength ranges, i.e. #1 ($\lambda = 0.365 \pm 0.010\,\mu$m), #3 ($0.484 \pm 0.006\,\mu$m), #5 ($0.62 \pm 0.06\,\mu$m) and #7 ($0.73 \pm 0.05\,\mu$m), we derive a ratio of $R(\lambda) = p(\lambda)/p(\lambda = 0.62\,\mu m)$. That is, at nearly $\alpha = 50°$, $0.75 \leqq R(\lambda = 0.37\,\mu\text{m}) \leqq 0.85$, $0.90 \leqq R(0.48\,\mu m) \leqq 1.0$, and $1.05 \leqq R(0.73\,\mu m) \leqq 1.15$.

If we set the complex refractive index at $\lambda = 0.62\,\mu$m to $n = 1.385$ and $k = 0.035$, which can give a good fit to the observed α-dependence of p at $\lambda = 0.62\,\mu$m, the values of $R(\lambda)$ lead to suitable regions of (n, k) at each wavelength range. Figure 3 shows the set of (n, k) for a reasonable value of $R(\lambda)$ as dashed curve. In addition, the condition of the α-dependence of p including $-10\% \leqq p \leqq 30\%$ at each wavelength range limits to the preferable values of (n, k) as shown by shaded areas in Fig. 3. Consequently, we reach the conclusion that the λ- and α-dependences of p computed from the cometary grains consisting of the materials with average complex refractive index within the shaded area in Fig. 3, agree with those observed in comets.

For reference, we computed λ-dependence of p at $\alpha = 50°$ and 64° by using the values of (n, k), i.e. (1.392, 0.024) at $\lambda = 0.365\,\mu$m, (1.387, 0.031) at $\lambda = 0.484\,\mu$m, (1.385, 0.035) at $\lambda = 0.62\,\mu$m and (1.383, 0.038) at $\lambda = 0.73\,\mu$m. A fairly good agreement

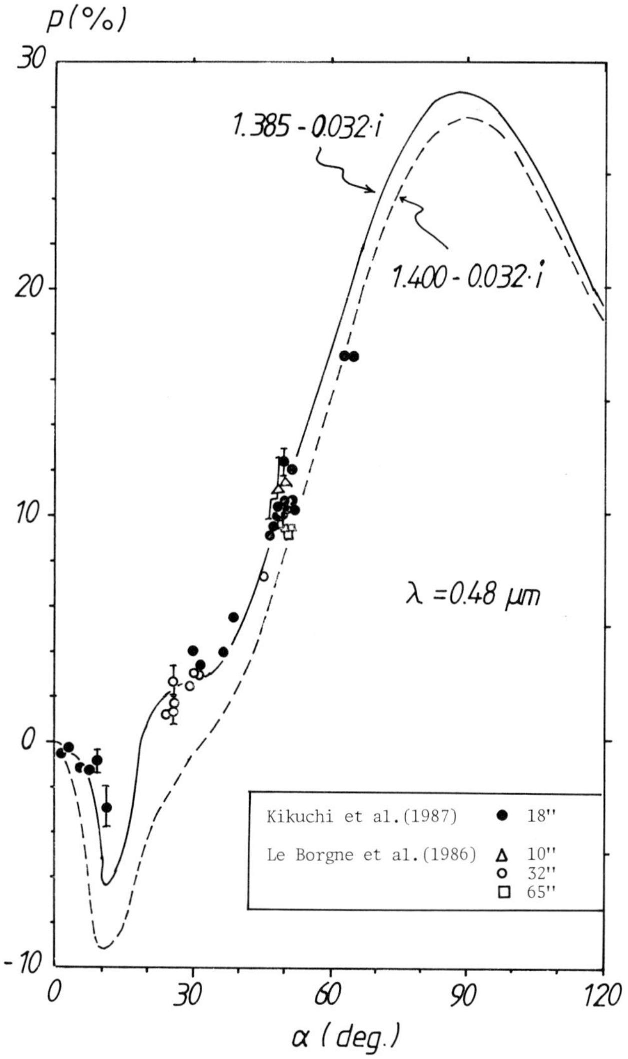

Fig. 2. Phase function of observed polarization p in comet Halley at a wavelength $0.48\,\mu$m with different aperture (diaphragm sizes are indicated in the figure), where α is a sun-comet-earth angle. Solid and dashed curves show the computed results for grain model with the complex refractive index of $1.385 - 0.032i$ and $1.400 - 0.032i$, respectively, assuming the size spectrum of grains reported in Mazets et al. (1986)

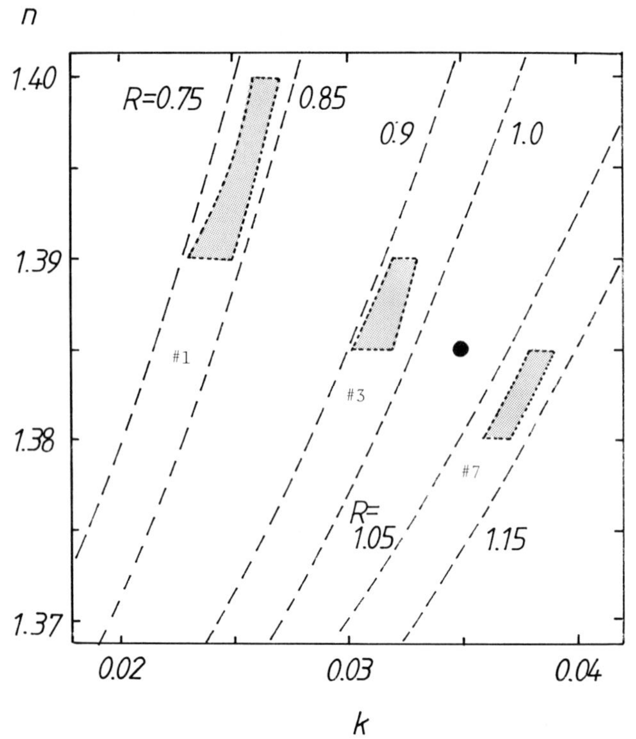

Fig. 3. Preferable region (shaded region) of the complex refractive index $m = n - i \cdot k$ of grain material, which can give an agreement with the observed phase angle- and wavelength-dependences of polarization p of comets at each wavelength range ($\#1 = 0.365\,\mu$m, $\#3 = 0.484\,\mu$m and $\#7 = 0.73\,\mu$m). R denotes a ratio of p in each wavelength range to that in $\lambda = 0.62\,\mu$m. Filled circle means a preferable set of (n, k) in $\lambda = 0.62\,\mu$m to explain the observed α-dependence of p

between the computed λ-dependence of p and those observed at nearly the same phase angle as shown in Fig. 4 suggests that the λ-dependence of (n, k) derived above seems to be reasonable.

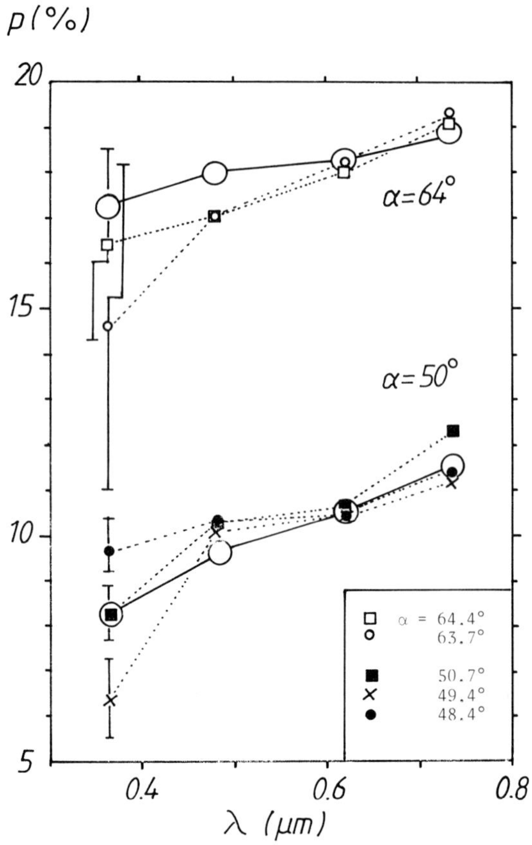

Fig. 4. Wavelength λ-dependence of the polarization p at various phase angle α. Large open circles connected by solid lines denote the computed values of p for the grains with the complex refractive index $m = n - i \cdot k$ of grain material, $(n, k) = (1.392, 0.024)$ at $\lambda = 0.365\,\mu$m, $(1.387, 0.031)$ at $\lambda = 0.484\,\mu$m, $(1.385, 0.035)$ at $\lambda = 0.62\,\mu$m and $(1.383, 0.038)$ at $\lambda = 0.73\,\mu$m, assuming the size spectrum of grains reported in Mazets et al. (1986)

5. Conclusion

Based on the Mie calculations for the grains with the size spectrum reported by the Vega missions, the phase angle- and wavelength-dependences of the visible polarization of comets have been examined to estimate the average complex refractive index of grain material. It is found that the real part of refractive index n gradually decreases from 1.392 at a wavelength $\lambda = 0.365\,\mu$m to 1.383 at $\lambda = 0.73\,\mu$m, whereas the absorption coefficient k slightly increases from 0.024 *at* $\lambda = 0.365\,\mu$m to 0.038 at $\lambda = 0.73\,\mu$m. These results might suggest that the major part of cometary grains consist of 'dirty' ice/silicate or some organic materials.

Acknowledgements. We would acknowledge the useful refereeing by J.F. Le Borgne. This work was partially supported by the Scientific Research Fund of Ministry of Education, Science, and Culture in Japan (60540163).

References

Bastien, P., Menard, F., Nadeau, R: 1986, *Monthly Notices Roy Astron. Soc.* **223**, 827

Brooke, T.Y., Knacke, R.F., Joyce, R.R.: 1986, 20th ESLAB Symp., ESA SP-250, Vol. **II**, 87

Campins, H., Hanner, M.S.: 1982, in *Comets*, ed. L.L. Wilkening, Univ. Arizona Press, p. 341

Dobrovolsky, O.V., Kiselev, N.N., Chernova, G.P.: 1986, *Earth, Moon, Planets* **34**, 189

Dollfus, A., Suchail, J-L., Crussaire, D., Killinger, R.: 1986, 20th ESLAB Symp., ESA SP-250, Vol. **II**, 41

Giese, R.H.: 1980, in *Solid Particles in the Solar System*, eds. I. Halliday, B.A. McIntosh, D. Reidel, p. 1

Isobe, S., Saito, K., Tomita, K., Maehara, H.: 1978, *Publ. Astron. Soc. Japan* **30**, 687

Kikuchi, S., Mikami, Y., Mukai, T., Mukai, S., Hough, J.H.: 1987 (in preparation)

Kiselev., N.N., Pushnin, P.A., Siklitsky, V.I., Chernova, G.P.: 1986, 20th ESLAB Symp., ESA SP-250, Vol. **III**, 29

Lamy, P.L.: 1986, in *Asteroids, Comets, Meteors II*, eds. C.-I. Lagerkvist, B.A. Lindblad, H. Lundstedt, H. Rickman, Uppsala Univ., Uppsala, p. 373

Le Borgne, J.F., Leroy, J.L., Arnaud, J.: 1986, 20th ESLAB Symp., ESA SP-250, Vol. **I**, 571

Mazets, E.P., Aptekar, R.L., Golenetskii, S.V., Guryan, Y.A., Dyachkov, A.V., Ilyinskii, V.N., Panov, V.N., Petrov, G.G., Savvin, A.V., Sagdeev, R.Z., Sokolov, I.A., Khavenson, N.G., Shapiro, V.D., Shevchenko, V.I.: 1986, *Nature* **321**, 276

Astron. Astrophys. 187, 653–660 (1987)

ASTRONOMY AND ASTROPHYSICS

Infrared emission from P/Halley's dust coma during March 1986

M.S. Hanner [1,]*, A.T. Tokunaga [2], W.F. Golisch [3], D.M. Griep [3], and C.D. Kaminski [3]

[1] Jet Propulsion Laboratory, California Institute of Technology, 4800 Oak Grove Drive, Pasadena, CA 91109, USA
[2] Institute for Astronomy, University of Hawaii, 2680 Woodlawn Drive, Honolulu, HI 96822, USA
[3] Infrared Telescope Facility, Institute for Astronomy P.O. Box 4729, Hilo, HI 96720, USA

Received February 13, accepted April 2, 1987

Summary. 2 to 20 μm photometry of the inner dust coma of comet Halley was obtained at the NASA IRTF on Mar 6.85, 12.8, 13.75, 17.7, and 24.8. Positions offset 10″ were measured as well as the central brightness. The strength of the 10 μm emission feature was observed to vary with location in the coma. The infrared emission is in general agreement with the dust size distribution measured from the Vega and Giotto spacecraft. Mar 6.8, 17.7, and 24.8 corresponded to strong dust jet activity. The strength of the 10 μm silicate emission is shown to be a sensitive indicator of grain size and thus of jet activity. Dust production rate on March 13.75, 6 h before Giotto encounter, was $\sim 10^7$ gm s^{-1}.

Key words: comet Halley – infrared – silicate

1. Introduction

Throughout comet Halley's apparition, we have monitored the infrared emission from the dust coma with the NASA 3-meter Infrared Telescope Facility (IRTF), Mauna Kea. The thermal infrared is particularly useful for determining the size, composition, and total amount of dust in the coma. The monitoring program and the pre-perihelion data have been described by Tokunaga et al. (1986; hereafter Paper I). A particular effort was made to obtain observations during March 1986, in order to compare with in situ measurements of the dust flux and mass distribution from the Giotto and Vega Spacecraft. In this paper, the March observations are presented and discussed in relation to the in situ data and other ground-based observations. We find the strength of the 10 micron silicate feature to be a sensitive indicator of grain size and thus of variable jet activity.

2. Observations

Five sets of data were acquired between March 6 and March 25 with the IRTF helium-cooled bolometer and 4 mm aperture (6″.8 FWHM). Because the solar elongation angle was only 45°–76°, the observations were carried out by day, centering on the maximum 10 micron signal. In order to investigate possible differences in grain properties across the coma, we also made measurements offset 10″ from the maximum brightness. The throw for the chopping secondary was 60″ N-S on March 6 and 24 and 120″ N-S on March 12, 13 and 17. Table 1 summarizes the observations.

The data reduction procedure follows that described in Paper I. The primary standard stars were α Lyr (0.0 mag at all λ) and γ Aql (Table 2). Magnitudes for γ Aql at M and 7.8–12.5 microns are the average of measurements versus α Lyr on Sept. 12, 1985 and March 12, 1986; we estimate the accuracy to be $\leq \pm 0.03$ mag. The N and Q magnitudes are from Tokunaga (1984). We used $K = -0.55$ (Koornneef, 1983) and interpolated $L = -0.72$ for γ Aql. BS 7754 (spectral type G2) was used as a secondary standard on March 17 because it was close to the comet; it was calibrated against α Lyr on March 12. March 6 and 12 were photometric, while March 13, 17, and 24 were wet, with some patchy clouds. The errors given in Table 1 include uncertainties in extinction and standard star measurements (the comet was brighter than the standards!). However, any errors in telescope tracking at the 10″ offset positions are not included. Corrections of a few percent were applied to the broadband M, N, and Q magnitudes to account for the difference between cometary and stellar spectral energy distribution (Hanner et al., 1984, Table VI).

Figure 1 presents the thermal emission spectra for the central coma. Blackbody curves have been drawn through the data points at 7.8 and 12.5 microns for illustration, although a single blackbody fit is clearly inadequate.

The data at 10″ offset are shown in Fig. 2 and the brightness distributions are compared in Table 3. During March, the comet-sun line was at $PA \simeq 80°$; thus our offset to the East is approximately in the sunward direction. Since the chopper throw was 120″ on March 12 and 17, no correction was made for contribution from the coma in the reference beam. On March 24, the chopper throw was 60″ N-S (30,000 km) implying a signal in the reference beam $\sim 17\%$ of the signal at 10″ if the outflow were uniform. We have used instead the asymmetric radial gradients derived by Green and Hughes (1986) from their CCD images to correct the brightness ratios in Table 3 for March 24.

The coma was always brightest to the East and faintest to the West, consistent with the space probe images depicting dust emission primarily from the sunward hemisphere and consistent also with 10.8 μm images of the coma on Mar 28.7–30.7 (Campins et al., 1986c). North, the "afternoon" quadrant, was brighter than

Send offprint requests to: M.S. Hanner

* Visiting Astronomer, Infrared Telescope Facility, operated by the University of Hawaii under contract from the National Aeronautics and Space Administration

Table 1. Observed magnitudes of comet Halley

UT Date Mar 1986	r (AU)	Δ (AU)	β (deg)	ε (deg)	Proj. FOV (km)		K λ 2.2	L 3.45	M 4.8	N 10.1	Q 20.0	7.8	8.7	9.8	10.3	11.6	12.5 μm
6.85	0.80	1.14	59	44	5620	Center	6.26 ±0.02	3.62 ±0.02	1.33 ±0.03	−2.60 ±0.03	−3.95 ±0.04	−1.14 ±0.05	−1.94 ±0.03	−2.63 ±0.03	−2.89 ±0.03	−2.93 ±0.03	−2.73 ±0.06
12.80	0.89	0.99	64	53	4900	Center	6.42 ±0.02	4.17 ±0.02	2.00 ±0.03	−2.07 ±0.02	−3.60 ±0.06	−0.62 ±0.06	−1.47 ±0.03	−2.24 ±0.03	−2.30 ±0.03	−2.59 ±0.03	−2.42 ±0.06
12.85						10″ E			3.83 ±0.03			+1.38 ±0.09	0.41 ±0.03		−0.38 ±0.03		−0.49 ±0.09
12.87						10″ W			4.28 ±0.04			1.84 ±0.22	0.87 ±0.03		+0.12 ±0.04		−0.24 ±0.10
13.75	0.90	0.97	64	55	4780	Center	6.98 ±0.02	4.69 ±0.03	2.37 ±0.04	−1.84 ±0.02		−0.38 ±0.04	−1.21 ±0.03	−1.89 ±0.03	−1.93 ±0.03	−2.27 ±0.03	−2.11 ±0.03
17.70	0.96	0.87	66	62	4290	Center	6.23 ±0.03	4.08 ±0.03	2.07 ±0.03	−2.36 ±0.02		−0.88 ±0.04	−1.66 ±0.03	−2.37 ±0.03	−2.55 ±0.03	−2.83 ±0.04	−2.76 ±0.05
17.75						10″ E	7.44 ±0.04	5.22 ±0.02	3.22 ±0.04	−1.03 ±0.02		0.62 ±0.04		−1.16 ±0.03			−1.47 ±0.05
17.76						10″ W	8.98 ±0.05	6.93 ±0.03	4.60 ±0.04	+0.31 ±0.02		1.84 ±0.06		0.50 ±0.03			−0.34 ±0.06
17.78						Center				−2.27 ±0.02							
17.78						10″ N	7.46 ±0.03	5.30 ±0.02	3.24 ±0.03	−1.04 ±0.02		0.66 ±0.04		−1.11 ±0.03			−1.46 ±0.05
17.79						10″ S	8.17 ±0.04	6.01 ±0.02	4.00 ±0.04	−0.33 ±0.02		1.27 ±0.07		−0.43 ±0.03			
24.77	1.07	0.69	65	76	3430	Center	6.35 ±0.02	4.34 ±0.02	2.20 ±0.03	−2.32 ±0.02		−0.70 ±0.04	−1.62 ±0.03	−2.42 ±0.03	−2.61 ±0.03	−2.87 ±0.03	−2.74 ±0.03
24.78						10″ N	8.24 ±0.04	6.13 ±0.02	4.02 ±0.03	−0.36 ±0.02		1.30 ±0.04		−0.46 ±0.03			−0.62 ±0.03
24.80						10″ S	8.66 ±0.05	6.61 ±0.02	4.48 ±0.04	−0.08 ±0.02		1.59 ±0.04		−0.18 ±0.03			−0.81 ±0.03
24.81						10″ E	7.86 ±0.03	5.83 ±0.03	3.76 ±0.03	−0.75 ±0.02		0.98 ±0.04		−0.82 ±0.03			−0.92 ±0.03
24.82						10″ W	9.38 ±0.13	7.46 ±0.04	5.03 ±0.04	+0.73 ±0.02		2.11 ±0.08		0.66 ±0.03			−0.25 ±0.03

Table 2. Magnitudes for γ Aql (BS 7525)

Bandpass	Magnitude
M (4.8 μm)	−0.50
N (10.1 μm)	−0.78t
Q (20.0 μm)	−0.82t
7.8 μm	−0.78
8.7 μm	−0.74
9.8 μm	−0.79
10.3 μm	−0.81
11.6 μm	−0.84
12.5 μm	−0.86

t: Tokunaga (1984)

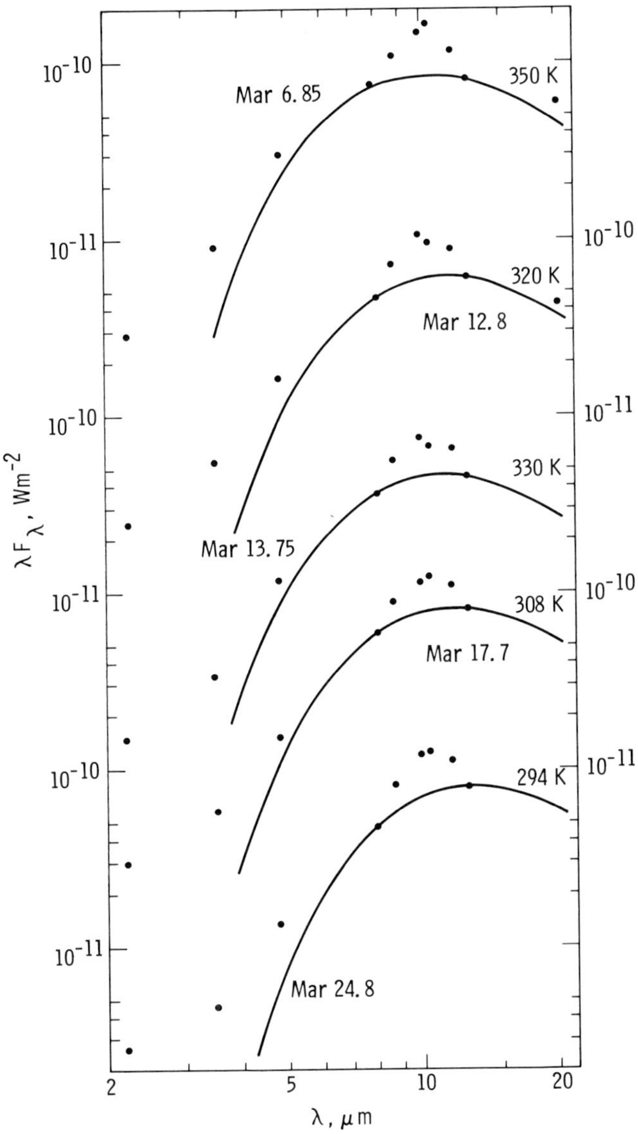

Fig. 1. Thermal emission from the central coma in March 1986, arranged chronologically. Each data set is offset by factor 10; the scales are shown on alternate sides of the figure. Blackbody curves have been fit through the 7.8 μm and 12.5 μm data points and the corresponding color temperatures are shown

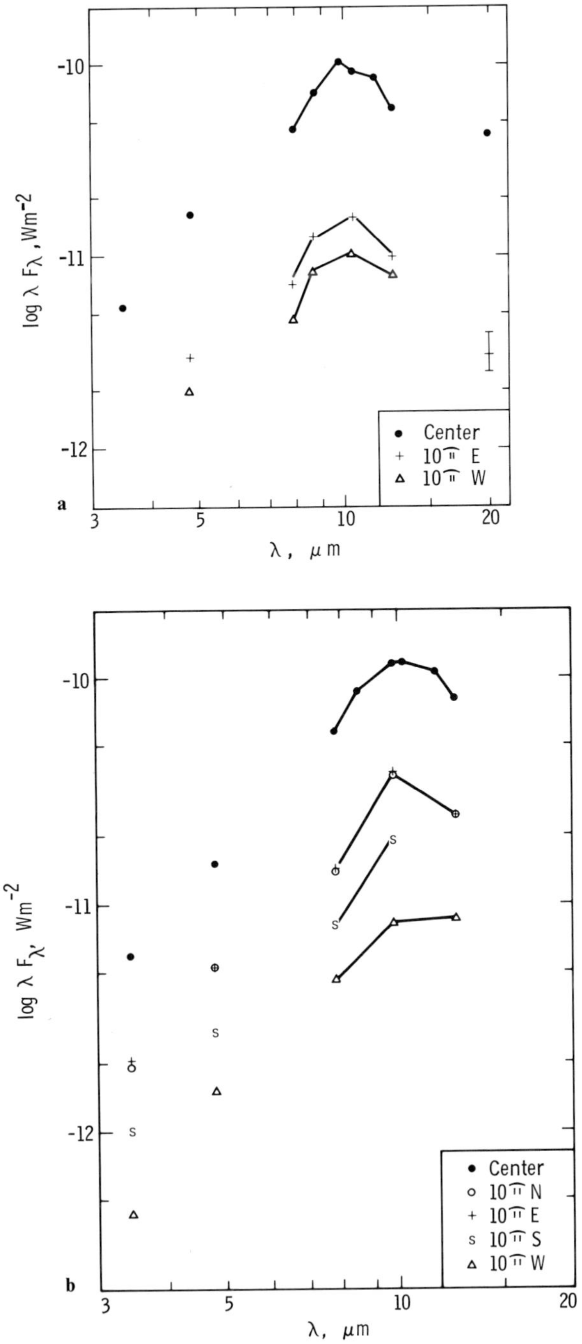

Fig. 2 a and b. Thermal emission spectra at positions offset 10″ from peak flux. **a** Mar 12.85; **b** Mar 17.75

South. If the dust were moving outward isotropically from the nucleus, the brightness at 10″ should be 0.17 that of the central coma for our 6″.8 beam. When jet activity was strong (Mar 17.7 and 24.8) the brightness at 10″ E exceeded this radial gradient.

The 12.5 micron flux was always relatively higher on the anti-sunward side, which we interpret to mean larger, cooler grains, and the silicate feature was weaker, again suggesting a lack of small grains.

Table 4 records the strength of the silicate emission feature, defined as the ratio of the observed 9.8 or 10.3 micron flux to the

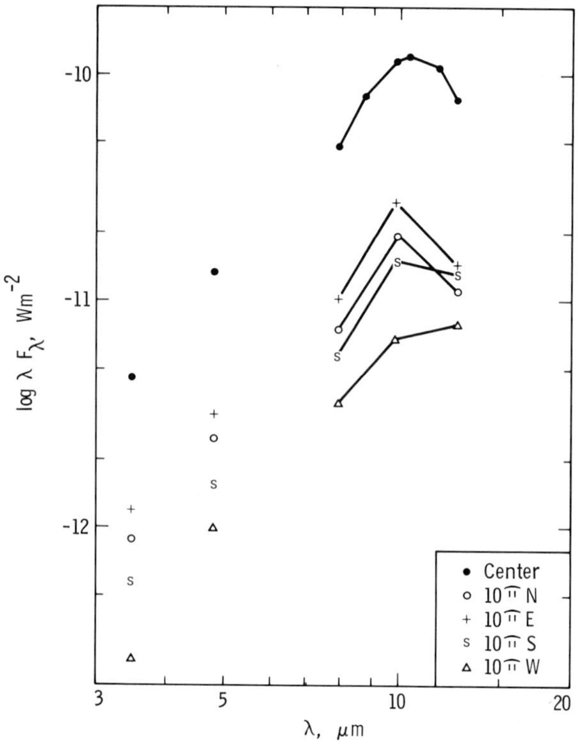

Fig. 2c. Thermal emission spectra at positions offset 10″ from peak flux on March 24.8

Table 3. Brightness distribution in coma

Date	λ	10″ E / Center	W/E	N/E	S/E
Mar 12	4.8	0.18	0.66		
	7.8	0.16	0.65		
	8.7	0.18	0.65		
	10.3	0.17	0.63		
	12.5	0.17	0.79		
Mar 17	2.2	0.33	0.24	0.98	0.51
	3.5	0.35	0.21	0.93	0.48
	4.8	0.35	0.28	0.99	0.51
	7.8	0.25	0.33	0.97	0.55
	9.8	0.33	0.22	0.97	0.52
	10.1 (N)	0.30	0.29	1.00	0.52
	12.5	0.30	0.36	1.00	–
Mar 24	2.2	0.27	0.28	0.72	0.51
	3.5	0.27	0.25	0.77	0.52
	4.8	0.25	0.35	0.80	0.55
	7.8	0.23	0.40	0.76	0.61
	9.8	0.24	0.29	0.73	0.59
	10.1 (N)	0.25	0.29	0.71	0.58
	12.5	0.20	0.62	0.77	0.96

continuum interpolated between 7.8 and 12.5 microns. Previous comet observations (e.g. Hanner et al., 1985; Ney, 1982) did not include the 7.8 micron filter; the continuum was generally drawn through the 8.7 and 12.5 micron points. Thus, the silicate emission may have been systematically underestimated and a weak feature,

Table 4. Strength of silicate emission

Date	Position	9.8/C [a]	10.3/C [a]
Jan 8.1	Center	1.9	2.2
Mar 6.8	Center	1.8	1.95
Mar 12.8	Center	1.7	1.6
	10″ E	–	1.6
	10″ W	–	1.4
Mar 13.8	Center	1.65	1.5
Mar 17.7	Center	1.5	1.5
	10″ E	1.8	–
	10″ W	1.2	–
	10″ N	1.8	–
	10″ N	1.75	–
Mar 24.8	Center	1.7	1.7
	10″ E	2.0	–
	10″ W	1.2	–
	10″ N	2.0	–
	10″ S	≧1.5	–

[a] C = continuum interpolated between 7.8 and 12.5 μm

such as that observed in Halley on Dec 13 at 1.3 AU pre-perihelion (Paper I), could have escaped detection. Campins et al. (1986) obtained 6–9 micron spectrophotometry of Halley from the KAO at 1.3 AU pre- and post-perihelion. When their data are fit with a dust model similar to those described in Sect. 4, the emission feature begins at ~8.2 microns (in contrast to Fig. 4 of Campins et al., where the data have been compared with a single 325 K blackbody).

3. Temporal variations

3.1. Silicate emission

The 10 micron emission feature was variable with time as well as with position. None of our March observations exhibited as strong a feature as we recorded on Jan 8.1 at 0.9 AU pre-perihelion. The average albedo of the grains was higher than usual on Jan 8, suggesting that smaller grains were present, and the central coma was ~30% brighter than on Jan. 9.0. On Mar 6.8 the feature was fairly strong. The very bright central coma (~30″) on CCD images taken Mar 7.4 indicated that an outburst had occurred about half a day earlier (West et al., 1986) and we must have detected it. Mar 24.8, which corresponded to a major outburst, also exhibited a very strong feature, particularly at 10″ E and N. This outburst began Mar 24.72 (Rettig et al., 1986). Thus, only the submicron grains, which are efficiently accelerated by the outflowing gas would have reached 10″ E and N. Although this outburst produced an anti-sunward streamer of small grains (Rettig et al.), we see no evidence of it in our 10″ W spectrum. On Mar 17.7 the silicate emission at the center was somewhat lower, although still strong in the coma. Apparently, an outburst of small grains had been in progress for several hours and was already subsiding. The strength of the 10 micron emission, then, can be a sensitive tracer of jet activity, with a strong feature indicating recent ejection of sub-micron grains.

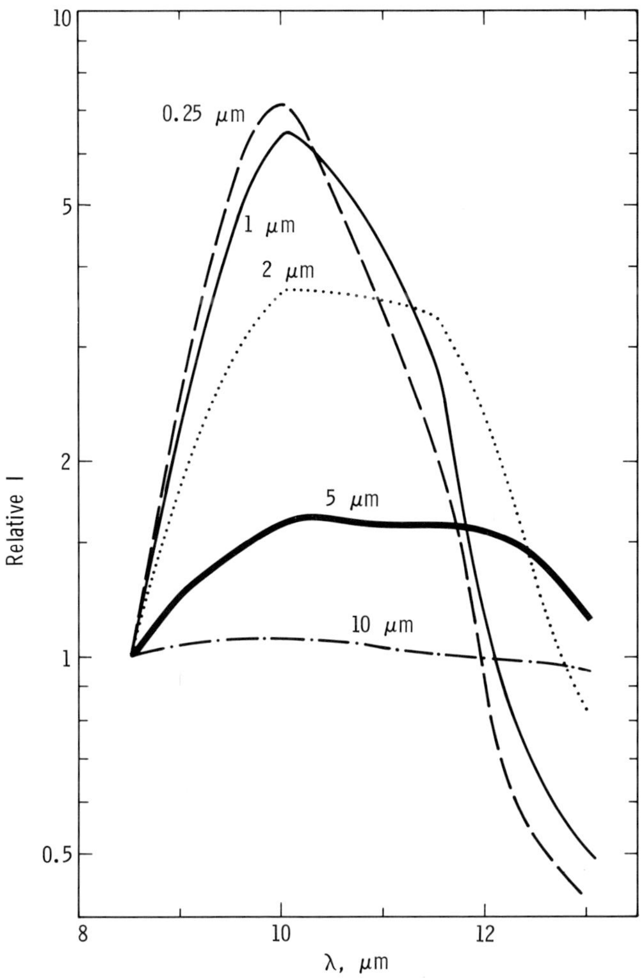

Fig. 3. Computed emission spectra for small silicate grains at $T = 330$ K, based on refractive indices of disordered olivine (Krätschmer and Huffman, 1979). ——— 0.25 μm; —— 1 μm; .... 2 μm; —— 5 μm; –·–· 10 μm. Spectra are normalized to 8.5 μm

Although there is a hint of structure in our spectra – and possible changes in shape with time – we caution against over-interpretation of these data; the shape of the emission feature is more appropriately studied with higher spectral resolution.

Temporal variability is further emphasized by comparison with other observations. While the feature was strong on Mar 24.8, it was weak or absent in a 14″ aperture on Mar 25.5, then moderately strong again in a 5″ beam on Mar 28.6 (Gehrz and Ney, 1986). The feature was absent in a 5″ aperture on Mar 31 (Green et al., 1986). (In these cases, the continuum was drawn from 8.7 to 12.5 microns, possibly masking a very weak feature). Campins et al. (1986c) imaged the coma at 10.8 μm and 12.8 μm on Mar 29.7. By comparing these images they find the silicate feature to be strongest at the center and decreasing in all directions, but more slowly to the East.

To illustrate the dependence of silicate emission on grain size, we present in Fig. 3 the computed emission for small silicate grains using refractive indices for disordered olivine (Krätschmer and Huffman, 1979). The feature is strong for grain radii $\leqq 1$ μm. However, for grains larger than 1 micron radius, the relative strength of the peak decreases, until no feature is evident in the emission from 10 μm grains. Also important for our discussion, the relative flux near 12.5 μm increases as $2\pi a/\lambda$ approaches unity. Changes in shape will be accentuated by differences in temperature with grain size. Although other silicate minerals may exhibit somewhat different 10 μm features, similar trends with particle size will apply.

The dust composition instrument on board Giotto and Vega detected very few pure silicate grains (Kissel et al., 1986a, b); generally the silicate was associated with carbonaceous material, as is also the case in the chondritic aggregate micrometeorites though to be of cometary origin (Brownlee, 1978).

3.2. *Level of activity vs. time*

In Fig. 4 we compare our observations with other measures of the comet's activity during March. The top panel presents the 10 μm (N) fluxes relative to Mar 12.8. These have been normalized to unit geocentric distance as Δ^{-1} and divided by the appropriate blackbody flux to account for the stronger emission from the grains at smaller heliocentric distance. The central dust coma was brighter on Mar 6.8, 17.7 and 24.8 than on Mar 12.8 or 13.8; the total range was less than a factor 2. Both Mar 6.8 and 24.8 correspond to outbursts detected optically. The H and L refer to far-infrared photometry obtained from the KAO Mar 15.7 and 16.7 with a 26″ beam (Campins et al., 1986). The comet was twice as bright on Mar 15 and the spectrum was steeper than on Mar 16. Comparison with panel d suggests that the total infrared variation from maximum to minimum would have been more than a factor 2. The count rate from the IUE Fine Error Sensor is displayed in the second panel (Festou et al., 1986; Feldman et al., 1986a, b). This sensor measured the visual brightness (mainly scattered light from dust) in the central 12″ × 12″. The count rates have not been corrected for distance since the product $r^2\Delta$ remained almost constant. The overlaps with the IRTF data are consistent in showing low to moderate brightness on Mar 12.8 and 13.8 and a maximum on Mar 24.8. The CS production rates derived from IUE spectra are plotted in the third panel (Festou et al.); they correlate with the trends in the FES count rates. Because CS has a relatively short lifetime, its abundance in the coma responds quickly to changes in nucleus activity. In the lowest panel we show the C_2 production rates from the photometry of Millis and Schleicher (1986), and Schleicher et al. (1986). The inferred times of maxima and minima are consistent with the IRTF photometry.

The dashes marked S_1 refer to the times of outbursts seen by the Lyman α camera on board Suisei (strong source 1) and the unmarked dashes to strong source 2 (Kaneda et al., 1986ab); these have been extrapolated backward from Mar 18 to Mar 5 using their 2.2 day period. The S_1 source corresponds to the high IRTF flux on Mar 24.8. Neither of our other maxima correspond to either S_1 or S_2; indeed, the extrapolated times of the S_1 source fall near minima in the C_2 production rate from Mar 9–16. Although the response time of the hydrogen emission will not necessarily coincide with the time of an outburst detected optically, one might expect a constant phase lag between them.

Millis and Schleicher have proposed that the rotation period of the nucleus is 7.4 days; a 7.4 day periodicity would match our data on Mar 17.7 and 24.8, if we assume that the outburst on Mar 17.7 had begun several hours earlier, as indicated by the high brightness and stronger 10 μm emission 10″ East and North of the nucleus. The same active area would than be responsible for the outburst on Mar 10.0 seen by FES. Clearly, the picture is complex, with numerous active areas on the nucleus, each responding to the pattern of insolation on a rotating, probably precessing object

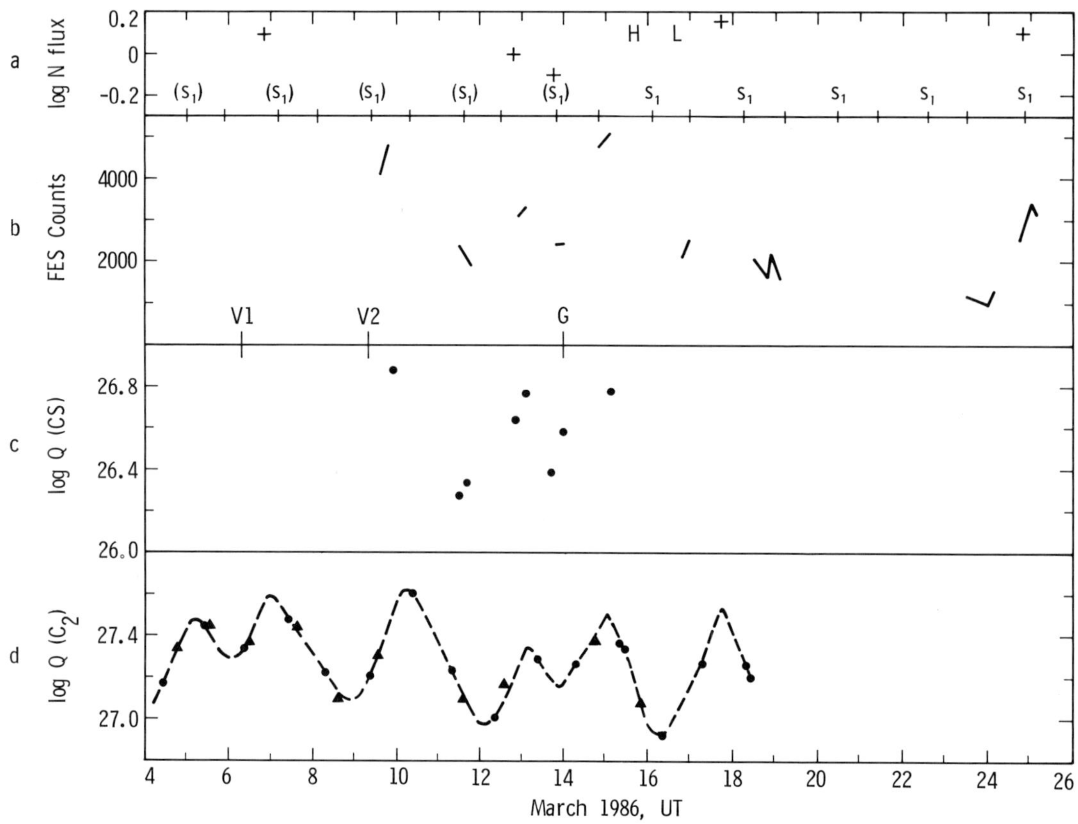

Fig. 4a–d. Variable activity of comet Halley in March 1986. **a** Infrared brightness +: IRTF relative 10 μm (N) flux (see text); H, L: High, Low far-infrared flux measured from KAO Mar 15.7, 16.7 (Campins et al., 1986); S_1: Strong source No. 1 detected by Lyman α camera on Suisei (Kaneda et al., 1986); (S_1): extrapolated times of S_1, using $P = 2.2$ days. **b** Count rate from IUE Fine Error Sensor (Festou et al., 1986; Feldman et al., 1986); V1: time of Vega 1 closest approach (9,000 km); V2: time of Vega 2 closest approach (8,000 km); G: time of Giotto closest approach (600 km). **c** Production rate of CS (mol s^{-1}) from IUE (Festou et al.). **d** Production rate of C_2 (mol s^{-1}) from ground-based filter photometry (Millis and Schleicher, 1986; Schleicher et al., 1986). Dashed curve is drawn to approximate the maxima and minima in the upper 3 panels

(e.g. Sekanina, 1987). Furthermore, measurements with different aperture sizes may not show maxima simultaneously.

4. Dust size distribution

The Vega and Giotto spacecraft carried detectors which measured the impact rate and mass distribution of dust particles along the trajectory (Mazets et al., 1986; McDonnell et al., 1986; Simpson et al., 1986; Vaisberg et al., 1986; Zarnecki et al., 1986). The data from all of these instruments are in general agreement, indicating an average integral mass distribution index $\alpha \approx 0.9$–1.1, with a turnover to shallower slope ~0.2 for small masses, although the location of the turnover differs in the various data sets. All of the instruments recorded changes in the mass distribution with position in the coma; the slope was usually steeper (more small particles) within jets. A high flux of small particles was detected, down to the sensor thresholds $\sim 10^{-16}$ g. (These particles, with their small emitting efficiency at infrared wavelengths, contribute negligibly to the integrated thermal emission from the dust.) The cumulative mass distribution recorded during the Giotto flyby is shown in Fig. 5 (McDonnell et al., 1986). Further discussion of the mass distribution functions is given by Divine and Newburn (1987).

We have computed the thermal emission spectra for these measured mass distributions, using the method and the temperatures for absorbing grains from Hanner (1983). A detailed discussion of these models will be published separately (Hanner, in preparation); an example is shown in Fig. 6 for March 12.8. To achieve a fit to the IRTF data, we find it is necessary to adjust somewhat the transition mass, M_t where the cumulative mass distribution changes slope. The model giving the best fit to the thermal emission on Mar 12 and 13 raises the flux of smallest particles by a factor of 2.5 for grain density 1 g cm^{-3} (Fig. 5), well within the uncertainties of effective area and dead time corrections in joining the two Giotto experiments (McDonnell et al.)[1]. This model fit implies a total dust mass of $\sim 2\ 10^{11}$ g within the 6.″8 beam on Mar 12.8 and $\sim 1.5\ 10^{11}$ g on Mar 13.75 (grain radii ≦1 mm). Adopting the velocity distribution for the dust grains from Gombosi (1986), we derive a dust production rate of 1.5 10^7 g s^{-1} on Mar 12.8 and 1.2 10^7 g s^{-1} on Mar 13.75, with an estimated uncertainty of a factor 2.

Taking into account the variability of the dust emission and associated changes in the size distribution, we conclude that our ground-based infrared data and the *in situ* measurements are indeed compatible.

Before the Halley encounters, our comet models incorporated a size distribution function derived from dynamical analysis of

[1] Recent analysis of PIA data suggests that the mass thresholds should be shifted toward larger mass, by up to a factor 10 (J. Kissel, private communication)

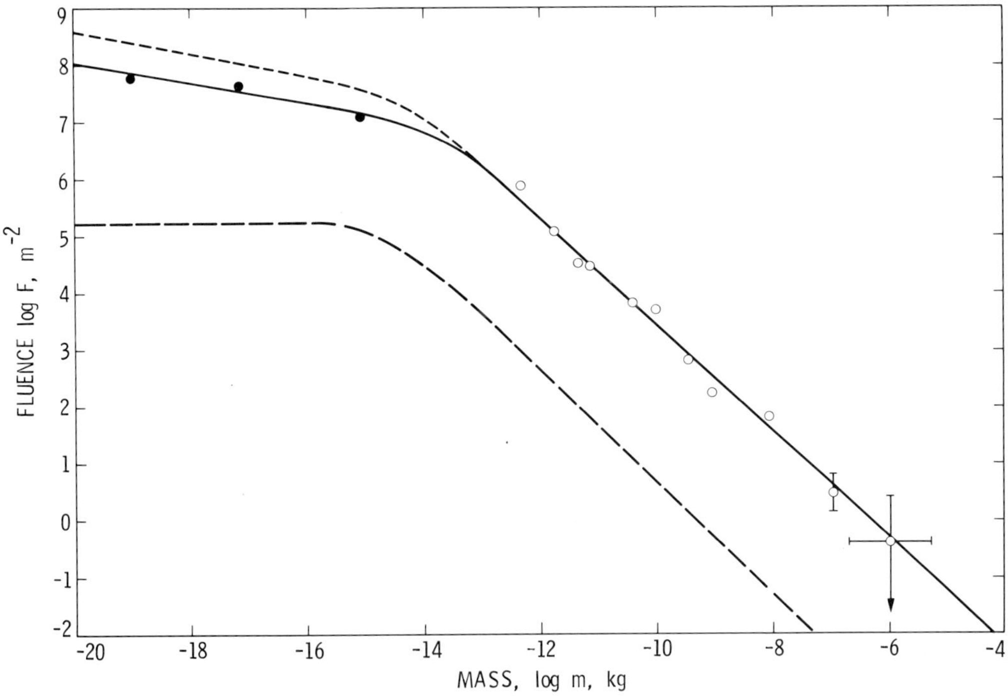

Fig. 5. Cumulative mass distribution integrated along the Giotto trajectory (McDonnell et al. 1986). ○: data from DIDSY experiment; ● data from PIA experiment. Solid line is analytical fit from Divine and Newburn (1987); short-dashed line is best fit to IRTF data Mar 12.8 and 13.75 if $\varrho = 1$ g cm^{-3}. Dashed line shows the model size distribution previously used (Divine et al. 1986; Hanner et al. 1985); normalization is arbitrary. Recent analysis of PIA data suggests that these mass thresholds should be shifted to larger mass, by up to a factor 10 (J. Kissel, private communication)

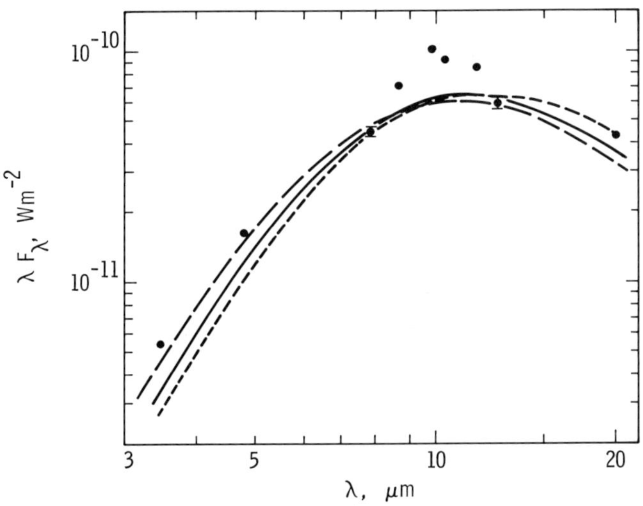

Fig. 6. Comparison of Mar 12.8 data with models. Solid line Giotto fluence, $m_t = 2\ 10^{-11}$ g; long dashes modified Giotto fluence, $m_t = 5\ 10^{-12}$ g (Fig. 5); short dashes Vega 2 fluence, $m_t = 1.6\ 10^{-10}$ g

dust tails, with a size parameter which was adjusted to match the thermal emission spectra (e.g. Hanner, 1983; Hanner et al., 1985; Divine et al., 1986). This size distribution is shown for comparison in Fig. 5. The slope $N = 3.7$ in those models corresponds to $\alpha = 0.9$, in agreement with the space probe data. A minimum radius of 0.1 μm ($4\ 10^{-15}$ g) was adopted, because smaller grains could not be detected readily in the integrated light from the dust coma. Particles <0.1 μm, although numerous in the Halley dust impact data, contribute <0.1 % to the thermal emission in the new models. Thus, our earlier models are not very different, and the dust production rates derived previously are not systematically inaccurate on this basis.

5. Concluding remarks

Halley's dust coma was never in a steady state during March 1986. Because the dust emission varied strongly with time, the size distribution of particles varied with both time and position in the coma, as documented by the in situ dust measurements. The infrared brightness varied in phase with the optical brightness and the production rates of CS and C_2, and the strength of the 10 μm silicate emission feature correlated with the level of jet activity. However, our data do not seem to have a constant phase relationship to the 2.2 day periodicity in Lyman α seen by Susei. Our observations give some support for a periodicity of 7.4 days in outburst activity during March. Neither Vega 1, Vega 2, nor Giotto apparently flew by the nucleus at a time of maximum activity. Further modeling of the dust emission, for comparison with infrared data, is in progress.

Acknowledgements. We thank the staff of the IRTF for their support, particularly George Kroenig and Lars Bergknut. ATT, WFG, DMG, and CDK acknowledge the support of NASA contract NASW-3159. A portion of this research was carried out at Jet Propulsion Laboratory, California Institute of Technology, under contract with the National Aeronautics and Space Administration, Planetary Astronomy Program.

References

Brownlee, D.E.: 1978, in *Cosmic Dust*, ed. J.A.M. McDonnell, J. Wiley, New York, p. 295

Campins, H., Bregman, J.D., Witteborn, F.C., Wooden, D.H., Rank, D.M., Allamandola, L.J., Cohen, M., Tielens, A.G.G.M.: 1986, 20th ESLAB Symp, ESA SP-250, Vol. **II**, p. 121

Campins, H., Joy, M., Harvey, P.M., Lester, D.F., Ellis, H.B.: 1986, 20th ESLAB Symp, ESA SP-250, Vol. **II**, p. 107

Campins, H., Telesco, C.M., Decher, R., Ramsey, B.D.: 1986c, 20th ESLAB Symp, ESA SP-250, Vol. **II**, p. 91

Divine, N., Fechtig, H., Gombosi, T.I., Hanner, M.S., Keller, H.U., Larson, S.M., Mendis, D.A., Newburn, R.L., Reinhard, R., Sekanina, Z., Yeomans, D.K.: 1986, *Space Sci. Rev.* **43**, 1

Divine, N., Newburn, R.L.: 1987, *Astron. Astrophys.* (this issue)

Feldman, P.D., A'Hearn, M.F., Festou, M.C., McFadden, L.A., Weaver, H.A., Woods, T.N.: 1986a, *Nature* **324**, 433

Feldman, P.D. et al.: 1986b, 20th ESLAB Symp, ESA SP-250, Vol. **I**, p. 325

Festou, M.C., Feldman, P.D., A'Hearn, M.F., Arpigny, C., Cosmovici, C.B., Danks, A.C., McFadden, L.A., Gilmozzi, R., Patriarchi, P., Tozzi, G.P., Wallis, M.K., Weaver, H.A.: 1986, *Nature* **321**, 361

Gehrz, R.D., Ney, E.P.: 1986, 20th ESLAB Symp, ESA SP-250, Vol. **II**, p. 101

Gombosi, T.I.: 1986, 20th ESLAB Symp, ESA SP-250, Vol. **II**, p. 167

Green, S.F., Hughes, D.W.: 1986, 20th ESLAB Symp, ESA SP-250, Vol. **II**, p. 157

Green, S.F., McDonnell, J.A.M., Pankiewicz, G.S.A., Zarnecki, J.C.: 1986, 20th ESLAB Symp, ESA SP-250, Vol. **II**, p. 81

Hanner, M.S.: 1983, in *Cometary Exploration*, ed. T.I. Gombosi, Hungar. Acad. Sci., Budapest, Vol. **II**, p. 1

Hanner, M.S., Tedesco, E., Tokunaga, A.T., Veeder, G.J., Lester, D.F., Witteborn, F.C., Bregman, J.D., Gradie, J., Lebofsky, L.: 1985, *Icarus* **64**, 11

Hanner, M.S., Tokunaga, A.T., Veeder, G.J., A'Hearn, M.F.: 1984, *Astron. J.* **89**, 162

Kaneda, E., Ashihara, O., Shimizu, M., Takagi, M., Hirao, K.: 1986a, *Nature* **321**, 297

Kaneda, E., Takagi, M., Hirao, K., Shimizu, M., Ashihara, O.: 1986b, 20th ESLAB Symp, ESA SP-250, Vol. **I**, p. 397

Kissel, J. et al.: 1986a, *Nature* **321**, 280

Kissel, J. et al.: 1986b, *Nature* **321**, 336

Koorneef, J.: 1983, *Astron. Astrophys. Suppl.* **51**, 489

Krätschmer, W., Huffman, D.R.: 1979, *Astrophys. Space Sci.* **61**, 195

Mazets, E.P. et al.: 1986, 20th ESLAB Symp, ESA SP-250, Vol. **II**, p. 3

McDonnell, J.A.M., Kissel, J., Grün, E., Grard, R.J.L., Langevin, Y., Olearczyk, R.E., Perry, C.H., Zarnecki, J.C.: 1986, 20th ESLAB Symp, ESA SP-250, Vol. **II**, p. 25

Millis, R.L., Schleicher, D.G.: 1986, *Nature* **324**, 646

Ney, E.P.: 1982, in *Comets*, ed. L.L. Wilkening, Univ. Arizona, Tucson, p. 323

Rettig, T.W., Kern, J.R., Ruchti, R., Baumbaugh, B., Baumbaugh, A., Knickerbocker, K., Dawe, J.: 1986, 20th ESLAB Symp, ESA SP-250, Vol. **III**, p. 93

Schleicher, D.G., Millis, R.L., Tholen, D., Lark, N., Birch, P.V., Martin, R., A'Hearn, M.F.: 1986, 20th ESLAB Symp, ESA SP-250, Vol. **I**, p. 565

Sekanina, Z.: 1987, *Nature* **325**, 326

Simpson, J.A., Rabinowitz, D., Tuzzolino, A.J., Ksanfomality, L.V., Sagdeev, R.Z.: 1986, 20th ESLAB Symp, ESA SP-250, Vol. **II**, p. 11

Tokunaga, A.T.: 1984, *Astron. J.* **89**, 172

Tokunaga, A.T., Golisch, W.F., Griep, D.M., Kaminski, C.D., Hanner, M.S.: 1986, *Astron. J.* **92**, 1183

Vaisberg, O., Smirnov, V., Omelchenko, A.: 1986, 20th ESLAB Symp, ESA SP-250, Vol. **II**, p. 17

West, R.M., Pedersen, H., Monderen, P., Vio, R., Grosbøl, P.: 1986, *Nature* **321**, 363

Zarnecki, J.C., McDonnell, J.A.M., Alexander, W.M., Burton, W.M., Hanner, M.S.: 1986, 20th ESLAB Symp, ESA SP-250, Vol. **II**, p. 185

Astron. Astrophys. 187, 661–664 (1987)

ASTRONOMY
AND
ASTROPHYSICS

The dust tail of comet P/Halley in April 1986 *

P.L. Lamy [1], **H. Pedersen** [2], **and R. Vio** [2]

[1] Max-Planck-Institut für Kernphysik, Postfach 103980, D-6900 Heidelberg, Federal Republic of Germany
[2] European Southern Observatory, La Silla, Chile

Received February 23, accepted April 2, 1987

Summary. Images of the dust tail of comet Halley were obtained with the ESO wide-field CCD camera during April 1986 with the Johnson *B*, *V* and *R* filters in polarized light. An analysis of selected images taken on April 6 and 11, processed for photometric analysis, is presented. Results are given for the absolute brightness, the color, the polarization and its variation with wavelength.

Key words: comets – photometry – polarization

1. Introduction

The accurate photometric analysis of cometary dust tails is a powerful means to investigate their physical properties. For instance, absolute brightness maps combined with a photometric model (e.g. Finson-Probstein method) allow to derive the size distribution function of the dust grains and the total dust production rate. The optical properties are best approached by the study of the color of the scattered solar light, the variation of polarization with scattering angle and with wavelength. However, this type of analysis presents numerous difficulties associated with the photometry of faint, extended objects. As a consequence, only a few comets have been studied so far, mainly those which displayed a broad and bright tail (e.g., comets Arend-Roland, Bennett, West).

The overall conditions for the large-scale observations of the dust tail of comet Halley were quite favorable (Lamy, 1985). The activity of the comet was sufficiently important and the viewing geometric conditions produced a good spreading of the tail and a net separation from the plasma tail. The main problem came during the first half of April 1986 from the interference of the Southern Milky Way and the rich stellar background.

2. Observations and images analysis

The post-perihelion observations of the tail of comet Halley were performed with a wide-field CCD camera (WFCC) set up at the European Southern Observatory (Lamy, 1985; West et al., 1986). This camera consists of a cooled, large-format CCD (630×1024 pixels) behind a Canon f/2.8, 100 mm lens and has the merit of combining a high sensitivity, wide-field ($5^\circ\!.0 \times 9^\circ$) and high resolution (1 pixel = 31″).

For the present study, 8 images have been selected which do not present too severe a problem of contamination by the Milky Way and which further illustrate the scientific information that can be obtained. For the first two observations, the heliocentric distance of the nucleus was $R = 1.26$ AU, its geocentric distance was $\Delta = 0.444$ AU and its phase angle was $45^\circ\!.5$. These parameters were respectively $R = 1.337$, $\Delta = 0.418$ and 31° for the other six observations. Additional information is presented in Table 1.

As discussed by Lamy (1986), the criteria for selecting the filters for observing cometary dust tails are entirely different from those imposed by comae. In the former case, the contamination comes from the plasma tail generally dominated by CO^+ and H_2O^+ emissions. When viewing geometric conditions result in a good separation of the two tails, this problem disappears except in the central region where the tails converge. Broad-band filters are therefore adequate and the selection of the Johnson system allows a direct calibration with field stars as discussed below. Finally, the orientation of the polarizor ($\perp$, $\|$) refers to the electric field vector with respect to the scattering plane.

Flat-field images of different levels where obtained at various levels of brightness through the lens and the different filters. Therefore, the vignetting function of the lens and the influence of the wavelength on this function and on the sensitivity of the CCD are taken care of by the flat-fielding procedure. It is worth mentioning that we first generated a new flat field adapted to the level of the science image by interpolation of the neighboring recorded flat-fields. Then the flat-fielding procedure with this new flat-field is performed removing all instrumental effects. The next step was to remove the stars which obviously perturb the photometry of the extended tail. We used the median filtering which replaces the pixel value by the median value of the distribution of the neighbouring pixel values, the neighbourhood or window 15×15 being chosen after trials. This way, pixel values deviating too much from the local mean such as stars or artefacts are rejected. This filtering has the extra advantage of improving the signal over noise ratio but is not totally efficient as the brightest stars still leave a contribution; this can be seen on the example (Fig. 1). A new procedure based on the local radii of curvature now under testing has demonstrated far superior results; it will be implemented for the final routine processing of all images.

Finally, for the present analysis and in particular for obtaining the polarization, a mean profile was generated for each image by taking the average in a band 15′ wide, centered onto the coma and oriented across that tail in the south-north direction (Fig. 1).

Send offprint requests to: P.L. Lamy

* Based on observations made at European Southern Observatory

Table 1. Parameters of the selected observations

Date UT Apr 86	Exposure (min)	Filter (polarizer)
6.341	3	$R(\parallel)$
6.350	3	$R(\perp)$
11.288	15	$B(\parallel)$
11.301	15	$B(\perp)$
11.332	6	$V(\perp)$
11.338	6	$V(\parallel)$
11.352	4	$R(\parallel)$
11.357	4	$R(\perp)$

3. Polarization

For the phase angles of the nucleus of 31° (Apr. 11) and 45°.5 (Apr. 6) corresponding to the present observations, the polarization is found positive. This can be considered as a "normal" situation of light scattered by dust grains in this range of phase angles. Detailed results for the April 11 observations are presented in Fig. 2: the polarization is plotted along the profile defined in Sect. 2 as a function of distance from the nucleus, in the northern direction, for the *B*, *V* and *R* filters. It is equal to 0.04 in the inner part (over 1°) and tends to increase slightly with distance. The decrease toward the innermost (coma) region, present only in the blue image, is probably a depolarization effect of the non-polarized CO^+ tail as it merges with the dust tail. An important conclusion is that the polarization appears independent of wavelength; at least no significant difference is detected at the present phase angle of 31°. An entirely different behavior was found for comet West as Lamy (1986) reported a strong increase of polarization with wavelength, namely $P_R/P_B = 1.6$ at a phase angle of $\approx 78°$. The fact that comet West was a new comet may possibly explain the difference as well as unusual circumstances (splitting of the nucleus and a striated tail) resulting in dust grains of (very?) different composition). There are conflicting results for $P(\lambda)$ of cometary comae, from a neutral behavior to an increase with wavelength (Lamy, 1986). The results obtained by Dollfus et al. (1987) on the coma of comet Halley indicate a very slight increase of the polarization with wavelength.

The observation of April 6 taken with the *R* filter yields a polarization of 0.075 at a phase angle of 45°.5. The two results are plotted in Fig. 3 along with results obtained *for the coma* by LeBorgne et al. (1987) and Mukai et al. (1985). Although the values are very close, the suggested slope is conspicuously different. As a consequence and when extrapolated, the present results indicate a phase angle corresponding to zero polarization

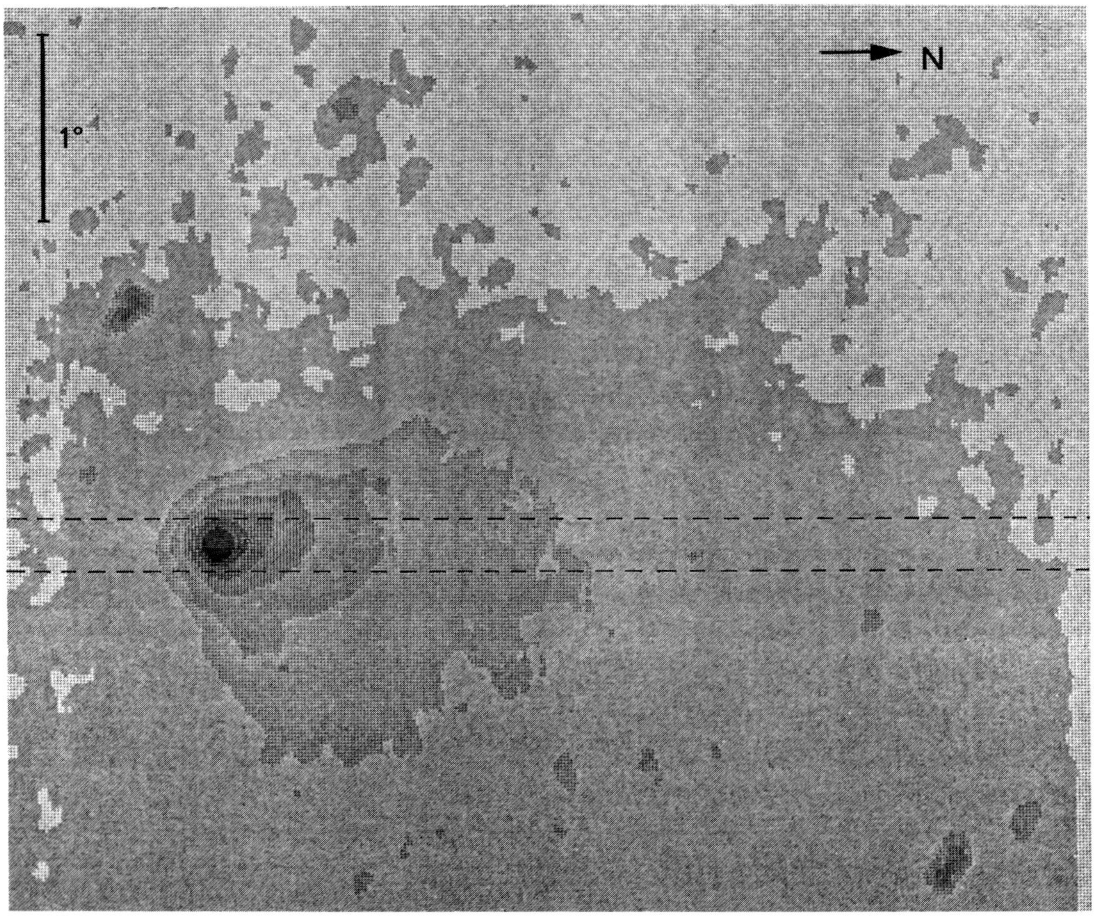

Fig. 1. Image of comet Halley on April 11.352, 1986 obtained with the *R* filter and the polarizor oriented in the praralled direction. The band which was used to generate the mean profiles is indicated

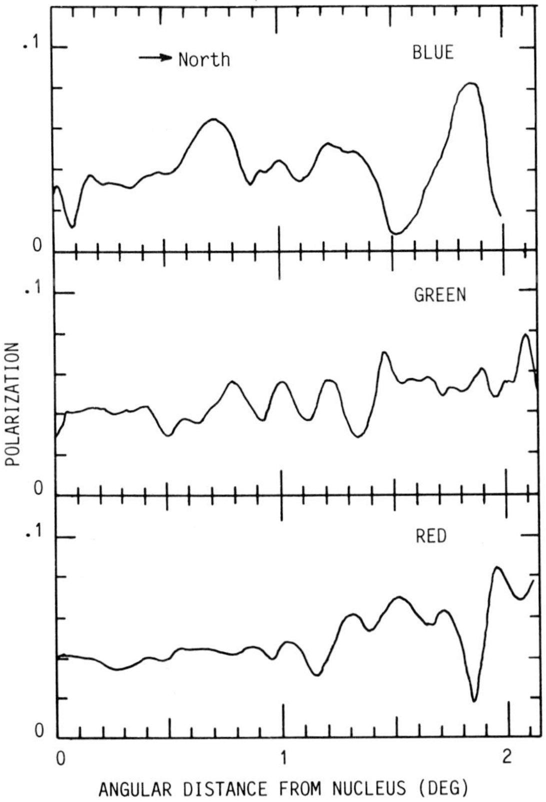

Fig. 2. Polarization versus angular distance from the nucleus for the three filters. The phase angle is 31°

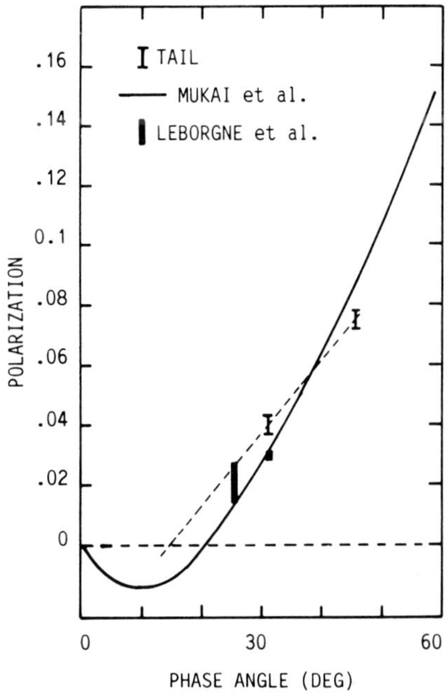

Fig. 3. Polarization versus phase angle. The present result for the tail (average of the V and R filters) are compared with those for the coma

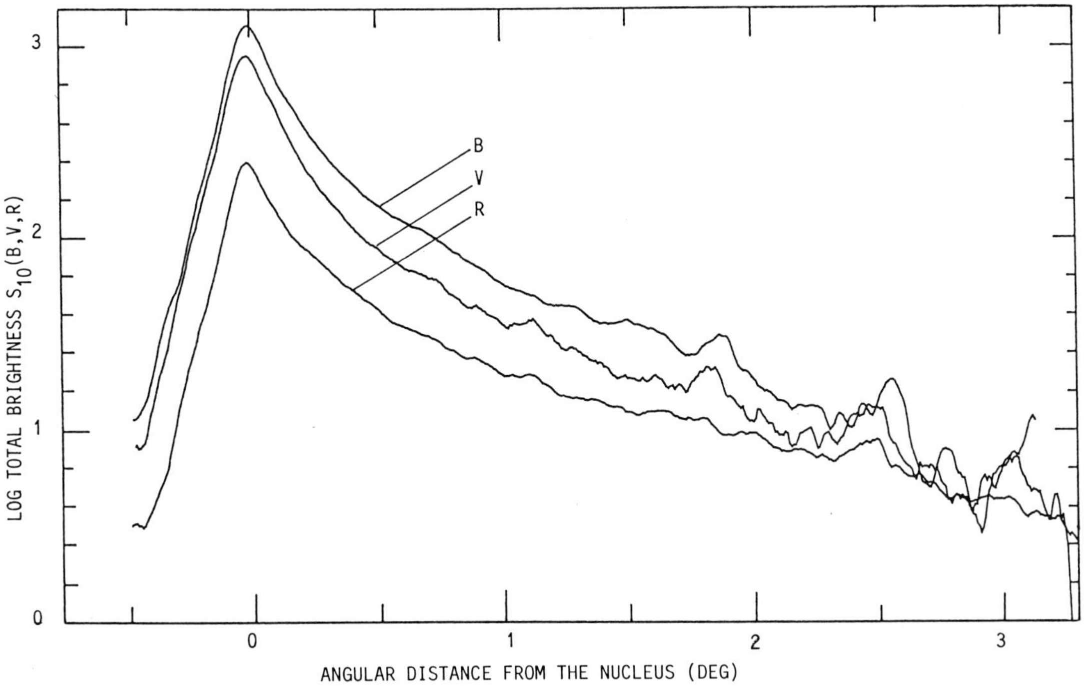

Fig. 4. Absolute total brightness versus angular distance from the nucleus on April 11, 1986 for the three filters

Table 2. Spectral type and *B*, *V*, *R* magnitude of the stars

HD	ST	*B*	*V*	*V-R*
127976	G8III	7.74	6.78	0.48
128582	F7V	6.58	6.07	0.28
130073	G8III	7.38	6.30	0.52
132301	F5V	7.06	6.58	0.21

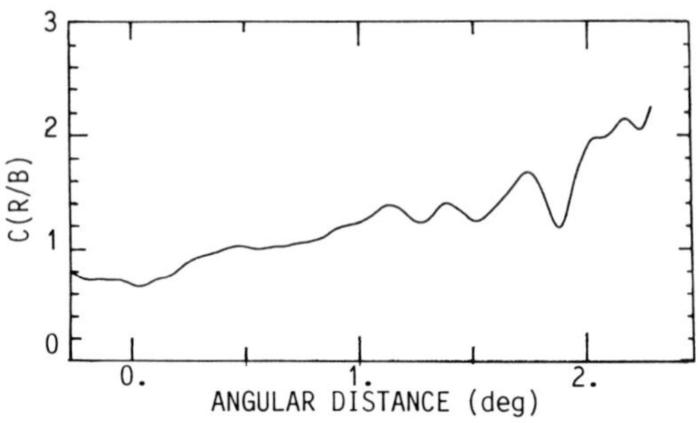

Fig. 5. The color index of the tail $C(R/B)$ normalized to that of the Sun versus angular distance from the nucleus (April 11, 1986)

(polarization reversal) of ≈15° for the tail and 20° for the coma. This result will have to be confirmed by the analysis of other images but it may reflect a true difference between the two dust populations.

4. Absolute photometry and color

The April 11 observations have been analyzed for absolute photometry using selected stars in the field whose spectral type is as close as possible to that of the Sun (G2V). The major problem comes from the limited number of measurements of *R* magnitudes. Indeed, this is lacking for the stars listed in Table 2 and the *V-R* color had to be deduced from the spectral type. The relationship between the pseudomagnitudes (measured on the images) and the true magnitudes was determined for each color. Then the flux per pixel was related to the number of stars of magnitude 10 per square degree $S_{10}(B, V, R)$. The absolute brightness along the profiles defined in Sect. 2 is plotted as a function of distance from the nucleus in the northern direction in Fig. 4.

Finally, the color of the dust tail was determined using the known color indices of the Sun (B-$V=0.65$, V-$R=0.52$). The ratio $C(R/B)$ of the *R* to the *V* brightness of the tail normalized to that of the Sun is plotted in Fig. 5 as a function of the distance from the nucleus. The fact that $C(R/B)$ is less than **1** in the inner part is readily understood by the presence of the CO^+ (blue) tail. Further away, when this tail is no longer present, $C(R/B)$ is larger than 1, indicating a dust tail redder than the Sun. This reddening effect tends to increase with increasing distances to the nucleus. At 1°.3, $(C(R/B))$ is equal to 1.3 approximately. A maximum value of the order of 1.8 seems indicated in Fig. 5. Such a reddening effect was also observed in the dust tail of comet West (Lamy, 1986) with a mean index of 1.68 (note a slight difference in the effective wavelength of the red channels, 0.65 μm in the case of comet West and 0.7 μm in the present case).

Acknowledgements. Image processing was performed at European Southern Observatory, Garching and at Max-Planck-Institut für Astronomie. P. Lamy expresses his gratitude to Prof. Elsässer for permission to use the facility at MPIA.

References

Dollfus, A., Suchail, J.L., Crussaire, D., Killinger, R.: 1986, 20th ESLAB Symp. ESA SP-250, Vol II, p. 41
Lamy, P.: 1985, *Adv. Space Res.* **5**, 317
Lamy, P.: 1986, in *Asteroids, Comets and Meteors II*, eds. C.I. Lagerkvist et al., Uppsala Universitet, Uppsala, p. 373
Le Borgne, J.F., Leroy, J.L., Arnaud, J.: 1987, 20th ESLAB Symp. ESA Sp-250 Vol I, p. 571
Mukai, S., Mukai, T., Kikuchi, S.: 1985, *Adv. Space Res.* **5**, 332
West, R.N., Pedersen, H., Monderen, P., Vio, R., Grosbol, P.: 1986, *Nature* **321**, 363

Astron. Astrophys. 187, 665–668 (1987)

ASTRONOMY
AND
ASTROPHYSICS

Albedo maps of comets P/Halley and P/Giacobini-Zinner

H.B. Hammel[1], C.M. Telesco[2]★, H. Campins[3]★, R. Decher[2]★, A.D. Storrs[1], and D.P. Cruikshank[1]★

[1] Institute for Astronomy, University of Hawaii, 2680 Woodlawn Drive, Honolulu, HI 96822, USA
[2] Space Science Laboratory, NASA/Marshall Space Flight Center, Huntsville, AL 35812, USA
[3] Planetary Science Institute, 2030 East Speedway, #201, Tucson, AZ 85719, USA

Received February 17, accepted April 14, 1987

Summary. Near-simultaneous infrared and visual images of P/Halley are combined to create a map of the spatial variation of average albedo (which is generally a factor of four larger than the geometric albedo defined by Hanner et al., 1981). The map is compared with a similar map of P/Giacobini-Zinner (P/G–Z) obtained previously. P/G–Z showed a minimum in albedo of 0.07 with an increase by a factor of 2 over an angular distance of 30″. The lowest albedos in P/G–Z were offset from the nucleus in the anti-sunward direction, coincident with a dust tail observed in the infrared. The P/Halley albedos are higher than those found for P/G–Z, ranging from 0.2–0.4, but both comets have relatively low albedo in the anti-sunward direction (along the tail). We attribute the albedo distribution to large, dark, fluffy grains confined to the orbital plane close to the nucleus. The high albedo values in P/Halley may be due to enhanced flux in the visual image because of the comet's very small phase angle.

Key words: comets – P/Halley, P/Giacobini-Zinner; comets – albedos

1. Introduction

In this paper we present pre-perihelion visual and infrared images of comet P/Halley. The visual image traces the distribution of light reflected by dust, while the IR image traces the dust's thermal emission. By combining the two data sets, we construct a map of the spatial distribution of average albedo in the inner coma. Our analysis of this map and a previously derived map of P/Giacobini-Zinner (hereafter P/G–Z; Telesco et al., 1986) provides a consistent picture accounting for both the albedo distribution in each comet and the difference in albedo between the comets.

2. Observations

2.1. P/Halley

2.1.1. Visual observations

Five CCD images were obtained with the University of Hawaii 2.24-m telescope (Mauna Kea Observatory) through an imaging quality filter distributed by the International Halley Watch (IHW). The filter samples the red continuum at 6840 Å. The image scale of the 500 × 500 array was 0″.138 pixel^{-1}. Two-by-two binning of the data was done at the telescope, yielding an image scale of 0″.276 pixel^{-1}. The field of each individual image is 69″ × 69″.

Four images were taken with the nucleus of P/Halley shifted into each of the four corners. The nucleus was centered in the fifth image. The position of the nucleus was determined to within a fraction of a pixel for each image. The images were then shifted appropriately and added to form a single composite image centered on the nucleus with a field of 125″ × 125″. The nuclear region of the composite was carefully compared with the single direct CCD image to verify that no spurious structure was created during the mosaic processing. All five 2.5-minute exposures were taken between 9:50 and 10:14 UT on 1985 November 18.

2.1.2. Infrared observations

The thermal IR image of P/Halley was made at the NASA Infrared Telescope Facility at Mauna Kea Observatory using the 20-pixel bolometer array developed at NASA Marshall Space Flight Center (described in Telesco et al., 1986). The wavelength was 10.8 μm with a $\Delta\lambda$(FWHM) equal to 5.3 μm. The pixel size was 4″.3 × 4″.3, with the center-to-center separation between pixels equal to 4″.5. The array field was 22″ × 18″ (RA by DEC). The total mapped field is about 80″ × 100″. The image was created by combining overlapping fields obtained over a period of 6.7 hours centered on 11:00 UT also on 18 November 1985. Figure 1a shows a contour map of the IR data.

2.1.3. Geometry

The geocentric distance of P/Halley was 0.7 AU on this date; the heliocentric distance was 1.7 AU. At the time of these observations, the comet was near opposition. The phase angle was 2° and changed by only about half a degree over the course of the observations. Thus, any significant dust tail should be approximately along the line of sight. The IR image does show a short dust tail extending toward the southeast (Fig. 1a); likewise, the visual image is asymmetric, with the lowest brightness levels extended toward the southeastern corner. The IR and visual observations were not simultaneous—the visual images were obtained 4 hours before the last sections of the IR image. Although the phase angle was nearly constant over this period, the position angle of the tail was changing rapidly, causing a clockwise rotation of the tail by 11° between the two data sets.

Send offprint requests to: H. Hammel

★ Visiting Astronomer, Infrared Telescope Facility, which is operated by the University of Hawaii, under contract with the National Aeronautics and Space Administration.

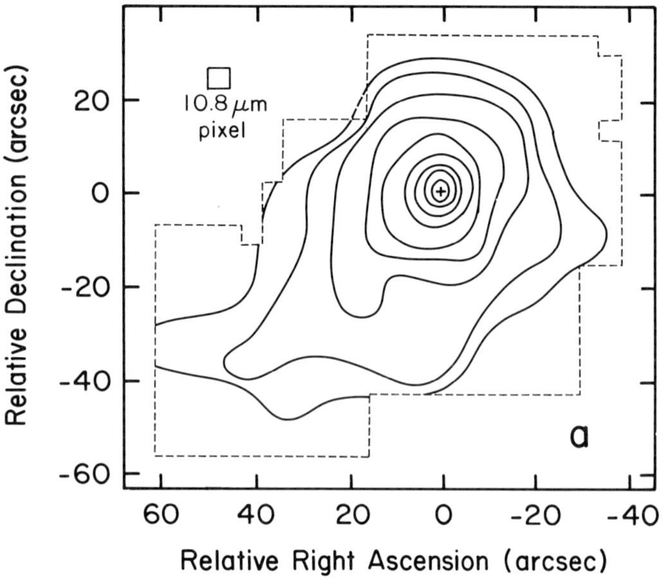

Fig. 1a. Map of P/Halley at 10.8 μm. The logarithmic-spaced contour levels are (Jy/pixel): 0.20, 0.30, 0.45, 0.68, 1.01, 1.52, 2.28, 3.42, and 5.13. The peak flux density (+) is 7.00 ± 0.26 Jy. The noise level is 0.10 Jy. The dotted line shows the extent of the mapped region

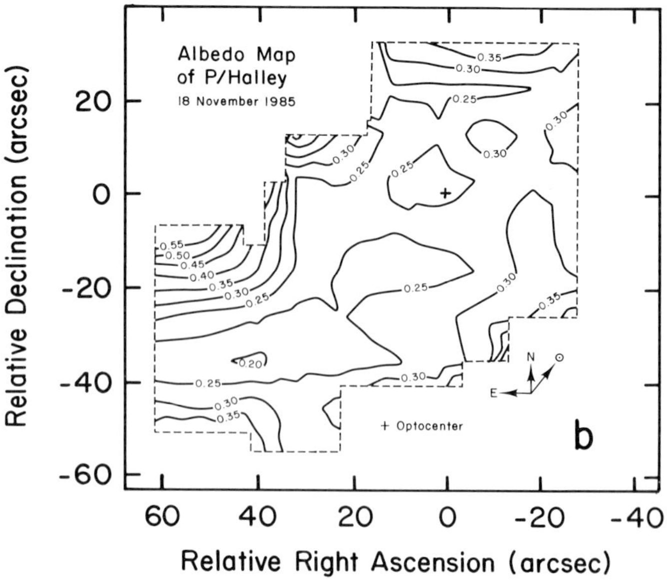

Fig. 1b. Albedo map of P/Halley. The contours are labeled with the average albedo (defined in the text). The location of the peak flux densities for both the visual and IR images is indicated (+). The dotted line shows the region of overlap between the visual and IR images

2.2. P/Giacobini-Zinner

2.2.1. Visual and IR observations

P/G–Z was observed from two sites on 4 August 1985. The visual image was taken with the same telescope, camera, and filter as the P/Halley observations. The IR image (Fig. 2a) was obtained with the same bolometer array, but at the Wyoming Infrared Observatory. The bandpass was also the same. Additional details about the observations can be found in Telesco et al. (1986).

2.2.2. Geometry

On 1985 August 4, the geocentric distance of P/G–Z was 0.59 AU and its heliocentric distance was 1.12 AU. The comet was at a moderate phase angle (64°) at the time of these observations, giving a favorable view of the tail structure. This was particularly evident in the IR thermal image, where there appears to be a large-grain tail superposed on the underlying coma and broader small-grain tail (Telesco et al., 1986). Because the IR observations bracket the visual observation, time variability of the comet should not be a problem.

3. Data reduction

3.1. Calibration

The visual P/Halley data were converted to standard fluxes using measurements of SAO star 076516 (HD 26736) in conjunction

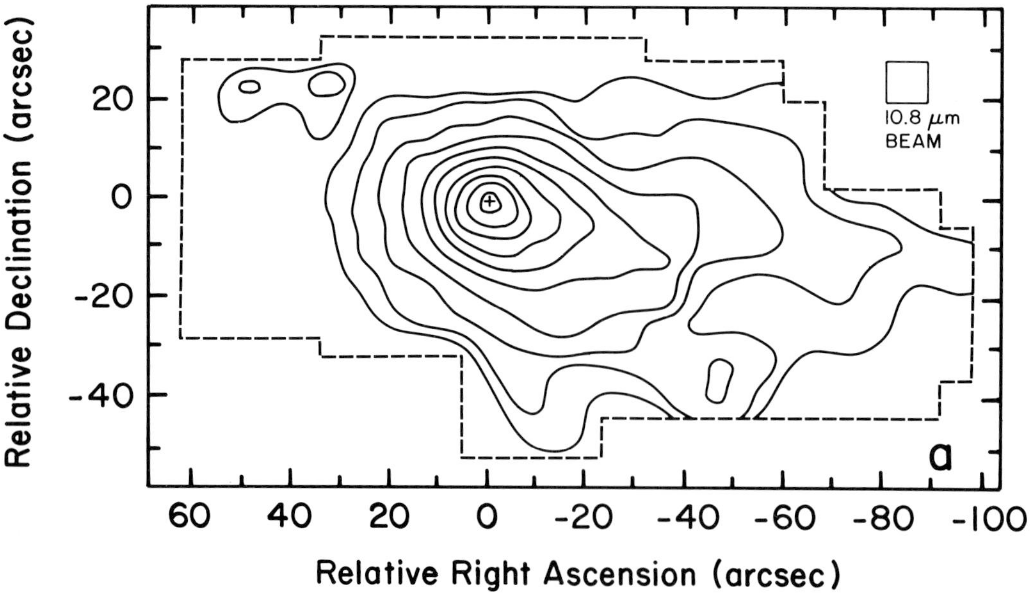

Fig. 2a. Map of P/Giacobini-Zinner at 10.8 μm (from Telesco et al., 1986). The logarithmic-spaced contour levels are 0.20, 0.30, 0.44, 0.67, 1.00, 1.50, 2.25, 3.38, 5.06, and 7.56 Jy per pixel. The peak flux density (+) is 9.3 ± 0.9 Jy. The average standard deviation of the measured mean for all points is ~0.1 Jy. The dotted line shows the extent of the mapped region

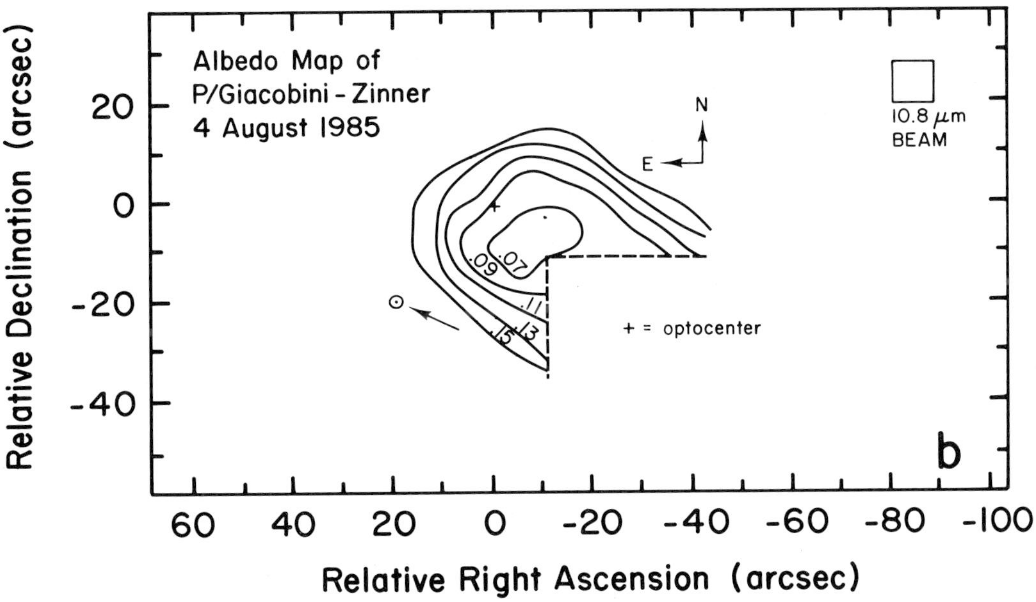

Fig. 2b. Albedo map of P/Giacobini-Zinner (from Telesco et al., 1986). The contours are labeled with the average albedo (defined in the text). The location of the peak flux densities for both the visual and IR images is indicated (+). The region to the lower right was contaminated by a star trail in the visual image. The albedo map was limited to the inner nucleus by the size of the single visual image

with the preliminary transformations for flux calibration outlined by IHW. The star α Herculis was used as a standard for the IR imaging. Its absolute calibration was provided by Tokunaga (1984). The visual data were rescaled and binned to match both the beam size (7.″5 × 7.″5) and sample spacing (8.″5 center-to-center) of the IR image. Care was taken to ensure that the images were properly aligned. Details of the calibration of the P/G–Z data can be found in Telesco et al. (1986).

3.2. Albedo calculation

The average albedo was calculated for each point in the two images by using the simplifying assumption (O'Dell, 1971; Ney, 1982) that the ratio of the total fluxes in the scattered solar and thermal emission energy distributions equals the ratio of the maximum values of λF_λ for each distribution. The ratio of λF_λ(visible)/λF_λ(infrared) is defined to be S. The albedo, g, is then equal to $S/(S+1)$. This average albedo (averaged over all wavelengths of incident solar radiation) is approximately a factor of four larger than the geometric albedo, $A_p(\theta)$, defined by Hanner et al. (1981). We correct for the fact that the maximum visual flux does not occur at 6840 Å by multiplying our visual data by a factor of 1.35 (Telesco et al., 1986). Albedos of other comets calculated with this method typically have values between 0.1 and 0.3 (Ney, 1982).

3.3. Possible sources of error

A possible source of error in determination of the albedo maps of both comets is misalignment of the two data sets used to calculate the final map. To estimate the magnitude and morphology of such an error, we deliberately offset the P/G–Z visual image from the corresponding IR image. Because the intensity distributions of both the visual and IR images are strongly peaked, even a very small error in alignment ($<3''$) manifested itself as an obvious distortion of albedo contours. The distortion caused by misalignment was significantly different in morphology from the albedo distribution derived from properly aligned images. Because the visual and IR images of P/Halley also show a strong central condensation, the same distortion would be observed. Misalignment of the data sets cannot cause the observed albedo distribution.

As discussed in Sect. 2.1.3, the position angle of P/Halley's tail changed by 11° between the observation times of the visual and IR images. However, the visual contours are very smooth over this range of angle, implying little or no change in the albedo map if the image were rotated.

Because the sky was photometric during the P/Halley observations, there should be no problem with the standard star calibration. Thin cirrus several hours before the P/G–Z observations motivated us to examine the effect of poor flux calibration. Simulated variations of the standard star flux by ±20% (an overestimate) of the observed value had no effect at all on the spatial distribution of albedo. The albedo values change by only ±0.01.

4. Discussion

The albedo map for P/Halley is shown in Fig. 1b; the P/G–Z map (from Telesco et al., 1986) is shown in Fig. 2b. The albedos (0.25–0.45) are systematically higher than those in the P/G–Z map (0.07–0.15; Telesco et al., 1986) and somewhat higher than those generally observed for other comets (0.1–0.3; Ney, 1982). The spatial distribution of albedo is similar to that of P/G–Z: the region of lowest albedo seems to be in the direction of the tail, i.e. the anti-sunward direction. The albedos increase radially from the nucleus, except along the tail, where the albedos are depressed. However, the range (or contrast) of albedo values is less in the P/Halley albedo map than it is in the P/G–Z map.

The difference in contrast between the P/Halley and P/G–Z maps may indicate a smaller concentration of large grains, which presumably cause the deep depression in the P/G–Z data. Interpretation of the IR image of P/G–Z alone indicated a tail composed of large ($>100\,\mu$m) grains spatially coincident with the region of lowest albedo (Telesco et al., 1986). The presence of large dark grains can account for both the low albedo (multiple scattering in large fluffy grains) and the location of low albedo relative to the nucleus and the large-grain tail. Large grains are expected to be closer to the nucleus and to the orbital plane because of slower ejection velocities. An analysis of the radial morphologies in IR images of P/Halley and P/G–Z has indicated a deficiency of large particles in P/Halley relative to P/G–Z (Campins et al., 1987). If the large grains are responsible for the albedo depression in the inner coma, P/Halley should show a smaller albedo variation because of this deficiency, just as we have seen. Thus, we have a consistent picture of grain-size distribution and albedo.

The overall higher albedos in P/Halley may be due to increased brightness in the visual image. Enhanced brightness at continuum wavelengths near opposition has been seen in other comets (e.g., A'Hearn et al., 1984; Millis et al., 1982). This has been interpreted as enhanced backscattering by dust, and it may produce a brightness increase as large as a factor of 3 at near-opposition phase angles (A'Hearn et al., 1984). Because the visual image of P/Halley was taken at very small phase angle, we might expect a backscattering peak. The albedo values in the Halley image would be comparable with typical cometary values if reduced by a factor of 2–3. Although a search for opposition brightening in P/Halley at shorter wavelengths did not appear to show any peak $>10\%$, substantial night-to-night variability was seen in the study (Meech and Jewitt, 1986). P/Halley may simply have a higher albedo. In any case, our general conclusions regarding the distribution and relative amount of large grains depend only on the morphology and contrast of the albedo distribution, not on its absolute value.

5. Conclusion

We have presented a map of the albedo distribution of comet P/Halley and compared it to a similar map of P/G–Z obtained earlier. The maps are derived from near-simultaneous visual and infrared imaging. Both comets show a similar distribution in albedo, with the albedo increasing radially from the nucleus except along the anti-sunward direction, where the albedos remain lower. The contrast in albedo is much more pronounced in the P/G–Z map. We interpret the albedo distribution as being produced by large grains concentrated near the nucleus and the orbital plane. The smaller contrast in the P/Halley map is consistent with IR observations that show a deficiency of large grains in P/Halley relative to P/G–Z.

The high albedos in P/Halley may be a result of increased brightness in the visual data produced by enhanced backscattering from the dust in the coma due to the near-opposition geometry. Backscattering peaks have been observed at small phase angles in other comets, typically causing an increase of a factor of 2–3 over the continuum values at moderate phase angles.

Acknowledgements. This research was partly funded by NASA grant NGL 12-001-057 to the University of Hawaii. H. Campins acknowledges NSF grant AST 84-14737 to the Planetary Science Institute, which is a division of Science Applications International Corporation. This is PSI contribution No. 231.

References

A'Hearn, M.F., Schleicher, D.G., Feldman, P.D., Millis, R.L., Thompson, D.T.: 1984, *Astron. J.* **89**, 579

Campins, H., Telesco, C.M., Decher, R., Ramsey, B.D.: 1987, *Astron. Astrophys.* (in press)

Hanner, M.S., Giese, R.H., Weiss, K., Zerrull, R.: 1981, *Astron. Astrophys.* **104**, 42

Meech, K.J., Jewitt, D.C.: 1986, in *Symp.* Paris, Volume I, p. 553

Millis, R.L., A'Hearn, M.J., Thompson, D.T.: 1982, *Astron. J.* **87**, 1310

Ney, E.P.: 1982, in *Comets*, ed. L.L. Wilkening, Tucson, University of Arizona Press, p. 323

O'Dell, C.R.: 1971, *Astrophys. J.* **66**, 675

Telesco, C.M., Decher, R., Baugher, C., Campins, H., Mozurkewich, D., Thronson, H.A., Cruikshank, D.P., Hammel, H.B., Larson, S., Sekanina, Z.: 1986, *Astrophys. J.* **310**, L61

Tokunaga, A.T.: 1984, *Astron. J.* **89**, 172

Astron. Astrophys. 187, 669–688 (1987)

ASTRONOMY AND ASTROPHYSICS

Polarimetry of grains in the coma of P/Halley

I. Observations

A. Dollfus[1] **and J.-L. Suchail**[2]

[1] Observatoire de Paris, Section de Meudon, F-92195 Meudon, France
[2] Dept. of Physics, Ecole Normale Supérieure, Paris, France

Received April 17, accepted June 16, 1987

Summary. With the new photopolarimeter that we built at the Observatoire de Paris, Section de Meudon, we analysed the polarization of the light in different areas over the coma of P/Halley, from October to December 1985 with the 100 cm telescope of Meudon, and then in greater detail in April 1986 with the 152 cm ESO telescope at La Silla, Chile. The data include 164 determinations of the linear polarization Q/I, 114 measurements of U/I and 74 values of the circular polarization V/I. The linear polarization is expressed as a function of the phase angle, separately for the dust recently ejected, for the oldest dust having travelled over at least 3000 km and for the bright envelope around the nucleus. Wavelength dependences are given. Deviations of the azimuth of polarization and circular polarizations are mapped over the coma. Interpretation will follow, for the characterizations of the coma grains.

Key words: comets – polarimetry – grains – interplanetary dust

1. Purpose

When a comet approaches the Sun, the heating of its nucleus produces a release of gas and solid particles, visible as the coma which surrounds the nucleus and as the tail. The solid grains are assumed to originate from the early phase of accretion which was responsible for the initial formation of the solar system. A period of gentle accumulation of grains and ice, with other volatile elements, produced first the proto-cometary nucleus, in which the solid grains remained embedded in ice. As such, they remained rather remarkably protected physically and chemically since the interior of the nucleus was never heated. Because of the subsequent gentle evaporation of ice, the pieces of pristine solid grains near the surface which have been reached by the removal of ice accumulated to form a crust. They were then exposed to "space weathering" agents, such as impacts, radiation and mechanical processes.

During the more recent episode when the comet is deflected toward the Sun, the increase of temperature, of radiation and of meteoroid impacts releases some gas and dust into space, often violently. The grains which are ejected from the interior are pristine, those blown out from the crust have been reworked. After release into the coma and tail, these grains eventually feed the reservoir of interplanetary dust.

Our purpose is to measure some physical properties of the grains, immediately after they have been released from the nucleus. We analyse, at the telescope, the state of polarization of the light which is scattered by the grains. Polarization produces information about albedo, absorption, texture, size, aggregation and shape for these grains.

In the present paper, which will be referenced as Part I in subsequent works, we present our observations. Other papers will assemble the other polarimetric data available, and interpretation will be given in terms of the nature of the cometary grains.

2. Observational method and data reduction

2.1. Polarization parameters

The state of polarization is conventionally described by the four Stokes parameters I, Q, U and V in which I is the photometric intensity. We denote as $+Q$ the flux which is linearly polarized with its electric vector in a given reference direction. It is $-Q$ when at $\pi/2$ from this azimuth. Q/I is the degree of polarization. $+U$ is the intensity for the linear polarization at $+45°$ from the reference, with $-U$ at $-45°$, $+V$ is the circularly polarized intensity with helicity left handed as seen by the observer, $-V$ is right handed and V/I is the degree of circular polarization.

In our case, the polarization is produced by a process of scattering which changes the direction of propagation; the reference azimuth is given by the electric vector perpendicular to the directions of incidence and of emergence. It is then expedient to express the linear polarization by another set of two parameters which are the degree of linear polarization P and the azimuth of the polarization θ, with $P=\sqrt{(Q/I)^2+(U/I)^2}$ and $\theta=1/2\,\mathrm{arc\,tg}\,(U/Q)$ reckoned trigonometrically from the direction perpendicular to the incident and emergent beams. The degree of circular polarization V/I is unchanged.

2.2. Telescopes

From October to the end of December 1985 we used the 100 cm diameter reflector at Meudon Observatory, France. From April 7 to 17, 1986, the 152 cm ESO telescope was made available to us at the European Southern Observatory (ESO), La Silla, Chile. Both

Send offprint requests to: A. Dollfus

these telescopes were tested on non-polarized stars to check the spurious polarization often produced by the aluminium coatings of the mirrors. These two telescopes produced no instrumental polarization within the accuracy of a few 10^{-4}, at any azimuth and for none of the wavelengths covered by our filters. No corrections were needed.

2.3. Photopolarimeter

The Meudon Observatory photoelectric polarimeter PPHR (Photo-Polarimètre à Haute Résolution) modulates the polarized component of the incident light at 40 Hz with a rotating achromatic half-wave plate in front of a Glen-Thomson polarizer prism. In order to avoid spurious polarization by edge effect, the field stop aperture is placed after the modulating device. The holes are drilled on metallic mirrors which reflect the image onto an eyeplece through which the observer can analyse the field, locate the hole on the image and secure fine guiding.

After a filter wheel, the light is collected by a Lallemand 20 dynodes photomultiplier tube, specially designed at the Observatoire de Paris, with an S11 cathode. Two lock-in amplifiers linked with the polarization modulator at phases 0° and 45° respectively produce simultaneously two rectified signals feeding the channels Q and U. The relative flux I is given by a D. C. channel.

In front of the modulating device, there is a polarization compensator made of a thin glass plate normal to the beam and which can be tilted of an angle α around an axis oriented at azimuths either 0° or 90° to produce a partial polarization along $+Q$ or $-Q$. It can also be oriented at +45° or −45° to compensate the polarization $\pm U$. The degree of linear polarization P produced was computed as a function of α, using Fresnel equations extended to the multiple reflexions between the two faces of the plate. The calibration curve $P(\alpha)$ so derived does not vary with time and has a small wavelength dependence.

The three output channels I, Q, and U, and the compensator angle α can be simultaneously recorded on a data acquisition system. For our comet Halley runs, we used pen recorders, as indicated below.

2.4. Holes and filters

Seven holes are available, the smallest of diameter 0.165 mm giving with the 152 cm ESO telescope an angular resolution of 2"1. The largest hole covers 107" in diameter.

The filter wheel introduces the 5 filters O, G, B, $UV(A)$, and $UV(B)$ routinely used in our polarimetric observations since 1960, plus a wide band filter covering the whole visible spectrum. Their spectral transmissions are given in Fig. 1. In addition, we used three IHW narrow band interference filters isolating the emissions C_2 (514 nm), CN (387 nm) and the continuum at 484 nm.

2.5. Method

First the whole photopolarimeter is positioned by rotation around its optical axis to set the azimuth $+Q$ at +90° from the azimuth θ of the Sun, computed from ephemeris with:

$$\cos\theta = [\cos\delta_\odot \sin(\alpha_\odot - \alpha)]/\sin\sigma$$

$$\cos\sigma = \sin\delta_\odot \sin\delta + \cos\delta_\odot \cos\delta \cos(\alpha_\odot - \alpha),$$

in which $\alpha_\odot$ and $\delta_\odot$ = coelestial coordinates of Sun, α and δ = coelestial coordinates of object and σ = Sun-Earth-object angle.

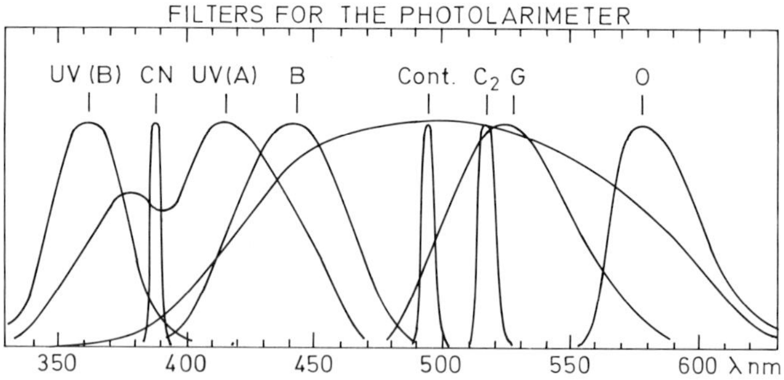

Fig. 1. Wavelength transmission of the filters used for the polarimetry of P/Halley. Filters CN, C_2 and Cont, are from the IHW set

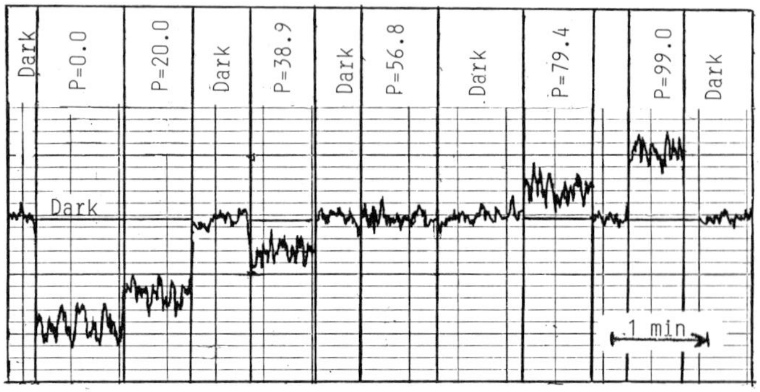

Fig. 2. Example of polarization measurements: Stokes parameter Q/I on April 8; 1986 at 02^h45^m, hole 7"3 centered on the P/Halley nucleus. Wide band filter. The sequences are labelled at top. P is the degree of polarization produced by the compensator. Dark is with the shutter closed (for analysis, see Fig. 3)

After a field area is selected for measurement, sequences of integrations are recorded with different values of the polarization compensator angle α. They are alternated with sequences of dark, with shutter closed. An example is given in Figure 2 which reproduces the integrations for the first Q/I measurement in Table 3 with the hole centered on the nucleus. We alternated sequences of dark and of measurements for 6 values of the compensation P, over a total of 8 min. of which 4.8 min. were on the comet. Then the deviations from dark are measured for each value of P and plotted in Figure 3. The regression line between the 6 deviations intersects the zero line at a point which corresponds to the exact compensation of the unknown polarization by the compensator polarization, giving in our case $Q/I = +(60.8 \pm 0.7)\ 10^{-3}$ with a slope $m = 2.66$ and a correlation $r = 0.997$. This method is independent of the electronic adjustment and of its variation. It enables the accumulation of successive integrations when the accuracy needed requires a long observation time.

The U/I signal is recorded simultaneously. Because it is usually small, the compensator is not needed, the deviations are translated into polarization with the same slope m used for Q/I, corrected for the gain ratio between the two Q and U amplifiers.

2.6. Circular polarization

V/I is analysed separately by introducing in front of the instrument a quarter wave birefringent plate, with its axis either at +45° or at −45° from the azimuth of Q. The circular polarization V is then transformed to a linear polarization, on channel U. Tilting the azimuth of the plate alternately from +45° to −45° and vice versa produces a modulation of amplitude 2V.

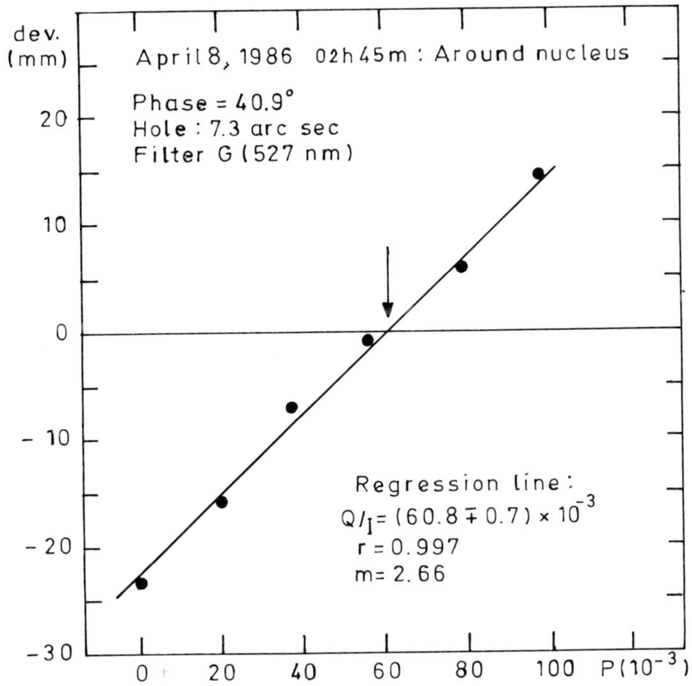

Fig. 3. Example of polarization record analysis. The data are from the pen recorder tracing of Fig. 2: The deflection from the dark signal is plotted as a function of the compensator degree of polarization P. The intersection of the regression line with the zero deviation gives the degree of linear polarization Q/I

For our comet Halley runs, for practical reasons, the quarter wave plate was set at only one position. The loss of sensitivity was more than a factor 2 because of the need to record precisely the dark current. The sign of the helicity escaped determination.

When the determination of V is made in presence of a linear polarization, as is the case for a comet, there is a danger that a significant fraction of the linear polarization might be transformed to circular, due to residual birefringence in the optics, or imperfection in the quarter wave plate or its misalignment. The result is the addition of a spurious V signal. Such was apparently the case for our runs. So we decided to correct the measurements for each night by the average of all the V values over the different areas of the coma. Thus, a permanent circular polarization all over the field, if it exists, is not recorded.

2.7. *Correction for sky brightness*

When the background sky luminance produced a non negligible flux, a correction was applied, namely, for the observations recorded at Meudon Observatory.

In the presence of a sky background, the object produces on channels I and Q the values I_n and Q_n recorded with the deviations $\delta(I_n)=k_1 I_n$ and $\delta(Q_n)=k_2 Q_n$. Then a polarization P is introduced with the compensator and new deviations are observed, respectively $\delta(I_p)$ and $\delta(I_b)$. On the sky background alone, we have $\delta(I_b)$ and $\delta(Q_b)$. Because the polarized components add their flux, the following equations apply:

$$I_n = I + I_b$$
$$\delta(Q_b) = k_2 (Q/I)_b I_b$$
$$\delta(Q_n) = k_2 [(Q/I)_b + (Q/I) I]$$
$$\delta(Q_p) = k_2 [(Q/I)_b + (Q/I) I + P \cdot I_n].$$

The correct value is derived:

$$\frac{Q}{I} = \frac{I_n}{I_n - I_b} P \frac{\delta(Q_p) - \delta(Q_b)}{\delta(Q_p) - \delta(Q_n)}.$$

2.8. *Correction for molecular emission*

The gas fluorescence produces several monochromatic emissions. Those for CN, C_2, and C_3 are partly included in the spectral transmission bands of some of our colour filters. They contaminate the data. A correction is needed.

$\tau(\lambda)$ is the transmission of the filter and $I_c(\lambda)$ is the luminance of the light scattered by the dust. The total flux transmitted by the dust continuum through the filter is $\theta_c = \int_0^\infty \tau(\lambda)\, I_c(\lambda)\, d\lambda$.

Each line has an intensity I_{em} expressed by the surface of its profile normalized to the local continuum I_{cont}. The global effect of all the emissions included in the spectral band transmitted by the filter is:

$$\theta_{em} = \sum \frac{I_{em}(\lambda_i)}{I_{cont}(\lambda_i)} I_c(\lambda_i)\, \tau(\lambda_i).$$

The polarization produced by the molecular emissions is

$$P_{em} = \frac{P_{max} \sin^2 V}{1 + P_{max} \cos^2 V}$$

in which V is the phase angle and P_{max} the polarization maximum. According to Le Borgne (1968), $P_{max} = 77\ 10^{-3}$.

From the addition of the polarized fluxes, we derive the polarization of the dust corrected for the emissions.

$$Q/I = (Q/I)_{obs} \left(1 + \frac{\theta_{em}}{\theta_c}\right) - \frac{\theta_{em}}{\theta_c} P_{em}.$$

During the period of our observations at the ESO station, the monochromatic emission intensities were monitored with spectra by Dr. Arpigny, at different distances form the nucleus. These spectra were generously made available to us by their author. One of these spectra is reproduced in Fig. 26. The ratio θ_{em}/θ_c found for each filter is reported in Table 1.

3. Morphology and photometry

3.1. *Envelope around the nucleus*

During all our polarimetric survey, the aspect of the coma was watched visually at the eyepiece of the photopolarimeter and

Table 1. P/Halley: Ratio of emission line to continuum intensities in April 1986

Filter	ϕ_{em}/ϕ_c around nucleus	ϕ_{em}/ϕ_c at 1000 km	ϕ_{em}/ϕ_c at 3000 km
O	0.08	0.14	0.29
G	0.15	0.28	0.54
B	0.05	0.09	0.18
UV (A)	0.11	0.19	0.40
UV (B)	0.00	0.00	0.00
Wide	0.11	0.19	0.40

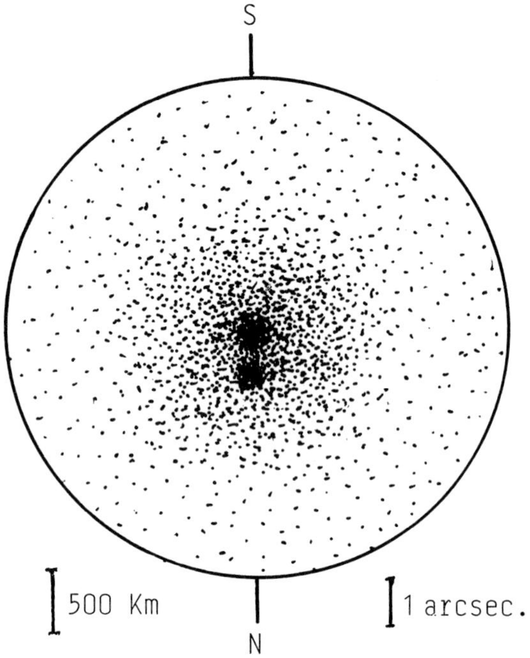

Fig. 4. Apulse of a star with the nuclear envelope of P/Halley. April 11, 1986, around 05^h30^m UT. Visual observation at the 1.52 m ESO telescope with a magnification of ×1000 by Audouin Dollfus. The cometary envelope is only slightly more fuzzy than the image of the star. See the scale in arcsec and in kilometers

sketched (see Figs. 5–12). A striking feature was the very bright point source seen at the nucleus, with an almost star-like appearance. In December 1985 and April 1986, this nuclear envelope was practically continuously noted. It gave the impression of having a greater extension before the perihelion passage. Its brightness was variable from night to night, depending on the amount of dust absorbing the light between the nucleus and the observer. On 1986 April 13, around 04^h UT, this point source was not seen at all (Fig. 9), because it was completely screened by a dust ejection in the direction of the Earth, dense enough to be opaque.

On April 11 at 05^h30^m UT, a star entered into the field and crossed the coma diametrally, to appulse the nucleus at only 1″. The star's magnitude was almost the same as the envelope and the seeing conditions were superb, with a magnification of ×1000. The image of the envelope was seen only slightly more fuzzy than the star (Fig. 4). The half width of the star image being of the order of 0.″3, the radius of the envelope at half brightness can be estimated to be around 50 – 100 km.

3.2. *Structure and photometry of the coma*

In April 1986, the ejection of dust toward the Sun, produced the classical fan-shaped structure and was continuously observed. Figures 5–7 show the details of the inner part of this feature, in a field of 25″ diameter. On April 8, a bright streamer at 170° dominated the other features. On April 9, the fan-like feature was symmetric. On April 10, it was still more developed and arcuated along its two edges, with additional small streamers.

The other observations, (see Figs. 8–12) refer to a larger field of 2′ diameter, showing the dust features at a wider scale. As the phase angle decreased from 31.°3 to 21.°6, the ejection cone was seen more and more from near its axis and the apparent angle of its

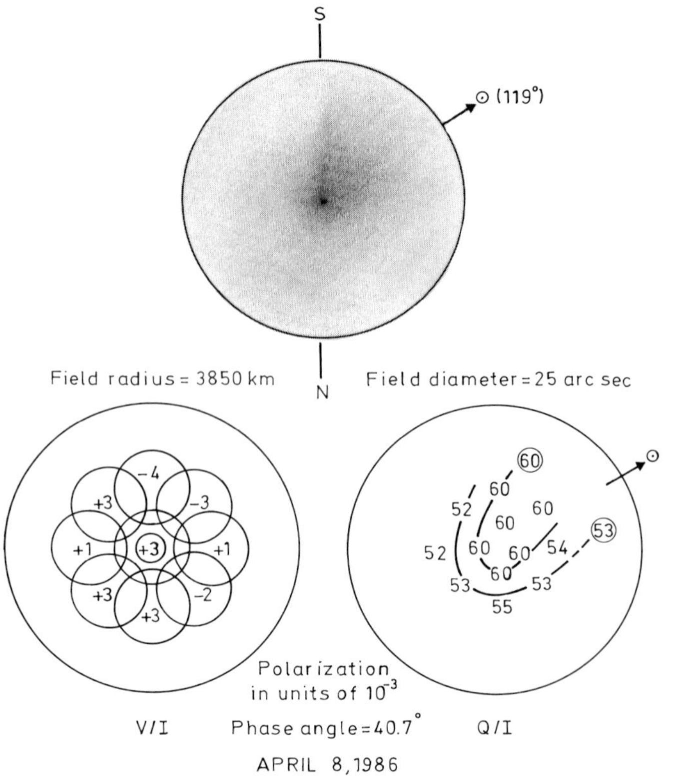

Fig. 5. Polarimetric analysis of P/Halley on April 8, 1986 from 05^h20^m to 06^h30^m UT, for the measurements around the nucleus. Wide band filter. The polarizations are not corrected from the emission bands (see Table 3). Field diameter: 25″ arcsec. Top: visual aspect of the inner coma and of the bright envelope, magnification ×1000. Bottom left: position and diameter of the holes, with values of V/I in units of 10^{-3}. Bottom right: values of Q/I in units of 10^{-3} and lines of equal degree of polarization

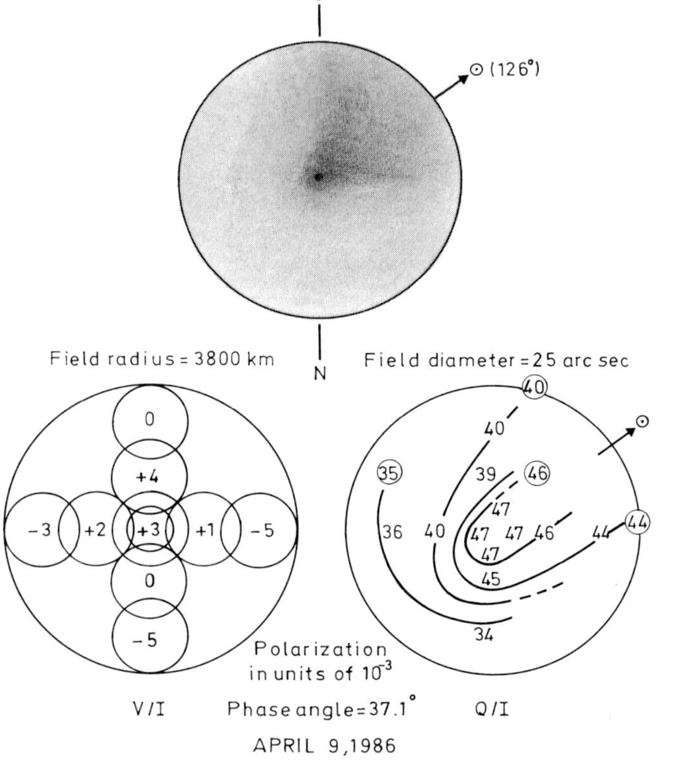

Fig. 6. Same as Fig. 5, but for April 9, 1986, from 05^h40^m to 06^h35^m UT and Table 4

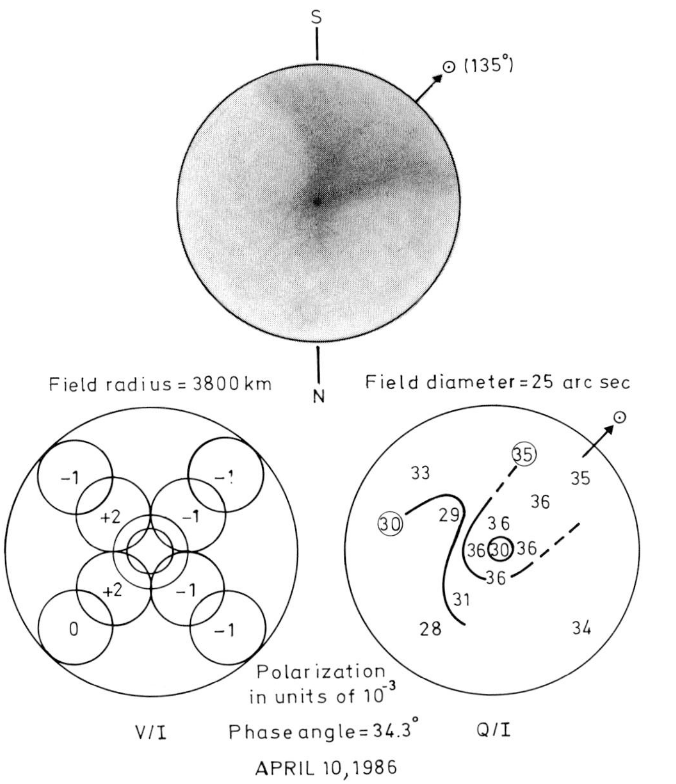

Fig. 7. Same as Fig. 5, but for April 10, 1986, from 03^h35^m to 05^h40^m UT, and Table 5

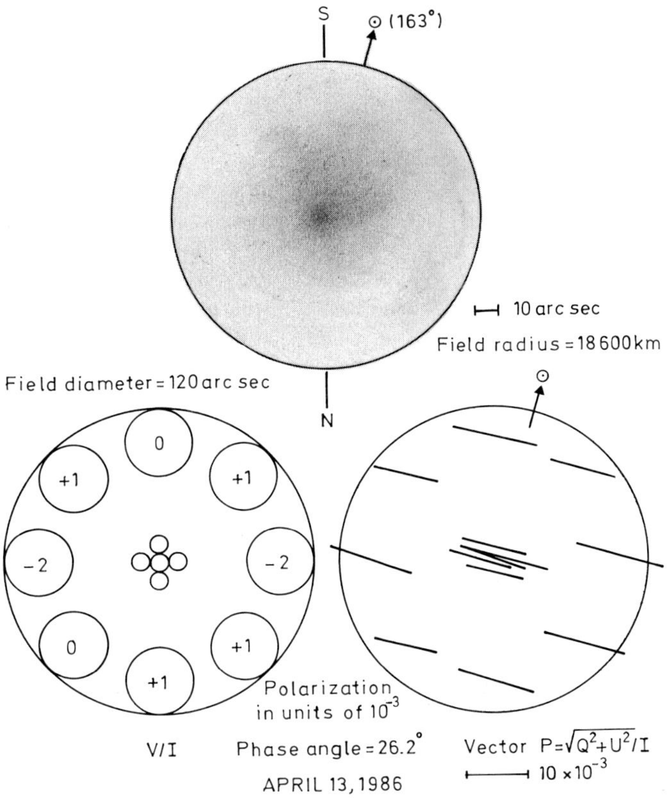

Fig. 9. Same as Fig. 8, but for April 13, 1986, from 02^h10^m to 06^h15^m UT, and Table 7. The aspect of the inner coma departs from its usual behaviour. There is no bright envelope seen around the nucleus. A dust streamer was ejected in the direction of the Earth and obscured the inner coma and envelope

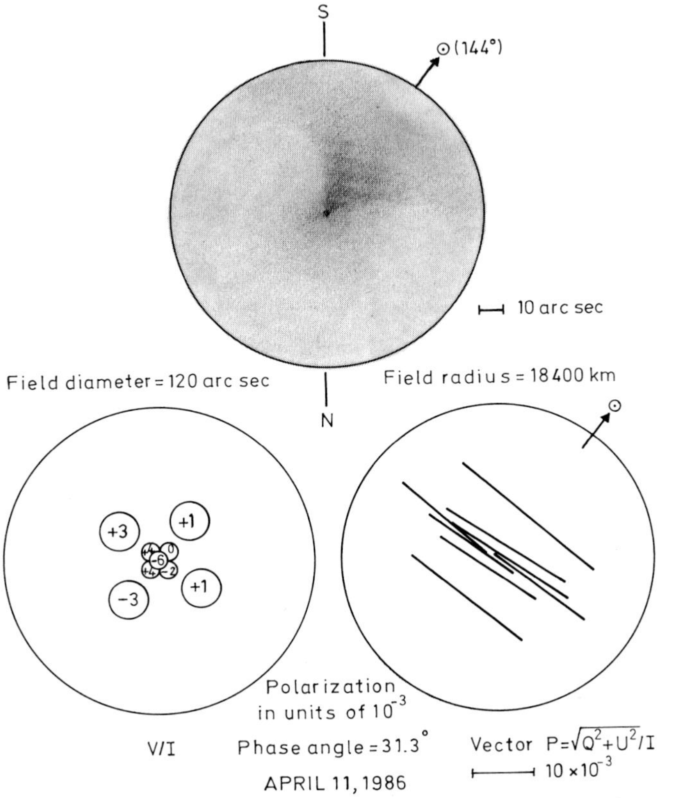

Fig. 8. Polarization analysis of P/Halley on April 11, 1986, from 03^h20^m to 07^h05^m. Same as Fig. 5, but in a field of 120″ diameter, and Table 6. At bottom right, the vectors give the degree P and the azimuth θ of the linear polarization

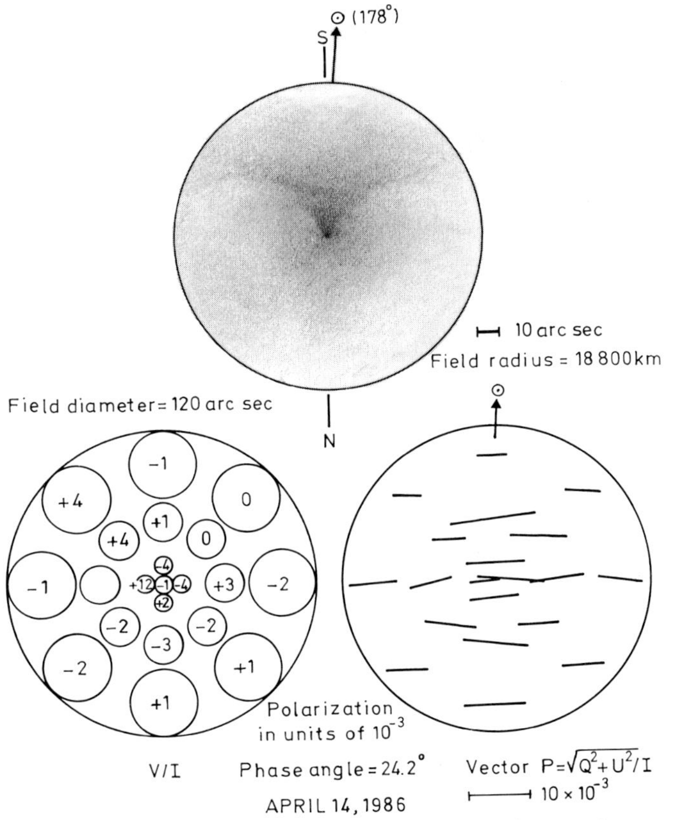

Fig. 10. Same as Figs. 8 and 9 but for April 14, 1986, from 02^h30^m to 07^h30^m and Table 8. The inner coma and envelope recovered their usual aspect, which was blurred in the previous night

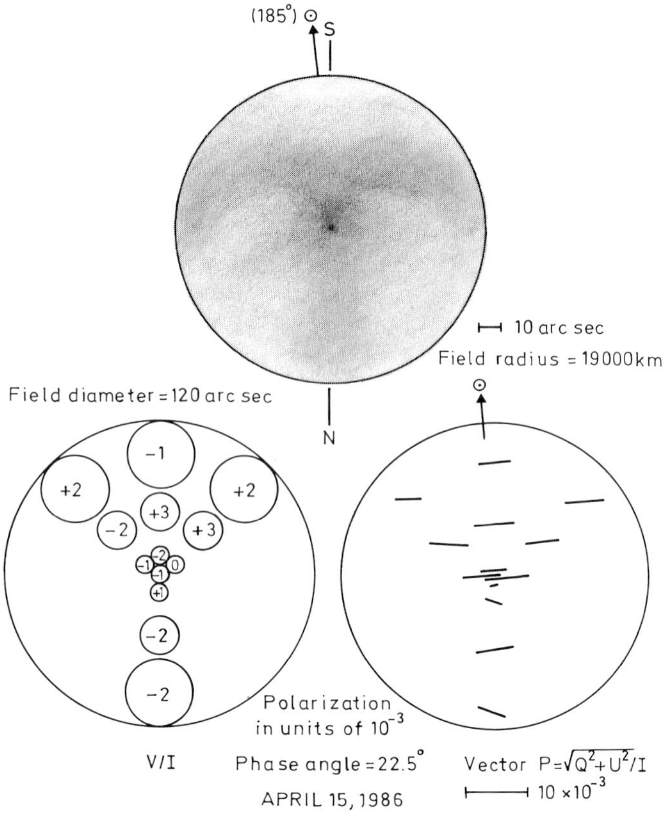

Fig. 11. Same as Fig. 8 but for April 15, 1986, from 03^h30^m to 06^h00^m, and Table 9

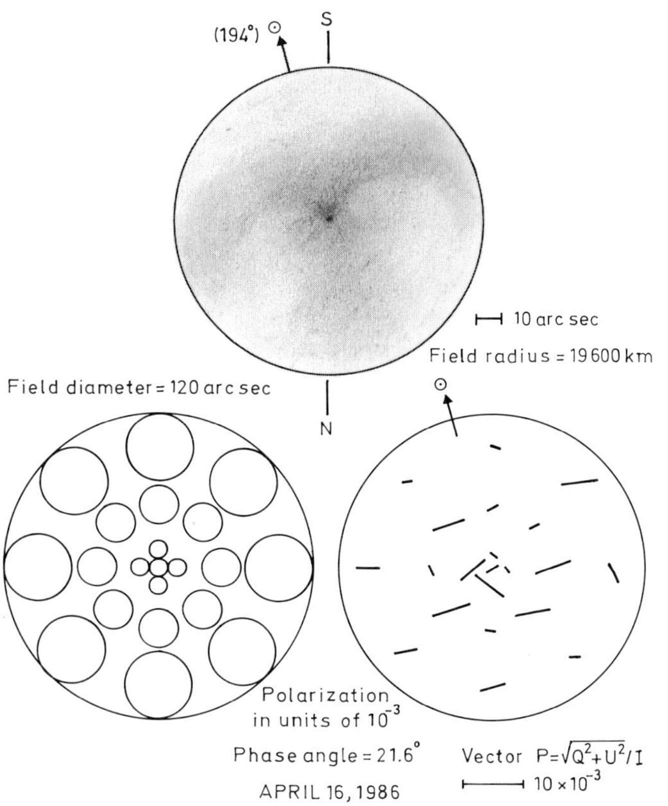

Fig. 12. Same as Fig. 8, but for April 16, 1986, from 03^h10^m to 06^h10^m, and Table 10

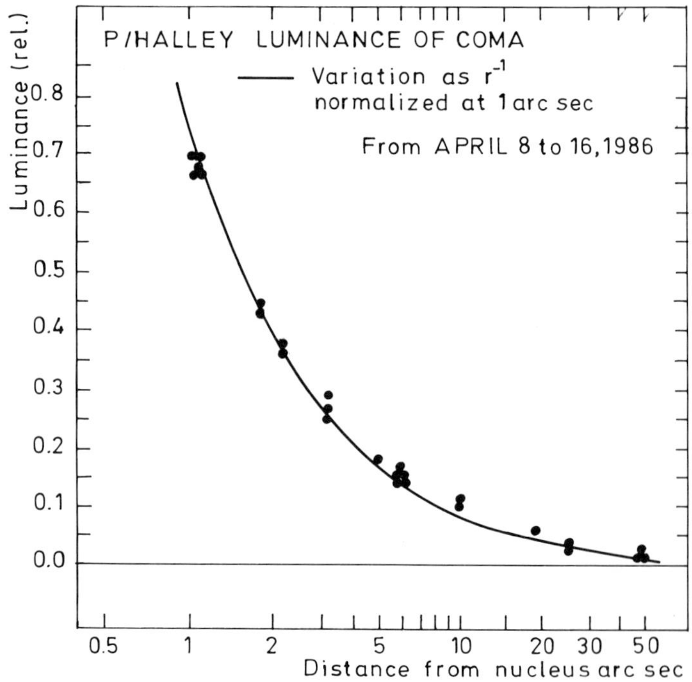

Fig. 13. Luminance of P/Halley coma averaged over circular zones, at increasing distance from the nucleus (in relative units). The sequences of measurements for April 8, 10, 13, 15, and 16 are normalized and added. The curve is the variation as r^{-1}

fan projection increased. On April 13, (Fig. 9) all the features were fuzzy and the central bright envelope was obscured as already noted, because of a dust ejection toward the Earth. On April 15, (Fig. 11), thin arcuated streamers were added to the fan which was more and more arched.

We analysed the increase of luminance of the coma toward the nucleus with the photopolarimeter. In Fig. 13, the relative luminance is expressed as a function of the distance r to the nucleus. The data are well fitted by a r^{-1} curve, indicative of a r^{-2} dust density decrease with distance, as expected if there is no change in the albedo of the grains with distance to the nucleus, or with time after ejection. There was almost no significant day to day variation, as shown in Fig. 14.

Star passages permitted an analysis of the optical absorption by the coma dust. The star which crossed the coma almost diametrally on April 11 (Fig. 4) showed, from the edge of the field and toward the nucleus, first an apparent increase of brightness due to the effect of the background coma luminance, then a progressive decrease: it reached exactly the magnitude of the nuclear envelope when at 1″ (Fig. 4) and then increased slightly in brightness when outbound. We estimated visually that, at minimum distance, the brightness decrease was of the order of 1 mag. Therefore, the coma along its diametral line of sight, had an optical transmission of around $40 \pm 10\%$ for visible light.

We observed several other stars crossing the field behind the coma. On April 10 at 23^h46^m, a star approached the nucleus at about 6″, or 2000 km, where it had approximately the same magnitude as for the nuclear envelope; it was seen about 15% brighter when at 15″. On April 13, the dust streamer ejected toward the Earth was completely opaque when seen along its axis and obscured completely the bright envelope; a star passing, at 35″ or about 10000 km from the nucleus (at 6^h54^m) showed a slight brightness decrease.

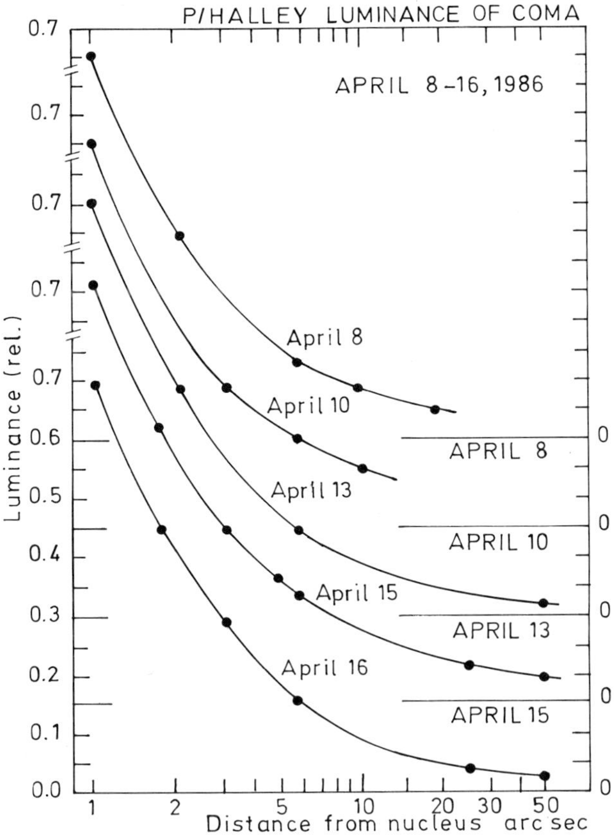

Fig. 14. Details of the five sequences of measurements added in the Fig. 13

Photometry of a star appulse on April 20–21, 1986 was reported by Kidger et al. (1987). At the minimum distance of 5″.3, or 2040 km, they observed no increase of magnitude within ±0.1 mag, at $\lambda = 1\ \mu$m.

Lecacheux et al. (1984) reported a stellar appulse photometry for comet IRAS-Iraki-Alcock, with no brightness variation larger than ±2.5% at the minimum distance of 420 km.

4. Linear polarization

Preliminary results have already been published by Dollfus et al. (1986) and by Dollfus and Suchail (1986).

4.1. Polarization maps of the coma

The measurements of the Stokes components over the coma are reported in Tables 2–10. Each measurement is localized by its distance to the nucleus expressed in arcsec. and its azimuth reckoned trigonometric from North. The fluxes Φ are in relative units and give the relative luminances L when divided by the surfaces of the hole. The degrees of polarization Q/I, U/I, and V/I are given in units of 10^{-3}, uncorrected from the effect of the fluorescence emissions. The azimuths θ of the linear polarizations are reckoned from the direction of the electric vector at +90° to the Sun. All the measurements in these tables are with the wide band filter centered at 500 nm.

The measurements from La Silla, Chile, are displayed in the eight polarization maps of the Figs. 5–12. Each figure shows at the top the visual aspect of the coma, at bottom left the field with the holes at their emplacements and diameters and, within the holes, the measurements of the circular polarization V/I, when available. At bottom right, the maps refer to the linear polarization.

The three first maps (Figs. 5–7) are over a narrow field of 12″.5 radius around the nucleus: Q/I is given for each hole (uncorrected values) and lines of equal polarization are derived. These lines are reminiscent of the visual aspect of the coma: the dust which is freshly ejected as streamers or in the cone oriented toward the Sun has a higher degree of polarization than the older dust elsewhere in the coma.

After April 10, we investigated the more distant coma and mapped it over a radius of 60″ around the nucleus (Figs. 8–12). The linear polarization decreased with the phase angle and deviations in the direction of the polarization became more significant: therefore, the linear polarization is presented by a vector proportional to the total polarization $P = \sqrt{(Q/I)^2 + (U/I)^2}$ and oriented in the proper azimuth θ.

4.2. Polarization of the inner coma and envelope

High angular resolution measurements centered on the nucleus were made with holes of increasing size, the smallest having 2″.1 in diameter. Between two holes I and $I+1$, the zone corresponds to a distance R_{zone} to the nucleus; it produces an average luminance L_{zone} and a polarization P_{zone}, with:

$$R_{zone} = R_i + \frac{R_{i+1} - R_i}{2}$$

$$L_{zone} = \frac{\theta_{i+1} - \theta_i}{S_{i+1} - S_i}$$

$$P_{zone} = \frac{(P_{i+1})(\theta_{i+1}) - (P_i)(\theta_i)}{\theta_{i+1} - \theta_i}.$$

Table 2. *1985 December 30.* Phase angle BETA = 53°.8. Position angle of Sun = 243°. Wide filter

Area	Time UT	Aperture diameter (arcsec)	Distance to nucleus (arcsec)	Azimuth around nucleus	Flux ϕ (relative)	Q/I (10^{-3})	U/I (10^{-3})	V/I (10^{-3})	θ (deg.)	U/I error	V/I error
N_4	17^h16^m	7.0	0	0	128	108 ∓ 5	–	–	–	–	–
A_1	17^h28^m	7.0	6	270	33	78 ∓ 10	–	–	–	–	–
A_2	17^h34^m	7.0	6	90	32	65 ∓ 12	–	–	–	–	–
A_3	17^h42^m	7.0	6	0	34	78 ∓ 10	–	–	–	–	–
A_4	17^h52^m	7.0	6	180	28	78 ∓ 10	–	–	–	–	–
A_5	18^h00^m	7.0	10	245	17	76 ∓ 20	–	–	–	–	–

Table 3. *1986 April 8.* Phase BETA = 40°.7. Position angle of Sun = 119°. Wide filter

Area	Time UT	Aperture diameter (arcsec)	Distance to nucleus (arcsec)	Azimuth around nucleus	Flux ϕ (relative)	Q/I (10^{-3})	U/I	V/I	θ (deg.)	U/I error	V/I error
N_4	02^h45^m	7.3	0	–	1080	+60.8∓0.7	−2.5∓1.2	+2.8∓2.4	−0.8∓0.3	2.0	1.0
A_1	05^h20^m	7.3	5	90	340	+53.5∓0.9	+1.3∓1.2	+1.2∓2.8	+1.0∓0.6	1.1	0.4
A_2		7.3	6	270	280	+51.6∓1.0	−0.6∓1.2	−0.8∓2.8	0.0∓0.6	0.5	0.3
A_3		7.3	6	0	270	+54.8∓1.0	−0.5∓1.2	+2.5∓2.8	0.0∓0.6	0.4	0.9
A_4		7.3	5	180	610	+60.3∓0.7	−2.6∓1.2	−3.8∓1.5	−1.0∓0.6	2.1	1.4
A_5		7.3	5	135	380	+60.0∓0.9	+2.8∓1.4	−2.8∓2.8	+1.8∓0.6	2.0	1.0
A_6		7.3	5	225	380	+52.1∓0.9	−1.2∓1.2	+2.8∓2.8	−0.4∓0.6	1.0	1.0
A_7		7.3	5	315	370	+52.8∓0.9	0.0∓1.3	+2.7∓2.8	+0.3∓0.6	0.0	0.9
A_8	06^h25^m	7.3	5	45	300	+52.8∓1.0	−2.0∓1.5	−1.8∓2.8	−0.8∓0.6	1.7	0.6
N_6	06^h30^m	2.1		–	180	+54.4∓1.2	–	–	–		

Table 4. *1986 April 9.* Phase angle BETA = 37°.1. Position angle of Sun = 126°. Wide filter

Area	Time UT	Aperture diameter (arcsec)	Distance to nucleus (arcsec)	Azimuth around nucleus	Flux ϕ (relative)	Q/I (10^{-3})	U/I	V/I	θ (deg.)	U/I error	V/I error
N_5	03^h20^m	5.1	0	–	220	+45.9∓1.2	–	–	–		
N_4		7.3	0	–	400	+46.8∓0.9	−5.7∓1.7	+2.5∓1.8	−3.2∓0.6	3.0	1.4
N_6	04^h05^m	2.1	0	–	48	+42.5∓2.6					
A_1	05^h40^m	7.3	5	90	240	+46.5∓1.1	−1.7∓1.7	+1.0∓1.8	−0.8∓1.3	1.0	0.6
A_2		7.3	5	270	150	+39.8∓1.1	+1.0∓1.7	+2.1∓1.8	+1.1∓1.3	0.6	1.2
A_3		7.3	5	180	220	+38.8∓1.1	+3.4∓1.7	+3.5∓1.8	+3.2∓1.3	2.0	1.9
A_4	06^h30^m	7.3	5	0	185	+44.8∓1.1	−1.7∓1.7	−0.4∓1.8	−1.8∓1.3	1.0	0.2
A_5	06^h35^m	7.3	10	90	180	+44.3∓1.1	+0.8∓1.7	−5.4∓2.0	−0.2∓1.3	0.5	2.7
A_6		7.3	10	270	100	+35.8∓1.1	+8.3∓1.7	+2.8∓1.8	+7.0∓1.3	4.9	1.5
A_7		7.3	10	180	165	+40.0∓1.1	+0.3∓1.7	0.0∓1.8	+0.6∓1.3	0.2	0.2
A_8		7.3	10	0	125	+33.5∓1.1	−0.4∓1.7	−5.2∓2.0	+0.3∓1.3	0.2	2.5
Scans	03^h00^m	16.2	20	N.S E.W	–	+40±3	–	–	–	–	–
Scans		16.2	40	N.S E.W	–	+40±3	–	–	–	–	–
Scans	03^h15^m	16.2	60	N.S E.W	–	+48±3	–	–	–	–	–

Table 5. *1986 April 10.* Phase BETA = 34°.3. Position angle of Sun = 135°. Wide filter

Area	Time UT	Aperture diameter (arcsec)	Distance to nucleus (arcsec)	Azimuth around nucleus	Flux ϕ (relative)	Q/I (10^{-3})	U/I	V/I	θ (deg.)	U/I error	V/I error
N_5	03^h35^m	5.1	0.0	–	144	+35.9∓1.6	–	–	–	–	–
N_4		7.3	0.0	–	240	+37.3∓1.3	−0.7∓1.9	–	−0.9∓1.7	0.4	–
N_6	03^h48^m	2.1	0.0	–	50	+30.0∓3	–	–		–	–
A_1	03^h50^m	7.3	5	135	150	+35.5∓1.8	−0.5∓1.9	−0.8∓2.1	−0.6∓1.7	0.3	0.4
A_2		7.3	5	315	105	+31.1∓1.8	+0.5∓1.9	+2.0∓2.1	+0.7∓1.7	0.3	1.0
A_3		7.3	5	225	90	+28.8∓1.8	+1.5∓1.9	+1.6∓2.1	+2.1∓1.7	0.8	0.8
A_4	04^h15^m	7.3	5	45	110	+32.0∓1.8	+4.6∓1.9	−0.6∓2.1	+4.1∓1.7	2.4	0.3
A_5	05^h15^m	7.3	10	135	90	+34.7∓1.6	+1.4∓1.9	−0.7∓2.1	+0.9∓1.7	0.7	0.3
A_6		7.3	10	315	70	+28.2∓1.6	−5.6∓1.9	+0.4∓2.1	−5.0∓1.7	2.9	0.2
A_7		7.3	10	225	65	+33.2∓1.6	+1.9∓1.9	−0.9∓2.1	−1.6∓1.7	1.0	0.4
A_8	05^h40^m	7.3	10	45	70	+34.4∓1.6	+0.4∓1.9	−1.3∓2.1	+0.2∓1.7	0.2	0.6

Table 6. *1986 April 11.* Phase BETA = 31°3. Position angle of Sun = 144°. Wide filter

Area	Time UT	Aperture diameter (arcsec)	Distance to nucleus (arcsec)	Azimuth around nucleus	Flux ϕ (relative)	Q/I (10^{-3})	U/I (10^{-3})	V/I (10^{-3})	θ (deg.)	U/I error	V/I error
N_5	03^h20^m	5.1	0.0	–	65	+21.3∓1.7	−1.0∓3.0	–	–	0.33	–
N_4	03^h30^m	7.3	0.0	–	120	+26.8∓0.8	−1.6∓2.0	−6.5∓3.9	−1.1∓3	0.80	1.6
A_1	03^h35^m	7.3	5.0	145	65	+22.6∓1.5	+1.3∓2.6	−0.2∓3.9	+1.6∓4	0.50	0.1
A_2	03^h45^m	7.3	5.0	325	55	+18.0∓1.5	+2.0∓2.6	+4.2∓3.9	+2.0∓4	0.77	1.1
A_3	04^h05^m	7.3	5.0	235	47	+21.3∓1.5	−4.8∓2.6	+3.8∓3.9	−4.4∓4	1.85	1.0
A_4	04^h15^m	7.3	5.0	55	60	+21.6∓1.5	+3.4∓2.6	−1.5∓3.9	+4.2∓4	1.30	0.4
A_5	06^h30^m	16.2	20.0	145	43	+26.5∓0.7	+1.2∓1.3	+1.1∓1.6	+1.8∓1.7	0.92	0.7
A_8	06^h45^m	16.2	20.0	55	143	+17.4∓0.7	+0.3∓1.3	−1.2∓1.6	−0.8∓1.7	0.23	0.8
A_6	07^h00^m	16.2	20.0	325	130	+20.0∓0.7	−0.3∓1.3	−1.5∓1.6	−0.1∓1.7	0.23	0.9
A_7	07^h05^m	16.2	20.0	235	117	+21.6∓0.7	−0.6∓1.6	−0.9∓1.7	−0.9∓1.7	0.46	0.5

Table 7. *1986 April 13.* Phase angle BETA = 26°2. Position angle of Sun = 163°. Wide filter

Area	Time UT	Aperture diameter (arcsec)	Distance to nucleus (arcsec)	Azimuth around nucleus	Flux ϕ (relative)	Q/I (10^{-3})	U/I (10^{-3})	V/I (10^{-3})	θ (deg.)	U/I error	V/I error
N_4	02^h10^m	7.3	0	–	607	+11.3∓0.8	−1.2∓1.0	–	−2.7∓1.5	1.2	–
N_6	02^h20^m	2.1	0	–	110	+13.7∓1.8	−3.5∓...	–	−5.2∓...	...	–
A_1	02^h30^m	7.3	5	180	280	+11.7∓1.3	−0.3∓1.1	–	−1.1∓2.0	0.3	–
A_2		7.3	5	0	252	+ 9.5∓1.3	+0.9∓1.1	–	+0.9∓2.0	0.8	–
A_3		7.3	5	270	233	+11.5∓1.3	−0.5∓1.1	–	−1.7∓2.0	0.5	–
A_4	03^h10^m	7.3	5	90	300	+13.4∓1.3	−0.2∓1.1	–	−0.3∓2.0	0.2	–
A_5	04^h40^m	28.1	50	0	430	+12.4∓1.1	−0.2∓1.0	+1.0∓1.0	−0.7∓1.8	0.2	1.0
A_6		28.1	50	180	370	+12.9∓1.1	−1.2∓1.0	+0.4∓1.0	+2.5∓1.8	1.2	0.4
A_7	05^h10^m	28.1	50	270	410	+13.6∓1.1	−0.9∓1.0	−1.1∓1.0	−1.7∓1.8	0.9	1.0
A_8		28.1	50	90	350	+13.1∓1.1	+0.8∓1.0	+0.8∓1.0	+1.8∓1.8	0.8	0.8
A_9		28.1	50	45	360	+14.5∓1.1	0.0∓1.0	+1.3∓1.0	+0.4∓1.8	0.0	1.3
A_{10}	05^h40^m	28.1	50	135	340	+11.8∓1.1	−0.7∓1.0	+0.7∓1.0	−1.9∓1.8	0.7	0.7
A_{11}	05^h55^m	28.1	50	225	370	+11.1∓1.1	+0.9∓1.0	+1.4∓1.0	+1.7∓1.8	0.9	1.4
A_{12}	06^h15^m	28.1	50	315	440	+11.0∓1.1	+1.7∓1.0	−0.2∓1.0	+3.7∓1.8	1.7	0.2
N_6	06^h25^m	2.1	0	–	150	+ 9.8∓2.0	–	–	–	–	–
N_6	07^h00^m	2.1	0	–	122	+12.4∓2.5	–	–	–	–	–

Other measurements with holes not centered at the nucleus but overlapping are corrected for their common surfaces.

The results are presented in Fig. 15–23. At the top of each figure the hole positions and sizes are given. Below, the degree of linear polarization P (uncorrected from emission) and the relative luminance L are plotted as a function of the distance to the nucleus, sunward, anti-sunward and along identified streamers.

The striking result is a drastic difference in the degree of polarization for the envelope, compared to the values for the coma. This polarization was higher on 1985 December 30, at phase angle 53°8 (Fig. 15), and smaller for all the phase angles between 40°7 and 22°8.

For the coma, a maximum polarization is observed at 1000 km from the nucleus where the dust was recently ejected. In the sunward direction, on the ejection cone which is facing the Sun, the recently ejected dust produces a high polarization, as demonstrated by the Figs. 18, 19, 21. On April 13, when the central envelope was obscured by a dense dust streamer recently ejected, the polarization was noted to be uniform all over the field and corresponded to the high value of the fresh dust (Fig. 20). On April 8, a streamer was observed southward (Fig. 5); its polarization kept the high value corresponding to fresh dust out to a large distance (dashed line in Fig. 16).

Conversely, in the anti-solar direction, where the dust was somewhat older, the polarization decreased and reached a lower value at 3000 km. Then it kept the same value out to the edge of the field (Fig. 18).

All these results, including other measurements at larger distance, are summarized in Fig. 24, with a logarithmic scale for the distance and with the P values corrected for the effect of the fluorescence emissions. The polarizations for the envelope, which are the asymptotes at left, are possibly slightly overestimated because the source is seen through the coma dust whose higher polarization is included in the hole.

Table 8. *1986 April 14.* Phase angle BETA = 24°.2. Position angle of Sun = 178°. Wide filter

Area	Time UT	Aperture diameter (arcsec)	Distance to nucleus (arcsec)	Azimuth around nucleus	Flux ϕ (relative)	Q/I (10^{-3})	U/I	V/I	θ (deg.)	U/I error	V/I error
N_5	02^h00^m	5.1	0	–	190	+ 4.8 ∓ 1.1	–	–	–	–	–
N_4		7.3	0	–	335	+ 9.2 ∓ 1.3	(+2.1 ∓ 0.9)	–	+ 5.2 ∓ 4	(2.3)	–
N_3	02^h30^m	16.2	0	–	900	+ 6.6 ∓ 0.8	−0.6 ∓ 0.5	− 0.9 ∓ 1.5	(− 2.6 ∓ 4)	1.2	0.6
A_1	02^h45^m	7.3	6	180	170	+ 9.6 ∓ 1.6	+2.4 ∓ 1.5	+ 4.2 ∓ 4.8	+ 5.6 ∓ 5	1.6	0.8
A_2	03^h15^m	7.3	6	0	120	+ 8.1 ∓ 1.6	+1.1 ∓ 1.5	+ 1.8 ∓ 4.8	+ 4.2 ∓ 5	0.8	0.4
A_3	03^h25^m	7.3	6	90	130	+10.0 ∓ 1.6	−1.0 ∓ 1.5	− 4.1 ∓ 4.8	− 5.4 ∓ 5	0.7	0.8
A_4	03^h32^m	7.3	6	270	105	+ 8.2 ∓ 1.6	+2.0 ∓ 1.5	+11.8 ∓ 4.8	+ 7.4 ∓ 5	1.3	2.5
A_5	03^h35^m	16.2	25	180	280	+14.2 ∓ 0.9	+4.2 ∓ 1.3	+ 1.0 ∓ 1.5	+ 3.6 ∓ 3.7	3.2	0.7
A_6		16.2	25	0	180	+11.5 ∓ 0.9	−0.8 ∓ 1.3	− 2.5 ∓ 1.5	− 5.1 ∓ 3.7	0.6	1.7
A_7	04^h20^m	16.2	25	270	190	+ 7.0 ∓ 0.9	+4.1 ∓ 1.3	− 1.9 ∓ 2.3	+13.4 ∓ 3.7	3.1	0.8
A_8		16.2	25	90	190	+ 8.4 ∓ 0.9	+2.3 ∓ 1.3	+ 2.8 ∓ 1.5	+ 7.3 ∓ 3.7	1.7	1.9
A_9	04^h50^m	16.2	25	315	175	+ 8.1 ∓ 0.9	−1.9 ∓ 1.3	− 1.8 ∓ 1.5	− 6.3 ∓ 3.7	1.5	1.2
A_{10}		16.2	25	135	260	+ 9.0 ∓ 0.9	−0.7 ∓ 1.3	− 0.4 ∓ 1.5	− 3.1 ∓ 3.7	0.5	0.3
A_{11}	05^h08^m	16.2	25	225	235	+ 5.4 ∓ 0.9	−0.8 ∓ 1.3	+ 3.9 ∓ 3.0	+ 2.0 ∓ 3.7	0.6	1.3
A_{12}	05^h23^m	16.2	25	45	180	+ 7.2 ∓ 0.9	−0.8 ∓ 1.3	− 1.9 ∓ 1.5	− 1.6 ∓ 3.7	0.6	1.3
A_{13}	06^h00^m	28.1	50	180	420	+ 4.5 ∓ 0.8	+0.1 ∓ 0.9	− 0.9 ± 1.5	+ 9.8 ∓ 4.4	0.1	0.6
A_{14}	06^h10^m	28.1	50	0	330	+ 8.4 ∓ 0.8	−0.3 ∓ 0.9	+0.6 ± 1.5	− 1.3 ∓ 4.4	0.3	0.4
A_{15}		28.1	50	225	370	+ 4.5 ∓ 0.8	−1.7 ∓ 0.9	–	− 1.2 ∓ 4.4	1.8	–
A_{16}	06^h30^m	28.1	50	135	360	+ 6.2 ∓ 0.8	−1.3 ∓ 0.9	− 0.1 ± 1.5	− 2.8 ∓ 4.4	1.4	0.0
A_{17}	07^h00^m	28.1	5.0	315	320	+ 7.8 ∓ 0.8	−0.4 ∓ 0.9	− 1.6 ± 1.5	− 0.7 ∓ 4.4	0.4	1.0
A_{18}		28.1	5.0	45	320	+ 7.6 ∓ 0.8	+0.5 ∓ 0.9	+ 0.7 ± 1.5	+ 3.0 ∓ 4.4	0.6	0.5
A_{19}		28.1	5.0	90	380	+ 7.8 ∓ 0.8	−2.0 ∓ 0.9	− 1.9 ± 1.5	− 6.5 ∓ 4.4	2.2	1.4
A_{20}	07^h30^m	28.1	50	270	320	+ 7.9 + 0.8	−0.7 + 0.9	− 0.5 + 1.5	+ 2.9 + 4.4	0.8	0.3

Table 9. *1986 April 15.* Phase angle BETA = 22°.5. Position angle of Sun = 185°. Wide filter

Area	Time UT	Aperture diameter (arcsec)	Distance to nucleus (arcsec)	Azimuth around nucleus	Flux ϕ (relative)	Q/I (10^{-3})	U/I	V/I	θ (deg.)	U/I error	V/I error
N_3	03^h00^m	16.2	0	–	660	+4.6 ∓ 0.7	+1.1 ∓ 0.6	−0.8 ∓ 0.6	+ 6.7 ∓ 2.0	1.8	−1.3
N_4		7.3	0	–	275	+2.8 ∓ 0.7	+0.3 ∓ 0.8	–	+ 3.0 ∓ 4.5	0.4	–
N_5		5.1	0	–	170	+2.6 ∓ 1.0	+1.5 ∓ 1.1	–	+15.0 ∓ 7.0	1.4	–
N_6	03^h30^m	2.1	0	–	45	+1.1 ∓ 3.0	–	–	–	–	–
A_1	03^h35^m	7.3	6	180	140	+4.4 ∓ 1.0	+0.3 ∓ 1.5	−2.0 ∓ 2.5	+ 1.9 ∓ 8.5	0.2	0.8
A_2		7.3	6	0	160	+3.8 ∓ 1.0	−3.4 ∓ 1.0	+0.7 ∓ 2.2	−21.0 ∓ 7.5	3.1	0.3
A_3		7.3	6	225	130	+7.3 ∓ 1.0	−0.1 ∓ 1.0	−1.0 ∓ 2.6	− 0.4 ∓ 9.0	0.1	0.4
A_4	04^h10^m	7.3	6	135	140	+7.1 ∓ 1.0	+0.1 ∓ 1.0	+0.6 ∓ 2.5	+ 0.4 ∓ 8.5	0.1	0.2
A_5	04^h15^m	16.2	25	180	165	+7.1 ∓ 1.1	−0.2 ∓ 1.0	+2.8 ∓ 2.1	− 0.8 ∓ 7.5	0.2	1.3
A_6		16.2	25	135	190	+5.3 ∓ 1.1	+1.2 ∓ 1.0	+2.8 ∓ 1.8	+ 6.3 ∓ 6.3	1.2	1.5
A_7		16.2	25	225	170	+7.2 ∓ 1.1	−2.2 ∓ 2.0	−2.0 ∓ 1.9	− 8.5 ∓ 7.0	1.1	0.9
A_8	05^h05^m	16.2	25	0	140	+6.6 ∓ 1.1	+0.6 ∓ 1.2	−2.0 ∓ 2.5	+ 2.6 ∓ 8.5	0.5	0.8
A_9		28.1	50	180	290	+4.4 ∓ 0.7	+0.8 ∓ 0.6	−1.2 ∓ 1.1	+ 5.1 ∓ 4.0	1.3	1.1
A_{10}		28.1	50	135	325	+6.9 ∓ 0.7	+0.1 ∓ 0.5	+1.6 ∓ 0.9	+ 0.4 ∓ 3.5	0.2	1.7
A_{11}		28.1	50	225	310	+4.0 ∓ 0.7	+2.0 ∓ 0.6	+2.1 ∓ 1.5	+13.2 ∓ 3.2	3.3	2.1
A_{12}	06^h00^m	28.1	50	0	280	+4.2 ∓ 0.7	−3.4 ∓ 0.6	−2.2 ∓ 1.1	−20.0 ∓ 4.2	5.6	1.1
A_{13}	06^h50^m	46.7	4′	0	380	+6.1 ∓ 0.7	+0.1 ∓ 0.5	–	+ 0.5 ∓ 3.0	0.2	–
A_{14}		46.7	7′	0	205	+5.7 ∓ 0.9	−0.2 ∓ 1.0	–	− 1.0 ∓ 6.0	0.2	–
A_{15}		46.7	10′	0	135	+7.0 ∓ 1.1	+1.1 ∓ 1.4	–	+ 4.5 ∓ 8.0	0.8	–
A_{16}		107	15′	0	610	+7.5 ∓ 0.0	−0.7 ∓ 0.4	–	− 2.7 ∓ 2.0	1.7	–
A_{17}		107	20′	0	330	+5.9 ∓ 0.7	−0.7 ∓ 0.6	–	− 3.4 ∓ 3.5	1.2	–
A_{18}	07^h30^m	107	25′	0	200	+7.3 ∓ 0.7	−0.1 ∓ 0.9	–	− 0.6 ∓ 6.0	0.1	–

Table 10. *1986 April 16*. Phase angle BETA = 21°.6. Position angle of Sun = 194°. Wide filter

Area	Time UT	Aperture diameter (arcsec)	Distance to nucleus (arcsec)	Azimuth around nucleus	Flux ϕ (relative)	Q/I (10^{-3})	U/I (10^{-3})	V/I (10^{-3})	θ (deg.)	U/I error	V/I error
N_6	03^h10^m	2.1	0	–	80	$+1.0\mp2.2$	–	–	–	–	–
N_5		5.1	0	–	275	$+1.9\mp0.7$	$+1.4\mp0.9$	–	$+18\mp10$	1.6	–
N_4		7.3	0	–	445	$+1.5\mp0.4$	$+1.5\mp2.5$	–	$+25\mp15$	2.5	–
A_1	03^h30^m	7.3	5	180	180	0.0 ∓ 1.0	-1.3 ∓ 1.4	–	-5 ∓ 8	0.9	–
A_2		7.3	5	0	170	-0.5 ∓ 1.0	-4.8 ∓ 1.4	–	-48 ∓ 8	3.4	–
A_3		7.3	5	90	145	0.0 ∓ 1.3	-0.4 ∓ 1.8	–	–	0.2	–
A_4		7.3	5	270	160	$+3.0\mp1.2$	$+4.6\mp1.7$	–	$+29\mp8$	2.7	–
A_5	03^h55^m	16.2	25	180	220	$+2.0\mp0.8$	$+1.4\mp1.1$	–	$+17\mp15$	1.3	–
A_6		16.2	25	0	185	$+2.7\mp1.0$	-1.6 ∓ 1.4	–	-15 ∓ 15	1.1	–
A_7		16.2	25	270	195	-1.2 ∓ 0.8	-0.4 ∓ 1.1	–	–	0.3	–
A_8		16.2	25	90	170	$+6.0\mp1.1$	$+2.4\mp1.5$	–	$+11\mp9$	1.6	–
A_9		16.2	25	315	200	$+6.9\mp0.8$	$+2.2\mp1.1$	–	$+9\mp5$	2.0	–
A_{10}		16.2	25	135	180	$+3.3\mp1.0$	$+2.1\mp1.4$	–	$+16\mp12$	1.5	–
A_{11}		16.2	25	225	230	$+4.7\mp0.9$	$+0.6\mp1.3$	–	$+4\mp9$	0.5	–
A_{12}	04^h45^m	16.2	25	45	170	$+6.1\mp1.1$	-0.8 ∓ 1.5	–	-4 ∓ 9	0.4	–
A_{13}	05^h10^m	28.1	50	180	330	$+0.9\mp0.6$	-2.8 ∓ 0.8	–	-36 ∓ 8	3.5	–
A_{14}		28.1	50	0	335	$+5.3\mp0.6$	$+0.5\mp0.8$	–	$+3\mp6$	0.6	–
A_{15}		28.1	50	225	365	$+1.7\mp0.6$	-0.4 ∓ 0.8	–	-6 ∓ 6	0.5	–
A_{16}		28.1	50	135	330	$+6.2\mp0.6$	$+0.6\mp0.8$	–	$+3\mp6$	0.7	–
A_{17}		28.1	50	315	360	$+4.5\mp0.5$	-0.2 ∓ 0.7	–	-1 ∓ 5	0.3	–
A_{18}		28.1	50	45	320	$+1.9\mp0.6$	-0.9 ∓ 0.8	–	-13 ∓ 10	1.1	–
A_{19}		28.1	50	270	320	$+4.1\mp0.6$	-2.2 ∓ 0.8	–	-14 ∓ 6	2.7	–
A_{20}	06^h10^m	28.1	50	90	340	-2.8 ∓ 0.6	-0.9 ∓ 0.8	–	$+99\mp9$	1.1	–

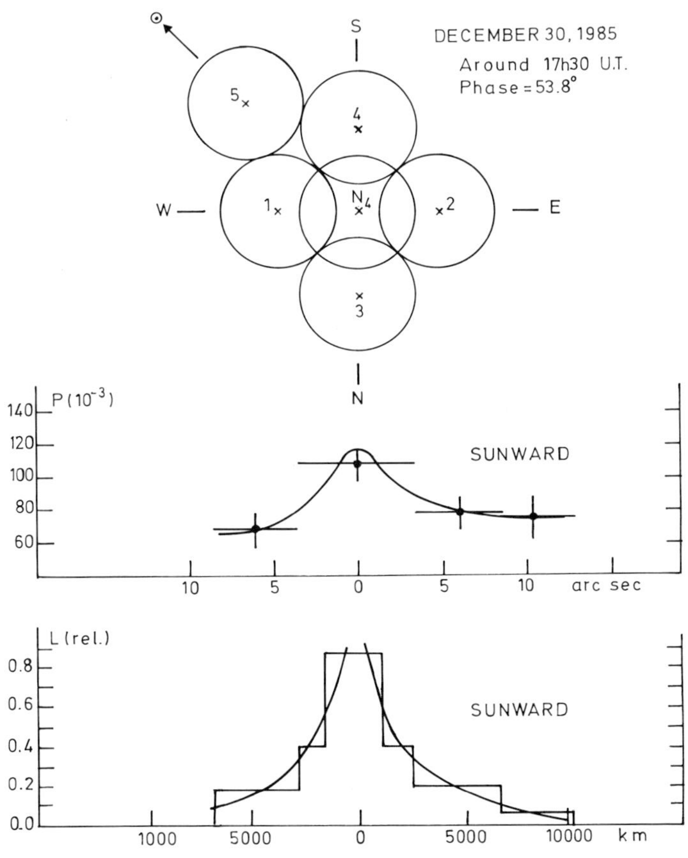

Fig. 15. Measurement around P/Halley nucleus on December 30, 1985, with the 100 cm Meudon Observatory telescope (see Table 2). At top, holes positions and sizes around the nucleus and identification number on Table 2. At bottom, degree of polarization (in units of 10^{-3} uncorrected for monochromatic emissions) and luminance (in relative units) as a function of distance to the nucleus, sunward and anti-sunward. The measurements are corrected from overlaps

4.3. Wavelength dependence of the linear polarization

We analysed the spectral variation of the polarization at phase angles 54°.2, 40°.7 and 26°.7. The results are given in Table 11. The values corrected for the effect of the fluorescense emissions are plotted in Figs. 25 and 26. There is no strong wavelength dependence. only a slight increase toward the longer wavelengths. This result contrasts with the strong decrease toward UV always observed on all silicaceous surface, in the laboratory and in the Solar System, on the Moon, Mercury and the asteroids.

4.4. Phase angle dependence of linear polarization

From all these data, plus the measurements recorded around the nucleus at Meudon in 1985 October and December, we derived the variation of the degree of polarization with phase angle and produced the relevant diagnostic "curves of polarization". The data are assembled in Table 12. The values corrected for the effect of the fluorescense emissions are plotted in Fig. 27. The envelope, the fresh dust at 1000 km from the nucleus and the older dust at more than 3000 km are separated. There is no strong variation with wavelength, as noted above.

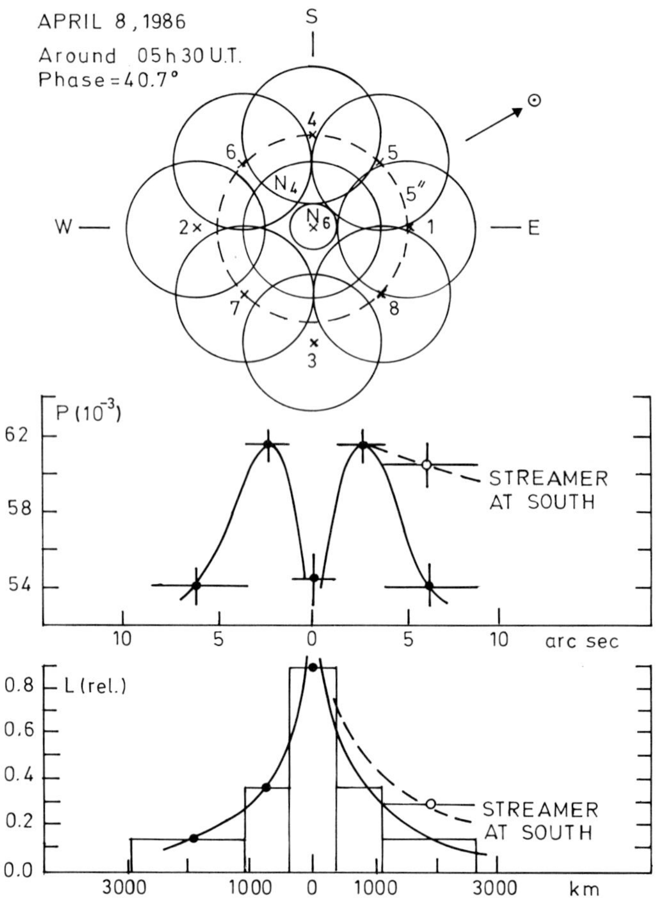

Fig. 16. Same as Fig. 15, but on April 8, 1986, with the 1.52 ESO telescope at La Silla, and Table 3. The concentric measurements around the nucleus are corrected from overlaps, as indicated in the text

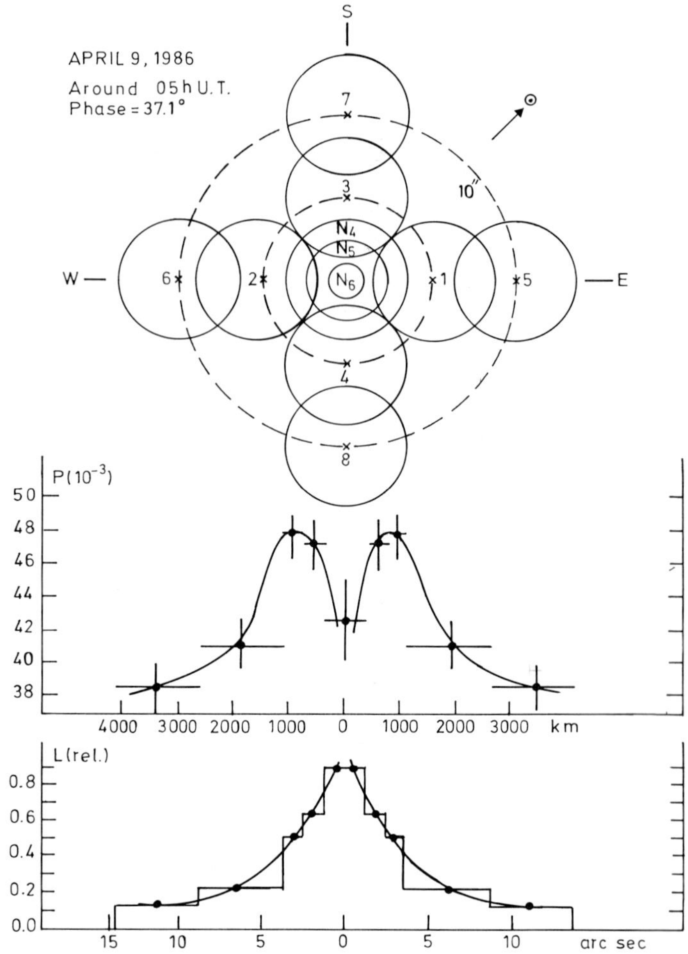

Fig. 17. Same as Fig. 16, but on April 9, 1986, Table 4 and measurements averaged for the N, E, S and W directions

Table 11. P/Halley – wavelength dependence of polarization. Data as observed, and corrected for emissions

Optical filter			Degree of polarization, in 10^{-3}					
Code	λ_{max} (nm)	$\Delta\lambda$ (nm)	Dec. 29, 1985–17^{h}30 Phase = 54°2		Apr. 8, 1986–04^{h}00 Phase = 40°7		Apr. 12, 1986–03^{h}30^m Phase = 26°	
			Obs.	Corr.	Obs.	Corr.	Obs.	Corr.
O	593	25	+100±15	+104±15	+65±1	+68±1	+13±1	+13±1
G	527	25	+ 77±10	+ 81±10	+61±0.7	+66±1	+12±1	+12±1
C_2 (IHW)	514	9	–	–	–	–	+11±2	+11±2
Wide band	490	170	+ 84±9	+ 93±10	+60±0.7	+67±1	+12±2	+13±2
Cont (IHW)	484	6.5	–	–	+61±0.5	+61±0.5	+16±2	+16±2
B	440	25	+ 73±10	+ 74±10	+61±1	+61±1	+14±1	+14±1
(A)	399	40	+ 82±12	+ 86±12	+50±0.6	+52±1	+12±2	+12±2
CN (IHW)	387	5	–	–	–	–	+ 9±2	+ 9±2
(B)	364	18	–	–	+54±1	+54±1	+16±2	+16±2

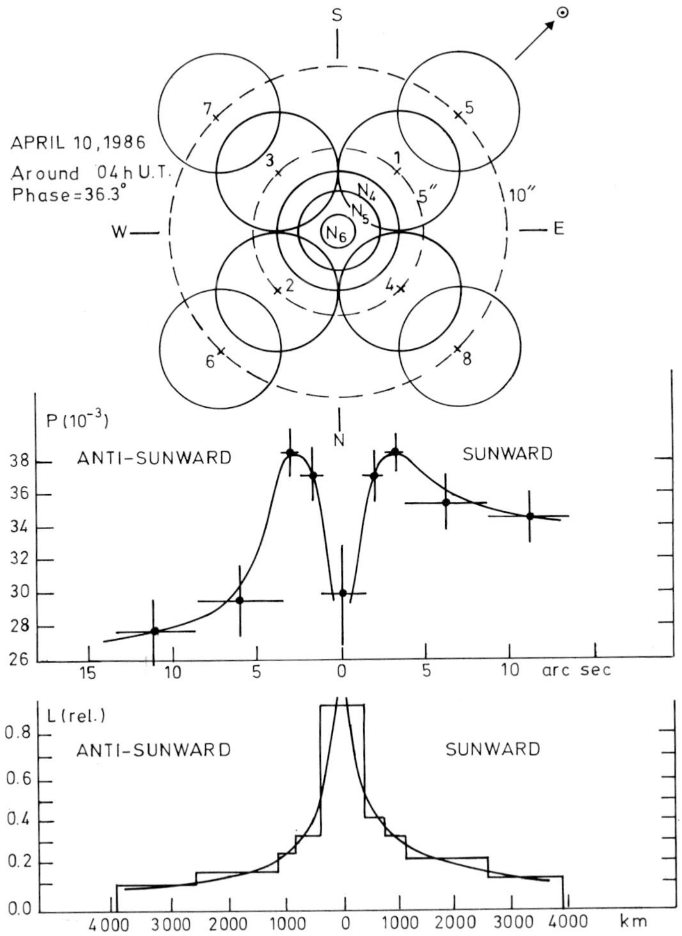

Fig. 18. Same as Fig. 16, but on April 10, 1986 and Table 5

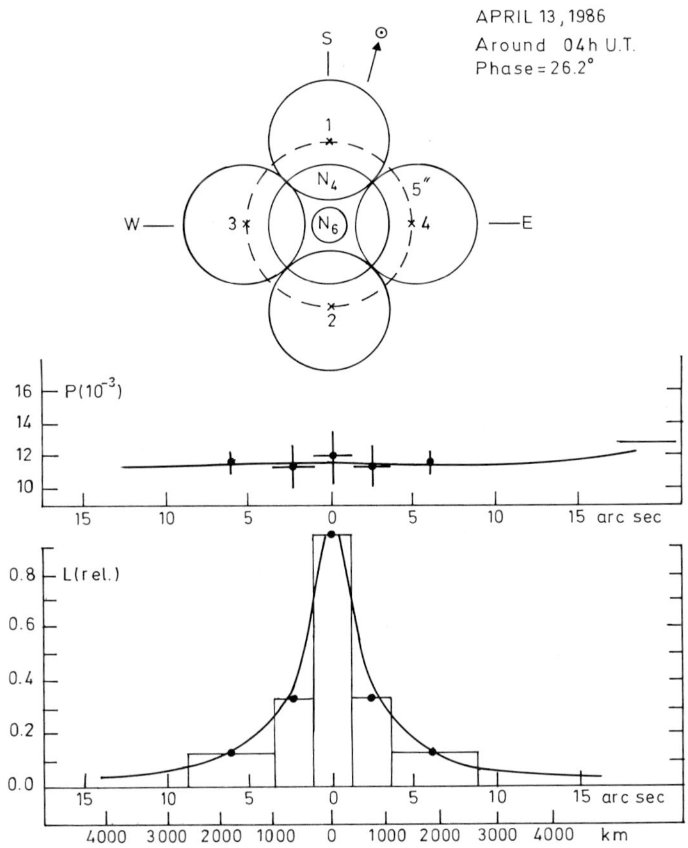

Fig. 20. Same as Fig. 16, but on April 13, 1986 and Table 7. The inner part of the coma and the envelope are obscured by a dust ejection toward the Earth; the polarization variation with distance is uniformized

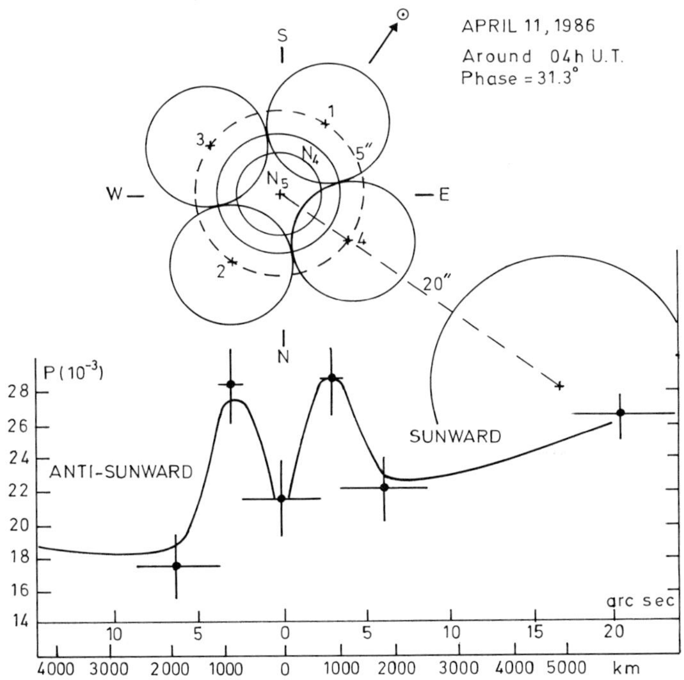

Fig. 19. Same as Fig. 16, but on April 11, 1986 and Table 6

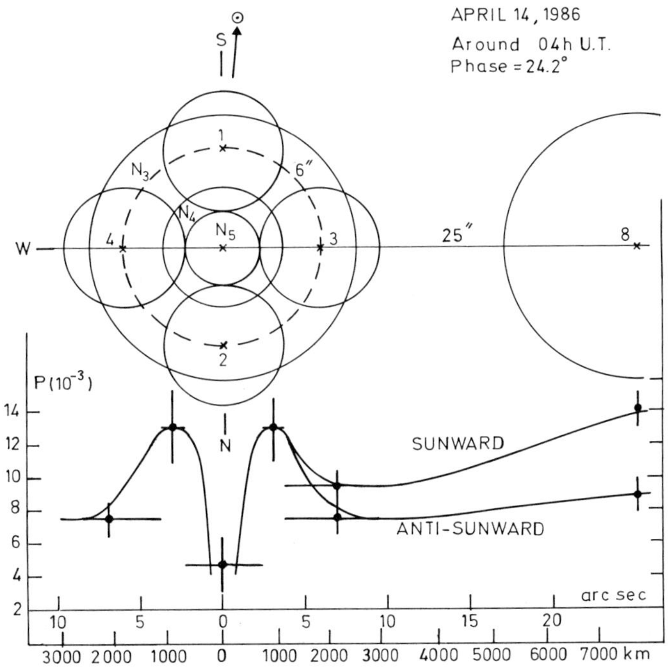

Fig. 21. Same as Fig. 16, but for April 14, 1986 and Table 8. The anti-sunward measurement is out of frame and not reproduced

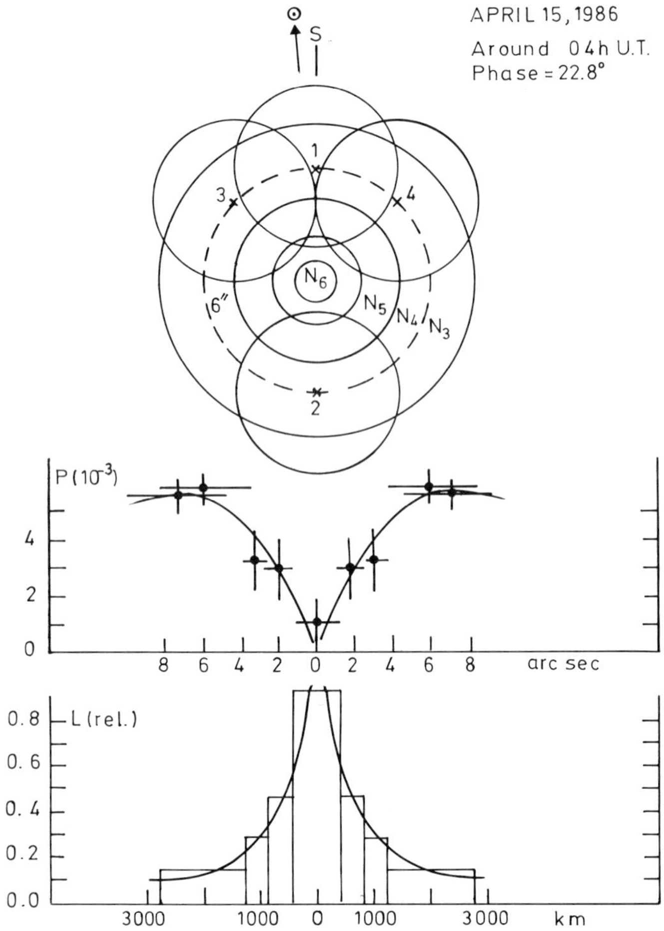

Fig. 22. Same as Fig. 16, but for April 15, 1986 and Table 9

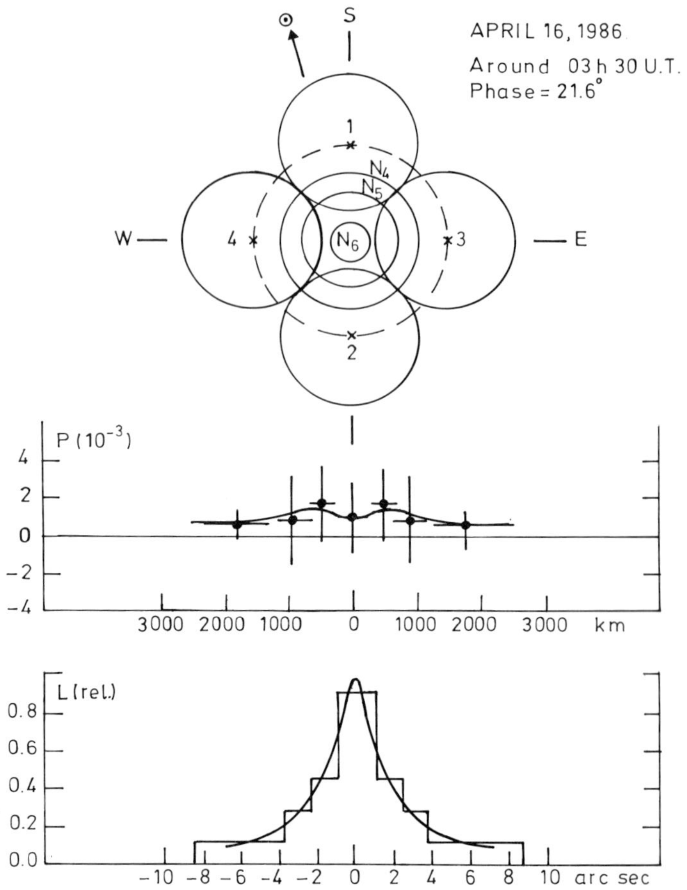

Fig. 23. Same as Fig. 16, but for April 16, 1986 and Table 10

Table 12. Polarization P/Halley: Continuum at 500 nm. Wide band filter Q/I, in units of 10^{-3}, as observed and corrected for emissions

Date	Time	Telescope	Phase BETA	Coma at 3000 km		Coma at 1000 km		Envelope	
				Obs.	Corr.	Obs.	Corr.	Obs.	Corr.
1985 Oct. 22	03^h30^m	T1 m Meudon	24.7	$+4\pm6$	$+2\pm6$	–	–	–	–
1985 Oct. 25	04^h00^m	T1 m Meudon	23.9	$+1\pm2$	0 ± 2	–	–	–	–
1985 Dec. 9	18^h40^m	T1 m Meudon	44.5	$+68\pm8$	$+73\pm10$	–	–	–	–
1985 Dec. 29	17^h30^m	T1 m Meudon	54.2	$+84\pm9$	$+93\pm10$	–	–	–	–
1985 Dec. 30	18^h30^m	T1 m Meudon	53.8	$+80\pm8$	$+90\pm10$	–	–	$+115\pm10$	$+122\pm12$
1986 April 8	03^h00^m	T152 ESO	40.7	$+54\pm2$	$+62\pm3$	$+61\pm1$	$+67\pm2$	$+54\pm3$	$+57\pm4$
1986 April 9	03^h30^m	T152 ESO	37.1	$+39\pm2$	$+44\pm3$	$+48\pm1$	$+52\pm2$	$+42\pm2$	$+43\pm3$
1986 April 10	03^h50^m	T152 ESO	34.2	$+28\pm2$	$+31\pm2$	$+38\pm1$	$+41\pm2$	$+30\pm3$	$+31\pm3$
1986 April 11	03^h20^m	T152 ESO	31.3	$+20\pm2$	$+20\pm2$	$+28\pm2$	$+29\pm2$	$+21\pm3$	$+21\pm3$
1986 April 13	02^h10^m	T152 ESO	26.2	–	–	$+12\pm2$	$+13\pm2$	–	–
1986 April 14	02^h30^m	T152 ESO	24.2	$+8\pm2$	$+7\pm2$	$+12\pm2$	$+12\pm2$	$+3\pm3$	$+2\pm3$
1986 April 15	03^h00^m	T152 ESO	22.6	$+2.5\pm2$	-1 ± 2	$+5.5\pm2$	$+5\pm2$	0 ± 2	-1 ± 2
1986 April 16	03^h00^m	T152 ESO	21.6	$+0.5\pm2$	-3 ± 2	$+1.5\pm1$	$+0\pm1$	$+1\pm2$	0 ± 2

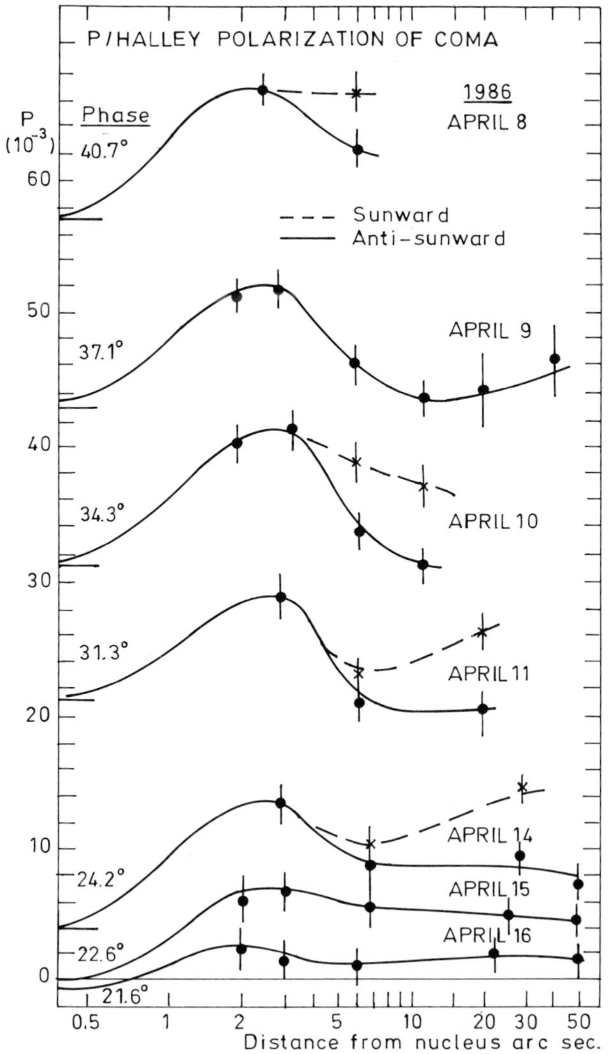

Fig. 24. Variation of the degree of polarization, corrected for the effect of the monochromatic emissions, as a function of the sunward and anti-sunward distances from the nucleus in a logarithmic scale. The phase angles are indicated. The curves are correctly positioned each other on the same vertical polarization scale

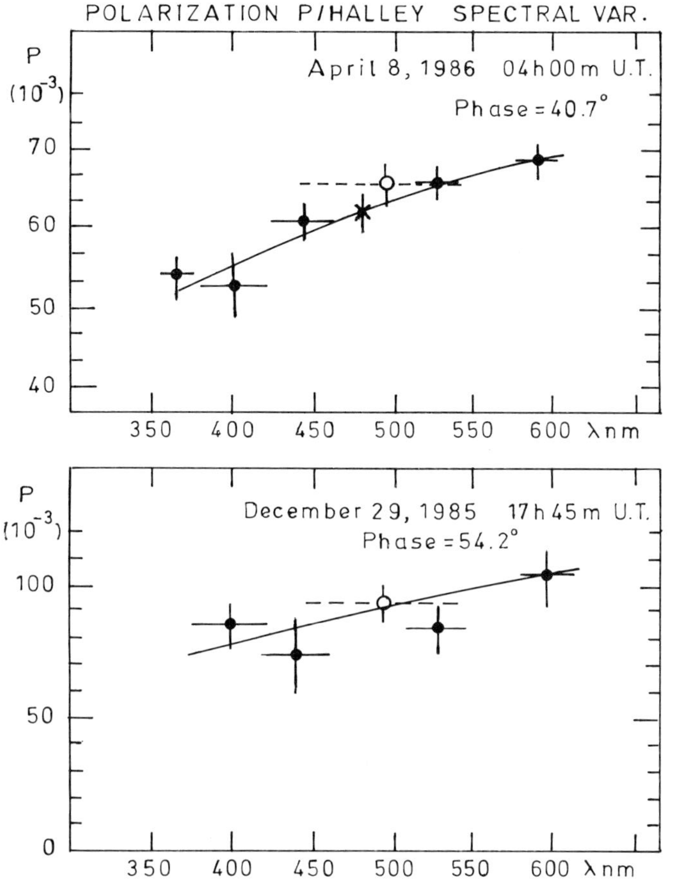

Fig. 25. Spectral variation of the polarization for P/Halley. Measurements centered on the nucleus (see Table 9) for phase angles 54°.2 and 40°.7. Values corrected for the effect of monochromatic emissions. Open circle is with the wide band filter. Cross is with the IHW continuum filter

4.5. Polarization along the tail

On April 15, when the phase angle of 22°.5 was near the value for which the polarization Q/I disappears, we measured the polarization along the tail, up to an apparent distance of 25′, or 0.48 10^6 km from the nucleus. The tail was seen at only 22°.5 from its axis and the foreshortening gave a real distance along the tail of 1.3 10^6 km, with a light path covering from 0.7 10^6 km to 1.8 10^6 km. The luminance and the linear polarization are shown in Fig. 28. There is no significant variation in the polarization along the portion of the tail which is observed, proving no change in the cross-over angle of the linear polarization (see Fig. 27), which is interpreted as no drastic modification in the microtexture of the cometary grains during their transportation from the nucleus out to the distance of 1.6 10^6 km.

For P/Halley, other polarization measurements along the tail have been reported by Lamy et al. (1986), on 1986 April 14, at phase angle 31°. They also observed no significant variation in the degree of polarization along our range of distances to the nucleus. But at larger distances, a variable and progressive increase of polarization is noted, from 38 10^{-3} at the nucleus to 60 10^{-3} at the apparent distance of 2.4 10^6 km (real distance 6 10^6 from nucleus). This last value is incompatible with a model of dust-aggregated fluffy grains (see Dollfus et al., 1986), but probably requires particles smaller than the wavelength.

5. Deviation of the azimuth of the linear polarization

5.1. Observations

125 measurements of U/I have been collected over 8 nights from April 8 to 16, 1986. They are reported in tables 2–10, column 8. Column 11 also gives the signal/noise ratio, which is the ratio of U/I to the error derived from the standard deviations between the sequences of integration. Column 10 gives the angle θ between the direction of polarization and the perpendicular to the direction of the sun.

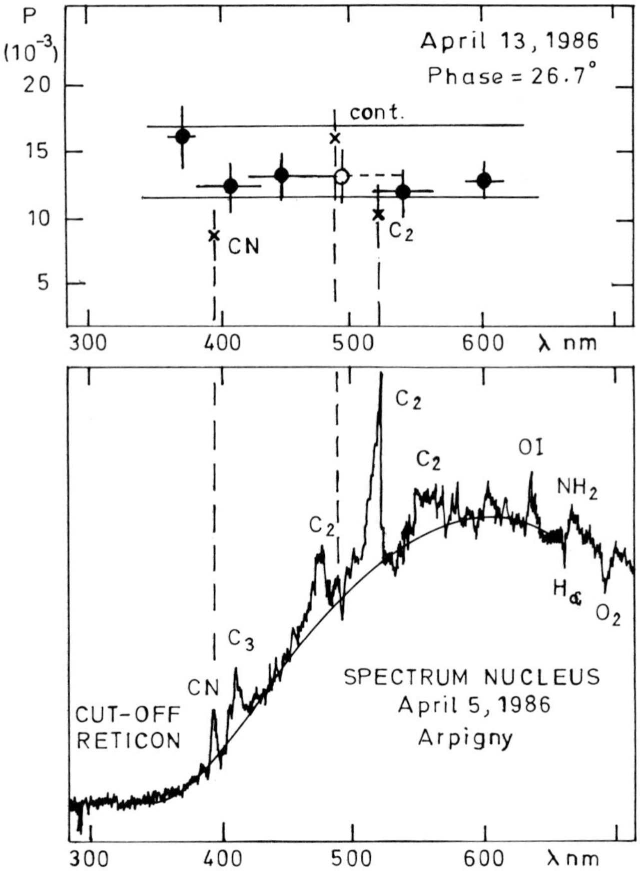

Fig. 26. Spectral variation as for Fig. 25 but at phase angle 26°.7, plus measurements with the IHW filters centered at the CN and C_2 molecular monochromatic emissions. At bottom, a spectrum of the coma with hole centered on the nucleus by Dr. Arpigny, taken with the same 1.52 m ESO telescope at La Silla, on April 5, 1986

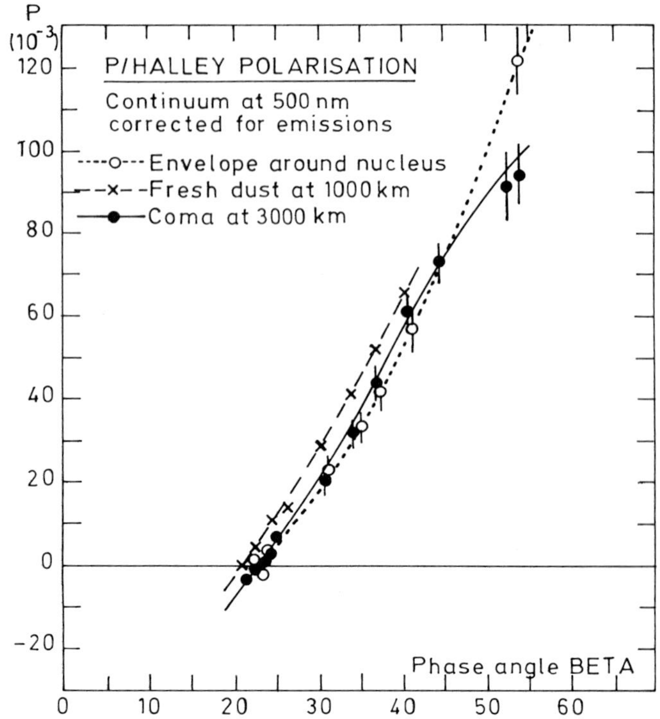

Fig. 27. Curves of polarization for P/Halley. The degree of polarization is corrected for the effects of monochromatic emissions and plotted as a function of phase angle, for the coma at 3000 km from nucleus (dots and solid line), for the inner coma at 1000 km (crosses and dashed line) and for the envelope around the nucleus (opened circles and dotted line). All observations by the authors with the Meudon photopolarimeter, at the 100 cm Meudon telescope and at the La Silla 1.52 m ESO telescope (see Table 11)

The observations on the envelope with hole centered on the nucleus, often give significant values of U/I compared to the error. The average (signal + noise)/noise ratio, designated $\frac{|U/I|}{\text{error}}$, is 1.4. The mean of all measurements is $U/I = (-2.0 \pm 1.0)\ 10^{-3}$ and $\theta = (-1.9 \pm 1.4)°$. It is not clear if this residual clockwise deviation is significant or due to misadjustments in the position angle of the instrument.

For the coma, we have 114 determinations of U/I. When compared to their accuracies, these data produce the histogram at the top of Fig. 29. The heavy curve is the best gaussian fit with $1/e$ at 1.0. Of these measurements, 60 % are smaller than errors, but 39 % have signal/noise between 1.0 and 3.6, one measurement is 5 times larger than the accuracy. In order to analyse the dependence of U/I on Q/I, we split the set into two groups. The histogram of figure 29b refers to the measurements when the linear polarization Q/I is larger than 14 10^{-3}; of these measurements, 35 % show significant deviation. The histogram of Fig. 29c with $Q/I < 10\ 10^{-3}$ produces 44 % of the measurements with significant deviations. We conclude that the deviations are not proportional to the degree of polarization; rather, they are due to the addition of a polarization which is not related to the direction of the sun.

This conclusion also emerges from Fig. 30 (a) in which the average value $\frac{\Sigma |U/I|}{n}$ is plotted for each date of observation, as a function of Q/I (bottom scale) or of the phase angle (top scale). The values remain around 1.3 10^{-3}, again with no clear dependence with the polarization Q/I produced by the isotropic scattering.

This point is also illustrated by the vector polarization maps in Figs. 8–12. For April 11 and 13 (Figs. 8 and 9) with $Q/I \gg U/I$, the effect of U/I on the orientation is negligible. On April 14 and 15; with Q/I reduced to 10 10^{-3} and to 4 10^{-5} respectively, the anisotropic polarization is now significant and produces clear deflections in the directions of polarization. On April 16, (Fig. 12), the phase angle of 21°.6 gives almost no average Q/I polarization; the anisotropic component U/I remains alone and displays a large variety of azimuths.

Similar effects have been observed at the telescope over the disk of Venus, for the phase angles at which the direct polarization disappears (Dollfus et al., 1979). Residual polarizations are observed at any azimuths and exhibit the presence of large scale anisotropies in the Venusian aerosol layer. The same is also true for the Saturn B ring (Dollfus, 1979) in which particles are subjected to strong anisotropies.

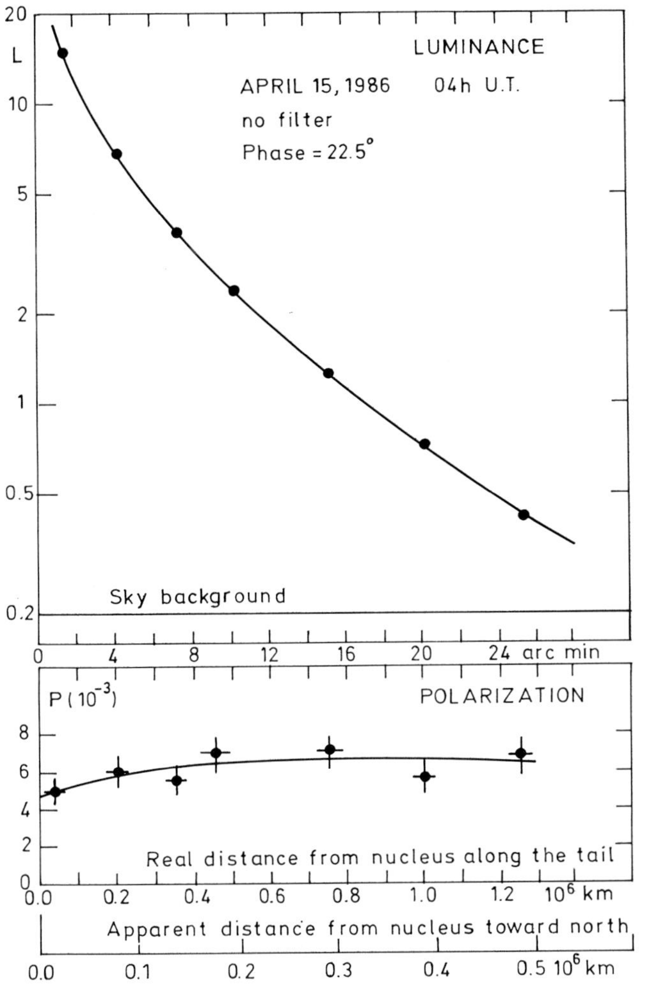

Fig. 28. Linear polarization and luminence along the tail, from the nucleus to an apparent distance of 0.48 10^6 km toward North, corresponding to a real distance along the tail of 1.25 10^6 km. April 15, 1986 at phase angle 22°.5 (cf. Table 9)

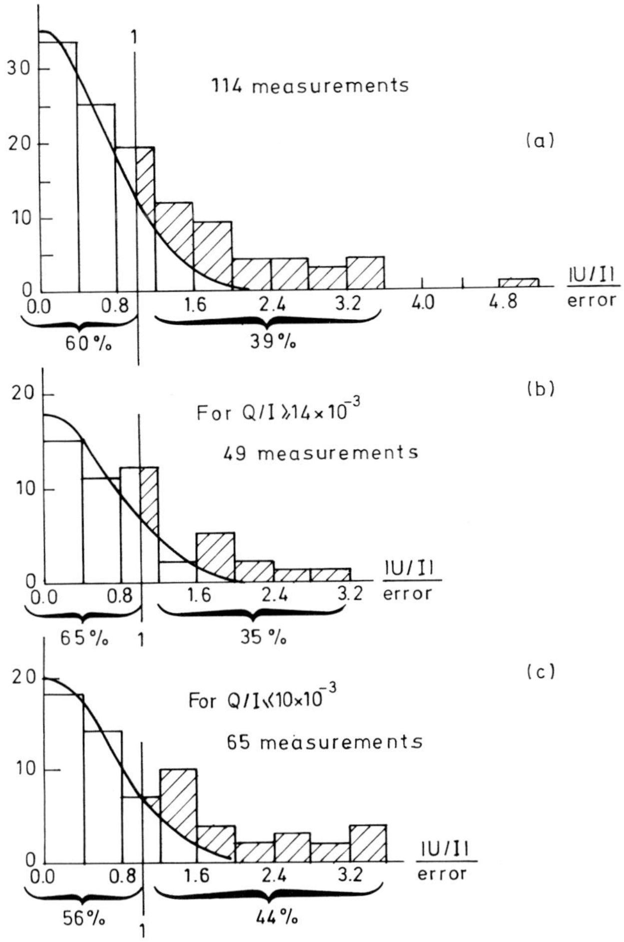

Fig. 29a–c. Histograms for the ratio of the absolute value of U/I to the accuracy, expressed by the error derived from the standard deviatons. The gaussian curves are with $1/e = 1$. **a** For all the measurements (cf. Tables 3–10). **b** For the measurements when $Q/I \geqq 14\ 10^{-3}$. **c** For the measurements when $Q/I \leqq 10\ 10^{-3}$

In Fig. 12, the azimuths do not draw characteristic features; no symmetries are detected and no obvious relationships are found with the dust streamer structures observed visually and reproduced aside.

These deviations of the azimuth of polarization can be understood in terms of anisotropies in the grain shapes, which may be oblate and oriented with their major axis along the same direction, in large areas of at least several hundred kilometers in the coma. The polarizations which are then produced are governed by the directions of the anisotropies.

5.2. *Comparison with other comets*

Deviations of the azimuth of the linear polarization have been analysed for other comets. The most extensive works are summarized in Table 13. All these measurements were obtained by the photographic method, which gives a good angular resolution but a low sensitivity. All of them correspond to phase angles of at least 40° where, in the circumstances of our measurements, no strong deviations are expected to be found.

For comet Mrkos, the extremely careful analysis of Martel (1960) over 678 areas in the coma, for 23 sequences of plates during 11 nights, shows almost no significant deviations, with only two exceptional occurrences of twice the value of the estimated error ±6°. The same is true for 11 areas on Giacobini-Zinner except for one area giving +20°, although a local photographic artifact may be considered. For comet West, a dense coverage of measurements by Isobe et al. (1978) produces almost no significant deviations in the coma, within the accuracy of ±6°. These results are consistent with our conclusions. However, between and around the four cores of the split nucleus after it broke into pieces, Isobe et al. (1978) found completely disturbed polarizations, in degree (from 50 10^{-3} to 450 10^{-3}, the coma giving 220 10^{-3}) and in azimuth (from −27° to +57°, the coma being 0°). Clearly the splitting produced an anomalous dusty environment, probably rich in very small elongated particles.

For comets Bennett (Clark, 1971) and Honda (Osherov, quoted by Dolginov et al. in 1975), the observers reported directions of polarization at all azimuths, in contrast with the rather clean organization found in the other comets. If these

Table 13. Comets: Deviations of the azimuth of linear polarization (reckoned from 90° to the direction of the Sun). Measurements by all the authors are photographic

Comet	Number of sequences	In the coma			Around the nucleus			Ref.
		Number of areas	Error	Extrema observed	Number of areas	Error	Extrema observed	
Mrkos (1957d)	23	678	±6°	−12° ± 13°	23	± 4°	− 7° + 6°	Martel (1960)
Giacobini-Zinner (1959b)	1	11	±6°	− 6° + 20°	–	–	–	Martel (1960)
West (1976 VI)	1	10^5 data points	±6°	− 7° + 8°	10^5 data points	±10°	−27° + 57°	Isobe et al. (1978)
Bennett (1969i)	1	23	?	All azimuths	–	–	–	Clarke (1971)
Honda (1969i)	1	121	?	All azimuths	–	–	–	Osherof, cited by Dolginov et al. (1975)

results are correct, they may indicate other types of physical conditions and dusty environments in these comets, which may involve very small elongated particles (see below). But the authors gave no indications about the accuracies of their measurements.

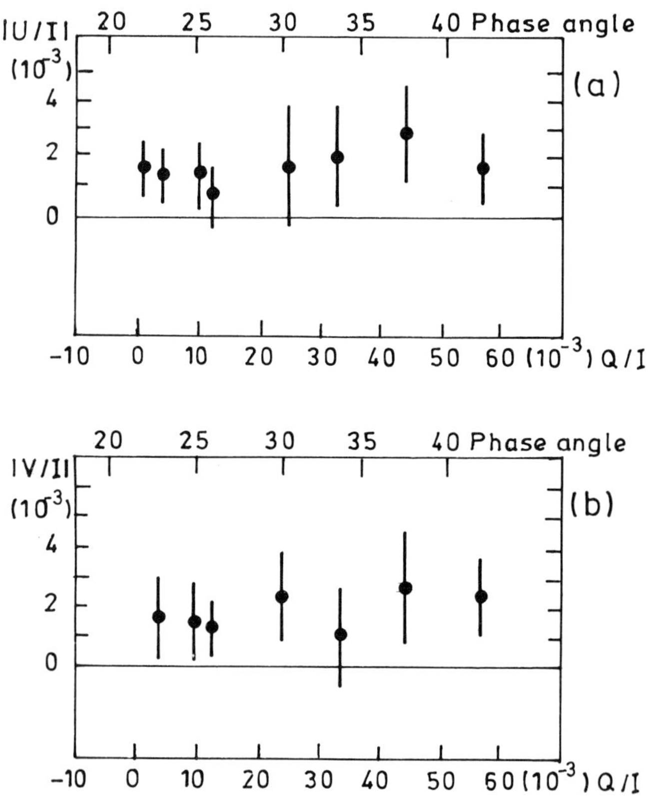

Fig. 30. *Top*: Mean absolute values of U/I as a function of Q/I (scale at bottom) or of the phase angle (scale at top). *Bottom*: same, but for the absolute value of V/I

6. Circular polarization

6.1. *Observations*

The measurements of V/I are listed in Tables 3–9, and mapped in Figs. 5–11. The best sequences of measurements produce an accuracy of $\pm 1.5\ 10^{-3}$. As for U/I, the values of V/I are compared to their errors in the histograms of Fig. 31. Out of 75 measurements, 43% of them exceed their errors, some of them by a factor 3 (Fig. 31a). When the linear polarization Q/I is smaller than $10\ 10^{-3}$, the V/I signals remain larger than the accuracy for 46% of the cases (Fig. 31c). Accordingly, the circular polarization is not directly related to the degree of linear polarization.

The plot of Fig. 30, at bottom, gives the average value $\frac{\Sigma|V/I|}{n}$ for each night of observations, as a function of the linear

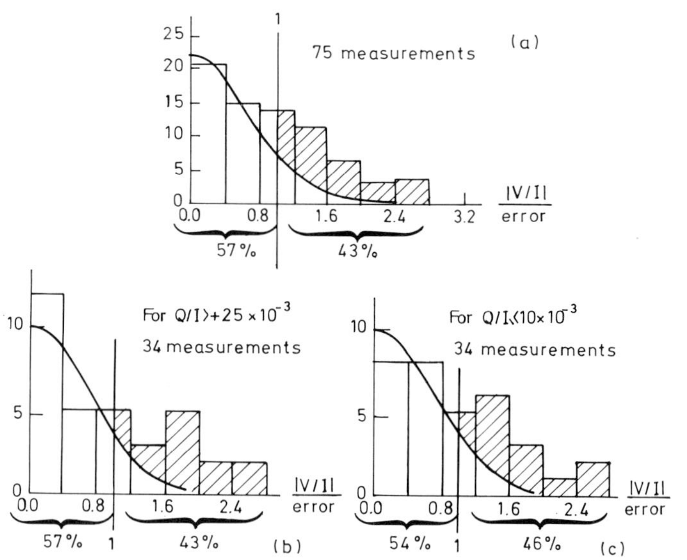

Fig. 31. Histograms for the ratio of the absolute value of V/I to the accuracy (cf. Tables 3–9). Same as for Fig. 29, with relevant values

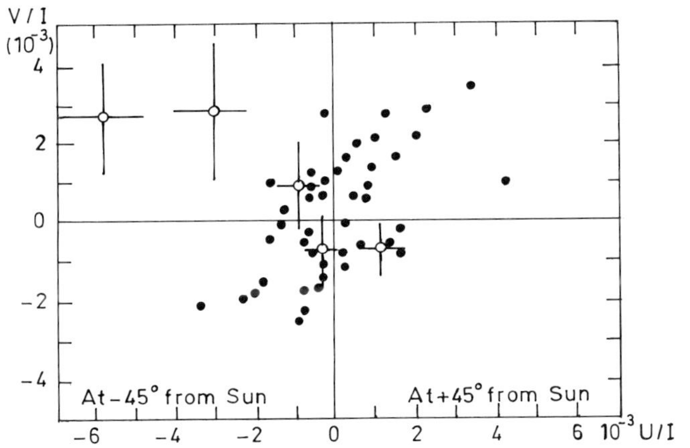

Fig. 32. Relationship between the circular polarization V/I and the deviated polarization U/I. *Dots*: measurements over the coma outside the nucleus. *Circles* with error bars: measurements with the hole centered on the nucleus

polarization Q/I and of the phase angle. These values remain around 1.7 10^{-3}, with a marginally significant increase with the phase angle. The effect which is suggested is the superimposition of circular polarizations, right handed or left handed, to the system of linear polarization produced by the ordinary scattering process.

There is a slight correlation between the degrees of circular polarization V/I and of linear polarization at 45° U/I (Fig. 32). When anisotropies produce a polarization at +45° from the direction of the sun, there is a tendency for a circular polarization with a certain helicity. When the anisotropies produce U/I at −45°, a circular polarization is associated with an opposite helicity.

For the measurements with the hole centered on the nucleus, there is also a correlation with U/I but with an opposite helicity (circles with errors bars in Fig. 32). The light from the nuclear envelope is then transmitted through the cloud of coma dust; circular and linear polarizations may be produced in the transmitted beam, in the presence of anisotropies.

The distributions of the circular polarization over the coma images which are displayed at bottom left in the 5 to 11 do not exhibit any clear trends, symmetries or associations with the coma features observed visually, other than the tendency reported above for association with the deviations of the linear polarization.

6.2. Comparison with other observations

There have been few other attempts to detect circular polarization in comets. Unsuccessful searches reported by several authors are reviewed by Dobrolewsky et al. (1986), including an inconclusive result by Wolff (1972) on comet Tago-Sato-Kosaki. More recently, Michalsky (1981) analysed three comets. For Kohoutek, he found $V/I \leqq 0.6\ 10^{-3}$, for Bradfield, $V/I \leqq 1.6\ 10^{-3}$ and for West, $V/I \leqq 0.5\ 10^{-3}$. All these negative results were collected with a hole centered on the nucleus.

For the coma outside the nucleus, there was a search by Häfner and Metz (1986) on P/Halley, using the photometer of the ESO 50 cm telescope at La Silla (Chile). They placed alternavely a right handed and a left handed circular analyser in front of the instrument. The accuracy was $\pm 5\ 10^{-3}$ only. Between 1986, March 16 and 20, for six areas over the coma, they found V/I values of the order of 5 to 9 10^{-3}, but always left handed. As a zero shift could be introduced in the signal by the linear polarization of the comet itself, the result may need a confirmation.

7. Conclusion

Our photopolarimetric measurements on P/Halley in 1985 at Meudon Observatory and in 1986 at the ESO station in Chile indicate that there are three types of results to be considered:

a) The cometary grains soon after their ejection from the nucleus, when still carried in new streamers, or in the cone of ejection facing the sun, or elsewhere at less than 1000 km from the nucleus. This fresh dust produces a dependence of the degree of linear polarization on phase angle with a cross-over at 21°.0 and a slope of 3.1 (10^{-3}/degree), as shown in Fig. 27, (dashed line and crosses). There is a small increase of the degree of linear polarization with wavelength, roughly proportional to the phase angle reckoned from the cross-over angle (Figs. 25 and 26). There are small deviations of the azimuth of the linear polarization, as if a field of complexly oriented small polarization vectors were added to the main polarization effect. This additional polarization does not depend on the phase angle and is mapped alone at the cross-over phase angle of 21° when the main polarization disappears. There is also a field of very faint circular polarization, left or right handed, added to the linear polarization and which does not depend on its degree. A relationship is suspected between this V/I circular polarization and the U/I component of circular polarization oriented at 45° from the scattering plane.

b) The cometary grains which have travelled for at least 3000 km outward from the nucleus and in the anti-solar direction. This older coma dust produces a linear polarization which behaves like that of the fresh dust but with a displacement of +1°.5 of the cross-over phase angle, followed by a similar slope (dark dots and heavy line in Fig. 27). There is also a field of additional small linear polarization related neither to the phase angle nor to the azimuth of the scattering plane, and a field of circular polarizaton, as for the fresh dust.

c) The bright envelope which surrounds the nucleus. This almost star-like feature had a radius not exceeding 100 km in April 1986 and was occasionally obscured by dust injection in the coma toward the earth. Its polarization, depart for the case of the coma grains, has a same cross-over angle of 22°. but a slope increasing with phase angle and the polarization reach 120 10^{-3} at phase angle 54° (circles and dotted line in Fig. 27).

In future papers, we plan to add to our analysis the polarization measurements made on P/Halley by other observers, in order to derive synthetic polarization curves extended to larger ranges of phase angles, wavelengths and time coverage. An interpretation will follow, intended to characterize the properties of the grains, for the fresh dust, the older dust and the envelope. Albedo, size distribution, texture, shape and aggregation will be considered.

Acknowledgements. The new photopolarimeter of Observatoire de Meudon was granted by CNRS and Institut National d'Astronomie et de Géophysique in France and constructed at the Observatory. The interference filters were loaned by international Halley Watch, thanks to Dr. M. A'Hearn. The instrument was used with the 100 cm telescope operated at Observatoire de Meudon, and in April 1986, at the European Southern Observatory, La Silla, Chile, with the 152 cm ESO telescope. We are indebted to ESO for having granted the whole mission from Paris to Chile, including the transportation of the polarimeter. At La Silla, we appreciated the help of MM. Le Saux, Miranda and Borguez for the operation of the telescope. Dr. Arpigny made available to us his cometary spectra, for our check, in real time, of the influence of the fluorescence emissions on our measurements. The program was set up, namely, by discussions with Prof. Giese and Dr. Zerull at Ruhr University, FRG, and with Prof. Levasseur-Regourd and Dr. Lamy in France. M.R. Killinger and D. Crussaire helped for the observations at Meudon. We thank the administration of Ecole Normale Supérieure in Paris for having authorized one of the authors to participate in the Chile observations.

References

Clark, D.: 1971, *Astron. Astrophys.* **14**, 90

Dobrolowsky, O.V., Kiselev, N.N., Chernova, G.P.: 1986, *Earth, Moon, Planets* **34**, 189

Dolginov, A.Z., Mitrofanov, I.G.: 1975, *Astron. Zh.* **52**, 1268 (in Russian) (transl.: *Sov. Astron.* **19**, 758)

Dollfus, A.: 1979, *Icarus* **37**, 404

Dollfus, A., Aurlère, M., Santer R.: 1979, *Astron. J.* **84**, 1410

Dollfus, A., Suchail, J.L., Crussaire, D., Killinger, R.: 1986, 20th ESLAB Symp. ESA-SP **250**, Vol. **II**. p. 41

Dollfus, A., Suchail, J.L.: 1987, *The Messenger*, Publ. European Southern Observatory, No. **47**, 39

Haefner, R., Metz, K.: 1986, *The Messenger*, Publ. European Southern Observatory, No. **44**, 16

Isobe, S., Saito, K., Tomita, K., Maehara H.: 1978, *Publ. Astron. Soc. Japan* **30**, 687

Kidger, M.R., Cepa, J., Keitch, G., Prieto, M., Collados, M.: 1987, Poster presentation at Symp. *Diversity and Similarity of Comets*, Brussels, Belgium, 6–9 April

Lamy, P., Pedersen, H., Vio, R.: 1986, 20th ESLAB Symposium, ESA SP-**250**, Vol. **II**. p. 69

Le Borgne, J.F., Leroy, J.L., Arnaud, J., 1986, 20th ESLAB Symp. ESA SP-**250**, Vol. **I**, p. 571

Lecacheux, J., Thuillot, W., Encrenaz, T., Laques, P., Rouan, D., Despiau, R.: 1984, *Icarus* **60**, 386

Martel, A.M.: 1960, *Ann. Astrophys.* **23**, 480

Michalsky, J.: 1981, *Icarus* **47**, 388

Wolff, G.W.: 1972, *Astron. J.* **77**, 576

Astron. Astrophys. 187, 689–692 (1987)

ASTRONOMY AND ASTROPHYSICS

Polarimetry of comet P/Halley

S. Kikuchi [1], **Y. Mikami** [1], **T. Mukai** [2], **S. Mukai** [2], **and J.H. Hough** [3]

[1] Tokyo Astronomical Observatory, University of Tokyo, Mitaka, Tokyo, 181 Japan
[2] Kanazawa Institute of Technology, Nonoichi, Ishikawa, 921 Japan
[3] Physics Department, Hatfield Polytechnic, PO Box 109, College Lane, Hatfield, AL 10 9 AB, England

Received April 7, accepted June 16, 1987

Summary. The linear polarization P of comet P/Halley was observed simultaneously at eight wavelengths ranging from 0.36 to 0.90 μm, for phase angles α (sun-comet-earth angle) between 2° and 65°. In addition, observations in the J and H bands were made for $\alpha = 30°$. In the continuum bands, we have obtained (i) negative polarization for $\alpha \lesssim 20°$, (ii) a quasi-linear increase of the polarization level with phase angle for $20° < \alpha < 50°$, and (iii) a slight increase of P with wavelength from the visible through to the near infrared.

Key words: polarization – comet – Halley

1. Observations

Observations of the linear polarization of the coma of comet Halley were made during 23 nights from 1985 November 11 to 1986 May 16, using an eight-channel polarimeter attached to the 91 cm telescope at the Dodaira Station (139° 12′E, +36° 00′, 879 m) of the Tokyo Astronomical Observatory. Phase angles of 1°.6–64°.4 were covered by these observations. Most measurements used an aperture of 18″.0, corresponding to an area around the nucleus of diameter varying between 8,000–17,000 km during the observing period. Particular attention was paid to the subtraction of the sky back-ground, since after January observations of the comet from the northern hemisphere were made at large zenith distances.

Details of the filter set are given in Table 1. Six of the eight channels were used for continuum observation. Among these six channels, were two interference filters supplied by the International Halley Watch (IHW) for channels Nos. 1 and 3. The fluxes by the molecular emission mainly due to NH_2 are contaminated in the results through the intermediate- and wide-band filters, but are considered to be less than few percents of the total flux through these filters except for channel No. 6. For channel No. 6, the observed contribution by the molecular emission was estimated to be 5–6% of the total flux which was larger than expected by a factor of 2–3. Besides these six channels for the study of the continuum light, we also used a narrow-band filter recommended by the IHW for channel No. 4 and a wide-band filter for channel No. 2. These filters were used in the molecular bands of C_2 and CN, respectively. Although the accurate photometric results for continuum and molecular bands are not at hand, the contribution of the continuum light in the last two bands seems to be 25–35% of the total fluxes, if we assume the Öhman's relation for the phase angle dependence of molecular emission (Öhman, 1941; Le Borgne et al., 1986, 1987).

The corrections for the instrumental polarization and the polarization efficiency, and also the determination of zero of position angles (θ) were made by the observation of polarimetric standard stars and by inserting a Glan-Taylor prism ahead of the rotating achromatic half-wave plate. We believe that any systematic uncertainties in P and θ are less than 0.1% and 1° respectively.

To investigate the wavelength dependence of the polarization in the visible and near infrared, polarization measurements in the *UBVRIJH* bands were made with the Hatfield polarimeter (Bailey and Hough, 1982) attached to the 3.89 m Anglo-Australian Telescope on 1986 May 3, and an aperture of 6″.0 was used. Polarimetric calibrations were made in the same way as those at Dodaira. Although observations in the *UBV* bands were affected by molecular emission from the comet, those in the *RIJH* bands were dominated almost completely by the continuum.

In Table 2 we give the nightly averaged polarization obtained with an 18″.0 aperture at Dodaira. Listed are the polarization levels. Also included are the phase angle of the observations, the geocentric and heliocentric distances.

Table 1. Effective wavelengths for observing bands

Channel No.	Effective wavelength (μm)	Band width [a] (μm)	Remark
1	0.365	0.010	Continuum
2	0.37	0.33–0.39	CN
3	0.484	0.006	Continuum
4	0.514	0.008	C_2
5	0.62	0.59–0.65	Continuum
6	0.67	0.66–0.69	Continuum
7	0.73	0.71–0.76	Continuum
8	0.90	0.85–1.10	Continuum

[a] We give full-widths for the half-maximum for narrow bands and the ranges of the efficiency of more than 10% of the maximum one in each band for other channels

Send offprint requests to: S. Kikuchi

Table 2. Polarization (in %) of the coma of comet Halley. Δ and r are the geocentric and the heliocentric distance of the comet, respectively. α denotes a sun-comet-earth angle (phase angle). λ is an effective wavelength of each of bands in μm

Date (UT)	Δ (AU)	r (AU)	α (deg)	Channel Nos. 1	2	3	4	5	6	7	8
Observation at Dodaira				$\lambda=0.365$ (μm)	0.37	0.484	0.514	0.62	0.67	0.73	0.90
1985 11 11.67	0.80	1.77	10.7		−1.5±0.4	−2.9±0.9	−1.3±0.8	−1.6±0.2	−1.7±1.9	−2.2±1.1	
12.70	0.78	1.75	9.2	−4.3±0.9	−1.2±0.2	−0.8±0.5	−1.4±0.3	−1.3±0.1	−2.8±0.7	−1.9±0.4	
13.72	0.76	1.74	7.7	−0.2±0.4	−0.7±0.1	−1.2±0.1	−1.1±0.1	−1.5±0.1	−1.1±0.2	−1.3±0.1	−1.4±0.6
14.72	0.74	1.73	6.1	−1.2±0.3	−0.5±0.1	−1.1±0.1	−0.6±0.1	−1.2±0.1	−1.2±0.2	−1.6±0.1	−2.5±0.7
17.68	0.69	1.68	1.6	−1.2±0.4	−0.4±0.1	−0.4±0.2	−0.4±0.1	−0.3±0.1	−0.6±0.2	−0.3±0.2	−2.6±0.6
18.70	0.68	1.67	2.0	−2.0±0.6	−0.6±0.1	−0.3±0.2	−0.9±0.1	−0.7±0.1	−0.5±0.3	−0.7±0.1	−3.1±0.9
12 04.56	0.66	1.43	36.7	4.7±0.9	3.4±0.2	4.0±0.2	4.1±0.2	4.9±0.2	6.0±0.2	6.1±0.3	9.8±1.7
05.52	0.67	1.41	38.5	4.3±0.6	4.1±0.1	5.5±0.2	3.9±0.1	5.7±0.1	5.4±0.2	5.9±0.2	9.3±0.7
10.49	0.74	1.34	46.1	8.6±0.7	6.2±0.1	9.1±0.2	5.4±0.1	8.8±0.1	8.0±0.2	9.3±0.2	9.3±0.7
12.45	0.78	1.31	48.4	9.7±0.4	7.0±0.1	10.4±0.1	6.7±0.1	10.4±0.1	9.9±0.2	11.4±0.1	11.7±0.5
13.51	0.79	1.29	49.4	6.4±0.9	6.9±0.1	10.1±0.3	6.4±0.1	10.5±0.1	9.3±0.3	11.2±0.3	10.6±1.0
14.52	0.81	1.28	50.3	10.8±1.2	6.3±0.2	10.6±0.5	5.8±0.2	10.6±0.1	8.2±0.5	11.2±0.3	13.1±1.6
15.50	0.83	1.26	51.0	8.5±0.9	6.2±0.2	10.6±0.4	6.2±0.1	10.7±0.1	10.1±0.4	11.1±0.4	11.6±1.3
1986 01 05.39	1.24	0.94	51.3	10.4±0.2	7.2±0.1	12.0±0.1	6.7±0.0	12.9±0.0	11.7±0.1	13.9±0.1	14.8±0.2
06.39	1.26	0.93	50.7	8.3±0.6	6.2±0.1	10.3±0.2	5.8±0.1	10.7±0.1	9.7±0.2	12.3±0.3	11.5±0.5
07.40	1.28	0.91	50.0	10.5±0.4	8.0±0.1	12.4±0.2	7.5±0.1	13.7±0.0	12.9±0.1	14.9±0.1	16.4±0.4
09.40	1.31	0.88	48.5	9.8±0.6	5.6±0.1	10.1±0.2	5.6±0.1	10.2±0.1	9.5±0.2	10.9±0.2	11.3±0.5
10.39	1.33	0.87	47.7	7.1±0.5	5.0±0.1	9.6±0.2	5.1±0.1	9.6±0.1	8.4±0.2	10.6±0.2	12.1±0.5
03 12.81	0.99	0.89	63.7	14.6±3.6	8.7±0.3	17.0±0.3	9.4±0.2	18.2±0.1	17.1±0.2	19.3±0.1	19.7±0.8
25.79	0.67	1.09	64.4	16.4±2.1	10.1±0.5	17.0±0.3	10.2±0.2	18.0±0.1	17.1±0.3	19.1±0.1	20.8±0.7
04 23.52	0.60	1.52	24.1	3.6±0.9	1.2±0.2	1.2±0.2	1.0±0.2	0.9±0.1	0.5±0.3	0.9±0.2	1.1±0.9
05 04.55	0.90	1.68	30.3	2.9±1.4	1.9±0.3	4.1±0.4	2.5±0.3	2.9±0.2	0.9±0.4	2.9±0.3	3.6±2.3
16.60	1.29	1.86	31.4		0.4±1.3	3.4±1.2	1.7±0.8	3.6±0.3	2.4±0.9	2.9±0.6	
Observation at AAT				U	B	V	R	I	J	H	
				$\lambda=0.36$	0.44	0.55	0.64	0.80	1.23	1.64	
1986 05 03.36	0.87	1.67	29.9	2.39±0.15	2.47±0.09	2.37±0.11	2.37±0.13	2.64±0.10	3.07±0.07	3.40±0.11	

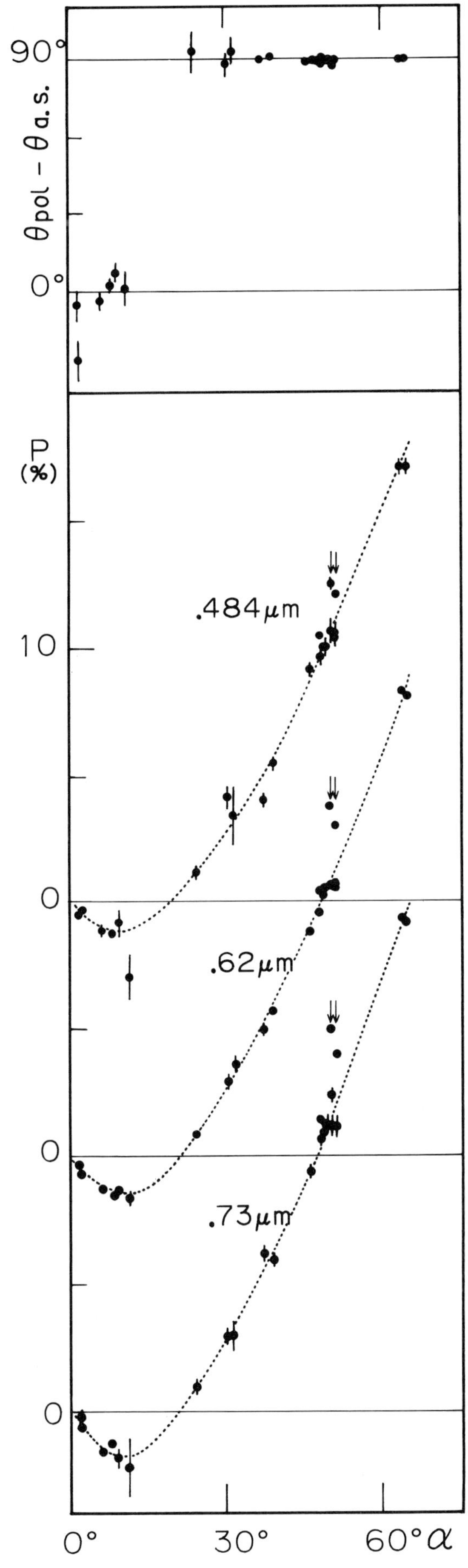

2. Phase angle dependence

Nearly all the polarization position angles are either parallel (for $\alpha < 20°$) or perpendicular (for $\alpha > 20°$) to the scattering plane (see Fig. 1). In accordance with the usual convention we will refer to such polarizations respectively as negative and positive. The variation of continuum polarization P with α is shown in Fig. 1. Displayed are the results obtained at wavelengths 0.484, 0.62, and 0.73 μm for which the polarization is considered to be almost completely due to the scattering of the solar continuum.

In general, the dependence of P on α is quite similar to that reported previously for other comets (e.g. Dobrovolsky et al., 1986; Lamy, 1986) and for comet Halley itself by Le Borgne et al. (1986), Dollfus et al. (1986), Kiselev et al. (1986), and Bastien et al. (1986).

We note the following features:

(i) Negative polarization is detected for phase angles less than 11°. Assuming a smooth interpolation of the data at $\lambda = 0.62$ μm between phase angles of 11° and 24°, we estimate the cross over between negative and positive polarizations occurs at a phase angle of $(22 \pm 2)°$. This value is typical of the behavior of other comets and agrees well with the values for comet Halley by Kiselev et al. (1986), Dollfus et al. (1986), and Bastien et al. (1986).

(ii) The maximum negative polarization at $\lambda = 0.62$ μm is estimated to be (-1.8 ± 0.2)% and to occur at a phase angle of $(12 \pm 2)°$. Again, both these values are typical for comets.

(iii) Between phase angles 25° and 50° the (positive) polarization increases linearly with phase angle with a slope of (0.38 ± 0.01) %/deg at $\lambda = 0.62$ μm.

(iv) We also measured the polarization in molecular bands with an IHW filter at $\lambda = 0.514$ μm (C_2) and with a wide-band filter at $\lambda = 0.37$ μm (CN). The results through these filters in the observing period are explained by the sum of the contributions due to the continuum and molecular emission. Especially, the negative polarization observed in molecular bands strongly suggests that the contamination of the continuum component is significant. Le Borgne et al. (1987) found that the polarization angles for CN deviated by at most 20° from the angle expected by the fluorescence scattering. However, our result shows no such deviation. It might be caused by the fact that our results have been obtained with the wide-band filter, and consequently they are rather insensitive for the measurement of the polarization due to molecular scattering.

3. Wavelength dependence

The observed wavelength dependence of polarization, $P(\lambda)$, at several values of α is shown in Fig. 2. As Dobrovolsky et al. (1986) and Lamy (1986) note, previous observations of $P(\lambda)$ for several comets give differing results. Neutral polarization was reported in the visible region for comets West, Austin and Churyumov-Gerasimenko, whereas an increase of P with λ was observed for comets Bennett and Kohoutek (see references in Lamy, 1986). The

Fig. 1. Phase angle dependence of polarization in the continuum bands at wavelengths of 0.484, 0.62, and 0.73 μm. $(\theta_{pol} - \theta_{a.s.})$ in the upper panel is the polarization angle for continuum measured from the position angle of the anti-solar direction at the comet. θ_{pol} is derived by averaging the results obtained at 0.484, 0.62, 0.67, and 0.73 μm. Arrows indicate the data obtained at the brightenings on 1986 January 5 (right) and 7 (left), respectively

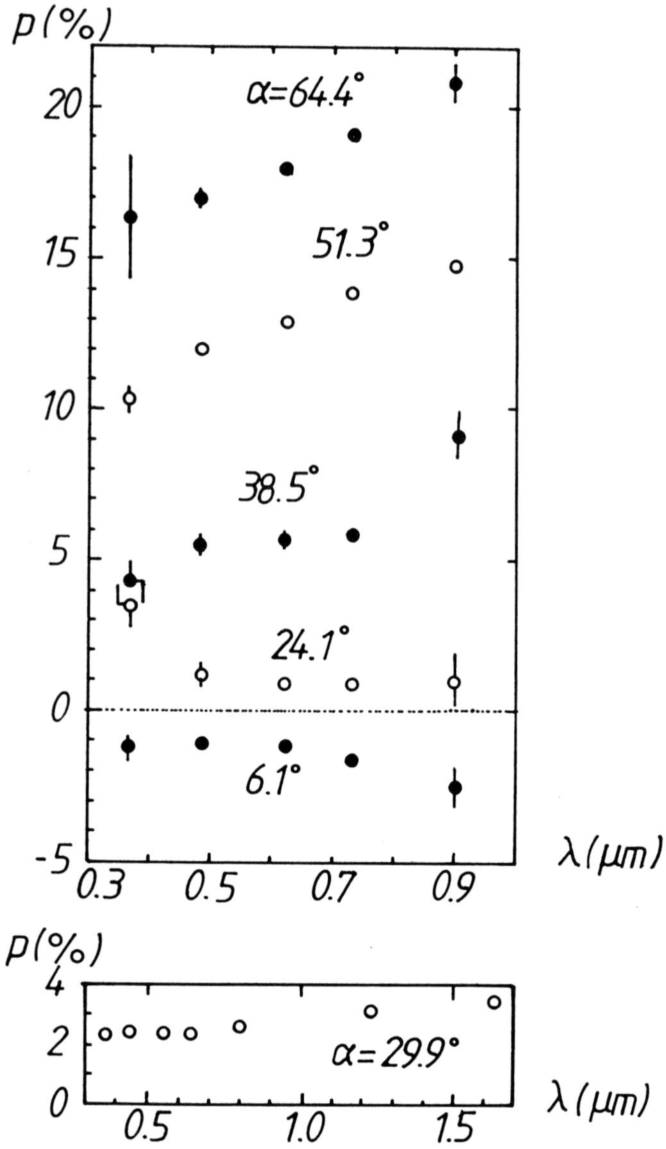

Fig. 2. Wavelength dependence of polarization at various phase angles

present observations for comet Halley suggest a small but significant increase of P with wavelength, extending through to the J and H bands for measurements made at $\alpha = 30°$. Brooke et al. (1986) also report an increase of P with λ for the J, H and K bands, measured at various phase angles.

4. Relation to the flux variation

Fluctuations in the level of polarization on a time-scale of a day are clearly present at some phase angles for the positive polarization branch. The variations are particularly pronounced on January 5–10. Polarization levels on January 5 and 7 enhanced between 2 and 3 % at $\lambda = 0.62$ μm, compared to those on January 6, 9, and 10. Similar increases were also observed at other wavelengths. We indicate the data on January 5 and 7 by arrows in Fig. 1. Our own estimates of brightness suggest that the comet brightened by at least 0.5 and 0.8 mag at $\lambda = 0.62$ μm on January 5 and 7, respectively. This would appear consistent with the increase in near infrared brightness reported by Tokunaga et al. (1986) for January 8 and 9. This suggests that higher polarization levels correlate with an increase of flux. It should also be noted that the higher polarizations occurred at Lyman-α flash time, as extrapolated from the measurements by the spacecraft Suisei (Kaneda et al., 1986). It might come from the differences in the size distribution of dust at the outbursts as Mukai et al. (1987) discussed.

5. Aperture dependence

A few polarization measurements were made with aperture sizes ranging from 13.″0 to 32.″6. For the continuum light, we found no systematic dependence of polarization on aperture size. Therefore, the features described above for polarization of the continuum light can be extended to regions out to about 10,000 km from the nucleus. We note that Dollfus et al. (1986) report a decrease in the levels of polarization for the central few arcsecs around the nucleus, whereas Bastien et al. (1986) found an increase at the inner part of the coma. In those bands for which molecular emission is important, we find that the observed polarization decreases with increasing aperture.

References

Bailey, J.A., Hough, J.H.: 1982, *Publ. Astron. Soc. Pacific* **94**, 618

Bastien, P., Menard, F., Nadeau, R.: 1986, *Monthly Notices Roy. Astron. Soc.* **223**, 827

Brooke, T.Y., Knacke, R.F., Joyce, R.R.: 1986, 20th ESLAB Symp., ESA SP-**250**, Vol. **II**, p. 87

Dobrovolsky, O.V., Kiselev, N.N., Chernova, G.P.: 1986, *Earth, Moon Planets* **34**, 189

Dollfus, A., Suchail, J.-L., Crussaire, D., Killinger, R.: 1986, 20th ESLAB Symp. ESA SP-**250**, Vol. **II**, p. 41

Kaneda, E., Hirao, K., Takagi, M., Ashihara, O., Itoh, T., Shimizu, M.: 1986, *Nature* **320**, 140

Kiselev, N.N., Pushnin, P.A., Siklitsky, V.I., Chernova, G.P.: 1986, 20th ESLAB Symp. ESA SP-**250**, Vol. **II**, p. 29

Lamy, P.: 1986, in *Asteroids, Comets and Meteors II*, eds. C.I. Lagerkulst et al., Uppsala Universitet, Uppsala, p. 373

Le Borgne, J.F., Leroy, J.L., Arnaud, J.: 1986, 20th ESLAB Symp. ESA SP-**250**, Vol. **I**, p. 571

Le Borgne, J.F., Leroy, J.L., Arnaud, J.: 1987, *Astron. Astrophys.* **173**, 180

Mukai, T., Mukai, S., Kikuchi, S.: 1987, ESA SP-**278** (in press)

Öhman, Y.: 1941, *Stockholm Obs. Annal.* **13**, No. 11

Tokunaga, A.T., Golisch, W.F., Griep, D.M., Kaminski, C.D., Hanner, M.S.: 1986, *Astron. J.* **92**, 1183

Astron. Astrophys. 187, 693–698 (1987)

ASTRONOMY AND ASTROPHYSICS

Calibration of the DIDSY-IPM dust detector and application to other impact ionisation detectors on board the P/Halley probes

J.R. Göller, E. Grün, and D. Maas

Max-Planck-Institut für Kernphysik, Postfach 10 39 80, D-6900 Heidelberg, Federal Republic of Germany

Received February 13, accepted April 2, 1987

Summary. Results from calibration tests with the DIDSY-IPM dust detector, flown on the Giotto mission, are reported and applied to the impact ionisation dust detectors SP-1 and SP-2, flown on the two Vega Halley probes. The DIDSY-IPM detector is the most sensitive subsystem of the Dust Impact Detection System on board the Giotto spacecraft. It consists of two impact ionisation detectors: sensor A is open, whereas sensor B is covered by a 2.5 μm thick aluminised Mylar film to measure the density of cometary particles. A geometrically very similar detector (SP-1) and an impact ionisation detector of different sensor geometry (SP-2) were flown on board the two Russian Vega spacecrafts. Impact experiments with the DIDSY-IPM detector were performed at the dust accelerators of the Max-Planck-Institut in Heidelberg and the Technical University in Munich. Iron, silicate and carbon particles with masses between 10^{-15} g and 10^{-6} g and speeds up to 70 km s^{-1} were used for sensor calibrations. At constant speeds a linear dependency between the impact charge and the particle mass was found. At an encounter speed of 69 km s^{-1} (Giotto) the charge yield Q/m is 700 C g^{-1} for iron particles and 3000 C g^{-1} for carbon and silicate particles. At 78 km s^{-1} (Vega) the yields are about twice as high. The penetration limit of the sensor B film is 6.5 10^{-15} g for iron particles, 1.3 10^{-14} g for silicate particles and 3.2 10^{-14} g for particles with a density of 1 g cm^{-3} at an encounter speed of 69 km s^{-1}. For particles with a density of 1 g cm^{-3} the penetration limits of the SP-1 films (with thicknesses of 0.6 μm and 2 μm) are 7.0 10^{-16} g and 1.4 10^{-14} g for Vega 1 and Vega 2 respectively. Due to the relatively fast risetimes ($<5\mu$s) of the charge signals the DIDSY-electronics has a nonlinear response if the impact charge is larger than 10^{-9} C.

Key words: comets – instruments – interplanetary medium – charge yield – penetration limit

1. Introduction

Several impact ionisation dust detectors were flown on the Halley probes to measure the dust flux, mass distribution and density of cometary particles in Halley's coma. The most sensitive subsystem of the Dust Impact Detection System (DIDSY, McDonnell et al., 1986a) on board Giotto is the Impact Plasma and Momentum detector (IPM). A schematic drawing of the DIDSY-IPM sensor is shown in Fig. 1. It consists of two impact ionisation detectors (IPM-P): sensor A is open, whereas sensor B is covered by a 2.5 μm thick aluminised Mylar film. The charges produced by particle impacts on the gold target are collected by two oppositely biased electrodes. The charge signals of both polarities are amplified, processed and digitized by the electronics. By help of the sensor calibrations introduced by Grün et al. (1984) and Göller et al. (1986) and extended in this paper the amplitudes of the charge signals provide information about the masses of the impacting particles. The penetration limit of the sensor B film depends on both the mass and density of the projectiles (Pailer and Grün, 1980). Since the impact speed during the Halley flyby was fixed at 69 km s^{-1}, the density of particles can be estimated by comparing the count rates on both sensors (Grün and McDonnell, 1983). A piezoelectric detector and a plasma probe also belong to DIDSY-IPM but shall not be discussed here (McDonnell et al., 1986a). DIDSY-IPM did not perform properly during the Halley encounter because of a malfunction of the instrument cover release mechanism (Lange et al., 1986). Results from the DIDSY experiment are presented by McDonnell et al. (1986b) and Zarnecki et al. (1986). Similar ionisation detectors (SP-1 and SP-2) were flown on the Vega spacecrafts (Vaisberg et al., 1986a and b, Mazets et al., 1986a and b, Grard et al., 1986, Smirnov et al., 1986).

Impact simulation experiments to calibrate the IPM-sensor were performed at the electrostatic dust accelerator of the Max-Planck-Institut in Heidelberg and with the plasma drag dust accelerator of the Technical University in Munich. Calibration tests in Heidelberg were made with the flight spare unit (FS) of DIDSY-IPM, whereas the experiments in Munich were performed with the engineering model (EM) of the sensor; all tests were made with the EM-electronics except some measurements in Munich which will be discussed later. With the flight unit (FU) only limited verification tests were performed at the dust accelerators in Heidelberg and Canterbury (Grün et al., 1984). The difference in calibration of the various units was found to be within the statistical uncertainties. The aim of this paper is to report the results of these calibrations and to apply them also to the impact ionisation detectors on the Vega spacecrafts.

Send offprint requests to: J.R. Göller

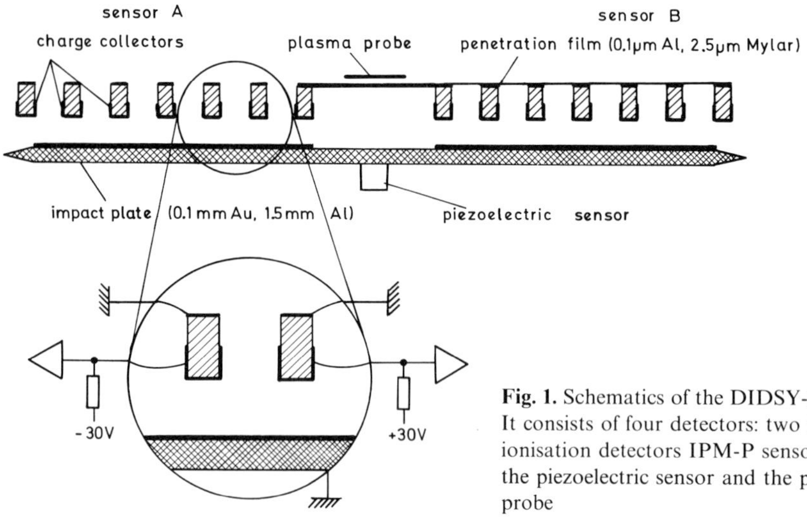

Fig. 1. Schematics of the DIDSY-IPM sensor. It consists of four detectors: two impact ionisation detectors IPM-P sensors A and B, the piezoelectric sensor and the plasma probe

2. Particle materials and parameters

The calibration experiments in Heidelberg were made with iron, carbon and silicate spherules. The acceleration process of the plasma drag accelerator requires particles with a low heat absorption coefficient and a high emission coefficient to avoid melting or even evaporation. Therefore silicate particles are the most suitable ones. As shown in Fig. 2 the electrostatic accelerator offers silicate particles with speeds between $2\,\mathrm{km\,s^{-1}}$ and $50\,\mathrm{km\,s^{-1}}$ in a mass range from 10^{-15}g to 10^{-10}g. For iron particles this range is somewhat larger, for carbon particles somewhat smaller. The projectiles of the plasma drag accelerator used in these tests are about 4 orders of magnitude larger in mass, but they cover a smaller speed interval of only $4\,\mathrm{km\,s^{-1}}$ to $12\,\mathrm{km\,s^{-1}}$ (see also Fig. 2).

3. Impact charge measurements (sensor A)

For a variety of small speed intervals and for all the projectile materials the mass dependency of the impact charge was analyzed. The data obtained in Heidelberg cover a mass range of typically one or two orders of magnitude in one speed interval. But in combining these data with those from the experiments in Munich the range can be enlarged. An example is shown in Fig. 3. In the speed interval from $10.0\,\mathrm{km\,s^{-1}}$ to $12.0\,\mathrm{km\,s^{-1}}$ a least-squares-fit in a double logarithmic diagram (impact charge versus particle mass) is shown. The slope of the fitted line is 1.02 ± 0.03. Hence the impact charge is proportional to the particle mass at constant speed. This result is also valid for other speed intervals and projectile materials. So it is suggestive to normalize the impact charge to the particle mass, leading to the so-called charge

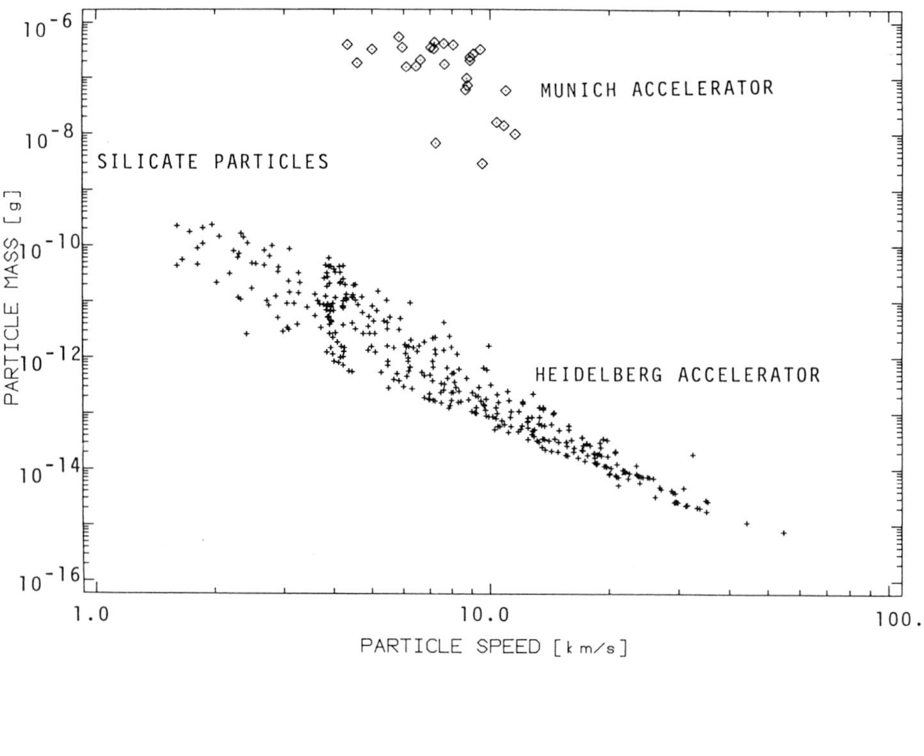

Fig. 2. Silicate particles shot onto DIDSY-IPM sensor A and used for sensor calibration

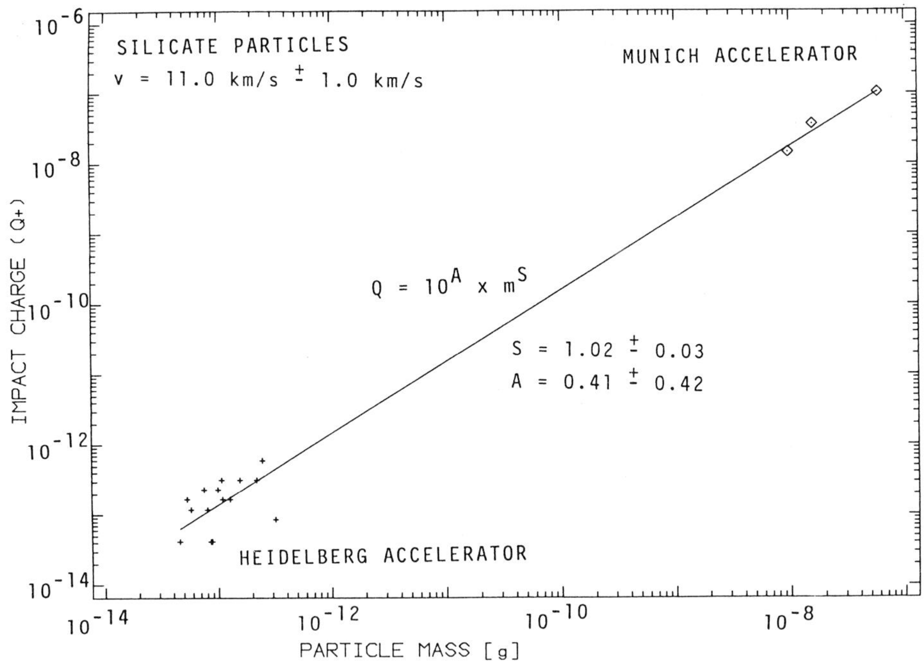

Fig. 3. Impact charge measured with IPM sensor A versus particle mass in a small impact speed interval of $11 \pm 1\,\mathrm{km\,s^{-1}}$. The slope of the fitted line gives a mass exponent of 1.02 ± 0.03

yield Q/m. The yield is speed dependent as shown in Fig. 4, again for silicate particles. For the other materials the yields are generally different as well as their speed dependencies. The crosses in Fig. 4 represent the mean value and the statistical deviations of all measurements in the speed interval indicated by the horizontal bars. It can be seen that the dependency is not given by a power law for the whole speed range. The encounter speeds of Giotto ($69\,\mathrm{km\,s^{-1}}$), Vega 1 ($79\,\mathrm{km\,s^{-1}}$) and Vega 2 ($77\,\mathrm{km\,s^{-1}}$) could not be reached in the calibration tests. So the yields had to be extrapolated from the high speed measurements ($20\,\mathrm{km\,s^{-1}}$ to $45\,\mathrm{km\,s^{-1}}$ for silicate particles). Such an extrapolation up to $69\,\mathrm{km\,s^{-1}}$ is also shown in Fig. 4, leading to a yield of $3000\,\mathrm{C\,g^{-1}}$ for silicate particles. The accuracy (1σ) is a factor of 3. The charge yields for the different encounter speeds of Vega 1 and 2 are not distinguished because they are equal within the errors. For iron particles measurements up to $70\,\mathrm{km\,s^{-1}}$ were possible. Thus the accuracy is better, it is about a factor of 2.5. Evaluating also the negative charge yields one obtains the same values within a deviation of a factor of 1.4. In the high speed region iron is an exception, because the yield of negative charges is about twice the yield of positive charges. All relevant values are summarized in Table 1.

From experiments with other detectors we know that the yields depend on the sensor geometry and electronic signal processing. For example the dust detectors for the Galileo and Ulysses missions have about a factor of 3 lower yields than

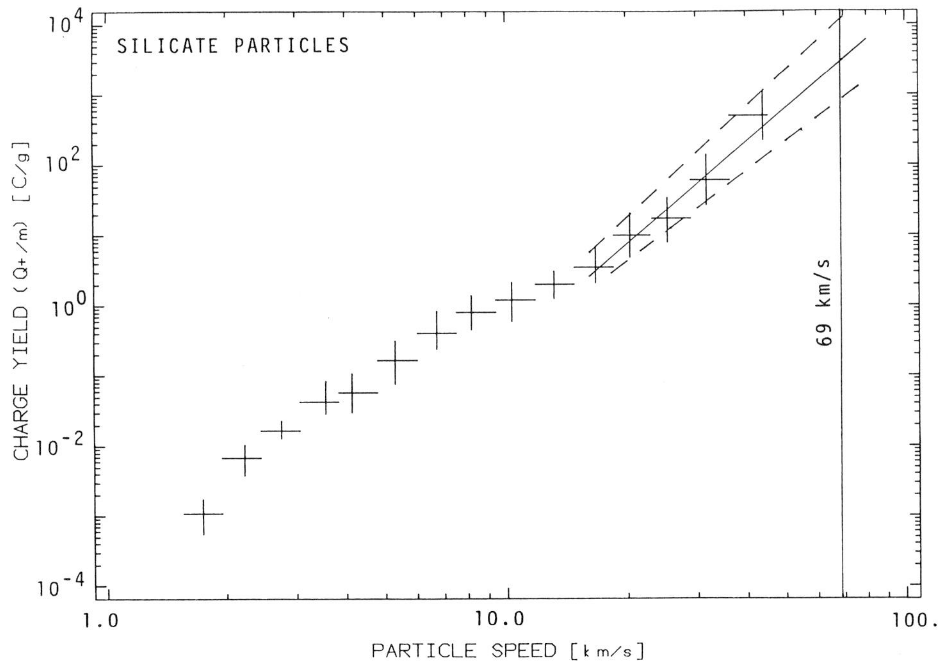

Fig. 4. Charge yield Q/m versus particle speed measured with IPM sensor A. High speed measurements were used to extrapolate the charge yields up to the encounter speeds of the various spacecrafts

Table 1. Positive charge yields for different projectile materials extrapolated to different encounter speeds derived from the calibration data of the DIDSY-IPM sensor A

Projectile material	Encounter speed of 69 km s^{-1}	Encounter speed of 78 km s^{-1}
Carbon	3000 C g^{-1}	6000 C g^{-1}
Silicate	3000 C g^{-1}	5500 C g^{-1}
Iron	700 C g^{-1}	1500 C g^{-1}

DIDSY-IPM-P at impact speeds larger than 20 km s^{-1} depending on the projectile material (Göller and Grün, 1984). SP-1 is quite similar in geometry to IPM-P but different in electronic signal processing (Grard et al., 1986). Thus for SP-1 the yields given in Table 1 are only valid within an increased uncertainty. Since the configuration of the SP-2 detector is different (Grard et al., 1986) from those mentioned above our results can be applied only with larger uncertainties of the order of one magnitude. Together with the charge detection limit (3 10^{-14} g) of the DIDSY-IPM-P electronics the yields of Table 1 correspond to detection thresholds of 10^{-17} g (carbon, silicate) and 4 10^{-17} g (iron).

4. Impact charge measurements (sensor B)

The detection threshold for the film covered sensor B is mainly given by the penetration limit of the mylar film. This depends on particle speed, mass and density. Pailer and Grün (1980) suggested a formula for the threshold mass:

$$m = cT^{2.5}v^{-2.2}\rho^{-0.825}$$

where T is the thickness of the film, v the velocity and ρ the density of the particle. Figure 5 shows the region where silicate particles were available at the accelerator and the particles that penetrated the foil. Obviously the measured speed dependency is consistent with the one given above. The same result was found for iron projectiles (Grün et al., 1984). So the formula was used to extrapolate the measurements to 69 km s^{-1}. The transmission (i.e. the number of particles penetrating the foil divided by the number of particles shot onto the foil) in a given mass interval was calculated. For silicate and iron projectiles this is shown in Fig. 6. The maximum transmission (66%) is given by the sensor geometry (sensitive area/total area) and indicated by a horizontal dashed line. The other dashed line indicates the penetration limit defined by a 10% transmission. Carbon particles used for the tests could not penetrate the foil at all, leading only to a lower limit. The SP-1 detectors were covered by films with thicknesses of 0.6 µm (Vega 1) and 2.0 µm (Vega 2, Smirnov et al., 1986). An extrapolation to these thicknesses and furthermore to a particle density of 1 g cm^{-3} at a mean speed of 78 km s^{-1} was made. All values are summarized in Table 2. Another effect of the film penetration is that the pulse shapes of the charge signals change. About a factor of 2 longer risetimes were observed for marginally penetrating particles.

5. Nonlinearity of the DIDSY electronics

The charge sensitive amplifiers (CSA) used in the DIDSY-IPM electronics were designed to measure impact charges between 10^{-14} C and 10^{-8} C to be sensitive for particles with masses from 10^{-17} g to 10^{-11} g at an impact speed of 69 km s^{-1}. But particles which produce larger charges were also expected. The dust accelerator of the Technical University in Munich provides the possibility to accelerate particles which produce impact charges between 10^{-10} C and 10^{-5} C (Maas et al., 1986). Thereby saturation effects could be studied. Two types of charge sensitive amplifiers were used:

a) CSA which are able to detect an impact charge between 10^{-10} C and 10^{-5} C with a linear response. They had been successfully used for similar impact experiments previously (Iglseder, 1986). A charge signal measured with these CSA is shown in Fig. 7. The risetime of the signals is typically 3 µs to 5 µs (10%

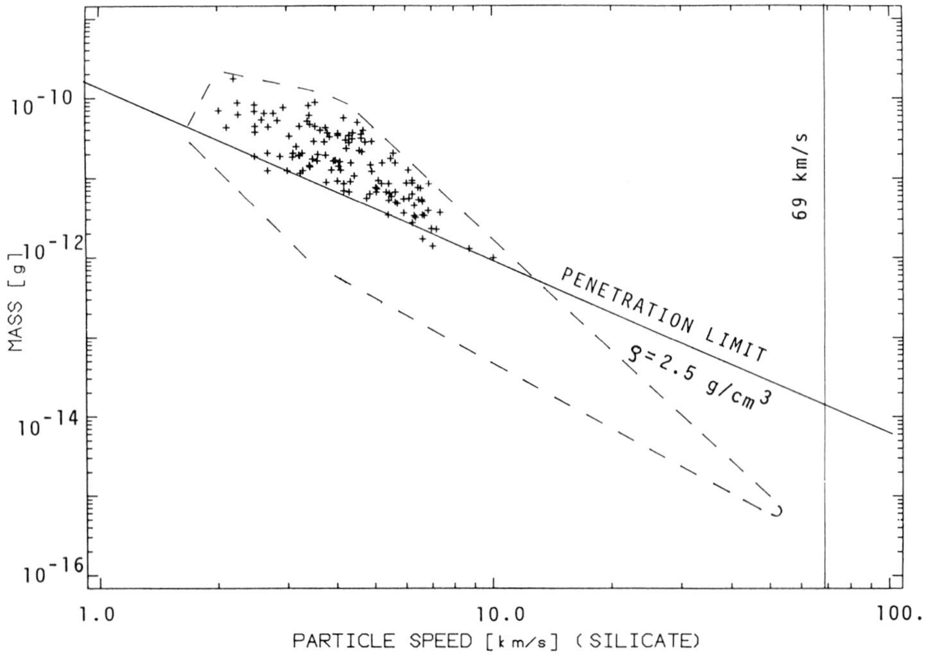

Fig. 5. The dust accelerator in Heidelberg offers particles of speeds and masses inside the dashed line. The crosses represent particles which could penetrate the sensor B film

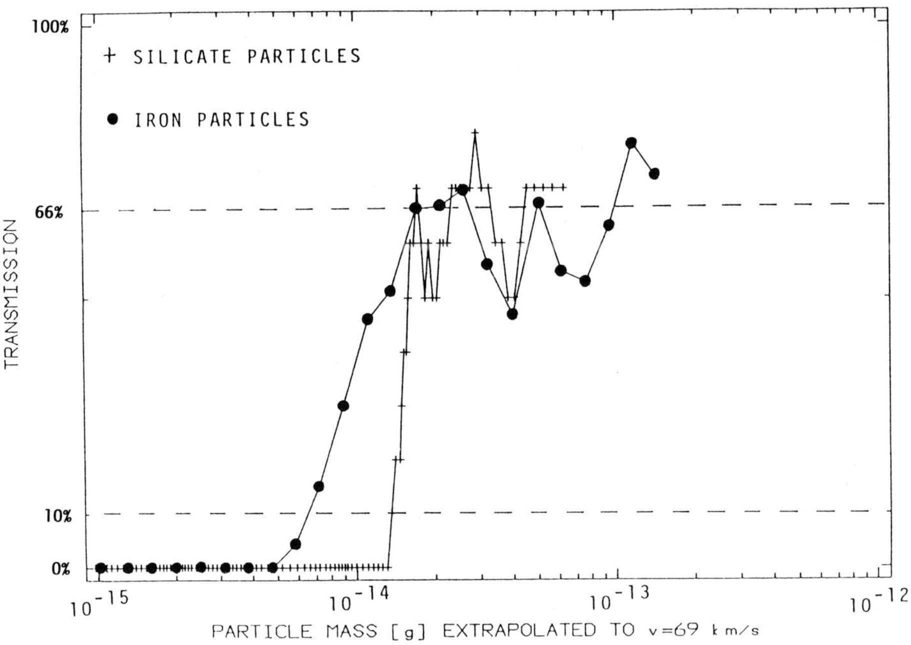

Fig. 6. Transmission of the B-sensor-film for iron and silicate particles at the encounter speed of Giotto. The ratio between sensitive area and total area of the sensor is 0.66

Table 2. Penetration limits of different films. All values are extrapolated to the impact speed v given in the first row. Additionally an extrapolation to the thickness T of the different films and to a particle density ρ of $1\,g\,cm^{-3}$ according to the Pailer-Grün-formula was made. The respective extrapolation is indicated by the letters behind each value

Particle material and density ρ	$v = 69\,km\,s^{-1}$ $T = 2.5\,\mu m$	$v = 78\,km\,s^{-1}$ $T = 2.0\,\mu m$	$v = 78\,km\,s^{-1}$ $T = 0.6\,\mu m$
Iron, $7.8\,g\,cm^{-3}$	$6.5\,10^{-15}\,g\,(v)$	$2.8\,10^{-15}\,g\,(v, T)$	$1.8\,10^{-16}\,g\,(v, T)$
Silicate, $2.5\,g\,cm^{-3}$	$1.3\,10^{-14}\,g\,(v)$	$5.6\,10^{-15}\,g\,(v, T)$	$3.7\,10^{-16}\,g\,(v, T)$
Carbon, $2.2\,g\,cm^{-3}$	$>1.8\,10^{-14}\,g\,(v)$	$>7.8\,10^{-15}\,g\,(v, T)$	$>5.1\,10^{-16}\,g\,(v, T)$
$1\,g\,cm^{-3}$	$3.2\,10^{-14}\,g\,(v, \rho)$	$1.4\,10^{-14}\,g\,(v, T, \rho)$	$7.0\,10^{-16}\,g\,(v, T, \rho)$

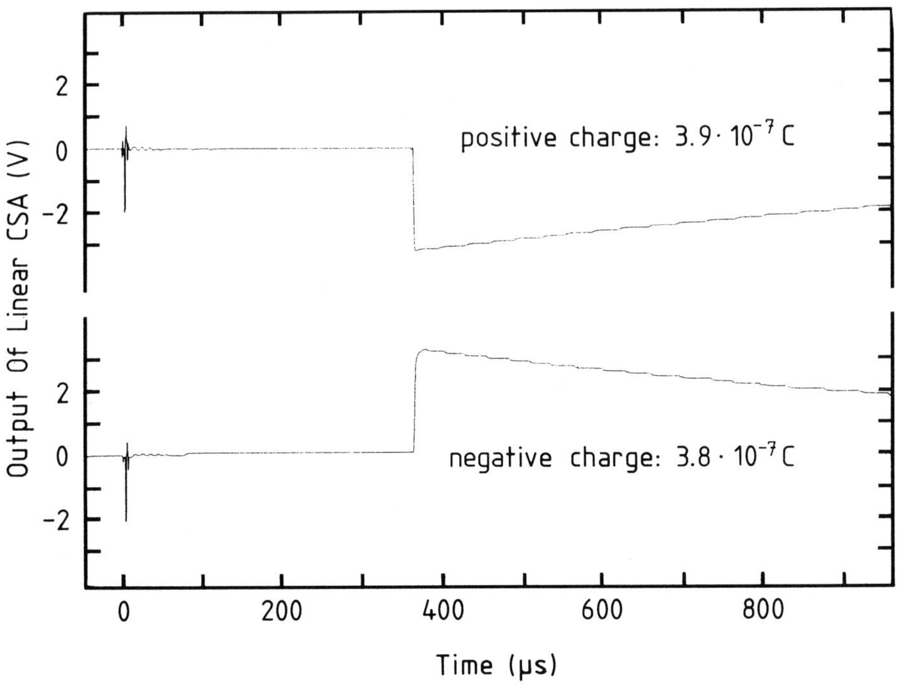

Fig. 7. Charge signal measured with charge sensitive amplifiers with a linear response. The shot was triggered at 0 μs. After 360 μs the projectile ($m = 5.5\,10^{-8}$ g, $v = 9.6\,km\,s^{-1}$ hit the target of the IPM sensor A and positive and negative impact charges were detected

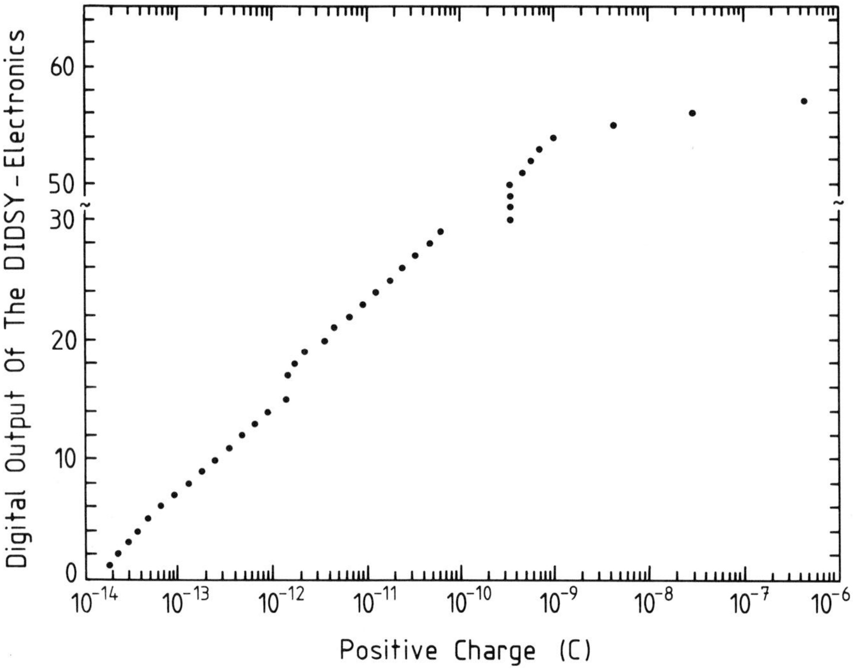

Fig. 8. Calibration of the DIDSY-IPM electronics. The IPM charge sensitive amplifiers have a nonlinear response if the impact charge is larger than 10^{-9} C. Electronic simulation tests showed, that for pulses with short risetimes the logarithmic charge intervals for digits 31, 49 and 50 are compressed. The digit 16 and the digits 32 to 48 were not used

to 90% of the amplitude) at impact speeds between $3\,\mathrm{km\,s^{-1}}$ and $12\,\mathrm{km\,s^{-1}}$

b) the CSA used in the DIDSY-IPM electronics (EM) which are of the same type as those of the flight model. With the results from impact simulation experiments and from electronic stimulation of the DIDSY-IPM electronics (EM and FS) it was found that the output of these CSA depends on the risetime of the charge signal. For charges lower than 10^{-10} C the conversion factor is independent of the risetime. For observed impact charge signals the response of the electronics (EM) is shown in Fig. 8. The nonlinear behaviour can be seen if the charge to be measured is larger than 10^{-9} C.

Acknowledgements. C. Nehls and G. Schäfer helped us a lot in processing the data and operating the accelerator in Heidelberg. The experiments in Munich were strongly supported by Prof. E. Igenbergs, Dr. U. Weishaupt, M. Rott and S. Aigner.

References

Göller, J.R., Grün, E.: 1985, in R.H. Giese, P. Lamy, eds., *Properties and Interaction of Interplanetary Dust*, Dordrecht, D. Reidel Publ. Co, p 113

Göller, J.R., Grün, E., Maas, D.: 1986, 20th ESLAB Symp. ESA SP-250, Vol. III, p 333

Grard, R.J.L., Gombosi, T., Sagdeev, R.Z.: 1986, ESA SP-1066, p 49

Grün, E., Bussoletti, E., Minafra, A., Kuczera, H., McDonnell, J.A.M.: 1984, *Adv. Space Res.* **4**, p 291

Grün, E., McDonnell, J.A.M.: 1983, *Adv. Space Res.* **2**, 183

Iglseder, H.: 1986, Doctoral thesis, Technical University Munich, FRG

Lange, G., Grün, E., Maas, D., Göller, J.R., Kissel, J., McDonnell, J.A.M., Zarnecki, J.C., Bussoletti, E., Grard, R.J.L.: 1986, 20th ESLAB Symp. ESA SP-250, Vol. II, p 191

Maas, D., Göller, J.R., Grün, E., Lange, G., Aigner, S., Igenbergs, E., Rott, M. and Weishaupt, U.: 1986, 20th ESLAB Symp., ESA SP-250, Vol. III, p 337

Mazets, E.P., Aptekar, R.L., Golenetskii, S.V., Guryan, Yu.A., Dyachkov, A.V., Ilyinskii, V.N., Panov, V.N., Petrov, G.G., Savvin, A.V., Sagdeev, R.Z., Sokolov, I.A., Khavenson, N.G., Shapiro, V.D., Shevchenko, V.I.: 1986a, *Nature* **321**, p 276

Mazets, E.P., Sagdeev, R.Z., Aptekar, R.L., Golenetskii, S.V., Guryan, Yu.A., Dyachkov, A.V., Ilyinskii, V.N., Panov, V.N., Petrov, G.G., Savvin, A.V., Sokolov, I.A., Frederiks, D.D., Khavenson, N.G., Shapiro, V.D., Shevchenko, V.I.: 1986b, 20th ESLAB Symp. ESA SP-250, Vol. II; p 3

McDonnell, J.A.M., Alexander, W.M., Burton, W.M., Bussoletti, E., Clark, D.H., Evans, G.C., Evans, S.T., Firth, J.G., Grard, R.J.L., Grün, E., Hanner, M.S., Hughes, D.W., Igenbergs, E., Kuczera, H., Lindblad, B.A., Mandeville, J.-C., Minafra, A., Reading, D., Ridgeley, A., Schwehm, G.H., Stevenson, T.J., Sekanina, Z., Turner, R.F., Wallis, M.K., Zarnecki, J.C.: 1986a, ESA SP-1077, p 85

McDonnell, J.A.M., Kissel, J., Grün, E., Grard, R.J.L., Langevin, Y., Olearczyk, R.E., Perry, C.H., Zarnecki, J.C.: 1986b, 20th ESLAB Symp. ESA SP-250, Vol. II p 25

Pailer, N., Grün, E.: 1980, *Planet. Space Sci.* **28**, p 321

Smirnov, V.N., Vaisberg, O.L., Anisimov, S.: 1986, 20th ESLAB Symp. ESA SP-250, Vol. II, p 195

Vaisberg, O.L., Smirnov, V.N., Gorn, L.S., Iovlev, M.V., Balikchin, M.A., Klimov, S.I., Savin, S.P., Shapiro, V.D., Shevchenko, V.I.: 1986a, *Nature* **321**, 274

Vaisberg, O., Smirnov, V., Omelchenko, A.: 1986b, 20th ESLAB Symp. ESA SP-250, Vol. II, p 17

Zarnecki, J.C., McDonnell, J.A.M., Alexander, W.M., Burton, W.M., Hanner, M.S.: 1986, 20th ESLAB Symp. ESA SP-250, Vol. II, p 185

Astron. Astrophys. 187, 699–706 (1987)

ASTRONOMY AND ASTROPHYSICS

Dust in comet P/Halley from Vega observations

E.P. Mazets[1], R.Z. Sagdeev[2], R.L. Aptekar[1], S.V. Golenetskii[1], Yu.A. Guryan[1], A.V. Dyachkov[2], V.N. Ilyinskii[1], V.N. Panov[1], G.G. Petrov[1], A.V. Savvin[1], I.A. Sokolov[1], D.D. Frederiks[1], N.G. Khavenson[2], V.D. Shapiro[2], and V.I. Shevchenko[2]

[1] A.F. Ioffe Physical-Technical Institute, Academy of Sciences, SU-194021 Leningrad, USSR

[2] Space Research Institute, Academy of Sciences, Profsoyuznaya 84/32, SU-117810 Moscow, USSR

Received February 11, accepted April 6, 1987

Summary. Direct measurements of the dust particle spatial and mass distributions in comet Halley have been carried out on Vega spacecraft over the mass range $\sim 10^{-16}$–10^{-6} g with the SP-2 dust particle detector consisting of a set of acoustic and impact plasma sensors. These measurements have revealed a large-scale, time-variable structure of the dust coma, the position and characteristics of its boundaries, and a strongly pronounced angular directivity of dust emission from the cometary nucleus. An unexpectedly high concentration of tiny dust particles with masses below 10^{-14} g and strong systematic variations of the shape of the particle mass distribution as a function of distance to the cometary nucleus have been observed. The mean dust production rate by the nucleus has been estimated at (10–13) 10^6 g s^{-1} (Vega 1) and (5–7) 10^6 g s^{-1} (Vega 2).

Key words: comet Halley – dust – spatial and mass distributions

1. Instrumentation and observations

The SP-2 experiment on Vega spacecraft was intended for studying the spatial and mass distributions of dust particles in the cometary coma over the mass range $\sim 10^{-16}$–10^{-6} g (Balebanov et al., 1985, p. 145; Mazets et al., 1986). Covering such a broad mass range was made possible by using sensors of two types, namely, impact plasma and acoustic, as well as by properly choosing their effective areas. To increase the measurement reliability, the sensors and the associated electronics were duplicated. Actually, the SP-2 instrument on board each of the Vega spacecraft comprises two independent impact plasma sensors, two independent acoustic sensors, and two identical electronic systems for signal processing, in-flight calibration, acquisition and storage of data, and their transfer to the spacecraft telemetry system.

The impact plasma sensors had four integral count rate channels each, uniformly spaced along a log mass scale and covering the ranges of $\sim 10^{-16}$–10^{-13} and $\sim 10^{-14}$–10^{-11} g, respectively. The acoustic sensors provided with 16 integral count rate channels each covered identical mass ranges of $\sim 3\,10^{-13}$–10^{-6} g. Each channel recorded the number of particle impacts per second with masses exceeding a threshold level, $N(>m)$.

Send offprint requests to: E.P. Mazets.

The flyby trajectory of the Vega 2 spacecraft through the comet in the cometocentric frame and the pattern of the dust particle flux distribution measured along the spacecraft trajectory are presented in Fig. 1. This trajectory can be specified by the coordinates of closest approach to the nucleus $R_0(R_{0x} = 7.10\,10^3$, $R_{0y} = 2.68\,10^3$, $R_{0z} = 2.63\,10^3$ km) and the components of the relative velocity vector $V = 76.8$ km s^{-1} ($V_x = 30.53$, $V_y = -69.44$, $V_z = -11.88$ km s^{-1}). The results of the dust flux measurements are referred to the time t reckoned from the instant of closest approach $t = 0$. The corresponding distances to the cometary nucleus can be found as $R = R_0(1 + (t/t_0)^2)^{1/2}$, where the parameter $t_0 = R_0/V \approx 105$ s. Vega 1 passed through the comet on a very similar trajectory ($R_{0x} = 7.83\,10^3$, $R_{0y} = 2.48\,10^3$, $R_{0z} = 3.40\,10^3$ km; $V_x = 28.64$, $V_y = -72.73$, $V_z = -12.95$ km s^{-1}; $t_0 \approx 112$ s). In the course of measurements, the number of particle impacts with masses exceeding the channel threshold is counted in each pulse height channel at 1 s-intervals along the trajectory:

$$N(>m) = F(>m)ST = n(>m)VST, \tag{1}$$

where $n(>m)$ is the number density, $F(>m) = n(>m)V$ the cumulative flux, S the effective sensor area, $T = \frac{63}{64}$ s the accumulation time in each measurement. For the acoustic sensors, dead time corrections were introduced at channel count rate $N(>m) \gtrsim 50\,\mathrm{s}^{-1}$. For the fast impact plasma sensors such corrections were not needed.

The impact plasma and acoustic sensors were calibrated in terms of collected charge Q and momentum P transferred in an impact, respectively. Calibration with hypervelocity particles (~ 80 km s^{-1}) was not carried out. The calibration relations used for preliminary mass m estimation from measured charge Q or momentum P were close to those accepted for the similar sensors on Giotto in accordance with McDonnell et al. (1981). Namely, the assumed ratio of the charge of the impact plasma to dust particle mass is $Q/m = 10^3$ C g^{-1}, and the momentum enhancement factor due to mass ejection at impact is $\varepsilon = 10$.

The pulse height channels of our instrument were arranged on the mass scale so as to provide simultaneous dust particle flux measurements by the sensors of the two types in overlapping mass regions. A possibility was thus provided to introduce appropriate corrections into the initial calibration of the impact plasma and acoustic sensors by comparing the measurements and thus to match with a sufficiently high accuracy their sensitivities.

Figure 2 illustrates dust particle mass spectra obtained on Vega 2 at different distances from the cometary nucleus. One

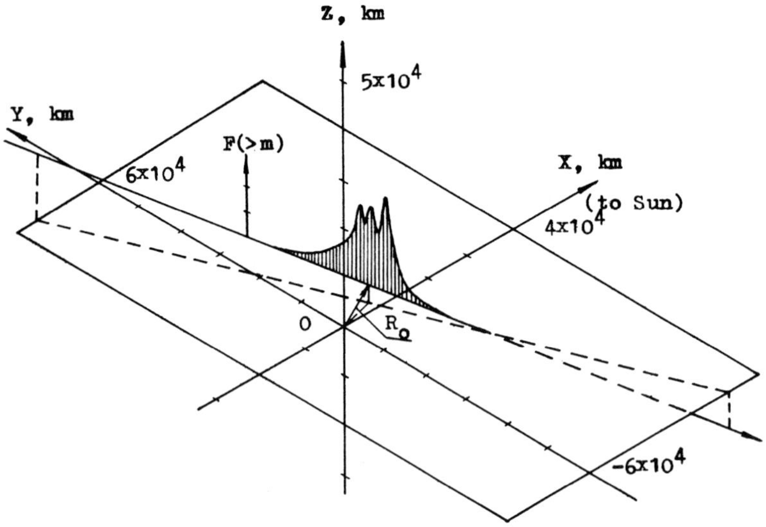

Fig. 1. Vega 2 trajectory through comet Halley and dust particle flux distribution along the path. Cometocentric frame is with X-axis directed towards the Sun and Z-axis antiparallel to orbital angular momentum

clearly sees a discrepancy to exist between the data of the impact plasma and acoustic sensors which is due to the accepted calibration and which results in an inconsistency between the two measured spectra. One has to introduce appropriate corrections to the calibration which would shift the mass scale channels so as to make the spectra recorded by the impact plasma and acoustic sensors merge continuously.

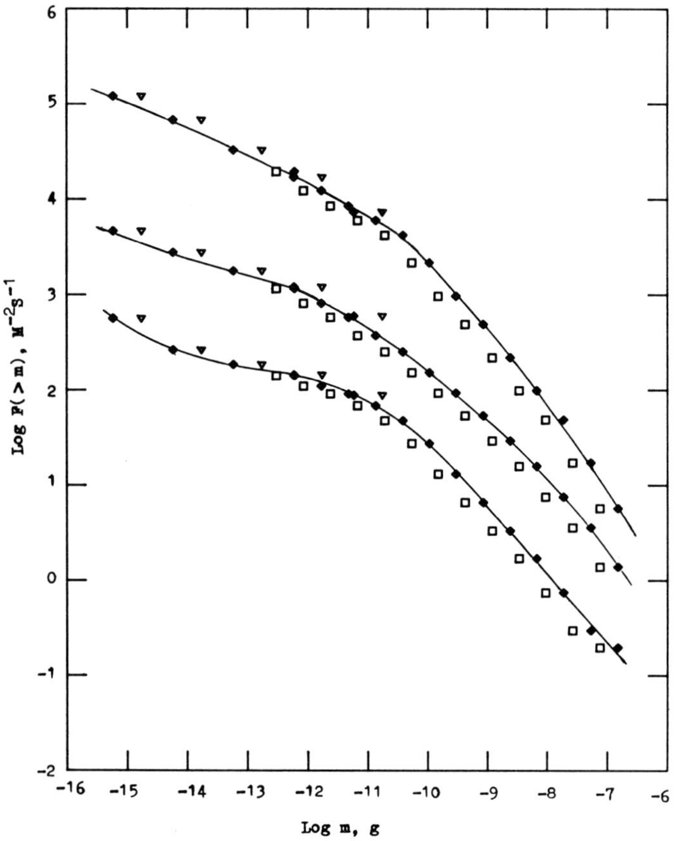

Fig. 2. Cumulative dust particle flux $F(>m)$ from Vega 2. Distances to the nucleus: 8.2 10^3 km (upper graph); 2.4 10^4 km (middle graph); 8.0 10^4 km (lower graph). Calibration corrections for the acoustic and impact plasma sensors required to match their sensitivities: $Q/m = 10^3\,\mathrm{C/g^{-1}}$, $\varepsilon = 10$ (open symbols); $Q/m = 3\,10^3\,\mathrm{C\,g^{-1}}$, $\varepsilon = 5$ (filled symbols)

The sensitivity of the acoustic sensors appears overestimated at $\varepsilon = 10$, and that of the impact plasma sensors, assuming $Q/m = 10^3\,\mathrm{C\,g^{-1}}$, underestimated. Therefore in the present stage of analysis, compromise values of $\varepsilon = 5$ and $Q/m = 3\,10^3\,\mathrm{C\,g^{-1}}$ have been accepted. As evident from Fig. 2, the corrected values of the calibration coefficients result in a good agreement between the two sets of data. This matching of calibration for the impact plasma and acoustic sensors is valid for all measurements on Vega 1 and Vega 2. Note, however, that these data provide a clear indication that our present day understanding of the dependences of the impact-induced plasma, light and mechanic effects on the particle and target parameters, particularly for hypervelocity impacts, is far from being complete. Therefore the calibration coefficients and the forced assumption of their relative and absolute values being constant over such a broad range of dust particle mass variation is certainly only a rough approximation.

Further consideration of our results requires mentioning certain malfunctions in the operation of the instruments during the mission. As evidenced by in-flight calibrations, up to the encounter with the comet both instruments functioned properly. On Vega 2, all the sensors fully maintained their sensitivity. On Vega 1, the sensitivity of the acoustic sensors decreased, their calibration shifting accordingly by one channel approximately. It is known that some instruments on the Vegas were damaged by impacting large dust particles during the encounter. On Vega 2, the SP-2 acoustic sensor plate suffered damage 251 s after closest approach to the nucleus, thus stopping transmission of useful information from these sensors. On Vega 1 acoustic sensors survived. However damage of other instruments on this spacecraft increased the level of electrical interference detectable by SP-2. As a result, starting with $t = 325$ s before closest approach the noise count rate in the first, highest sensitivity channels of the acoustic sensors reached abruptly saturation. Due to specific features of the electronic circuit of dead time formation which is common for the first four acoustic channels, this part of information (obtained in these four channels) was lost. The level of interference in the first, most sensitive channel of the impact plasma sensors likewise turned out to be enhanced. The analysis

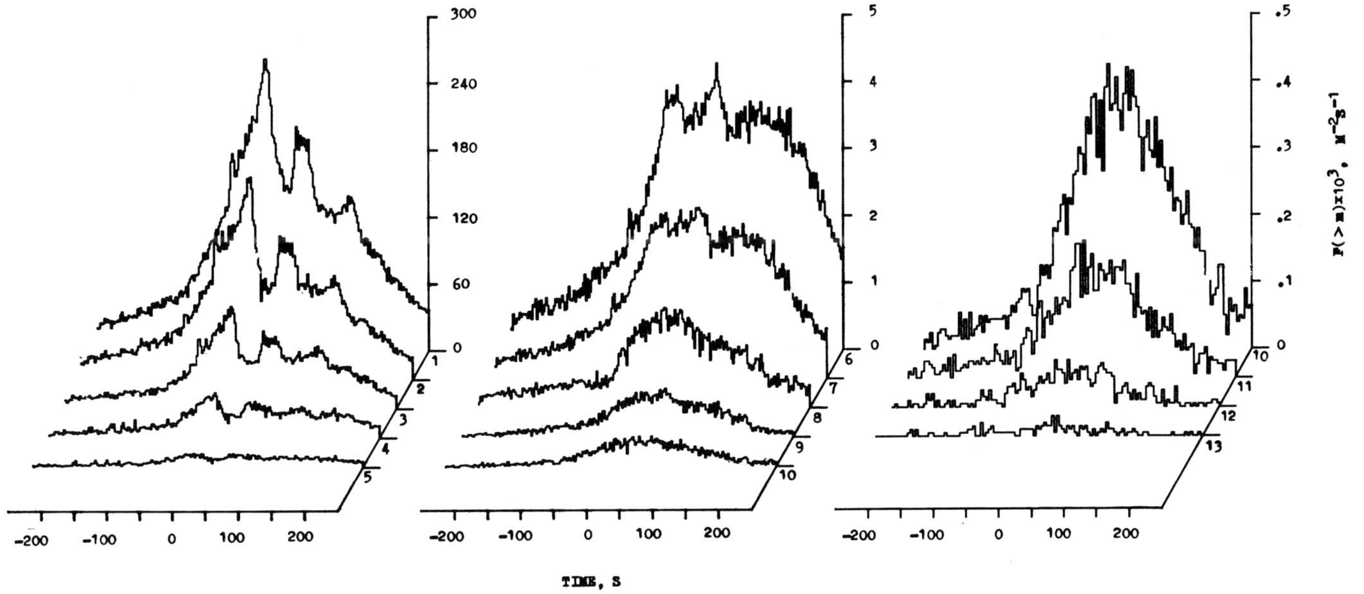

Fig. 3. Cumulative dust flux distribution in the inner dust coma, $R < 2\,10^4$ km, Vega 1. Curves 1–13: $m > 6.0\,10^{-16}$, $6.0\,10^{-15}$, $6.0\,10^{-14}$, $6.0\,10^{-13}$, $6.0\,10^{-12}$, $1.2\,10^{-10}$, $3.3\,10^{-10}$, $9.2\,10^{-10}$, $2.6\,10^{-9}$, $7.3\,10^{-9}$, $2.1\,10^{-8}$, $5.8\,10^{-8}$, and $1.7\,10^{-7}$ g, respectively

of the corresponding records has not yet been completed, and the information obtained in these channels is not considered here.

By their nature and the time of onset, however, all these malfunctions did not affect noticeably the volume and quality of the information obtained on the dust coma.

2. Results

Figure 3 shows the time profiles of the cumulative flux $F(>m)$ measured on the Vega 1 in the inner region of the dust coma. The data obtained over a more extended part of the trajectory starting from distances to the cometary nucleus of $\sim 10^5$ km are presented in Fig. 4. The results of the measurements were averaged over intervals determined by the count rate statistics, particularly for the larger particles and greater separations from the nucleus. For the constant velocity of the impacting particles, $V = \text{const}$, these plots characterized also integral spatial number density distributions, $n(>m)$. The set of $F(>m)$ data for a given moment of time provides the cumulative mass distribution for the corresponding point of the spacecraft trajectory. Some examples of such spectra are given in Fig. 5. Note that if a part of the spectrum is approximated by a power law, $F(>m) \propto m^{-\alpha}$, then these relationships will characterize also the differential distribution of dust particles as a function of $\ln m$:

$$dN/d(\ln m) = \alpha n(>m) = \frac{\alpha}{V} F(>m)\,. \tag{2}$$

Analogous data obtained during the Vega 2 encounter with comet are shown in Figs. 6 to 8.

3. Discussion

It is widely accepted that dust particles are carried away from the nuclear surface by the gas flux and accelerated by it to a terminal velocity depending on the size a and density ρ_d of the particles, $V_d \propto (\rho_d a)^{-1/2}$ (Gombosi et al., 1983). The dust number density drops with increasing distance from the nucleus R as R^{-2}, and its distribution along the trajectory should scale as $\propto (1 + t/t_0)^2)^{-1}$. This free expansion of the dust particles is, however, counteracted by solar radiation pressure which drives the particles in the antisolar direction with an acceleration

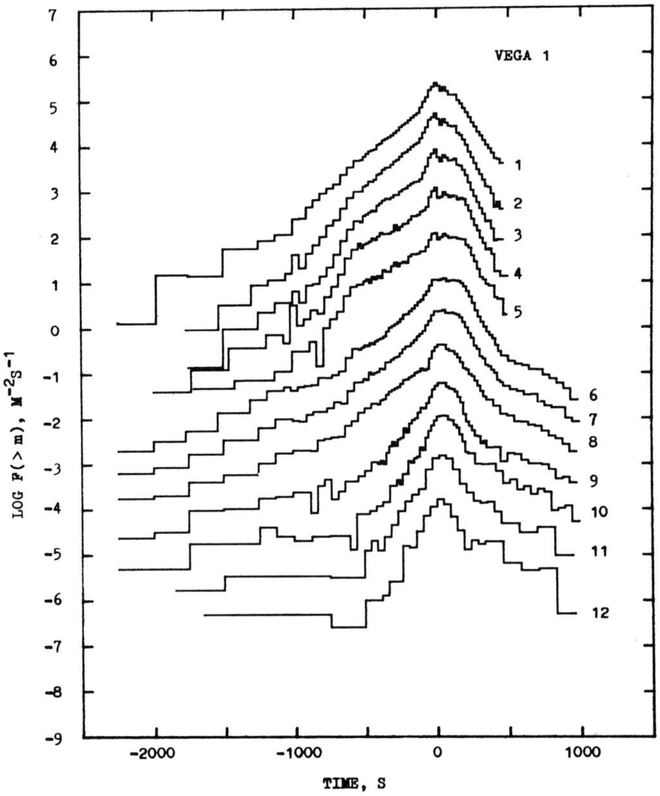

Fig. 4. $F(>m)$ distribution along Vega 1 trajectory. Curves 1–12: $m > 6.0\,10^{-16}$, $6.0\,10^{-15}$, $6.0\,10^{-14}$, $6.0\,10^{-13}$, $6.0\,10^{-12}$, $1.2\,10^{-10}$, $3.3\,10^{-10}$, $9.2\,10^{-10}$, $2.6\,10^{-9}$, $7.3\,10^{-9}$, $2.1\,10^{-8}$, and $5.8\,10^{-8}$ g, respectively. Vertical scale refers to curve 1, the others being shifted downward half-an-order of magnitude each

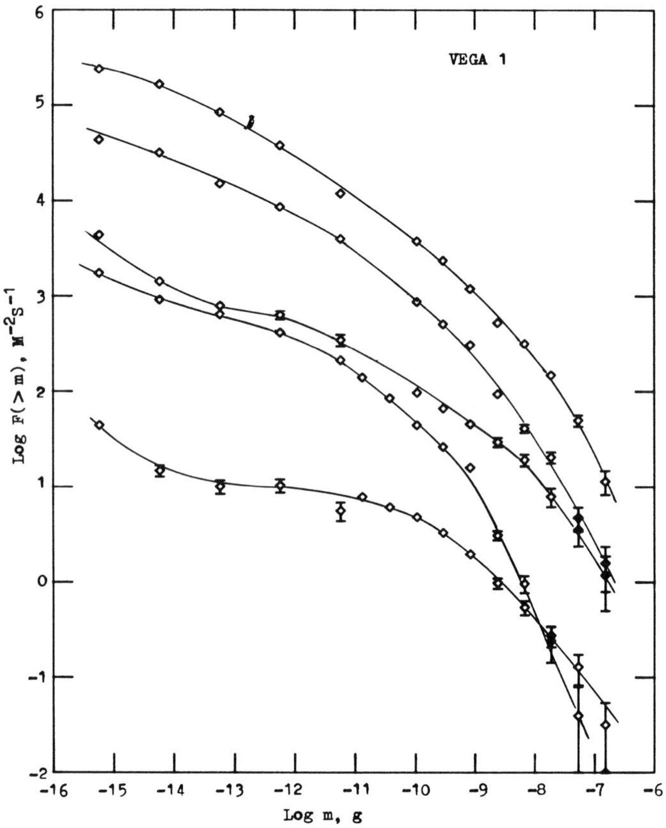

Fig. 5. Cumulative flux distribution $F(>m)$ at different distances from the nucleus, Vega 1 (from top downward): 8.9 10^3 km; 1.2 10^4 km; 3.7 10^4 km (after the closest approach; 8.0 to 4.0 10^4 km; 1.8 10^5 to 8.0 10^4 km

$$g = \frac{3LQ_{\rm pr}}{16\pi c R_s^2 \rho_d a}, \tag{3}$$

where L is the solar luminosity, R_s the comet-to-Sun distance, $Q_{\rm pr}$ the radiation pressure efficiency, and c the velocity of light.

As a result, the dust particles are driven into the cometary tail, the trajectories of their motion filling a paraboloid of revolution $y^2 + z^2 = p^2 - 2px$ with the parameter $p = V_d^2/g$. The apex of the paraboloid lies at a distance $x_0 = p/2$ from the cometary nucleus.

The light scattering cross section depends in a complicated way on the size of the dust particles, their optical characteristics and wavelength λ. When averaged over the solar radiation spectrum, the efficiencies $Q_{\rm pr}$ for particles of different composition have a broad maximum at sizes $a \sim \lambda$ (Balebanov et al., 1985, p. 13) or, accordingly, at masses $m \sim 10^{-14}$–10^{-11} g. In this region the dependence of parameter p on particle size a becomes weak suggesting that the boundaries of the dust coma for particles of different mass should be about the same.

Generally, the data obtained agree quite satisfactorily with this simple model. One also finds, however, remarkable features which reflect finer details in the mechanisms of dust particle emission and motion.

In Figs. 4 and 7 the envelope of the dust paraboloid manifests itself clearly in the spatial distributions of the dust flux number density as steep humps at distances of a few tens of thousand km. For particles with a mass $\sim 10^{-14}$–10^{-11} g whose size is comparable with the wavelengths of solar radiation the boundaries of the dust coma do indeed coincide very closely. Within a relatively short distance interval, $\Delta R/R \approx 0.2$, the spatial number density of such particles drops by a factor $\sim 10^1$–10^2. Part of the particles escape from the coma. The number of such particles is small, the count rate statistics providing only a rough estimate of the dependence of their number density on cometocentric distance. These features wash out gradually as one moves towards both larger and smaller particles. The number density of the smallest and larger particles varies more smoothly with distance from the nucleus. Accordingly, at large distances from the cometary nucleus the particle mass distribution acquires a peculiar shape (Figs. 5 and 8) depleted in particles of intermediate mass, 10^{-14}–10^{-10} g.

This spatial distribution of the particle flux reflects undoubtedly a decrease of the radiation pressure efficiency $Q_{\rm pr}$ and

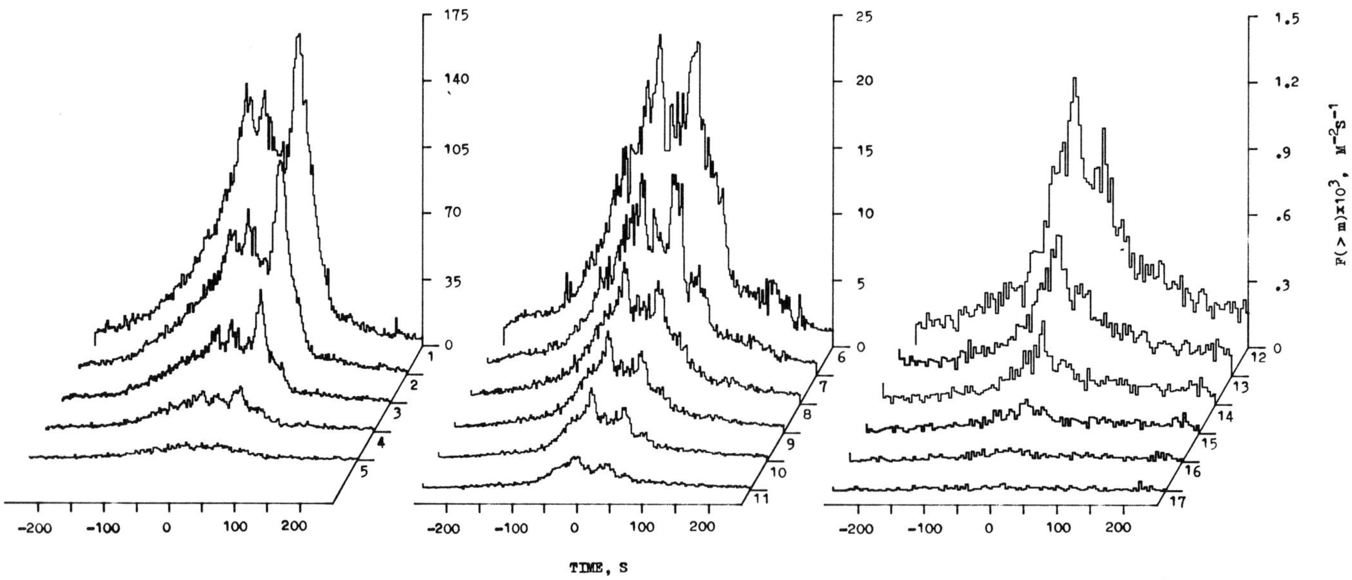

Fig. 6. Cumulative dust flux distribution in the inner dust coma, $R < 2\,10^4$ km, Vega 2. Curves 1–17: $m >$ 6.0 10^{-16}, 6.0 10^{-15}, 6.0 10^{-14}, 6.0 10^{-13}, 6.0 10^{-12}, 6.5 10^{-13}, 1.8 10^{-12}, 5.2 10^{-12}, 1.5 10^{-11}, 4.1 10^{-11}, 1.2 10^{-10}, 3.3 10^{-10}, 9.2 10^{-10}, 2.6 10^{-9}, 7.3 10^{-9}, 2.1 10^{-8}, and 5.8 10^{-8} g, respectively

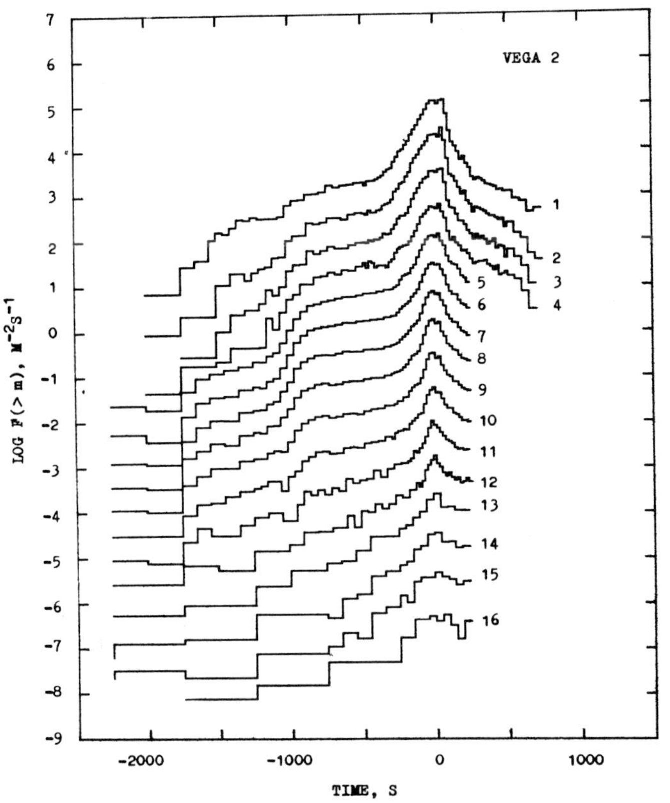

Fig. 7. $F(>m)$ distribution along Vega 2 trajectory. Curves 1–16: $m > 6.0\,10^{-16}$, $6.0\,10^{-15}$, $6.0\,10^{-14}$, $6.0\,10^{-13}$, $6.5\,10^{-13}$, $1.8\,10^{-12}$, $5.2\,10^{-12}$, $1.5\,10^{-11}$, $4.1\,10^{-11}$, $1.2\,10^{-10}$, $3.3\,10^{-10}$, $9.2\,10^{-10}$, $2.6\,10^{-9}$, $7.3\,10^{-9}$, $2.1\,10^{-8}$, and $5.8\,10^{-8}$ g, respectively. Vertical scale refers to curve 1, the others being shifted downward half-an-order of magnitude each

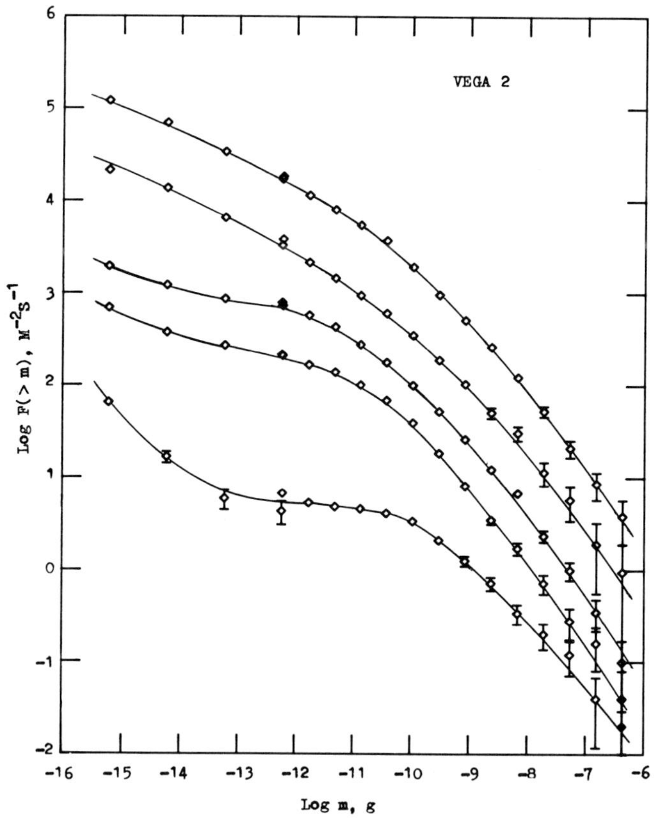

Fig. 8. Cumulative flux distribution $F(>m)$ at different distances from the nucleus, Vega 2 (from top downward): $8.1\,10^3$ km; $1.4\,10^4$ km; 5.7 to $2.7\,10^4$ km; 9.5 to $5.7\,10^4$ km; $1.7\,10^5$ to $9.5\,10^4$ km

its strong dependence on size a of the smallest and largest particles. For such particles the magnitude of parameter p does not remain constant resulting in a washing out of the clearly pronounced paraboloid boundary.

The washing out of the coma boundary within $\Delta R/R \approx 0.2$ is most probably associated with a spread in the terminal velocities to which the particles are accelerated by the gases sublimating from the nucleus.

In the inner regions of the coma, the dust particle number density varies with cometocentric distance R substantially stronger than as R^{-2}. This argues persuasively for the dust emission from the nucleus being anisotropic. The most active are apparently those regions of the nuclear surface where the direction to the Sun is close to local zenith.

The expected distribution of the dust particle number density as a function of the observer's coordinates (x, R) can be calculated under different assumptions concerning the angular dependence of dust emission from the nucleus, $f(\theta)$, and the velocity V_d spread which is accounted for by the distribution function $f_1(p)$. Examples of such a calculation for $f(\theta) = f(\cos\theta)$ and

$$f_1(p) = \frac{1}{(2\pi)^{-1/2}\,\Delta p} \exp\left[-(p - p_0)^2/2(\Delta p)^2\right] \tag{4}$$

are presented in Fig. 9 demonstrating a good agreement with the observational data. For the Vega 1 data obtained in a narrow mass interval Δm from $6.0\,10^{-13}$ to $6.0\,10^{-12}$ g, the calculation was carried out with $f(\theta) = 0.5 + 0.5\cos\theta$, $p_0 = 3.7\,10^4$ km, $\Delta p = 6.5\,10^3$ km. For the Vega 2 data, for mass interval from $4.1\,10^{-11}$ to $1.2\,10^{-10}$ g and on the unbound trajectory until closest approach, $t < 0$, the calculated curve corresponds to $f(\theta) = 0.2 + 0.8\cos\theta$, $p_0 = 5.0\,10^4$ km, $\Delta p = 6.0\,10^3$ km. For the outbound part of the trajectory, $t > 0$, $f(\theta) = 0.2 + 0.8\cos^2\theta$, $p_0 = 6.5\,10^4$ km, $\Delta p = 1.0\,10^4$ km.

It should be stressed that these examples are not a rigorous solution of the inverse problem, i.e. of determination of the source parameters from the distribution of the emission. They clearly show, however, that the data obtained permit formulation of such a problem, and its solution can yield valuable information on the sources and mechanisms of dust emission.

It is not surprising that the above estimates of the coma boundaries and the shape of $f(\theta)$ are somewhat different for different parts of the trajectory, for particles of different mass and on different days. The mean value and scatter of the dust particle terminal velocities, as well as the particle angular distribution should undoubtedly depend strongly on the orientation and illumination by the Sun of a cometary nucleus of such an irregular shape (Sagdeev et al., 1986). One should bear in mind that an instrument traversing the coma intercepts particles emitted from the nucleus at different times and at different levels of activity. Differences in particle age may be as high as a few days, which

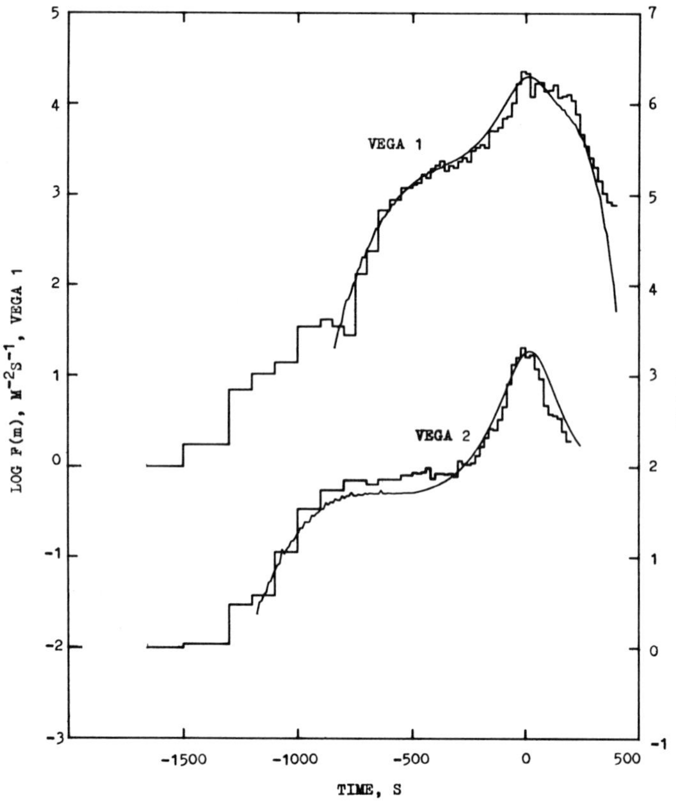

Fig. 9. $F(\Delta m)$ fluxes for narrow mass intervals $\Delta m = 6.0\,10^{-13}$–$6.0\,10^{-12}$ g (Vega 1) and $4.1\,10^{-11}$–$1.2\,10^{-10}$ g (Vega 2). Smooth curves: calculations (see text)

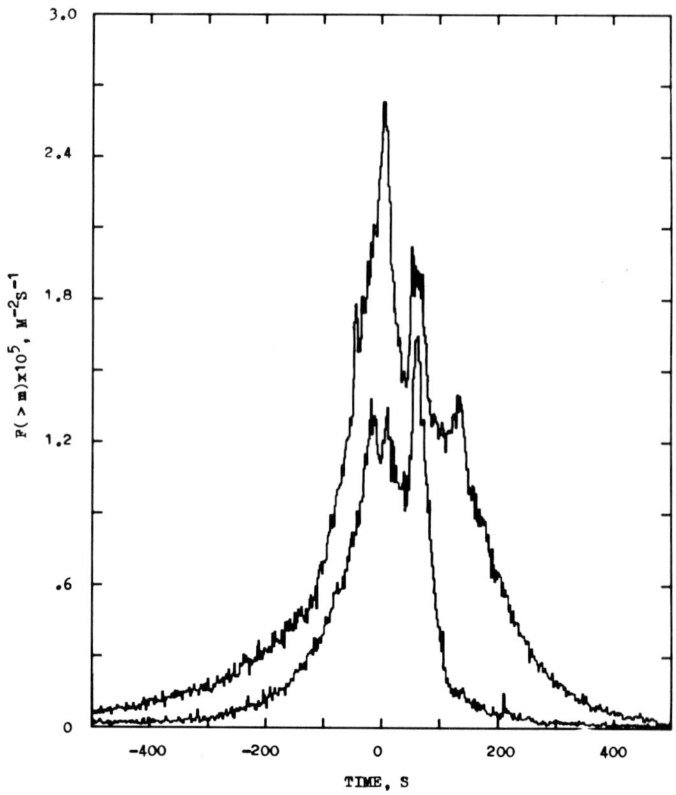

Fig. 10. $F(>m)$ distributions along Vega 1 (upper curve) and Vega 2 (lower curve) trajectories, $m > 6\,10^{-16}$ g

exceeds the nuclear spin period.

On the whole, the above estimates of the dust coma size correspond to the expected value of parameter p_0 for $R_S = 0.8$ AU, $Q_{pr} \sim 1$ and values of V_d calculated by Gombosi et al. (1983). They agree also quite well with the results of ground based Halley observations performed during this period (Sekanina and Larson, 1986; West et al., 1986).

Measurements carried out close to the cometary nucleus reveal a large scale structure of the inner dust coma. Figure 10 compares dust flux distributions sampled along the Vega 1 and Vega 2 trajectories. As already pointed out, in the cometocentric frame the two trajectories are very close. Therefore to the same times t correspond similar angular positions of the two spacecraft relative to the comet-Sun line. The rotation of the nucleus is believed to be prograde (Sagdeev et al., 1986). As follows from Fig. 10, in the period preceding the Vega 1 flyby the evening part of the nuclear surface was substantially more active. The intense peaks in the time profiles evidence the existence on the nuclear surface of strong local dust sources. Particularly remarkable is the very strong peak at $t \approx 50$ s observed during the Vega 2 flyby. In Fig. 11 the Vega 2 data for the inner region of the coma are presented for individual particle mass intervals as obtained from differences of the count rates in neighboring integral channels. As the dust particle mass increases, this maximum is seen to shift to the left, towards smaller distances from the nucleus. This effect, as already suggested by Vaisberg et al. (1986), is most probably associated with a velocity dispersion of the dust particles emitted in a narrow jet from a small active spot on the surface of the rotating nucleus. However the position of the other

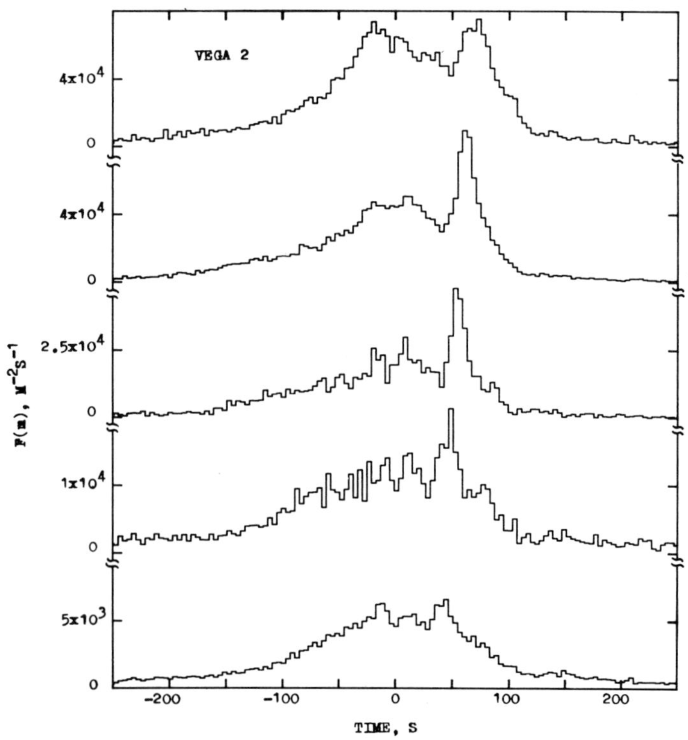

Fig. 11. Shift of the dust jet position as observed from Vega 2 in different particle mass intervals (from top downward): $6.0\,10^{-16}$–$6.0\,10^{-15}$; $6.0\,10^{-15}$–$6.0\,10^{-14}$; $6.0\,10^{-14}$–$6.0\,10^{-13}$; $6.0\,10^{-13}$–$5.2\,10^{-12}$; $5.2\,10^{-12}$–$4.1\,10^{-11}$ g

peaks recorded in the flybys of both Vega 1 and Vega 2 does not exhibit such dependence on particle mass. One does not observe here this mass-dependent effect of particle velocity dispersion. These peaks are possibly associated with explosive dust ejections for which the assumptions of particle acceleration by a stationary gas flux are not valid.

The dust particle mass distributions vary considerably along the flyby trajectory (Figs. 5 and 8). The least variable is the distribution of large particles of $m > 10^{-10}$ g. Variations in the spectra are characteristic of smaller particles. At small distances from the nucleus the relative number density of small dust particles of $m \sim 10^{-15}$–10^{-12} g is high. The relative contribution of such particles to the mass distribution decreases with increasing distance from the nucleus. The slope of the spectrum decreases in the region of small masses. It is possibly due to a faster vaporization of small particles consisting of volatiles. Near the boundary of the main dust coma the shape of the dust particle mass spectrum, as already pointed out, is strongly affected by differences in the velocity V_d and acceleration g and, hence, by the variability of parameter $p = V_d^2/g$ for the largest and smallest particles. At large ($R \approx 10^5$ km) distances from the nucleus one observes a clearly pronounced separation of the dust into two groups, namely, of the smallest ($m < 10^{-14}$ g) and fairly large ($m > 10^{-10}$ g) particles.

During the Vega 1 flyby, the shape of the mass spectrum exhibits the gradual variation with distance from the nucleus. Within the interval 250 to 1000 s before closest approach the spectrum reveals a considerable drop in the number density of particles of $m > 10^{-10}$ g (Figs. 4 and 5). This feature reflects apparently variations in the source spectrum during the dust ejection, a few days before the observations.

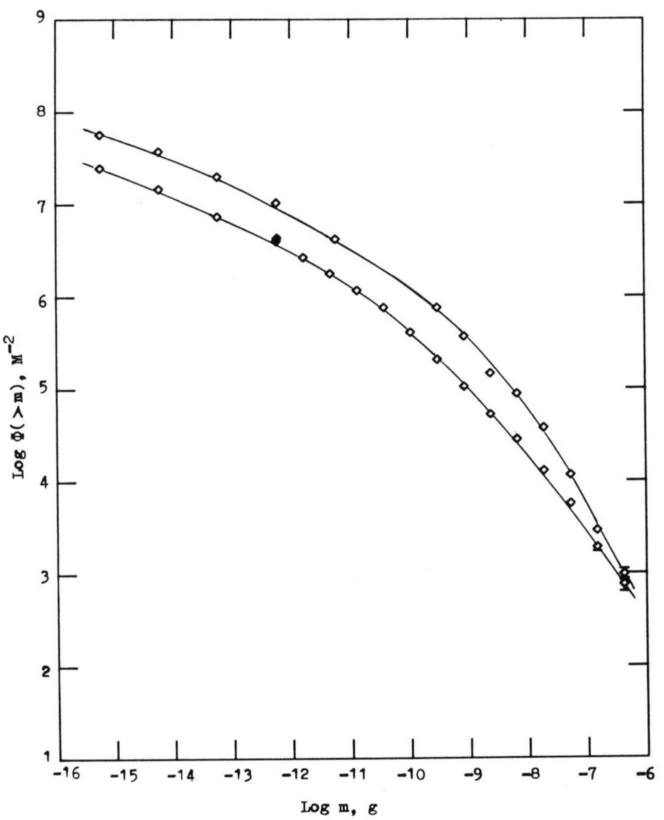

Fig. 12. Dust particle fluences from Vega 1 (upper curve) and Vega 2 (lower curve)

Figure 12 displays net particle mass distribution obtained on Vega 1 and Vega 2 for the inner coma. Actually, these distributions characterize the total number of particles of different mass intercepted by a unit area during the spacecraft flybys. The mean mass spectra of the dust particles will hopefully be useful in the analysis of photometric observations of the comet made in the reflected solar light continuum.

Determination of the angular pattern of dust emission from the nucleus from the dust number density distribution along the flyby trajectory has provided a sufficiently reliable estimate of the mean dust production rate by the nucleus, $\dot{Q}_d$. These estimates support our preliminary data (Mazets et al., 1986) and yield $\approx (10$–$13)\,10^6\,\mathrm{g\,s^{-1}}$ (Vega 1) and $\approx (5$–$7)\,10^6\,\mathrm{g\,s^{-1}}$ (Vega 2).

4. Conclusion

On the whole, direct studies of comet Halley's dust environment support the concept developed by Whipple (1950) of a cometary nucleus being an aggregate of ices and solid particles, as well as the major guidelines of the cometary models constructed when preparing the missions (Balebanov et al., 1985, pp. 9, 13 and 21; Divine et al., 1986).

The principal unexpected results of measurements on Vegas have been the discovery of a high number density of the smallest dust particles, of a very narrow angular pattern in the dust emission from the nucleus, and of the variability in the dustt particle mass distribution at different distances from the nucleus. Further analysis of the data obtained may yield valuable data on the relative content of particles differing in composition and characterized by essentially different radiation pressure efficiencies Q_{pr}, on the process of dust acceleration by the gas stream and the dependence of the terminal velocities on the gas to dust mass ratio, on possible explosive processes of dust ejection, and on the characteristics of strong local sources on the surface of the cometary nucleus.

References

Balebanov, V.M., Skuridin, G.A., Vorontsova, E.V., Bassolo, V.S. (eds.): 1985, Venus-Halley Mission, Louis-Jean, Gap

Divine, N., Fechtig, H., Gombosi, T.I., Hanner, M.S., Keller, H.U., Larson, S.M., Mendis, D.A., Newburn Jr., R.L., Reinhard, R., Sekanina, Z., Yeomans, D.K.: 1986, *Space Sci. Rev.* **43**, 1

Gombosi, T.I., Szegö, K., Gribov, B.E., Sagdeev, R.Z., Shapiro, V.D., Shevchenko, V.I., Cravens, T.E.: 1983, Central Res. Inst. Physics, preprint No KFRI-1983-70, Budapest

Mazets, E.P., Aptekar, R.L., Golenetskii, S.V., Guryan, Y.A., Dyachkov, A.V., Ilyinskii, V.N., Panov, V.N., Petrov, G.G., Savvin, A.V., Sagdeev, R.Z., Sokolov, I.A., Khavenson, N.G., Shapiro, V.D., Shevchenko, V.I.: 1986, *Nature* **321**, 276

McDonnell, J.A.M., Grün, E., Evans, G.C., Turner, R.F., Firth, J.G., Carey, W.C., Kuczera, H., Alexander, W.M., Clark, D.H., Grard, R.J.L., Hanner, M.S., Hughes, D.W., Igenbergs, E., Lindblad, B.A., Mandeville, J.C., Schwehm, G., Sekanina, Z.: 1981, Proc. Int. Meeting on the Giotto Mission, Noordwijkerhout, Netherlands (ESA SP-169), p. 61

Sagdeev, R.Z., Szabó, F., Avanesov, G.A., Cruvellier, P., Szabó, L., Szegö, K., Abergel, A., Balazs, A., Barinov, I.V., Bertaux, L.-J., Blamont, J., Detaille, M., Demarelis, E., Dul'nev, G.N.,

Endröszy, G., Gardos, M., Kanyo, M., Kostenko, V.I., Krasikov, V.A., Nguen-Trong, T., Nyitrai, Z., Reny, I., Rusznyak, P., Shamis, V.A., et al.: 1986, *Nature* **321**, 262
Sekanina, Z., Larson, S.M.: 1986, *Nature* **321**, 357
Vaisberg, O.L., Smirnov, V.N., Gorn, L.S., Iovlev, M.V., Balikchin, M.A., Klimov, S.I., Savin, S.P., Shapiro, V.D., Shevchenko, V.I.: 1986, *Nature* **321**, 274
West, R.M., Pedersen, H., Monderen, P., Vio, R., Grosbøl, P.: 1986, *Nature* **321**, 363
Whipple, F.L.: 1950, *Astrophys. J.* **111**, 375

Astron. Astrophys. 187, 707–711 (1987)

ASTRONOMY AND ASTROPHYSICS

Properties of dust in comet P/Halley measured by the Vega-2 three-channel spectrometer

V. A. Krasnopolsky [1], V. I. Moroz [1], A. A. Krysko [1], A. Y. Tkachuk [1], G. Moreels [2], J. Clairemidi [2], J. P. Parisot [2], M. Gogoshev [3], and T. Gogosheva [3]

[1] Space Research Institute, Profsoyusnaya 84/32, SU-117810 Moscow GSP-7, USSR
[2] University of Besançon, F-25000 Besançon, France
[3] Central Laboratory for Space Research, Observatory Stara Zagora, BG-6000 Stara Zagora, Bulgaria

Received February 12, accepted May 18, 1987

Summary. Optical properties of cometary dust exhibit no substantial variations at distances from 100 to 15 000 km from the nucleus. Dust ejection can be approximated by $\cos\varphi$ where φ is the angle with the sun-nucleus axis. Phase functions of dust scattering at 10 wavelengths from 328 nm to 1.4 µm are obtained and discussed. Dust reflectivity is nearly constant in the range 0.3 µm to 1.8 µm. Vega-2 in situ measurements are analysed and compared with optical data. Phase functions calculated for Mie and Fresnel spheres fail to reproduce adequately the obtained data. More promising seems to be the approximation of rough spheres. Dust density was evaluated as 0.3 g cm^{-3} from comparison of the in situ data with our measurements. Dust velocities are approximated by the expression $V_g/V_d = 1 + (\varrho r/b)^{1/2}$ with $b = 4$ g µm cm^{-3} obtained in our measurements. Total dust production rate equals 10 tons^{-1} and $Q_d/Q_g \approx 0.3$ on 1986 March 9.

Key words: comet P/Halley – dust

1. Introduction

The experimental set-up and some preliminary results have been discussed by Krasnopolsky et al. (1986a–c) and Moreels et al. (1986). The measured spectra, especially in the inner coma, exhibit very strong background radiation caused by dust scattering. We will consider this effect in detail.

Figure 1 reproduces in-flight sensitivities of both channels of the instrument with wavelength ranges of 0.95–1.9 µm and 270–710 nm. The instrument was calibrated in Toulouse at five selected wavelengths (Hg 313, 405, 436, 546 and Zn 636 nm) against a photomultiplier with known sensitivity and in Moscow against a tungsten filament lamp with known brightness and a screen of fresh magnesium oxide. Both calibrations in the visible range coincide fairly well. To control the in-flight sensitivity, spectra of a filament iodine lamp inside the instrument were measured before every scanning sequence. These spectra show a decrease in sensitivity of 270–710 nm channel by a factor of 1.6 as compared with the preflight calibration and an increase for the 0.95–1.9 µm channel by a factor 10.5 (due to cooling of the detector).

Send offprint requests to: V. A. Krasnopolsky

One of the difficulties encountered in the interpretation of the experimental data was a deviation of the instrument optical axis from the initial direction parallel to the TV camera. In the previous papers (Krasnopolsky et al., 1986b, c) we tried to find both the deviation and the spatial distribution of dust in the comet starting with our infrared data only. Better accuracy can be achieved with TV images (Sagdeev et al., 1986). Figure 2 shows an isophote with the brightness factor $R = \pi I/I_\odot = 1.75\ 10^{-4}$ (I is the measured brightness and $I_\odot$ is the solar flux) measured with the 750–850 nm filter and a 10-s exposure at the Vega 2 flyby session in 1986 March 9. The image was obtained at a distance of $4.15\ 10^5$ km from the nucleus at a phase angle of 110°. The filter range is suitable for separation of dust scattering. Among the isophotes measured during the Vega-2 flyby session the one shown in Fig. 2 is of maximal size. Only the narrow angle camera with the image size of $26' \times 35'$ and 512×512 pixels was used in this session, and a mode of 128×128 pixels was utilized at its beginning. Our preliminary analysis (Krasnopolsky et al., 1986b) showed that angular distribution of dust ejection can be approximated by $\cos\varphi$, where φ is the angle with the sun-nucleus axis. This gives a circle-shaped isophote at a phase angle close to 90°. Actually the mean deviation of the measured isophote from the circle equals 6% for φ varying between 0° and 75°, which covers our observations. The better approximation of the isophote is given by

$$f(\varphi) = [1 + 0.13\sin(x - 0.5)]\exp[-((x + 0.6)/4.94)^2] + 0.09\,,$$

where $x = \varphi/15°$, which provided an accuracy of 2% in the whole range of φ, $\pm 180°$. Assuming that the spatial distribution of the brightness factor at 0.8 µm is given by $R = 0.088\,f(\varphi)/r$, where r is the distance in kilometers between a point of observation and the nucleus, and taking into account the phase dependence described below we have found by the least square technique an angular deviation of the visible channel optical axis of 70′ from the nucleus in sunward direction and 30′ in the transverse direction. These results are used for further interpretation.

Both our experiment and the TV camera give similar isophotes measured at different wavelengths. Contribution of emission bands does not exceed several percents at distances less than 500 km from the nucleus for four main filters of the TV camera centered at 550, 700, 800, and 920 nm. Therefore the similarity of isophotes presumes similarity of optical properties of dust ejected at different angles up to $\varphi = 75°$. We do not consider here small-scale variations.

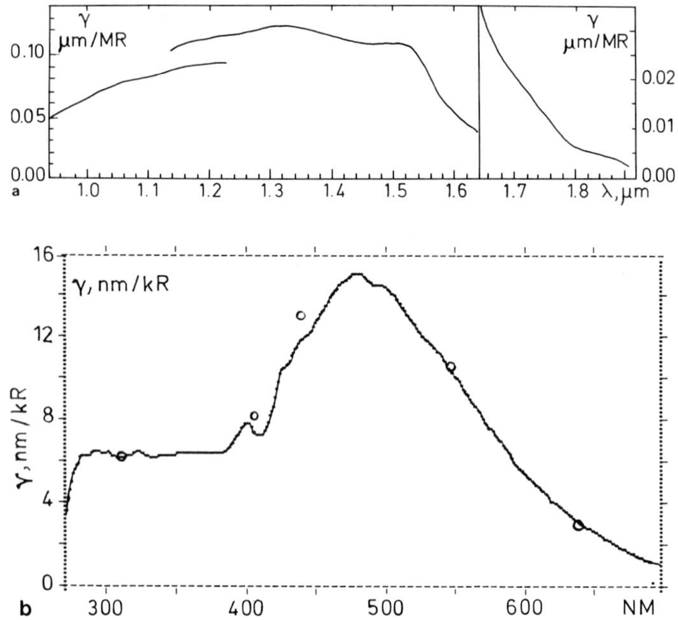

Fig. 1. In-flight sensitivities of **a** the 0.9–1.9 μm and **b** 270–710 nm channels based on calibrations made in Moscow (curves) and in Toulouse (circles). Sensitivity γ per telemetric unit is shown along the Y-axis. The wavelength range 1.13 to 1.23 μm was measured twice by different parts of the circular variable filter

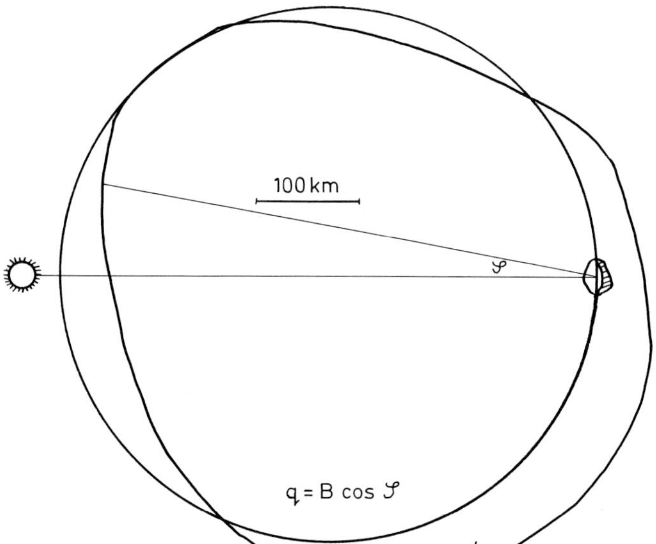

Fig. 2. Shape of the isophote with a brightness factor of 1.75 10^{-4} at $\lambda = 0.8$ μm measured by the Vega-2 TV camera on March 9, 1986. Circle is the representation of the isophote based on preliminary interpretation of our infrared measurements. Mean deviation of the measured isophote from the circle equals 6% for φ varying between 0° and 75°

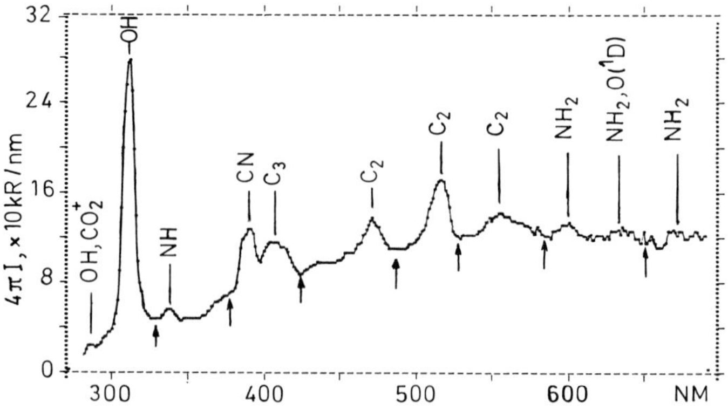

Fig. 3. Spectrum of comet Halley measured at a distance of 1000 km from the nucleus at an angle of 38° with the sun-nucleus axis. Arrows indicate the wavelengths chosen for an analysis of dust scattering

2. Phase functions of dust scattering

A typical spectrum in the range 270–700 nm measured at a distance of 1000 km from the nucleus is shown in Fig. 3. The mean near infrared spectrum was given by Krasnopolsky et al. (1986a).

First of all, we choose wavelengths where dust scattering dominates and contribution of gaseous emissions is negligible: 328, 375, 425, 525, 585, 650 nm in the 270–710 nm channel and almost the whole range of the infrared channel. Then 30 spectra measured at a spacecraft-comet distance of 5.1 10^5 km and a phase angle of 110° are compared with those at 6 10^4 km and 106°.5. The optical axis impact parameter equals 15 000 km and 1600 km respectively, while the phase angles are almost the same. With the exception of 328 nm, intensity ratios at the selected wavelengths prove to be almost the same at all wavelengths with a mean deviation of 4.5%. Larger deviation at 328 nm may be attributed to contamination by gaseous emissions. Thus optical properties of cometary dust do not vary up to 16 000 km. Results discussed below refer to the data obtained at distances of 100–1500 km from the nucleus, i.e. for a spacecraft-nucleus distance smaller than 6 10^4 km.

The Vega-2 flyby observations covered the phase angles Φ (sun-nucleus-spacecraft angle) from 110° to 14°. Phase dependence of scattering is known to be an important characteristic of dust. To obtain these dependences at different wavelengths the measured intensities were corrected for the slit size and its position relative to the nucleus, for the total amount of dust along the optical axis which is proportional to $(1 + \sin \Phi)$, for the angular distribution of the dust ejection given by $\cos \varphi$, and the position of the optical axis in the sun-nucleus-spacecraft plane. The derived phase functions at 1.4 and 0.49 μm are shown in Fig. 4. The scatter of points is about 10% and 2.5% at 1.4 and 0.49 μm, respectively, and some oscillations seen may be real. These functions are given as the products of volume scattering coefficient at a distance of 1000 km from the nucleus in the sunward direction and phase function $\kappa(\theta)$ with $\int \kappa \, d\Omega = 4\pi$ (Ω is the solid angle, the integral is taken over the sphere).

In spite of a three-fold difference in the wavelengths, the phase functions obtained nearly coincide in absolute values, show very low variations (~10%) and some common features which exceed the error bars: the minimum near $\Phi \simeq 80°$ and the increase by a factor of 1.25 towards 95°. Such functions for other wavelengths can be derived using intensity ratios measured with much better accuracy. They are shown in Fig. 5. Most of the oscillations of these curves are real, especially those measured by the visible channel of the instrument which provides simultaneous measurements of the whole spectrum with high signal-to-noise ratio.

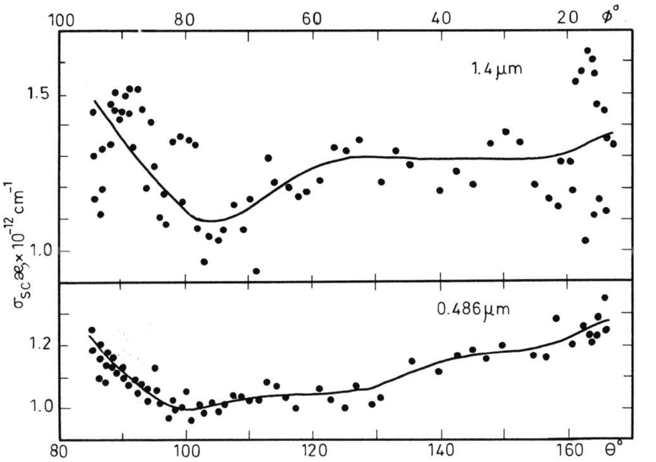

Fig. 4. Phase function of dust scattering at 1.4 µm and 486 nm; σ_{sc} is the volume scattering coefficient at 1000 km from the nucleus in the direction of the Sun and κ is the phase function normalized to 4π. θ is the scattering angle and $\varphi = 180° - \theta$ is the phase angle

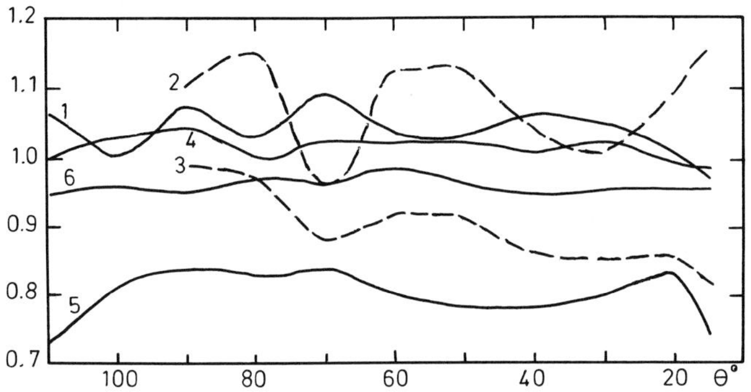

Fig. 5. Mean intensity ratios versus scattering angle θ: 1.4/1.0 µm (*1*), 1.0/0.65 µm (*2*), 1.4/0.328 µm (*3*), 0.65/0.486 µm (*4*), 0.65/0.328 µm (*5*), 0.486/0.425 µm (*6*)

3. Spectrum of dust scattering

Low phase variations of the intensity ratios made it possible to use their mean values to form the mean spectrum of dust scattering in the comet shown in Fig. 6. For the sake of comparison, we also present in the figure the ground-based *J-H-K* photometry (Moroz and Taranova, 1986; Hanner, 1986) and the IUE data (Feldman et al., 1986). *J-H-K* colors were translated into brightness factors using the usual technique with subsequent normalization. The measured spectrum exhibits a nearly constant reflectivity of cometary dust from 0.3 to 1.7 µm, Accuracy of our calibration is about 5 to 10%, therefore some of the features of the spectrum may not be real. Jewitt and Meech (1986) found a decrease of 20% in the dust reflectivity from 650 to 450 nm in Halley's comet at 1.6 AU. Our measurements a 0.83 AU do not show this behavior.

Spectra of dust scattering are influenced by the size distribution of dust particles, the wavelength dependence of their complex refractive index $n - ik$, and the shape and roughness of their surface. If $n - ik$ is constant in the spectral range under consideration, shape and surface roughness are independent of size, and a particle sizes spectrum is given by $n = br^{-u}$ (for $0 < r < \infty$) where n is the number density of particles with a

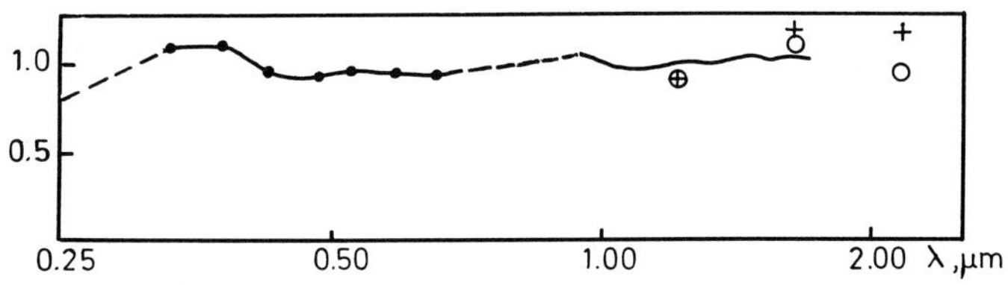

Fig. 6. Spectrum of dust scattering in the comet (solid line, our measurements). Also given are the data of *J-H-K* photometry (Moroz and Taranova, 1986; Hanner, 1986, open circles and crosses) and the IUE measurements (Feldman et al., 1986) at 250–300 nm. Our measurements are averaged for phase angles from 14 to 110°

radius exceeding r; then the reflectivity of dust medium is given by $R = A\lambda^{-u+2}$ (Krasnopolsky and Krysko, 1979). If scattering by very large particles which is independent of wavelength dominates, then contributions of different size ranges to scattered radiation are proportional to $r^2 (dn/d\ln r) \sim r^{2-u}$, thus $u \simeq 2$ in ranges giving maximal contributions. This can be easily checked, using the particle mass spectrum measured in situ (Sect. 4). Some small features of the spectra can be caused either by variation of $n - ik$ or by interference of light scattered by surface inhomogeneities (Schiffer, 1985).

4. Comparison with in situ measurements

Out of the three instruments onboard Vega-2 that measuremd mass distribution of dust particles only one (Masets et al., 1986) gives data in the mass range from 10^{-11} to 10^{-6} g which is the main contributor to light scattering. To compare with the optical measurements, we chose a high-quality mass spectrum measured at a distance of 8100 km from the nucleus. This spectrum was measured at a position closest to the sunward direction ($\varphi \simeq 30°$) and to the nucleus, thus minimizing the effect of solar radiation deceleration. We approximated this spectrum in the log-log coordinates by the five-order polynomial. The subsequent differentiation of this polynomial gives the spectrum shown in Fig. 7. It is extrapolated from 10^{-6} to 10^{-3} g with $\alpha = u/3 = 1$ according to the Giotto data (McDonnel et al., 1986).

Figure 8 shows contribution of different mass ranges to the total intensity of scattered light

$$I(m_r, \varrho) = \int_0^x i(x) \frac{dn}{dx} dx \Big/ \int_0^\infty i(x) \frac{dn}{dx} dx .$$

Here ϱ is the dust density, $x = 2\pi a/\lambda$, a is the particle radius, $i(x)$ is the intensity of scattering given by the Mie formulas for a refractive index of $m_r = 1.4 - 0.05\,i$. This figure shows that the mass range from 10^{-12} to 10^{-6} g is responsible for 80% of the scattered light. The slope of the mass spectrum in this range is $\alpha = 2/3$ (Fig. 7), as discussed in Sect. 3.

5. Simulation of the measured phase functions

Our attempts to reproduce the measured phase functions by an appropriate choice of the refractive index and the dust density fail to give satisfactory results. Figure 9 shows the results of calculations with the refractive index $1.39 - 0.035\,i$ recommended by Mukai et al. (1986) on the basis of their polarization measurements of comet Halley. Large particles with $x \simeq 100$ yield

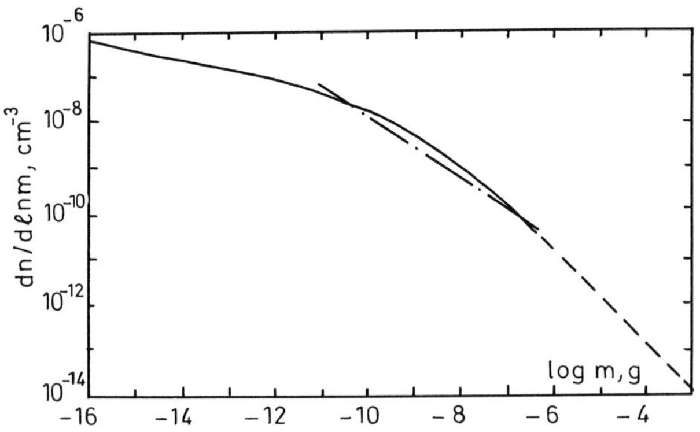

Fig. 7. Differential mass spectrum of dust particles derived from the Vega-2 in situ measurements by Mazets et al. (1986) at 8100 km. Dashes is the extrapolation with the slope $\alpha = 1$ given by the Giotto measurements (McDonnel et al., 1986). Dot-dash line shows a slope of the spectrum $\alpha = 2/3$ in the range of maximal optical response

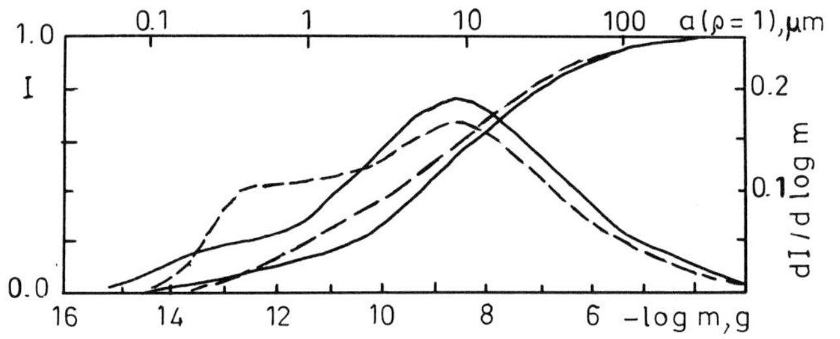

Fig. 8. Integral and differential contributions of different mass ranges to scattered radiation:

$I(m_r, \varrho) = \int_0^x i(x) \frac{dn}{dx} dx \int_0^\infty i(x) \frac{dn}{dx} dx$

calculations are made for $\lambda = 0.65\,\mu m$, $\varphi = 70°$, complex reflective index 1.4–0.05i, dust density $\varrho = 0.1$ and 1 g/cm^{-3} (solid and dashed curves)

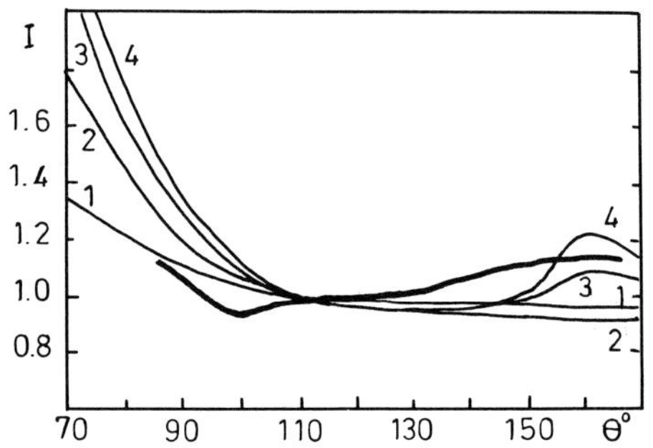

Fig. 9. Comparison of the measured scattering function of the cometary dust at 0.65 μm (solid line) with functions calculated by the Mie formulas with the refractive index $m_r = 1.39 - 0.035i$ and $\varrho = 0.1$ g cm^{-3} (3) and 1 g cm^{-3} (4), with that for Fresnell spheres with the same m_r (2) and with $m_r = 1.8$ and 1.6–0.5i (the results coincide, curve *1*)

the maximal scattering effect (Fig. 7). It follows that the representation of dust by Fresnel reflecting spheres is similar to that by Mie formulas with the exception of a small peak at $\varphi \simeq 20°$. It is possible to choose the value of the refractive index which provides better fit to the measured curve, but such a value would disagree with polarization measurements by Mukai et al. (1986) and our, less accurate, measurements, the latter giving, nevertheless, practically the same value of the refractive index. Evidently, cometary dust particles are not spherical and neither the Fresnel nor Mie formulas could be used to describe the results of measurements. A better representation is given by rough spheres (Mukai and Mukai, 1985; see also Perrin and Lamy, 1985) with the roughness parameter $\sigma \simeq 2$. Comparison of our results with the laboratory studies of particles scattering by microwave analog and laser techniques (Zerull, 1985; Weiss-Wrana, 1983; Hanner et al., 1981) gives the best coincidence for low absorbing compact irregular particles with $x = 19$, refractive index 1.45 $- 0.05i$ (Hanner et al., 1981). This coincidence may be accidental due to the large variation of x in our observations, but it should be noted that scattering by large particles is independent of x.

6. Density of dust particles

Let us now derive the density of dust particles from the known mass spectrum (Fig. 7), reflectivity (Fig. 5) and optical properties. This will be done in the approximation of very large particles with a size-independent phase-function (see Sect. 2):

$$\sigma\kappa = \kappa\pi \int_0^\infty r^2 \frac{dn}{dr} dr = \kappa\pi \varrho^{-2/3} I(\varrho = 1),$$

where $I(\varrho = 1)$ is the integral evaluated with $\varrho = 1$ g cm^{-3}. The dust paraboloid apex is 2.5 10^4 km (Mazets et al., 1986) and solar radiation deceleration is accounted for by a factor of 0.82 at 8100 km and the fact that $\varphi \simeq 30°$ by a factor $(\cos 30°)^{-1} = 1.15$. Numerical substitution gives $\varrho \simeq 5\kappa^{3/2}$ g cm^3. Laboratory measurements of fluffy particles with $x \simeq 33$ and refractive index of $1.45 - 0.05i$ and $1.65 - 0.25i$ gave $\kappa \simeq 0.15$ (Hanner et al., 1981), thus $\varrho \simeq 0.3$ g cm^{-3}. The accuracy of this value is hardly better than a factor of 2 due to uncertainties in κ and in the in situ data.

7. Velocities of dust particles

Detection of the same jet at the water vapor 1.38 μm band and in the in situ dust observations and the analysis of the time delay between the appearances of the jet make it to possible derive gas and dust velocities, V_g nad V_d, as functions of their ratio (Krasnopolsky et al., 1986b, c). This analysis shows that V_d is almost independent of the adopted value of V_d/V_g and varies from 0.53 to 0.63 km s^{-1} for V_d/V_g varying from 1.2 to 2 respectively. Dust measurements (Mazets et al., 1986) show that the dust jet was most pronounced for masses of particles $m = 10^{-10\pm2}$ g. This jet does not reveal velocity dispersion of particles with different masses while some other jets observed do and the theory and common sense predict. Following Probstein (1968) we assume

$$V_g/V_d = 1 + (\varrho r/b)^{1/2}.$$

Then our calculations of $\varrho \simeq 0.3$ g cm^{-3} and $m \simeq 10^{-10}$ g favor $b = 4$ g cm^{-3} μm. This formula is in fair agreement with the more sophisticated calculations of Gombosi et al. (1986) which are based on hydrodynamic modelling. We adopt $V_g = 0.8$ km s^{-1} according to Krankowsky et al. (1986) and Krasnopolsky et al. (1986b, c).

8. Total production rate of dust

It is possible to calculate the total dust production rate of the comet with the known mass spectrum of dust at 8100 km (Fig. 6), dust velocities and spatial distribution of dust ejection approximated by $\cos\varphi$. In this approximation the number density of particles and their total production rate are given by

$$n = B \cos\varphi / VR^2; \quad Q = \pi B$$

(Krasnopolsky et al., 1986b). Similar expressions in the case of dust mass spectrum yield

$$Q = \pi R^2 \alpha \cos^{-1}\varphi \int_0^\infty \frac{dn}{d\ln m} V(m)\, dm.$$

Here $R = 8.1\ 10^8$ cm is the distance from the nucleus, $\alpha = 0.82$ is the reduction factor to account for the solar radiation deceleration. For $m > 10^{-6}$ g it is assumed that $dn/d\ln m \sim m^{-1}$ according to the Giotto results (McDonnel et al., 1986). A change of the slope to $dn/d\ln m \sim m^{-0.5}$ for $m > 10^{-5}$ g proposed by McDonnel et al. (1986) is based on a preliminary estimate of a total effective mass impact on Giotto equal to 1.9 g and should be thus revised with a corrected value of 0.32 g (Edenhofer et al., 1986). The integral does not diverge with velocity decreasing with m, and total dust production rate of comet Halley was equal to 10 ton/s during the Vega-2 flyby (March 9, 1986).

9. Discussion

The simultaneous optical and in situ measurements of cometary dust provided good opportunity to study its properties. As a result, spatial distribution of dust, its phase functions, spectra of scattered light, density, velocities and total production rate have been derived to complement mass spectra and some other in situ data. It should be noted, however, that the interpretation of some data, e.g. phase functions, is more difficult than in the case of Mie scattering. This is due, first, to difficulty in describing scattering by iregular particles. Second, cometary dust is a mixture of several kinds of particles with different properties, and in such cases mean values are not always a good representation.

References

Edenhofer, P., Buschert, H., Porsche, H., Bird, M.K., Volland, H., Brenkle, J.P., Kursinsky, E.R., Mottinger, N.A., Stelzried, S.T.: 1986, 20th ESLAB Symp., ESA SP-250, Vol. **II**, p. 215

Feldman, P.D., Festou, M.C., A'Hearn, M.F., Arpigny, C., Butterworth, P.S., Gosmovici, C.B., Danks, A.V., Grilmozzi, R., Jackson, W.M., McFadden, L.A., Patriarchi, P., Schleicher, D.C., Tozzi, G.P., Wallis, M.K., Weaver, H.A., Woods, T.N.: 1986, paper presented at the Symposium "New Insights in Astrophysics", July 1986, London

Gombosi, T.I., Nagy, A.F., Cravens, T.E.: 1986, *Rev. Geophys.* **20**, 667

Hanner, N.S.: 1986, *Adv. Space Res.* **5**

Hanner, M.S., Giese, R.H., Weiss, K., Zerull, R.: 1981, *Astron. Astrophys.* **104**, 42

Jewitt, D.C., Meech, K.J.: 1986, 20th ESLAB Symp., ESA SP-250, Vol. **II**, p. 47

Krankowsky, D., Lämmerzahl, P., Herworth, X., Weweries, J., Eberhardt, P., Dolder, U., Norrman, U., Schulte, W., Borthelier, J.J., Illiano, J.M., Hodges, R.R., Hoffman, J.H.: 1986, *Nature* **321**, 326

Krasnopolsky, V.A., Krysko, A.A.: 1979, *Planetary Space Sci.* **27**, 951

Krasnopolsky, V.A., M. Gogoshev, M., Moreels, G., Moroz, V.I., Krysko, A.A., Gogosheva, T., Palasov, K., Sargoichev, S., Clairemidi, J., Vincent, M., Bertaux, J.L., Blamont, J.E., Troshin, V.S., Valnichek, B.: 1986a, *Nature* **321**, 269

Krasnopolsky, V.A., Gogoshev, M., Moreels, G., Moroz, V.I., Krysko, A.A., Barke, V.V., Jegulev, V.S., Parshev, V.A., Sanko, N.F., Tomashova, G.V., Tkachuk, A.Y., Troshin, V.S., Novikov, B.S., Perminov, V.G., Sulakov, I.I., Fedorov, O.G., Gogosheva, T., Sargoichev, S., Palasov, K., Georgiev, A., Nedkov, I., Kanev, K., Clairemidi, J., Vincent, M. et al.: 1986b, *Adv. Space Res.* **5**, No. **12**, 143

Krasnopolsky, V.A., Moroz, V.I., Krysko, A.A., Jegulev, V.S., Tkachuk, A.Y., Gogoshev, M., Moreels, G., Gogosheva, T., Clairemidi, J., Parisot, J.P.: 1986c, 20th ESLAB Symp., ESA SP-250, Vol. **I**, p. 459

Mazets, N.P., Sagdeev, R.Z., Aptecar, R.L., Golenetskii, S.V., Guryan, Y.A., Dychkov, A.V., Ilyinskii, V.N., Panov, V.N., Petrov, G.G., Shavvin, A.V., Sokolov, I.A., Frederiks, D.D., Khavenson, N.G., Shapiro, V.D., Shevchenko, V.I.: 1986, 20th ESLAB Symp., ESA SP-250, Vol. **II**, p. 3

McDonnel, J.A.M., Kissel, J., Grün, E., Grard, R.J.L., Langevin, Y., Olearczyk, R.E., Perry, C.H., Zarnecki, J.C.: 1986, 20th ESLAB Symp., ESA SP-250, Vol. **II**, p. 25

Moroz, V.I., Taranova, O.G.: 1986 (private communication)

Moreels, G., Gogoshev, M., Krasnopolsky, V.A., Clairemidi, J., Vincent, M., Parisot, J.P., Bertaux, J.L., Blamont, J.E., Festou, M.C., Gogoshev, T., Sargoichev, S., Palasov, K., Moroz, V.I., Vanysek, V.: 1986, *Nature* **321**, 271

Mukai, T., Mukai, S., Kikuchi, S.: 1986, 20th ESLAB Symp., ESA SP-250, Vol. **II**, p. 59

Mukai, S., Mukai, T.: 1985, in *Properties and Interaction of Interplanetary Dust*, eds. R.H. Giese, P. Lamy, Reidel, Dordrecht, p. 235

Perrin, J.M., Lamy, P.L.: 1985, in *Properties and Interaction of Interplanetary Dust*, eds. R.H. Giese, P. Lamy, Reidel, Dordrecht, p. 245

Probstein, R.F.: 1968, in *Problems of Hydrodynamics and Continuum Mechanics*, Philadelphia, p. 568

Sagdeev, R.Z., Szabo, F., Avanesov, G.A., Cruvelier, P., Szabo, L., Szego, K., Abergel, A., Balazs, A., Barinov, I.V., Bertaux, J.L., Blamont, J., Detaile, M., Demarelis, E., Dul'nev, G.N., Endröczy, G., Gardos, M., Kanyo, M., Kostenko, V.I., Krasikov, V.A., Nguyen-Trong, T., Nyitrai, Z., Reny, I., Rusznyak, P., Shamis, V.A., et al.: 1986, *Nature* **321**, 262

Schiffer, R.: 1985, *Astron. Astrophys.* **148**, 265

Weiss-Wrana, K.: 1983, *Astron. Astrophys.* **126**, 240

Zerull, N.N.: 1985, in *Properties and Interaction of Interplanetary Dust*, eds. R.H. Giese, P. Lamy, Reidel, Dordrecht, p. 197

Astron. Astrophys. 187, 712–718 (1987)

ASTRONOMY AND ASTROPHYSICS

Dust distribution of comet P/Halley's inner coma determined from the Giotto Radio-Science Experiment

P. Edenhofer [1], M.K. Bird [2], J.P. Brenkle [3], H. Buschert [1], E.R. Kursinski [3], N.A. Mottinger [3], H. Porsche [4], C.T. Stelzried [3], and H. Volland [2]

[1] Institut für Hoch- und Höchstfrequenztechnik, Universität Bochum, D-4630 Bochum, Federal Republic of Germany
[2] Radioastronomisches Institut, Universität Bonn, D-5300 Bonn, Federal Republic of Germany
[3] Jet Propulsion Laboratory, Pasadena, CA 91109, USA
[4] Deutsche Forschungs- und Versuchsanstalt für Luft- und Raumfahrt (DFVLR), D-8031 Oberpfaffenhofen, Federal Republic of Germany

Received February 16, accepted July 10, 1987

Summary. Measurements of the Giotto Radio-Science Experiment, consisting of the Doppler frequency shift and the intensity level of the X-band downlink signal of the Giotto spacecraft during Halley encounter, are analysed and interpreted. Continuous radio-science data were recorded throughout the encounter. The Doppler shift observed over a time interval of about 100 s is attributed to a drag effect in the cometary atmosphere causing a deceleration of the spacecraft. The total change of velocity of Giotto was $23.05 \pm 0.05\,\mathrm{cm\,s^{-1}}$, yielding a best estimate for the total cometary mass impacting the spacecraft of 0.32 g. The cometary mass fluence and mean mass density along the Giotto trajectory was found to be $1.2\ 10^{-4}\,\mathrm{kg\,m^{-2}}$ and $1.8\ 10^{-11}\,\mathrm{kg\,m^{-3}}$, respectively. Within a time interval ± 45 s about closest approach, at least four well-established, sharply confined dust jets are distinguished, for which the characteristic geometries and production rates can be inferred. The total dust production rate is estimated to be $1.5\ 10^3\,\mathrm{kg\,s^{-1}}$. The comet's jet structure derived from these Doppler data is shown to correlate with measurements on board Giotto.

Key words: comets – occultations – radio science

1. Introduction

The carrier signal of the Giotto spacecraft was monitored in amplitude and phase (frequency) during the encounter with comet Halley on 13/14 March 1986. The scientific objectives of this Giotto Radio-Science Experiment (GRE) were to determine the electron content of the cometary ionosphere along the signal's ray path to Earth and the mass fluence (dust and gas) of the cometary atmosphere along the spacecraft's trajectory (Edenhofer et al., 1986a, and prior references therein). A series of interim reports on the data analysis and interpretation have already appeared in the literature (Edenhofer et al., 1986b–d). This paper presents the results of the work on GRE to date. Further studies will be necessary, however, before a more comprehensive and unambiguous interpretation of the encounter scenario can be formulated from the radio metric and supplementary spacecraft housekeeping data.

Send offprint requests to: P. Edenhofer

2. Observations

Giotto X-band (8.429 GHz) radio metric data for GRE were recorded during Halley encounter at the Australian 64-m Deep Space Network (DSN) station DSS 43 at Tidbinbilla (near Canberra) using two independent receiver systems, referred to as "closed-loop" and "open-loop", respectively. Virtually identical results were obtained from these two systems. The closed-loop recordings were generated at the maximum possible sampling rate (10 Hz) from the primary phase-locked loop (PLL) receivers with a fixed PLL-bandpass. The raw open-loop data are digitized signal voltages (wave form samples) recorded at high sampling rate (50 kHz) with specially designed radio science receivers (bandwidth controlled by on-line processing of received signal spectrum). The data are filtered, decimated and Fourier transformed to provide measurements of signal amplitude and frequency. The receiver and data processing parameters such as effective bandwidth, integration time, etc., can be varied to improve signal-to-noise ratios and provide optimum recording of the data.

Figure 1 shows the open-loop radio metric data (integration time 0.128 s) of the Giotto carrier signal for a 150-s interval about the time of closest approach to the comet (00:11:02.7 $\pm$ 0.4 s UT from Keller et al., 1986b). The upper panel shows the frequency residuals in Hz, i.e. the deviation from the expected downlink frequency. Corrections were applied for the known orbital Doppler shift and a linear drift of the Giotto oscillator (ca. $+1\,\mathrm{Hz\,min^{-1}}$) that persisted throughout the encounter period. The lower panel of Fig. 1 shows the corresponding signal level in dB (pre-encounter signal level -129.6 dBm). Very similar encounter profiles of the signal amplitude and frequency were recorded at the 64-m CSIRO radio telescope at Parkes, which served as the primary ESA ground receive facility (Hall, 1986).

Simultaneous GRE recordings of the Giotto downlink at S-band (2.299 GHz) were planned for the encounter in order to determine the cometary electron content from dispersive Doppler shift at two frequencies (Edenhofer et al., 1986a). In preparation for this part of the experiment, the S-band signal of the nearby

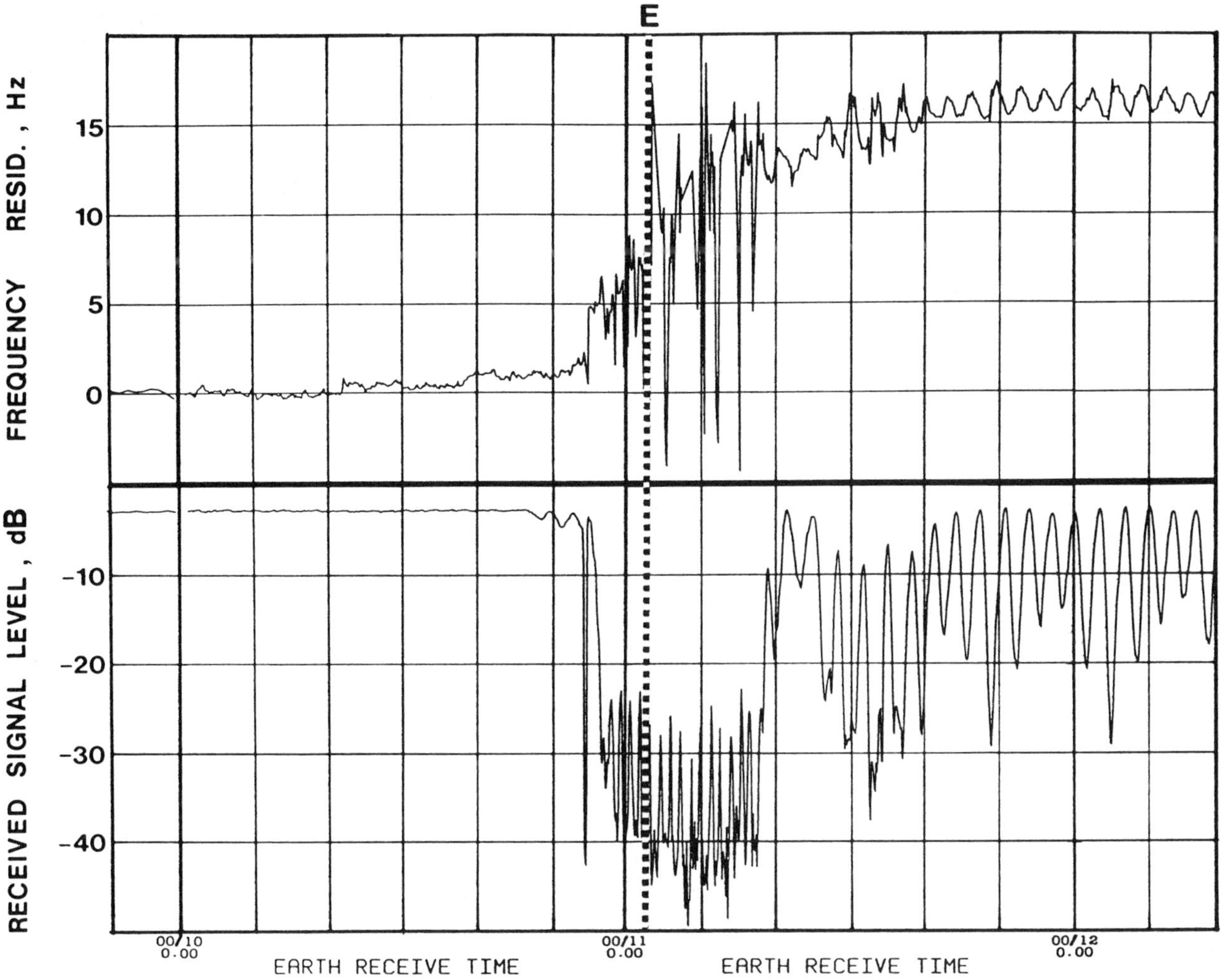

Fig. 1. Giotto X-band carrier signal parameters during Halley flyby. The residual frequency in Hz (upper panel) and the received signal level in dB (lower panel) are plotted for the Earth receive time interval of 150 s centered approximately at the time of nearest encounter (vertical dotted line marked "E") at 00:11:02.7 ± 0.4 UT. These data, recorded with open-loop radio science receivers at the Australian DSN station DSS 43, are shown here with a time resolution of 0.128 s

Sakigake probe was tracked during Giotto encounter at the 34-m station DSS 42 in order to provide a calibration for noncometary (mainly interplanetary) plasmas. Contributions from the terrestrial ionosphere were further monitored at DSS 43 by taking VHF Faraday rotation measurements from the geostationary ETS 2 satellite. Unfortunately, the final decision of the Giotto Project Management was to transmit only at X-band during encounter. Subsequent simulation studies (Edenhofer et al., 1986c) using cometary electron density models based on in situ observations (Krankowsky et al., 1986; Balsiger et al., 1986), performed with actual S- and X-band Doppler data from the cruise phase of the Giotto mission, have shown that:

(a) a highly structured, easily detectable cometary electron content profile could have been derived from a dual-frequency Doppler experiment, and

(b) extraction of the cometary electron content from the actual X-band Doppler profiles will be extremely difficult, if not impossible.

The feasibility of the dual-frequency experiment was underscored by the successful detection of Halley's ionosphere by the two Vega probes using coherent radio signals at 0.9 and 6 GHz (Savich et al., 1986). The peak electron densities at closest approach (flyby distances 8890 and 8030 km) inferred from these data were 2500 el cm^{-3} (Vega-1) and 1500 el cm^{-3} (Vega-2), respectively. Extrapolating these values to the Giotto flyby distance, one obtains electron content levels which are very consistent with the above mentioned Giotto simulation (Edenhofer et al., 1986c).

A more detailed interpretation of the features seen in Fig. 1 will be deferred to Sect. 4. It is appropriate at this point, however, to provide an overview of all radiometric "facts" as they are presently understood.

The total decrease in velocity of the Giotto spacecraft due to drag deceleration in Halley's coma was first determined to be $\Delta v \simeq 23.2$ cm s^{-1}, directed along the relative velocity vector between comet and spacecraft (Edenhofer et al., 1986b; Morley and Fertig, 1986). This estimate was made from highly accurate, two-way Doppler and ranging measurements taken before and after encounter, treating the comet interaction as an "involuntary" spacecraft maneuver. These two-way data, the phase stability of which is controlled by a highly stable uplink generated at the ground station, are not subject to the intrinsic frequency instability of the on-board crystal oscillator. A more thorough subsequent navigational analysis (Morley, private communication from unpublished results), using all calibrated pre- and post-encounter tracking passes, has yielded a final result for the total velocity decrease given by:

$$\Delta v = 23.05 \pm 0.05\ \mathrm{cm\,s^{-1}}. \tag{1}$$

This value is consistent with the independent determinations from both ranging and Doppler data and serves as a "boundary condition" for the asymptotic behavior of the one-way Doppler data during encounter. No significant secular accelerations of the spacecraft could be detected in the post-encounter tracking data. The entire deceleration is thus presumed to have occurred over an interval of approximately $\tau_0 = 100$ s from ca. 00:10:20 UT to 00:12:00 UT (Fig. 1).

The total X-band blue shift recorded one-way at encounter was found to be $\Delta f = 16.9 \pm 0.2$ Hz, compared with the value $\Delta f = 4.64 \pm 0.01$ Hz derived from the two-way pre- and post-encounter data. So far, two possible explanations have been given for this discrepancy. The first is that the Doppler variation at encounter is enlarged by an on-board oscillator drift that is just proportional to the deceleration of the spacecraft, i.e. proportional to the applied drag force. The effect on the oscillator would presumably occur via acoustic, electrical, or thermal energy deposition. The net result would be a frequency curve which qualitatively displays the spacecraft deceleration, but with an incorrect scaling. The second possibility is that the frequency data are correct up to the time when the nutation starts or the time of the sharp drop in signal amplitude and concurrent discontinuity in frequency at 00:10:54.4 UT (Fig. 1). This loss of signal for about 0.6 s is thought to be the switchover of the X-band travelling wave tube amplifier from the primary unit TWT1 to the redundant hot standby TWT2. This TWT switchover may have been induced by the large-amplitude electrical transient that terminated the NMS experiment at this time (Krankowsky et al., 1986). If the on-board crystal oscillator were also subjected to a large transient pulse, whether of acoustic, electrical or thermal origin, it is conceivable that an unforeseen frequency drift of some 12 Hz could occur over a time scale of 10–20 s. In this case, the spacecraft's drag deceleration profile after the anomaly could only be obtained with knowledge of the "pseudo-Doppler" contributed by the on-board oscillator. The transient response of the radio system was not monitored during pre-launch tests of TWT switchovers. Since no other TWT switchovers or high-voltage pulses occured during the Giotto mission, the behavior of the downlink frequency and amplitude under these unique circumstances is unknown.

The real drag decelerations attributed to various features in the measured Doppler profile (Fig. 1, upper panel), and therefore the associated distribution of mass in the comet's atmosphere, are obviously dependent upon which interpretation is relevant. Since it is presently impossible to eliminate either of the above mentioned candidates, it will be assumed, subject to further investigation, that the first interpretation is valid. This means that the Doppler profile need only be rescaled to attain its asymptotic value of $\Delta v = 23.05\ \mathrm{cm\,s^{-1}}$, corresponding to $\Delta f = 4.64$ Hz, at times well after the encounter. If the alternate scenario eventually turns out to be correct, a quantitative correction to the deceleration profile would become necessary. Neither the qualitative structure of the inner coma, i.e. the locations of the most obvious concentrations of cometary matter, nor the value of the total cometary mass fluence would be affected by this correction.

3. Determination of mass fluence and dust production rate

The encounter geometry for the flyby of the Giotto spacecraft at comet Halley is shown in Fig. 3. During encounter the Giotto radio signal received on Earth is subject to a Doppler frequency shift, measured as the time derivative of phase shift

$$\Delta f(t) = \Delta f_{\mathrm{d}} + \Delta f_{\mathrm{p}} = -\frac{f\,\Delta v_{\mathrm{r}}(t)}{c} + \frac{40.3\,\dot{I}(t)}{cf}. \tag{2}$$

The first term Δf_{d} corresponds to an increase in signal frequency (blue shift) caused by a decrease in spacecraft velocity due to drag effects in the atmosphere of comet Halley. The second term Δf_{p} represents the rate of change of the electron content $\dot{I}$ (integrated electron density along the signal ray path) in the cometary ionosphere (c = velocity of light, MKS units). For the X-band downlink ($f = 8.429$ GHz) the drag-induced spacecraft deceleration is dominant over the electron content effect by about two orders of magnitude (Bird et al., 1983). This can be verified by taking the maximum rate of change of the total electron content $\dot{I} \simeq 10^{14}\ \mathrm{el\,m^{-2}\,s^{-1}}$ as measured by the two Vega probes (Savich et al., 1986), extrapolated to the Giotto flyby distance to yield a plasma-induced X-band Doppler frequency shift $\Delta f_{\mathrm{p}} \simeq 0.02$ Hz from (2). The relative radial velocity Δv_{r} is taken along the line-of-sight Giotto-Earth and is related to the residual velocity along the Giotto trajectory by $\Delta v_{\mathrm{r}} = \Delta v \cos(44.3°)$.

The change in velocity Δv is caused by cometary particles of total mass m_0 (dust and gas) impacting the Giotto spacecraft and accumulating the cometary mass fluence along the Giotto trajectory (Fig. 3). The quantity m_0 can be determined by applying the equations for conservation of momentum and energy (Edenhofer et al., 1986c):

$$m_0 v_0 = M \Delta v - mv, \tag{3}$$

$$\frac{m_0 v_0^2}{2} = \frac{M \Delta v^2}{2} + \frac{mv^2}{2} + q, \tag{4}$$

where $v_0 = 68.37\ \mathrm{km\,s^{-1}}$ is the relative encounter velocity; $M = 573.7$ kg is the mass of the Giotto spacecraft at encounter; m and v are the mass and velocity of recoil particles; and the quanity q encompasses energy losses due to heating, ionization, etc. From Eqs. (3) and (4) Δv can be related to m_0 as follows:

$$\Delta v = v_0 \frac{m_0}{M}\left(1 + \sqrt{\frac{m}{m_0}\left(1 - \frac{2q}{m_0 v_0^2}\right)}\right) = v_0 \frac{m_0}{M}(1 + \varepsilon), \tag{5}$$

where the momentum enhancement factor

$$\varepsilon = mv/(m_0 v_0) \tag{6}$$

has been introduced to account for the increase in momentum transferred by the spectrum of cometary mass particles impacting at hypervelocity conditions ($\varepsilon = 0$ or 1 for inelastic and elastic collisions, respectively). The value of ε for light cometary particles is assumed to be of the order of 10 or more (McDonnell et al., 1984), corresponding to effects such as explosive mass release and ionization on the front side of the Giotto composite bumper shield. The momentum enhancement factor decreases for heavier particles that penetrate the first shield. Using a constant value of $\varepsilon = 11$ for nonpenetrating particles of mass less than $m_{\mathrm{c}} = 1\ \mu$g and a decrease in ε for heavier particles m_0 according to $(m_{\mathrm{c}}/m_0)^{2/3}$ (McDonnell et al., 1986a), a rough estimate may be obtained for

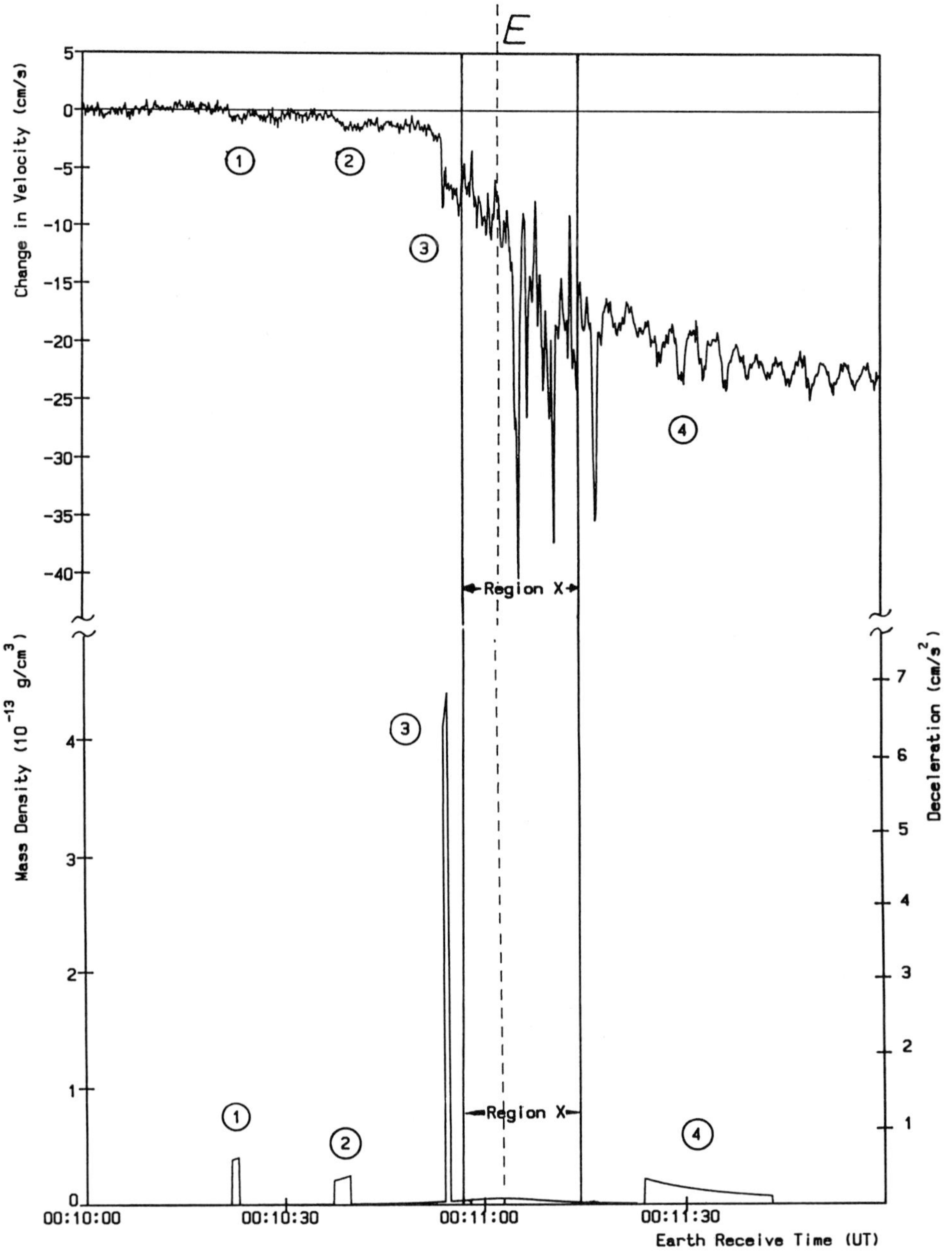

Fig. 2. Velocity and deceleration residuals of the Giotto spacecraft deduced from closed-loop X-band Doppler data. The velocity decrements (upper panel) are used to derive the mass density of the impacting cometary matter, which is proportional to the spacecraft deceleration (lower panel). A deceleration of $1\ \mathrm{cm\,s^{-2}}$ is equivalent to an impacting mass density of ca. $8\ 10^{-14}\ \mathrm{g\,cm^{-3}}$. Regions (1)–(4) are considered to be distinct dust jets. Region X has not been analysed due to the uncertainty of the frequency measurements. The slow decrease in deceleration associated with jet (4) can be explained by an approximately inverse-square decrease in mass density as Giotto moves out along the jet

the mean momentum enhancement factor $\langle\varepsilon\rangle$, weighted over an appropriate cometary mass fluence distribution (Divine et al., 1986). Depending on the choice of the upper mass cutoff, the higher values of mass fluence in the spectral range of heavier particles reduce the effective value of $\langle\varepsilon\rangle$ to the order of 3 to 6. For the time being, a value of $\langle\varepsilon\rangle = 5$ has been selected to determine the cometary mass m_0 from (5). The value Δv from (1) yields a total cometary mass (dust and gas) of $m_0 = 320$ mg impacting the Giotto spacecraft upon an effective cross-sectional area $A = 2.64\ \mathrm{m}^2$ (Edenhofer et al., 1986a). The cometary mass fluence is thus derived to be $F = m_0/A = 1.2\ 10^{-4}\ \mathrm{kg\,m^{-2}}$ and the mean mass density of the cometary atmosphere along the Giotto trajectory is found to be

$$\langle N\rangle = m_0/(A v_0 \tau_0) = 1.8\ 10^{-11}\ \mathrm{kg\,m^{-3}}$$

(Edenhofer et al., 1986d). This total mass density yields a mean mass flux at encounter of

$$\langle N\rangle v_0 = 1.2\ 10^{-6}\ \mathrm{kg\,m^{-2}\,s^{-1}},$$

a mean momentum flux of

$$\langle N\rangle v_0^2 = 0.083\ \mathrm{kg\,m^{-1}\,s^{-2}},$$

and a mean drag force on the spacecraft of

$$\langle N\rangle v_0^2 A (1 + \langle\varepsilon\rangle) = 1.3\ N.$$

Using an appropriate model for the sources of dust and gas, it is possible to calculate a mass production rate. The spatial density

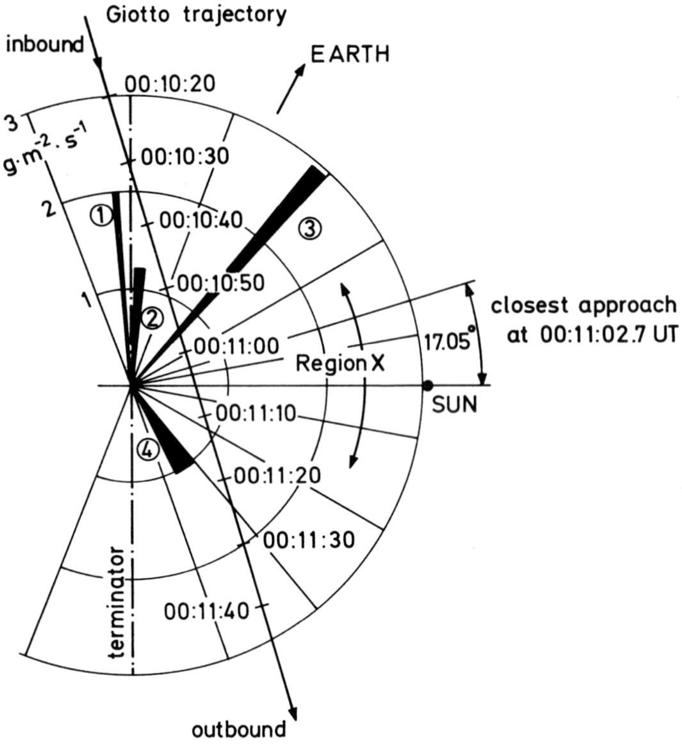

Fig. 3. Dust jet structure of the inner coma of comet Halley during Giotto flyby. The cometary nucleus is in the center of the coordinate system. The Giotto trajectory moves from upper left to lower right. The Sun is to the right. The angle between the Giotto trajectory and the direction to Earth is 44°3. The lengths of the four jets ($g\,m^{-2}\,s^{-1}$) are a measure of the dust production rate, normalized to a constant radius of 7 km. The uncertain region X extends over about 60° of the subsolar cometary coma

distributions N_d for each dust source and N_g for gas are assumed to be given by

$$N_d(r) = \frac{Q_d}{\pi\,\alpha^2\,u_d\,r^2}; \quad N_g(r) = \frac{Q_d}{4\pi\,u_g\,r^2}\exp\left(\frac{-r}{\tau_g\,u_g}\right), \tag{7}$$

where $Q_{d,g}$ is the dust/gas production rate, and $u_{d,g}$ is the terminal (constant) outflow velocity for dust/gas. Spherical symmetry was assumed for the distribution of gas and a model with distinct jet-like sources was used for the dust. τ_g is the mean molecular lifetime against dissociation/ionization, and α is the (small) angular half-width of each jet source. By defining the entry and exit times for each jet, one may determine the angular width 2α. The jet's angular extent in the dimension perpendicular to the plane of Fig. 3 was assumed to be of the same order 2α (axial symmetry). The gas production rate $Q_g = 6.9\ 10^{29}$ molecules s^{-1} ($2.3\ 10^4\ kg\,s^{-1}$) and an outflow velocity of $u_g = 900\ m\,s^{-1}$ were used (Krankowsky et al., 1986). A value of $u_d = 400\ m\,s^{-1}$ (Edenhofer et al., 1986d) was assumed for the mean dust outflow velocity at the spacecraft, again weighted over the dust fluence distribution. The mass impacting the spacecraft in a specific time interval Δt is given by

$$m_{d,g}(t) = N_{d,g}\,A\,v_0\,\Delta t. \tag{8}$$

A calculation of the mass density and production rate was performed for the entire encounter time frame from 00:09:00 UT to 00:13:00 UT. Figure 2 (upper panel) shows the velocity profile of the spacecraft during encounter (closed-loop data at sampling interval 0.1 s, in contrast to the open-loop data of Fig. 1). Figure 2 has been normalized to the known asymptotic value of $23.05\ cm\,s^{-1}$. The lower panel of Fig. 2 shows the cometary mass density and spacecraft deceleration as derived from the upper panel. If the region X is also included in the analysis, as many as 11 jet-like structures can be distinguished in the time interval ± 45 s about closest approach. Eight of these structures are encountered within ± 10 s of closest approach. The greatest contribution ($17.2\ cm\,s^{-1}$, about 75%) to the total spacecraft deceleration occured in this interval.

The total dust production rate Q_d from all dust sources included in the above calculation was $Q_d = 1.5\ 10^3\ kg\,s^{-1}$. This is about seven times smaller than the production rate of an earlier calculation (Edenhofer et al., 1986c) which was based on isotropic dust outflow over a solid angle of 4π. Comparing the dust production rate with the gas production rate, the dust-to-gas ratio is found to be 0.07. McDonnell et al. (1986b) calculated a total dust production rate $Q_d = 3.3\ 10^3\ kg\,s^{-1}$ by integrating the emission dust spectrum up to a maximum of 1 g.

4. Characteristics of individual dust jets

Cometary matter impacting the spacecraft can only cause a deceleration. Knowing the encounter geometry, this deceleration is observed as a blue shift of the downlink signal. Those data segments which show a red shift are thus interpreted as noise, most probably originating from spacecraft nutation or anomalous phase drift from the on-board oscillator. Only those data can be interpreted as real spacecraft deceleration where an irreversible blue shift of the frequency occurs. These events are denoted as (1)–(4) in Figs. 2 and 3. The greatest uncertainties are associated with the region X. In this region the noise is particularly high because of the low signal level and the spacecraft nutation, which both make an interpretation of the data difficult. Moreover, it is unclear which proportion of the frequency increase in this region can be attributed to deceleration in several distinct steps or merely to an anomalous frequency shift of the on-board oscillator. The region X is thus not included in the following discussion of the distinguishable dust jets.

The images of the innermost coma taken by the Giotto camera HMC (Keller et al., 1986a, b) show that the dust is emitted from the nucleus in strictly confined jets. It is known from earth-bound observations that the cometary jet structure stretches far out from the nucleus (Sekanina and Larson, 1986; Larson et al., 1986; Chakaveh et al., 1986). A strong deviation from spherical symmetry in the drag deceleration of the spacecraft would thus be one effect expected from these jets. The times the spacecraft traversed these jets, characterized by the increase in downlink frequency and decrease in residual velocity, are indicated as (1)–(4) in Fig. 2.

The rotation of the spacecraft was perturbed by a nutation (characteristic period: 3.2 s) that became measurable starting about 00:10:45 UT (Fig. 1). The spin period also increased from 3.998 s to 4.010 s over the encounter. An imbalance of the impinging cometary dust flux on Giotto's front bumper shield, culminating in the probable impact of a fairly large single particle (mass >0.1 g) at 00:10:54.4 UT, caused a spacecraft tilt that resulted in a depointing of the downlink (3 dB X-band beamwidth: 1°8) away from the Earth. The signal level thus decreased rapidly over a time comparable to one spacecraft rotation period and stayed low for about 20 seconds (from 00:10:59 to 00:11:18 UT)

Table 1. Some characteristics of the four well-established dust jets

Dust jet code number	(1)	(2)	(3)	(4)
Spacecraft change in velocity Δv (cm s^{-1})	0.6±0.1	0.8±0.2	4.1±0.5	3.2±0.4
Flight time in jet Δt (s)	1.2±0.6	2.4±0.6	0.6±0.2	19±5
Spacecraft deceleration $\Delta \dot{v}$ (cm s^{-2})	0.5	0.3	6.8	0.17
Spacecraft drag force (N)	3	2	39	1
Power onto spacecraft (MW)	0.2	0.13	2.7	0.07
Flight distance in jet (km)	80	160	40	1300
Mean distance from nucleus (km)	2820	1720	820	2260
Angular width of jet (deg)	0.4	1.9	2.1	8.9
Source size at nucleus, $r_0 = 7$ km (m)	50	230	260	1100
Mass impacted on spacecraft (mg)	8	11	57	45
Mean density in jet at S/C (10^{-11} kg m^{-3})	3.7	2.6	53	1.3
Mass production rate of jet (kg s^{-1})	4	25	150	500
Mass flux density of jet (g m^{-2} s^{-1})	2.1	0.6	2.9	0.5

Notes:
Assumed values for construction of the above table:
Time of closest approach: 00:11:02.7±0.4 UT; Distance of closest approach: 605±8 km
Spacecraft mass at encounter: $M = 573.7$ kg; Relative flyby velocity: $v_0 = 68.37$ km s^{-1}
Momentum enhancement factor: $\langle \varepsilon \rangle = 5$; Dust outflow velocity: $u_d = 400$ m s^{-1}
Dust sources: axial symmetric cones, radial outflow with mass density falling off as r^{-2}

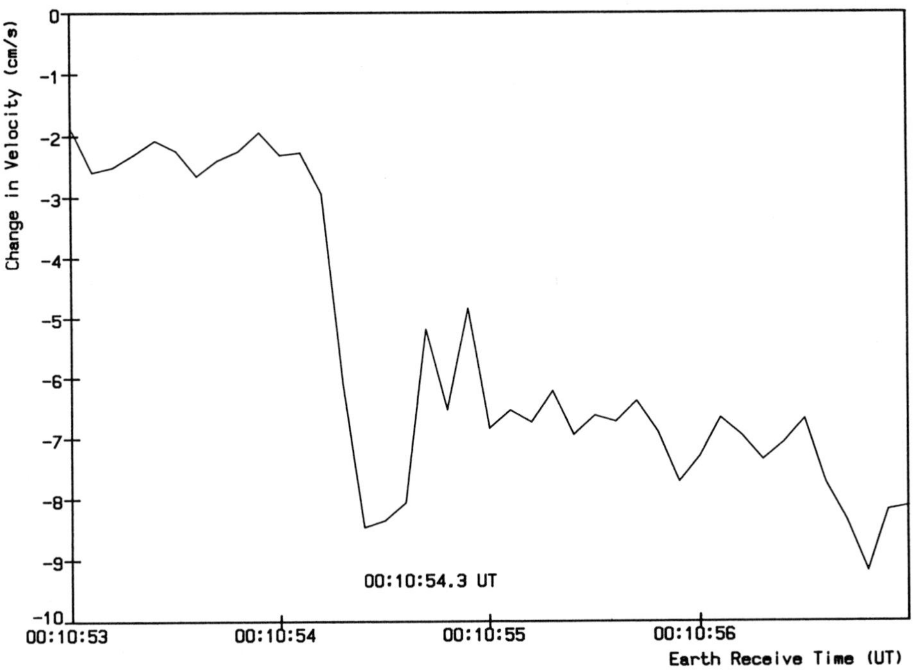

Fig. 4. Close-up of the temporal variations of the velocity decrease showing detailed time signature from dust jet (3) of Fig. 2 (upper panel). In order to reduce the noise level on the data sampled at the maximum rate of 10 per second, a running mean average algorithm was applied (average over three samples, successively incrementing by one sample). The dust jet is centered at the Earth receive time of 00:10:54.6 UT

until the despin motor was adjusted via the on-board attitude control system to the new phase and spin rate. As mentioned earlier, the noisy data in this 20 s interval (Region X) cannot be unambiguously interpreted. So far only the dust jets denoted (1)–(4) can be described in more detail. Some characteristics of these four jets, which contributed about 40% of the total spacecraft deceleration, are summarized in Table 1.

Transferring into a comet-centered coordinate system, one derives from Fig. 2 the jet structure as given in Fig. 3, which is a projection onto the plane containing the Giotto trajectory and the cometary nucleus. The radial lengths of the four jet sources are proportional to the specific dust production rate (mass per unit area and per unit time). They have been normalized to a given distance of $r_0 = 7$ km from the nucleus by applying an inverse-square law correction for the density.

A velocity decrease of $\Delta v = 0.6 \pm 0.1$ cm s^{-1} could be measured, denoted by (1) in the figures, between 00:10:21.8±0.3 UT and 00:10:23.0±0.3 UT. This decrease was significantly higher than the otherwise typical fluctuations arising from oscillator instabilities. This change in velocity yields a spacecraft deceleration of $\langle \dot{\Delta v} \rangle = 0.5 \pm 0.3$ cm s^{-2}. The corresponding drag force acting on the spacecraft was 3±2 N. Certainly, this deceleration must be considered marginal. However, this interpretation is supported by a simultaneous slight increase recorded by the photometer of the OPE experiment on board Giotto (Levasseur-Regourd et al., 1986).

Another pre-encounter jet (2) was traversed between 00:10:37.9 ± 0.3 UT and 00:10:40.3 ± 0.3 UT causing an increase in velocity of $\Delta v = 0.8 \pm 0.2$ cm s^{-1}. This results in a deceleration $(\dot{\Delta v}) = 0.3 \pm 0.2$ cm s^{-2} associated with a drag force of about 2 ± 1 N. Unfortunately, so far no reliable correlation of this feature can be found with the Giotto in situ experiments (apart from the general increase in intensity observed by OPE up to the termination of operations at 00:10:45 UT due to the increasing spacecraft nutation).

No correlating in situ data exist for the most prominent dust event (3) seen by GRE in the 0.6 s period from 00:10:54.3 ± 0.1 UT to 00:10:54.9 ± 0.1 UT. Figure 4 shows a close-up of this dust event from Fig. 2 (upper panel) with enhanced time resolution. The velocity decrease during this event was determined to be 4.1 ± 0.5 cm s^{-1}, yielding $(\dot{\Delta v}) = 6.8 \pm 3.1$ cm s^{-2}. The drag force reached its maximum value of 39 ± 18 N during this short-duration dust barrage, which also delivered 2.7 ± 1.2 MW of power to Giotto's bumper shield.

All events of region X cannot be individually analysed due to the uncertainty of measurements in this time interval (data reconstruction still in progress). A generally linear increase in frequency of about 10–12 Hz was recorded during the time of low signal level.

In spite of the nutation of the spacecraft, which modulates the downlink carrier signal, a reliable determination of the average frequency increase in the period (4) is possible. A spectral analysis of the post-encounter amplitude and frequency out to 00:20 UT clearly reveals the spacecraft inertial spin period (4.0 s) and inertial nutation period (3.2 s) that combine to produce a beat modulation at a period close to 16.4 s. The corresponding dust event responsible for the deceleration started at 00:11:23 UT and ended roughly at or slightly before 00:12:00 UT. The overall Doppler shift for jet (4) between the times 00:11:24.8 and 00:11:44.1 UT is equivalent to a velocity change of 3.2 ± 0.4 cm s^{-1}. The deceleration decreased monotonically with time from a maximum value $(\dot{\Delta v}) \simeq 0.26$ cm s^{-2} at the beginning of the interval. Mean values for these and other derived quantities are found in Table 1. The enhanced dust-flux responsible for this deceleration was confirmed by the DID investigation on board Giotto (Zarnecki et al., 1986).

Acknowledgements. We wish to express our appreciation for the many contributions to GRE from our collegues at JPL (Pasadena) and the DSN tracking station Tidbinbilla. We would like to extend thanks to Trevor Morley (ESOC) for a special reprocessing of the Giotto navigational data.

References

Balsiger, H., Altwegg, K., Bühler, F., Geiss, J., Ghielmetti, A.G., Goldstein, B.E., Goldstein, R., Huntress, W.T., Ip, W.-H., Lazerus, A.J., Meier, A., Neugebauer, M., Rettenmund, U., Rosenbauer, H., Schwenn, R., Sharp, R.D., Shelley, E.G., Ungstrup, E., Young, D.T.: 1986, *Nature* **321**, 330

Bird, M.K., Volland, H., Porsche, H., Edenhofer, P.: 1983, *Mitt. Astron. Ges.* **58**, 157

Chakaveh, S.C., Green, S.F., Ridley, J.K., McDonnell, J.A.M., Hughes, D.W.: 1986, in B. Battrick et al. (eds.), 20th ESLAB Symp., ESA SP-250, Vol. **II**, p. 163

Divine, N., Fechtig, H., Gombosi, T.I., Hanner, M.S., Keller, H.U., Larson, S.M., Mendis, D.A., Newburn, R.L., Reinhard, R., Sekanina, Z., Yeomans, D.K.: 1986, *Space Sci. Rev.* **43**, 1

Edenhofer, P., Bird, M.K., Buschert, H., Esposito, P.B., Porsche, H., Volland, H.: 1986a, in R. Reinhard, B. Battrick (eds), ESA SP-1077, p. 173

Edenhofer, P., Bird, M.K., Brenkle, J.P., Buschert, H., Esposito, P.B., Porsche, H., Volland, H.: 1986b, *Nature* **321**, 355

Edenhofer, P., Bird, M.K., Esposito, P.B., Porsche, H., Volland, H.: 1986c, *Adv. Space Res.* **5**, No. 12, 201

Edenhofer, P., Buschert, H., Bird, M.K., Volland, H., Porsche, H., Brenkle, J.P., Kursinski, E.R., Mottinger, N.A., Stelzried, C.T.: 1986d, in B. Battrick et al. (eds.), 20th ESLAB Symp., ESA SP-250, Vol. **II**, p. 215

Hall, P.J., 1986, *Proc. ASA* **6** (3), 298

Keller, H.U., Arpigny, C., Barbieri, C., Bonnet, R.M., Cazes, S., Coradini, M., Cosmovici, C.B., Delamere, W.A., Huebner, W.F., Hughes, D.W., Jamar, C., Malaise, D., Reitsema, H.J., Schmidt, H.U., Schmidt, W.K.H., Seige, P., Whipple, F.L., Wilhelm, K.: 1986a, *Nature* **321**, 320

Keller, H.U., Arpigny, C., Barbieri, C., Bonnet, R.M., Cazes, S., Coradini, M., Cosmovici, C.B., Curdt, W., Delamere, W.A., Huebner, W.F., Hughes, D.W., Jamar, C., Kramm, R., Malaise, D., Reitsema, H., Schmidt, H.U., Schmidt, K., Schmidt, W.K.H., Seige, P., Whipple, F.L., Wilhelm, K.: 1986b, in B. Battrick et al. (eds.), 20th ESLAB Symp., ESA SP-250, Vol. **II**, p. 347

Krankowsky, D., Lämmerzahl, P., Herrwerth, I., Woweries, J., Eberhardt, P., Dolder, U., Herrmann, U., Schulte, W., Berthelier, J.J., Illiano, J.M., Hodges, R.R., Hoffman, J.H.: 1986, *Nature* **321**, 326

Larson, S., Sekanina, Z., Levy, D., Tapia, S., Senay, M., 1986, in B. Battrick et al. (eds.), 20th ESLAB Symp., ESA SP-250, Vol. **II**, p. 45

Levasseur-Regourd, A.C., Bertaux, J.L., Dumont, R., Festou, M., Giese, R.H., Giovane, F., Lamy, P., Le Blanc, J.M., Llebaria, A., Weinberg, J.L.: 1986, *Nature* **321**, 341

McDonnell, J.A.M., Alexander, M., Lyons, D., Tanner, W., Anz, P., Hyde, T., Chen, A.-L., Stevenson, T.J., Evans, S.T.: 1984, *Adv. Space Res.* **4**, No. 9, 297

McDonnell, J.A.M., Alexander, W.M., Burton, W.M., Bussoletti, E., Clark, D.H., Grard, R.J.L., Grün, E., Hanner, M.S., Hughes, D.W., Igenbergs, E., Kuczera, H., Lindblad, B.A., Mandeville, J.-C., Minafra, A., Schwehm, G.H., Sekanina, Z., Wallis, M.K., Zarnecki, J.C., Chakaveh, S.C., Evans, G.C., Evans, S.T., Firth, J.G., Littler, A.N., Massonne, L., Olearczyk, R.E., Pankiewicz, G.S., Stevenson, T.J., Turner, R.F.: 1986a, *Nature* **321**, 338

McDonnell, J.A.M., Kissel, J., Grün, E., Grard, R.J.L., Langevin, Y., Olearczyk, R.E., Perry, C.H., Zarnecki, J.C.: 1986b, in B. Battrick et al. (eds.), 20th ESLAB Symp., ESA SP-250, Vol. **II**, p. 25

Morley, T.A., Fertig, J.: 1986, ESA Bulletin No. 46, p. 71

Savich, N.A., Altunin, V.I., Andreev, V.E., Basos, I.F., Boguslavskaia, N.-E.E., Vyshlov, A.S., Gavrik, A.L., Dorospchuk, O.N., Kruchkovich, M.M., Liaskovskaia, V.I., Molotov, E.P., Morskoy, I.M., Nabatov, A.S., Ovsiannikova, M.A., Seleznev, V.V., Sukhanov, K.G., Sheverdiaev, A.S.: 1986, *Pisma Astron. Zh.* **12**, 675

Sekanina, Z., Larson, S.M.: 1986, *Nature* **321**, 357

Zarnecki, J.C., McDonnell, J.A.M., Alexander, W.M., Burton, W.M., Hanner, M.S.: 1986, in B. Battrick et al. (eds.), 20th ESLAB Symp., ESA SP-250, Vol. **II**, p. 185

Astron. Astrophys. 187, 719–741 (1987)

ASTRONOMY AND ASTROPHYSICS

The dust distribution within the inner coma of comet P/Halley 1982i: encounter by Giotto's impact detectors

J.A.M. McDonnell [1], W.M. Alexander [2], W.M. Burton [3], E. Bussoletti [4], G.C. Evans [1], S.T. Evans [1], J.G. Firth [3], R.J.L. Grard [5], S.F. Green [1], E. Grun [6], M.S. Hanner [7], D.W. Hughes [8], E. Igenbergs [9], J. Kissel [6], H. Kuczera [10], B.A. Lindblad [11], Y. Langevin [12], J.-C. Mandeville [13], S. Nappo [1], G.S.A. Pankiewicz [1], C.H. Perry [1], G.H. Schwehm [5], Z. Sekanina [7], T.J. Stevenson [1], R.F. Turner [3], U. Weishaupt [9], M.K. Wallis [14], and J.C. Zarnecki [1]

[1] Unit for Space Sciences, University of Kent at Canterbury, Canterbury, Kent CT2 7NR, UK
[2] Institute for Environmental Studies, Baylor University, Waco, TX 76703, USA
[3] Rutherford Appleton Laboratory, Chilton Didcot, Oxon OX11 OQX, UK
[4] Instituto Universitario Navale, Via Acton 38, I-80133 Naples, Italy
[5] ESA Space Science Department, ESTEC, Postbus 299, NL-2200 AG Noordwijk, The Netherlands
[6] Max-Planck-Institut für Kernphysik, Postfach 103980, D-6900 Heidelberg, Federal Republic of Germany
[7] Jet Propulsion Laboratory, Pasadena, CA 91103, USA
[8] Physics Department, University of Sheffield, Sheffield S3 7RH, UK
[9] Technische Universität München, D-8000 München 2, Federal Republic of Germany
[10] Messerschmidt-Bölkow-Blohm GmbH, Space Division, UR-RA 40, D-8000 München 80, Federal Republic of Germany
[11] Institute for Astronomy, Lund Observatory, S-22100 Lund, Sweden
[12] Laboratoire Rene-Bernas, F-91406 Orsay, France
[13] ONERA/CERT-DERTS, BP 4045, F-31055 Toulouse, France
[14] Department of Applied Mathematics, University College Cardiff, Cardiff, CF1 1XL, UK

Received February 16, accepted June 23, 1987

Summary. Analysis of the data from Giotto's Dust Impact Detection System experiment (DIDSY) is presented. These data represent measurement of the size of dust grains incident on the Giotto dust shield along its trajectory through the coma of comet P/Halley on 1986 March 13/14. First detection occurred at some 287000 km distance from the nucleus on the inbound leg; the majority of the DIDSY subsystems remained operational after closest approach (604 km) yielding the last detection at about 202000 km from the nucleus.

In order to improve the data coverage (and especially for the smallest grains, to approximately 10^{-19} kg particle mass), data from the PIA instrument has been combined with DIDSY data. Flux profiles are presented for the various mass channels showing, to a first approximation, a $1/R^2$ flux dependence, where R is the distance of the detection point from the cometary nucleus, although significant differences are noted. Deviations from this dependence are observed, particularly close to the nucleus. From the flux profiles, mass and geometrical area distributions for the dust grains are derived for the trajectory through the coma. Groundbased CCD imaging of the dust continuum in the inner coma at the time of encounter is also used to derive the area of grains intercepted by Giotto. The results are consistent with the area functions derived by Giotto data and the low albedo of the grains deduced from infrared emission.

For the close encounter period (-5 min to $+5$ min), the cumulative mass distribution function has been investigated, initially in 20 second periods; there is strong evidence from the data for a steepening of the index of the mass distribution for masses greater than 10^{-13} kg during passage through dust jets which is not within the error limits of statistical uncertainty. The fluence for dust grains along the entire trajectory is calculated; it is found that extrapolation of the spectrum determined at intermediate masses (cumulative mass index $\alpha = 0.85$) is not able to account for the spacecraft deceleration as observed by the Giotto Radio Science Experiment and by ESOC tracking operations. Data at large masses ($>10^{-8}$ kg) recently analysed from the DIDSY data set show clear evidence of a decrease in the mass distribution index at these masses within the coma, and it is shown that such a value of the mass index can provide sufficient mass for consistency with the observed deceleration. The total particulate mass output from the nucleus of comet P/Halley at the time of encounter would be dependent on the maximum mass emitted if this change in slope observed in the coma were also applicable to the emission from the nucleus; this matter is discussed in the text.

The flux time profiles have been converted through a simple approach to modelling of the particle trajectories to yield an indication of nucleus surface activity. There is indication of an enhancement in flux at $t \approx -29$ s corresponding to crossing of the dawn terminator, but the flux detected prior to crossing of the dawn terminator is shown to be higher than predicted by simple modelling. Further enhancements corresponding to jet activity are detected around $+190$ s and $+270$ s.

Key words: comets – interplanetary medium

1. Introduction

At 0004h UT on 1986 March 14 the European Space Agency's Giotto spacecraft passed at a distance of 604 km from the nucleus

Send offprint requests to: J.A.M. McDonnell

of comet Halley. The Dust Impact Detection System DIDSY recorded the impacts of particulates from 287000 km pre-encounter to 202000 km post-encounter. Mass sensitive detectors comprising piezoelectric systems, thin film capacitor and impact charge systems, determined the intercepted particulate flux at $68\,km\,s^{-1}$ enabling the mass distribution to be determined. The Particulate Impact Analyser (PIA), recorded numerous impact spectra and also provided flux rates at the closer regions of encounter; the flux rates are especially valuable at a time close to encounter when DIDSY data was showing anomalies on the IPM sensor. A full description of the DIDSY and PIA instruments can be found in McDonnell et al. (1986a) and Kissel (1986).

Preliminary results for some of the pre-encounter data have been presented by McDonnell et al. (1986b). With the recovery of most of the post-encounter data, a more firm assessment of the instrument performance has been achieved. Methods of converting observed impact rates to true fluxes are described which allow derivation of mass distribution indices as a function of time. The total mass and area of grains intercepted by Giotto can then be calculated and compared respectively with the spacecraft deceleration as measured by the Radio Science Experiment (Edenhofer et al., 1986a) and groundbased imaging data at the same epoch. The measured mass distribution represents dust grains actually encountered along the spacecraft coma trajectory, whereas the quantities of more fundamental astrophysical interest are related to the emission from the nucleus and must await better modelling before conclusions can be firmly drawn.

2. Data processing and instrument performance

2.1. Dust Impact Detection System (DIDSY)

Data from the DIDSY experiment are available from two types of source, namely digital and analogue. Analogue data comprise an integrated measure of the event rate and event amplitude and are sampled infrequently compared with the digital data. Such analogue data are transmitted by the digital microprocessor without modification and are useful in several ways: (i) as an average measure of front-end detector activity to assess noise levels, (ii) to confirm that digital samples are representative of the actual impact activity, and (iii) to give a measure of average activity when impact rates saturate the digital processor. Except in the case of the plasma sensor in DID 1 (IPM-A), such analogue data is not presented here although they have been shown to exhibit similar behaviour to the digital records presented. Digital data are acquired whenever any one of several sensors is stimulated. All outputs are then sampled within relevant subgroups over a period of milliseconds. The logarithmically compressed amplitude set is then operated on via software algorithms to increment counters (bins) or to be transmitted as *discrete data* representing the actual set. Various priorities operate, dependent on the data and its generation rate. For all the piezo-electric sensors, when a measurement is made, a voltage is fed back to "back-off" that channel. The back-off voltage decrements according to the acoustical decay constant of that sensor to avoid echoes or multiple reads. A feature of this processing is that high event rates are compressed but it should be noted that the larger impact events in the higher mass binned channels are never masked because they exceed previous back-off signals from smaller events; large events generally have priority but not in the further selection of the smaller set of discrete data, where the selection is unbiased as long as the appropriate selection criteria are satisfied. The feedback process is deterministic in operation, and hence from the actual transmitted data for each data gathering interval (DGI), the probable input flux rates can be deduced for each channel. The discrete data, which are limited in availability, are subject to random selection within each category, namely the last such event. In the interim data presented herein, discrete data are used only for the large mass dust shield events; analysis of the discrete data has, however, been used to confirm the data binned in the various channels used here and will be the subject of more detailed future presentations. A more complete system description is given by McDonnell et al. (1986a) and McDonnell (1987).

2.2. Particulate Impact Analyser (PIA)

The Particulate Impact Analyser (PIA) experiment on Giotto is a time-of-flight mass spectrometer and is therefore primarily designed for compositional analysis of dust grains with submicron dimensions (Kissel, 1986). As part of its system the experiment includes three impact rate counters, two of which detect the incident number of small and large particles on the experiment, while the third counts the number of events which produce spectra. Due to the unexpected operation of the DIDSY IPM sensor, little information is available on the low mass ($<10^{-13}$ kg) flux. The PIA impact rate counters are therefore invaluable in defining the flux within this gap. Impact rates are determined from the three front end channels of the experiment, namely the target and acceleration grid (via charge sensitive amplifiers) and a photomultiplier to detect the flash. The channels can be switched between two sensitivities, each having a dynamic range of 1000 and being separated by a factor of 100. The outputs of the front end channels are sent to three single channel discriminators. The impact rate counter is incremented by triggers from the lowest discriminator threshold, corresponding to a limiting mass (threshold sensitivity) of some 10^{-19} kg in high sensitivity and 100 times greater in low sensitivity. The large signal impact counter is triggered by the upper discrimination threshold and corresponds to 10^2 and 10^4 times threshold sensitivity in high and low modes respectively. The final impact rate counter is triggered only by a simultaneous detection by one or more of the front end detectors and does not directly correspond to a limiting mass. The number of coincident channels required depends on the telemetry rate and the relative number of large particle impacts to particles which produce spectra. The sensitivity mode of the channels is similarly determined from the telemetry and impact rates, and switchover is checked after 4 s of high-sensitivity time. Once selected, low-sensitivity lasts for 4 s before automatically reverting to high sensitivity. This provides the equivalent of three different mass channels, one at threshold, two at 10^2 times threshold and one at 10^4 times threshold near encounter when switching between modes occurs. Assessment of the in-flight PIA thteshold sensitivity is currently under review; it should be appreciated in this context that measurement of flux rates is not the primary purpose of the PIA instrument, and that further refinement of the threshold may be possible by e.g. comparing fluxes from the similar instrument, PUMA, with the other onboard dust detectors on Vega 1 to reduce calibration uncertainties.

2.3. DIDSY instrument performance

At encounter all impact detection systems were operating within specified temperatures and voltages. No events had been detected

Table 1. Giotto DIDSY performance summary. The instrument normally cycles through modes requiring DID (2 or 3), DID 2, DID 3, DID (2 and 3) every format, but the latter was selected by a transient at encounter. DIDs 2, 3 and 4 fluxes are combined where appropriate in the analysis. Area of sensors [a] is dependent on a weighting according to the mass distribution. Figures shown are for $\alpha = 1.0$. Effective mass sensitivity assumes momentum enhancement of 11 for non-penetrating particles

Sensor sub-system	Limiting mass (kg)	Geometric area (m^2)	Effective area	First detection (time and distance)	Last detection
DID 1 (IPM-M)	$4\ 10^{-13}$	$1.19\ 10^{-2}$	$1.19\ 10^{-2}$	-24^m30^s (101 10^3 km)	$+41^m49^s$ (172 10^3 km)
DID 1 (IPM-PA) (sensor open)	10^{-20}	$5.96\ 10^{-3}$	$5.96\ 10^{-3}$	-61^m45^s (253 10^3 km)	Under analysis
DID 1 (IPM-PB) (sensor + foil)	10^{-20} (+2.5 μm foil)	$5.96\ 10^{-3}$	$5.96\ 10^{-3}$	-36^m09^s (148 10^3 km)	Under analysis
DID 2 or 3 (mode 3)	$4\ 10^{-12}$	1.94	$6.17\ 10^{-1}$ [a]	-64^m41^s (265 10^3 km)	
DID 2 and 3 (mode 0)	$1.0\ 10^{-10}$	1.94	$2.80\ 10^{-1}$ [a]	-7^m14^s (30 10^3 km)	$+7^m49^s$ (32 10^3 km)
DID 4	$4\ 10^{-12}$	$2.2\ 10^{-1}$	$7.88\ 10^{-2}$ [a]	-70^m00^s (287 10^3 km)	$+49^m11^s$ (202 10^3 km)
Mode 3 was deselected at around closest approach and therefore did not operate on outward leg.					
DID 5 (RSM)	10^{-9}	2.2	~0.2	-3^m00^s (12 10^3 km)	$+1^m05^s$ (4.5 10^3 km)
DID 7 (CIS)	$3\ 10^{-13}$	10^{-1}	10^{-1}	-33^m59^s (139 10^3 km)	-9^s (malfunction) (862 km)

[a] Depends on incident mass spectrum (values quoted are for a cumulative mass index of 1.0)

during the cruise phase, representing (as expected) many days of normal interplanetary operation. The IPM-A plasma sensor unit on DID 1 was not showing its intended voltage but this was typical of its status throughout cruise and it had been concluded that this unit had suffered a malfunction at launch. Following the first impact event, on DID 4 at $t = -70^m00^s$ (287000 km), all sensors registered data increasing through to the last minute of approach, when several anomalies relating to the impact scene were observed: 1) the DID 6 processor block counter lost synchronism with the expected position in the data stream, 2) an uncommanded DIDSY mode switch was effected and the spacecraft interface showed signs of intermittent operation, 3) the IPM-P unit gave a burst of zeros and then became "noisy", 4) the IPM-A output, which had been quiescent, became noisy and then rose towards saturation for a short period. Data was then received intermittently with some periods of loss of signal; full data coverage was not regained until 32 min after encounter. Table 1 shows a summary of performance, while Table 2 gives details of mass sensitivities for the different sensors and appropriate mass channels.

Subsequently the best records from three Reed-Solomon decoded tapes from the Parkes Receiving Station, Australia and from NASA's DSN were merged. Even these records were found to have substantial gaps in the data very close to encounter, where activity was high and system performance atypical; partial data frames have, however, been able to be extracted very much more effectively "by hand", forming a record where the longest gap without any data at all is now only 22 s, and 30 s for binned data (centred on encounter). First analyses confirmed nominal behaviour on all piezo-electric systems, both pre- and post-encounter, with the exception of the higher sensitivity DID 1 IPM-M sensor, which showed a varying ratio when compared with the shield sensor DID 4.

Records of the IPM-P sensor were analysed prior to the onset of noise only and IPM-A records were not recovered until August 1986. At that time, flux rates from the PIA sensor became available for the last 5 min of encounter; in the very short period available when PIA yielded flux data and before IPM-P became noisy, PIA showed an excess flux of up to two orders of magnitude compared with the IPM-P sensor rates at an equivalent mass.

A critical and full appraisal of all IPM-P, −M and −A data was effected and presented to the DIDSY Co-Investigators at the University of Lecce (September, 1986). In order to protect this IPM unit during the solid propellant firing for injection of Giotto into its Halley intercept trajectory, a thin cover required to be of almost "zero" weight, had been provided as a late modification to the unit. Release of this cover was commanded on February 10th, 1986 and the activating current monitored successfully through

Table 2. Mass thresholds used in this interim assessment for DIDSY subsystem channels and for the Particulate Impact Analyser (PIA)

Sensor	Channel	Mass thresholds (kg)
DID 1-M	1	$4\ 10^{-13}$
	2	$1\ 10^{-12}$
(IPM-M)	3	$5\ 10^{-12}$
	4	$4\ 10^{-11}$
DID 2+3	1	$4\ 10^{-12}$
	2	$5\ 10^{-11}$
	3	$7\ 10^{-10}$
DID 2.3	1	$1\ 10^{-10}$
DID 4	1	$4\ 10^{-12}$
	2	$6\ 10^{-11}$
	3	$4\ 10^{-10}$
DID 5	1	$1\ 10^{-9}$ (shield penetration)
PIA	1	ca. 10^{-19}
	2	ca. 10^{-17}
	3	ca. 10^{-15}

telemetry records. The absence of a signal on IPM-A, which might have then (but only marginally) seen a photoelectric current, was disturbing and several more extended firing pulses were transmitted subsequently during cruise. Only at encounter, however, would the true test occur, namely the observation of flux differences between the IPM-P A and IPM-P B sensors, since the B sensor incorporates a 2.5 µm foil and would filter out the smallest particles. If the cover were in position, then its 22 µm thickness would largely mask the effect of an extra 2.5 µm. In the early pre-encounter period, vastly different flux rates were indeed observed at the A and B sensors in line with cover retraction; in the total absence of knowledge of what the size distribution from Halley would be along the trajectory, approaching the nucleus from the dark pre-dawn side, such data was interpreted as being indicative of the true in-situ distribution.

Re-examination of the data set available from the decoded tapes after encounter showed that the cover must have remained in place, although it had never failed on test and had been carefully scrutinised at Kourou by the Principal Investigator and Project Manager. A more detailed explanation of the experimental tests subsequently conducted is offered by Maas et al. (1986); this shows how the IPM-A and B flux differences are due to the ability of the thin foil to stop the *plasma* which is generated after the thin cover foils acted (as meteoroid bumper) to vapourise the particles. The IPM-A "noisy" sensor records are now clearly interpreted (Fig. 1) in terms of the abrasion of this foil by dust; the ratio of the flux rates of the IPM-M sensor to those deduced from DID 4 and PIA data is also used to demonstrate this abrasion process (Figs. 2a and 2b) which commences at −70 s and is near-complete by −10 s. This ratio would be expected to be unity in the absence of a cover and this is indeed the post-encounter behaviour. A possible under-estimation of the sensitivity of the IPM-M channel 1 sensor might be suggested which is unlikely to arise from any remains of the cover. We have concluded therefore beyond doubt that post-encounter the IPM-M data is valid, with the possible reservation concerning the assigned sensitivity level of channel 1; pre-encounter records used for processing have therefore been purged of such IPM-M and IPM-P data presented at this time. IPM-P data is now being re-assessed to assign the various events detected to the larger masses which are required to penetrate this cover before its abrasion. Flux profiles for some of the validated DIDSY channels are shown in Fig. 3.

Design and development of the IPM cover was a Principal Investigator responsibility and effected at the Rutherford Appleton Laboratory. Prior to flight, it had met all test requirements and performed faultlessly in the qualification programme. The hot soak experienced by Giotto at temperatures of up to 150 °C for several months following launch was a factor which could not, of course, be simulated on the flight spacecraft. One suggestion for failure is that this led to a suspected deformation of the stressed release heater element, thus eliminating the contact with the fusible nylon loop.

Data from the CIS sensor yields a valuable independent validation of the pre-encounter flux since it operates by a penetration technique whose calibration is relatively straightforward and quite independent of the momentum and impact plasma sensors (McDonnell et al., 1986a). This sensor operated correctly until about −16 mins before encounter at which time an impact-related malfunction occurred. However at −40 s prior encounter, correct operation resumed until closest approach. After this time, no further impacts were registered despite a correctly indicating monitor voltage, implying a failure due to the very high impact rate at closest approach. During this 40 s period prior to closest approach, CIS data can be compared with that from the other sensors; over the period −40 s to −20 s this shows reasonable agreement between data from different sensors though it should be appreciated that the IPM cover may not have been fully abraded by this time. All data on fluxes is normalised relative to the percentage telemetry coverage with the exception of the close encounter loss where for the fluence a $1/R^2$ cometocentric dependence is used for interpolation. This does not account for the possibility of very short duration jets in the period, but the existence of these is not supported by the PIA flux rates which bridge this period and have a cumulative spectrum event identifier which is read after the telemetry gaps. Closure of the PIA cover is found to be only some 20 % for a period of 4.5 s near to encounter, thus arguing against abnormally high flux rates.

2.4. PIA instrument performance

Spectra were obtained over a wide range of sensitivities. From the event counter it was seen that the event priority switching was operating to require higher levels of signal coincidence as the impact rate increased. System sensitivity and priority level changes also switched in accordance with the increasing flux of larger particles; only within the last minute was a signal to restrict the aperture shutter detected, and the overall flux rate thereby restricted by some 20 % for a matter of 0.5 min. Analysis of the impact rate on PIA at three mass levels (nominally 10^{-19} kg, 10^{-17} kg and 10^{-15} kg) has provided the time profiles shown in Figs. 4 and 5. These have been combined with the DIDSY flux. Comparison of the similar PUMA sensor on Vega with the SP1 and SP2 data (Mazets, 1986; Vaisberg, 1986) indicates good agreement between these different sensing techniques and relative calibrations.

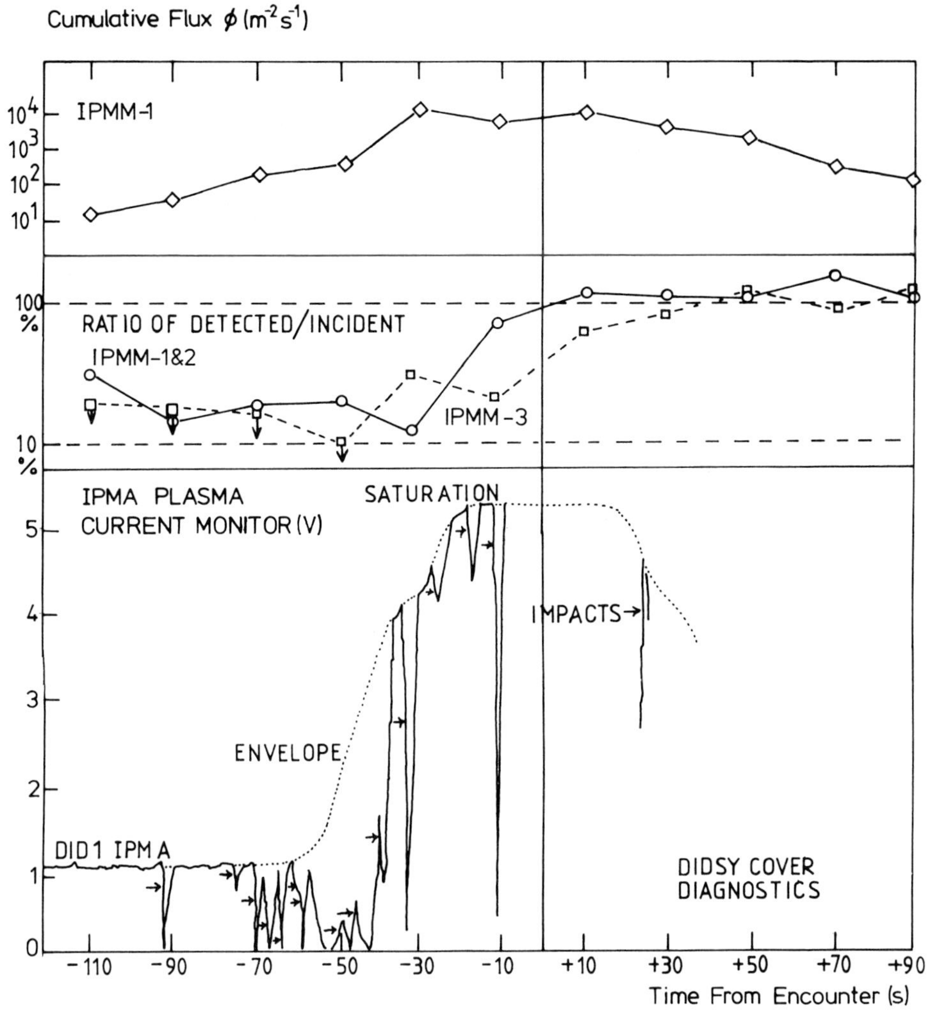

Fig. 1. DIDSY IPM diagnostics. Top: flux rate incident on IPM-unit throughout −110 s to +90 s (note logarithmic scale) deduced from DID 4 and PIA flux data. Centre: flux rates from IPM-M sensor behind cover relative to incident flux. Post-encounter rates are consistent with total erosion of the cover. Lower: plasma current monitored by IPM-A sensor behind cover (Grard et al., 1986). Impacts, many coincident with IPM-P events, are seen as a result of electron collection from cover penetrations. The envelope shows a rapid approach to saturation as the cover is abraded at approx. −35 s to −10 s. Recovery to the quiescent status on IPM-A is not completed after encounter due to contamination or damage

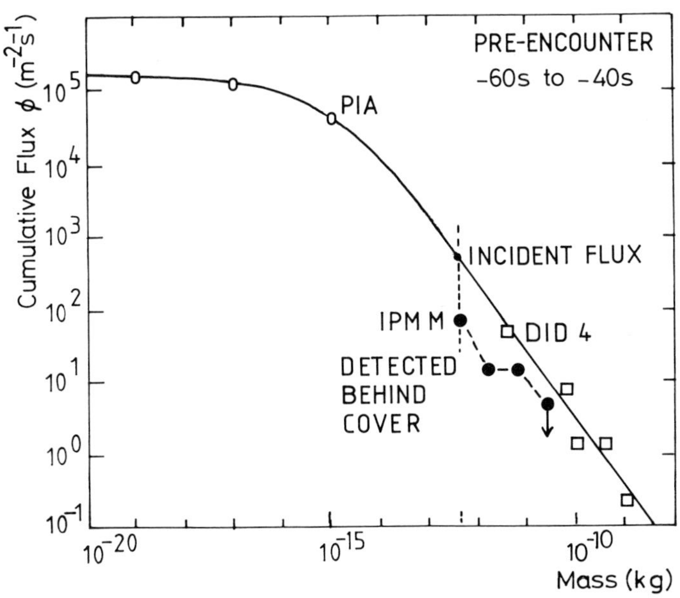

Fig. 2a. DIDSY and PIA fluxes at −50 s showing incident flux a IPM-M channel 1 mass threshold derived from interpolation between sensors without flight cover. A reduction in flux corresponding to a mass threshold for perforation of approx. 10^{-10} kg is seen

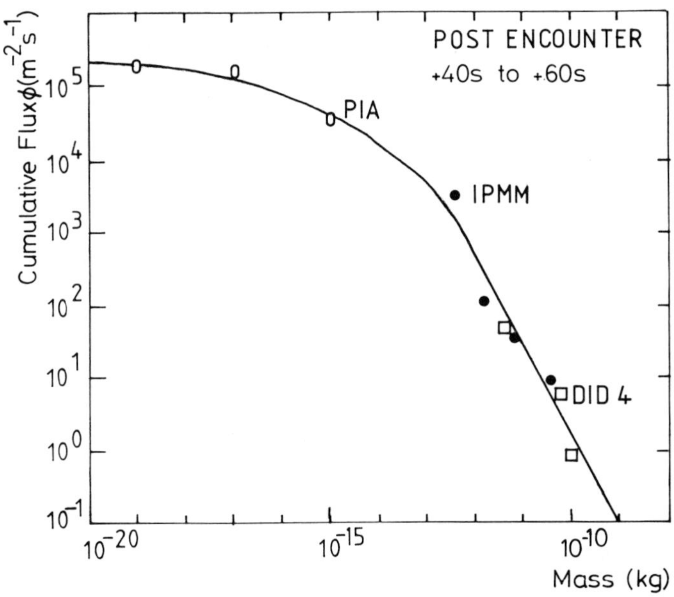

Fig. 2b. At +50 s the cover is effectively abraded by multiple perforations and flux levels are nominal for IPM-M

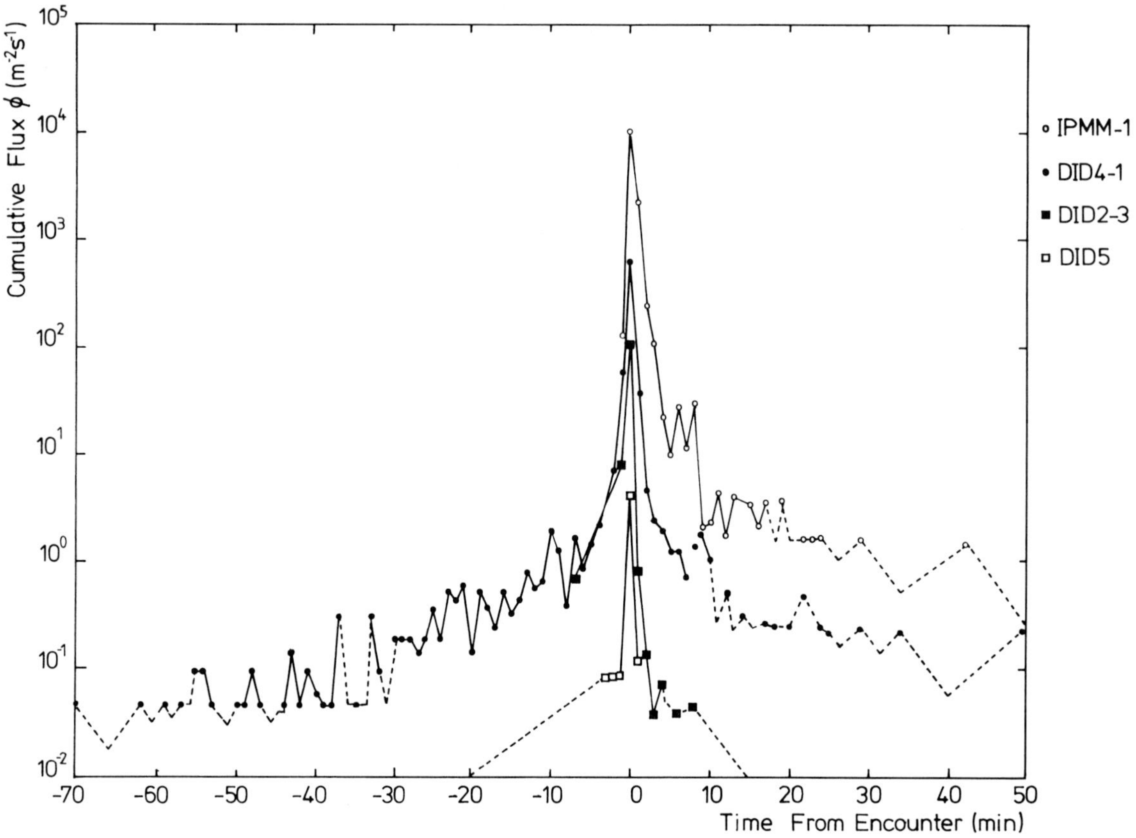

Fig. 3. Flux rates for sample channels of DIDSY sensors throughout encounter in 1 min intervals. First and last detection by DID 4 (which is combined pre-encounter with DID 2 + 3 in this plot) yields an envelope corresponding to an apex distance of approx. 120000 km. Where no data are shown, an upper limit corresponding to one event is sketched (dashed line). Expanded data are shown in Fig. 4 and 5. Mass sensitivities are shown in Table 2

2.5. DIDSY and PIA flux rates

Impact rates very close to encounter for PIA and DIDSY (plotted from −5 min to +5 min and −1 min to +1 min in Figs. 4 and 5); comparison is offered with an idealised isotropic flux distribution varying inversely as the square of the nucleus distance for the close encounter data. The pre-encounter IPM-M data is indeed suppressed relative to the form of the other sensors because of the cover remaining at this time and this data is not used subsequently; pre-encounter data at this mass is derived from interpolation between the PIA flux rate and the DID 4 sensors. Questions regarding the total mass intercepted by the spacecraft area of 2.9 m^2 deduced from the flux measurements of PIA and DIDSY should be considered in the context of the form of the flux data very close to encounter; such peak values could dominate, even in some 30 s, the total mass encountered elsewhere along the trajectory. The significance of this will be discussed below but at this stage it might at least be noted that a $1/R^2$ functional interpolation over the missing 30 s would not appear to be grossly inappropriate; the data does show a somewhat steeper dependence which is being analysed in terms of an additional angular dependence from the nucleus and subsolar point. Extreme fluctuations in fluxes during the 30 s missing from the DIDSY data are constrained by the profile of the cumulative PIA data at $m \leq 10^{-15}$ kg which is more complete throughout this period. Saturation in the lower mass PIA channels could be significant in this period but not for the PIA threshold of 10^{-15} kg.

2.6. Data processing

Analysis is performed for DIDSY in the first instance within instrument cycles; corrections for "dead-times" following an impact are applied but this is required only within one minute of closest approach. Where records are incomplete within an averaging period, the flux is normalised to the data coverage. Close to encounter this selection effect of missing data is significant for two periods, at +10 s and −10 s for the 20 s averages for DIDSY. Fluxes from different sensors are plotted as cumulative with respect to mass, although in analysing detailed behaviour of sensors or mass indices, the actual sensor counts within channel boundaries may be used.

When required, a least squares fitting of equations is performed which are either single power laws or polynomials in the log flux/log mass plane: these equations may be used to bridge gaps between sensor channels but where available, actual flux counts are used. For PIA, the area is 4.7 cm^2 except for a 20% reduction in a period of 4.5 s near encounter. For the DIDSY dust shield sensors, the area is usually an *effective one*, depending on the slope of the distribution. This has been computed using detailed shield mapping measurements, and is described below (Sect. 2.7). For the IPM-M sensor unit, the area is a geometric one and is independent of the mass distribution (119 cm^2).

Discrete multiple detection data from DIDSY's shield detectors DID 2, 3, 4, and 5 (generally the large mass data) is corrected for a reduction of the signal due to penetration.

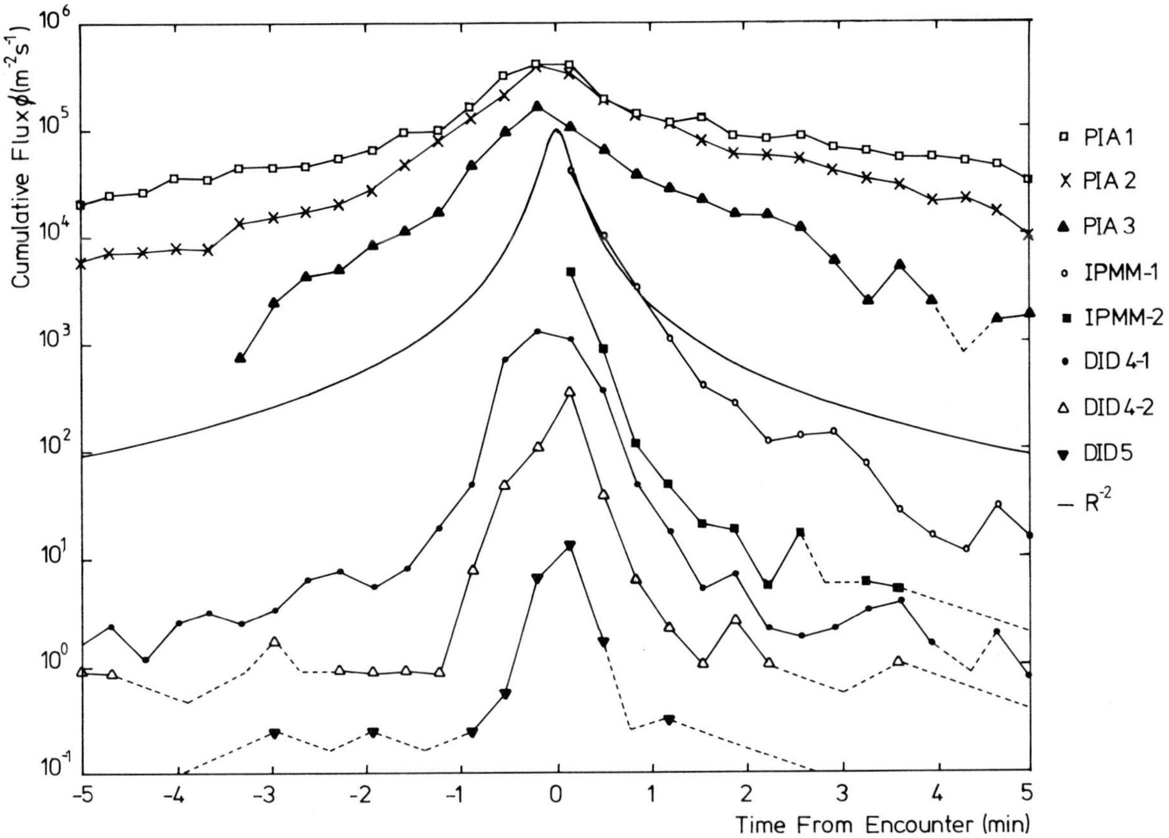

Fig. 4. DIDSY and PIA fluxes from -5 min to $+5$ min averaged in 20 s periods. A $1/R^2$ Halleycentric flux dependence is shown for comparison. Where no lines are shown, no events were detected and upper limits apply (dashed lines). Mass sensitivities are shown in Table 2

Momentum sensitive detection is shown to be appropriate (McDonnell, 1969; McDonnell et al., 1984); above the penetration limiting mass m_{pen}, the detected signal on the 1 mm shield (but not the total momentum) is derated to a value of $11.(m_{pen}/m)^{\gamma}$ at incident mass m. This function is derived by extrapolation to $68.4\ \mathrm{km\,s^{-1}}$ from lower velocity measurements and we assign a probable value of $\gamma = 2/3$ (Wallis, 1986). In this analysis we perform the calculations for a range of values of γ, in order to assess the sensitivity of the implications to this assumption. Especial attention is focussed on processing of the close encounter period -5 min to $+5$ min which contains some 95% of detected mass. For these large events (multiple detection) the effective sensor area approaches the actual geometric shield area and hence there is less dependence either on calibration or on the incident mass distribution although the data set is smaller and must be corrected for selection effect during close encounter (below).

2.7. *Sub-system effective areas*

Determination of incident flux rates requires division of the raw impact rates by the area-integration time product. For the piezoelectric microphone sensors mounted on the Giotto front dust field (DID 2, DID 3 and DID 4), the determination of the effective area requires considerable care since the entire geometric area of the front shield (approx. $2\ \mathrm{m}^2$) is not effective in detecting incident dust particles except for the larger masses; this results from the attenuation of acoustic waves in the shield during propagation towards the appropriate sensor. The overall effect of this attenuation is to produce a large variation in sensitivity over the dust shield and so the effective area for any sensor or sensor mass channel is determined by convolution with various values of the incident mass distribution function. This is illustrated in Fig. 6, where the effective area for various of the DID 4 channels is plotted as a function of the mass distribution slope, assuming a power law function of the form

$$N(>m) = Km^{-\alpha}. \tag{1}$$

These values are sensitive to the value assumed for the penetration limit of the front dust shield (taken as 10^{-9} kg, McDonnell, 1979). Uncertainty in this value may lead to an error in the determination of α of approximately 0.05. Effective areas used in this analysis are shown in Table 3; for IPM-M the geometric area is used because of uniform sensitivity. A different assumption for the penetration limit would affect channels corresponding to higher masses more critically and, additionally, such a change would be a function of mass index. For example, if one assumes a penetration limit of $4\ 10^{-9}$ kg, the effective areas for DID^4 channel 3 (limiting mass $4\ 10^{-10}$ kg) would be higher by about 45% for $\alpha = 0.5$ and about 65% for $\alpha = 1.5$. A distinction between the marginal threshold for perforation and the critical penetrating mass for transmission of forward momentum (i.e. loss of detected momentum) should be made. We assume $5\ 10^{-9}$ kg for the latter in line with simulations (Arnadeau et al., 1984). Absolute sensitivity is determined from hypervelocity impacts on the 2 MV electrostatic accelerator at Canterbury.

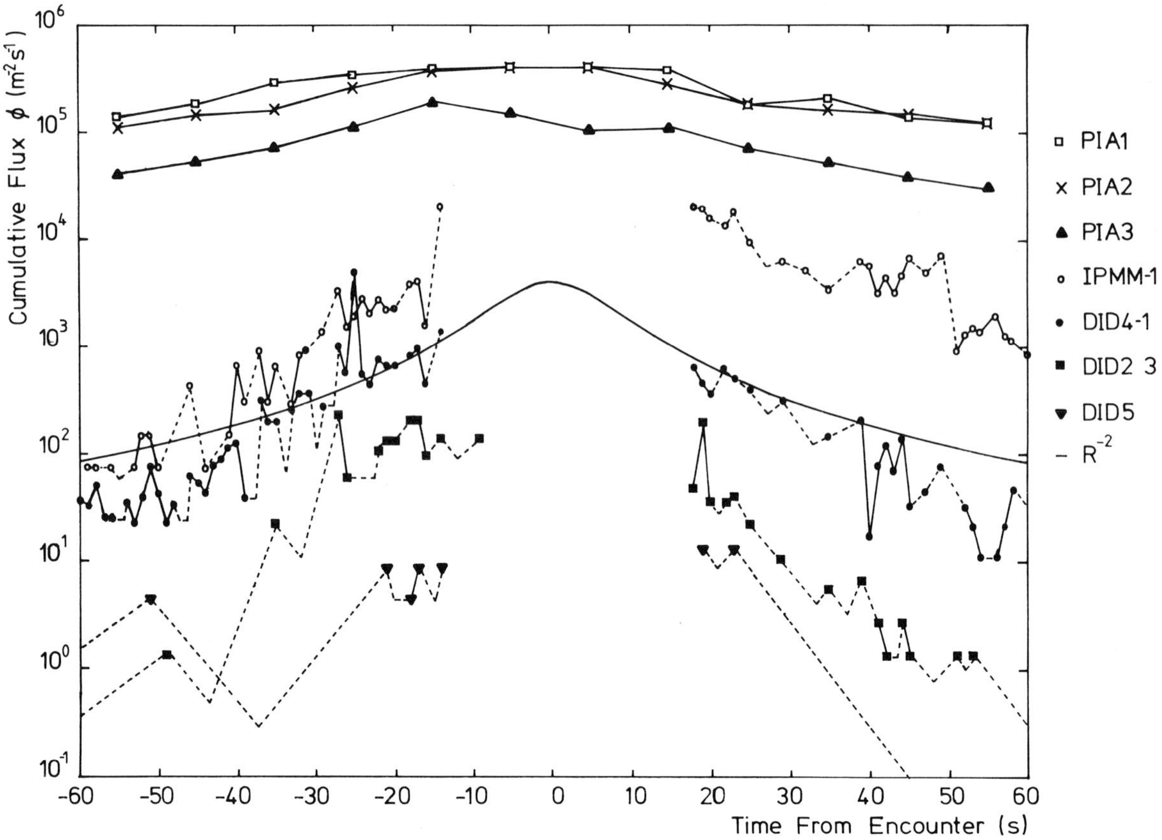

Fig. 5. DIDSY and PIA fluxes in the interval −1 min to +1 min. Note the steep rise of IPM-M flux as the cover erodes pre-encounter; the IPM flux before encounter is not used to derive fluences. Sensitivities are shown in Table 2. Saturation in channel 1 PIA data is present at this epoch

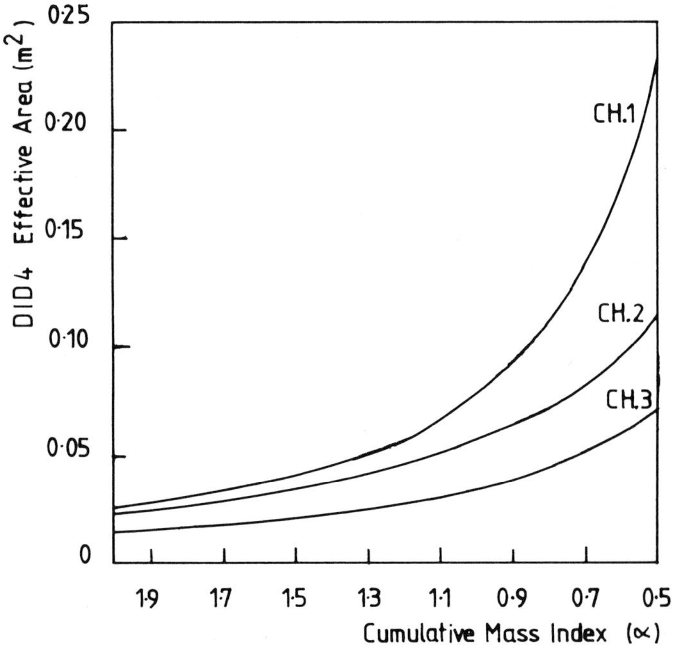

Fig. 6. Effective area as a function of cumulative mass distribution index (assuming a power law form) for the DID 4 detection subsystem. Three mass channels are shown corresponding respectively to: channel 1: $m > 4\ 10^{-12}$ kg; channel 2: $m > 6\ 10^{-11}$ kg; channel 3: $m > 4\ 10^{-10}$ kg

Table 3. Effective area in cm^2 as a function of cumulative mass index (α) for the various sensors used in this analysis. For limiting masses for the sensor channels, see Table 2

Sensor	Cumulative mass index (α)				
	0.5	0.8	1.0	1.2	1.5
DID 4 Channel 1	2324	1097	788	580	403
Channel 2	1152	722	570	460	342
Channel 3	707	440	339	271	203
DID 2 or DID 3					
Channel 1	8085	4124	2798	1987	1275
DID 2 and DID 3					
Channel 1	10650	7628	6168	5044	3810

3. Flux time profiles

Data presented from PIA and DIDSY in Figs. 3–5 show a distribution not inconsistent with pre-encounter modelling of the dust environment (Divine et al., 1986). An approximate inverse square dependence is observed close to the nucleus but with significant deviations:

(i) The traversing of jets post-encounter is identified and changes of mass index (Sect. 4).

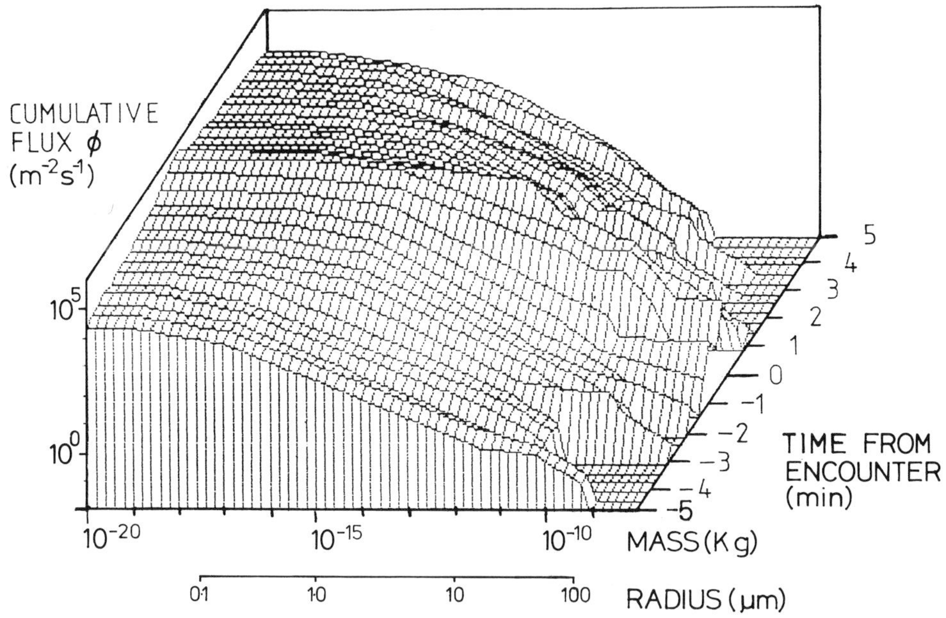

Fig. 7. Cumulative flux distribution for DIDSY and PIA from -5 min to $+5$ min for masses $m < 10^{-8}$ kg. Approximate particle sizes using the Halley environment density model are shown

(ii) The pre-encounter flux is lower than a $1/R^2$ dependence and appears to be associated with an approach to the nucleus from the dark side; a dawn enhancement is clearly seen in the data at around -30 seconds but not the sharp transition anticipated (Sect. 7.3).

(iii) Although PIA flux data may demonstrate saturation at highest sensitivity and flux rates, the flux profiles at 10^{-15} kg show evidence of a relative depletion at closer than 4000 km. This would be consistent with a generation of these small particles from the fragmentation (or less probably ablation) of larger particles *after* liberation from the nucleus. If this is accepted then it would be quite reasonable to attribute the "flat" spatial distribution of the smaller mass channels of about 10^{-19} kg as due to the generation within a distributed region of some 10000 km of the nucleus. This would imply a low local velocity for these femtogram particles.

(iv) Intermediate and large mass profiles show a steeper than $1/R^2$ flux dependence at >2000 km nucleus distance in the post terminator region which is associated with an enhanced sunward emission. Fluctuations appear to be less than one magnitude, and typically a factor of ± 2 about the "mean" level applicable at any distance. At larger distances, fluctuations are statistical in nature.

(v) The envelope of the global dust coma distribution may be estimated e.g. at mass 4 10^{-12} kg (DID 4). It yields a width along Giotto's trajectory close to the first and last detection (at -70 and $+50$ min) which implies an apex distance of some 120000 km for this mass. Pre-encounter modelling (Divine et al., 1986) had suggested figures less than half this value. Two factors could be used to account for this, namely (i) higher ejection velocities perhaps resulting from the dominance of jets as the principal ejection mechanism or (ii) optical properties differing from the model parameters.

Figure 7 shows the cumulative flux distribution throughout the 10 min encounter period for masses $<10^{-8}$ kg. This is processed firstly to yield a derivative in mass so that differential fluxes may be used to calculate e.g. the area distribution (see Fig. 13) or alternatively integrated in time to yield fluences. The fluence offers a better measure of the average mass distribution at Giotto and further derivatives for the nucleus emission are offered (Sect. 5).

4. Mass distribution

4.1. Fitting of mass distribution function

For the cumulative mass distribution function, we assume for the purpose of this analysis, a function of the form of Eq. (1). A weighted least squares fit has been carried out for the log cumulative flux from 20 s integration periods (or several such periods combined when the count rate is low). Each cumulative flux data point is weighted according to the inverse square of the standard deviation of that particular measurement. This is important as the standard deviation in a set of flux data points can typically vary from a few percent to 50 %. The uncertainty is made up of two contributions, namely (a) a $\sqrt{N}$ uncertainty in the incident count N and (b) an uncertainty in the dead time correction factor, K. Under low impact rate conditions, the incident flux is obtained simply by dividing the number of impacts for a particular sensor by the appropriate area time product. However, for a relatively short period around closest approach, the impact rate is sufficiently high to cause some impacts to fail to be detected by the relevant detection sub-system. Such a correction is necessary for certain sub-system channels over the period -60 s to $+60$ s. In the most extreme case the correction factor, K, by which the detected count rate must be multiplied to obtain the true incident count rate, amounts to 2.44 averaged over the appropriate 20 s period. We have taken the uncertainty in the correction factor to be ± 50 %. Generally, it is the statistical uncertainty that dominates except in the few instances where the deadtime correction is large.

This analysis is based on binned data only. The discrete data, which can potentially offer greater accuracy in mass determination, is currently being examined to confirm and extend the analyis to date on the mass spectrum and its variation.

4.2. Results

The best fitting power law index, α, was derived from the cumulative mass distribution function for single (or multiple) 20 s periods from -300 s to $+220$ s. There are not sufficient data in the

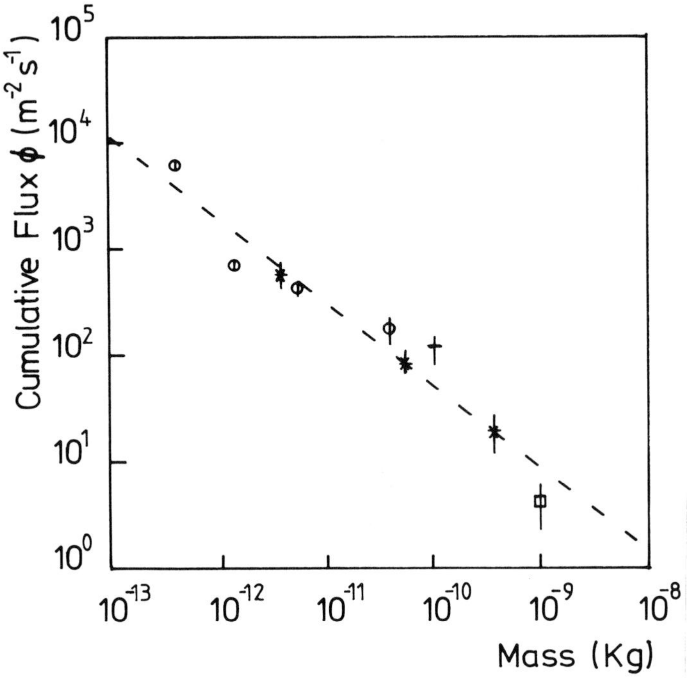

Fig. 8. Cumulative mass distribution for the period $t = -20$ s to $t = -14$ s. Data points are for the following sensor subsystems: ϕ IPM-M; ∗ DID 4; + DID 2/3; ϕ DID 5. Errors are 1 sigma error bars based on count rate and dead time correction uncertainties

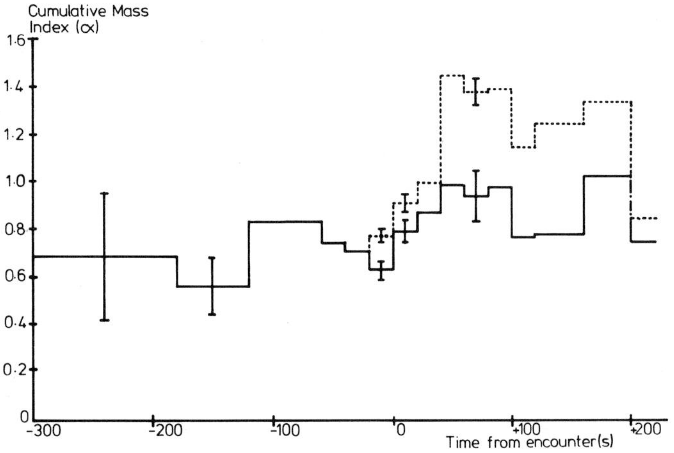

Fig. 9. Cumulative mass index as a function of time in 20 s intervals. Up to -20 s the relevant mass range is $m > 4\ 10^{-12}$ kg. From -20 s, two values are shown corresponding to $m > 4\ 10^{-13}$ kg (dotted line) and $m > 1.4\ 10^{-12}$ kg (solid line)

period $+220$ s to $+300$ s to allow an accurate fit to be attempted. A typical spectrum is shown in Fig. 8, while the variation of α with time is shown in Fig. 9. For the period from -10 s, two values for α have been plotted depending on whether the most sensitive mass channel (IPMM-1, $m > 4\ 10^{-13}$ kg) has been included. This is because this one flux point does appear to lie marginally but consistently above the extrapolated spectrum from the other flux points. At present we are unable to determine if this is a genuine effect resulting from a change in spectral slope in this mass region or whether it results from a systematic error in the calibration of this particular mass channel. It should be noted that no such effect is noted for any other channels, which have been calibrated by similar techniques. Typical error limits in the determination of α are also shown.

In order to assess whether we are justified in using a power law form to fit the cumulative mass distribution function, the linear correlation coefficient has been determined for all the data considered here. This can then be used to determine quantitatively whether the data points could represent a sample derived from an uncorrelated parent population or whether the variables are related by the assumed power law form. Of 25 spectral fits performed all but 2 are found to satisfy a power law relationship at the 90% confidence level and in many cases at 95% or better. The two cases which do not satisfy this criterion correspond to the data periods -300 s to -200 s and $+120$ s to $+160$ s (lowest mass channel excluded). This may well be due to the relatively low number of impacts in these periods resulting in poor statistical precision. The generally high degree of correlation is taken to be a justification for using the power law form and it certainly provides, therefore, a very good indication of the slope of the cumulative distribution function. However, it does not follow that the correct functional form of this distribution is necessarily a power law.

There is a clear indication of a significant change in the spectral slope starting at around $+20$ s compared with that over the period -300 s to $+20$ s. This change is apparent from the data both with and without the lowest mass channel but is clearly more marked if this channel is considered. Subsequently it appears that the spectral slope returns to its previous lower value at around $+200$ s. This change of slope, to a higher value, in the immediate post-encounter period compared with the pre-encouter value had already been noted in the initial publication of DIDSY data (footnote to Table 2 in McDonnell et al., 1986b).

Considering the values of α derived excluding the lowest mass channel (shown by the solid line in Fig. 9 after 0 s), the progressive increase in its value from about $+20$ s is regarded as meaningful bearing in mind the relatively high precision in determining α around this period (indicated by the appropriate error bars in Fig. 9). It might appear that the increase in α should be described as starting at closest approach to Halley (i.e. 0 s). However, due to the data loss around closest approach (described in Sect. 2.3), the data period 0 s to $+20$ s contains only data over the approximate period $+18$ s to $+20$ s; for this reason, the change in slope is therefore referred to as beginning at around $+20$ s.

One method of studying the significance of the change in slope of the mass distribution function is to look for correlation with the absolute fluxes measured by the DIDSY sensors. Because the overall variation of flux detected follows a dependence of approximately $1/R^2$ and similar behaviour was observed during the two Vega encounters (Vaisberg, 1986b; Mazets et al., 1986; Simpson et al., 1986), intrinsic variations in the spatial density of the dust can easily be masked. To remove this variation the detected flux can be normalised by the appropriate $1/R^2$ factor; such data is presented in Fig. 10. When displayed in this form, completely uniform emission from the nucleus would appear as a flat distribution. What is clear from this data is that considerable relative increase in dust efflux was detected by the Giotto DIDSY sensors at around $+20$ s. Similar behaviour is seen in most of the other mass channels (not all displayed in Fig. 10). It is only in the channels corresponding to the highest masses where the effect is not necessarily apparent due, at least in part, to the relatively low fluxes at these high dust particle masses.

Further information on this local enhancement in dust density can be gained from the Giotto Radio Science Experiment (GRE, Edenhofer et al., 1986b, c). The latest analysis of data from this experiment (Edenhofer et al., 1986a) indicates those time periods during the Halley encounter phase when Giotto experienced greatest deceleration due to impacts on or through the front dust shield. Ten distinct periods when the deceleration is most significant have been identified, but most fall within the 30 s period when experiment data from the spacecraft was lost. However, the last of these periods of significant deceleration does fall within the period when data from DIDSY are available. It corresponds to the period from +20 s to at least +60 s although the exact cut-off time is not easy to define, yielding a duration of at the very least 40 s. The clearest interpretation of these events is to identify them with Giotto's passage through discrete dust jets which are now a well established feature of comet Halley (Keller et al., 1986; Sekanina and Larson, 1986a; Larson and Sekanina, 1984).

The flux enhancement shown in Fig. 10 clearly appears to be a manifestation of the same phenomenon – where GRE measures the resultant deceleration on the spacecraft, DIDSY measures the dust impacts on the spacecraft directly. The bottom profile in Fig. 10, which represents the normalised flux in the most sensitive mass channel considered in this analysis, seems to indicate that the dust enhancement or jet is very broad, extending from +20 s to perhaps +100 s. In fact there is indication of mass dispersion in this jet whereby larger mass particles are encountered before the smaller ones. Such an effect has indeed been observed by the SP-1 detectors on the Vega spacecraft (Vaisberg et al., 1986a, 1986b). This would arise from the combined effect of a rotating nucleus and the differing velocities of dust particles of different masses. Detailed analysis of such effects in the DIDSY data will be published later.

Other analyses of the DIDSY data indicate passage of Giotto through a dust jet at this time. Figure 18 shows the absolute flux as a function of time and a broad enhancement over a $1/R^2$ dependence is apparent. Figure 10 also shows some evidence of passage through another jet at around +140 s, supported by Fig. 18 which shows a fairly sharp enhancement occuring at a distance of about +10000 km from the nucleus. It is tempting to associate a second increase in α, evident in Fig. 9 at around +160 s, with passage through this jet also.

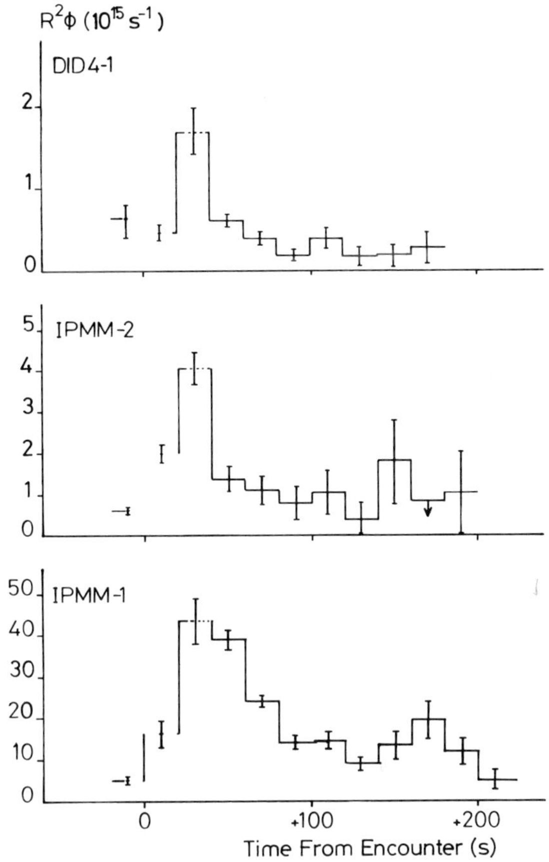

Fig. 10. Flux profile for the period around and just after closest approach, normalised with a $1/R^2$ dependence to remove the effect of varying distance of the measurement point from the cometary nucleus. Three mass ranges are shown corresponding to, reading upwards, $4\ 10^{-13}$ to $1\ 10^{-12}$ kg, $1\ 10^{-12}$ to $5\ 10^{-12}$ kg and $4\ 10^{-12}$ to $6\ 10^{-11}$ kg. The data are summed in 20 s integration periods and the ordinate is in relative flux units (linear). Loss of telemetry around closest approach results in loss of data in some of the 20 s integration periods. Where this loss amounts to more than 10% in a 20 s period, the time when data is collected is shown with a solid line. Loss of data is significant only over the period −20 s to +40 s

4.3. Discussion

We have considered three particular time spans here, namely (i) −300 s to −40 s, (ii) +20 s to +120 s, and (iii) +200 s to +300 s. These we identify with "background" (or quiescent), jet and "background" dust respectively, and have computed the average cumulative mass index for each period. In the case of the jet index, two averages have been determined corresponding to an average taken over time and flux respectively (Table 4). Because of the relatively low impact rate in the period +200 s to +300 s, only a relatively coarse determination can be made. Also quoted are values of α derived from large particles impacts ($m > 10^{-8}$ kg). These are from the DIDSY discrete data (Sect. 2.1 and McDonnell et al., 1986a) whereby for a small number of impacts, amplitude information from all sensors is recorded allowing a more accurate determination of incident mass to be made. The values derived indicate a much flatter spectrum in that mass range, which has significance in determination of total mass encountered by Giotto along its trajectory. Also of note is the value derived from the DIDSY data summed along the entire trajectory, namely 0.85 ± 0.05 (Sect. 5) for median masses. This value is intermediate between the low value of a corresponding to the "background" dust and the higher value corresponding to the jet dust derivel in this analysis. Thus, this may represent the weight mean value from these two separate sources or the average of two different coma distributions. Note that the value quoted from the model of Divine (Divine, 1981) has been converted from a value at the cometary source (1.07) to that tabulated which refers to the Giotto trajectory and takes account of the mass dispersion of the emitted grains due to their differing terminal velocities.

Also shown for comparison are values from the Vega dust detectors which tend to show a relatively wide range of values depending on the detector and mass range considered.

5. Fluence and nucleus emission rates

5.1. Fluences

The fluence $F(m, T)$ is the integral of the cumulative flux

$$F(m, T) = \int_0^T \phi(m, t)\, dt. \qquad (2)$$

Table 4. Cumulative mass indices derived in this work and compared with values from other sources

	Cumulative mass index (α)	Comments	Ref.
This work	0.69 ± 0.15	$t = -300$ s to -40 s	This work
		$m > 4\ 10^{-12}$ kg	"Background" dust
		$t = +20$ s to $+120$ s	Jet dust
	0.91 ± 0.12	$m > 1\ 10^{-12}$ kg	Time average
	1.27 ± 0.07	$m > 4\ 10^{-13}$ kg	
	0.90 ± 0.10	$m > 1\ 10^{-12}$ kg	Flux average
	1.13 ± 0.09	$m > 4\ 10^{-13}$ kg	
	$\sim 0.7-0.8$	$m > 4\ 10^{-13}$ kg	"Background" dust
	0.30 ± 0.09	$t = -300$ s to -40 s	"Large" mass ($m > 10^{-9}$ kg)
	0.46 ± 0.13	$t = -40$ s to $+40$ s	
Other sources	0.85 ± 0.05	$m > 4\ 10^{-13}$ kg	Summed over trajectory (this paper)
	0.66 ± 0.05	$t = -45$ s to -16 s	$m > 10^{-13}$ kg (McDonnell et al., 1986b)
	0.45	$10^{-16} < m < 10^{-14}$ kg	Vega SP-1 (Vaisberg et al., 1986a)
	0.9 to 2.0	$10^{-16} < m < 10^{-13}$ kg	Vega 1 and 2 DUCMA (Simpson et al., 1986)
	0.80	$m > 10^{-12}$ kg	Vega 2 SP-2 (Mazets et al., 1986)
	0.50 to 0.67	$10^{-15} < m < 10^{-12}$ kg	
	0.71	$m < 10^{-4}$ kg	Halley meteor streams (Hughes, 1986)
	1.27	$m > 10^{-4}$ kg	
	0.90	$m > 10^{-13}$ kg	Model prediction (Divine, 1981)

It yields the most significant measure of the average mass distribution from comet Halley's nucleus. $F(m, T)$ represents here the total number of particles at $>$mass m per meter^2 of Giotto's area. The source data include compensation for DIDSY data samples missing close to encounter, but on an assumed $1/R^2$ dependence fitted to the nearest samples, and are presented in Figs. 3–5. Evidence does indicate that this could be underestimated somewhat but any enhancements must be transient jet crossings of less than 30 s in duration and centred on the closest encounter period. From the integral

$$\int F(m, t)\, dm \tag{3}$$

we find the total mass incident on Giotto per meter^2 and for the total 2.9 m^2 spacecraft area we would derive a total incident mass of 55 mg assuming an upper limit of integration to an incident mass of 5.7 mg. This calculation is considerably lower than that required to decelerate the spacecraft on present understanding, even including momentum enhancement, and previously posed a problem. The large particle mass spectrum, now presented from DIDSY data, shows that this slope is not applicable to all intercepted masses (see Fig. 22) where the pre- and post-encounter fluences are shown for various values of assumed parameters γ (Sect. 2.6).

5.2. Area distributions

From the differentiation of the cumulative flux $\phi(m,t)$ with respect to m (performed in logarithmic intervals at 5 points per decade), we find $\oint(m, t)$ the particle differential (logarithmic) flux. The mass distribution of the flux may be obtained as

$$M(m, t) = \oint(m, t)\, m \tag{4}$$

and also the cross-sectional area

$$A(m, t) = \oint(m, t)\, \pi \left(\frac{3m}{4\pi\varrho}\right)^{2/3} \tag{5}$$

assuming a particle density ϱ (Fig. 13).

We take here for ϱ, values from the Halley environmental model (Divine et al., 1986) varying from $\varrho = 3000$ kg m^{-3} at small masses to $\varrho = 800$ km m^{-3} at large masses. Though ϱ may vary with mass more sharply, unless it is a strong function of mass, the area distribution (Fig. 13) provides a very useful measure of the changes in distribution. Pre-encounter at 20000 km the large particles (of mass $\gtrsim 10^{-9}$ kg, diameter $\gtrsim 100$ µm) are seen to dominate, but the situation developes to a flat distribution with particles of 1 µm to 100 µm diameter equally important at about -4000 km. At encounter and thereafter we find a peak developing in the region of 10 µm particle diameter. Though such cross-sectional areas must be weighted by optical efficiencies to yield scattering functions the dimensions of the dominent particles are generally larger compared with pre-encounter models and would suggest a considerable reddening of the scattered light pre-encounter although, as with the fluence, remote observations will be weighted by the higher jet fluxes of some 5 µm diameter particles. Significant contributions from large particles are implied from the fluence data though the optical effect is reduced by lower scattering efficiencies.

5.3. Dust distribution at source

From the coma fluence distribution we present the number, mass and area distributions based on detected particles up to mass 10^{-8} kg (Fig. 14). Above this mass, the distributions do not

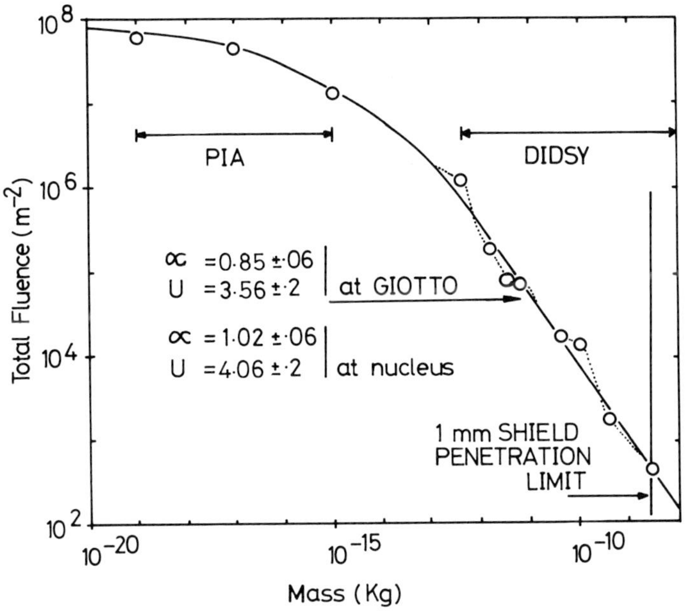

Fig. 11. Fluences derived from -5 min to $+5$ min from DIDSY and PIA sensors. Discussions in the text show the actual fluence above 10^{-8} kg as being grossly underestimated due to a change in mass index α

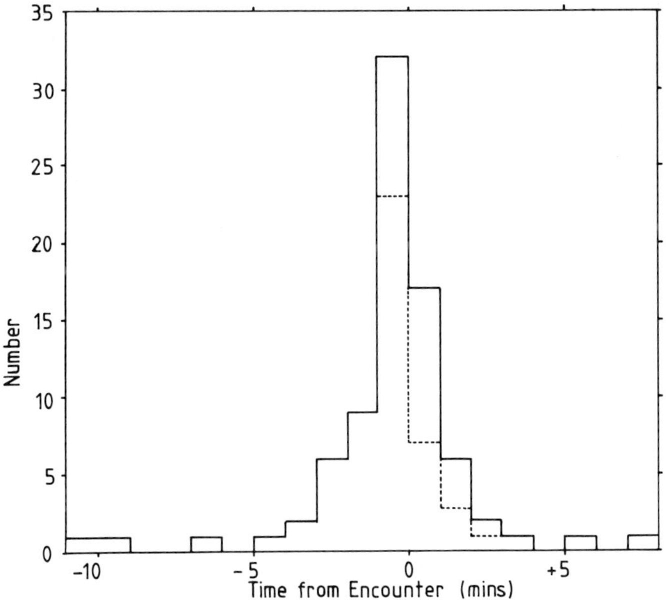

Fig. 12. A time profile in one minute intervals of the number of particles detected in the discrete data with mass $m > 10^{-9}$ kg. The actual data (dotted line) has been enhanced near encounter (solid line) to take account of telemetry loss

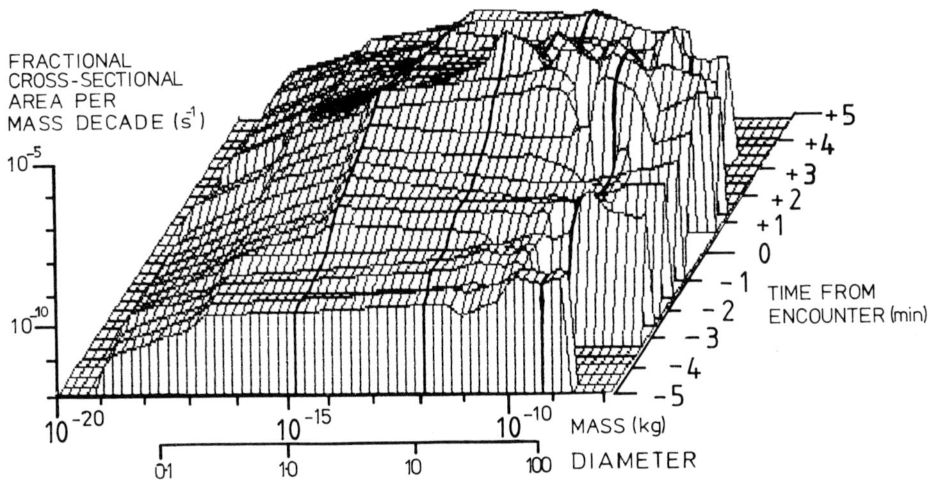

Fig. 13. Area distribution of the close encounter period. Note the dominance of large particles (greater than 200 microns) at -5 min (approx. 20000 km), a trend which continues beyond the few large particles detected at that time. This is followed by a flat distribution from 1 micron to 100 micron. At encounter and thereafter, as jets are traversed, particles in the range 1 micron to 10 microns dominate

continue monotonically; the area distribution is found to rise representing the larger detected particles (below).

Giotto's PIA and DIDSY sensors detected in the coma a nucleus distribution transformed by:

1) the emission process releasing grains from the icy matrix,
2) the subsequent acceleration by gas,
3) the subsequent deceleration by solar radiation pressure, and
4) possible fragmentation or evaporation of grains.

We are able, under relatively crude assumptions of, e.g. a $1/R^2$ radial flux scaling, and a velocity scaling according to particle size (Gombosi, 1986) to estimate the mass distributions at the nucleus. Post-encounter rates are consistent with a mean of $7\ 10^{-6}\ \mathrm{kg\,m^{-2}\,s^{-1}}$ integrated to mass 10^{-8} kg. The coma mass index of $\alpha = 0.85$ transforms at large masses to $\alpha = 1.02$ at the nucleus and hence the integral to large masses would (just) be bound at this index. An emission rate of $8\ 10^{-6}\ \mathrm{kg\,m^{-2}\,s^{-1}}$ would be implied and for an active area of πr^2 where r is the nuclear radius ($= 4$ km, Keller et al., 1986), we would deduce a total emission of $\approx 380\ \mathrm{kg\,s^{-1}}$. The mass index of the large particle coma flux (though not yet related quantitatively to the source distribution) leads to significant differences, however. In addition to uncertainties in the dynamics of particles, the time varying nucleus emission rate complicates the scene and could well account for the observed low index at high masses in the situation of a post-perihelion decline in the emission rate. Larger particles are more abundant because they diffuse away from the nucleus slowly and represent a period of higher intrinsic activity.

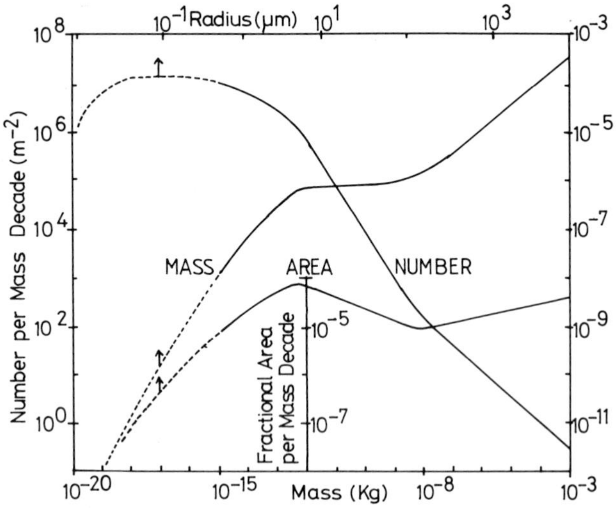

Fig. 14. Fluence distribution functions per mass decade applicable to Giotto trajectory. The low mass index applicable to particles $m > 10^{-8}$ kg implies an increase of the relative contributions to both the mass and geometric area functions

5.4. *Total mass and momentum exchange at Giotto*

From the flux rates for each sensor as a function of time we calculate the incident mass rate and total incident mass having made appropriate corrections for telemetry loss. This is calculated assuming the stated momentum enhancements for penetrating and non-penetrating particles (McDonnell et al., 1984). The *effective* mass relevant to spacecraft deceleration is then calculated. It is seen that with a cumulative mass index of $\alpha = 0.85$, (implying a maximum probable intercepted mass of 5.7 mg) that the non-penetrating particles must dominate, but only because of the range of integration and enhancement of the impulse for the smaller non-penetrating particles. A total of 235 mg effective mass is thereby deduced to which is added the *gas* contribution of ~20 mg (Krankowsky et al., 1986), yielding 255 mg compared with 1.95 g required. The basis for this calculation is independent of the momentum enhancement ε; this arises because the incident mass which is deduced from the signal voltage V by

$$m_i = kV/\varepsilon \tag{6}$$

and hence the effective momentum transfer is

$$\varepsilon m_i v_i = kVv_i \tag{7}$$

thus independent of ε. For large masses, ε must approach 1 as the shield traps the momentum inelastically.

We examined (McDonnell et al., 1986) the possibility on empirical grounds that the flux of very large particles, e.g. $> 10^{-9}$ kg might not be characterised by the intermediate mass particle slope. It was found that we require a value of $\alpha = 0.54$ implying a maximum mass (single particle) of $M = 0.8$ g in order to account for the spacecraft deceleration. This requirement, indicated by integral relationships, has now been largely overtaken by analysis of DIDSY's large particle mass distribution (Fig. 22) which demonstrates clearly the existence of an index in this range. Though data is missing from this set, the selection is unbiased concerning signal magnitude and shows that such a slope does pertain to the large mass fluence detected in the coma at this epoch. A fuller analysis of the large particle distribution will be presented but initially we present the data without absolute flux calibration to demonstrate index changes.

Within the period −40 s to +40 s all such large mass events have been decoded and the cumulative mass distribution index (α) determined (see Sect. 4.1 for details of fitting procedure). This yields a slope of $\alpha = 0.46 \pm 0.13$. The relatively large error probably reflects the counting statistics rather than intrinsic scatter in the data and does not include any possible systematic effects due to, for example, the uncertainty in the momentum derating factor for penetrating particles (Sect. 2.6 and Fig. 22). Other time periods have been investigated in the same way and yield similar, and consistent, low values for the slope of the cumulative mass distribution function. We have performed this analysis for derating of the momentum for penetrating particles for values of $\gamma = 0.40$, 0.55, and 0.7. γ (defined in Sect. 2.6) is considered to be near two-thirds (Wallis, 1986).

In the period closest to encounter where a flux of very large particles would be anticipated, only a selection of discrete events is available in the data. Examining such a distribution expected, with a mass distribution index of $\alpha = 0.5$, we find that the higher flux of particles of mass $\sim 10^{-9}$ kg would indeed preclude the appearance in this set of masses $\gtrsim 10^{-5}$ kg, i.e. 10 mg. Thus the absence of the larger masses in this discrete data set is not indicative of their non-existence, and the large mass slope must be presumed to apply to the 10 mg mass region during encounter. A maximum mass of between 1 and 10 mg is indeed deduced in practice entirely consistent with the change in observed slope. We examine below consequences of this low slope in the coma distribution above mass 10^{-9} kg.

Implications for the fluence and estimation of Halley's mass output could be substantial. Although DIDSY cannot *determine* the index α at masses above values of approximately 1 mg, the global mass balance for Giotto requires that this low index should extend to intercepted masses up to a value of ~1 g. Though statistical variations preclude our saying this mass was actually encountered the arguments certainly call for a maximum single particle mass in the region of 1 g and a concentration of the majority of the mass in the range 100 mg to 1 g, with little momentum enhancement implied. By contrast the smaller optically dominant particulates contribute only 200 mg and even this contribution is derived with considerable momentum enhancement. The fluence derived from the PIA and DIDSY data is compatible in form with the fluxes derived from the Vega data (Vaisberg et al., 1986a; Mazets et al., 1986) although the Vega data do not extend to such large masses by virtue of their larger miss-distances and smaller areas.

The fluence deduced for the Giotto encounter of Halley on March 13th and 14th 1986 is described by the form:

$$\log \phi(m) = k(\log m + 12.4) + k' |(\log m + 12.4)|^{1.45} + c, \tag{8}$$

where for

$m < 10^{-8}$ kg,	$k = -0.854$	$c = 5.903$
$m > 10^{-8}$ kg,	$k = -0.54$	$c = 4.521$
$m < 4\ 10^{-13}$ kg,	$k' = -0.239$	
$m > 4\ 10^{-13}$ kg,	$k' = 0.0$.	

5.5. *Nucleus emission rates*

Implications for the total emission rates from Halley resulting from the observed "bimodal coma population" affect significantly the derived gas-to-dust ratio deduced to apply at the source. From the emission rate integrated up to mass 10^{-8} kg, namely $7\ 10^{-6}\,\mathrm{kg\,m^{-2}\,s^{-1}}$, we could tentatively extend the integration up to a maximum mass m_{max}, using the index of $\alpha = 0.54$ at Giotto [which transforms to a nucleus source index of $\alpha = 0.71$ assuming a grain velocity dependence of $1/m^{1/6}$ (Divine et al., 1986)]. Using the cumulative emission flux ϕ at $m = 10^{-8}$ kg derived from the DIDSY flux, we find a total dust production of $2900\,\mathrm{kg\,s^{-1}}$ if integrated up to $m_{max} = 1$ g. Compared with a total gas emission rate of $2.1\ 10^4\,\mathrm{kg\,s^{-1}}$ (Krankowsky et al., 1986), we find the gas-to-dust ratio of 7:1. If the emission distribution were to continue however, to a maximum mass of $m_{max} = 1$ kg with the same index, an output of 7 times this mass production rate would result and the gas to dust ratio would hence be deduced as 1:1. Clearly the maximum mass lifted from Halley is critical in determining this parameter and its finer definition must now await a more detailed modelling of the large particle fluxes and time variant nucleus production rates. We refer to modelling by Fertig and Schwehm (1984) and note especially the concentration of very large masses emitted from the sunlit hemisphere which drift towards the pre-encounter approach path Giotto; the mimicking of this in the Giotto discrete data (Fig. 12) certainly supports their model predictions on a qualitative basis. Some special points do arise from the detection of the mass index change in the coma:

1) Such dust, probably contributing the dominant output in terms of mass from Halley, would be optically significant though not necessarily dominant; it should be expected to be seen in remote observations, such as far infrared and radar measurements at the time of Giotto encounter and perhaps other comets post-perihelion.

2) Such dust *could* provide relatively massive inputs to the Zodiacal cloud from comets if typical of emission at such masses, inputs which could have escaped detection to date by remote optical sensing and calculations which assumed the perpetuation of a high size index of $U \approx 4.2$ ($\alpha = 1.07$ at nucleus or $\alpha = 0.90$ in coma) to masses of ~1 kg. Whether this peak in the tail of the mass distribution does reflect itself in a true peak in the nucleus emission distribution or is a feature of the coma dynamics of these large masses or temporal flux changes will have to be resolved.

3) Such a mass output, at dimensions generally greater than the optically detected components (and even generally the dust detectors) could provide a very rich reservoir for comminution and fragmentation within the coma and meteor streams.

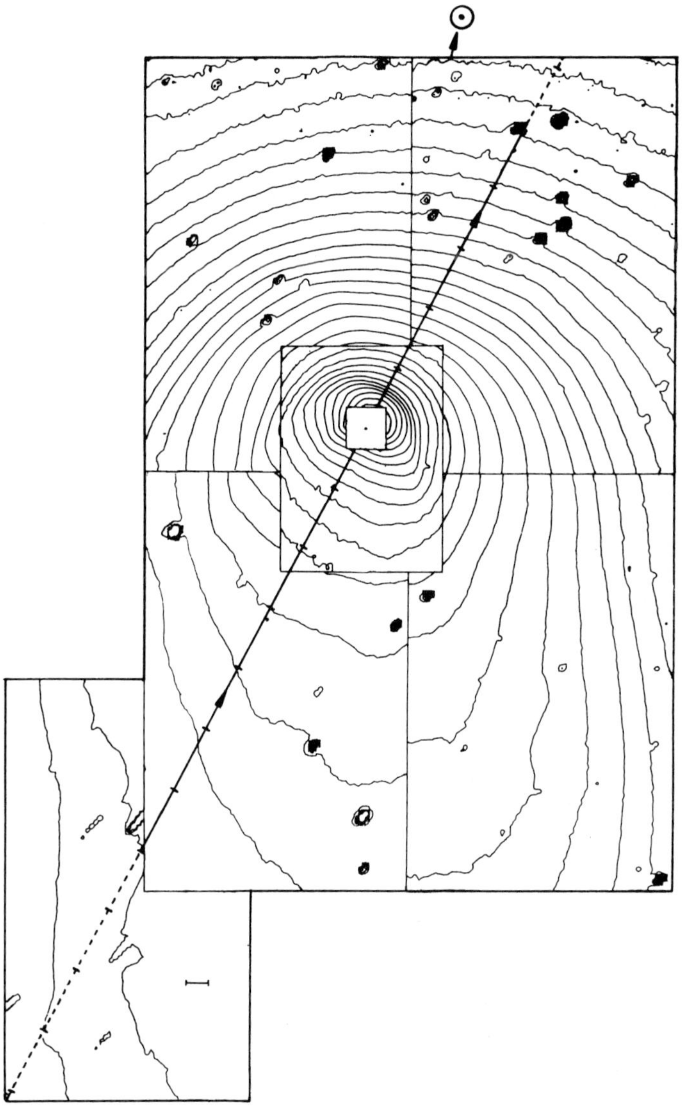

Fig. 15. Mosaic of six CCD images (13/461-6) taken with the red continuum (730 nm) filter between 1825 and 1850 UT on 13 March 1986. North is to the right and East is up. Isophotes are at intervals of 20%. No attempt has been made to correct for residual differences at the edge of each image, which are approximately 20%. The projected path of Giotto is shown as a solid line for the period within which DIDSY detected impacts, with tick marks at 10 minute intervals. The central window is shown enlarged in Fig. 16

6. Comparison of DIDSY area distribution with ground-based CCD images

6.1. Observations

Almost 200 CCD images were taken between March 3–24 1986 using the 3.9 m Anglo-Australian Telescope and filters identical to those incorporated in the Giotto Halley Multicolour Camera. The filter set used prior to the CCD detector contained narrow-band filters to isolate C_2 (570 nm), C_3 (410 nm), and OH (320 nm) emissions and the dust continuum at 450 nm and 730 nm, as well as several wide band filters. Each 512 × 320 pixel image covered an area approx. 180000 × 110000 km with seeing limited resolution of approx. 400–800 km. The inner coma structure changed dramatically from night to night and the dust images exhibited a number of jets. A series of these taken only 5 hours before Giotto encounter have been used to estimate the dust density along the trajectory for comparison with data from the Dust Impact Detection System.

6.2. March 13 dust images

Figure 15 shows a mosaic of six images taken with the red continuum filter between 1825 and 1850 UT on 1986 March 13. The central part of the image is shown enlarged in Fig. 16. Scale bars are 10000 km and the projected path of Giotto is marked with times from closest approach.

Points to note are the strong asymmetry in the central part of the image, irregularities in the inner contours indicative of the presence of dust jets, a higher dust concentration in the northern part of the tail, and a sharper intensity gradient in the sunward (Easterly) direction. These features can be qualitatively understood in terms of the effect of solar radiation pressure on the dust grains, the sources of emission from the nuclear surface, and its rotation. To the East (almost the sunward direction) of the nucleus it follows a dependence of the form $D^{-0.94}$ up to a projected distance from the nucleus of $D = 32$ pixels or 18000 km.

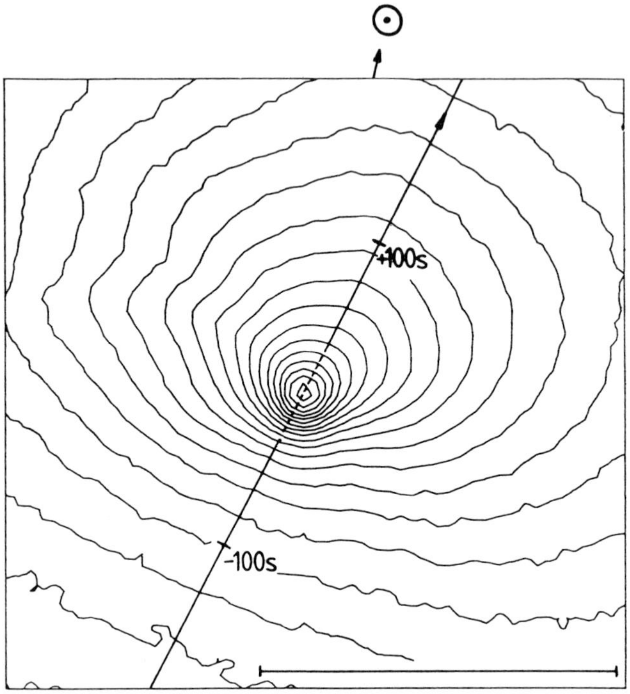

Fig. 16. Central region of image 13/463. The projected Giotto trajectory is shown with times from closest approach

In the other directions the brightness does not follow a simple power law. This is not surprising since the spacecraft images and the near nuclear regions of these images show strong sunward emission. The observed intensity distribution would, in simple modelling, follow a roughly D^{-1} gradient in the solar direction up to a point where the majority of observable particles are "reflected" back by radiation pressure. In the anti-sunward direction, the initial intensity is lower due to the lower surface activity but becomes relatively larger due to the "reflected" particles from other parts of the surface. The apex distance at which particles are forced back depends on grain size, density and ejection velocity, and to a lesser extent, composition. Jets seen on March 11 and 12 probably produced the concentration of dust observed at approx. 18000 km.

There is evidence for a considerable quantity of dust at large distances from the nucleus as measured by DIDSY and from impacts on the Japanese probe Suisei (Hirao and Itoh, 1986) at 170000 km.

6.3. *Area of dust particles encountered by Giotto*

An independent estimate of the area of particles encountered by Giotto can be obtained from the CCD data. This is achieved by relating the observed intensity from a column along the line of sight, to the particle cross sectional area at a point along the column, through assumptions about the distribution of particles and their scattering properties.

From DIDSY and PIA data the area of grains in the mass range 10^{-19} to 10^{-8} kg between -300 s and $+300$ s (see Fig. 13) has been calculated in 20 s intervals. When no impacts were detected in the larger mass bins, a mass index of $\alpha = 0.9$ was used for extrapolation. Figure 17 shows the area of grains encountered (A in $m^2\,m^{-3}$) as a function of distance from the nucleus (R) compared with an inverse square relation. If there are $N = \int_{a_{min}}^{a_{max}} n(a)\,da$ grains per cubic meter with geometric cross-

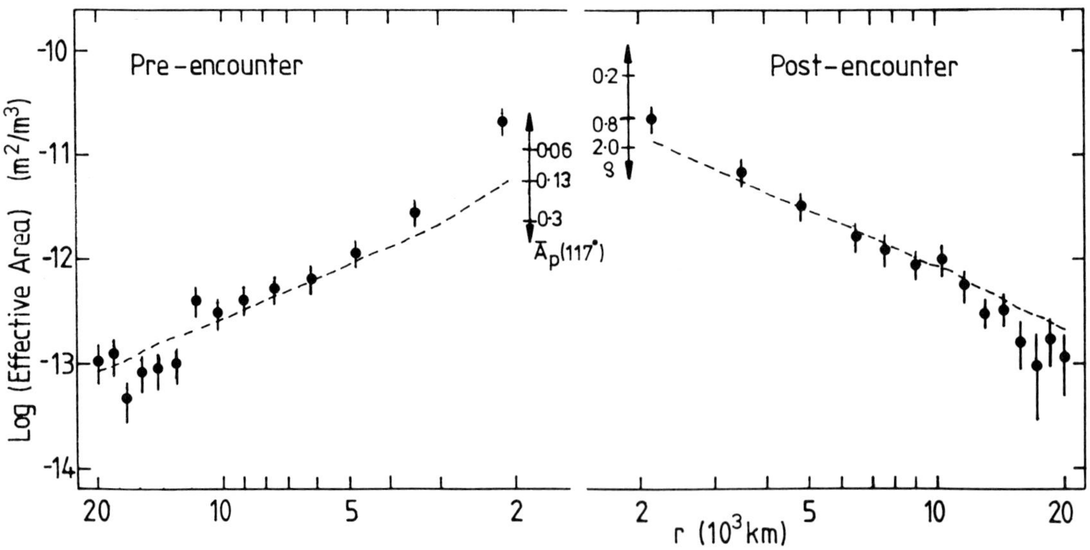

Fig. 17. Effective area of grains per cubic metre along the Giotto trajectory calculated from Giotto DIDSY and PIA data (circles) and from ground-based CCD data (dashed line). The area function A is calculated assuming a density $\varrho = 3 - 2.2\,(a/(a + 2.0))$ (Divine et al., 1986) where a is the grain radius in microns, yielding densities between 3.0 and 0.8 g cm^{-3}. The effect on the calculated areas of changing the density for large particles from 0.8 g cm^{-3} is shown, as is the effect of changing the geometric albedo $\bar{A}_p$ (117°) from the best fit value of 0.13. Error bars for DIDSY data correspond to statistical uncertainties in the area and do not include any possible systematic errors

sectional area $G(a)$ at distance R, and the dust space density follows an inverse square relationship, the observed flux at the Earth from scattered sunlight is given by

$$F_\lambda = \frac{d^2}{\Delta^2}\left(\frac{F_{\lambda\odot}}{r^2}\right)\frac{R^2}{R_0}\,\pi\,\frac{A}{N}\int_{a_{\min}}^{a_{\max}}\frac{\sigma(\theta,a)\,n(a)}{G(a)}\,da\,, \tag{9}$$

where d is pixel dimension, Δ is geocentric distance, $F_{\lambda\odot}$ is the solar constant at wavelength λ, r is heliocentric distance in AU, R_0 is the projected distance from the nucleus and $\sigma(\theta, a)$ is the differential scattering cross section. Since the "geometric" albedo for a single grain is defined as

$$A_p(\theta) = \pi\sigma(\theta, a)/G\,, \tag{10}$$

(Hanner, 1981), we can define a mean geometric albedo for the grain population,

$$\bar{A}_p(\theta) = \frac{\pi}{N}\int_{a_{\min}}^{a_{\max}}\frac{\sigma(\theta,a)\,n(a)}{G(a)}\,da\,. \tag{11}$$

A CCD image at a wavelength of 0.73 µm taken on March 13 gives the flux density $F_\lambda(\mathrm{Wm^{-2}\,\mu m^{-1}})$ observed at the Earth from a column through the coma with cross section of one pixel. One CCD pixel = 345 km at the distance of the comet (0.9 AU) so on the Giotto trajectory, inclined at an angle of 45°.7 to the plane of the sky, 1 pixel = 494 km. The solar constant at 0.73 µm is taken as 1336 $\mathrm{Wm^{-2}\,\mu m^{-1}}$ (Labs and Neckel, 1970). In Fig. 17 the best fit between the areas A calculated from the CCD data and from Giotto is found for $\bar{A}_p(117°) = 0.13$. This is a crude estimate for the following reasons:

i) The true dust distribution in the coma does not follow an inverse square law even in the inner coma where the surface brightness varies approximately as R^{-1}. This is due to jets and non-isotropic emission from the nucleus.

ii) The DIDSY area distribution is dependent on the large particle cut-off where the sampling is generally poor. Using a mass index $\alpha = 0.54$ for masses from 10^{-8} to 10^{-3} kg causes a mean increase of ~60% in the pre-encounter DIDSY areas but only ~10% post encounter. Since the true mass index for large particles probably also varies with time, they have not been included in the calculation, so the mean albedo will be an overestimate.

iii) The density used for the derivation of areas from Giotto data (Divine et al., 1986) may be incorrect. The effect of changing the large grain density (0.8 $\mathrm{g\,cm^{-3}}$) is shown in Fig. 17.

Considering these effects the mean geometric albedo upper limit of 0.13 is not out of context compared with the values of about 0.05 derived from infrared observations (e.g. Hanner, 1986). The higher value of albedo would lend support however to the contention that there is a significant contribution from large grains.

7. Modelling nucleus emission characteristics

7.1. Source fluxes

Dust impact data from Giotto, Vega 1 and Vega 2 can be used to derive information on the distribution of emission over the surface of the nucleus. Dust particles travel on parabolic trajectories after acceleration to a terminal velocity within a few nuclear radii and reach apex distances of the order of 10^5 km (Divine, 1981) where solar radiation pressure is the dominant force in retardation and acceleration into the tail. We consider here only such particles relatively close to the source, so that their trajectories are approximately described by straight lines between the emission point the interception point at the spacecraft. The dust flux can then be transformed back to a given position on the nucleus, according to the mass of particle arriving at the spacecraft, since the acceleration here is radial with respect to the comet nucleus (Divine et al., 1986). This assumption will be considered more fully later, in the light of recent data, but will be taken to be valid for the discussion below.

We now discuss mass distributions and fluxes obtained from the spacecraft, how they relate to the assumptions made on transformations to the nucleus, and comparisons of their data transformed back to the nuclear surface using these assumptions.

To transform the fluxes back to a Halleycentric nucleus co-ordinate system, the sub-spacecraft track along the surface of the nucleus must be obtained as a function of time. This requires a particular spin model for the nucleus, and our early work used an argument ϕ of 300° and obliquity I of 30° for the north pole of a simply rotating nucleus, following the analysis of Sekanina and Larson (1984) derived from image-processed photographs of the 1910 Halley apparition (Sekanina and Larson, 1986b).

Grün et al. (1986) also determined a rotation axis position differing by 29°–50° from the 1910 solution of Sekanina, and interpretation of a sunward spike seen in May 1986 (Sekanina, 1986) gives a spin axis inclined more nearly perpendicular to the comet orbit plane. However, since the photometric observations of Halley by Millis and Schleicher (1986) in March and April 1986, a second nuclear period of 7.4 days has been discovered. This has been interpreted by Sekanina (1987) as a true rotation of an ellipsoidally shaped, torque-free nucleus, with axes in the ratio 1.9:1:1 (Sagdeev at al., 1986), so that the 2.2 day period determined earlier would agree to a precessional motion of such a body with a precessional angle of 77° (Sekanina, 1987).

If this first-approximation model linking these two observed rotational periods is correct, then it appears that the originally observed "spin axis" is in fact a precessional axis, and the observed north pole is therefore much nearer to the nuclear equator.

We consider a particle of mass m striking a spacecraft at time t after closest approach. In all flyby cases, the particle would have been emitted at some earlier time, dependent on the velocity of the particle, $v(a)$, in a direction to the south of the sub-spacecraft track, since the displacement due do precession is about 3 times as large as that due to spin.

If θ_S and ϕ_S are the sub-spacecraft co-ordinate at time t, then the emission co-ordinates are given by:

$$\theta_E = \theta_S + \delta\theta_{PR} - \delta\theta_{SP}\,, \tag{12}$$

$$\phi_E = \phi_S - \delta\phi_{PR}\,, \tag{13}$$

where $\delta\theta_{SP}$ is a small change in longitude due to the spin of the nucleus and $\delta\theta_{PR}$ and $\delta\phi_{PR}$ are the changes in longitude and latitude due to the precession of the nucleus.

If δ is the change in the co-ordinate required, then generally we have

$$\delta \propto \frac{R}{v(a)}\cdot\frac{1}{P}\,, \tag{14}$$

where

$$R = (R_0^2 + v_S^2 t^2)^{1/2} \tag{15}$$

is the distance from comet nucleus, P = rotation period of comet, R_0 = closest approach distance, v_S = spacecraft velocity and t = time after encounter.

Table 5. Angular displacement from sub-spacecraft track for a rotation period of 52 hours, for a point in a flyby 1200 km from the nucleus

Particle radius (m)	$\theta_S - \theta_E$	$v_S (ms^{-1})$
10^{-9}	3°37	680
10^{-8}	4°16	550
10^{-7}	6°54	350
10^{-6}	13°47	170
10^{-5}	38°18	60

The particle velocity, $v(a)$, for a particle of radius a, is obtained from the parametric form below, using computational solutions from *T.* Gombosi (personal communication, 1985):

$$v(a) = 1 \cdot 8\,\overline{V}\,\frac{\left(\frac{a_H}{a}\right)^{1/2}}{1+\left(\frac{a_H}{a}\right)^{1/2}}, \tag{16}$$

where $a_H = a_p (0.59/r_\odot)^{2.2}$, $\overline{V}$ is the molecular gas velocity, and is a function of the nuclear surface temperature, $r_\odot$ is the heliocentric distance, and for Halley near encounter, we take the corner radius, $a_p = 2\,\mu m$, distinguishing small particles travelling at close to the gas velocity, and larger particles with much smaller terminal velocities. More recent data (Gombosi, 1986) has suggested a value of a_p nearer 10 µm.

The radii of particles are obtained by considering them to be spheres of density derived as below (Divine et al., 1986):

$$\varrho = \varrho_0 - \sigma_0 \left(\frac{a}{a+a_p}\right). \tag{17}$$

Table 5 gives an indication of the displacement of the emission point from the sub-spacecraft track, 15 s before the Giotto closest approach, for different particle radii, with a spin period of 52 hours.

7.2. Comparison of data at the spacecraft

Fluxes in the mass interval 10^{-15} to 10^{-14} kg are shown in Fig. 18, together with an R^{-2} comparison. A notable feature of the Giotto flux profile occurs on the in-bound phase at about 1500 km from the nucleus near the crossing of the dawn terminator at $t \approx -29$ s, and on the out-bound phase at 12000 km (+170 s), and 18000 km (+270 s).

The form of the surge in flux seen close to the dawn terminator in the Giotto data is, of course, closely linked to the angular source distribution relative to the sub-solar point. A question arises, however, as to whether dark-side emission was seen before Giotto crossed the dawn terminator, as the Halley Environment Working Group reference model suggested (Divine et al., 1986).

We have therefore compared a simple bright side emission model with data from the DIDSY experiments on Giotto. Such a model predicts a flux component before 29 s to encounter as due to particles reflected at the dust envelope from the bright side. The flux at the spacecraft, ϕ_S, is given by

$$\phi_S \propto 1/R_R^2 \tag{18}$$

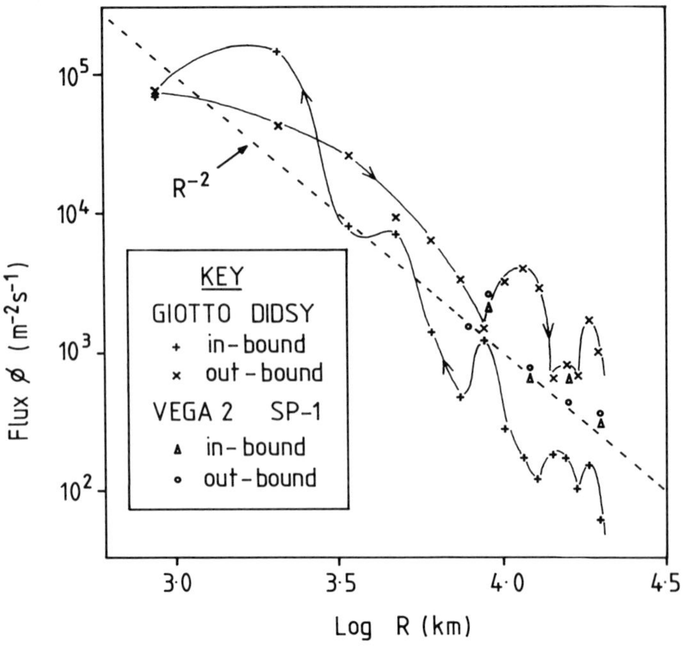

Fig. 18. Dust fluxes in the mass channel 10^{-15} to 10^{-14} kg are shown as a function of distance from the comet, for Giotto and Vega (SP-2), on their inbound and outbound trajectories. A $1/R^2$ dependence is also shown.

where R_R is the distance travelled by the "reflected" particle. This is given approximately by using twice the particle apex distance, E, (Divine, 1981) where

$$E = v(a)^2\, \varrho\, a\, 8\pi r_\odot^2 c/3 L_\odot \tag{19}$$

at distances close to encounter according to pre-encounter modelling theory. The particle trajectory distance only by about 2% from $t = -300$ s and $t = +300$ s from the value of $R = 2E$.

After crossing the dawn terminator, there is both a direct and reflected component in flux measured by the spacecraft, and

$$\phi_S \propto \frac{1}{R_R^2} + \frac{1}{R^2} \tag{20}$$

which is almost entirely obtained from the direct component near encounter, since $R_R \gg R$.

A comparison of these fluxes has been made for particles of 10^{-13} kg, where we find pre-encounter modelling leads to an expected apex distance of 38750 km using Eqs. (16), (17), and (19).

Figure 19 shows 20 s averaged flux data interpolated at 10^{-13} kg from the DID 2/3/4 experiments, from -300 s to $+300$ s along the Giotto trajectory. A bright side emission only model discussed above has been fitted to this to give closest agreement post-encounter.

This model for emission from the bright side only is based on the fountain model for dust particle trajectories. Another approach (Fertig and Schwehm, 1984) has been to attribute parabolic trajectories to small particles (less than 10^{-11} kg) which are strongly accelerated by solar radiation, and to look in more detail at the trajectories of larger particles, which remain in the near nucleus region for a longer time.

For bright side emission only, their model predicts symmetrical parabolic trajectories for small particles but highly asymmetri-

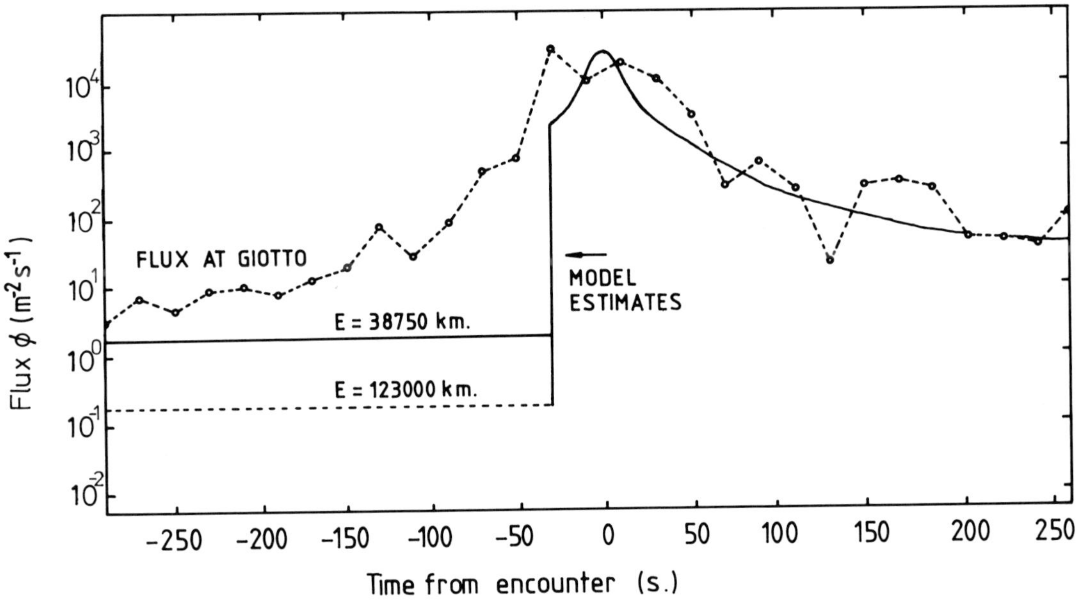

Fig. 19. Dust fluxes are plotted as a function of time for the Giotto trajectory, interpolated at a mass channel of 10^{-13} to 10^{-12} kg from 3 DIDSY sensors. A model curve is fitted to the post-encounter data for a nucleus with bright side emission only, and is drawn for apex distances of 38750 km and 123000 km

cal trajectories for particles of mass greater than 10^{-11} kg, such that the Giotto spacecraft would meet nearly all of the larger particles before encounter. The Giotto flyby did show that the largest events (e.g. a few milligrams) occurred before encounter. Such a model would give an increasing flux prior to encounter, and a particle apex, as such, would be difficult to distinguish.

Measured fluxes are generally higher than predicted from about -200 s pre-encounter; four possible reasons for this are:

i) Non-radial acceleration of dust grains, producing a dawn terminator "jet" which would be directed back into the night-side by as much as 15°;

ii) Dark-side emission is taking place;

iii) A non-spherical nucleus: regions nominally on the dark side for a spherical nucleus could thus be illuminated;

iv) Non-parabolic trajectories, giving rise to higher pre-encounter fluxes of larger particles, even close to the nucleus ($R < 20000$ km).

We have obtained an apex distance of 123000 km for particles of 10^{-13} kg, much larger than the model estimate. This would lower the pre-dawn terminator flux estimates by about a magnitude, and a simple reflected component would be quite insufficient. Halley Multicolour Camera photographs (Keller et al., 1986) would also indicate that Halley does have a highly asymmetrical nucleus. The image taken at 267 s (18270 km) before closest approach in fact shows a nuclear feature on the night side with an elevation that extends it into the sunlit region. It is therefore not so easy to ascribe such a simple dawn terminator situation to the nucleus of Halley.

7.3. *Comparison of the data at source*

To compare data seen by the spacecraft with Halleycentric emission co-ordinates, it was decided to obtain a flat source background by a method different from direct comparison with a pre-encounter dust model (Divine et al., 1986). This enables data for very small mass particles to be included, as such dust models took smallest dust masses of about 3.5 10^{-17} kg, whereas particles as small as 10^{-19} kg were seen by Giotto (Kissel et al., 1986) and Vega 1 and Vega 2 (Mazets et al., 1986; Vaisberg et al., 1986b) during the encounters.

The data was flat-fielded by normalizing to the source, and then by using the following equation to flatten the data in magnitude mass bins. If ϕ_N is the flux normalized to the source by R^{-2}, where R is the distance from the comet nucleus, as a function of mass and time, then we define the relative activity function to be:

$$A_R(m,t) = \frac{\phi_N(m,t)}{\frac{1}{T}\sum_{t=1,T} \phi_N(m,t)} \tag{21}$$

where T is the number of time steps over which the fluxes were measured. In this way, one would expect a flat distribution for a comet nucleus with a perfectly homogeneous source function.

If the source activity suddenly increases near the dawn terminator of the nucleus, an increase in the function A_R would therefore be expected.

Such maps have been obtained from the Vega 2 spacecraft and for the Giotto spacecraft and are shown in Figs. 20 and 21.

The relative activity function, A_R, is seen in a mass-time space from about 2000 km before encounter, through closest approach, to about 2000 km after encounter, for both spacecraft.

For the Vega 2 map, data from Vaisberg et al. (1986b) was used, averaged over one minute intervals, and for the Giotto map, data was averaged over 20 s intervals. The Vega 2 map shows a relative decrease in activity towards closest approach for particles between 10^{-19} and 10^{-14} kg, but shows a large increase at around $t = +60$ s, corresponding to the jet examined by Vaisberg et al. (1986b).

For dust particles of masses 10^{-20} to 10^{-15} kg in the Giotti map, and over a 200 s period around encounter, data interpolated from the PIA experiment used here only at mass 10^{-19} kg must be

Table 6. Summary of results

On 2.9 m^2 spacecraft

Total mass $<10^{-8}$ kg	55 mg (actual) 235 mg (effective)
Inferred total mass	1.7 g (actual) 1.9 g (effective)

Local mass index for fluence

$10^{-13} < m < 10^{-8}$ kg, $\alpha = 0.85$
$m > 10^{-8}$ kg, $\alpha = 0.54$

Shield perforations:
Max. rate inferred 53 s^{-1} at encounter ± 10 s.
Total number: 1,500

Apex distance for mass 10^{-12} kg: $\sim$120000 km

Shield abrasion

Microabrasion
(assuming crater diameter = 10 × particle diameter)

To encounter	$<10^{-8}$ kg	0.9%
Total	$<10^{-8}$ kg	2.6%

Total geometric cross-section of particulates in coma

Calculation assumes $1/R^2$ dependence and coma radius 5 10^4 km plus particle density from the Halley Environment Model (Divine et al., 1986) yielding a cross-section of 1.9 10^{14} A_G m^2 where A_G is the particle cross-section per unit area integrated through Giotto's trajectory.

Integration range	A_G	Area (km^2)
0–10^{-8} kg	2.6 10^{-4}	5.0 10^4
10^{-8} kg–1 g	1.4 10^{-4}	2.6 10^4
1 g–1 kg	2.6 10^{-4}	4.8 10^4

Nucleus area (4 km radius): 50 km^2

At nucleus

Mass index	10^{-13} kg $< m < 10^{-8}$ kg,	$\alpha = 1.02$
	10^{-8} kg $< m$,	$\alpha = 0.71$

Surface emission rate per square meter

$<10^{-8}$ kg:	6.7 10^{-6} kg m^{-2} s^{-1}
<1 g:	5.7 10^{-5} kg m^{-2} s^{-1}
<1 kg:	4.2 10^{-4} kg m^{-2} s^{-1}

Upper mass limit	Diameter	Total surface emission rate to upper mass limit (kg s^{-1} for nucleus radius 4 km)
1.58 10^{-18} kg	0.1 micron	7.5 10^{-4}
1.60 10^{-15} kg	1.0 micron	7.8 10^{-1}
6.31 10^{-13} kg	10.0 micron	37
10^{-8} kg	0.62 mm	340
1 g	1.34 cm	2900
10 g	2.88 cm	5500
100 g	6.2 cm	11000
1 kg	13.4 cm	21000

Gas-to-dust ratio

Assuming gas production = 2.1 10^4 kg s^{-1} (Krankowsky et al., 1986)

to mass 10^{-8} kg:	62:1	
to mass 1 g:	7:1	likely range
to mass 1 kg:	1:1	likely range

Total perihelion passage ablation

Assuming $1/R^2$ heliocentric dependence of emission rate:

		Average	On active areas (10%)
Integrated to mass limit	1 g	0.25 m	2.5 m
	1 kg	1.8 m	18 m

considered with some caution, as saturation effects may have reduced the overall activity function here (Langevin, personal communication, 1986). However, a general high increase is observed near $t = -30$ s, which would correspond to the crossings of the dawn terminator and a rise in emission there. Two broad peaks are also seen at $t = +190$ s and $t = +270$ s, spread over a large mass range, and correspond to the two jets seen at distances of 12000 km and 18000 km from the nucleus.

Of course, dust may be detected which originated from the same spot on the nucleus but which was ejected during one or more earlier rotation periods for large masses or at larger distances. Two relatively active regions do appear at the same time but at different points across the mass spectrum, each exhibiting about 4 times the average amount of activity defined by A_R. These points are located at -290 s in time and 10^{-20} to 10^{-9} kg in mass, respectively. If we assume that these points do represent particles from the same spot on the nucleus, then we require a velocity distribution that gives the correct time of flight for particles of such masses, the times differing by one rotation period only. Although other solutions might be found, using the form of Eq. (19) and a rotation period of 52 h, these flight times can be satisfied if the particle corner radius, a_p; is 3 μm, a value not in conflict with present particle velocity distributions.

Giotto's dust impact detectors have established, at the time of encounter on March 13th and 14th, the spatial density and mass distribution of particulates within the inner coma of comet Halley above a mass of 10^{-19} kg. Spatial densities encountered are significantly lower than results from the Vega encounters,

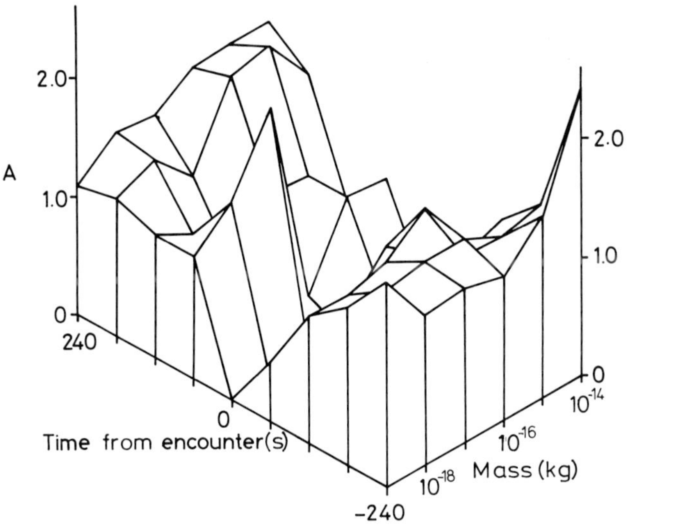

Fig. 20. Activity function A (see Sect. 7.3) is plotted in a mass-time space for the Vega 2 and Giotto flybys. Increased activity is seen in the Giotto data at $t=-30$ s (terminator), $t=+190$ s and $t=+270$ s

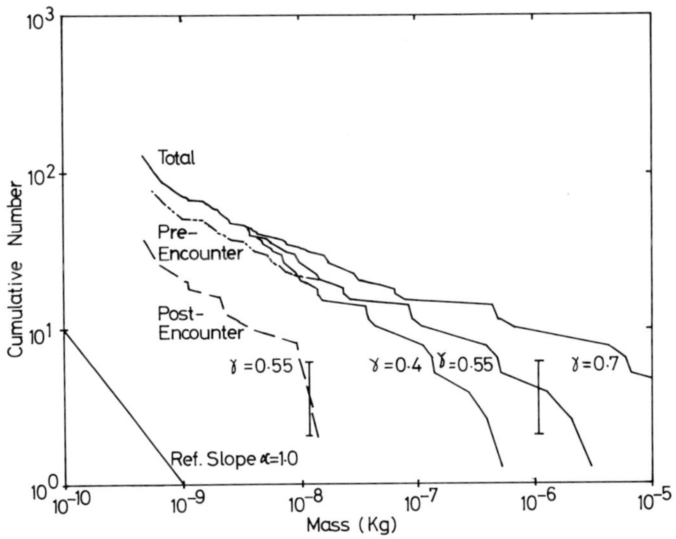

Fig. 22. DIDSY's large particle mass distribution, corresponding to multiple sensor detections on the dust shield. The perforation limit of the shield is taken as $m_{\text{pen}}=3\ 10^{-9}$ kg; cumulative number refers to the total event set and is not yet converted to flux rate because of undetermined data selection effects. The distribution index α is seen to be markedly different from the intermediate mass index of $\alpha=0.96$ for all three test values of the impulse derating exponent γ applied to all masses above m_{pen} (Sect. 2.6). The onset of index change is seen to occur as a mass lower than this penetration limit

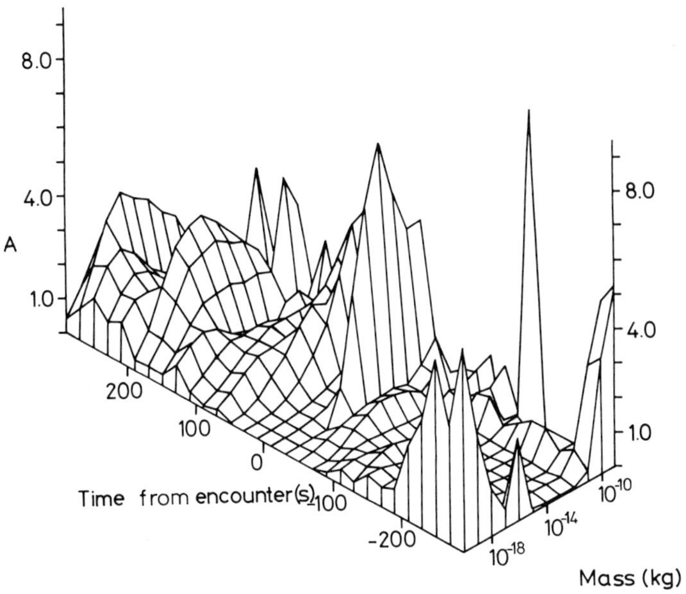

Fig. 21. Same as Fig. 20

demonstrating a substantial decrease of dust emission. Resulting from the combination of Giotto's large area for sensing (the 2 m^2 dust shield) and the very close encounter distance, the mass distribution has been measured to especially large masses, namely some 10 mg; a significant change in the slope is measured in the coma dust distribution which may be partly associated with reduced activity and also the complicated dynamics of the larger particles. Nevertheless, this detected large particle "excess" has significant implications for the cross-sectional area which is of relevance to in-situ and remote optical and infrared scattering experiments.

Derivation of the mass distribution of the nucleus must wait further modelling as also must the proper evaluation of the gas-to-dust ratio but nevertheless it is clear that very significant mass is contained within the range of mass $>10^{-8}$ kg and that calculations of the dust production at the time of Giotto's encounter must take full account of the changing index of the large mass flux. The total mass encountered by Giotto, and responsible for its deceleration, is certainly dominated by particles of mass above 10^{-9} kg which penetrate the dust shield with little impulse enhancement; it is likely to be concentrated in a relatively small number of particles encountered near closest approach extending up to a mass of approximately 1 g.

Acknowledgements. The contribution of the entire DIDSY team is gratefully acknowledged, as well as the fincial support of the Science and Engineering Research Council (U.K.) for the DIDSY experiment at the University of Kent and for management and technical support through the Rutherford Appleton Laboratory.

The filters used for the A. A. T. observations were generously provided by the H. M. C. team (P. I. H. U. Keller). We would like to thank the staff at the A. A. T. for their assistance with scheduling and conducting the observations. The processing was performed on the University College London and University of Leicester Starlink nodes.

Acknowledgement is also due to the Bundesminister für Forschung und Technologie (D), CNES (F), and the Piano Spaziale Nazionale, Consiglio Nazionale delle Recherche (I) for support of flight hardware development and to the national space agencies of Sweden and the United States (NASA) for co-investigator support. The success of the Giotto mission is an indication of the dedication of the project teams at the European Space Agency, British Aerospace (prime contractors) and the sub-contractors throughout Europe.

The authors thank N. Divine for very valuable comments and corrections during review.

References

Arnadeau, F., Winkelmuller, G., de Rouvray: 1984, Engineering Systems International, ESA/ESTEC, contract No. 4637/81, NL/AB-CCNO 7

Chakaveh, S.C., Green, S.F., McDonnell, J.A.M., Hughes, D.W.: 1986, 20th *ESLAB Symp.* ESA SP-250, Vol. **II**, p. 163

Cosmovici, C.B., Green, S.F., Hughes, D.W., Keller, H.U., Mack, P., Moreno-Insertis, F., Schmidt, H.U.: 1986, *20th ESLAB Symp.*, ESA SP-250, Vol **II**, p. 375

Divine, N.: 1981, *The Comet Halley Gas and Dust Environment*, ESA SP-174, 25

Divine, N., Fechtig, H., Gombosi, T.I., Hanner, M.S., Keller, H.U., Larson, S.M., Mendis, D.A., Newburn, R.L., Jr., Reinhard, R., Sekanina, Z., Yeomans, D.K.: 1986, *Space Sci. Rev.* **43**, 1

Edenhofer, P., Buschert, H., Bird, M.K., Volland, H., Porsche, H., Brenkle, J.P., Kursinsky, E.R., Mottinger, N.A., Stelreid, C.T.: 1986a, *20th ESLAB Symp.*, ESA SP-250, Vol. **II**, p. 215

Edenhofer, P., Bird, M.K., Brenkle, J.P., Buschert, H., Esposito, P.B., Porsche, H., Volland, H.: 1986b, *Nature* **321**, 355

Edenhofer, P., Bird, M.K., Buschert, H., Esposito, P.B., Porsche, H., Volland, H.: 1986c, *The Giotto Mission – Its Scientific Investigations*, ESA SP-1077, 173

Fertig, J., Schwehm, G.H.: 1984, *Adv. Space Res.* **4**, 9, 213

Gombosi, T.I.: 1986, *20th ESLAB Symp.*, ESA SP-250, Vol. **II**, p. 167

Grard, R.J.L., McDonnell, J.A.M., Grun, E., Gringauz, K.: 1986, *20th ESLAB Symp.*, ESA SP-250, Vol. **III**, Poster B.327

Green, S.F., Hughes, D.W.: 1986, *20th ESLAB Symp.*, ESA SP-250, Vol. **II**, p. 157

Grun, E., Graser, U., Kohoutek, L., Thiele, U., Massone, L., Schwehm, G.: 1986, *Nature* **321**, 144

Hanner, M.S., Giese, R.H., Weiss, K., Zerrull, R.: 1981, *Astron. Astrophys.* **104**, 42

Hanner, M.S.: 1986, *Adv. Space Res.* **4**, 9, 189

Hirao, K., Itoh, T.: 1986, *Nature* **321**, 294

Hughes, D.W.: 1986, *Proc. ESA Workshop Comet Nucleus Sample Return*, ESA SP-249, 173

Keller, H.U., Arpigny, C., Barbieri, C., Bonnet, R.M., Cazes, S., Coradini, M., Cosmovici, C.B., Delamere, W.A., Huebner, W.F., Hughes, D.W., Jamar, C., Malaise, D., Reitsma, H.J., Schmidt, H.U., Schmidt, W.K.H., Seige, P., Whipple, F.L., Wilhelm, K.: 1986, *Nature* **321**, 320

Kissel, J.: 1986, in *The Giotto Mission – its Scientific Investigations*, ESA SP-1077, 67

Kissel, J., Sagdeev, R.Z., Bertaux, J.L., Angrov, V.N., Audouze, J.E., Blamont, J.E., Buchler, K., Evlanov, E.N., Fechtig, H., Fomenkova, M.N., von Hoerner, H., Inogamov, N.A., Khromov, V.N., Knabe, W., Krueger, F.R., Langevin, Y., Leonas, V.B., Levasseur-Regourd, A.C., Managadaze, G.G., Podkolzin, S.N. Shapiro, V.D., Tabaldyev, S.R., Zubkov, B.V.: 1986, *Nature* **321**, 280

Krankowsky, D., Lämmerzahl, P., Herrwerth, I., Woweries, J., Eberhardt, P., Dolder, U., Herrmann, U., Schulte, W., Berthelier, J.J., Illiano, J.M., Hodges, R.R., Hoffmann, J.H.: 1986, *Nature* **321**, 326

Labs, D., Neckel, H.: 1970, *Solar Phys.* **15**, 79

Larson, S.M., Sekanina, Z.: 1984, *Asstron. J.* **89**, 571 and 600

Larson, S., Sekanina, Z., Levy, D., Tapia, S., Senay, M.: 1986, *20th ESLAB Symp.*, ESA-SP 250, Vol **II**, p. 145

Maas, D., Aigner, S., Goller, J.R., Grun, E., Igenbergs, E., Rott, M., Weishaupt, U.: 1986, *20th ESLAB Symp.*, ESA SP-250, Vol. **III**, Poster B.337

Mazets, E.P., Aptekar, R.L., Golenetskii, S.V., Guryan, Yu.A., Dyachkov, A.V., Illyinski, V.N., Panov, V.N., Petrov, G.G., Savin, A.V., Sagdeev, R.Z., Sokolov, I.A., Khavenson, N.G., Shapiro, V.D., Shevchenko, V.I.: 1986, *Nature* **321**, 276

McDonnell, J.A.M.: 1969, *J. Sci. Instr. (J. Phys. E)* **2**, 2, 1026

McDonnell, J.A.M.: 1979, *Proc. Workshop Comet Halley Probe Design*, ESA SP-153, 93

McDonnell, J.A.M.: 1987, *J. Phys. E* **20**, 741

McDonnell, J.A.M., Alexander, W.M., Lyons, D., Tanner, W., Anz, P., Hyde, T., Chen, A.-I., Stevenson, T.J., Evans, S.T.: 1984, *Adv. Space Res.* **4**, 9, 297

McDonnell, J.A.M., Alexander, W.M., Burton, W.M., Bussoletti, E., Clark, D.H., Evans, G.C., Evans, S.T., Firth, J.G., Grard, R.J.L., Grun, E., Hanner, M.S., Hughes, D.W., Igenbergs, E., Kuczera, H., Lindblad, B.A., Mandeville, J.-C., Minafra, A., Reading, D., Ridgeley, A., Schwehm, G.H., Stevenson, T.J., Sekanina, Z., Turner, R.F., Wallis, M.K., Zarnecki, J.C.: 1986a, *The Giotto Mission – Its Scientific Investigations*, ESA SP-1077, 85

McDonnell, J.A.M., Alexander, W.M., Burton, W.M., Bussoletti, E., Clark, D.H., Firth, J.G., Grard, R.J.L., Grun, E., Hanner, MS., Hughes, D.W., Igenbergs, E., Kuczera, H., Lindblad, B.A., Mandeville, J.-C., Minafra, A., Schwehm, G.H., Sekanina, Z., Wallis, M.K., Zarnecki, J.C., Chakaveh, S.C., Evans, G.C., Evans, S.T., Firth, J.G., Littler, A.N., Massonne, L., Olearczyk, R.E., Pankiewicz, G.S.A., Stevenson, T.J., Turner, R.F.: 1986b, *Nature* **321**, 338

Millis, R.L., Schleicher, O.G.: 1986, *Nature* **324**, 646

Morley, T.A., Fertig, J.: 1986, *ESA Bull.* **46**, 71

Pettengill, G.H.: 1984, paper presented at General Assembly of U.R.S.I., Florence

Sagdeev, R.Z., Szabo, F., Avenesov, G.A., Cruvellier, P., Szabo, L., Szego, K., Abergel, A., Balacz, A., Barinov, I.V., Bertaux, J.-L., Blamont, J., Detaille, M., Demorelis, E., Dul'nev, G.N., Endroczny, G., Gordos, M., Kanyo, M., Kostenko, V.I., Krasikov, V.A., Nguyen-Trong, T., Nyitrai, Z., Reny, I., Rusznyak, P., Shamis, V.A., et al.: 1986, *Nature* **321**, 262

Schmidt, W.K.H., Keller, H.U., Wilhelm, K., Arpigny, C., Barbieri, C., Biermann, L., Bonnet, R.M., Cazes, S., Cosmovici, C.B., Delamere, W.A., Huebner, W.F., Hughes, D.W., Jamar, C., Malaise, D., D., D., Reitsma, H., Seige, P., Whipple, F.L.: 1986, *The Giotto Mission – Its Scientific Investigations*, ESA SP-1077, 149

Sekanina, Z.: 1986, *Adv. Space Res.* **5**, 12, 307

Sekanina, Z.: 1987, *Nature* **325**, 326

Sekanina, Z., Larson, S.M.: 1984, *Astron. J.* **89**, 1408

Sekanina, Z., Larson, S.M.: 1986a, *Nature* **321**, 357

Sekanina, Z., Larson, S.M.: 1986b, *Astron. J.* **92**, 462

Simpson, J.A., Sagdeev, R.Z., Tuzzolino, A.J., Perkins, M.A., Ksanfomality, L.V., Rabinowitz, D., Lentz, G., Afonin, V.V., Ero, J., Keppler, E., Kosorokov, J., Petrova, E., Szabo, L., Umlauft, G.: 1986, *Nature* **321**, 278

Stone, R.P., Baldwin, J.A.: 1983, *Monthly Notices Roy. Astron. Soc.* **204**, 347

Vaisberg, O.L., Smirnov, V., Omelchenko, A.: 1986a, *20th ESLAB Symp.*, ESA SP-250, Vol. **II**, 17

Vaisberg, O.L., Smirnov, V.N., Gorn, L.S., Iovlev, M.V., Balikchin, M.A., Slimov, S.I., Savin, S.P., Shapiro, V.D., Shevchenko, V.I.: 1986b, *Nature* **321**, 274

Astron. Astrophys. 187, 742–752 (1987)

ASTRONOMY AND ASTROPHYSICS

The dust coma of comet P/Halley: measurements on the Vega-1 and Vega-2 spacecraft

J.A. Simpson[1], **D. Rabinowitz**[1], **A.J. Tuzzolino**[1], **L.V. Ksanfomality**[2], **and R.Z. Sagdeev**[2]

[1] Enrico Fermi Institute, University of Chicago, Chicago, IL 60637, USA

[2] Space Research Institute, Profsoyuznaya 88, 117810 Moscow, V-485, USSR

Received February 17, accepted May 25, 1987

Summary. The spatial, temporal and mass distributions of coma dust particles for masses $> 10^{-13}$ g, which we measured with the DUCMA instruments and reported earlier (Simpson et al., 1986a), have been confirmed and extended to reveal additional properties of the dust coma. We have analyzed the inverse square dependence ($\propto R^{-2}$) of flux with distance R from the nucleus and show that there are well-defined envelope boundaries for the measured range of masses. Regions of enhanced fluxes above these R^{-2} 'baselines' inside these boundaries are identified as dust jets. Of special interest is the giant flux enhancement near closest approach for Vega-1, for which we show that the mass spectrum is similar to the pre-encounter spectra and that there is a smooth rate of rise of intensity over more than four seconds leading into the peak. This is a flux gradient similar to that found for several other much smaller flux enhancements inbound and outbound. The mass dependence of terminal velocities for some jets is consistent with recent predictions, and recurring jets observed by both spacecraft—when traced back to the nucleus—correlate with the active regions predicted by Sekanina and Larson (1986). We have discovered particles with a wide range of masses arriving in clusters or 'packets' (i.e., a non-Poisson distribution of dust particles) throughout the regions beyond the envelope boundaries – namely, the fringe regions of the dust coma. These observations suggest the emission from the nucleus of large conglomerates of small particles which gently disintegrate as they travel outward to account for the 10^{-13} g particles observed beyond the envelope boundaries. Also, dust fluences over the missions have been determined. Finally, we report that both the DUCMA electronics and dust sensors maintained their calibrations and performance throughout the encounters and during the post-encounter interplanetary missions.

Key words: comet P/Halley – comet dust – dust mass spectra – dust jets – dust clusters – dust envelopes

1. Introduction

During the past several months since our publication of the initial results of the coma measurements obtained with the Dust Counter and Mass Analyzer (DUCMA), carried on the Vega-1 and Vega-2 spacecraft (Simpson et al., 1986a), we have re-examined our initial work and extended our research based on these measurements. Both the measurements and the conclusions we reached in this initial report remain unchanged. In addition to the observations of all the Vega dust instruments reported in *Nature* (1986) which are in agreement, there have been some significant differences which we discuss in this report. In particular, we have included in Sect. 4 detailed information on the large flux increase measured by DUCMA near Vega-1 closest approach. These and other new results, which we report here, were presented recently at the International Cometary Dust Workshop (Simpson et al., 1986b) – hereafter referred to as ICDW 86, the ESLAB Symposium on Halley's comet and summarized in ESA publication SP-250 (Simpson et al., 1986c).

Send offprint requests to: J.A. Simpson

2. DUCMA detector calibrations

The theory and details of detector operation have been published (Simpson and Tuzzolino, 1985; Perkins et al., 1985) and we shall only briefly review key points here. The basic detection element is the dust particle mass detector, called the M detector, consisting of a 28 μm thick film composed of the electrically polarized polymer polyvinylidene fluoride (PVDF), whose area is 75 cm^2. A high-velocity dust particle impacting the M detector produces very rapid local depolarization in the foil volume destroyed by the dust particle. This local volume depolarization results in a fast (nanosecond range) current pulse at the input to the electronics, as shown in Fig. 1a. The total charge carried by the current pulse is determined by the mass, m, and relative velocity, v, of the impacting particle.

Prelaunch calibrations of the M detectors were carried out at the Heidelberg dust accelerator, using dust particles over the mass range $\sim 10^{-10}$ g to 10^{-13} g and over the velocity range ~ 1 to 12 km s^{-1}. For this mass and velocity range, all particles had a penetration depth $l < 6\,\mu$m, which was small compared to the M detector thickness $L = 28\,\mu$m. For these stopping particles ($l \ll L$), the output signal, N, was found to vary as $m^a v^b$, with the exponents being $a = 1.3$, $b = 3.0$, as shown in Fig. 1b (Simpson and Tuzzolino, 1985). Since calibrations at velocities ~ 80 km s^{-1} were not possible, particle masses corresponding to the four DUCMA electronic thresholds (Sect. 3) were determined by extrapolating our stopping particle calibrations (exponents $a = 1.3$, $b = 3.0$) to the encounter velocities, with all of our reported encounter mass spectra (Simpson et al., 1986a,b,c) being

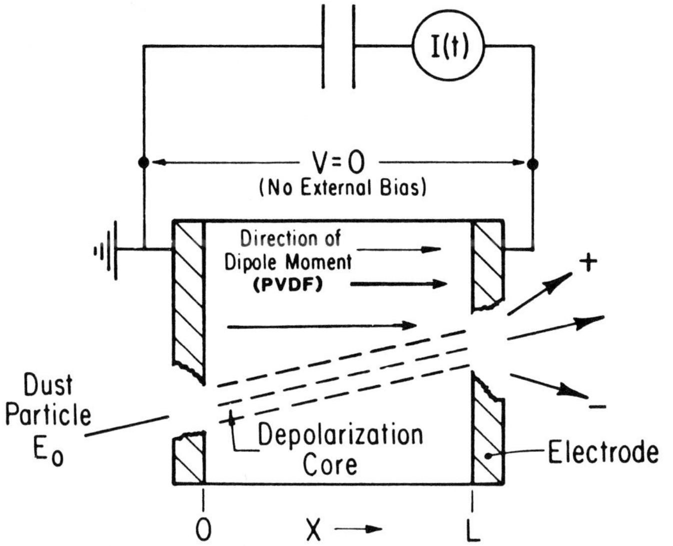

Fig. 1a. Schematic diagram illustrating the PVDF dust detection principle.

based on the particle mass thresholds so determined (Table 1). However, our Heidelberg measurements carried out on a PVDF detector under conditions where the particles penetrated the detector ($l > L$) showed that, for penetrating particles, the exponents a and b had the values $a = 0.6 \pm 0.2$, $b = 1.8 \pm 0.2$ (Simpson and Tuzzolino, 1985; see Eq. (12) and Fig. 2) so that, for penetrating events, the dependence of PVDF detector signal on particle mass is much flatter than that for stopping dust particles.

Based on the discussion and experimental results given in this earlier work, we conclude that the Halley encounter dust particles which fired the higher DUCMA electronic thresholds (Sect. 3) could have masses significantly greater than the values given in Table 1, which were deduced from extrapolation of our extensive stopping particle calibrations to the encounter velocities. However, from physical considerations (to be published) we believe that our lowest mass threshold continues to be correct.

The possibility that our detector signal dependence on particle mass may be closer to the dependence we measured for penetrating particles and the fact that our reported encounter mass spectra are significantly steeper than those reported by the other Vega dust experiments (Mazets et al., 1986a,b; Vaisberg et al., 1986a,b) led us to re-examine our encounter mass spectra on the basis of an assumed dependence of M detector signal on particle mass which varies as $m^{0.8}$. We emphasize that we have as yet no dust calibration data for the DUCMA M detector to justify this assumption. This choice of $a = 0.8$ was arrived at on the basis of our stopping and penetrating particle studies on thick and thin PVDF detectors mentioned above. In Sect. 5, we show that DUCMA encounter mass spectral shapes, based on this assumption, are in good agreement with the spectral shapes

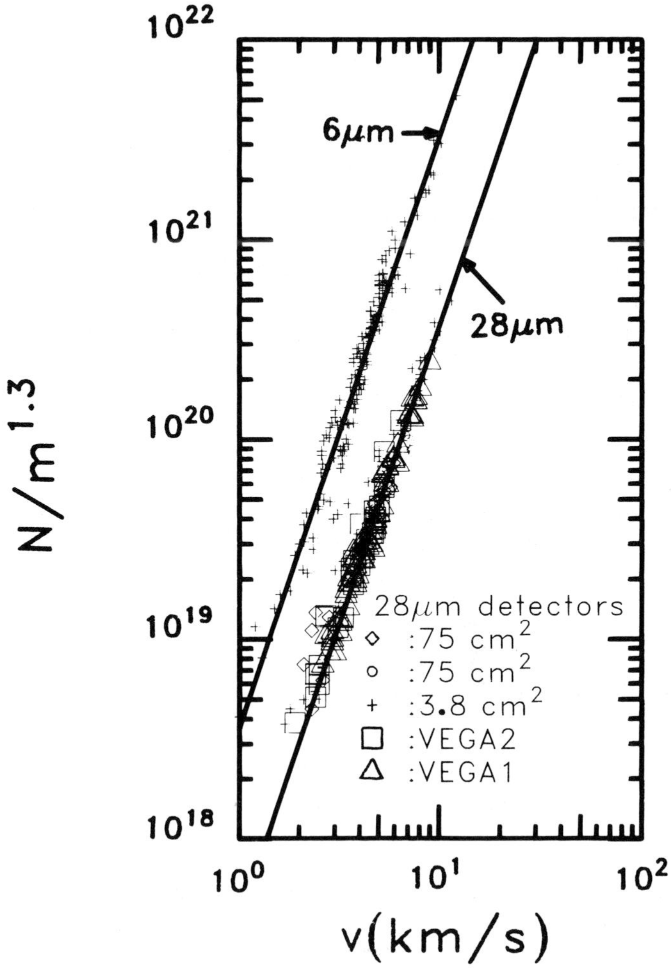

Fig. 1b. Plot of the ratio $N/m^{1.3}$ *vs.* measured projectile velocity v (km/s), where N (electrons) is the measured detector signal and m(g) is the measured projectile mass. Data were obtained at the Heidelberg dust accelerator

obtained from SP-2 (Mazets et al., 1986b) for Vega-2. Spectra discussed elsewhere in this report are based on our previously reported mass assignments (Table 1).

3. Instrument response

The DUCMA sensor system was mounted on the spacecraft so that the M detector was not exposed to possible ejecta from massive dust particles striking the solar arrays or other spacecraft structures. The exception is a small portion of the guard cone (of area $\sim 100\,\text{cm}^2$) which retained the surrounding thermal blanket (Perkins et al., 1985) and ICDW 86, page 24). Electronic thresholds, M1 through M4 for the M detector, define the

Table 1. DUCMA counting rate channels

Channel	M1, M1NV	M2, M2NV	M3, M3NV	M4, M4NV	VL, VH
Particle mass threshold (g)	$\geq 1.5\,10^{-13}$	$\geq 9\,10^{-13}$	$\geq 9\,10^{-12}$	$\geq 9\,10^{-11}$	—

four dust particle mass thresholds listed in Table 1 (see also ICDW 86, page 26).

The fast response of the M detector, combined with fast electronics (3 μs pulse shaping, 4 μs discriminator width), permitted a high counting rate capability of up to ~2 10^4 random impacts per second without dead time corrections. The 19-bit counters for each independent channel excluded the possibility of counter overflow throughout the two encounters.

As described in our earlier reports (Simpson and Tuzzolino, 1985; Perkins et al., 1985), PVDF detectors also have a secondary but minor mode of pulse response due to their piezoelectric properties. Therefore, to ensure that no signals from the M detector could be attributed to acoustic events, special precautions were taken both in the detector mounting and in the addition of a second detector which provided information on any acoustic events occurring on the spacecraft that could interfere with the dust measurements. Throughout the encounters, background acoustic events were found to be of negligible importance.

Each DUCMA instrument contains an electronic In-Flight-Calibrator (IFC) which periodically sends a sequence of pulses to the DUCMA electronics (automatic calibration of DUCMA occurs once each 24.3 days) such that a rigorous verification of electronic performance is obtained (Perkins et al., 1985). The stability of the four DUCMA particle mass thresholds listed in Table 1 is illustrated in Fig. 2, and the DUCMA instruments continuously measure the integral counts above these mass thresholds. Throughout the encounters the data accumulation times for each of the ten channels were two seconds for Vega-1 and one second for Vega-2.

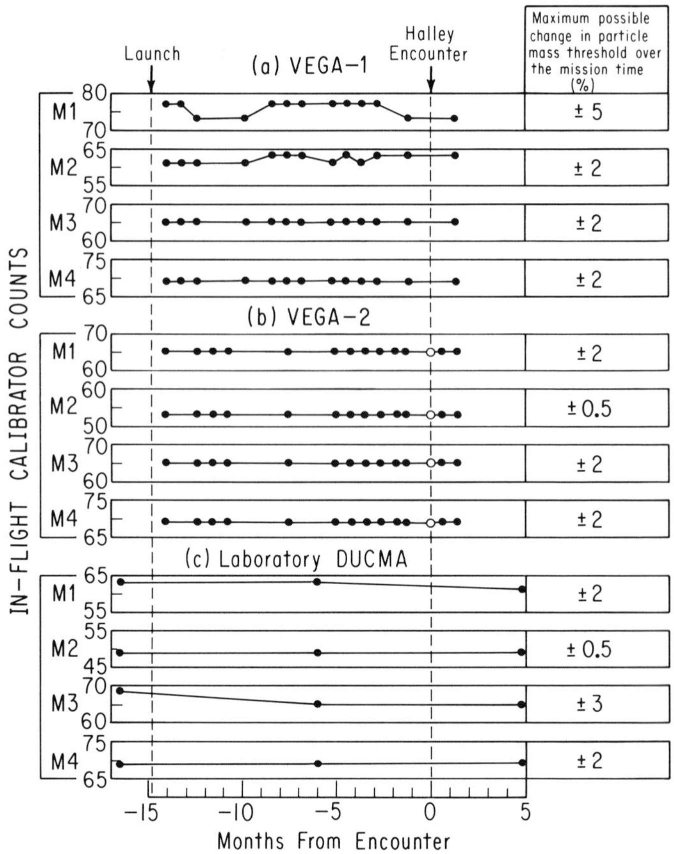

Fig. 2. Stability of the four DUCMA particle mass thresholds listed in Table 1 *versus* time for the DUCMA instruments on Vega-1 and Vega-2 and for a flight back-up instrument. The offset vertical scales give the number of firings of each discriminator resulting from the DUCMA In-Flight-Calibrator (IFC). The number of discriminator firings are a function of the discriminator threshold level. From these data and the known characteristics of the counters and IFC, the maximum possible changes in particle mass thresholds over the time interval shown may be determined and are as indicated. For Vega-2, the IFC was activated during the Vega-2 encounter (~10 minutes before closest approach), with this encounter IFC data (open circles) being identical with the immediate pre- and post-encounter IFC data. Over the ~20 month time interval shown, all thresholds for the three DUCMA instruments are seen to be stable to better than ±2%

4. Flux profiles and mass spectra

4.1. Overview within ± 500s of closest approach

The flux measurements for Vega-1 and Vega-2 have been averaged over six second intervals, as shown in Figs. 3 and 4 for all dust particles with mass greater than the mass thresholds indicated. Up until the onset of the large Vega-1 flux peak at closest approach, shown in Fig. 3, the flux curves for 10^{-13} g particles are similar for the two pre-closest-approach intervals including the rapid fluctuations of dust intensities, as reported earlier (Simpson et al., 1986a). However, at and after closest approach, the flux curves for Vega-1 and Vega-2 show dramatic differences, with the Vega-1 lower mass fluxes showing an intense first peak at closest approach, followed by four additional flux enhancements extending out to ~200 s (~ 1.8 10^4 km). In contrast, the Vega-2 outbound flux profile is only slightly different from the inbound profile. These results are consistent with our observations and those of other investigators on the Vega missions that, for the Vega-1 encounter, the active side of the nucleus faced the Sun, whereas, for the Vega-2 encounter, the relatively inactive side faced the Sun. We note that there is no measurable time dispersion between the Vega-1 flux enhancements measured at and within ~ 200 s after closest approach for the four DUCMA mass thresholds.

4.2. Flux profile and mass spectra near closest approach for Vega-1

We have analyzed the period of outstanding and rapid increase of flux near Vega-1 closest approach (Fig. 3) in detail and report here our confirmation of these data.

a) In Fig. 5 we show that the flux data averaged over two second intervals within ±20 s of closest approach reveal a progressive flux increase with time for this event – a factor ~15 increase in the first two seconds, followed by a further factor 3 increase in the succeeding two seconds, with the flux peak occupying the interval 0 to 4 s. This rise time corresponds to a spatial distance ~ 300 to 500 km, over a dust mass range exceeding a factor of ~10^3. This flux gradient is frequently observed on a much smaller amplitude scale on both Vega-1 and Vega-2 (e.g., see insert Fig. 2a, Simpson et al., 1986a) and is not uniquely limited to the large flux increase shown in Fig. 5.

b) The integral mass spectrum varies continuously before, during and after the peak intensity period. To illustrate this time variation, we assume that the integral flux varies as $m^{-\alpha}$, and determine a value for α from the integral fluxes corresponding to our two lowest mass thresholds. Although a simple power law does not, in general, describe the measured mass distribution

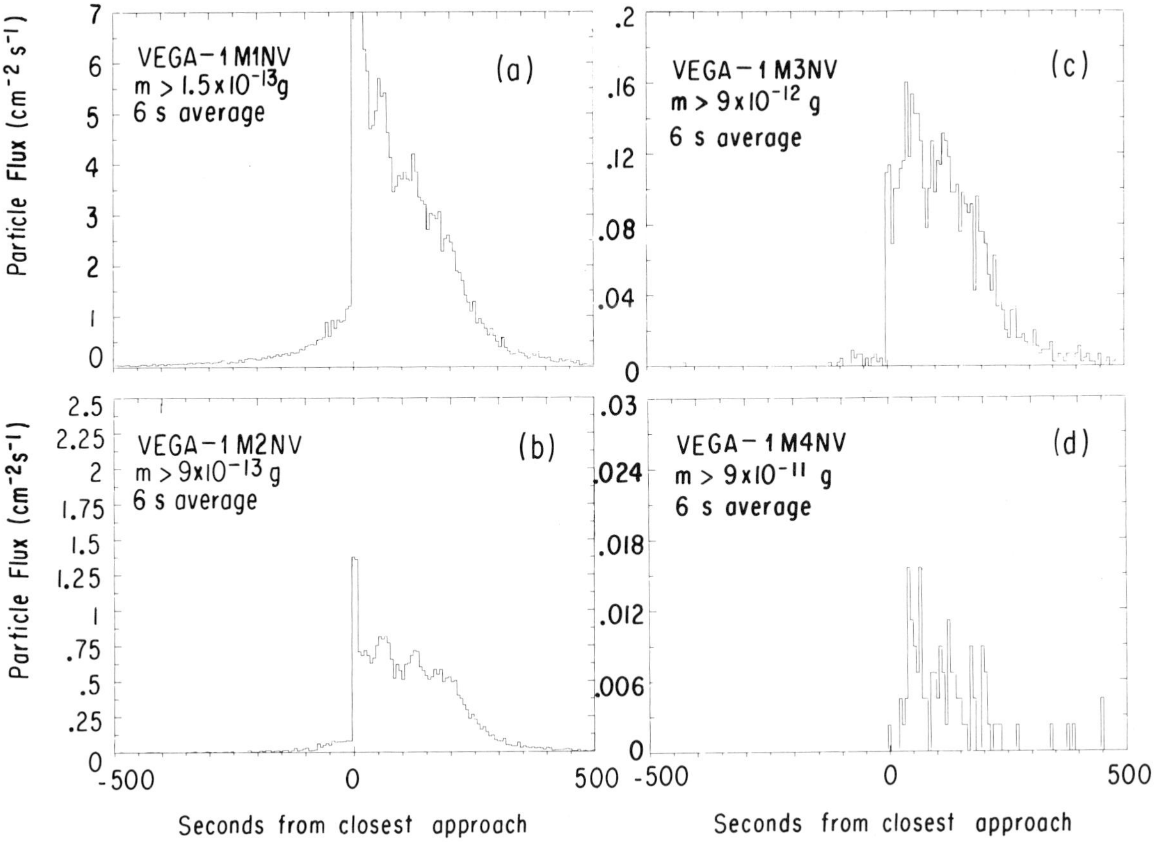

Fig. 3. Time profile of the integral particle fluxes corresponding to the four DUCMA mass thresholds listed in Table 1 for the Vega-1 encounter. Six second averaging intervals are used in each case. Note that in panel (a) the peak flux is off scale (see Fig. 5)

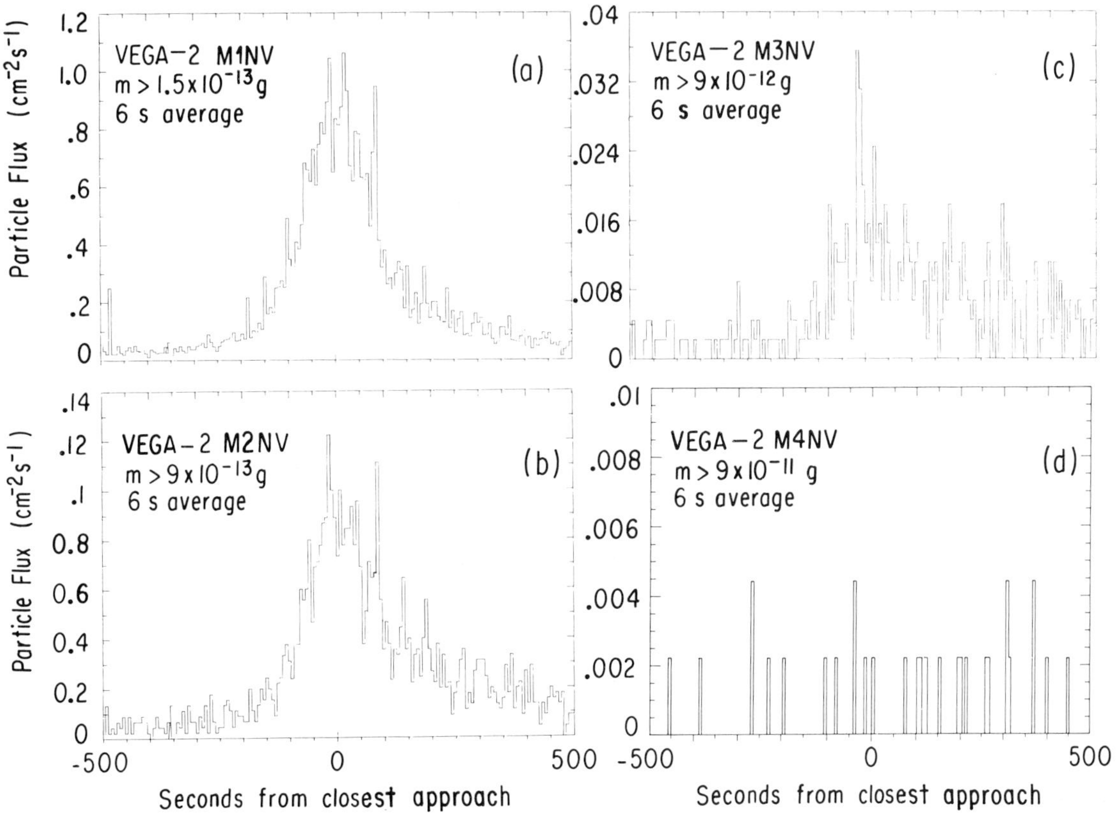

Fig. 4. Time profiles of the integral particle fluxes corresponding to the four DUCMA mass thresholds listed in Table 1 for the Vega-2 encounter. Six second averaging intervals are used in each case

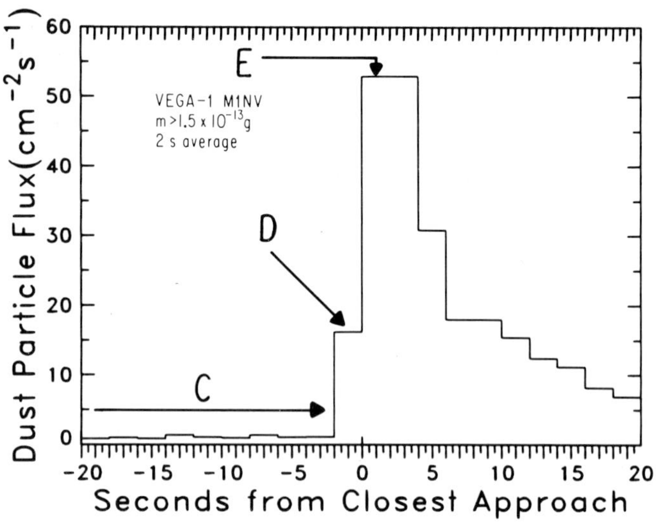

Fig. 5. Time profile of the integral flux of dust particles of mass $\geq 1.5\,10^{-13}$ g near Vega-1 closest approach (2 s averages)

over the entire DUCMA mass interval (either for our work or the reports of the other investigators), the value of α, as computed, provides a useful measure of the particle mass spectrum, and changes in α provide a valid measure of the spectral changes throughout the coma. The time dependence of α is shown in Fig. 6 for the Vega-1 and Vega-2 encounters. We note that both Vega-1 and Vega-2 show higher α values inbound than outbound, and that α for Vega-1 shows considerable variation, ranging from a maximum of 1.9 before and at closest approach to a minimum of 0.8 after closest approach. Spectra for intervals preceding and including the peak intensity are shown in Fig. 7 (and listed in Table 2), where it is apparent that the spectral slopes near closest approach are not greatly different from the pre-encounter spectra (e.g., compare spectra A and E), even though the integral intensities are greatly different. Furthermore, there is a monotonic increase in α up to the value at the peak intensity – namely, α begins to increase $\sim$ 20 s before any rapid increase in the flux.

c) We have analyzed integral fluxes measured at large distances from the nucleus ($\approx 10^5$ km) and find that—for each spacecraft—the inbound integral fluxes for our lowest mass thresholds (M1NV and M2NV) differ from the corresponding outbound fluxes by less than 50%. Further, at these great distances, the Vega-2 and Vega-1 fluxes are within a factor of two of each other. This overall consistency demonstrates that there was no significant change in M detector sensitivity for either Vega-1 or Vega-2 as a result of the encounters, especially with regard to the large flux increase near Vega-1 closest approach.

d) We have carefully examined all ten of the DUCMA counting rates, as well as other DUCMA parameters (Perkins et al., 1985) obtained for each two second interval during the Vega-1 closest approach, and find that the instrument functioned properly throughout the period. Furthermore, the IFC data presented in Fig. 2 show that the DUCMA instruments remained in calibration both during and after the encounter. Finally, we have examined the post-encounter interplanetary dust fluxes measured in the period March 1986 – October 1986 and find that

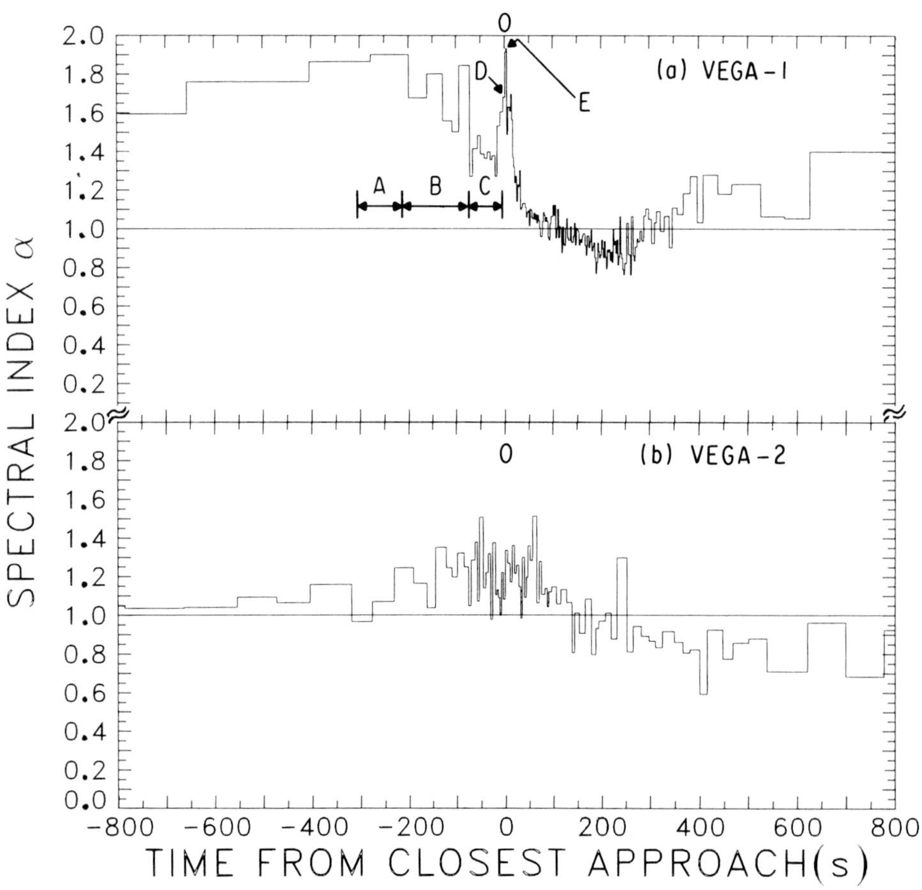

Fig. 6. Spectral index α versus time from closest approach for the Vega-1 and Vega-2 encounters. α is computed from the integral fluxes corresponding to the two lowest mass thresholds listed in Table 1. For all time intervals over which α is computed, the uncertainty in α is <0.1. The horizontal line at $\alpha = 1.0$ serves as a reference

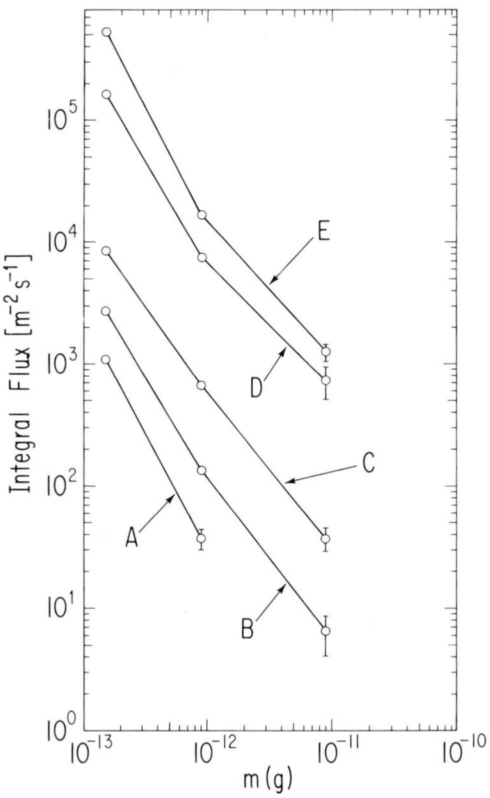

Fig. 7. Integral mass spectra for time intervals preceding and including Vega-1 closest approach. Table 2 lists corresponding values of α assuming the integral flux to vary as $m^{-\alpha}$ for each of the two mass intervals. The time intervals corresponding to spectra A through E are indicated in Fig. 6. Those corresponding to spectra C through E are also indicated in Fig. 5

the fluxes measured by the Vega-1 and Vega-2 DUCMA sensors are the same as for the interplanetary pre-encounter levels of 1985. Thus, both sensors and electronics performed nominally before, during and after encounter.

We conclude that the intense flux enhancements measured by DUCMA at and after Vega-1 closest approach result from dust jets as discussed below in Sect. 6.

4.3. Comparison of DUCMA measurements with other Vega dust measurements

We had noted earlier (Simpson et al., 1986c) what appeared to be a discrepancy between the time profiles of the dust flux measured by DUCMA and by SP-2 as reported in *Nature* (Mazets et al., 1986a, Fig. 1) for ~ 10^{-13} g particles near Vega-1 closest approach. Specifically, there were discrepancies in the times of occurrence of the intense flux enhancements measured by the two instruments near Vega-1 closest approach, which could not be accounted for by mass dispersion effects. However, a comparison of the subsequent results reported by SP-2 (Mazets et al., 1986b, Fig. 10) with the DUCMA results, shown in Fig. 8, now shows that the time profiles of the dust fluxes are in substantial agreement. At this time we do not have any explanation for the remaining differences in the results from the different Vega dust instruments, which include differences in the absolute fluxes measured and differences in the spectral slope of the mass spectra, which will be analyzed in a later publication. However, in Sect. 5 we illustrate a possible reason for the differences in the spectral slopes of the reported mass spectra.

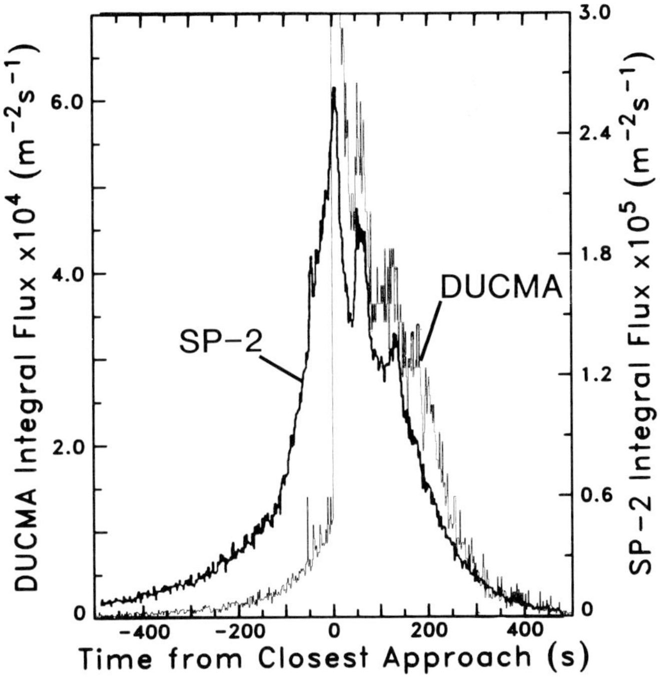

Fig. 8. Comparison of time profiles of integral fluxes for the Vega-1 encounter measured by DUCMA ($m > 1.5\,10^{-13}$ g) and SP-2 ($m > 6\,10^{-16}$ g). The SP-2 flux curve is a photo reproduction of Fig. 10 in Mazets et al. (1986b). Two second averaging intervals are used for the DUCMA data. Note that the flux scales are different for the two sets of data

Table 2. Vega-1 integral mass spectral index α

Spectrum	Time interval	α $1.5\,10^{-13}\,\mathrm{g} < m < 9\,10^{-13}\,\mathrm{g}$	α $9\,10^{-13}\,\mathrm{g} < m < 9\,10^{-12}\,\mathrm{g}$
A	−304 to −214 s	1.90 ± 0.1	—
B	−214 to −70 s	1.68 ± 0.05	1.30 ± 0.2
C	−70 to −2 s	1.42 ± 0.03	1.25 ± 0.1
D	−2 to 0 s	1.71 ± 0.05	1.01 ± 0.16
E	0 to 4 s	1.92 ± 0.03	1.12 ± 0.08

5. Integral fluences

The integral fluences (i.e., the integral over time of the integral fluxes) provide an important measure of the dust mass distribution from the comet nucleus. The time profiles of the integral fluences for the two encounters are shown in Fig. 9. For Vega-1, the first and second dust events ($1.5\,10^{-13}\,\mathrm{g} \le m < 9\,10^{-13}\,\mathrm{g}$) were recorded at -8068 s ($6.4\,10^5$ km) and -4208 s ($3.3\,10^5$ km), respectively, with the subsequent buildup of events as shown in Fig. 9. For Vega-2, the first event was recorded at -3313 s ($2.5\,10^5$ km). A comparison of the inbound data for the two encounters shows that heavy particles ($m > 10^{-11}$ g) are measured at greater distances from the nucleus for Vega-2 than for Vega-1.

In Fig. 10a, the Vega-2 fluence reported by SP-2 (Mazets et al., 1986b) is compared with the Vega-2 DUCMA fluence. The DUCMA spectrum labeled A is based on the extrapolations of our stopping particle calibrations, as discussed in Sect. 2 and reported earlier (Simpson et al., 1986a,b,c). DUCMA spectrum B is obtained by shifting the second, third and fourth points for spectrum A to higher masses consistent with the assumption discussed in Sect. 2. DUCMA Spectrum C is obtained by arbitrarily shifting the first DUCMA point in Spectrum A to coincide with the SP-2 data, and then shifting the remaining DUCMA points by amounts again consistent with the assumption discussed in Sect. 2. It is clear that DUCMA spectra B and C have spectral shapes in good agreement with that of the SP-2 spectrum. However, should any modification of our earlier reported spectra prove to be necessary, we believe the modifications will result in spectra much closer to spectrum B than spectrum C because of physical constraints on the mass and velocity dependence of our detector signal. Thus, a discrepancy in intensity would still exist between the DUCMA and SP-2 spectra.

In Fig. 10b, we show the DUCMA fluence data for both encounters. The spectra labeled A are based on the extrapolations of our calibration data, and those labeled B are obtained as for Fig. 10a. The best-fit spectral indices are indicated, with spectra B being significantly flatter than the corresponding A spectra.

If future dust particle calibrations on PVDF detectors should verify our assumption (Sect. 2), then much of the discrepancy between the reported Vega mass spectral shapes would be removed, with the remaining discrepancy being that of intensity.

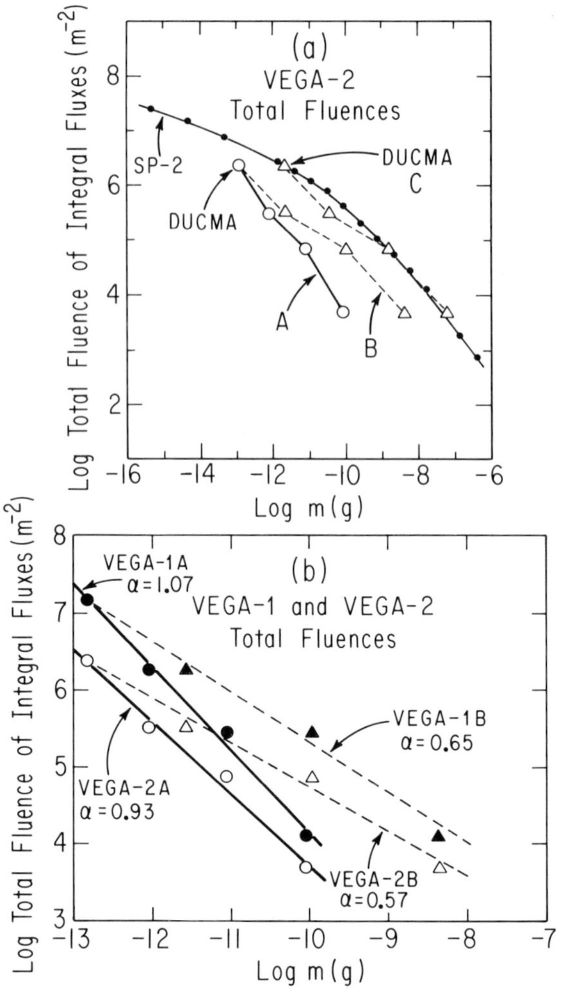

Fig. 10. **a** Comparison of the mass dependence of the total integral fluence for the Vega-2 encounter measured by DUCMA and SP-2. The SP-2 curve is a photoreproduction of Fig. 12 in Mazets et al. (1986b). DUCMA spectrum A is based on extrapolation of dust calibration data (Simpson and Tuzzolino, 1985). DUCMA spectra B and C are obtained from A, as described in Sect. 5. **b** Total integral fluence measured by DUCMA for the Vega-1 and Vega-2 encounters. Spectra A are based on extrapolation of our calibration data, and spectra B are obtained from A, as described in Sect. 5. The spectral indices corresponding to the best-fit lines are indicated

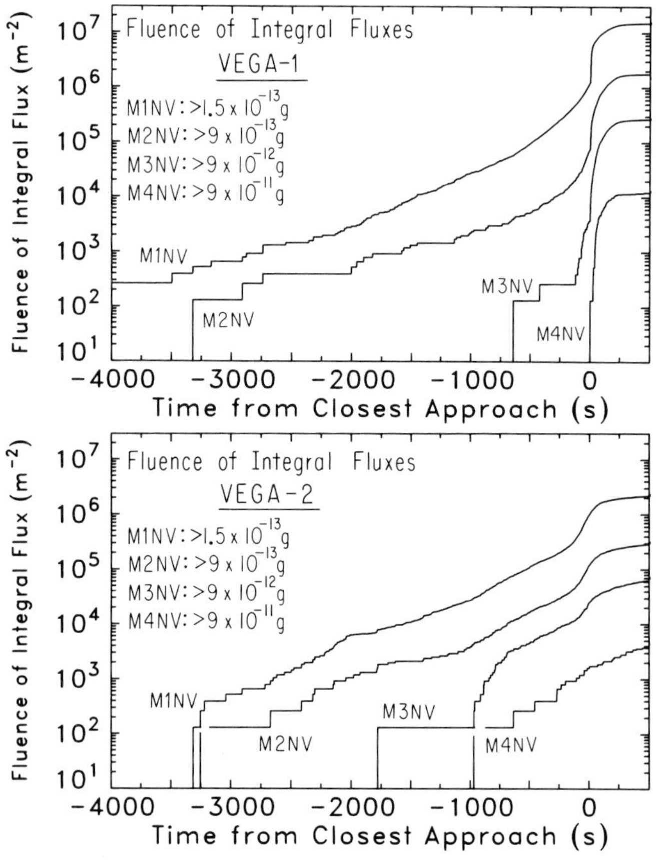

Fig. 9. Fluence of integral fluxes *vs.* time for the Vega-1 and Vega-2 encounters

6. Jet source mapping and comparison with the fountain model

6.1. R^{-2} dependent fluxes and dust envelopes

According to the fountain model and common assumptions concerning the properties of cometary dust particles, dust envelopes in the mass range 10^{-13} to 10^{-11} g are expected at distances of 10^4 to 10^5 km from the nucleus (Divine et al., 1986). Inside these boundaries, fluxes are expected to fall off with the inverse square of distance (R) to the nucleus. Fig. 11 (ICDW 86, page 35) shows *differential* fluxes ($\Delta m1$, $\Delta m2$, and $\Delta m3$) corresponding to differences in the integral mass channels – i.e., M1NV-M2NV, M2NV-M3NV, M3NV-M4NV. These curves have been normalized by $(R/R_0)^2$, with R_0 the closest approach distance for each spacecraft. Note the R^{-2} dependence inbound on Vega-1, and both inbound and outbound on Vega-2. The persistence of this R^{-2} dependence suggests a background dust production at the nucleus which is isotropic and quasi steady-state. Distinct envelope crossings appear where the curves suddenly steepen. The three most distinct $\Delta m1$ boundaries (panels a, b and d) correspond to a parabolic envelope apex (E) at $5(\pm 1) \times 10^4$ km, in agreement with the results of Vaisberg et al. (1986a,b).

6.2. Terminal velocity dispersion

Another feature predicted by the fountain model is that the terminal velocities (v_0) to which particles are radially accelerated by subliming gasses should increase with decreasing particle size. As a result, a discrete jet active on a rotating nucleus will generate in the coma a wide stratified spiral, with an inner edge characterized by slowly moving, heavy particles and an outer edge of fast, light particles. Fig. 12 (ICDW 86, page 36) is a plot of $(R/R_0)^2$ normalized fluxes from Vega-1 plotted against time. There appear two peaks, labelled 'a' and 'b', which move to later times as the particle mass decreases. This is an indication of two discrete sources on the nucleus, jet 'a' and jet 'b' with each jet generating a stratified spiral. Assuming a 52.1h rotation period and setting $v_0 = 0.40\,\mathrm{km\,s^{-1}}$ for $\Delta m1$ particles, values for v_0 of $0.33\,(\pm 0.01)\mathrm{km\,s^{-1}}$ and $0.30\,(\pm 0.01)\mathrm{km\,s^{-1}}$ are obtained for the

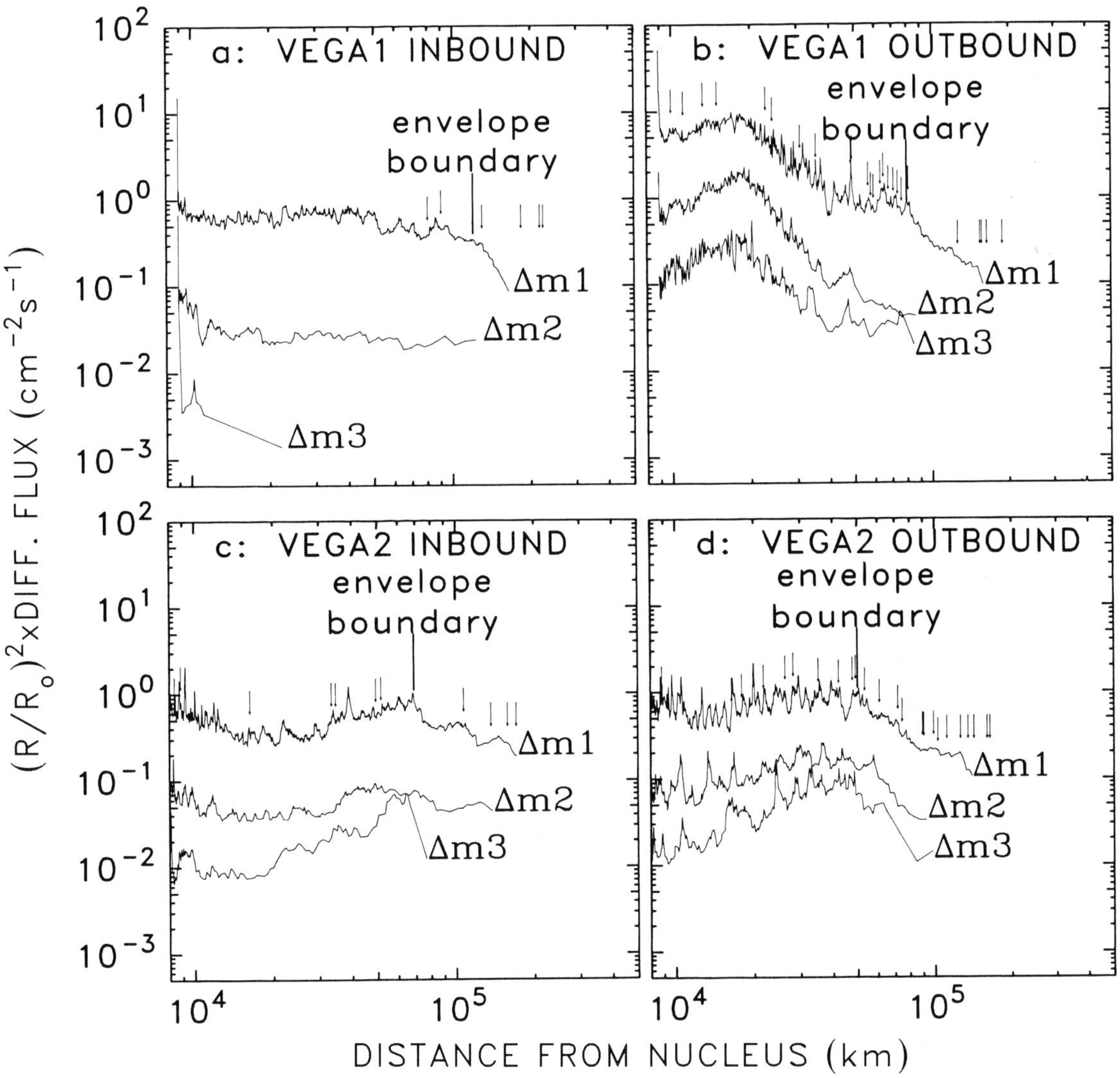

Fig. 11. Differential fluxes normalized by $(R/R_0)^2$ *versus* distance R from the nucleus, where R_0 is the closest approach distance for each spacecraft. See Fig. 14 for identification of vertical tick marks

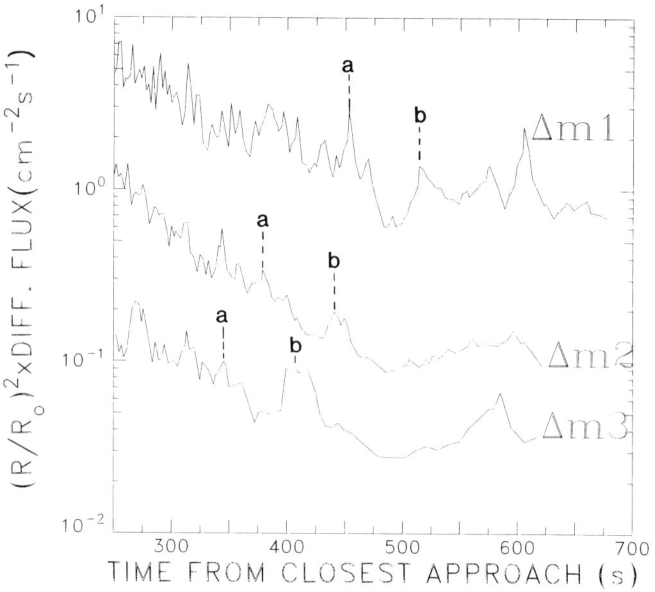

Fig. 12. $(R/R_0)^2$ normalized differential fluxes for Vega-1 *versus* time (outbound) showing jets 'a' and 'b'

particles of mass $\Delta m2$ and $\Delta m3$, respectively, in these jets. These velocities are consistent with theoretical calculations based on an assumed dust to gas mass ratio of 0.15 (Gombosi in Divine et al., 1986) and agree with those velocities used in a similar analysis of Vega-2 dust measurements reported earlier (Vaisberg et al., 1986a, b). Velocity dispersion effects have also been observed by Mazets et al. (1986b).

6.3. *Jet source mapping*

Encouraged by this correspondence between the fountain model and the measured $\Delta m1$ flux profiles, a simple fountain theory was used to map observed $\Delta m1$ flux enhancements to jet source locations on the nucleus. All $\Delta m1$ particles were assumed to follow parabolic trajectories, parameterized by the same E and v_0. Preliminary results of this analysis were reported (Simpson et al., 1986c; ICDW 86, pages 37–39).

Figure 13 shows source departure points for dust which reaches Vega-1 and Vega-2, assuming particles with $v_0 = 0.42$ and $0.38\ \mathrm{km\,s^{-1}}$, respectively, that form an envelope with $E = 5\,10^4$ km. Major ticks mark source locations mapped from spacecraft positions separated by 100s of flight time. The coordinate system, rotation period (52.1 hours), and spin vector orientation for the nucleus are as specified by Sekanina and Larson (1986), except that the subsolar point at 1986 perihelion defines 0° longitude. For each spacecraft, respective times of envelope penetration (−1810 and −2000 s), closest approach ($t = 0$), and envelope departure (840 and 860 s) are indicated next to the corresponding source points. Sources track east to west until about −150s, at which point they reverse direction. Notice that each source track circles the nucleus three times inbound (segments 1, 2, and 3) and twice outbound (segments 4 and 5). A single source, active over

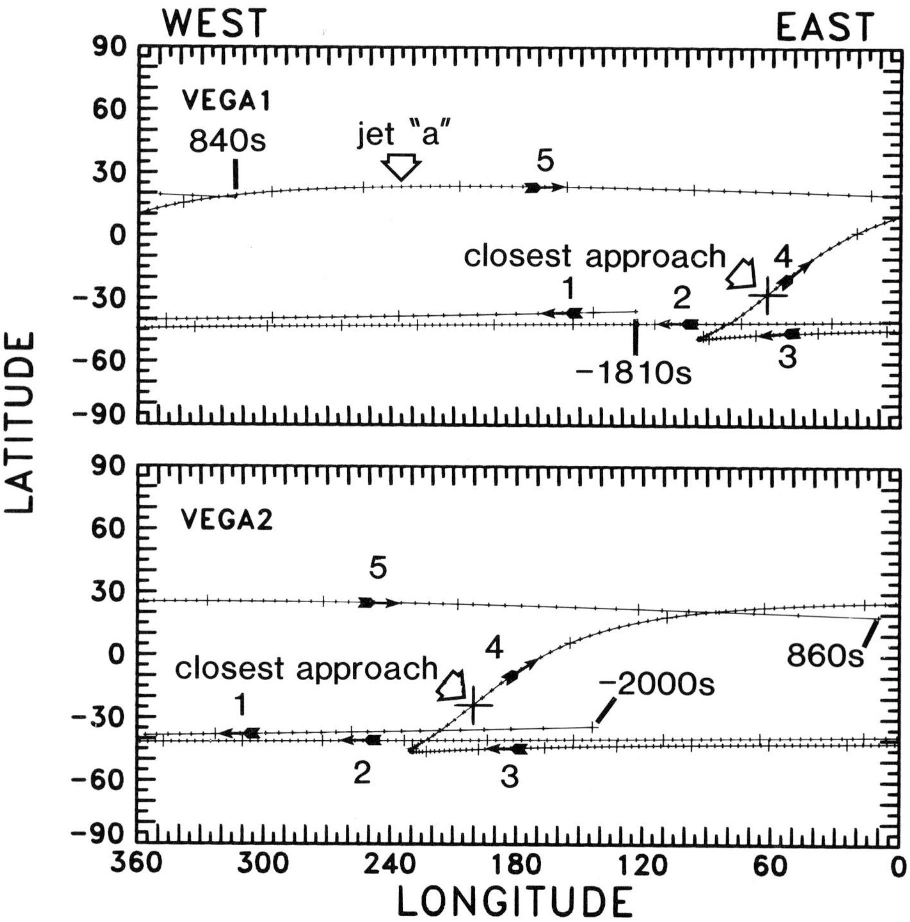

Fig. 13. Dust source departure points for Vega-1 and Vega-2, assuming particles with $v_0 = 0.42$ and $0.38\ \mathrm{km\,s^{-1}}$, respectively, and $E = 5\,10^4$ km. The coordinate system, spin orientation, and rotation period (52.1h) are as specified by Sekanina and Larson (1986) except that 0° is subsolar at 1986 perihelion

many rotation periods, may appear in the measured flux profile each time the spacecraft source track passes near that source's coordinates (ICDW 86, pages 38–39). These encounter times correspond to places where the spacecraft crosses the spiral-shaped band of dust unwinding from the source. Pankiewicz and McDonnell (1986) have made similar maps for the Giotto and Vega-2 trajectories.

6.4. Source map results

The results we obtain by extending this analysis (to be published later) show that recurring enhancements observed by both spacecraft map to four discrete regions of local source activity on the nucleus when the source departure points shown in Fig. 13 are used. The regions are spaced by roughly 90° in longitude and are consistent (see ICDW 86, page 39) with the 1910 observations (Sekanina and Larson, 1986), assuming that the prime meridian of 1910 has moved to the 1986 longitude of 105°. The region of enhanced jet activity seen outbound on Vega-1 ($8\,10^3$ to $4\,10^4$ km, Fig. 11b, and 0 to 500 s, Fig. 3a) maps from longitude 65° eastward to longitude $-150° = 210°$. Overall, we confirm the existence of an active side on the comet nucleus, as well as discrete regions of jet activity.

7. Clusters and packets of dust particles in the outer coma

7.1. Low mass particles beyond the dust envelopes

One of the surprises reported in our initial dust measurements was the existence of very low mass particles at great distances from the nucleus – i.e., in the fringes of the coma (Simpson et al., 1986a, c). For example, in Fig. 11 we see that beyond the envelope boundaries, in general, there remains an extended region containing low mass particles at relatively high intensity. By careful analysis of our flux *vs.* time profiles in this region, we have discovered that dust particles arrive in sequences of events preceded and followed by relatively long time intervals (time gaps) during which no dust particles are detected. Furthermore, an examination of these particle clusters shows that there are 'packets' within a single DUCMA readout period containing substantial numbers of low mass events that can not be accounted for by the average intensity preceding and following these readouts. Tick marks have been added in Fig. 11 showing where we have obtained detailed analysis of these clusters of particles. We further conclude that this phenomenon persists inward towards closest approach, although it is more difficult to observe due to the high dust background.

7.2. Non-Poisson distributions

Both the time gaps between sequences of events and the packets of events within a single readout are observed inbound and outbound on Vega-1 and Vega-2 as illustrated in Fig. 14, where A through F indicate dust packets containing several particles over a wide mass range. In some cases there are a few particles in preceding and following readouts associated with the packet (e.g., 'F', outbound Vega-1). By comparison with the expected gap distribution assuming Poisson statistics (Simpson et al., 1986c), we find that the observed distribution is totally different from that predicted on the basis of random arrival of events.

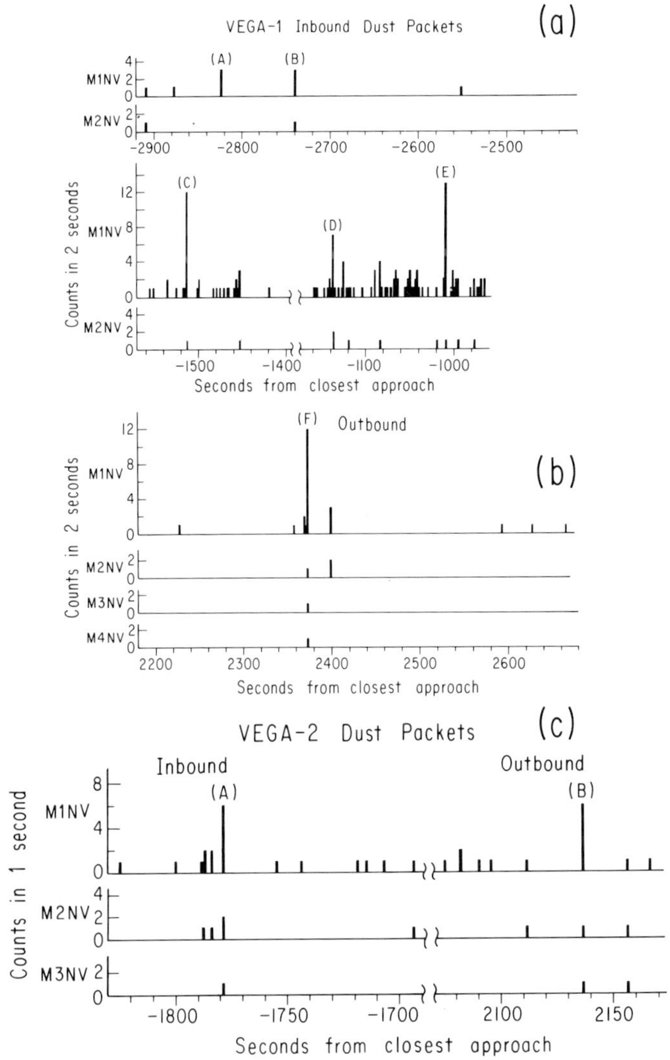

Fig. 14. Time profiles of dust events for selected portions of the Vega-1 ([a] and [b]) and Vega-2 ([c]) encounters illustrating dust clusters and dust packets (A through F). The locations of measured clusters and packets are indicated in Fig. 11 by vertical tick marks. See Table 1 for identification of mass thresholds

8. Conclusions on low mass particles beyond the envelope boundaries

From this study we reach the following conclusions and implications:

a) We have searched for possible artifacts and found none – including local spallation effects.

b) Clusters of particles have very small velocity and spatial dispersion on the average.

c) The data so far suggest emission from the nucleus of large conglomerates of small particles which gently disintegrate so that a cluster of smaller particles all continue to travel in the distant coma with approximately the same velocity.

d) The packets, or clusters, appear to be intermixed with the overall dust distributions inside the envelope boundaries.

e) These results could account for the small mass particles observed at great distances from the nucleus, as reported earlier (Simpson et al., 1986a).

Acknowledgements. We gratefully acknowledge the continuing cooperation with the Max-Planck Institute (Lindau), the Central Research Institute (Hungary), and the support both of the Space Research Institute, Moscow (L.V. Ksanfomality and R.Z. Sagdeev) and the NASA Contract NAS W-3959 (J.A. Simpson, and A.J. Tuzzolino). J.A. Simpson thanks E. Grun and J.A.M. McDonnell for useful discussions and D. Rabinowitz thanks NASA for a Predoctoral Fellowship.

References

Divine, N., Fechtig, H., Gombosi, T.I., Hanner, M.S., Keller, H.U., Larson, S.M., Mendis, D.A., Newburn, R.L., Reinhard, R., Sekanina, Z., Yeomans, D.K.: 1986, *Space Sci. Rev.* **43**, 1

Mazets, E.P., Aptekar, R.L., Golenetskii, S.V., Guryan, Yu A., Dyachkov, A.V., Llyinskii, V.N., Panov, V.N., Petrov, G.G., Savvin, A.V., Sagdeev, R.Z., Sokolov, I.A., Khavenson, N.G., Shapiro, V.D., Shevchenko, V.I.: 1986a, *Nature* **321**, 276

Mazets, E.P., Sagdeev, R.Z., Aptekar, R.L., Golenetskii, S.V., Guryan, Yu A., Dyachkov, A.V., Llyinskii, V.N., Panov, V.N., Petrov, G.G., Savvin, R.Z., Sokolov, I.A., Fredericks, D.D., Khavenson, N.G., Shapiro, V.D., Shevchenko, V.I.: 1986b, *20th ESLAB Symp., ESA SP-250 Vol. II*, 3

Pankiewicz, G.S., McDonnell, J.A.M.: 1986, *20th ESLAB Symp.*, ESA SP-*250 Vol. II*, 201

Perkins, M.A., Simpson, J.A., Tuzzolino, A.J.: 1985, *Nucl. Instrum. Meth.* **A239**, 310

Sekanina, Z., Larson, S.M.: 1986, *Astron. J.* **92**, 462

Simpson, J.A., Tuzzolino, A.J.: 1985, *Nucl. Instrum. Meth.* **A236**, 187

Simpson, J.A., Sagdeev, R.Z., Tuzzolino, A.J., Perkins, M.A., Ksanfomality, L.V., Rabinowitz, D., Lentz, G.A., Afonin, V.V., Ero, J., Keppler, E., Kosorokov, J., Petrova, E., Szabo, L., Umlauft, G.: 1986a, *Nature* **321**, 278

Simpson, J.A., Rabinowitz, D., Tuzzolino, A.J., Ksanfomality, L.V., Sagdeev, R.Z.: 1986b, Preliminary Report, International Cometary Dust Workshop, 23–25 October, Heidelberg, FRG, ed. E. Grun, (unpublished)

Simpson, J.A., Rabinowitz, D., Tuzzolino, A.J., Ksanfomality, L.V., Sagdeev, R.Z.: 1986c, *20th ESLAB Symp., ESA SP-250 Vol. II*, 11

Vaisberg, A.L., Smirnov, V.N., Gorn, L.S., Iovlev, M.V., Balikchin, M.A., Klimov, S.I., Savin, S.P., Shapiro, V.D., Shevchenko, V.I.: 1986a, *Nature* **321**, 274

Vaisberg, O., Smirnov, V., Omelchenko, A.: 1986b, *20th ESLAB Symp., ESA SP-250 Vol. II*, 17

Astron. Astrophys. 187, 753–760 (1987)

ASTRONOMY AND ASTROPHYSICS

Spatial and mass distribution of low-mass dust particles ($m < 10^{-10}$ g) in comet P/Halley's coma

O.L. Vaisberg, V. Smirnov, A. Omelchenko, L. Gorn, and M. Iovlev

Space Research Institute, USSR Academy of Sciences, Profsoyuznaya 84/32, SU-117810 Moscow GSP-7, USSR

Received February 2, accepted July 4, 1987

Summary. The spatial distribution of cometary dust particles in the mass range $\sim 10^{-10}$ to $\sim 3\ 10^{-17}$ g was obtained with the plasma impact detectors SP-1 on Vega-1 and Vega-2. The mass spectrum continues to rise in the range 10^{-14}–$3\ 10^{-17}$ g. The evolution of the mass spectrum with distance suggests radiation pressure separation and some other effects. Several dust shells are easily distinguished by spatial gradients and by different properties of the grains. The spatial mass dispersion in the prominent jet observed on Vega-2 is used to evaluate the rotation of the nucleus and to calculate the velocity dispersion of the dust.

Key words: comets – dust mass spectrum – dust spatial distribution

1. Introduction

First direct measurements of dust in comet Halley's coma showed some differences with respect to the models developed from astronomical observations (Divine et al., 1986). A large number of small particles, $m < 10^{-14}$ g, with rising mass spectrum down at least to $\sim 10^{-16}$ g, was found (Vaisberg et al., 1986; Mazets et al., 1986; McDonnell et al., 1986). Moderate depletion of particles in the mass range 10^{-14}–10^{-12} g was observed and tentatively attributed to radiation pressure (Vaisberg et al., 1986), as proposed in theoretical models (Divine et al., 1986; Bertaux et al., 1982). This paper is a more systematic discription of the first results of an experiment with the plasma impact detector SP-1 on the Vega-1 and -2 spacecraft (Vaisberg et al., 1986).

2. Instrumentation

The SP-1 detector (Gorn et al., 1987) is based on the measurements of the charge of a plasma cloud originating at hypervelocity impact of grains on a thick target (McDonnell, 1978). The total charge in the plasma cloud is approximately proportional to the grain mass (Grün et al., 1984) constituting $\sim 10^3$ Coulombs per gram at impact velocity $\sim 80\ \mathrm{km\,s^{-1}}$, near the Vega's encounter velocity. With this conversion factor the decade mass-ranges m_n of the SP-1 detector are:

$$m_n = (10^{-n-9}\text{–}10^{-n-10})\ \mathrm{g} \qquad (1)$$

for $n = 1, \ldots 6$, and $M_7 = (10^{-16}$–$3\ 10^{-17})$ g. Charge thresholds were set to predetermined values with an accuracy of about 5%. Stability of thresholds determined by temporal and thermal stability of resistors was about 10%. The target was made of gold and had an open area of 81 cm^2 oriented normal to the encounter velocity vector.

The collector system consists of gold coated bars suspended above the target and forming elongated "topless" boxes with approximately square cross section of 0.5 cm × 0.5 cm dimension. The collectors are shielded from direct impacts. One of two identical sensors is shielded by a thin plastic foil. The collectors are biased in a geometric succession by +30 V and −30 V. Charge pulses were amplitude-separated, accumulated during 2 sec in decade counters, and transmitted through the memory of the APV-N plasma wave instrument to the spacecraft telemetry.

Data from SP-1 were obtained on two Vega spacecraft. Near the closest approach (CA) both instruments suffered temporary overloadings that apparently resulted in a change of some thresholds of SP-1 on Vega-1. A minor fault of the SP-1 electronics on Vega-1 led to the summing of counts of channels 1 and 3, 4 and 6, 5 and 7. Instrumental noise before and after the encounter with the comet has not been observed.

3. Spatial and mass distribution

The first dust particles were recorded at $\sim 260{,}000$ km on Vega-1 and at $\sim 320{,}000$ km on Vega-2, both in the lower mass channels. Profiles of the dust counting rates are shown in Fig. 1. Only inbound data are given for Vega-1 in a different format due to the specific performance of the instrument. Dust counting rates were ~ 2.5 times higher for Vega-1 (March 6, 1986) than for Vega-2 (March 9, 1986) at comparable distances, but the dust envelope during the Vega-1 passage was more compact. Radial profiles seem to show similar and identifiable features with boundaries P_i indicated in Fig. 1 and in Table 1. These boundaries were identified by stronger gradients in the counting rate, coinciding with changes in the mass spectra (see below). Some of these boundaries coincide for a wide mass range. The radial profiles show strong differences varying from $\sim R^{-1}$ dependence (between P_2 and P_3) to $\sim R^{-4.5}$ dependence (beyond P_3). Small particles are observed well beyond the outer identifiable boundary P_1.

Differences in the dust envelopes are also seen in the mass spectra. Typical averaged spectra for Vega-2 are shown in Fig. 2 for different dust shells (indicated in Fig. 1). Spectrum A was obtained from fluence counts beyond approximately the P_1

Send offprint requests to: O. Vaisberg

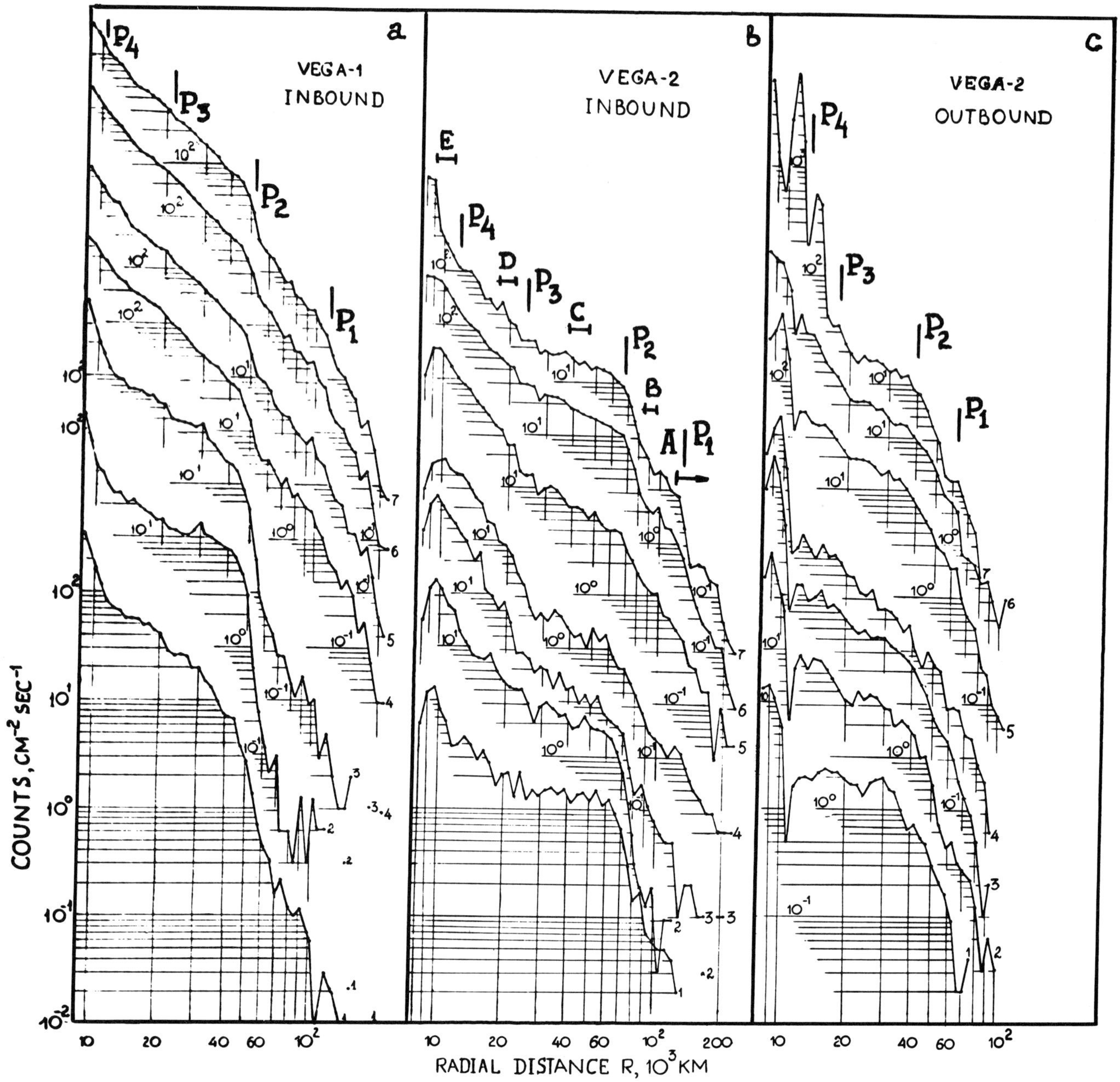

Fig. 1a–c. Radial distributions of dust grains (averaged) obtained with SP-1: **a** Vega-1 inbound, **b** Vega-2 inbound, **c** Vega-2 outbound. The Vega-1 profiles are: 1. channels (1 + 3), 2. channel (2), 3. (1 + 2 + 3), 4. (4 + 6), 5. (5 + 7), 6. (4 + 5 + 6 + 7), 7. (1 + 2 + 3 + 4 + 5 + 6 + 7); numbers on the Vega-2 profiles correspond to channel numbers. The conversion of channel numbers into mass ranges is given by Eq. (1)

boundary. The error bar corresponds to the error of the average obtained from a set of 10 sec averaged spectra for certain distance intervals. This error is usually significantly larger than the $\sqrt{N}/N$ value where N is the total number of counts for the radial interval considered. The most important difference between the measured spectra and preflight models (Divine et al., 1986) is the large number of small particles with $m \lesssim 10^{-14}$ g. Assuming that the mass spectrum continues to rise below $\sim 10^{-16}$ g, the counts in the lowest mass range 7, having a width of only half a decade, were corrected to match the other full decade ranges. Another interesting feature is the depletion in channels 2, 3 and 4. Depletion in channel 4 (10^{-14}–10^{-13} g) is already seen at $\sim$10,000 km. It slightly increases with increasing R. Depletion in

Table 1. Apexes of dust shells

S/c	Leg	P_1 $\times 10^3$ km	P_2 $\times 10^3$ km	P_3 $\times 10^3$ km	P_4 $\times 10^3$ km
V-1	IN	41.4	21.3	10.8	8.4
V-2	IN	42.7	24.5	11.2	7.9
V-2	OUT	45.3	30.6	16.5	10.7

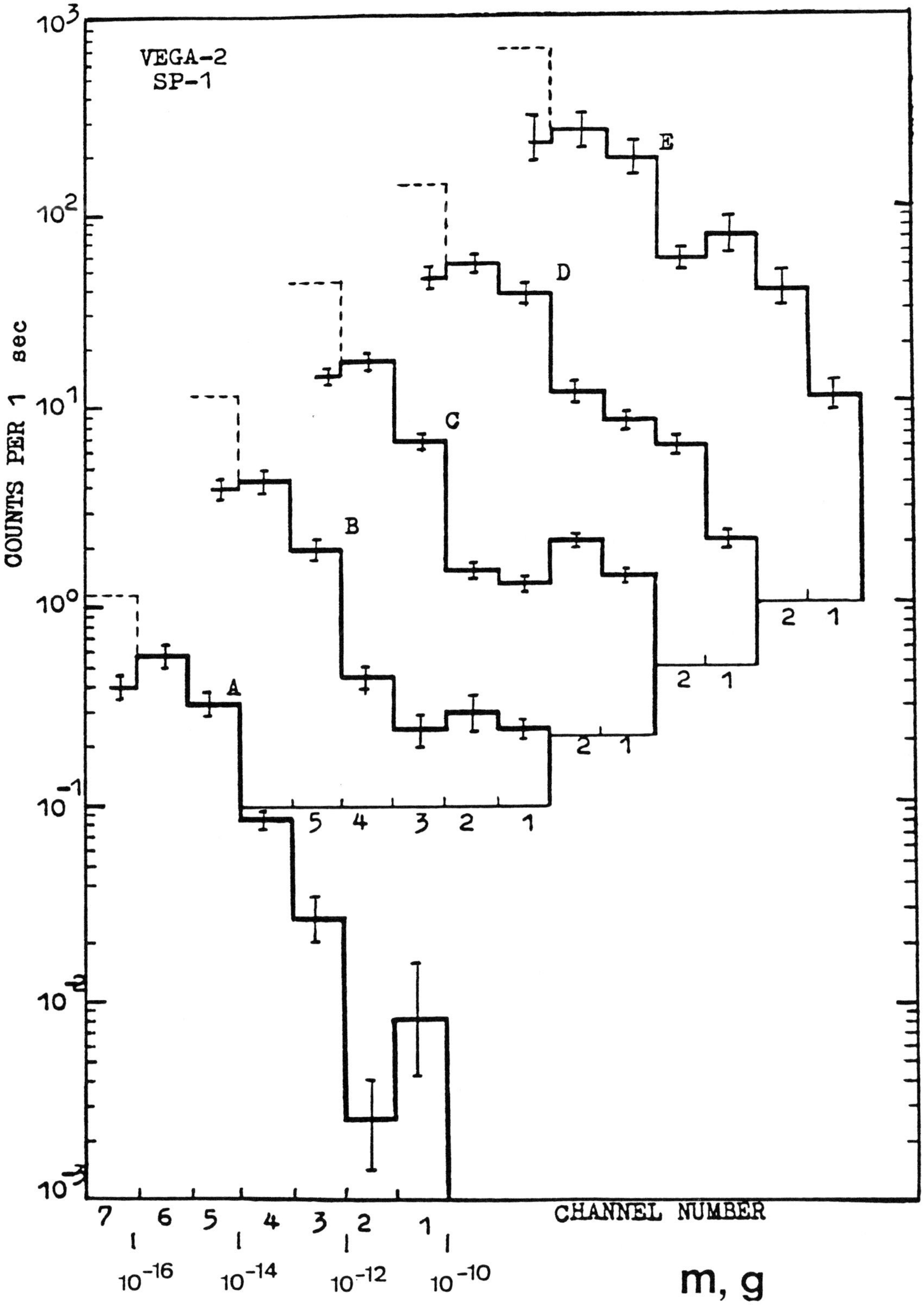

Fig. 2. Mass spectra in decade mass ranges averaged over different radial intervals (Vega-2 inbound; see Fig. 1 b for reference). The counting rate in the semidecade channel 7 is corrected (dashed lines) to match the other mass intervals (see text)

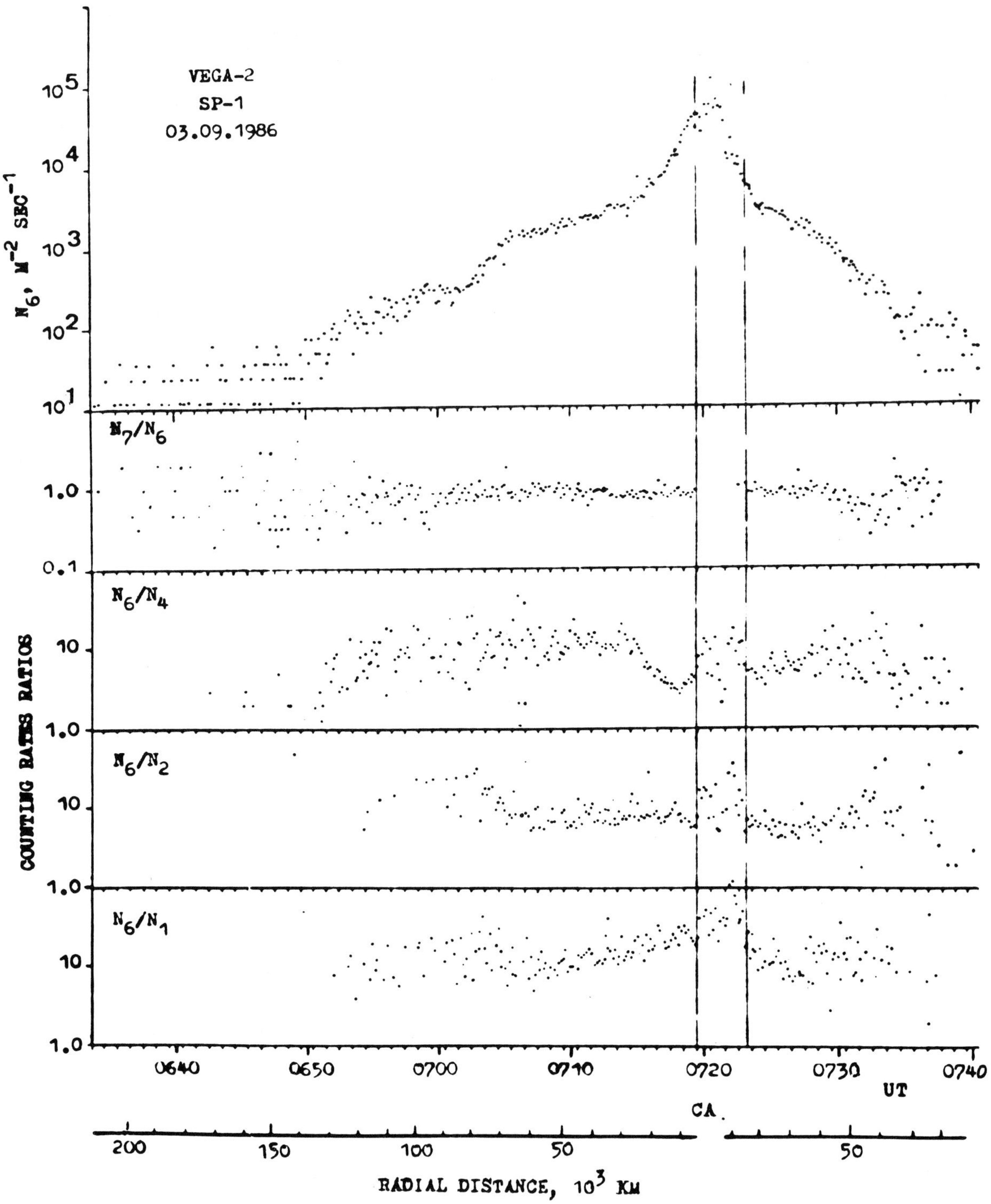

Fig. 3. The Vega-2 count rate profile of grains in channel 6 $(10^{-16}$–10^{-15} g) (above) and count rate ratios in selected mass channels, shown in logarithmic scale (below). The lower index in ^0N_i denotes the channel number. Vertical bars mark the sunward dust cone

channel 3 $(10^{-13}$–10^{-12} g) developes at larger distances, and relative depletion of even larger masses, 10^{-12}–10^{-11} g (channel 2) is suggested in the outer coma. Depletion in channels 3, 4 in the SP-1 mass spectra although not as strong as in theoretical models (Divine et al., 1986; Bertaux et al., 1982; Gombosi et al., 1982), was attributed to radiation pressure. No single exponent can be attributed to the differential mass spectra $dN/dm = m^{-\alpha}$ in the 10^{-16}–10^{10} g range at ~10,000 km, where the measured mass distribution is closer to the source spectrum, and up to ~20,000 km the lowest mass range $(10^{-16}$–10^{-14} g) has $\alpha \approx 1.2$ in the mass power law spectrum, whereas the highest mass range $(10^{-13}$–10^{-10} g) has $\alpha \approx 1.5$.

Some features of the spatial distribution of grains are seen from the ratios of the counting rates in the SP-1 mass channels (Fig. 3). The spatial distribution seen in the lowest mass ranges (6 and 7) is nearly identical. This and the high counting rate in

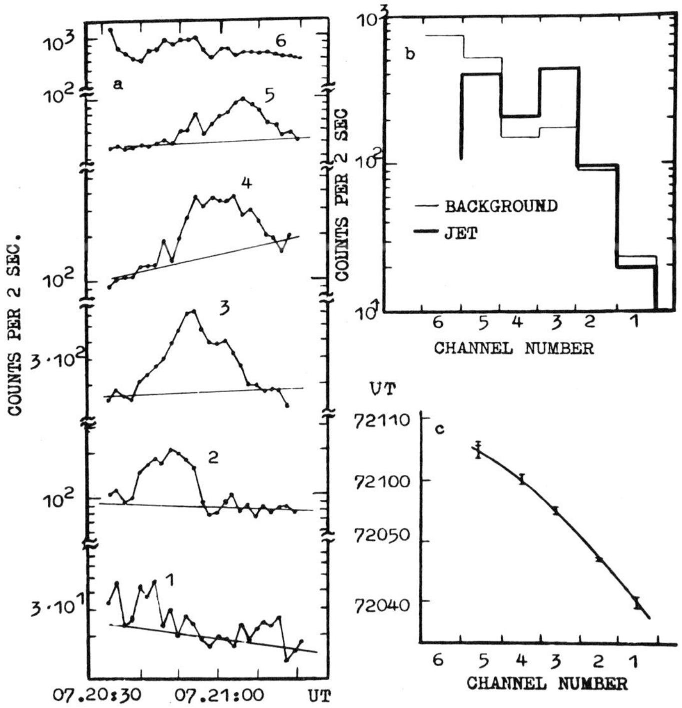

Fig. 4a–c. Count rate profiles in different mass ranges showing a dispersive jet (**a**), **b** mass spectra of the jet and of background dust, and **c** spatial displacement of the center of the jet for different mass intervals

channel 6 allowed us to use it as a reference. The ratio of counting rates of the 6th and 4th channels increases between ~12,000 km and 25,000 km showing the already mentioned radiation pressure influence that leads to a redistribution of grains. The boundaries P_i mentioned earlier are clearly seen in the ratios of the counting rates.

4. Narrow dispersive jet

Velocity dispersion (Gombosi et al., 1982; Divine et al., 1982) is second to radiation pressure as a factor leading to spatial separation of dust. A prominent jet with mass-dispersion was found in the Vega-2 data and explained in terms of velocity dispersion (Vaisberg et al., 1986). This jet was observed at ~9,000 km from the nucleus near the subsolar point. Temporal counting rate enhancements by a factor of 2–4 were observed in the higher-mass channels of SP-1 (Fig. 4a), no sign of the jet was found in channels 6 and 7 ($m < 10^{-15}$ g). The different locations of the counting rate enhancements on the spacecraft trajectory allows one to infer a mass-dispersion curve. The center of the jet in every mass range was determined in the following way. Background lines were drawn through adjacent parts of the counting rate curves. Excess above this background was considered as a jet. The mass spectrum of the jet (Fig. 4b) was obtained as the maximum excess counting rate above the background, while the background mass spectrum was drawn through the background counting rate at the respective maximum. The location of the jet's center in every mass range was determined in two ways: (1) as a center of gravity of the excess counting rates n on the time scale, $t_i = \sum_k t_k \cdot n_{ik} / \sum_k n_{ik}$, where i is the mass channel number, and k the time interval, and (2) as the middle of the line drawn through semimaximum counting rates in the jet. The mass dispersion curve (Fig. 4c) was drawn through the bars determined by the described procedure. The background mass spectrum (Fig. 4b) is very similar to spectrum E in Fig. 2, while the mass spectrum of the jet is markedly different, the jet spectrum being narrower and showing a low-mass cut-off. Both spectra show depletion in channel 4 (10^{-14}–10^{-13} g).

5. Discussion

The use of a proper charge to mass conversion factor is important for the interpretation. A factor as high as 3000 C/g was obtained in the laboratory for silicate grains at an impact velocity of 69 km s^{-1} (Göller et al., 1986). Laboratory calibration of SP-1 was made with shaped charge pulses identical to the ones obtained during accelerator tests of the Giotto impact detector that was the prototype of SP-1. Yet the differences in electronics and other technical details of the two instruments as well as differences between cometary grains and laboratory projectiles leave an uncertainty of a factor of ~3.

A crude test of the mass scale can be made assuming that the observed depletion in channel 4 (10^{-14}–10^{-13} g) is associated with radiation pressure. It is known that the radiation pressure maximum is at 0.2–0.3 μm radius for nonconducting spheres (Burns et al., 1979). The particles of these dimensions with density 0.3–1.0 g cm^{-3} will fit the middle of channel 4. This simple consideration suggests that our mass scale may be relatively correct.

Yet the displacement of the depletion in the mass spectrum from channel 4 through 3 to 2 with increasing distance (Fig. 2) shows that things are not so simple. To fit the middle of channel 3 (10^{-13}–10^{-12} g) nonconducting particles suffering maximum radiation pressure should have the density ~3–10 g cm^{-2}. Conducting particles in this mass range and any particles in channel 2 (10^{-12}–10^{-11} g) should have unreasonably high density to show pronounced radiation pressure effects. Another disagreement with the simple picture of radiation pressure effects lies in the fact that the depletion already seen in channel 4 at ~10,000 km is not complete at ~50,000 km. So the systematic changes in the mass spectra with radial distance suggest that other factors are important and that an increasing depletion in the mass spectrum is filled in by "new" particles.

It is believed that the comets are the main sources of interplanetary dust (Fechtig, 1981). Detailed statistics on interplanetary dust was obtained from lunar microcraters (Fechtig, 1982). The size spectrum of microcraters has a slope αD^{-2} for crater diameters $D < 2$ μm and αD^{-1} for $D > 2$ μm. The diameter of craters at hypervelocity impact is 4–10 times the projectile diameter (Fechtig, 1981; Sagdeev et al., 1983). So a change of the spectral slope at $D \sim 2$ μm corresponds to $m \sim (4\ 10^{-15}$–$6\ 10^{-14}) \cdot \varrho g$, where ϱ is the projectile's density. It appears that the outer part of the coma could be the source for comparatively long-living interplanetary particles, while short-lived particles and particles strongly influenced by radiation pressure should be concentrated to the inner coma. In the outer part of comet Halley's coma (shells B and C, Fig. 2) the change of the slope α of the mass spectrum $m^{-\alpha}$ occurs in channel 4 (10^{-14}–10^{-13} g), being 1.5 ± 0.15 for small masses and 1.0 ± 0.1 for larger masses. Converting the indexes of the mass spectra α to the ones of the size spectra u, we obtain $u = 2.5 \pm 0.45$ and $u = 1 \pm 0.3$, respectively, for the lower and higher ranges in the size spectra, D^{-u}. So if $\varrho \sim 1$ g cm^{-3}, the similar changes of the slope of

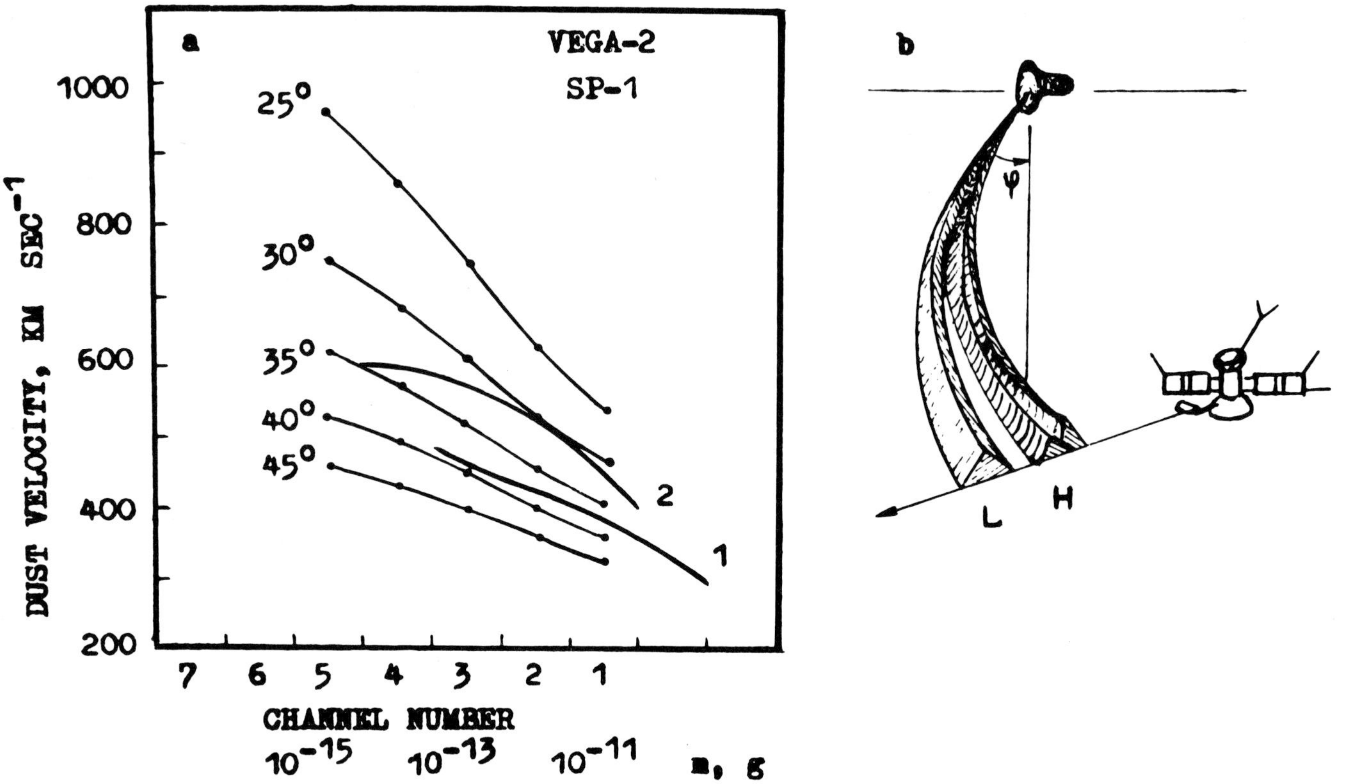

Fig. 5a and b. Interpretation of the dispersive jet as velocity-dependent rotating structure (**a**), and **b** velocity dispersion curves obtained from the spatial dispersion of the jet for different locations φ_0 of the source. Theoretical dispersion curves are: 1) from Divine and Newburn (1982), and 2) from Gombosi et al. (1982)

the mass-spectrum inferred from lunar microcraters occurs at a mass of about 1.5 10^{-14} g, compared to 10^{-14}–10^{-13} g in the outer part of comet Halley's coma. This reasonable agreement of the shape size spectrum of lunar microcraters with the spectrum of cometary dust supports the correctness of our mass scale within a factor of 3.

Rapid changes of the structure of the dust coma of comet Halley (Larson et al., 1986) make the interpretation of the observed dust profile quite difficult. The more compact linear scale of the Vega-1 profile compared to the Vega-2 one (by factor ~1.2 in the location of the boundaries and of the first recorded particles) suggest an expansion velocity for time interval several days preceding March 6 which is smaller by ~10% as compared to one preceding March 9. Higher by factor ~2.5 counting rate of dust during the Vega-1 encounter compared to Vega-2 can be used to roughly estimate the dust production rate change between the two periods separated by 3 days. Our estimate of the dust production rate M in the $m \leq 10^{-10}$ g range is ~200 kg s^{-1} for the Vega-2 encounter, assuming that dust emission occurs only on the sunlit hemisphere. Assuming that M_i per decade mass range for $m > 10^{-10}$ g is changing in the same way as in our channels 3, 2 and 1, namely $dM_i/\text{dlgm} = -0.15$, we can extrapolate M_i to higher masses. That gives a total dust production rate of ~3.2 tons s^{-1} for the Vega-2 encounter. The corresponding value is ~8 tons s^{-1} for Vega-1.

The complicated structure of dust coma with several boundaries (Fig. 1) prevents an easy discrimination of large-scale envelopes from the dust paraboloid. The outermost boundaries P_1 for Vega-2 yield the best agreement of the apexes (~45 10^3 km), determined from the inbound and outbound locations of P_1. Other boundaries extrapolated to the subsolar point along the parabolas give less consistent values for the apexes, outbound boundaries always give higher figures than inbound ones. This suggests a higher dust velocity on the "evening" side of the comet and the influence of thermal inertia of the nucleus' surface. P_2 boundaries limiting the most pronounced dust shell for smaller and higher mass ranges give apex values of ~24,500 km for the inbound pass and ~30,600 km for the outbound one, or about 27,500 km as a mean. The existence of several dust shells (or paraboloids) with different mass spectra (Figs. 1–3) suggests that they may consist of different kinds of particles.

The innermost part of the dust envelope E ($R \leq 12{,}000$ km) shows a very steep radial gradient. Large azimuthal motion of the spacecraft in this region suggests that this increase is associated mainly with angular asymmetry of dust ejection. It was noticed (Vaisberg et al., 1986) that strong dust emission with jet-like structures in an approximately Sun-centered cone of 70°–80° width has an optical counterpart evident of TV-images (Sagdeev et al., 1986; Keller et al., 1986) as bright Sun-directed cone. This kind of fan-shaped coma where most of the dust (and gas) release takes place was found earlier in astronomical data (Sekanina, 1979).

The narrow dispersive jet in the Vega-2 data was interpreted as time-of-flight separated feature emitted from a rotating source (Fig. 5a) (Vaisberg et al., 1986). With a dust velocity of $V \sim 350$ m s^{-1} for heavier particles and $V \sim 500$ m s^{-1} for smaller ones, it is easy to obtain a rough estimate of the rotation period of the nucleus of $T \sim 50$ hours and the correct sense of rotation in agreement with the analysis of the Vega TV images (Sagdeev et al., 1986). If we adopt the period found by these authors, $T = 2.2$ d, we can try to locate the origin of the jet and to obtain the velocity dispersion curve from

$$V_i = \frac{360^\circ}{\varphi - \varphi_0} \cdot \frac{R_i}{T}, \qquad (2)$$

where V_i is the dust velosity. φ_i and R_i are the azimuth and radial distance of the observed jet, φ_0 is the azimuth of the source, and i is the mass range. Varying φ_0 we obtain the family of velocity dispersion curves fitting the observations (Fig. 5a). Theoretical curves are shown for comparison: (1) obtained in (Divine et al., 1982) for a terminal gas expansion velocity of 754 m s^{-1} and (2) obtained in (Gombosi et al., 1982) for a gas velocity of 650 m s^{-1}. Recent estimates of gas expansion velocity give 900 ± 200 m s^{-1} (Krankowsky et al., 1986) and 2/3 km s^{-1} (A'Hearn et al., 1986), justifying the application of these models. Better agreement of experimentally obtained dispersion curves with theoretical ones (Gombosi et al., 1982; Divine et al., 1982) is obtained for an azimuth equal to $35° \pm 5°$ for the source of the jet at the time of the Vega-2 passage. Vega-2 TV-pictures reveal the crossing of subprobe trace on the nucleus with a linear jet at $\varphi_0 \sim 45°$ (Szegö, 1986). The edge-on view of the jet was also noticed at an azimuth close to 35° (Smith, 1986). So proposed explanation of a dispersive jet is therefore not inconsistent with the available observations.

There are several pieces of evidences for an evolution of the mass spectrum and the supply of new particles in the low-mass range while dust propagates from the source. Among them is the existence of common boundaries for light and heavier grains differing in mass by several orders of magnitude (Fig. 1). The second one is the relatively small dependence of (radiation pressure) depletion in the mass spectrum on distance (Fig. 2), suggesting the supply of new particles in this mass range. Thirdly the smallest particles (channels 6 and 7) behave similarly throughout the coma (Fig. 3), suggesting that all particles in the mass range $\lesssim 10^{-15}$ g may be secondary ones. Fourth is the absence of small ($\lesssim 10^{-15}$ g) particles in the narrow jet's mass spectrum (Fig. 4) that may be explained either as due to their lacking in the original spectrum, or by their lateral expansion after birth from mother particles. The fifth one is the existence of clusters of small particles (about 5–10 events recorded in 2 s at a background rate of ~ 1 event/2 s) observed at large distances in agreement with other observations (Mazets, 1986; Simpson et al., 1986). Finally, evidence for activity of cometary grains was recently obtained from ground-based observations of CN jets (Divine et al., 1986).

6. Conclusion

The analysis of SP-1 plasma impact detector measurements on the Vega-1 and Vega-2 spacecraft show that:

– the mass spectra continue to rise between $\sim 10^{-14}$ g and $\sim 10^{-16}$ g;

– an evolution of the mass spectrum with distance is observed that may be partly due to radiation pressure;

– the dust coma consists of several envelopes or paraboloids with different characteristics suggesting the existence of different kinds of particles;

– small grains are observed well beyond the main paraboloid(s) (with apex 26,500 km and/or $\sim 45{,}000$ km);

– observations of a mass dispersive jet were associated with the period and direction of nucleus' rotation and were used to estimate the velocity dispersion curve in the 10^{-15}–10^{-10} g mass range;

– several pieces of evidence of an evolution (including splitting) of the grains during their motion in the cometary coma were found.

Acknowledgements. The authors are grateful to R.Z. Sagdeev, A.A. Galeev, V.D. Shapiro, E.P. Mazets, R.L. Aptekar for helpful discussions and to R.A. Isaeva and A.M. Kuznetsova for help in the preparation of this manuscript.

References

A'Hearn, M.F., Hoban, S., Birch, P.V., Bowers, C., Martin, R., Klinglesmith III, D.A.: 1986, 20th ESLAB Symp. ESA SP-250, Vol. **I**, p. 483

Bertaux, J.L., Cot, C.: 1982, Cometary Exploration, II, p. 113

Burns, J.A., Lamy, Ph.L., Soter, S.: 1979, *Icarus* **40**, 1

Divine, N., Fechtig, H., Gombosi, T.I., Hanner, M.S., Keller, H.U., Larson, S.M., Mendis, D.A., Newburn, R.L., Reinhard, R., Sekanina, Z., Yeomans, D.K.: 1986, *Space Sci. Rev.* **43**, 1

Divine, N., Newburn, R.L., Jr.: 1982, Cometary Exploration, II, p. 81

Fechtig, H.: 1981, Proceedings of Alpbach Summer School, ESA SP-164, 203

Fechtig, H.: 1982, Cometary Exploration, II, p. 203

Göller, J.R., Grün, E., Maas, D.: 1986, 20th ESLAB Symp. ESA SP-250, Vol. **III**, p. 333

Gombosi, T.I., Szegö, K., Gribov, B.E., Sagdeev, R.Z., Shapiro, V.D., Shevchenko, V.I., Cravens, T.E.: 1982, Cometary Exploration, II, p. 99

Gorn, L.S., Iovlev, M.V., Shifrin, A.V., Vaisberg, O.L., Smirnov, V.N.: 1987, Problems of nuclear science and techniques, Atomizdat (in press) (in Russian)

Grün, E., Bussoletti, E., Minafra, A., Kuczera, H., McDonnell, J.A.M.: 1984, *Adv. Space Res.* **4**, N9, 291

Keller, H.U., Arpigny, C., Barbieri, C., Bonnet, R.M., Cazes, S., Coradini, M., Cosmovici, C.B., Delamero, W.A., Huebner, W.F., Hughes, D.W., Jamar, C., Malaise, D., Reitsema, H.J., Schmidt, H.U., Schmidt, W.K.H., Seige, P., Whipple, F.L., Wilhelm, K.: 1986, *Nature* **321**, 320

Krankowsky, D., Lämmerzahl, P., Herrwerth, I., Wowers, I., Eberhardt, P., Dolder, U., Herrmann, U., Schulte, W., Berthelier, J.J., Illiano, J.M., Holdes, R.R., Hoffmann, J.H.: 1986, *Nature* **321**, 326

Larson, S., Sekanina, Z., Levy, D., Tapia, S., Senay, M.: 1986, 20th ESLAB Symp. ESA SP-250, Vol. **II**, p. 145

Mazets, E.P.: 1986 (private communication)

Mazets, E.P., Aptekar, R.L., Golenetskii, S.V., Guryan, Yu.A., Dyachkov, A.V., Ilyinski, V.N., Panov, V.N., Petrov, G.G., Savvin, A.V., Sagdeev, R.Z., Sokolov, I.A., Knavenson, N.G., Shapiro, V.D., Shevchenko, V.I.: 1986, *Nature* **321**, 276

McDonnell, J.A.M., Alexander, W.M., Burton, W.M., Bussoletti, E., Clark, D.H., Grard, R.J.L., Grün, E., Hanner, M.S., Hughes, D.W., Igenbergs, E., Kuczera, H., Lindblad, B.A., Mandeville, J.-C., Minafra, A., Schwehm, G.H., Sekanina, Z., Wallis, M.K., Zarnecki, J.C., Chakaveh, S.C., Evans, G.C., Evans, S.T., Firth, J.G., Littler, A.N., Massonne, L., et al.: 1986, *Nature* **321**, 338

McDonnell, J.A.M.: 1978, *Cosmic Dust*, Wiley, Chichester, New York, p. 337

Sagdeev, R.Z., Anisimov, S.I., Galeev, A.A., Shapiro, V.D., Shevchenko, V.I.: 1983, *Adv. Space Res.* **2**, N12, p. 133
Sagdeev, R.Z., Szabo, F., Avanesov, G.A., Cruvellier, P., Szabo, L., Szegö, K., Abergel, A., Balazs, A., Barinov, I.V., Bertaux, J.L., Blamont, J., Detaille, M., Demarelis, E., Dul'nev, G.N., Endröczy, G., Gardos, M., Kanyo, M., Kostenko, V.I., Krasikov, V.A.., Nguyen-Trong, T., Nyitrai, Z., Reny, I., Rusznyak, P., et al.: 1986b, *Nature* **321**, 262
Sekanina, Z.: 1979, *Icarus*, **37**, 420
Simpson, J.A., Rabinowitz, D., Tuzzoling, A.J., Ksanfomality, L.V., Sagdeev, R.Z.: 1986, 20th ESLAB Symp. ESA SP-250, Vol. **II**, p. 11
Smith, B.: 1986 (private communication)
Szegö, K.: 1986 (private communication)
Vaisberg, O.L., Smirnov, V.N., Gorn, L.S., Iovlev, M.V., Balikhin, M.A., Klimov, S.I., Savin, S.P., Shapiro, V.D., Shevchenko, V.I.: 1986, *Nature* **321**, 274

Astron. Astrophys. 187, 761–766 (1987)

ASTRONOMY AND ASTROPHYSICS

First statistical analysis of 5000 mass spectra of cometary grains obtained by PUMA 1 (Vega 1) and PIA (Giotto) impact ionization mass spectrometers in the compressed modes

Y. Langevin[1], **J. Kissel**[2], **J-L. Bertaux**[3], **and E. Chassefière**[3]

[1] Laboratoire René-Bernas, Bat. 108, F-91406 Orsay, France
[2] Max-Planck Institut für Kernphysik, D-6900 Heidelberg, Federal Republic of Germany
[3] Service d'aéronomie, F-91370 Verrieres-le-Buisson, France

Received January 22, accepted April 3, 1987

Summary. Impact ionization mass spectrometers have been flown on board the space vehicles Giotto (instrument PIA), Vega 1 and Vega 2 (instruments PUMA 1 and 2). All three instruments obtained a large collection of mass spectra of individual cometary dust particles in the mass range $3\,10^{-16}$–$3\,10^{-10}$ g. Spectra from PUMA 2 are still being recovered from data transmission problems, and only results from PUMA 1 and PIA are presented. A large fraction of particles ($\sim 30\%$) are dominated by light elements (H, C, O, and N, or "CHON"). $\sim 35\%$ of particles can be interpreted as minerals with a compositional range similar to that encountered in carbonaceous chondrites. The remaining 35% appear to be mixtures in varying amounts of these two major components. Statistically significant variations have been observed as a function of time.

Key words: cometary grains – Giotto – Vega 1 and 2 – chemical composition

1. Introduction

More than 5000 mass spectra of cometary dust particles have been obtained by three instruments flown on board Giotto (PIA), Vega 1 and and Vega 2 (PUMA-1 and 2) spacecrafts (J. Kissel, P.I; Kissel et al., 1986a, 1986b). Upon impact of a dust particle at 69 km s^{-1} (Giotto) or 78 km s^{-2} (Vega 1 and 2), a significant fraction of the grain is ionized and analyzed by a time of flight mass spectrometer. The encounter with comet P/Halley occurred on March 6 (Vega 1), March 9 (Vega 2) and March 14 1986 (Giotto). The closest approach of the Giotto spacecraft (600 km) was more than ten times smaller than that of the Vega 1 and 2 spacecrafts (8000 km and 8900 km respectively), and the activity of comet P/Halley was a factor of 2 higher during the Vega 1 encounter. Results obtained from these three instruments should therefore allow to follow the time evolution of dust emitted from the comet during 8 days, at different stages of its activity cycle. Comparing Giotto and Vega data should yield information on possible changes in the composition of the grains as a function of their distance to the nucleus, hence the duration of their exposure to solar radiation.

The physical processes responsible for ion emission during hypervelocity impacts are difficult to model and to simulate. Saturation effects are also not excluded for relatively large grains, and much work is still needed to obtain quantitative results on elemental abundances. However, ion abundance ratios and comparisons between grains are much less model dependent and should provide information on the typology of cometary grains, as a function of the distance ot the nucleus, the activity of the comet, and the phase angle.

2. Experimentals

The instruments PIA, PUMA 1 and PUMA 2 have the same general design: ions formed by the hypervelocity impact of a dust particle on the metallic target are accelerated by a grid, then analyzed in a time of flight spectrometer, which uses a reflectron to correct for the wide initial energy distribution of ions. Ion current is converted in a logarithmic amplifier which nominally provides 5 decades for the full scale of 0 to 127. Three entry channels can trigger the time of flight sequence: a photomultiplier, which detects the light flash induced by the impact, and two charge collectors. The triggering levels can be increased by a nominal factor of 100, which provides a high and a low sensitivity mode. When the flux was large, near closest approach, these modes were commuted every 4 s. At either sensitivity setting, the number of impacts with at least one channel above level 1 (all triggered events) and level 100 ("large" events) were recorded every s. The mass range of dust particles was $\sim 3\,10^{-16}$ g to $\sim 3\,10^{-10}$ g. These impact rates provide a flux estimate across the 5 cm^2 entry window for 4 dust particle classes, which can be compared with the other dust detectors (DIDSY, McDonnell et al., 1986; DUCMA, Simpson et al., 1986; SP-1, Vaisberg et al., 1986; SP-2, Mazets et al., 1986).

The mass resolution achieved on most of the spectra was >50, which allows to separate the mass lines up to Fe, but obviously not the isobars. Out of 25 spectra, one is uncompressed, consisting of up to 800 times samples covering the mass range 1 to 140. The 24 other spectra are sent in 3 increasingly com-

Send offprint requests to: Y. Langevin

pressed modes, depending on the data acquisition to transmission ratio. In these modes, the amplitude and the output of a t^2 generator are transmitted for each peak of the ion current, as well as few regularly spaced amplitudes and t^2 outputs. These spectra in compressed modes, although more difficult to interpret, provide the broadest statistical basis for the qualitative classification of dust grains from Halley presented in this paper.

Each spectrum was assigned a class on the basis of the number of coincidences in entry channels (1 to 3), the number and amplitude of mass lines corresponding to major elements. All spectra in uncompressed modes, as well as every other spectrum in compressed modes, were selected for transmission by opening a time window and waiting for the first impact to occur. These spectra therefore provide a representative statistical sample of the grain population. For the other half of spectra in compressed modes, the spectrum with the highest class was selected whenever several spectra were ready to be transmitted.

Specific characteristics: These three instruments present specific characteristics which must be taken into account before attempting to compare the statistical information on grain composition:

– target: the target on PIA is a platinum foil doped with 10% silver, whereas that of PUMA-2 is a pure silver foil, and that of PUMA-1 consists in a series of small "steps" making an angle of $\sim 60°$ with the direction of incoming dust particles.

– PUMA 1 and 2 operate either in a wide energy window mode similar to that of PIA, or in a narrow energy window mode which increases the mass resolution, and enhances the transmission of heavy ions as compared to light ions (< 20 a.m.u.). PUMA 1 and 2 switched from one mode to the other every 30 s.

3. Operations

All three instruments obtained a large collection of useful spectra, although several problems where encountered:

– On PUMA 2, the large power drawn by the operation of the scanning platform resulted in a reset of the high voltage to a low back-up level. This resulted in a reduction in gain for each dynode of the logarithmic amplifier, and the nominal dynamic range for PUMA 2 is 3.5 decades. Furthermore, transmission problems were encountered during the Vega 2 encounter. In particular, the header was modified for part of the spectra. Work is still in progress to solve these problems, but it already appears that more than 1000 spectra should be recovered. In this preliminary statistical analysis, we will therefore discuss only data from PIA and PUMA 1.

– the low sensitivity setting of PUMA 1 and 2 corresponds to an increase by a factor of ~ 7 of the triggering levels of the front channels, as compared to the nominal value of 100 which was obtained on PIA

– The logarithmic amplifier on PUMA 1 operated nominally. On PIA, the 0 level seems to correspond to a higher ion current than for PUMA 1, as all peaks are lower, and the numerous small peaks corresponding to ion noise are missing. The distribution of peak amplitudes presents a strong peak near 20. These two features are presently interpreted as a failure of dynodes 1 and 3 (lowest and 3rd lowest) among the 6 dynodes of the logarithmic amplifier. The increase in the baseline decreased by a factor 3 to 4 the data required by each compressed spectrum, and the number of spectra obtained by PIA is larger than that obtained by PUMA 1, although the dust fluence was lower. The likely loss of dynode 3 presents serious problems for interpretation: a tentative corrected scale has been used, but 0.8 decades of information have been lost.

Only compressed spectra with at least one significant peak corresponding to a major element where selected. At the present stage, the resulting data base consists in 2783 spectra for PIA and 2284 spectra for PUMA 1. We show below that the narrow energy window mode of PUMA 1 (or "long" mode, from the length of uncompressed spectra) strongly favors elements with $Z > 20$ over elements with $Z < 20$, when compared to the wide energy window mode (or "short", from the length of uncompressed spectra). A reliable relative calibration of the short and long operating modes will require extensive simulation and modeling (relevant results are already presented by Sagdeev et al. in these special issues). It is therefore necessary to analyze separately the 1220 spectra obtained in the short mode and the 1064 spectra obtained in the long mode.

4. Results and discussion

4.1. Impact rates

The four monitored fluxes correspond to entry channel levels of 1, 100, 100 and 10,000 for PIA, as compared to 1, ~ 10, 100 and ~ 1000 for PUMA 1. For PUMA-1, there is very good agreement with SP-2 data (Mazets et al., 1986). The large peak in DUCMA data (Simpson et al., 1986) was not observed. The comparison with PIA data is not straightforward, and is discussed by McDonnell et al. (1987). The distribution for PIA is more narrow than for PUMA-1. This is consistent with the lower activity observed during VEGA-1 encounter. The peak fluxes for PIA are of course higher due to the smaller miss distance. PUMA-1 observed at least 3 peaks which are most likely related to jets. The proportion of large particles increases as a function of time.

4.2. Interpretation of compressed spectra

The uncompressed spectra show much less mass contamination than SIMS or other laboratory mass spectrometry techniques: isotopic ratios compatible with solar system values have been observed for elements such as carbon ($^{12}C/^{13}C \sim 100$), magnesium ($^{24}Mg/^{26}Mg \sim 8$), silicon or iron ($^{56}Fe/^{54}Fe \sim 15$), within uncertainties (~ 2 on the peak ratios). Peaks corresponding to doubly charged ions are also very low. It is therefore valid as a first approximation to consider that each peak corresponds mainly to the singly charged monoatomic ion. Due to jitter in the t^2 generator, each mass correspond to a range of t^2 value, and it is difficult to separate masses closer than 4 amu. C, N, and O could however be sorted out in the low noise PIA spectra. The background level for PUMA 1 was evaluated from peak values in regions where no large mass peak are expected. Figure 1 shows that the discrimination of significant and background peaks in the mass 1 region is quite straightforward. For each spectrum, peak values were evaluated for H, C, O, Na, Mg, Si, S, Ca, Fe and Ag (target) for PUMA 1, and the same elements with nitrogen replacing sodium for PIA. Assuming the nominal dynamics of 25 levels per decade, it was then possible to derive the number of ions produced by each particle.

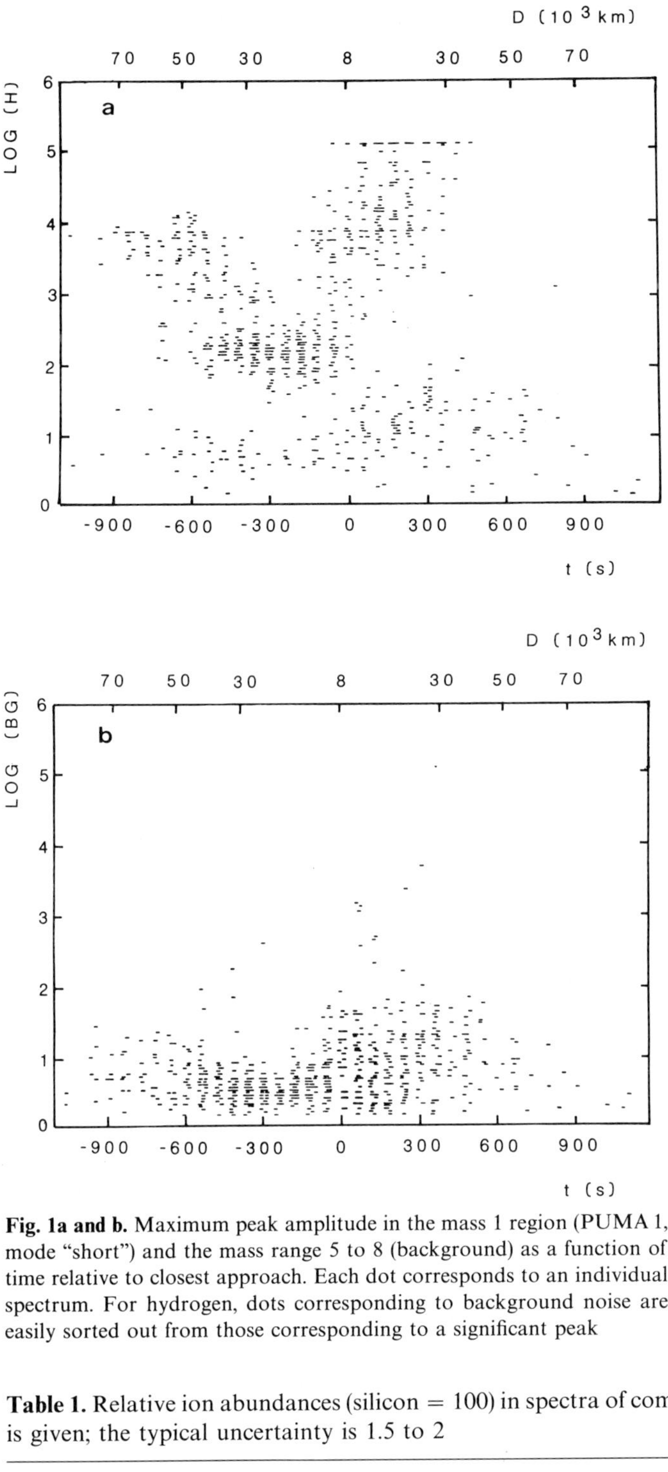

Fig. 1a and b. Maximum peak amplitude in the mass 1 region (PUMA 1, mode "short") and the mass range 5 to 8 (background) as a function of time relative to closest approach. Each dot corresponds to an individual spectrum. For hydrogen, dots corresponding to background noise are easily sorted out from those corresponding to a significant peak

The physical processes responsible for ion formation during the very high velocity impacts of Halley's dust particles are difficult to model and to simulate: only very small grains can reach velocities of several 10 km/s; the ultraclean conditions during the encounter with Halley (15 months of outgassing + intense neutral and ion sputtering of the metallic target) cannot easily be duplicated in the laboratory. Much more work is therefore required before attempting to derive accurate elemental abundances for each individual particle, which requires reliable information on the ionization efficiencies. We will therefore discuss here only the ion "abundances" measured. An average *ion* composition of cometary particles is reported in Table 1 for PIA, and PUMA 1 in both the "short" and "long" modes of operation. It was obtained by considering 5 independent evaluations: total number of ions collected per mass, mean peak value, and peak values below which lie 50% (median), 75% (quartile) and 90% of spectra. The listed standard deviation on the ratio with silicon shows that these evaluations are consistent within a factor of 2.

For all elements with $Z > 20$, the spread of ion abundance ratios is less than 1.5. The two modes of PUMA 1 yield very similar results within these uncertainties. On the contrary, light elements show quite large variations (typically by a factor of 2), and the short mode abundances are consistently a factor of 8 higher than the long mode abundances. This most likely results from a change in the transmission factor between the two modes. Apart from hydrogen, PIA values are closer to the short mode values, which is to be expected as this instrument also operated with a wide energy window. The significant differences observed are probably due in part to the different targets (platinum doped with silver for PIA, as compared to pure silver for PUMA 1 and 2), and to information loss induced by dynode 3 of the amplifier. The comparison of these results with *elemental* abundances in carbonaceous chondrites is quite striking: with a few exceptions (the very high abundance of carbon, the small excess of magnesium, and the relatively low abundance of iron), these *ion* abundances could be interpreted as a carbonaceous chondritic composition, with ionization efficiencies similar for all elements.

Where does the large carbon increase come from? if we consider a diagram showing the ratio of silicon to carbon as a function of the magnesium to carbon ratio (Fig. 2), it clearly appears that silicon and magnesium are correlated, while carbon is provided by another component. Indeed, a large fraction of the spectra (~30%) are dominated by light elements (CHON), while another fraction (~35%) are dominated by mineral forming elements (Na, Mg, Si, Ca, Fe), the other grains representing a mixture between these two major components.

Table 1. Relative ion abundances (silicon = 100) in spectra of cometary particles; the log standard deviation between five evaluations is given; the typical uncertainty is 1.5 to 2

	H	C	N	O	Na	Mg	Si	S	Ca	Fe
PIA	500	300	15	410		210	100	16	5	20
Log standard deviation	0.30	0.20	.10	0.17		0.14		0.15	0.17	0.17
PUMA 1 (short)	2000	800		460	12	180	100	30	17	33
	0.33	0.11		0.11	0.11	0.09		0.04	0.02	0.15
PUMA 1 (long)	400	100		70	10	250	100	25	18	28
	0.70	0.30		0.30	0.18	0.04		0.12	0.12	0.04
Carbonaceous chondrites (elemental abundances)	590	76	6	730	12	103	100	48	7	86

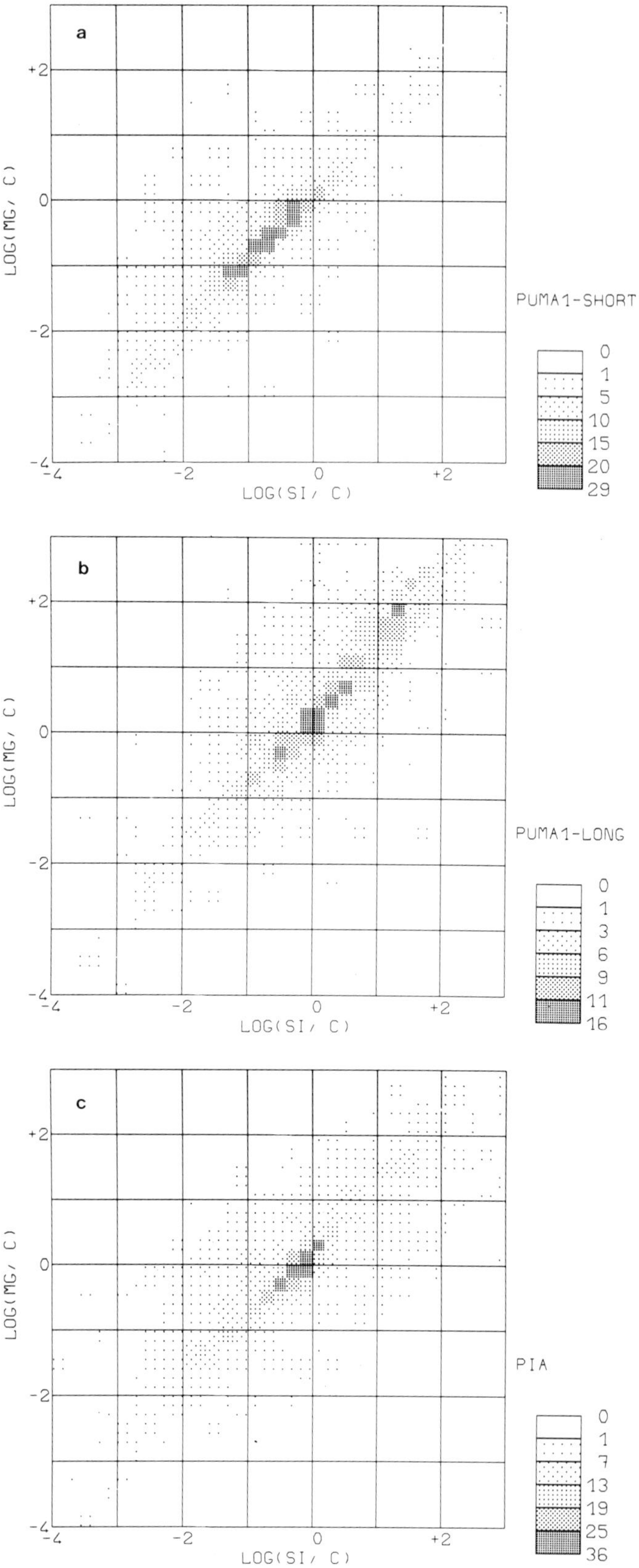

Fig. 2a–c. Correlation plot between Mg/C and Si/C abundance ratios (log scale on both axes). The degree of shading indicates the number of spectra in squares of 0.2 × 0.2 magnitudes. The clustering of most spectra near and slightly above the diagonal indicates a strong correlation between Mg and Si over more than 5 orders of magnitude

The CHON component could be similar to the very small "tar balls" observed in stratospheric particles. Due to the nearly complete break-up of molecular species induced by the impact ionization process, it is however not possible to discriminate between complex organic compounds and simpler compounds (down to a mixture of CO_2, H_2O and NH_3, which is however unlikely to survive long at 0.8 au from the sun). Other contributions (Clark et al., 1987) discuss in detail this CHON component, and we will discuss mainly the mineral component and the possible variations in grain composition with time:

– Mg rich silicates dominate the mineral fraction (2/3 of all mineral grains). The Fe to Ca ratio varies by more than a factor 100 (Fig. 3a, 3b, 3c). The possible silicate compositions therefore cover the full range of meteoritic minerals, from olivines and low Ca pyroxenes to calcium-rich pyroxenes, hydrated minerals and feldspars. The relatively tight grouping of spectra near the 1, 1 point for PIA (Fig. 3c) is most likely linked to the truncation of peaks higher than 21 and lower than 42, in which lie many Ca and Fe amplitudes.

– Sulfur appears both in CHON rich grains and mineral grains. The correlation diagram between sulfur, iron and carbon (Fig. 4a, 4b, 4c) shows that sulfur is slightly more correlated with iron than with carbon. 3 to 4% of grains, with $S/C > 100$ and $S/Fe \gtrsim 1$, can be interpreted as iron sulfide particles. Such mineral particles with low Si content (sulfides, carbides, oxydes) constitute up to 25% of all mineral grains.

– In order to bring out possible variations of the dust composition along the trajectory of the spacecrafts, we plotted the ratio of all "mineral" ions (Na, Mg, Si, Ca, Fe) to carbon ions as a function of time. (Fig. 5a, 5b, 5c). The spread of compositions is very wide at any given time relative to closest approach (often 3 order of magnitude for the mineral to carbon ratio). However, for PUMA 1 in short mode, the CHON grains are more numerous (42%) between −600 s and closest approach than either before (33%) or after (19%). The statistical error for the two last periods is ~3%, so this variation can be considered significant. The same trend is observed for PUMA 1 in long mode (14%, 24%, 17%). It can be seen that most "pure" mineral grains are indeed observed either long before or after closest approach. For PIA, there are already mineral grains near closest approach. If we select the period from −500 s to −50 s, the proportion of CHON grains varies from 38% before to 45% during and 27% after this period.

5. Conclusion

The impact ionization mass spectrometers on board Giotto and Vega spacecraft allowed a first qualitative characterization of cometary dust on a large statistical basis. There is an extreme variety of composition at the submicron scale, which is consistent with results obtained on stratospheric dust particles (Fraundorf et al., 1979). A large fraction of particles are dominated by light elements. Part of this large CHON component could be lost either during transit to Earth or during atmospheric entry. The mineral fraction shows a range of composition similar to the range of meteoritic minerals. The derivation of quantitative elemental abundances requires a sustained simulation and modeling program so as to better understand ion formation processes during hypervelocity impacts.

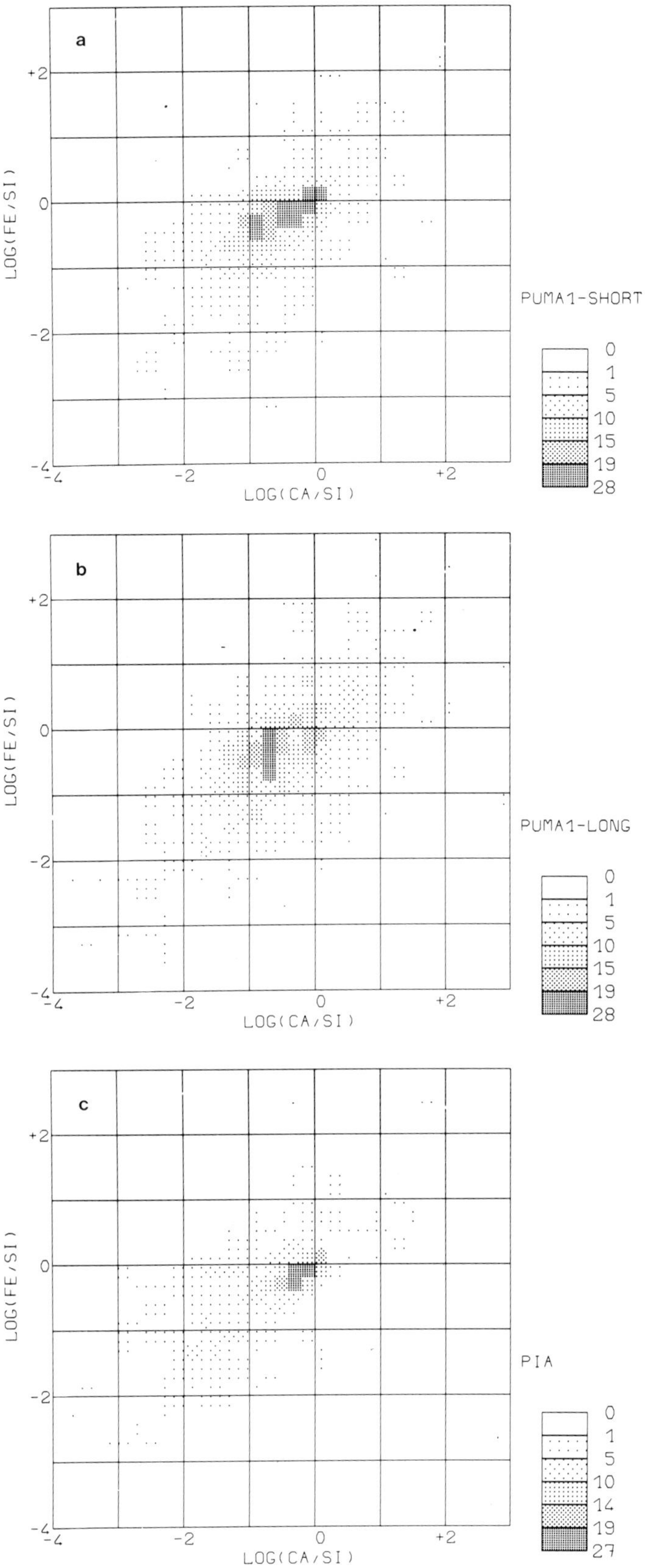

Fig. 3a–c. Correlation plot between Fe/Si and Ca/Si abundance ratios for PUMA 1, mode "short" (**a**), PUMA 1, mode "long" (**b**), and PIA (**c**) A wide spread in silicate composition is observed, with variations of the Ca/Fe ratio by more than a factor 100. For PIA, the high density of spectra near the (1, 1) point (approximately equal amounts of Si, Ca and Fe) is probably linked to the amplifier malfunction

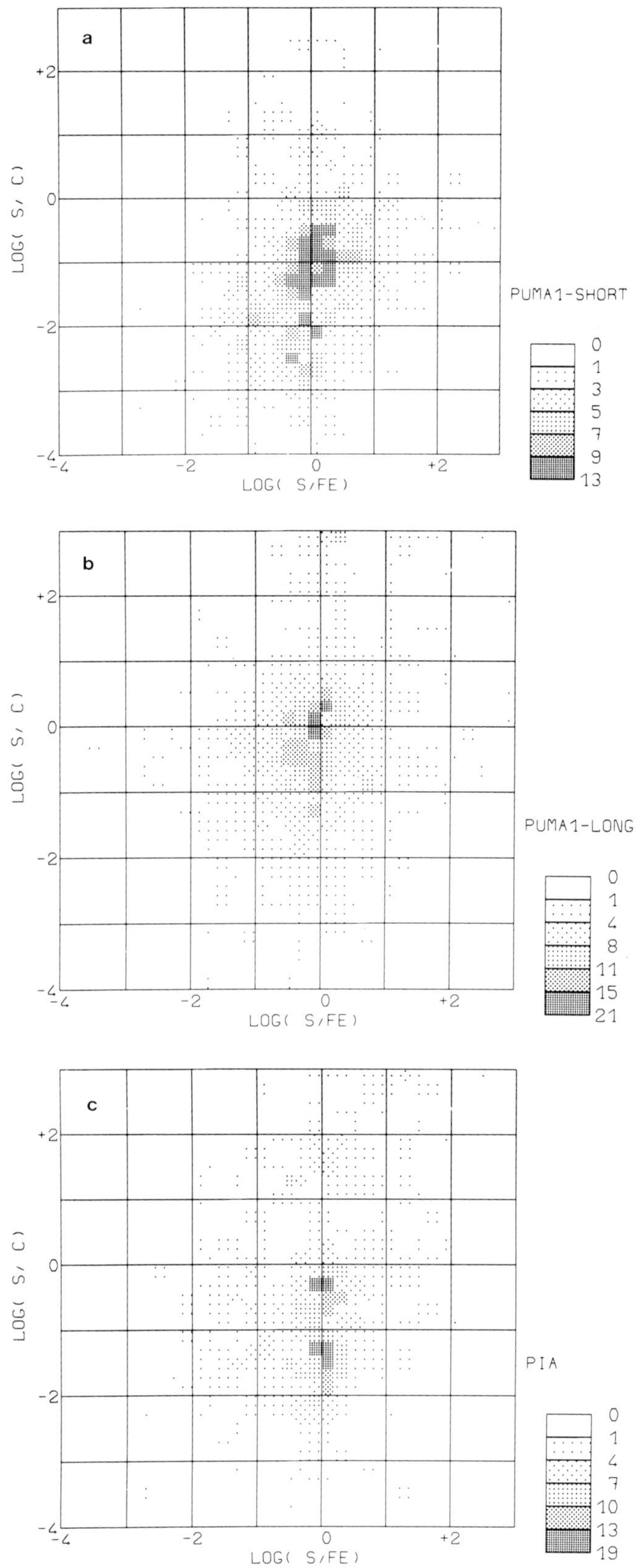

Fig. 4a–c. Correlation plot between S/Fe and S/C abundance ratios for PUMA 1, mode "short" (**a**), PUMA 1, mode "long" (**b**), and PIA (**c**). The wide spread observed along both directions suggests that sulfur can either be linked to carbon (most likely in organic material), or to iron (in iron sulfides). Correlation with iron is slightly stronger than with carbon. Grains near the top of the figure (S/C > 100) are most likely rich in iron sulfides

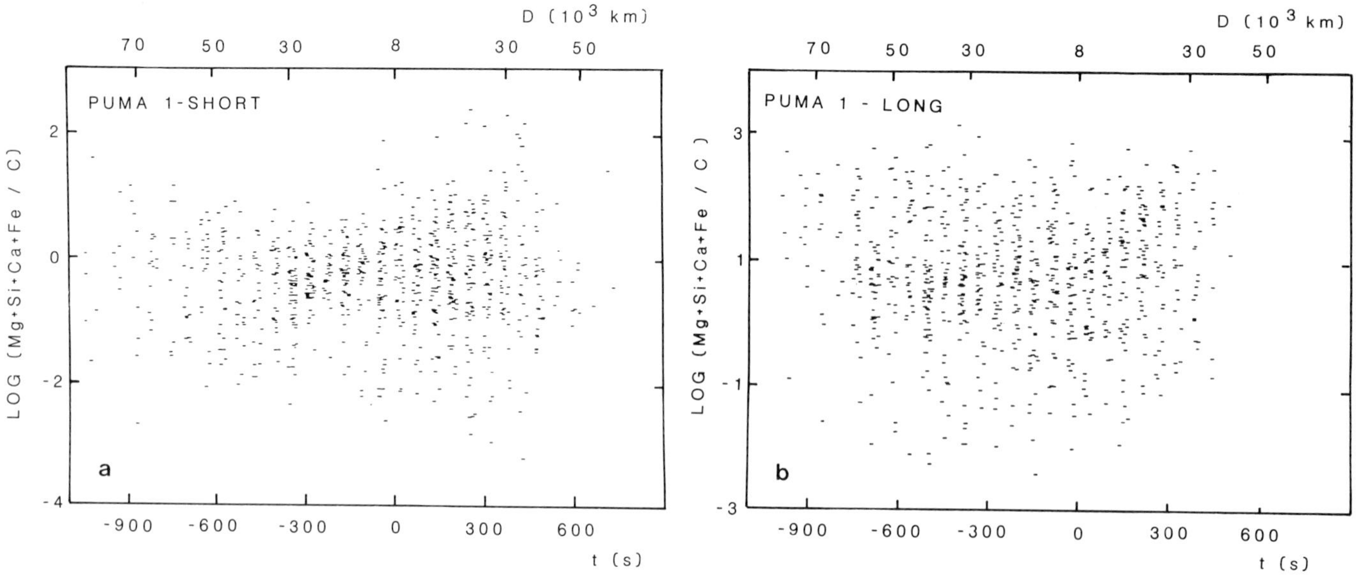

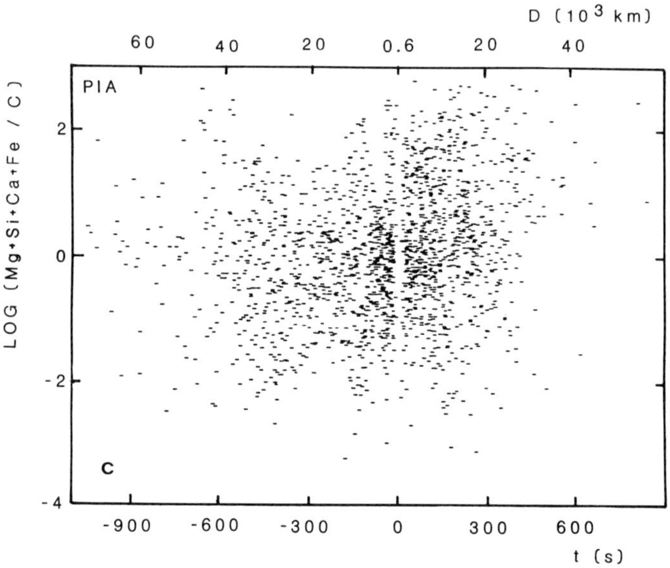

Fig. 5a–c. Abundance ratio of "mineral" ions (Na, Mg, Si, Ca, Fe) to carbon ions (log scale) as a function of time to closest approach, for PUMA 1, mode "short" (**a**), PUMA 1, mode "long" (**b**), and PIA (**c**). The distance to the comet is also reported. The very wide spread in composition at any given time is clearly apparent; the trend favoring carbon-rich grains before closest approach, and mineral grains far before, or after closest approach, can also be observed

References

Clark, B., Mason, L.W., Kissel, J.: 1987, Proc 20th ESLAB Symp., ESA SP-250 Vol. **III**, 353

Fraundorf, P., Shirk, J.: 1979, *Proc., 10th Lunar Planet. Sci. Conf.* p. 951

Kissel, J., Brownlee, D.E., Büchler, K., Clark, B.C., Fechtig, H., Grün, E., Hornung, K., Igenbergs, E.B., Jessberger, E.K., Krueger, F.R., Kucsera, H., McDonnell, J.A.M., Morfill, G.M., Rahe, J., Schwehm, G.H., Sekanina, Z., Utterback, N.G., Völk, H.J., Zook, H.A.: 1986a, *Nature* **321**, 336

Kissel, J., Sagdeev, R.Z., Bertaux, J.L., Angarov, V.N., Audouze, J., Blamont, J.E., Büchler, K., Evlanov, E.N., Fechtig, H., Fomenkova, M.N., Von Hoerner, H., Inogamov, N.A., Khromo, V.N., Knabe, W., Krueger, F.R., Langevin, Y., Leonas, V.B., Levasseur-Regourd, A.C., Managadze, G.G., Podkolzin, S.N., Shapiro, V.D., Tabaldyev, S.R., Zubkov, B.V.: 1986b, *Nature* **321**, 280

Mazets, E.P., Aptekar, R.L., Golenetskii, S.V., Guryan, Y.A., Dyachkov, A.V., Ilyinskii, V.N., Panov, V.N., Petrov, G.G., Savvin, A.V., Sagdeev, R.Z., Sokolov, I.A., Khavenson, N.G., Shapiro, V.D., Shevchenko, V.I.: 1986, *Nature* **321**, 276

McDonnell, J.A.M., Alexander, W.M., Burton, W.M., Bussoletti, E., Clark, D.H., Grard, R.J.L., Grün, E., Hanner, M.S., Hughes, D.W., Igenbergs, E., Kuczera, H., Lindblad, B.A., Mandeville, J-C., Minagra, A., Schwehm, G.H., Sekanina, Z., Wallis, M.K., Zarnecki, J.C., Chakaveh, S.C., Evans, G.C., Evans, S.T., Firth, J.G., Littler, A.N., Massonne, L., Olearczyk, R.E., Pankiewicz, G.S., Stevenson, T.J., Turner, R.F.: 1986, *Nature* **321**, 338

McDonnell, J.A.M., Kissel, J., Grün, E., Grard, R.J.L., Langevin, Y., Olearczyk, R.E., Perry, C.H., Zarnecki, J.C.: 1987, Proc. 20th ESLAB Symp., ESA SP-250 Vol. **II**, 25

Simpson, J.A., Sagdeev, R.Z., Tuzzolino, A.J., Perkins, M.A., Ksanfomality, L.V., Rabinowitz, D., Lentz, G.A., Afonin, V.V., Erö, J., Keppler, E., Kosorokov, J., Petrova, E., Szabö, L., Umlauft, G.: 1986, *Nature* **321**, 278

Vaisberg, O.L., Smirnov, V.N., Gorn, L.S., Iovlev, M.V., Blaikchin, M.A., Klimov, S.I., Savin, S.P., Shapiro, V.D., Shevchenko, V.I.: 1986, *Nature* **321**, 274

Astron. Astrophys. 187, 767–773 (1987)

ASTRONOMY AND ASTROPHYSICS

Comet P/Halley: implications of the mass distribution function for the photopolarimetric properties of the dust coma

P.L. Lamy [1], E. Grün [1], and J.M. Perrin [2]

[1] Max-Planck-Institut für Kernphysik, Postfach 103980, D-6900 Heidelberg, Federal Republic of Germany
[2] Laboratoire d'Astronomie Spatiale, Les Trois Lucs, F-13012 Marseille, France

Received March 10, accepted May 11, 1987

Summary. The dust particle fluences measured aboard the Vega (SP-1 and SP-2 impact sensors) and Giotto probes DIDSY impact sensors) are analyzed to obtain the differential size distribution function of the dust in the coma of comet Halley. The brightness integral is then calculated for perfectly spherical (Mie scattering) and rough grains of various compositions. It is shown that the photopolarimetric observations rule out the dominating presence of weakly absorbing silicates such as olivine but are compatible with rough moderately absorbing silicate grains having a density decreasing with radius and with rough graphite grains. A mixture of the two types gives in fact the best agreement with the observations.

Key words: comet Halley – dust – photometry – polarization

1. Introduction

The size distribution function of cometary grains has always been indirectly determined from either their visible scattering light or their thermal emission. Many assumptions are of course required in this process such as the optical properties of the grains. For the first time, the space missions to comet Halley gave direct access to this function thanks to the various impact detectors. Once this distribution is established, an important question is to check the compatibility of the measured dust flux with the optical observations and, more generally, to investigate its implications for the physical and optical properties of the dust grains. We have therefore calculated the brightness integral and the polarization under various assumptions for the composition, the density and the shape of the grains. A comparison with the observations allow to place constraints on these parameters and help characterizing the properties of the grains.

2. From the mass fluence to the differential spatial density

2.1. *Discussion of the in situ measurements*

Each of the Halley space probes (Vega 1, 2 and Giotto) carried on board several dust detectors in order to determine the dust flux over a wide mass range. The most sensitive detectors are impact ionization detectors which measure the electrical charge release upon impact of fast particles onto the sensor. They typically record particles in the mass range 10^{-17} g to 10^{-10} g at impact speeds of 70 to 80 km s^{-1}. The IPM-P sensor of the DIDSY (McDonnell et al., 1986a) experiment on Giotto as well as the SP-1 (Balebanov et al., 1985, p. 145) and part of the SP-2 (Balebanov et al., 1985, p. 150) experiments on Vega 1 and 2 are of this type. Also the particulate impact analysers on Giotto PIA (Kissel et al., 1986) and on Vega 1 and 2 PUMA (Balebanov et al., 1985, p. 88) use the charge released to detect particle impacts. However, these instruments have been optimized to obtain chemical analysis of the dust particles. Therefore an absolute impact rate is not easy to extract from the data and hence the results on the impact rate are still preliminary. The chemical analysis of cometary dust, however, yielded already the following important results (Jessberger et al., 1986): The average abundance of the major elements is chondritic within a factor of two except that carbon is enriched by a factor of 8 compared to C1 chondrites. Halley's dust is composed of chondritic silicates and refractory carbonaceous material in varying proportions.

Intermediate in sensitivity are penetration detectors. Particles are recorded after they have penetrated a certain depth of material. The sensitivity of these detectors is determined by the thickness of the material which has to be penetrated. The thickness and corresponding mass sensitivities range from a few μm (10^{-14} to 10^{-13} g) in the case of the IPM-P sensor of DIDSY, the SP-1 and DUCMA (Simpson et al., 1986) experiments, to 50–100 μm ($\sim 10^{-10}$ g) for the CIS sensor of DIDSY and the FOTON experiments (Balebanov et al., 1985, p. 155) of Vega 1 and 2 and up to 1000 μm (10^{-6} g) for the RSM sensor of DIDSY. Note that this latter sensor, being mounted of the rear shield, requires that all detected particles have penetrated the 1 mm thick front shield. The largest particles (10^{-12} to 10^{-3} g) are measured by acoustic microphone detectors. Different from the SP-2 and DIDSY-4 microphones which use fixed sensitive areas, the DIDSY 2 and 3 microphones use most of the Giotto impact shield as sensitive surface and hence, the magnitude of the impact determines the effective area of the detector. Because of the very large sensitive area (~ 2 m^2) for big particles and because of closest proximity of Giotto to the nucleus at closest approach, the largest particles ($m \sim 10^{-3}$ g) were recorded by the DIDSY experiment.

Besides the technical differences among the detectors, one has to consider characteristics of the missions as well as the performances of the different instruments during the flight. A major disturbance occurred within the Giotto flight when, from

Send offprint requests to: P.L. Lamy

minus 10 s to about plus 20 s to closest approach, the communication link was lost and only gradually recovered afterwards (McDonnell et al., 1986b). This way, the major part of dust flux measurements by Giotto above the sunlit hemisphere of the nucleus was lost. In addition the sensor cover of DIDSY-IPM did malfunction and therefore no data on the smallest particles was obtained from this impact ionisation detector (Lange et al., 1986). Because of these technical problems, McDonnell et al. (1986b) interpolated the impact rate for the time interval around closest approach according to a r^{-2} law (r = distance from the nucleus). As a consequence, any enhancement of the flux from the central sunlit portion of the nucleus was neglected. Preliminary flux data from PIA supplemented the DIDSY small particle data. Further analysis of the PIA data (J. Kissel and F. Krueger, private communication) indicates a shift of the flux values given by McDonnell et al. (1986b) to larger masses by about one order of magnitude. In our analysis we have taken into account this shift in the small particle data.

The mission profile of both Vega missions was much smoother and no major disturbances occurred. The widest mass range ($6\ 10^{-16}$ g to $5\ 10^{-7}$ g) was covered by the impact ionization and acoustic detectors of the SP-2 experiment (Mazets et al., 1986). Both data were obtained from the inbound and outband legs of the Vega trajectories. Although the actual flight instruments have not been calibrated at a hypervelocity dust accelerator, the underlying measurement principle is well understood and extensively tested and therefore, the accuracy of the mass determination should be better than an order of magnitude (Göller et al., 1987). The flux values measured by Vega 1 are generelly a factor 3 higher than those on Vega 2 due to an enhanced activity during the earlier period. Since the Vega 2 measurements published by Mazets et al. (1986) are represented by a few more data points than the Vega 1 measurements we will refer in the following to the Vega 2 SP-2 measurements. The SP-1 data (Vaisberg et al., 1986a) agree quite well with the SP-2 fluxes and extend the range down to masses as small as $3\ 10^{-17}$ g.

DUCMA data (Simpson et al., 1986) cover only a small mass range from $1.5\ 10^{-13}$ g to 10^{-10} g. They are not considered here. The considerable differences between the DUCMA and SP-1, SP-2 data are attributed by us to the new measurement principle employed in the DUCMA experiment. The calibration of this experiment is based on tests with very small particles at the bottom of the dynamic range and therefore involves major uncertainties especially for the larger particles.

The statistically best data of the mass distribution are the fluence data because they include all measurements along the trajectory. The largest contribution to the data comes from measurements taken during closest approach which occurred close to the noon meridian on the nucleus where the activity is expected to be highest. This averaging of the particle flux along the trajectory smoothes out dynamical effects on the mass distribution in the distant parts of the trajectory (Massonne and Grün, 1986) as well as the dispersive effect of rotation in the near nucleus part of the trajectory (Vaisberg et al., 1986b). Above all, this averaging process to obtain fluence values correspond directly to the averaging along a line-of-sight in groundbased coma observations.

2.2. Derivation of the differential spatial density

Let $N_c(r, m)$ be the cumulative spatial density (in m^{-3}), that is the total number of grains with masses up to m, per unit volume at a distance r from the nucleus.

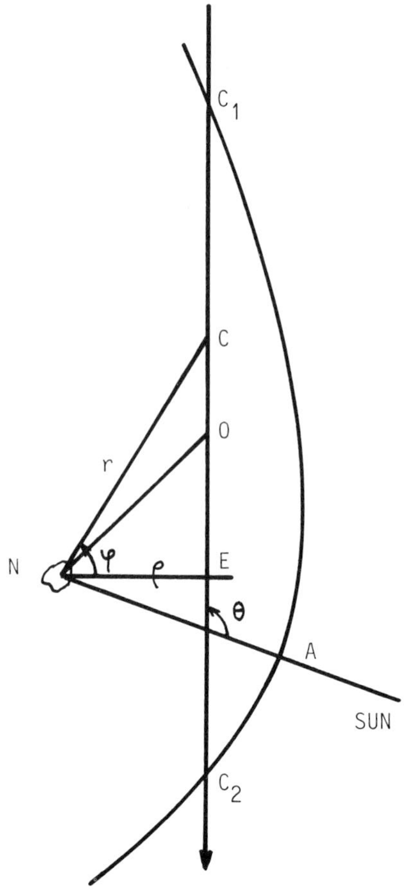

Fig. 1. The geometry of the comet flyby. The angle φ is positive for direct rotation

Using the geometry and the notations of Fig. 1. The fluence (in m^{-2}), which is the integral of $N_c(r, m)$ over the trajectory is written

$$\Phi(m) = \int_{C_1}^{C_2} N_c(r, m)\, dl$$

$$= \int_{r_1}^{r_2} N_c(r, m)\, (r^2 - \varrho^2)^{-1/2}\, r\, dr\,, \tag{1}$$

where ϱ is the miss-distance of the trajectory.

It is of course impossible to retrieve $N_c(r, m)$ from $\Phi(m)$. Having in mind that we are going to calculate a brightness integral corresponding to optical measurements averaged over a line-of-sight, we may look for an averaged cumulative spatial density independent of r, $N_c(m)$. The radial variation is then assumed to follow the $(r_0/r)^2$ law of a simple radial-outflow model; then

$$\Phi(m) = -r_0^2\, \varrho^{-1} \int_{\varphi_1}^{\varphi_2} N_c(m)\, d\varphi\,. \tag{2}$$

Using a reference distance of $r_0 = 10^6$ m, we obtain

$$N_c(m) = 10^{-12}\, \varrho\, (\varphi_1 + \varphi_2)^{-1}\, \Phi(m)\,. \tag{3}$$

$(\varphi_1 + \varphi_2)$ is the total angle (in rad) subtented at the nucleus by the segment C_1C_2 of the trajectory which lies inside the envelope of

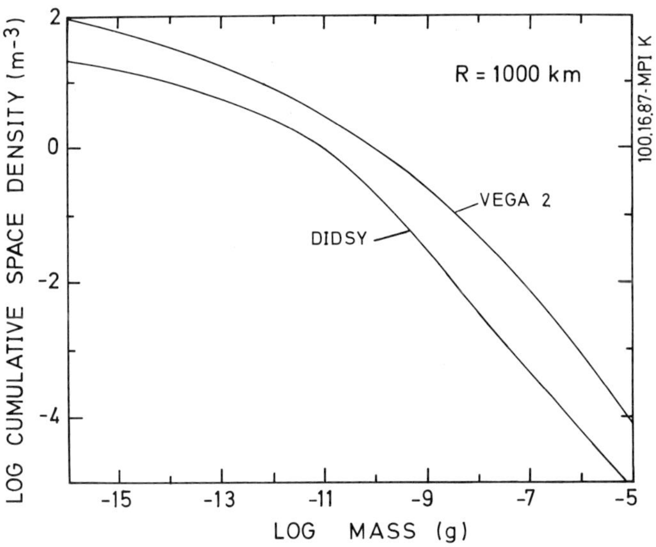

Fig. 2. The cumulative spatial density as a function of the mass of the grains deduced from the fluence measured by Vega 2 SP1, SP2 and Giotto DIDSY impact detectors

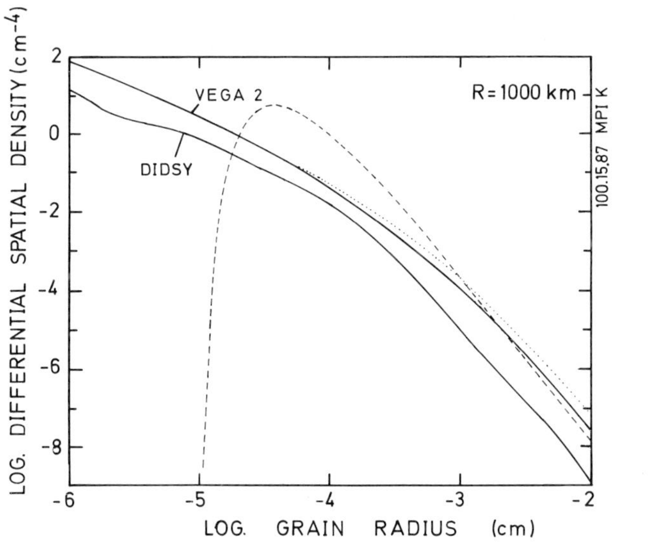

Fig. 3. The differential spatial density as a function of radius of the grains deduced from the cumulative densities of Fig. 2. The solid curves correspond to a constant bulk density of 2.2 g cm^{-3} while the dotted line illustrates the influence of a variable density (see text) for the Vega 2 case. The model predicted by Divine et al. (1986) is shown for comparison (broken line)

Table 1. The coefficients a_i of the best fit polynomial to the Vega 2-SP1-SP2 data

A0 $= -0.1204042413\ 10^2$
A1 $= -0.2063823315\ 10^1$
A2 $= -0.1058154891$
A3 $= -0.1975492189\ 10^{-2}$

the dust coma. For a parabolic boundary having a parameter of $2\ 10^7$ m, we obtain distances EC_1 and EC_2 (E is the point of closest approach) in good agreement with the first events detected by the sensors leading to $\varphi_1 + \varphi_2 = 2.98$ rad for Vega 2 and 3.12 rad for Giotto.

We have implicitly assumed that the trajectories cross the nucleus – Sun line (i.e., the axis of the paraboloid). This is not strictly true but the "off-axis" always remains small and the effect is negligible for the present study.

Best fit polynomials to the [$\log N_c$, $\log m$] Vega 2 (SP-1, SP-2) and DIDSY data have been determined and are illustrated in Fig. 2. From

$$\log N_c(m) = \sum_{i=0}^{m} a_i (\log m)^i \tag{4}$$

it is straightforward to obtain the differential spatial density $N(s)$ (in m^{-4}) as a function of grain radius s:

$$N(s) = -3\, N_c s \sum_{i=1}^{m} i\, a_i \log m)^{i-1} \tag{5}$$

at the reference distance $r_0 = 10^6$ m. The mass and radius are simply related by

$$m = (4/3)\, \pi\, \delta\, s^3\,, \tag{6}$$

where δ is the density of the grains.

Table 2. Bulk density and complex index of refraction at $\lambda = 0.58\ \mu$m of the selected materials. For graphite, the index refers to $E_\perp$. Alternative densities for chondrite and the silicate A have also been considered (see text)

Material	Density (g cm^{-3})	Index of refraction	
		n	k
Olivine	3.3	1.66	1.0 10^{-4}
Chondrite	2.2	1.90	1.9 10^{-3}
Tholin	1.45	1.7	2.0 10^{-2}
Silicate A	2.2	1.72	3.0 10^{-2}
Silicate B	2.2	1.4	3.0 10^{-2}
Magnetite	2.16	2.56	6.0 10^{-1}
Graphite	5.2	2.7	1.4

We give the coefficients a_i for the reliable Vega 2 results (the DIDSY curve is still preliminary) in Table 1, where N_c is in m^{-3} and m in g. The deduced differential spatial densities are displayed in Fig. 3 and compared with the model of Divine et al. (1986). The curve derived by Lamy and Perrin (1986) from the Vega 2 SP-2 cumulative flux measured at $r = 8.2\ 10^6$ m (see Mazets et al., 1986, Fig. 2) was ill-determined below $s \sim 0.05\ \mu$m so that the difference with the present model is not meaningful (it will be further shown later that these grains negligibly contribute to the brightness); however, there is a significant difference in the size range 0.1 to 1 μm which indicates that, as observed, grains of this size are more numerous near the nucleus.

The density of the grains δ is a key parameter. Table 2 lists the bulk density of materials selected for the present calculation. However, it appears plausible that cometary grains are porous aggregates and have therefore lower effective densities; a few examples were introduced in the calculation. As the elementary grains making the aggregates have their normal bulk density, a variable density, decreasing with increasing radius is probably a better description. We have retained the law

$$\delta\,(\mathrm{g\,cm^{-3}}) = 2.2 - 1.4\,s/(s_0 + s) \tag{7}$$

with $s_0 = 1\,\mu\mathrm{m}$, for the cases where a bulk density of $2.2\,\mathrm{g\,cm^{-3}}$ applies.

In a $\log N$ versus $\log s$ representation, the slope of the Vega 2 curve (assuming constant bulk density of $2.2\,\mathrm{g\,cm^{-3}}$) is -4.08 at $100\,\mu\mathrm{m}$, -3.04 at $10\,\mu\mathrm{m}$, -2.23 at $1\,\mu\mathrm{m}$ and -1.54 at $0.1\,\mu\mathrm{m}$.

3. The brightness integral

3.1. Basic expression

Let us consider the geometry for an observer O located in the dust coma (Fig. 1) and assume that:

i) the dust coma is bounded by a parabola,

ii) the line-of-sight crosses the nucleus-Sun line or the off-axis is small,

iii) all grains along the line-of-sight scatter at the scattering angle θ,

iv) the differential spatial density may be written

$$N(s)\,(r_0/r)^2 \tag{8}$$

with $N(s)$ given by Eq. (5).

The $(r_0/r)^2$ law allows a straightforward integration along the line-of-sight in terms of the angle φ and the observer O (angle φ_0) will measure a brightness I at wavelength

$$I(\lambda) = \pi\, B_{\odot}(\lambda)\, \frac{R_{\odot}^2}{R^2}\, \frac{r_0^2}{\varrho}\, (\varphi_1 - \varphi_0)\, \psi(\lambda, \theta)\,, \tag{9}$$

where $R_{\odot}$ = radius of the Sun, R = heliocentric distance, $B_{\odot}(\lambda)$ = spectral brightness of the Sun, ϱ = miss-distance of the line-of-sight, r_0 = reference distance (10^6 m), and, $\varphi_1 - \varphi_0$ = angle subtented at the nucleus by the segment C_1O.

The quantity ψ is known as the volume scattering function

$$\psi(\lambda, \theta) = \left(\frac{\lambda}{2\pi}\right)^2 \int_{s_1}^{s_2} \frac{i_1(\lambda,\theta,s) + i_2(\lambda,\theta,s)}{2}\, N(s)\, ds \tag{10}$$

and i_1 and i_2 are the squares of the amplitude functions for the two directions of polarization; they are further functions of the composition of the grains (via the complex index of refraction).

The generalization to an observer located outside the dust coma is straightforward and, in the above expression of $I(\lambda)$, $(\varphi_1 - \varphi_0)$ is replaced by $(\varphi_1 + \varphi_2)$, the angle subtented by the segment C_1C_2.

3.2. Scope of the calculations

We limit the present investigation to the brightness and polarization of the scattered light by the dust coma, essentially in the visible (the thermal emission is not considered for the moment). The calculation of the quantities i_1 and i_2 are carried out along two lines:

i) the Mie scattering theory which assumes perfect spheres,

ii) the vectorial description in the eikonal picture recently proposed by Perrin and Lamy (1986) and which applies to large rough spheres.

This latter restriction imposes a minimum radius of about $3\,\mu\mathrm{m}$ (for visible wavelengths) and therefore the second approach is in fact a combination of the Mie model for grains smaller than $3\,\mu\mathrm{m}$ and the Perrin-Lamy model for larger grains.

The materials whose index of refraction is given in Table 2 were selected to be compatible with the results of the mass spectrometric particle analysers. Four are basically of the silicates family ranging from slightly to strongly absorbing. The one denoted "silicate A" is the "astronomical silicate" whose dielectric function has been constructed by Draine (1985) so as to be consistent with astronomical observations. The material denoted "silicate B" is an hypothetical silicate whose optical constants are in the range determined by Mukai et al. (1986) to fit the observed polarization. As the presence of organic material is supported by various evidences, we have retained tholin as representative of this class of grains. Although processed under the conditions of the Titan atmosphere (Khare et al., 1984), we note that the imaginary part $k = 0.02$ is similar to that inferred for the "refractory residue" produced in the laboratory under interstellar conditions (Greenberg, private communication). Surprisingly, there is only a slight difference with the optical constants of the "astronomical silicate", implying that either case may represent a composite grain with a silicate core and a tholin mantle.

4. Analysis of the results

4.1. General remarks

The preliminary analysis of Lamy and Perrin (1986) and of Mukai et al. (1986) outlined the major implication of the observed mass distribution function, namely that about half of the scattered light is produced by submicronic particles. The direct consequence is that polarization predicted by Mie theory for moderately absorbing silicates is negative at intermediate scattering angles and positive at backscattering angles, in contradiction with the observations. Rough, strongly absorbing (e.g., magnetite) grains can reproduce the observed visible polarization but their domination is unlikely.

The new size distribution function derived above alleviates these problems to some extent and we now discuss several aspects.

4.2. Volume scattering function

The brightness will be discussed in terms of the volume scattering function $\psi(\theta)$ which is independent of the observational conditions except for the scattering angle θ. We first consider $\theta = 107°$, an angle apropriate to the Giotto optical probe experiment and to the post-encounter Vega 2 television observations.

The integral in the expression of ψ has generally been calculated in the size range 0.01 to $100\,\mu\mathrm{m}$. Figure 4 shows ψ as a function of the upper limit of the integral s_2 for the case of the astronomical silicate (cumulative representation) and illustrates how the various size intervals contribute to the total value; for $s < 0.1\,\mu\mathrm{m}$; the contribution is negligible. The integral reaches half its final value at $s \sim 3\,\mu\mathrm{m}$ when a constant density is assumed and at $s \sim 8\,\mu\mathrm{m}$ when the density is variable; in this latter case, note that the integration extends to $300\,\mu\mathrm{m}$ as the 100–300 µm interval contributes about 5% to the total value.

Figure 5 displays the volume scattering function $\psi(\theta = 107°)$ and the associated polarization for the selected materials under various conditions. Basically, ψ ranges from 0.8 to $4.3\ 10^{-14}\,\mathrm{cm^{-1}\,sr^{-1}}$. This result is very similar to that already

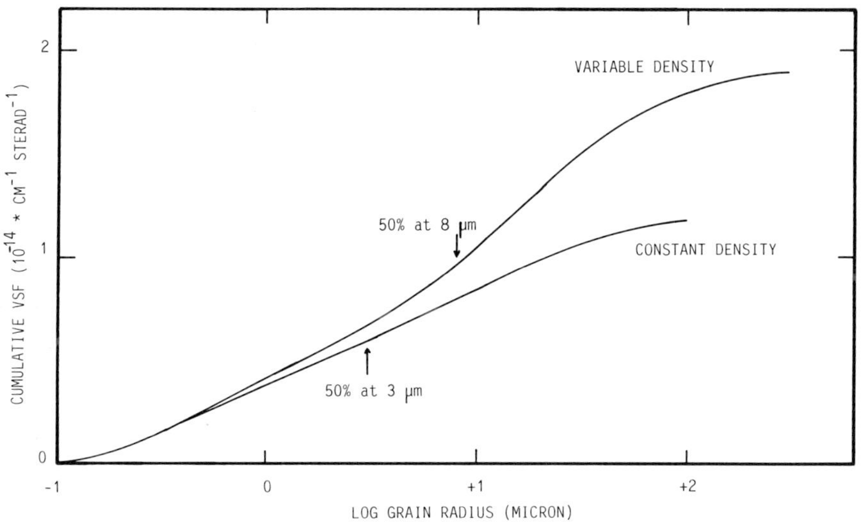

Fig. 4. Cumulative representation of the volume scattering function for the case of the astronomical silicate and Mie scattering

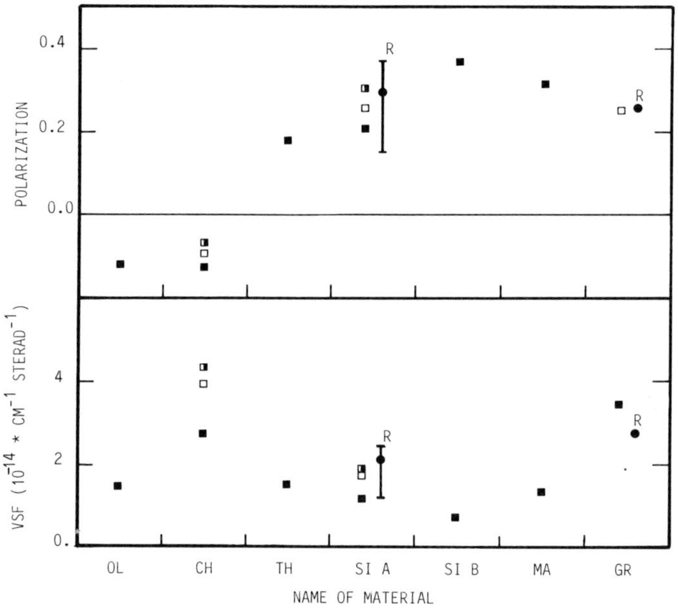

Fig. 5. The volume scattering function and the polarization at $\theta = 107°$ for the 7 selected materials: OL = olivine, CH = chondrite, TH = tholin, SI A = silicate A (astronomical), SI B = silicate B, MA = magnetite, GR = graphite. The squares corresponds to perfect spheres (Mie theory) of various densities: $2.2\,\mathrm{g\,cm^{-3}}$ (■), $1\,\mathrm{g\,cm^{-3}}$ (□) and variable density (◧). The filled circles correspond to rough grains and are labelled "R". For the astronomical silicate, the vertical bars show the range of values for different roughness

obtained by Lamy and Perrin (1986). On one hand, the new size distribution function does not much affect ψ. In the case of chondrite and the silicate, the variable density results in an increase by a factor ~ 1.6 which, in fact, is close, to assuming a constant density of $1\,\mathrm{g\,cm^{-3}}$. On the other hand, the new size distribution function appreciably shifts the polarization to larger values. The variable density also increases the polarization, by as much as 0.1 for the silicate.

The effect of grain roughness has been investigated in two cases. For the astronomical silicate, with variable density, we illustrate in Fig. 5 the range of values corresponding to various roughness, either constant or proportional to the radius of the grains. The solution singled out in the graph corresponds to a "proportional" roughness with $\Delta s/s = 0.15$ and an opening angle of the asperities $w = 120°$ (see Perrin and Lamy, 1983 for these definitions) and yields $\psi = 2.13\ 10^{-14}\,\mathrm{cm^{-1}\,sr^{-1}}$ and $p \sim 0.3$. For the graphite, we obtained $\psi = 2.82\ 10^{-14}\,\mathrm{cm^{-1}\,sr^{-1}}$ and $p = 0.26$ for $\Delta s/s = 0.1$ and the same w.

A preliminary calibration of an image obtained by the Vega 2 camera one day after encounter (Abergel and Bertaux, 1987) indicates that the sunward brightness profile allows the ϱ^{-1} law and yields a volume scattering function

$$\psi \simeq 2.5\ 10^{-14}\,\mathrm{cm^{-1}\,sr^{-1}}$$

in satisfactory agreement with the above calculated results.

Turning now to the Giotto optical probe experiment (Levasseur-Regourd et al., 1986) at a distance of $r = 5000$ km where there is no deviation from the r^{-2} law, preliminary calibration yields

$$\psi \simeq 7.4\ 10^{-15}\,\mathrm{cm^{-1}\,sr^{-1}}.$$

If we retain a value of 2 to 2.5 10^{-14} for the Vega 2 case, we conclude from the above result that the spatial density (i.e., the dust production rate) was reduced by a factor ~ 3 at the time of the Giotto encounter. Direct calculations using the DIDSY differential spatial density for the case of the astronomical silicate with variable density gives a larger factor of approximately 5.6. It has been reported that the dust activity was minimal at the time of the Giotto encounter (e.g., Festou et al., 1986; Cosmovici et al., 1986) but the latter factor is probably an upper limit inherent to the present procedure of retrieving the DIDSY fluence.

4.3. Polarization

The polarization p at $\theta = 107°$ (Fig. 5) exhibit a large variation from -0.13 to $+0.37$. The observations suggest that p lies between 0.20 and 0.24 (see below for further details) ruling out several of the selected materials. In fact, one sees that overall, the astronomical silicate, tholin (or optically equivalent materials as already pointed out) and graphite, produce acceptable results.

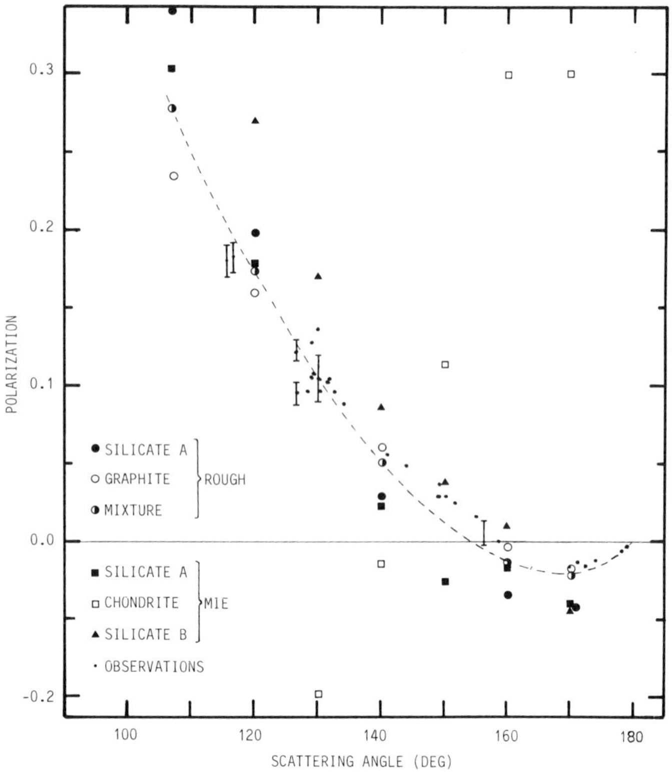

Fig. 6. Polarization at large scattering angles. The dots are the observational data from Mukai et al. (1986), Dollfus et al. (1986) and LeBorgne et al. (private communication) with typical error bars. Only a few examples of model calculations are displayed for clarity but they all assume a variable bulk density of the grains. The "mixture" (broken line) is a combination of 50% of astronomical silicate and 50% or graphite by grain number

Figure 6 regroups the polarization data obtained by several groups for scattering angles larger than about 100°. The polarisation behaviour, characterized by a reversal at an angle of about 160° and a negative branch is similar to that reported for other comets (Lamy, 1986). The calculated values confirm that olivine and chondrite are not acceptable. Worth mentioning is the large difference between Mukai et al. (1986) and our present result for the same "silicate B" (for instance, -0.1 against -0.04 at $\theta = 170°$) probably emphasizing the importance of the size distribution function and of the density.

One cannot expect, at this early stage of the investigation, a perfect agreement with the observational data. However, we notice that the trends presented by the rough grains of astronomical silicate with variable density and by the rough grains of graphite are the most promising. In the first case, the roughness smoothes the negative branch already existing with Mie theory and which results from the value of the complex index of refraction. In the second case (graphite), the roughness induces a negative branch which does not exist for perfect spheres. Both solutions exhibit furthermore a strong backscattering enhancement of the volume scattering function, by a factor for 3 approximately, in agreement with the interpretation by Hammel et al. (1986) of their albedo maps of comet Halley.

4.4. The color and the albedo of the dust

From photometric observations with the International Halley Watch filters, Tholen et al. (1986) deduced a neutral color of the dust coma over the wavelength range 0.3 to 0.7 µm. But these authors have called attention to contamination from the wings of emission bands. High resolution spectrophotometry observations are far superior to tackle this question since they allow to select clean windows to measure the continuum. Both the IUE spectra (Feldman et al., 1986) and those obtained by Jewitt and Meech (1986) indicate a substantial reddening, 7% per 0.01 µm in the ultraviolet and 10% per 0.1 µm in the visible. This latter result is quite similar to those obtained on comets P/Stephan-Oterma, P/Neujmin I, P/Churyumov-Gerasimenko and P/Arend-Rigaux (Tholen et al., 1986).

We performed a preliminary investigation of the color for the case of the astronomical silicate with variable density and found indeed a reddening in the interval 0.4–0.7 µm but of only 3.5% for 0.1 µm. This question requires further analysis.

The definition of an albedo for small particles always poses problems discussed by Hanner et al. (1981) who proposed an analogous of the geometric albedo at any angle in terms of the backscattering by a white Lambert disc of equal cross-section:

$$A_p(\theta) = \frac{\sigma(\theta)}{s^2} = \left(\frac{\lambda}{2\pi s}\right)^2 \frac{i_1(\lambda,\theta,s) + i_2(\lambda,\theta,s)}{2}. \tag{11}$$

In the case of a size distribution, we can generalize this definition by calculating an average in a given size interval $[s_1, s_2]$ viz.

$$\bar{A}_p(\theta) = \left(\frac{\lambda}{2\pi}\right)^2 \int_{s_1}^{s_2} \frac{i_1 + i_2}{2} \frac{N(s)}{s^2}\, ds \Big/ \int_{s_1}^{s_2} N(s)\, ds. \tag{12}$$

For the case of the astronomical silicate with variable density, we calculated $\bar{A}_p$ at $\theta = 107°$ integrated over the whole size distribution and found $\bar{A}_p = 0.036$. This compares favorably with the infrared albedos of 0.032 at 1.25 µm and of 0.045 at 1.65 µm measured by Tokunaga et al. (1986). However, the albedo calculated in restricted size intervals (e.g., 1 µm) decreases with increasing size being typically 0.035 at $s \sim 1$ µm and 0.02 at $s \sim 10$ µm. For the interval 1–30 µm. Because of the backscattering enhancement noted in Sect. 4.3, these values should be multiplied by a factor 3 approximately to obtain $\bar{A}_p$ at $\theta = 180°$ and to make a sound comparison with the geometric albedo of large objects.

5. Conclusion

The size distribution function of the grains present in the coma of comet Halley has been derived from the fluence measurements obtained with the Vega 2 SP1 and SP2 instruments. The power index varies from -4 at $s \sim 100$ µm to -1.54 at $s \sim 0.1$ µm.

The brightness calculated with this function is shown to be compatible with a preliminary calibrated observation with the Vega 2 camera obtained on March 10, 1986. A still preliminary comparison with the results from the Giotto optical probe indicates a decrease by a factor 3 approximately of the dust spatial density, a trend supported by independent observations. Direct calculations based on the fluence curve estimated from the DIDSY-PIA data yields a larger decrease (5–6) which is probably too large and points to persisting problems with these data.

Visible polarization data impose severe constraints on the nature of the dust grains. Assuming that the bulk density of the grains decreases with increasing size, and that the grains are rough, the astronomical silicate constructed by Draine (1985) offers an acceptable trend. Graphite, intended here to represent the overabundant carbon observed in the grains, also provides an acceptable solution in the case of rough grains. In fact, a combination of these two materials may eventually prove to be most appropriate. Indeed the simple mixture of 50 % silicate, 50 % graphite grains by number (but probably not totally realistic) yields $\psi(\theta=107°)=2.5\ 10^{-14}\,\mathrm{cm}^{-1}\,\mathrm{sr}^{-1}$ and a polarization curve in nice agreement with the available data (Fig. 5).

The fact that the astronomical silicate provides an acceptable trend is of particular importance since it will directly explain the 10 µm emission band. Also a limited calculation of the polarization at $\lambda=1.65\,\mu\mathrm{m}$ showed a satisfactory agreement with the data of Brooke et al. (1986). Here again, it is worth emphasizing that both conditions on the variable bulk density and on the roughness of the grains are required.

References

Abergel, A., Bertaux, J.L.: 1987, *Astron. Astrophys.* (submitted)

Balebanov, V.M., Skuridin, G.A., Vorontzova, E.V., Bassolo, V.S.: 1985, Venus-Halley Mission, Imprimerie Louis-Jean, Gap

Brooke, T.Y., Knacke, R.F., Joyce, R.R.: 1986, 20th ESLAB Symp., ESA SP-250, Vol. **II**, p. 87

Cosmovici, C.B., Mack, P., Craubner, H., Schwarz, G.: 1986, 20th ESLAB Symp., ESA SP-250, Vol. **II**, p. 151

Divine, N., Fechtig, H., Gombosi, T.I., Hanner, M.S., Keller, H.U., Larson, M.S., Mendis, D.A., Newburn, R.L., Reinhard, R., Sekanina, Z., Yeomans, D.K.: 1986, *Space Sci. Rev.* **43**, 1

Dollfus, A., Suchail, J.-L., Crussaire, D.: 1986, 20th ESLAB Symp., ESA SP-250, Vol. **II**, p. 41

Draine, B.T.,: 1985, *Astrophys. J. Suppl.* **57**, 587

Feldman, P.D., Festou, M.C., A'Hearn, M.F., Arpigny, C., Butterworth, P.S., Cosmovici, C.B., Danks, A.C., Crilmozzi, R., Jackson, W.M., McFadden L.A., Patriardi, P., Schleicher, D.G., Tozzi, G.P., Wallis, M.K., Weaver, H.A., Woods, T.N.: 1986, ESA SP-263, p. 39

Festou, M.C., Feldman, P.D., A'Hearn, M.F., Arpigny, C., Cosmovici, C.B., Danks, A.C., McFadden, L.A., Gilmozzi, R., Patriarchi, P., Tozzi, G.P., Wallis, M.K., Weaver, H.A.: 1986, *Nature* **321**, 361

Göller, J.R., Grün, E., Maas, D.: 1987, *Astron. Astrophys.* (submitted)

Hammel, H.B., Storrs, A.D., Cruikshank, D.P., Telesco, C.M., Decker, R.M., Campins, H.: 1986, 20th ESLAB Symp., ESA SP-250, Vol. **III**, p. 503

Hanner, M.S., Giese, R.H., Weiss, K., Zerull, R.: 1981, *Astron. Astrophys.* **104**, 42

Jessberger, E.K., Kissel, J., Fechtig, H., Krueger, F.R.: 1986, ESA SP-249, p. 27

Jewitt, D.C., Meech, K.J.: 1986, 20th ESLAB Symp., ESA SP-250, Vol. **II**, p. 47

Khare, B.N., Sagan, C., Arakawa, E.T., Suits, F., Callcot, T.A., Williams, M.W.: 1984, *Icarus* **60**, 127

Kissel, J.: 1986, ESA SP-1077, p. 67

Lamy, P.: 1986, in *Asteroids, Comet and Meteors* **II**, eds. I. Lagerkvist et al.

Lamy, P.L., Perrin, J.M.: 1986, 20th ESLAB Symp., ESA SP-250, Vol. **II**, p. 65

Lange, G., Grün, E., Maas, D., Göller, J.R., Kissel, J., McDonnell, J.A.M., Zarnecki, J.C., Bussoletti, E., Grard, R.J.L.: 1986, 20th ESLAB Symp., ESA SP-250, Vol. **II**, p. 191

Levasseur-Regourd, A.-L., Weinberg, J.L., Bertaux, J.-L., Dumont, R., Festou, M., Giese, R.H., Giovane, F., Lamy, P., Le Blanc, J.M., Llebaria, A.: 1986, *Adv. Space Res.* **5**, 12, 197

Massonne, L., Grün, E.: 1986, 20th ESLAB Symp., ESA SP-250, Vol. **II**, p. 319

Mazets, E.P., Sagdeev, R.Z., Aptekar, R.L., Golenitskii, S.V., Guryan, Yu.A., Dyachkov, A.V., Ilyinskii, Y.N., Panov, V.N., Petrov, G.G., Savvin, A.V., Sokolov, I.A., Frederiks, D.D., Khavenson, N.G., Shapiro, V.D., Shevchenko, V.I.: 1986, 20th ESLAB Symp. ESA SP-250, Vol. **II**, p. 3

McDonnell, J.A.M., Alexander, W.M., Burton, W.M., Bussoletti, E., Clark, D.H., Evans, G.C., Evans, S.T., Firth, J.G., Grard, R.J.L., Grün, E., Hanner, M.S., Hughes, D.W., Igenbergs, E., Kuczera, H., Lindblad, B.A., Mandeville, J.-C., Minafra, A., Reading, D., Ridgeley, A., Schwehm, G.H., Stevenson, T.J., Sekanina, Z., Turner, R.F., Wallis, M.K., Zarnecki, J.C.: 1986a, ESA SP-1077, p. 85

McDonnell, J.A.M., Kissel, J., Grün, E., Grard, R.J.L., Langevin, Y., Olearczyk, R.E., Perry, C.H., Zarnecki, J.C.: 1986b, 20th ESLAB Symp., ESA SP-250, Vol. **II**, p. 25

Mukai, T., Mukai, S., Kikuchi, S.: 1986, 20th ESLAB Symp., ESA SP-250, Vol. **II**, p. 59

Perrin, J.M., Lamy, P.L.: 1983, *Optica Acta* **30**, 1223

Perrin, J.M., Lamy, P.L.: 1986, *Optica Acta* **33**, 1001

Simpson, J.A., Rabinowitz, D., Tuzzolino, A.J., Ksanfomality, L.V., Sagdeev, R.Z.: 1986, 20th ESLAB Symp., ESA SP-250, Vol. **II**, p. 11

Tokunaga, A.T., Golish, W.F., Griep, D.M., Kaminski, C.D., Hanner, M.S.: 1986, *Astron. J.* **92**, 1183

Tholen, D.J., Cruikshank, D.P., Hammel, H., Piscitelli, J.R., Hartmann, W.K., Lark, N.: 1986, 20th ESLAB Symp., ESA SP-250, Vol. **II**, p. 73

Vaisberg, O.L., Smirnov, V.N., Omelchenko, A.: 1986a, 20th ESLAB Symp., ESA SP-250, Vol. **II**, p. 17

Vaisberg, O.L., Smirnov, V.N., Gorn, L.S., Iovlev, M.V., Balikchin, M.A., Klimov, S.I., Savin, S.P., Shapiro, V.D., Shevchenko, V.I.: 1986b, *Nature* **321**, 274

Astron. Astrophys. 187, 774–778 (1987)

ASTRONOMY AND ASTROPHYSICS

An attempt to evaluate the structure of cometary dust particles

V.N. Smirnov [1], **O.L. Vaisberg** [1] **and S. Anisimov** [2]

[1] Space Research Institute, USSR Academy of Sciences, Profsoyuznaya 84/32, SU-117810 Moscow GSP-7, USSR
[2] Landau Institute of Theoretical Physics, Moscow, USSR

Received February 4, accepted June 19, 1987

Summary. A comparison of the counting rates of two sensors of the SP-1 plasma impact detector, one open and one shielded by a thin foil is used to roughly estimate the mean density of dust particles. Some inconsistencies between the current understanding of the penetration of film by particles and the observed behaviour are found. A mass density of $\lesssim 1\,\mathrm{g\,cm^{-3}}$ does not contradict to measurements in the low mass range ($\lesssim 10^{-14}$ g). A lower density value is suggested for heavier particles ($m \gtrsim 10^{-12}$ g), implying a fluffy structure of these particles.

Key words: cometary dust – dust particle's density

1. Introduction

The knowledge of the physical properties of dust particles including their density is essential in understanding the cometary nucleus. The studies of micrometeoroids believed to be the remnants of cometary particles have given densities between 0.01 and $1\,\mathrm{g\,cm^{-3}}$ (Fechtig, 1981; Brownlee, 1978).

Thin films were used previously to study interplanetary dust (Grün et al., 1980), as the thickness of the film that can be penetrated by fast grains depends on the mass and on the density of the particles. McDonnell et al. (1981) suggested the DIDSY dust experiment for the Giotto mission. One of the DIDSY detectors was used as a prototype of the SP-1 plasma impact detector on the Vega spacecraft (Gorn et al., 1985; Vega Project Description, 1985).

2. The experiment

SP-1 consists of two identical detectors. Each one consists of a thick (0.1 mm) golden target with negatively and positively biased collector systems suspended above the target. The charge pulse of the plasma cloud originating at the impact of grains on the target is used to measure the particles' mass. Six decade counters and one semidecade counter accumulated amplitude-separated pulses during 2 s. Conversion of the amplitude range to a mass-range was made with a charge-to-mass conversion factor of 10^3 Coulombs per gram (Grün et al., 1984) for an impact velocity of $\sim 80\,\mathrm{km\,s^{-1}}$. The transformation of the SP-1 channel number to mass range is made with

$$\Delta m_n = (10^{-9-n} \div 10^{-10-n})\,\mathrm{g} \qquad (1)$$

for $n = 1, 2, \dots 6$ and $\Delta m_7 = (10^{-16} \div 3\;10^{-17})$ g. One of two detectors of SP-1 was shielded by a foil made of lavsan, $(OCC_6H_5OOCH_2CH_2O) \cdot N$, with a density of $\sim 1.4\,\mathrm{g\,cm^{-3}}$. The foil thickness was chosen as 0.6 μm for the detector on Vega-1, and was equal to 2.0 μm on Vega-2. The conversion factor for the shielded detector may be different from (1) due to differences in the processes during foil penetration by a projectile as compared to the impact on a semi-infinite target.

3. The measurements

Profiles of Vega-1 dust counting rates for two wide mass ranges are shown in Fig. 1 along with the ratios of the counting rates observed with the unshielded detector and the shielded one (0.6 μm foil). Representation chosen was influenced by minor fault of the SP-1 electronics on Vega-1 leading to a summing of the counting rates of channels 7 and 5, 6 and 4, and 3 and 1. We assumed that the mass spectra of dust and their change with cometocentric distance were the same for the Vega-1 fly-by as measured during the Vega-2 fly-by. Subsequent analysis of Vega-1 data (that consists mostly of combined rates of pairs of mass-channels) was made using SP-1 measurements on Vega-2 (Vaisberg et al., 1986, 1987) that provided the following average ratios of count rates: $N_7/N_5 \sim 2$, approaching 1 closer to the nucleus, $N_6/N_4 \sim 10$, dropping to 3 closer to nucleus, and $N_3/N_1 \sim 0.8$ rising to 5 closer to the nucleus. The dust profiles and ratios of counting rates displayed in Fig. 1 show the existence of several regions. Counting rates beyond the most remote dust boundary ($R \sim 100{,}000$ km on the inbound trajectory) are too low, and this region will be not considered here. Within the outer dust coma (100,000–50,000 km) the ratio of the counts from the $(7+5)$ channels approximately corresponds to a complete rejection of particles in the 7th channel ($3\;10^{-17} - 10^{-16}$ g) by the foil. Channel 5 ($10^{-15} - 10^{-11}$ g) seems also partially suffering the influence of the foil. The ratio of the counting rates of channels $(6+4)$ in the same region also corresponds to complete rejection of particles in the 6th channel ($10^{-16} - 10^{-15}$ g) by the foil. At smaller radial distances ($\lesssim 40{,}000$ km) the ratios in the smaller masses' channels decrease monotonically down to $R \sim 10{,}000$ km.

In mass channels 1 to 3 ($10^{-13} - 10^{-10}$ g) the ratios are higher than 1 at $R \sim 70{,}000$–$40{,}000$ km and quite close to 1 at smaller distances.

Measurements with the 2 μm foil on Vega-2 differ drastically from the Vega-1 data. Vega-2 measurements are shown in Fig. 2

Send offprint requests to: O.L. Vaisberg

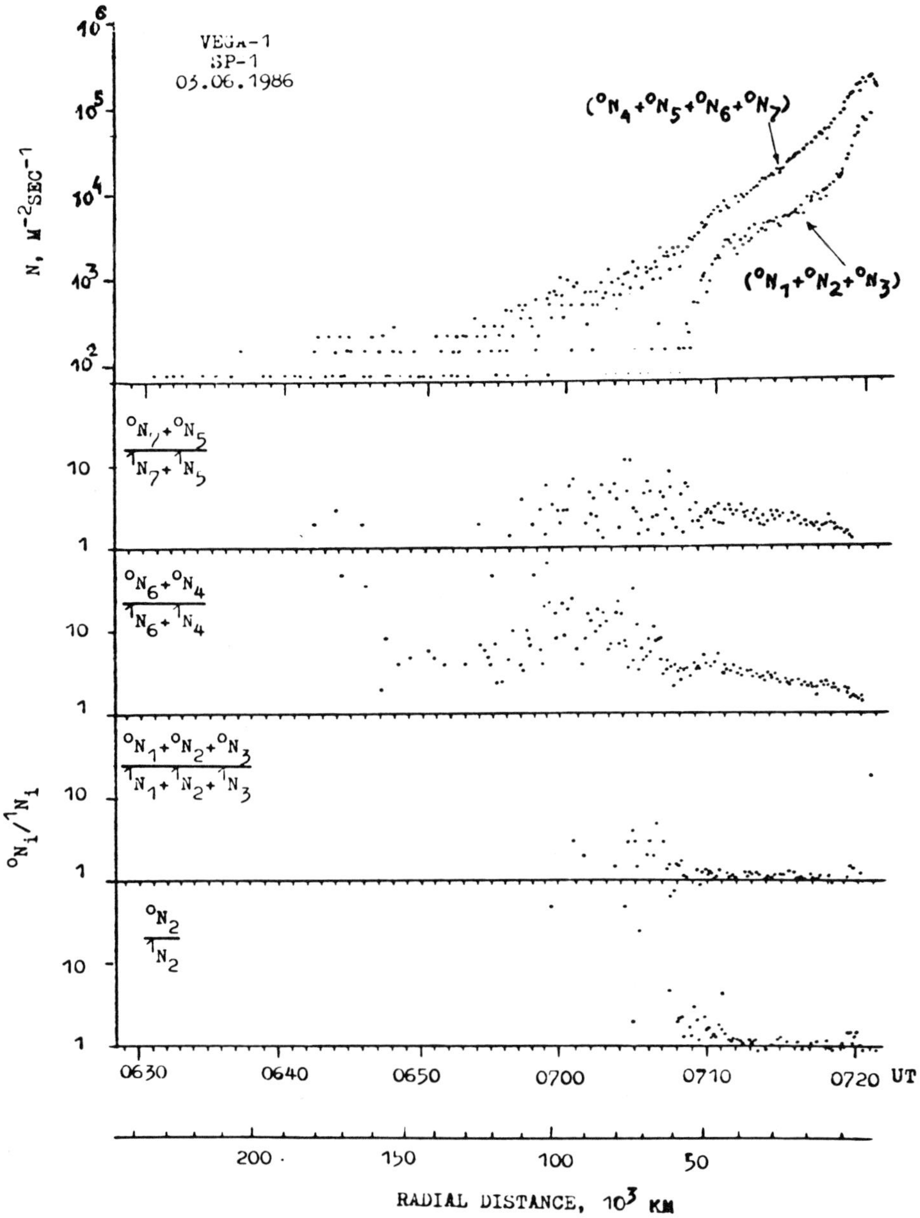

Fig. 1. Profiles of the Vega-1 dust counting rates for the mass ranges 3 10^{-17}–10^{-13} g and 10^{-13}–10^{-10} g (upper panel). Below: ratios of the counting rates of the open detector and those of the shielded detector in different mass channels. In $^{k}N_{n}$ the index $k=0$ stands for the open detector and $k=1$ stands for the shielded detector. n is the channel number. All data are averaged for 10 s. Note the logarithmic scale for the counting rate ratios

with dust profiles in similar form as in Fig. 1 for Vega-1. The sum of the counting rates channels 5, 6 and 7 of the detector shielded by a 2.0 μm foil is shown in addition.

What is evident is that the values of the sum of the counting rates and the dust profiles of channels $5+6+7$ of the shielded detector and channels $1+2+3$ of the unshielded detector are nearly identical. At the same time the counting rate of the shielded detector and its profile in the low-mass (low charge amplitude) range are strongly different from ones in the low-mass range of the unshielded detector. This is more clearly seen in the ratios (two lower panels on Fig. 2). Ratios of counting rates of channels 1 and 2 of the unshielded detector and channels 6 and 7 of the shielded detector, respectively, are close to 1 in the entire dust envelope where the counting rate could be measured. Countrary to that, the counting rates in the same channels (by number) of the shielded detector and the unshielded one are very different (by about order of magnitude) and change throughout the coma.

4. Identification of mass range of the shielded detector

The similarity of the counting rates and dust profiles of channels 6 and 7 of the detector shielded by a 2.0 μm foil and the counting rates and profiles of channels 1 and 2 of the unshielded detector, respectively, suggests that the "low-mass" channels of the shielded detector are recording much heavier dust particles.

One more test of that was made when a dust jet was crossed by Vega-2 (Vaisberg et al., 1987). This mass-dispersed jet was recorded in several channels of the open detector and of the shielded one (Fig. 3). It is easily seen that the time profiles both for the jet and for the background are very close in pairs of SP-1 channels: (7 shielded – 2 open) and (6 shielded – 1 open). A more precise correlation of the channels of the shielded detector to the mass scale of the open detector was made with the use of the spatial dispersion curve (Fig. 3c). The center of the jet in the different channels of the shielded detector is located on the mass

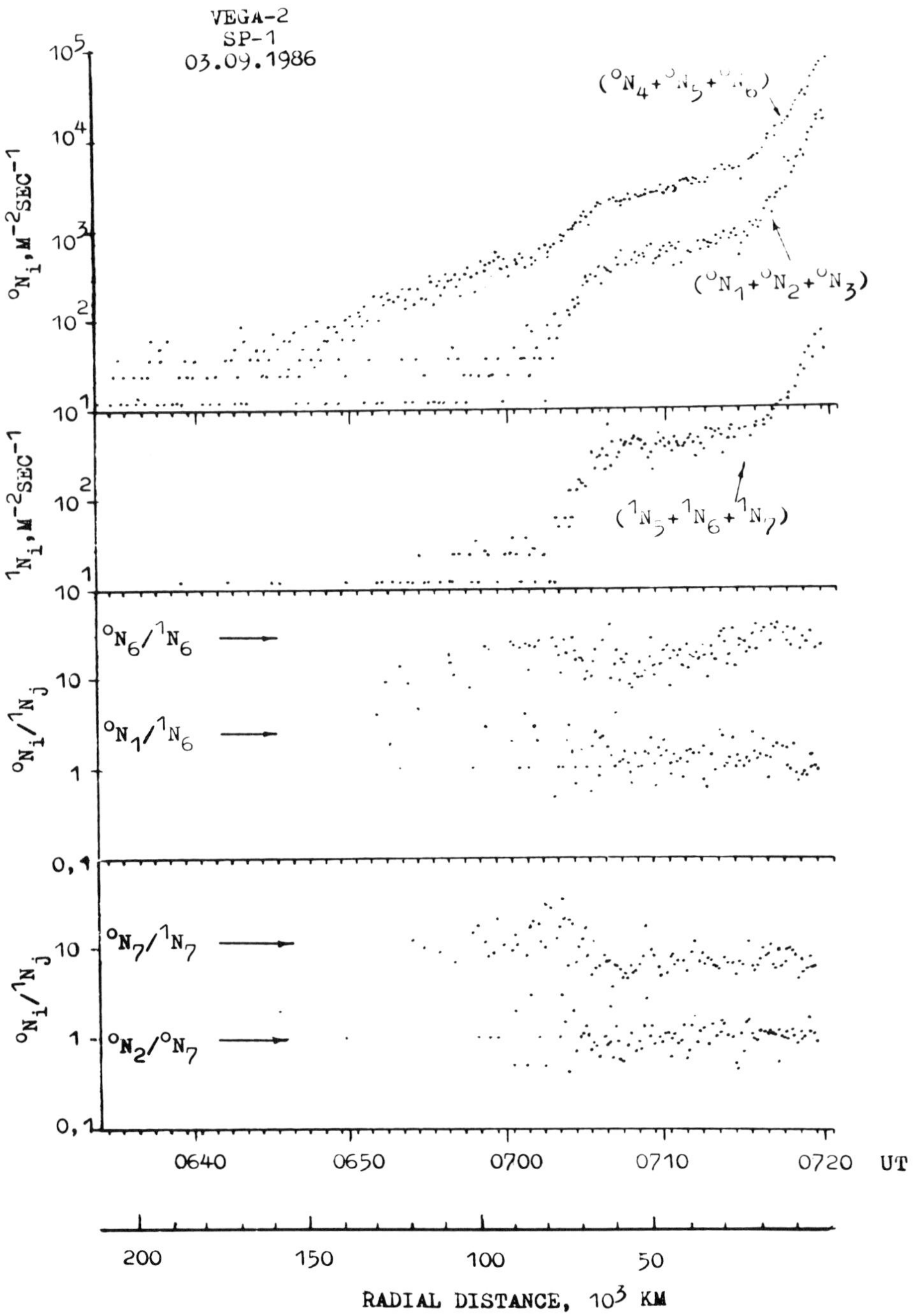

Fig. 2. Profiles of the Vega-2 dust counting rates for the mass ranges 10^{-16}–10^{-13} g and 10^{-13}–10^{-10} g (upper panel), and profile of the counting rate of the shielded (2.0 μm thick foil) detector in channels 5 to 7 (second panel from above). Two lower panels: ratios of the counting rates from the open and the shielded detectors. For the explanation of the symbols see caption of Fig. 1

scale of the open detector in the following way: 7th – between 1st and 2nd mass ranges, 6th – beyond the 1st mass range, and 5th – at significantly higher masses outside the SP-1 mass range.

The counting rates in the other channels of the shielded detector of SP-1 on Vega-2 are very low. In both the 1st and 4th channel only one event per channel was registered during fly-by. Several tens of events were recorded in channels 2 and 3. Fluence spectra for the shielded and open detectors are shown in Fig. 4. It appears that the "spectrum" measured with the shielded detector consists of two parts. The "low-mass" range of it was discussed above. The meaning of the "high-mass" range will be considered in the next paragraph.

5. Discussion

Conventionally, the expression for the thickness of the foil marginally penetrated by a projectile is (Pailer and Grün, 1980):

$$T = c \cdot m^{0.4} \cdot V^{0.88} \cdot \varrho^{0.33}, \tag{2}$$

where T is the thickness in cm, c is a coefficient ~ 1.5, m, V, and ϱ are the mass, velocity and density of dust particle. This formula was obtained with different projectiles but for a relatively limited range of velocities. According to (2) the limiting mass of a particle with density $\sim 1\,\text{g cm}^{-3}$ penetrating a 0.6 μm foil is $\sim 7\;10^{-16}$ g and one penetrating a 2.0 μm foil is $\sim 2\;10^{-14}$ g for the Vega

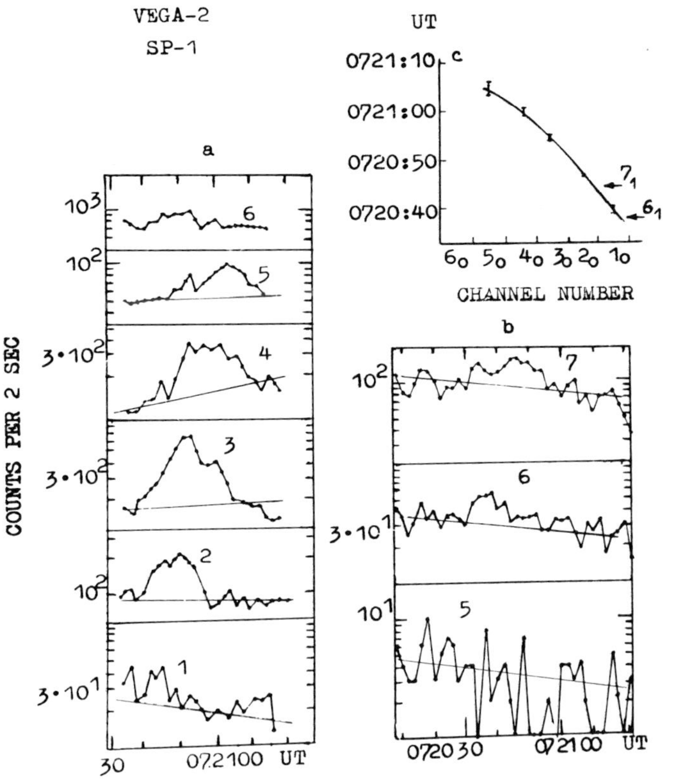

Fig. 3a–c. Counting rate profiles in 6 channels of the open detector (**a**) and in 3 channels of the detector shielded by a 2.0 μm thick foil (**b**) of SP-1 on Vega-2 are shown for the part of the trajectory where a dispersive jet was crossed. The background is approximated by a straight line. The spatial dispersion curve (**c**) is drawn through the maxima of the jet as observed in different channels of the open detector. The locations of the jet as recorded by the shielded detector are indicated by numbered arrows

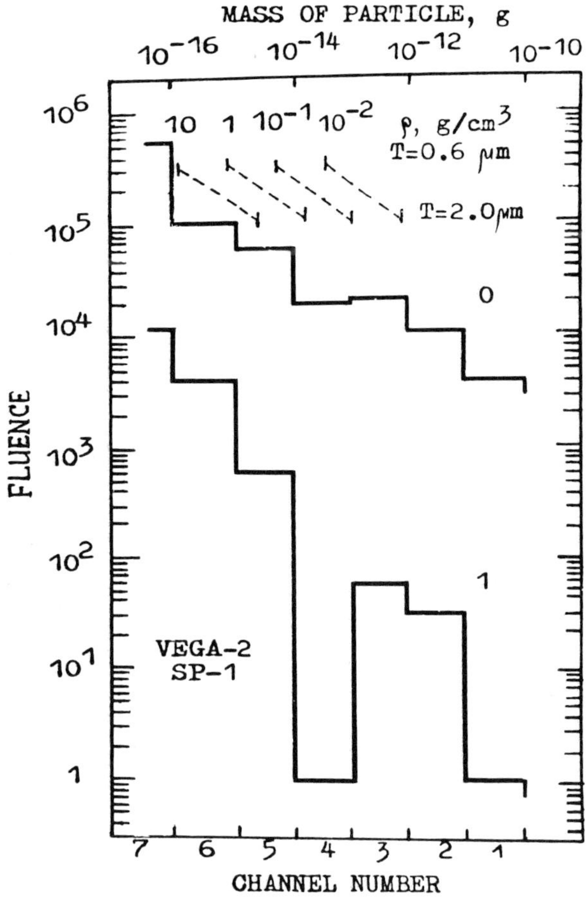

Fig. 4. Fluence spectra from the open detector (0) and the shielded detector (1) of SP-1 on Vega-2. The number of counts in channel 7 of the open detector was influenced by overloading near CA and should be corrected

encounter velocity ($\sim 80\ \mathrm{km\ s^{-1}}$). The validity of (2) was tested in the configuration of the DIDSY detector, which is very similar to the SP-1 detector, with iron projectiles and proved to be correct at least to a velocity $16\ \mathrm{km\ s^{-1}}$ (Grün et al., 1984). It was stated that particles with a mass above the penetration limit (2) give healthy pulses. Penetration limits for two foils and for projectiles of different densities calculated from (2) are shown in Fig. 4.

Measurements performed on Vega-1 with the 0.6 μm foil seem not to contradict the penetration limit (2). Channels 5, 6, 7 will suffer from the influence of the 0.6 μm foil, as measurements show, if the density is ~ 0.1–$0.3\ \mathrm{g\ cm^{-3}}$. The decrease of the ratios $({}^0N_7 + {}^0N_5)/({}^1N_7 + {}^1N_5)$ and $({}^0N_6 + {}^0N_4)/({}^1N_6 + {}^1N_4)$ with decreasing R at $R \lesssim 40{,}000$ km may reflect the increase of the number of denser particles in the mass range $m \lesssim 10^{-14}$ g.

The influence of the thin foil on the measurements on Vega-2 in channels 1, 2, 3 (Figs. 2 and 4) is not easy to understand. Because the exponents of m and ϱ in (2) are close to each other, the drop of ϱ to $\sim 0.01\ \mathrm{g\ cm^3}$ will move the penetration limit (2) inside the 4th channel, and thus will not change the ratios $({}^0N_1 + {}^0N_2 + {}^0N_3)/({}^1N_1 + {}^1N_2 + {}^1N_3)$ and ${}^0N_2/{}^1N_2$, contrary to observations (Fig. 1). It may suggest that formula (2) is not applicable to particles in the mass range $m \gtrsim 10^{-13}$ g. That is supported by measurements with the 2.0 μm foil. The observed behaviour contradicts any penetration theory and experiment, as the largest projectiles should show the least effect after penetration of a thin film.

It was shown above that measurements in channels 5, 6, and 7 of the SP-1 shielded detector on Vega-2 (2.0 μm foil) correspond to particles with masses about 10^5 times heavier. The influence of dead-time effects on the instrument response should be ruled out as the dead-time of SP-1 is 10 μs and the maximum counting rate was $\sim 10^3\ \mathrm{s^{-1}}$ (in the lower mass channels). Besides that the influence of the 2 μm foil was the same even at lower counting rates.

Calibrations of the electronics were repeated during the flight of the Vega space crafts with current pulses of known amplitudes and one performed within 1 day after the Vega-2 fly-by does not reveal any evidence for a change of thresholds of SP-1. So the change of the mass scale of the shielded detector can result from the change of the ionization efficiency and/or the shape of the charge pulse for foil penetration compared to the processes studied with solid targets. Laboratory calibration of electronics was made with current pulses with ~ 0.1 μs risetime and ~ 0.3 μs decay time. The conversion factor was taken equal to 10^3 Coulombs per gram. The product of two factors, namely the ionization efficiency and the reciprocal pulse duration, determine the change of the mass scale. Without detailed models and/or laboratory modelling with particles of different structure it is difficult to decide what is the relative contribution of these two factors.

In any case, the measurements using a 2.0 μm foil contradict the penetration limit (2). The only reasonable explanation which we can suggest now is that particles registered in channels 1, 2, and 3 of the open detector have very low density or specific structure for which formula (2) is not applicable. If cometary particles are made in a complicated manner as suggested by Greenberg (1982) or if they are similar to Brownlee particles (Brownlee, 1978), it is possible that penetration processes are different for this kind of projectiles. It is possible that these particles will be destroyed by the foil and will give very low recorded charge amplitudes. Particles registered in channels 2 and 3 of the detector shielded by the 2.0 μm thick foil (Fig. 4) should be either yet heavier particles or particles in the proper mass range but with larger densities. A decrease of the particles' density with mass would be in general agreement with models (Divine, 1981).

The question remains whether the change of the count rates of the unshielded and the shielded detectors with time (or distance) may be caused by a damage of the foil by dust or gas. The upper limit on the total area of holes is $\sim 0.4\,\mathrm{cm}^2$ while the total area of the detector is 81 cm^2. The limit is obtained by assuming that the diameter of the holes corresponds to 10 times the diameter of the projectiles and that the fluence mass spectrum F (Fig. 4) continues in the mass range $10^{-10}\,\mathrm{g} < m < 10^{-5}\,\mathrm{g}$ with $d\log F/d\log m = -0.46$. (The slope of the fluence spectrum for channels 1 and 2.) The total number of molecules hitting the foil during the fly-by is $\pi n_0 R_0 = 4\ 10^{15}\,\mathrm{cm}^{-2}$ (with $n_0 R_0^2 = 10^{24}\,\mathrm{cm}^{-1}$ and $R_0 \approx 8\ 10^8$ cm) or about 10^{-7} g for water molecules. This is a negligible mass even for a foil with an area density of 0.6 μm.

6. Conclusion

In order to get a better understanding of the measurements performed with foils it is necessary to make a detailed analysis of the processes occurring during the penetration of the foil by different particles. Measurements with the SP-1 detector on the Vega spacecraft suggest that:

– cometary particles in the mass range below $\sim 10^{-14}$ g have a density of ~ 0.1–$0.3\,\mathrm{g\,cm}^{-3}$ in the outer coma and closer to $\sim 1\,\mathrm{g\,cm}^{-3}$ in the inner coma;

– cometary particles in the mass range $\gtrsim 10^{-12}$ g are predominantly particles of very low density or of complicated structure.

Acknowledgements. The authors are indebted to R.Z. Sagdeev, A.A. Galeev, V.D. Shapiro, and V.C. Lichtenstein for useful discussions and to A.M. Kuznetsova and R.A. Isaeva for help in the preparation of this paper.

References

Brownlee, D.E.: 1978, in *Cosmic Dust*, ed. J.A.M. McDonnell, Wiley, New York, p. 295

Divine, N.: 1981, *The Comet Halley Dust and Gas Environment*, ESA SP-**174**, p. 25

Fechtig, H.: 1981, *Proc. Intern. Meeting on the Giotto Mission*, ESA SP-**169**, p. 47

Gorn, L.S., Iovlev, M.V., Shifrin, A.V., Vaisberg, O.L., Smirnov, V.N.: 1987, *Problems of Nuclear Science and Techniques*, Atomizdat, in press (in russian)

Greenberg, J.M.: 1982, *Proc. Intern. Conf. Cometary Exploration*, ed. T.I. Gombosi, p. 23

Gringauz, K.I., Gombosi, T.I., Remizov, A.P., Apáthy, I., Szemerey, I., Verigin, M.I., Denchikova, L.I., Dyachkov, A.V., Keppler, E., Klimenko, I.N., Richter, A.K., Somogyi, A.J., Szegö, K., Szendrö, S., Tátrallyay, M., Varga, A., Vladimirova, G.A.: 1986, *Nature* **321**, 282

Grün, E., Pailer, M., Fechtig, H., Kissel, J.: 1980, *Planet. Space Sci.* **28**, 333

Grün, E., Bussoletti, E., Minafra, A., Kuczera, H., McDonnell, J.A.M.: 1984, *Adv. Space Res.* **4**, 291

McDonnell, J.A.M., Grün, E., Evans, G.E., Turner, R.F., Firth, J.G., Carey, W.O., Kuczera, H., Alexander, W.M., Clark, D.H., Grard, R.J.L., Hanner, M., Hughes, D.W., Igenbergs, E., Lindblad, B.A., Manderville, J.O., Schwehm, G., Sekanina, Z.: 1981, *Proc. Intern. Meeting on the Giotto Mission*, ESA SP-**169**, p. 61

Pailer, N., Grün, E.: 1980, *Planet. Space Sci.* **28**, 321

Sagdeev, R.Z., Anisimov, S.I., Galeev, A.A., Shapiro, V.D., Shevchenko, V.I.: 1983, *Adv. Space Res.* **2**, 12, 133

Vaisberg, O.L., Smirnov, V.N., Gorn, L.S., Iovlev, M.V., Balikhin, M.A., Klimov, S.I., Savin, S.P., Shapiro, V.D., Shevchenko, V.I.: 1986, *Nature* **321**, 274

Vaisberg, O.L., Smirnov, V.N., Omelchenko, A.: 1987, *Astron. Astrophys.* **187**, 753 (also Vega Project. Description, Moscow, 1985)

Astron. Astrophys. 187, 779–784 (1987)

ASTRONOMY AND ASTROPHYSICS

Systematics of the "CHON" and other light-element particle populations in comet P/Halley

B.C. Clark [1], L.W. Mason [1], and J. Kissel [2]

[1] Planetary Sciences Laboratory-0560, Martin Marietta Denver Aerospace, Denver, CO 80201, USA
[2] Max-Planck-Institut für Kernphysik, PO Box 103980, D-6900 Heidelberg, Federal Republic of Germany

Received February 17, accepted June 2, 1987

Summary. Based on chemical signatures measured by the PIA experiment during the Giotto flyby of comet P/Halley, a number of particle classifications have been designated. In addition to silicate-like grains and particles of mixed (cosmic) composition, there appear to be several light-element rich populations, including the CHON, (H,C), (H,C,O), and (H,C,N) particle types. In some cases, these compositional classes are further distinguished by differences in mass distributions, a density indicator, and variations in relative abundance within the coma. These particle populations are evidence for chemical heterogeneity in the surface of the cometary nucleus. Some particles found mainly in the inner coma may be volatile icy grains. Most of the N of the comet may be found in up to three different populations of grains; one or more of these may be responsible for the observation of cyanojets.

Key words: CHON – cyanojets – comet – Halley – Giotto

1. Introduction

It has long been known that the cometary nucleus is a complex mixture of "ices" and "dust" grains. Water, organics, and silicates had all been inferred from ground-based spectroscopic observations of various comets. From micro-compositional studies of meteorites and dust particles collected in the Earth's stratosphere, it has been possible to speculate about the variety of grain types that may exist in the cometary nucleus and coma (Anders, 1981). Of special interest have been the interplanetary dust particles (IDP) discovered in the stratospheric particle collection (Brownlee, 1978), with the archetypical putative cometary particle being the chondritic porous aggregate, containing elements from C to Fe in proportions not unlike the composition of the primitive solar nebular matter, as inferred from the composition of carbonaceous meteorites and the sun.

Early results from the Giotto PIA instrument and Vega-1 PUMA instrument reported the predominance of particles in the coma of Halley's comet that are similar to, but not necessarily identical with C1 carbonaceous chondrite composition (Kissel, 1986a, b). Subsequent, preliminary analysis has indicated the average composition of these particles, relative to silicon, may be compatible to within a factor of two with this reference composition, except for excess abundances in the elements carbon (Jessberber, 1986) and possibly magnesium (Sagdeev, 1986) and nitrogen (Jessberger, 1986). In this report we shall refer to these particles as being of "mixed" composition, reflecting the presence of significant levels of C, O, Mg, Si, and often S, Ca, Fe, Na, and Al simultaneously in the same particle. In addition to the mixed particles, the rare occurrence of grains of apparent pure silicate composition (Si, O, Mg/Fe) has been noted (Kissel, 1986a, b).

In the present study, interest is focused on particles composed predominantly or solely of one or more of the light elements: hydrogen, carbon, nitrogen and oxygen. As seen in the PIA data, a number of particles exist that appear to contain all four of these elements in major proportions (Kissel, 1986b); we have dubbed this the "CHON" classification. Although this type is the second most abundant particle detected by the PIA during the Giotto flyby, a number of additional particle types containing three or less of these light elements seem also to occur. In this early communication, we report tentative identifications of these additional classes of light-element particles and examine some systematics of their occurrence and nature.

2. Analysis methodology

The PIA instrument, a time-of-flight mass spectrometer for analyzing the elemental composition of individual grains during flythrough of the Halley coma, has been previously described (Kissel, 1986c). To optimize the number of particles that could be analyzed, yet achieve compatibility with limited data transmission, spectra were taken in several different modes during the flyby. Mode 0 spectra, which contain the most detailed information but represent only 3% of the spectra, were analyzed first to examine mass line shapes, understand instrument response, and identify particle types. It should be noted that because of difficulties in reliably accelerating particles to the 68 km s^{-1} velocity of the flyby, precise calibration and extensive study of instrument response has not been possible in the laboratory. In addition to various instrumental factors, the physical process of impact of irregular grains onto a target, which itself may be somewhat micro-heterogeneous, appears to affect the yield of ions from each constituent element in a stochastic manner. No estimates of relative abundances of the elements within each particle classification will be made at this time. It has been empirically determined, however, that the mass peaks are mainly, and perhaps exclusively, a result of singly-charged cations of the elements of interest, and not to molecular or multiply-charged ions.

Send offprint requests to: B. Clark

An interim set of mode 1, 2, and 3 spectra comprise the data upon which the present results are based. Time is estimated from the PIA internal clock, assuming a nominal timing interval of 118 ms. Algorithms have been developed for computer selection of the mass scale for each spectrum and for classifying the major element composition of each particle detection event in the following manner. A search is first made for peaks in the region expected for carbon and silver (target material) ions. If both peaks are found, the mass scale is adjusted accordingly. If only carbon is present, the correction is made based on that carbon time of flight and on the global average location of silver (byte 246 in the PIA time scale). If neither occur, an average time scale, which is byte 21 for carbon and 246 for silver, is assumed. Suitable time windows have been established for each element, with the highest amplitude reading that occurs within this window being selected. Spectra containing significant line intensities in mass regions of dubious geochemical validity (chiefly, amu = 70 to 100) have been eliminated from consideration at this time. These approaches have been developed as a result of manual inspection and classification of over 3000 individual spectra. Only maximum peak heights are extracted from each selected mass range and used in subsequent analyses; conversion to numerical values has been made both with and without preliminary amplifier in-flight performance corrections. Preliminary ion yield factors (same as Jessberger, 1986) have also been applied to allow estimates of the relative mass of each particle.

3. Particle population characteristics

Particles can be classified on the basis of several different criteria, including elemental composition, total signal amplitude (relative mass), presence of target ions in the spectra, location/time of occurrence, and "quality" of the spectrum. To the extent that one or more of these characteristics can be shown to be common to a group of particles, such a group can be termed a population of a given particle type. For the reasons given above, it is emphasized that the following results are preliminary in nature.

3.1. Chemical classification

The mixed, silicate, and CHON classifications have already been noted. There are in principle 15 classes formed from the elements H, C, O, and N when taken 1, 2, 3, or 4 at a time. The CHON class is but one of these. We have searched the data for these classes and found that several do appear to constitute particle populations at a statistically significant level. For example, a type of particle containing the elements hydrogen, carbon, and oxygen has been identified. We refer to such particles as the (H, C, O) classification. This nomenclature of listing of the elements in increasing atomic number, separated by commas, is intended to avoid any inferences of relative elemental stoichiometry of chemical structure as it is premature to attempt such assignments at this time (see Discussion, Sect. 4).

Examples of classifications that have appreciable occurrences include (H,C) and (H,C,N). Rarer cases of (C,O), (H,N), and (H,O) have also been found. A preliminary summary of the classifications made on the interim data set is presented in Table 1. Other important classes discovered, such as (C,O,Mg) and (C,S,Fe), will not be discussed at this time. An example of a light-element dominated, but very rare case is that of (H,C,N,S).

Misclassification is a potential problem. For example, the (H,C,N) could be a subset of the CHON particles, with low amplitude (below detectable signal) for O. Likewise, any given (C,O) could be a case of an (H,C,O) with a missing detection of H. Improper peak assignments can likewise cause misinterpretations. For example, cases of (H,N) could be misidentified as (H,C), and vice-versa, because C and N occupy adjacent mass lines. Further analysis on a case by case basis may be necessary to distinguish between these two possibilities.

Table 1. Particle compositional classification. Summary (preliminary)

Particle class	Number of events detected	
	Exclusive [a]	Total
Mixed	350	103
CHON	115	420
(H,C)	114	150
(H,C,O)	48	111
(H,C,N)	88	127
(H,O)	17	31
(C,O)	10	28

[a] No other elements detected

Composite particles are another possibility. Many cases are known of IDPs with two or more compositionally-distinct grains or regions within a single particle. In our classification, a combination of (H,C) grains and (H,O) grains would produce the (H,C,O) particle type. Likewise, the mixed, CHON, and silicate particle populations could be gradations of a single population of variable organic/inorganic mixing proportions, as suggested by Langevin et al. (1986). On chemical composition grounds, it is presently possible to neither verify nor reject such a hypothesis. It is indeed noteworthy that particles we classify as CHON often contain Na and/or Mg at low levels.

3.2. Particle mass distributions

In principle, the relative mass of each detected particle can be determined by integrating the product of the measured ion signal amplitudes and atomic mass, corrected by the ion yield factor for each atomic species. Neglecting, for the time being, additional possible complicating effects such as particle density and instrumental response, we have calculated integral mass distribution functions for those particle populations with significant event frequencies. These preliminary calculations are shown in Fig. 1. Most noteworthy is the apparent larger average mass for the mixed particles than for CHON or other light-element particles. Also apparent in this figure are the "humps" at the lower end of the CHON and mixed mass distributions, indicating the existance of a few large AWMW (amplitude weighted molecular weight) particles in these populations.

3.3. Particle density

Based upon laboratory experiments, albeit at lower projectile velocities, the occurrence of target ions in the ion mass spectra is characteristic of compact, relatively high density grains. Silver ions from the 90% Pt/10% Ag target foil are detected in many spectra (the instrument is not designed to analyse for Pt), at the frequencies quoted in Table 2. The contrast between particles

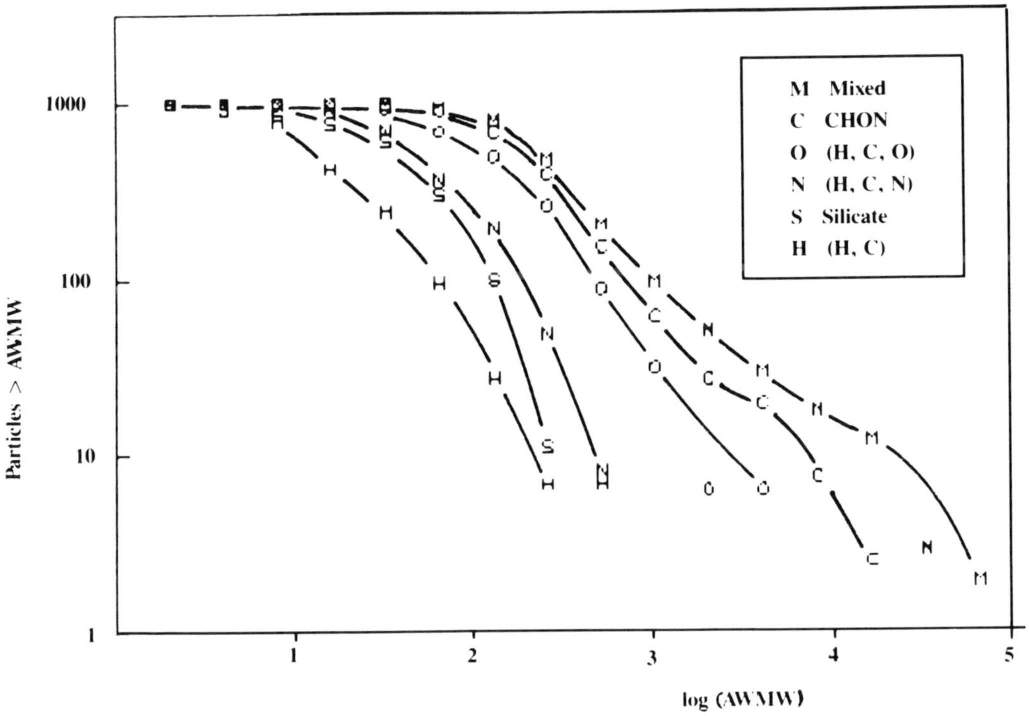

Fig. 1. Integral relative mass distribution functions (preliminary)

Table 2. Target material occurrence (preliminary)

Particle Class	% of spectra with Ag ion
Mixed	62
CHON	12
Silicate	40
All other (H,C,N,O) classes	12–19
[Fe,(H/C/S)]	60

classified as silicates and the particles rich in light elements is quite striking and statistically secure.

3.4. Class relative abundance variations

The Giotto flight path sampled the coma at nucleocentric distances of from 40,000 km to 600 km in a time span of less than 10 minutes. Because of the approach angle and the nature of dust grain trajectories, the sampling of the coma was in both space and time in the sense that particles "ages" (time since release from the surface of the nucleus) ranged from 10^3 to 10^5 s or more. Figure 2 is a plot by class of individual particle occurrences, with the amplitude representing the calculated relative mass of each particle (AWMW). The abcissa is the spectrum number of the particle impact events and is related to sequential sampling through the coma. However, because the instrument automatically buffers and prioritizes spectra before data transmission, this plot is not perfectly time-ordered, but does correctly represent the time sequence of events on the >10 s time scale.

The relative abundances of CHON and mixed particles does not remain constant, even allowing for statistical variations. This is further emphasized in Fig. 3, where the ratio of CHON/(CHON + Mixed) is displayed in 48- and 256-spectra running averages (neglecting all other spectra). The mean value of this ratio is 0.3 over the total measurement path. During this period, however, the ratio varies strongly, approaching its theoretical limits of 0 and 100% several times. There are, therefore, regions within the coma presumably reflecting specific areas of differing dominance on the cometary surface by CHON and mixed particles.

Refering to Fig. 2, the CHON particles exhibit a moderate decrease centered around closest approach, complemented by an increase in (H,C,O) particles. The latter group clusters in the "inner coma" region, commonly taken as 10^4 km from the nucleus. Grouping of the (H,O) and (C,O) particles types is also noted in this region (possibly a testament to relatedness of these three classes).

The (H,C,N) particles are detected predominantly on the inbound portion of the trajectory, with some significant reoccurrence late in the outbound leg. The (H,C) group also are more populous in the portion of the coma sampled first, and occur with dramatically decreased frequency after closest approach.

4. Discussion

Comets have long been known to exhibit dynamic variability in their coma and tail structures. The Vega and Giotto direct images of the nucleus show it to be topographically irregular and heterogeneous in both the surface active sites and the direction of their emissions. Combined with the knowledge that the coma contains gases, ions, and radicals of considerable chemical variety, these results point to the possibility of chemical heterogeneity in the grains as well. Study of meteorites and IDPs (Brownlee, 1978) also leads to such expectations. The ubiquitous "mixed" particles found in the coma could, when combined with the ices, have

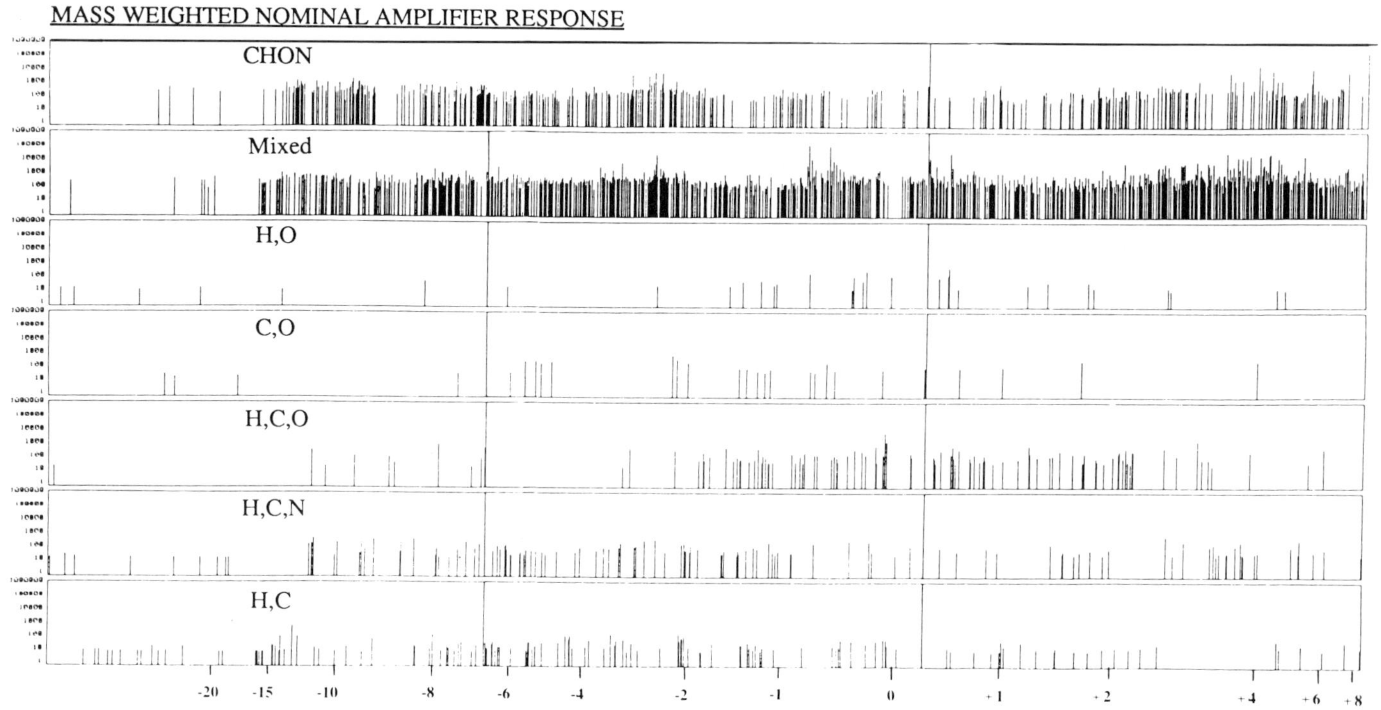

Fig. 2. Particle occurrances plotted as a function of sequential data record number. Approximate time (min) from closest approach is indicated by tic marks on the abscissa. The ordinate shows AWMW on a logarithmic scale spanning six orders of magnitude

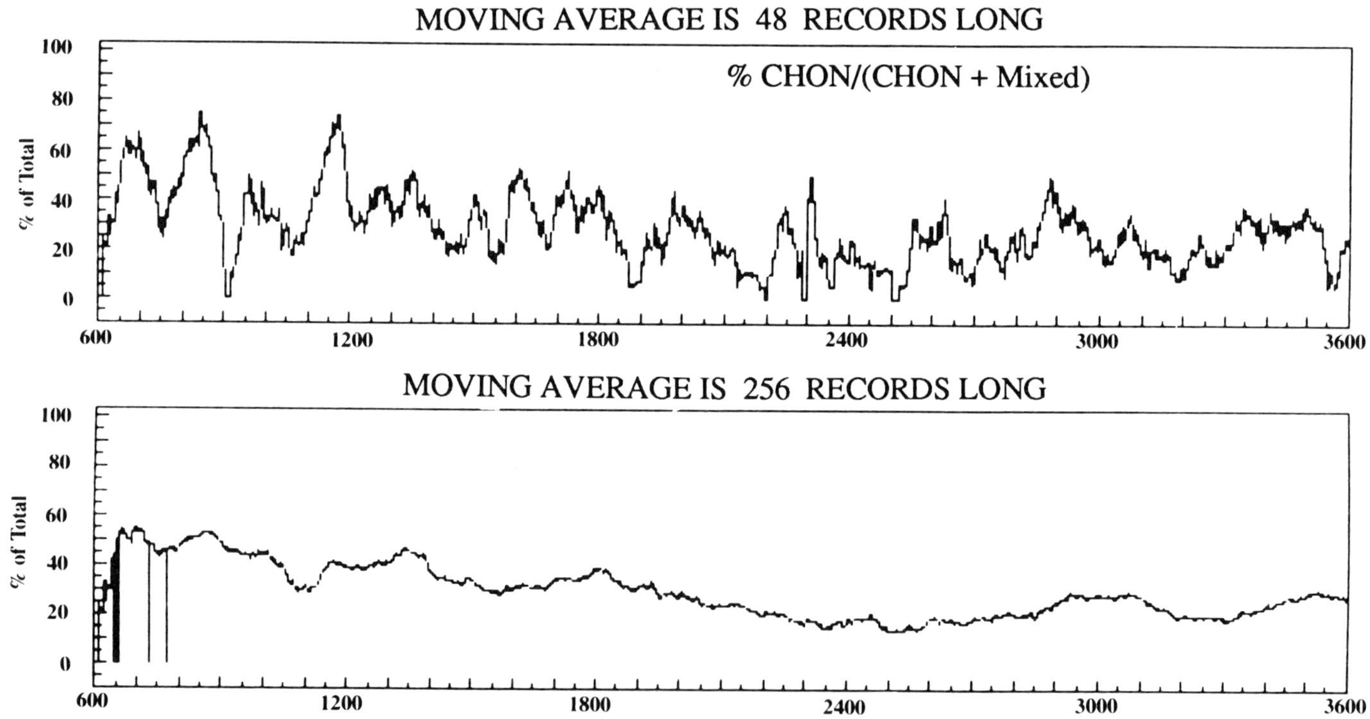

Fig. 3. Ratio of CHON/(CHON + Mixed) particles plotted as a function of sequential data record

allowed the bulk comet to correspond to the average "cosmic" abundances of the elements. Concepts of the cometary nucleus as an agglomerate of many individual cometesimals also permit chemical heterogeneity, especially if the cometestimals themselves or subunits therein originated from different formation locations within the solar nebula, analogous to the mixing of condensation products observed in certain meteorites (Boynton, 1978).

The data presented in Sect. 3 show there may be several classes of particles of distinctive chemical composition, density attributes, and mass and size distribution functions. The fact that the particles so classified occur, in several cases, mainly in specific regions is further evidence that at least some of these classes are unique. It is also evidence of the chemical heterogeneity of the surface of the nucleus. During the Giotto flyby, the spacecraft sampled particles in both direct and indirect trajectories from the surface. The subspacecraft point swept across the sunward face, the source of direct particles, in a matter of ± 50 s (Sekanina, personal communication 1986) around closest approach. Further analysis will have to be done to determine loci on the surface of the nucleus that contributed the particles measured.

The search for evidence of icy grains in cometary comae has been a long and continuing endeavor, especially by Hanner and coworkers (Hanner, 1986; Sekanina, 1986). The (H, C, O) class of particles definitely occur with much greater frequency inside the inner coma, and are symmetrically distributed around the point of closest approach, allowing speculation of this as a candidate class for icy grains. If this clustering resulted from gradual particle volatilization, then it might be expected that both the event frequency and relative mass of such particles would continously increase as the spacecraft approached the nucleus. The results are not clear on this point, although the statistics and heterogeneities in surface activity could be masking this expectation. It is clear from these data, however, that the CHON particles follow none of the expected patterns for highly volatile constituents.

In view of the parent ices of H_2O, CO, and CO_2, icy grains containing O, plus H and/or C, would be possible. Detection of apparent particles of (H, O) and (C, O), also mostly occurring in the inner coma, is noteworthy in this regard. On the other hand, the (H, C, O) group could just as easily be representative of a less-volatile organic compound, such as formaldehyde or polymers thereof. Formaldehyde in the gas phase has indeed been reported in the coma of Halley by the Vega IKS experiment (Combes, 1986).

Spectroscopic mapping of the coma has revealed evidence of sporadic activity with chemical signatures, such as the CO_2^+ emission accompanying a dust outburst, observed by Feldman et al. (1986). The dynamics of this ion, possibly derived from CO_2 grains, and the "puzzle" of the distribution of atomic carbon in the inner coma (Woods, 1986) may both be related to the existence of (C, O) and/or (H, C, O) grains. Likewise, the high abundance of CH inferred from in situ measurements is surprising in view of its short lifetime (Festou, 1986), and could derive from one or more of the particle classes discussed here. Polycyclic aromatic hydrocarbons (PAH) and quenched carbonaceous composites (QCC) have been suggested as constituents of interstellar molecules or grains (Sakata, 1984; Allamandola, 1985), and either may easily relate to particle classes discussed here.

Of considerable interest are the CN jets discovered by A'Hearn et al. (1986), which they interpret as evidence of grains that slowly produce the CN radical. This could be result from sublimation or outgassing of molecular species such as HCN, HC_3N, and CH_3CN. The first of these has been detected in Halley's coma and only relatively high upper limits have been established for the latter (Brockelée-Morvan, 1986). More likely, argue A'Hearn et al. (1986), the observed CN radical is released directly from the grain surface presumably by photodesorption or energetic ion sputtering. For example, Matthews and Ludicky (1986) have predicted the probable presence of a black $(HCN)_n$ polymer produced by radiolysis of frozen volatiles and subsequent condensed-phase polymerization of HCN. The minority classes corresponding to this are the CHON and (H, C, N) particles. Both exhibit structure within the coma, but at different locations (Fig. 2), opening the possibility that correlations with other observational data may help decide whether either or both are responsible for the patterns observed.

It should be pointed out that in addition to the CHON and (H, C, N) classes, the mixed particles contain the element nitrogen. Particulates thus carry significant quantities of N and may compensate for the observed low abundance of N (N/O is an order of magnitude below cosmic) in the gas phase (Balsiger, 1986). Further data analysis will be necessary to quantitatively estimate the partitioning of N among these three classes, although the CHON particles definitely contain much, if not the majority, of the cometary nitrogen.

5. Summary and conclusion

Different particle types (alternatively, major fluctuations toward various end-members in a continuous series of particles) are present in Halley's coma. These particles can be distinguished, to first order, by their chemical composition as measured in their constitutent elements. They also form classes that are distinct with respect to their mass distributions and competence in producing target material ionization during hypervelocity impact. The relative abundance of particle types is a function of location within the coma. Certain classes, notably that of the (H, C, O) particles, are found at much lower relative frequency outside the inner coma, possibly as a result of destruction by sublimation of the particles during their outward movement. Other changes in particle relative frequency appear to be evidence for significant compositional heterogeneity in the particle source regions, a fact that could have major implications for theories of the origin of comets, was well as the sampling strategies for a comet nucleus sample return mission.

Acknowledgements. This work was supported, in part, by NASA under contract JPL-956214. We wish to thank P. Demuth for processing and providing the interim data tapes. We also thank M. F. A'Hearn for access to recent results prior to publication.

References

A'Hearn, M. F., Hoban, S., Birch, P. V., Bowers, C., Martin, R., Klinglesmith D. A.: 1986, *Nature* **324**, 649

Allamandola, L. J., Tielens, G. G. M., Barker, J. R.: 1985, *Astrophys. J.* **290**, L25

Anders, E.: 1981, *Meteoritics* **16**, 289

Balsiger, H. et al.: 1986, *Nature* **321**, 330

Bockelee-Morvan, D. et al.: 1986, 20th ESLAB Symp., ESA SP-250, Vol. **I**, p. 365

Boynton, W. V.: 1978, in *Protostars and Planets*, ed. T. Gehrels, Univ. Arizona Press, p. 427

Brownlee, D. E.: 1978, in *Protostars and Planets*, ed. T. Gehrels, Univ. Arizona Press, p. 134

Combes, M. et al.: 1986, 20th ESLAB Symp., ESA SP-250, Vol. **I**, p. 353

Feldman, P. D., A'Hearn, M. F., Festou, M. C., McFadden, L. A., Weaver, H. A., Woods, T. N.: 1986, *Nature* **324**, 433

Festou, M. C.: 1986, 20th ESLAB Symp., SP-250, Vol. **I**, p. 55

Hanner, M. S.: 1981, *Icarus* **47**, 342

Jessberger, E. K., Kissel, J., Fechtig, H., Krueger, F. R.: 1986, in *Proc. Workshop on Comet Nucleus Sample Return*, ESA SP-249, p. 27

Kissel, J. et al.: 1986a, *Nature* **321**, 280
Kissel, J. et al.: 1986b, *Nature* **321**, 336
Kissel, J.: 1986c, ESA SP **1077**, 67
Langevin, Y., Kissel, J., Bertaux, J.L.: 1986, 20th ESLAB Symp. (Abstracts), Heidelberg
Matthews, C.N., Ludicky, R.: 1986, 20th ESLAB Symp., ESA SP-250, Vol. **II**, p. 273
Sagdeev, R.Z. et al.: 1986, 20th ESLAB Symp., ESA SP-250, Vol. **III**, p. 349
Sakata, A., Wada, S., Tanabe, T., Onaka, T.: 1984, *Astrophys. J.* **287**, L51
Sekanina, Z.: 1986, 20th ESLAB Symp., ESA SP-250, Vol. **II**, p. 131
Woods, T.N., Feldman, P.D., Dymond, K.: 1986, 20th ESLAB Symp., ESA SP-250, Vol. **I**, p. 431

Astron. Astrophys. 187, 785–788 (1987)

ASTRONOMY AND ASTROPHYSICS

Secondary electron emission induced by gas and dust impacts on Giotto, Vega-1 and Vega-2 in the environment of comet P/Halley

R.J.L. Grard [1], J.A.M. McDonnell [2], E. Grün [3], and K.I. Gringauz [4]

[1] Space Science Department, ESA/ESTEC, Keplerlaan, NL-2200 AG Noordwijk, The Netherlands
[2] Unit for Space Sciences, University of Kent at Canterbury, Canterbury CT2 7NR, United Kingdom
[3] Max-Planck-Institut für Kernphysik, Postfach 103980, D-6900 Heidelberg, Federal Republic of Germany
[4] Space Research Institute, Profsoyuznaya 84/32, SU-117810 Moscow GSP-7, USSR

Received February 18, accepted April 30, 1987

Summary. Giotto, Vega-1 and Vega-2 have been bombarded by a flow of gas and dust particles during their flyby of comet Halley. The emission of secondary electrons and sputtered ions caused by the large impact velocities (70–80 km s^{-1}) perturbed the plasma density in the spacecraft vicinity and was a possible source of interference for electric field and plasma measurements. Identical impact plasma detectors were mounted on the three space probes; the saturation currents of secondary electrons emitted from gold targets were measured once per second. The results obtained during the three flybys are presented and compared. Information about the gas density profile and nucleus gas production rate can be derived from the measurements.

Key words: Giotto – Vega – comet Halley – cometary atmosphere – dust impacts – secondary electron emission – spacecraft charging

1. Introduction

The interaction between a spacecraft and a cometary environment during a high relative velocity flyby has been the subject of numerous studies and review papers prior to the encounters with comet Halley in March 1986 (Grard, 1982, 1983; Grard et al., 1981; Reinhard, 1981, 1984; Pedersen, 1983; Young, 1983, 1986).

In short, the impact of molecules and dust particles causes the emission of sputtered ions and secondary electrons from the fore section of the spacecraft surface. This mechanism is responsible for the existence of an artificial plasma cloud around the space probe and the possible perturbation of the surface potential during the flyby.

This situation can be simulated numerically (Parker and Holeman, 1981; Thieman et al., 1986a, b) but requires input information about the flux and energy of the emitted charged species. Some of these parameters can be obtained from laboratory measurements (Arends and Schmidt, 1983; Schmidt and Arends, 1984, 1985) but it seemed nevertheless essential to measure in-situ the most important quantity, namely the secondary electron flux. Identical detectors were therefore flown on Giotto, Vega-1 and Vega-2. The results obtained from the Vega missions have already been presented elsewhere (Grard et al., 1986a; Grard and Gringauz, 1986); this paper focuses more on the Giotto data and the consistency of the three sets of measurements.

The Giotto instrument was mounted behind a protective cover which was intended to protect another instrument from contamination during the operation of the orbit insertion motor, but it became subsequently doubtful whether the cover opening command has been properly executed. The analysis of the impact plasma detector data brings a crucial contribution to this controversial issue.

2. The instrument

The sensor consists of a circular target surrounded by a guard ring and polarized at a potential of −17 V with respect to the spacecraft structure. The probe is planar, gold plated and mounted in a plane perpendicular to the flow direction (Fig. 1). The saturation electron current emitted from the central electrode is measured once per second with a nominal 8 bit resolution and a time constant of the order of 1 s. More detail about the instrument is given elsewhere (Grard, 1984; Grard et al., 1986b; McDonnell et al., 1986).

The Giotto sensor is protected at launch with a cover made of a thermal blanket which consists of three 7 μm thick insulating layers. This cover is loaded with three flat springs and is rolled up to one side of the aperture when released.

3. The measurements

3.1. The Vega results

The measurements performed during the flybys of Vega-1 and Vega-2 are shown in Figs. 2 and 3, respectively. Time is measured relative to closest approach; the scales for current and time are identical on both plots.

The Vega-1 and Vega-2 profiles are similar, but the latter is incomplete because of an experimental anomaly which interrupted the data flow for nearly 27 minutes.

Limited resolution is responsible for the step-like discontinuities observed in both figures. The current consists of two components: a background signal and a series of superimposed discrete spikes.

Send offprint requests to: R.J.L. Grard

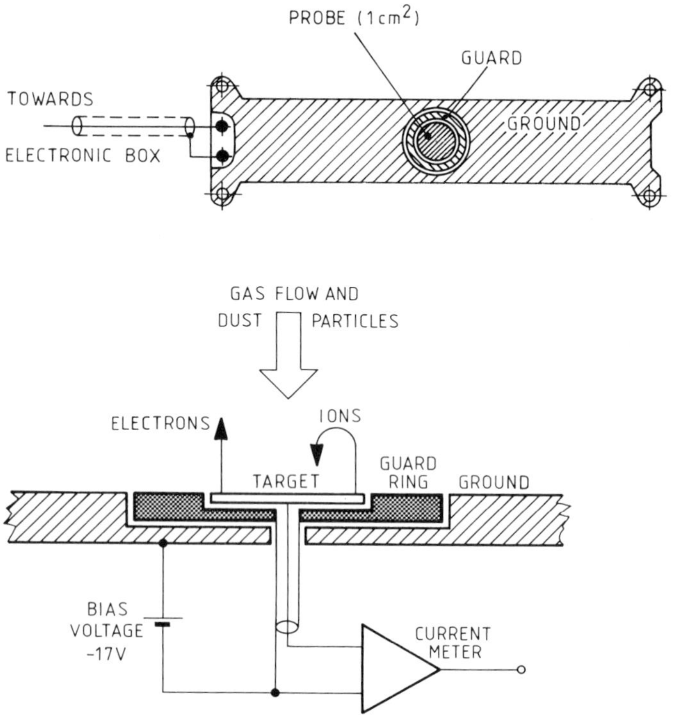

Fig. 1. Target configuration and simplified diagram of sensor and electronics

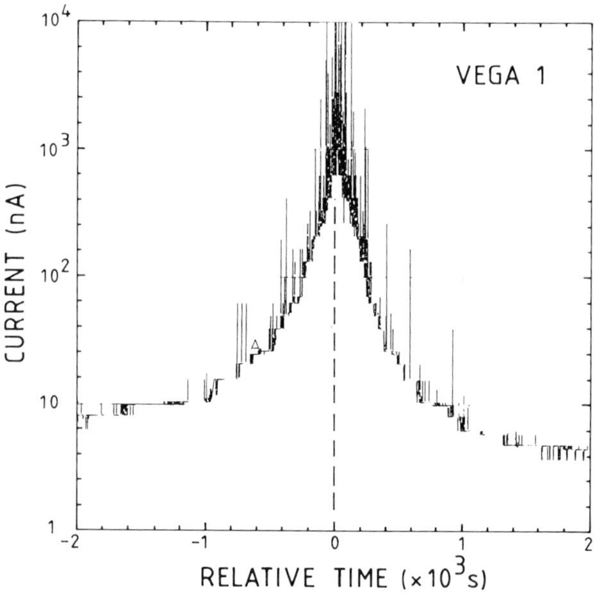

Fig. 2. Secondary electron current emitted by the target during the Vega-1 flyby on 6 March 1986; the closest approach occurred at 07:20:06 CET (cometary event time), at a distance of 8889 km

One fraction of the background current is associated with secondary electron emission due to molecular bombardment during flyby; it may also include the effect of the incoming flow of small dust particles which is integrated by the electronic circuitry (time constant of the order of 1 s). The target emits also a current of photoelectrons which is constant at large distances from the nucleus; the magnitude of this current is about 8 nA before encounter and is subsequently reduced by half. This phenomenon

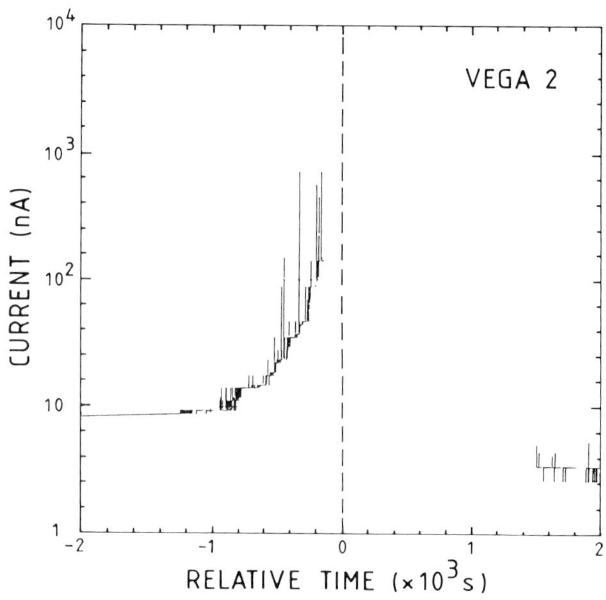

Fig. 3. Secondary electron current emitted by the target during the Vega-2 flyby on 9 March 1986; the closest approach occurred ad 07:20:00 CET, at a distance of 8030 km. The data flow is interrupted for about 27 min due to an instrumental anomaly

is observed identically on Vega-1 and Vega-2 and can be explained by surface degradation during the flyby due to erosion or contamination.

The discrete spikes are caused by the random impact of relatively larger particles; their amplitudes can exceed 10^4 nA near closest approach as shown in Fig. 2, but they never reach the upper limit of the dynamic range of the instrument which is approximately 10^5 nA.

3.2. The Giotto results

The Giotto results are illustrated in Fig. 4. The current scale is the same as in Figs. 2 and 3, but the time scale has been expanded to improve legibility. Before encounter, the average current intensity equals 4 nA, typically half the value observed on the Vega spacecraft under similar circumstances (solar distance and aspect angle), which leads to assume that the protective cover is still closed and that the measurements should rather be explained in terms of current leakage or experimental anomaly (erroneous calibration or instrumental drift). This hypothesis is all the more convincing as the dark current did not exhibit the expected variation when the cover release was ordered by telecommand on 10 February 1986; it has indeed been predicted that photoemission caused by exposure to sunlight should have given rise to a measurable current increase.

Following a series of negative impulsive events such as those seen as discrete spikes at −93 s and −75 s but which are probably overlaping in the interval −70 to −39 s, the intensity of the current suddenly jumps by more than three orders of magnitude up to 10^4 nA. Thereafter the current increases and reaches the upper limit of the telemetry dynamic range at −20 s; superimposed negative spikes are still observed at −33, −26, −19, and −11 s.

The negative pulses observed before −60 s are observed simultaneously on another instrument which is also protected by the same protective cover before opening. These events are

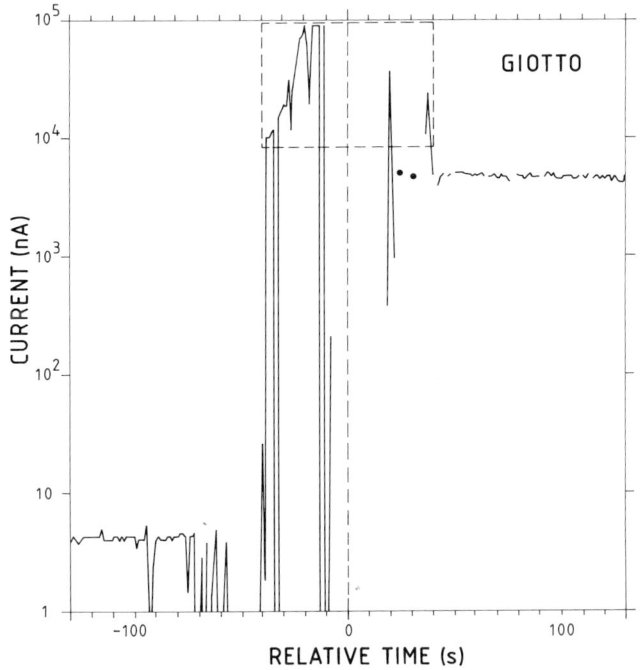

Fig. 4. Secondary electron current emitted by the target during the Giotto flyby on 14 March 1986; the closest approach occurred at 00:03:02 CET, at a distance of 605 km. The protective cover was opened 40 s before transit through pericenter. The data contained within the dashed rectangle are plotted, in a different format, in Fig. 5

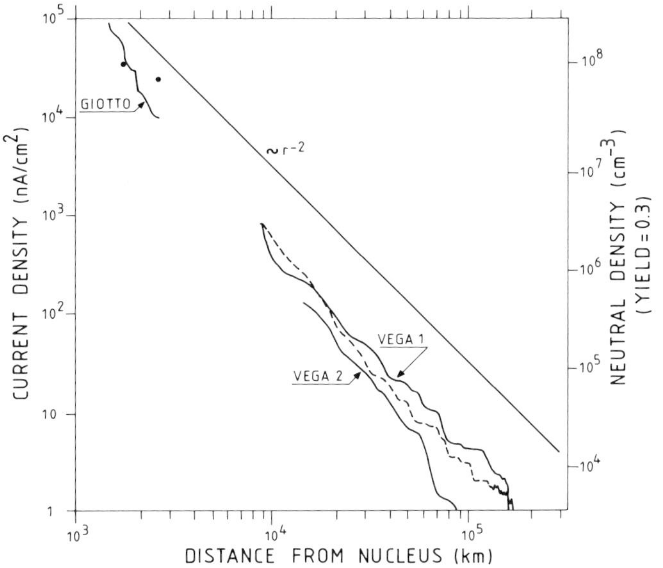

Fig. 6. The envelopes of the secondary electron emission measured during the three flybys plotted against the distance from nucleus, *r*, during the approach (full line) and exit (dashed line for Vega-1 or dots for Giotto). The left ordinate axis gives the current density; the right ordinate axis give the equivalent neutral density evaluated for a yield of 0.3 and a flyby velocity of 78 km s^{-1}. The slope r^{-2} is given for reference

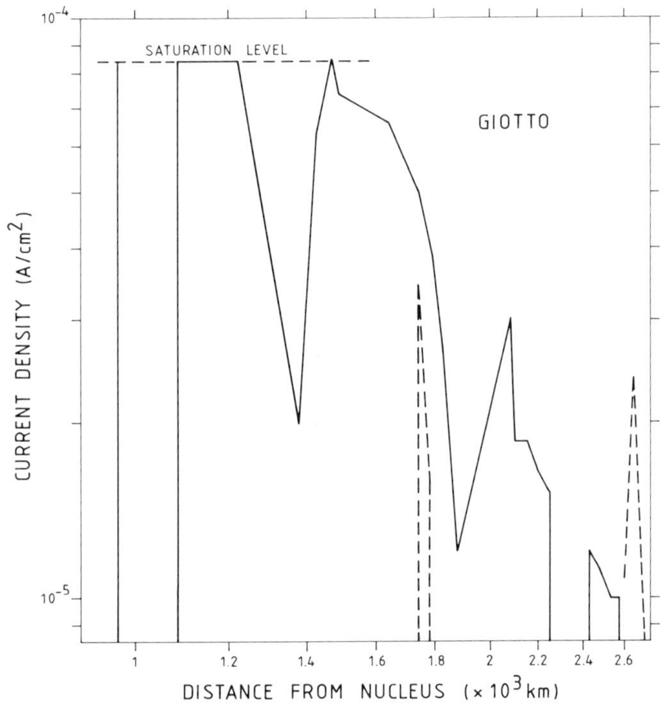

Fig. 5. The secondary electron current emitted from the Giotto target plotted against distance from nucleus during the approach (full line) and exit (dashed line)

believed to be linked to the collection of electrons with energies larger than 17 eV generated by the perforation of the closed cover by dust particles. The occurrence of a few additional events after −40 s leads us to assume that the cover was not completely released at that time but progressively eroded.

The absence of positive spikes after −40 s, such as those seen superimposed on the Vega background signals in Figs. 2 and 3 may simply be due to the combined effect of the high dust impact rate and the long time constant of the instrument (~1 s). Giotto was hit by a relatively large particle at −14 s; the impact caused the spacedraft to perform a nutation with an amplitude such that the telemetry link could not be maintained continuously. After +40 s, the average current level varies very little, decreasing very slowly from 4.7 10^3 nA down to 3.5 10^3 nA in about 25 min. The latter data are erroneous since they are probably dominated by a stray current due to sensor damage or electronic failure.

The information contained within the dashed rectangle is possibly valid as it was taken before the instrument failure and after the cover was totally or partially opened, whether it was released or abraded. This information is plotted in a different format, current density vs. distance from the nucleus, in Fig. 5; the full and dashed lines link the measurements taken during the approach and the exit, respectively.

4. Data comparison and consistency

The data shown in Figs. 2, 3, and 5 are plotted in Fig. 6 against distance from nucleus, *r*; the discrete spikes associated with dust impacts have been subtracted. Measurements performed during the outbound leg of the trajectories are indicated by dots (Giotto) or a dashed line (Vega-1). The slope r^{-2} is shown for reference.

The three sets of data exhibit a reasonable degree of compatibility; the Vega-1 and Vega-2 profiles differ by an amount which can be easily explained on terms of variation in nucleus activity and the Giotto results appear as an acceptable extrapolation of the Vega measurements. It can therefore be inferred that the Giotto impact plasma detector was exposed during the approach to the incoming flow of dust and gas at a distance of 2600 km from the nucleus.

5. Conclusion

We shall conclude with the following remarks:

(1) Information about the neutral density profile can be derived from the secondary electron flux, F, through the relation $n = F/YV$. The right axis of Fig. 6 has been graduated for a yield $Y = 0.3$ and a relative gas flow velocity $V = 78\ \text{km s}^{-1}$ (Grard and Gringauz, 1986). Departure from a r^{-2} profile is caused by nucleus activity fluctuation and target surface degradation (Grard et al., 1986a).

(2) The Giotto cover was certainly not released before the encounter; it was opened or, at least, abraded and partially torn a few seconds before closest approach.

(3) It is possible to estimate, by extrapolation, that a secondary electron current of the order of 27 A was emitted from the Giotto shield during its flyby at the cometary pericenter (600 km).

References

Arends, H., Schmidt, R.: 1983, ESA SP-198, p. 133

Grard, R. (ed.): 1982, *The Giotto spacecraft gas and plasma environments*, ESA SP-187

Grard, R.: 1983, *Adv. Space Res.* **2**, 167

Grard, R.: 1984, ESA SP-224, p. 87

Grard, R., Apathy, I., Gringaux, K., Grün, E., McDonnell, J., Thiemann, H.: 1986, *Adv. Space Res.* **5**, 149

Grard, R., Gombosi, T., Sagdeev, R.: 1986, ESA SP-1066, p. 49

Grard, R., Gringaux, K.: 1986, *Geophys. Res. Letters* **13**, 877

Grard, R., Parker, L., Young, D.: 1981, *Adv. Space Res.* **1**, 403

Krankowski, D., Lämmerzahl, P., Herrwerth, I., Woweries, J., Eberhardt, P., Dolder, U., Herrmann, U., Schulte, W., Berthelier, J.J., Illiano, J.M., Hodges, R., Hoffman, J.: 1986, *Nature* **321**, 326

McDonnell, J., Alexander, W., Burton, W., Bussoletti, E., Clark, D., Evans, G., Evans, S., Firth, J., Grard, R., Grün, E., Hanner, M., Hughes, D., Igenbergs, E., Kuczera, H., Lindblad, B., Mandeville, J.-C., Minafra, A., Reading, D., Ridgeley, A., Schwehm, G., Stevenson, T., Sekanina, Z., Turner, R., Wallis, M., Zarnecki, J.: 1986, ESA SP-1077, p. 85

Parker, L., Holeman, E.: 1981, ESA SP-155, p. 67

Pedersen, A. (ed.): 1983, *Spacecraft/plasma interactions and their influence on field and particle measurements*, ESA SP-198

Reinhard, R. (ed.): 1981, *The comet Halley probe plasma environment*, ESA SP-155

Reinhard, R. (ed.): 1984, *Proc. of the Giotto PEWG Meeting*, 10–11 April 1984, ESA SP-224

Schmidt, R., Arends, H.: 1984, ESA SP-224, p. 15

Schmidt, R., Arends, H.: 1985, *Planet. Space Sci.* **33**, 667

Thieman, H., Grard, R., Singh, N., Schunk, R.: 1986a, *Adv. Space Res.* **5**, 227

Thieman, H., Singh, N., Schunk, R., Grard, R.: 1986b, *J. Geophys. Res.* **93** (A3), 2989

Young, D.: 1983, in *Cometary exploration, Proc. International Conference on Cometary Exploration*, 1982, November 15–19, ed. T. Gombosi, Hungarian Academy of Sciences, Vol. **3**, p. 197

Young, D.: 1986, *Impact-induced spacecraft charging and related effects during the Halley encounters*, ESA SP-1066, p. 85

Astron. Astrophys. 187, 789–795 (1987)

ASTRONOMY AND ASTROPHYSICS

Dust environment of comet P/Halley: a review

Z. Sekanina

MS 183-401, Jet Propulsion Laboratory, California Institute of Technology, 4800 Oak Grove Drive, Pasadena, CA 91109, USA

Received February 2, accepted May 13, 1987

Summary. Recent major advances in the understanding of comet Halley's dust environment are highlighted and a wide range of related problems reviewed. Addressed individually are selected spacecraft results, namely the dust composition, the detection of subfemtogram grains, and the particle-mass distribution; jets, the dust-emission pattern, and the existence of discrete sources on the nucleus surface; CN and C_2 jets; large-scale phenomena, including an antitail, streamers, and a sunward spike; infrared observations, highlighted by recent detections of the 3.4 μm emission feature, first applications of array detectors, and rare far-infrared data; the chemical modification of low-temperature ices by UV and cosmic-ray irradiation processes, their laboratory simulations, and applications to the evolution of comets; the "interstellar-connection" models of comet origin; and the formation of a mantle made of refractory organic material on the nucleus surface.

Key words: comet Halley – dust emission – chemical composition – jet evolution – UV and cosmic-ray irradiation

1. Introduction

The current apparition of comet P/Halley has triggered an unprecedented interest in the study of comets and an avalanche of new results on their origin, nature, and behavior. Many major events observed in Halley involved dust. Conventionally, this term was used in cometary science to describe the refractory component of solids emitted by expanding gas from the nucleus into the coma and tail. The dust is now perceived as a regulator of cometary activity and the bearer of information on some of the most fundamental aspects of the origin and evolution of comets and the solar system.

2. Highlights of spacecraft dust experiments

2.1. Dust chemistry

One of the most exciting discoveries made by the in situ experiments is the detection of extremely small grains ($\sim 10^{-16}$ g in mass), most of them rich in carbon, hydrogen, oxygen, and nitrogen (Kissel et al., 1986a, b) and referred to as the CHON (or low-Z) particles. Under prevailing observing conditions they are not detected by Earth-based techniques. It is therefore hardly surprising that they had never been included in comet dust models (e.g., Divine et al., 1986).

The CHON particles are of major importance for the understanding of the evolution of comets. They offer strong evidence for the presence of organic matter in comets and a solution to the problem of the "missing" carbon, and also lend support to the "interstellar-connection" models (Greenberg, 1986a). A classification of the CHON particles was proposed by Clark et al. (1986), who believe that these grains carry much of the comet's nitrogen supply and are not very volatile. Besides the CHONs, grains were detected with silicate, mineral, mixed, and various exotic compositions.

The major compositional heterogeneity on a submicron scale suggests that Halley's nucleus is the product of a structured accretion process, consistent with Greenberg's (1982) concept of aggregation of comets from interstellar grains. The preponderance of the CHON particles also provides evidence for radiation and other possible processing of cometary material (Sect. 6).

2.2. Subfemtogram grains and the mass distribution

The extension of the dust population to subfemtogram grains entails a number of problems that await solutions. From the ratio of target ions to projectile ions in the mass spectra (Kissel et al., 1986a), the bulk densities of the grains come out to be extremely low. The required high porosity on such a small scale could perhaps be understood as the result of fairy-castle-type structure consisting of even smaller grains (or giant molecules?).

Motions of subfemtogram grains may be affected by a variety of forces, but probably very little by solar radiation pressure, especially if they are slightly absorbing or nearly dielectric (Hanner, 1986; Sekanina, 1986a; Sekanina et al., 1987). Even though subfemtogram grains are by far the most numerous in the coma, they contribute little mass to the comet's total dust output. The results from the dust-impact experiments on Giotto (McDonnell et al., 1986; Zarnecki et al., 1986) indicate that the index s of the cumulative particle-mass distribution function m^{-s} attains values near unity at masses greater than $\sim 10^{-10}$ g but is much lower at 10^{-16} g. There is a suggestion that the mass index may have been higher after closest approach. On the average, these mass indices appear to be lower than those determined by Simpson et al. (1986) from the earlier flight data in the particle mass range down to 10^{-13} g (the above values of s refer to space density; they require a correction of $+0.17$ for large particles – but practically none for very small grains – to describe the mass distribution in the dust flux from the nucleus).

2.3. Dust emission pattern

There are several lines of evidence from spacecraft experiments on Halley's highly anisotropic pattern of dust emission: (1) direct imaging of the environment of the nucleus; (2) Doppler measurements of the time profile of the momentum transferred by particle impacts to the spacecraft; (3) major deviations, in the particle-impact curve, from the inverse-square power law of distance from the nucleus; (4) optical probing of the ejecta from within the coma.

The appearance of the nucleus on the sequence of images taken with the Giotto camera (Keller et al., 1986) is unquestionably by far the most dramatic and impressive documentation of Halley's pattern of dust emission, dominated by several discrete jets on the sunlit hemisphere, with the rest of the surface essentially, or entirely, inactive. The high degree of collimation exhibited by the jets on the spacecraft images explains the detection of the ejecta as fairly sharp-edge features on Earth-based images for days after ejection. The cause of the collimation is not known with certainty, but a possible explanation is a simple funnel-type flow from sources that are depressed with respect to the surrounding terrain.

The Doppler-shift measurements of Giotto's downlink radio signal indicate that the probe was decelerated because of momentum transfer from impacting ejecta (primarily dust), the velocity change amounting to about $23\,\mathrm{cm\,s^{-1}}$ (Edenhofer et al., 1986a). The refined, but still preliminary, results show several distinct discontinuities along the spacecraft path, the most prominent event taking place a few seconds prior to closest approach and attributed to Giotto's encounter with a major jet (Edenhofer et al., 1986b).

Additional evidence on Halley's emission anisotropy is furnished by the in situ dust-impact experiments (cf. McDonnell et al., 1986; Simpson et al., 1986). In general, the measured particle-impact profiles are inconsistent with the inverse-square power law of distance from the nucleus expected for a steady-state, isotropic emission. In an extreme case, a sudden increase in the impact rate by a factor of ~ 40 was recorded by Simpson et al. (1986).

Levasseur-Regourd et al. (1986) report an apparent signature of a jet in their data from the Halley optical probe experiment. Their sampling interval of four seconds may be short enough to provide additional information from the brightness variation along the spacecraft path. Further analysis appears to be warranted.

3. Jets

3.1. *Brief historical account*

Discrete jets are now universally recognized as the principal mode of nucleus activity of Halley and probably many other comets. A prominent curved jet in a drawing of comet Halley made by Hevelius (1682) on 1682 September 8 clearly illustrates that observers have been aware of these phenomena for centuries. During the comet's 1835/36 apparition it was noticed that the basic structure of its coma was changing dramatically from day to day and sometimes on an hourly basis (cf. Bessel, 1836; Herschel, 1847). During the return of 1910 the expansion rates of coma features (which include various jets, arcs, envelopes, etc.) were measured on high-resolution images and the projected velocities determined (e.g., Bobrovnikoff, 1931, and references therein). These turned out to be in a general range of 0.1–$1\,\mathrm{km\,s^{-1}}$, coinciding with the velocities predicted by theory for ejection of microscopic dust particles entrained in the flow of expanding gas (e.g., Probstein, 1969). The dust nature of these features is further corroborated by their comparative sharpness, which is indicative of highly organized particle motions but inconsistent with the semi-chaotic distribution of molecular motions.

Finally, since lately it has been possible to follow the evolution of distinct features through the coma from their early phase of rapidly developing spiral structure to the eventual disappearance of gradually fading, expanding envelopes; to model this process quantitatively; and to interpret the observations in terms of continuous ejection of dust from discrete sources located on the sunlit hemisphere of a rotating nucleus (Sekanina and Larson, 1983, 1984, 1986a, b; Larson and Sekanina, 1984, 1985).

3.2. *Concept of discrete dust-emission sources*

The simplest relevant case of continuous emission of dust from a discrete source is that of a point-like area (small vent) on a spherical nucleus. Particles are assumed to be ejected along the normal to the surface and subjected to accelerations by solar radiation pressure that depend upon their size, density, and optical properties. The ejection velocity is assumed to be a parametric function of the acceleration which is determined in the process of model fitting. The comet is rotating about an axis whose position is described by two angles defined in Sekanina (1981a). The period of rotation and the equatorial coordinates of the emission point complete the list of unknowns. This model can readily be upgraded to apply to a precessing nonspherical nucleus and to incorporate extended emission sources.

The model equations based on this concept have been found to "generate" features that in the projection onto the sky bear a strong resemblance to the actual features on high-resolution images of Halley's coma. The computer simulation of the day-to-day evolution of features has also mimicked the observed patterns. Besides this intuitively plausible scenario for the development of Halley's spiral jets into expanding envelopes, the postulated concept of continuous dust emission from discrete sources explains in a natural manner other phenomena, such as the major deviations in the outlines of Halley's envelopes from parabolic shape and their slight slant with respect to the radius vector, which Bobrovnikoff (1931) detected but could not understand on the basis of the traditional fountain model of isotropic emission. The new model further explains, as a necessary consequence of differences among comets in the spin rate, why spiral jets are observed in some comets and semi-circular halos in others.

3.3. *Computer processing of Halley's images*

Jets are usually features of low surface brightness, so their positional reduction necessitates contrast enhancement. Digital processing of high-resolution images becomes, consequently, an integral part of modern jet evolution studies. A powerful intensity-derivative algorithm that invoves both radial and rotational shift with respect to the emission center (nucleus) substantially reduces intensity gradients of a spatial frequency greater than the applied shift, but offers a considerable contrast enhancement of intensity variations on the scales that are smaller than the shift (Larson and Sekanina, 1984, 1985). Hence, it is the outlines and expansion rate of the outer boundary of the discrete feature that the quantitative modeling is involved with.

Many high-resolution images of comet Halley in the current apparition have been obtained with charge coupled device (CCD) arrays and are therefore available in digital from (Larson et al., 1986). Whenever photographic images are involved, such as is the case with the Halley data from 1910, the necessary digitization must be performed with a pixel size that does not compromise structural detail, but at the same time compensates for the noise due to photographic grain (Larson and Sekanina, 1984, 1985).

3.4. *Halley's dust jet activity*

While expressions of amazement abound at the degree to which Halley's nucleus activity is concentrated in jets on the Giotto-camera images, this very fact had been established from jet photometry (Larson and Sekanina, 1984) and dynamical analysis of the ejecta (Sekanina and Larson, 1984), performed on the comet's Earth-based images from 1910. For example, the minimum production rate of dust per unit surface area of an extended emission region was found to be 40–60 times the rate averaged over the nucleus surface.

The recently completed analysis of dust features in 1910 (Sekanina and Larson, 1986b) has demonstrated that the assumption of point-like sources fails for most of the jets, leading to a concept of linear sources. Although its application proved very successful, showing that the observations of jets are always satisfied on the assumption that dust emission was terminated at local sunset, the question of solution multiplicity has remained unanswered. For example, a "linear" source may consist of a number of small isolated vents lined up near one another on the nucleus surface. With the spatial resolution usually available from Earth, the individual microjets emanating from such minor vents would coalesce to yield a seemingly single, massive outburst.

The discrete features on the 1986 images exhibit a remarkable qualitative similarity to those identified on the 1910 images and include pinwheels, envelopes, arcs, and straight, antisunward jets. While work continues on high-resolution images of the comet's dust coma, obtained in a collaborative effort coordinated by the International Halley Watch's Near Nucleus Studies Network (Larson et al., 1986), future developments depend in part on the progress in our understanding of the comet's rotation. Millis and Schleicher's (1986) detection of a second recurrence period in Halley's activity variations and the subsequent models of a precessing nucleus (e.g., Sekanina, 1986b, 1987; Julian, 1987) indicate that the comet's rotation is apparently much more complex than thought at the time of modeling the 1910 jets. In addition to any free precession, there are also effects of torques due to outgassing, deviations from the assumed spheroidal figure, the probably nonuniform mass distribution in the nucleus interior, and the energy dissipation due to the nonrigidity of the nucleus.

3.5. *Cyano and C_2 jets*

A'Hearn et al. (1986a) made systematic observations of pinwheel-shaped features in the CN (0,0) bandpass, which were generally not detected in the continuum bandpasses. Extending to more than 60,000 km from the nucleus and most prominent in late April 1986, this type of structure persisted for weeks with varying brightness. The cyano jets were confirmed by Larson et al. (1986) who commented on the lack of correspondence between the positions of these jets and the usual dust jets on images from the same nights. Larson et al. were also able to identify these features on several preperihelion images of Halley taken in January 1986. A'Hearn et al. (1986b) subsequently detected the same jets in the emission of C_2.

A'Hearn et al. (1986a, b) argue convincingly that the radicals in the spiral structures are not emitted directly from the nucleus but must be produced in situ, that is, most probably released from dust grains in the coma, possibly from the CHON particles (Sect. 2.1). Still open is the question of the release mechanism involved, either dissociation of very short-lived parent molecules or direct emission from the grains. The fact that the jets are not detected in the continuum implies that the parent grains are virtually transparent to the visible light which means that their dimensions are substantially smaller than 0.1 μm (Sekanina, 1986a). This model is consistent with the inferred grain velocity, estimated at $0.5\,\mathrm{km\,s^{-1}}$ or more (Samarasinha et al., 1986).

4. Large scale dust phenomena

4.1. *Tail*

The dominating force that drives tail particles is of course solar radiation pressure. Since the dust tail is a relatively thin layer of ejecta extending in the orbit plane, confined to the sector between the prolonged radius vector and the vector of negative orbital velocity, its appearance to the Earth-based observer is affected by the geometric conditions. With these being radically different in 1910 and 1985/86, the tail likewise presented itself in very diverse projections at the two apparitions.

Halley's large-scale images obtained throughout 1985/86 indicate that the gross properties of its dust tail were very much as predicted from by now routine particle-trajectory analysis (Sekanina, 1983). In particular, the tail had been very narrow for most of the time before perihelion but broadened substantially by mid-February 1986, when the comet reappeared from behind the Sun. Tentative results by Lamy (1986) indicate that the peak particle acceleration from solar radiation pressure in the main tail amounted to ~2.5 times the solar attraction, consistent with the earlier determination based on observed tail lengths about the time of Earth's transit across the comet's orbit plane in May 1910 (Sekanina, 1981b).

As expected (Sekanina, 1983), Halley did display an antitail after perihelion. The feature became the most prominent in the second half of February 1986 but was rapidly fading away during early March. According to Lamy (1986) the antitail's sharp boundary indicated that appreciable emission of dust had commenced when the comet approached the Sun to ~2 AU and that typical particle accelerations were on the order of 0.005 the solar attraction, implying the presence of submillimeter-size particles.

The antitail was only a part of the system of straight streamers stretching from the northeast to the west-northwest for up to a few degrees from Halley's nucleus condensaton. The streamers were unquestionably signatures in the tail of the episodic dust-jet activity from discrete sources on the nucleus surface; the independent determinations of the ejection times by Sekanina (1986c) and by Lamy (1986) consistently show that the events recurred with a period of ~2.2 days or a multiple thereof.

4.2. *Sunward spike*

Virtually all dust ejecta in comets are confined to distances smaller than ~100,000 km from the nucleus on its sunward side because of the relationship between ejection velocity and solar radiation pressure acceleration that control the apex distances of particle trajectories. A major exception to this rule is the population of extremely small, subfemtogram grains which were for the first time identified by the in situ experiments on the Halley missions

(Sect. 2.2). Sekanina (1986a) argues that these grains also made up the faint sunward spike whose projected length at the time of discovery in late April-early May 1986 was reported to be up to three quarters of a million of kilometers. The spike's variations in the position angle have now been measured to provide a data base for further investigations and the most likely position of the plane normal to the comet's angular-momentum vector was determined (Sekanina et al., 1987).

In retrospect, Herschel's (1847) and Maclear's (1838) careful descriptions and systematic measurements of Halley's halo which, subsequent to a huge outburst in late January 1836, was observed for a few weeks to expand to its maximum measurable projected dimension of 460,000 km toward the Sun, is of considerable interest.

5. Infrared observations of dust

An excellent introductory review of infrared observations of comet Halley has recently been presented by Hanner (1986). Only some of the most interesting topics discussed by her are mentioned below.

5.1. Near infrared

Of considerable interest is the postperihelion detection, by several independent groups of observers, of an emission feature near 3.4 μm, believed to be indicative of organic grains or molecules. Hanner summarizes these observations and calculates thermal-emission spectra for one particular organic residue produced in the laboratory, which, in sharp contrast to graphite or amorphous carbon, is only slightly absorbing, as are the organic mantles of interstellar grains (Greenberg, 1982).

A number of times from early December 1985 till mid-April 1986 near-infrared arrays were used to map the spatial distribution of dust near the nucleus in passbands between 1 and 4 μm. Halley is the first comet to which this technique has ever been applied.

5.2. Thermal infrared

The character of the thermal emission was similar to that of most other comets. The color temperature in the 8–20 μm region exceeded the black-body temperature by several percent throughout Halley's apparition. The NASA's Kuiper Airborne Observatory was flown on several occasions between December 1985 and April 1986 to observe the comet in the spectral region between 5 and 30 μm.

A mean albedo of Halley's dust virtually coincided with the albedos of periodic comets Kopff and Giacobini-Zinner, decreasing from 0.03 at phase angle 20° to 0.02 at 50°. The 10 μm silicate signature was first detected at 1.3 AU inbound and became prominent at heliocentric distances smaller than 1 AU.

5.3. Far infrared

On several flights of the NASA's Kuiper Airbone Observatory in December 1985 and March–April 1986 measurements were obtained at wavelengths greater than 30 μm. These are especially important for studies of dust particles whose dimensions exceed ~10 μm. The results of a 40–160 μm photometry reported by Harvey et al. (1986) indicate a variation by nearly a factor of two on two consecutive days in mid-March and a flux falloff with increasing wavelength that is characteristic of particles with sizes comparable to wavelengths of the passbands used. A less steep flux distribution on the date of lower brightness is consistent with smaller, faster dust particles' more rapid evacuation of the near-nucleus space (relative to larger, slower particles) following a major emission event that must have taken place more than 24 h before.

6. Irradiation effects and evolution of comets

6.1. Dust mantle

Concepts of the formation and evolution of a mantle of solid material on the surface of a comet nucleus are currently undergoing a major review. Independent studies have demonstrated the plausibility of a scenario, based on Whipple's (1950) icy conglomerate model, in which the mantle is a by-product of the sublimation process: dust particles entrained in the ice mixture that are too heavy to be dragged away by subliming gas fall back on the surface and build up an insulating layer. Quantitative models (Brin and Mendis, 1979; Brin, 1980; Fanale and Salvail, 1984; to mention a few) have shown that the formation of such a layer is dictated by certain, physically plausible conditions at the nucleus surface. While estimates of a maximum thickness of this dust layer are rather uncertain, they are always $\ll 1$ m and usually in the mm-cm range. Some of the models even predict that the nucleus should be "bare" (that is, with the exposed pristine "dirty" ice) at heliocentric distances smaller than ~1 AU, so that the mantle is generally perceived as a transient phenomenon.

6.2. Proton irradiation of ice mixtures

Recent major advances in the laboratory simulation of long-term exposure of comets to cosmic radiation, especially energetic protons, in the Oort cloud have stimulated fundamentally new ideas on the nature of the outer layers of a cometary nucleus. A number of experiments (e.g., Brown et al., 1982; Johnson et al., 1982; Moore and Donn, 1982; Pirronello et al., 1982; Johnson et al., 1983; Moore et al., 1983; Strazzulla et al., 1983; Foti et al., 1984; Lanzerotti et al., 1985; Strazzulla, 1985) have demonstrated that a low-temperature mixture of ices rich in hydrocarbons is subjected to a number of thermochemical processes when irradiated by protons in the keV-MeV range and then gradually warmed up. New volatile species are synthesized at very low temperatures due to irradiation. The subsequent warmup is observed to trigger a sequence of discrete chemiluminescent events (mainly at temperatures 20–40 K) that are accompanied by concurrent pressure pulses. As the warmup continues, the less volatile fractions are gradually polymerized into an amorphous organic residue which is stable at room temperature.

Very recently, Strazzulla (1986) has reported that continuing proton irradiation intensifies gradual depletion of primarily molecular hydrogen from the low-density residue, transforming it into a dark, carbon-like substance. Similar experimental results have also been obtained by Sakata et al. (1983, 1984) by synthesizing a quenched carbonaceous composite of ionized methane gas discharged in a fused quartz tube. In addition, solid-state hot atom chemistry may also be of some importance (Rössler, 1986).

6.3. UV irradiation and interstellar grain evolution

The processes triggered by proton irradiation of the low-temperature ice mixtures have been, to a degree, duplicated in the laboratory by UV irradiation, with the exception that UV light penetrates only microscopic layers of substrates (Greenberg, 1982, 1983, 1986b; d'Hendecourt et al., 1986; and a large number of related papers). Processing of submicron-size grains in interstellar space and the presolar nebula is in fact dominated by UV radiation. Greenberg's prescription for making comets is the process of coagulation of interstellar grains first into micron-size bird's nest particles (Greenberg and Gustafson, 1981) and then gradually larger clumps until they eventually grow into kilometer-size objects during final gravitational collapse of the solar nebula or its satellite. What counts most in this scenario is the kind of processing that the grains had undergone prior to their aggregation. Greenberg argues that if cometesimals accreted at $>10^4$ AU from the center of the nebula, their ingredients were not subjected to temperatures higher than 20 K, once the silicate grains had cooled down subsequent to their expulsion from the atmospheres of red stars. UV photolysis of condensibles accreting on the silicate grains causes the formation of new species and a gradual buildup of organic refractory mantles which survive the conditions in low-density diffuse clouds. During their passes through the dense molecular clouds, the core-mantle grains accrete an outer shell of condensibles rich in H_2O ice.

Considering that all interstellar material is eventually recycled, Greenberg was able to come up with the prediction for the distribution of the principal chemical components in a primordial comet consisting of aggregated interstellar grains.

6.4. *Effects or irradiation on comets*

The synthesis of new chemical species triggered by UV and cosmic-ray irradiation of low-temperature ice mixtures and the formation of complex organic residues are documented by major changes in the infrared spectra that were obtained in the course of irradiation and during the subsequent warmup of the samples (Greenberg, 1982; Moore and Donn, 1982; Moore et al., 1983; Strazzulla et al., 1984; Calcagno et al., 1985; d'Hendecourt et al., 1986).

The penetrating power of cosmic rays increases rapidly with their energy. However, for relativistic particles this power is determined primarily by induced nuclear reactions (Cooper, 1983; Draganic et al., 1984) and Strazzulla's (1986) estimate that during a comet's 4.6 billion year long stay in the Oort cloud virtually all carbon to depths of a few tens of meters below the nucleus surface would be polymerized must be scaled down accordingly. On the other hand, since the penetration path is measured by the columnar mass encountered, the very low bulk density attributed nowadays to cometary nuclei (Greenberg, 1986c; Rickman, 1986) implies polymerization depths exceeding considerably the often quoted 1 meter (e.g., Donn, 1976) and amounting to perhaps some 5–10 meters on the average.

The effects of irradiation are interesting in that they result in the formation of both highly reactive species and a very refractory residue and that the dramatic changes in the composition that they bring about tend to trigger observable events only after a warmup. Depending upon the conditions during their formation and subsequent evolution, comets may never have been exposed to temperatures higher than ~ 20 K until their first journey toward the Sun.

The significance of cosmic ray irradiation of ices for the explanation of the extraordinary activity of new comets at large heliocentric distances along the inbound leg of their orbits has been recognized by Whipple (1977). The following sequence of developments was proposed, which should lead to a virtually permanent insulating mantle over much of the nucleus surface (Sekanina, 1986d). A virtually identical scenario was independently described by Johnson et al. (1986). If the approach to the Sun of a new comet does indeed represent the first time for its surface temperature to rise above a few tens of degrees K, then the warmup at heliocentric distances of about 100 AU should already be sufficient for the reactive species formed by irradiation to recombine exothermically to give rise to major comet activity (Moore et al., 1983). Concurrently, as the volatile fraction is being released, the polymer-like organic residue, synthesized from the frozen hydrocarbons in the ice mixture, should be gradually building up at the nucleus surface, trapping – and being reinforced by – silicate, organic, and other refractory grains already present in the agglomerated comet. Some of the mantle debris is stripped off and carried away as "dust" particles by the escaping gases. Before sublimation by solar radiation can even commence, a reservoir of more volatile species at greater depths has already been sealed off over perhaps much of the surface. By the time the object is discovered from Earth, a possible thick insulating mantle of inert material has spread over at least some parts of the nucleus. The comet exhibits a dust tail that testifies to an activity long before perihelion (Sekanina, 1973, 1975, 1982), but the performance of the comet around and especially after the time of perihelion passage is disappointing, with only limited activity persisting from unmantled portions of the surface. Indeed, since the surface-temperature distribution of a new comet during its journey to the Sun is spin dependent and since the initial surface distribution of carbon-carrying species is, in general, variable, the growth of the mantle over the surface should be quite uneven. If the comet is subsequently captured into a periodic orbit with a fairly small aphelion distance, the process of mantle growth should speed up, being assisted by the thermal wave penetrating ever deeper into the interior of the nucleus. However, with reactive species trapped beneath the surface, there exists the possibility of outbursts that can destroy or heavily damage the mantle on a local scale. This disintegration process is assisted by the mantle's low conductivity, which implies a steep temperature gradient in the uppermost layer of the mantle at smaller heliocentric distances.

In summary, it appears plausible to speculate that discrete emission vents on the nuclei of comets with highly evolved orbits, such as comet Halley, are the areas where the structure of the organic residue was too weak to survive. This scenario further suggests that in principle there are two kinds of insulating mantle on cometary nuclei:

1. *Irradiation-driven mantle.* Its development is related to the comet's early evolution. It may have grown quite thick in the areas of the nucleus with favorable conditions for the polymerization process but much less so elsewhere. On Halley it seems to extend over most of the surface, making the nucleus generally dark and largely inactive. This mantle is likely to remain basically unaffected by the comet's activity, but it is subjected to thermal stresses as mentioned above.

2. *Sublimation-driven mantle.* This is a relatively thin, short-lived, and possibly cyclic layer, which is functionally confined to the emission vents, that is, to the regions not covered by the irradiation-driven mantle or where the latter is structurally weak. The formation and erosion of the sublimation-driven mantle on a local scale may accentuate variable activity of comets.

7. Future work

Cometary science is at crossroads, as new ideas are put forward to accommodate the avalanche of data on Halley from experiments on board of the spacecraft intercepting the comet, from many other spaceborne and airborne observations, and from very impressive Earth-based observations. This information offers some very significant correlations with the results of recent sophisticated laboratory experiments on low-temperature ice mixtures.

In the area of dust activity, the most urgently needed data concern: (1) the diurnal evolution of jets, from activation in the morning till termination near sunset; and (2) the nature, lifetimes, and properties of the emission vents. Although the Halley missions have resulted in qualitatively new information, it is restricted to a few "snapshots" – an inherent limitation of every flyby mission. Further major leaps in understanding comets can now be accomplished only with rendezvous missions, which offer sufficient time for necessary spacecraft maneuvering in the vicinity of the target and for implementation of a "smart" approach to data acquisition. Before a first rendezvous mission becomes a reality, the most productive way of studying the jet activity is by correlating the close-range snapshot views of Halley with its systematic Earth-based imaging observations.

There exist no measurements of the thickness of the mantle that covers much of Halley's surface. It is, however, argued that in general it is considerable because a thin mantle would easily get perforated at many more points than observed on the Giotto images, considering the elevation variations detected over the surface.

Some interesting conclusions about the properties of the surface mantle can be deduced from the generally accepted inferences that its porosity is high. This implies a low thermal conductivity (Smoluchowski, 1982, 1985) which, in turn, leads to a high temperature gradient at the surface upon external heating near the Sun and brings about considerable thermal stresses, exceeding by orders of magnitude the mechanical strength of fragile structure of organic aggregates.

The results of the in situ dust composition experiments on the Halley missions provide strong evidence for organic matter in the comet's nucleus and lend support to Greenberg's interstellar-connection theory of cometary origin. High porosity, low albedo, and the omnipresence of subfemtogram grains provide additional evidence. Processes have been singled out by Fechtig and Mukai (1985), by Johnson (1985), by Strazzulla et al. (1985), and by Johnson and Lanzerotti (1986) that could transform the highly porous, bird's nest particles into more compact ones, such as Brownlee's particles that have been collected in the Earth's atmosphere (Brownlee, 1978, 1985).

Various lines of evidence appear to corroborate the general perception that the conventional discrimination of comet ejecta between gas and dust is artificial and that there is no major gap between the two, either in the size of a basic "building block" or in the volatility. In other words, there may not be any difference between very tiny grains and very complex molecules, and there certainly are species in comets that are less volatile than water ice, yet less refractory than silicate particles. The exploration of these new frontiers presents a major challenge to cometary science in the post-Halley era.

Acknowledgement. The research described in this paper was carried out by the Jet Propulsion Laboratory, California Institute of Technology, under contract with the National Aeronautics and Space Administration.

References

A'Hearn, M.F., Hoban, S., Birch, P.V., Bowers, C., Martin, R., Klinglesmith, D.A.: 1986a, *Nature* **324**, 649
A'Hearn, M.F., Hoban, S., Birch, P.V., Bowers, C., Martin, R., Klinglesmith, D.A.: 1986b, 20th ESLAB Symp., ESA SP-250, Vol. **I**, p. 483
Bessel, F.W.: 1836, *Astron. Nachr.* **13**, 185
Bobrovnikoff, N.T.: 1931, *Publ. Lick Obs.* **17**, 307
Brin, G.D.: 1980, *Astrophys. J.* **237**, 265
Brin, G.D., Mendis, D.A.: 1979, *Astrophys. J.* **229**, 402
Brown, W.L., Lanzerotti, L.J., Johnson, R.E.: 1982, *Science* **218**, 525
Brownlee, D.E.: 1978, in *Cosmic Dust*, ed. J.A.M. McDonnell, Wiley, Chichester, England, p. 295
Brownlee, D.E.: 1985, *Ann. Rev. Earth Planet. Sci.* **13**, 147
Calcagno, L., Foti, G., Torrisi, L., Strazzulla, G.: 1985, *Icarus* **63**, 31
Clark, B.C., Mason, L.W., Kissel, J.: 1986, 20th ESLAB Symp., ESA SP-250, Vol. **III**, p. 353
Cooper, J.F.: 1983, *J. Geophys. Res.* **88**, 39
d'Hendecourt, L.B., Allamandola, L.J., Grim, R.J.A., Greenberg, J.M.: 1986, *Astron. Astrophys.* **158**, 119
Divine, N., Fechtig, H., Gombosi, T.I., Hanner, M.S., Keller, H.U., Larson, S.M., Mendis, D.A., Newburn, R.L., Reinhard, R., Sekanina, Z., Yeomans, D.K.: 1986, *Space Sci. Rev.* **43**, 1
Donn, B.: 1976, NASA SP-393, p. 611
Draganic, I.G., Draganic, Z.D., Vujosevic, S.: 1984, *Icarus* **60**, 464
Edenhofer, P., Bird, M.K., Brenkle, J.P., Buschert, H., Esposito, P.B., Porsche, H., Volland, H.: 1986a, *Nature* **321**, 355
Edenhofer, P., Buschert, H., Bird, M.K., Volland, H., Porsche, H., Brenkle, J.P., Kursinsky, E.R., Mottinger, N.A., Stelzried, C.T.: 1986b, 20th ESLAB Symp., ESA SP-250, Vol. **II**, p. 215
Fanale, F.P., Salvail, J.R.: 1984, *Icarus* **60**, 476
Fechtig, H., Mukai, T.: 1985, in *Ices in the Solar System*, eds. J. Klinger, D. Benest, A. Dollfus, R. Smoluchowski, Reidel, Dordrecht, p. 251
Foti, G., Calcagno, L., Sheng, K.L., Strazzulla, G.: 1984, *Nature* **310**, 126
Greenberg, J.M.: 1982, in *Comets*, ed. L.L. Wilkening, University of Arizona, Tucson, p. 131
Greenberg, J.M.: 1983, in *Cometary Exploration II*, ed. T.I. Gombosi, Hung. Acad. Sci., Budapest, p. 23
Greenberg, J.M.: 1986a, ESA SP-249, p. 47
Greenberg, J.M.: 1986b, in *Galaxy and the Solar System*, ed. R. Smoluchowski, University of Arizona, Tucson, p. 103
Greenberg, J.M.: 1986c, in *Asteroids, Comets, Meteors II*, eds. C.-I. Lagerkvist, B.A. Lindblad, H. Lundstedt, H. Rickman, Uppsala University, Uppsala, p. 221
Greenberg, J.M., Gustafson, B.A.S.: 1981, *Astron. Astrophys.* **93**, 35
Hanner, M.S.: 1986, *Adv. Space. Res.* **5**, No. 12, p. 325
Harvey, P.M., Campins, H., Ellis, H.B., Jr., Joy, M., Lester, D.F.: 1986, *Adv. Space Res.* **5**, No. 12, p. 335
Herschel, J.F.W.: 1847, *Results of Astronomical Observations at the Cape of Good Hope*, Smith, Elder, London, p. 393

Hevelius, J.: 1682, *Acta erudit.*, Dec., p. 389
Johnson, R. E.: 1985, in *Ices in the Solar System*, eds. J. Klinger, D. Benest, A. Dollfus, R. Smoluchowski, Reidel, Dordrecht, p. 337
Johnson, R. E., Cooper, J. F., Lanzerotti, L. J.: 1986, 20th ESLAB Symp., ESA SP-250, Vol. **II**, p. 269
Johnson, R. E., Lanzerotti, L. J.: 1986, *Icarus* **66**, 619
Johnson, R. E., Lanzerotti, L. J., Brown, W. L.: 1982, *Nucl. Instr. Meth.* **198**, 147
Johnson, R. E., Lanzerotti, L. J., Brown, W. L., Augustyniak, W. M., Mussil, C.: 1983, *Astron. Astrophys.* **123**, 343
Julian, W. H.: 1987, *Nature* **326**, 57
Keller, H. U., Arpigny, C., Barbieri, C., Bonnet, R. M., Casez, S., Coradini, M., Cosmovici, C. B., Delamare, W. A., Huebner, W. F., Hughes, D. W., Jamar, C., Malaise, D., Reitsema, H. J., Schmidt, H. U., Schmidt, W. K. H., Seige, P., Whipple, F. L., Wilhelm, K.: 1986, *Nature* **321**, 320
Kissel, J., Brownlee, D. E., Buchler, K., Clark, B. C., Fechtig, H., Grun, E., Hornung, K., Igenbergs, E. B., Jessberger, E. K., Krueger, F. R., Kuczera, H., McDonnell, J. A. M., Morfill, G. M., Rahe, J., Schwehm, G. H., Sekanina, Z., Utterback, N. G., Volk, H. J., Zook, H. A.: 1986a, *Nature* **321**, 336
Kissel, J., Sagdeev, R. Z., Bertaux, J. L., Angarov, V. N., Audouze, J., Blamont, J. E., Buchler, K., Evlanov, E. N., Fechtig, H., Fomenkova, M. N., von Hoerner, H., Inogamov, N. A., Khromov, V. N., Knabe, W., Krueger, F. R., Langevin, Y., Leonas, V. B., Levasseur-Regourd, A. C., Managadze, G. G., Podkolzin, S. N., Shapiro, V. D., Tabaldyev, S. R., Zubkov, B. V.: 1986b, *Nature* **321**, 280
Lamy, P.: 1986, *Adv. Space Res.* **5**, No. 12, p. 317
Lanzerotti, L. J., Brown, W. L., Johnson, R. E.: 1985, in *Ices in the Solar System*, eds. J. Klinger, D. Benest, A. Dollfus, R. Smoluchowski, Reidel, Dordrecht, p. 317
Larson, S. M., Sekanina, Z.: 1984, *Astron. J.* **89**, 571, 600
Larson, S. M., Sekanina, Z.: 1985, *Astron. J.* **90**, 823, 917
Larson, S. M., Sekanina Z., Levy, D., Tapia, S., Senay, M.: 1986, 20th ESLAB Symp., ESA SP-250, Vol. **II**, p. 145
Levasseur-Regourd, A. C., Bertraux, J. L., Dumont, R., Festou, M., Giese, R. H., Giovane, F., Lamy, P., Le Blanc, J. M., Llebaria, A., Weinberg, J. L.: 1986, *Nature* **321**, 341
Maclear, T.: 1838, *Mem. Roy Astron. Soc.* **10**, 91
McDonnell, J. A. M., Kissel, J., Grün, E., Grard, R. J. L., Langevin, Y., Olearczyk, R. E., Perry, C. H., Zarnecki, J. C.: 1986, 20th ESLAB Symp., ESA SP-250, Vol. **II**, p. 25
Millis, R. L., Schleicher, D. G.: 1986, *Nature* **324**, 646
Moore, M. H., Donn, B.: 1982, *Astrophys. J.* **257**, L 47
Moore, M. H., Donn, B., Khanna, R., A'Hearn, M. F.: 1983, *Icarus* **54**, 388
Pirronello, V., Brown, W. L., Lanzerotti, L. J., Marcantonio, K. J., Simmons, E.: 1982, *Astrophys. J.* **262**, 636
Probstein, R. F.: 1969, in *Problems of Hydrodynamics and Continuum Mechanics*, eds. F. Bisshopp et al., Society of Industrial and Applied Mathematics, Philadelphia, p. 568
Rickman, H.: 1986, ESA SP-249, p. 195
Rössler, K.: 1986, ESA SP-249, p.161
Sakata, A., Wada, S., Okutsu, Y., Shintani, H., Nakada, Y.: 1983, *Nature* **301**, 493
Sakata, A., Wada, S., Tanabe, T., Onaka, T.: 1984, *Astrophys. J.* **287**, L 51
Samarasinha, N. H., A'Hearn, M. F., Hoban, S., Klinglesmith, D. A.: 1986, ESA SP-250, Vol. **I**, p. 487
Sekanina, Z.: 1973, *Astrophys. Letters* **14**, 175
Sekanina, Z.: 1975, *Icarus* **25**, 218
Sekanina, Z.: 1981a, *Ann. Rev. Earth Planet. Sci.* **9**, 113
Sekanina, Z.: 1981b, ESA SP-174, p. 55
Sekanina, Z.: 1982, *Astron. J.* **87**, 161
Sekanina, Z.: 1983, in *The Comet Halley Handbook*, 2nd ed., Jet Propulsion Laboratory, Pasadena, CA, p. 14
Sekanina, Z.: 1986a, *Adv. Space Res.* **5**, No. 12, p. 307
Sekanina, Z.: 1986b, IAU *Circ.* No. **4273**
Sekanina, Z.: 1986c, IAU *Circ.* No. **4187**
Sekanina, Z.: 1986d, 20th ESLAB Symp., ESA SP-250, Vol. **II**, p. 131
Sekanina, Z.: 1987, *Nature* **325**, 326
Sekanina, Z., Larson, S. M.: 1983, *Bull. Am. Astron. Soc.* **15**, 651
Sekanina, Z., Larson, S. M.: 1984, *Astron. J.* **89**, 1408, 1446
Sekanina, Z., Larson, S. M.: 1986a, *Nature* **321**, 357
Sekanina, Z., Larson, S. M.: 1986b, *Astron. J.* **92**, 462
Sekanina, Z., Larson, S. M., Emerson, G., Helin, E. F., Schmidt, R. E.: 1987, *Astron. Astrophys.* (in press)
Simpson, J. A., Sagdeev, R. Z., Tuzzolino, A. J., Perkins, M. A., Ksanformality, L. V., Rabinowitz, D., Lentz, G. A., Afonin, V. V., Ero, J., Keppler, E., Kosorokov, J., Petrova, E., Szabo, L., Umlauft, G.: 1986, *Nature* **321**, 278
Smoöuchowski, R.: 1982, *J. Geophys. Res.* **87**, A422
Smoluchowski, R.: 1985, in *Ices in the Solar System*, eds. J. Klinger, D. Benest, A. Dollfus, R. Smoluchowski, Reidel, Dordrecht, p. 397
Strazzulla, G.: 1985, *Icarus* **61**, 48
Strazzulla, G.: 1986, *Icarus* **67**, 63
Strazzulla, G., Calcagno, L., Foti, G., Sheng, K. L.: 1985, in *Ices in the Solar System*, eds. J. Klinger, D. Benest, A. Dollfus, R. Smoluchowski, Reidel, Dordrecht, p. 273
Strazzulla, G., Cataliotti, R. S., Calcagno, L., Foti, G.: 1984, *Astron. Astrophys.* **133**, 77
Strazzulla, G., Pirronello, V., Foti, G.: 1983, *Astron. Astrophys.* **123**, 93
Whipple, F. L.: 1950, *Astrophys. J.* **111**, 375
Whipple, F. L.: 1977, in *Comets, Asteroids, Meteorites: Interrelations, Evolution, and Origins*, ed. A. H. Delsemme, University of Toledo, Toledo, OH, p. 25
Zarnecki, J. C., McDonnell, J. A. M., Alexander, W. M., Burton, W. M., Hanner, M. S.: 1986, 20th ESLAB Symp., ESA SP-250, Vol. **II**, p. 185

Astron. Astrophys. 187, 796–800 (1987)

ASTRONOMY AND ASTROPHYSICS

Charging of dust particles in comets and in interplanetary space

P. Notni and H. Tiersch

Zentralinstitut für Astrophysik der AdW der DDR, Sternwarte Babelsberg, Rosa-Luxemburg-Strasse 17a, DDR-1591 Potsdam, German Democratic Republic

Received January 27, accepted June 16, 1987

Summary. We calculate the electric potential which is to be expected on dust particles in a plasma, covering a wide range of plasma temperature, density and relative velocity between plasma and dust particle, two plasma compositions (H_2O^+ and H^+) and some selected values of photoemission and secondary electron emission coefficients. An error concerning the high temperature region in the graphs of our paper (Notni and Tiersch, 1986) is corrected. Systematic trends in the results are discussed, and a crude estimate is given of the influence of a charge on the motion of dust in comets.

Key words: comets – dust – charge on dust

1. Introduction

The charge to be expected on a dust particle in a cometary or interplanetary environment is of interest mainly for two reasons:

1. The motion of a charged particle is influenced by electromagnetic fields of any kind in the plasma (Coulomb interaction, large scale magnetic fields etc.).
2. The particle may explode due to electrostatic forces.

Since the charge cannot yet be measured directly, theoretical calculations have to be made (Notni, 1964; Wyatt, 1969; Drain and Salpeter, 1979; Millet et al., 1980; Lafon et al., 1981; Mukai, 1981; Wallis and Hassan, 1983; Böhnhardt, 1976; and others). Most of the cited papers are concerned with some special aspects of the problem. We thought it useful, therefore, to have at hand a set of theoretical data for free interpolation into parameter combinations which normally occur in interplanetary space or cometary comae. Regrettably, an undetected code error in the computer programme influenced our results in the first version of the conference proceedings (Notni and Tiersch, 1986), where positive values of the grain potential are plotted too high in the region of significant secondary emission. We present the corrected version of these calculations in the following, slightly expanded paper.

We used the theory of charging as developed by Notni (1964), updating in the calculations the values and sometimes the functional dependences of the grain parameters. One can refer to the extensive paper of Böhnhardt (1985, also 1986) for a similar discussion of the input parameters. Böhnhardt, however, used an incorrect way to derive his fundamental equations of interaction between a thermalized plasma and a dust grain; his results differ from ours, therefore, and have to be used with great caution. (He assumed a Maxwellian velocity distribution of the plasma particles *near the surface of the charged grains*. This is incorrect, since the mean free path of the plasma particles is much greater ($10^{10\pm}$) than the Debye screening distance of the grain. The plasma is not in thermal equilibrium near the grain's surface, therefore, and a free-particle approximation has to be used.)

Send offprint requests to: P. Notni

2. Input data

The physical parameters of the plasma as well as grain parameters are expected to have a wide range. We try to cover a useful range of the following quantities by the calculations:

a) In the plasma:
 Electron and ion temperature, T_e, T_i; mostly $T_e = T_i$.
 plasma number density, $n = n_e = n_i$,
 plasma bulk velocity with respect to the grain, v,
 plasma composition (H^+-electron and H_2O^+-electron plasma).

b) Grain parameters (independent of diameter):
 Photoelectric yield, $c(\chi)$,
 secondary electron emission (SEE-) yield, δ.

We considered 3 model types of grains, spanning the probable range of the coefficents: silicate, graphite and "aluminium", the latter being a "model material" for (organic?) grains which have a low SEE yield.

We did not consider the following effects which have been discussed by other authors but are of mostly minor importance within our range of parameters:

– Non-Maxwellian electron or ion energy distribution (Lafon et al., 1981) which may occur near shock fronts (Feldman et al., 1983) and which may lead to multivalued potentials (Meyer-Vernet, 1982),

– a very small grain radius, which influences the photoelectric yield (Mukai, 1981) and, at high energies, the values of the sticking and secondary electron emission coefficients (Draine and Salpeter, 1979; Böhnhardt, 1985).

– effects of thermoionic emission which may be important at small distances from the sun (Millet et al., 1980),

– field emission (Drain and Salpeter, 1979; Meyer-Vernet, 1984),

– sputtering,

– effects of stochastic charge fluctuations (Gail and Sedlmayr, 1975) which are of some significance only for the smallest cometary particles (0.01 μm) at very low plasma temperatures,

– and the time dependence of the charging process.

2.1. The photoelectric yield

The dependence of the photoelectric yield on photon energy has been approximated by the formulas given by Mukai (1981). To characterize to *total* photoelectric yield under solar irradiation, we introduce the parameter $c(\chi)$, the value of the integral photoelectric yield under solar irradiation referred to that of silicate at 1 AU (calculated to be 1.255 10^{10} photoelectrons $cm^{-2}\,s^{-1}$). χ stands for the kind of the material (the "chemical composition"). For our three materials, viz. Si, C, Al, $c(\chi)$ equals 1, 0.12, and 0.82, resp., at 1 AU. Also, effects of grain radius r (Mukai, 1981) or those of varying distance from the sun R are easily incorporated into c, writing $c(\chi, r, R, \ldots)$. It is easily seen that in the low-temperature range (no SEE) the potential on the grain depends only on the *combined* parameter $n/c(\chi)$, n being the plasma number density. Materials of widely differing photoelectric properties as well as the influence of different distances from the sun or shadowing effects, and different plasma densities, are thus described by a single set of curves.

The distribution of photoelectrons on energy has been taken from Mukai (1981).

2.2. Secondary emission

The secondary electron emission yield, δ, depends on the primary's energy and reaches a peak value δ_m at an energy E_m. We used $E_m = 420$, 250, 420 eV and $\delta_m = 2.5$, 1.0, 0.95 for silicate, graphite and "aluminium", respectively.

Since we will restrict ourselves to plasma temperatures $< 10^7$ K, we can still use an approximation for the dependence of δ on E as given by Notni (1964), $\delta = \delta_m(1 - \exp(-4.3\,E/E_m))$. The formulas of Drain et al. (1979) or of Sternglass (1972) would have been a better choice for $E > E_m$; we did not use these expressions simply because Notni's formula can be integrated analytically to especially simple expressions which are of sufficient accuracy at the moderate energies considered. The error introduced by this is negligible for $kT < E_m$ and of little significance ($< 30\%$) up to $kT < 3\,E_m$, that means $T < (1 \ldots 2)\ 10^7$ K. One should bear in mind, however, that the near constancy of the calculated grain potential at $T > 10^7$ K is caused in part by this assumption. At very high temperatures, $T >$ a few 10^7 K, the potential of real grains will swich back to negative values. The use of Notni's expression implies further that formally no multivalued solutions occur, even for non-Maxwellian plasmas, in contrast to the situation for a δ-curve descending after E_m (Meyer-Vernet, 1982).

The energy distribution function of the secondary electrons has been taken from Drain and Salpeter (1979); its unintended omission is the reason for the strongly biased results of Notni and Tiersch (1986) at positive potentials in the high temperature range.

Secondary electrons from ions have been considered using $E_m = 50$ keV, $\delta_m = 3$.

We want to stress the fact, that all calculated positive potentials suffer strongly from the uncertainties of the experimental energy distribution functions of photoelectrons and secondary electrons. Errors of up to, say, 30% seem easily possible.

3. Results

Figures 1–4 give an overview of our results. A cometary plasma with a "high density" ($n > 1000\ cm^{-3}$) and a relative velocity between dust and plasma near zero is assumed in Fig. 1. The comparison of the potentials acquired by the grains in a H_2O^+-plasma ($eU = -3.66\,kT$ at low T) with that in a H^+-plasma ($eU = -2.50\,kT$ at low T) shows that the kind of ions (and thus also a mixture of different ions) is of minor importance as long as no secondary electron emission sets in. At higher temperatures the coefficient for SEE is the most important parameter. If δ is noticeably greater than 1 the equilibrium between the electrons and their secondaries determines the potential, whereas at $\delta < 1$ the equilibrium between the incident ions and the small number of sticking electrons may become most delicate – see the curves for $\delta_m = 0.95$ ("Al").

Figures 2 and 3 show for the high secondary electron emitter silicate the electrostatic potential under physical conditions typical for an interplanetary environment at 1 AU, i.e. at high bulk velocities and small plasma densities, also contrasted to velocity zero and high density. Positive potentials are expected in interplanetary space at nearly all instances ($n/c(\chi) < 100\ cm^{-3}$, $v > 400\ km\,s^{-1}$). At low densities ($n/c < 1$) the photoeffect, in equilibrium with the plasma electrons, determines the potential nearly independently of bulk velocity (Fig. 2), SEE-coefficient (Fig. 3) or temperature. At high solar wind velocities the SEE causes a positive potential already at small temperatures even at high densities, the plasma density plays but a minor role then (Fig. 3).

Note that the results can be used for $R \neq 1$ AU if $c(\chi)$ is put equal to $c(\chi) * (1/R^2\,[\mathrm{AU}])$.

To check what happens if electrons and ions are not in thermal equilibrium, Fig. 4 exhibits a comparison of results obtained for $T_i = 0$ and $T_i = T_e$. At low plasma densities the "missing" of the ions is not felt (photoelectrons outnumber the ions), whereas at high densities the potential is dramatically driven to high negative values at low temperatures (in the limit of very high density the equilibrium is determined by the electrons and their secondaries solely. For low secondary electron emitters the potential is unlimited in the envisaged situation).

4. Discussion and conclusions

The results presented here are in accord with smilar calculations made over the years by other authors in the order-of-magnitude of the expected potentials (papers cited in the introduction). The basic conclusions, reached by these authors regarding the disintegration of grains by electrostatic forces or the influence of the charge on their motion remain unaltered, therefore (see for inst. Mendis et al., 1985, and references therein). The inclusion of a material of low SEE shows, as expected, that such materials can reach very high negative potentials in a high-temperature environment. Organic compounds, nowadays supposed to occur in comets (Clark et al., 1986; Sekanina, 1986 and references therein) belong to this class (Ono and Kanaya, 1979). This may make such materials especially suitable for plasma-grain coupling.

Interesting effects may occur if the plasma parameters are near to an "instability region" of the solutions, i.e. near the sudden jumps in the dependence on T or, less dramatically, near the quick changes of the potential depending on $n/c(\chi)$ (between 10 and 1000). Small environmental changes may have great effects there, as discussed by many authors. For instance, passage through a shock wave may switch the potential from one sign to the other (Cravens et al., 1984), or different shadowing of the grains in optically thick environments [changing $c(\chi)$ in our graphs, cf. Sect. 2.1] may have similar effects (Meyer-Vernet, 1984).

Semiquantitative interpolations can be made into the set of curves shown. This should be especially useful if in-situ

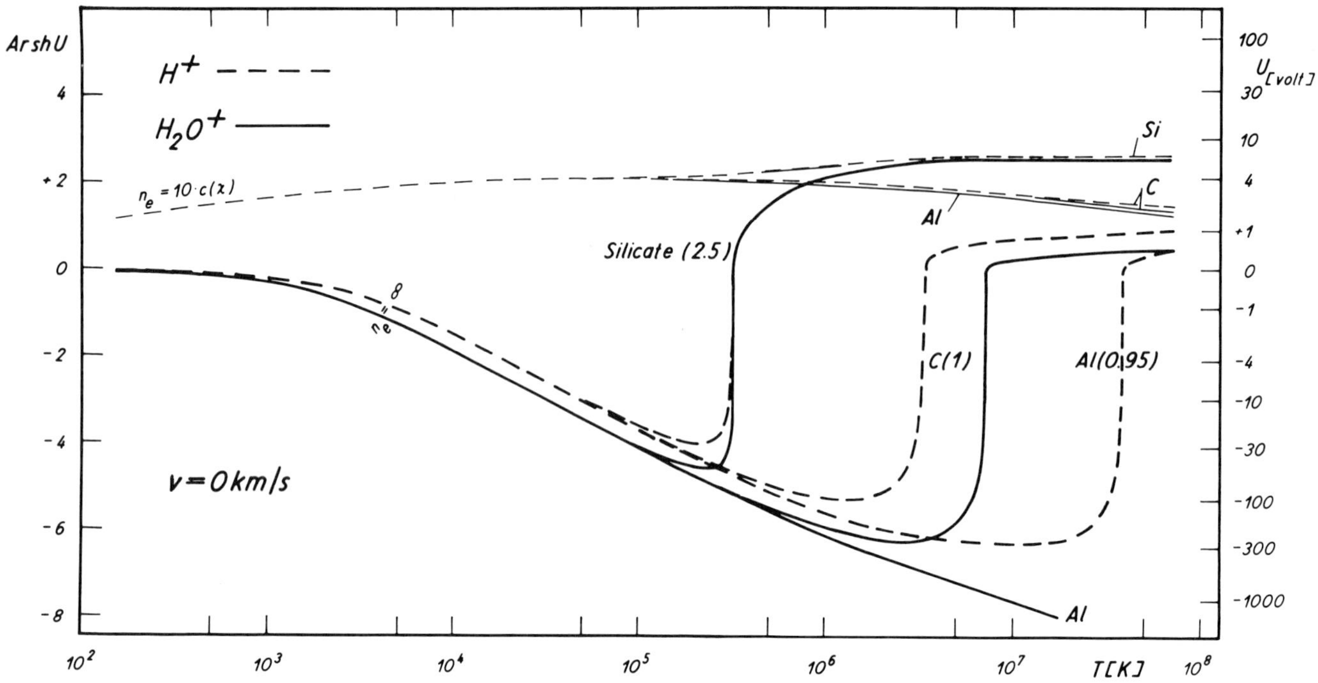

Fig. 1. Electric potential U on dust grains of different composition in dense plasma (H_2O^+-resp. H^+-electron) at a relative velocity near zero (thick lines). The maximum secondary electron emission (SEE) yield, which determines the behaviour at high temperatures, is given in parentheses behind the material names for quick orientation. For comparison, the (upper) thin-lined set of curves gives the results for a low-density plasma at $n/c(\chi) = 10\,\mathrm{cm}^{-3}$ where the photocurrent is the most important contributor. $c(\chi)$ is the total photoelectric yield under solar irradiation referred to that of silicate at 1 AU distance from the sun (for graphite, e.g., $c \approx 0.1$ at 1 AU, and the curves shown refer to $n \approx 1\,\mathrm{cm}^{-3}$). Ordinate is the inverse hyperbolic sine, $\operatorname{ar\,sh} U = \ln(U + \sqrt{U^2+1})$, which is linear at small U and logarithmic for $|U| \gg 1$; see right hand scale for U

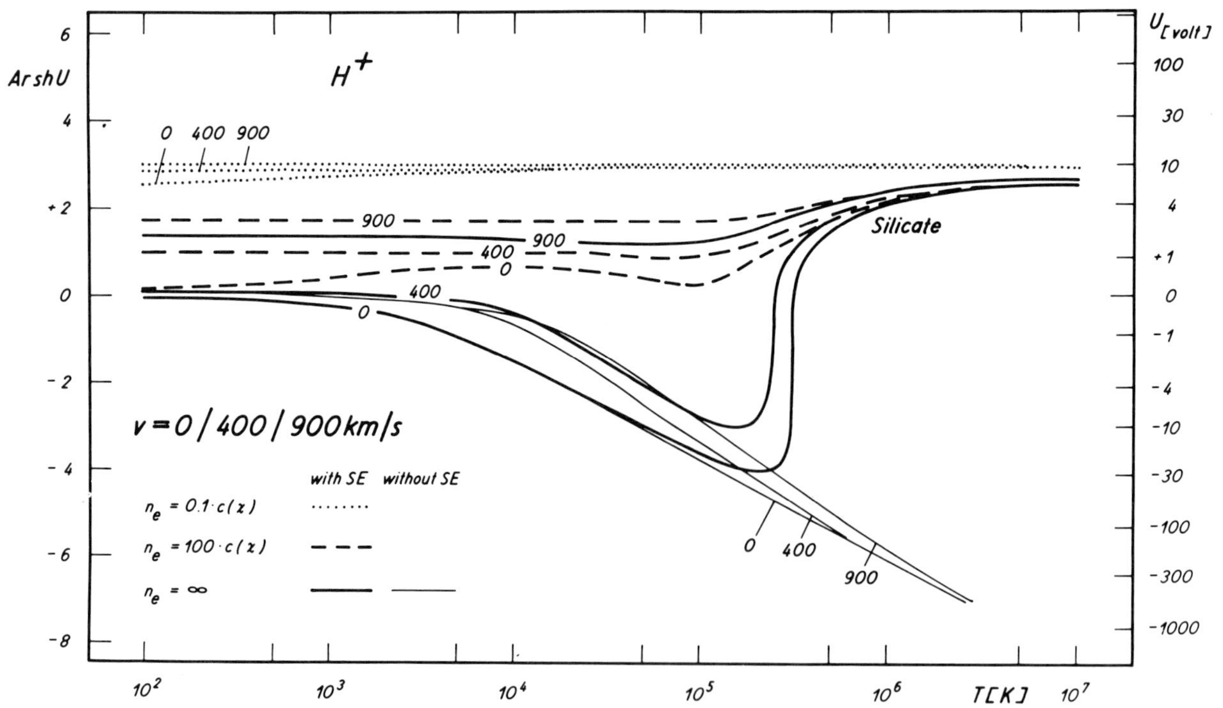

Fig. 2. Electric potential U for silicate material in a H^+-electron-plasma at different bulk velocities of the plasma and three densities. In the low density limit, SEE due to the bulk motion becomes negligible even for the high secondary electron emitter silicate – the photoeffect prevails (dotted curves). – Results for a material with an assumed SEE yield of zero are shown for the high density limit for comparison (thin lines at negative U). Note that these curves stay at $U < 0$ always! $c(\chi) = 1$, see legend to Fig. 1 and text

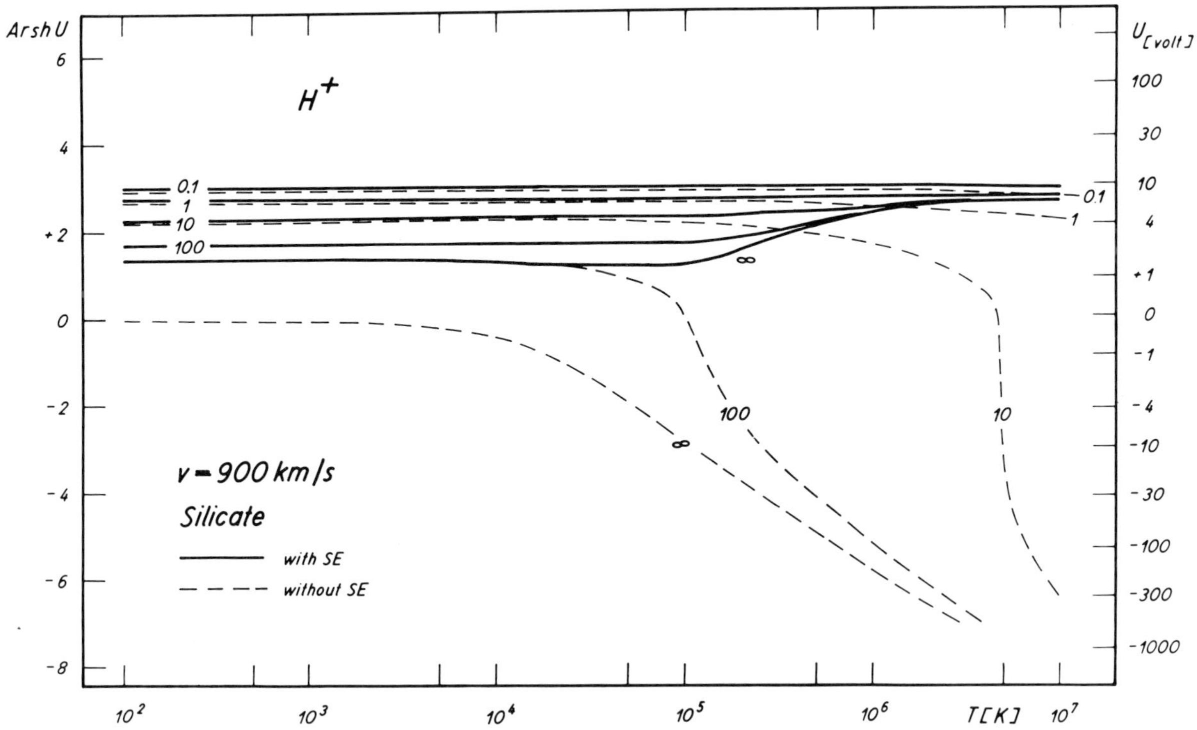

Fig. 3. Electric potential U on silicate dust grains at high relative velocity in a H^+-electron-plasma, at various densities. Results using the high SEE coefficient of silicate are compared to those for zero secondary electron emission. Numbers at the curves are plasma densities in cm^{-3}. ($c(\chi) = 1$)

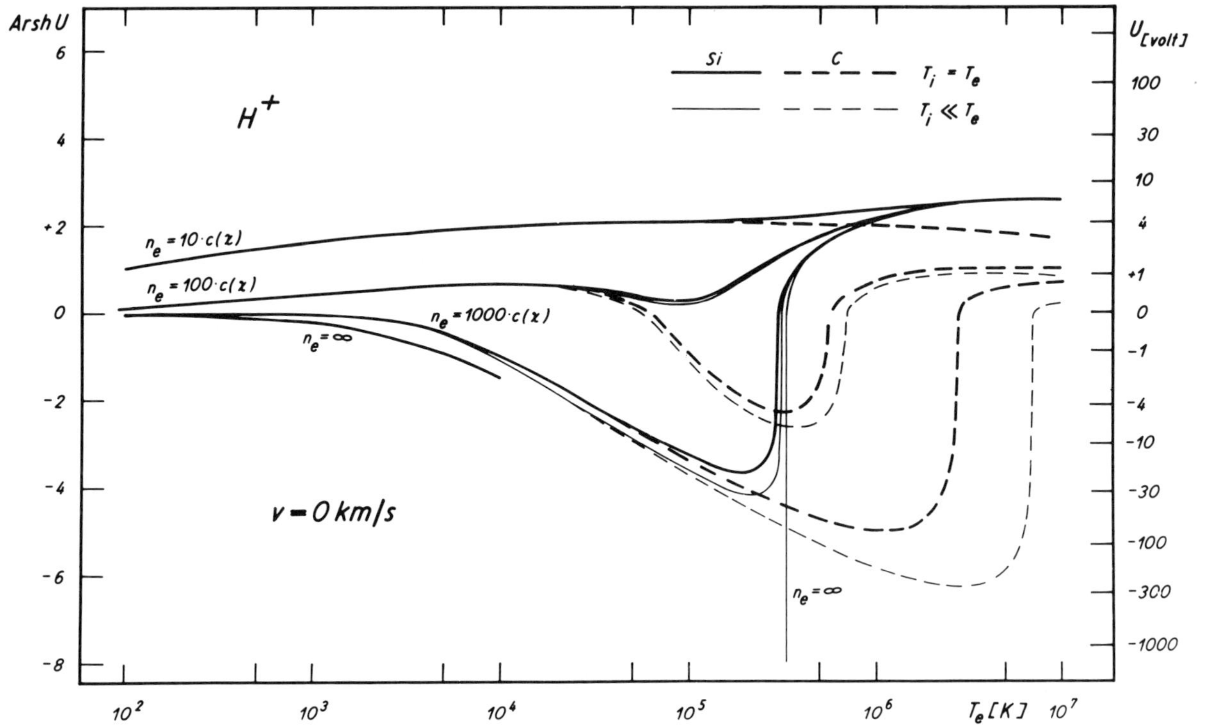

Fig. 4. The influence of non-equilibrium ion and electron temperatures on the electric potential U of silicate and graphite dust grains in a H^+-electron-plasma at zero bulk velocity. Thick lines – $T_i = T_e$, thin lines – $T_i = 0$. SEE has been taken into account. The curve showing the high density limit at $T_i = T_e$ has not been drawn fully for clarity, comp. Fig. 1. For $c(\chi)$, see text

measurements of plasma and grain parameters become available. In the dense central parts of the magnetotail of P/Giacobini-Zinner, for instance (compiled among others in Galeev, 1986, references therein), at $T=2\ 10^4$ K, $n=600\ \mathrm{cm}^{-3}$ the potential reached by Si, C and Al will be -2.2, -4.7, -3.1 Volts resp. (H_2O^+-plasma). The plasma density is yet insufficient for any significant coupling by coulomb interaction save for the lightest particles (acceleration $5\ 10^{-2}\ \mathrm{cm\,s}^{-2}$ for $r\approx 0.01\ \mu$m; one can use the calculations given in Notni (1964, Fig. 15) for this order-of-magnitude estimate. Spherical grains and a density $\varrho=1\ \mathrm{g\,cm}^{-3}$ is assumed). Magnetic interaction at $B=50$ nT, on the other hand, results in a gyration-radius of only $L=1.5\ 10^3$ km for $r\approx 0.01\ \mu$m and $U=-5$ V if the particles have the low speed of $v_\perp=100\ \mathrm{m\,s}^{-1}$. Very small and slow particles will thus not be able to leave the magnetotail orthogonal to its direction (see also Cravens et al., 1984).

In more active comets densities and field strengths will be higher; all materials will reach about the limiting potential for $n\to\infty$ (if $n>1000\ \mathrm{cm}^{-3}$), that is $U\,[\mathrm{Volts}]=-3.66\,kT\,(\mathrm{eV})$ in a H_2O^+-plasma. We expect a charge between -5 and -10 V at temperatures of some 10^4 K. Taking the optimistic value $U=-10$ V, and $n=10^4\ \mathrm{cm}^{-3}$, Coulomb interaction results in accelerations above $0.1\ \mathrm{cm\,s}^{-2}$ for particles below $0.1\ \mu$m (solar attraction is $0.5\ \mathrm{cm\,s}^{-2}$ at 1 AU). The magnetic interaction is expected to be larger than in P/Giacobini-Zinner, likewise. Thus, small particles will be influenced by the plasma-stream for some time downstream in the tail in very active comets; this must be considered in addition to the action of the solar radiation pressure.

Acknowledgements. We thank our referee, Dr. N. Meyer-Vernet, for her valuable remarks which helped us much to improve the presentation of our paper.

References

Böhnhardt, H.: 1985, Diss. Univ. Erlangen-Nürnberg, Nat. Wiss. Fak. I

Böhnhardt, H.: 1986, 20th ESLAB Symp., ESA SP-250, Vol. **II**, p. 207

Clark, B., Mason, L., Kissel, I.: 1986, 20th ESLAB Symp., ESA SP-250, Vol. **III**, p. 353

Cravens, T.E., Gombosi, T.I., Gribov, B.E., Horànyi, M., Kecskemèty, K., Kőrösmezey, A., Marconi, M.L., Mendis, D.A., Nagy, A.F., Sagdeev, R.Z., Shevchenko, V.I., Shapiro, V.D., Szegő, K.: 1984, Hungarian Academy of Sciences, Central Research Institute for Physics, Budapest, Publ. KFKI – 1984, p. 41

Draine, B.T., Salpeter, E.E.: 1979, *Astrophys. J.* **231**, 77

Feldman, W.C., Anderson, R.C., Bame, S.J., Gary, S.P., Gosling, J.T., McComas, D.J., Thomsen, M.F.: 1983, *J. Geophys. Res.* **88**, 96

Gail, H.-P., Sedlmayr, E.: 1975, *Astron. Astrophys.* **41**, 359

Galeev, A.A.: 1986, 20th ESLAB Symp., ESA SP-250, Vol. **I**, p. 3

Lafon, J.P.J., Lamy, P.L., Millet, J.: 1981, *Astron. Astrophys.* **95**, 295

Mendis, D.A., Houpis, H.L.F., Marconi, M.L.: 1985, *Fund. Cosmic Phys.* **10**, 1

Meyer-Vernet, N.: 1982, *Astron. Astrophys.* **105**, 98

Meyer-Vernet, N.: 1984, *Icarus* **57**, 422

Millet, J., Lafon, J.P.J., Lamy, P.L.: 1980, *Astron. Astrophys.* **92**, 6

Mukai, T.: 1981, *Astron. Astrophys.* **99**, 1

Notni, P.: 1964, *Veröff. Sternwarte Babelsberg* **XV**, H. 1

Notni, P., Tiersch, H.: 1986, 20th ESLAB Symp., ESA SP-250, Vol. **III**, p. 385

Ono, S., Kanaya, K.: 1979, *J. Phys.* **D 12**, 619

Sekanina, Z.: 1986, 20th ESLAB Symp., ESA SP-250, Vol. **II**, p. 131

Sternglass, E.J.: 1972, Westinghouse Res. Lab. Sci. Paper

Wallis, M.K., Hassan, M.H.A.: 1983, *Astron. Astrophys.* **121**, 10

Wyatt, S.P.: 1969, *Planet. Space Sci.* **17**, 155

Astron. Astrophys. 187, 801–806 (1987)

ASTRONOMY AND ASTROPHYSICS

Evaporating grains in P/Halley's coma

M.K. Wallis, R. Rabilizirov, and N.C. Wickramasinghe

Department of Applied Mathematics and Astronomy, University College, P.O. Box 78, Cardiff CF1 1XL, Wales, UK

Received January 29, accepted May 11, 1987

Summary. Evidence is presented for the hypothesis that the organic grains streaming from a cometary nucleus lose substantial volatile constituents within 10^5 s. To develop the hypothesis, we study grain properties appropriate to both the comet-probe encounters at 0.8–0.9 AU and to comas observed at 4–5 AU. While submicron-sized grains of ice contaminated by 1 % or more absorptive graphite can explain the strong OH coma far from the Sun, as in comet Bowell at 4.6 AU, the dynamical lifetime of the grain coma implies mm-sized and therefore cooler grains, with weaker H_2O or OH attachment than ice at about 10 kcal/mole. Grains in comet Halley, if composed of similar organic material must also contain fractions of more strongly bound components. In this study we adopt kerogen as a prototype organic complex having a range of bond energies. Temperatures and evaporation rates of kerogen grains are calculated using Arrhenius relations and Mie-Güttler radiative scattering theory with complex refractive index as measured for two cases: organic material of biological origin and UV-processed methane ice ("tholin"). A time-scale of 10^4 s at 0.9 AU corresponds to submicron grains with rather strongly-bound organics of ~ 40 kcal/mole, identified with bridges of ether, ketone, sulphide etc., bonds. The steeply increasing strength of H and O inside 1 AU ($\propto r^{-4}$), unlike OH, corresponds to increasing breakdown following rupture of the chemical bridges. The 10 μm IR feature can be explained by grains of an organo-siliceous material or organic mantle on a siliceous core, a few μm in size and about 20 kcal/mole bond strength.

Key words: OH-coma – grain sublimation – organic composition – comet Bowell – comet P/Halley

1. Introduction

That cometary comas contain icy or ice-coated dust grains is a proposal of many years standing (Huebner and Weigert, 1966). In offering a larger surface area to the solar radiation, such grains on evaporation can even dominate over the nucleus as sources of gas and change the hydrodynamics of the coma. We now know from Vega and Giotto that small grains (femtogramme and smaller) are abundant, though it is unknown whether they arise directly from the nuclear material or condense out of the super-saturated gas near the nucleus (as in laboratory simulation of sublimating mixtures; see Dobrovolsky and Kaimakov, 1977). Optical studies indicate a new composition (Moroz, 1986), not studied hitherto. We know from in situ analysis (Kissel et al., 1986) and spectroscopy of the 3–4 μm waveband (Wickramasinghe and Allen, 1986; Knacke et al., 1986) that much of the dust is organic rather than mineral in composition. High ratios of radiation pressure to solar gravity at $\beta \simeq 2.5$, inferred from tail structure (Lamy, 1986), are consistent with ~ 0.1 μm grains of graphite-like absorptive material being present in the inner coma and relatively abundant close to perihelion.

An organic composition for the grains was first proposed by Vanýsek and Wickramasinghe (1975). They hypothesised polyoxymethelene or polyformaldehyde compounds, whose radiation scattering properties are generally suitable. Such grains would release appropriate OH, CN, HCO, etc., under the solar radiation.

Evidence for the evaporation of grains as the gas and dust mixture expands from the comet comes from various directions. The comet-probes showed that the fraction of grains with Si:C exceeding 50 % is lower close in to the nucleus (Kissel et al., 1986) while there are substantial amounts of C, CH, and C^+ in the gas and plasma phases (Balsiger et al., 1986; Krankowsky, 1986). Molecular CO increases outwards with a 10^4 km scale (Eberhardt et al., 1986) while CN increases both on this scale and a short (600 km) scale (Krasnopolsky et al., 1986). The report of radial structures in images covering the 388 nm CN band (A'Hearn et al., 1986) convinced many skeptics that grains are a source of some CN and other cometary molecules. The relative dominance of CN in comae at large heliocentric distance, compared with C_2 etc. (Wurm and Balazs, 1963) suggests that CN is more loosely bound and released at lower temperature.

The major H_2O component probably also comes in part from grains. An OH coma was marginally detected in Halley at 4.4 AU pre-perihelion (Festou et al., 1986), compatible with O I 630 nm observed six weeks earlier and of usual strength relative to CN; at such distance, sublimation from contaminated crystalline ice at ~ 130 K is too slow (cf. Sect. 2), so less strongly bound H_2O may be inferred. The H and O abundances, judged from H α and L α strengths (Craven et al., 1986) and from O I 630 nm intensities (Sherb et al., 1986), increased more strongly than OH (Feldman et al., 1986) with closeness to the Sun, especially inside 1 AU where the OH strength was roughly constant but with strong fluctuations. Suisei's L α observations (Kaneda et al., 1986) imply a source of H, perhaps CH, with faster photo-production than from H_2O. The OH if it all came from evaporation of H_2O-ice from the nucleus surface, requires an emitting surface area of some 20 km^2 at the time of the Giotto encounter (5 10^{29} molecules/s, Feldman et al., 1986) from a surface normal to the solar radiation at 0.9 AU), which is 3–4 times the projected area of the "jet source", that is itself structured (Keller et al., 1986; Huebner et al., 1986a). The R^{-2} variation of H_2O density, obeyed quite closely along the

Send offprint requests to: M.K. Wallis

probe trajectories even on the approach from behind the nucleus (Curtis et al., 1986), is understandable if much of the H_2O issues from the grains despite strongly anisotropic initial emission from the sunlit surface. Although the grains move sunward at $\lesssim 0.5\,\mathrm{km\,s^{-1}}$, H_2O molecules evaporating at higher speed and isotropically from individual grains can fill the tail-side coma, resulting in much reduced anisotropy.

2. Grains at several AU from the Sun

2.1. *What size icy grains in comet Bowell?*

Comet Bowell 1980b had a well-developed particle coma more than a year prior to perihelion at 3.4 AU. In early 1981, it showed high OH strength over $r = 5.3$–4.6 AU (for $>10^7$ s) but lower OH a year later around perihelion. If this arose from large ice grains composing the long-lived coma (the favoured explanation – A'Hearn et al., 1984; Sekanina, 1982; Jewitt, 1984), then the grains must be large enough for a long dynamical lifetime, which for a grain radius a in mm and $D = 50''$ diameter coma (Sekanina, 1982) under radiative acceleration g is

$$\sqrt{(D/g)} = 5\ a\,[\mathrm{mm}]\ 10^7\ [\mathrm{s}]. \tag{1}$$

Evidently this implies mm-sized grains in the 1981 coma. However, contaminated icy grains would be too cold to emit significant H_2O at 4.6 AU (Hanner and Campins 1986), being around the 130 K black-body temperature [their total surface area being inferred from the 20 µm IR flux (Hanner and Campins, 1986)]. If icy grains are to supply H_2O for the production of OH, they have to be warmer, as are micron-sized grains.

2.2. *Analysis of micron-sized dirty-ice grains*

Taking various fractions of contaminant absorbers embedded in the ice, we have calculated the grain temperature on the basis of Mie scattering theory for spherical grains. We could use the Maxwell-Garnett mixing rule for spherical inclusions (Bohren and Wickramasinghe, 1977) but such complexity makes a difference only for a large imaginary part to the refractive index near the solar spectrum peak. Mukai (1986) who does use the mixing rule also assumes a single size to the inclusions, but results are very sensitive to that. So we employed the simple linear combination

$$m = (1 - w_g)\, m_{\mathrm{ice}} + w_g\, m_{\mathrm{graphite}} \tag{2}$$

for fractional contamination (by volume) w_g, and adopted the tabulated values of refractive indices m for ice and graphite. We used a Planckian spectrum for the solar radiation as giving adequate accuracy. The temperature for slightly contaminated 5 µm grains agrees closely with Hanner (1981) for a real spectral distribution. We allowed for the energy lost in sublimation and calculated the grain lifetime against evaporation according to the half-size formula

$$\tfrac{1}{2}a/|\dot{a}| \quad \text{with} \quad |\dot{a}| = (p/\varrho)\,(m/2\pi kT)^{0.5}. \tag{3}$$

Figures 1 depict results for various values of w_g, showing that the evaporation lifetime of 5 µm grains at 4.6 AU is 10^7 s if $w_g \simeq 1\%$ (more precisely, 0.25–1.5%), but longer for higher or lower contaminant fractions than this. Smaller grains have generally much shorter lifetimes (as the 0.3 and 0.1 µm grains in Fig. 1b, excepting the uncontaminated 0.3 µm example). We have compared results for the absorbing graphitic proportion confined to

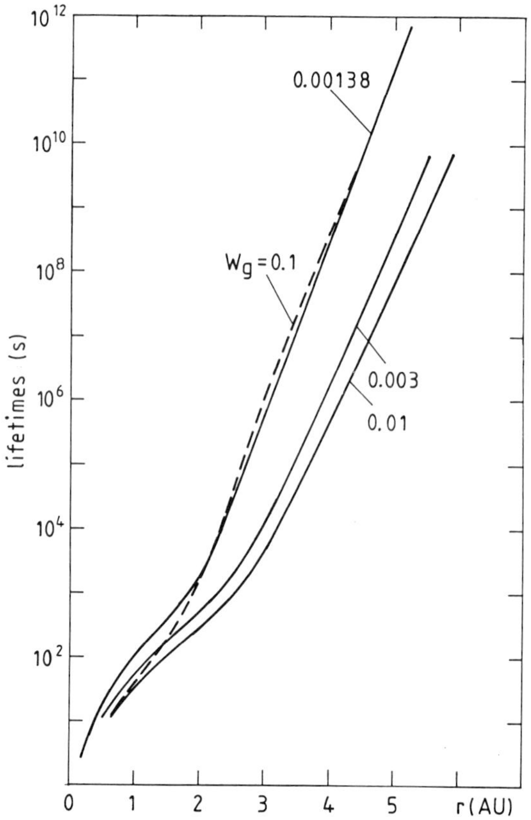

Fig. 1a. Lifetimes of 5 µm radius ice grains contaminated by small fractions of graphite w_g (by volume). The "lifetimes" given here are really half-lives, being the time to decrease to half the initial radius, by when the scattering properties and graphite fraction have changed drastically. The value $w_g = 0.00138$ is chosen to give the absorption coefficient $k = 0.002$ adopted by Hanner (1981). The fraction $w_g = 0.01$ gives hotter and faster sublimating grains than larger or smaller fractions for the 5 µm grains

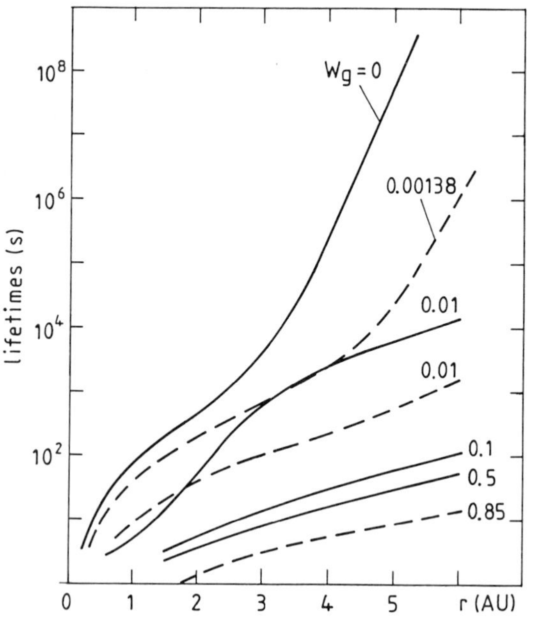

Fig. 1b. Lifetimes of submicron ice grains: solid lines – 0.3 µm radius; broken lines – 0.1 µm radius, with fractional contamination of dispersed graphite w_g as labelled

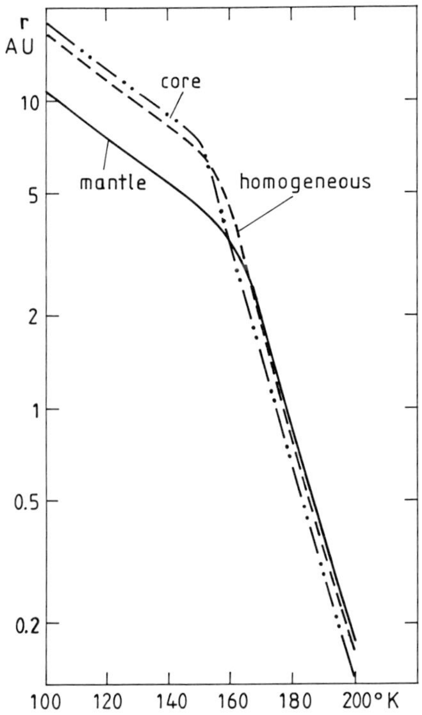

Fig. 2. Solutions of the sublimation equation for 0.1 μm ice grains containing 1% ($w_g = 0.01$) graphite, showing heliocentric distance where a given temperature is attained, for cases with the graphite in a mantle, in a central core, or homogeneously distributed as in Figs. 1. There is a marked change in slope where sublimation starts to become important in the energy balance (at 155 K). The 10 K spread in temperature at the 4.6 AU distance of comet Bowell is significant, showing that a contaminated grain that starts evaporating but develops a mantle of the non-volatile graphite will have its lifetime lengthened by several times

a central core, or to an outer mantle. Figures 2 give results for $w_g = 1\%$ graphite. The case with it concentrated in a core is close to the uniform distribution, T differing by under 5 K. A graphite-mantled grain, however, radiates efficiently and is substantially cooler beyond a few AU. The cooling is much stronger in the larger example shown (0.4 μm, Fig. 2b; cf. 0.1 μm of Fig. 2a). At 4.6 AU, mantled grains of 0.4 μm and larger are close to the 130 K black-body temperature and evaporate too slowly. Evidently grains of a few μm and initially uniform contamination could sublimate at 4.6 AU to produce the observed OH, but as the contaminating graphite accumulates at their surfaces they cool and sublimation is stopped. In any case, the residence time (1) of such grains in the coma is $<10^6$ s, so continual emission of such grains in early 1981 is implied, but lower emission around perihelion one year later. The small grain hypothesis would not explain the relatively sharp coma boundary reported (Jewitt, 1984), nor would small grains alone give the 20 μm IR flux and the 10 μm:20 μm ratio observed (Hanner and Campins, 1986).

2.3. Alternative to ice grains

Our preferred explanation is for large grains of primarily organic composition (Hoyle et al., 1985) with temperature similar to a black body (130 K at 4.6 AU) that emit H_2O/OH on the 10^7 s time scale. This requires slightly weaker binding energy than ice, at about $E = (130/155) \times L[H_2O] \simeq 10$ kcal/mole. On loss of adsorbed H_2O, CO_2 and weakly bonded side-chains (e.g. CH_2, NH, H, CN), these grains would shrivel and become more absorbant (C=C bonds), giving in particular the dark grains observed postperihelion at 3.5 AU (1982 June) with temperature (Hanner and Campins, 1986) 15% above that of a black-body (i.e. IR absorptivity <0.25).

3. Comet Halley's grains

The infra-red spectra of comet Halley imply that the grains probably consist of a complex mixture of organic compounds (Knacke et al., 1986). If of similar initial oranic composition to those in comet Bowell, the submicron femtogram component would shed its loosely-bonded H_2O/OH, H, CN etc., within 100 s in free space at 0.9 AU (Fig. 1). Near the nucleus, however, IR radiation is strongly blanketed by gaseous H_2O, CO_2, etc., (Bockelée-Morvan and Crovisier, 1986; Wallis, 1986) and also recondensation on the grains may occur (Kitamura and Yamamoto, 1986; Crifo, 1986). Still, the loosely-bonded (10 kcal/mole) fraction has surely left the grain surfaces within a few thousand km (Crifo, 1986). Rupture of C=C bonds at 70–80 kcal/mole would not however proceed. Between these extreme bond strengths, various chemical components do come off.

3.1. Grains composed of kerogen

In the absence of specific composition data, we take as prototype the material classified as *kerogen* – the insoluble components of organic residues on Earth (Tissot and Welte, 1978) and thought to resemble that in carbonaceous chondrites and interstellar grains (Pelet, 1980). We assume the bonds of strength E_i are disrupted at rates given by the Arrhenius formula

$$d(\ln n/dt) = A_i \exp(-E_i/RT) \quad (4)$$

which simulates independent rupturing of bonds in a frozen lattice (Tissot and Welte, 1978 – Part V, Sect. 4.2). We adopt the bond strengths and proportions given for Type II terrestrial kerogen (reproduced in Table 1), but expect the weakly-bonded ($E \lesssim 30$) fractions to be higher in space organics. The radiative absorptivity of organic grains is uncertain, for the graphitic fraction found in terrestrial kerogens may not be prevalent in space. Organics are largely dielectric plus pigments, typified by chlorophyll with absorption peaks in the visible. Using measurements of terrestrial grass extracts and micro-organisms (Al-Mufti, 1985), we take the real part of the refractive index as 1.4 and imaginary part dominated by peaks around 433 and 660 nm and with significant value in the IR (details in Table 2). From polarization data, Mukai et al. (1986) derived the real refractive index as 1.4, while Krasnopolsky et al. (1986) give 1.35, but the difference is not significant for the present analysis.

The incident energy going to rupturing the bonds in the submicron grains is tiny, so the temperature given by ignoring that component can be calculated via Mie theory for spherical grains of various sizes. By setting the timescale for transformation at 10^4 s appropriate to submicron grains' residence time in the inner 10^4 km coma, we calculate from (3) the heliocentric distances where the various bonds E_i suffer rupture. The results for Type II kerogen are given in Table 1: the grain temperatures are numerically close to $14 E_i$ [in K for E in kcal/mole]. Evidently, bonds up to 50 kcal/mole rupture in Halley's grains near perihelion ($q = 0.6$ AU) where temperatures of 700 K are attained.

Table 1. Heliocentric distance r_{crit} for a time scale of 10^4 s for breaking bonds of strength E in three sizes of organic grains. The critical temperatures vary little: $T_{crit} \simeq 14E$. The E-data is a representation of experimental results on Type II kerogen (Tissot and Welte 1978, V-4.2); the constants A_i given there are not used as they apply to transformations in liquid rather than the solid lattice

E (kcal/mole)		10	30	50	60	70	80
f–fraction of bonds		0.022	0.034	0.251	0.152	0.116	0.120
$a = 0.05$ μm	T_{crit} (K)	140	420	670	820	965	1105
	r_{crit} AU	15	1.6	0.64	0.44	0.33	0.26
$a = 0.1$ μm	T_{crit} (K)	143	430	680	835	985	1125
	r_{crit} AU	16	1.6	0.64	0.44	0.33	0.26
$a = 0.3$ μm	T_{crit} (K)	148	445	710	860	1010	1160
	r_{crit} AU	13	1.25	0.50	0.34	0.25	0.20

Table 2. Complex refractive index of model organic grains. We employ $m = 1.4–ik$ with $k(\lambda)$ based on experimental results for biological organic material (Hoover et al., 1986): a) in the UV over $\lambda = 0.2–0.3$ μm, Fig. 7.6 and Table 7.17 of Al-Mufti (1985) measured for a mixture of micro-organisms; b) at visual $\lambda = 0.4–0.9$ μm, values of absorbance for grass extracts; c) in the IR, $\lambda = 2.5–16$ μm, transmittance measured for *E. coli* bacteria (Fig. 4.7 of Al-Mufti). In the remaining wavebands, we have adopted constant values for the mass absorption coefficient K. At $\lambda < 0.2$ μm, $K = 25\,000$; at $\lambda > 20$ μm, $K = 500$ and $k = 1.2\ 10^{-5} K$ [$cm^2 s^{-1}$]. The value $\mathrm{Re}(m) = 1.4$ corresponds to compact material and has been used to explain polarization of the scattered radiation (Mukai et al., 1986): lower values $\lesssim 1.2$ would better simulate porous or fluffy grains, but present results are not sensitive to it

λ (μm)	0.21	0.23	0.25	0.35	0.45	0.55	0.65	0.75	0.85	0.95
k	0.08	0.077	0.375	0.075	0.19	0.01	0.156	0.0014	0.0015	0.0017
λ (μm)	1.05	2.05	3.05	4.05	5.05	6.05	7.05	8.05	9.05	10.05
k	0.0019	0.0037	0.035	0.015	0.018	0.10	0.06	0.08	0.125	0.94

Figure 3 presents calculations of the evaporation lifetimes in fuller detail, including the energy absorbed in breaking the bonds (analogous to sublimation energy) and using the relative fractions of bonds f from Tissot and Welte (1978). We find, for example, that at 0.9 AU the 30 kcal/mole bonds have a very short 10–100 s lifetime (so diffusion of radicals through the grain matrix may in practice limit the escape rate), while 40 kcal/mole bonds last 4 10^3 s in 0.1 μm grains but 4 10^5 s in 0.3 μm grains (and ten times shorter at 0.8 AU).

3.2. *Grains composed of synthesised organics*

An alternative propotype for organic grain material is the "tholin" obtained by irradiating frozen methane and water (Khare et al., 1984). Mie calculations using their tabulated refractive index give lower temperatures (by ~100 K) and longer lifetimes, because the material lacks the absorption peaks in the visible of our biological organic. However, grains larger than 0.1 μm are still quite hot enough to evaporate and undergo chemical "cooking" (Hanner, 1986). The latter author shows the smallest grains of tholin (0.01 μm) to be very cold, but we find them to be hot (e.g. ~500 K at 0.9 AU) so still liable to rapid cooking. Figure 2 presents comparable calculations for $E = 30$ kcal/mole bonds in spherical grains of tholin – showing lifetimes up to twice as long but nowhere strongly different from the biological organics. Figure 2 shows also sample lifetimes for ice grains, including the contamination fraction $w_g = 0.00138$ corresponding to absorptive $k = 0.002$ in the visual adopted by Hanner (1981). At 0.9 AU, life-times are under 100 s even for pure ice of $a = 0.3$ μm.

4. Discussion

The correspondence between bond strengths and particular chemical components is not very close, because of the complex chemical linkages and "cages". Breaking one bond induces rearrangement of neighbouring ones, so modelling by independent breakages [Eq. (4)] may not be adequate. The 10 kcal/mole strength corresponds to physical/chemical absorption and some H/OH bonding. Carbonyl, carboxy and methoxy side-chains tend to be superficially attached with 20–30 kcal/mole. Ether (–O–), ketone (–CO–) and sulphide (–S– or –S–S–) bonds tend to serve as bridges with 40–50 kcal/mole, while carbon bonds are as strong as 60–80 kcal/mole.

Both the biological and the "tholin" organic prototypes (Sects. 3.1, 3.2) have significant absorption in the visible waveband, so grains of either substance are strongly heated in the outstreaming coma. If contaminant graphite were present as common in terrestrial kerogen, they would be still hotter.

Evidently the carbon bonds stay intact in Halley's submicron grains but, with rupture of bonds up to 50 kcal/mole, substantial

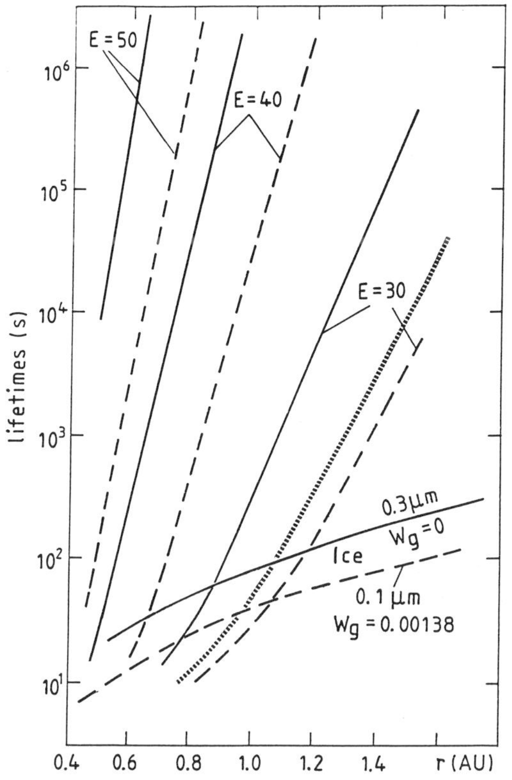

Fig. 3. Lifetimes of biological organic grains of 0.1 μm (broken lines) and 0.3 μm radius (solid lines) to evaporation or destruction of chemical bonds of various strengths E (kcal/mole) as indicated. Evidently for comet Halley at 0.8–0.9 AU, 0.1 μm grains quickly lose their $E=40$ components, but 0.3 μm grains retain them for $\sim 10^5$ s. $E=50$ components are hardly affected within the 10^5 km coma. Results for a laboratory organic "tholin" 0.1 μm grain with $E=30$ are shown dotted, being some twice as long due to lower absorptivity. Sample ice-grain lifetimes are plotted for comparison, being under 100 s for the Halley spacecraft encounters, so evaporation competes with condensation (Crifo, 1986) in the very inner coma

carbonisation may be achieved within 0.7 AU (explaining the high radiative repulsion indicated by the $\beta \simeq 2.5$ syndyne – Lamy, 1986 – see introduction). The time-scale at 0.9 AU for breaking the chemical "bridges" is sensitive to grain size, but comparable to outstreaming times (0.5 km s^{-1} speeds). As bridge-breaking exposes interior CH_2 etc., this can explain the decreasing fraction of C with distance from the nucleus (Kissel et al., 1986 – correlation with grain size would check this). The bridge-breaking also allows increasingly efficient release of H (and O) at smaller heliocentric distance, unlike OH if it comes mainly from H_2O. Molecular H_2O and CO_2 do come off immature kerogen that is heated to <600 K (but at lower temperature in the presence of clay catalysts), so these would come off among the first components from fresh submicron grains released around 1 AU.

Why does comet Halley's 10 μm "silicate" feature appear only inside 1.3 AU, becoming prominant at 1 AU? Such behaviour is known in other comets, and an evaporating organic coating has been proposed in explanation (Hanner, 1986). Do the grains have silicate cores (Mukai, 1986) or is –SiO– a strongly bound structural component? As grains larger than a few microns are required, we have calculated the evaporation times for material with the refractive index of our organic mixture. Being cooler than the submicron grains, time scales of 10^4–10^5 s around 1 AU are found for radii of 3–5 μm if $E=20$ kcal/mole. Since dielectric siliceous inclusions within a bulk of such weakly-bound organic material would not change significantly the absorption properties, such a composite would be suitable to explain the heliocentric distance dependence of the 10 μm feature. An alternative would be the organo-siliceous mixture, used previously in grain modelling (Hoyle et al., 1985), which has a biological analogue in siliceous diatoms (Hoover et al., 1986).

5. Conclusion

The discovery that "CHON" grains (A'Hearn et al., 1986) composed of the light elements form an abundant component of comet Halley's dust (Kissel et al., 1986) has confirmed the earlier hypothesis that complex organics are a significant constituent, which partially evaporates from the grains and contributes to the gas coma. This is one part of the new cometary paradigm, that is superseding the icy conglomerate theory.

We have shown that micron and submicron grains composed of biological "kerogen" or synthetic "tholin" have appropriate radiation emission properties and suitable evaporation lifetimes. Specific studies of the gas emission propeties of these classes of organics, and comparison with gas and dust distributions in the coma, to be detailed from the comet-probe and astronomical observations, should allow refinement of the new theory.

Acknowledgement. Support for RR under an SERC research award is acknowledged.

References

A'Hearn, M.F., Schleicher, D.G., Feldman, P.D., Millis, R.L., Thompson, D.T.: 1984, *Astron. J.* **89**, 579

A'Hearn, M.F., Hoban, S., Birch, P.V., Bowers, C., Martin, R., Klinglesmith, D.A.: 1986, 20th ESLAB Symp., ESA SP-250, Vol. **1**, p. 483

Al-Mufti, S.: 1985, Ph. D. Thesis, Univ. College Cardiff, Wales, UK

Balsiger, H. et al.: 1986, *Nature* **321**, 330

Bockelée-Morvan, D., Crovisier, J. 1986, *Asteroids Comets Meteors* **II**, ed. C.-I. Lagerkvist et al., Uppsala Univ., p. 279

Bohren, C.F., Wickramasinghe, N.C.: 1977, *Astrophys. Space Sci.* **50**, 461

Craven, J.D., Frank, L.A., Rairden, R.L., Dvorsky, M.R.: 1986, *Geophys. Res. Letters* **13**, 873

Crifo, J.F., 1986, 20th ESLAB Symp., SP-250, Vol. **1**, 533

Curtis, C.C. et al., 1986, 20th ESLAB Symp., ESA SP-250, Vol. **1**, 391

Dobrovolsky, O.V., Kaimakov, E.A.: 1977, *Comets, Asteroids, Meteroids*, ed. A.H. Delsemme, Univ. Toledo, p. 37

Eberhardt, P. et al., 1986, 20th ESLAB Symp., ESA SP-250, Vol. **1**, p. 383

Feldman, P.D. et al., 1986, ESA SP-263, 39

Festou, M.C. et al.: 1986, *Astron. Astrophys.* **155**, L 17

Hanner, M.S., Campins, H.: 1986, *Icarus* **67**, 51

Hanner, M.S.: 1981, *Icarus* **47**, 342

Hanner, M.S.: 1986, *Adv. Space Res.* **5**, 12, 325
Hoover, R.B., Hoyle, F., Wickramasinghe, N.C., Hoover, M.J., Al-Mufti, S.: 1986, *Earth Moon Planets* **35**, 19
Hoyle, F., Wickramasinghe, N.C., Wallis, M.K.: 1985, *Earth, Moon Planets* **33**, 179
Huebner, W.F., Weigert, A.: 1966, *Z. Astrophys* **64**, 185
Jewitt, D., 1984, *Icarus* **60**, 373
Kaneda, E., Takagi, M., Hirao, K., Ashihara, O., Shimizu, M.: 1986, 20th ESLAB Symp., ESA SP-250, Vol. **1**, p. 397
Keller, H.U. et al.: 1986, *Nature* **321**, 320
Khare, B.N., Sagan, C., Arakawa, E.T., Suits, F., Callcott, T.A., Williams, M.W.: 1984, *Icarus* **60**, 127
Kissel, J. et al.: 1986, *Nature* **321**, 280
Kitamura, Y., Yamamoto, T.: 1986, *Icarus* **68**, 266
Knacke, R.F., Brooke, T.Y., Joyce, R.R.: 1986, 20th ESLAB Symp. ESA SP-250, Vol. **2**, p. 95
Krankowsky, D., 1986 (personal communication to MKW)
Krasnopolsky, V.A. et al.: 1986 20th ESLAB Sym. ESA SP-250, Vol. **1**, p. 459
Lamy, P.: 1986, *Adv. Space Res.* **5**, 12, 317
Moroz, V.I.: 1986, Review paper delivered at Heidelberg, 20th ESLAB Symp. (unpublished)
Mukai, T., Mukai, S., Kikuchi, S.: 1986, 20th ESLAB Sym., ESA SP-250, Vol. **2**, p. 59
Mukai, T.: 1986, *Astron. Astrophys.* **164**, 397
Pelet, R.: 1980, in *Kerogen*, ch. 15, ed. B. Durand, Ed. Technip, Paris
Sekanina, Z.: 1982, *Astron. J.* **87**, 161
Sherb, F., Roesler, F.L., Magee, K., Harlander, J., Reynolds, R.J.: 1986, *Adv. Space Res.* **5**, 12, 275
Tissot, B.P., Welte, D.H.: 1978, *Petroleum Formation and Occurance*, Sects. II-4.7, II-5.2, V-4.2, Springer
Vanýsek, V., Wickramasinghe, N.C.: 1975, *Astrophys. Space Sci.* **33**, L19
Wallis, M.K.: 1986, *Asteroids Comets Meteors II*, ed. C-I. Lagerkvist et al., Uppsala Univ., p. 363
Wurm, K., Balazs, B.: 1963, *Icarus* **2**, 334
Wickramasinghe, D.T., Allen, D.A.: 1986, *Nature* **323**, 44

Astron. Astrophys. 187, 807–823 (1987)

ASTRONOMY AND ASTROPHYSICS

Comet P/Halley's nucleus and its activity

H. U. Keller [1], W. A. Delamere [2], W. F. Huebner [3], H. J. Reitsema [2], H. U. Schmidt [4], F. L. Whipple [5], K. Wilhelm [1], W. Curdt [1], R. Kramm [1], N. Thomas [1], C. Arpigny [6], C. Barbieri [7], R. M. Bonnet [8], S. Cazes [9], M. Coradini [8], C. B. Cosmovici [10], D. W. Hughes [11], C. Jamar [6], D. Malaise [6], K. Schmidt [12], W. K. H. Schmidt [1], and P. Seige [13]

[1] Max-Planck-Institut für Aeronomie, Postfach 20, D-3411 Katlenburg-Lindau, Federal Republic of Germany
[2] Ball Aerospace Systems Division, Boulder, CO, USA
[3] Southwest Research Institute, San Antonio, TX, USA
[4] Max-Planck-Institut für Physik und Astrophysik, D-8046 Garching, Karl-Schwarzschild-Straße 1, Federal Republic of Germany
[5] Smithsonian Astrophysical Observatory, Cambridge, MA, USA
[6] Institut d'Astrophysique, Université de Liège, Belgium
[7] Istituto di Astronomia, Universita di Padova, Padova, Italy
[8] European Space Agency, Paris, France
[9] Laboratoire de Physique Stellaire et Planétaire, Verrières-le-Buisson, France
[10] Istituto Fisica Spazio Interplanetario, Frascati, Italy
[11] University of Sheffield, Sheffield, U.K.
[12] Deutsche Forschungs- und Versuchsanstalt für Luft- und Raumfahrt, Oberpfaffenhofen, Federal Republic of Germany

Received February 23, accepted April 28, 1987

Summary. The Halley Multicolour Camera on board ESA's Giotto spacecraft observed the nucleus of comet P/Halley and its environment and returned more than 2000 images. The observations are summarized, their calibration is described, the status of the analysis and the results are discussed. Topographic features on the nuclear surface and areas of activity are identified. The optical thickness of the dust produced in jet-like features is estimated. The impact and constraints of the observations on cometary nucleus models are discussed.

Key words: comet – nucleus – dust – imaging

1. Introduction

The primary scientific goal of the Halley Multicolour Camera (HMC) was the detection and characterization of the nucleus of comet Halley during the flyby of ESA's spacecraft Giotto. Images taken during the last half hour before closest approach (CA) have revealed the nucleus clearly. The immediate results – the large size of the nucleus, its extreme darkness, its irregular, roughly elliptical shape, the existence of a non-volatile crust of high temperature, and its highly asymmetric and localized activity – have already become part of our common understanding of cometary nuclei.

It was particularly difficult to achieve high resolution images from the spinning spacecraft. The comet could only be observed for a few milliseconds at a certain phase during each spin period. It could not be followed continuously. The position and exact moment for the next image had to be predicted by a complicated tracking programme. The exposure times were set during the design of the instrument by the chosen resolution and depended only on the offset angle of the comet from the spacecraft spin axis. The spacecraft rotation (period 4s) and the high flyby velocity required completely autonomous operation of the instrument. Schmidt et al. (1986) and Keller et al. (1987) give descriptions of the camera and its performance.

The Halley Multicolour Camera telescope was a modified Ritchey-Chrétien design with correcting lens and a periscope mirror in front of the entrance pupil. Its aperture was 160 mm and it had a focal length of 998 mm and an effective speed of f/7.7. The instrument operated in line scan mode, making use of the image motion introduced by the spin of the spacecraft. HMC used two uniphase charged coupled devices (CCDs) 584 lines of 390 pixels each at its focal plane. Each CCD was divided into two sections which were clocked independently. They were converted into analogue storage areas by covering them with metal plates with narrow slits on top of each section. When the object moved across the slit, the charge packets were moved from line to line with the same speed. Three of the sections were exposed through fixed filters [broad band red (B), blue (D), and clear (E)] and the fourth section (C), through a filter wheel with 11 filters and polarizers. The effective exposure times were defined by the line clocking time between 1 ms and 14 μs (for a separation angle, ϱ, of 90°) times the slit width measured in pixels (between 4 and 8, depending on the CCD section). The geometric resolution was 22.4 μrad × 22.4 μrad per pixel. One section corresponded to a field of view (FOV) of $0\overset{\circ}{.}5 \times 0\overset{\circ}{.}36$. The CCD signals were digitized to 12 bits with a gain switch for an additional factor of 4. The information was square-root-encoded, giving a dynamic range equivalent to a 14-bit word by transmitting only an 8-bit word to ground.

The communication with the Giotto spacecraft failed and the camera operation was interrupted by a power reset shortly before closest approach. Therefore, the aspect angle of the cometary nucleus varied by only a few degrees. The HMC could only reveal one side of the nucleus. A programme has been started to determine the 3-dimensional shape of the nucleus in co-operation with scientists from the Vega imaging team. Two Vega spacecraft passed the nucleus of comet Halley several days earlier than Giotto, observing the nuclear region from three sides at different

Send offprint requests to: H. U. Keller

phases of nuclear rotation (Sagdeev et al., 1986a). The combination of images will also lead to a more precise determination of the rotation period and position of the axis. Some of the presented interpretations and results will be based on a preliminary investigation (Wilhelm et al., 1986).

Early images during Single Detector Mode (SDM) of large size (full frame 390 by 292 and 196 by 196 pixels) show the inner coma out to about 4000 km in projection. The field of view therefore included the point of closest approach of the spacecraft so that the trajectory can be correlated with the observed dust jets. The HMC observations are described in Sect. 2.

Light levels observed during the encounter were much lower than anticipated because the reflectivities of the nucleus and the dust were lower than expected when the camera was designed in early 1981. Therefore, the dark current signals have to be subtracted particularly carefully. The calibration of images produced by detector section C using the clear filter has been essentially completed. Calibration of the other detectors and filters is in progress and comprises a major effort of the HMC team. Details will be given in Sect. 3.

Section 4 will present the results concerning the nucleus and its surface. Many details of surface features (morphology) can be seen in the later images during Multi Detector Mode (MDM). The reflectivity of the surface is determined in areas of low activity. Dust jets can be traced down to specific features on the nucleus.

Dust jets are very prominent in most images. The activity of the nucleus was concentrated towards the sun. The distribution of activity and dust is described and globally analyzed in Sect. 5. The gas and dust production rates are estimated as well as the optical depth of the dust column.

The constraints of the HMC observations on the nature of comet Halley's nucleus and its physics are discussed along present models in Sect. 6. These observations are also correlated with results of other experiments on board Giotto, Vega and from ground-based experiments.

With one exception, images presented in this paper are not enchanced by image processing methods; they represent calibrated data from which coherent noise has been removed. The low contrast of many parts of the images, mainly caused by the dust and the extremely low-light levels on one hand and the high dynamic range of up to 1000 in a single image on the other hand, make a representation of the images in print rather difficult.

2. Observations

The HMC took the first images under realistic conditions when the instrument was pointed towards Jupiter in September 1985, two months after launch. The observations of the earth from a distance of more than 21 Gm in October of 1985 in MDM proved the HMC to be fully operational (Keller et al., 1985). Several stars were observed during cruise. These images are useful for absolute calibration of the instrument and for determination of the point spread function (PSF) (Wilhelm, 1986). In total the camera was operated for 220 h; 36 kbyte of commands had to be uplinked (for details of the operations see Curdt and Specht, 1986). The search for comet Halley started on February 15, 1986. Only when the offset angle became small enough was the comet unambiguously detected on March 4. The Giotto spacecraft was still 59 Gm away. The comet was observed again on March 6, 8, 9, 12, and early 13, 1987.

The HMC was switched on at 4:12 h before CA on March 13 during the encounter. The first image was taken at 3:06 h before CA, when the distance to the nucleus was more than 773,000 km. The inner coma was observed using all 11 filters and polarizers during Single Detector Mode. The transmission curves of the filters (except polarizers) are illustrated in Fig. 1 together with the sensitivity of the camera (approximate numbers). Images of various small sizes were sent back to earth in order to reduce transmission time. Table 1 lists the different formats. In addition, pixels could be summed up on the CCD to exchange resolution for improved sensitivity and FOV. Five "superpixel formats" (SPF) were used. SPF 0 corresponds to the full resolution. In addition, the integration time could be extended to 150 ms and 1.5 s by suppressing the image clocking. This photometer mode was used for the narrow band filters giving up resolution. Altogether 2043 images were taken in SDM before the camera switched to the Multi Detector Mode, 6 min before CA. The first full frame image (390 by 292 px) is No. 0704, taken at a distance of 767,000 km. Its width corresponds to 6530 km. The comet is in the centre and the coma can be traced to the edges of the image on the sunward side (see Fig. 2). It took about 12 spins to transmit a full frame image. Therefore, the large images had to be interleaved with smaller ones, generally with lower resolution and narrow band filters. The maximum image size was reduced to 196 by 196 px at about 1/2 h before CA. The last full frame image was No. 3104 (blue filter), the

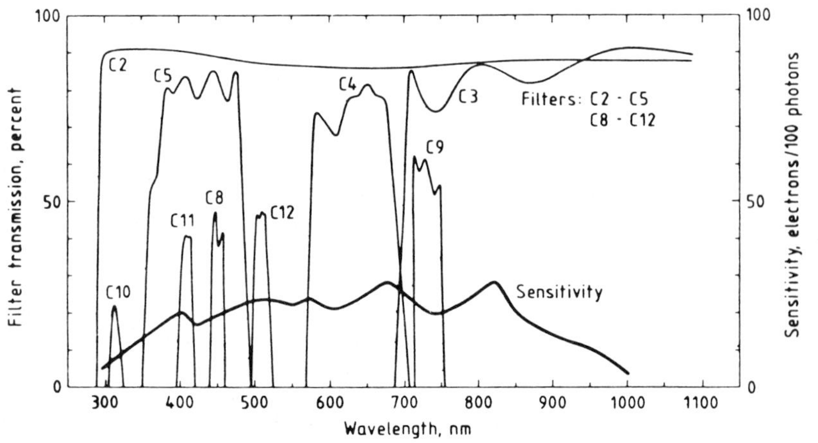

Fig. 1. Transmission (left ordinate) of the various HMC filters (excluding polarizers) are displayed versus wavelength. C2 is the clear filter. The overall sensitivity of the camera is also shown (right hand ordinate)

Table 1. Images in Single Detector Mode. Column 1 notes the type of filter, column 2 SPF is the superpixel format SPF 0:1 × 1 pixel (full resolution), SPF 1:2 × 2, SPF 2:4 × 4, SPF 3:8 × 8, SPF 4:15 × 16, SPF 5:4 × 3. Column 3 size of images [in super pixels]. Column 4 total number of images in SDM. In photometer mode clocking of the CCD is suppressed giving integration times within the slit of 150 ms and 1.5 s

Filter	SPF	Size	Number
Blue	0	390 × 292	32
	0	196 × 196	9
	3	49 × 36	33
	5	98 × 97	33
Red	0	390 × 292	33
	0	196 × 196	10
	3	49 × 36	33
	5	98 × 97	33
Orange	0	390 × 292	33
	0	196 × 196	10
	3	49 × 36	33
	5	98 × 97	33
Clear	0	390 × 292	33
	0	368 × 26	210
	0	98 × 98	99
	0	36 × 36	132
	0	34 × 276	211
	1	37 × 37	71
	3	49 × 36	33
	5	98 × 97	38
OH		Photometer mode	100
	2	98 × 73	13
	3	49 × 36	13
	4	25 × 18	50
	5	98 × 97	5
C-2		Photometer mode	82
	2	98 × 73	12
	3	49 × 36	13
	4	25 × 18	41
	5	98 × 97	8
C-3		Photometer mode	82
	2	98 × 73	13
	3	49 × 36	13
	4	25 × 18	41
	5	98 × 97	8
Cont. 1	4	25 × 18	132
	5	98 × 97	41
Cont. 2	4	25 × 18	132
	5	98 × 97	42
P⊥	5	98 × 97	42
P ∥	5	98 × 97	43

last large image before switchover to MDM was No. 3432 (orange filter). Image No. 3416 taken at a distance of 25,680 km was the last SDM image using the clear filter. The exposure time was determined by the design of the camera and the offset angle given by the trajectory. It varied from 6 ms down to 265 μm for the last image.

The first image (No. 3436) in MDM was taken 295 s before CA at a distance of 20,200 km when the offset angle of the comet increased beyond 1°.6. The images then passed across all four detector sections. Sets of four images each were taken almost simultaneously through the red, blue and clear wide-band filters in SPF 1 (2 × 2 px), and in full resolution through the clear filter on detector section C. 74 by 74 pixels in full resolution and 37 by 37 px in half resolution were transmitted every spin before the next image was taken. This corresponds to a FOV of 0°.1. Only 2% of the data on the CCDs could be transmitted back to earth (rate 20 kbit s^{-1}). Transmission was interrupted shortly before CA. The last transmitted frame was taken 15 s before CA. However, the last useful image (No. 3504) showing details of the nucleus was taken from a slant distance of 1680 km. The operation of HMC was interrupted by a reset of one of the microprocessors. The sequence of MDM images comprises 69 sets of images. The last image showing the full nucleus (No. 3461) was taken from a distance of 13,340 km. The camera FOV centred on the brightest spot, the activity region on the northern subsolar limb. This area is shown with steadily increasing resolution, down to 40 m per pixel, with a corresponding decrease in field of view. The phase angle of the cometary nucleus varied from 107° during approach to 90° for image No. 3504 (see Fig. 3). The small range of phase angle variation and the smaller and smaller cutout of the nuclear surface prevent the derivation of physical data based on the changes of scattering and/or phase angles.

The flyby parameters of the spacecraft were determined in real time by the on-board software analyzing the images. The following values were found: encounter distance 602 km, time of closest approach March 14, 1986 00:03:02.3 UT and an angle α south of the sun-comet-spacecraft velocity plane of 30°.8 (see Fig. 3). Ground-based analysis based on preliminary knowledge of the spacecraft attitude provided the following values: distance 591.5 km ± 2 km, time 00:03:01.9 ± 0.2 s and $\alpha = 30°.7 \pm 0°.3$.

3. Calibration

The task of calibrating the images has turned out to be more difficult than anticipated. The relatively low signal to noise ratio of the images caused an elaborate and detailed investigation. The temperature of the detectors was considerably higher (about −5° C in MDM) than the design requirement −20 °C). Therefore the accumulated dark current during image taking and read out constitutes a major portion of the signal for the images taken with the narrow filters. Quantitative data reduction is impaired and may be impossible in some cases (e. g., for the low signal level images taken with the OH filter).

Thermal background calibration has been based essentially on dark images taken shortly before the encounter. However, the temperature range was generally higher than during the encounter. The camera was passively cooled and the spacecraft temperature had been decreasing with increasing heliocentric distance of the spacecraft and with solar aspect angle. Therefore, dark images have been extrapolated to the correct temperature.

A second source contributing to the noise was caused by the clocking of the CCDs during image taking and read out ("spurious charge") and, therefore, depended on the offset angle and on the location on the detector (pixel). A two dimensional grid of dark matrices had to be constructed as function of ϱ and T. Full frame SDM images taken with the blue filter were used from which the relatively weak signal had been extracted by comparison with the corresponding clear filter image. These images were

Fig. 2. Image No. 704 was taken from a distance of 767,000 km. The length of the central vertical line corresponds to 6530 km. North is up. The direction towards the sun is indicated. The sun is 17° behind the image plane. The line labeled ROT indicates the approximate direction of the rotation axis. The projected point of closest approach (CA) is indicated as a cross and labelled ENC. The isophotes are labelled in units of [$mW\,m^{-2}\,sr^{-1}$]

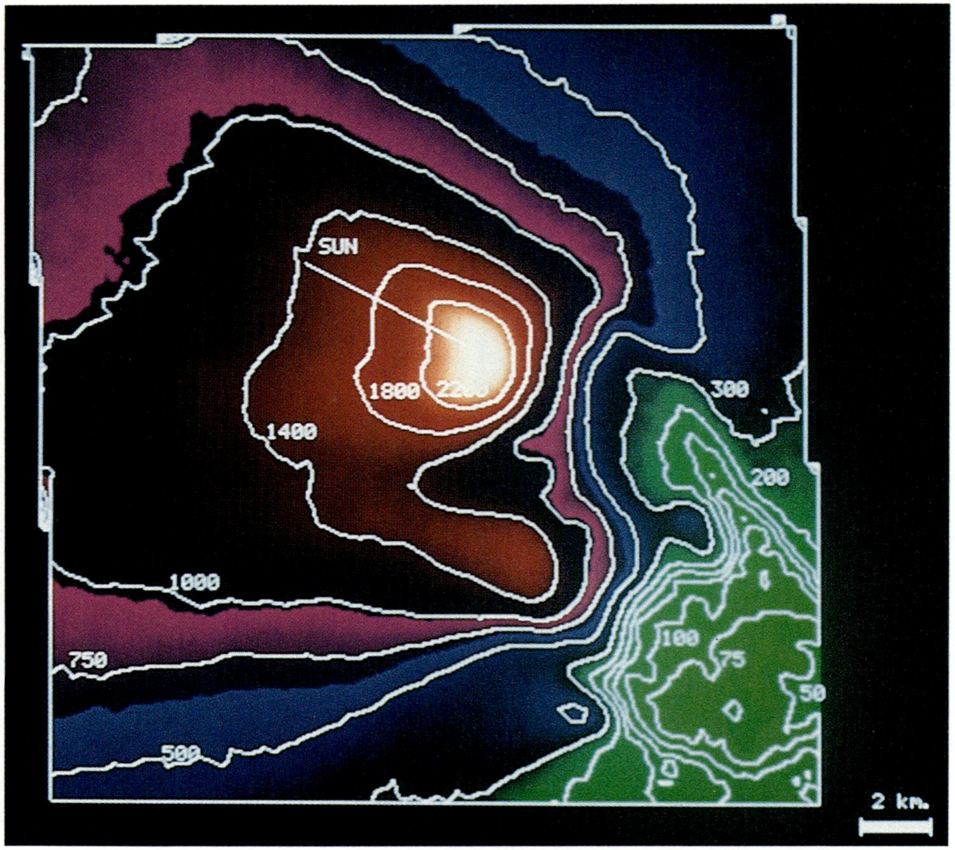

Fig. 16. Image No. 3461 was taken from a distance of 13,390 km (frame size 22 × 22 km). North is up. The sun is 15° behind the image plane. The isophotes are labelled in units of [$mW\,m^{-2}\,sr^{-1}$]

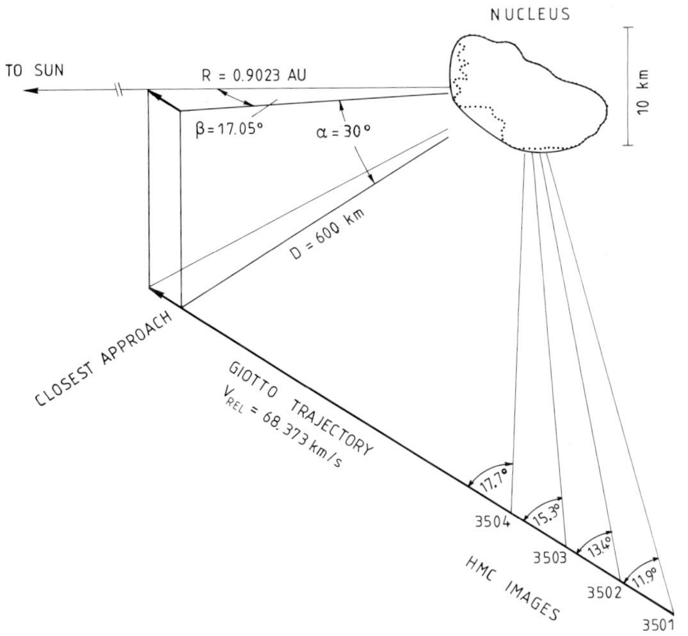

Fig. 3. The Giotto flyby trajectory relative to the cometary nucleus and sun is shown, including offset angles and positions for the last 4 useful HMC images

combined with the mosaic of dark images taken in the MDM covering most of the CCD. This method could only be applied to images taken with section C. Equivalent SDM images for the other detector sections (B, D, E) are of course not available. This is the main reason why the status of calibration of SPF1 images of the MDM is still rudimentary.

Most images show a slight component of electronic interference (only 2 to 3 DU peak to peak). Smaller frame sizes in SDM seem to be more affected. As a first step, all images have been investigated for coherent noise. The main components of the frequency spectrum have been removed in the frequency domain by Fourier transformations. The remaining noise after dark current subtraction is about 2 to 3 digital units for 1σ rms, comparable to the noise introduced by the digitisation of the signal. A few noise spikes remain; they are caused by minor defects of some pixels. These have been eliminated by substituting signals from neighbouring pixels.

The responsivity matrix degenerates to one dimension for the line scanning operation, simplifying this step of calibration considerably. All values have been averaged along columns. The corrective terms had been determined on ground. The variation is of minor importance (a few percent) along the line.

The second step in restoration of the images is the geometric correction. The detector slit swept over an annular region of the sky due to spacecraft rotation. During the exposure period, the slit covered a segment of the annulus that was stored in the rectangular detector line by line. The parts of the image further away were undersampled, those closer to the spin axis were oversampled. The segment has to be restored resulting in the characteristic "pie shape" of the geometrically corrected HMC images (see Fig. 2). The distortions of the last images ($\varrho > 5°$) are minor for section C but are more pronounced for the other sections because the image did not pass their scanning lines perpendicularly. The on-board programme of the HMC determined the position of the spin axis and the changes of ϱ. However, the erratic changes due to nutation of the spacecraft caused by the dust impacts could not be taken into account during the final approach. The actual position of the spin axis can be determined from the operations protocol and the movement of the comet within the CCD sections. Figure 4 shows the derived residual error in the offset angle caused by the nutation of the spacecraft, triggered by dust impacts. Overlaying of images, e.g., with different colours, requires subpixel co-registration accuracy.

The predictions of the on-board tracking programme showed some systematic deviations from the nominal values of ϱ as long as ϱ was less than 2°. Figure 5 displays the predicted values and the values derived from the analysis of the trajectory. The clocking of the CCD was derived from the predicted value and therefore was faster than required by the apparent comet motion. This led to an essentially linear distortion (stretching) of the images along the direction of image movement, up to 20% around $\varrho \lesssim 1$. The loss in resolution is minor, less than 1 pixel in the worst case (see below). The analysis is still in progress and will also lead to an improvement of the flyby parameters.

After geometric correction, scaling, and absolute registration, images can be combined to improve signal to noise, to look for variations with time and with colour, and to superimpose the MDM images. Later images can be inserted into earlier ones replacing the portions which are covered with higher resolution (see Fig. 6). This "composition" technique allows one to judge the ever increasing resolution of the later images in the frame (and location) of the whole nucleus. Care must be exercised in the interpretation, because the variations of resolution in a composite image cause artifacts such as changes in brightness and in sharpness of features running across image boundaries.

A further step to improve the contrast of features can be achieved by considering the blurring caused by the PSF. The PSF has been determined by evaluating the star images taken during cruise. Its full width at half maximum (FWHM) of intensity is slightly less than two pixels (Kramm et al., 1987). The deconvolution is done in the frequency domain after radiometric correction. The quality of the processed images depends strongly on the noise and therefore the method has not been applied to the very last images with shortest exposures. Additional information can be gained in areas of the nucleus which are not in the vicinity of the sources of the northern active jets, and which are therefore excluded from the close up images because of the restricted FOV.

Further improvements of the contrast and clarity of the images can be gained by various filter techniques which reduce high frequency noise and methods to enhance the contrast. These have occasionally been used in order to overcome the losses due to the printing process. Care has to be taken regarding artifacts introduced by these enhancement procedures. The quantitative analyses of this paper are not based on the additional enhancement results since more experience has to be accumulated.

Absolute calibration of the HMC has been derived from observations of the earth cloud patterns in October 1985 from a distance of 21 Gm. The signal received per pixel per second is

$$\text{signal} = k \int_{\lambda_1}^{\lambda_2} R_\lambda Q F T \, d\lambda . \tag{1}$$

The quantities $Q(\lambda)$ (quantum efficiency) and $T(\lambda)$ (filter transmission) describe properties of the camera and k is a constant determined by the electronics. $F(\lambda) = F_0(\lambda)/r_h^2$ is the solar radiation flux scaled to the heliocentric distance, r_h. $R_\lambda(\lambda)$ is the reflectivity of the imaged surface. The signal is measured in digital

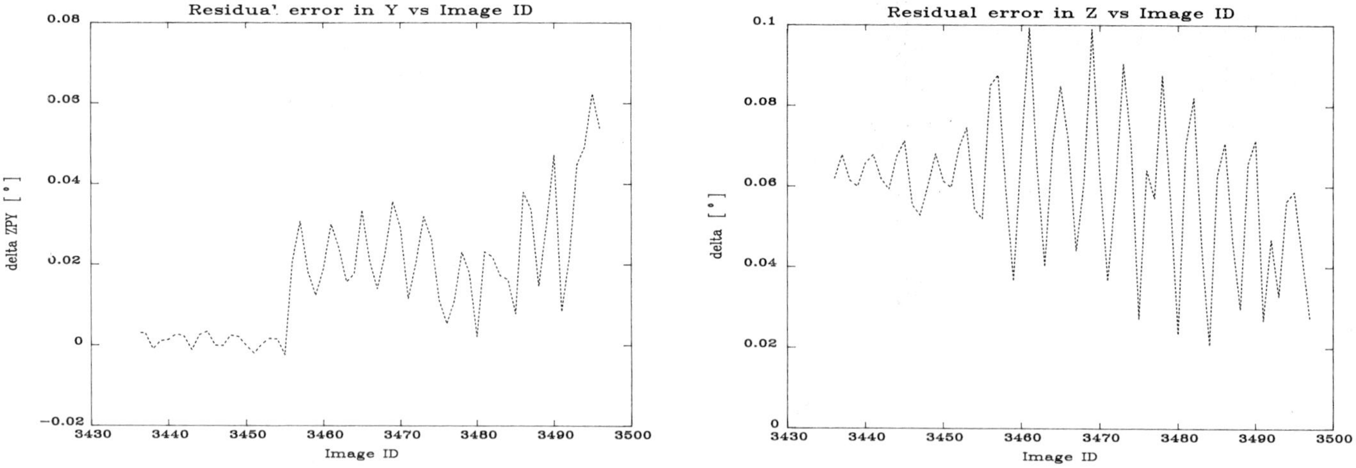

Fig. 4. Near encounter image positioning. Residual error of the cometary centre of brightness in HMC image plane coordinates; after image No. 3457 the spin axis is tilted by 0°.02 and starts nutating with an 0°.03 amplitude

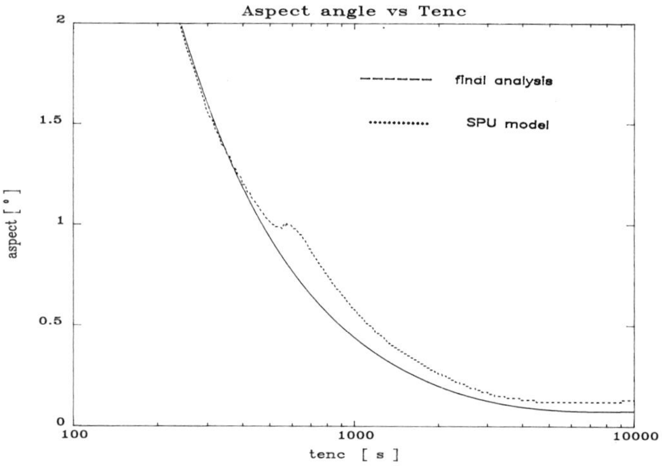

Fig. 5. Aspect angle ϱ versus time to encounter dotted: onboard predictions solid: actual aspect from trajectory

units (DU/second). If one considers the reflectivity to be independent of λ, the camera properties can be combined leading to

$$\text{signal} = k\,R \int_{\lambda_1}^{\lambda_2} Q\,F\,T\,d\lambda = k_c\,R, \tag{2}$$

connecting the signal to the image brightness by the proportionality constant k_c and the reflectivity R. The value of k_c for observations with the clear filter and detector C during the encounter at $r_h = 0.9$ AU is

$$k_c = 1.5\ 10^8\,. \tag{3}$$

The error is estimated to be about $\pm 20\,\%$.

The reflectivity can also be defined by comparing the imaged surface to a Lambertian surface

$$R_L = \pi\,R\,. \tag{4}$$

The reflected intensity within the passband of the clear filter is related to R_L by

$$\Pi I = F_0^c\,R_L\,. \tag{5}$$

F_0^c is the solar flux at the comet within the effective passband of the clear filter to 374 nm and takes the value of 1.08 10^3 Wm^{-2}.

4. Nucleus

During the late hours of the Giotto encounter it became obvious that the nucleus of comet Halley is a single solid body of roughly elliptical shape with a projected size larger than anticipated (15 × 8 km), implying a very dark albedo. Highly asymmetric and localized activity was observed. The nucleus of comet Halley is visible on several hundred images of the HMC. The last 60 images have been used to produce the composite image shown in Fig. 6. About 25 % of the surface turned towards the camera is seen to be illuminated by the sun and shows a wealth of surface features. The morning terminator is well defined and gives an indication of the irregularity of the surface. The outline of the nighttime region stands out as a dark silhouette against the background dust.

4.1. Limb and terminator

The position of the limb in Fig. 6 is easily visible over most of the circumference of the nucleus and is shown in the diagram in Fig. 7. The nucleus projection viewed by Giotto is roughly elliptical with dimensions 14.9 × 8.2 km (Reitsema et al., 1986). There remain regions of slight uncertainty due to dust contamination in the vicinity of the dust jets but repeated measurements agreed within 100 to 200 metres of each other. Figure 8 shows that the local gradients do not materially assist in defining the limb in these areas.

The morning terminator divides the night and sunlit portions of the nucleus. Most structural details are visible near the northern tip which was contained in the later, high resolution images. However, the terminator gives evidence that the surface in the middle region is generally smoother than that at either of the ends with the notable exception of the "mountain" feature.

4.2. Surface features

The roughness of the nucleus shape can be judged from the limb profile, the abundance of surface features and the angularity of the terminator. There are many discrete features of several different

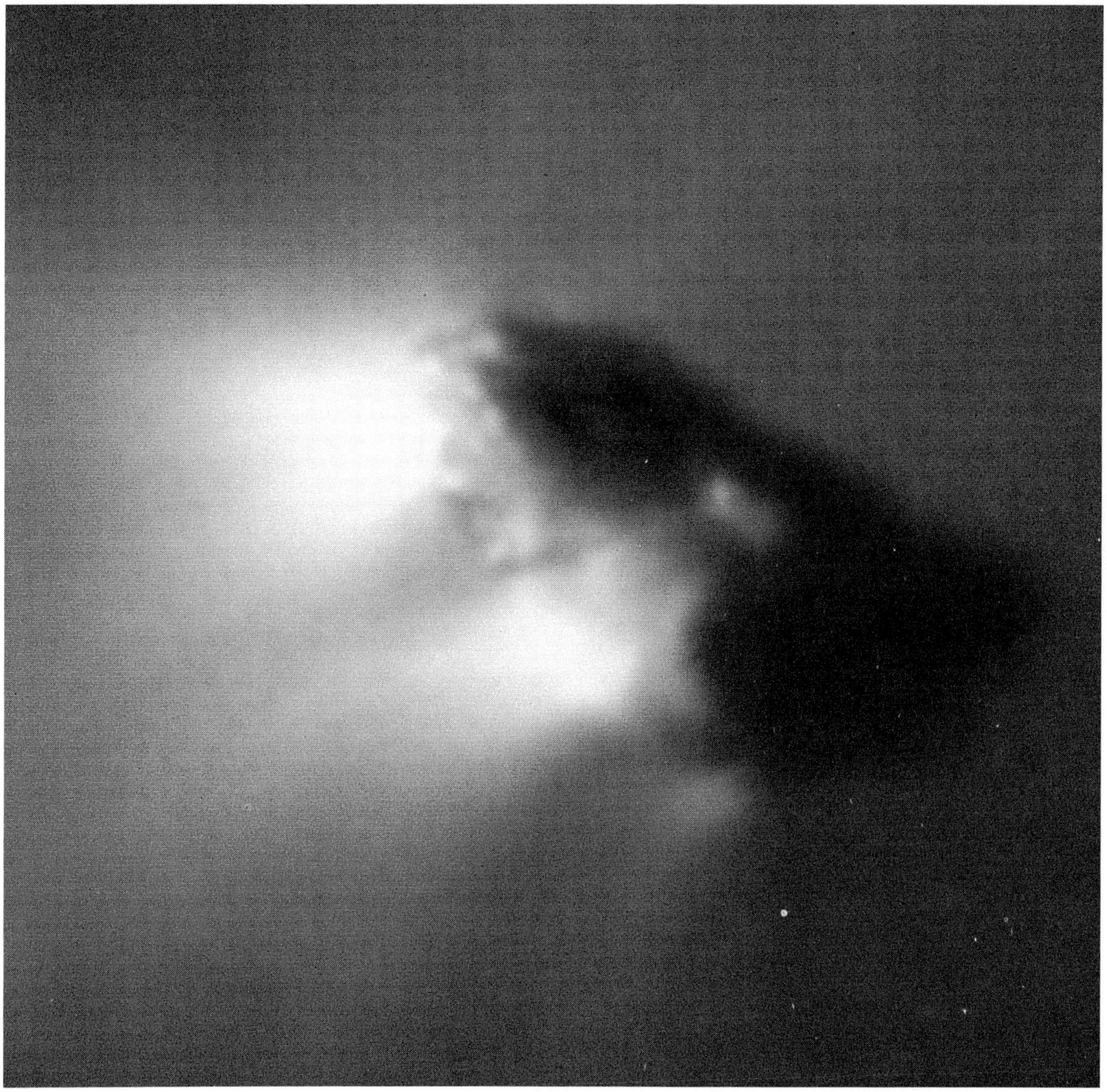

Fig. 6. Composite image of the nucleus of comet Halley. North is up and the sun is to the left and 15° below the image plane. This photograph is a composite of 60 HMC images and has resolution varying from 800 m in the lower right to 100 m in the vicinity of the northern tip. Image processing steps required to achieve this composite include subtraction of thermal background charge, Fourier filtering of coherent noise, detector responsivity correction, geometric rectification, and image registration. (Reitsema et al., 1986)

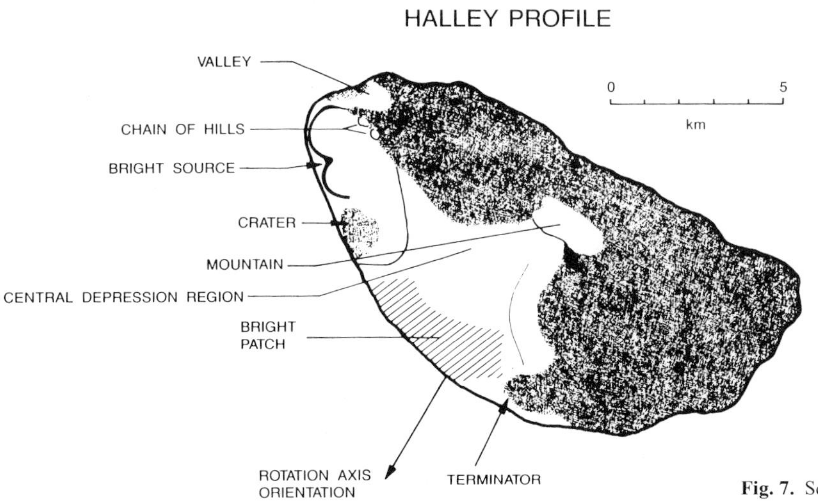

Fig. 7. Schematic drawing of major nuclear features as seen by the HMC

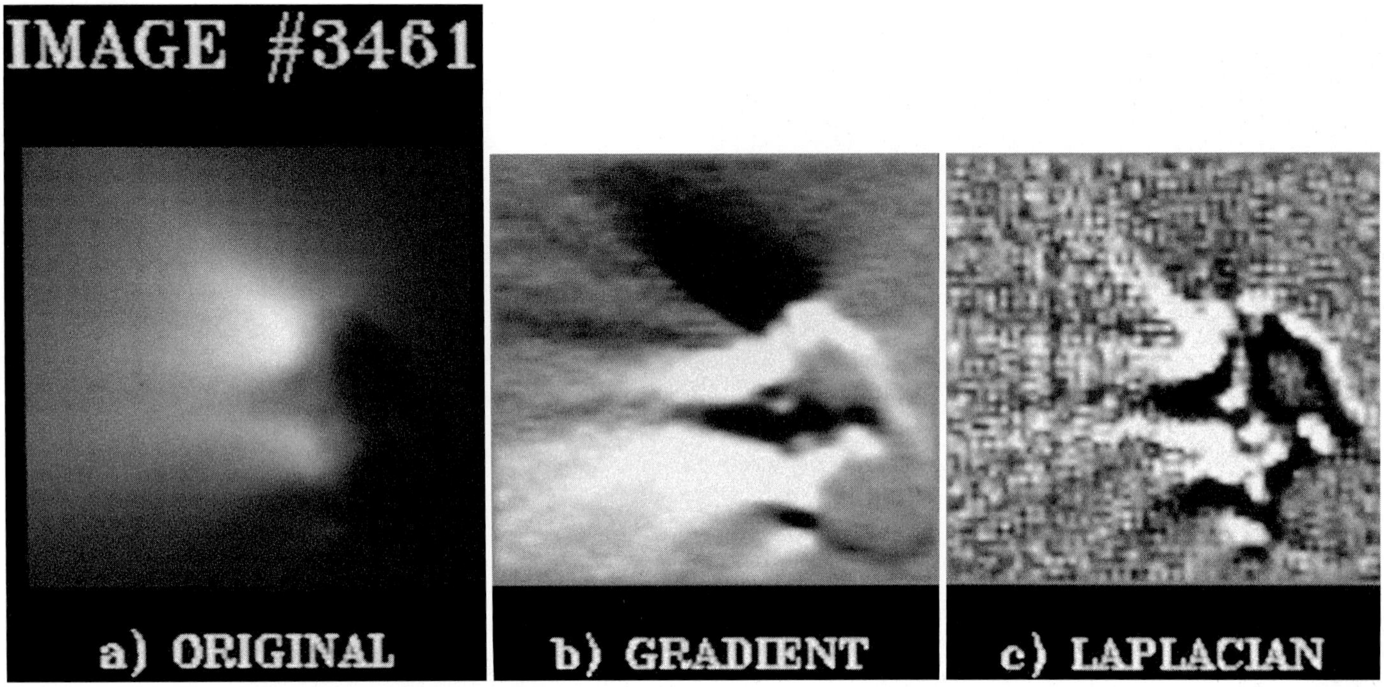

Fig. 8a–c. Masks calculating the local gradient (**b**) and the second derivative (Laplacian) (**c**) are applied to image 3461 (**a**). The changes in gradient are particularly obvious in the area of activity. A semi-regular structure is visible next to the "bright patch" in the central depression. One of the strongest intensity changes occurs right at the northern edge of the "mountain"

characteristics. These include features such as "craters", "valleys", and "mountains", and apparently coherent structures such as chains of "hills". Some features are related to jet activity.

Interpretation of the nucleus images is complicated by the need to distinguish among several processes which can produce bright regions in the data. Surface topography, surface albedo and scattering properties, and overlying dust can all produce brightness contrast in the images; identifying the specific factor is more difficult because of the limited range of phase angles during HMC observations (107°–89°). The topographic origin of many of the bright features is confirmed through their association with structure in the terminator or by the existence of corresponding dark features. In the vicinity of the bright jet, a number of bright "upslope" features with adjacent "down slopes" to their right are seen.

The largest discrete feature observed is the circular shape ("crater") just north of of the central depression on the sunward surface. It appears to be a depression with a projected diameter of 1600 m. The contour of the rim is not smooth and fully circular. The dark inner rim does not seem to be caused by shadow, only by a "downslope". Photoclinometry has given a depth for the feature of as little as 100 m (Schwarz et al., 1987). Dust obviously contributed to the signal so that the contrast of light from the surface may be larger than assumed, making the depth of the depression deeper. Higher resolution images reveal more and more globular substructures in the centre part. South of this "crater" there may be another circular feature.

The "mountain" is a bright region near the centre of the observed nucleus. Although it appears to be a high relief feature, it need not be more than 0.5 km elevation. It could in fact be a cliff marking the end of the central depression. The illuminated surface of the "mountain" covers an area of 2×1 km and it is brightest and narrowest at the northern end. Its orientation is almost parallel to the dark limb (see scan of Fig. 9).

The most prominent coherent features are the bright ridges which form the northern and southern ends of the terminator. A dark region runs parallel to the southern ridge at ~ 1 km distance. The northern ridges show 4 or 5 almost periodic structures typically 500 m in size ("chain of hills"). Lumpy morphology seems to be more common in this region, although this may be caused by the increased resolution achieved by the camera at the northern tip.

A smooth "central depression" tapers towards the "mountain" from the sunward limb. Photoclinometry shows that the general decrease in brightness from the limb to terminator is produced by the change in solar elevation angle associated with the curvature of a body of 5 km radius (Delamere et al., 1986).

Another example of large scale structure is the dark ridge at the very northern tip of the nucleus, enclosing a triangular-shaped bright feature. This triangular notch-shaped valley suggests a depth of several hundred metres.

Two distinct bright spots connected by a bright arc can be identified near the northern limb. They are located near the base of the dust jet activity and seem to be the source region for the jets. This increased brightness could be caused by the concentration of dust or by higher reflectivity of the surface. Other nearby bright patches have not been associated with dust jets.

The source region for the strong middle jet is a bright patch which is approximately 50% brighter than the surrounding surface. It is quite uniform in brightness and appears to be bounded by a narrow dark band. The sunward boundary indicates that this is a well-defined surface area with different optical properties rather than a diffuse dust cloud.

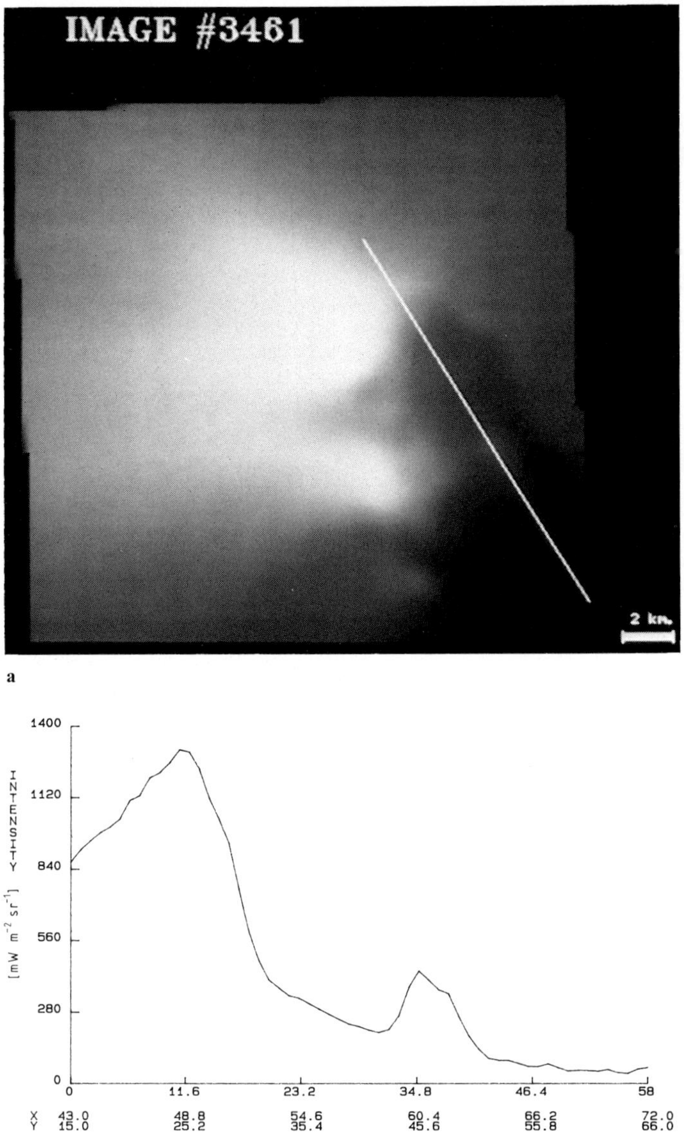

Fig. 9. **a** Image No. 3461 was taken at a distance of 13,390 km (frame size 22 × 22 km). North is up. The sun is 27° above the horizontal on the left and 15° behind the image plane. This image is the last one showing the whole nucleus. The white line indicates the scan represented in **b** which shows intensity within the passband of the clear filter

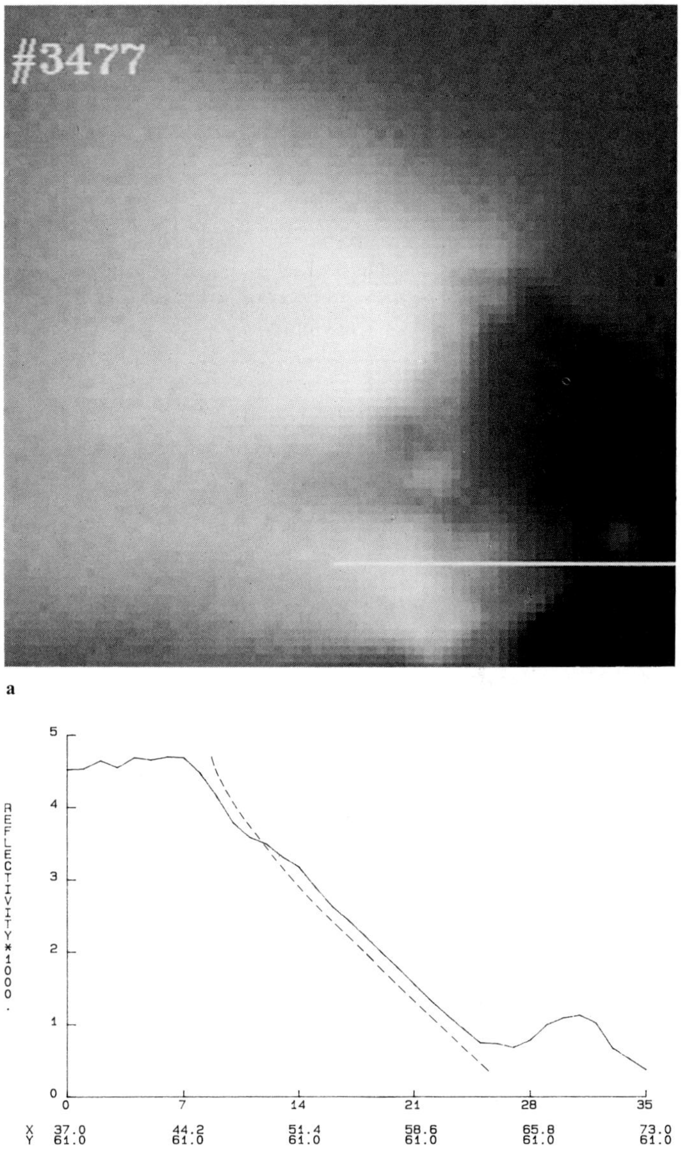

Fig. 10. **a** Image No. 3477 taken from 9030 km distance. The white lines indicate the scan (**b**) shown along line 61. The length of the scan corresponds to 7.3 km. R_L, the reflectivity relative to a Lambertian surface with inclination angle zero is given as ordinate. Also indicated, dashed, is the signal decrease from a cylinder of radius 7 km with a Lambertian surface. Note this figure has been revised since the Heidelberg meeting (Delamere et al., 1986)

As was immediately obvious from the large size of the nucleus, the surface of comet Halley is extremely dark (Keller et al., 1986a). The reflected intensity (normalized by the response of a Lambertian surface with $i = 0$) along a line of the image No. 3477 is shown in Fig. 10. The peak signal corresponds to a reflectivity [defined in Eq. (4)] of less than 0.005 at the limb of the nucleus, consistent with a geometric albedo of 4%. The albedo near the bright limb is thus less than 1/8 of the full-phase geometric albedo. This is not unreasonable since, for example, in the case of the moon the ratio of the average surface brightness at phase angle 107° divided by that at full-phase has a similar value.

The reflectivity of other areas is found by dividing R with the cosine of the corresponding angle of incidence, i. Photoclinometric methods can be applied to determine the 3-dimensional structure of the bright surface.

4.3. The 3-dimensional shape

The 3-dimensional shape of the nucleus can be derived from the combination of Vega and Giotto images. In the HMC images, the long axis of the nucleus appears somewhat foreshortened. The northern tip of the nucleus is tilted out of the image plane towards the observer by about 20 to 30° (Wilhelm et al., 1986). A self-consistent model has been produced using an ellipsoid with 8.2, 8.5, and 16 km for the three axes (Wilhelm, 1987) in order to

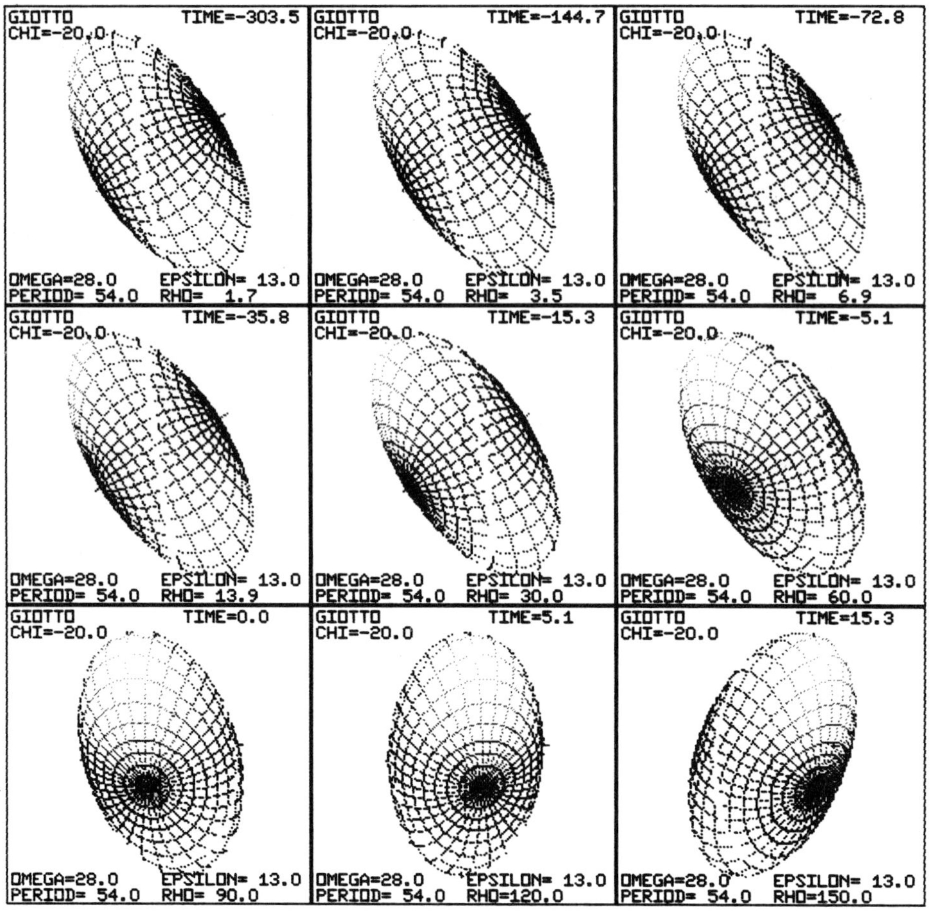

Fig. 11. Aspects of the nucleus of comet Halley schematically represented by an ellipsoid are shown for the Giotto flyby. The co-ordinate system is centred on the rotation poles. North is approximately up. The terminator is shown by suppression of the co-ordinate grid points. RHO corresponds to the HMC offset angle. The light points correspond to the sub-solar point

investigate the orientation of the body, its rotation axis and period. The spin vector is indicated in Fig. 7. It points about 60° south of the comet to sun direction and is about perpendicular to the apparent long axis of the nucleus, 13° out of the image plane away from the approaching spacecraft. The resulting obliquity is about 25°. A spin period of 54 ± 1 h matches the data from the three spacecraft. The merging of the terminator with the sunward limb on the southern end of the nucleus (see Fig. 11) is in agreement with calculations of the rotation of the nucleus based on this simplified tri-axial ellipsoid model.

The acceleration of the concentrated dust jet activity leads to limited nutation (free precession) of a few degrees (Wilhelm, 1987). The larger end of the nucleus (the southern end on the HMC images) was oriented towards the sun during the Vega-1 flyby on 6 March 1986 while the orientation for Vega-2 was almost the same as for Giotto (see also Fig. 12). The best images of Vega-2 taken during closest approach from an almost subsolar view point show the opposite side of the nucleus.

5. Dust and source activity

The HMC images have revealed that the activity of comet Halley's nucleus was concentrated towards the sun (see Fig. 13). The variability of the comet during MDM was seen to be small, although it is apparent that some changes in the activity occurred during the 3 h encounter. Near the nucleus many small jet-like structures have been observed. It is clear that only the largest scale jet features observed by the HMC could have been seen by earth-based observers. Using the HMC, the dust distribution out to 3,000 km in projection can be well-defined.

5.1. Overall dust distribution

On the HMC images taken from large distances the observed dust is confined to a sector (as seen in projection) of about 70° extending approximately from the sunward direction southwards (see Keller et al., 1986a). Beyond a few 100 km, the hemisphere on the nightside was devoid of dust within the sensitivity of HMC ($\sim 4\,\mathrm{mW\,m^{-2}\,sr^{-1}}$) (see image No. 704, Fig. 2). The overall appearance of the dust activity in Fig. 13, when compared to image No. 1186 of Vega-2 (see Sagdeev et al., 1986c, Fig. 9e) shows that the dust distribution near the nucleus two rotations apart is remarkably similar. The influence of solar radiation pressure on the visible dust is negligible at this distance from the nucleus, as this is only of significance at about several 10^4 km. Particles on reversed trajectories could not be detected. Ground-based observations covering the period of the Giotto encounter are consistent with the overall features as seen, for example, in image No. 704 (Fig. 2) (Cosmovici et al., 1986). 3-dimensional calculations are required for a detailed comparison. The orientation of the dust emission activity outside the image plane (essentially the Giotto target plane) can only be indirectly assessed

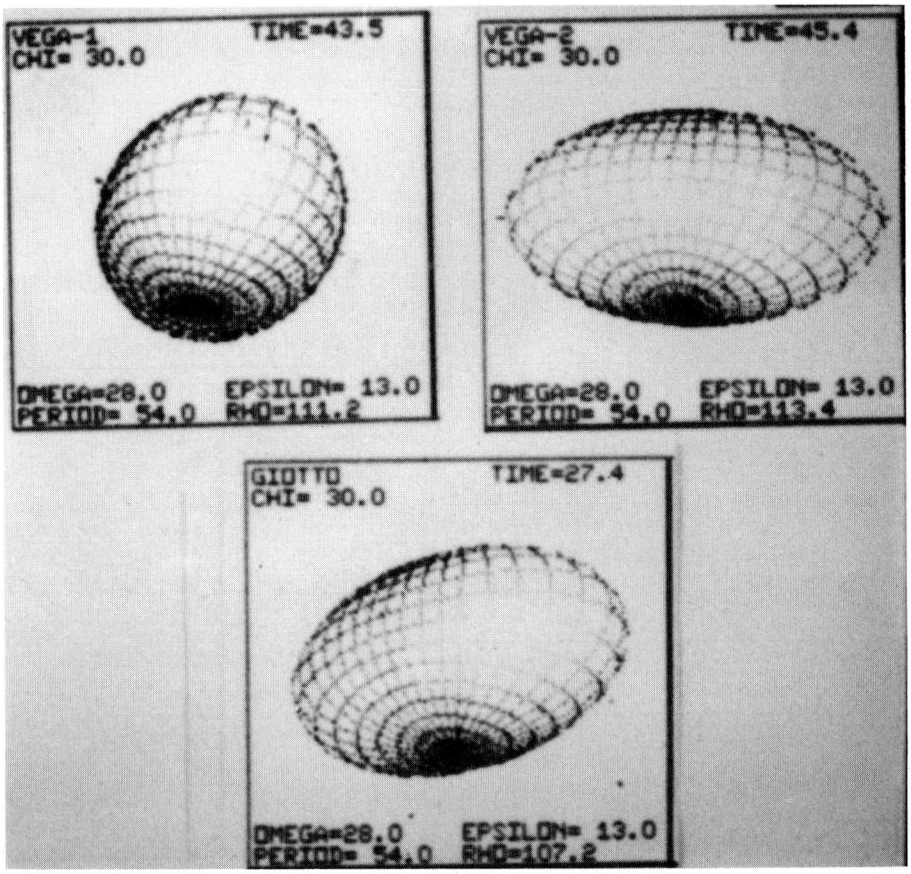

Fig. 12. This figure shows the projected surfaces of the nucleus as seen from the Sun during the a) Vega-1, b) Vega-2, c) Giotto closest approaches. The nucleus is approximated by an ellipsoid. Vega-1 looks at the "big end". The illuminated hemispheres are almost identical for Vega-2 and Giotto

from the HMC observations alone. An extension of the HMC observations into the outer coma seems feasible.

The radial extent of the isophotes can be used as a measure for the average activity in a certain direction (at distances $\gg$ size of the nucleus). As seen in Fig. 2, the strongest emission was not directed towards the sun. It seems that the projected directions of the rotation axis and the strongest dust emission were approximately coincident. This emission was about 1.6 times stronger than that roughly directed towards the sun. Small particles with radii of about 1 μm leaving the nucleus at a velocity of 0.5 km s^{-1} would reach the limit of the FOV of HMC in about 2 h if they travel close to the image plane.

Isophotes on later close-up images indicate pronounced activity directed ~13° north of the solar direction, emanating from the visible northern subsolar activity area of the nucleus (cf. image No. 3461). A slightly stronger activity was directed 35° south of the sunward direction. The direction of the maximum extent of the isophotes changes from 52° for the outer isophotes (distance >1000 km) to 35° for the inner isophotes (<100 km). The apparent change of direction is visible on the intermediate image No. 3256 of Fig. 1 in Keller et al. (1986a). The following interpretation is suggested: a strong emission of dust, directed away from the observer (HMC), originated from the hemisphere not visible to HMC on which the illuminated pole of rotation was located (see Fig. 12). The direction of emission came close to the mean normal of the insolated surface. This activity produced most of the background dust seen on the antisunward side of the nucleus.

The shape of the isophotes is controlled by the emission from the northern tip of the nucleus. Part of this active area was observed by HMC. The diameter of this jet is about 3 km. With a rotation period of 54 h, the limb is about 5 h away from the morning terminator. The activity of this region should have been increasing during and after the approach of Giotto. This relative increase in production towards the sun is confirmed by the ground based observations.

A second major activity area was observed near the "central depression" associated with the "bright patch". Its dimension is comparable to that of the northern activity area, however, the production of dust maybe only about 2/3 as judged by the intensity ratios. A small dust activity can be found at the southern end of the nucleus the source of which is located just beyond the limb. Its view may be partly obscured by outcroppings of the nucleus.

At least 13 faint, narrow jets can be identified on HMC images. There are three faint jets, two pointing northerly and one southerly, in projection away from the sun, only visible on the latest calibrated images (No. 3416). These are clearly discernible on Fig. 14 which shows azimuthal gradients by subtraction of a rotated image. These faint jets could still originate from the afternoon (dusk) part of the nucleus located within 17° of the night terminator directed almost parallel to the line of sight. Their apparent orientation is caused by projection.

5.2. *The northern area of activity*

A bright region of activity was observed near the north sunward limb (just north of the "crater"). The width of the dust emission area measured just at the limb is about 3 km. At least 4 individual sub-jets, narrow almost collimated beams of less than 10° full-

Fig. 13. Image No. 3416. An example of a 196 × 196 image taken at the end of Single Detector Mode, 376 seconds before CA from a distance of 25,780 km. The Sun is 16° behind the image plane and 26° above the horizontal. The horizontal middle line corresponds to 113 km

cone angle, can be identified. Their typical diameter is 0.5 km. Three of them can be traced to two of the bright spots on the surface discussed in Sect. 4 (Fig. 15). The jets cross each other with angles up to 20°. The terrain has to be rather rough showing slope variations if they emanate perpendicularly to the surface. It is difficult to assess whether all the surface within the area of activity emitted dust or whether the dust was ejected out of numerous small spots of which only the strongest could be distinguished. The individual emissions remaining visible at larger distances as faint jets carried only a minor part of the total dust production from activity regions. Whether the increased brightness of the foot points and their enhanced activity was due to a less inhibited sublimation or caused by chemical inhomogenity remains an open question (see also Sect. 6).

5.3. Dust-gas interaction

In considering this topic, it should be noted that the HMC observed scattered radiation from dust particles through the broad band filters. The gas environment can be inferred from the observations of the dust since the particles are entrained by the gas near the surface. Investigations of the radial intensity profiles beyond a distance of 1 or 2 km from the surface on medium resolution images have not revealed a zone of dust grain acceleration. A steepening above the $1/r$ profile due to acceleration cannot be expected since the hydrodynamical scalelength for particles smaller than about 1 μm is of this order (Hellmich, 1981; Gombosi et al., 1985). Investigations of the high resolution images reveal bright, narrow subjets indicating differences in the strength of gas activity. The physics of the collimation of the jets has still to be explained.

Gas produced in active regions must spread across and around the nuclear surface to equalize the pressure. The finer dust is more likely to be entrained in this surface breeze. Evidence for the spreading of dust across the surface of the nucleus has been found from the gradient of dust columns between the dark side of the nucleus and the observer. The observed column density decreases

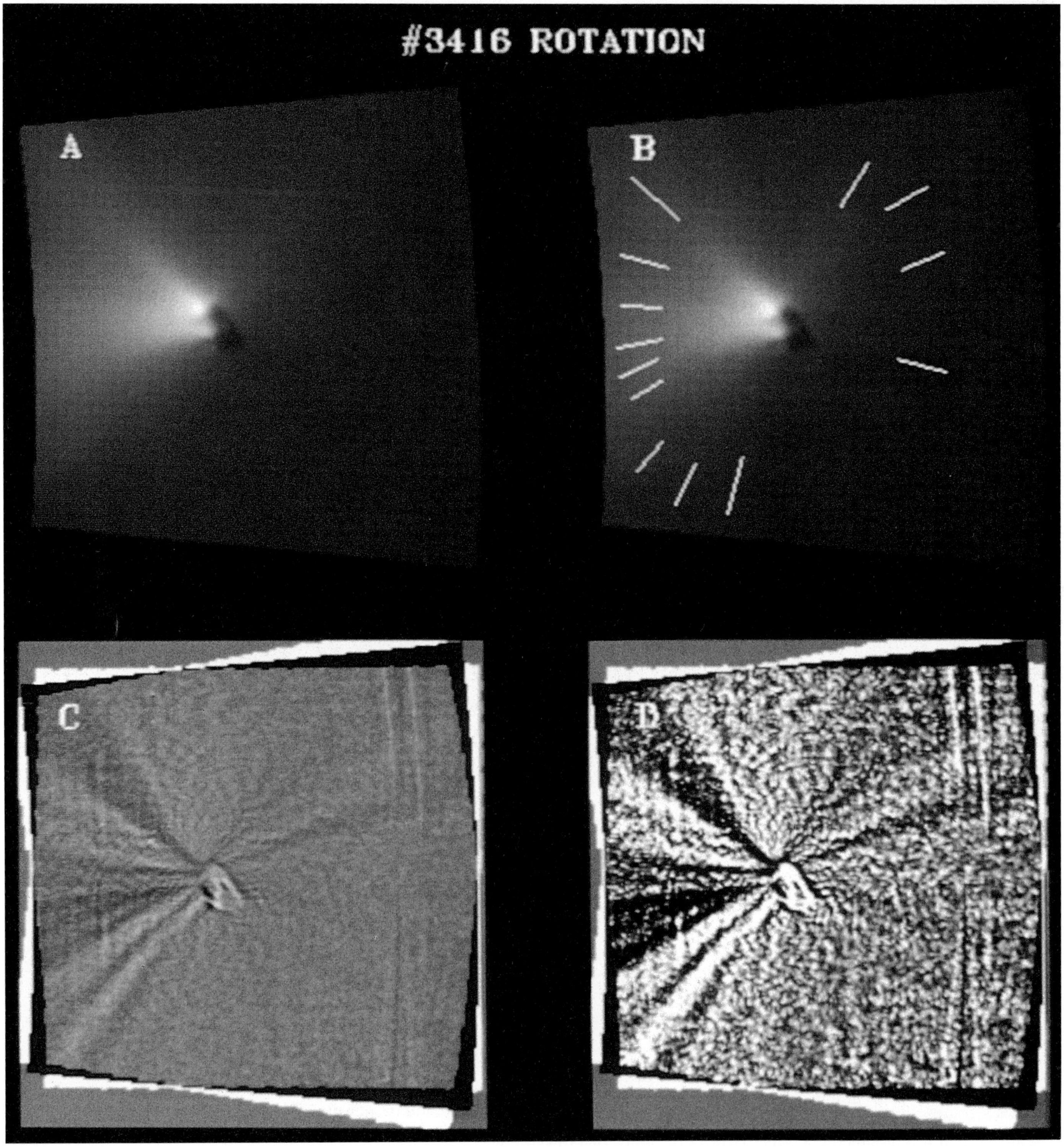

Fig. 14a–d. In image No. 3416 (**a**) some jet structures are obvious. However, the azimuth gradient technique (**c**) and (**d**) reveals more jet-like features near the nucleus. 13 features have been noted and marked in **b**

by about a factor 3 from the north to the south end (Fig. 9). The scalelength of the dust distributions is small compared to the size of the nucleus indicating that the dust was close to the surface and stemmed from the area of activity on the northern bright limb (Fig. 16). The minimum intensity on the nucleus is about $50\ \mathrm{mW\ m^{-2}\ sr^{-1}}$. This value is about a factor 2.5 less than the intensity ($123\ \mathrm{mW\ m^{-2}\ sr^{-1}}$) from the dust just off the nucleus on the nightside, indicating again that the amount of dust in the background was larger than that between the camera and the nucleus.

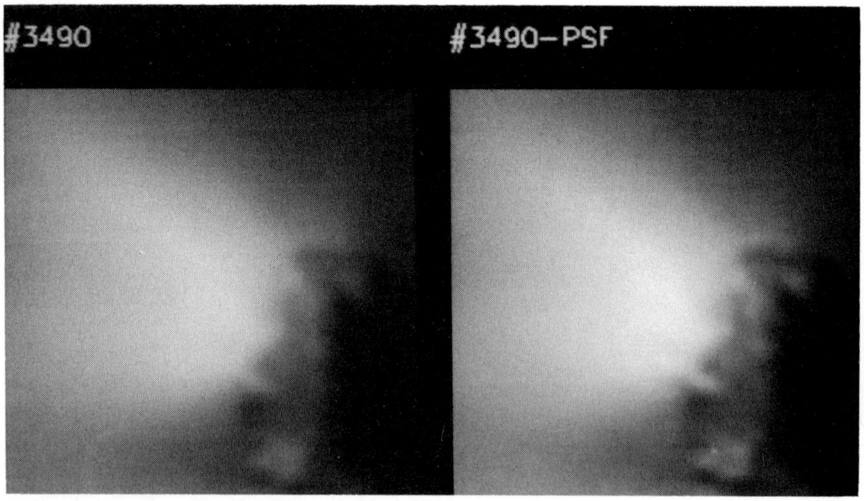

Fig. 15. Image No. 3490 is taken from a distance of 5490 km. The sun is 29°.5 above the horizontal and 5°.3 below the image plane. This is the last image showing the "crater" as a whole. a) radiometrically corrected image b) additional filtering to suppress high frequency noise and deconvolution of the point spread function are applied. Resolution and contrast are improved

5.4. Total area of activity

The active area on the northern limb visible to the HMC contributes only a fraction of the overall dust production. Following the arguments given above, the strongest sources were located on the afternoon side of the nucleus.

Sublimation of an icy mixture consisting of 80% water and 20% more volatile compounds (Krankowski et al., 1986) at the heliocentric distance of comet Halley during the Giotto encounter and for an albedo of 0.04 would yield a flux of $Z = 1.9\ 10^{18}$ molecules $cm^{-2}\,s^{-1}$ for insolation with normal incidence angle. The active area of 3 km diameter (if assumed to be circular) would have produced $1.3\ 10^{29}$ molecules s^{-1}.

The other observed major source (the bright patch) had similar dimensions. A production rate of $1.2\ 10^{29}$ molecules s^{-1} can be estimated. The contribution of the source on the afternoon side, estimated to be 1.6 times stronger, should then produce $4.0\ 10^{29}$ molecules s^{-1} yielding a total gas production of $6.5\ 10^{29}$ molecules s^{-1}, in good agreement with the value of $6.9\ 10^{29}$ molecules s^{-1} determined during the Giotto flyby (Krankowsky et al., 1986). Because of the uncertainty of the size of the active areas beyond the limb, the associated error is about 25%. This error is obtained from the relative intensity (somewhat less than a factor of 2) of the two observed jets.

The total active area on the nucleus is therefore $7+7+1.6 \times 14 = 36$ in [km^2]. This is about 30 to 40% of the projected surface and 10% of the total surface.

5.5. Dust column density

Densities and column densities of the dust cannot be calculated based on HMC data alone. In addition, the particle size distribution and optical properties of the dust are needed. For a first assessment it is assumed that dust particles of radius of 1 μm were optically most active.

The energy scattered in direction θ by a particle with a differential scattering cross section $\sigma(\theta)$ is given by

$$j(\theta) = F_0^c \cdot \sigma(\theta)\,[\mathrm{W\,sr^{-1}}], \tag{6}$$

where F_0^c is the solar flux at the particle and θ is the scattering angle. The intensity produced by a dust column along the line-of-sight, l, is

$$I(\theta) = \int j(\theta)\, n\, dl, \tag{7}$$

where n is the particle density. The differential scattering cross-section of the particles can be expressed in terms of the "phase dependent" geometric albedo of a particle (see Divine et al., 1986).

$$\sigma(\theta) = \frac{A_p(\theta)}{\pi} \cdot q = \frac{p}{\pi} \cdot \frac{\phi(\theta)}{\phi(180°)} \cdot q, \tag{8}$$

where ϕ is the scattering function [sr^{-1}], q is the geometrical cross section ($\pi a^2 = q$), and $A_p(180°) = p$, the geometric albedo. The intensity in direction of θ is then given by

$$I(\theta) = F_0^c \cdot \frac{A_p(\theta)}{\pi} \cdot q \int n\, dl. \tag{9}$$

The geometric albedo of the dust particles is rather low, $p \simeq 0.03$ (Divine et al., 1986). The scattering angle for the Giotto approach was 73°. The scattering function of Divine et al. (1986, Fig. 20) yields $\phi(73°)/\phi(180°) \simeq 1$. Therefore,

$$I(73°) = F_0^c \frac{p}{\pi} \cdot q \int n\, dl. \tag{10}$$

The solar flux within the pass band of the HMC clear filter at 0.9 AU (see Wilhelm, 1986) was

$$F_0^c = 1.08\ 10^3\ \mathrm{W m^{-2}}. \tag{11}$$

The peak signal coming from the northern dust activity area was

$$I_{max} = 2.8\ \mathrm{W m^{-2}\, sr^{-1}}. \tag{12}$$

On substitution into Eq. (10), the column density is found to be

$$N = \int n\, dl = 8.7\ 10^{10}\ \mathrm{m^{-2}} \tag{13}$$

for $a = 1$ μm particles.

This column density can be used to derive an order-of-magnitude estimate for the maximum optical thickness of the dust corresponding to the peak signal. If the optical extinction cross section is equated to the geometrical cross section, which is valid within a factor 2 for absorbing particles in the size range of 1 μm, the optical depth becomes

$$\tau_{max} = q \int n\, dl = 0.28. \tag{14}$$

The optical depth, τ, is inversely proportional to p. Values for $p \ll 0.03$ have not been observed. Therefore, it can be assumed that the optical thickness of the dust seen by the HMC was <1 everywhere.

This conclusion is also corroborated by the intensity distribution of the dust near the brightest areas around the base of the northern most jet. Intensity profiles generally follow a $1/r$ law (Keller et al., 1986b). Flattening of the observed intensity profile near the nucleus, indicating high optical depth, is not observed.

There are also no deviations in the opposite direction, that is, to steeper gradients than the $1/r$ law. Such gradients would indicate regions of acceleration of the dust particles on scale-lengths larger than 1 km. Of course it cannot be excluded that the flattening due to optical thickness and the steepening due to acceleration of dust accidentally counter-balance.

5.6. Visibility of the nuclear surface near dust activities

The reflectivity of the nuclear surface at the bright limb relative to that of the dust can be estimated following the approach of Hellmich and Keller (1981). In the single scattering approximation the signal from the surface attenuated by $e^{-\tau/2}$ is comparable to that of the column of dust in the line of sight tangential to the surface $[\sim(1-e^{-\tau})]$ if the reflectivity R_L is about 0.3 of the dust particle albedo, i.e.,

$$R_L/A_p = \frac{(1-e^{-\tau})}{e^{-\tau/2}} = 0.3\,. \tag{15}$$

The transition from the surface to the dust jets at places of highest activity is hardly discernible in the HMC images taken with the clear filter. Therefore, the condition $R_L \approx 1/3\, A_p$ at the limb, under phase angle of 107°, should hold. This is in good agreement with the direct measurement which yields $R < 1\%$ (see Sect. 4).

6. Discussion

The immediate results of the HMC images confirmed the existence of a solid cometary nucleus as predicted by the icy conglomerate model of Whipple (1950). The activity of gas production was strongly oriented towards the sun, coming from isolated areas of the surface. The surface has to be widely covered by a crust or mantle that prevents free sublimation of the icy conglomerate.

6.1. Shape

The shape of the nucleus showed considerable deviation from a sphere. Its 2:1 elongation makes it improbable that the nucleus was originally formed as a spherical body. This ellipsoidal shape could not be formed by preferential sublimation over long time intervals. The axis of rotation would not be stable within the body if the moments of inertia are changed so drastically. Reorientation to rotation around the axis of maximum moment of inertia would occur. One exception would be the preferential loss of mass at the poles of rotation. Then the spherical nucleus would slowly form a flattened sphere but not an elongated ellipsoid. A possible way to sustain non-uniform sublimation and to reconcile the present shape with an initially more spherical body could be the suppression of sublimation by a crust on certain parts of the surface. In the case of comet Halley the sublimation on both ends should be strongly suppressed. This, however, is in clear disagreement with the strong activity level during the Vega 1 flyby when an end pointed to the sun (Sagdeev et al., 1986). The large amount of subliming ice required to reduce the diameter from 16 to 8 km (a 4 km thick layer) would require at least a thousand revolutions within the inner solar system (see below) and make comet Halley an extremely "old" object.

Consequently, the overall shape of the nucleus was not caused by non-uniform sublimation. It had to be formed during the creation process. The nucleus could consist of a few rather large subnuclei having coalesced during collisions. Alternatively, the irregular shape might result from the break up of a larger body already in the early period of formation. Break up by collision is plausible during formation but practically impossible later. The break up caused by tensions within the rather fragile body, frequently observed as splitting of cometary nuclei, is more probable. This splitting is also facilitated by large-scale heterogeneity. The irregular elongated shape of the nucleus of comet Halley suggests that the body consists of subnuclei up to a few km in diameter. If these subnuclei differ somewhat in composition the inferred variability and inhomogeneity of the cometary coma and even the variability in activity could be naturally explained. The CN jets did not coincide with the dust observed in the continuum (Larson et al., 1986; A'Hearn et al., 1986) and the corresponding grains may originate from different places of the nuclear surface. Some outbursts, manifested only in enhanced gas and ion production, and other outbursts, manifested in general activity, indicate a large scale inhomogeneity of the nucleus (Feldman et al., 1986).

6.2. Size

The 3-dimensional shape of the nucleus is being investigated in collaboration with the Vega imaging team. The overall approximation by an ellipsoid of 4.1, 4.2, and 8 km for the semimajor axes yields a surface of about 400 km² and a volume of 580 km². These values can be compared to the "best guess" before the spacecraft encounters (Divine et al., 1986), a sphere with $R = 3$ km, a geometic albedo $p = 0.06$ and a density of 1 g cm^{-3}. This small radius yields a surface of 113 km² (projected area 28 km²) and a volume of 113 km³. The larger, observed surface goes well with the activity being restricted to about 30 to 40 % of the projected area (10 % of the overall surface). It is obvious that free sublimation of ice can take place only on a fraction of the surface, otherwise the total production rate would be much too high. The alternative of a restricted outflow (Horanyi and Kecskeméty, 1983) through a thin mantle over most of the illuminated or total surface can be ruled out by the HMC observations. The increase in projected area varying somewhere between the possible minimum of 50 km² (if the obliquity of the rotation axis is zero) and the maximum around 100 km² reduces the geometric albedo as determined from the total visual magnitude of the comet at large heliocentric distances by the ratios of the earlier and present projected surface areas. A more recent assessment of the observations (Whipple, 1986; Delamere et al., 1986) leads to $p = 0.04$. The obliquity of the rotation axis has to be large enough (~20–30°, Sagdeev et al., 1986c; Wilhelm et al., 1986) so that the minimum – maximum variation is considerably less than a factor of 2. Otherwise the rotation period would have been easily determined from preperihelion observations.

The reflectivity of active areas was somewhat higher than that of the surrounding surface but still very small. The ice must be "dirty" and dark. It did not differ much from the crust, if the ice was directly visible in the bright spots. It has been pointed out (Whipple, 1986) that the total production of gas and dust during comet Halley's 1986 apparition was slightly higher (by less than a factor of 2) than predicted. The acceleration by non-gravitational forces was estimated by Yeomans (1977, 1984) and others, limiting the mass of comet Halley's nucleus to about 10^{14} kg. For a recent thorough analysis see Rickman (1986). He obtained a mean density for the nucleus (outlined by the ellipsoid) of less than $0.2\,g\,cm^{-3}$.

The discovery of the small albedo and the low density of the nucleus of comet Halley are a direct consequence of the observational results of the Halley Multicolour Camera. The low albedo value is in line with recent infrared determinations (for example, Hartmann et al., 1985). The inferred low density of about $0.2\,g\,cm^{-3}$ implies that cometary nuclei are extremely fragile and are the lowest density bodies in the solar system. This is consistent with the very low tensile strength determined from the splitting of comets (particularly comet West 1976 VI, Sekanina and Farrell, 1978).

It seems difficult to imagine how such a large body of such low density could be formed, particularly if collisions of subbodies are involved. The compaction of large parts of the volume may be unavoidable, but if the material is brittle, then compaction may not occur. Laboratory experiments with such low density icy materials and extremely low temperatures are urgently needed. These exotic nuclei must have been formed at the outer fringes of the planetary system or beyond. Many results of the gas and dust analyses point to a strong connection with interstellar material. It is expected that material of such low density has a low heat conductivity and capacity. This, in conjunction with the long spin period of the nucleus of 54 h, should prevent activity from continuing on the night side. However, a minor activity centre on the night side seems visible on the Vega-2 images (Smith et al., 1986) and some of the faint jets on the HMC images could originate from the early evening part of the nucleus (see above). Energy may be stored locally. For example, pockets of more volatile ice could explain erratic and short-term activity (times-cales 1 h, Feldman et al. 1986) and sustain activity beyond the evening terminator.

A surface crust or mantle that prevents free sublimation will also be a good overall insulator (low thermal conductivity), preventing the subsurface of the nucleus exceeding the mean temperature determined by the heat absorption averaged over the cometary orbit. On the other hand, the surface temperature of the crust will reach almost that of a black body.

6.3. *Surface features*

Dust from jets obscured the signal from the near-by inactive parts of the illuminated surface near the morning terminator. This contribution from the dust has to be carefully assessed before photoclinometric results based also on the three-dimensional model of the nucleus can be expected.

The large roundish feature, the "crater", could be an active surface not yet triggered by the rising sun. Some smaller similar basins appear to be located south of the "crater". Sublimation of ice at its centre may therefore have been responsible for the formation of the "crater". If the "erosion" is controlled by the sublimation heat of a 80:20 water to more volatile ice mixture (Krankowsky et al., 1986; Huebner et al., 1986) an evaporation flux of $Z = 2\ 10^{18}\,cm^{-2}\,s^{-1}$ can be estimated for high solar incident angles at the heliocentric distance $r_h = 0.9$ AU. This activity level could be considered typical for the 3 months of activity around perihelion, with a turn-on fraction of 1/3. Then $5\ 10^{24}$ molecules per cm^2 are sublimated per orbital revolution. This converts to a lost layer of about 6 m assuming a density of the material of $0.25\,g\,cm^{-3}$. Over the known time of short periodicity (30 revolutions) of comet Halley this amounts to less than 200 m, a value comparable to an estimated depth of this feature (Schwarz et al., 1987). It could have already existed before the comet reached its present orbit. Anyway, if comet Halley is young in terms of orbits through the inner solar system (let us say 30 out of less than 1000 possible revolutions) the present shape and features could be traced back to its pre-active time. If comet Halley has passed its "youth", having orbited the inner solar system many hundred times, then its small and medium scale (0.5 to 1 km) topographic features could be completely formed or altered by the sublimation process.

It is hard to imagine that the crust or mantle is only a few centimetres thick, covering a terrain with valleys and hills of up to the order of a kilometre evenly enough to prevent sublimation. 1 % of dust particles from the sublimated layer would already lead to metres of debris on the surface.

The observed asymmetry of the dust emission relative to the comet-sun line indicates that the strongest emission does not come from the subsolar point but rather from the afternoon hemisphere. The near coincidence of the strongest dust production with the rotation axis (in projection) may indicate that the region around the continuously illuminated pole of rotation contributes a sizable fraction of the dust production. This conclusion is supported by the fact that the same asymmetry is also observed by Vega 1 and 2. All three spacecraft approached the comet with similar phase angles. In particular, the images taken by Vega 1 (Sagdeev et al., 1986c) more than 7 days earlier from $4\ 10^5$ km at a different rotation phase of the nucleus show a very similar cone opening of less than 90° south of the comet-sun direction. These should be compared with HMC image No. 3056 taken at a distance of $1.2\ 10^5$ km (Keller et al., 1986a, Fig. 1).

One could speculate that near the pole the continuous insolation increases the production flux compared to that at lower latitudes where the insolation is modulated by rotation. Thus the sub-surface temperature at the illuminated pole may be higher than at the sub-solar regions, causing enhanced activity. It is also possible that the largest active area is located near the pole. In any case it is obvious that the major production is not directed towards the sun with a small lag angle during the observational period of the Vega and Giotto spacecrafts. If this is generalized for the orbital position of comet Halley, it is in conflict with the general assumption of calculations of the non-gravitational forces. The effective "lag" angle seems to be much greater than the upper limit of 10° which is generally assumed. This increases the effect of the non-gravitational forces and thus increases the calculated mass and density of the nucleus.

In this contribution only some of the consequences of the observations of the nucleus of comet Halley could be discussed. More facts will be established in future analysis of the data. There is no doubt that the HMC images are leading to a new, detailed, picture of cometary nuclei, forming the basis of cometary physics.

Acknowledgements. The authors thank the many scientists and engineers involved in the design and the development of the instrument for their dedicated work during the course of the project. We should also like to acknowledge the contributions of the ESA Project Team and the Project Scientist to the successful completion of the HMC experiment within the Giotto Mission. The experiment was funded by BMFT, MPG, DFVLR (FRG), CNES (France), SPPS (Belgium), CNR/PSN (Italy), and NASA (USA).

References

A'Hearn, M.F., Birch, P.V., Klinglesmith, D.A.: 1986, 20th ESLAB Symp., ESA SP-250, Vol. **I**, p. 483

Cosmovici, C.B., Green, S.F., Hughes, D.W., Keller, H.U., Mack, P., Moreno-Insertis, F., Schmidt, H.U.: 1986, 20th ESLAB Symp., ESA SP-250, Vol. **II**, p. 375

Curdt, W., Specht, H.: 1986, Internal Report Max-Planck-Institut für Aeronomie, MPAE-W-55-86-29

Delamere, W.A., Huebner, W.F., Keller, H.U., Reitsema, H.J., Schmidt, H.U., Schmidt, W.K.H., Whipple, F.L., Wilhelm, K.: 1986, 20th ESLAB Symp., ESA SP-250, Vol. **II**, p. 355

Divine, N., Fechtig, H., Gombosi, T.I., Hanner, M.S., Keller, H.U., Larson, S.M., Mendis, D.A., Newburn Jr., R.L., Reinhard, R., Sekanina, Z., Yeomans, D.K.: 1986, Space Sci. Rev. **43**, 1

Edenhofer, P., Bird, M.K., Brenkle, J.P., Buschert, H., Kursinsky, E.R., Mottinger, N.A., Porsche, H., Stelzried, C.T., Volland, H.: 1986, 20th ESLAB Symp., ESA SP-250, Vol. **II**, p. 215

Feldman, P.D., Festou, M.C., A'Hearn, M.F., Arpigny, C., Butterworth, P.S., Cosmovici, C.B., Danks, A.C., Gilmozzi, R., Jackson, W.M., McFadden, L.A., Patriarchi, P., Schleicher, D.G., Tozzi, G.P., Wallis, M.K., Weaver, H.A., Woods, T.N.: 1986, 20th ESLAB Symp., ESA SP-250, Vol. **I**, p. 325

Gombosi, T.I., Cravens, T.E., Nagy, A.F.: 1985, *Astrophys. J.* **293**, 328

Hartmann, W.K., D.P., Tholen, D.J.: 1985, *Ices in the Solar System*, eds. J. Klinger et al., Reidel, Dordrecht, p. 169

Hellmich, R.: 1981, *Astron. Astrophys.* **93**, 341

Hellmich, R., Keller, H.U.: 1981, *Icarus* **47**, 325

Horanyi, M., Kecskeméty, K.: 1983, *Asteroids, Comets, Meteors*, ed. Lagerkvist and Rickmann, p. 269

Huebner, W.F., Delamere, W.A., Keller, H.U., Reitsema, H.J., Schmidt, H.U., Whipple, F.L., Wilhelm, K.: 1986, 20th ESLAB Symp., ESA SP-250, Vol. **II**, p. 363

Keller, H.U., Schmidt, W.K.H., Wilhelm, K.: 1985, *ESA Bulletin* **44**, 96

Keller, H.U., Arpigny, C., Barbieri, C., Bonnet, R.M., Cazes, S., Coradini, M., Cosmovici, C.B., Delamere, W.A., Huebner, W.F., Hughes, D.W., Jamar, C., Malaise, D., Reitsema, H.J., Schmidt, H.U., Schmidt, W.K.H., Seige, P., Whipple, F.L., Wilhelm, K.: 1986a, *Nature* **321**, 320

Keller, H.U., Delamere, W.A., Huebner, W.F., Reitsema, H., Schmidt, H.U., Schmidt, W.K.H., Whipple, F.L., Wilhelm, K.: 1986b, 20th ESLAB Symp., ESA SP-250, Vol. **II**, p. 359

Keller, H.U., Angrilli, F., Becker, C., Bianchini, G., Bouryries, P., Brunello, P., Cazes, S., Cucchiaro, A., Curdt, W., Delamere, W.A., Engelhardt, W., Fanti, G., Gliem, F., Hartwig, H., Jamar, C., Krahn, E., Kramm, J.R., Meyer, H.J., Reitsema, H., Schmidt, H.P., Schmidt, R., Schmidt, W.K.H., Schwarz, G., Turner, J.J., Wilhelm, K.: 1987, *J. Phys. E: Sci Instrum.* **20**, 807

Kramm, J.R., Möhring, W., Keller, H.U.: 1987, *Geophys. Res. Lett.* **14**, 677

Krankowsky, D., Berthelier, J.J., Dolder, U., Eberhardt, P., Herrmann, U., Herrwerth, I., Hodges, R.R., Hoffman, J.H., Illiano, J.M., Lämmerzahl, P., Schulte, W., Woweries, J.: 1986, *Nature* **321**, 326

Larson, S., Sekanina, Z., Levy, D., Tapia, S., Senay, M.: 1986, 20th ESLAB Symp., ESA SP-250, Vol. **II**, p. 145

Reitsema, H.J., Delamere, W.A., Huebner, W.F., Keller, H.U., Schmidt, H.U., Schmidt, W.K.H., Whipple, F.L., Wilhelm, K.: 1986, 20th ESLAB Symp., ESA SP-250, Vol. **II**, p. 351

Rickman, H.: 1986, ESA Workshop on *Comet Nucleus Return Mission*, Canterbury

Sagdeev, R.Z., Szabó, F., Avanesov, G.A., Cruvellier, P., Szabo, L., Szegö, K., Abergel, A., Balazs, A., Barinov, I.V., Berteaux, J.-L., Blamont, J., Detaille, M., Demarelis, E., Dul'nev, G.N., Endröczy, G., Gardos, M., Kanyo, M., Kostenko, V.I., Krasikov, V.A., Nguyen-Trong, T., Nyitrai, Z., Reny, I., Rusznyak, P., Shamis, V.A., Smith, B., Sukhanov, K.G., Szabó, F., Szalai, S., Tarnopolsky, V.I., Toth, I., Tsukanova, G., Valníček, B.I., Varhalmi, L., Zaiko, Y.K., Zatsepin, S.I., Ziman, Y.L., Zsenei, M., Zhukov, B.S.: 1986a, *Nature* **321**, 262

Sagdeev, R.Z., Krasikov, V.A., Larson, S., Merenyi, E., Shamis, V.A., Smith, B., Szegö, K., Tarnopolski, V.I., Toth, I.: 1986b, 20th ESLAB Symp., ESA SP-250, Vol. **II**, p. 335

Sagdeev, R.Z., Avanesov, G.A., Moroz, V.I., Shamis, V.A., Smith, B., Tarnopolski, V.I., Toth, I., Ziman, Ya L., Zhukov, B.S.: 1986c, 20th ESLAB Symp., ESA SP-250, Vol. **II**, p. 317

Schmidt, W.K.H., Keller, H.U., Wilhelm, K., Arpigny, C., Barbieri, C., Biermann, L., Bonnet, R.M., Cazes, S., Cosmovici, C.B., Delamere, W.A., Huebner, W.F., Hughes, D.W., Jamar, C., Malaise, D., Reitsema, H., Seige, P., Whipple, F.L.: 1986, ESA SP-**1077**, p. 149

Schwarz, G., Craubner, H., Delamere, A., Goebel, M., Gonano, M., Huebner, W.F., Keller, H.U., Kramm, R., Mikusch, E., Reitsema, H., Whipple, F.L., Wilhelm, K.: 1987, *Astron. Astrophys.* (this volume)

Sekanina, Z., Farrell, J.A.: 1978, *Astron. J.* **83**, 1675–1689 (Paper I)

Sekanina, Z., Larson, S.M.: 1983, *Bull. Am. Astron. Soc.* **15**, 651

Smith, B., Szegö, K., Larson, S., Merényi, E., Toth, I., Sagdeev, R.Z., Avanesov, G.A., Krasikov, V.A., Shamis, V.A., Tarnopolsky, V.I.: 1986, 20th ESLAB Symp., ESA SP-250, Vol. **II**, p. 327

Whipple, F.L.: 1950, *Astrophys. J.* **111**, 375

Whipple, F.L.: 1986, 20th ESLAB Symp., ESA SP-250, Vol. **II**, p. 281

Wilhelm, K.: 1986, Internal Report Max-Planck-Institut für Aeronomie, MPAE-W-55-86-06

Wilhelm, K.: 1987, *Nature* **327**, 27

Wilhelm, K., Cosmovici, C.B., Delamere, W.A., Huebner, W.F., Keller, H.U., Reitsema, H., Schmidt, H.U., Whipple, F.L.: 1986, 20th ESLAB Symp., ESA SP-250, Vol. **II**, p. 367

Yeomans, D.K.: 1977, *Astron. J.* **82**, 435

Yeomans, D.K.: 1984, *Cometary Astronomy*, eds. D.K. Yeomans, R.M. West, R.S. Harrington, B.G. Marsden, p. 167

Astron. Astrophys. 187, 824–828 (1987)

ASTRONOMY AND ASTROPHYSICS

Electrostatic charging and fragmentation of dust near P/Giacobini-Zinner and P/Halley

H. Boehnhardt [1] **and H. Fechtig** [2]

[1] Dr. Remeis-Sternwarte, Astronomisches Institut der Universität Erlangen-Nürnberg, Sternwartstrasse 7, D-8600 Bamberg, Federal Republic of Germany
[2] Max-Planck-Institut für Kernphysik, Saupfercheckweg, D-6900 Heidelberg, Federal Republic of Germany

Received February 17, accepted May 27, 1987

Summary. In the coma of a comet, the dust grains emitted by the cometary nucleus are embedded in a surrounding plasma and exposed to the sunlight. Several effects (impinging charged plasma particles, photo-, secondary, thermionic, and field emission of electrons, sputtering of ions) might cause the electrostatic charging and possibly the fragmentation of some of these grains. Using in-situ-measurements of the comets P/Giacobini-Zinner (P/G-Z) and P/Halley, the electrostatic charging and fragmentation of spherical dust particles of carbon and silicate in the cometary comae is investigated theoretically. The results are compared with in-situ-measurements of dust in the coma of P/Halley.

Key words: comets – dust – electrostatic charging – fragmentation

1. Introduction

During the fly-by of the Vega-1, Vega-2, and Giotto spacecrafts at comet P/Halley some indications of fragmentation and/or ablation processes, affecting the dust particles in the coma, are observed. Vaisberg et al. (1986) mention changes in the mass distribution of the dust particles which cannot be exclusively explained by the effect of the radiation pressure of the sunlight. McDonnell et al. (1986) suggest a region extending 10 000 km from the nucleus, in which additional dust particles are produced. Simpson et al. (1986) detected clusters of grains inside and also outside the cometopause.

In the following we describe in detail one of the suggested fragmentation processes, i.e. electrostatic fragmentation of dust within the coma of P/G-Z and P/Halley. We shortly discuss other processes of fragmentation and ablation which might affect the dust of these comets.

2. Electrostatic charging and fragmentation of cometary dust

2.1. Cometary dust

It is widely accepted (Fechtig, 1982; Fraundorf et al., 1982), that cometary dust is very similar to the fluffy-like Brownlee particles (Brownlee et al., 1976) which are collected from interplanetary space by the earth's atmosphere. These dust grains, mostly irregular in shape, are conglomerates consisting of a great number of small building blocks (diameter $<0.1\,\mu$m). In-situ-measurements near comet P/Halley (Smirnov et al., 1986) and laboratory experiments with artificial cometary nuclei (Ibadinov, 1982; Ibadinov et al., 1984) indicate low density ϱ ($\varrho < 1\,\mathrm{g\,cm^{-3}}$) and low tensile strength σ_T ($\sigma_T = 5\ 10^4$ to $2\ 10^6\,\mathrm{dyn\,cm^{-2}}$) of the cometary dust and of the crust of the nucleus. Carbon and silicate were found to be two of the basic materials of cometary dust (Kissel et al., 1986).

2.2. Electrostatic fragmentation

We compare the stress $\boldsymbol{\sigma}_c$ of the electrostatic field $\boldsymbol{F}$ on a surface element $\boldsymbol{dA}$ of a charged grain with the tensile strength σ_T of the dust material in order to derive a theoretical criterion for the electrostatic fragmentation of the dust particles. $\boldsymbol{\sigma}_c$ can be calculated from the Maxwellian stress tensor $\mathbf{T}$ (Boehnhardt, 1985)

$$\boldsymbol{\sigma}_c = \mathbf{T}\cdot\frac{\boldsymbol{dA}}{|\boldsymbol{dA}|} = \frac{1}{2}\frac{\varepsilon_0}{\varepsilon}\boldsymbol{F}^2\begin{pmatrix}\sin 2\vartheta\\ 0\\ \cos 2\vartheta\end{pmatrix} \tag{1a}$$

or

$$\sigma_c = |\boldsymbol{\sigma}_c| = \frac{1}{2}\frac{\varepsilon_0}{\varepsilon}\boldsymbol{F}^2 \tag{1b}$$

with ϑ being the angle between $\boldsymbol{F}$ and $\boldsymbol{dA}$, $\varepsilon_0, \varepsilon$ being the dielectric constants of the vacuum and of the dust material, respectively.

Using Eqs. (1a) and (1b) it is easy to show that a critical surface potential φ_T, given by

$$\varphi_T = r(2\,\varepsilon\,\sigma_T/\varepsilon_0)^{1/2} \tag{2}$$

might initiate the fragmentation of a homogenously charged spherical dust grain of radius r and tensile strength σ_T. In the following we use Eq. (2) to estimate the possibility of the electrostatic fragmentation of cometary dust charged up in the surrounding cometary plasma. Certainly only a small fraction of the cometary dust particles is spherically shaped as assumed for the derivation of Eq. (2). But from Eqs. (1a) and (1b) one can expect that very irregular mother grains might loose smaller daughter grains from edges and surface peaks, where the electrostatic field strength is high, even more efficiently than spherical grains of comparable size. Also inhomogeneities of the

Send offprint requests to: H. Fechtig

charge distribution in/on the grains might highly influence the electrostatic fragmentation of the particles, either supporting it or reducing it. So far no experimental measurements of the electrostatic fragmentation of small dust grains are published which could be used to model this effect on cometary dust.

For numerical reasons the electrostatic field strength of dust grains, irregular in shape and/or inhomogeneously charged, cannot be easily calculated. In addition the charging currents of these grains due to the surrounding plasma cannot be modelled analytically.

Therefore in this paper we calculate the electrostatic charging and fragmentation of homogeneously charged, spherical dust grains of carbon and silicate in the cometary coma. For the dust material a tensile strength σ_T of 10^3 to 10^6 dyn cm^{-2} is used. A surface potential of about 6 and of about 10 V is necessary to disrupt fluffy particles of carbon and silicate with radii of 0.4 µm and with tensile strength of 10^4 dyn cm^{-2}.

2.3. Electrostatic charging of dust grains

In the cases of the comets P/G-Z and P/Halley only the currents of the thermal plasma electrons and of the photoelectrons are found to be of major importance for the charging of dust grains moving in the coma plasma. Therefore we give in this paper only the expressions of these two charging currents, though our program code can also model the effects of monoenergetic plasma flows, secondary electron emission and field emission. For a detailed description of all theoretical equations used in the program we refer to Boehnhardt (1985, 1986). Thermionic and sputtering currents are not important for dust grains at about 1 AU solar distance.

The current J_{TH} of a thermal plasma onto a spherical dust grain with a surface potential φ can be calculated by

$$J_{TH} = \pi r^2 q N \left(\frac{8E}{\pi m}\right)^{1/2} \exp\left(-\frac{q\varphi}{E}\right)(1-\alpha)(1-\kappa), \quad (3),$$

where N and E are the density and the mean thermal energy, q and m denote the charge and the mass of the plasma particles. α and κ mean the numbers of particles to be reflected (α) and penetrating (κ) the grain. A discussion of these less important effects for the charging of the cometary dust is also included in Boehnhardt (1985, 1986).

Following the experimental and the theoretical work of Feuerbacher et al. (1972a, b) and Mukai (1981), one can express the photoelectron current J_{PH} onto spherical dust grains with radii greater than 0.1 µm, when exposed to sunlight, by

$$J_{PH} = \pi r^2 \Gamma \left(1 + \frac{\max(0, q\varphi)}{E_{smax}}\right) \exp\left(-\frac{\max(0, q\varphi)}{E_{smax}}\right), \quad (4)$$

where Γ denotes the current density and E_{smax} the energy of the maximum in the energy distribution of the photoelectrons (E_{smax} is of the order of 2.5 eV for silicate and carbon). We use Γ-values given by Mukai (1981) for 1 AU (Γ is 2.0 10^{10} cm^{-2} s^{-1} for silicate and 2.5 10^9 cm^{-2} s^{-1} for carbon); these values were scaled to the solar distances of the comets considered and corrected for opacity effects in the solar Lyman α line due to the extended neutral hydrogen cloud around the comae (2 to 4% of Γ, Feuerbacher et al., 1972a, b).

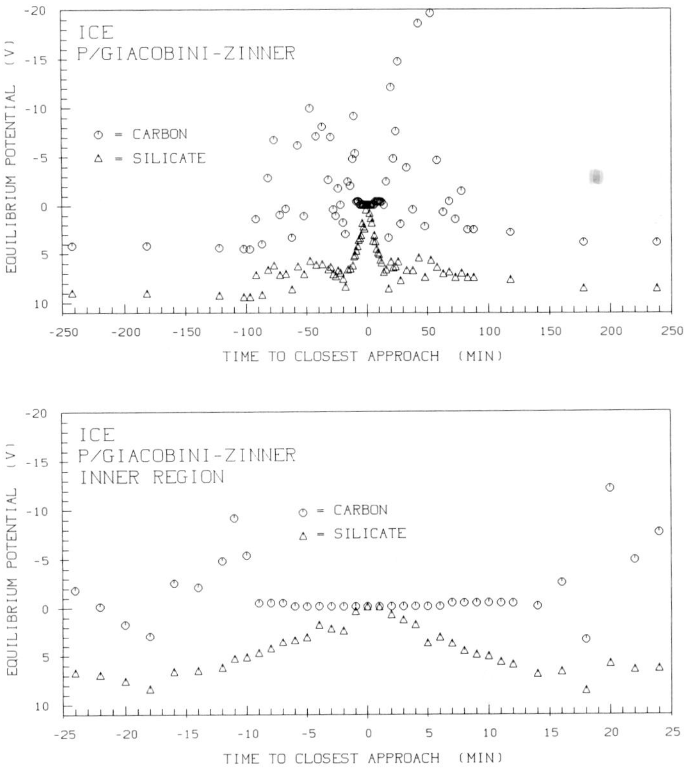

Fig. 1a and b. Equilibrium potential of carbon and silicate dust grains during the fly-by of ICE at P/Giacobini-Zinner. **a** Overview, **b** inner region

3. Results for P/G-Z and P/Halley

As plasma input data for our computer calculations we use the *in-situ*-measurements of the International Cometary Explorer (ICE) spacecraft at P/G-Z (Bame et al., 1986; Meyer-Vernet et al., 1986; Ogilvie et al., 1986) and of the Giotto spacecraft at P/Halley (Rème et al., 1986; Johnstone et al., 1986; Korth et al., 1986). For the equilibrium potentials (ELP) of silicate and carbon grains only unique solutions are derived, which show no significant dependence (<0.1 V) from the grain radius r ($r \geqq 0.1$ µm). Positive ELPs are found in regions of the comae, where the photoelectron current dominates that of the plasma electrons, and negative ELPs vice versa. From the calculated characteristic time of charging (CTC), which gives the time for charging an uncharged grain up to 95% of its ELP (Boehnhardt, 1986), it can be concluded that all dust particles adapt potentials closely to their ELPs nearly instantly compared with the time for crossing through the respective plasma region. Using Eq. (2), the critical radii (CR) for electrostatic fragmentation are derived for a fixed set of tensile strengths ($\sigma_T = 10^3, 10^4, 10^5, 10^6$ dyn cm^{-2}).

3.1. ICE fly-by at P/G-Z

For silicate grains the ELPs (Fig. 1a) vary between +10 and +6 V within the bow shock (±90 min to encounter) and outside the cometopause (±10 min to encounter); within the cometopause (Fig. 1b), the ELPs are steadily decreasing to slightly negative numbers (−0.1 V) due to the high density of the low-energy plasma electrons. Very weak spherical grains ($\sigma_T = 10^3$ dyn cm^{-2}) of up to 1 µm radii can fragmentate outside

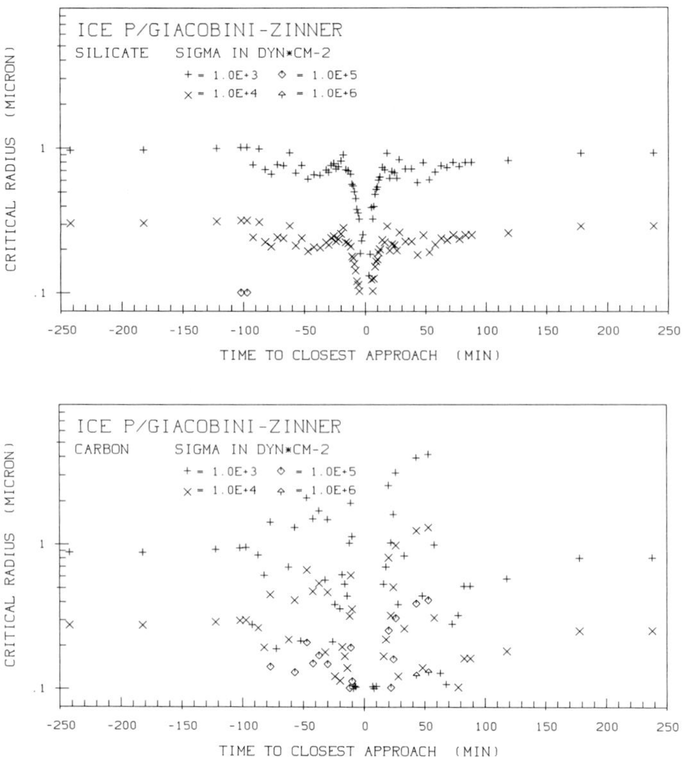

Fig. 2a and b. Critical radii for the fragmentation of dust grains during the fly-by of ICE at P/Giacobini-Zinner. **a** Silicate grains, **b** carbon grains

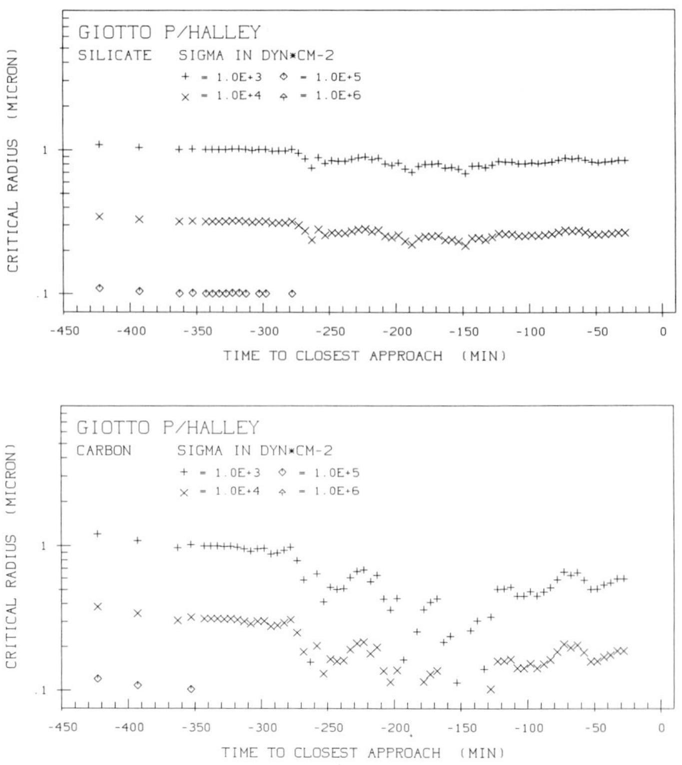

Fig. 4a and b. Critical radii for the fragmentation of dust grains during the fly-by of Giotto at P/Halley. **a** Silicate grains, **b** carbon grains

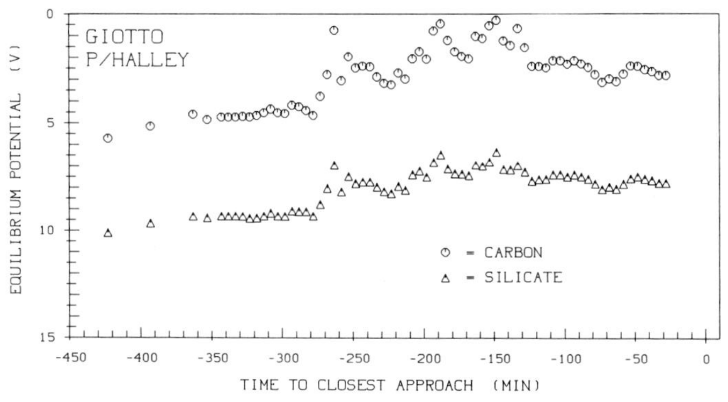

Fig. 3. Equilibrium potential of carbon and silicate dust grains during the fly-by of Giotto at P/Halley

the cometopause (Fig. 2a), inside however, the conditions are becoming worse with decreasing distance of the grains from the nucleus. Within roughly 2 min to closest approach no fragmentation is indicated by the calculations.

For carbon grains the ELPs (Fig. 1a) alternate between +5 and −19 V in dependence of the changing parameters of the surrounding plasma outside the cometopause. Inside of it, they quickly achieve slightly negative potentials (−0.1 V). Therefore electrostatic fragmentation is excluded within this region of the coma even for the weakest grains considered, but might start instantly outside of this region for spherical grains of up to 4 µm radii and tensile strength of 10^3 dyn cm^{-2} (Fig. 2b).

3.2. Giotto fly-by at P/Halley

During the fly-by of Giotto at P/Halley the ELPs of the grains of both materials considered are dominated by the current of the photoelectrons, at least outside the cometopause, which is located 24 min before closest approach (Fig. 3). Silicate grains carry ELPs of +10 to +6 V, the ELPs of carbon grains are lower by about 4 to 5 V in the same plasma region. The CRs are comparable to those calculated for P/G-Z, which means that very weak grains ($\sigma_T = 10^3$ dyn cm^{-2}) of up to 1 µm radii can fragmentate (Fig. 4a and b). The electron experiment onboard Giotto was not able to detect the very dense low-energy electron plasma ($E < 10$ eV) inside the cometopause, so the ELPs of the grains in this region cannot be modelled exactly. But using the results of the ion density measurements (Korth et al., 1986) and a theoretical expression for the thermal energy E of the electron plasma in order to produce a negative ELP of the dust grains (Boehnhardt, 1985) – and assuming an electrically neutral coma plasma –

$$E \gtrsim \frac{\pi}{8} m \, (\Gamma/\alpha N)^2 \tag{5}$$

it seems to be reasonable to expect a similar general behaviour of the ELPs of dust grains inside the cometopause of P/Halley as in the case of P/G-Z (Fig. 5). If Eq. (5) is fulfilled by the electron plasma, the ELPs to be carried by the dust grains are of the same order like the thermal energy of the electron plasma (Boehnhardt, 1985).

Theoretical calculations by Körösmezey et al. (1986) show that the thermal energy E of the very dense ($N = 2000$ cm^{-3}) electron plasma inside the cometopause amounts to about 1 eV at a distance of 10000 km from the nucleus of P/Halley. Farther

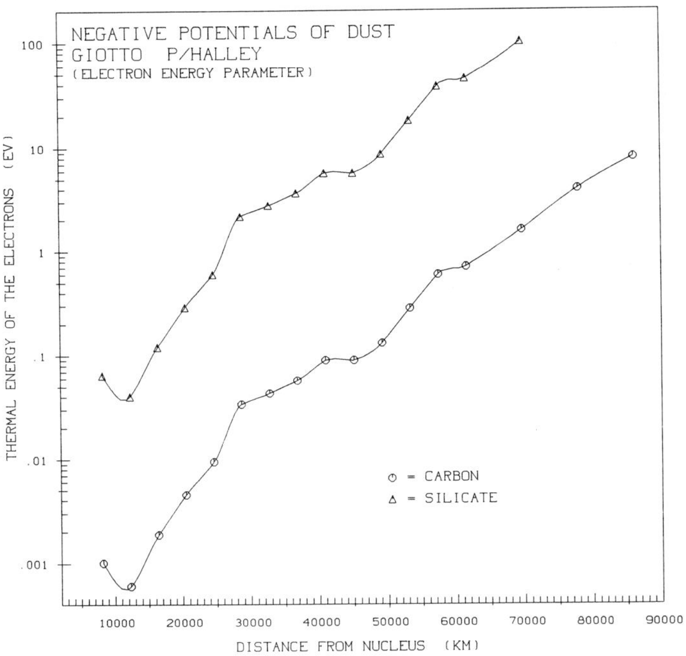

Fig. 5. Lower limit of the thermal energy of the electron plasma to produce negative grain potentials within the cometopause of P/Halley

outside E should increase steadily, while N decreases until the plasma parameters of the cometopause are achieved at 130 000 km distance as measured by Rème et al. (1986). This extrapolated behaviour of the electron plasma parameters E and N is not only indicated by the plasma measurements of ICE near P/G-Z, but is also supported by the measurements of the respective parameters of the ion plasma inside the cometopause as reported by Korth et al. (1986); a thermal energy of about 5 eV was detected at 50 000 km distance from the nucleus of P/Halley. Therefore we expect electrostatic fragmentation for very weak ($\sigma_T = 10^3$ dyn cm^{-2}) and very small ($r = 0.1$ μm) grains starting at about 10 000 km distance from the nucleus with the tendency of increasing radii ($r = 0.5$ μm) and slightly heavier grains farther away from the nucleus in the inner coma. For silicate grains this fragmentation process ceases at the distance of about 50 000 km where the ELPs are expected to be changed from negative to positive sign due to the decreasing density of the plasma electrons. For carbon this change of sign certainly happens closer to the cometopause.

4. Discussion

Electrostatic fragmentation was found to be effective only for small ($r = 0.1$ to 1 μm) and very weak ($\sigma_T = 10^3$ to 10^4 dyn cm^{-2}) spherical grains. From the discussion in Sect. 2.2 we expect a similar efficiency of this fragmentation process also for irregular grains of comparable size. For the coma plasma of P/G-Z the possibility of electrostatic fragmentation of dust grains can be modelled in detail, but no in-situ-measurements of the mass distribution of the dust were made by the ICE mission. For P/Halley this effect might contribute to the changes in the mass distribution of dust, at least outside in the cometosphere (> 100 000 km). Some evidence exists for the presence of a very fluffy dust component (Smirnov et al., 1986). Clusters of dust grains (Simpson et al., 1986) and changes in the mass distribution (Vaisberg et al., 1986) were observed which can be interpreted as fragmentation phenomenon. But one had to be cautions to argue for electrostatic fragmentation of grains with σ_T of the order of 10^3 dyn cm^{-2} from these experimental results. From laboratory measurements (Ibadinov, 1982; Ibadinov et al., 1984) the tensile strength of the dust crust of an artificial cometary nucleus was found to be in the order of 10^4 to 10^6 dyn cm^{-2}. Our calculations show that the electrostatic fragmentation seems to work properly outside the cometopause of P/Halley on dust grains of 0.3 μm radii with tensile strength of 10^4 dyn cm^{-2}, whereas for even heavier grains ($\sigma_T = 10^5$ dyn cm^{-2}) this process is restricted to grains of about 0.1 μm radii. However, we expect no influence by this effect on the dust within 10 000 km of the nucleus. Whether it can contribute to the observed changes of the grain distribution inbetween this region and the cometopause, cannot be calculated now because the plasma parameters, necessary for modelling, are not yet available. This holds also for the effects on the grain potentials by the plasma components of hot electrons (keV) as observed by Vega-2 (Gringauz et al., 1986), which might drastically increase the negative charging of the larger grains, if the density of these electrons is sufficiently high.

Using the equations given by Rhee et al. (1976), it can be shown that rotational bursting caused by the radiation pressure of the sun may not account for the changes in the spectra of the dust masses mentioned above. The characteristic time for bursting is found to be of the order of 5 days in the case of P/Halley.

Wallis et al. (1986) have argued for an evaporation of a volatile mantle around the cores of the grains, which should efficiently reduce (within 10 000 s) the radii of 0.1 to 0.3 μm grains freshly emitted by the nucleus. The observed CN-jets (A'Hearn et al., 1986) and the increase of the density of CO gas within a distance of 20 000 km from the nucleus (Eberhardt et al., 1986) confirm this explanation. On the other hand, the observed clusters of grains contradict this model, unless an evaporation break-up of larger grains is postulated.

In contrast to simple ablation, fragmentation increases the overall number of dust particles in the cometary coma. An experimental detection of this different behaviour encounters some difficulties because the mass distribution, observed by the spacecrafts, are strongly influenced by the radiation pressure inducing the spatial separation of dust grains of different size and material.

Finally we like to point out that different processes, for instance ablation by sublimation of a volatile ice mantle around the dust grains and fragmentation caused by evaporation of frozen gas enclosed in the grain matrix or caused by electrostatic charging, might influence different dust components present in the coma or might work together in producing the many femtogram particles of dust observed near P/Halley.

Acknowledgements. We thank J. Kissel and E. Grün, MPI für Kernphysik, Heidelberg, for helpful discussions. The research was supported by the DFG grant Ral36/13-1.

References

A'Hearn, M.F., Birch, P.V., Klinglesmith III, D.A., Bowers, C., Martin, R.: 1986, 20th ESLAB Symp., ESA-SP 250, Vol. **I**, p. 483

Bame, S.J., Anderson, R.C., Asbridge, J.R., Baker, D.N., Feldman, W.C., Fuselier, S.A., Gosling, J.T., McComas, D.J., Thomsen, M.F., Young, D.T., Zwickl, R.D.: 1986, *Science* **232**, 356

Boehnhardt, H.: 1985, PhD thesis, Univ. Erlangen-Nürnberg

Boehnhardt, H.: 1986, 20th ESLAB Symp., ESA-SP 250, Vol. **II**, p. 207

Brownlee, D.E., Tomandl, D., Blanchard, M.B., Ferry, G.V., Kyte, F.T.: 1976, NASA TMX-**73**, 152

Eberhardt, P., Krankowsky, D., Schulte, W., Dolder, U., Lämmerzahl, P., Berthelier, J.J., Woweries, J., Stubbemann, U., Hodges, R.R., Hoffman, J.H., Illiona, J.M.: 1986, 20th ESLAB Symp., ESA-SP 250, Vol. **I**, p. 383

Feuerbacher, B., Anderegg, M., Fitton, B., Laude, L.D., Willis, R.F., Grard, R.J.L.: 1972a, *Geochim. Cosmochim. Acta Suppl.* **3**, 2655

Fechtig, H.: 1982, in *Comets*, ed. L.L. Wilkening, Univ. Arizona Press, Tucson, USA, p. 370

Feuerbacher, B., Fitton, B.: 1972b, *J. Appl. Phys.* **43**, 1563

Fraundorf, P., Brownlee, D.E., Walker, R.M.: 1982, in *Comets*, ed. L.L. Wilkening, Univ. Arizona Press, Tucson, USA, p. 383

Gringauz, K.I., Remizov, A.P., Verigin, M.I., Richter, A.K., Tatrallyay, M., Szegö, K., Klimenko, I.N., Apathy, I., Gombosi, T.I., Szemerey, T.: 1986, 20th ESLAB Symp., ESA-SP-250, Vol. **I**, p. 195

Ibadinov, K.I.: 1982, *Komety i Meteory* **34**, 19

Ibadinov, K.I., Aliev, S., Rakhmonov, A.A.: 1984, *Cometniy Circ.* **33**, 3

Johnstone, A.D., Coates, A.J., Heath, J., Thomsen, M.F., Wilken, B., Jockers, K., Formisano, V., Amata, E., Winningham, J.D., Borg, H., Bryant, D.A.: 1986, 20th ESLAB Symp., ESA-SP 250, Vol. **I**, p. 25

Kissel, J., Brownlee, D.E., Büchler, K., Clark, B.C., Fechtig, H., Grün, E., Hornung, K., Igenbergs, E.B., Jessberger, E.K., Krueger, F.R., Kuczera, H., McDonnell, J.A.M., Morfill, G.M., Rahe, J., Schwehm, G.H., Sekanina, Z., Utterback, N.G., Völk, H.J., Zook, H.A.: 1986, *Nature* **321**, 336

Körösmezey, A., Cravens, T.E., Gombosi, T.I., Nagy, A.F., Mendis, D.A., Szegö, K., Gribov, B.E., Sagdeev, R.Z., Shapiro, V.D., Shevchenko, V.I.: 1986, 20th ESLAB Symp., ESA-SP 250, Vol. **I**, p. 235

Korth, A., Richter, A.K., Mendis, D.A., Anderson, K.A., Carlson, W., Curtis, D.W., Lin, R.P., Mitchell, D.L., Reme, H., Sauvaud, J.A., d'Uston, C.: 1986, 20th ESLAB Symp., ESA-SP 250, Vol. **I**, p. 199

McDonnell, J.A.M., Kissel, J., Grün, E., Grard, R.J.L., Langevin, Y., Olearczyk, R.E., Perry, C.H., Zarnecki, J.C.: 1986, 20th ESLAB Symp., ESA-SP 250, Vol. **II**, p. 25

Meyer-Vernet, N., Couturier, P., Hoang, S., Perche, C., Steinberg, J.L., Fainberg, J., Meetre, C.: 1986, *Science* **232**, 370

Mukai, T.: 1981, *Astron. Astrophys.* **99**, 1

Ogilvie, K.W., Coplan, M.A., Bochsler, P., Geiss, J.: 1986, *Science* **232**, 374

Rème, H., Sauvaud, J.A., d'Uston, C., Cros, A., Anderson, K.A., Carlson, C.W., Curtis, D.W., Lin, R.P., Korth, A., Richter, A.K., Mendis, D.A.: 1986, 20th ESLAB Symp., ESA-SP 250, Vol. **I**, p. 29

Rhee, J.W., Paddack, S.J.: 1976, in *Lecture Notes Phys.* **48**, 453

Simpson, J.A., Rabinowitz, D., Tuzzolino, A.J., Ksanfomality, L.V., Sagdeev, R.Z.: 1986, 20th ESLAB Symp., ESA-SP 250, Vol. **II**, p. 11

Smirnov, V., Vaisberg, O., Anisimov, S.: 1986, 20th ESLAB Symp., ESA-SP 250, Vol. **II**, p. 195

Vaisberg, O., Smirnov, V., Omelchenko, A.: 1986, 20th ESLAB Symp., ESA-SP 250, Vol. **II**, p. 17

Wallis, M.K., Rabilizirov, R., Wickramasinghe, N.C.: 1986, 20th ESLAB Symp., ESA-SP 250, Vol. **II**, p. 251

Astron. Astrophys. 187, 829–834 (1987)

ASTRONOMY
AND
ASTROPHYSICS

Evolution of comet P/Halley in early march 1986 as observed from Vega pictures

A. Abergel and J. L. Bertaux

Service d'Aéronomie du CNRS, F-91371 Verrières-le-Buisson, France

Received February 5, accepted June 25, 1987

Summary. From 4 to 10 march 1986, the internal dust coma of comet P/Halley was observed every day with Vega-1 and 2 narrow angle cameras of TVS system. The distribution of dust was monitored within 30,000 km from the nucleus. Strong variations are observed from one day to the other, indicating corresponding variations of the nucleus activity. These variations affect both the overall dust number density and the shape of its distribution around the nucleus. Dust intensities are given in absolute units of brightness.

Key words: comet Halley – dust coma evolution

1. Introduction

From 4 to 10 March 1986, Vega-1 and Vega-2 Narrow Angle Cameras of the TVS system have observed comet P/Halley for two hours every day. The first results obtained by these instruments were described by Sagdeev et al. (1986). Since the closest approach occurred on March 6th for Vega-1 and March 9th for Vega-2, we have data far from the nucleus as well as close to it. The purpose of this paper is to present the evolution of the shape and intensity brightness of the dust coma from 4 to 10. This should allow useful comparisons with ground-based observations in two respects. First, the absolute calibration of the instruments provides a good data base for brightness evolution along this period of time, which could help to understand better the time-variation activity pattern of comet Halley. Secondly, our data provide a unique opportunity of performing 3-D analysis of some features of the coma brightness. Indeed, the viewing angle form the Vega probes and those from the ground-based observations are different.

2. Observations

Far from the nucleus, on March 4, 5, and 7 for Vega-1 and on March 7 and 10 for Vega-2, the distribution of dust was observed during 2 hours. The Narrow Angle Camera has a telescope whose focal length is equal to 1200 mm. Images are registered by a CCD detector. The dimensions of images sent to the earth are 512×512 pixels (whole frame) or 128×128 pixels (floating window around the brightest part of the image), to save the telemetry. The pixels are rectangular and their angular size is equal to $3'' \times 4''$.

A great number of images were sent to the Earth during each stage, with several filters and several exposure times. The distance from the nucleus was near 14×10^6 km two days before each encounter, and $7\ 10^6$ km one day before and after. The phase angle was about 109° before encounters and 66° after. The precise values are given in Table 1, at 07:20 UT. As the spacecraft were far from the nucleus, the geometrical parameters did not change considerably during these observation stages. Since the general shape of the comet looks quite similar for all filters, we have used for this report only images obtained with the glass filter, whose spectral range goes from 400 to 1000 nm. We will analyze and compare at a later time images obtained with the other filters, in order to attempt to enhance some dust or gas structures. Parameters defining each image are summarized in Table 2.

All images presented in this paper were processed to remove the noise. The periodic and coherent parasitic signals were subtracted by Fourier filtering. The high frequency noise was removed by a 5×5 median filter. Isophotes are given in relative unit of brightness. Images have been geometrically corrected to take into account the rectangular shape of the pixels.

We next present the evolution of the dust distribution day by day, starting on March 4.

March 4: The first image sent to the Earth by the TVS system of Vega-1 was a full frame image (512×512 pixels) with the glass filter (Fig. 1). It is the first cometary image obtained by an interplanetary probe. The image, whose total dimensions are equal to $105{,}000 \times 140{,}000$ km, shows the beginning of the tail in the anti-solar direction. The apparent diameter of the coma is equal to 100,000 km. However, the real diameter is certainly greater because of a threshold effect which cuts all the signals under a fixed "negative pedestal" of one sixth of the dynamic range. This effect has affected all the images obtained by Vega-1. To take it into account, a constant level has been added to the measured signal to obtain the isophotes. The innermost isophotes are almost symmetrical. The photometrical profile obtained along the solar axis (see Fig. 6) shows that the brightness varies inversely with the distance from the nucleus.

Green and Hughes (1986) have observed comet Halley with the 3.9 m Anglo-Austrian Telescope on March 3 and 4, at 19:00 UT. It is interesting to note that these ground-based observations display the same overall shape of the distribution. The isophote of images that they obtained with a red filter centered on 730 nm (HMC filter) on March 4 has the same overall shape that the one obtained by Vega-1 twelve hours. earlier.

March 5: Figure 2a and b correspond to the same position. The first is over-exposed near the nucleus. It is a full frame image, which displays the coma far from the nucleus. The tail looks

Send offprint request to: A. Abergel

Table 1. The observation stages far from the nucleus

	Vega-1			Vega-2	
Date (March 1986)	4	5	7	7	10
Distance from the nucleus at 07:20 UT (10^6 km)	13.69	6.85	6.85	13.27	6.63
Phase angle at 07:20 UT (degrees)	107.74	109.44	67.10	110.31	65.02
Beginning (UT)	05:30	06:30	07:50	03:30	05:30
End (UT)	07:30	08:30	09:30	05:30	07:30

Table 2. Characteristics of pictures

Figure no.	Spacecraft	Day (1986)	Type of picture	Exposure time (sec)	Pixel size (km × km)	Image size (km × km)
1	Vega-1	March 4	Full frame (1)	40.96	205 274	105,000 140,000
2a	Vega-1	March 5	Full frame	20.48	102 137	52,500 70,000
2b	Vega-1	March 5	Window (2)	5.12	102 137	13,125 17,500
3	Vega-1	March 7	Window	5.12	102 137	13,125 17,500
4	Vega-2	March 7	1/4 window (3)	2.56	198 264	12,700 16,900
5	Vega-2	March 10	Window	10,24	99 132	12,700 16,900

(1) Full frame: 512 × 512 pixels
(2) Window: 128 × 128 pixels
(3) 1/4 window: 64 × 64 pixels

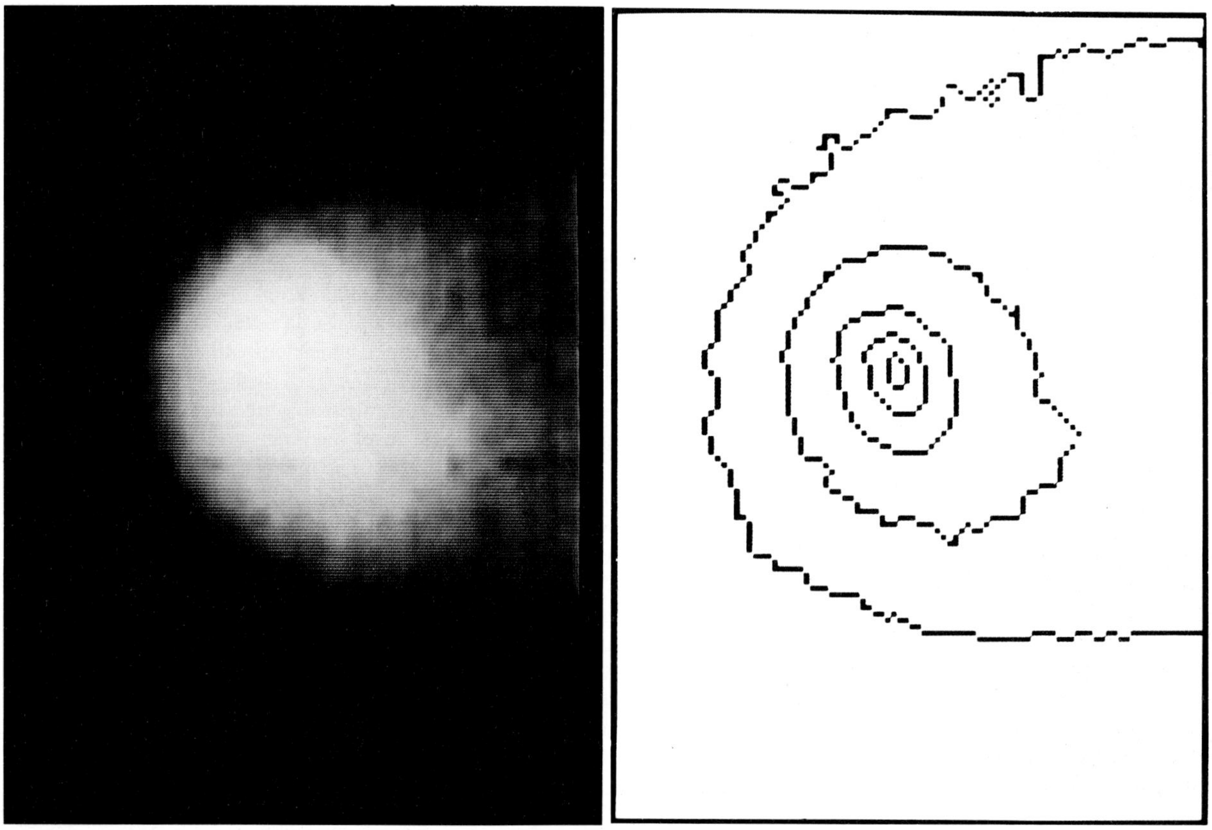

Fig. 1. Full frame image (512 × 512 pixels) obtained by Vega-1 two days before the encounter (March 4). The sun in exactly to the left. – Isophotes: 1., 0.8, 0.6, 0.4, 0.2 (arbitrary units)

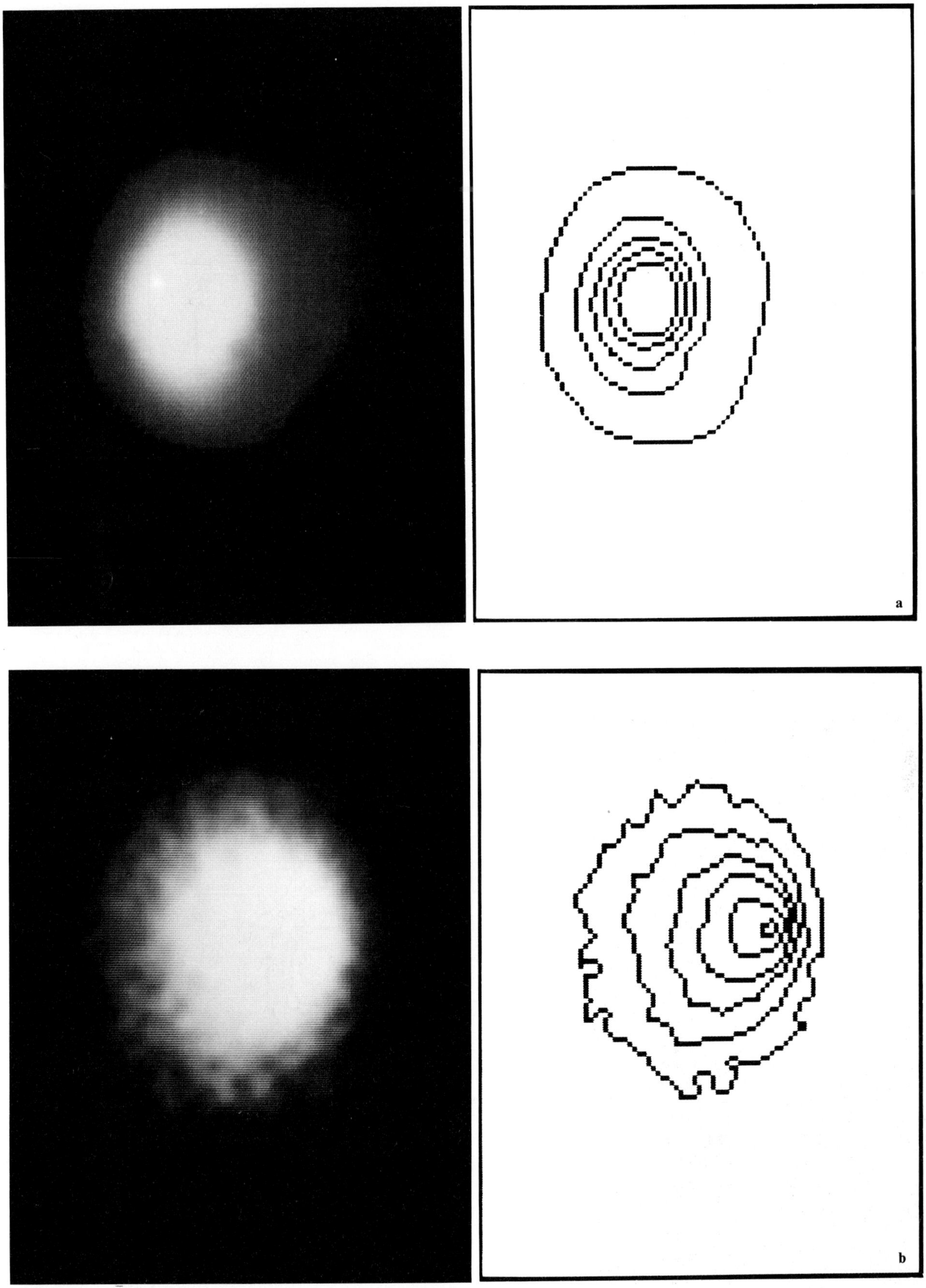

Fig. 2a and b. Images obtained by Vega-1 one day before the encounter (March 5). The sun is again exactly to the left. (**a**) Full frame image (512 × 512 pixels). – Isophotes: 1, 0.8, 0.6, 0.4, 0.2. (**b**) Window image (128 × 128 pixels). – Isophotes: 1, 0.8, 0.6, 0.5, 0.4, 0.2

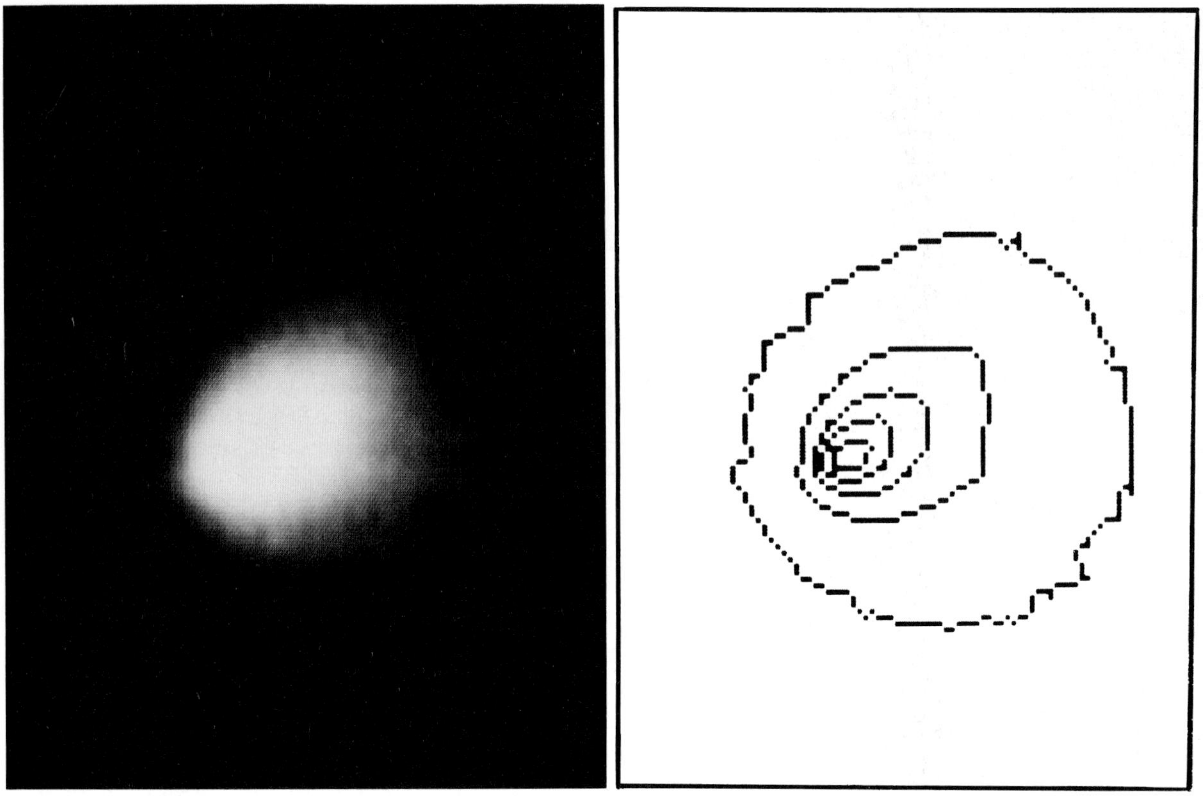

Fig. 3. Window image (128 × 128 pixels) obtained by Vega-1 one day after the encounter (March 7). The sun is exactly to right. – Isophotes: 1, 0.8, 0.6, 0.4, 0.2

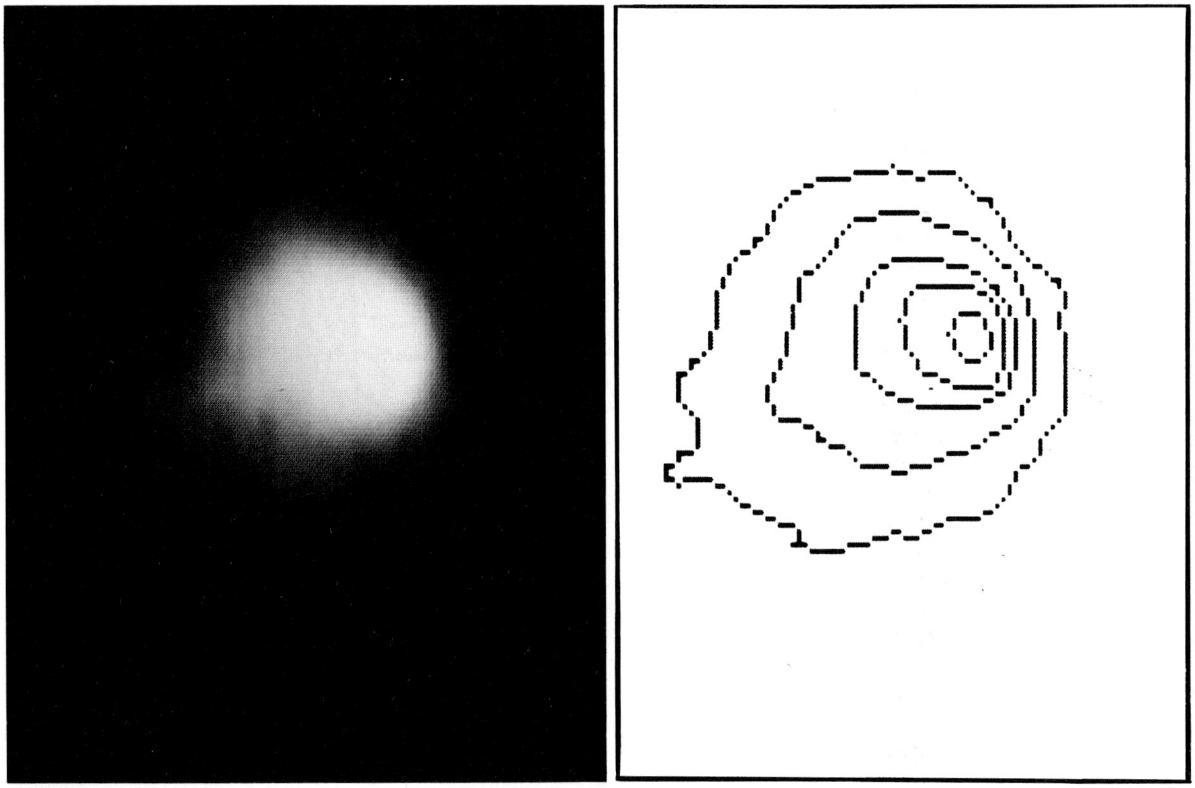

Fig. 4. A quarter of window image (64 × 64 pixels) obtained by Vega-2 two days before the encounter (March 7). The sun is to the left. – Isophotes: 1, 0.8, 0.6, 0.4, 0.2

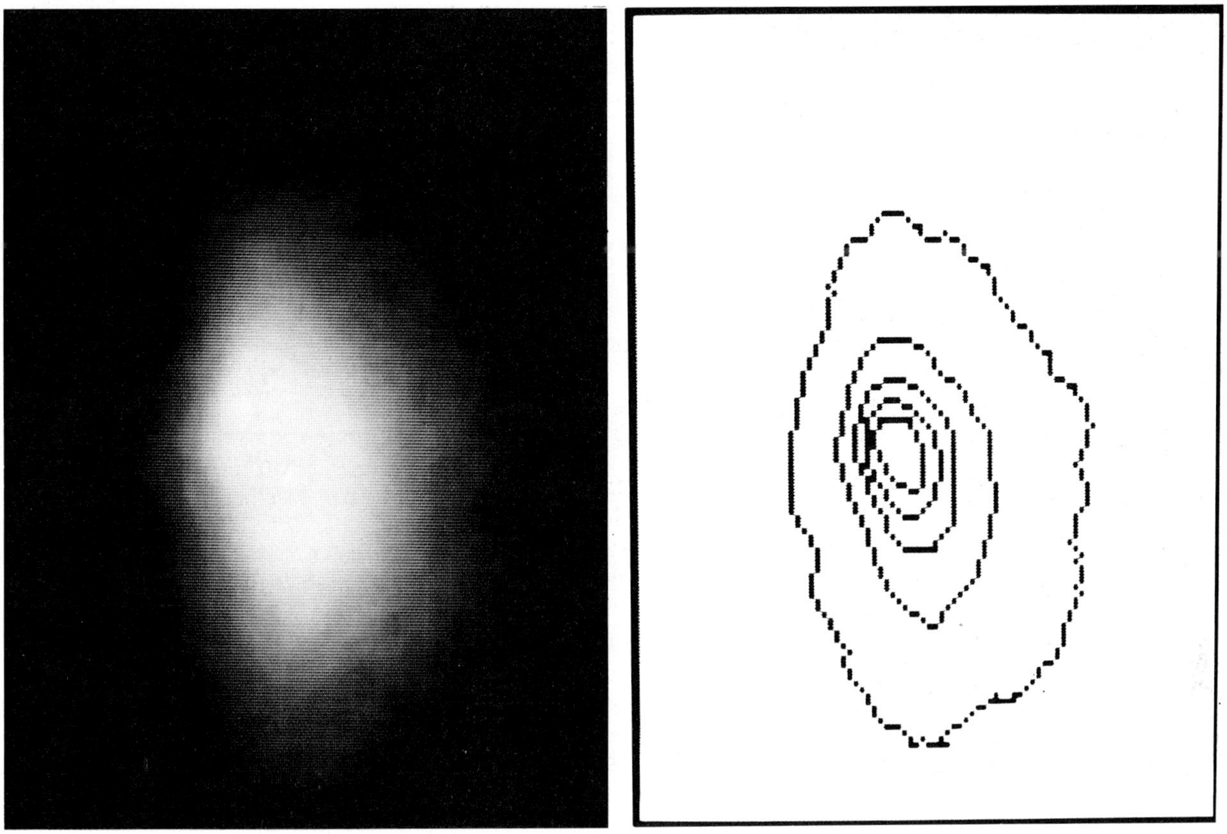

Fig. 5. Window image (128 × 128 pixels) obtained by Vega-2 one day after the encounter (March 10). The sun is exactly to the right. - Isophotes: 1, 0.8, 0.6, 0.4, 0.2

smaller than on March 4 most likely because of the threshold effect.

The brightest part of Fig. 2b is just under the saturation level, so it shows the dust distribution near the nucleus. In this region, the brightness of the coma is considerably stronger than on March 4 (the exposure time of the window image of Fig. 2b is shorter than the one of Fig. 1 by a factor of 8), and the isophotes are non circular. The coma seems to be symmetric with an axis of symmetry which coincides with the solar direction. Thus, the observed distribution is a perfect demonstration of the simple "fountain" model of dust ejection, when it is restricted to the sunlit hemisphere. To compare the brightness of the comet on March 4 and 5, we have plotted on Fig. 6 the photometric profiles, with the same relative units. The brightness in the central part was multiplied by a factor 8 between March 4 and 5. Thus an outburst probably occurred in the 24 h before the observations of March 5. During that day, the observations were continuously performed during 2 h, from 06:30 to 08:30 UT, but the received images were all identical. It shows that a steady state of dust ejection was probably reached at this time.

March 7: The comet was observed nearly simultaneously by Vega-1 and Vega-2 (Figs. 3 and 4 respectively). The Vega-1 image signal shows that the global brightness of the comet was almost the same as on March 5. This time, the isophote distribution, which is still non-circular, is strongly elongated near the nucleus toward a direction at about 50 degrees from the sun axis (towards Ecliptic North). Four hours before the Vega-1 observations (Table 1), Vega-2 was sending its first images, with a geometrical position opposite to that of Vega-1, and a distance from the nucleus twice larger. Differences between images can only be due to the difference of phase angle (Table 1). In the future, comparative analysis of the optical and photometrical properties of the two cameras will allow us to compare quantitatively the images obtained by the two spacecraft.

March 10: The coma (Fig. 5) is weaker than on March 7, and it looks like a broad fan extended to the sun direction.

3. Discussion

It is interesting to compare our results with the images obtained by West et al. (1986) at the European Southern Observatory (ESO), La Silla, between 1986 March 5 and 14. Our images give details on the so-called "Central Condensation". We confirm the rotation of the elongation axis of the dust coma between March 5 and 7 as reported by these authors. Moreover, the comet has been observed on March 5 and 7 almost at the same time by Vega-1 (about 08:00 UT) and by West (08:30–09:30 UT). It is thus possible to combine the two observations to calculate the direction of the elongation axis in the ecliptic coordinates system (J. 1950). We find:

March 5: elongation axis $l=-25^\circ\!.2$, $b=25^\circ\!.6$, angular coordinates of the sun as seen from the comet: $l=66^\circ\!.3$, $b=-2^\circ\!.45$.

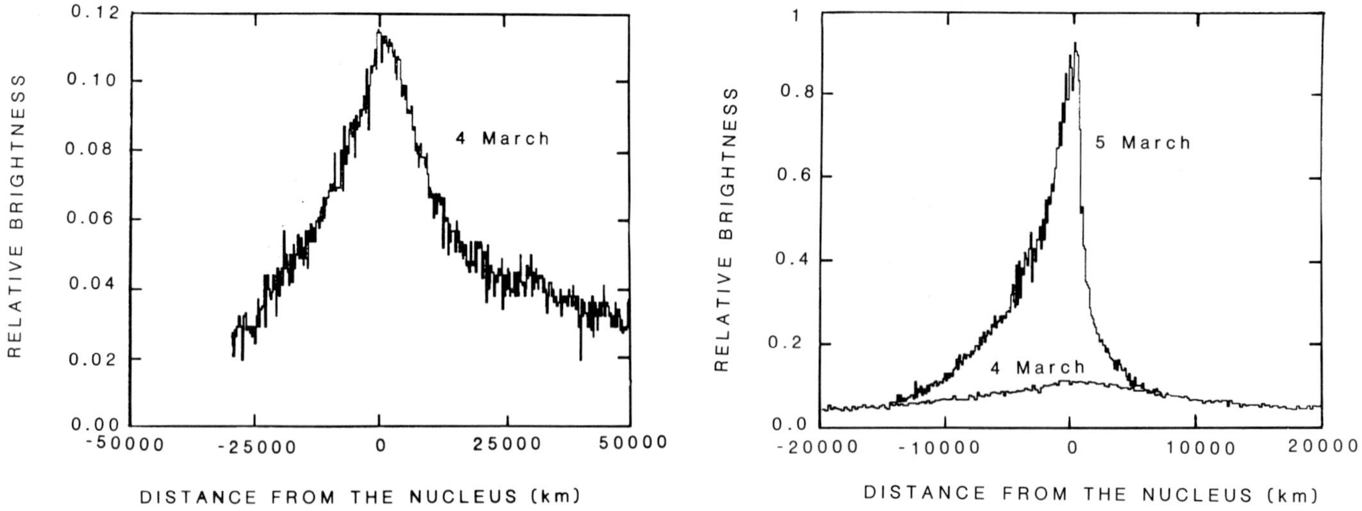

Fig. 6. Photometric profiles of the comet along the Sun-comet axis, obtained from images received on March 4 (Fig. 1) and on March 5 (Fig. 2a and b). The sun is on the left. The unit of brightness is arbitrarely chosen, and is the same for the two pictures

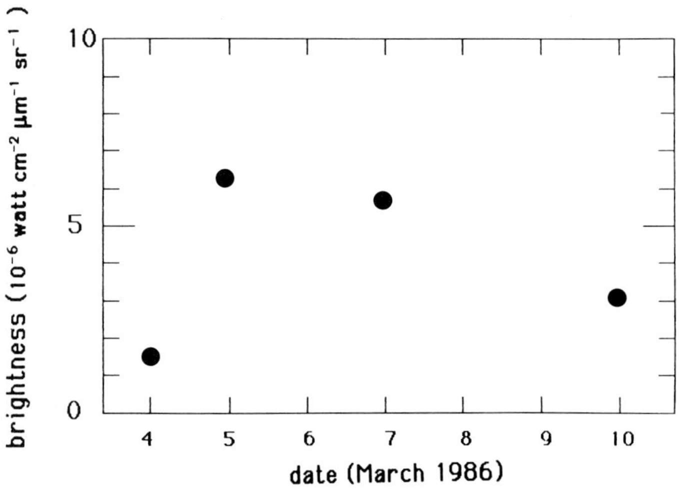

Fig. 7. Variation of the average brightness of the central part of the comet, between March 4 and 10. The brightness is averaged within the brightest surface whose area is equal to 10^7 km^2

March 7: elongation axis: $l = -8^\circ\!.5$, $b = 71^\circ\!.1$, angular coordinates of the sun: $l = 63^\circ\!.1$, $b = -1^\circ\!.44$

The orientation of the elongation axis does not coincide by far with the orientation of the comet-sun axis.

March 5: elongation axis $l = -25^\circ\!.2$, $b = 25^\circ\!.2$, angular coordinates of the sun as seen from the comet: $l = 66^\circ\!.3$, $b = -2^\circ\!.45$.

March 7: elongation axis: $l = -8^\circ\!.5$, $b = 71^\circ\!.1$, angular coordinates of the sun: $l = 63^\circ\!.1$, $b = -1^\circ\!.44$
The orientation of the elongation axis does not coincide by far with the orientation of the comet-sun axis.

The photometric properties of the cameras are described by Abergel et al. (1987). Using the values of the Vega-2 spectral sensitivity for the clear filter, we can convert the digital number measured by pixels in to absolute brightness. The variations with time of the average brightness of the brightest part of the comet are plotted on Fig. 7. The area of this part is equal to 3 10^7 km^2. The Vega-1 observations are normalized with the Vega-2 data with the simultaneous observations obtained on March 7. So the phase effect is neglected, and the phase angle assumed every day equal to about 110°.

Our series of observations indicate that an event occurred between March 4, at 07:30 UT, and March 5 at 06:30 UT. In the same way the observations of West et al (1986), show that an outburst of dust activity should have occurred 24 h previous March 7, at 09:00 UT. Therefore two outbursts have successively occurred. They have to be linked with the maxima of the mean light curve of comet Halley given by Festou et al (1987) from several independant ground-based observations.

The next step of our work will be to give all images in absolute units of brightness and to compare the images received with different filters, in order to derive photometric properties of the dust paricles.

Acknowledgements. We wish to thank M. Dimarellis for developing image processing software, and all devoted personel who helped to carry out the TVS imaging system investigation on Vega-1 and Vega-2. Image processing was performed on CIRCE computer at Orsay.

References

Abergel, A., Bertaux, J.L., Avanessov, G.A., Tarnopolsky, V.I., Zhukov, B.S.: 1987, *Appl. Opt.* (submitted)

Festou, MC., Drossar, P., Lecacheux, J., Encrenaz, T., Puel, F., Kohl-Moreira, J.L.: 1987, *Astron. Astrophys.* (submitted)

Green, S.F., Hughes, D.W.: 1986, 20th ESLAB Symp., ESA SP-**250**, Vol. **II**, p. 157

Sagdeev, R.Z., et al.: 1986, *Nature* **321**, 262

West, R.M., Pedersen, H., Monderen, P., Vio, R., Grosbl, P.: 1986, *Nature* **321**, 363

Astron. Astrophys. 187, 835–838 (1987)

ASTRONOMY
AND
ASTROPHYSICS

The spatial distribution of dust jets seen during the Vega 2 flyby

R. Z. Sagdeev[3], **B. Smith**[2], **K. Szegö**[1], **S. Larson**[2], **I. Tóth**[1], **E. Merényi**[1,2], **G. A. Avanesov**[3], **V. A. Krasikov**[3], **V. A. Shamis**[3], **and V. I. Tarnapolski**[3]

[1] Central Research Institute of Physics of the Hungarian Academy of Sciences, P.O. Box 49, H-1525 Budapest 114, Hungary
[2] Lunar and Planetary Laboratory, University of Arizona, Tucson, AZ 85721, USA
[3] Space Research Institute, USSR Academy of Sciences, Profsoyuznaya Str. 88, SU-117810 Moscow, USSR

Received January 30, accepted May 11, 1987

Summary. One important physical characteristic of comet P/Halley is dust jet activity, seen from Earth as changing coma patterns. At spacecraft resolution it becomes possible to precisly locate the sources of active jets on an irregularly shaped nucleus. The data set employed in this analysis consists of 11 full-format (512 × 512 px) Vega-2 images taken over a 15 min interval starting approximately 370 s before closest approach. The distance to the nucleus was always less then 45,000 km and became as close as 8030 km. Most of the active zones formed at least one quasi-linear structure on the surface. We found evidence of the activity on the dark hemisphere, at least one of the faint but active jet sources is there.

Key words: comet – nucleus – jets

1. Introduction

One important physical characteristic of comet Halley is dust jet activity, seen from Earth as changing coma patterns when the comet is within 2 AU of the Sun. Hevelius made one of the earliest observations of a jet on comet Halley in 1692, and jets were widely observed during the 1835–36 apparition by Herschel and Struve (1847). The application of photography during the 1910 apparition provided the first objective data on jets. Modern digital image processing and computer modeling techniques have been recently applied to the 1910 photographs to derive quantitative information on the jet mechanism, nuclear spin vector, and distribution of jet sources on the nucleus by Larson and showed the observes patterns are consistent with continous emission of dust from discrete natur of these sources suggested an inhomogeneous surface, possibly a crust depleted in volatiles, but with cracks or vents allowing outgassing from subsurface volatile reserves.

At groundbased resolution the jets, diffused and curved by the effects of solar radiation pressure on a distribution of particle sizes, serve well as tracers of nuclear rotation, but can yield only approximate location for their sources on the nuclear surface. At spacecraft resolution, however, the effects of radiation pressure are negligible, and it now becomes possible to precisely locate the sources of active jets on an irregularly shaped nucleus.

Send offprint requests to: K. Szegö

In the following sections we shall discuss the unique Vega 2 data set, the methods of reduction and analysis and, finally our conclusions.

2. Vega 2 observations

The data set employed in this analysis consists of 11 full-format (512 × 512 px) Vega 2 imaging frames taken over a 15 min interval starting approximately 370 s before closest approach to the nucleus and continuing until 560 s afterwards. The distance to the nucleus was always less than 45,000 km and became as close as 8030 km. The pertinent data for each of these images is summarized in Table 1. Image 1174 was the first of the full-format frame made necessary by the loss of the primary tracking system; image 1214 was the final frame used in our analysis, because by this time the relative trajectory between the comet and spacecraft was such that the viewing geometry was changing only very slowly. Figure 1 shows the subspacecraft and subsolar coordinates projected onto a sphere centered on the Halley nucleus; the points at which the 11 analysis images were shuttered is indicated. The preliminary analysis of the dust jets (Sagdeev et al., 1985) made use of unprocessed images. The analysis reported here uses photometrically corrected images where the quasi-coherent noise has been removed by Fourier filtering, i.e. by suppression of the offending spatial frequencies in the Fourier domain. The image processing and data analysis have taken place both at CRIP in Budapest an LPL in Tucson.

3. Reduction an analysis

The method employed in evaluating the spatial orientation of the dust jets as they existed during the Vega 2 encounter is straightforward in concept. Each jet or jet complex can be thought of as an organized set of spatial vectors. On any given image, the vector sets are seen projected onto the image plane which is always normal to the spacecraft-comet vector, and always has its horizontal, (row) direction aligned parallel to the ecliptic plane. Thus measurements of the same feature seen on different images, together with knowledge of the changing spatial relationship between the spacecraft and comet, yield the absolute spatial orientation of feature. With the additional knowledge of the 3-dimensional shape and orientation of the Halley nucleus, it becomes possible to map each measured feature back onto the

Table 1

Image number	Time [a]	Range	Resolution [b]	Field of view [b]	Phase angle [c]	Sun angle [c]	Filter [d]
1174	−370.5 s	29 540 km	1.03 km	265 km	98.4	89.0	NIR
1178	−282.9 s	23 140 km	0.81 km	207 km	94.0	89.3	VIS
1182	−197.0 s	17 120 km	0.60 km	153 km	86.6	89.6	RED
1186	−101.7 s	11 200 km	0.39 km	100 km	69.3	69.3	GLS
1190	− 1.5 s	8 030 km	0.28 km	72 km	28.7	28.7	NIR
1194	+ 98.7 s	11 040 km	0.39 km	99 km	23.5	23.5	VIS
1198	+187.3 s	16 460 km	0.58 km	147 km	38.7	38.7	RED
1202	+280.5 s	22 970 km	0.80 km	206 km	46.8	46.8	GLS
1206	+372.7 s	29 700 km	1.04 km	266 km	51.3	51.3	NIR
1210	+467.9 s	36 790 km	1.29 km	330 km	54.3	54.3	VIS
1214	+558.1 s	43 570 km	1.53 km	390 km	56.2	56.2	RED

[a] Time is relative to closest approach, which occurred at 0720.0 UT
[b] The pixels in the Vega CCD detectors are rectangular, measuring 18 × 24 μm. This results in differing resolution in the vertical and horizontal directions and in a rectangular field of view. The values given here represent an average of the two orthogonal directions. Resolution is defined as a line pair
[c] Phase angle is the sun-comet-spacecraft angle. Sun angle is the direction of the sun in each frame, measured counter-clockwise from the top, which is also ecliptic north
[d] GLS = clear glass: VIS = Visual; RED = Red; NIR = Near Infrared. See Sagdeev et al. (1986a)

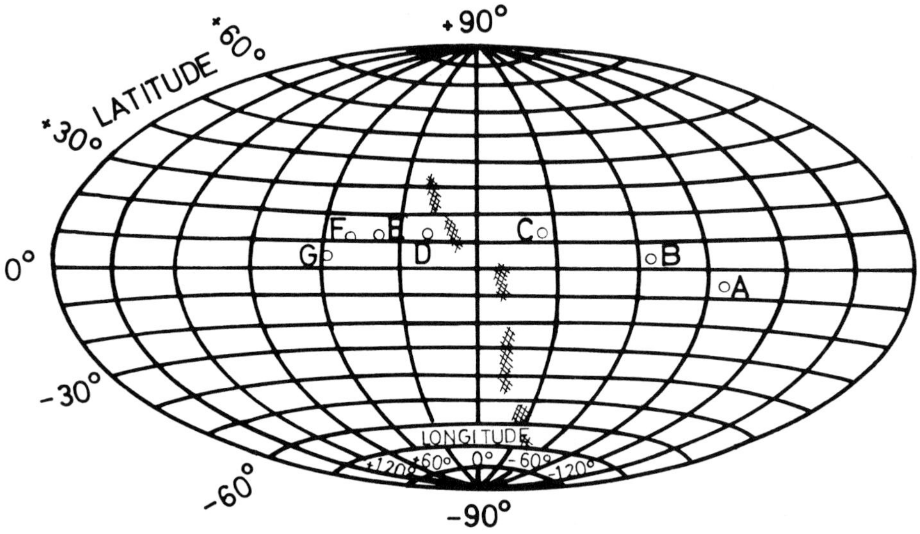

Fig. 1. The path of Vega 2 projected onto a sphere centered on the nucleus. North is ecliptic north on the figure, the subsolar point defines the zero longitude. On the path, dots indicate the recording of the images; e.g. we list for images A-G the recording time relative to encounter (E) and distance from the nucleus. (A: E-370.5 s, 29539 km; B: E-101.7 s, 11197 km; C: E-1.5 s, 8030 km; D: E+98.7 s, 11037 km; E: E+187.3 s, 16462 km; F: E+280.5 s, 22971 km; G: E+558.1 s, 43567 km). The identified line-like jet sources are shaded on the sphere

nuclear surface as a dust jet source. As we defined it here, a feature may be thought of in the simplest case as a single collimated jet originating from a small, point-like source or as a more complex system of jets such as a sheet or fan originating from a linear or curvilinear source.

Prior to the reduction of the data, it was necessary to determine the precise spatial direction and distance of the Halley nucleus from the Vega 2 spacecraft at the moment of exposure for each of the analysis images. Using heliocentric velocity vectors for comet Halley and Vega 2 together with unpublished impact parameters provided by Vega project, we derived a solution which yields the ecliptic latitude and longitude and distance of the Halley nucleus, along with the phase angle and the image-plane projection of comet-sun vector for each image in the analysis. Slight modification to the Vega project impact parameters ($<0.^\circ2$) were needed to preserve geometrical symmetry with respect to the point of closest approach to the nucleus.

The projection of various jet features on the image plane were measured by two different methods, measurement of high-quality photogaphic prints and the use of a videodisplay monitor with special graphics capability. The final product for each image was a set of position angles (measured counter-clockwise from ecliptic north) for all measurable dust-jet features. If the feature was a narrowly confined dust jet, we measured only the core of the feature. Wider fans were measured at their extremities.

Several of the more prominent jets could easily be identified by casual inspection as features common to several of the image (Fig. 2). Others were much more difficult, with candidate solution

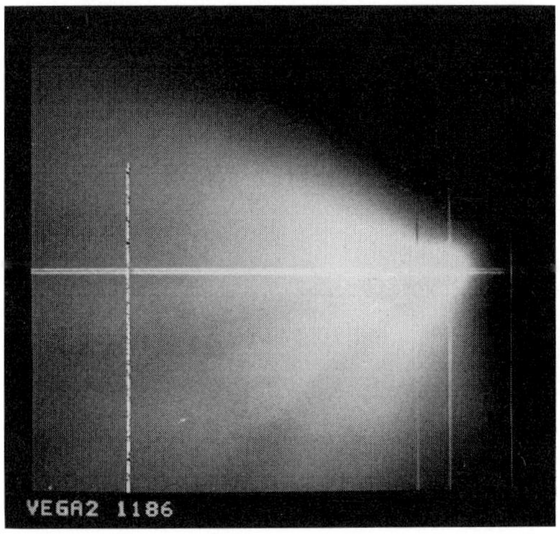

Fig. 2. Image 1186 of Vega 2

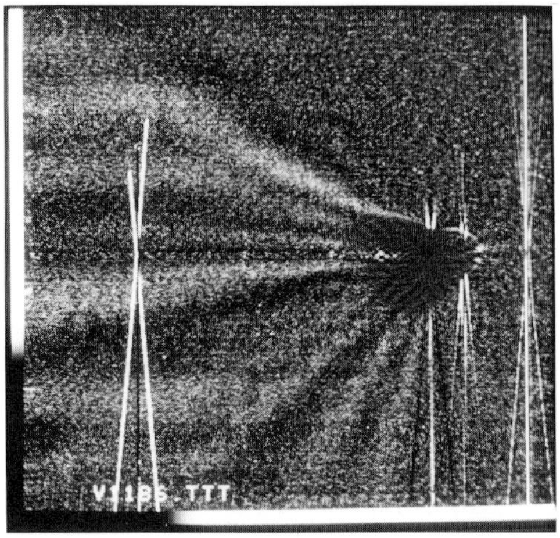

Fig. 3. Shift-differentiated version of image 1186

frequently predicting unconfirmable positions in other images. Several trials for a solution were often necessary before one could be found that would satisfy the observed features in all of the analysis images. To enhance the less-obvious features, we used a shift-differentiation algorithm developed earlier by Larson and Sekanina (1985). Because all jet boundaries were seen as straight, radial lines in our images (with a field of view, of only a few hundred kilometers, the curvature of the jet boundaries is negligible), only azimuthal differentiation was used to detect the faint linear structure. Radial differentation, however, was used to pinpoint the jet sources on the nucleus. With this method, the jets would vanish when the radial center was placed at the jets. Figure 3 shows the azimuthally shift-differentiated version of Figure 2. Some features that appeared broad an without structure on images taken far from the nucleus, were resolved as a set of smaller, but still not point-like jets at closer range.

The most viable solutions are shown in Fig. 1, as they would appear on a spherical nucleus. Such is certainly not the case for comet Halley; however, even with a spherically projected distribution, it is possible to observe serval noteworthy characteristics: The most prominent sources are far from the terminator, with the exeption of the strong source near $-70°$. At the time of the Vega 2 encounter, there is no pronounced solar phase lag; the sources seen to be fairly evenly distributed between late morning and early afternoon. The sources are not randomly distributed, but seem to show some degree of organization into quasilinear patterns, some of which are clearly discontinous with others appearing to have continuous segments, at least at Vega resolution. From these observations we may conclude that dust jets are produced by discrete sources, some of them relatively isolated and others seemingly connected in curvilinear patterns, similar to the general configurations found in 1910 (Sekanina and Larson, 1986). Thus, the volatiles that expel the dust appear to be coming from individual pits and cracks in the surface of the nucleus and the sources, therefore, can become active only when sunlight reaches these ices that lie deep within the pits and cracks. To understand their true distribution, we must first determine the approximate shape of the Halley nucleus.

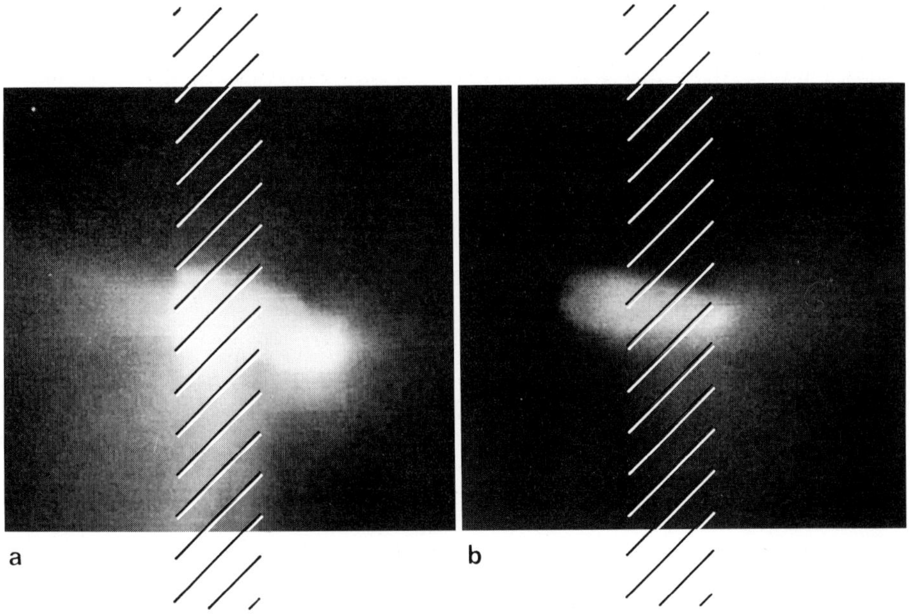

Fig. 4a and b. Images 1190 and 1194 of Vega 2. The shaded regions contain the most prominent sources

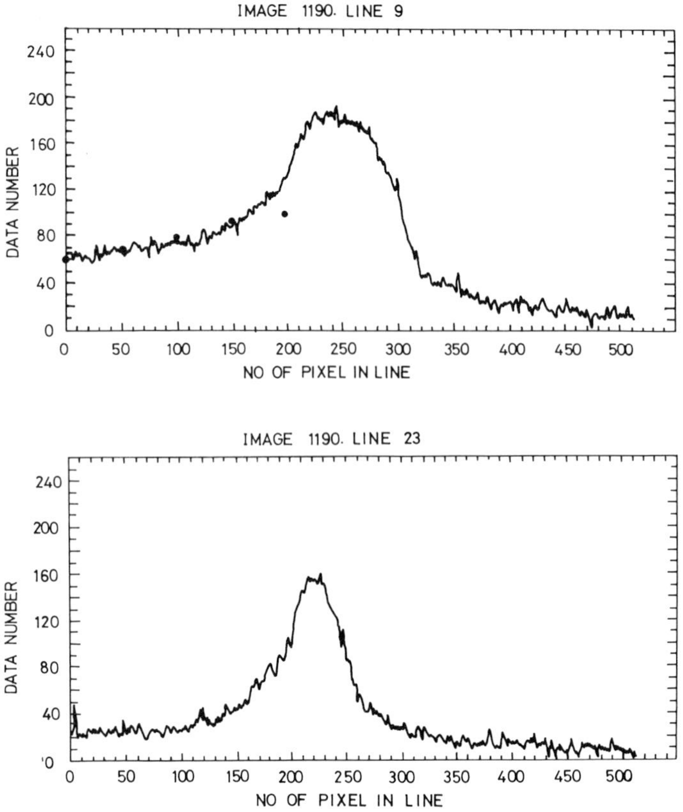

Fig. 5a and b. Intensity distribution an image 1190 along two lines (line 9 and 23) perpendicular to the nucleus axis. The first perpendicular line starts at the left-hand side of the nucleus as seen in Fig. 4a, the last line, line 25, is at his right hand side. The distances between the lines are equal. In Fig. 5a dots indicate the $1/r$ behaviour.

To locate the jet sources on the non-spherical nucleus we have tried to conciliate a number of different observations. We started with the generals appearance of Vega 2 images 1190 and 1194 (Fig. 4). On 1190 the right side of the nucleus is seemingly free of jets, and even the "waist" and much of the left side of the dumbell-like structure is not heavily screened by jet features. Photometric scans confirm these visual observations. Brightness profiles along lines parallel to and perpendicular to the longest axis of the nucleus enable us to discriminate between domains having strong jet contributions (showing an approximately $1/r$ behaviour in brightness well away from the nucleus) and those which are relatively free of dust, where one sees an abrupt dropoff in brightness. Figure 5 exhibits examples of jet-free and dusty cross sections of the nucleus.

A similar analysis of 1194 reveals the left side to be nearly free of external dust. The "waist" and right side of the dumbell-like form, however, is somewhat screened by jets emitted toward the right of the image. Radial shift-differentation of the images from appropriately chosen centers causes the jets to almost vanish; this does not happen with centers chosen on the limb, far from the assumed position of the sources. The shaded region in Fig. 4 contains the most prominent sources on the nucleus. To locate them better will be possible only after having an improved 3-dimensional model to work with. Ever since the arrival of the various spacecraft at Halley, it somehow became a new paradigm that comets are active only on their sunlit hemispheres. It must be said that we too contributed to its foundation. However, in the process of shift-differentiating the images, there appeared several faint features we had not recognized previously. One of these could be followed through images 1174, 1178, 1186, 1202, 1206, and 1210. This feature turned out to be a jet emitted from a source directed 130° from the Sun, that is, toward the antisolar hemisphere. On those images where it is not apparent, it is masked by other features. At present we interpret this feature as a jet coming from the dark hemisphere, from a source still active after sunset.

References

Herschel, J.F.W.: 1847, Results of Astronomical Obeservations at the Cape of Good Hope, Smith, Elder, and Co., London, 393
Hevelius, J.: 1685, Annus Climactericus, Rhetius, Danzig
Larson, S.M., Sekanina, Z.: 1984, *Astron. J.* **89**, 571
Larson, S.M., Sekanina, Z.: 1985, *Astron. J.* **90**, 823
Sagdeev, R.Z. et al.: 1985, *Adv. Space Res.* **5**, 12, 95
Sekanina, Z., Larson, S.M.: 1984 *Astron. J.* **89**,1408
Sekanina, Z., Larson, S.M.: 1986, *Astron. J.* **92**, 462
Struve, F.G.W.: 1839, Beobachtungen des Halleyschen Cometen, Kaiserliche Akademie der Wissenschaften, St. Petersburg, 132

Astron. Astrophys. 187, 839–842 (1987)

ASTRONOMY AND ASTROPHYSICS

Temperature and size of the nucleus of comet P/Halley deduced from IKS infrared Vega 1 measurements

C. Emerich[1], J.M. Lamarre[1], V.I. Moroz[2], M. Combes[3], N.F. Sanko[2], Y.V. Nikolsky[2], F. Rocard[4], R. Gispert[1], N. Coron[1], J.P. Bibring[4], T. Encrenaz[3], and J. Crovisier[3]

[1] Laboratoire de Physique Stellaire et Planétaire, B.P. 10, F-91371 Verrières le Buisson Cedex, France
[2] IKI, Profsoyuznaya 84/32, 117810 Moscow, GSP 7, USSR
[3] Observatoire de Paris-Meudon, F-92190 Meudon, France
[4] Laboratoire René Bernas, F-91406 Orsay, France

Received February 2, accepted April 14, 1987

Summary. The infrared thermal emission of the nucleus of Halley's comet was measured by the imaging channel of the IKS spectrometer in two wavelengths bands (7–10 μm and 9–14 μm) and in two perpendicular directions of analysis. The effective dimensions of the infrared emissive region in these directions are estimated from a simulation of the flyby configuration, and compared to observations made in the visible range. A colour temperature larger than 360 K was deduced using on-ground preflight calibrations. A temperature of the same order is obtained independently, comparing the absolute value of the measured fluxes with model simulations. Such temperatures, significantly higher than the sublimating temperature of ices, and near the blackbody temperatures at 0.8 au, suggest the existence of an essentially inactive dark crust covering a significant part of the surface of the nucleus.

Key words: comet P/Halley – nucleus – infrared – temperature

1. Introduction

1.1. Principle of the experiment

The imaging channel of "IKS" (Infra Krasnoi Spectrometer, or Infrared Spectrometer) was fully described in different papers (Arduini et al., 1983; Lamarre et al., 1984; Crifo, 1981). It analyses only one angular frequency of the image of the comet, with a period of about one minute of arc, in two infrared bands: 9–14 μm for channels I and II, and 7–10 μm for channels III and IV. The grid direction of channels I and IV is roughly parallel to the plane of the comet orbit, and roughly perpendicular for channels II and III. This situation is summarized in Table 1. The signals are proportional to the infrared flux emitted by the comet and to a shape factor (positive and smaller than one) which represents the proportion of the flux detected at the angular frequency of the analysing grid.

Send offprint requests to: C. Emerich

Table 1

Orientation of the grid relative to cometary orbital plane	//	⊥
7–10 μm	IV	III
9–14 μm	I	II

1.2. First results

The signals acquired during the flyby of P/Halley by IKS on VEGA 1 have been presented with a preliminary analysis (Combes et al., 1986; Lamarre et al., 1986) and are represented in Fig. 1. The colour temperature of the source, deduced from comparison of channels with identical directions of analysis and different bandpaths is 420 K ± 60 K. The signal proved to be strongly dependent on the phase angle and roughly proportional to the illumination by the sun of the sub-spacecraft point of the nucleus. In this paper, using new model simulations of the measurements, the time dependence (i.e. the shape) and the absolute values of the measured signals (in IKS units) are discussed, and an interpretation is outlined.

2. Model simulations

2.1. Models of nucleus temperature distribution

The simplest nucleus models initially used to forecast the time variation of the measurements were models with a uniform temperature of about 200 K, because very weak diurnal variations of the nucleus surface temperatures were expected for an icy conglomerate nucleus model (Weissman and Kieffer, 1981). In fact, such models proved to be unable to explain the measured signals, neither their absolute levels nor their variations with time.

For example, the dashed-dotted curve in Fig. 2 shows that a temperature of about 320 K is necessary to account for the maximum signal of about 70 units measured near the closest approach. This temperature is far from the previously expected value of 200 K for the sublimating ice (which would have only

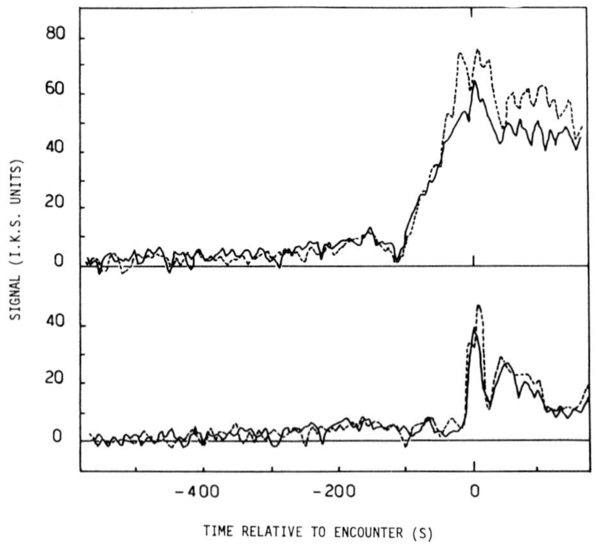

Fig. 1. Measured cometary signals as a function of the time relative to the encounter. The solid lines correspond to the 9–14 μm bandpass. The dashed lines to the 7–10 μm bandpass. On the top, the signals correspond to the grid direction parallel to the cometary orbital plane (i.e. to an analysis direction perpendicular to this plane). At the bottom, the signals correspond to the grid direction perpendicular to the cometary plane (i.e. to an analysing direction parallel to this plane)

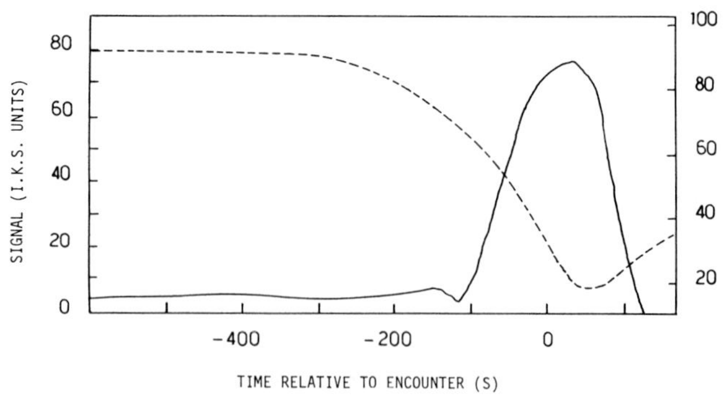

Fig. 3. In solid line, the cometary signal is simulated for a grid parallel to the comet orbital plane, with a nucleus diameter corresponding to the TV camera data, but with an heated emissive region limited to an angle $\pm 30°$ around the subsolar point. The dashed line represents the phase angle variation during the flyby

given a maximum signal of about 5 units at closest approach). It is rather close to the mean equilibrium temperature of a fast rotating blackbody at 0.8 au of the Sun. On the other hand, such a uniform model predicts a ratio of about 0.65 instead of the measured ratio of about 0.1 between the two successive maxima observed by the channels I and IV, respectively around -150 s and 0 s relative to the closest approach (see Fig. 1). This means that a uniform temperature distribution does not reflect the very sharp variation of the phase angle which induces variation of the sun illumination conditions of the nucleus, and then of the temperature distribution on the nucleus surface as seen from the spacecraft (see Fig. 3). As these variations are expected to induce steep changes of the shape factor (see definition in 1.1) associated with the distribution of the thermal emission of the nucleus, the phase angle variation was introduced in a simple model of the temperature distribution on the surface of the nucleus. In this model, in order to obtain reasonable computer time consumption, the nucleus is spherical with a radius RN. The temperature distribution results from the equilibrium between one part of the solar radiation and the thermal emission, the other part being converted into sublimation heat. For an icy conglomerate, the thermal emission was shown to be negligible at distances from the sun lower than 1 au (Delsemme, 1982).

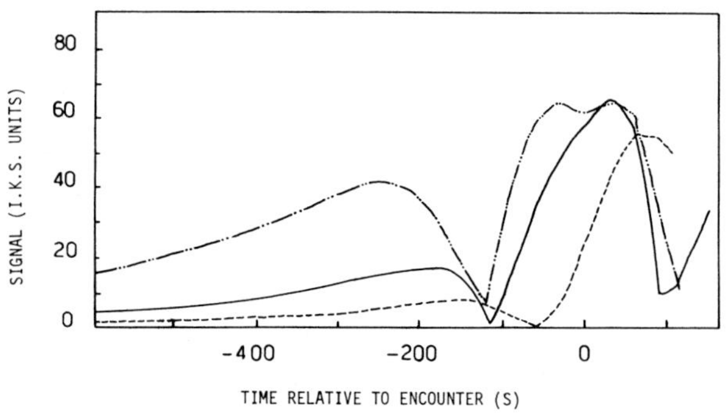

Fig. 2. Simulated signals as a function of the time relative to the encounter. The dashed-dotted line corresponds to a uniform disc with a 5 km diameter, a temperature of 320 K and an infrared emissivity of 0.8. The solid line corresponds to a grid direction parallel to the comet orbit plane and to a nucleus temperature distribution depending on the solar illumination (see text). The model parameters are $RN = 2.5$ km, $\Delta l = 20°$, $T_{MAX} = 390$ K. The dashed line corresponds to a grid direction perpendicular to the comet orbit plane and to model parameters $RN = 1.5$ km, $\Delta l = 20°$, $T_{MAX} = 360$ K

In the models presented here, in order to explain the large signals measured near the closest approach and to account for the large values of the measured color temperatures, we adopted the opposite viewpoint, assuming that for most of the nucleus surface, the solar energy absorbed in the visible part of the spectrum is essentially converted into thermal emission following the Stefan-Boltzmann law. This is probably the case in the inactive regions of the nucleus covered with a refractory black mantle, and this also seems to be justified by the fact that the active regions, where ices are sublimated, represent only some 10% of the total nucleus surface (Keller et al., 1986). Therefore, in the solar illuminated part of the nucleus, the temperature distribution adopted is given by:

$$T = T_{MAX}(\cos\theta)^{1/4} \tag{1}$$

T_{MAX} is the highest temperature on the nucleus surface, θ is the angle between the local zenith and the direction corresponding to the maximum temperature:

$$\theta = \arccos\left[(RN^2 - y^2 - z^2)^{1/2}\cos(\phi + \Delta l)/RN - y\sin(\phi + \Delta l)/RN\right] \tag{2}$$

y and z are two of the coordinates of a point of the nucleus surface, in a cometocentric reference system in which the x axis is parallel to the sight direction and the (x, y) plane contains the sun direction. ϕ is the phase angle and Δl the possible angular lag between the warmest point of the nucleus surface and the subsolar point. The IR surface emissivity is an adjustable parameter and a minimum temperature lower than 200 K was assumed both in the dayside and in the nightside of the nucleus.

2.2. *Instrument model*

The modelisation of the transfer function of the instrument was made using the on-ground preflight calibration data providing the spectral responsivity and the signal variation as a function of a blackbody known source (Bibring et al., 1984). It provides the shape factor associated with each measurement.

2.3. *Results*

An example of the signals obtained using such a model is represented in Fig. 2 and compared to the previously used uniform model. The ratio of the two successive maxima measured during the flyby (around -150 s and 0 s) can be roughly explained by the new temperature distribution, confirming the predominent effect of the phase angle, that is to say of the large temperature gradients on the nucleus surface. The most likely explanation of these facts is that the main infrared flux measured by the experiment comes from the nucleus itself and not from a dust cloud around it (Lamarre et al., 1986). Concerning the influence of the model parameters, the calculations show that the radius RN, the angular lag Δl and the temperature distribution essentially govern the shape of the signal and particularly the position in time of the successive maxima and minima during the flyby. The maximum temperature T_{MAX}, rather deals with the absolute value of the simulated signals, i.e. the total infrared flux measured for each position of the spacecraft.

3. Discussion

3.1. *Time dependence of the signals*

In order to roughly fit the time variation of the measurements of channels I and IV with the model presented, one has to take a diameter of about 5 km and an angular lag of about 20°, as can be seen comparing Figs. 1 and 2. This in fact allows to reproduce the position in time of the measured minima and maxima. Concerning the two other channels, a diameter of about 3 km and an angular lag of the same order are necessary to account for the time position of the extrema. In fact, it is not possible to completely fit the time function variation of the measured signals, especially those of channels II and III and the signal variation observed after the encounter time. The uncertainties relative to the time scale, the phase angle variation and the directions of analysis of the grid in a reference frame linked to the nucleus certainly are not negligible. But they cannot explain the discrepancy between the dimensions deduced from the IR measurements and those measured by the TV camera (Sagdeev et al., 1986a, b). It appears that the dimensions of the effective emissive region measured in the infrared are significantly smaller than the nucleus as observed in the visible range. However, it has to be noted that the infrared experiment is above all sensitive to contrasted features. Therefore, the observed discrepancy can be explained by the fact that the real temperature distribution, which is one of the main factors governing the time signal variation, may be more complex than is assumed in the model used where active regions and non homogeneous surface of the nucleus are not taken into account. This hypothesis was checked by simulating such possible inhomogeneities: using a diameter of 10 km, with a heated area of the nucleus limited to a certain angle around subsolar direction, it is possible to reproduce the same kind of signal time variation as measured (Fig. 3). These calculations suggest a possible explanation which could reconcile infrared and visible observations by the existence of hot spots or heated regions covering some percentage of the nucleus surface (up to about 25%). The rest of the nucleus (taken as a whole) has to be at temperatures lower than about 270 K in order not to create secondary maxima which were not detected. Another explanation that cannot be discarded a priori (but not checked by our models) is the possible detection of dust jets which would have been projected into the direction of the probe around the encounter time. In that case, these jets would have sufficient I.R. optical depths to be detected. In fact, such hot dust jets seem not to have been detected by other experiments at the time of the flyby. Anyway, in order to reproduce the measured time variation of the signal, it is necessary to assume large temperature gradients on the nucleus surface.

The simple model presented here does not pretend to reproduce the details of the time variation of the signals. However it explains the main characteristics correctly.

3.2. *Absolute value of the signals*

Note that in the three figures shown in this paper, the signals are only scaled in "IKS units" because the conversion into the incident flux (in watt) depends on each measurement point through the variation of the shape factor (see Sect. 1.1). After having grossly explained the time variation of the signal and its general shape, it is necessary to account for the absolute scale of the measurements -which is obviously related to the received total infrared flux.

A first determination of a brightness temperature was already indicated in Sect. 2.1, using a constant temperature of 320 K to explain the maximum signal measured near the encounter. Another check was made using a constant temperature on the dayside part of the nucleus and 200 K on the nightside part. The best fit for the maximum signal near the encounter is 320 K for a heated source diameter of 5 km as well as for a source diameter of 10 km. Such a temperature appears then to correspond to a mean equivalent temperature of the sunlit part of the nucleus.

But, in order to fit both the maximum observed near the closest approach and the rest of the curves, it was shown previously that a temperature distribution on the nucleus given by Eq. (1) has to be used. The value of T_{MAX} then is adjusted to provide the correct absolute scale: a maximum temperature between 350 and 400 K is necessary, depending on the details of the temperature distribution chosen and on the different hypotheses concerning the dimension of the hot area of the nucleus. Nevertheless, calculations show that these absolute values do not depend much on the assumed diameter of the observed scene, because the shape factor decreases when the diameter increases, compensating then the increase of the total flux. Consequently, the maximum brightness temperatures deduced from this method do not depend much on the assumed diameter of the infrared source.

These temperatures are slightly lower than the colour temperature of 420 K $\pm$ 60 K deduced from comparison of the ratio of the signals to prefligth on-ground calibration curves (Combes et al., 1986; Lamarre et al., 1986). It is however satisfactory that these two completely independent and different methods lead to temperature determinations with overlapping error bars, giving the most probable maximum temperature between 360 K and 400 K.

4. Conclusion

The more reliable and the newest result concerns the temperature of the nucleus, or at least a part of it, the implication of which deals with the physical nature of its surface. The temperature was obtained with two distinct methods, one depending on models, the other not, both leading to a similar large value of about 320 K in mean value and 400 K in maximum value. These high temperatures justify a posteriori the neglect of dust effects in the simulations (Crifo, 1983). Comparison with previous physical models of nucleus seems then to rule out an homogeneous bare model which leads to a temperature around 200 K (Weissman and Kieffer, 1981). The higher measured temperatures and the very low albedo of the nucleus surface (Sagdeev et al., 1986a) suggested that it is covered with some black thermo insulating layer. A porous structure of this layer is then necessary to provide a low thermoconductivity and to allow the gas molecules to penetrate this layer without a substantial attenuation of the gas production rate (Sagdeev et al., 1986b). This layer might consist of weakly bounded dust grains. Different models of such a surface layer were discussed during the last years (for example, the "friable sponge model" (Brin and Mendis, 1979; Horanyi et al., 1984)), but these models failed in explaining the observed high gas production rates. More recently, taking into account the Vega results, Ip and Rickman (1986) presented a new version of the friable sponge model which seems to be free from this contradiction. However, a possible problem raised by such models is that a porous layer would probably be efficiently cooled by the ices sublimating beneath the porous mantle, and this would be in conflict with the large temperatures measured by the experiment. In that case, an alternative explanation could be the coexistence on the nucleus surface of two distinct kinds of regions: an active one, relatively cool, in which the sublimating ices exposed to the solar flux would produce both the gases and the dust jets, and an inactive one, consisting in a hot refractory dark crust. The infrared experiment would then have measured the thermal emission and the temperature of the hottest part of this essentially inactive crust, when it was directly exposed to the solar flux. This crust, heated to blackbody temperatures, represented about 25% of the projected nucleus area, in the conditions of the VEGA 1 flyby. In any case, the high surface temperatures deduced from the observations presented here bring a new constraint for any model of the cometary nucleus to be developed in the future.

Acknowledgements. We thank the French and the Soviet technical teams for having made this experiment possible. Financial support was provided by the Centre National d'Etudes Spatiales.

References

Arduini, M., Bibring, J.P., Cazes, S., Combes, M., Crifo, J.F., Encrenaz, T., Gispert, R., Harduin, D., Lamarre, J.M.: 1983, *Adv. Space Res.* **2**, 113

Bibring, J.P., Cazes, S., Charra, J., Combes, M., Coron, N., Cougrand, B., Crifo, J.F., Crovisier, J., Emerich, C., Encrenaz, T., Gispert, R., Gondet, B., Guyot, G., Harduin, D., Lamarre, J.M., Levanti, G., Maurel, C., Parisot, D., Rocard, F., Salvetat, P., Soufflot, A.: 1984, *Adv. Space Res.* **4**, 273

Brin, G.D., Mendis, D.A.: 1979, Astr. J., 229, 402.

Combes, M., Moroz, V.I., Crifo, J.F., Lamarre, J.M., Charra, J., Sanko, N.F., Soufflot, A., Bibring, J.P., Cazes, S., Coron, N., Crovisier, J., Emerich, C., Encrenaz, T., Gispert, R., Grigoryev, A.V., Guyot, G., Kranopolsky, V.A., Nikolsky, Yu.V., Rocard, F.: 1986, *Nature* **321**, 266

Crifo, J.F.: 1981, Proc. Alpbach Summer School 27 July–7 August 1981, ESA SP-164, 229

Crifo, J.F.: 1983, *Adv. Space Res.* **2**, 203

Delsemme, A.H.: 1982, in *Comets*, ed. L.L. Wilkening, Univ. Arizona Press, p. 85

Horanyi, M., Gombosi, T.I., Cravens, T.E., Korosmezey, A., Kecskemety, K., Nagy, A.F., Szego, K.: 1984, *Astron. J.* **278**, 449

Ip, W.H., Rickman, H.: 1986, in *Proc. Workshop Comet Nucleus Sample Return Mission*, Canterbury 15–17 July 1986, (to appear)

Keller, H.U., Arpigny, C., Barbieri, C., Bonnet, R.M., Cazes, S., Coradini, M., Cosmovici, C.B., Delamere, W.A., Huebner, W.F., Hughes, D.W., Jamar, C., Malaise, D., Reitsema, H.J., Schmidt, H.U., Schmidt, W.K.H., Seige, P., Whipple, F.L., Wilhelm, K.: 1986, *Nature* **321**, 320

Lamarre, J.M., Gondet, B., Gispert, R., Emerich, C., Rocard, F.: 1984, *Adv. Space Res.* **4**, 277

Lamarre, J.M., Emerich, C., Moroz, V.I., Combes, M., Sanko, N.F., Rocard, F., Gispert, R., Nikolsky, Yu.V., Bibring, J.P., Coron, N., Crovisier, J., Encrenaz, T.: 1986, *Adv. Space Res.* **5**, 123

Sagdeev, R.Z., Blamont, J., Galeev, A.A., Moroz, V.I., Shapiro, V.D., Shevchenko, V.I., Szegö, K.: 1986a, *Nature* **321**, 259

Sagdeev, R.Z., Szabo, F., Avanesov, G.A., Cruvellier, P., Szabo, L., Szego, K., Abergel, A., Balazs, A., Barinov, I.V., Bertaux, J.L., Blamont, J., Detaille, M., Dimarellis, E., Dul'nev, G.N., Endröczy, G., Gardos, M., Kanyo, M., Kostenko, V.I., Krasikov, V.A., Nguyen-Trong, T., Nyitrai, Z., Reny, I., Rusznyak, P. et al.: 1986b, *Nature* **321**, 262

Weissman, P.R., Kieffer, H.H.: 1981, *Icarus* **47**, 302

Astron. Astrophys. 187, 843–846 (1987)

ASTRONOMY AND ASTROPHYSICS

Fine dust structures in the emission of comet P/Halley observed by the Halley Multicolour Camera on board Giotto

N. Thomas and H. U. Keller

Max-Planck-Institut für Aeronomie, Postfach 20, D-3411 Katlenburg-Lindau 3, Federal Republic of Germany

Received May 12, accepted June 25, 1987

Summary: Halley Multicolour Camera (HMC) observations on board the ESA Giotto spacecraft provided images of comet P/Halley with resolution 4 orders of magnitude better than from the ground in March 1986. The fine structure seen in the emission from the cometary surface has led to some confusion over the nomenclature of jets and jet-like features streaming from the nucleus. In this letter, emission features are classified according to directions with respect to the sun and to positions with respect to the nucleus. In total 17 fine dust structures (filaments) are identified, more than doubling the number presented earlier (Keller et al., 1986). This will assist in correlation with other spacecraft and ground-based observations.

Key words: comet p/Halley – HMC observations – dust

The fine structure seen in HMC images of comet P/Halley's dust emission has led to a number of names such as "sub-jets", "mini-jets" and "jetties" being mentioned in an effort to distinguish these structures from the "jets" described by Larson and Sekanina (1984) in their analysis of observations of the 1910 apparition of the comet. Following the suggestion of Huebner (personal communication), we propose that the almost collimated features in the emission be referred to as "dust filaments".

HMC images have revealed 17 filaments, although improved image processing may increase this number. Ground-based observations appear to show the presence of 3 jets over the observing period on encounter night (Cosmovici et al., 1986). It is therefore desirable to classify the filaments in such a way that clear identification of major activity is also apparent. In this regard, the filaments can be split conveniently into three groups, each associated with an area of activity on the surface plus a fourth covering the remaining filaments. Figure 1a shows the last clear filter image taken in single detector mode (Keller et al., 1987) from a distance of 25000 km. A schematic diagram of the same image showing the four classification groups is shown in Fig. 1b. "N" filaments are associated with the northern active region and of these filaments, the majority appear to be on the side of the nucleus observed by HMC. The "C" filaments come from the region near the central depression (Keller et al., 1987) while the "S" filaments are associated with the southern area of activity

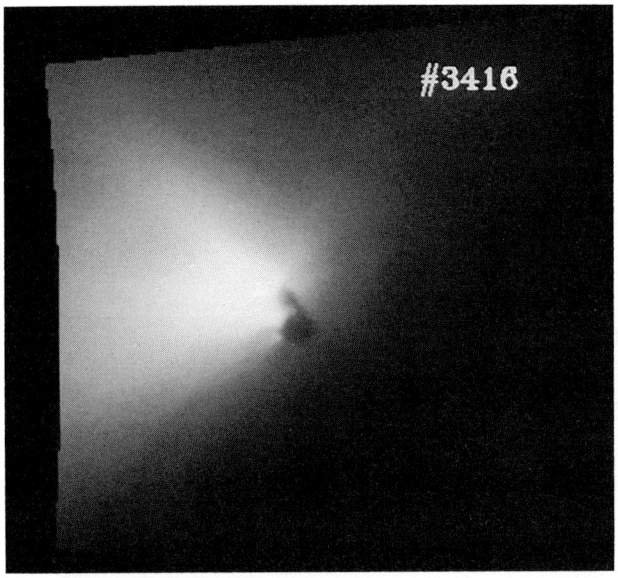

Fig. 1a. Image No. 3416. Last clear image in single detector mode (SDM). Taken from a distance of 25 000 km giving a frame size of $\sim 113 \times 113\,km^2$ at a resolution of 580 meters/pixel. The solar aspect angle is 105°.9. The image shows the presence of fine structure within the emission from the cometary nucleus.

which is at present believed to be on the side of the nucleus not observed by HMC. "U" (unidentified) filaments are those which do not fall into the above categories. Their sources are not observed and their relationship to active regions is either speculative or unknown.

There are two factors which must be taken into account when completing the identification of each dust filament. Firstly, as Giotto moved towards closest approach, the restricted field of view imposed by the limited telemetry allowed only earlier, comparatively poor resolution images of the "S" and "U" filaments. The consequence of this was that three reference images were used to identify filaments. Secondly, the last images transmitted before the communication failure show a small but significant change in solar aspect angle. Investigation of whether this change is sufficient to give three-dimensional information is currently in progress. Only images where the change in aspect angle remains small were considered for this note, thereby ruling

Send offprint requests to: N. Thomas

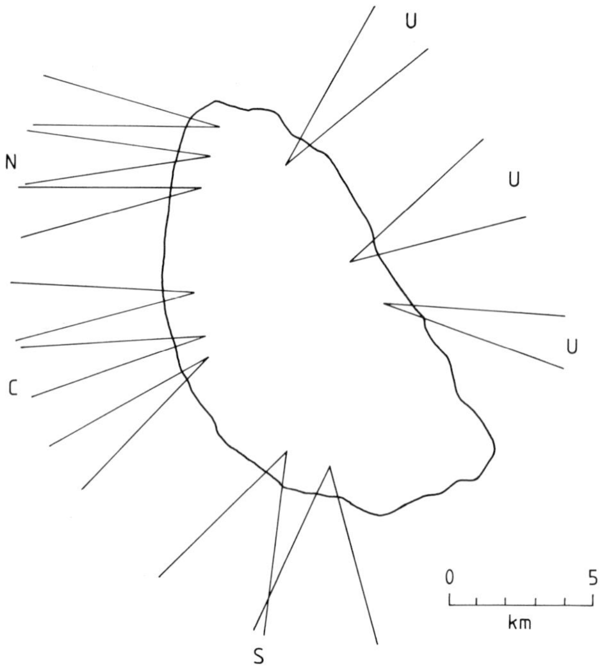

Fig. 1b. Schematic diagram of image No. 3416 indicating the four proposed classification groups

Fig. 2. Image No. 3475. The second reference image used in the classification was taken from a distance of 9500 km giving a frame size of $\sim 17 \times 17\ \mathrm{km}^2$ at a resolution of 220 meters/pixel. The solar aspect angle is 104°.0

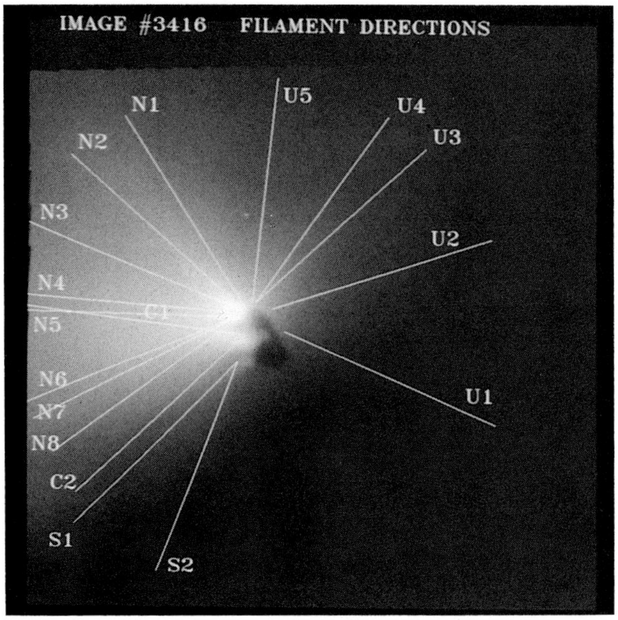

Fig. 1c. Filament lines and their names superimposed on image No. 3416

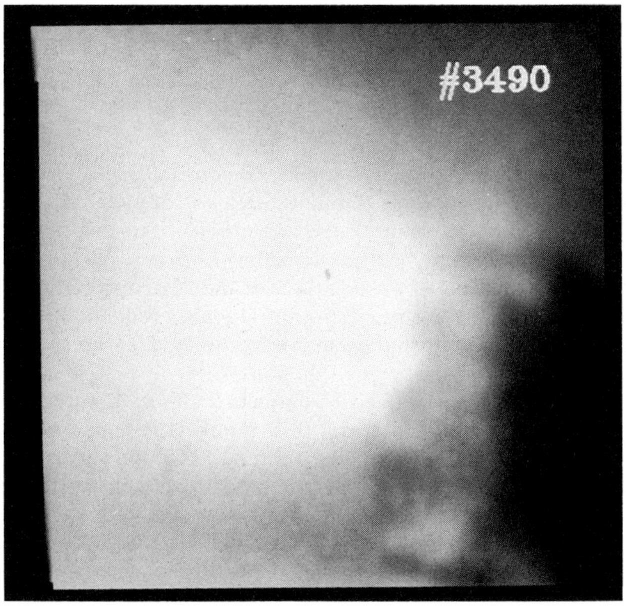

Fig. 3. Image No. 3490. The third reference image used in the classification was taken from a distance of 5400 km giving a frame size of $\sim 10 \times 10\ \mathrm{km}^2$ at a resolution of 130 meters/pixel. The solar aspect angle is 101°.7

out the last, highest resolution, images. This, however, allowed precise identification of a dust filament according to its angle of emission in the image plane relative to the apparent sun direction, and by using registered images, allowed the identification of a point on the line of the filament relative to a clearly identifiable feature on the nuclear surface, which then uniquely identifies each filament in the two-dimensional image plane.

The feature on the nucleus chosen as a reference point was the brightest point on the "mountain". This feature is associated with a steep gradient on its northern margin and is the best defined "point" on HMC images of the nucleus. The images chosen to define the filament positions were No. 3416 (Fig. 1a), No. 3475 (Fig. 2), and No. 3490 (Fig. 3). The mountain feature is not visible on image No. 3490 although accurate geometrical rectification of

Table 1

HMC image number	Filament	Distance of emission from mountain [a] (km)	Angle from mountain-Sun line [a] (deg)	Angle of emission (deg)	Dist. of mountain from jet along line perpendicular to Sun-comet direction. (km)
3416	U1	4.1 ± 1.2	146 ± 6	174.3 ± 1.5	2.0
3416	U2	5.4 ± 1.2	77 ± 6	132.8 ± 1.5	6.7
3475	U2	5.1 ± 0.3	56 ± 4	133.0 ± 4.0	7.3
3416	U3	6.8 ± 1.2	44 ± 6	108.1 ± 1.5	19.7
3475	U3	6.7 ± 0.3	39 ± 4	111.0 ± 4.0 [b]	17.8
3416	U4	6.8 ± 1.2	44 ± 6	95.7 ± 1.5	53.7
3475	U4	6.7 ± 0.3	39 ± 4	99.0 ± 4.0	37.1
3475	U5	6.7 ± 0.3	37 ± 4	67.0 ± 4.0 [c]	− 8.6
3490	U5	6.7 ± 0.1	38 ± 1	64.0 ± 4.0	− 6.7
3416	N1	6.6 ± 1.2	19 ± 6	26.9 ± 1.5	− 1.0
3475	N1	5.8 ± 0.3	23 ± 4	29.0 ± 4.0	− 0.7
3490	N1	6.4 ± 0.1	26 ± 1	24.0 ± 4.0	0.2
3416	N2	6.6 ± 1.2	19 ± 6	12.9 ± 1.5	0.7
3475	N2	6.6 ± 0.3	16 ± 4	10.0 ± 4.0	0.7
3490	N2	6.9 ± 0.1	19 ± 1	8.0 ± 4.0	1.3
3416	N3	6.6 ± 1.2	19 ± 6	354.2 ± 1.5	2.8
3475	N3	6.7 ± 0.3	17 ± 4	350.4 ± 4.0	3.0
3490	N3	6.5 ± 0.1	18 ± 1	353.0 ± 4.0	2.8
3475	N4	6.5 ± 0.3	9 ± 4	332.0 ± 4.0	4.4
3490	N4	6.0 ± 0.1	16 ± 1	335.0 ± 4.0	4.3
3475	N5	6.0 ± 0.3	2 ± 4	329.0 ± 4.0	3.8
3490	N5	5.8 ± 0.1	7 ± 1	332.0 ± 4.0	3.8
3475	N6	6.1 ± 0.3	11 ± 4	308.0 ± 4.0	8.8
3490	N6	6.7 ± 0.1	16 ± 1	309.0 ± 4.0	9.8
3416	N7	6.3 ± 1.2	8 ± 6	303.5 ± 1.5 [d]	15.2
3475	N8	6.2 ± 0.3	6 ± 4	294.0 ± 4.0	14.5
3490	N8	6.5 ± 0.1	18 ± 1	296.0 ± 4.0	14.7
3416	C1	4.1 ± 1.2	326 ± 6	337.8 ± 1.5	− 0.9
3475	C1	4.9 ± 0.3	325 ± 4	337.0 ± 4.0	− 1.1
3416	C2	4.1 ± 1.2	326 ± 6	286.7 ± 1.5 [e]	9.0
3416	S1	8.3 ± 1.2	284 ± 6	284.7 ± 1.5 [d]	− 0.4
3416	S2	8.3 ± 1.2	284 ± 6	260.8 ± 1.5	−20.5

Notes: The error bars for the angles of emission are based on a possible uncertainty of 2 pixels in identifying the start and finish position. This value is approximately equal to the HMC point spread function. The error bars for the points on the filament refer to a 2 pixel error on the unscaled images

[a] To identify each filament uniquely, the distance and angle from a point on the filament to the "mountain" has been found. In most cases, the point used was the visible start point of the filament. However, this was not always the case because of the complex structure of the emission near the source regions. This explains the rather good agreement of the distances for the same filament in different images in some cases but not in others

[b] Questionable identification

[c] Not clearly identified in image 3416

[d] Confirmed in images 3438 and 3450

[e] Possibly from northern area rather than central

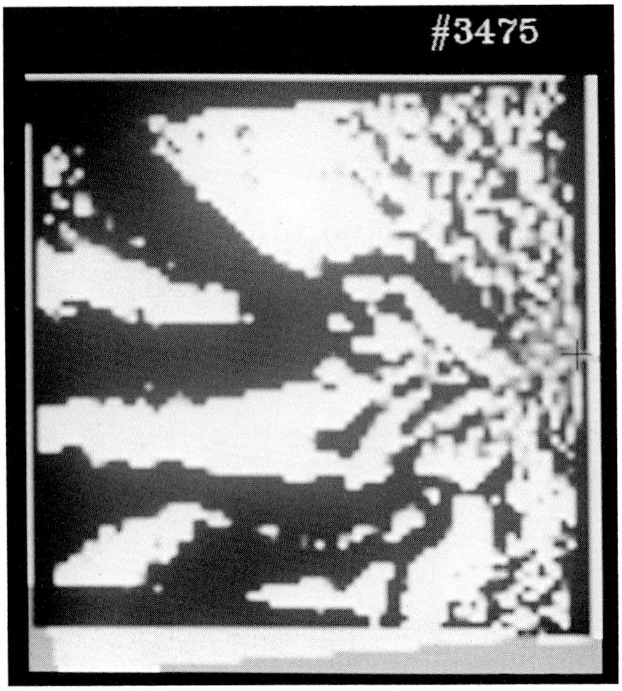

Fig. 4. Image No. 3475 processed using the rotational gradient technique (see text) which enhances the dust filaments. The centre of rotation is indicated by the cross on the right of the image. Moving the centre of rotation accentuates different filaments so that care must be taken interpreting this type of diagram

images and the assumption that the change in aspect angle is of only minor effect for all but the last images give the position of the mountain within the frame of image 3490 accurately. Self-consistency checks have been made, however, to ensure the accuracy of the determinations.

Rotational and shift gradients were used to show the dust filaments more clearly. However, any filament not visible on the original image must be considered suspicious. Cross-checks between the original images (radiometrically and geometrically corrected) and the images produced by gradient methods were repeatedly carried out. An example of the rotation method for image No. 3475, is shown in Fig. 4. To produce this display, the equation:

$$I_r = I(5) + I(-5) - 2\,I(0)$$

has been used where I_r is the resultant image and $I(x)$ is the original image rotated by the angle x (in degrees clockwise being positive). The centre of rotation in this example is inedicated by the cross although this can be varied to accentuate specific features. In addition to these rotational gradients, shift differences and Laplacian gradient masks were also used. The result of this analysis, with associated errors, is listed in Table 1 and the full identification of the filaments shown, with reference to image No. 3416, in Fig. 1c.

In conclusion, this nomenclature describes each dust filament and jet accurately and can easily be extended upon the discovery of new filaments or upon improved analysis of the existing dust structures. It also forms a basis for the analysis of HMC images taken well before closest approach to check for variabilty of the cometary emission over the observing period and for the analysis of the last HMC images in which the changes of the aspect angle are significant.

Acknowledgements. The authors acknowledge useful conversations with W. Huebner and NT acknowledges support from MPG.

References

Cosmovici, C.B., Green, S.F., Hughes, D.W., Keller, H.U., Mack, P., Moreno-Insertis, F., Schmidt, H.U.: 1986, *20th ESLAB Symp.*, ESA SP-250, Vol. **II**, p. 375

Keller, H.U., Arpigny, C., Barbieri, C., Bonnet, R.M., Cazes, S., Coradini, M., Cosmovici, C.B., Curdt, W., Delamere, W.A., Huebner, W.F., Hughes, D.W., Jamar, C., Malaise, D., Reitsema, H.J., Schmidt, H.U., Schmidt, W.K.H., Seige, P., Whipple, F.L., Wilhelm, K.: 1986, *Nature* **321**, 320

Keller, H.U., Schmidt, W.K.H., Wilhelm, K., Becker, C., Curdt, W., Hartwig, H., Kramm, J.R., Meyer, H.J., Schmidt, R., Gliem, F., Krahn, E., Schmidt, H.P., Schwarz, G., Turner, J.J., Bouyries, P., Cazes, S., Angrilli, F., Bianchini, G., Fanti, G., Brunello, P., Delamere, A., Reitsema, H., Jamar, C., Cucchiaro, A.: 1987, *J. Phys. E.*

Keller, H.U., Delamere, W.A., Huebner, W.F., Reitsema, H.J., Schmidt, H.U., Whipple, F.L., Wilhelm, K., Curdt, W., Kramm, R., Thomas, N., Arpigny, C., Barbieri, C., Bonnet, R.M., Cazes, S., Coradini, M., Cosmovici, C.B., Hughes, D.W., Jamar, C., Malaise, D., Schmidt, K., Schmidt, W.K.H., Seige, P.: 1987, *Astron. Astrophys.* (in press)

Larson, S., Sekanina, Z.: 1984, *Astron. J.* **89**, 571

Astron. Astrophys. 187, 847–851 (1987)

ASTRONOMY AND ASTROPHYSICS

Detailed analysis of a surface feature on comet P/Halley

G. Schwarz [1], H. Craubner [1], A. Delamere [2], M. Göbel [1], M. Gonano [1], W. F. Huebner [3], H. U. Keller [4], R. Kramm [4], E. Mikusch [1], H. Reitsema [2], F. L. Whipple [5], and K. Wilhelm [4]

[1] DFVLR, Hauptabteilung Angewandte Datentechnik, D-8031 Oberpfaffenhofen, Federal Republic of Germany
[2] Ball Aerospace Systems Division, Boulder, CO, USA
[3] Los Alamos Scientific Laboratory, Los Alamos, NM, USA
[4] Max-Planck-Institut für Aeronomie, D-3411 Katlenburg-Lindau, Federal Republic of Germany
[5] Smithsonian Astrophysical Observatory, Cambridge, MA, USA

Received February 5, accepted June 10, 1987

Summary. The surface of the nucleus of comet P/Halley is visible on many images taken by the Halley Multicolour Camera on board the Giotto spacecraft. A number of structural features like "mountains" or "surface dips" are clearly discernible. This paper outlines the methods that can be applied to analyze a surface feature with the aim of obtaining its characteristics.

A relatively large craterlike surface feature is selected for detailed analysis and an attempt is made to determine its size and shape as well as to assess the attainable accuracy. The surface topography is computed using a photoclinometric approach, while dimensions are determined based on apparent object sizes and an estimate of aspect angle.

Key words: comets – P/Halley – data analysis – image processing – photometry

1. Introduction

The Halley Multicolour Camera on board the Giotto spacecraft provided impressive pictures, which clearly show the shape of the cometary nucleus together with a number of morphological details. The first imaging results (Keller et al., 1986) outline the dimensions of the nucleus and its basic shape, while more recent publications (Wilhelm et al., 1986; Reitsema et al., 1986) provide details of all dimensions and describe the surface morphology.

2. Observations

On the surface of comet Halley, as shown in Fig. 1, a number of morphological details are clearly visible. One of the prominent surface details, located on the upper left side below the bright jet, appears as a craterlike surface dip. This feature appears to be real since it can be identified on all images taken in the Nucleus Mode of the Halley Multicolour Camera. Figure 2 contains a series of radiometrically corrected images on which the craterlike feature is clearly visible. A visual analysis of this surface detail on an image

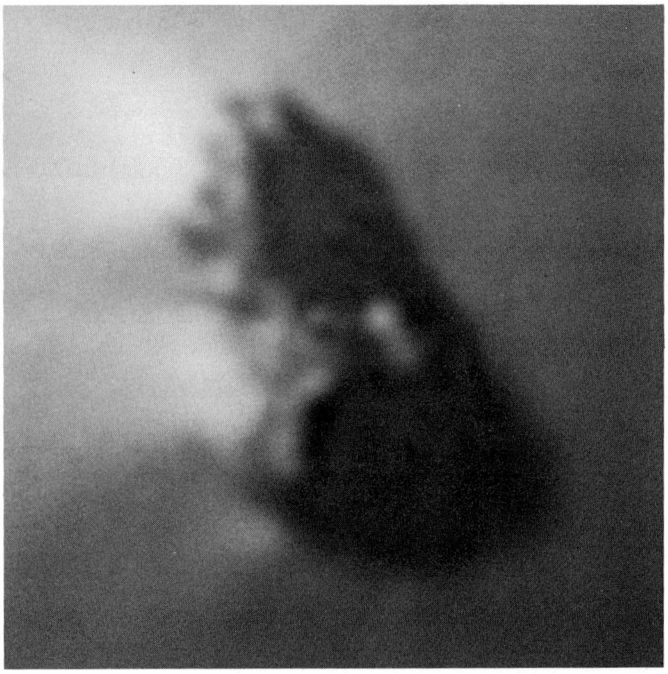

Fig. 1. The nucleus of comet Halley (composite image, Sun from upper left direction)

shows that a dark basin is surrounded by a brighter rim. The diameter of what appears to the eye to be the darker interior corresponds to a projected distance of about 1600 m. If an attempt is made to locate a definite rim by looking at separate pixel amplitudes it becomes evident that a high degree of uncertainty is involved. The brightness variations along the suspected rim are sometimes clearly evident and sometimes uncertain. This uncertainty results mainly from the amplitude quantization during transmission to the ground, which gives step sizes comparable to the variations in signal amplitude along the contour line. Even when corresponding pixel amplitudes of subsequent images are compared, a crisp continuous rim line cannot be determined. This is partly due to apparent substructures which appear in the images taken from closer distances.

Send offprint requests to: G. Schwarz

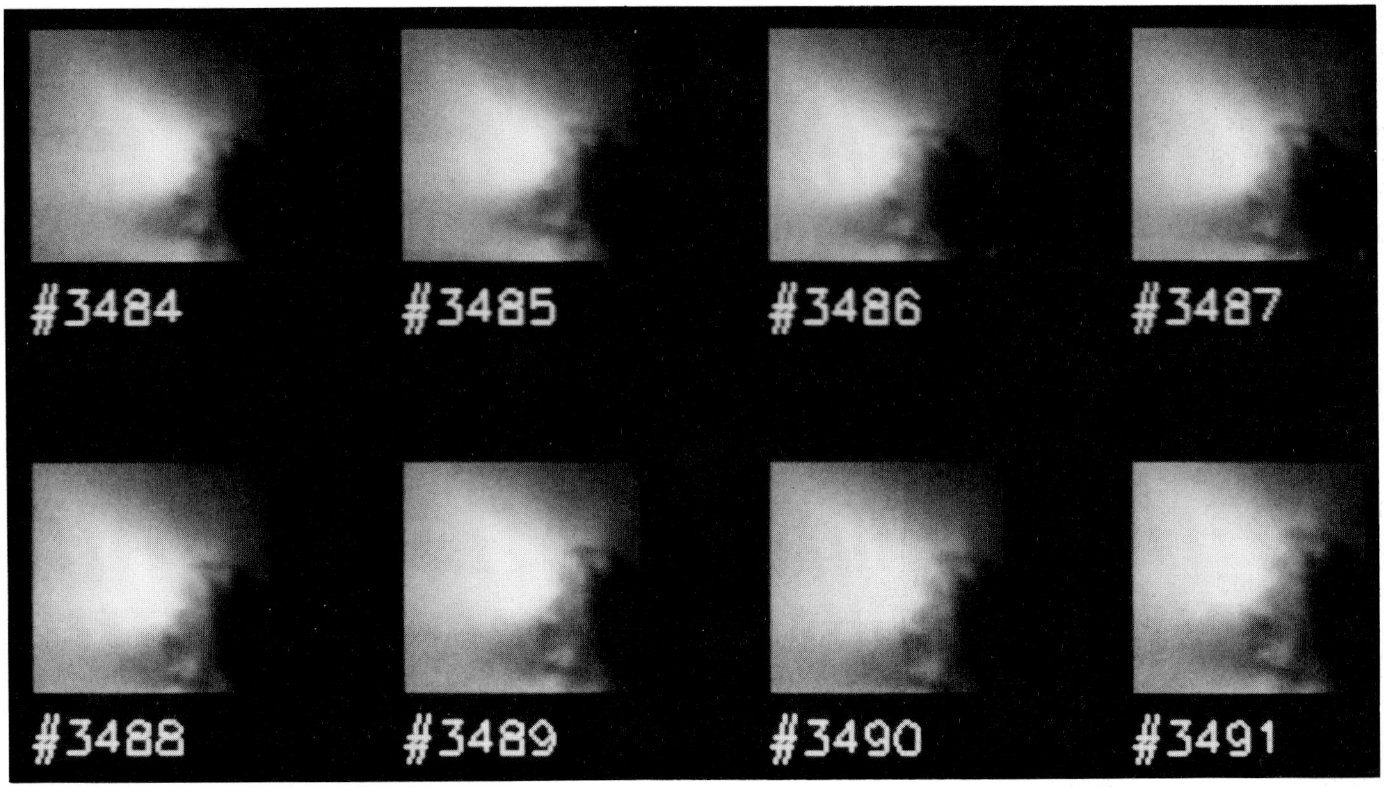

Fig. 2. Surface feature on a series of images

Table 1. Pixel amplitudes of the surface feature

	40	41	42	43	44	45	46	47	48	49	50	51
52:	157	141	127	119	122	123	124	117	115	107	109	107
53:	143	131	116	111	103	98	100	104	105	106	108	104
54:	132	117	108	101	93	88	92	94	96	99	103	103
55:	131	120	105	93	89	87	87	89	97	95	103	104
56:	126	115	101	95	92	91	90	89	94	95	103	101
57:	121	112	101	96	96	97	99	100	96	99	103	106
58:	120	106	102	99	101	100	100	106	99	99	106	108
59:	118	112	103	100	102	106	103	107	105	107	113	114
60:	114	109	102	97	99	103	104	108	114	120	120	121
61:	114	111	103	100	104	109	108	114	122	129	133	127
62:	106	111	106	106	119	119	125	127	130	135	138	135
63:	107	109	114	110	123	130	135	135	129	132	134	134

Nevertheless, the pixel amplitudes indicate a dark basin with two minima surrounded by an irregularly shaped brighter contour. Table 1 gives an example of pixel amplitudes.

3. Analysis

Even when no clear edges, shadow casts or stereoscopic views are available, the topography of a surface can be determined using photoclinometric methods. Photoclinometry delivered results for a number of surface studies of planets and their satellites (e.g. Rindfleisch, 1966; Hapke et al., 1975; Squyres, 1981; Mouginis-Mack et al., 1981; Howard et al., 1982; Jantunen et al., 1983).

Photoclinometry uses the photometric function of the surface, i.e. how the reflected and observed light intensity depends on the angles of light incidence, emission and phase. The immediate results of photoclinometric computations are slopes of surface elements. These slope elements can be combined to derive global topographic surface profiles. The computational complexity is minimized if photoclinometric profiles are taken along lines that lie perpendicular to topographic strike. Then the brightness variations depend mainly on the surface slope angles along the track and are negligible across the track direction. In the following we will derive such profiles from simple two-dimensional configurations. These profiles have to be brought to absolute scale. A more detailed description of photoclinometry can be found e.g. in Squyres (1981).

Photoclinometry presupposes some conditions: the optical reflectivity (albedo) of the entire surface must be constant, the photometric function of the surface must be known and picture elements may not lie in the shadow.

For the crater area the constancy of the surface reflectivity is assumed. At present, however, no proof exists, that this assumption is justified.

The derivation of a completely defined photometric function is a classical astronomical task (Hapke, 1966). In our case the phase angle of the near nucleus images is approximately constant; the comet is not a big spherical planet allowing determination of definite latitude/longitude dependencies of reflected intensities. Therefore, the analytical derivation of a full photometric function is difficult and an attempt is made to use classical formulas to model the light reflection. Besides the Minnaert relationship (Howard et al., 1982) there are two fundamental formulas used in astronomy, namely Lambert's law (1) or Lommel-Seeliger's reflection law (2):

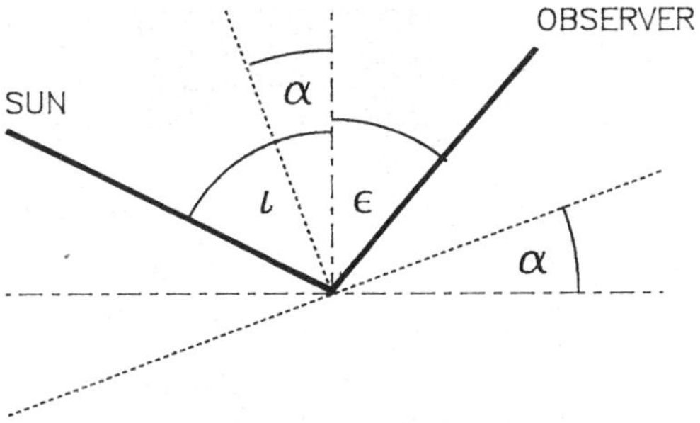

Fig. 3. Geometrical relationships for photoclinometry

$$I(\iota) = C\cos(\iota) \tag{1}$$

resp.

$$I(\iota, \varepsilon) = C\frac{\cos(\iota)}{\cos(\iota) + \cos(\varepsilon)}, \tag{2}$$

where I denotes the reflected intensity, ι is the angle of incidence, ε is the angle of emergence (both angles refer to the surface normal) and C is a constant, which has to be computed for each profile as it includes the projection from three onto two dimensions. In our case the phase angle is considered to be constant. Figure 3 shows an example within a plane. A tilt of the surface about an angle α yields $\iota' = \iota - \alpha$ and $\varepsilon' = \varepsilon + \alpha$. Solving the equations for α determines the local surface slope.

Both laws will be used to model the surface topology. The basic validity of a Lambertian approach is tested by Delamere et al. (1986). In general, the reflection of bright surfaces can be approximated by Lambert's law, while lunar-like and darker surfaces, as in our case, are better approximated by the Lommel-Seeliger relationship. Therefore preference should be given to the latter law.

The additional restriction for photoclinometry, that picture elements may not lie in the shadow, can be determined easily. The near nucleus images show a mountain ridge running from the northern tip southward. On the sunward side surface structures are clearly discernible, on the night side the surface appears to be dark. The amplitudes of this dark area are used as a comparison if parts of the analyzed area lie in the shadow.

The determination of topographic profiles by photoclinometry depends on the photometric accuracy of the picture elements. The noise along a profile line can be reduced by averaging several picture elements. This can be done either by averaging coregistered images with corresponding scale, mean amplitude level and orientation, or by averaging e.g. two pixels perpendicular to the analyzed profile within one image. Both solutions induce a loss of fine detail but this is tolerable, as the general profile is the topic to be analyzed. For reasons of computational simplicity, the second alternative was chosen.

When images, which show the entire cometary nucleus are analyzed, then it becomes apparent that the northern part of the comet is surrounded by atmospheric dust which originates from the jets. This dust makes the northern part appear brighter than the southern part. For the photoclinometric analysis we assume that the intensity reflected by the cometary surface can be determined by subtracting a spatially varying dust profile from the pixel amplitudes. A first approximation of the slope of the dust profile can be found by analyzing the pixel amplitudes on the night side of the comet. The night side profile can be extrapolated to cover the area to be analyzed. Furthermore, we assume that the atmospheric dust does not cause considerable blurring of the signal reflected by the cometary surface and that the direct sunlight is the main source of illumination.

4. Interpretation

As a first step, an image line from image 3489 was selected that runs parallel to the direction of the sun and crosses the darkest area of the basin. Two parallel lines of pixels were averaged to yield a profile. This profile and its relation to the surface of the nucleus is depicted in Fig. 4. In order to assess the accuracy of this

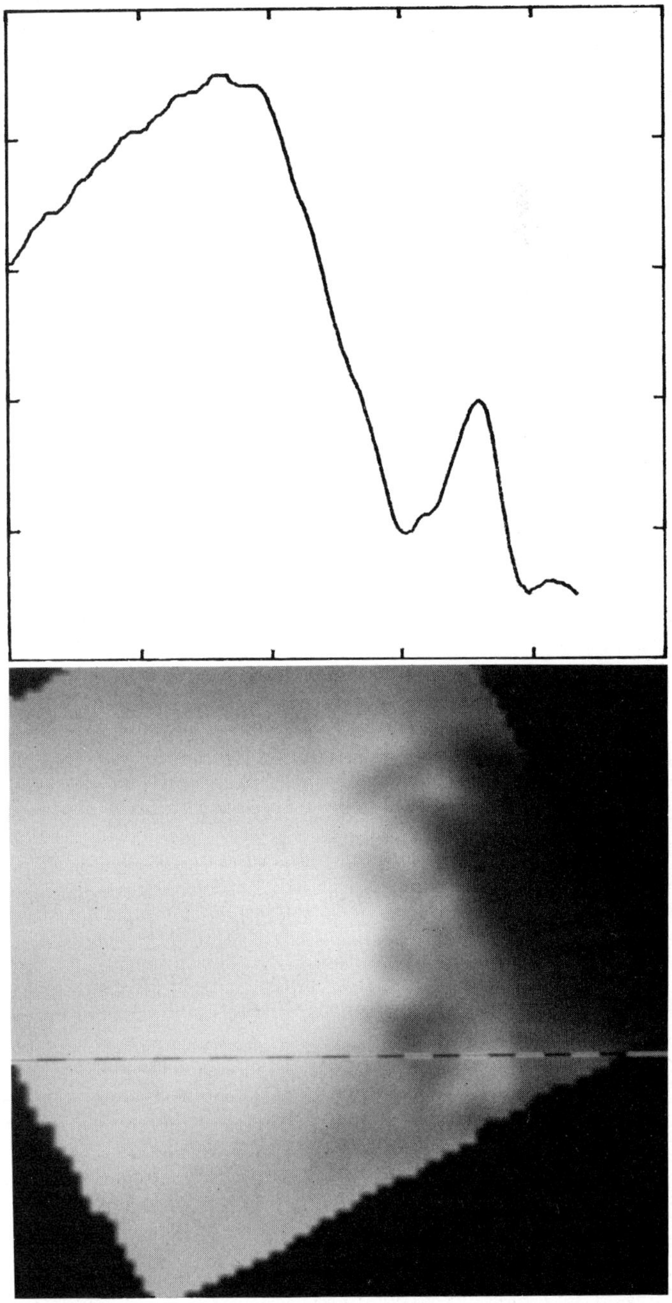

Fig. 4. Profile across the surface feature

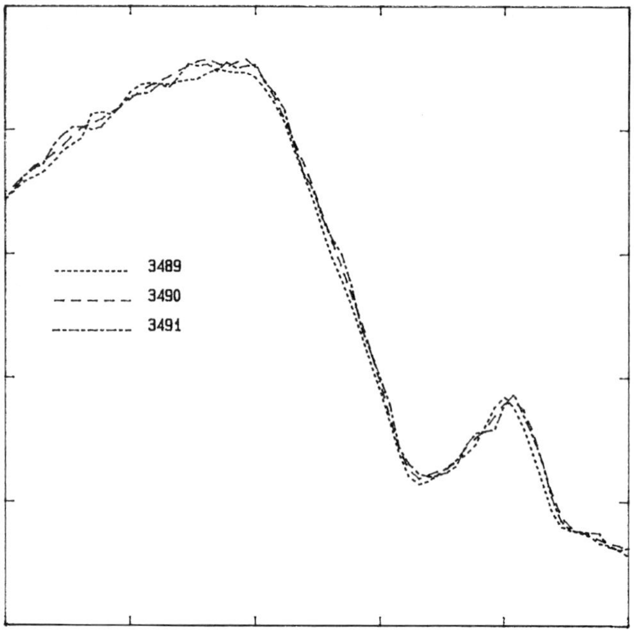

Fig. 5. Surface feature in different images

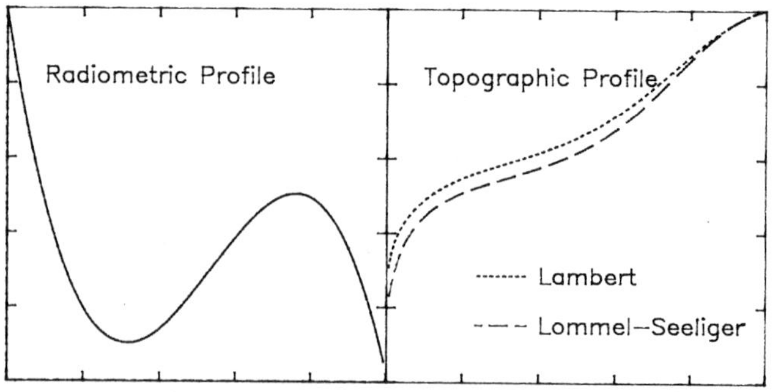

Fig. 6. Relationship between radiometric and geometric profiles

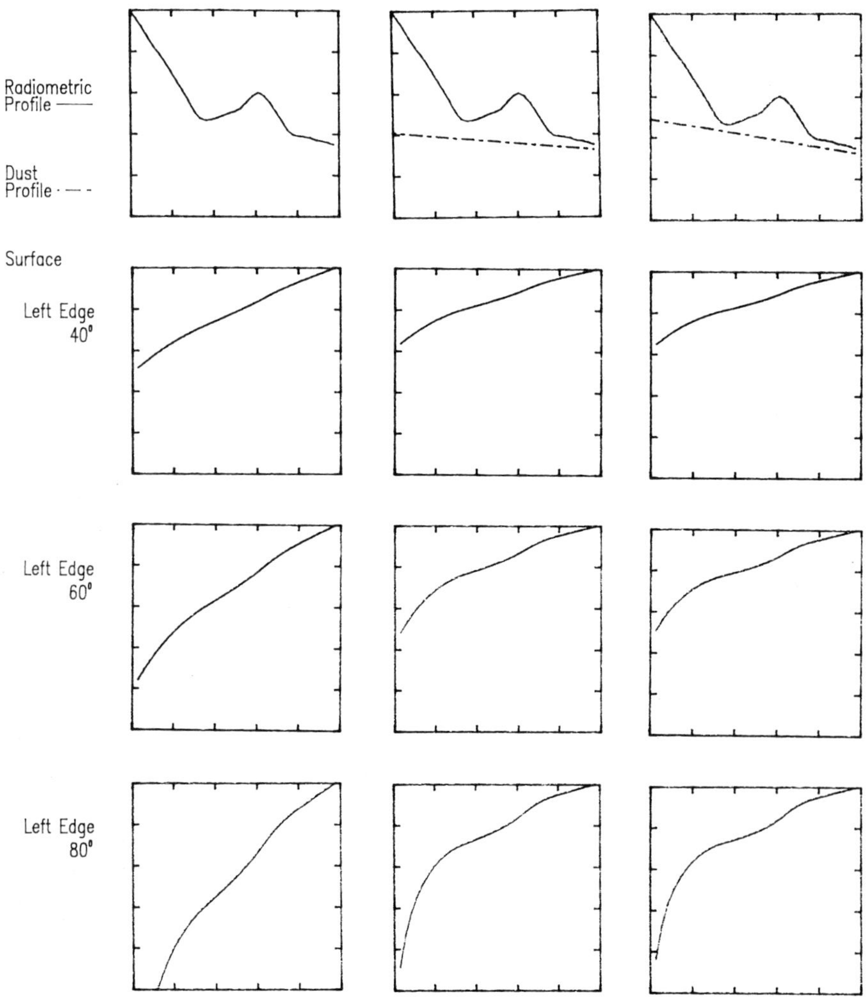

Fig. 7. Effects of parameter variations

	Dip depth (m)	Rim to rim distance (m)
40°	25, 45, 60	2250, 1990, 2050
60°	40, 70, 90	2190, 2020, 2070
80°	80, 110, 130	2570, 2100, 2160

profile, the same procedure was applied to the images 3490 and 3491. These images had to be corrected in scale, mean brightness level and orientation. All resulting profiles are very similar. Details are shown in Fig. 5.

The interpretation of brightness variations as depicted in Fig. 4 means that the peak in the brightness curve corresponds to the steepest slope of the topographic profile. The relation of such a radiometric profile and the corresponding topographic profile is depicted in Fig. 6. A significant problem is the fact that the comet limb position cannot be seen directly. The visible amplitude at the assumed limb position is, as described before, the sum of the atmospheric dust and the surface signals. Therefore, the surface angle at a definite location close to the assumed limb position was estimated, together with the dust signal which was subtracted.

These estimates which are combinations of parameters called for a simulation programme which handled all parameter variations. Simulation results are contained in Fig. 7 showing some selected parameter variations and their effect upon the surface topography. From the data in Fig. 7 the general shape of the surface dip can be determined. Both the rim to rim distance (due to the slope projection) as well as the depth of the dip, are affected by a variation of the parameters.

5. Conclusions

While Fig. 7 contains some simulated examples of computed dip depths and diameters, additional physical conditions lead to a more precise determination of results. The dust profile in the right column of Fig. 7 is the extrapolation of the atmospheric dust amplitudes observed on the dark side of the nucleus. The subtraction of this profile results in slopes near the morning terminator, which are very close to 12 degrees. This value corresponds to what is expected from the viewing geometry. Therefore we assume that this is the correct dust profile. If the overall shape of the nucleus resembles an ellipsoid, then the surface slope at the point where the profile plots start is approximately 70 degrees. These parameters result in a depth of around 110 m and a diameter of 2100 m.

Acknowledgements. The authors are grateful to the entire camera design, development and data evaluation team involved in this ambitious project. Their dedicated work during all phases of the project generated the data which enabled the publication of results.

References

Delamere, W. A., Huebner, W. F., Keller, H. U., Reitsema, H. J., Schmidt, H. U., Schmidt, W. K. H., Whipple, F. L., Wilhelm, K.: 1986, 20th ESLAB Symp., ESA SP-250, Vol. **II**, p. 355

Hapke, B.: 1966, *Astron. J.* **71**, 333

Hapke, B., Danielson, E., Klaasen, K., Wilson, L.: 1975, *J. Geophys. Res.* **80**, 2431

Howard, A. D., Blasius, K. R., Cutts, J. A.: 1982, *Icarus* **50**, 245

Jantunen, H., Raitala, J.: 1983, *Moon Planets* **29**, 7

Keller, H. U., Arpigny, C., Barbieri, C., Bonnet, R. M., Cazes, S., Coradini, M., Cosmovici, C. B., Delamere, W. A., Huebner, W. F., Hughes, D. W., Jamar, C., Malaise, D., Reitsema, H. J., Schmidt, H. U., Schmidt, W. K. H., Seige, P., Whipple, F. L., Wilhelm, K.: 1986, *Nature* **321**, 320

Mouginis-Mack, P. J., Wilson, L.: 1981, *Comp. Geosci.* **7**, 35

Reitsema, H. J., Delamere, W. A., Huebner, W. F., Keller, H. U., Schmidt, H. U., Schmidt, W. K. H., Whipple, F. L., Wilhelm, K.: 1986, 20th ESLAB Symp., ESA SP-250, Vol. **II**, p. 351

Rindfleisch, T.: 1966, Photogram. Eng. **32**, 262

Squyres, S. W., Veverka, J.: 1981, *Icarus* **46**, 137

Squyres, S. W.: 1981, *Icarus* **46**, 156

Wilhelm, K., Cosmovici, C. B., Delamere, W. A., Huebner, W. F., Keller, H. U., Reitsema, H., Schmidt, H. U., Whipple, F. L.: 1986, 20th ESLAB Symp. ESA SP-250, Vol. **II**, p. 367

Astron. Astrophys. 187, 852–858 (1987)

ASTRONOMY AND ASTROPHYSICS

The cometary nucleus: current concepts

F. L. Whipple

Smithsonian Astrophysical Observatory, 60 Garden Street, MS 7, Cambridge, MA 02138, USA

Received February 11, accepted July 20, 1987

Summary. This paper deals with current ideas concerning the nature of cometary nuclei as modified by observations of recent years, particularly by those of Halley's comet during the 1986 apparition. The major trend has been to increase the postulated dimensions of the nuclei and reduce their albedos. The Vega and Giotto space missions have established invaluable check points with regard to comet nuclei. Striking is the low geometric albedo of 0.044 deduced from the directly observed dimensions of the nucleus and the brightness at large solar distances. The dimensions, the observed nongravitational motions and the jet activity combine to suggest an extremely low density for the nucleus of Halley's comet – in the range of 0.1 to 0.5 $\mathrm{g\,cm^{-3}}$.

The gross composition of the nucleus appears to conform roughly to abundances of the elements and isotope ratios expected if comets were frozen out of a Solar-System mix, supporting theories of cometary origin concurrent with the Sun, planets and asteroids. The extremely low albedo and the prevalent "CHON" composition of the dust, the newly observed dust particles of mass less than 10^{-17} g, and evidence of formation at low temperatures support accumulation processes for comets from material like interstellar grains imbedded in a proto-planetary nebula.

The life expectances of comets and their calculated contributions to the interplanetary complex are little changed by the new results. Two "rotation" periods for Halley's comet require further investigation.

Key words: comet nuclei – albedo – density – nature – origin

1. Introduction

The 1986 apparition of Halley's comet has sparked a unique resurgence of interest in comets. This has inspired massive observational programs. Most ground based telescopes now boast superb CCD imaging systems and infrared sensors while the groundbased observers, both professional and amateur, have been encouraged by the International Halley Watch, which promises to provide a central storehouse of cometary observations. Radio techniques add another facet to the observational program. In addition, observations from air and space internationally have been encouraged by the Interagency Consultative Group. Five space-craft converged on the comet during the interval March 6–14 1986, to provide close-up observations of many kinds never before possible in all history. These observational programs have then inspired theoretical activity concerning comets: their nature, origin and significance. At the present time, on the crest of the wave of analysis of the diverse observations, only a progress report is possible, including, nevertheless, a few definitive, if qualitative, results and a few suggestions of results to come. In the spirit of reporting rather than analyzing the manifold of P/Halley activities, this paper will attempt to discuss some of the recent results and how they are affecting concepts of the cometary nucleus.

2. Nucleus: dimensions and albedo

Direct observations of the nucleus of P/Halley by the two Vega and the Giotto spacecraft have immediately established new and significant facts about the nature of comets. The orientation of cometary activity generally towards the Sun has completely supported Bessel's "fountain" theory deduced from his 1835 observations.

The brilliance of the nucleus has put to rest any doubts as to whether the apparent brightest point of light and center of gravity of a periodic comet differ in astrometric terms sufficiently to affect the calculated values of non-gravitational forces. The problem is thoroughly discussed by Devine et al. (1985). Granted that the finite size of an "ideal" cometary image observed in large telescope generally amounts to several hundred to a few thousand kilometers, the center measured from short exposures or with short integration times will nearly coincide with the nucleus itself, and the deviation should be calculable if needed, almost entirely caused by the dust jets. This conclusion is probably secure for all, or almost all, periodic comets because P/Halley is one of the most active and in 1910, according to Donn (1977) and recently by many observers, has shown at least a "medium" if not strong dust component compared to other comets. In the ultraviolet Festou et al. (1986) find that comet Halley differs from others observed by the IUE satellite in that the continuum emission is very strong relative to the gaseous emissions. Quite possibly some of the "new" comets, recently acquired from the Öpik-Oort Cloud, might show more activity, involving a coma that outshines the nuclear concentration of light.

In any case, the clear-cut outline of the nucleus of P/Halley observed directly near the time of its maximum acitvity by Giotto and Vega space-craft forces a revision in certain theories of comet activity in which the sunlight scattered by the dust provides significant heat and therefore sublimation on the "dark" portion of the nucleus. This subject will be discussed in a later section.

The direct observations of the nucleus provide for the first time a knowledge of the dimensions of a comet nucleus. The values are roughly an apparent crossectional area of $\sim 80\,\mathrm{km^2}$ (Sagdeev et al., 1986; Keller et al., 1986) an elongated irregular nucleus with maximum length about 16 km and minimum width some 7.5–8.0 km. An oppositon magnitude at great solar distances when the comet was inactive then provides a measure of the geometric albedo (Allen, 1973). A number of such measures were made before P/Halley clearly became active at about 6 AU from the Sun in the fall of 1984. These magnitude determinations over the period from October 1982 through February 1984, however, show variations in the opposition absolute magnitude, $V(1,0)$ at 1 AU from Earth and Sun (see e.g. Sekanina, 1985; Delamere et al., this publication, for discussions). The mean value of $V(1,0)$ may range over more than about 0.5 mag. depending on whether the apparent variations are: a) a combination of observational errors and variations caused by rotation, b) intrinsic variations produced by dust production, or c) a combination of all of these. If (c) is adopted, a reasonable value of $V(1,0) = 14^{\mathrm{m}}.0$ leads to a geometric albedo of $p_v = 0.044$ if the apparent surface area is $80\,\mathrm{km}^2$. Considering that for the Moon $p_v = 0.112$ and the Bond albedo $= 0.067$, the total reflectivity of the nucleus of Halley's comet is only 0.027 if its reflection variation with phase angle is similar to that of the Moon.

New values of the radius for other comets, R(km), with $p_v = 0.044$ from the visual opposition extreme magnitude $V(1,0)$, can be estimated to be the order of

$$\log R\,(\mathrm{km}) = 3.50 - 0.2\,V(1,0). \tag{1}$$

Equation (1) will increase the radius from, for example, Kresák's 1973 estimate where $p_v = 0.3$, by a factor of about 2.6. This decrease in albedo indicated by the spacecraft direct observation of the nucleus of Halley's comet, although an extremely valuable result, is not surprising in view of the improvement of infrared observing during recent years. In their infrared observations of comets and asteroid, Hartmann, Cruikshank and Tholen (1985) have compiled albedo measures of averages (17 comets in all) leading to a mean value (reduced to the visual system) of $p_v = 0.051 \pm 0.010$ (see also Hanner et al., 1985; Tokunage and Hanner, 1985) The extreme values are 0.01 and 0.13, with $\sigma = \pm 0.037$ for one determination. Dusty comas and measuring errors can account for much of the scatter. Their result is in complete accord with that from direct measures of Halley's comet and shows, as they empasize, that cometary soils are very similar in their radiation scattering properties to the dark soils on the surfaces of many small bodies in the outer solar system, most closely resembling those of the D-class Trojan asteroids.

3. Mass loss and density

The "official" expectation of the total icy mass to be lost by P/Halley in its entire appariton centered on 1986 was probably that by Devine et al (1985) involving eleven eminent comet scientists (see also Hughes, 1985). The expected total ice-gas loss amounted to about $2.2\ 10^{11}$ kg or $7\ 10^{10}$ kg on the inward leg and $1.5\ 10^{11}$ kg after perihelion, of which the ratio of the total number molecules compared to that of H_2O was 1.29 and the mean molecular weight of all was 22.3.

The actual icy mass loss by the comet cannot yet be determined in a definitive manner. The H_2O loss measured by the satellite *International Ultraviolet Explorer* (IUE) and reported by Feldman (1986) was used by Rickman (1986) and here by the author to extrapolate a total loss of some $2.4\ 10^{11}$ kg of H_2O molecules alone. If we increase this value by a factor of 1.20 (22.3/18) = 1.49, in accordance with Devine et al. and Krankowsky et al. (1986), the total icy loss becomes about $3.5\ 10^{11}$ kg, $1.5\ 10^{11}$ kg inbound and $2.0\ 10^{11}$ outbound.

The dust/gas ratio is required to complete our estimate of the total mass loss by P/Halley during this appariton. On the Vega spacecraft the dust counter and mass analyzer (DUCMA) of Simpson et al. (1986) and SP-2 detector of Mazets et al. (1986) and on Giotto, the dust impact detection system (DIDSY) of McDonnell et al. (1986) and the dust particle analyzers of Kissel et al. (1986a, b) on the (PUMA) and Giotto (PIA) all registered a surprisingly large number of extremely small particles, $\ll 10^{-17}$ kg in mass. The cumulative mass index (for the distribution function m-x), however, gave $\alpha < 1.0$ such as to make the integral in mass depend upon the upper limit in mass, a quantity not really determinable. McDonnell et al. (1986) on the basis of some assumptions, suggest a dust to gas production ratio of $\sim 1/5$ not inconsistent with many other estimates. Hanner (1986), from routine infrared monitoring of comet Halley at several observatories, places the dust production rate at $\sim 10^4\,\mathrm{kg\,s^{-1}}$ at 0.9 AU where the H_2O production rate was 5 to 10 $10^{29}\,\mathrm{s^{-1}}$. Thus with the mass factor of 1.49 used above the dust/gas production ratio would be 1/4.5 to 1/2.2. Let us adopt 0.3 for this ratio, leading to the rough estimate of a total mass loss for P/Halley of $\sim 5\ 10^{11}$ kg during the current apparition.

All the mass losses above are subject to considerable correction as more data become available, but they indicate that P/Halley from the 1986 apparition data has exceeded its total expected mass loss by no more than two times, perhaps by no more than a factor of 1.6. Generally, then earlier estimates of cometary mass contributions to the interplanetary complex, related to the Zodiacal Light and the Gegenschein are little affected by the increases dimensions now attributed to cometary nuclei. Only in the realm of extremely fine particles of masses below $\sim 10^{-17}$ kg is the cometary contribution significantly larger than anticipated before the Halley missions.

On the other hand, the modest (if real) increase in estimated mass loss coupled with the order-of-magnitude increase in the volume of the nucleus as compared to pre-mission estimates creates a proven and striking concept regarding cometary density. The only observational basis for estimating cometary masses and hence density has been from non-gravitational motions (Whipple, 1950). These observed deviations for several comets have generally been reasonably well explained by the jet action of sublimating ices within the framework of the measured mass losses and nuclear masses based on dimensions calculated with relatively high albedos and a density $\sim 1\,\mathrm{g\,cm^{-3}}$. An order-of-magnitude increase in volume, resulting from the lower albedos now clearly applicable to cometary nuclei, forces a reduction in the estimated density. In other words, accounting for the nongravitational forces determines the ratio of gas loss to total mass while the gas loss is measured observationally. Any increase in the assumed volume forces a corresponding decrease in the calculated density. Following Belton's (1985) pre-perihelion calculations, Rickman (1986) has analyzed this situation thoroughly for P/Halley. From the Halley's missions he adopts a nucleus volume of $5\ 10^{11}\,\mathrm{m}^3$ and the mass losses of H_2O plus 15% CO (by number) mentioned earlier. He then carries out a wide spectrum of calculations using the non-gravitational accelerations compiled by Yeomans (1977 and 1984) and Kiang (1972), $A_1 = 0.28\ 10^{-8}$ AU/d^2, $A_2 = 0.0159\ 10^{-8}$ AU/d^2, and the delay in perihelion passage of $4^{\mathrm{d}}.1$ per revolution. Depending upon his assumption concerning lag angle and gas production asymmetry before and after perihelion he finds a mass for the nucleus in the range 5 to 13 10^{13} kg or a mean density in the range 0.08 to $0.24\,\mathrm{g\,cm^{-3}}$.

Rickmann prefers the range in density of 0.1 to $0.2\,g\,cm^{-3}$, perhaps a mean of $0.15\,g\,cm^{-3}$. By assuming a large asymmetry of sublimation around perihelion, as comes from the Devine et al. (1985) model, he shows that the density might be calculated to reach a maximum value of $0.3\,g\,cm^{-3}$. From my own rough calculations I agree with this maximum value. Because periodic comets statistically are brighter post perihelion compared to pre-perihelion (Whipple, 1978) it is tempting to suspect that a large asymmetry in sublimation actually produced the increase in period for P/Halley and should bias the A_1 and A_2 accelerations towards positive values for other periodic comets. Marsden's (1986) recent compilation of non-gravitation accelerations for 45 periodic comets favors positive values of A_1 by a factor of about 6 times. The problem needs more study as negative A_2 values, contrary to those for P/Halley, are favored by a ratio of 24 to 14 in Marsden's catalogue.

It is interesting that the first investigators to derive low densities for cometary nuclei on the basis of non-gravitational accelerations were Wallis and Macpherson (1981). They based their argument largely on their conclusion that earlier investigators had over estimated the jet force by more than a factor of two by assuming approximately radial outgassing on the sunward side of the nucleus. They postulated "pressure-driven flow to the shaded side, where condensation partially compensates for radiative heat losses". The Vega and the Giotto images of the Halley comet nucleus rather clearly indicate that the jets occur almost exclusively on the sunward side of the nucleus. Thus various theories that derive extremely high opacities near the nucleus to produce both radiative and gas warming on the night side do not apply to P/Halley and perhaps not to any of the periodic comets, which are generally less active than P/Halley.

The concept that comets are truly like terrestrial snow in density is highly attractive in explaining 1) the apparent disappearance of several comets while under observation (Sekanina, 1984), 2) a number that simply could not be found on expected returns, 3) twenty odd that have split into two or more components, 4) the occurrence of short-lived sometimes star-like condensations near the nuclei of bright comets such as P/Halley and 5) many cometary bursts of a magnitude or more in brightness. It is extremely difficult to visualize the physics of the separation of rocky or glacial type formations from small comet nuclei by sublimation processes or even by activity of warmed amorphous ices (note Smoluchowski, 1984; Klinger, 1984; or Greenberg, 1982). Low-density snowy material of variable physical and chemical composition could be forced away much more easily than rocky material, permafrost or dirty ice. In addition, increasing the radius of a comet's spherical nucleus by a factor, f, reduces the gravitational velocity of escape by a factor of $f^{1/2}$ whereas the centrifugal force at the equator for a given period of rotation is reduced by a factor of f. For nuclei that are irregularly shaped, either by accumulation processes or sublimation of heterogeneous material, these factors favor fracture and escape of protuberances or appendages.

A low mean density for comets strongly favors a very low heat conductivity throughout the cometary mass, compared to that for a glacial or permafrost type of material. Thus a very low primitive temperature might be maintained near the surface.

Correspondingly, the high temperature measured on part of the nucleus by the Vega 1, infrared group of Combes et al. (1986) led them to suggest that "A surface layer of a porous dark substance with a low thermal conductivity, covering the ice, could explain the high measured temperatures." The low mean density of the nucleus would, of course, suggest that an outer layer of dusty debris of cometary crust that had lost its icy component might well exhibit voids exceeding 30 to 90% of the volume. It would, therefore, exhibit an excessively low thermal conductivity. A formula for the mass of a comet as a function of its absolute magnitude, $V(1,0)$, is almost certainly far less likely to predict successfully than one for the radius. We still have no reliable benchmark value of mass or density and it is possible, if not probable, that comet densities may vary widely from one to another and with depth from the surface for large comets. Calculated values of comet lifetimes are not much affected because the ratio of mass loss per period to the total mass is little changed, the only measure of this ratio being derived from non-gravitational accelerations. The adopted mass loss of $5\ 10^{11}$ kg per revolution of P/Halley over an area of some $320\,km^2$ corresponds to a surface loss of $1.6/\varrho$m, where ϱ is the density in $g\,cm^{-3}$. This suggests a life expectancy of many fewer than 1000 revolutions for P/Halley in its present orbit.

The small mean density for Halley's comet supports the supposition that no large-scale radioactive heating of the nucleus occurred. Any such melting would have increased the density of the inner regions to the values of 1.2 to $1.3\,g\,cm^{-3}$ expected for a re-frozen cometary mix. It is difficult to believe that huge voids could exist in cometary nuclei even though the cohesive gravitational forces are small.

4. Rotation and activity of the nucleus

Nuclear activity appears to be confined largely to broad-based jets on the nucleus of Halley's comet as illustrated by the direct close-up images obtained by the two Vegas and the Giotto spacecraft (Sagdeev et al., 1986; Keller et al., 1986). A simple calculation shows that an area of some $25\,km^2$ of black H_2O ice perpendicular to solar radiation at $r = 0.9$ AU would provide the H_2O sublimation rate of $5\ 10^{29}\,mol\,s^{-1}$ or 15 tons s^{-1}, observed near the time of the Giotto close approach. This calculated area is roughly 1/4 to 1/3 the area of the nucleus exposed normally to the Sun's radiation. If then, H_2O ice with its high latent heat of vaporization is the only sublimating ice, perhaps 0.3 to 0.5 of the surface is active, allowing for various efficiencies on the process. More likely, some 20% or so of the icy mixture is more volatile than H_2O, reducing the active areas to the observed 15 per cent or so of the irradiated surface (Keller et al., 1986). Thus most of the area is covered with an insulating or partially insulating blanket, which in sunlight can provide the high surface temperature observed in the infrared.

Sekanina and Larson (1984) from analysis of near-nucleus images of P/Halley in 1910 came to a similar conclusion concerning dust sources, coupled with the fact that on a subsequent revolution an active area had usually become inactive. They derived a rotation period of 52.1 h, confirmed by "breathing" of the hydrogen Lyman-α emission measured by the Suisei spacecraft (Kaneda et al., 1986). The rotation was prograde with respect to the orbital motion as expected from the non-gravitational increase in period. Observations by Gehrz and Ney (1987), Leibowitz and Brosch (1987) also favor a rotation period of ~ 53.5 h. On the other hand ground-based photometric observations by Millis et al. (1987) and by Schleicher et al. (1987) fail to support the 53.5-hour period but yield one of 7.4 days. The resolution of this discrepancy will require more observations, analyses and discussions, beyond the scope of this paper. Sekanina, for example, has suggested a solution in which both periods are involved in a tumbling motion via the well-known Euler equations for a rigid rotator. The pole has not yet been accurately located in the 1986 apparition according to the writer's sources, except that the obliquity appears not to be huge (B. Smith, private communication).

Sekanina and Larson concluded that one hemisphere is relatively dusty and active "*either* made up of a complex maze of mostly linear sources...; *or* it represents an essentially single, vast reservoir of dust which is highly structured horizontally and probably also vertically". The Giotto images (Keller et al., 1986) definitely disprove the first possibility because the sun-lit areas appear much like the terminator on the Moon, suggesting a createred of lacelike surface structure. Just how such a structure came into being is a fascinating and important problem of cometary physics, chemistry and history for future study.

At great solar distances (>7 AU Flammer et al., 1986) have found support for the idea of Mendis et al. (1981) that electrostatic effects of the solar wind can raise dust on the night side of a cometary nucleus. The former identified high-speed solar streams on the Sun's H-α synoptic charts that coincided with observed intervals of increased brightness of P/Halley. On the other hand a number of investigators including Smoluchowski (1984), Klinger (1984), Greenberg (1982), Rickman et al. (1984) and others have accredited comet activity at great solar distances (e.g. P. Schwassman-Wachmann 1, up to 7 AU) to the transition of amorphous H_2O ice to the ordinary crystalline phase. Among other possible candidates are, of course, CO_2, NH_3, CO, and CH_4 although the latter two have such high vapor pressures that it is difficult to see how they could have been frozen in comet nuclei.

In terms of general theory the current P/Halley data have not yet formed the basis for new concepts. The lack of a high dust opacity and scattering around the nucleus and the low mean density of the nucleus are consistent in reducing the estimates of heat conductivity within the nucleus and also heat transport by scattering to the night side of the nucleus. The core-mantle concept of layers produced by temperature-sublimation near the surface of the nucleus as envisaged by Mendis and Brin (1977, 1978) with modification by Weissman and Kieffer (1981) and others (see e.g. Horanyi et al., 1982) seems flexible enough to meet the new demand of the P/Halley observations. The result will be chiefly to tighten the constraints on numerical values. The variable activity on individual areas of the nucleus from rotation to rotation, however, is now well established. This demand may lead to important new concepts of nucleus activity (see e.g. Brin and Mendis, 1979; Smoluchowski, 1982; Fanale and Salvail, 1984; Froeschle and Rickman, 1984; Bar-Nun et al., 1985).

Kurth has investigated thermal stresses on ices in and near the surfaces of cometary nuclei induced by rotation about their axes and revolution about the Sun. He shows "that thermal stresses can give a plausible explanation for cometary outbursts and splits". However, the extent to which low-density highly porous conglomerate ices are cracked by such stresses probably demands laboratory study. Personally, still without a satisfactory theory for the storage of energy near the surface, I (Whipple, 1980) have favored the process of particle fallback from a small active source as a means of propagating activity over large regions that, for some reason, have become prepared for active sublimation.

5. Nucleus structure and origin

The low albedo, low density, variable jet activity from relatively small areas on the surface and the chemistry of the gas and dust observed for the nucleus of P/Halley add strong restraints to theories of cometary structure and origin. The surprising quantity of extremely fine dust and the fact that "most of the particles are rich in light elements such as H, C, N, and O" (Kissel et al., 1986a, b), strongly support their origin in cold interstellar type clouds, whether in the outskirts of the primitive Solar System or elsewhere. The general nature of the particles is consistent with Greenberg's (1982) theory that they are indeed made by interstellar dust and that "$\sim$20% of comets consists of complex nonvolatile organic molecules of prebiotic type".

Two pieces of data, although not definitive, strongly support the idea that the main body of a comet has remained extremely cold since its formation. The discovery of the unlikely S_2 molecule by A'Hearn et al. (1983) in the spectrum of comet IRAS-Araki-Alcock led A'Hearn and Feldman (1985) to argue "that the presence of S_2 implies that the ice in most comets arrived there in solid form directly from the interstellar medium". The extremely low mean density suggests, as mentioned above, that no widespread internal heating has occurred.

Some striking inhomogeneities in the composition of dust (if not the ices?) is indicated by the observations of CN emission but not in the continuum from jets on several days in April and May of 1986 by Larson (1986) and A'Hearn et al. (1986). Whether or not these variations and other such variations probably to be observed in the spectrum of Halley's comet really represent basic differences in composition with position on the comet may be in doubt. Because more volatile substances may be depleted near the surface, producing the layered effect discussed by Mendis and Brin (1977, 1978), certain areas sublimating from the surface or from near the surface may vary in volatility and therefore composition. On the other hand, comets vary greatly one to the other in their spectra at similar solar distances suggesting that the place of origin or the physical conditions (such as temperature and pressure) at the time of the origin are widely different for individual comets and possibly for different parts of a given comet.

These arguments bear heavily on the question of whether the comets originated in the outer planetary regions of the early solar system or in the infalling interstellar cloud that formed the "Laplace" nebula or, ideed, elsewhere. One might expect greater heterogeneity among and within comets that accumulated in regions where large planets were forming than among or within comets formed directly from interstellar material away from disturbing proto-planetary masses.

An issue here that gave me a personal bias towards higher mean densities in cometary nuclei ($\sim 1\,\mathrm{g\,mc^{-3}}$) concerns the effects of encounters as comets accumulate from dust to dimensions of kilometers. It seemed to me that the encounters would produce compression that would compact the "snow" to higher and higher densities increasing with comet dimensions. Now that the evidence is clear that the mean densities are well below that of solid ice, one asks major questions (really only faced by Donn, 1963). The first question concerns the velocity distribution among proto-comets, if, indeed, they formed, say in the region of proto-Uranus and proto-Neptune. How could they be thrown into the Öpik-Ort cloud without colliding at velocities nearing the limit for H_2O ice, 2.3 km s^{-1} (see e.g. Safronov, 1972)? Note that those with excessive collisional velocities would, of course, volatilize and disappear as entities.

The second question concerns the effect of low-velocity collisions on the mean densities of dirty snowballs at temperatures of the order of 30 K. The latter question is probably a laboratory subject and the former a theoretical one. Possibly neither of these questions is relevant to comets forming in the interstellar medium well away from planetary masses, but both questions should be addressed seriously. Probably our experience with snow near its melting point has prejudiced our thinking with regard to increased density resulting from collisional impact. At extremely low temperatures the low-density snow probably shatters on impact with little if any increase in density. Laboratory experiments would help guide our thinking regarding this phenomenon.

The observations by Krankowsky et al. (1986) with the neutral mass spectrometer on Giotto of the ratio of $^{34}S^+$ to $^{32}S^+$ as 0.045 ± 0.010, compared to the terrestrial ratio of 0.044 suggests an origin from a source of material in common with that of the Solar System. The conclusion is supported by ground-based spectral measures of the isotope ratios $^{12}C/^{13}C$ and $^{14}N/^{15}N$ by Wehinger et al. (1986) who find values in reasonable accordance with those of the solar system. The gross abundances of the elements in comets also appear to be similar to Solar System abundances with an apparent depletion of C in the ices (Delsemme, 1982). When analyses of the dust are added to those of the ice, the P/Halley data suggest that no such deficiency exists although definitive calculations are yet lacking (see e.g.: Kissel et al., 1986a, b; Festou et al., 1986; Krankowsky et al., 1986; Balsiger et al., 1986). Even though no age determinations are yet possible from cometary material, such isotopic data may also well establish or strongly support a common origin in material and, therefore in time, with the Solar System. If so, answers to the two questions above may differentiate between comets as the left-over building blocks of the outer planets or else comets as independently formed entities resulting from the collapse of the "Laplace" interstellar cloud. At the moment, the huge heterogeneity among comets in icy composition and in dust size distribution suggest that they may well have accumulated in various parts of the outer planetary systems where conditions varied greatly with solar distances.

Several models of cometary nuclei have been pictured by various inverstigators. Donn (1981) and Donn and Rahe (1982) picture a low-density porous nucleus formed from interstellar type grains. Their papers and references should be read carefully by the serious student. Of interest is the "fractal model" of a comet by Donn and Hughes (1986), involving the mathematics of random accretion. An "icy-glue" model has been submitted by Gombosi and Houpis (1986) in which the porous boulders are "glued" together by an accumulated matrix of ice-dust grains. The percolation theory of cometary nuclei by Horanyi et al. and Kecskemety (1982) applies a mathematical packing formula for dust grains of various sizes, leading to a mean volume density or concentration of grains equal to 0.31. Weissman's (1986) comet nucleus as a "primordial rubble pile" probable summarizes best and most simply the ideas of many cometary students as to the nature of typical comets.

With regard to the future of comets, except for dissolution, ejection from the Solar System, or collision with Solar System bodies, there remains the question as to whether some may attain asteroidal-type orbits to become inactive, indistinguisable members of the asteroid population. So far the P/Halley observations have helped answer this question by proving the low albedo of the nucleus and by demonstrating its very low mean density. Both characteristics tend to suggest that the very small near-Earth asteroids of Aten-Apollo-Amor group are probably not old comets (see e.g. Whipple, 1986; Hartmann et al., 1985, for a discussion of their colors). Hartmann et al. (1985) do find, however, that a moderate fraction of the asteroids now moving in orbits that suggest a cometary origin do fall in color classes that are cometary. Most of these bodies, however, are very much larger than the nucleus of Halley's comet so that pressure and radioactive heat may well have increased their densities to icy values like some of the icy satellites of the giant planets. Several could well be remnants of early Solar System formation.

6. Conclusion

A discrete nucleus is clearly evident in the Giotto and Vega images of P/Halley. The center of light is nearly, if not exactly on the nucleus, effectively at the center of gravity, so that this long standing question is solved except for observational problems. The nucleus is larger than expected, with dimensions the order of $16 \times 8 \times 7.5$ km. The corresponding geometric albedo is ~ 0.05 and Bond albedo ~ 0.03 if the distanct oppositon magnitude at 1 AU form Sun and Earth is taken to be 14.0. The effect of this measurement is roughly to double the linear dimensions of the nuclei of periodic comets compared with earlier estimates.

The total rate of gas and dust loss by P/Halley is only a small factor larger than predicted. As a result, the jet forces to produce non-gravitational motion of the comet require no significant change in the calculated mass of the nucleus. The calculated mean density, therefore, decreases to a surprisingly small value, in the range 0.1 to 0.24 g cm^{-3} according to calculations by Rickman with upper limits ~ 0.4 to 0.5 g cm^{-3} by myself and Morose.

The expected life expectancies of comets are little altered by the new observations nor are the expected contributions of comets to the Zodiacal Cloud.

Several lines of evidence suggest a 53.5-h rotation period for the nucleus of P/Halley although a few observations suggest a 7.5-h period. The pole is not yet well located but appears to be in the general neighbourhood of the pole of the orbit plane (rotation and revolution in the same sense.

The activity of the comet is a) almost all if not entirely from the sunlit side of the nucleus, b) highly variable in time and position on the nucleus, c) mostly from one "hemisphere" of the nucleus, d) evidenced by dust jets from a small fraction ($\sim 15\%$) of the surface of the nucleus, and e) highly variable in the composition of sources (e.g. C_2, CH and dust). The inactive areas in sunlight are very hot, above black-body temperatures. These facts support a "mantle" theory or layered theory of the near surface of the nucleus, a low dust opacity around the nucleus and probably a very low heat conductivity for cometary solids.

The dust contains a sizeable but hitherto unobservable component of extremely fine size, $< 10^{-17}$ kg in mass. The dust contains a major fraction of the so-called CHON basic elements. The overall composition shows no lack of C as earlier suspected, relative to comparative solar values. Also the iosopic ratios, $^{13}C/^{12}C$, $^{15}N/^{14}N$, and the newly observed $^{34}S/^{32}S$, although not precisely measured, appear to be typical of Solar-System values. The CO molecule appears to be an order of magnitude more abundant than CO_2, while the abundances of NH_2 and CH_4 can only be given as upper limits H_2O being the major ice ($\sim 80\%$ of ices). The low density of the nucleus suggests that no serious radioactive heating has occurred, particularly not by ^{26}Al.

This all supports the formation of comets from and on interstellar grains as proposed by M. Greenberg. The formation seems to have taken place in the Solar Nebula. The lack of homogeneity in the nucleus suggests formation in the neighborhood of the planets as does the battered appearance of the nucleus. The low density favors formation in gravitationally less active regions.

Clearly we need cometary rendezvous missions and sample-return missions to gain a deeper understanding of cometary structure, composition activity and origin.

For quite different viewpoints concerning the nuclei of comets the reader is referred to discussions by Wood (1986) and Sekanina (1986).

Acknowledgements. I am indebted to Zdenek Sekanina and Bradford Smith for discussions and to a number of contributors who sent me preprints of their current publications in press. This research has been supported by a grant from the Planetary Division of the National Aeronautics and Space Administration.

References

A'Hearn, M.F., Hoban, S., Birch, P., Bowers, C., Martin, R.: 1986, see Sekanina, 1986

A'Hearn, M.F., Feldman, P.D.: 1985, in *Ices in the Solar System*, eds. J. Klinger, D. Benest, A. Dollfus, R. Smoluchowski, Reidel, Dordrecht, p. 463

A'Hearn, M.F., Schleicher, D.G., Feldman, P.D.: 1983, *Astrophys. J.* **274**, L99–L103

Allen, C.W.: 1973, *Astrophysical Quantities*, The Athlone Press, London, pp. 142 and 162

Balsiger, H., Altwegg, Bühler, F., Geiss, J., Ghielmetti, A.G., Goldstein, B.E., Goldstein, R., Huntress, W.T., Ip, W-H., Lazarus, A.J., Meier, A., Neugebauer, M., Rettenmund, U., Rosenbauer, H., Schwenn, R., Sharp, R.D., Shelley, E.G., Ungstrup, E., Young, D.T.: 1986, *Nature* **321**, 330

Bar-Nun, A., Herman, G., Laufer, D., Rappaport, M.L.: 1985, *Icarus* **63**, 317

Belton, M.J.S.: 1985, *Science* **230**, 1229

Belton, M.J.S., Wehinger, P., Spinrad, H., Wyckoff, S.: 1987, 20th ESLAB Symp., Vol. I, p. 599

Brin, G.D., Mendis, D.A.: 1979, *Astrophys. J.* **229**, 402

Combes, M., Moroz, V.I., Crifo, J.F., Lamarre, J.M., Charra, J., Sanko, N.F., Soufflot, A., Bibring, J.P., Cazes, S., Coron, N., Crovisier, J., Encrenaz, T., Gispert, R., Grigoryev, A.V., Guyot, G., Krasnopolsky, V.A., Kikolsky, Yu V., Rocard, F.: 1986, *Nature* **321**, 266

Delamere, W.A., Huebner, W.F., Keller, H.U., Reitsema, H.J., Schmidt, H.U., Schmidt, W.K.H., Whipple, F.L., Wilhelm, K.: 1987, 20th ESLAB Symp., Vol. II, p. 355

Delsemme, A.H.: 1982, *Comets*, ed. L.L. Wilkening, Univ. Arizona Press, Tucson, p. 85

Devine, N., Fechting, H., Gombosi, H., Hanner, T.I., Keller, M.S., Keller, H.U., Larson, S.M., Mendis, D.A., Newburn, Jr., R.L., Reinhard, R., Sekanina, Z., Yeomans, D.K.: 1985, *Space Sci. Rev.* **43**, 1

Donn, B.: 1963, *Icarus*, 2, 396

Donn, B.: 1977, in *Comets Asteroids Meteorites*, ed. A.H. Delsemme, Univ. of Toledo, p. 15

Donn, B.: 1981, in *Comets and the Origin of Life*, ed. C. Ponnamperuma, Reidel, Dordrecht, p. 21

Donn, B., Hughes, D.: 1986, 20th ESLAB Symp., Vol. III, p. 523

Donn, B., Rahe, J.: 1982, in *Comets*, ed. L.L. Wilkening, Univ. Arizona Press, Tucson, p. 203

Fanale, F.P., Salvail, J.R.: 1984, *Icarus*, 60, 476

Feldman, P.D.: 1986, *Pub. Am. Phil. Soc.* (in press)

Festou, M.C., Feldman, P.D., A'Hearn, M.F., Arpigny, C., Cosmovici, C.B., Danks, A.C., McFadden, L.A., Gilmozzi, R., Patriarchi, P., Tozzi, G.P., Wallis, M.K., Weaver, H.A.: 1986, *Nature* **321**, 361

Flammer, K.R., Jackson, B., Mendis, D.A.: 1986, *Earth, Moon, Planets* **35**, 203

Froeschlé, C., Rickman, H.: 1984, *Asteroids, Comets, Meteors*, eds. E.I. Lagerkvist, H. Rickmann, Uppsala Univ.

Gehrz, R.D., Ney, E.P.: 1987, 20th ESLAB Symp., Vol. III, p. 101

Gombosi, T.I., Houpis, H.L.F.: 1986, *Nature* **324**, 43

Greenberg, J.M.: 1982, in *Comets*, ed. L.L. Wilkening, Univ. Arizona, Tucson, p. 131

Hanner, MS.: 1986, "A preliminary look at the dust in Comet Halley" in Symposium No. 12, XXVI Plenary Meeting *Cospar*, Toulouse, France (in press)

Hanner, M.S., Aitken, D.K., Knacke, R., McCorkle, S., Roche, P.F., Tokunaga, A.T.: 1985, *Icarus* **62**, 97

Hartmann, W.K., Cruishank, D.P., Tholen, D.J.: 1985, in *Ices in the Solar System*, eds. J. Klinger et al., Reidel, Dordrecht, p. 169

Horanyi, M., Gombosi, T.I., Cravens, T.E., Kecsremity, K., Nagy, A.F., Szegö, K.: 1982, in *Cometary Exploration*, Vol. 1, ed. T.I. Gombosi, Hungarian Acad. Sci., p. 59

Hughes, D.W.: 1985, *Monthly Notices Roy. Astron. Soc.* **213**, 103

Kaneda, E., Asihara, O., Shimizu, M., Takagi, M., Hirio, K.: 1986, *Nature* **321**, 297

Keller, H.U., Arpigny, C., Barbieri, C., Bonnet, R.M., Cazes, S., Coradini, M., Cosmovici, C.B., Delamere, W.A., Huebner, W.F., Hughes, D.W., Jamar, C., Malaise, D., Reitsema, H.J., Schmidt, H.U., Schmidt, W.K.H., Seige, P., Whipple, F.L., Wilhelm, K.: 1986, *Nature* **321**, 320

Kiang, T.: 1972, *Monthly Notices Roy. Astron. Soc.* **76**, 27

Kissel, J., Sagdeev, E.Z., Bertaux, J.L., Angarov, V.N., Audouze, J.B., Blamont, J.E., Büchler, K., Evlanov, E.N., Fechting, H., Fomenkova, M.N., von Hoerner, H., Inogamov, N.A., Khromov, V.N., Knabe, W., Frueger, F.R., Langevin, Y., Leonas, V.B., Levasseur-Regourd, A.C., Managadze, G.G., Podkozin, S.N., Shapiro, V.D., Tabaldyev, S.R., Zubkov, B.V.: 1986, *Nature* **321**, 280

Kissel, J., Brownlee, D.E., Büchler, K., Clark, B.C., Fechting, H., Hornung, K., Igenbergs, E.B., Jessberger, E.K., Krueger, F.R., Kuczera, H., McDonnell, J.A.M., Morfill, G.M., Rahe, J., Schwehm, G.H., Sekanina, Z., Utterback, N.G., Volk, H.J., Zook, H.A.: 1986, *Nature* **321**, 336

Klinger, J.: 1984, in *Ices in the Solar System*, eds. J. Klinger, B. Benest, Al Dolfus, R. Smoluchowski, Reidel, Dordrecht, 407

Krankowsky, D.: 1986, *Nature* **321**, 326

Kresák, L.: 1973, *Bull. Astron. Inst. Czech.* **24**, 264

Kürth, E.: 1984, *Icarus* **60**, 512

Larson, S.M.: 1986, see Sekanina, 1986

Leibowitz, E.M., Brosch, N.: 1987, 20th ESLAB Symp., Vol. I, 605

McDonnell, J.A.M., Alexander, W.M., Burton, W.M., Bussoletti, W.M., Clark, D.H., Grard, R.J.L., Grün, E., Hanner, M.S., Hughes, D.W., Igenbergs, E., Kuczera, H., Lindblad, B.A., Mandeville, J.C., Minagra, A., Schwehn, G.H., Sekanina, Z., Wallis, M.K., Zarnecki, J.C., Chakaveh, S.C., Evans, G.C., Evans, S.T., Firth, J.G., Littler, A.N., Massonne, L., Olearczyk, R.E., Pankiewicz, G.S., Stevenson, T.J., Turner, R.F.: 1986, *Nautre* **321** 338

Marsden, B.J.: 1986, *Catalogue of Cometary Orbits*, 5th Edn., Smithsonian Astrophys. Obs., pp. 65-66

Mazets, E.P., Aptekar, R.L., Golenetskii, S.V., Guryan, Yu.A., Dyachkov, A.V., Ilyinskii, V.N., Panov, V.N., Petrov, G.G., Savvin, A.V., Sagdeev, R.Z., Sokolov, I.A., Khavenson, N.G., Shapiro, V.D., Shevchenko, V.I.: 1986, *Nature* **32**, 276

Mendis, D.A., Brin, G.D.: 1977, *The Moon* **17**, 353

Mendis, D.A., Brin, G.D.: 1978, *The Moon and the Planets* **18**, 77

Mendis, D.A., Hill, J.R., Houpis, H.L.F., Whipple, E.C., Jr.: 1981, *Astrophys. J.* **249**, 787

Richman, H., Froeschle, C., Klinger, J.: 1984, in *Ices in the Solar System*, eds. J. Klinger, B. Benest, A. Dolfus, R. Smoluchowski, Reidel, Dordrecht, p. 419

Rickman, H.: 1986, ESA Workshop on Comet Nucleus Return Mission, Canterbury (in press)

Safronov, V.S.: 1972, Israel Program for Scientific Translations, Jerusalem

Sagdeev, R.Z., Szabo, F., Avanesov, G.A., Cruvellier, P., Szabó, L., Szegö, K., Abergel, A., Balazs, A., Barinov, I.V., Bertaux, J-L., Blamont, J., Detaille, M., Demarelis, E., Dul'nev, G.N., Endroczy, G., Gardos, M., Kanyo, M., Kostenko, V.I., Krasikov, V.A., Nguyen-Trong, T., Nyitrai, Z., Reny, I., Rusznyak, P., Shamis, V.A., Smith, B., Sukhanov, K.G., Szabo, F., Szalai, S., Tarnopolsky, V.I., Toth, I., Tsukanova, G. et al.: 1986, *Nature* **321**, 262

Schleicher, D.G., Millis, R.H., Tholen, D., Lark, N., A'Hearn, M.F., Hoban, S., Birch, P.V., Martin, R.: 1986, 20th ESLAB Symp., Vol. I, 565

Sekanina, Z., Larson, S.M.: 1984, *Astron. J.* **89**, 1408

Sekanina, Z., Yeomans, D.K.: 1985, *Astron. J.* **90**, 2335

Sekanina, Z.: 1986, *Advances in Space Research*, (in press)

Sekanina, Z.: 1987, *Nature* **325**, 326

Simpson, J.A., Sagdeev, R.Z., Tuzzolino, A.J., Perkins, M.A., Ksanfomality, L.V., Rabinowitz, D., Lentz, G.A., Afonin, V.V., Ero, J., Keppler, E., Kosorokov, J., Petrova, E., Szabo, L., Umlauft, G.: 1986, *Nature* **321**, 278

Smoluchowski, R.: 1982, *J. Geophys. Res.* **87**, Suppl. 422

Smoluchowski, R.: 1984, in *Ices in the Solar System*, eds. J. Klinger, B. Benest, A. Dolfus, R. Smoluchowski, Reidel, Dordrecht, P. 397

Tokunaga, A.T., Hanner, M.S.: 1985, *Astrophys. J.* **296**, L13

Wehinger, P.A., Peterson, B.A., Wychoff, S., Lindholm, E., Festou, M.C.: 1986, 20th ESLAB Symp., Vol. III, 413

Weissman, P.R.: 1986, *Nature* **320**, 242

Weissman, P.R., Kieffer, H.H.: 1981, *Icarus* **47**, 302

Wallis, M.K., Macpherson, A.K.: 1981, *Astron. Astrophys.* **98**, 45

Wood, J.A.: 1986, *Proc. E.S.A. Conf.* (in press)

Whipple, F.L.: 1950, *Astrophys. J.* **111**, 779

Whipple, F.L.: 1978, *Moon Planets* **18**, 343

Whipple, F.L.: 1980, *Astron. J.* **85**, 305

Whipple, F.L.: 1986, *Pub. Roy. Soc.* (in press)

Yeomans, D.K.: 1977, *Astron. J.* **82**, 435

Yeomans, D.K.: 1985, in *Cometary Astronomy*, eds. E.K. Yeomans, R.M. West, R.S. Harrington, B.G. Marsden, p. 167

Astron. Astrophys. 187, 859–866 (1987)

ASTRONOMY
AND
ASTROPHYSICS

Composition measurements and the history of cometary matter

J. Geiss

Physikalisches Institut, University of Bern, Sidlerstrasse 5, CH-3012 Bern, Switzerland

Received February 23, accepted May 22, 1987

Summary. Composition data show that production from many nucleosynthetic sites and a sequence of condensation processes characterize the history of matter found in the smaller objects of the solar system. We present estimates for elemental abundances in the material emitted by Halley's comet. They are based on in-situ measurements on the gas and dust in the coma. It appears that Halley's nucleus is not depleted in the volatile elements C and O, whereas N is deficient. The abundance data show that a large fraction of the material in Halley's nucleus condensed at very low temperature. Comparison of the results published so far by the Giotto and Vega groups on molecular, elemental and isotopic abundances with data from meteorites, meteors, and the interstellar gas confirm that comets are regular members of the solar system which have preserved the original characteristics of the condensed and accreted matter better than other bodies in this system.

Key words: nucleosynthesis – interstellar medium: clouds: general – interstellar medium: dust – solar system: general – comets – meteors, meteorites

1. Introduction

Solar system matter is derived from a variety of nucleosynthetic processes and sources, i.e. from thermonuclear reactions in the big bang and in stars at various stages of evolution and from nuclear spallation in the interstellar medium. In the sun and planets, all the products from these sources were irreversibly mixed before or during accretion and by indigenous activity. However, in the smaller objects of the solar system traces of specific nucleosynthetic processes have been preserved. We have evidence for this in the isotopic anomalies that are found in components of meteorites which themselves are probably fragments of asteroids.

Chemical compositions and isotopic abundances indicate that terrestrial planets, asteroids and comets formed almost exclusively by accretion of condensed matter. Even most of the material which is now contained in the atmospheres and hydrospheres of the terrestrial planets was probably accreted in the form of grains (cf. Anders and Owen, 1977). Thus nucleosynthesis and condensation are the processes that above all determined the composition of the smaller objects in the solar system.

In Table 1 we have schematically summarized important epochs in the evolution of solar system material. Three stages of condensation are distinguished in this simplified list: (1) Condensation following stellar explosions and in expanding atmospheres of stars. The grains formed should contain source specific isotopic abundances, perhaps diluted by an admixture of ambient interplanetary matter. (2) Condensation from the interstellar gas. Probably most significant was the condensation in the (hypothetical) dense and cold interstellar molecular cloud (IMC) from which the solar nebula and then the solar system formed. However, residues of condensates from earlier IMCs may have survived destruction during the intervening period in the diffuse interstellar medium (Greenberg, 1982). The ambient UV radiation field could then have produced the high degree of polymerization that is suggested by the high abundances of volatile elements in the dust of the coma of P/Halley (Kissel et al., 1986; Jessberger et al., 1986; Greenberg, 1986). (3) Condensation in the solar nebula. The collapse that produced the nebula caused its matter to be heated. The subsequent cooling phase probably produced most of the condensed matter that went into the terrestrial planets and asteroids.

About 25 years ago, it was widely assumed that heating during the formation of the solar nebula was severe enough to evaporate all grains that existed before. This view rested on (a) the remarkable uniformity of isotopic ratios in different classes of meteorites and in the earth; and (b) the rather high temperatures predicted by models of the solar nebula. The picture has changed during the last two decades. Modern model calculations for the collapse and contraction of the solar nebula yield lower temperatures (Morfill et al., 1985), and a variety of isotopic inhomogeneities were found in meteorites, at least some of which reflect condensation at the nucleosynthetic source.

Some of the most clear-cut indicators for the preservation of condensates predating the formation of the solar system are given in Table 2. The existence of a neon component consisting of pure ^{22}Ne in some meteorites can satisfactorily be explained only by condensation of short-lived ^{22}Na near a nucleosynthetic source, followed by decay to ^{22}Ne (Eberhardt et al., 1981). Similarly, the high ^{13}C enrichments found in meteoritic concentrates (Swart et al., 1983) and grains (Niederer et al., 1985; Zinner and Epstein, 1987) has to be explained by condensation from a ^{13}C-rich gas rather than by isotopic fractionation. There are other observations in meteorites pointing to condensation in a stellar source region, e.g. the isotopic anomalies of several refractory elements (Wasserburg et al., 1977, 1980; Begemann, 1980) or the s-process component of meteoritic xenon (Srinivasan and Anders, 1978).

When it was realized that deuterium is about eight times more abundant on earth and in meteorites than it was in the solar nebula, it became clear that such a strong enrichment can only be due to chemical reactions at low temperature (Geiss and Reeves, 1972). The still greater, up to 40-fold, deuterium enrichments in

Table 1. Stages in the history of solar system matter (times are approximate)

	TIME (10^9 YRS)	EPOCH	CONDENSATION (time-scales)
	18	Big-Bang: Quarks > Antiquarks → H, D and He isotopes → Birth of galaxy	None
	18 to 4.8	Synthesis of CNO..U in stars and delivery into interstellar space by stellar winds and explosions	Condensation at source (≤1 year) (source specific isotopes)
	4.8 to 4.6	Our Molecular Cloud: Low gas temperature Local star formation "Late nucleosynthesis"	Condensation at low temp (10^7 yrs) Condensation at sources (≤1 year).
	4.6	Solar Nebula → Birth of solar system	(partial) reevaporation of pre-existing grains. Condensation in cooling nebula (10^3 yrs)
	4.6 to 0	Evolution of planets	≈ None

Table 2. Indicators for condensation in the solar nebula, in the IMC preceding the nebula, and near nucleosynthetic sources

CONDENSATION: WHERE?

	SOLAR NEBULA	IMC	STELLAR SOURCE
EARTH	● a)	● D/H b)	
METEORITES	● a)	● D/H c)	● ^{22}Ne, ^{13}C e)
COMETS		● D/H d)	

a) probably dominant
b) average
c) high D/H in organics
d) stratospheric particles
e) rare grains

organic compounds of meteorites (Robert et al., 1979; Yang and Epstein, 1983) apparently require temperatures well below 200 K, and at such low temperatures, conventional chemical reactions (with appreciable activation energies) are implausibly slow. Only ion-molecule reactions, which have no activation energy, would be adequately fast at very low temperatures. Such reactions apparently occur in interstellar molecular clouds (IMC; $T < 50$ K, $n_H \sim 10^4$–10^6 cm^{-3}), and are thought to be responsible for the very high D enrichments (10^2–10^5) of interstellar molecules. Thus, the D/H observations in meteorites imply that some organic matter from an IMC survived the heating in the solar nebula and was accreted into meteorite parent bodies (Kolodny et al., 1980; Geiss and Reeves, 1981).

The high D/H ratios found in stratospheric particles (Fahey et al., 1985), some of which are likely to be of cometary origin (Fraundorf et al., 1982), confirm this view. It is remarkable that the deuterium-rich material was not lost and has resisted dilution or exchange with atmospheric hydrogen.

In terrestrial planets and meteorites, the history of chemical processes and condensation is even harder to read than the nuclear history, because much of the record was destroyed by heating in the solar nebula, during accretion and by indigenous processes after accretion. Cometary material was probably privileged over meteoritic/asteroidal material on all counts. According to Oort's model comets formed between ~20 and ~30 AU (cf. Oort, 1986; Yamamoto, 1985; Weissman, 1986) where solar nebula temperatures remained lower than in the asteroidal belt (Morfill et al., 1985). The accretion energy was small (10^{-2} joule g^{-1} for a 20 km sized body), external heating was negligible after accretion, and indigenous activity should have remained on a lower level than in asteroids. Of course, the full range of analytical techniques can only be applied to cometary matter after return of cryogenically preserved samples. However, the fly-by missions to comet Halley have already yielded results on the nature and composition of comets that compare quite well with first missions to other solar system bodies. The advantage is that unlike planets, moons, or asteroids, comets freely give away material characteristic of the body as a whole, and this material was analyzed in the coma of P/Halley with a set of instruments having remarkable sensitivities and resolution.

2. In-situ measurements

Photodissociation times of molecules at 0.89 AU solar distance are of the order of 10^3 to 10^5 s. Thus, considering the gas expansion speed of $800 \, m \, s^{-1}$ (Lämmerzahl et al., 1986), Giotto came close enough to allow a fairly unaltered mix of "parent molecules" to be analyzed by the Neutral Mass Spectrometer (NMS, Krankowsky et al., 1986). However, even with the good resolution of the NMS and its capability to exploit differences in ionization potential, the identification of some important parent molecules and radicals remains difficult, because of overlaps in molecular weight. The pairs CO and N_2, or NH_3 and OH are examples. The problem can be alleviated by looking at the compounds containing the rarer isotopes, by comparison with optical observations, and by investigating the radial evolution of neutral and ionized fragments.

Whereas the mass spectrometer data on the neutral gas near the nucleus yield the most direct information on the nature of *parent molecules*, radial profiles of the neutral and ion composition are needed in addition to reveal the abundances of *elements* in the cometary gas. Solar UV gradually breaks up the original molecules into smaller radicals, atoms, and ions until eventually all components are dissected into atomic ions that can be more easily identified in a mass spectrum. Therefore, extending radial profiles of ion mass spectra to distances of $\sim 3 \; 10^5$ km (Balsiger et al., 1986b) helps to derive the elemental composition in the volatile material of the comet. First estimates of element abundances in the gas phase have been made by Balsiger et al. (1986a) who found a C/O ratio (0.33) that is about half the solar ratio, a comparably low N/O ratio (≤ 0.01), and an S/O ratio (0.03) that is close to the solar value. A relatively low abundance of N is consistent with recent abundance estimates for nitrogen bearing molecules. Bockelée-Morvan et al. (1986) and Schloerb et al. (1986) reported $HCN/H_2O = 10^{-3}$ and Allen et al. (1987) derived $NH_3/H_2O \approx 0.013$.

The gas streaming away from the cometary nucleus carries along fine solid particles. Contrary to recent cometary dust models, the observed size distribution was found to extend down to 10^{-17} g (Mazets et al., 1986; McDonnell et al., 1986), and there are indications that even finer grains are present. The composition of a large number of dust particles was investigated by time-of-flight mass spectrometers onboard the Vega and Giotto spacecraft (Kissel et al., 1986). The high encounter velocity with P/Halley (of $68-79 \, km \, s^{-1}$) – otherwise quite a headache – is a virtue when interpreting the dust mass spectra. The impact energy of several hundred electron volts per atom causes the grains to dissociate almost completely into atomic species. Moreover, the ion yields are high, so that the composition of particles with masses as low as 10^{-15} g can be studied. Even a cursory look at the mass spectra reveals a strong variability in composition (Kissel et al., 1986; Jessberger et al., 1986). The relative abundances among the "stony", non-volatile elements vary, but most remarkable is the high abundance of the volatile elements HCNO, and the extreme variations in the relative proportion of the volatile and non-volatile components (Kissel et al., 1986). Some grains, called CHON-particles, are composed almost entirely of volatile materials, presumably in polymerized form. Preliminary average elemental abundances in the dust have been given by Jessberger et al., (1986). They find the average carbon content in the grains of Halley to be an order of magnitude higher than in C1 meteorites. However, the C/N ratios in the Halley grains and in C1 chondrites are similar.

The mass spectrometer data on cometary composition obtained in the coma of P/Halley should turn out to be most valuable in providing "ground truth" calibration for optical or UV spectroscopic observations of comets. Improving the interpretation of optical data might help to decide whether there exist significant intrinsic compositional differences among the comets, differences that could arise if comets formed over a wide range of solar distances (cf. Delsemme, 1982; A'Hearn et al., 1985; Yamamoto, 1985; Wood, 1986). Similarly, the in-situ data on dust composition and texture can be used to identify which of the particles collected in the stratosphere have a cometary origin. If such an identification could be unambiguously made, the full power of laboratory methods could be applied to the less volatile component of cometary material.

3. Bulk composition of comet Halley

Determination of the average chemical composition of the Halley nucleus at the time of formation is a most important but distant cosmochemical goal. It may be as difficult to attain as it is in the case of other bodies in the solar system, the earth not excluded. An extrapolation over several steps is needed to translate coma composition into the composition of the original nucleus, because at the present time, we cannot exclude a number of fractionating events. During its early life, the nucleus – perhaps aided by indigenous energy – may have lost a fraction of its most volatile components. Late in its life, the frequent approaches to the sun could have led to a systematic depletion of volatiles in parts of the nucleus that feed the coma. Also, there may be short-time changes, for instance in the relative proportion of emitted gas and dust. To be sure, there are good reasons to hope that these effects will turn out not to be overwhelmingly important. Thus, a determination of the average element composition in the material given off by P/Halley at the time of the Vega and Giotto encounters is a worthwhile effort. For such a determination, the gas and dust compositions as well as the gas/dust ratio are needed. This ratio is still difficult to judge. A gas/dust ratio of 2 (at the source) with an uncertainty of a factor of 2 or 3 is the best present estimate (McDonnell et al., 1986).

For the gas composition, we adopt the C/O ratio given by Balsiger et al. (1986a) and estimate a ratio $H/O \approx 2$ from the measured molecule abundances (Krankowsky et al., 1986; Eberhardt et al., 1986a) and ion spectra (Balsiger et al. 1986a, b). Nitrogen is more uncertain than the other volatile elements, and therefore, we give two abundance values, N (low) and N (high). Based on the estimates of Balsiger et al. (1986a) and Allen et al. (1987) discussed in Sect. 2, we adopt $N/O = 0.01$ as the low value for the gas. For the dust, we use the preliminary average composition data of Jessberger et al. (1986). In the case of nitrogen, we use their ratio $N/C = 1/29$ as the N (low) value. For obtaining N (high), we double the N/O ratio in the gas and the N/C ratio in the dust.

The relative proportions in the gas and the dust for the elements HCNOSi obtained are given in Table 3. The high proportion of C and N in the dust is remarkable, because the grains were analyzed (Kissel et al., 1986; Jessberger et al., 1986) after they had been heated by the sun to temperatures of up to a few hundred degree Celsius (Hanner, 1982) for a few hours. Probably low H/C ratios and polymer bonds prevent a more complete evaporation of the volatile substances (see also Khare et al., 1981; Greenberg, 1982, 1986; Anders, 1986).

There are several indications that some evaporation from the grains takes place even several hours after their release from the nucleus. Eberhardt et al. (1986a) observed an unusual radial profile at mass 28 in the neutral spectrum and interpret this as continued release from the grains of CO or a short-lived progenitor. A'Hearn et al. (1986) concluded that submicron

Table 3. The distribution of a few key elements between grains and gas in the material released by P/Halley. The arrow indicates that some evaporation from the grains continues in the coma

ELEMENTS (%) RELEASED BY P/HALLEY

	GRAINS ------→	GAS
H	1	99
O	13	87
C	47	53
N	50	50
Si	100	0

ASSUMED IS GAS/DUST = 2:1 (BY WEIGHT)

Table 4. Relative element abundances of the material released by P/Halley. The estimates are based on a gas/dust ratio of two at the source. The element abundance ratios given are uncertain by a factor of about two (excluding the variability and uncertainty in the gas/dust ratio). N is more uncertain than the other elements, and therefore, two estimates are given. For comparison, the solar system abundances as given by Anders and Ebihara (1982) and the solar abundances by Grevesse (1984) are included

	"Halley"	Solar system	Sun
H/Si	42	27200	28200
C/Si	12.9	12.1	13.8
O/Si	24.1	20.1	23.4
$(N/Si)_{high}$	0.84	2.48	2.75
$(N/Si)_{low}$	0.42		
$(N/C)_{high}$	0.065	0.205	0.199
$(N/C)_{low}$	0.033		

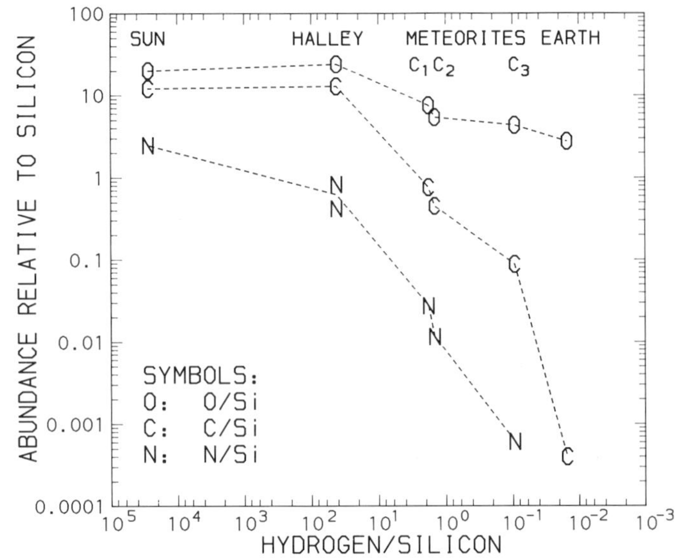

Fig. 1. Relative abundances of elements in the material released by P/Halley. A gas/dust ratio of two at the source was assumed. For comparison, the abundances in the solar nebula (Anders and Ebihara, 1982) and in carbonaceous meteorites (Kerridge, 1985) are given. Estimates for the abundance on earth (crust plus mantle) are also included (Larimer, 1971; Wänke, 1981; Wänke et al., 1984)

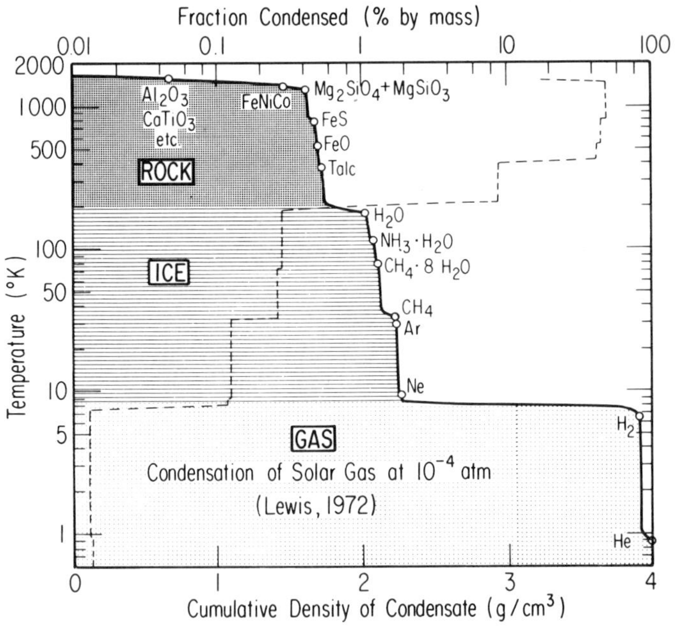

Fig. 2. Condensation under equilibrium conditions in a gas of solar composition at 10^{-4} atm (after Lewis, 1972; Anders, 1986). The dashed line is the cumulative density of the condensate

particles in jets release CN. Balsiger et al. (1986a) suggested that the high abundance and radial distribution of C^+ indicate release of highly unsaturated carbon compounds from grains. Simpson et al. (1986) and Vaisberg et al. (1986) found clear evidence for disintegration of conglomerates of grains in the coma which could be explained by the loss of "glue" that held them together.

The elemental abundances (gas plus grains in 2:1 proportion) of the material released by P/Halley at the time of the encounter are given in Table 4 and shown in Fig. 1. H, C, N, and O are related to silicon which represents a typical non-volatile element. For comparison, the solar abundances as well as the average composition of C1, C2, and C3 chondrites, and of the earth are included in Fig. 1. The P/Halley abundances are partly based on preliminary data, and as emphasized above the matter emitted into the coma does not necessarily have a composition identical to the average composition of the nucleus. Notwithstanding these reservations, the data shown in Fig. 1 allow us to draw some conclusions and to make comparisons between cometary abundances and the composition of other objects in the solar system.

Figure 1 demonstrates that the content of volatiles in Halley is more than an order of magnitude higher than even in C1 condrites. The C/Si and O/Si ratios in the derived Halley composition are close to the solar ratios, and the hydrogen content is as high as it can be expected in solid material (if $T \gtrsim 50$ K). Since the N/C ratios are similar in the gas and the dust (cf. Table 3), the N/C estimate for the released total material is virtually independent of the gas/dust ratio. Thus, the difference between the cometary and solar N/C ratios given in Table 4 indicates that the nitrogen underabundance in Halley is significant.

For discussing the Halley abundances in terms of condensation conditions, we show in Fig. 2 the condensation sequence that would result under equilibrium condition in a cooling gas of solar composition at a pressure of 10^{-4} atm (Lewis, 1972). Chemical abundances, mineral chemistry, and petrology of meteorites are roughly consistent with condensation under equilibrium conditions down to temperatures of ~ 400 K (Larimer and

Anders, 1967; Lewis, 1972; Larimer, 1973; Laul et al., 1973; Krähenbühl et al., 1973). Somewhere below 400 K, many chemical reactions become so slow that thermodynamic equilibrium brakes down. Kinetic effects in gas reactions, incomplete catalytic synthesis, or ionization by external radiation become important. In fact, even in meteorites there are clear signs of non-equilibrium chemistry. The high proportion of straight-chain isomers among the hydrocarbons in carbonaceous chondrites (Hayatsu and Anders, 1981) and the ^{13}C enrichment in the carbonates of these meteorites are best explained by kinetic effects (Lancet and Anders, 1970; Anders, 1986). The high D/H ratios found in organic molecules of meteorites point to ion-molecule reactions at very low temperatures (cf. Sects. 1 and 4).

The molecular forms produced in ion-molecule reactions are quite different from those prevailing in equilibrium. The dominant forms of C are CO and related species, not CH_4, and the dominant form of N is N_2, not NH_3 (cf. Mitchell et al., 1981; Anders, 1986, Fig. 5). Condensation at low temperature of the products of ion-molecule reactions appears to be roughly consistent with the Halley abundances of the volatile elements presented in Table 4 and Fig. 1. Oxygen and carbon are higher than in C1 meteorites, because water ice, CO_2 and CO are condensed and retained in the cometary nucleus. The nitrogen abundance is higher than in the carbonaceous meteorites, but it is below the solar abundance. This implies that a considerable fraction of nitrogen remained uncondensed, probably because it was present as N_2. Why is then the similarly volatile CO found in Halley? Unlike molecular nitrogen, CO makes weak bonds with metals and other materials. In fact, the observations of Eberhardt et al. (1986a) of an extended CO source indicates that most of the comet's CO is not condensed in the ice phase, but it is loosely bound in the grains.

The low nitrogen abundance, the high C/N ratio, and the near solar C/O ratio – if substantiated by further data analysis – are strong points on which to build models of the condensation history of cometary matter. The expected chemical behaviour of nitrogen is sketched in Table 5. Under most IMC conditions, N_2 ought to be the preponderant form of nitrogen (Mitchell et al., 1981). Under equilibrium conditions in the solar nebula, N would be present as ammonia (Lewis, 1972). However, Lewis and Prinn (1980) have pointed out that the transition from N_2 to NH_3 is not likely to occur under the conditions that are inferred for the solar nebula. Thus, a nitrogen deficiency in the condensed phase could result in both these environments. Some N_2 might be contained in the cometary ices if condensation temperatures were very low.

Table 5. Theoretically inferred forms of nitrogen in IMCs and in the solar nebula

IMC	(N_2)		Mitchell, Prasad, Huntress, 1981
Solar Nebula (equil)	$N_2 \rightarrow$	(NH_3)	Lewis 1972
Solar Nebula (kinetic)	(N_2)	NH_3	Lewis + Prinn 1980

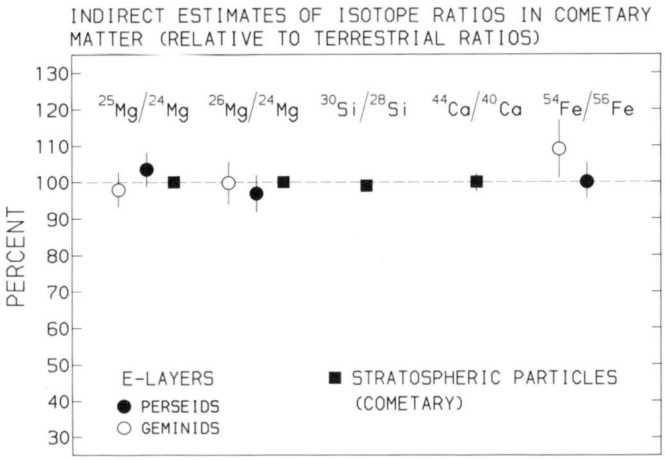

Fig. 3. Isotopic ratios in material derived from comets (relative to terrestrial ratios). Data by Zbinden et al. (1975), Herrmann (1979) for E-layers; by Esat et al. (1979) and McKeegan et al. (1985) for stratospheric particles

4. Isotopes

Prior to the in-situ investigations of P/Halley, some isotopic ratios, in particular $^{13}C/^{12}C$, had been investigated in comets by optical spectroscopy (cf. Vanysek and Rahe, 1978). Also, the metallic component in E-layers during meteor showers and dust particles collected in the stratosphere was investigated with the aim of obtaining isotopic abundances in cometary material. The most recent data on the isotopes of Mg, Si, Ca, and Fe are reproduced in Fig. 3. Precise isotopic ratios were obtained in these studies, but the identification with cometary material remains somewhat conjectural. The debris of comets certainly supplies metals to the E-layers in the upper atmosphere during meteor showers. However, it is difficult to exclude a contribution from sporadic meteors, a fraction of which may be asteroidal in origin. Thus, the E-layer data represent isotopic ratios in cometary matter, with a possible asteroidal admixture. On the basis of their texture, a fraction of the particles from stratospheric collection were tentatively identified as cometary in origin (Fraundorf et al., 1982). So far, the investigations on the dust in the coma of P/Halley have neither greatly strengthened nor weakened these identifications. The data given in Fig. 3 are from $\sim 10\,\mu m$ particles, much larger than any particle studied in the P/Halley in-situ investigations (Kissel et al., 1986; Jessberger et al., 1986). Simpson et al. (1986) found that larger Halley particles disintegrate spontaneously into smaller grains after several hours in sunlight. Of course, other large particles may be more stable, but are they stable enough to survive atmospheric entry? Thus, with some reservation concerning the correct identification of the stratospheric particles investigated, we note that Fig. 3 shows a remarkable agreement between the isotopic abundances in cometary and terrestrial matter, an agreement that extends to all meteorite classes. This finding supports the view (Oort, 1986) that comets and the other bodies in the planetary system formed from the same reservoir 4.6 10^9 yr ago and that the Oort cloud was not replenished since then from sources alien to the solar system.

So far, in-situ determinations of isotopic abundances have been published for cometary sulfur by Krankowsky et al. (1986) and for hydrogen and oxygen in cometary water by Eberhardt et al. (1986b). More isotopic data are likely to result from the continuing data evaluation efforts (cf. Jessberger and Kissel, 1987).

Whereas *average* isotopic abundances in comets and in other bodies of the solar system appear to be very similar, isotopic ratios in meteorites are found to vary if small enough samples are analyzed. The carbon isotopes are an important example as is

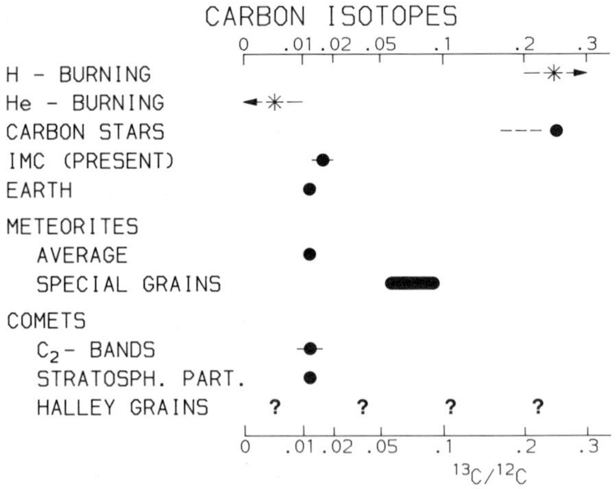

Fig. 4. Isotopic composition of carbon in the galaxy and in bodies of the solar system. Data from Tomkin et al. (1976), Wannier (1980), Niederer et al. (1985), Zinner and Epstein (1987), Vanysek and Rahe (1978), Pillinger (1986), and references quoted therein

Table 6. The deuteration of polar molecules in dense IMCs. H_2O might condense faster than some other constituents, possibly limiting the deuterium enrichment in water ice

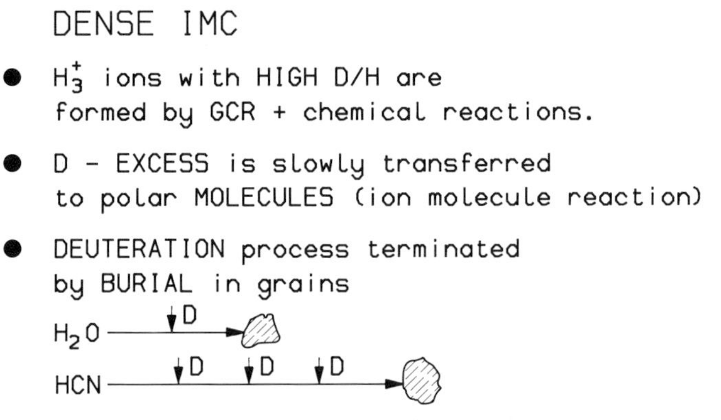

demonstrated by Fig. 4. The average $^{12}C/^{13}C$ abundance ratio is the same on earth, in meteorites and, within larger limits of error, in cometary material. ^{12}C and ^{13}C are formed by totally different nuclear processes (He-burning versus warm H-burning; cf. Audouze, 1977), and therefore, the isotopic signatures of this element could be strongly source specific, if condensates from stellar sources have survived. Large ^{13}C enhancements were indeed found in special mineral concentrates (Swart et al., 1983) and in metal oxide-rich grains (Niederer et al., 1985; Zinner and Epstein, 1987) of carbonaceous meteorites. The in-situ investigations of P/Halley offer an opportunity to learn whether grains with even higher $^{13}C/^{12}C$ ratios entered the solar system intact. Submicron grains have preserved their individuality, and the high impact energy produces enough ions to permit analyses of these grains that have 10^3 to 10^4 times less mass than the smallest meteoritic grains analyzed so far for carbon isotopes. In a recent progress report on their data analyses, Jessberger and Kissel (1987) show some grains with high 13/12 peak ratios, but the problem is to find conclusive criteria to correct for CH^+ on the $^{13}C^+$ peak.

Isotopic analyses in the gas phase are hampered by the large variety of hydrides occurring among the molecules, radicals and ions. The H_2DO^+/H_3O^+ ratio investigated by Eberhardt et al. (1986b) is exceptionally suited because H_3O^+ has a high abundance in the innermost part of the coma visited by Giotto, and the position of H_2DO^+ on the mass scale is relatively free of other ions. Still, the coincidence of $H_2{}^{18}O^+$ with H_2DO^+ is a major source of uncertainty. The authors obtain for D/H in the water of Halley $0.6\ 10^{-4} \leq D/H \leq 4.8\ 10^{-4}$, and they expect further data analyses to give a much more precise value.

As Eberhardt et al. (1986b) point out, their brackets for the deuterium abundance in cometary water cover the range of D/H ratios in many hydrogen-poor bodies of the solar system, i.e. the earth, silicates in meteorites (Yang and Epstein, 1983), or CH_4 in Titan (Owen et al., 1986). The D abundance in Halley water is lower than in organic compounds of some meteorites (Robert et al., 1979; Yang and Epstein, 1983; Pillinger, 1984) and much lower than in the HNC, HCN, and HCO^+ molecules of IMCs (cf. Wannier, 1980). However, this latter observation does not necessarily speak against an IMC origin of cometary water (cf. Table 6). It is believed that the high deuterium abundance in the polar molecules of IMCs is due to the transfer of deuterium from H_3^+ by ion molecule reactions (Watson, 1976). There is no a priori reason to assume that H_2O vapor would not participate in this process. However, deuteration is terminated by condensation which could occur faster for H_2O than for other molecules, such as HCN. Thus, a better understanding of the deuteration process of H_2O in IMCs is needed before the D/H ratio observed in the water ice of P/Halley can be fully exploited for delimiting the condensation conditions of cometary water.

5. Conclusion

The grains which accreted into Halley's nucleus $\sim 4.6\ 10^9$ yr ago did not form in a single environment at a single time. Isotopic anomalies, as yet much better documented for meteoritic than for cometary material, demonstrate the survival of solids that condensed near a stellar source. The close association of volatile and non-volatile elements and the evidence for polymerization in the extremely fine Halley grains is consistent with interstellar condensation. The high deuterium abundance in gas-poor objects of the solar system (earth, meteorites, comets) and in particular the D/H enhancements in organic substances of meteorites result from low temperature ion-molecule reactions which occur in IMCs. In analogy to contemporary observations of star formation, the solar system is thought to have formed out of an IMC by gravitational collapse. Whether the interstellar grains or grain remnants which accreted into comets condensed solely in this IMC or whether these grains have ages of up to several billion years as advocated by Greenberg (1982, 1986) is as yet difficult to judge on compositional grounds. Long lifetimes in interstellar space would provide ample UV energy for polymerization, but would also produce a high degree of nuclear spallation by cosmic rays. A large fraction of spallation products would have been lost from the fine grains (Geiss and Reeves, 1981), but heavy species like xenon could have been retained. Spallation xenon levels corresponding to $\gtrsim 10^9$ yr of cosmic ray exposure in grains of stony meteorites have not been reported, but probably have not been systematically searched for. The prospects for in-situ detection of spallation products in Halley grains are very small, the limitation being the dynamic range of the instruments.

A variety of observations point to interstellar grains as important constituents of comets (cf. also Weissman, 1986; Whipple, 1986). Their survival would be consistent with the low temperatures inferred for the trans-Jovian part of the solar nebula (Morfill et al., 1985). However, solar nebula condensates are probably not negligible in comets. Appreciable metal abundances in the interstellar gas (Snow, 1975) and the occurrence of metal atoms and ions in cometary comae (Oppenheimer, 1980; Barbieri et al., 1986; Gringauz et al., 1986; cf. Geiss et al., 1986) are indicators for non-equilibrium and incomplete condensation which could then accelerate in the dense solar nebula. Furthermore, grains forming closer to the sun ought to contain a higher proportion of nebula condensates, and some such material was probably transported outward and mixed in with the material in the comet forming region (cf. Morfill et al., 1985; Anders, 1986; Weissman, 1986). Thus, the conclusion at the present time is that comets contain solids that condensed near nucleosynthetic sources, in interstellar space and in the solar nebula. The mixing ratio is not yet well determined. In particular, we cannot yet be certain where all the water condensed which is the most abundant substance in Halley's nucleus.

Acknowledgements. The author wishes to thank E. Anders, H. Balsiger, P. Eberhardt, H. Fechtig, E. Grün, W.T. Huntress, E.K. Jessberger, J. Kissel, and H. Völk for discussions, E. Anders and W.T. Huntress for many suggestions and comments concerning the conclusions in this paper, Mrs. G. Troxler for typing the manuscript, and J. Fischer for preparing the figures. This work was supported in part by the Swiss National Science Foundation.

References

A'Hearn, M.F., Birch, P.V., Feldman, P.D., Millis, R.L.: 1985, *Icarus* **64**, 1

A'Hearn, M.F., Hoban, S., Birch, P.V., Bowers, C., Martin, R., Klinglesmith, D.A.: 1986, *Nature* (submitted)

Allen, M., Delitsky, M., Huntress, W.T., Yung, Y., Ip, W.-I., Schwenn, R., Balsiger, H., Geiss, J.: 1987 (this volume)

Anders, E.: 1986, in *The Comet Nucleus Sample Return Mission Proc. Workshop, Canterbury, UK, 15–17 July 1986*, ESA SP-249, p. 31

Anders, E., Owen, T.: 1977, *Science* **198**, 453

Anders, E., Ebihara, M.: 1982, *Geochim. Cosmochim. Acta* **46**, 2363

Audouze, J. (ed.): 1977, *CNO Isotopes in Astrophysics*, Reidel, Dordrecht

Balsiger, H., Altwegg, K., Bühler, F., Geiss, J., Ghielmetti, A.G., Goldstein, B.E., Goldstein, R., Huntress, W.T., Ip, W.-H., Lazarus, A.J., Meier, A., Neugebauer, M., Rettenmund, U., Rosenbauer, H., Schwenn, R., Sharp, R.D., Shelley, E.G., Ungstrup, E., Young, D.T.: 1986a, *Nature* **321**, 330

Balsiger, H., Altwegg, K., Bühler, F., Fuselier, S.A., Geiss, J., Goldstein, R., Lazarus, A.J., Meier, A., Neugebauer, M., Rettenmund, U., Rosenbauer, H., Schwenn, R., Shelley, E.G., Ungstrup, E., Young, D.T.: 1986b, 20th ESLAB Symp. ESA SP-250, Vol. **I**, p. 99

Barbieri, C., Iijima, T., Sabbadin, F.: 1986, IAUC 4169

Begemann, F.: 1980, *Rep. Prog. Phys.* **43**, 1309

Bockelée-Morvan, D., Crovisier, J., Despois, D., Forveille, T., Gérard, E., Schraml, J., Thum, C.: 1986, 20th ESLAB Symp. ESA SP-250, Vol. **I**, p. 365

Delsemme, A.H.: 1982, in *Comets*, ed. L.L. Wilkening, University of Arizona Press, Tucson, Arizona, p. 85

Eberhardt, P., Jungck, M.H.A., Meier, F.O., Niederer, F.R.: 1981, *Geochim. Cosmochim. Acta* **45**, 1515

Eberhardt, P., Krankowsky, D., Schulte, W., Dolder, U., Lämmerzahl, P., Berthelier, J.J., Woweries, J., Stubbemann, U., Hodges, R.R., Hoffman, J.H., Illiano, J.M.: 1986a, 20th ESLAB Symp. ESA SP-250, Vol. **I**, p. 383

Eberhardt, P., Dolder, U., Schulte, W., Krankowsky, D., Lämmerzahl, P., Hoffman, J.H., Hodges, R.R., Berthelier, J.J., Illiano, J.M.: 1986b, 20th ESLAB Symp., ESA SP-250, Vol. **I**, p. 539

Esat, T.M., Brownlee, D.E., Papanastassiou, D.A., Wasserburg, G.J.: 1979, *Science* **206**, 190

Fahey, A., McKeegan, K.D., Sandford, S.A., Walker, R.M., Wopenka, B., Zinner, E.: 1985, in *Properties and Interactions of Interplanetary Dust*, eds. R.H. Giese, P. Lamy, Reidel, Dordrecht, p. 149

Fraundorf, P., Brownlee, D.E., Walker, R.M.: 1982, in *Comets*, ed. L.L. Wilkening, University of Arizona Press, Tucson, Arizona, p. 383

Geiss, J., Reeves, H.: 1972, *Astron. Astrophys.* **18**, 126

Geiss, J., Reeves, H.: 1981, *Astron. Astrophys.* **93**, 189

Geiss, J., Bochsler, P., Ogilvie, K.W., Coplan, M.A.: 1986, *Astron. Astrophys.* **166**, L1

Greenberg, J.M.: 1982, in *Comets*, ed. L.L. Wilkening, University of Arizona Press, Tucson, Arizona, p. 131

Greenberg, J.M.: 1986, in *The Comet Nucleus Sample Return Mission Proc. Workshop, Canterbury, UK, 15–17 July 1986*, ESA SP-249, p. 47

Grevesse, N.: 1984, *Physica Scripta T8*, 49

Gringauz, K.I., Gombosi, T.I., Remizov, A.P., Apáthy, I., Szemerey, I., Verigin, M.I., Denchikova, L.I., Dyachkov, A.V., Keppler, E., Klimenko, I.N., Richter, A.K., Somogyi, A.J., Szegö, K., Szendrö, S., Tátrallyay, M., Varga, A., Vladimirova, G.A.: 1986, *Nature* **321**, 282

Hanner, M.S.: 1982, in *Cometary Exploration II*, ed. T.I. Gombosi, Hungarian Acad. of Sciences, p. 1

Hayatsu, R., Anders, E.: 1981, *Topics Current Chem.* **99**, 1

Herrmann, U.P.: 1979, *Ph.D. Thesis*, University of Bern

Jessberger, E.K., Kissel, J., Fechtig, H., Krueger, F.R.: 1986, in *The Comet Nucleus Sample Return Mission Proc. Workshop, Canterbury, UK, 15–17 July 1986*, ESA SP-249, p. 27

Jessberger, E.K., Kissel, J.: 1987, in *Lunar Planet. Sci. XVIII* (in press)

Kerridge, J.F.: 1985, *Geochim. Cosmochim. Acta* **49**, 1707

Khare, B.N., Sagan, C., Zumberge, J.E., Sklarew, D.S., Nagy, B.: 1981, *Icarus* **48**, 290

Kissel, J., Sagdeev, R.Z., Bertaux, J.L., Angarov, V.N., Audouze, J., Blamont, J.E., Büchler, K., Evlanov, E.N., Fechtig, H., Fomenkova, M.N., von Hoerner, H., Inogamov, N.A., Khromov, V.N., Knabe, W., Krueger, F.R., Langevin, Y., Leonas, V.B., Levasseur-Regourd, A.C., Managadze, G.G., Podkolzin, S.N., Shapiro, V.D., Tabaldyev, S.R., Zubkov, B.V.: 1986, *Nature* **321**, 280

Kolodny, Y., Kerridge, J.F., Kaplan, I.R.: 1980, *Earth Planet. Sci. Letters* **46**, 149

Krähenbühl, U., Morgan, J.W., Ganapathy, R., Anders, E.: 1973, *Geochim. Cosmochim. Acta* **37**, 1353

Krankowsky, D., Lämmerzahl, P., Herrwerth, I., Woweries, J., Eberhardt, P., Dolder, U., Herrmann, U., Schulte, W., Berthelier, J.J., Illiano, J.M., Hodges, R.R., Hoffman, J.H.: 1986, *Nature* **321**, 326

Lämmerzahl, P., Krankowsky, D., Hodges, R.R., Stubbemann, U., Woweries, J., Herrwerth, I., Berthelier, J.J., Illiano, J.M., Eberhardt, P., Dolder, U., Schulte, W., Hoffman, J.H.: 1986, 20th ESLAB Symp., ESA SP-250, Vol. **I**, p. 179

Lancet, M.S., Anders, E.: 1970, *Science* **170**, 980

Larimer, J.W.: 1971, *Geochim. Cosmochim. Acta* **35**, 769

Larimer, J.W.: 1973, *Geochim. Cosmochim. Acta* **37**, 1603

Larimer, J.W., Anders, E.: 1967, *Geochim. Cosmochim. Acta* **31**, 1239

Laul, J.C., Ganapathy, R., Anders, E., Morgan, J.W.: 1973, *Geochim. Cosmochim. Acta* **37**, 329

Lewis, J.S.: 1972, *Icarus* **16**, 241

Lewis, J.S., Prinn, R.G.: 1980, *Astrophys. J.* **238**, 357

Mazets, E.P., Sagdeev, R.Z., Aptekar, R.L., Golenetskii, S.V., Guryan, Yu. A., Dyachkov, A.V., Ilyinskii, V.N., Panov, V.N., Petrov, G.G., Savvin, A.V., Sokolov, I.A., Frederiks, D.D., Khavenson, N.G., Shapiro, V.D., Shevchenko, V.I.: 1986, 20th ESLAB Symp. ESA SP-250, Vol. **II**, p. 3

McDonnell, J.A.M., Kissel, J., Grün, E., Grard, R.J.L., Langevin, Y., Olearczyk, R.E., Perry, C.H., Zarnecki, J.C.: 1986, 20th ESLAB Symp. ESA SP-250, Vol. **II**, p. 25

McKeegan, K.D., Walker, R.M., Zinner, E.: 1985, *Geochim. Cosmochim. Acta* **49**, 1971

Mitchell, G.F., Huntress, W.T., Prasad, S.S.: 1981, *Astrophys. J.* **244**, 1087

Morfill, G.E., Tscharnuter, W., Völk, H.J.: 1985, in *Protostars and Planets II*, eds. D.C. Black, M.S. Matthews, University of Arizona Press, Tucson, Arizona, p. 493

Niederer, F.R., Eberhardt, P., Geiss, J.: 1985, *Meteoritics* **20**, 716

Oort, J.H.: 1986, The origin and dissolution of comets, Halley Lecture for 1986, delivered at Oxford, May 6, 1986

Oppenheimer, M.: 1980, *Astrophys. J.* **240**, 923

Owen, T., Lutz, B.L., de Bergh, C.: 1986, *Nature* **320**, 244

Pillinger, C.T.: 1984, *Geochim. Cosmochim. Acta* **48**, 2739

Pillinger, C.T.: 1986, *Proc. Roy. Soc.* (submitted)

Robert, F., Merlivat, L., Javoy, M.: 1979, *Nature* **282**, 785

Schloerb, F.P., Kinzel, W.M., Swade, D.A., Irvine, W.M.: 1986, 20th ESLAB Symp. ESA SP-250, Vol. **I**, p. 577

Simpson, J.A., Rabinowitz, D., Tuzzolino, A.J., Ksanfomality, L.V., Sagdeev, R.Z.: 1986, 20th ESLAB Symp. ESA SP-250, Vol. **II**, p. 11

Snow, T.P., Jr.: 1975, *Astrophys. J.* **202**, L87

Srinivasan, B., Anders, E.: 1978, *Science* **201**, 51

Swart, P.K., Grady, M.M., Pillinger, C.T., Lewis, R.S., Anders, E.: 1983, *Science* **220**, 406

Tomkin, J., Luck, R.E., Lambert, D.L.: 1976, *Astrophys. J.* **210**, 694

Vaisberg, O., Smirnov, V., Omelchenko, A.: 1986, 20th ESLAB Symp. ESA SP-250, Vol. **II**, p. 17

Vanysek, V., Rahe J.: 1978, *Moon Planets* **18**, 441

Wänke, H.: 1981, *Phil. Trans. Roy. Soc. Lond. A* **303**, 287

Wänke, H., Dreibus, G., Jagoutz, E.: 1984, in *Archaean Geochemistry*, eds. A. Kröner et al., Springer, Berlin, Heidelberg, New York, p. 1

Wannier, P.G.: 1980, *Ann. Rev. Astron. Astrophys.* **18**, 399

Wasserburg, G.J., Lee, T., Papanastassiou, D.A.: 1977, *Geophys. Res. Letters* **4**, 299

Wasserburg, G.J., Papanastassiou, D.A., Lee, T.: 1980, in *Early Solar System Processes and the Present Solar System*, ed. D. Lal, North-Holland, Amsterdam, p. 144

Watson, W.D.: 1976, *Rev. Mod. Phys.* **48**, 513

Weissman, P.R.: 1986, in *The Comet Nucleus Sample Return Mission Proc. Workshop, Canterbury, UK, 15–17 July 1986*, ESA Sp-249, p. 15

Whipple, F.L.: 1986, 20th ESLAB Symp. ESA SP-250, Vol. **II**, p. 281

Wood, J.A.: 1986, in *The Comet Nucleus Sample Return Mission Proc. Workshop, Canterbury, UK, 15–17 July 1986*, ESA SP-249, p. 123

Yamamoto, T.: 1985, *Astron. Astrophys.* **142**, 31

Yang, J., Epstein, S.: 1983, *Geochim. Cosmochim. Acta* **47**, 2199

Zbinden, P.A., Hidalgo, M.A., Eberhardt, P., Geiss, J.: 1975, *Planet. Space Sci.* **23**, 1621

Zinner, E., Epstein, S.: 1987, *Earth Planet. Sci. Letters* (to be published)

Astron. Astrophys. 187, 867–872 (1987)

ASTRONOMY AND ASTROPHYSICS

Modeling Halley before and after the encounters

N. Divine and R. L. Newburn, Jr.

Jet Propulsion Laboratory, California Institute of Technology, 4800 Oak Grove Drive, Pasadena, CA 91109, USA

Received February 13, accepted June 12, 1987

Summary. Numerical models developed prior to the 1986 spacecraft flybys at comet P/Halley described anticipated dust and gas environments. Predicted values for gas production, for dynamic ranges of gas and dust experiments, and for likely spacecraft effects were matched during the flybys well within expected margins; in particular the dust-impact-generated attitude disturbance experienced by Giotto several seconds before closest approach was within the envelope of expectations from the models. We have identified three major areas in which 1986 results for Halley might be used to improve the models, namely (1) gas production values from Earth-based and in-situ observations, (2) dust flux and fluence values in the coma from three spacecraft, and (3) dust size distributions from the in-situ data which require the presence of numerous particles at masses less than 10^{-17} kg.

Key words: comet – cometary gas – cometary dust

1. Introduction

In preparation for the spacecraft missions to comet P/Halley, numerical models for the gas and dust environments, relying mainly on earth-based observations made prior to Halley's current apparition, were developed by a Working Group on the Halley Environment. These models, as described by Divine et al. (1986), contain numerous features which were either confirmed or contradicted by the in-situ data gathered on March 6, 9, and 14, 1986. Quantitative comparisons between the model predictions and the data interpretations are useful to assess the validity of the considerations which form the bases for the model, and to support the development of improved environmental models for future cometary exploration spacecraft and mission design. For this purpose we concentrate below on three quantitative aspects of the model results, namely gas production rates, dust fluxes, and dust size distributions.

2. Gas production

Models of P/Halley prepared before the availability of water production data were based on a semi-empirical photometric theory (Newburn, 1981) tying dust and gas production to the visual magnitude as a function of time (the light curve, normalized to 1 AU from Earth). The accuracy of such models depends on the quality of the photometric data, on the extent to which the light curve repeats from apparition to apparition, and on the applicability to Halley of the values for dust-to-gas ratio (by light) and for resonance fluorescence efficiency (selected from data for several recent comets used to calibrate the theory). For Halley, the model developed by Divine et al. (1986) anticipates that the predicted gas production, as shown by the solid line in Fig. 1, would be correct to within a factor $4^{\pm 1}$ for Halley within a few months of its 1986 perihelion passage. The gas production predictions were tied to the 1910 Halley light curve, which itself was somewhat uncertain. In 1985 Halley was brighter between 2.0 and 1.0 AU preperihelion than predicted by most modelers, and therefore the predictions were low at that time. The fault lay more with an inadequate 1910 light curve than with the theory. The scatter in the gas production values in 1985 and 1986 reflects primarily the intrinsic variations in the comet from day to day, and secondarily some inadequacies in the theories used to reduce data from diverse sources.

The Earth-based data (represented by open symbols) in Fig. 1 represent estimates of water production (whose ratio to gas production is very close to unity) from several sources (none direct H_2O measurements, and none in print at this writing). The data provided by Feldman et al. (1986) are H_2O values derived from the OH emission band at wavelength 309 nm, as observed with IUE (International Ultraviolet Explorer). Those provided by Millis et al. (1986) are ground-based photometric data for OH, multiplied by 15/14, since roughly one fifteenth of the water dissociates into H_2 and $O[^1D]$ at times near sunspot minimum. Those provided by Spinrad et al. (1986) are $O\,[^1D]$ values derived from its emission line at wavelength 630 nm, and have been multiplied by 15 (see above). Those provided by Gerard et al. (1986) are OH values derived from its emission lines near wavelength 18 cm, and have been multiplied by a factor 5 (as well as by 15/14, as above) to correct for inadequacies in the theory of OH line formation (as discussed at the 1986 NRAO Cometary Radio Astronomy Workshop).

The in-situ spacecraft data (the remaining symbols) in Fig. 1 have been reported by Gringauz et al. (1986; Vega-1 Plasmag), by Krasnopolsky et al. (1986; Vega-2 TKS), and by Krankowsky et al. (1986; Giotto NMS). The several values for Vega-2 arise from the use of data from emission lines at different wavelengths and from different modeling techniques for data reduction. The Earth-based and in-situ data are evidence for the variability of Halley's gas production on time scales of days or less, as supported by the variable Lyman-alpha observations from Suisei (Kaneda et al., 1986). All of the data in Fig. 1 are within a factor 3 of the model prediction, although the latter is clearly too low preperihelion; at the time of the spacecraft encounters the agreement is excellent, suggesting that the calibration using median values for several

Send offprint requests to: N. Divine

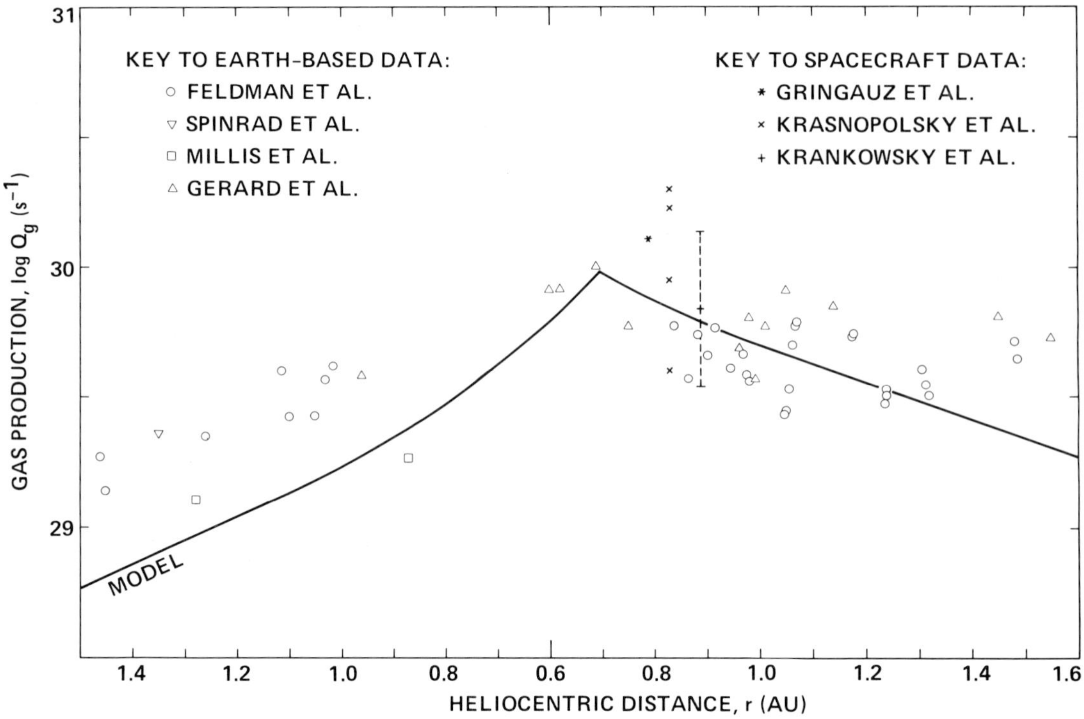

Fig. 1. Gas production as a function of heliocentric distance (preperihelion, left half; and postperihelion, right half) for Halley in 1985 and 1986. As explained in the text, the open symbols represent water production rates derived from Earth-based observations, whereas the other symbols and lines represent total gas production rates derived from encounter spacecraft data and from pre-encounter models, respectively. Pre-encounter models were based on the Halley light curve in 1910

recently observed comets shows that Halley's gas production is quite normal.

3. Dust flux and fluence

The Giotto and two Vega spacecraft were well inside the envelopes [of the order of 10^8 m, as suggested by the fountain model; cf. Eq. (7.9) of Divine et al. (1986)] for a few thousand seconds. For most of this time interval the predicted dust particle concentrations and fluxes are proportional to R^{-2}, where R is the cometocentric distance. To compare the absolute values of the dust number flux J detected by the several spacecraft-borne dust detectors it is useful to construct the scaled flux J_s, comparable to the dust flux at the surface of the comet nucleus, in the form

$$J_s = \frac{R^2 u_T}{R_N^2 V_s} J \quad (1)$$

for dust particles exceeding some size threshold. Here R_N is the nuclear radius, u_T is the terminal outflow velocity of the gas, and V_s is the spacecraft velocity relative to the nucleus. Figure 2 displays constant values at four mass thresholds, as obtained by applying Eq. (1) to the peak flux predicted for Giotto by Divine et al. (1986), and sample profiles from one experiment at three mass thresholds, for J_s (using $u_T/R_N^2 = 0.02\ \mathrm{km}^{-1}\ \mathrm{s}^{-1}$). In this particular comparison the ratio of the observed to model-predicted fluxes is near unity before time $t = -80$ s and near 3 at $t = +150$ s for mass $m > 10^{-14}$ kg, between 0.4 and 1.3 throughout the encounter for mass $m > 10^{-16}$ kg, and somewhat larger than unity for mass $m > 10^{-18}$ kg. Using similar comparisons from all available in-situ data (as published in the journal issue containing Mazets et al., 1986a), we conclude that (1) the model-predicted fluxes are close to the midrange of the in-situ data set, except at masses below 10^{-16} kg, where particles unanticipated by the models are present (see Sect. 4 below); (2) the absolute calibrations of the several experiments, even on different spacecraft, are generally consistent with one another (commonly within a factor 0.3 to 3, but there are a few exceptions to be investigated elsewhere); and (3) the time (or position) variations of scaled dust fluxes at fixed mass thresholds, as revealed by a single spacecraft detector, commonly range over a factor ten, and in some cases a factor near 100 (e.g., for DUCMA on Vega-1 near closest approach; Simpson et al., 1986), on scales greater than about 1 s (or about 100 km). The last of these three items results directly from the existence of real jets (i.e., significant variations in the dust and gas activity with time, insolation, and position on the nucleus surface), as observed in both ground-based and spacecraft images of Halley near the encounter times.

Some cometary dust particles were detected on board the several spacecraft at distances much greater then predicted by the model. For example, at $m = 10^{-16}$ kg, the fountain model apex distances are near 10^4 and 2 10^4 km for the two kinds of particles anticipated by Divine et al. (1986), corresponding to limiting cometocentric distances near 2 10^4 and 4 10^4 km (on the flanks of the paraboloidal envelopes) for the approach and departure legs of the flyby trajectories. At this mass Simpson et al. (1986) report detections well beyond 10^5 km, and they suggest that the radiation pressure efficiencies may be smaller than anticipated, and/or that large particles may be releasing smaller ones many hours after their departure from the nucleus. Later detailed modeling may permit assessment of these alternatives, and, in the latter case, whether simple sublimation of icy glue interior to the larger

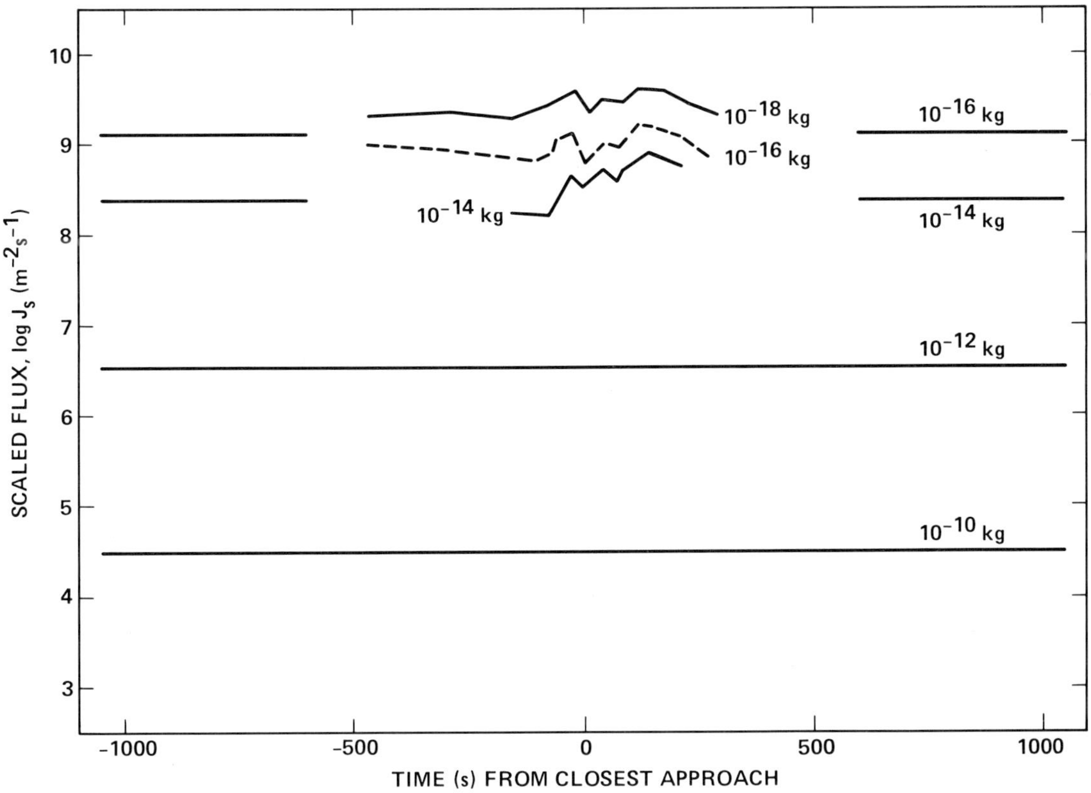

Fig. 2. Scaled flux [cf. Eq. (1)] at Halley as a function of time, for several mass thresholds, from Vega-2 SP-2 data (Fig. 2 of Mazets et al., 1986a; inner profiles), and for the model prediction (Divine et al., 1986; horizontal lines, omitted in data region)

particles, or possibly electrostatic fragmentation, might be important here.

A noteworthy event in the Giotto flyby is the attitude perturbation starting at $t = -14$ s (Reinhard, 1986). This event has been interpreted as the consequence of an off-center impact of a dust particle massive enough to induce a temporary spacecraft nutation of amplitude 0.9 deg. If the responsible dust particle had a mass of 0.1 g (cf. Sect. 10.2 of Divine et al., 1986), the probability of impact, using the nominal model prediction, is 5%, or somewhat larger if the gas production rate is increased (cf. Sect. 2 above). This probability was sufficiently large that the spacecraft and operations designs were intended to ensure survival and to minimize data loss (from interruption of downlink telemetry, actually experienced as a consequence of antenna attitude perturbation) in case of such an impact. Thus the flyby events, including telemetry recovery, validated the engineering analysis and design.

We note also that the deceleration inferred from Doppler data by the Radio Science Experiment aboard Giotto (Edenhofer et al., 1986) requires a total mass fluence near 0.32 g, larger than the (dust portion of the) nominal model prediction by a factor 9. If the major portion of this mass fluence were attributed to a single dust particle of mass >0.1 g, the model-predicted probability of impact is $<5\%$ (cf. Sect. 10.1 of Divine et al., 1986). One possible resolution is that the larger particles, whose outflow velocity (following release from the nucleus) is small, can remain within several thousand km of the nucleus for several days, and may thus be remnants from a prior interval of higher gas and dust activity. Alternatively, as suggested by McDonnell et al. (1986b), the dust size distribution exponent α (see Sect. 4 below) may decrease for particles larger (i.e., $>10^{-8}$ kg) than those registered in the DIDSY analog data. Not only is the fluence reported from DIDSY (excepting the discrete data for which α is near 0.5; McDonnell et al., 1986b) much smaller than required for the deceleration, it is also comparable to that from Vega-2 (cf. Figs. 3 and 4), whereas one would have expected it to be about 13 times larger (using the ratio of closest approach distances). One explanation may be that the dominant contribution to the mass fluence occurred in those dust jets encountered during the telemetry outages (for which McDonnell et al., 1986b, have simply scaled the flux by R^{-2}), as seems reasonable from the analysis by Edenhofer et al. (1986, Fig. 3).

In contrast, the Vega cumulative fluence data exceed the model prediction by factors between 1.0 and 10 at masses above 10^{-13} kg, and are smaller than the model prediction by factors between 0.3 and 1.0 at smaller masses.

4. Dust size distribution

The in-situ dust data at Halley suggest that power law representations of the particle distributions with mass require small (but non-zero) exponent values at masses $<10^{-15}$ kg and larger ones at masses $>10^{-12}$ kg (McDonnell et al., 1986b; Mazets et al., 1986a, b). To model these distributions we introduce a new function of dust particle mass m in which the dummy variable

$$x = (m/m_t)^{1/\gamma} \tag{2}$$

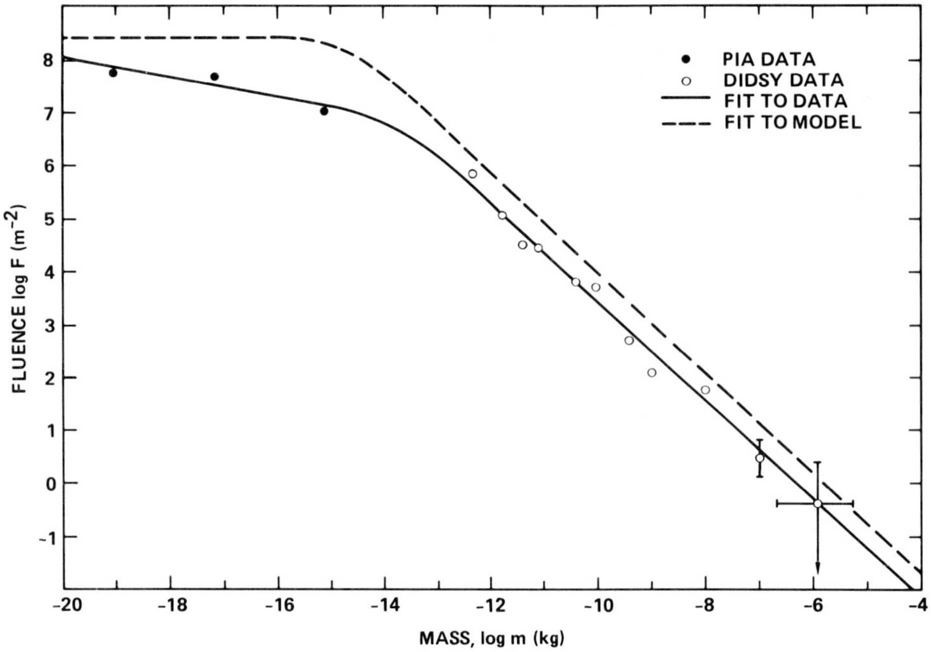

Fig. 3. Sample mass distributions from model (Divine et al., 1986) and as fitted to fluence data points from Giotto (Fig. 11 of McDonnell et al., 1986b). Parameter values for the dashed and solid lines, respectively, appear in the first and second rows of Table 1

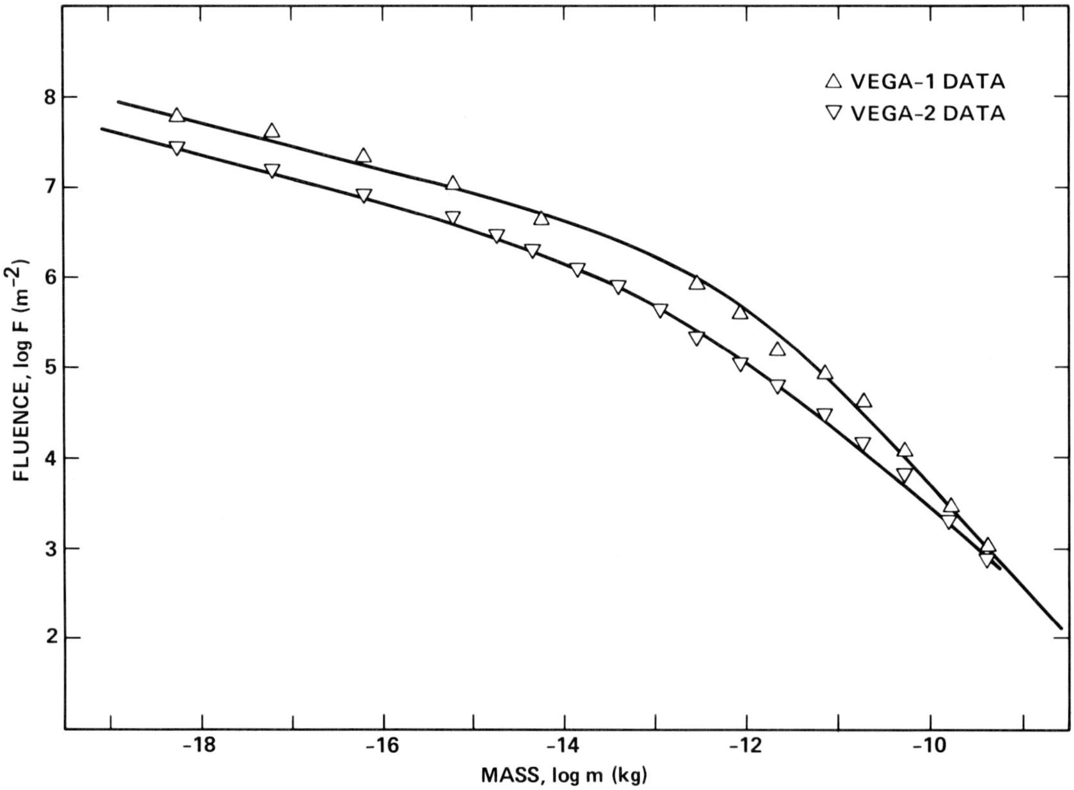

Fig. 4. Sample mass distributions fitted to fluence data points from Vega-1 and Vega-2 SP-2 data (Fig. 12 of Mazets et al., 1986b). Parameter values appear in the third and fourth rows of Table 1

is used in the form

$$J = J_t \left[\frac{(1+x)^{\beta-1}}{x^\beta} \right]^{\alpha\gamma} \tag{3}$$

for the cumulative flux (for fluence replace J by $F = \int J\,dt$). Here the transition between the behavior at masses $m < m_t$ and that at masses $m > m_t$, where m_t is the transition mass, is sharp for small $\gamma > 0$, and gradual for large γ. For masses large compared to m_t (i.e., for $x \gg 1$) the approximate negative exponent of mass m is α for the cumulative distribution J and $\alpha + 1$ for the incremental distribution $-dJ/dm$, whereas that of particle radius a is 3α for the cumulative distribution J and $3\alpha + 1$ for the incremental distribution $-dJ/da$, when $m = (4\pi/3)\varrho a^3$ for spherical particles

Table 1. Parameter values [cf. Eq. (3)] for fits to cometary dust mass distributions (cf. Figs. 3 and 4)

Figure Number	Fluence Curve	α	β	γ	m_t (kg)	F_t (m^{-2})
3	Model	0.955	0.0	1.0	2.13×10^{-15}	2.9×10^{8}
3	*Giotto*	0.94	0.19	1.0	2.0×10^{-14}	8.07×10^{6}
4	VEGA-1	1.19	0.218	1.9	1.0×10^{-12}	1.43×10^{6}
4	VEGA-2	0.90	0.29	2.16	1.6×10^{-13}	9.8×10^{5}

of constant density ϱ. Thus α has the same meaning as used by McDonnell et al. (1986b) and by Zarnecki et al. (1986); and for small masses ($x \ll 1$) the product $\alpha\beta$ substitutes for α in these exponent expressions.

The distributions specified by Eqs. (2) and (3) are used to model the local concentration, flux or fluence of dust particles in the coma, using particle mass m as the independent variable, and thus contrast with those suggest by Divine et al. (1986) and by Gombosi (1986), whose distributions model the production (at the nucleus surface) in terms of particle radius a. In the models cited, the local concentrations result from a convolution of the production distribution with the outflow velocity (a function of particle size and other properties) and with the time history of the dust production. Nevertheless it is possible to select parameters values for Eqs. (2) and (3) such that the local flux distribution mimics that derived in the prior models. The fluence distribution predicted for Giotto by Divine et al. (1986) is approximated by the entries in the first row of Table 1 [where F_t replaces J_t in Eq. (3), for fluence]. Here the fit to the model convolution results yields $\alpha = 0.955$, rather than 0.90 (which would be expected from an analytic approximation to the model, and which is good for $10^{-9} < m < 10^{-4}$ kg, but poor below 10^{-10} kg). Note that, as shown in Fig. 3, there are very few particles at masses less than 10^{-15} kg in the model distribution, as a result of the fact that they are "invisible", i.e., their scattering cross sections are too small to have contributed significantly to the Earthbased observations from which the models were derived.

The parameters in Eqs. (2) and (3) have also been fitted to several flux and fluence data sets from the three flyby spacecraft. Table 1 and Figs. 3 and 4 display the resulting values for the three fluence distributions. The value of the exponent α which fits the model for masses above m_t (see Table 1, first row) matches those for the data quite well for two (Giotto and Vega-2) of the three fluence cases, whereas it is too small for the third (Vega-1); this ratio (between satisfactory and divergent comparisons) is typical for the several data sets (both flux and fluence) we have fitted. Below the transition mass m_t (whose value ranges from 10^{-15} to 10^{-12} kg in the fits) the model predicts an absence of particles (note that $\beta = 0$ in the first row of Table 1), whereas both the data and the fits require that the abundance of particles continue to increase toward smaller masses (typical values of $\alpha\beta$ are 0.2; cf. Table 1). The relative contributions to the dust infrared cross section as a function of wavelength, and the shape of infrared synthetic spectra, are quite sensitive to the values used for α and m_t with these fits. In particular Hanner et al. (1987) find that ground-based infrared data (obtained near the times of the spacecraft encounters) can be fitted using parameter values similar to those in the last three rows of Table 1, with $0.9 < \alpha < 1.0$, and β near 0.2, but with somewhat smaller m_t, of the order of $5\ 10^{-15}$ to 10^{-14} kg.

We anticipate that additional comparisons, for example, among the model and data within and beyond dust jets, will be useful when more complete data compilations from the several sensors aboard the three flyby spacecraft are analysed and published.

Acknowledgements. We thank several of the authors cited for sharing their data with us in advance of publication. The research described in this paper was carried out by the Jet Propulsion Laboratory, California Institute of Technology, under contract with the National Aeronautics and Space Administration.

References

Divine, N., Fechtig, H., Gombosi, T.I., Hanner, M.S., Keller, H.U., Larson, S.M., Mendis, D.A., Newburn, R.L., Reinhard, R., Sekanina, Z., Yeomans, D.K.: 1986, *Space Sci. Rev.* **43**, 1

Edenhofer, P., Buschert, H., Bird, M.K., Volland, H., Porsche, H., Brenkle, J.P., Kursinsky, E.R., Mottinger, N.A., Stelzried, C.T.: 1986, 20th ESLAB Symp., ESA SP-250, Vol. **II**, p. 215

Feldman, P.D., Festou, M.C., A'Hearn, M.F., Arpigny, C., Butterworth, P.S., Cosmovici, C.B., Danks, A.C., Gilmozzi, R., Jackson, W.M., McFadden, L.A., Patriarchi, P., Schleicher, D.G., Tozzi, G.P., Wallis, M.K., Weaver, H.A., Woods, T.N.: 1986, paper presented at symposium *New Insights in Astrophysics*, 8 *Years of UV Astronomy with IUE*, London

Gerard, E., Bockelée-Morvan, D., Bourgois, G., Colom, P., Crovisier, J.: 1986: 20th ESLAB Symp., ESA SP-250, Vol. **I**, p. 589

Gringauz, K.I., Gombosi, T.I., Remizov, A.P., Apathy, I., Szemerey, I., Verigin, M.I., Denchikova, L.I., Dyachkov, A.V., Keppler, E., Klimenko, I.N., Richter, A.K., Somogyi, A.J., Szego, K., Szendro, S., Vladimirova, G.A.: 1986, *Nature* **321**, 282

Hanner, M.S., Tokunaga, A.T., Golisch, W.F., Griep, D.M., Kaminski, C.D.: 1987, *Astron. Astrophys.* (this issue)

Kaneda, E., Ashihara, O., Shimizu, M., Takagi, M., Hirao, K.: 1986, *Nature* **321**, 297

Krankowsky, D., Lämmerzahl, P., Herrwerth, I., Woweries, J., Eberhardt, P., Dolder, U., Herrmann, U., Schulte, W., Berthelier, J.J., Illiano, J.M., Hodges, R.R., Hoffman, J.H.: 1986, *Nature* **321**, 326

Krasnopolsky, V.A., Gogoshev, M., Moreels, G., Moroz, V.I., Krysko, A.A., Gogosheva, T., Palazov, K., Sargoichev, S., Clairemidi, J., Vincent, M., Bertaux, J.L., Blamont, J.E., Troshin, V.S., Valnicek, B.: 1986, *Nature* **321**, 269

Mazets, E.P., Aptekar, R.L., Golenetskii, S.V., Guryan, Y.A., Dyachkov, A.V., Ilyinskii, V.N., Panov, V.N., Petrov, G.G., Savvin, A.V., Sagdeev, R.Z., Sokolov, I.A., Khavenson, N.G., Shapiro, V.D., Shevchenko, V.I.: 1986a, *Nature* **321**, 276

Mazets, E.P., Sagdeev, R.Z., Aptekar, R.L., Golenetskii, S.V., Guryan, Y.A., Dyatchkov, A.V., Ilyinskii, V.N., Panov, V.N., Petrov, G.G., Savvin, A.V., Sokolov, I.A., Frederiks, D.D., Khavenson, N.G., Shapiro, V.D., Shevchenko, V.I.: 1986b, 20th ESLAB Symp., ESA SP-250, Vol. **II**, p. 3

McDonnell, J. A. M., Alexander, W. M., Burton, W. M., Bussoletti, E., Clark, D. H., Grard, R. J. L., Grun, E., Hanner, M. S., Hughes, D. W., Igenbergs, E., Kuczera, H., Lindblad, B. A., Mandeville, J.-C., Minafra, A., Schwehm, G. H., Sekanina, Z., Wallis, M. K., Zarnecki, J. C., Chakaveh, S. C., Evans, G. C., Evans, S. T., Firth, J. G., Littler, A. N., Massonne, L., et al.: 1986a, *Nature* **321**, 338

McDonnell, J. A. M., Kissel, J., Grun, E., Grard, R. J. L., Langevin, Y., Olearczyk, R. E., Perry, C. H., Zarnecki, J. C.: 1986b, 20th ESLAB Symp., ESA SP-250, Vol. **II**, p. 25

Millis, R. L., Schleicher, D. G., Thompson, D. T., Bus, S., Martin, L., Jansen, S., Birch, P. V., Martin, R., A'Hearn, M. F.: 1986, 20th ESLAB Symp., ESA SP-250, *Abstracts*, p. 219

Newburn, R. L.: 1981, ESA SP-**174**, 3

Reinhard, R.: 1986, *Nature* **321**, 313

Simpson, J. A., Sagdeev, R. Z., Tuzzolino, A. J., Perkins, M. A., Ksanfomality, L. V., Rabinowitz, D., Lentz, G. A., Afonin, V. V., Ero, J., Keppler, E., Kosorokov, J., Petrova, E., Szabo, L., Umlauft, G.: 1986, *Nature* **321**, 278

Spinrad, H., McCarthy, P. J., Strauss, M. A.: 1986 (private communication)

Zarnecki, J. C., McDonnell, J. A. M., Alexander, W. M., Burton, W. M., Hanner, M. S.: 1986, 20th ESLAB Symp., ESA SP-250, Vol. **II**, p. 185

Astron. Astrophys. 187, 873–878 (1987)

ASTRONOMY AND ASTROPHYSICS

Post-perihelion brightening of comet P/Halley: Springtime for Halley

P.R. Weissman

Earth and Space Sciences Division, Jet Propulsion Laboratory, Pasadena, CA 91109, USA

Received February 2, accepted April 21, 1987

Summary. Increased brightness and gas production rates for Halley's comet after perihelion can be explained as a result of seasonal effects on an obliquely rotating nucleus. The highly eccentric cometary orbit causes a rapid change in solar declination as the comet rounds perihelion, resulting in drastic changes in the insolation reaching the northern and southern hemispheres of the nucleus. The rapid heating of the northern hemisphere post-perihelion likely results in substantial cracking of the non-volatile surface crust due to thermal stresses, exposing areas of fresh volatile ices. The orientation of the triaxial ellipsoid nucleus may also play a role in exposing more surface area to continuous sunlight and sublimation after perihelion. Post-perihelion brightening models based on heat flow and storage in sub-surface layers of the nucleus pre-perihelion are likely not viable because of the low thermal conductivities of porous, low density cometary surface materials.

Key words: Halley's comet – post-perihelion brightening – obliquity – thermal conductivity – nucleus modeling – thermal stress – cracking

1. Introduction

Halley's comet has typically been brighter following perihelion passage, as compared with before (Newburn, 1981). This was again true during the 1985–86 apparition. Post-perihelion brightening is common, though not universal among short-period comets (Sekanina, 1979), and is in contrast to dynamically new long-period comets which are typically far brighter pre-perihelion, particularly at large solar distances (Oort and Schmidt, 1951).

No good explanation of post-perihelion brightening existed prior to the Halley spacecraft encounters in 1986. Some past work on thermal modeling of icy conglomerate nuclei (Kührt, 1984; Smoluchowski, 1986) suggested that heat flow into the nucleus surface on the pre-perihelion leg of the orbit could act as an additional energy source post-perihelion as the comet moved away from the sun and the nucleus surface cooled. However, the low thermal conductivities suggested for porous cometary surfaces, and indicated by the small lag angles observed for non-gravitational forces, would not allow sufficient energy to be deposited in the cometary surface layers. Also, such a model cannot explain those long and short-period comets which are brighter pre-perihelion.

2. Nucleus thermal modeling

A detailed thermal model (Weissman and Kieffer, 1981, 1984a, 1984b) of the behavior of icy conglomerate spheres in cometary orbits has been developed and used to study a variety of sublimation related processes on cometary nuclei. The model includes the effects of surface heat flow and thermal inertia, rotation period and pole orientation, and energy feedback from multiple scattering and thermal emission by dust in the cometary coma. The computer model was rerun following the Halley spacecraft encounters using a revised set of nucleus parameters based on results from the 1985–86 apparition. These parameters are summarized in Table 1. The nucleus radius was chosen to give a total surface area equal to that for the observed triaxial ellipsoid shape of Halley's nucleus (Sagdeev et al., 1986). It appears in spacecraft images that the entire nucleus surface area was not actively sublimating, but rather that a large fraction of it, on the order of 80% or more, was covered by an insulating non-volatile crust. However, 100% sublimation was assumed in the nucleus thermal model, so as to be able to use the model results to estimate the fraction of surface area which was, in fact, active.

The rotation period and pole orientation were taken from Wilhelm et al. (1986). Considerable doubts have now been raised as to the validity of the 2.2 day rotation period (Millis and Schleicher, 1986). However, the 2.2 day period is still sufficiently long for the Halley nucleus to be considered a "slow rotator." Results for running the comet nucleus thermal model with a 7.4 day period were virtually identical to that described below.

Complex precession models have been proposed for the Halley nucleus (Lissauer, 1986; Sekanina, 1987) in an attempt to accommodate both the 2.2 and 7.4 day periods. It is not yet clear which, if any, of the precession models is correct. In fact, it is not clear yet if such complex models are required as one of the observed periods may be shown to be spurious. Given the current

Table 1. Halley nucleus thermal model parameters

Nucleus radius	5.5	km
Surface albedo	0.04	
Surface ice density	0.5	$\mathrm{g\,cm^{-3}}$
Thermal inertia	0.003	$\mathrm{cal\,cm^{-2}\,s^{-1/2}\,K^{-1}}$
Rotation period	2.2, 7.4	days
Rotation pole obliquity	30.	degrees
Dust to gas ratio	0.2	
Dust particle density	0.8	$\mathrm{g\,cm^{-3}}$

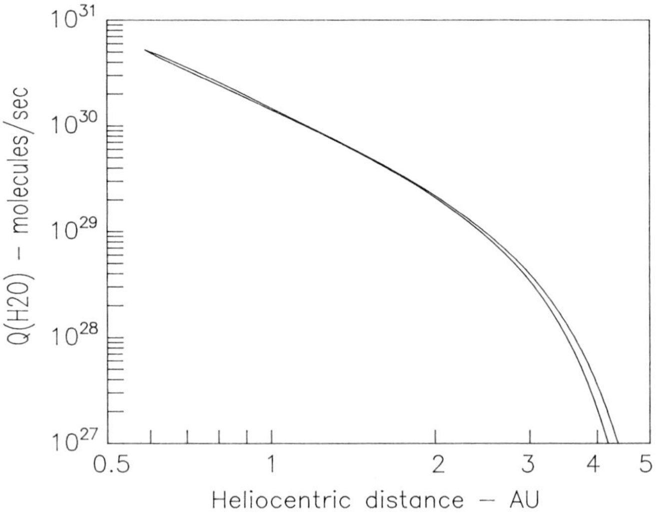

Fig. 1. Predicted gas production rate for Halley's comet as a function of heliocentric distance, for the nominal nucleus model parameters in Table 1

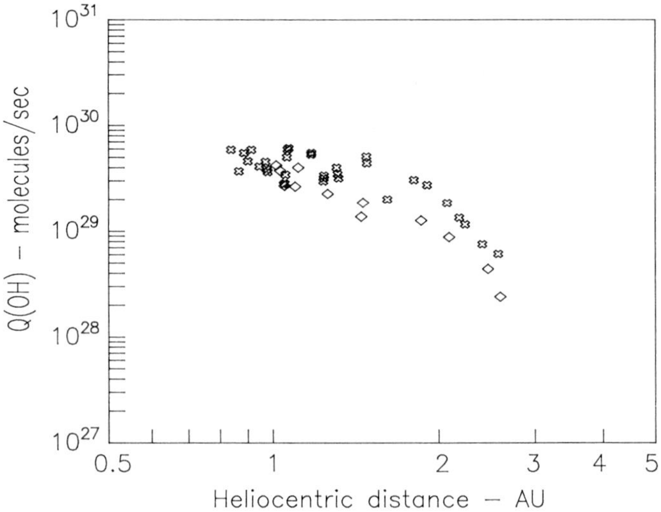

Fig. 3. Observations of OH production rate for Halley's comet in 1985–86 by the International Ultraviolet Explorer. Open diamonds are pre-perihelion observations; open crosses are post-perihelion

uncertainty, the approach followed in this paper will be to assume a simply rotating, non-precessing nucleus.

The surface thermal inertia was chosen to be that for a compacted but still porous snow, about one order of magnitude less than that expected for solid water ice, but much higher than measured values for the icy surfaces of the Galilean satellites (Morrison, 1977). The effects of higher and lower values of thermal inertia were also investigated.

Predicted water production rates for the nominal nucleus model are shown in Fig. 1. The pre- and post-perihelion legs of the orbit are very nearly identical, with the post-perihelion rates being very slightly greater, particularly at large heliocentric distances. For comparison, observed OH production rates for Halley's comet in 1985–86 are shown in Figs. 2 and 3 for measurements made by Stewart (1987) with the Pioneer Venus (PV) spacecraft, and by Feldman et al. (1986) with the International Ultraviolet Explorer (IUE) spacecraft, respectively. In

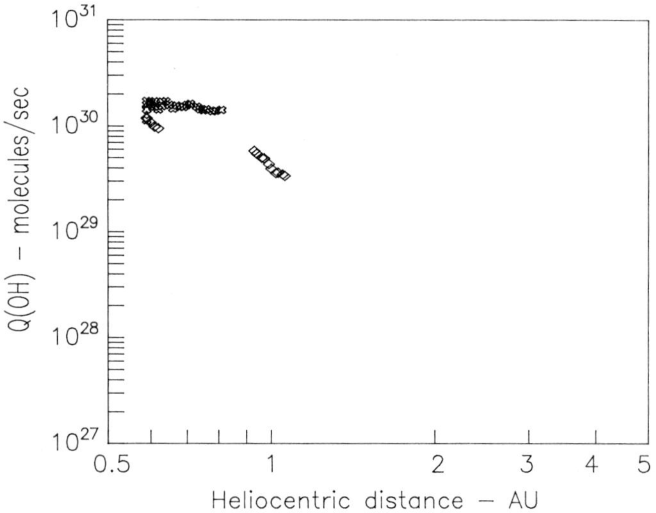

Fig. 2. Observations of OH production rate for Halley's comet in 1985–86 by the Pioneer Venus spaccecraft. Open diamonds are pre-perihelion observations; open crosses are post-perihelion

each figure, pre-perihelion observations are shown by open diamonds, while post-perihelion measurements are indicated by open crosses. The much higher post-perihelion gas production rates are clearly evident in each dataset. Also, it is seen that the comet's gas production rate did not follow an approximately inverse square dependence with heliocentric distance while within about 2.0 AU of the sun, as would be expected from energy arguments.

Comparison of the predicted gas production rate curve with the PV and IUE observations shows that the fraction of area apparently active on the nucleus surface varied considerably throughout the 1985–86 apparition. Initially, on the inbound leg, it is about 40% at 2 AU, but then drops to 30% and eventually 20% shortly before perihelion. At perihelion of 0.5871 AU the gas production rate rises rapidly, to about 35% of the average surface area being active. The gas production rate as found by Pioneer Venus remains fairly constant even as the comet began to recede from the sun. This could be interpreted to indicate that the average fraction of active area continued to increase following perihelion as the sun continued to move northward in declination.

There is a discontinuity in the match between the PV and IUE data at about 0.8 AU outbound, the OH production rate appearing to drop by somewhat more than a factor of two. If this is real it would represent a sudden decrease in the fraction of active nucleus surface area from about 58% to 27%. More likely, the problem is observational. Note also that the IUE data shows greater fluctuations between successive data points than the PV data, probably indicative of the shorter integration times for the IUE instrument, and thus its greater ability to see short-term variations in the comet's brightness.

At the time of the Giotto flyby the estimated fraction of active surface area (comparing the thermal model with the IUE observations) is 25%. This compares well with a previous estimate of 15% to 30% by Ip and Rickman (1986) using a similar thermal model. At about 1.0 AU the OH production rates observed by IUE increase back to about 50% of the surface area being active, and then jump again at 1.4 AU to about 90%, before dropping down to about 70% beyond 2.0 AU.

The estimated fraction of active area on the Halley nucleus based on Giotto imaging is about 10% (Keller et al., 1986). Given the estimates above, it would seem that at least some of the gas production must additionally come from the apparently "inactive" areas on the nucleus. This would be the result of evolving gas from buried dirty ice layers diffusing through a porous non-volatile crust. It also would seem to indicate that the crust does not serve to completely insulate the dirty ice mix below it, and therefore the non-volatile layer is likely fairly thin. On the other hand, the Giotto estimate of the fraction of active surface area, based on an oblique view of the nucleus's morning hemisphere, may be lower than the average for the entire sunlit hemisphere of the nucleus.

The above discussion has not yet considered the question of the non-spherical shape of the Halley nucleus, and the actual changing aspect presented to the sun by the triaxial ellipsoid nucleus as the comet moved around its highly eccentric orbit. Also, as already noted by observers (Festou et al., 1986) comet Halley displayed rapid fluctuations in gas production rate and brightness. These have been interpreted as being due to rotation, as "active" and "inactive" hemispheres on the rotating nucleus became illuminated by the sun. More complex patterns of surface activity have been suggested by observations which indicate that the rotation period is not 2.2 days, but rather 7.4 days (Millis and Schleicher, 1986) with a double-maximum light curve. This double-maximum may reflect a complex combination of effects from the large variation in illuminated surface area as the triaxial ellipsoid nucleus rotated (and precessed?) about some axis, and some asymmetric distribution of active areas on the nucleus surface. Still, these fluctuations do not explain the gross differences in the overall gas production rates for the nucleus pre- and post-perihelion.

To examine the effect of increased surface heat flow acting as an additional energy source post-perihelion, additional cases of the computer thermal model were run with both higher and lower values of the surface thermal inertia. The resulting predicted gas production curves are shown in Figs. 4 and 5 for thermal inertias of 0.03 (10 times the nominal value) and 0.0001 ($\frac{1}{30}$th the nominal value), respectively. In each case, the gas production

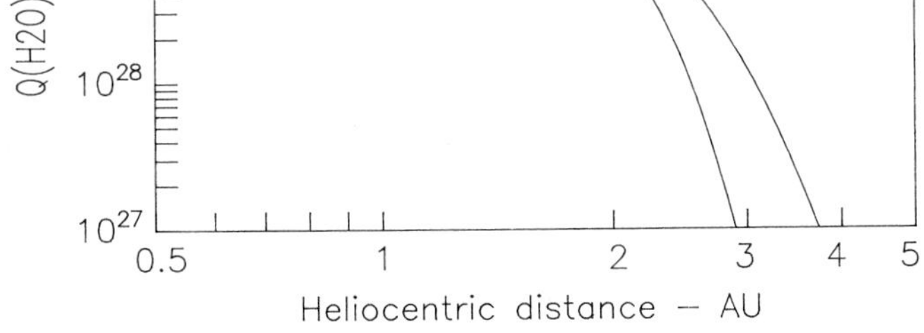

Fig. 4. Predicted gas production rate for Halley's comet assuming a surface thermal inertia of 0.03, equal to that for solid water ice. Note the large variation in gas production rate at large solar distances

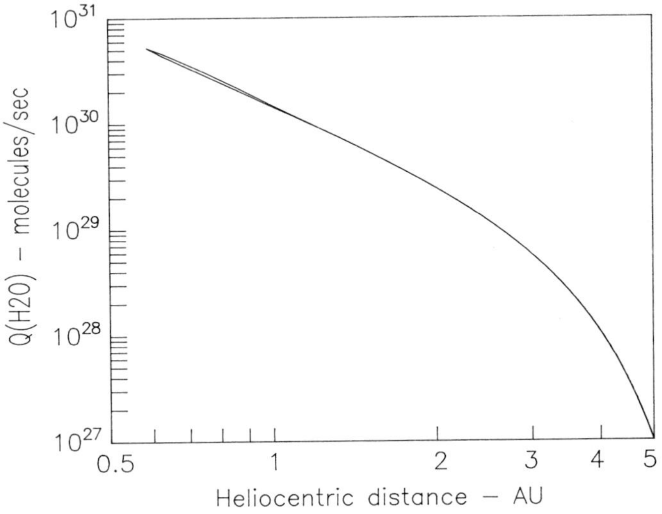

Fig. 5. Predicted gas production rate for Halley's comet assuming a surface thermal inertia of 0.0001, a very low value. The pre- and post-perihelion legs of the orbit are virtually identical

rates close to the sun are not appreciably changed, and there is very little pre- versus post-perihelion asymmetry. On the other hand, the behavior of the nucleus at large solar distances is quite different. For the very low thermal inertia value illustrated in Fig. 5, the nucleus gas production rate remains fairly high out to 5 AU or more, and identical on the inbound and outbound legs of the orbit. But for the very high value of surface thermal inertia in Fig. 4, typical of solid water ice, the comet's gas production rate beyond 3 AU is greatly lowered on the inbound leg of the orbit, and also substantially decreased on the outbound leg.

These sharp differences mean that the observed behavior of Halley's comet at large solar distances can be used to bound the average thermal inertia of the nucleus surface material. Observations at 4 to 6 AU are critical in this respect. Such observations for Comet Halley on the inbound leg of its orbit are shown in Fig. 6 (Wyckoff et al., 1985), along with predictions of the comet's brightness by M. Belton. The slowly rising dotted curve is the predicted brightness of the bare cometary nucleus based on its recovery observations in 1982. The more sharply rising dashed curve at the right of the figure is a prediction of the total coma brightness as water ice began to sublimate from the nucleus, based on the comet thermal model of Weissman and Kieffer (1984a,b). It is seen that the observations of comet Halley followed this prediction quite well. The prediction was based on a cometary nucleus with a radius of 3 km and an albedo of 0.15. Given the current thermal model parameters in Table 1, the same gas production rates would be achieved assuming that 26% of the average surface area was active. Interestingly, at coma "turn on" in early 1985 at about 5.8 AU, the predicted gas production rate for the entire Halley nucleus was only about one gram/second.

Given that the estimate of active surface area above is in the range of estimates for the inbound leg of the orbit from the PV and IUE observations, then the assumed value of 0.003 for the mean thermal inertia is likely close to the correct value. Observations of Halley outbound at these same solar distances will be extremely valuable in additionally tying down this critical number.

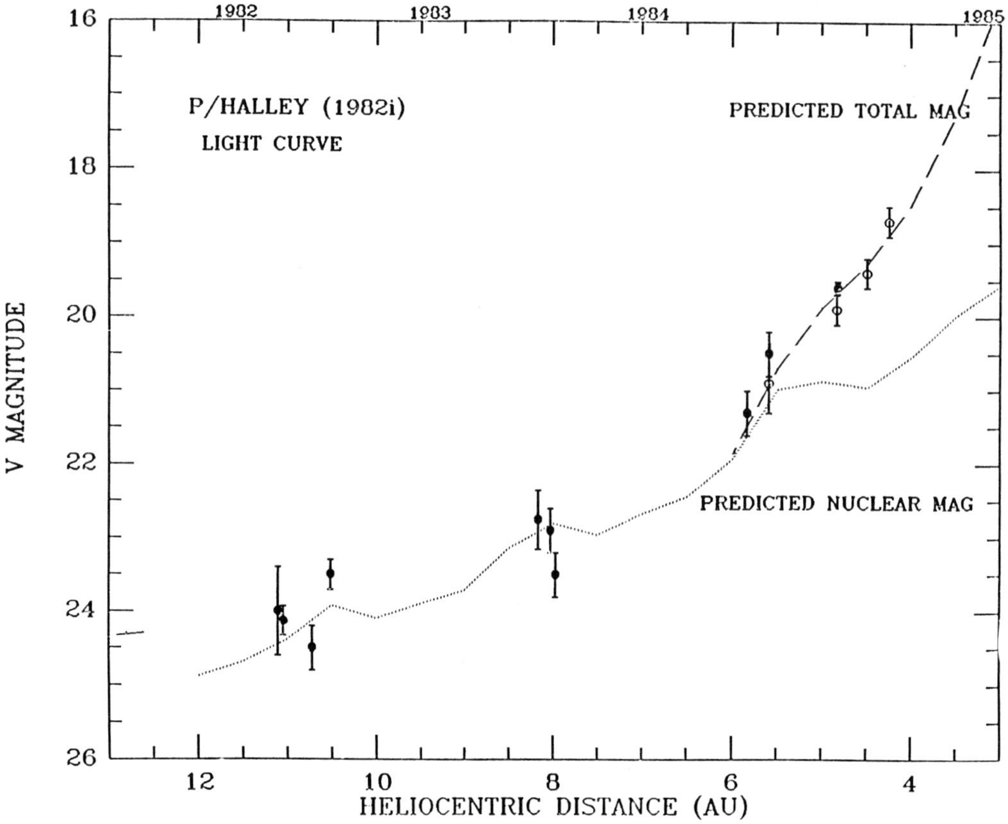

Fig. 6. Observations of Halley's comet at large solar distances inbound in 1982–85. The dotted curve is the predicted nucleus magnitude based on its brightness at recovery in 1982. The dashed curve is the predicted total magnitude (by M. Belton) based on results of the comet thermal modeling program. From Wyckoff et al. (1985)

3. Seasonal variations

The thermal modeling results above still do not provide any explanation of the substantially different pre-perihelion and post-perihelion behavior of comet Halley. This is because the state of knowledge of cometary phenomena has now passed beyond many of the simplifying assumptions used in the thermal model. The weakest of these appears to be the assumption of a nucleus surface with uniform composition and thermal properties. The Vega and Giotto images (Sagdeev et al., 1986; Keller et al., 1986) as well as analyses of Halley activity in 1910 (Sekanina and Larson, 1986) show that the nucleus is divided into discrete active and inactive areas. The distribution of these active areas appears to be random over the nucleus surface, but there does not yet exist enough data to establish that fact clearly. Also, no conclusions can be drawn about the temporal persistence of the active areas, though there is at least some evidence that they can last for several nucleus rotation periods or more.

Some insight into a possible mechanism for explaining the post-perihelion behavior of comet Halley can be obtained by considering the latitude of the sub-solar point on the Halley nucleus as a function of time from perihelion. A plot of solar declination is shown in Fig. 7 for three different solutions for the Halley rotation pole orientation (Wilhelm et al., 1986; Grün et al., 1986; Sekanina and Larson, 1986). In each case the solar declination is at its southernmost minimum shortly before perihelion, moves rapidly northward as the comet rounds perihelion, and remains near its maximum value as the comet moves away from the sun. The direction "north" is defined by the direction of the nucleus's rotational angular momentum vector. The very rapid change in solar declination is a result of the large change in true anomaly as the comet rounds perihelion on its highly eccentric orbit.

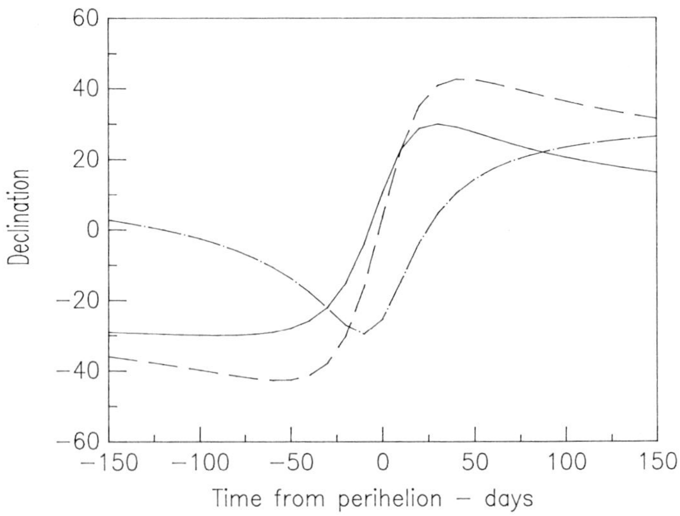

Fig. 7. Declination of the sub-solar point on the Halley nucleus for three possible solutions for the rotation pole orientation: solid curve, Wilhelm et al. (1986); dashed curve, Grün et al. (1986); dot-dash curve, Sekanina and Larson (1986). In each case the sun moves rapidly northward as the comet rounds perihelion

In effect, what is happening is that the Halley nucleus is going through very rapid seasonal changes, with the spring equinox for the northern hemisphere occurring very close to perihelion. The period following perihelion corresponds to spring and then summer on the northern hemisphere of the Halley nucleus. The strong southern declination of the sun for at least 50 days prior to perihelion means that the northern hemisphere is poorly illuminated, with a polar cap of 30 to 45 degrees in latitude in total darkness. Average surface temperatures on the northern hemisphere can be expected to be very low with the northern hemisphere contributing very little to the comet's total gas production. As the comet rounds perihelion the northern hemisphere will be suddenly and completely illuminated, with the solar zenith angle continuing to decrease for 50 to 100 days after perihelion. This all occurs when the comet is closest to the sun, so the sudden heating of the northern hemisphere of the nucleus is quite extreme. The effect is likely increased even further when one considers the triaxial ellipsoid shape of the Halley nucleus.

Theoretical studies of the behavior of volatile ices have shown that the typical thermal stresses expected in comets will result in substantial cracking (Kührt, 1984; Green, 1986). The sudden heating described above is even more extreme than the gradients considered in those studies, and will result in extensive cracking and break-up of any insulating non-volatile crust on the nucleus surface, thus creating large new active areas in the northern hemisphere. As the comet continues along its orbit and the sun continues to move northward in declination, the decreased solar zenith angle for northern latitudes serves to compensate for the increasing solar distance, keeping the gas production rates high.

The gas production from the southern hemisphere of the Halley nucleus pre-perihelion is not as great as from the northern hemisphere post-perihelion for several reasons. The southern hemisphere is heated more gradually as the comet moves in towards the sun, thus allowing the thermal stresses to be more easily accommodated. Also, the slower heating would allow for a gradual build-up of a lag deposit of larger non-volatile particles on the nucleus surface, actually decreasing the effective fraction of the surface area available for sublimation. This is, in fact, what was seen for the fit between the IUE and PV observations and the comet nucleus thermal modeling results in Sect. 2, above.

Because of the high eccentricity of the Halley orbit, the length of the respective "seasons" in the northern and southern hemispheres are quite different. Depending on the orbit rotation pole used, summer in the southern hemisphere can either be quite short, on the order of only about 100 days, or very long, on the order of tens of years. On the other hand, summer in the northern hemisphere will occupy either a large fraction or virtually all of the comet's 76 year orbital period. After Halley's many orbits around the sun, the effects of these very different heating profiles could have led to profound differences in the evolution and thermal properties of the surfaces in the respective hemispheres.

4. Discussion

The mechanism proposed here to explain post-perihelion brightening of Halley's comet follows as a natural consequence of what has been learned about comet Halley from the spacecraft encounters and from Earth-based observations in 1985–86. It is now known that the comet has an irregularly shaped nucleus, slowly rotating with an obliquity of about 30°, and covered with some irregular pattern of active and inactive areas. Sublimation from the comet is clearly controlled by water ice but the fraction of active surface area seems to vary with heliocentric distance and may be related to the evolution of non-volatile surface crusts and the changing solar declination as the comet moves around its orbit. Very rapid changes in solar declination as the comet rounded perihelion suddenly illuminated a northern hemisphere that had been poorly illuminated during the inbound leg of the orbit, resulting in sudden heating and a large increase in the fraction of nucleus surface area available for sublimation. This led to a very high level of activity which was reflected in much higher observed gas production rates post-perihelion.

This mechanism can be generalized to all comets and may help to explain the great range of cometary behavior that is observed. It is expected that the orientation of cometary rotation poles is random over the celestial sphere, and likely unrelated to the orbital planes of the comets. Thus, each comet would have its own unique distribution of "seasons" around its orbit. While highly eccentric orbits would still be expected to have very large "seasonal" changes near perihelion, the change might be such as to take a comet from spring to fall in the relevant hemisphere, rather than the Halley case where it appears to go from winter to summer in the northern hemisphere. For such a case, the comet's behavior might appear quite symmetric around perihelion. Similarly, the orientation of the nucleus might be such that a brief intense "summer" for a previously unilluminated hemisphere occurs just before perihelion, rather than after. Thus, the nucleus orientation appears to be a free parameter that could be used to explain any cometary light curve. However, this cannot be entirely true since certain trends in cometary behavior are observed as a function of orbital type. This was referred to in Sect. 1, above. Dynamically new comets are typically brighter pre-perihelion, particularly beyond 3 AU, and the larger fraction of short-period comets do show some post-perihelion brightening, though usually not as great as for Halley. Pre-perihelion brightening of dynamically new comets has usually been explained in terms of a pristine layer of more volatile ices and trapped free radicals which is sublimated away on the first perihelion passage (Weissman, 1986). On the other hand, the behavior of short-period comets may be as a result of the substantial thermal evolution they have undergone after so many orbits close to the sun.

As already noted, alternative explanations of post-perihelion brightening which invoke heat flow into the cometary crust are likely not viable because of the low conductivity of the cometary surface materials. Another mechanism, proposed by Rickman et al. (1984), involves heat release from the amorphous to crystalline ice transition, presumably occurring at some buried interface, and not triggered until the perihelion thermal wave penetrates to the depth of the interface. However, Prialnik and Bar-Nun (1987) have modeled just such transitions for Halley and showed that they occur in brief spurts, followed by long quiescent periods lasting tens or even hundreds of orbits. In addition, they found that the thermal wave did not penetrate to the required depth, about 15 meters, until the comet was 5 AU from the sun on its outbound leg. For Halley, the brightening is evident right at perihelion.

Clearly, the behavior of any comet on any given orbit is the result of a great number of factors which include its original composition and formation history, its past thermo-physical evolution, its current orbit and orientation, and its detailed nucleus shape and physical parameters. The Halley spacecraft encounters in 1986 have given us a fantastically enticing, but unfortunately brief and incomplete view of a cometary nucleus. They have helped to explain, but have not provided the full information needed to understand the complete range of cometary phenomena observed for comet Halley and other comets. That information will have to come from future spacecraft missions to comets, in particular, the Comet Rendezvous/Asteroid Flyby mission, and the Comet Nucleus Sample Return mission.

Acknowledgments. The author is very grateful to Hugh Kieffer for his collaboration on the comet nucleus thermal model. Also, the author thanks Ian Stewart and Paul Feldman for the use of their published and unpublished Halley observations, and Susan Wyckoff for the figure from her Nature paper. This work was supported by the NASA Planetary Geology and Geophysics Program, and was performed at the Jet Propulsion Laboratory under contract with the National Aeronautics and Space Administration.

References

Feldman, P.D., Festou, M.C., A'Hearn, M.F., Arpigny, C., Butterworth, P.S., Cosmovici, C.B., Danks, A.C., Gilmozzi, R., Jackson, W.M. McFadden, L.A., Patriarchi, P., Schleicher, D.G., Tozzi, G.P., Wallis, M.K., Weaver, H.A., Woods, T.N.: 1986, *20th ESLAB Symp.* ESA SP-250 Vol. 1, p. 325

Festou, M.C., Feldman, P.D., A'Hearn, M.F., Arpigny, C., Cosmovici, C.B., Danks, A.C., McFadden, L.A., Gilmozzi, R., Patriarchi, P., Tozzi, G.P., Wallis, M.K., Weaver, H.A.: 1986, *Nature* **321**, 361

Green, J.R.: 1986, *Bull. Amer. Astron. Soc.* **18**, 800

Grün, E., Grasser, U., Kohoutek, L., Thiele, U., Massonne, L., Schwehm, G.: 1986, *Nature* **321**, 144

Ip, W-H., Rickman, H.: 1986, in *The Comet Nucleus Sample Return Mission*, ESA SP-249, p. 181

Keller, H.U., Arpigny, C., Barbieri, C., Bonnet, R.M., Cazes, S., Coradini, M., Cosmovici, C.B., Delamere, W.A., Huebner, W.F., Hughes, D.W., Jamar, C., Malaise, D., Reitsema, H.J., Schmidt, H.U., Schmidt, W.K.H., Seige, P., Whipple, F.L., Wilhelm, K.: 1986, *Nature* **321**, 320

Kührt, E.: 1984, *Icarus* **60**, 512

Lissauer, J.: 1986, talk presented at the *20th ESLAB Symp.* Heidelberg, FRG

Millis, R.L., Schleicher, D.G.: 1986, *Nature* **324**, 646

Morrison, D.: 1977, in *Planetary Satellites*, ed. J.A. Burns, Univ. Arizona Press, Tucson, p. 269

Newburn, R.L.: 1981, in *The Comet Halley Gas and Dust Environment*, ESA SP-174, p. 3

Oort, J.H., Schmidt, M.: 1951, *Bull. Astron. Inst. Netherlands* **12**, 259

Prialnik, D., Bar-Nun, A.: 1987, *Astrophys. J.* **313**, 893

Rickman, H., Froeschle, C., Klinger, J.: 1984, in *Ices in the Solar System*, eds. J. Klinger, D. Benest, A. Dolfus, R. Smoluchowski, p. 419

Sagdeev, R.Z., Blamont, J., Galeev, A.A., Moroz, V.I., Shapiro, V.D., Shevchenko, V.I., Szego, K.: 1986, *Nature* **321**, 259

Sekanina, Z.: 1979, *Icarus* **37**, 420

Sekanina, Z.: 1987, *Nature* **325**, 326

Sekanina, Z., Larson, S.M.: 1986, *Astron. J.* **92**, 462

Smoluchowski, R.: 1986, in *Asteroids, Comets, Meteors II*, eds. C.I. Lagerkvist et al., Uppsala University, p. 305

Stewart, A.I.F.: 1987, *Astron. Astrophys.* (in press)

Weissman, P.R., Kieffer, H.H.: 1981, *Icarus* **47**, 302

Weissman, P.R., Kieffer, H.H.: 1984a, *J. Geophys. Res.* **89**, C358

Weissman, P.R., Kieffer, H.H.: 1984b, *Adv. Space Res.* **4**, 221

Weissman, P.R.: 1986, in *The Comet Nucleus Sample Return Mission*, ESA SP-249, p. 15

Wilhelm, K., Cosmovici, C.B., Delamere, W.A., Huebner, W.F., Keller, H.U., Reitsema, H., Schmidt, H.U., Whipple, F.L.: 1986, *20th ESLAB Symp.*, ESA SP-250, Vol. 2, p. 367

Wyckoff, S., Wagner, R.M., Wehinger, P.A., Schleicher, D.G., Festou, M.C.: 1985, *Nature* **316**, 241

Astron. Astrophys. 187, 879–888 (1987)

ASTRONOMY
AND
ASTROPHYSICS

P/Halley dust characteristics: a comparison between Orionid and Eta Aquarid meteor observations and those from the flyby spacecraft

D.W. Hughes

Department of Physics, The University, Sheffield S3 7RH, UK

Received February 3, accepted April 21, 1987

Summary. Dust from P/Halley was detected by the Giotto and Vega spacecrafts in mid March 1986 and can also be observed from Earth every May and October in the form of the Eta Aquarid and Orionid meteor showers. Unfortunately, the two techniques measure dust in dramatically different places (with respect to the parent comet) and over non-overlapping mass ranges. This paper compares and contrasts the fluxes and mass distribution indices obtained.

Key words: dust – meteoroid stream – mass index

1. Introduction

Comet Halley has passed through the inner solar system many times. Hughes (1985) estimated that the recent apparition was of the order of its 3000th visit. Each time it moves through perihelion the dirty snowball nucleus loses mass and some fraction of this mass, that made up of dust particles more massive than about 10^{-9} g, goes into the meteoroid stream. These dust particles have orbits which are similar but not exactly the same as that of the parent comet and soon (i.e. in a few dozen cometary orbital periods) an annulus of dust is formed around the cometary orbit, an annulus which is thinnest at perihelion.

The geometry is shown in Fig. 1. When Earth passes through the stream in April and May it is, at its closest, 0.065 AU (9,800,000 km) away from the comet's orbit. In October when it again passes through the stream, it is 0.18 AU (28,000,000 km) away. Both these values should be compared with the Giotto flyby of 1986 March 14 when the spacecraft got to within 605 km of the active cometary nucleus at its descending node. The orbital parameters are given in Table 1.

The dust observed by the Giotto spacecraft was made up of three components,

(1) dust that has recently (hours, days) been emitted by the nucleus of P/Halley,

(2) dust in Halley's meteoroid stream that was emitted at a previous perihelion passage,

(3) sporadic dust which just happens to be in orbits which collide with Giotto during its recording phase.

The ratio between the components is mass dependent but for the Giotto mass range (10^{-17} g – 10^{-3} g) component (1) is by far the most dominant. Dust detected from Earth in the two Halley meteoroid streams is of type (2). It has taken a long time for the dust to get from the comet orbit to the meteoroid stream orbit (see Table 1). Hughes (1985, 1986) calculated that the comet was first captured into its present inner solar system orbit about 3000 perihelion passages ago (i.e. ~200,000 years). At that time its dirty snowball nucleus had an equivalent radius of about 9.6 km. At present this radius is 5.6 km, and the mass loss is equivalent to a 140 cm thick layer being removed from all over the surface at each apparition. As time progresses the mass loss at each apparition decreases. (It is obvious from the Halley Multicolour Camera images that the activity of the nucleus is patchy, in fact temporal changes in the light curve of the comet can be interpreted in terms of a *single* active region on a spinning nucleus, and so this simple model of the nucleus losing an 'onion skin' layer at each apparition is a crude approximation. The general smoothness of the 'avocado pear' shaped nucleus does indicate however that mass loss over many apparitions leads to a rounding off of the surface contours as opposed to a production of deep pits and crevasses. So the 'uniform layer loss' assumption seems reasonable in the long run.)

The cometary decay has been modelled by Hughes (1985) and is applied specifically to P/Halley in Table 2. As is to be expected, the majority of the dust in the stream comes from the outer regions of the original nucleus. The relative velocities between the individual dust particles do however, lead to the stream being well mixed, a tendency that will in time produce a homogeneity throughout the volume. We must stress the words 'in time'. The velocity of a dust particle with respect to the parent nucleus, when it has stopped being accelerated by gas pressure (i.e. when it is further than about 10 R_c from the nucleus where R_c is the equivalent radius of that nucleus) is proportional to $(R_c/r_m\rho_m r^{9/4})^{1/2}$ where r_m and ρ_m are the radius and density of the dust particle and r is the heliocentric distance of the nucleus at the time of emission (see Hughes, 1977). Visual meteoroids (mass ~ 10^{-1} g) will spread round the cometary orbit much slower than radio meteoroids (mass ~ 10^{-5} g) and very much slower than satellite meteoroids (say mass ~ 10^{-11} g). So the smoothness of the stream spatial density is both time and mass dependent. *If* the source was switch off, one would expect the spatial density of P/Halley's radio meteoroids to be reasonably uniform after about 1500 years (*uniform* not *constant*, the density decreases gently from perihelion to aphelion). But the source is not switched off. Every 76 years an additional short pulse of dust is added to the stream. In fact this pulse is also modulated by the

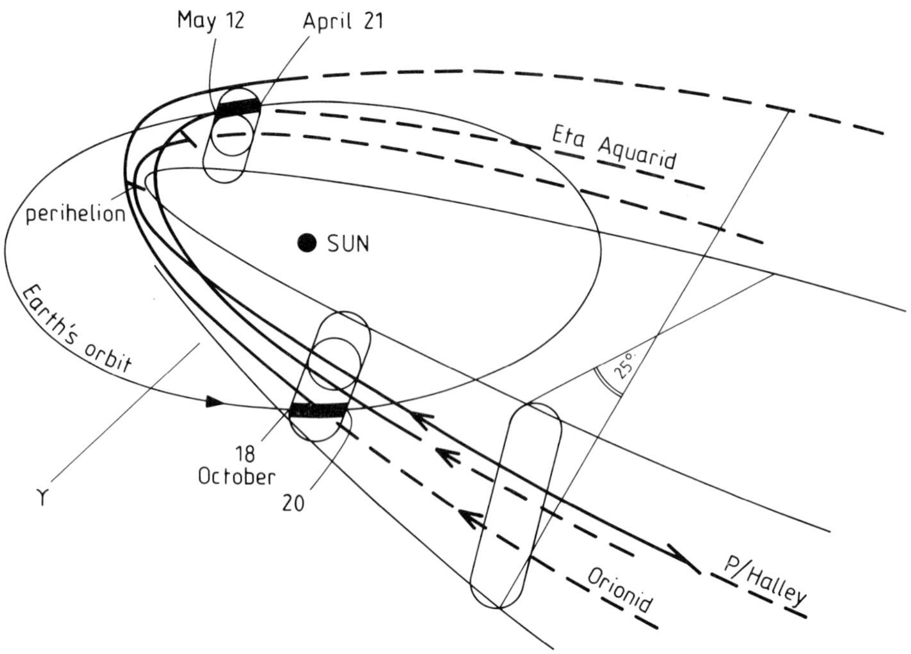

Fig. 1. The geometry of the intersection of the Halley meteoroid stream and the Earth's orbit. Two showers result, the Eta Aquarids in May and the Orionids in October. The orbit of the comet is librating and this develops a flattened tube of dust which contains both the comet and the stream. Following Kozai (1979) the majority of the tube subtends an angle of about 25° along the major axis of the orbit

Table 1. Orbital parameters of the comet and its meteoroid showers

	Ω°	ω°	e	i°	a(AU)	q(AU)
Halley (1982i)	58.15	111.85	0.967	162.24	17.94	0.587
Eta Aquarids	42.4	95.2	0.958	163.9	13.3	0.560
Orionids	28.0	82.5	0.962	163.9	15.03	0.571

spin of the nucleus. So the stream is expected to have a uniform component due to dust emitted many cometary orbits ago with an added filamentary structure due to dust emitted recently from the limited active regions on the nucleus. Collisional fragmentation will tend, in the case of low interparticle relative velocities, to increase the cross-section of the stream and decrease the variations in the spatial density. High velocity collisions will remove meteoroids from the stream altogether, leading to an enhancement of the sporadic background (see Hughes, 1978). The Orionid and Eta Aquarid showers are produced by meteoroids which find themselves at differing distances from the present orbit of the comet and thus could have been emitted at different times during the comet's evolution. In considering the relative positions of the comet and the meteoroid stream we must be mindful of the fact that the cometary orbit is not fixed in the solar system and has suffered considerable perturbation during its history (Yeomans and Kiang, 1981). The orbit is gradually precessing with respect to a fixed first point of Aries. For example $(\omega - \Omega)$ was 53°.70 in 1986 and 57°.46 in 87 BC. Not only is the source of the meteoroid stream moving, the stream itself will be precessing in a similar fashion.

Table 2. The decay of P/Halley

Apparition number	Radius of nucleus (km)	Mass of dust in meteoroid stream (g)	
0	9.6	0	0%
500	8.9	$1.27\,10^{17}$	25%
1000	8.2	$2.35\,10^{17}$	47%
1500	7.5	$3.26\,10^{17}$	65%
2000	6.8	$4.02\,10^{17}$	80%
2500	6.1	$4.64\,10^{17}$	93%
3000 (The present)	5.4	$5.00\,10^{17}$	100%

2. The meteoroid stream

When a meteoroid hits the Earth's upper atmosphere it ablates leaving behind a trail of ionized and excited atoms and molecules. It is this train that produces the cylindrical reflector of radar waves and the brief streak of light – the visual meteor. Verniani (1973) observed underdense radio meteors ($\alpha < 2\,10^{14}\,\mathrm{m}^{-1}$) and found empirically that the number of electrons per metre, α, left behind by the retarding meteoroid was given by

$$\alpha = 7.7\,10^{-10} m^{0.92} V_G^{3.91} \tag{1}$$

where m (g) is the instantaneous mass of the meteoroid and V_G (cm s^{-1}) is its geocentric velocity. The electron line density, α, can be related to the absolute magnitude, M, of the meteor (its zenithal magnitude at a height of 100 km) by

$$M = 40 - 2.5 \log \alpha\,. \tag{2}$$

Table 3. Meteor magnitudes and meteoroid masses for the Halley showers and the comet dust

Meteor absolute magnitude	30	27.5	25	22.5	20	17.5	15	12.5	10	7.5	5	2.5	0	−2.5	−5	−7.5
Meteoroid mass (g)	2.0 10^{-15}	2.4 10^{-14}	3.0 10^{-13}	3.6 10^{-12}	4.4 10^{-11}	5.4 10^{-11}	6.6 10^{-9}	8.1 10^{-8}	9.9 10^{-7}	1.2 10^{-5}	1.5 10^{-4}	1.8 10^{-3}	2.2 10^{-2}	0.27	3.3	40
												←—— Naked eye meteors ——				
											←—— Telescopic and Super Schmidt ——					
								←—— Television with image intensification ——								
							←—— Radar meteors ——									
	—— Giotto, DIDSY ——→															
	—— Vega, DUCMA ——→															
	—— Vega, SP-2 ——→															

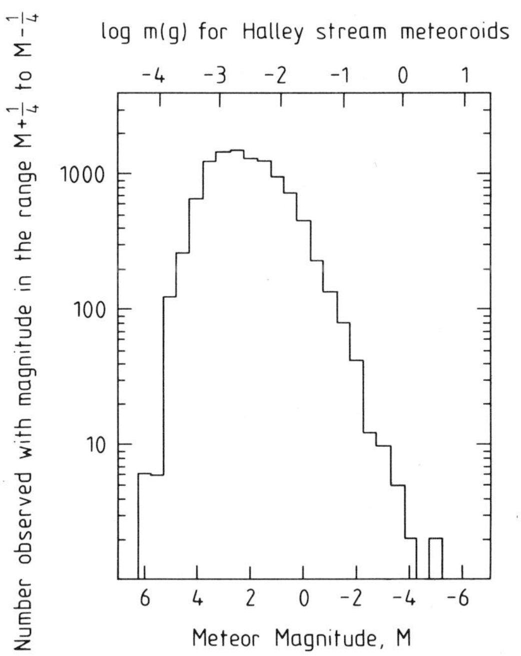

Fig. 2. 10500 sporadic visual meteors have been grouped according to their absolute magnitude, M. The number observed in each interval $M + 0.25$ to $M - 0.25$ is plotted as a function of M

Halley's comet and the two meteor showers associated with it are both retrograde and this results in a V_G of $65\,10^5\,\mathrm{cm\,s^{-1}}$. Meteoroid mass can be related to absolute magnitude by using this V_G value in conjunction with Eqs. 1 and 2, giving

$$M = -3.81 - 2.3 \log m. \tag{3}$$

Particles with masses below the micrometeoroid limit can radiate heat away from their surfaces faster than they receive energy from the impinging air molecules. For the Halley stream this limit is $\sim 9\,10^{-13}$ g, and all meteoroids more massive than this produce a train. Magnitudes and masses, together with observing techniques are given in Table 3.

Most meteor stream observing campaigns essentially make only two measurements. When a meteor or dust particle is observed one records a) the time of occurrence and b) the mass of the meteoroid. Different techniques have different mass sensitivities. This can be illustrated in terms of visual data. The time-honoured method of observing meteors – standing outside on a clear moonless night and gazing intently at the sky – still provides many useful results. Figure 2 shows the observed distribution of *sporadic* meteors as a function of their visual magnitude (data comes from Stohl and Millman, 1973). The field of view of the normal eye is a cone of apex angle around 120°, the sensitivity to streaking lights (i.e. shooting stars) being strongly dependent on the position of the meteor in the field of view. From Fig. 2 it can be seen that the most probable observed magnitude is +2.5. Only 15% of the meteors are brighter than $+0^{m}.75$ and 75% have magnitudes in the range $3^{m}.75 > M > 0^{m}.75$. The limiting magnitude was $+6^{m}$. No meteors were seen brighter than -5^{m} simply because these are so rare that none occured during the observing session. The curve can be divided into two portions. The linear region for magnitudes less than 0^{m} is not markedly affected by observational selection and the gradient of this curve can be converted into a magnitude distribution index, a, or a

Table 4. Detection probability factors $P(M)$ for visual meteors of magnitude M

M	6	5	4	3
$P(M)$	$7\,10^{-4}$	$8\,10^{-3}$	0.064	0.232
M	2	1	0	−1
$P(M)$	0.343	0.420	0.48	0.57
M	−2	−3	−4	−5
$P(M)$	0.73	0.87	0.97	0.98
M	−6	−7		
$P(M)$	1.0	1.0		

mass distribution index, s (The number of meteors, N, with magnitudes less than (i.e. brighter than) M is assumed to be given by $N \propto a^M$; the number of meteoroids with masses greater than m is given by $N \propto m^{1-s}$). It can be seen from Fig. 2 that only about 9% of the visual meteors can be used to obtain the magnitude or mass distribution indices. For magnitudes greater than 0^m the histogram progressively falls away from the straight line as fewer and fewer meteors are detected. The value $+6^m$ is often quoted as the normal limit for naked eye observations. Very few of these faint meteors are actually seen and these lucky ones have to be very close to the most sensitive regions of the eye's field of view.

Most meteor detection systems suffer from some degree of observational bias. The naked eye meteor problem was tackled by Kresáková (1966) who introduced a factor, $P(M)$, defined as being the probability of noting, in the field of view, a meteor of magnitude M. Values for this factor are given in Table 4. Unfortunately, the easiest way of calculating a series of $P(M)$ values is to measure the index a for bright meteors and then to extrapolate to fainter magnitudes using a fixed a value. This usually masks variations in a, these variations often being the main aim of the search.

A distribution curve for the Halley showers (as opposed to the sporadic meteors) would have a different gradient for the $M < 0$ section. The other features would, however, be essentially similar to those shown in Fig. 2. Equation 3 has been used to provide the upper abscissa in Fig. 2, and this relationship between magnitude and mass is only applicable to the Halley stream. Zero magnitude is equivalent to a mass of $2.2\,10^{-2}$ g, so visual measurements of a and s only apply to particles with masses greater than this value.

Shower measurements can yield values for the flux and for the two indices. The flux can be converted into a meteoroid stream spatial density by using intersection geometrical factors and the relative velocities between the Earth and the meteoroids. Figure 3 is a guide to the stream-Earth intersection geometry. Remember (see Fig. 1) that the comet orbit is not perpendicular to the Earth's path. Typical visual flux curves for both showers are shown in Fig. 4.

Radar fluxes for the showers are shown in Fig. 5. The Orionid results (Hajduk 1983) are an amalgam of data from many radar systems working at frequencies around 40 MHz. Usually only echoes with durations greater than 1 s are counted, a value that is equivalent to a limiting magnitude of around 8^m. Values obtained by Jones (1983) for the 1980 and 1981 Orionids are shown in Fig. 6.

The following general conclusions can be drawn (see Hajduk, 1970, 1973).

(1) There is a considerable variation in the flux and in the position of maximum activity from year to year. This indicates that the density of the stream varies not only along the orbit, i.e. as a function of heliocentric distance but is also 'patchy' and 'filamentary' and varies as a function of time at any specific spot. These density variations can reach an amplitude of 4:1. No fixed periodicity is found in the occurrence frequencies of maximum or minimum activity. Remember that a meteoroid stream is a very dynamic entity. New dust is being added and, with time, the structure should become less pronounced, but dust is also continually being lost from the stream. The ratio 4:1 is obviously

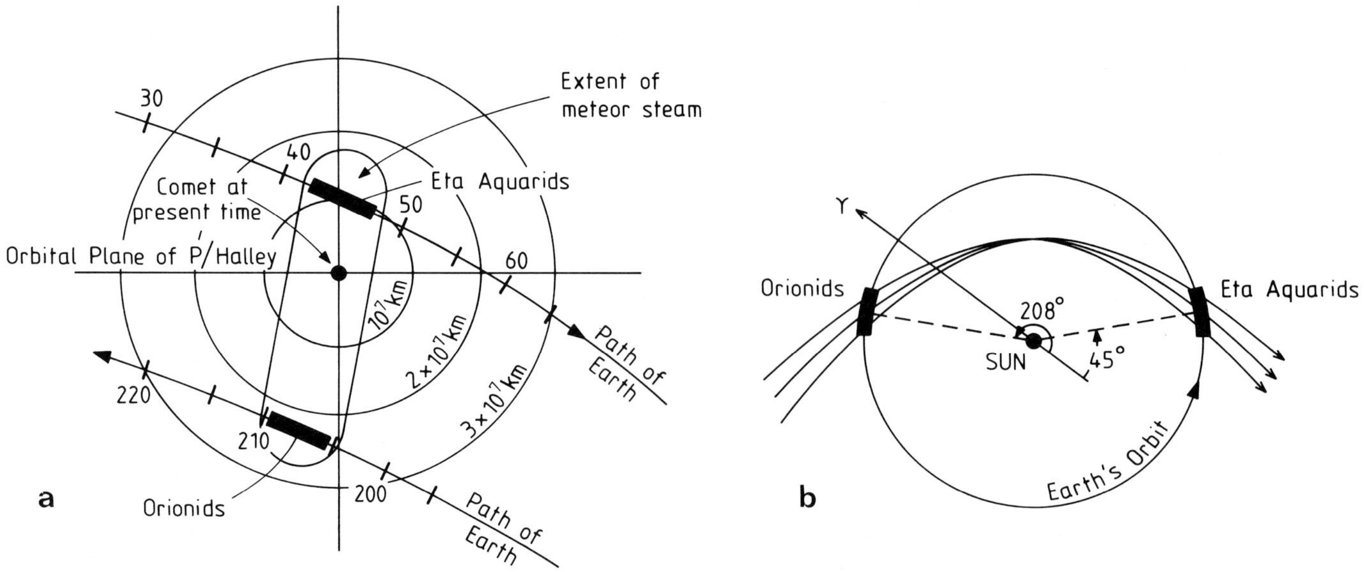

Fig. 3a and b. The stream-Earth orbit intersection geometry. The Earth crosses the stream twice. In **a** the numbers along the Eta Aquarid and Orionid lines represent the solar longitude (see Fig. 1). The black bars along the trajectories represent the regions of maximum meteor activity. **b** gives some impression as to the volumetric extent of the stream (after Hajduk, 1980)

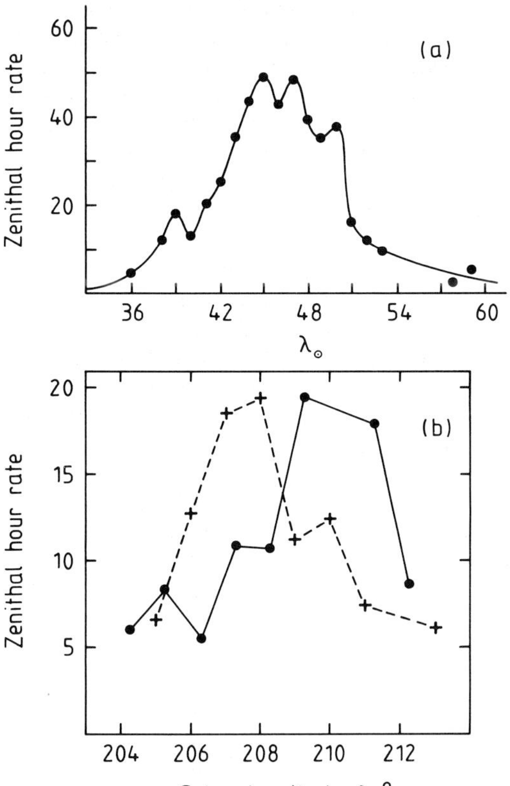

Fig. 4a. The visual Eta Aquarid rate as a function of solar longitude. 2100 Eta Aquarid meteors were obtained from 121 man hours of observing. The observing rate f_0 has been converted to a zenith rate f_z by using $f_z = f_0$ cosec(ε + 3) where ε is the radiant elevation (after Hajduk and Buhagiar, 1982). **b** Visual Orionid rates ● from Meteor News, No. 48 (Jan. 1980), + from Prentice (1938), see Lovell (1954)

a function of the time scales of these separate processes.

(2) In a long series of observations, certain sequences of years show a regular progression in the movement of the position of maximum activity (usually by around 1° in solar longitude per year). This gradual displacement of the activity peak can be explained by successive encounters with a specific stream filament.

(3) The strength of this filamentary structure is more noticeable in the radar (low mass) data than it is in the visual data. The enhanced density in the filaments could be due to variations in the activity of specific regions on the cometary nucleus or to variations in the general activity from one apparition to the next. The degree of variability is probably a function of the mass of the emitted dust particles.

(4) The Eta Aquarids are slightly more constant in activity than the Orionids, their radio maximum varying between 300 and 460 hr^{-1} whereas the Orionid maximum varies between 240 and 560 hr^{-1}.

3. Fluxes

There are two approaches to the representation of meteoroid fluxes. One is to follow Jones (1983). His plot of the number of Orionids that hit every km^2 of the Earth's upper atmosphere each hour, at the time of maximum activity is shown in Fig. 7. the relevance of this is easily visualised from Figs. 3a and b in which the Earth is seen ploughing through the stream with the meteoroids striking its leading surface. Figure 7 is a cumulative flux curve, each point representing the number of meteors that are observed having masses *greater than* a specific mass. Curves can easily be plotted in a non-cumulative fashion in which the number of meteoroids in a specific mass range m to $m + dm$ is represented on the ordinate. Figure 2 is a typical example. A non-cumulative representation invariably requires more and better data. Cumulative data produces an easily calculable a and s value but suffers from the fact that the graphical points are not independent, a low mass point, showing how many particles there are that are larger than the specified mass, containing an accumulation of *all* the higher mass data points. Standard deviations of a and s values are thus difficult to access from a cumulative graph.

The second approach is to consider percentages of the total mass of dust in the stream. Figure 8 has been taken from Hughes (1975) in which it is assumed that a typical meteoroid stream contains a total of 10^{15} g of dust. A logarithmic mass interval, m^n to m^{n+1} g can then be selected and the mass of dust made up of individual particles with masses in that interval can then be plotted as a function of mass. Figure 8 uses mean values for the mass distribution index, s, in meteoroid streams. The index has been found to be ~ 1.71 in the region $10^{-13} < m < 10^{-1}$ g and ~ 2.27 in the region $10^{-1} < m < 10^{+6}$ g (see Hughes, 1974). It can be seen that about 48% of the stream mass is in the form of particles in the 10^{-2} to 1 g mass range, 73% in the 10^{-3} to 10 g range and 86% in the 10^{-4} to 100 g range.

This conclusion is supported by Hajduk (1982) who finds, for the Halley stream, that the relative contribution to the total mass of the stream reaches a maximum for particles of masses of $\sim 10^{-2}$ g. The gradients of the histogram in Fig. 8 fortunately make the mass limits rather unimportant. The lower limit can be set by a radiation pressure cut off such that everything less massive than about 10^{-15} g simply gets speedily blown out of the stream (Kresák, 1976). The upper limit is governed by the particle which can just escape from the cometary nucleus as it is accelerated away by momentum transfer from the sublimating gases. Whipple (1951) suggests a mass value of about 10^5 g for this upper limit. This would mean that there are no meteors brighter than magnitude −15 in the Halley stream.

Using meteoroid stream fluxes and mass distribution index values, McIntosh and Hajduk (1983) have calculated that the total mass of the Halley stream is $5\,10^{17}$ g (see also Hajduk and Cevolani, 1981).

4. Mass distribution index

The mass distribution index, s, is defined such that the number of meteoroids having individual masses between m and $m + dm$ is proportional to m^{-s}. The cumulative number of particles with masses greater than m will be proportional to m^{1-s} (see for example, Hughes, 1972). So the gradient of the curve in Fig. 7 is $(1 - s)$ and the gradients of the histograms in Fig. 8 are $(2 - s)$.

A summary of the s values obtained for Eta Aquarid and Orionid meteors is given in Table 5 and Fig. 9. How do these results compare with the expectations hinted at in Fig. 8? Notice that the mean solar system meteoroid stream, had a change in s values at about 0.1 g. For lower masses s was less than 2.0, for higher masses s was greater than 2.0. From equation 3 it can be seen that Halley stream dust particles of masses 0.01, 0.1 and 1 g

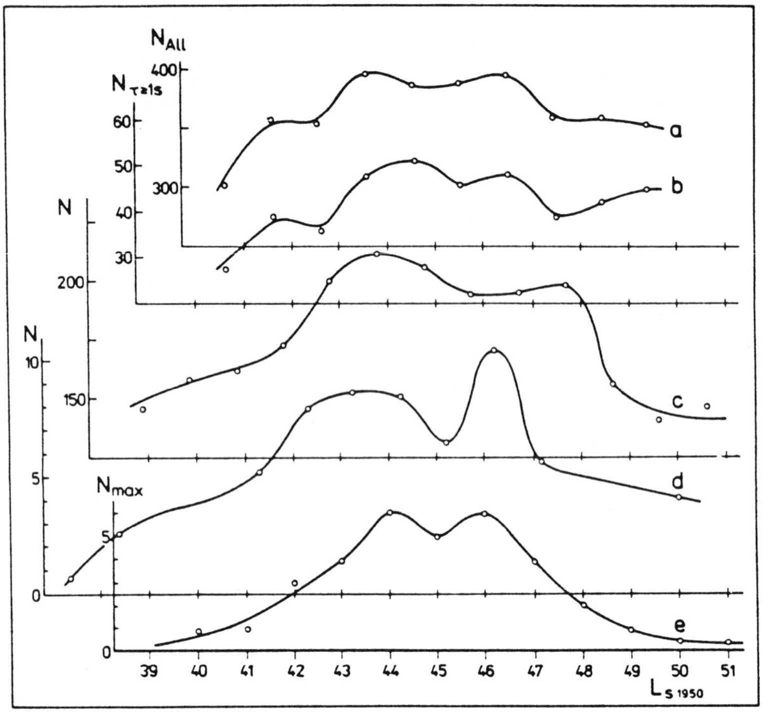

Fig. 5a. Mean meteor hourly rates during the Eta Aquarid shower period: a) Ottawa radar data (1958–1967) – all echoes; b) Ottawa radar data (1958–1967) – echoes with duration, $\tau \geq 1$ s; c) New Zealand radar data (1960, 1963–65); d) New Zealand visual data (1928–33); e) peak rates from visual and radar data (1910–1971) (from Hajduk, 1980)

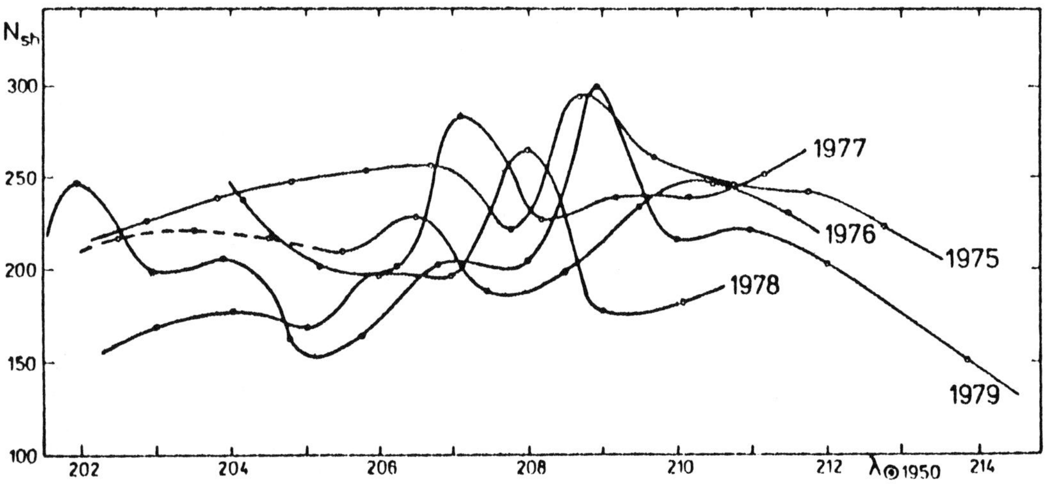

Fig. 5b. Relative radar meteor echo rates during the Orionid shower periods in the years 1975 to 1979 (from Hajduk, 1983)

have equivalent meteor magnitudes of 0.79, -1.51 and -3.81 respectively. The kindest thing that can be said about the data given in Table 5 and Fig. 9 is that the expected variation of s with mass is not obvious. The scatter in the data is horrendous, especially when one remembers that s is a power. It is also disheartening when one considers the significance of s. Values of s greater than 2.0 indicate that mass is concentrated in the small particles and values of s less than 2.0 indicate the opposite, i.e. that mass is concentrated in the large particles.

The data hints that the Orionids have a higher s value than the Eta Aquarids and this could be interpreted as indicating that the s value decreases as one approaches the present core of the stream, ie to-day's orbit of Halley's comet. Whether this is due to some temporal variation in stream parameters caused by mass dependent perturbation effects (i.e. Ponyting-Robertson) or collision probabilities or whether the s of the dust varies as a function of depth in the nucleus will be left to a further paper. From distance considerations alone it is tempting to suggest that the Orionids were released from the comet some time before the Eta Aquarids.

The Giotto and Vega data should help solve this problem. Unfortunately (from a meteor scientist's standpoint) the leading surface area of Giotto was so small that it detected very few large ($m > 10^{-3}$ g, $M < 3.1$) particles (the largest impacting particle had a mass of $1.1\,10^{-3}$ g). In fact a statistically significant sample was only possible for dust particles with masses less than 10^{-5} g. Also the four hour pre-encounter data-take only encompassed a distance of 10^6 km which, as can be seen from Fig. 3a,

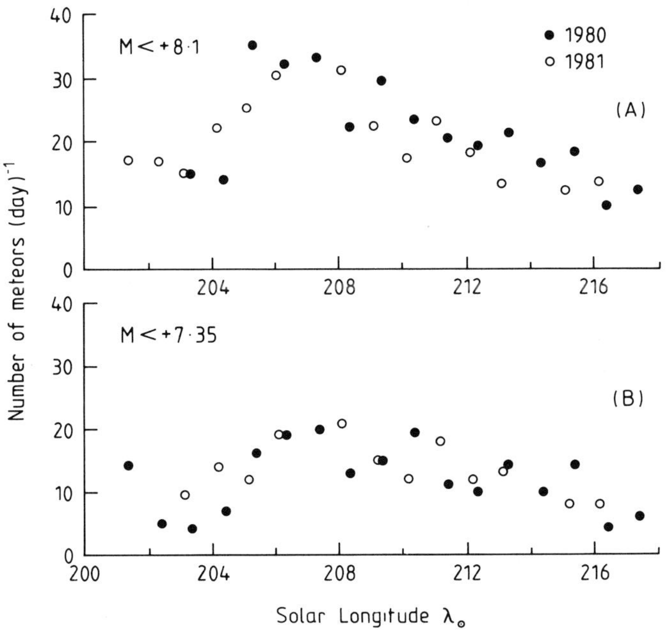

Fig. 6. Radar meteor echo rates for the Orionids taken from Jones (1983)

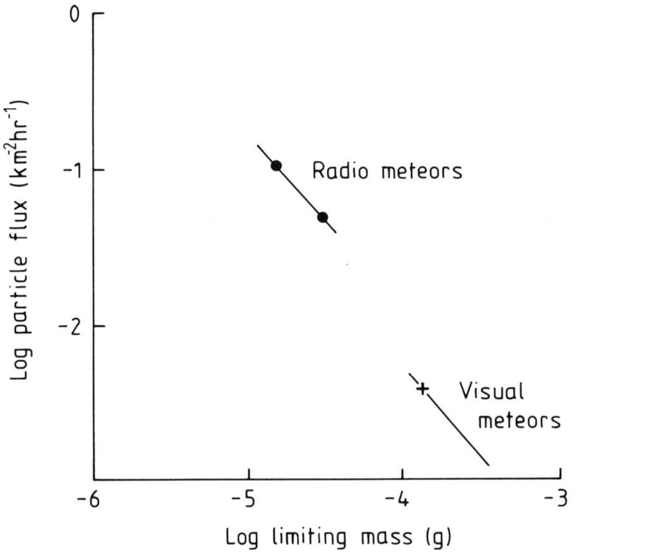

Fig. 7. A typical flux curve, this one being for the Orionid shower. The flux to each km^2 of the 'front' of the Earth's atmosphere, of meteoroids having a mass greater than a specific mass, m, is plotted as a function of that mass. The number of meteoroids with individual masses between m and $m + dm$ is taken as proportional to m^{-s}. Cumulative curves, as shown above, have a gradient of $(1 - s)$

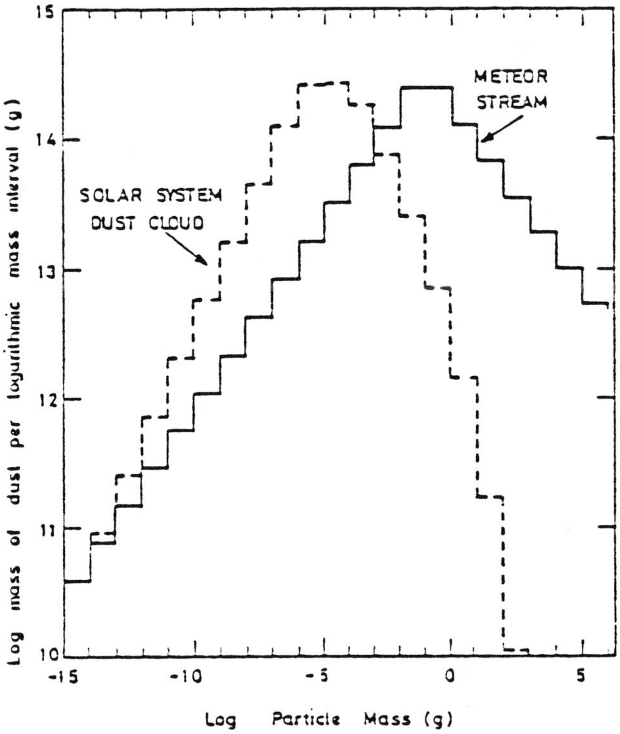

Fig. 8. The bold histogram represents the mass of dust in each logarithmic mass interval of the 10^{15} g in an average meteor stream. It has been assumed that the stream particles have a mass distribution index of 1.71 in the region $10^{-13}\,\text{g} < m < 10^{-1}\,\text{g}$ and 2.27 in the region $10^{-1}\,\text{g} < m < 10^{+6}\,\text{g}$; these values being mean stream s values in these mass ranges (see Hughes, 1975). The dashed histogram represents the distribution of the dust after it has been perturbed into the solar system dust cloud

is only a small fraction of the stream radius. The Giotto spacecraft did, however, sample the small dust particles close to their source.

5. The spacecraft results

Halley's comet was visited by three dust detecting spacecraft in March 1986. Brief details are given in Table 6. The spatial density of the dust and thus the flux to the detectors increased as the spacecraft approached the nucleus. A typical result is shown in Fig. 10, these data being taken from the Vega 2, DUCMA experiment (Simpson, 1986). As can be seen, the flux to the spacecraft varies approximately as R^{-2} where R is the distance to the cometary nucleus. The variability of the flux, by a factor of about 50% about the mean, is due to filamentary structures of width 200 to 300 km, possibly caused by the discrete nature of the dust emission process and the fact that only a small area of the surface of the nucleus is active.

The picture is even more complicated because the mass distribution index of the dust changes as a function of R. So particles of different masses have differing spatial density variation characteristics. This is illustrated in the Giotto DIDSY data (Fig. 11) which clearly shows the flux increasing as the nucleus is approached and the mass index changing (McDonnell et al., 1986a). Similar results are obtained by the SP-2 experiment on Vega 1 and 2 (Mazets, 1986), see Fig. 12, and the DUCMA experiment on the two Vega spacecraft, see Fig. 13 (Simpson, 1986). This variation of s with R is only to be expected, mass fractionation being caused by the variation in emission velocity and radiation pressure effects (see Massonne, 1985). For the total flux to the front shield of the Giotto spacecraft McDonnell et al. (1986b) quoted an s value 'at Giotto' of 1.96 ± 0.06 this converting to a (model dependent) value of 2.13 ± 0.06 at the surface of the nucleus.

We have already criticised the meteor astronomers for the lack of agreement between their values for the mass distribution index, s. Unfortunately space physicists seem to fare little better.

Table 5. Mass distribution indices, *s*

Magnitude range	*s*	Technique	Ref.
Eta Aquarids			
$-1.7 < M_R < 0.5$	2.0	Radar echo duration	Weiss (1961)
$0.5 \leq M_R \leq 1.75$	1.4	Radar echo duration	Weiss (1961)
$5 \leq M_R \leq 7.5$	1.61 ± 0.1	Radar echo amplitude	Weiss (1961)
$(M_V < 1)$	1.90	Visual	Levin (1953)
$-3.5 < M_V \leq 2.5$	1.725	Visual $P(M) = 1.0$	Kresáková (1966)
$-3.5 < M_V \leq 2.5$	1.95	Visual, variable $P(M)$	Kresáková (1966)
	1.94	Radar echoes	Hajduk (private communication)
Orionids			
$-3.5 \leq M_V \leq 2.5$	2.10	Visual, $P(M) = 1.0$	Kresáková (1966)
$-3.5 \leq M_V \leq 2.5$	2.15	Visual, variable $P(M)$	Kresáková (1966)
	2.51	Visual	Levin (1953, 1955)
$-3 < M_V < 2$	1.85	Visual	Krisciunas (1980)
$-1 < M_V < 10$	2.39 ± 0.09	Visual and telescopic	Znojil (1968)
$+5 < M_R < +7.5$	2.08	Radar echo amplitude (u.d.)	Hajduk (1968)
$+1 < M_R < +5$	1.95	Radar echo amplitude (o.d.)	Hajduk (1968)
$+1 < M_R < +5$	2.4	Radar echo duration (o.d.)	Hajduk and Simek (1979)
	1.90	Radar echoes	Hajduk (private communication)
$7.35 < M_R < 8.1$	2.07	Radar echoes	Jones (1983)

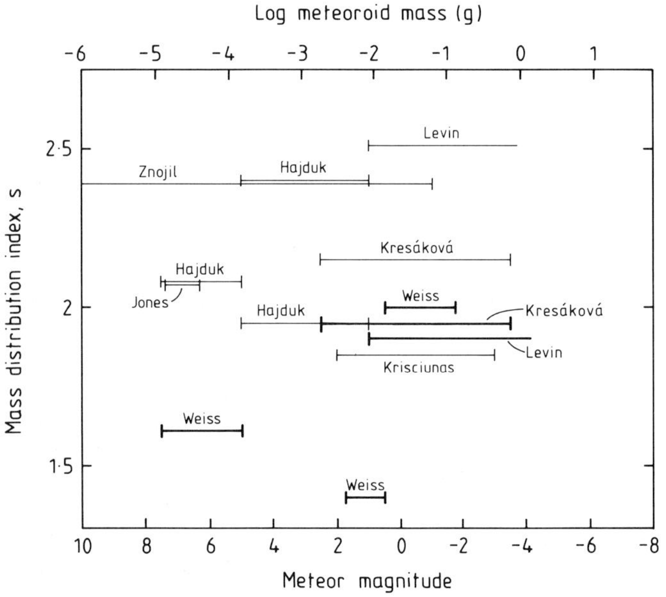

Fig. 9. Orionid (——) and Eta Aquarid (——) mass distribution indices as a function of magnitude and mass

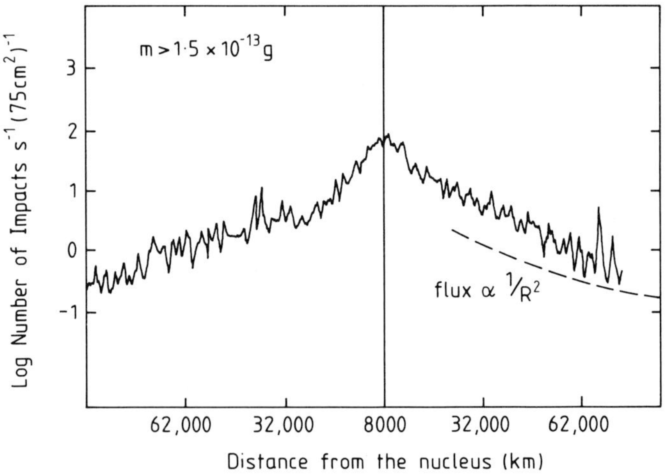

Fig. 10. The logarithm of the number of impacts of mass $\geq 1.5\,10^{-13}$ g particles, per second, is plotted as a function of distance from the cometary nucleus for the DUCMA experiment on board Vega 2 (see Simpson, 1986)

Table 6. Spacecraft flyby parameters at Halley encounter

Spacecraft	Closest approach to comet (km)	Time of closest approach	Speed (km s^{-1})
Giotto	605	March 14, $00^h 03^m$	68.37
Vega 1	8890	March 6, $7^h 20^m$	79.2
Vega 2	8030	March 9, $7^h 20^m$	76.8

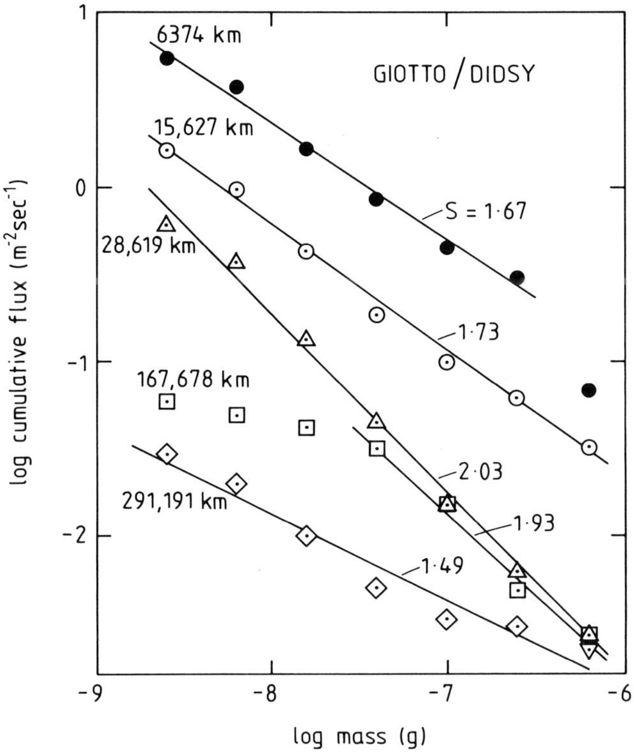

Fig. 11. Logarithmic flux versus mass curves for the Giotto/DIDSY data, as a function of distance from the cometary nucleus (see McDonnell, 1986a)

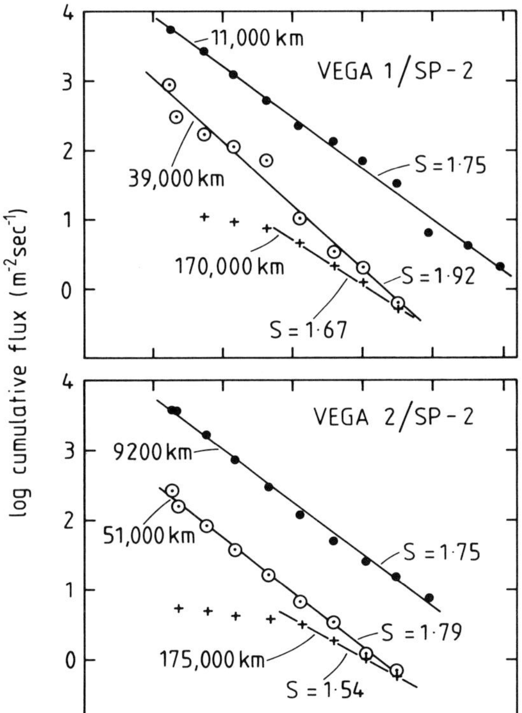

Fig. 12. Logarithmic flux versus mass curves for the Vega SP-2 dust detectors (see Mazets, 1986)

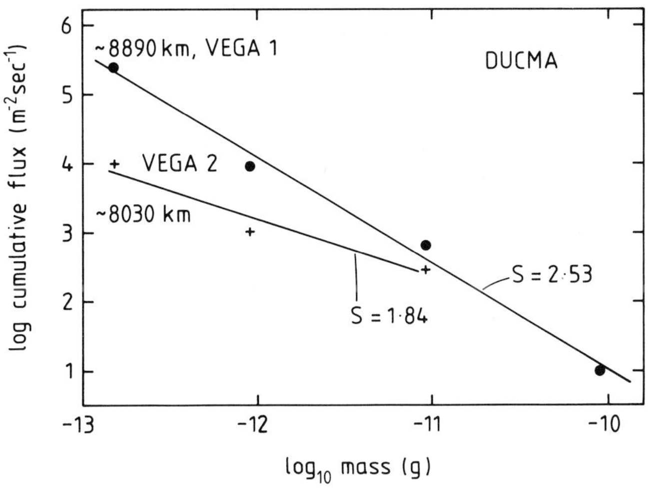

Fig. 13. Logarithmic flux versus mass curves for the Vega DUCMA dust detectors (see Simpson, 1986)

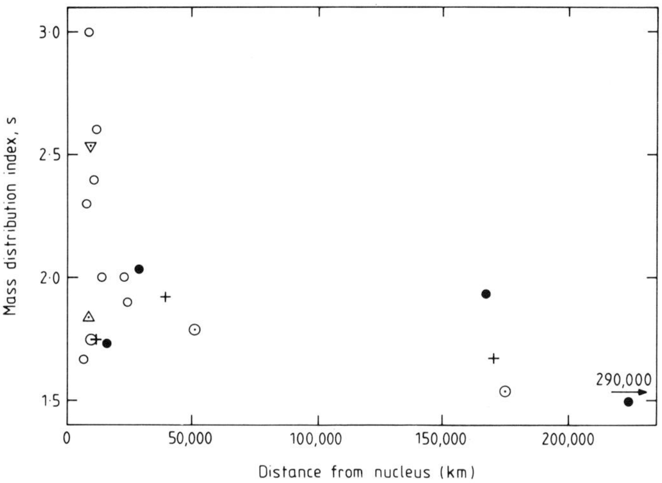

Fig. 14. Mass distribution indices obtained from spacecraft observations, as a function of distance from the cometary nucleus. ● Giotto DIDSY, + Vega 1 SP-2, ⊙ Vega 2 SP-2, ▽ Vega 1 DUCMA, △ Vega 2 DUCMA, ○ DUCMA (see Table 2, Simpson 1986)

Figure 14 shows the s values taken from the DIDSY, SP-2 and DUCMA results plotted as a function of distance from the cometary nucleus. The large majority of DIDSY (Giotto) and SP-2 (Vega 1 and 2) results lead to s values below 2.0. The DUCMA experiment (Vega 1 and 2) leads to results in the $3.0 > s > 1.84$ range. It should be pointed out that there are much more limited test data for the DUCMA experiment in comparison to the others, so that the difference in results might be due to instrumental effects. The uncertainties in the individual s values and the scatter in the data means that, at present, the relationship between mass distribution index and distance is somewhat uncertain.

Intercomparisons between ground based and spacecraft observations of Halley dust are bedevilled by two main problems. The major one is that sampling took place over only slightly overlapping mass ranges. The majority of present day meteor detecting systems are only sensitive to meteors produced by meteoroids more massive than about 10^{-5} g. The results published to date for the Vega SP-2 experiment do not extend to particles more massive than 10^{-6} g. The reason for this is only too clear. The top curve in Fig. 12A, for the region around 11,000 km from the nucleus, indicates that the cumulative flux, $F(\mathrm{m}^{-2}\,\mathrm{s}^{-1})$ was given by

$$\log F = -4.24 - 0.746 \log m .$$

The detector had a surface area of 500 cm^2, so at that distance was only recording ~ 5 particles min^{-1} with masses greater than 10^{-6} g. Even with Giotto, the close-flyby spacecraft, the most massive single particle detected, had a mass of $1.1\,10^{-3}$ g. In the total dust flux to the spacecraft there were only 8 particles with $m > 10^{-4}$ g and about 200 with $m > 7.5\,10^{-6}$ g (see McDonnell et al., 1986b).

The second problem concerns the distances between the detectors and the cometary nucleus. The DIDSY detector produced results over the $3\,10^5$ km to $6\,10^2$ km range. Vega's results were for the region $1.7\,10^5$ km to $9\,10^3$ km. Both measurements were made when the comet was near its descending node. It is difficult to compare these results with those from the Eta Aquarid and Orionid observations. Here the position of the comet in its orbit is of little significance, the annual flux of meteoroids being reasonably constant, indicating that the large particles are evenly distributed around the orbit. At the time of the maximum activities of these showers the Earth based observer is sampling regions $9.8\,10^6$ and $2.7\,10^6$ km from the *present* stream orbit. It is, however, extremely difficult to convert these activities into fluxes in the vicinity of the nucleus. The radio results presented by Jones (1983) for the Orionid shower follow the formula

$$\log F(\mathrm{m}^{-2}\,\mathrm{s}^{-1}) = -15.67 - 1.07 \log m$$

The zero mass intercepts of the two flux curves quoted above, one for the close vicinity of the nucleus, the other for the near Earth stream differ by a factor of $2.5\,10^{-11}$!

6. Conclusions

Ground based meteor and spacecraft dust experiments both produce important results concerning the flux of particles that originated from the nucleus of P/Halley. Unfortunately their intercomparison is extremely difficult because they are measuring dust in completely different places and, at present, over different mass ranges. An easy solution to the second of these problems is to extend the meteor range to lower and lower masses by using more sensitive detectors.

It should be mentioned that the Infrared Astronomy Satellite (IRAS) data reveals annuli of dust around comet orbits. The analysis of these results should provide valuable clues to the size of meteoroid streams. Unfortunately the calculation of the particle size responsible for scattering the radiation is still somewhat model dependent.

References

Hajduk, A.: 1968, *Bull. Astron. Inst. Czech.* **19**, 338

Hajduk, A.: 1970, *Bull. Astron. Inst. Czech.* **21**, 37

Hajduk, A.: 1973, *Bull. Astron. Inst. Czech.* **24**, 9

Hajduk, A.: 1980, in *Solid Particles in the Solar System*, eds. I. Halliday, B.A. McIntosh, D. Reidel, p. 149

Hajduk, A.: 1982, in *Sun and Planetary System*, ed. W. Fricke, G. Teleki, D. Reidel, p. 335

Hajduk, A.: 1983, in *Asteroids, Comets, Meteors*, eds. C.-I. Lagerkvist, H. Rickman, Astronomical Observatory, University of Uppsala, p. 425

Hajduk, A., Buhagiar, M.: 1982, *Bull. Astron. Inst. Czech.* **33**, 262

Hajduk, A., Cevolani, G.: 1981, *Bull. Astron. Inst. Czech.* **32**, 304

Hajduk, A., Simek, M.: 1979, *Contr. Astron. Obs. Skalnaté Pleso.* **8**, 81

Hughes, D.W.: 1972, *Planet Space Sci.* **20**, 1949

Hughes, D.W.: 1974, Space Res. **14**, 789

Hughes, D.W.: 1975, Space Res. **15**, 565

Hughes, D.W.: 1977, Space Res. **17**, 565

Hughes, D.W.: 1978, in *Cosmic Dust*, ed J.A.M. McDonnell, John Wiley and Sons, p. 123

Hughes, D.W.: 1985, *Monthly Notices Roy. Astron. Soc.* **213**, 103

Hughes, D.W.: 1987, *Phil. Trans. Roy. Soc. London*, (in press)

Jones, J.: 1983, *Monthly Notices Roy. Astron. Soc.* **204**, 765

Kresáková, M.: 1966, *Contr. Astron. Obs. Skalnaté Pleso* **3**, 75

Kresák, Ľ.: 1976, *Bull. Astron. Inst. Czech.* **27**, 35

Krisciunas, K.: 1980, *Icarus* **43**, 381

Kozai, Y.: 1979, in *Dynamics of the Solar System*, IAU Symp. **81** ed. R.L. Duncombe, D. Reidel, Dordrecht, p. 231

Levin, B.: 1953, *Doklady A.N.* **90**, 513

Levin, B.: 1955, in *Meteors*, ed. T.R. Kaiser, Pergamon Press, p. 131

Lovell, A.C.B.: 1954, *Meteor Astronomy*, Oxford Univ. Press, p. 289

Massonne, L.: 1985, Adv. Space Res. **5**, 187

Mazets, E.P. et al.: 1986, *Nature* **321**, 276

McIntosh, B.A. Hajduk, A.: 1983, *Monthly Notices Roy. Astron. Soc.* **205**, 931

McDonnell, J.A.M. et al.: 1986a, *Nature* **321**, 338

McDonnell, J.A.M. et al.: 1986b, 20th ESLAB Symp. ESA SP 250, Vol. **II**, p. 25

Simpson, J.A. et al.: 1986, *Nature* **321**, 278

Stohl, J., Millman, P.M.: 1973, *Bull. Astron. Inst. Czech.* **24**, 321

Verniani, F.: 1973, *J. Geophys. Res.* **78**, 8429

Weiss, A.A.: 1961, *Australian J. Phys.* **14**, 102

Whipple, F.L.: 1951, *Astrophys. J.* **113**, 464

Yeomans. D.K., Kiang, T.: 1981, *Monthly Notices Roy. Astron. Soc.* **197**, 633

Znojil, V.: 1968, *Bull. Astron. Inst. Czech.* **19**, 306

Astron. Astrophys. 187, 889–892 (1987)

ASTRONOMY AND ASTROPHYSICS

Radiation formation of a non-volatile comet crust

R.E. Johnson [1], **J.F. Cooper** [2], **L.J. Lanzerotti** [3], **and G. Strazzulla** [4]

[1] Dept. of Nuclear Eng. and Eng. Physics., Univ. of Virginia, Charlottesville, VA 22901, USA
[2] California Institute of Technology, Pasadena, CA 91125, USA
[3] AT&T Bell Laboratories, Murray Hill, NJ 07974, USA
[4] Instituto di Astronomia, Citta Universitaria, I-95125 Catania, Italy

Received February 9, accepted June 29, 1987

Summary. Ion irradiation of the outer meters of a cometary surface can produce new molecular species in the solid state. Because of the interfaces with the interplanetary vacuum these species segregate in an irreversible way into a nonvolatile residue and new very volatile species. The latter are ejected directly or lost when the comet enters the inner solar system. Therefore, a comet exposed to background particle radiations in the Oort cloud obtains an outer web of non-volatile material ($\sim 10^2\,\mathrm{g\,cm^{-2}}$) which will lead to the formation of a substantial "crust". When a new comet enters the inner solar system there will be early activity, initial fizzures in the crust and the break-off of unstable pieces of the crust, due to warming of subsurface species. If this comet enters a periodic orbit in the inner solar system the remaining mantle should be continuously hardened due, primarily, to thermal processing. There will also be permanently active regions on such a comet which were initially shaded from the cosmic ray radiation when the comet was in the Oort cloud or which subsequently lost their crust.

Key words: comet crust – radiation modification

1. Introduction

Results from the missions to Uranus and Halley clearly indicate that the solid matter on/in those objects (satellite and ring surfaces, comet nucleus and dust) contains radiation processed ices or organic materials (e.g. Smith et al., 1986; Krimigis et al., 1986; Kissel et al., 1986a, b). These findings have drawn attention to the results obtained from laboratory experiments studying the production of organic residues from carbon-containing gases or frozen targets irradiated by electrical discharges (e.g. Sagan and Khare, 1979; Sagan et al., 1984), by UV light (e.g. Allamandola et al., 1980; Greenberg, 1982) and by energetic ions (e.g. Cheng and Lanzerotti, 1978; Foti et al. 1984; Lanzerotti et al., 1985). Such experiments can help answer a number of important questions about these dark outer solar system materials (i) What are their properties (chemical composition, physical structure)? (ii) When and where have the materials been formed and processed: in the presolar interstellar medium, in the early stages (e.g. T-Tau phase) of the solar evolution, or in "situ" after the formation of the various objects? (iii) are the differences these materials exhibit in different objects due to different formation scenarios or do they only reflect different evolutive stages of the same process?

In this paper we describe a particular in situ radiation processing effect, the formation of a cometary "crust" activated by cosmic-ray irradiation in the Oort cloud. The recent observations of the nucleus of P/Halley indicate that it has a very low albedo and that the gas and dust primarily effuse from very localized regions. If the volatile material is assumed to have been distributed fairly uniformly throughout the nucleus when the comet was formed, then these observations would imply that there are large regions in which the nucleus has aquired a thermally stable "crust" with a tensile strength sufficient to contain the internal subliming gas pressures. Alternatively, the nucleus may be very heterogeneous with the non-active regions containing primarily rocky, meteoritic materials (e.g. Mendis, 1985). Here we consider the first possibility, and show that radiation-induced changes in the outer layers can create a thermally stable "crust" prior to entry into the inner solar system.

The organic nature of the crust and the existence in the comet of radiation-processed materials is suggested by the results from dust particle analyzer experiments (PUMA, PIA) on Vega and Giotto (e.g. Kissel et al., 1986a, b). These experiments found that a large fraction of the carbon is in cometary dust particles and a large number of these particles are composed of light elements (CHON). In addition, infrared emission bands observed both from Vega-1 at ~ 3.3 and $\sim 7.5\,\mu\mathrm{m}$ (Combes et al., 1986) and from the ground at $3.4\,\mu\mathrm{m}$ (Wickramasinghe and Allen, 1986) have been attributed to the vibrations of $C-H$ and $C=O$ groups in the cometary grains. These results would seem to support the model proposed by Greenberg (1982) in which comets are an aggregate of pre-cometary (interstellar) grains (Table 1). These consist of a silicate core plus an inner organic-refractory mantle formed by photon irradiation of condensed molecules in the interstellar clouds with, possibly, a contribution from cosmic-ray irradiation (Johnson et al., 1983; Strazzulla et al., 1983; Lanzerotti et al., 1987). Finally the grains have an outer "icy" mantel of molecules condensed prior to solar system formation.

This model of the nucleus does not directly account for the inhomogeneous, involatile crust observed. Also it is not clear that the interstellar grains as described can survive the early solar system processing (e.g. Strazzula, 1986), although this is debated (Grim and Greenberg, 1987: Moore and Donn, 1987). However, because the grain forming process (condensation of refractories followed by formation of organics and condensation of volatiles) can also occur in the early solar system we use the model of a pre-

Send offprint requests to: R.E. Johnson

Table 1. Grain composition[a]

Component	Mass fraction
Silicates	0.23
Carbon	0.06
Organic refractory	0.19
H_2O	0.20
CO	0.03
CO_2	0.04
(H_2CO, HCN, CN, H_2, HCO, ...)	0.27

[a] From J.M. Greenberg (1986), two grain types contributing

cometary grain in Table 1 in order to describe the formation of a stable cometary crust. Below we review the relevant laboratory data and then apply these to the mantle formation process (Johnson et al., 1986).

2. Radiation modifications

It is well established that ionizing radiations (UV, X-rays, γ-rays, electrons, fast ions) alter low temperature molecular solids such as condensed gases and organics. The irradiations initially produce radicals (bond breaking) during the electronic relaxation processes. At the low temperatures (10–20 K) in the Oort cloud individual radicals can, in principle, be stored for considerable periods. Radical recombination is then induced in very narrow temperature regions during the warm-up of such materials. This appears as an enhanced volatility in the ices (Moore et al., 1983; Johnson et al., 1983; Greenberg, 1982). Therefore Whipple (1977), Donn (1976), Shullman (1972) and others have proposed that, if the outer layers of a comet were modified by ionizing radiations, a comet newly introduced into the inner solar system could become active at rather larger than expected distances from the sun, consistent with observations of many comets (e.g. Sekanina, 1982). The storage of a high density of radicals requires extremely small diffusion coefficients and/or large recombination barriers as the radical and its parent are generally not displaced many lattice distances. Therefore, the possibility of annealing the broken bonds needs to be examined for Oort cloud temperatures and those temperatures in the interior of an evolved comet when considering transient releases of material in the inner solar system.

On the other hand, if the *local* density of the excitations is high, permanent alterations are directly produced in the material without warming. This occurs along the "track" of a fast ion, in the "spurs" associated with secondary electrons produced by energetic electrons, ions or photons, or when the irradiation dose is high. These alterations have been shown to contain, in addition to radicals, both very volatile and involatile components if the materials contain carbon or sulfur. For example, irradiation of organics results in the formation of carbon residues *and* the production of H_2, O_2, CO, etc. (Venkatesan, 1984; Brown et al., 1987). Recent experiments have roughly quantified the radiation precipitation of residues. Moore et al. (1983) have shown that an interstellar ice mixture (H_2O, NH_3, CH_4) forms a residue (of the order of 1 %) after an average dose of only 10 eV/molecule. In pure materials, such as CO and SO_2, a small fraction (of the order of a few percent) of the material is converted into a residue (Chrisey et al., 1985; Haring et al., 1984; Moore, 1984) after somewhat higher doses, whereas in H_2S, CS_2 (Boring et al., 1985) and CH_4 (Foti et al., 1984; Lanzerotti et al., 1985) most of the material is converted into a residue (after ~200 eV/mol). This is especially the case for very energetic protons. Lanzerotti et al. (1987) have also studied the sample thickness dependence and ionization density effect on residue formation in condensed CD_4.

Besides condensed gases, all organics are readily carbonized while releasing volatiles and producing black residues (Venkatesan, 1984; Strazzulla et al., 1984; Calcagno et al., 1985). Such processes occur efficiently due, in part, to the enhanced diffusion initiated by the incident ions. This produces a segregation in the material, after long exposure, with the involatile species accreting into a porous condensed material and the volatile species brought to the vacuum interface or stored in the large defects and pores in the solid. Such materials can also contain unreacted radicals, depending on the thermal history. Therefore, ion irradiation causes progressive evolution in the target as the time of exposure to the ions (the fluence) increases. This affects both the chemical composition and the optical properties of the bombarded layers, as seen in Table 2 for a frozen methane target bombarded by MeV protons. The original target evolves from clear to dark black as the H to C ratio decreases with increasing fluence. In pure water ice, although the material becomes porous upon irradiation (Johnson et al., 1985), residue formation does not proceed. Therefore, at very low temperatures, the irradiated H_2O remains a stable mix of radicals and small molecules.

The organics produced by ion irradiation exhibit, in their IR spectra, bands due to C−H and C=C groups (Strazzulla et al., 1984), with band depths which evolve with ion fluence (Calcagno et al., 1985). In particular, they exhibit a complex feature around 3.4 μm (CH_2 and CH_3 stretching vibrations) similar to the one observed on many astronomical objects including Halley (Combes et al., 1986; Wickramasinghe and Allen, 1986). In fact, the 3.4 μm feature is also produced in such materials by other ionizing radiations, such as UV (Greenberg, 1986). Such spectra also indicate that the new organics produced by ion bombardment "forget" the chemical nature of their original parents. They respond primarily to the average H/C ratio exhibiting similar spectral properties roughly independent of the original target and bombarding ion and energy (Calcagno et al., 1985). Because the above results for ions are obtained whatever the original form of the carbon, they can be applied to many astrophysical situations, including comets, in which there is considerable H_2O.

3. Comet surface alteration

A number of authors (Draganic et al., 1984; Moore et al., 1983; Johnson et al., 1983; Pirronello et al., 1982; Whipple, 1977) have discussed the implications of galactic cosmic ray irradiation in the Oort cloud. This irradiation acts on a fluffy porous object which may be composed of grains that are already highly processed (see Table 1). The cosmic ray ions lose their energy in these grains by direct ionization and by inducing nuclear reactions.

Due to the absence of solar modulation effects, cosmic ray ions in the Oort cloud should have significantly higher intensities at energies below a few GeV/nucleon than those presently measurable within the heliospheric boundary at 50–100 AU from the Sun (Fisk, 1974). Protons at energies less than 300 MeV would deposit energy by direct ionization in cometary surface depths less than 100 grams/cm^2. At higher energies the secondary products of nuclear interactions in cometary ices (i.e., H_2O) would deposit

Table 2. Summary of the evolution of frozen methane under bombardment by MeV protons at 10 K

H^+/cm^2	Dose (eV/mol)	C/H [a]	Resulting material [b]	Color
0	0	1/4	Frozen methane (~10 K)	Clear
2 10^{16}	100	1/3(1/1)	Org. residue 50% (~300 K)	Yellow
5 10^{16}	250	1/2(1/0.7)	Org. residue 75% (~300 K)	Brown
1 10^{17}	500	1/1.5(1/0.5)	Carbonaceous 90% (~300 K)	Black

[a] Brackets are the resulting C/H ratios after heating to (~300 K)
[b] Percents are approximate fractions of original carbon in residue after warming

Table 3. Cosmic ray ion doses [a]

Depth (m)		Dose (4.5 10^9 yr)	
1 g cm^{-3}	0.2 g cm^{-3}	Mrad	(eV/molecule) [b]
0.1	0.5	3 10^5	600
1	5	1 10^5	200
5	25	1 10^4	~ 20
10	50	3 10^3	~ 2

[a] Data from Ryan and Draganic (1986); data in Strazzulla (1986) exclude the important nuclear interations
[b] Based on methane equivalent

much of the initial proton energy at depths of order 100 g/cm^2 (Cooper, 1983). On the other hand the large fluxes of protons measured at energies below 30 MeV within the solar system (McDonald et al., 1974) would penetrate to depths of order of one gram/cm^2. Locally accelerated ions at 100 keV energies in the cometary magnetosphere would affect only a micron-thick outer layer. Comparison of predicted differential fluxes for protons and electrons in interstellar space (Fisk, 1974) indicates that the electron fluxes may be significantly larger at energies below 100 MeV. In this case the electrons might also make a significant, perhaps even dominant contribution to irradiation of the outer 100 g cm^{-2} layers of cometary nuclei. According to Ryan and Draganic (1986) at a depth of one meter in a non-porous icy object (therefore many meters in a porous aggregation of grains) every molecule would have received an average of ~200 eV of energy in 4.5 10^9 yr. This dose increases rapidly with decreasing depth (Table 3) although the total dose in the outer layers depends on the energy spectrum assumed for the low energy ionizing particles. Draganic et al. (1984) also estimate the dose of radiation due to radioactive decay which occurs throughout the comet. These are much smaller (~3 eV/molecule), altering only a few percent of the material. Therefore, the principle alternation occurs in the surface layers.

In carbon and sulfur containing materials significant residue formation can occur at the cosmic ray doses in Table 3. Using the composition in Table 1 and the results in Table 2, at a depth ~100 g cm^{-2} the overlayers on the grains are depleted in volatiles (e.g. Johnson et al., 1983) and the organic molecules and the organic mantles are highly carbonized. The residues obtained on warming have been shown to be black porous, locally compacted, filamentary materials that are stable at room temperature (Moore et al., 1983; Calcagno et al., 1985). At the low temperature of the Oort cloud, the external and internal surfaces of the porous residues, made from the irradiated organics and ices, would be coated with volatile molecular species. Most H_2 presumably would diffuse out of the porous surface at these temperatures (Brown et al., 1987), with some trapped in defects and bubbles. These materials would also contain some radiacals. The levels of ion irradiation in Table 3 also enhance adhesion between the new non-volatile residues, the modified organics, and the silicate grain cores (Strazzulla et al., 1985; Johnson et al., 1985; Johnson and Lanzerotti, 1986). Therefore, cosmic rays produce highly processed organic material (about ~40% of the original by mass) and bind together grain cores (~27% by mass) at depths of the order of 100 g cm^{-2}, so that, after a long residence period in the outer solar system the comet has acuired a "crust" with a web-like construction of refractory materials. Because the original comet is not a smooth sphere, but rather a locally irregular aggregation of a weekly compacted material we imagine this "crust" is not uniform. That is, there are numerous large cracks, crevices, and shadowed surfaces with relatively pristine material not irradiated by the cosmic rays.

On warming this object residual radicals will react and the very volatile gases in the porous "crust" will be lost leaving a dark residue depleted in volatiles which covers a more volatile underlying material. The final state is much like that described by Fanale and Savail (1984) in which a thin layer is produced by thermally enhanced adhesion and volatile depletion. In the present model a much thicker volatile-depleted mantle is formed by the radiation alteration processes described above. In the active "surface" regions, which were not fully exposed to the cosmic ray flux, thermally produced crusts will temporarily form after each release of material.

Subsequent warming of the dark "crust" will continue the depeltion of volatiles. Heat will only slowly be conducted into the underlying regions partially depleting the volatiles below the porous "crust". The net effect would be the formation of a relatively thick outer mantle ($\gtrsim$100 g cm^{-2}) so that gas ejection would come preferentially from regions lacking such a mantle. However, with a sufficiently rapid thermal increase on close approach, the radiation produced mantle may be broken, releasing volatiles and giving a cometary burst which would also contain large amounts of dark refractory debris and possibly, reactive material depending on the thermal history of the interior. In those regions in which the mantel survives, the continuous thermal compaction and additional ionizing radiations will act to harden the "crust". For the grains ejected from active regions, subsequent radiation modification occurs due to the locally-produced cometary plasma and, eventually, solar particles (Strazzulla, 1985; Johnson and Lanzerotti, 1986).

4. Conclusion

That some radiation processing of a cometary material has occurred is suggested by the depletion of H and the significant percent of carbon and sulfur in a refractory form. Further, this refractory material is often fragile and underdense, consistent with the expectations for a radiation processed organic material. This processing may have been produced in the grains prior to accretion due to UV photon and cosmic ray bombardment, a shock in the interstellar medium, or in the early period of the formation of the solar system.

The radiation-induced alteration of the outer meters of a comet in the Oort cloud, however, is certain to occur. This results in the formation of more volatile molecular species and the precipitation of new refractory materials when carbon or sulfur are present as organics or condensed molecules. This can account for the enhanced activity of new comets, the formation of a dark, thermally stable "crust", and the existence of locally active regions (see also Sekanina, these proceedings). Clearly there may be other scenarios for producing the observed mantle as the aggregate of grains is initially porous and , possibly, dark (Greenberg, 1986). The principle question that needs to be answered is the degree of completion of the mantle forming process described here and the ability of this mantle to survive the heating produced on entering the inner solar system. However, this problem pertains to *all models* that begin with a mixture of ices and refractory materials. In the present model, the cosmic ray irradiation process significantly enhances the ability to form a thick mantle of refractory material.

Acknowledgements. We acknowledge helpful comments from Z. Sekanina, and J.M. Greenberg. REJ acknowledges the support of the NSF Astronomy Division under grant AST-85-11391

References

Allamandola, L.J., Greenberg, J.M., Norman, C.A., Hagen, W.: 1980, *IAU Symp.* **87**, 373

Boring, J.W., Zhao Nansheng, Chrisey, D.B., O'Shaunessy, D.J., Phipps, J.A., Johnson, R.E.: 1985, in *Asteroids, Comet and Meteors*, Univ. Uppsala Press, p. 229

Brown, W.L., Foti, G., Lanzerotti, L.J., Bower, J.E., Johnson, R.E.: 1987, *Nucl. Instr. Methods* (in press)

Calcagno, L., Foti, G., Torrisi, L., Strazzulla, G.: 1985, *Icarus* **63**, 31

Cheng, A.F., Lanzerotti, L.J.: 1978, *J. Geophys. Res* **83**, 2597

Chrisey, D.B., Boring, J.W., Phipps, J.A., Johnson, R.E., Brown, W.L.: 1985, *Nucl. Instr. Methods* **B13**, 360

Combes, M. et al.: 1986, *Nature* **321**, 266

Cooper, J.F.: 1983, *J. Geophys. Res.* **88**, 39

Donn, B.: 1976, in *The Study of Comets*, NASA SP-393, p. 611

Draganic, I.G., Draganic, Z.D., Vujosevic, S.: 1984, *Icarus* **60**, 464

Draganic, I.G., Ryanh, M.P., Draganic, Z.D.: 1987, *Adv Space Sci.* (in press)

Fanale, F.R., Salvail, J.R.: 1984, *Icarus* **60**, 476

Fisk, L.A.: 1974, in *High Energy Particles and Quanta in Astrophysics*, eds. F.B. McDonald, C.E. Fitchell, MIT Press, Cambridge, MA, p. 120

Foti, G., Calcagno, L., Sheng, K.L., Strazzulla, G.: 1984, *Nature* **310**, 126

Greenberg, J.M.: 1982, in *Comets*, L.L. Wilkening, ed., Univ. of Arizona Press, Tucson, p. 131

Greenberg, J.M.: 1986, *Nature* **321**, 385

Grim, J., Greenberg, J.M.: 1987, *Astron. Astrophys.* (this issue)

Haring, R.A., Pedrys, R., Haring, A., de Vries, A.E.: 1984, *Nucl. Instr. Methods* **135**, 476

Johnson, R.E. et al.: 1985, in *Ices in the Solar System*, eds. J. Klinger et al., Reidel, Dordrecht, p. 301

Johnson, R.E., Cooper, J.F., Lanzerotti, L.J.: 1986, ESA Publication: Comet Halley Data

Johnson, R.E., Lanzerotti, L.J.: 1986, *Icarus* **66**, 619

Johnson, R.E., Lanzerotti, L.J., Brown, W.L., Augustiniak, W.M., Mussil, C.: 1983, *Astron. Astrophys.* **123**, 343

Kissel, J. et al.: 1986a, *Nature* **321**, 280

Kissel, J. et al.: 1986b, *Nature* **321**, 336

Krimigis, S.M. et al.: 1986, *Science* **233**, 97

Lanzerotti, L.J., Brown, W.L., Johnson, R.E.: 1985, in *Ices in the Solar System*, eds. J. Klinger et al., Reidel, Dordrecht, p. 317

Lanzerotti, L.J., Brown, W.L., Marcantonio, K.J.: 1987, *Astrophys. J.* (in press)

Mendis, D.A.: 1985, in *Ices in the Solar System*, eds. J. Klinger et al., Reidel, Dordrecht, p. 317

McDonald, F.B., Fitchel, C.E., Fisk, L.A.: 1974, in *High Energy Particles and Quanta in Astrophysics*, eds F.B. McDonald, C.E. Fitchel, MIT Press Cambridge, MA, p. 212

Moore, M.H., Donn, B., Khanna, R., A'Hearn, M.F.: 1983, *Icarus* **388**

Moore, M.H.: 1984, *Icarus* **59**, 114

Moore, M.H., Donn, B.: 1987, *Icarus* (in press)

Pirronello, V.: 1982, *Astrophys. J.* **262**, 636

Ryan, H.P., Draganic, I.G.: 1986, *Astron. Space Sci.* **125**, 49

Sagan, C., Khare, B.N.: 1979, *Nature* **277**, 102

Sagan, C., Khare, B.N., Lewis, J.S.: 1984, in *Saturn*, eds. T. Gehrels, M.S. Matthews, Univ. of Arizona Press, Tucson, p. 788

Sekanina, Z.: 1982, *Astrophys. J. Letters* **14**, 175

Shullman, L.M.: 1972, *IAU Symp.* **45**, 265

Smith, B.A. et al.: 1986, *Science* **233**, 43

Strazzulla, G.: 1985, *Icarus* **61**, 48

Strazzulla, G.: 1986, *Icarus* **67**, 63

Strazzulla, G., Calcagno, L., Foti, G.: 1983, *Monthly Notices Roy. Astron. Soc.* **204**, 59

Strazzulla, G., Cataliotti, R.S., Calcagno, L., Foti, G.: 1984, *Astron. Astrophys.* **133**, 77

Strazzulla, G., Calcagno, L., Foti, G., Sheng, K.L.: 1985, in *Ices in the Solar System*, eds J. Klinger et al., Reidel, Dordrecht, p. 273

Venkatesan, T.: 1984, *Nucl. Instr. Methods.* **B1**, 286

Whipple, F.L.: 1977, in *Comets, Asteroids, Meteorites, Interrelation, Evolution and Origins*, ed. A.H. Delsemme, Univ. of Toleda, p. 25

Wickramasinghe, D.T., Allen, D.A.: 1986, *Nature* **232**, 44

Astron. Astrophys. 187, 893–895 (1987)

ASTRONOMY AND ASTROPHYSICS

Astrometric positions of comet P/Halley

C. Barbieri [1], A. Kranjc [2], M. Scardia [2], and G. Cremonese [1]

[1] Dipartimento di Astronomia, Università di Padova, Vicolo dell'Osservatorio 5, I-35122 Padova, Italy
[2] Osservatorio Astronomico di Brera-Merate, Via E. Bianchi 46, I-22055 Merate (Como), Italy

Received February 27, accepted March 10, 1987

Summary. One hundred astrometric positions of comet P/Halley (mostly in the pre-perihelion phase) have been derived from images taken at the Asiago and Merate Observatories. These positions have been used to determine a set of orbital elements for the comet that, although in good agreement with those of IHW Orbit Nr. 54, give smaller residuals.

Key words: astrometry – comets

In the course of the astrometric campaign promoted by the International Halley Watch in preparation for the encounters between the ICE and Giotto satellites and the Giacobini-Zinner (G–Z) and Halley comets respectively, much data was collected at the Observatories of Asiago and Merate. Twenty-nine accurate positions of comet G–Z have already been published in Barbieri et al., (1987) and that paper gives further details on our reduction procedure. Here we report on 100 positions obtained for comet Halley. Three telescopes have been used: at Asiago the 67–92 cm Schmidt, at Asiago-Cima Ekar the 182-cm at its *f*/9 Cassegrain focus (both photography and CCD images); at Merate the ZEN Astrograph of 20 cm aperture (focal length 2077 mm). The geographical coordinates of the three sites are given in Table 1. The different configurations have been coded in Table 2 as: S for Schmidt, C and CCD for the 182-cm and M for Merate. For completeness we mention that in addition to those 100 data, we rejected 9 plates with badly discordant results. Several other Schmidt plates and CCD frames could not be used for astrometric purposes, especially those obtained in the December–January period.

The reference stars were taken from a magnetic tape version of the SAO Catalog. For the reduction of the Schmidt plates we could use an average of 19 reference stars, that gave mean residuals of 0".37 in RA and 0".34 in DEC. On the astrographic plates, the average number of SAO stars was 12, with mean residuals of 0".31 and 0".29 respectively. The procedure for the 182-cm telescope images was more elaborate: the small field of the plates and especially of the CCD frames forced a two-step procedure with an intermediate reference provided by faint stars visible both on our images and on the POSS prints. Larger systematic and casual errors are therefore possible. At any rate only a small number of CCD frames could be processed for astrometric purposes. In the images obtained in December and in January, the comet fills the entire active area and no useful reference stars could be found.

Our topocentric positions are presented in Table 2. Col. 1 gives the date of the observation (in UT), Cols. 2 and 3 the RA and DEC at 1950.0, Cols. 4 and 5 the residuals (O–C) in both coordinates, Col. 6 the instrument, coded as already explained.

The (O–C)'s were calculated with our astrometric program using the elements of the IHW orbit number 54 (Yeomans, 1987 private communication). Our program takes into account also the planetary perturbations. The agreement between calculations and observations is reasonably good: the average residual is 1".8 in RA and 1".3 in DEC. This accuracy is sufficiently high to attempt to improve on the original orbital elements, yielding the slightly different elements in Table 3. Using these elements the residuals in RA become equal to those in DEC, namely 1".1 thus suggesting that our elements are somewhat better than those of the IHW orbit 54. Actually, the agreement with the IHW orbits decreased systematically from orbit Nr. 33 to orbit Nr. 48 and finally to orbit Nr. 54. One reason might be that our program did not include non-gravitational perturbations (small, however, for comet Halley), or more likely that the displacement between the photocenter and the center of mass became increasingly important.

Table 1. Geographical coordinates of the observatories

Observatory	Longitude east	Latitude	Elevation
Asiago Schmidt	00^{h}46^{m}06^s.4	+45°51′40″	1060 m
Asiago 182-cm	46 17.13	50 36.2	1350
Merate	37 42.83	41 54.1	334

Send offprints requests to: C. Barbieri

Table 2. Topocentric positions and (O–C)

DATE (UT)	RA (1950)	DEC	(O-C) RA	(O-C) DE	Tel
	h m s	° ′ ″	″	″	
1984 12 23.1032	5 55 24.84	11 57 12.1	- 1.9	- 0.5	CCD
1985 2 20.8268	4 59 50.12	13 17 11.9	- 1.1	- 1.2	CCD
1985 8 21.1036	6 3 39.16	19 11 24.9	- 0.1	0.5	CCD
1985 8 21.1077	6 3 39.32	19 11 25.2	0.4	0.6	CCD
1985 8 21.1119	6 3 39.42	19 11 25.1	0.0	0.3	CCD
1985 8 22.0675	6 4 8.18	19 12 15.0	1.1	0.3	CCD
1985 8 22.0741	6 4 8.36	19 12 15.2	0.9	0.2	CCD
1985 8 22.0826	6 4 8.62	19 12 16.2	1.0	0.7	CCD
1985 8 22.0913	6 4 8.89	19 12 16.5	1.2	0.5	CCD
1985 8 22.0983	6 4 9.07	19 12 17.0	0.8	0.6	CCD
1985 9 12.1236	6 12 0.40	19 32 16.4	- 0.6	- 0.2	M
1985 9 13.0994	6 12 12.51	19 33 24.4	0.7	1.5	CCD
1985 9 13.1032	6 12 12.57	19 33 24.7	0.9	1.5	CCD
1985 9 13.1180	6 12 12.73	19 33 25.7	0.8	1.4	CCD
1985 9 13.1225	6 12 12.75	19 33 26.0	0.5	1.4	CCD
1985 9 13.1278	6 12 12.89	19 33 26.2	1.6	1.2	CCD
1985 9 13.1355	6 12 12.94	19 33 26.5	1.1	1.0	CCD
1985 9 18.0146	6 12 54.21	19 39 22.7	- 1.0	- 1.1	C
1985 9 18.0938	6 12 54.59	19 39 28.9	- 0.7	- 1.5	S
1985 9 18.1087	6 12 54.74	19 39 30.1	0.5	- 1.5	M
1985 9 20.0139	6 13 1.49	19 42 5.1	2.0	- 0.7	S
1985 10 11.0920	6 5 24.34	20 25 9.1	- 0.1	- 1.5	M
1985 10 12.0056	6 4 32.92	20 27 50.6	1.2	- 1.8	S
1985 11 8.0354	4 48 35.47	22 12 17.3	0.4	0.7	C
1985 11 8.9195	4 43 13.32	22 13 28.7	2.2	0.2	M
1985 11 8.9354	4 43 7.00	22 13 28.7	- 2.1	- 0.9	M
1985 11 8.9524	4 43 0.77	22 13 30.5	0.6	- 0.2	M
1985 11 8.9722	4 42 53.34	22 13 31.5	1.4	- 0.5	M
1985 11 14.9807	3 57 56.03	21 55 55.7	- 1.4	0.3	M
1985 11 14.9898	3 57 51.22	21 55 52.7	- 1.7	1.8	M
1985 11 15.0016	3 57 44.97	21 55 46.7	- 1.6	1.5	M
1985 11 15.0113	3 57 39.83	21 55 41.7	- 1.5	1.3	M
1985 11 15.0224	3 57 33.98	21 55 36.7	- 1.2	1.8	M
1985 11 16.9017	3 40 21.00	21 35 57.7	- 1.3	- 0.7	M
1985 11 16.9142	3 40 13.80	21 35 48.3	- 1.4	- 0.5	M
1985 11 16.9281	3 40 5.67	21 35 38.5	- 2.7	0.3	M
1985 11 16.9385	3 39 59.72	21 35 29.8	- 1.8	- 0.4	M
1985 11 16.9503	3 39 52.93	21 35 20.6	- 1.2	- 0.5	M
1985 12 2.7050	0 48 1.11	12 14 49.0	- 2.3	- 0.3	M
1985 12 2.7168	0 47 54.05	12 14 14.8	- 2.2	0.5	M
1985 12 2.7300	0 47 46.19	12 13 35.6	- 2.2	0.3	M
1985 12 2.7425	0 47 38.69	12 12 58.2	- 2.6	- 0.1	M
1985 12 2.7578	0 47 29.54	12 12 13.0	- 2.8	0.1	M
1985 12 2.7703	0 47 22 14	12 11 35.1	- 1.9	- 0.7	M
1985 12 12.7197	23 28 47.30	4 54 57.3	- 0.6	- 1.4	M
1985 12 12.7349	23 28 41.71	4 54 25.5	- 1.5	- 0.4	M
1985 12 12.7425	23 28 38.83	4 54 7.0	- 3.0	- 2.5	M
1985 12 12.7509	23 28 35.93	4 53 50.5	- 0.5	- 0.8	M
1985 12 12.7592	23 28 32.81	4 53 31.7	- 2.1	- 1.8	M
1985 12 13.7335	23 22 50.61	4 19 20.5	- 1.8	- 0.4	M
1985 12 13.7453	23 22 46.41	4 18 55.8	- 3.5	- 0.8	M
1985 12 13.7585	23 22 41.84	4 18 26.7	- 4.0	- 2.8	M
1985 12 13.7738	23 22 36.69	4 17 55.5	- 2.1	- 2.5	M
1985 12 13.7891	23 22 31.33	4 17 26.6	- 3.4	- 0.0	M
1985 12 14.7057	23 17 26.87	3 46 47.6	- 1.6	- 0.7	M
1985 12 14.7675	23 17 6.71	3 44 46.9	- 1.0	- 0.4	M
1985 12 15.7127	23 12 8.97	3 14 42.9	- 3.1	- 0.8	M
1985 12 15.7328	23 12 2.69	3 14 5.7	- 3.7	- 0.5	M
1985 12 15.7953	23 11 43.45	3 12 10.3	- 2.6	0.3	M
1985 12 15.8043	23 11 40.69	3 11 52.3	- 2.5	- 1.0	M
1985 12 16.7252	23 7 6.44	2 44 4.8	- 2.8	- 0.1	M
1985 12 19.7273	22 53 39.67	1 22 0.0	- 2.1	- 0.2	M
1985 12 19.7405	22 53 36.30	1 21 37.9	- 3.1	- 2.3	M
1985 12 19.7578	22 53 31.95	1 21 12.4	- 3.5	- 1.6	M
1985 12 20.7335	22 49 36.20	0 57 10.8	- 1.7	- 0.1	M
1985 12 20.7543	22 49 31.25	0 56 39.8	- 1.6	- 1.1	M
1985 12 20.7752	22 49 26.33	0 56 10.1	- 0.9	- 0.7	M
1985 12 22.7064	22 42 12.82	0 11 59.4	- 3.6	- 0.1	M
1985 12 29.7078	22 20 47.95	- 1 58 23.8	- 0.5	- 0.6	M
1985 12 29.7155	22 20 46.76	- 1 58 30.5	0.1	0.1	M
1986 1 4.7064	22 6 33.05	- 3 24 10.4	- 1.1	- 1.1	M
1986 1 4.7161	22 6 31.77	- 3 24 17.0	- 1.3	- 0.3	M
1986 1 4.7335	22 6 29.43	- 3 24 32.5	- 2.2	- 2.3	M
1986 1 9.7083	21 56 24.28	- 4 24 48.1	- 2.2	0.4	M
1986 1 9.7125	21 56 23.73	- 4 24 52.2	- 3.2	- 0.8	M
1986 1 12.7123	21 50 46.95	- 4 58 28.9	- 0.5	- 1.8	M
1986 1 12.7318	21 50 44.69	- 4 58 40.8	- 1.9	- 0.9	M
1986 1 12.7401	21 50 43.69	- 4 58 46.1	- 3.0	- 0.7	M
1986 1 13.7179	21 48 56.98	- 5 9 28.4	- 3.1	- 2.7	M
1986 1 13.7297	21 48 55.81	- 5 9 33.5	- 1.3	- 0.1	M
1986 1 14.7165	21 47 9.24	- 5 20 15.7	- 1.9	- 1.6	M
1986 1 14.7311	21 47 7.83	- 5 20 22.4	0.6	1.1	M
1986 1 15.7137	21 45 22.79	- 5 30 59.9	1.4	- 2.4	M
1986 1 15.7325	21 45 20.65	- 5 31 10.6	- 0.5	- 1.0	M
1986 1 16.7200	21 43 35.90	- 5 41 46.6	0.5	- 2.5	M
1986 1 16.7408	21 43 33.64	- 5 41 57.3	- 0.1	0.2	M
1986 1 17.7241	21 41 49.84	- 5 52 30.7	- 0.8	- 2.4	M
1986 1 17.7422	21 41 48.00	- 5 52 41.2	0.5	- 1.3	M
1986 1 20.7283	21 36 34.82	- 6 24 46.6	- 2.6	- 3.6	M
1986 5 8.8608	10 36 37.49	- 12 58 23.6	2.3	0.1	M
1986 5 8.8715	10 36 36.34	- 12 58 3.7	0.3	0.1	M
1986 5 9.8375	10 35 11.01	- 12 29 2.7	- 1.9	1.2	M
1986 5 9.8503	10 35 9.90	- 12 28 41.5	- 1.9	0.1	M
1986 5 9.8628	10 35 8.74	- 12 28 19.1	- 3.1	0.6	M
1986 5 9.8632	10 35 8.91	- 12 28 17.3	0.1	1.7	S
1986 5 10.8364	10 33 50.94	- 12 0 51.8	0.9	2.5	M
1986 5 10.8490	10 33 49.86	- 12 0 31.1	- 0.4	2.6	M
1986 5 10.8628	10 33 48.80	- 12 0 7.9	- 0.0	3.1	M
1986 6 4.8660	10 24 11.73	- 6 6 44.9	- 0.7	- 2.1	M
1986 6 4.8833	10 24 11.89	- 6 6 37.1	- 0.1	- 0.7	S

Table 3. Orbital elements of comet P/Halley

	IHW orbit 54	This paper
q (AU)	0.5870943	0.5870947
e	0.9673512	0.9673485
i	162.2425	162.2427
long. asc. node	58.1640	58.1641
Arg. of perihelion	111.8720	111.8721
Perihelion time	2446470.9799	2446470.9796
Osc. epoch	2446800.5	2446800.5

Acknowledgement. This research has been partly supported by CNR/PSN in the frame of the Giotto Mission. We thank C. Bonoli and M. D'Alessandro for the in kind help with the CCD data.

References

Barbieri, C., Kranjc, A., Scardia, M.: 1987, *Astron. Astrophys. Suppl. Ser.* **67**, 507

Yeomans, D.K., 1987 (private communication)

Astron. Astrophys. 187, 896–898 (1987)

ASTRONOMY AND ASTROPHYSICS

A new approach to investigations of the long-term motion of comet P/Halley

G. Sitarski and K. Ziolkowski

Space Research Centre, Bartycka 18, PL-00-716 Warsaw, Poland

Received January 30, accepted March 3, 1987

Summary. A new method, based on the moments of perihelion passages as observational data to the orbit improvement, was used to link all the observed apparitions of comet Halley in 240 BC–1986 AD by one system of orbital elements. Preliminary results are presented. The problem of secular variations in the nongravitational parameters of the comet's motion, influencing the results of the orbit improvement, is discussed. It is shown that a successful modelling of the motion of comet Halley over the last two millennia is also possible if we take into account a variability of the nongravitational effects with time.

Key words: comet Halley – orbit determination – nongravitational effects

1. Introduction

Among several results of investigations of the motion of comet Halley in the long past, two of them were based on Marsden's method of including nongravitational effects (Marsden et al., 1973): Yeomans and Kiang (1981) used the nongravitational parameters A_1 and A_2 as constant values with time while Landgraf (1984, 1986) considered those parameters as some linear function of time. In both works the orbit was improved by the least squares method using the positional observations from few last apparitions of the comet. The equations of motion were then integrated backwards over a time-span of some thousands of years. However, to obtain good agreement of computed perihelion times with the observed ones, the authors had to make some subjective changes in the orbital elements for 837 AD when the comet closely approached the Earth to within 0.03 AU. Therefore, those results cannot be regarded as the successful modelling of Halley's motion since they do not link all the observed returns of the comet by one system of orbital elements.

We applied an entirely new procedure of orbit improvement (Sitarski, 1987), based on the least squares method, too, and enabling the use of the full observational material of comet Halley since 240 BC. Our approach is to use the instants T of perihelion passages of the comet as observational data for composing the observational equations in the form:

$$\sum_{i=1}^{6} \frac{\partial T}{\partial E_i} \Delta E_i + \frac{\partial T}{\partial A_1} \Delta A_1 + \frac{\partial T}{\partial A_2} \Delta A_2 = \Delta T. \tag{1}$$

Send offprint requests to: G. Sitarski

The set of Eqs. (1) should be solved by the least squares method to find corrections to the six orbital elements E_i and to the nongravitational parameters A_1 and A_2. In practical computations we used the mean anomaly M computed for the moment T, and Eq. (1) was transformed by putting M instead of T. Thus, we could join the set of Eqs. (1) to the observational equations resulting from the positional observations of the comet.

In our method of orbit improvement one can use in a uniform way both modern positional observations as well as inaccurate ancient observations known from the historical records, from which the dates of perihelion passages of the comet have been deduced (Kiang, 1972; Yeomans and Kiang, 1981).

We applied this method to link all the apparitions of comet Halley by one system of dynamical parameters of motion, treating the nongravitational parameters either as being constant with time as it has been assumed by Yeomans and Kiang (1981) or used their secular variation according to Landgraf (1984).

2. Computations and results

In order to improve the orbit of comet Halley we selected the 250 best positional observations from 1835–1986 and used 28 observed perihelion times from the interval 1835 AD–240 BC. We solved the observational equations in two cases corresponding to the procedures by Yeomans and Kiang and by Landgraf respectively:

(I) A_1 and A_2 were taken as constant values with time,

(II) A_1 was constant but it was assumed that $A_2 = A_{02}$ $(1 - B \cdot t)$ where the value of B was taken from Landgraf (1984).

The process of orbit improvement required many iterations. The equations of motion as well as special differential equations to obtain values of the partial derivatives in Eq. (1) were integrated by recurrent power series (Sitarski, 1984) including perturbations due to all the planets and also relativistic effects (Sitarski, 1983). In both cases we determined the corrections to six orbital elements and corrections to A_1, A_2 in case (I), and to A_1, A_{02} in case (II).

In cases (I) and (II) we improved the orbit in two variants:

(a) we took the observational data from the period 1986 AD–12 BC and then used the perihelion times in 87 BC, 164 BC and 240 BC to test the prediction based on the resulting orbital elements,

(b) we used the observational data from the interval 1986 AD–240 BC to link all the observed apparitions by one system of elements.

As a result we obtained the four sets of orbital elements and nongravitational parameters given in Table 1. Starting from the

Table 1

	Ia	Ib	IIa	IIb
$T = 1835$	Nov. $16\overset{d}{.}43950$	Nov. $16\overset{d}{.}43966$	Nov. $16\overset{d}{.}44227$	Nov. $16\overset{d}{.}44229$
q au	0.58657764	0.58657485	0.58649656	0.58649647
e	0.96737632	0.96736810	0.96733665	0.96733665
ω	$110\overset{\circ}{.}68930$	$110\overset{\circ}{.}68999$	$110\overset{\circ}{.}68747$	$110\overset{\circ}{.}68751$
Ω	56.80698	56.80718	56.79382	56.79382
i	162.26606	162.26661	162.24413	162.24421
$10^8 A_1$	+ 0.4479	+ 0.6251	+ 1.3756	+ 1.3347
$10^8 A_{02}$	+ 0.01521	+ 0.01519	+ 0.01604	+ 0.01604
	$A_2 = A_{02}$		$A_2 = A_{02}(1 - B \cdot t)$ $10^4 B = -0.00495$, t in days from 1986 Feb. 19.0 ET	

Epoch 1835 Nov. 18.0 ET, equinox 1950.0

Table 2

No.	T observed	Residuals O–C Ia	Ib	IIa	IIb
1	1986 Feb. $9\overset{d}{.}46$	$0\overset{d}{.}00$	$0\overset{d}{.}00$	$0\overset{d}{.}00$	$0\overset{d}{.}00$
2	1910 Apr. 20.18	0.00	0.00	0.00	0.00
3	1835 Nov. 16.44	0.00	0.00	0.00	0.00
4	1759 Mar. 13.05	+ 0.45	+ 0.49	− 0.19	− 0.18
5	1682 Sep. 15.27	+ 0.48	+ 0.53	− 0.03	− 0.02
6	1607 Oct. 27.56	+ 1.49	+ 1.57	− 0.73	− 0.73
7	1531 Aug. 25.8	+ 1.04	+ 1.11	− 1.32	− 1.32
8	1456 June 9.1	+ 0.49	+ 0.55	− 0.88	− 0.88
9	1378 Nov. 9.02	− 0.90	− 0.75	− 1.89	− 1.89
10	1301 Oct. 24.53	− 0.51	− 0.28	− 0.53	− 0.52
11	1222 Sep. 30.8	+ 1.62	+ 1.72	+ 3.08	+ 3.05
12	1145 Apr. 21.25	+ 1.45	+ 1.42	+ 4.57	+ 4.57
13	1066 Mar. 23.5	+ 1.39	+ 1.29	+ 4.98	+ 4.93
14	989 Sep. 9.0	+ 1.70	+ 1.40	+ 4.89	+ 4.89
15	(912 July 9.5)	(− 9.73)	(− 9.92)	(− 8.40)	(− 8.40)
16	837 Feb. 28.27	− 0.10	− 0.06	− 0.10	− 0.10
17	760 May 22.5	+ 1.93	+ 1.88	+ 1.98	+ 2.02
18	684 Sep. 28.5	− 3.96	− 4.10	− 2.81	− 2.68
19	607 Mar. 13.0	− 2.11	− 2.05	+ 0.16	+ 0.23
20	530 Sep. 26.7	− 0.82	− 0.54	+ 1.39	+ 1.29
21	451 June 24.5	− 4.11	− 3.64	− 2.53	− 2.73
22	374 Feb. 17.4	+ 0.81	+ 1.25	+ 2.34	+ 2.16
23	295 Apr. 20.5	+ 0.56	− 0.03	+ 0.14	+ 0.40
24	218 May 17.5	+ 0.43	− 2.14	− 1.70	− 0.67
25	141 Mar. 22.35	+ 0.21	− 4.36	− 1.63	+ 0.24
26	66 Jan. 26.5	+ 0.02	− 6.20	+ 0.31	+ 2.79
27	− 11 Oct. 5.5	− 6.57	− 9.23	− 2.49	− 1.27
28	− 86 Aug. 2.5	(− 4.59)	+ 0.39	(+ 1.46)	+ 0.49
29	(−163 Nov. 17)?	(+ 1.02)	(+23.28)	(+27.27)	(+20.39)
30	−239 Mar. 30.5	(−55.93)	+ 5.68	(+ 6.28)	− 1.52

The data in brackets were not used for the orbit improvement. The perihelion time in −163 is given between November 9 to 26 (Stephenson et al., 1985); we took the middle moment.

initial data taken either from the solution (I) or from solution (II), the equations of motion of the comet were integrated backwards. A comparison of the results of integration with the observed perihelion times is given in Table 2.

3. Discussion

A comparison of the residuals O–C given in Table 2 shows that solutions (I) and (II) are quite different although we applied the same computational method and used the same observational data for the improvement of the orbit. The secular change in the nongravitational parameter A_2, included in the process of orbit improvement, resulted in different values of improved elements in cases (I) and (II).

The prediction of the perihelion time in 240 BC is distinctly better in solution (II a) than in solution (I a). Moreover, if we use the perihelion time in 240 BC as the observational data for the orbit improvement, we obtain a good result in case (II b) whereas we hardly may regard the result in case (I b) as a satisfactory one. An analysis of residuals O–C in case (I) could suggest that the moment of perihelion passage in 240 BC is erroneously deduced from the Chinese observations, and probably Yeomans has therefore omitted that date when comparing his computations with observed perihelion times of the comet (Yeomans and Kiang, 1981, p. 644). However, if we include the secular variation in the nongravitational parameter we are able to find orbital elements representing all the observational data well.

The moment of perihelion passage in 164 BC, given in Table 2, cannot be considered to be equivalent to other observations because Stephenson et al. (1985) have estimated that the perihelion passage in that year would have happened only between November 9 to 26.

We should mention that the representation of the observational data by our computations in case (I a) is in a good agreement with results by Yeomans and Kiang (1981). This means that their computations were correct although some empirical adjustment had to be made to the orbital elements during the process of numerical integration by those authors.

The value of B accepted by us after Landgraf (1984) made it possible to obtain a different and in general terms no less satisfactory representation of all the perihelion times. However, we think that the linear approximation of the parameter A_2 is too simple for modelling the real variability of nongravitational effects over such long periods of time. An indication of this may be that the residuals during 989–1222 show a distinct maximum in case (II) while case (I) gives a better representation of these apparitions. Taking the value of $B = -0.002955\ 10^{-4}$ from a recent paper by Landgraf (1986) we have obtained another solution, which is in good agreement with Landgraf's results, and similar to our presented solution (I a) when assuming a constant value of A_2.

4. Conclusions

The new method for the orbit improvement used in this work made it possible to reexamine the modelling of the motion of comet Halley over a long time interval. Our preliminary results presented in this paper show that the secular changes of nongravitational effects in comet Halley's motion play an important role in the process of orbit determination and should be carefully investigated. Therefore, although Yeomans' dynamic model of the motion of comet Halley represents the ancient observations successfully, one should be cautious about conclusions based on the assumption that the nongravitational forces remained constant over two millennia (e.g. Yeomans, 1986).

The method applied in this work allows us to look for variability of nongravitational effects in a comet's motion as is successfully shown in case of the periodic comet Encke (Sitarski, 1987). We are working on the determination of changes of the nongravitational parameters for comet Halley to elaborate the new theory of its motion.

Acknowledgements. We are indebted to Dr. T. Jopek, Poznan Astronomical Observatory, for making available his long-term ephemeris of the Solar System as well as for the subroutines for converting calendar dates to Julian dates and vice versa. The computations were made with the R-60 and R-40 computers at the Warsaw University. The work reported here was supported by the Polish Academy of Sciences in CPBP 01.20.

References

Kiang, T.: 1972, *Mem. Roy. Astron. Soc.* **76**, 45
Landgraf, W.: 1984, ESTEC EP/14.7/6184 Final Report
Landgraf, W.: 1986, *Astron. Astrophys.* **163**, 246
Marsden, B.G., Sekanina, Z., Yeomans, D.K.: 1973, *Astron. J.* **78**, 211
Sitarski, G.: 1983–7, *Acta Astron.* **33**, 295, **34**, 53, **37**, in press
Stephenson, F.R., Yau, K.K.C., Hunger, H.: 1985, *Nature* **314**, 587
Yeomans, D.K.: 1986, Proc. 20th ESLAB Symp. ESA SP-250, Vol. II, p. 419
Yeomans, D.K., Kiang, T.: 1981, *Monthly Notices Roy. Astron. Soc.* **197**, 633

Astron. Astrophys. 187, 899–905 (1987)

ASTRONOMY AND ASTROPHYSICS

High-order librations of Halley-type comets

A. Carusi [1], L. Kresák [2], E. Perozzi [1], and G.B. Valsecchi [1]

[1] IAS – Reparto di Planetologia, Viale dell' Università 11, I-00185 Rome, Italy
[2] Astronomical Institute, SAV, Bratislava, Czechoslovakia

Received February 18, accepted July 6, 1987

Summary. Halley-type comets, usually referred to as those with revolution periods of 20 to 200 yr, can be dynamically defined by the values of their Tisserand invariants with respect to Jupiter, which account for the processes that brought these comets into the present orbits. Long-term integrations of their motion show oscillations of their heliocentric elements when the comets are far from the Sun and the centre of their motion is the barycentre of the solar system. The shift from barycentric to heliocentric motion is also responsible for the major changes of orbital elements within a single revolution which, in the case of direct orbits, may follow regular libration patterns with typical periods of 350–400 yr. This is the reason for the concentration of periods around 70 yr. Some examples of this behaviour and a short discussion of its importance for the dynamical history of Halley-type comets are presented.

Key words: comets – Halley-type comets – resonances

1. Introduction

A periodic comet may be classified as of Halley-type (HT) if its orbital period P is comprised between 20 and 200 yr.

On a dynamical basis, most HT comets behave differently from typical short-period Jupiter family comets, and there is evidence that the past history of these objects has also been essentially different; dynamical considerations should then be used to find a satisfactory definition.

Kresák (1972) showed that all periodic comets with $P > 20$ yr have their Tisserand invariants with respect to Jupiter (J), defined by

$$J = a_j/a + 2\sqrt{a/a_j(1-e^2)}\cos i$$

smaller than 2 (where a, e, i are the comet's semimajor axis, eccentricity and inclination, and a_j is the semimajor axis of Jupiter). A definition of HT comets as those with $J < 2$ therefore seems plausible (see Fig. 1). However, there are comets with $P < 20$ yr (P/Tuttle, P/IRAS and the newly discovered P/Machholz) which in that case should be included among the HT comets, something that could seem rather disputable especially in the case of P/Machholz, whose period is about 5.2 yr and whose J is 1.94. An important argument in favour of identifying HT comets by means of their Tisserand invariants, rather tan by the revolution periods alone, is that the latter criterion holds only for those comets which are bright enough for the present search techniques. The long-term integrations by Carusi et al. (1985a) have shown that 10% of all known comets of Jupiter's family had, or will have, periods between 20 and 200 yr sometimes within ± 400 yr from now. At that time their perihelia were, or will be, situated close to the orbit of Jupiter. These comets have very high Tisserand invariants not only with respect to Jupiter, but also with respect to Saturn – as another signature of their stepwise capture. There we are dealing with a transient evolutionary stage which can continue with an orbit characteristic for the observable population of the Jupiter family, the dynamical history of which essentially differs from that of HT comets. A definition based on the values of J then highlights the main dynamical characteristic of HT comets, i.e. the fact that their capture into a periodic orbit has apparently taken place as a result of a single major encounter with Jupiter (Everhart, 1973; Kresák, 1981; Carusi and Valsecchi, 1985). Exceptions are of course possible, as in the case of P/IRAS, characterized by $J = 1.96$, $P = 13.2$ yr, whose present motion is controlled by Saturn (its nodes being far from Jupiter's orbit, see Carusi et al., 1985b), and is therefore to be considered an interloper among the short-period comets with $J < 2$.

Perhaps the best way to put things is to say that dynamically one can consider as of HT those comets whose capture has been mainly due to a single close encounter with Jupiter, and that comets with $J < 2$ are very probably of HT, while those with $J < 1.5$ are almost certainly of HT. Including P/Machholz there are 22 known short-period comets with $J < 2$ and $P < 200$ yr (see Table 1); however that comet is not taken into consideration in our study because of the uncertainty in its orbital elements (it is still under observation) and to the lack of a good long-term integration of its motion. P/IRAS is not included for the reasons already explained.

2. Barycentic and heliocentric motion

The long-term dynamical evolution of HT comets is mainly controlled by the indirect perturbations by the major planets, and in particular by Jupiter. These rather weak perturbations, which do not involve close encounters with any planet, may be explained quite easily by the repeated exchange of the centre of motion between the Sun and the solar system barycentre, as the comets approach or leave perihelion. Close encounters with Jupiter are essential for the capture of HT comets into short-period orbits, but less important for the subsequent evolution. A rough estimate of the time during which a heliocentic orbit may well represent the

Send offprint requests to: A. Carusi

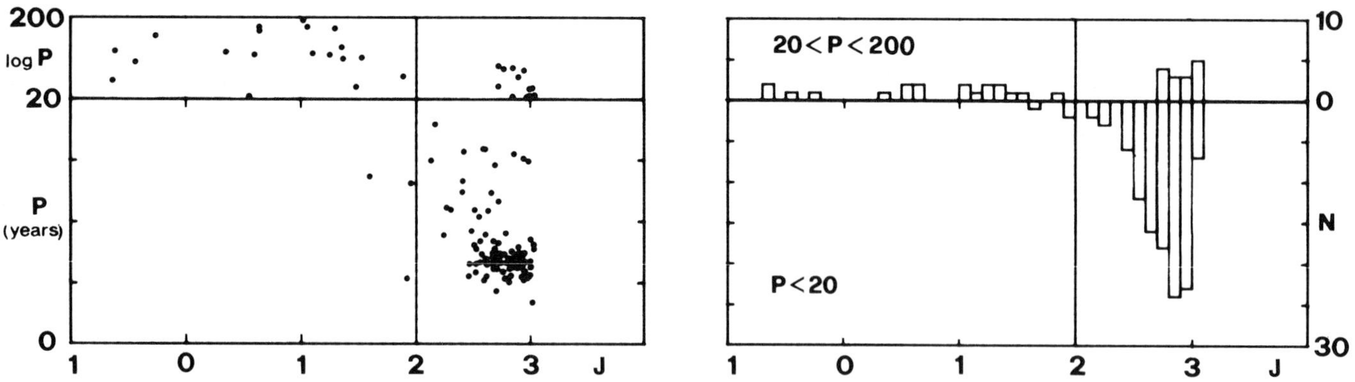

Fig. 1. **a** Distribution and **b** frequency histogram of short-period comets as a function of J and P. The horizontal lines separate comets with $P > 20$ yr from the others. Points in the upper right parts refer to pre-capture and post-ejection orbits of comets presently with $P < 20$ yr. The three dots on the lower left refer (from left to right) to P/Tuttle, P/Machholz, and P/IRAS

Table 1. Orbital data of short-period comets with $J < 2$

N [a]	Name	J	P	t	i(°)	a(AU)	e
2	P/Tempel-Tuttle	−0.64	32.92	1366,1965	162.71	10.27	0.904
1	P/Halley	−0.61	75.99	−239,1986	162.24	17.94	0.967
505	P/Pons-Gambart	−0.45	57.46	1827	136.45	14.89	0.946
507	P/Swift-Tuttle	−0.27	119.98	1862	113.56	24.33	0.960
506	P/de Vico	0.37	76.30	1846	85.11	17.99	0.963
547	P/Hartley-IRAS	0.54	21.46	1984	95.73	7.72	0.834
14	P/Pons-Brooks	0.60	70.92	1812,1954	74.18	17.13	0.955
517	P/Mellish	0.64	145.37	1917	32.68	27.65	0.993
38	P/Herschel-Rigollet	0.64	154.91	1788,1939	64.20	28.84	0.974
520	P/Wilk	1.03	187.38	1937	26.02	32.74	0.981
548	P/Bradfield	1.07	151.41	1984	51.79	28.41	0.952
27	P/Brorsen-Metcalf	1.11	71.97	1847,1919	19.21	17.30	0.972
15	P/Olbers	1.25	69.58	1815,1956	44.61	16.92	0.930
510	P/Barnard 2	1.29	145.36	1889	31.25	27.65	0.960
521	P/Väisälä 2	1.36	85.42	1942	38.00	19.40	0.934
25	P/Westphal	1.36	61.87	1852,1913	40.88	15.64	0.920
11	P/Crommelin	1.48	27.41	1818,1984	29.10	9.09	0.919
519	P/Dubiago	1.52	62.35	1921	22.32	15.72	0.929
8	P/Tuttle	1.60	13.69	1790,1980	54.46	5.72	0.823
41	P/Stephan-Oterma	1.89	37.71	1867,1980	17.98	11.25	0.860
560	P/Machholz	1.94	5.25	1986	59.99	3.02	0.958
546	P/IRAS	1.96	13.16	1983	46.18	5.57	0.696

[a] The numbering order is the one introduced in Carusi et al., 1985a. t is the time from the first to the last apparition (in case of more than one apparition)

motion of HT comets is given by the time spent inside Jupiter's orbit. Because of the generally high eccentricities, this time span is about two years for all comets of this type, which is a small fraction of their revolution periods. As a consequence, the regular oscillations of the heliocentric orbital elements far from perihelion are only apparent, and may be removed simply by computing the barycentric orbital elements; a sampling in time of the heliocentric elements may not then be representative of the real changes in the motion of these comets (see Fig. 2). On the other hand, the perturbations due to the shift of the centre of motion undergone by a comet once per revolution are recognizable by sudden changes of its barycentric elements. Their extent depends both on the characteristics of individual orbits (being larger for orbits of larger semimajor axes) and on the geometrical configuration of the major planets when the comet passes perihelion. A qualitative view of the distribution of these perturbations is shown in Fig. 3, where the barycentric semimajor axes of the 20 HT comets, as a function of their corresponding perihelion distances, are plotted at equispaced intervals of 5000 days from 1585 to 2406. The data are taken from the extended integrations by Carusi et al. (1985a). In

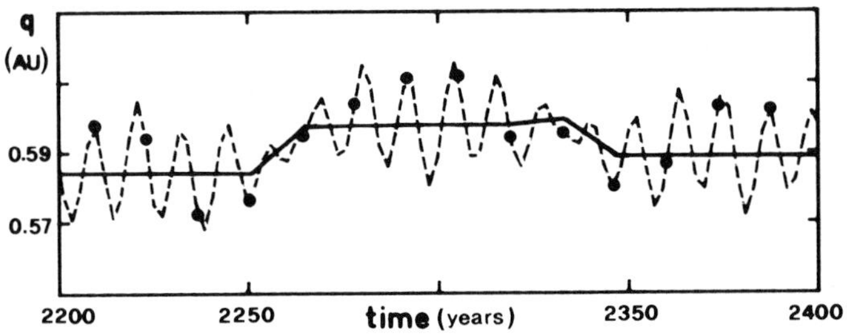

Fig. 2. Time evolution of the perihelion distance of P/Halley from 2200 to 2400. Heliocentric (dashed) and barycentric (solid line) evolutions of q are shown. Dots represent a sampling of heliocentric q at equispaced intervals of 5000 days

the case of direct orbits, a general tendency to follow constant eccentricity lines is recognizable, while retrograde orbits tend to move perpendicularly with respect to these lines.

Table 2. Energy levels (AU^{-1}) and differences for some resonances with Jupiter of the type n:1 and n:2

Resonance	Energy level	Energy difference
3:1	92399 ($\times 10^{-6}$)	9024 ($\times 10^{-6}$)
7:2	83375	7102
4:1	76273	5760
9:2	70513	4783
5:1	65730	4046
11:2	61684	3477
6:1	58207	3024
13:2	55183	2660
7:1	52523	2361
15:2	50162	2113
8:1	48049	1903
17:2	46146	1725
9:1	44421	

3. Resonant motions

The widely spread opinion about the concentration of aphelia of HT comets towards the orbits of major planets (the Saturn, Uranus, and Neptune families) lacks in reasoning. In fact, the dynamics of these comets is mainly under the control of Jupiter, and their aphelion distribution does not significantly differ from randomness. The number of known HT comets remains approximately constant over uniform intervals of $Q^{1/2}$ (or $a^{1/2}$, $P^{1/3}$), and local irregularities are as large as predicted with a 13% probability for a random distribution. This is too high over the conventional rejection level, and does not demand searching for any physical explanation. The same result of 13% is obtained when the checks are made within superimposed intervals between exact n:1 commensurabilities with the period of Jupiter, using the osculating periods. Unfortunately, here we are limited to only 13

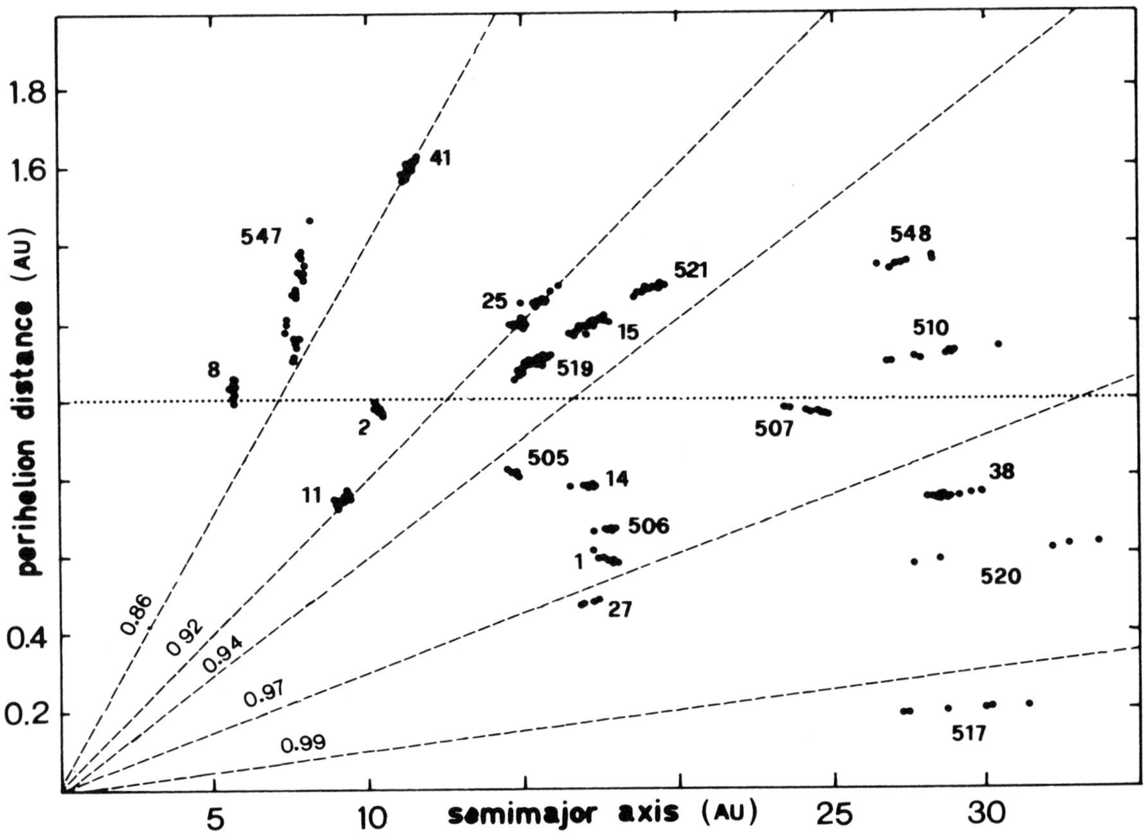

Fig. 3. Plot of the jovicentric semimajor axes of 20 Halley-type comets as a function of their corresponding perihelion distances at equispaced intervals of 5,000 days from 1585 to 2406. Lines of constant eccentricity (dashed lines) are also shown

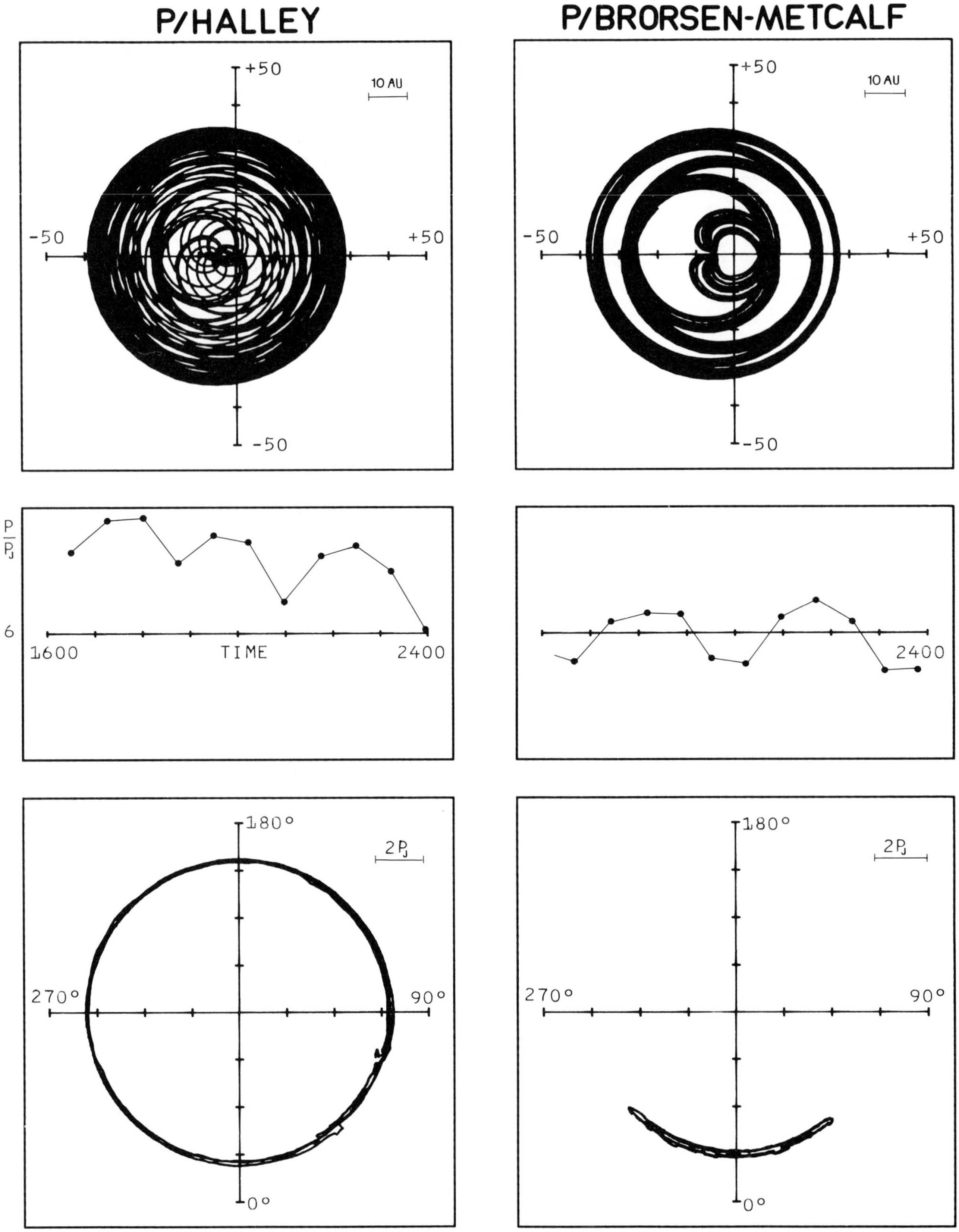

Fig. 4. Orbital evolution of comets P/Halley and P/Brorsen-Metcalf from 1585 to 2406. Top: comets' trajectories in a frame centred on Jupiter and rotating with it around the Sun. Centre: plots of the osculating periods (in units of Jupiter's period); the plots extend from -0.5 to $+0.5$ around the exact resonance. Bottom: polar diagrams of the osculating periods (radius vectors) and the heliocentric elongation of Jupiter from the comet (at perihelion of the latter, position angles)

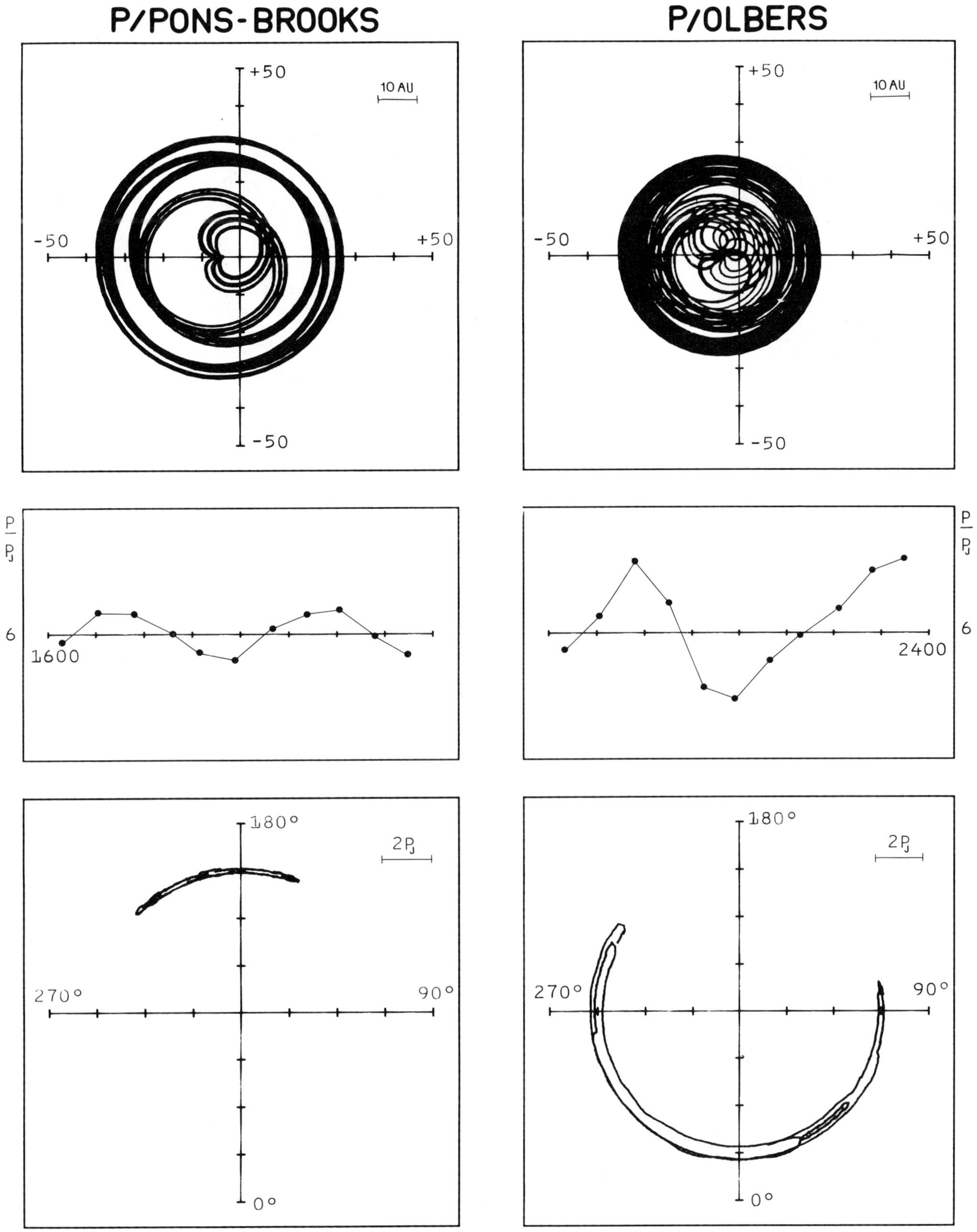

Fig. 5. Same as Fig. 4 for P/Pons-Brooks and P/Olbers, both librating around the 6:1 resonance with Jupiter

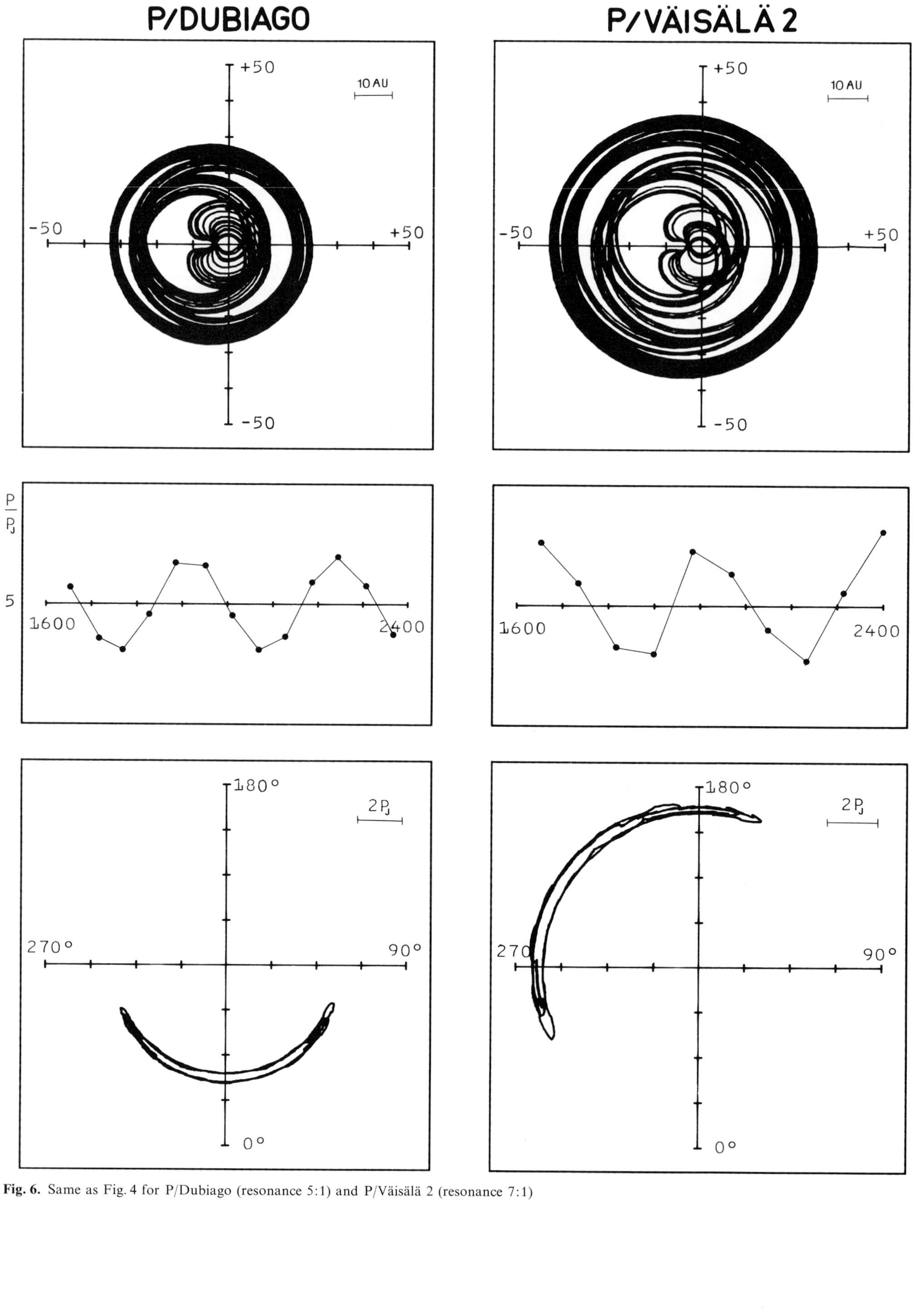

Fig. 6. Same as Fig. 4 for P/Dubiago (resonance 5:1) and P/Väisälä 2 (resonance 7:1)

objects with revolution periods between 20 and 100 yr, since otherwise n becomes too high. Moreover, the periods of some of these comets are known with poor accuracy: for P/de Vico, P/Dubiago, and P/Väisälä 2 the estimated error of P exceeds 10% of Jupiter's period, and for P/Pons-Gambart is nearly the whole Jupiter's period.

If we take the whole set of data obtained from integrations over 800 yr (Carusi et al., 1985a), i.e. 10 to 40 revolutions of each of these 13 comets, some concentration towards resonances with Jupiter becomes evident. This is due to rather regular librations of some HT comets around integer multiples of Jupiter's period, with a typical cycle of 350–400 yr, or 5 to 6 revolutions. These comets are P/Brorsen-Metcalf, P/Pons-Brooks and P/Olbers (all in resonance 6:1 and with well determined orbits), and also P/Dubiago (5:1) and P/Väisälä 2 (7:1), as far as their one-apparition orbits can be relied upon. All of them move in direct orbits with periods of 60 to 90 yr; and inversely, all comets which do not exhibit this libration pattern have either periods outside these limits or retrograde orbits (like P/Halley), with the exception of P/de Vico ($i = 85°$) and P/Westphal. The barycentric shift connected with the comet's entry into, and exit from, the sphere around the Sun of radius equal to Jupiter's semiaxis can help to understand this librating behaviour. As a matter of fact, at these moments the total perturbations due to the displacement of the barycentre are of opposite sign according to whether Jupiter is on one or the other side of the line of apses of the comet's orbit, and vanish only if Jupiter and the comet are in conjunction at the time of perihelion passage of the latter. Displacements from this equilibrium configuration lead to temporary librations for a direct comet, while for a retrograde orbit they tend to repel the comet from the commensurability.

The fact that this kind of libration is limited to the simplest resonance ratios of the type n:1, with $4 < n < 8$, can be tentatively explained as follows. With longer periods and with more complicated ratios of n:2, n:3 etc., the differences in energy at the neighbouring resonances become dangerously close to typical perturbations by the above mechanism: for HT comets with $P > 60$ yr, the r.m.s variation of $1/a$ (barycentric) in one orbit, taken from the integrations of Carusi et al. (1985a), is about $1070\ 10^{-6}\ \mathrm{AU}^{-1}$. This value should be compared with the energy differences listed in Table 2: if the perturbations are too large, the comets may pass across the resonance without being captured. With shorter periods, the comets spend a too large fraction of each revolution within the planetary region, and the mechanism becomes less effective as compared with the direct perturbations by the planets. If this interpretation is correct, then the HT librations, although temporary, should increase the number of comets within the interval of periods in which they occur. There is indeed an excess of comets with P around 70 yr, and it is only the statistics of small numbers which makes the statistical tests inconvincing.

Figure 4 illustrates the different behaviours of two of the comets whose orbital period is not far from six times that of Jupiter: P/Halley and P/Brorsen-Metcalf. The top plots show the trajectories of these comets in a frame centred on Jupiter and rotating with it so that the Sun is always on the negative x-axis. The central plots show the changes in the revolution period between subsequent perihelion passages, the horizontal line indicating the exact commensurability. Finally, the bottom plots are polar diagrams where the radius vector is the osculating period, in units of Jupiter's period, and the position angle is the heliocentric elongation of Jupiter from the comet when this is at perihelion. P/Halley shows no regular pattern in its long-term behaviour. On the contrary, the libration of P/Brorsen-Metcalf reflects in all three plots. The jovicentric pattern is locked so that the jovicentric longitude of the comet at its perihelion passages remains almost constant, and the perturbations experienced in the inward and outward branches nearly compensate, making the motion very stable. The period undergoes regular oscillations depicted in the second plot. The extrema of the elongation in the lowest plot coincide with the passages of the comet's period through the exact commensurability, and the zero elongation points correspond to the largest amplitudes in the second plot.

Two more HT comets with periods around the 6:1 resonance with Jupiter are shown in Fig. 5. While the libration pattern of P/Pons-Brooks seems satisfactorily regular, that of P/Olbers is markedly irregular and an extension of the integration period seems necessary in order better to understand the long-term dynamics of this comet which, anyway, does not circulate as – for instance – P/Halley.

Figure 6 contains two more comets: P/Dubiago, with a very regular libration around the 5:1 resonance with Jupiter, and P/Väilsälä 2, whose libration pattern, around the 7:1 resonance with Jupiter, is recognizable but irregular like that of P/Olbers.

In the long-term dynamical evolution the resonances can increase the stability of motion, making the dynamical lifetimes of HT comets longer than indicated by simplified statistical modelling. Libration of the argumet of perihelion, as described by Kozai (1979), would act in the same direction. All the integrations currently available are still too short compared with the libration period, and the nongravitational effects, which may become considerable at these small perihelion distances, would not a priori allow a definite conclusion even on the basis of much longer integrations.

Acknowledgements. This work has been conducted in the framework of the scientific agreement between CNR (Italy) and SAV (Czechoslovakia). The two Academies have provided grants to facilitate the exchange of visits in both Countries. The authors are indebted to the Referee, B.G. Marsden, for helpful comments on the manuscript.

References

Carusi, A., Kresák, L., Perozzi, E., Valsecchi, G.B.: 1985a, *Long-term Evolution of Short-period Comets*, A. Hilger, Bristol

Carusi, A., Kresák, L., Perozzi, E., Valsecchi, G.B.: 1985b, in *Dynamics of Comets: their Origin and Evolution*, eds. A. Carusi, G.B. Valsecchi, Reidel, Dordrecht, p. 319

Carusi, A., Valsecchi, G.B.: 1985, in *Dynamics of Comets: their Origin and Evolution*, eds. A. Carusi, G.B. Valsecchi, Reidel, Dordrecht, p. 261

Everhart, E.: 1973, *Astron. J.* **78**, 329

Kozai, Y.: 1979, in *Dynamics of the Solar System*, ed. R.L. Duncombe, Reidel, Dordrecht, p. 231

Kresák, L.: 1972, *Bull. Astron. Inst. Czechosl.* **23**, 1

Kresák, L.: 1981, *Bull. Astron. Inst. Czechosl.* **32**, 321

Astron. Astrophys. 187, 906–908 (1987)

ASTRONOMY
AND
ASTROPHYSICS

Dormant phases in the aging of periodic comets

L. Kresák

Astronomical Institute, Slovak Academy of Sciences, 84228 Bratislava, Czechoslovakia

Received February 10, accepted March 13, 1987

Summary. Implications of the observations of P/Halley for the problem of comet aging are discussed. It is suggested that the active lifetimes of short-period comets are intermitted by dormant phases during which they remain completely, or almost completely, inactive. Such phases are tentatively identified for several short-period comets by the absence of observations of their prediscovery apparitions at which they should have been very bright and favourably placed.

Key words: comets – brightness changes – evolution – lifetimes

1. Introduction

Among the numerous exciting results obtained by the Halley missions, two are of particular significance for the problem of cometary evolution: the unexpectedly large size of the nucleus and the high heterogeneity of its surface. The comet's gas and dust production was found to come from a limited number of discrete sources, with the rest of the surface completely, or almost completely, inactive. The sources seem to persist over several revolutions (Sekanina and Larson, 1986), but must get exhausted after some time, being replaced by new ones. Owing to their limited number, it appears reasonable to assume that there can be periods during which none of them is active. The existence of such intermissions, which would be of fundamental importance for the comet's lifetime, can only be demonstrated in cases when a periodic comet has escaped detection in the past, although it was favourably placed, and should have been very bright according to its later observations.

The identification of such cases poses several problems. In view of the improving observing techniques and increasing sky coverage, they can only be found among short-period comets of small perihelion distance, which exhibit major changes of their apparent brightness with the varying solar elongation at individual perihelion passages. Fortunately, it is exactly these comets which evolve most rapidly, and among which definite extinctions have been observed. Unfortunately, such comets exhibit the largest, and often erratic, non-gravitational effects in their motion, which limits the reliability of the backward integrations of their present orbits. The situation becomes critical when the comet passes close to Jupiter, and its real motion starts to diverge rapidly from the computed one due to the differential perturbations.

2. The search for missing observations

A search for missing observations of comets, indicative of a dormant phase in their evolution, was based on the backward integrations of the motion of all known short-period comets by Belyaev et al. (1986), including constant non-gravitational parameters. The divergence of orbits was checked with the integrations by Carusi et al. (1985), which represent purely gravitational solutions. The brightness estimates were mainly taken from the catalogue of Vsekhsvyatskij (1958) for the period prior to 1954, from its supplements for the period 1954–1977, and from the International Comet Quarterly from 1978 onwards. New values of absolute magnitudes were derived using only observations obtained with visual telescopes of low magnifying power, under very favourable conditions near the perihelion. For each suspect case positional and magnitude ephemerides were computed, with the time of perihelion passage varied within reasonable limits, the interference of twilight and moonlight being also taken into account. Then, the missing apparitions were checked with the records of inadequately observed and uncatalogized comets for which no orbit computation was possible (Galle, 1894; Baldet, 1949 and the original sources quoted therein). In this way P/Grigg-Skjellerup was removed from the preliminary list by being identified with comet 1808 III, observed solely by Pons on two nights (Kresák, 1987).

This screening has resulted in the list of missed comet apparitions given in Table 1. For each comet, the following data are listed: the years of the first and last observation, the dates of missing observations, the maximum apparent brightness which the comet should have reached at that time, $M_1 - M_2$, and the number of nights for which the brightness should have remained within 1 mag of this maximum for the northern and southern observers, N_N/N_S. The periods N_N and N_S are represented by their net values, sometimes composed of several shorter intervals, and refer to the geographic latitudes of $+40°$ and $-30°$, respectively. Subtracted are the periods when the comet was at an altitude of less than 10° at the time of astronomical twilight, and when the moon was near its full phase or near the comet; the results are rounded off to the nearest 5 nights. The magnitudes M_1 and M_2 correspond to the extremes of the total absolute magnitude reached by these comets during all later apparitions

Table 1

Comet	Observed	Missed	$M_1 - M_2$	N_N/N_S
P/Biela	1772–1852	1778 Nov.–Dec.	3.5–5.0	35/40
P/Brorsen	1846–1879	1835 Jan.	3.0–5.0	10/10
P/Finlay	1886–1981	1827 Sep.–Oct.	3.5–5.5	30/35
P/Honda-Mrkos-Pajdušáková	1948–1985	1878 Sep.–Nov.	5.5–7.0	15/20
		1889 Sep.	5.0–6.5	10/10
		1917 Jan.	5.5–7.0	10/15
		1927 Nov.–Dec.	6.0–7.5	15/20
P/Boethin	1975–1986	1919 Oct.–1920 Feb.	6.5–8.5	75/75
P/Hartley-IRAS	1983–1984	1962 July–Sep.	5.5–9.0	50/10
P/Denning-Fujikawa	1881–1978	1829 Dec.	1.5–6.0	5/0
		1881 July–Aug.	4.0–8.5	5/15
		1916 Sep.–Oct.	5.0–9.5	10/15
		1960 July–Sep.	5.0–9.5	15/35
		1969 Aug.–Sep.	5.0–9.5	10/30

of favourable observing geometry. No secular brightness decrease is assumed, because this is believed to be quite insignificant.

3. Discussion of individual cases

Interestingly enough, the two comets at the beginning of the list are exactly those for which a later complete extinction appears almost certain. The 1778 perihelion passage of P/Biela followed immediately after its discovery apparition of 1772. While it may be conjectured that comet searches were just starting at that time, it is difficult to believe that the comet could have escaped detection, as a very easy naked-eye object visible for several weeks at large solar elongations, ten times brighter than in 1772. After having split into two components, last observed in 1852, P/Biela was definitely extinct in February 1866, when it should have been again a favourably situated naked-eye object, and on several similar occasions since then. P/Brorsen should have been in 1835 an equally bright naked-eye object but less favourably placed, observable around its conjunction with the sun. The missed perihelion passage precedes the discovery by two revolutions, which would normally guarantee a good ephemeris. Unfortunately, there was an encounter with Jupiter in 1842, which could degrade the accuracy of the extrapolation; however, the observing conditions do not change appreciably when T is varied by ± 10 days, with the maximum brightness remaining within 0.5 mag of that of Table 1, and a good observability from both hemispheres. P/Brorsen was definitely extinct in March 1890 when it should have been a favourably placed naked-eye object again, and on several similar occasions since then.

The next two comets of the table, P/Finlay and P/Honda-Mrkos-Pajdušáková, have been under almost continuous observation for 15 and 8 perihelion passages, respectively, with only a few returns missed due to unfavourable observing geometry. Their orbits and nongravitational effects are well determined. P/Finlay should have been a bright naked-eye object in 1827, observable for several weeks from both hemispheres not far from the opposition. The only questionable point is the series of approaches to Jupiter in 1862, 1850 and 1839 which, combined with wild nongravitational effects, could have produced some deviation in the backward integrations. However, for deviations of ± 10 days the maximum remains within 1–2 mag of the tabulated value, and larger deviations do not seem probable. For P/Honda-Mrkos-Pajdušáková the situation is different in that the past observing opportunities were not so outstanding, but there were four of them. Any of these perihelion passages could have been missed quite easily, but for all of them it does not seem likely. There was a close encounter with Jupiter in 1935, which makes the situation similar to that with P/Brorsen. Another similar encounter occurred in 1876, and extrapolation over it would be certainly unsound; it may just be noted that, accordingly, the comet should have reached $M_1 = 4.5$, $M_2 = 6.0$ in 1837.

The next two comets were only discovered quite recently. P/Boethin was observed at two apparitions, and the linked-up orbit is of ample accuracy for extrapolating five revolutions back, where an observing opportunity of long duration was identified. Although the comet should have not been a naked-eye object, it could have hardly escaped detection with an absolute brightness comparable to that of 1975 and 1986. In 1986 it was 2 mag brighter than in 1975, which may indicate a gradual increase of activity from a totally or partially dormant phase. P/Hartley-IRAS was observed only once, but the accuracy of its osculating orbit makes the extrapolation one revolution back quite safe. The observing opportunity of 1962 was of long duration, and the comet was circumpolar for the northern observers for nearly two months. With the same absolute lightcurve as in 1984, it should have reached naked-eye visibility in the second half of August 1962, close to the north pole. However, the brightness variations in 1984 were quite unusual, indicating a slow and long brightness surge. The absolute brightness increased from ~ 8 in January (M_2 of Table 1) to ~ 5 at the end of February (M_1), and dropped again to ~ 7 by the end of May.

For illustrating the difficulties involved in the interpretation of this kind of data, P/Denning-Fujikawa can serve as the best example. It was observed at two apparitions, in 1881 and 1978, and its absolute brightness was 4.5 mag lower at the latter. The comet should have passed near the earth in 1829, without any closer approach to Jupiter in the meantime. Now, interpreting the 1978-1881 difference as a secular brightness decrease (at a rate comparable to that found by some authors for other comets), we obtain for 1829 a maximum apparent brightness of -1 mag! Assuming the absolute magnitude of 1881 as a constant value, we still obtain for 1829 $M_1 = 1.5$, which is definitely impossible. Moreover, the same assumption would predict $M_1 = 4$ two months before the comet's first discovery, at a large solar elongation in the southern hemisphere, and $M_1 = 5$ at the three missed returns in 1916, 1960, and 1969. Assuming the absolute magnitude of 1978 as a constant value, it is still difficult to believe that the comet could have remained undiscovered in the 60's, with $M_2 = 9.5$ for several weeks. Hence, the only plausible explanation is that the comet was entirely, or almost entirely, inactive for a number of revolutions, and experienced a temporary brightness burst after the 1881 perihelion passage.

The recently discovered comets P/Takamizawa, P/Hartley 2 and P/Machholz promise additional examples of irregular behaviour in the past, as soon as more accurate orbits and their backward integrations become available.

4. Conclusions

The presence of dormant phases in the evolution of periodic comets has an important impact on the determination of their lifetimes. While the values found previously with the use of different approaches (Dobrovol'skij, 1972; Kresák, 1981; Fernández, 1985; Rickman, 1986) are in fair mutual agreement, they can also represent transient periods of activity spaced by dormant phases, or their sums. It appears that there are not too many such phases in the lifetime of a typical comet, but their duration cannot be estimated from the observations available. Therefore, efforts should be made to rediscover also those comets for which the number of missed returns suggests a complete disappearance, and to monitor the asteroidal objects moving in cometlike orbits, most of which have been discovered quite recently.

References

Baldet, F.: 1949, *Liste Générale des Comètes de l'Origine a 1948*, Gauthier-Villars, Paris

Belyaev, N.A., Kresák, L., Pittich, E.M., Pushkarev, N.A.: 1986, *Catalogue of Short-period Comets*, Veda, Bratislava

Carusi, A., Kresák, L., Perozzi, E., Valsecchi, G.B.: 1985, *The Long-term Evolution of Short-period Comets*, Hilger, Bristol

Dobrovol'skij, O.V.: 1972, *IAU Symp.* **45**, 352

Fernández, J.A.: 1985, *Icarus* **64**, 308

Galle, J.G.: 1894, *Verzeichniss der Elemente der bisher berechneten Cometenbahnen*, Engelmann, Leipzig

Kresák, L.: 1981, *Bull. Astron. Inst. Czech.* **32**, 321

Kresák, L.: 1987, *Bull. Astron. Inst. Czech.* **38**, 65

Rickman, H.: 1986, ESA SP-249, 195

Sekanina, Z., Larson, S.M.: 1986, *Nature* **321**, 357

Vsekhsvyatskij, S.K.: 1958, *Fizicheskie Kharakteristiki Komet*, Gos. Izd. Fiz. Mat. Literatury, Moskva

Astron. Astrophys. 187, 909–912 (1987)

ASTRONOMY AND ASTROPHYSICS

The dynamical lifetime of comet P/Halley

D. I. Olsson-Steel

Lund Observatory, Box 43, S-22100 Lund, Sweden

Received June 10, accepted June 25, 1987

Summary. Using an extension of Öpik's techniques the stability of the orbit of Halley's comet under close encounters with the planets is investigated, and it is found that the orbital changes due to such encounters expected over a long time-base are of the same order as the effects of more distant planetary perturbations and non-gravitational effects. For the present orbit the half-life of the comet against ejection from the solar system as a result of a close encounter is 68,000 apparitions; the half-life against planetary impacts or catastrophic disruption in a close approach is about a million apparitions: dynamical loss is therefore unlikely. For a past lifetime of 2000–3000 perihelion passages there is a probability of about 50% of a single large deflection into an orbit of ($q < 0.5$ or > 0.7 AU), ($a < 15$ or > 25 AU), or ($i < 150°$ or $> 170°$). During this period the perihelion distance has most probably remained between 0.5 and 0.7 AU, so that physical models based upon sensibly constant rates of material loss from the surface through its lifetime appear to be justified.

Key words: comets – Halley's comet – meteors – meteoroid streams – solar system dynamics

1. Introduction

One problem in developing models for the past and future physical evolution of Halley's comet is our lack of knowledge of its past and future orbital evolution. Recently-developed models for the long-term mass-loss rate by the nucleus and hence the supply of material to the meteoroid stream (McIntosh and Hajduk, 1983; Hajduk, 1985, 1986; Hughes, 1985, 1986; Jones and McIntosh, 1986) have tacitly assumed that the size and shape of the comet's orbit are sensibly constant over the age of the comet and meteoroid stream in the inner solar system. Since the mass-loss rate of a comet depends critically upon its orbital parameters, and in particular its perihelion distance (q), it is necessary to know whether the orbit of the comet might have been substantially different (e.g. $q < 0.5$ AU or $q > 0.7$ AU, compared to the present value of 0.587 AU) during the past lifetime of the comet, or indeed during the future lifetime. Thus, although generally the physical lifetimes of short-period comets are much less than the dynamical lifetimes of idealized particles on similar orbits (Kresák, 1981, 1986), it is necessary to have some idea of the comet's orbital history if its physical history is to be reconstructed. It is the purpose of this paper to address this question.

From physical considerations and the mass of the meteoroid stream associated with it, the age of P/Halley (the "age" being understood to imply the length of time that the comet has been in its present basic orbit of period ~ 75 yr and $q \sim 0.6$ AU) has been estimated as being a few thousand perihelion passages, with a similar number to come before the nucleus is exhausted (Hajduk, 1985, 1986; Hughes, 1985, 1986); however Wallis and Wickramasinghe (1985) and Whipple (1986) have favoured a much shorter life expectancy. For present purposes it will be assumed that the characteristic time-span which is of interest is 2000–3000 apparitions, or about $2\ 10^5$ yr.

Unfortunately, it is not possible to numerically integrate the orbit of the comet either backwards or forwards in time over more than a few tens of apparitions. Yeomans and Kiang (1981) found that the orbit prior to close approaches to the Earth in 837, 607, 374, and 141 A.D. could only be rectified by using observations from prior apparitions; since there are no observations available from before a close approach in 1404 B.C., they were unable to integrate the orbit backwards beyond that time. The discordance between the results of different orbital integrations has been investigated by Sitarski and Ziolkowski (1986, 1987) and the form of the corrections applied for non-gravitational effects shown to be crucial. Generally the accuracy of numerical integrations is lost after a close approach with a planet, particularly Jupiter (Carusi et al., 1985a, 1986) although for P/Halley the high-velocity shallow encounters are not very effective in changing its orbit (Öpik, 1976, p. 84; Carusi et al., 1985b). Nevertheless we expect the close encounters to cause the largest orbital changes, and hence their frequency and effect are of interest.

In order to obtain statistical results long-term integrations with slightly different starting parameters, like those carried out for Chiron by Oikawa and Everhart (1979), would be of utility; however, that type of simulation is extremly expensive in terms of computer time and an alternative approach is used here.

2. Brief description of method

In a recent paper Olsson-Steel (1987) has described an extension of the techniques of Öpik (1976) which allows statistical results to be derived, and this method has been applied here to P/Halley. The main drawback of this technique is that the close encounter is modelled as two separate two-body events: prior to entry to the planetary sphere of influence the orbit is taken to be a heliocentric ellipse, and within this sphere the cometary trajectory is taken to be a planetocentric hyperbola. In fact for P/Halley, with its high-

Send offprint requests to: D.I. Olsson-Steel, Department of Physics, University of Adelaide, GPO Box 498, Adelaide, SA 5001, Australia

Table 1. Minimum and maximum encounter velocities and resultant angular deflections for encounters between P/Halley and each of the planets

	Venus	Earth	Mars	Jupiter	Saturn	Uranus	Neptune
V_{min} (km/s)	80.1	65.8	47.7	23.8	16.2	10.0	6.9
- mean deflection	0°.006	0°.008	0°.002	0°.51	0°.27	0°.10	0°.15
- RMS deflection	0°.011	0°.013	0°.004	0°.88	0°.48	0°.19	0°.29
V_{max} (km/s)	82.0	67.6	56.0	25.9	18.0	11.1	7.1
- mean deflection	0°.006	0°.007	0°.001	0°.43	0°.22	0°.08	0°.15
- RMS deflection	0°.010	0°.013	0°.003	0°.75	0°.39	0°.16	0°.28

velocity encounters and hence small angular deviations from its unperturbed path (cf. Tisserand parameter = −0.6), this model is a reasonably good approximation. Presented in Table 1 are the minimum and maximum encounter velocities with each of the planets, along with the resultant mean and root-mean-square angular deflections (for a complete description of the derivation of these quantities, see Weidenschilling, 1975, and Olsson-Steel, 1986, 1987). As can be seen, the angular deflections and hence the changes in the orbit are rather small, as was noted by Öpik (1976).

The encounter probability method used assumes that the argument of perihelion and the longitude of the node are random on a time-scale short compared to the dynamical lifetime. Kozai (1979) studied the secular perturbations of P/Halley and found that the argument of perihelion may librate between 47° and 133°; however Kozai's model assumes that (i) only the four giant planets perturb the comet, whereas close approaches to the Earth have been shown to significantly alter its orbit (Yeomans and Kiang, 1981); (ii) the semi-major axis of the comet does not change appreciably, whereas Yeomans and Kiang (1981) find variations from 16.7 to 18.4 AU over just the last 45 apparitions; and (iii) the four disturbing planets all have circular orbits in the same plane as Jupiter. Thus Kozai (1979) himself pointed out that a small change or error in the calculations could lead to P/Halley no longer being found to be a librator; in any case at the extremes of the libration cycle approaches to within 0.25 AU of Jupiter would be possible, which would again lead to (ii) being violated. Thus, it is not clear that the comet should be regarded as a librator (hence avoiding very close approaches to the giant planets) in the long-term. Yeomans (1985, 1986) pointed out that if the line of apsides continued to rotate at the same rate as found for the 45 apparitions integrated by Yeomans and Kiang (1981) then P/Halley would have had its node close to the orbit of Jupiter about 16,000 years ago, which is therefore a minimum age for the comet if it is assumed that P/Halley was captured into its present general orbit by a close encounter with that planet. A full rotation of the argument of perihelion, if it continued at the same rate, would take about 400 apparitions which is less than 20% of the estimated age of the comet. Randomness of the argument of perihelion (and the node relative to each planet) is a condition of the method used in this paper, and this should be borne in mind should later rigorous studies prove P/Halley to be a librator. Based upon observations of the Halleyid meteor showers Hajduk (1985, 1986) concluded that the maximum period since the comet entered the inner solar system is about 2 10^5 yr; this calculation was founded upon the assumption that the cometary orbit librates in the same way as described by Kozai (1979).

The probability distributions for new orbital elements for P/Halley were therefore calculated here by the method described by Olsson-Steel (1987), using the following initial elements for the comet (Marsden, 1986):

$q = 0.587$ AU;

$e = 0.967$;

$i = 162°.2$;

from which

$a = 17.79$ AU;

$P = 75.04$ yr;

and

$Q = 35.0$ AU.

The results are not critically dependent upon the initial elements [cf. Yeomans and Kiang (1981) find variations in q from 0.572 to 0.634 AU, in e from 0.962 to 0.969, in i from 162°.2 to 163°.7, and in P from 68.15 to 79.29 yr, all within the last 45 apparitions]. The resultant probabilities for different events are given in Table 2.

Apart from the probabilities per year for each event occurring as a result of encounters with specific planets, in the final column of Table 2 are given the half-lives against each event. This parameter is defined as:

$$t_{1/2} = -\ln 2 \,/\, 75.04 \ln (1 - \Sigma p),$$

where Σp is the sum of the individual probabilities from the first seven columns, and the factor of 75.04 means that $t_{1/2}$ is expressed as a number of perihelion passages, or apparitions. Thus the half-life against loss in a physical impact upon a planet is 5.5 million apparitions, and hence is an unlikely fate for the comet.

Rickman (1986) gives probable density limits for P/Halley of 100–200 kg m^{-3}: if the true density is close to the upper limit then the chance of the comet being catastrophically disrupted in a passage through the Roche radius of Jupiter is about five times the collision probability, and at the lower limit, seven times; the half-life in either case is about a million apparitions. Similarly, the half-life against ejection from the solar system on an hyperbolic heliocentric orbit is 68,000 apparitions, but in such a length of time not only would the comet be physically exhausted but also the orbit would have changed substantially due to a random walk caused by many small orbital changes: a deflection (or transit through one of the planetary spheres of influence) occurs about every 18 apparitions. Next a consideration of the likely effects of such deflections is necessary.

Table 2. Probability distributions and half-lives for events resulting from planetary close encounters by P/Halley

	Venus (yr^{-1})	Earth (yr^{-1})	Mars (yr^{-1})	Jupiter (yr^{-1})	Saturn (yr^{-1})	Uranus (yr^{-1})	Neptune (yr^{-1})	Half-life (passages)
Collision	1.7E-10	6.5E-11	6.3E-12	1.2E-09	2.0E-10	9.7E-12	1.3E-11	5.5E+06
Deflection	1.3E-05	9.6E-06	1.7E-06	2.8E-04	1.2E-04	3.8E-05	6.4E-05	18
Ejection				1.3E-07	4.7E-09	4.2E-11	5.9E-11	6.8E+04
q < 0.1 AU					1.0E-09	1.0E-10	1.9E-10	7.2E+06
< 0.2					4.8E-09	3.2E-10	6.6E-10	1.6E+06
< 0.3				2.3E-08	1.6E-08	8.9E-10	2.0E-09	2.2E+05
< 0.4				2.1E-07	5.7E-08	3.1E-09	6.0E-09	3.3E+04
< 0.5				2.1E-06	4.6E-07	2.7E-08	5.8E-08	3500
< 0.6	1.3E-05	9.6E-06	1.7E-06	1.6E-04	7.2E-05	2.3E-05	3.8E-05	29
> 0.6	1.8E-12	1.0E-10	6.0E-13	1.2E-04	5.0E-05	1.5E-05	2.6E-05	44
> 0.7				1.4E-06	2.6E-07	1.8E-08	6.6E-08	5300
> 0.8				5.0E-07	1.0E-07	5.7E-09	1.8E-08	1.5E+04
> 0.9				2.6E-07	5.0E-08	3.1E-09	9.9E-09	2.9E+04
> 1.0				1.6E-07	3.1E-08	2.0E-09	6.5E-09	4.6E+04
> 1.2				8.3E-08	1.6E-08	1.1E-09	3.6E-09	8.9E+04
> 1.5				4.1E-08	8.3E-09	5.8E-10	2.1E-09	1.8E+05
> 2.0				1.9E-08			1.2E-09	4.6E+05
a < 5 AU				3.1E-09				3.0E+06
< 6				1.1E-08				8.4E+05
< 7				2.4E-08	9.0E-11			3.8E+05
< 8				4.5E-08	6.2E-10			2.0E+05
< 9				8.0E-08	1.5E-09			1.1E+05
<10		2.8E-12		1.3E-07	3.1E-09			6.9E+04
<12	6.3E-12	3.4E-11		4.1E-07	1.1E-08	1.8E-11		2.2E+04
<14	1.2E-10	1.6E-10		1.2E-06	4.1E-08	1.8E-10		7400
<16	1.0E-09	1.1E-09	4.8E-12	5.6E-06	2.1E-07	1.6E-09	1.5E-10	1600
<18	1.3E-05	9.5E-06	1.7E-06	1.5E-04	7.1E-05	3.1E-05	6.4E-05	27
>18	7.8E-08	1.1E-07	6.2E-10	1.3E-04	5.1E-05	7.5E-06	9.6E-08	49
>19	3.0E-09	3.4E-09	2.0E-11	8.6E-05	8.8E-07	6.5E-09	4.4E-09	106
>20	1.0E-09	1.1E-09	5.1E-12	1.2E-05	2.7E-07	2.1E-09	1.8E-09	750
>25	1.1E-10	1.5E-10		5.0E-07	4.5E-08	1.8E-10	3.6E-10	1.7E+04
Q <10 AU				1.1E-08	1.3E-10			8.3E+05
<15				5.0E-08	1.4E-09			1.8E+05
<20		5.0E-12		1.8E-07	5.2E-09	1.5E-11		5.0E+04
<25	2.9E-11	6.5E-11		6.2E-07	1.4E-08	1.3E-10		1.5E+04
<30	4.3E-10	5.1E-10	1.3E-12	4.1E-06	1.4E-07	9.2E-10	5.1E-11	2200
$q < Q_{Merc}$				8.4E-07	1.6E-07	8.4E-09	1.7E-08	9000
$q > Q_{Venus}$			5.0E-13	1.4E-06	2.3E-07	1.8E-08	6.0E-08	5400
$q > Q_{Earth}$				1.5E-07	3.1E-08	2.0E-09	6.4E-09	4.9E+04
$q > Q_{Mars}$				3.1E-08	6.3E-09	5.2E-10	1.8E-09	2.3E+05
$q > Q_{Jup}$					4.6E-10	5.3E-11	2.8E-10	1.2E+07
$q > Q_{Sat}$						1.7E-11	1.2E-10	6.7E+07
$Q < q_{Sat}$				9.6E-09				9.6E+05
$Q < q_{Uran}$		5.0E-15		1.3E-07	4.0E-09			6.9E+04
$Q < q_{Nep}$	4.4E-10	5.2E-10	1.3E-12	4.0E-06	1.4E-07	9.2E-10		2200
$i < 150^0$				4.0E-07	1.4E-07	6.6E-09	4.0E-08	1.6E+04
$< 155^0$	4.4E-11			2.0E-06	3.5E-06	2.4E-08	1.4E-06	1300
$< 160^0$	5.5E-06	1.3E-11	5.1E-07	7.0E-05	4.0E-05	7.7E-06	1.8E-05	65
$< 165^0$	7.5E-06	9.6E-06	1.3E-06	2.2E-04	9.1E-05	3.3E-05	4.7E-05	23
$> 165^0$	5.5E-06	2.9E-12	4.0E-07	5.4E-05	3.1E-05	5.3E-06	1.6E-05	82
$> 170^0$	3.9E-12			1.7E-06	7.2E-07	2.1E-08	3.2E-07	3300

3. Results of close encounters

For physical models the most significant alterations to the orbit are those to the perihelion distance. Table 2 shows that the most likely effect of a close encounter is that q is not much changed, with about (18/29) or 60% of new orbits having $q < 0.6$ AU and about (18/44) or 40% having $q > 0.6$ AU. The half-lives for single deflections into orbits of $q < 0.5$ AU or > 0.7 AU are 3500 and 5300 apparition respectively, so if the age of the comet is about 2300 apparitions (Hajduk, 1985; Hughes, 1985) then there is less than a 50% chance that in that time a single large deflection into such an orbit will occur. The change in q per close encounter is of the order of 0.03 AU, equivalent to about 0.0015 AU per apparition; this is half of the mean change in q from apparition to apparition found by Yeomans and Kiang (1981) as a result of planetary perturbations and non-gravitational effects.

It appears unlikely that the semi-major axis of P/Halley has been or will be altered substantially as the result of a close encounter; deflections into orbits of $a < 15$ or > 25 AU have less than a 50% probability over a time-span equivalent to the putative age of the comet. The figures are consistent with a remaining within about 3 AU of its present value.

Changes in the planets crossed by P/Halley in the past or future do not occur frequently. Diversion into a Mercury-crossing orbit has a half-life of 9000 apparitions. Reduction of the aphelion distance to $Q < 30$ AU, or within the orbit of Neptune, occurs with a half-life of 2200 apparitions if only close-encounters are considered, as herein. The exact integrations of Yeomans and Kiang (1981) and Carusi et al. (1985b) show variations in Q of the order of 0.5 AU from apparition to apparition; thus it appears that the aphelion distance may change largely as the result of a random walk caused by planetary perturbations and non-gravitational effects.

Finally in Table 2, the possible changes in the inclination are given. Since the other orbital parameters are not affected much by close encounters, the variations in the inclination are of especial significance because the probability of subsequent encounters varies as $(1/\sin i)$. Thus, an increase in the inclination to $i > 170°$ would mean that thereafter close encounters occur about twice as frequently. In the long-term, with only the inclination changing appreciably, the planetary encounter rate of P/Halley would be between once per 10 and once per 25 perihelion passages.

4. Conclusions

The results of this investigation can be summarized as follows. If the period since P/Halley entered its present orbit, probably as the result of a close encounter with Jupiter, is a few thousand perihelion passages (i.e. about 2 10^5 yr) then there is less than a 50% chance that during that time there have been any close planetary encounters which have altered the orbit from a substantially different form. By "substantially different" an orbit of ($q < 0.5$ or > 0.7 AU), ($a < 15$ or > 25 AU), ($i < 150°$ or $> 170°$) is implied. Rather, random walks in these orbital parameters with much smaller changes are expected. Comparison with precise numerical integrations (Yeomans and Kiang, 1981; Carusi et al., 1985b) show that the variations in the elements brought about by planetary perturbations and non-gravitational effects are of the same order as those occurring as the result of close encounters, so that a long-term monotonic change in the orbit might occur if the non-gravitational forces continue to act in the same sense over a period of tens or hundreds of orbits.

For physical models of the comet's evolution knowledge of the history of the perihelion distance is of importance since the total insolation per apparition depends critically upon q; here, large variations in q are *not* found on a time-scale of the same order as the estimated age of the comet, so that the implied assumption of a quasi-constant q in many physical evolution models appears to be justified.

Acknowledgements. The author is the recipient of a European Space Agency Fellowship at the Lund Observatory. Computing costs associated with the work reported in this paper were borne by the Swedish Board for Space Activities. The helpful comments of the referees, Drs. D. W. Hughes and A. Hajduk, are gratefully acknowledged.

References

Carusi, A., Perozzi, E., Pittich, E. M., Valsecchi, G. B.: 1985a, in *Dynamics of Comets: Their Origin and Evolution, IAU Coll.* **83**, eds. A. Carusi, G. B. Valsecchi, Reidel, Dordrecht, p. 227

Carusi, A., Kresák, L., Perozzi, E., Valsecchi, G. B.: 1985b, in *Long-term Evolution of Short-period Comets*, Hilger, Bristol, Boston, p. 1

Carusi, A., Kresák, L., Perozzi, E., Valsecchi, G. B.: 1986, 20th ESLAB Symp., ESA SP-**250**, Vol. **II**, p. 413

Hajduk, A.: 1985, in *Dynamics of Comets: Their Origin and Evolution, IAU Coll.* **83**, eds. A. Carusi, G. B. Valsecchi, Reidel, Dordrecht, p. 399

Hajduk, A.: 1986, 20th ESLAB Symp., ESA SP-**250**, Vol. **II**, p. 239

Hughes, D. W.: 1985, *Monthly Notices Roy. Astron. Soc.* **213**, 103

Hughes, D. W.: 1986, in *Proc. Comet Sample Return Workshop*, ESA SP-**249**, p. 173

Jones, J., McIntosh, B. A.: 1986, 20th ESLAB Symp., ESA SP-**250**, Vol. **II**, p. 233

Kozai, Y.: 1979, in *Dynamics of the Solar System, IAU Symp.* **81**, ed. R. L. Duncombe, Reidel, Dordrecht, p. 231

Kresák, L.: 1981, *Bull. Astron. Inst. Czech.* **32**, 321

Kresák, L.: 1986, 20th ESLAB Symp., ESA SP-**250**, Vol. **II**, p. 433

Marsden, B. G.: 1986, *Catalogue of Cometary Orbits*, IAU, Minor Planet Center, Cambridge, MA, USA

McIntosh, B. A., Hajduk, A.: 1983, *Monthly Notices Roy. Astron. Soc.* **205**, 931

Oikawa, S., Everhart, E.: 1979, *Astron. J.* **84**, 134

Olsson-Steel, D.: 1986, *Monthly Notices Roy. Astron. Soc.* **219**, 47

Olsson-Steel, D.: 1987, *Icarus* **69**, 51

Öpik, E. J.: 1976, *Interplanetary Encounters*, Elsevier, Amsterdam, New York

Rickman, H.: 1986, in *Proc. Comet Sample Return Workshop*, ESA SP-**249**, p. 195

Sitarski, G., Ziolkowski, K.: 1986, 20th ESLAB Symp., ESA SP-**250**, Vol. **III**, p. 299

Sitarski, G., Ziolkowski, K.: 1987, paper presented at the *Symp. Diversity and Similarity of Comets*, Brussels, April 6–9

Wallis, M. K., Wickramasinghe, N. C.: 1985, *Monthly Notices Roy. Astron. Soc.* **216**, 453

Weidenschilling, S. J.: 1975, *Astron. J.* **80**, 145

Whipple, F. L.: 1986, 20th ESLAB Symp., ESA SP-**250**, Vol. **II**, p. 281

Yeomans, D. K.: 1985, in *Dynamics of Comets: Their Origin and Evolution, IAU Coll.* **83**, eds. A. Carusi, G. B. Valsecchi, Reidel, Dordrecht, p. 389

Yeomans, D. K.: 1986, 20th ESLAB Symp., ESA SP-**250**, Vol. **II**, p. 419

Yeomans, D. K., Kiang, T.: 1981, *Monthly Notices Roy. Astron. Soc.* **197**, 633

Astron. Astrophys. 187, 913–918 (1987)

ASTRONOMY AND ASTROPHYSICS

Galactic tides affect the Oort cloud: an observational confirmation

A.H. Delsemme

Department of Physics and Astronomy, The University of Toledo, Toledo, Ohio 43606, USA

Received February 16, accepted March 16, 1987

Summary. There are 152 known original orbits of comets with a period larger than ten thousand years. Their aphelia are shown here to avoid *three* zones of the celestial sphere, namely the two galactic polar caps and a strip along the galactic equator. Such an axial symmetry of the aphelia distribution in respect to the plane of the galaxy cannot be explained by the traditional mechanism proposed to bring the Oort cloud comets into visibility – namely perturbations from random stellar passages. It cannot be explained either by any other *random* mechanism – like the passage of molecular clouds, a more recent variation of the previous idea. Finally, none of the known observational biases could introduce such a specific symmetry with respect to the galaxy. Standing in contrast, the observed symmetry is predicted by the "vertical" galactic tide, namely the differential attraction, on the sun and on each comet of the Oort cloud respectively, that comes from the mass distribution within the thick galactic disk.

Key words: comets – the galaxy: disk – the galaxy: solar neighbourhood

1. Introduction

It will be shown here that the cometary aphelia of dynamically young comets do not seem to be distributed at random on the celestial sphere, but seem to display the axisymmetrical features expected from the galactic vertical tide. The present paper presents therefore the first observational confirmation that this tidal mechanism is likely to be *the* major cause bringing the Oort cloud comets into visibility.

Since Eddington (1913), the distribution over the celestial sphere of the aphelia of "new" and long period comets, as well as their irregular clustering, have repeatedly puzzled many authors all over the years.

Early determination (Tyror 1957, Hurnik 1959) have considered the largest possible number of long period comets, using the traditional limit of a 200-year period as the cutoff between long and short period comets. Such a sample where osculating and not original orbits are used, contains many orbits which have been drastically and repeatedly perturbed by the planets.

Oja (1975) prefers to limit himself to the *original* orbits of those 73 comets with the greatest major axes ($1/a < 10^{-4}\,\mathrm{AU}^{-1}$). The rationale is that those comets have not yet been much influenced by the solar system. He computes the distribution ellipsoid of all aphelia, but recognizes that observational selection may change their distribution. Unable to choose among several hypotheses, his major achievement is probably to establish that the chances of this clustering to be random is only 0.1%. Although the sample he used is too small to draw definite conclusions, it is interesting in hindsight to see that his data already suggest the existence of those three zones of avoidance that we are going to speak about.

Bogart and Noerdlinger (1982) extend the study to three orbital elements specifying the spatial distribution of the orbits: the orbital plane normal $\hat{R}$, the perihelion direction $\hat{P}$, and the direction of the velocity at perihelion $\hat{Q}$. Using χ^2 tests, they find that the velocity $\hat{Q}$ is rather consistent with randomness, whereas the deviations of $\hat{R}$ and $\hat{P}$ from randomness are large, confirming with greater certainty the conclusion of Tyror (1957) and Hasegawa (1976). Unfortunately, they use for basic sample the 545 comet orbits listed in Marsden's (1979) catalog as "long period", mixing all good and dubious orbits (285 orbits are parabolas, most only because of insufficient observations to be able to distinguish a short-period from a long-period orbit). In particular, the fact that their orbital planes tend to prefer the ecliptic plane is clearly an artifact of the presence of many short period comets in the sample. Bogart and Noerdlinger conclude inconclusively: "any inferences about preferred theories are probably premature..."

Biermann et al. (1983) bring at last clarity in the confusion of the previous literature. Using 80 "new" and 59 "intermediate" comets with the best "original" orbits available, they confirm a large observed deviation from randomness that they call "aphelion clustering"; they interpret the clustering effect as coming from the recent passage of a very small number of stars perturbing the Oort cloud; in particular, they analyze and identify one conspicuous star track, showing the alignment of 17 aphelia; when the effect of this star track is removed, they find that the "antapex concentration of aphelia," reported by several authors, disappears in the sense that it is only slightly larger than the expected statistical fluctuation. Other star tracks cannot be identified as easily; however, they assume by analogy that the rest of the clustering comes from the same basic effect. Although their sample is not large enough to be conclusive, our three zones of avoidance are already visible, in hindsight, in their Fig. 1.

Lüst (1984) finds the need to extend the previous sample. This had become possible thanks to the additional orbits contained in the 1982 edition of Marsden's Catalog of Cometary Orbits. She classifies 89 comets as "new", 69 as "intermediate" and 65 as "old". She confirms the absence of clustering around the solar antapex, as well as the existence of a strong north-south asymmetry that she explains quite reasonably by an observational bias due to the geographic distribution of the observers. She discovers the existence of a "zone of avoidance," a $\pm 10^\circ$ belt along the galactic equator. She also mentions two further gaps around the poles; however, presumably concerned by the marginal results of the southern cap, she mentions only the north polar cap in the conclusions summarizing the paper. She notes: "though these large deviations point very strongly to a real effect, we did not find yet any plausible explanation."

Independently, the present author (Delsemme, 1985, 1986) has studied the aphelion distribution and the orbital angular momenta of a set of 126 dynamically young comets. His purpose was the possible detection of the passage of a massive body through the Oort cloud. However, he mentions in a note added in proof that his data seem to contain the first observational detection of the vertical galactic tide on the Oort cloud, because the two polar galactic caps and the zone close to the galactic equator seem to display a rather large depletion in aphelia. The present work brings substance to this assertion and extends the work to 152 dynamically young comets.

2. Data

The 152 comets of Table 1 are all comets with periods longer than 10,000 years, whose original orbits are known with accuracy. The comets names are taken straight from Marsden and Roemer's (1982) table. They are in the same order of increasing $(1/a)$; the galactic coordinates of their perihelia have been added.

Although the period cutoff is rather arbitrary, it has been demonstrated by Lüst (1984) that such a limit is safe in conserving the original aphelion clusterings detected for "new" comets. It is also easy to verify that our results do not change when the cutoff is changed, within reasonable limits; only our statistical results have been somewhat improved with 152 comets instead of 126.

In order to study the aphelia distribution, we divide the galactic coordinates into 24 zones of equal areas of $\pi/6$ steradians, all parallel to the galactic equator. The zone limits are given by galactic parallels, whose galactic latitudes are:

0.00° $\pm 4.78^\circ$ $\pm 9.59^\circ$ $\pm 14.47^\circ$ $\pm 19.47^\circ$ $\pm 24.62^\circ$
$\pm 30.00^\circ$ $\pm 35.68^\circ$ $\pm 41.81^\circ$ $\pm 48.59^\circ$ $\pm 56.44^\circ$ $\pm 66.44^\circ$

Table 1 lists perihelia instead of aphelia for convenience, since statistics can be applied indifferently.

The number of aphelia in each zone can be immediately deduced from Table 1. Although the three zones of avoidance can already be seen, the average number of aphelia per zone is too small, therefore statistical fluctuations are too large. We wish to diminish the total number of zones. However, the depletion of each polar cap does not go much beyond the $23^\circ 1/2$ radius of the cap. In order not to loose the significance of the polar cap's depletion, and to keep intact all symmetry properties, we fold the southern galactic hemisphere onto the northern galactic hemisphere, and add two by two the zones of same galactic longitude. This reduces the number of zones from 24 to 12, and increases the average number of aphelia per zone by a factor of two, improving the statistical significance of the results.

The results are given by the lower line on Fig. 1. The two equatorial zones contain *six* and *six* aphelia respectively. The sum of the two polar caps contains *four* aphelia. The rest of the sky contains 136 aphelia for nine zones, that is an average of more than 15 aphelia per zone.

3. Correction for observational bias

Let's now consider whether observational biases could affect the results. Biermann et al's (1883) work has established that the antapex concentration of aphelia is not a major bias. The only major bias to take into account is therefore the geographical distribution of perihelia. There is a distinct depletion of comets' discoveries in the southern hemisphere, assumedly because of the depletion of the southern observers. Since comets become brightest near perihelion, there must therefore be a smaller number of cometary perihelia in the southern hemisphere which is an artifact of the observers' depletion.

In order to assess this geographical bias numerically, we have divided the sky into 24, 12 or 6 zones of equal areas, this time all parallel to the celestial equator. Figure 2 shows the distribution of the 152 perihelia into six zones classified by declinations. It is apparent that approximate declination -20° divides the sky into two very different regions. The sky north of -20° does not show any evidence of depletion, whereas the sky south of -20° shows a large and almost uniform depletion which is, for our sample, by a factor of about 1.9. This declination limit of -20° seems to come from the average limit of easy and extended access of the southern sky for those numerous observers at the average

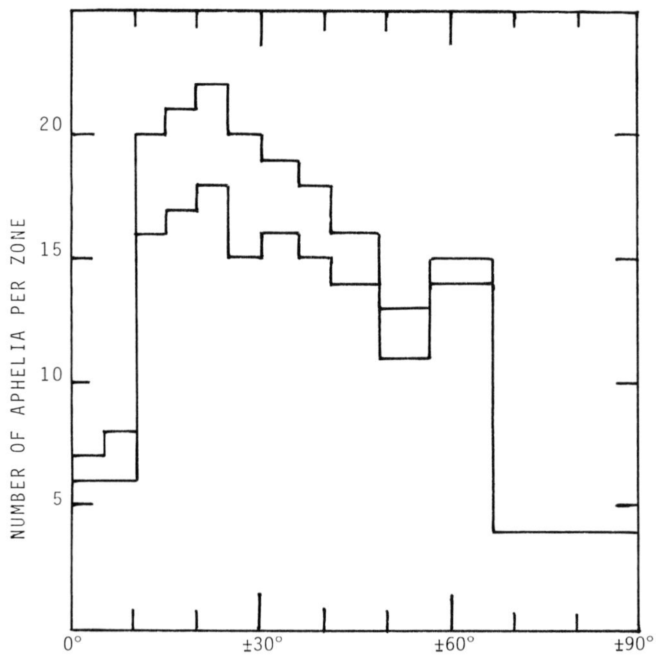

Fig. 1. Distribution of cometary aphelia in 12 zones of equal areas, symmetrical in respect to the plane of the galaxy. The zones of equal galactic latitude have been added by folding the south galactic hemisphere onto the north galactic hemisphere. This preserves the axial symmetry of the distribution. Only the equator zones and the two polar caps contain a number of aphelia smaller than 11 per zone. The lower curve is without correction. The upper curve is corrected for the geographical bias mentioned in Fig. 2

Table 1. Galactic coordinates of the set of 152 cometary perihelia

Comet	ℓ	b	Comet	ℓ	b	Comet	ℓ	b
1976 I	221.3	-10.4	1903 IV	140.6	34.1	1950 I	5.1	78.5
1955 V	52.3	-65.7	1979 VII	170.9	43.7	1970 XV	57.8	17.3
1959 III	16.5	-19.8	1948 V	70.1	38.6	1977 X	169.0	-41.1
1895 IV	211.7	-49.3	1947 VIII	144.0	-17.8	1973 II	187.7	-30.0
1971 V	40.7	-5.8	1925 VI	245.6	40.9	1886 II	61.5	40.3
1960 II	8.2	-37.4	1975 VIII	320.2	-21.7	1910 IV	40.8	27.5
1953 II	264.7	-15.7	1954 XII	293.2	-26.3	1972 XII	251.7	-41.9
1940 III	243.4	31.0	1976 IX	130.0	30.5	1949 I	331.7	-11.8
1899 I	127.4	-50.8	1951 I	317.6	43.8	1969 IX	275.3	-44.1
1957 III	319.6	-5.3	1956 I	134.0	23.6	1863 I	111.3	40.5
1968 VI	112.1	-21.3	1925 I	58.3	-10.6	1930 IV	180.3	41.2
1904 II	36.7	62.3	1959 X	40.9	12.8	1948 IV	254.5	41.9
1911 IV	116.1	22.1	1979 VI	27.0	-14.1	1973 X	36.6	61.7
1898 VIII	190.5	10.9	1955 VI	156.3	18.4	1880 II	164.8	21.5
1932 VII	70.6	25.4	1946 VI	306.6	20.0	1970 III	136.0	38.8
1975 XI	5.1	-1.5	1976 XIII	360.0	-5.7	1959 IV	51.1	0.2
1942 VIII	198.8	24.6	1932 VI	326.4	14.6	1900 II	62.2	-42.7
1892 VI	251.4	32.2	1912 II	45.4	-12.5	1927 IV	344.9	51.6
1849 II	17.1	56.5	1886 IX	108.6	24.4	1905 VI	72.2	63.1
1886 I	62.5	59.1	1889 I	133.7	-59.5	1966 II	43.9	-19.5
1946 I	133.1	60.3	1954 VIII	207.2	27.6	1974 III	228.6	18.4
1947 I	167.1	-49.0	1967 II	223.5	58.9	1949 IV	82.5	41.0
1941 I	221.6	24.0	1972 VIII	286.2	50.7	1975 IX	129.2	30.1
1952 VI	53.5	9.4	1975 II	300.9	44.5	1844 II	290.0	36.0
1897 I	12.2	-1.6	1900 I	124.7	-42.4	1968 I	67.9	69.7
1974 XII	128.8	-28.0	1937 IV	34.5	39.3	1892 II	135.3	13.8
1853 III	43.0	-34.4	1898 VII	323.8	-28.3	1980u	113.9	39.1
1863 VI	108.5	39.0	1972 IX	55.3	40.4	1890 VI	27.9	-3.7
1942 IV	262.9	17.2	1959 IX	301.2	78.7	1889 II	191.9	-17.5
1917 III	45.9	-17.9	1954 X	293.5	47.1	1893 II	31.1	-12.2
1957 VI	359.9	25.4	1973 IX	212.5	-48.7	1980t	201.3	22.3
1921 II	39.5	59.5	1915 II	324.9	-46.6	1864 III	283.1	11.4
1944 IV	304.3	31.8	1959 I	127.5	18.5	1903 I	190.4	54.9
1936 I	68.0	-11.8	1941 VIII	91.9	35.2	1924 I	264.1	-44.6
1973 XII	27.3	-17.0	1966 V	47.6	-27.4	1892 I	15.7	14.2
1919 V	22.2	-33.1	1954 V	53.2	29.8	1963 III	22.8	55.1
1922 II	90.6	-16.4	1890 II	55.7	24.1	1948 XI	54.8	-25.3
1975 V	235.1	-42.3	1980b	357.0	18.1	1873 V	331.0	-57.7
1925 VIII	143.7	21.2	1937 V	47.5	-7.5	1952 I	263.1	31.4
1948 I	5.4	-10.9	1914 II	26.6	10.4	1863 II	0.3	17.8
1962 III	38.0	-27.0	1910 I	223.9	11.9	1969 I	147.2	24.2
1907 I	232.9	20.0	1976 XII	176.1	57.3	1973 VII	136.7	30.8
1903 II	200.4	30.4	1882 I	191.3	-59.0	1863 IV	193.1	6.3
1914 III	20.8	1.4	1908 III	21.8	-15.2	1976 VI	202.0	23.3
1902 III	8.6	17.4	1847 II	171.2	80.8	1971 I	2.2	13.9
1948 II	65.3	39.4	1847 VI	297.6	-25.6	1847 III	89.6	33.3
1905 IV	217.4	44.4	1904 I	36.6	41.4	1927 IX	141.6	17.2
1978 XXI	304.4	20.7	1947 VI	37.8	-28.5	1975 XII	250.0	-25.1
1914 V	102.9	52.4	1977 XIV	84.1	-47.1	1957 V	140.2	2.3
1977 IX	322.2	31.3	1958 III	248.2	61.8	1941 IV	222.8	29.5
						1877 III	123.7	51.2
						1863 III	56.0	30.4

northern latitudes; whatever the reason, the depletion factor only, is relevant to the following considerations.

To correct the geographical bias in Fig. 1, we now assign a weight of 2 (a slight overcorrection) to each of the 32 perihelia located at declinations smaller than $-20°$. Now each of the corresponding aphelia is therefore counted twice in its relevant galactic zone. This gives the upper curve of Fig. 1; this curve shows that the three depletion zones do not change very much. The correction we used is a valid statistical substitute for the missing comets; indeed, it substitutes to the missing perihelia, a sample whose

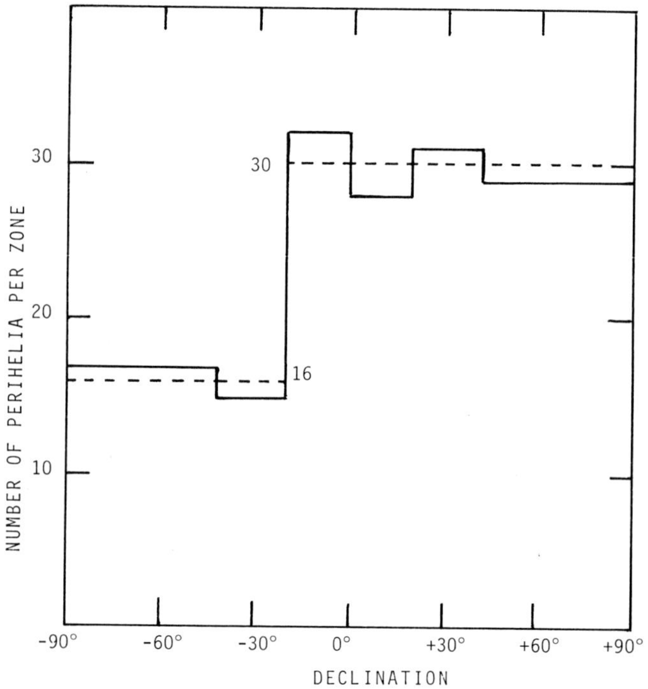

Fig. 2. Apparent depletion of cometary perihelia in the southern hemisphere. The sky has been divided into six zones of equal areas. The dashed lines define the *mean* number of cometary perihelia per zone for declinations smaller than $-20°$ (16 perihelia) and for declinations larger than $-20°$ (30 perihelia). The ratio of these two numbers gives the depletion factor, close to 1.9 in the zones south of declination $-20°$. No depletion is apparent north of declination $-20°$. Since comets become brightest near their perihelion, discoveries are influenced by the geographical distribution of the observers

distribution in declination is the same as that of the observed perihelia – a quite natural assumption if one considers Fig. 2. By necessity (since the added perihelia are the same) the sample has also the same distribution in galactic latitude, so that the basic statistical symmetry properties of Fig. 1 are not modified by the process. Of course the depletion zones, while remaining at the same locations, could have been drastically diminished by the correction of the bias. They were not, as this is understandable, mostly because the celestial polar axis is almost at 60° from the galactic polar axis.

4. Statistical significance

The number of comets perturbed into visibility is of course an extremely small fraction of the total number N perturbed in the Oort cloud; their individual probability p of becoming visible is also extremely small, but the point is that the product $\lambda = Np$ is an observable number; as a matter of fact, with 152 comets distributed in 12 zones, $\lambda = 152/12 = 12.67$ comets per zone. If their distribution over the sky is random, this is a typical case where Poisson's approximation is excellent. Poisson's equation is $P_n(\lambda) = \lambda^n e^{-\lambda}/n!$, where $P_n(\lambda)$ is the probability of finding n aphelia in one zone, for a given λ (the mean value of n).

The probability we are interested in here, is that of a random distribution that would simulate the symmetry properties of the galactic vertical tide *at least* as well as the observations, namely the three depletions observed at the galactic poles and the galactic equator. In particular, we would also simulate a galactic tide with *any smaller* number of aphelia in anyone of the three depleted zones. The sum of the probabilities of the 245 combinations of the three sectors containing at most 4, 6 and 6 perihelia, is given by:

$$\left[\sum_{n=0}^{6} P_n(\lambda)\right]^2 \left[\sum_{n=0}^{4} P_n(\lambda)\right] = 4.8\,10^{-6}$$

In spite of the large number of possible combinations, the probability that random processes duplicate the observed depletions in the zones predicted by the vertical galactic tide theory remains extremely small.

However, a fraction of the result may be due to observational bias. We must repeat the same type of computations with those aphelia that are missing in our uncorrected statistics. However, the use of 48 more aphelia would change λ; it would become $\lambda = (152 + 48)/12 = 16.67$ aphelia per zone, a fictitious value that would diminish Poisson's probabilities and paradoxically enhance our results! To remain on the safe side, we keep the previous λ (total of 152 observed comets) but we assume that three comets were missing in the two depleted zones near the galactic equator (see Fig. 1). Then one has:

$$\left[\sum_{n=0}^{8} P_n\right]\left[\sum_{n=0}^{7} P_n\right]\left[\sum_{n=0}^{4} P_n\right] = 3.6\,10^{-5},$$

a conservative result. After correction of the geographical bias, the degree of confidence that the galactic tides explain the observations remains therefore better than 99.996%, or four times the standard deviation σ.

5. Galactic tides

Since galactic tides seem the proper answer to the observations, it seems reasonable to say a few words about them. The tidal force towards the center of the galaxy was considered first by Byl (1983). However, it soon became apparent that the tidal field component perpendicular to the galactic plane is overwhelmingly more important than that to the center; in short, we call it "vertical" tidal force (with the figure of speech that the galactic plane is "horizontal"). Harrington (1985) used it but only to refute van Flandern's (1977, 1978) theory on the origin of comets; Morris and Muller (1986), Heisler and Tremaine (1986), Torbett (1986) developed the theory of the vertical tidal force as a possible source for the "new" comets coming from the Oort cloud.

The tidal force is a non-central force that changes the angular momentum of comets in the Oort cloud. Its component bringing comets into visibility is proportional to sin $2b$ (b galactic latitude); therefore, it is maximum for $b = 45°$ and nil for $b = \pm 90°$ and $b = 0°$, predicting three zones of avoidance.

If it is true that the local galactic density is nearly uniform and is likely to consist (for one half to two thirds of the mass) of a very large number of small objects, typically brown dwarfs (Bahcall, 1984; Bahcall et al., 1985), the proposed mechanism might be competitive with the traditional random perturbations of passing stars, or more recently, of molecular clouds.

The present observational confirmation of the principal role of the galactic tides removes the uncertainties in the numerical parameters and confirms Bahcall's estimate of the local galactic density.

6. Aphelion clustering

The present results do not imply that the traditional mechanism (Oort, 1950) is inoperative, but only that it is not strong enough to completely hide the depletions due to the galactic tidal field. In particular, it is likely that the large observed deviations from randomness recognized by previous authors and called "aphelion clustering" by Biermann et al. (1983) come indeed from the recent passage of a very small number of stars that add their tidal fields to the general galactic field. These authors' "star track" of 17 aphelia is very convincing.

This track goes across the depleted equatorial belt, inadvertently adding 8 aphelia to it. The depletion in the equatorial belt is therefore much more pronounced if these aphelia are removed; indeed only 4 aphelia are found in the rest of the equatorial belt (covering 305° in longitude). We will avoid the temptation of doing all over again the computation of the statistical significance with less aphelia (at most 3 and 3) for the two double equatorial zones; this would amount to an overkill.

Lüst (1984) mentions another "huge bulk" produced by an elongated cluster of aphelia, larger and wider than (and distinct from) the "star track" mentioned before. Delsemme (1986) had independently studied this "huge bulk" that he recognizes as the largest aphelion clustering in the sky, containing 29 aphelia. He identifies it as the major source of anisotropy in the orbital angular momenta of cometary orbits: 21 of its 29 comets (or 72%) seem to have received an impulse in the same direction as the major axis of this elongated sector. He shows that an extremely slow perturbing body is needed, moving typically at $200\,\mathrm{m\,s^{-1}}$ and of the order of 30 times the mass of Jupiter, in order to drag so many comets in the *same* direction.

He shows in particular that fast ($20\,\mathrm{km\,s^{-1}}$) stars or molecular clouds could not produce any detectable anisotropy in the direction of the perturbation.

The need of such a small velocity to induce the observed anisotropy suggests first the existence of a body bound to the solar system. When the periodicity of the mass extinctions of species on Earth (Raup and Sepkowski, 1984) was attributed to the action of "Nemesis", a small stellar companion of the sun (Davis et al., 1984; Whitmire and Jackson, 1984), it was tempting to identify it with Nemesis.

Since the search for Nemesis has not yet brought any positive result, and since the geologists themselves dispute the quasi-periodicity of the mass extinctions, the possibility of an unbound body cannot be rejected. As a matter of fact, the very detection of the galactic tidal field may be used as a support for Bahcall's contention that the number of small brown or black dwarfs is much larger than believed before, making more probable that many have gone through the Oort cloud. Because of their small mass, those that move with typical relative velocities of 20–30 $\mathrm{km\,s^{-1}}$ would have practically no effect on the Oort cloud of comets. Only those that would move extremely slowly could have a very large influence on the cometary cloud. The most recent of these objects may have produced the largest clustering in cometary aphelia.

Let's assess a crude probability for such a slow velocity passage through the Oort cloud. If we assume a solar velocity of $20\,\mathrm{km\,s^{-1}}$ with respect to the neighboring stars and a gaussian dispersion of stellar velocities of $\sigma = 20\,\mathrm{km\,s^{-1}}$, only 0.24% of the stars will approach the sun with a velocity at infinity not larger than $\pm 100\,\mathrm{m\,s^{-1}}$ (corresponding to a perihelion velocity of at most $300\,\mathrm{m\,s^{-1}}$). We can assume with Weissman (1986) that one star per million years passes within 10^5 AU from the sun; then one slow velocity star ($\pm 100\,\mathrm{m\,s^{-1}}$) passes per 400 million years, making a recent passage rather implausible. This result is rather insensitive to the value of the dispersion σ; for instance, if $\sigma = 30\,\mathrm{km\,s^{-1}}$, 0.32% of the stars will be within $\pm 100\,\mathrm{m\,s^{-1}}$ of the sun's velocity. If, however, we accept Bahcall's interpretation of the mass distribution in the galactic disk, the small individual mass of the black dwarfs (typically $0.03\,M_\odot$ for our brown dwarf) implies that their number may be 60 times as large as that of ordinary stars. We conclude that such a slow passage of a brown dwarf must occur every 6.7 million years on the average; this is a duration of the same order as the relaxation time of the observed cometary orbits against diffusion (1 to 20 Myr, depending on perihelia; Delsemme 1986); therefore observing the effect of a recent slow passage of at least *one* brown dwarf becomes the norm and not the exception. A final remark: in contrast to fast passages, the perihelion distance of the slow intruder will be much smaller than its impact parameter (assumed to be 10^5 AU), a focusing effect due to the quasi-parabolic trajectory, as opposed to the quasi-straight hyperbolic branch of a fast intruder.

7. Conclusion

Among the 152 comets used in this analysis, about 32 have probably been brought into visibility by only 2 discrete stellar passages: the fast passage of an ordinary star, identified by Biermann et al. (1983) and the slow passage of a brown dwarf, that has produced the orbital anisotropy of the large cluster studied by Delsemme (1985, 1986). The rest, that is about 120 comets, seems to have been brought into visibility mostly by the vertical galactic tide.

Acknowledgements. Mrs Rhea Lüst, as a referee, has clarified a misunderstanding. B. Soonthorntum and M. Patmiou have helped in the numerical computations. The research was undertaken under NSF grant AST-8611734 and a NASA grant under the Planetary Atmospheres Program.

References

Bahcall, J.N.: 1984, *Astrophys. J.* **276**, 169

Bahcall, J.N., Bahcall, S.: 1984, *Nature* **316**, 706

Biermann, L., Huebner, W.F. Lüst, Rh.: 1983, *Proc. Natl. Acad. Sci. USA* **80**, 5151

Bogart, R.S., Noerdlinger, P.D.: (1982), *Astron. J.* **87**, 911

Byl, J.: 1983, *Moon Planets* **29**, 121

Davis, M., Hut, P., Muller, R.A.: 1984, *Nature* **308**, 715

Delsemme, A.H.: 1985 in *Dynamics of Comets: Their Origin and their Evolution* A. Carusi, G.B. Valsecchi, eds., Reidel, p. 71

Delsemme, A.H.: 1986 in *The Galaxy and the Solar System*, R. Smoluchovski, J.N. Bahcall, M.S. Matthews, eds., U. Arizona Press, Tucson, p. 173

Eddington, A.S.: 1913 *Observatory* **36**, 142

Everhart, E., Raghavan, N. 1970, *Astron. J.* **75**, 258

Harrington, R.S. 1985, *Icarus* **61**, 60

Hasegawa, I.: 1976, *Publ. Astron. Soc. Japan* **28**, 259

Heisler, J., Tremaine, S.: 1986, *Icarus* **65**, 13

Hurnik H.: 1959, *Acta Astron.* **9**, 207

Lüst, R. 1984, *Astron. Astrophys.* **141**, 94

Marsden, B.: 1979, Catalog of Cometary Orbits, 3rd edit., Cambridge

Marsden, B.: 1982, Catalog of Cometary Orbits, 4th edit., Cambridge

Marsden, B., Roemer, E.: 1982 in *Comets* ed. L.L. Wilkening, U. Arizona Press, Tucson, p. 720

Morris, D.E., Muller R.A.: 1986, *Icarus* **65**, 1

Oja, H.: 1975, *Astron. Astrophys.* **43**, 317

Oort, J.H.: 1950, *Bull. Astron. Inst. Netherlands* **11**, 91

Raup, D.M., Sepkoski, J.J.: 1984, *Proc. Natl. Acad. Sci.* USA **81**, 801

Torbett, M.V.: 1986 in *The Galaxy and the Solar System*, R. Smoluchovski, J.N. Bahcall, M.S. Matthews, eds., U. Arizona Press, Tucson, p. 147

Tyror, J.G.: 1957, *Monthly Notices Roy. Astron. Soc.* **117**, 370

van Flandern, T.C.: 1977 in *Comets, Asteroids and Meteorites*, ed. A.H. Delsemme, U. Toledo, Ohio, p. 475

van Flandern, T.C.: 1978, *Icarus* **36**, 51

Weissman, P.R.: 1986, in *The Galaxy and the Solar System*, R. Smoluchovski, J.N. Bahcall, M.S. Matthews, eds., U. Arizona Press, Tucson, p. 204

Whitmire, D.P., Jackson, A.A.: 1984 *Nature*, **308**, 713

Astron. Astrophys. 187, 919–920 (1987)

ASTRONOMY AND ASTROPHYSICS

The P/Halley meteor showers in 1985–1986

M. Hajduková[1], **A. Hajduk**[2], **G. Cevolani**[3], **and C. Formiggini**[3]

[1] Department of Astronomy, Comenius University, CS-84215 Bratislava, Czechoslovakia
[2] Astronomical Institute, Slovak Academy of Sciences, CS-84228 Bratislava, Czechoslovakia
[3] FISBAT Laboratory, CNR, I-40126 Bologna, Italy

Received February 10, accepted May 12, 1987

Summary. The main characteristics of the meteor streams associated with comet P/Halley, Orionids and Eta Aquarids, have been deduced from simultaneous radar observations at Ondřejov (Czechoslovakia) and Budrio (Italy), during the shower periods in 1985 and 1986. The flux of particles with masses $m \geqq 10^{-6}$ kg was found to be $F = 1.5\ 10^{-12}\,\mathrm{m}^{-2}\,\mathrm{s}^{-1}$ for the central parts of the streams. This value is less than one third of the maximum values observed within the stream. The differential mass index, $s = 1.85$, derived for the last three returns of the showers, is compared with the results from 1980–1983, $s = 2.2$. The difference suggests the presence of two populations of particles in the stream.

Key words: Halley stream – radar meteors – mass index – particle flux

1. The Halley meteor showers and the approach of the comet

One of the most discussed questions concerning the meteor showers associated with comet Halley is whether or not the activity of showers will be influenced by the approach of the comet to the Earth's orbit in 1985–1986. Theoretical calculations show that the minimum distance between the present orbit of the comet and the Earth's orbit (0.154 AU and 0.065 AU for the Orionid and Eta Aquarid meteor shower periods, respectively) is too large to observe fresh ejecta at the limit of visual observations (i.e. for $m \geqq 10^{-5}$ kg) or at the limit of most radar equipments (i.e. for $m \geqq 10^{-6}$ kg). However, a higher shower activity could also be caused by an expanding dust cloud in the comet's orbit, formed during previous returns. Some predictions for the detection of higher activity have been published for the dates of the closed distance of the Earth to the comet's orbit (Yeomans, 1977; Buhagiar, 1984). However, the observations show that the showers do not follow this criterion very precisely (Fig. 1.).

2. Observations in 1985–1986 and comparisons with previous years

Simultaneous observations of the Orionid and Eta Aquarid meteor showers have been made in 1985–1986 at the Ondřejov Astronomical Observatory (14°47′ E, 49°55′ N) and at the Budrio station of the CNR FISBAT Laboratory (11°38′ E, 44°33′ N). The observational program has been extended to cover the periods proposed by the International Halley Watch program (Babadzhanov et al., 1985). The meteor radars at Ondřejov and Budrio have been transmitting at frequencies of 37.5 MHz and 42.7 MHz, respectively, with peak powers of 25 kW and 200 kW, respectively. Equipment parameters as well as data reduction methods have been described in an earlier paper (Hajduk and Cevolani, 1981). A total of about 10000 meteor echoes, recorded during the periods Oct. 16–29, 1985 and May 1–14, 1986 have been analyzed. The echoes were recorded during two 2-h intervals symmetric with respect to the time of meridian transit of the shower radiant. Meteor activity, expressed as the shower to sporadic background ratio of echoes, is plotted in Fig. 1 versus solar longitude L_S. The activity level is for both Orionid and Eta Aquarid showers less than the average activity in previous years. The activity variations are typical as usually observed in this shower. As reported in a series of papers, describing the results of simultaneous observations of Orionids and Eta Aquarids (Hajduk, 1970, 1973, 1981; Hajduk et Cevolani, 1981; Hajduk et al., 1983), no fixed positions of the activity maxima in these two showers exist. The main conclusions as to the stream structure have been summarized by McIntosh and Hajduk (1983). The

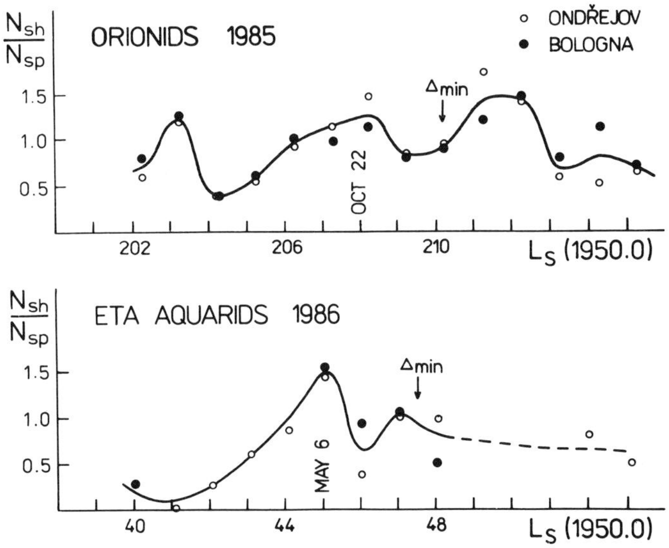

Fig. 1. Shower activity curves from simultaneous observations at Ondřejov and Budrio (expressed as shower to sporadic background ratio) versus solar longitude L_S. $\Delta_{\min}$ denotes the position of the minimum distance between the orbit of comet Halley and the Earth

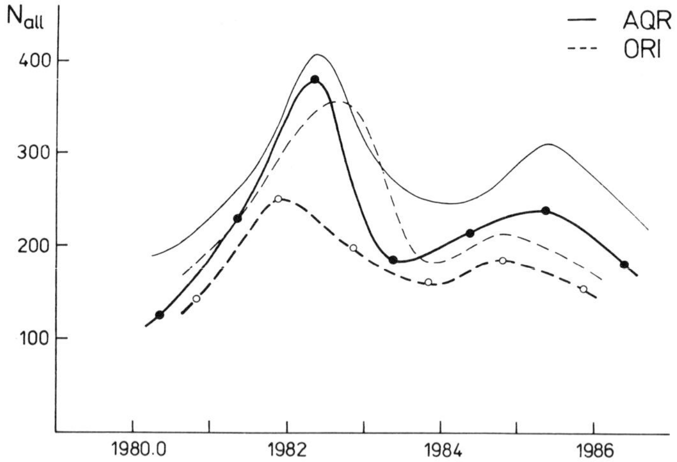

Fig. 2. Mean echo hourly rates for the central part of the Eta Aquarid and Orionid showers (heavy lines) and peak rates (light lines) from observations 1980–86 at Ondřejov Observatory

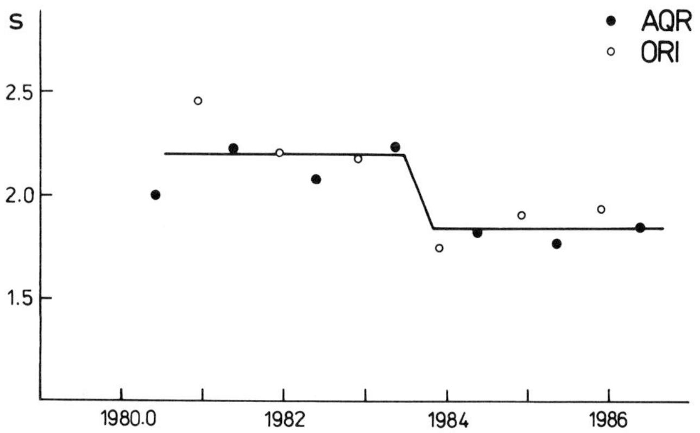

Fig. 3. The differential mass index s in different sections of the stream

1985–1986 data show that for particle $m \geqq 10^{-6}$ kg, and for distances between 0.065 AU and 0.154 AU from the comet orbit, the volume density of stream particles, did not increase at the approach of the parent comet. On the contrary, the mean activity of the central parts of the two showers (between $44° < L < 48°$ and $206° < L < 210°$, respectively, as well as peak values for the whole shower period, are considerably lower than in previous years, mainly in 1982 (Fig. 2) – and lower than the average values from the quoted series of observations. However, the proportion of long duration echoes was larger in 1984-86. Hence the observed, recent decrease of meteoroid flux is caused mainly by an absence of small particles.

3. Mass distribution and flux

Few results have been published to date concerning the mass distribution of the Orionids – and even fewer about the Eta Aquarids. Moreover, published values of the differential mass index s differ considerably (from $s = 1.4$ to $s = 2.5$). Most of the recent reports give an s-value of about 2.0. Our results in Fig. 3 have been deduced for masses of 10^{-5} and 10^{-6} kg. These masses correspond to the thresholds of overdense and underdense ionization trails, respectively, recorded by the Ondřejov meteor radar (5.0 and 7.5 apparent magnitude). The concentration of values at two separate levels $s = 2.2$ and $s = 1.85$, with an abrupt change in 1984 is surprising.

The spatial mass distribution within the stream depends critically on the mass index s – resulting in large particles for $s < 2$ and vice versa (Hajduk, 1987). It follows that we encounter at least two different meteoroid populations in the stream. These may be differently mixed in different parts of the stream. It is therefore difficult to give a representative mean flux value for particles in the stream. It is also difficult to evaluate the contribution of particles in particular mass ranges to the total mass of the stream. In the central belt of the stream (corresponding to the oribital section observed in 1985–1986) the heliocentric flux F of particles with masses $m \geqq 10^{-6}$ kg is $F = 1.5\ 10^{-11}\,\mathrm{m}^{-2}\,\mathrm{s}^{-1}$. This is about one third of the maximum flux observed in the stream for the same particle masses. For particles with masses $m \geqq 10^{-5}$ kg the flux variation in different sections along the stream is less, reaching twice the value of the 1986 section, which ist about $2\ 10^{-12}\,\mathrm{m}^{-2}\,\mathrm{s}^{-1}$. The flux of large particles in the central belt of the meteoroid stream appears to be relatively stable in comparison with other parts of the stream. This indicates a very small diffusion process as a consequence of very low ejection velocities of these particles from the comet.

References

Babadzhanov, P.B., Hajduk, A., Lindblad, B.A., McIntosh, B.A.: 1985, *IHW Newsletter* **7**, 12

Buhagiar, M.: 1984, *IHW Amateur Obs. Bull. No.* **5**, 3

Hajduk, A.: 1970, *Bull. Astron. Inst. Czechosl.* **21**, 37

Hajduk, A.: 1973, *Bull. Astron. Inst. Czechosl.* **24**, 9

Hajduk, A.: 1980, in *Solid Particles in the Solar System*, eds. I. Halliday, B.A., McIntosh, Reidel, Dordrecht, p. 149

Hajduk, A.: 1981, *Contr. Astron. Obs. Skalnaté Pleso* **10**, 129

Hajduk, A., Cevolani, G.: 1981, *Bull. Astron. Inst. Czechosl.* **32**, 304

Hajduk, A., Hajduková, M., Babdzhanov, P.B., Chebotarev, R.P.: 1983, *Acta Astron. Geophys. Univ. Comen.* 8–9, 39

Hajduk, A.: 1987 (this volume)

McIntosh, B.A., Hajduk, A.: 1983, *Monthly Notices Roy Astron. Soc.* **205**, 931

Yeomans, D.K.: 1977, in *Comets, Asteroids, Meteorites*, ed. A.H. Delsemme, Toledo, p. 61

Astron. Astrophys. 187, 921–924 (1987)

ASTRONOMY AND ASTROPHYSICS

The spectra of meteors from Halley's comet

I. Halliday

Herzberg Institute of Astrophysics, National Research Council of Canada, Ottawa, Ontario, Canada, K1A OR6

Received January 26, accepted March 16, 1987

Summary. The spectra of 16 bright Orionid meteors are studied in the spectral range from 3440 to 8700 Å. The brightest members of the group have peak absolute magnitudes near −8, indicating masses of a few grams and providing clear evidence that large particles from comet Halley survive to encounter the Earth's atmosphere. Spectral lines from six neutral atomic species and three singly-ionized species are observed. The Orionid spectra are very similar to Perseid meteor spectra, suggesting any differences in abundances in the parent comets appear to be marginal for the elements observed.

Key words: meteors – comets – P/Halley – spectroscopy

1. Introduction

An extensive program of meteor spectroscopy was conducted at the Meanook and Newbrook Meteor Observatories, Alberta, Canada, from 1952 to 1969 (Millman, 1959). As many as 22 slitless meteor spectrographs with transmission diffraction gratings were used on all suitable clear nights, with first-order dispersions in the range from 100 to 2000 Å mm^{-1} (Halliday, 1958). Considerable efforts were made in Alberta and in other Canadian observing programs to secure spectra of meteors in the near infrared region (Millman and Halliday, 1961) and also in the ultraviolet (Halliday, 1969). Spectra of 16 Orionid meteors were obtained between 1958 and 1968 at Meanook and Newbrook but none of these was produced by an exceptionally bright meteor and little attention was devoted to them at that time.

2. Observational material

The Orionid and Eta-Aquarid meteor showers are produced by debris from comet Halley. The Orionids are a modest annual shower with peak activity near October 22 while the Eta-Aquarids are mainly a daytime shower early in May. The complex features of the stream have been explained recently by McIntosh and Hajduk (1983) and studied further by Jones and McIntosh (1987).

The 16 Orionids observed spectroscopically were observed between solar longitudes of 204°.8 and 211°.5 with a mean value of 207°.7. There is a strong concentration of nine meteors between 207°.0 and 207°.6, which is not a time-of-night selection effect since the meteors are distributed in phase of the leap-year cycle. These nine have a mean value of 207°.3. McIntosh and Hajduk showed that for radar meteors the main shower is between solar longitudes of 206° and 210° with two significant maxima near 207°.1 and 209°.1 and three lesser concentrations, due to libration cycles of the shower orbit. Jones and McIntosh find that 1-gram particles are likely to be more strongly concentrated to the earlier maximum (207°.1) than are small particles of mass 10^{-4} g. Although the numbers are small, the large particles responsible for the spectra do seem to cluster strongly near the earlier maximum, as predicted by Jones and McIntosh. The first columns of Table 1 list the time of appearance and the corresponding solar longitude for the 16 meteors with the duration of the trail recorded by the spectrographs shown in the next column. The durations range from 0.02 s to 0.59 s, with a mean value of 0.26 s.

3. Heights

The observed height interval was determined for some meteors in Table 1 in the following manner, The angular velocity at one point on the trail was determined for the five meteors photographed through a rotating shutter that displayed at least three segments of the trail. Assuming a shower velocity of 66 km s^{-1} and a radiant position for the date of each meteor as given by Cook (1973), the heights were determined from:

$$H = (V \sin \theta_R \sin h)/\omega \tag{1}$$

where V is the velocity, θ_R the angular distance from the radiant, h the elevation of the meteor point above the horizon and ω the observed angular velocity. The beginning and end heights shown in Table 1, H_B and H_E, follow from the known time relationships along the trail and the vertical component of velocity, derived from the elevation of the radiant. Due to uncertainties in the times of appearance, which affect the elevation of the radiant, and in the precise shutter velocities, the heights could be in error by as much as 4%.

The mean beginning height is 101 km and the mean end height is 90 km. With one exception (meteor 11) the heights are based on zero-order meteor trails photographed through a transmission diffraction grating which limits detection to quite bright meteors, brighter than about −3 magnitude. Meteors of the Orionid velocity photographed with conventional small cameras normally appear between heights of 108 and 90 km (Millman and McKinley, 1963). Although the meteors in Table 1 are brighter than the small-camera meteors, it is not surprising that the spectrographs fail to record the faint upper portions of the trails, but it is perhaps surprising that the meteors fail to persist to lower heights in spite of their larger masses. This may be explained by their tendency to crumble in a terminal flare.

Bright, fast meteors frequently exhibit one or more flares late in the trajectory, often a terminal burst attributed to complete

Table 1. Data for 16 Orionid meteors

No.	y	Oct. (UT)	$\lambda_{\odot}$ °	Duration s	Heights H_B km	H_E km	M_{pan}	Mass g	
1	58	21.39	207.5	0.59			-7.4	5.8	
2	58	21.47	207.5	0.20			-6.3	3.5	
3	58	21.53	207.6	0.28			-6.3	1.8	
4	61	18.42	204.8	0.04					
5	62	22.29	208.4	0.19	95	90			
6	62	22.30	208.4	0.06			-5.7	0.4	a
7	62	25.36	211.5	0.02					
8	63	20.47	206.3	0.12					
9	63	21.37	207.2	0.30	102	90	-5.2	1.1	b
10	63	21.39	207.3	0.33					
11	63	21.47	207.3	0.44	108	86	-6.3	1.0	
12	63	23.37	209.2	0.27	103	92	-7.8	8.6	
13	66	23.39	209.5	0.51					
14	68	20.33	207.0	0.31	99	90	-6.3	3.1	c
15	68	20.35	207.0	0.33					
16	68	20.44	207.1	0.19					

Notes: a-Photometry based on a blue emulsion
b-Scattered cloud, brightness and mass might be greater than shown
c-Photometry based on an infrared emulsion

crumbling of a fragile meteoroid as it encounters increasingly dense atmosphere. A flare may be the only portion of the path sufficiently bright to produce a useful spectrum. The Orionids are typical in this respect with five that end at peak brightness with a terminal flare while another three end with a flare somewhat less intense than an earlier maximum. Four others might have had terminal flares that were not recorded due to the phasing of the occulting shutter. The remaining four appear to have smooth light curves but are generally the faintest members of the entire group.

4. Meteoroid masses

Mass estimates for some of the bright Orionids may be obtained by constructing light curves and converting the total luminosity to a mass. For eight of the meteors there were suitable zero-order images to compare with zero-order star trails. By applying corrections for atmospheric extinction, vignetting of the lens, relative trailing rates of the meteor and star images on the film, range of the meteor from the camera and the effect of the occulting shutter on the star trails, light curves were constructed for these meteors. The absolute magnitude of the meteor (instantaneous brightness if it had been observed in the zenith at a range of 100 km) for a panchromatic spectral range, M_{pan}, was obtained for several points on each trail. Differences in intensity of the meteor and star images caused by their locations in the field of the grating are believed to be small and no corrections have been applied. Also, no corrections were attempted for color variations although red stars were avoided in the comparisons since the color index of fast meteors is rather blue.

To convert the total luminosity to a mass estimate requires a value for the luminous efficiency, the fraction of the kinetic energy converted to light. Laboratory experiments indicate values near 0.01 or less for the blue spectral region (Friichtenicht and Becker, 1973; Savage and Boitnott, 1973) whereas observations of the fall of the Innisfree meteorite gave values as high as 0.08 for a panchromatic spectral range at effective heights of about 35 km (Halliday et al., 1981). For fast Orionid meteors near heights of 95 km we adopt a panchromatic luminous efficiency of 0.02, which is equivalent to the ablation of 0.041 g of material to maintain a luminosity of 0 mag for one second. The peak panchromatic magnitudes and the masses are shown in the final columns of Table 1. Errors in the photometry are small compared to the uncertainty in the appropriate values of the luminous efficiency, where an uncertainty as large as a factor of 3 may remain. Nevertheless, there is no doubt that meteoroids as large as several grams are required, i.e. fragments of comet Halley in this mass range survive and regularly encounter the Earth's atmosphere as Orionid meteors.

Table 2. Multiplets identified

Atom or Ion	Multiplets
N I	1, 2, 3, 8
O I	1, 4, 10, 3F
Na I	1
Mg I	2, 3
Mg II	4
Si II	1, 2, 3, 5
Ca I	2
Ca II	1, 2
Fe I	2, 4, 5, 6, 15, 20, 21, 23, 41, 42, 43, 45, 152, 326

5. The Orionid spectra

Some meteors were observed by several spectrographs so a total of 24 spectrograms exist for the 16 Orionids, but many of these are of little value due to low intensity, poor inclination of the dispersion to the direction of the trail, inferior focus, etc. There are five spectrograms on infrared film and two on blue-sensitive plates from the single spectrograph with quartz optics for ultraviolet studies.

The better spectrograms were measured for wavelength identifications, normally at the position of peak intensity. Since the appearance of cometary meteor spectra depends primarily on velocity, the spectra of Orionid meteors resemble closely the spectra of Perseid meteors (60 km s^{-1}). The Perseid shower has produced more strong spectra than any other shower so the features measured in the Orionids are most conveniently identified by comparison with a study of some 230 features identified in Perseid spectra (Halliday, 1961) supplemented to some extent in the ultraviolet by a Lyrid spectrum (42 km s^{-1}) which provided the best detail at wavelengths below 3700 A (Halliday, 1969).

Table 2 lists the multiplets identified in the Orionid spectra, using Moore's (1945) multiplet numbers. In general, only those multiplets that are considered major contributors to an observed feature are listed, hence the list contains many fewer multiplets of Fe I or Ca I than the Perseid list, a result of the lower intensity and inferior quality of the Orionid spectra, not an indication of any fundamental difference between the showers.

The two Orionid spectra shown in Fig. 1 are meteor 2 on a panchromatic emulsion from a spectrograph equipped with a

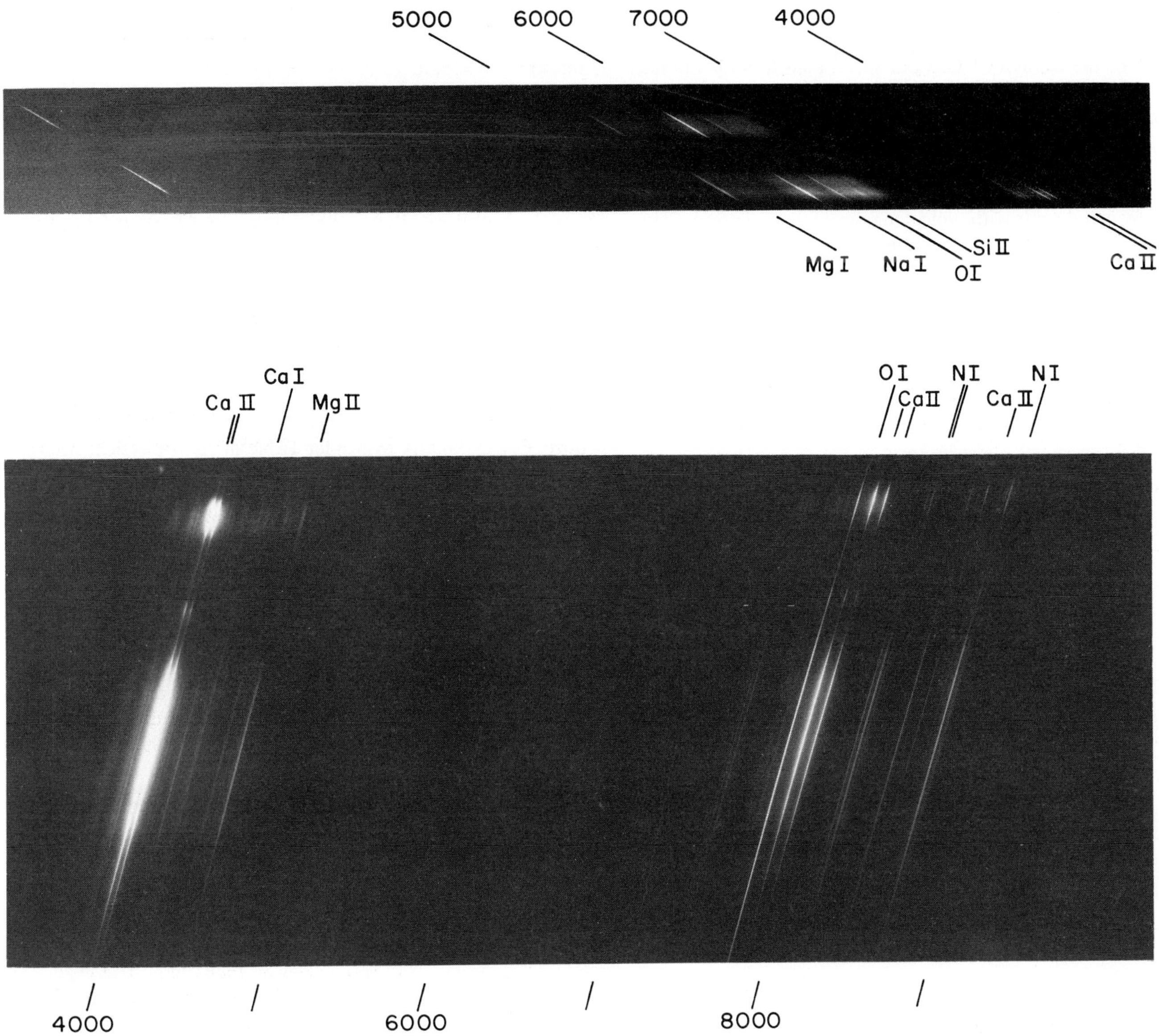

Fig. 1. Spectra of meteor 2 (above) and meteor 12 (below). Meteor 2 was moving down and meteor 12 was moving up as shown in the figure. Meteor 2 was photographed on a panchromatic emulsion and shows the zero-order image at the left plus strong lines in the first order and the second-order H and K lines of Ca II. Meteor 12 was on infrared film and the strong lines are in the first-order blue and infrared regions except for the second-order H and K lines. Some of the strong features are identified in both spectra and wavelength scales are indicated

rotating shutter and a spectrum of meteor 12 on infrared film, with high sensitivity in the blue and infrared regions. No. 12 was apparently the most massive of the meteoroids. The H and K lines of Ca II are very strong in meteor 12 and in fact all the Orionids tend to exhibit strong lines from multiplets of singly-ionized magnesium, silicon and calcium. The normally strong line of neutral calcium at 4226 Å is present but not prominent, possibly an indication that most of the calcium has been ionized in these very fast meteors. There is no clear evidence for the presence of the Hα line of hydrogen although it probably would appear if the spectrum of a strong flare had appeared with suitable dispersion on a film with good sensitivity at this wavelength.

The spectrograms reveal all nine infrared multiplets previously identified in meteor spectra, due to N I, O I and Ca II. The strong oxygen line at 7774 Å appears at greater heights and its intensity is less influenced by flares than lines of Ca II, as in other fast meteors. The relative intensities of multiplet 2 of Ca II and the nearby multiplets O I, 4 and N I, 1, show the increasing prominence of Ca II with increasing meteor luminosity as observed in Perseid spectra (Millman and Halliday, 1961).

In the visible spectrum region, three of the Orionids show the forbidden auroral green line of neutral oxygen at 5577 Å. Meteor 15 was photographed with one of the jumping-film meteor spectrographs (Halliday and Griffin, 1963) and shows six images of the green line recorded during at least 1.8 s of its decay. The lines of neutral magnesium near 5176 and 3833 Å are sometimes surprisingly strong. In the final segment of meteor 2 (too weak to be shown in Fig. 1) they are both present, along with 4226 of Ca I, the H and K lines of Ca II and multiplet 10 of O I at 6157 Å as the only other features. The previously intense Na D lines have disappeared completely, a phenomenon observed once before by Millman (1972) in a Giacobinid spectrum (velocity 23 km s^{-1})

and attributed to rapid baking out of the sodium after complete fragmentation of the meteoroid.

Neither of the ultraviolet Orionid spectra shows much detail although they add a few multiiplets of Fe I to Table 2. A more intense spectrogram would presumably have shown lines due to Ti, Cr and Ni previously identified in the best ultraviolet meteor spectra.

In summary, this group of Orionid meteor spectra was produced by meteoroids with typical masses of several grams. The spectra are very similar to those of other fast meteors, particularly the Perseids which are produced by fragments from P/Swift-Tuttle (1862 III). Although some subtle differences may exist between the showers, such as a possible higher abundance of magnesium in the Orionids, it is safe to conclude that the composition of P/Halley is essentially similar to the composition of other comets of comparable period.

References

Cook, A.F.: 1973, in *Evolutionary and Physical Properties of Meteoroids*, NASA SP-319, p. 183

Friichtenicht, J.F., Becker, D.G.: 1973, in *Evolutionary and Physical Properties of Meteoroids*, NASA SP-319, p. 53

Halliday, I.: 1958, *J. Roy. Astron. Soc. Canada* **52**, 169

Halliday, I.: 1961, *Publ. Dominion Obs. Ottawa* **25**, 1

Halliday, I.: 1969, *Publ. Dominion Obs. Ottawa* **25**, 313

Halliday, I., Griffin, A.A.: 1963, J. Sci. Instruments **40**, 187

Halliday, I., Griffin, A.A., Blackwell, A.T.: 1981, *Meteoritics* **16**, 153

Jones, J., McIntosh, B.A.: 1987, 20th *ESLAB Symp.* (in press)

McIntosh, B.A., Hajduk, A.: 1983, *Monthly Notices Roy. Astron. Soc.* **205**, 931

Millman, P.M.: 1959, *J. Roy. Astron. Soc. Canada* **53**, 16

Millman, P.M.: 1972, *J. Roy. Astron. Soc. Canada* **66**, 201

Millman, P.M., Halliday, I.: 1961, *Planetary and Space Sci.* **5**, 137

Millman, P.M., McKinley, D.W.R.: 1963, in *The Moon, Meteorites and Comets*, ed. B.M. Middlehurst, G.P. Kuiper, University of Chicago Press, p. 674

Moore, C.E.: 1945, A Multiplet Table of Astrophysical Interest, *Contr. Princeton University Observatory*, No. 20

Savage, H.F., Boitnott, C.A.: 1973, in *Evolutionary and Physical Properties of Meteoroids*, NASA SP-319, p. 83

Astron. Astrophys. 187, 925–927 (1987)

ASTRONOMY AND ASTROPHYSICS

Meteoroids from comet P/Halley. The comet's mass production and age

A. Hajduk

Astronomical Institute, Slovak Academy of Sciences, CS-84228 Bratislava, Czechoslovakia

Received February 10, accepted June 2, 1987

Summary. The flux of meteoroid size particles ejected from comet P/Halley in the mass range from 10^{-9} to 10^{-1} kg as observed by radar and optical techniques is compared with the flux of dust particles in the mass range from 10^{-20} to 10^{-8} kg as measured by space probes. The theory of the orbital motion of the comet and of the shell model of the meteor stream give within the observational errors of particle fluxes an initial mass of the comet of about $3.2\ 10^{15}$ kg and an effective diameter of 21 km. An estimated age of about 2300 revolutions is derived.

Key words: meteoroids – mass production – mass distribution – comet's age

1. Comet P/Halley meteor showers

The association between the Orionid and Eta Aquarid meteor showers and comet Halley is now generally accepted. The only objection is based on the relatively small values of the semimajor axis of the meteor orbits, which are derived from photographic and radar recordings. This discrepancy can be explained by the uncertainty in the geocentric velocity determination, on which the resulting value of the meteor's semimajor axis critically depends. The observed position in solar longitude of both Orionid and Eta Aquarid showers, their width, activity level and common structural features, in combination with studies of the orbital history of the parent comet, led McIntosh and Hajduk (1983) to a proposed shell model for the stream. The model, in comparison with the old toroidal one, complies much better with the requirements of the theory of an association between the comet and its stream, and with the observational data.

2. Mass distribution of meteoroids and dust particles

The correct determination of the mass distribution index for different mass ranges is the key problem in any study of the relation between cometary mass loss and the mass contribution of a meteor stream. Here we study the differential mass index s defined by the equation $\log N = k - (s-1)\log m$, where N is the observed number of particles with the mass m or larger than m. [The frequently used integrated mass index is $S = 1 + s$ (Millman, 1970).] For two mass categories differing by one order of magnitude, the number of particles increases 10 times, when $s = 2$. The derived mass contribution of particles of different mass categories is very sensitive to the s-value. When $s < 2$ for a mass m_0, then the mass contribution of particles increases with $m > m_0$ a vice versa. Unfortunately, most showers have s-values very close to 2. The Halley associated meteor showers are no exception. As shown by Hajduková et al. (1986), two different levels of s have been derived from the most recent radar observations ($s = 1.85$ and $s = 2.2$). These s-values may characterize different components of the meteoroid mass flow within the stream. The results are shown in the right part of the mass-flux diagram in Fig. 1. Meteoroids of smaller masses are very soon expelled from the stream as a consequence of larger ejection velocities, radiation pressure and destructive forces, while the supply of larger bodies is limited by the ejection cut-off. It follows that the total mass contribution from different mass ranges peaks at about 10^{-5} kg for meteor showers in general and at about $3\ 10^{-4}$ kg for the Halley meteor showers, as indicated in Fig. 1. The particles expelled from meteor streams contribute to the background, increasing the sporadic flux, which reaches its maximum mass contribution at 10^{-8} kg. This mass value corresponds also to the maximum of the total volume density of all solid particles (McDonnell, 1978). The mass distribution index s remains smaller than 2 for $m < 10^{-9}$ kg (Fechtig, 1973).

In order to study the interrelation between the mass loss of the comet and the mass contribution to the stream it is interesting to derive the mass index s of fresh cometary ejecta as observed by space probes to Halley. Local values of the differential mass index, derived from the dust sensor on Giotto are $s = 1.96$ for $10^{-13} < m < 10^{-8}$ kg and $s = 1.50$ for $m > 10^{-8}$ kg; at the nucleus are the values higher, with $s = 2.13$ and 1.67 respectively (McDonnell et al., 1986). The mass index, which is characterized by the slope of the flux-mass dependence, varies with the mass and with the distance from nucleus. This is shown in the upper left part of Fig. 1 for the selected flux data from Giotto, as well as from Vega spaceprobes (Mazets et al., 1986). The flux measured at larger cometary distances gives much smaller s-values especially for particles with $m < 10^{-12}$ kg. This is confirmed by data from the Vega 1 and Vega 2 dust sensors. Figure 1 suggests te existence of two different components of the ejected particles, differing strongly in the mass distribution. The mean value of the mass index s thus results from a mixture of two dust components. An explanation may be proposed based on two different mechanisms of ejection: in jets and in gas pressure. Most of the particles in cometary jets have large initial velocities upon leaving the surface of the nuclus (therefore small particles occur preferentially in jets). For the larger particles gas pressure is necessary to overcome gradually the escape velocity from the nucleus. For typical gas production rates measured by space probes – between $7\ 10^{29}$ – $2\ 10^{30}$ molecules s^{-1} – and a gas expansion velocity of about

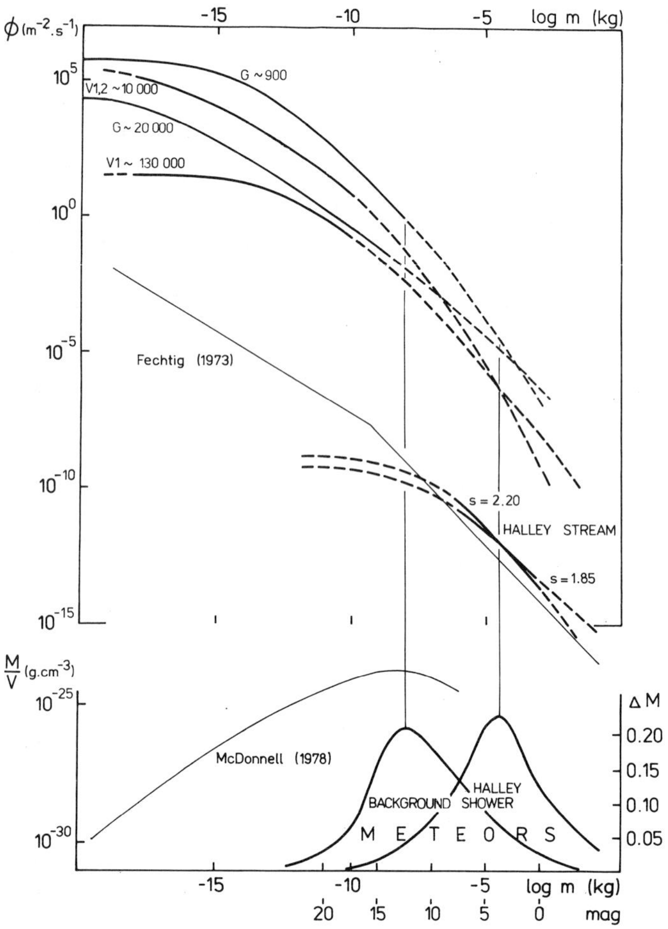

Fig. 1. Mass-flux relation obtained from space probes to comet Halley and ground-bases meteor oberservations. Symbols: G = Giotto, V = Vega (numbers = distance from nucleus in km), s = differential mass index, ΔM = relative mass contribution, M/V = space density

$1\ \mathrm{km\,s^{-1}}$ we obtain outflow velocities of a few meteors per second for meteoroid particles of about 1 g. These values are just above the escape limit.

Larger particles may leave the nucleus due to the gas pressure, forcing the particles gradually to larger distances with smaller escape velocities. This mechanism may produce a halo of larger particles extending to very large distances from the cometary nucleus and very slowly spreading near the comet's orbit. This halo component seems to be responsible for the low mass distribution index observed in meteor streams. The higher contribution of large pariticles derived from meteor observations (assuming small ejection velocities) led Hajduk (1986) to predict the impact on the Giotto spacecraft of two particles of masses of about 0.1 g during the 30 seconds of closest approach to the nucleus. The small mass index s, especially for the halo component, shown in Fig. 1, for large distances, suggest that the impact of two particles on Giotto (0.1 g particle and another somewhat smaller particle) causing two gaps in radiometric data (Edenhofer et al., 1986), corresponds to the real flux of large particles. The extrapolation of space probe measured s-values in the mass range $m < 10^{-8}$ kg towards meteoroid size particles (and vice versa the extrapolation of s-values determined from meteor observations towards the range of small dust particles shown the evolution of the parameter s: The dashed curves in Fig. 1 show the gradual change of s from smaller values corresponding to the halo around the nucleus to the larger ones, corresponding to the cloud of fresh cometary particles in the orbit of the comet, and finally to the largest values, corresponding to the meteor stream.

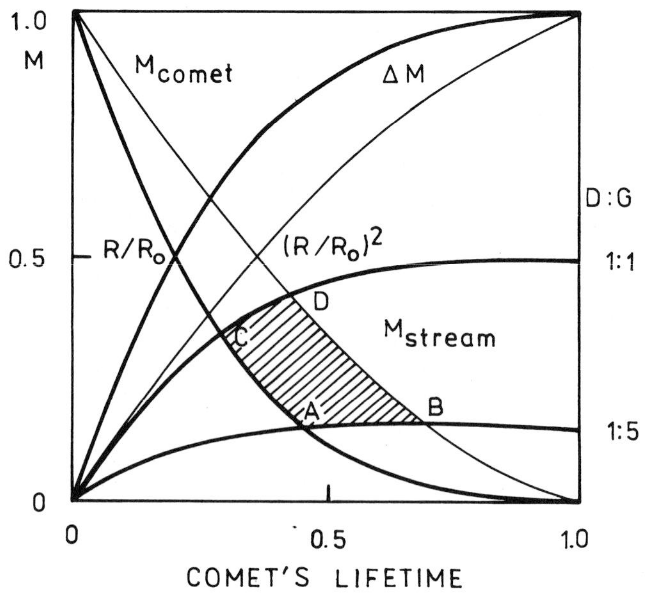

Fig. 2. Relation between the mass of comet Halley and associated meteor streams for different dust gas production rates during the comet's lifetime.

3. Meteoroid flux, mass of the stream and the comet's mass loss and age

The flux of particles within the stream, as derived from meteor observations, shows considerable variations along and across the stream. The mean flux $\phi = 2.5\ 10^{-11}\ \mathrm{m^{-2}\,s^{-1}}$ of 10^{-6} kg particles (7.25 mag) as obtained from a long series of radar observations (Hajduk, 1982), combined with the mass distribution shown in Fig. 1, led to a mass flux $F = 2\ 10^{-16}\ \mathrm{kg\,m^{-2}\,s^{-1}}$ for the central belt of the Halley associated streams. The total mass of the stream is $M_s = 5\ 10^{14}$ kg (McIntosh and Hajduk, 1983). Much lower flux was obtained by Jones (1983) from a two year observational series. However, his values, corrected for 98% of undetected meteor echoes and based on a different mass-magnitude relation may be significantly influenced by the method used.

The mass of comet Halley as determined from space probes is approximately the same as the mass of the associated meteor streams. The total dust production rate is $3.1\ 10^{3}\ \mathrm{kg\,s^{-1}}$ (Reinhard, 1986) and the dust to gas ratio is estimated at 1:6, for masses $m < 10^{-3}$ kg. However, including masses up to 1 kg this ratio may be 1:0.6 (McDonnell et al., 1986). Hence the total mass loss of P/Halley during one revolution as determined from space experiments (about $1.5\ 10^{11}$ kg) is one half of the mean meteoroid production per revolution (through the whole history of the comet), as determined from meteor observations. The cometary mass loss rate, of course varies with the comet's age. When we postulate that the cometary mass changes proportionally to the radius (R/R_0) as suggested by Drummond (1981), or proportionally to the surface $(R/R_0)^2$, we obtain for different dust to gas ratios the cometary lifetimes shown in Fig. 2 (dashed area).

For a dust gas ratio of 1:5 the comet has spent between 0.47–0.7 of its lifetime. At point A in Fig. 2, we find the initial cometary parameters $M_0 = 7\,M_c$, $R_0 = 1.9\,R$ and $(R/R_0)^2 = 3.6$. At point D in the diagram we have $M_0 = 2.3\,M_c$, $R_0 = 2.7\,R$ and $(R/R_0)^2 = 7.3$. Assuming the values $M_c = 5\ 10^{14}$ kg and $\Delta M = 1.5\ 10^{11}$ kg per revolution we obtain a required time of 12300 revolutions for point A and 2300 revolutions for point D. The last value corresponds exactly to the 8 Kozai cycles needed for the capture of the comet by Jupiter (Hajduk, 1984). The same number of revolutions can be obtained by increasing the estimated value of the total mass loss of the comet and instead decreasing the dust gas ratio. As the dust production rate determined from space experiments was obtained for particle masses below the maximum space density of the dust background, it is likely that the dust production rate is considerably higher than $3.1\ 10^3\ \mathrm{kg\,s^{-1}}$.

We thus conclude that the theory of the orbital history of Halley's comet and that of the shell model of the meteoroid stream agree within the observational errors with recent space probe data and data from ground based observations. We conclude that the limits of the comet's initial mass are about 6–7 times that of its present mass, with a corresponding effective diameter of about 20–22 km. The comet's age is close to 175000 years (2300 revolutions) which is about one half of its lifetime. An age of 2300 revolutions, as suggested by Hajduk (1984), was derived by Hughes (1985) from slightly different assumptions. It is interesting to note that the same basic problem arises in our study of comet Halley as in the general study of interplanetary matter, namely insufficient mass loss of the parent body (Kresák 1980). The discrepancy is diminished by accepting a predominant mass contribution from large particles in the Halley associated meteoroid stream produced by the halo component of the comet's nucleus. The dust gas ratios will then be correspondingly higher.

References

Drummond, J.D.: 1981, *Icarus* **47**, 500

Edenhofer, P., Bird, M.K., Brenkle, J.P., Buschert, H., Esposito, P.B., Porsche, H., Volland, H.: 1986, *Nature* **321**, 355

Fechtig, H.: 1973, in *Evolutionary and Physical Properties of Meteoroids*, eds. C.L. Hemenway, P.M. Millman, A.F. Cook, Washington D.C., NASA SP-319, p. 209

Hajduk, A.: 1982, in *Sun and Planetary System*, eds. W. Fricke, G. Teleki, Reidel, Dordrecht, p. 335

Hajduk, A.: 1984 in *Dynamics of Comets: Their Origin and Evolution*, eds. A. Carusi, G.B. Valsecchi, Reidel, Dordrecht, p. 399

Hajduk, A.: 1986, in *Asteroids, Comets, Meteors II*, eds. C.-I. Lagerkvist, B.A. Lindblad, H. Lundstedt, H. Rickman, Uppsala, p. 497

Hajduková, M., Hajduk, A., Cevolani, G., Formiggini, C.: 1987 (this volume)

Hughes, D.W.: 1985, *Monthly Notices Roy. Astron. Soc.* **213**, 103

Jones, J.: 1983, *Monthly Notices Roy. Astron. Soc.* **204**, 765

Kresák, L.: 1980, in *Solid Paricles in the Solar System*, eds. I. Halliday, B.A. McIntosh, Reidel, Dordrecht, p. 211

Mazets, E.P.: 1986, 20th ESLAB Symp., ESA SP-250, Vol. **II**, p. 3

McDonnell, J.A.M.: 1978, in *Cosmic Dust*, Wiley Publ., New York, p. 337

McDonnell, J.A.M. et al.: 1986, 20th ESLAB Symp., ESA SP-250, Vol. **II**, p. 25

McIntosh, B.A. and Hajduk, A.: 1983, *Monthly Notices Roy. Astron. Soc.* **205**, 931

Millman, P.M.: 1970, *Space Res.* **X**, 260

Reinhard, R.: 1986, *Nature* **321**, 313

Astron. Astrophys. 187, 928–930 (1987)

ASTRONOMY AND ASTROPHYSICS

The 1985 return of the Giacobinid meteor stream

B. A. Lindblad

Lund Observatory, Box 43, S-221 00 Lund, Sweden

Received February 14, accepted May 21, 1987

Summary. Radar observations of the 1985 return of the Giacobinid meteor shower were carried out at the Onsala Space Observatory in Sweden. The observations showed that the shower was of approximately 3^h duration, rising steeply to a maximum at 09^h35^m UT on 1985 October 8. The meteor shower thus showed the same short period characteristics as the intense showers of 1933 and 1946. Peak meteor echo rates of 75 occurred in the ten-minute interval $09^h30^m - 09^h40^m$ UT. The corresponding shower zenithal rate at maximum after removal of background activity was 484 echoes per hour. The longitude of the descending node of the shower at maximum was 194°.55 (equinox 1950). The plane of the central part of the 1985 shower was displaced by 0°.15 in longitude with respect to the orbital plane of the comet.

Key words: Giacobinid meteors 1985 – Draconid meteors 1985

1. Introduction

Comet P/Giacobini-Zinner (1900 III) was discovered by Giacobini in 1900 and re-discovered by Zinner in 1913. The comet is particularly interesting because it is responsible for the Giacobinid (Draconid) meteor shower. The orbital geometry is such that Giacobini-Zinner comes close to the Earth every thirteen years. The possibility that the comet could give rise to a meteor shower was first suggested by Davidson (1915). Intense meteoric storms with rates of several thousand Giacobinid meteors per hour were observed in 1933 and 1946, when the Earth passed very near the comet's orbit and reached the nodal point 80 and 15 days, respectively, after the comet (de Roy, 1933; Watson, 1934; Lovell, 1954). Weaker displays were observed in 1926 and 1952, when the Earth reached the node 70 and 193 days, respectively, before the comet (Denning, 1926; Davies and Lovell, 1955). No shower was observed in 1939 when the Earth passed the node 136 days earlier than the comet. Very few Giacobinid meteors were observed in the period 1953–1984.

Since P/Giacoboni-Zinner is a member of the Jupiter comet family the comet's orbit undergoes continuous changes. The past and future orbit of Giacobini-Zinner has beed studied in detail by Yevdokimov and Yevdokimov (1974). Close approaches to Jupiter occurred in 1958 and 1969 (Yeomans, 1971; Yevdokimov, 1972a). The 1969 approach (Δ(min) = 0.58 AU) changed the perihelion distance of the comet from 0.93 to 0.99 AU thus increasing the likelihood of a Draconid meteor display in 1972. McIntosh (1972) discusses the predictions for a 1972 Draconid display. It turned out, however, that the 1972 shower was extremely weak in the visual magnitude range with a peak rate of two to three visual meteors per hour (Harvey, 1973). Only a feeble Draconid shower was observed by radar (Hughes and Thompson, 1973; Simek, 1974). After a new approach to Jupiter in 1981 (Δ(min) = 1.6 AU) the perihelion distance increased to 1.03 AU and the orbit has once again come into close touch with the Earth's orbit (Yeomans, 1985; Yeomans and Brandt, 1985; Yevdokimov, 1985).

2. Conditions for a 1985 display

The longitude of the descending node of the comet is 194°.706 (equinox 1950.0). The Earth reached the nodal point on 1985 October 8.549 UT, passing 0.033 AU inside the comet's orbit at the descending node 26.5 days after the comet (Fig. 1). Assuming that the meteor stream from comet Giacobini-Zinner is at least 0.033 AU wide and that the Earth's arrival time at the descending node is also the time of shower maximum, a Giacobinid meteor display centered on Oct. 8, 13^h10^m UT was predicted. The theoretical radiant point for meteoroids moving in orbits similar to the comet was $\alpha = 260°.6$, $\delta = 57°.1$.

3. Observations

3.1. Observing schedule

Radar observations at the time of the Giacobinid meteor shower were carried out at the Onsala Space Observatory (57°.4 N, 12°.0 E) in Sweden between 00.00–15.00 h UT on 1985 Oct. 8 and 9. Additional observations were made using the same observational schedule on 1986 Oct. 7 and 8. The Onsala meteor radar operates at a frequency of 32.6 MHz. Meteor echoes are presented on range-time and amplitude-time displays, which are photographed on continuously moving film (film speed 10 m h^{-1}). The range interval recorded is 0–380 km. The transmitting antenna is a half-wave dipole. The receiving antenna is a steerable array of four 3-element Yagis. The combined beam has a half-power width of about 45° in the vertical and 40° in the horizontal plane. Further details of the radar equipment are given elsewhere (Lindblad, 1956; Lindblad, 1963). During the Giacobinid observations the antenna beam was directed to azimuth 60° W and continuously changed in elevation between 25° and 50°, so as to maintain a 90° distance from the predicted radiant.

3.2. Observed hourly rate

Analysis of the 1985 recordings are presented here. Meteor echoes were counted from the films down to the limiting echo

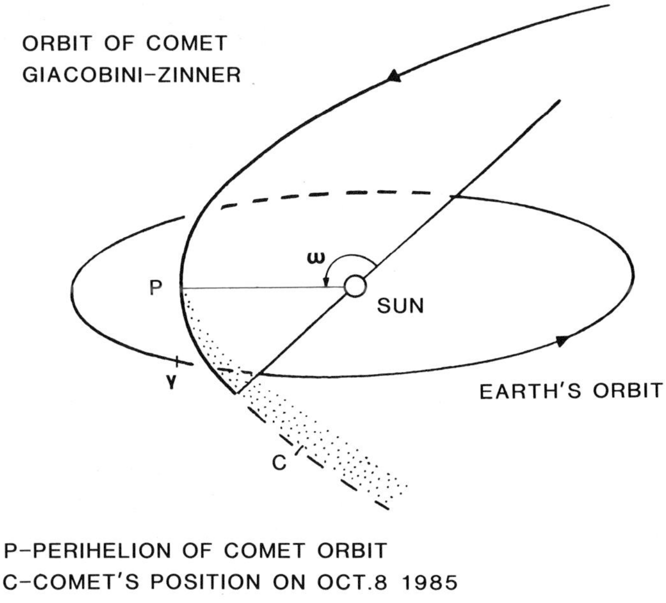

Fig. 1. Schematic orbit of comet Giacobini-Zinner

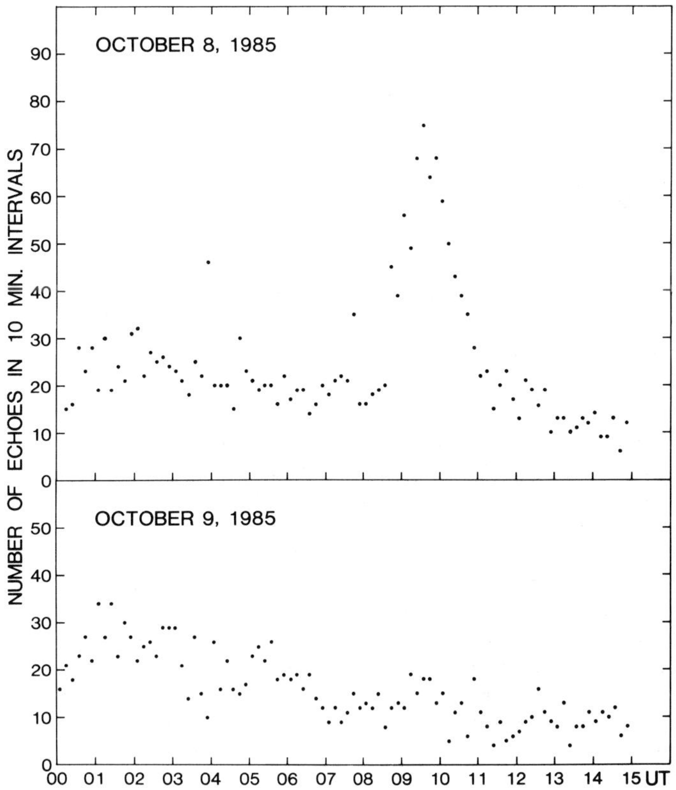

Fig. 2. a Radar recordings of Giacobinid meteors on 1985 Oct. 8. **b** Radar recordings of sporadic background on 1985 Oct. 9

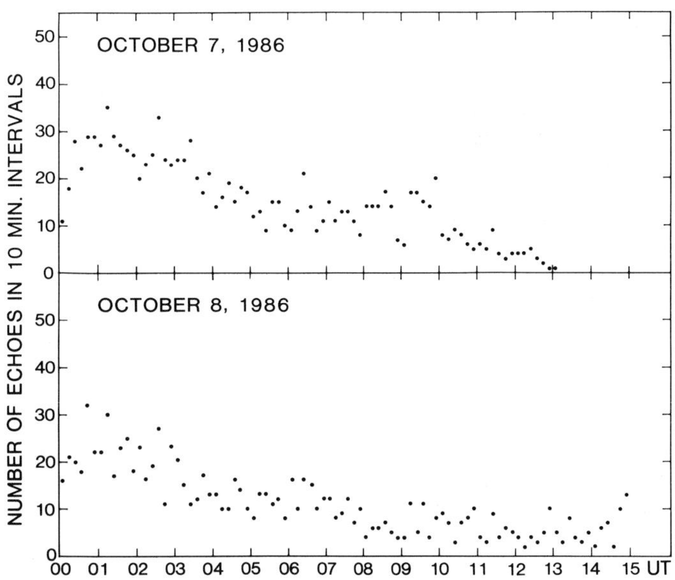

Fig. 3. a Radar recordings of sporadic background on 1986 Oct. 7. **b** Radar recordings of sporadic background on 1986 Oct. 8

signal detectable. Previous studies show that the Onsala radar detects meteors down to absolute visual magnitude $5^{\mathrm{m}}.5$ (Lindblad, 1963). The number of meteor radar echoes observed in ten-minute intervals on 1985 Oct. 8, is depicted in Fig. 2a. The background level as determined from observations with identical antenna settings on 1985 Oct. 9, is shown in Fig. 2b. From Fig. 2a we observe that Giacobinid shower activity rose significantly above the background level at about $08^{\mathrm{h}}40^{\mathrm{m}}$ UT. Peak rates occurred in the ten-minute interval $09^{\mathrm{h}}30^{\mathrm{m}}-09^{\mathrm{h}}40^{\mathrm{m}}$ (five-minute interval $09^{\mathrm{h}}30^{\mathrm{m}}-09^{\mathrm{h}}35^{\mathrm{m}}$). By 11.10 UT shower activity had subsided. It is evident from Fig. 2b that no meteor shower was observed on 1985 Oct. 9. Meteor echo rates recorded on 1986 Oct. 7 and 8 are shown in Fig. 3. We note that there is no indication of a 1986 recurrence of the Giacobinid meteor shower. A comparison of Figs. 2 and 3 shows that the 1985 and 1986 non-shower rates are very similar.

The Giacobinid peak in Fig. 2a is slightly asymmetric. Owing to this skewness the time of shower maximum was determined from a weighted mean of the five rate measurements in the interval $09^{\mathrm{h}}10^{\mathrm{m}}$ to $10^{\mathrm{h}}00^{\mathrm{m}}$. This procedure placed the maximum at $09^{\mathrm{h}}35^{\mathrm{m}} \pm 2^{\mathrm{m}}$ UT.

A comparision of the echo range distributuions recorded on Oct. 8 and 9 showed that the peak in meteor echoes centered on $09^{\mathrm{h}}35^{\mathrm{m}}$ UT Oct. 8 exihibited the range distribution that one would expect from a meteor radiant at $\alpha = 261°$, $\delta = 57°$ with shower members ablating at a mean atmospheric height of 93.8 km (Jacchia et al., 1950). In Fig. 2a we note a secondary peak consisting of 46 meteors observed between $03^{\mathrm{h}}50^{\mathrm{m}}-04^{\mathrm{h}}00^{\mathrm{m}}$. The range distribution of these echoes showed a wide dispersion in range, indicating that this spike is caused by random fluctuations in the sporadic background rate. The range distributions on Oct. 9 as well as time periods far from the Oct. 8 peak also exihibited a wide dispersion in radar range.

3.3. *Giacobinid hourly rates*

The true Giacobinid flux was estimated by subtracting from the total echo rates on Oct. 8 the sporadic meteor rate observed on Oct. 9. The sporadic data were smoothed by using a running mean of five 10-min. intervals.

One difficulty in computing the peak shower rate is that the variation in echo rates near maximum was so rapid that the mean hourly rates, determined by averaging over short time intervals, are considerably affected by the length of the interval chosen. In the present study the zenithal rate at maximum was determined by averaging over a ten-minute period centered on 09^h35^m. After background subtraction a maximum observed hourly rate of 348 echoes was determined. Assuming a sec z_R law, we deduce a zenithal hourly rate of 484. We note that if we instead had chosen the five-minute interval 09^h30^m–09^h35^m we observed 48 echoes; after background subtraction we derive an observed hourly rate of 480 and a zenithal rate of 667.

The extremely low ionizing efficiency of Giacobinid meteors (geocentric velocity = 24 km s^{-1}) makes a direct comparison with the hourly rates of other meteor showers uncertain. However, it is interesting to note that the computed zenithal rate of 484 echoes per hour is almost twice the highest meteor echo rates ever observed during the Perseid shower maximum at the Onsala station in over thirty years of Perseid observations.

4. Discussion

4.1. Comparision with other observations

An intense Giacobinid meteor shower was observed all over Japan on the evening of 1985 Oct. 8. TV-observations during the declining phase of the shower have been reported by Nagasawa and Kanda (1986). From 7 Draconid video trails a mean radiant position corrected for zenithal attraction at $\alpha = 262.^\circ2$, $\delta = 55.^\circ3$ was derived. Visual recordings of the meteor shower have been reported by Nakano (1985). Although these observations were severely affected by sky conditions, the results indicate a maximum at (or before) 09^h40^m UT. Simek (1986) infers from radar observations a maximum at about 10^h00^m. However, owing to equipment problems no film recordings were obtained between 05^h30^m and 09^h55^m UT and it is obvious that the main maximum was missed. Mason (1986) presents an hourly rate curve based on 70.3 MHz radio forward scatter measurements. The activity curve indicates a maximum in the five-minute interval 09^h30^m–09^h35^m, in excellent agreement with our results.

5. Conclusions

The Onsala radar recordings indicate that shower maximum in 1985 occured on Oct. 8, 09^h35^m UT, corresponding to solar longitude $L_0 = 194.^\circ55$ (equinox 1950). Somewhat surprisingly we thus note that the 1985 shower maximum did not occur at the descending node of the parent comet, but 03^h35^m earlier. The orbital plane of the cometary debris observed in 1985 was thus displaced in nodal longitude by $0.^\circ15$ with respect to the orbital plane of the comet. This result has been reported previously (Lindblad, 1986).

The observations of the 1933, 1946, 1952, and 1985 returns are remarkably consistent in the estimates of shower duration. In all four cases the shower had a duration of less than 3 h, with 75 % of the meteoroids appearing within a 1 h period. Hence the cometary debris must be concentrated in a plane with a lateral thickness of not more than 300000 km.

It is obvious that the planetary pertubations on the stream are at present not well understood. The meteorid particles causing the 1985 display must have been released at an earlier apparition of the comet and Jupiter must have shifted the plane of the particle orbits with respect to the comet. In this connection it is interesting to note that Yevdokimov (1985) predicted a shower maximum at about 05^h UT on 1985 Oct. 8.

The fact that Giacobinid meteors are only observed in close proximity to the comet indicates that the meteor shower must be of recent origin. Several authors have attempted to estimate ejection velocities and times of ejection of the cometary debris. Yevdokimov (1963a, b, 1972b) found by numerical simulations that the 1946 Giacobinid shower was the result of particles ejected from the comet in 1940, with a tangential component of velocity of less than 13.7 m s^{-1}. Similar numerical computations, carried out for the other observed Giacobinid returns, should greatly improve our knowledge of the mass loss of comets and the manner in which cometary debris is dispersed into meteoriod streams. Continued observations over many decades of the Giacobinid meteoroid stream and other young streams are necessary in order to increase the data base for these studies.

Acknowledgements. The meteor radar program at the Onsala Space Observatory has been supported by grants from the Swedish Natural Science Research Council and Alice and Knut Wallenbergs Stiftelse. The data analysis has been supported by grants from Crafoordska Stiftelsen.

References

Davidson, M.: 1915, *J. Brit. Astron. Assoc.* **25**, 292

Davies, J.G., Lovell, A.C.B.: 1955, *Monthly Notices Roy. Astron. Soc.* **115**, 23

Denning, W.F.: 1926, *Monthly Notices Roy. Astron. Soc.* **87**, 104

Harvey, G.A.: 1973, *Publ. Astron. Soc. Pacific.* **83**, 474

Hughes, D.W., Thompson, D.A.: 1973, *Monthly Notices Roy. Astron. Soc.* **163**, 3

Jacchia, L.G., Kopal, Z., Millman, P.M.: 1950, *Astrophys. J.* **111**,104

Lindblad, B.A.: 1956, *Meddel. Lunds Astron. Obs., Ser.* **I**, No. 189

Lindblad, B.A.: 1963, *Smithsonian Contr. Astrophys.* **7**, 27

Lindblad, B.A.: 1986, *IHW Newsletter* **8**, 25

Lovell, B.: 1954, *Meteor Astronomy*, Oxford Univ. Press

McIntosh, B.A.: 1972, *J. Roy. Soc. Canada* **66**, 149

Mason, J.W.: 1986, Proc. 20th ESLAB Symp., ESA SP-250, p. 379

Nagasawa. K., Kanda, T.: 1986, *Tokyo Astron. Bull., Second Ser.* **274**, 3159

Nakano, S.: 1985, *IAU Circ. No.* **4124**

de Roy, F.: 1933, *Gazette Astron.* **20**, 165

Simek, M.: 1974, *Bull. Astron. Inst. Czech.* **25**, 66

Simek, M.: 1986, *Bull. Astron. Inst. Czech.* **37**, 246

Watson, F.: 1934, *Bull. Harv. Coll. Obs. No.* **895**, 9

Yeomans, D.K.: 1971, *Astron. J.* **76**, 83

Yeomans, D.K.: 1985, *IHW Newsletter* **6**, 41

Yeomans, D.K., Brandt, J.C.: 1985, *Giacobini-Zinner Handbook*, Jet Propulsion Lab., Pasadena

Yevdokimov, Y.V.: 1963a, *Smithson. Contr. Astrophys.* **7**, 297

Yevdokimov, Y.V.: 1963b, *Sov. Astron. J.* **7**, 415

Yevdokimov, Y.V.: 1972a, *IAU Circ. No.* **2372**

Yevdokimov, Y.V.: 1972b, in *The Motion, Evolution of Orbits, and Origin of Comets*, eds. G. Cheborarev, E. Kazimirchak-Polonskaya, B. Marsden, Reidel, Dordrecht, p. 173

Yevdokimov, Y.V.: 1985, *Komet Zirk. No.* **340**, Kiev

Yevdokimov, Y.V., Yevdokimov, I.Y.: 1974, *Publ. Kazan Gorod. Astron. Obs.* **40**, 47

Astron. Astrophys. 187, 931–932 (1987)

ASTRONOMY
AND
ASTROPHYSICS

The meteor stream associated with comet P/Grigg-Skjellerup

B.A. Lindblad

Lund Observatory, Box 43, S-221 00 Lund, Sweden

Received February 14, accepted April 6, 1987

Summary. A 1964 Jupiter approach of P/Grigg-Skjellerup perturbed the comet's orbit so that very close approaches to the Earth's orbit now occur. A recently observed southern hemisphere meteor shower, the σ Puppids, is associated with P/Grigg-Skjellerup. Observations of this meteor shower now provide us with a unique opportunity to observe the birth and evolution of a meteoroid stream.

Key words: comet Grigg-Skjellerup meteors – Puppid meteor stream

1. Introduction

It is now fairly well-established that meteor showers occur when the Earth passes near the orbits of comets, which have expelled a part of their dust content. Comet Grigg-Skjellerup has long been recognised as a possible progenitor of meteors (Hoffmeister, 1948; Porter, 1952; Hasegawa, 1958; Drummond, 1981). The comet is included in Porter's list of cometary meteor radiants arising from elliptical orbits that approach the Earth's orbit to within 0.10 AU. Porter predicted a meteor shower on April 26 with a theoretical radiant in Puppis at $\alpha = 109°$, $\delta = -35°$. However, until recently no observations of this Puppid meteor shower have been reported. This is not surprising in view of the paucity of meteor observations in the southern hemisphere and the extremely low geocentric velocity of the meteors ($V_{\text{geoc}} = 15\,\text{km s}^{-1}$). The low velocity implies that only large meteoroid masses will produce sufficient luminosity for detection. The low velocity also implies a diffuse radiant making it difficult for visual observers to recognize the Puppid meteors.

The orbit of P/Grigg-Skjellerup has recently been perturbed into closer contact with the Earth's orbit, and the prospects for detection of an associated meteor stream have now drastically improved. This note summarizes the information available on the σ Puppid meteor stream.

2. Orbit

Comet Grigg-Skjellerup (1902 II) was discovered by Grigg in 1902 and rediscovered by Skjellerup in 1922 (for a previous observation by Pons in 1808, see Kresak, 1986). Since 1922 the comet has been observed at every apparition. The aphelion point of the orbit is close to Jupiter's orbit; hence the comet's orbit undergoes continuous changes. The orbital evolution of P/Grigg-Skjellerup over the time period 1947–1977 has been studied by Sitarski (1964). In particular a 1964 Jupiter approach perturbed the comet's orbit so that very close approaches to the Earth's orbit now occur. Since the comet's ascending node and perihelion almost coincide, these approaches occur at every perihelion passage. Sitarski predicted that meteor showers might be observed during the 1967, 1972 and 1977 perihelion passages. The distance between the Earth's orbit and that of the comet at the nodal point was 0.0027, 0.0044 and 0.0123 AU, respectively, the Earth passing through the nodal point 97, 51 and 12 days after the comet. A theoretical radiant at $\alpha = 109°$, $\delta = -45°$ on April 23 and a geocentric velocity of $V_{\text{geoc}} = 15\,\text{km s}^{-1}$ was predicted. After correction for earth attraction we deduce an observed velocity of $V_{\text{obs}} = 18.5\,\text{km s}^{-1}$.

3. Observations

It followed from Sitarski's work that the best conditions for a meteoric display occurred in 1972 and 1977 (IAU Circs. 2371, 3055). Radar observations in 1972 of a meteor shower associated with P/Grigg-Skjellerup have been described by Baggaley (1973). Excess hourly rates were observed on the Christchurch meteor radar at the predicted time of the shower. The excess rate was, however, not large enough to determine a reliable radiant position. It is therefore somewhat difficult to assess the significance of the 1972 Christchurch observations. Visual observers in Western Australia reported only minor activity on April 23, 1972 from a radiant positioned at $\alpha = 109°$, $\delta = -49°$ (near σ Puppis).

Visual observations in 1977 have been reported by several groups. Observations in the northern hemisphere were hampered by the low elevation of the radiant. Visual observations in the southern hemisphere were, however, very successful (IAU Circs. 3069, 3092). Wood (1986), director of the Western Australian Meteor Society (now Meteor Section of National Association of Planetary Observers), reports a Grigg-Skjellerup meteor shower with maximum activity occurring on April 23. The meteors were characterized as "slow and long". A zenithal peak shower hourly rate on April 23.60 UT of about 75 meteors was derived by Buhagiar (1977). Since few Grigg-Skjellerup meteors were reported in 1977 by observers in New Zealand and Brazil it is likely that the 1977 shower was of very short duration.

Since 1977 Grigg-Skjellerup meteors have been monitored on an annual basis by Australian meteor observers. Few, if any shower meteors were reported in the period 1978–1981. In 1982 a σ Puppid watch was carried out in the period April 16–25. A sharp maximum in σ Puppid activity was recorded with a peak zenithal hourly rate (hr) of 23 on April 23.49 UT. Shower rates on adjacent nights were less than 2 per hour. Observations in 1983 indicated a peak on April 23.87 UT with a zenithal hr of 12.7, while observations in 1984 gave an hr of 3.3 at peak activity.

Based on 10 radiants observed by Wood and coworkers in 1977–83 (Wood, 1987) a mean apparent radiant position at maximum shower activity of $\alpha = 110\overset{\circ}{.}3$, $\delta = -42\overset{\circ}{.}6$ was derived. This position agrees closely with the theoretical radiants given by Sitarski and Drummond.

4. Discussion

The UT times of peak shower activity for 1977, 1982 and 1983 were converted to solar longitude and compared. We note that maximum shower activity in 1977 occurred at solar longitude $32\overset{\circ}{.}94$ (equinox 1950.0). Maximum activity in 1982 and 1983 occurred at solar longitudes $32\overset{\circ}{.}56$ and $32\overset{\circ}{.}68$, respectively, which is very nearly the time of passage of the Earth through the comet's orbital plane (comet at descending node at solar longitude $32\overset{\circ}{.}63$, equinox 1950.0). It follows that the cometary debris is gradually being dispersed along the cometary orbit, but is still largely confined to the orbital plane of the comet. It is evident that continued obsservations of Grigg-Skjellerup meteors over the next few years will provide unique information on the rate at which cometary material is dispersed into a meteoroid stream.

From a total of 582 apparent magnitudes of stream meteors recorded in 1982–84 by members of the Western Australian Meteor Society we deduce a slope $\log_{10} r = 0.30$ for the log cumulative numbers versus magnitude law. The derived value $r = 2.0$ is rather low, indicating a lack of faint meteors in the stream. This result is rather unexpected in view of the youth of the meteoroid stream. It is possible that observational selection in combination with the very low geocentric velocity of the meteoroids has influenced this result.

In view of the extremely low velocity of the σ Puppid meteors the observed peak hourly rate of 75 in 1977 is quite remarkable. To illustrate this point let us assume a v^3-law for the luminous intensity. Members of the σ Puppid shower are then 3.8 stellar magnitudes fainter than Perseid meteors of comparable masses. If the observed visual rate is scaled up to a velocity of 60 km s^{-1} and if we assume r-values in the interval 2.0–2.9 for the log cumulative number versus magnitude law we deduce a corrected zenithal rate of some 1000–4000 meteors per hour.

Future studies of the σ Puppid stream will depend on the efforts of visual and radar meteor observers in the southern hemisphere. The short duration of the shower implies that maximum activity may occur in daylight over a considerable area of the Earth. Radar observations of the σ Puppid meteors should therefore be encouraged. Since meteor ionization and luminosity follow the same velocity dependence one would expect the σ Puppid meteors to have a very low ionizing efficiency. Hence, with conventional, low power, meteor radars the prospects for high meteor echo counts are not very good. However, a short duration, well defined maximum should be detectable. The radar rate profile of this shower will probably be quite similar to that of the 1985 Draconid meteor stream (Lindblad, 1987).

Acknowledgements. The author is indebted to P. Jenniskens and G. Spalding for directing his attention to the southern hemisphere σ Puppid observations and to J. Wood for supplying detailed information on visual rates, magnitudes and radiants of the σ Puppid meteors.

References

Baggaley, J.: 1973, *Observatory* **93**, 23
Buhagiar, M.: 1977, *Western Australian Meteor Soc. Bull.* **88**, (mimeographed report)
Drummond, J.: 1981, *Icarus* **47**, 500
Hasegawa, I.: 1958, *J. Observateurs*, **11**
Hoffmeister, C.: 1948, *Meteorströme*, J. Ambrosius Barth Verlag, Leipzig
Kresak, L.: 1986, *IAU Circular* **4234**
Lindblad, B.A.: 1987, *Astron. Astrophys.* **187**, 928
Porter, J.: 1952, *Comets and Meteor Streams*, Chapman and Hall, London
Sitarski, G.: 1964, *Acta Astron.* **14**, 25
Wood, J.: 1986, (private communication. See also *Western Australian Meteor Soc. Bull.* **180**, mimeographed report)
Wood, J.: 1987, (private communication)

Astron. Astrophys. 187, 933–934 (1987)

ASTRONOMY AND ASTROPHYSICS

Meteor contribution by short-period comets

J. Štohl

Astronomical Institute, Slovak Academy of Sciences, CS-84228 Bratislava, Czechoslovakia

Received February 12, accepted April 25, 1987

Summary. The contribution of P/Halley and other comets to the meteor streams are discussed. Particular attention is given to the meteoric production by P/Encke as the potential major contributor to the meteoric complex in the inner part of the solar system. The density and total mass of the meteoric stream associated with P/Encke is derived, taking into account the whole complex both of the showers and of the very broad and diffuse stream of seemingly sporadic meteoroids, produced by P/Encke in its long history via its showers.

Key words: P/Encke's stream – meteoric production by comets – meteoric influx

1. Introduction

In the course of their decay short-period comets make an important contribution to the meteoric complex in the solar system. By studying meteor streams, we can obtain information about the meteoric production of their parent comets in the past. Estimates, both theoretical and observational, of how large is the actual meteoric contribution by a particular comet vary considerably.

Among the short-period comets, P/Halley together with P/Encke seem to be the major contributors to the meteoric complex in the inner part of the solar system. The meteoric production rate by the present population of the active short-period comets, however, seems to be too low to ensure a steady state of the meteoric complex undergoing various dissipative processes (Whipple, 1967, 1976; Delsemme, 1976; Kresák, 1980). It was the very low rate of the total meteoric contribution by comets which led Delsemme (1976) to propose the hypothesis, supported further by Kresák (1980) and also by Clube (1983) and Napier (1983), that the meteoric complex is not in a permanent equilibrium, most of its mass being supplied by rare, exceptionally bright comets, captured occasionally into a short-period orbit. Such a comet could have been P/Encke, with its exceptionally large and dispersed meteoric complex.

2. Meteoric complex associated with P/Encke

There are several major and minor meteor showers whose association with P/Encke can be taken as confirmed or at least possible. The whole complex of such showers consists primarily of the Nothern Taurids, Southern Taurids and β Taurids (Drummond, 1973), but very probably also of the Southern Arietids (identical with the Northern Piscids) and ζ Perseids, i.e. of the showers comprising the Sekanina's Taurus Cluster, which may also include several minor showers (Sekanina, 1973). It should be noted that even within the same shower there are relatively large differences between the mean orbital elements as derived from different sets of data obtained by various methods of observations and in different periods of the shower's activity, which is especially true for the showers of such a long duration as in the case of e.g. the Taurids. One should therefore be careful in giving too much confidence to spurious relations and, on the other hand, not to reject beforehand possible real associations between showers derived from the analysis of their orbits.

Though association of some of the minor showers with the Taurus Cluster suggested by Sekanina may be taken as spurious, it is still possible to confirm the existence of an extremely broad and rich complex of showers associated with P/Encke, whose preperihelion and postperihelion encounters with the Earth occur in the periods from September till December, and from April till July, respectively. To this complex also belongs a large number of the background meteoroids with apparently sporadic orbits which fail to be recognized as belonging to the complex through the usual method of the D-criterion, but reveal themselves by higher strengths of the helion and antihelion radiant sources in the corresponding periods of the year (Štohl, 1986).

3. Meteor contribution by P/Encke

To estimate the influx of meteoric material and of the total mass of meteoroids in the streams associated with P/Encke, we can make use of the very homogeneous set of data on hourly rates of meteors from the Skalnaté Pleso Observatory, where visual observations had been carried out in 1944–1950 during whole periods of the Taurid complex activity. At the maximum activity of the Taurid shower, occurring around the solar longitude of about 226°, the peak hourly rate of shower meteors recorded by one observer was about $n_{\max}=20$, very close to the value $n_{\max}=17$, derived from visual observations in India by Lokanadham (1980). For the limiting magnitude of the visual meteors, a value $+5^{\mathrm{m}}$ can be accepted, to which the lower mass limit $8\ 10^{-3}$ g corresponds in Öpik's scale. For the upper mass limit of the sample we can accept the value $8\ 10^{1}$ g, to which a meteor of magnitude -5^{m} corresponds.

The hourly rates of meteors observed visually are biased by the perception abilities of the observers. To correct the observed rates, we must know both the magnitude distribution of meteors and the corresponding perception coefficients, which are strongly dependent on the magnitude of meteors. While the perception coefficients for the Skalnaté Pleso data have been estimated very carefully by Kresáková (1966), there is no direct determination of the magnitude distribution for the Taurids available on the basis of the Skalnaté Pleso observations. With good reason, however, we can accept the value $r = 3.0$ as the first approximation to the magnitude distribution index of the Taurids for the whole range of magnitudes involved, i.e. from -5^m to $+5^m$.

After correcting the rate $n_{max} = 20$ by the proper perception coefficients we obtain the value $N_{max} = 393$ for the total number of meteors which could be seen by an observer in one hour over his observing area of the sky. This value, integrated over the whole hemisphere, gives then the flux $2.2\ 10^{-11}\,m^{-2}\,s^{-1}$ for the meteoroids of the Taurid stream with masses $M > 8\ 10^{-3}$ g in the period of the stream's maximum activity. The density of the stream can be calculated by the method used by Hughes to calculate the dust influx for the Perseids, Quadrantids and Geminids (Hughes, 1974). Accepting the differential mass index $s = 2.2$ (corresponding to the magnitude distribution index $r = 3.0$) for meteors with $m < +5^m$, and $s = 1.9$ for meteors with $m > +5^m$ and summing up the masses derived for each magnitude interval between extrapolated limiting masses from 10^{-6} to 10^{+6} g, we obtain the total incident mass of meteoric material in the Taurid stream at its maximum activity of $7.8\ 10^{-14}\,g\,m^{-2}\,s^{-1}$.

For the equivalent transit width of the whole Taurid stream, defined as the time the shower would have to last at maximum rate to give the same meteoric influx as observed, we can assume the value 25 days, with respect also to the stream activity curve derived by Lokanadham (1980). By multiplying the incident mass of meteoroids at the shower maximum by the equivalent stream width and the proper cross-sectional area of the Earth, we obtain the total mass of meteoric material incident on the Earth during its passage through the stream, which amounts to $4.7\ 10^{8}$ g, i.e. to 473 tons/per passage. The corresponding maximum spatial density of the stream then is $\varrho_{max} = 2.6\ 10^{-24}\,g\,cm^{-3}$, from which we derive the total mass of meteoroids in the stream within the investigated mass limits $M_{tot} = 1.6\ 10^{16}$ g.

In deriving the above values of ϱ_{max} and M_{tot} only shower meteors have been taken into account, while the large part of meteors belonging to the Taurid complex but dispersed in the sporadic background were left out in the calculations. Taking into account also this "sporadic stream", the values of the influx rate, and also of the spatial density and of the total mass should be multiplied by a factor of approximately 2. For the whole complex of meteoroids associated with P/Encke through the Taurid complex we thus have $\varrho_{max} = 5\ 10^{-24}\,g\,cm^{-3}$ and $M_{tot} = 3\ 10^{16}$ g.

4. Discussion

The values derived above for the spatial density and for the total mass of the meteoric complex associated with P/Encke need further verification by the corresponding radio observations, which give information not only on the preperihelion, but also on the postperihelion encounter of the complex with the Earth. In spite of this it is possible to suggest that the above values may be taken as lower limits of the actual values of the stream.

Table 1

Stream	Maximum density ϱ_{max} (g cm^{-3})	Total mass M_{tot} (g)
P/Encke's	$5\ 10^{-24}$	$3\ 10^{16}$
P/Halley's	$5\ 10^{-24}$	$5\ 10^{17}$
Quadrantids	$8\ 10^{-24}$	$6\ 10^{13}$
Perseids	$5\ 10^{-25}$	$2\ 10^{15}$
Geminids	$7\ 10^{-24}$	$9\ 10^{14}$

The density $\varrho_{max} = 5\ 10^{-24}\,g\,cm^{-3}$ and $M_{tot} = 3\ 10^{16}$ g are to be compared with corresponding values obtained for the P/Halley's stream by McIntosh and Hajduk (1983) and for some other streams by Hughes (1974). All these values are presented in Table 1.

It is seen that while the total mass of the streams presented in Table 1 differ substantially, by as much as 5 orders of magnitude, the differences in their spatial densities are surprisingly small, not exceeding 1 order of magnitude. The differences in the mass of the streams are evidently the result of different rates and durations of the meteoric production by the parent bodies of the streams. It is also evident that the comets P/Halley and P/Encke are by far the most effective producers of the meteoric material. As compared to P/Halley with its very large intermediate-period orbit, P/Encke with its orbit of the shortest period appears to be major contributor to the meteoric material in the inner part of the solar system. The extremely large dispersion of P/Encke's meteoric complex, including the semi-sporadic debris, together with the large concentrations of orbits identified as its major and minor showers, do not necessarily imply a continuous and uniform mass loss by P/Encke. On the contrary, a few very active periods of meteoric production, connected presumably with some kind of catastrophes seem much more likely.

References

Clube, S.V.M.: 1983, in *Asteroids, Comets, Meteors*, eds. C.I. Lagerkvist, H. Rickman, Uppsala Universitet, p. 369

Delsemme, A.H.: 1976, *IAU Coll.* **31**, 31

Drummond, J.D.: 1973, *Icarus* **45**, 545

Hughes, D.W.: 1974, *Space Res.* **14**, 709

Kresák, Ľ.: 1980, *IAU Symp.* **90**, 211

Kresáková, M.: 1966, *Contr. Astron. Obs. Skalnaté Pleso* **3**, 75

Lokanadham, B.: 1980, *IAU Symp.* **90**, 163

McIntosh, B.A., Hajduk, A.: 1983, *Monthly Notices Roy. Astron. Soc.* **205**, 931

Napier, W.M.: 1983, in *Asteroids, Comets, Meteors*, eds. C.I. Lagerkvist, H. Rickman, Uppsala Universitet, p. 391

Sekanina, Z.: 1973, *Icarus* **18**, 253

Štohl, J.: 1986, in *Asteroids, Comets, Meteors* **II**, eds. C.I. Lagerkvist, B.A. Lindblad, H. Lundstedt, H. Rickman, Uppsala Universitet, p. 565

Whipple, F.: 1967, NASA SP-150, 409

Whipple, F.: 1976, *IAU Coll.* **31**, 403

Astron. Astrophys. 187, 935–936 (1987)

ASTRONOMY AND ASTROPHYSICS

Associations between ancient comets and meteor showers

M. Kresáková

Astronomical Institute, Slovak Academy of Sciences, 84228 Bratislava, Czechoslovakia

Received February 10, accepted April 3, 1987

Summary. Association criteria for comets and meteor showers are reviewed. Their application to the records in the Far East chronicles suggests three or four such associations, the most probable pairs being the comets and meteor showers of 461, 647, 684, 1063, and 1539. Most of them seem to be of low inclination and short revolution period. The number of meteor events associated with comet Halley implies its uniqueness in meteoroid production.

Key words: comets and meteor showers – associations

1. Introduction

While there exist historical records of meteor showers covering more than 25 centuries (Imoto and Hasegawa, 1958), we have clear evidence of their associations with individual parent objects only for the last 150 to 200 years. For this period we know the parent objects of about two thirds of the major permanent showers, and essentially of all the strongest temporary showers (Kresák, 1980).

The permanent showers, observable annually for several days to several weeks, are associated with parent objects of higher orbital stability: the Taurids and β Taurids with P/Encke (period 3.3 yr), the Ursids with P/Tuttle (14 yr), the Leonids with P/Tempel-Tuttle (33 yr), the Orionids and η Aquarids with P/Halley (76 yr), the Perseids with P/Swift-Tuttle (120 yr), the Lyrids with comet 1861 I Thatcher (415 yr), and the Geminids with the Apollo object 3200 Phaethon (1.4 yr). The temporary showers, with higher meteor rates but durations of not more than a few hours, are mostly associated with typical comets of the Jupiter family: the Andromedids with P/Biela, the Draconids with P/Giacobini-Zinner, the Bootids with P/Pons-Winnecke, and the τ Herculids with P/Schwassmann-Wachmann 3, all with periods between 5.4 and 6.6 yr. The orbits of these comets are rather unstable due to frequent encounters with Jupiter, and the resulting perturbations prevent a continuous ring of detectable particle density from being formed (see Carusi et al., 1982). Moreover, the lower geocentric velocities corresponding to the lower inclinations and eccentricities of these comets make the meteor phenomena less intense, larger particles being necessary to produce an equivalent brightness and ionization.

Under these circumstances, spectacular meteor showers attracting general attention are mainly produced by fresh cometary ejecta. They appear when a cometary orbit passes close to the earth, and when the comet and the earth arrive at the crossing point approximately at the same time. Hence, the appearance of a shower would often coincide with that of its parent comet. Indeed, the extraordinary Perseid display of 1862 occurred at the time when P/Swift-Tuttle was a very bright naked-eye object, and P/Schwassmann-Wachmann 3 and P/Giacobini-Zinner were marginal naked-eye objects one week before their showers of 1930 and 1946. This fact makes it feasible to search for associations between comets and meteor showers in historical records, even when the information included is insufficient for orbit determination.

2. Criteria for comet-meteor associations

A summary of the criteria for comet-meteor associations, depending on the kind of data available and ordered by their decreasing degree of reliability, is presented in Table 1. The meaning of Criterion I and II is self-explanatory, with the Southworth parameter D collecting all the orbital elements into one suitable function. Criterion III identifies the position of the comet's node – its heliocentric longitude and distance – with the position of the earth at the time of the shower, without the possibility of comparing the directions of motion. Criterion IV identifies the geocentric longitude of the point where the apparent path of the comet crosses the ecliptic, λ_C, with the position of the earth at the time of the shower; $\odot_C$ and $\odot_M$ are the solar longitudes at the time of these two events. The derivation of this criterion is explained in more detail elsewhere (Kresáková, 1987). Criterion V is of purely statistical nature, based on the deviations of the real distribution of time intervals between the meteor showers and the nearest comet apparitions, from a random distribution (Kresáková, 1986). It allows to estimate the total number of associations within a sufficiently large sample, and to identify the pairs of highest association probability to be checked by circumstantial evidence.

For unexpected strong meteor showers, Criterion I can only be used quite exceptionally, because it requires special observing techniques – double-station photographic observations with a rotating shutter, or special radio or TV observations. Criteria II and III, with less requirements on the meteor observations, can only be used when the comet's orbit is well determined. As a rule, this condition is not satisfied by the old historical records, giving only rough indication of the comet's apparent position and motion.

Table 1

Comet data	Meteor data	Criterion
I Orbital elements	Time radiant velocity	Approximate equality of all elements except T D-criterion of Southworth and Hawkins (1963)
II Orbital elements	Time radiant	The same, assuming equal geocentric velocities
III Orbital elements	Time	$\Omega_M \simeq \Omega_C$, $q_C(1 + e_C) \simeq 1 + e_C \cos \omega_C$
IV Time longitude of crossing point	Time	$\lambda_C \simeq \frac{1}{2}(\odot_C + \odot_M) \pm 90°$
V Time	Time	Correlation in time distribution (statistical)

3. Individual associations and conclusions

With a single, not quite convincing exception, all the associations which can be found in the ancient and medieval records are restricted to the application of criteria IV and V. Criterion V, when applied to the 62 showers of unidentified origin listed by Imoto and Hasegawa (1958) and to 950 comet apparitions prior to 1600 as listed by Hasegawa (1980), gives an estimate of about four real associations (Kresáková, 1986). Then, the application of Criterion IV to the original quotations of comet observations (Ho, 1962) allows to specify the most probable pairs, as identified below by the dates of appearance and the numbers from the catalogues of Hasegawa (H) and Imoto and Hasegawa (IH):

461 April (H 313 and IH 14)
647 September (H 385 and IH 60)
684 November–685 January (H 402 and IH 94)
1063 August (H 591 and IH 51)
1539 April–June (H 906 and IH 20)

For more details on individual cases see Kresáková (1987). An important point is that for most of the suggested associations a low orbital inclination is indicated, which is characteristic for short-period comets. Imoto and Hasegawa have already pointed out three cases of clustering of the showers around certain solar longitudes, suggesting their origin in the same periodic comet. The shower of 1063, with other suspect apparitions in 1037, 1548 and 1560, is just one of them. Accordingly, its parent comet should have a revolution period of 12–13 yr, or one half of this, which would make it a typical member of the Jupiter family.

In general, these results are in good agreement with the current ideas on the evolution of comets and dispersion of meteor streams. The system of broad meteor streams crossing the orbit of the earth has not undergone substantial changes during the last two millennia. On the other hand, the narrow streams of recent origin, producing strong temporary showers of short duration, were different from those which we observe now. The varying pattern is mainly due to a much higher sensitivity to planetary perturbations at the small cross sections of these streams. The statistical difference between the revolution periods of permanent and temporary meteor showers, as mentioned in the Introduction, contributes to this difference, too, because the temporary showers are mostly produced by comets with aphelia at the distance of Jupiter. The orbits of such comets are changing on a much shorter time scale. And finally, some of the comets which have produced the ancient meteor showers may already be extinct and inactive, which is again more probable for those of very short revolution period.

As regards P/Halley, these results present additional evidence of its uniqueness in meteor production. 40% of the old meteor shower records have already been identified by Imoto and Hasegawa (1958) with the permanent showers still observed at present. 14 of them belong to P/Tempel-Tuttle, 14 to P/Swift-Tuttle, 10 to P/Halley, and 3 to 1861 I. The orbits of all of these comets except P/Halley are very favourably situated with respect to earth crossing, aproaching to the earth orbit deep within 0.01 AU in the vicinity of their perihelia. Extrapolation of the particle flux from the crossing point to the orbit of the parent comet implies a much larger mass of the meteor stream of P/Halley than of those of the other comets, with a possible exception of P/Encke.

References

Carusi, A., Kresáková, M., Valsecchi, G.B.: 1982, *Astron. Astrophys.* **116**, 201

Hasegawa, I.: 1980, *Vistas Astron.* **24**, 59

Ho, Peng-Yoke: 1962, *Vistas Astron.* **5**, 127

Imoto, S., Hasegawa, I.: 1958, *Smithsonian Contr. Astrophys.* **2**, 131

Kresák, L.: 1980, *IAU Symp.* **90**, 211

Kresáková, M.: 1986, *Bull. Astron. Inst. Czech.* **37**, 339

Kresáková, M.: 1987, *Bull. Astron. Inst. Czech.* **38**, 75

Southworth, R.B., Hawkins, G.S.: 1963, *Smithsonian Contr. Astrophys.* **7**, 261

Supplement

Science Summary
Space Missions To Halley's Comet
International Halley Watch

EXPLORATION OF HALLEY's COMET

M. Grewing · F. Praderie
R. Reinhard (Eds.)

Exploration of Halley's Comet: symposium summary

D. A. Mendis

Department of Electrical and Computer Engineering and Center for Astrophysics and Space Science, University of California, San Diego, La Jolla, CA 92093, USA

Summary. The recent missions to comets P/Giacobini-Zinner and P/Halley, together with the coordinated ground-based observations, have led to a quantum leap forward in our knowledge of comets. Careful analysis of the accumulated data in conjunction with more sophisticated theoretical modelling will no doubt lead to a much deeper and more secure understanding of these cosmogonically highly significant solar system bodies. This process is already well under way. In this symposium summary, which makes no claims to being comprehensive, some of the most important of these new findings are highlighted. An attempt is also made to briefly summarize our pre- encounter understanding of comets, so that these new discoveries may be seen in perspective. In conclusion, the question of what needs to be done in the immediate future is briefly addressed.

Key words: Comets

1. Introduction

There have been many important cometary symposia in the past which contributed to our gradual increase in the knowledge and understanding of these fascinating solar system bodies. But the Heidelberg Symposium on the "Exploration of Halley's Comet is the culmination of the most significant year for cometary studies, which started with the first fly-by of a spacecraft through the tail of a comet in September of last year, and reached a climax with the encounter of Halley's Comet by five spacecraft within a week in March of this year.

While the in situ observations by spacecraft undoubtedly produced the most dramatic discoveries, highlighted by the first images of a cometary nucleus, Earth-based observations also contributed very significantly. The worldwide ground-based observational program organized by the International Halley Watch, as well as observations by Earth-orbiting satellites, rockets and high-flying planes, have led to important discoveries that complement the in situ ones. Also, as evidenced by the size of this meeting, there has been a dramatic increase in interest in comets within the scientific community, which has led to the acceleration of the rate of significant theoretical contributions too.

Clearly, within the space allocated for my summary, I cannot do justice to all these new discoveries; there are just too many. Consequently, I will confine myself to a partial list of those which, in my opinion, have had the most immediate impact on our understanding of comets.

2. The nucleus

The dominant, but not universal, view of a discrete cohesive nucleus of volatile solids bearing essentially all the mass of the comet is by no means new. It dates back at least to Laplace (1813) and later to Bessel (1836). However, the subsequent association of meteor streams with individual comets, starting about 1860, had such an impact that comets themselves began to be regarded as the densest part of such streams. This model of a comet as a "flying gravel bank" held sway for almost a century. In a seminal paper in 1950 Whipple resurrected the idea of a discrete nucleus, with the description and analysis of his so-called "icy conglomerate" model wherein the nucleus is considered as an aggregate of ices (e.g., H_2O, CO_2, CO, CH_4, NH_3, etc.), and meteoritic dust. It must be emphasized that Whipple did not merely resurrect Laplace's early idea, he modified, and most importantly, quantified it in such a way that it not only explained dynamical effects on cometary orbits (the so-called nongravitational recoil effects associated with non-isotropic outgassing) but also (with a few modifications) the essential features of all cometary observations. It became the basis of all subsequent work on the dynamics, physics and chemistry of comets and in so doing put the subject, which had begun to lose its way over a century ago, firmly back on track.

Therefore, the detection of the nucleus of Halley's Comet by the cameras on board the Vega 1 and 2 and Giotto spacecraft (Sagdeev et al., 1986a; Keller et al., 1986) came as no surprise. Nor did its irregular shape, variously described as a potato, peanut and an avocado, which was expected for such a small body, although it was generally approximated by a sphere in modelling.

Its very low geometric albedo (0.04, Keller et al., 1987), which puts it among the darkest objects in the solar system, while being much smaller than earlier assumptions, was close to those estimated more recently [e.g., 0.06 (Divine et al., 1986), 0.04 (Cruikshank et al., 1985)]. Its presently estimated volume ($\approx 500\,km^3$) while being about a factor of 5 larger than that estimated by Divine et al. ($R_n \approx 3$ km; Divine et al., 1986), is about a factor of 4 smaller than that estimated by Cruikshank et al. (1985).

Independent estimates of the mass of comet P/Halley, based on non-gravitational effects of its orbit, put it in the range of $5 \cdot 10^{16} \sim 10^{17}$ g (Rickman, 1986). This relatively small mass, despite its larger size, would make its inferred bulk density $\bar{\varrho}\,(0.1 \sim 0.2\,g\,cm^{-3})$ much smaller than generally assumed (e.g., $\varrho \sim 1.0\,g\,cm^{-3}$; Divine et al., 1986). However, the estimate of the mass, based on the rather uncertain radial non-gravitational parameter, must be taken with some caution (D.K. Yeomans, private communication), and at present our uncertainty of the value of $\bar{\varrho}$ should perhaps be bracketed by the range $0.1 \sim 1.0\,\mathrm{g\,cm^{-3}}$.

Regarding the chemical composition of the nucleus, Delsemme and Swings (1952) made a significant contribution by pointing out that while H_2O constitutes the dominant volatile, all other volatiles (e.g., CO) would be trapped within the H_2O-ice lattice as "guests" to form what is commonly called a clathrate hydrate. Since at most one guest molecule can be trapped for every six molecules of H_2O, this model predicts that the production rate of all other species $\lesssim 1/6$ that of H_2O. Furthermore, since the sublimation of all species, whatever their individual volatility, will now be controlled by that of the clathrate (which is close to that of pure crystalline H_2O-ice), this model anticipates that all significant coma activity begins typically at a distance of 2–3 AU, where the H_2O-ice sublimation "turns on." This model is consistent with most though not all cometary observations. More recently Houpis et al. (1986) have pointed out that even if initially there were more of the volatile species (e.g., CO or CO_2) than can be trapped in the clathrate hydrate lattice, chemical differentiation of the upper layers of periodic comets will cause them to behave as if they were H_2O dominated. This model overcomes some of the difficulties of the purely clathrate model by showing that $\dot{Q}$ (guest) could be $\gtrsim \frac{1}{6}\, \dot{Q}(H_2O)$ in periodic comets, although not by much. It also anticipates the activity of "new" comets coming into the inner solar system for the first time, at large heliocentric distances.

In this connection it must also be pointed out that several authors (e.g., Klinger, 1980; Smoluchowski, 1981) have argued that the H_2O-ice in a new comet must be amorphous and that the transition from amorphous to crystalline as the cometary nucleus heats up to a critical temperature could be the cause of sudden flare-ups between about 3 and 6 AU.

A so-called "average heuristic model" for the elemental abundance of the cometary nucleus was constructed by Delsemme in 1982. Here the H:C:O:S ratio was constructed rather directly from the UV observations of atomic species, while the ratios of several of the other elements were deduced less directly. While H was depleted by a factor of about 10^3 vis-à-vis the so-called "cosmic abundance", which is not surprising, C appeared to be depleted by a factor of about 4 from the cosmic abundance. Delsemme pointed out that on the one hand C could have been trapped in the very volatile CH_4 at cosmogonic times, which would imply that the cometary material condensed at relatively high temperatures, or on the other hand, that C is trapped in the cometary dust, as was first suggested by Vanysek and Wickramasinghe (1975).

The chemical composition of the cometary dust as determined by the dust impact analyzer aboard Vega (Kissel et al., 1986) seems to support the latter possibility: namely, that much of the C is trapped in grains. I will discuss this in greater detail in the next section.

The view that H_2O is the dominant volatile species is also supported both by the in situ as well as by earth-based observations. Indeed, the production rate of H_2O (by number) appears to be $\gtrsim 80\%$ of the total production rate, whereas CO appears to be the next most abundant volatile species, with a production rate of $\gtrsim 10 \sim 15\%$ [on the basis of the IUE observations, Festou et al. (1986)]. The observed and predicted production rate of H_2O (e.g., Divine et al., 1986) are compatible, but the observed production rate is highly variable, which is not surprising in view of the observation that the nuclear activity, both in gas and dust, appears to be largely associated with discrete jets, mainly on the sunward side of the spinning nucleus.

With regard to the overall structure of the surface, it was already anticipated by Whipple (1950) that it would be covered by the dust mantle. The possibility of forming dust mantles due to cosmic ray processing was discussed by Shul'man (1972), Donn (1976), and Whipple (1977). The last author estimated a one-meter deep "outer frosting" which would probably be blown off at the first entry of a new comet into the inner solar system. The possible formation and disruption of such mantles due to solar radiation processing of the upper layers was considered by Brin and Mendis (1979) in order to explain the observed heliocentric brightness asymmetry of comets. Horanyi et al. (1984) refined the model of Brin and Mendis in several ways; most importantly by assuming that the dust mantle is "friable" and that the dust flux is β times the momentum flux of the gas. By taking various values of β they were able to anticipate various types of mantle evolution. One case (large β) corresponded to the periodic buildup and purging of the dust mantle obtained earlier by Brin and Mendis. However, for smaller values of β mantles were continuously built up. The diffusion of gases through the mantle was treated in detail by Fanale and Salvail (1984) and the buildup of multiple mantles (both dust and clathrate plus dust) over a more volatile core was considered by Houpis et al. (1985, 1986). The last model predicted a dust mantle $\lesssim 1$ cm and a surface temperature of about 320 K at a distance of about 0.83 AU. The infrared spectrometer (IKS) on board Vega 2 estimated a surface temperature of ≈ 300–400 K. While warm circumnuclear dust could contaminate this measurement somewhat, the nucleus surface temperature was certainly warmer than 190 K, the temperature of an evaporating surface of dirty ice at 0.83 AU. It is therefore reasonable to assume that what we are observing is not a bare icy nucleus but rather a nucleus covered by a layer of dark, warm dust.

The morphology of certain asymmetric dust features, particularly sunward jets, was exploited by Sekanina and Whipple to determine the rotation vectors of a number of comets (1979; see also Sekanina, 1979a). More recently Sekanina and Larson (1984) used the morphology of the spiral jets seen in the processed images of comet P/Halley in its 1910 apparition to obtain its rotation period ($T \approx 52$ h) and spin axis. Assuming that these spiral jets arise from "active regions" that are turned on at sunrise and turned off at sunset, Sekanina and Larson have also been able to map these regions on the cometary surface. The surface locations of a number of dust jets observed by the Vega missions (Sagdeev et al., 1986b) are in reasonable agreement with Sekanina and Larson's predictions. Also, support for the spin period deduced by Sekanina and Larson comes from both the Vega 1 and 2 observations of the nucleus (53.5 h), as well as the Suisei UV images which showed a periodic variability of the brightness of the Ly-α images with a periodic of about 52–53 h (Kaneda et al., 1986). Further support for this period is provided by the analysis of observed brightness variations from the ground (Belton et al., 1986).

Up to the time of the Heidelberg conference this 2.2 day period appeared to be one of the well-established parameters of the cometary nucleus. It was therefore a surprise when several other observations claimed a different period of 7.3–7.4 d. Spectrophotometric observations of Halley in March and April, 1986 show a pronounced brightness variation with a period of 7.4 d (Millis and Schleicher, 1986).

Julian (1987) has attempted to reconcile these different inferences of the rotation period of Halley's nucleus by regarding it as a rigid, homogeneous asymmetric top (specifically as an ellipsoid with axis ratios $l_1 : l_2 : l_3 = 16:8:7$), in Eulerian free precession. As he points out, two qualitatively distinct free precessional modes are allowed for torque free rotation about its center of mass, each exhibiting both the 2.2 d and 7.4 d periodicities. In one mode, the terminus of the angular momentum vector traces a closed curve (called the polhode about the long axis (l_1) with a period of 7.4 d. In the other, it traces a similar curve about the short axis l_3. (That there is no such mode about the intermediate axis l_2 is associated with the fact that rotation about that axis is unstable, while rotations about l_1 and l_3 are stable).

The long axis mode is essentially the same as the nuclear motion proposed by Sekanina (1987), who assumed that the two short axes l_2 and l_3 are equal (symmetrial top). In that case, only the long axis mode is present; and Sekanina identified motion of $\boldsymbol{\Omega}$ around the long axis as the "spin" of the nucleus about this axis. Also, in this case, the angle, θ_1 between the long axis, l_1, and the spatially fixed angular momentum vector $\boldsymbol{M} \approx 77°$, and the precession period of l_3 about $\boldsymbol{M}$ (incidentally l_3, $\boldsymbol{\Omega}$ and $\boldsymbol{M}$ are always coplanar) ≈ 2.2 d.

This mode of precession however, appears to be in conflict with the combined observations of VEGA's 1 and 2 and Giotto, as has been pointed out recently by Smith et al. (1987). Basically, the problem is that the profiles of the nucleus observed by VEGA 2 and Giotto, 4.7 d apart can only allow an integral number of rotations about the long axis, during that interval. As Julian points out, his second short-axis mode is not ruled-out by these observations. This corresponds to a rotation period $(2\pi/\boldsymbol{\Omega}) \approx 2.08$ d [which incidentally is very close to the 50 h period derived by Vaisberg et al. (1986) from the shape of the dust jets encountered during the VEGA encounters] about an axis that is inclined at only a small angle ($\omega_3 \approx 11°$) to the short axis l_3. As stated before, the precession period of $\boldsymbol{\Omega}$ about l_3 is 7.4 d. This motion may also be described approximately as a "spin" about the short axis (l_3) with a period ≈ 2.2 d, and a precession of this axis with a period ≈ 7.4 d about an axis inclined to it at $\approx 11°$.

The main drawbacks of this model are firstly that the actual shape of the nucleus appears to be markedly different from an ellipsoid and secondly, that prevalent asymmetric jet activity would lead to substantial torques about the center of the mass. Clearly, these complications need to be addressed. However, Julian's "short-axis mode" may represent a reasonable approximation to the actual rotation of the nucleus.

3. Dust

The study of the dynamics of dust tails of comets also has a long history, dating back about 150 years. In an attempt to explain the morphology of the tail of comet P/Halley in its 1835 apparition, Bessel (1936) derived the equations of motion of dust particles emitted from the nucleus and driven away from the Sun by some repulsive force, which like gravitation varied inversely with the heliocentric distance. The physical nature of the postulated force was recognized to be radiation pressure only much later (Arrhenius, 1900; Schwarzschild, 1901). Significant contributions to the theory of dust tail dynamics have been made subsequently by a number of authors, and particularly by Probstein (1968) and Finson and Probstein (1968). For a detailed review see Mendis et al. (1985).

The two basic questions about the cometary dust concern the chemical composition and physical structure. Among the latter are the size distribution, shape, bulk density, and single scattering albedo. Before the encounters with Halley's Comet, information about the first came mainly from the presence of broad emission bands in the IR around 10 and 20 μm. Information about the second came from various lines of investigation. One line was the dynamical studies of dust in the tail, following Finson and Probstein. The other line came from optical and IR data on the continuum radiation by the dust. The relevant observations here were brightness and polarization measurements of scattered sunlight as functions of scattering angle and wavelength, and the variation of the thermal emission as a function of the heliocentric distance (and hence dust temperature). Two less direct sources of information pertaining to these questions came from the interaction of meteor streams with the earth's upper atmosphere and from the so-called Brownlee particles collected in the Earth's stratosphere, which are believed by some authors to be of cometary origin.

The observed 10 μm and 20 μm features were generally attributed to silicate grains, perhaps in the form of amorphous olivine (Hanner, 1980), although Vanysek and Wickramasinghe (1975), Mendis and Wickramasinghe (1975) had attributed these to organics such as polymerized formaldehyde grains. The coexistence of an absorbing component (e.g. magnetite) had also been invoked (e.g., Divine et al., 1986) to explain the thermal reradiation. That cometary grains may have a composite core mantle structure with a silicate core and an organic mantle had also been speculated (Greenberg, 1982).

The dust composition analyzers on board the Vega spacecraft have indicated the existence of at least three broad classes of grains: one class composed mainly of low-z elements (mainly CHON), a second class similar to CI chondrites but enriched in C, and a third class, similar to the second but more enriched in hydrogen.

That small (sub-micron) grains of the CHON type may be the "parents" of the observed spiral CN and C_2 jets in Halley's Comet has also been suggested. Some support for this view comes from the observations of Millis et al. (1986) which indicate that the production rate of CN is systematically larger than the production rate of HCN, although the latter molecule is generally believed to be the main (gas-phase) parent of CN.

Lastly, it has been suggested (Ogilvie et al., 1986a) that C-clusters in the form of rings or chains may be the parent of the observed C_2 radicals. If either these C-clusters or the CHON-grains are the parents of the observed C_3, it would remove an existing problem in atmospheric modelling-namely that of under-producing C_3 without assuming the existence of a very large quantity of complex hydrocarbons (Mitchell et al., 1981). What the present observations indicate is that the "parents" of some of the observed radicals need not be in the gas phase.

Indeed the concept that the parents of at least some of the radicals observed in the cometary atmosphere could be small dust grains rather than stable molecules is a new one that has emerged during the Heidelberg symposium.

On the basis of several lines of argument, a dust size distribution had been constructed and used for estimating the differential dust fluxes and fluences during the encounter (e.g., Divine et al., 1986). This was given by

$$n(a)\,da = g_0 \left(1 - \frac{a_0}{a}\right)^M \left(\frac{a_0}{a}\right)^N da; \qquad a \geqq a_0$$
$$= 0 \qquad\qquad a < a_0$$

with $N(=4.2)$ defining the slope of the larger grains, as determined by Sekanina (1979b) from the study of anti-tails of comets, and M taken equal to 12. Also, a_0 was taken equal to 0.1 μm. Neither the optical nor the IR data give any reliable information about particles of sizes $\lesssim 0.1$ μm. The reason that the cut-off radius a_0 was taken as 0.1 μm, however, was due to the expectation that cometary grains were aggregates of the Brownlee type, which had its smallest crystalline building blocks of about that size.

What is clear from the various dust detectors on both the Vega and the Giotto spacecraft is the existence of large quantities of very small particles. Instead of a cutoff at $a_0 = 0.1$ μm the dust mass spectrum continues to increase towards the smallest sizes that are detected; which may be as small as $a = 0.01$ μm for reasonable bulk densities (≈ 1 g cm^{-3}) for these smallest particles.

Another interesting discovery of the Vega dust detectors was the spatial distributions of the dust particles of various sizes. The smallest particles were detected much farther away than expected. They were well outside their bounding paraboloids defined by the classical theory based on radiation pressure deceleration, if they were even slightly absorbing (e.g., Vaisberg et al., 1986). Several possibilities have been suggested to explain this anomalous spatial distribution. One possibility, of course, is that the grains are indeed exotic (e.g., perfect dielectrics), another is that larger grains leaving the nucleus get broken up at large cometocentric distances (Simpson et al., 1986), and finally, that these grains, which are necessarily electrically charged, are electrically accelerated by the motional electric fields they experience in the magnetized plasma environment (Horanyi and Mendis, 1986). If indeed the larger grains do get broken up at large distances from the nucleus, the question arises as to the cause of this breakup. Do the aggregates come unglued due to the sublimation of the "icy glue"? Or are they electrostatically disrupted? In the latter connection, it must be recognized that small grains take a long time to reach their equilibrium electrostatic potentials.

While the spacecraft observations have provided us with a wealth of new information about the dust particles, including their elemental abundance and their size and spatial distributions, these observations are complemented by Earth-based observations. Of particular interest are the IR photometric and spectrophotometric observations carried out by the Kuiper Airborne Observatory (KAO) over a wavelength range from a few μm all the way to about 160 μm (Campins, 1986). Besides the earlier observed features at 10 and 20 μm, there is now a newly observed, unidentified feature around 28.4 μm. These observations also provide us with information about the relatively large ($\gtrsim 10$ μm) grains. Indeed it is already being claimed, on the basis of IR maps, that Halley is deficient in such grains vis-à-vis Giacobini-Zinner (Campins et al., 1986).

4. Gas: neutrals and ions

At the time of the last apparition of Halley's Comet in 1910, spectroscopy was still in its infancy. Even so, several chemical species, both neutral and ionized (CH, C_2, C_3, CN, Na, CO^+, and N_2^+) were identified in the head and plasma tail (Bobrovnikoff, 1931). Advances in spectrophotometric techniques, together with the extension of the spectral range both shortward (UV) and longward (IR and radio) of optical wavelengths have resulted in the detection of many more chemical species in subsequent comets. A composite list of all the identifications is given in Table 1. While the atomic species H, C, O, and S were detected via their resonance lines in the vacuum UV, the chemically stable molecules HCN and CH_3CN were detected at radio wavelengths.

A strong motivation for the study of cometary atmospheres arises from the fact that at present it is the only available method of deducing the chemical compositon of the volatile component of the nucleus. However, as is clear from Table 1, most of the

Table 1. Chemical species identified in cometary spectra prior to the encounter (tentative identifications in parenthesis)

Atome	Molecules	Ions
H	C_2	C^+
O		Ca^+
C	CH	CO^+
S	CN	CH^+
Na	CO	CN^+
K	CS	N_2^+
Ca	NH	CO_2^+
V	OH	H_2O^+
Mn	C_3	H_2S^+
Fe	NH_2	
Co	(H_2O)	
Ni	HCN	
Cu	CH_3CN	
	S_2	
	HCO	
	NH_3	
	(H_2CO)	
	(NH_4)	

observed species are chemically reactive photo-destruction fragments of presumably more stable molecules that are stored in the nucleus. As pointed out earlier, there were theoretical grounds and indirect observational evidence to suppose that H_2O is the dominant parent molecule. But the claim of the direct detection of H_2O in the radio range at about 1.35 cm was considered uncertain.

It was also recognized since the pioneering work of Oppenheimer (1975) that fast, generally exothermic ion-molecule reactions involving mass transfer play a crucial role in the chemical processing of the inner collision-dominated coma. Indeed, all the atmospheric/ionospheric models constructed in recent times had predicted that the stable ion H_3O^+ would be the dominant ion in the inner coma (unless the percentage of NH_3 or Na were substantial). However, the spectrum of this species was unknown.

One of the major discoveries of the ion mass spectrometers aboard the ICE spacecraft to Giacobini-Zinner (Ogilvie et al., 1986b) and the Vega and Giotto spacecraft to Halley (e.g., Gringauz et al., 1986; Balsiger et al., 1986; Korth et al., 1986) was the detection of this ion, as predicted, thereby establishing the general validity of the atmosphere/ionosphere models.

The neutral and ion mass spectrometers on board Giotto and Vega spacecraft, as well as the 3-channel and IR remote sensing photon spectrometers aboard the Vega spacecraft, have detected most of the species identified earlier, and a number of new species. The Vega 1 IKS spectrometer (Combes et al., 1986) made a definite detection of H_2O at 2.7 μm and also CO_2 (for the first time) at 4.24 μm. The stretching band of C-H at 3.4 μm was also detected, while the possible detection of H_2CO at 3.6 μm is less certain due to the noise level. Also, H_2O was detected at 1.38 μm with the Vega 2 TKS spectrometer (Krasnopolsky et al., 1986).

But undoubtedly the most convincing detection of H_2O was done even earlier, using the high-resolution IR spectroscope on board KAO, which identified 12 rotational lines of the 2.7 μm band at 1 AU pre-perihelion and subsequently 15 lines of the same band post-perihelion at the same distance (Mumma et al., 1986). This observation has led to the determination of the ortho-para ratio of H_2O vapor in comet P/Halley.

The ion mass spectrometers on board the Vega and Giotto spacecraft have clearly identified the dominant H_3O^+, H_2O^+, OH^+, and O^+ ions of the H_2O-group. A clear feature at $m/q = 12$ is almost certainly C^+. A composite list of other possible identifications includes CO_2^+, CO^+, CS_2^+, S_2^+, CS^+, S^+, CH^+, Fe^+, NH_4^+, CH_3^+, and Na^+. So far unidentified ions with $m/q \approx 100$ are also observed (Korth et al., 1986). The Giotto neutral mass spectrometer also detected the dominant neutral species H_2O, O, and OH, and probably also CO_2.

Recently Huebner (1987) has shown that the peaks observed at 45, 61, 75, 91, and 105 amu with the PICCA ion analyzer of Korth et al. (1986) on board GIOTTO is consistent with these being various ion fragments broken off a polyoxymethylene, $[(H_2CO)_n]$ chain. He also argues that this polymer may be associated with the "CHON" dust grains discussed earlier. If this identification is correct, this would be the first polymer identified in space although earlier Vanysek and Wickramasinghe (1975) and Mendis and Wickramasinghe (1975) had used IR data as well as inferred physical data from earlier comets to propose that cometary dust was largely composed of $(H_2CO)_n$. Based on the analysis of the widths of the peaks at 45, 61, 75, 91, and 105 amu and knowledge of the ion temperature, Mitchell et al. (1987) point out that each of these peaks are associated with more than just a single ion. So while the $(H_2CO)_n$ fragments may indeed be present, there are contributions from other as yet unknown ions to each of the peaks. Indeed it is likely that $(H_2CO)_n$ is not the only polymer associated with the CHON grains. Mixed polymers also including $(HCN)_n$ are more likely. In that case the CN jets, which are also presumably associated with CHON grains may be a dissociation product of such a mixed polymer.

As exciting as these in situ observations are the earth-based observations. We have already mentioned the discoveries of the KAO in the near and far IR. The extremely high-resolution spectroscopy of the CN (0–0) violet system of Halley (Wehinger et al., 1986) has produced the first reliable estimates of two isotope ratios of a comet. They obtain $C^{12}/C^{13} = 80 \pm 20$, $N^{14}/N^{15} = 250 \pm 100$ which within the error bars are compatible with their so-called cosmic values (89 in the former case and 270 in the latter).

Before the encounters only column densities and total abundances of various species were known. Data on temperatures and velocities (except down the ion tail) were virtually unknown. After the encounters we have information on the number densities, translational temperatures and velocities along the spacecraft path (e.g., Balsiger et al., 1986; Lämmerzahl et al., 1987). This constitutes the basic observational input required for testing the atmospheric/ionospheric models. Indeed the observations inside the ionopause (Lämmerzahl et al., 1987) appears to be in reasonable agreement with the predictions of the latest models (e.g., Korozmezey et al., 1987; Marconi and Mendis, 1986) although the neutral temperatures deduced are only upper limits. Also, the high-resolution KAO near-IR observations of H_2O have the potential of providing the rotational temperature of this molecule. This information is vital in understanding the radiation coupling between the dust and the gas (Marconi and Mendis, 1986).

The role that dust played both in the dynamics (Probstein, 1968) and the thermodynamics (Marconi and Mendis, 1986) of the atmosphere was appreciated prior to the present apparition of Halley's Comet. What was was not appreciated was the important role that they would also play in the chemistry of the atmosphere. The observations of the spiral jets of CN by A'Hearn et al. (1986) discussed earlier, and the detection of the low-z CHON dust grains, suggest that some dust grains themselves may be the long sought after parents for some of the cometary radicals. Clearly this aspect will have to be considered in future atmospheric modelling. A comprehensive review of the pre-encounter atmospheric/ionospheric modelling is given in Mendis et al. (1986).

5. The solar wind interaction

Ever since Biermann (1951) used the observed aberration of plasma tails of comets to infer the continuous outflow of plasma from the sun (since referred to as the solar wind) comets have been used to good advantage to delineate the average global flow parameters of the solar wind (e.g., Brandt and Rother, 1986). Also, dramatic changes observed in the plasma tail, such as its sudden total disconnection, have more recently been associated with discontinuous changes in the solar wind flow past the comet, such as the passage of magnetic sector boundaries (e.g., Niedner and Brandt, 1978; Niedner, 1986). These authors have developed a model which invokes magnetic reconnection in the head during sector boundary traversals. The possibility of reconnection in the tail, as well as the possibility of the rapid growth of the flute instability at the head during encounters with high-speed streams, has also been suggested as alternative mechanisms for the observed tail disconnection events.

The basic nature of the solar wind interaction with a comet was discussed in several pioneering papers. Alfvén (1957) predicted the draping of the interplanetary magnetic field around the cometary ionosphere to form a cometary magneto-tail. Axford (1964) predicted the existence of a stationary bow shock upstream of the comet, while other authors had considered the possibility of travelling shock waves earlier. The structure of this shock was first calculated by Biermann et al. (1967) using a hydrodynamic model, and its basic nature was clarified by Wallis (1973). MHD models of the comet-solar wind interaction have been constructed subsequently by several authors (e.g., Schmidt and Wegmann, 1982) and they all predicted a weak ($M = 2$) shock.

The existence of a tangential discontinuity surface (loosely called a contact surface or ionopause) was already anticipated in the early work of Biermann et al. (1967) but the basic mechanism for its formation, viz., by the balance of the electromagnetic ($\boldsymbol{j} \times \boldsymbol{B}$) force by the neutral drag on the plasma ahead of it, was proposed by Ip and Axford (1982).

The possible existence of an inner shock, where the supersonically outflowing cometary ions are decelerated and diverted into the tail, was proposed by Wallis and Dryer (1976), and its basic structure was calculated by Houpis and Mendis (1980). While Marconi and Mendis (1983) have subsequently argued that this shock would be very weak due to plasma heating in the inner coma, the possibility of an inner shock of an electrostatic nature, due to the sudden heating of the electrons following their thermal decoupling from the ions has also been suggested recently (Körösmezey et al., 1987). Finally, the existence of a so-called "collisionopause" ahead of the tangential discontinuity surface due to rapid increase of the collision rate of the contaminated solar wind ions with the outflowing neutrals had been suggested (e.g., Flammer et al., 1986).

The global nature of the solar wind interaction with the well-developed atmosphere of a comet is shown in Fig. 1. One of the major contributions of the missions to comets P/Giacobini Zinner and P/Halley was in the advances made in our understanding of the nature of the comet-solar wind interaction.

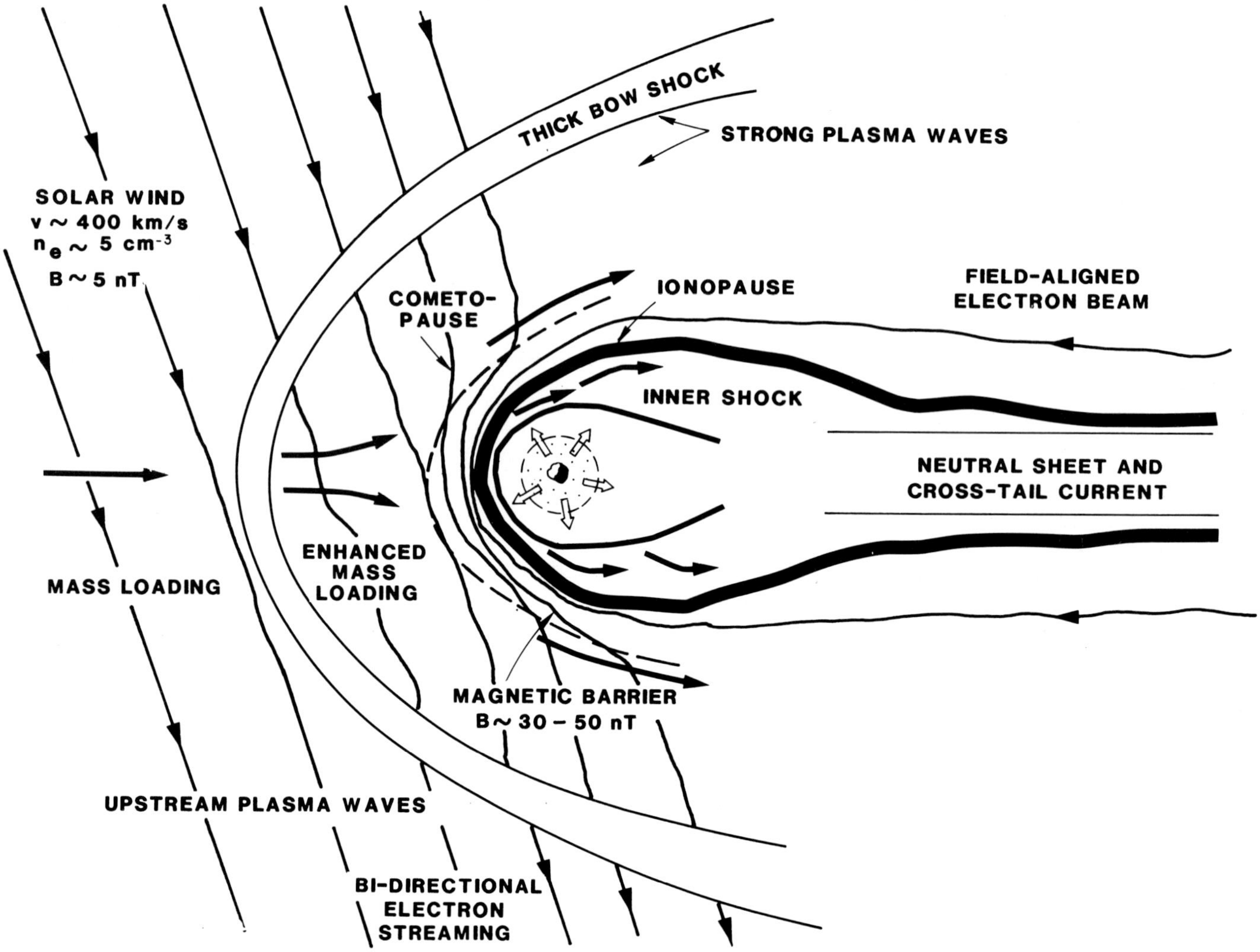

Fig. 1. Schematic representation of the global morphology of the solarwind interaction with a well developed cometary atmosphere

The magnetometer on board the NASA ICE spacecraft [Smith et al. (1986) spectacularly confirmed the magnetotail model of Alfvén (1957)] as it flew through the plasma tail of comet P/Giacobini-Zinner. Two magnetic lobes of opposite polarity separated by a current-carrying neutral sheet were observed as expected. The magnetometer as well as the plasma instruments also observed a feature which could be associated with the weak bow shock more or less at the anticipated distance (e.g., Mendis et al., 1986). However, possibly due to the decay of the shock wave in the flanks, as predicted by Wallis and Dryer (1985), this feature appeared to be intermediate to a compression wave and a fully developed shock. The existence of a weak collisionless shock on the sunward hemisphere of Halley's Comet was clearly seen by magnetometers and plasma analyzers on board the Vega, Giotto and Suisei.

Another dramatic discovery was of the "ionopause" of comet P/Halley by the magnetometer on board Giotto (Neubauer et al., 1986). An essentially magnetic field-free cavity, containing purely cometary ions separated from the inflowing contaminated solar wind by a sharp boundary, was observed at a cometocentric distance ≈ 4700 km inbound and at about 3800 km outbound. The average distance of the ionopause is entirely consistent with the prediction of the Ip and Axford model (1982) if the neutral drag coefficient on the ions $\approx 10^{-9}\,cm^3\,s^{-1}$. Although a well-defined ionopause is not seen down the tail of comet P/Giacobini-Zinner, a region of relatively low magnetic field was observed within about 1000 km of the axis (Smith et al., 1986). Ershkovich et al. (1986) have studied the stability of the cometary ionopause and have shown that both the exclusion of the magnetic field in the case of Halley and the partial penetration of the magnetic field in the case of Giacobini-Zinner can be explained in terms of the relative production rates of the molecules and the respective solar wind conditions.

The collisionopause, whose position more or less corresponds to the beginning of the sudden enhancement of the magnetic field (the "magnetic barrier") is at the expected distance both in the case of P/Giacobini-Zinner (Mendis et al., 1986) and in the case of comet P/Halley (Reme et al., 1986). However, there is no general agreement at present as to the mechanism responsible for its formation, and in order not to prejudge the issue, it is now often referred to as the "cometopause" (e.g., Gringauz et al., 1986).

Another interesting feature is the one called the "ion pile-up region" just ahead of the ionopause. The ion density reaches a local maximum around 12000 km from the nucleus and drops sharply by about a factor of 3, by about 8000 km from the nucleus, before increasing again (e.g., Balsiger et al., 1986). The description of this feature as a "pile-up" is a misnomer. It is really a region of ion depletion, and appears to be largely due to the rapid decrease of plasma temperature, which increases the loss rate of ions by dissociative recombinations, (Baumgärtel and Sauer, 1987). A similar effect, in reverse, was observed in the numerical simulations of the cometary ionsphere (Marconi and Mendis, 1983). Around $2\text{–}3 \cdot 10^3$ km from the nucleus the electrons thermally decouple from the outflowing ions and rapidly increase in temperature. This in turn leads to a decrease in the dissociative recombination rate and a consequent increase in the electron and ion density by about a factor of two.

Two other sharp surfaces of discontinuity whose nature is not presently understood (referred to as "mysterious boundaries") also exist in the particles and fields data of comet P/Halley (e.g., Galeev, 1986).

The large-scale (global) structure of the comet-solar wind interaction region observed by the missions to comets P/Giacobini-Zinner and P/Halley were to a rather surprising degree close to predictions, considering that the models were either hydrodynamic or ideal MHD, about whose applicability there had always been considerable doubt. What was surprising, however, was the high level and large spatial extent of the hydromagnetic turbulence. Besides large amplitude variations of $|\boldsymbol{B}|$ and plasma parameters with a time scale of minutes to tens of minutes, a variety of plasma waves, including the ion cyclotron, lower hybrid, whistlers, electron cyclotron and electron acoustic, were observed. These have now strongly focussed our attention on the micro-physics associated with the comet-solar wind interaction. It is clear that central to all this activity is the pick-up of heavy cometary ions by the inflowing solar wind over distance scales exceeding by 10^6 the linear dimensions of the nucleus.

It was pointed out by Sagdeev et al. (1986c) that when the solar wind flow is oblique to the direction of the interplanetary magnetic field, the velocity space distribution of the newly picked up ions in the solar wind frame of reference is a ring perpendicular to the magnetic field and a beam parallel to it. It was also shown by these authors that the beaming of these ions along the magnetic field would lead to the generation of Alfvén waves and that the inelastic scattering of these ions on these waves will in turn lead to their being rather rapidly isotropized into a shell distribution. This is consistent with the in-situ measurements of several instruments at Halley's Comet (Neugebauer et al., 1986; Mukai et al., 1986), which indicates a partially filled shell.

To the second order, this scattering can also lead to the further energization of the pick-up ions.

Large fluxes of energetic ions ($E \sim 250\text{–}400$ keV) (which are significantly more energetic than what is expected for newly picked up O^+ or H_2O^+ ions, which have typical energies $\lesssim 70$ keV) have been observed at comets P/Giacobini-Zinner and P/Halley (e.g. Hynds et al., 1986; McKenna-Lawlor et al., 1986). Using a simple model, Ip and Axford (1986) have argued that the so called second order Fermi acceleration mechanism (which is essentially a stochastic acceleration due to interaction with propagating Alfvén waves) is responsible for these energetic ions. More detailed calculations, using quasilinear theory (Isenberg, 1987; Gribov, 1986) appears to support this view.

Noting the observation of lower hybrid waves in the ion pick-up regions of both comets P/Giacobini-Zinner and P/Halley, Buti and Lakhina (1987) have shown that the turbulence generated by these waves leads to electric fields that are strong enough to stochastically accelerate the pick-up ions to the observed high energies even if only a fraction of the stochastic electric fields are used for the acceleration. Clearly this is an alternative to the second order Fermi process, in the energization of the pick-up ions to the observed values.

The role of collective plasma processes, particularly the possible contribution of the Alfvén critical velocity ionization mechanism at Halley's Comet, has also been discussed (e.g., Galeev et al., 1986).

As a result of all these considerations, the theoretical modelling of the comet-solar wind interaction is moving away from the fluid approach to a more kinetic approach. While partial models have been or are being developed to explain various aspects such as the structure of the collisionless shock, we are still very far from a complete kinetic description of the comet-solar wind interaction.

6. Conclusion

The multitude of missions to comets P/Giacobini-Zinner and P/Halley, together with the coordinated earth-based observations, have led to a quantum leap during the last year in our knowledge of comets. We are only beginning to appreciate the implications of all the new discoveries and there is no doubt that in time this will lead to a much deeper understanding of the nature of comets and their interaction with the solar radiation and the solar wind.

Clearly, much remains to be done. While we continue to analyze the existing data and build more sophisticated models to explain them, we should not forget the need for future missions to comets, to provide answers to some old unanswered questions as well as new ones that have arisen.

The view that comets constitute the most pristine material in the solar system was one of the strongest motivations for missions to comets. It was appreciated that a knowledge of the chemical composition of the cometary nucleus could lead to far-reaching inferences regarding the chemical composition and physical state of the primordial dusty nebula in which comets and all other solar system bodies formed. Also, a knowledge of a comet's internal physical structure (which was not expected to change much since its birth) could lead to an understanding of the important physical processes that were involved in its formation. However, as we pointed out earlier, what was observed of Halley's nucleus was not the evaporating surface but rather a dust mantle or crust that largely insulates the ices lying underneath, with the upper layers of these ices also possibly chemically differentiated up to several decimeters or more. Clearly what is required in a future mission is to sample the subsurface material which is still undifferentiated. Ideally this is accomplished by using a penetrator, as has been included in the proposed NASA-CRAF mission. Such a rendezvous mission with a short-period comet would clearly pick up from where the missions to comets P/Giacobini-Zinner and P/Halley left off. Also, by observing the cometary nucleus more closely at large heliocentric distances, when it is inactive, one would obtain a much better knowledge of its topography.

From the point of view of the solar wind interaction, what we have learned pertains to the interaction with a well-developed atmosphere when a comet is fairly close to the sun ($d \approx 1$ AU). What makes comets particularly intersting from the point of view of comparative planetology is that they produce not one but rather a whole range of atmospheres, from a fledgling one around $d \approx 3$ AU to a dense one closer to the Sun ($d \lesssim 1.5$ AU). A mission such as CRAF would enable us to study the interaction of the solar wind with the entire range of atmospheres, as well as its interaction with the bare nucleus, which may lead to interesting electrostatic effects such as electrostatic levitation and blow-off of fine surface dust.

Acknowledgements. The author wishes to acknowledge support from the following grants; NASA NGR 05-009-110 of the NASA Planetary Atmospheres Program and NSF AST 85-05921 of the NSF Solary System Astronomy program.

References

A'Hearn, M.L., Hoban, S., Birch, P.V., Bowers, C., Martin, R., Klinglesmith III, D.A.: 1986, *Proc. 20th ESLAB Symposium on the Exploration of Halley's Comet*, eds. B. Battrick, E.J. Rolfe, R. Reinhard, ESA SP-250, Vol. I, p. 483

Alfvén, H.: 1957, *Tellus* **9**, 92

Arrhenius, S.A.: 1900. *Phys. Z.* **2**, 81

Axford, W.I.: 1964, *Planet. Space Sci.* **12**, 719

Balsiger, H., Altwegg, K., Bühler, F., Geiss, J., Ghielmetti, A.G., Goldstein, B.E., Goldstein, R., Huntress, W.T., Ip, W.-H., Lazarus, A.J., Meier, A., Neugebauer, M., Rettenmund, U., Rosenbauer, H., Schwenn, R., Sharp, R.D., Shelley, E.G., Ungstrup, E., Young, D.T.: 1986, *Nature* **321**, 330

Baumgärtel, K., Sauer, K.: 1987, Fluid Simulation of Halley's Ionosphere, *Astron. Astrophys.* **187**, 307

Belton, M.J.S., Wehinger, P., Spinrad, H., Wyckoff, S.: 1986, *Proc. 20th ESLAB Symposium on the Exploration of Halley's Comet*, eds. B. Battrick, E.J. Rolfe, R. Reinhard, ESA SP-250, Vol, I, p. 599

Bessel, F.W.: 1836, *Astron. Nachr.* **13**, 185

Biermann, L.: 1951, *Z. Astrophys.* **29**, 274

Biermann, L., Brosowski, B., Schmidt, H.U.: 1967, *Solar Phys.* **1**, 254

Bobrovnikoff, N.T.: 1931, *Pub. Lick Observ.* **17**, Part II, 309

Brandt, J.C., Rothe, E.D.: 1976, in *Study of Comets*, eds. B. Donn et al., NASA SP-393, p. 878

Brin, G.D., Mendis, D.A.: 1979, *Astrophys. J.* **229**, 402

Buti, B., Lakhina, G.S.: 1987, Stochastic acceleration of ions in the pre-shock region of comet Halley, *Geophys. Res. Lett.* **14**, 107

Campins, H., Telesco, C.M., Decher, R., Ramsey, B.D.: 1986, *Proc. 20th ESLAB Symposium on the Exploration of Halley's Comet*, eds. B. Battrick, E.J. Rolfe, R. Reinhard, ESA SP-250, Vol. II, p. 91

Combes, M., Moroz, V.I., Crifo, J.F., Lamarre, J.M., Charra, J., Sanko, N.F., Soufflot, A., Bibring, J.P., Cazes, S., Coron, N., Crovisier, J., Emerich, C., Encrenaz, T., Gispert, R., Grigoryev, A.V., Guyot, G., Krasonopolsky, V.A., Nikolsky, Yu.V., Rocard, F.: 1986, *Nature* **321**, 266

Cruikshank, D.P., Hartmann, W.K., Tholen, D.J.: 1985, *Nature* **315**, 122

Delsemme, A.H.: 1982, in *Comets*, ed. L.L. Wilkening, Univ. of Arizona Press, Tucson, Arizona, p. 85

Delsemme, A.H., Swings, P.: 1952, *Ann. Astrophys.* **15**, 1

Divine, N., Fechtig, H., Gombosi, T.I., Hanner, M.S., Keller, H.U., Larson, S.M., Mendis, D.A., Newburn, R.L., Jr., Reinhard, R., Sekanina, Z., Yeomans, D.K.: 1986, *Space Sci. Rev.* **43**, 1

Donn, B.: 1976, in *Study of Comets*, Part 2, eds. B. Donn, M. Mumma, W. Jackson, M. A'Hearn, R. Harrington, NASA SP-373, p. 611

Ershkovich, A.I., Flammer, K.R., Mendis, D.A.: 1986, *Astrophys. J.* **311**, 1031

Fanale, F.P., Salvail, J.R.: 1984, *Icarus* **60**, 476

Festou, M.C., Feldman, P.D., A'Hearn, M.F., Arpigny, C., Cosmovici, C.B., Danks, A.C., McFadden, L.A., Gilmozzi, R., Patriarchi, P., Tossi, G.P., Wellis, M.K., Weaver, H.A.: 1986, *Nature* **321**, 361

Finson, M.L., Probstein, R.F.: 1968, *Astrophys. J.* **154**, 327

Flammer, K.R., Mendis, D.A., Houpis, H.L.F.: 1986, in *Proc. STIP Symposium on Retrospective Analysis and Future Coordinated Intervals*, ed. M.A. Shea, Reidel, Dordrecht, in press)

Galeev, A.: 1986, *Proc. 20th ESLAB Symposium on the Exploration of Halley's Comet*, eds. B. Battrick, E.J. Rolfe, R. Reinhard, ESA SP-250, Vol. I, p. 3

Galeev, A., Gringauz, K.I., Klimov, S.I., Remizov, A.P., Sagdeev, R.Z., Savin, S.P., Sokolov, A.Yu., Verigin, M.I.: 1986, *Geophys. Res. Lett.* **13**, 845

Greenberg, J.M.: 1982, in *Comets*, eds. L.L. Wilkening, Univ. of Arizona Press, Tucson Arizona, p. 131

Gribov, B.E., Kecskeméty, K., Sagdeev, R.Z., Shapiro, V.D., Shevchenko, V.I., Somogyi, A.J., Szegö, K., Erdos, G., Eroshenko, E.G., Gringunz, K.I., Keppler, E., Marsdan, R., Remizov, A.P., Richter, A.K., Riedler, W., Schwingenschuh, K., Wenzel, K.P.: 1986, *Proc. 20th ESLAB Symposium on the Exploration of Halley's Comet*, eds. B. Battrick, E.J. Rolfe, R. Reinhard, ESA SP-250, Vol. I, p. 211

Gringauz, K.I., Gombosi, T.I., Remizov, A.P., Apáthy, I., Szemerey, I., Verigin, M.I., Denchikova, L.I., Dyachkov, A.V., Keppler, E., Klimenko, I.N., Richter, A.K., Somogyi, A.J., Szegö, K., Szendro, S., Tátrallyay, M., Varga, A., Vladimirova, G.A.: 1986, *Nature* **321**, 282

Hanner, M.: 1980, in *Solid Particles in the Solar System*, eds. I. Halliday, B.M. McIntosh, Reidel, Dordrecht, p. 233

Horanyi, M., Gombosi, T.I., Cravens, T.E., Körösmezey, T.E., Kecskemety, K., Nagy, A.F., Szegö, K.: 1984, *Astrophys. J.* **278**, 449

Horanyi, M., Mendis, D.A.: 1986, *Astrophys. J.* **307**, 800

Houpis, H.L.F., Ip, W.-H., Mendis, D.A.: 1986, *Astrophys. J.* **295**, 654

Houpis, H.L.F., Mendis, D.A.: 1980, *Astrophys. J.* **239**, 1107

Houpis, H.L.F., Mendis, D.A., Ip, W.-H.: 1985, in *Asteroids Comets Meteors II: Exploration and Theoretical Modelling*, eds. C.I. Lagerkvist et al., Uppsala Univ. Press, Uppsala, p. 341

Huebner, W.F.: 1987, *Science* **237**, 628

Hynds, R.J., Cowley, S.W.H., Sanderson, T.R., Wenzel, K.P., Van Rooijer, J.J.: 1986, *Science* **232**, 361

Isenberg, P.A.: 1987, *J. Geophys. Res.* (in press)

Ip, W.-H., Axford, W.I.: 1982, in *Comets*, ed. L.L. Wilkening, Univ. of Arizona Press, Tucson, p. 588

Ip, W.-H., Axford, W.I.: 1986, *Planet Space Sci* **34**, 1061

Julian, W.H.: 1987, *Nature* **326**, 57

Kaneda, E., Ashihara, O., Shimizu, M., Takagai, M., Hiro, K.: 1986, *Nature* **321**, 297

Keller, H.U., Arpigny, C., Barbieri, C., Bonnet, R.M., Cazes, S., Coradini, M., Cocmovici, C.B., Delamere, W.A., Huebner, W.F., Hughes, D.W., Jamar, C., Malaise, D., Reitsema, H.J., Schmidt, H.U., Schmidt, W.K.H., Seige, P., Whipple, F.I., Wilhelm, K.: 1986, *Nature* **321**, 320

Keller, H.U., Delamere, W.A., Huebner, W.F., Reitsema, H.J., Schmidt, H.U., Whipple, F.L., Wilhelm, K., Curdt, W., Kramm, R., Thomas, N., Arpigny, C., Barbieri, C., Bonnet, R.M., Cazes, S., Coradini, M., Cosmovici, C.B., Hughes, D.W., Jamar, C., Malaise, D., Schmidt, K., Schmidt, W.K.H., Seige, P.: 1987, *Astron. Astrophys.* **187**, 807

Kissel, J., Sagdeev, R.Z., Bertaux, J.L., Angarov, V.N., Audouze, J., Blamont, J.E., Büchler, K., Evlanov, E.N., Fechtig, H., Fomenkova, M.N., von Hoerner, H., Inogamov, N.A., Khromov, V.N., Knabe, W., Krueger, F.R., Langevin, Y., Leonas, V.B., Levasseur-Regourd, A.C., Managadze, G.G., Podkolzin, S.N., Shapiro, V.D., Tabaldyev, S.R., Zubkov, B.V.: 1986, *Nature* **321**, 280

Klinger, J.: 1980, *Science* **209**, 271

Körösmezey, A., Cravens, T.E., Gombosi, T.I., Nagy, A.F., Mendis, D.A., Szegö, K., Gribov, B.E., Sagdeev, R.Z., Shapiro, V.D., Schevchenko, V.I.: 1987, *J. Geophys. Res.* **92**, 7331

Korth, A., Richter, A.K., Loidl, A., Anderson, K.A., Carlson, C.W., Curtis, D.W., Lin, R.P., Reme, H., Sauvaud, J.A., d'Uston, C., Cotin, F., Cros, A., Mendis, D.A.: 1986, *Nature* **321**, 335

Krasnopolsky, V.A., Gogoshev, M., Moreels, G., Moroz, V.I., Krysko, A.A., Gogosheva, Ts., Palozov, K., Sargoichev, S., Clairemide, J., Vincent, M., Bertaux, J.L., Blamont, J.E., Troshin, V.S., Valnićek, B.: 1986, *Nature* **321**, 269

Lämmerzahl, P., Krankowsky, D., Hodges, R.R., Stubbeman, U., Woweries, J., Herrwerth, I., Berthelier, J.J., Illiana, J.M., Eberhardt, P., Dolder, U., Schulte, W., Hoffman, J.H.: 1987, *Astron. Astrophys.* **187**, 169

Laplace, P.S.: 1813, *Exposition du Systeme du Monde* (4th edn.), Paris, p. 130

Marconi, M.L., Mendis, D.A.: 1983, *Astrophys. J.* **273**, 381

Marconi, M.L., Mendis, D.A.: 1986, *Earth Moon, Planets* **36**, 249

McKenna-Lawlor, S., Kirsch, E., O'Sullivan, D., Thompson, A., Wenzel, K.P.: 1986, *Nature* **321**, 347

Mendis, D.A., Wickramasinghe, N.C.: 1975, *Astrophys. and Space Sci.* **37**, L13

Mendis, D.A., Houpis, H.L.F., Marconi, M.L.: 1985, *Fundamentals of Cosmic Physics* **10**, 1

Mendis, D.A., Smith, E.J., Tsurutani, B.T., Slavin, J.A., Jones, D.E., Siscoe, G.L.: 1986, *Geophys. Res. Lett.* **13**, 239

Millis, S.L., Schleicher, D.G.: 1986, *Nature* **324**, 646

Millis, R.L., Schleicher, D.G., Thompson, D.T., Bus, S., Martin, L., Jansen, S., Birch, P.V., Martin, R., A'Hearn, M.F.: 1986, paper red at the *20th ESLAB Symposium: Exploration of Halley's Comet* (Heidelberg, October, 1986)

Mitchell, G.F., Prasad, S.A., Huntress, W.T.: 1981, *Astrophys. J.* **244**, 1087

Mitchell, D.L., Lin, R.P., Anderson, K.A., Carlson, C.W., Curtis, D.W., Korth, A., Reme, H., Sauvand, J.A., D'Uston, C., Mendis, D.A.: 1987, *Science* **237**, 626

Mukai, T., Miyake, N., Terasawa, T., Ketayama, M., Hiro, K.: 1986, *Geophys. Res. Lett.* **13**, 829

Mumma, M.J., Weaver, H.A., Larson, H.P.: 1986, *Proc. 20th ESLAB Symposium on the Exploration of Halley's Comet*, eds. B. Battrick, E.J. Rolfe, R. Reinhard, ESA SP-250, Vol. I, p. 341

Neugebauer, M., Lazarus, A.J., Altwegg, K., Balsiger, H., Goldstein, B.E., Goldstein, R., Neubauer, F.M., Rosenbauer, H., Schwenn, R., Shelley, E.G., Ungstrup, E.: 1986, *Proc. 20th ESLAB Symposium on the Exploration of Halley's Comet*, eds. B. Battrick, E.J. Rolfe, R. Reinhard, ESA SP-250, Vol. I, p. 341

Neubauer, F.M., Glassmeier, K.H., Pohl, M., Raeder, J., Acuna, M.H., Burlaga, L.F.D., Ness, N.F., Musmann, G., Mariani, F., Wallis, M.K., Ungstrup, E., Schmidt, H.U.: 1986, *Nature* **321**, 352

Niedner, M.B., Jr., Brandt, J.C.: 1978, *Astrophys. J.* **223**, 655

Niedner, M.B., Jr.: 1986, "Dynamics of cometary plasma tails", talk given at the *Solar Wind Interaction Symposium (COSPAR)*, Toulouse, France, July, 1986

Ogilvie, K.M., Coplan, M.A., A'Hearn, M.F.: 1986a, paper read at this meeting

Ogilvie, K.W., Coplan, M.A., Bochsler, P., Geiss, J.:1986b, *Science* **232**, 374

Oppenheimer, M.: 1975, *Astrophys. J.* **196**, 251

Probstein, R.F.: 1968, in *Problems of Hydrodynamics and Continuum Mechanics, Soc. Industr. Appl. Math.* 568

Reme, H., Sauvaud, J.A., d'Uston, C., Cotin, F., Cros, A., Anderson, K.A., Carlson, C.W., Curtis, D.W., Lin, R.P., Mendis, D.A., Korth, A., Richter, A.K.: 1986, *Nature* **321**, 349

Rickman, H.: 1986, *Uppsala Preprints in Astron.*, No. 8

Sagdeev, R.Z., Avansov, G.A., Shamis, V.A., Ziman, Y.L., Krasikov, V.A., Tarnopolsky, V.A., Kuzmin, A.A., Szegö, K., Merenyi, E., Smith, B.A.: 1986a, *Nature* **321**, 262

Sagdeev, R.Z., Avansov, G.A., Shamis, V.A., Ziman, Ya.L., Krasikov, V.A., Tarnopolsky, V.A., Kuzmin, A.A., Szegö, K., Merenyi, E., Smith, B.A.: 1986b, *Proc. ESLAB Symposium on the Exploration of Halley's Comet*, eds. B. Battrick, E.J. Rolfe, R. Reinhard, ESA SP-250, Vol. II, p. 295

Sagdeev, R.Z., Shapiro, V.D., Shevchenko, V.I., Szegö, K.: 1986c, *Geophys. Res. Lett.* **13**, 85

Schmidt, H.U., Wegmann, R.: 1982, in *Comets*, ed. L.L. Wilkening, Univ. of Arizona Press, Tucson, p, 538

Schwarzschild, K.: 1901, *Sitz Bayer. Akad. Wiss. München*, 293

Sekanina, Z.: 1987, *Nature* **325**, 326

Sekanina, Z.: 1979a, *Icarus* **37**, 420

Sekanina, Z., 1979b, *ESA SP-153*, 125

Sekanina, Z., Larson, S.: 1984, *Astrophys. J.* **89**, 1408

Shul'man, E.M.: 1972, in *The Motion Evolution of Orbits and Origin of Comets*, eds. G.A. Chebotarev, E.I. Kazimirchak-Polonkaya, G.B. Marsden, Reidel, Dordrecht, p. 271

Simpson, J.A., Rabinowitz, D., Tuzzolino, A.J., Ksanfomality, L.V., Sagdeev, R.Z.: 1986, paper read at the *ESLAB Symposium on the Exploration of Halley's Comet*, eds. B. Battrick, E.J. Rolfe, R. Reinhard, ESA SP-250, Vol. II, 11

Smith, B.A., Larson, S.H., Szegö, K., Sagdeev, R.Z.: 1987, *Nature* **326**, 573

Smith, E.J., Tsurutani, B.T., Slavin, J.A., Jones, D.E., Siscoe, G.L., Mendis, D.A.: 1986, *Science* **232**, 382

Smoluchowski, R.: 1981, *Astrophys. J.* **244**, L31

Vaisberg, O.L., Smirnov, V.N., Gom, L.S., Iovlev, M.V., Balikchin, M.A., Klimov, S.I., Savin, S.P., Shapiro, V.D., Shevchenko, V.I.: 1986, *Nature* **321**, 274

Vanysek, V., Wickramasinghe, N.C.: 1975, *Astrophys. Space Sci.* **33**, L19

Wallis, M.K.: 1973, *Planet. Space Sci.* **21**, 1647

Wallis, M.K., Dryer, M.: 1976, *Astrophys. J.* **205**, 895

Wallis, M.K., Dryer, M.: 1985, *Nature* **318**, 646

Wehinger, P.A., Peterson, B.A., Wyckoff, S., Lindblom, E., Festou, M.: 1986, *Proc. 20th ESLAB Symposium Exploration of Halley's Comet*, eds. B. Battrick, E.J. Rolfe, R. Reinhard, ESA SP-250, Vol. III, 473

Whipple, F.W.: 1950, *Astron. J.* **111**, 375

Whipple, F.L.: 1977, in *Comets Asteroids Meteorites*, ed. A.H. Delsemme, Univ. of Toledo Press, Toledo, Ohio, p. 25

Whipple, F.L., Sekanina, Z.: 1979, *Astrophys. J.* **84**, 1894

EXPLORATION OF HALLEY's COMET

M. Grewing · F. Praderie
R. Reinhard (Eds.)

The Giotto mission to Halley's Comet

R. Reinhard

Space Science Department of ESA, ESTEC, Noordwijk, The Netherlands

Introduction

Halley's Comet is the only one among 750 different recorded comets which has both a well-known orbit and high gas and dust production rates. The encounter with Halley could take place in March 1986 when it was still close to the Sun (0.89 AU) and therefore highly active and also observable from the Earth which is important for correlating in-situ observations with remote ground-based observations. Furthermore, a mission to Halley requires one of the lowest launch energies of all possible cometary missions.

Halley's Comet is thus an outstanding target for a first cometary mission and, as it returns only every 76 yr, the Science Programme Committee of the European Space Agency (ESA) approved a mission to this comet in July 1980. The mission was named "Giotto" after the Italian painter Giotto di Bondone who in 1304 depicted Comet Halley as the "Star of Bethlehem" in one of the frescoes in the Scrovegni chapel in Padua.

The Giotto mission to Comet Halley was ESA.s first interplanetary mission, and Giotto was ESA's first scientific spacecraft to be launched by an Ariane rocket.

The spacecraft

The Giotto spacecraft (Fig. 1) is spin-stabilised, nominally at 15 rpm. Its diameter is 1.86 m, and the height from the tip of the tripod to the bottom adaptor ring is 2.85 m. At launch Giotto weighted 960 kg, reducing to 573.7 kg during the Halley flyby, after the solid-propellant kick motor used to inject the spacecraft from a geostationary transfer orbit into a heliocentric orbit had burnt out, and 9 kg of the available 69 kg of hydrazine had been used for three orbit correction manoeuvres and a large number of attitude corrections.

Because Halley's orbit is retrograde, that is, of opposite sense to those of the Earth and the spacecraft, the relative encounter velocity was very high (68.373 km s^{-1}). Compared with this, the speed of the cometary gas (1 km s^{-1}) and dust particles (a few hundred m s^{-1}) was very low. Giotto thus saw an essentially static situation as it traversed the coma, and all the particles struck the spacecraft from the forward direction. During the Halley encounter the spacecraft spin axis was aligned with the relative velocity vector, that is, the cometary particle streaming was from below in Fig. 1. A dust particle with a mass of only 0.1 g and a velocity of 68 km s^{-1} can penetrate an aluminium sheet 8 cm thick. Protection of the spacecraft from high-velocity dust particle impacts was thus essential for the spacecraft to approach to within a few hundred kilometres of the nucleus where the dust fluxes are high. Giotto carries a dual-sheet bumper shield, consisting of a thin front sheet (1 mm Al) and a thick rear sheet (7.5 mm epoxy Kevlar, 5 mm polyurethane foam, 2 mm epoxy Kevlar) separated by 23 cm (Fig. 1, bottom). A dust particle striking the thin front sheet is completely vaporized. The vapour cloud then expands into the empty space between the two sheets and strikes the rear sheet, where its energy is distributed over a large area. It had been calculated that the shield which weighs only 50 kg can withstand 70 km s^{-1} impacts of dust particles up to 0.1 g in mass or even higher.

At the other end, well protected from dust particle impacts, Giotto carries a high-gain antenna with an effective reflector diameter of 1.47 m. This antenna can be operated in either the S-band (receiving at 2.1 GHz, transmitting at 2.3 GHz) or the X-band (transmitting at 8.4 GHz). The antenna beam is inclined 44°.3 with respect to the spacecraft spin axis, and the antenna itself is despun, so that the beam could point continuously at the Earth during the encounter. The pointing requirements in the X-band are rather stringent: if the antenna beam were to be misaligned by more than 1° the telemetry link to the ground receiving station would be lost. It was calculated before the mission that the impact of a 0.1 g dust particle on the rim of the bumper shield could change the spacecraft attitude and hence the antenna pointing direction by more than 1°.

The solar cell array provided 196 W of power during the encounter which was not quite sufficient to power the spacecraft and all experiments. Giotto, therefore, carries four Si-Cd batteries in addition.

Giotto's scientific payload comprises 10 hardware experiments (Table 1). All experiments are mounted on the "experiment platform" of the spacecraft which is on top of the rear sheet of the bumper shield (Fig. 2). The experiment sensors protrude up to 17 cm (this limit was imposed by the fairing of the Ariane 1 launch vehicle) from the spacecraft side wall to allow measurements in the undisturbed flow of cometary particles. The sensors of the Dust Impact Detection System (DID) are mounted on the outer face of the front bumper shield, the two magnetometer sensors (MAG) are mounted on the carbon fibre tripod as far away from the spacecraft's magnetic field sources as possible, and the Optical Probe Experiment (OPE) is mounted on the upper platform inside the spacecraft, looking rearward. All other experiments look in the forward direction (downward in Fig. 1), apart from some plasma experiments, which look to the side to observe the changing characteristics of the solar wind plasma. The camera can be rotated by 180°. This, together with the spacecraft spin allows viewing in all directions.

Giotto has no onboard data storage capability, all data are transmitted in real time. Two experiments (MAG and EPA;

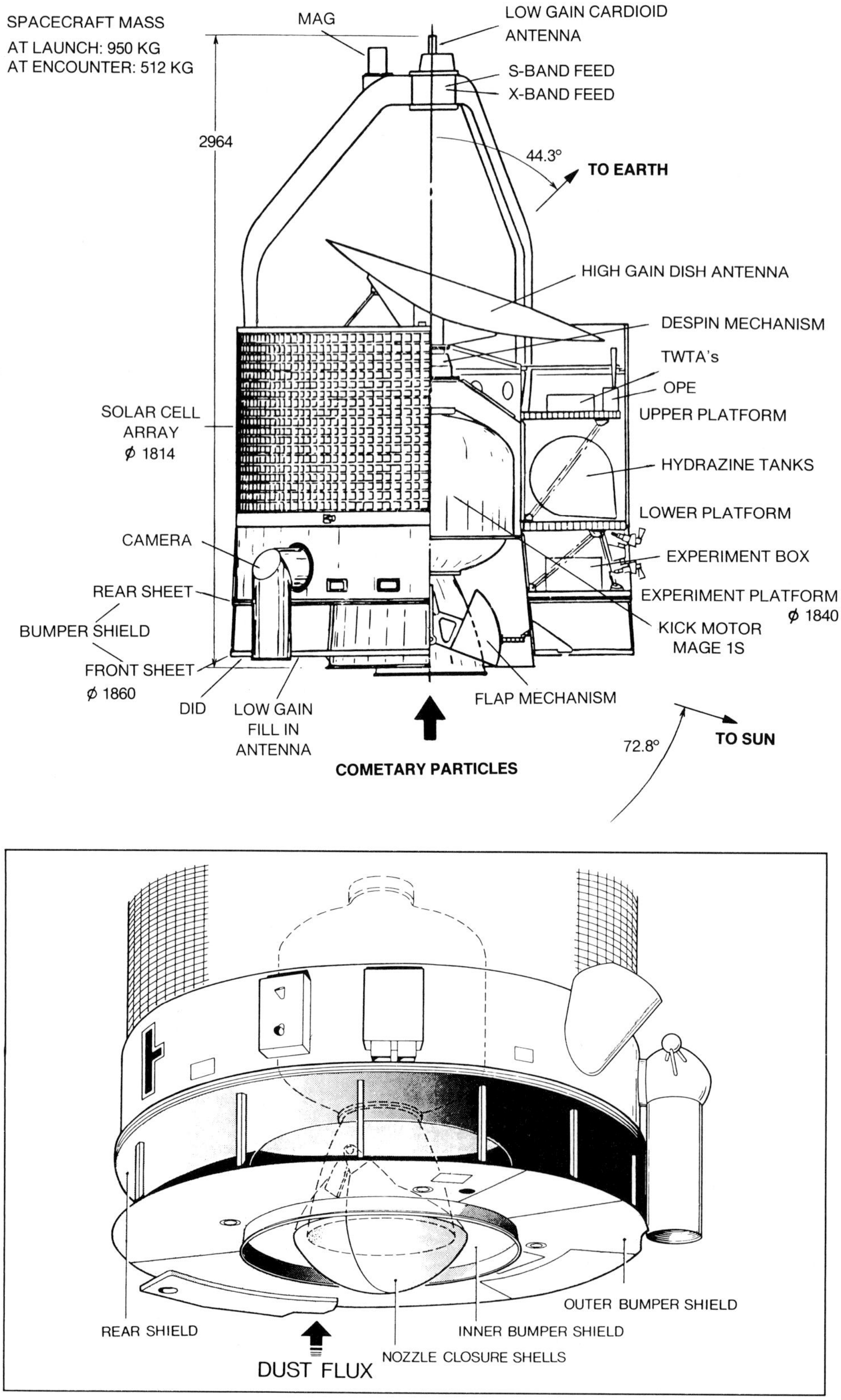

Fig. 1. *Top:* Cross-section through the Giotto spacecraft. *Bottom:* Giotto's dust impact protection system, with thin front sheet, thick rear sheet, and nozzle-closure shells

Table 1. Giotto scientific payload

Experiment (acronym)	Mass [kg]	Data rate (bit/s)			Goal and instrument parameters
		F1	F2	F3	
Camera (HMC)	13.51	20058	20058	723	Inner coma and nucleus imaging, CCD[a] narrow-angle camera, 11 m resolution from 500 km
Neutral mass spectrometer (NMS) Mass analyser Energy analyser	12.70	4156	4156		Neutral gas composition Mass analyser: 1–36 AMU Energy analyser: 1–57, 9–89 AMU
Ion mass spectrometer (IMS) High-energy-range spectrometer (HERS) High-intensity-range spectrometer (HIS)	9.00	3253	3253	1084	Ion composition HERS: 1–35 AMU/q HIS: 12–57 AMU/q
Dust mass spectrometer (PIA)	9.89	2891	5782		Dust particle flux and composition, 1–110 AMU, 3 10^{-16}–5 10^{-11} g
Dust impact detection system (DID) Meteoroid shield momentum sensor (MSM/RSM) Impact plasma and momentum sensor (IPM) Capacitor impact sensor (CIS)	2.26	361	903		Dust particle flux and mass distribution IPM: 6 10^{-17}–6 10^{-11} g CIS: $>10^{-10}$ g MSM/RSM: 10^{-10}–10^{-1} g
Plasma analysis 1 (JPA) Fast ion sensor (FIS) Implanted ion sensor (IIS)	4.70	3975	1265	1355	FIS: 3-dimensional ion velocity distributions, 10 eV–20 keV IIS: ion flux, mass and velocity distributions, 90 eV–90 keV, 1–45 AMU/q
Plasma analysis 2 (RPA) Electron electrostatic analyser (EESA) Positive ion cluster composition analyser (PICCA)	3.21	2530	1807	904	EESA: 3-dimensional electron velocity distributions, 10 eV–30 keV PICCA: composition of cold ions, 10–50 AMU, 50–203 AMU
Energetic particle analyser (EPA)	0.95	181	181	181	Energy and flux of electrons and accelerated ions, $\geqq 20$ keV
Magnetometer (MAG)	1.36	1265	1265	407	Magnetic field, 0.004–65536 nT, up to 25.4 vectors/s
Optical probe experiment (OPE)	1.32	723	723		Coma brightness in 4 continuum bands (dust), 4 discrete emissions (OH, CN, CO^+, C_2)
Radio-science experiment (GRE)					Dust and gas column densities in the coma
Total	58.90	39393	39393	4654	

[a] Charge-coupled device

acronyms are defined in Table 1) have small memories to bridge gaps in ground-station contact during the cruise phase.

As the spacecraft traversed the coma it was decelerated by dust and gas impacts (atmospheric drag); a radio science experiment was designed to measure this deceleration by observing the Doppler frequency shift of the Giotto radio signal, and to derive from it the dust column density along the Giotto trajectory. Detailed experiment, spacecraft and mission descriptions can be found in the ESA Special Publication No. 1077 (1986), in Lo Galbo (1984) and Reinhard (1987a).

The encounter

Giotto was launched on 2 July 1985 by an Ariane 1 from Centre Spatial Guyanais (CSG) in Kourou (Fig. 3). Eight months later, on 13/14 March 1986, Giotto encountered Halley's Comet. The interplanetary trajectory (Fig. 4) was in the ecliptic with a closest approach to the Sun of 0.72 AU, reached on 31 December 1985.

For the routine operation of the spacecraft in the S-band, ESA's 15 m ground station at Carnarvon, Australia, was used, while the 64 m ground station at Parkes, Australia was used for

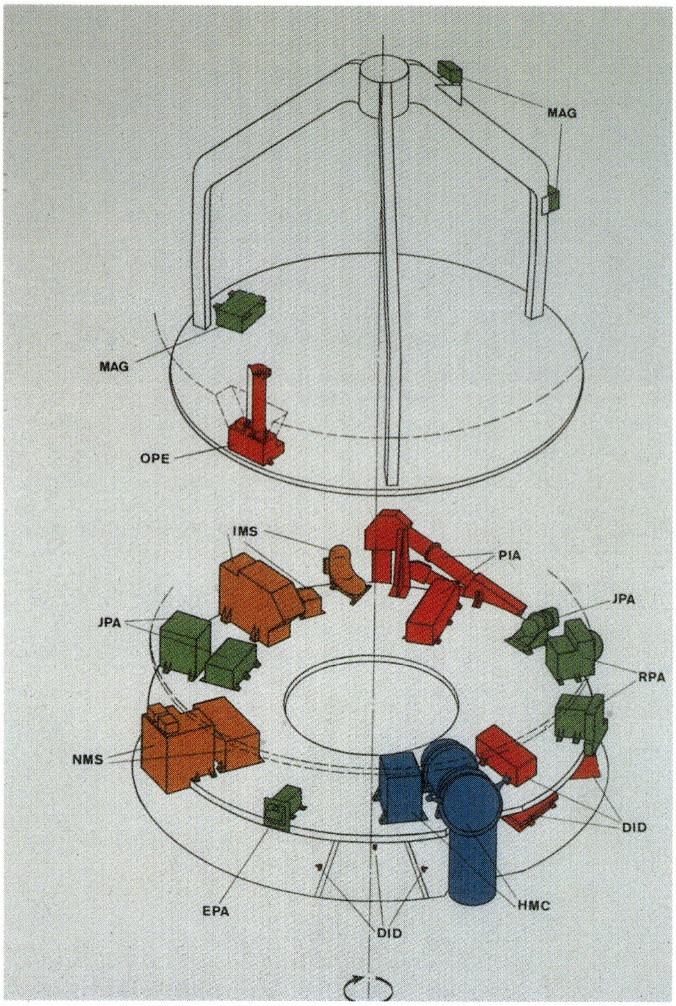

Fig. 2. Positioning of Giotto's experiments on the spacecraft structure (experiment acronyms explained in Table 1). They can be grouped in four categories: imaging (blue), gas mass spectroscopy (orange), dust experiments (red), and plasma experiments (green)

high-rate transmission of scientific data (Format 1 or 2, Table 1). The Parkes antenna is owned and operated by the Commonwealth Scientific and Industrial Research Organisation (CSIRO), and is normally used for radioastronomy. During the cruise phase, a few times per week scientific data were transmitted to Carnarvon and the 30 m station at Weilheim, W. Germany at low rate (Format 3). During the encounter, continuous data coverage for 70 h before and 30 h after closest approach in high-data rate mode was provided by the Parkes ground station together with the NASA 64 m Deep Space Network (ground stations at Goldstone. Madrid, Canberra). To save power, not all the experiments were operated throughout that period. Also, during orbit and attitude manoeuvres, the payload was switched off (except for MAG and EPA which stayed on continuously). The periods for which encounter science data are available are given for each experiment in Table 2.

The geometry of the Giotto encounter with Comet Halley is shown in Fig. 5, together with the encounter geometries of the other five Halley flyby spacecraft. The phase angles and flyby velocities were all very similar. Giotto was the last spacecraft to encounter Comet Halley and the closest one.

Giotto was targeted to fly past the nucleus at a closest distance of 540 km on the sunward side. This distance was chosen as a compromise between the partially conflicting requirements of three groups of experiments. The camera experimenters wanted to fly past ideally at 1000 km but no closer than 500 km because at distances <500 km the camera could not rotate fast enough to follow the apparent motion of the nucleus during the flyby. A second group of experimenters (NMS, IMS, OPE, MAG) wanted to approach as close as possible to the nucleus, even if the spacecraft would not survive, and a third group (PIA, DID, JPA, RPA) also wanted to approach as close as possible, while maintaining a high survival probability. The distance of 500 km was acceptable to all three groups, however the combined 1σ uncertainty of the spacecraft and Halley nucleus positions of 40 km had to be added to this distance.

Targeting the spacecraft to this preselected point was a difficult task, as the nucleus is too small to be observed from the Earth even by the largest telescopes, and, moreover, is masked by the coma dust and gas. Also, due to the asymmetric outgassing – the dust and gas is mostly emitted into the sunward hemisphere – the nucleus is accelerated away from the Sun (non-gravitational forces). Compared with the purely Keplerian orbit that it follows at large heliocentric distances, Halley's orbit is extended by four days. This outgassing effet is irregular and therefore difficult to model. Also, because of the asymmetric outgassing, there is an offset between the centre of light (assumed to be the comet nucleus position in the astrometric observations) and the centre of mass (actual nucleus position). This offset can be considered either as a bias towards the Sun or as an uncertainty in the radial direction. Between Halley's recovery on 16 October 1982 and mid January 1986, 4000 astrometric observations were made of Halley. They were coordinated via the Astrometry Net of the International Halley Watch. From these observations, in combination with a model which includes the effects of non-gravitational perturbations, the nucleus position could be determined for the time of the Giotto encounter with a 3σ uncertainty of 750 km by 220 km in the Giotto target plane and 4700 km downtrack. These uncertainties were calculated assuming a $1''$ observational error in right ascension and declination and, in addition, a 1000 km uncertainty in the radial direction. The downtrack error is of little importance as it reflects only the uncertainty in arrival time (± 68 s). To these values has to be added the Giotto spacecraft position uncertainty of about hundred kilometres (3σ) to give the overall targeting uncertainty. Compared with the intended flyby distance of 540 km this is relatively large. Fortunately, Giotto was the last of the spacecraft to encounter the comet (Fig. 5), and the earlier arriving Vega 1 and Vega 2 spacecraft, having located the nucleus, could pass this information on to Giotto. This was the so-called "pathfinder concept", the principle of which is illustrated in Fig. 6.

The uncertainty in spacecraft position (30–50 km) is shown in Fig. 6 as a small circle around the Giotto path, at the time when the last orbit correction manoeuvre was made. The large circle around the path of the comet reflects the relatively large 1σ uncertainty that could be achieved via the Astrometry Net. Vega located the comet nucleus during its flyby on 6 March 1986 with an uncertainty given by the angular uncertainties in spacecraft attitude and pointing direction of the platform on which the camera was mounted. The Vega position uncertainty also contributed to the error. Using only conventional (6 GHz) ranging and Doppler, Soviet experts expected a geocentric Vega position uncertainty of several hundred kilometres. It was estimated that the Vega position uncertainty could be reduced to ~ 40 km by

Fig. 3. Left: The Giotto spacecraft mounted on top of the Ariane 1 launch vehicle. The camera (HMC) is seen on the left side of the spacecraft, looking downward (this is the viewing direction during the approach to the nucleus), the dust mass spectrometer (PIA) is seen on the right side of the spacecraft, enclosed by black multilayer insulation. In the background, one half of the Ariane fairing is visible. *Right:* Lift-off of Ariane Flight 14 carrying Giotto at 11:23 UT on 2 July 1985

using Very Long Baseline Interferometry (VLBI) techniques. This was achieved by NASA using the widely separated tracking stations of its Deep Space Network and precise L-band VLBI. After processing all the data, the position of the comet nucleus was known with much better accuracy, represented by the small circle around the Halley path. Giotto was targeted to the centre of the small circle projected to the intersection of the comet and the spacecraft paths. Between the time of the nucleus detection by Vega and the Giotto encounter, the uncertainty grew slightly due to the non-gravitational forces, which cannot be modelled precisely. In fact, both Vega 1 and 2 were used as pathfinders (Fig. 6 shows one Vega for simplicity). The pathfinder concept reduced the targeting uncertainty to 40 km (Fig. 7).

The flyby distance that was actually achieved was 596 ± 2 km, as derived by Curdt et al. (1988) from camera observations (variation of the camera offset angle with time). All relevant encounter parameters are summarised in Table 3. The time of closest approach was determined as 00:03:01.84 (± 0.20 s) UT on 14 March. This time is given as "spacecraft event time" (SCET); however, some experimenters prefer to use "ground station received time" (GSRT). The two are related by GSRT = SCET + 8 min 0.1 s.

At 00:10:54.3 UT (GSRT), i.e. 7.6 s before closest approach, Giotto was hit by a "large" dust particle. The impact caused the spacecraft angular momentum vector to shift by 1° and the spacecraft performed a nutation around the new axis with a period

Table 2. Periods of encounter science data availability after the last orbit correction manoeuvre (completed 12 Mar, 05:00); (Intervals of intermittent data transmission due to spacecraft nutation around closest approach are omitted)

Exp		Data available		Damage assessment
		From (SCET)[a]	Until (SCET)[a]	
HMC		12 Mar, 20:38 13 Mar, 06:19 13 Mar, 08:30 13 Mar, 19:44	13 Mar, 00:33 13 Mar, 06:41 13 Mar, 09:46 14 Mar, 00:02:52.8	Mirror degradation Baffle loss
NMS		13 Mar, 20:18	14 Mar, 00:02:53.4	CCDs of both detectors dead
IMS	HERS	12 Mar, 05:15 13 Mar, 05:00	13 Mar, 03:57 14 Mar, 00:02:41	HV damage
	HIS	13 Mar, 05:07	14 Mar, 02:21	No damage
PIA		13 Mar, 19:23	14 Mar, 03:49:42	Mass spectrum channel: 30% damage
DID		13 Mar, 19:25:17	14 Mar, 03:50:11	MSM: no damage CIS: indeterminate IPM: shows anomalous behaviour
JPA	IIS	12 Mar, 05:16:52 13 Mar, 04:54:14 14 Mar, 00:33:03	13 Mar, 04:00:40 14 Mar, 00:02:59 15 Mar, 02:33:53	No damage
	FIS	12 Mar, 05:15:48 13 Mar, 04:54:06	13 Mar, 04:00:59 14 Mar, 01:15:52	HV drop of 400 V
RPA	EESA	12 Mar, 05:35 13 Mar, 04:53	13 Mar, 03:51 14 Mar, 00:02:29	Partial damage
	PICCA	13 Mar, 19:03:09	14 Mar, 00:02:41	HV damage
EPA		<12 Mar, 05:00 (continuous data available)	15 Mar, 02:47:39	No damage
MAG		<12 Mar, 05:00 (continuous data available)	15 Mar, 02:51:26	No damage
OPE		13 Mar, 20:06	14 Mar, 03:31	No damage

Format 2 from 13 Mar, 22:38:30 until 14 Mar, 01:04:00; other times Format 1

[a] SCET: spacecraft event time

of 3.2 s and an amplitude of 1° (Fig. 8). Thus the maximum deviation from the desired attitude was 2° and, as had been expected in such an event, the telecommunications link to Earth could no longer be maintained.

The effect of the dust impact is clearly seen in the strength of the Giotto radio signal received at the Parkes ground station around the time of closest approach (Fig. 9). The signal strength dropped by almost 40 dB at 00:10:54.3 and recovered within 0.6 s. This short-duration loss of signal is due to the switch-over from travelling wave tube 1 to travelling wave tube 2, which is thought to be caused by a large amplitude electrical transient resulting from the formation of a plasma cloud after the impact of the "large" dust particle. About one second later the signal strength dropped again, this time caused by the change in spacecraft attitude due to the dust impact induced nutation. At about the same time the spin period increased from 3.998 s to 4.010 s. This change in spin period caused a 1° misalignment of the antenna beam, in addition to the misalignment due to nutation. The signal level thus decreased rapidly and stayed low for the next 21.75 s. By that time the antenna despin motor was adjusted via the onboard attitude control system to the new phase and spin rate. After 00:11:19 the signal strength fluctuated with a period of 3.2 s, which is the spacecraft inertial nutation period. This is the period with which the despun antenna dish nutates. Superimposed on the 3.2 s period in the received radio signal strength is a 16.4 s period which is the onboard nutation period. This is the period with which all components, such as the Sun sensor, which are fixed on the spinning spacecraft nutate. The sun sensor provides a see-Sun pulse which is used to control the antenna despin motor (closed loop). If the see-Sun pulse is slightly delayed or advanced due to the nutation of the Sun sensor, the antenna is not precisely despun and does not point precisely at the Earth. In this way the

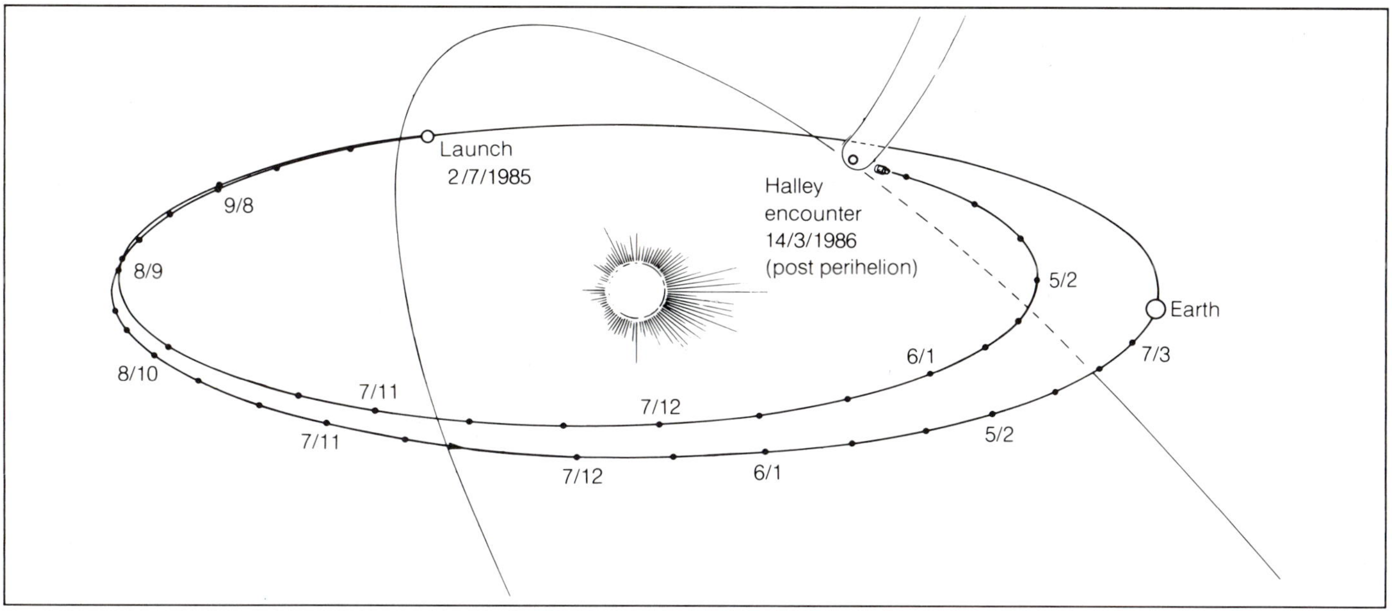

Fig. 4. Interplanetary trajectory for Giotto from launch on 2 July 1985 until post-perihelion encounter with Halley on 13/14 March 1986. Giotto's orbit is the inner circle and lies completely in the ecliptic. Halley's orbit is retrograde and inclined at 18° with respect to the ecliptic

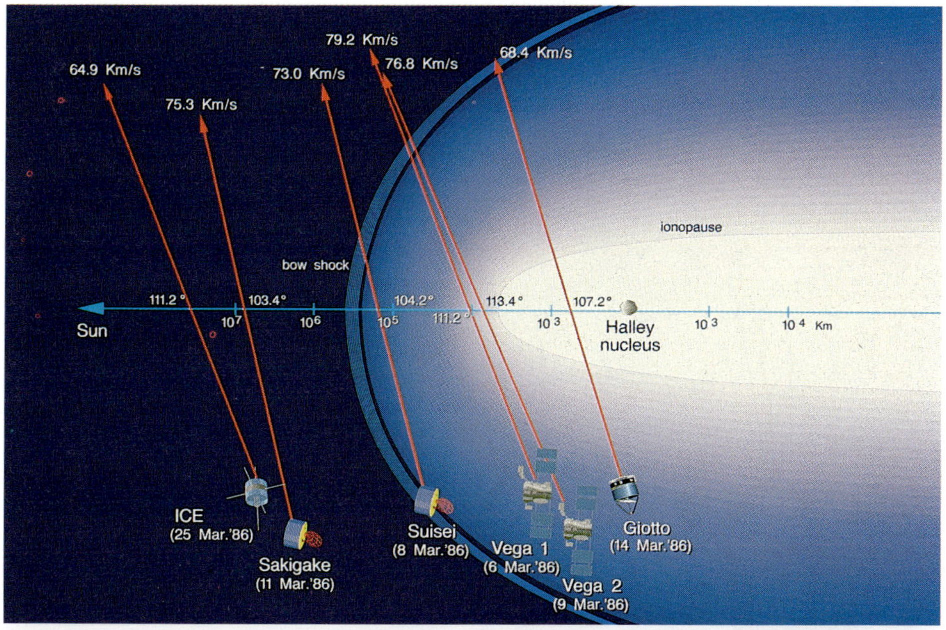

Fig. 5. Flyby geometry of the six Halley spacecraft at encounter. The Sun is to the left; the distance scale is logarithmic. For each mission the flyby dates are given at the bottom, flyby phase angles in the centre and flyby speeds at the top

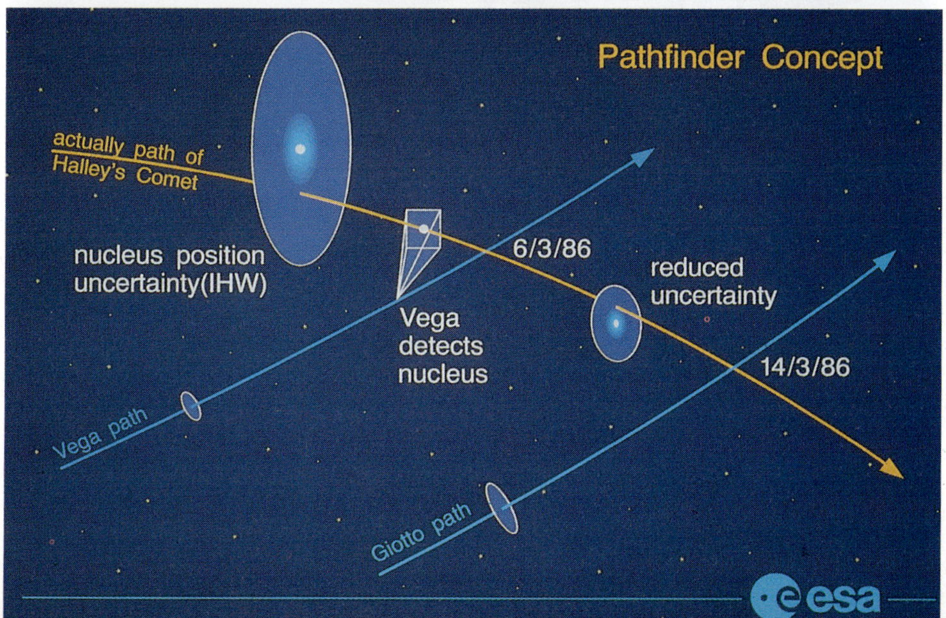

Fig. 6. Principle of the "pathfinder concept" (courtesy J.F. Jordan)

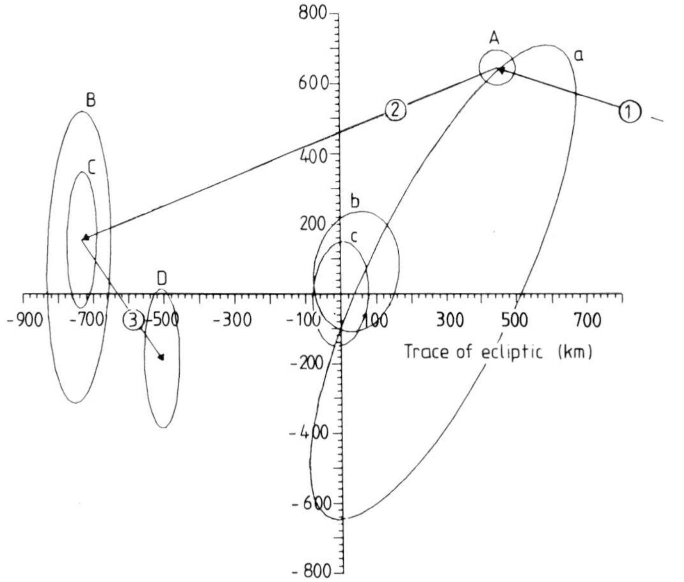

Fig. 7. 3σ uncertainty ellipses for Comet Halley and Giotto. The ellipses shown are projections of the three-dimensional position uncertainty ellipsoids onto the Giotto target plane; the x axis is the trace of the ecliptic, the y axis points north. Ellipses a–c refer to the comet location: a, from ground-based observations; b, from Vega 1 observations; c, from Vega 1 and Vega 2 observations. Ellipses A–D refer to the Giotto position at different times: A, on 27 August 1985; B, on 5 March 1986; C, on 12 March 1986; D, on 14 March 1986 – projected forward to the time of closest approach. The lines identified by circled numbers show the effect of the three orbit correction manoeuvres performed: 1, on 26 August 1985; 2, on 12 February 1986; 3, on 12 March 1986 – the last one being based on pathfinder results (courtesy R. Münch)

Table 3. Parameters at closest approach

Distance to comet nucleus	596 ± 2 km
Distance to Earth	$1.44\ 10^8$ km
Distance to Sun	$1.35\ 10^8$ km
Distance below the ecliptic	$3\ 10^6$ km
Time of closest approach (SCET) [a]	$00{:}03{:}01.84 \pm 0.20$ s
Phase angle (angle between the spacecraft velocity vector in the comet frame of reference and the Sun-comet line)	107°05
Angle between the comet-spacecraft line and the plane defined by the Sun-comet line and the spacecraft velocity vector in the comet frame of reference	30°6 ± 0°1 (south)
Spacecraft velocity in the comet frame of reference	68.373 km s^{-1}
Time of large dust impact (SCET) [a]	$00{:}02{:}54.2 \pm 0.1$ s
Spacecraft spin period	
– before dust impact	3.998 s
– after dust impact	4.010 s

[a] SCET: spacecraft event time

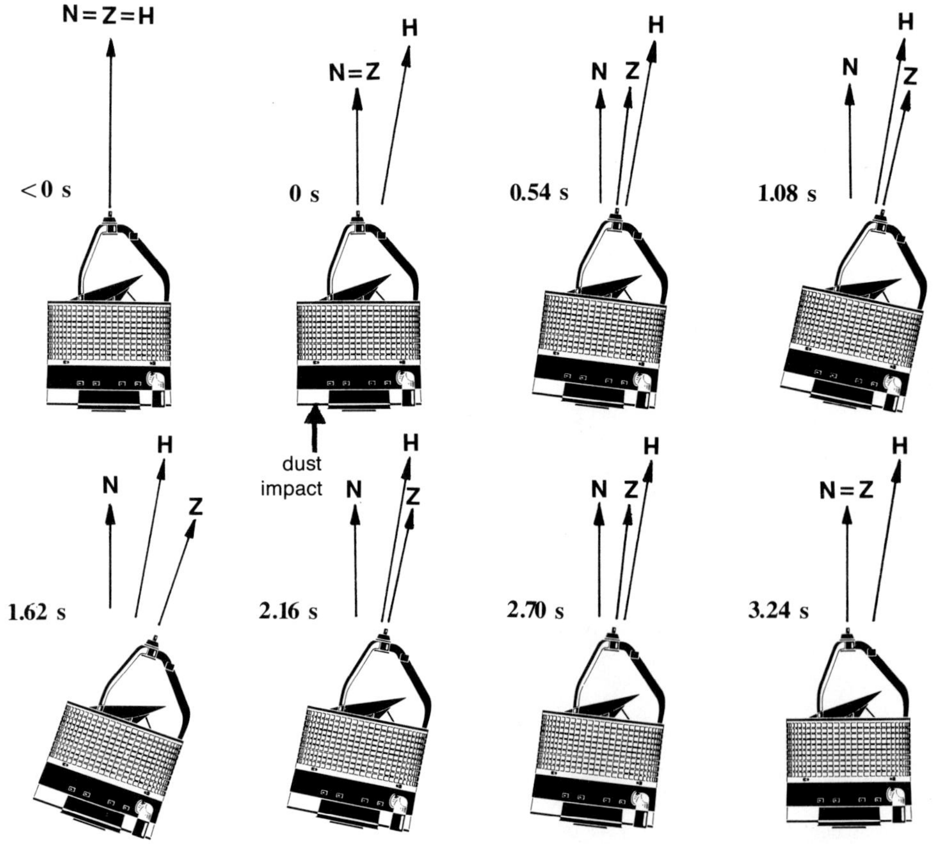

Fig. 8. Giotto attitude perturbation after dust particle impact, 7.6 s before closest approach. N is the undisturbed spin axis inertial direction, Z the spacecraft spin axis and H the spacecraft angular momentum. The effect is shown a factor 10 exaggerated, the actual angle between N and H was only 1°

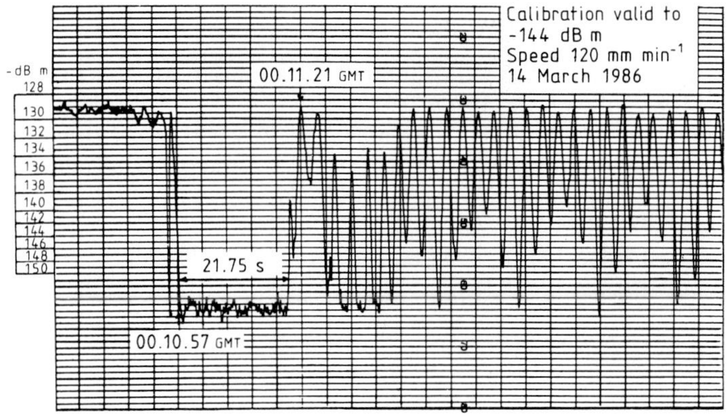

Fig. 9. Giotto rado signal strength recording received at the Parkes ground station from 00:10:32 to 00:12:47 (GSRT). Closest approach was at 00:11:01.94. The vertical lines are 5 s apart (courtesy D. Wilkins)

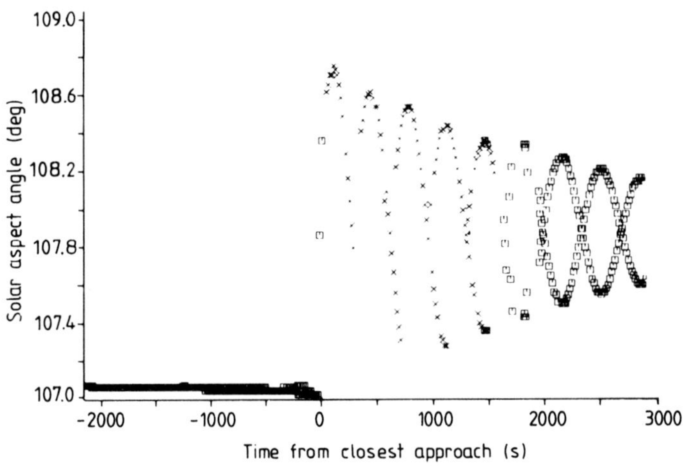

Fig. 10. Solar aspect angle (angle between Giotto's spin axis and the direction to the Sun) as a function of time around closest approach (one tick corresponds to 500 s). Data represented by squares have been obtained from standard computer processing requiring a stable telemetry link, data shown as crosses were reconstructed from raw telemetry. Due to dust particle impact 7.6 s before closest approach the spacecraft started to nutate with an amplitude of 1°. The actual oscillation is between the upper and lower envelopes and is much more frequent (16.4 s period) than is apparent. The seemingly lower frequency is a beat phenomenon caused by sampling of the data points. The nutation was damped and stable telemetry was recovered after 1800 s (courtesy T. Morley and J. Fertig)

16.4 s Sun sensor nutation is reflected in the received radio signal strength.

For the next 32 min scientific data were received only intermittently. By that time the onboard nutation dampers (two 60 cm long tubes filled with a viscous liquid) had reduced the maximum deviation from the Earth-pointing direction to $\leq 1°$ and continuous data were again received (Fig. 10).

From the Doppler frequency shift at the time of the "large" dust particle impact the spacecraft deceleration was found to be 4.1 cm s^{-1}. This deceleration occurred over a period of 0.6 s, corresponding to a spatial extent of 40 km. It is concluded (Edenhofer et al., 1987) that at that time Giotto traversed a region with a high dust concentration (dust "jet"). From the deceleration, assuming an average momentum enhancement factor $\varepsilon = 1$ as for elastic collision, the total mass of the impacting dust particles in this dust jet is found to be ~ 0.17 g. Values for ε are not very well known at high impact velocities; they are of the order of 10 or more for small particles and approach the inelastic case ($\varepsilon = 0$) for particles $> 10^{-6}$ g which penetrate the front shield. The total impacting mass of ~ 0.17 g also sets the upper limit of the mass of the impacting single "large" dust particle in this jet which caused the spacecraft nutation. A dust particle of that mass can cause the observed nutation amplitude of 1° only if it impacted on the outer edge of the spacecraft bumper shield or on one of the spacecraft or experiment sensors protruding beyond the spacecraft side wall.

A sudden increase in the spin period is indicative of a dust particle impact or dust particle impacts on an oblique surface which is inclined opposite to the spin direction. Until 9 s before closest approach the camera was still operational. By that time the camera rotation, required to follow the apparent motion of the comet nucleus during the flyby, had reached 40°. It was at that time the only large oblique area in the forward direction and it was inclined opposite to the spin direction. Presumably, the spacecraft spin was slowed by the impacts of a large number of dust particles on the oblique baffle around the time of closest approach.

A few hours after closest approach the camera was rotated twice to 150°. In this position the camera baffle should have cast a shadow on the solar cell array, causing a decrease in the solar cell array current. In contrast to similar pre-encounter operations such a decrease was, however, not observed and it must be concluded that the camera baffle or at least a major part of it disappeared during the encounter. Analysis of the effects of post-encounter rotations of the camera, which is no longer properly balanced, on the dynamics of the spacecraft body (small nutations and spin-rate changes) also led to the conclusion that the camera lost its baffle during the Halley flyby.

Since the encounter the spacecraft has a permanent wobble of 0°.2, i.e. it is no longer properly balanced. This is explained by a loss of a few hundred grammes of material from the periphery. The loss of the camera baffle (330 g) could account for this effect.

It must be assumed that the numerous dust particles that impacted on the camera baffle while it was in its oblique position not only slowed the spacecraft spin-rate, but also gradually eroded the baffle itself.

The last camera image that still shows details and was completely transmitted was taken at a distance of 1930 km from the nucleus.

About half of the experiments worked flawlessly throughout the encounter, while the other half suffered damage due to dust impacts (Table 2). The last experiment was switched off $26\frac{1}{2}$ h after closest approach.

The spacecraft also suffered some damage during the encounter but it was possible to redirect it to the Earth before it was put in a hibernation configuration on 2 April 1986. This manoeuvre was carried out in 3 steps on 19/20 March using ~ 35 kg of hydrazine. Presently 25.45 kg (corresponding to ~ 85 m s^{-1} in ΔV) of hydrazine are left. Giotto will swing by the Earth on 2 July 1990 at a distance of 22,000 km. Using the assistance of the Earth's gravity it would be possible to encounter another comet. The short-period comet Grigg-Skjellerup has been identified as the most suitable target.

It is an "old", relative inactive comet, which is favourable for imaging, and would also provided an interesting contrast to Halley. Its orbit is very well known; the flyby would occur on 10 July 1992. Details of the Giotto extended mission are described in Reinhard (1987b).

It is presently planned to switch-on and check-out the spacecraft and the experiments in early 1990. The Giotto mission extension will then be decided upon, weighing spacecraft and experiment performance against mission cost.

Acknowledgement. It is a pleasure to thank the 11 Giotto Principal Investigators and their experiment teams, the ESTEC Giotto Project team and the ESOC Giotto operations and flight dynamics team for many discussions and for the close collaboration through all phases of the mission.

References

Curdt, W., Wilhelm, K., Craubner, A., Krahn, E., Keller, H.U.: 1988, *Astron. Astrophys.* **191**, L1

Edenhofer, P., Bird, M.K., Brenkle, J.P., Buschert, H., Kursinski, E.R., Mottinger, N.A., Porsche, H., Stelzried, C.T., Volland, H.: 1987, *Astron. Astrophys.* **187**, 712

Lo Galbo, P.: 1984, *ESA J.* **8**, 215

Reinhard, R.: 1987a, *J. Phys. E: Sci. Instrum.* **20**, 700

Reinhard, R.: 1987b, ESA SP-278, 523

EXPLORATION OF HALLEY's COMET

M. Grewing · F. Praderie
R. Reinhard (Eds.)

The Vega mission to Halley's Comet

R. Z. Sagdeev

Space Research Institute, Academy of Sciences of the USSR, Moscow, USSR

1. Introduction

A unique opportunity to combine missions to Venus and comet Halley was available in 1985/86 by employing a two-element space vehicle consisting of a Venus lander and a Halley flyby probe. The mission was called Vega, a contraction of the Russian words Venera (Venus) and Gallei (Halley) and was conducted by the USSR in cooperation with a number of other countries.

Two spacecraft, Vega-1 and Vega-2, have been launched by Proton rockets from the cosmoport Baykonur on 15 and 21 December 1984, respectively. The two spacecraft are identical and the redundancy is aimed at increasing the overall reliability of the mission. The basic design of the spacecraft chosen for this mission is the same as has been used many times to deliver Soviet landers and orbiters to Venus. On 11 and 15 June 1985, the two Vega spacecraft successfully delivered the first balloons into the Venus atmosphere (each carrying 4 scientific experiments) as well as delivering landers with 9 experiments to the surface of the planet. The results of this part of the mission are published in Sagdeev et al. (1986a–c). The flyby modules made a gravitational manoeuvre near Venus, and finally Vega 1 and Vega 2 encountered comet Halley on 6 and 9 March 1986.

The Vega project was an international one. Although the spacecraft themselves were controlled by the Soviet Union, the scientific programme and payload were coordinated by the International Science and Technical Committee (CIST), representing scientific institutions from nine countries. The CIST designed the Vega mission to be complementary to the European Giotto and the Japanese Suisei cometary missions.

2. The spacecraft

The Vega spacecraft was composed of the Halley flyby probe and a Venus descent module; the whole system weighed about 4.5 t at launch. The Halley probe is shown in Fig. 1 in its nominal Halley flyby configuration. On its trajectory the probe was still surmounted by the Venus descent module (not shown in Fig. 1) which was a spherical object with a diameter of 2.5 m and a mass of approximately 2 t. The spacecraft is 3-axis stabilised with a wingspan of the order of 10 m. It carries 14 experiments (Table 1), among them a TV system for imaging the inner coma and comet nucleus.

Successful imaging of the comet from the spacecraft required a steerable platform which could be automatically pointed with great accuracy at the cometary nucleus. The most difficult problem was to locate the nucleus with its very low albedo, against the background of the very bright coma with gas and dust jets. The accurate pointing achieved allowed the examination of the innermost parts of the coma and near-nucleus region by means of television cameras and infrared and optical spectrometers placed on the steerable platform. The platform is shown in the lower part of Fig. 1, it has a mass of 82 kg and carries 64 kg of instrumentation. It can scan an angular sector of 110° in the ecliptic plane and 60° in a plane perpendicular to the ecliptic. The pointing platform could track the nucleus with an accuracy of the order of 2.′5.

The Vega spacecraft (Fig. 1) is derived from the Venera series of spacecraft. A number of modifications improve the reliability of the probe, for example, 5 m^2 of shield have been added in order to protect the most essential subsystems against the bombardment of dust particles with masses ≤ 0.01 g. A dual-sheet bumper shield has been adopted, composed of a thin metallic front sheet (0.4 mm) and a thicker rear sheet, separated by several centimeters.

The spacecraft structure resembles a cylindrical body connected to two conical skirts. The lower skirt houses a motor for orbital manoeuvres and a toroidal pressurised utility instrument bay. The cylindrical compartment contains the fuel tanks and the upper skirt is the interface that held the Venus lander. Two pairs of deployable solar panels are mounted on each side of the cylindrical section; the solar array has a total area of nearly 10 m^2. Three-axis stabilisation with 1° accuracy in flight is achieved by a gyroscopic system and a number of gas nozzles, most of which are mounted on the solar panels.

The telemetry system consists of a high-data-rate channel (BRL) and a low-data-rate channel (BTM). The BRL channel is used for real-time transmission only. Its telemetry rate of 65536 bps can be reduced by half if required by propagation conditions. The telemetry rate of the BTM channel is 3072 bps. Scientific data can also be stored by onboard magnetic tape recorders with 5 Mbit capacity and subsequently telemetered through the BTM channel. This was done every 20 days during the interplanetary cruise phase and once every 20 min during the cometary flyby. The spacecraft orientation must be such that the high gain antenna is pointed at the Earth during high data rate transmission periods.

3. Mission and encounter

The interplanetary orbit of Vega-1 from launch to Venus flyby (Fig. 2, left) and from Venus flyby to Halley flyby (Fig. 2, right) is entirely in the ecliptic. The plane of the Halley orbit is inclined at 18° with respect to the ecliptic, Halley is above the ecliptic from the ascending to the descending node. The orbit for Vega-2 is almost

Table 1. VEGA sientific payload

Acronym	Experiment	Mass [kg]	Power [W]	Direct telemetry [bit/s]	Recorded telemetry [bit/20 min]	Goal and instrument parameters	Collaborating institutes (Principal Investigators)
TVS	Television System	32	50	32768	–	Inner coma and nucleus imaging. Two CCD* cameras (fields of view, $0\overset{\circ}{.}43 \times 0\overset{\circ}{.}57$ and $3\overset{\circ}{.}5 \times 5\overset{\circ}{.}3$)	LAS, Marseille, France (P. Cruvellier) Central Research Institute for Physics, Budapest, Hungary (L. Szabo) IKI, Moscow, USSR (G. Avanesov)
IKS	Infrared Spectro-meter	18	18	2048	4320	Detection of infrared emissions of coma and thermal radiation of nucleus ($2.5 < \lambda < 12\,\mu m$)	Observatoire de Meudon, France (M. Combes) IKI, Moscow, USSR
TKS	Three-Channel Spectro-meter	14	30	12288	–	Spectral mapping of coma emissions in the range $0.12 < \lambda < 1.9\,\mu m$	Observatoire de Besançon, France (G. Moreels) IKI, Moscow, USSR (V. Krasnopolskii) Bulgaria (M. Gogoshev)
PHOTON	Shield Penetration Detector	2	4	108	–	Large dust particle detection (under anti-dust shield)	USSR
DUCMA	Dust Particle Detector	3	2	100	100	Dust particle flux and mass spectrum ($m > 1.5\ 10^{-13}$ g)	University of Chicago, USA (J. Simpson) MPI, Lindau, W. Germany IKI, Moscow, USSR Central Research Institute for Physics, Budapest
SP-2	Dust Particle Detector	4	4	1024	2160	Dust particle flux and mass spectrum ($m > 10^{-16}$ g)	IKI, Leningrad, USSR (E. Mazets)
SP-1	Dust Particle Detector	2	1	150	2160	Dust particle flux and mass spectrum ($m > 10^{-16}$ g)	IKI, Moscow, USSR (O. Vaisberg)
PUMA	Dust Mass Spectro-meter	19	31	10240	–	Dust particle elemental composition	MPI, Heidelberg, W. Germany (J. Kissel) Service d'Aéronomie, Verrières, France (J.-L. Bertaux) IKI, Moscow, USSR (R. Sagdeev)
ING	Neutral Gas Mass Spectro-meter	7	8	1024	1080	Neutral gas composition	MPI, Lindau. W. Germany (E. Keppler) Central Research Institute for Physics, Budapest IKI, Moscow, USSR University of Arizona, USA

Table 1. (continued)

Acronym	Experiment	Mass [kg]	Power [W]	Direct telemetry [bit/s]	Recorded telemetry [bit/20 min]	Goal and instrument parameters	Collaborating institutes (Principal Investigators)
PLASMAG	Plasma Energy Analyser	9	8	2048	15120	Ion flux composition, energy spectra of ions and electrons	IKI, Moscow, USSR (K. Gringauz) Central Research Institute for Physics, Budapest MPI, Lindau, W. Germany ESA Space Science Dept., ESTEC, Netherlands
TUNDE	Energetic Particle Analyser	5	6	512	6480	Energy and flux of accelerated cometary ions	Central Research Institute for Physics, Budapest, Hungary (A. Somogyi) IKI, Moscow, USSR MPI, Lindau, W. Germany ESA Space Science Dept., ESTEC, Netherlands Nuclear Research Institute, Moscow, USSR
MISCHA	Magneto-meter	4	6	512	2160	Magnetic field	Space Research Institute, Graz, Austria (W. Riedler) Izmiran, Troitsk, USSR
APV-N	Wave and Plasma Analyser	5	8	2048	28080	Plasma waves, 0.01–1,000 Hz, plasma ion flux fluctuations	IKI, Moscow, USSR (S. Klimov) Aviation Institute, Warsaw, Poland Geophysical Scientific Institute, Prague
APV-V	Wave and Plasma Analyser	3	2	512	15120	Plasma waves, 0–300 kHz, plasma density and temperature	ESA Space Science Dept., ESTEC (R. Grard) LPCE, Orleans, France Izmiran, Troitsk, USSR

* Charge – coupled device

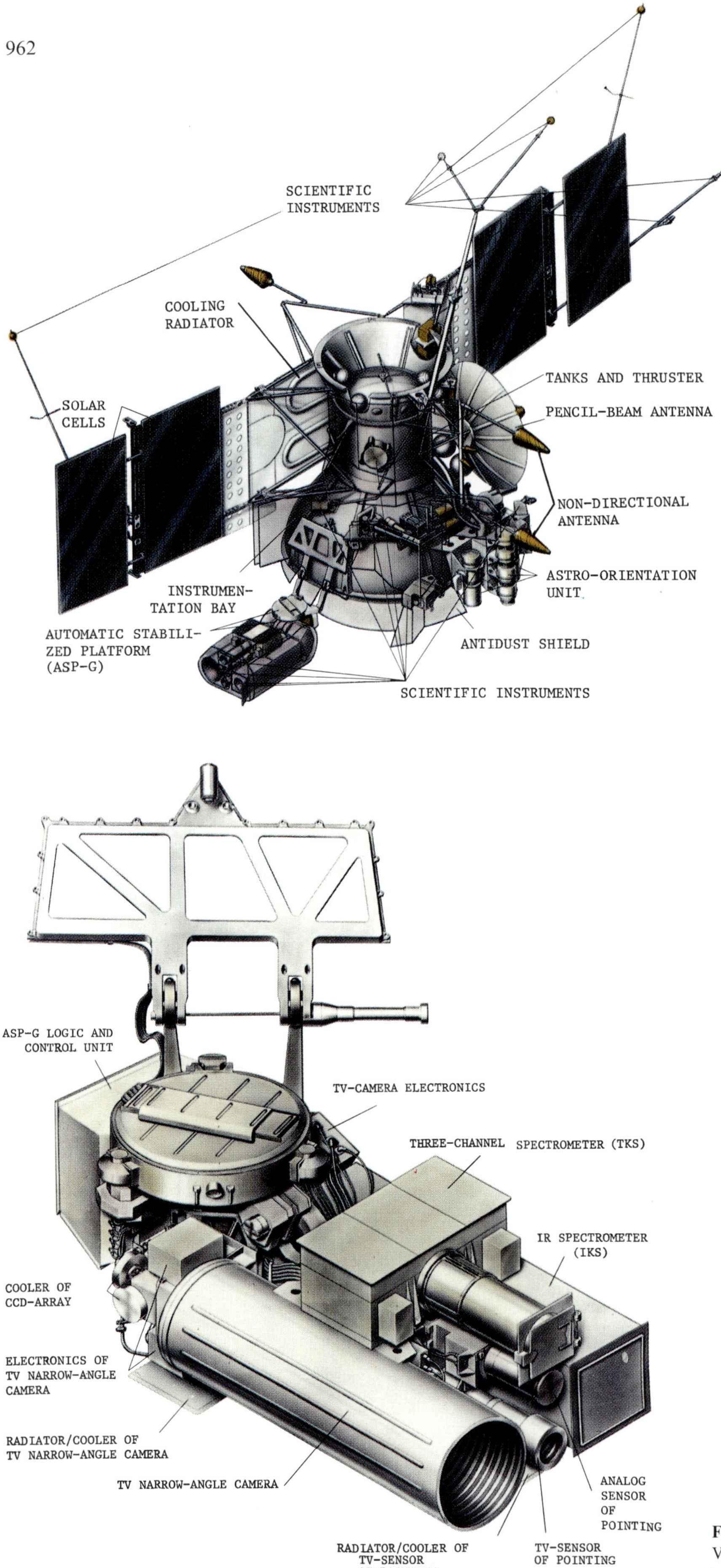

Fig. 1. The Halley probe of the Vega spacecraft (Vega 1 and Vega 2 were identical), showing the locations of scientific instruments

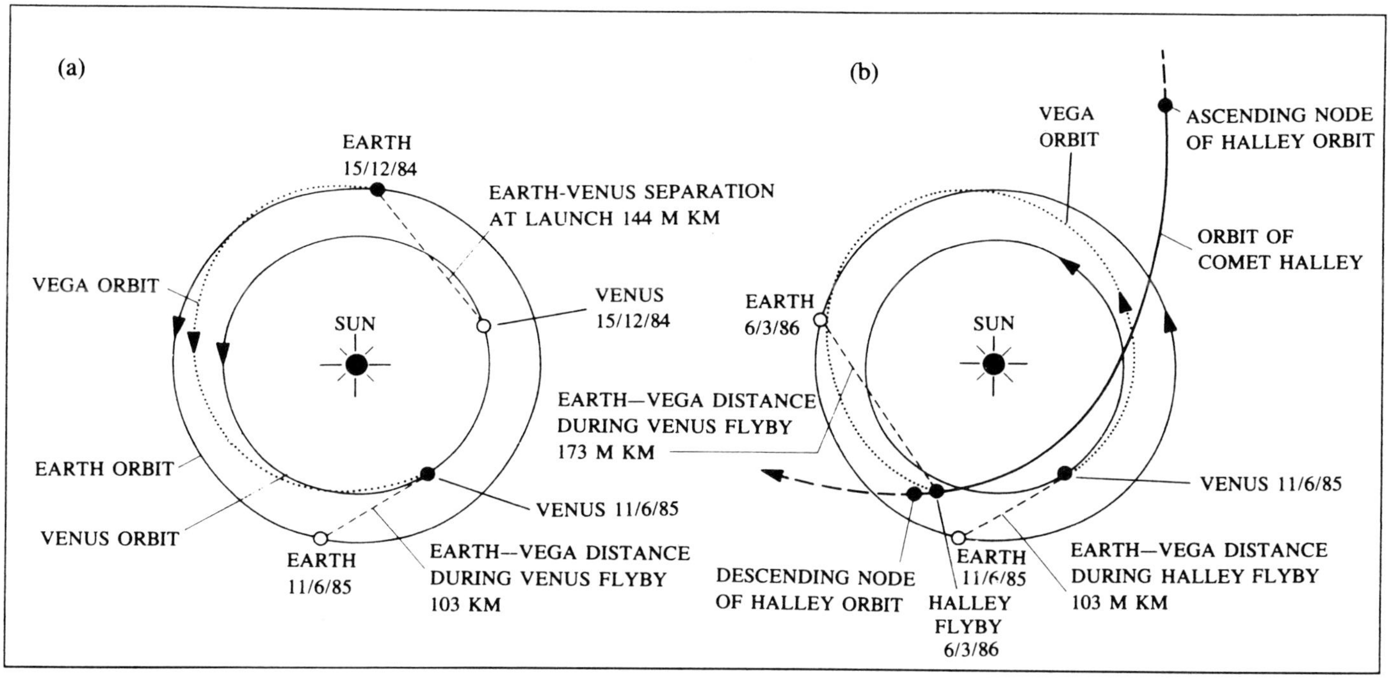

Fig. 2. The path of the Vega probes (a) from Earth to Venus, and (b) from Venus to Halley. The projection of the inner portion of the trajectory of Halley on the plane of the ecliptic is also shown

Table 2. Parameters of the Vega mission flights

	Vega 1	Vega 2
Launch	15 December 1984	21 December 1984
Venus fly-by, descent probe landing, balloon insertion into the Venus atmosphere	11 June 1985	15 June 1985
Passage through the comet Halley coma	7:20:06[a] 6 March 1986	7:20:00[a] 9 March 1986
Nearest distance to the nucleus (km)	8,890	8,030
Encounter velocity (km s^{-1})	79.2	76.8
Distance of the encounter point from the Sun (AU)	0.7923	0.8341

[a] Time of closest approach to the comet nucleus (UT)

identical except for slightly different flyby dates. The Halley flyby speeds for Vega-1 and Vega-2 were extremely high (Table 2) because Halley's orbit sense is retrograde, i.e. directed opposite to that of the Earth and thus of the spacecraft. During the encounter dust particles impacted on the spacecraft at ~ 80 km s^{-1} and caused damage to some experiments and the unprotected solar panels.

The relative position of the spacecraft to the comet nucleus during the flyby is shown schematically in Fig. 3 for four different times around closest approach. As the encounters of both spacecraft took place before 10 March (descending node), the spacecraft and the nucleus were above the ecliptic during the flyby. The spacecraft attitude during the encounter was determined by the requirements to point the high gain antenna at the Earth and to obtain maximum power from the solar panels. Consequently, the solar power panels could not be aligned with the relative velocity vector (Fig. 4) and damage to the panels was inevitable. After the encounters the power from the solar panels decreased by almost 50%.

The flight operations centre was located in Evpatoria (Crimea), but most investigators were in Moscow at the Space Research Institute, where they were able to obtain all data in real time. Deep-space antennae in Evpatoria (70 m) and Medvezy Ozera (64 m), near Moscow, received the telemetry.

All experiments were switched on two days before closest approach. The direct high speed telemetry was transmitted from -48 to -46 h, from -24 to -22 h, from -3 to $+1$ h, from $+21$ to $+24$ h, and from $+45$ to $+48$ h during the cometary flyby. A limited amount of data were also stored on the onboard tape recorder and transmitted every 20 min in order to cover the 20-hour gaps between high-speed telemetry intervals.

The Vega mission, in particular the experiments, are described in a comprehensive document entitled "Venus-Halley mission, Experiment description and scientific objectives of the inter-

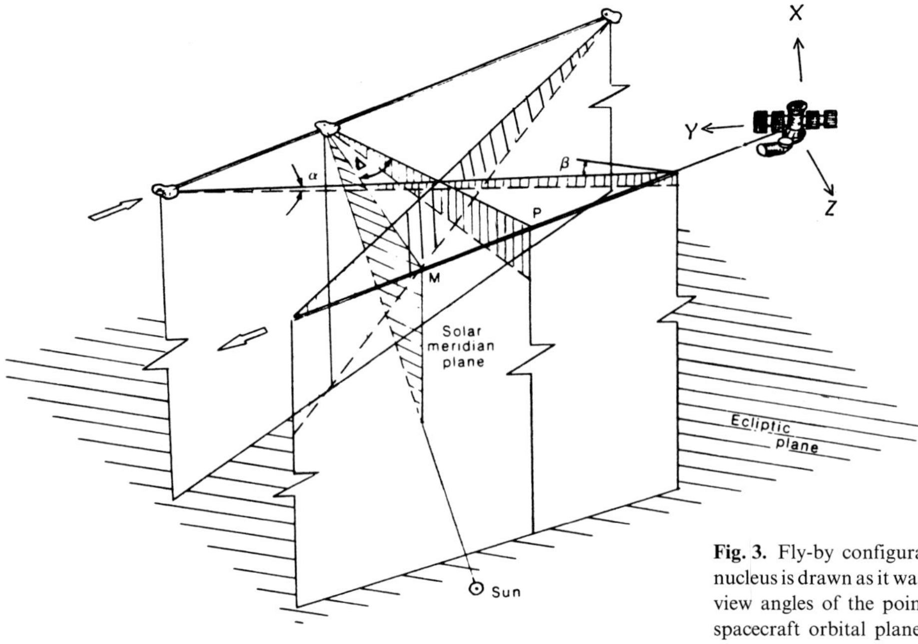

Fig. 3. Fly-by configuration of the Vega spacecraft and comet nucleus. The nucleus is drawn as it was during the Vega 1 encounter. The angles *β* and *α* are the view angles of the pointing platform (measured in and perpendicular to the spacecraft orbital plane) and *Δ* is the phase angle. *P* is the point of closest approach, *M* is the point where the spacecraft crossed the solar meridian plane

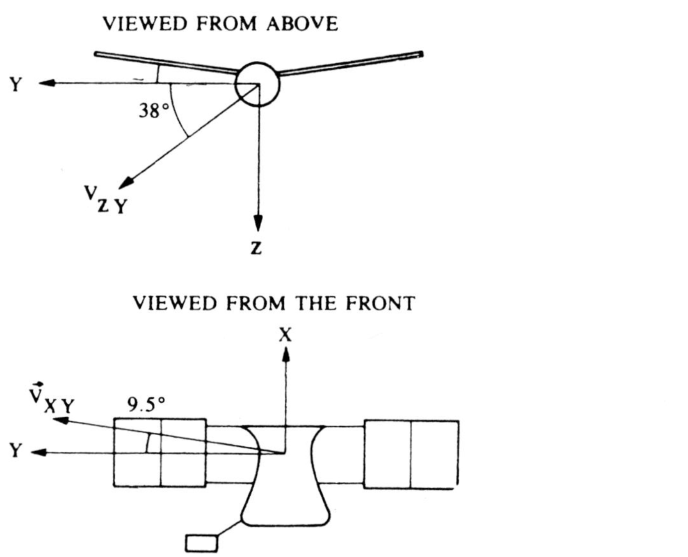

Fig. 4. Orientation of the Vega spacecraft during the Halley flyby in the spacecraft coordinate system

national project VEGA (1984–1986)", published by Imprimerie Louis-Jean, 05002 Gap, France, in May 1985. A shortened version can be found in Grard et al. (1986).

References

Sagdeev, R.Z., Moroz, V.I.: 1986a, *Astron. Z. Letters* **12**, 5

Sagdeev, R.Z. et al.: 1986b, *Astron. Z. Letters* **12**, 10

Sagdeev, R.Z., Linkin, V.M., Blamont, J.E., Preston, R.A.: 1986c, *Science* **231**, 1407

Grard, R., Gombosi, T.I., Sagdeev, R.Z.: 1986, ESA SP-**1066**, 49

EXPLORATION OF HALLEY's COMET

M. Grewing · F. Praderie
R. Reinhard (Eds.)

The Sakigake/Suisei missions to Halley's Comet

K. Hirao[1] **and T. Itoh**[2]

[1] Department of Aeronautics and Astronautics, Tokai University, 1117, Kitakaname, Hiratsuka, 259-12 Japan
[2] The Institute of Space and Astronautical Science 3-1-1, Yoshino-dai, Sagamihara, 229, Japan

1. Introduction

Japan's Institute of Space and Astronatical Science (ISAS) decided in 1979 to send a spacecraft to comet Halley. This decision was supported both by Japanese scientists who were anxious to carry out an interplanetary mission and by those institute engineers who wanted to demonstrate the capabilities of a new launch vehicle. Comet Halley was considered a scientifically meaningful target because it has never been explored previously from close-range, it displays a wide range of interesting phenomena, and it returns only every 76 yr.

The Japanese mission to comet Halley involved two spacecraft, a test spacecraft called "Sakigake" (japanese for forerunner, this spacecraft was called MS-T5 until launch) and a dedicated comet Halley encounter spacecraft called "Suisei" (japanese for comet, this spacecraft was called Planet-A until launch).

2. The spacecraft

Sakigaki and Suisei are almost identical spacecraft (Fig. 1) except for the scientific experiments they carry. The spin-stabilised spacecraft are cylindrical in shape with a diameter of 140 cm and a height of 70 cm. An off-set despun parabolic antenna of 70 cm diameter is mounted on top of the cylinder. At the bottom there is a low gain, cross dipole antenna and a medium-gain antenna of the co-linear type. The total height of the spacecraft, including antennas, is ~ 2.5 m. Around 70% of the outside of the cylinder is covered with solar cells which produce about 100 W during cruise when the spin axis of the spacecraft is kept perpendicular to the ecliptic plane. The Sakigake and Suisei spacecraft weigh 138.1 and 139.5 kg, respectively, including 10 kg of hydrazine propellant in each case. The attitude and orbit control system is composed of a two-thruster complex mounted diagonally at the top of the cylindrical body, the thrusting force of each thruster being 3 N. The spacecraft is usually spin-stabilised at 6 rpm which can be reduced to 0.2 rpm in the case of Suisei by means of a momentum wheel when the ultraviolet imager of Suisei is in operation. Information on the spacecraft attitude is provided by a Sun sensor and a star scanner.

The communications subsystem operates in S-band using a 5 W transmitter. The spacecraft have two data rates, a "high bit rate" mode (2048 bps) and a "low bit rate" mode (64 bps). The bit rate used depended on the communication distance. The ground station for the mission was built at Usuda, about 170 km northwest of Tokyo. The station's 64 m dish antenna (Fig. 2) has a gain of approximately 63 dB. During the Halley encounter data were received at Usuda at a rate of 64 bps. A 1 Mbit magnetic bubble memory onboard the spacecraft recorded data obtained during periods of non-contact with the Usuda Deep Space Station.

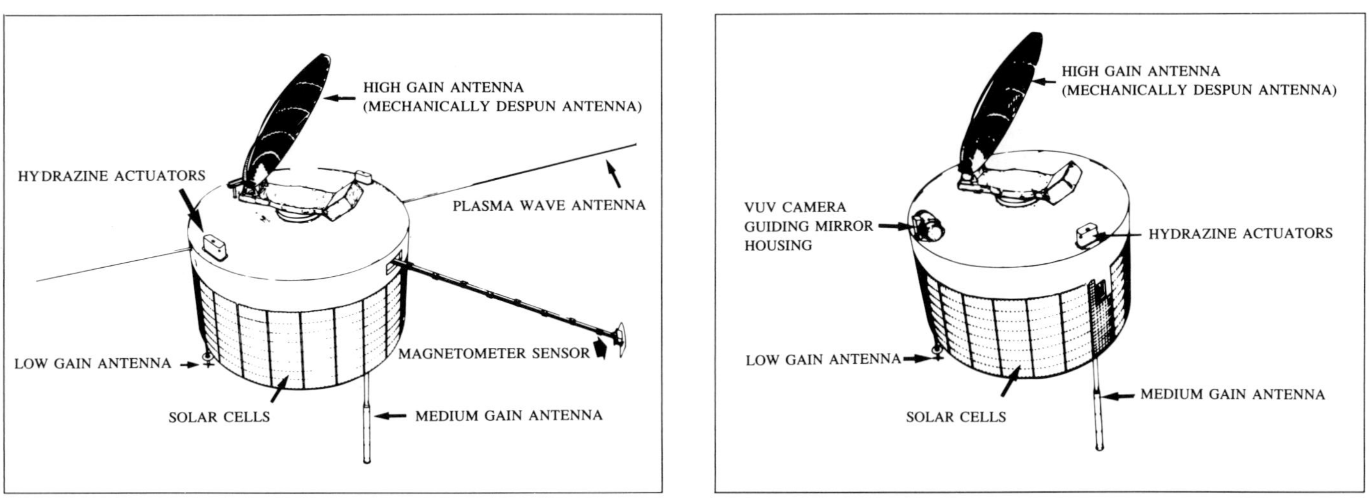

Fig. 1. The Sakigake (*right*) and Suisei (*left*) spacecraft

Fig. 2. The 64-m antenna of the Usuda Deep Space Center

The scientific payload onboard the two spacecraft is shown in Tables 1 and 2 and described in more detail in Hirao and Itoh (this volume).

3. The mission

Both spacecraft were launched by M-3S II launchers. This launcher is a three-stage solid propellant rocket with a solid kick-stage motor attached. Both spacecraft were injected directly into a heliocentric orbit.

Sakigake was launched at 19.26 UT on 8 January 1985 and Suisei was launched at 23.33 UT on 18 August 1985 (Fig. 3). The interplanetary orbit is shown in Fig. 4, the geometry during the encounter with comet Halley in Fig. 5. The key encounter parameters are given in Table 3.

All experiments onboard Sakigake were switched on at the end of February 1985 and both experiments onboard Suisei were switched on in September 1985. UV images of the comet's hydrogen corona have been recorded continuously from 26 November 1985 until 14 April 1986 except for the period 11 January to 8 February when the elongation angle of the comet precluded observation. At specific times during and a few days before the encounter data were obtained in the photometry mode.

During the encounter all experiments on both spacecraft worked flawlessly. They could not be operated continuously

Table 1. The Sakigake scientific payload

Experiment	Mass [kg]	Power [W]	Data rate Format 1/2/3/4	Principal investigator
PWP	5.2 [a]	1.7	H 512/768/768/0 L 32/48/48/0	H. Oya, Faculty of Science, Tohoku Univ., Sendai
SOW	2.0	1.1	H 512/768/0/768 L 32/48/0/48	K. Oyama, Res. Div. Planetary Science, ISAS, Tokyo
IMF	5.4 [b]	2.	H 400/0/768/768 L 25/0/48/48	T. Saito, Faculty of Science, Tohoku Univ., Sendai

Format				
1	PWP	SPW	IMF	H = High bit rate
2	PWP	SOW		L = Low bit rate
3	PWP	IMF		
4	SOW	IMF		

[a] Includes dipole antenna, search coil and electronics
[b] Includes boom

Table 2. The Suisei scientific payload

Experiment	Mass [kg]	Power [kg]	Data rate [bit/s] Format 1/2	Principal investigator
UVI	7.5	8.9	H 1568/0 L 98/0	E. Kaneda, Geophys. Res. Lab., Faculty of Science, Univ. of Tokyo
ESP	4.7	4.9	H 0/1408 L 0/88	T. Mukai, Res. Div. Planetary Science, ISAS, Tokyo

Format		
1	UVI and UVI check	H = High bit rate
2	ESP	L = Low bit rate

Table 3. Flight parameters of the Suisei/Sakigake missions

	Sakigake	Suisei
Launch	8 January 1985, UT	18 August 1985, UT
Time of closest approach	11 March 1986, 04:17:51 UT	8 March 1986, 13:05:49 UT
Heliocentric distance	0.86 AU	0.82 AU
Flyby distance	6.99 10^6 km	151 000 km
Flyby speed	75.3 km s^{-1}	73.0 km s^{-1}
Phase angle	109°.4	104°.2

Fig. 3. (*left*) The Suisei spacecraft on top of the third stage of the M-3S II launcher. For this launch a small kick stage was added between the third stage and the spacecraft (*right*) Launching of Suisei from the Kagoshima Space Center at 08.33 on 19 August 1986, local time

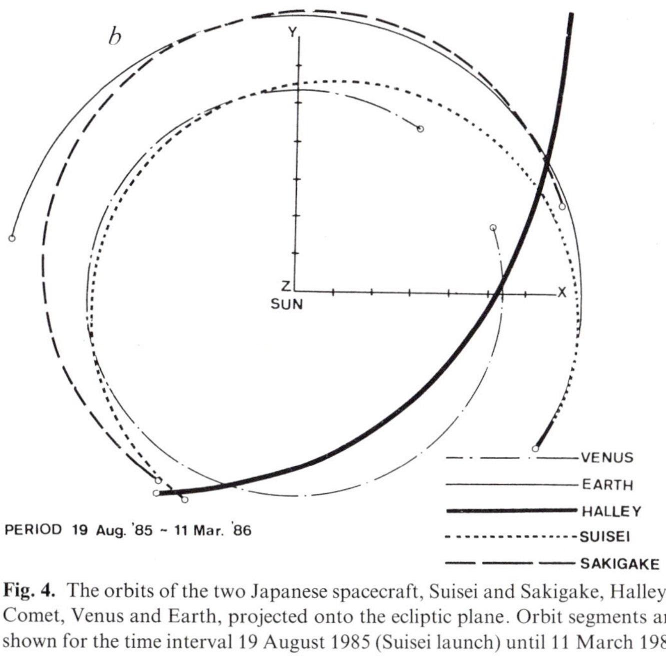

Fig. 4. The orbits of the two Japanese spacecraft, Suisei and Sakigake, Halley's Comet, Venus and Earth, projected onto the ecliptic plane. Orbit segments are shown for the time interval 19 August 1985 (Suisei launch) until 11 March 1986

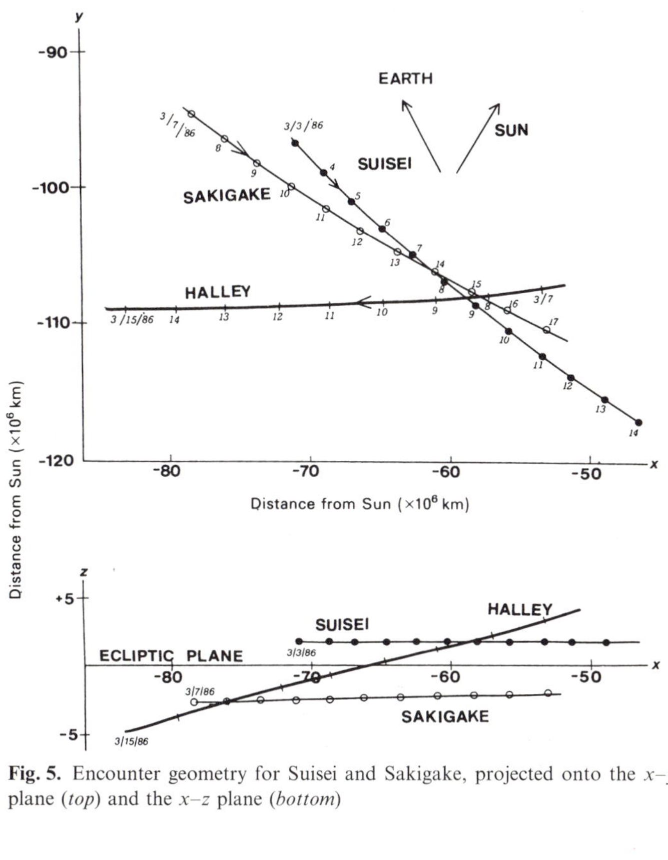

Fig. 5. Encounter geometry for Suisei and Sakigake, projected onto the *x–y* plane (*top*) and the *x–z* plane (*bottom*)

Table 4. Observation periods for experiments onboard Sakigake during the Halley encounter (no observations were made on 8, 9, and 14 March)

Exp.	Data available		Mode
	From (UT)	Until (UT)	
IMF PWP SOW	10 Mar, 02.46	10 Mar, 03.44	R
	10 Mar, 19.40	11 Mar, 03.47	R
	11 Mar, 04.07	11 Mar, 08.39	S
	11 Mar, 19.37	12 Mar, 03.49	R
	12 Mar, 19.36	13 Mar, 02.45	R

R = real time
S = storage

because of the limited visibility from the Usuda ground station. The intervals around closest approach for which science data are available from the experiments on both spacecraft are given in Tables 4 and 5.

Acknowledgements. The Authors express their thanks to all members of The Institute of Space and Astronautical Science supervised by the Director-General, Prof. M. Oda. The project was financially supported by the Ministry of Education, Science and Culture, and technically supported by the engineers of the Institute as well as those of many private industries in Japan. The tracking of spacecraft was supported by NASA, and also by NASDA during the initial phase. The data acquisition of Suisei at the time of its closest approach to the comet was supported by Goldstone and Madrid Deep Space Network Stations of NASA. Key parameters and Comet Halley dust model data were supplied by the members of IACG and its working groups.

Reference

Hirao, K., Itoh, T. 1987, *Astron. Astrophys.* **187**, 39

Table 5. Observation periods for experiments onboard Suisei (UVI: all, ESP: during the Halley encounter)

Exp.	Data available		Mode	Remarks
	From (UT)	Until (UT)		
UVI	14 Nov, 22.00	15 Nov, 01.00	Imaging	
	26 Nov, 23.00	11 Jan, 03.00	Imaging	10 days of no obs.
	9 Feb, 16.00	15 Apr, 01.00	Imaging	17 days of no obs.
	8 Mar, 05.00	8 Mar, 12.20	Photometry	
	21 Mar, 15.00	23 Mar, 13.00	Photometry	
ESP	7 Mar, 21.48	8 Mar, 04.00	R	
	8 Mar, 12.32	8 Mar, 15.29	S	
	8 Mar, 15.29	8 Mar, 19.40	R	Madrid, Goldstone
	8 Mar, 19.40	9 Mar, 04.02	R	

R = real time
S = storage

EXPLORATION OF HALLEY's COMET

M. Grewing · F. Praderie
R. Reinhard (Eds.)

The International Cometary Explorer

J.C. Brandt[1], **R.W. Farquhar**[2], **S.P. Maran**[2], **M.B. Niedner**[2], **Jr., and T.T. von Rosenvinge**[2]

[1] University of Colorado, Boulder, CO 80309, USA
[2] NASA/Goddard Space Flight Center, Greenbelt, MD 20771, USA

Summary. The International Cometary Explorer (ICE) spacecraft passed through the plasma tail of comet Giacobini-Zinner (G/Z) on September 11, 1985 and made in situ measurements of particles, fields, and waves. The results establish comets and their interaction with the solar wind as a rich source of plasma physics phenomena. Time of closest approach to the comet was approximately 11:02 UT and occurred at a tailward distance of 7,800 km. The spacecraft was in the strong interaction region for roughly $3\frac{1}{2}$ h, and spent about 10 min in the central tail region.

The scientific results for comet Giacobini-Zinner indicate:

(1) that the magnetic field capture and draping model, originated by Alfvén is correct. Oppositely-polarized magnetic tail lobes, and a current sheet separating them, were observed,

(2) that the comet/solar-wind interaction produced energetic ions, probably by the "pick-up" process,

(3) the presence of intense plasma wave activity,

(4) that the central plasma tail is dense and cold,

(5) that the principal ions are in the $H_2O^+ - H_3O^+$ group,

(6) that the bow wave, as seen on the flanks, is not a conventional shock, but resembles an extended interaction region,

(7) that impacts of micron-sized dust particles were detected.

The ICE spacecraft survived the encounter with comet G/Z relatively unscathed and passed approximately 0.2 AU upstream of Halley's Comet in late March 1986 with closest approach on the 25th. The scientific results for comet P/Halley indicate:

(1) an apparent detection of the comet via plasma wave and energetic particle measurements over the nuclear distance range 28,000,000 to 35,000,000 km.

(2) that the total interaction region for a Halley-sized comet can be quite large.

Key words: plasma tail, plasma physics, comet Giacobini-Zinner

1. Introduction and history

The Third International Sun-Earth Explorer (ISEE-3) was launched on August 12, 1978, as part of a three-spacecraft mission to study the solar-wind interaction with the Earth's magnetosphere (see Ogilvie et al., 1977). The spacecraft monitored the solar wind input from a "halo orbit" around the L_1 libration point. Four years of uninterrupted operation were completed at that location.

Among the opportunities to use ISEE-3 in an extended mission phase was exploration of the distant geotail and an intercept of comet Giacobini-Zinner in September 1985. To explore the comet option, a subsommittee of the ISEE Science Working Team (SWT) was formed at the request of ISEE-3 Project Scientist, author T. von Rosenvinge. The subcommittee, chaired by E. Smith of JPL, issued a report entitled "Intercept of Giacobini-Zinner by ISEE-3" in June 1982 (revised in May 1983). The necessary maneuvers began on June 10, 1982 for the geotail part of the option, and the full option was ultimately approved by NASA Headquarters. The final lunar encounter took place on December 22, 1983; see discussion below.

The first extensive survey of the distant geotail was completed by ISEE-3 in 1983. Distances as great as 237 Earth radii in the tail were achieved. The scientific experience obtained in the tail – data on particles, fields, and waves – provided an excellent rehearsal for experimental operations during the G/Z encounter. Extensive results were obtained from the geotail excursion and these were presented in the October 1984 issue (Vol. 11, No. 10) of *Geophysical Research Letters*.

When the spacecraft completed its last lunar encounter, its mission was irrevocably changed to the study of the comet, and accordingly, the name was changed to the *International Cometary Explorer* (ICE). The primary ICE scientific objective was to study the interaction between the solar wind and a cometary atmosphere by passing through the plasma tail of comet G/Z on September 11, 1985. Clearly the instrumentation onboard the spacecraft was well-suited to this objective.

Generally, the physical features or model that were expected to be proved were derived from Alfvén's (1957) pioneering paper on the plasma tails of comets. Briefly, the ionization of molecules in the cometary atmosphere creates an obstacle for the magnetized solar wind. The solar-wind magnetic field lines wrap or drape around the comet to form, in the terminal case, the two-lobed configuration. A neutral or current sheet was expected to separate the lobes of opposite magnetic polarity. In addition, because the solar-wind flow is supersonic and super-Alfvénic, a bow shock was expected. Except for the visible plasma tail, none of these features had been observed. Hence, the passage of the ICE spacecraft through the tail of G/Z would provide a critical test of our concepts.

Pre-encounter descriptions of the ICE Mission are given by Farquhar and Dunham (1983), Brandt et al. (1985a), and Brandt (1986a). A preliminary post-encounter description is found in Brandt et al. (1985b).

2. The trajectory

The maneuvers necessary to produce the trajectory that would send the spacecraft into the distant geomagnetic tail and to

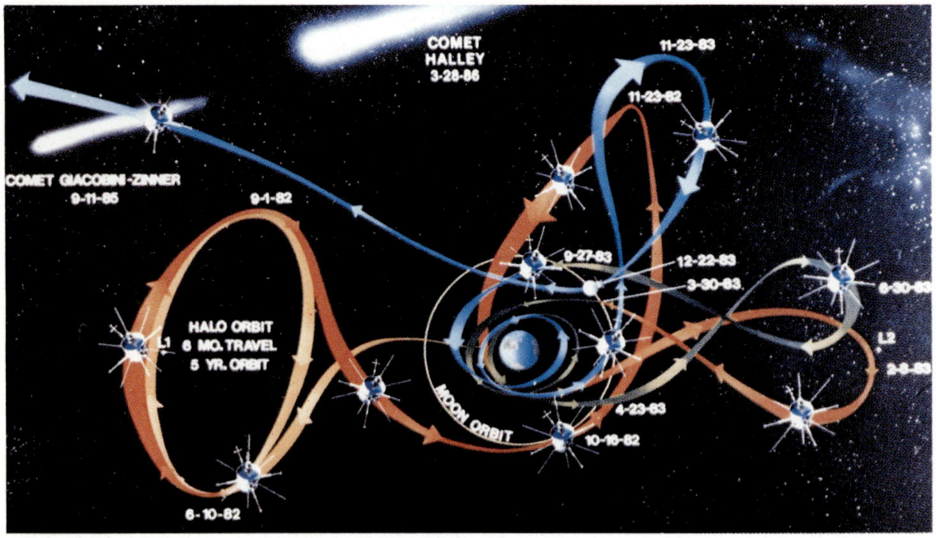

Fig. 1. A schematic diagram showing the trajectory of the ISEE-3/ICE spacecraft. Note that the spacecraft returns to the vicinity of Earth in the year A.D. 2014. If retrieved, it might provide samples of cometary dust for analysis

Fig. 2. A bipolar plot, with the Sun-Earth line held fixed, of the ICE trajectory showing the encounter with G/Z and the radial alignments with Halley's Comet

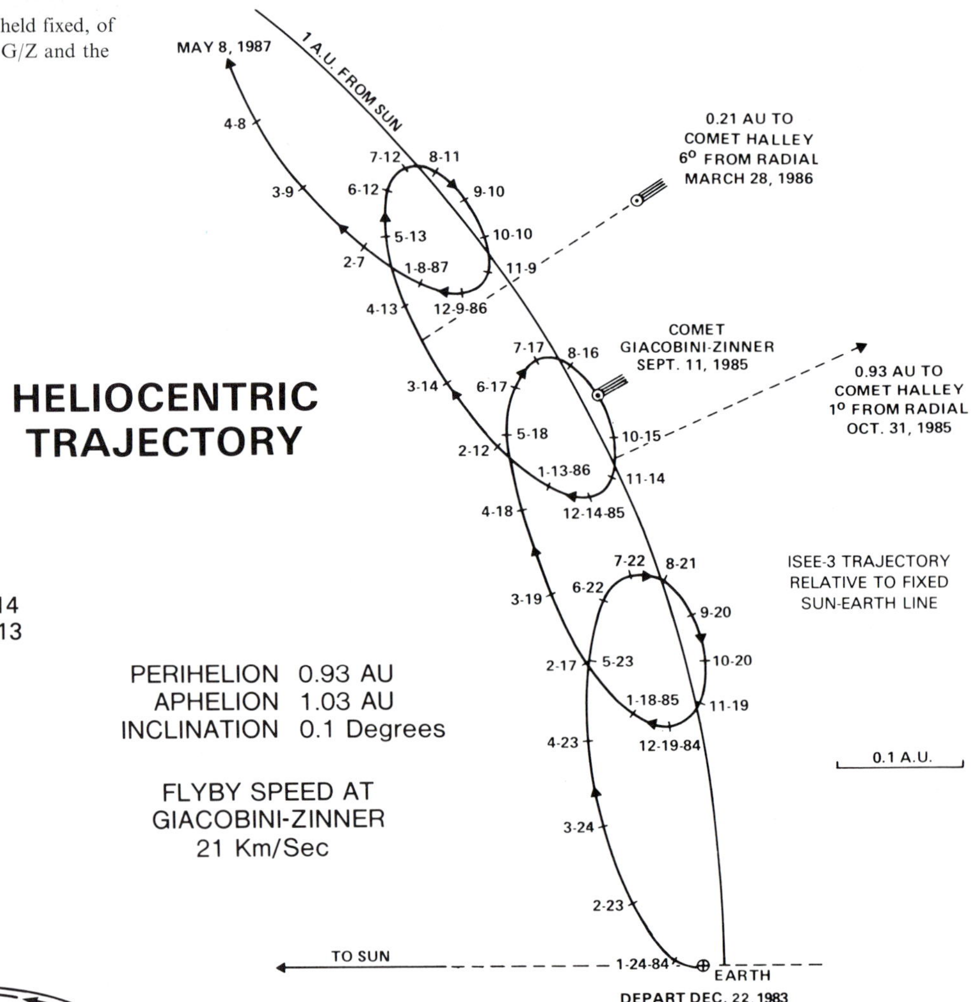

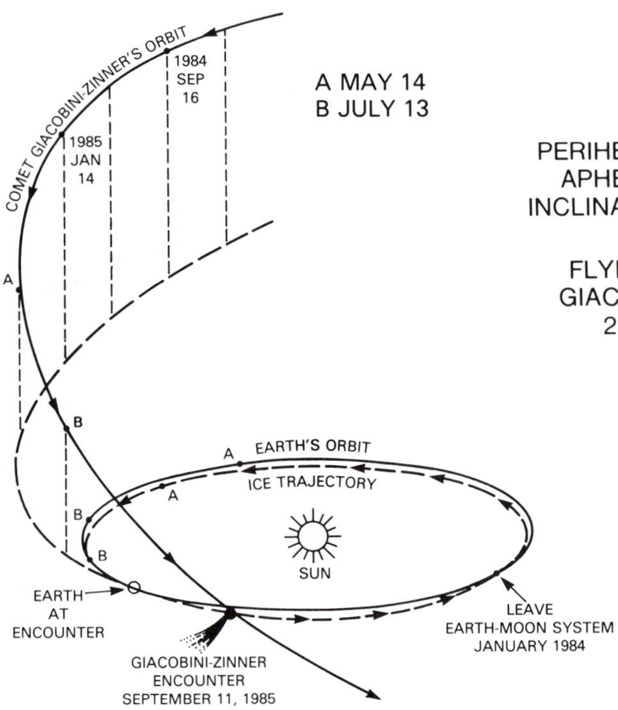

Fig. 3. Trajectory of the ICE spacecraft with the orbits of Earth and comet Giacobini-Zinner

intercept G/Z were not simple. They were the brainchild of author R. Farquhar, Flight Director for ISEE-3/ICE. Five gravitational encounters with the moon were required. The last encounter took place on December 22, 1983 when the spacecraft made a close swingby, passing only 120 km above the moon's surface. This last maneuver effectively "launched" the spacecraft from the Earth-moon system. A schematic summary of the orbital maneuvers is shown in Fig. 1.

A plot to scale of the ICE orbit to G/Z is shown in Fig. 2. At intercept, ICE and G/Z were 0.47 AU from Earth and 1.03 AU from the Sun. The encounter occurred near the plane of the ecliptic. The relative speed was 21 km s^{-1}. As viewed from the comet, the spacecraft traversed the tail moving south to north roughly perpendicular to the ecliptic. A perspective view of the encounter is shown in Fig. 3.

3. The ICE spacecraft and experiments

The spacecraft is shown in Figs. 4 and 5. The body is a 16-sided cylinder, 1.74 m in diameter and 1.6 m high. Solar arrays mounted on the faces of the cylinder provide power. A tower elevates the telemetry antenna above the spacecraft body. The spacecraft body supports a total of 10 appendages.

The radio system consists of two redundant S band transponders operating at 2217 and 2270 MHz. There are two low-gain antennas and a telemetry antenna having medium gain and a fan beam of $\pm 6°$. The standard telemetry rate in halo orbit was 2048 information bits per second, and other (lower) rates are available.

The spacecraft is spin-stabilized at a rate of 19.75 rev per min. Attitude information and control is provided by two Fine Sun Sensors and a Panoramic Attitude Sensor. The spin axis is maintained perpendicular to the ecliptic plane to within one degree.

Propulsion and attitude control use monopropellant hydrazine. A large amount of propellant was carried to launch as a safeguard against potentially large errors associated with the launch vehicle. As a result of a nominal launch, gas usage has been much lower than anticipated, resulting in a substantial amount of gas being available.

The distance at encounter of 0.47 AU greatly exceeded the distance of 0.01 AU while in the halo orbit. Hence, a lower bit rate (1024 bits per second) and tracking by NASA's Deep Space Network (DSN) was requited. Outfitting antennas in the DSN and at Arecibo, Puerto Rico to operate at the ICE frequencies was a major effort of the mission. At the time of the encounter, data were collected by Arecibo, Madrid, and Goldstone. Later in the day, data were collected by Canberra and Usuda, Japan.

The principal experiments – out of 13 total – thought capable of producing useful data on the comet were:

(1) Solar wind plasma experiment. Measures speed, density, and temperature of electrons. Scans in 16 steps covering the energy range 5 eV to 1500 eV. Sampling takes 24 s.

Principal Investigator: S. Bame, Los Alamos National Laboratory

(2) Magnetometer experiment. Obtains three, high-accuracy, triaxial measurements of magnetic field strength per second in 8 ranges from 4 nT (lowest full range) to 1.4 10^5 nT (highest full range) and a sensitivity of 1/256 of each full range.

Principal Investigator: E. Smith, Jet Propulsion Laboratory

(3) Radio wave experiment. Samples the range 30 kHz to 1 MHz in 12 steps with $\Delta f = 3$ kHz. Samples the range 40 kHz to 2 MHz in 12 steps with $\Delta f = 10$ kHz. Each scan takes 56 s.

Principal Investigator: J.-L. Steinberg, Observatoire de Meudon

(4) Plasma wave experiment. Samples 16 channels in electric field in one second in the frequency range 20 Hz to 10^5 Hz. Obtains magnetic field data from a search coil.

Principal Investigator: F. Scarf, TRW

(5) Ion composition experiment. Experiment reprogrammed for the cometary encounter. Determines flow speeds from alpha particles and ion composition for the mass range 14 to 34 for speeds of 80 to 224 km per sec. Each scan takes 20 min.

Principal Investigator: K. Ogilvie, NASA-Goddard Space Flight Center

(6) Low energy cosmic ray experiment. Measures the elemental abundances, charge state composition, energy spectra and angular distributions of ions over parts of the range 2 keV/charge to 80 MeV/nucleon and of electrons between 75 and 1300 keV.

Principal Investigator: D. Hovestadt, Max-Planck-Institut für Physik and Astrophysik.

(7) Energetic proton experiment. Three telescopes aligned at different angles measure particles in eight channels between 35 keV and 1.6 MeV. Full measurement takes 32 s.

Principal Investigator: R. Hynds, Imperial College

4. Comet P/Giacobini-Zinner

The comet was discovered by M. Giacobini at Nice in 1900 and rediscovered in 1913 by E. Zinner at Bamberg. The period is 6.5 yr and the perihelion distance is 1.03 AU. The orbital eccentricity and inclination are 0.71 and 31°.9, respectively. The comet has been observed at previous apparitions. Figure 6, taken in 1959, shows some typical dimensions. The coma (or atmosphere) has a diameter of about 50,000 km and the plasma tail can reach lengths of 500,000 km or more. Photometry at past apparitions established a total gas production rate at perihelion of about 2.3 10^{28} molecules per second. The comet has been known to be a dust producer from its spectrum and polarization.

Comet G/Z is associated with the Draconid (Giacobinid) meteor showers. These produced spectacular meteor showers in 1933 and 1946. The dust generated by G/Z represented a possible hazard to the ICE spacecraft. The principal concern was dust-impact degradation of the solar power arrays, which could reduce the power available to operate the spacecraft and payload during the cometary intercept. A G/Z dust model was prepared by N. Divine, NASA/Jet Propulsion Laboratory, and it was used to implement a conservative approach to power management.

The targeting strategy was based on data from previous apparitions and considered the following elements. (1) The dust hazard just mentioned. (2) The possibility that a fully-developed plasma tail shown in Fig. 6 might not exist too close to the nucleus. (3) The flaring of the tail increases with increasing distance from the head of the comet Ershkovich, Niedner, and Brandt, 1982). (4) The position of the tail wags back and forth in response to changing solar wind velocity. The scientific objective was to pass through the plasma tail. Targeting accuracy was estimated to be 600 km. This figure was based on the ability to maneuver the

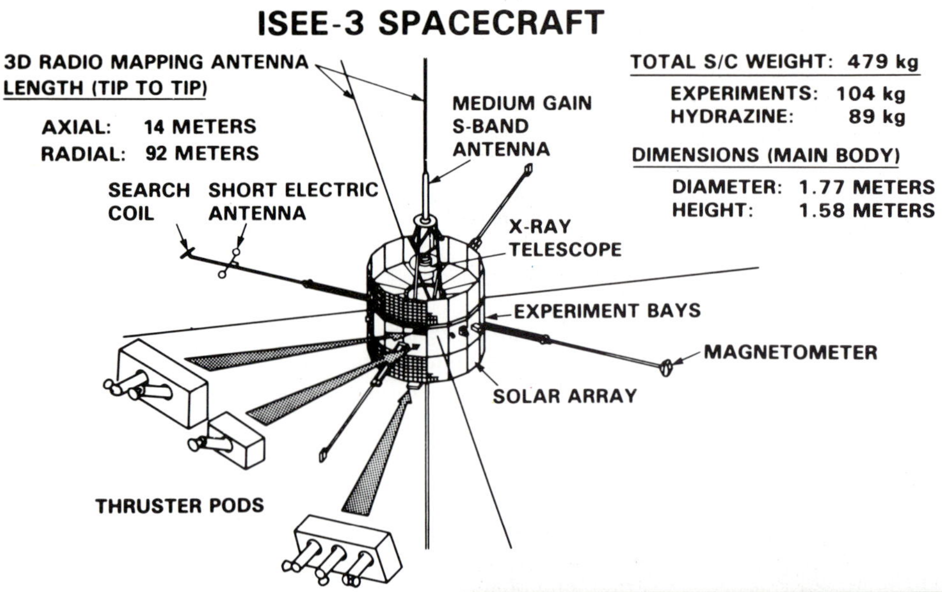

Fig. 4. Schematic diagram of the ICE spacecraft as built under the name of ISEE-3

Fig. 5. Artist's conception of the ICE spacecraft close to encounter with comet Giacobini-Zinner

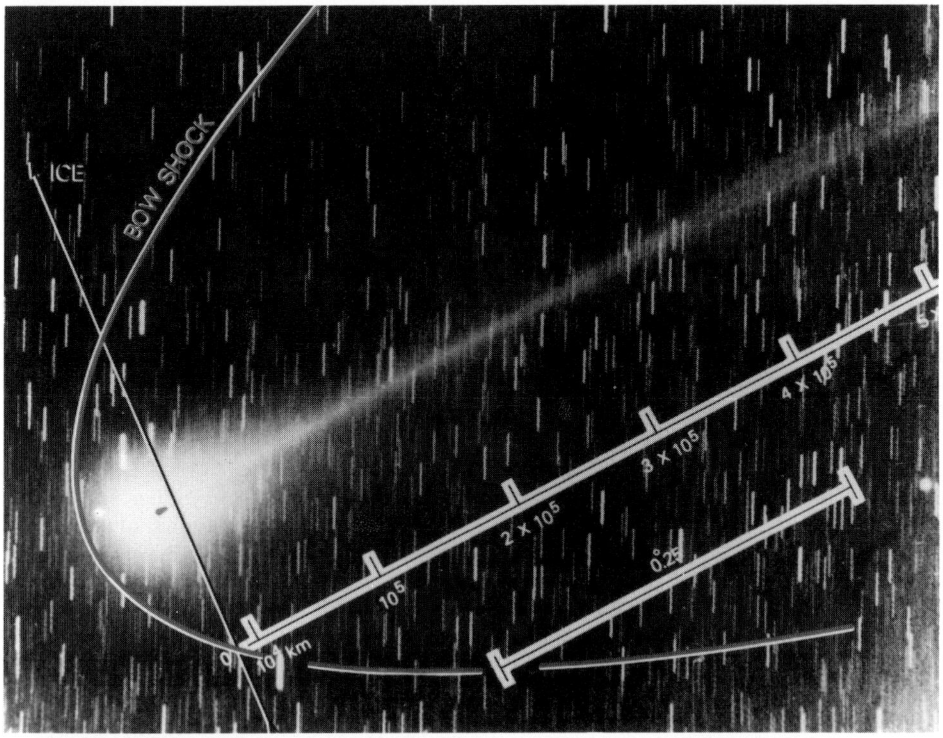

Fig. 6. Photograph of comet Giacobini-Zinner, obtained by E. Roemer on October 26, 1959. (Official US Navy photograph). The scales have been added as have the estimated position of the bow shock and the ICE trajectory, here shown with closest approach distance of 10,000 km. Actual distance was 7,800 km

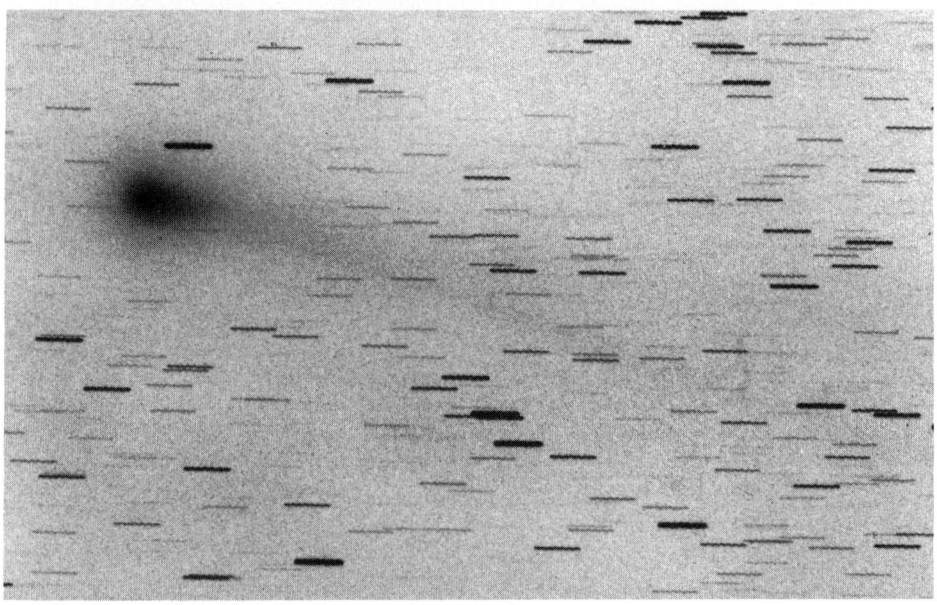

Fig. 7. Photograph of comet Giacobini-Zinner obtained by F. Dossin at the Haute-Provence Observatory at Saint Michel, France on July 13, 1972

spacecraft and on the estimated precision of the comet's location as supplied by D.K. Yeomans, NASA/Jet Propulsion Laboratory, who carried out the astrometric and orbital work for the encounter. Elements (1), (2), and (3) argue for an encounter away from the head, while element (4) argued for a close encounter. A model was set up by author M.B. Niedner to estimate the probabilities of intersecting the tail and a target distance in the range 5,000 to 15,000 km was found to satisfy $P \geqq 95\%$; the value of 10,000 km was adopted. Comet G/Z is known to have important and variable non-gravitational forces which significantly affect its orbit. At the time of the last scheduled orbital maneuver of the spacecraft, the orbit gave an intercept distance of approximately 7,800 km at a point several hundred km off the tail axis. The off-axis component was corrected on September 8, while the intercept distance was left at 7,800 km.

The comet at this apparition was recovered on April 3, 1984 at the Kitt Peak National Observatory. The recoverers were S. Djorgovski and H. Spinrad, University of California, Berkeley, and G. Will and M.J.S. Belton, Kitt Peak National Observatory. The recovery used an advanced CCD detector and the comet was at 23rd magnitude.

The comet developed toward the expected state illustrated in Fig. 6, although a more realistic photographic appearance was that shown in Fig. 7 at the 1972 apparition. By August 1985, most of the major spectral emission had been observed from a combination of ground-based observatories and the International

Ultraviolet Explorer (IUE). Of particular interest was the report of production of H_2O^+; CO^+ was also reported but at a very low intensity.

The development of the comet probably peaked shortly before the encounter as evidenced by gas production rates estimated at over 10^{29} molecules per second. At the time of encounter, the comet had already begun to fade and the gas production rate was estimated roughly at 5 10^{28} molecules per second. From this value and our models, on the day before the encounter we estimated roughly 3 to 4 h in the comet (strictly, the shock-to-shock time) and roughly 10 min in the main tail.

Establishing the correspondence between the traditional picture of a comet (e.g., Brandt, 1982) and the new, in situ data is a major challenge. Ground-based coverage, focused on the month of September 1985, has been obtained from the *International Halley Watch* (IHW).

5. The encounter of comet P/Giacobini-Zinner and scientific results

The first scientific results from the passage of ICE through the plasma tail of comet Giacobini-Zinner have appeared in *Science* (Vol. 232, 1986), and further findings of the mission are contained in the March and April 1986 issues of *Geophysical Research Letters*, *Advances in Space Research*. (Vol. 5, 1985) entitled *Comets Halley and Giacobini-Zinner*, and the *Proceedings of the 20th ESLAB Symposium on the Exploration of Halley's Comet* (1986). We provide here a brief overview of the results to data, including discussion of observations made by remote observers. Note that results continue to appear in the literature.

A. In situ results

Although the spacecraft was inside the bow wave for only ~ 3 h, the large-scale structure of the comet was in fact well-determined and found to be in general agreement with several long-standing theories. In particular, the field draping model of Alfvén (1957), in which the cometary capture of interplanetary magnetic fields (IMF) from the solar wind leads to the formation of a tail with oppositely-polarized magnetic lobes separated by a current sheet, was confirmed. Both the lobes and the current sheet were directly observed, and very importantly, the polarity sense of the lobes and the orientation of the current sheet were consistent with the local geometry of the IMF (Smith et al., 1986a; Slavin et al., 1986a, b).

An extended upstream mass-loading region was observed via turbulence seen in the magnetometer data (Tsurutani and Smith, 1986a, b) as well as in the plasma wave and electron measurements (Scarf et al., 1986a, c; Gosling et al., 1986; Scarf, 1986). Much of this turbulence has been proposed to be due to instabilities associated with mass-loading of the solar wind, primarily by the water-group ions (Tsurutani and Smith, 1986a, b; Tsurutani et al., 1986b). The results are in agreement with the early model of Biermann et al. (1967), as well as with others which have followed, which predict a comet/solar wind interaction region extending over distances $\sim 10^6$ km.

The ICE electron plasma and magnetic field observations do not show a conventional sharp bow shock such as is observed at Earth, Venus, Jupiter, Saturn and Uranus (Bame et al., 1986; Thomsen et al., 1986), although the recent analysis by Smith et al. (1986b) suggests an $M = 1.5$ shock was encountered inbound and an $M = 0.73$ expansion wave outbound. Rather, the conventional shock is replaced with a broader, more extended "transition region" whose characteristic (inbound, outbound) thicknesses are $\approx 20\%$ and $\approx 38\%$ of the (inbound, outbound) outer bow wave distances (Jones et al., 1986). In the transition region, solar-wind electron plasma is compressed and heated by approximate factors of 5 and 2, respectively. There are several possible reasons for the lack of a classical shock. It could be that the pickup of heavy ions has inhibited the formation of a shock, as was suggested some years ago by Wallis (1973). Or perhaps there is a shock at the subsolar point which decays to an expansion wave on the flanks (Wallis and Dryer, 1985; Wallis, 1986). Another alternative is that the proper scale length for the shock is the gyroradius of the pickup ions in the solar wind frame ($\sim 8{,}000$ km), in which case the entire transition region might be properly identified as the shock. The much larger gyroradius for pickup ions is due to the large mass of water ions relative to protons ($m_p/m_p = 18$) and to the fact that pickup ions are created in the solar wind frame at large velocities with respect to the IMF.

Sporadic bursts of activity associated with energetic plasma, in particular reverse electron heat fluxes and the enhanced electron plasma oscillations with which they are correlated, appear to be associated with the magnetic connection/disconnection of the spacecraft with the bow wave and transition region (Bame et al., 1986; Fuselier et al., 1986). Fuselier et al. (1986) propose that the electron plasma oscillations derive their energy from the back streaming electrons when there is connection with the transition region. Using the times of contact of 25 observed connections with the transition region, they fitted a paraboloid to the data to derive a sub-solar standoff distance ($\approx$ bow wave distance at the nose) of 4 10^4 km. This distance and the theory of Schmidt and Wegmann (1982) then lead to a neutral water molecule production rate of 3 10^{28} molecules per second, in excellent agreement with independent, remote observations (e.g., Combi et al., 1986; and McFadden et al., 1987).

The properties of the sheath (the outer envelope of which is the transition region) are as follows. The sheath is characterized by magnetic fields which are becoming more and more draped the deeper into the sheath the spacecraft traveled (Smith et al., 1986a), decreasing solar-wind bulk velocities, and, surprisingly, decreasing electron densities and temperatures (with respect to the maxima of these quantities in the transition region; Bame et al., 1986). Small-scale electron density enhancements were observed in the sheath which may be due to instabilities associated with the ion pickup process (Bame et al., 1986).

The tail itself was composed of cold ($\sim 50{,}000\,^\circ$K), dense ($N_e \sim 200\,\mathrm{cm}^{-3}$), and slowly-moving ($\sim 30\,\mathrm{km\,s}^{-1}$) plasma (according to the solar-wind electron experiment; cf. Bame et al., 1986). Although the solar-wind electron, magnetometer, and radio science experiments essentially all saw strong evidence of a plasma tail corresponding in approximate dimensions and properties to that seen in the remote imaging, there were differences in the specific properties assigned. For example, the radio science experiment measured a temperature and density in mid-tail passage of 16,000 °K and 650 cm^{-3}, respectively, figures significantly different from the electron experiment results quoted above (Meyer-Vernet, 1985; Meyer-Vernet et al., 1986). The magnetometer observed a magnetotail of approximate width 10,000 km (Slavin et al., 1986a, b), about three times as wide as that measured by the electron experiment, and twice that of the radio science experiment.

It is interesting to compare these figures. Specific transverse dimensions have been quoted for electron densities higher than a specified value and these are collected in Table 1.

In addition, Scarf et al. (1986a) quote a dimension of 32,000 km for a full tail width and several other experimenters

Table 1. Measured tail widths in comet P/Giacobini-Zinner

Minimum N_e [cm^{-3}]	Dimension [km]	Experimenter
10	30,000	Steinberg
30	15,000	Scarf
100	4,000	Steinberg
120	4,000	Scarf
200	3,000	Bame

(All figures from *Science* articles)

(Hynds, Ogilvie, and Smith) quote or present data indicating a typical value of 10,000 km. The optical value of 5600 km (see below) corresponds to a minimum density in the range 80 to 90 per cm^3. Clearly, all of these figures are consistent with a typical tail width value of 10,000 km. If a more precise definition is required, the general variation is shown in Table 1 and the original papers should be consulted.

The peak observed magnitude of the tail magnetic field was 60 nT, which, considering the rather close encounter distance (7800 km) and the large 20° flaring angle of the lobe fields (Slavin et al., 1986a), is consistent with the view advocated for years by Ershkovich and his colleagues (Ershkovich, 1980; Ershkovich, Niedner, and Brandt, 1982) that the field magnitude in the distant tail ($>10^6$ km) must be comparable to the interplanetary field. This viewpoint has also essentially been advocated by Slavin et al. (1986a).

Specific properties of the mass-loading process were provided by the Low Energy Cosmic Ray Experiment (Ipavich et al., 1986) and the Energetic Proton Experiment (Hynds et al., 1985, 1986; Sanderson et al., 1986a; Richardson et al., 1986). In the outer mass-loading region ($\sim 10^6$ km from the nucleus), these experiments observed ion energies greater than 300 keV, energies well in excess of those predicted for water ions captured onto field lines of the undisturbed solar wind, and then "heated" by adiabatic compression of the magnetic field. Clearly, some acceleration process is needed to produce the observed high energies.

As far as the actual composition of the cometary ions is concerned, the Ion Composition Experiment indicated the presence of mostly water group ions, H_2O^+, H_3O^+; probably CO^+ and HCO^+, and an as-yet-unknown ion whose mass is 23 or 24 (Ogilvie, 1985; Ogilvie et al., 1986). The predominance of water ions is considered evidence in favor of Whipple's dirty snowball model of cometary nuclei, in which water is the predominant ice.

During the ICE flyby, dust particle impacts with the spacecraft were detected by the plasma wave instrument. The temporal profile of the impact rate over the encounter suggests an r^{-2} fall-off (Gurnett et al., 1986), which is consistent with the nucleus being the central and only source of cometary dust. Gurnett et al. (1986) calculate that the masses of typical impacting dust particles was in the range 10^{-10}–10^{-12} gm.

A simplified schematic overview of the scientific findings for comet Giacobini-Zinner from the ICE encounter is given in Fig. 8. The ICE trajectory is also shown schematically.

B. Remote-sensing results

Our knowledge of the general state of the comet on September 11, 1985 from ground-based observations is not as detailed as expected. A major contributing factor was the poor weather over much of the southwestern United States during September. In addition, the distinct plasma structure seen quite clearly on Roemer's photographs from the 1959 apparition appears to be absent. Rather, the overall appearance is similar to the apparitions of 1946 and 1972 (see Yeomans and Brandt, 1985). We return below to the question of CO^+ in comet P/Giacobini-Zinner. As data continue to come into the various disciplines of the International Halley Watch, we expect an improvement in our knowledge. The *International Halley Watch* plans to collect results for comet Giacobini-Zinner into one volume of its *Archive*.

The total gas production rate as estimated from ultraviolet and radio observations peaked in late August at values in excess of

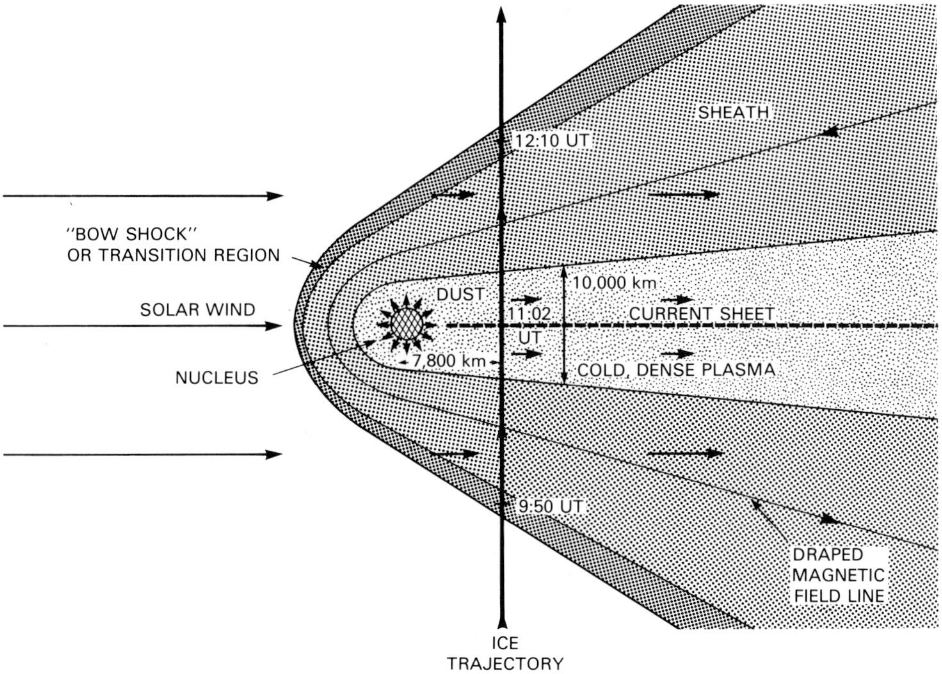

Fig. 8. Summary schematic of results for comet Giacobini-Zinner. Times (UT) along the trajectory refer to the date of the encounter, September 11, 1985

1 10^{29} molecules per second. The rate steadily declined to a value of approximately 5 10^{28} molecules per second at encounter. Note that this value is consistent with the cometary dimensions (e.g., pseudo bow shock location) found by the ICE spacecraft. The dust production rate has been estimated by Hanner et al. (1985) as approximately 3 10^5 grams per second. These two rates yield a gas-to-dust ratio by mass for comet Giacobini-Zinner near perihelion of 5. This high value is consistent with the historical view of comet Giacobini-Zinner as a low dust comet. In addition, Hanner et al. (1985) estimate the size of the predominant dust particles as being a few tenths of a micron.

Spectroscopy from the Earth's surface and from above the Earth's atmosphere have established the definitive presence of the following species: H, OH, NH, CN, C_3, C_2, CH, NH_2, [O I], and H_2O^+. The situation with respect to CO^+ is unclear, but this species is very weak if detectably present at all.

Dimensions of cometary structures are as follows. The hydrogen cloud had a diameter of 10 million km for emission above the background as determined by Lyman-alpha observations taken by the Pioneer Venus Orbiter (Stewart et al., 1985). The coma is asymmetrical as shown in Fig. 9. The chord along the trajectory across the visible coma corresponds to approximately 75,000 km. This dimension closely matches the length of significant dust impact detection by the plasma wave experiment, a dimension of 1 h at 21 km s^{-1}. Thus, the 75,000 km dimension refers to both gas and dust. An estimate of tail length comes from observation on August 17 and September 12 (Figs. 10 and 11). Measurements show a featureless tail stretching over 10^6 km on August 17 and a tail at least 20′ or 450,000 km in length on September 12. We have only limited estimates of tail width. At 10,000 km tailward in H_2O^+, the width (FWHM) was 5600 km (Spinrad et al., 1985).

6. The encounter of Halley's Comet

Shortly after the ICE encounter with comet G/Z, comets Halley and G/Z were within 2° of each other as seen on the sky. This historic "changing-of-the-guard" circumstance is shown in Fig. 12.

The ICE spacecraft contributed to the exploration of Halley's Comet by providing first indications of spacecraft problems in a cometary environment, specifically dust impact and spacecraft charging. At Halley's Comet, ICE would, at a minimum, serve as a valuable upstream monitor of solar-wind conditions.

However, the detection of streaming electrons and ion beams by the ICE spacecraft millions of kilometers from the nucleus of comet Giacobini-Zinner indicated the possibility of the direct detection of comet Halley by the spacecraft. The dimensions of plasma-associated features (e.g., "bow shock" and contact surface) vary approximately linearly with total gas production rate (Schmidt and Wegmann, 1982; also see Yeomans and Brandt, 1985) in the range 10^{28} to 10^{30} molecules per second. Thus, detection of comet Halley was possible because a total gas productuon rate of ten times the rate for comet Giacobini-Zinner was expected. At closest approach on March 25, ICE was 28 million km upstream of Halley's Comet; on March 27, ICE made its closest approach to the Sun-comet line at distance of 14 million km, while the comet's total distance was 30 million km. A few days either side of these dates were optimum for detection of cometary structures. If the only detection possible would be streaming electrons or ion beams, the optimum time was in the first week of April 1986 when the comet and the spacecraft were best connected by the solar-wind magnetic field lines.

Detections of Halley's Comet were reported in late March 1986 by the Plasma Wave Experiment (Scarf et al., 1986b; also see Brandt, 1986b) and the Energetic Proton Experiment (Sanderson et al., 1986b). However, negative results have been reported by Tsurutani et al. (1986a).

The physical situation is believed to be as follows. Heavy ions (probably from the water group, i.e., H_2O^+, OH^+, or O^+) having their origin in neutrals from Halley's Comet are produced upstream from the ICE spacecraft, are measured by the Energetic Proton Experiment (Wenzel et al., 1986), and cause the waves (Fig. 13) measured by the Plasma Wave Experiment (Scarf et al., 1986c). The conditions necessary for neutrals to reach the ICE spacecraft from comet P/Halley have been discussed by Daly (1987).

The current situation (September 1987) is as follows. The detections cited in the previous paragraph are still supported by the respective experiment teams. Waves might have been detected by the Magnetometer Experiment (see Tsurutani et al., 1986a) and particle intensity increases by Energetic Proton Experiment. The magnetometer team feels that the phenomena seen near the closest approach to comet P/Halley are not related to the comet. Thus, the association of phenomena measured 0.2 AU upstream of comet P/Halley by ICE is reasonable but equivocal. Nevertheless, it is fair to say that ICE became the first spacecraft to directly investigate two comets.

7. Fate of the ICE spacecraft

Spacecraft maneuvers and tracking have continued since the encounters. The current bit rate is 256 and the last orbital maneuver was made on 7 April 1986. The hydrazine system provided a ΔV of 39 m s^{-1}. Since that time, there have been only attitude maneuvers, and no further maneuvers are currently planned. The orbit should pass close to the moon on 10 August, 2014. If the spacecraft were subsequently brought back to Earth, it would make a splendid addition to the Smithsonian Air and Space Museum, Washington, DC, USA.

8. Conclusion

In addition to the scientific results, the success of the ISEE-3/ICE mission illustrates the utility of extended mission phases for spacecraft and the viability of multiple lunar swingbys to make major orbital changes. By becoming the sixth spacecraft to directly investigate Halley's Comet, ICE became the first spacecraft to investigate two comets. The first chapter of the exploration of comets by spacecraft was opened and closed by the International Cometary Explorer.

Acknowledgements. The scientific results presented here and elsewhere were made possible by the effort of many individuals associated with the International Cometary Explorer Project at the NASA/Goddard Space Flight Center. In particular, we note the efforts of the Project Manager, Mr. Patrick Corrigan.

Fig. 9. The coma region of comet G/Z on August 22, 1985 as recorded by U. Fink et al., University of Arizona

Fig. 10. The plasma tail of comet G/Z as observed on August 17, 1985. The tail is relatively featureless, but extends for at least 1,000,000 km. Photograph obtained by E. Moore at the Joint Observatory for Cometary Research (JOCR) operated by the Laboratory for Astronomy and Solar Physics, NASA/Goddard Space Flight Center and the New Mexico Institute of Mining and Technology

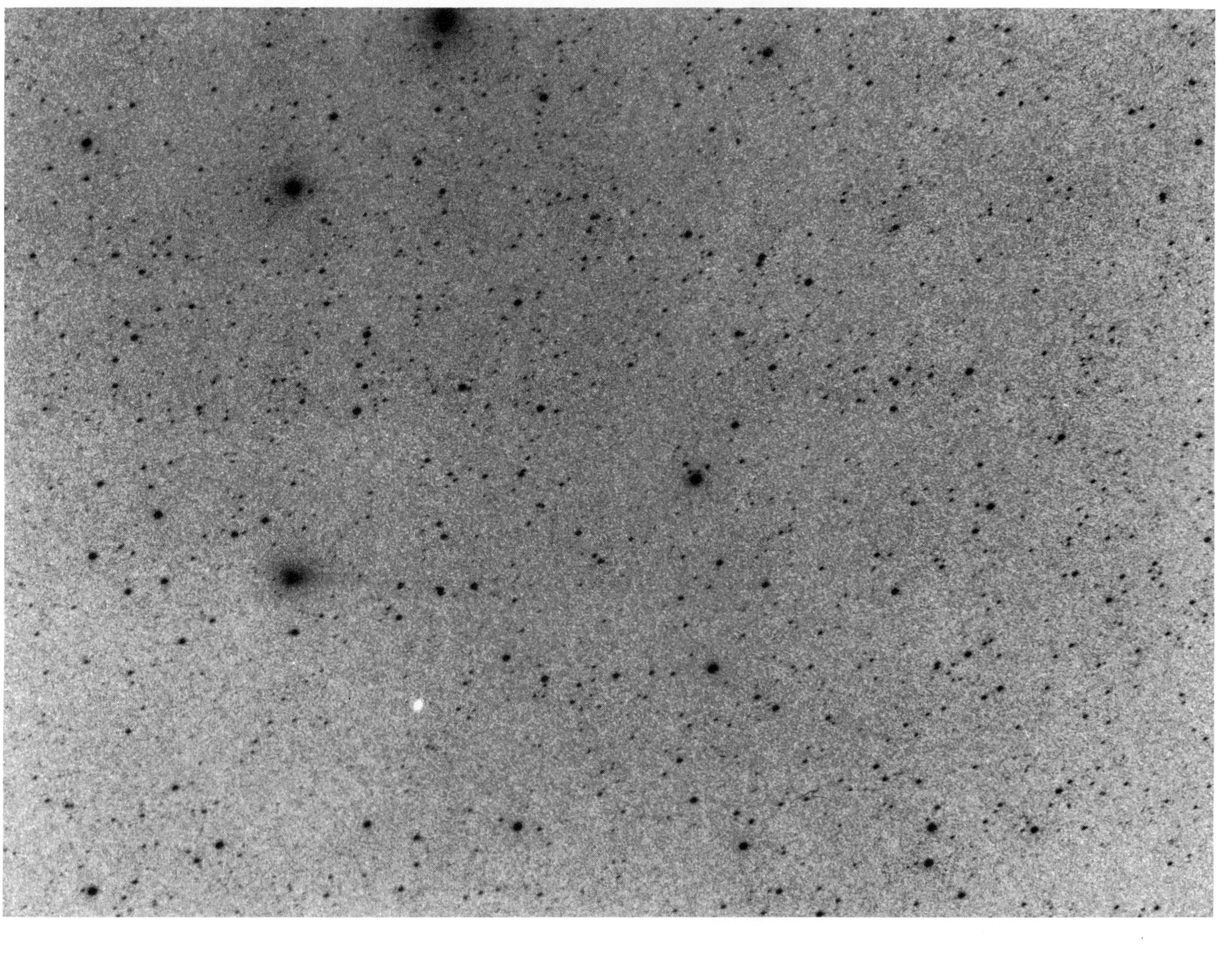

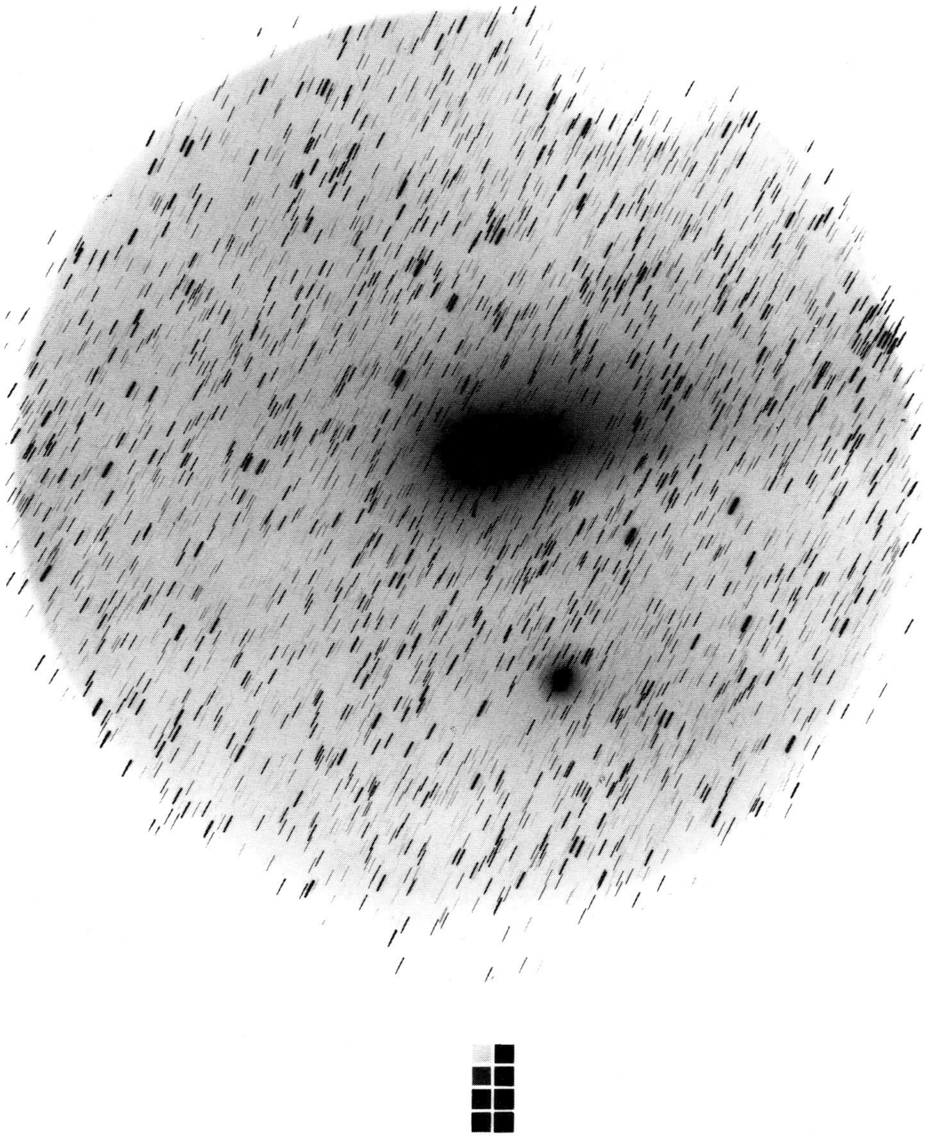

Fig. 11. Comet Giacobini-Zinner on September 12, 1985 clearly showing a tail extending over 450,000 km. (NOAO photograph taken by Richard Kron and Steven Majesky, Yerkes Observatory, University of Chicago)

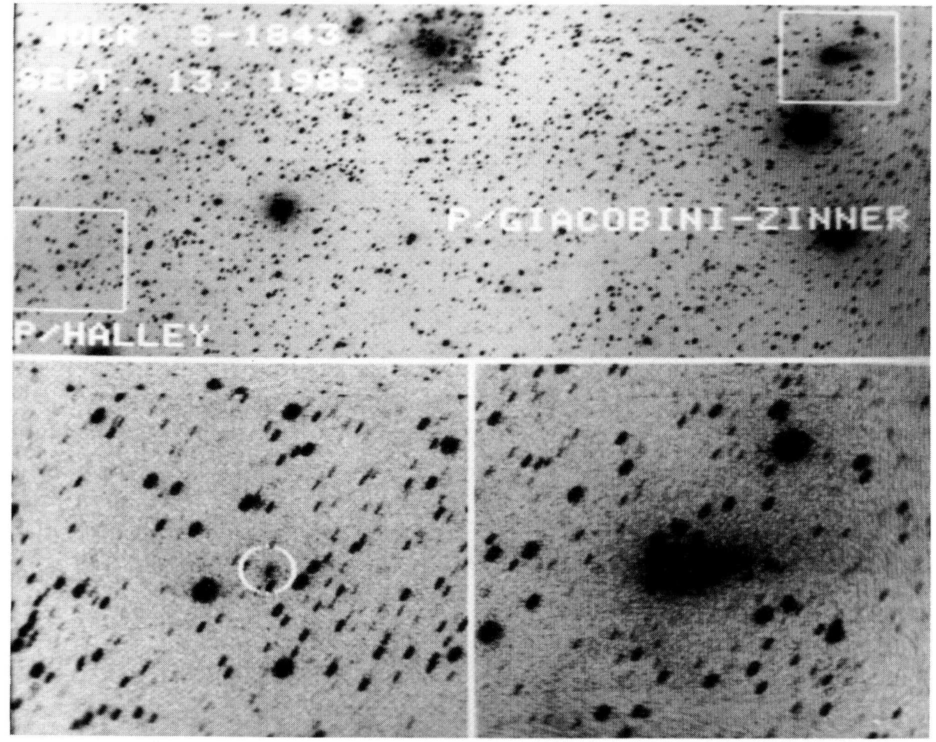

Fig. 12. Comets Giacobini-Zinner and Halley on the same wide field image, September 13, 1985. The two comets are shown enlarged at the bottom of the figure. Photographed by E. Moore at the JOCR

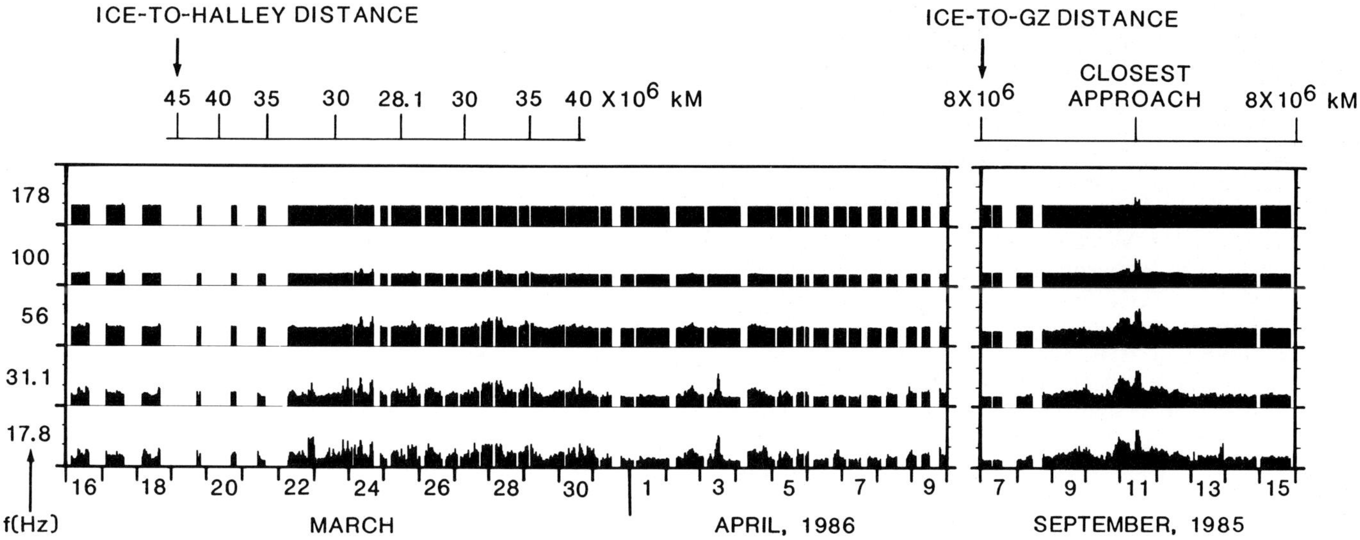

Fig. 13. Low frequency *B*-field waves observed during closest approach to comets Halley and Giacobini-Zinner. (From Scarf et al., 1986c)

References

Alfvén, H.: 1975, *Tellus* **9**, 92

Bame, S., Anderson, R., Asbridge, J., Baker, D., Feldman, W., Fuselier, S., Gosling J., McComas, D., Thomsen, M., Young, D., Zwickl, R.: 1986, *Science* **232**, 356

Biermann, L., Browsowski, B., Schmidt, H.U.: 1967, *Solar Physics* **1**, 254

Brandt, J.C.: 1982, Observations and dynamics of plasma tails, in *Comets*, ed. L.L. Wilkening (U. Arizona Press, Tucson) pp. 519–537

Brandt, J.C.: 1986a, The International Cometary Explorer (ICE) Mission to Comets Giacobini-Zinner and Halley, in *Space Missions to Halley's Comet*, eds. R. Reinhard and B. Battrick, ESA SP-1066, pp. 99–115

Brandt, J.C.: 1986b, *Nature* **321**, 391

Brandt, J.C., Farquhar, R.W., Maran, S.P., Niedner, M.B., von Rosenvinge, T.: 1985a, The International Cometary Explorer (ICE) mission to comet Giacobini-Zinner (G/Z), in *Dynamics of Comets: Their Origin and Evolution*, eds. A. Carusi and G.B. Valsecchi (Reidel, Dordrecht), pp. 405–414

Brandt, J.C., Niedner, M.B., von Rosenvinge, T.: 1985b, *Adv. Space Res.* **5**, 3

Combi, M.R., Stewart, A.I.F., Smyth, W.H.: 1986, *Geophys. Res. Letters* **13**, 385

Daly, P.W.: 1987, *Geophys. Res. Letters* **14**, 648

Ershkovich, A.I.: 1980, *Space Science Reviews* **25**, 3

Ershkovich, A.I., Niedner, M.B., Brandt, J.C.: 1982, *Astrophys. J.* **262**, 396

Farquhar, R.W., Dunham, D.W.: 1983, *Astronautics and Aeronautics* **21**, ♯9, 50

Fuselier, S.A., Feldman, W.C., Bame, S.J., Smith, E.J., Scarf, F.L.: 1986, *Geophys. Res. Letters* **13**, 247

Gosling, J.T., Asbridge, J.R., Bame, S.J., Thomsen, M.F., Zwickl, R.D.: 1986, *Geophys. Res. Letters* **13**, 267

Gurnett, D.A., Averkamp, T.F., Scarf, F.L., Grün, E.: 1986, *Geophys. Res. Letters* **13**, 291

Hanner, M., Veeder, G., Johnson, T., Matson, D., Morrison, D.: 1985, *Bull. Am. Astron. Soc.* **17**, 687

Hynds, R.J., Cowley, S.W.H., Richardson, I.G., Sanderson, T.R., Tranquille, C., Wenzel, K.-P., Daly, P.W.: 1985, *Adv. Space Res.* **5**, 17

Hynds, R.J., Cowley, S.W.H., Sanderson, T.R., Wenzel, K.-P., Van Rooijen, J.J.: 1986, *Science* **232**, 361

Ipavich, F.M., Galvin, A.B., Gloeckler, G., Hovestadt, D., Klecker, B., Scholer, M.: 1986, *Science* **232**, 366

Jones, D.E., Smith, E.J., Slavin, J.A., Tsurutani, B.T., Siscoe, G.L., Mendis, D.A.: 1986, *Geophys. Res. Letters* **13**, 243

McFadden, L.A., A'Hearn, M.F., Feldman, P.D., Böhnhardt, H., Rahe, J., Festou, M.C., Brandt, J.C., Maran, S.P., Niedner, M.B., Smith, A.M., Schleicher, D.G.: 1987, *Icarus* **69**, 329

Meyer-Vernet, N.: 1985, *Adv. Space Res.* **5**, 37

Meyer-Vernet, N., Couturier, P., Hoang, S., Perche, C., Steinberg, J.-L.: 1986, *Geophys. Res. Letters* **13**, 279

Ogilvie, K.W., von Rosenvinge, T., Durney, A.C.: 1977, *Science* **198**, 131

Ogilvie, K.W.: 1985, *Adv. Space Res.* **5**, 27

Ogilvie, K.W., Coplan, M.A., Bochsler, P., Geiss, J.: 1986, *Science* **232**, 374

Richardson, I.G., Cowley, S.W.H., Hynds, R.J., Sanderson, T.R., Wenzel, K.-P., Daly, P.W.: 1986, *Adv. Space Res.* **6**, 213

Sanderson, T.R., Wenzel, K.-P., Daly, P.W., Cowley, S.W.H., Hynds, R.J., Smith, E.J., Bame, S.J., Zwickl, R.D.: 1986a, *Adv. Space Res.* **6**, 209

Sanderson, T.R., Wenzel, K.-P., Daly, P.W., Cowley, S.W.H., Hynds, R.J., Richarson, I.G., Smith,E.J., Bame, S.J., Zwickl, R.D.: 1986b, Observations of Heavy Energetic Ions far Upstream from Comet Halley, in *20th ESLAB Symposium on the Exploration of Halley's Comet*, ESA SP-250, **1**, pp. 105–108

Scarf, F.L.: 1986, *Adv. Space Res.* **6**, 329

Scarf, F.L., Coroniti, F.V., Kennel, C.F., Gurnett, D.A., Ip, W.-H., Smith, E.J.: 1986a, *Science* **232**, 377

Scarf, F.L., Coroniti, F.V., Kennel, C.F., Sanderson, T.R., Wenzel, K.-P., Hynds, R.J., Smith, E.J., Bame, S.J., Zwickl, R.D.: 1986b, *Geophys. Res. Letters* **13**, 857

Scarf, F.L., Coroniti, F.V., Kennel, C.F., Gurnett, D.A., Ip, W.-H., Smith, E.J.: 1986c, Observations of Cometary Plasma

Wave Phenomena, in *20th Eslab Symposium on the Exploration of Halley's Comet*, ESA SP-250, **1**, pp. 163–168

Schmidt, H.U., Wegmann, R.: 1982, Plasma flow and magnetic field in comets, in *Comets*, ed. L.L. Wilkening, (U. Arizona Press, Tucson), pp. 538–560

Slavin, J.A., Smith, E.J., Tsurutani, B.T., Siscoe, G.L., Jones, D.E., Mendis, D.A.: 1986a, *Geophys. Res. Letters* **13**, 283

Slavin, J.A., Smith, E.J., Daly, P.W., Flammer, K.R., Gloeckler, G., Goldberg, B.A., McComas, D.J., Scarf, F.L., Steinberg, J.-L.: 1986b, The P/Giacobini-Zinner Magnetotail, in *20th Eslab Symposium on the Exploration of Halley's Comet*, ESA SP-250, **1**, pp. 81–87

Smith, E.J., Tsurutani, B.T., Slavin, J.A., Jones, D.E., Siscoe, G.L., Mendis, D.A.: 1986a, *Science* **232**, 382

Smith, E.J., Slavin, J.A., Bame, S.J., Thomsen, M.F., Cowley, S.W.H., Richardson, I.G., Hovestadt, D., Ipavich, F.M., Ogilvie, K.W., Coplan, M.A., Sanderson, T.R., Wenzel, K.-P., Scarf, F.L., Vinas, A.F., Scudder, J.D.: 1986b, Analysis of the Giacobini-Zinner bow wave, in *20th Eslab Symposium on the Exploration of Halley's Comet*, ESA SP-250, **3**, pp. 461–465

Spinrad, H., McCarthy, P.J., Strauss, M.A.: 1985, *Bull. Am. Astron. Soc.* **17**, 686

Stewart, A.I.F., Combi, M.R., Smyth, W.H.: 1985, *Bull. Am. Astron. Soc.* **17**, 686

Thomsen, M.F., Bame, S.J., Feldman, W.C., Gosling, J.T., McComas, D.J., Young, D.T.: 1986, *Geophys. Res. Letters* **13**, 393

Tsurutani, B.T., Smith, E.J.: 1986a, *Geophys. Res. Letters* **131**, 259

Tsurutani, B.T., Smith, E.J.: 1986b, *Geophys. Res. Letters* **13**, 263

Tsurutani, B.T., Brinca, A.L., Smith, E.J., Thorne, R.M., Scarf, F.L., Gosling, J.T., Ipavich, F.M.: 1986a, MHD Waves detected by ICE at distances $\geqq 28 \times 10^6$ km from Comet Halley: Cometary or Solar-wind Origin?, in *20th Eslab Symposium on the Exploration of Halley's Comet*, ESA SP-250, **3**, pp, 451–456

Tsurutani, B.T., Smith, E.J., Thorne, R.M., Gosling, J.T., Matsumoto, H.: 1986b, Steepened Magnetosonic waves in the High β plasma surrounding Comet Giacobini-Zinner, in *20th Eslab Symposium on the Exploration of Halley's Comet*, ESA SP-250, **3**, pp. 457–460

Wallis, M.K.: 1973, *Planet. Space Sci.* **21**, 1647

Wallis, M.K.: 1986, *Adv. Space Res.* **6**, 239

Wallis, M.K., Dryer, M.: 1985, *Nature* **318**, 646

Wenzel, K.-P., Sanderson, T.R., Richardson, I.G., Cowley, S.W.H., Hynds, R.J., Bame, S.J.,Zwickl, R.D., Smith, E.J., Tsurutani, B.T.: 1986, *Geophys. Res. Letters* **13**, 861

Yeomans, D.K., Brandt, J.C.: 1985, *The Comet Giacobini-Zinner Handbook*, NASA/JPL Publication (JPL 400–254)

EXPLORATION OF HALLEY's COMET

M. Grewing · F. Praderie
R. Reinhard (Eds.)

The International Halley Watch

Stephen J. Edberg[1], **Ray L. Newburn, Jr.**[1], **and Jurgen Rahe**[2]

[1] International Halley Watch, Jet Propulsion Laboratory, California Institute of Technology, 4800 Oak Grove Drive, Pasadena, CA 91109, USA
[2] Dr.-Remeis-Sternwarte, Sternwartstrasse 7, D-8600 Bamberg, Federal Republic of Germany

Summary. The International Halley Watch has received the spirited cooperation of astronomers, space agencies, and governments since it began. With much of the data acquisition complete, efforts now turn to the production of the archive of observations. The history, current status, and plans for the future are outlined.

Key words: International Halley Watch – comet P/Halley – comet P/Giacobini-Zinner – comet P/Crommelin – comets – archive

1. Purposes

Advocacy

The IHW has encouraged and supported any scientifically valid means of studying Comet Halley.

Coordination

The IHW has coordinated activities among the groundbased disciplines and with flight projects in order to maximize the scientific value of the entire body of observations.

Standardization

The scientists in each IHW discipline have set useful standards for observations in that discipline. There was no absolute requirement that such standards be followed for IHW participation, however.

Archiving

All properly documented data (reduced data, not interpretations) will be published in a Halley archive. This archive will complement the usual interpretive papers in the open literature, not supplant them.

P/Giacobini-Zinner

All data are being treated in the same manner as those on P/Halley, in support of the ICE spacecraft encounter with P/Giacobini-Zinner.

2. History and organization

The International Halley Watch (IHW) was conceived in 1979 by Louis Friedman, then at the Jet Propulsion Laboratory (JPL). His basic idea was to promote international cooperation in all phases of the study of Halley's Comet. The U.S. National Aeronautics and Space Administration (NASA) supported a JPL study of Friedman's ideas. In 1980 a science working group was organized by NASA to consider JPL's proposal to establish the IHW (Brandt et al., 1980). Following the working group's recommendation to proceed, "lead centers" (LC) for the IHW were created in 1981 in Pasadena, California at JPL and in Bamberg, Federal Republic of Germany (FRG) at the University of Erlangen-Nürnberg's Astronomical Institute. The centers receive support from NASA and the FRG, respectively, and agreed to split international communications responsibilities: the Bamberg LC would serve as a focus for Europe, Africa, and Asia (except Japan) and the Pasadena LC would serve North and South America and astronomers around the Pacific Ocean.

Even as this external foundation was being designed, the internal structure was being generated. A steering group, international in composition, whose members were widely recognized experts in the field of cometary studies was convened to advise on the methods and goals of the IHW. Ultimately 27 astronomers from 13 countries constituted the steering group (Table 1).

A call for proposals for discipline specialists (DSs) in seven areas of cometary studies went out. The disciplines were specified on the basis of their investigation technology:

1. Astrometry
2. Infrared spectroscopy and radiometry
3. Large-scale phenomena (tail studies)
4. Near-nucleus studies
5. Photometry and polarimetry
6. Radio studies
7. Spectroscopy and spectrophotometry.

An eighth discipline, meteor studies, was added later (Table 2). An amateur observation network was organized directly by the lead centers.

1981 also saw the presentation of the IHW idea to the Interagency Consultative Group (IACG), composed of the European Space Agency, the Intercosmos Council of the Soviet Union, Japan's Institute of Space and Astronautical Science (ISAS), and NASA. These organizations were making plans to cooperate in spacecraft missions to Comet Halley and the groundbased efforts of the IHW would be very useful in the navigation of the spacecraft and in placing the brief spacecraft encounters in the context of the whole apparition. The next year the IACG (which meets annually) accepted the IHW as an independent, non-voting member.

Late in 1981 the international DS teams were selected by the steering group. At the International Astronomical Union (IAU)

Table 1. International steering group

W.I. Axford	New Zealand	R. Lust	FRG
M.J.S. Belton	USA	A. Massevitch	USSR
J. Blamont	France	C.R. O'Dell	USA
G. Briggs	USA	V. Pankonin	USA
W. Brunk	USA	R. Reinhard	The Netherlands
A. Delsemme	USA	H.E. Schuster	Chile
B. Donn	USA	K.R. Sivaraman	India
H. Fechtig	FRG	V. Vanysek	Czechoslovakia
L. Friedman	USA	J.F. Veverka	USA
S.M. Gong	China	G. Wetherill	USA
I. Halliday	Canada	F.L. Whipple	USA
G. Herbig	USA	I.P. Williams	Great Britain
L. Kresak	Czechoslovakia	Ya.S. Yatskiv	USSR
Y. Kozai	Japan		

Table 2. Discipline specialist teams

Astrometry	
D.K. Yeomans	Jet Propulsion Laboratory
R.M. West	European Southern Observatory
R.S. Harrington	U.S. Naval Observatory
B.G. Marsden	Smithsonian Astrophysical Observatory
Infrared Spectroscopy and Radiometry	
R.F. Knacke	SUNY (Stony Brook)
T. Encrenaz	Observatoire de Paris
Large-scale phenomena	
J.C. Brandt	University of Colorado
M.B. Niedner	Goddard Space Flight Center
J. Rahe	Astronomical Inst., U. Erlangen-Nürnberg
Meteor studies	
P.B. Babadzhanov	Academy of Sciences of the Tajik SSR
A. Hajduk	Astronomical Inst. SAV
B.A. Lindblad	Lund Observatory
B.A. McIntosh	Herzberg Institute of Astrophysics
Near-nucleus studies	
S. Larson	University of Arizona
Z. Sekanina	Jet Propulsion Laboratory
J. Rahe	Astronomical Inst., U. Erlangen-Nürnberg
Photometry and Polarimetry	
M. A'Hearn	University of Maryland
V. Vanysek	Charles University
Radio studies	
W. Irvine	University of Massachusetts
F.P. Schloerb	University of Massachusetts
E. Gerard	Observatoire de Paris
R.D. Brown	Monash University
P. Godfrey	Monash University
Spectroscopy and spectrophotometry	
S. Wyckoff	Arizona State University
P. Wehinger	Arizona State University
M.C. Festou	Observatoire de Besançon

General Assembly in 1982 the DS teams made presentations of their plans and began organizing their worldwide networks of observers. It was at the IAU that the IHW was recognized as the sole coordinator of worldwide groundbased Halley observations (Fig. 1). Comet Halley was recovered in October, more than three years before perihelion (IAU Circular 3737). In space, NASA began the process that, in three years time, would lead the International Sun-Earth Explorer No. 3 spacecraft, renamed the International Comet Explorer (ICE), through the tail of Comet P/Giacobini-Zinner in 1985 and near Halley in 1986.

As the networks of observers were organized, the planning for a trial run on intermediate period comet P/Crommelin advanced. Held early in 1984, the trial run provided a useful lesson in end-to-end data handling problems and in the preparation of an archive (Sekanina and Aronsson, 1985).

P/Giacobini-Zinner was recovered in 1984 (IAU Circular 3937). As a subprogram of the IHW, a Giacobini-Zinner Watch was instituted in support of ICE in 1985, with the G-Z archive projected for publication in late 1988.

By the time the prime Halley observing period opened in the fall of 1985 over 1000 astronomers in 51 countries had registered their intentions to observe Halley, and they did so with unprecedented enthusiasm throughout 1986. The printed archive of all these observations is projected for publication in 1989.

3. Status – August 1987

All the IHW observation networks attained or exceeded their original goals for data acquisition. In additon, all the networks involved in space-flight-project support met the requirements placed on them. For both the professional and amateur branches of the IHW, participation exceeded 100% of the observers registered in advance. Table 3 lists present estimates of quantities of data acquired. The biggest problem remaining is to get all of these results transmitted for inclusion in the archive by March 1989.

4. The archive

The Halley and G-Z archives will be produced in three forms, two of them widely available. Included in digital versions will be the data acquired by the spacecraft that studied these comets.

The primary archive will be housed on compact optical disks (CDs). It will include all valid digital data, including imagery and spacecraft results. A printed archive will contain entries for each observation. Numerical data will appear in tables or graphically and selected imagery will be produced as halftones. The CD and printed archives will be made available to all data contributors who request them and to research libraries.

The IHW agreed that each participating space agency would receive a full set of archive magnetic tapes. These sets will include the groundbased and flight-acquired data and will be housed in Pasadena, Bamberg, Paris, Moscow, and Tokyo.

Finally, the archives will be accessible through the planetary data system (PDS) in the U.S. and through similar systems in

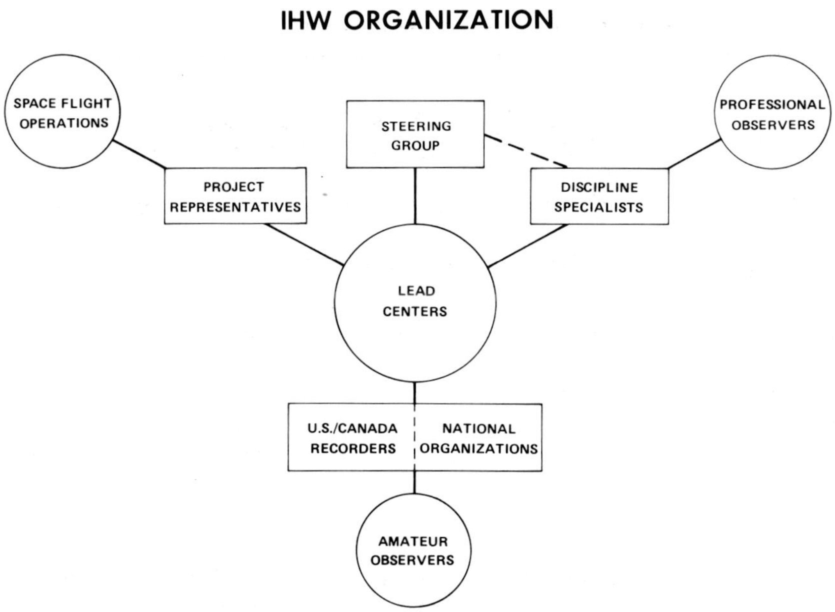

Fig. 1. The organization and lines of communication of the International Halley Watch

Table 3. Data acquired: approximate numbers of observations

Net	P/Halley		P/Giacobini-Zinner	
	Logged observations	Anticipated	Logged observations	Anticipated
Astrometry	6655	7000	1127	1200
Infrared Photometry	1000	1700	30	150
– Spectra	15	100	2	2
– Polarimetry	40	50	0	0
– Images	3	60	29	29
Large-scale phenomena	2000	7000	50	> 50
Meteor studies – Radar	5000	8500	100	100
– Visual	3000	5000	60	60
Near-nucleus studies	5000	6000	200	350
Photometry and polarimetry	41000	55000	1000	1200
Radio studies	1900	2200	275	400
Spectroscopy and spectrophotometry	9600	10000	20	250
Amateur observations – Magnitude	6000	10000	1000	1000
– Drawings	360	700	53	50
– Photographs	1400	2000	42	40
– Spectra	52	52	0	0

other nations that choose to create them. The collection of amateur drawings and photographs will be housed at the cometary node of the PDS for use by researchers.

5. The future

With completion of the publication of the Halley archive in 1990 the IHW will have finished its work. A proposal has been made to begin an International Comet Watch (ICW) in 1988 as an international extension of the PDS comet node. The ICW would primarily be a passive organization, serving to receive data on all comets and make them available to any scientist through a PDS-type system. It would promote astrometric programs to improve ephemerides and serve as a resource to all research programs involving comets. The ICW would be ready to actively promote and coordinate research efforts when the "comet of the century" finally arrives.

Acknowledgements. The IHW would not have worked without the spirited cooperation of space agencies and of professional and amateur astronomers and their governments and universities around the world.

The research described in this paper was carried out in part by the Jet Propulsion Laboratory, California Institute of Technology, under contract with the National Aeronautics and Space Administration.

References

Brandt, J.C., Friedman, L.D., Newburn, R.L., Yeomans, D.K. (eds.): 1980, *The International Halley Watch – Report of the Science Working Group*, NASA-JPL Publication TM 82181 (400-88), Pasadena

Sekanina, Z., Aronsson, M. (eds.): 1985, *Archive of Observations of Periodic Comet Crommelin*, NASA-JPL Publication 86-2, Pasadena